Precalculus

Graphical, Numerical, Algebraic

Ninth Edition

Franklin D. Demana
The Ohio State University

Bert K. Waits
The Ohio State University

Gregory D. Foley
Ohio University

Daniel Kennedy
Baylor School

David E. Bock
Ithaca High School

PEARSON

Boston Columbus Indianapolis New York San Francisco Upper Saddle River
Amsterdam Cape Town Dubai London Madrid Milan Munich Paris Montréal Toronto
Delhi Mexico City São Paulo Sydney Hong Kong Seoul Singapore Taipei Tokyo

Editor in Chief	Anne Kelly
Senior Content Editor	Rachel Reeve
Editorial Assistant	Judith Garber
Senior Managing Editor	Karen Wernholm
Associate Managing Editor	Tamela Ambush
Senior Production Project Manager	Peggy McMahon
Digital Assets Manager	Marianne Groth
Project Manager, MyMathLab	Kristina Evans
QA Manager, Assessment Content	Marty Wright
Marketing Manager, AP&E Math	Jackie Flynn
Senior Author Support/ Technology Specialist	Joe Vetere
Procurement Manager	Vincent Scelta
Procurement Specialist	Debbie Rossi
Image Manager	Rachel Youdelman
Liaison Manager, Permissions	Joseph Croscup
Cover and Text Design, Production Coordination, Composition, and Illustrations	Cenveo® Publisher Services
Cover photo	*Geometric Glass Façade,* ER_09/Shutterstock

For permission to use copyrighted material, grateful acknowledgment is made to the copyright holders listed on page 951, which is hereby made part of this copyright page.

Library of Congress Cataloging-in-Publication Data

Demana, Franklin D., 1938-
Precalculus : graphical, numerical, algebraic / Franklin D. Demana, The Ohio State University, Bert K. Waits, The Ohio State University, Gregory D. Foley, Ohio University, Daniel Kennedy, Baylor School, David E. Bock. — 9th edition.
pages cm
Previous title: Precalculus. Boston : Addison Wesley, c2011.
ISBN 0-13-351845-0 — ISBN 0-13-351879-5
I. Waits, Bert K. II. Foley, Gregory D. III. Kennedy, Daniel, 1946- IV. Bock, David E. V. Title.
QA154.3.P74 2013
512'.13—dc23
2013035546

8 17

PEARSON

www.PearsonSchool.com/Advanced

ISBN-13: 978-0-13-351845-0
ISBN-10: 0-13-351845-0
(High School binding)

FOREWORD

We are proud of the fact that earlier editions of *Precalculus: Graphical, Numerical, Algebraic* were among the first to recognize the potential of hand-held graphers for helping students understand function behavior. The power of visualization eventually transformed the teaching and learning of calculus on the college level and in the AP® program, then led to reforms in the high school curriculum articulated in the NCTM *Principles and Standards for School Mathematics* and now in the Common Core State Standards. All along the way, this textbook has kept current with the best practices while continuing to pioneer new ideas in exploration and pedagogy that enhance student learning (for example, the study of function behavior based on the Twelve Basic Functions, an idea that has gained widespread acceptance in the textbook world). For those students continuing to a calculus course, this precalculus textbook concludes with a chapter that prepares students for the two central themes of calculus: instantaneous rate of change and continuous accumulation. This intuitively appealing preview of calculus is both more useful and more reasonable than the traditional, unmotivated foray into the computation of limits, and it is more in keeping with the stated goals and objectives of the AP courses and their emphasis on depth of knowledge.

Recognizing that precalculus is a capstone course for many students, we include *quantitative literacy* topics such as probability, statistics, and the mathematics of finance and integrate the use of data and modeling throughout the text. Our goal is to provide students with the critical-thinking skills and mathematical know-how needed to succeed in college, career, or any endeavor.

Continuing in the spirit of the eight earlier editions, we have integrated graphing technology throughout the course, not as an additional topic but as an essential tool for both mathematical discovery and effective problem solving. Graphing technology enables students to study a full catalog of basic functions at the beginning of the course, thereby giving them insights into function properties that are not seen in many books until later chapters. By connecting the algebra of functions to the visualization of their graphs, we are even able to introduce students to parametric equations, piecewise-defined functions, limit notation, and an intuitive understanding of continuity as early as Chapter 1. However, the advances in technology and increased familiarity with calculators have blurred some of the distinctions between solving problems and supporting solutions that we had once assumed to be apparent. Therefore, we ask that some exercises be solved without calculators. (See the "Technology and Exercises" section.)

Once students are comfortable with the language of functions, the text guides them through a more traditional exploration of twelve basic functions and their algebraic properties, always reinforcing the connections among their algebraic, graphical, and numerical representations. This book uses a consistent approach to modeling, emphasizing the use of particular types of functions to model behavior in the real world. Modeling is a fundamental aspect of our problem-solving process that is introduced in Section 1.1 and used throughout the book. The text has a wealth of data and range of applications to illustrate how mathematics and statistics connect to every facet of modern life.

This textbook has faithfully incorporated not only the teaching strategies that have made *Calculus: Graphical, Numerical, Algebraic* so popular, but also some of the strategies from the popular Pearson high school algebra series, and thus has produced a seamless pedagogical transition from prealgebra through calculus for students. Although this book can certainly be appreciated on its own merits, teachers who seek coherence and vertical alignment in their mathematics sequence might consider this pedagogical approach to be an additional asset of *Precalculus: Graphical, Numerical, Algebraic.*

This textbook is written to address current and emerging state curriculum standards. In particular, we embrace NCTM's *Focus in High School Mathematics: Reasoning and Sense Making* and its emphasis on the importance of helping students make sense of and reason using mathematics. The NCTM's *Principles and Standards for School Mathematics* identified five "Process Standards" that should be fundamental in mathematics education. The first of these standards was Problem Solving. Since then, the emphasis on problem solving has continued to grow, to the point that it is now integral to the instructional process in many mathematics classrooms. When the Common Core State Standards in Mathematics detailed eight "Standards for Mathematical Practice" that should be fundamental in mathematics education, again the first of these addressed problem solving. Individual states have also released their own standards over the years, and problem solving is invariably front and center as a fundamental objective. Problem solving, reasoning, sense making, and the related processes and practices of mathematics are central to the approach we use in *Precalculus: Graphical, Numerical, Algebraic.*

With this new edition we embrace the growing importance and wide applicability of Statistics. Because Statistics is increasingly used in college coursework, the workplace, and everyday life, we have added a new Chapter 10 to help students see that statistical analysis is an investigative process that turns loosely formed ideas into scientific studies. Our five sections on data analysis, probability, and statistical literacy are aligned with the *GAISE* Report published by the American Statistical Association, the College Board's AP® Statistics curriculum, and the Common Core State Standards. Chapter 10 is not intended as a course in statistics but rather as an introduction to set the stage for possible further study.

CONTENTS

CHAPTER P

Prerequisites

P.1 Real Numbers 2

Representing Real Numbers • Order and Interval Notation • Basic Properties of Algebra • Integer Exponents • Scientific Notation

P.2 Cartesian Coordinate System 12

Cartesian Plane • Absolute Value of a Real Number • Distance Formulas • Midpoint Formulas • Equations of Circles • Applications

P.3 Linear Equations and Inequalities 21

Equations • Solving Equations • Linear Equations in One Variable • Linear Inequalities in One Variable

P.4 Lines in the Plane 28

Slope of a Line • Point-Slope Form Equation of a Line • Slope-Intercept Form Equation of a Line • Graphing Linear Equations in Two Variables • Parallel and Perpendicular Lines • Applying Linear Equations in Two Variables

P.5 Solving Equations Graphically, Numerically, and Algebraically 40

Solving Equations Graphically • Solving Quadratic Equations • Approximating Solutions of Equations Graphically • Approximating Solutions of Equations Numerically with Tables • Solving Equations by Finding Intersections

P.6 Complex Numbers 49

Complex Numbers • Operations with Complex Numbers • Complex Conjugates and Division • Complex Solutions of Quadratic Equations

P.7 Solving Inequalities Algebraically and Graphically 54

Solving Absolute Value Inequalities • Solving Quadratic Inequalities • Approximating Solutions to Inequalities • Projectile Motion

Key Ideas 59

Review Exercises 60

CHAPTER 1

Functions and Graphs

1.1 Modeling and Equation Solving 64

Numerical Models • Algebraic Models • Graphical Models • The Zero Factor Property • Problem Solving • Grapher Failure and Hidden Behavior • A Word About Proof

1.2 Functions and Their Properties 80
Function Definition and Notation • Domain and Range • Continuity • Increasing and Decreasing Functions • Boundedness • Local and Absolute Extrema • Symmetry • Asymptotes • End Behavior

1.3 Twelve Basic Functions 99
What Graphs Can Tell Us • Twelve Basic Functions • Analyzing Functions Graphically

1.4 Building Functions from Functions 110
Combining Functions Algebraically • Composition of Functions • Relations and Implicitly Defined Functions

1.5 Parametric Relations and Inverses 119
Relations Defined Parametrically • Inverse Relations and Inverse Functions

1.6 Graphical Transformations 129
Transformations • Vertical and Horizontal Translations • Reflections Across Axes • Vertical and Horizontal Stretches and Shrinks • Combining Transformations

1.7 Modeling with Functions 140
Functions from Formulas • Functions from Graphs • Functions from Verbal Descriptions • Functions from Data

Key Ideas 152
Review Exercises 152
Chapter Project 156

CHAPTER 2

Polynomial, Power, and Rational Functions

2.1 Linear and Quadratic Functions and Modeling 158
Polynomial Functions • Linear Functions and Their Graphs • Average Rate of Change • Association, Correlation, and Linear Modeling • Quadratic Functions and Their Graphs • Applications of Quadratic Functions

2.2 Power Functions with Modeling 174
Power Functions and Variation • Monomial Functions and Their Graphs • Graphs of Power Functions • Modeling with Power Functions

2.3 Polynomial Functions of Higher Degree with Modeling 185
Graphs of Polynomial Functions • End Behavior of Polynomial Functions • Zeros of Polynomial Functions • Intermediate Value Theorem • Modeling

2.4 Real Zeros of Polynomial Functions 197
Long Division and the Division Algorithm • Remainder and Factor Theorems • Synthetic Division • Rational Zeros Theorem • Upper and Lower Bounds

2.5 Complex Zeros and the Fundamental Theorem of Algebra 210
Two Major Theorems • Complex Conjugate Zeros • Factoring with Real Number Coefficients

2.6 Graphs of Rational Functions 218
Rational Functions • Transformations of the Reciprocal Function • Limits and Asymptotes • Analyzing Graphs of Rational Functions • Exploring Relative Humidity

2.7 Solving Equations in One Variable 228
Solving Rational Equations • Extraneous Solutions • Applications

2.8 Solving Inequalities in One Variable 236
Polynomial Inequalities • Rational Inequalities • Other Inequalities • Applications

Key Ideas 245
Review Exercises 246
Chapter Project 250

CHAPTER 3

Exponential, Logistic, and Logarithmic Functions

3.1 Exponential and Logistic Functions 252
Exponential Functions and Their Graphs • The Natural Base e • Logistic Functions and Their Graphs • Population Models

3.2 Exponential and Logistic Modeling 265
Constant Percentage Rate and Exponential Functions • Exponential Growth and Decay Models • Using Regression to Model Population • Other Logistic Models

3.3 Logarithmic Functions and Their Graphs 274
Inverses of Exponential Functions • Common Logarithms—Base 10 • Natural Logarithms—Base e • Graphs of Logarithmic Functions • Measuring Sound Using Decibels

3.4 Properties of Logarithmic Functions 283
Properties of Logarithms • Change of Base • Graphs of Logarithmic Functions with Base b • Re-expressing Data

3.5 Equation Solving and Modeling 292
Solving Exponential Equations • Solving Logarithmic Equations • Orders of Magnitude and Logarithmic Models • Newton's Law of Cooling • Logarithmic Re-expression

3.6 Mathematics of Finance 304
Simple and Compound Interest • Interest Compounded k Times per Year • Interest Compounded Continuously • Annual Percentage Yield • Annuities—Future Value • Loans and Mortgages—Present Value

Key Ideas 313
Review Exercises 314
Chapter Project 318

CHAPTER 4

Trigonometric Functions

4.1 Angles and Their Measures 320
The Problem of Angular Measure • Degrees and Radians • Circular Arc Length • Angular and Linear Motion

4.2 Trigonometric Functions of Acute Angles 329
Right Triangle Trigonometry • Two Famous Triangles • Evaluating Trigonometric Functions with a Calculator • Common Calculator Errors when Evaluating Trig Functions • Applications of Right Triangle Trigonometry

4.3 Trigonometry Extended: The Circular Functions 338
Trigonometric Functions of Any Angle • Trigonometric Functions of Real Numbers • Periodic Functions • The 16-Point Unit Circle

4.4 Graphs of Sine and Cosine: Sinusoids 350
The Basic Waves Revisited • Sinusoids and Transformations • Modeling Periodic Behavior with Sinusoids

4.5 Graphs of Tangent, Cotangent, Secant, and Cosecant 361
The Tangent Function • The Cotangent Function • The Secant Function • The Cosecant Function

4.6 Graphs of Composite Trigonometric Functions 369
Combining Trigonometric and Algebraic Functions • Sums and Differences of Sinusoids • Damped Oscillation

4.7 Inverse Trigonometric Functions 378
Inverse Sine Function • Inverse Cosine and Tangent Functions • Composing Trigonometric and Inverse Trigonometric Functions • Applications of Inverse Trigonometric Functions

4.8 Solving Problems with Trigonometry 388
More Right Triangle Problems • Simple Harmonic Motion

Key Ideas 399
Review Exercises 399
Chapter Project 402

CHAPTER 5

Analytic Trigonometry

5.1 Fundamental Identities 404

Identities • Basic Trigonometric Identities • Pythagorean Identities • Cofunction Identities • Odd-Even Identities • Simplifying Trigonometric Expressions • Solving Trigonometric Equations

5.2 Proving Trigonometric Identities 413

A Proof Strategy • Proving Identities • Disproving Non-Identities • Identities in Calculus

5.3 Sum and Difference Identities 421

Cosine of a Difference • Cosine of a Sum • Sine of a Difference or Sum • Tangent of a Difference or Sum • Verifying a Sinusoid Algebraically

5.4 Multiple-Angle Identities 428

Double-Angle Identities • Power-Reducing Identities • Half-Angle Identities • Solving Trigonometric Equations

5.5 The Law of Sines 434

Deriving the Law of Sines • Solving Triangles (AAS, ASA) • The Ambiguous Case (SSA) • Applications

5.6 The Law of Cosines 442

Deriving the Law of Cosines • Solving Triangles (SAS, SSS) • Triangle Area and Heron's Formula • Applications

Key Ideas 450

Review Exercises 450

Chapter Project 454

CHAPTER 6

Applications of Trigonometry

6.1 Vectors in the Plane 456

Two-Dimensional Vectors • Vector Operations • Unit Vectors • Direction Angles • Applications of Vectors

6.2 Dot Product of Vectors 467

The Dot Product • Angle Between Vectors • Projecting One Vector onto Another • Work

6.3 Parametric Equations and Motion 475

Parametric Equations • Parametric Curves • Eliminating the Parameter • Lines and Line Segments • Simulating Motion with a Grapher

6.4 Polar Coordinates 487

Polar Coordinate System • Coordinate Conversion • Equation Conversion • Finding Distance Using Polar Coordinates

6.5 Graphs of Polar Equations 494
Polar Curves and Parametric Curves • Symmetry • Analyzing Polar Graphs • Rose Curves • Limaçon Curves • Other Polar Curves

6.6 De Moivre's Theorem and *n*th Roots 503
The Complex Plane • Polar Form of Complex Numbers • Multiplication and Division of Complex Numbers • Powers of Complex Numbers • Roots of Complex Numbers

Key Ideas 513
Review Exercises 514
Chapter Project 517

CHAPTER 7

Systems and Matrices

7.1 Solving Systems of Two Equations 519
The Method of Substitution • Solving Systems Graphically • The Method of Elimination • Applications

7.2 Matrix Algebra 529
Matrices • Matrix Addition and Subtraction • Matrix Multiplication • Identity and Inverse Matrices • Determinant of a Square Matrix • Applications

7.3 Multivariate Linear Systems and Row Operations 543
Triangular Form for Linear Systems • Gaussian Elimination • Elementary Row Operations and Row Echelon Form • Reduced Row Echelon Form • Solving Systems Using Inverse Matrices • Partial Fraction Decomposition • Other Applications

7.4 Systems of Inequalities in Two Variables 557
Graph of an Inequality • Systems of Inequalities • Linear Programming

Key Ideas 565
Review Exercises 565
Chapter Project 569

CHAPTER 8

Analytic Geometry in Two and Three Dimensions

8.1 Conic Sections and a New Look at Parabolas 571
Conic Sections • Geometry of a Parabola • Translations of Parabolas • Reflective Property of a Parabola

8.2 Circles and Ellipses 582
Transforming the Unit Circle • Geometry of an Ellipse • Translations of Ellipses • Orbits and Eccentricity • Reflective Property of an Ellipse

8.3 Hyperbolas 593
Geometry of a Hyperbola • Translations of Hyperbolas • Eccentricity and Orbits • Reflective Property of a Hyperbola • Long-Range Navigation

8.4 Quadratic Equations with *xy* Terms 603
Quadratic Equations Revisited • Axis Rotation Formulas • Discriminant Test

8.5 Polar Equations of Conics 612
Eccentricity Revisited • Writing Polar Equations for Conics • Analyzing Polar Equations of Conics • Orbits Revisited

8.6 Three-Dimensional Cartesian Coordinate System 621
Three-Dimensional Cartesian Coordinates • Distance and Midpoint Formulas • Equation of a Sphere • Planes and Other Surfaces • Vectors in Space • Lines in Space

Key Ideas 629
Review Exercises 630
Chapter Project 632

CHAPTER 9

Discrete Mathematics

9.1 Basic Combinatorics 634
Discrete Versus Continuous • The Importance of Counting • The Multiplication Principle of Counting • Permutations • Combinations • Subsets of an *n*-Set

9.2 Binomial Theorem 644
Powers of Binomials • Pascal's Triangle • Binomial Theorem • Factorial Identities

9.3 Sequences 650
Infinite Sequences • Limits of Infinite Sequences • Arithmetic and Geometric Sequences • Sequences and Technology

9.4 Series 658
Summation Notation • Sums of Arithmetic and Geometric Sequences • Infinite Series • Convergence of Geometric Series

9.5 Mathematical Induction 667
Tower of Hanoi Problem • Principle of Mathematical Induction • Induction and Deduction

Key Ideas 673
Review Exercises 673
Chapter Project 675

CHAPTER 10

Statistics and Probability

10.1 Probability 677

Sample Spaces and Probability Functions • Determining Probabilities • Venn Diagrams • Tree Diagrams • Conditional Probability

10.2 Statistics (Graphical) 691

Statistics • Categorical Data • Quantitative Data: Stemplots • Frequency Tables • Histograms • Describing Distributions: Shape • Time Plots

10.3 Statistics (Numerical) 704

Parameters and Statistics • Describing and Comparing Distributions • Five-Number Summary • Boxplots • The Mean (and When to Use It) • Variance and Standard Deviation • Normal Distributions

10.4 Random Variables and Probability Models 717

Probability Models and Expected Values • Binomial Probability Models • Normal Model • Normal Approximation for Binomial Distributions

10.5 Statistical Literacy 732

Uses and Misuses of Statistics • Correlation Revisited • Importance of Randomness • Samples, Surveys, and Observational Studies • Experimental Design • Using Randomness • Simulations

Key Ideas 747

Review Exercises 747

Chapter Project 752

CHAPTER 11

An Introduction to Calculus: Limits, Derivatives, and Integrals

11.1 Limits and Motion: The Tangent Problem 754

Average Velocity • Instantaneous Velocity • Limits Revisited • The Connection to Tangent Lines • The Derivative

11.2 Limits and Motion: The Area Problem 765

Distance from a Constant Velocity • Distance from a Changing Velocity • Limits at Infinity • The Connection to Areas • The Definite Integral

11.3 More on Limits 773

A Little History • Defining a Limit Informally • Properties of Limits • Limits of Continuous Functions • One-Sided and Two-Sided Limits • Limits Involving Infinity

11.4 Numerical Derivatives and Integrals 784

Derivatives on a Calculator • Definite Integrals on a Calculator • Computing a Derivative from Data • Computing a Definite Integral from Data

Key Ideas 793
Review Exercises 793
Chapter Project 795

APPENDIX A

Algebra Review

A.1 Radicals and Rational Exponents 797
Radicals • Simplifying Radical Expressions • Rationalizing the Denominator • Rational Exponents

A.2 Polynomials and Factoring 802
Adding, Subtracting, and Multiplying Polynomials • Special Products • Factoring Polynomials Using Special Products • Factoring Trinomials • Factoring by Grouping

A.3 Fractional Expressions 809
Algebraic Expressions and Their Domains • Reducing Rational Expressions • Operations with Rational Expressions • Compound Rational Expressions

APPENDIX B

Logic

B.1 Logic: An Introduction 814
Statements • Compound Statements

B.2 Conditionals and Biconditionals 820
Forms of Statements • Valid Reasoning

APPENDIX C

Key Formulas

C.1 Formulas from Algebra 827
C.2 Formulas from Geometry 828
C.3 Formulas from Trigonometry 828
C.4 Formulas from Analytic Geometry 830
C.5 Gallery of Basic Functions 831

Bibliography 832
Glossary 833
Selected Answers 851
Applications Index 953
Index 957

ABOUT THE AUTHORS

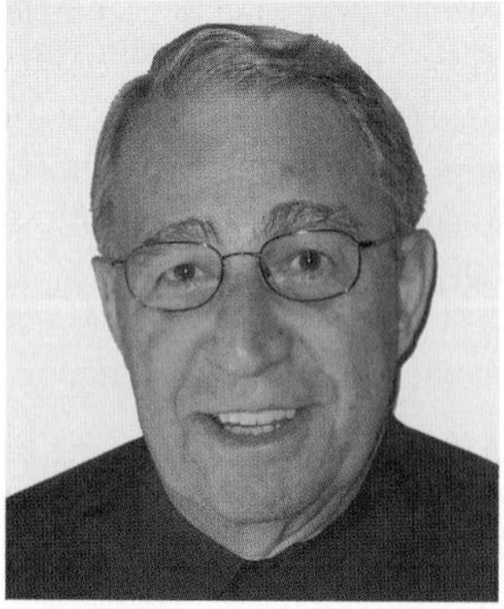

Franklin D. Demana

Frank Demana received his master's degree in mathematics and his Ph.D. from Michigan State University. Currently, he is Professor Emeritus of Mathematics at The Ohio State University. As an active supporter of the use of technology to teach and learn mathematics, he is cofounder of the international Teachers Teaching with Technology (T^3) professional development program. He has been the director and codirector of more than $10 million of National Science Foundation (NSF) and foundational grant activities, including a $3 million grant from the U.S. Department of Education Mathematics and Science Educational Research program awarded to The Ohio State University. Along with frequent presentations at professional meetings, he has published a variety of articles in the areas of computer- and calculator-enhanced mathematics instruction. Dr. Demana is also cofounder (with Bert Waits) of the annual International Conference on Technology in Collegiate Mathematics (ICTCM). He is co-recipient of the 1997 Glenn Gilbert National Leadership Award presented by the National Council of Supervisors of Mathematics, and co-recipient of the 1998 Christofferson-Fawcett Mathematics Education Award presented by the Ohio Council of Teachers of Mathematics.

Dr. Demana coauthored *Calculus: Graphical, Numerical, Algebraic; Essential Algebra: A Calculator Approach; Transition to College Mathematics; College Algebra and Trigonometry: A Graphing Approach; College Algebra: A Graphing Approach; Precalculus: Functions and Graphs;* and *Intermediate Algebra: A Graphing Approach.*

Bert K. Waits

Bert Waits received his Ph.D. from The Ohio State University and is currently Professor Emeritus of Mathematics there. Dr. Waits is cofounder of the international Teachers Teaching with Technology (T^3) professional development program, and has been codirector or principal investigator on several large National Science Foundation projects. Dr. Waits has published articles in more than 70 nationally recognized professional journals. He frequently has given invited lectures, workshops, and minicourses at national meetings of the MAA and the National Council of Teachers of Mathematics (NCTM) on how to use computer technology to enhance the teaching and learning of mathematics. Dr. Waits is co-recipient of the 1997 Glenn Gilbert National Leadership Award presented by the National Council of Supervisors of Mathematics, and is the cofounder (with Frank Demana) of the ICTCM. He is also co-recipient of the 1998 Christofferson-Fawcett Mathematics Education Award presented by the Ohio Council of Teachers of Mathematics. Dr. Waits was one of the six authors of the high school portion of the groundbreaking 1989 *NCTM Standards.*

Dr. Waits coauthored *Calculus: Graphical, Numerical, Algebraic; College Algebra and Trigonometry: A Graphing Approach; College Algebra: A Graphing Approach; Precalculus: Functions and Graphs;* and *Intermediate Algebra: A Graphing Approach.*

Gregory D. Foley

Greg Foley received B.A. and M.A. degrees in mathematics and a Ph.D. in mathematics education from The University of Texas at Austin. He is the Robert L. Morton Professor of Mathematics Education at Ohio University. Dr. Foley has taught elementary arithmetic through graduate-level mathematics, as well as upper division and graduate-level mathematics education classes. He has held full-time faculty positions at North Harris County College, Austin Community College, The Ohio State University, Sam Houston State University, and Appalachian State University, and served as Director of the Liberal Arts and Science Academy and as Senior Scientist for Secondary School Mathematics Improvement for the Austin Independent School District in Austin, Texas. Dr. Foley has presented over 300 lectures, workshops, and institutes throughout the United States and, internationally, has directed or codirected more than 50 funded projects totaling some $5 million; he has published over 40 scholarly works. Active in numerous professional organizations, Dr. Foley serves as chair of the editorial panel for the *Mathematics Teacher*, published by the National Council of Teachers of Mathematics. In 1998, he received the biennial American Mathematical Association of Two-Year Colleges (AMATYC) Award for Mathematics Excellence; in 2005, the annual Teachers Teaching with Technology (T^3) Leadership Award; and in 2013, Ohio University's Patton College award for distinguished graduate teaching.

Dr. Foley coauthored *Precalculus: A Graphing Approach*; *Precalculus: Functions and Graphs*; and *Advanced Quantitative Reasoning: Mathematics for the World Around Us.*

Daniel Kennedy

Dan Kennedy received his undergraduate degree from the College of the Holy Cross and his master's degree and Ph.D. in mathematics from the University of North Carolina at Chapel Hill. Since 1973 he has taught mathematics at the Baylor School in Chattanooga, Tennessee, where he holds the Cartter Lupton Distinguished Professorship. Dr. Kennedy joined the Advanced Placement® Calculus Test Development Committee in 1986, then in 1990 became the first high school teacher in 35 years to chair that committee. It was during his tenure as chair that the program moved to require graphing calculators and laid the early groundwork for the 1998 reform of the Advanced Placement Calculus curriculum. The author of the 1997 *Teacher's Guide—AP® Calculus*, Dr. Kennedy has conducted more than 50 workshops and institutes for high school calculus teachers. His articles on mathematics teaching have appeared in the *Mathematics Teacher* and the *American Mathematical Monthly*, and he is a frequent speaker on education reform at professional and civic meetings. Dr. Kennedy was named a Tandy Technology Scholar in 1992 and a Presidential Award winner in 1995.

Dr. Kennedy coauthored *Calculus: Graphical, Numerical, Algebraic; Prentice Hall Algebra I; Prentice Hall Geometry;* and *Prentice Hall Algebra 2.*

David E. Bock

Dave Bock holds degrees from the University at Albany (NY) in mathematics (B.A.) and statistics/education (M.S.). Mr. Bock taught mathematics at Ithaca High School for 35 years, including both BC Calculus and AP Statistics. He also taught Statistics at Tompkins-Cortland Community College, Ithaca College, and Cornell University, where he recently served as K–12 Education and Outreach Coordinator and Senior Lecturer for the Mathematics Department. Mr. Bock serves as a Statistics consultant to the College Board, leading numerous workshops and institutes for AP Statistics teachers. He has been a reader for the AP Calculus exam and both a reader and a table leader for the AP Statistics exam. During his career Mr. Bock won numerous teaching awards, including the MAA's Edyth May Sliffe Award for Distinguished High School Mathematics Teaching (twice) and Cornell University's Outstanding Educator Award (three times), and was also a finalist for New York State Teacher of the Year.

Mr. Bock coauthored the AP Statistics textbook *Stats: Modeling the World*, the non-AP text *Stats in Your World*, Barron's *AP Calculus* review book, and Barron's *AP Calculus Flash Cards.*

PREFACE

Our Approach

The Rule of Four—A Balanced Approach

A principal feature of this text is the balance among the algebraic, numerical, graphical, and verbal methods of representing problems: the rule of four. For instance, we obtain solutions algebraically when that is the most appropriate technique to use, and we obtain solutions graphically or numerically when algebra is difficult to use. We urge students to solve problems by one method and then support or confirm their solutions by using another method. We believe that students must learn the value of each of these methods or representations and must learn to choose the one most appropriate for solving the particular problem under consideration. This approach reinforces the idea that to understand a problem fully, students need to understand it algebraically as well as graphically and numerically.

Problem-Solving Approach

Systematic problem solving is emphasized in the examples throughout the text, using the following variation of Polya's problem-solving process:

- *understand* the problem,
- *develop* a mathematical model,
- *solve* the mathematical model and support or confirm the solutions, and
- *interpret* the solution.

Students are encouraged to use this process throughout the text.

Twelve Basic Functions

Twelve basic functions are emphasized throughout the book as a major theme and focus. These functions are

- The Identity Function
- The Squaring Function
- The Cubing Function
- The Reciprocal Function
- The Square Root Function
- The Exponential Function
- The Natural Logarithm Function
- The Sine Function
- The Cosine Function
- The Absolute Value Function
- The Greatest Integer Function
- The Logistic Function

One of the most distinctive features of this textbook is that it introduces students to the full vocabulary of functions early in the course. Students meet the twelve basic functions graphically in Chapter 1 and are able to compare and contrast them as they learn about concepts like domain, range, symmetry, continuity, end behavior, asymptotes, extrema, and even periodicity—concepts that are difficult to appreciate when the only examples a teacher can refer to are polynomials. With this book, students are able to characterize functions by their behavior within the first month of classes. Once students have a comfortable understanding of functions in general, the rest of the course consists of studying the various types of functions in greater depth, particularly with respect to their algebraic properties and modeling applications.

These functions are used to develop the fundamental analytic skills that are needed in calculus and advanced mathematics courses. A complete gallery of basic functions is included in Appendix C and inside the back cover of the book for easy reference.

Applications and Real Data

The majority of the applications in the text are based on real data from cited sources, and their presentations are self-contained. As they work through the applications, students are exposed to functions as mechanisms for modeling real-life problems. They learn to analyze and model data, represent data graphically, interpret from graphs, and fit curves. Additionally, the tabular representation of data presented in this text highlights the concept that a function is a correspondence between numerical variables. This helps students build the connection between numerical quantities and graphs and recognize the importance of a full graphical, numerical, and algebraic understanding of a problem. For a complete listing of applications, please see the Applications Index on page 1003.

Technology and Exercises

The authors of this textbook have encouraged the use of technology in mathematics education for three decades. Our approach to problem solving (pages 70–71) distinguishes between **solving** the problem and **supporting** or **confirming** the solution, and how technology figures into each of those processes.

We have come to realize, however, that advances in technology and increased familiarity with calculators have gradually blurred some of the distinctions between solving and supporting that we had once assumed to be apparent. We do not want to retreat in any way from our support of modern technology, but we feel that the time has come to provide more guidance about the intent of the various exercises in our textbook.

Therefore, as a service to teachers and students alike, exercises in this textbook that **should be solved without calculators** are identified with gray ovals around the exercise numbers. These usually are exercises that demonstrate how various functions behave algebraically or how algebraic representations reflect graphical behavior and vice versa. Application problems usually have no restrictions, in keeping with our emphasis on **modeling** and on bringing **all representations** to bear when confronting real-world problems.

Incidentally, we continue to encourage the use of calculators to **support** answers graphically or numerically after the problems have been solved with pencil and paper. Any time students can make connections among the graphical, analytical, and numerical representations, they are doing good mathematics.

As a final note, we will freely admit that different teachers use our textbook in different ways, and some will probably override our no-calculator recommendations to fit with their pedagogical strategies. In the end, the teachers know what is best for their students, and we are just here to help.

Content Changes to This Edition

Mindful of the need to keep the applications of mathematics relevant to our students, we have changed many of the examples and exercises throughout the book to include the most current data available to us at the time of publication. Additionally, calculator screens were updated to conform to the enhanced capabilities of modern graphers. We have also added more student and teacher notes. The importance of statistics and probability for college, career, and everyday life has grown to the point that we now have a separate chapter titled Statistics and Probability. The Common Core Edition is built upon the prior editions of this textbook.

In **Chapter P** the use of the point-slope form of a line has been integrated into the solution of more examples. In **Chapter 1** references to calculator regression models were reworded to avoid giving the wrong signals about how statisticians actually operate. The discussion of linear correlation was revised in **Chapter 2** to complement this edition's more extensive treatment of Statistics. The section on financial mathematics in **Chapter 3** was updated, and simple interest was included. Additionally, a predator-prey application was added.

In **Chapter 6**, several significant textual changes have been made in order to tie the topics of this chapter (vectors, parametric equations, and polar graphing) more directly to the topics in the preceding chapters, particularly the unifying concepts of functions and their graphs in the Cartesian plane. The material on partial fractions has been incorporated into Section 7.3 to streamline **Chapter 7**.

Within **Chapter 8** the treatment of conic sections has been changed to emphasize that they are extensions of previously studied topics, even if their graphs do not pass the vertical line test. Explorations have been added to allow students to make the connections with earlier topics; for example, rotation of axes is introduced by prompting students to treat one of their twelve basic functions (the reciprocal function) as a hyperbola and find its vertices and foci.

The new **Chapter 10** expands the discussion of Statistics and probability. Section 10.1 opens the chapter with a discussion of basic probability concepts, including sample spaces, determining probabilities of compound events, Venn diagrams, tree diagrams, and conditional probability. Section 10.2 examines the creation and interpretation of graphical displays of data, including pie charts and bar charts of categorical data, stemplots and histograms for quantitative data, and time plots. Section 10.3 presents numerical summaries of center and spread for describing and comparing distributions, including the five-number summary, mean, and standard deviation, and introduces both boxplots and the Normal curve. Section 10.4 expands the discussion of probability to include random variables and probability models, including expected value, binomial probabilities, and Normal probabilities, and links these models to data and decision making by introducing the concept of statistical significance. Section 10.5 closes the chapter with a broad look at statistical literacy, the design of statistical studies, the important role of randomness, and the use of simulations to estimate probabilities and assess statistical significance. Throughout the chapter the emphasis is on proper statistical terminology and practice, attention to applications, and statistical thinking.

Features

Chapter Openers include a general description of an application that can be solved with the concepts learned in the chapter. The application is revisited later in the chapter via a specific problem that is solved.

A **Chapter Overview** begins each chapter to give students a sense of what they are going to learn. This overview provides a roadmap of the chapter, as well as tells how the topics in the chapter are connected under one big idea. It is always helpful to remember that mathematics isn't modular, but interconnected, and that the skills and concepts learned throughout the course build on one another to help students understand more complicated processes and relationships. Similarly, the **What you'll learn about . . . and why** feature presents the big ideas in each section and explains their purpose.

Throughout the book, **Vocabulary** is highlighted in yellow for easy reference. Additionally, **Properties**, **Definitions**, and **Theorems** are boxed in blue, and **Procedures** in purple, so that they can be easily found. The **Web/Real Data** icon marks the examples and exercises that use real cited data.

Each example ends with a suggestion to **Now Try** a related exercise. Working the suggested exercise is an easy way for students to check their comprehension of the material while reading each section. In the *Annotated Teacher's Edition*, various examples are marked for the teacher with the icon. Alternates are provided for these examples in the PowerPoint Slides.

Explorations appear throughout the text and provide students with the perfect opportunity to become active learners and to discover mathematics on their own. This will help hone critical-thinking and problem-solving skills. Some are technology-based and others involve exploring mathematical ideas and connections.

Margin Notes and Tips on various topics appear throughout the text. *Tips* offer practical advice on using the grapher to obtain the best, most accurate results. *Margin notes* include historical information and hints about examples, and provide additional insight to help students avoid common pitfalls and errors.

The **Looking Ahead to Calculus** Σ icon is found throughout the text next to many examples and topics to point out concepts that students will encounter again in calculus. Ideas that foreshadow calculus, such as limits, maximum and minimum, asymptotes, and continuity, are highlighted. Some calculus notation and language are introduced in the early chapters and used throughout the text to establish familiarity.

The **Chapter Review** material at the end of each chapter consists of sections dedicated to helping students review the chapter concepts. **Key Ideas** are broken into parts: Properties, Theorems, and Formulas; Procedures; and Gallery of Functions. The **Review Exercises** represent the full range of exercises covered in the chapter and give additional practice with the ideas developed in the chapter. The exercises with red numbers indicate problems that would make up a good chapter test. **Chapter Projects** conclude each chapter and require students to analyze data. They can be assigned as either individual or group work. Each project expands upon concepts and ideas taught in the chapter, and many projects refer to the Web for further investigation of real data.

Exercise Sets

Each exercise set begins with a **Quick Review** to help students review skills needed in the exercise set and references others sections students can go to for help. Some exercises are designed to be solved *without* a *calculator*; the numbers of these exercises are printed within a gray oval. Students are urged to **support** the answers to these (and all) exercises graphically or numerically, but only after they have solved them with pencil and paper.

There are over 6000 exercises, including 720 Quick Review Exercises. The section exercises have been carefully graded from routine to challenging. The following types of skills are tested in each exercise set:

- Algebraic and analytic manipulation
- Connecting algebra to geometry
- Interpretation of graphs
- Graphical and numerical representations of functions
- Data analysis

Also included in the exercise sets are thought-provoking exercises:

- **Standardized Test Questions** include two true-false problems with justifications and four multiple-choice questions.
- **Explorations** are opportunities for students to discover mathematics on their own or in groups. These exercises often require the use of critical thinking to explore the ideas.

- **Writing to Learn** exercises give students practice at communicating about mathematics and opportunities to demonstrate understanding of important ideas.
- **Group Activity** exercises ask students to work on the problems in groups or solve them as individual or group projects.
- **Extending the Ideas** exercises go beyond what is presented in the textbook. These exercises are challenging extensions of the book's material.

This variety of exercises provides sufficient flexibility to emphasize the skills most needed for each student or class.

Technology Resources

The following supplements are available for purchase:

MathXL® for School (optional, for purchase only)—access code required www.mathxlforschool.com

MathXL for School is a powerful online homework, tutorial, and assessment supplement that aligns to Pearson Education's textbooks in mathematics or statistics. MathXL for School is the homework and assessment engine that runs MyMathLab for School. With MathXL for School, teachers can do the following:

- Create, edit, and assign auto-graded online homework and tests correlated at the objective level to the textbook.
- Utilize automatic grading to rapidly assess students' understanding.
- Track both student and group performance in an online gradebook.
- Prepare students for high-stakes testing, including aligning assignments to state and Common Core State Standards, where available.
- Deliver quality, effective instruction regardless of experience level.

With MathXL for School, students can do the following:

- Complete their homework and receive immediate feedback.
- Get self-paced assistance on problems in a variety of ways (guided solutions, step-by-step examples, video clips, and animations).
- Choose from a large number of practice problems, helping them master a topic.
- Receive personalized study plans and homework based on test results.

For more information and to purchase student access codes after the first year, visit our Web site at www.mathxlforschool.com or contact your Pearson School Sales Representative.

MyMathLab® for School Online Course (optional, for purchase only)—access code required

MyMathLab for School delivers **proven results** in helping individual students succeed. It provides **engaging experiences** that personalize, stimulate, and measure learning for each student. And it comes from a **trusted partner** with educational expertise and an eye on the future. To learn more about how MyMathLab combines proven learning applications with powerful assessment, visit **www.mymathlabforschool.com** or contact your Pearson School Sales Representative. In this **MyMathLab® for School** course, you have access to the most cutting-edge, innovative study solutions proven to increase students' success.

Additional Teacher Resources

Most of the teacher supplements and resources available for this text are available electronically for download at the Instructor Resource Center (IRC). Please go to www.PearsonSchool.com/Access_Request *and select "we need IRC (Instructor Resource Access)." You will be required to complete a one-time registration subject to verification before being emailed access information for download materials. Once logged into the IRC, enter the Student Edition ISBN in the* ***Search our Catalog*** *box to locate your resources.*

The following supplements are available to qualified adopters:

Annotated Teacher's Edition

- Provides answers in the margins next to the corresponding problem for almost all exercises, including sample answers for writing exercises.
- Various examples marked with the icon indicate that alternative examples are provided in the PowerPoint Slides.
- Provides notes written specifically for the teacher. These notes include chapter and section objectives, suggested assignments, lesson guides, and teaching tips.

Online Solutions Manual (Download Only)

Provides complete solutions to all exercises, including Quick Reviews, Exercises, Explorations, and Chapter Reviews.

Online Resource Manual (Download Only)

Provides Major Concepts Review, Group Activity Worksheets, Sample Chapter Tests, Standardized Test Preparation Questions, Contest Problems.

Online Tests and Quizzes (Download Only)

Provides two parallel tests per chapter, two quizzes for every three to four sections, two parallel midterm tests covering Chapters P–5, and two parallel end-of-year tests, covering Chapters 6–10.

TestGen® (Download Only)

TestGen enables teachers to build, edit, print, and administer tests using a computerized bank of questions developed to cover all the objectives of the text. TestGen is algorithmically based, allowing teachers to create multiple but equivalent versions of the same question or test with the click of a button. Teachers can also modify test bank questions or add new questions. Tests can be printed or administered online.

PowerPoint Slides (Download Only)

Features presentations written and designed specifically for this text, including figures, alternate examples, definitions, and key concepts.

Web Site

Our Web site, www.pearsonhighered.com/demana, provides dynamic resources for teachers and students. Some of the resources include TI graphing calculator downloads, online quizzing, teaching tips, study tips, Explorations, end-of-chapter projects, and more.

Common Core: Student Practice and Review Guide (Download Only)

This resource provides complete daily support for every lesson. The following resources are included: Problem Solving, Practice and Standardized Test Prep.

Common Core: Implementation Guide (Download Only)

All the support a teacher needs to make the transition to a Common Core curriculum. Includes:

- Overview of the Common Core State Standards
- Standards for Mathematical Practice Observational Protocol
- Common Core Correlations
- Common Core assessment resources

ACKNOWLEDGMENTS

We wish to express our gratitude to the reviewers of this and previous editions who provided such invaluable insight and comment.

Judy Ackerman
Montgomery College

Ignacio Alarcon
Santa Barbara City College

Ray Barton
Olympus High School

Nicholas G. Belloit
Florida Community College at Jacksonville

Margaret A. Blumberg
University of Southwestern Louisiana

Ray Cannon
Baylor University

Marilyn P. Carlson
Arizona State University

Edward Champy
Northern Essex Community College

Janis M. Cimperman
Saint Cloud State University

Wil Clarke
La Sierra University

Marilyn Cobb
Lake Travis High School

Donna Costello
Plano Senior High School

Gerry Cox
Lake Michigan College

Deborah A. Crocker
Appalachian State University

Marian J. Ellison
University of Wisconsin—Stout

Donna H. Foss
University of Central Arkansas

Betty Givan
Eastern Kentucky University

Brian Gray
Howard Community College

Daniel Harned
Michigan State University

Vahack Haroutunian
Fresno City College

Celeste Hernandez
Richland College

Rich Hoelter
Raritan Valley Community College

Dwight H. Horan
Wentworth Institute of Technology

Margaret Hovde
Grossmont College

Miles Hubbard
Saint Cloud State University

Sally Jackman
Richland College

T. J. Johnson
Hendrickson High School

Stephen C. King
University of South Carolina—Aiken

Jeanne Kirk
William Howard Taft High School

Georgianna Klein
Grand Valley State University

Fred Koenig
Walnut Ridge High School

Deborah L. Kruschwitz-List
University of Wisconsin—Stout

Carlton A. Lane
Hillsborough Community College

James Larson
Lake Michigan University

Edward D. Laughbaum
Columbus State Community College

Ron Marshall
Western Carolina University

Janet Martin
Lubbock High School

Beverly K. Michael
University of Pittsburgh

Paul Mlakar
St. Mark's School of Texas

John W. Petro
Western Michigan University

Cynthia M. Piez
University of Idaho

Debra Poese
Montgomery College

Jack Porter
University of Kansas

Antonio R. Quesada
The University of Akron

Hilary Risser
Plano West Senior High

Thomas H. Rousseau
Siena College

David K. Ruch
Sam Houston State University

Sid Saks
Cuyahoga Community College

Mary Margaret Shoaf-Grubbs
College of New Rochelle

Malcolm Soule
California State University, Northridge

Sandy Spears
Jefferson Community College

Shirley R. Stavros
Saint Cloud State University

Stuart Thomas
University of Oregon

Janina Udrys
Schoolcraft College

Mary Voxman
University of Idaho

Eddie Warren
University of Texas at Arlington

Steven J. Wilson
Johnson County Community College

Gordon Woodward
University of Nebraska

Cathleen Zucco-Teveloff
Trinity College

Consultants

We would like to extend a special thank you to the following consultants for their guidance and invaluable insight in the development of recent editions.

Jane Nordquist
Ida S. Baker High School, Florida

Sudeepa Pathak
Williamston High School, North Carolina

Laura Reddington
Forest Hill High School, Florida

James Timmons
Heide Trask High School, North Carolina

Jill Weitz
The G-Star School of the Arts, Florida

We express our gratitude to Chris Brueningsen, Linda Antinone, and Bill Bower for their work on the Chapter Projects. We greatly appreciate Jennifer Blue, Nathan Kidwell, Brianna Kurtz, and James Lapp for their meticulous accuracy checking and Lisa Collette for her careful proofreading. We are grateful to Cenveo, who pulled off an amazing job on composition, and wish to offer special thanks to project manager John Orr, who kept us on track throughout the project. Our thanks as well are extended to the professional and remarkable staff at Pearson. We wish to thank our families for their support, patience, and understanding throughout the process. We dedicate this edition to them!

—F. D. D.
—B. K. W.
—G. D. F.
—D. K.
—D. E. B.

CHAPTER **P**

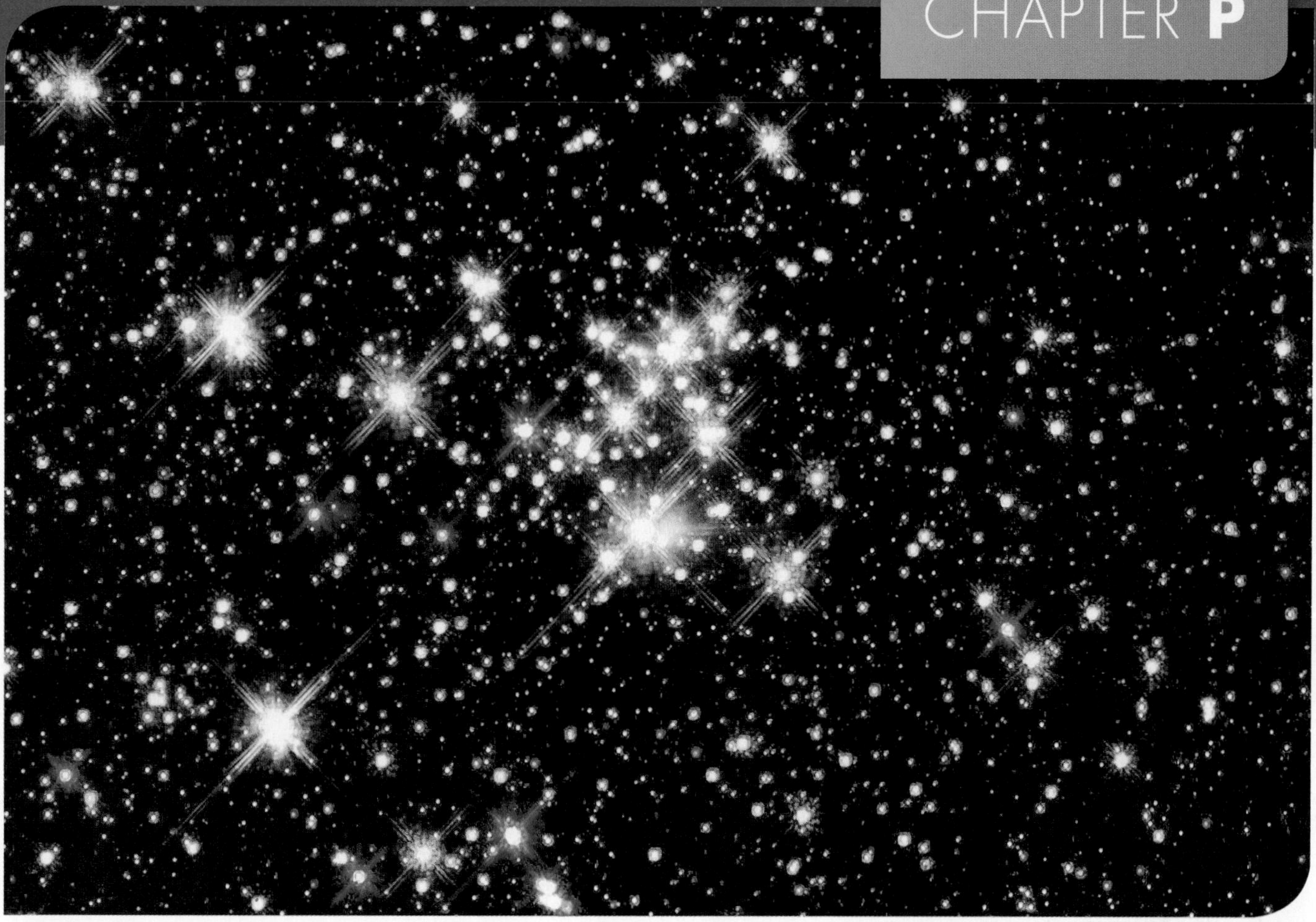

P.1 Real Numbers

P.2 Cartesian Coordinate System

P.3 Linear Equations and Inequalities

P.4 Lines in the Plane

P.5 Solving Equations Graphically, Numerically, and Algebraically

P.6 Complex Numbers

P.7 Solving Inequalities Algebraically and Graphically

Prerequisites

Large distances are measured in *light years*, the distance that light travels in one year. Scientists use the speed of light, which is roughly 299,800 km/sec, to approximate distances within the solar system. See page 35 for examples.

Chapter P Overview

Historically, algebra was used to represent problems with symbols (algebraic models) and solve them by reducing the solution to algebraic manipulation of symbols. This technique is still important today. In addition, graphing calculators are now used to represent problems with graphs (graphical models) and solve them with the numerical and graphical techniques of technology.

We begin with basic properties of real numbers and introduce absolute value, distance formulas, midpoint formulas, and equations of circles. Slope of a line is used to write standard equations for lines, and applications involving linear equations are discussed. The basics of complex numbers are explained. Equations and inequalities are solved using both algebraic and graphical techniques.

P.1 Real Numbers

What you'll learn about

- Representing Real Numbers
- Order and Interval Notation
- Basic Properties of Algebra
- Integer Exponents
- Scientific Notation

... and why

These topics are fundamental in the study of mathematics and science.

Representing Real Numbers

A **real number** is any number that can be written as a decimal. Real numbers are represented by symbols such as $-8, 0, 1.75, 2.333...,\ 0.\overline{36},\ 8/5,\ \sqrt{3},\ \sqrt[3]{16},\ e$, and π.

The set of real numbers contains several important subsets:

The **natural (or counting) numbers**:	$\{1, 2, 3, \ldots\}$
The **whole numbers**:	$\{0, 1, 2, 3, \ldots\}$
The **integers**:	$\{\ldots, -3, -2, -1, 0, 1, 2, 3, \ldots\}$

Braces { } are used to enclose the **elements**, or **objects**, of a set. The rational numbers are another important subset of the real numbers. A **rational number** is any number that can be written as a ratio a/b of two integers, where $b \neq 0$. We can use **set-builder notation** to define the rational numbers:

$$\left\{\frac{a}{b}\,\middle|\, a, b \text{ are integers, and } b \neq 0\right\}$$

The vertical bar that follows a/b is read "such that."

The decimal form of a rational number either **terminates** like $7/4 = 1.75$, or is **infinitely repeating** like $4/11 = 0.363636... = 0.\overline{36}$. The bar over the 36 indicates the block of digits that repeats. A real number is **irrational** if it is *not* rational. The decimal form of an irrational number is infinitely nonrepeating. For example, $\sqrt{3} = 1.7320508...$ and $\pi = 3.14159265....$

A real number can be approximated by giving a few of its digits. Sometimes we can find the decimal form of rational numbers with calculators, but not very often.

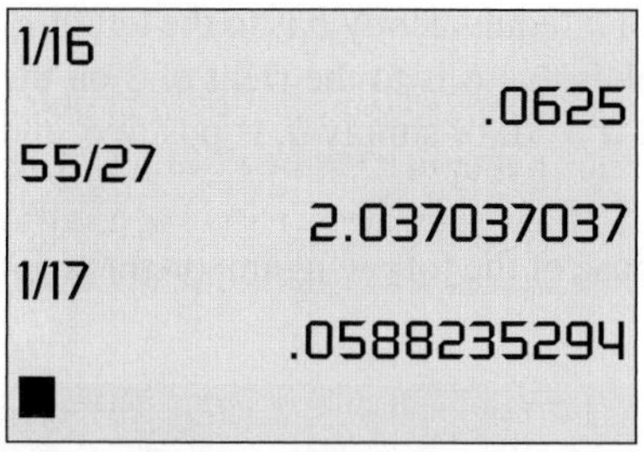

FIGURE P.1 Calculator decimal representations of 1/16, 55/27, and 1/17 with the calculator set in floating decimal mode. (Example 1)

EXAMPLE 1 Examining Decimal Forms of Rational Numbers

Determine the decimal form of 1/16, 55/27, and 1/17.

SOLUTION Figure P.1 suggests that the decimal form of 1/16 terminates and that of 55/27 repeats in blocks of 037.

$$\frac{1}{16} = 0.0625 \quad \text{and} \quad \frac{55}{27} = 2.\overline{037}$$

We cannot predict the *exact* decimal form of 1/17 from Figure P.1; however, we can say that $1/17 \approx 0.0588235294$. The symbol $\approx$ is read "*is approximately equal to*." We can use long division (see Exercise 66) to prove that

$$\frac{1}{17} = 0.\overline{0588235294117647}.$$

Now try Exercise 3.

The real numbers and the points of a line can be matched *one-to-one* to form a **real number line**. We start with a horizontal line and match the real number zero with a point O, the **origin**. **Positive numbers** are assigned to the right of the origin, and **negative numbers** to the left, as shown in Figure P.2.

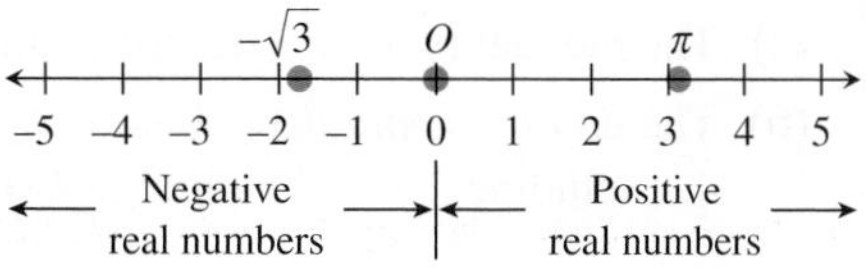

FIGURE P.2 The real number line.

Every real number corresponds to one and only one point on the real number line, and every point on the real number line corresponds to one and only one real number. Between every pair of real numbers on the number line there are infinitely many more real numbers.

The number associated with a point is **the coordinate of the point**. As long as the context is clear, we will follow the standard convention of using the real number for both the name of the point and its coordinate.

Order and Interval Notation

The set of real numbers is **ordered**. This means that we can use inequalities to compare any two real numbers that are not equal and say that one is "less than" or "greater than" the other.

Unordered Systems

Not all number systems are ordered. For example, the complex number system, to be introduced in Section P.6, has no natural ordering.

Order of Real Numbers

Let a and b be any two real numbers.

Symbol	Definition	Read
$a > b$	$a - b$ is positive	a is greater than b
$a < b$	$a - b$ is negative	a is less than b
$a \geq b$	$a - b$ is positive or zero	a is greater than or equal to b
$a \leq b$	$a - b$ is negative or zero	a is less than or equal to b

The symbols $>$, $<$, $\geq$, and $\leq$ are **inequality symbols**.

Opposites and Number Line

$a < 0 \Leftrightarrow -a > 0$

If $a < 0$, then a is to the left of 0 on the real number line, and its opposite, $-a$, is to the right of 0. Thus, $-a > 0$. This logic can be reversed: If $-a > 0$, then $a < 0$.

Geometrically, $a > b$ means that a is to the right of b (equivalently b is to the left of a) on the real number line. For example, $6 > 3$ implies that 6 is to the right of 3 on the real number line. Note also that $a > 0$ means that $a - 0$, or simply a, is positive, and $a < 0$ means that a is negative.

We are able to compare any two real numbers because of the following important property of the real numbers.

Trichotomy Property

Let a and b be any two real numbers. Exactly one of the following is true:

$$a < b, \quad a = b, \quad \text{or} \quad a > b$$

Inequalities can be used to describe **intervals** of real numbers, as illustrated in Example 2.

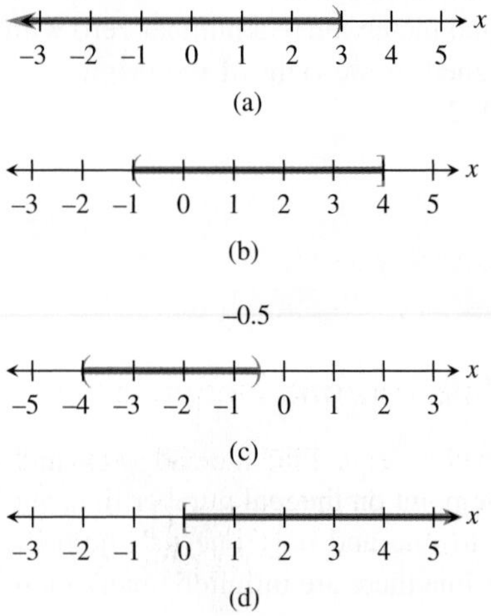

FIGURE P.3 In graphs of inequalities, parentheses correspond to $<$ and $>$ and brackets to $\le$ and $\ge$. (Examples 2 and 3)

EXAMPLE 2 Interpreting Inequalities

Describe and graph the interval of real numbers for the inequality.

(a) $x < 3$ **(b)** $-1 < x \le 4$

SOLUTION

(a) The inequality $x < 3$ describes all real numbers less than 3 (Figure P.3a).

(b) The *double inequality* $-1 < x \le 4$ represents all real numbers between -1 and 4, excluding -1 and including 4 (Figure P.3b). **Now try Exercise 5.**

EXAMPLE 3 Writing Inequalities

Write an interval of real numbers using an inequality and draw its graph.

(a) The real numbers between -4 and -0.5

(b) The real numbers greater than or equal to zero

SOLUTION

(a) $-4 < x < -0.5$ (Figure P.3c)

(b) $x \ge 0$ (Figure P.3d) **Now try Exercise 15.**

As shown in Example 2, inequalities define *intervals* on the real number line. We often use $[2, 5]$ to describe the *bounded interval* determined by $2 \le x \le 5$. This interval is **closed** because it contains its *endpoints* 2 and 5. There are four types of **bounded intervals**.

Bounded Intervals of Real Numbers

Let a and b be real numbers with $a < b$.

Interval Notation	Interval Type	Inequality Notation	Graph
$[a, b]$	Closed	$a \le x \le b$	
(a, b)	Open	$a < x < b$	
$[a, b)$	Half-open	$a \le x < b$	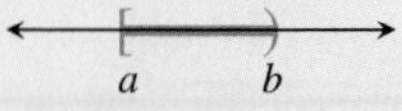
$(a, b]$	Half-open	$a < x \le b$	

The numbers a and b are the **endpoints** of each interval.

We use the interval notation $(-\infty, \infty)$ to represent the entire set of real numbers. The symbols $-\infty$ (*negative infinity*) and ∞ (*positive infinity*) allow us to use interval notation for unbounded intervals and are *not* real numbers.

Interval Notation at $\pm\infty$

Because $-\infty$ is *not* a real number, we use $(-\infty, 2)$ instead of $[-\infty, 2)$ to describe $x < 2$. Similarly, we use $[-1, \infty)$ instead of $[-1, \infty]$ to describe $x \geq -1$.

The interval of real numbers determined by the inequality $x < 2$ can be described by the *unbounded interval* $(-\infty, 2)$. This interval is **open** because it does *not* contain its endpoint 2. In addition to $(-\infty, \infty)$, there are four types of **unbounded intervals**.

Unbounded Intervals of Real Numbers

Let a and b be real numbers.

Interval Notation	Interval Type	Inequality Notation	Graph
$[a, \infty)$	Closed	$x \geq a$	
(a, ∞)	Open	$x > a$	
$(-\infty, b]$	Closed	$x \leq b$	
$(-\infty, b)$	Open	$x < b$	

Each of these intervals has exactly one endpoint, namely a or b.

EXAMPLE 4 Converting Between Intervals and Inequalities

Convert interval notation to inequality notation or vice versa. Find the endpoints and state whether the interval is bounded, its type, and graph the interval.

(a) $[-6, 3)$ **(b)** $(-\infty, -1)$ **(c)** $-2 \leq x \leq 3$

SOLUTION

(a) The interval $[-6, 3)$ corresponds to $-6 \leq x < 3$ and is bounded and half-open (Figure P.4a). The endpoints are -6 and 3.

(b) The interval $(-\infty, -1)$ corresponds to $x < -1$ and is unbounded and open (Figure P.4b). The only endpoint is -1.

(c) The inequality $-2 \leq x \leq 3$ corresponds to the closed, bounded interval $[-2, 3]$ (Figure P.4c). The endpoints are -2 and 3. Now try Exercise 29.

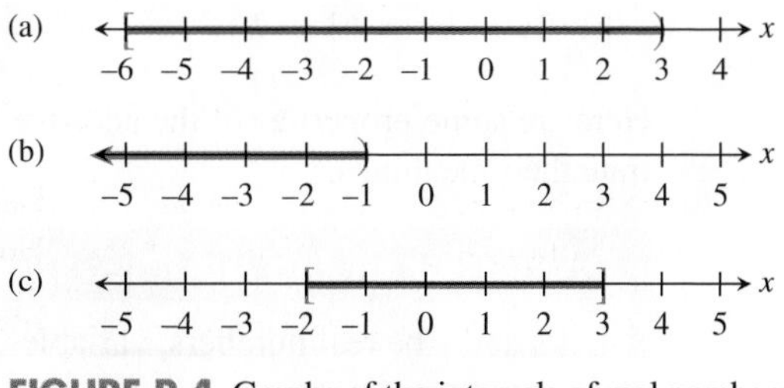

FIGURE P.4 Graphs of the intervals of real numbers in Example 4.

Basic Properties of Algebra

Algebra involves the use of letters and other symbols to represent real numbers. A **variable** is a letter or symbol (for example, x, y, t, θ) that represents an unspecified real number. A **constant** is a letter or symbol (for example, -2, 0, $\sqrt{3}$, π) that represents a specific real number. An **algebraic expression** is a combination of variables and constants involving addition, subtraction, multiplication, division, powers, and roots.

Subtraction vs. Negative Numbers

On many calculators, there are two "−" keys, one for subtraction and one for negative numbers or opposites. Be sure you know how to use both keys correctly. Misuse can lead to incorrect results.

We state some of the properties of the arithmetic operations of addition, subtraction, multiplication, and division, represented by the symbols $+, -, \times$ (or $\cdot$) and $\div$ (or /), respectively. Addition and multiplication are the primary operations. Subtraction and division are defined in terms of addition and multiplication.

Subtraction: $a - b = a + (-b)$

Division: $\frac{a}{b} = a\left(\frac{1}{b}\right), b \neq 0$

In the above definitions, $-b$ is the **additive inverse** or **opposite** of b, and $1/b$ is the **multiplicative inverse** or **reciprocal** of b. Perhaps surprisingly, additive inverses are not always negative numbers. The additive inverse of 5 is the negative number -5. However, the additive inverse of -3 is the positive number 3.

The following properties hold for real numbers, variables, and algebraic expressions.

Properties of Algebra

Let u, v, and w be real numbers, variables, or algebraic expressions.

1. Commutative properties
Addition: $u + v = v + u$
Multiplication: $uv = vu$

2. Associative properties
Addition:
$(u + v) + w = u + (v + w)$
Multiplication: $(uv)w = u(vw)$

3. Identity properties
Addition: $u + 0 = u$
Multiplication: $u \cdot 1 = u$

4. Inverse properties
Addition: $u + (-u) = 0$
Multiplication: $u \cdot \frac{1}{u} = 1, u \neq 0$

5. Distributive properties
Multiplication over addition:
$u(v + w) = uv + uw$
$(u + v)w = uw + vw$

Multiplication over subtraction:
$u(v - w) = uv - uw$
$(u - v)w = uw - vw$

The left-hand sides of the equations for the distributive property show the **factored form** of the algebraic expressions, and the right-hand sides show the **expanded form**.

EXAMPLE 5 Using the Distributive Property

(a) Write the expanded form of $(a + 2)x$.
(b) Write the factored form of $3y - by$.

SOLUTION

(a) $(a + 2)x = ax + 2x$
(b) $3y - by = (3 - b)y$

Now try Exercise 37.

Here are some properties of the additive inverse together with examples that help illustrate their meanings.

Properties of the Additive Inverse

Let u and v be real numbers, variables, or algebraic expressions.

Property	Example
1. $-(-u) = u$	$-(-3) = 3$
2. $(-u)v = u(-v) = -(uv)$	$(-4)3 = 4(-3) = -(4 \cdot 3) = -12$
3. $(-u)(-v) = uv$	$(-6)(-7) = 6 \cdot 7 = 42$
4. $(-1)u = -u$	$(-1)5 = -5$
5. $-(u + v) = (-u) + (-v)$	$-(7 + 9) = (-7) + (-9) = -16$

Integer Exponents

Exponential notation is used to shorten products of factors that repeat. For example,

$$(-3)(-3)(-3)(-3) = (-3)^4 \quad \text{and} \quad (2x + 1)(2x + 1) = (2x + 1)^2.$$

Exponential Notation

Let a be a real number, variable, or algebraic expression and n a positive integer. Then

$$a^n = \underbrace{a \cdot a \cdot \cdots \cdot a}_{n \text{ factors}},$$

where n is the **exponent**, a is the **base**, and a^n is the ***n*th power of *a***, read as "a to the nth power."

The two exponential expressions in Example 6 have the same value but have different bases. Be sure you understand the difference.

Understanding Notation

$(-3)^2 = 9$

$-3^2 = -9$

Be careful!

EXAMPLE 6 Identifying the Base

(a) In $(-3)^5$, the base is -3.

(b) In -3^5, the base is 3.

Now try Exercise 43.

Here are the basic properties of exponents together with examples that help illustrate their meanings.

Properties of Exponents

Let u and v be real numbers, variables, or algebraic expressions and m and n be integers. All bases are assumed to be nonzero.

Property	Example
1. $u^m u^n = u^{m+n}$	$5^3 \cdot 5^4 = 5^{3+4} = 5^7$
2. $\dfrac{u^m}{u^n} = u^{m-n}$	$\dfrac{x^9}{x^4} = x^{9-4} = x^5$
3. $u^0 = 1$	$8^0 = 1$
4. $u^{-n} = \dfrac{1}{u^n}$	$y^{-3} = \dfrac{1}{y^3}$
5. $(uv)^m = u^m v^m$	$(2z)^5 = 2^5 z^5 = 32z^5$
6. $(u^m)^n = u^{mn}$	$(x^2)^3 = x^{2 \cdot 3} = x^6$
7. $\left(\dfrac{u}{v}\right)^m = \dfrac{u^m}{v^m}$	$\left(\dfrac{a}{b}\right)^7 = \dfrac{a^7}{b^7}$

To simplify an expression involving powers means to rewrite it so that each factor appears only once, all exponents are positive, and exponents and constants are combined as much as possible.

Moving Factors

Be sure you understand how exponent property 4 permits us to move factors from the numerator to the denominator and vice versa:

$$\frac{v^{-m}}{u^{-n}} = \frac{u^n}{v^m}$$

EXAMPLE 7 Simplifying Expressions Involving Powers

(a) $(2ab^3)(5a^2b^5) = 10(aa^2)(b^3b^5) = 10a^3b^8$

(b) $\dfrac{u^2v^{-2}}{u^{-1}v^3} = \dfrac{u^2u^1}{v^2v^3} = \dfrac{u^3}{v^5}$

(c) $\left(\dfrac{x^2}{2}\right)^{-3} = \left(\dfrac{2}{x^2}\right)^3 = \dfrac{2^3}{(x^2)^3} = \dfrac{8}{x^6}$

Now try Exercise 47.

Scientific Notation

Any positive number can be written in **scientific notation**,

$$c \times 10^m, \text{ where } 1 \le c < 10 \text{ and } m \text{ is an integer.}$$

This notation provides a way to work with very large and very small numbers. For example, the distance between the Earth and the Sun is about 93,000,000 miles. In scientific notation,

$$93{,}000{,}000 \text{ mi} = 9.3 \times 10^7 \text{ mi.}$$

The *positive exponent* 7 indicates that moving the decimal point in 9.3 to the right 7 places produces the decimal form of the number.

The mass of an oxygen molecule is about

$$0.000\,000\,000\,000\,000\,000\,000\,054 \text{ g.}$$

In scientific notation,

$$0.000\,000\,000\,000\,000\,000\,000\,054 \text{ g} = 5.4 \times 10^{-23} \text{ g.}$$

The *negative exponent* -23 indicates that moving the decimal point in 5.4 to the left 23 places produces the decimal form of the number.

EXAMPLE 8 Converting to and from Scientific Notation

(a) $2.375 \times 10^8 = 237{,}500{,}000$

(b) $0.000000349 = 3.49 \times 10^{-7}$

Now try Exercises 57 and 59.

EXAMPLE 9 Using Scientific Notation

Simplify $\dfrac{(360{,}000)(4{,}500{,}000{,}000)}{18{,}000}$.

SOLUTION

$$\begin{aligned}\frac{(360{,}000)(4{,}500{,}000{,}000)}{18{,}000} &= \frac{(3.6 \times 10^5)(4.5 \times 10^9)}{1.8 \times 10^4}\\ &= \frac{(3.6)(4.5)}{1.8} \times 10^{5+9-4}\\ &= 9 \times 10^{10}\\ &= 90{,}000{,}000{,}000\end{aligned}$$

Now try Exercise 63.

Using a Calculator Figure P.5 shows two ways to perform the computation. In the first, the numbers are entered in decimal form. In the second, the numbers are entered in scientific notation. The calculator uses "9E10" to stand for 9×10^{10}.

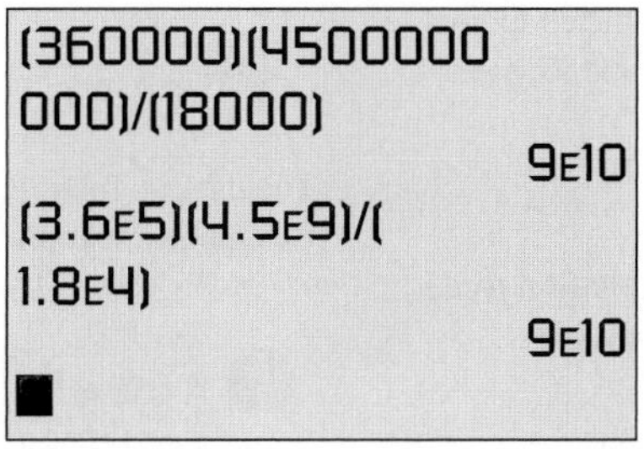

FIGURE P.5 Be sure you understand how your calculator displays scientific notation. (Example 9)

QUICK REVIEW P.1

1. List the positive integers between -3 and 7.

2. List the integers between -3 and 7.

3. List all negative integers greater than -4.

4. List all positive integers less than 5.

In Exercises 5 and 6, use a calculator to evaluate the expression. Round the value to two decimal places.

5. (a) $4(-3.1)^3 - (-4.2)^5$ **(b)** $\dfrac{2(-5.5) - 6}{7.4 - 3.8}$

6. (a) $5[3(-1.1)^2 - 4(-0.5)^3]$ **(b)** $5^{-2} + 2^{-4}$

In Exercises 7 and 8, evaluate the algebraic expression for the given values of the variables.

7. $x^3 - 2x + 1, x = -2, 1.5$

8. $a^2 + ab + b^2, a = -3, b = 2$

In Exercises 9 and 10, list the possible remainders.

9. When the positive integer n is divided by 7

10. When the positive integer n is divided by 13

SECTION P.1 Exercises

Exercise numbers with a gray background indicate problems that the authors have designed to be solved *without a calculator*.

In Exercises 1–4, find the decimal form for the rational number. State whether it repeats or terminates.

1. $-37/8$ **2.** $15/99$

3. $-13/6$ **4.** $5/37$

In Exercises 5–10, describe and graph the interval of real numbers.

5. $x \leq 2$ **6.** $-2 \leq x < 5$

7. $(-\infty, 7)$ **8.** $[-3, 3]$

9. x is negative.

10. x is greater than or equal to 2 and less than or equal to 6.

In Exercises 11–16, use an inequality to describe the interval of real numbers.

11. $[-1, 1)$ **12.** $(-\infty, 4]$

13.

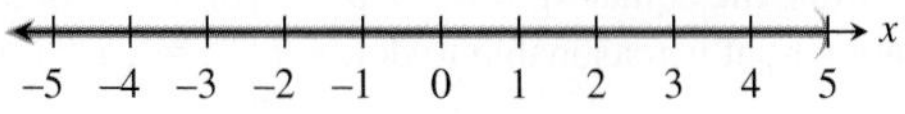

14.

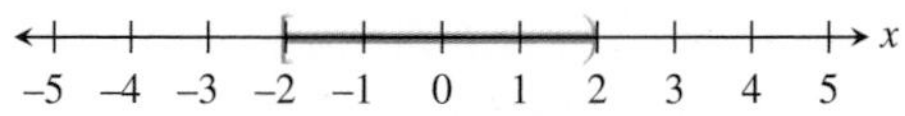

15. x is between -1 and 2.

16. x is greater than or equal to 5.

In Exercises 17–22, use interval notation to describe the interval of real numbers.

17. $x > -3$

18. $-7 < x < -2$

19.

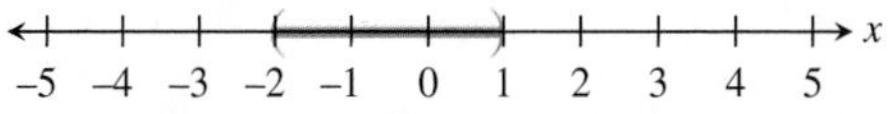

20.

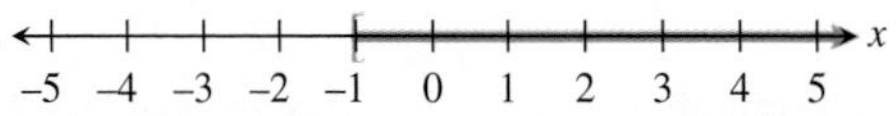

21. x is greater than -3 and less than or equal to 4.

22. x is positive.

In Exercises 23–28, use words to describe the interval of real numbers.

23. $4 < x \leq 9$ **24.** $x \geq -1$

25. $[-3, \infty)$ **26.** $(-5, 7)$

27.

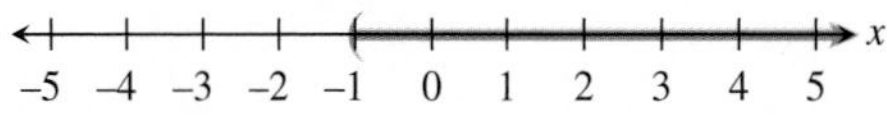

28.

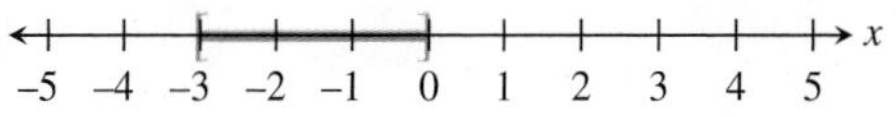

In Exercises 29–32, convert to inequality notation. Find the endpoints and state whether the interval is bounded or unbounded and its type.

29. $(-3, 4]$ **30.** $(-3, -1)$

31. $(-\infty, 5)$ **32.** $[-6, \infty)$

In Exercises 33–36, use both inequality and interval notation to describe the set of numbers. State the meaning of any variables you use.

33. Writing to Learn Bill is at least 29 years old.

34. Writing to Learn No item at Sarah's Variety Store costs more than $2.00.

35. Writing to Learn The price of a gallon of gasoline varies from $3.099 to $4.399.

36. Writing to Learn Salary raises at California State University at Chico will be between 2% and 6.5% this year.

In Exercises 37–40, use the distributive property to write the factored form or the expanded form of the given expression.

37. $a(x^2 + b)$ **38.** $(y - z^3)c$

39. $ax^2 + dx^2$ **40.** $a^3z + a^3w$

In Exercises 41 and 42, find the additive inverse of the number.

41. $6 - \pi$ **42.** -7

In Exercises 43 and 44, identify the base of the exponential expression.

43. -5^2 **44.** $(-2)^7$

45. Group Activity Discuss which algebraic property or properties are illustrated by the equation. Try to reach a consensus.

(a) $(3x)y = 3(xy)$ **(b)** $a^2b = ba^2$

(c) $a^2b + (-a^2b) = 0$ **(d)** $(x + 3)^2 + 0 = (x + 3)^2$

(e) $a(x + y) = ax + ay$

46. Group Activity Discuss which algebraic property or properties are illustrated by the equation. Try to reach a consensus.

(a) $(x + 2)\dfrac{1}{x + 2} = 1$ **(b)** $1 \cdot (x + y) = x + y$

(c) $2(x - y) = 2x - 2y$

(d) $2x + (y - z) = 2x + (y + (-z))$
$= (2x + y) + (-z) =$
$(2x + y) - z$

(e) $\dfrac{1}{a}(ab) = \left(\dfrac{1}{a}a\right)b = 1 \cdot b = b$

In Exercises 47–52, simplify the expression. Assume that the variables in the denominators are nonzero.

47. $\dfrac{x^4y^3}{x^2y^5}$ **48.** $\dfrac{(3x^2)^2y^4}{3y^2}$

49. $\left(\dfrac{4}{x^2}\right)^2$ **50.** $\left(\dfrac{2}{xy}\right)^{-3}$

51. $\dfrac{(x^{-3}y^2)^{-4}}{(y^6x^{-4})^{-2}}$ **52.** $\left(\dfrac{4a^3b}{a^2b^3}\right)\left(\dfrac{3b^2}{2a^2b^4}\right)$

The data in Table P.1 give the expenditures in millions of dollars for U.S. public schools for the 2008–2009 school year.

Table P.1 U.S. Public School Expenditures

Category	Amount (millions of $)
Current expenditures	518,997
Capital outlay	65,882
Interest on school debt	16,691
Total	610,110

Source: National Center for Education Statistics, U.S. Department of Education, as reported in The World Almanac and Book of Facts 2012.

In Exercises 53–56, write the amount of expenditures in dollars obtained from the category in scientific notation.

53. Current expenditures

54. Capital outlay

55. Interest on school debt

56. Total

In Exercises 57 and 58, write the number in scientific notation.

57. The mean distance from Jupiter to the Sun is about 483,900,000 miles.

58. The electric charge, in coulombs, of an electron is about −0.000 000 000 000 000 000 16.

In Exercises 59–62, write the number in decimal form.

59. 3.33×10^{-8} **60.** 6.73×10^{11}

61. The distance that light travels in 1 year (*one light year*) is about 5.87×10^{12} mi.

62. The mass of a neutron is about 1.6747×10^{-24} g.

In Exercises 63 and 64, use scientific notation to simplify.

63. $\dfrac{(1.3 \times 10^{-7})(2.4 \times 10^8)}{1.3 \times 10^9}$ without using a calculator

64. $\dfrac{(3.7 \times 10^{-7})(4.3 \times 10^6)}{2.5 \times 10^7}$

Explorations

65. Investigating Exponents For positive integers m and n, we can use the definition to show that $a^m a^n = a^{m+n}$.

(a) Examine the equation $a^m a^n = a^{m+n}$ for $n = 0$ and explain why it is reasonable to define $a^0 = 1$ for $a \neq 0$.

(b) Examine the equation $a^m a^n = a^{m+n}$ for $n = -m$ and explain why it is reasonable to define $a^{-m} = 1/a^m$ for $a \neq 0$.

66. **Decimal Forms of Rational Numbers** Here is the third step when we divide 1 by 17. (The first two steps are not shown, because the quotient is 0 in both cases.)

$$\begin{array}{r} 0.05 \\ 17\overline{)1.00} \\ \underline{85} \\ 15 \end{array}$$

By convention we say that 1 is the first remainder in the long division process, 10 is the second, and 15 is the third remainder.

(a) Continue this long division process until a remainder is repeated, and complete the following table:

Step	Quotient	Remainder
1	0	1
2	0	10
3	5	15
⋮	⋮	⋮

(b) Explain why the digits that occur in the quotient between the pair of repeating remainders determine the infinitely repeating portion of the decimal representation. In this case

$$\frac{1}{17} = 0.\overline{0588235294117647}.$$

(c) Explain why this procedure will always determine the infinitely repeating portion of a rational number whose decimal representation does not terminate.

Standardized Test Questions

67. **True or False** The additive inverse of a real number must be negative. Justify your answer.

68. **True or False** The reciprocal of a positive real number must be less than 1. Justify your answer.

In Exercises 69–72, solve these problems without using a calculator.

69. **Multiple Choice** Which of the following inequalities corresponds to the interval $[-2, 1)$?

(A) $x \leq -2$ **(B)** $-2 \leq x \leq 1$

(C) $-2 < x < 1$ **(D)** $-2 < x \leq 1$

(E) $-2 \leq x < 1$

70. **Multiple Choice** What is the value of $(-2)^4$?

(A) 16 **(B)** 8

(C) 6 **(D)** −8

(E) −16

71. **Multiple Choice** What is the base of the exponential expression -7^2?

(A) −7 **(B)** 7

(C) −2 **(D)** 2

(E) 1

72. **Multiple Choice** Which of the following is the simplified form of $\frac{x^6}{x^2}$, $x \neq 0$?

(A) x^{-4} **(B)** x^2

(C) x^3 **(D)** x^4

(E) x^8

Extending the Ideas

The **magnitude** of a real number is its distance from the origin.

73. List the whole numbers whose magnitudes are less than 7.

74. List the natural numbers whose magnitudes are less than 7.

75. List the integers whose magnitudes are less than 7.

76. **Writing to Learn Combining Rational and Irrational Numbers** In each case, write an explanation to justify your answer.

(a) When two rational numbers are added, is the sum a rational number?

(b) When two rational numbers are multiplied, is the product a rational number?

(c) When a rational number and an irrational number are added, is the sum a rational number?

(d) When a *nonzero* rational number and an irrational number are multiplied, is the product a rational number?

P.2 Cartesian Coordinate System

What you'll learn about

- Cartesian Plane
- Absolute Value of a Real Number
- Distance Formulas
- Midpoint Formulas
- Equations of Circles
- Applications

... and why

These topics provide the foundation for the material that will be addressed in this textbook.

Cartesian Plane

The points in a plane correspond to ordered pairs of real numbers, just as the points on a line are associated with individual real numbers. This correspondence creates the **Cartesian plane**, or the **rectangular coordinate system**, in the plane.

To construct a rectangular coordinate system, or a Cartesian plane, draw a pair of perpendicular real number lines, one horizontal and the other vertical, with the lines intersecting at their respective 0-points (Figure P.6). The horizontal line is usually the ***x*-axis** and the vertical line is usually the ***y*-axis**. The positive direction on the *x*-axis is to the right, and the positive direction on the *y*-axis is up. Their point of intersection, *O*, is the **origin of the Cartesian plane**.

Each point *P* of the plane is associated with an **ordered pair (*x*, *y*)** of real numbers, the **(Cartesian) coordinates of the point**. The ***x*-coordinate** represents the intersection of the *x*-axis with the vertical line from *P*, and the ***y*-coordinate** represents the intersection of the *y*-axis with the horizontal line from *P*. Figure P.6 shows the points *P* and *Q* with coordinates $(4, 2)$ and $(-6, -4)$, respectively. We use the ordered pair (a, b) for both the name of the point and its coordinates.

The coordinate axes divide the Cartesian plane into four **quadrants**, as shown in Figure P.7.

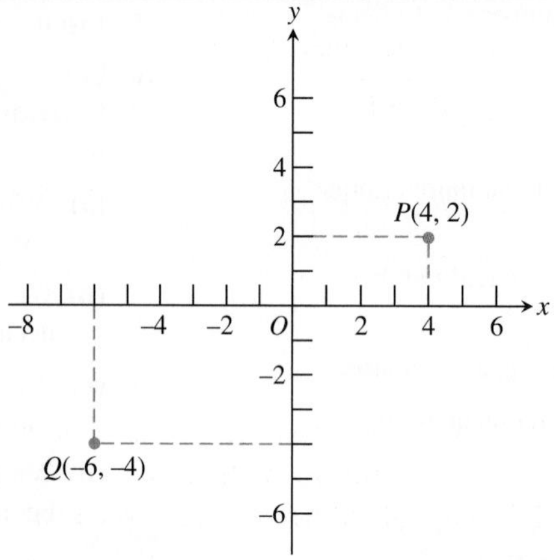

FIGURE P.6 The Cartesian coordinate plane.

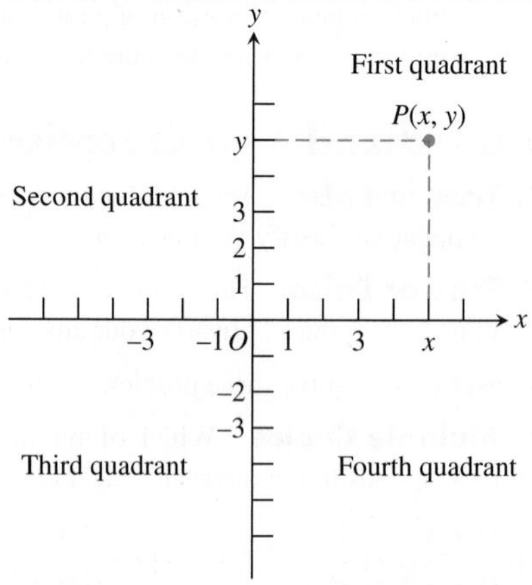

FIGURE P.7 The four quadrants. Points on the *x*- or *y*-axis are not in any quadrant.

Table P.2 U.S. Exports to Mexico

Year	U.S. Exports (billions of dollars)
2003	97.4
2004	110.8
2005	120.4
2006	134.0
2007	136.0
2008	151.2
2009	128.9
2010	163.5

Source: U.S. Census Bureau, The World Almanac and Book of Facts 2012.

EXAMPLE 1 Plotting Data on U.S. Exports to Mexico

The values in billions of dollars of U.S. exports to Mexico from 2003 to 2010 are given in Table P.2. Plot the (year, export value) ordered pairs on a rectangular coordinate system.

SOLUTION The points are plotted in Figure P.8 on page 13. **Now try Exercise 31.**

A **scatter plot** is a graph of the (x, y) data pairs on a Cartesian plane. Figure P.8 is a scatter plot of the data from Table P.2.

Absolute Value of a Real Number

The *absolute value of a real number* suggests its **magnitude** (size). For example, the absolute value of 3 is 3, and the absolute value of -5 is 5.

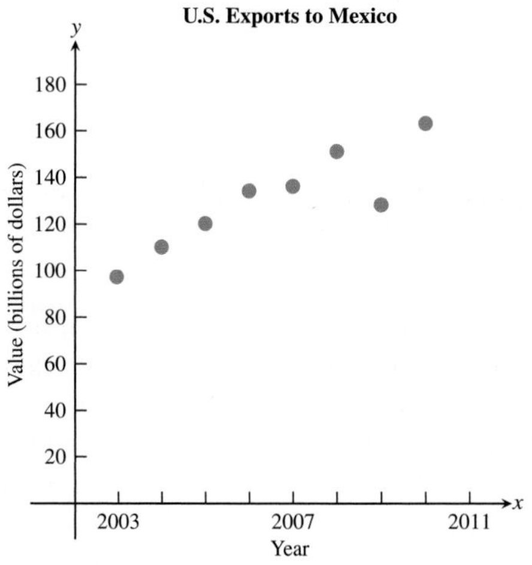

FIGURE P.8 The graph for Example 1.

DEFINITION Absolute Value of a Real Number

The **absolute value of a real number *a*** is

$$|a| = \begin{cases} a, \text{ if } a > 0 \\ -a, \text{ if } a < 0 \\ 0, \text{ if } a = 0. \end{cases}$$

EXAMPLE 2 Using the Definition of Absolute Value

Evaluate:

(a) $|-4|$ **(b)** $|\pi - 6|$

SOLUTION

(a) Because $-4 < 0$, $|-4| = -(-4) = 4$.

(b) Because $\pi \approx 3.14$, $\pi - 6$ is negative, so $\pi - 6 < 0$. Thus, $|\pi - 6| = -(\pi - 6) = 6 - \pi \approx 2.858$.

Now try Exercise 9.

Here is a summary of some important properties of absolute value.

Properties of Absolute Value

Let a and b be real numbers.

1. $|a| \geq 0$

2. $|-a| = |a|$

3. $|ab| = |a||b|$

4. $\left|\frac{a}{b}\right| = \frac{|a|}{|b|}, b \neq 0$

Distance Formulas

The *distance* between -1 and 4 on the number line is 5 (Figure P.9). This distance may be found by subtracting the smaller number from the larger: $4 - (-1) = 5$. If we use absolute value, the order of subtraction does not matter:

$$|4 - (-1)| = |-1 - 4| = 5$$

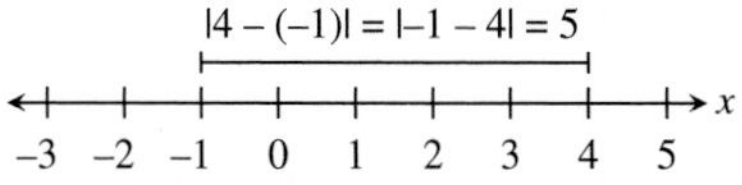

FIGURE P.9 Finding the distance between -1 and 4.

Distance Formula (Number Line)

Let a and b be real numbers. The **distance between *a* and *b*** is

$$|a - b|.$$

Note that $|a - b| = |b - a|$.

Absolute Value and Distance

If we let $b = 0$ in the distance formula, we see that the distance between a and 0 is $|a|$. Thus, the absolute value of a number is its distance from zero.

To find the *distance* between two points that lie on the same horizontal or vertical line in the Cartesian plane, we use the distance formula for points on a number line. For example, the distance between points x_1 and x_2 on the x-axis is $|x_1 - x_2| = |x_2 - x_1|$ and the distance between points y_1 and y_2 on the y-axis is $|y_1 - y_2| = |y_2 - y_1|$.

To find the distance between two points $P(x_1, y_1)$ and $Q(x_2, y_2)$ that do not lie on the same horizontal or vertical line, we form the right triangle determined by P, Q, and $R(x_2, y_1)$ (Figure P.10).

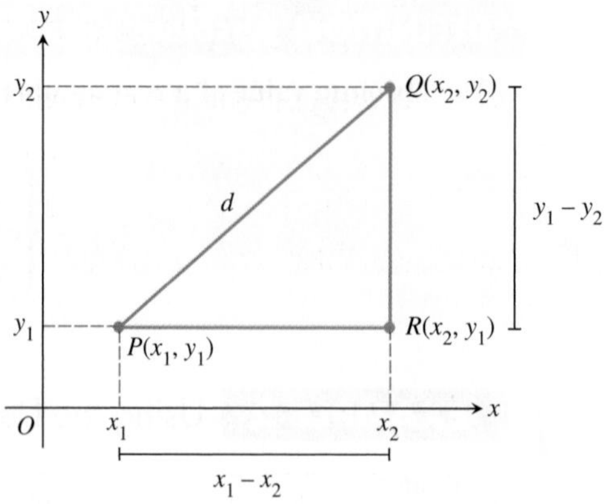

FIGURE P.10 Forming a right triangle with hypotenuse $\overline{PQ}$.

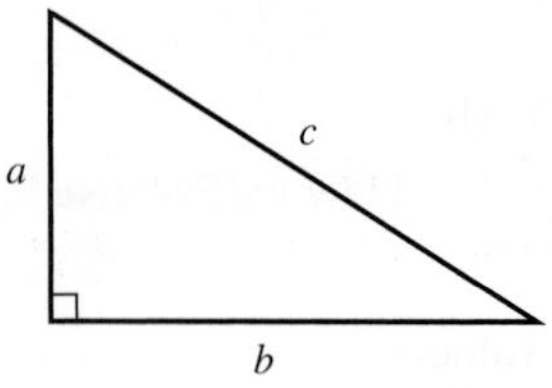

FIGURE P.11 The Pythagorean Theorem: $c^2 = a^2 + b^2$.

The distance from P to R is $|x_1 - x_2|$, and the distance from R to Q is $|y_1 - y_2|$. By the **Pythagorean Theorem** (Figure P.11), the distance d between P and Q is

$$d = \sqrt{|x_1 - x_2|^2 + |y_1 - y_2|^2}.$$

Because $|x_1 - x_2|^2 = (x_1 - x_2)^2$ and $|y_1 - y_2|^2 = (y_1 - y_2)^2$, we obtain the following formula.

Distance Formula (Coordinate Plane)

The **distance d between points $P(x_1, y_1)$ and $Q(x_2, y_2)$** in the coordinate plane is

$$d = \sqrt{(x_1 - x_2)^2 + (y_1 - y_2)^2}.$$

EXAMPLE 3 Finding the Distance Between Two Points

Find the distance d between the points $(1, 5)$ and $(6, 2)$.

SOLUTION

$$\begin{aligned} d &= \sqrt{(1 - 6)^2 + (5 - 2)^2} && \text{The distance formula} \\ &= \sqrt{(-5)^2 + 3^2} \\ &= \sqrt{25 + 9} \\ &= \sqrt{34} \approx 5.831 && \text{Using a calculator} \end{aligned}$$

Now try Exercise 11.

Midpoint Formulas

When the endpoints of a segment on a number line are known, we take the average of their coordinates to find the midpoint of the segment.

Midpoint Formula (Number Line)

The **midpoint of the line segment with endpoints a and b** is

$$\frac{a + b}{2}.$$

EXAMPLE 4 Finding the Midpoint of a Line Segment

The midpoint of the line segment with endpoints -9 and 3 on a number line is

$$\frac{(-9) + 3}{2} = \frac{-6}{2} = -3.$$

See Figure P.12. Now try Exercise 23.

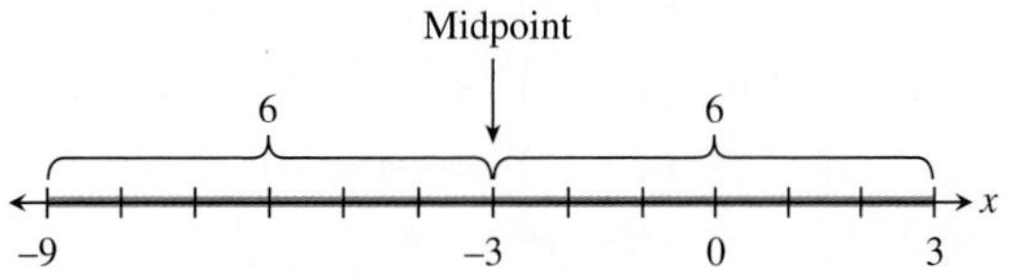

FIGURE P.12 Notice that the distance from the midpoint, -3, to 3 or to -9 is 6. (Example 4)

Just as with number lines, the midpoint of a line segment in the coordinate plane is determined by its endpoints. Each coordinate of the midpoint is the average of the corresponding coordinates of its endpoints.

Midpoint Formula (Coordinate Plane)

The **midpoint of the line segment with endpoints (a, b) and (c, d)** is

$$\left(\frac{a+c}{2}, \frac{b+d}{2}\right).$$

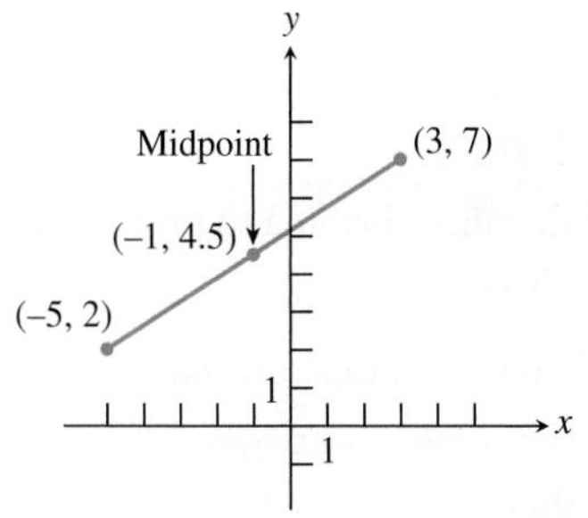

FIGURE P.13 Midpoint of a line segment (Example 5).

EXAMPLE 5 Finding the Midpoint of a Line Segment

The midpoint of the line segment with endpoints $(-5, 2)$ and $(3, 7)$ is

$$(x, y) = \left(\frac{-5 + 3}{2}, \frac{2 + 7}{2}\right) = (-1, 4.5).$$

See Figure P.13. Now try Exercise 25.

Equations of Circles

A **circle** is the set of points in a plane at a fixed distance (**radius**) from a fixed point (**center**). Figure P.14 shows the circle with center (h, k) and radius r. If (x, y) is any point on the circle, the distance formula gives

$$\sqrt{(x - h)^2 + (y - k)^2} = r.$$

Squaring both sides, we obtain the following equation for a circle.

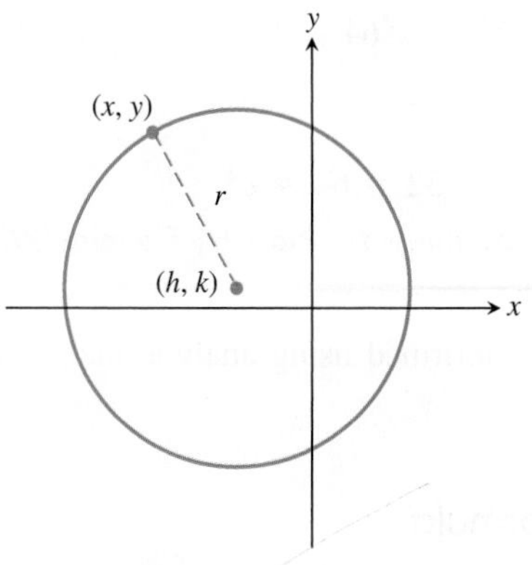

FIGURE P.14 The circle with center (h, k) and radius r.

DEFINITION Standard Form Equation of a Circle

The **standard form equation of a circle** with center (h, k) and radius r is

$$(x - h)^2 + (y - k)^2 = r^2.$$

EXAMPLE 6 Finding Standard Form Equations of Circles

Find the standard form equation of the circle.

(a) Center $(-4, 1)$, radius 8 **(b)** Center $(0, 0)$, radius 5

SOLUTION

(a) $(x - h)^2 + (y - k)^2 = r^2$ Standard form equation

$(x - (-4))^2 + (y - 1)^2 = 8^2$ Substitute $h = -4, k = 1, r = 8$.

$(x + 4)^2 + (y - 1)^2 = 64$

(b) $(x - h)^2 + (y - k)^2 = r^2$ Standard form equation

$(x - 0)^2 + (y - 0)^2 = 5^2$ Substitute $h = 0, k = 0, r = 5$.

$x^2 + y^2 = 25$

Now try Exercise 41.

Applications

EXAMPLE 7 Using an Inequality to Express Distance

We can state that "the distance between x and -3 is less than 9" using the inequality

$$|x - (-3)| < 9 \quad \text{or} \quad |x + 3| < 9.$$

Now try Exercise 51.

The converse of the Pythagorean Theorem is true. That is, if the sum of squares of the lengths of the two sides of a triangle equals the square of the length of the third side, then the triangle is a right triangle.

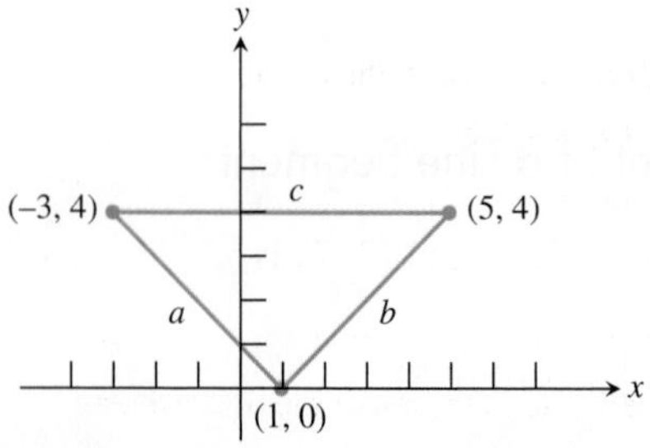

FIGURE P.15 The triangle in Example 8.

EXAMPLE 8 Verifying Right Triangles

Use the converse of the Pythagorean Theorem and the distance formula to prove that the points $(-3, 4)$, $(1, 0)$, and $(5, 4)$ determine a right triangle.

SOLUTION The three points are plotted in Figure P.15. We need to show that the lengths of the sides of the triangle satisfy the Pythagorean relationship $a^2 + b^2 = c^2$. Applying the distance formula we find that

$$a = \sqrt{(-3 - 1)^2 + (4 - 0)^2} = \sqrt{32}$$

$$b = \sqrt{(1 - 5)^2 + (0 - 4)^2} = \sqrt{32}$$

$$c = \sqrt{(-3 - 5)^2 + (4 - 4)^2} = \sqrt{64}$$

The triangle is a right triangle because

$$a^2 + b^2 = (\sqrt{32})^2 + (\sqrt{32})^2 = 32 + 32 = 64 = c^2.$$

Now try Exercise 39.

Properties of geometric figures can sometimes be confirmed using analytic methods such as the midpoint formulas.

EXAMPLE 9 Using the Midpoint Formula

It is a fact from geometry that the diagonals of a parallelogram bisect each other. Prove this with a midpoint formula.

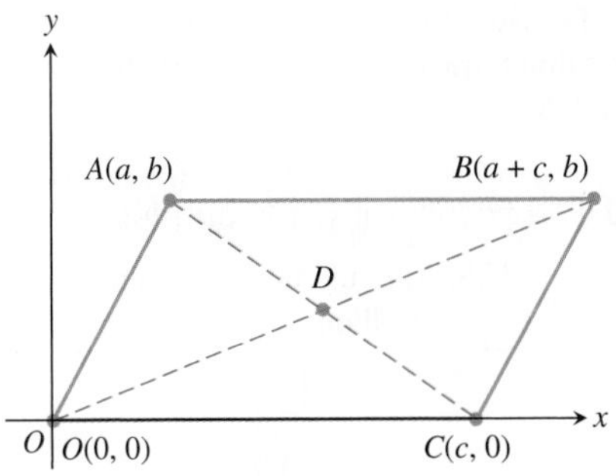

FIGURE P.16 The coordinates of B must be $(a + c, b)$ in order for CB to be parallel to OA. (Example 9)

SOLUTION We can position a parallelogram in the rectangular coordinate plane as shown in Figure P.16. Applying the midpoint formula for the coordinate plane to segments OB and AC, we find that

$$\text{midpoint of segment } OB = \left(\frac{0 + a + c}{2}, \frac{0 + b}{2}\right) = \left(\frac{a + c}{2}, \frac{b}{2}\right)$$

$$\text{midpoint of segment } AC = \left(\frac{a + c}{2}, \frac{b + 0}{2}\right) = \left(\frac{a + c}{2}, \frac{b}{2}\right)$$

The midpoints of segments OA and AC are the same, so the diagonals of the parallelogram $OABC$ meet at their midpoints and thus bisect each other.

Now try Exercise 37.

QUICK REVIEW P.2

In Exercises 1 and 2, plot the two numbers on a number line. Then find the distance between them.

1. $\sqrt{7}, \sqrt{2}$ **2.** $-\frac{5}{3}, -\frac{9}{5}$

In Exercises 3 and 4, plot the real numbers on a number line.

3. $-3, 4, 2.5, 0, -1.5$ **4.** $-\frac{5}{2}, -\frac{1}{2}, \frac{2}{3}, 0, -1$

In Exercises 5 and 6, plot the points.

5. $A(3, 5), B(-2, 4), C(3, 0), D(0, -3)$

6. $A(-3, -5), B(2, -4), C(0, 5), D(-4, 0)$

In Exercises 7–10, use a calculator to evaluate the expression. Round your answer to two decimal places.

7. $\frac{-17 + 28}{2}$ **8.** $\sqrt{13^2 + 17^2}$

9. $\sqrt{6^2 + 8^2}$ **10.** $\sqrt{(17 - 3)^2 + (-4 - 8)^2}$

SECTION P.2 Exercises

Exercise numbers with a gray background indicate problems that the authors have designed to be solved *without a calculator*.

In Exercises 1 and 2, estimate the coordinates of the points.

1.

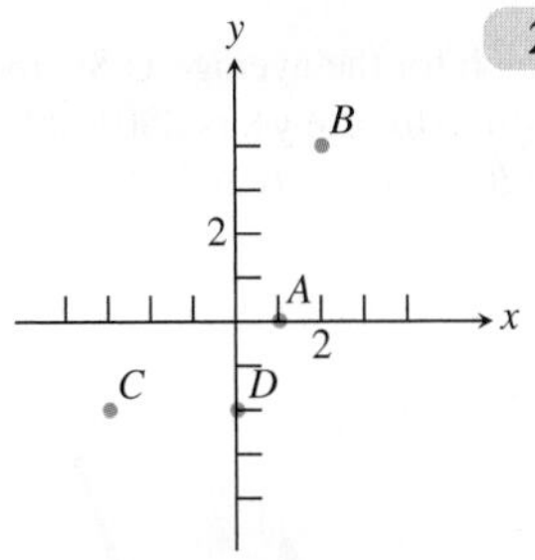

2.

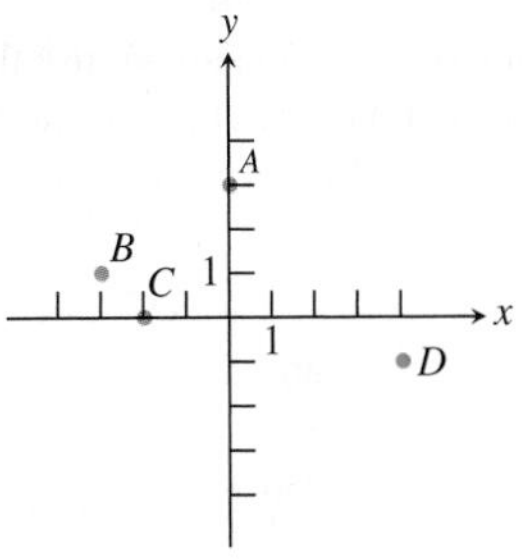

In Exercises 3 and 4, name the quadrants containing the points.

3. **(a)** $(2, 4)$ **(b)** $(0, 3)$ **(c)** $(-2, 3)$ **(d)** $(-1, -4)$

4. **(a)** $\left(\frac{1}{2}, \frac{3}{2}\right)$ **(b)** $(-2, 0)$ **(c)** $(-1, -2)$ **(d)** $\left(-\frac{3}{2}, -\frac{7}{3}\right)$

In Exercises 5–8, evaluate the expression.

5. $3 + |-3|$ **6.** $2 - |-2|$

7. $|(-2)3|$ **8.** $\frac{-2}{|-2|}$

In Exercises 9 and 10, rewrite the expression without using absolute value symbols.

9. $|\pi - 4|$ **10.** $|\sqrt{5} - 5/2|$

In Exercises 11–18, find the distance between the points.

11. $-9.3, 10.6$ **12.** $-5, -17$

13. $(-3, -1), (5, -1)$ **14.** $(-4, -3), (1, 1)$

15. $(0, 0), (3, 4)$ **16.** $(-1, 2), (2, -3)$

17. $(-2, 0), (5, 0)$ **18.** $(0, -8), (0, -1)$

In Exercises 19–22, find the perimeter and area of the figure determined by the points.

19. $(-5, 3), (0, -1), (4, 4)$

20. $(-2, -2), (-2, 2), (2, 2), (2, -2)$

21. $(-3, -1), (-1, 3), (7, 3), (5, -1)$

22. $(-2, 1), (-2, 6), (4, 6), (4, 1)$

In Exercises 23–28, find the midpoint of the line segment with the given endpoints.

23. $-9.3, 10.6$ **24.** $-5, -17$

25. $(-1, 3), (5, 9)$

26. $(3, \sqrt{2}), (6, 2)$

27. $(-7/3, 3/4), (5/3, -9/4)$

28. $(5, -2), (-1, -4)$

In Exercises 29–34, draw a scatter plot of the data given in the table.

29. U.S. Motor Vehicle Production The total number of motor vehicles in thousands (y) produced by the United States each year from 2004 through 2010 is given in the table. *(Source: Automotive News Data Center and R. L. Polk Marketing Systems as reported in The World Almanac and Book of Facts 2012.)*

x	2004	2005	2006	2007	2008	2009	2010
y	12,021	12,018	11,351	10,611	8,503	5,591	7,632

30. World Motor Vehicle Production The total number of motor vehicles in thousands (y) produced in the world each year from 2004 through 2010 is given in the table. *(Source: American Automobile Manufacturers Association as reported in The World Almanac and Book of Facts 2012.)*

x	2004	2005	2006	2007	2008	2009	2010
y	65,654	67,892	70,992	74,647	67,602	59,096	73,311

31. U.S. Imports from Mexico The total in billions of dollars of U.S. imports from Mexico from 2003 through 2010 is given in Table P.3.

Table P.3 U.S. Imports from Mexico

Year	U.S. Imports (billions of dollars)
2003	138.1
2004	155.9
2005	170.1
2006	198.3
2007	210.7
2008	215.9
2009	176.7
2010	229.9

Source: U.S. Census Bureau, The World Almanac and Book of Facts 2012.

32. U.S. Agricultural Exports The total in billions of dollars of U.S. agricultural exports from 2003 through 2010 is given in Table P.4.

Table P.4 U.S. Agricultural Exports

Year	U.S. Agricultural Exports (billions of dollars)
2003	59.4
2004	61.4
2005	63.2
2006	70.9
2007	90.0
2008	114.8
2009	98.5
2010	115.8

Source: U.S. Department of Agriculture, The World Almanac and Book of Facts 2012.

33. U.S. Agricultural Trade Surplus The total in billions of dollars of U.S. agricultural trade surplus from 2003 through 2010 is given in Table P.5.

Table P.5 U.S. Agricultural Trade Surplus

Year	U.S. Agricultural Trade Surplus (billions of dollars)
2003	12.0
2004	7.4
2005	3.9
2006	5.6
2007	18.1
2008	34.3
2009	26.8
2010	34.0

Source: U.S. Department of Agriculture, The World Almanac and Book of Facts 2012.

34. U.S. Exports to Canada The total in billions of dollars of U.S. exports to Canada from 2003 through 2010 is given in Table P.6.

Table P.6 U.S. Exports to Canada

Year	U.S. Exports (billions of dollars)
2003	169.9
2004	189.9
2005	211.9
2006	230.7
2007	248.9
2008	261.2
2009	204.7
2010	249.1

Source: U.S. Census Bureau, The World Almanac and Book of Facts 2012.

In Exercises 35 and 36, use the graph for the average U.S. city prices (in cents) of regular unleaded gasoline for the years 2001–2011. *(Source: The World Almanac and Book of Facts 2012.)*

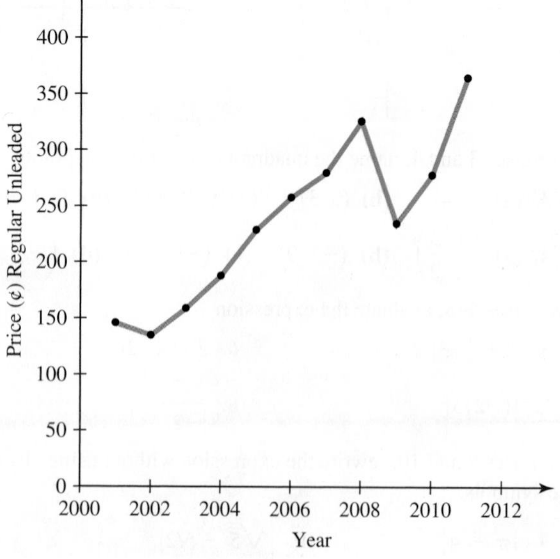

35. Reading from Graphs Estimate the price of gasoline (in dollars) for

(a) 2002 **(b)** 2007 **(c)** 2010

36. Percent Increase Estimate the percent increase in the price of gasoline from

(a) 2001 to 2006 **(b)** 2009 to 2011

37. Prove that the figure determined by the points is an isosceles triangle: $(1, 3)$, $(4, 7)$, $(8, 4)$

38. Group Activity Prove that the diagonals of the figure determined by the points bisect each other.

(a) Square $(-7, -1)$, $(-2, 4)$, $(3, -1)$, $(-2, -6)$

(b) Parallelogram $(-2, -3)$, $(0, 1)$, $(6, 7)$, $(4, 3)$

39. (a) Find the lengths of the sides of the triangle in the figure.

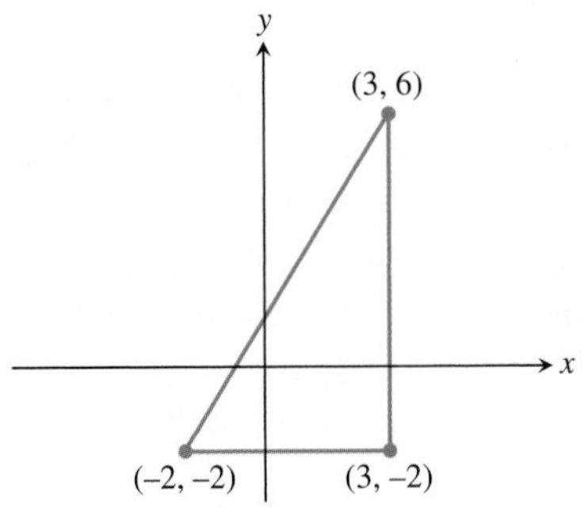

(b) Writing to Learn Prove that the triangle is a right triangle.

40. (a) Find the lengths of the sides of the triangle in the figure.

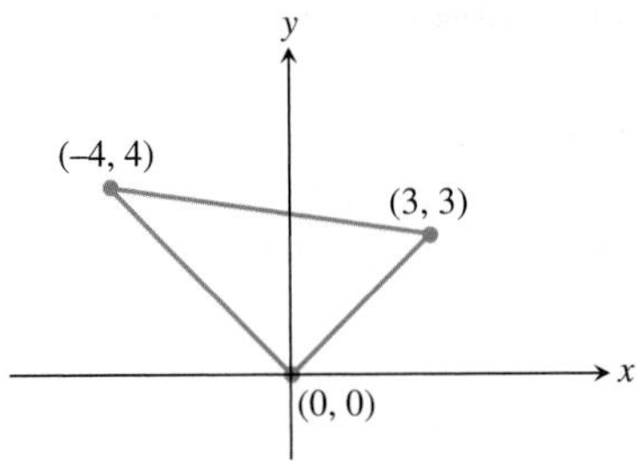

(b) Writing to Learn Prove that the triangle is a right triangle.

In Exercises 41–44, find the standard form equation for the circle.

41. Center $(1, 2)$, radius 5

42. Center $(-3, 2)$, radius 1

43. Center $(-1, -4)$, radius 3

44. Center $(0, 0)$, radius $\sqrt{3}$

In Exercises 45–48, find the center and radius of the circle.

45. $(x - 3)^2 + (y - 1)^2 = 36$

46. $(x + 4)^2 + (y - 2)^2 = 121$

47. $x^2 + y^2 = 5$

48. $(x - 2)^2 + (y + 6)^2 = 25$

In Exercises 49–52, write the statement using absolute value notation.

49. The distance between x and 4 is 3.

50. The distance between y and -2 is greater than or equal to 4.

51. The distance between x and c is less than d units.

52. y is more than d units from c.

53. Let $(4, 4)$ be the midpoint of the line segment determined by the points $(1, 2)$ and (a, b). Determine a and b.

54. Writing to Learn Isosceles but Not Equilateral Prove that the triangle determined by the points $(3, 0)$, $(-1, 2)$, and $(5, 4)$ is isosceles but not equilateral.

55. Writing to Learn Equidistant Point Prove that the midpoint of the hypotenuse of the right triangle with vertices $(0, 0)$, $(5, 0)$, and $(0, 7)$ is equidistant from the three vertices.

56. Writing to Learn Describe the set of real numbers that satisfy $|x - 2| < 3$.

57. Writing to Learn Describe the set of real numbers that satisfy $|x + 3| \geq 5$.

Standardized Test Questions

58. True or False If a is a real number, then $|a| \geq 0$. Justify your answer.

59. True or False Let ΔABC and $\Delta AMM'$ be right triangles as shown in the figure. If M is the midpoint of segment AB, then M' is the midpoint of segment AC. Justify your answer.

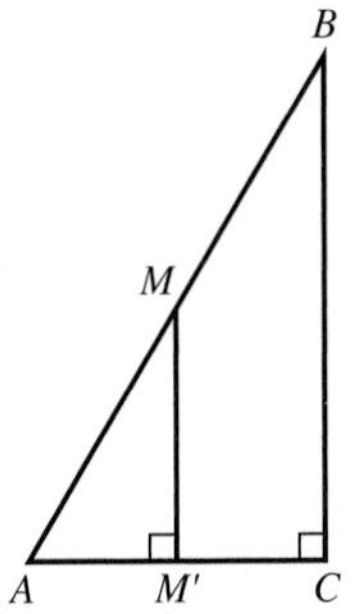

In Exercises 60–63, solve these problems without using a calculator.

60. Multiple Choice Which of the following is equal to $|1 - \sqrt{3}|$?

(A) $1 - \sqrt{3}$ **(B)** $\sqrt{3} - 1$

(C) $(1 - \sqrt{3})^2$ **(D)** $\sqrt{2}$

(E) $\sqrt{1/3}$

61. Multiple Choice Which of the following is the midpoint of the line segment with endpoints -3 and 2?

(A) 5/2 **(B)** 1

(C) $-1/2$ **(D)** -1

(E) $-5/2$

62. Multiple Choice Which of the following is the center of the circle $(x - 3)^2 + (y + 4)^2 = 2$?

(A) $(3, -4)$ **(B)** $(-3, 4)$
(C) $(4, -3)$ **(D)** $(-4, 3)$
(E) $(3/2, -2)$

63. Multiple Choice Which of the following points is in the third quadrant?

(A) $(0, -3)$ **(B)** $(-1, 0)$
(C) $(2, -1)$ **(D)** $(-1, 2)$
(E) $(-2, -3)$

Explorations

64. Dividing a Line Segment into Thirds

(a) Find the coordinates of the points one-third and two-thirds of the way from $a = 2$ to $b = 8$ on a number line.

(b) Repeat (a) for $a = -3$ and $b = 7$.

(c) Find the coordinates of the points one-third and two-thirds of the way from a to b on a number line.

(d) Find the coordinates of the points one-third and two-thirds of the way from the point $(1, 2)$ to the point $(7, 11)$ in the coordinate plane.

(e) Find the coordinates of the points one-third and two-thirds of the way from the point (a, b) to the point (c, d) in the coordinate plane.

Extending the Ideas

65. Writing to Learn Equidistant Point from Vertices of a Right Triangle Prove that the midpoint of the hypotenuse of any right triangle is equidistant from the three vertices.

66. Comparing Areas Consider the four points $A(0, 0)$, $B(0, a)$, $C(a, a)$, and $D(a, 0)$. Let P be the midpoint of the line segment CD and Q the point one-fourth of the way from A to D on segment AD.

(a) Find the area of triangle BPQ.

(b) Compare the area of triangle BPQ with the area of square $ABCD$.

In Exercises 67–69, let $P(a, b)$ be a point in the first quadrant.

67. Find the coordinates of the point Q in the fourth quadrant so that the x-axis is the perpendicular bisector of PQ.

68. Find the coordinates of the point Q in the second quadrant so that the y-axis is the perpendicular bisector of PQ.

69. Find the coordinates of the point Q in the third quadrant so that the origin is the midpoint of the segment PQ.

70. Writing to Learn Prove that the distance formula for the number line is a special case of the distance formula for the Cartesian plane.

P.3 Linear Equations and Inequalities

What you'll learn about

- Equations
- Solving Equations
- Linear Equations in One Variable
- Linear Inequalities in One Variable

... and why

These topics provide the foundation for algebraic techniques needed throughout this textbook.

Equations

An **equation** is a statement of equality between two expressions. Here are some properties of equality that we use to solve equations algebraically.

Properties of Equality

Let u, v, w, and z be real numbers, variables, or algebraic expressions.

1. **Reflexive** $u = u$
2. **Symmetric** If $u = v$, then $v = u$.
3. **Transitive** If $u = v$, and $v = w$, then $u = w$.
4. **Addition** If $u = v$ and $w = z$, then $u + w = v + z$.
5. **Multiplication** If $u = v$ and $w = z$, then $uw = vz$.

Solving Equations

A **solution of an equation in *x*** is a value of x for which the equation is true. To **solve an equation in *x*** means to find all values of x for which the equation is true, that is, to find all solutions of the equation.

EXAMPLE 1 Confirming a Solution

Prove that $x = -2$ is a solution of the equation $x^3 - x + 6 = 0$.

SOLUTION

$$(-2)^3 - (-2) + 6 \stackrel{?}{=} 0$$
$$-8 + 2 + 6 \stackrel{?}{=} 0$$
$$0 = 0$$

Now try Exercise 1.

Linear Equations in One Variable

The most basic equation in algebra is a *linear equation*.

DEFINITION Linear Equation in *x*

A **linear equation in *x*** is one that can be written in the form

$$ax + b = 0,$$

where a and b are real numbers and $a \neq 0$.

The equation $2z - 4 = 0$ is linear in the variable z. The equation $3u^2 - 12 = 0$ is *not* linear in the variable u. A linear equation in one variable has exactly one solution. We solve such an equation by transforming it into an *equivalent equation* whose solution is obvious. Two or more equations are **equivalent** if they have the same solutions. For example, the equations $2z - 4 = 0$, $2z = 4$, and $z = 2$ are all equivalent. Here are operations that produce equivalent equations.

Operations for Equivalent Equations

An equivalent equation is obtained if one or more of the following operations are performed.

Operation	Given Equation	Equivalent Equation
1. Combine like terms, reduce fractions, and remove grouping symbols.	$2x + x = \frac{3}{9}$	$3x = \frac{1}{3}$
2. Perform the same operation on both sides.		
(a) Add (-3).	$x + 3 = 7$	$x = 4$
(b) Subtract $(2x)$.	$5x = 2x + 4$	$3x = 4$
(c) Multiply by a nonzero constant (1/3).	$3x = 12$	$x = 4$
(d) Divide by a nonzero constant (3).	$3x = 12$	$x = 4$

The next two examples illustrate how to use equivalent equations to solve linear equations.

EXAMPLE 2 Solving a Linear Equation

Solve $2(2x - 3) + 3(x + 1) = 5x + 2$. Support the result with a calculator.

SOLUTION

$$2(2x - 3) + 3(x + 1) = 5x + 2$$
$$4x - 6 + 3x + 3 = 5x + 2 \quad \text{Distributive property}$$
$$7x - 3 = 5x + 2 \quad \text{Combine like terms.}$$
$$2x = 5 \quad \text{Add 3, and subtract } 5x.$$
$$x = 2.5 \quad \text{Divide by 2.}$$

To support our algebraic work we can evaluate the expressions in the original equation for $x = 2.5$. Figure P.17 shows that each side of the original equation is equal to 14.5 if $x = 2.5$.

Now try Exercise 23.

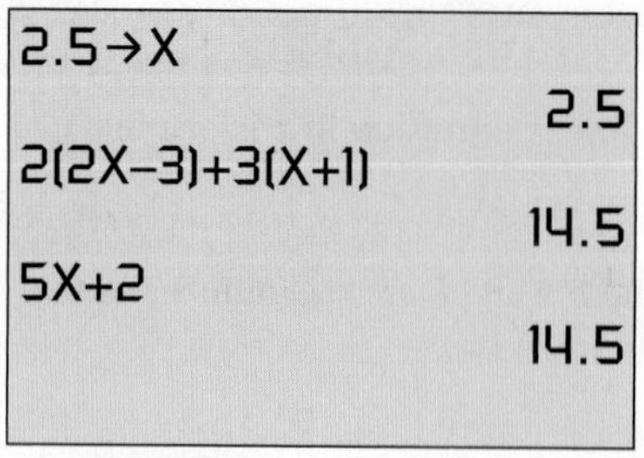

FIGURE P.17 The top line stores the number 2.5 into the variable x. (Example 2)

If an equation involves fractions, find the least common denominator (LCD) of the fractions and multiply both sides by the LCD. This is sometimes referred to as *clearing the equation of fractions*. Example 3 illustrates.

Integers and Fractions

Notice in Example 3 that $2 = \frac{2}{1}$.

EXAMPLE 3 Solving a Linear Equation Involving Fractions

Solve

$$\frac{5y - 2}{8} = 2 + \frac{y}{4}.$$

SOLUTION The denominators are 8, 1, and 4. The LCD of the fractions is 8. (See Appendix A.3 if necessary.)

$$\begin{aligned} \frac{5y - 2}{8} &= 2 + \frac{y}{4} \\ 8\left(\frac{5y - 2}{8}\right) &= 8\left(2 + \frac{y}{4}\right) && \text{Multiply by the LCD 8.} \\ 8 \cdot \frac{5y - 2}{8} &= 8 \cdot 2 + 8 \cdot \frac{y}{4} && \text{Distributive property} \\ 5y - 2 &= 16 + 2y && \text{Simplify.} \\ 5y &= 18 + 2y && \text{Add 2.} \\ 3y &= 18 && \text{Subtract } 2y. \\ y &= 6 && \text{Divide by 3.} \end{aligned}$$

We leave it to you to check the solution using either paper and pencil or a calculator.

Now try Exercise 25.

Linear Inequalities in One Variable

We used inequalities to describe order on the number line in Section P.1. For example, if x is to the left of 2 on the number line, or if x is any real number less than 2, we write $x < 2$. The most basic inequality in algebra is a *linear inequality*.

DEFINITION Linear Inequality in x

A **linear inequality in x** is one that can be written in the form

$$ax + b < 0, \quad ax + b \le 0, \quad ax + b > 0, \quad \text{or} \quad ax + b \ge 0,$$

where a and b are real numbers and $a \neq 0$.

To **solve an inequality in x** means to find all values of x for which the inequality is true. A **solution of an inequality in x** is a value of x for which the inequality is true. The set of all solutions of an inequality is the **solution set** of the inequality. We **solve an inequality** by finding its solution set. Here is a list of properties we use to solve inequalities.

Direction of an Inequality

Multiplying (or dividing) an inequality by a positive number preserves the direction of the inequality. Multiplying (or dividing) an inequality by a negative number reverses the direction.

Properties of Inequalities

Let u, v, w, and z be real numbers, variables, or algebraic expressions, and c a real number.

1. Transitive	If $u < v$ and $v < w$, then $u < w$.
2. Addition	If $u < v$, then $u + w < v + w$. If $u < v$ and $w < z$, then $u + w < v + z$.
3. Multiplication	If $u < v$ and $c > 0$, then $uc < vc$. If $u < v$ and $c < 0$, then $uc > vc$.

The above properties are true if $<$ is replaced by $\le$. There are similar properties for $>$ and $\ge$.

The set of solutions of a linear inequality in one variable forms an interval of real numbers. Just as with linear equations, we solve a linear inequality by transforming it into an *equivalent inequality* whose solutions are obvious. Two or more inequalities are **equivalent** if they have the same solution set. The properties of inequalities listed above describe operations that transform an inequality into an equivalent one.

EXAMPLE 4 Solving a Linear Inequality

Solve $3(x - 1) + 2 \le 5x + 6$.

SOLUTION

$$
\begin{aligned}
3(x - 1) + 2 &\le 5x + 6 && \\
3x - 3 + 2 &\le 5x + 6 && \text{Distributive property} \\
3x - 1 &\le 5x + 6 && \text{Combine like terms.} \\
3x &\le 5x + 7 && \text{Add 1.} \\
-2x &\le 7 && \text{Subtract } 5x. \\
\left(-\frac{1}{2}\right) \cdot -2x &\ge \left(-\frac{1}{2}\right) \cdot 7 && \text{Multiply by } -1/2. \text{ (The inequality reverses.)} \\
x &\ge -3.5 &&
\end{aligned}
$$

The solution set of the inequality is the set of all real numbers greater than or equal to -3.5. In interval notation, the solution set is $[-3.5, \infty)$.

Now try Exercise 41.

Because the solution set of a linear inequality is an interval of real numbers, we can display the solution set with a number line graph as illustrated in Example 5.

EXAMPLE 5 Solving a Linear Inequality Involving Fractions

Solve the inequality, and graph its solution set.

$$\frac{x}{3} + \frac{1}{2} > \frac{x}{4} + \frac{1}{3}$$

SOLUTION The LCD of the fractions is 12.

$$
\begin{aligned}
\frac{x}{3} + \frac{1}{2} &> \frac{x}{4} + \frac{1}{3} && \\
12 \cdot \left(\frac{x}{3} + \frac{1}{2}\right) &> 12 \cdot \left(\frac{x}{4} + \frac{1}{3}\right) && \text{Multiply by the LCD 12.} \\
4x + 6 &> 3x + 4 && \text{Simplify.} \\
x + 6 &> 4 && \text{Subtract } 3x. \\
x &> -2 && \text{Subtract 6.}
\end{aligned}
$$

The solution set is the interval $(-2, \infty)$. Its graph is shown in Figure P.18.

Now try Exercise 43.

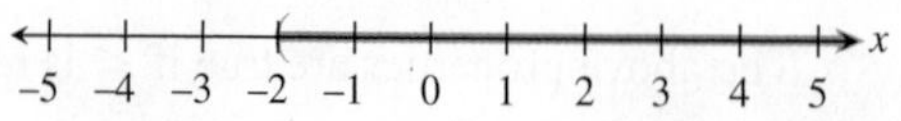

FIGURE P.18 The graph of the solution set of the inequality in Example 5.

Sometimes two inequalities are combined in a *double inequality,* which is solved by isolating x as the middle expression. Example 6 illustrates this.

EXAMPLE 6 Solving a Double Inequality

Solve the inequality, and graph its solution set.

$$-3 < \frac{2x + 5}{3} \leq 5$$

SOLUTION

$$-3 < \frac{2x + 5}{3} \leq 5$$

$$-9 < 2x + 5 \leq 15 \quad \text{Multiply by 3.}$$

$$-14 < 2x \leq 10 \quad \text{Subtract 5.}$$

$$-7 < x \leq 5 \quad \text{Divide by 2.}$$

The solution set is the set of all real numbers greater than -7 and less than or equal to 5. In interval notation, the solution is set $(-7, 5]$. Its graph is shown in Figure P.19.

Now try Exercise 47.

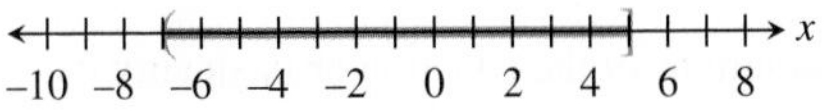

FIGURE P.19 The graph of the solution set of the double inequality in Example 6.

QUICK REVIEW P.3

In Exercises 1 and 2, simplify the expression by combining like terms.

1. $2x + 5x + 7 + y - 3x + 4y + 2$
2. $4 + 2x - 3z + 5y - x + 2y - z - 2$

In Exercises 3 and 4, use the distributive property to expand the products. Simplify the resulting expression by combining like terms.

3. $3(2x - y) + 4(y - x) + x + y$
4. $5(2x + y - 1) + 4(y - 3x + 2) + 1$

In Exercises 5–10, use the LCD to combine the fractions. Simplify the resulting fraction.

5. $\frac{2}{y} + \frac{3}{y}$
6. $\frac{1}{y - 1} + \frac{3}{y - 2}$
7. $2 + \frac{1}{x}$
8. $\frac{1}{x} + \frac{1}{y} - x$
9. $\frac{x + 4}{2} + \frac{3x - 1}{5}$
10. $\frac{x}{3} + \frac{x}{4}$

SECTION P.3 Exercises

Exercise numbers with a gray background indicate problems that the authors have designed to be solved *without a calculator.*

In Exercises 1–4, which values of x are solutions of the equation?

1. $2x^2 + 5x = 3$
   (a) $x = -3$ **(b)** $x = -\frac{1}{2}$ **(c)** $x = \frac{1}{2}$
2. $\frac{x}{2} + \frac{1}{6} = \frac{x}{3}$
   (a) $x = -1$ **(b)** $x = 0$ **(c)** $x = 1$
3. $\sqrt{1 - x^2} + 2 = 3$
   (a) $x = -2$ **(b)** $x = 0$ **(c)** $x = 2$
4. $(x - 2)^{1/3} = 2$
   (a) $x = -6$ **(b)** $x = 8$ **(c)** $x = 10$

In Exercises 5–10, determine whether the equation is linear in x.

5. $5 - 3x = 0$
6. $5 = 10/2$
7. $x + 3 = x - 5$
8. $x - 3 = x^2$
9. $2\sqrt{x} + 5 = 10$
10. $x + \frac{1}{x} = 1$

In Exercises 11–24, solve the equation without using a calculator.

11. $3x = 24$
12. $4x = -16$
13. $3t - 4 = 8$
14. $2t - 9 = 3$
15. $2x - 3 = 4x - 5$
16. $4 - 2x = 3x - 6$
17. $4 - 3y = 2(y + 4)$
18. $4(y - 2) = 5y$
19. $\frac{1}{2}x = \frac{7}{8}$
20. $\frac{2}{3}x = \frac{4}{5}$

21. $\frac{1}{2}x + \frac{1}{3} = 1$

22. $\frac{1}{3}x + \frac{1}{4} = 1$

23. $2(3 - 4z) - 5(2z + 3) = z - 17$

24. $3(5z - 3) - 4(2z + 1) = 5z - 2$

In Exercises 25–28, solve the equation. Support your answer with a calculator.

25. $\frac{2x - 3}{4} + 5 = 3x$

26. $2x - 4 = \frac{4x - 5}{3}$

27. $\frac{t + 5}{8} - \frac{t - 2}{2} = \frac{1}{3}$

28. $\frac{t - 1}{3} + \frac{t + 5}{4} = \frac{1}{2}$

29. Writing to Learn Write a statement about a solution of an equation suggested by the computations in the figure.

(a)
```
-2→X
                -2
2X²+X-6
                 0
```

(b)
```
3/2→X
               1.5
2X²+X-6
                 0
```

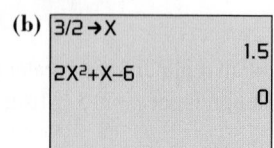

30. Writing to Learn Write a statement about a solution of an equation suggested by the computations in the figure.

(a)
```
2→X
                 2
7X+5
                19
4X-7
                 1
```

(b)
```
-4→X
                -4
7X+5
               -23
4X-7
               -23
```

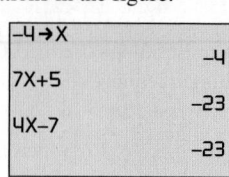

In Exercises 31–34, which values of x are solutions of the inequality?

31. $2x - 3 < 7$

(a) $x = 0$ (b) $x = 5$ (c) $x = 6$

32. $3x - 4 \geq 5$

(a) $x = 0$ (b) $x = 3$ (c) $x = 4$

33. $-1 < 4x - 1 \leq 11$

(a) $x = 0$ (b) $x = 2$ (c) $x = 3$

34. $-3 \leq 1 - 2x \leq 3$

(a) $x = -1$ (b) $x = 0$ (c) $x = 2$

In Exercises 35–42, solve the inequality, and draw a number line graph of the solution set.

35. $x - 4 < 2$

36. $x + 3 > 5$

37. $2x - 1 \leq 4x + 3$

38. $3x - 1 \geq 6x + 8$

39. $2 \leq x + 6 < 9$

40. $-1 \leq 3x - 2 < 7$

41. $2(5 - 3x) + 3(2x - 1) \leq 2x + 1$

42. $4(1 - x) + 5(1 + x) > 3x - 1$

In Exercises 43–54, solve the inequality.

43. $\frac{5x + 7}{4} \leq -3$

44. $\frac{3x - 2}{5} > -1$

45. $4 \geq \frac{2y - 5}{3} \geq -2$

46. $1 > \frac{3y - 1}{4} > -1$

47. $0 \leq 2z + 5 < 8$

48. $-6 < 5t - 1 < 0$

49. $\frac{x - 5}{4} + \frac{3 - 2x}{3} < -2$

50. $\frac{3 - x}{2} + \frac{5x - 2}{3} < -1$

51. $\frac{2y - 3}{2} + \frac{3y - 1}{5} < y - 1$

52. $\frac{3 - 4y}{6} - \frac{2y - 3}{8} \geq 2 - y$

53. $\frac{1}{2}(x - 4) - 2x \leq 5(3 - x)$

54. $\frac{1}{2}(x + 3) + 2(x - 4) < \frac{1}{3}(x - 3)$

In Exercises 55–58, find the solutions of the equation or inequality that are displayed in Figure P.20.

55. $x^2 - 2x < 0$

56. $x^2 - 2x = 0$

57. $x^2 - 2x > 0$

58. $x^2 - 2x \leq 0$

X	Y1	
0	0	
1	-1	
2	0	
3	3	
4	8	
5	15	
6	24	

Y1 = X²–2X

FIGURE P.20 The second column gives values of $y_1 = x^2 - 2x$ for $x = 0, 1, 2, 3, 4, 5,$ and 6.

59. Writing to Learn Explain how the second equation was obtained from the first.

$$x - 3 = 2x + 3, \quad 2x - 6 = 4x + 6$$

60. Writing to Learn Explain how the second equation was obtained from the first.

$$2x - 1 = 2x - 4, \quad x - \frac{1}{2} = x - 2$$

61. Group Activity Determine whether the two equations are equivalent.

(a) $3x = 6x + 9, \quad x = 2x + 9$

(b) $6x + 2 = 4x + 10, \quad 3x + 1 = 2x + 5$

62. Group Activity Determine whether the two equations are equivalent.

(a) $3x + 2 = 5x - 7, \quad -2x + 2 = -7$

(b) $2x + 5 = x - 7, \quad 2x = x - 7$

Standardized Test Questions

63. **True or False** $-6 > -2$. Justify your answer.

64. **True or False** $2 \leq \frac{6}{3}$. Justify your answer.

In Exercises 65–68, you may use a graphing calculator to solve these problems.

65. **Multiple Choice** Which of the following equations is equivalent to the equation $3x + 5 = 2x + 1$?

(A) $3x = 2x$ **(B)** $3x = 2x + 4$

(C) $\frac{3}{2}x + \frac{5}{2} = x + 1$ **(D)** $3x + 6 = 2x$

(E) $3x = 2x - 4$

66. **Multiple Choice** Which of the following inequalities is equivalent to the inequality $-3x < 6$?

(A) $3x < -6$ **(B)** $x < 10$

(C) $x > -2$ **(D)** $x > 2$

(E) $x > 3$

67. **Multiple Choice** Which of the following is the solution to the equation $x(x + 1) = 0$?

(A) $x = 0$ or $x = -1$ **(B)** $x = 0$ or $x = 1$

(C) Only $x = -1$ **(D)** Only $x = 0$

(E) Only $x = 1$

68. **Multiple Choice** Which of the following represents an equation equivalent to the equation

$$\frac{2x}{3} + \frac{1}{2} = \frac{x}{4} - \frac{1}{3}$$

that is cleared of fractions?

(A) $2x + 1 = x - 1$ **(B)** $8x + 6 = 3x - 4$

(C) $4x + 3 = \frac{3}{2}x - 2$ **(D)** $4x + 3 = 3x - 4$

(E) $4x + 6 = 3x - 4$

Explorations

69. **Testing Inequalities on a Calculator**

(a) The calculator we use indicates that the statement $2 < 3$ is true by returning the value 1 (for true) when $2 < 3$ is entered. Try it with your calculator.

(b) The calculator we use indicates that the statement $2 < 1$ is false by returning the value 0 (for false) when $2 < 1$ is entered. Try it with your calculator.

(c) Use your calculator to test which of these two numbers is larger: 799/800, 800/801.

(d) Use your calculator to test which of these two numbers is larger: $-102/101$, $-103/102$.

(e) If your calculator returns 0 when you enter $2x + 1 < 4$, what can you conclude about the value stored in x?

Extending the Ideas

70. **Perimeter of a Rectangle** The formula for the perimeter P of a rectangle is

$$P = 2(L + W).$$

Solve this equation for W.

71. **Area of a Trapezoid** The formula for the area A of a trapezoid is

$$A = \frac{1}{2}h(b_1 + b_2).$$

Solve this equation for b_1.

72. **Volume of a Sphere** The formula for the volume V of a sphere is

$$V = \frac{4}{3}\pi r^3.$$

Solve this equation for r.

73. **Celsius and Fahrenheit** The formula for Celsius temperature in terms of Fahrenheit temperature is

$$C = \frac{5}{9}(F - 32).$$

Solve the equation for F.

P.4 Lines in the Plane

What you'll learn about

- Slope of a Line
- Point-Slope Form Equation of a Line
- Slope-Intercept Form Equation of a Line
- Graphing Linear Equations in Two Variables
- Parallel and Perpendicular Lines
- Applying Linear Equations in Two Variables

... and why

Linear equations are used extensively in applications involving business and behavioral science.

Slope of a Line

The slope of a nonvertical line is the ratio of the amount of vertical change to the amount of horizontal change between any two points on the line. For the points (x_1, y_1) and (x_2, y_2), the vertical change is $\Delta y = y_2 - y_1$ and the horizontal change is $\Delta x = x_2 - x_1$. Δy is read "delta" y. See Figure P.21.

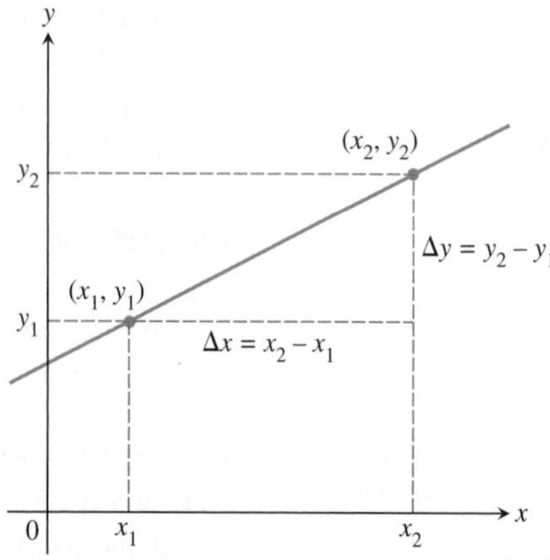

FIGURE P.21 The slope of a nonvertical line can be found from the coordinates of any two points of the line.

DEFINITION Slope of a Line

The **slope** of a nonvertical line through the points (x_1, y_1) and (x_2, y_2) is

$$m = \frac{\Delta y}{\Delta x} = \frac{y_2 - y_1}{x_2 - x_1}.$$

If the line is vertical, then $x_1 = x_2$ and the slope is undefined.

EXAMPLE 1 Finding the Slope of a Line

Find the slope of the line through the two points. Sketch a graph of the line.

(a) $(-1, 2)$ and $(4, -2)$ **(b)** $(1, 1)$ and $(3, 4)$

SOLUTION

(a) The two points are $(x_1, y_1) = (-1, 2)$ and $(x_2, y_2) = (4, -2)$. Thus,

$$m = \frac{y_2 - y_1}{x_2 - x_1} = \frac{(-2) - 2}{4 - (-1)} = -\frac{4}{5}.$$

(b) The two points are $(x_1, y_1) = (1, 1)$ and $(x_2, y_2) = (3, 4)$. Thus,

$$m = \frac{y_2 - y_1}{x_2 - x_1} = \frac{4 - 1}{3 - 1} = \frac{3}{2}.$$

The graphs of these two lines are shown in Figure P.22.

Now try Exercise 3.

Slope Formula

The slope does not depend on the order of the points. We could use $(x_1, y_1) = (4, -2)$ and $(x_2, y_2) = (-1, 2)$ in Example 1a. Check it out.

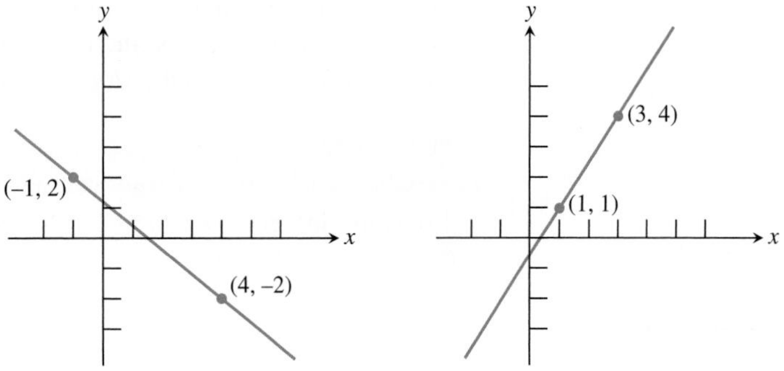

FIGURE P.22 The graphs of the two lines in Example 1.

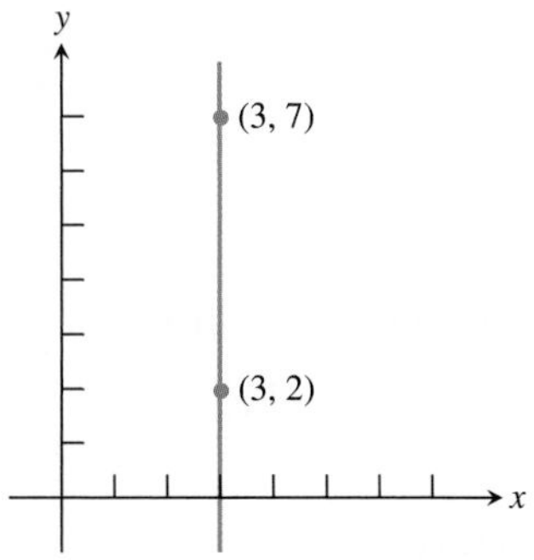

FIGURE P.23 Applying the slope formula to this vertical line gives $m = 5/0$, which is not defined. Thus, the slope of a vertical line is undefined.

Figure P.23 shows a vertical line through the points $(3, 2)$ and $(3, 7)$. If we try to calculate its slope using the slope formula $(y_2 - y_1)/(x_2 - x_1)$, we get zero in the denominator. So, it makes sense to say that a vertical line does not have a slope, or that its slope is undefined.

Point-Slope Form Equation of a Line

If we know the coordinates of one point on a line and the slope of the line, then we can find an equation for that line. For example, the line in Figure P.24 passes through the point (x_1, y_1) and has slope m. If (x, y) is any other point on this line, the definition of the slope yields the equation

$$m = \frac{y - y_1}{x - x_1} \quad \text{or} \quad y - y_1 = m(x - x_1).$$

An equation written this way is in the *point-slope form*.

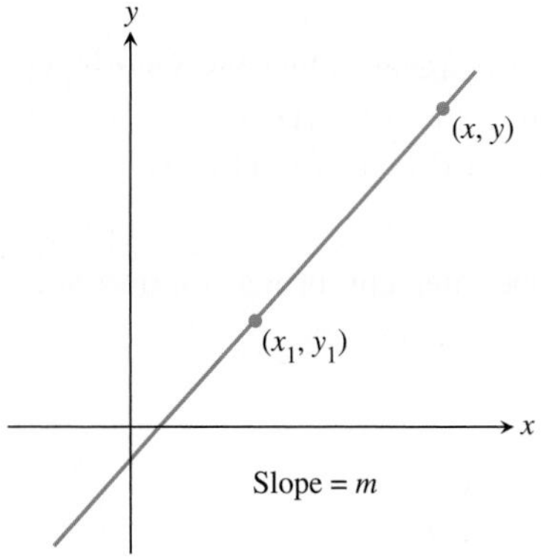

FIGURE P.24 The line through (x_1, y_1) with slope m.

DEFINITION Point-Slope Form of an Equation of a Line

The **point-slope form** of an equation of a line that passes through the point (x_1, y_1) and has slope m is

$$y - y_1 = m(x - x_1).$$

EXAMPLE 2 Using the Point-Slope Form

Use the point-slope form to find an equation of the line that passes through the point $(-3, -4)$ and has slope 2.

SOLUTION Substitute $x_1 = -3$, $y_1 = -4$, and $m = 2$ into the point-slope form, and simplify the resulting equation.

$$\begin{aligned} y - y_1 &= m(x - x_1) && \text{Point-slope form} \\ y - (-4) &= 2(x - (-3)) && x_1 = -3,\ y_1 = -4,\ m = 2 \\ y + 4 &= 2(x + 3) && \text{Simplify.} \end{aligned}$$

For graphing purposes, this equation can be written as $y = 2(x + 3) - 4$ or as $y = 2x + 2$.

Now try Exercise 11.

y-Intercept

The b in $y = mx + b$ is often referred to as "the y-intercept" instead of "the y-coordinate of the y-intercept."

Slope-Intercept Form Equation of a Line

The **y-intercept** of a nonvertical line is the point where the line intersects the y-axis. If we know the y-intercept and the slope of the line, we can apply the point-slope form to find an equation of the line.

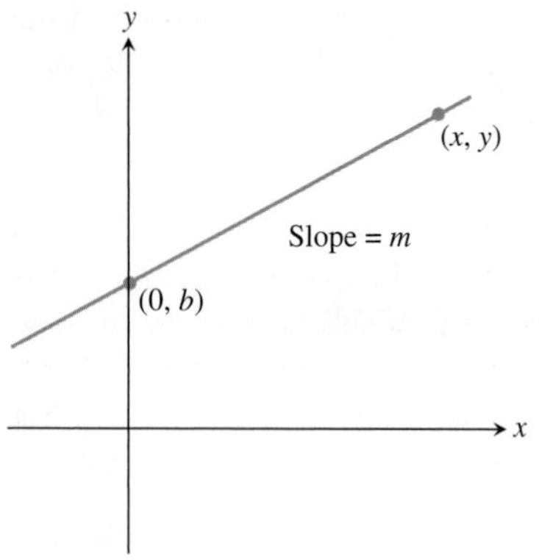

FIGURE P.25 The line with slope m and y-intercept b.

Figure P.25 shows a line with slope m and y-intercept $(0, b)$, or b for short. A point-slope form equation for this line is $y - b = m(x - 0)$. By rewriting this equation we obtain the form known as the *slope-intercept form*.

DEFINITION Slope-Intercept Form of an Equation of a Line

The **slope-intercept form** of an equation of a line with slope m and y-intercept $(0, b)$ is

$$y = mx + b.$$

EXAMPLE 3 Using the Slope-Intercept Form

Using the slope-intercept form, write an equation of the line with slope 3 that passes through the point $(-1, 6)$.

SOLUTION

$y = mx + b$	Slope-intercept form
$y = 3x + b$	$m = 3$
$6 = 3(-1) + b$	$y = 6$ when $x = -1$
$b = 9$	

The slope-intercept form of the equation is $y = 3x + 9$. **Now try Exercise 21.**

Alternative Solution

You could solve Example 3 using the point-slope form:

$$y - 6 = 3(x - (-1))$$
$$y = 3(x + 1) + 6$$
$$y = 3x + 3 + 6$$
$$y = 3x + 9$$

We should not use the phrase "*the* equation of a line" because each line has many equations. Every line has an equation that can be written in the form $Ax + By + C = 0$ where A and B are not both zero. This form is the **general form** for an equation of a line.

If $B \neq 0$, the general form can be changed to the slope-intercept form as follows:

$$Ax + By + C = 0$$
$$By = -Ax - C$$
$$y = \underbrace{-\frac{A}{B}}_{\text{slope}}x + \underbrace{\left(-\frac{C}{B}\right)}_{y\text{-intercept}}$$

Forms of Equations of Lines

General form:	$Ax + By + C = 0$, A and B not both zero
Slope-intercept form:	$y = mx + b$
Point-slope form:	$y - y_1 = m(x - x_1)$
Vertical line:	$x = a$
Horizontal line:	$y = b$

Graphing Linear Equations in Two Variables

A **linear equation in x and y** is one that can be written in the form

$$Ax + By = C,$$

where A and B are not both zero. Rewriting the equation as $Ax + By - C = 0$ we see that it is closely related to the general form. If $B = 0$, the line is vertical, and if $A = 0$, the line is horizontal.

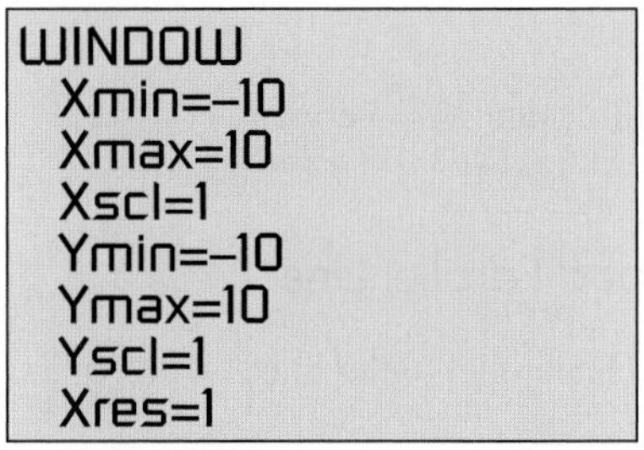

FIGURE P.26 The window dimensions for the *standard window*. The notation "$[-10, 10]$ by $[-10, 10]$" is used to represent window dimensions like these.

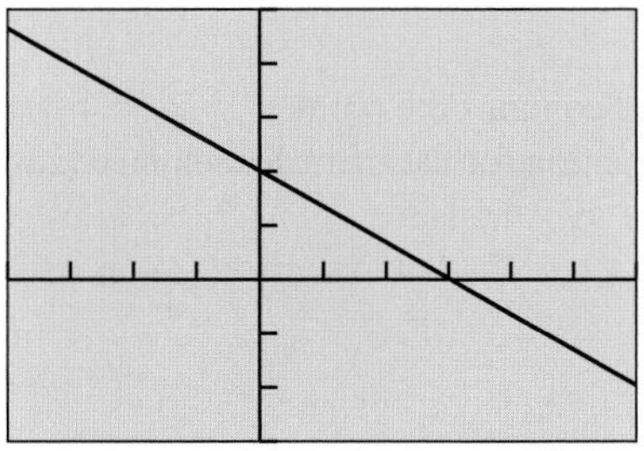

[–4, 6] by [–3, 5]

FIGURE P.27 The graph of $2x + 3y = 6$. The points $(0, 2)$ (y-intercept) and $(3, 0)$ (x-intercept) appear to lie on the graph and, as pairs, are solutions of the equation, providing visual support that the graph is correct. (Example 4)

Viewing Window

The **viewing window** $[-4, 6]$ by $[-3, 5]$ in Figure P.27 means $-4 \le x \le 6$ and $-3 \le y \le 5$.

Square Viewing Window

A **square viewing window** on a grapher is one in which angles appear to be true. For example, the line $y = x$ will appear to make a 45° angle with the positive x-axis. Furthermore, a distance of 1 on the x- and y-axes will appear to be the same. That is, if Xscl = Yscl, the distance between consecutive tick marks on the x- and y-axes will appear to be the same.

The **graph** of an equation in x and y consists of all pairs (x, y) that are solutions of the equation. For example, $(1, 2)$ is a **solution** of the equation $2x + 3y = 8$ because substituting $x = 1$ and $y = 2$ into the equation leads to the true statement $8 = 8$. The pairs $(-2, 4)$ and $(2, 4/3)$ are also solutions.

Because the graph of a linear equation in x and y is a straight line, find two solutions and then connect them with a straight line to draw the graph. If a line is neither horizontal nor vertical, then two easy points to find are its x-intercept and the y-intercept. The **x-intercept** is the point $(x', 0)$ where the graph intersects the x-axis. Set $y = 0$ and solve for x to find the x-intercept. The coordinates of the y-intercept are $(0, y')$. Set $x = 0$ and solve for y to find the y-intercept.

Graphing with a Graphing Utility

To draw a graph of an equation using a grapher:

1. Rewrite the equation in the form $y =$ (an expression in x).
2. Enter the expression into the grapher.
3. Select an appropriate **viewing window** (see Figure P.26).
4. Press the "graph" key.

A graphing utility, or *grapher*, computes y-values for a select set of x-values between Xmin and Xmax and plots the corresponding (x, y) points.

EXAMPLE 4 Using a Graphing Utility

Draw the graph of $2x + 3y = 6$.

SOLUTION First we solve for y.

$$2x + 3y = 6$$

$$3y = -2x + 6 \quad \text{Solve for } y.$$

$$y = -\frac{2}{3}x + 2 \quad \text{Divide by 3.}$$

Figure P.27 shows the graph of $y = -(2/3)x + 2$, or equivalently, the graph of the linear equation $2x + 3y = 6$ in the $[-4, 6]$ by $[-3, 5]$ viewing window.

Now try Exercise 27.

Parallel and Perpendicular Lines

EXPLORATION 1 Investigating Graphs of Linear Equations

1. What do the graphs of $y = mx + b$ and $y = mx + c$, $b \ne c$, have in common? How are they different?
2. Graph $y = 2x$ and $y = -(1/2)x$ in a *square viewing window* (see margin note). On the calculator we use, the "decimal window" is square. Estimate the angle between the two lines.
3. Repeat part 2 for $y = mx$ and $y = -(1/m)x$ with $m = 1, 3, 4,$ and 5.

Parallel lines and perpendicular lines were involved in Exploration 1. Using a grapher to decide whether lines are parallel or perpendicular is risky. Here is an algebraic test to determine whether two lines are parallel or perpendicular.

Parallel and Perpendicular Lines

1. Two nonvertical lines are parallel if and only if their slopes are equal. Any two distinct vertical lines are parallel.
2. Two nonvertical lines are perpendicular if and only if their slopes m_1 and m_2 are opposite reciprocals, that is, if and only if

$$m_1 = -\frac{1}{m_2}.$$

A vertical line is perpendicular to a horizontal line, and vice versa.

EXAMPLE 5 Finding an Equation of a Parallel Line

Find an equation of the line through $P(1, -2)$ that is parallel to the line L with equation $3x - 2y = 1$.

SOLUTION We find the slope of L by writing its equation in slope-intercept form.

$$3x - 2y = 1 \quad \text{Equation for } L$$
$$-2y = -3x + 1 \quad \text{Subtract } 3x.$$
$$y = \frac{3}{2}x - \frac{1}{2} \quad \text{Divide by } -2.$$

The slope of L is $3/2$.

The line whose equation we seek has slope $3/2$ and contains the point $(x_1, y_1) = (1, -2)$. Thus, the point-slope form equation for the line we seek is

$$y + 2 = \frac{3}{2}(x - 1),$$

which can be written in slope-intercept form as

$$y = \frac{3}{2}x - \frac{7}{2}.$$

Now try Exercise 41(a).

EXAMPLE 6 Finding an Equation of a Perpendicular Line

Find an equation of the line through $P(2, -3)$ that is perpendicular to the line L with equation $4x + y = 3$. Support the result with a grapher.

SOLUTION We find the slope of L by writing its equation in slope-intercept form.

$$4x + y = 3 \quad \text{Equation for } L$$
$$y = -4x + 3 \quad \text{Subtract } 4x.$$

The slope of L is -4.

The line whose equation we seek has slope $-1/(-4) = 1/4$ and passes through the point $(x_1, y_1) = (2, -3)$. Thus, the point-slope form equation for the line we seek is

$$y - (-3) = \frac{1}{4}(x - 2)$$
$$y + 3 = \frac{1}{4}x - \frac{1}{2} \quad \text{Distributive property}$$
$$y = \frac{1}{4}x - \frac{7}{2}$$

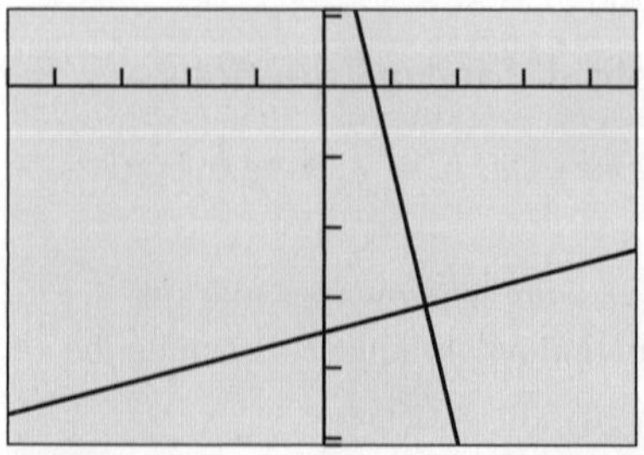

[–4.7, 4.7] by [–5.1, 1.1]

FIGURE P.28 The graphs of $y = -4x + 3$ and $y = (1/4)x - 7/2$ in this square viewing window appear to intersect at a right angle. (Example 6)

Figure P.28 shows the graphs of the two equations in a square viewing window and suggests that the graphs are indeed perpendicular. **Now try Exercise 43(b).**

Applying Linear Equations in Two Variables

Linear equations and their graphs occur frequently in applications. Algebraic solutions to these application problems often require finding an equation of a line and solving a linear equation in one variable. Grapher techniques complement algebraic ones.

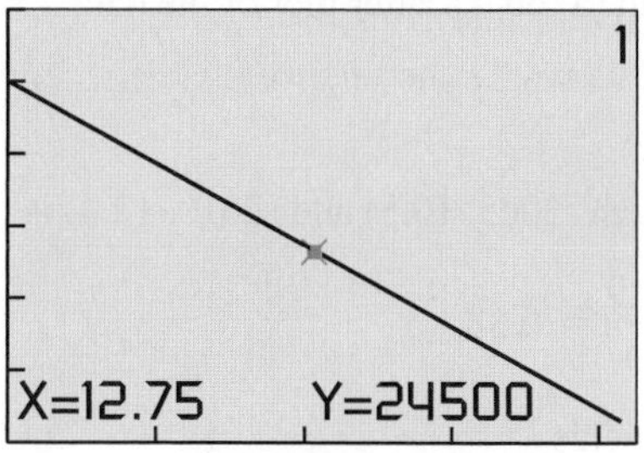

[0, 23.5] by [0, 60000]

(a)

X	Y_1	
12	26000	
12.25	25500	
12.5	25000	
12.75	24500	
13	24000	
13.25	23500	
13.5	23000	
Y_1 = −2000X+50000		

(b)

FIGURE P.29 A (a) graph and (b) table of values for $y = -2000x + 50{,}000$. (Example 7)

EXAMPLE 7 Finding the Depreciation of Real Estate

Camelot Apartments purchased a \$50,000 building. For tax purposes, its value depreciates \$2000 per year over a 25-year period.

(a) Write a linear equation giving the value y of the building in terms of the years x after the purchase.

(b) In how many years will the value of the building be \$24,500?

SOLUTION

(a) We need to determine the value of m and b so that $y = mx + b$, where $0 \le x \le 25$. We know that $y = 50{,}000$ when $x = 0$, so the line has y-intercept $(0, 50{,}000)$ and $b = 50{,}000$. One year after purchase $(x = 1)$, the value of the building is $50{,}000 - 2000 = 48{,}000$. So when $x = 1$, $y = 48{,}000$. Using algebra, we find

$$\begin{aligned} y &= mx + b \\ 48{,}000 &= m \cdot 1 + 50{,}000 && y = 48{,}000 \text{ when } x = 1 \\ -2000 &= m && \text{Subtract } 50{,}000. \end{aligned}$$

The value y of the building x years after its purchase is

$$y = -2000x + 50{,}000.$$

(b) We need to find the value of x when $y = 24{,}500$.

$$y = -2000x + 50{,}000$$

Using algebra again, we find

$$\begin{aligned} 24{,}500 &= -2000x + 50{,}000 && \text{Set } y = 24{,}500. \\ -25{,}500 &= -2000x && \text{Subtract } 50{,}000. \\ 12.75 &= x && \text{Divide by } -2000. \end{aligned}$$

The depreciated value of the building will be \$24,500 exactly 12.75 years, or 12 years 9 months, after the building was purchased by Camelot Apartments. We can support our algebraic work both graphically and numerically. The trace coordinates in Figure P.29a show graphically that $(12.75, 24{,}500)$ is a solution of $y = -2000x + 50{,}000$. This means that $y = 24{,}500$ when $x = 12.75$.

Figure P.29b is a table of values for $y = -2000x + 50{,}000$ for a few values of x. The fourth line of the table shows numerically that $y = 24{,}500$ when $x = 12.75$.

Now try Exercise 45.

Figure P.30 on page 34 shows Americans' income from 2005 to 2010 in trillions of dollars and a corresponding scatter plot of the data. In Example 8, we model the data in Figure P.30 with a linear equation.

Year	Amount (trillions of dollars)
2005	10.5
2006	11.3
2007	11.9
2008	12.5
2009	11.9
2010	12.4

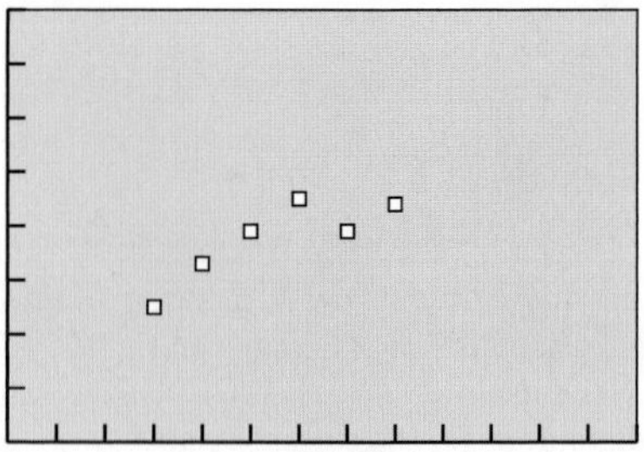

[2002, 2015] by [8, 16]

FIGURE P.30 Americans' personal income. (Example 8)
Source: U.S. Census Bureau, The World Almanac and Book of Facts 2012.

EXAMPLE 8 Finding a Linear Model for Americans' Personal Income over Time

Americans' total personal income in trillions of dollars is given in Figure P.30.

(a) Write a linear equation for Americans' income y in terms of the year x using the points (2005, 10.5) and (2010, 12.4).

(b) Use the equation in (a) to estimate Americans' income in 2008.

(c) Use the equation in (a) to predict Americans' income in 2015.

(d) Superimpose a graph of the linear equation in (a) on a scatter plot of the data like the one shown in Figure P.30.

SOLUTION

(a) Let $y = mx + b$. The slope of the line through (2005, 10.5) and (2010, 12.4) is

$$m = \frac{12.4 - 10.5}{2010 - 2005} = \frac{1.9}{5} = 0.38.$$

Using this slope and the point (2005, 10.5) yields

$$y = 0.38(x - 2005) + 10.5.$$

(b) To estimate Americans' total personal income for the year 2008, we let $x = 2008$ in the equation found in part (a):

$$\begin{aligned} y &= 0.38(2008 - 2005) + 10.5 \\ &= 0.38 \cdot 3 + 10.5 \\ &= 1.14 + 10.5 \\ &= 11.64 \end{aligned}$$

This value of roughly \$11.6 trillion underestimates the actual total income of \$12.5 trillion.

(c) To predict Americans' total personal income for 2015, we let $x = 2015$:

$$\begin{aligned} y &= 0.38(2015 - 2005) + 10.5 \\ &= 0.38 \cdot 10 + 10.5 \\ &= 3.8 + 10.5 \\ &= 14.3 \end{aligned}$$

Our linear model projects that the total personal income of all Americans in 2015 should be about \$14.3 trillion.

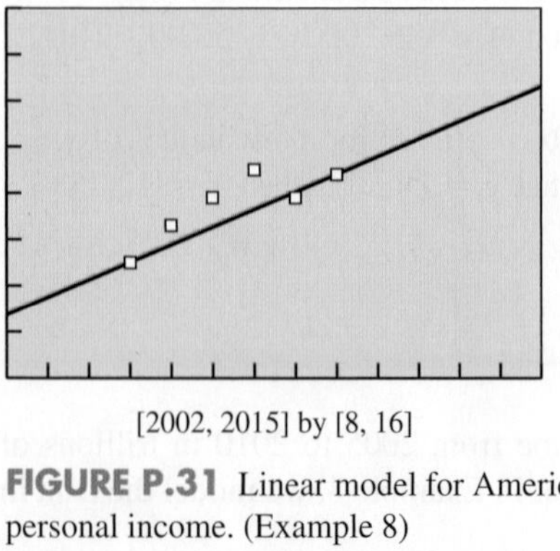

[2002, 2015] by [8, 16]

FIGURE P.31 Linear model for Americans' personal income. (Example 8)

(d) Figure P.31 shows the scatter plot as well as the graph of the line we used to make our estimation and prediction. **Now try Exercise 51.**

The Speed of Light

Whether light traveled instantaneously or actually took some time was an open question for thousands of years. Galileo Galilei (1564–1642) was one of the first to approximate the speed of light. Jean Bernard Léon Foucault (1819–1868) established a modern estimate for light's speed. The speed of light played an important role in Einstein's development of special relativity and continues to be important in physics and astronomy. In 2013, whether the speed of light in a vacuum is truly a universal constant was called into question.

CHAPTER OPENER **Problem** (from page 1)

Problem: Assume that the speed of light is about 299,800 km/sec (186,300 mi/sec).

(a) If light travels from the Sun to Earth in 8.32 min, approximate the distance between Earth and the Sun.

(b) If the distance from the Moon to Earth is roughly 384,400 km, approximate the time required for light to travel from the Moon to Earth.

(c) If light travels on average from the Sun to Pluto in about 5 hr 28 min, approximate the average distance between the Sun and Pluto.

Solution: We use the linear equation $d = r \cdot t$ (distance = rate × time) and the given rate $r = 299{,}800$ km/sec.

(a) Because $t = 8.32$ min $= 499.2$ sec,

$$d = r \cdot t = 299{,}800 \text{ km/sec} \times 499.2 \text{ sec} \approx 150{,}000{,}000 \text{ km}.$$

The distance from the Sun to Earth is about 150 million kilometers (93 million miles).

(b) Because $d = 384{,}400$ km,

$$t = \frac{d}{r} = \frac{384{,}400 \text{ km}}{299{,}800 \text{ km/sec}} \approx 1.282 \text{ sec}.$$

The time it takes light to travel from the Moon to Earth is about 1.282 sec.

(c) Because $t = 5$ hr 28 min $= 328$ min $= 19{,}680$ sec,

$$d = r \cdot t = 299{,}800 \text{ km/sec} \times 19{,}680 \text{ sec} = 5{,}900{,}064{,}000 \text{ km}.$$

The average distance from the Sun to Pluto is about 5.9×10^9 km.

QUICK REVIEW **P.4**

Exercise numbers with a gray background indicate problems that the authors have designed to be solved *without a calculator*.

In Exercises 1–4, solve for x.

1. $-75x + 25 = 200$

2. $400 - 50x = 150$

3. $3(1 - 2x) + 4(2x - 5) = 7$

4. $2(7x + 1) = 5(1 - 3x)$

In Exercises 5–8, solve for y.

5. $2x - 5y = 21$

6. $\frac{1}{3}x + \frac{1}{4}y = 2$

7. $2x + y = 17 + 2(x - 2y)$

8. $x^2 + y = 3x - 2y$

In Exercises 9 and 10, simplify the fraction.

9. $\dfrac{9 - 5}{-2 - (-8)}$

10. $\dfrac{-4 - 6}{-14 - (-2)}$

SECTION P.4 Exercises

Exercise numbers with a gray background indicate problems that the authors have designed to be solved *without a calculator*.

In Exercises 1 and 2, estimate the slope of the line.

1.

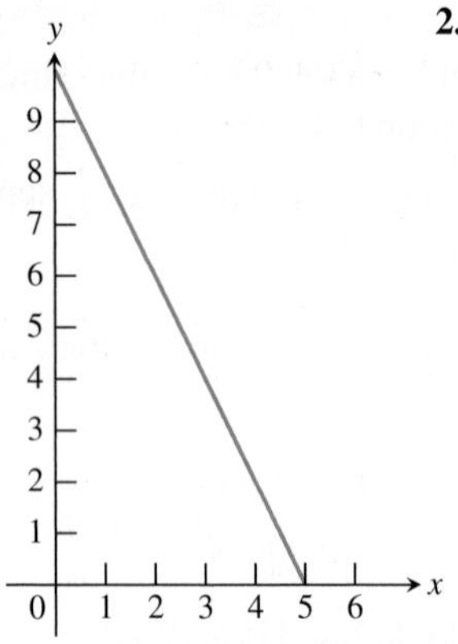

2.

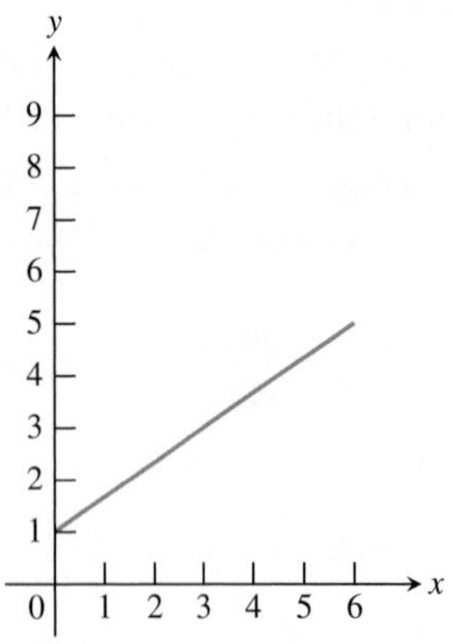

In Exercises 3–6, find the slope of the line through the pair of points.

3. $(-3, 5)$ and $(4, 9)$ **4.** $(-2, 1)$ and $(5, -3)$

5. $(-2, -5)$ and $(-1, 3)$ **6.** $(5, -3)$ and $(-4, 12)$

In Exercises 7–10, find the value of x or y so that the line through the pair of points has the given slope.

Points	Slope
7. $(x, 3)$ and $(5, 9)$	$m = 2$
8. $(-2, 3)$ and $(4, y)$	$m = -3$
9. $(-3, -5)$ and $(4, y)$	$m = 3$
10. $(-8, -2)$ and $(x, 2)$	$m = 1/2$

In Exercises 11–14, find a point-slope form equation for the line through the point with given slope.

Point	Slope	Point	Slope
11. $(1, 4)$	$m = 2$	**12.** $(-4, 3)$	$m = -2/3$
13. $(5, -4)$	$m = -2$	**14.** $(-3, 4)$	$m = 3$

In Exercises 15–20, find a general form equation for the line through the pair of points.

15. $(-7, -2)$ and $(1, 6)$ **16.** $(-3, -8)$ and $(4, -1)$

17. $(1, -3)$ and $(5, -3)$ **18.** $(-1, -5)$ and $(-4, -2)$

19. $(-1, 2)$ and $(2, 5)$ **20.** $(4, -1)$ and $(4, 5)$

In Exercises 21–26, find a slope-intercept form equation for the line.

21. The line through $(0, 5)$ with slope $m = -3$

22. The line through $(1, 2)$ with slope $m = 1/2$

23. The line through the points $(-4, 5)$ and $(4, 3)$

24. The line through the points $(4, 2)$ and $(-3, 1)$

25. The line $2x + 5y = 12$

26. The line $7x - 12y = 96$

In Exercises 27–30, graph the linear equation on a grapher. Choose a viewing window that shows the line intersecting both the x- and y-axes.

27. $8x + y = 49$ **28.** $2x + y = 35$

29. $123x + 7y = 429$ **30.** $2100x + 12y = 3540$

In Exercises 31 and 32, the line contains the origin and the point in the upper right corner of the grapher screen.

31. Writing to Learn Which line shown here has the greater slope? Explain.

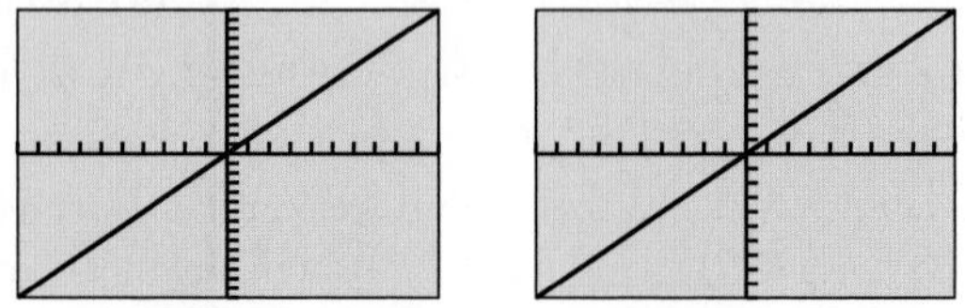

[–10, 10] by [–15, 15]
(a)

[–10, 10] by [–10, 10]
(b)

32. Writing to Learn Which line shown here has the greater slope? Explain.

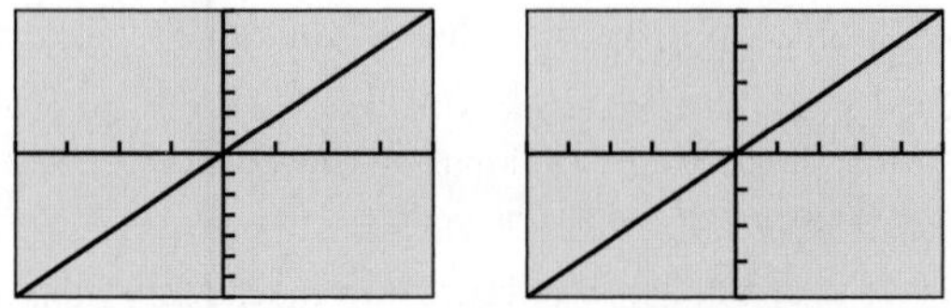

[–20, 20] by [–35, 35]
(a)

[–5, 5] by [–20, 20]
(b)

In Exercises 33–36, find the value of x and the value of y for which $(x, 14)$ and $(18, y)$ are points on the graph.

33. $y = 0.5x + 12$ **34.** $y = -2x + 18$

35. $3x + 4y = 26$ **36.** $3x - 2y = 14$

In Exercises 37–40, find the values for Ymin, Ymax, and Yscl that will make the graph of the line appear in the viewing window as shown here.

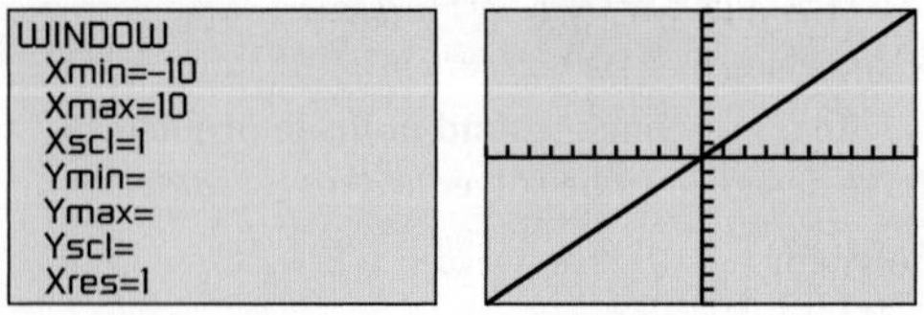

37. $y = 3x$ **38.** $y = 5x$

39. $y = \frac{2}{3}x$ **40.** $y = \frac{5}{4}x$

In Exercises 41–44, (a) find an equation for the line passing through the point and parallel to the given line, and (b) find an equation for the line passing through the point and perpendicular to the given line. Support your work graphically.

	Point	**Line**
41.	$(1, 2)$	$y = 3x - 2$
42.	$(-2, 3)$	$y = -2x + 4$
43.	$(3, 1)$	$2x + 3y = 12$
44.	$(6, 1)$	$3x - 5y = 15$

45. Real Estate Appreciation Bob Michaels purchased a house 8 years ago for $42,000. This year it was appraised at $67,500.

(a) A linear equation $V = mt + b, 0 \leq t \leq 15$, represents the value V of the house for 15 years after it was purchased. Determine m and b.

(b) Graph the equation and trace to estimate in how many years after purchase this house will be worth $72,500.

(c) Write and solve an equation algebraically to determine how many years after purchase this house will be worth $74,000.

(d) Determine how many years after purchase this house will be worth $80,250.

46. Investment Planning Mary Ellen plans to invest $18,000, putting part of the money x into a savings account that pays 5% annually and the rest into an account that pays 8% annually.

(a) What are the possible values of x in this situation?

(b) If Mary Ellen invests x dollars at 5%, write an equation that describes the total interest I received from both accounts at the end of one year.

(c) Graph and trace to estimate how much Mary Ellen invested at 5% if she earned $1020 in total interest at the end of the first year.

(d) Use your grapher to generate a table of values for I to find out how much Mary Ellen should invest at 5% to earn $1185 in total interest in one year.

47. Navigation A commercial jet airplane climbs at takeoff with slope $m = 3/8$. How far in the horizontal direction will the airplane fly to reach an altitude of 12,000 ft above the takeoff point?

48. Grade of a Highway Interstate 70 west of Denver, Colorado, has a section posted as a 6% grade. This means that for a horizontal change of 100 ft there is a 6-ft vertical change.

6% grade

(a) Find the slope of this section of the highway.

(b) On a highway with a 6% grade what is the horizontal distance required to climb 250 ft?

(c) A sign along the highway says 6% grade for the next 7 mi. Estimate how many feet of vertical change there are along those next 7 mi. (There are 5280 ft in 1 mile.)

49. Writing to Learn Building Specifications Asphalt shingles do not meet code specifications on a roof that has less than a 4-12 pitch. A 4-12 pitch means there are 4 ft of vertical change in 12 ft of horizontal change. A certain roof has slope $m = 3/8$. Could asphalt shingles be used on that roof? Explain.

50. Revisiting Example 8 Use the linear equation found in Example 8 to estimate Americans' income in 2004, 2006, 2007 displayed in Figure P.30.

51. Americans' Spending The (x, y) table shows total personal consumption expenditures (y) in the United States in trillions of dollars for selected years (x). (*Source: U.S. Bureau of Economic Analysis, The World Almanac and Book of Facts 2012.*)

x	1990	1995	2000	2005	2010
y	3.8	5.0	6.8	8.8	10.2

(a) Write a linear equation for Americans' spending (y) in terms of the year (x), using the points (1990, 3.8) and (2010, 10.2).

(b) Use the equation in (a) to estimate Americans' expenditures in 2005.

(c) Use the equation in (a) to predict Americans' expenditures in 2015.

(d) Superimpose a graph of the linear equation in (a) on a scatter plot of the data.

52. U.S. Imports from Mexico The (x, y) table shows total U.S. imports from Mexico (y) in billions of dollars for selected years (x). (*Source: U.S. Census Bureau, The World Almanac and Book of Facts 2012.*)

x	2003	2004	2005	2006	2007	2008	2009	2010
y	138.1	155.9	170.1	198.3	210.7	215.9	176.7	229.9

(a) Use the pairs (2003, 138.1) and (2010, 229.9) to write a linear equation for x and y.

(b) Superimpose the graph of the linear equation in (a) on a scatter plot of the data.

(c) Use the equation in (a) to predict the total U.S. imports from Mexico in 2015.

53. World Population Table P.7 shows the midyear worldwide human population for selected years.

Table P.7 World Population

Year	Population (millions)
1980	4453
1990	5282
2000	6085
2010	6972
2013	7130

Source: U.S. Census Bureau.

(a) Let $x = 0$ represent 1980, $x = 1$ represent 1981, and so forth. Draw a scatter plot of the data.

(b) Use the 1980 and 2010 data to write a linear equation for the population y (in millions) in terms of the year x. Superimpose the graph of the linear equation on the scatter plot of the data.

(c) Use the graph in (b) to predict the midyear world population in 2020.

54. U.S. Exports to Canada Table P.8 shows the total exports from the United States to Canada in billions of dollars for selected years.

Table P.8 U.S. Exports to Canada

Year	U.S. Exports (billions of dollars)
2003	169.9
2004	189.9
2005	211.9
2006	230.7
2007	248.9
2008	261.2
2009	204.7
2010	249.1

Source: U.S. Census Bureau, The World Almanac and Book of Facts 2012.

(a) Draw a scatter plot of the data.

(b) Use the 2003 and 2010 data to write a linear equation for the U.S. exports y in terms of the year x. Superimpose the graph of the equation on the scatter plot.

(c) Use the equation in (b) to predict the U.S. exports to Canada for 2015.

In Exercises 55 and 56, determine a so that the line segments AB and CD are parallel.

55.

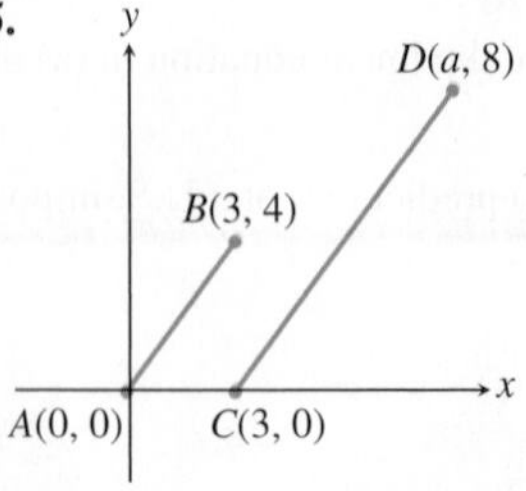

56.

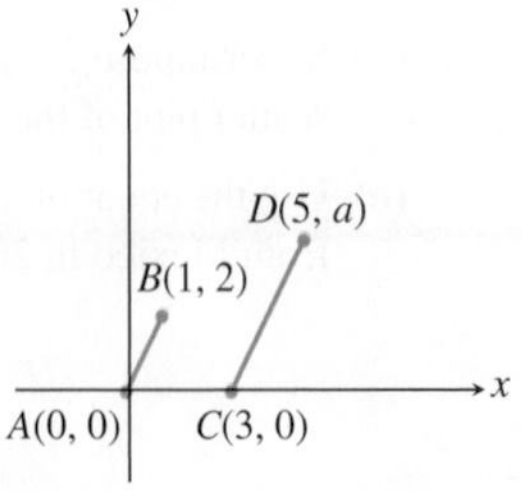

In Exercises 57 and 58, determine a and b so that figure $ABCD$ is a parallelogram.

57.

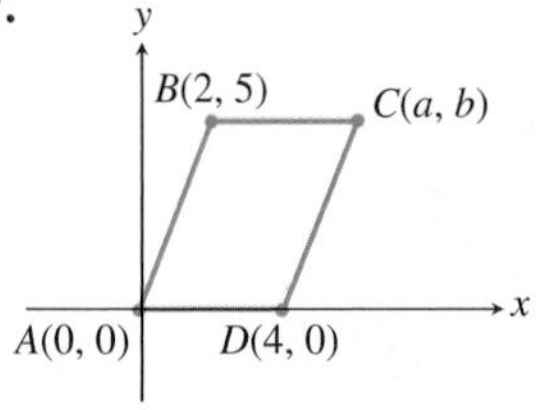

58.

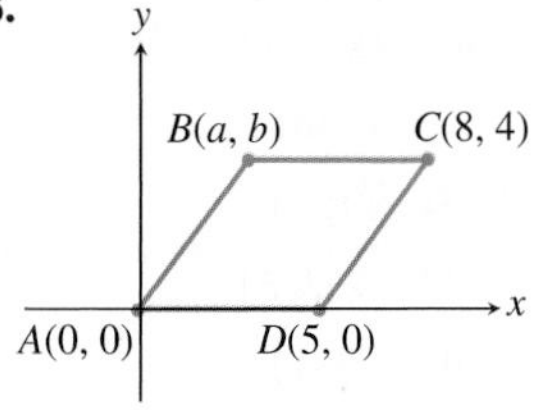

59. Writing to Learn Perpendicular Lines

(a) Is it possible for two lines with positive slopes to be perpendicular? Explain.

(b) Is it possible for two lines with negative slopes to be perpendicular? Explain.

60. Group Activity Parallel and Perpendicular Lines

(a) Assume that $c \neq d$ and a and b are not both zero. Show that $ax + by = c$ and $ax + by = d$ are parallel lines. Explain why the restrictions on a, b, c, and d are necessary.

(b) Assume that a and b are not both zero. Show that $ax + by = c$ and $bx - ay = d$ are perpendicular lines. Explain why the restrictions on a and b are necessary.

Standardized Test Questions

61. True or False The slope of a vertical line is zero. Justify your answer.

62. True or False The graph of any equation of the form $ax + by = c$, where a and b are not both zero, is always a line. Justify your answer.

In Exercises 63–66, you may use a graphing calculator to solve these problems.

63. Multiple Choice Which of the following is an equation of the line through the point $(-2, 3)$ with slope 4?

(A) $y - 3 = 4(x + 2)$ **(B)** $y + 3 = 4(x - 2)$

(C) $x - 3 = 4(y + 2)$ **(D)** $x + 3 = 4(y - 2)$

(E) $y + 2 = 4(x - 3)$

64. Multiple Choice Which of the following is an equation of the line with slope 3 and y-intercept -2?

(A) $y = 3x + 2$ **(B)** $y = 3x - 2$

(C) $y = -2x + 3$ **(D)** $x = 3y - 2$

(E) $x = 3y + 2$

65. Multiple Choice Which of the following lines is perpendicular to the line $y = -2x + 5$?

(A) $y = 2x + 1$ **(B)** $y = -2x - \frac{1}{5}$

(C) $y = -\frac{1}{2}x + \frac{1}{3}$ **(D)** $y = -\frac{1}{2}x + 3$

(E) $y = \frac{1}{2}x - 3$

66. Multiple Choice Which of the following is the slope of the line through the two points $(-2, 1)$ and $(1, -4)$?

(A) $-\frac{3}{5}$ **(B)** $\frac{3}{5}$

(C) $-\frac{5}{3}$ **(D)** $\frac{5}{3}$

(E) -3

Explorations

67. Exploring the Graph of $\frac{x}{a} + \frac{y}{b} = c,\ a \neq 0,\ b \neq 0$

Let $c = 1$.

(a) Draw the graph for $a = 3, b = -2$.

(b) Draw the graph for $a = -2, b = -3$.

(c) Draw the graph for $a = 5, b = 3$.

(d) Use your graphs in (a), (b), (c) to conjecture what a and b represent when $c = 1$. Prove your conjecture.

(e) Repeat (a)–(d) for $c = 2$. What do a and b represent when $c = 2$?

(f) If $c = -1$, what do a and b represent?

68. Investigating Graphs of Linear Equations

(a) Graph $y = mx$ for $m = -3, -2, -1, 1, 2, 3$ in the window $[-8, 8]$ by $[-5, 5]$. What do these graphs have in common? How are they different?

(b) If $m > 0$, what do the graphs of $y = mx$ and $y = -mx$ have in common? How are they different?

(c) Graph $y = 0.3x + b$ for $b = -3, -2, -1, 0, 1, 2, 3$ in $[-8, 8]$ by $[-5, 5]$. What do these graphs have in common? How are they different?

Extending the Ideas

69. Connecting Algebra and Geometry Show that if the midpoints of consecutive sides of any quadrilateral (see figure) are connected, the result is a parallelogram.

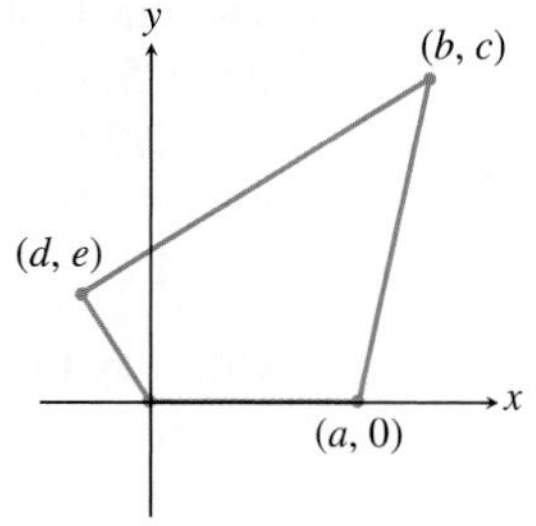

Art for Exercise 69

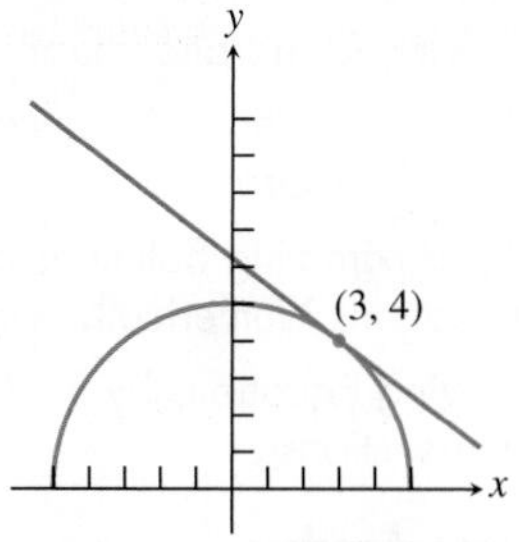

Art for Exercise 70

70. Connecting Algebra and Geometry Consider the semicircle of radius 5 centered at $(0, 0)$ as shown in the figure. Find an equation of the line tangent to the semicircle at the point $(3, 4)$. (*Hint:* A line tangent to a circle is perpendicular to the radius at the point of tangency.)

71. Connecting Algebra and Geometry Show that in any triangle (see figure), the line segment joining the midpoints of two sides is parallel to the third side and is half as long.

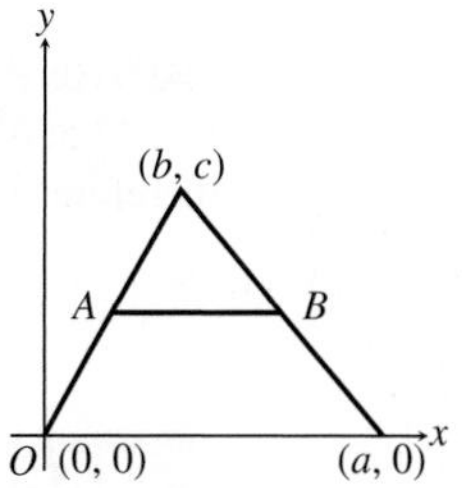

P.5 Solving Equations Graphically, Numerically, and Algebraically

What you'll learn about

- Solving Equations Graphically
- Solving Quadratic Equations
- Approximating Solutions of Equations Graphically
- Approximating Solutions of Equations Numerically with Tables
- Solving Equations by Finding Intersections

... and why

These basic techniques are involved in using a graphing utility to solve equations in this textbook.

Solving Equations Graphically

The graph of the equation $y = 2x - 5$ (in x and y) can be used to solve the equation $2x - 5 = 0$ (in x). Using the techniques of Section P.3, we can show algebraically that $x = 5/2$ is a solution of $2x - 5 = 0$. Therefore, the ordered pair $(5/2, 0)$ is a solution of $y = 2x - 5$. Figure P.32 suggests that the x-intercept of the graph of the line $y = 2x - 5$ is the point $(5/2, 0)$ as it should be.

One way to solve an equation graphically is to find all its x-intercepts. There are many graphical techniques that can be used to find x-intercepts.

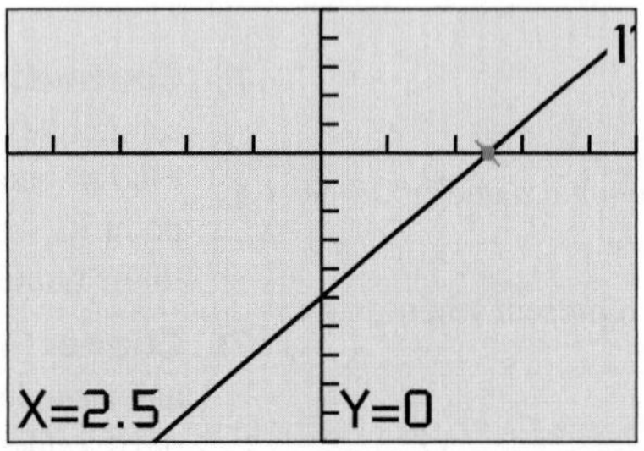

[–4.7, 4.7] by [–10, 5]

FIGURE P.32 Using the TRACE feature of a grapher, we see that $(2.5, 0)$ is an x-intercept of the graph of $y = 2x - 5$ and, therefore, $x = 2.5$ is a solution of the equation $2x - 5 = 0$.

EXAMPLE 1 Solving by Finding x-Intercepts

Solve the equation $2x^2 - 3x - 2 = 0$ graphically. Confirm algebraically.

SOLUTION

Solve Graphically Find the x-intercepts of the graph of $y = 2x^2 - 3x - 2$ (Figure P.33). We use TRACE to see that $(-0.5, 0)$ and $(2, 0)$ are x-intercepts of this graph. Thus, the solutions of this equation are $x = -0.5$ and $x = 2$. Answers obtained graphically are really approximations, although in general they are very good approximations.

Confirm Algebraically In this case, we can use factoring to find exact values.

$$2x^2 - 3x - 2 = 0$$
$$(2x + 1)(x - 2) = 0 \quad \text{Factor.}$$

We can conclude that

$$2x + 1 = 0 \quad \text{or} \quad x - 2 = 0$$
$$x = -1/2 \quad \text{or} \quad x = 2$$

So, $x = -1/2$ and $x = 2$ are the exact solutions of the original equation.

Now try Exercise 1.

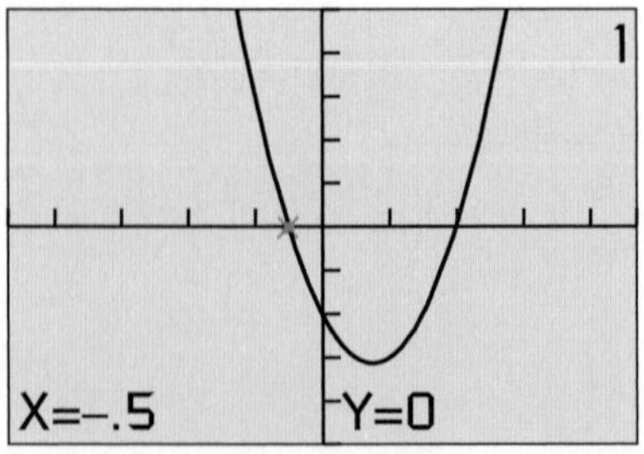

[–4.7, 4.7] by [–5, 5]

FIGURE P.33 It appears that $(-0.5, 0)$ and $(2, 0)$ are x-intercepts of the graph of $y = 2x^2 - 3x - 2$. (Example 1)

The algebraic solution procedure used in Example 1 is a special case of the following important property.

Zero Factor Property

Let a and b be real numbers.

If $ab = 0$, then $a = 0$ or $b = 0$.

Solving Quadratic Equations

Linear equations $(ax + b = 0)$ and *quadratic equations* are two members of the family of *polynomial equations*, which will be studied in detail in Chapter 2.

DEFINITION Quadratic Equation in x

A **quadratic equation in x** is one that can be written in the form

$$ax^2 + bx + c = 0,$$

where a, b, and c are real numbers and $a \neq 0$.

We review some of the basic algebraic techniques for solving quadratic equations. One algebraic technique that we have already used in Example 1 is *factoring*.

Quadratic equations of the form $(ax + b)^2 = c$ are fairly easy to solve as illustrated in Example 2.

Finding Square Roots

If $t^2 = k > 0$, then $t = \sqrt{k}$ or $t = -\sqrt{k}$.

EXAMPLE 2 Solving by Extracting Square Roots

Solve $(2x - 1)^2 = 9$ algebraically.

SOLUTION

$$(2x - 1)^2 = 9$$

$$2x - 1 = \pm 3 \qquad \text{Extract square roots.}$$

$$2x = 4 \quad \text{or} \quad 2x = -2 \qquad \text{Add 1.}$$

$$x = 2 \quad \text{or} \quad x = -1 \qquad \text{Divide by 2.}$$

Now try Exercise 9.

The technique of Example 2 is more general than you might think because every quadratic equation can be written in the form $(x + b)^2 = c$. The procedure we need to accomplish this is *completing the square*.

Completing the Square

To solve $x^2 + bx = c$ by **completing the square**, add $(b/2)^2$ to both sides of the equation and factor the left side of the new equation.

$$x^2 + bx + \left(\frac{b}{2}\right)^2 = c + \left(\frac{b}{2}\right)^2$$

$$\left(x + \frac{b}{2}\right)^2 = c + \frac{b^2}{4}$$

In general, to solve a quadratic equation by completing the square, first we divide both sides of the equation by the coefficient of x^2, and then we complete the square as illustrated in Example 3.

EXAMPLE 3 Solving by Completing the Square

Solve $4x^2 - 20x + 17 = 0$ by completing the square.

SOLUTION

$$4x^2 - 20x + 17 = 0$$

$$x^2 - 5x + \frac{17}{4} = 0 \qquad \text{Divide by 4.}$$

$$x^2 - 5x = -\frac{17}{4} \qquad \text{Subtract } \frac{17}{4}.$$

Completing the square for the equation above we obtain

$$x^2 - 5x + \left(-\frac{5}{2}\right)^2 = -\frac{17}{4} + \left(-\frac{5}{2}\right)^2 \qquad \text{Add } \left(-\frac{5}{2}\right)^2.$$

$$\left(x - \frac{5}{2}\right)^2 = 2 \qquad \text{Factor and simplify.}$$

$$x - \frac{5}{2} = \pm\sqrt{2} \qquad \text{Extract square roots.}$$

$$x = \frac{5}{2} \pm \sqrt{2}$$

$$x = \frac{5}{2} + \sqrt{2} \approx 3.91 \text{ or } x = \frac{5}{2} - \sqrt{2} \approx 1.09$$

Now try Exercise 13.

The procedure of Example 3 can be used to solve the general quadratic equation $ax^2 + bx + c = 0$, producing the following formula. (See Exercise 68.)

Quadratic Formula

The solutions of the quadratic equation $ax^2 + bx + c = 0$, where $a \neq 0$, are given by the **quadratic formula**

$$x = \frac{-b \pm \sqrt{b^2 - 4ac}}{2a}.$$

EXAMPLE 4 Using the Quadratic Formula

Solve the equation $3x^2 - 6x = 5$.

SOLUTION First we subtract 5 from both sides of the equation to put it in the form $ax^2 + bx + c = 0$: $3x^2 - 6x - 5 = 0$. We can see that $a = 3$, $b = -6$, and $c = -5$.

$$x = \frac{-b \pm \sqrt{b^2 - 4ac}}{2a} \qquad \text{Quadratic formula}$$

$$x = \frac{-(-6) \pm \sqrt{(-6)^2 - 4(3)(-5)}}{2(3)} \qquad a = 3, b = -6, c = -5$$

$$x = \frac{6 \pm \sqrt{96}}{6} \qquad \text{Simplify.}$$

$$x = \frac{6 + \sqrt{96}}{6} \approx 2.63 \quad \text{or} \quad x = \frac{6 - \sqrt{96}}{6} \approx -0.63$$

The graph of $y = 3x^2 - 6x - 5$ in Figure P.34 supports that the x-intercepts are approximately -0.63 and 2.63.

Now try Exercise 19.

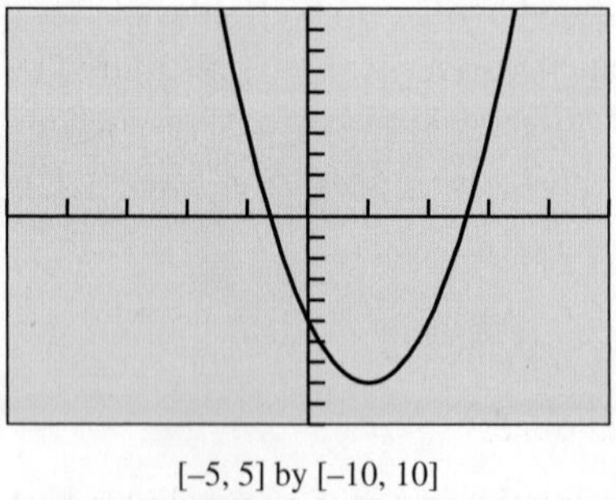

[–5, 5] by [–10, 10]

FIGURE P.34 The graph of $y = 3x^2 - 6x - 5$. (Example 4)

Solving Quadratic Equations Algebraically

There are four basic ways to solve quadratic equations algebraically.

1. **Factoring** (see Example 1)
2. **Extracting Square Roots** (see Example 2)
3. **Completing the Square** (see Example 3)
4. **Using the Quadratic Formula** (see Example 4)

Approximating Solutions of Equations Graphically

A solution of the equation $x^3 - x - 1 = 0$ is a value of x that makes the value of $y = x^3 - x - 1$ equal to zero. Example 5 illustrates a built-in procedure on graphing calculators to find such values of x.

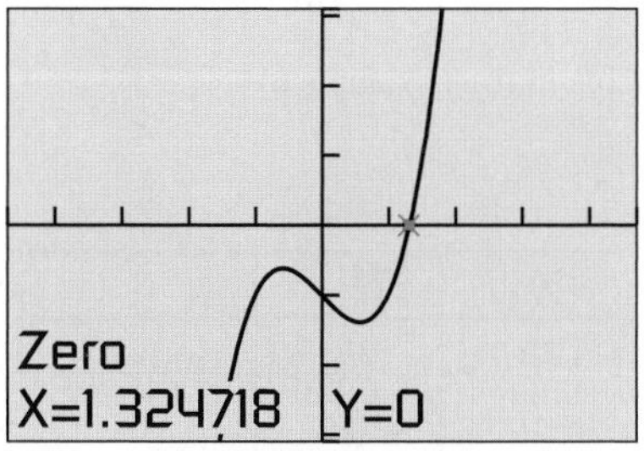

[–4.7, 4.7] by [–3.1, 3.1]

(a)

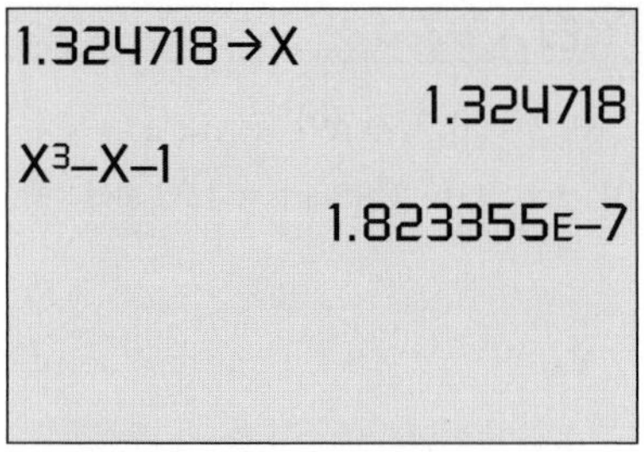

(b)

FIGURE P.35 The graph of $y = x^3 - x - 1$. (a) shows that $(1.324718, 0)$ is an approximation to the x-intercept of the graph. (b) supports this conclusion. (Example 5)

EXAMPLE 5 Solving Graphically

Solve the equation $x^3 - x - 1 = 0$ graphically.

SOLUTION Figure P.35a suggests that $x \approx 1.324718$ is the solution we seek. Figure P.35b provides numerical support that $x = 1.324718$ is a close approximation to the solution because, when $x = 1.324718$, $x^3 - x - 1 \approx 1.82 \times 10^{-7}$, which is nearly zero.

Now try Exercise 31.

When solving equations graphically, we usually get approximate solutions and not exact solutions. We will use the following agreement about accuracy in this book.

Agreement About Approximate Solutions

For applications, round to a value that is reasonable for the context of the problem. For all other problems, round to two decimal places unless directed otherwise.

With this accuracy agreement, we would report the approximate solution found in Example 5 as 1.32.

Approximating Solutions of Equations Numerically with Tables

The table feature on graphing calculators provides a numerical *zoom-in procedure* that we can use to find accurate solutions of equations. We illustrate this procedure in Example 6 with the same equation used in Example 5.

EXAMPLE 6 Solving Using Tables

Solve the equation $x^3 - x - 1 = 0$ using grapher tables.

SOLUTION From Figure P.35a, we know that the solution we seek is between $x = 1$ and $x = 2$. Figure P.36a sets the starting point of the table $(\text{TblStart} = 1)$ at $x = 1$ and the input increments in the table $(\Delta\text{Tbl} = 0.1)$ at 0.1. Figure P.36b shows that the zero of $x^3 - x - 1$ lies between $x = 1.3$ and $x = 1.4$.

The next two steps in this process are shown in Figure P.37.

From Figure P.37a, we can read that the zero is between $x = 1.32$ and $x = 1.33$; from Figure P.37b, we can read that the zero is between $x = 1.324$ and $x = 1.325$. Thus, we conclude that the zero is approximately 1.32, according to our accuracy agreement. **Now try Exercise 37.**

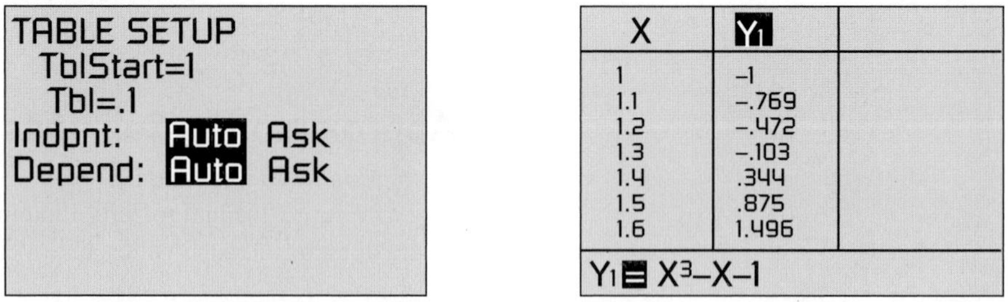
TABLE SETUP
TblStart=1
Tbl=.1
Indpnt: Auto Ask
Depend: Auto Ask
X
Y1
1
1.1
1.2
1.3
1.4
1.5
1.6
-1
-.769
-.472
-.103
.344
.875
1.496
Y1 = X3-X-1

(a) (b)

FIGURE P.36 (a) gives the setup that produces the table in (b). (Example 6)

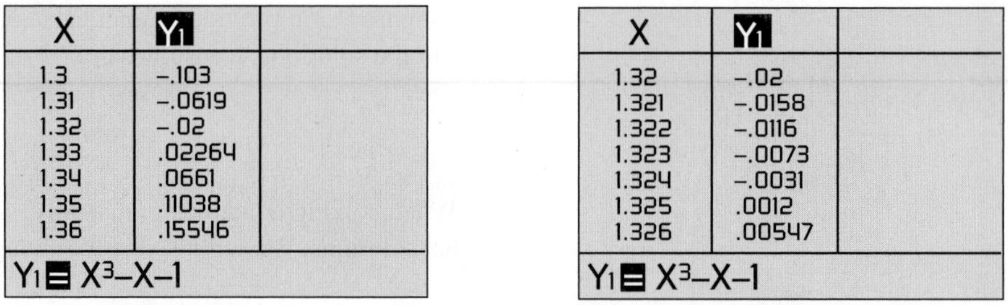
X
Y1
1.3
1.31
1.32
1.33
1.34
1.35
1.36
-.103
-.0619
-.02
.02264
.0661
.11038
.15546
Y1 = X3-X-1
X
Y1
1.32
1.321
1.322
1.323
1.324
1.325
1.326
-.02
-.0158
-.0116
-.0073
-.0031
.0012
.00547
Y1 = X3-X-1

(a) (b)

FIGURE P.37 In (a) TblStart $= 1.3$ and ΔTbl $= 0.01$, and in (b) TblStart $= 1.32$ and ΔTbl $= 0.001$. (Example 6)

EXPLORATION 1 Finding Real Zeros of Equations

Consider the equation $4x^2 - 12x + 7 = 0$.

1. Use a graph to show that this equation has two real solutions, one between 0 and 1 and the other between 2 and 3.
2. Use the numerical zoom-in procedure illustrated in Example 6 to find each zero accurate to two decimal places.
3. Use the built-in zero finder (see Example 5) to find the two solutions. Then round them to two decimal places.
4. If you are familiar with the graphical zoom-in process, use it to find each solution accurate to two decimal places.
5. Compare the numbers obtained in parts 2, 3, and 4.
6. Support the results obtained in parts 2, 3, and 4 numerically.
7. Use the numerical zoom-in procedure illustrated in Example 6 to find each zero accurate to six decimal places. Compare with the answer found in part 3 with the zero finder.

Solving Equations by Finding Intersections

Sometimes we can rewrite an equation and solve it graphically by finding the *points of intersection* of two graphs. A point (a, b) is a **point of intersection** of two graphs if it lies on both graphs.

We illustrate this procedure with the absolute value equation in Example 7.

EXAMPLE 7 Solving by Finding Intersections

Solve the equation $|2x - 1| = 6$.

SOLUTION Figure P.38 suggests that the V-shaped graph of $y = |2x - 1|$ intersects the graph of the horizontal line $y = 6$ twice. We can use TRACE or the intersection feature of our grapher to see that the two points of intersection have coordinates $(-2.5, 6)$ and $(3.5, 6)$. This means that the original equation has two solutions: -2.5 and 3.5.

We can use algebra to find the exact solutions. The only two real numbers with absolute value 6 are 6 itself and -6. So, if $|2x - 1| = 6$, then

$$2x - 1 = 6 \quad \text{or} \quad 2x - 1 = -6$$

$$x = \frac{7}{2} = 3.5 \quad \text{or} \quad x = -\frac{5}{2} = -2.5$$

Now try Exercise 39.

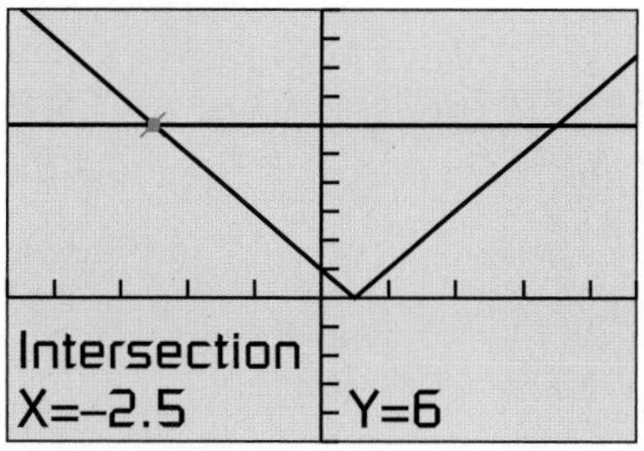

[–4.7, 4.7] by [–5, 10]

FIGURE P.38 The graphs of $y = |2x - 1|$ and $y = 6$ intersect at $(-2.5, 6)$ and $(3.5, 6)$. (Example 7)

QUICK REVIEW P.5

In Exercises 1–4, expand the product.

1. $(3x - 4)^2$
2. $(2x + 3)^2$
3. $(2x + 1)(3x - 5)$
4. $(3y - 1)(5y + 4)$

In Exercises 5–8, factor the expression.

5. $25x^2 - 20x + 4$
6. $15x^3 - 22x^2 + 8x$
7. $3x^3 + x^2 - 15x - 5$
8. $y^4 - 13y^2 + 36$

In Exercises 9 and 10, combine the fractions and reduce the resulting fraction to lowest terms.

9. $\dfrac{x}{2x + 1} - \dfrac{2}{x + 3}$

10. $\dfrac{x + 1}{x^2 - 5x + 6} - \dfrac{3x + 11}{x^2 - x - 6}$

SECTION P.5 Exercises

In Exercises 1–6, solve the equation graphically by finding x-intercepts. Confirm by using factoring to solve the equation.

1. $x^2 - x - 20 = 0$
2. $2x^2 + 5x - 3 = 0$
3. $4x^2 - 8x + 3 = 0$
4. $x^2 - 8x = -15$
5. $x(3x - 7) = 6$
6. $x(3x + 11) = 20$

In Exercises 7–12, solve the equation by extracting square roots.

7. $4x^2 = 25$
8. $2(x - 5)^2 = 17$
9. $3(x + 4)^2 = 8$
10. $4(u + 1)^2 = 18$
11. $2y^2 - 8 = 6 - 2y^2$
12. $(2x + 3)^2 = 169$

In Exercises 13–18, solve the equation by completing the square.

13. $x^2 + 6x = 7$
14. $x^2 + 5x - 9 = 0$
15. $x^2 - 7x + \frac{5}{4} = 0$
16. $4 - 6x = x^2$
17. $2x^2 - 7x + 9 = (x - 3)(x + 1) + 3x$
18. $3x^2 - 6x - 7 = x^2 + 3x - x(x + 1) + 3$

In Exercises 19–24, solve the equation using the quadratic formula.

19. $x^2 + 8x - 2 = 0$
20. $2x^2 - 3x + 1 = 0$
21. $3x + 4 = x^2$
22. $x^2 - 5 = \sqrt{3}x$
23. $x(x + 5) = 12$
24. $x^2 - 2x + 6 = 2x^2 - 6x - 26$

In Exercises 25–28, estimate any x- and y-intercepts that are shown in the graph.

25.

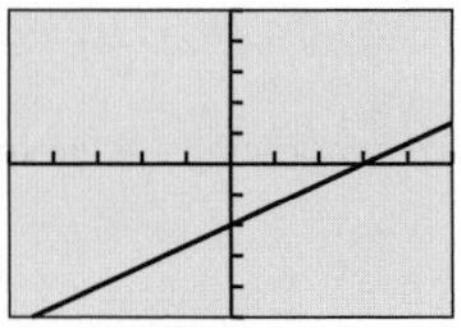

[–5, 5] by [–5, 5]

26.

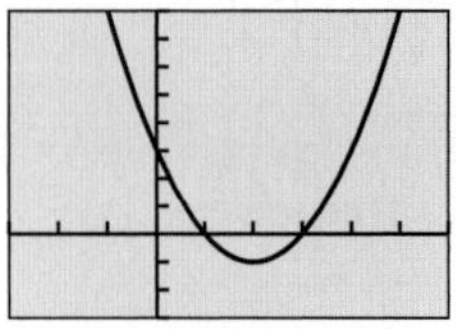

[–3, 6] by [–3, 8]

27

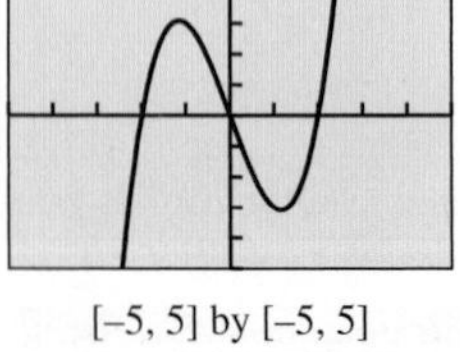

[–5, 5] by [–5, 5]

28

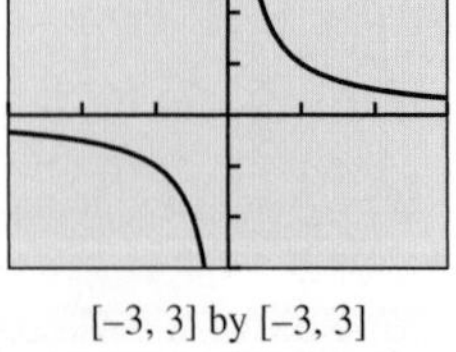

[–3, 3] by [–3, 3]

In Exercises 29–34, solve the equation graphically by finding x-intercepts.

29. $x^2 + x - 1 = 0$
30. $4x^2 + 20x + 23 = 0$
31. $x^3 + x^2 + 2x - 3 = 0$
32. $x^3 - 4x + 2 = 0$
33. $x^2 + 4 = 4x$
34. $x^2 + 2x = -2$

In Exercises 35 and 36, the table permits you to estimate a zero of an expression. State the expression and give the zero as accurately as can be read from the table.

35.

X	Y1
.4	–.04
.41	–.0119
.42	.0164
.43	.0449
.44	.0736
.45	.1025
.46	.1316

Y1 = X²+2X–1

36.

X	Y1
–1.735	–.0177
–1.734	–.0117
–1.733	–.0057
–1.732	3E–4
–1.731	.0063
–1.73	.01228
–1.729	.01826

Y1 = X³–3X

In Exercises 37 and 38, use tables to find the indicated number of solutions of the equation accurate to two decimal places.

37. Two solutions of $x^2 - x - 1 = 0$

38. One solution of $-x^3 + x + 1 = 0$

In Exercises 39–44, solve the equation graphically by finding intersections. Confirm your answer algebraically.

39. $|t - 8| = 2$
40. $|x + 1| = 4$
41. $|2x + 5| = 7$
42. $|3 - 5x| = 4$
43. $|2x - 3| = x^2$
44. $|x + 1| = 2x - 3$

45. Interpreting Graphs The graphs in the two viewing windows shown here can be used to solve the equation $3\sqrt{x + 4} = x^2 - 1$ graphically.

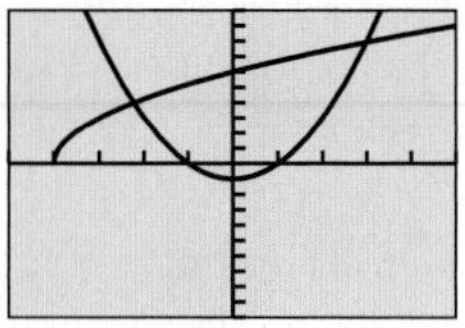

[–5, 5] by [–10, 10]
(a)

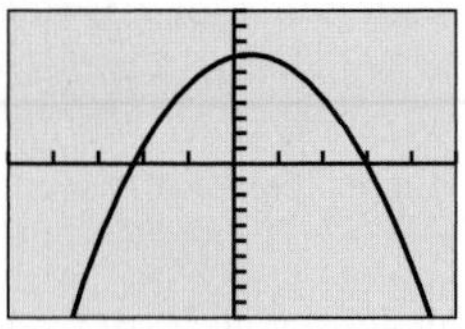

[–5, 5] by [–10, 10]
(b)

(a) The viewing window in (a) illustrates the intersection method for solving. Identify the two equations that are graphed.

(b) The viewing window in (b) illustrates the x-intercept method for solving. Identify the equation that is graphed.

(c) Writing to Learn How are the intersection points in (a) related to the x-intercepts in (b)?

46. Writing to Learn Revisiting Example 6 Explain why all real numbers x that satisfy $1.324 < x < 1.325$ round to 1.32.

In Exercises 47–56, use a method of your choice to solve the equation.

47. $x^2 + x - 2 = 0$
48. $x^2 - 3x = 12 - 3(x - 2)$
49. $|2x - 1| = 5$
50. $x + 2 - 2\sqrt{x + 3} = 0$
51. $x^3 + 4x^2 - 3x - 2 = 0$
52. $x^3 - 4x + 2 = 0$
53. $|x^2 + 4x - 1| = 7$
54. $|x + 5| = |x - 3|$
55. $|0.5x + 3| = x^2 - 4$
56. $\sqrt{x + 7} = -x^2 + 5$

57. Group Activity Discriminant of a Quadratic The radicand $b^2 - 4ac$ in the quadratic formula is called the **discriminant** of the quadratic polynomial $ax^2 + bx + c$ because it can be used to describe the nature of its zeros.

(a) **Writing to Learn** If $b^2 - 4ac > 0$, what can you say about the zeros of the quadratic polynomial $ax^2 + bx + c$? Explain your answer.

(b) **Writing to Learn** If $b^2 - 4ac = 0$, what can you say about the zeros of the quadratic polynomial $ax^2 + bx + c$? Explain your answer.

(c) **Writing to Learn** If $b^2 - 4ac < 0$, what can you say about the zeros of the quadratic polynomial $ax^2 + bx + c$? Explain your answer.

58. Group Activity Discriminant of a Quadratic Use the information learned in Exercise 57 to create a quadratic polynomial with the following numbers of real zeros. Support your answers graphically.

(a) Two real zeros?

(b) Exactly one real zero

(c) No real zeros

59. Size of a Soccer Field Several of the World Cup '94 soccer matches were played in Stanford University's stadium in Menlo Park, California. The field is 30 yd longer than it is wide, and the area of the field is 8800 yd^2. What are the dimensions of this soccer field?

60. Height of a Ladder John's paint crew knows from experience that its 18-ft ladder is particularly stable when the distance from the ground to the top of the ladder is 5 ft more than the distance from the building to the base of the ladder as shown in the figure. In this position, how far up the building does the ladder reach?

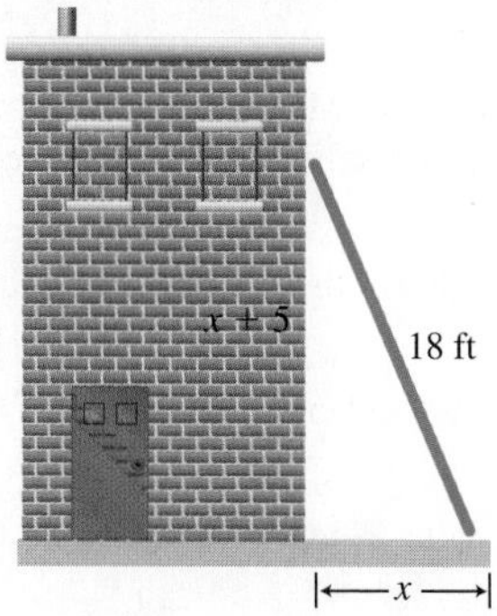

61. Finding the Dimensions of a Norman Window A Norman window has the shape of a square with a semicircle mounted on it. Find the width of the window if the total area of the square and the semicircle is to be 200 ft^2.

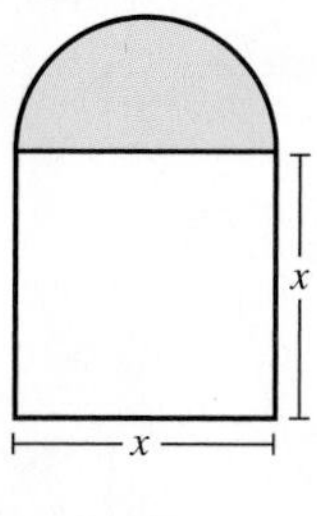

Standardized Test Questions

62. True or False If 2 is an x-intercept of the graph of $y = ax^2 + bx + c$, then 2 is a solution of the equation $ax^2 + bx + c = 0$. Justify your answer.

63. True or False If $2x^2 = 18$, then x must be equal to 3. Justify your answer.

In Exercises 64–67, you may use a graphing calculator to solve these problems.

64. Multiple Choice Which of the following are the solutions of the equation $x(x - 3) = 0$?

(A) Only $x = 3$ (B) Only $x = -3$

(C) $x = 0$ and $x = -3$ (D) $x = 0$ and $x = 3$

(E) There are no solutions.

65. Multiple Choice Which of the following replacements for ? make $x^2 - 5x + ?$ a perfect square?

(A) $-\frac{5}{2}$ (B) $\left(-\frac{5}{2}\right)^2$

(C) $(-5)^2$ (D) $\left(-\frac{2}{5}\right)^2$

(E) -6

66. Multiple Choice Which of the following are the solutions of the equation $2x^2 - 3x - 1 = 0$?

(A) $\frac{3}{4} \pm \sqrt{17}$ (B) $\frac{3 \pm \sqrt{17}}{4}$

(C) $\frac{3 \pm \sqrt{17}}{2}$ (D) $\frac{-3 \pm \sqrt{17}}{4}$

(E) $\frac{3 \pm 1}{4}$

67. Multiple Choice Which of the following are the solutions of the equation $|x - 1| = -3$?

(A) Only $x = 4$ (B) Only $x = -2$

(C) Only $x = 2$ (D) $x = 4$ and $x = -2$

(E) There are no solutions.

Explorations

68. Deriving the Quadratic Formula Follow these steps to use completing the square to solve $ax^2 + bx + c = 0$, $a \neq 0$.

(a) Subtract c from both sides of the original equation and divide both sides of the resulting equation by a to obtain

$$x^2 + \frac{b}{a}x = -\frac{c}{a}.$$

(b) Add the square of one-half of the coefficient of x in (a) to both sides and simplify to obtain

$$\left(x + \frac{b}{2a}\right)^2 = \frac{b^2 - 4ac}{4a^2}.$$

(c) Extract square roots in (b) and solve for x to obtain the quadratic formula

$$x = \frac{-b \pm \sqrt{b^2 - 4ac}}{2a}.$$

Extending the Ideas

69. Finding Number of Solutions Consider the equation $|x^2 - 4| = c$.

(a) Find a value of c for which this equation has four solutions. (There are many such values.)

(b) Find a value of c for which this equation has three solutions. (There is only one such value.)

(c) Find a value of c for which this equation has two solutions. (There are many such values.)

(d) Find a value of c for which this equation has no solutions. (There are many such values.)

(e) Writing to Learn Are there any other possible numbers of solutions of this equation? Explain.

70. Sums and Products of Solutions of $ax^2 + bx + c = 0$, $a \neq 0$ Suppose that $b^2 - 4ac > 0$.

(a) Show that the sum of the two solutions of this equation is $-b/a$.

(b) Show that the product of the two solutions of this equation is c/a.

71. Exercise 70 Continued The equation $2x^2 + bx + c = 0$ has two solutions x_1 and x_2. If $x_1 + x_2 = 5$ and $x_1 \cdot x_2 = 3$, find the two solutions.

P.6 Complex Numbers

What you'll learn about

- Complex Numbers
- Operations with Complex Numbers
- Complex Conjugates and Division
- Complex Solutions of Quadratic Equations

... and why

The zeros of polynomials are complex numbers.

Complex Numbers

Figure P.39 shows that the function $f(x) = x^2 + 1$ has no real zeros, so $x^2 + 1 = 0$ has no real-number solutions. To remedy this situation, mathematicians in the 17th century extended the definition of $\sqrt{a}$ to include negative real numbers a. First the number $i = \sqrt{-1}$ is defined as a solution of the equation $i^2 + 1 = 0$ and is the **imaginary unit**. Then for any negative real number $\sqrt{a} = \sqrt{|a|} \cdot i$.

The extended system of numbers, called the *complex numbers*, consists of all real numbers and sums of real numbers and real number multiples of i. The following are all examples of complex numbers:

$$-6, \quad 5i, \quad \sqrt{5}, \quad -7i, \quad \frac{5}{2}i + \frac{2}{3}, \quad -2 + 3i, \quad 5 - 3i, \quad \frac{1}{3} + \frac{4}{5}i$$

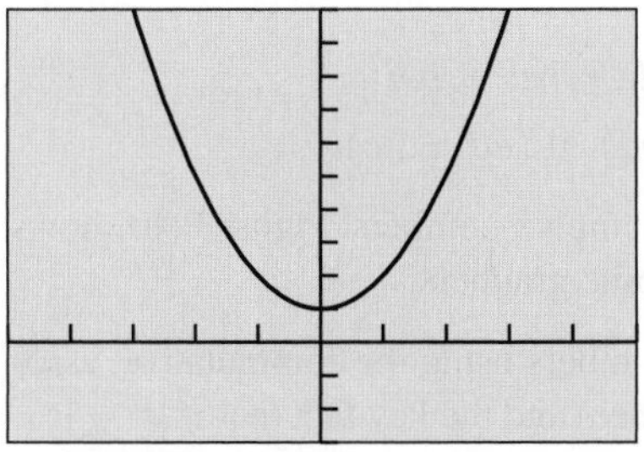

[–5, 5] by [–3, 10]

FIGURE P.39 The graph of $f(x) = x^2 + 1$ has no x-intercepts.

DEFINITION Complex Number

A **complex number** is any number that can be written in the form

$$a + bi,$$

where a and b are real numbers. The real number a is the **real part**, the real number b is the **imaginary part**, and $a + bi$ is the **standard form**.

A real number a is the complex number $a + 0i$, so *all real numbers are also complex numbers*. If $a = 0$ and $b \neq 0$, then $a + bi$ becomes bi, and is an **imaginary number**. For instance, $5i$ and $-7i$ are imaginary numbers.

Two complex numbers are **equal** if and only if their real and imaginary parts are equal. For example,

$$x + yi = 2 + 5i \quad \text{if and only if} \quad x = 2 \text{ and } y = 5.$$

Historical Note

René Descartes (1596–1650) coined the term imaginary in a time when negative solutions to equations were considered *false*. Carl Friedrich Gauss (1777–1855) gave us the term complex number and the symbol i for $\sqrt{-1}$. Today practical applications of complex numbers abound.

Operations with Complex Numbers

Adding complex numbers is done by adding their real and imaginary parts separately. Subtracting complex numbers is also done using the same parts.

DEFINITION Addition and Subtraction of Complex Numbers

If $a + bi$ and $c + di$ are two complex numbers, then

Sum: $(a + bi) + (c + di) = (a + c) + (b + d)i,$

Difference: $(a + bi) - (c + di) = (a - c) + (b - d)i.$

EXAMPLE 1 Adding and Subtracting Complex Numbers

(a) $(7 - 3i) + (4 + 5i) = (7 + 4) + (-3 + 5)i = 11 + 2i$

(b) $(2 - i) - (8 + 3i) = (2 - 8) + (-1 - 3)i = -6 - 4i$

Now try Exercise 3.

The **additive identity** for the complex numbers is $0 = 0 + 0i$. The **additive inverse** of $a + bi$ is $-(a + bi) = -a - bi$ because

$$(a + bi) + (-a - bi) = 0 + 0i = 0.$$

Many of the properties of real numbers also hold for complex numbers. These include

- *Commutative* properties of addition and multiplication,
- *Associative* properties of addition and multiplication, and
- *Distributive* properties of multiplication over addition and subtraction.

Using these properties and the fact that $i^2 = -1$, complex numbers can be multiplied by treating them as algebraic expressions.

EXAMPLE 2 Multiplying Complex Numbers

$$\begin{aligned}(2 + 3i)\cdot(5 - i) &= 2(5 - i) + 3i(5 - i) && \text{Distributive property}\\ &= 10 - 2i + 15i - 3i^2 && \text{Distributive property}\\ &= 10 + 13i - 3(-1) && i^2 = -1\\ &= 13 + 13i\end{aligned}$$

Now try Exercise 9.

We can generalize Example 2 as follows:

$$\begin{aligned}(a + bi)(c + di) &= ac + adi + bci + bdi^2\\ &= (ac - bd) + (ad + bc)i\end{aligned}$$

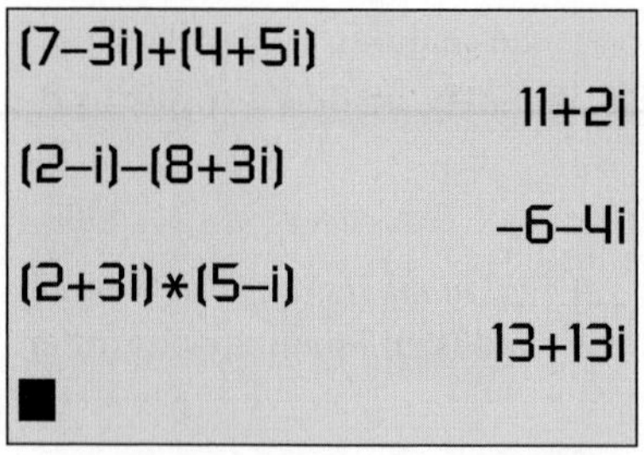

FIGURE P.40 Complex number operations on a grapher. (Examples 1 and 2)

Many graphers can perform basic calculations on complex numbers. Figure P.40 shows how the operations of Examples 1 and 2 look on some graphers.

We compute positive integer powers of complex numbers using the commutative, associative, and distributive properties of complex numbers and the key fact that $i^2 = -1$.

EXAMPLE 3 Raising a Complex Number to a Power

If $z = \frac{1}{2} + \frac{\sqrt{3}}{2}i$, find z^2 and z^3.

SOLUTION

$$\begin{aligned}z^2 &= \left(\frac{1}{2} + \frac{\sqrt{3}}{2}i\right)\left(\frac{1}{2} + \frac{\sqrt{3}}{2}i\right)\\ &= \frac{1}{4} + \frac{\sqrt{3}}{4}i + \frac{\sqrt{3}}{4}i + \frac{3}{4}i^2\\ &= \frac{1}{4} + \frac{2\sqrt{3}}{4}i + \frac{3}{4}(-1)\\ &= -\frac{1}{2} + \frac{\sqrt{3}}{2}i\end{aligned}$$

$$\begin{aligned}z^3 = z^2\cdot z &= \left(-\frac{1}{2} + \frac{\sqrt{3}}{2}i\right)\left(\frac{1}{2} + \frac{\sqrt{3}}{2}i\right)\\ &= -\frac{1}{4} - \frac{\sqrt{3}}{4}i + \frac{\sqrt{3}}{4}i + \frac{3}{4}i^2\\ &= -\frac{1}{4} + 0i + \frac{3}{4}(-1)\\ &= -1\end{aligned}$$

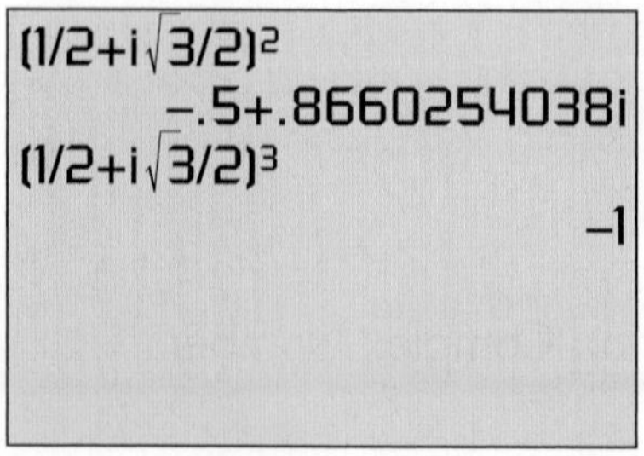

FIGURE P.41 The square and cube of a complex number. (Example 3)

Figure P.41 supports these results numerically.

Now try Exercise 27.

Example 3 demonstrates that $1/2 + (\sqrt{3}/2)i$ is a cube root of -1 and a solution of $x^3 + 1 = 0$. In Section 2.5, we will explore complex zeros of polynomial functions in depth. Section 6.6 delves into the geometry of complex numbers.

Complex Conjugates and Division

The product of the complex numbers $a + bi$ and $a - bi$ is a positive real number:

$$(a + bi) \cdot (a - bi) = a^2 - (bi)^2 = a^2 + b^2$$

We introduce the following definition to describe this special relationship.

DEFINITION Complex Conjugate

The **complex conjugate** of the complex number $z = a + bi$ is

$$\bar{z} = \overline{a + bi} = a - bi.$$

The **multiplicative identity** for the complex numbers is $1 = 1 + 0i$. The **multiplicative inverse**, or **reciprocal**, of $z = a + bi$ is

$$z^{-1} = \frac{1}{z} = \frac{1}{a + bi} = \frac{1}{a + bi} \cdot \frac{a - bi}{a - bi} = \frac{a}{a^2 + b^2} - \frac{b}{a^2 + b^2} i.$$

In general, a quotient of two complex numbers, written in fraction form, can be simplified as we just simplified $1/z$—by multiplying the numerator and denominator of the fraction by the complex conjugate of the denominator.

EXAMPLE 4 Dividing Complex Numbers

Write the complex number in standard form.

(a) $\dfrac{2}{3 - i}$ **(b)** $\dfrac{5 + i}{2 - 3i}$

SOLUTION Multiply the numerator and denominator by the complex conjugate of the denominator.

(a)
$$\frac{2}{3 - i} = \frac{2}{3 - i} \cdot \frac{3 + i}{3 + i} = \frac{6 + 2i}{3^2 + 1^2} = \frac{6}{10} + \frac{2}{10} i = \frac{3}{5} + \frac{1}{5} i$$

(b)
$$\frac{5 + i}{2 - 3i} = \frac{5 + i}{2 - 3i} \cdot \frac{2 + 3i}{2 + 3i} = \frac{10 + 15i + 2i + 3i^2}{2^2 + 3^2} = \frac{7 + 17i}{13} = \frac{7}{13} + \frac{17}{13} i$$

Now try Exercise 33.

Complex Solutions of Quadratic Equations

Recall that the solutions of the quadratic equation $ax^2 + bx + c = 0$, where a, b, and c are real numbers and $a \neq 0$, are given by the quadratic formula

$$x = \frac{-b \pm \sqrt{b^2 - 4ac}}{2a}.$$

The radicand $b^2 - 4ac$ is the **discriminant**, and tells us whether the solutions are real numbers. In particular, if $b^2 - 4ac < 0$, the solutions involve the square root of a negative number and thus lead to complex-number solutions. In all, there are three cases, which we now summarize:

Discriminant of a Quadratic Equation

For a quadratic equation $ax^2 + bx + c = 0$, where a, b, and c are real numbers and $a \neq 0$,

- If $b^2 - 4ac > 0$, there are two distinct real solutions.
- If $b^2 - 4ac = 0$, there is one (repeated) real solution.
- If $b^2 - 4ac < 0$, there is a complex conjugate pair of solutions.

EXAMPLE 5 Solving a Quadratic Equation

Solve $x^2 + x + 1 = 0$.

SOLUTION

Solve Algebraically Using the quadratic formula with $a = b = c = 1$, we obtain

$$x = \frac{-(1) \pm \sqrt{(1)^2 - 4(1)(1)}}{2(1)} = \frac{-(1) \pm \sqrt{-3}}{2} = -\frac{1}{2} \pm \frac{\sqrt{3}}{2}i.$$

So the solutions are $-1/2 + \left(\sqrt{3}/2\right)i$ and $-1/2 - \left(\sqrt{3}/2\right)i$, a complex conjugate pair.

Confirm Numerically Substituting $-1/2 + \left(\sqrt{3}/2\right)i$ into the original equation, we obtain

$$\left(-\frac{1}{2} + \frac{\sqrt{3}}{2}i\right)^2 + \left(-\frac{1}{2} + \frac{\sqrt{3}}{2}i\right) + 1$$
$$= \left(-\frac{1}{2} - \frac{\sqrt{3}}{2}i\right) + \left(-\frac{1}{2} + \frac{\sqrt{3}}{2}i\right) + 1 = 0.$$

By a similar computation we can confirm the second solution.

Now try Exercise 41.

QUICK REVIEW P.6

In Exercises 1–4, add or subtract, and simplify.

1. $(2x + 3) + (-x + 6)$ **2.** $(3y - x) + (2x - y)$

3. $(2a + 4d) - (a + 2d)$ **4.** $(6z - 1) - (z + 3)$

In Exercises 5–10, multiply and simplify.

5. $(x - 3)(x + 2)$

6. $(2x - 1)(x + 3)$

7. $(x - \sqrt{2})(x + \sqrt{2})$

8. $(x + 2\sqrt{3})(x - 2\sqrt{3})$

9. $[x - (1 + \sqrt{2})][x - (1 - \sqrt{2})]$

10. $[x - (2 + \sqrt{3})][x - (2 - \sqrt{3})]$

SECTION P.6 Exercises

Exercise numbers with a gray background indicate problems that the authors have designed to be solved *without a calculator.*

In Exercises 1–8, write the sum or difference in the standard form $a + bi$ without using a calculator.

1. $(2 - 3i) + (6 + 5i)$ **2.** $(2 - 3i) + (3 - 4i)$

3. $(7 - 3i) + (6 - i)$ **4.** $(2 + i) - (9i - 3)$

5. $(2 - i) + (3 - \sqrt{-3})$

6. $(\sqrt{5} - 3i) + (-2 + \sqrt{-9})$

7. $(i^2 + 3) - (7 + i^3)$

8. $(\sqrt{7} + i^2) - (6 - \sqrt{-81})$

In Exercises 9–16, write the product in standard form without using a calculator.

9. $(2 + 3i)(2 - i)$ **10.** $(2 - i)(1 + 3i)$

11. $(1 - 4i)(3 - 2i)$ **12.** $(5i - 3)(2i + 1)$

13. $(7i - 3)(2 + 6i)$ **14.** $(\sqrt{-4} + i)(6 - 5i)$

15. $(-3 - 4i)(1 + 2i)$ **16.** $(\sqrt{-2} + 2i)(6 + 5i)$

In Exercises 17–20, write the expression in the form bi, where b is a real number.

17. $\sqrt{-16}$ **18.** $\sqrt{-25}$

19. $\sqrt{-3}$ **20.** $\sqrt{-5}$

In Exercises 21–24, find the real numbers x and y that make the equation true.

21. $2 + 3i = x + yi$ **22.** $3 + yi = x - 7i$

23. $(5 - 2i) - 7 = x - (3 + yi)$

24. $(x + 6i) = (3 - i) + (4 - 2yi)$

In Exercises 25–28, write the complex number in standard form.

25. $(3 + 2i)^2$ **26.** $(1 - i)^3$

27. $\left(\frac{\sqrt{2}}{2} + \frac{\sqrt{2}}{2}i\right)^4$ **28.** $\left(\frac{\sqrt{3}}{2} + \frac{1}{2}i\right)^3$

In Exercises 29–32, find the product of the complex number and its conjugate.

29. $2 - 3i$ **30.** $5 - 6i$

31. $-3 + 4i$ **32.** $-1 - \sqrt{2}i$

In Exercises 33–40, write the expression in standard form without using a calculator.

33. $\frac{1}{2 + i}$ **34.** $\frac{i}{2 - i}$

35. $\frac{2 + i}{2 - i}$ **36.** $\frac{2 + i}{3i}$

37. $\frac{(2 + i)^2(-i)}{1 + i}$ **38.** $\frac{(2 - i)(1 + 2i)}{5 + 2i}$

39. $\frac{(1 - i)(2 - i)}{1 - 2i}$ **40.** $\frac{(1 - \sqrt{2}i)(1 + i)}{(1 + \sqrt{2}i)}$

In Exercises 41–44, solve the equation.

41. $x^2 + 2x + 5 = 0$ **42.** $3x^2 + x + 2 = 0$

43. $4x^2 - 6x + 5 = x + 1$ **44.** $x^2 + x + 11 = 5x - 8$

Standardized Test Questions

45. True or False There are no complex numbers z satisfying $z = -\bar{z}$. Justify your answer.

46. True or False For the complex number i, $i + i^2 + i^3 + i^4 = 0$. Justify your answer.

In Exercises 47–50, solve the problem without using a calculator.

47. Multiple Choice Which of the following is the standard form for the product $(2 + 3i)(2 - 3i)$?

(A) $-5 + 12i$ **(B)** $4 - 9i$ **(C)** $13 - 3i$

(D) -5 **(E)** $13 + 0i$

48. Multiple Choice Which of the following is the standard form for the quotient $\frac{1}{i}$?

(A) 1 **(B)** -1 **(C)** i **(D)** $-1/i$ **(E)** $0 - i$

49. Multiple Choice Assume that $2 - 3i$ is a solution of $ax^2 + bx + c = 0$, where a, b, c are real numbers. Which of the following is also a solution of the equation?

(A) $2 + 3i$ **(B)** $-2 - 3i$ **(C)** $-2 + 3i$

(D) $3 + 2i$ **(E)** $\frac{1}{2 - 3i}$

50. Multiple Choice Which of the following is the standard form for the power $(1 - i)^3$?

(A) $-4i$ **(B)** $-2 + 2i$ **(C)** $-2 - 2i$ **(D)** $2 + 2i$ **(E)** $2 - 2i$

Explorations

51. Group Activity The Powers of *i*

(a) Simplify the complex numbers $i, i^2, \ldots, i^8$ by evaluating each one.

(b) Simplify the complex numbers $i^{-1}, i^{-2}, \ldots, i^{-8}$ by evaluating each one.

(c) Evaluate i^0.

(d) Writing to Learn Discuss your results from (a)–(c) with the members of your group, and write a summary statement about the integer powers of i.

52. Writing to Learn Describe the nature of the graph of $f(x) = ax^2 + bx + c$ when a, b, and c are real numbers and the equation $ax^2 + bx + c = 0$ has nonreal complex solutions.

Extending the Ideas

53. Prove that the difference between a complex number and its conjugate is a complex number whose real part is 0.

54. Prove that the product of a complex number and its complex conjugate is a complex number whose imaginary part is zero.

55. Prove that the complex conjugate of a product of two complex numbers is the product of their complex conjugates.

56. Prove that the complex conjugate of a sum of two complex numbers is the sum of their complex conjugates.

57. Writing to Learn Explain why $-i$ is a solution of $x^2 - ix + 2 = 0$ but i is not.

P.7 Solving Inequalities Algebraically and Graphically

What you'll learn about

- Solving Absolute Value Inequalities
- Solving Quadratic Inequalities
- Approximating Solutions to Inequalities
- Projectile Motion

... and why

These techniques are involved in using a graphing utility to solve inequalities in this textbook.

Solving Absolute Value Inequalities

We now extend the methods for solving inequalities introduced in Section P.3. We start with two basic rules for solving absolute value inequalities.

Solving Absolute Value Inequalities

Let u be an algebraic expression in x, and let a be a real number with $a \geq 0$.

1. If $|u| < a$, then u is in the interval $(-a, a)$. That is,

$$|u| < a \quad \text{if and only if} \quad -a < u < a.$$

2. If $|u| > a$, then u is in the interval $(-\infty, -a)$ or (a, ∞), that is,

$$|u| > a \quad \text{if and only if} \quad u < -a \text{ or } u > a.$$

The inequalities $<$ and $>$ can be replaced with $\leq$ and $\geq$, respectively. See Figure P.42.

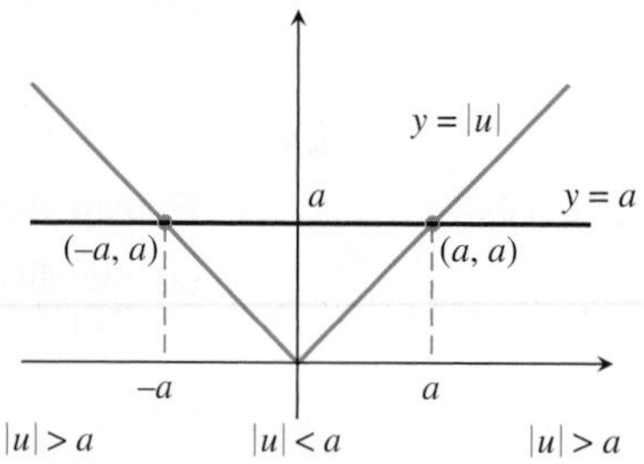

FIGURE P.42 The solution of $|u| < a$ is represented by the portion of the number line for which the graph of $y = |u|$ is below the graph of $y = a$. The solution of $|u| > a$ is represented by the portion of the number line where the graph of $y = |u|$ is above the graph of $y = a$.

EXAMPLE 1 Solving an Absolute Value Inequality

Solve $|x - 4| < 8$.

SOLUTION

$\lvert x - 4\rvert < 8$	Original inequality
$-8 < x - 4 < 8$	Equivalent double inequality
$-4 < x < 12$	Add 4.

As an interval the solution is $(-4, 12)$.

Figure P.43 shows that points on the graph of $y = |x - 4|$ are below the points on the graph of $y = 8$ for values of x between -4 and 12.

Now try Exercise 3.

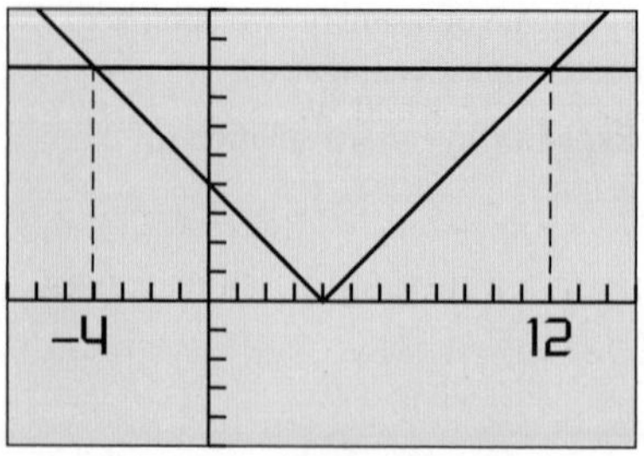

[−7, 15] by [−5, 10]

FIGURE P.43 The graphs of $y = |x - 4|$ and $y = 8$. (Example 1)

EXAMPLE 2 Solving Another Absolute Value Inequality

Solve $|3x - 2| \geq 5$.

SOLUTION The solution of this absolute value inequality consists of the solutions of both of these inequalities.

$3x - 2 \leq -5$	or	$3x - 2 \geq 5$	
$3x \leq -3$	or	$3x \geq 7$	Add 2.
$x \leq -1$	or	$x \geq \frac{7}{3}$	Divide by 3.

The solution consists of all numbers that are in either one of the two intervals $(-\infty, -1]$ or $[7/3, \infty)$, which may be written as $(-\infty, -1] \cup [7/3, \infty)$. The notation "$\cup$" is read as "union." (See the margin note.)

Union of Two Sets

The **union of two sets** A **and** B, denoted by $A \cup B$, is the set of all objects that belong to A or B or both.

Figure P.44 shows that points on the graph of $y = |3x - 2|$ are above or on the points on the graph of $y = 5$ for values of x to the left of and including -1 and to the right of and including 7/3. **Now try Exercise 7.**

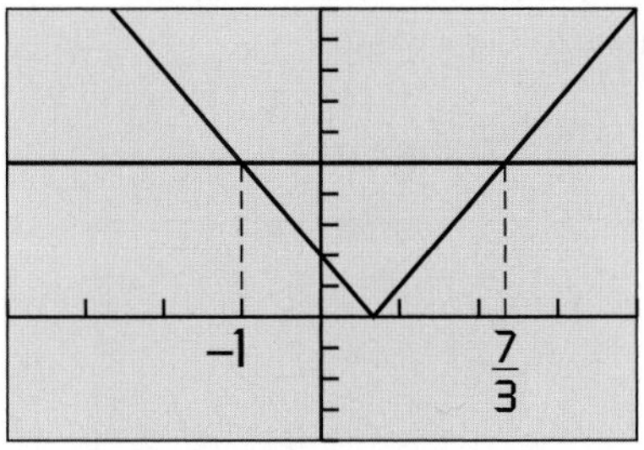

[–4, 4] by [–4, 10]

FIGURE P.44 The graphs of $y = |3x - 2|$ and $y = 5$. (Example 2)

Solving Quadratic Inequalities

To solve a quadratic inequality such as $x^2 - x - 12 > 0$ we begin by solving the corresponding quadratic equation $x^2 - x - 12 = 0$. Then we determine the values of x for which the graph of $y = x^2 - x - 12$ lies above the x-axis.

EXAMPLE 3 Solving a Quadratic Inequality

Solve $x^2 - x - 12 > 0$.

SOLUTION First we solve the corresponding equation $x^2 - x - 12 = 0$.

$$x^2 - x - 12 = 0$$
$$(x - 4)(x + 3) = 0 \quad \text{Factor.}$$
$$x - 4 = 0 \quad \text{or} \quad x + 3 = 0 \quad ab = 0 \Rightarrow a = 0 \text{ or } b = 0$$
$$x = 4 \quad \text{or} \quad x = -3 \quad \text{Solve for } x.$$

The solutions of the corresponding quadratic equation are -3 and 4, and they are not solutions of the original inequality because $0 > 0$ is false. Figure P.45 shows that the points on the graph of $y = x^2 - x - 12$ are above the x-axis for values of x to the left of -3 and to the right of 4.

The solution of the original inequality is $(-\infty, -3) \cup (4, \infty)$.

Now try Exercise 11.

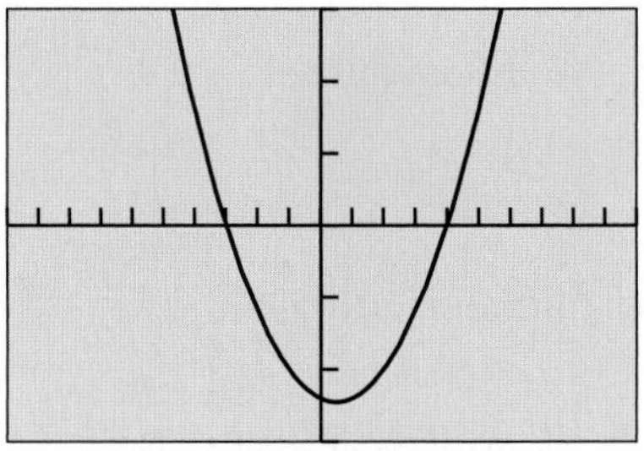

[–10, 10] by [–15, 15]

FIGURE P.45 The graph of $y = x^2 - x - 12$ appears to cross the x-axis at $x = -3$ and $x = 4$. (Example 3)

In Example 4, the quadratic inequality involves the symbol $\le$. In this case, the solutions of the corresponding quadratic equation are also solutions of the inequality.

EXAMPLE 4 Solving Another Quadratic Inequality

Solve $2x^2 + 3x \le 20$.

SOLUTION First we subtract 20 from both sides of the inequality to obtain $2x^2 + 3x - 20 \le 0$. Next, we solve the corresponding quadratic equation $2x^2 + 3x - 20 = 0$.

$$2x^2 + 3x - 20 = 0$$
$$(x + 4)(2x - 5) = 0 \quad \text{Factor.}$$
$$x + 4 = 0 \quad \text{or} \quad 2x - 5 = 0 \quad ab = 0 \Rightarrow a = 0 \text{ or } b = 0$$
$$x = -4 \quad \text{or} \quad x = \frac{5}{2} \quad \text{Solve for } x.$$

The solutions of the corresponding quadratic equation are -4 and $5/2 = 2.5$. You can check that they are also solutions of the inequality.

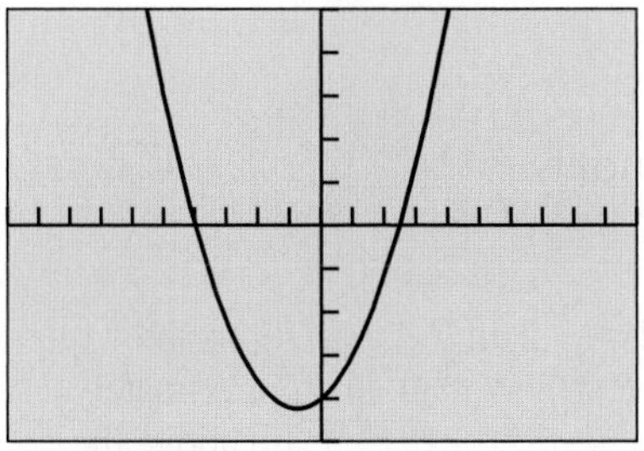

[–10, 10] by [–25, 25]

FIGURE P.46 The graph of $y = 2x^2 + 3x - 20$ appears to be below the x-axis for $-4 < x < 2.5$. (Example 4)

Figure P.46 shows that the points on the graph of $y = 2x^2 + 3x - 20$ are below the x-axis for values of x between -4 and 2.5. The solution of the original inequality is $[-4, 2.5]$. We use square brackets because the endpoints -4 and 2.5 are solutions of the inequality. **Now try Exercise 9.**

In Examples 3 and 4 the corresponding quadratic equation factored. If this doesn't happen we will need to approximate the zeros of the quadratic equation if it has any. Then we use our accuracy agreement from Section P.5 and write the endpoints of any intervals accurate to two decimal places as illustrated in Example 5.

EXAMPLE 5 Solving a Quadratic Inequality Graphically

Solve $x^2 - 4x + 1 \geq 0$ graphically.

SOLUTION We can use the graph of $y = x^2 - 4x + 1$ in Figure P.47 to determine that the solutions of the equation $x^2 - 4x + 1 = 0$ are about 0.27 and 3.73. Thus, the solution of the original inequality is roughly $(-\infty, 0.27] \cup [3.73, \infty)$. We use square brackets because the zeros of the quadratic equation are solutions of the inequality even though we only have approximations to their values.

Now try Exercise 21.

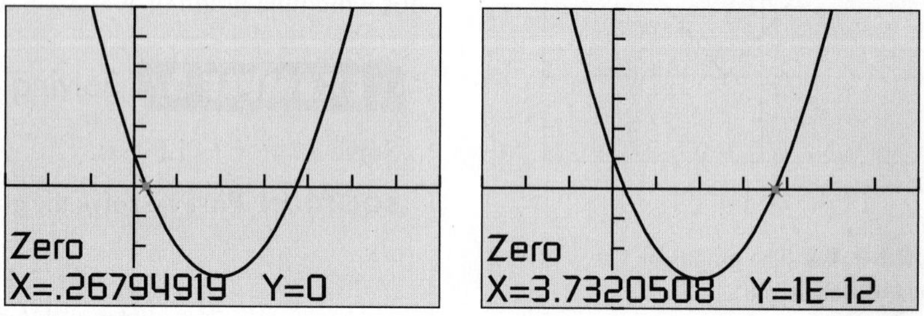

[–3, 7] by [–4, 6]   [–3, 7] by [–4, 6]

FIGURE P.47 This figure suggests that $y = x^2 - 4x + 1$ is zero for $x \approx 0.27$ and $x \approx 3.73$. (Example 5)

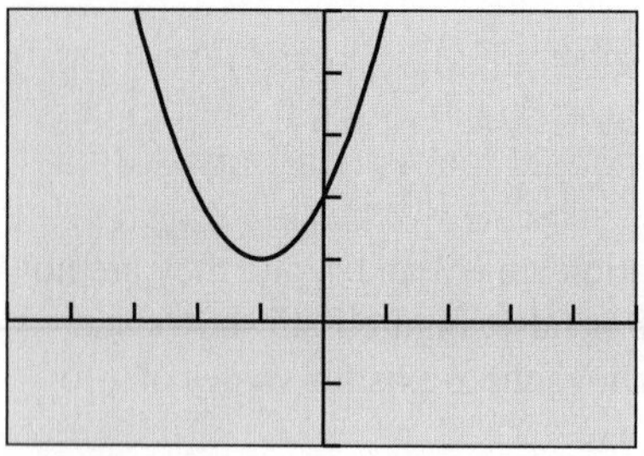

[–5, 5] by [–2, 5]

FIGURE P.48 The values of $y = x^2 + 2x + 2$ are never negative. (Example 6)

EXAMPLE 6 Showing That a Quadratic Inequality Has No Solution

Solve $x^2 + 2x + 2 < 0$.

SOLUTION Figure P.48 shows that the graph of $y = x^2 + 2x + 2$ lies above the x-axis for all values for x. Thus, the inequality $x^2 + 2x + 2 < 0$ has *no* solution.

Now try Exercise 25.

Figure P.48 also shows that the solution of the inequality $x^2 + 2x + 2 > 0$ is the set of all real numbers or, in interval notation, $(-\infty, \infty)$. A quadratic inequality can also have exactly one solution (see Exercise 31).

Approximating Solutions to Inequalities

To solve the inequality in Example 7 we approximate the zeros of the corresponding graph. Then we determine the values of x for which the corresponding graph is above or on the x-axis.

EXAMPLE 7 Solving a Cubic Inequality

Solve $x^3 + 2x^2 - 1 \geq 0$ graphically.

SOLUTION We can use the graph of $y = x^3 + 2x^2 - 1$ in Figure P.49 to show that the solutions of the corresponding equation $x^3 + 2x^2 - 1 = 0$ are approximately -1.62, -1, and 0.62. The points on the graph of $y = x^3 + 2x^2 - 1$ are above the x-axis for values of x between -1.62 and -1, and for values of x to the right of 0.62.

The solution of the inequality is $[-1.62, -1] \cup [0.62, \infty)$. We use square brackets because the zeros of $y = x^3 + 2x^2 - 1$ are solutions of the inequality.

Now try Exercise 27.

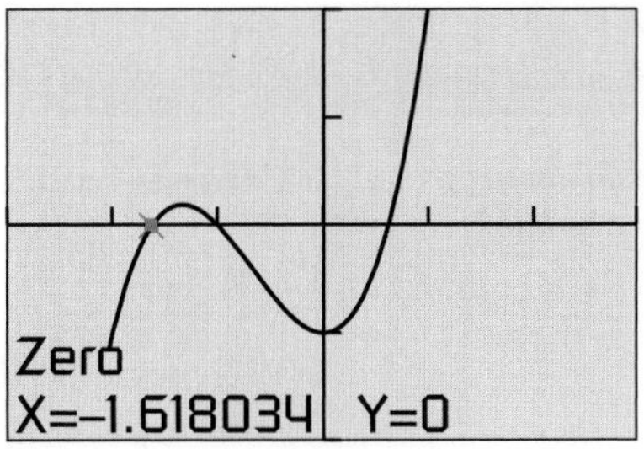

[–3, 3] by [–2, 2]

FIGURE P.49 The graph of $y = x^3 + 2x^2 - 1$ appears to be above the x-axis between the two negative x-intercepts and to the right of the positive x-intercept. (Example 7)

Projectile Motion

The movement of an object that is propelled vertically, but then subject only to the force of gravity, is an example of **projectile motion**.

Projectile Motion

Suppose an object is launched vertically from a point s_0 feet above the ground with an initial velocity of v_0 feet per second. The vertical position s (in feet) of the object t seconds after it is launched is

$$s = -16t^2 + v_0t + s_0.$$

EXAMPLE 8 Finding the Height of a Projectile

A projectile is launched straight up from ground level with an initial velocity of 288 ft/sec.

(a) When will the projectile's height above ground be 1152 ft?

(b) When will the projectile's height above ground be at least 1152 ft?

SOLUTION Here $s_0 = 0$ and $v_0 = 288$. So, the projectile's height is $s = -16t^2 + 288t$.

(a) We need to determine when $s = 1152$.

$$s = -16t^2 + 288t$$
$$1152 = -16t^2 + 288t \quad \text{Substitute } s = 1152.$$
$$16t^2 - 288t + 1152 = 0 \quad \text{Add } 16t^2 - 288t.$$
$$t^2 - 18t + 72 = 0 \quad \text{Divide by 16.}$$
$$(t - 6)(t - 12) = 0 \quad \text{Factor.}$$
$$t = 6 \quad \text{or} \quad t = 12 \quad \text{Solve for } t.$$

The projectile is 1152 ft above ground twice; the first time at $t = 6$ sec on the way up, and the second time at $t = 12$ sec on the way down (Figure P.50).

(b) The projectile will be at least 1152 ft above ground when $s \geq 1152$. We can see from Figure P.50 together with the algebraic work in (a) that the solution is $[6, 12]$. This means that the projectile is at least 1152 ft above ground for times between $t = 6$ sec and $t = 12$ sec, including 6 and 12 sec.

In Exercise 32 we ask you to use algebra to solve the inequality $s = -16t^2 + 288t \geq 1152$.

Now try Exercise 33.

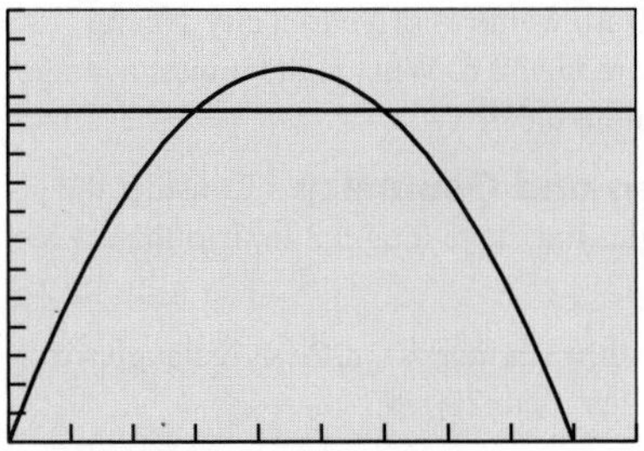

[0, 20] by [0, 1500]

FIGURE P.50 The graphs of $s = -16t^2 + 288t$ and $s = 1152$. We know from Example 8a that the two graphs intersect at (6, 1152) and (12, 1152).

QUICK REVIEW P.7

Exercise numbers with a gray background indicate problems that the authors have designed to be solved *without a calculator*.

In Exercises 1–3, solve for x.

1. $-7 < 2x - 3 < 7$
2. $5x - 2 \geq 7x + 4$
3. $|x + 2| = 3$

In Exercises 4–6, factor the expression completely.

4. $4x^2 - 9$
5. $x^3 - 4x$
6. $9x^2 - 16y^2$

In Exercises 7 and 8, reduce the fraction to lowest terms.

7. $\dfrac{z^2 - 25}{z^2 - 5z}$
8. $\dfrac{x^2 + 2x - 35}{x^2 - 10x + 25}$

In Exercises 9 and 10, add the fractions and simplify.

9. $\dfrac{x}{x - 1} + \dfrac{x + 1}{3x - 4}$
10. $\dfrac{2x - 1}{x^2 - x - 2} + \dfrac{x - 3}{x^2 - 3x + 2}$

SECTION P.7 Exercises

In Exercises 1–8, solve the inequality algebraically. Write the solution in interval notation and draw its number line graph.

1. $|x + 4| \geq 5$
2. $|2x - 1| > 3.6$
3. $|x - 3| < 2$
4. $|x + 3| \leq 5$
5. $|4 - 3x| - 2 < 4$
6. $|3 - 2x| + 2 > 5$
7. $\left|\dfrac{x + 2}{3}\right| \geq 3$
8. $\left|\dfrac{x - 5}{4}\right| \leq 6$

In Exercises 9–16, solve the inequality. Use algebra to solve the corresponding equation.

9. $2x^2 + 17x + 21 \leq 0$
10. $6x^2 - 13x + 6 \geq 0$
11. $2x^2 + 7x > 15$
12. $4x^2 + 2 < 9x$
13. $2 - 5x - 3x^2 < 0$
14. $21 + 4x - x^2 > 0$
15. $x^3 - x \geq 0$
16. $x^3 - x^2 - 30x \leq 0$

In Exercises 17–26, solve the inequality graphically.

17. $x^2 - 4x < 1$
18. $12x^2 - 25x + 12 \geq 0$
19. $6x^2 - 5x - 4 > 0$
20. $4x^2 - 1 \leq 0$
21. $9x^2 + 12x - 1 \geq 0$
22. $4x^2 - 12x + 7 < 0$
23. $4x^2 + 1 > 4x$
24. $x^2 + 9 \leq 6x$
25. $x^2 - 8x + 16 < 0$
26. $9x^2 + 12x + 4 \geq 0$

In Exercises 27–30, solve the cubic inequality graphically.

27. $3x^3 - 12x + 2 \geq 0$
28. $8x - 2x^3 - 1 < 0$
29. $2x^3 + 2x > 5$
30. $4 \leq 2x^3 + 8x$

31. **Group Activity** Give an example of a quadratic inequality with the indicated solution.
    (a) All real numbers
    (b) No solution
    (c) Exactly one solution
    (d) $[-2, 5]$
    (e) $(-\infty, -1) \cup (4, \infty)$
    (f) $(-\infty, 0] \cup [4, \infty)$

32. **Revisiting Example 8** Solve the inequality $-16t^2 + 288t \geq 1152$ algebraically and compare your answer with the result obtained in Example 10.

33. **Projectile Motion** A projectile is launched straight up from ground level with an initial velocity of 256 ft/sec.
    (a) When will the projectile's height above ground be 768 ft?
    (b) When will the projectile's height above ground be at least 768 ft?
    (c) When will the projectile's height above ground be less than or equal to 768 ft?

34. **Projectile Motion** A projectile is launched straight up from ground level with an initial velocity of 272 ft/sec.
    (a) When will the projectile's height above ground be 960 ft?
    (b) When will the projectile's height above ground be more than 960 ft?
    (c) When will the projectile's height above ground be less than or equal to 960 ft?

35. **Writing to Learn** Explain the role of equation solving in the process of solving an inequality. Give an example.

36. **Travel Planning** Barb wants to drive to a city 105 mi from her home in no more than 2 h. What is the lowest average speed she must maintain on the drive?

37. **Connecting Algebra and Geometry** Consider the collection of all rectangles that have length 2 in. less than twice their width.
    (a) Find the possible widths (in inches) of these rectangles if their perimeters are less than 200 in.
    (b) Find the possible widths (in inches) of these rectangles if their areas are less than or equal to 1200 in.^{2}.

38. **Boyle's Law** For a certain gas, $P = 400/V$, where P is pressure and V is volume. If $20 \leq V \leq 40$, what is the corresponding range for P?

39. **Cash-Flow Planning** A company has current assets (cash, property, inventory, and accounts receivable) of \$200,000 and current liabilities (taxes, loans, and accounts payable) of \$50,000. How much can it borrow if it wants its ratio of assets to liabilities to be no less than 2? Assume the amount borrowed is added to both current assets and current liabilities.

Standardized Test Questions

40. True or False The absolute value inequality $|x - a| < b$, where a and b are real numbers, always has at least one solution. Justify your answer.

41. True or False Every real number is a solution of the absolute value inequality $|x - a| \geq 0$, where a is a real number. Justify your answer.

In Exercises 42–45, solve these problems without using a calculator.

42. Multiple Choice Which of the following is the solution to $|x - 2| < 3$?

(A) $x = -1$ or $x = 5$ **(B)** $[-1, 5)$

(C) $[-1, 5]$ **(D)** $(-\infty, -1) \cup (5, \infty)$

(E) $(-1, 5)$

43. Multiple Choice Which of the following is the solution to $x^2 - 2x + 2 \geq 0$?

(A) $[0, 2]$ **(B)** $(-\infty, 0) \cup (2, \infty)$

(C) $(-\infty, 0] \cup [2, \infty)$ **(D)** All real numbers

(E) There is no solution.

44. Multiple Choice Which of the following is the solution to $x^2 > x$?

(A) $(-\infty, 0) \cup (1, \infty)$ **(B)** $(-\infty, 0] \cup [1, \infty))$

(C) $(1, \infty)$ **(D)** $(0, \infty)$

(E) There is no solution.

45. Multiple Choice Which of the following is the solution to $x^2 \leq 1$?

(A) $(-\infty, 1]$ **(B)** $(-1, 1)$

(C) $[1, \infty)$ **(D)** $[-1, 1]$

(E) There is no solution.

Explorations

46. Constructing a Box with No Top An open box is formed by cutting squares from the corners of a regular piece of cardboard (see figure) and folding up the flaps.

(a) What size corner squares should be cut to yield a box with a volume of 125 in.^{3}?

(b) What size corner squares should be cut to yield a box with a volume more than 125 in.^{3}?

(c) What size corner squares should be cut to yield a box with a volume of at most 125 in.^{3}?

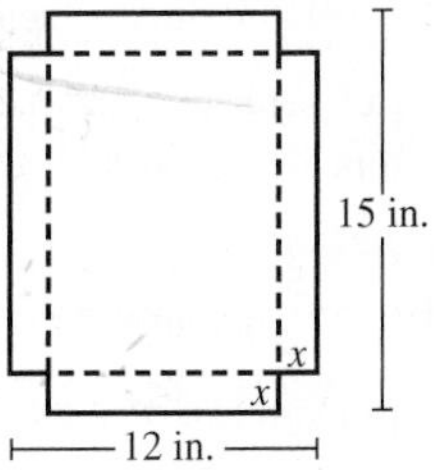

Extending the Ideas

In Exercises 47 and 48, use a combination of algebraic and graphical techniques to solve the inequalities.

47. $|2x^2 + 7x - 15| < 10$ **48.** $|2x^2 + 3x - 20| \geq 10$

CHAPTER P Key Ideas

Properties, Theorems, and Formulas

Trichotomy Property 4
Properties of Algebra 6
Distance Formulas 13, 14
Midpoint Formulas 14, 15
Equations of a Circle 15
Properties of Equality 21
Operations for Equivalent Equations 22
Properties of Inequalities 23
Equations of a Line 30
Zero Factor Property 41
Quadratic Formula 42
Discriminant of a Quadratic Equation 52
Projectile Motion 57

Procedures

Graphing with a Graphing Utility 31
Completing the Square 41
Solving Quadratic Equations Algebraically 43
Agreement About Approximate Solutions 43
Solving Absolute Value Inequalities 54

CHAPTER P Review Exercises

Exercise numbers with a gray background indicate problems that the authors have designed to be solved *without a calculator*.

The collection of exercises marked in red could be used as a chapter test.

In Exercises 1 and 2, find the endpoints and state whether the interval is bounded or unbounded.

1. $[0, 5]$ **2.** $(2, \infty)$

3. Distributive Property Use the distributive property to write the expanded form of $2(x^2 - x)$.

4. Distributive Property Use the distributive property to write the factored form of $2x^3 + 4x^2$.

In Exercises 5 and 6, simplify the expression. Assume that denominators are not zero.

5. $\frac{(uv^2)^3}{v^2u^3}$ **6.** $(3x^2y^3)^{-2}$

In Exercises 7 and 8, write the number in scientific notation.

7. The mean distance from Pluto to the Sun is about 3,680,000,000 miles.

8. The diameter of a red blood corpuscle is about 0.000007 meter.

In Exercises 9 and 10, write the number in decimal form.

9. Our solar system is about 5×10^9 years old.

10. The mass of an electron is about 9.1094×10^{-28} g (gram).

11. The data in Table P.9 give the Fiscal 2009 final budget for some Department of Education programs. Using scientific notation and no calculator, write the amount in dollars for the programs.

Table P.9 Fiscal 2009 Budget

Program	Amount
Title 1 district grants	\$14.5 billion
Title 1 school improvement grants	\$545.6 million
IDEA (Individuals with Disabilities Education Act) state grants	\$11.5 billion
Teacher Incentive Fund	\$97 million
Head Start	\$7.1 billion

Source: U.S. Departments of Education, Health and Human Services as reported in Education Week, May 13, 2009.

(a) Title 1 district grants

(b) Title 1 school improvement grants

(c) IDEA state grants

(d) Teacher Incentive Fund

(e) Head Start

12. Decimal Form Find the decimal form for $-5/11$. State whether it repeats or terminates.

In Exercises 13 and 14, find **(a)** the distance between the points and **(b)** the midpoint of the line segment determined by the points.

13. -5 and 14 **14.** $(-4, 3)$ and $(5, -1)$

In Exercises 15 and 16, show that the figure determined by the points is the indicated type.

15. Right triangle: $(-2, 1)$, $(3, 11)$, $(7, 9)$

16. Equilateral triangle: $(0, 1)$, $(4, 1)$, $(2, 1 - 2\sqrt{3})$

In Exercises 17 and 18, find the standard form equation for the circle.

17. Center $(0, 0)$, radius 2

18. Center $(5, -3)$, radius 4

In Exercises 19 and 20, find the center and radius of the circle.

19. $(x + 5)^2 + (y + 4)^2 = 9$

20. $x^2 + y^2 = 1$

21. (a) Find the length of the sides of the triangle in the figure.

(b) Writing to Learn Show that the triangle is a right triangle.

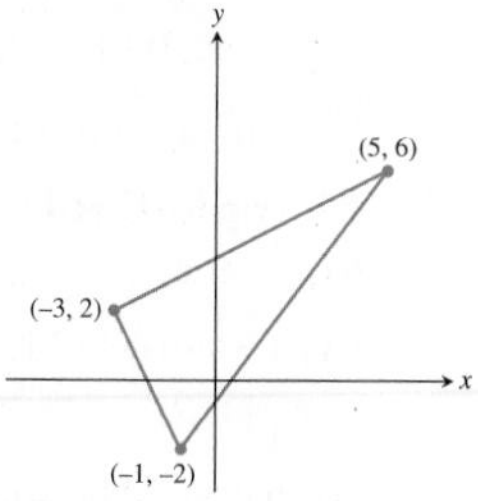

22. Distance and Absolute Value Use absolute value notation to write the statement that the distance between z and -3 is less than or equal to 1.

23. Finding a Line Segment with Given Midpoint Let (3, 5) be the midpoint of the line segment with endpoints $(-1, 1)$ and (a, b). Determine a and b.

24. Finding Slope Find the slope of the line through the points $(-1, -2)$ and $(4, -5)$.

25. Finding Point-Slope Form Equation Find an equation in point-slope form for the line through the point $(2, -1)$ with slope $m = -2/3$.

26. Find an equation of the line through the points $(-5, 4)$ and $(2, -5)$ in the general form $Ax + By + C = 0$.

In Exercises 27–32, find an equation in slope-intercept form for the line.

27. The line through $(3, -2)$ with slope $m = 4/5$

28. The line through the points $(-1, -4)$ and $(3, 2)$

29. The line through $(-2, 4)$ with slope $m = 0$

30. The line $3x - 4y = 7$

31. The line through $(2, -3)$ and parallel to the line $2x + 5y = 3$

32. The line through $(2, -3)$ and perpendicular to the line $2x + 5y = 3$

33. SAT Math Scores Scores on each part of the SAT are on a scale of 200–800. Table P.10 shows the average SAT math score for selected years.

Table P.10 Average SAT Math Scores

Year	Scaled Scores
1995	506
2000	514
2005	520
2006	518
2007	515
2008	515
2009	515
2010	516
2011	514

Source: The College Board, The World Almanac and Book of Facts 2012.

(a) Draw a scatter plot of the data.

(b) Use the 1995 and 2005 data to write a linear equation for the average SAT math score y in terms of the year x. Superimpose the graph of the equation on the scatter plot.

(c) Use the equation in (b) to predict the average SAT math score for 2015.

(d) **Writing to Learn** Do you think the prediction in (c) is valid? Explain.

34. Consider the point $(-6, 3)$ and Line L: $4x - 3y = 5$. Write an equation **(a)** for the line passing through this point and parallel to L, and **(b)** for the line passing through this point and perpendicular to L. Support your work graphically.

In Exercises 35 and 36, assume that each graph contains the origin and the upper right-hand corner of the viewing window.

35. Find the slope of the line in the figure.

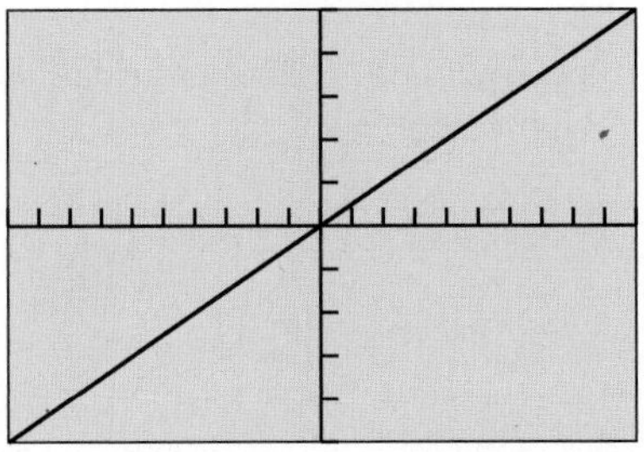

[–10, 10] by [–25, 25]

36. Writing to Learn Which line has the greater slope? Explain.

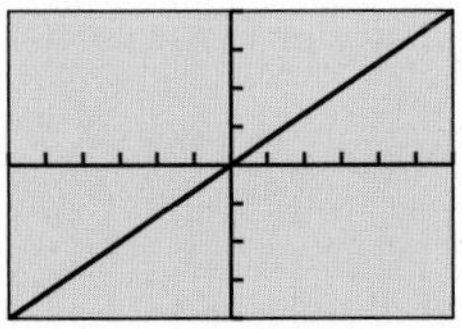

[–6, 6] by [–4, 4]

(a)

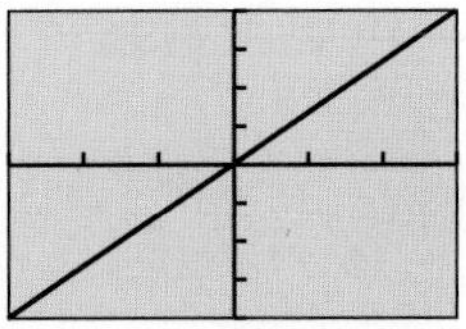

[–15, 15] by [–12, 12]

(b)

In Exercises 37–52, solve the equation algebraically without using a calculator.

37. $3x - 4 = 6x + 5$

38. $\dfrac{x-2}{3} + \dfrac{x+5}{2} = \dfrac{1}{3}$

39. $2(5 - 2y) - 3(1 - y) = y + 1$

40. $3(3x - 1)^2 = 21$

41. $x^2 - 4x - 3 = 0$

42. $16x^2 - 24x + 7 = 0$

43. $6x^2 + 7x = 3$

44. $2x^2 + 8x = 0$

45. $x(2x + 5) = 4(x + 7)$

46. $|4x + 1| = 3$

47. $4x^2 - 20x + 25 = 0$

48. $-9x^2 + 12x - 4 = 0$

49. $x^2 = 3x$

50. $4x^2 - 4x + 2 = 0$

51. $x^2 - 6x + 13 = 0$

52. $x^2 - 2x + 4 = 0$

53. Completing the Square Use completing the square to solve the equation $2x^2 - 3x - 1 = 0$.

54. Quadratic Formula Use the quadratic formula to solve the equation $3x^2 + 4x - 1 = 0$.

In Exercises 55–58, solve the equation graphically.

55. $3x^3 - 19x^2 - 14x = 0$

56. $x^3 + 2x^2 - 4x - 8 = 0$

57. $x^3 - 2x^2 - 2 = 0$

58. $|2x - 1| = 4 - x^2$

In Exercises 59 and 60, solve the inequality and draw a number line graph of the solution.

59. $-2 < x + 4 \le 7$

60. $5x + 1 \ge 2x - 4$

In Exercises 61–72, solve the inequality.

61. $\dfrac{3x - 5}{4} \le -1$

62. $|2x - 5| < 7$

63. $|3x + 4| \ge 2$

64. $4x^2 + 3x > 10$

65. $2x^2 - 2x - 1 > 0$

66. $9x^2 - 12x - 1 \le 0$

67. $x^3 - 9x \le 3$

68. $4x^3 - 9x + 2 > 0$

69. $\left|\frac{x + 7}{5}\right| > 2$

70. $2x^2 + 3x - 35 < 0$

71. $4x^2 + 12x + 9 \geq 0$

72. $x^2 - 6x + 9 < 0$

In Exercises 73–80, perform the indicated operation, and write the result in the standard form $a + bi$ without using a calculator.

73. $(3 - 2i) + (-2 + 5i)$

74. $(5 - 7i) - (3 - 2i)$

75. $(1 + 2i)(3 - 2i)$

76. $(1 + i)^3$

77. $(1 + 2i)^2(1 - 2i)^2$

78. i^{29}

79. $\sqrt{-16}$

80. $\frac{2 + 3i}{1 - 5i}$

81. **Projectile Motion** A projectile is launched straight up from ground level with an initial velocity of 320 ft/sec.

(a) When will the projectile's height above ground be 1538 ft?

(b) When will the projectile's height above ground be at most 1538 ft?

(c) When will the projectile's height above ground be greater than or equal to 1538 ft?

82. **Navigation** A commercial jet airplane climbs at takeoff with slope $m = 4/9$. How far in the horizontal direction will the airplane fly to reach an altitude of 20,000 ft above the takeoff point?

83. **Connecting Algebra and Geometry** Consider the collection of all rectangles that have length 1 cm more than three times their width w.

(a) Find the possible widths (in cm) of these rectangles if their perimeters are less than or equal to 150 cm.

(b) Find the possible widths (in cm) of these rectangles if their areas are greater than 1500 cm^2.

1.1 Modeling and Equation Solving

1.2 Functions and Their Properties

1.3 Twelve Basic Functions

1.4 Building Functions from Functions

1.5 Parametric Relations and Inverses

1.6 Graphical Transformations

1.7 Modeling with Functions

Functions and Graphs

One of the central principles of economics is that the value of money is not constant; it is a function of time. Since fortunes are made and lost by people attempting to predict the future value of money, much attention is paid to quantitative measures like the Consumer Price Index, a basic measure of inflation in various sectors of the economy. See page 146 for a look at how the Consumer Price Index for housing has behaved over time.

Chapter 1 Overview

In this chapter we begin the study of functions that will continue throughout the book. Your previous courses have introduced you to some basic functions. These functions can be visualized using a graphing calculator, and their properties can be described using the notation and terminology that will be introduced in this chapter. A familiarity with this terminology will serve you well in later chapters when we explore properties of functions in greater depth.

1.1 Modeling and Equation Solving

What you'll learn about

- Numerical Models
- Algebraic Models
- Graphical Models
- The Zero Factor Property
- Problem Solving
- Grapher Failure and Hidden Behavior
- A Word About Proof

... and why

Numerical, algebraic, and graphical models provide different methods to visualize, analyze, and understand data.

Numerical Models

Scientists and engineers have always used mathematics to model the real world and thereby to unravel its mysteries. A **mathematical model** is a mathematical structure that approximates phenomena for the purpose of studying or predicting their behavior. Thanks to advances in computer technology, the process of devising mathematical models is now a rich field of study itself, **mathematical modeling**.

We will be concerned primarily with three types of mathematical models in this book: *numerical models*, *algebraic models*, and *graphical models*. Each type of model gives insight into real-world problems, but the best insights are often gained by switching from one kind of model to another. Developing the ability to do that will be one of the goals of this course.

Perhaps the most basic kind of mathematical model is the **numerical model**, in which numbers (or *data*) are analyzed to gain insights into phenomena. A numerical model can be as simple as the major league baseball standings or as complicated as the network of interrelated numbers that measure the global economy.

Table 1.1 The Minimum Hourly Wage

Year	Minimum Hourly Wage (MHW)	Purchasing Power in 1996 Dollars
1955	0.75	4.39
1960	1.00	5.30
1965	1.25	6.23
1970	1.60	6.47
1975	2.10	6.12
1980	3.10	5.90
1985	3.35	4.88
1990	3.80	4.56
1995	4.25	4.38
2000	5.15	4.69
2005	5.15	4.14
2010	7.25	5.22

Source: www.infoplease.com

EXAMPLE 1 Tracking the Minimum Wage

The numbers in Table 1.1 show the growth of the minimum hourly wage (MHW) from 1955 to 2010. The table also shows the MHW adjusted to the purchasing power of 1996 dollars (using the CPI-U, the Consumer Price Index for all Urban Consumers). Answer the following questions using only the data in the table.

(a) In what five-year period did the actual MHW increase the most?

(b) In what year did a worker earning the MHW enjoy the greatest purchasing power?

(c) A worker on minimum wage in 1980 was earning nearly twice as much as a worker on minimum wage in 1970, and yet there was great pressure to raise the minimum wage again. Why?

SOLUTION

(a) In the period 2005 to 2010 it increased by \$2.10. Notice that the minimum wage never goes down, so we can tell that there were no other increases of this magnitude even though the table does not give data from every year.

(b) In 1970.

(c) Although the MHW increased from \$1.60 to \$3.10 in that period, the purchasing power actually dropped by \$0.57 (in 1996 dollars). This is one way inflation can affect the economy.

Now try Exercise 11.

The numbers in Table 1.1 provide a numerical model for one aspect of the U.S. economy by using another numerical model, the urban Consumer Price Index (CPI-U), to adjust the data. Working with large numerical models is standard operating procedure in business and industry, where computers are relied upon to provide fast and accurate data processing.

Table 1.2 U.S. Prison Population (thousands)

Year	Total	Male	Female
1980	329	316	13
1985	502	479	23
1990	774	730	44
1995	1125	1057	68
2000	1391	1298	93
2005	1528	1420	108
2010	1612	1499	113

Source: U.S. Justice Department.

Table 1.3 Female Percentage of U.S. Prison Population

Year	Female
1980	3.9
1985	4.6
1990	5.7
1995	6
2000	6.7
2005	7.1
2010	7.0

Source: U.S. Justice Department.

EXAMPLE 2 Analyzing Prison Populations

Table 1.2 shows the growth in the number of prisoners incarcerated in state and federal prisons at year's end for selected years between 1980 and 2010. Is the proportion of female prisoners increasing over the years?

SOLUTION The *number* of female prisoners over the years is certainly increasing, but so is the total number of prisoners, so it is difficult to discern from the data whether the *proportion* of female prisoners is increasing. What we need is another column of numbers showing the ratio of female prisoners to total prisoners.

We could compute all the ratios separately, but it is easier to do this kind of repetitive calculation with a single command on a computer spreadsheet. You can also do this on a graphing calculator by manipulating lists (see Exercise 19). Table 1.3 shows the percentage of the total population each year that consists of female prisoners. With these data to extend our numerical model, it is clear that the proportion of female prisoners has been increasing (with a slight decline between 2005 and 2010).

Now try Exercise 19.

Algebraic Models

An **algebraic model** uses formulas to relate variable quantities associated with the phenomena being studied. The added power of an algebraic model over a numerical model is that it can be used to generate numerical values of unknown quantities by relating them to known quantities.

EXAMPLE 3 Comparing Pizzas

A pizzeria sells a rectangular 18 in. by 24 in. pizza for the same price as its large round pizza 24-in. diameter). If both pizzas are of the same thickness, which option gives the most pizza for the money?

SOLUTION We need to compare the *areas* of the pizzas. Fortunately, geometry has provided algebraic models that allow us to compute the areas from the given information.

For the rectangular pizza:

$$Area = l \times w = 18 \times 24 = 432 \text{ in.}^2$$

For the circular pizza:

$$Area = \pi r^2 = \pi\left(\frac{24}{2}\right)^2 = 144\pi \approx 452.4 \text{ in.}^2$$

The round pizza is larger and therefore gives more for the money.

Now try Exercise 21.

The algebraic models in Example 3 come from geometry, but you have probably encountered algebraic models from many other sources in your algebra and science courses.

EXPLORATION 1 **Designing an Algebraic Model**

A department store is having a sale in which everything is discounted 25% off the marked price. The discount is taken at the sales counter, and then a state sales tax of 6.5% and a local sales tax of 0.5% are added on.

1. The discount price d is related to the marked price m by the formula $d = km$, where k is a certain constant. What is k?
2. The actual sale price s is related to the discount price d by the formula $s = d + td$, where t is a constant related to the combined total sales tax. What is t?
3. Using the answers from steps 1 and 2 you can find a constant p that relates s directly to m by the formula $s = pm$. What is p?
4. If you only have \$30, can you afford to buy a shirt marked \$36.99?
5. If you have a credit card but are determined to spend no more than \$100, what is the maximum total value of your marked purchases before you present them at the sales counter?

The ability to generate numbers from formulas makes an algebraic model far more useful as a predictor of behavior than a numerical model. Indeed, one optimistic goal of scientists and mathematicians when modeling phenomena is to fit an algebraic model to numerical data and then (even more optimistically) to analyze why it works. Not all models can be used to make accurate predictions. For example, nobody has ever devised a successful formula for predicting the ups and downs of the stock market as a function of time, although that does not stop investors from trying.

If numerical data do behave reasonably enough to suggest that an algebraic model might be found, it is often helpful to look at a picture first. That brings us to graphical models.

Graphical Models

A **graphical model** is a visible representation of a numerical model or an algebraic model that gives insight into the relationships between variable quantities. Learning to interpret and use graphs is a major goal of this book.

EXAMPLE 4 Visualizing Galileo's Gravity Experiments

Galileo Galilei (1564–1642) spent a good deal of time rolling balls down inclined planes, carefully recording the distance they traveled as a function of elapsed time. His experiments are commonly repeated in physics classes today, so it is easy to reproduce a typical table of Galilean data.

Elapsed time (sec)	0	1	2	3	4	5	6	7	8
Distance traveled (in.)	0	0.75	3	6.75	12	18.75	27	36.75	48

What graphical model fits the data? Can you find an algebraic model that fits?

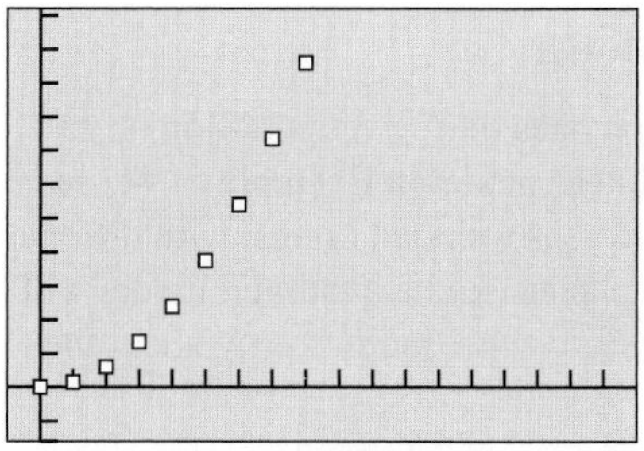

[–1, 18] by [–8, 56]

FIGURE 1.1 A scatter plot of the data from a Galileo gravity experiment. (Example 4)

SOLUTION A scatter plot of the data is shown in Figure 1.1.

Galileo's experience with quadratic functions suggested to him that this figure was a parabola with its vertex at the origin; he therefore modeled the effect of gravity as a quadratic function:

$$d = kt^2.$$

Because the ordered pair $(1, 0.75)$ must satisfy the equation, it follows that $k = 0.75$, yielding the equation

$$d = 0.75t^2.$$

You can verify numerically that this algebraic model correctly predicts the rest of the data points. We will have much more to say about parabolas in Chapter 2.

Now try Exercise 23.

This insight led Galileo to discover several basic laws of motion that would eventually be named after Isaac Newton. Although Galileo had found the algebraic model to describe the path of the ball, it would take Newton's calculus to explain why it worked.

EXAMPLE 5 Fitting a Curve to Data

Table 1.4 shows the average life expectancy for persons born in the United States in each given year. The data are drawn from census years between 1950 and 2010.

Table 1.4 Years of Life Expected at Birth in the United States, 1950–2010

Years After 1950	0	10	20	30	40	50	60
Life Expectancy	68.2	69.7	70.8	73.7	75.4	76.8	78.7

Source: National Center for Health Statistics, U.S. Department of Health and Human Services.

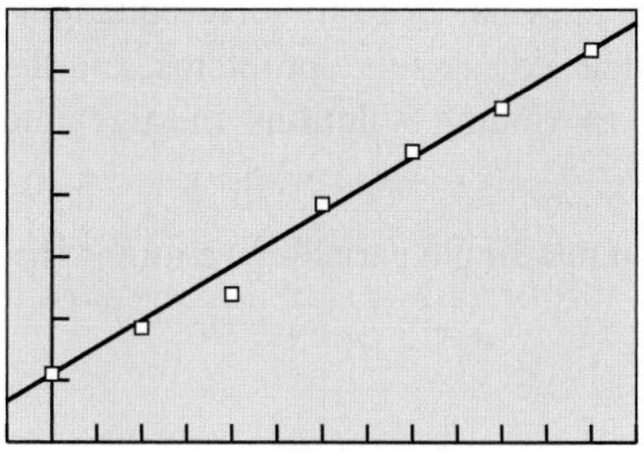

[–5, 65] by [66, 80]

FIGURE 1.2 The line with equation $y = 0.175x + 68.2$ is a good model for the life expectancy data in Table 1.4. (Example 5)

Model the trend graphically and use the graphical model to suggest an algebraic model.

SOLUTION A scatter plot of the data is shown in Figure 1.2. Since the points show a linear pattern, a linear equation is an appropriate algebraic model. We will describe in Chapter 2 how a statistician would find the best line to fit the data, but we can get a pretty good fit for now by putting the line through $(0, 68.2)$ and $(60, 78.7)$.

The slope is $(78.7 - 68.2)/(60 - 0) = 0.175$ and the y-intercept is 68.2.

You can see that the line $y = 0.175x + 68.2$ does a very nice job of modeling the data.

Now try Exercises 13 and 15.

EXPLORATION 2 Interpreting the Model

The parabola in Example 4 arose from a law of physics that governs falling objects, which should inspire more confidence than the linear model in Example 5. We can repeat Galileo's experiment many times with differently sloped ramps, with different units of measurement, and even on different planets, and a quadratic model will fit it every time. The purpose of this Exploration is to think more deeply about the linear model for the life expectancy data.

1. The linear model we found will not continue to predict life expectancy indefinitely. Why must it eventually fail?
2. Do you think that our linear model will give an accurate estimate for the life expectancy of a person born in the United States in 2015? Why or why not?
3. The linear model is such a good fit that it actually calls our attention to the unusually low point in 1970. Statisticians might look for some unusual circumstance in 1970 that might explain the temporary dip in life expectancy. Can you think of one? As a hint, consider the scatter plot in Figure 1.3, which shows the life expectancy for *males* in the United States for the same time period.

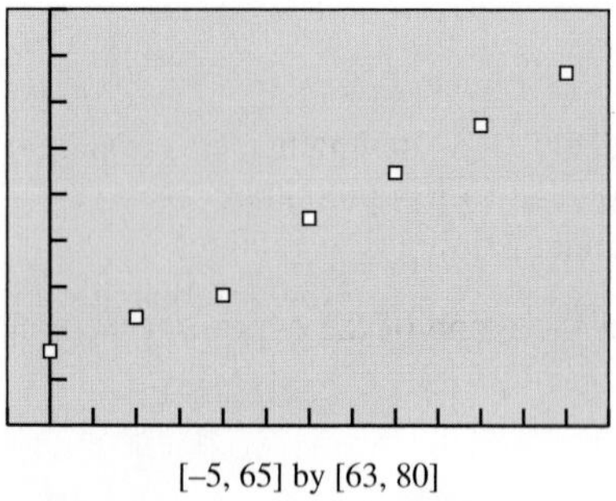

[–5, 65] by [63, 80]

FIGURE 1.3 A scatter plot showing the life expectancy of U.S. *males* for the same years as shown in Figure 1.2. The low point in 1970 is even more apparent.

There are other ways of graphing numerical data that are particularly useful for statistical studies. We will treat some of them in Chapter 10. The scatter plot will be our choice of data graph for the time being, as it provides the closest connection to graphs of functions in the Cartesian plane.

The Zero Factor Property

The main reason for studying algebra through the ages has been to solve equations. We develop algebraic models for phenomena so that we can solve problems, and the solutions to the problems usually come down to finding solutions of algebraic equations.

If we are fortunate enough to be solving an equation in a single variable, we might proceed as in the following example.

EXAMPLE 6 Solving an Equation Algebraically

Find all real numbers x for which $6x^3 = 11x^2 + 10x$.

SOLUTION We begin by changing the form of the equation to $6x^3 - 11x^2 - 10x = 0$.

We can then solve this equation algebraically by factoring:

$$6x^3 - 11x^2 - 10x = 0$$
$$x(6x^2 - 11x - 10) = 0$$
$$x(2x - 5)(3x + 2) = 0$$
$$x = 0 \quad \text{or} \quad 2x - 5 = 0 \quad \text{or} \quad 3x + 2 = 0$$
$$x = 0 \quad \text{or} \quad x = \frac{5}{2} \quad \text{or} \quad x = -\frac{2}{3}$$

Now try Exercise 31.

In Example 6, we used the important Zero Factor Property of real numbers.

Zero Factor Property

A product of real numbers is zero if and only if at least one of the factors in the product is zero.

It is this property that algebra students use to solve equations in which an expression is set equal to zero. Modern problem solvers are fortunate to have an alternative way to find such solutions.

If we graph the expression, then the x-intercepts of the graph of the expression will be the values for which the expression equals 0.

EXAMPLE 7 Solving an Equation: Comparing Methods

Solve the equation $x^2 = 10 - 4x$.

SOLUTION

Solve Algebraically The given equation is equivalent to $x^2 + 4x - 10 = 0$. This quadratic equation has irrational solutions that can be found by the quadratic formula.

$$x = \frac{-4 + \sqrt{16 + 40}}{2} \approx 1.7416574$$

and

$$x = \frac{-4 - \sqrt{16 + 40}}{2} \approx -5.7416574$$

Although the decimal answers are certainly accurate enough for all practical purposes, it is important to note that only the expressions found by the quadratic formula give the *exact* real-number answers. The tidiness of exact answers is a worthy mathematical goal. Realistically, however, exact answers are often impossible to obtain, even with the most sophisticated mathematical tools.

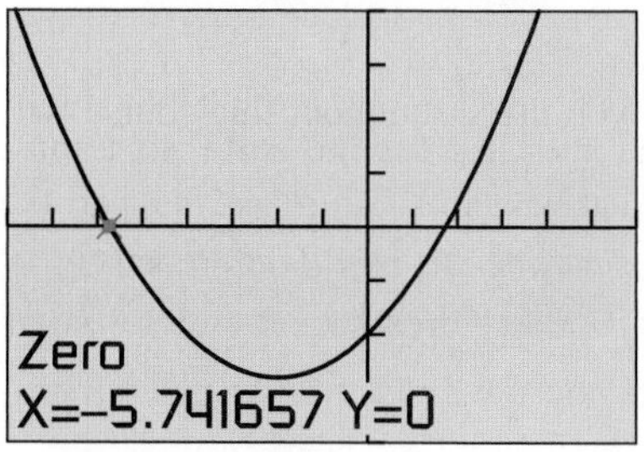

[–8, 6] by [–20, 20]

FIGURE 1.4 The graph of $y = x^2 + 4x - 10$. (Example 7)

Solve Graphically We first find an equivalent equation with 0 on the right-hand side: $x^2 + 4x - 10 = 0$. We next graph the equation $y = x^2 + 4x - 10$, as shown in Figure 1.4.

We then use the grapher to locate the x-intercepts of the graph:

$$x \approx 1.7416574 \text{ and } x \approx -5.7416574$$

Now try Exercise 35.

Solving Equations with Technology

Example 7 shows one method of solving an equation with technology. Some graphers could also solve the equation in Example 7 by finding the *intersection* of the graphs of $y = x^2$ and $y = 10 - 4x$. Some graphers have built-in equation solvers. Each method has its advantages and disadvantages, but we recommend the "finding the x-intercepts" technique for now because it most closely parallels the classical algebraic techniques for finding roots of equations, and makes the connection between the algebraic and graphical models easier to follow and appreciate.

We used the graphing utility of the calculator to **solve graphically** in Example 7. Most calculators also have solvers that would enable us to **solve numerically** for the same decimal approximations without considering the graph. Some calculators have computer algebra systems that will solve numerically to produce exact answers in certain cases. In this book we will distinguish between these two technological methods and the traditional pencil-and-paper methods used to **solve algebraically**.

Every method of solving an equation usually comes down to finding where an expression equals zero. If we use $f(x)$ to denote an algebraic expression in the variable x, the connections are as follows:

Fundamental Connection

If a is a real number that solves the equation $f(x) = 0$, then these three statements are equivalent:

1. The number a is a **root** (or **solution**) of the **equation** $f(x) = 0$.
2. The number a is a **zero** of $y = f(x)$.
3. The number a is an **x-intercept** of the **graph** of $y = f(x)$. (Sometimes the point $(a, 0)$ is referred to as an x-intercept.)

Problem Solving

George Pólya (1887–1985) is sometimes called the father of modern problem solving, not only because he was good at it (as he certainly was) but also because he published the most famous analysis of the problem-solving process: *How to Solve It: A New Aspect of Mathematical Method*. His "four steps" are well known to most mathematicians:

Pólya's Four Problem-Solving Steps

1. Understand the problem.
2. Devise a plan.
3. Carry out the plan.
4. Look back.

The problem-solving process that we recommend you use throughout this course will be the following version of Pólya's four steps.

Problem-Solving Process

Step 1—Understand the problem.

- Read the problem as stated, several times if necessary.
- Be sure you understand the meaning of each term used.
- Restate the problem in your own words. Discuss the problem with others if you can.
- Identify clearly the information that you need to solve the problem.
- Find the information you need from the given data.

Step 2—Develop a mathematical model of the problem.

- Draw a picture to visualize the problem situation. It usually helps.
- Introduce a variable to represent the quantity you seek. (In some cases there may be more than one.)
- Use the statement of the problem to find an equation or inequality that relates the variables you seek to quantities that you know.

Step 3—Solve the mathematical model and support or confirm the solution.

- **Solve algebraically** using traditional algebraic methods and **support graphically or support numerically** using a graphing utility.

Problem Solving, Process, and Practice

NCTM's *Principles and Standards for School Mathematics* (2000) identified five "Process Standards" that should be fundamental in mathematics education. The first of these standards was Problem Solving. The Common Core State Standards in Mathematics identified eight "Standards for Mathematical Practice" that should be fundamental in mathematics education. Again, the first of these was about Problem Solving. Some individual states have opted for their own standards, but Problem Solving is invariably front and center as a fundamental objective. (For example, the state of Texas has a list of Process Standards for Precalculus, the first two of which are about Problem Solving. In fact, the second one specifically recommends a problem-solving model that is similar to the one we recommend here.)

- **Solve graphically or numerically** using a graphing utility and **confirm algebraically** using traditional algebraic methods.
- **Solve graphically or numerically** because there is no other way possible.

Step 4—Interpret the solution in the problem setting.

- Translate your mathematical result into the problem setting and decide whether the result makes sense.

EXAMPLE 8 Applying the Problem-Solving Process

The engineers at an auto manufacturer pay students \$0.08 per mile plus \$25 per day to road test their new vehicles.

(a) How much did the auto manufacturer pay Sally to drive 440 mi in one day?

(b) John earned \$93 test-driving a new car in one day. How far did he drive?

SOLUTION

Model A picture of a car or of Sally or John would not be helpful, so we go directly to designing the model. Both John and Sally earned \$25 for one day, plus \$0.08 per mile. Multiply dollars/mile by miles to get dollars.

So if p represents the pay for driving x miles in one day, our algebraic model is

$$p = 25 + 0.08x.$$

Solve Algebraically

(a) To get Sally's pay we let $x = 440$ and solve for p:

$$\begin{aligned} p &= 25 + 0.08(440) \\ &= 60.20 \end{aligned}$$

(b) To get John's mileage we let $p = 93$ and solve for x:

$$\begin{aligned} 93 &= 25 + 0.08x \\ 68 &= 0.08x \\ x &= \frac{68}{0.08} \\ x &= 850 \end{aligned}$$

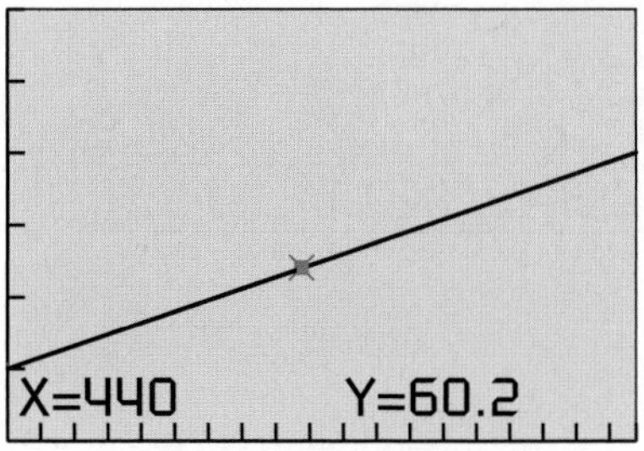

[0, 940] by [0, 150]

(a)

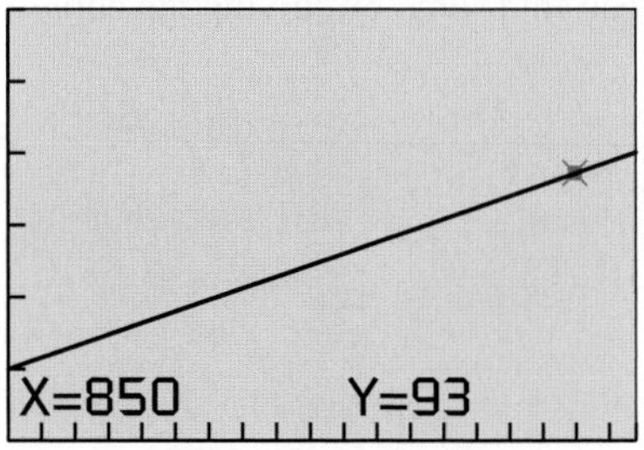

[0, 940] by [0, 150]

(b)

FIGURE 1.5 Graphical support for the algebraic solutions in Example 8.

Support Graphically Figure 1.5a shows that the point $(440, 60.20)$ is on the graph of $y = 25 + 0.08x$, supporting our answer to (a). Figure 1.5b shows that the point $(850, 93)$ is on the graph of $y = 25 + 0.08x$, supporting our answer to (b). (We could also have **supported** our answer **numerically** by simply substituting in for each x and confirming the value of p.)

Interpret Sally earned \$60.20 for driving 440 mi in one day. John drove 850 mi in one day to earn \$93.00.

Now try Exercise 47.

It is not really necessary to *show* written support as part of an algebraic solution, but it is good practice to support answers wherever possible simply to reduce the chance for error. We will often show written support of our solutions in this book in order to highlight the connections among the algebraic, graphical, and numerical models.

Grapher Failure and Hidden Behavior

Although the graphs produced by computers and graphing calculators are wonderful tools for understanding algebraic models and their behavior, it is important to keep in mind that machines have limitations. Occasionally they can produce graphical models that misrepresent the phenomena we wish to study, a problem we call **grapher failure**. Sometimes the viewing window will be too large, obscuring details of the graph, which we call **hidden behavior**. We will give an example of each just to illustrate what can happen, but rest assured that these difficulties rarely occur with graphical models that arise from real-world problems.

Technology Note

One way to get the table in Figure 1.6b is to use the "Ask" feature of your graphing calculator and enter each x-value separately.

EXAMPLE 9 Seeing Grapher Failure

Look at the graph of $y = 3 - \dfrac{1}{\sqrt{x^2 - 1}}$ in the ZDecimal window on a graphing calculator. Are there any x-intercepts?

SOLUTION The graph is shown in Figure 1.6a.

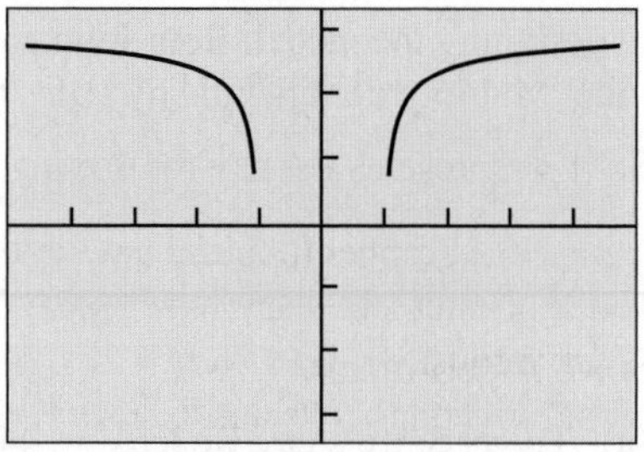

[–4.7, 4.7] by [–3.1, 3.1]

(a)

X	Y_1	
.8	ERROR	
.9	ERROR	
1	ERROR	
1.1	.81782	
1.2	1.4924	
1.3	1.7961	
1.4	1.9794	

Y_1 = 3–1/√(X²–1)

(b)

FIGURE 1.6 (a) A graph with no apparent intercepts. (b) The function $y = 3 - 1/\sqrt{x^2 - 1}$ is undefined when $|x| \le 1$.

The graph seems to have no x-intercepts, yet we can find some by solving the equation $0 = 3 - 1/\sqrt{x^2 - 1}$ algebraically:

$$\begin{aligned} 0 &= 3 - 1/\sqrt{x^2 - 1} \\ 1/\sqrt{x^2 - 1} &= 3 \\ \sqrt{x^2 - 1} &= 1/3 \\ x^2 - 1 &= 1/9 \\ x^2 &= 10/9 \\ x &= \pm\sqrt{10/9} \approx \pm 1.054 \end{aligned}$$

There should be x-intercepts at about ± 1.054. What went wrong?

The answer is a simple form of grapher failure. As the table shows, the function is undefined for the sampled x-values until $x = 1.1$, at which point the graph "turns on," beginning with the pixel at $(1.1, 0.81782)$ and continuing from there to the right. Similarly, the graph coming from the left "turns off" at $x = -1$, before it gets to the x-axis. The x-intercepts might well appear in other windows, but for this particular function in this particular window, the behavior we expect to see is not there.

Now try Exercise 49.

EXAMPLE 10 Not Seeing Hidden Behavior

Solve graphically: $x^3 - 1.1x^2 - 65.4x + 229.5 = 0$.

SOLUTION Figure 1.7a shows the graph in the standard $[-10, 10]$ by $[-10, 10]$ window, an inadequate choice because too much of the graph is off the screen. Our horizontal dimensions look fine, so we adjust our vertical dimensions to $[-500, 500]$, yielding the graph in Figure 1.7b.

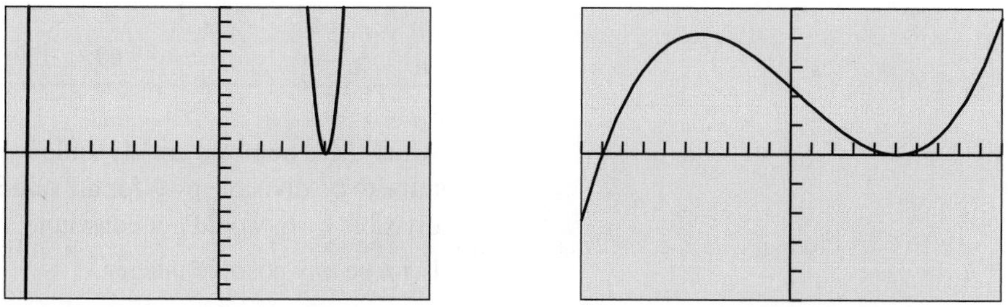

[–10, 10] by [–10, 10]
(a)

[–10, 10] by [–500, 500]
(b)

FIGURE 1.7 The graph of $y = x^3 - 1.1x^2 - 65.4x + 229.5$ in two viewing windows. (Example 10)

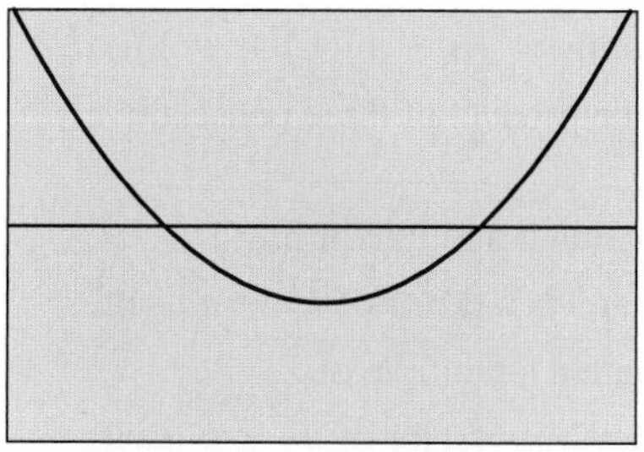

[4.95, 5.15] by [–0.1, 0.1]

FIGURE 1.8 A closer look at the graph of $y = x^3 - 1.1x^2 - 65.4x + 229.5$. (Example 10)

We use the grapher to locate an x-intercept near -9 (which we find to be -9) and then an x-intercept near 5 (which we find to be 5). The graph leads us to believe that we have finished. However, if we zoom in closer to observe the behavior near $x = 5$, the graph tells a new story (Figure 1.8).

In this graph we see that there are actually *two* x-intercepts near 5 (which we find to be 5 and 5.1). There are therefore three roots (or zeros) of the equation $x^3 - 1.1x^2 - 65.4x + 229.5 = 0$: $x = -9$, $x = 5$, and $x = 5.1$. **Now try Exercise 51.**

You might wonder if there could be still *more* hidden x-intercepts in Example 10! We will learn in Chapter 2 how the *Fundamental Theorem of Algebra* guarantees that there are not.

A Word About Proof

While Example 10 is still fresh in our minds, let us point out a subtle, but very important, consideration about our solution.

We *solved graphically* to find two solutions, then eventually three solutions, to the given equation. Although we did not show the steps, it is easy to *confirm numerically* that the three numbers found are actually solutions by substituting them into the equation. But the problem asked us to find *all* solutions. Although we could explore that equation graphically in a hundred more viewing windows and never find another solution, our failure to find them would not *prove* that they are not out there somewhere. That is why the Fundamental Theorem of Algebra is so important. It tells us that there can be at most three real solutions to *any* cubic equation, so we know for a fact that there are no more.

Exploration is encouraged throughout this book because it is how mathematical progress is made. Mathematicians are never satisfied, however, until they have *proved* their results. We will show you proofs in later chapters and we will ask you to produce proofs occasionally in the exercises. That will be a time for you to set the technology aside, get out a pencil, and show in a logical sequence of algebraic steps that something is undeniably and universally true. This process is called **deductive reasoning**.

EXAMPLE 11 Proving a Peculiar Number Fact

Prove that 6 is a factor of $n^3 - n$ for every positive integer n.

SOLUTION You can explore this expression for various values of n on your calculator. Table 1.5 shows it for the first 12 values of n.

Table 1.5 The First 12 Values of $n^3 - n$

n	1	2	3	4	5	6	7	8	9	10	11	12
$n^3 - n$	0	6	24	60	120	210	336	504	720	990	1320	1716

All of these numbers are divisible by 6, but that does not prove that they will continue to be divisible by 6 for all values of n. In fact, a table with a billion values, all divisible by 6, would not constitute a proof. Here is a proof:

Let n be *any* positive integer.

- We can factor $n^3 - n$ as the product of three numbers: $(n - 1)(n)(n + 1)$.
- The factorization shows that $n^3 - n$ is always the product of three consecutive integers.
- Every set of three consecutive integers must contain a multiple of 3.
- Since 3 divides a factor of $n^3 - n$, it follows that 3 is a factor of $n^3 - n$ itself.
- Every set of three consecutive integers must contain a multiple of 2.
- Since 2 divides a factor of $n^3 - n$, it follows that 2 is a factor of $n^3 - n$ itself.
- Since both 2 and 3 are factors of $n^3 - n$, we know that 6 is a factor of $n^3 - n$.

End of proof!

Now try Exercise 53.

QUICK REVIEW 1.1 *(For help, go to Section A.2.)*

Exercise numbers with a gray background indicate problems that the authors have designed to be solved *without a calculator*.

Factor the following expressions completely over the real numbers.

1. $x^2 - 16$

2. $x^2 + 10x + 25$

3. $81y^2 - 4$

4. $3x^3 - 15x^2 + 18x$

5. $16h^4 - 81$

6. $x^2 + 2xh + h^2$

7. $x^2 + 3x - 4$

8. $x^2 - 3x + 4$

9. $2x^2 - 11x + 5$

10. $x^4 + x^2 - 20$

SECTION 1.1 Exercises

In Exercises 1–10, match the numerical model to the corresponding graphical model (*a–j*) and algebraic model (*k–t*).

1.

x	3	5	7	9	12	15
y	6	10	14	18	24	30

2.

x	0	1	2	3	4	5
y	2	3	6	11	18	27

3.

x	2	4	6	8	10	12
y	4	10	16	22	28	34

4.

x	5	10	15	20	25	30
y	90	80	70	60	50	40

5.

x	1	2	3	4	5	6
y	39	36	31	24	15	4

6.

x	1	2	3	4	5	6
y	5	7	9	11	13	15

7.

x	5	7	9	11	13	15
y	1	2	3	4	5	6

8.

x	4	8	12	14	18	24
y	20	72	156	210	342	600

9.

x	3	4	5	6	7	8
y	8	15	24	35	48	63

10.

x	4	7	12	19	28	39
y	1	2	3	4	5	6

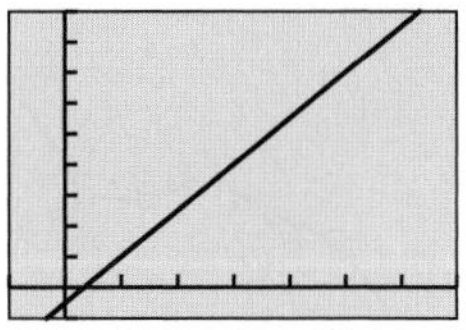

[–2, 14] by [–4, 36]

(a)

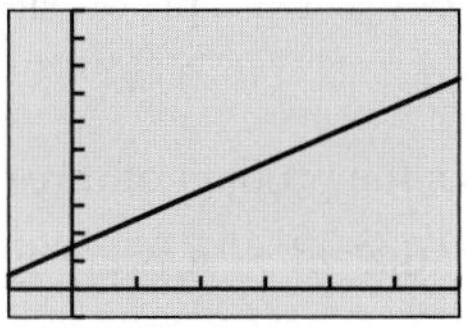

[–1, 6] by [–2, 20]

(b)

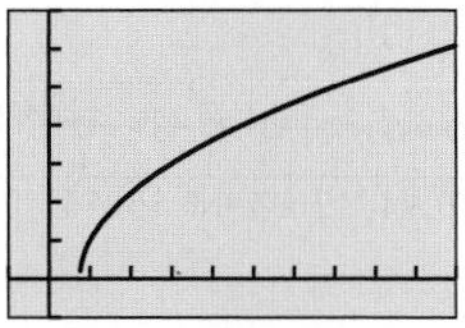

[–4, 40] by [–1, 7]

(c)

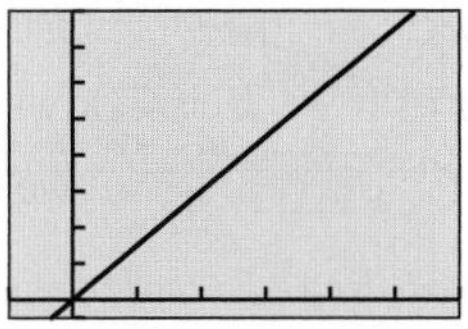

[–3, 18] by [–2, 32]

(d)

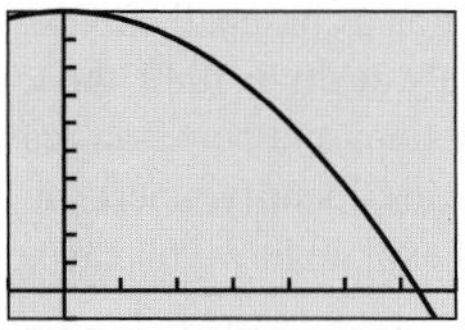

[–1, 7] by [–4, 40]

(e)

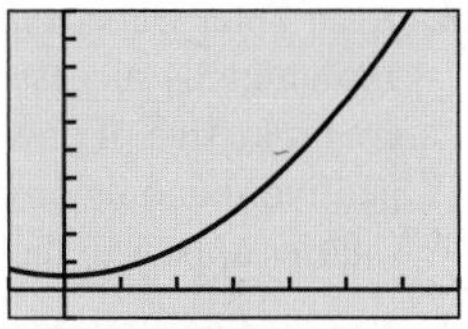

[–1, 7] by [–4, 40]

(f)

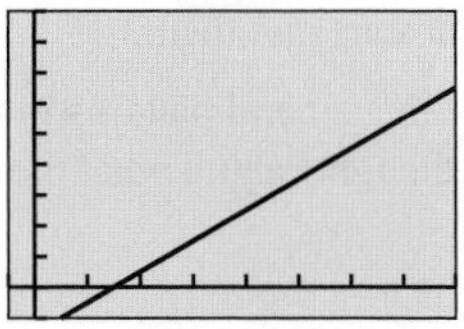

[–1, 16] by [–1, 9]

(g)

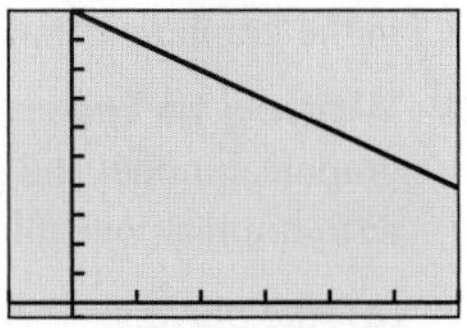

[–5, 30] by [–5, 100]

(h)

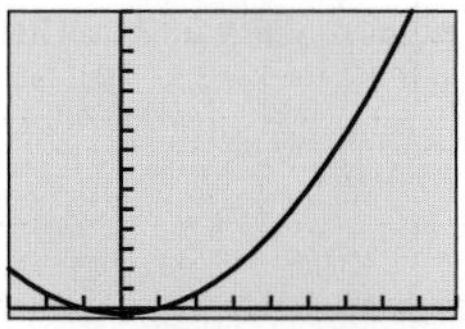

[–3, 9] by [–2, 60]

(i)

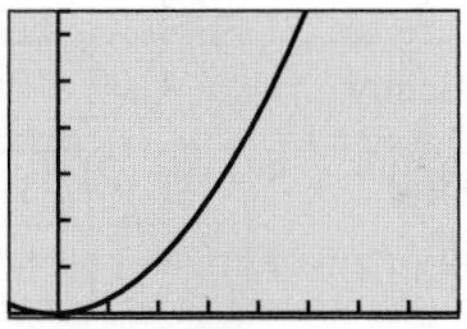

[–5, 40] by [–10, 650]

(j)

(k) $y = x^2 + x$

(l) $y = 40 - x^2$

(m) $y = (x + 1)(x - 1$

(n) $y = \sqrt{x - 3}$

(o) $y = 100 - 2x$

(p) $y = 3x - 2$

(q) $y = 2x$

(r) $y = x^2 + 2$

(s) $y = 2x + 3$

(t) $y = \dfrac{x - 3}{2}$

Exercises 11–18 refer to the data in Table 1.6 below, showing the participation rate of the male and female population in the United States civilian work force in selected years from 1960 to 2010 (seasonally adjusted):

Table 1.6 Labor Force Participation Rate, Age 16 and Over

Year	Female	Male
1960	37.7	83.4
1965	39.3	80.5
1970	43.3	79.6
1975	46.3	77.2
1980	51.5	77.0
1985	54.5	76.1
1990	57.5	76.3
1995	58.9	74.6
2000	59.9	74.7
2005	59.3	73.2
2010	58.6	70.7

Source: U.S. Bureau of Labor Statistics (www.bls.gov).

11. (a) According to the numerical model, what has been the trend in the percentage of females participating in the civilian work force since 1960?

(b) In what 5-year interval did the percentage of female participation change the most?

12. (a) According to the numerical model, what has been the trend in the percentage of males participating in the civilian work force since 1960?

(b) In what 5-year interval did the percentage of male participation change the most?

13. Model the data graphically with two scatter plots on the same graph, one showing the percentage of females employed as a function of time and the other showing the same for males. Measure time in years since 1960.

14. Over the 50-year period, have the male percentages been falling faster than the female percentages are rising, or vice versa?

15. Model the male and female data algebraically with separate linear equations of the form $y = mx + b$. [Use the 1960 and 2010 ordered pairs (with x-coordinates 0 and 50) to compute the slopes.]

16. If the percentages follow the linear models found in Exercise 15, at what point will the lines intersect? What is the significance of the intersection point in terms of employment participation?

17. Which data in Table 1.6 (male or female) appear to be less likely to follow a linear model? On the basis of social and historical circumstances, can you explain the nonlinear behavior in that model?

18. **Writing to Learn** Explain why the percentages cannot continue indefinitely to follow the linear models that you wrote in Exercise 15.

19. **Doing Arithmetic with Lists** Enter the data from the "Total" column of Table 1.2 of Example 2 into list L_1 in your calculator. Enter the data from the "Female" column into list L_2. Check a few computations to see that either one of the procedures in (a) and (b) causes the calculator to divide each element of L_2 by the corresponding entry in L_1, multiply it by 100, and store the resulting list of percentages in L_3.

 (a) On the home screen, enter the command: $100 \times L_2/L_1 \rightarrow L_3$.

 (b) Go to the top of list L_3 and enter $L_3 = 100(L_2/L_1)$.

20. **Comparing Cakes** A bakery sells a 9″ by 13″ cake for the same price as an 8″ diameter round cake. If the round cake is twice the height of the rectangular cake, which option gives the most cake for the money?

21. **Stepping Stones** A garden shop sells 12″ by 12″ square stepping stones for the same price as 13″ round stones. If all of the stepping stones are the same thickness, which option gives the most rock for the money?

22. **Free Fall of a Smoke Bomb** At the Oshkosh, WI, air show, Jake Trouper drops a smoke bomb to signal the official beginning of the show. Ignoring air resistance, an object in free fall will fall d feet in t seconds, where d and t are related by the algebraic model $d = 16t^2$.

 (a) How long will it take the bomb to fall 180 ft?

 (b) If the smoke bomb is in free fall for 12.5 sec after it is dropped, how high was the airplane when the smoke bomb was dropped?

23. **Physics Equipment** A physics student obtains the following data involving a ball rolling down an inclined plane, where t is the elapsed time in seconds and y is the distance traveled in inches.

t	0	1	2	3	4	5
y	0	1.2	4.8	10.8	19.2	30

Find an algebraic model that fits the data.

24. **U.S. Air Travel** The number of revenue passengers enplaned in the United States over the 14-year period from 1994 to 2007 is shown in Table 1.7.

Table 1.7 U.S. Air Travel

Year	Passengers (millions)	Year	Passengers (millions)
1994	528.8	2001	622.1
1995	547.8	2002	614.1
1996	581.2	2003	646.5
1997	594.7	2004	702.9
1998	612.9	2005	738.3
1999	636.0	2006	744.2
2000	666.2	2007	769.2

Source: www.airlines.org

 (a) Graph a scatter plot of the data. Let x be the number of years since 1994.

 (b) From 1994 to 2000 the data seem to follow a linear model. Use the 1994 and 2000 points to find an equation of the line and superimpose the line on the scatter plot.

 (c) According to the linear model, in what year did the number of passengers seem destined to reach 900 million?

 (d) What happened to disrupt the linear model?

Exercises 25–28 refer to the graph below, which shows the *minimum* salaries in major league baseball over a recent 18-year period and the *average* salaries in major league baseball over the same period. Salaries are measured in dollars and time is measured after the starting year (year 0).

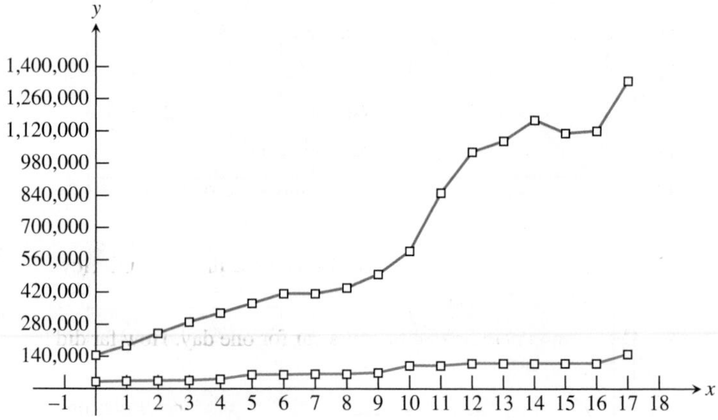

Source: Major League Baseball Players Association.

25. Which line is which, and how do you know?

26. After Peter Ueberroth's resignation as baseball commissioner in 1988 and his successor's untimely death in 1989, the team owners broke free of previous restrictions and began an era of competitive spending on player salaries. Identify where the 1990 salaries appear in the graph and explain how you can spot them.

27. The owners attempted to halt the uncontrolled spending by proposing a salary cap, which prompted a players' strike in 1994. The strike caused the 1995 season to be shortened and left many fans angry. Identify where the 1995 salaries appear in the graph and explain how you can spot them.

28. **Writing to Learn** Analyze the general patterns in the graphical model and give your thoughts about what the long-term implications might be for

 (a) the players;

 (b) the team owners;

 (c) the baseball fans.

In Exercises 29–38, solve the equation algebraically and confirm graphically.

29. $v^2 - 5 = 8 - 2v^2$

30. $(x + 11)^2 = 121$

31. $2x^2 - 5x + 2 = (x - 3)(x - 2) + 3x$

32. $x^2 - 7x - \frac{3}{4} = 0$

33. $x(2x - 5) = 12$

34. $x(2x - 1) = 10$

35. $x(x + 7) = 14$

36. $x^2 - 3x + 4 = 2x^2 - 7x - 8$

37. $x + 1 - 2\sqrt{x + 4} = 0$

38. $\sqrt{x} + x = 1$

In Exercises 39–46, solve the equation graphically by converting it to an equivalent equation with 0 on the right-hand side and then finding the x-intercepts.

39. $2x - 5 = \sqrt{x + 4}$ **40.** $|3x - 2| = 2\sqrt{x + 8}$

41. $|2x - 5| = 4 - |x - 3|$ **42.** $\sqrt{x + 6} = 6 - 2\sqrt{5 - x}$

43. $2x - 3 = x^3 - 5$ **44.** $x + 1 = x^3 - 2x - 5$

45. $(x + 1)^{-1} = x^{-1} + x$ **46.** $x^2 = |x|$

47. Swan Auto Rental charges \$32 per day plus \$0.18 per mile for an automobile rental.

(a) Elaine rented a car for one day and she drove 83 mi. How much did she pay?

(b) Ramon paid \$69.80 to rent a car for one day. How far did he drive?

48. Connecting Graphs and Equations The curves on the graph below are the graphs of the three curves given by

$y_1 = 4x + 5$

$y_2 = x^3 + 2x^2 - x + 3$

$y_3 = -x^3 - 2x^2 + 5x + 2.$

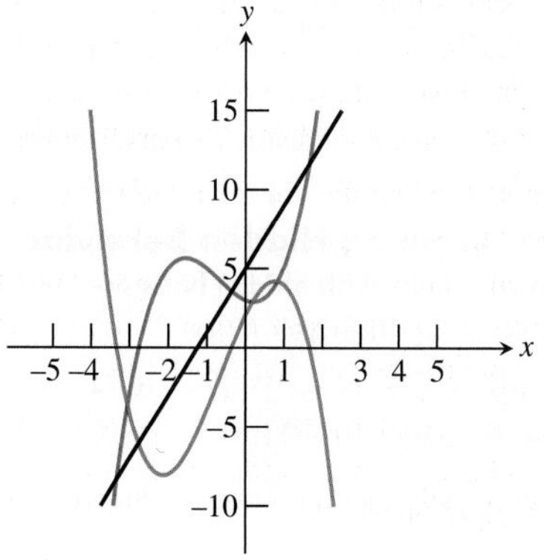

(a) Write an equation that can be solved to find the points of intersection of the graphs of y_1 and y_2.

(b) Write an equation that can be solved to find the x-intercepts of the graph of y_3.

(c) Writing to Learn How does the graphical model reflect the fact that the answers to (a) and (b) are equivalent algebraically?

(d) Confirm numerically that the x-intercepts of y_3 give the same values when substituted into the expressions for y_1 and y_2.

49. Exploring Grapher Failure Let $y = (x^{200})^{1/200}$.

(a) Explain algebraically why $y = x$ for all $x \geq 0$.

(b) Graph the equation $y = (x^{200})^{1/200}$ in the window $[0, 1]$ by $[0, 1]$.

(c) Is the graph different from the graph of $y = x$?

(d) Can you explain why the grapher failed?

50. Connecting Algebra and Geometry Explain how the algebraic equation $(x + b)^2 = x^2 + 2bx + b^2$ models the areas of the regions in the geometric figure shown below on the left:

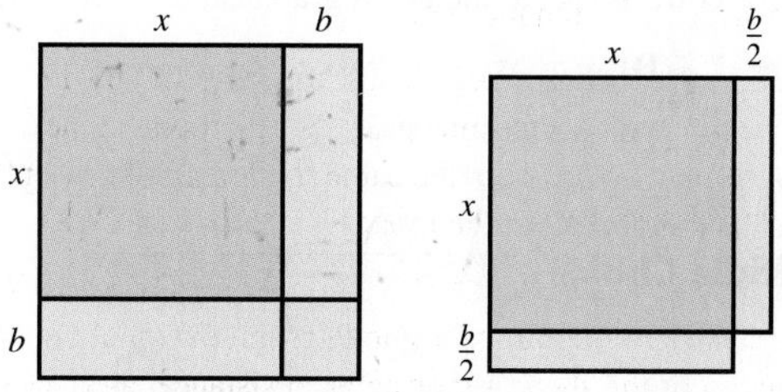

(Ex. 50) (Ex. 52)

51. Exploring Hidden Behavior Solving graphically, find all real solutions to the following equations. Watch out for hidden behavior.

(a) $y = 10x^3 + 7.5x^2 - 54.85x + 37.95$

(b) $y = x^3 + x^2 - 4.99x + 3.03$

52. Connecting Algebra and Geometry The geometric figure shown on the right above is a large square with a small square missing.

(a) Find the area of the figure.

(b) What area must be added to complete the large square?

(c) Explain how the algebraic formula for completing the square models the completing of the square in (b).

53. Proving a Theorem Prove that if n is a positive integer, then $n^2 + 2n$ is either odd or a multiple of 4. Compare your proof with those of your classmates.

54. **Writing to Learn** The graph below shows the distance from home against time for a jogger. Using information from the graph, write a paragraph describing the jogger's workout.

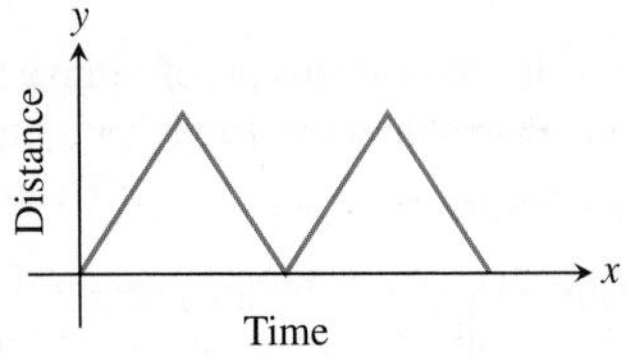

Standardized Test Questions

55. **True or False** A product of real numbers is zero if and only if every factor in the product is zero. Justify your answer.

56. **True or False** An algebraic model can always be used to make accurate predictions.

In Exercises 57–60, you may use a graphing calculator to decide which algebraic model corresponds to the given graphical or numerical model.

(A) $y = 2x + 3$ **(B)** $y = x^2 + 5$

(C) $y = 12 - 3x$ **(D)** $y = 4x + 3$

(E) $y = \sqrt{8 - x}$

57. **Multiple Choice**

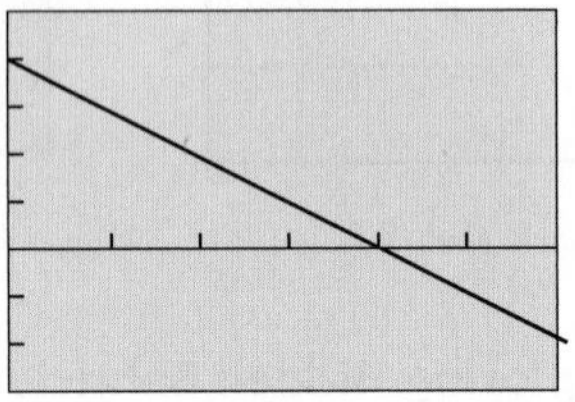

[0, 6] by [−9, 15]

58. **Multiple Choice**

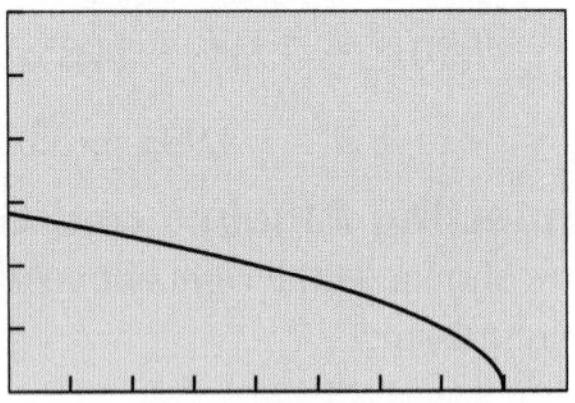

[0, 9] by [0, 6]

59. **Multiple Choice**

x	1	2	3	4	5	6
y	6	9	14	21	30	41

60. **Multiple Choice**

x	0	2	4	6	8	10
y	3	7	11	15	19	23

Explorations

61. **Analyzing the Market** Both Ahmad and LaToya watch the stock market throughout the year for stocks that make significant jumps from one month to another. When they spot one, each buys 100 shares. Ahmad's rule is to sell the stock if it fails to perform well for three months in a row. LaToya's rule is to sell in December if the stock has failed to perform well since its purchase.

The graph below shows the monthly performance in dollars (Jan.–Dec.) of a stock that both Ahmad and LaToya have been watching.

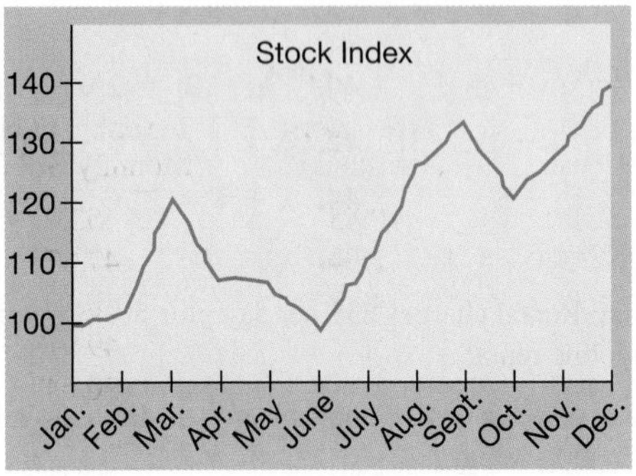

(a) Both Ahmad and LaToya bought the stock early in the year. In which month?

(b) At approximately what price did they buy the stock?

(c) When did Ahmad sell the stock?

(d) How much did Ahmad lose on the stock?

(e) **Writing to Learn** Explain why LaToya's strategy was better than Ahmad's for this particular stock in this particular year.

(f) Sketch a 12-month graph of a stock's performance that would favor Ahmad's strategy over LaToya's.

62. **Group Activity Creating Hidden Behavior** You can create your own graphs with hidden behavior. Working in groups of two or three, try this exploration.

(a) Graph the equation $y = (x + 2)(x^2 - 4x + 4)$ in the window $[-4, 4]$ by $[-10, 10]$.

(b) Confirm algebraically that this function has zeros only at $x = -2$ and $x = 2$.

(c) Graph the equation $y = (x + 2)(x^2 - 4x + 4.01)$ in the window $[-4, 4]$ by $[-10, 10]$

(d) Confirm algebraically that this function has only one zero, at $x = -2$. (Use the discriminant.)

(e) **Graph the Equation** $(x + 2)(x^2 - 4x + 3.99)$ in the window $[-4, 4]$ by $[-10, 10]$.

(f) Confirm algebraically that this function has three zeros. (Use the discriminant.)

Extending the Ideas

63. The Proliferation of Cell Phones Table 1.8 shows the number of cellular phone subscribers in the United States and their average monthly bill in the years from 2000 to 2010.

Table 1.8 Cellular Phone Subscribers

Year	Subscribers (millions)	Average Local Monthly Bill ($)
2000	109.5	45.27
2001	128.4	47.37
2002	140.8	48.40
2003	158.7	49.91
2004	182.1	50.64
2005	207.9	49.98
2006	233.0	50.56
2007	255.4	49.79
2008	262.7	50.07
2009	276.6	48.16
2010	300.5	47.21

Source: Cellular Telecommunication and Internet Association.

(a) Graph the scatter plots of the number of subscribers and the average local monthly bill as functions of time, letting time $t =$ the number of years after 1990.

(b) One of the scatter plots clearly suggests a linear model. Use the points at $t = 10$ and $t = 20$ to find a model in the form $y = mx + b$.

(c) Superimpose the graph of the linear model onto the scatter plot. Does the fit appear to be good?

(d) Why does a linear model seem inappropriate for the other scatter plot? Can you think of a curve that might fit the data better?

(e) In 1995 There were 33.8 million subscribers with an average local monthly bill of $51.00. Add these points to the scatter plots. Do the 1995 points match well with the trends in the rest of the data?

(f) **Writing to Learn** Give a plausible explanation for the unusual position of the 1995 point in the local monthly bill scatter plot.

64. Group Activity (Continuation of Exercise 63) Discuss the economic forces suggested by the data in Exercise 63 and speculate about the future by analyzing the scatter plots.

1.2 Functions and Their Properties

What you'll learn about

- Function Definition and Notation
- Domain and Range
- Continuity
- Increasing and Decreasing Functions
- Boundedness
- Local and Absolute Extrema
- Symmetry
- Asymptotes
- End Behavior

... and why

Functions and graphs form the basis for understanding the mathematics and applications you will see both in your workplace and in course-work in college.

In this section we will introduce the terminology that is used to describe functions throughout this book. Feel free to skim over parts with which you are already familiar, but take the time to become comfortable with concepts that might be new to you (like continuity and symmetry). Even if it takes several days to cover this section, it will be precalculus time well spent.

Function Definition and Notation

Mathematics and its applications abound with examples of formulas by which quantitative variables are related to each other. The language and notation of functions is ideal for that purpose. A function is actually a simple concept; if it were not, history would have replaced it with a simpler one by now. Here is the definition.

DEFINITION Function, Domain, and Range

A function from a set D to a set R is a rule that assigns to every element in D a unique element in R. The set D of all input values is the **domain** of the function, and the set R of all output values is the **range** of the function.

There are many ways to look at functions. One of the most intuitively helpful is the "machine" concept (Figure 1.9), in which values of the domain (x) are fed into the machine (the function f) to produce range values (y). To indicate that y comes from the function acting on x, we use Euler's elegant **function notation** $y = f(x)$ (which we read as "**y equals f of x**" or "**the value of f at x**"). Here x is the **independent variable** and y is the **dependent variable**.

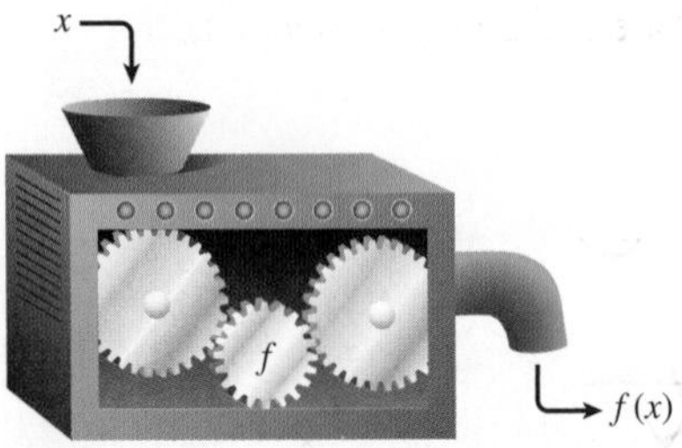

FIGURE 1.9 A "machine" diagram for a function.

A function can also be viewed as a **mapping** of the elements of the domain onto the elements of the range. Figure 1.10a shows a function that maps elements from the domain X onto elements of the range Y. Figure 1.10b shows another such mapping, but *this one is not a function*, since the rule does not assign the element x_1 to a *unique* element of Y.

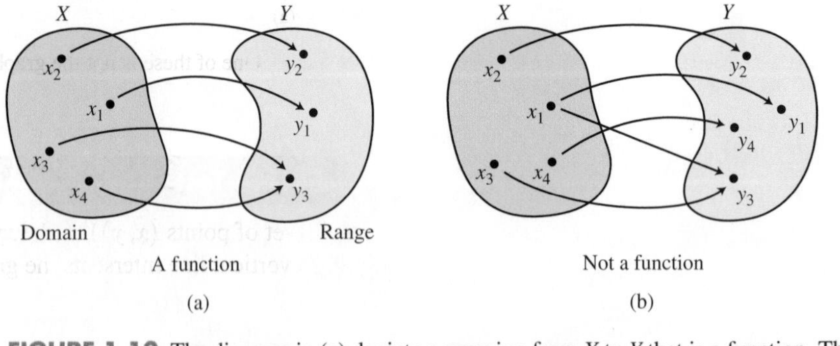

FIGURE 1.10 The diagram in (a) depicts a mapping from X to Y that is a function. The diagram in (b) depicts a mapping from X to Y that is not a function.

A Bit of History

The word *function* in its mathematical sense is generally attributed to Gottfried Leibniz (1646–1716), one of the pioneers in the methods of calculus. His attention to clarity of notation is one of his greatest contributions to scientific progress, which is why we still use his notation in calculus courses today. Ironically, it was not Leibniz but Leonhard Euler (1707–1783) who introduced the familiar notation $f(x)$.

This uniqueness of the range value is very important to us as we study function behavior. Knowing that $f(2) = 8$ tells us something about f, and that understanding would be contradicted if we were to discover later that $f(2) = 4$. That is why you will never see a function defined by an ambiguous formula like $f(x) = 3x \pm 2$.

EXAMPLE 1 Defining a Function

Does the formula $y = x^2$ define y as a function of x?

SOLUTION Yes, y is a function of x. In fact, we can write the formula in function notation: $f(x) = x^2$. When a number x is substituted into the function, the square of x will be the output, and there is no ambiguity about what the square of x is.

Now try Exercise 3.

Another useful way to look at functions is graphically. The **graph of the function $y = f(x)$** is the set of all points $(x, f(x))$, x in the domain of f. We match domain values along the x-axis with their range values along the y-axis to get the ordered pairs that yield the graph of $y = f(x)$.

EXAMPLE 2 Seeing a Function Graphically

Of the three graphs shown in Figure 1.11, which is *not* the graph of a function? How can you tell?

SOLUTION The graph in (c) is not the graph of a function. There are three points on the graph with x-coordinate 0, so the graph does not assign a *unique* value to 0. (Indeed, we can see that there are plenty of numbers between -2 and 2 to which the graph assigns multiple values.) The other two graphs do not have a comparable problem because no vertical line intersects either of the other graphs in more than one point. Graphs that pass this *vertical line test* are the graphs of functions.

Now try Exercise 5.

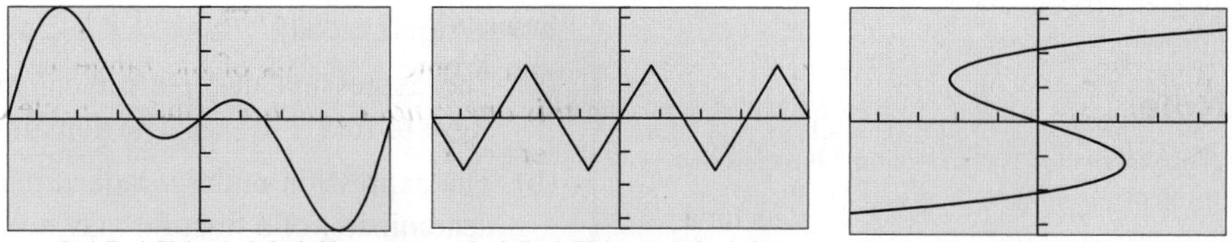

[–4.7, 4.7] by [–3.3, 3.3]
(a)

[–4.7, 4.7] by [–3.3, 3.3]
(b)

[–4.7, 4.7] by [–3.3, 3.3]
(c)

FIGURE 1.11 One of these is not the graph of a function. (Example 2)

Vertical Line Test

A graph (set of points (x, y)) in the xy-plane defines y as a function of x if and only if no vertical line intersects the graph in more than one point.

Domain and Range

We will usually define functions algebraically, giving the rule explicitly in terms of the domain variable. The rule, however, does not tell the complete story without some consideration of what the domain actually is.

What About Data?

When moving from a numerical model to an algebraic model we will often use a function to approximate data pairs that by themselves violate our definition. In Figure 1.12 we can see that several pairs of data points fail the vertical line test, and yet the linear function approximates the data quite well.

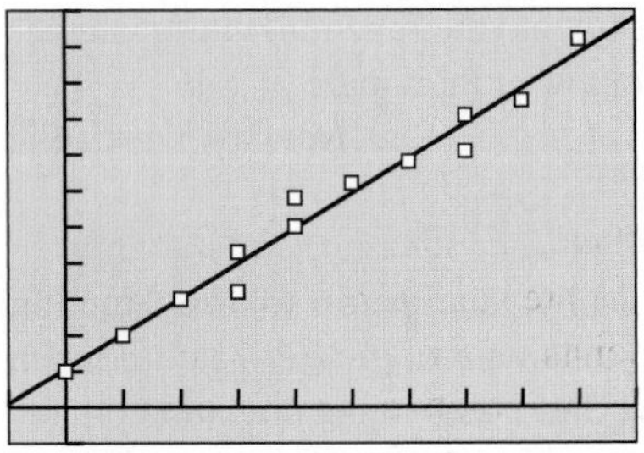

[–1, 10] by [–1, 11]

FIGURE 1.12 The data points fail the vertical line test but are nicely approximated by a linear function.

For example, we can define the volume of a sphere as a function of its radius by the formula

$$V(r) = \frac{4}{3}\pi r^3 \text{ (Note that this is "}V\text{ of }r\text{"—not "}V \cdot r\text{").}$$

This *formula* is defined for all real numbers, but the volume *function* is not defined for negative r-values. So, if our intention were to study the volume function, we would restrict the domain to be all $r \geq 0$.

Agreement About Domain

Unless we are dealing with a model (like volume) that necessitates a restricted domain, we will assume that the domain of a function defined by an algebraic expression is the same as the domain of the algebraic expression, the **implied domain**. For models, we will use a domain that fits the situation, the **relevant domain**.

EXAMPLE 3 Finding the Domain of a Function

Find the domain of each of these functions:

(a) $f(x) = \sqrt{x + 3}$

(b) $g(x) = \dfrac{\sqrt{x}}{x - 5}$

(c) $A(s) = (\sqrt{3}/4)s^2$, where $A(s)$ is the area of an equilateral triangle with sides of length s.

SOLUTION

Solve Algebraically

(a) The expression under a radical may not be negative. We set $x + 3 \geq 0$ and solve to find $x \geq -3$. The domain of f is the interval $[-3, \infty)$.

(b) The expression under a radical may not be negative; therefore $x \geq 0$. Also, the denominator of a fraction may not be zero; therefore, $x \neq 5$. The domain of g is the interval $[0, \infty)$ with the number 5 removed, which we can write as the *union* of two intervals: $[0, 5) \cup (5, \infty)$.

(c) The algebraic expression has domain $(-\infty, \infty)$, but the behavior being modeled restricts s from being negative. The domain of A is the interval $[0, \infty)$.

Support Graphically We can support our answers in (a) and (b) graphically, as the calculator should not plot points where the function is undefined.

(a) Notice that the graph of $y = \sqrt{x + 3}$ (Figure 1.13a) shows points only for $x \geq -3$, as expected.

(b) The graph of $y = \sqrt{x}/(x - 5)$ (Figure 1.13b) shows points only for $x \geq 0$, as expected. Some calculators might show an unexpected line through the x-axis at $x = 5$. This line, another form of grapher failure, should not be there. Ignoring it, we see that 5, as expected, is not in the domain.

(c) The graph of $y = (\sqrt{3}/4)s^2$ (Figure 1.13c) shows the unrestricted domain of the algebraic expression: all real numbers. The calculator has no way of knowing that s is the length of a side of a triangle. **Now try Exercise 11.**

Note

The symbol "∪" is read "union." It means that the elements of the two sets are combined to form one set.

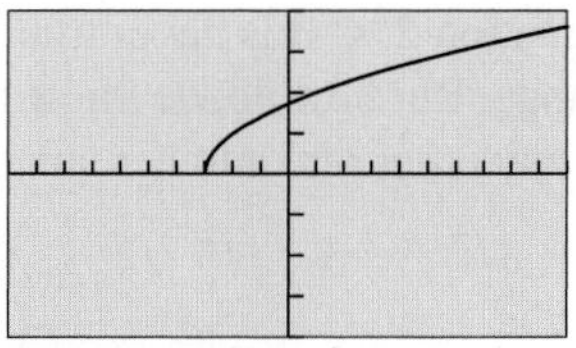

[–10, 10] by [–4, 4]
(a)

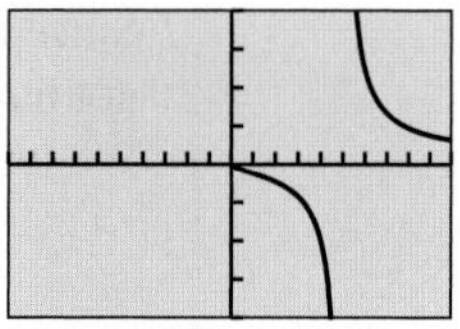

[–10, 10] by [–4, 4]
(b)

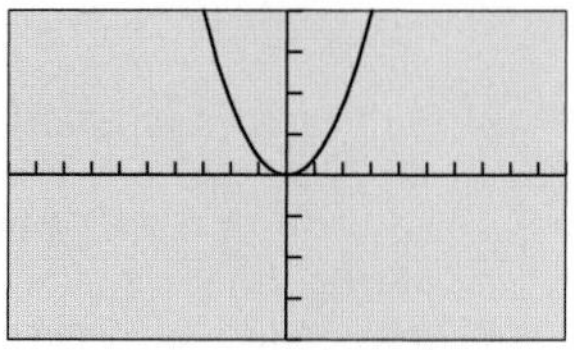

[–10, 10] by [–4, 4]
(c)

FIGURE 1.13 Graphical support of the algebraic solutions in Example 3. The points in (c) with negative x-coordinates should be ignored because the calculator does not know that x is a length (but we do).

Finding the range of a function algebraically is often much harder than finding the domain, although graphically the things we look for are similar: To find the *domain* we look for all *x-coordinates* that correspond to points on the graph, and to find the *range* we look for all *y-coordinates* that correspond to points on the graph. A good approach is to use graphical and algebraic approaches simultaneously, as we show in Example 4.

EXAMPLE 4 Finding the Range of a Function

Find the range of the function $f(x) = \frac{2}{x}$.

SOLUTION

Solve Graphically The graph of $y = \frac{2}{x}$ is shown in Figure 1.14.

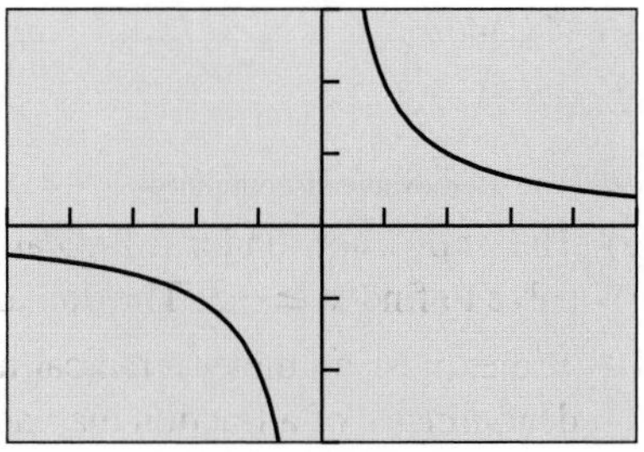

[–5, 5] by [–3, 3]

FIGURE 1.14 The graph of $y = 2/x$. Is $y = 0$ in the range?

It appears that $x = 0$ is not in the domain (as expected, because a denominator cannot be zero). It also appears that the range consists of all real numbers except 0.

Confirm Algebraically We confirm that 0 is not in the range by trying to solve $2/x = 0$:

$$\frac{2}{x} = 0$$
$$2 = 0 \cdot x$$
$$2 = 0$$

(continued)

Function Notation

A grapher typically does not use function notation. So the function $f(x) = x^2 + 1$ is entered as $y_1 = x^2 + 1$. On some graphers you can evaluate f at $x = 3$ by entering $y_1(3)$ on the home screen. On the other hand, on other graphers $y_1(3)$ means $y_1 * 3$.

Since the equation $2 = 0$ is never true, $2/x = 0$ has no solutions, and so $y = 0$ is not in the range. But how do we know that all other real numbers are in the range? We let k be any other real number and try to solve $2/x = k$:

$$\frac{2}{x} = k$$
$$2 = k \cdot x$$
$$x = \frac{2}{k}$$

As you can see, there was no problem finding an x this time, so 0 is the only number not in the range of f. We write the range $(-\infty, 0) \cup (0, \infty)$.

Now try Exercise 17.

You can see that this is considerably more involved than finding a domain, but we are hampered at this point by not having many tools with which to analyze function behavior. We will revisit the problem of finding ranges in Exercise 86, after having developed the tools that will simplify the analysis.

Continuity

One of the most important properties of the majority of functions that model real-world behavior is that they are *continuous*. Graphically speaking, a function is continuous at a point if the graph does not come apart at that point. We can illustrate the concept with a few graphs (Figure 1.15):

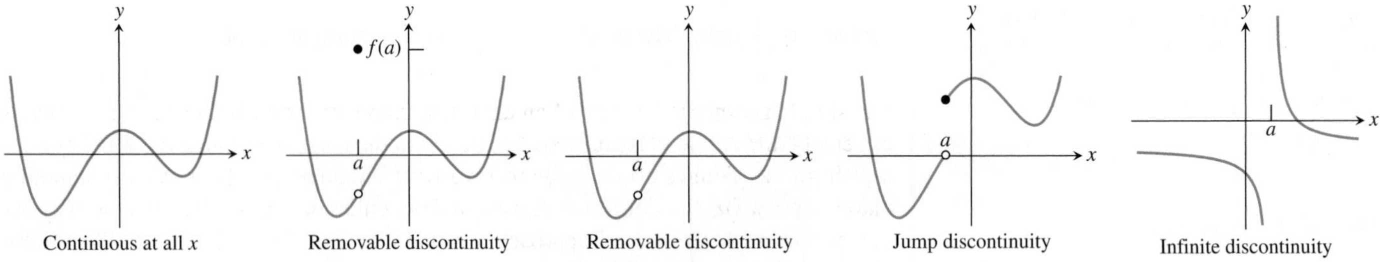

FIGURE 1.15 Some points of discontinuity.

Let's look at these cases individually.

This graph is continuous everywhere. Notice that the graph has no breaks. This means that if we are studying the behavior of the function f for x values close to any particular real number a, we can be assured that the $f(x)$ values will be close to $f(a)$.

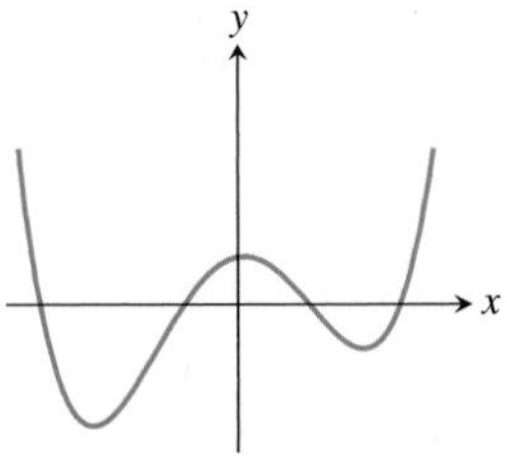

Continuous at all x

This graph is continuous everywhere except for the "hole" at $x = a$. If we are studying the behavior of this function f for x values close to a, we *cannot* be assured that the $f(x)$ values will be close to $f(a)$. In this case, $f(x)$ is smaller than $f(a)$ for x near a. This is called a **removable discontinuity** because it can be patched by redefining $f(a)$ so as to plug the hole.

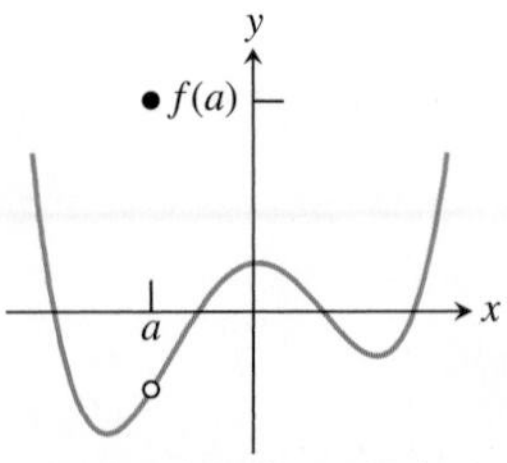

Removable discontinuity

This graph also has a **removable discontinuity** at $x = a$. If we are studying the behavior of this function f for x values close to a, we are still not assured that the $f(x)$ values will be close to $f(a)$, because in this case $f(a)$ doesn't even exist. It is removable because we could define $f(a)$ in such a way as to plug the hole and make f continuous at a.

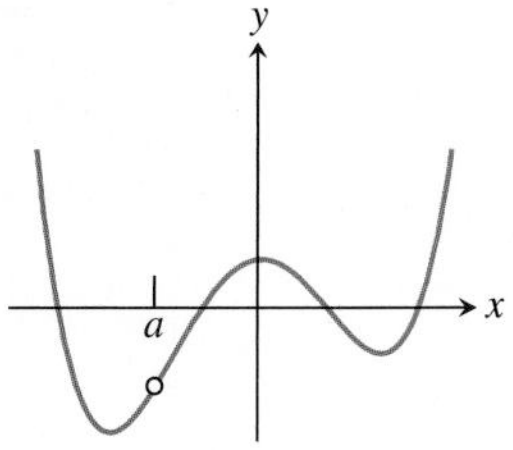

Removable discontinuity

Here is a discontinuity that is not removable. It is a **jump discontinuity** because there is more than just a hole at $x = a$; there is a *jump* in function values that makes the gap impossible to plug with a single point $(a, f(a))$, no matter how we try to redefine $f(a)$.

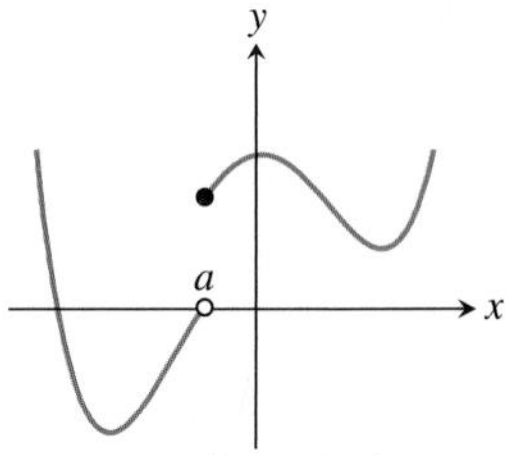

Jump discontinuity

This is a function with an **infinite discontinuity** at $x = a$. It is definitely not removable.

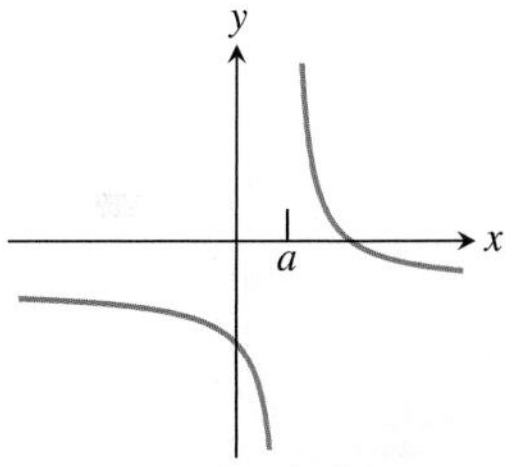

Infinite discontinuity

A Limited Use of Limits

Although the *notation* of limits is easy to understand, the algebraic *definition* of a limit can be a little intimidating and is best left to future courses. We will have more to say about limits in Chapter 11. For now, if you understand the statement $\lim_{x \to 5}(x^2 - 1) = 24$, you are where you need to be.

The simple geometric concept of an unbroken graph at a point is a visual notion that is extremely difficult to communicate accurately in the language of algebra. The key concept from the pictures seems to be that we want the point $(x, f(x))$ to slide smoothly onto the point $(a, f(a))$ without missing it from either direction. We say that $f(x)$ approaches $f(a)$ as a *limit* as x approaches a, and we write $\lim_{x \to a} f(x) = f(a)$. This "limit notation" reflects graphical behavior so naturally that we will use it throughout this book as an efficient way to describe function behavior, beginning with this definition of continuity. A function f is **continuous at $x = a$** if $\lim_{x \to a} f(x) = f(a)$. A function is **discontinuous at $x = a$** if it is not continuous at $x = a$.

EXAMPLE 5 Identifying Points of Discontinuity

Judging from the graphs, which of the following figures shows functions that are discontinuous at $x = 2$? Are any of the discontinuities removable?

SOLUTION Figure 1.16 shows a function that is undefined at $x = 2$ and hence not continuous there. The discontinuity at $x = 2$ is not removable.

The function graphed in Figure 1.17 is a quadratic polynomial whose graph is a parabola, a graph that has no breaks because its domain includes all real numbers. It is continuous for all x.

The function graphed in Figure 1.18 is not defined at $x = 2$ and so cannot be continuous there. The graph looks like the graph of the line $y = x + 2$, except that there is a hole where the point $(2, 4)$ should be. This is a removable discontinuity.

Now try Exercise 21.

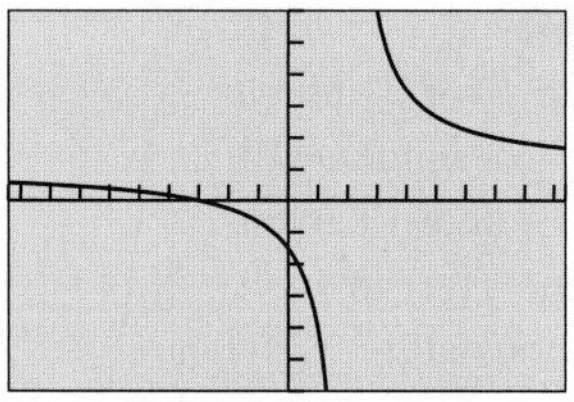

[–9.4, 9.4] by [–6, 6]

FIGURE 1.16 $f(x) = \dfrac{x + 3}{x - 2}$

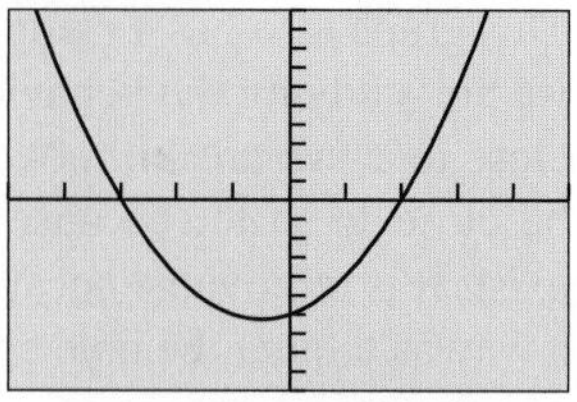

[–5, 5] by [–10, 10]

FIGURE 1.17 $g(x) = (x + 3)(x - 2)$

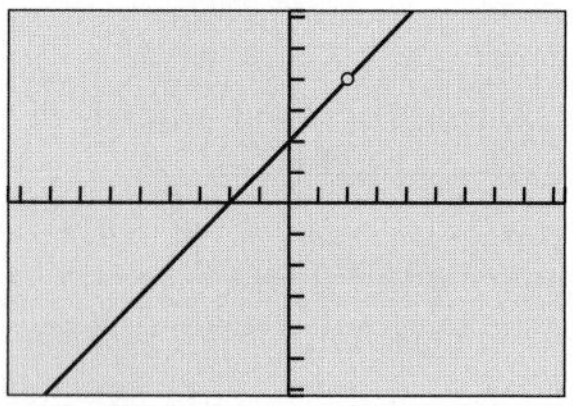

[–9.4, 9.4] by [–6.2, 6.2]

FIGURE 1.18 $h(x) = \dfrac{x^2 - 4}{x - 2}$

Increasing and Decreasing Functions

Another function concept that is easy to understand graphically is the property of being increasing, decreasing, or constant on an interval. We illustrate the concept with a few graphs (Figure 1.19):

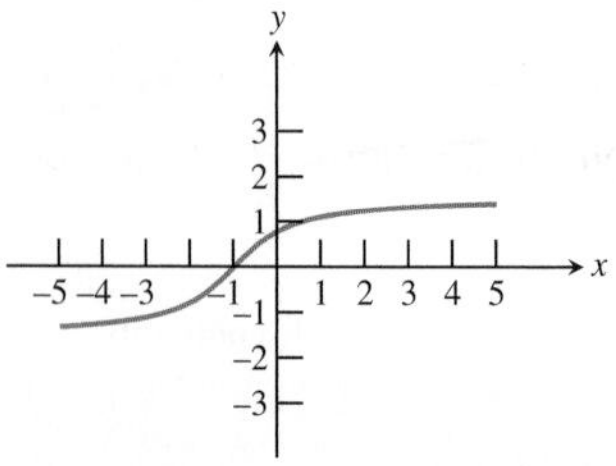

Increasing

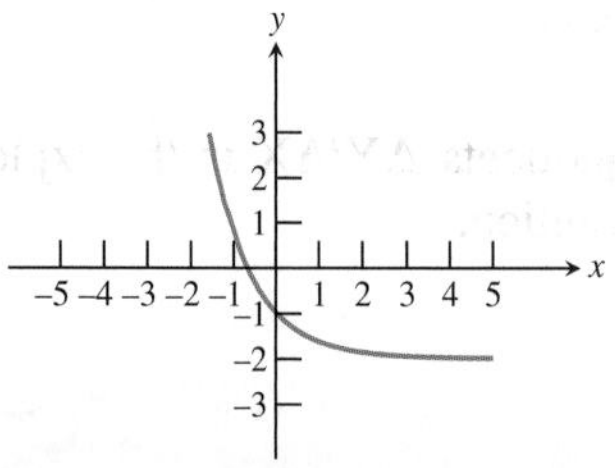

Decreasing

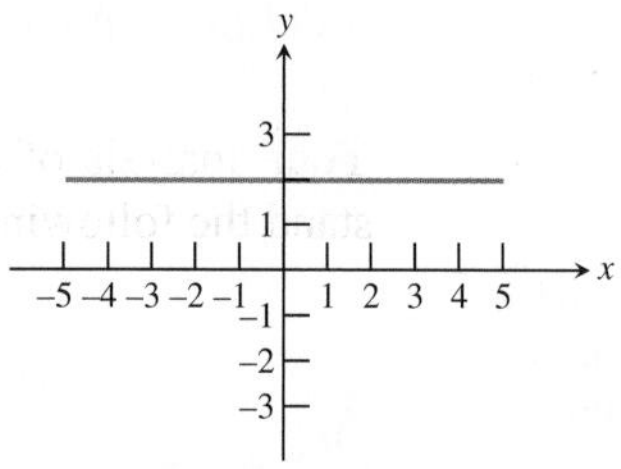

Constant

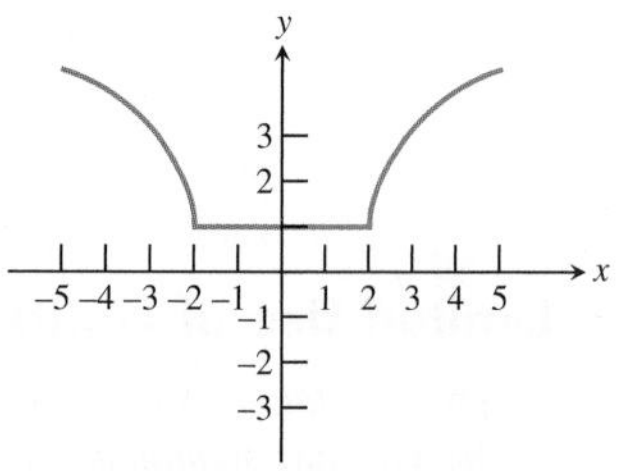

Decreasing on $(-\infty, -2]$
Constant on $[-2, 2]$
Increasing on $[2, \infty)$

FIGURE 1.19 Examples of increasing, decreasing, or constant on an interval.

Once again the idea is easy to communicate graphically, but how can we identify these properties of functions algebraically? Exploration 1 will help to set the stage for the algebraic definition.

EXPLORATION 1 Increasing, Decreasing, and Constant Data

1. Of the three tables of numerical data below, which would be modeled by a function that is (a) increasing, (b) decreasing, (c) constant?

X	Y1
–2	12
–1	12
0	12
1	12
3	12
7	12

X	Y2
–2	3
–1	1
0	0
1	–2
3	–6
7	–12

X	Y3
–2	–5
–1	–3
0	–1
1	1
3	4
7	10

ΔList on a Calculator

Your calculator might be able to help you with the numbers in Exploration 1. Some calculators have a "ΔList" operation that will calculate the changes as you move down a list. For example, the command "ΔList (L1) → L3" will store the differences from L1 into L3. Note that ΔList (L1) is always one entry shorter than L1 itself.

2. Make a list of ΔY1, the *change* in Y1 values as you move down the list. As you move from Y1 = a to Y1 = b, the change is ΔY1 = $b - a$. Do the same for the values of Y2 and Y3.

X moves from	ΔX	ΔY1
−2 to −1	1	
−1 to 0	1	
0 to 1	1	
1 to 3	2	
3 to 7	4	

X moves from	ΔX	ΔY2
−2 to −1	1	
−1 to 0	1	
0 to 1	1	
1 to 3	2	
3 to 7	4	

X moves from	ΔX	ΔY3
−2 to −1	1	
−1 to 0	1	
0 to 1	1	
1 to 3	2	
3 to 7	4	

3. What is true about the quotients ΔY/ΔX for an increasing function? For a decreasing function? For a constant function?

4. Where else have you seen the quotient ΔY/ΔX? Does this reinforce your answers in part 3?

Your analysis of the quotients ΔY/ΔX in the exploration should help you to understand the following definition.

DEFINITION Increasing, Decreasing, and Constant Function on an Interval

A function f is **increasing** on an interval if, for any two points in the interval, a positive change in x results in a positive change in $f(x)$.

A function f is **decreasing** on an interval if, for any two points in the interval, a positive change in x results in a negative change in $f(x)$.

A function f is **constant** on an interval if, for any two points in the interval, a positive change in x results in a zero change in $f(x)$.

EXAMPLE 6 Analyzing a Function for Increasing-Decreasing Behavior

For each function, tell the intervals on which it is increasing and the intervals on which it is decreasing.

(a) $f(x) = (x + 2)^2$ **(b)** $g(x) = \dfrac{x^2}{x^2 - 1}$

SOLUTION

Solve Graphically

(a) We see from the graph in Figure 1.20 that f is decreasing on $(-\infty, -2]$ and increasing on $[-2, \infty)$. (Notice that we include -2 in both intervals. Don't worry that this sets up some contradiction about what happens *at* -2, because we only talk about functions increasing or decreasing on *intervals*, and -2 is not an interval.)

(continued)

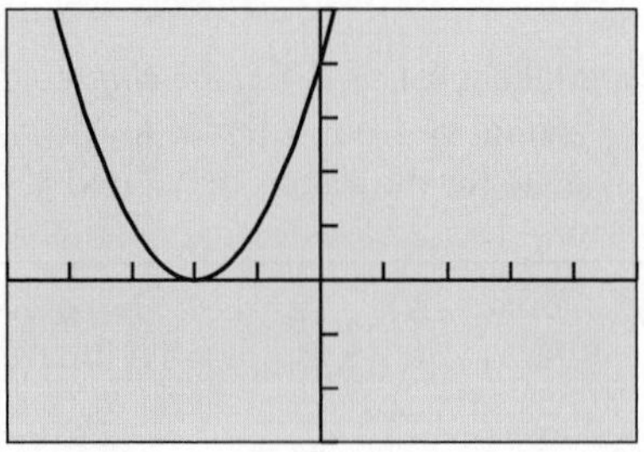

FIGURE 1.20 The function $f(x) = (x + 2)^2$ decreases on $(-\infty, -2]$ and increases on $[-2, \infty)$. (Example 6)

(b) We see from the graph in Figure 1.21 that g is increasing on $(-\infty, -1)$, increasing again on $(-1, 0]$, decreasing on $[0, 1)$, and decreasing again on $(1, \infty)$.

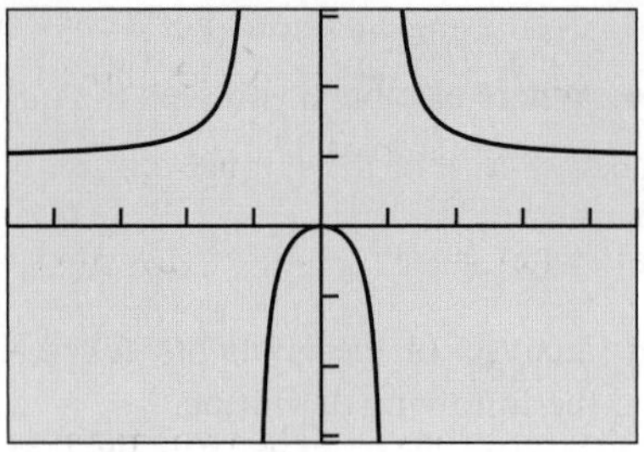

FIGURE 1.21 The function $g(x) = x^2/(x^2 - 1)$ increases on $(-\infty, -1)$ and $(-1, 0]$; the function decreases on $[0, 1)$ and $(1, \infty)$. (Example 6)

Now try Exercise 33.

You may have noticed that we are making some assumptions about the graphs. How do we know that they don't turn around somewhere off the screen? We will develop some ways to answer that question later in the book, but the most powerful methods will await you when you study calculus.

Boundedness

The concept of *boundedness* is fairly simple to understand both graphically and algebraically. We will move directly to the algebraic definition after motivating the concept with some typical graphs (Figure 1.22).

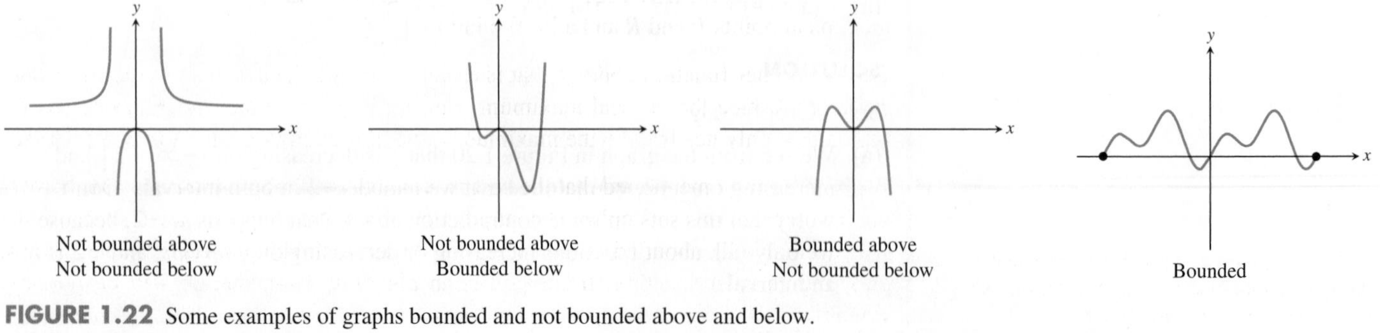

FIGURE 1.22 Some examples of graphs bounded and not bounded above and below.

DEFINITION Lower Bound, Upper Bound, and Bounded

A function f is **bounded below** if there is some number b that is less than or equal to every number in the range of f. Any such number b is called a **lower bound** of f.

A function f is **bounded above** if there is some number B that is greater than or equal to every number in the range of f. Any such number B is called an **upper bound** of f.

A function f is **bounded** if it is bounded both above and below.

We can extend the above definition to the idea of **bounded on an interval** by restricting the domain of consideration in each part of the definition to the interval we wish to consider. For example, the function $f(x) = 1/x$ is bounded above on the interval $(-\infty, 0)$ and bounded below on the interval $(0, \infty)$.

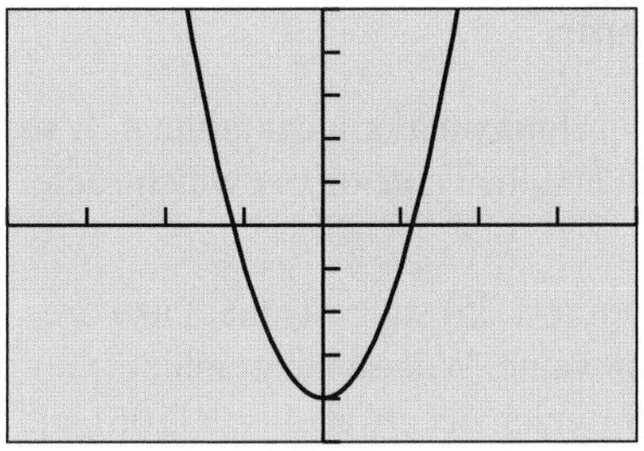

[-4, 4] by [-5, 5]
(a)

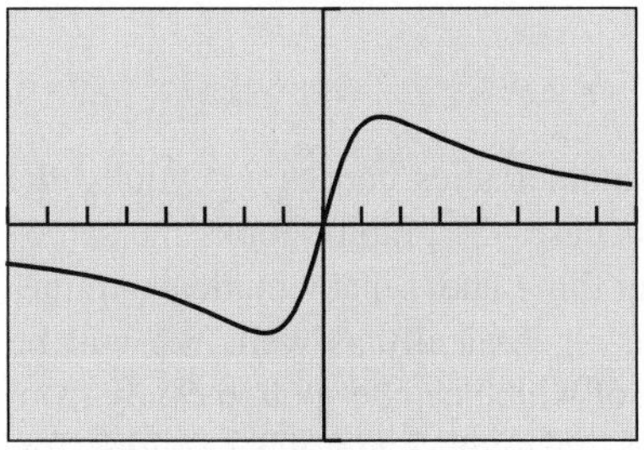

[-8, 8] by [-1, 1]
(b)

FIGURE 1.23 The graphs for Example 7. Which are bounded where?

EXAMPLE 7 Checking Boundedness

Identify each of these functions as bounded below, bounded above, or bounded.

(a) $w(x) = 3x^2 - 4$ **(b)** $p(x) = \dfrac{x}{1 + x^2}$

SOLUTION

Solve Graphically The two graphs are shown in Figure 1.23. It appears that w is bounded below, and p is bounded.

Confirm Algebraically We can confirm that w is bounded below by finding a lower bound, as follows:

$$x^2 \geq 0 \quad x^2 \text{ is nonnegative.}$$
$$3x^2 \geq 0 \quad \text{Multiply by 3.}$$
$$3x^2 - 4 \geq 0 - 4 \quad \text{Subtract 4.}$$
$$3x^2 - 4 \geq -4$$

Thus, -4 is a lower bound for $w(x) = 3x^2 - 4$.

We leave the verification that p is bounded as an exercise (Exercise 77).

Now try Exercise 37.

Local and Absolute Extrema

Many graphs are characterized by peaks and valleys where they change from increasing to decreasing and vice versa. The extreme values of the function (or *local extrema*) can be characterized as either *local maxima* or *local minima*. The distinction can be easily seen graphically. Figure 1.24 shows a graph with three local extrema: local maxima at points P and R and a local minimum at Q.

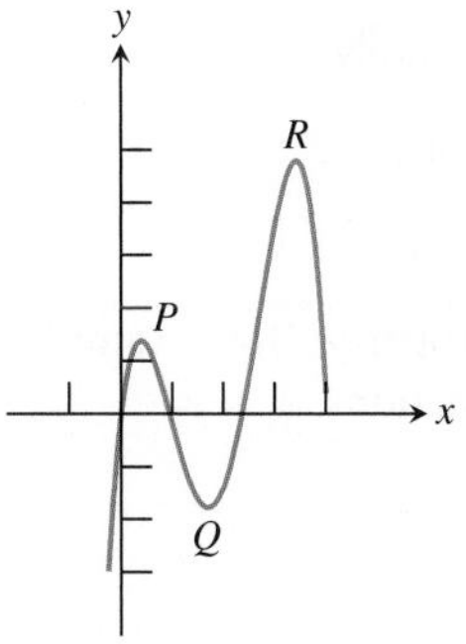

FIGURE 1.24 The graph suggests that f has a local maximum at P, a local minimum at Q, and a local maximum at R.

This is another function concept that is easier to see graphically than to describe algebraically. Notice that a local maximum does not have to be *the* maximum value of a function; it only needs to be the maximum value of the function on *some* tiny interval.

We have already mentioned that the best method for analyzing increasing and decreasing behavior involves calculus. Not surprisingly, the same is true for local extrema. We will generally be satisfied in this course with approximating local extrema using a graphing calculator, although sometimes an algebraic confirmation will be possible when we learn more about specific functions.

DEFINITION Local and Absolute Extrema

A **local maximum** of a function f is a value $f(c)$ that is greater than or equal to all range values of f on some open interval containing c. If $f(c)$ is greater than or equal to all range values of f, then $f(c)$ is the **maximum** (or **absolute maximum**) value of f.

A **local minimum** of a function f is a value $f(c)$ that is less than or equal to all range values of f on some open interval containing c. If $f(c)$ is less than or equal to all range values of f, then $f(c)$ is the **minimum** (or **absolute minimum**) value of f.

Local extrema are also called **relative extrema**.

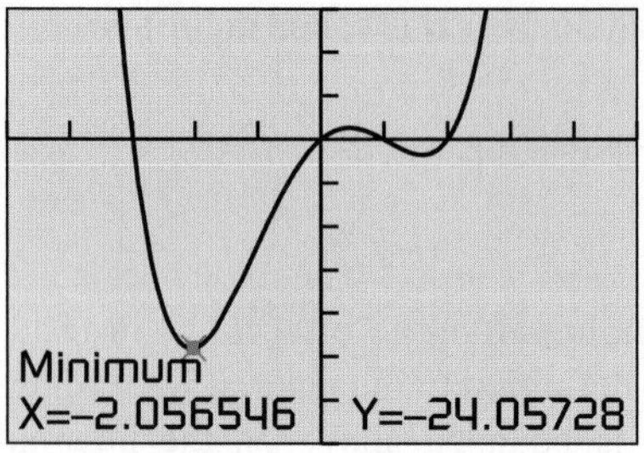

[–5, 5] by [–35, 15]

FIGURE 1.25 A graph of $y = x^4 - 7x^2 + 6x$. (Example 8)

EXAMPLE 8 Identifying Local Extrema

Decide whether $f(x) = x^4 - 7x^2 + 6x$ has any local maxima or local minima. If so, find each local maximum value or minimum value and the value of x at which each occurs.

SOLUTION The graph of $y = x^4 - 7x^2 + 6x$ (Figure 1.25) suggests that there are two local minimum values and one local maximum value. We use the graphing calculator to approximate local minima as -24.06 (which occurs at $x \approx -2.06$) and -1.77 (which occurs at $x \approx 1.60$). Similarly, we identify the (approximate) local maximum as 1.32 (which occurs at $x \approx 0.46$).

Now try Exercise 41.

Using a Grapher to Find Local Extrema

Most modern graphers have built-in "maximum" and "minimum" finders that identify local extrema by looking for sign changes in Δy. It is not easy to find local extrema by zooming in on them, as the graphs tend to flatten out and hide the very behavior you are looking for. If you use this method, keep narrowing the vertical window to maintain some curve in the graph.

Symmetry

In the graphical sense, the word "symmetry" in mathematics carries essentially the same meaning as it does in art: The picture (in this case, the graph) "looks the same" when viewed in more than one way. The interesting thing about mathematical symmetry is that it can be characterized numerically and algebraically as well. We will be looking at three particular types of symmetry, each of which can be spotted easily from a graph, a table of values, or an algebraic formula, once you know what to look for. Since it is the connections among the three models (graphical, numerical, and algebraic) that we need to emphasize in this section, we will illustrate the various symmetries in all three ways, side-by-side.

Symmetry with respect to the y-axis

Example: $f(x) = x^2$

Graphically

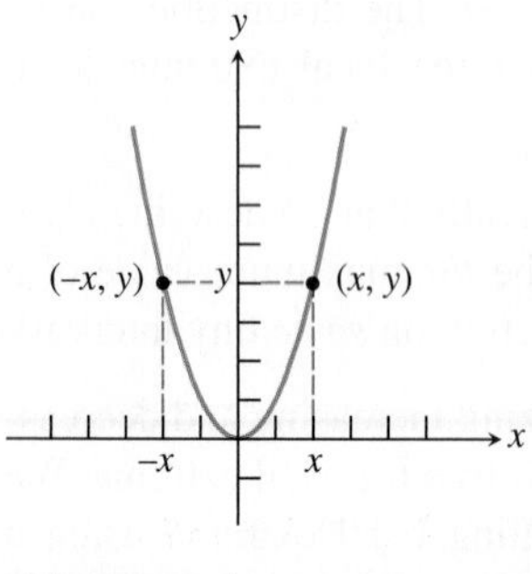

FIGURE 1.26 The graph looks the same to the left of the y-axis as it does to the right of it.

Numerically

x	$f(x)$
-3	9
-2	4
-1	1
1	1
2	4
3	9

Algebraically

For all x in the domain of f,

$$f(-x) = f(x).$$

Functions with this property (for example, x^n, n even) are **even** functions.

Symmetry with respect to the x-axis
Example: $x = y^2$

Graphically

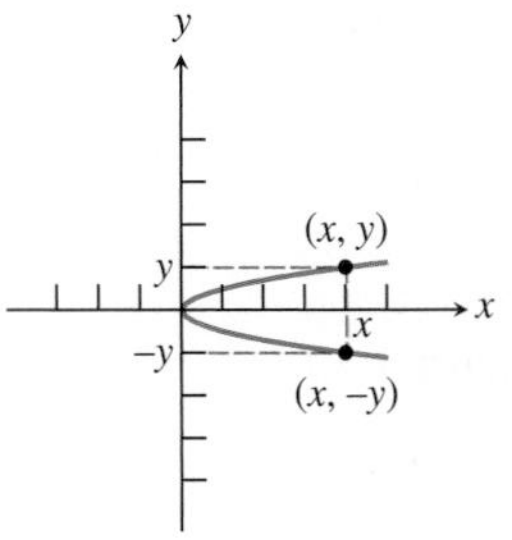

FIGURE 1.27 The graph looks the same above the x-axis as it does below it.

Numerically

x	y
9	−3
4	−2
1	−1
1	1
4	2
9	3

Algebraically

Graphs with this kind of symmetry are not functions (except the zero function), but we can say that $(x, -y)$ is on the graph whenever (x, y) is on the graph.

Symmetry with respect to the origin
Example: $f(x) = x^3$

Graphically

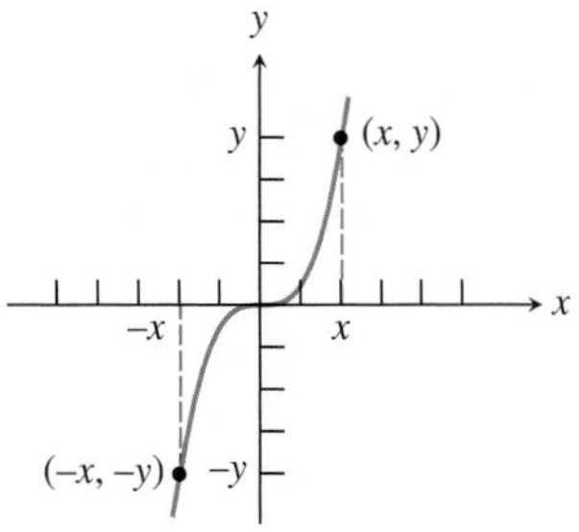

FIGURE 1.28 The graph looks the same upside-down as it does rightside-up.

Numerically

x	y
−3	−27
−2	−8
−1	−1
1	1
2	8
3	27

Algebraically

For all x in the domain of f,

$$f(-x) = -f(x).$$

Functions with this property (for example, x^n, n odd) are **odd** functions.

EXAMPLE 9 Checking Functions for Symmetry

Tell whether each of the following functions is odd, even, or neither.

(a) $f(x) = x^2 - 3$ **(b)** $g(x) = x^2 - 2x - 2$ **(c)** $h(x) = \dfrac{x^3}{4 - x^2}$

SOLUTION

(a) Solve Graphically The graphical solution is shown in Figure 1.29.

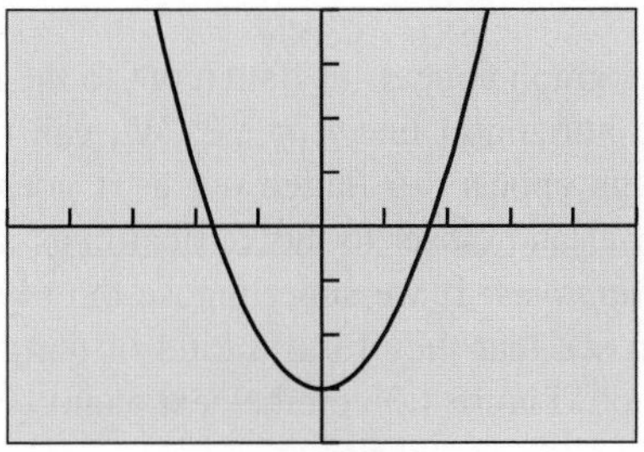

[–5, 5] by [–4, 4]

FIGURE 1.29 This graph appears to be symmetric with respect to the y-axis, so we conjecture that f is an even function.

(continued)

Confirm Algebraically We need to verify that

$$f(-x) = f(x)$$

for all x in the domain of f.

$$\begin{aligned} f(-x) &= (-x)^2 - 3 = x^2 - 3 \\ &= f(x) \end{aligned}$$

Since this identity is true for all x, the function f is indeed even.

(b) Solve Graphically The graphical solution is shown in Figure 1.30.

Confirm Algebraically We need to verify that

$$\begin{aligned} g(-x) &\neq g(x) \text{ and } g(-x) \neq -g(x). \\ g(-x) &= (-x)^2 - 2(-x) - 2 \\ &= x^2 + 2x - 2 \\ g(x) &= x^2 - 2x - 2 \\ -g(x) &= -x^2 + 2x + 2 \end{aligned}$$

So $g(-x) \neq g(x)$ and $g(-x) \neq -g(x)$.

We conclude that g is neither odd nor even.

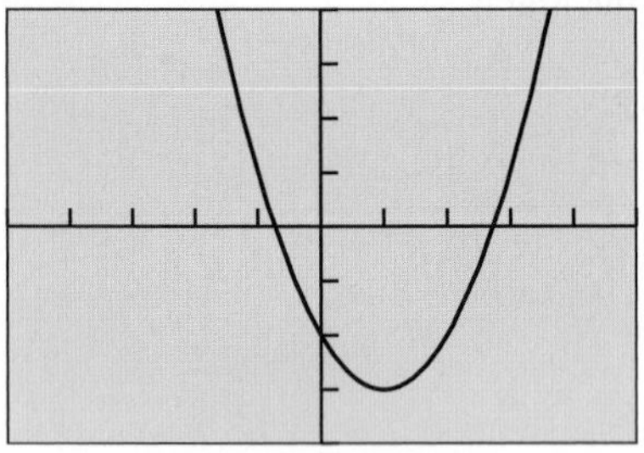

[–5, 5] by [–4, 4]

FIGURE 1.30 This graph does not appear to be symmetric with respect to either the y-axis or the origin, so we conjecture that g is neither even nor odd.

(c) Solve Graphically The graphical solution is shown in Figure 1.31.

Confirm Algebraically We need to verify that

$$h(-x) = -h(x)$$

for all x in the domain of h.

$$\begin{aligned} h(-x) &= \frac{(-x)^3}{4 - (-x)^2} = \frac{-x^3}{4 - x^2} \\ &= -h(x) \end{aligned}$$

Since this identity is true for all x except ± 2 (which are not in the domain of h), the function h is odd.

Now try Exercise 49.

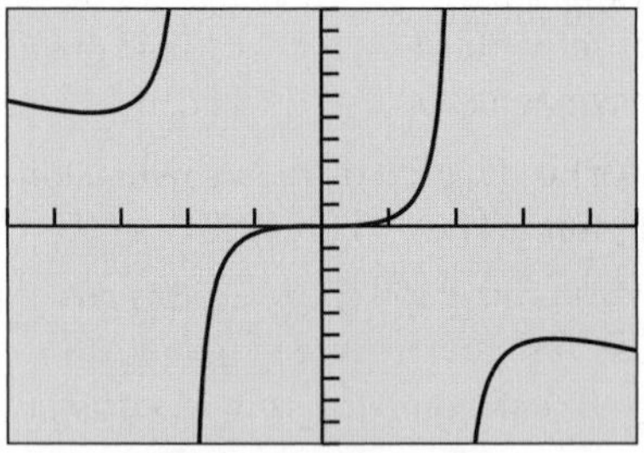

[–4.7, 4.7] by [–10, 10]

FIGURE 1.31 This graph appears to be symmetric with respect to the origin, so we conjecture that h is an odd function.

Asymptotes

Consider the graph of the function $f(x) = \dfrac{2x^2}{4 - x^2}$ in Figure 1.32.

The graph appears to flatten out to the right and to the left, getting closer and closer to the horizontal line $y = -2$. We call this line a *horizontal asymptote*. Similarly, the graph appears to flatten out as it goes off the top and bottom of the screen, getting closer and closer to the vertical lines $x = -2$ and $x = 2$. We call these lines *vertical asymptotes*. If we superimpose the asymptotes onto Figure 1.32 as dashed lines, you can see that they form a kind of template that describes the limiting behavior of the graph (Figure 1.33 on the next page).

Since asymptotes describe the behavior of the graph at its horizontal or vertical extremities, the definition of an asymptote can best be stated with limit notation. In this definition, note that $x \to a^-$ means "x approaches a from the left," and $x \to a^+$ means "x approaches a from the right."

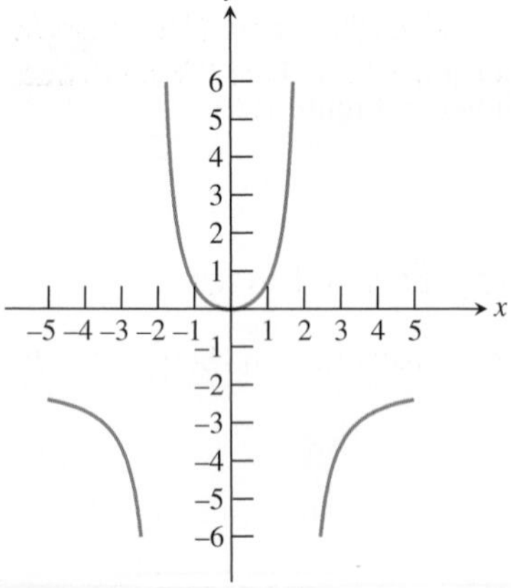

FIGURE 1.32 The graph of $f(x) = 2x^2/(4 - x^2)$ has two vertical asymptotes and one horizontal asymptote.

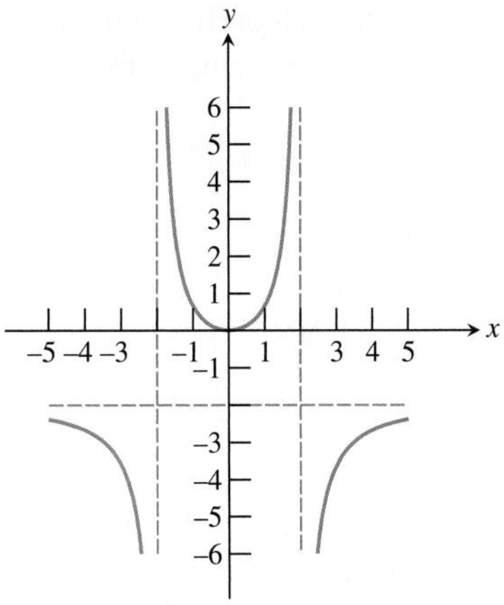

FIGURE 1.33 The graph of $f(x) = 2x^2/(4 - x^2)$ with the asymptotes shown as dashed lines.

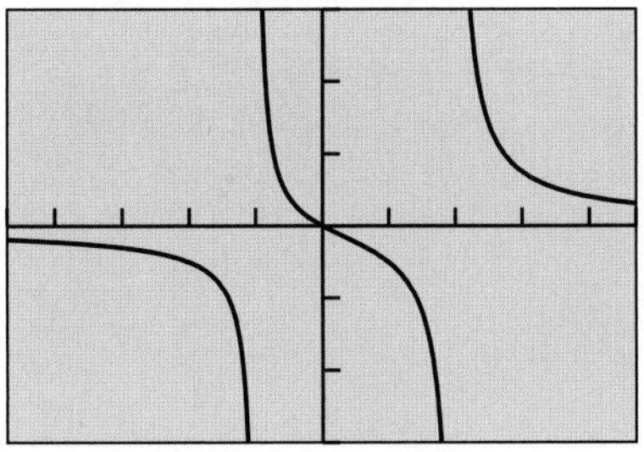

[–4.7, 4.7] by [–3, 3]

FIGURE 1.34 The graph of $y = x/(x^2 - x - 2)$ has vertical asymptotes of $x = -1$ and $x = 2$ and a horizontal asymptote of $y = 0$. (Example 10)

DEFINITION Horizontal and Vertical Asymptotes

The line $y = b$ is a **horizontal asymptote** of the graph of a function $y = f(x)$ if $f(x)$ approaches a limit of b as x approaches $+\infty$ or $-\infty$.

In limit notation:

$$\lim_{x \to -\infty} f(x) = b \quad \text{or} \quad \lim_{x \to +\infty} f(x) = b$$

The line $x = a$ is a **vertical asymptote** of the graph of a function $y = f(x)$ if $f(x)$ approaches a limit of $+\infty$ or $-\infty$ as x approaches a from either direction.

In limit notation:

$$\lim_{x \to a^-} f(x) = \pm\infty \quad \text{or} \quad \lim_{x \to a^+} f(x) = \pm\infty$$

EXAMPLE 10 Identifying the Asymptotes of a Graph

Identify any horizontal or vertical asymptotes of the graph of

$$y = \frac{x}{x^2 - x - 2}.$$

SOLUTION The quotient $x/(x^2 - x - 2) = x/((x + 1)(x - 2))$ is undefined at $x = -1$ and $x = 2$, which makes them likely sites for vertical asymptotes. The graph (Figure 1.34) provides support, showing vertical asymptotes of $x = -1$ and $x = 2$.

For large values of x, the numerator (a large number) is dwarfed by the denominator (a *product* of *two* large numbers), suggesting that $\lim_{x \to \infty} x/((x + 1)(x - 2)) = 0$. This would indicate a horizontal asymptote of $y = 0$. The graph (Figure 1.34) provides support, showing a horizontal asymptote of $y = 0$ as $x \to \infty$. Similar logic suggests that $\lim_{x \to -\infty} x/((x + 1)(x - 2)) = -0 = 0$, indicating the same horizontal asymptote as $x \to -\infty$. Again, the graph provides support for this.

Now try Exercise 57.

End Behavior

A horizontal asymptote gives one kind of end behavior for a function because it shows how the function behaves as it goes off toward either "end" of the x-axis. Not all graphs approach lines, but it is helpful to consider what *does* happen "out there." We illustrate with a few examples.

EXAMPLE 11 Matching Functions Using End Behavior

Match the functions with the graphs in Figure 1.35 by considering end behavior. All graphs are shown in the same viewing window.

(a) $y = \dfrac{3x}{x^2 + 1}$ **(b)** $y = \dfrac{3x^2}{x^2 + 1}$

(c) $y = \dfrac{3x^3}{x^2 + 1}$ **(d)** $y = \dfrac{3x^4}{x^2 + 1}$

(continued)

Tips on Zooming

Zooming out is often a good way to investigate end behavior with a graphing calculator. Here are some useful zooming tips:

- Start with a "square" window.
- Set Xscl and Yscl to zero to avoid fuzzy axes.
- Be sure the zoom factors are both the same. (They will be unless you change them.)

SOLUTION When x is very large, the denominator $x^2 + 1$ in each of these functions is almost the same number as x^2. If we replace $x^2 + 1$ in each denominator by x^2 and then reduce the fractions, we get the simpler functions

(a) $y = \frac{3}{x}$ (close to $y = 0$ for large x) **(b)** $y = 3$

(c) $y = 3x$ **(d)** $y = 3x^2$.

So, we look for functions that have end behavior resembling, respectively, the functions

(a) $y = 0$ **(b)** $y = 3$ **(c)** $y = 3x$ **(d)** $y = 3x^2$.

Graph (iv) approaches the line $y = 0$. Graph (iii) approaches the line $y = 3$. Graph (ii) approaches the line $y = 3x$. Graph (i) approaches the parabola $y = 3x^2$. So, (a) matches (iv), (b) matches (iii), (c) matches (ii), and (d) matches (i).

Now try Exercise 65.

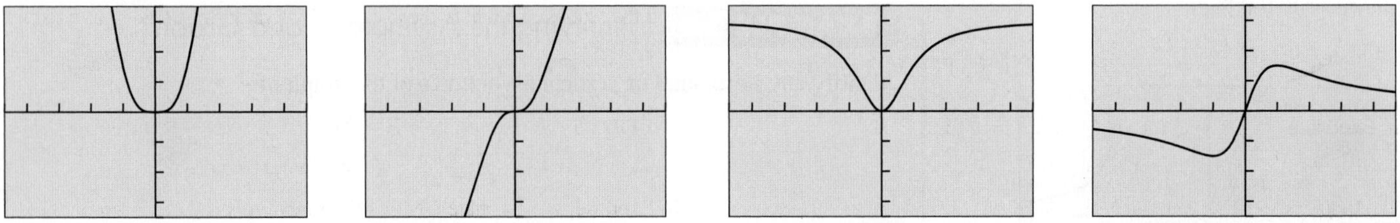

[–4.7, 4.7] by [–3.5, 3.5]
(i)

[–4.7, 4.7] by [–3.5, 3.5]
(ii)

[–4.7, 4.7] by [–3.5, 3.5]
(iii)

[–4.7, 4.7] by [–3.5, 3.5]
(iv)

FIGURE 1.35 Match the graphs with the functions in Example 11.

For more complicated functions we are often content with knowing whether the end behavior is bounded or unbounded in either direction.

QUICK REVIEW 1.2 *(For help, go to Sections A.3, P.3, and P.5.)*

Exercise numbers with a gray background indicate problems that the authors have designed to be solved *without a calculator*.

In Exercises 1–4, solve the equation or inequality.

1. $x^2 - 16 = 0$ **2.** $9 - x^2 = 0$

3. $x - 10 < 0$ **4.** $5 - x \leq 0$

In Exercises 5–10, find all values of x algebraically for which the algebraic expression is *not* defined. Support your answer graphically.

5. $\frac{x}{x - 16}$ **6.** $\frac{x}{x^2 - 16}$

7. $\sqrt{x - 16}$ **8.** $\frac{\sqrt{x^2 + 1}}{x^2 - 1}$

9. $\frac{\sqrt{x + 2}}{\sqrt{3 - x}}$

10. $\frac{x^2 - 2x}{x^2 - 4}$

SECTION 1.2 Exercises

In Exercises 1–4, determine whether the formula determines y as a function of x. If not, explain why not.

1. $y = \sqrt{x - 4}$ **2.** $y = x^2 \pm 3$

3. $x = 2y^2$ **4.** $x = 12 - y$

In Exercises 5–8, use the vertical line test to determine whether the curve is the graph of a function.

5.

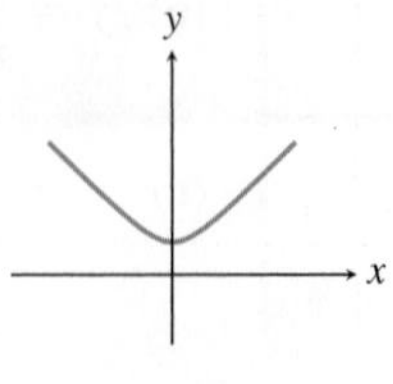

6.

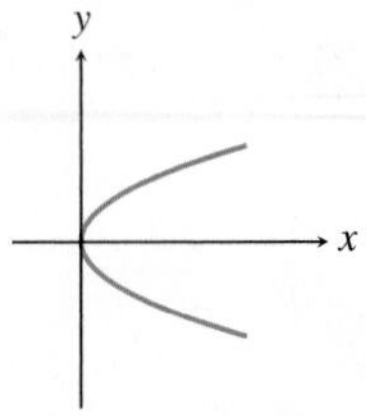

7. 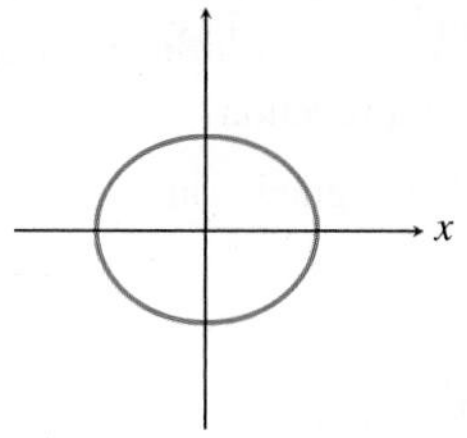

8. 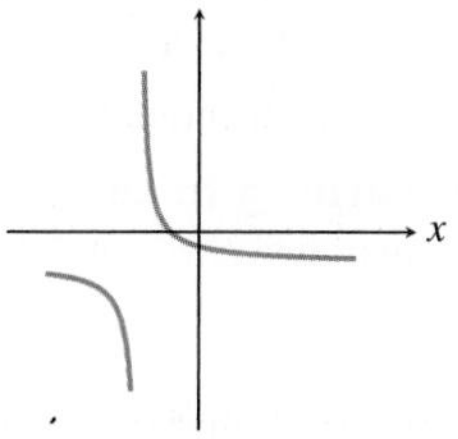

In Exercises 9–16, find the domain of the function algebraically.

9. $f(x) = x^2 + 4$

10. $h(x) = \dfrac{5}{x - 3}$

11. $f(x) = \dfrac{3x - 1}{(x + 3)(x - 1)}$

12. $f(x) = \dfrac{1}{x} + \dfrac{5}{x - 3}$

13. $g(x) = \dfrac{x}{x^2 - 5x}$

14. $h(x) = \dfrac{\sqrt{4 - x^2}}{x - 3}$

15. $h(x) = \dfrac{\sqrt{4 - x}}{(x + 1)(x^2 + 1)}$

16. $f(x) = \sqrt{x^4 - 16x^2}$

In Exercises 17–20, find the range of the function.

17. $f(x) = 10 - x^2$

18. $g(x) = 5 + \sqrt{4 - x}$

19. $f(x) = \dfrac{x^2}{1 - x^2}$

20. $g(x) = \dfrac{3 + x^2}{4 - x^2}$

In Exercises 21–24, graph the function and tell whether or not it has a point of discontinuity at $x = 0$. If there is a discontinuity, tell whether it is removable or nonremovable.

21. $g(x) = \dfrac{3}{x}$

22. $h(x) = \dfrac{x^3 + x}{x}$

23. $f(x) = \dfrac{|x|}{x}$

24. $g(x) = \dfrac{x}{x - 2}$

In Exercises 25–28, state whether each labeled point identifies a local minimum, a local maximum, or neither. Identify intervals on which the function is decreasing and increasing.

25. 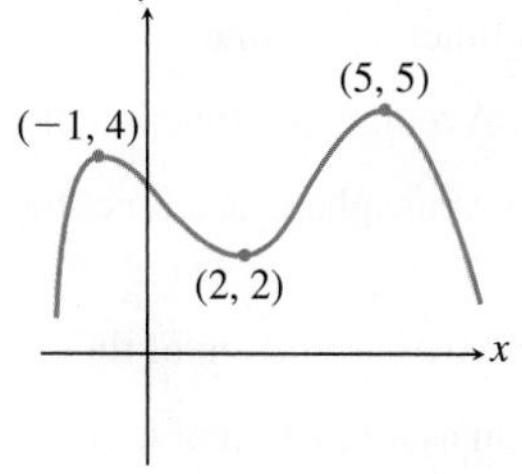

26. 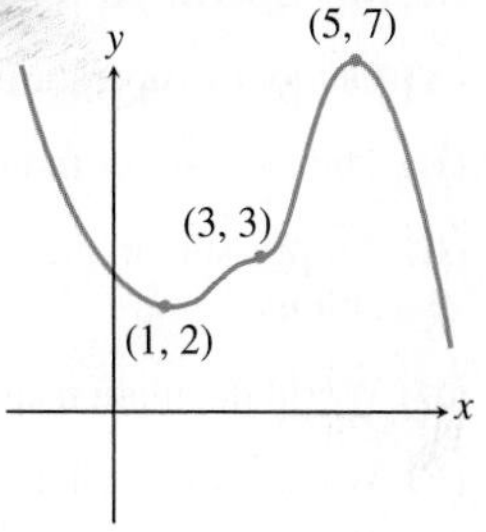

27. 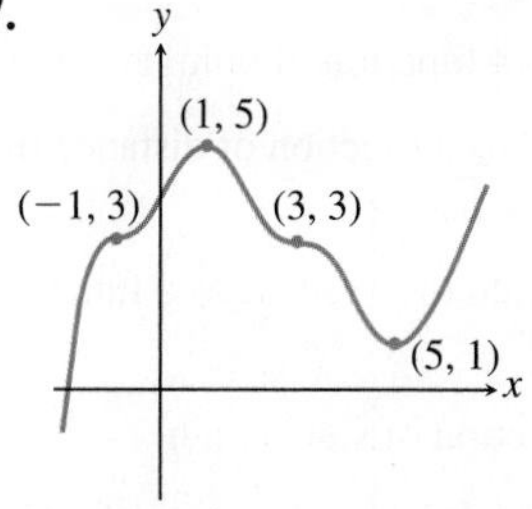

28. 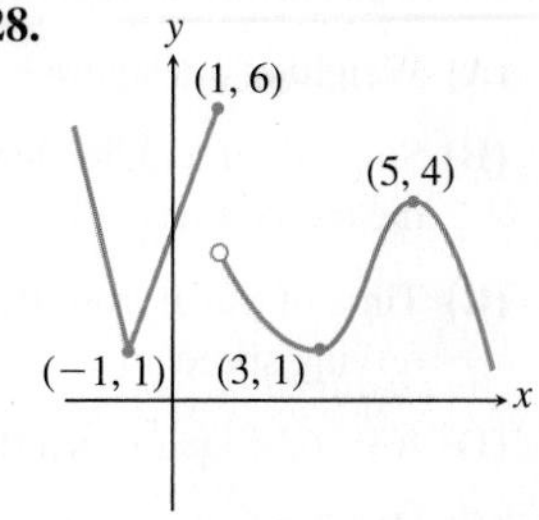

In Exercises 29–34, graph the function and identify intervals on which the function is increasing, decreasing, or constant.

29. $f(x) = |x + 2| - 1$

30. $f(x) = |x + 1| + |x - 1| - 3$

31. $g(x) = |x + 2| + |x - 1| - 2$

32. $h(x) = 0.5(x + 2)^2 - 1$

33. $g(x) = 3 - (x - 1)^2$

34. $f(x) = x^3 - x^2 - 2x$

In Exercises 35–40, determine whether the function is bounded above, bounded below, or bounded on its domain.

35. $y = 32$

36. $y = 2 - x^2$

37. $y = 2^x$

38. $y = 2^{-x}$

39. $y = \sqrt{1 - x^2}$

40. $y = x - x^3$

In Exercises 41–46, use a grapher to find all local maxima and minima and the values of x where they occur. Give values rounded to two decimal places.

41. $f(x) = 4 - x + x^2$

42. $g(x) = x^3 - 4x + 1$

43. $h(x) = -x^3 + 2x - 3$

44. $f(x) = (x + 3)(x - 1)^2$

45. $h(x) = x^2\sqrt{x + 4}$

46. $g(x) = x|2x + 5|$

In Exercises 47–54, state whether the function is odd, even, or neither.

47. $f(x) = 2x^4$

48. $g(x) = x^3$

49. $f(x) = \sqrt{x^2 + 2}$

50. $g(x) = \dfrac{3}{1 + x^2}$

51. $f(x) = -x^2 + 0.03x + 5$

52. $f(x) = x^3 + 0.04x^2 + 3$

53. $g(x) = 2x^3 - 3x$

54. $h(x) = \dfrac{1}{x}$

In Exercises 55–62, use a method of your choice to find all horizontal and vertical asymptotes of the function.

55. $f(x) = \dfrac{x}{x - 1}$

56. $q(x) = \dfrac{x - 1}{x}$

57. $g(x) = \dfrac{x + 2}{3 - x}$

58. $q(x) = 1.5^x$

59. $f(x) = \dfrac{x^2 + 2}{x^2 - 1}$

60. $p(x) = \dfrac{4}{x^2 + 1}$

61. $g(x) = \dfrac{4x - 4}{x^3 - 8}$

62. $h(x) = \dfrac{2x - 4}{x^2 - 4}$

In Exercises 63–66, match the function with the corresponding graph by considering end behavior and asymptotes. All graphs are shown in the same viewing window.

63. $y = \dfrac{x + 2}{2x + 1}$

64. $y = \dfrac{x^2 + 2}{2x + 1}$

65. $y = \dfrac{x + 2}{2x^2 + 1}$

66. $y = \dfrac{x^3 + 2}{2x^2 + 1}$

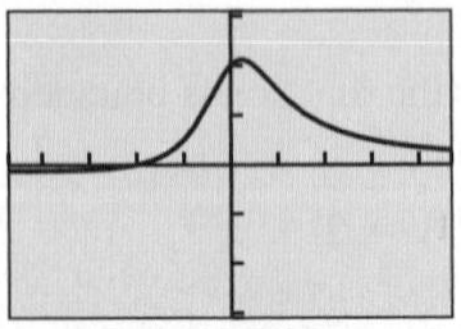

[–4.7, 4.7] by [–3.1, 3.1]
(a)

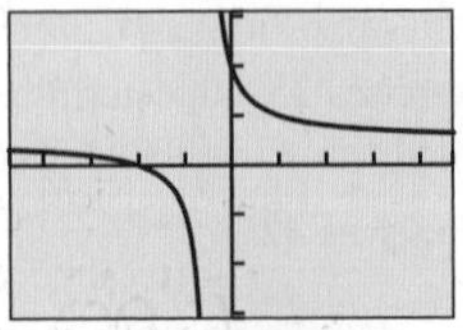

[–4.7, 4.7] by [–3.1, 3.1]
(b)

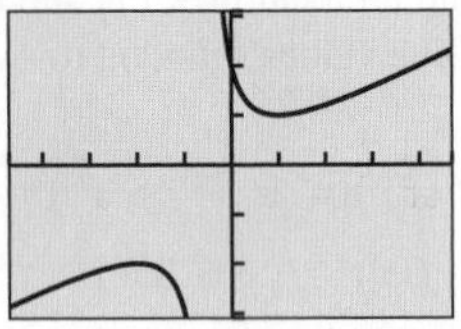

[–4.7, 4.7] by [–3.1, 3.1]
(c)

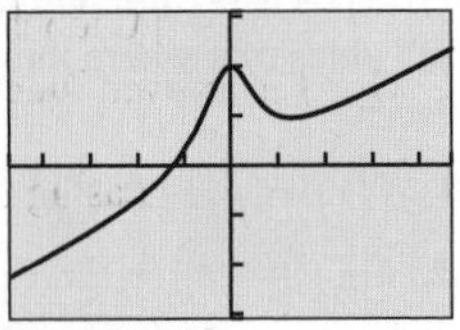

[–4.7, 4.7] by [–3.1, 3.1]
(d)

67. Can a Graph Cross Its Own Asymptote? The Greek roots of the word "asymptote" mean "not meeting," since graphs tend to approach, but not meet, their asymptotes. Which of the following functions have graphs that *do* intersect their horizontal asymptotes?

(a) $f(x) = \dfrac{x}{x^2 - 1}$

(b) $g(x) = \dfrac{x}{x^2 + 1}$

(c) $h(x) = \dfrac{x^2}{x^3 + 1}$

68. Can a Graph Have Two Horizontal Asymptotes? Although most graphs have at most one horizontal asymptote, it is possible for a graph to have more than one. Which of the following functions have graphs with more than one horizontal asymptote?

(a) $f(x) = \dfrac{|x^3 + 1|}{8 - x^3}$

(b) $g(x) = \dfrac{|x - 1|}{x^2 - 4}$

(c) $h(x) = \dfrac{x}{\sqrt{x^2 - 4}}$

69. Can a Graph Intersect Its Own Vertical Asymptote? Graph the function $f(x) = \dfrac{x - |x|}{x^2} + 1$.

(a) The graph of this function does not intersect its vertical asymptote. Explain why it does not.

(b) Show how you can add a single point to the graph of f and get a graph that *does* intersect its vertical asymptote.

(c) Is the graph in (b) the graph of a function?

70. Writing to Learn Explain why a graph cannot have more than two horizontal asymptotes.

Standardized Test Questions

71. True or False The graph of function f is defined as the set of all points $(x, f(x))$, where x is in the domain of f. Justify your answer.

72. True or False A relation that is symmetric with respect to the x-axis cannot be a function. Justify your answer.

In Exercises 73–76, answer the question without using a calculator.

73. Multiple Choice Which function is continuous?

(A) Number of children enrolled in a particular school as a function of time

(B) Outdoor temperature as a function of time

(C) Cost of U.S. postage as a function of the weight of the letter

(D) Price of a stock as a function of time

(E) Number of soft drinks sold at a ballpark as a function of outdoor temperature

74. Multiple Choice Which function is *not* continuous?

(A) Your altitude as a function of time while flying from Reno to Dallas

(B) Time of travel from Miami to Pensacola as a function of driving speed

(C) Number of balls that can fit completely inside a particular box as a function of the radius of the balls

(D) Area of a circle as a function of radius

(E) Weight of a particular baby as a function of time after birth

75. Multiple Choice Which function is decreasing?

(A) Outdoor temperature as a function of time

(B) The Dow Jones Industrial Average as a function of time

(C) Air pressure in the Earth's atmosphere as a function of altitude

(D) World population since 1900 as a function of time

(E) Water pressure in the ocean as a function of depth

76. Multiple Choice Which function cannot be classified as either increasing or decreasing?

(A) Weight of a lead brick as a function of volume

(B) Strength of a radio signal as a function of distance from the transmitter

(C) Time of travel from Buffalo to Syracuse as a function of driving speed

(D) Area of a square as a function of side length

(E) Height of a swinging pendulum as a function of time

Explorations

77. Bounded Functions As promised in Example 7 of this section, we will give you a chance to prove algebraically that $p(x) = x/(1 + x^2)$ is bounded.

(a) Graph the function and find the smallest integer k that appears to be an upper bound.

(b) Verify that $x/(1 + x^2) < k$ by proving the equivalent inequality $kx^2 - x + k > 0$. (Use the quadratic formula to show that the quadratic has no real zeros.)

(c) From the graph, find the greatest integer k that appears to be a lower bound.

(d) Verify that $x/(1 + x^2) > k$ by proving the equivalent inequality $kx^2 - x + k < 0$.

78. Baylor School Grade Point Averages Baylor School uses a sliding scale to convert the percentage grades on its transcripts to grade point averages (GPAs). Table 1.9 shows the GPA equivalents for selected grades.

Table 1.9 Converting Grades

Grade (x)	GPA (y)
60	0.00
65	1.00
70	2.05
75	2.57
80	3.00
85	3.36
90	3.69
95	4.00
100	4.28

Source: Baylor School College Counselor.

(a) Considering GPA (y) as a function of percentage grade (x), is it increasing, decreasing, constant, or none of these?

(b) Make a table showing the *change* (Δy) in GPA as you move down the list. (See Exploration 1.)

(c) Make a table showing the change in Δy as you move down the list. (This is $\Delta\Delta y$.) Considering the *change* (Δy) in GPA as a function of percentage grade (x), is it increasing, decreasing, constant, or none of these?

(d) In general, what can you say about the shape of the graph if y is an increasing function of x and Δy is a decreasing function of x?

(e) Sketch the graph of a function y of x such that y is a decreasing function of x and Δy is an increasing function of x.

79. Group Activity Sketch (freehand) a graph of a function f with domain all real numbers that satisfies all of the following conditions:

(a) f is continuous for all x;

(b) f is increasing on $(-\infty, 0]$ and on $[3, 5]$;

(c) f is decreasing on $[0, 3]$ and on $[5, \infty)$;

(d) $f(0) = f(5) = 2$;

(e) $f(3) = 0$.

80. Group Activity Sketch (freehand) a graph of a function f with domain all real numbers that satisfies all of the following conditions:

(a) f is decreasing on $(-\infty, 0)$ and decreasing on $(0, \infty)$;

(b) f has a nonremovable point of discontinuity at $x = 0$;

(c) f has a horizontal asymptote at $y = 1$;

(d) $f(0) = 0$;

(e) f has a vertical asymptote at $x = 0$.

81. Group Activity Sketch (freehand) a graph of a function f with domain all real numbers that satisfies all of the following conditions:

(a) f is continuous for all x;

(b) f is an even function;

(c) f is increasing on $[0, 2]$ and decreasing on $[2, \infty)$;

(d) $f(2) = 3$.

82. Group Activity Get together with your classmates in groups of two or three. Sketch a graph of a function, but do not show it to the other members of your group. Using the language of functions (as in Exercises 79–81), describe your function as completely as you can. Exchange descriptions with the others in your group and see if you can reproduce each other's graphs.

Extending the Ideas

83. A function that is bounded above has an infinite number of upper bounds, but there is always a *least upper bound*, i.e., an upper bound that is less than all the others. This least upper bound may or may not be in the range of f. For each of the following functions, find the least upper bound and tell whether or not it is in the range of the function.

(a) $f(x) = 2 - 0.8x^2$

(b) $g(x) = \dfrac{3x^2}{3 + x^2}$

(c) $h(x) = \dfrac{1 - x}{x^2}$

(d) $p(x) = 2 \sin (x)$

(e) $q(x) = \dfrac{4x}{x^2 + 2x + 1}$

84. Writing to Learn A continuous function f has domain all real numbers. If $f(-1) = 5$ and $f(1) = -5$, explain why f must have at least one zero in the interval $[-1, 1]$. (This generalizes to a property of continuous functions known as the Intermediate Value Theorem.)

85. Proving a Theorem Prove that the graph of every odd function with domain all real numbers must pass through the origin.

86. Finding the Range Graph the function $f(x) = \dfrac{3x^2 - 1}{2x^2 + 1}$ in the window $[-6, 6]$ by $[-2, 2]$.

(a) What is the apparent horizontal asymptote of the graph?

(b) Based on your graph, determine the apparent range of f.

(c) Show algebraically that $-1 \le \dfrac{3x^2 - 1}{2x^2 + 1} < 1.5$ for all x, thus confirming your conjecture in part (b).

87. Looking Ahead to Calculus A key theorem in calculus, the Extreme Value Theorem, states, if a function f is continuous on a closed interval $[a, b]$ then f has both a maximum value and a minimum value on the interval. For each of the following functions, verify that the function is continuous on the given interval and find the maximum and minimum values of the function and the x values at which these extrema occur.

(a) $f(x) = x^2 - 3$, $[-2, 4]$

(b) $f(x) = \dfrac{1}{x}$, $[1, 5]$

(c) $f(x) = |x + 1| + 2$, $[-4, 1]$

(d) $f(x) = \sqrt{x^2 + 9}$, $[-4, 4]$

1.3 Twelve Basic Functions

What you'll learn about

- What Graphs Can Tell Us
- Twelve Basic Functions
- Analyzing Functions Graphically

... and why

As you continue to study mathematics, you will find that the twelve basic functions presented here will come up again and again. By knowing their basic properties, you will recognize them when you see them.

What Graphs Can Tell Us

The preceding section has given us a vocabulary for talking about functions and their properties. We have an entire book ahead of us to study these functions in depth, but in this section we want to set the scene by just *looking* at the graphs of twelve "basic" functions that are available on your graphing calculator.

You will find that function attributes such as domain, range, continuity, asymptotes, extrema, increasingness, decreasingness, and end behavior are every bit as graphical as they are algebraic. Moreover, the visual cues are often much easier to spot than the algebraic ones.

In future chapters you will learn more about the algebraic properties that make these functions behave as they do. Only then will you able to *prove* what is visually apparent in these graphs.

Twelve Basic Functions

The Identity Function

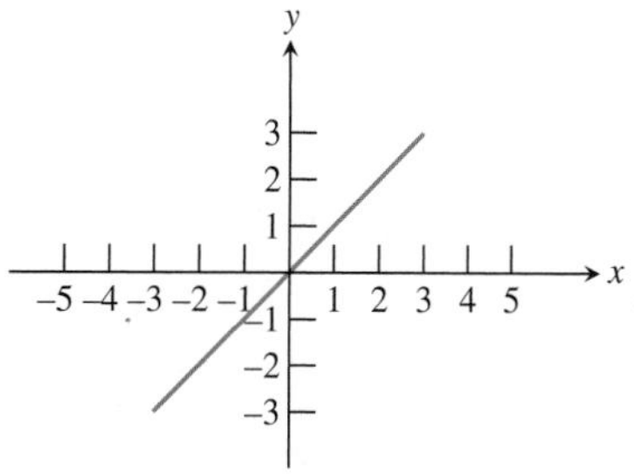

$f(x) = x$

Interesting fact: This is the only function that acts on every real number by leaving it alone.

FIGURE 1.36

The Squaring Function

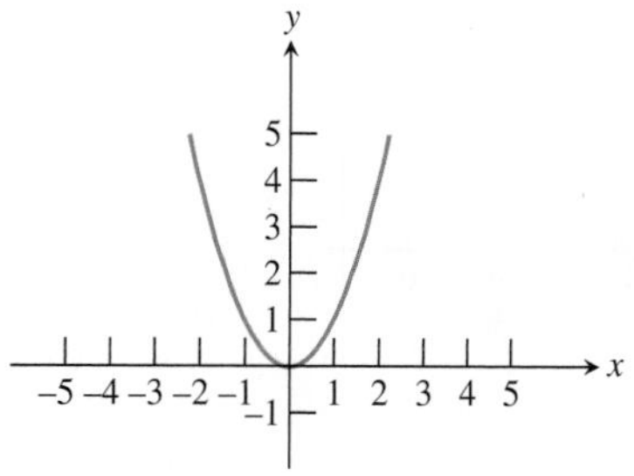

$f(x) = x^2$

Interesting fact: The graph of this function, called a parabola, has a reflection property that is useful in making flashlights and satellite dishes.

FIGURE 1.37

The Cubing Function

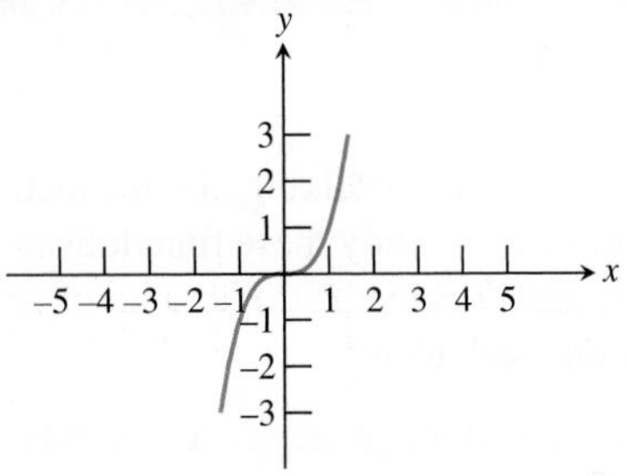

$f(x) = x^3$

Interesting fact: The origin is called a "point of inflection" for this curve because the graph changes curvature at that point.

FIGURE 1.38

The Reciprocal Function

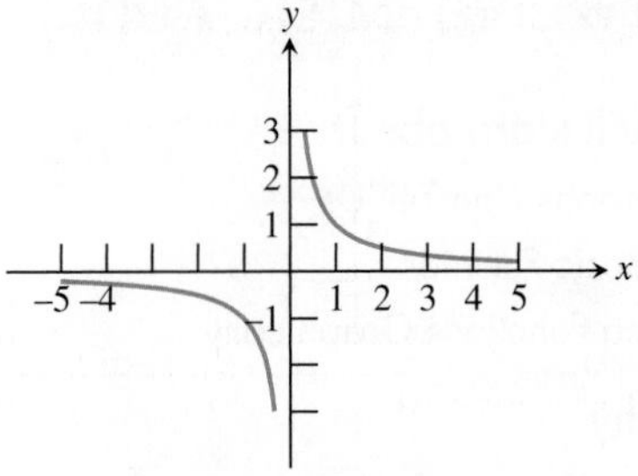

$f(x) = \frac{1}{x}$

Interesting fact: This curve, called a hyperbola, also has a reflection property that is useful in satellite dishes.

FIGURE 1.39

The Square Root Function

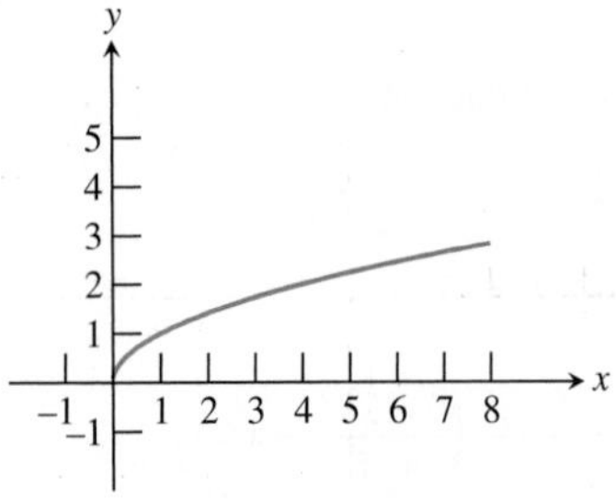

$f(x) = \sqrt{x}$

Interesting fact: Put any positive number into your calculator. Take the square root. Then take the square root again. Then take the square root again, and so on. Eventually you will always get 1.

FIGURE 1.40

The Exponential Function

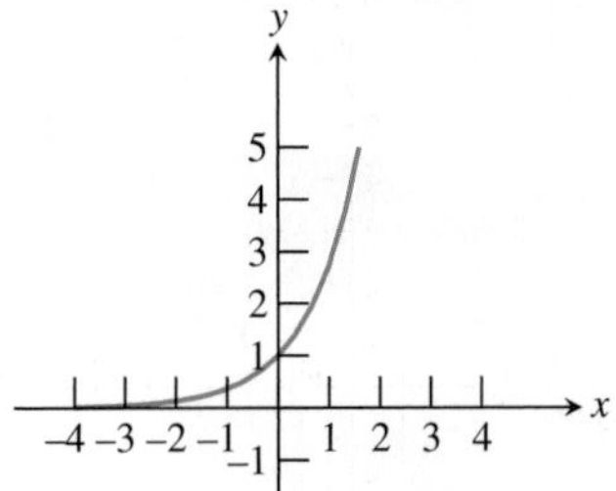

$f(x) = e^x$

Interesting fact: The number e is an irrational number (like π) that shows up in a variety of applications. The symbols e and π were both brought into popular use by the great Swiss mathematician Leonhard Euler (1707–1783).

FIGURE 1.41

The Natural Logarithm Function

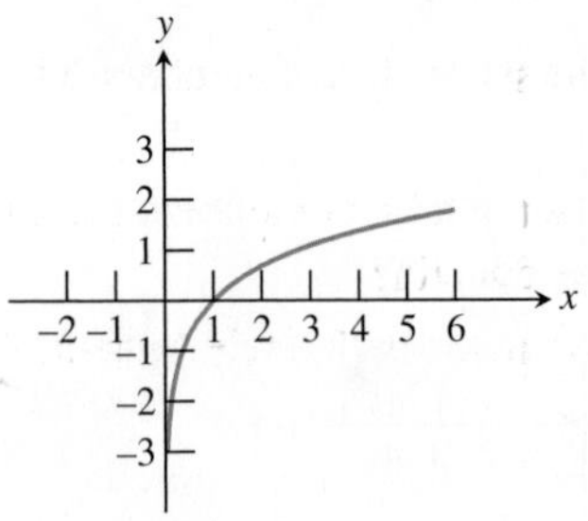

$f(x) = \ln x$

Interesting fact: This function increases very slowly. If the x-axis and y-axis were both scaled with unit lengths of 1 in., you would have to travel more than 2.5 mi along the curve just to get 1 ft above the x-axis.

FIGURE 1.42

The Sine Function

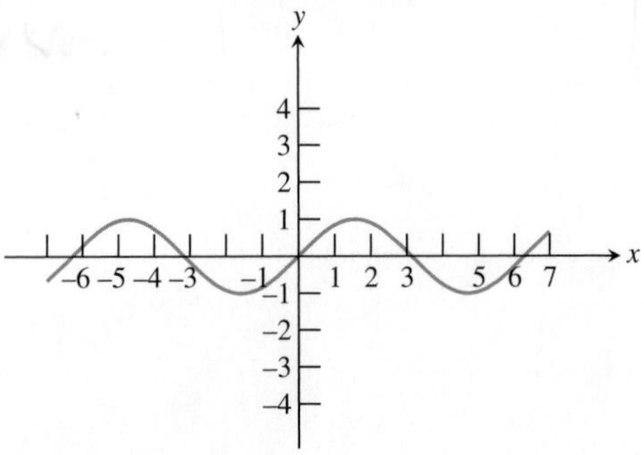

$f(x) = \sin x$

Interesting fact: This function and the sinus cavities in your head derive their names from a common root: the Latin word for "bay." This is due to a 12th-century mistake made by Robert of Chester, who translated a word incorrectly from an Arabic manuscript.

FIGURE 1.43

The Cosine Function

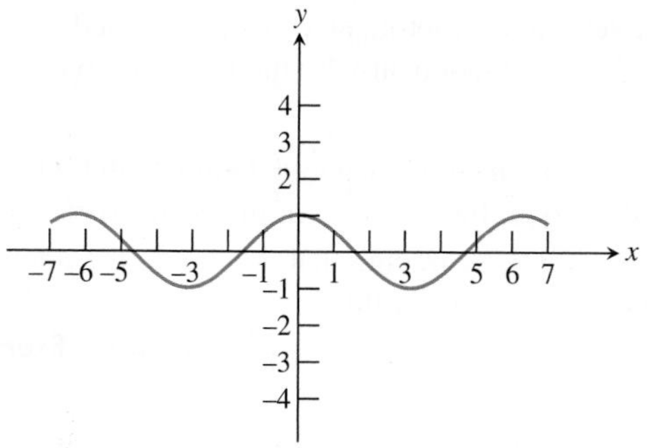

$f(x) = \cos x$

Interesting fact: The local extrema of the cosine function occur exactly at the zeros of the sine function, and vice versa.

FIGURE 1.44

The Absolute Value Function

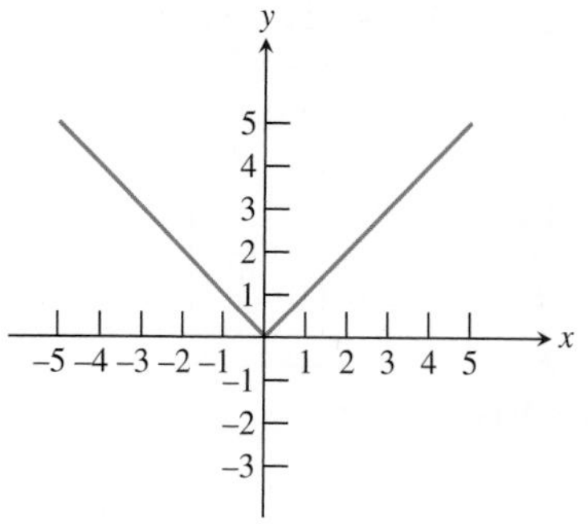

$f(x) = |x| = \text{abs}\,(x)$

Interesting fact: This function has an abrupt change of direction (a "corner") at the origin, while our other functions are all "smooth" on their domains.

FIGURE 1.45

The Greatest Integer Function

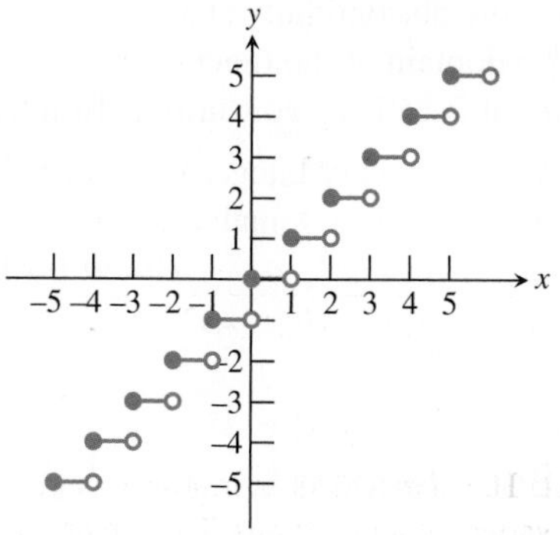

$f(x) = \text{int}\,(x)$

Interesting fact: This function has a jump discontinuity at every integer value of x. Similar-looking functions are called *step functions*.

FIGURE 1.46

The Logistic Function

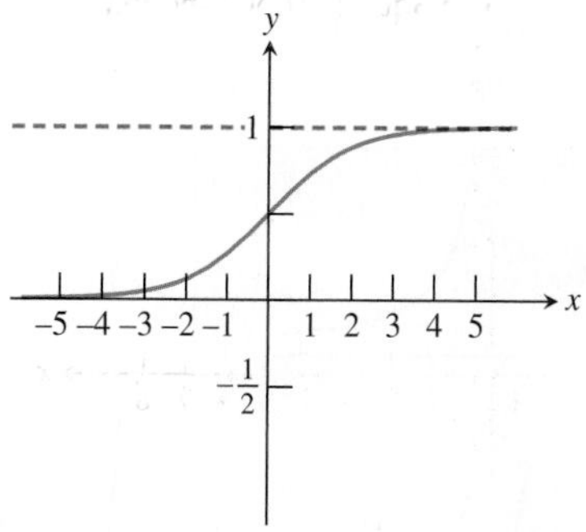

$$f(x) = \frac{1}{1 + e^{-x}}$$

Interesting fact: There are two horizontal asymptotes, the x-axis and the line $y = 1$. This function provides a model for many applications in biology and business.

FIGURE 1.47

EXAMPLE 1 Looking for Domains

(a) Nine of the functions have domain the set of all real numbers. Which three do not?

(b) One of the functions has domain the set of all real numbers except 0. Which function is it, and why isn't zero in its domain?

(c) Which two functions have no negative numbers in their domains? Of these two, which one is defined at zero?

SOLUTION

(a) Imagine dragging a vertical line along the x-axis. If the function has domain the set of all real numbers, then the line will always intersect the graph. The intersection might occur off screen, but the TRACE function on the calculator will show the y-coordinate if there is one. Looking at the graphs in Figures 1.39, 1.40, and 1.42, we conjecture that there are vertical lines that do not intersect

(continued)

the curve. A TRACE at the suspected x-coordinates confirms our conjecture (Figure 1.48). The functions are $y = 1/x$, $y = \sqrt{x}$, and $y = \ln x$.

(b) The function $y = 1/x$, with a vertical asymptote at $x = 0$, is defined for all real numbers except 0. This is explained algebraically by the fact that division by zero is undefined.

(c) The functions $y = \sqrt{x}$ and $y = \ln x$ have no negative numbers in their domains. (We already knew that about the square root function.) Although 0 is in the domain of $y = \sqrt{x}$, we can see by tracing that it is not in the domain of $y = \ln x$. We will see the algebraic reason for this in Chapter 3.

Now try Exercise 13.

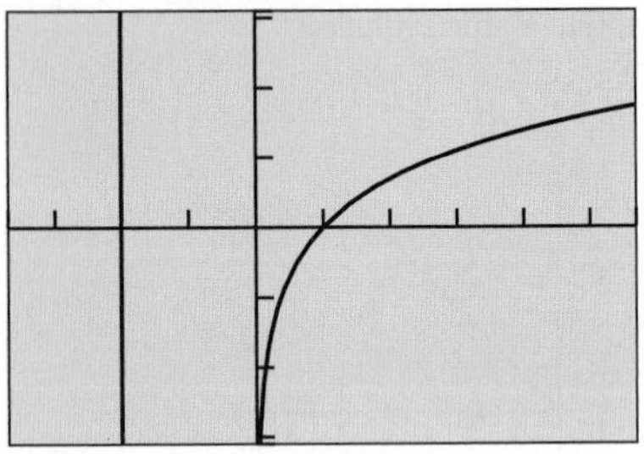

[–3.7, 5.7] by [–3.1, 3.1]

(a)

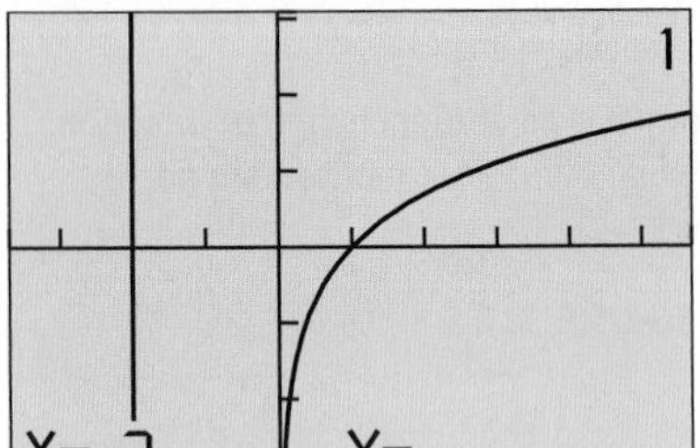

[–3.7, 5.7] by [–3.1, 3.1]

(b)

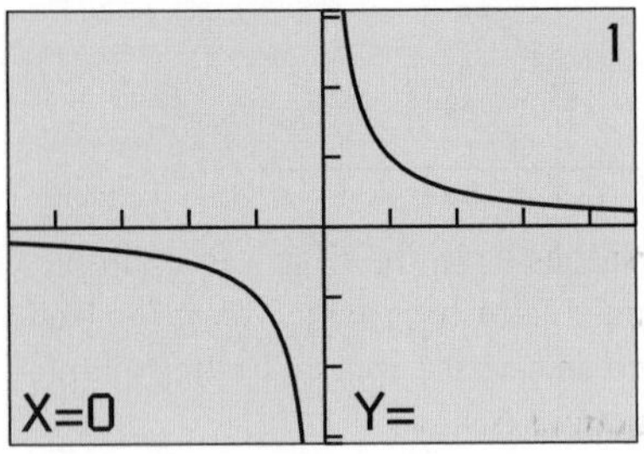

[–4.7, 4.7] by [–3.1, 3.1]

(c)

FIGURE 1.48 (a) A vertical line through -2 on the x-axis appears to miss the graph of $y = \ln x$. (b) A TRACE confirms that -2 is not in the domain. (c) A TRACE at $x = 0$ confirms that 0 is not in the domain of $y = 1/x$. (Example 1)

EXAMPLE 2 Looking for Continuity

Only two of twelve functions have points of discontinuity. Are these points in the domain of the function?

SOLUTION All of the functions have continuous, unbroken graphs except for $y = 1/x$, and $y = \text{int}(x)$.

The graph of $y = 1/x$ clearly has an infinite discontinuity at $x = 0$ (Figure 1.39). We saw in Example 1 that 0 is not in the domain of the function. Since $y = 1/x$ is continuous for every point *in its domain*, it is called a **continuous function**.

The graph of $y = \text{int}(x)$ has a discontinuity at every integer value of x (Figure 1.46). Since this function has discontinuities at points in its domain, it is *not* a continuous function.

Now try Exercise 15.

EXAMPLE 3 Looking for Boundedness

Only three of the twelve basic functions are bounded (above and below). Which three?

SOLUTION A function that is bounded must have a graph that lies entirely between two horizontal lines. The sine, cosine, and logistic functions have this property (Figure 1.49). It looks like the graph of $y = \sqrt{x}$ might also have this property, but we know that the end behavior of the square root function is unbounded: $\lim_{x \to \infty} \sqrt{x} = \infty$, so it is really only bounded below. You will learn in Chapter 4 why the sine and cosine functions are bounded.

Now try Exercise 17.

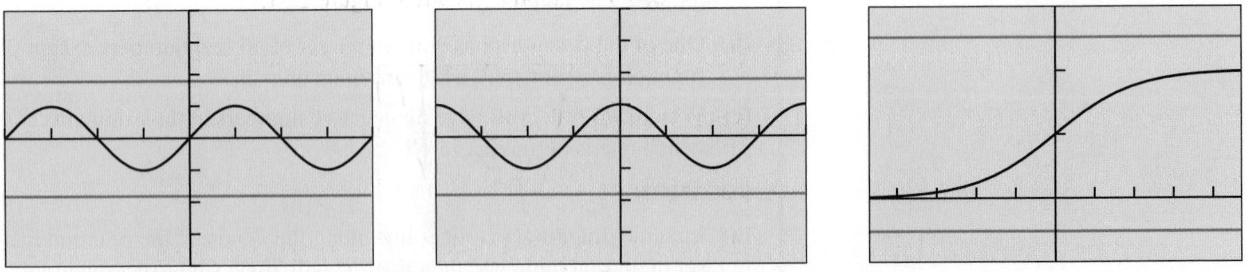

[–2π, 2π] by [–4, 4] (a) | [–2π, 2π] by [–4, 4] (b) | [–4.7, 4.7] by [–0.5, 1.5] (c)

FIGURE 1.49 The graphs of $y = \sin x$, $y = \cos x$, and $y = 1/(1 + e^{-x})$ lie entirely between two horizontal lines and are therefore graphs of bounded functions. (Example 3)

EXAMPLE 4 Looking for Symmetry

Three of the twelve basic functions are even. Which are they?

SOLUTION Recall that the graph of an even function is symmetric with respect to the y-axis. Three of the functions exhibit the required symmetry: $y = x^2$, $y = \cos x$, and $y = |x|$ (Figure 1.50). **Now try Exercise 19.**

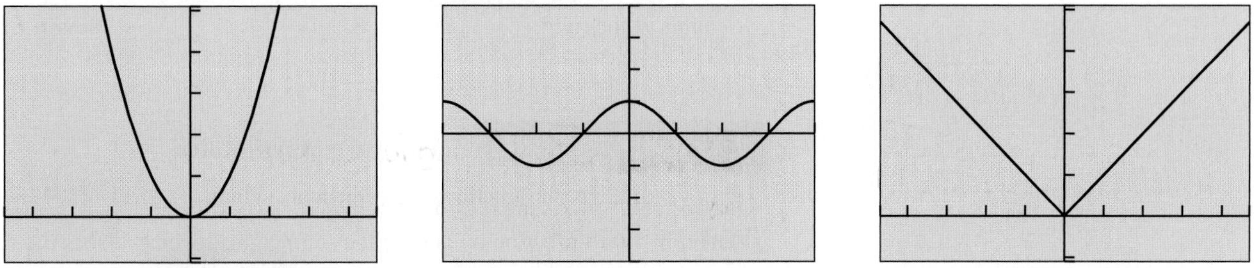

$[-4.7, 4.7]$ by $[-1.1, 5.1]$
(a)

$[-2\pi, 2\pi]$ by $[-4, 4]$
(b)

$[-4.7, 4.7]$ by $[-1.1, 5.1]$
(c)

FIGURE 1.50 The graphs of $y = x^2$, $y = \cos x$, and $y = |x|$ are symmetric with respect to the y-axis, indicating that the functions are even. (Example 4)

Analyzing Functions Graphically

We could continue to explore the twelve basic functions as in the first four examples, but we also want to make the point that there is no need to restrict ourselves to the basic twelve. We can alter the basic functions slightly and see what happens to their graphs, thereby gaining further visual insights into how functions behave.

EXAMPLE 5 Analyzing a Function Graphically

Graph the function $y = (x - 2)^2$. Then answer the following questions:

(a) On what interval is the function increasing? On what interval is it decreasing?

(b) Is the function odd, even, or neither?

(c) Does the function have any extrema?

(d) How does the graph relate to the graph of the basic function $y = x^2$?

SOLUTION The graph is shown in Figure 1.51.

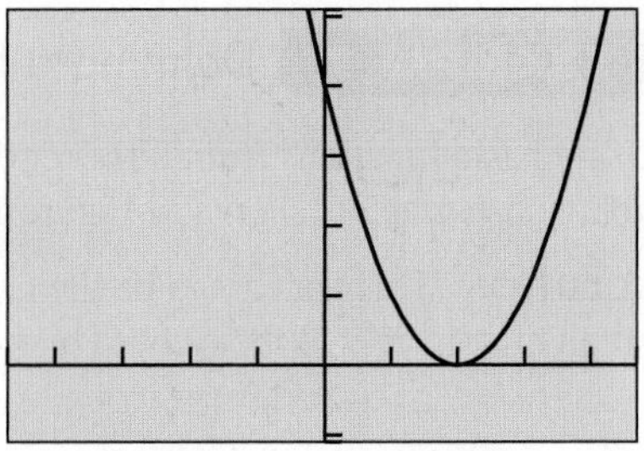

$[-4.7, 4.7]$ by $[-1.1, 5.1]$

FIGURE 1.51 The graph of $y = (x - 2)^2$. (Example 5)

(continued)

(a) The function is increasing if its graph is headed upward as it moves from left to right. We see that it is increasing on the interval $[2, \infty)$. The function is decreasing if its graph is headed downward as it moves from left to right. We see that it is decreasing on the interval $(-\infty, 2]$.

(b) The graph is not symmetric with respect to the y-axis, nor is it symmetric with respect to the origin. The function is neither.

(c) Yes, we see that the function has a minimum value of 0 at $x = 2$. (This is easily confirmed by the algebraic fact that $(x - 2)^2 \geq 0$ for all x.)

(d) We see that the graph of $y = (x - 2)^2$ is just the graph of $y = x^2$ moved two units to the right. **Now try Exercise 35.**

EXPLORATION 1 Looking for Asymptotes

1. Two of the basic functions have vertical asymptotes at $x = 0$. Which two?
2. Form a new function by adding these functions together. Does the new function have a vertical asymptote at $x = 0$?
3. Three of the basic functions have horizontal asymptotes at $y = 0$. Which three?
4. Form a new function by adding these functions together. Does the new function have a horizontal asymptote at $y = 0$?
5. Graph $f(x) = 1/x$, $g(x) = 1/(2x^2 - x)$, and $h(x) = f(x) + g(x)$. Does $h(x)$ have a vertical asymptote at $x = 0$?

EXAMPLE 6 Identifying a Piecewise-Defined Function

Which of the twelve basic functions has the following **piecewise** definition over separate intervals of its domain?

$$f(x) = \begin{cases} x & \text{if } x \geq 0 \\ -x & \text{if } x < 0 \end{cases}$$

SOLUTION You may recognize this as the definition of the absolute value function (Chapter P). Or, you can reason that the graph of this function must look just like the line $y = x$ to the right of the y-axis, but just like the graph of the line $y = -x$ to the left of the y-axis. That is a perfect description of the absolute value graph in Figure 1.45. Either way, we recognize this as a piecewise definition of $f(x) = |x|$. **Now try Exercise 45.**

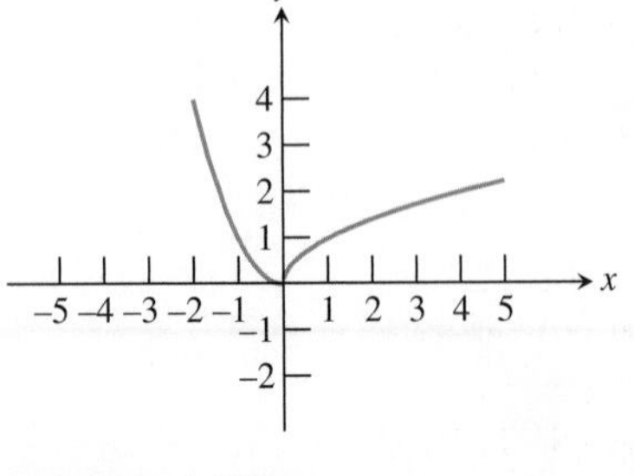

FIGURE 1.52 A piecewise-defined function. (Example 7)

EXAMPLE 7 Defining a Function Piecewise

Using basic functions from this section, construct a piecewise definition for the function whose graph is shown in Figure 1.52. Is your function continuous?

SOLUTION This appears to be the graph of $y = x^2$ to the left of $x = 0$ and the graph of $y = \sqrt{x}$ to the right of $x = 0$. We can therefore define it piecewise as

$$f(x) = \begin{cases} x^2 & \text{if } x \leq 0 \\ \sqrt{x} & \text{if } x > 0. \end{cases}$$

The function is continuous. **Now try Exercise 47.**

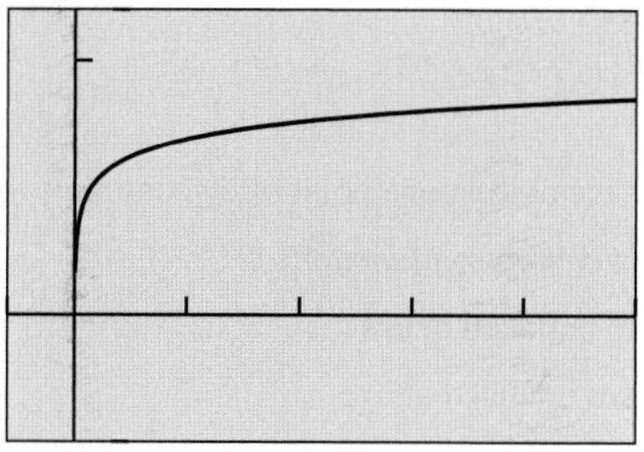

[–600, 5000] by [–5, 12]

FIGURE 1.53 The graph of $y = \ln x$ still appears to have a horizontal asymptote, despite the much larger window than in Figure 1.42. (Example 8)

You can go a long way toward understanding a function's behavior by looking at its graph. We will continue that theme in the exercises and then revisit it throughout the book. However, you can't go *all* the way toward understanding a function by looking at its graph, as Example 8 shows.

EXAMPLE 8 Looking for a Horizontal Asymptote

Does the graph of $y = \ln x$ (see Figure 1.42) have a horizontal asymptote?

SOLUTION In Figure 1.42 it certainly *looks* like there is a horizontal asymptote that the graph is approaching from below. If we choose a much larger window (Figure 1.53), it still looks that way. In fact, we could zoom out on this function all day long and it would *always* look like it is approaching some horizontal asymptote—but it is not. We will show algebraically in Chapter 3 that the end behavior of this function is $\lim_{x\to\infty} \ln x = \infty$, so its graph must eventually rise above the level of any horizontal line.That rules out any horizontal asymptote, even though there is no *visual* evidence of that fact that we can see by looking at its graph.

Now try Exercise 55.

EXAMPLE 9 Analyzing a Function

Give a complete analysis of the basic function $f(x) = |x|$.

SOLUTION

BASIC FUNCTION The Absolute Value Function

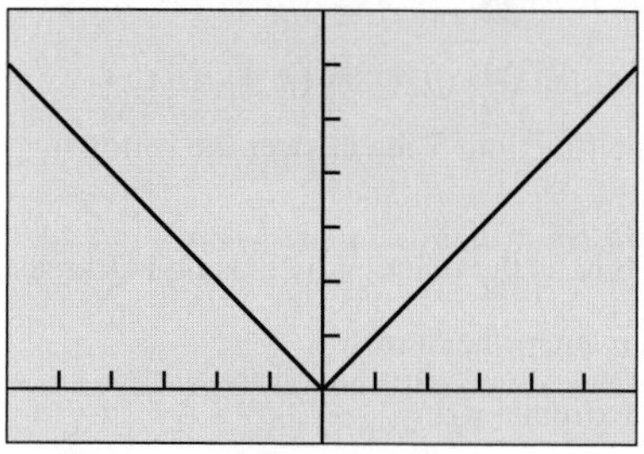

[–6, 6] by [–1, 7]

FIGURE 1.54 The graph of $f(x) = |x|$.

$f(x) = |x|$
Domain: $(-\infty, \infty)$
Range: $[0, \infty)$
Continuous
Decreasing on $(-\infty, 0]$; increasing on $[0, \infty)$
Symmetric with respect to the y-axis (an even function)
Bounded below
Local minimum at $(0, 0)$
No horizontal asymptotes
No vertical asymptotes
End behavior: $\lim_{x\to-\infty} |x| = \infty$ and $\lim_{x\to\infty} |x| = \infty$

Now try Exercise 67.

QUICK REVIEW 1.3 *(For help, go to Sections P.1, P.2, 3.1, and 3.3.)*

Exercise numbers with a gray background indicate problems that the authors have designed to be solved *without a calculator.*

In Exercises 1–10, evaluate the expression without using a calculator.

1. $|-59.34|$

2. $|5 - \pi|$

3. $|\pi - 7|$

4. $\sqrt{(-3)^2}$

5. $\ln(1)$

6. e^0

7. $(\sqrt[3]{3})^3$

8. $\sqrt[3]{(-15)^3}$

9. $\sqrt[3]{-8^2}$

10. $|1 - \pi| - \pi$

SECTION 1.3 Exercises

In Exercises 1–12, each graph is a slight variation on the graph of one of the twelve basic functions described in this section. Match the graph to one of the twelve functions (a)–(l) and then support your answer by checking the graph on your calculator. (All graphs are shown in the window $[-4.7, 4.7]$ by $[-3.1, 3.1]$.)

(a) $y = -\sin x$ **(b)** $y = \cos x + 1$ **(c)** $y = e^x - 2$

(d) $y = (x + 2)^3$ **(e)** $y = x^3 + 1$ **(f)** $y = (x - 1)^2$

(g) $y = |x| - 2$ **(h)** $y = -1/x$ **(i)** $y = -x$

(j) $y = -\sqrt{x}$ **(k)** $y = \text{int}(x + 1)$

(l) $y = 2 - 4/(1 + e^{-x})$

1.
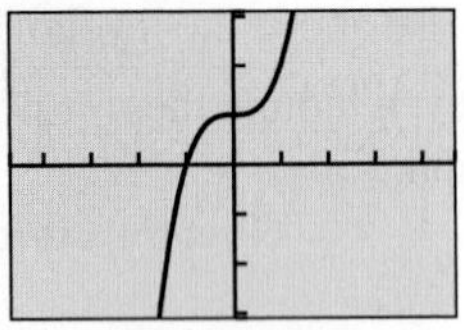

2.
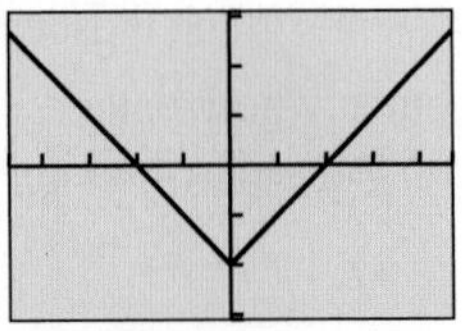

3.
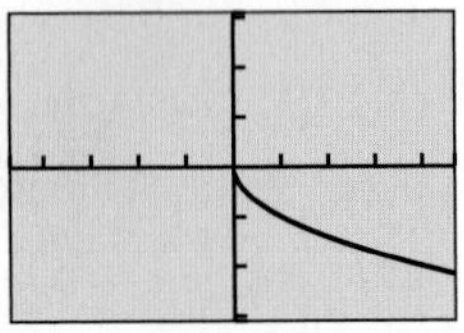

4.
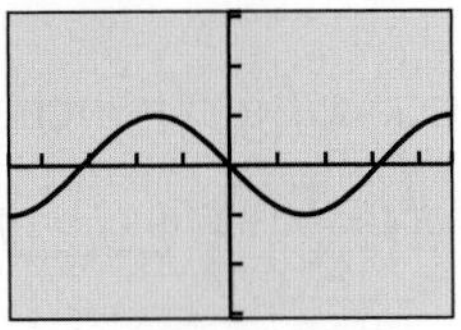

5.
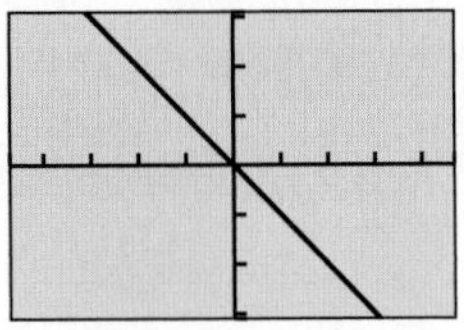

6.
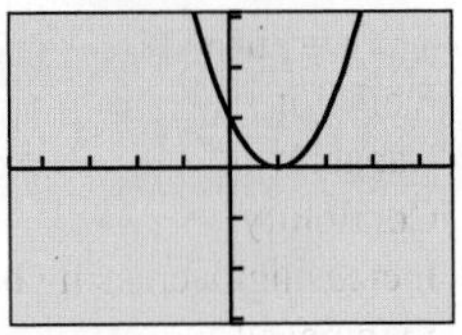

7.
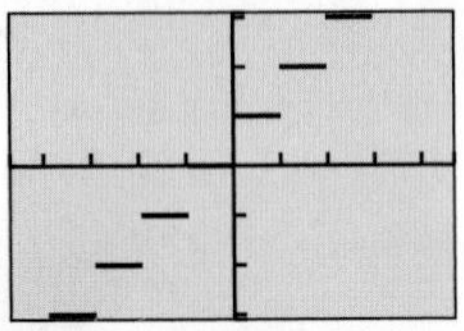

8.
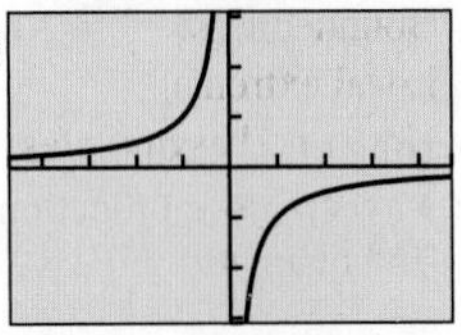

9.
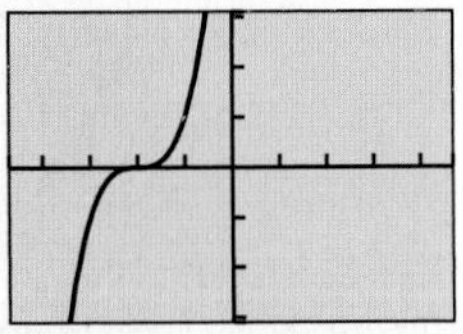

10.
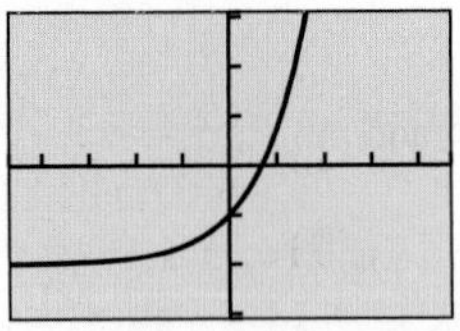

11.
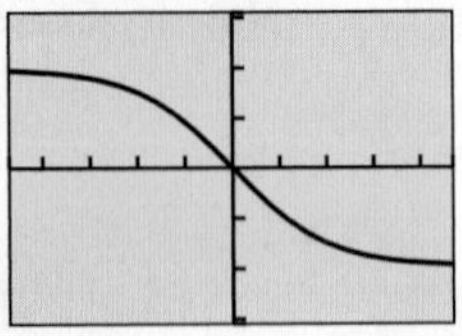

12.
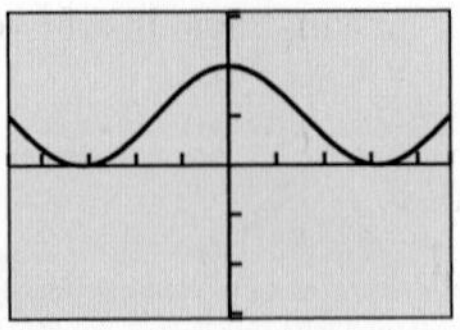

In Exercises 13–18, identify which of Exercises 1–12 display functions that fit the description given.

13. The function whose domain excludes zero

14. The function whose domain consists of all nonnegative real numbers

15. The two functions that have at least one point of discontinuity

16. The function that is not a *continuous function*

17. The six functions that are bounded below

18. The four functions that are bounded above

In Exercises 19–28, identify which of the twelve basic functions fit the description given.

19. The four functions that are odd

20. The six functions that are increasing on their entire domains

21. The three functions that are decreasing on the interval $(-\infty, 0)$

22. The three functions with infinitely many local extrema

23. The three functions with no zeros

24. The three functions with range {all real numbers}

25. The four functions that do *not* have end behavior $\lim_{x \to +\infty} f(x) = +\infty$

26. The three functions with end behavior $\lim_{x \to -\infty} f(x) = -\infty$

27. The four functions whose graphs look the same when turned upside-down and flipped about the *y*-axis

28. The two functions whose graphs are identical except for a horizontal shift

In Exercises 29–34, use your graphing calculator to produce a graph of the function. Then determine the domain and range of the function by looking at its graph.

29. $f(x) = x^2 - 5$

30. $g(x) = |x - 4|$

31. $h(x) = \ln(x + 6)$

32. $k(x) = 1/x + 3$

33. $s(x) = \text{int}(x/2)$

34. $p(x) = (x + 3)^2$

In Exercises 35–42, graph the function. Then answer the following questions:

(a) On what interval, if any, is the function increasing? Decreasing?

(b) Is the function odd, even, or neither?

(c) Give the function's extrema, if any.

(d) How does the graph relate to a graph of one of the twelve basic functions?

35. $r(x) = \sqrt{x - 10}$

36. $f(x) = \sin(x) + 5$

37. $f(x) = 3/(1 + e^{-x})$

38. $q(x) = e^x + 2$

39. $h(x) = |x| - 10$

40. $g(x) = 4\cos(x)$

41. $s(x) = |x - 2|$

42. $f(x) = 5 - \text{abs}(x)$

43. Find the horizontal asymptotes for the graph shown in Exercise 11.

44. Find the horizontal asymptotes for the graph of $f(x)$ in Exercise 37.

In Exercises 45–52, sketch the graph of the piecewise-defined function. (Try doing it without a calculator.) In each case, give any points of discontinuity.

45. $f(x) = \begin{cases} x & \text{if } x \le 0 \\ x^2 & \text{if } x > 0 \end{cases}$

46. $g(x) = \begin{cases} x^3 & \text{if } x \le 0 \\ e^x & \text{if } x > 0 \end{cases}$

47. $h(x) = \begin{cases} |x| & \text{if } x < 0 \\ \sin x & \text{if } x \ge 0 \end{cases}$

48. $w(x) = \begin{cases} 1/x & \text{if } x < 0 \\ \sqrt{x} & \text{if } x \ge 0 \end{cases}$

49. $f(x) = \begin{cases} \cos x & \text{if } x \le 0 \\ e^x & \text{if } x > 0 \end{cases}$

50. $g(x) = \begin{cases} |x| & \text{if } x < 0 \\ x^2 & \text{if } x \ge 0 \end{cases}$

51. $f(x) = \begin{cases} -3 - x & \text{if } x \le 0 \\ 1 & \text{if } 0 < x < 1 \\ x^2 & \text{if } x \ge 1 \end{cases}$

52. $f(x) = \begin{cases} x^2 & \text{if } x < -1 \\ |x| & \text{if } -1 \le x < 1 \\ \text{int}(x) & \text{if } x \ge 1 \end{cases}$

53. Writing to Learn The function $f(x) = \sqrt{x^2}$ is one of our twelve basic functions written in another form.

(a) Graph the function and identify which basic function it is.

(b) Explain algebraically why the two functions are equal.

54. Uncovering Hidden Behavior The function $g(x) = \sqrt{x^2 + 0.0001} - 0.01$ is *not* one of our twelve basic functions written in another form.

(a) Graph the function and identify which basic function it appears to be.

(b) Verify numerically that it is not the basic function that it appears to be.

55. Writing to Learn The function $f(x) = \ln(e^x)$ is one of our twelve basic functions written in another form.

(a) Graph the function and identify which basic function it is.

(b) Explain how the equivalence of the two functions in (a) shows that the natural logarithm function is *not* bounded above (even though it *appears* to be bounded above in Figure 1.42).

56. Writing to Learn Let $f(x)$ be the function that gives the cost, in cents, to mail a first-class package that weighs x ounces. In 2013, the cost was \$2.07 for a package that weighed up to 3 ounces, plus 17 cents for each additional ounce or portion thereof (up to 13 ounces). (*Source: United States Postal Service.*)

(a) Sketch a graph of $f(x)$.

(b) How is this function similar to the greatest integer function? How is it different?

Packages

Weight Not Over	Price
3 ounces	\$2.07
4 ounces	\$2.24
5 ounces	\$2.41
6 ounces	\$2.58
7 ounces	\$2.75
8 ounces	\$2.92
9 ounces	\$3.09
10 ounces	\$3.26
11 ounces	\$3.43
12 ounces	\$3.60
13 ounces	\$3.77

57 Analyzing a Function Set your calculator to DOT mode and graph the greatest integer function, $y = \text{int}(x)$, in the window $[-4.7, 4.7]$ by $[-3.1, 3.1]$. Then complete the following analysis.

BASIC FUNCTION

The Greatest Integer Function

$f(x) = \text{int } x$
Domain:
Range:
Continuity:
Increasing/decreasing behavior:
Symmetry:
Boundedness:
Local extrema:
Horizontal asymptotes:
Vertical asymptotes:
End behavior:

Standardized Test Questions

58. True or False The greatest integer function has an inverse function. Justify your answer.

59. True or False The logistic function has two horizontal asymptotes. Justify your answer.

In Exercises 60–63, you may use a graphing calculator to answer the question.

60. Multiple Choice Which function has range {all real numbers}?

(A) $f(x) = 4 + \ln x$

(B) $f(x) = 3 - 1/x$

(C) $f(x) = 5/(1 + e^{-x})$

(D) $f(x) = \text{int}(x - 2)$

(E) $f(x) = 4 \cos x$

61. **Multiple Choice** Which function is bounded both above and below?

(A) $f(x) = x^2 - 4$

(B) $f(x) = (x - 3)^3$

(C) $f(x) = 3e^x$

(D) $f(x) = 3 + 1/(1 + e^{-x})$

(E) $f(x) = 4 - |x|$

62. **Multiple Choice** Which of the following is the same as the restricted-domain function $f(x) = \text{int}(x), 0 \le x < 2$?

(A) $f(x) = \begin{cases} 0 & \text{if } 0 \le x < 1 \\ 1 & \text{if } x = 1 \\ 2 & \text{if } 1 < x < 2 \end{cases}$

(B) $f(x) = \begin{cases} 0 & \text{if } x = 0 \\ 1 & \text{if } 0 < x \le 1 \\ 2 & \text{if } 1 < x < 2 \end{cases}$

(C) $f(x) = \begin{cases} 0 & \text{if } 0 \le x < 1 \\ 1 & \text{if } 1 \le x < 2 \end{cases}$

(D) $f(x) = \begin{cases} 1 & \text{if } 0 \le x < 1 \\ 2 & \text{if } 1 \le x < 2 \end{cases}$

(E) $f(x) = \begin{cases} x & \text{if } 0 \le x < 1 \\ 1 + x & \text{if } 1 \le x < 2 \end{cases}$

63. **Multiple Choice** Which function is increasing on the interval $(-\infty, \infty)$?

(A) $f(x) = \sqrt{3 + x}$

(B) $f(x) = \text{int}(x)$

(C) $f(x) = 2x^2$

(D) $f(x) = \sin x$

(E) $f(x) = 3/(1 + e^{-x})$

Explorations

64. **Which Is Bigger?** For positive values of x, we wish to compare the values of the basic functions x^2, x, and $\sqrt{x}$.

(a) How would you order them from least to greatest?

(b) Graph the three functions in the viewing window $[0, 30]$ by $[0, 20]$. Does the graph confirm your response in (a)?

(c) Now graph the three functions in the viewing window $[0, 2]$ by $[0, 1.5]$.

(d) Write a careful response to the question in (a) that accounts for all positive values of x.

65. **Odds and Evens** There are four odd functions and three even functions in the gallery of twelve basic functions. After multiplying these functions together pairwise in different combinations and exploring the graphs of the products, make a conjecture about the symmetry of:

(a) a product of two odd functions;

(b) a product of two even functions;

(c) a product of an odd function and an even function.

66. **Group Activity** Assign to each student in the class the name of one of the twelve basic functions, but secretly so that no student knows the "name" of another. (The same function name could be given to several students, but all the functions should be used at least once.) Let each student make a one-sentence self-introduction to the class that reveals something personal "about who I am that really identifies me." The rest of the students then write down their guess as to the function's identity. Hints should be subtle and cleverly anthropomorphic. (For example, the absolute value function saying "I have a very sharp smile" is subtle and clever, while "I am absolutely valuable" is not very subtle at all.)

67. **Pepperoni Pizzas** For a statistics project, a student counted the number of pepperoni slices on pizzas of various sizes at a local pizzeria, compiling the following table:

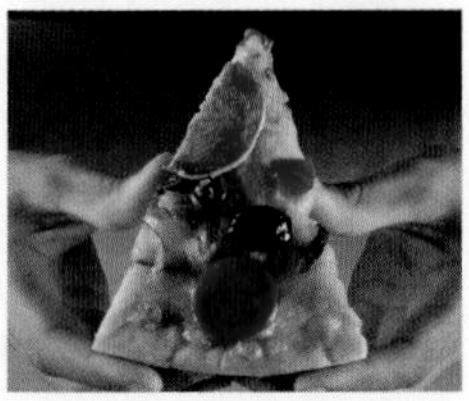

Table 1.10

Type of Pizza	Radius	Pepperoni Count
Personal	4″	12
Medium	6″	27
Large	7″	37
Extra large	8″	48

(a) Explain why the pepperoni count (P) ought to be proportional to the square of the radius (r).

(b) Assuming that $P = k \cdot r^2$, use the data pair $(4, 12)$ to find the value of k.

(c) Does the algebraic model fit the rest of the data well?

(d) Some pizza places have charts showing their kitchen staff how much of each topping should be put on each size of pizza. Do you think this pizzeria uses such a chart? Explain.

Extending the Ideas

68. **Inverse Functions** Two functions are said to be *inverses* of each other if the graph of one can be obtained from the graph of the other by reflecting it across the line $y = x$. For example, the functions with the graphs shown below are inverses of each other:

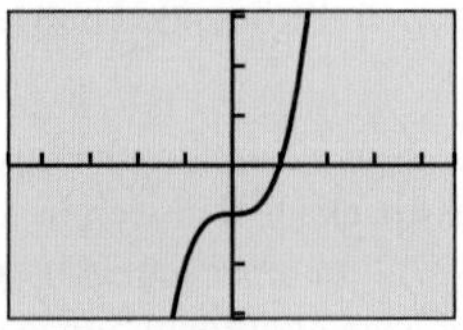

[–4.7, 4.7] by [–3.1, 3.1]
(a)

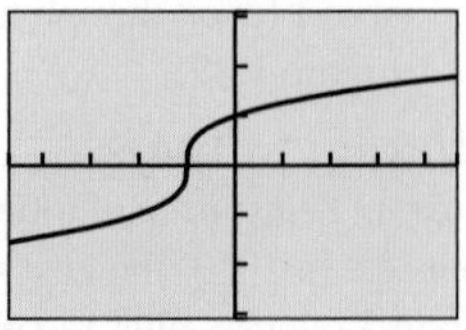

[–4.7, 4.7] by [–3.1, 3.1]
(b)

(a) Two of the twelve basic functions in this section are inverses of each other. Which are they?

(b) Two of the twelve basic functions in this section are their own inverses. Which are they?

(c) If you restrict the domain of one of the twelve basic functions to $[0, \infty)$, it becomes the inverse of another one. Which are they?

69. Identifying a Function by Its Properties

(a) Seven of the twelve basic functions have the property that $f(0) = 0$. Which five do not?

(b) Only one of the twelve basic functions has the property that $f(x + y) = f(x) + f(y)$ for all x and y in its domain. Which one is it?

(c) One of the twelve basic functions has the property that $f(x + y) = f(x)f(y)$ for all x and y in its domain. Which one is it?

(d) One of the twelve basic functions has the property that $f(xy) = f(x) + f(y)$ for all x and y in its domain. Which one is it?

(e) Four of the twelve basic functions have the property that $f(x) + f(-x) = 0$ for all x in their domains. Which four are they?

1.4 Building Functions from Functions

What you'll learn about

- Combining Functions Algebraically
- Composition of Functions
- Relations and Implicitly Defined Functions

... and why

Most of the functions that you will encounter in calculus and in real life can be created by combining or modifying other functions.

Combining Functions Algebraically

Knowing how a function is "put together" is an important first step when applying the tools of calculus. Functions have their own algebra based on the same operations we apply to real numbers (addition, subtraction, multiplication, and division). One way to build new functions is to apply these operations, using the following definitions.

DEFINITION Sum, Difference, Product, and Quotient of Functions

Let f and g be two functions with intersecting domains. Then for all values of x in the intersection, the algebraic combinations of f and g are defined by the following rules:

Sum: $(f + g)(x) = f(x) + g(x)$

Difference: $(f - g)(x) = f(x) - g(x)$

Product: $(fg)(x) = f(x)g(x)$

Quotient: $\left(\frac{f}{g}\right)(x) = \frac{f(x)}{g(x)}$, provided $g(x) \neq 0$

In each case, the domain of the new function consists of all numbers that belong to both the domain of f and the domain of g. As noted, the zeros of the denominator are excluded from the domain of the quotient.

Euler's function notation works so well in the above definitions that it almost obscures what is really going on. The "+" in the expression "$(f + g)(x)$" stands for a brand new operation called *function addition*. It builds a new function, $f + g$, from the given functions f and g. Like any function, $f + g$ is defined by what it does: It takes a domain value x and returns a range value $f(x) + g(x)$. Note that the "+" sign in "$f(x) + g(x)$" *does* stand for the familiar operation of real-number addition. So, with the same symbol taking on different roles on either side of the equal sign, there is more to the above definitions than first meets the eye.

Fortunately, the definitions are easy to apply.

EXAMPLE 1 Defining New Functions Algebraically

Let $f(x) = x^2$ and $g(x) = \sqrt{x + 1}$.

Find formulas for the functions $f + g$, $f - g$, fg, f/g, and gg. Give the domain of each.

SOLUTION We first determine that f has domain all real numbers and that g has domain $[-1, \infty)$. These domains overlap, the intersection being the interval $[-1, \infty)$. So:

$$(f + g)(x) = f(x) + g(x) = x^2 + \sqrt{x + 1} \quad \text{with domain } [-1, \infty).$$

$$(f - g)(x) = f(x) - g(x) = x^2 - \sqrt{x + 1} \quad \text{with domain } [-1, \infty).$$

$$(fg)(x) = f(x)g(x) = x^2\sqrt{x + 1} \quad \text{with domain } [-1, \infty).$$

$$\left(\frac{f}{g}\right)(x) = \frac{f(x)}{g(x)} = \frac{x^2}{\sqrt{x + 1}} \quad \text{with domain } (-1, \infty).$$

$$(gg)(x) = g(x)g(x) = (\sqrt{x + 1})^2 \quad \text{with domain } [-1, \infty).$$

Note that we could express $(gg)(x)$ more simply as $x + 1$. That would be fine, but the simplification would not change the fact that the domain of gg is (by definition) the interval $[-1, \infty)$. Under other circumstances the function $h(x) = x + 1$ would have domain all real numbers, but under these circumstances it cannot; it is a product of two functions with restricted domains. **Now try Exercise 3.**

Composition of Functions

It is not hard to see that the function $\sin(x^2)$ is built from the basic functions $\sin x$ and x^2, but the functions are not put together by addition, subtraction, multiplication, or division. Instead, the two functions are combined by simply applying them in order—first the squaring function, then the sine function. This operation for combining functions, which has no counterpart in the algebra of real numbers, is called *function composition.*

DEFINITION Composition of Functions

Let f and g be two functions such that the domain of f intersects the range of g. The **composition f of g**, denoted $f \circ g$, is defined by the rule

$$(f \circ g)(x) = f(g(x)).$$

The domain of $f \circ g$ consists of all x-values in the domain of g that map to $G(x)$ values in the domain of f. (See Figure 1.55.)

The composition g of f, denoted $g \circ f$, is defined similarly. In most cases $g \circ f$ and $f \circ g$ are different functions. (In the language of algebra, "function composition is not commutative.")

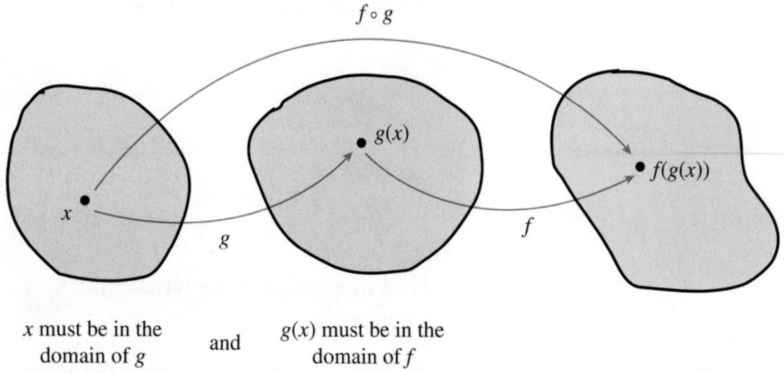

FIGURE 1.55 In the composition $f \circ g$, the function g is applied first and then f. This is the reverse of the order in which we read the symbols.

EXAMPLE 2 Composing Functions

Let $f(x) = e^x$ and $g(x) = \sqrt{x}$. Find $(f \circ g)(x)$ and $(g \circ f)(x)$ and verify numerically that the functions $f \circ g$ and $g \circ f$ are not the same.

SOLUTION

$(f \circ g)(x) = f(g(x)) = f(\sqrt{x}) = e^{\sqrt{x}}$

$(g \circ f)(x) = g(f(x)) = g(e^x) = \sqrt{e^x}$

One verification that these functions are not the same is that they have different domains: $f \circ g$ is defined only for $x \geq 0$, but $g \circ f$ is defined for all real numbers. We could also consider their graphs (Figure 1.56), which agree only at $x = 0$ and $x = 4$.

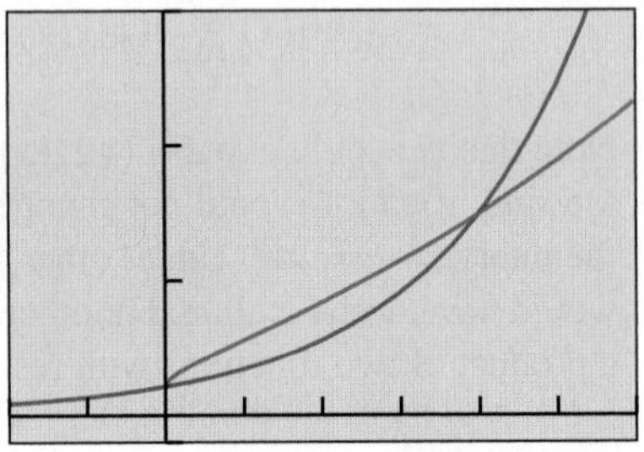

[–2, 6] by [–1, 15]

FIGURE 1.56 The graphs of $y = e^{\sqrt{x}}$ and $y = \sqrt{e^x}$ are not the same. (Example 2)

Finally, the graphs suggest a numerical verification: Find a single value of x for which $f(g(x))$ and $g(f(x))$ give different values. For example, $f(g(1)) = e$ and $g(f(1)) = \sqrt{e}$. The graph helps us to make a judicious choice of x. You do not want to check the functions at $x = 0$ and $x = 4$ and conclude that they are the same!

Now try Exercise 15.

EXPLORATION 1 **Composition Calisthenics**

One of the f functions in column B can be composed with one of the g functions in column C to yield each of the basic $f \circ g$ functions in column A. Can you match the columns successfully without a graphing calculator? If you are having trouble, try it with a graphing calculator.

A	B	C
$f \circ g$	f	g
x	$x - 3$	$x^{0.6}$
x^2	$2x - 3$	x^2
$\lvert x \rvert$	$\sqrt{x}$	$\dfrac{(x-2)(x+2)}{2}$
x^3	x^5	$\ln(e^3 x)$
$\ln x$	$\lvert 2x + 4 \rvert$	$\dfrac{x}{2}$
$\sin x$	$1 - 2x^2$	$\dfrac{x+3}{2}$
$\cos x$	$2 \sin x \cos x$	$\sin\left(\dfrac{x}{2}\right)$

EXAMPLE 3 Finding the Domain of a Composition

Let $f(x) = x^2 - 1$ and let $g(x) = \sqrt{x}$. Find the domains of the composite functions

(a) $g \circ f$ **(b)** $f \circ g$.

SOLUTION

(a) We compose the functions in the order specified:

$$(g \circ f)(x) = g(f(x))$$
$$= \sqrt{x^2 - 1}$$

For x to be in the domain of $g \circ f$, we must first find $f(x) = x^2 - 1$, which we can do for all real x. Then we must take the square root of the result, which we can only do for nonnegative values of $x^2 - 1$.

Therefore, the domain of $g \circ f$ consists of all real numbers for which $x^2 - 1 \geq 0$, namely the union $(-\infty, -1] \cup [1, \infty)$.

(b) Again, we compose the functions in the order specified:

$$(f \circ g)(x) = f(g(x))$$
$$= (\sqrt{x})^2 - 1$$

For x to be in the domain of $f \circ g$, we must first be able to find $g(x) = \sqrt{x}$, which we can only do for nonnegative values of x. Then we must be able to square the result and subtract 1, which we can do for all real numbers. Therefore, the domain of $f \circ g$ consists of the interval $[0, \infty)$.

Support Graphically We can graph the composition functions to see if the grapher respects the domain restrictions. The screen to the left of each graph shows the setup in the "Y =" editor. Figure 1.57b shows the graph of $y = (g \circ f)(x)$, and Figure 1.57d shows the graph of $y = (f \circ g)(x)$. The graphs support our algebraic work quite nicely.

Now try Exercise 17.

Caution

We might choose to express $(f \circ g)$ more simply as $x - 1$. However, you must remember that the composition is restricted to the domain of $g(x) = \sqrt{x}$, or $[0, \infty]$. The domain of $x - 1$ is all real numbers. It is a good idea to work out the domain of a composition before you simplify the expression for $f(g(x))$. One way to simplify and maintain the restriction on the domain in Example 3 is to write $(f \circ g)(x) = x - 1$, $x \geq 0$.

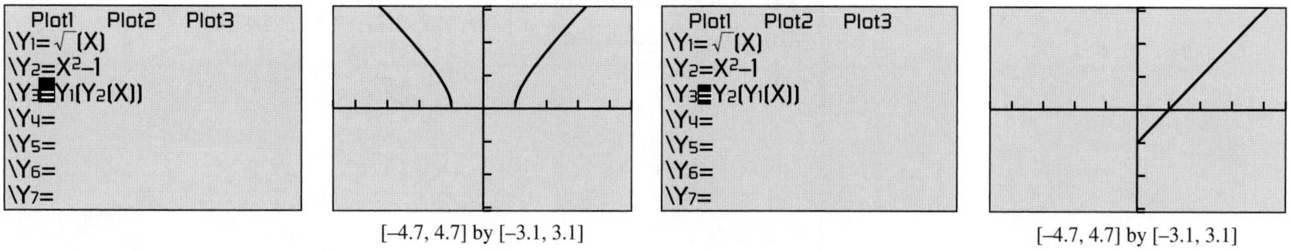

(a) (b) (c) (d)

FIGURE 1.57 The functions Y1 and Y2 are composed to get the graphs of $y = (g \circ f)(x)$ and $y = (f \circ g)(x)$, respectively. The graphs support our conclusions about the domains of the two composite functions. (Example 3)

In Examples 2 and 3 two functions were *composed* to form new functions. There are times in calculus when we need to reverse the process. That is, we may begin with a function h and *decompose* it by finding functions whose composition is h.

EXAMPLE 4 Decomposing Functions

For each function h, find functions f and g such that $h(x) = f(g(x))$.

(a) $h(x) = (x + 1)^2 - 3(x + 1) + 4$

(b) $h(x) = \sqrt{x^3 + 1}$

(continued)

SOLUTION

(a) We can see that h is quadratic in $x + 1$. Let $f(x) = x^2 - 3x + 4$ and let $g(x) = x + 1$. Then

$$h(x) = f(g(x)) = f(x + 1) = (x + 1)^2 - 3(x + 1) + 4.$$

(b) We can see that h is the square root of the function $x^3 + 1$. Let $f(x) = \sqrt{x}$ and let $g(x) = x^3 + 1$. Then

$$h(x) = f(g(x)) = f(x^3 + 1) = \sqrt{x^3 + 1}.$$

Now try Exercise 25.

There is often more than one way to decompose a function. For example, an alternate way to decompose $h(x) = \sqrt{x^3 + 1}$ in Example 4b is to let $f(x) = \sqrt{x + 1}$ and let $g(x) = x^3$. Then $h(x) = f(g(x)) = f(x^3) = \sqrt{x^3 + 1}$.

EXAMPLE 5 Modeling with Function Composition

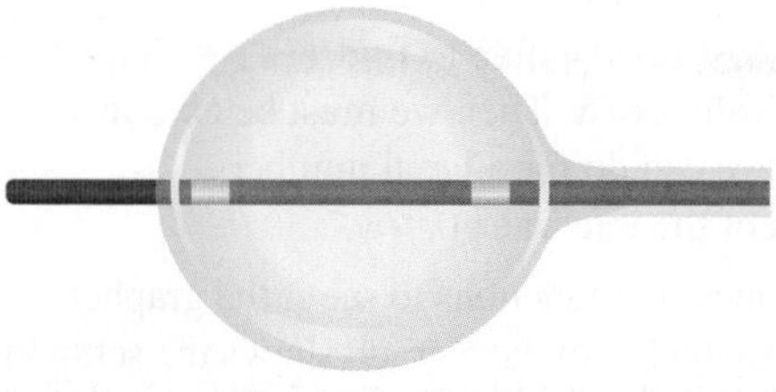

FIGURE 1.58 (Example 5)

In the medical procedure known as angioplasty, doctors insert a catheter into a heart vein (through a large peripheral vein) and inflate a small, spherical balloon on the tip of the catheter. Suppose the balloon is inflated at a constant rate of 44 cubic millimeters per second (Figure 1.58).

(a) Find the volume after t seconds.

(b) When the volume is V, what is the radius r?

(c) Write an equation that gives the radius r as a function of the time. What is the radius after 5 sec?

SOLUTION

(a) After t seconds, the volume will be $44t$.

(b) Solve Algebraically

$$\frac{4}{3}\pi r^3 = V$$

$$r^3 = \frac{3V}{4\pi}$$

$$r = \sqrt[3]{\frac{3V}{4\pi}}$$

(c) Substituting $44t$ for V gives $r = \sqrt[3]{\frac{3 \cdot 44t}{4\pi}}$ or $r = \sqrt[3]{\frac{33t}{\pi}}$. After 5 sec, the radius will be $r = \sqrt[3]{\frac{33 \cdot 5}{\pi}} \approx 3.74$ mm.

Now try Exercise 31.

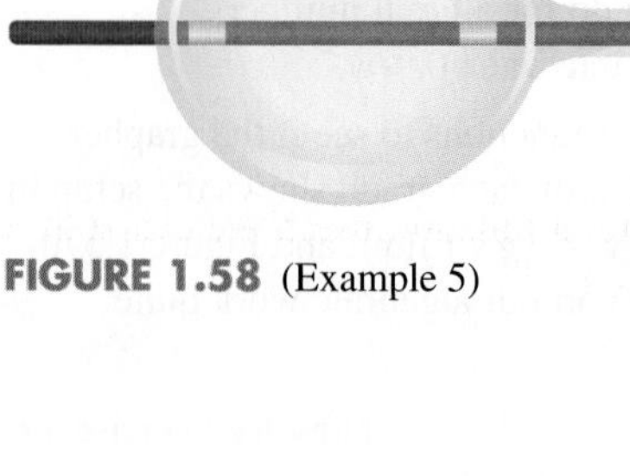

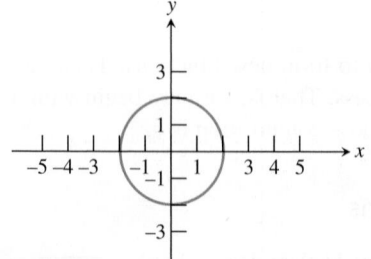

FIGURE 1.59 A circle of radius 2 centered at the origin. This set of ordered pairs (x, y) defines a *relation* that is not a function, because the graph fails the vertical line test.

Relations and Implicitly Defined Functions

There are many useful curves in mathematics that fail the vertical line test and therefore are not graphs of functions. One such curve is the circle in Figure 1.59. Even though y is not related to x as a function in this instance, there is certainly some sort of relationship going on. In fact, not only does the shape of the graph show a significant *geometric* relationship among the points, but the ordered pairs (x, y) exhibit a significant *algebraic* relationship as well: They consist exactly of the solutions to the equation $x^2 + y^2 = 4$.

Graphing Relations

Relations that are not functions are often not easy to graph. We will study some special cases later in the course (circles, ellipses, etc.), but some simple-looking relations like the one in Example 6 can be difficult. If you are curious to see what this one looks like, see Exercise 55.

The general term for a set of ordered pairs (x, y) is a **relation**. If the relation happens to relate a *single* value of y to each value of x, then the relation is also a function and its graph will pass the vertical line test. In the case of the circle with equation $x^2 + y^2 = 4$, both $(0, 2)$ and $(0, -2)$ are in the relation, so y is not a function of x.

EXAMPLE 6 Verifying Pairs in a Relation

Determine which of the ordered pairs $(2, -5)$, $(1, 3)$, and $(2, 1)$ are in the relation defined by $x^2y + y^2 = 5$. Is the relation a function?

SOLUTION We simply substitute the x- and y-coordinates of the ordered pairs into $x^2y + y^2$ and see if we get 5.

$(2, -5)$:	$(2)^2(-5) + (-5)^2 = 5$	Substitute $x = 2, y = -5$.
$(1, 3)$:	$(1)^2(3) + (3)^2 = 12 \neq 5$	Substitute $x = 1, y = 3$.
$(2, 1)$:	$(2)^2(1) + (1)^2 = 5$	Substitute $x = 2, y = 1$.

So, $(2, -5)$ and $(2, 1)$ are in the relation, but $(1, 3)$ is not.

Since the equation relates two different y values (-5 and 1) to the same x value (2), the relation cannot be a function. **Now try Exercise 35.**

Let us revisit the circle $x^2 + y^2 = 4$. While it is not a function itself, we can split it into two equations that *do* define functions, as follows:

$$x^2 + y^2 = 4$$
$$y^2 = 4 - x^2$$
$$y = +\sqrt{4 - x^2} \text{ or } y = -\sqrt{4 - x^2}$$

The graphs of these two functions are, respectively, the upper and lower semicircles of the circle in Figure 1.59. They are shown in Figure 1.60. Since all the ordered pairs in either of these functions satisfy the equation $x^2 + y^2 = 4$, we say that the relation given by the equation defines the two functions **implicitly**.

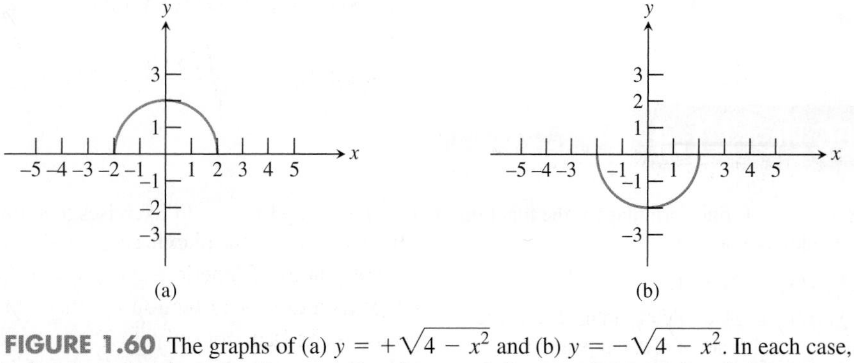

FIGURE 1.60 The graphs of (a) $y = +\sqrt{4 - x^2}$ and (b) $y = -\sqrt{4 - x^2}$. In each case, y is defined as a function of x. These two functions are defined *implicitly* by the relation $x^2 + y^2 = 4$.

EXAMPLE 7 Using Implicitly Defined Functions

Describe the graph of the relation $x^2 + 2xy + y^2 = 1$.

SOLUTION This looks like a difficult task at first, but notice that the expression on the left of the equal sign is a factorable trinomial. This enables us to split the relation into two implicitly defined functions as follows:

$$x^2 + 2xy + y^2 = 1$$

$$(x + y)^2 = 1 \quad \text{Factor.}$$

$$x + y = \pm 1 \quad \text{Extract square roots.}$$

$$x + y = 1 \text{ or } x + y = -1$$

$$y = -x + 1 \text{ or } y = -x - 1 \quad \text{Solve for } y.$$

The graph consists of two parallel lines (Figure 1.61), each the graph of one of the implicitly defined functions. **Now try Exercise 37.**

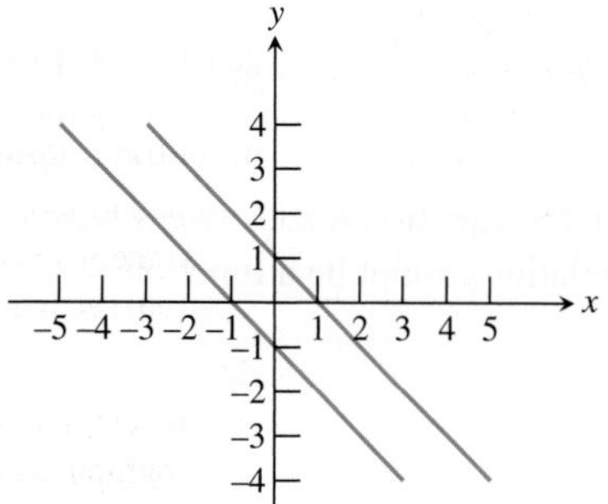

FIGURE 1.61 The graph of the relation $x^2 + 2xy + y^2 = 1$. (Example 7)

QUICK REVIEW 1.4 *(For help, go to Sections P.1, 1.2, and 1.3.)*

Exercise numbers with a gray background indicate problems that the authors have designed to be solved *without a calculator*.

In Exercises 1–10, find the domain of the function and express it in interval notation.

1. $f(x) = \dfrac{x - 2}{x + 3}$
2. $g(x) = \ln(x - 1)$
3. $f(t) = \sqrt{5 - t}$
4. $g(x) = \dfrac{3}{\sqrt{2x - 1}}$
5. $f(x) = \sqrt{\ln(x)}$
6. $h(x) = \sqrt{1 - x^2}$
7. $f(t) = \dfrac{t + 5}{t^2 + 1}$
8. $g(t) = \ln(|t|)$
9. $f(x) = \dfrac{1}{\sqrt{1 - x^2}}$
10. $g(x) = 2$

SECTION 1.4 Exercises

In Exercises 1–4, find formulas for the functions $f + g$, $f - g$, and fg. Give the domain of each.

1. $f(x) = 2x - 1$; $g(x) = x^2$
2. $f(x) = (x - 1)^2$; $g(x) = 3 - x$
3. $f(x) = \sqrt{x}$; $g(x) = \sin x$
4. $f(x) = \sqrt{x + 5}$; $g(x) = |x + 3|$

In Exercises 5–8, find formulas for f/g and g/f. Give the domain of each.

5. $f(x) = \sqrt{x + 3}$; $g(x) = x^2$
6. $f(x) = \sqrt{x - 2}$; $g(x) = \sqrt{x + 4}$
7. $f(x) = x^2$; $g(x) = \sqrt{1 - x^2}$
8. $f(x) = x^3$; $g(x) = \sqrt[3]{1 - x^3}$

9. $f(x) = x^2$ and $g(x) = 1/x$ are shown below in the viewing window $[0, 5]$ by $[0, 5]$. Sketch the graph of the sum $(f + g)(x)$ by adding the y-coordinates directly from the graphs. Then graph the sum on your calculator and see how close you came.

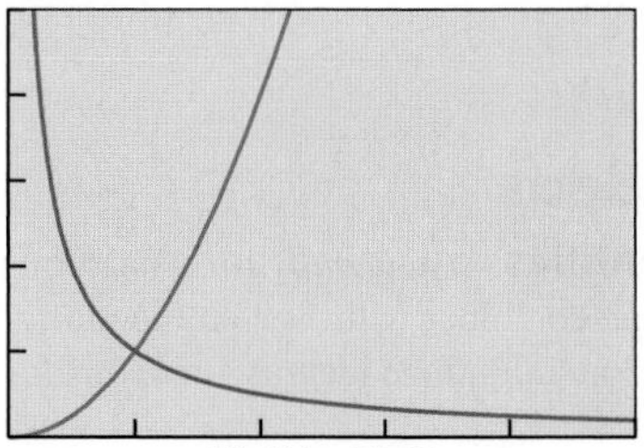

[0, 5] by [0, 5]

10. The graphs of $f(x) = x^2$ and $g(x) = 4 - 3x$ are shown in the viewing window $[-5, 5]$ by $[-10, 25]$. Sketch the graph of the difference $(f - g)(x)$ by subtracting the y-coordinates directly from the graphs. Then graph the difference on your calculator and see how close you came.

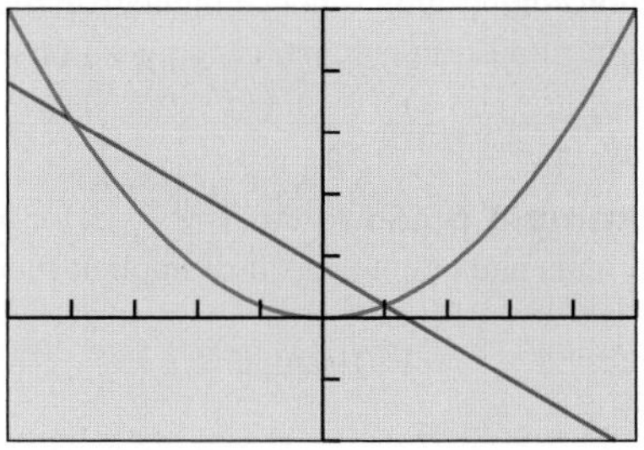

[–5, 5] by [–10, 25]

In Exercises 11–14, find $(f \circ g)(3)$ and $(g \circ f)(-2)$.

11. $f(x) = 2x - 3;\ g(x) = x + 1$

12. $f(x) = x^2 - 1;\ g(x) = 2x - 3$

13. $f(x) = x^2 + 4;\ g(x) = \sqrt{x + 1}$

14. $f(x) = \dfrac{x}{x + 1};\ g(x) = 9 - x^2$

In Exercises 15–22, find $f(g(x))$ and $g(f(x))$. State the domain of each.

15. $f(x) = 3x + 2;\ g(x) = x - 1$

16. $f(x) = x^2 - 1;\ g(x) = \dfrac{1}{x - 1}$

17. $f(x) = x^2 - 2;\ g(x) = \sqrt{x + 1}$

18. $f(x) = \dfrac{1}{x - 1};\ g(x) = \sqrt{x}$

19. $f(x) = x^2;\ g(x) = \sqrt{1 - x^2}$

20. $f(x) = x^3;\ g(x) = \sqrt[3]{1 - x^3}$

21. $f(x) = \dfrac{1}{2x};\ g(x) = \dfrac{1}{3x}$

22. $f(x) = \dfrac{1}{x + 1};\ g(x) = \dfrac{1}{x - 1}$

In Exercises 23–30, find $f(x)$ and $g(x)$ so that the function can be described as $y = f(g(x))$. (There may be more than one possible decomposition.)

23. $y = \sqrt{x^2 - 5x}$

24. $y = (x^3 + 1)^2$

25. $y = |3x - 2|$

26. $y = \dfrac{1}{x^3 - 5x + 3}$

27. $y = (x - 3)^5 + 2$

28. $y = e^{\sin x}$

29. $y = \cos(\sqrt{x})$

30. $y = (\tan x)^2 + 1$

31. Weather Balloons A high-altitude spherical weather balloon expands as it rises due to the drop in atmospheric pressure. Suppose that the radius r increases at the rate of 0.03 in./sec and that $r = 48$ in. at time $t = 0$. Determine an equation that models the volume V of the balloon at time t and find the volume when $t = 300$ sec.

32. A Snowball's Chance Jake stores a small cache of 4-inch-diameter snowballs in the basement freezer, unaware that the freezer's self-defrosting feature will cause each snowball to lose about 1 cubic inch of volume every 40 days. He remembers them a year later (call it 360 days) and goes to retrieve them. What is their diameter then?

33. Satellite Photography A satellite camera takes a rectangle-shaped picture. The smallest region that can be photographed is a 5-km by 7-km rectangle. As the camera zooms out, the length l and width w of the rectangle increase at a rate of 2 km/sec. How long does it take for the area A to be at least 5 times its original size?

34. Computer Imaging New Age Special Effects, Inc., prepares computer software based on specifications prepared by film directors. To simulate an approaching vehicle, they begin with a computer image of a 5-cm by 7-cm by 3-cm box. The program increases each dimension at a rate of 2 cm/sec. How long does it take for the volume V of the box to be at least 5 times its initial size?

35. Which of the ordered pairs $(1, 1)$, $(4, -2)$, and $(3, -1)$ are in the relation given by $3x + 4y = 5$?

36. Which of the ordered pairs $(5, 1)$, $(3, 4)$, and $(0, -5)$ are in the relation given by $x^2 + y^2 = 25$?

In Exercises 37–44, find two functions defined implicitly by the given relation.

37. $x^2 + y^2 = 25$

38. $x + y^2 = 25$

39. $x^2 - y^2 = 25$

40. $3x^2 - y^2 = 25$

41. $x + |y| = 1$

42. $x - |y| = 1$

43. $y^2 = x^2$

44. $y^2 = x$

Standardized Test Questions

45. True or False The domain of the quotient function $(f/g)(x)$ consists of all numbers that belong to both the domain of f and the domain of g. Justify your answer.

46. True or False The domain of the product function $(fg)(x)$ consists of all numbers that belong to either the domain of f or the domain of g. Justify your answer.

You may use a graphing calculator when solving Exercises 47–50.

47. Multiple Choice Suppose f and g are functions with domain all real numbers. Which of the following statements is *not* necessarily true?

(A) $(f+g)(x) = (g+f)(x)$ **(B)** $(fg)(x) = (gf)(x)$

(C) $f(g(x)) = g(f(x))$ **(D)** $(f-g)(x) = -(g-f)(x)$

(E) $(f \circ g)(x) = f(g(x))$

48. Multiple Choice If $f(x) = x - 7$ and $g(x) = \sqrt{4-x}$, what is the domain of the function f/g?

(A) $(-\infty, 4)$ **(B)** $(-\infty, 4]$ **(C)** $(4, \infty)$

(D) $[4, \infty)$ **(E)** $(4, 7) \cup (7, \infty)$

49. Multiple Choice If $f(x) = x^2 + 1$, then $(f \circ f)(x) =$

(A) $2x^2 + 2$ **(B)** $2x^2 + 1$ **(C)** $x^4 + 1$

(D) $x^4 + 2x^2 + 1$ **(E)** $x^4 + 2x^2 + 2$

50. Multiple Choice Which of the following relations defines the function $y = |x|$ implicitly?

(A) $y = x$ **(B)** $y^2 = x^2$ **(C)** $y^3 = x^3$

(D) $x^2 + y^2 = 1$ **(E)** $x = |y|$

Explorations

51. Three on a Match Match each function f with a function g and a domain D so that $(f \circ g)(x) = x^2$ with domain D.

f	g	D
e^x	$\sqrt{2-x}$	$x \neq 0$
$(x^2+2)^2$	$x+1$	$x \neq 1$
$(x^2-2)^2$	$2 \ln x$	$(0, \infty)$
$\dfrac{1}{(x-1)^2}$	$\dfrac{1}{x-1}$	$[2, \infty)$
$x^2 - 2x + 1$	$\sqrt{x-2}$	$(-\infty, 2]$
$\left(\dfrac{x+1}{x}\right)^2$	$\dfrac{x+1}{x}$	$(-\infty, \infty)$

52. Be a g Whiz Let $f(x) = x^2 + 1$. Find a function g so that

(a) $(fg)(x) = x^4 - 1$

(b) $(f+g)(x) = 3x^2$

(c) $(f/g)(x) = 1$

(d) $f(g(x)) = 9x^4 + 1$

(e) $g(f(x)) = 9x^4 + 1$

Extending the Ideas

53. Identifying Identities An *identity* for a function operation is a function that combines with a given function f to return the same function f. Find the identity functions for the following operations:

(a) Function addition. That is, find a function g such that $(f+g)(x) = (g+f)(x) = f(x)$.

(b) Function multiplication. That is, find a function g such that $(fg)(x) = (gf)(x) = f(x)$.

(c) Function composition. That is, find a function g such that $(f \circ g)(x) = (g \circ f)(x) = f(x)$.

54. Is Function Composition Associative? You already know that function composition is not commutative; that is, $(f \circ g)(x) \neq (g \circ f)(x)$. But is function composition associative? That is, does $(f \circ (g \circ h))(x) = ((f \circ g) \circ h))(x)$? Explain your answer.

55. Revisiting Example 6 Solve $x^2y + y^2 = 5$ for y using the quadratic formula and graph the pair of implicit functions.

1.5 Parametric Relations and Inverses

What you'll learn about

- Relations Defined Parametrically
- Inverse Relations and Inverse Functions

... and why

Some functions and graphs can best be defined parametrically, while some others can be best understood as inverses of functions we already know.

Relations Defined Parametrically

Another natural way to define functions or, more generally, relations, is to define *both* elements of the ordered pair (x, y) in terms of another variable t, called a **parameter**. We illustrate with an example.

EXAMPLE 1 Defining a Function Parametrically

Consider the set of all ordered pairs (x, y) defined by the equations

$$x = t + 1$$
$$y = t^2 + 2t$$

where t is any real number.

(a) Find the points determined by $t = -3, -2, -1, 0, 1, 2,$ and 3.

(b) Find an algebraic relationship between x and y. (This is often called "eliminating the parameter.") Is y a function of x?

(c) Graph the relation in the (x, y) plane.

SOLUTION

(a) Substitute each value of t into the formulas for x and y to find the point that it determines parametrically:

t	$x = t + 1$	$y = t^2 + 2t$	(x, y)
-3	-2	3	$(-2, 3)$
-2	-1	0	$(-1, 0)$
-1	0	-1	$(0, -1)$
0	1	0	$(1, 0)$
1	2	3	$(2, 3)$
2	3	8	$(3, 8)$
3	4	15	$(4, 15)$

(b) We can find the relationship between x and y algebraically by the method of substitution. First solve for t in terms of x to obtain $t = x - 1$.

$$y = t^2 + 2t \quad \text{Given}$$
$$y = (x - 1)^2 + 2(x - 1) \quad t = x - 1$$
$$= x^2 - 2x + 1 + 2x - 2 \quad \text{Expand.}$$
$$= x^2 - 1 \quad \text{Simplify.}$$

This is consistent with the ordered pairs we had found in the table. As t varies over all real numbers, we will get all the ordered pairs in the relation $y = x^2 - 1$, which does indeed define y as a function of x.

(c) Since the parametrically defined relation consists of all ordered pairs in the relation $y = x^2 - 1$, we can get the graph by simply graphing the parabola $y = x^2 - 1$. See Figure 1.62.

Now try Exercise 5.

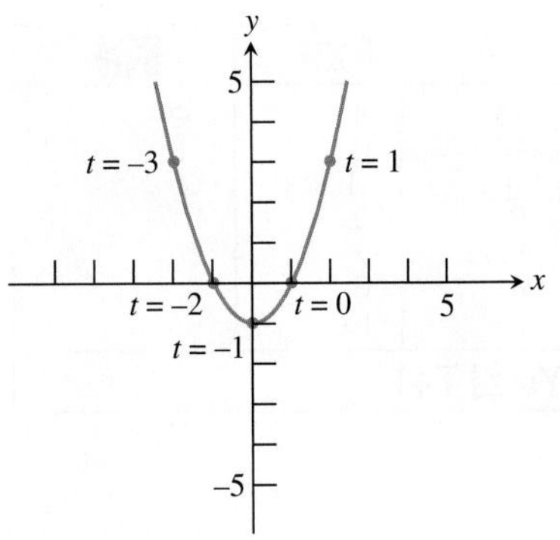

FIGURE 1.62 (Example 1)

EXAMPLE 2 Using a Graphing Calculator in Parametric Mode

Consider the set of all ordered pairs (x, y) defined by the equations

$$x = t^2 + 2t$$
$$y = t + 1$$

where t is any real number.

(a) Use a graphing calculator to find the points determined by $t = -3, 2, -1, 0, 1, 2,$ and 3.

(b) Use a graphing calculator to graph the relation in the (x, y) plane.

(c) Is y a *function* of x?

(d) Find an algebraic relationship between x and y.

t	(x, y)
−3	(3, −2)
−2	(0, −1)
−1	(−1, 0)
0	(0, 1)
1	(3, 2)
2	(8, 3)
3	(15, 4)

SOLUTION

(a) When the calculator is in *parametric mode*, the "Y =" screen provides a space to enter both X and Y as functions of the parameter T (Figure 1.63a). After entering the functions, use the table setup in Figure 1.63b to obtain the table shown in Figure 1.63c. The table shows, for example, that when T = −3 we have X1T = 3 and Y1T = −2, so the ordered pair corresponding to $t = -3$ is $(3, -2)$.

(b) In parametric mode, the "WINDOW" screen contains the usual x-axis information, as well as "Tmin," "Tmax," and "Tstep" (Figure 1.64a). To include most of the points listed in part (a), we set Xmin = −5, Xmax = 5, Ymin = −3, and Ymax = 3. Since $t = y - 1$, we set Tmin and Tmax to values one less than those for Ymin and Ymax.

The value of Tstep determines how far the grapher will go from one value of t to the next as it computes the ordered pairs. With Tmax − Tmin = 6 and Tstep = 0.1, the grapher will compute 60 points, which is sufficient. (The more points, the smoother the graph. See Exploration 1.) The graph is shown in Figure 1.64b. Use TRACE to find some of the points found in (a).

(c) No, y is not a function of x. We can see this from part (a) because $(0, -1)$ and $(0, 1)$ have the same x-value but different y-values. Alternatively, notice that the graph in (b) fails the vertical line test.

(d) We can use the same algebraic steps as in Example 1 to get the relation in terms of x and y: $x = y^2 - 1$.

Now try Exercise 7.

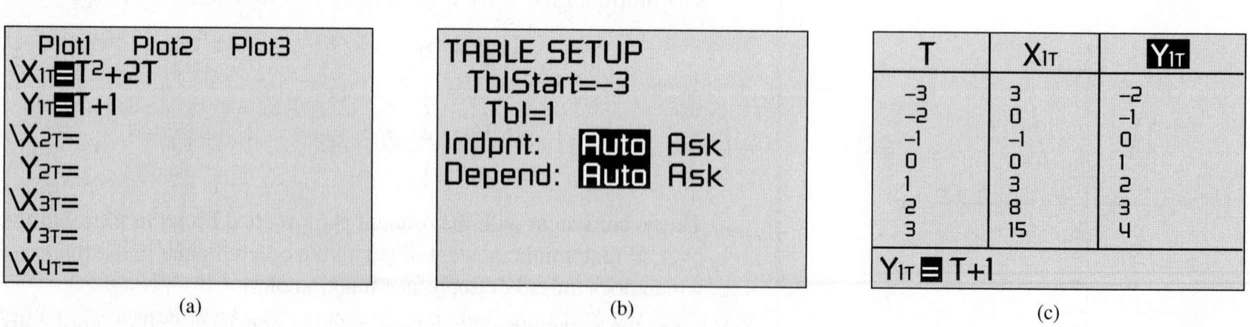

(a) (b) (c)

FIGURE 1.63 Using the table feature of a grapher set in parametric mode. (Example 2)

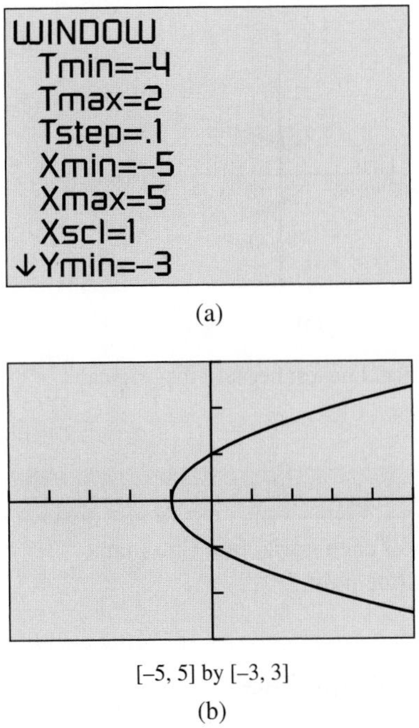

FIGURE 1.64 The graph of a parabola in parametric mode on a graphing calculator. (Example 2)

EXPLORATION 1 Watching Your Tstep

1. Graph the parabola in Example 2 in parametric mode as described in the solution. Press TRACE and observe the values of T, X, and Y. At what value of T does the calculator begin tracing? What point on the parabola results? (It's off the screen.) At what value of T does it stop tracing? What point on the parabola results? How many points are computed as you TRACE from start to finish?
2. Leave everything else the same and change the Tstep to 0.01. Do you get a smoother graph? Why or why not?
3. Leave everything else the same and change the Tstep to 1. Do you get a smoother graph? Why or why not?
4. What effect does the Tstep have on the speed of the grapher? Is this easily explained?
5. Now change the Tstep to 2. Why does the left portion of the parabola disappear? (It may help to TRACE along the curve.)
6. Change the Tstep back to 0.1 and change the Tmin to -1. Why does the bottom side of the parabola disappear? (Again, it may help to TRACE.)
7. Make a change to the window that will cause the grapher to show the bottom side of the parabola but not the top.

Inverse Relations and Inverse Functions

What happens when we reverse the coordinates of all the ordered pairs in a relation? We obviously get another relation, as it is another set of ordered pairs, but does it bear any resemblance to the original relation? If the original relation happens to be a function, will the new relation also be a function?

We can get some idea of what happens by examining Examples 1 and 2. The ordered pairs in Example 2 can be obtained by simply reversing the coordinates of the ordered pairs in Example 1. This is because we set up Example 2 by switching the parametric equations for x and y that we used in Example 1. We say that the relation in Example 2 is the *inverse relation* of the relation in Example 1.

DEFINITION Inverse Relation

The ordered pair (a, b) is in a relation if and only if the ordered pair (b, a) is in the **inverse relation.**

We will study the connection between a relation and its inverse. We will be most interested in inverse relations that happen to be *functions*. Notice that the graph of the inverse relation in Example 2 (Figure 1.64) fails the vertical line test and is therefore not the graph of a function. Can we predict this failure by considering the graph of the original relation in Example 1? Figure 1.65 suggests that we can.

The inverse graph in Figure 1.65b fails the vertical line test because two different y values have been paired with the same x value. This is a direct consequence of the fact that the original relation in Figure 1.65a paired two different x values with the same y value. The inverse graph fails the *vertical* line test precisely because the original graph fails the *horizontal* line test. This gives us a test for relations whose inverses are functions.

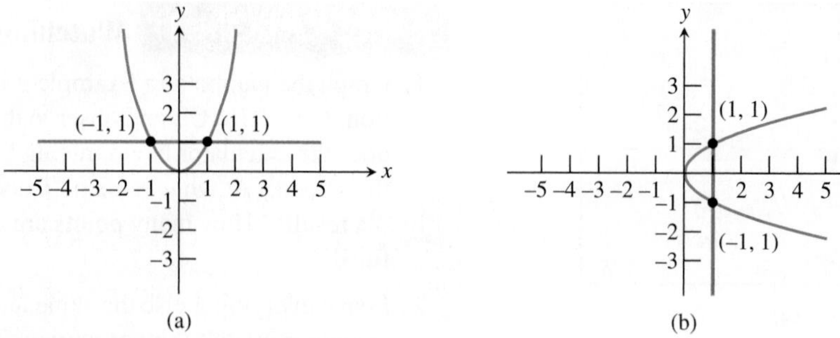

FIGURE 1.65 The inverse relation in (b) fails the vertical line test because the original relation in (a) fails the horizontal line test.

Horizontal Line Test

The inverse of a relation is a function if and only if each horizontal line intersects the graph of the original relation in at most one point.

EXAMPLE 3 Applying the Horizontal Line Test

Which of the graphs (1)–(4) in Figure 1.66 are graphs of

(a) relations that are functions?

(b) relations that have inverses that are functions?

SOLUTION

(a) Graphs (1) and (4) are graphs of functions because these graphs pass the vertical line test. Graphs (2) and (3) are not graphs of functions because these graphs fail the vertical line test.

(b) Graphs (1) and (2) are graphs of relations whose inverses are functions because these graphs pass the horizontal line test. Graphs (3) and (4) fail the horizontal line test so their inverse relations are not functions. **Now try Exercise 9.**

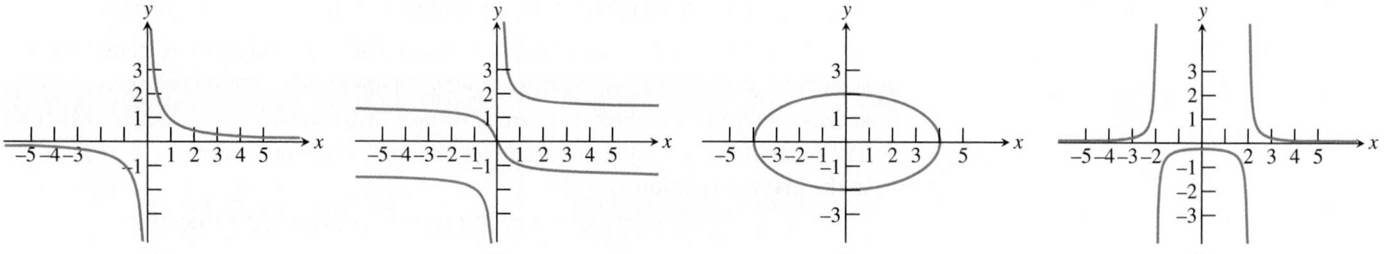

FIGURE 1.66 (Example 3)

A *function* whose inverse is a function has a graph that passes both the horizontal and vertical line tests (such as graph (1) in Example 3). Such a function is **one-to-one**, since every x is paired with a unique y and every y is paired with a unique x.

Caution About Function Notation

The symbol f^{-1} is read "f inverse" and should never be confused with the reciprocal of f. If f is a function, the symbol f^{-1} can *only* mean f inverse. The reciprocal of f must be written as $1/f$.

DEFINITION Inverse Function

If f is a one-to-one function with domain D and range R, then the **inverse function of f**, denoted f^{-1}, is the function with domain R and range D defined by

$$f^{-1}(b) = a \quad \text{if and only if} \quad f(a) = b.$$

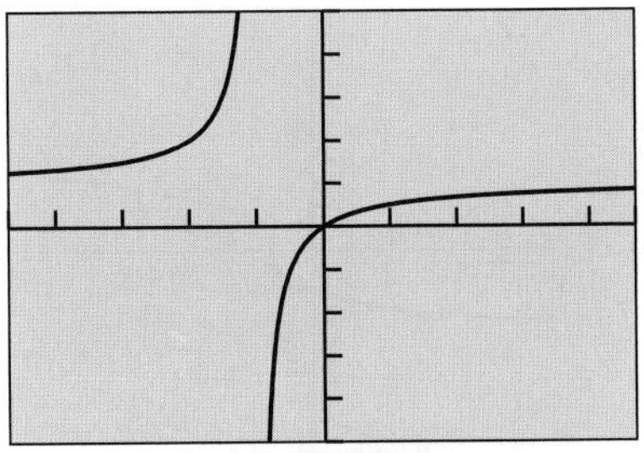

[–4.7, 4.7] by [–5, 5]

FIGURE 1.67 The graph of $f(x) = x/(x + 1)$. (Example 4)

EXAMPLE 4 Finding an Inverse Function Algebraically

Find an equation for $f^{-1}(x)$ if $f(x) = x/(x + 1)$.

SOLUTION The graph of f in Figure 1.67 suggests that f is one-to-one. The original function satisfies the equation $y = x/(x + 1)$. If f truly is one-to-one, the inverse function f^{-1} will satisfy the equation $x = y/(y + 1)$. (Note that we just switch the x and the y.)

If we solve this new equation for y we will have a formula for $f^{-1}(x)$:

$$x = \frac{y}{y + 1}$$

$$x(y + 1) = y \quad \text{Multiply by } y + 1.$$

$$xy + x = y \quad \text{Distributive property}$$

$$xy - y = -x \quad \text{Isolate the } y \text{ terms.}$$

$$y(x - 1) = -x \quad \text{Factor out } y.$$

$$y = \frac{-x}{x - 1} \quad \text{Divide by } x - 1.$$

$$y = \frac{x}{1 - x} \quad \text{Multiply numerator and denominator by } -1.$$

Therefore $f^{-1}(x) = x/(1 - x)$. **Now try Exercise 15.**

Let us candidly admit two things regarding Example 4 before moving on to a graphical model for finding inverses. First, many functions are not one-to-one and so do not have inverse functions. Second, the algebra involved in finding an inverse function in the manner of Example 4 can be extremely difficult. We will actually find very few inverses this way. As you will learn in future chapters, we will usually rely on our understanding of how f maps x to y to understand how f^{-1} maps y to x.

It is possible to use the graph of f to produce a graph of f^{-1} without doing any algebra at all, thanks to the following geometric reflection property:

Inverse Reflection Principle

The points (a, b) and (b, a) in the coordinate plane are symmetric with respect to the line $y = x$. The points (a, b) and (b, a) are **reflections** of each other across the line $y = x$.

EXAMPLE 5 Finding an Inverse Function Graphically

The graph of a function $y = f(x)$ is shown in Figure 1.68. Sketch a graph of the function $y = f^{-1}(x)$. Is f a one-to-one function?

SOLUTION We need not find a formula for $f^{-1}(x)$. All we need to do is to find the reflection of the given graph across the line $y = x$. This can be done geometrically.

Imagine a mirror along the line $y = x$ and draw the reflection of the given graph in the mirror (Figure 1.69).

Another way to visualize this process is to imagine the graph to be drawn on a large pane of glass. Imagine the glass rotating around the line $y = x$ so that the *positive* x-axis switches places with the *positive* y-axis. (The back of the glass must be rotated to the front for this to occur.) The graph of f will then become the graph of f^{-1}.

Since the inverse of f has a graph that passes the horizontal and vertical line test, f is a one-to-one function. **Now try Exercise 23.**

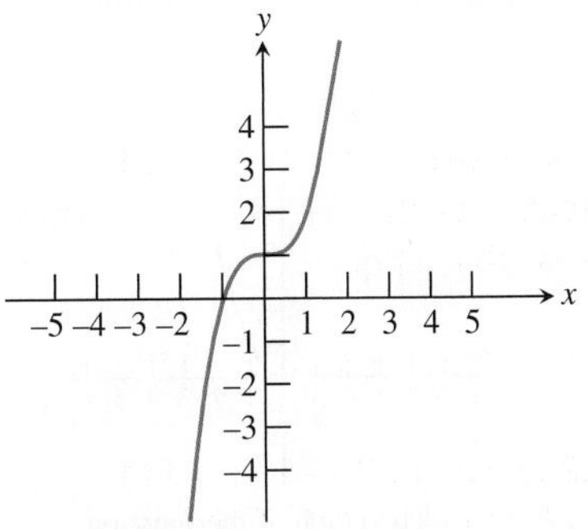

FIGURE 1.68 The graph of a one-to-one function. (Example 5)

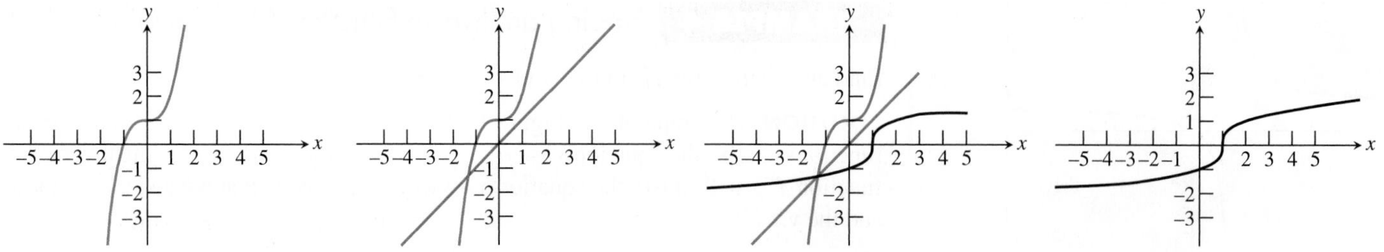

The graph of f | The mirror $y = x$ | The reflection | The graph of f^{-1}

FIGURE 1.69 The mirror method. The graph of f is reflected in an imaginary mirror along the line $y = x$ to produce the graph of f^{-1}. (Example 5)

There is a natural connection between inverses and function composition that gives further insight into what an inverse actually does: It "undoes" the action of the original function. This leads to the following rule:

Inverse Composition Rule

A function f is one-to-one with inverse function g if and only if

$$f(g(x)) = x \text{ for every } x \text{ in the domain of } g, \text{ and}$$
$$g(f(x)) = x \text{ for every } x \text{ in the domain of } f.$$

EXAMPLE 6 Verifying Inverse Functions

Show algebraically that $f(x) = x^3 + 1$ and $g(x) = \sqrt[3]{x - 1}$ are inverse functions.

SOLUTION We use the Inverse Composition Rule.

$$f(g(x)) = f\left(\sqrt[3]{x-1}\right) = \left(\sqrt[3]{x-1}\right)^3 + 1 = x - 1 + 1 = x$$
$$g(f(x)) = g(x^3 + 1) = \sqrt[3]{(x^3 + 1) - 1} = \sqrt[3]{x^3} = x$$

Since these equations are true for all x, the Inverse Composition Rule guarantees that f and g are inverses.

You do not have far to go to find graphical support of this algebraic verification, since these are the functions whose graphs are shown in Example 5!

Now try Exercise 27.

Some functions are so important that we need to study their inverses even though they are not one-to-one. A good example is the square root function, which is the "inverse" of the square function. It is not the inverse of the *entire* squaring function, because the full parabola fails the horizontal line test. Figure 1.70 shows that the function $y = \sqrt{x}$ is really the inverse of a "restricted-domain" version of $y = x^2$ defined only for $x \geq 0$.

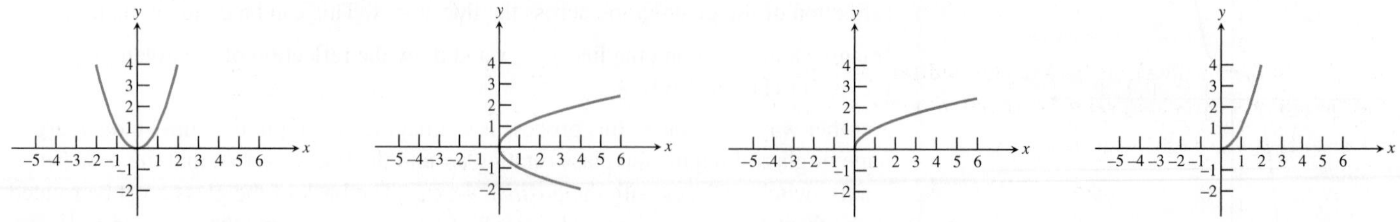

The graph of $y = x^2$ (not one-to-one) | The inverse relation of $y = x^2$ (not a function) | The graph of $y = \sqrt{x}$ (a function) | The graph of the function whose inverse is $y = \sqrt{x}$

FIGURE 1.70 The function $y = x^2$ has no inverse function, but $y = \sqrt{x}$ is the inverse function of $y = x^2$ on the restricted domain $[0, \infty)$.

The consideration of domains adds a refinement to the algebraic inverse-finding method of Example 4, which we now summarize:

How to Find an Inverse Function Algebraically

Given a formula for a function f, proceed as follows to find a formula for f^{-1}.

1. Determine that there is a function f^{-1} by checking that f is one-to-one. State any restrictions on the domain of f. (Note that it might be necessary to impose some to get a one-to-one version of f.)
2. Switch x and y in the formula $y = f(x)$.
3. Solve for y to get the formula $y = f^{-1}(x)$. State any restrictions on the domain of f^{-1}.

EXAMPLE 7 Finding an Inverse Function

Show that $f(x) = \sqrt{x + 3}$ has an inverse function and find a rule for $f^{-1}(x)$. State any restrictions on the domains of f and f^{-1}.

SOLUTION

Solve Algebraically The graph of f passes the horizontal line test, so f has an inverse function (Figure 1.71). Note that f has domain $[-3, \infty)$ and range $[0, \infty)$.

To find f^{-1} we write

$y = \sqrt{x + 3}$	where $x \geq -3, y \geq 0$	
$x = \sqrt{y + 3}$	where $y \geq -3, x \geq 0$	Interchange x and y.
$x^2 = y + 3$	where $y \geq -3, x \geq 0$	Square.
$y = x^2 - 3$	where $y \geq -3, x \geq 0$	Solve for y.

Thus $f^{-1}(x) = x^2 - 3$, with an "inherited" domain restriction of $x \geq 0$. Figure 1.71 shows the two functions. Note the domain restriction of $x \geq 0$ imposed on the parabola $y = x^2 - 3$.

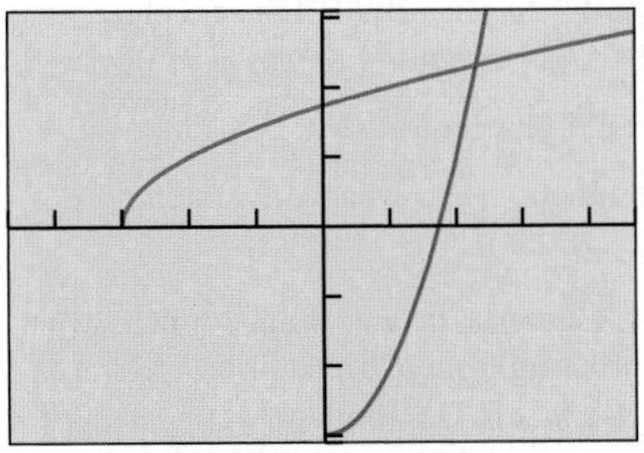

[–4.7, 4.7] by [–3.1, 3.1]

FIGURE 1.71 The graph of $f(x) = \sqrt{x + 3}$ and its inverse, a restricted version of $y = x^2 - 3$. (Example 7)

Support Graphically Use a grapher in parametric mode and compare the graphs of the two sets of parametric equations with Figure 1.71:

$$x = t \qquad\qquad x = \sqrt{t + 3}$$
$$y = \sqrt{t + 3} \quad \text{and} \quad y = t$$

Now try Exercise 17.

QUICK REVIEW 1.5 *(For help, go to Sections P.3 and P.4.)*

Exercise numbers with a gray background indicate problems that the authors have designed to be solved *without a calculator.*

In Exercises 1–10, solve the equation for y.

1. $x = 3y - 6$
2. $x = 0.5y + 1$
3. $x = y^2 + 4$
4. $x = y^2 - 6$
5. $x = \dfrac{y - 2}{y + 3}$
6. $x = \dfrac{3y - 1}{y + 2}$
7. $x = \dfrac{2y + 1}{y - 4}$
8. $x = \dfrac{4y + 3}{3y - 1}$
9. $x = \sqrt{y + 3}, y \geq -3$
10. $x = \sqrt{y - 2}, y \geq 2$

SECTION 1.5 Exercises

In Exercises 1–4, find the (x, y) pair for the value of the parameter.

1. $x = 3t$ and $y = t^2 + 5$ for $t = 2$

2. $x = 5t - 7$ and $y = 17 - 3t$ for $t = -2$

3. $x = t^3 - 4t$ and $y = \sqrt{t + 1}$ for $t = 3$

4. $x = |t + 3|$ and $y = 1/t$ for $t = -8$

In Exercises 5–8, complete the following. **(a)** Find the points determined by $t = -3, -2, -1, 0, 1, 2,$ and 3. **(b)** Find a direct algebraic relationship between x and y and determine whether the parametric equations determine y as a function of x. **(c)** Graph the relationship in the xy-plane.

5. $x = 2t$ and $y = 3t - 1$

6. $x = t + 1$ and $y = t^2 - 2t$

7. $x = t^2$ and $y = t - 2$

8. $x = \sqrt{t}$ and $y = 2t - 5$

In Exercises 9–12, the graph of a relation is shown. **(a)** Is the relation a function? **(b)** Does the relation have an inverse that is a function?

9.

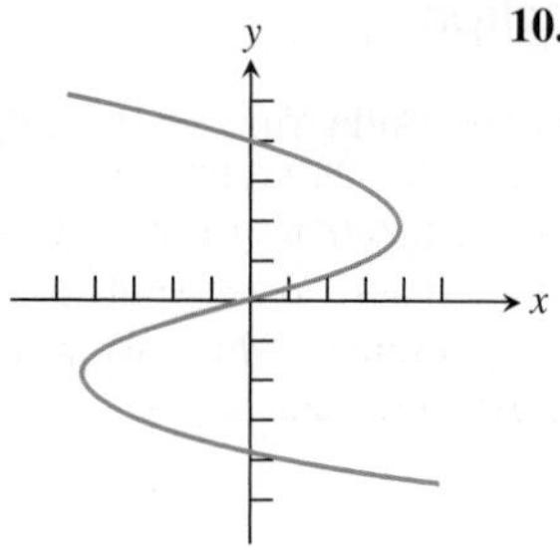

10.

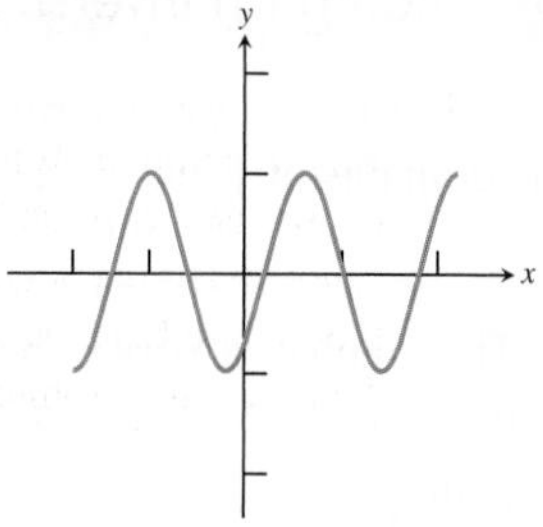

11.

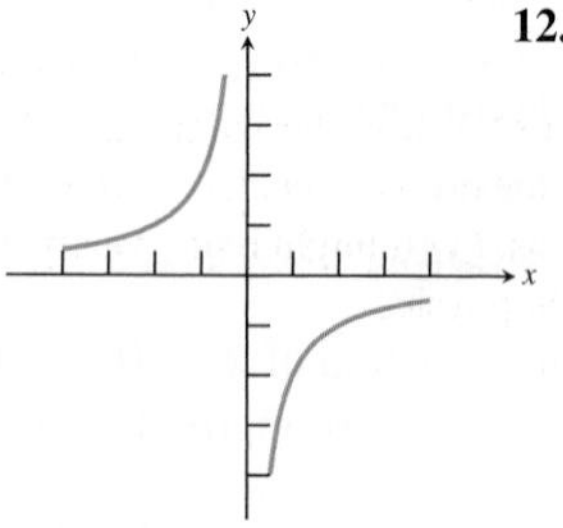

12.

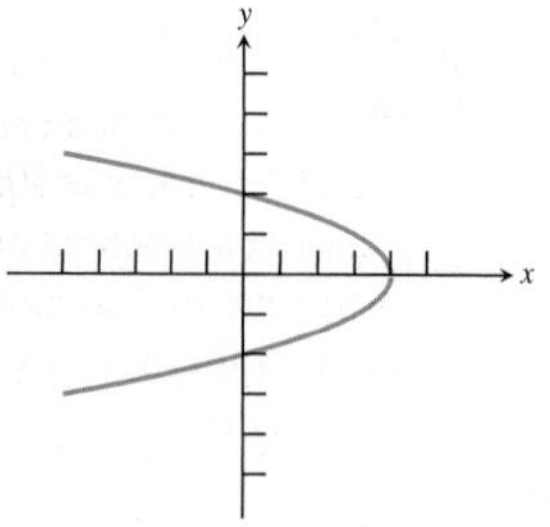

In Exercises 13–22, find a formula for $f^{-1}(x)$. Give the domain of f^{-1}, including any restrictions "inherited" from f.

13. $f(x) = 3x - 6$

14. $f(x) = 2x + 5$

15. $f(x) = \dfrac{2x - 3}{x + 1}$

16. $f(x) = \dfrac{x + 3}{x - 2}$

17. $f(x) = \sqrt{x - 3}$

18. $f(x) = \sqrt{x + 2}$

19. $f(x) = x^3$

20. $f(x) = \sqrt{x^3 + 5}$

21. $f(x) = \sqrt[3]{x + 5}$

22. $f(x) = \sqrt[3]{x - 2}$

In Exercises 23–26, determine whether the function is one-to-one. If it is one-to-one, sketch the graph of the inverse.

23.

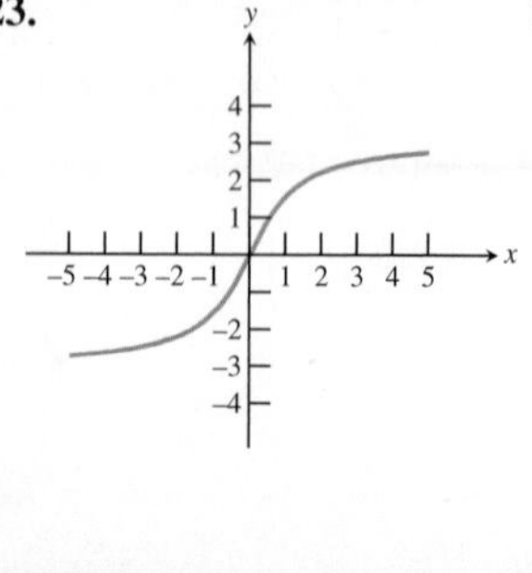

24.

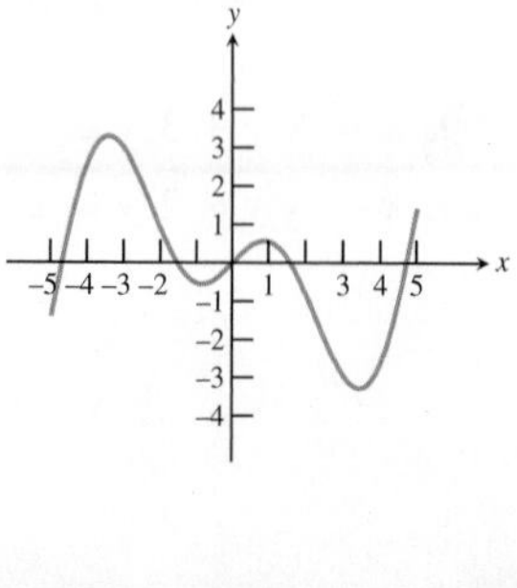

25.

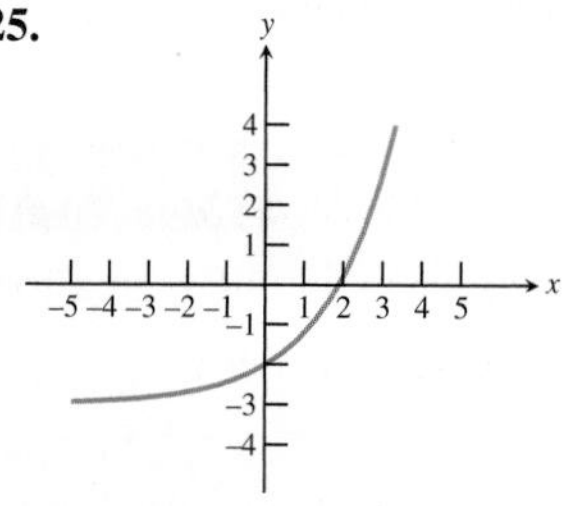

26.

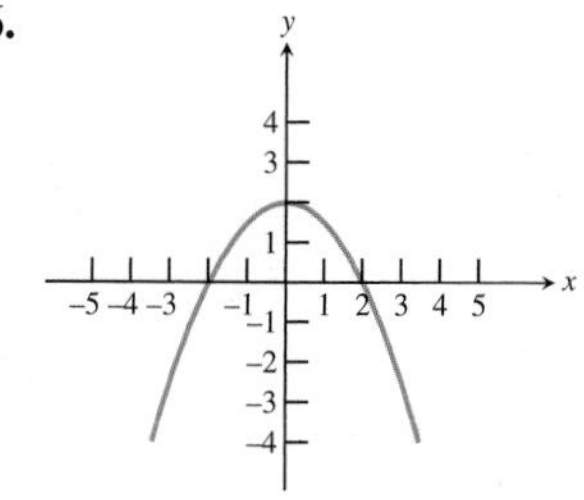

In Exercises 27–32, confirm that f and g are inverses by showing that $f(g(x)) = x$ and $g(f(x)) = x$.

27. $f(x) = 3x - 2$ and $g(x) = \dfrac{x + 2}{3}$

28. $f(x) = \dfrac{x + 3}{4}$ and $g(x) = 4x - 3$

29. $f(x) = x^3 + 1$ and $g(x) = \sqrt[3]{x - 1}$

30. $f(x) = \dfrac{7}{x}$ and $g(x) = \dfrac{7}{x}$

31. $f(x) = \dfrac{x + 1}{x}$ and $g(x) = \dfrac{1}{x - 1}$

32. $f(x) = \dfrac{x + 3}{x - 2}$ and $g(x) = \dfrac{2x + 3}{x - 1}$

33. Currency Conversion In May of 2013 the exchange rate for converting U.S. dollars (x) to euros (y) was $y = 0.76x$.

(a) How many euros could you get for $100 U.S.?

(b) What is the inverse function, and what conversion does it represent?

(c) In the spring of 2013, a tourist had an elegant lunch in Provence, France, ordering from a "fixed price" 48-euro menu. How much was that in U.S. dollars?

34. Temperature Conversion The formula for converting Celsius temperature (x) to Kelvin temperature is $k(x) = x + 273.16$. The formula for converting Fahrenheit temperature (x) to Celsius temperature is $c(x) = (5/9)(x - 32)$.

(a) Find a formula for $c^{-1}(x)$. What is this formula used for?

(b) Find $(k \circ c)(x)$. What is this formula used for?

35. Which pairs of basic functions (Section 1.3) are inverses of each other?

36. Which basic functions (Section 1.3) are their own inverses?

37. Which basic function can be defined parametrically as follows?

$$x = t^3 \text{ and } y = \sqrt{t^6} \text{ for } -\infty < t < \infty$$

38. Which basic function can be defined parametrically as follows?

$$x = 8t^3 \text{ and } y = (2t)^3 \text{ for } -\infty < t < \infty$$

Standardized Test Questions

39. True or False If f is a one-to-one function with domain D and range R, then f^{-1} is a one-to-one function with domain R and range D. Justify your answer.

40. True or False The set of points $(t + 1, 2t + 3)$ for all real numbers t form a line with slope 2. Justify your answer.

In Exercises 41–44, answer the questions without using a calculator.

41. Multiple Choice Which ordered pair is in the *inverse* of the relation given by $x^2y + 5y = 9$?

(A) $(2, 1)$ **(B)** $(-2, 1)$ **(C)** $(-1, 2)$ **(D)** $(2, -1)$

(E) $(1, -2)$

42. Multiple Choice Which ordered pair is not in the *inverse* of the relation given by $xy^2 - 3x = 12$?

(A) $(0, -4)$ **(B)** $(4, 1)$ **(C)** $(3, 2)$ **(D)** $(2, 12)$

(E) $(1, -6)$

43. Multiple Choice Which function is the *inverse* of the function $f(x) = 3x - 2$?

(A) $g(x) = \frac{x}{3} + 2$

(B) $g(x) = 2 - 3x$

(C) $g(x) = \frac{x + 2}{3}$

(D) $g(x) = \frac{x - 3}{2}$

(E) $g(x) = \frac{x - 2}{3}$

44. Multiple Choice Which function is the *inverse* of the function $f(x) = x^3 + 1$?

(A) $g(x) = \sqrt{x - 1}$

(B) $g(x) = \sqrt[3]{x} - 1$

(C) $g(x) = x^3 - 1$

(D) $g(x) = \sqrt[3]{x + 1}$

(E) $g(x) = 1 - x^3$

Explorations

45. Function Properties Inherited by Inverses There are some properties of functions that are automatically shared by inverse functions (when they exist) and some that are not. Suppose that f has an inverse function f^{-1}. Give an algebraic or graphical argument (not a rigorous formal proof) to show that each of these properties of f must necessarily be shared by f^{-1}.

(a) f is continuous.

(b) f is one-to-one.

(c) f is odd (graphically, symmetric with respect to the origin).

(d) f is increasing.

46. Function Properties Not Inherited by Inverses There are some properties of functions that are not necessarily shared by inverse functions, even if the inverses exist. Suppose that f has an inverse function f^{-1}. For each of the following properties, give an example to show that f can have the property while f^{-1} does not.

(a) f has a graph with a horizontal asymptote.

(b) f has domain all real numbers.

(c) f has a graph that is bounded above.

(d) f has a removable discontinuity at $x = 5$.

47. Scaling Algebra Grades A teacher gives a challenging algebra test to her class. The lowest score is 52, which she decides to scale to 70. The highest score is 88, which she decides to scale to 97.

(a) Using the points (52, 70) and (88, 97), find a linear equation that can be used to convert raw scores to scaled grades.

(b) Find the inverse of the function defined by this linear equation. What does the inverse function do?

48. Writing to Learn (Continuation of Exercise 47) Explain why it is important for fairness that the scaling function used by the teacher be an *increasing* function. (*Caution:* It is *not* because "everyone's grade must go up." What would the scaling function in Exercise 47 do for a student who does enough "extra credit" problems to get a raw score of 136?)

Extending the Ideas

49. Modeling a Fly Ball Parametrically A baseball that leaves the bat at an angle of 60° from horizontal traveling 110 ft/sec follows a path that can be modeled by the following pair of parametric equations. (You might enjoy verifying this if you have studied motion in physics.)

$$x = 110(t)\cos(60°)$$

$$y = 110(t)\sin(60°) - 16t^2$$

You can simulate the flight of the ball on a grapher. Set your grapher to parametric mode and put the functions above in for X2T and Y2T. Set X1T = 325 and Y1T = 5T to draw a 30-ft fence 325 ft from home plate. Set Tmin = 0, Tmax = 6, Tstep = 0.1, Xmin = 0, Xmax = 350, Xscl = 0, Ymin = 0, Ymax = 300, and Yscl = 0.

(a) Now graph the function. Does the fly ball clear the fence?

(b) Change the angle to 30° and run the simulation again. Does the ball clear the fence?

(c) What angle is optimal for hitting the ball? Does it clear the fence when hit at that angle?

50. The Baylor GPA Scale Revisited *(See Problem 78 in Section 1.2.)* The function used to convert Baylor School percentage grades to GPAs on a 4-point scale is

$$y = \left(\frac{3^{1.7}}{30}(x - 65)\right)^{\frac{1}{1.7}} + 1.$$

The function has domain $[65, 100]$. Anything below 65 is a failure and automatically converts to a GPA of 0.

(a) Find the inverse function algebraically. What can the inverse function be used for?

(b) Does the inverse function have any domain restrictions?

(c) Verify with a graphing calculator that the function found in (a) and the given function are really inverses.

51. Group Activity (Continuation of Exercise 50) The number 1.7 that appears in two places in the GPA scaling formula is called the scaling factor (k). The value of k can be changed to alter the curvature of the graph while keeping the points $(65, 1)$ and $(95, 4)$ fixed. It was felt that the lowest D (65) needed to be scaled to 1.0, while the middle A (95) needed to be scaled to 4.0. The faculty's Academic Council considered several values of k before settling on 1.7 as the number that gives the "fairest" GPAs for the other percentage grades.

Try changing k to other values between 1 and 2. What kind of scaling curve do you get when $k = 1$? Do you agree with the Baylor decision that $k = 1.7$ gives the fairest GPAs?

1.6 Graphical Transformations

What you'll learn about

- Transformations
- Vertical and Horizontal Translations
- Reflections Across Axes
- Vertical and Horizontal Stretches and Shrinks
- Combining Transformations

... and why

Studying transformations will help you to understand the relationships between graphs that have similarities but are not the same.

Transformations

The following functions are all different:

$$y = x^2$$
$$y = (x - 3)^2$$
$$y = 1 - x^2$$
$$y = x^2 - 4x + 5$$

However, a look at their graphs shows that, while no two are exactly the same, all four have the same identical *shape* and *size*. Understanding how algebraic alterations change the shapes, sizes, positions, and orientations of graphs is helpful for understanding the connection between algebraic and graphical models of functions.

In this section we relate graphs using (geometric) **transformations**, which are functions that map points to points. By acting on the x-coordinates and y-coordinates of points, transformations change graphs in predictable ways. **Rigid transformations**, which leave the size and shape of a graph unchanged, include horizontal translations, vertical translations, reflections, or any combination of these. **Nonrigid transformations**, which generally distort the shape of a graph, include horizontal or vertical stretches and shrinks.

Vertical and Horizontal Translations

A **vertical translation** of the graph of $y = f(x)$ is a shift of the graph up or down in the coordinate plane. A **horizontal translation** is a shift of the graph to the left or the right. The following exploration will give you a good feel for what translations are and how they occur.

EXPLORATION 1 Introducing Translations

Set your viewing window to $[-5, 5]$ by $[-5, 15]$ and your graphing mode to sequential as opposed to simultaneous.

1. Graph the functions

$$y_1 = x^2 \qquad y_4 = y_1(x) - 2 = x^2 - 2$$
$$y_2 = y_1(x) + 3 = x^2 + 3 \qquad y_5 = y_1(x) - 4 = x^2 - 4$$
$$y_3 = y_1(x) + 1 = x^2 + 1$$

on the same screen. What effect do the $+3$, $+1$, -2, and -4 seem to have?

2. Graph the functions

$$y_1 = x^2 \qquad y_4 = y_1(x - 2) = (x - 2)^2$$
$$y_2 = y_1(x + 3) = (x + 3)^2 \qquad y_5 = y_1(x - 4) = (x - 4)^2$$
$$y_3 = y_1(x + 1) = (x + 1)^2$$

on the same screen. What effect do the $+3$, $+1$, -2, and -4 seem to have?

3. Repeat steps 1 and 2 for the functions $y_1 = x^3$, $y_1 = |x|$, and $y_1 = \sqrt{x}$. Do your observations agree with those you made after steps 1 and 2?

Technology Alert

In Exploration 1, the notation $y_1(x + 3)$ means the function y_1, evaluated at $x + 3$. It does not mean multiplication.

In general, *replacing x* by $x - c$ shifts the graph horizontally c units. Similarly, *replacing y* by $y - c$ shifts the graph vertically c units. If c is positive the shift is to the right or up; if c is negative the shift is to the left or down.

This is a nice, consistent rule that unfortunately gets complicated by the fact that the c for a vertical shift rarely shows up being subtracted from y. Instead, it usually shows up on the other side of the equal sign being *added* to $f(x)$. That leads us to the following rule, which only *appears* to be different for horizontal and vertical shifts:

Translations of Graphs

Let c be a positive real number. Then the following transformations result in translations of the graph of $y = f(x)$:

Horizontal Translations

$y = f(x - c)$ a translation to the right by c units

$y = f(x + c)$ a translation to the left by c units

Vertical Translations

$y = f(x) + c$ a translation up by c units

$y = f(x) - c$ a translation down by c units

EXAMPLE 1 Vertical and Horizontal Translations

Describe how the graph of $y = |x|$ can be transformed to the graph of the given equation.

(a) $y = |x| - 4$

(b) $y = |x + 2|$

SOLUTION

(a) The equation is in the form $y = f(x) - 4$, a translation down by 4 units. See Figure 1.72.

(b) The equation is in the form $y = f(x + 2)$, a translation left by 2 units. See Figure 1.73.

Now try Exercise 3.

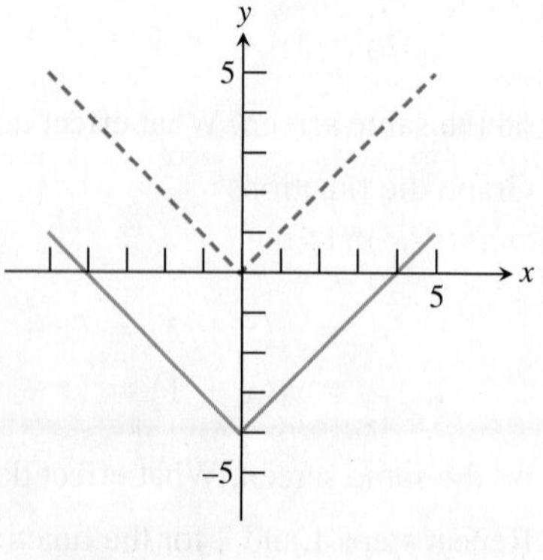

FIGURE 1.72 $y = |x| - 4$. (Example 1)

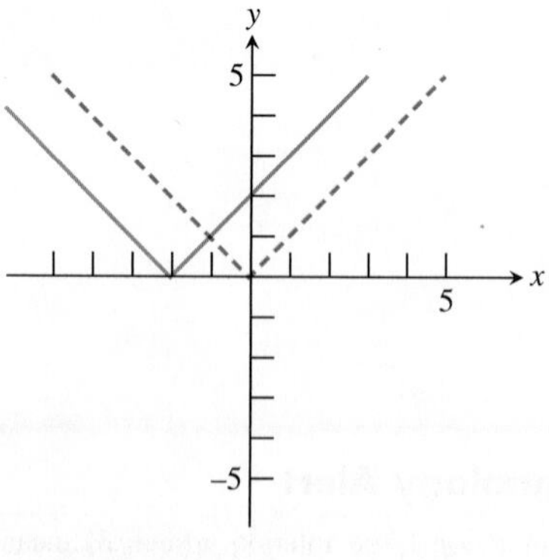

FIGURE 1.73 $y = |x + 2|$. (Example 1)

EXAMPLE 2 Finding Equations for Translations

Each view in Figure 1.74 shows the graph of $y_1 = x^3$ and a vertical or horizontal translation y_2. Write an equation for y_2 as shown in each graph.

SOLUTION

(a) $y_2 = x^3 - 3 = y_1(x) - 3$ (a vertical translation down by 3 units)

(b) $y_2 = (x + 2)^3 = y_1(x + 2)$ (a horizontal translation left by 2 units)

(c) $y_2 = (x - 3)^3 = y_1(x - 3)$ (a horizontal translation right by 3 units)

Now try Exercise 25.

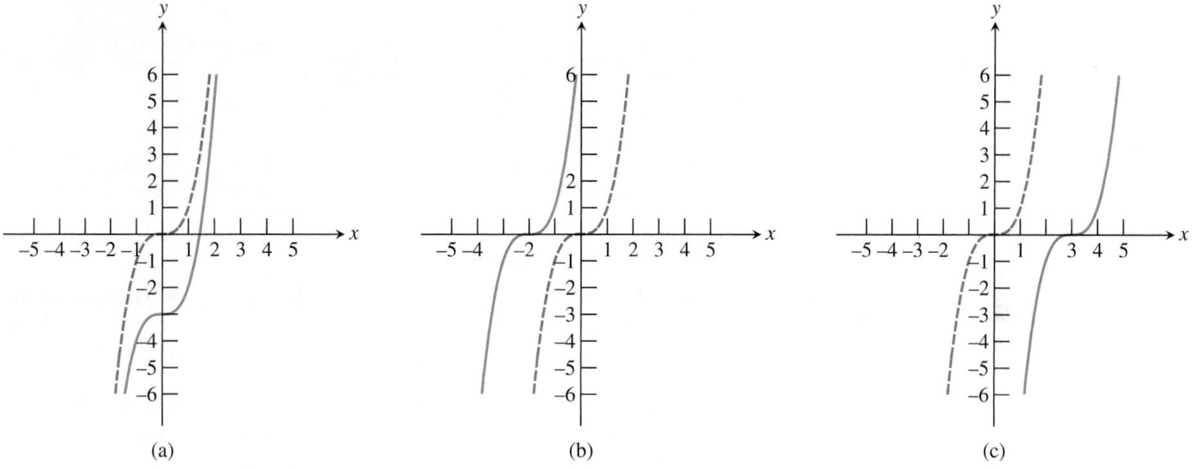

FIGURE 1.74 Translations of $y_1 = x^3$. (Example 2)

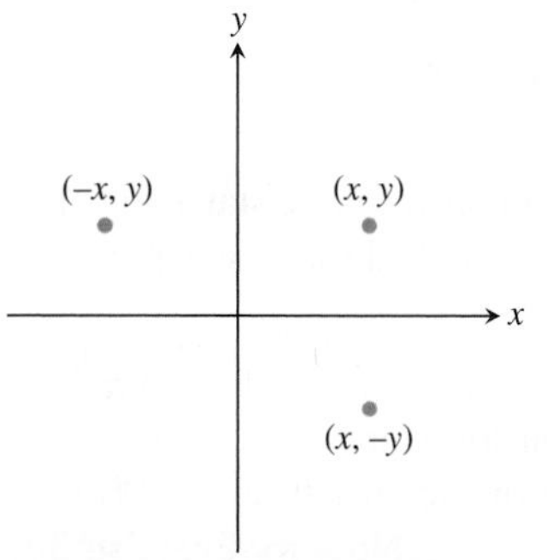

FIGURE 1.75 The point (x, y) and its reflections across the x- and y-axes.

Reflections Across Axes

Points (x, y) and $(x, -y)$ are **reflections of each other across the x-axis**. Points (x, y) and $(-x, y)$ are **reflections of each other across the y-axis**. (See Figure 1.75.) Two points (or graphs) that are symmetric with respect to a line are **reflections of each other across that line**.

Figure 1.75 suggests that a reflection across the x-axis results when y is replaced by $-y$, and a reflection across the y-axis results when x is replaced by $-x$.

Reflections of Graphs

The following transformations result in reflections of the graph of $y = f(x)$:

Across the x-axis
$y = -f(x)$

Across the y-axis
$y = f(-x)$

Through the origin
$y = -f(-x)$

Double Reflection

Note that a reflection through the origin is the result of reflections in both axes, performed in either order.

EXAMPLE 3 Finding Equations for Reflections

Find an equation for the reflection of $f(x) = \dfrac{5x - 9}{x^2 + 3}$ across each axis.

(continued)

SOLUTION

Solve Algebraically

Across the x-axis: $y = -f(x) = -\dfrac{5x - 9}{x^2 + 3} = \dfrac{9 - 5x}{x^2 + 3}$

Across the y-axis: $y = f(-x) = \dfrac{5(-x) - 9}{(-x)^2 + 3} = \dfrac{-5x - 9}{x^2 + 3}$

Support Graphically The graphs in Figure 1.76 support our algebraic work.

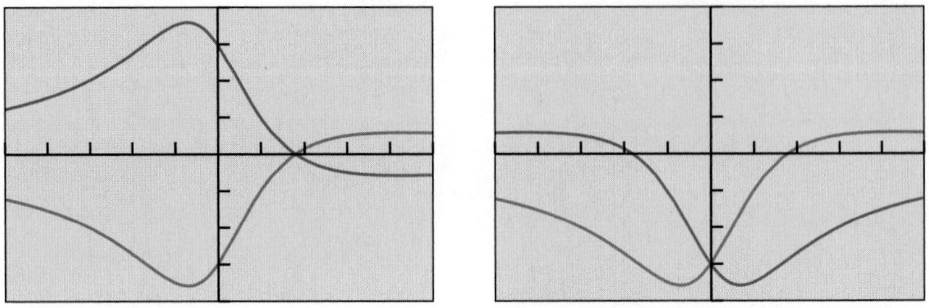

FIGURE 1.76 Reflections of $f(x) = (5x - 9)/(x^2 + 3)$ across (a) the x-axis and (b) the y-axis. (Example 3)

Now try Exercise 29.

You might expect that odd and even functions, whose graphs already possess special symmetries, would exhibit special behavior when reflected across the axes. They do, as shown by Example 4 and Exercises 33 and 34.

EXAMPLE 4 Reflecting Even Functions

Prove that the graph of an even function remains unchanged when it is reflected across the y-axis.

SOLUTION Note that we can get plenty of graphical support for these statements by reflecting the graphs of various even functions, but what is called for here is **proof**, which will require algebra.

Let f be an even function; that is, $f(-x) = f(x)$ for all x in the domain of f. To reflect the graph of $y = f(x)$ across the y-axis, we make the transformation $y = f(-x)$. But $f(-x) = f(x)$ for all x in the domain of f, so this transformation results in $y = f(x)$. The graph of f therefore remains unchanged.

Now try Exercise 33.

Graphing Absolute Value Compositions

Given the graph of $y = f(x)$,

- the graph of $y = |f(x)|$ can be obtained by reflecting only the portion of the graph below the x-axis across the x-axis, leaving the portion above the x-axis unchanged;
- the graph of $y = f(|x|)$ can be obtained by *replacing* the portion of the graph to the left of the y-axis by a reflection of the portion to the right of the y-axis across the y-axis, leaving the portion to the right of the y-axis unchanged. (The result will show even symmetry.)

Function compositions with absolute value can be realized graphically by reflecting portions of graphs, as you will see in the following Exploration.

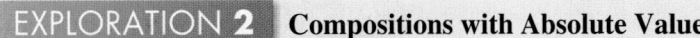

EXPLORATION 2 Compositions with Absolute Value

The graph of $y = f(x)$ is shown at the right. Match each of the graphs below with one of the following equations and use the language of function reflection to defend your match. Note that two of the graphs will not be used.

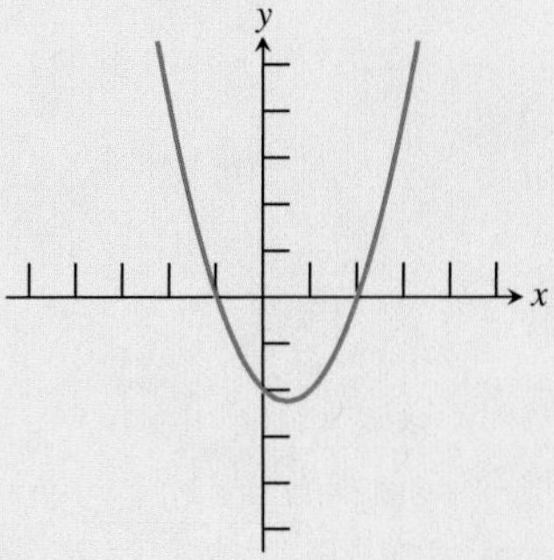

1. $y = |f(x)|$
2. $y = f(|x|)$
3. $y = -|f(x)|$
4. $y = |f(|x|)|$

(A)

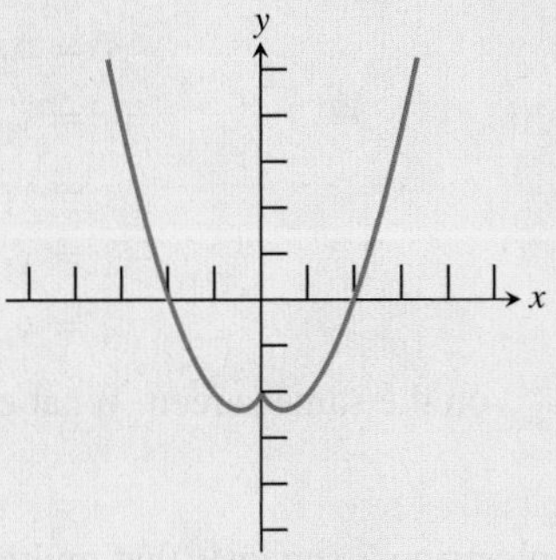

(B)

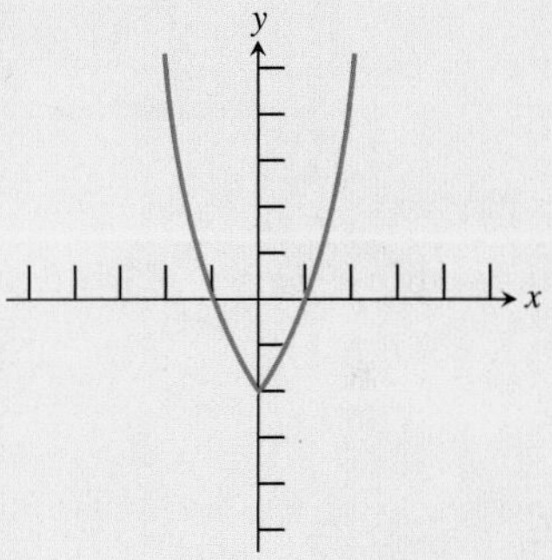

(C)

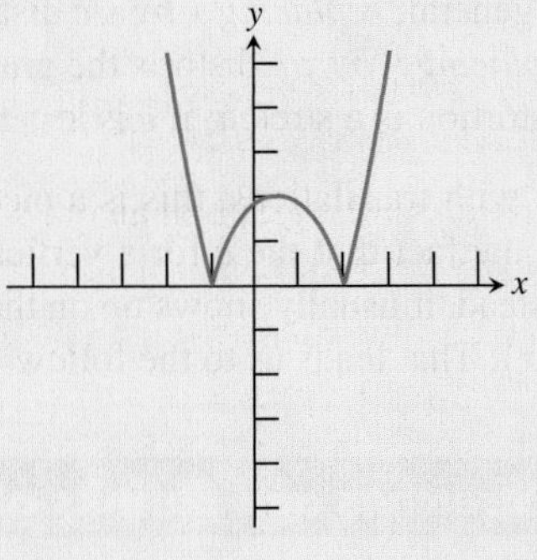

(D)

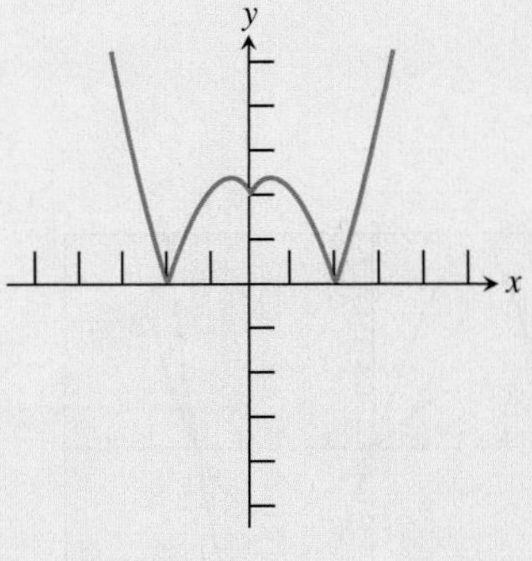

(E)

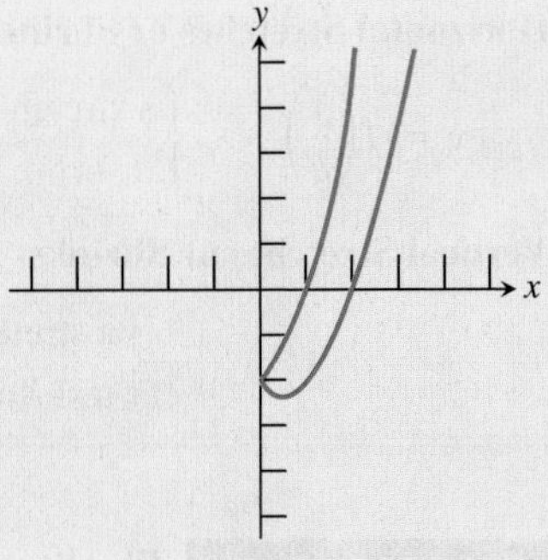

(F)

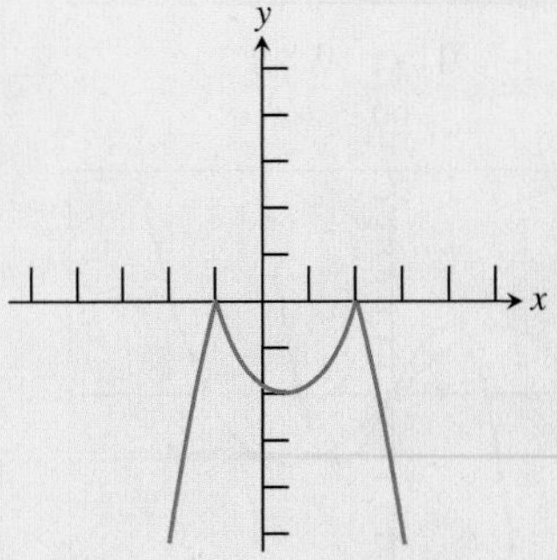

Vertical and Horizontal Stretches and Shrinks

We now investigate what happens when we multiply all the y-coordinates (or all the x-coordinates) of a graph by a fixed real number.

EXPLORATION 3 Introducing Stretches and Shrinks

Set your viewing window to $[-4.7, 4.7]$ by $[-1.1, 5.1]$ and your graphing mode to sequential as opposed to simultaneous.

1. Graph the functions

$$y_1 = \sqrt{4 - x^2}$$
$$y_2 = 1.5y_1(x) = 1.5\sqrt{4 - x^2}$$
$$y_3 = 2y_1(x) = 2\sqrt{4 - x^2}$$
$$y_4 = 0.5y_1(x) = 0.5\sqrt{4 - x^2}$$
$$y_5 = 0.25y_1(x) = 0.25\sqrt{4 - x^2}$$

on the same screen. What effect do the 1.5, 2, 0.5, and 0.25 seem to have?

2. Graph the functions

$$y_1 = \sqrt{4 - x^2}$$
$$y_2 = y_1(1.5x) = \sqrt{4 - (1.5x)^2}$$
$$y_3 = y_1(2x) = \sqrt{4 - (2x)^2}$$
$$y_4 = y_1(0.5x) = \sqrt{4 - (0.5x)^2}$$
$$y_5 = y_1(0.25x) = \sqrt{4 - (0.25x)^2}$$

on the same screen. What effect do the 1.5, 2, 0.5, and 0.25 seem to have?

Exploration 3 suggests that multiplication of x or y by a constant c results in a horizontal or vertical stretching or shrinking of the graph.

In general, *replacing x by x/c* distorts the graph horizontally by a factor of c. Similarly, *replacing y by y/c* distorts the graph vertically by a factor of c. If c is greater than 1 the distortion is a stretch; if c is less than 1 the distortion is a shrink.

As with translations, this is a nice, consistent rule that unfortunately gets complicated by the fact that the c for a vertical stretch or shrink rarely shows up as a divisor of y. Instead, it usually shows up on the other side of the equal sign as a *factor* multiplied by $f(x)$. That leads us to the following rule:

Stretches and Shrinks of Graphs

Let c be a positive real number. Then the following transformations result in stretches or shrinks of the graph of $y = f(x)$:

Horizontal Stretches or Shrinks

$$y = f\left(\frac{x}{c}\right) \quad \begin{cases} \text{a stretch by a factor of } c & \text{if } c > 1 \\ \text{a shrink by a factor of } c & \text{if } c < 1 \end{cases}$$

Vertical Stretches or Shrinks

$$y = c \cdot f(x) \quad \begin{cases} \text{a stretch by a factor of } c & \text{if } c > 1 \\ \text{a shrink by a factor of } c & \text{if } c < 1 \end{cases}$$

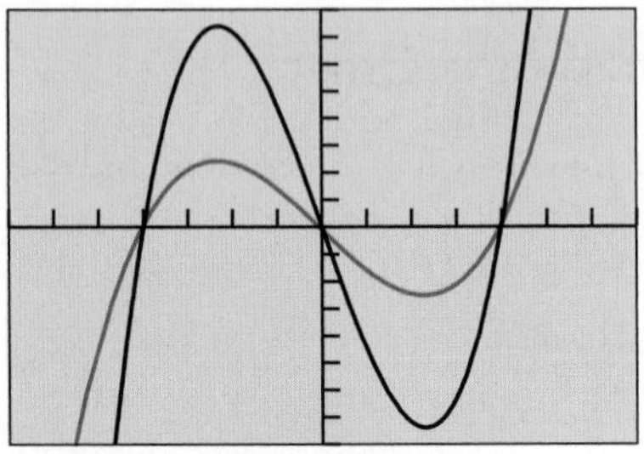

[–7, 7] by [–80, 80]
(a)

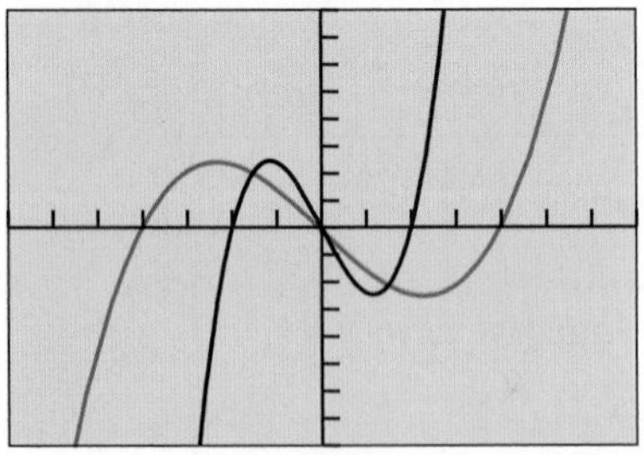

[–7, 7] by [–80, 80]
(b)

FIGURE 1.77 The graph of $y_1 = f(x) = x^3 - 16x$, shown with (a) a vertical stretch and (b) a horizontal shrink. (Example 5)

EXAMPLE 5 Finding Equations for Stretches and Shrinks

Let C_1 be the curve defined by $y_1 = f(x) = x^3 - 16x$. Find equations for the following nonrigid transformations of C_1:

(a) C_2 is a vertical stretch of C_1 by a factor of 3.

(b) C_3 is a horizontal shrink of C_1 by a factor of 1/2.

SOLUTION

Solve Algebraically

(a) Denote the equation for C_2 by y_2. Then

$$\begin{aligned} y_2 &= 3 \cdot f(x) \\ &= 3(x^3 - 16x) \\ &= 3x^3 - 48x \end{aligned}$$

(b) Denote the equation for C_3 by y_3. Then

$$\begin{aligned} y_3 &= f\left(\frac{x}{1/2}\right) \\ &= f(2x) \\ &= (2x)^3 - 16(2x) \\ &= 8x^3 - 32x \end{aligned}$$

Support Graphically The graphs in Figure 1.77 support our algebraic work.

Now try Exercise 39.

Combining Transformations

Transformations may be performed in succession—one after another. If the transformations include stretches, shrinks, or reflections, the order in which the transformations are performed may make a difference. In those cases, be sure to pay particular attention to order.

EXAMPLE 6 Combining Transformations in Order

(a) The graph of $y = x^2$ undergoes the following transformations, in order. Find the equation of the graph that results.

- a horizontal shift 2 units to the right
- a vertical stretch by a factor of 3
- a vertical translation 5 units up

(b) Apply the transformations in (a) in the opposite order and find the equation of the graph that results.

SOLUTION

(a) Applying the transformations in order, we have

$$x^2 \Rightarrow (x - 2)^2 \Rightarrow 3(x - 2)^2 \Rightarrow 3(x - 2)^2 + 5.$$

Expanding the final expression, we get the function $y = 3x^2 - 12x + 17$.

(b) Applying the transformations in the opposite order, we have

$$x^2 \Rightarrow x^2 + 5 \Rightarrow 3(x^2 + 5) \Rightarrow 3((x - 2)^2 + 5).$$

Expanding the final expression, we get the function $y = 3x^2 - 12x + 27$.

The second graph is ten units higher than the first graph because the vertical stretch lengthens the vertical translation when the translation occurs first. Order often matters when stretches, shrinks, or reflections are involved.

Now try Exercise 47.

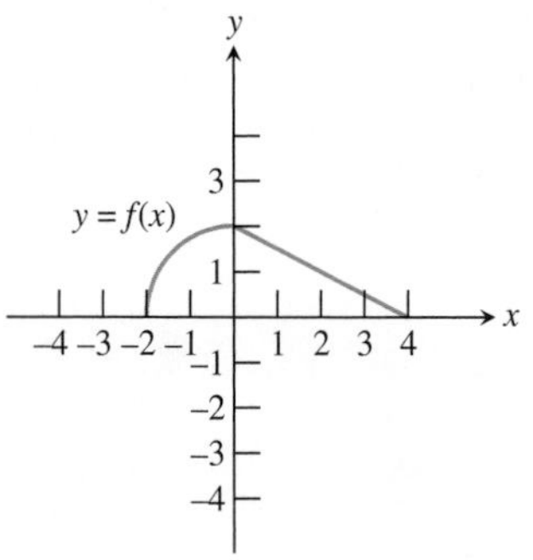

FIGURE 1.78 The graph of the function $y = f(x)$ in Example 7.

EXAMPLE 7 Transforming a Graph Geometrically

The graph of $y = f(x)$ is shown in Figure 1.78. Determine the graph of the composite function $y = 2f(x + 1) - 3$ by showing the effect of a sequence of transformations on the graph of $y = f(x)$.

SOLUTION

The graph of $y = 2f(x + 1) - 3$ can be obtained from the graph of $y = f(x)$ by the following sequence of transformations:

(a) a vertical stretch by a factor of 2 to get $y = 2f(x)$ (Figure 1.79a)

(b) a horizontal translation 1 unit to the left to get $y = 2f(x + 1)$ (Figure 1.79b)

(c) a vertical translation 3 units down to get $y = 2f(x + 1) - 3$ (Figure 1.79c)

(The order of the first two transformations can be reversed without changing the final graph.)

Now try Exercise 51.

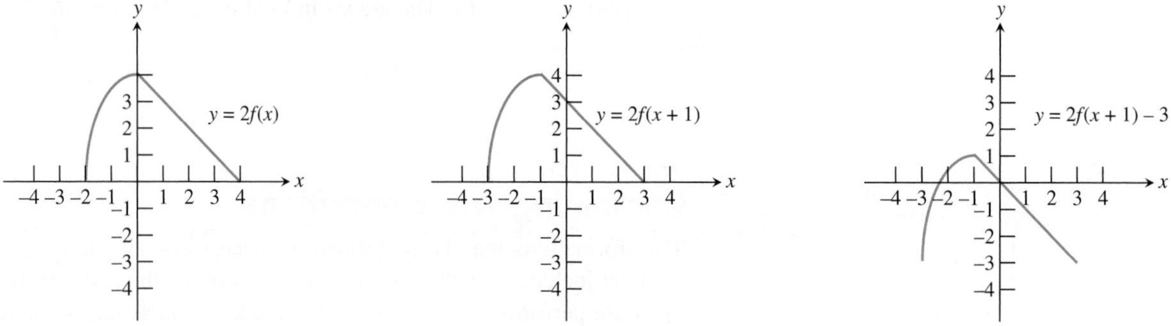

Vertical stretch of factor 2 (a)

Horizontal translation left 1 unit (b)

Vertical translation down 3 units (c)

FIGURE 1.79 Transforming the graph of $y = f(x)$ in Figure 1.78 to get the graph of $y = 2f(x + 1) - 3$. (Example 7)

QUICK REVIEW 1.6 *(For help, go to Section A.2.)*

Exercise numbers with a gray background indicate problems that the authors have designed to be solved *without a calculator*.

In Exercises 1–6, write the expression as a binomial squared.

1. $x^2 + 2x + 1$

2. $x^2 - 6x + 9$

3. $x^2 + 12x + 36$

4. $4x^2 + 4x + 1$

5. $x^2 - 5x + \dfrac{25}{4}$

6. $4x^2 - 20x + 25$

In Exercises 7–10, perform the indicated operations and simplify.

7. $(x - 2)^2 + 3(x - 2) + 4$

8. $2(x + 3)^2 - 5(x + 3) - 2$

9. $(x - 1)^3 + 3(x - 1)^2 - 3(x - 1)$

10. $2(x + 1)^3 - 6(x + 1)^2 + 6(x + 1) - 2$

SECTION 1.6 Exercises

In Exercises 1–8, describe how the graph of $y = x^2$ can be transformed to the graph of the given equation.

1. $y = x^2 - 3$

2. $y = x^2 + 5.2$

3. $y = (x + 4)^2$

4. $y = (x - 3)^2$

5. $y = (100 - x)^2$

6. $y = x^2 - 100$

7. $y = (x - 1)^2 + 3$

8. $y = (x + 50)^2 - 279$

In Exercises 9–12, describe how the graph of $y = \sqrt{x}$ can be transformed to the graph of the given equation.

9. $y = -\sqrt{x}$

10. $y = \sqrt{x - 5}$

11. $y = \sqrt{-x}$

12. $y = \sqrt{3 - x}$

In Exercises 13–16, describe how the graph of $y = x^3$ can be transformed to the graph of the given equation.

13. $y = 2x^3$

14. $y = (2x)^3$

15. $y = (0.2x)^3$

16. $y = 0.3x^3$

In Exercises 17–20, describe how to transform the graph of f into the graph of g.

17. $f(x) = \sqrt{x + 2}$ and $g(x) = \sqrt{x - 4}$

18. $f(x) = (x - 1)^2$ and $g(x) = -(x + 3)^2$

19. $f(x) = (x - 2)^3$ and $g(x) = -(x + 2)^3$

20. $f(x) = |2x|$ and $g(x) = 4|x|$

In Exercises 21–24, sketch the graphs of f, g, and h by hand. Support your answers with a grapher.

21. $f(x) = (x + 2)^2$
$g(x) = 3x^2 - 2$
$h(x) = -2(x - 3)^2$

22. $f(x) = x^3 - 2$
$g(x) = (x + 4)^3 - 1$
$h(x) = 2(x - 1)^3$

23. $f(x) = \sqrt[3]{x + 1}$
$g(x) = 2\sqrt[3]{x} - 2$
$h(x) = -\sqrt[3]{x - 3}$

24. $f(x) = -2|x| - 3$
$g(x) = 3|x + 5| + 4$
$h(x) = |3x|$

In Exercises 25–28, the graph is that of a function $y = f(x)$ that can be obtained by transforming the graph of $y = \sqrt{x}$. Write a formula for the function f.

25.

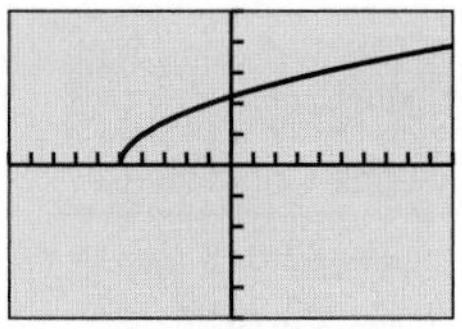

[–10, 10] by [–5, 5]

26.

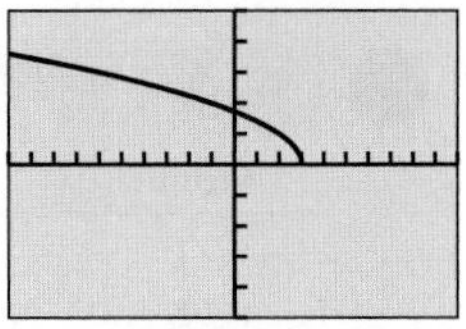

[–10, 10] by [–5, 5]

27.

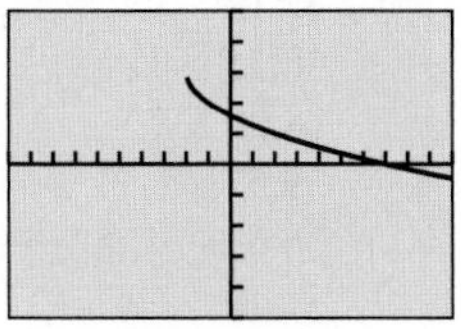

[–10, 10] by [–5, 5]

28.

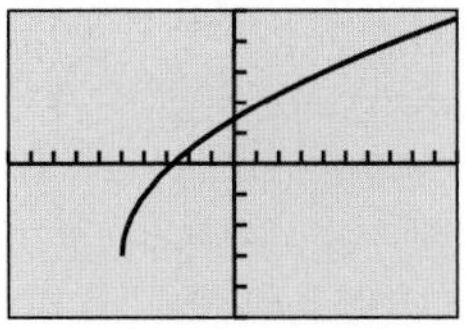

[–10, 10] by [–5, 5]
Vertical stretch = 2

In Exercises 29–32, find the equation of the reflection of f across **(a)** the x-axis and **(b)** the y-axis.

29. $f(x) = x^3 - 5x^2 - 3x + 2$

30. $f(x) = 2\sqrt{x + 3} - 4$

31. $f(x) = \sqrt[3]{8x}$

32. $f(x) = 3|x + 5|$

33. Reflecting Odd Functions Prove that the graph of an odd function is the same when reflected across the x-axis as it is when reflected across the y-axis.

34. Reflecting Odd Functions Prove that if an odd function is reflected about the y-axis and then reflected again about the x-axis, the result is the original function.

Exercises 35–38 refer to the graph of $y = f(x)$ shown at the top of the next column. In each case, sketch a graph of the new function.

35. $y = |f(x)|$

36. $y = f(|x|)$

37. $y = -f(|x|)$

38. $y = |f(|x|)|$

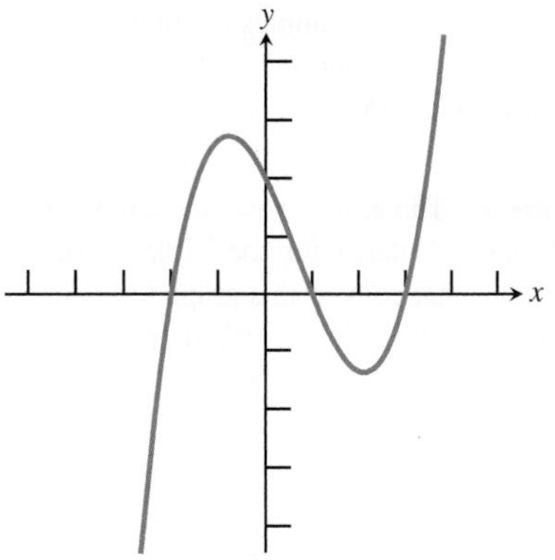

In Exercises 39–42, transform the given function by **(a)** a vertical stretch by a factor of 2, and **(b)** a horizontal shrink by a factor of 1/3.

39. $f(x) = x^3 - 4x$

40. $f(x) = |x + 2|$

41. $f(x) = x^2 + x - 2$

42. $f(x) = \dfrac{1}{x + 2}$

In Exercises 43–46, describe a basic graph and a sequence of transformations that can be used to produce a graph of the given function.

43. $y = 2(x - 3)^2 - 4$

44. $y = -3\sqrt{x + 1}$

45. $y = (3x)^2 - 4$

46. $y = -2|x + 4| + 1$

In Exercises 47–50, a graph G is obtained from a graph of y by the sequence of transformations indicated. Write an equation whose graph is G.

47. $y = x^2$: a vertical stretch by a factor of 3, then a shift right 4 units.

48. $y = x^2$: a shift right 4 units, then a vertical stretch by a factor of 3.

49. $y = |x|$: a shift left 2 units, then a vertical stretch by a factor of 2, and finally a shift down 4 units.

50. $y = |x|$: a shift left 2 units, then a horizontal shrink by a factor of 1/2, and finally a shift down 4 units.

Exercises 51–54 refer to the function f whose graph is shown below.

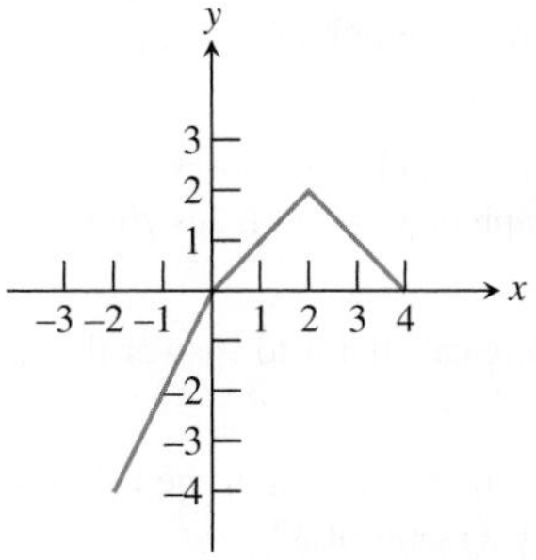

51. Sketch the graph of $y = 2 + 3f(x + 1)$.

52. Sketch the graph of $y = -f(x + 1) + 1$.

53. Sketch the graph of $y = f(2x)$.

54. Sketch the graph of $y = 2f(x - 1) + 2$.

55. Writing to Learn Graph some examples to convince yourself that a reflection and a translation can have a different effect when combined in one order than when combined in the opposite order. Then explain in your own words why this can happen.

56. Writing to Learn Graph some examples to convince yourself that vertical stretches and shrinks do not affect a graph's x-intercepts. Then explain in your own words why this is so.

57. Celsius vs. Fahrenheit The graph shows the temperature in degrees Celsius in Windsor, Ontario, for one 24-hr period. Describe the transformations that convert this graph to one showing degrees Fahrenheit. [*Hint*: $F(t) = (9/5)C(t) + 32$.]

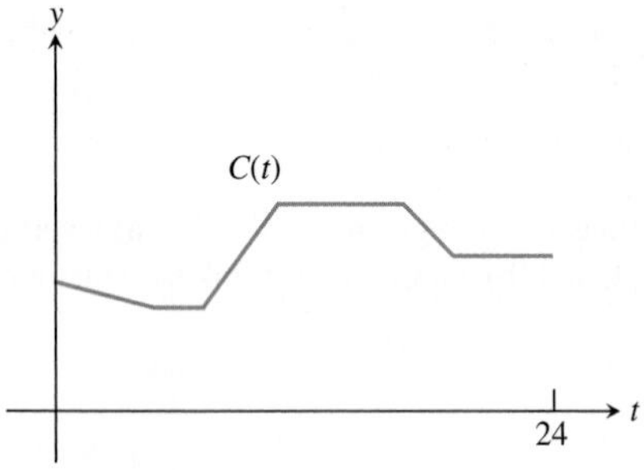

58. Fahrenheit vs. Celsius The graph shows the temperature in degrees Fahrenheit in Mt. Clemens, Michigan, for one 24-hr period. Describe the transformations that convert this graph to one showing degrees Celsius. [*Hint*: $F(t) = (9/5)C(t) + 32$.]

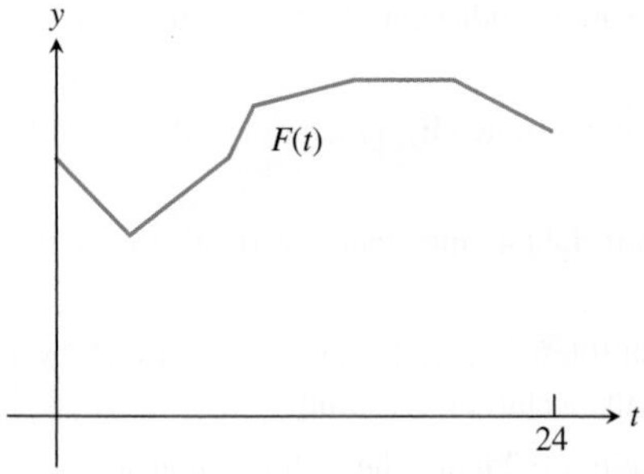

Standardized Test Questions

59. True or False The function $y = f(x + 3)$ represents a translation to the right by 3 units of the graph of $y = f(x)$. Justify your answer.

60. True or False The function $y = f(x) - 4$ represents a translation down 4 units of the graph of $y = f(x)$. Justify your answer.

In Exercises 61–64, you may use a graphing calculator to answer the question.

61. Multiple Choice Given a function f, which of the following represents a vertical stretch by a factor of 3?

(A) $y = f(3x)$ **(B)** $y = f(x/3)$

(C) $y = 3f(x)$ **(D)** $y = f(x)/3$

(E) $y = f(x) + 3$

62. Multiple Choice Given a function f, which of the following represents a horizontal translation of 4 units to the right?

(A) $y = f(x) + 4$ **(B)** $y = f(x) - 4$

(C) $y = f(x + 4)$ **(D)** $y = f(x - 4)$

(E) $y = 4f(x)$

63. Multiple Choice Given a function f, which of the following represents a vertical translation of 2 units upward, followed by a reflection across the y-axis?

(A) $y = f(-x) + 2$ **(B)** $y = 2 - f(x)$

(C) $y = f(2 - x)$ **(D)** $y = -f(x - 2)$

(E) $y = f(x) - 2$

64. Multiple Choice Given a function f, which of the following represents reflection across the x-axis, followed by a horizontal shrink by a factor of 1/2?

(A) $y = -2f(x)$ **(B)** $y = -f(x)/2$

(C) $y = f(-2x)$ **(D)** $y = -f(x/2)$

(E) $y = -f(2x)$

Explorations

65. International Finance Table 1.11 shows the (adjusted closing) price of a share of stock in the Coca-Cola company for the first trading day of each month of 2012.

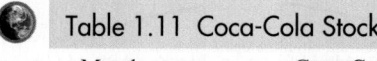

Table 1.11 Coca-Cola Stock

Month	Coca-Cola Price
1	32.61
2	33.73
3	36.00
4	37.12
5	36.35
6	38.29
7	39.57
8	36.63
9	37.40
10	36.66
11	37.65
12	37.39

Source: Yahoo! Finance.

(a) Graph price (y) as a function of month (x) as a line graph, connecting the points to make a continuous graph.

(b) Explain what transformation you would apply to this graph to produce a graph showing the price of the stock in Japanese yen.

66. Group Activity Get with a friend and graph the function $y = x^2$ on both your graphers. Apply a horizontal or vertical stretch or shrink to the function on one of the graphers. Then change the *window* of that grapher to make the two graphs look the same. Can you formulate a general rule for how to find the window?

Extending the Ideas

67. The Absolute Value Transformation Graph the function $f(x) = x^4 - 5x^3 + 4x^2 + 3x + 2$ in the viewing window $[-5, 5]$ by $[-10, 10]$. (Put the equation in Y1.)

(a) Study the graph and try to predict what the graph of $y = |f(x)|$ will look like. Then turn Y1 off and graph Y2 = abs (Y1). Did you predict correctly?

(b) Study the original graph again and try to predict what the graph of $y = f(|x|)$ will look like. Then turn Y1 off and graph Y2 = Y1(abs (X)). Did you predict correctly?

(c) Given the graph of $y = g(x)$ shown below, sketch a graph of $y = |g(x)|$.

(d) Given the graph of $y = g(x)$ shown below, sketch a graph of $y = g(|x|)$.

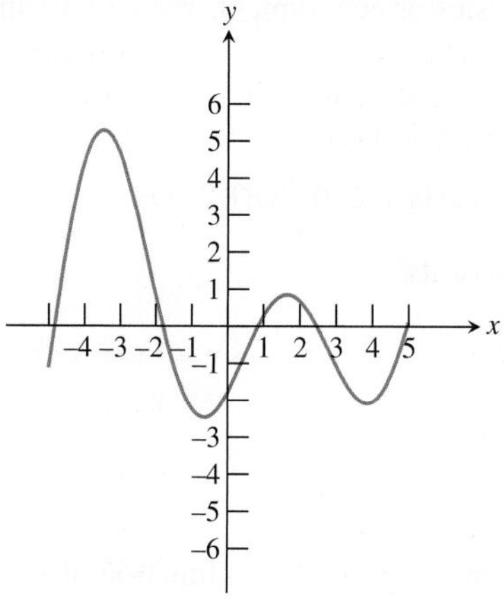

68. Parametric Circles and Ellipses Set your grapher to parametric and radian mode and your window as follows:

Tmin = 0, Tmax = 7, Tstep = 0.1

Xmin = −4.7, Xmax = 4.7, Xscl = 1

Ymin = −3.1, Ymax = 3.1, Yscl = 1

(a) Graph the parametric equations $x = \cos t$ and $y = \sin t$. You should get a circle of radius 1.

(b) Use a transformation of the parametric function of x to produce the graph of an ellipse that is 4 units wide and 2 units tall.

(c) Use a transformation of both parametric functions to produce a circle of radius 3.

(d) Use a transformation of both functions to produce an ellipse that is 8 units wide and 4 units tall.

(You will learn more about ellipses in Chapter 8.)

1.7 Modeling with Functions

What you'll learn about

- Functions from Formulas
- Functions from Graphs
- Functions from Verbal Descriptions
- Functions from Data

... and why

Using a function to model a variable under observation in terms of another variable often allows one to make predictions in practical situations, such as predicting the future growth of a business based on known data.

Functions from Formulas

Now that you have learned more about what functions are and how they behave, we want to return to the modeling theme of Section 1.1. In that section we stressed that one of the goals of this course was to become adept at using numerical, algebraic, and graphical models of the real world in order to solve problems. We now want to focus your attention more precisely on modeling with *functions.*

You have already seen quite a few formulas in the course of your education. Formulas involving two variable quantities always relate those variables implicitly, and quite often the formulas can be solved to give one variable explicitly as a function of the other. In this book we will use a variety of formulas to pose and solve problems algebraically, although we will not assume prior familiarity with those formulas that we borrow from other subject areas (like physics or economics). We *will* assume familiarity with certain key formulas from mathematics.

EXAMPLE 1 Forming Functions from Formulas

Write the area A of a circle as a function of its

(a) radius r.

(b) diameter d.

(c) circumference C.

SOLUTION

(a) The familiar area formula from geometry gives A as a function of r:

$$A = \pi r^2.$$

(b) This formula is not so familiar. However, we know that $r = d/2$, so we can substitute that expression for r in the area formula:

$$A = \pi r^2 = \pi(d/2)^2 = (\pi/4)d^2.$$

(c) Since $C = 2\pi r$, we can solve for r to get $r = C/(2\pi)$. Then substitute to get A: $A = \pi r^2 = \pi(C/(2\pi))^2 = \pi C^2/(4\pi^2) = C^2/(4\pi)$. **Now try Exercise 19.**

EXAMPLE 2 A Maximum Value Problem

A square of side x inches is cut out of each corner of an 8 in. by 15 in. piece of cardboard and the sides are folded up to form an open-topped box (Figure 1.80).

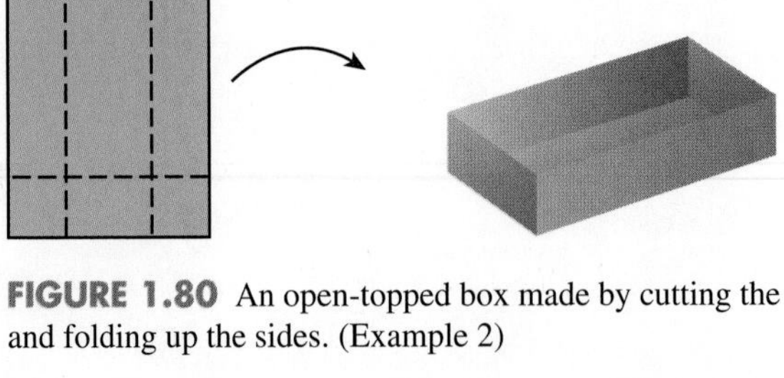

FIGURE 1.80 An open-topped box made by cutting the corners from a piece of cardboard and folding up the sides. (Example 2)

(a) Write the volume V of the box as a function of x.

(b) Find the domain of V as a function of x. (Note that the model imposes restrictions on x.)

(c) Graph V as a function of x over the domain found in part (b) and use the maximum finder on your grapher to determine the maximum volume such a box can hold.

(d) How big should the cut-out squares be in order to produce the box of maximum volume?

SOLUTION

(a) The box will have a base with sides of width $8 - 2x$ and length $15 - 2x$. The depth of the box will be x when the sides are folded up. Therefore $V = x(8 - 2x)(15 - 2x)$.

(b) The formula for V is a polynomial with domain $(-\infty, \infty)$. However, the depth x must be nonnegative, as must the width of the base, $8 - 2x$. Together, these two restrictions yield a domain of $[0, 4]$. (The endpoints give a box with no volume, which is as mathematically feasible as other zero concepts.)

(c) The graph is shown in Figure 1.81. The maximum finder shows that the maximum occurs at the point $(5/3, 90.74)$. The maximum volume is about 90.74 in.^{3}.

(d) Each square should have sides of 5/3 in.

Now try Exercise 33.

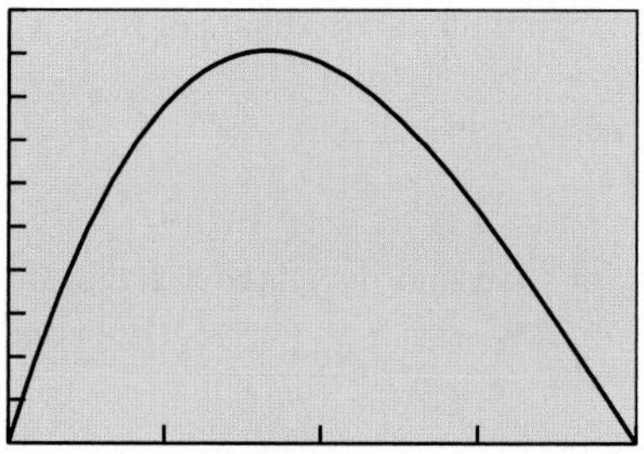

[0, 4] by [0, 100]
(a)

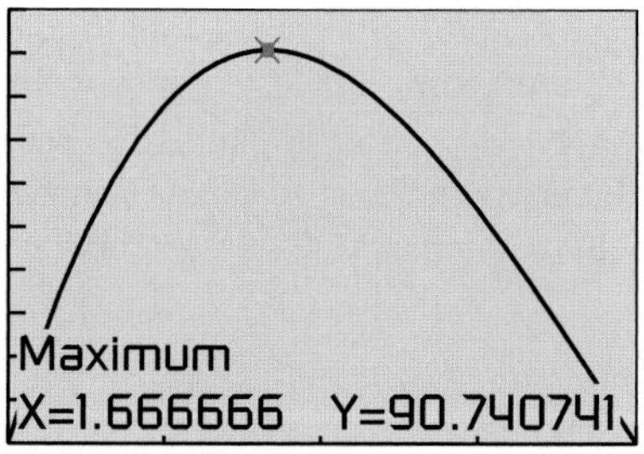

[0, 4] by [0, 100]
(b)

FIGURE 1.81 The graph of the volume of the box in Example 2.

Functions from Graphs

When "thinking graphically" becomes a genuine part of your problem-solving strategy, it is sometimes actually easier to start with the graphical model than it is to go straight to the algebraic formula. The graph provides valuable information about the function.

EXAMPLE 3 Protecting an Antenna

A small satellite dish is packaged with a cardboard cylinder for protection. The parabolic dish is 24 in. in diameter and 6 in. deep, and the diameter of the cardboard cylinder is 12 in. How tall must the cylinder be to fit in the middle of the dish and be flush with the top of the dish? (See Figure 1.82.)

SOLUTION

Solve Algebraically The diagram in Figure 1.82a showing the cross section of this 3-dimensional problem is also a 2-dimensional graph of a quadratic function. We can transform our basic function $y = x^2$ with a vertical shrink so that it goes through the points $(12, 6)$ and $(-12, 6)$, thereby producing a graph of the parabola in the coordinate plane (Figure 1.82b).

$$y = kx^2 \quad \text{Vertical shrink}$$
$$6 = k(\pm 12)^2 \quad \text{Substitute } x = \pm 12,\ y = 6.$$
$$k = \frac{6}{144} = \frac{1}{24} \quad \text{Solve for } k.$$

Thus, $y = \frac{1}{24}x^2$.

(continued)

To find the height of the cardboard cylinder, we first find the y-coordinate of the parabola 6 in. from the center, that is, when $x = 6$:

$$y = \frac{1}{24}(6)^2 = 1.5.$$

From that point to the top of the dish is $6 - 1.5 = 4.5$ in. **Now try Exercise 35.**

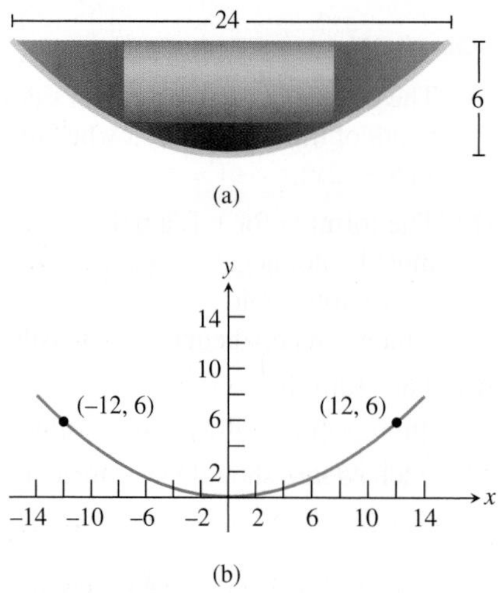

FIGURE 1.82 (a) A parabolic satellite dish with a protective cardboard cylinder in the middle for packaging. (b) The parabola in the coordinate plane. (Example 3)

Although Example 3 serves nicely as a "functions from graphs" example, it is also an example of a function that must be constructed by gathering relevant information from a verbal description and putting it together in the right way. People who do mathematics for a living are accustomed to confronting that challenge regularly as a necessary first step in modeling the real world. In honor of its importance, we have saved it until last to close out this chapter in style.

Functions from Verbal Descriptions

There is no fail-safe way to form a function from a verbal description. It can be hard work, frequently a good deal harder than the mathematics required to solve the problem once the function has been found. The 4-step problem-solving process in Section 1.1 gives you several valuable tips, perhaps the most important of which is to *read* the problem carefully. Understanding what the words say is critical if you hope to model the situation they describe.

EXAMPLE 4 Finding the Model and Solving

Grain is leaking through a hole in a storage bin at a constant rate of 8 cubic inches per minute. The grain forms a cone-shaped pile on the ground below. As it grows, the height of the cone always remains equal to its radius. If the cone is 1 ft tall now, how tall will it be in 1 hr?

SOLUTION

Reading the problem carefully, we realize that the formula for the volume of the cone is needed (Figure 1.83). From memory or by looking it up, we get the formula $V = (1/3)\pi r^2 h$. A careful reading also reveals that the height and the radius are always equal, so we can get volume directly as a function of height: $V = (1/3)\pi h^3$. When $h = 12$ in., the volume is $V = (\pi/3)(12)^3 = 576\pi$ in.^{3}.

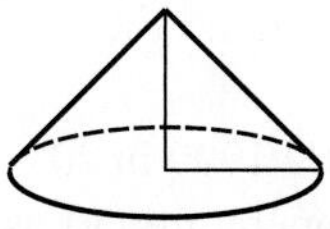

FIGURE 1.83 A cone with equal height and radius. (Example 4)

One hour later, the volume will have grown by $(60 \text{ min})(8 \text{ in.}^3/\text{min}) = 480 \text{ in.}^3$. The total volume of the pile at that point will be $(576\pi + 480) \text{ in.}^3$. Finally, we use the volume formula once again to solve for h:

$$\frac{1}{3}\pi h^3 = 576\pi + 480$$

$$h^3 = \frac{3(576\pi + 480)}{\pi}$$

$$h = \sqrt[3]{\frac{3(576\pi + 480)}{\pi}}$$

$$h \approx 12.98 \text{ in.}$$

Now try Exercise 37.

EXAMPLE 5 Letting Units Work for You

How many rotations does a 15-in. (radius) tire make per second on a sport utility vehicle traveling 70 mph?

SOLUTION It is the perimeter of the tire that comes in contact with the road, so we first find the circumference of the tire:

$$C = 2\pi r = 2\pi(15) = 30\pi \text{ in.}$$

This means that 1 rotation $= 30\pi$ in. From this point we proceed by converting "miles per hour" to "rotations per second" by a series of **conversion factors** that are really factors of 1:

$$\frac{70 \text{ miles}}{1 \text{ hour}} \times \frac{1 \text{ hour}}{60 \text{ min}} \times \frac{1 \text{ min}}{60 \text{ sec}} \times \frac{5280 \text{ feet}}{1 \text{ mile}} \times \frac{12 \text{ inches}}{1 \text{ foot}} \times \frac{1 \text{ rotation}}{30\pi \text{ inches}}$$

$$= \frac{70 \times 5280 \times 12 \text{ rotations}}{60 \times 60 \times 30\pi \text{ sec}} \approx 13.07 \text{ rotations per second}$$

Now try Exercise 39.

Functions from Data

In this course we will use the following 3-step strategy to construct functions from data.

Constructing a Function from Data

Given a set of data points of the form (x, y), to construct a formula that approximates y as a function of x:

1. Make a scatter plot of the data points. The points do not need to pass the vertical line test.
2. Determine from the shape of the plot whether the points seem to follow the graph of a familiar type of function (line, parabola, cubic, sine curve, etc.).
3. Transform a basic function of that type to fit the points as closely as possible.

Step 3 might seem like a lot of work, and for earlier generations it certainly was; it required all of the tricks of Section 1.6 and then some. We, however, will gratefully use technology to do this "curve-fitting" step for us, as shown in Example 6.

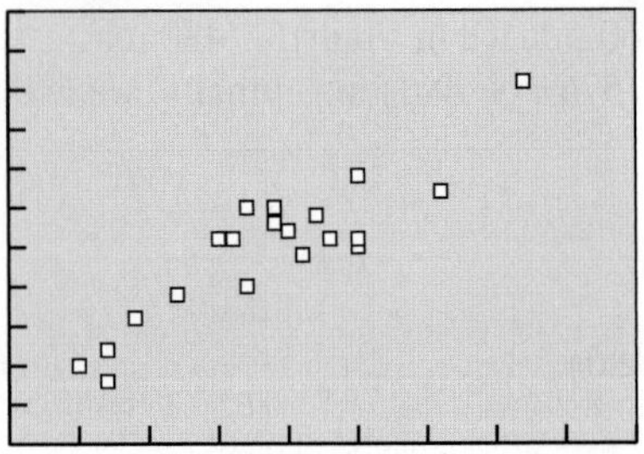

[45, 90] by [60, 115]

FIGURE 1.84 The scatter plot of the temperature data in Example 6.

EXAMPLE 6 Curve Fitting with Technology

Table 1.12 records the low and high daily temperatures observed on 9/9/1999 in 20 major American cities. Find a function that approximates the high temperature (y) as a function of the low temperature (x). Use this function to predict the high temperature that day for Madison, WI, given that the low was 46.

SOLUTION The scatter plot is shown in Figure 1.84.

Table 1.12 Temperature on 9/9/99

City	Low	High	City	Low	High
New York, NY	70	86	Miami, FL	76	92
Los Angeles, CA	62	80	Honolulu, HI	70	85
Chicago, IL	52	72	Seattle, WA	50	70
Houston, TX	70	94	Jacksonville, FL	67	89
Philadelphia, PA	68	86	Baltimore, MD	64	88
Albuquerque, NM	61	86	St. Louis, MO	57	79
Phoenix, AZ	82	106	El Paso, TX	62	90
Atlanta, GA	64	90	Memphis, TN	60	86
Dallas, TX	65	87	Milwaukee, WI	52	68
Detroit, MI	54	76	Wilmington, DE	66	84

Source: AccuWeather, Inc.

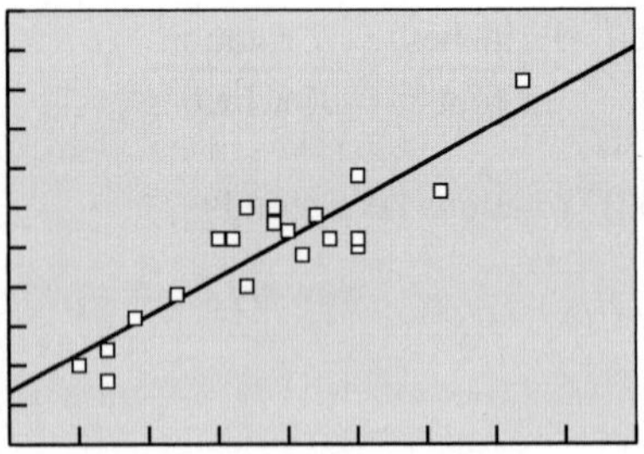

[45, 90] by [60, 115]

FIGURE 1.85 The temperature scatter plot with the regression line shown. (Example 6)

Notice that the points do not fall neatly along a well-known curve, but they do seem to fall *near* an upwardly sloping line. We therefore choose to model the data with a function whose graph is a line. We could fit the line by sight (as we did in Example 5 in Section 1.1), but this time we will use the calculator to find the line of "best fit," called the **regression line**. (See your grapher's owner's manual for how to do this.) The regression line is found to be approximately $y = 0.97x + 23$. As Figure 1.85 shows, the line fits the data as well as can be expected.

If we use this function to predict the high temperature for the day in Madison, WI, we get $y = 0.97(46) + 23 = 67.62$. (For the record, the high that day was 67.)

Now try Exercise 47, parts (a) and (b).

Professional statisticians would be quick to point out that this function should not be trusted as a model for all cities, despite the fairly successful prediction for Madison. (For example, the prediction for San Francisco, with a low of 54 and a high of 64, is off by more than 11 degrees.) *The effectiveness of a data-based model is highly dependent on the number of data points and on the way they were selected.* The functions we construct from data in this book should be analyzed for how well they model the data, not for how well they model the larger population from which the data came.

In addition to lines, we can model scatter plots with several other curves by choosing the appropriate regression option on a calculator or computer. The options to which we will refer in this book (and the chapters in which we will study them) are shown in the following table:

Regression Type	Equation	Graph	Applications
Linear (Chapter 2)	$y = ax + b$	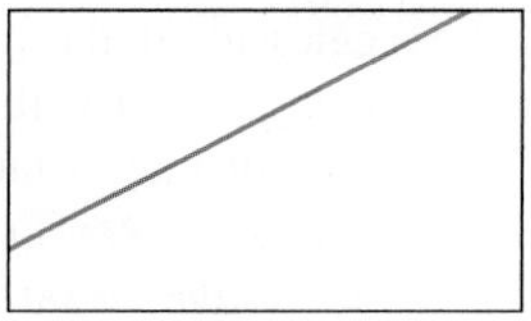	Fixed cost plus variable cost, linear growth, free-fall velocity, simple interest, linear depreciation, many others
Quadratic (Chapter 2)	$y = ax^2 + bx + c$ (requires at least 3 points)	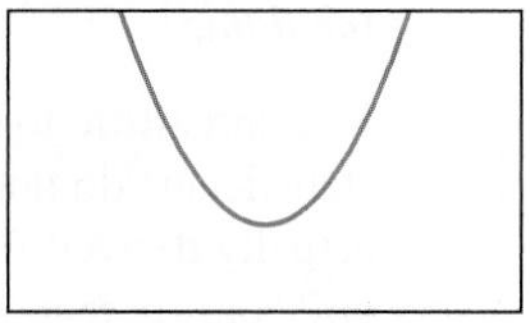	Position during free fall, projectile motion, parabolic reflectors, area as a function of linear dimension, quadratic growth, etc.
Cubic (Chapter 2)	$y = ax^3 + bx^2 + cx + d$ (requires at least 4 points)	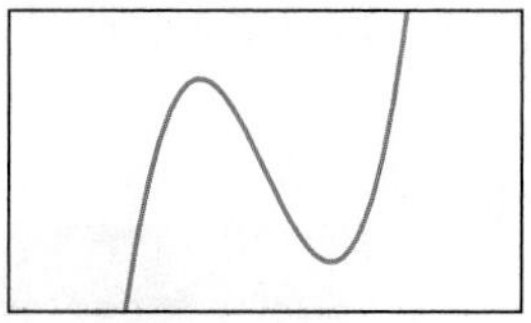	Volume as a function of linear dimension, cubic growth, miscellaneous applications where quadratic regression does not give a good fit
Quartic (Chapter 2)	$y = ax^4 + bx^3 + cx^2 + dx + e$ (requires at least 5 points)	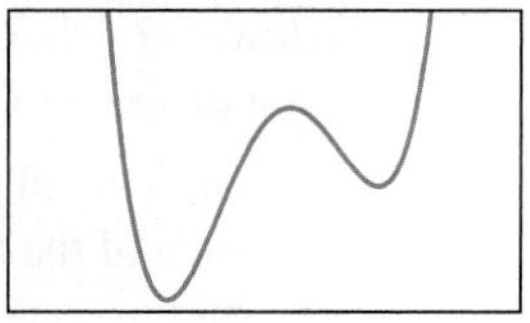	Quartic growth, miscellaneous applications where quadratic and cubic regression do not give a good fit
Natural logarithmic (ln) (Chapter 3)	$y = a + b \ln x$ (requires $x > 0$)	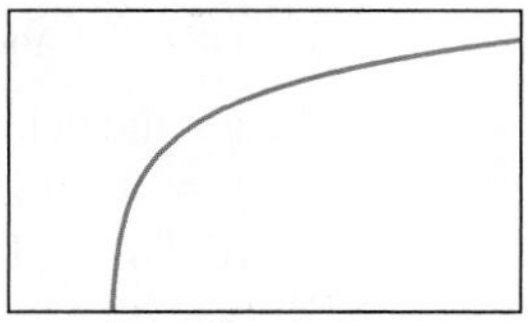	Logarithmic growth, decibels (sound), Richter scale (earthquakes), inverse exponential models
Exponential ($b > 1$) (Chapter 3)	$y = a \cdot b^x$	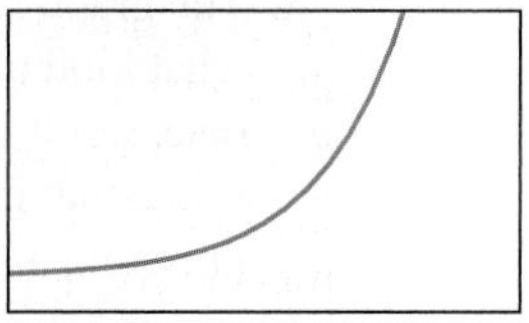	Exponential growth, compound interest, population models
Exponential ($0 < b < 1$) (Chapter 3)	$y = a \cdot b^x$	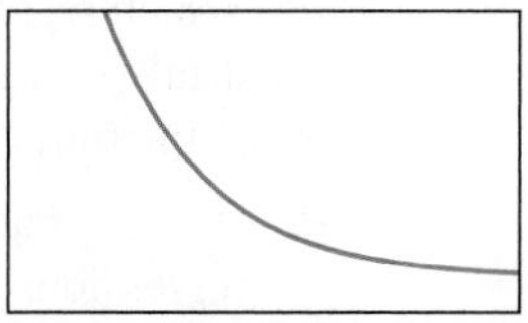	Exponential decay, depreciation, temperature loss of a cooling body, etc.
Power (requires $x, y > 0$) (Chapter 2)	$y = a \cdot x^b$	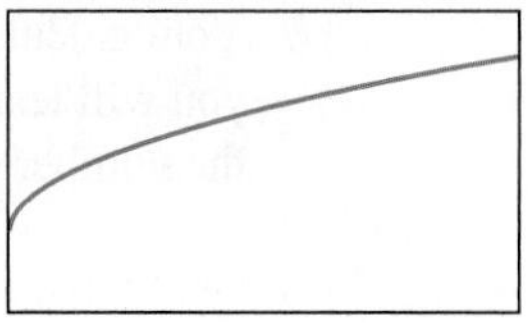	Inverse-square laws, Kepler's Third Law
Logistic (Chapter 3)	$y = \dfrac{c}{1 + a \cdot e^{-bx}}$	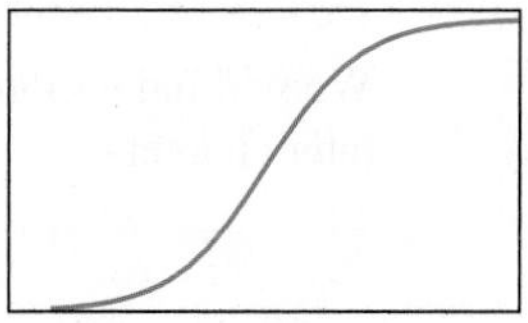	Logistic growth: spread of a rumor, population models
Sinusoidal (Chapter 4)	$y = a \sin (bx + c) + d$	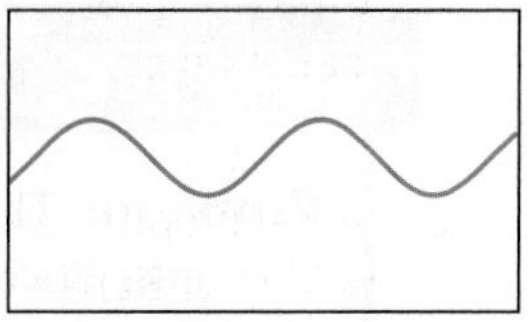	Periodic behavior: harmonic motion, waves, circular motion, etc.

Displaying Diagnostics

If your grapher is giving regression formulas without displaying the values of r or r^2 or R^2, you may be able to fix that. If your MODE display has a second page, choose "ON" for "STATDIAGNOSTICS." Otherwise, choose "DiagnosticOn" from the CATALOG menu. ENTER the command on the home screen and see the reply "Done." Your next regression should display the diagnostic values.

These graphs are only examples, as they can vary in shape and orientation. (For example, any of the curves could appear upside-down.) The grapher uses various strategies to fit these curves to the data, most of them based on combining function composition with linear regression. Depending on the regression type, the grapher may display a number r called the **correlation coefficient** or a number r^2 or R^2 called the **coefficient of determination**. In either case, a useful "rule of thumb" is that *the closer the absolute value of this number is to 1, the better the curve fits the data.*

It is unrealistic to expect a regression curve to fit data from the real world perfectly, so statisticians do not expect to see r^2 or R^2 equal to 1. However, if the ordered pairs are actually derived from a function defined by one of the regression types, the calculator will confirm that. As an example, you will use a calculator to find a geometric formula in Exploration 1.

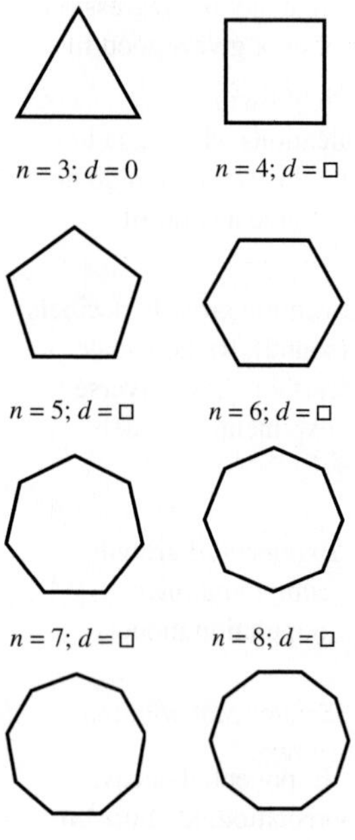

FIGURE 1.86 Some polygons. (Exploration 1)

EXPLORATION 1 Diagonals of a Regular Polygon

How many diagonals does a regular polygon have? Can the number be expressed as a function of the number of sides? Try this Exploration.

1. Draw in all the diagonals (i.e., segments connecting nonadjacent points) in each of the regular polygons shown in Figure 1.86 and fill in the number (d) of diagonals in the space below the figure. The values of d for the triangle ($n = 3$) and the decagon ($n = 10$) are filled in for you.
2. Put the values of n in list L1, *beginning with* $n = 4$. (We want to avoid that $y = 0$ value for some of our regressions later.) Put the corresponding values of d in list L2. Display a scatter plot of the ordered pairs.
3. The graph shows an increasing function with some curvature, but it is not clear what kind of function it is. Try these regressions in order and record which ones result in a value of r^2 or R^2 equal to 1: power, quadratic, cubic, quartic. (Note that it is not worth trying linear, logarithmic, logistic, or sinusoidal regression, as the curvature is not right for those functions.)
4. Which regressions result in a perfect fit? What kind of function matches the data? (It might appear at first that there are multiple kinds, but look more closely at the functions you get.)
5. Looking back, could you have predicted the results of the cubic and quartic regressions after seeing the result of the quadratic regression?
6. Your calculator has found the formula for d as a function of n. (In Chapter 9 you will learn how to derive this formula for yourself.) Use the formula to find the number of diagonals in a 128-gon.

We will have more to say about curve fitting as we study the various function types in later chapters.

CHAPTER OPENER Problem (from page 63)

Problem: The table at the top of the next page shows the growth in the Consumer Price Index (CPI) for housing for selected years between 1990 and 2007 (based on 1983 dollars). How can we construct a function to predict the housing CPI for the years 2008–2015?

Consumer Price Index (Housing)

Year	Housing CPI
1990	128.5
1995	148.5
2000	169.6
2002	180.3
2003	184.8
2004	189.5
2005	195.7
2006	203.2
2007	209.6

Source: Bureau of Labor Statistics, quoted in The World Almanac and Book of Facts 2013.

Solution: A scatter plot of the data is shown in Figure 1.87, where x is the number of years since 1990. A linear model would work pretty well, but the slight upward curve of the scatter plot suggests that a quadratic model might work better. Using a calculator to compute the quadratic regression curve, we find its equation to be

$$y = 0.089x^2 + 3.17x + 129.$$

As Figure 1.88 shows, the parabola fits the data impressively well.

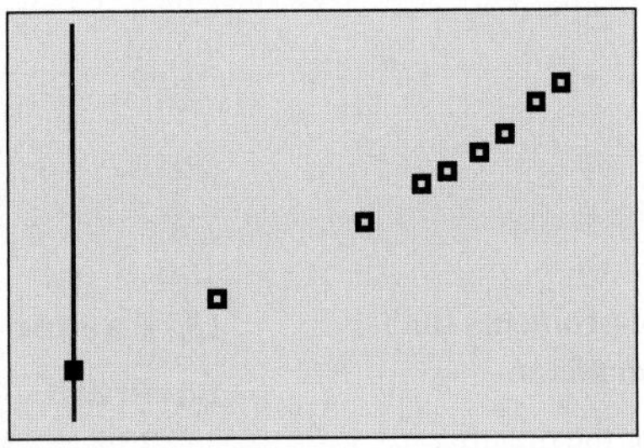

[–2, 19] by [115, 225]

FIGURE 1.87 Scatter plot of the data for the housing CPI.

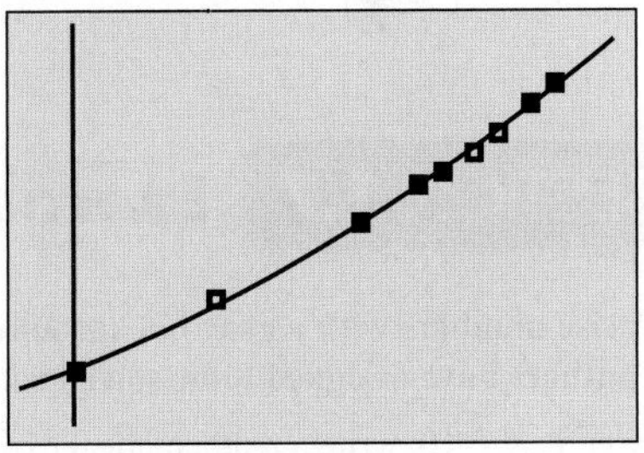

[–2, 19] by [115, 225]

FIGURE 1.88 Scatter plot with the regression curve shown.

To predict the housing CPI for 2008, use $x = 18$ in the regression equation. Similarly, we can predict the housing CPI for each of the years 2008–2015, as shown below:

Predicted CPI (Housing)

Year	Predicted Housing CPI	Actual Housing CPI*
2008	214.9	216.2
2009	221.4	217.1
2010	228.0	216.3
2011	234.8	219.1
2012	241.8	
2013	249.0	
2014	256.3	
2015	263.9	

**Source: Bureau of Labor Statistics, quoted in The World Almanac and Book of Facts 2013.*

As you can see, even with a regression curve that fits the data as beautifully as in Figure 1.88, it is risky to predict this far beyond the data set. Statistics like the CPI are dependent on many volatile factors that can quickly render any mathematical model obsolete. In fact, the mortgage model that fueled the housing growth up to 2007 proved to be unsustainable, and when it broke down it took many well-behaved economic curves (like this one) down with it. In light of that fact, you might enjoy comparing the rest of these "predictions" with the actual housing CPI numbers as the years go by!

QUICK REVIEW 1.7 *(For help, go to Sections P.3 and P.4.)*

In Exercises 1–10, solve the given formula for the given variable.

1. **Area of a Triangle** Solve for h: $A = \frac{1}{2}bh$
2. **Area of a Trapezoid** Solve for h: $A = \frac{1}{2}(b_1 + b_2)h$
3. **Volume of a Right Circular Cylinder** Solve for h: $V = \pi r^2 h$
4. **Volume of a Right Circular Cone** Solve for h: $V = \frac{1}{3}\pi r^2 h$
5. **Volume of a Sphere** Solve for r: $V = \frac{4}{3}\pi r^3$
6. **Surface Area of a Sphere** Solve for r: $A = 4\pi r^2$
7. **Surface Area of a Right Circular Cylinder** Solve for h: $A = 2\pi rh + 2\pi r^2$
8. **Simple Interest** Solve for t: $I = Prt$
9. **Compound Interest** Solve for P: $A = P\left(1 + \frac{r}{n}\right)^{nt}$
10. **Free Fall from Height H** Solve for t: $s = H - \frac{1}{2}gt^2$

SECTION 1.7 Exercises

Exercise numbers with a gray background indicate problems that the authors have designed to be solved *without a calculator.*

In Exercises 1–10, write a mathematical expression for the quantity described verbally.

1. Five more than three times a number x
2. A number x increased by 5 and then tripled
3. Seventeen percent of a number x
4. Four more than 5% of a number x
5. **Area of a Rectangle** The area of a rectangle whose length is 12 more than its width x
6. **Area of a Triangle** The area of a triangle whose altitude is 2 more than its base length x
7. **Salary Increase** A salary after a 4.5% increase, if the original salary is x dollars
8. **Income Loss** Income after a 3% drop in the current income of x dollars
9. **Sale Price** Sale price of an item marked x dollars, if 40% is discounted from the marked price
10. **Including Tax** Actual cost of an item selling for x dollars if the sales tax rate is 8.75%

In Exercises 11–14, choose a variable and write a mathematical expression for the quantity described verbally.

11. **Total Cost** The total cost is \$34,500 plus \$5.75 for each item produced.
12. **Total Cost** The total cost is \$28,000 increased by 9% plus \$19.85 for each item produced.
13. **Revenue** The revenue when each item sells for \$3.75
14. **Profit** The profit consists of a franchise fee of \$200,000 plus 12% of all sales.

In Exercises 15–20, write the specified quantity as a function of the specified variable. It will help in each case to draw a picture.

15. The height of a right circular cylinder equals its diameter. Write the volume of the cylinder as a function of its radius.
16. One leg of a right triangle is twice as long as the other. Write the length of the hypotenuse as a function of the length of the shorter leg.
17. The base of an isosceles triangle is half as long as the two equal sides. Write the area of the triangle as a function of the length of the base.
18. A square is inscribed in a circle. Write the area of the square as a function of the radius of the circle.
19. A sphere is contained in a cube, tangent to all six faces. Find the surface area of the cube as a function of the radius of the sphere.
20. An isosceles triangle has its base along the x-axis with one base vertex at the origin and its vertex in the first quadrant on the graph of $y = 6 - x^2$. Write the area of the triangle as a function of the length of the base.

In Exercises 21–36, write an equation for the problem and solve the problem.

21. One positive number is 4 times another positive number. The sum of the two numbers is 620. Find the two numbers.

22. When a number is added to its double and its triple, the sum is 714. Find the three numbers.

23. **Salary Increase** Mark received a 3.5% salary increase. His salary after the raise was \$36,432. What was his salary before the raise?

24. **Consumer Price Index** The Consumer Price Index for food and beverages in 2007 was 203.3 after a hefty 3.9% increase from the previous year. What was the Consumer Price Index for food and beverages in 2006? (*Source: U.S. Bureau of Labor Statistics*)

25. **Travel Time** A traveler averaged 52 mph on a 182-mi trip. How many hours were spent on the trip?

26. **Travel Time** On their 560-mi trip, the Bruins basketball team spent two more hours on the interstate highway than they did on local highways. They averaged 45 mph on local highways and 55 mph on the interstate highways. How many hours did they spend driving on local highways?

27. **Sale Prices** At a shirt sale, Jackson sees two shirts that he likes equally well. Which is the better bargain, and why?

28. **Job Offers** Ruth is weighing two job offers from the sales departments of two competing companies. One offers a base salary of \$25,000 plus 5% of gross sales; the other offers a base salary of \$20,000 plus 7% of gross sales. What would Ruth's gross sales total need to be to make the second job offer more attractive than the first?

29. **Cell Phone Antennas** From December 2006 to December 2007, the number of cell phone antennas in the United States grew from 195,613 to 213,299. What was the percentage increase in U.S. cell phone antennas in that one-year period? (*Source: CTIA, quoted in The World Almanac and Book of Facts 2009*)

30. **Cell Phone Antennas** From December 1996 to December 1997, the number of cell phone antennas in the United States grew from 30,045 to 51,600. What was the percentage increase in U.S. cell phone antennas in that one-year period? (*Source: CTIA, quoted in The World Almanac and Book of Facts 2009*)

31. **Mixing Solutions** How much 10% solution and how much 45% solution should be mixed together to make 100 gal of 25% solution?

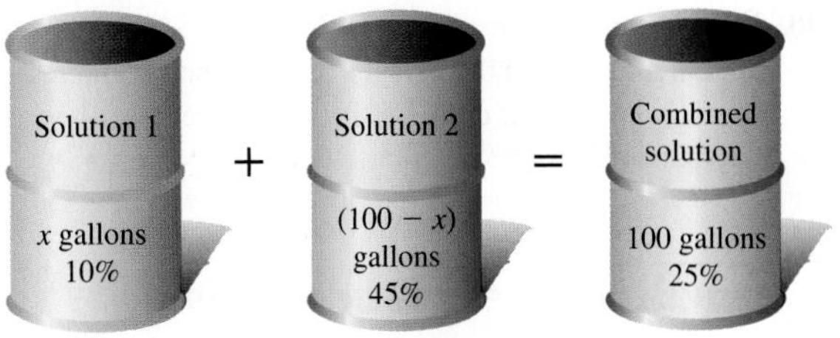

(a) Write an equation that models this problem.

(b) Solve the equation graphically.

32. **Mixing Solutions** The chemistry lab at the University of Hardwoods keeps two acid solutions on hand. One is 20% acid and the other is 35% acid. How much 20% acid solution and how much 35% acid solution should be used to prepare 25 L of a 26% acid solution?

33. **Maximum Value Problem** A square of side x inches is cut out of each corner of a 10-in. by 18-in. piece of cardboard and the sides are folded up to form an open-topped box.

(a) Write the volume V of the box as a function of x.

(b) Find the domain of your function, taking into account the restrictions that the model imposes in x.

(c) Use your graphing calculator to determine the dimensions of the cut-out squares that will produce the box of maximum volume.

34. **Residential Construction** DDL Construction is building a rectangular house that is 16 ft longer than it is wide. A rain gutter is to be installed in four sections around the 136-ft perimeter of the house. What lengths should be cut for the four sections?

35. **Protecting an Antenna** In Example 3, suppose the parabolic dish has a 32-in. diameter and is 8 in. deep, and the radius of the cardboard cylinder is 8 in. Now how tall must the cylinder be to fit in the middle of the dish and be flush with the top of the dish?

36. **Interior Design** Renée's Decorating Service recommends putting a border around the top of the four walls in a dining room that is 3 ft longer than it is wide. Find the dimensions of the room if the total length of the border is 54 ft.

37. **Finding the Model and Solving** Water is stored in a conical tank with a faucet at the bottom. The tank has depth 24 in. and radius 9 in., and it is filled to the brim. If the faucet is opened to allow the water to flow at a rate of 5 cubic inches per second, what will the depth of the water be after 2 min?

38. **Investment Returns** Reggie invests \$12,000, part at 7% annual interest and part at 8.5% annual interest. How much is invested at each rate if Reggie's total annual interest is \$900?

39. **Unit Conversion** A tire of a moving bicycle has radius 16 in. If the tire is making 2 rotations per second, find the bicycle's speed in miles per hour.

40. Investment Returns Jackie invests \$25,000, part at 5.5% annual interest and the balance at 8.3% annual interest. How much is invested at each rate if Jackie receives a 1-year interest payment of \$1571?

Standardized Test Questions

41. True or False A correlation coefficient gives an indication of how closely a regression line or curve fits a set of data. Justify your answer.

42. True or False Linear regression is useful for modeling the position of an object in free fall. Justify your answer.

In Exercises 43–46, tell which type of regression is likely to give the most accurate model for the scatter plot shown without using a calculator.

(A) Linear regression

(B) Quadratic regression

(C) Cubic regression

(D) Exponential regression

(E) Sinusoidal regression

43. Multiple Choice

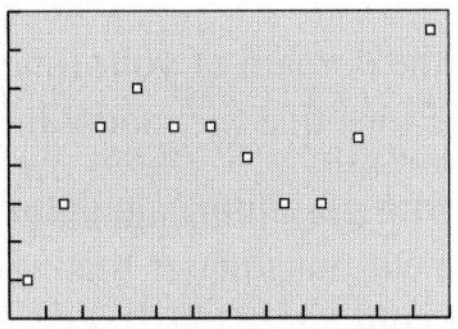

[0, 12] by [0, 8]

44. Multiple Choice

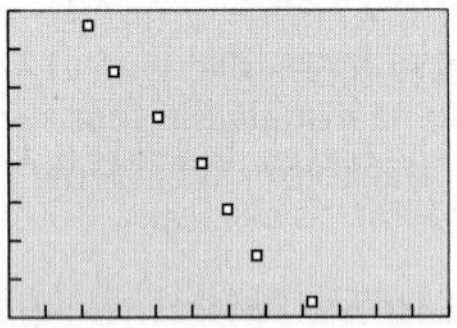

[0, 12] by [0, 8]

45. Multiple Choice

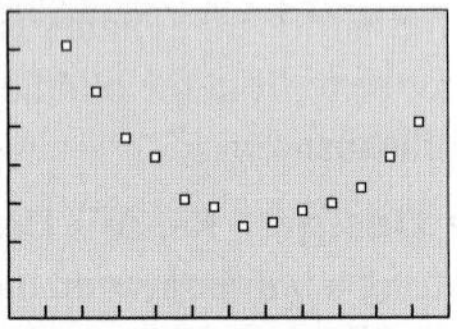

[0, 12] by [0, 8]

46. Multiple Choice

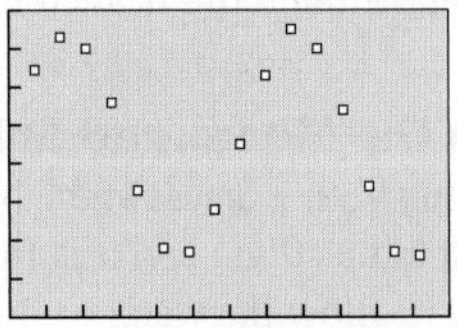

[0, 12] by [0, 8]

Exploration

47. Manufacturing The Buster Green Shoe Company determines that the annual cost C of making x pairs of one type of shoe is \$30 per pair plus \$100,000 in fixed overhead costs. Each pair of shoes that is manufactured is sold wholesale for \$50.

(a) Find an equation that models the cost of producing x pairs of shoes.

(b) Find an equation that models the revenue produced from selling x pairs of shoes.

(c) Find how many pairs of shoes must be made and sold in order to break even.

(d) Graph the equations in (a) and (b). What is the graphical interpretation of the answer in (c)?

48. Employee Benefits John's company issues employees a contract that identifies salary and the company's contributions to pension, health insurance premiums, and disability insurance. The company uses the following formulas to calculate these values.

Salary	x (dollars)
Pension	12% of salary
Health insurance	3% of salary
Disability insurance	0.4% of salary

If John's total contract with benefits is worth \$48,814.20, what is his salary?

49. Manufacturing Queen, Inc., a tennis racket manufacturer, determines that the annual cost C of making x rackets is \$23 per racket plus \$125,000 in fixed overhead costs. It costs the company \$8 to string a racket.

(a) Find a function $y_1 = u(x)$ that models the cost of producing x unstrung rackets.

(b) Find a function $y_2 = s(x)$ that models the cost of producing x strung rackets.

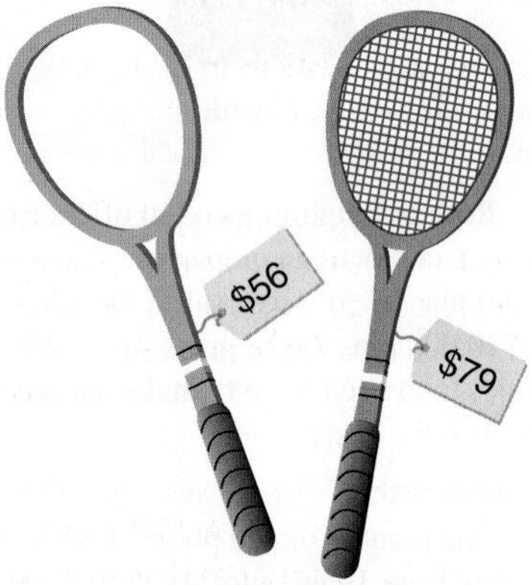

(c) Find a function $y_3 = R_u(x)$ that models the revenue generated by selling x unstrung rackets.

(d) Find a function $y_4 = R_s(x)$ that models the revenue generated by selling x rackets.

(e) Graph y_1, y_2, y_3, and y_4 simultaneously in the window $[0, 10{,}000]$ by $[0, 500{,}000]$.

(f) Writing to Learn Write a report to the company recommending how it should manufacture its rackets, strung or unstrung. Assume that you can include the viewing window in (e) as a graph in the report, and use it to support your recommendation.

50. Carbon Dioxide Levels Table 1.13 shows the atmospheric concentration of carbon dioxide (in parts per million) recorded by the Mauna Loa Observatory in Hawaii for selected years between 1960 and 2011.

Table 1.13 Atmospheric CO_2

Year	CO_2 (ppm)
1960	317
1970	326
1980	339
1990	354
2000	369
2005	380
2006	382
2007	384
2008	386
2009	387
2010	390
2011	392

Source: Carbon Dioxide Information Analysis Center, U.S. Dept. of Energy, as reported in The World Almanac and Book of Facts 2013.

(a) Produce a scatter plot of CO_2 concentration (y) as a function of years since 1950 (x).

(b) Find the linear regression equation for the years 1960 to 2011. Round the coefficients to the nearest hundredth.

(c) Find the linear regression equation for the years 2000 to 2011. Round the coefficients to the nearest hundredth.

(d) Use both regression equations to estimate the CO_2 concentration for the year 2015. Which do you think is a safer prediction?

(e) Writing to Learn Use the results of parts (a)–(d) to explain why it is risky to predict y values for x values that are not very close to the data points, even when the regression plot fits the data points quite well.

Extending the Ideas

51. Newton's Law of Cooling A 190° cup of coffee is placed on a desk in a 72° room. According to Newton's Law of Cooling, the temperature T of the coffee after t minutes will be $T = (190 - 72)b^t + 72$, where b is a constant that depends on how easily the cooling substance loses heat. The data in Table 1.14 are from a simulated experiment of gathering temperature readings from a cup of coffee in a 72° room at 20 one-minute intervals:

Table 1.14 Cooling a Cup of Coffee

Time	Temp	Time	Temp
1	184.3	11	140.0
2	178.5	12	136.1
3	173.5	13	133.5
4	168.6	14	130.5
5	164.0	15	127.9
6	159.2	16	125.0
7	155.1	17	122.8
8	151.8	18	119.9
9	147.0	19	117.2
10	143.7	20	115.2

(a) Make a scatter plot of the data, with the times in list L1 and the temperatures in list L2.

(b) Store L2 − 72 in list L3. The values in L3 should now be an exponential function $(y = a \times b^x)$ of the values in L1.

(c) Find the exponential regression equation for L3 as a function of L1. How well does it fit the data?

52. Group Activity Newton's Law of Cooling If you have access to laboratory equipment (such as a CBL or CBR unit for your grapher), gather experimental data such as in Exercise 51 from a cooling cup of coffee. Proceed as follows:

(a) First, use the temperature probe to record the temperature of the room. It is a good idea to turn off fans and air conditioners that might affect the temperature of the room during the experiment. It should be a constant.

(b) Heat the coffee. It need not be boiling, but it should be at least 160°. (It also need not be coffee.)

(c) Make a new list consisting of the temperature values minus the room temperature. Make a scatter plot of this list (y) against the time values (x). It should appear to approach the x-axis as an asymptote.

(d) Find the equation of the exponential regression curve. How well does it fit the data?

(e) What is the equation predicted by Newton's Law of Cooling? (Substitute your initial coffee temperature and the temperature of your room for the 190 and 72 in the equation in Exercise 51.)

(f) Discussion What sort of factors would affect the value of b in Newton's Law of Cooling? Discuss your ideas with the group.

CHAPTER 1 Key Ideas

Properties, Theorems, and Formulas

Zero Factor Property 69
Fundamental Connection 70
Vertical Line Test 81
Horizontal Line Test 122
Inverse Reflection Principle 123
Inverse Composition Rule 124
Translations of Graphs 130
Reflections of Graphs 131
Stretches and Shrinks of Graphs 134

Procedures

Pólya's Four Problem-Solving Steps 70
Problem-Solving Process 70–71
Agreement About Domain 82
How to Find an Inverse Function Algebraically 125
Graphing Absolute Value Compositions 132
Constructing a Function from Data 143

CHAPTER 1 Review Exercises

The collection of exercises marked in red could be used as a chapter test.

In Exercises 1–10, match the graph with the corresponding function (a)–(j) from the list below. Use your knowledge of function behavior, *not* your grapher.

(a) $f(x) = x^2 - 1$ **(b)** $f(x) = x^2 + 1$
(c) $f(x) = (x - 2)^2$ **(d)** $f(x) = (x + 2)^2$
(e) $f(x) = \dfrac{x - 1}{2}$ **(f)** $f(x) = |x - 2|$
(g) $f(x) = |x + 2|$ **(h)** $f(x) = -\sin x$
(i) $f(x) = e^x - 1$ **(j)** $f(x) = 1 + \cos x$

1.
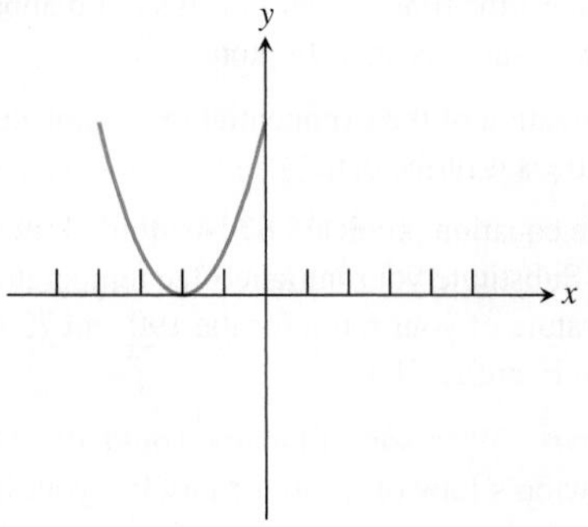

2.
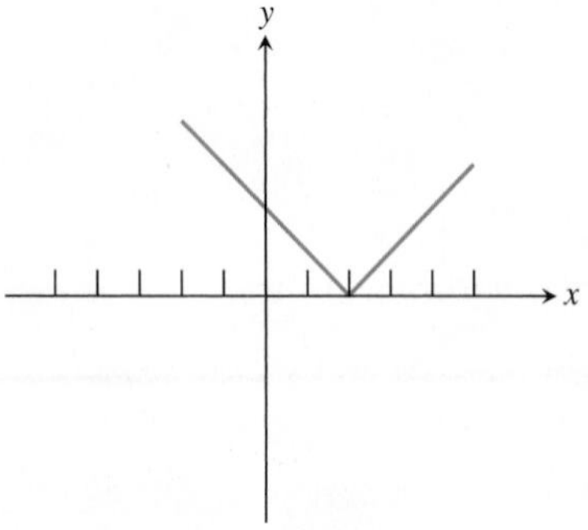

3.
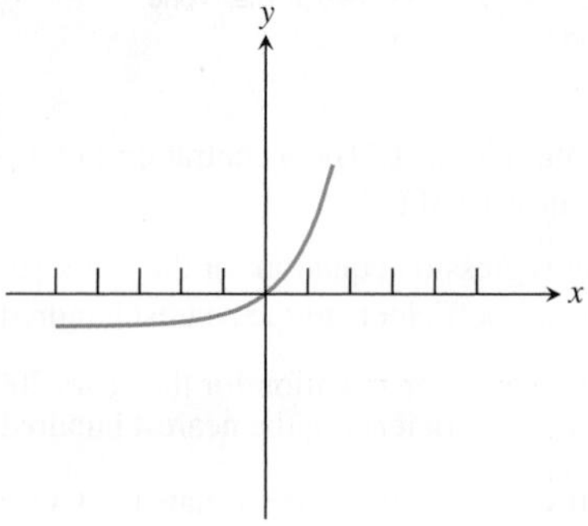

4.
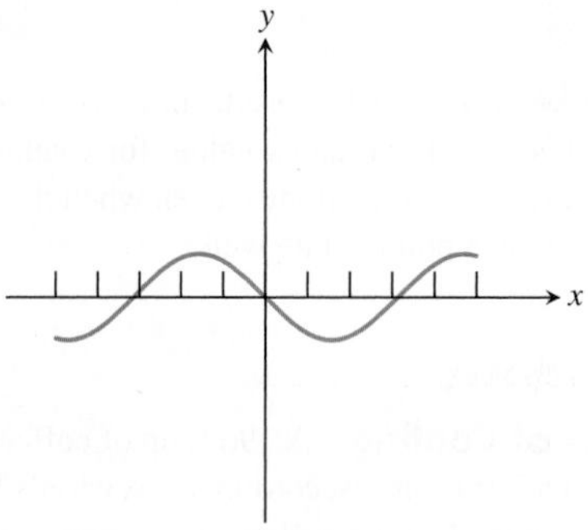

5.
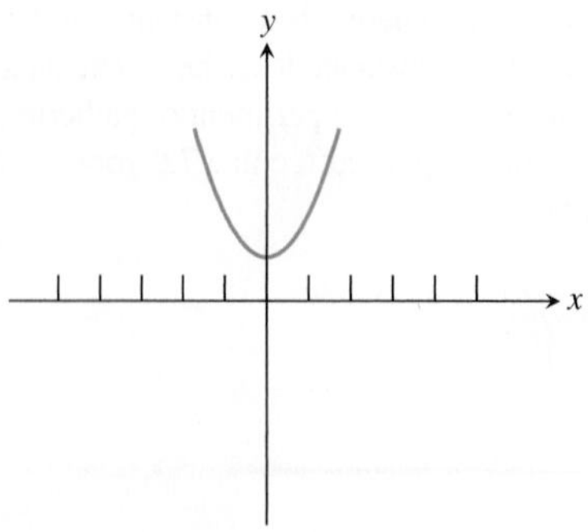

6.

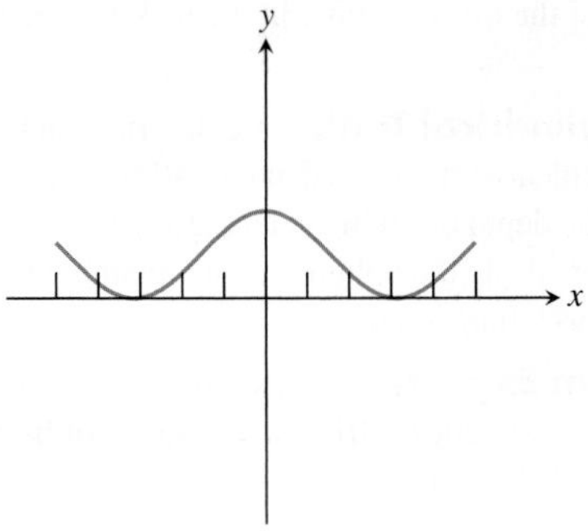

7.

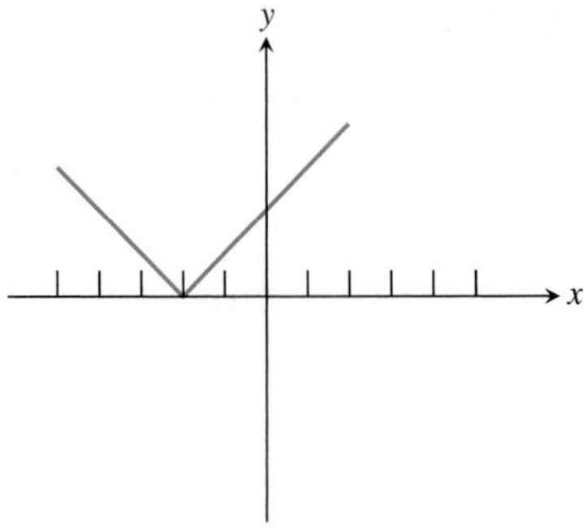

8.

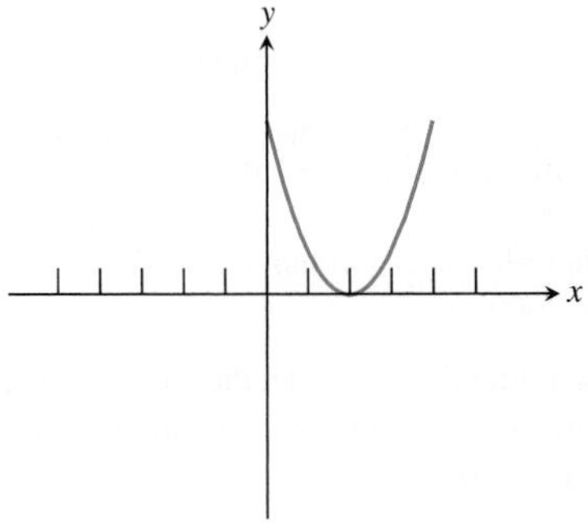

9.

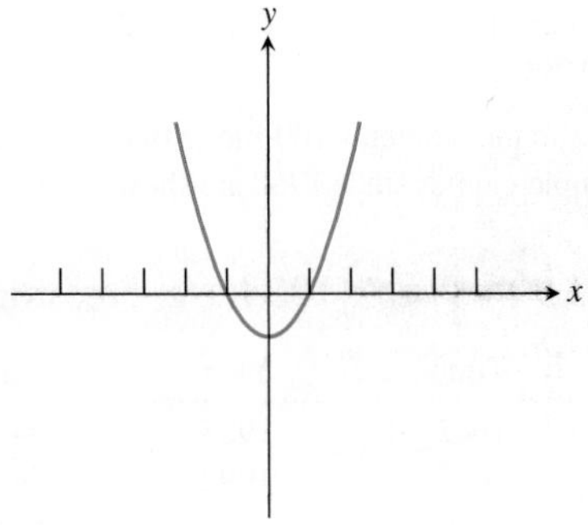

10.

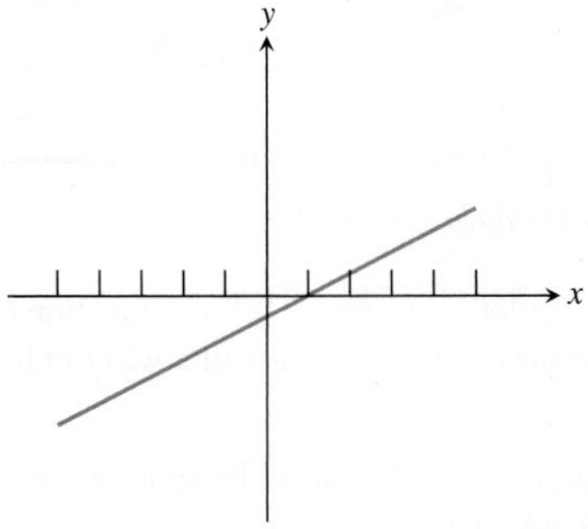

In Exercises 11–18, find (a) the domain and (b) the range of the function.

11. $g(x) = x^3$ **12.** $f(x) = 35x - 602$

13. $g(x) = x^2 + 2x + 1$ **14.** $h(x) = (x - 2)^2 + 5$

15. $g(x) = 3|x| + 8$ **16.** $k(x) = -\sqrt{4 - x^2}$

17. $f(x) = \dfrac{x}{x^2 - 2x}$ **18.** $k(x) = \dfrac{1}{\sqrt{9 - x^2}}$

In Exercises 19 and 20, graph the function, and state whether the function is continuous at $x = 0$. If it is discontinuous, state whether the discontinuity is removable or nonremovable.

19. $f(x) = \dfrac{x^2 - 3}{x + 2}$ **20.** $k(x) = \begin{cases} 2x + 3 \text{ if } x > 0 \\ 3 - x^2 \text{ if } x \le 0 \end{cases}$

In Exercises 21–24, find all (a) vertical asymptotes and (b) horizontal asymptotes of the graph of the function. Be sure to state your answers as equations of lines.

21. $y = \dfrac{5}{x^2 - 5x}$ **22.** $y = \dfrac{3x}{x - 4}$

23. $y = \dfrac{7x}{\sqrt{x^2 + 10}}$ **24.** $y = \dfrac{|x|}{x + 1}$

In Exercises 25–28, graph the function and state the intervals on which the function is *increasing*.

25. $y = \dfrac{x^3}{6}$ **26.** $y = 2 + |x - 1|$

27. $y = \dfrac{x}{1 - x^2}$ **28.** $y = \dfrac{x^2 - 1}{x^2 - 4}$

In Exercises 29–32, graph the function and tell whether the function is bounded above, bounded below, or bounded.

29. $f(x) = x + \sin x$ **30.** $g(x) = \dfrac{6x}{x^2 + 1}$

31. $h(x) = 5 - e^x$ **32.** $k(x) = 1000 + \dfrac{x}{1000}$

In Exercises 33–36, use a grapher to find all (a) relative maximum values and (b) relative minimum values of the function. Also state the value of x at which each relative extremum occurs.

33. $y = (x + 1)^2 - 7$ **34.** $y = x^3 - 3x$

35. $y = \dfrac{x^2 + 4}{x^2 - 4}$ **36.** $y = \dfrac{4x}{x^2 + 4}$

In Exercises 37–40, graph the function and state whether the function is odd, even, or neither.

37. $y = 3x^2 - 4|x|$ **38.** $y = \sin x - x^3$

39. $y = \dfrac{x}{e^x}$ **40.** $y = x \cos(x)$

In Exercises 41–44, find a formula for $f^{-1}(x)$.

41. $f(x) = 2x + 3$ **42.** $f(x) = \sqrt[3]{x - 8}$

43. $f(x) = \dfrac{2}{x}$ **44.** $f(x) = \dfrac{6}{x + 4}$

Exercises 45–52 refer to the function $y = f(x)$ whose graph is given below.

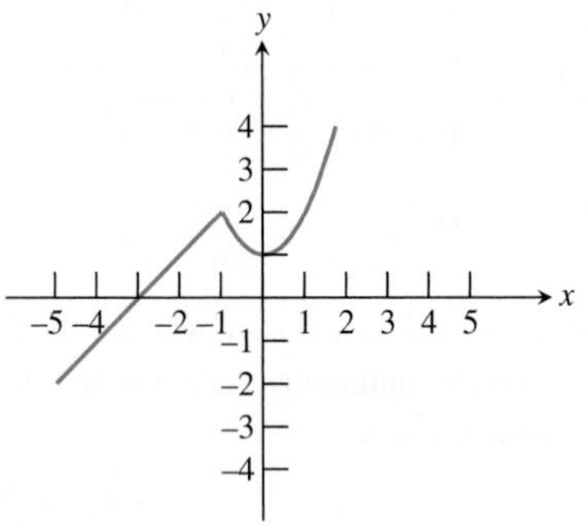

45. Sketch the graph of $y = f(x) - 1$.

46. Sketch the graph of $y = f(x - 1)$.

47. Sketch the graph of $y = f(-x)$.

48. Sketch the graph of $y = -f(x)$.

49. Sketch a graph of the inverse relation.

50. Does the inverse relation define y as a function of x?

51. Sketch a graph of $y = |f(x)|$.

52. Define f algebraically as a piecewise function. (*Hint*: the pieces are translations of two of our "basic" functions.)

In Exercises 53–58, let $f(x) = \sqrt{x}$ and let $g(x) = x^2 - 4$.

53. Find an expression for $(f \circ g)(x)$ and give its domain.

54. Find an expression for $(g \circ f)(x)$ and give its domain.

55. Find an expression for $(fg)(x)$ and give its domain.

56. Find an expression for $\left(\frac{f}{g}\right)(x)$ and give its domain.

57. Describe the end behavior of the graph of $y = f(x)$.

58. Describe the end behavior of the graph of $y = f(g(x))$.

In Exercises 59–64, write the specified quantity as a function of the specified variable. Remember that drawing a picture will help.

59. Square Inscribed in a Circle A square of side s is inscribed in a circle. Write the area of the circle as a function of s.

60. Circle Inscribed in a Square A circle is inscribed in a square of side s. Write the area of the circle as a function of s.

61. Volume of a Cylindrical Tank A cylindrical tank with diameter 20 ft is partially filled with oil to a depth of h feet. Write the volume of oil in the tank as a function of h.

62. Draining a Cylindrical Tank A cylindrical tank with diameter 20 ft is filled with oil to a depth of 40 ft. The oil begins draining at a constant rate of 2 cubic feet per second. Write the volume of the oil remaining in the tank t seconds later as a function of t.

63. Draining a Cylindrical Tank A cylindrical tank with diameter 20 ft is filled with oil to a depth of 40 ft. The oil begins draining at a constant rate of 2 cubic feet per second. Write the depth of the oil remaining in the tank t seconds later as a function of t.

64. Draining a Cylindrical Tank A cylindrical tank with diameter 20 ft is filled with oil to a depth of 40 ft. The oil begins draining so that the depth of oil in the tank decreases at a constant rate of 2 ft/hr. Write the volume of oil remaining in the tank t hours later as a function of t.

65. U.S. Petroleum Exports The average U.S. exports of petroleum for the years 2001–2011, in thousands of barrels per day, are shown in Table 1.15.

Table 1.15 Daily U.S. Oil Exports

Year	Exports (1000 bls./day)
2001	971
2002	984
2003	1027
2004	1048
2005	1165
2006	1317
2007	1433
2008	1802
2009	2024
2010	2353
2011	2924

Source: Monthly Energy Review, Aug. 2012, as reported in The World Almanac and Book of Facts 2013.

(a) Sketch a scatter plot of export numbers (y) as a function of years since 2000 (x).

(b) Use quadratic regression to model the data with a quadratic function and superimpose the parabolic curve on the scatter plot. Is it a good fit?

(c) Based on the quadratic model, approximately how many thousands of barrels of oil would the United States export per day in 2016?

66. The winning times in the women's 100-meter freestyle event at the Summer Olympic Games since 1960 are shown in Table 1.16.

Table 1.16 Women's 100-Meter Freestyle

Year	Time	Year	Time
1960	61.2	1988	54.93
1964	59.5	1992	54.64
1968	60.0	1996	54.50
1972	58.59	2000	53.83
1976	55.65	2004	53.84
1980	54.79	2008	53.12
1984	55.92	2012	53.00

Source: The World Almanac and Book of Facts 2013.

(a) Sketch a scatter plot of the times (y) as a function of the years (x) beyond 1900. (The values of x will run from 60 to 112.)

(b) Explain why a linear model cannot be appropriate for these times over the long term.

(c) The points appear to be approaching a horizontal asymptote of $y = 52$. What would this mean about the times in this Olympic event?

(d) Subtract 52 from all the times so that they will approach an asymptote of $y = 0$. Redo the scatter plot with the new y-values. Now find the *exponential* regression curve and superimpose its graph on the vertically shifted scatter plot.

(e) According to the regression curve, what will be the winning time in the women's 100-meter freestyle event at the 2020 Olympics? (Don't forget to compensate for the 52-sec data shift in (d).)

67. Inscribing a Cylinder Inside a Sphere A right circular cylinder of radius r and height h is inscribed inside a sphere of radius $\sqrt{3}$ in.

(a) Use the Pythagorean Theorem to write h as a function of r.

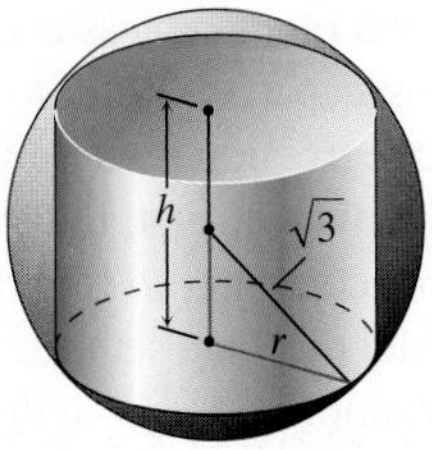

(b) Write the volume V of the cylinder as a function of r.

(c) What values of r are in the domain of V?

(d) Sketch a graph of $V(r)$ over the domain $[0, \sqrt{3}]$.

(e) Use your grapher to find the maximum volume that such a cylinder can have.

68. Inscribing a Rectangle Under a Parabola A rectangle is inscribed between the x-axis and the parabola $y = 36 - x^2$ with one side along the x-axis, as shown in the figure below.

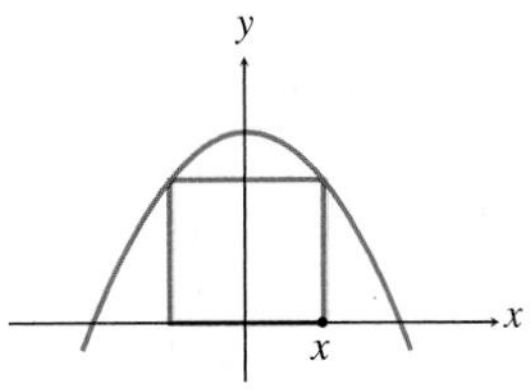

(a) Let x denote the x-coordinate of the point denoted in the figure. Write the area A of the rectangle as a function of x.

(b) What values of x are in the domain of A?

(c) Sketch a graph of $A(x)$ over the domain.

(d) Use your grapher to find the maximum area that such a rectangle can have.

CHAPTER 1 Project

Modeling the Growth of a Business

In 1971, Starbucks Coffee opened its first location in Pike Place Market—Seattle's legendary open-air farmer's market. By 1987, the number of Starbucks stores had grown to 17 and by 2005 there were 10,241 locations. The data in the table below (obtained from Starbucks Coffee's Web site, www.starbucks.com) summarize the growth of this company from 1987 through 2005.

Year	Number of Locations
1987	17
1989	55
1991	116
1993	272
1995	676
1997	1412
1999	2498
2001	4709
2003	7225
2005	10,241

Explorations

1. Enter the data in the table into your grapher or computer. (Let $t = 0$ represent 1987.) Draw a scatter plot for the data.
2. Refer to page 145 in this chapter. Look at the types of graphs displayed and the associated regression types. Notice that the exponential regression model with $b > 1$ seems to most closely match the plotted data. Use your grapher or computer to find an exponential regression equation to model this data set (see your grapher's guidebook for instructions on how to do this).
3. Use the exponential model you just found to predict the total number of Starbucks locations for 2007 and 2008.
4. There were 15,011 Starbucks locations in 2007 and 16,680 locations in 2008. (You can verify these numbers and find more up-to-date information in the investors' section of the Starbucks Web site.) Why is there such a big difference between your predicted values and the actual number of Starbucks locations? What real-world feature of business growth was not accounted for in the exponential growth model?
5. You need to model the data set with an equation that takes into account the fact that growth of a business eventually levels out or reaches a carrying capacity. Refer to page 145 again. Notice that the logistic regression modeling graph appears to show exponential growth at first, but eventually levels out. Use your grapher or computer to find the logistic regression equation to model this data set (see your grapher's guidebook for instructions on how to do this).
6. Use the logistic model you just found to predict the total number of Starbucks locations for 2007 and 2008. How do your predictions compare with the actual number of locations for 2007 and 2008? How many locations do you think there will be in the year 2020?

CHAPTER 2

Polynomial, Power, and Rational Functions

2.1 Linear and Quadratic Functions and Modeling

2.2 Power Functions with Modeling

2.3 Polynomial Functions of Higher Degree with Modeling

2.4 Real Zeros of Polynomial Functions

2.5 Complex Zeros and the Fundamental Theorem of Algebra

2.6 Graphs of Rational Functions

2.7 Solving Equations in One Variable

2.8 Solving Inequalities in One Variable

Humidity and relative humidity are measures used by weather forecasters. Humidity affects our comfort and our health. If humidity is too low, our skin can become dry and cracked, and viruses can live longer. If humidity is too high, it can make warm temperatures feel even warmer, and mold, fungi, and dust mites can live longer. See page 224 to learn how relative humidity is modeled as a rational function.

Chapter 2 Overview

Chapter 1 laid a foundation of the general characteristics of functions, equations, and graphs. In this chapter and the next two, we will explore the theory and applications of specific families of functions. We begin this exploration by studying three interrelated families of functions: polynomial, power, and rational functions. These three families of functions are used in the social, behavioral, and natural sciences.

This chapter includes a thorough study of the theory of polynomial equations. We investigate algebraic methods for finding both real- and complex-number solutions of such equations and relate these methods to the graphical behavior of polynomial and rational functions. The chapter closes by extending these methods to inequalities in one variable.

2.1 Linear and Quadratic Functions and Modeling

What you'll learn about

- Polynomial Functions
- Linear Functions and Their Graphs
- Average Rate of Change
- Association, Correlation, and Linear Modeling
- Quadratic Functions and Their Graphs
- Applications of Quadratic Functions

... and why

Many business and economic problems are modeled by linear functions. Quadratic and higher-degree polynomial functions are used in science and manufacturing applications.

Polynomial Functions

Polynomial functions are among the most familiar of all functions.

DEFINITION Polynomial Function

Let n be a nonnegative integer and let $a_0, a_1, a_2, \ldots, a_{n-1}, a_n$ be real numbers with $a_n \neq 0$. The function given by

$$f(x) = a_n x^n + a_{n-1}x^{n-1} + \cdots + a_2x^2 + a_1x + a_0$$

is a **polynomial function of degree n**. The **leading coefficient** is a_n.
The zero function $f(x) = 0$ is a polynomial function. It has no degree and no leading coefficient.

Polynomial functions are defined and continuous on all real numbers. It is important to recognize whether a function is a polynomial function.

EXAMPLE 1 Identifying Polynomial Functions

Which of the following are polynomial functions? For those that are polynomial functions, state the degree and leading coefficient. For those that are not, explain why not.

(a) $f(x) = 4x^3 - 5x - \frac{1}{2}$ **(b)** $g(x) = 6x^{-4} + 7$

(c) $h(x) = \sqrt{9x^4 + 16x^2}$ **(d)** $k(x) = 15x - 2x^4$

SOLUTION

(a) f is a polynomial function of degree 3 with leading coefficient 4.

(b) g is not a polynomial function because of the exponent -4.

(c) h is not a polynomial function because it cannot be simplified into polynomial form. Notice that $\sqrt{9x^4 + 16x^2} \neq 3x^2 + 4x$.

(d) k is a polynomial function of degree 4 with leading coefficient -2.

Now try Exercise 1.

The zero function and all constant functions are polynomial functions. Some other familiar functions are also polynomial functions, as shown below.

Polynomial Functions of No and Low Degree

Name	Form	Degree
Zero function	$f(x) = 0$	Undefined
Constant function	$f(x) = a \ (a \neq 0)$	0
Linear function	$f(x) = ax + b \ (a \neq 0)$	1
Quadratic function	$f(x) = ax^2 + bx + c \ (a \neq 0)$	2

We study polynomial functions of degree 3 and higher in Section 2.3. For the remainder of this section, we turn our attention to the nature and uses of linear and quadratic polynomial functions.

Linear Functions and Their Graphs

Linear equations and graphs of lines were reviewed in Sections P.3 and P.4, and some of the examples in Chapter 1 involved linear functions. We now take a closer look at the properties of linear functions.

A **linear function** is a polynomial function of degree 1 and so has the form

$$f(x) = ax + b, \text{ where } a \text{ and } b \text{ are constants and } a \neq 0.$$

If we use m for the leading coefficient instead of a and let $y = f(x)$, then this equation becomes the familiar slope-intercept form of a line:

$$y = mx + b.$$

Vertical lines are not graphs of functions because they fail the vertical line test, and horizontal lines are graphs of constant functions. A line in the Cartesian plane is the graph of a linear function if and only if it is a **slant line**, that is, neither horizontal nor vertical. We can apply the formulas and methods of Section P.4 to problems involving linear functions.

Surprising Fact

Not all lines in the Cartesian plane are graphs of linear functions.

EXAMPLE 2 Finding an Equation of a Linear Function

Write an equation for the linear function f such that $f(-1) = 2$ and $f(3) = -2$.

SOLUTION

Solve Algebraically We seek a line through the points $(-1, 2)$ and $(3, -2)$. The slope is

$$m = \frac{y_2 - y_1}{x_2 - x_1} = \frac{(-2) - 2}{3 - (-1)} = \frac{-4}{4} = -1.$$

Using this slope and the coordinates of $(-1, 2)$ with the point-slope formula, we have

$$\begin{aligned} y - y_1 &= m(x - x_1) \\ y - 2 &= -1(x - (-1)) \\ y - 2 &= -x - 1 \\ y &= -x + 1 \end{aligned}$$

Converting to function notation gives us the desired form:

$$f(x) = -x + 1$$

(continued)

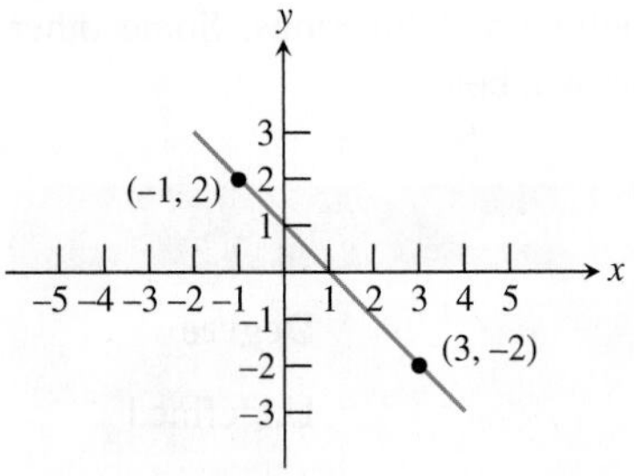

FIGURE 2.1 The graph of $y = -x + 1$ passes through $(-1, 2)$ and $(3, -2)$. (Example 2)

Support Graphically We can graph $y = -x + 1$ and see that it includes the points $(-1, 2)$ and $(3, -2)$ (Figure 2.1).

Confirm Numerically Using $f(x) = -x + 1$ we prove that $f(-1) = 2$ and $f(3) = -2$:

$$f(-1) = -(-1) + 1 = 1 + 1 = 2, \text{ and } f(3) = -(3) + 1 = -3 + 1 = -2$$

Now try Exercise 7.

Average Rate of Change

Another property that characterizes a linear function is its *rate of change*. The **average rate of change** of a function $y = f(x)$ between $x = a$ and $x = b$, $a \neq b$, is

$$\frac{f(b) - f(a)}{b - a}.$$

You are asked to prove the following theorem in Exercise 85.

THEOREM Constant Rate of Change

A function defined on all real numbers is a linear function if and only if it has a constant nonzero average rate of change between any two points on its graph.

Constant Functions

The rate of change for any constant function is 0.

Because the average rate of change of a linear function is constant, it is called simply the **rate of change** of the linear function. The slope m in the formula $f(x) = mx + b$ is the rate of change of the linear function. In Exploration 1, we revisit Example 7 of Section P.4 in light of the rate of change concept.

EXPLORATION 1 Modeling Depreciation with a Linear Function

Camelot Apartments bought a $50,000 building and for tax purposes are depreciating it $2000 per year over a 25-yr period using straight-line depreciation.

1. What is the rate of change of the value of the building?
2. Write an equation for the value $v(t)$ of the building as a linear function of the time t since the building was placed in service.
3. Evaluate $v(0)$ and $v(16)$.
4. Solve $v(t) = 39{,}000$.

Rate and Ratio

All rates are ratios, whether expressed as miles per hour, dollars per year, or even rise over run.

The rate of change of a linear function is the signed ratio of the corresponding line's rise over run. That is, for a linear function $f(x) = mx + b$,

$$\text{rate of change} = \text{slope} = m = \frac{\text{rise}}{\text{run}} = \frac{\text{change in } y}{\text{change in } x} = \frac{\Delta y}{\Delta x}.$$

This formula allows us to interpret the slope, or rate of change, of a linear function numerically. For instance, in Exploration 1 the value of the apartment building fell from $50,000 to $0 over a 25-yr period. In Table 2.1 we compute $\Delta y/\Delta x$ for the apartment building's value (in dollars) as a function of time (in years). Because the average rate of change $\Delta y/\Delta x$ is the nonzero constant -2000, the building's value is a linear function of time decreasing at a rate of $2000/yr.

Table 2.1 Rate of Change of the Value of the Apartment Building in Exploration 1: $y = -2000x + 50{,}000$

x (time)	y (value)	Δx	Δy	$\Delta y/\Delta x$
0	50,000			
		1	−2000	−2000
1	48,000			
		1	−2000	−2000
2	46,000			
		1	−2000	−2000
3	44,000			
		1	−2000	−2000
4	42,000			

In Exploration 1, as in other applications of linear functions, the constant term represents the value of the function for an input of 0. In general, for any function f, $f(0)$ is the **initial value of f**. So for a linear function $f(x) = mx + b$, the constant term b is the initial value of the function. For any polynomial function $f(x) = a_n x^n + \cdots + a_1 x + a_0$, the **constant term** $f(0) = a_0$ is the function's initial value. Finally, the initial value of any function—polynomial or otherwise—is the y-intercept of its graph.

We now summarize what we have learned about linear functions.

Characterizing the Nature of a Linear Function

Point of View	Characterization
Verbal	polynomial of degree 1
Algebraic	$f(x) = mx + b \ (m \neq 0)$
Graphical	slant line with slope m and y-intercept b
Analytical	function with constant nonzero rate of change m: f is increasing if $m > 0$, decreasing if $m < 0$; initial value of the function $= f(0) = b$

Association, Correlation, and Linear Modeling

In Section 1.7 we approached modeling from several points of view. Along the way you learned how to use a grapher to create a scatter plot, compute a regression line for a data set, and overlay a regression line on a scatter plot. We touched on the notion of correlation coefficient. We now go deeper into these modeling and regression concepts.

Figure 2.2 on page 162 shows five types of scatter plots. When the points of a scatter plot are clustered along a line, we say there is a **linear association** between the quantities represented by the data. When an oval is drawn around the points in the scatter plot, generally speaking, the narrower the oval, the stronger the linear association.

When the oval tilts like a line with positive slope (as in Figure 2.2a and b), the data have a **positive linear association**. On the other hand, when it tilts like a line with negative slope (as in Figure 2.2d and e), the data have a **negative linear association**. Some scatter plots exhibit little or no linear association (as in Figure 2.2c), or have nonlinear patterns.

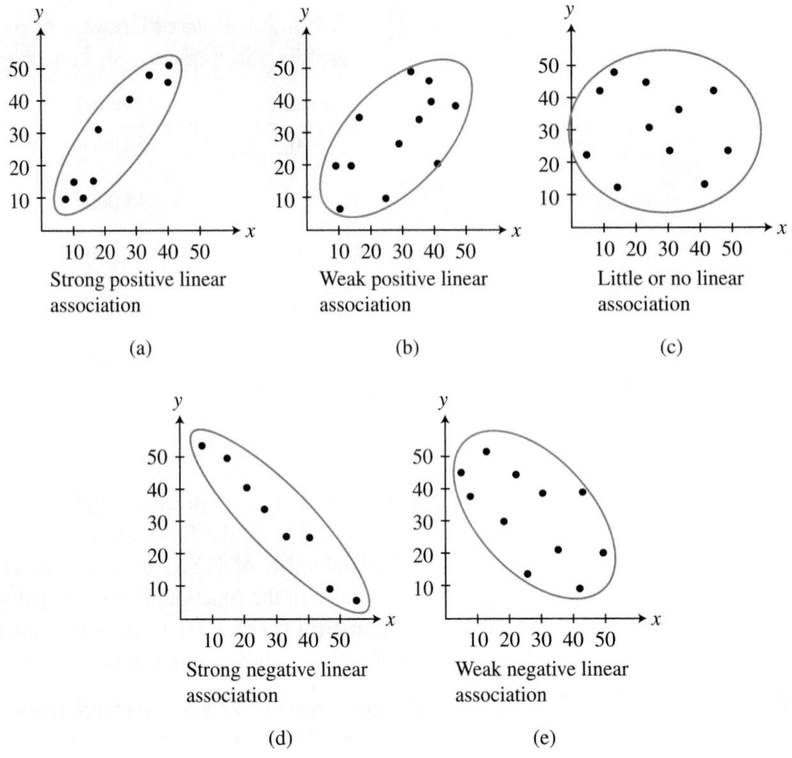

FIGURE 2.2 Five scatter plots and the types of linear association they suggest.

When a scatter plot shows evidence of a linear association in the data set, we measure the strength and direction of the association with the **correlation coefficient, r.**

Note that *correlation* describes linear associations only. If the scatter plot shows a curved relationship, do not calculate r.

Correlation vs. Causation

Correlation does not imply causation. Two variables can be strongly correlated, but that does not necessarily mean that one *causes* the other.

For example, the strong correlation between shoe size and reading comprehension among elementary school children does not imply that reading ability resides in one's feet. As kids get taller, their feet grow and they become better readers.

Properties of the Correlation Coefficient, r

1. $-1 \le r \le 1$.
2. When there is a positive linear association, $r > 0$.
3. When there is a negative linear association, $r < 0$.
4. When there is a strong linear association, $|r| \approx 1$.
5. When $r \approx 0$ there is weak or no linear association.

Correlation informs the linear modeling process by giving us a measure of goodness of fit. Good modeling practice, however, demands that we have a theoretical reason for selecting a model. In business, for example, fixed cost is modeled by a constant function. (Otherwise, the cost would not be fixed.)

In economics, a linear model is often used for the demand for a product as a function of its price. For instance, suppose Twin Pixie, a large supermarket chain, conducts a market analysis on its store brand of doughnut-shaped oat breakfast cereal. The chain sets various prices for its 15-oz box at its different stores over a period of time. Then, using these data, the Twin Pixie researchers predict the weekly sales at the entire chain of stores for each price and obtain the data shown in Table 2.2.

Table 2.2 Weekly Sales Data Based on Marketing Research

Price per Box	Boxes Sold
\$2.40	38,320
\$2.60	33,710
\$2.80	28,280
\$3.00	26,550
\$3.20	25,530
\$3.40	22,170
\$3.60	18,260

EXAMPLE 3 Modeling and Predicting Demand

Use the data in Table 2.2 to write a linear model for demand (in boxes sold per week) as a function of the price per box (in dollars). Describe the strength and direction of the linear association. Then use the model to predict weekly cereal sales if the price is dropped to \$2.00 or raised to \$4.00 per box.

SOLUTION

Model We enter the data and obtain the scatter plot shown in Figure 2.3a. It appears that the data have a strong negative linear association.

We then find the linear regression model to be approximately

$$y = -15{,}358.93x + 73{,}622.50,$$

where x is the price per box of cereal and y the number of boxes sold.

Figure 2.3b shows the scatter plot for Table 2.2 together with a graph of the regression line. You can see that the line fits the data fairly well. The correlation coefficient of $r \approx -0.98$ supports this visual evidence.

Solve Graphically Our goal is to predict the weekly sales for prices of \$2.00 and \$4.00 per box. Using the value feature of the grapher, as shown in Figure 2.3c, we see that y is about 42,900 when x is 2. In a similar manner we could find that $y \approx 12{,}190$ when x is 4.

Interpret If Twin Pixie drops the price for its store brand of doughnut-shaped oat breakfast cereal to \$2.00 per box, the linear model predicts demand will rise to about 42,900 boxes per week. On the other hand, if they raise the price to \$4.00 per box, the model predicts demand will drop to around 12,190 boxes per week.

Now try Exercise 49.

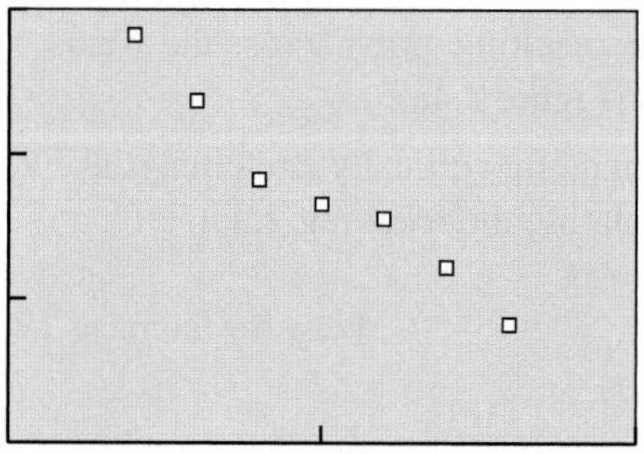

[2, 4] by [10000, 40000]

(a)

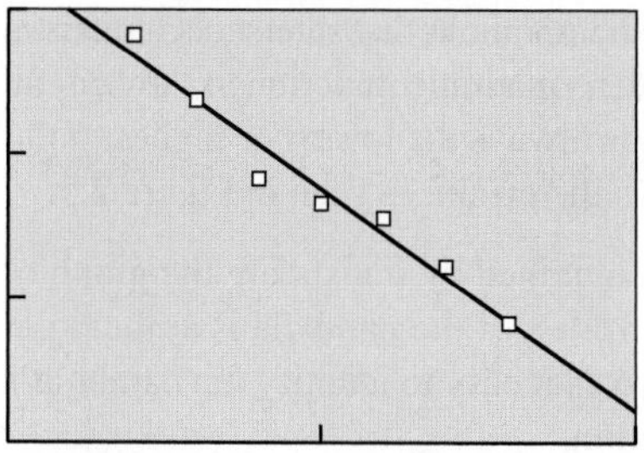

[2, 4] by [10000, 40000]

(b)

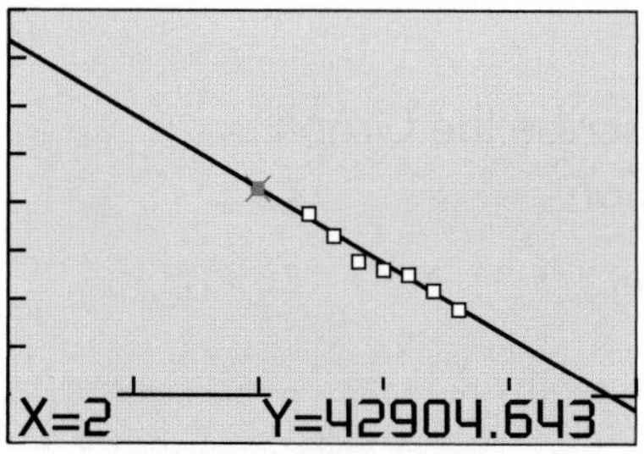

[0, 5] by [–10000, 80000]

(c)

FIGURE 2.3 Scatter plot and regression line graphs for Example 3.

We summarize for future reference the analysis used in Example 3.

Regression Analysis

1. Enter and plot the data (scatter plot). Check for evidence of linearity.
2. Find the regression model that fits the problem situation.
3. Superimpose the graph of the regression model on the scatter plot, and observe the fit.
4. Use the regression model to make the predictions called for in the problem.

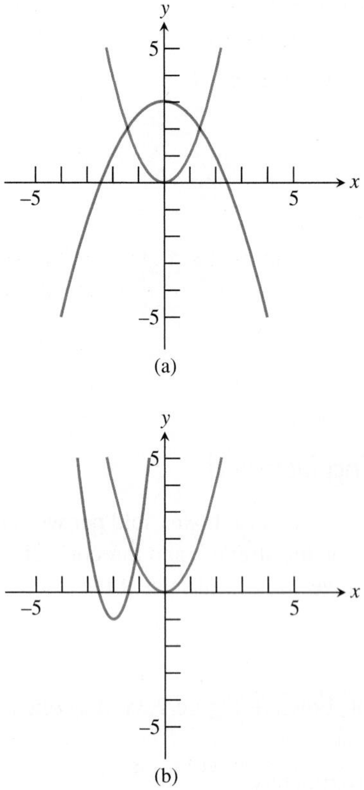

FIGURE 2.4 The graph of $f(x) = x^2$ (blue) shown with
(a) $g(x) = -(1/2)x^2 + 3$ and
(b) $h(x) = 3(x + 2)^2 - 1$. (Example 4)

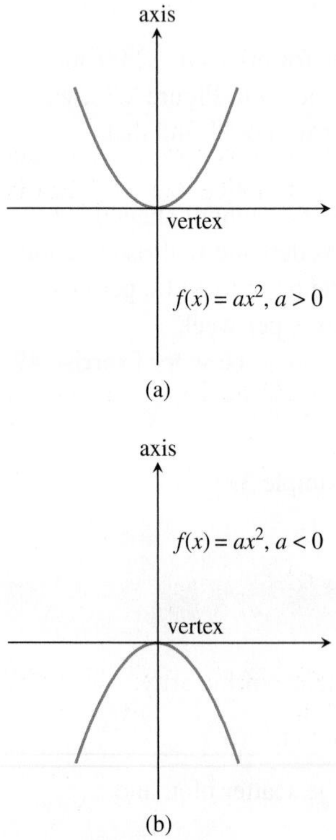

FIGURE 2.5 The graph $f(x) = ax^2$ for (a) $a > 0$ and (b) $a < 0$.

Quadratic Functions and Their Graphs

A **quadratic function** is a polynomial function of degree 2. Recall from Section 1.3 that the graph of the squaring function $f(x) = x^2$ is a parabola. We will see that the graph of every quadratic function is an upward- or downward-opening parabola. This is because the graph of any quadratic function can be obtained from the graph of the squaring function $f(x) = x^2$ by a sequence of translations, reflections, stretches, and shrinks.

EXAMPLE 4 Transforming the Squaring Function

Describe how to transform the graph of $f(x) = x^2$ into the graph of the given function. Sketch its graph by hand.

(a) $g(x) = -(1/2)x^2 + 3$ **(b)** $h(x) = 3(x + 2)^2 - 1$

SOLUTION

(a) The graph of $g(x) = -(1/2)x^2 + 3$ is obtained by vertically shrinking the graph of $f(x) = x^2$ by a factor of 1/2, reflecting the resulting graph across the x-axis, and translating the reflected graph up 3 units (Figure 2.4a).

(b) The graph of $h(x) = 3(x + 2)^2 - 1$ is obtained by vertically stretching the graph of $f(x) = x^2$ by a factor of 3 and translating the resulting graph left 2 units and down 1 unit (Figure 2.4b).

Now try Exercise 19.

The graph of $f(x) = ax^2$, $a > 0$, is an upward-opening parabola. When $a < 0$, its graph is a downward-opening parabola. Regardless of the sign of a, the y-axis is the line of symmetry for the graph of $f(x) = ax^2$. The line of symmetry for a parabola is its **axis of symmetry**, or **axis** for short. The point on the parabola that intersects its axis is the **vertex** of the parabola. Because the graph of a quadratic function is always an upward- or downward-opening parabola, its vertex is always the lowest or highest point of the parabola. The vertex of $f(x) = ax^2$ is always the origin, as seen in Figure 2.5.

In general, the graph of $f(x) = a(x - h)^2 + k$ is obtained by translating the graph of $f(x) = ax^2$ horizontally and vertically so that the vertex of the parabola is at the point (h, k). This *vertex form* for a quadratic function makes it easy to identify the parabola's vertex and axis and to sketch the graph of the function.

Any function written in **standard quadratic form** $y = ax^2 + bx + c$ can be rewritten in vertex form by completing the square.

EXAMPLE 5 Using Algebra to Describe the Graph of a Quadratic Function

Use completing the square to describe the graph of $f(x) = 3x^2 + 12x + 11$. Support your answer graphically.

SOLUTION

Solve Algebraically

$$
\begin{aligned}
f(x) &= 3x^2 + 12x + 11 \\
&= 3(x^2 + 4x) + 11 && \text{Factor 3 from the x-terms.} \\
&= 3(x^2 + 4x + (\ ) - (\ )) + 11 && \text{Prepare to complete the square.} \\
&= 3(x^2 + 4x + (2^2) - (2^2)) + 11 && \text{Complete the square.} \\
&= 3(x^2 + 4x + 4) - 3(4) + 11 && \text{Distribute the 3.} \\
&= 3(x + 2)^2 - 1
\end{aligned}
$$

The graph of f is an upward-opening parabola with vertex $(-2, -1)$, axis of symmetry $x = -2$. (The x-intercepts are $x = -2 \pm \sqrt{3}/3$, or about -2.577 and -1.423.)

Support Graphically The graph in Figure 2.6 supports these results.

Now try Exercise 33.

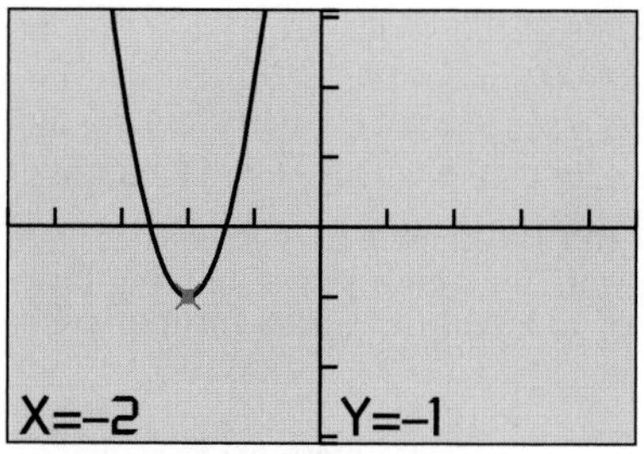

[–4.7, 4.7] by [–3.1, 3.1]

FIGURE 2.6 The graphs of $f(x) = 3x^2 + 12x + 11$ and $y = 3(x + 2)^2 - 1$ appear to be identical. The vertex $(-2, -1)$ is highlighted. (Example 5)

Expanding $f(x) = a(x - h)^2 + k$ and comparing the resulting coefficients with the standard quadratic form $ax^2 + bx + c$, where the powers of x are arranged in descending order, we can obtain formulas for h and k.

$$\begin{aligned} f(x) &= a(x - h)^2 + k \\ &= a(x^2 - 2hx + h^2) + k && \text{Expand } (x - h)^2. \\ &= ax^2 + (-2ah)x + (ah^2 + k) && \text{Distributive property} \\ &= ax^2 + bx + c && \text{Let } b = -2ah \text{ and } c = ah^2 + k. \end{aligned}$$

Because $b = -2ah$ and $c = ah^2 + k$ in the last line above, $h = -b/(2a)$ and $k = c - ah^2$.

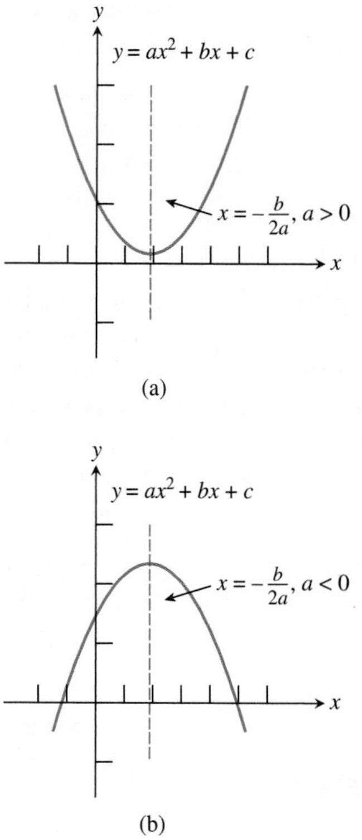

FIGURE 2.7 The vertex is at $x = -b/(2a)$, which therefore also describes the axis of symmetry. (a) When $a > 0$, the parabola opens upward. (b) When $a < 0$, the parabola opens downward.

Vertex Form of a Quadratic Function

Any quadratic function $f(x) = ax^2 + bx + c$, $a \neq 0$, can be written in the **vertex form**

$$f(x) = a(x - h)^2 + k.$$

The graph of f is a parabola with vertex (h, k) and axis $x = h$, where $h = -b/(2a)$. If $a > 0$, the parabola opens upward, and if $a < 0$, it opens downward (Figure 2.7).

The formula $h = -b/(2a)$ is useful for locating the vertex and axis of the parabola associated with a quadratic function. To help you remember it, notice that $-b/(2a)$ is part of the quadratic formula

$$x = \frac{-b \pm \sqrt{b^2 - 4ac}}{2a}.$$

(Cover the radical term.) You need not remember $k = c - ah^2$ because you can use $k = f(h)$ instead, as illustrated in Example 6.

EXAMPLE 6 Finding the Vertex and Axis of a Quadratic Function

Use the vertex form of a quadratic function to find the vertex and axis of the graph of $f(x) = 6x - 3x^2 - 5$. Rewrite the equation in vertex form.

SOLUTION

Solve Algebraically The standard polynomial form of f is

$$f(x) = -3x^2 + 6x - 5.$$

(continued)

So $a = -3$, $b = 6$, and $c = -5$, and the coordinates of the vertex are

$$h = -\frac{b}{2a} = -\frac{6}{2(-3)} = 1 \text{ and}$$
$$k = f(h) = f(1) = -3(1)^2 + 6(1) - 5 = -2.$$

The equation of the axis is $x = 1$, the vertex is $(1, -2)$, and the vertex form of f is

$$f(x) = -3(x - 1)^2 + (-2).$$

Now try Exercise 27.

We now summarize what we know about quadratic functions.

Characterizing the Nature of a Quadratic Function

Point of View	Characterization
Verbal	polynomial of degree 2
Algebraic	$f(x) = ax^2 + bx + c$ or $f(x) = a(x - h)^2 + k \ (a \neq 0)$
Graphical	parabola with vertex (h, k) and axis $x = h$; opens upward if $a > 0$, opens downward if $a < 0$; initial value = y-intercept = $f(0) = c$; $x\text{-intercepts} = \dfrac{-b \pm \sqrt{b^2 - 4ac}}{2a}$

Applications of Quadratic Functions

In economics, when demand is linear, revenue is quadratic. Example 7 illustrates this by extending the Twin Pixie model of Example 3.

EXAMPLE 7 Predicting Maximum Revenue

Use the model $y = -15{,}358.93x + 73{,}622.50$ from Example 3 to develop a model for the weekly revenue generated by doughnut-shaped oat breakfast cereal sales. Determine the maximum revenue and how to achieve it.

SOLUTION

Model Revenue can be found by multiplying the price per box, x, by the number of boxes sold, y. So the revenue is given by

$$R(x) = x \cdot y = -15{,}358.93x^2 + 73{,}622.50x,$$

a quadratic model.

Solve Graphically In Figure 2.8, we find a maximum of about 88,227 occurs when x is about 2.40.

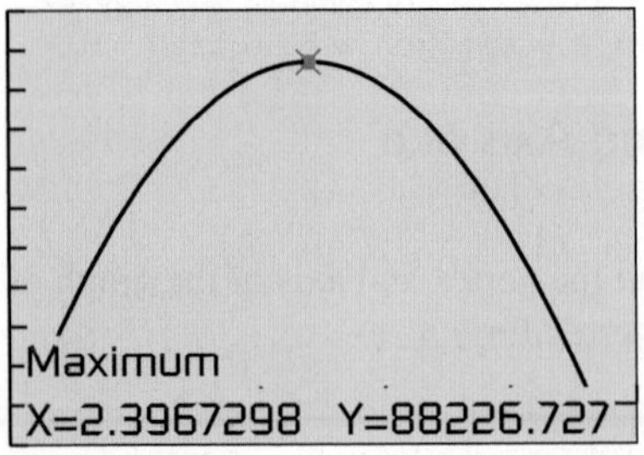

[0, 5] by [–10000, 100000]

FIGURE 2.8 The revenue model for Example 7.

Interpret To maximize revenue, Twin Pixie should set the price for its store brand of doughnut-shaped oat breakfast cereal at \$2.40 per box. Based on the model, this will yield a weekly revenue of about \$88,227.

Now try Exercise 55.

Recall that the average rate of change of a linear function is constant. In Exercise 78 you will see that the average rate of change of a quadratic function is not constant.

In calculus you will study not only average rate of change but also *instantaneous rate of change*. Such instantaneous rates include velocity and acceleration, which we now begin to investigate.

Since the time of Galileo Galilei (1564–1642) and Isaac Newton (1642–1727), the vertical motion of a body in free fall has been well understood. The *vertical velocity* and *vertical position* (*height*) of a free-falling body (as functions of time) are classical applications of linear and quadratic functions.

Vertical Free-Fall Motion

The **height** s and **vertical velocity** v of an object in free fall are given by

$$s(t) = -\frac{1}{2}gt^2 + v_0t + s_0 \quad \text{and} \quad v(t) = -gt + v_0,$$

where t is time (in seconds), $g \approx 32 \text{ ft/sec}^2 \approx 9.8 \text{ m/sec}^2$ is the **acceleration due to gravity**, v_0 is the *initial vertical velocity* of the object, and s_0 is its *initial height.*

These formulas disregard air resistance, and the two values given for g are valid at sea level. We apply these formulas in Example 8, and will use them from time to time throughout the rest of the book.

The data in Table 2.3 were collected in Boone, North Carolina (about 1 km above sea level), using a Calculator-Based Ranger™ (CBR™) and a 15-cm rubber air-filled ball. The CBR™ was placed on the floor face up. The ball was thrown upward above the CBR™, and it landed directly on the face of the device.

EXAMPLE 8 Modeling Vertical Free-Fall Motion

Use the data in Table 2.3 to write models for the height and vertical velocity of the rubber ball. Then use these models to predict the maximum height of the ball and its vertical velocity when it hits the face of the CBR™.

SOLUTION

Model First we make a scatter plot of the data, as shown in Figure 2.9a. We see an association that is curved and strong. Using quadratic regression, we find the model for the height of the ball to be about

$$s(t) = -4.676t^2 + 3.758t + 1.045,$$

with $R^2 \approx 0.999$, indicating an excellent fit.

Our free-fall theory says the leading coefficient of -4.676 is $-g/2$, giving us a value for $g \approx 9.352 \text{ m/sec}^2$, which is a bit less than the theoretical value of 9.8 m/sec^2. We also obtain $v_0 \approx 3.758$ m/sec. So the model for vertical velocity becomes

$$v(t) = -gt + v_0 \approx -9.352t + 3.758.$$

(continued)

Table 2.3 Rubber Ball Data from CBR™

Time (sec)	Height (m)
0.0000	1.03754
0.1080	1.40205
0.2150	1.63806
0.3225	1.77412
0.4300	1.80392
0.5375	1.71522
0.6450	1.50942
0.7525	1.21410
0.8600	0.83173

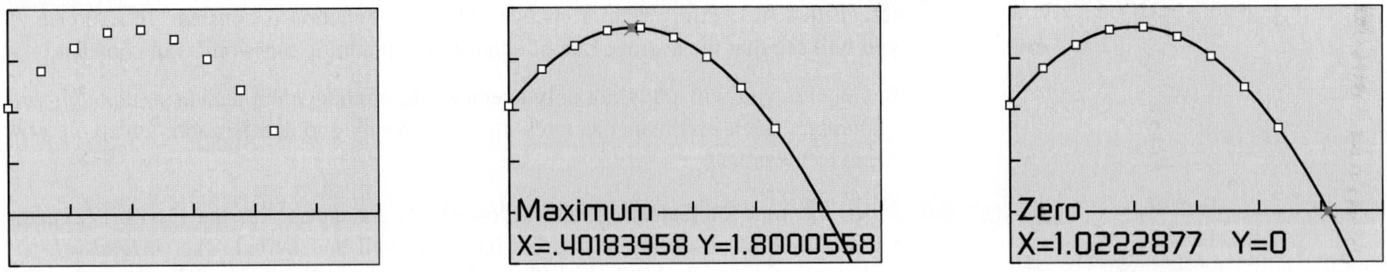

[0, 1.2] by [−0.5, 2.0]
(a)

[0, 1.2] by [−0.5, 2.0]
(b)

[0, 1.2] by [−0.5, 2.0]
(c)

FIGURE 2.9 Scatter plot and graph of height versus time for Example 8.

Solve Graphically and Numerically The maximum height is the maximum value of $s(t)$, which occurs at the vertex of its graph. We can see from Figure 2.9b that the vertex has coordinates of about $(0.402, 1.800)$.

In Figure 2.9c, to determine when the ball hits the face of the CBR™, we calculate the positive-valued zero of the height function, which is $t \approx 1.022$. We turn to our linear model to compute the vertical velocity at impact:

$$v(1.022) = -9.352(1.022) + 3.758 \approx -5.80 \text{ m/sec}$$

Interpret The maximum height the ball achieved was about 1.80 m above the face of the CBR™. The ball's downward rate is about 5.80 m/sec when it hits the CBR™.

The curve in Figure 2.9b appears to fit the data extremely well, and $R^2 \approx 0.999$. You may have noticed, however, that Table 2.3 contains the ordered pair $(0.4300, 1.80392)$ and that $1.80392 > 1.800$, which is the maximum shown in Figure 2.9b. So, even though our model is theoretically based and an excellent fit to the data, like all models, it is not perfect. Despite its imperfections, the model provides accurate and reliable predictions about the CBR™ experiment. **Now try Exercise 63.**

Reminder

Recall from Section 1.7 that R^2 is the coefficient of determination, which measures goodness of fit.

QUICK REVIEW 2.1 *(For help, go to Sections A.2. and P.4)*

Exercise numbers with a gray background indicate problems that the authors have designed to be solved *without a calculator*.

In Exercises 1–2, write an equation in slope-intercept form for a line with the given slope m and y-intercept b.

1. $m = 8, \quad b = 3.6$ **2.** $m = -1.8, \quad b = -2$

In Exercises 3–4, write an equation for the line containing the given points. Graph the line and points.

3. $(-2, 4)$ and $(3, 1)$ **4.** $(1, 5)$ and $(-2, -3)$

In Exercises 5–8, expand the expression.

5. $(x + 3)^2$ **6.** $(x - 4)^2$

7. $3(x - 6)^2$ **8.** $-3(x + 7)^2$

In Exercises 9–10, factor the trinomial.

9. $2x^2 - 4x + 2$ **10.** $3x^2 + 12x + 12$

SECTION 2.1 Exercises

In Exercises 1–6, determine which are polynomial functions. For those that are, state the degree and leading coefficient. For those that are not, explain why not.

1. $f(x) = 3x^{-5} + 17$ **2.** $f(x) = -9 + 2x$

3. $f(x) = 2x^5 - \frac{1}{2}x + 9$ **4.** $f(x) = 13$

5. $h(x) = \sqrt[3]{27x^3 + 8x^6}$ **6.** $k(x) = 4x - 5x^2$

In Exercises 7–12, write an equation for the linear function f satisfying the given conditions. Graph $y = f(x)$.

7. $f(-5) = -1$ and $f(2) = 4$

8. $f(-3) = 5$ and $f(6) = -2$

9. $f(-4) = 6$ and $f(-1) = 2$

10. $f(1) = 2$ and $f(5) = 7$

11. $f(0) = 3$ and $f(3) = 0$

12. $f(-4) = 0$ and $f(0) = 2$

In Exercises 13–18, match a graph to the function. Explain your choice.

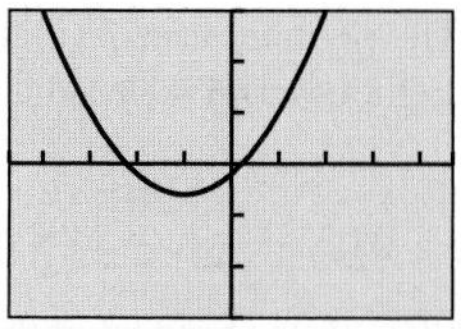
(a)

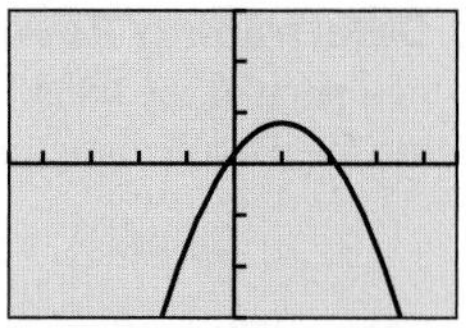
(b)

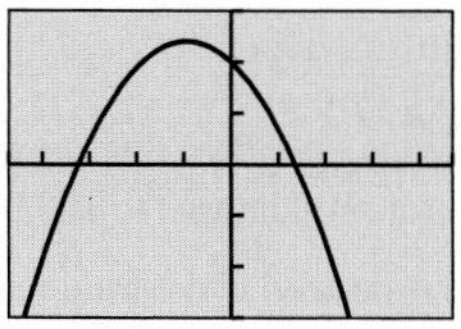
(c)

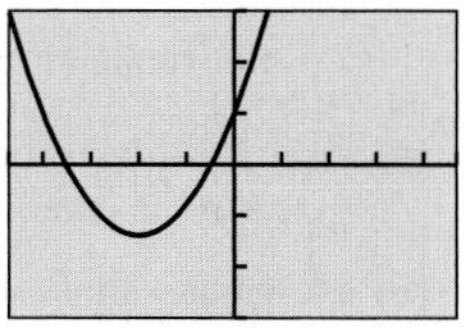
(d)

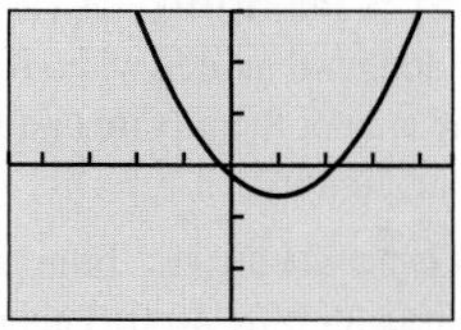
(e)

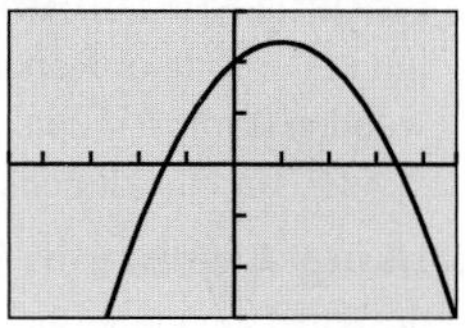
(f)

13. $f(x) = 2(x + 1)^2 - 3$ **14.** $f(x) = 3(x + 2)^2 - 7$

15. $f(x) = 4 - 3(x - 1)^2$ **16.** $f(x) = 12 - 2(x - 1)^2$

17. $f(x) = 2(x - 1)^2 - 3$ **18.** $f(x) = 12 - 2(x + 1)^2$

In Exercises 19–22, describe how to transform the graph of $f(x) = x^2$ into the graph of the given function. Sketch each graph by hand.

19. $g(x) = (x - 3)^2 - 2$ **20.** $h(x) = \frac{1}{4}x^2 - 1$

21. $g(x) = \frac{1}{2}(x + 2)^2 - 3$ **22.** $h(x) = -3x^2 + 2$

In Exercises 23–26, find the vertex and axis of the graph of the function.

23. $f(x) = 3(x - 1)^2 + 5$ **24.** $g(x) = -3(x + 2)^2 - 1$

25. $f(x) = 5(x - 1)^2 - 7$ **26.** $g(x) = 2(x - \sqrt{3})^2 + 4$

In Exercises 27–32, find the vertex and axis of the graph of the function. Rewrite the equation for the function in vertex form.

27. $f(x) = 3x^2 + 5x - 4$ **28.** $f(x) = -2x^2 + 7x - 3$

29. $f(x) = 8x - x^2 + 3$ **30.** $f(x) = 6 - 2x + 4x^2$

31. $g(x) = 5x^2 + 4 - 6x$ **32.** $h(x) = -2x^2 - 7x - 4$

In Exercises 33–38, use completing the square to describe the graph of each function. Support your answers graphically.

33. $f(x) = x^2 - 4x + 6$ **34.** $g(x) = x^2 - 6x + 12$

35. $f(x) = 10 - 16x - x^2$ **36.** $h(x) = 8 + 2x - x^2$

37. $f(x) = 2x^2 + 6x + 7$ **38.** $g(x) = 5x^2 - 25x + 12$

In Exercises 39–42, write an equation for the parabola shown, using the fact that one of the given points is the vertex.

39.

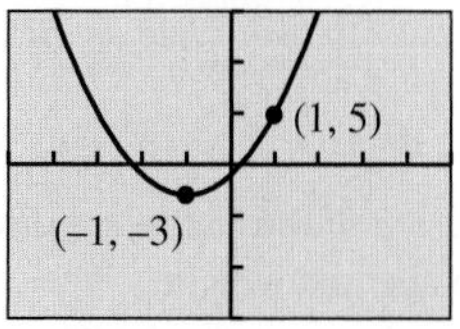

[–5, 5] by [–15, 15]

40.

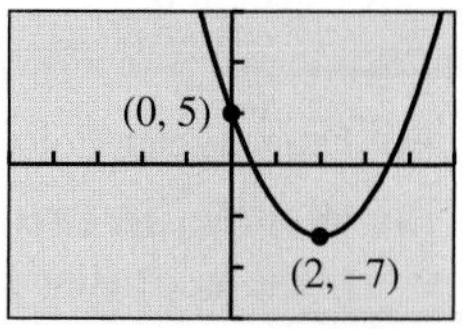

[–5, 5] by [–15, 15]

41.

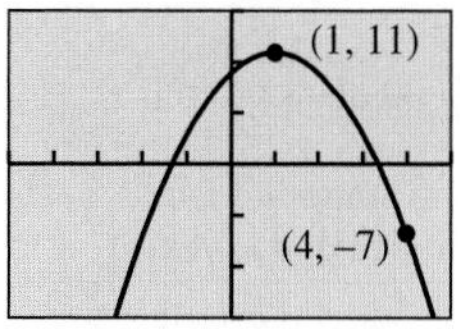

[–5, 5] by [–15, 15]

42.

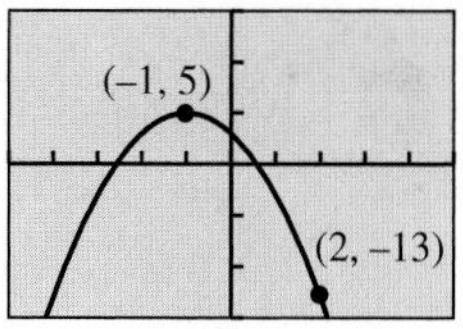

[–5, 5] by [–15, 15]

In Exercises 43 and 44, write an equation for the quadratic function whose graph contains the given vertex and point.

43. Vertex $(1, 3)$, point $(0, 5)$

44. Vertex $(-2, -5)$, point $(-4, -27)$

In Exercises 45–48, describe the strength and direction of the linear correlation.

45.

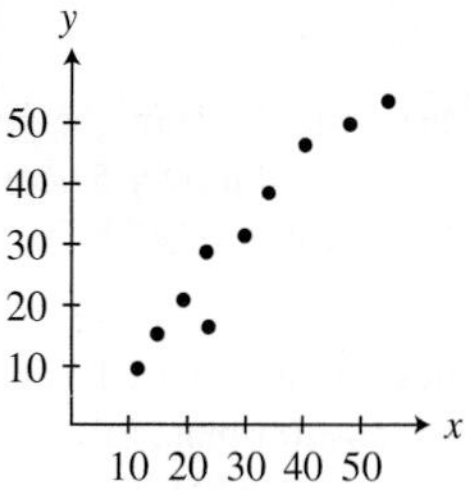

46.

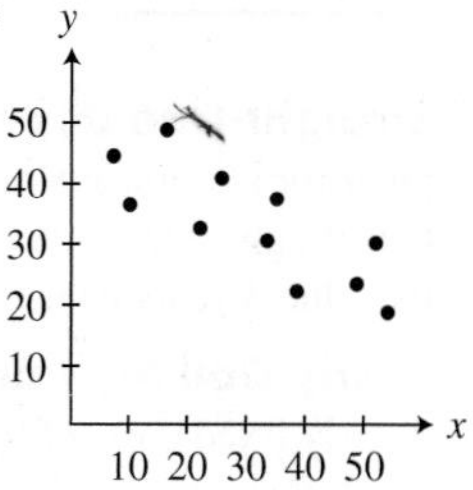

47.

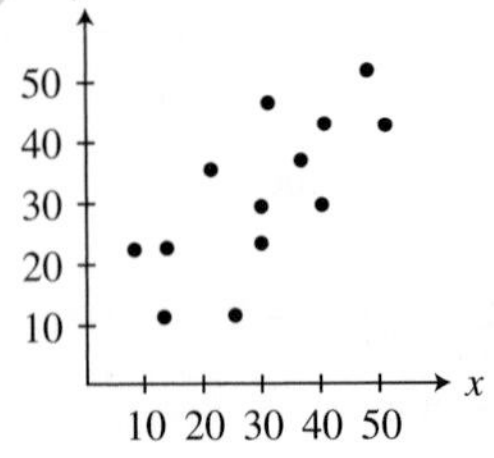

48.

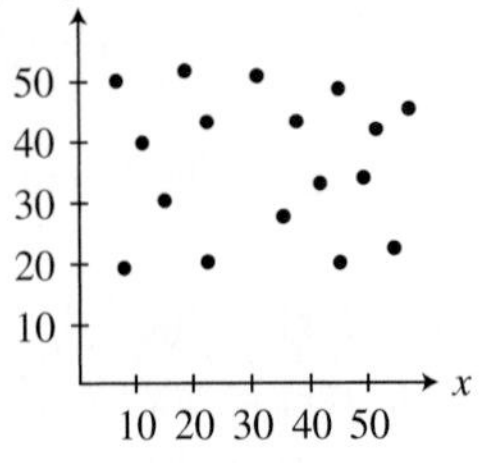

49. Comparing Age and Weight A group of male children were weighed. Their ages and weights are recorded in Table 2.4.

Table 2.4 Children's Age and Weight

Age (months)	Weight (pounds)
18	23
20	25
24	24
26	32
27	33
29	29
34	35
39	39
42	44

(a) Draw a scatter plot of these data.

(b) Writing to Learn Describe the strength and direction of the association between age and weight.

50. Life Expectancy Table 2.5 shows the average number of additional years a U.S. citizen is expected to live for various ages.

Table 2.5 U.S. Life Expectancy

Age (years)	Remaining Life Expectancy (years)
10	68.8
20	59.0
30	49.5
40	40.1
50	31.0
60	22.7
70	15.2

Source: United States Life Tables, 2008, National Vital Statistics Reports, September, 2012.

(a) Draw a scatter plot of these data.

(b) Writing to Learn Describe the strength and direction of the association between age and life expectancy.

51. Straight-Line Depreciation Mai Lee bought a computer for her home office and depreciated it over 5 years using the straight-line method. If its initial value was \$2350, what is its value 3 years later?

52. Costly Doll Making Patrick's doll-making business has weekly fixed costs of \$350. If the cost for materials is \$4.70 per doll and his total weekly costs average \$500, about how many dolls does Patrick make each week?

53. Fuel Economy Table 2.6 shows the average U.S. fuel economy for "light duty" vehicles (passenger cars and small trucks) for several years. Let x be the number of years since 1985, so that $x = 5$ stands for 1990 and so forth.

Table 2.6 Light Duty Vehicles

Year	Fuel Economy (mpg)
1990	20.3
1995	21.1
2000	21.9
2005	22.1
2010	23.5

Source: National Transportation Statistics 2012, U.S. Department of Transportation.

(a) Writing to Learn Find the linear regression model for these data. What does the slope in the regression model represent?

(b) Use the linear regression model to predict the average U.S. fuel economy for light duty vehicles in the year 2015.

54. Finding Maximum Area Among all the rectangles whose perimeters are 100 ft, find the dimensions of the one with maximum area.

55. Determining Revenue The per unit price p (in dollars) of a popular toy when x units (in thousands) are produced is modeled by the function

$$\text{price} = p = 12 - 0.025x.$$

The revenue (in thousands of dollars) is the product of the price per unit and the number of units (in thousands) produced. That is,

$$\text{revenue} = xp = x(12 - 0.025x).$$

(a) State the dimensions of a viewing window that shows a graph of the revenue model for producing 0 to 100,000 units.

(b) How many units should be produced if the total revenue is to be \$1,000,000?

56. Finding the Dimensions of a Painting A large painting in the style of Rubens is 3 ft longer than it is wide. If the wooden frame is 12 in. wide, the area of the picture and frame is 208 ft^2, find the dimensions of the painting.

57. Using Algebra in Landscape Design Julie Stone designed a rectangular patio that is 25 ft by 40 ft. This patio is surrounded by a terraced strip of uniform width planted with small trees and shrubs. If the area A of this terraced strip is 504 ft^2, find the width x of the strip.

58. Management Planning The Welcome Home apartment rental company has 1600 units available, of which 800 are currently rented at \$300 per month. A market survey indicates that each \$5 decrease in monthly rent will result in 20 new leases.

(a) Determine a function $R(x)$ that models the total rental income realized by Welcome Home, where x is the number of \$5 decreases in monthly rent.

(b) Find a graph of $R(x)$ for rent levels between \$175 and \$300 (that is, $0 \le x \le 25$) that clearly shows a maximum for $R(x)$.

(c) What rent will yield Welcome Home the maximum monthly income?

59. Group Activity Beverage Business The Sweet Drip Beverage Co. sells cans of soda pop in machines. It finds that sales average 26,000 cans per month when the cans sell for 50¢ each. For each nickel increase in the price, the sales per month drop by 1000 cans.

(a) Determine a function $R(x)$ that models the total revenue realized by Sweet Drip, where x is the number of \$0.05 increases in the price of a can.

(b) Find a graph of $R(x)$ that clearly shows a maximum for $R(x)$.

(c) How much should Sweet Drip charge per can to realize the maximum revenue? What is the maximum revenue?

60. Group Activity Sales Manager Planning Jack was named District Manager of the Month at the Athens Wire Co. due to his hiring study. It shows that each of the 30 salespersons he supervises average \$50,000 in sales each month, and that for each additional salesperson he would hire, the average sales would decrease \$1000 per month. Jack concluded his study by suggesting a number of salespersons that he should hire to maximize sales. What was that number?

61. Free-Fall Motion As a promotion for the Houston Astros downtown ballpark, a competition is held to see who can throw a baseball the highest from the front row of the upper deck of seats, 83 ft above field level. The winner throws the ball with an initial vertical velocity of 92 ft/sec and it lands on the infield grass.

(a) Find the maximum height of the baseball.

(b) How much time is the ball in the air?

(c) Determine its vertical velocity when it hits the ground.

62. Baseball Throwing Machine The Sandusky Little League uses a baseball throwing machine to help train 10-year-old players to catch high pop-ups. It throws the baseball straight up with an initial velocity of 48 ft/sec from a height of 3.5 ft.

(a) Find an equation that models the height of the ball t seconds after it is thrown.

(b) What is the maximum height the baseball will reach? How many seconds will it take to reach that height?

63. Fireworks Planning At the Bakersville Fourth of July celebration, fireworks are shot by remote control into the air from a pit that is 10 ft below the earth's surface.

(a) Find an equation that models the height of an aerial bomb t seconds after it is shot upward with an initial velocity of 80 ft/sec. Graph the equation.

(b) What is the maximum height above ground level that the aerial bomb will reach? How many seconds will it take to reach that height?

64. Landscape Engineering In her first project after being employed by Land Scapes International, Becky designs a decorative water fountain that will shoot water to a maximum height of 48 ft. What should be the initial velocity of each drop of water to achieve this maximum height? (*Hint*: Use a grapher and a guess-and-check strategy.)

65. Patent Applications Create a quadratic regression model using the data in Table 2.7, letting $x = 0$ stand for 1980, $x = 10$ for 1990, and so on. In what year does this model predict the number of patent applications to have first exceeded 350,000?

Table 2.7 U.S. Patent Applications

Year	Applications (thousands)
1980	112.4
1990	176.3
1995	228.2
2000	315.0
2005	417.5
2008	485.3
2010	520.3
2012	576.8

Source: U.S. Patent Statistics Table, U.S. Patent and Trademark Office, 2012.

66. Highway Engineering Interstate 70 west of Denver, Colorado, has a section posted as a 6% grade. This means that for a horizontal change of 100 ft there is a 6-ft vertical change.

(a) Find the slope of this section of the highway.

(b) On a highway with a 6% grade what is the horizontal distance required to climb 250 ft?

(c) A sign along the highway says 6% grade for the next 7 mi. Estimate how many feet of vertical change there are along those 7 mi. (There are 5280 ft in 1 mi.)

67. A group of female children were weighed. Their ages and weights are recorded in Table 2.8.

Table 2.8 Children's Ages and Weights

Age (months)	Weight (pounds)
19	22
21	23
24	25
27	28
29	31
31	28
34	32
38	34
43	39

(a) Draw a scatter plot of the data.

(b) Find the linear regression model.

(c) Interpret the slope of the linear regression equation.

(d) Superimpose the regression line on the scatter plot.

(e) Use the regression model to predict the weight of a 30-month-old girl.

68. Table 2.9 shows the median U.S. income of women (in 2011 dollars) for selected years. Let x be the number of years since 1940.

Table 2.9 Median Income of Women in the United States (in 2011 dollars)

Year	Median Income ($)
1950	7,771
1960	8,380
1970	11,563
1980	12,786
1990	16,799
2000	21,000
2010	21,430

Source: Historical Income Tables, U.S. Census Bureau, 2012.

(a) Find the linear regression model for the data.

(b) Use it to predict the median U.S. female income in 2020.

In Exercises 69–70, complete the analysis for the given Basic Function.

69. Analyzing a Function

BASIC FUNCTION

The Identity Function

$f(x) = x$
Domain:
Range:
Continuity:
Increasing-decreasing behavior:
Symmetry:
Boundedness:
Local extrema:
Horizontal asymptotes:
Vertical asymptotes:
End behavior:

70. Analyzing a Function

BASIC FUNCTION

The Squaring Function

$f(x) = x^2$
Domain:
Range:
Continuity:
Increasing-decreasing behavior:
Symmetry:
Boundedness:
Local extrema:
Horizontal asymptotes:
Vertical asymptotes:
End behavior:

Standardized Test Questions

71. True or False The initial value of $f(x) = 3x^2 + 2x - 3$ is 0. Justify your answer.

72. True or False The graph of the function $f(x) = x^2 - x + 1$ has no x-intercepts. Justify your answer.

In Exercises 73–76, you may use a graphing calculator to solve the problem.

In Exercises 73 and 74, $f(x) = mx + b$, $f(-2) = 3$, and $f(4) = 1$.

73. Multiple Choice What is the value of m?

(A) 3 **(B)** -3 **(C)** -1 **(D)** 1/3 **(E)** $-1/3$

74. Multiple Choice What is the value of b?

(A) 4 **(B)** 11/3 **(C)** 7/3 **(D)** 1 **(E)** $-1/3$

In Exercises 75 and 76, let $f(x) = 2(x + 3)^2 - 5$.

75. Multiple Choice What is the axis of symmetry of the graph of f?

(A) $x = 3$ **(B)** $x = -3$ **(C)** $y = 5$
(D) $y = -5$ **(E)** $y = 0$

76. Multiple Choice What is the vertex of f?

(A) $(0, 0)$ **(B)** $(3, 5)$ **(C)** $(3, -5)$
(D) $(-3, 5)$ **(E)** $(-3, -5)$

Explorations

77. Writing to Learn Identifying Graphs of Linear Functions

(a) Which of the lines graphed on the next page are graphs of linear functions? Explain.

(b) Which of the lines graphed on the next page are graphs of functions? Explain.

(c) Which of the lines graphed on the next page are not graphs of functions? Explain.

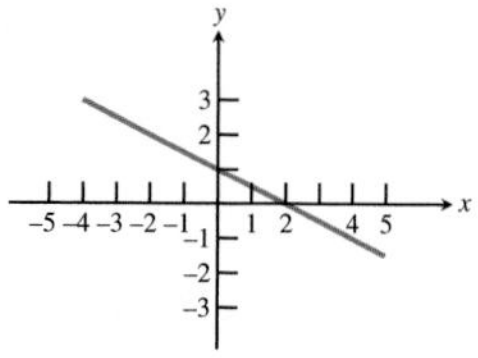

(i)

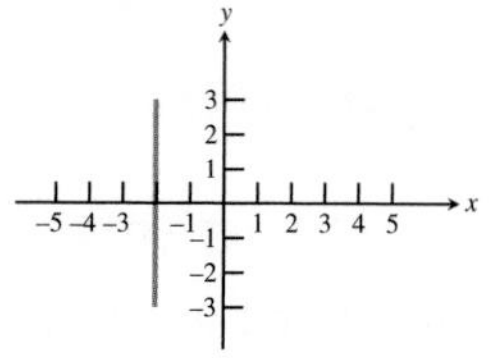

(ii)

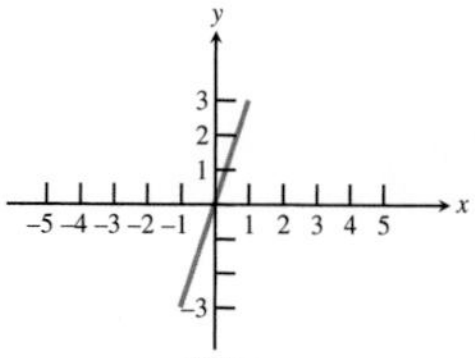

(iii)

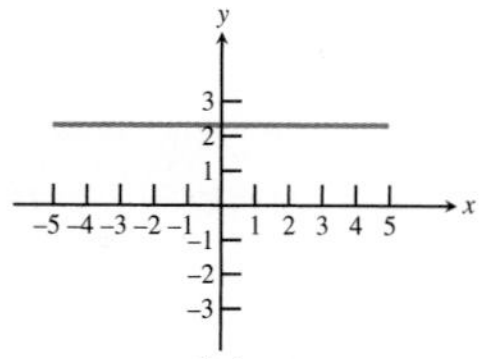

(iv)

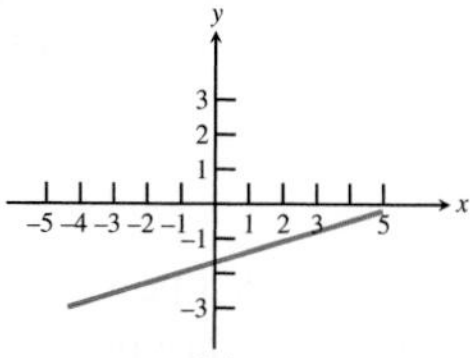

(v)

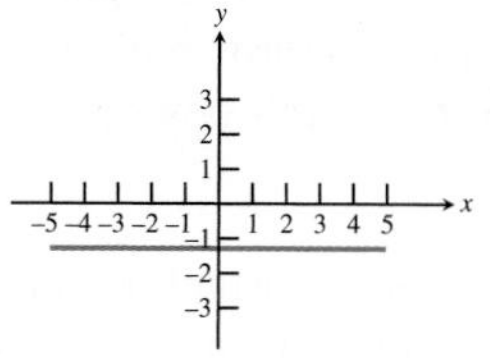

(vi)

78. Average Rate of Change Let $f(x) = x^2$, $g(x) = 3x + 2$, $h(x) = 7x - 3$, $k(x) = mx + b$, and $l(x) = x^3$.

(a) Compute the average rate of change of f from $x = 1$ to $x = 3$.

(b) Compute the average rate of change of f from $x = 2$ to $x = 5$.

(c) Compute the average rate of change of f from $x = a$ to $x = c$.

(d) Compute the average rate of change of g from $x = 1$ to $x = 3$.

(e) Compute the average rate of change of g from $x = 1$ to $x = 4$.

(f) Compute the average rate of change of g from $x = a$ to $x = c$.

(g) Compute the average rate of change of h from $x = a$ to $x = c$.

(h) Compute the average rate of change of k from $x = a$ to $x = c$.

(i) Compute the average rate of change of l from $x = a$ to $x = c$.

Extending the Ideas

79. Minimizing Sums of Squares The linear regression line is often called the **least-square lines** because it minimizes the sum of the squares of the **residuals**, the differences between actual y values and predicted y values:

$$residual = y_i - (ax_i + b),$$

where (x_i, y_i) are the given data pairs and $y = ax + b$ is the regression equation, as shown in the figure.

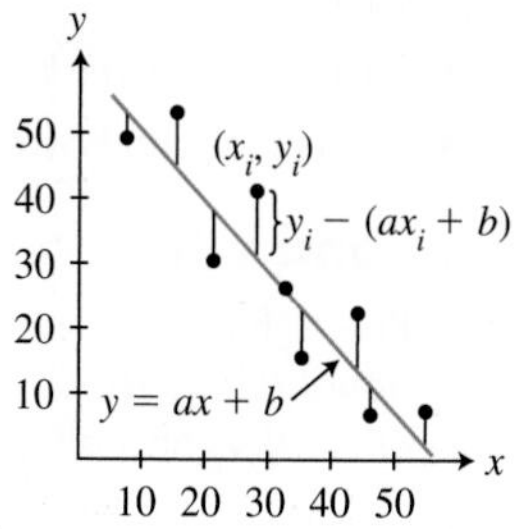

Use these definitions to explain why the regression line obtained from reversing the ordered pairs in Table 2.2 is not the inverse of the function obtained in Example 3.

80. Median-Median Line Read about the median-median line by going to the Internet, your grapher owner's manual, or a library. Then use the following data set to complete this problem.

$$\{(2, 8), (3, 6), (5, 9), (6, 8), (8, 11), (10, 13), (12, 14), (15, 4)\}$$

(a) Draw a scatter plot of the data.

(b) Find the linear regression equation and graph it.

(c) Find the median-median line equation and graph it.

(d) Writing to Learn For these data, which of the two lines appears to be the line of better fit? Why?

81. Suppose $b^2 - 4ac > 0$ for the equation $ax^2 + bx + c = 0$.

(a) Prove that the sum of the two solutions of this equation is $-b/a$.

(b) Prove that the product of the two solutions of this equation is c/a.

82. Connecting Algebra and Geometry Prove that the axis of the graph of $f(x) = (x - a)(x - b)$ is $x = (a + b)/2$, where a and b are real numbers.

83. Connecting Algebra and Geometry Identify the vertex of the graph of $f(x) = (x - a)(x - b)$, where a and b are any real numbers.

84. Connecting Algebra and Geometry Prove that if x_1 and x_2 are real numbers and are zeros of the quadratic function $f(x) = ax^2 + bx + c$, then the axis of the graph of f is $x = (x_1 + x_2)/2$.

85. Prove the Constant Rate of Change Theorem, which is stated on page 160.

2.2 Power Functions with Modeling

What you'll learn about

- Power Functions and Variation
- Monomial Functions and Their Graphs
- Graphs of Power Functions
- Modeling with Power Functions

... and why

Power functions specify the proportional relationships of geometry, chemistry, and physics.

Power Functions and Variation

Five of the basic functions introduced in Section 1.3 were power functions. Power functions are an important family of functions in their own right and are important building blocks for other functions.

DEFINITION Power Function

Any function that can be written in the form

$$f(x) = k \cdot x^a, \text{ where } k \text{ and } a \text{ are nonzero constants,}$$

is a **power function**. The constant a is the **power**, and k is the **constant of variation**, or **constant of proportion**. We say $f(x)$ **varies as** the ath power of x, or $f(x)$ **is proportional to** the ath power of x.

In general, if $y = f(x)$ varies as a constant power of x, then y is a power function of x. Many of the most common formulas from geometry and science are power functions.

Name	Formula	Power	Constant of Variation
Circumference	$C = 2\pi r$	1	2π
Area of a circle	$A = \pi r^2$	2	π
Force of gravity	$F = k/d^2$	-2	k
Boyle's Law	$V = k/P$	-1	k

These four power functions involve relationships that can be expressed in the language of *variation and proportion*:

- The circumference of a circle varies directly as its radius.
- The area enclosed by a circle is directly proportional to the square of its radius.
- The force of gravity acting on an object is inversely proportional to the square of the distance from the object to the center of the Earth.
- Boyle's Law states that the volume of an enclosed gas (at a constant temperature) varies inversely as the applied pressure.

The power function formulas with positive powers are statements of **direct variation**, and power function formulas with negative powers are statements of **inverse variation**. Unless the word *inversely* is included in a variation statement, the variation is assumed to be direct, as in Example 1.

EXAMPLE 1 Writing a Power Function Formula

From empirical evidence and the laws of physics it has been found that the period of time T for the full swing of a pendulum varies as the square root of the pendulum's length l, provided that the swing is small relative to the length of the pendulum. Express this relationship as a power function.

SOLUTION Because it does not state otherwise, the variation is direct. So the power is positive. The wording tells us that T is a function of l. Using k as the constant of variation gives us

$$T(l) = k\sqrt{l} = k \cdot l^{1/2}.$$

Now try Exercise 17.

Section 1.3 introduced five basic power functions:

$$x, x^2, x^3, x^{-1} = \frac{1}{x}, \quad \text{and} \quad x^{1/2} = \sqrt{x}$$

Example 2 describes two other power functions: the *cube root function* and the *inverse-square* function.

EXAMPLE 2 Analyzing Power Functions

State the power and constant of variation for the function, graph it, and analyze it.

(a) $f(x) = \sqrt[3]{x}$ **(b)** $g(x) = \dfrac{1}{x^2}$

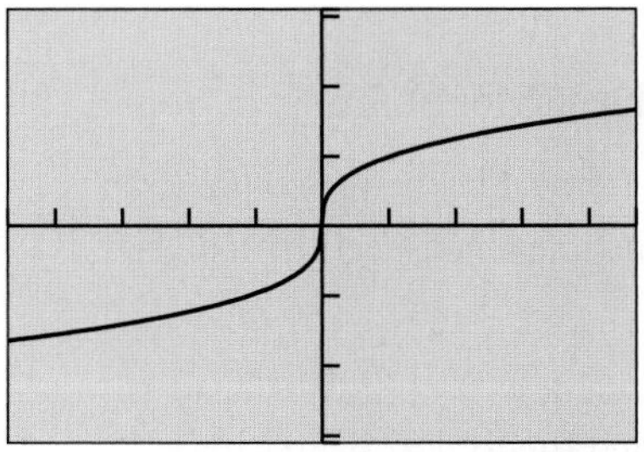

[–4.7, 4.7] by [–3.1, 3.1]
(a)

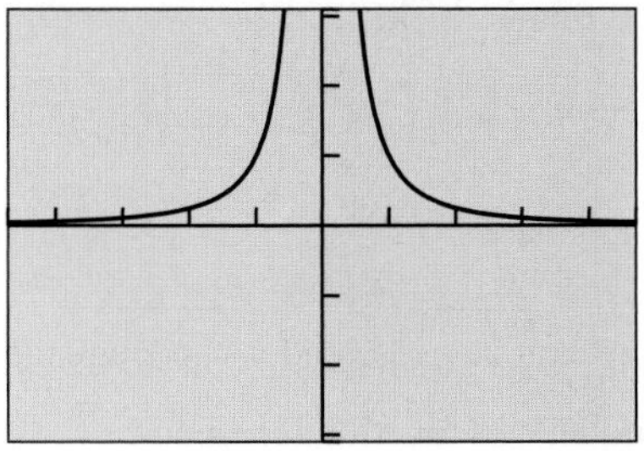

[–4.7, 4.7] by [–3.1, 3.1]
(b)

FIGURE 2.10 The graphs of (a) $f(x) = \sqrt[3]{x} = x^{1/3}$ and (b) $g(x) = 1/x^2 = x^{-2}$. (Example 2)

SOLUTION

(a) Because $f(x) = \sqrt[3]{x} = x^{1/3} = 1 \cdot x^{1/3}$, its power is 1/3, and its constant of variation is 1. The graph of f is shown in Figure 2.10a.

Domain: $(-\infty, \infty)$
Range: $(-\infty, \infty)$
Continuous
Increasing for all x
Symmetric with respect to the origin (an odd function)
Not bounded above or below
No local extrema
No asymptotes
End behavior: $\lim_{x\to-\infty} \sqrt[3]{x} = -\infty$ and $\lim_{x\to\infty} \sqrt[3]{x} = \infty$
Interesting fact: The cube root function $f(x) = \sqrt[3]{x}$ is the inverse of the cubing function.

(b) Because $g(x) = 1/x^2 = x^{-2} = 1 \cdot x^{-2}$, its power is -2, and its constant of variation is 1. The graph of g is shown in Figure 2.10b.

Domain: $(-\infty, 0) \cup (0, \infty)$
Range: $(0, \infty)$
Continuous on its domain; discontinuous at $x = 0$
Increasing on $(-\infty, 0)$; decreasing on $(0, \infty)$
Symmetric with respect to the y-axis (an even function)
Bounded below, but not above
No local extrema
Horizontal asymptote: $y = 0$; vertical asymptote: $x = 0$
End behavior: $\lim_{x\to-\infty} (1/x^2) = 0$ and $\lim_{x\to\infty} (1/x^2) = 0$
Interesting fact: $g(x) = 1/x^2$ is the basis of scientific *inverse-square laws*, for example, the inverse-square gravitational principle $F = k/d^2$ mentioned above. So $g(x) = 1/x^2$ is sometimes called the inverse-square function, but it is *not* the inverse of the squaring function but rather its *multiplicative* inverse.

Now try Exercise 27.

Monomial Functions and Their Graphs

A single-term polynomial function is a *monomial function.*

DEFINITION Monomial Function

Any function that can be written as

$$f(x) = k \text{ or } f(x) = k \cdot x^n,$$

where k is a constant and n is a positive integer, is a **monomial function**.

So the zero function and constant functions are monomial functions, but the more typical monomial function is a power function with a positive integer power, which is the degree of the monomial. For example, the basic functions x, x^2, and x^3 are typical monomial functions. It is important to understand the graphs of monomial functions because every polynomial function is either a monomial function or a sum of monomial functions.

In Exploration 1, we take a close look at six basic monomial functions. They have the form x^n for $n = 1, 2, \ldots, 6$. We group them by even and odd powers.

EXPLORATION 1 Comparing Graphs of Monomial Functions

Graph the triplets of functions in the stated windows and explain how the graphs are alike and how they are different. Consider the relevant aspects of analysis from Example 2. Which ordered pairs do all three graphs have in common?

1. $f(x) = x$, $g(x) = x^3$, and $h(x) = x^5$ in the window $[-2.35, 2.35]$ by $[-1.5, 1.5]$, then $[-5, 5]$ by $[-15, 15]$, and finally $[-20, 20]$ by $[-200, 200]$.
2. $f(x) = x^2$, $g(x) = x^4$, and $h(x) = x^6$ in the window $[-1.5, 1.5]$ by $[-0.5, 1.5]$, then $[-5, 5]$ by $[-5, 25]$, and finally $[-15, 15]$ by $[-50, 400]$.

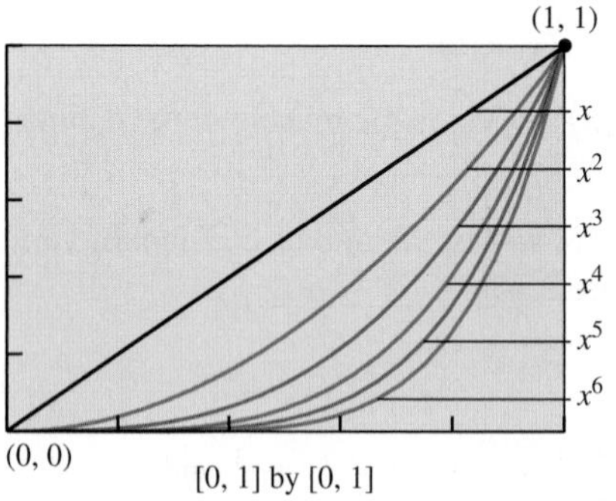

[0, 1] by [0, 1]

FIGURE 2.11 The graphs of $f(x) = x^n$, $0 \le x \le 1$, for $n = 1, 2, \ldots, 6$.

From Exploration 1 we see that

$$f(x) = x^n \text{ is an even function if } n \text{ is even and an odd function if } n \text{ is odd.}$$

Because of this symmetry, it is enough to know the first quadrant behavior of $f(x) = x^n$. Figure 2.11 shows the graphs of $f(x) = x^n$ for $n = 1, 2, \ldots, 6$ in the first quadrant near the origin.

The following conclusions about the basic function $f(x) = x^3$ can be drawn from your investigations in Exploration 1.

BASIC FUNCTION The Cubing Function

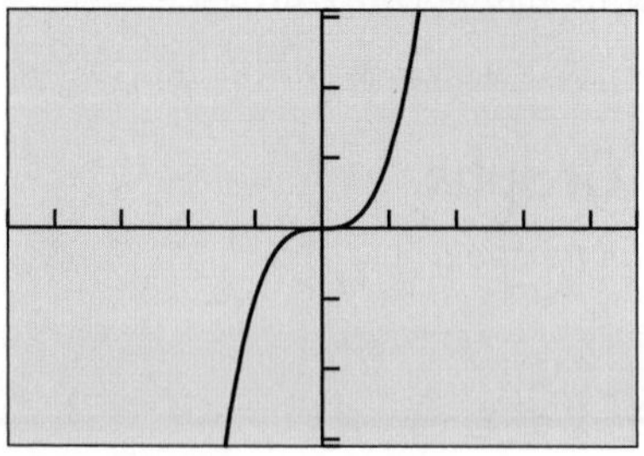

$[-4.7, 4.7]$ by $[-3.1, 3.1]$

FIGURE 2.12 The graph of $f(x) = x^3$.

$f(x) = x^3$
Domain: $(-\infty, \infty)$
Range: $(-\infty, \infty)$
Continuous
Increasing for all x
Symmetric with respect to the origin (an odd function)
Not bounded above or below
No local extrema
No horizontal asymptotes
No vertical asymptotes
End behavior: $\lim_{x \to -\infty} x^3 = -\infty$ and $\lim_{x \to \infty} x^3 = \infty$

EXAMPLE 3 Graphing Monomial Functions

Describe how to obtain the graph of the given function from the graph of $g(x) = x^n$ with the same power n. Sketch the graph by hand and support your answer with a grapher.

(a) $f(x) = 2x^3$ **(b)** $f(x) = -\dfrac{2}{3}x^4$

SOLUTION

(a) We obtain the graph of $f(x) = 2x^3$ by vertically stretching the graph of $g(x) = x^3$ by a factor of 2. Both are odd functions (Figure 2.13a).

(b) We obtain the graph of $f(x) = -(2/3)x^4$ by vertically shrinking the graph of $g(x) = x^4$ by a factor of 2/3 and then reflecting it across the x-axis. Both are even functions (Figure 2.13b).

Now try Exercise 31.

We ask you to explore the graphical behavior of power functions of the form x^{-n} and $x^{1/n}$, where n is a positive integer, in Exercise 65.

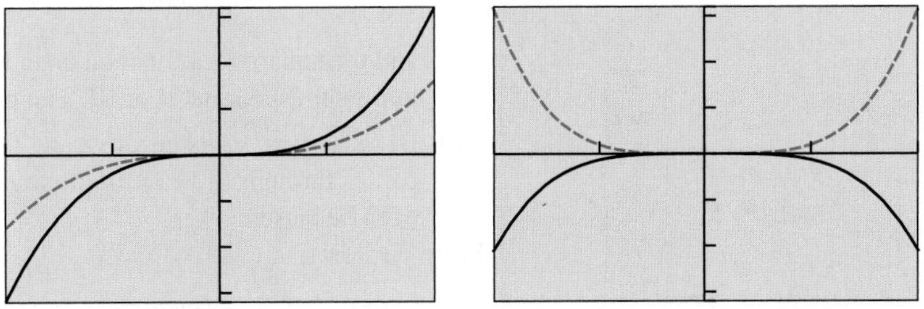

[–2, 2] by [–16, 16]
(a)

[–2, 2] by [–16, 16]
(b)

FIGURE 2.13 The graphs of (a) $f(x) = 2x^3$ with basic monomial $g(x) = x^3$, and (b) $f(x) = -(2/3)x^4$ with basic monomial $g(x) = x^4$. (Example 3)

Graphs of Power Functions

The graphs in Figure 2.14 represent the four shapes that are possible for general power functions of the form $f(x) = kx^a$ for $x \geq 0$. In every case, the graph of f contains $(1, k)$. Those with positive powers also pass through $(0, 0)$. Those with negative exponents are asymptotic to both axes.

When $k > 0$, the graph lies in Quadrant I, but when $k < 0$, the graph is in Quadrant IV.

In general, for any power function $f(x) = k \cdot x^a$, one of three following things happens when $x < 0$.

- f is undefined for $x < 0$, as is the case for $f(x) = x^{1/2}$ and $f(x) = x^{\pi}$.
- f is an even function, so f is symmetric about the y-axis, as is the case for $f(x) = x^{-2}$ and $f(x) = x^{2/3}$.
- f is an odd function, so f is symmetric about the origin, as is the case for $f(x) = x^{-1}$ and $f(x) = x^{7/3}$.

Predicting the general shape of the graph of a power function is a two-step process as illustrated in Example 4.

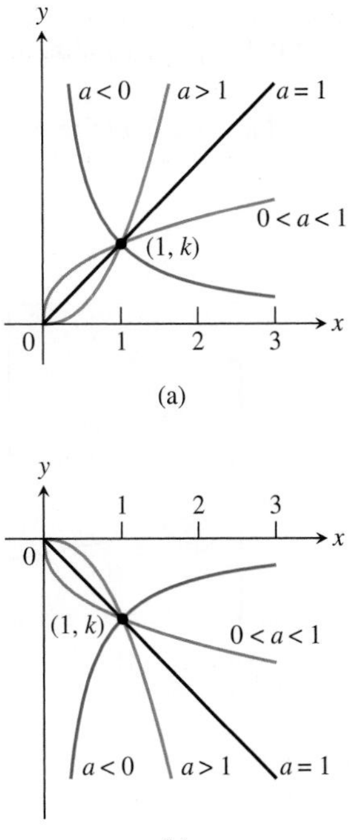

FIGURE 2.14 The graphs of $f(x) = k \cdot x^a$ for $x \geq 0$. (a) $k > 0$, (b) $k < 0$.

EXAMPLE 4 Graphing Power Functions $f(x) = k \cdot x^a$

State the values of the constants k and a. Describe the portion of the curve that lies in Quadrant I or IV. Determine whether f is even, odd, or undefined for $x < 0$. Describe the rest of the curve if any. Graph the function to see whether it matches the description.

(a) $f(x) = 2x^{-3}$ **(b)** $f(x) = -0.4x^{1.5}$ **(c)** $f(x) = -x^{0.4}$

SOLUTION

(a) Because $k = 2$ is positive and the power $a = -3$ is negative, the graph passes through $(1, 2)$ and is asymptotic to both axes. The graph is decreasing in the first quadrant. The function f is odd because

$$f(-x) = 2(-x)^{-3} = \frac{2}{(-x)^3} = -\frac{2}{x^3} = -2x^{-3} = -f(x).$$

So its graph is symmetric about the origin. The graph in Figure 2.15a supports all aspects of the description.

(b) Because $k = -0.4$ is negative and the power $a = 1.5 > 1$, the graph contains $(0, 0)$ and passes through $(1, -0.4)$. In the fourth quadrant, it is decreasing. The function f is undefined for $x < 0$ because

$$f(x) = -0.4x^{1.5} = -\frac{2}{5}x^{3/2} = -\frac{2}{5}(\sqrt{x})^3,$$

and the square root function is undefined for $x < 0$. So the graph of f has no points in Quadrants II or III. The graph in Figure 2.15b matches the description.

(c) Because $k = -1$ is negative and $0 < a < 1$, the graph contains (0, 0) and passes through $(1, -1)$. In the fourth quadrant, it is decreasing. The function f is even because

$$\begin{aligned} f(-x) &= -(-x)^{0.4} = -(-x)^{2/5} = -(\sqrt[5]{-x})^2 = -(-\sqrt[5]{x})^2 \\ &= -(\sqrt[5]{x})^2 = -x^{0.4} = f(x). \end{aligned}$$

So the graph of f is symmetric about the y-axis. The graph in Figure 2.15c fits the description.

Now try Exercise 43.

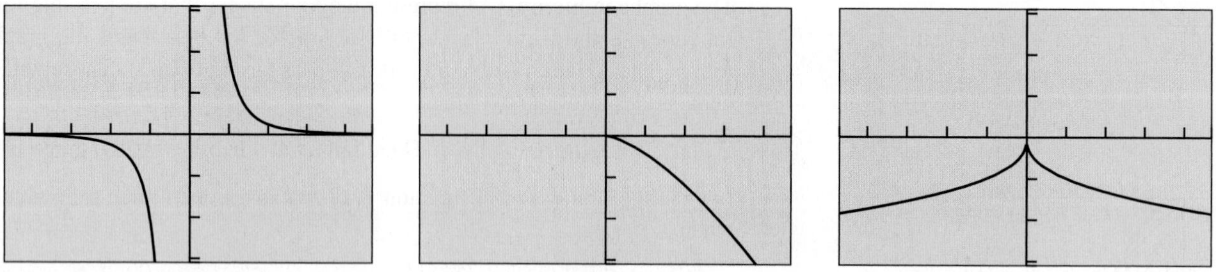

[–4.7, 4.7] by [–3.1, 3.1]
(a)

[–4.7, 4.7] by [–3.1, 3.1]
(b)

[–4.7, 4.7] by [–3.1, 3.1]
(c)

FIGURE 2.15 The graphs of (a) $f(x) = 2x^{-3}$, (b) $f(x) = -0.4x^{1.5}$, and (c) $f(x) = -x^{0.4}$. (Example 4)

The following information about the basic function $f(x) = \sqrt{x}$ follows from the investigation in Exercise 65.

BASIC FUNCTION The Square Root Function

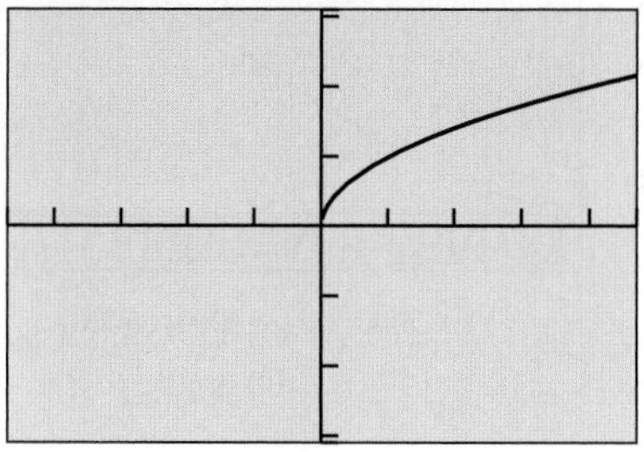

[–4.7, 4.7] by [–3.1, 3.1]

FIGURE 2.16 The graph of $f(x) = \sqrt{x}$.

$f(x) = \sqrt{x}$
Domain: $[0, \infty)$
Range: $[0, \infty)$
Continuous on $[0, \infty)$
Increasing on $[0, \infty)$
No symmetry
Bounded below but not above
Local minimum at $x = 0$
No horizontal asymptotes
No vertical asymptotes
End behavior: $\lim_{x \to \infty} \sqrt{x} = \infty$

Modeling with Power Functions

Noted astronomer Johannes Kepler (1571–1630) developed three laws of planetary motion that are used to this day. Kepler's Third Law states that the square of the period of orbit T (the time required for one full revolution around the Sun) for each planet is proportional to the cube of its average distance a from the Sun. Table 2.10 gives the relevant data for the six planets that were known in Kepler's time. The distances are given in millions of kilometers, or gigameters (Gm).

Table 2.10 Average Distances and Orbital Periods for the Six Innermost Planets

Planet	Average Distance from Sun (Gm)	Period of Orbit (days)
Mercury	57.9	88
Venus	108.2	225
Earth	149.6	365.2
Mars	227.9	687
Jupiter	778.3	4332
Saturn	1427	10,760

Source: Shupe, Dorr, Payne, Hunsiker, et al., National Geographic Atlas of the World (rev. 6th ed.). Washington, DC: National Geographic Society, 1992, plate 116.

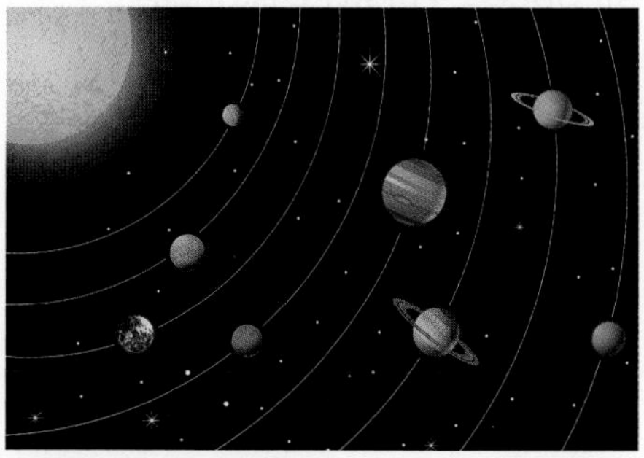

A Bit of History

Example 5 shows the predictive power of a well-founded model. Exercise 67 asks you to find Kepler's elegant form of the equation, $T^2 = a^3$, which he reported in *The Harmony of the World* in 1619.

EXAMPLE 5 Modeling Planetary Data with a Power Function

Use the data in Table 2.10 to obtain a power function model for orbital period as a function of average distance from the Sun. Then use the model to predict the orbital period for Neptune, which is 4497 Gm from the Sun on average.

SOLUTION

Model First we make a scatter plot of the data, as shown in Figure 2.17a on page 180. The association appears strong and curved. Using power regression, we find the model for the orbital period to be about

$$T(a) \approx 0.20a^{1.5} = 0.20a^{3/2} = 0.20\sqrt{a^3}.$$

Figure 2.17b shows the scatter plot for Table 2.10 together with a graph of the power regression model just found. You can see that the curve fits the data quite well. The coefficient of determination is $r^2 \approx 0.999999912$, indicating an amazingly close fit and supporting the visual evidence.

(continued)

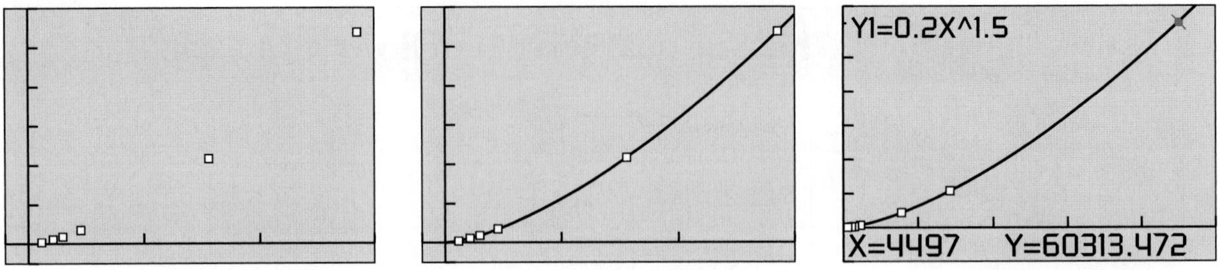

[–100, 1500] by [–1000, 12000]
(a)

[–100, 1500] by [–1000, 12000]
(b)

[0, 5000] by [–10000, 65000]
(c)

FIGURE 2.17 Scatter plot and graphs for Example 5.

Solve Numerically To predict the orbit period for Neptune we substitute its average distance from the Sun in the power regression model:

$$T(4497) \approx 0.2(4497)^{1.5} \approx 60{,}313$$

Interpret It takes Neptune about 60,313 days to orbit the Sun, or about 165 years, which is the value given in the *National Geographic Atlas of the World.*

Figure 2.17c reports this result and gives some indication of the relative distances involved. Neptune is much farther from the Sun than the six innermost planets and especially the four closest to the Sun—Mercury, Venus, Earth, and Mars.

Now try Exercise 55.

In Example 6, we return to free-fall motion, with a new twist. The data in the table come from the same CBR™ experiment referenced in Example 8 of Section 2.1. This time we are looking at the downward distance (in meters) the ball has traveled since reaching its peak height and its downward speed (in meters per second). It can be shown (see Exercise 68) that free-fall speed is proportional to a power of the distance traveled.

EXAMPLE 6 Modeling Free-Fall Speed Versus Distance

Use the data in Table 2.11 to obtain a power function model for speed p versus distance traveled d. Then use the model to predict the speed of the ball at impact, given that impact occurs when $d \approx 1.80$ m.

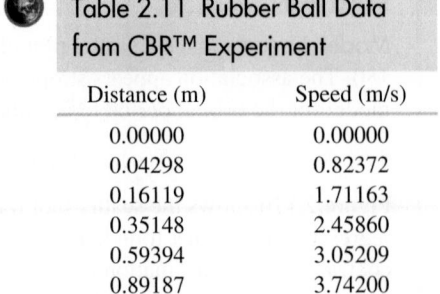

Table 2.11 Rubber Ball Data from CBR™ Experiment

Distance (m)	Speed (m/s)
0.00000	0.00000
0.04298	0.82372
0.16119	1.71163
0.35148	2.45860
0.59394	3.05209
0.89187	3.74200
1.25557	4.49558

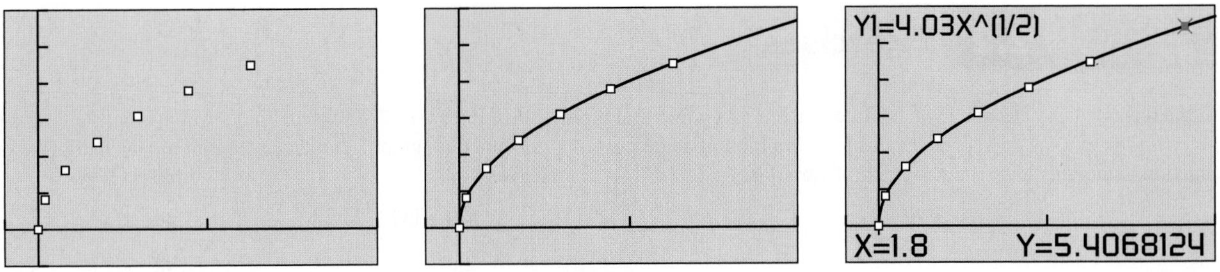

[–0.2, 2] by [–1, 6]
(a)

[–0.2, 2] by [–1, 6]
(b)

[–0.2, 2] by [–1, 6]
(c)

FIGURE 2.18 Scatter plot and graphs for Example 6.

Why p?

We use p for speed to distinguish it from velocity v. Recall that speed is the *absolute value* of velocity.

A Word of Warning

The regression routine traditionally used to compute power function models involves taking logarithms of the data, and therefore, all of the data must be strictly positive numbers. So we must leave out $(0, 0)$ to compute the power regression equation.

SOLUTION

Model Figure 2.18a is a scatter plot of the data. Using power regression, we find the model for speed p versus distance d to be about

$$p(d) \approx 4.03d^{0.5} = 4.03d^{1/2} = 4.03\sqrt{d}.$$

(See margin notes.) Figure 2.18b shows the scatter plot for Table 2.11 together with a graph of the power regression equation just found. You can see that the curve fits the data nicely. The coefficient of determination is $r^2 \approx 0.99770$, indicating a close fit and supporting the visual evidence.

Solve Numerically To predict the speed at impact, we substitute $d \approx 1.80$ into the obtained power regression model:

$$p(1.80) \approx 5.4$$

See Figure 2.18c.

Interpret The speed at impact is about 5.4 m/sec. This is slightly less than the value obtained in Example 8 of Section 2.1, using a different modeling process for the same experiment.

Now try Exercise 57.

QUICK REVIEW 2.2 *(For help, go to Section A.1.)*

Exercise numbers with a gray background indicate problems that the authors have designed to be solved *without a calculator*.

In Exercises 1–6, write the following expressions using only positive integer powers.

1. $x^{2/3}$
2. $p^{5/2}$
3. d^{-2}
4. x^{-7}
5. $q^{-4/5}$
6. $m^{-1.5}$

In Exercises 7–10, write the following expressions in the form $k \cdot x^a$ using a single rational number for the power a.

7. $\sqrt{9x^3}$
8. $\sqrt[3]{8x^5}$
9. $\sqrt[3]{\dfrac{5}{x^4}}$
10. $\dfrac{4x}{\sqrt{32x^3}}$

SECTION 2.2 Exercises

In Exercises 1–10, determine whether the function is a power function, given that c, g, k, and π represent constants. For those that are power functions, state the power and constant of variation.

1. $f(x) = -\frac{1}{2}x^5$

2. $f(x) = 9x^{5/3}$

3. $f(x) = 3 \cdot 2^x$

4. $f(x) = 13$

5. $E(m) = mc^2$

6. $KE(v) = \frac{1}{2}kv^5$

7. $d = \frac{1}{2}gt^2$

8. $V = \frac{4}{3}\pi r^3$

9. $I = \frac{k}{d^2}$

10. $F(a) = ma$

In Exercises 11–16, determine whether the function is a monomial function, given that l and π represent constants. For those that are monomial functions state the degree and leading coefficient. For those that are not, explain why not.

11. $f(x) = -4$

12. $f(x) = 3x^{-5}$

13. $y = -6x^7$

14. $y = -2 \cdot 5^x$

15. $S = 4\pi r^2$

16. $A = lw$

In Exercises 17–22, write the statement as a power function equation. Use k for the constant of variation if one is not given.

17. The area A of an equilateral triangle varies directly as the square of the length s of its sides.

18. The volume V of a circular cylinder with fixed height is proportional to the square of its radius r.

19. The current I in an electrical circuit is inversely proportional to the resistance R, with constant of variation V.

20. Charles's Law states the volume V of an enclosed ideal gas at a constant pressure varies directly as the absolute temperature T.

21. The energy E produced in a nuclear reaction is proportional to the mass m, with the constant of variation being c^2, the square of the speed of light.

22. The speed p of a free-falling object that has been dropped from rest varies as the square root of the distance traveled d, with a constant of variation $k = \sqrt{2g}$.

In Exercises 23–26, write a sentence that expresses the relationship in the formula, using the language of variation or proportion.

23. $w = mg$, where w and m are the weight and mass of an object and g is the constant acceleration due to gravity.

24. $C = \pi D$, where C and D are the circumference and diameter of a circle and π is the usual mathematical constant.

25. $n = c/v$, where n is the refractive index of a medium, v is the velocity of light in the medium, and c is the constant velocity of light in free space.

26. $d = p^2/(2g)$, where d is the distance traveled by a free-falling object dropped from rest, p is the speed of the object, and g is the constant acceleration due to gravity.

In Exercises 27–30, state the power and constant of variation for the function, graph it, and analyze it in the manner of Example 2 of this section.

27. $f(x) = 2x^4$

28. $f(x) = -3x^3$

29. $f(x) = \frac{1}{2}\sqrt[4]{x}$

30. $f(x) = -2x^{-3}$

In Exercises 31–36, describe how to obtain the graph of the given monomial function from the graph of $g(x) = x^n$ with the same power n. State whether f is even or odd. Sketch the graph by hand and support your answer with a grapher.

31. $f(x) = \frac{2}{3}x^4$

32. $f(x) = 5x^3$

33. $f(x) = -1.5x^5$

34. $f(x) = -2x^6$

35. $f(x) = \frac{1}{4}x^8$

36. $f(x) = \frac{1}{8}x^7$

In Exercises 37–42, match the equation to one of the curves labeled in the figure.

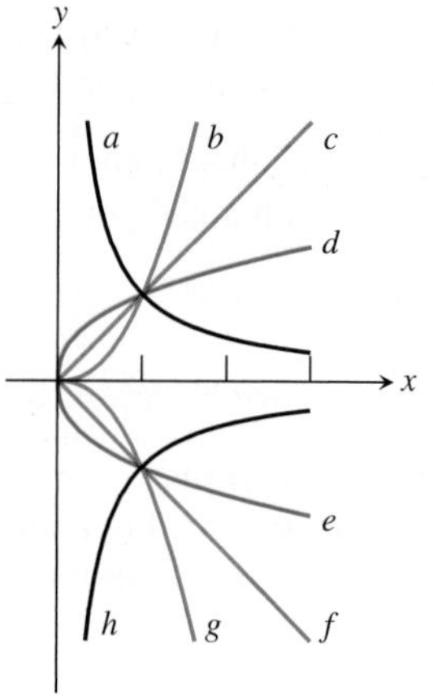

37. $f(x) = -\frac{2}{3}x^4$

38. $f(x) = \frac{1}{2}x^{-5}$

39. $f(x) = 2x^{1/4}$

40. $f(x) = -x^{5/3}$

41. $f(x) = -2x^{-2}$

42. $f(x) = 1.7x^{2/3}$

In Exercises 43–48, state the values of the constants k and a for the function $f(x) = k \cdot x^a$. Before using a grapher, describe the portion of the curve that lies in Quadrant I or IV. Determine whether f is even, odd, or undefined for $x < 0$. Describe the rest of the curve if any. Graph the function to see whether it matches the description.

43. $f(x) = 3x^{1/4}$

44. $f(x) = -4x^{2/3}$

45. $f(x) = -2x^{4/3}$

46. $f(x) = \frac{2}{5}x^{5/2}$

47. $f(x) = \frac{1}{2}x^{-3}$

48. $f(x) = -x^{-4}$

In Exercises 49 and 50, data are given for y as a power function of x. Write an equation for the power function, and state its power and constant of variation.

49.

x	2	4	6	8	10
y	2	0.5	0.222...	0.125	0.08

50.

x	1	4	9	16	25
y	−2	−4	−6	−8	−10

51. Boyle's Law The volume of an enclosed gas (at a constant temperature) varies inversely as the pressure. If the pressure of a 3.46-L sample of neon gas at a temperature of 302 K is 0.926 atm, what would the volume be at a pressure of 1.452 atm if the temperature does not change?

52. Charles's Law The volume of an enclosed gas (at a constant pressure) varies directly as the absolute temperature. If the pressure of a 3.46-L sample of neon gas at a temperature of 302 K is 0.926 atm, what would the volume be at a temperature of 338 K if the pressure does not change?

53. Diamond Refraction Diamonds have the extremely high refraction index of $n = 2.42$ on average over the range of visible light. Use the formula from Exercise 25 and the fact that $c \approx 3.00 \times 10^8$ m/sec to determine the speed of light through a diamond.

54. Windmill Power The power P (in watts) produced by a windmill is proportional to the cube of the wind speed v (in mph). If a wind of 10 mph generates 15 watts of power, how much power is generated by winds of 20, 40, and 80 mph? Make a table and explain the pattern.

55. Keeping Warm For mammals and other warm-blooded animals to stay warm requires quite a bit of energy. Temperature loss is related to surface area, which is related to body weight, and temperature gain is related to circulation, which is related to pulse rate. In the final analysis, scientists have concluded that the pulse rate r of mammals is a power function of their body weight w.

(a) Draw a scatter plot of the data in Table 2.12.

(b) Find the power regression model.

(c) Superimpose the regression curve on the scatter plot.

Table 2.12 Weight and Pulse Rate of Selected Mammals

Mammal	Body Weight (kg)	Pulse Rate (beats/min)
Rat	0.2	420
Guinea pig	0.3	300
Rabbit	2	205
Small dog	5	120
Large dog	30	85
Sheep	50	70
Human	70	72

Source: A. J. Clark, Comparative Physiology of the Heart. New York: Macmillan, 1927.

(d) Use the regression model to predict the pulse rate for a 450-kg horse. Is the result close to the 38 beats/min reported by A. J. Clark in 1927?

56. Even and Odd Functions If n is an integer, $n \geq 1$, prove that $f(x) = x^n$ is an odd function if n is odd and is an even function if n is even.

57. Light Intensity Velma and Reggie gathered the data in Table 2.13 using a 100-watt light bulb and a Calculator-Based Laboratory™ (CBL™) with a light-intensity probe.

(a) Draw a scatter plot of the data in Table 2.13

(b) Find the power regression model. Is the power close to the theoretical value of $a = -2$?

(c) Superimpose the regression curve on the scatter plot.

(d) Use the regression model to predict the light intensity at distances of 1.7 m and 3.4 m.

Table 2.13 Light-Intensity Data for a 100-W Light Bulb

Distance (m)	Intensity (W/m^2)
1.0	7.95
1.5	3.53
2.0	2.01
2.5	1.27
3.0	0.90

Standardized Test Questions

58. True or False The function $f(x) = x^{-2/3}$ is even. Justify your answer.

59. True or False The graph $f(x) = x^{1/3}$ is symmetric about the y-axis. Justify your answer.

In Exercises 60–63, solve the problem without using a calculator.

60. Multiple Choice Let $f(x) = 2x^{-1/2}$. What is the value of $f(4)$?

(A) 1 **(B)** −1 **(C)** $2\sqrt{2}$ **(D)** $\dfrac{1}{2\sqrt{2}}$ **(E)** 4

61. Multiple Choice Let $f(x) = -3x^{-1/3}$. Which of the following statements is true?

(A) $f(0) = 0$ **(B)** $f(-1) = -3$ **(C)** $f(1) = 1$

(D) $f(3) = 3$ **(E)** $f(0)$ is undefined.

62. Multiple Choice Let $f(x) = x^{2/3}$. Which of the following statements is true?

(A) f is an odd function.

(B) f is an even function.

(C) f is neither an even nor an odd function.

(D) The graph f is symmetric with respect to the x-axis.

(E) The graph f is symmetric with respect to the origin.

63. Multiple Choice Which of the following is the domain of the function $f(x) = x^{3/2}$?

(A) $(-\infty, \infty)$ **(B)** $[0, \infty)$ **(C)** $(0, \infty)$

(D) $(-\infty, 0)$ **(E)** $(-\infty, 0) \cup (0, \infty)$

Explorations

64. Group Activity Rational Powers Working in a group of three students, investigate the behavior of power functions of the form $f(x) = k \cdot x^{m/n}$, where m and n are positive with no factors in common. Have one group member investigate each of the following cases:

- n is even
- n is odd and m is even
- n is odd and m is odd

For each case, decide whether f is even, f is odd, or f is undefined for $x < 0$. Solve graphically and confirm algebraically in a way to convince the rest of your group and your entire class.

65. Comparing the Graphs of Power Functions Graph the functions in the stated windows and explain how the graphs are alike and how they are different. Consider the relevant aspects of analysis from Example 2. Which ordered pairs do all four graphs have in common?

(a) $f(x) = x^{-1}$, $g(x) = x^{-2}$, $h(x) = x^{-3}$, and $k(x) = x^{-4}$ in the windows $[0, 1]$ by $[0, 5]$, $[0, 3]$ by $[0, 3]$, and $[-2, 2]$ by $[-2, 2]$.

(b) $f(x) = x^{1/2}$, $g(x) = x^{1/3}$, $h(x) = x^{1/4}$, and $k(x) = x^{1/5}$ in the windows $[0, 1]$ by $[0, 1]$, $[0, 3]$ by $[0, 2]$, and $[-3, 3]$ by $[-2, 2]$.

Extending the Ideas

66. Writing to Learn Irrational Powers A negative number to an irrational power is undefined. Analyze the graphs of $f(x) = x^{\pi}, x^{1/\pi}, x^{-\pi}, -x^{\pi}, -x^{1/\pi}$, and $-x^{-\pi}$. Prepare a sketch of all six graphs on one set of axes, labeling each of the curves. Write an explanation for why each graph is positioned and shaped as it is.

67. Planetary Motion Revisited Convert the time and distance units in Table 2.10 to the Earth-based units of years and astronomical units using

$$1 \text{ yr} = 365.2 \text{ days and } 1 \text{ AU} = 149.6 \text{ Gm}.$$

Use this "re-expressed" data to obtain a power function model. Show algebraically that this model closely approximates Kepler's equation $T^2 = a^3$.

68. Free Fall Revisited The **speed** p of an object is the absolute value of its velocity v. The distance traveled d by an object dropped from an initial height s_0 with a current height s is given by

$$d = s_0 - s$$

until it hits the ground. Use this information and the free-fall motion formulas from Section 2.1 to prove that

$$d = \frac{1}{2}gt^2,\ p = gt, \text{ and therefore } p = \sqrt{2gd}.$$

Do the results of Example 6 approximate this last formula?

69. Prove that $g(x) = 1/f(x)$ is even if and only if $f(x)$ is even and that $g(x) = 1/f(x)$ is odd if and only if $f(x)$ is odd.

70. Use the results in Exercise 69 to prove that $g(x) = x^{-a}$ is even if and only if $f(x) = x^a$ is even and that $g(x) = x^{-a}$ is odd if and only if $f(x) = x^a$ is odd.

71. Joint Variation If a variable z varies as the product of the variables x and y, we say z **varies jointly** as x and y, and we write $z = k \cdot x \cdot y$, where k is the constant of variation. Write a sentence that expresses the relationship in each of the following formulas, using the language of joint variation.

(a) $F = m \cdot a$, where F and a are the force and acceleration acting on an object of mass m.

(b) $KE = (1/2)m \cdot v^2$, where KE and v are the kinetic energy and velocity of an object of mass m.

(c) $F = G \cdot m_1 \cdot m_2/r^2$, where F is the force of gravity acting on objects of masses m_1 and m_2 with a distance r between their centers and G is the universal gravitational constant.

2.3 Polynomial Functions of Higher Degree with Modeling

What you'll learn about

- Graphs of Polynomial Functions
- End Behavior of Polynomial Functions
- Zeros of Polynomial Functions
- Intermediate Value Theorem
- Modeling

... and why

These topics are important in modeling and can be used to provide approximations to more complicated functions, as you will see if you study calculus.

Graphs of Polynomial Functions

As we saw in Section 2.1, a polynomial function of degree 0 is a constant function and graphs as a horizontal line. A polynomial function of degree 1 is a linear function; its graph is a slant line. A polynomial function of degree 2 is a quadratic function; its graph is a parabola.

We now consider polynomial functions of higher degree. These include **cubic functions** (polynomials of degree 3) and **quartic functions** (polynomials of degree 4). Recall that a polynomial function of degree n can be written in the form

$$p(x) = a_n x^n + a_{n-1}x^{n-1} + \cdots + a_2x^2 + a_1x + a_0,\ a_n \neq 0.$$

Here are some important definitions associated with polynomial functions and this equation.

DEFINITION The Vocabulary of Polynomials

- Each monomial in this sum—$a_nx^n, a_{n-1}x^{n-1}, \ldots, a_0$—is a **term** of the polynomial.
- A polynomial function written in this way, with terms in descending degree, is written in **standard form**.
- The constants $a_n, a_{n-1}, \ldots, a_0$ are the **coefficients** of the polynomial.
- The term a_nx^n is the **leading term**, and a_0 is the constant term.

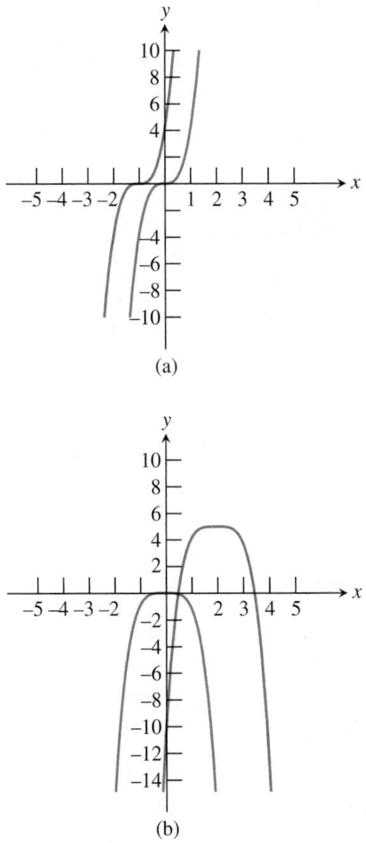

FIGURE 2.19 (a) The graphs of $g(x) = 4(x + 1)^3$ and $f(x) = 4x^3$. (b) The graphs of $h(x) = -(x - 2)^4 + 5$ and $f(x) = -x^4$. (Example 1)

In Example 1 we use the fact from Section 2.1 that the constant term a_0 of a polynomial function p is both the initial value of the function $p(0)$ and the y-intercept of the graph to provide a quick and easy check of the transformed graphs.

EXAMPLE 1 Graphing Transformations of Monomial Functions

Describe how to transform the graph of an appropriate monomial function $f(x) = a_nx^n$ into the graph of the given function. Sketch the transformed graph by hand and support your answer with a grapher. Compute the location of the y-intercept as a check on the transformed graph.

(a) $g(x) = 4(x + 1)^3$ **(b)** $h(x) = -(x - 2)^4 + 5$

SOLUTION

(a) You can obtain the graph of $g(x) = 4(x + 1)^3$ by shifting the graph of $f(x) = 4x^3$ one unit to the left, as shown in Figure 2.19a. The y-intercept of the graph of g is $g(0) = 4(0 + 1)^3 = 4$, which appears to agree with the transformed graph.

(b) You can obtain the graph of $h(x) = -(x - 2)^4 + 5$ by shifting the graph of $f(x) = -x^4$ two units to the right and five units up, as shown in Figure 2.19b. The y-intercept of the graph of h is $h(0) = -(0 - 2)^4 + 5 = -16 + 5 = -11$, which appears to agree with the transformed graph.

Now try Exercise 1.

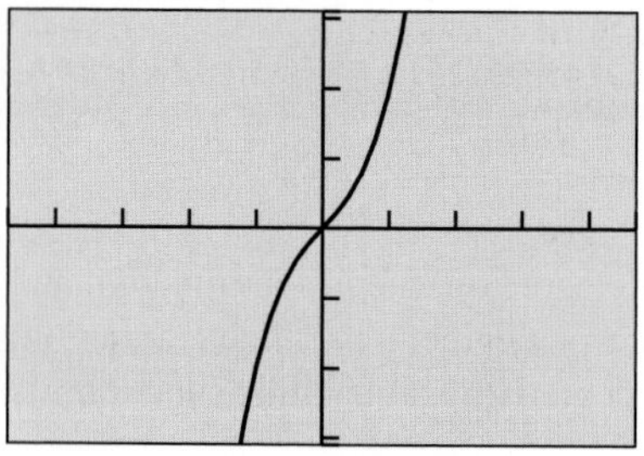

[–4.7, 4.7] by [–3.1, 3.1]
(a)

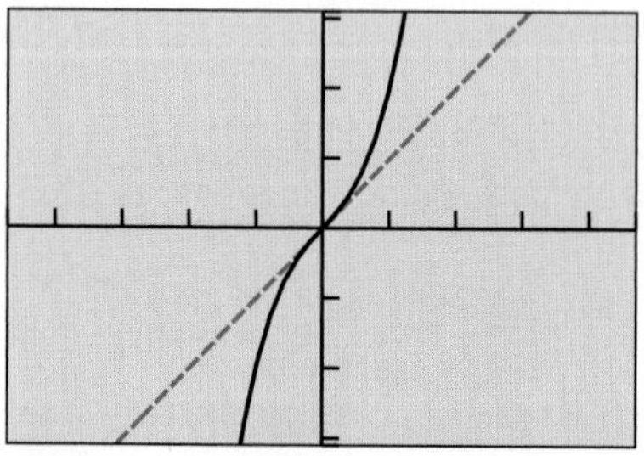

[–4.7, 4.7] by [–3.1, 3.1]
(b)

FIGURE 2.20 The graph of $f(x) = x^3 + x$ (a) by itself and (b) with $y = x$. (Example 2a)

Example 2 shows what can happen when simple monomial functions are combined to obtain polynomial functions. The resulting polynomials are *not* mere translations of monomials.

EXAMPLE 2 Graphing Combinations of Monomial Functions

Graph the polynomial function, locate its extrema and zeros, and explain how it is related to the monomials from which it is built.

(a) $f(x) = x^3 + x$ **(b)** $g(x) = x^3 - x$

SOLUTION

(a) The graph of $f(x) = x^3 + x$ is shown in Figure 2.20a. The function f is increasing on $(-\infty, \infty)$, with no extrema. The function factors as $f(x) = x(x^2 + 1)$ and has one zero at $x = 0$.

The general shape of the graph is much like the graph of its leading term x^3, but near the origin f behaves much like its other term x, as shown in Figure 2.20b. The function f is odd, just like its two building block monomials.

(b) The graph of $g(x) = x^3 - x$ is shown in Figure 2.21a. The function g has a local maximum of about 0.38 at $x \approx -0.58$ and a local minimum of about -0.38 at $x \approx 0.58$. The function factors as $g(x) = x(x + 1)(x - 1)$ and has zeros located at $x = -1$, $x = 0$, and $x = 1$.

The general shape of the graph is much like the graph of its leading term x^3, but near the origin g behaves much like its other term $-x$, as shown in Figure 2.21b. The function g is odd, just like its two building block monomials.

Now try Exercise 7.

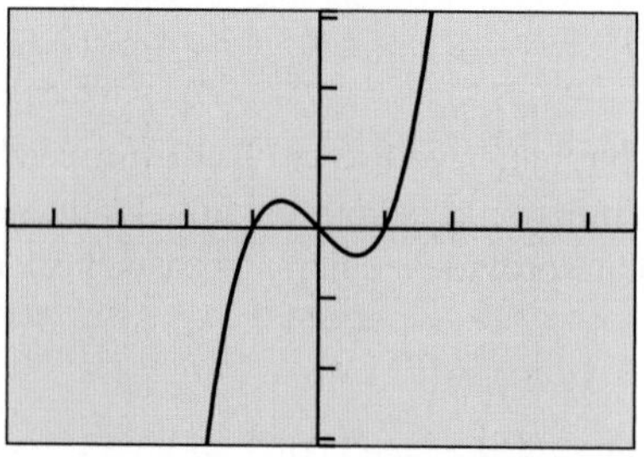

[–4.7, 4.7] by [–3.1, 3.1]
(a)

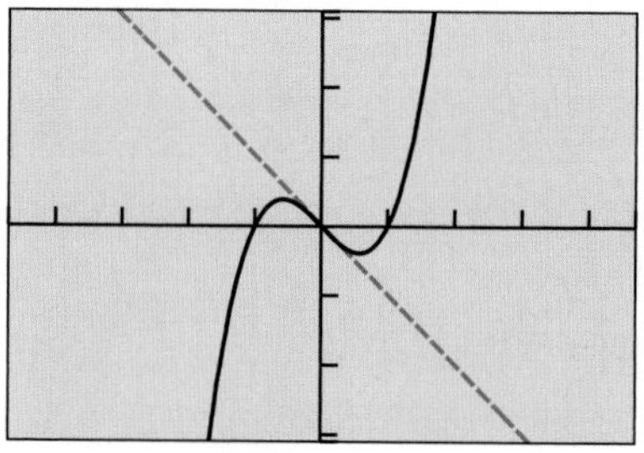

[–4.7, 4.7] by [–3.1, 3.1]
(b)

FIGURE 2.21 The graph of $g(x) = x^3 - x$ (a) by itself and (b) with $y = -x$. (Example 2b)

We have now seen a few examples of graphs of polynomial functions, but are these typical? What do graphs of polynomials look like in general?

To begin our answer, let's first recall that every polynomial function is defined and continuous for all real numbers. Not only are graphs of polynomials unbroken without jumps or holes, but they are *smooth*, unbroken lines or curves, with no sharp corners or cusps. Typical graphs of cubic and quartic functions are shown in Figures 2.22 and 2.23.

Imagine horizontal lines passing through the graphs in Figures 2.22 and 2.23, acting as x-axes. Each intersection would be an x-intercept that would correspond to a zero of the function. From this mental experiment, we see that cubic functions have at most three zeros and quartic functions have at most four zeros. Focusing on the high and low points in Figures 2.22 and 2.23, we see that cubic functions have at most two local extrema and quartic functions have at most three local extrema. These observations generalize in the following way:

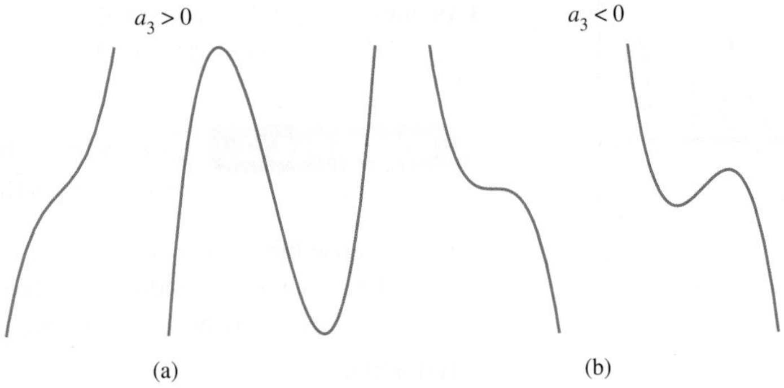

FIGURE 2.22 Graphs of four typical cubic functions: (a) two with positive and (b) two with negative leading coefficients.

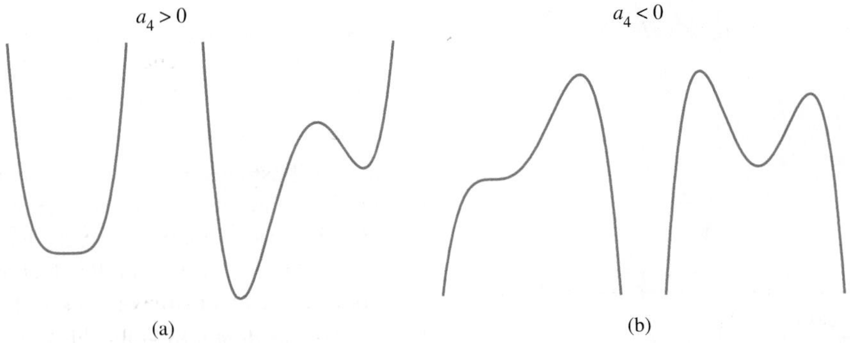

FIGURE 2.23 Graphs of four typical quartic functions: (a) two with positive and (b) two with negative leading coefficients.

THEOREM Local Extrema and Zeros of Polynomial Functions

A polynomial function of degree n has at most $n - 1$ local extrema and at most n zeros.

Technology Note

For a cubic, when you change the horizontal window by a factor of k, it usually is a good idea to change the vertical window by a factor of k^3. Similar statements can be made about polynomials of other degrees.

End Behavior of Polynomial Functions

An important characteristic of polynomial functions is their end behavior. As we shall see, the end behavior of a polynomial is closely related to the end behavior of its leading term. Exploration 1 examines the end behavior of monomial functions, which are potential leading terms for polynomial functions.

EXPLORATION 1 Investigating the End Behavior of $f(x) = a_n x^n$

Graph each function in the window $[-5, 5]$ by $[-15, 15]$. Describe the end behavior using $\lim_{x\to\infty} f(x)$ and $\lim_{x\to-\infty} f(x)$.

1. **(a)** $f(x) = 2x^3$ **(b)** $f(x) = -x^3$
   (c) $f(x) = x^5$ **(d)** $f(x) = -0.5x^7$
2. **(a)** $f(x) = -3x^4$ **(b)** $f(x) = 0.6x^4$
   (c) $f(x) = 2x^6$ **(d)** $f(x) = -0.5x^2$
3. **(a)** $f(x) = -0.3x^5$ **(b)** $f(x) = -2x^2$
   (c) $f(x) = 3x^4$ **(d)** $f(x) = 2.5x^3$

Describe the patterns you observe. In particular, how do the values of the coefficient a_n and the degree n affect the end behavior of $f(x) = a_n x^n$?

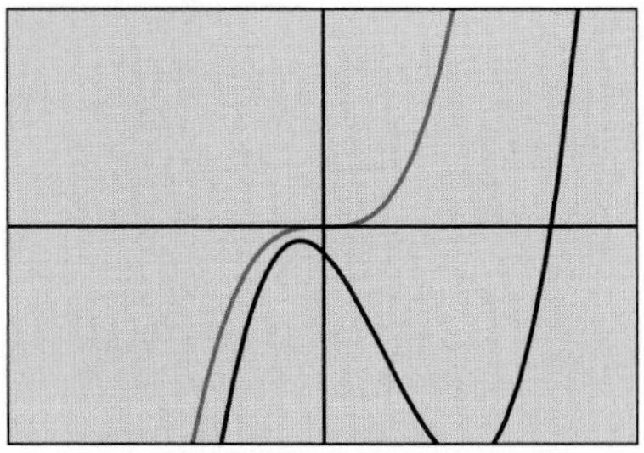

[–7, 7] by [–25, 25]

(a)

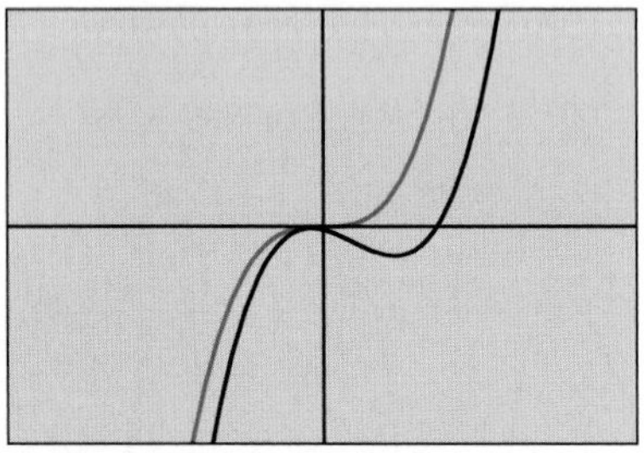

[–14, 14] by [–200, 200]

(b)

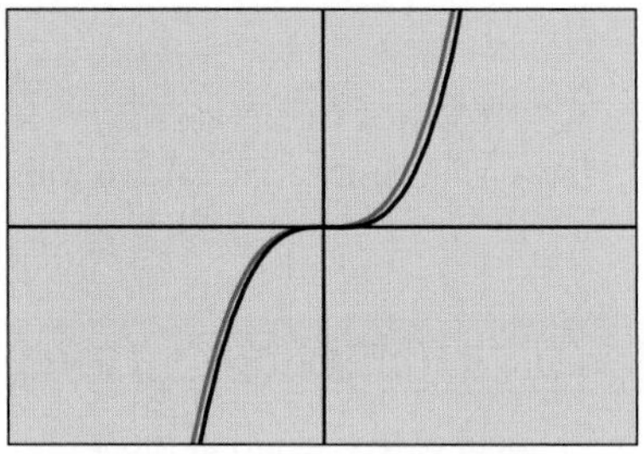

[–56, 56] by [–12800, 12800]

(c)

FIGURE 2.24 As the viewing window gets larger, the graphs of $f(x) = x^3 - 4x^2 - 5x - 3$ and $g(x) = x^3$ look more and more alike. (Example 3)

Example 3 illustrates the link between the end behavior of a polynomial $f(x) = a_n x^n + \cdots + a_1 x + a_0$ and its leading term $a_n x^n$.

EXAMPLE 3 Comparing the Graphs of a Polynomial and Its Leading Term

Superimpose the graphs of $f(x) = x^3 - 4x^2 - 5x - 3$ and $g(x) = x^3$ in successively larger viewing windows, a process called **zoom out**. Continue zooming out until the graphs look nearly identical.

SOLUTION

Figure 2.24 shows three views of the graphs of $f(x) = x^3 - 4x^2 - 5x - 3$ and $g(x) = x^3$ in progressively larger viewing windows. As the dimensions of the window increase, it gets harder to tell them apart. Moreover,

$$\lim_{x\to\infty} f(x) = \lim_{x\to\infty} g(x) = \infty \quad \text{and} \quad \lim_{x\to-\infty} f(x) = \lim_{x\to-\infty} g(x) = -\infty.$$

Now try Exercise 13.

Example 3 illustrates something that is true for all polynomials: *In sufficiently large viewing windows, the graph of a polynomial and the graph of its leading term appear to be identical.* Said another way, the leading term *dominates* the behavior of the polynomial as $|x| \to \infty$. Based on this fact and what we have seen in Exploration 1, there are four possible end behavior patterns for a polynomial function. The power and coefficient of the leading term tell us which one of the four patterns occurs.

Leading Term Test for Polynomial End Behavior

For any polynomial function $f(x) = a_n x^n + \cdots + a_1 x + a_0$, the limits $\lim_{x\to\infty} f(x)$ and $\lim_{x\to-\infty} f(x)$ are determined by the degree n of the polynomial and its leading coefficient a_n:

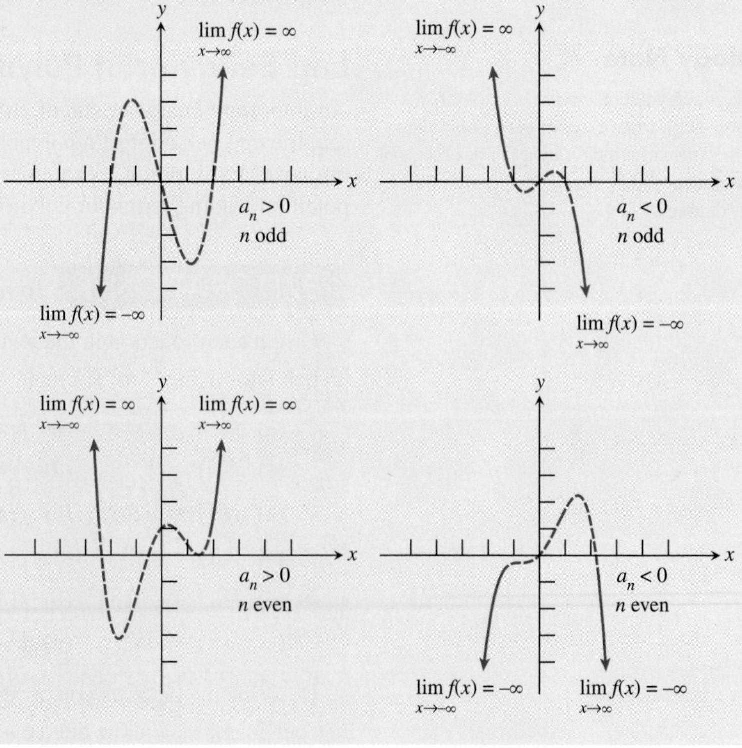

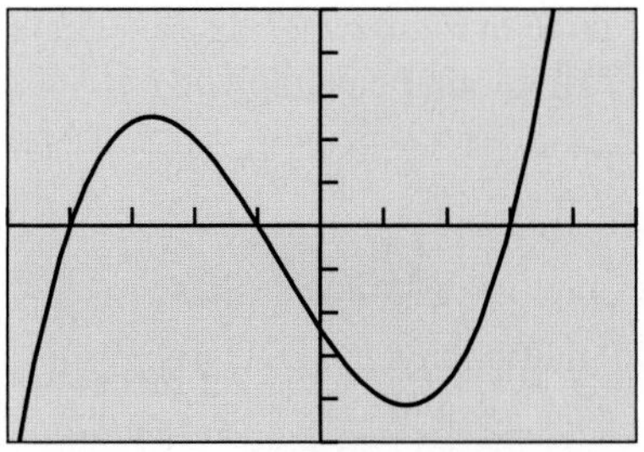

[–5, 5] by [–25, 25]

(a)

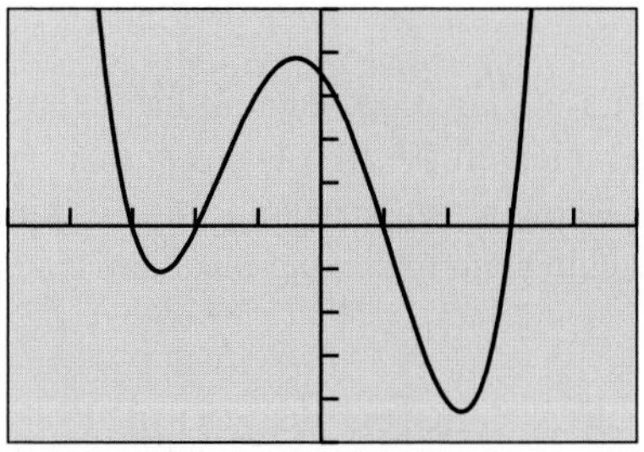

[–5, 5] by [–50, 50]

(b)

FIGURE 2.25 Example 4
(a) $f(x) = x^3 + 2x^2 - 11x - 12$,
(b) $g(x) = 2x^4 + 2x^3 - 22x^2 - 18x + 35$.

EXAMPLE 4 Applying Polynomial Theory

Graph the polynomial in a window showing its extrema and zeros and its end behavior. Describe the end behavior using limits.

(a) $f(x) = x^3 + 2x^2 - 11x - 12$

(b) $g(x) = 2x^4 + 2x^3 - 22x^2 - 18x + 35$

SOLUTION

(a) The graph of $f(x) = x^3 + 2x^2 - 11x - 12$ is shown in Figure 2.25a. The function f has 2 extrema and 3 zeros, the maximum number possible for a cubic. $\lim_{x\to\infty} f(x) = \infty$ and $\lim_{x\to-\infty} f(x) = -\infty$.

(b) The graph of $g(x) = 2x^4 + 2x^3 - 22x^2 - 18x + 35$ is shown in Figure 2.25b. The function g has 3 extrema and 4 zeros, the maximum number possible for a quartic. $\lim_{x\to\infty} g(x) = \infty$ and $\lim_{x\to-\infty} g(x) = \infty$. **Now try Exercise 19.**

Zeros of Polynomial Functions

Recall that finding the real number zeros of a function f is equivalent to finding the x-intercepts of the graph of $y = f(x)$ or the solutions to the equation $f(x) = 0$. Example 5 illustrates that *factoring* a polynomial function makes solving these three related problems an easy matter.

EXAMPLE 5 Finding the Zeros of a Polynomial Function

Find the zeros of $f(x) = x^3 - x^2 - 6x$.

SOLUTION

Solve Algebraically We solve the related equation $f(x) = 0$ by factoring:

$$x^3 - x^2 - 6x = 0$$

$$x(x^2 - x - 6) = 0 \quad \text{Remove common factor x.}$$

$$x(x - 3)(x + 2) = 0 \quad \text{Factor quadratic.}$$

$$x = 0, \quad x - 3 = 0, \quad \text{or} \quad x + 2 = 0 \quad \text{Zero factor property}$$

$$x = 0, \quad x = 3, \quad \text{or} \quad x = -2$$

So the zeros of f are 0, 3, and -2.

Support Graphically Use the features of your calculator to approximate the zeros of f. Figure 2.26 shows that there are three values. Based on our algebraic solution we can be sure that these values are exact. **Now try Exercise 33.**

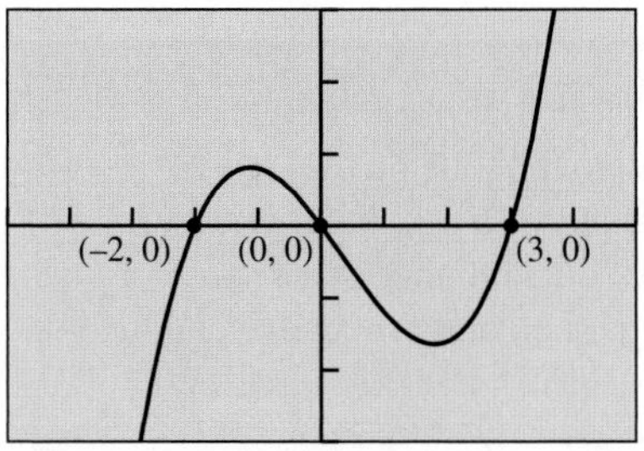

[–5, 5] by [–15, 15]

FIGURE 2.26 The graph of $y = x^3 - x^2 - 6x$, showing the three x-intercepts. (Example 5)

From Example 5, we see that if a polynomial function f is presented in factored form, each factor $(x - k)$ corresponds to a zero $x = k$, and if k is a real number, $(k, 0)$ is an x-intercept of the graph of $y = f(x)$.

When a factor is repeated, as in $f(x) = (x - 2)^3(x + 1)^2$, we say the polynomial function has a *repeated zero*. The function f has two repeated zeros. Because the factor $x - 2$ occurs three times, 2 is a zero of *multiplicity* 3. Similarly, -1 is a zero of multiplicity 2. The following definition generalizes this concept.

DEFINITION Multiplicity of a Zero of a Polynomial Function

If f is a polynomial function and $(x - c)^m$ is a factor of f but $(x - c)^{m+1}$ is not, then c is a zero of **multiplicity m** of f.

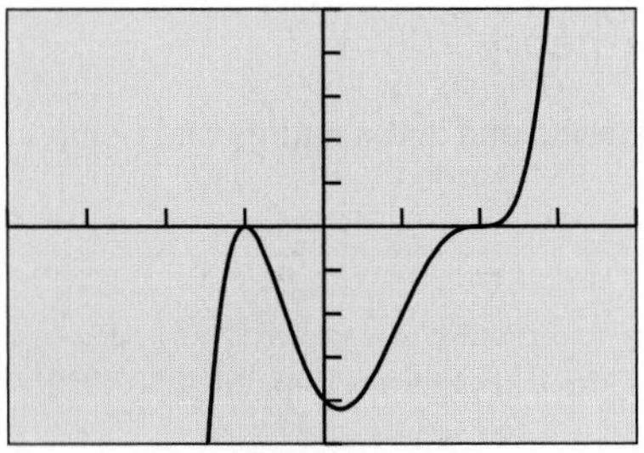

[–4, 4] by [–10, 10]

FIGURE 2.27 The graph of $f(x) = (x - 2)^3(x + 1)^2$, showing the x-intercepts.

A zero of multiplicity $m \geq 2$ is a **repeated zero**. Notice in Figure 2.27 that the graph of f just *kisses* the x-axis without crossing it at $(-1, 0)$, but that the graph of f crosses the x-axis at $(2, 0)$. This too can be generalized.

Zeros of Odd and Even Multiplicity

If a polynomial function f has a real zero c of odd multiplicity, then the graph of f crosses the x-axis at $(c, 0)$ and the value of f changes sign at $x = c$.
If a polynomial function f has a real zero c of even multiplicity, then the graph of f does not cross the x-axis at $(c, 0)$ and the value of f does not change sign at $x = c$.

In Example 5 none of the zeros were repeated. Because a nonrepeated zero has multiplicity 1, and 1 is odd, the graph of a polynomial function crosses the x-axis and has a sign change at every nonrepeated zero (Figure 2.26). Knowing where a graph crosses the x-axis and where it doesn't is important in curve sketching and in solving inequalities.

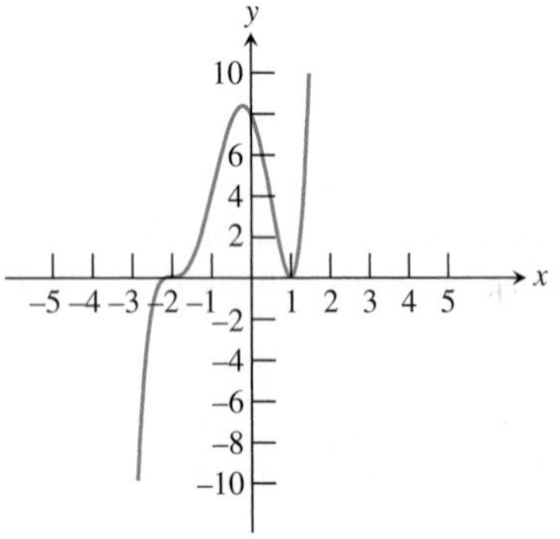

FIGURE 2.28 A sketch of the graph of $f(x) = (x + 2)^3(x - 1)^2$ showing the x-intercepts.

EXAMPLE 6 Sketching the Graph of a Factored Polynomial

State the degree and list the zeros of the function $f(x) = (x + 2)^3(x - 1)^2$. State the multiplicity of each zero and whether the graph crosses the x-axis at the corresponding x-intercept. Then sketch the graph of f by hand.

SOLUTION The degree of f is 5 and the zeros are $x = -2$ and $x = 1$. The graph crosses the x-axis at $x = -2$ because the multiplicity 3 is odd. The graph does not cross the x-axis at $x = 1$ because the multiplicity 2 is even. Notice that values of f are positive for $x > 1$, positive for $-2 < x < 1$, and negative for $x < -2$. Figure 2.28 shows a sketch of the graph of f. **Now try Exercise 39.**

Intermediate Value Theorem

The *Intermediate Value Theorem* tells us that a sign change implies a real zero.

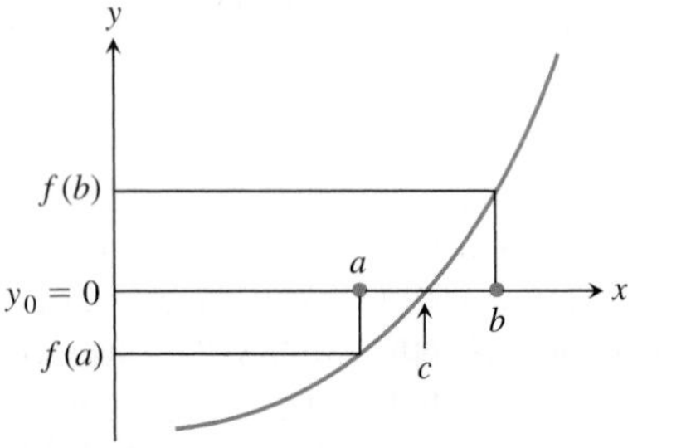

FIGURE 2.29 If $f(a) < 0 < f(b)$ and f is continuous on $[a, b]$, then there is a zero $x = c$ between a and b.

THEOREM Intermediate Value Theorem

If a and b are real numbers with $a < b$ and if f is continuous on the interval $[a, b]$, then f takes on every value between $f(a)$ and $f(b)$. In other words, if y_0 is between $f(a)$ and $f(b)$, then $y_0 = f(c)$ for some number c in $[a, b]$.
In particular, if $f(a)$ and $f(b)$ have opposite signs (i.e., one is negative and the other is positive), then $f(c) = 0$ for some number c in $[a, b]$ (Figure 2.29).

EXAMPLE 7 Using the Intermediate Value Theorem

Explain why a polynomial function of odd degree has at least one real zero.

SOLUTION Let f be a polynomial function of odd degree. Because f is odd, the leading term test tells us that $\lim_{x \to \infty} f(x) = -\lim_{x \to -\infty} f(x)$. So there exist real numbers a and b with $a < b$ and such that $f(a)$ and $f(b)$ have opposite signs. Because every polynomial function is defined and continuous for all real numbers, f is continuous on the interval $[a, b]$. Therefore, by the Intermediate Value Theorem, $f(c) = 0$ for some number c in $[a, b]$, and thus c is a real zero of f. **Now try Exercise 61.**

In practice, the Intermediate Value Theorem is used in combination with our other mathematical knowledge and technological know-how.

Exact vs. Approximate

In Example 8, note that $x = 0.50$ is an exact answer; the others are approximate. Use by-hand substitution to confirm that $x = 1/2$ is an exact real zero.

EXAMPLE 8 Zooming to Uncover Hidden Behavior

Find all of the real zeros of $f(x) = x^4 + 0.1x^3 - 6.5x^2 + 7.9x - 2.4$.

SOLUTION

Solve Graphically Because f is of degree 4, there are at most four zeros. The graph in Figure 2.30a suggests a single zero (multiplicity 1) around $x = -3$ and a triple zero (multiplicity 3) around $x = 1$. Closer inspection around $x = 1$ in Figure 2.30b reveals three separate zeros. Using the grapher, we find the four zeros to be $x \approx 1.37$, $x \approx 1.13$, $x = 0.50$, and $x \approx -3.10$. (See the margin note.)

Now try Exercise 75.

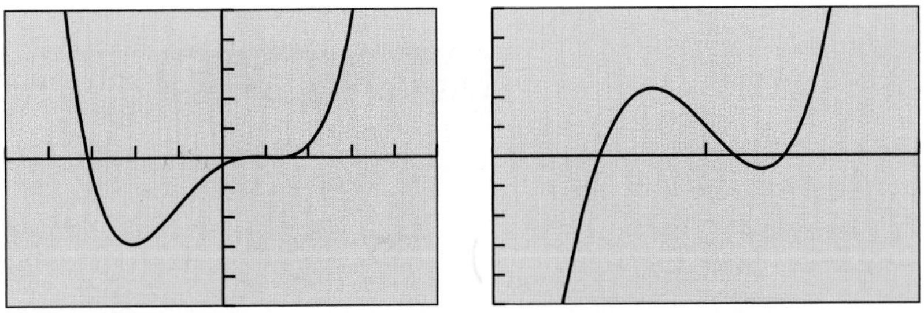

[–5, 5] by [–50, 50]
(a)

[0, 2] by [–0.5, 0.5]
(b)

FIGURE 2.30 Two views of $f(x) = x^4 + 0.1x^3 - 6.5x^2 + 7.9x - 2.4$. (Example 8)

Modeling

In the problem-solving process presented in Section 1.1, step 2 is to develop a mathematical model of the problem. When the model being developed is a polynomial function of higher degree, the algebraic and geometric thinking required can be rather involved. In solving Example 9 you may find it helpful to make a physical model out of paper or cardboard.

X	Y1	
1	414	
2	672	
3	798	
4	816	
5	750	
6	624	
7	462	

Y1 = X(20–2X)(25–...

FIGURE 2.32 A table to get a feel for the volume values in Example 9.

EXAMPLE 9 Designing a Box

Dixie Packaging Company has contracted to make boxes with a volume of approximately 484 in.^{3}. Squares are to be cut from the corners of a 20-in. by 25-in. piece of cardboard, and the flaps folded up to make an open box. (See Figure 2.31.) What size squares should be cut from the cardboard?

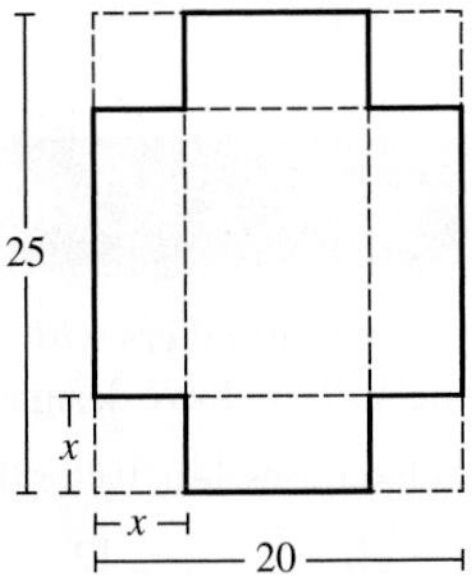

FIGURE 2.31

SOLUTION

Model We know that the volume $V = \text{height} \times \text{length} \times \text{width}$. So let

$$x = \text{edge of cut-out square (height of box)}$$
$$25 - 2x = \text{length of the box}$$
$$20 - 2x = \text{width of the box}$$
$$V = x(25 - 2x)(20 - 2x)$$

Solve Numerically and Graphically For a volume of 484, we solve the equation $x(25 - 2x)(20 - 2x) = 484$. Because the width of the cardboard is 20 in., $0 \le x \le 10$. We use the table in Figure 2.32 to get a sense of the volume values to set the window for the graph in Figure 2.33. The cubic volume function intersects the constant volume of 484 at $x \approx 1.22$ and $x \approx 6.87$.

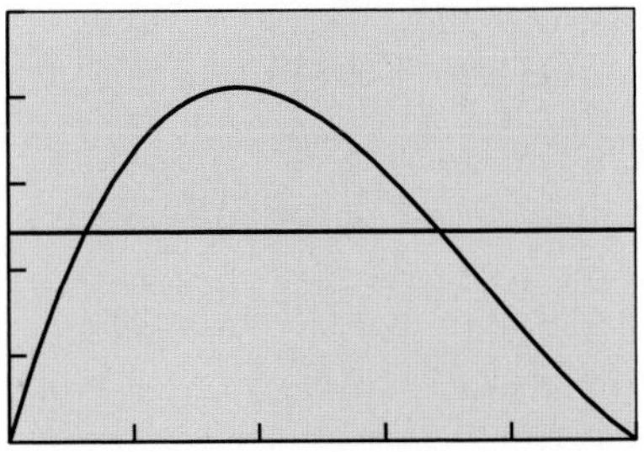

[0, 10] by [0, 1000]

FIGURE 2.33 $y_1 = x(25 - 2x)(20 - 2x)$ and $y_2 = 484$. (Example 9)

(continued)

Interpret Squares with lengths of approximately 1.22 in. or 6.87 in. should be cut from the cardboard to produce a box with a volume of 484 in.^{3}. **Now try Exercise 67.**

Just as any two points in the Cartesian plane with different x values and different y values determine a unique slant line and its related linear function, any three noncollinear points with different x values determine a quadratic function. In general, $(n + 1)$ points positioned with sufficient generality determine a polynomial function of degree n. The process of fitting a polynomial of degree n to $(n + 1)$ points is **polynomial interpolation**. Exploration 2 involves two polynomial interpolation problems.

EXPLORATION 2 Interpolating Points with a Polynomial

1. Use cubic regression to fit a curve through the four points given in the table.

x	-2	1	3	8
y	2	0.5	-0.2	1.25

2. Use quartic regression to fit a curve through the five points given in the table.

x	3	4	5	6	8
y	-2	-4	-1	8	3

How good is the fit in each case? Why?

Generally we want a reason beyond "it fits well" to choose a model for genuine data. However, when no theoretical basis exists for picking a model, a balance between goodness of fit and simplicity of model is sought. For polynomials, we try to pick a model with the lowest possible degree that has a reasonably good fit. Collected data usually exhibit small-scale variability that is essentially meaningless, so we should not try to create a complicated curve that meanders through every data point. Our objective is to find a simple curve that describes the underlying relationship between the variables.

QUICK REVIEW 2.3 *(For help, go to Sections A.2. and P.5.)*

Exercise numbers with a gray background indicate problems that the authors have designed to be solved *without a calculator*.

In Exercises 1–6, factor the polynomial into linear factors.

1. $x^2 - x - 12$

2. $x^2 - 11x + 28$

3. $3x^2 - 11x + 6$

4. $6x^2 - 5x + 1$

5. $3x^3 - 5x^2 + 2x$

6. $6x^3 - 22x^2 + 12x$

In Exercises 7–10, solve the equation *mentally*.

7. $x(x - 1) = 0$

8. $x(x + 2)(x - 5) = 0$

9. $(x + 6)^3(x + 3)(x - 1.5) = 0$

10. $(x + 6)^2(x + 4)^4(x - 5)^3 = 0$

SECTION 2.3 Exercises

In Exercises 1–6, describe how to transform the graph of an appropriate monomial function $f(x) = x^n$ into the graph of the given polynomial function. Sketch the transformed graph by hand and support your answer with a grapher. Compute the location of the y-intercept as a check on the transformed graph.

1. $g(x) = 2(x - 3)^3$ **2.** $g(x) = -(x + 5)^3$

3. $g(x) = -\frac{1}{2}(x + 1)^3 + 2$ **4.** $g(x) = \frac{2}{3}(x - 3)^3 + 1$

5. $g(x) = -2(x + 2)^4 - 3$ **6.** $g(x) = 3(x - 1)^4 - 2$

In Exercises 7 and 8, graph the polynomial function, locate its extrema and zeros, and explain how it is related to the monomials from which it is built.

7. $f(x) = -x^4 + 2x$ **8.** $g(x) = 2x^4 - 5x^2$

In Exercises 9–12, match the polynomial function with its graph. Explain your choice. Do not use a graphing calculator.

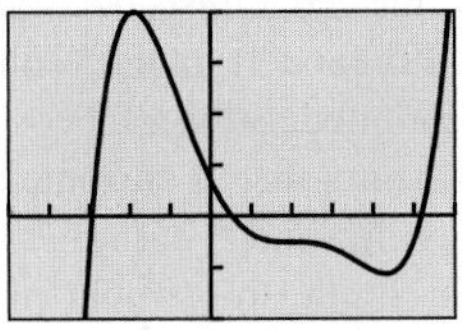
[–5, 6] by [–200, 400]
(a)

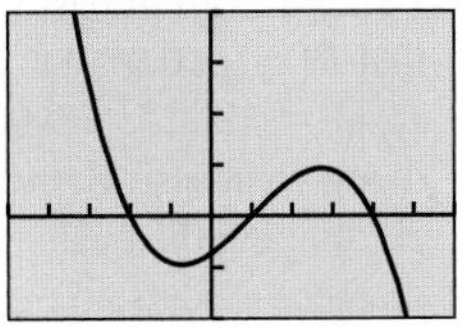
[–5, 6] by [–200, 400]
(b)

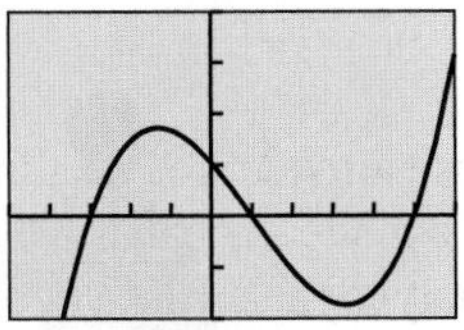
[–5, 6] by [–200, 400]
(c)

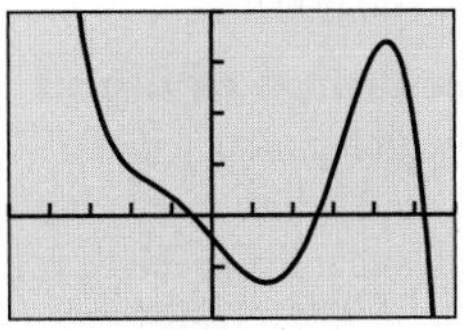
[–5, 6] by [–200, 400]
(d)

9. $f(x) = 7x^3 - 21x^2 - 91x + 104$

10. $f(x) = -9x^3 + 27x^2 + 54x - 73$

11. $f(x) = x^5 - 8x^4 + 9x^3 + 58x^2 - 164x + 69$

12. $f(x) = -x^5 + 3x^4 + 16x^3 - 2x^2 - 95x - 44$

In Exercises 13–16, graph the function pairs in the same series of viewing windows. Zoom out until the two graphs look nearly identical and state your final viewing window.

13. $f(x) = x^3 - 4x^2 - 5x - 3$ and $g(x) = x^3$

14. $f(x) = x^3 + 2x^2 - x + 5$ and $g(x) = x^3$

15. $f(x) = 2x^3 + 3x^2 - 6x - 15$ and $g(x) = 2x^3$

16. $f(x) = 3x^3 - 12x + 17$ and $g(x) = 3x^3$

In Exercises 17–24, graph the function in a viewing window that shows all of its extrema and x-intercepts. Describe the end behavior using limits.

17. $f(x) = (x - 1)(x + 2)(x + 3)$

18. $f(x) = (2x - 3)(4 - x)(x + 1)$

19. $f(x) = -x^3 + 4x^2 + 31x - 70$

20. $f(x) = x^3 - 2x^2 - 41x + 42$

21. $f(x) = (x - 2)^2(x + 1)(x - 3)$

22. $f(x) = (2x + 1)(x - 4)^3$

23. $f(x) = 2x^4 - 5x^3 - 17x^2 + 14x + 41$

24. $f(x) = -3x^4 - 5x^3 + 15x^2 - 5x + 19$

In Exercises 25–28, describe the end behavior of the polynomial function using $\lim_{x\to\infty} f(x)$ and $\lim_{x\to-\infty} f(x)$.

25. $f(x) = 3x^4 - 5x^2 + 3$

26. $f(x) = -x^3 + 7x^2 - 4x + 3$

27. $f(x) = 7x^2 - x^3 + 3x - 4$

28. $f(x) = x^3 - x^4 + 3x^2 - 2x + 7$

In Exercises 29–32, match the polynomial function with its graph. Approximate all of the real zeros of the function.

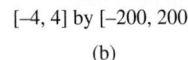

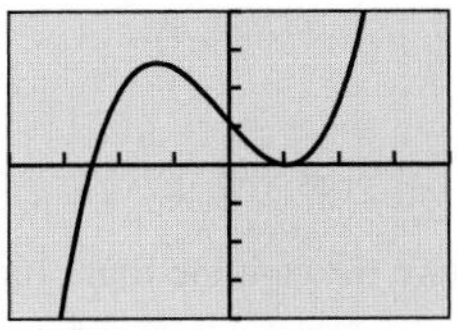
[–4, 4] by [–200, 200]
(a)

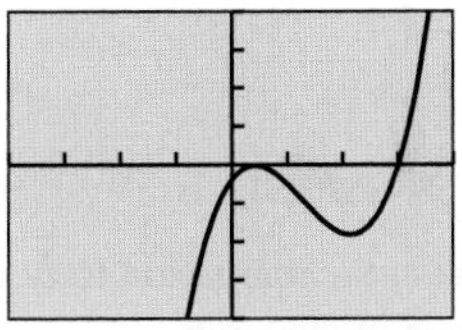
[–4, 4] by [–200, 200]
(b)

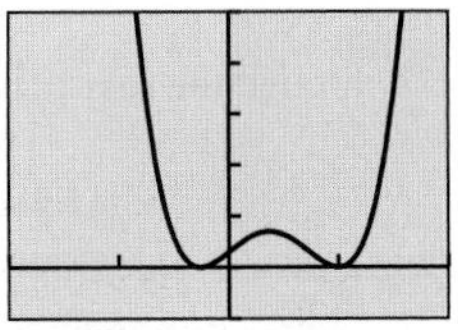
[–2, 2] by [–10, 50]
(c)

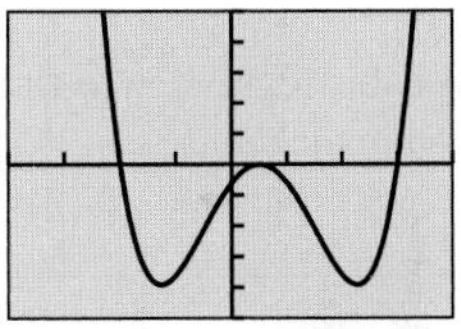
[–4, 4] by [–50, 50]
(d)

29. $f(x) = 20x^3 + 8x^2 - 83x + 55$

30. $f(x) = 35x^3 - 134x^2 + 93x - 18$

31. $f(x) = 44x^4 - 65x^3 + x^2 + 17x + 3$

32. $f(x) = 4x^4 - 8x^3 - 19x^2 + 23x - 6$

In Exercises 33–38, find the zeros of the function algebraically.

33. $f(x) = x^2 + 2x - 8$ **34.** $f(x) = 3x^2 + 4x - 4$

35. $f(x) = 9x^2 - 3x - 2$ **36.** $f(x) = x^3 - 25x$

37. $f(x) = 3x^3 - x^2 - 2x$ **38.** $f(x) = 5x^3 - 5x^2 - 10x$

In Exercises 39–42, state the degree and list the zeros of the polynomial function. State the multiplicity of each zero and whether the graph crosses the x-axis at the corresponding x-intercept. Then sketch the graph of the polynomial function by hand.

39. $f(x) = x(x - 3)^2$

40. $f(x) = -x^3(x - 2)$

41. $f(x) = (x - 1)^3(x + 2)^2$

42. $f(x) = 7(x - 3)^2(x + 5)^4$

In Exercises 43–48, graph the function in a viewing window that shows all of its x-intercepts and approximate all of its zeros.

43. $f(x) = 2x^3 + 3x^2 - 7x - 6$

44. $f(x) = -x^3 + 3x^2 + 7x - 2$

45. $f(x) = x^3 + 2x^2 - 4x - 7$

46. $f(x) = -x^4 - 3x^3 + 7x^2 + 2x + 8$

47. $f(x) = x^4 + 3x^3 - 9x^2 + 2x + 3$

48. $f(x) = 2x^5 - 11x^4 + 4x^3 + 47x^2 - 42x - 8$

In Exercises 49–52, find the zeros of the function algebraically or graphically.

49. $f(x) = x^3 - 36x$

50. $f(x) = x^3 + 2x^2 - 109x - 110$

51. $f(x) = x^3 - 7x^2 - 49x + 55$

52. $f(x) = x^3 - 4x^2 - 44x + 96$

In Exercises 53–56, using only algebra, find a cubic function with the given zeros. Support by graphing your answer.

53. $3, -4, 6$

54. $-2, 3, -5$

55. $\sqrt{3}, -\sqrt{3}, 4$ **56.** $1, 1 + \sqrt{2}, 1 - \sqrt{2}$

57. Use cubic regression to fit a curve through the four points given in the table.

x	-3	-1	1	3
y	22	25	12	-5

58. Use cubic regression to fit a curve through the four points given in the table.

x	-2	1	4	7
y	2	5	9	26

59. Use quartic regression to fit a curve through the five points given in the table.

x	3	4	5	6	8
y	-7	-4	-11	8	3

60. Use quartic regression to fit a curve through the five points given in the table.

x	0	4	5	7	13
y	-21	-19	-12	8	3

In Exercises 61–62, explain why the function has at least one real zero.

61. Writing to Learn $f(x) = x^7 + x + 100$

62. Writing to Learn $f(x) = x^9 - x + 50$

63. Stopping Distance A state highway patrol safety division collected the data on stopping distances in Table 2.14 in the next column.

(a) Draw a scatter plot of the data.

(b) Find the quadratic regression model.

(c) Superimpose the regression curve on the scatter plot.

(d) Use the regression model to predict the stopping distance for a vehicle traveling at 25 mph.

(e) For what speed would this regression model predict the stopping distance to be 300 ft?

Table 2.14 Highway Safety Division

Speed (mph)	Stopping Distance (ft)
10	15.1
20	39.9
30	75.2
40	120.5
50	175.9

64. Analyzing Profit Economists for Smith Brothers, Inc., find the company profit P by using the formula $P = R - C$, where R is the total revenue generated by the business and C is the total cost of operating the business.

(a) Using data from past years, the economists determined that $R(x) = 0.0125x^2 + 412x$ models total revenue, and $C(x) = 12{,}225 + 0.00135x^3$ models the total cost of doing business, where x is the number of customers patronizing the business. For how many customers do these models predict that Smith Bros. would be profitable?

(b) For how many customers do these models predict that Smith Bros. would realize an annual profit of \$60,000?

65. Circulation of Blood Research conducted at a national health research project shows that the speed at which a blood cell travels in an artery depends on its distance from the center of the artery. The function $v = 1.19 - 1.87r^2$ models the velocity (in centimeters per second) of a cell that is r centimeters from the center of an artery.

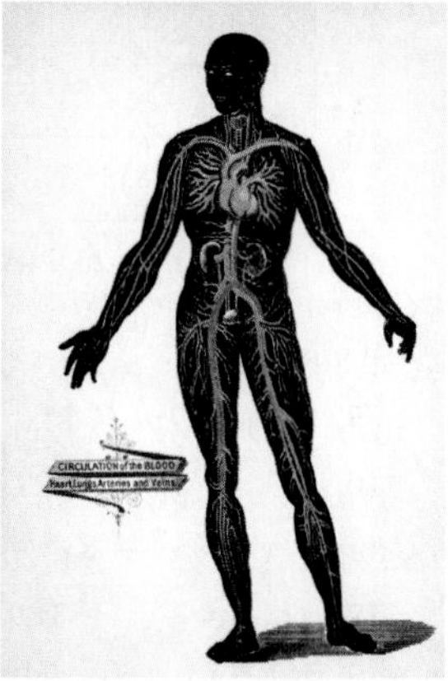

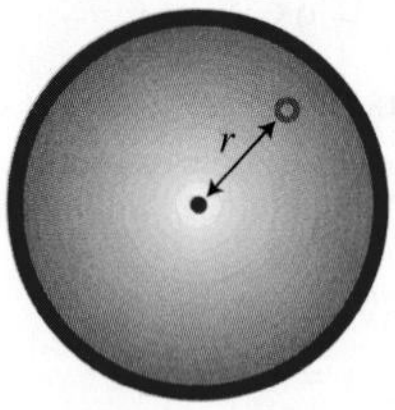

(a) Find a graph of v that reflects values of v appropriate for this problem. Record the viewing-window dimensions.

(b) At what distance from the center of the artery does this model predict that a blood cell travels at 0.975 cm/sec?

66. Volume of a Box Dixie Packaging Co. has contracted to manufacture a box with no top that is to be made by removing squares of width x from the corners of a 15-in. by 60-in. piece of cardboard.

(a) Show that the volume of the box is modeled by $V(x) = x(60 - 2x)(15 - 2x)$.

(b) Determine x so that the volume of the box is at least 450 in.^{3}

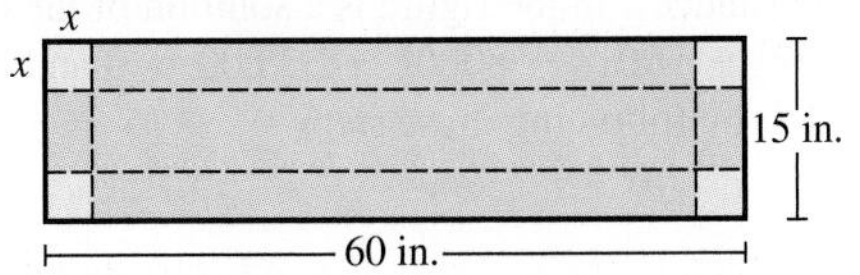

67. Volume of a Box Squares of width x are removed from a 10-cm by 25-cm piece of cardboard, and the resulting edges are folded up to form a box with no top. Determine all values of x so that the volume of the resulting box is at most 175 cm^3.

68. Volume of a Box The function $V = 2666x - 210x^2 + 4x^3$ represents the volume of a box that has been made by removing squares of width x from each corner of a rectangular sheet of material and then folding up the sides. What values are possible for x?

Standardized Test Questions

69. True or False The graph of $f(x) = x^3 - x^2 - 2$ crosses the x-axis between $x = 1$ and $x = 2$. Justify your answer.

70. True or False If the graph of $g(x) = (x + a)^2$ is obtained by translating the graph of $f(x) = x^2$ to the right, then a must be positive. Justify your answer.

In Exercises 71–74, solve the problem without using a calculator.

71. Multiple Choice What is the y-intercept of the graph of $f(x) = 2(x - 1)^3 + 5$?

(A) 7 **(B)** 5 **(C)** 3 **(D)** 2 **(E)** 1

72. Multiple Choice What is the multiplicity of the zero $x = 2$ in $f(x) = (x - 2)^2(x + 2)^3(x + 3)^7$?

(A) 1 **(B)** 2 **(C)** 3 **(D)** 5 **(E)** 7

In Exercises 73 and 74, which of the specified functions might have the given graph?

73. Multiple Choice

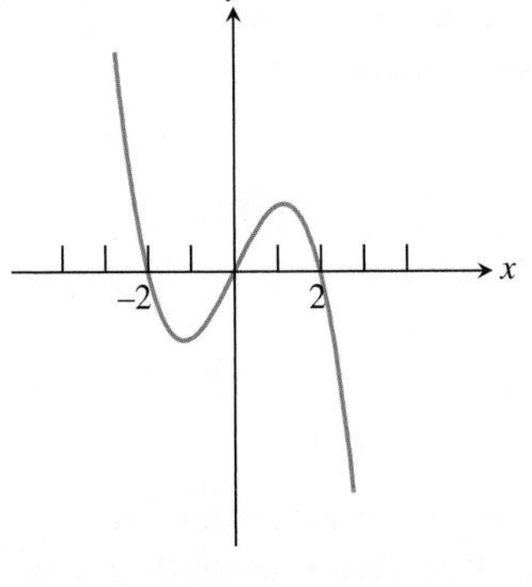

(A) $f(x) = -x(x + 2)(2 - x)$

(B) $f(x) = -x(x + 2)(x - 2)$

(C) $f(x) = -x^2(x + 2)(x - 2)$

(D) $f(x) = -x(x + 2)^2(x - 2)$

(E) $f(x) = -x(x + 2)(x - 2)^2$

74. Multiple Choice

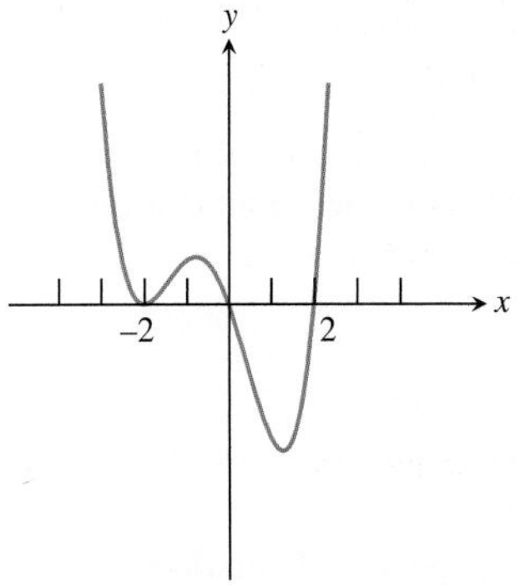

(A) $f(x) = x(x + 2)^2(x - 2)$

(B) $f(x) = x(x + 2)^2(2 - x)$

(C) $f(x) = x^2(x + 2)(x - 2)$

(D) $f(x) = x(x + 2)(x - 2)^2$

(E) $f(x) = x^2(x + 2)(x - 2)^2$

Explorations

In Exercises 75 and 76, two views of the function are given.

75. Writing to Learn Describe why each view of the function

$$f(x) = x^5 - 10x^4 + 2x^3 + 64x^2 - 3x - 55,$$

by itself, may be considered inadequate.

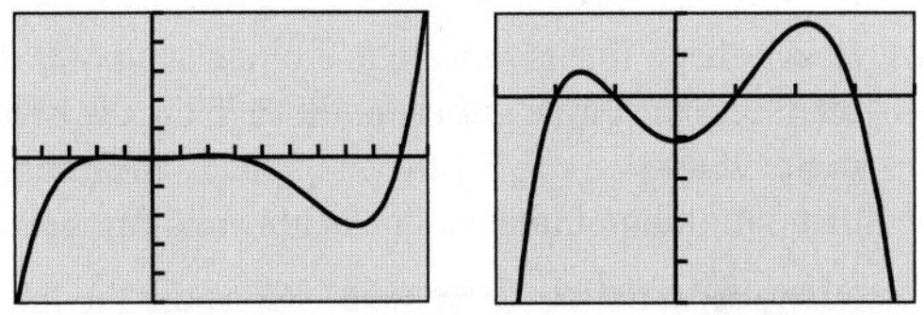

[–5, 10] by [–7500, 7500]
(a)

[–3, 4] by [–250, 100]
(b)

76. Writing to Learn Describe why each view of the function

$$f(x) = 10x^4 + 19x^3 - 121x^2 + 143x - 51,$$

by itself, may be considered inadequate.

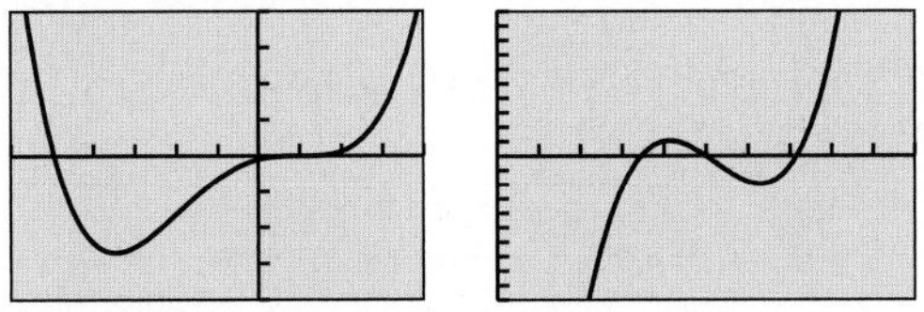

[–6, 4] by [–2000, 2000]
(a)

[0.5, 1.5] by [–1, 1]
(b)

In Exercises 77–80, the function has hidden behavior when viewed in the window $[-10, 10]$ by $[-10, 10]$. Describe what behavior is hidden, and state the dimensions of a viewing window that reveals the hidden behavior.

77. $f(x) = 10x^3 - 40x^2 + 50x - 20$

78. $f(x) = 0.5(x^3 - 8x^2 + 12.99x - 5.94)$

79. $f(x) = 11x^3 - 10x^2 + 3x + 5$

80. $f(x) = 33x^3 - 100x^2 + 101x - 40$

Extending the Ideas

81. Graph the left side of the equation

$$3(x^3 - x) = a(x - b)^3 + c.$$

Then explain why there are no real numbers a, b, and c that make the equation true. (*Hint*: Use your knowledge of $y = x^3$ and transformations.)

82. Graph the left side of the equation

$$x^4 + 3x^3 - 2x - 3 = a(x - b)^4 + c.$$

Then explain why there are no real numbers a, b, and c that make the equation true.

83. Looking Ahead to Calculus The figure shows a graph of both $f(x) = -x^3 + 2x^2 + 9x - 11$ and the line L defined by $y = 5(x - 2) + 7$.

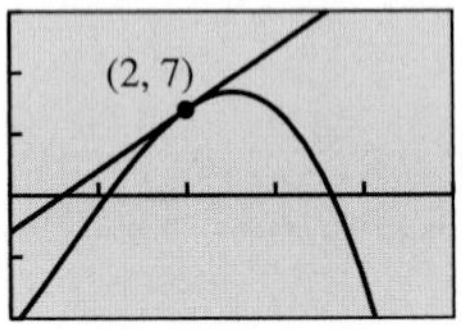

[0, 5] by [–10, 15]

(a) Confirm that the point $Q(2, 7)$ is a point of intersection of the two graphs.

(b) Zoom in at point Q to develop a visual understanding that $y = 5(x - 2) + 7$ is a *linear approximation* for $y = f(x)$ near $x = 2$.

(c) Recall that a line is *tangent* to a circle at a point P if it intersects the circle only at point P. View the two graphs in the window $[-5, 5]$ by $[-25, 25]$, and explain why that definition of tangent line is not valid for the graph of f.

84. Looking Ahead to Calculus Consider the function $f(x) = x^n$ where n is an odd integer.

(a) Suppose that a is a positive number. Show that the slope of the line through the points $P(a, f(a))$ and $Q(-a, f(-a))$ is a^{n-1}.

(b) Let $x_0 = a^{1/(n-1)}$. Find an equation of the line through point $(x_0, f(x_0))$ with the slope a^{n-1}.

(c) Consider the special case $n = 3$ and $a = 3$. Show both the graph of f and the line from part b in the window $[-5, 5]$ by $[-30, 30]$.

85. Derive an Algebraic Model of a Problem Show that the distance x in the figure is a solution of the equation $x^4 - 16x^3 + 500x^2 - 8000x + 32{,}000 = 0$ and find the value of D by following these steps.

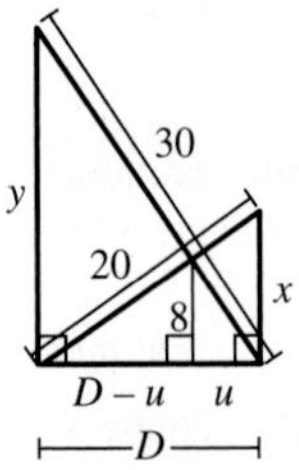

(a) Use the similar triangles in the diagram and the properties of proportions learned in geometry to show that

$$\frac{8}{x} = \frac{y - 8}{y}.$$

(b) Show that $y = \dfrac{8x}{x - 8}$.

(c) Show that $y^2 - x^2 = 500$. Then substitute for y, and simplify to obtain the desired degree 4 equation in x.

(d) Find the distance D.

86. Group Activity Consider functions of the form $f(x) = x^3 + bx^2 + x + 1$ where b is a nonzero real number.

(a) Discuss as a group how the value of b affects the graph of the function.

(b) After completing (a), have each member of the group (individually) predict what the graphs of $f(x) = x^3 + 15x^2 + x + 1$ and $g(x) = x^3 - 15x^2 + x + 1$ will look like.

(c) Compare your predictions with each other. Confirm whether they are correct.

2.4 Real Zeros of Polynomial Functions

What you'll learn about

- Long Division and the Division Algorithm
- Remainder and Factor Theorems
- Synthetic Division
- Rational Zeros Theorem
- Upper and Lower Bounds

... and why

These topics help identify and locate the real zeros of polynomial functions.

Long Division and the Division Algorithm

We have seen that factoring a polynomial reveals its zeros and much about its graph. Polynomial division gives us new and better ways to factor polynomials. First we observe that the division of polynomials closely resembles the division of integers:

112	$1x^2 + 1x + 2$	← Quotient
$32\overline{)3587}$	$3x + 2\overline{)3x^3 + 5x^2 + 8x + 7}$	← Dividend
$\underline{32}$	$\underline{3x^3 + 2x^2}$	← Multiply: $1x^2 \cdot (3x + 2)$
387	$3x^2 + 8x + 7$	← Subtract
$\underline{32}$	$\underline{3x^2 + 2x}$	← Multiply: $1x \cdot (3x + 2)$
67	$6x + 7$	← Subtract
$\underline{64}$	$\underline{6x + 4}$	← Multiply: $2 \cdot (3x + 2)$
3	3	← Remainder

Division, whether integer or polynomial, involves a *dividend* divided by a *divisor* to obtain a *quotient* and a *remainder*. We can check and summarize our result with an equation of the form

$$(\text{Divisor})(\text{Quotient}) + \text{Remainder} = \text{Dividend}.$$

For instance, to check or summarize the long divisions shown above we could write

$$32 \times 112 + 3 = 3587 \qquad (3x + 2)(x^2 + x + 2) + 3 = 3x^3 + 5x^2 + 8x + 7.$$

The *division algorithm* contains such a summary *polynomial equation*, but with the dividend written on the left side of the equation.

Division Algorithm for Polynomials

Let $f(x)$ and $d(x)$ be polynomials with the degree of f greater than or equal to the degree of d, and $d(x) \neq 0$. Then there are unique polynomials $q(x)$ and $r(x)$, called the **quotient** and **remainder**, such that

$$f(x) = d(x) \cdot q(x) + r(x), \tag{1}$$

where either $r(x) = 0$ or the degree of r is less than the degree of d.

The function $f(x)$ in the division algorithm is the **dividend**, and $d(x)$ is the **divisor**. If $r(x) = 0$, we say $d(x)$ **divides evenly** into $f(x)$.

The summary statement (1) is sometimes written in *fraction form* as follows:

$$\frac{f(x)}{d(x)} = q(x) + \frac{r(x)}{d(x)} \tag{2}$$

For instance, to summarize the polynomial division example above we could write

$$\frac{3x^3 + 5x^2 + 8x + 7}{3x + 2} = x^2 + x + 2 + \frac{3}{3x + 2}.$$

EXAMPLE 1 Using Polynomial Long Division

Use long division to find the quotient and remainder when $2x^4 - x^3 - 2$ is divided by $2x^2 + x + 1$. Write a summary statement in both polynomial and fraction form.

(continued)

SOLUTION

Solve Algebraically

$$\begin{array}{rl}
 & x^2 - x \qquad \leftarrow \text{Quotient} \\
2x^2 + x + 1 \,\big)\!\!\! & \overline{2x^4 - x^3 + 0x^2 + 0x - 2} \\
 & \underline{2x^4 + x^3 + x^2} \\
 & -2x^3 - x^2 + 0x - 2 \\
 & \underline{-2x^3 - x^2 - x} \\
 & x - 2 \qquad \leftarrow \text{Remainder}
\end{array}$$

The division algorithm yields the polynomial form

$$2x^4 - x^3 - 2 = (2x^2 + x + 1)(x^2 - x) + (x - 2).$$

Using equation (2), we obtain the fraction form

$$\frac{2x^4 - x^3 - 2}{2x^2 + x + 1} = x^2 - x + \frac{x - 2}{2x^2 + x + 1}.$$

Support Graphically Figure 2.34 supports the polynomial form of the summary statement.

Now try Exercise 1.

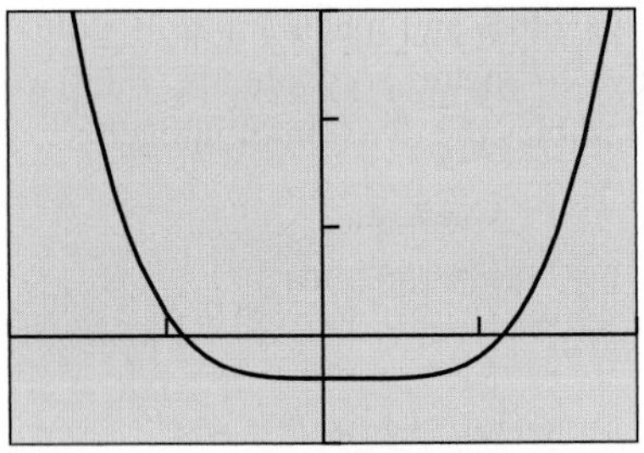

[–2, 2] by [–5, 15]

FIGURE 2.34 The graphs of $y_1 = 2x^4 - x^3 - 2$ and $y_2 = (2x^2 + x + 1)(x^2 - x) + (x - 2)$ are a perfect match. (Example 1)

Remainder and Factor Theorems

An important special case of the division algorithm occurs when the divisor is of the form $d(x) = x - k$, where k is a real number. Because the degree of $d(x) = x - k$ is 1, the remainder is a real number. We obtain the following simplified summary statement for the division algorithm:

$$f(x) = (x - k)q(x) + r \tag{3}$$

We use this special case of the division algorithm throughout the rest of the section.

Using equation (3), we evaluate the polynomial $f(x)$ at $x = k$:

$$f(k) = (k - k)q(k) + r = 0 \cdot q(k) + r = 0 + r = r$$

So $f(k) = r$, which is the remainder. This reasoning yields the following theorem.

THEOREM Remainder Theorem

If a polynomial $f(x)$ is divided by $x - k$, then the remainder is $r = f(k)$.

Example 2 shows a clever use of the Remainder Theorem that gives information about the factors, zeros, and x-intercepts.

EXAMPLE 2 Using the Remainder Theorem

Find the remainder when $f(x) = 3x^2 + 7x - 20$ is divided by

(a) $x - 2$ **(b)** $x + 1$ **(c)** $x + 4$.

SOLUTION

Solve Numerically (by hand)

(a) We can find the remainder without doing long division! Using the Remainder Theorem with $k = 2$ we find that

$$r = f(2) = 3(2)^2 + 7(2) - 20 = 12 + 14 - 20 = 6.$$

X	Y1	
−4	0	
−3	−14	
−2	−22	
−1	−24	
0	−20	
1	−10	
2	6	
Y1 = 3X^2+7X−20		

FIGURE 2.35 Table for $f(x) = 3x^2 + 7x - 20$ showing the remainders obtained when $f(x)$ is divided by $x - k$, for $k = -4, -3, \ldots, 1, 2$.

(b) $r = f(-1) = 3(-1)^2 + 7(-1) - 20 = 3 - 7 - 20 = -24$

(c) $r = f(-4) = 3(-4)^2 + 7(-4) - 20 = 48 - 28 - 20 = 0$

Interpret Because the remainder in part (c) is 0, $x + 4$ *divides evenly* into $f(x) = 3x^2 + 7x - 20$. So, $x + 4$ is a factor of $f(x) = 3x^2 + 7x - 20$, -4 is a solution of $3x^2 + 7x - 20 = 0$, and -4 is an x-intercept of the graph of $y = 3x^2 + 7x - 20$. We know all of this without ever dividing, factoring, or graphing!

Support Numerically (using a grapher) We can find the remainders of several division problems at once using the table feature of a grapher (Figure 2.35).

Now try Exercise 13.

Our interpretation of Example 2c leads us to the following theorem.

THEOREM Factor Theorem

A polynomial function $f(x)$ has a factor $x - k$ if and only if $f(k) = 0$.

Proof of the Factor Theorem

If $f(x)$ has a factor $x - k$, there is a polynomial $g(x)$ such that

$$f(x) = (x - k)g(x) = (x - k)g(x) + 0.$$

By the uniqueness condition of the division algorithm, $g(x)$ is the quotient and 0 is the remainder, and by the Remainder Theorem, $f(k) = 0$.

Conversely, if $f(k) = 0$, the remainder $r = 0$, $x - k$ divides evenly into $f(x)$, and $x - k$ is a factor of $f(x)$.

Applying the ideas of the Factor Theorem to Example 2, we can factor $f(x) = 3x^2 + 7x - 20$ by dividing it by the known factor $x + 4$.

$$\begin{array}{r@{}l}
 & 3x - 5 \\
x + 4 \,\big) & 3x^2 + 7x - 20 \\
 & 3x^2 + 12x \\
 & \quad\; -5x - 20 \\
 & \quad\; -5x - 20 \\
 & \qquad\qquad 0
\end{array}$$

So, $f(x) = 3x^2 + 7x - 20 = (x + 4)(3x - 5)$. In this case, there really is no need to use long division or fancy theorems; traditional factoring methods can do the job. However, for polynomials of degree 3 and higher, these sophisticated methods can be quite helpful in solving equations and finding factors, zeros, and x-intercepts. Indeed, the Factor Theorem ties in nicely with earlier connections we have made in the following way.

Fundamental Connections for Polynomial Functions

For a polynomial function f and a real number k, the following statements are equivalent:

1. $x = k$ is a solution (or root) of the equation $f(x) = 0$.
2. k is a zero of the function f.
3. k is an x-intercept of the graph of $y = f(x)$.
4. $x - k$ is a factor of $f(x)$.

Synthetic Division

We continue with the important special case of polynomial division with the divisor $x - k$. The Remainder Theorem gave us a way to find remainders in this case without long division. We now learn a method for finding both quotients and remainders for division by $x - k$ without long division. This shortcut method for the division of a polynomial by a linear divisor $x - k$ is **synthetic division**.

We illustrate the evolution of this method on the next page, progressing from long division through two intermediate stages to synthetic division.

Moving from stage to stage, focus on the coefficients and their relative positions. Moving from stage 1 to stage 2, we suppress the variable x and the powers of x, and then from stage 2 to stage 3, we eliminate unneeded duplications and collapse vertically.

Stage 1
Long Division

$$\begin{array}{r}
2x^2 + 3x + 4 \\
x - 3\overline{)2x^3 - 3x^2 - 5x - 12} \\
\underline{2x^3 - 6x^2} \qquad\qquad \\
3x^2 - 5x - 12 \\
\underline{3x^2 - 9x} \qquad \\
4x - 12 \\
\underline{4x - 12} \\
0
\end{array}$$

Stage 2
Variables Suppressed

$$\begin{array}{r}
2 \quad 3 \quad 4 \\
-3\overline{)2 \quad -3 \quad -5 \quad -12} \\
\underline{2 \quad -6} \qquad\qquad\quad \\
3 \quad -5 \quad -12 \\
\underline{3 \quad -9} \qquad \\
4 \quad -12 \\
\underline{4 \quad -12} \\
0
\end{array}$$

Stage 3
Collapsed Vertically

$$\begin{array}{r|rrrrl}
-3 & 2 & -3 & -5 & -12 & \text{Dividend} \\
 & & -6 & -9 & -12 & \\
\hline
 & 2 & 3 & 4 & 0 & \text{Quotient, remainder}
\end{array}$$

Finally, from stage 3 to stage 4, we change the sign of the number representing the divisor and the signs of the numbers on the second line of our division scheme. These sign changes yield two advantages:

- The number standing for the divisor $x - k$ is now k, its zero.
- Changing the signs in the second line allows us to add rather than subtract.

Stage 4
Synthetic Division

$$\begin{array}{lr|rrrrl}
\text{Zero of divisor} \rightarrow & 3 & 2 & -3 & -5 & -12 & \text{Dividend} \\
 & & & 6 & 9 & 12 & \\
\hline
 & & 2 & 3 & 4 & 0 & \text{Quotient, remainder}
\end{array}$$

With stage 4 we have achieved our goal of synthetic division, a highly streamlined version of dividing a polynomial by $x - k$. How does this "bare bones" division work? Example 3 explains the steps.

EXAMPLE 3 Using Synthetic Division

Divide $2x^3 - 3x^2 - 5x - 12$ by $x - 3$ using synthetic division and write a summary statement in fraction form.

SOLUTION

Set Up The zero of the divisor $x - 3$ is 3, which we put in the divisor position. Because the dividend is in standard form, we write its coefficients in order in the dividend position, *making sure to use a zero as a placeholder for any missing term.* We leave space for the line for products and draw a horizontal line below the space. (See below.)

Calculate

- Because the leading coefficient of the dividend must be the leading coefficient of the quotient, copy the 2 into the first quotient position.

$$\begin{array}{lr|rrrrl}
\textbf{Zero of Divisor} & 3 & 2 & -3 & -5 & -12 & \textbf{Dividend} \\
\text{Line for products} & & & & & & \\
\hline
 & & 2 & & & &
\end{array}$$

- Multiply the zero of the divisor (3) by the most recently determined coefficient of the quotient (2). Write the product above the line and one column to the right.

- Add the next coefficient of the dividend to the product just found and record the sum below the line in the same column.
- Repeat the "multiply" and "add" steps until the last row is completed.

Zero of Divisor	3	2	−3	−5	−12	**Dividend**
Line for products			6	9	12	
Line for sums		2	3	4	0	**Remainder**
		Quotient				

Interpret The numbers in the last line are the coefficients of the quotient polynomial and the remainder. The quotient must be a quadratic function. (Why?) So the quotient is $2x^2 + 3x + 4$ and the remainder is 0. We conclude that

$$\frac{2x^3 - 3x^2 - 5x - 12}{x - 3} = 2x^2 + 3x + 4, x \neq 3.$$

Now try Exercise 7.

Rational Zeros Theorem

Real zeros of polynomial functions are either **rational zeros**—zeros that are rational numbers—or **irrational zeros**—zeros that are irrational numbers. For example,

$$f(x) = 4x^2 - 9 = (2x + 3)(2x - 3)$$

has the rational zeros $-3/2$ and $3/2$, and

$$f(x) = x^2 - 2 = \left(x + \sqrt{2}\right)\left(x - \sqrt{2}\right)$$

has the irrational zeros $-\sqrt{2}$ and $\sqrt{2}$.

The Rational Zeros Theorem tells us how to make a list of all potential rational zeros for a polynomial function with integer coefficients.

THEOREM Rational Zeros Theorem

Suppose f is a polynomial function of degree $n \geq 1$ of the form

$$f(x) = a_n x^n + a_{n-1}x^{n-1} + \cdots + a_0,$$

with every coefficient an integer and $a_0 \neq 0$. If $x = p/q$ is a rational zero of f, where p and q have no common integer factors other than ± 1, then

- p is an integer factor of the constant coefficient a_0, and
- q is an integer factor of the leading coefficient a_n.

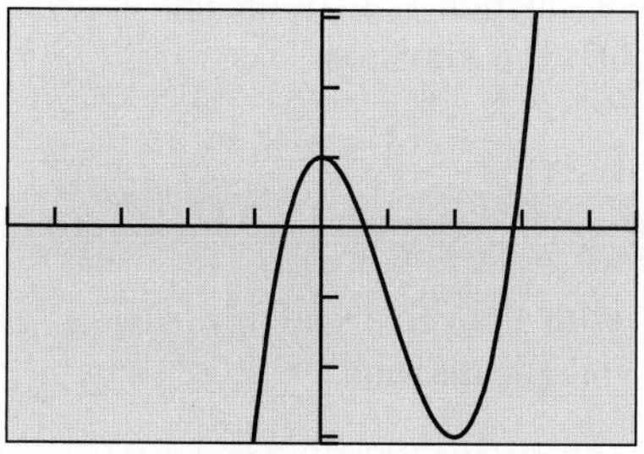

[–4.7, 4.7] by [–3.1, 3.1]

FIGURE 2.36 The function $f(x) = x^3 - 3x^2 + 1$ has three real zeros. (Example 4)

EXAMPLE 4 Finding the Rational Zeros

Find the rational zeros of $f(x) = x^3 - 3x^2 + 1$.

SOLUTION Because the leading and constant coefficients are both 1, according to the Rational Zeros Theorem, the only potential rational zeros of f are 1 and -1. We check to see whether they are in fact zeros of f:

$$f(1) = (1)^3 - 3(1)^2 + 1 = -1 \neq 0$$
$$f(-1) = (-1)^3 - 3(-1)^2 + 1 = -3 \neq 0$$

So f has no rational zeros. Figure 2.36 shows that the graph of f has three x-intercepts. Therefore, f has three real zeros. All three must be irrational numbers.

Now try Exercise 33.

In Example 4 the Rational Zeros Theorem gave us only two candidates for rational zeros, neither of which "checked out." Often this theorem suggests many candidates, as we see in Example 5. In such a case, we use technology and a variety of algebraic methods to locate the rational zeros.

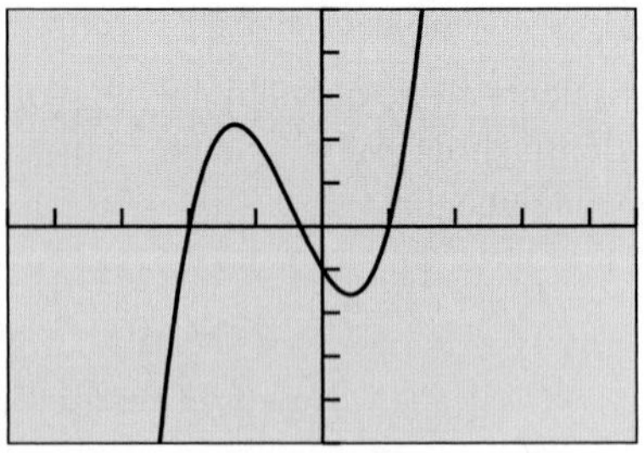

[–4.7, 4.7] by [–10, 10]

FIGURE 2.37 The function $f(x) = 3x^3 + 4x^2 - 5x - 2$ has three real zeros. (Example 5)

EXAMPLE 5 Finding the Rational Zeros

Find the rational zeros of $f(x) = 3x^3 + 4x^2 - 5x - 2$.

SOLUTION Because the leading coefficient is 3 and constant coefficient is -2, the Rational Zeros Theorem yields several potential rational zeros of f. We take an organized approach to our solution.

Potential Rational Zeros:

$$\frac{\text{Factors of } -2}{\text{Factors of } 3} : \frac{\pm 1, \pm 2}{\pm 1, \pm 3} : \pm 1, \pm 2, \pm\frac{1}{3}, \pm\frac{2}{3}$$

Figure 2.37 suggests that, among our candidates, 1, -2, and possibly $-1/3$ or $-2/3$ are the most likely to be rational zeros. We use synthetic division because it tells us whether a number is a zero and, if so, how to factor the polynomial. To see whether 1 is a zero of f, we synthetically divide $f(x)$ by $x - 1$:

$$\begin{array}{r|rrrr} 1 & 3 & 4 & -5 & -2 \\ & & 3 & 7 & 2 \\ \hline & 3 & 7 & 2 & 0 \end{array}$$

Zero of Divisor: 1; Dividend: 3 4 −5 −2; Quotient: 3 7 2; Remainder: 0

Because the remainder is 0, $x - 1$ is a factor of $f(x)$ and 1 is a zero of f. By the division algorithm and factoring, we conclude

$$\begin{aligned} f(x) &= 3x^3 + 4x^2 - 5x - 2 \\ &= (x - 1)(3x^2 + 7x + 2) \\ &= (x - 1)(3x + 1)(x + 2) \end{aligned}$$

Therefore, the rational zeros of f are 1, $-1/3$, and -2. **Now try Exercise 35.**

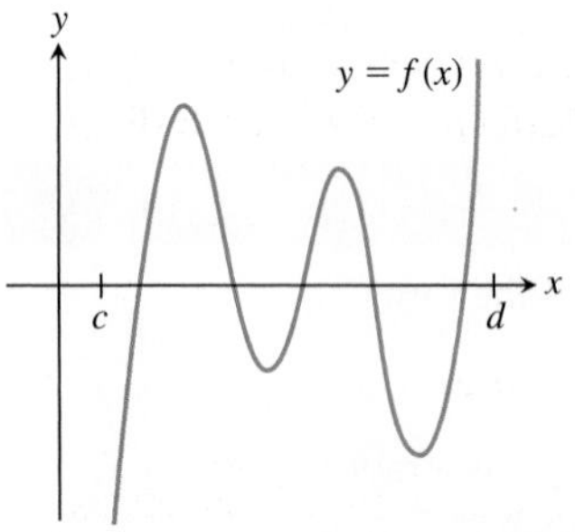

FIGURE 2.38 c is a lower bound and d is an upper bound for the real zeros of f.

Upper and Lower Bounds

We narrow our search for real zeros by using a test that identifies upper and lower bounds for real zeros. A number k is an **upper bound for the real zeros** of f if $f(x)$ is never zero when x is greater than k. On the other hand, a number k is a **lower bound for the real zeros** of f if $f(x)$ is never zero when x is less than k. So if c is a lower bound and d is an upper bound for the real zeros of a function f, all of the real zeros of f must lie in the interval $[c, d]$. Figure 2.38 illustrates this situation.

Upper and Lower Bound Tests for Real Zeros

Let f be a polynomial function of degree $n \geq 1$ with a positive leading coefficient. Suppose $f(x)$ is divided by $x - k$ using synthetic division.

- If $k \geq 0$ and every number in the last line is nonnegative (positive or zero), then k is an *upper bound* for the real zeros of f.
- If $k \leq 0$ and the numbers in the last line are alternately nonnegative and nonpositive, then k is a *lower bound* for the real zeros of f.

EXAMPLE 6 Establishing Bounds for Real Zeros

Prove that all of the real zeros of $f(x) = 2x^4 - 7x^3 - 8x^2 + 14x + 8$ must lie in the interval $[-2, 5]$.

SOLUTION We must prove that 5 is an upper bound and -2 is a lower bound on the real zeros of f. The function f has a positive leading coefficient, so we employ the upper and lower bound tests, and use synthetic division:

$$\begin{array}{r|rrrrrl} \underline{5} & 2 & -7 & -8 & 14 & 8 & \\ & & 10 & 15 & 35 & 245 & \\ \hline & 2 & 3 & 7 & 49 & 253 & \text{Last line} \end{array}$$

$$\begin{array}{r|rrrrrl} \underline{-2} & 2 & -7 & -8 & 14 & 8 & \\ & & -4 & 22 & -28 & 28 & \\ \hline & 2 & -11 & 14 & -14 & 36 & \text{Last line} \end{array}$$

Because the last line in the first division scheme consists of all positive numbers, 5 is an upper bound. Because the last line in the second division consists of numbers of alternating signs, -2 is a lower bound. All of the real zeros of f must therefore lie in the closed interval $[-2, 5]$. **Now try Exercise 37.**

EXAMPLE 7 Finding the Real Zeros of a Polynomial Function

Find all of the real zeros of $f(x) = 2x^4 - 7x^3 - 8x^2 + 14x + 8$.

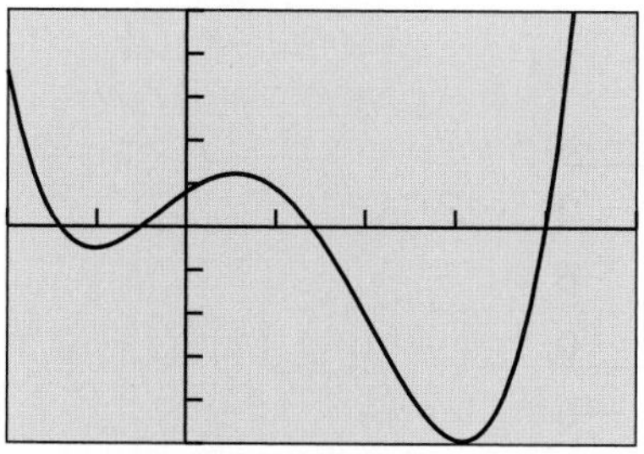

[–2, 5] by [–50, 50]

FIGURE 2.39 (Example 7) $f(x) = 2x^4 - 7x^3 - 8x^2 + 14x + 8$ has all of its real zeros in $[-2, 5]$.

SOLUTION From Example 6 we know that all of the real zeros of f must lie in the closed interval $[-2, 5]$. So in Figure 2.39 we set our Xmin and Xmax accordingly.

Next we use the Rational Zeros Theorem.

Potential Rational Zeros:

$$\frac{\text{Factors of } 8}{\text{Factors of } 2} : \frac{\pm 1, \pm 2, \pm 4, \pm 8}{\pm 1, \pm 2} : \pm 1, \pm 2, \pm 4, \pm 8, \pm \frac{1}{2}$$

We compare the x-intercepts of the graph in Figure 2.39 and our list of candidates, and decide 4 and $-1/2$ are the only potential rational zeros worth pursuing.

$$\begin{array}{r|rrrrr} \underline{4} & 2 & -7 & -8 & 14 & 8 \\ & & 8 & 4 & -16 & -8 \\ \hline & 2 & 1 & -4 & -2 & 0 \end{array}$$

From this first synthetic division we conclude

$$\begin{aligned} f(x) &= 2x^4 - 7x^3 - 8x^2 + 14x + 8 \\ &= (x - 4)(2x^3 + x^2 - 4x - 2) \end{aligned}$$

and we now divide the cubic factor $2x^3 + x^2 - 4x - 2$ by $x + 1/2$:

$$\begin{array}{r|rrrr} \underline{-1/2} & 2 & 1 & -4 & -2 \\ & & -1 & 0 & 2 \\ \hline & 2 & 0 & -4 & 0 \end{array}$$

(continued)

This second synthetic division allows us to complete the factoring of $f(x)$.

$$\begin{aligned} f(x) &= (x-4)(2x^3 + x^2 - 4x - 2) \\ &= (x-4)\left(x + \frac{1}{2}\right)(2x^2 - 4) \\ &= 2(x-4)\left(x + \frac{1}{2}\right)(x^2 - 2) \\ &= (x-4)(2x+1)(x+\sqrt{2})(x-\sqrt{2}) \end{aligned}$$

The zeros of f are the rational numbers 4 and $-1/2$ and the irrational numbers $-\sqrt{2}$ and $\sqrt{2}$. **Now try Exercise 49.**

A polynomial function cannot have more real zeros than its degree, but it can have fewer. When a polynomial has fewer real zeros than its degree, the upper and lower bound tests help us know that we have found them all, as illustrated by Example 8.

EXAMPLE 8 Finding the Real Zeros of a Polynomial Function

Prove that all of the real zeros of $f(x) = 10x^5 - 3x^2 + x - 6$ lie in the interval $[0, 1]$, and find them.

SOLUTION We first prove that 1 is an upper bound and 0 is a lower bound for the real zeros of f. The function f has a positive leading coefficient, so we use synthetic division and the upper and lower bound tests:

$$\begin{array}{r|rrrrrrl} 1 & 10 & 0 & 0 & -3 & 1 & -6 & \\ & & 10 & 10 & 10 & 7 & 8 & \\ \hline & 10 & 10 & 10 & 7 & 8 & 2 & \text{Last line} \\ 0 & 10 & 0 & 0 & -3 & 1 & -6 & \\ & & 0 & 0 & 0 & 0 & 0 & \\ \hline & 10 & 0 & 0 & -3 & 1 & -6 & \text{Last line} \end{array}$$

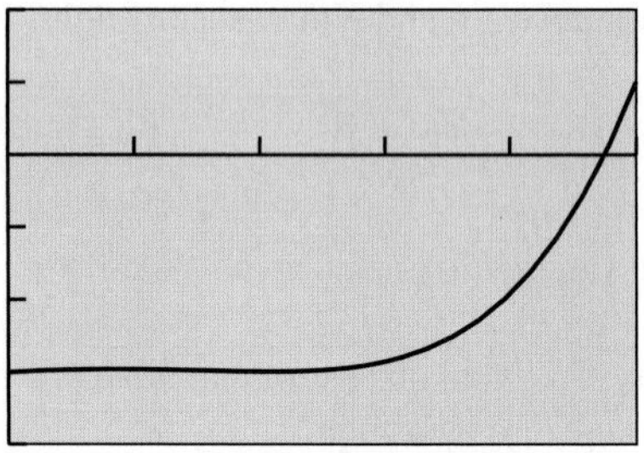

[0, 1] by [–8, 4]

FIGURE 2.40 $y = 10x^5 - 3x^2 + x - 6$. (Example 8)

Because the last line in the first division scheme consists of all nonnegative numbers, 1 is an upper bound. Because the last line in the second division consists of numbers that are alternately nonnegative and nonpositive, 0 is a lower bound. All of the real zeros of f must therefore lie in the closed interval $[0, 1]$. So in Figure 2.40 we set our Xmin and Xmax accordingly.

Next we use the Rational Zeros Theorem.

Potential Rational Zeros:

$$\frac{\text{Factors of } -6}{\text{Factors of } 10} : \frac{\pm 1, \pm 2, \pm 3, \pm 6}{\pm 1, \pm 2, \pm 5, \pm 10} :$$

$$\pm 1, \pm 2, \pm 3, \pm 6, \pm\frac{1}{2}, \pm\frac{3}{2}, \pm\frac{1}{5}, \pm\frac{2}{5}, \pm\frac{3}{5}, \pm\frac{6}{5}, \pm\frac{1}{10}, \pm\frac{3}{10}$$

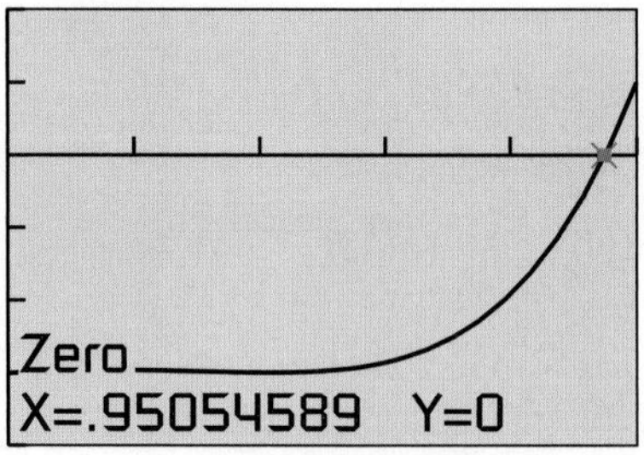

[0, 1] by [–8, 4]

FIGURE 2.41 An approximation for the irrational zero of $f(x) = 10x^5 - 3x^2 + x - 6$. (Example 8)

We compare the x-intercepts of the graph in Figure 2.40 and our list of candidates, and decide f has no rational zeros. From Figure 2.40 we see that f changes sign on the interval $[0.8, 1]$. Thus by the Intermediate Value Theorem, f must have a real zero on this interval. Because it is not rational, we conclude that it is irrational. Figure 2.41 shows that this lone real zero of f is approximately 0.95. **Now try Exercise 55.**

QUICK REVIEW 2.4 *(For help, go to Sections A.2. and A.3.)*

Exercise numbers with a gray background indicate problems that the authors have designed to be solved *without a calculator*.

In Exercises 1–4, rewrite the expression as a polynomial in standard form.

1. $\dfrac{x^3 - 4x^2 + 7x}{x}$

2. $\dfrac{2x^3 - 5x^2 - 6x}{2x}$

3. $\dfrac{x^4 - 3x^2 + 7x^5}{x^2}$

4. $\dfrac{6x^4 - 2x^3 + 7x^2}{3x^2}$

In Exercises 5–10, factor the polynomial into linear factors.

5. $x^3 - 4x$

6. $6x^2 - 54$

7. $4x^2 + 8x - 60$

8. $15x^3 - 22x^2 + 8x$

9. $x^3 + 2x^2 - x - 2$

10. $x^4 + x^3 - 9x^2 - 9x$

SECTION 2.4 Exercises

In Exercises 1–6, divide $f(x)$ by $d(x)$, and write a summary statement in polynomial form and fraction form.

1. $f(x) = x^2 - 2x + 3;\ d(x) = x - 1$

2. $f(x) = x^3 - 1;\ d(x) = x + 1$

3. $f(x) = x^3 + 4x^2 + 7x - 9;\ d(x) = x + 3$

4. $f(x) = 4x^3 - 8x^2 + 2x - 1;\ d(x) = 2x + 1$

5. $f(x) = x^4 - 2x^3 + 3x^2 - 4x + 6;\ d(x) = x^2 + 2x - 1$

6. $f(x) = x^4 - 3x^3 + 6x^2 - 3x + 5;\ d(x) = x^2 + 1$

In Exercises 7–12, divide using synthetic division, and write a summary statement in fraction form.

7. $\dfrac{x^3 - 5x^2 + 3x - 2}{x + 1}$

8. $\dfrac{2x^4 - 5x^3 + 7x^2 - 3x + 1}{x - 3}$

9. $\dfrac{9x^3 + 7x^2 - 3x}{x - 10}$

10. $\dfrac{3x^4 + x^3 - 4x^2 + 9x - 3}{x + 5}$

11. $\dfrac{5x^4 - 3x + 1}{4 - x}$

12. $\dfrac{x^8 - 1}{x + 2}$

In Exercises 13–18, use the Remainder Theorem to find the remainder when $f(x)$ is divided by $x - k$.

13. $f(x) = 2x^2 - 3x + 1;\ k = 2$

14. $f(x) = x^4 - 5;\ k = 1$

15. $f(x) = x^3 - x^2 + 2x - 1;\ k = -3$

16. $f(x) = x^3 - 3x + 4;\ k = -2$

17. $f(x) = 2x^3 - 3x^2 + 4x - 7;\ k = 2$

18. $f(x) = x^5 - 2x^4 + 3x^2 - 20x + 3;\ k = -1$

In Exercises 19–24, use the Factor Theorem to determine whether the first polynomial is a factor of the second polynomial.

19. $x - 1;\ x^3 - x^2 + x - 1$

20. $x - 3;\ x^3 - x^2 - x - 15$

21. $x - 2;\ x^3 + 3x - 4$

22. $x - 2;\ x^3 - 3x - 2$

23. $x + 2;\ 4x^3 + 9x^2 - 3x - 10$

24. $x + 1;\ 2x^{10} - x^9 + x^8 + x^7 + 2x^6 - 3$

In Exercises 25 and 26, use the graph to guess possible linear factors of $f(x)$. Then completely factor $f(x)$ with the aid of synthetic division.

25. $f(x) = 5x^3 - 7x^2 - 49x + 51$

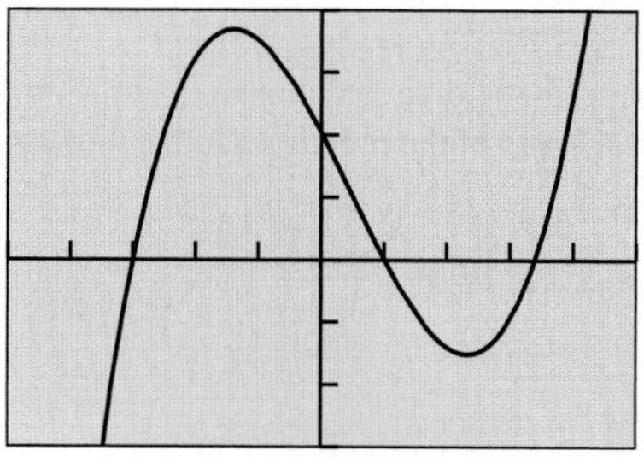

[–5, 5] by [–75, 100]

26. $f(x) = 5x^3 - 12x^2 - 23x + 42$

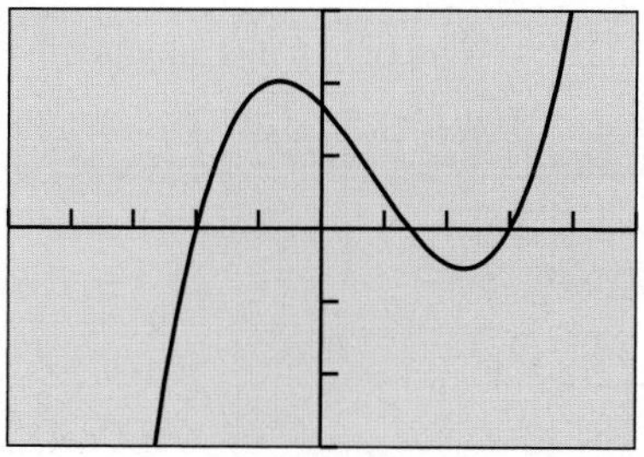

[–5, 5] by [–75, 75]

In Exercises 27–30, find the polynomial function with leading coefficient 2 that has the given degree and zeros.

27. Degree 3, with -2, 1, and 4 as zeros

28. Degree 3, with -1, 3, and -5 as zeros

29. Degree 3, with 2, $\frac{1}{2}$, and $\frac{3}{2}$ as zeros

30. Degree 4, with -3, -1, 0, and $\frac{5}{2}$ as zeros

In Exercises 31 and 32, using only algebraic methods, find the cubic function with the given table of values. Check with a grapher.

31.

x	-4	0	3	5
$f(x)$	0	180	0	0

32.

x	-2	-1	1	5
$f(x)$	0	24	0	0

In Exercises 33–36, use the Rational Zeros Theorem to write a list of all potential rational zeros. Then determine which ones, if any, are zeros.

33. $f(x) = 6x^3 - 5x - 1$

34. $f(x) = 3x^3 - 7x^2 + 6x - 14$

35. $f(x) = 2x^3 - x^2 - 9x + 9$

36. $f(x) = 6x^4 - x^3 - 6x^2 - x - 12$

In Exercises 37–40, use synthetic division to prove that the number k is an upper bound for the real zeros of the function f.

37. $k = 3$; $f(x) = 2x^3 - 4x^2 + x - 2$

38. $k = 5$; $f(x) = 2x^3 - 5x^2 - 5x - 1$

39. $k = 2$; $f(x) = x^4 - x^3 + x^2 + x - 12$

40. $k = 3$; $f(x) = 4x^4 - 6x^3 - 7x^2 + 9x + 2$

In Exercises 41–44, use synthetic division to prove that the number k is a lower bound for the real zeros of the function f.

41. $k = -1$; $f(x) = 3x^3 - 4x^2 + x + 3$

42. $k = -3$; $f(x) = x^3 + 2x^2 + 2x + 5$

43. $k = 0$; $f(x) = x^3 - 4x^2 + 7x - 2$

44. $k = -4$; $f(x) = 3x^3 - x^2 - 5x - 3$

In Exercises 45–48, use the upper and lower bound tests to decide whether there could be real zeros for the function outside the window shown. If so, check for additional zeros.

45. $f(x) = 6x^4 - 11x^3 - 7x^2 + 8x - 34$

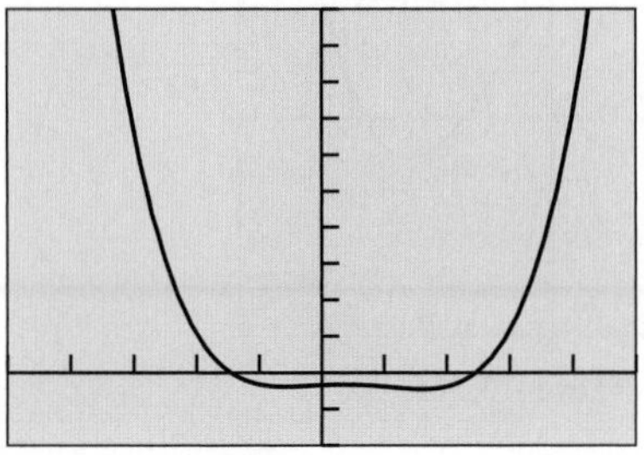

[–5, 5] by [–200, 1000]

46. $f(x) = x^5 - x^4 + 21x^2 + 19x - 3$

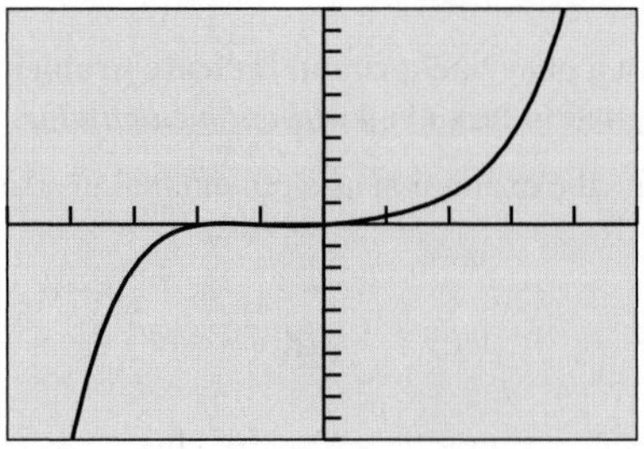

[–5, 5] by [–1000, 1000]

47. $f(x) = x^5 - 4x^4 - 129x^3 + 396x^2 - 8x + 3$

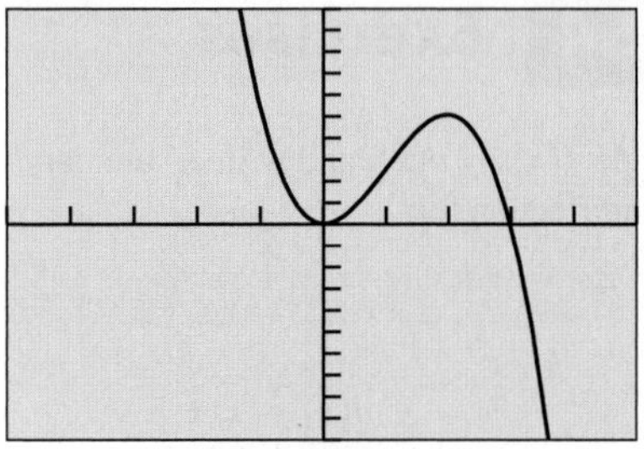

[–5, 5] by [–1000, 1000]

48. $f(x) = 2x^5 - 5x^4 - 141x^3 + 216x^2 - 91x + 25$

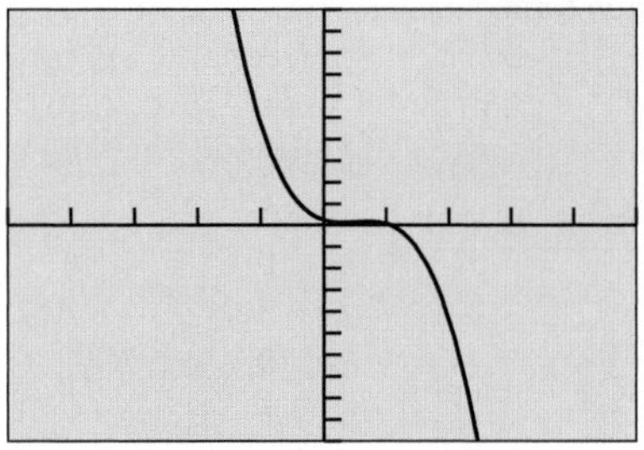

[–5, 5] by [–1000, 1000]

In Exercises 49–56, find all of the real zeros of the function, finding exact values whenever possible. Identify each zero as rational or irrational.

49. $f(x) = 2x^3 - 3x^2 - 4x + 6$

50. $f(x) = x^3 + 3x^2 - 3x - 9$

51. $f(x) = x^3 + x^2 - 8x - 6$

52. $f(x) = x^3 - 6x^2 + 7x + 4$

53. $f(x) = x^4 - 3x^3 - 6x^2 + 6x + 8$

54. $f(x) = x^4 - x^3 - 7x^2 + 5x + 10$

55. $f(x) = 2x^4 - 7x^3 - 2x^2 - 7x - 4$

56. $f(x) = 3x^4 - 2x^3 + 3x^2 + x - 2$

57. Setting Production Schedules The Sunspot Small Appliance Co. determines that the supply function for their EverCurl hair dryer is $S(p) = 6 + 0.001p^3$ and that its demand function is $D(p) = 80 - 0.02p^2$, where p is the price. Determine the price for which the supply equals the demand and the number of hair dryers corresponding to this equilibrium price.

58. Setting Production Schedules The Pentkon Camera Co. determines that the supply and demand functions for their 35 mm–70 mm zoom lens are $S(p) = 200 - p + 0.000007p^4$ and $D(p) = 1500 - 0.0004p^3$, where p is the price. Determine the price for which the supply equals the demand and the number of zoom lenses corresponding to this equilibrium price.

59. Find the remainder when $x^{40} - 3$ is divided by $x + 1$.

60. Find the remainder when $x^{63} - 17$ is divided by $x - 1$.

61. Let $f(x) = x^4 + 2x^3 - 11x^2 - 13x + 38$.

(a) Use the upper and lower bound tests to prove that all of the real zeros of f lie on the interval $[-5, 4]$.

(b) Find all of the rational zeros of f.

(c) Factor $f(x)$ using the rational zero(s) found in (b).

(d) Approximate all of the irrational zeros of f.

(e) Use synthetic division and the irrational zero(s) found in (d) to continue the factorization of $f(x)$ begun in (c).

62. Lewis's distance D from a motion detector is given by the data in Table 2.15.

Table 2.15 Motion Detector Data

t (sec)	D (m)	t (sec)	D (m)
0.0	1.00	4.5	0.99
0.5	1.46	5.0	0.84
1.0	1.99	5.5	1.28
1.5	2.57	6.0	1.87
2.0	3.02	6.5	2.58
2.5	3.34	7.0	3.23
3.0	2.91	7.5	3.78
3.5	2.31	8.0	4.40
4.0	1.57		

(a) Find a cubic regression model, and graph it together with a scatter plot of the data.

(b) Use the cubic regression model to estimate how far Lewis is from the motion detector initially.

(c) Use the cubic regression model to estimate when Lewis changes direction. How far from the motion detector is he when he changes direction?

Standardized Test Questions

63. True or False The polynomial function $f(x)$ has a factor $x + 2$ if and only if $f(2) = 0$. Justify your answer.

64. True or False If $f(x) = (x - 1)(2x^2 - x + 1) + 3$, then the remainder when $f(x)$ is divided by $x - 1$ is 3. Justify your answer.

In Exercises 65–68, you may use a graphing calculator to solve the problem.

65. Multiple Choice Let f be a polynomial function with $f(3) = 0$. Which of the following statements is not true?

(A) $x + 3$ is a factor of $f(x)$. **(B)** $x - 3$ is a factor of $f(x)$.

(C) $x = 3$ is a zero of $f(x)$. **(D)** 3 is an x-intercept of $f(x)$.

(E) The remainder when $f(x)$ is divided by $x - 3$ is zero.

66. Multiple Choice Let $f(x) = 2x^3 + 7x^2 + 2x - 3$. Which of the following is not a possible rational root of f?

(A) -3 **(B)** -1 **(C)** 1 **(D)** 1/2 **(E)** 2/3

67. Multiple Choice Let $f(x) = (x + 2)(x^2 + x - 1) - 3$. Which of the following statements is not true?

(A) The remainder when $f(x)$ is divided by $x + 2$ is -3.

(B) The remainder when $f(x)$ is divided by $x - 2$ is -3.

(C) The remainder when $f(x)$ is divided by $x^2 + x - 1$ is -3.

(D) $x + 2$ is not a factor of $f(x)$.

(E) $f(x)$ is not evenly divisible by $x + 2$.

68. Multiple Choice Let $f(x) = (x^2 + 1)(x - 2) + 7$. Which of the following statements is not true?

(A) The remainder when $f(x)$ is divided by $x^2 + 1$ is 7.

(B) The remainder when $f(x)$ is divided by $x - 2$ is 7.

(C) $f(2) = 7$ **(D)** $f(0) = 5$

(E) f does not have a real root.

Explorations

69. Archimedes' Principle A spherical buoy has a radius of 1 m and a density one-fourth that of seawater. By Archimedes' Principle, the weight of the displaced water will equal the weight of the buoy.

- Let x = the depth to which the buoy sinks.
- Let d = the density of seawater.
- Let r = the radius of the circle formed where buoy, air, and water meet. See the figure below.

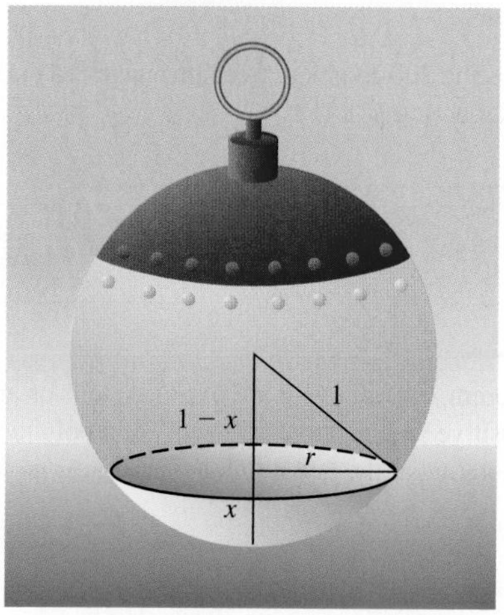

Notice in the figure that $r^2 = 1 - (1 - x)^2 = 2x - x^2$, and recall from geometry that the volume of the submerged *spherical cap* is $V = \frac{\pi x}{6} \cdot (3r^2 + x^2)$.

(a) Verify that the volume of the buoy is $4\pi/3$.

(b) Use your result from (a) to establish the weight of the buoy as $\pi d/3$.

(c) Prove the weight of the displaced water is $\pi d \cdot x(3r^2 + x^2)/6$.

(d) Approximate the depth to which the buoy will sink.

70. Archimedes' Principle Using the scenario of Exercise 69, find the depth to which the buoy will sink if its density is one-fifth that of seawater.

71. Biological Research Stephanie, a biologist who does research for the poultry industry, models the population P of wild turkeys, t days after being left to reproduce, with the function

$$P(t) = -0.00001t^3 + 0.002t^2 + 1.5t + 100.$$

(a) Graph the function $y = P(t)$ for appropriate values of t.

(b) Find what the maximum turkey population is and when it occurs.

(c) Assuming that this model continues to be accurate, when will this turkey population become extinct?

(d) Writing to Learn Create a scenario that could explain the growth exhibited by this turkey population.

72. Architectural Engineering Dave, an engineer at the Trumbauer Group, Inc., an architectural firm, completes structural specifications for a 172-ft-long steel beam, anchored at one end to a piling 20 ft above the ground. He knows that when a 200-lb object is placed d feet from the anchored end, the beam bends s feet where

$$s = (3 \times 10^{-7})d^2(550 - d).$$

(a) What is the independent variable in this polynomial function?

(b) What are the dimensions of a viewing window that shows a graph for the values that make sense in this problem situation?

(c) How far is the 200-lb object from the anchored end if the vertical deflection is 1.25 ft?

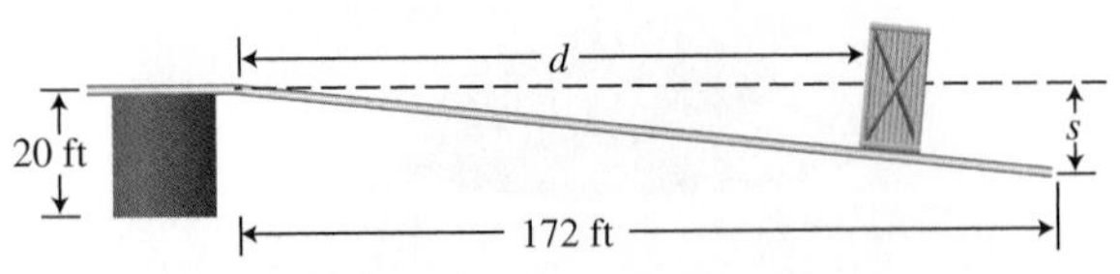

73. A classic theorem, **Descartes' Rule of Signs**, tells us about the number of positive and negative real zeros of a polynomial function, by looking at the polynomial's variations in sign.

A *variation in sign* occurs when consecutive coefficients (in standard form) have opposite signs.

If $f(x) = a_n x^n + \cdots + a_0$ is a polynomial of degree n, then

- The number of positive real zeros of f is equal to the number of variations in sign of $f(x)$, or that number less some even number.
- The number of negative real zeros of f is equal to the number of variations in sign of $f(-x)$, or that number less some even number.

Use Descartes' Rule of Signs to determine the possible numbers of positive and negative real zeros of the function.

(a) $f(x) = x^3 + x^2 - x + 1$

(b) $f(x) = x^3 + x^2 + x + 1$

(c) $f(x) = 2x^3 + x - 3$

(d) $g(x) = 5x^4 + x^2 - 3x - 2$

Extending the Ideas

74. Writing to Learn Graph each side of the Example 3 summary equation:

$$f(x) = \frac{2x^3 - 3x^2 - 5x - 12}{x - 3} \text{ and}$$

$$g(x) = 2x^2 + 3x + 4, \quad x \neq 3$$

How are these functions related? Include a discussion of the domain and continuity of each function.

75. Writing to Learn Explain how to carry out the following division using synthetic division. Work through the steps with complete explanations. Interpret and check your result.

$$\frac{4x^3 - 5x^2 + 3x + 1}{2x - 1}$$

76. Writing to Learn The figure shows a graph of $f(x) = x^4 + 0.1x^3 - 6.5x^2 + 7.9x - 2.4$. Explain how to use a grapher to justify the statement.

$$f(x) = x^4 + 0.1x^3 - 6.5x^2 + 7.9x - 2.4$$
$$\approx (x + 3.10)(x - 0.5)(x - 1.13)(x - 1.37)$$

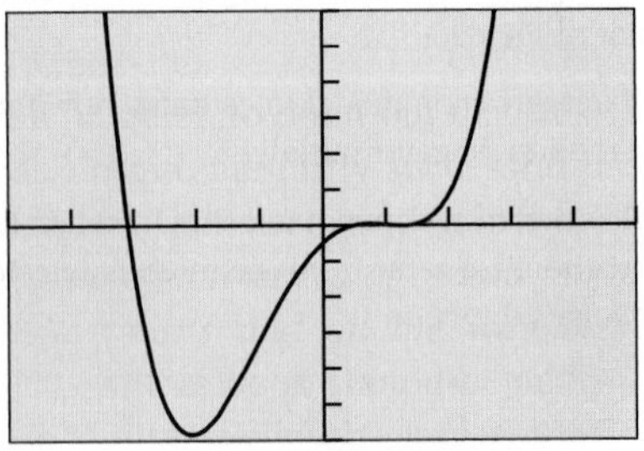

[–5, 5] by [–30, 30]

77. (a) **Writing to Learn** Write a paragraph that describes how the zeros of $f(x) = (1/3)x^3 + x^2 + 2x - 3$ are related to the zeros of $g(x) = x^3 + 3x^2 + 6x - 9$. In what ways does this example illustrate how the Rational Zeros Theorem can be applied to find the zeros of a polynomial with *rational* number coefficients?

(b) Find the rational zeros of $f(x) = x^3 - \frac{7}{6}x^2 - \frac{20}{3}x + \frac{7}{2}$.

(c) Find the rational zeros of $f(x) = x^3 - \frac{5}{2}x^2 - \frac{37}{12}x + \frac{5}{2}$.

78. Use the Rational Zeros Theorem to prove $\sqrt{2}$ is irrational.

79. **Group Activity** *Work in groups of three.* Graph $f(x) = x^4 + x^3 - 8x^2 - 2x + 7$.

(a) Use grapher methods to find approximate real number zeros.

(b) Identify a list of four linear factors whose product could be called an *approximate factorization of* $f(x)$.

(c) Discuss what graphical and numerical methods you could use to show that the factorization from part (b) is reasonable.

2.5 Complex Zeros and the Fundamental Theorem of Algebra

What you'll learn about

- Two Major Theorems
- Complex Conjugate Zeros
- Factoring with Real Number Coefficients

... and why

These topics provide the complete story about the zeros and factors of polynomials with real number coefficients.

Two Major Theorems

In Section 2.3 we learned that a polynomial function of degree n has at most n real zeros. Figure 2.42 shows that the polynomial function $f(x) = x^2 + 2x + 5$ of degree 2 has no real zeros. (Why?) A little arithmetic, however, shows that the complex number $-1 + 2i$ is a zero of f:

$$\begin{aligned} f(-1 + 2i) &= (-1 + 2i)^2 + 2(-1 + 2i) + 5 \\ &= (-3 - 4i) + (-2 + 4i) + 5 \\ &= 0 + 0i \\ &= 0 \end{aligned}$$

The quadratic formula shows that $-1 \pm 2i$ are the two zeros of f and can be used to find the complex zeros for any polynomial function of degree 2. In this section we will learn about complex zeros of polynomial functions of higher degree and how to use these zeros to factor polynomial expressions.

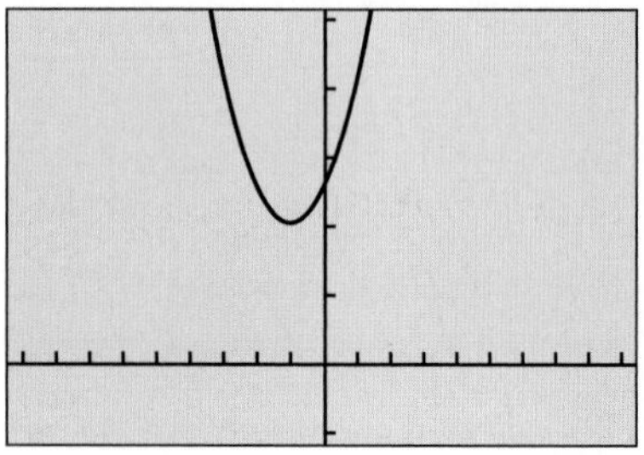

[–9.4, 9.4] by [–2, 10]

FIGURE 2.42 The graph of $f(x) = x^2 + 2x + 5$ has no x-intercepts, so f has no real zeros.

THEOREM Fundamental Theorem of Algebra

A polynomial function of degree n has n complex zeros (real and nonreal). Some of these zeros may be repeated.

The Factor Theorem extends to the complex zeros of a polynomial function. Thus, k is a complex zero of a polynomial if and only if $x - k$ is a factor of the polynomial, even if k is not a real number. We combine this fact with the Fundamental Theorem of Algebra to obtain the following theorem.

THEOREM Linear Factorization Theorem

If $f(x)$ is a polynomial function of degree $n > 0$, then $f(x)$ has precisely n linear factors and

$$f(x) = a(x - z_1)(x - z_2) \cdots (x - z_n)$$

where a is the leading coefficient of $f(x)$ and $z_1, z_2, \ldots, z_n$ are the complex zeros of $f(x)$. The z_i are not necessarily distinct numbers; some may be repeated.

The Fundamental Theorem of Algebra and the Linear Factorization Theorem are *existence theorems*. They tell us of the existence of zeros and linear factors, but not how to find them.

One connection is lost going from real zeros to complex zeros. If k is a *nonreal* complex zero of a polynomial function $f(x)$, then k is *not* an x-intercept of the graph of f. The other connections hold whether k is real or nonreal:

Fundamental Polynomial Connections in the Complex Case

The following statements about a polynomial function f are equivalent if k is a complex number:

1. $x = k$ is a solution (or root) of the equation $f(x) = 0$.

2. k is a zero of the function f.

3. $x - k$ is a factor of $f(x)$.

EXAMPLE 1 Exploring Fundamental Polynomial Connections

Write the polynomial function in standard form, and identify the zeros of the function and the x-intercepts of its graph.

(a) $f(x) = (x - 2i)(x + 2i)$
(b) $f(x) = (x - 5)(x - \sqrt{2}i)(x + \sqrt{2}i)$
(c) $f(x) = (x - 3)(x - 3)(x - i)(x + i)$

SOLUTION

(a) The quadratic function $f(x) = (x - 2i)(x + 2i) = x^2 + 4$ has two zeros: $x = 2i$ and $x = -2i$. Because the zeros are not real, the graph of f has no x-intercepts.

(b) The cubic function

$$\begin{aligned} f(x) &= (x - 5)(x - \sqrt{2}i)(x + \sqrt{2}i) \\ &= (x - 5)(x^2 + 2) \\ &= x^3 - 5x^2 + 2x - 10 \end{aligned}$$

has three zeros: $x = 5$, $x = \sqrt{2}i$, and $x = -\sqrt{2}i$. Of the three, only $x = 5$ is an x-intercept.

(c) The quartic function

$$\begin{aligned} f(x) &= (x - 3)(x - 3)(x - i)(x + i) \\ &= (x^2 - 6x + 9)(x^2 + 1) \\ &= x^4 - 6x^3 + 10x^2 - 6x + 9 \end{aligned}$$

has four zeros: $x = 3$, $x = 3$, $x = i$, and $x = -i$. There are only three distinct zeros. The real zero $x = 3$ is a repeated zero of multiplicity two. Due to this even multiplicity, the graph of f touches but does not cross the x-axis at $x = 3$, the only x-intercept.

Figure 2.43 supports our conclusions regarding x-intercepts.

Now try Exercise 1.

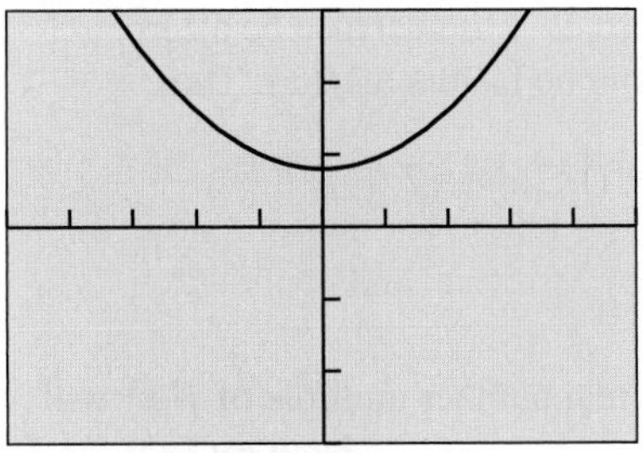

[–5, 5] by [–15, 15]
(a)

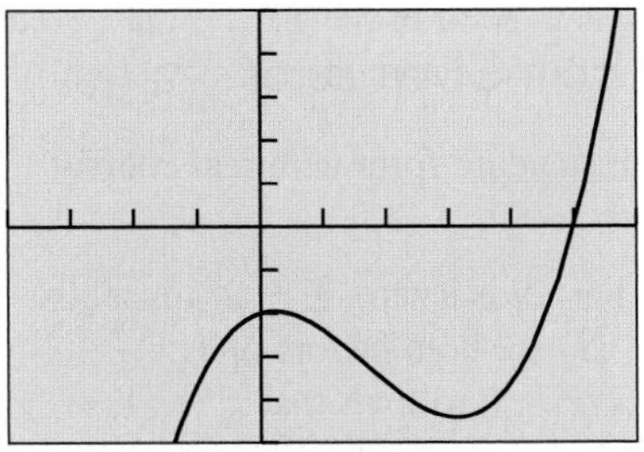

[–4, 6] by [–25, 25]
(b)

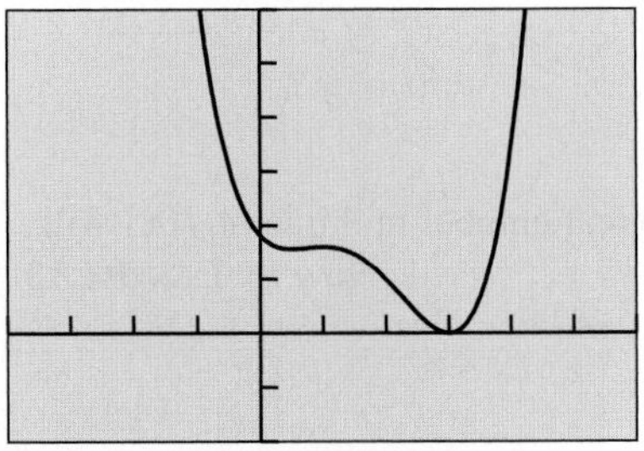

[–4, 6] by [–10, 30]
(c)

FIGURE 2.43 The graphs of (a) $y = x^2 + 4$, (b) $y = x^3 - 5x^2 + 2x - 10$, and (c) $y = x^4 - 6x^3 + 10x^2 - 6x + 9$. (Example 1)

Complex Conjugate Zeros

In Section P.6 we saw that, for quadratic equations $ax^2 + bx + c = 0$ with real coefficients, if the discriminant $b^2 - 4ac$ is negative, the solutions are a conjugate pair of complex numbers. This relationship generalizes to polynomial functions of higher degree in the following way:

THEOREM Complex Conjugate Zeros

Suppose that $f(x)$ is a polynomial function with *real coefficients*. If a and b are real numbers with $b \neq 0$ and $a + bi$ is a zero of $f(x)$, then its complex conjugate $a - bi$ is also a zero of $f(x)$.

EXPLORATION 1 What Can Happen if the Coefficients Are Not Real?

1. Use substitution to verify that $x = 2i$ and $x = -i$ are zeros of $f(x) = x^2 - ix + 2$. Are the conjugates of $2i$ and $-i$ also zeros of $f(x)$?
2. Use substitution to verify that $x = i$ and $x = 1 - i$ are zeros of $g(x) = x^2 - x + (1 + i)$. Are the conjugates of i and $1 - i$ also zeros of $g(x)$?
3. What conclusions can you draw from parts 1 and 2? Do your results contradict the theorem about complex conjugate zeros?

EXAMPLE 2 Finding a Polynomial from Given Zeros

Write a polynomial function of minimum degree in standard form with real coefficients whose zeros include -3, 4, and $2 - i$.

SOLUTION Because -3 and 4 are real zeros, $x + 3$ and $x - 4$ must be factors. Because the coefficients are real and $2 - i$ is a zero, $2 + i$ must also be a zero. Therefore, $x - (2 - i)$ and $x - (2 + i)$ must both be factors of $f(x)$. Thus,

$$\begin{aligned} f(x) &= (x + 3)(x - 4)[x - (2 - i)][x - (2 + i)] \\ &= (x^2 - x - 12)(x^2 - 4x + 5) \\ &= x^4 - 5x^3 - 3x^2 + 43x - 60 \end{aligned}$$

is a polynomial of the type we seek. Any nonzero real number multiple of $f(x)$ will also be such a polynomial. **Now try Exercise 7.**

EXAMPLE 3 Finding a Polynomial from Given Zeros

Write a polynomial function of minimum degree in standard form with real coefficients whose zeros include $x = 1, x = 1 + 2i, x = 1 - i$.

SOLUTION Because the coefficients are real and $1 + 2i$ is a zero, $1 - 2i$ must also be a zero. Therefore, $x - (1 + 2i)$ and $x - (1 - 2i)$ are both factors of $f(x)$. Likewise, because $1 - i$ is a zero, $1 + i$ must be a zero. It follows that $x - (1 - i)$ and $x - (1 + i)$ are both factors of $f(x)$. Therefore,

$$\begin{aligned} f(x) &= (x - 1)[x - (1 + 2i)][x - (1 - 2i)][x - (1 + i)][x - (1 - i)] \\ &= (x - 1)(x^2 - 2x + 5)(x^2 - 2x + 2) \\ &= (x^3 - 3x^2 + 7x - 5)(x^2 - 2x + 2) \\ &= x^5 - 5x^4 + 15x^3 - 25x^2 + 24x - 10 \end{aligned}$$

is a polynomial of the type we seek. Any nonzero real number multiple of $f(x)$ will also be such a polynomial. **Now try Exercise 13.**

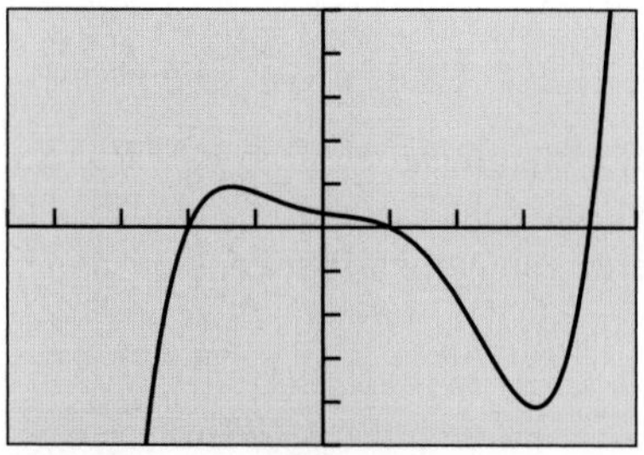

[−4.7, 4.7] by [−125, 125]

FIGURE 2.44 $f(x) = x^5 - 3x^4 - 5x^3 + 5x^2 - 6x + 8$ has three real zeros. (Example 4)

EXAMPLE 4 Factoring a Polynomial with Complex Zeros

Find all zeros of $f(x) = x^5 - 3x^4 - 5x^3 + 5x^2 - 6x + 8$, and write $f(x)$ in its linear factorization.

SOLUTION Figure 2.44 suggests that the real zeros of f are $x = -2$, $x = 1$, and $x = 4$.

Using synthetic division we can verify these zeros and show that $x^2 + 1$ is a factor of f. So $x = i$ and $x = -i$ are also zeros. Therefore,

$$\begin{aligned} f(x) &= x^5 - 3x^4 - 5x^3 + 5x^2 - 6x + 8 \\ &= (x + 2)(x - 1)(x - 4)(x^2 + 1) \\ &= (x + 2)(x - 1)(x - 4)(x - i)(x + i) \end{aligned}$$

Now try Exercise 29.

Synthetic division can be used with complex number divisors in the same way it is used with real number divisors.

EXAMPLE 5 Finding Complex Zeros

The complex number $z = 1 - 2i$ is a zero of $f(x) = 4x^4 + 17x^2 + 14x + 65$. Find the remaining zeros of $f(x)$, and write it in its linear factorization.

SOLUTION We use synthetic division to show that $f(1 - 2i) = 0$:

$$\begin{array}{r|rrrrr} 1-2i & 4 & 0 & 17 & 14 & 65 \\ & & 4-8i & -12-16i & -27-26i & -65 \\ \hline & 4 & 4-8i & 5-16i & -13-26i & 0 \end{array}$$

Thus $1 - 2i$ is a zero of $f(x)$. The conjugate $1 + 2i$ must also be a zero. We use synthetic division on the quotient found above to find the remaining quadratic factor:

$$\begin{array}{r|rrrr} 1+2i & 4 & 4-8i & 5-16i & -13-26i \\ & & 4+8i & 8+16i & 13+26i \\ \hline & 4 & 8 & 13 & 0 \end{array}$$

Finally, we use the quadratic formula to find the two zeros of $4x^2 + 8x + 13$:

$$\begin{aligned} x &= \frac{-8 \pm \sqrt{64 - 208}}{8} \\ &= \frac{-8 \pm \sqrt{-144}}{8} \\ &= \frac{-8 \pm 12i}{8} \\ &= -1 \pm \frac{3}{2}i \end{aligned}$$

Thus the four zeros of $f(x)$ are $1 - 2i$, $1 + 2i$, $-1 + (3/2)i$, and $-1 - (3/2)i$. Because the leading coefficient of $f(x)$ is 4, we obtain

$$f(x) = 4[x - (1 - 2i)][x - (1 + 2i)]\left[x - \left(-1 + \tfrac{3}{2}i\right)\right]\left[x - \left(-1 - \tfrac{3}{2}i\right)\right].$$

If we wish to remove fractions in the factors, we can distribute the 4 to get

$$f(x) = [x - (1 - 2i)][x - (1 + 2i)][2x - (-2 + 3i)][2x - (-2 - 3i)].$$

Now try Exercise 33.

Factoring with Real Number Coefficients

Let $f(x)$ be a polynomial function with real coefficients. The Linear Factorization Theorem tells us that $f(x)$ can be factored into the form

$$f(x) = a(x - z_1)(x - z_2)\cdots(x - z_n),$$

where z_i are complex numbers. Recall, however, that nonreal complex zeros occur in conjugate pairs. The product of $x - (a + bi)$ and $x - (a - bi)$ is

$$\begin{aligned}[x - (a + bi)][x - (a - bi)] &= x^2 - (a - bi)x - (a + bi)x + (a + bi)(a - bi)\\ &= x^2 - 2ax + (a^2 + b^2)\end{aligned}$$

So the quadratic expression $x^2 - 2ax + (a^2 + b^2)$, whose coefficients are real numbers, is a factor of $f(x)$. Such a quadratic expression with real coefficients but no real zeros is **irreducible over the reals**. In other words, if we require that the factors of a polynomial have real coefficients, the factorization can be accomplished with linear factors and irreducible quadratic factors.

Factors of a Polynomial with Real Coefficients

Every polynomial function with real coefficients can be written as a product of linear factors and irreducible quadratic factors, each with real coefficients.

EXAMPLE 6 Factoring a Polynomial

Write $f(x) = 3x^5 - 2x^4 + 6x^3 - 4x^2 - 24x + 16$ as a product of linear and irreducible quadratic factors, each with real coefficients.

SOLUTION The Rational Zeros Theorem provides the candidates for the rational zeros of f. The graph of f in Figure 2.45 suggests which candidates to try first. Using synthetic division, we find that $x = 2/3$ is a zero. Thus,

$$\begin{aligned}f(x) &= \left(x - \frac{2}{3}\right)(3x^4 + 6x^2 - 24)\\ &= \left(x - \frac{2}{3}\right)(3)(x^4 + 2x^2 - 8)\\ &= (3x - 2)(x^2 - 2)(x^2 + 4)\\ &= (3x - 2)(x - \sqrt{2})(x + \sqrt{2})(x^2 + 4)\end{aligned}$$

Because the zeros of $x^2 + 4$ are complex, any further factorization would introduce nonreal complex coefficients. We have taken the factorization of f as far as possible, subject to the condition that *each factor has real coefficients*.

Now try Exercise 37.

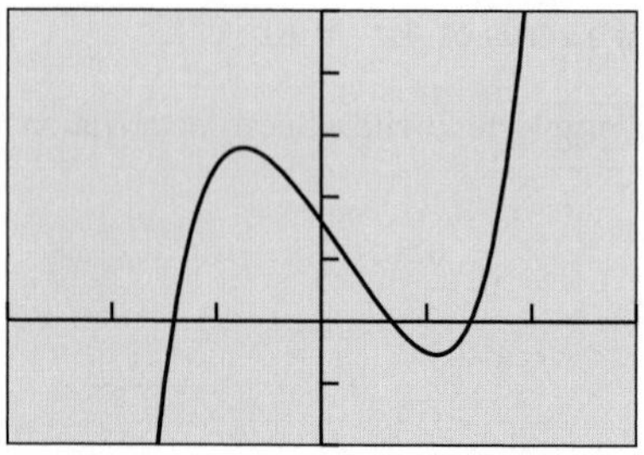

[–3, 3] by [–20, 50]

FIGURE 2.45 (Example 6) $f(x) = 3x^5 - 2x^4 + 6x^3 - 4x^2 - 24x + 16$ has three real zeros.

We have seen that if a polynomial function has real coefficients, then its nonreal complex zeros occur in conjugate pairs. *Because a polynomial of odd degree has an odd number of zeros, it must have at least one zero that is real.* This confirms Example 7 of Section 2.3 in light of complex numbers.

Polynomial Function of Odd Degree

Every polynomial function of odd degree with real coefficients has at least one real zero.

The function $f(x) = 3x^5 - 2x^4 + 6x^3 - 4x^2 - 24x + 16$ in Example 6 fits the conditions of this theorem, so we know immediately that we are on the right track in searching for at least one real zero.

QUICK REVIEW 2.5 *(For help, go to Sections P.5, P.6, and 2.4.)*

Exercise numbers with a gray background indicate problems that the authors have designed to be solved *without a calculator*.

In Exercises 1–4, perform the indicated operation, and write the result in the form $a + bi$.

1. $(3 - 2i) + (-2 + 5i)$

2. $(5 - 7i) - (3 - 2i)$

3. $(1 + 2i)(3 - 2i)$

4. $\dfrac{2 + 3i}{1 - 5i}$

In Exercises 5 and 6, factor the quadratic expression.

5. $2x^2 - x - 3$

6. $6x^2 - 13x - 5$

In Exercises 7 and 8, solve the quadratic equation.

7. $x^2 - 5x + 11 = 0$

8. $2x^2 + 3x + 7 = 0$

In Exercises 9 and 10, list all potential rational zeros.

9. $3x^4 - 5x^3 + 3x^2 - 7x + 2$

10. $4x^5 - 7x^2 + x^3 + 13x - 3$

SECTION 2.5 Exercises

In Exercises 1–4, write the polynomial in standard form, and identify the zeros of the function and the x-intercepts of its graph.

1. $f(x) = (x - 3i)(x + 3i)$

2. $f(x) = (x + 2)(x - \sqrt{3}i)(x + \sqrt{3}i)$

3. $f(x) = (x - 1)(x - 1)(x + 2i)(x - 2i)$

4. $f(x) = x(x - 1)(x - 1 - i)(x - 1 + i)$

In Exercises 5–12, write a polynomial function of minimum degree in standard form with real coefficients whose zeros include those listed.

5. i and $-i$

6. $1 - 2i$ and $1 + 2i$

7. 1, $3i$, and $-3i$

8. -4, $1 - i$, and $1 + i$

9. 2, 3, and i

10. -1, 2, and $1 - i$

11. 5 and $3 + 2i$

12. -2 and $1 + 2i$

In Exercises 13–16, write a polynomial function of minimum degree in standard form with real coefficients whose zeros and their multiplicities include those listed.

13. 1 (multiplicity 2), -2 (multiplicity 3)

14. -1 (multiplicity 3), 3 (multiplicity 1)

15. 2 (multiplicity 2), $3 + i$ (multiplicity 1)

16. -1 (multiplicity 2), $-2 - i$ (multiplicity 1)

In Exercises 17–20, match the polynomial function graph to the given zeros and multiplicities.

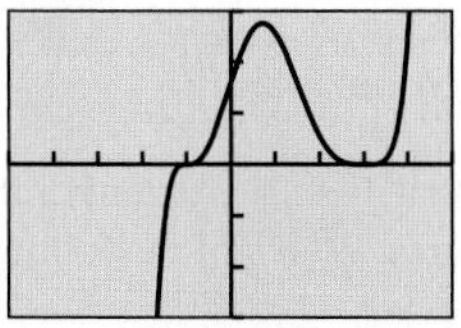

[–5, 5] by [–150, 150]
(a)

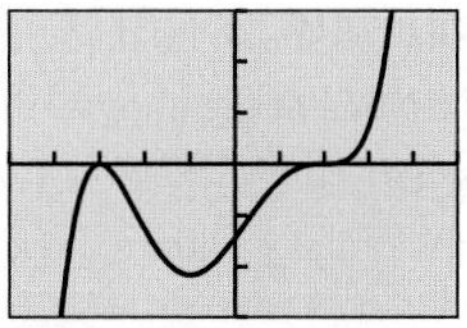

[–5, 5] by [–150, 150]
(b)

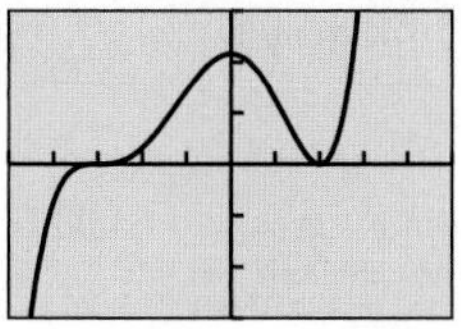

[–5, 5] by [–150, 150]
(c)

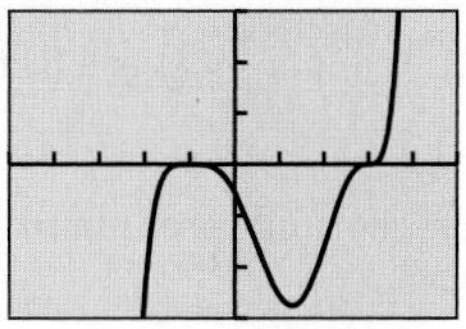

[–5, 5] by [–150, 150]
(d)

17. -3 (multiplicity 2), 2 (multiplicity 3)

18. -3 (multiplicity 3), 2 (multiplicity 2)

19. -1 (multiplicity 4), 3 (multiplicity 3)

20. -1 (multiplicity 3), 3 (multiplicity 4)

In Exercises 21–26, state how many complex and real zeros the function has.

21. $f(x) = x^2 - 2x + 7$

22. $f(x) = x^3 - 3x^2 + x + 1$

23. $f(x) = x^3 - x + 3$

24. $f(x) = x^4 - 2x^2 + 3x - 4$

25. $f(x) = x^4 - 5x^3 + x^2 - 3x + 6$

26. $f(x) = x^5 - 2x^2 - 3x + 6$

In Exercises 27–32, find all of the zeros and write a linear factorization of the function.

27. $f(x) = x^3 + 4x - 5$

28. $f(x) = x^3 - 10x^2 + 44x - 69$

29. $f(x) = x^4 + x^3 + 5x^2 - x - 6$

30. $f(x) = 3x^4 + 8x^3 + 6x^2 + 3x - 2$

31. $f(x) = 6x^4 - 7x^3 - x^2 + 67x - 105$

32. $f(x) = 20x^4 - 148x^3 + 269x^2 - 106x - 195$

In Exercises 33–36, using the given zero, find all of the zeros and write a linear factorization of $f(x)$.

33. $1 + i$ is a zero of $f(x) = x^4 - 2x^3 - x^2 + 6x - 6$.

34. $4i$ is a zero of $f(x) = x^4 + 13x^2 - 48$.

35. $3 - 2i$ is a zero of $f(x) = x^4 - 6x^3 + 11x^2 + 12x - 26$.

36. $1 + 3i$ is a zero of $f(x) = x^4 - 2x^3 + 5x^2 + 10x - 50$.

In Exercises 37–42, write the function as a product of linear and irreducible quadratic factors, all with real coefficients.

37. $f(x) = x^3 - x^2 - x - 2$

38. $f(x) = x^3 - x^2 + x - 6$

39. $f(x) = 2x^3 - x^2 + 2x - 3$

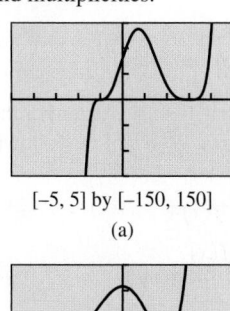

40. $f(x) = 3x^3 - 2x^2 + x - 2$

41. $f(x) = x^4 + 3x^3 - 3x^2 + 3x - 4$

42. $f(x) = x^4 - 2x^3 + x^2 - 8x - 12$

In Exercises 43 and 44, use *Archimedes' Principle,* which states that when a sphere of radius r with density d_S is placed in a liquid of density $d_L = 62.5$ lb/ft^3, it will sink to a depth h where

$$\frac{\pi}{3}(3rh^2 - h^3)d_L = \frac{4}{3}\pi r^3 d_S.$$

Find an approximate value for h if:

43. $r = 5$ ft and $d_S = 20$ lb/ft^3.

44. $r = 5$ ft and $d_S = 45$ lb/ft^3.

In Exercises 45–48, answer yes or no. If yes, include an example. If no, give a reason.

45. Writing to Learn Is it possible to find a polynomial of degree 3 with real number coefficients that has -2 as its only real zero?

46. Writing to Learn Is it possible to find a polynomial of degree 3 with real coefficients that has $2i$ as its only nonreal zero?

47. Writing to Learn Is it possible to find a polynomial $f(x)$ of degree 4 with real coefficients that has zeros $-3, 1 + 2i$, and $1 - i$?

48. Writing to Learn Is it possible to find a polynomial $f(x)$ of degree 4 with real coefficients that has zeros $1 + 3i$ and $1 - i$?

In Exercises 49 and 50, find the unique polynomial with real coefficients that meets these conditions.

49. Degree 4; zeros at $x = 3, x = -1$, and $x = 2 - i$; $f(0) = 30$

50. Degree 4; zeros at $x = 1 - 2i$ and $x = 1 + i$; $f(0) = 20$

51. Sally's distance D from a motion detector is given by the data in Table 2.16.

(a) Find a cubic regression model, and graph it together with a scatter plot of the data.

(b) Describe Sally's motion.

(c) Use the cubic regression model to estimate when Sally changes direction. How far is she from the motion detector when she changes direction?

Table 2.16 Motion Detector Data

t (sec)	D (m)	t (sec)	D (m)
0.0	3.36	4.5	3.59
0.5	2.61	5.0	4.15
1.0	1.86	5.5	3.99
1.5	1.27	6.0	3.37
2.0	0.91	6.5	2.58
2.5	1.14	7.0	1.93
3.0	1.69	7.5	1.25
3.5	2.37	8.0	0.67
4.0	3.01		

52. Jacob's distance D from a motion detector is given by the data in Table 2.17.

(a) Find a quadratic regression model, and graph it together with a scatter plot of the data.

(b) Describe Jacob's motion.

(c) Use the quadratic regression model to estimate when Jacob changes direction. How far is he from the motion detector when he changes direction?

Table 2.17 Motion Detector Data

t (sec)	D (m)	t (sec)	D (m)
0.0	4.59	4.5	1.70
0.5	3.92	5.0	2.25
1.0	3.14	5.5	2.84
1.5	2.41	6.0	3.39
2.0	1.73	6.5	4.02
2.5	1.21	7.0	4.54
3.0	0.90	7.5	5.04
3.5	0.99	8.0	5.59
4.0	1.31		

Standardized Test Questions

53. True or False There is at least one polynomial with real coefficients with $1 - 2i$ as its only nonreal zero. Justify your answer.

54. True or False A polynomial of degree 3 with real coefficients must have two nonreal zeros. Justify your answer.

In Exercises 55–58, you may use a graphing calculator to solve the problem.

55. Multiple Choice Let z be a nonreal complex number and $\bar{z}$ its complex conjugate. Which of the following is not a real number?

(A) $z + \bar{z}$ **(B)** $z\bar{z}$ **(C)** $(z + \bar{z})^2$ **(D)** $(z\bar{z})^2$ **(E)** z^2

56. Multiple Choice Which of the following cannot be the number of real zeros of a polynomial of degree 5 with real coefficients?

(A) 0 **(B)** 1 **(C)** 2 **(D)** 3 **(E)** 4

57. Multiple Choice Which of the following cannot be the number of nonreal zeros of a polynomial of degree 5 with real coefficients?

(A) 0 **(B)** 2 **(C)** 3 **(D)** 4

(E) None of the above

58. Multiple Choice Assume that $1 + 2i$ is a zero of the polynomial f with real coefficients. Which of the following statements is not true?

(A) $x - (1 + 2i)$ is a factor of $f(x)$.

(B) $x^2 - 2x + 5$ is a factor of $f(x)$.

(C) $x - (1 - 2i)$ is a factor of $f(x)$.

(D) $1 - 2i$ is a zero of f.

(E) The number of nonreal complex zeros of f could be 1.

Explorations

59. Group Activity The Powers of $1 + i$

(a) Selected powers of $1 + i$ are displayed in Table 2.18. Find a pattern in the data, and use it to extend the table to powers 7, 8, 9, and 10.

(b) Compute $(1 + i)^7$, $(1 + i)^8$, $(1 + i)^9$, and $(1 + i)^{10}$ using the fact that $(1 + i)^6 = -8i$.

(c) Compare your results from parts (a) and (b) and reconcile, if needed.

Table 2.18 Powers of $1 + i$

Power	Real Part	Imaginary Part
0	1	0
1	1	1
2	0	2
3	−2	2
4	−4	0
5	−4	−4
6	0	−8

60. Group Activity The Square Roots of i
Let a and b be real numbers such that $(a + bi)^2 = i$.

(a) Expand the left-hand side of the given equation.

(b) Think of the right-hand side of the equation as $0 + 1i$, and separate the real and imaginary parts of the equation to obtain two equations.

(c) Solve for a and b.

(d) Check your answer by substituting them in the original equation.

(e) What are the two square roots of i?

61. Verify that the complex number i is a zero of the polynomial $f(x) = x^3 - ix^2 + 2ix + 2$.

62. Verify that the complex number $-2i$ is a zero of the polynomial $f(x) = x^3 - (2 - i)x^2 + (2 - 2i)x - 4$.

Extending the Ideas

In Exercises 63 and 64, verify that $g(x)$ is a factor of $f(x)$. Then find $h(x)$ so that $f = g \cdot h$.

63. $g(x) = x - i$; $f(x) = x^3 + (3 - i)x^2 - 4ix - 1$

64. $g(x) = x - 1 - i$; $f(x) = x^3 - (1 + i)x^2 + x - 1 - i$

65. Find the three cube roots of 8 by solving $x^3 = 8$.

66. Find the three cube roots of -64 by solving $x^3 = -64$.

2.6 Graphs of Rational Functions

What you'll learn about

- Rational Functions
- Transformations of the Reciprocal Function
- Limits and Asymptotes
- Analyzing Graphs of Rational Functions
- Exploring Relative Humidity

... and why

Rational functions are used in calculus and in scientific applications such as inverse proportions.

Rational Functions

Rational functions are ratios (or quotients) of polynomial functions.

DEFINITION Rational Functions

Let f and g be polynomial functions with $g(x) \neq 0$. Then the function given by

$$r(x) = \frac{f(x)}{g(x)}$$

is a **rational function**.

The domain of a rational function is the set of all real numbers except the zeros of its denominator. Every rational function is continuous on its domain.

EXAMPLE 1 Finding the Domain of a Rational Function

Find the domain of f and use limits to describe its behavior at value(s) of x not in its domain.

$$f(x) = \frac{1}{x - 2}$$

SOLUTION The domain of f is all real numbers $x \neq 2$. The graph in Figure 2.46 strongly suggests that f has a vertical asymptote at $x = 2$. As x approaches 2 from the left, the values of f decrease without bound. As x approaches 2 from the right, the values of f increase without bound. Using the notation introduced in Section 1.2 on page 92 we write

$$\lim_{x \to 2^-} f(x) = -\infty \quad \text{and} \quad \lim_{x \to 2^+} f(x) = \infty.$$

The tables in Figure 2.47 support this visual evidence numerically.

Now try Exercise 1.

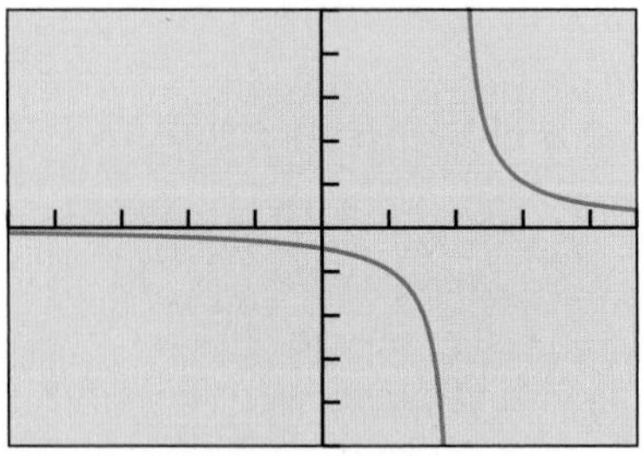

[–4.7, 4.7] by [–5, 5]

FIGURE 2.46 The graph of $f(x) = 1/(x - 2)$. (Example 1)

X	Y1	
2	ERROR	
2.01	100	
2.02	50	
2.03	33.333	
2.04	25	
2.05	20	
2.06	16.667	

Y1 = 1/(X–2)

(a)

X	Y1	
2	ERROR	
1.99	–100	
1.98	–50	
1.97	–33.33	
1.96	–25	
1.95	–20	
1.94	–16.67	

Y1 = 1/(X–2)

(b)

FIGURE 2.47 Table of values for $f(x) = 1/(x - 2)$ for values of x (a) to the right of 2, and (b) to the left of 2. (Example 1)

In Chapter 1 we defined horizontal and vertical asymptotes of the graph of a function $y = f(x)$. The line $y = b$ is a *horizontal asymptote* of the graph of f if

$$\lim_{x \to -\infty} f(x) = b \quad \text{or} \quad \lim_{x \to \infty} f(x) = b.$$

The line $x = a$ is a *vertical asymptote* of the graph of f if

$$\lim_{x \to a^-} f(x) = \pm\infty \quad \text{or} \quad \lim_{x \to a^+} f(x) = \pm\infty.$$

We can see from Figure 2.46 that $\lim_{x \to -\infty} 1/(x - 2) = \lim_{x \to \infty} 1/(x - 2) = 0$, so the line $y = 0$ is a horizontal asymptote of the graph of $f(x) = 1/(x - 2)$. Because $\lim_{x \to 2^-} f(x) = -\infty$ and $\lim_{x \to 2^+} f(x) = \infty$, the line $x = 2$ is a vertical asymptote of $f(x) = 1/(x - 2)$.

Transformations of the Reciprocal Function

One of the simplest rational functions is the reciprocal function

$$f(x) = \frac{1}{x},$$

which is one of the basic functions introduced in Chapter 1. It can be used to generate many other rational functions.

Here is what we know about the reciprocal function.

BASIC FUNCTION The Reciprocal Function

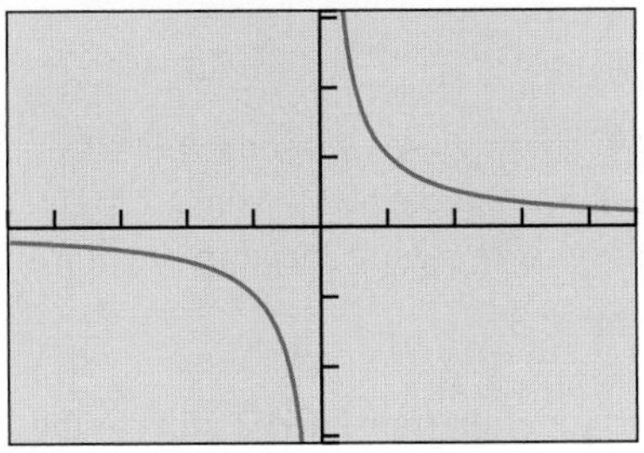

[–4.7, 4.7] by [–3.1, 3.1]

FIGURE 2.48 The graph of $f(x) = 1/x$.

$f(x) = \dfrac{1}{x}$
Domain: $(-\infty, 0) \cup (0, \infty)$
Range: $(-\infty, 0) \cup (0, \infty)$
Continuity: All $x \neq 0$
Decreasing on $(-\infty, 0)$ and $(0, \infty)$
Symmetric with respect to origin (an odd function)
Unbounded
No local extrema
Horizontal asymptote: $y = 0$
Vertical asymptote: $x = 0$
End behavior: $\lim_{x \to -\infty} f(x) = \lim_{x \to \infty} f(x) = 0$

EXPLORATION 1 Comparing Graphs of Rational Functions

1. Sketch the graph and find an equation for the function g whose graph is obtained from the reciprocal function $f(x) = 1/x$ by a translation of 2 units to the right.
2. Sketch the graph and find an equation for the function h whose graph is obtained from the reciprocal function $f(x) = 1/x$ by a translation of 5 units to the right, followed by a reflection across the x-axis.
3. Sketch the graph and find an equation for the function k whose graph is obtained from the reciprocal function $f(x) = 1/x$ by a translation of 4 units to the left, followed by a vertical stretch by a factor of 3, and finally a translation 2 units down.

The graph of any nonzero rational function of the form

$$g(x) = \frac{ax + b}{cx + d},\ c \neq 0$$

can be obtained through transformations of the graph of the reciprocal function. If the degree of the numerator is greater than or equal to the degree of the denominator, we can use polynomial division to rewrite the rational function.

EXAMPLE 2 Transforming the Reciprocal Function

Describe how the graph of the given function can be obtained by transforming the graph of the reciprocal function $f(x) = 1/x$. Identify the horizontal and vertical asymptotes and use limits to describe the corresponding behavior. Sketch the graph of the function.

(a) $g(x) = \dfrac{2}{x + 3}$ **(b)** $h(x) = \dfrac{3x - 7}{x - 2}$

SOLUTION

(a) $g(x) = \dfrac{2}{x + 3} = 2\left(\dfrac{1}{x + 3}\right) = 2f(x + 3)$

The graph of g is the graph of the reciprocal function shifted left 3 units and then stretched vertically by a factor of 2. So the lines $x = -3$ and $y = 0$ are vertical and horizontal asymptotes, respectively. Using limits we have $\lim_{x \to \infty} g(x) = \lim_{x \to -\infty} g(x) = 0$, $\lim_{x \to -3^+} g(x) = \infty$, and $\lim_{x \to -3^-} g(x) = -\infty$. The graph is shown in Figure 2.49a.

(b) We begin with polynomial division:

$$\begin{array}{r} 3 \\ x - 2\overline{)3x - 7} \\ \underline{3x - 6} \\ -1 \end{array}$$

So, $h(x) = \dfrac{3x - 7}{x - 2} = 3 - \dfrac{1}{x - 2} = -f(x - 2) + 3.$

Thus the graph of h is the graph of the reciprocal function translated 2 units to the right, followed by a reflection across the x-axis, and then translated 3 units up. (Note that the reflection must be executed before the vertical translation.) So the lines $x = 2$ and $y = 3$ are vertical and horizontal asymptotes, respectively. Using limits we have $\lim_{x \to \infty} h(x) = \lim_{x \to -\infty} h(x) = 3$, $\lim_{x \to 2^+} g(x) = -\infty$, and $\lim_{x \to 2^-} g(x) = \infty$. The graph is shown in Figure 2.49b. **Now try Exercise 5.**

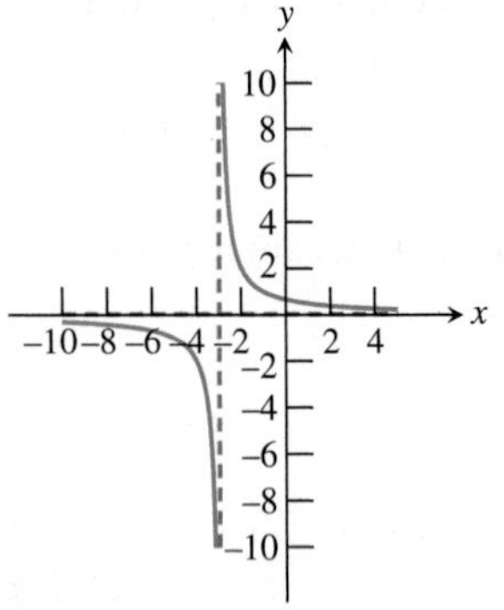

(a)

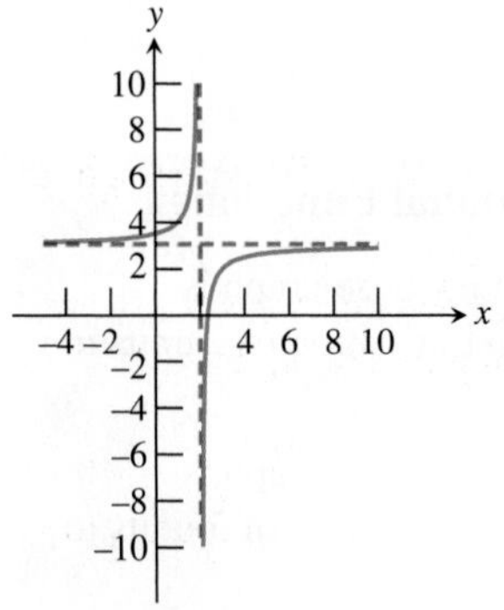

(b)

FIGURE 2.49 The graphs of (a) $g(x) = 2/(x + 3)$ and (b) $h(x) = (3x - 7)/(x - 2)$, with asymptotes shown in red.

Limits and Asymptotes

In Example 2 we found asymptotes by translating the known asymptotes of the reciprocal function. In Example 3, we use graphing and algebra to find an asymptote.

EXAMPLE 3 Finding Asymptotes

Find the horizontal and vertical asymptotes of $f(x) = (x^2 + 2)/(x^2 + 1)$. Use limits to describe the corresponding behavior of f.

SOLUTION

Solve Graphically The graph of f in Figure 2.50 suggests that

$$\lim_{x \to \infty} f(x) = \lim_{x \to -\infty} f(x) = 1$$

and that there are no vertical asymptotes. The horizontal asymptote is $y = 1$.

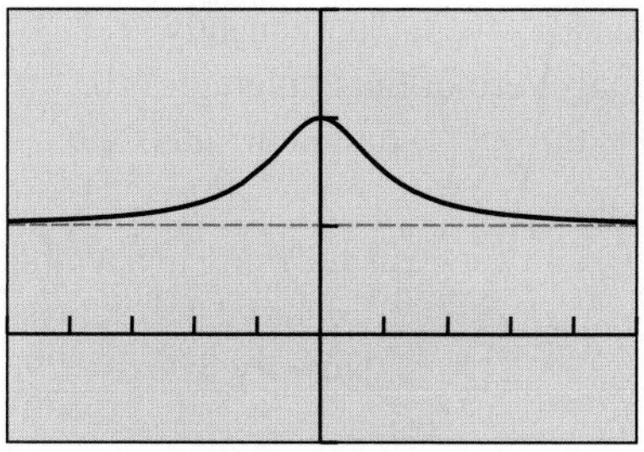

[–5, 5] by [–1, 3]

FIGURE 2.50 The graph of $f(x) = (x^2 + 2)/(x^2 + 1)$ with its horizontal asymptote $y = 1$.

Solve Algebraically Because the denominator $x^2 + 1 > 0$, the domain of f is all real numbers. So there are no vertical asymptotes. Using polynomial long division, we find that

$$f(x) = \frac{x^2 + 2}{x^2 + 1} = 1 + \frac{1}{x^2 + 1}.$$

When the value of $|x|$ is large, the denominator $x^2 + 1$ is a large positive number, and $1/(x^2 + 1)$ is a small positive number, getting closer to zero as $|x|$ increases. Therefore,

$$\lim_{x \to \infty} f(x) = \lim_{x \to -\infty} f(x) = 1,$$

so $y = 1$ is indeed a horizontal asymptote. **Now try Exercise 19.**

Example 3 shows the connection between the end behavior of a rational function and its horizontal asymptote. We now generalize this relationship and summarize other features of the graph of a rational function:

Graph of a Rational Function

The graph of $y = f(x)/g(x) = (a_n x^n + \cdots)/(b_m x^m + \cdots)$ has the following characteristics:

1. **End behavior asymptote:**
   If $n < m$, the end behavior asymptote is the horizontal asymptote $y = 0$.
   If $n = m$, the end behavior asymptote is the horizontal asymptote $y = a_n/b_m$.
   If $n > m$, the end behavior asymptote is the quotient polynomial function $y = q(x)$, where $f(x) = g(x)q(x) + r(x)$. There is no horizontal asymptote.
2. **Vertical asymptotes:** These occur at the real zeros of the denominator, provided that the zeros are not also zeros of the numerator of equal or greater multiplicity.
3. ***x*-intercepts:** These occur at the real zeros of the numerator that are not also zeros of the denominator.
4. ***y*-intercept:** This is the value of $f(0)$, if defined.

It is a good idea to find all of the asymptotes and intercepts when graphing a rational function. If the end behavior asymptote of a rational function is a slant line, we call it a **slant asymptote**, as illustrated in Example 4.

EXAMPLE 4 Graphing a Rational Function

Find the asymptotes and intercepts of the function $f(x) = x^3/(x^2 - 9)$ and graph the function.

SOLUTION The degree of the numerator is greater than the degree of the denominator, so the end behavior asymptote is the quotient polynomial. Using polynomial long division, we obtain

$$f(x) = \frac{x^3}{x^2 - 9} = x + \frac{9x}{x^2 - 9}.$$

So the quotient polynomial is $q(x) = x$, a slant asymptote. Factoring the denominator,

$$x^2 - 9 = (x - 3)(x + 3),$$

(continued)

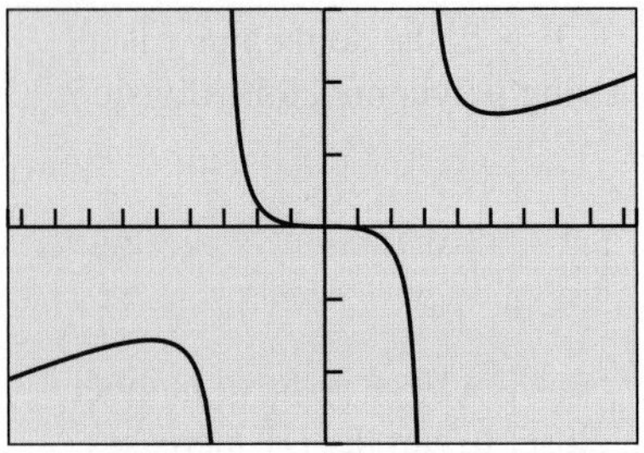

[–9.4, 9.4] by [–15, 15]
(a)

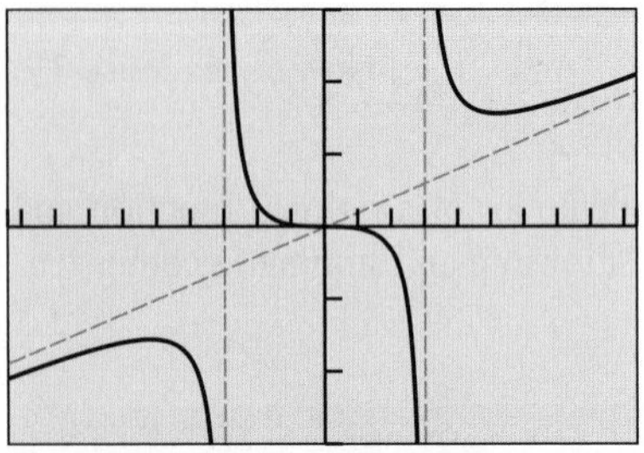

[–9.4, 9.4] by [–15, 15]
(b)

FIGURE 2.51 The graph of $f(x) = x^3/(x^2 - 9)$ (a) by itself and (b) with its asymptotes. (Example 4)

shows that the zeros of the denominator are $x = 3$ and $x = -3$. Consequently, $x = 3$ and $x = -3$ are the vertical asymptotes of f. The only zero of the numerator is 0, so $f(0) = 0$, and thus we see that the point (0, 0) is the only x-intercept and is the y-intercept of the graph of f.

The graph of f in Figure 2.51a passes through (0, 0) and suggests the vertical asymptotes $x = 3$ and $x = -3$ and the slant asymptote $y = q(x) = x$. Figure 2.51b shows the graph of f with its asymptotes overlaid. **Now try Exercise 29.**

Analyzing Graphs of Rational Functions

Because the degree of the numerator of the rational function in Example 5 is less than the degree of the denominator, we know that the graph of the function has $y = 0$ as a horizontal asymptote.

EXAMPLE 5 Analyzing the Graph of a Rational Function

Find the intercepts and asymptotes, use limits to describe the behavior at the vertical asymptotes, and analyze and draw the graph of the rational function

$$f(x) = \frac{x - 1}{x^2 - x - 6}.$$

SOLUTION The numerator is zero when $x = 1$, so the x-intercept is 1. Because $f(0) = 1/6$, the y-intercept is 1/6. The denominator factors as

$$x^2 - x - 6 = (x - 3)(x + 2),$$

so there are vertical asymptotes at $x = -2$ and $x = 3$. From the comment preceding this example we know that the horizontal asymptote is $y = 0$. Figure 2.52 supports this information and allows us to conclude that

$$\lim_{x \to -2^-} f(x) = -\infty, \lim_{x \to -2^+} f(x) = \infty, \lim_{x \to 3^-} f(x) = -\infty, \textit{ and } \lim_{x \to 3^+} f(x) = \infty.$$

Domain: $(-\infty, -2) \cup (-2, 3) \cup (3, \infty)$
Range: $(-\infty, \infty)$
Continuity: All $x \neq -2, 3$
Decreasing on $(-\infty, -2)$, $(-2, 3)$, and $(3, \infty)$
Not symmetric
Unbounded
No local extrema
Horizontal asymptotes: $y = 0$
Vertical asymptotes: $x = -2, x = 3$
End behavior: $\lim_{x \to -\infty} f(x) = \lim_{x \to \infty} f(x) = 0$ **Now try Exercise 39.**

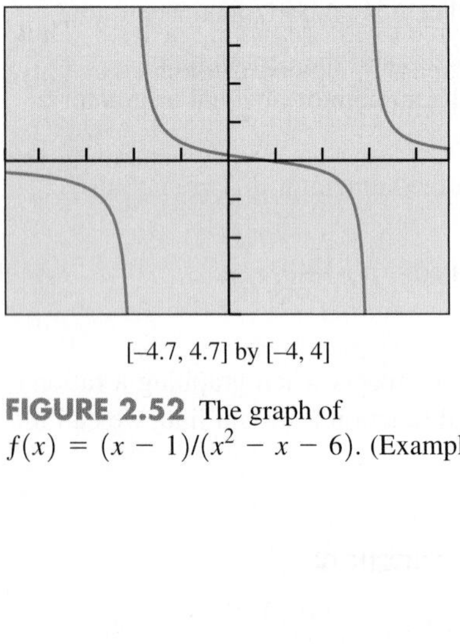

[–4.7, 4.7] by [–4, 4]

FIGURE 2.52 The graph of $f(x) = (x - 1)/(x^2 - x - 6)$. (Example 5)

The degrees of the numerator and denominator of the rational function in Example 6 are equal. Thus, we know that the graph of the function has $y = 2$, the quotient of the leading coefficients, as its end behavior asymptote.

EXAMPLE 6 Analyzing the Graph of a Rational Function

Find the intercepts, analyze, and draw the graph of the rational function

$$f(x) = \frac{2x^2 - 2}{x^2 - 4}.$$

SOLUTION The numerator factors as

$$2x^2 - 2 = 2(x^2 - 1) = 2(x + 1)(x - 1),$$

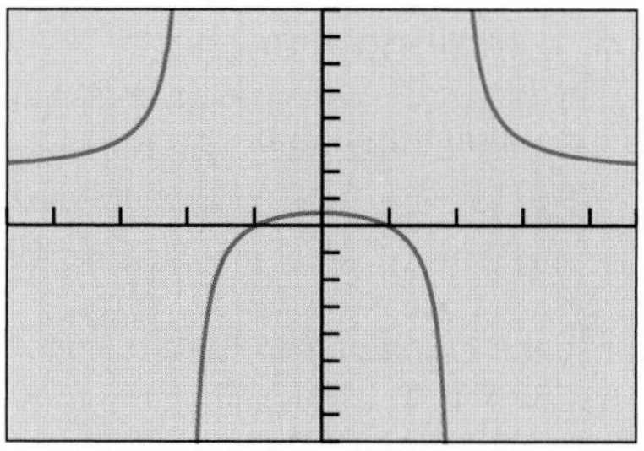

[–4.7, 4.7] by [–8, 8]

FIGURE 2.53 The graph of $f(x) = (2x^2 - 2)/(x^2 - 4)$. It can be shown that f takes on no value between 1/2, the y-intercept, and 2, the horizontal asymptote. (Example 6)

so the x-intercepts are -1 and 1. The y-intercept is $f(0) = 1/2$. The denominator factors as

$$x^2 - 4 = (x + 2)(x - 2),$$

so the vertical asymptotes are $x = -2$ and $x = 2$. From the comment preceding this example we know that $y = 2$ is the horizontal asymptote. Figure 2.53 supports this information and allows us to conclude that

$$\lim_{x \to -2^-} f(x) = \infty, \quad \lim_{x \to -2^+} f(x) = -\infty, \quad \lim_{x \to 2^-} f(x) = -\infty, \quad \lim_{x \to 2^+} f(x) = \infty.$$

Domain: $(-\infty, -2) \cup (-2, 2) \cup (2, \infty)$
Range: $(-\infty, 1/2] \cup (2, \infty)$
Continuity: All $x \neq -2, 2$
Increasing on $(-\infty, -2)$ and $(-2, 0]$; decreasing on $[0, 2)$ and $(2, \infty)$
Symmetric with respect to the y-axis (an even function)
Unbounded
Local maximum of 1/2 at $x = 0$
Horizontal asymptotes: $y = 2$
Vertical asymptotes: $x = -2, x = 2$
End behavior: $\lim_{x \to -\infty} f(x) = \lim_{x \to \infty} f(x) = 2$

Now try Exercise 41.

In Examples 7 and 8 we will investigate the rational function

$$f(x) = \frac{x^3 - 3x^2 + 3x + 1}{x - 1}.$$

The degree of the numerator of f exceeds the degree of the denominator by 2. Thus, there is no horizontal asymptote. We will see that the end behavior asymptote is a polynomial of degree 2.

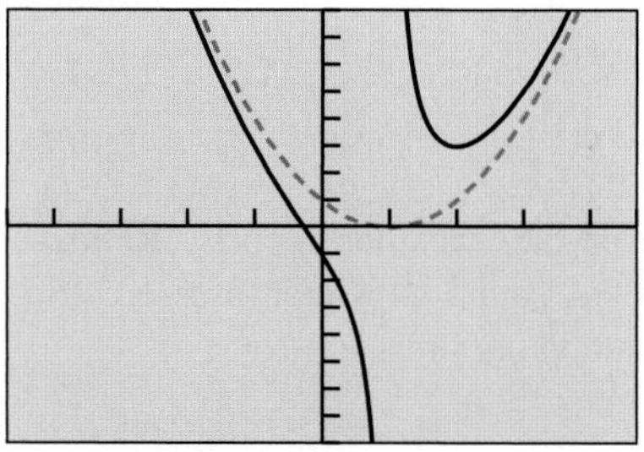

[–4.7, 4.7] by [–8, 8]

FIGURE 2.54 The graph of $f(x) = (x^3 - 3x^2 + 3x + 1)/(x - 1)$ as a solid black line and its end behavior asymptote $y = x^2 - 2x + 1$ as a dashed blue line. (Examples 7 and 8)

EXAMPLE 7 Finding an End Behavior Asymptote

Find the end behavior asymptote of

$$f(x) = \frac{x^3 - 3x^2 + 3x + 1}{x - 1}$$

and graph it together with f in two windows:

(a) one showing the details around the vertical asymptote of f,

(b) one showing a graph of f that resembles its end behavior asymptote.

SOLUTION The graph of f has a vertical asymptote at $x = 1$. Divide $x^3 - 3x^2 + 3x + 1$ by $x - 1$ to show that

$$f(x) = \frac{x^3 - 3x^2 + 3x + 1}{x - 1} = x^2 - 2x + 1 + \frac{2}{x - 1}.$$

The end behavior asymptote of f is $y = x^2 - 2x + 1$.

(a) The graph of f in Figure 2.54 shows the details around the vertical asymptote. We have also overlaid the graph of its end behavior asymptote as a dashed line.

(b) If we draw the graph of $f(x) = (x^3 - 3x^2 + 3x + 1)/(x - 1)$ and its end behavior asymptote $y = x^2 - 2x + 1$ in a large enough viewing window, the two graphs will appear to be identical (Figure 2.55).

Now try Exercise 47.

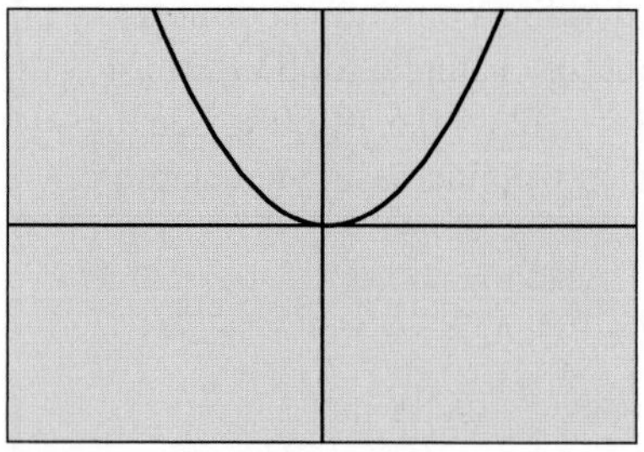

[–40, 40] by [–500, 500]

FIGURE 2.55 The graphs of $f(x) = (x^3 - 3x^2 + 3x + 1)/(x - 1)$ and its end behavior asymptote $y = x^2 - 2x + 1$ appear to be identical. (Example 7)

EXAMPLE 8 Analyzing the Graph of a Rational Function

Find the intercepts, analyze, and draw the graph of the rational function

$$f(x) = \frac{x^3 - 3x^2 + 3x + 1}{x - 1}.$$

SOLUTION f has only one x-intercept and we can use the graph of f in Figure 2.54 to show that it is about -0.26. The y-intercept is $f(0) = -1$. The vertical asymptote is $x = 1$ as we have seen. We know that the graph of f does not have a horizontal asymptote, and from Example 7 we know that the end behavior asymptote is $y = x^2 - 2x + 1$. We can also use Figure 2.54 to show that f has a local minimum of 3 at $x = 2$. Figure 2.54 supports this information and allows us to conclude that

$$\lim_{x \to 1^-} f(x) = -\infty \quad \text{and} \quad \lim_{x \to 1^+} f(x) = \infty.$$

Domain: $(-\infty, 1) \cup (1, \infty)$
Range: $(-\infty, \infty)$
Continuity: All $x \neq 1$
Decreasing on $(-\infty, 1)$ and $(1, 2]$; increasing on $[2, \infty)$
Not symmetric
Unbounded
Local minimum of 3 at $x = 2$
No horizontal asymptotes; end behavior asymptote: $y = x^2 - 2x + 1$
Vertical asymptote: $x = 1$
End behavior: $\lim_{x \to -\infty} f(x) = \lim_{x \to \infty} f(x) = \infty$

Now try Exercise 55.

Exploring Relative Humidity

The phrase *relative humidity* is familiar from everyday weather reports. **Relative humidity** is the ratio of constant vapor pressure to saturated vapor pressure. So, relative humidity is inversely proportional to saturated vapor pressure.

CHAPTER OPENER Problem (from page 157)

Problem: Determine the relative humidity values that correspond to the saturated vapor pressures of 12, 24, 36, 48, and 60 millibars, at a constant vapor pressure of 12 millibars. (In practice, saturated vapor pressure increases as the temperature increases.)

Solution: Relative humidity (RH) is found by dividing constant vapor pressure (CVP) by saturated vapor pressure (SVP). So, for example, for SVP = 24 millibars and CVP = 12 millibars, RH = 12/24 = 1/2 = 0.5 = 50%. See the table below, which is based on CVP = 12 millibars with increasing temperature.

SVP (millibars)	RH (%)
12	100
24	50
36	$33.\overline{3}$
48	25
60	20

QUICK REVIEW 2.6 *(For help, go to Section 2.4.)*

Exercise numbers with a gray background indicate problems that the authors have designed to be solved *without a calculator*.

In Exercises 1–6, use factoring to find the real zeros of the function.

1. $f(x) = 2x^2 + 5x - 3$
2. $f(x) = 3x^2 - 2x - 8$
3. $g(x) = x^2 - 4$
4. $g(x) = x^2 - 1$
5. $h(x) = x^3 - 1$
6. $h(x) = x^2 + 1$

In Exercises 7–10, find the quotient and remainder when $f(x)$ is divided by $d(x)$.

7. $f(x) = 2x + 1, d(x) = x - 3$
8. $f(x) = 4x + 3, d(x) = 2x - 1$
9. $f(x) = 3x - 5, d(x) = x$
10. $f(x) = 5x - 1, d(x) = 2x$

SECTION 2.6 Exercises

In Exercises 1–4, find the domain of the function f. Use limits to describe the behavior of f at value(s) of x not in its domain.

1. $f(x) = \dfrac{1}{x + 3}$
2. $f(x) = \dfrac{-3}{x - 1}$
3. $f(x) = \dfrac{-1}{x^2 - 4}$
4. $f(x) = \dfrac{2}{x^2 - 1}$

In Exercises 5–10, describe how the graph of the given function can be obtained by transforming the graph of the reciprocal function $g(x) = 1/x$. Identify the horizontal and vertical asymptotes and use limits to describe the corresponding behavior. Sketch the graph of the function.

5. $f(x) = \dfrac{1}{x - 3}$
6. $f(x) = -\dfrac{2}{x + 5}$
7. $f(x) = \dfrac{2x - 1}{x + 3}$
8. $f(x) = \dfrac{3x - 2}{x - 1}$
9. $f(x) = \dfrac{5 - 2x}{x + 4}$
10. $f(x) = \dfrac{4 - 3x}{x - 5}$

In Exercises 11–14, evaluate the limit based on the graph of f shown.

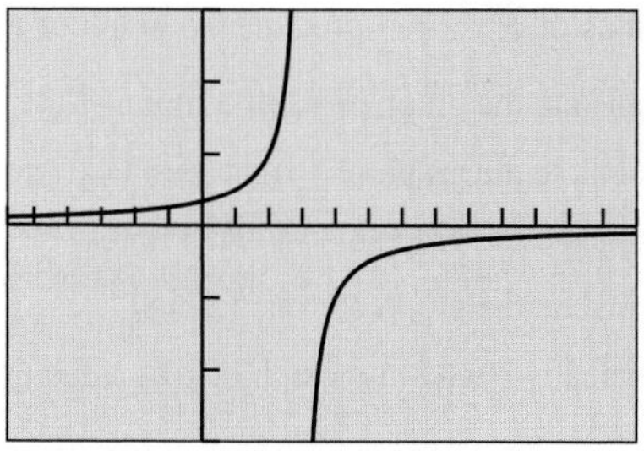

[–5.8, 13] by [–3, 3]

11. $\lim_{x \to 3^-} f(x)$
12. $\lim_{x \to 3^+} f(x)$
13. $\lim_{x \to \infty} f(x)$
14. $\lim_{x \to -\infty} f(x)$

In Exercises 15–18, evaluate the limit based on the graph of f shown.

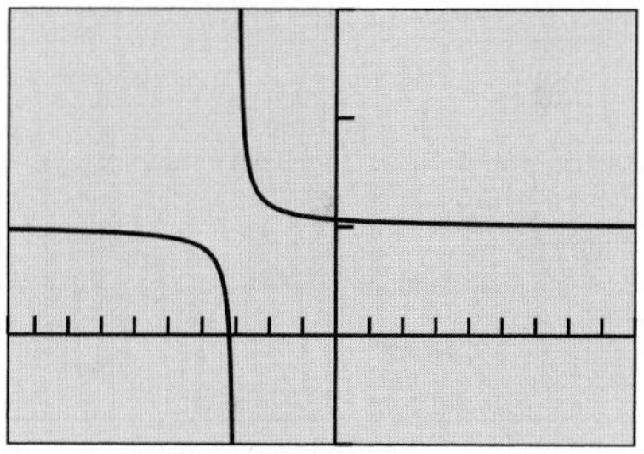

[–9.8, 9] by [–5, 15]

15. $\lim_{x \to -3^+} f(x)$
16. $\lim_{x \to -3^-} f(x)$
17. $\lim_{x \to -\infty} f(x)$
18. $\lim_{x \to \infty} f(x)$

In Exercises 19–22, find the horizontal and vertical asymptotes of $f(x)$. Use limits to describe the corresponding behavior.

19. $f(x) = \dfrac{2x^2 - 1}{x^2 + 3}$
20. $f(x) = \dfrac{3x^2}{x^2 + 1}$
21. $f(x) = \dfrac{2x + 1}{x^2 - x}$
22. $f(x) = \dfrac{x - 3}{x^2 + 3x}$

In Exercises 23–30, find the asymptotes and intercepts of the function, and graph the function.

23. $g(x) = \dfrac{x - 2}{x^2 - 2x - 3}$
24. $g(x) = \dfrac{x + 2}{x^2 + 2x - 3}$
25. $h(x) = \dfrac{2}{x^3 - x}$
26. $h(x) = \dfrac{3}{x^3 - 4x}$
27. $f(x) = \dfrac{2x^2 + x - 2}{x^2 - 1}$
28. $g(x) = \dfrac{-3x^2 + x + 12}{x^2 - 4}$
29. $f(x) = \dfrac{x^2 - 2x + 3}{x + 2}$
30. $g(x) = \dfrac{x^2 - 3x - 7}{x + 3}$

In Exercises 31–36, match the rational function with its graph. Identify the viewing window and the scale used on each axis.

31. $f(x) = \dfrac{1}{x-4}$ **32.** $f(x) = -\dfrac{1}{x+3}$

33. $f(x) = 2 + \dfrac{3}{x-1}$ **34.** $f(x) = 1 + \dfrac{1}{x+3}$

35. $f(x) = -1 + \dfrac{1}{4-x}$ **36.** $f(x) = 3 - \dfrac{2}{x-1}$

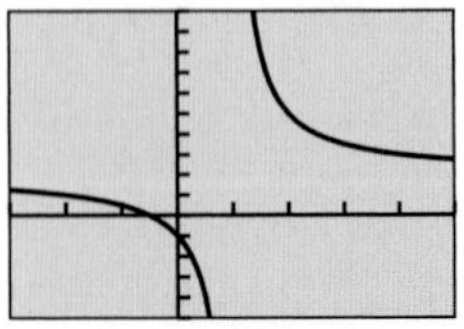
(a)

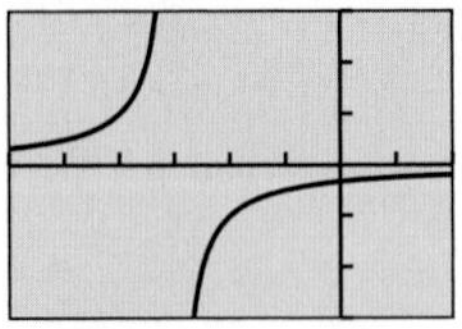
(b)

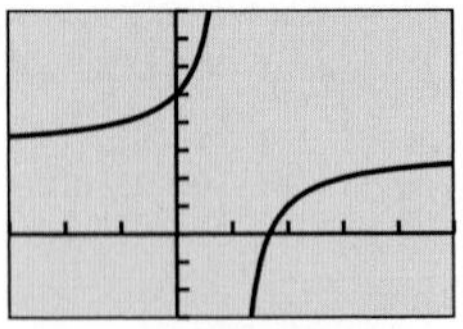
(c)

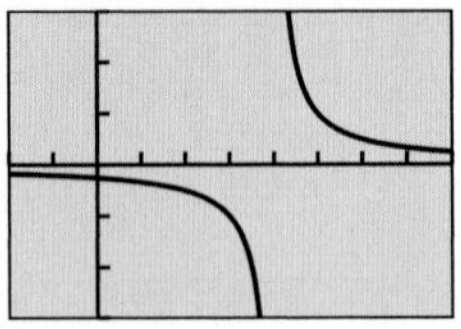
(d)

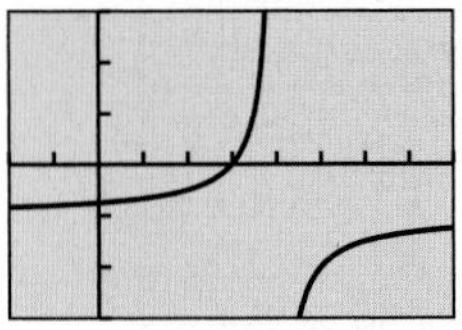
(e)

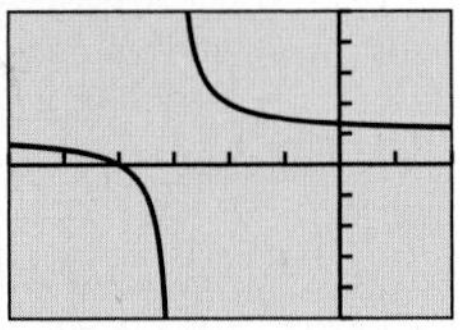
(f)

In Exercises 37–44, find the intercepts and asymptotes, use limits to describe the behavior at the vertical asymptotes, and analyze and draw the graph of the given rational function.

37. $f(x) = \dfrac{2}{2x^2 - x - 3}$ **38.** $g(x) = \dfrac{2}{x^2 + 4x + 3}$

39. $h(x) = \dfrac{x-1}{x^2 - x - 12}$ **40.** $k(x) = \dfrac{x+1}{x^2 - 3x - 10}$

41. $f(x) = \dfrac{x^2 + x - 2}{x^2 - 9}$ **42.** $g(x) = \dfrac{x^2 - x - 2}{x^2 - 2x - 8}$

43. $h(x) = \dfrac{x^2 + 2x - 3}{x+2}$ **44.** $k(x) = \dfrac{x^2 - x - 2}{x-3}$

In Exercises 45–50, find the end behavior asymptote of the given rational function f and graph it together with f in two windows:

(a) One showing the details around the vertical asymptote(s) of f.
(b) One showing a graph of f that resembles its end behavior asymptote.

45. $f(x) = \dfrac{x^2 - 2x + 3}{x-5}$ **46.** $f(x) = \dfrac{2x^2 + 2x - 3}{x+3}$

47. $f(x) = \dfrac{x^3 - x^2 + 1}{x+2}$ **48.** $f(x) = \dfrac{x^3 + 1}{x-1}$

49. $f(x) = \dfrac{x^4 - 2x + 1}{x-2}$ **50.** $f(x) = \dfrac{x^5 + 1}{x^2 + 1}$

In Exercises 51–56, find the intercepts, analyze, and graph the given rational function.

51. $f(x) = \dfrac{3x^2 - 2x + 4}{x^2 - 4x + 5}$ **52.** $g(x) = \dfrac{4x^2 + 2x}{x^2 - 4x + 8}$

53. $h(x) = \dfrac{x^3 - 1}{x-2}$ **54.** $k(x) = \dfrac{x^3 - 2}{x+2}$

55. $f(x) = \dfrac{x^3 - 2x^2 + x - 1}{2x - 1}$ **56.** $g(x) = \dfrac{2x^3 - 2x^2 - x + 5}{x-2}$

In Exercises 57–62, find the intercepts, vertical asymptotes, and end behavior asymptote, and graph the function together with its end behavior asymptote.

57. $h(x) = \dfrac{x^4 + 1}{x+1}$ **58.** $k(x) = \dfrac{2x^5 + x^2 - x + 1}{x^2 - 1}$

59. $f(x) = \dfrac{x^5 - 1}{x+2}$ **60.** $g(x) = \dfrac{x^5 + 1}{x-1}$

61. $h(x) = \dfrac{2x^3 - 3x + 2}{x^3 - 1}$ **62.** $k(x) = \dfrac{3x^3 + x - 4}{x^3 + 1}$

Standardized Test Questions

63. True or False A rational function must have a vertical asymptote. Justify your answer.

64. True or False $f(x) = \dfrac{x^2 - x}{\sqrt{x^2 + 4}}$ is a rational function. Justify your answer.

In Exercises 65–68, you may use a graphing calculator to solve the problem.

65. Multiple Choice Let $f(x) = \dfrac{-2}{x^2 + 3x}$. What values of x have to be excluded from the domain of f?

(A) Only 0 **(B)** Only 3 **(C)** Only -3
(D) Only 0, 3 **(E)** Only 0, -3

66. Multiple Choice Let $g(x) = \dfrac{2}{x+3}$. Which of the transformations of $f(x) = \dfrac{2}{x}$ produce the graph of g?

(A) Translate the graph of f left 3 units.
(B) Translate the graph of f right 3 units.
(C) Translate the graph of f down 3 units.
(D) Translate the graph of f up 3 units.
(E) Vertically stretch the graph of f by a factor of 2.

67. Multiple Choice Let $f(x) = \dfrac{x^2}{x+5}$. Which of the following statements is true about the graph of f?

(A) There is no vertical asymptote.
(B) There is a horizontal asymptote but no vertical asymptote.
(C) There is a slant asymptote but no vertical asymptote.
(D) There is a vertical asymptote and a slant asymptote.
(E) There is a vertical and horizontal asymptote.

68. Multiple Choice What is the degree of the end behavior asymptote of $f(x) = \dfrac{x^8 + 1}{x^4 + 1}$?

(A) 0 **(B)** 1 **(C)** 2 **(D)** 3 **(E)** 4

Explorations

69. Group Activity *Work in groups of two.* Compare the functions $f(x) = \dfrac{x^2 - 9}{x - 3}$ and $g(x) = x + 3$.

(a) Are the domains equal?

(b) Does f have a vertical asymptote? Explain.

(c) Explain why the graphs appear to be identical.

(d) Are the functions identical?

70. Group Activity Explain why the functions are identical or not. Include the graphs and a comparison of the functions' asymptotes, intercepts, and domain.

(a) $f(x) = \dfrac{x^2 + x - 2}{x - 1}$ and $g(x) = x + 2$

(b) $f(x) = \dfrac{x^2 - 1}{x + 1}$ and $g(x) = x - 1$

(c) $f(x) = \dfrac{x^2 - 1}{x^3 - x^2 - x + 1}$ and $g(x) = \dfrac{1}{x - 1}$

(d) $f(x) = \dfrac{x - 1}{x^2 + x - 2}$ and $g(x) = \dfrac{1}{x + 2}$

71. Boyle's Law This ideal gas law states that the volume of an enclosed gas at a fixed temperature varies inversely as the pressure.

(a) Writing to Learn Explain why Boyle's Law yields both a rational function model and a power function model.

(b) Which power functions are also rational functions?

(c) If the pressure of a 2.59-L sample of nitrogen gas at a temperature of 291 K is 0.866 atm, what would the volume be at a pressure of 0.532 atm if the temperature does not change?

72. Light Intensity Aileen and Malachy gathered the data in Table 2.19 using a 75-watt lightbulb and a Calculator-Based Laboratory™ (CBL™) with a light-intensity probe.

(a) Draw a scatter plot of the data in Table 2.19.

(b) Find an equation for the data, assuming it has the form $f(x) = k/x^2$ for some constant k. Explain your method for choosing k.

(c) Superimpose the regression curve on the scatter plot.

(d) Use the regression model to predict the light intensity at distances of 2.2 m and 4.4 m.

Table 2.19 Light-Intensity Data for a 75-W Lightbulb

Distance (m)	Intensity (W/m^2)
1.0	6.09
1.5	2.51
2.0	1.56
2.5	1.08
3.0	0.74

Extending the Ideas

In Exercises 73–76, graph the function. Express the function as a piecewise-defined function without absolute value, and use the result to confirm the graph's asymptotes and intercepts algebraically.

73. $h(x) = \dfrac{2x - 3}{|x| + 2}$

74. $h(x) = \dfrac{3x + 5}{|x| + 3}$

75. $f(x) = \dfrac{5 - 3x}{|x| + 4}$

76. $f(x) = \dfrac{2 - 2x}{|x| + 1}$

77. Describe how the graph of a nonzero rational function

$$f(x) = \frac{ax + b}{cx + d},\ c \neq 0$$

can be obtained from the graph of $y = 1/x$. (*Hint*: Use long division.)

78. Writing to Learn Let $f(x) = 1 + 1/(x - 1/x)$ and $g(x) = (x^3 + x^2 - x)/(x^3 - x)$. Does $f = g$? Support your answer by making a comparative analysis of all of the features of f and g, including asymptotes, intercepts, and domain.

2.7 Solving Equations in One Variable

What you'll learn about

- Solving Rational Equations
- Extraneous Solutions
- Applications

... and why

Applications involving rational functions as models often require that an equation involving the model be solved.

Solving Rational Equations

Equations involving rational expressions or fractions (see Appendix A.3) are **rational equations**. Every rational equation can be written in the form

$$\frac{f(x)}{g(x)} = 0.$$

If $f(x)$ and $g(x)$ are polynomial functions with no common factors, then the zeros of $f(x)$ are the solutions of the equation.

Usually it is not necessary to put a rational equation in the form of $f(x)/g(x)$. To solve an equation involving fractional expressions we begin by finding the LCD (least common denominator) of all the terms of the equation. Then we clear the equation of fractions by multiplying each side of the equation by the LCD. Sometimes the LCD contains variables.

When we multiply or divide an equation by an expression containing variables, the resulting equation may have solutions that are *not* solutions of the original equation. These are **extraneous solutions**. For this reason we must check each solution of the resulting equation in the original equation.

EXAMPLE 1 Solving by Clearing Fractions

Solve $x + \frac{3}{x} = 4$.

SOLUTION

Solve Algebraically The LCD is x.

$$x + \frac{3}{x} = 4$$

$$x^2 + 3 = 4x \quad \text{Multiply by } x.$$

$$x^2 - 4x + 3 = 0 \quad \text{Subtract } 4x.$$

$$(x - 1)(x - 3) = 0 \quad \text{Factor.}$$

$$x - 1 = 0 \quad \text{or} \quad x - 3 = 0 \quad \text{Zero factor property}$$

$$x = 1 \quad \text{or} \quad x = 3$$

Confirm Numerically

For $x = 1, x + \frac{3}{x} = 1 + \frac{3}{1} = 4$, and for $x = 3, x + \frac{3}{x} = 3 + \frac{3}{3} = 4$.

Each value is a solution of the original equation. **Now try Exercise 1.**

When the fractions in Example 2 are cleared, we obtain a quadratic equation.

EXAMPLE 2 Solving a Rational Equation

Solve $x + \dfrac{1}{x-4} = 0.$

SOLUTION

Solve Algebraically The LCD is $x - 4$.

$$x + \frac{1}{x-4} = 0$$

$$x(x-4) + \frac{x-4}{x-4} = 0 \quad \text{Multiply by } x - 4.$$

$$x^2 - 4x + 1 = 0 \quad \text{Distributive property}$$

$$x = \frac{4 \pm \sqrt{(-4)^2 - 4(1)(1)}}{2(1)} \quad \text{Quadratic formula}$$

$$x = \frac{4 \pm 2\sqrt{3}}{2} \quad \text{Simplify.}$$

$$x = 2 \pm \sqrt{3} \quad \text{Simplify.}$$

$$x \approx 0.268, 3.732$$

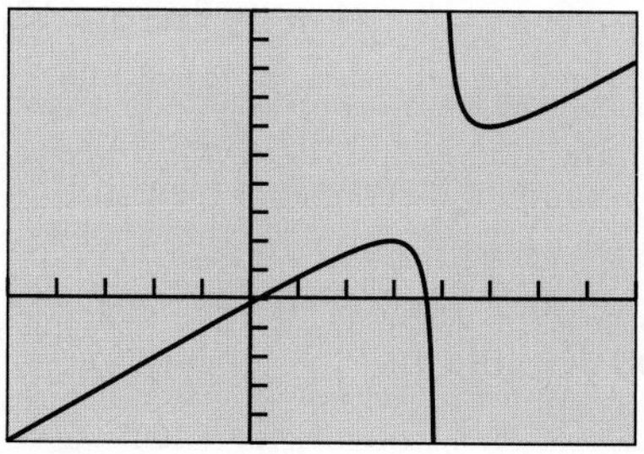

[–5, 8] by [–5, 10]

FIGURE 2.56 The graph of $y = x + 1/(x - 4)$. (Example 2)

Support Graphically The graph in Figure 2.56 suggests that the function $y = x + 1/(x - 4)$ has two zeros. We can use the graph to find that the zeros are about 0.268 and 3.732, agreeing with the values found algebraically.

Now try Exercise 7.

Extraneous Solutions

We will find extraneous solutions in Examples 3 and 4.

EXAMPLE 3 Eliminating Extraneous Solutions

Solve the equation

$$\frac{2x}{x-1} + \frac{1}{x-3} = \frac{2}{x^2 - 4x + 3}.$$

SOLUTION

Solve Algebraically The denominator of the right-hand side, $x^2 - 4x + 3$, factors into $(x - 1)(x - 3)$. So the least common denominator (LCD) of the equation is $(x - 1)(x - 3)$, and we multiply both sides of the equation by this LCD:

$$(x-1)(x-3)\left(\frac{2x}{x-1} + \frac{1}{x-3}\right) = (x-1)(x-3)\left(\frac{2}{x^2 - 4x + 3}\right)$$

$$2x(x-3) + (x-1) = 2 \quad \text{Distributive property}$$

$$2x^2 - 5x - 3 = 0 \quad \text{Distributive property}$$

$$(2x+1)(x-3) = 0 \quad \text{Factor.}$$

$$x = -\frac{1}{2} \quad \text{or} \quad x = 3$$

Confirm Numerically We replace x by $-1/2$ in the original equation:

$$\frac{2(-1/2)}{(-1/2) - 1} + \frac{1}{(-1/2) - 3} \stackrel{?}{=} \frac{2}{(-1/2)^2 - 4(-1/2) + 3}$$

$$\frac{2}{3} - \frac{2}{7} \stackrel{?}{=} \frac{8}{21}$$

(continued)

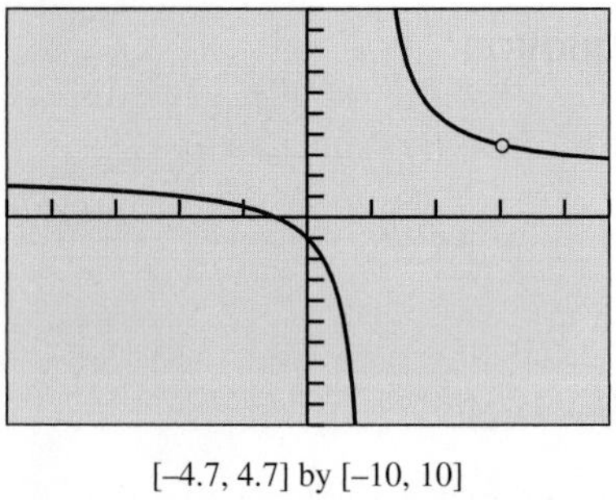

[–4.7, 4.7] by [–10, 10]

FIGURE 2.57 The graph of $f(x) = 2x/(x - 1) + 1/(x - 3) - 2/(x^2 - 4x + 3)$, with the missing point at $(3, 3.5)$ emphasized. Note that some graphers may not display such points. (Example 3)

The equation is true, so $x = -1/2$ is a valid solution. The original equation is not defined for $x = 3$, so $x = 3$ is an extraneous solution.

Support Graphically The graph of

$$f(x) = \frac{2x}{x - 1} + \frac{1}{x - 3} - \frac{2}{x^2 - 4x + 3}$$

in Figure 2.57 suggests that $x = -1/2$ is an x-intercept and $x = 3$ is not.

Interpret Only $x = -1/2$ is a solution. **Now try Exercise 13.**

We will see that Example 4 has no solutions.

EXAMPLE 4 Eliminating Extraneous Solutions

Solve

$$\frac{x - 3}{x} + \frac{3}{x + 2} + \frac{6}{x^2 + 2x} = 0.$$

SOLUTION The LCD is $x(x + 2)$.

$$\frac{x - 3}{x} + \frac{3}{x + 2} + \frac{6}{x^2 + 2x} = 0$$

$$(x - 3)(x + 2) + 3x + 6 = 0 \quad \text{Multiply by } x(x + 2).$$

$$x^2 - x - 6 + 3x + 6 = 0 \quad \text{Expand.}$$

$$x^2 + 2x = 0 \quad \text{Simplify.}$$

$$x(x + 2) = 0 \quad \text{Factor.}$$

$$x = 0 \quad \text{or} \quad x = -2$$

Substituting $x = 0$ or $x = -2$ into the original equation results in division by zero. So both of these numbers are extraneous solutions and the original equation has no solution. **Now try Exercise 17.**

Applications

EXAMPLE 5 Calculating Acid Mixtures

How much pure acid must be added to 50 mL of a 35% acid solution to produce a mixture that is 75% acid? (See Figure 2.58.)

SOLUTION

Model

Word statement: $\frac{\text{mL of pure acid}}{\text{mL of mixture}} = \text{concentration of acid}$

0.35×50 or 17.5 = mL of pure acid in 35% solution

x = mL of acid added

$x + 17.5$ = mL of pure acid in resulting mixture

$x + 50$ = mL of the resulting mixture

$\frac{x + 17.5}{x + 50}$ = concentration of acid

Solve Graphically

$$\frac{x + 17.5}{x + 50} = 0.75 \quad \text{Equation to be solved}$$

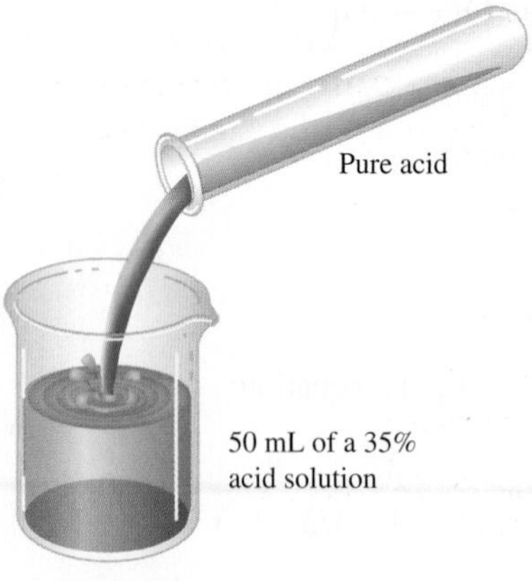

FIGURE 2.58 Mixing solutions. (Example 5)

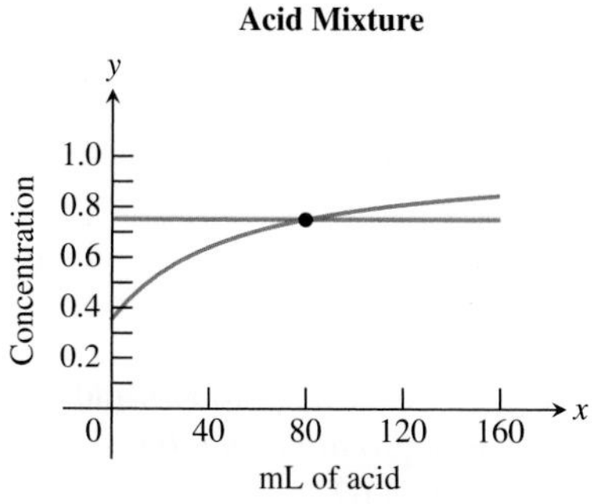

Intersection: $x = 80$; $y = .75$

FIGURE 2.59 The graphs of $f(x) = (x + 17.5)/(x + 50)$ and $g(x) = 0.75$. (Example 5)

Figure 2.59 shows graphs of $f(x) = (x + 17.5)/(x + 50)$ and $g(x) = 0.75$. The point of intersection is (80, 0.75).

Interpret We need to add 80 mL of pure acid to the 35% acid solution to make a solution that is 75% acid. **Now try Exercise 31.**

EXAMPLE 6 Finding a Minimum Perimeter

Find the dimensions of the rectangle with minimum perimeter if its area is 200 square meters. Find this least perimeter.

SOLUTION

Model Draw the diagram in Figure 2.60.

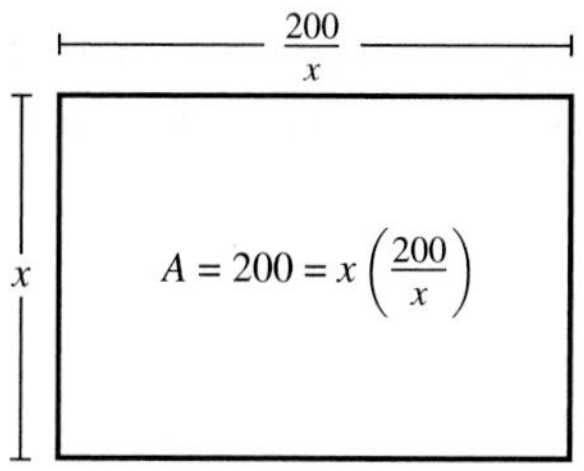

FIGURE 2.60 A rectangle with area 200 m^2. (Example 6)

Word statement: Perimeter $= 2 \times$ length $+ 2 \times$ width

$$x = \text{width in meters}$$

$$\frac{200}{x} = \frac{\text{area}}{\text{width}} = \text{length in meters}$$

Function to be minimized: $P(x) = 2x + 2\left(\frac{200}{x}\right) = 2x + \frac{400}{x}$

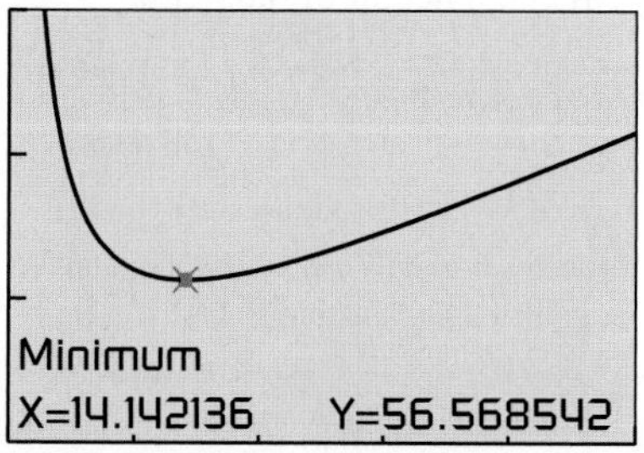

[0, 50] by [0, 150]

FIGURE 2.61 A graph of $P(x) = 2x + 400/x$. (Example 6)

Solve Graphically The graph of P in Figure 2.61 shows a minimum of approximately 56.57, occurring when $x \approx 14.14$.

Interpret A width of about 14.14 m produces the minimum perimeter of about 56.57 m. Because $200/14.14 \approx 14.14$, the dimensions of the rectangle with minimum perimeter are 14.14 m by 14.14 m, a square. **Now try Exercise 35.**

FIGURE 2.62 A tomato juice can. (Example 7)

EXAMPLE 7 Designing a Juice Can

Stewart Cannery packages tomato juice in 2-L cylindrical cans. Find the radius and height of the cans if the cans have a surface area of 1000 cm^2. (See Figure 2.62.)

SOLUTION

Model

$$S = \text{surface area of can in cm}^2$$
$$r = \text{radius of can in centimeters}$$
$$h = \text{height of can in centimeters}$$

Using volume (V) and surface area (S) formulas and the fact that 1 L $= 1000$ cm^3, we conclude that

$$V = \pi r^2 h = 2000 \quad \text{and} \quad S = 2\pi r^2 + 2\pi r h = 1000.$$

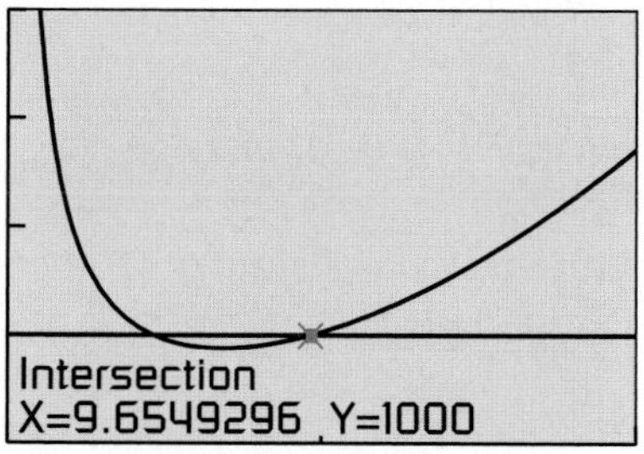

[0, 20] by [0, 4000]

FIGURE 2.63 Two points of intersection (Example 7)

(continued)

So

$$2\pi r^2 + 2\pi rh = 1000$$

$$2\pi r^2 + 2\pi r\left(\frac{2000}{\pi r^2}\right) = 1000 \quad \text{Substitute } h = 2000/(\pi r^2).$$

$$2\pi r^2 + \frac{4000}{r} = 1000 \quad \text{Equation to be solved}$$

Solve Graphically Figure 2.63 shows the graphs of $f(x) = 2\pi r^2 + 4000/r$ and $g(x) = 1000$. One point of intersection occurs when r is approximately 9.65. A second point of intersection occurs when r is approximately 4.62.

Because $h = 2000/(\pi r^2)$, the corresponding values for h are

$$h = \frac{2000}{\pi(4.619\ldots)^2} \approx 29.83 \quad \text{and} \quad h = \frac{2000}{\pi(9.654\ldots)^2} \approx 6.83.$$

Interpret With a surface area of 1000 cm^2, the cans either have a radius of 4.62 cm and a height of 29.83 cm or have a radius of 9.65 cm and a height of 6.83 cm.

Now try Exercise 37.

QUICK REVIEW 2.7 *(For help, go to Sections A.3. and P.5.)*

Exercise numbers with a gray background indicate problems that the authors have designed to be solved *without a calculator*.

In Exercises 1 and 2, find the missing numerator or denominator.

1. $\dfrac{2x}{x-3} = \dfrac{?}{x^2+x-12}$

2. $\dfrac{x-1}{x+1} = \dfrac{x^2-1}{?}$

In Exercises 3–6, find the LCD and rewrite the expression as a single fraction reduced to lowest terms.

3. $\dfrac{5}{12} + \dfrac{7}{18} - \dfrac{5}{6}$

4. $\dfrac{3}{x-1} - \dfrac{1}{x}$

5. $\dfrac{x}{2x+1} - \dfrac{2}{x-3}$

6. $\dfrac{x+1}{x^2-5x+6} - \dfrac{3x+11}{x^2-x-6}$

In Exercises 7–10, use the quadratic formula to find the zeros of the quadratic polynomials.

7. $2x^2 - 3x - 1$

8. $2x^2 - 5x - 1$

9. $3x^2 + 2x - 2$

10. $x^2 - 3x - 9$

SECTION 2.7 Exercises

In Exercises 1–6, solve the equation algebraically. Support your answer numerically and identify any extraneous solutions.

1. $\dfrac{x-2}{3} + \dfrac{x+5}{3} = \dfrac{1}{3}$

2. $x + 2 = \dfrac{15}{x}$

3. $x + 5 = \dfrac{14}{x}$

4. $\dfrac{1}{x} - \dfrac{2}{x-3} = 4$

5. $x + \dfrac{4x}{x-3} = \dfrac{12}{x-3}$

6. $\dfrac{3}{x-1} + \dfrac{2}{x} = 8$

In Exercises 7–12, solve the equation algebraically and graphically. Check for extraneous solutions.

7. $x + \dfrac{10}{x} = 7$

8. $x + 2 = \dfrac{15}{x}$

9. $x + \dfrac{12}{x} = 7$

10. $x + \dfrac{6}{x} = -7$

11. $2 - \dfrac{1}{x+1} = \dfrac{1}{x^2+x}$

12. $2 - \dfrac{3}{x+4} = \dfrac{12}{x^2+4x}$

In Exercises 13–18, solve the equation algebraically. Check for extraneous solutions. Support your answer graphically.

13. $\dfrac{3x}{x+5} + \dfrac{1}{x-2} = \dfrac{7}{x^2+3x-10}$

14. $\dfrac{4x}{x+4} + \dfrac{3}{x-1} = \dfrac{15}{x^2+3x-4}$

15. $\dfrac{x-3}{x} - \dfrac{3}{x+1} + \dfrac{3}{x^2+x} = 0$

16. $\dfrac{x+2}{x} - \dfrac{4}{x-1} + \dfrac{2}{x^2-x} = 0$

17. $\dfrac{3}{x+2} + \dfrac{6}{x^2+2x} = \dfrac{3-x}{x}$

18. $\dfrac{x+3}{x} - \dfrac{2}{x+3} = \dfrac{6}{x^2+3x}$

In Exercises 19–22, two possible solutions to the equation $f(x) = 0$ are listed. Use the given graph of $y = f(x)$ to decide which, if any, are extraneous.

19. $x = -5$ or $x = -2$

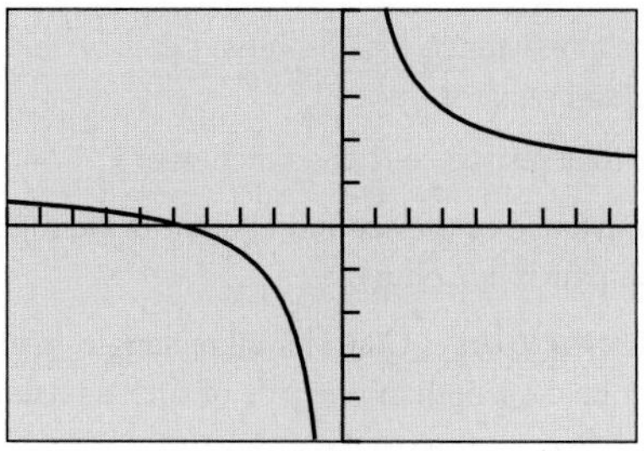

[–10, 8.8] by [–5, 5]

20. $x = -2$ or $x = 3$

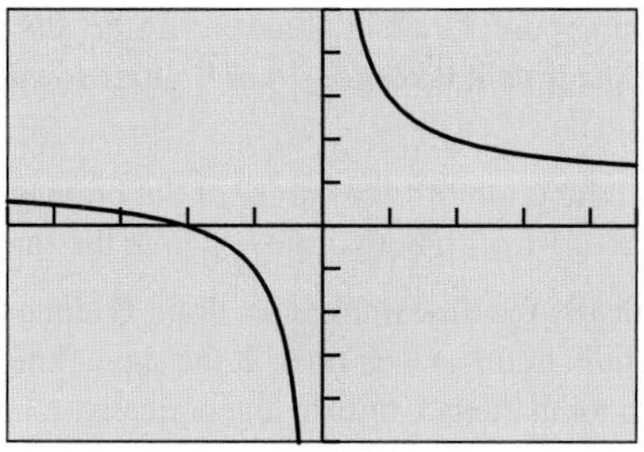

[–4.7, 4.7] by [–5, 5]

21. $x = -2$ or $x = 2$

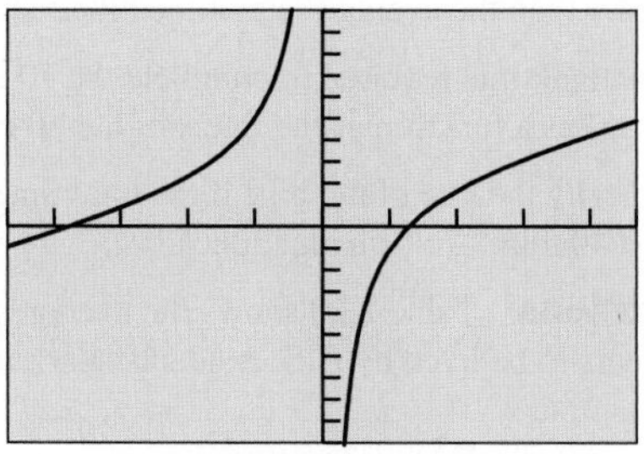

[–4.7, 4.7] by [–10, 10]

22. $x = 0$ or $x = 3$

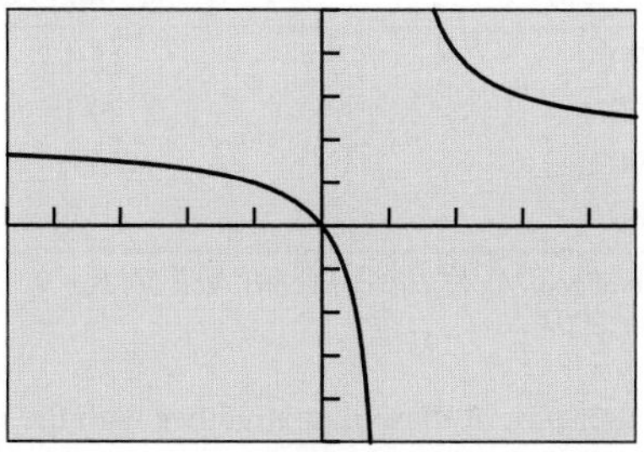

[–4.7, 4.7] by [–5, 5]

In Exercises 23–30, solve the equation.

23. $\dfrac{2}{x-1} + x = 5$

24. $\dfrac{x^2 - 6x + 5}{x^2 - 2} = 3$

25. $\dfrac{x^2 - 2x + 1}{x + 5} = 0$

26. $\dfrac{3x}{x+2} + \dfrac{2}{x-1} = \dfrac{5}{x^2 + x - 2}$

27. $\dfrac{4x}{x+4} + \dfrac{5}{x-1} = \dfrac{15}{x^2 + 3x - 4}$

28. $\dfrac{3x}{x+1} + \dfrac{5}{x-2} = \dfrac{15}{x^2 - x - 2}$

29. $x^2 + \dfrac{5}{x} = 8$

30. $x^2 - \dfrac{3}{x} = 7$

31. Acid Mixture Suppose that x mL of pure acid are added to 125 mL of a 60% acid solution. How many mL of pure acid must be added to obtain a solution of 83% acid?

(a) Explain why the concentration $C(x)$ of the new mixture is

$$C(x) = \frac{x + 0.6(125)}{x + 125}.$$

(b) Suppose the viewing window in the figure below is used to find a solution to the problem. What is the equation of the horizontal line?

(c) Writing to Learn Write and solve an equation that answers the question of this problem. Explain your answer.

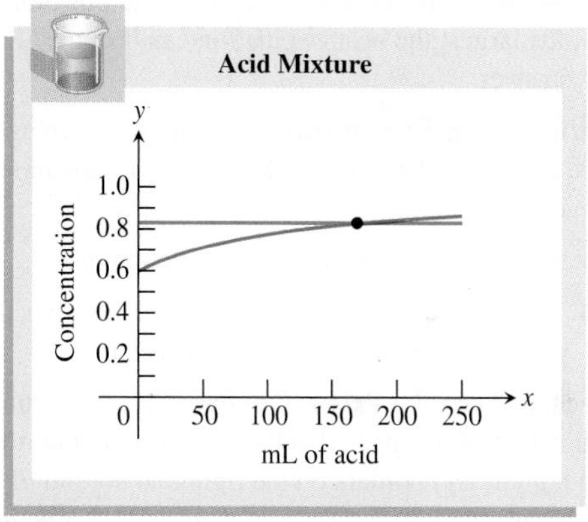

32. Acid Mixture Suppose that x mL of pure acid are added to 100 mL of a 35% acid solution.

(a) Express the concentration $C(x)$ of the new mixture as a function of x.

(b) Use a graph to determine how much pure acid should be added to the 35% solution to produce a new solution that is 75% acid.

(c) Solve (b) algebraically.

33. Breaking Even Mid Town Sports Apparel, Inc., has found that it needs to sell golf hats for \$2.75 each in order to be competitive. It costs \$2.12 to produce each hat, and the business has weekly overhead costs of \$3000.

(a) Let x be the number of hats produced each week. Express the average cost (including overhead costs) of producing one hat as a function of x.

(b) Solve algebraically to find the number of golf hats that must be sold each week to make a profit. Support your answer graphically.

(c) Writing to Learn How many golf hats must be sold to make a profit of \$1000 in 1 week? Explain your answer.

34. Bear Population The number of bears at any time t (in years) in a federal game reserve is given by

$$P(t) = \frac{500 + 250t}{10 + 0.5t}.$$

(a) Find the population of bears when the value of t is 10, 40, and 100.

(b) Does the graph of the bear population have a horizontal asymptote? If so, what is it? If not, why not?

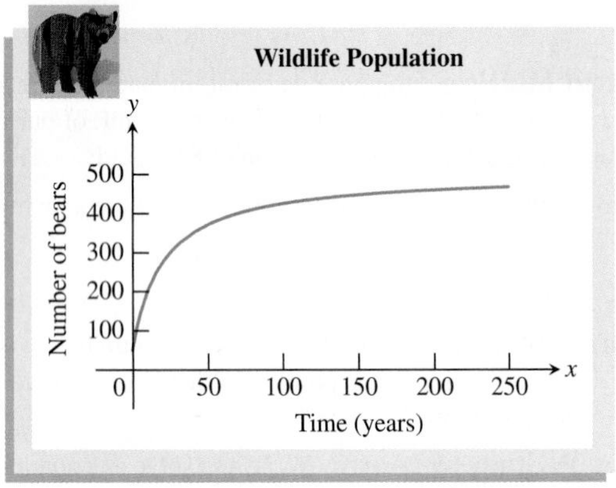

(c) Writing to Learn According to this model, what is the largest the bear population can become? Explain your answer.

35. Minimizing Perimeter Consider all rectangles with an area of 182 ft^2. Let x be the length of one side of such a rectangle.

(a) Express the perimeter P as a function of x.

(b) Find the dimensions of the rectangle that has the least perimeter. What is the least perimeter?

36. Group Activity Page Design Hendrix Publishing Co. wants to design a page that has a 0.75-in. left border, a 1.5-in. top border, and borders on the right and bottom of 1 in. They are to surround 40 in.^{2} of print material. Let x be the width of the print material.

(a) Express the area of the page as a function of x.

(b) Find the dimensions of the page that has the least area. What is the least area?

37. Industrial Design Drake Cannery will pack peaches in 0.5-L cylindrical cans. Let x be the radius of the can in cm.

(a) Express the surface area S of the can as a function of x.

(b) Find the radius and height of the can if the surface area is 900 cm^2.

38. Group Activity Designing a Swimming Pool Thompson Recreation, Inc., wants to build a rectangular swimming pool with the top of the pool having surface area 1000 ft^2. The pool is required to have a walk of uniform width 2 ft surrounding it. Let x be the length of one side of the pool.

(a) Express the area of the plot of land needed for the pool and surrounding sidewalk as a function of x.

(b) Find the dimensions of the plot of land that has the least area. What is the least area?

39. Resistors The total electrical resistance R of two resistors connected in parallel with resistances R_1 and R_2 is given by

$$\frac{1}{R} = \frac{1}{R_1} + \frac{1}{R_2}.$$

One resistor has a resistance of 2.3 ohms. Let x be the resistance of the second resistor.

(a) Express the total resistance R as a function of x.

(b) Find the resistance of the second resistor if the total resistance of the pair is 1.7 ohms.

40. Designing Rectangles Consider all rectangles with an area of 200 m^2. Let x be the length of one side of such a rectangle.

(a) Express the perimeter P as a function of x.

(b) Find the dimensions of a rectangle whose perimeter is 70 m.

41. Swimming Pool Drainage Drains A and B are used to empty a swimming pool. Drain A alone can empty the pool in 4.75 h. Let t be the time it takes for drain B alone to empty the pool.

(a) Express as a function of t the part D of the drainage that can be done in 1 h with both drains open at the same time.

(b) Find graphically the time it takes for drain B alone to empty the pool if both drains, when open at the same time, can empty the pool in 2.6 h. Confirm algebraically.

42. Time-Rate Problem Josh rode his bike 17 mi from his home to Columbus, and then traveled 53 mi by car from Columbus to Dayton. Assume that the average rate of the car was 43 mph faster than the average rate of the bike.

(a) Express the total time required to complete the 70-mi trip (bike and car) as a function of the rate x of the bike.

(b) Find graphically the rate of the bike if the total time of the trip was 1 h 40 min. Confirm algebraically.

43. Late Expectations Table 2.20 shows the average number of remaining years to be lived by U.S. residents surviving to particular ages.

Table 2.20 Expectation for Remaining Life

Age (years)	Remaining Years
70	15.1
80	9.1
90	5.0
100	2.6

Source: National Vital Statistics Reports, Vol. 56, No. 9, December 2007.

(a) Draw a scatter plot of these data together with the model

$$E(a) = \frac{170}{a - 58},$$

where a is a person's age and E is the expected years remaining in the person's life.

(b) Use the model to predict how much longer the average U.S. 74-year-old will live.

44. Fuel Economy The weight of a vehicle is a major factor in its fuel consumption. Table 2.21 lists the weights and gas mileage performances for four 2012 automobiles.

Table 2.21 2012 Automobiles

Weight (pounds)	Fuel Economy (mpg)
2190	30.5
2585	26.5
2910	21.9
3620	18.6

Source: Consumer Reports, 2012.

A model for these data is given by $y = \dfrac{133{,}000}{2x + 1}$.

(a) Graph the model together with a scatter plot of the data.

(b) Use the model to estimate the improvement in fuel economy a manufacturer could achieve by finding a way to reduce the weight of a current 3200-lb car by 100 lb.

Standardized Test Questions

45. True or False An extraneous solution of a rational equation is also a solution of the equation. Justify your answer.

46. True or False The equation $1/(x^2 - 4) = 0$ has no solution. Justify your answer.

In Exercises 47–50, solve the problem without using a calculator.

47. Multiple Choice Which of the following are the solutions of the equation $x - \dfrac{3x}{x + 2} = \dfrac{6}{x + 2}$?

(A) $x = -2$ or $x = 3$

(B) $x = -1$ or $x = 3$

(C) Only $x = -2$

(D) Only $x = 3$

(E) There are no solutions.

48. Multiple Choice Which of the following are the solutions of the equation $1 - \dfrac{3}{x} = \dfrac{6}{x^2 + 2x}$?

(A) $x = -2$ or $x = 4$

(B) $x = -3$ or $x = 0$

(C) $x = -3$ or $x = 4$

(D) Only $x = -3$

(E) There are no solutions.

49. Multiple Choice Which of the following are the solutions of the equation $\dfrac{x}{x + 2} + \dfrac{2}{x - 5} = \dfrac{14}{x^2 - 3x - 10}$?

(A) $x = -5$ or $x = 3$

(B) $x = -2$ or $x = 5$

(C) Only $x = 3$

(D) Only $x = -5$

(E) There are no solutions.

50. Multiple Choice Ten liters of a 20% acid solution are mixed with 30 L of a 30% acid solution. Which of the following is the percent of acid in the final mixture?

(A) 21% **(B)** 22.5% **(C)** 25% **(D)** 27.5% **(E)** 28%

Explorations

51. Revisiting Example 4 Consider the following equation, which we solved in Example 4.

$$\frac{x - 3}{x} + \frac{3}{x + 2} + \frac{6}{x^2 + 2x} = 0$$

Let $f(x) = \dfrac{x - 3}{x} + \dfrac{3}{x + 2} + \dfrac{6}{x^2 + 2x}$.

(a) Combine the fractions in $f(x)$ but do *not* reduce to lowest terms.

(b) What is the domain of f?

(c) Write f as a piecewise-defined function.

(d) Writing to Learn Graph f and explain how the graph supports your answers in (b) and (c).

Extending the Ideas

In Exercises 52–55, solve for x.

52. $y = 1 + \dfrac{1}{1 + x}$

53. $y = 1 - \dfrac{1}{1 - x}$

54. $y = 1 + \dfrac{1}{1 + \dfrac{1}{x}}$

55. $y = 1 + \dfrac{1}{1 + \dfrac{1}{1 - x}}$

2.8 Solving Inequalities in One Variable

What you'll learn about

- Polynomial Inequalities
- Rational Inequalities
- Other Inequalities
- Applications

... and why

Designing containers, as well as other types of applications, often requires that an inequality be solved.

Polynomial Inequalities

A polynomial inequality takes the form $f(x) > 0$, $f(x) \geq 0$, $f(x) < 0$, $f(x) \leq 0$, or $f(x) \neq 0$, where $f(x)$ is a polynomial. There is a fundamental connection between inequalities and the positive or negative sign of the corresponding expression $f(x)$:

- To solve $f(x) > 0$ is to find the values of x that make $f(x)$ *positive.*
- To solve $f(x) < 0$ is to find the values of x that make $f(x)$ *negative.*

If the expression $f(x)$ is a product, we can determine its sign by determining the sign of each of its factors. Example 1 illustrates that a polynomial function changes sign only at its real zeros of odd multiplicity.

EXAMPLE 1 Finding Where a Polynomial Is Zero, Positive, or Negative

Let $f(x) = (x + 3)(x^2 + 1)(x - 4)^2$. Determine the real number values of x that cause $f(x)$ to be **(a)** zero, **(b)** positive, **(c)** negative.

SOLUTION We begin by verbalizing our analysis of the problem:

(a) The real zeros of $f(x)$ are -3 (with multiplicity 1) and 4 (with multiplicity 2). So $f(x)$ is zero if $x = -3$ or $x = 4$.

(b) The factor $x^2 + 1$ is positive for all real numbers x. The factor $(x - 4)^2$ is positive for all real numbers x except $x = 4$, which makes $(x - 4)^2 = 0$. The factor $x + 3$ is positive if and only if $x > -3$. So $f(x)$ is positive if $x > -3$ and $x \neq 4$.

(c) By process of elimination, $f(x)$ is negative if $x < -3$.

This verbal reasoning process is aided by the following **sign chart**, which shows the x-axis as a number line with the real zeros displayed as the locations of potential sign change and the factors displayed with their sign value in the corresponding interval:

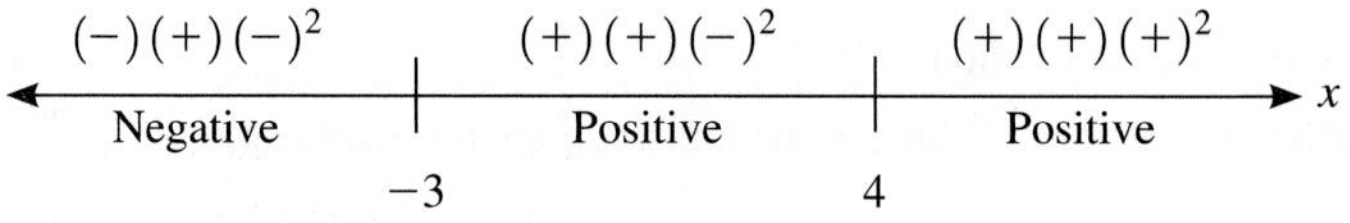

Figure 2.64 supports our findings because the graph of f is above the x-axis for x in $(-3, 4)$ or $(4, \infty)$, is on the x-axis for $x = -3$ or $x = 4$, and is below the x-axis for x in $(-\infty, -3)$. *Now try Exercise 1.*

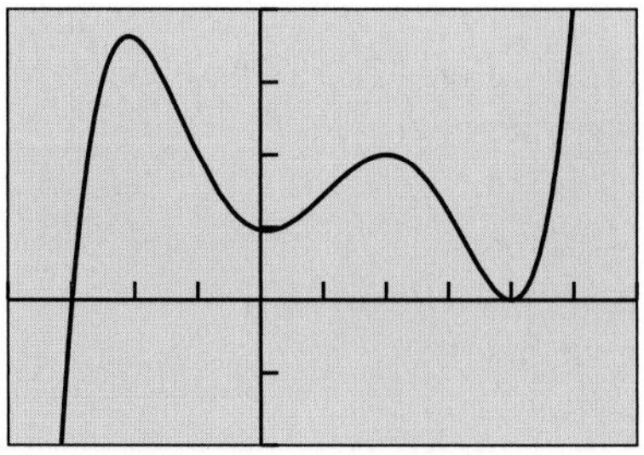

[–4, 6] by [–100, 200]

FIGURE 2.64 The graph of $f(x) = (x + 3)(x^2 + 1)(x - 4)^2$. (Example 1)

Our work in Example 1 allows us to report the solutions of four polynomial inequalities:

- The solution of $(x + 3)(x^2 + 1)(x - 4)^2 > 0$ is $(-3, 4) \cup (4, \infty)$.
- The solution of $(x + 3)(x^2 + 1)(x - 4)^2 \geq 0$ is $[-3, \infty)$.
- The solution of $(x + 3)(x^2 + 1)(x - 4)^2 < 0$ is $(-\infty, -3)$.
- The solution of $(x + 3)(x^2 + 1)(x - 4)^2 \leq 0$ is $(-\infty, -3] \cup \{4\}$.

Example 1 illustrates some important general characteristics of polynomial functions and polynomial inequalities. The polynomial function $f(x) = (x + 3)(x^2 + 1)(x - 4)^2$ in Example 1 and Figure 2.64:

- changes sign at its real zero of odd multiplicity $(x = -3)$;
- touches the x-axis but does not change sign at its real zero of even multiplicity $(x = 4)$;
- has no x-intercepts or sign changes at its nonreal complex zeros associated with the irreducible quadratic factor $(x^2 + 1)$.

This is consistent with what we learned about the relationships between zeros and graphs of polynomial functions in Sections 2.3–2.5. The real zeros and their multiplicity together with the end behavior of a polynomial function give us sufficient information about the polynomial to sketch its graph well enough to obtain a correct sign chart, as shown in Figure 2.65.

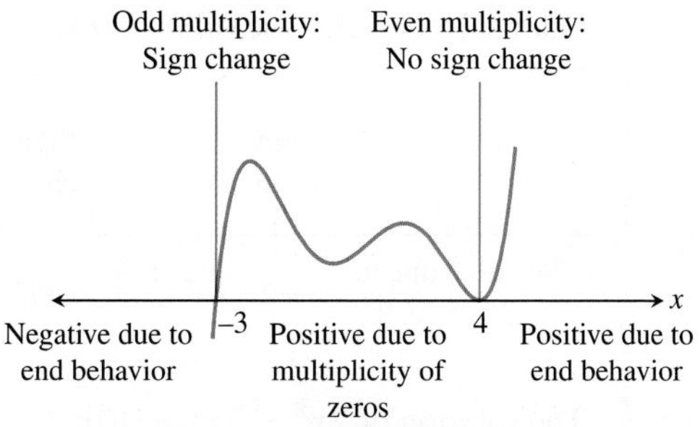

FIGURE 2.65 The sign chart and graph of $f(x) = (x + 3)(x^2 + 1)(x - 4)^2$ overlaid.

EXPLORATION 1 Sketching a Graph of a Polynomial from Its Sign Chart

Use your knowledge of end behavior and multiplicity of real zeros to create a sign chart and sketch the graph of the function. Check your sign chart using the factor method of Example 1. Then check your sketch using a grapher.

1. $f(x) = 2(x - 2)^3(x + 3)^2$
2. $f(x) = -(x + 2)^4(x + 1)(2x^2 + x + 1)$
3. $f(x) = 3(x - 2)^2(x + 4)^3(-x^2 - 2)$

So far in this section all of the polynomials have been presented in factored form and all of the inequalities have had zero on one of the sides. Examples 2 and 3 show us how to solve polynomial inequalities when they are not given in such a convenient form.

Worth Trying

You may wish to make a table or graph for the function f in Example 2 to support the analytical approach used.

EXAMPLE 2 Solving a Polynomial Inequality Analytically

Solve $2x^3 - 7x^2 - 10x + 24 > 0$ analytically.

SOLUTION Let $f(x) = 2x^3 - 7x^2 - 10x + 24$. The Rational Zeros Theorem yields several possible rational zeros of f for factoring purposes:

$$\pm 1, \pm 2, \pm 3, \pm 4, \pm 6, \pm 8, \pm 12, \pm 24, \pm\frac{1}{2}, \pm\frac{3}{2}$$

A table or graph of f can suggest which of these candidates to try. In this case, $x = 4$ is a rational zero of f, as the following synthetic division shows:

$$\begin{array}{r|rrrr} 4 & 2 & -7 & -10 & 24 \\ & & 8 & 4 & -24 \\ \hline & 2 & 1 & -6 & 0 \end{array}$$

(continued)

The synthetic division lets us start the factoring process, which can then be completed using basic factoring methods:

$$\begin{aligned} f(x) &= 2x^3 - 7x^2 - 10x + 24 \\ &= (x - 4)(2x^2 + x - 6) \\ &= (x - 4)(2x - 3)(x + 2) \end{aligned}$$

So the zeros of f are 4, 3/2, and -2. They are all real and all of multiplicity 1, so each will yield a sign change in $f(x)$. Because the degree of f is odd and its leading coefficient is positive, the end behavior of f is given by

$$\lim_{x\to\infty} f(x) = \infty \quad \text{and} \quad \lim_{x\to-\infty} f(x) = -\infty.$$

Our analysis yields the following sign chart:

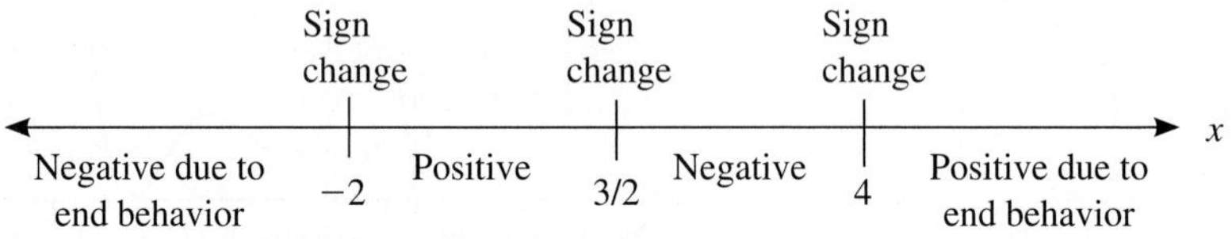

The solution of $2x^3 - 7x^2 - 10x + 24 > 0$ is $(-2, 3/2) \cup (4, \infty)$.

Now try Exercise 11.

EXAMPLE 3 Solving a Polynomial Inequality Graphically

Solve $x^3 - 6x^2 \le 2 - 8x$ graphically.

SOLUTION First we rewrite the inequality as $x^3 - 6x^2 + 8x - 2 \le 0$. Then we let $f(x) = x^3 - 6x^2 + 8x - 2$ and find the real zeros of f graphically as shown in Figure 2.66. The three real zeros are approximately 0.32, 1.46, and 4.21. The solution consists of the x-values for which the graph is on or below the x-axis. So the solution of $x^3 - 6x^2 \le 2 - 8x$ is approximately $(-\infty, 0.32] \cup [1.46, 4.21]$.

The end points of these intervals are accurate to two decimal places. We use square brackets because the zeros of the polynomial are solutions of the inequality even though we only have approximations of their values.

Now try Exercise 13.

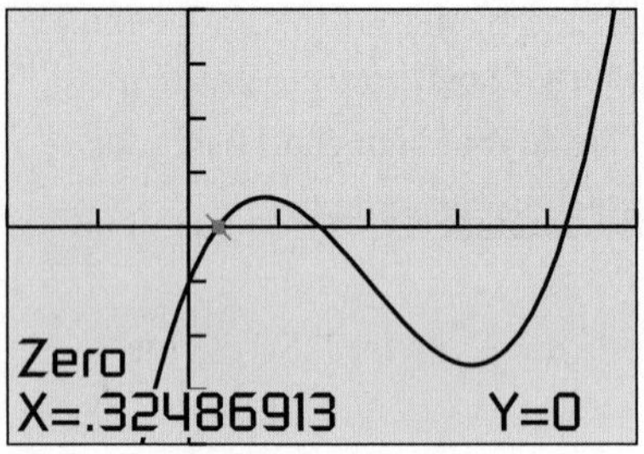

[–2, 5] by [–8, 8]

FIGURE 2.66 The graph of $f(x) = x^3 - 6x^2 + 8x - 2$, with one of three real zeros highlighted. (Example 3)

When a polynomial function has no sign changes, the solutions of the associated inequalities can look a bit unusual, as illustrated in Example 4.

EXAMPLE 4 Solving a Polynomial Inequality with Unusual Answers

(a) The inequalities associated with the strictly positive polynomial function $f(x) = (x^2 + 7)(2x^2 + 1)$ have unusual solution sets. We use Figure 2.67a as a guide to solving the inequalities:

- The solution of $(x^2 + 7)(2x^2 + 1) > 0$ is $(-\infty, \infty)$, all real numbers.
- The solution of $(x^2 + 7)(2x^2 + 1) \ge 0$ is also $(-\infty, \infty)$.
- The solution set of $(x^2 + 7)(2x^2 + 1) < 0$ is empty. We say an inequality of this sort has no solution.
- The solution set of $(x^2 + 7)(2x^2 + 1) \le 0$ is also empty, so the inequality has no solution.

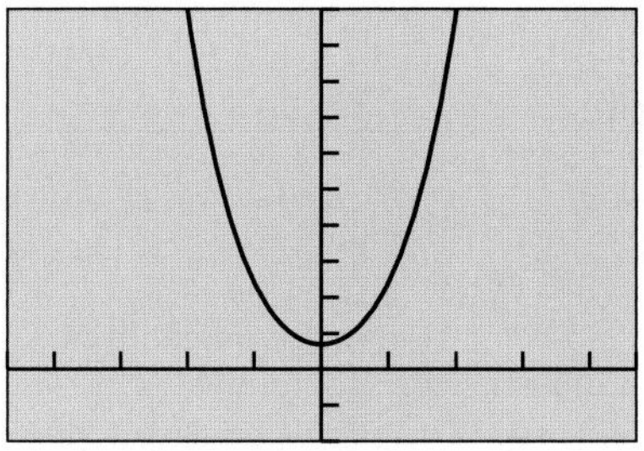

[–4.7, 4.7] by [–20, 100]

(a)

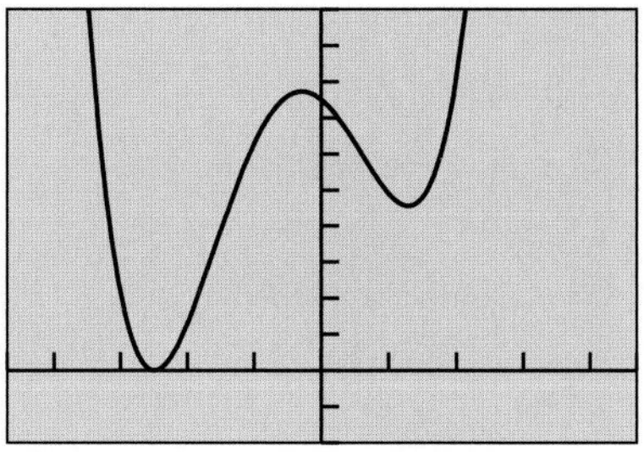

[–4.7, 4.7] by [–20, 100]

(b)

FIGURE 2.67 The graphs of (a) $f(x) = (x^2 + 7)(2x^2 + 1)$ and (b) $g(x) = (x^2 - 3x + 3)(2x + 5)^2$. (Example 4)

(b) The inequalities associated with the nonnegative polynomial function $g(x) = (x^2 - 3x + 3)(2x + 5)^2$ also have unusual solution sets. We use Figure 2.67b as a guide to solving the inequalities:

- The solution of $(x^2 - 3x + 3)(2x + 5)^2 > 0$ is $(-\infty, -5/2) \cup (-5/2, \infty)$, all real numbers except $x = -5/2$, the lone real zero of g.
- The solution of $(x^2 - 3x + 3)(2x + 5)^2 \geq 0$ is $(-\infty, \infty)$, all real numbers.
- The inequality $(x^2 - 3x + 3)(2x + 5)^2 < 0$ has no solution.
- The solution of $(x^2 - 3x + 3)(2x + 5)^2 \leq 0$ is the single number $x = -5/2$.

Now try Exercise 21.

Rational Inequalities

A polynomial function $p(x)$ is positive, negative, or zero for all real numbers x, but a rational function $r(x)$ can be positive, negative, zero, or *undefined*. In particular, a rational function is undefined at the zeros of its denominator. Thus when solving rational inequalities we modify the kind of sign chart used in Example 1 to include the real zeros of both the numerator and the denominator as locations of potential sign change.

EXAMPLE 5 Creating a Sign Chart for a Rational Function

Let $r(x) = (2x + 1)/((x + 3)(x - 1))$. Determine the real number values of x that cause $r(x)$ to be **(a)** zero, **(b)** undefined. Then make a sign chart to determine the real number values of x that cause $r(x)$ to be **(c)** positive, **(d)** negative.

SOLUTION

(a) The real zeros of $r(x)$ are the real zeros of the numerator $2x + 1$. So $r(x)$ is zero if $x = -1/2$.

(b) $r(x)$ is undefined when the denominator $(x + 3)(x - 1) = 0$. So $r(x)$ is undefined if $x = -3$ or $x = 1$.

These findings lead to the following sign chart, with three points of potential sign change:

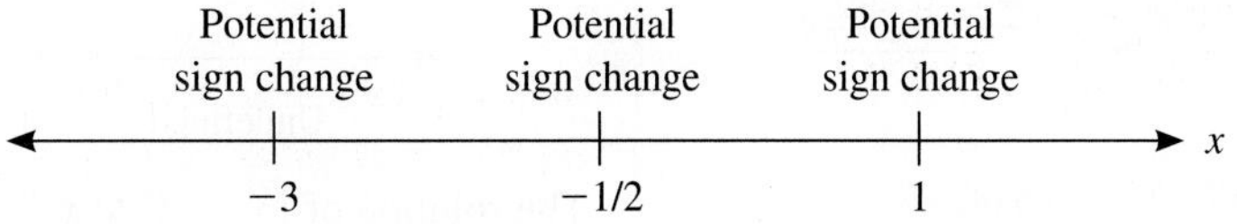

Analyzing the factors of the numerator and denominator yields:

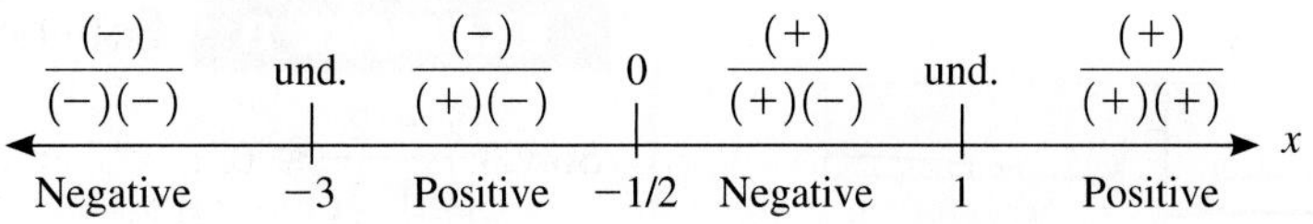

(c) So $r(x)$ is positive if $-3 < x < -1/2$ or $x > 1$, and the solution of $(2x + 1)/((x + 3)(x - 1)) > 0$ is $(-3, -1/2) \cup (1, \infty)$.

(d) Similarly, $r(x)$ is negative if $x < -3$ or $-1/2 < x < 1$, and the solution of $(2x + 1)/((x + 3)(x - 1)) < 0$ is $(-\infty, -3) \cup (-1/2, 1)$.

Figure 2.68 supports our findings because the graph of r is above the x-axis for x in $(-3, -1/2) \cup (1, \infty)$ and is below the x-axis for x in $(-\infty, -3) \cup (-1/2, 1)$.

Now try Exercise 25.

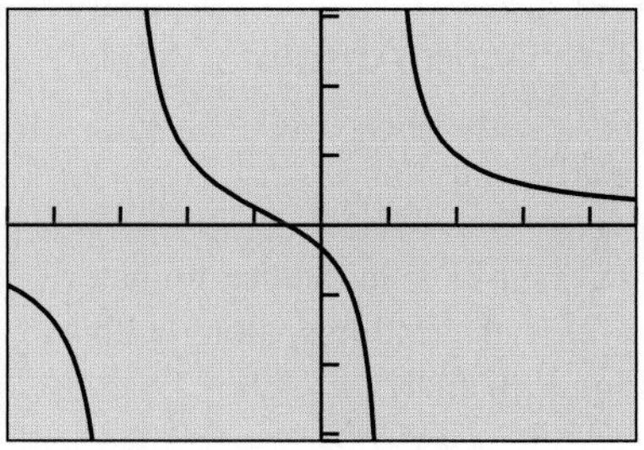

[–4.7, 4.7] by [–3.1, 3.1]

FIGURE 2.68 The graph of $r(x) = (2x + 1)/((x + 3)(x - 1))$. (Example 5)

EXAMPLE 6 Solving a Rational Inequality by Combining Fractions

Solve $\dfrac{5}{x+3} + \dfrac{3}{x-1} < 0.$

SOLUTION We combine the two fractions on the left-hand side of the inequality using the least common denominator $(x + 3)(x - 1)$:

$$\frac{5}{x+3} + \frac{3}{x-1} < 0 \qquad \text{Original inequality}$$

$$\frac{5(x-1)}{(x+3)(x-1)} + \frac{3(x+3)}{(x+3)(x-1)} < 0 \qquad \text{Use LCD to rewrite fractions.}$$

$$\frac{5(x-1) + 3(x+3)}{(x+3)(x-1)} < 0 \qquad \text{Add fractions.}$$

$$\frac{5x - 5 + 3x + 9}{(x+3)(x-1)} < 0 \qquad \text{Distributive property}$$

$$\frac{8x + 4}{(x+3)(x-1)} < 0 \qquad \text{Simplify.}$$

$$\frac{2x + 1}{(x+3)(x-1)} < 0 \qquad \text{Divide both sides by 4.}$$

This inequality matches Example 5d. The solution is $(-\infty, -3) \cup (-1/2, 1)$.

Now try Exercise 49.

Other Inequalities

The sign chart method can be adapted to solve other types of inequalities, and we can support our solutions graphically as needed or desired.

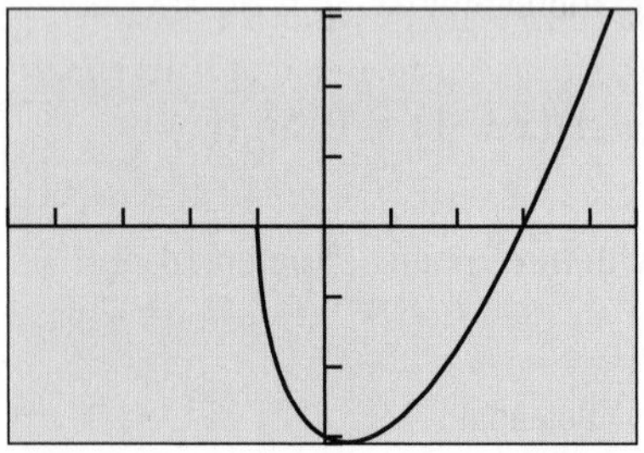

[-4.7, 4.7] by [-3.1, 3.1]

FIGURE 2.69 The graph of $f(x) = (x - 3)\sqrt{x + 1}$. (Example 7)

EXAMPLE 7 Solving an Inequality Involving a Radical

Solve $(x - 3)\sqrt{x + 1} \geq 0$.

SOLUTION Let $f(x) = (x - 3)\sqrt{x + 1}$. Because of the factor $\sqrt{x + 1}$, $f(x)$ is undefined if $x < -1$. The zeros of f are 3 and -1. These findings, along with a sign analysis of the two factors, lead to the following sign chart:

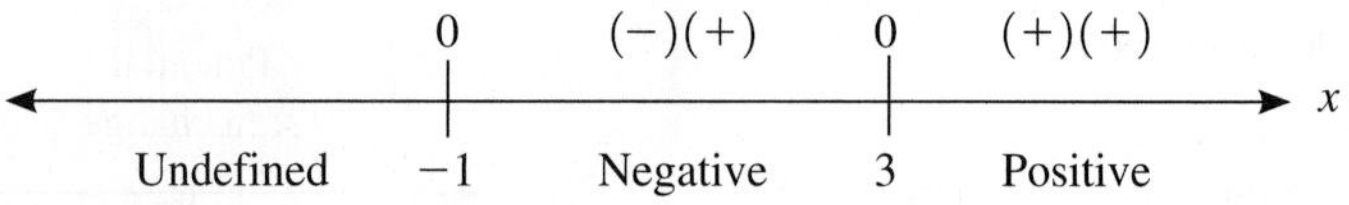

The solution of $(x - 3)\sqrt{x + 1} \geq 0$ is $\{-1\} \cup [3, \infty)$. The graph of f in Figure 2.69 supports this solution.

Now try Exercise 43.

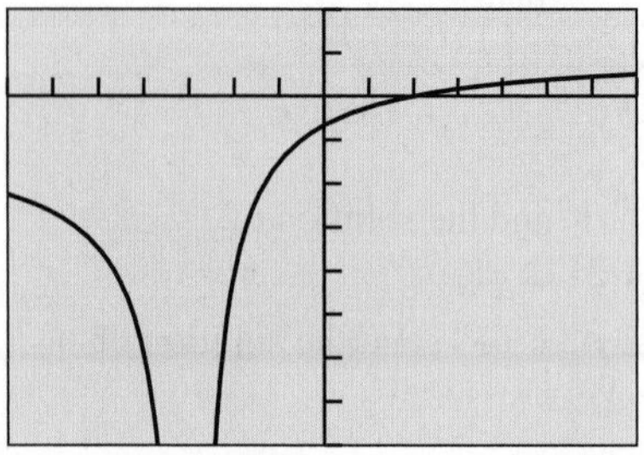

[-7, 7] by [-8, 2]

FIGURE 2.70 The graph of $f(x) = (x - 3)/|x + 3|$. (Example 8)

EXAMPLE 8 Solving an Inequality Involving Absolute Value

Solve $\dfrac{x - 2}{|x + 3|} \leq 0.$

SOLUTION Let $f(x) = (x - 2)/|x + 3|$. Because $|x + 3|$ is in the denominator, $f(x)$ is undefined if $x = -3$. The only zero of f is 2. These findings, along with a sign analysis of the two factors, lead to the following sign chart:

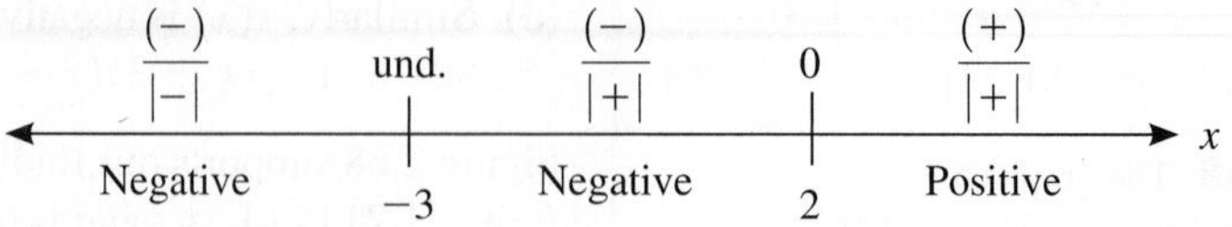

The solution of $(x - 2)/|x + 3| \leq 0$ is $(-\infty, -3) \cup (-3, 2]$. The graph of f in Figure 2.70 supports this solution.

Now try Exercise 53.

Applications

EXAMPLE 9 Designing a Box—Revisited

Dixie Packaging Company has contracted with another firm to design boxes with a volume of *at least* 600 in.^{3}. Squares are still to be cut from the corners of a 20-in. by 25-in. piece of cardboard, with the flaps folded up to make an open box. What size squares should be cut from the cardboard? (See Example 9 of Section 2.3 and Figure 2.31.)

SOLUTION

Model Recall that the volume V of the box is given by

$$V(x) = x(25 - 2x)(20 - 2x),$$

where x represents both the side length of the removed squares and the height of the box. To obtain a volume of at least 600 in.^{3}, we solve the inequality

$$x(25 - 2x)(20 - 2x) \geq 600.$$

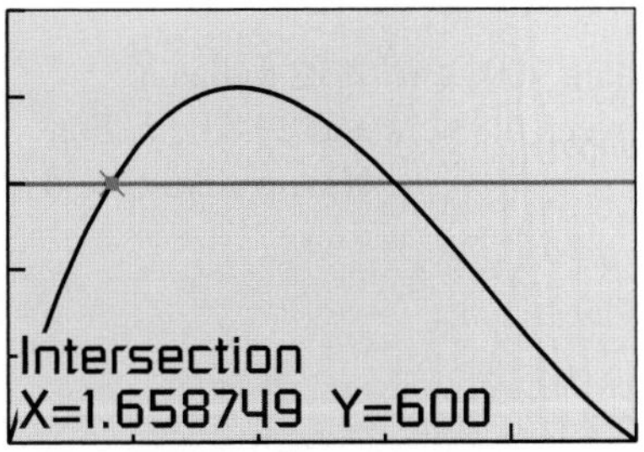

[0, 10] by [0, 1000]

FIGURE 2.71 The graphs of $y_1 = x(25 - 2x)(20 - 2x)$ and $y_2 = 600$. (Example 9)

Solve Graphically Because the width of the cardboard is 20 in., $0 \leq x \leq 10$, and we set our window accordingly. In Figure 2.71, we find the values of x for which the cubic function is on or above the horizontal line. The solution is the interval $[1.66, 6.16]$.

Interpret Squares with side lengths between 1.66 in. and 6.16 in., inclusive, should be cut from the cardboard to produce a box with a volume of at least 600 in.^{3}.

Now try Exercise 59.

EXAMPLE 10 Designing a Juice Can—Revisited

Stewart Cannery will package tomato juice in 2-L (2000 cm^3) cylindrical cans. Find the radius and height of the cans if the cans have a surface area that is less than 1000 cm^2. (See Example 7 of Section 2.7 and Figure 2.63.)

SOLUTION

Model Recall that the surface area S is given by

$$S(r) = 2\pi r^2 + \frac{4000}{r}.$$

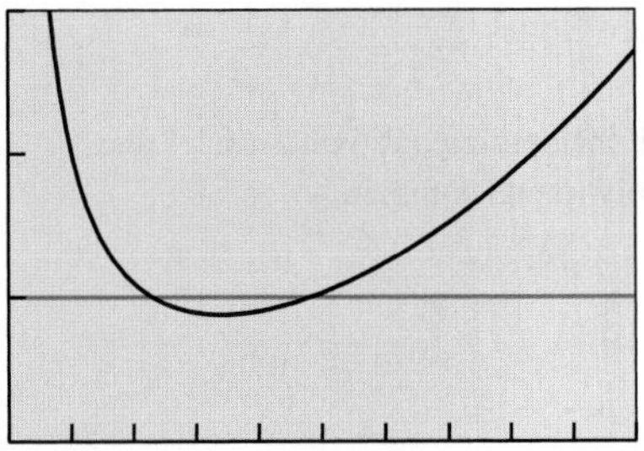

[0, 20] by [0, 3000]

FIGURE 2.72 The graphs of $y_1 = 2\pi x^2 + 4000/x$ and $y_2 = 1000$. (Example 10)

The inequality to be solved is

$$2\pi r^2 + \frac{4000}{r} < 1000.$$

Solve Graphically Figure 2.72 shows the graphs of $y_1 = S(r) = 2\pi r^2 + 4000/r$ and $y_2 = 1000$. Using grapher methods we find that the two curves intersect at approximately $r \approx 4.619\ldots$ and $r \approx 9.654\ldots$. (We carry all the extra decimal places for greater accuracy in a computation below.) So the surface area is less than 1000 cm^3 if

$$4.62 < r < 9.65.$$

The volume of a cylindrical can is $V = \pi r^2 h$ and $V = 2000$. Using substitution we see that $h = 2000/(\pi r^2)$. To find the values for h we build a double inequality for $2000/(\pi r^2)$.

(continued)

$4.62 < r < 9.65$	Original inequality
$4.62^2 < r^2 < 9.65^2$	$0 < a < b \Rightarrow a^2 < b^2$.
$\pi \cdot 4.62^2 < \pi r^2 < \pi \cdot 9.65^2$	Multiply by π.
$\frac{1}{\pi \cdot 4.62^2} > \frac{1}{\pi r^2} > \frac{1}{\pi \cdot 9.65^2}$	$0 < a < b \Rightarrow \frac{1}{a} > \frac{1}{b}$.
$\frac{2000}{\pi \cdot 4.62^2} > \frac{2000}{\pi r^2} > \frac{2000}{\pi \cdot 9.65^2}$	Multiply by 2000.
$\frac{2000}{\pi(4.619\ldots)^2} > h > \frac{2000}{\pi(9.654\ldots)^2}$	Use the extra decimal places now.
$29.83 > h > 6.83$	Compute.

Interpret The surface area of the can will be less than 1000 cm^3 if its radius is between 4.62 cm and 9.65 cm and its height is between 6.83 cm and 29.83 cm. For any particular can, h must equal $2000/(\pi r^2)$. **Now try Exercise 61.**

QUICK REVIEW 2.8 *(For help, go to Sections A.2, A.3, and 2.3.)*

Exercise numbers with a gray background indicate problems that the authors have designed to be solved *without a calculator.*

In Exercises 1–4, use limits to state the end behavior of the function.

1. $f(x) = 2x^3 + 3x^2 - 2x + 1$
2. $f(x) = -3x^4 - 3x^3 + x^2 - 1$
3. $g(x) = \dfrac{x^3 - 2x^2 + 1}{x - 2}$
4. $g(x) = \dfrac{2x^2 - 3x + 1}{x + 1}$

In Exercises 5–8, combine the fractions and reduce your answer to lowest terms.

5. $x^2 + \dfrac{5}{x}$
6. $x^2 - \dfrac{3}{x}$
7. $\dfrac{x}{2x + 1} - \dfrac{2}{x - 3}$
8. $\dfrac{x}{x - 1} + \dfrac{x + 1}{3x - 4}$

In Exercises 9 and 10, **(a)** list all the possible rational zeros of the polynomial and **(b)** factor the polynomial completely.

9. $2x^3 + x^2 - 4x - 3$
10. $3x^3 - x^2 - 10x + 8$

SECTION 2.8 Exercises

In Exercises 1–6, determine the x values that cause the polynomial function to be **(a)** zero, **(b)** positive, and **(c)** negative.

1. $f(x) = (x + 2)(x + 1)(x - 5)$
2. $f(x) = (x - 7)(3x + 1)(x + 4)$
3. $f(x) = (x + 7)(x + 4)(x - 6)^2$
4. $f(x) = (5x + 3)(x^2 + 6)(x - 1)$
5. $f(x) = (2x^2 + 5)(x - 8)^2(x + 1)^3$
6. $f(x) = (x + 2)^3(4x^2 + 1)(x - 9)^4$

In Exercises 7–12, complete the factoring if needed, and solve the polynomial inequality using a sign chart. Support graphically.

7. $(x + 1)(x - 3)^2 > 0$
8. $(2x + 1)(x - 2)(3x - 4) \le 0$
9. $(x + 1)(x^2 - 3x + 2) < 0$
10. $(2x - 7)(x^2 - 4x + 4) > 0$
11. $2x^3 - 3x^2 - 11x + 6 \ge 0$
12. $x^3 - 4x^2 + x + 6 \le 0$

In Exercises 13–20, solve the polynomial inequality graphically.

13. $x^3 - x^2 - 2x \ge 0$
14. $2x^3 - 5x^2 + 3x < 0$
15. $2x^3 - 5x^2 - x + 6 > 0$
16. $x^3 - 4x^2 - x + 4 \le 0$
17. $3x^3 - 2x^2 - x + 6 \ge 0$
18. $-x^3 - 3x^2 - 9x + 4 < 0$
19. $2x^4 - 3x^3 - 6x^2 + 5x + 6 < 0$
20. $3x^4 - 5x^3 - 12x^2 + 12x + 16 \ge 0$

In Exercises 21–24, solve the following inequalities for the given function $f(x)$.

(a) $f(x) > 0$ **(b)** $f(x) \geq 0$ **(c)** $f(x) < 0$ **(d)** $f(x) \leq 0$

21. $f(x) = (x^2 + 4)(2x^2 + 3)$

22. $f(x) = (x^2 + 1)(-2 - 3x^2)$

23. $f(x) = (2x^2 - 2x + 5)(3x - 4)^2$

24. $f(x) = (x^2 + 4)(3 - 2x)^2$

In Exercises 25–32, determine the real values of x that cause the function to be **(a)** zero, **(b)** undefined, **(c)** positive, and **(d)** negative.

25. $f(x) = \dfrac{x - 1}{((2x + 3)(x - 4))}$

26. $f(x) = \dfrac{(2x - 7)(x + 1)}{(x + 5)}$

27. $f(x) = x\sqrt{x + 3}$

28. $f(x) = x^2|2x + 9|$

29. $f(x) = \dfrac{\sqrt{x + 5}}{(2x + 1)(x - 1)}$

30. $f(x) = \dfrac{x - 1}{(x - 4)\sqrt{x + 2}}$

31. $f(x) = \dfrac{(2x + 5)\sqrt{x - 3}}{(x - 4)^2}$

32. $f(x) = \dfrac{3x - 1}{(x + 3)\sqrt{x - 5}}$

In Exercises 33–44, solve the inequality using a sign chart. Support graphically.

33. $\dfrac{x - 1}{x^2 - 4} < 0$

34. $\dfrac{x + 2}{x^2 - 9} < 0$

35. $\dfrac{x^2 - 1}{x^2 + 1} \leq 0$

36. $\dfrac{x^2 - 4}{x^2 + 4} > 0$

37. $\dfrac{x^2 + x - 12}{x^2 - 4x + 4} > 0$

38. $\dfrac{x^2 + 3x - 10}{x^2 - 6x + 9} < 0$

39. $\dfrac{x^3 - x}{x^2 + 1} \geq 0$

40. $\dfrac{x^3 - 4x}{x^2 + 2} \leq 0$

41. $x|x - 2| > 0$

42. $\dfrac{x - 3}{|x + 2|} < 0$

43. $(2x - 1)\sqrt{x + 4} < 0$

44. $(3x - 4)\sqrt{2x + 1} \geq 0$

In Exercises 45–54, solve the inequality.

45. $\dfrac{x^3(x - 2)}{(x + 3)^2} < 0$

46. $\dfrac{(x - 5)^4}{x(x + 3)} \geq 0$

47. $x^2 - \dfrac{2}{x} > 0$

48. $x^2 + \dfrac{4}{x} \geq 0$

49. $\dfrac{1}{x + 1} + \dfrac{1}{x - 3} \leq 0$

50. $\dfrac{1}{x + 2} - \dfrac{2}{x - 1} > 0$

51. $(x + 3)|x - 1| \geq 0$

52. $(3x + 5)^2|x - 2| < 0$

53. $\dfrac{(x - 5)|x - 2|}{\sqrt{2x - 3}} \geq 0$

54. $\dfrac{x^2(x - 4)^3}{\sqrt{x + 1}} < 0$

55. Writing to Learn Write a paragraph that explains two ways to solve the inequality $3(x - 1) + 2 \leq 5x + 6$.

56. Company Wages Pederson Electric charges \$25 per service call plus \$18/hr for repair work. How long did an electrician work if the charge was less than \$100? Assume the electrician rounds the time to the nearest quarter hour.

57. Connecting Algebra and Geometry Consider the collection of all rectangles that have lengths 2 in. less than twice their widths. Find the possible widths (in inches) of these rectangles if their perimeters are less than 200 in.

58. Planning for Profit The Grovenor Candy Co. finds that the cost of making a certain candy bar is \$0.13 per bar. Fixed costs amount to \$2000 per week. If each bar sells for \$0.35, find the minimum number of candy bars that will earn the company a profit.

59. Designing a Cardboard Box Picaro's Packaging Plant wishes to design boxes with a volume of *not more than* 100 in.^{3}. Squares are to be cut from the corners of a 12-in. by 15-in. piece of cardboard (see figure), with the flaps folded up to make an open box. What size squares should be cut from the cardboard?

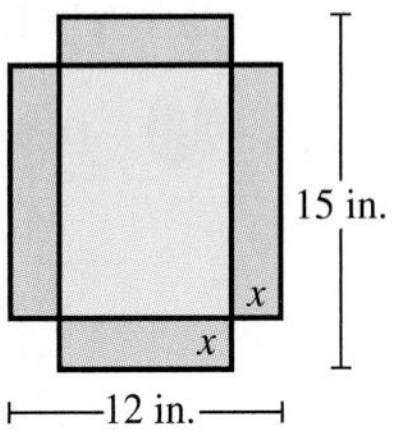

60. Cone Problem Beginning with a circular piece of paper with a 4-in. radius, as shown in (a), cut out a sector with an arc of length x. Join the two radial edges of the remaining portion of the paper to form a cone with radius r and height h, as shown in (b). What length of arc will produce a cone with a volume greater than 21 in.^{3}?

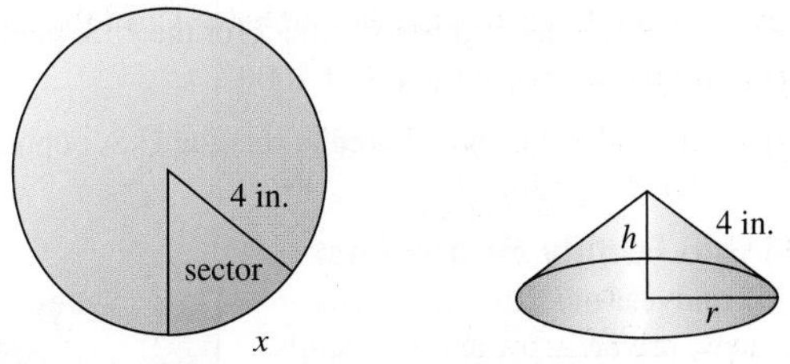

(a) (b)

61. Design a Juice Can Flannery Cannery packs peaches in 0.5-L cylindrical cans.

(a) Express the surface area S of the can as a function of the radius x (in cm).

(b) Find the dimensions of the can if the surface is less than 900 cm^2.

(c) Find the least possible surface area of the can.

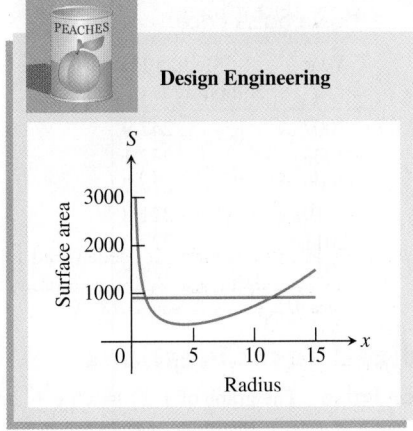

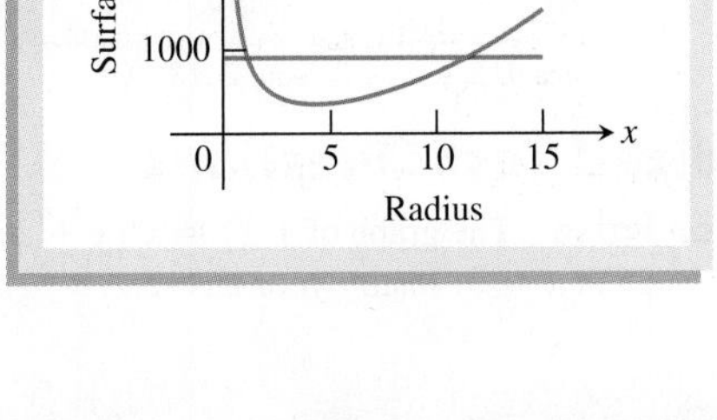

62. Resistors The total electrical resistance R of two resistors connected in parallel with resistances R_1 and R_2 is given by

$$\frac{1}{R} = \frac{1}{R_1} + \frac{1}{R_2}.$$

One resistor has a resistance of 2.3 ohms. Let x be the resistance of the second resistor.

(a) Express the total resistance R as a function of x.

(b) Find the resistance in the second resistor if the total resistance of the pair is at least 1.7 ohms.

63. The Growing of America Table 2.22 shows the midyear (July 1) U.S. population estimates in millions of persons for the years 2004 through 2011. Let x be the number of years since July 1, 2000.

Table 2.22 U.S. Population

Year	Population (in millions)
2004	292.8
2005	295.5
2006	298.4
2007	301.2
2008	304.1
2009	306.8
2010	309.3
2011	311.6

Source: U.S. Census Bureau, 2012.

(a) Find the linear regression model for the U.S. population in millions since the middle of 2000.

(b) When does this model predict that the U.S. population will reach 320 million?

64. Single-Family House Cost The midyear median sales prices of new, privately owned one-family houses sold in the United States are given for selected years in Table 2.23. Let x be the number of years since July 1, 2000.

(a) Find the quadratic regression model for the data.

(b) When does this model predict that the median price for a new home returned to its July 1, 2007, level?

Table 2.23 Midyear New Home Sales Price

Year	Median Price (in thousands of dollars)
2007	247.9
2008	232.1
2009	216.7
2010	221.8
2011	227.2
2012	239.8

Source: U.S. Census Bureau, 2013.

Standardized Test Questions

65. True or False The graph of $f(x) = x^4(x + 3)^2(x - 1)^3$ changes sign at $x = 0$. Justify your answer.

66. True or False The graph $r(x) = \dfrac{2x - 1}{(x + 2)(x - 1)}$ changes sign at $x = -2$. Justify your answer.

In Exercises 67–70, solve the problem without using a calculator.

67. Multiple Choice Which of the following is the solution to $x^2 < x$?

(A) $(0, \infty)$ **(B)** $(1, \infty)$ **(C)** $(0, 1)$
(D) $(-\infty, 1)$ **(E)** $(-\infty, 0) \cup (1, \infty)$

68. Multiple Choice Which of the following is the solution to $\dfrac{1}{(x + 2)^2} \geq 0$?

(A) $(-2, \infty)$ **(B)** All $x \neq -2$ **(C)** All $x \neq 2$
(D) All real numbers **(E)** There are no solutions.

69. Multiple Choice Which of the following is the solution to $\dfrac{x^2}{x - 3} < 0$?

(A) $(-\infty, 3)$ **(B)** $(-\infty, 3]$ **(C)** $(-\infty, 0] \cup (0, 3)$
(D) $(-\infty, 0) \cup (0, 3)$ **(E)** There are no solutions.

70. Multiple Choice Which of the following is the solution to $(x^2 - 1)^2 \leq 0$?

(A) $\{-1, 1\}$ **(B)** $\{1\}$ **(C)** $[-1, 1]$
(D) $[0, 1]$ **(E)** There are no solutions.

Explorations

In Exercises 71 and 72, find the vertical asymptotes and intercepts of the rational function. Then use a sign chart and a table of values to sketch the function by hand. Support your result using a grapher. (*Hint*: You may need to graph the function in more than one window to see different parts of the overall graph.)

71. $f(x) = \dfrac{(x - 1)(x + 2)^2}{(x - 3)(x + 1)}$ **72.** $g(x) = \dfrac{(x - 3)^4}{x^2 + 4x}$

Extending the Ideas

73. Group Activity Looking Ahead to Calculus Let $f(x) = 3x - 5$.

(a) Assume x is in the interval defined by $|x - 3| < 1/3$. Give a convincing argument that $|f(x) - 4| < 1$.

(b) Writing to Learn Explain how (a) is modeled by the figure below.

(c) Show how the algebra used in (a) can be modified to show that if $|x - 3| < 0.01$, then $|f(x) - 4| < 0.03$. How would the figure below change to reflect these inequalities?

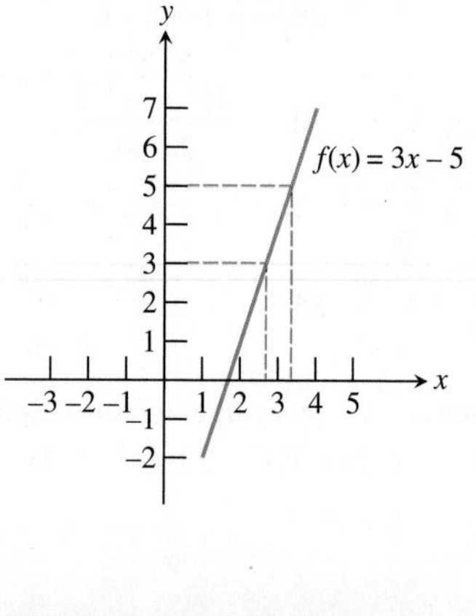

74. Writing to Learn Boolean Operators The Test menu of many graphers contains inequality symbols that can be used to construct inequality statements, as shown in (a). An answer of 1 indicates the statement is true, and 0 indicates the statement is false. In (b), the graph of $Y_1 = (x^2 - 4 \geq 0)$ is shown using Dot mode and the window $[-4.7, 4.7]$ by $[-3.1, 3.1]$. Experiment with the Test menu, and then write a paragraph explaining how to interpret the graph in (b).

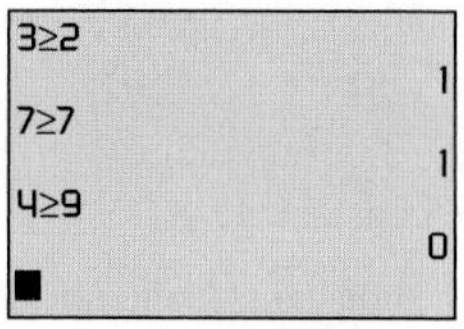

(a)

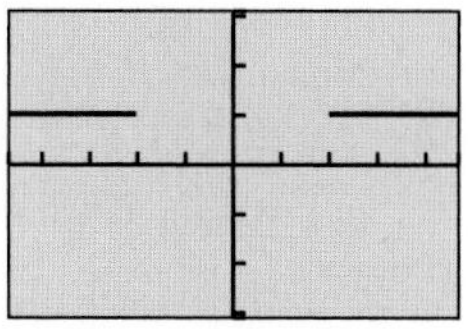

[–4.7, 4.7] by [–3.1, 3.1]

(b)

In Exercises 75 and 76, use the properties of inequality from Chapter P to prove the statement.

75. If $0 < a < b$, then $a^2 < b^2$.

76. If $0 < a < b$, then $\frac{1}{a} > \frac{1}{b}$.

CHAPTER 2 Key Ideas

Properties, Theorems, and Formulas

- Constant Rate of Change Theorem 160
- Properties of the Correlation Coefficient, *r* 162
- Vertex Form of a Quadratic Function 165
- Vertical Free-Fall Motion 167
- Local Extrema and Zeros of Polynomial Functions 187
- Leading Term Test for Polynomial End Behavior 188
- Zeros of Odd and Even Multiplicity 190
- Intermediate Value Theorem 190
- Division Algorithm for Polynomials 197
- Remainder Theorem 198
- Factor Theorem 199
- Fundamental Connections for Polynomial Functions 199
- Rational Zeros Theorem 201
- Upper and Lower Bound Test for Real Zeros 202
- Fundamental Theorem of Algebra 210
- Linear Factorization Theorem 210
- Fundamental Polynomial Connections in the Complex Case 211
- Complex Conjugate Zeros Theorem 211
- Factors of a Polynomial with Real Coefficients 214
- Polynomial Function of Odd Degree 214
- Graph of a Rational Function 221

Procedures

- Regression Analysis 163
- Synthetic Division 200–201
- Solving Inequalities Using Sign Charts 236–240

Gallery of Functions

Identity

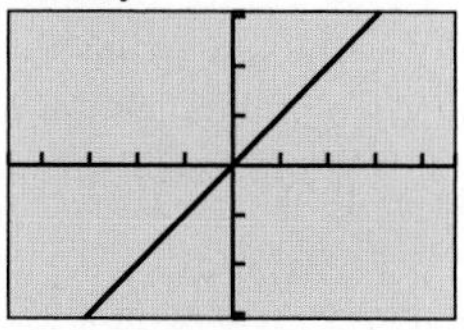

[–4.7, 4.7] by [–3.1, 3.1]

$f(x) = x$

Squaring

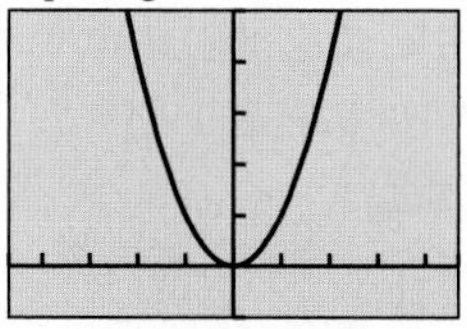

[–4.7, 4.7] by [–1, 5]

$f(x) = x^2$

Cubing

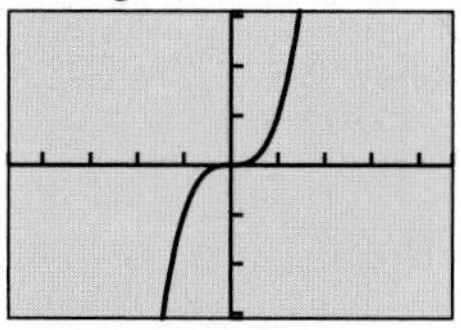

[–4.7, 4.7] by [–3.1, 3.1]

$f(x) = x^3$

Reciprocal

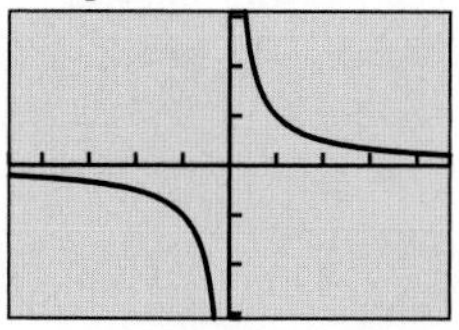

[–4.7, 4.7] by [–3.1, 3.1]

$f(x) = 1/x = x^{-1}$

Square Root

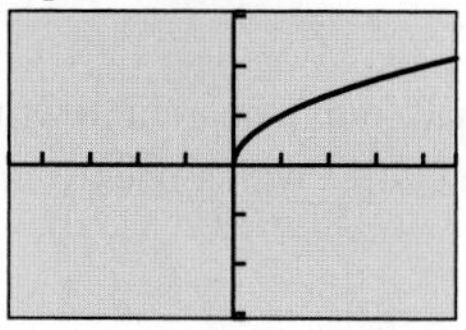

[–4.7, 4.7] by [–3.1, 3.1]

$f(x) = \sqrt{x} = x^{1/2}$

Cube Root

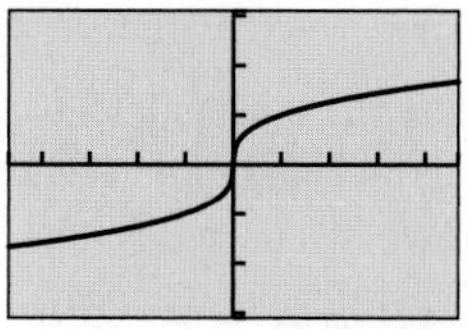

[–4.7, 4.7] by [–3.1, 3.1]

$f(x) = \sqrt[3]{x} = x^{1/3}$

Inverse-Square

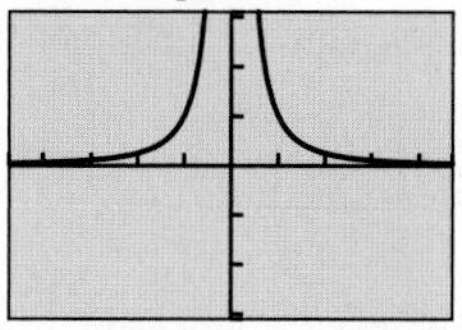

[–4.7, 4.7] by [–3.1, 3.1]

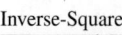

$f(x) = 1/x^2 = x^{-2}$

CHAPTER 2 Review Exercises

Exercise numbers with a gray background indicate problems that the authors have designed to be solved *without a calculator*.

The collection of exercises marked in red could be used as a chapter test.

In Exercises 1 and 2, write an equation for the linear function f satisfying the given conditions. Graph $y = f(x)$.

1. $f(-3) = -2$ and $f(4) = -9$
2. $f(-3) = 6$ and $f(1) = -2$

In Exercises 3 and 4, describe how to transform the graph of $f(x) = x^2$ into the graph of the given function. Sketch the graph by hand and support your answer with a grapher.

3. $h(x) = 3(x - 2)^2 + 4$
4. $g(x) = -(x + 3)^2 + 1$

In Exercises 5–8, find the vertex and axis of the graph of the function. Support your answer graphically.

5. $f(x) = -2(x + 3)^2 + 5$
6. $g(x) = 4(x - 5)^2 - 7$
7. $f(x) = -2x^2 - 16x - 31$
8. $g(x) = 3x^2 - 6x + 2$

In Exercises 9 and 10, write an equation for the quadratic function whose graph contains the given vertex and point.

9. Vertex $(-2, -3)$, point $(1, 2)$
10. Vertex $(-1, 1)$, point $(3, -2)$

In Exercises 11 and 12, write an equation for the quadratic function with graph shown, given one of the labeled points is the vertex of the parabola.

11.

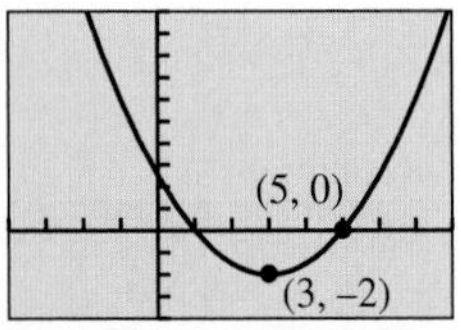

[–4, 8] by [–4, 10]

12.

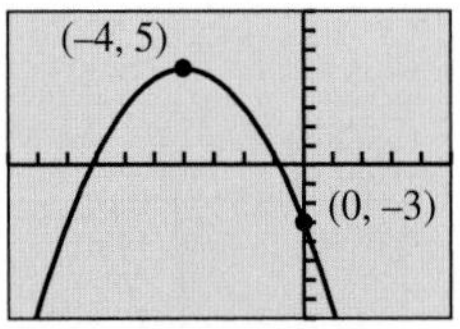

[–10, 5] by [–8, 8]

In Exercises 13–16, graph the function in a viewing window that shows all of its extrema and x-intercepts.

13. $f(x) = x^2 + 3x - 40$
14. $f(x) = -8x^2 + 16x - 19$
15. $f(x) = x^3 + x^2 + x + 5$
16. $f(x) = x^3 - x^2 - 20x - 2$

In Exercises 17 and 18, write the statement as a power function equation. Let k be the constant of variation.

17. The surface area S of a sphere varies directly as the square of the radius r.
18. The force of gravity F acting on an object is inversely proportional to the square of the distance d from the object to the center of the earth.

In Exercises 19 and 20, write a sentence that expresses the relationship in the formula, using the language of variation or proportion.

19. $F = kx$, where F is the force it takes to stretch a spring x units from its unstressed length and k is the spring's force constant.
20. $A = \pi \cdot r^2$, where A and r are the area and radius of a circle and π is the usual mathematical constant.

In Exercises 21–24, state the values of the constants k and a for the function $f(x) = k \cdot x^a$. Describe the portion of the curve that lies in Quadrant I or IV. Determine whether f is even, odd, or undefined for $x < 0$. Describe the rest of the curve if any. Graph the function to see whether it matches the description.

21. $f(x) = 4x^{1/3}$
22. $f(x) = -2x^{3/4}$
23. $f(x) = -2x^{-3}$
24. $f(x) = (2/3)x^{-4}$

In Exercises 25–28, divide $f(x)$ by $d(x)$, and write a summary statement in polynomial form.

25. $f(x) = 2x^3 - 7x^2 + 4x - 5$; $d(x) = x - 3$
26. $f(x) = x^4 + 3x^3 + x^2 - 3x + 3$; $d(x) = x + 2$
27. $f(x) = 2x^4 - 3x^3 + 9x^2 - 14x + 7$; $d(x) = x^2 + 4$
28. $f(x) = 3x^4 - 5x^3 - 2x^2 + 3x - 6$; $d(x) = 3x + 1$

In Exercises 29 and 30, use the Remainder Theorem to find the remainder when $f(x)$ is divided by $x - k$. Check by using synthetic division.

29. $f(x) = 3x^3 - 2x^2 + x - 5$; $k = -2$
30. $f(x) = -x^2 + 4x - 5$; $k = 3$

In Exercises 31 and 32, use the Factor Theorem to determine whether the first polynomial is a factor of the second polynomial.

31. $x - 2$; $x^3 - 4x^2 + 8x - 8$
32. $x + 3$; $x^3 + 2x^2 - 4x - 2$

In Exercises 33 and 34, use synthetic division to prove that the number k is an upper bound for the real zeros of the function f.

33. $k = 5$; $f(x) = x^3 - 5x^2 + 3x + 4$
34. $k = 4$; $f(x) = 4x^4 - 16x^3 + 8x^2 + 16x - 12$

In Exercises 35 and 36, use synthetic division to prove that the number k is a lower bound for the real zeros of the function f.

35. $k = -3$; $f(x) = 4x^4 + 4x^3 - 15x^2 - 17x - 2$
36. $k = -3$; $f(x) = 2x^3 + 6x^2 + x - 6$

In Exercises 37 and 38, use the Rational Zeros Theorem to write a list of all potential rational zeros. Then determine which ones, if any, are zeros.

37. $f(x) = 2x^4 - x^3 - 4x^2 - x - 6$
38. $f(x) = 6x^3 - 20x^2 + 11x + 7$

In Exercises 39–42, perform the indicated operation, and write the result in the form $a + bi$.

39. $(1 + i)^3$ **40.** $(1 + 2i)^2(1 - 2i)^2$

41. i^{29} **42.** $\sqrt{-16}$

In Exercises 43 and 44, solve the equation.

43. $x^2 - 6x + 13 = 0$ **44.** $x^2 - 2x + 4 = 0$

In Exercises 45–48, match the polynomial function with its graph. Explain your choice.

45. $f(x) = (x - 2)^2$ **46.** $f(x) = (x - 2)^3$

47. $f(x) = (x - 2)^4$ **48.** $f(x) = (x - 2)^5$

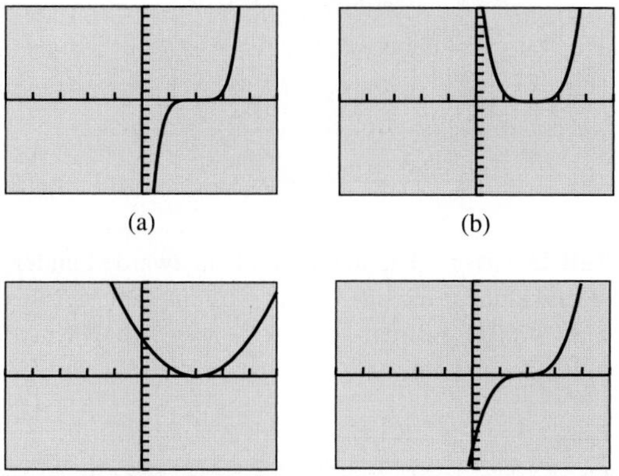

(a) (b) (c) (d)

In Exercises 49–52, find all of the real zeros of the function, finding exact values whenever possible. Identify each zero as rational or irrational. State the number of nonreal complex zeros.

49. $f(x) = x^4 - 10x^3 + 23x^2$

50. $k(t) = t^4 - 7t^2 + 12$

51. $h(x) = x^3 - 2x^2 - 8x + 5$

52. $k(x) = x^4 - x^3 - 14x^2 + 24x + 5$

In Exercises 53–56, find all of the zeros and write a linear factorization of the function.

53. $f(x) = 2x^3 - 9x^2 + 2x + 30$

54. $f(x) = 5x^3 - 24x^2 + x + 12$

55. $f(x) = 6x^4 + 11x^3 - 16x^2 - 11x + 10$

56. $f(x) = x^4 - 8x^3 + 27x^2 - 50x + 50$, given that $1 + 2i$ is a zero.

In Exercises 57–60, write the function as a product of linear and irreducible quadratic factors all with real coefficients.

57. $f(x) = x^3 - x^2 - x - 2$

58. $f(x) = 9x^3 - 3x^2 - 13x - 1$

59. $f(x) = 2x^4 - 9x^3 + 23x^2 - 31x + 15$

60. $f(x) = 3x^4 - 7x^3 - 3x^2 + 17x + 10$

In Exercises 61–66, write a polynomial function with real coefficients whose zeros and their multiplicities include those listed.

61. Degree 3; zeros: $\sqrt{5}, -\sqrt{5}, 3$

62. Degree 2; -3 only real zero

63. Degree 4; zeros: $3, -2, 1/3, -1/2$

64. Degree 3; zeros: $1 + i, 2$

65. Degree 4; zeros: -2(multiplicity 2), 4(multiplicity 2)

66. Degree 3; zeros: $2 - i, -1$, and $f(2) = 6$

In Exercises 67 and 68, describe how the graph of the given function can be obtained by transforming the graph of the reciprocal function $f(x) = 1/x$. Identify the horizontal and vertical asymptotes.

67. $f(x) = \dfrac{-x + 7}{x - 5}$

68. $f(x) = \dfrac{3x + 5}{x + 2}$

In Exercises 69–72, find the asymptotes and intercepts of the function, and graph it.

69. $f(x) = \dfrac{x^2 + x + 1}{x^2 - 1}$ **70.** $f(x) = \dfrac{2x^2 + 7}{x^2 + x - 6}$

71. $f(x) = \dfrac{x^2 - 4x + 5}{x + 3}$ **72.** $g(x) = \dfrac{x^2 - 3x - 7}{x + 3}$

In Exercises 73–74, find the intercepts, analyze, and graph the given rational function.

73. $f(x) = \dfrac{x^3 + x^2 - 2x + 5}{x + 2}$

74. $f(x) = \dfrac{-x^4 + x^2 + 1}{x - 1}$

In Exercises 75–82, solve the equation or inequality algebraically, and support graphically.

75. $2x + \dfrac{12}{x} = 11$

76. $\dfrac{x}{x + 2} + \dfrac{5}{x - 3} = \dfrac{25}{x^2 - x - 6}$

77. $2x^3 + 3x^2 - 17x - 30 < 0$

78. $3x^4 + x^3 - 36x^2 + 36x + 16 \ge 0$

79. $\dfrac{x + 3}{x^2 - 4} \ge 0$

80. $\dfrac{x^2 - 7}{x^2 - x - 6} < 1$

81. $(2x - 1)^2|x + 3| \le 0$

82. $\dfrac{(x - 1)|x - 4|}{\sqrt{x + 3}} > 0$

83. Writing to Learn Determine whether

$$f(x) = x^5 - 10x^4 - 3x^3 + 28x^2 + 20x - 2$$

has a zero outside the viewing window. Explain. (See graph.)

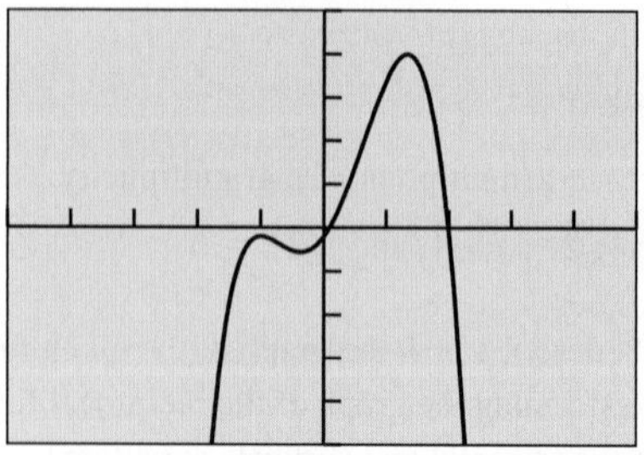

[–5, 5] by [–50, 50]

84. Launching a Rock Larry uses a slingshot to launch a rock straight up from a point 6 ft above level ground with an initial velocity of 170 ft/sec.

(a) Find an equation that models the height of the rock t seconds after it is launched and graph the equation. (See Example 8 in Section 2.1.)

(b) What is the maximum height of the rock? When will it reach that height?

(c) When will the rock hit the ground?

85. Volume of a Box Villareal Paper Co. has contracted to manufacture a box with no top that is to be made by removing squares of width x from the corners of a 30-in. by 70-in. piece of cardboard.

(a) Find an equation that models the volume of the box.

(b) Determine x so that the box has a volume of 5800 in.^{3}.

86. Architectural Engineering DeShanna, an engineer at J. P. Cook, Inc., completes structural specifications for a 255-ft-long steel beam anchored between two pilings 50 ft above ground, as shown in the figure. She knows that when a 250-lb object is placed d feet from the west piling, the beam bends s feet where

$$s = (8.5 \times 10^{-7})d^2(255 - d).$$

(a) Graph the function s.

(b) What are the dimensions of a viewing window that shows a graph for the values that make sense in this problem situation?

(c) What is the greatest amount of vertical deflection s, and where does it occur?

(d) Writing to Learn Give a possible scenario explaining why the solution to part (c) does not occur at the halfway point.

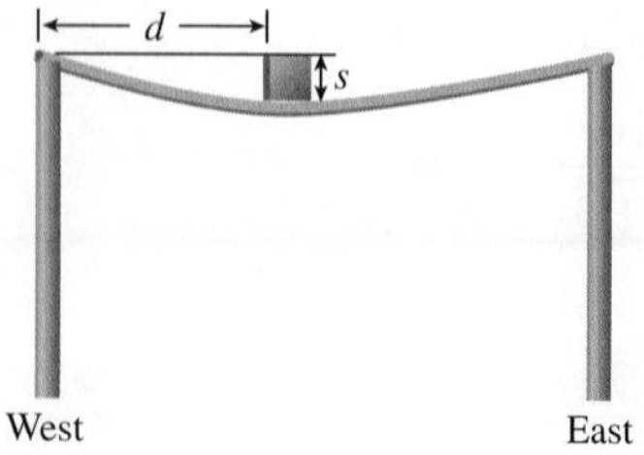

87. Storage Container A liquid storage container on a truck is in the shape of a cylinder with hemispheres on each end as shown in the figure. The cylinder and hemispheres have the same radius. The total length of the container is 140 ft.

(a) Determine the volume V of the container as a function of the radius x.

(b) Graph the function $y = V(x)$.

(c) What is the radius of the container with the largest possible volume? What is the volume?

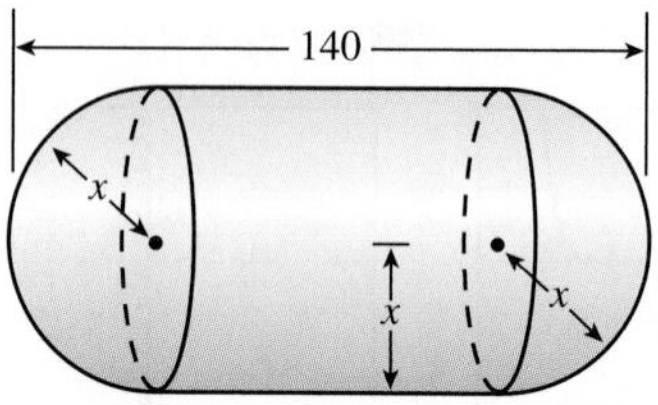

88. Pell Grants The maximum loan awarded under the federal student-aid program is given in Table 2.24 for several years. Let $x = 0$ represent 1975, $x = 1$ represent 1976, and so forth.

Table 2.24 Maximum Pell Grant Awards

Year	Amount (dollars)
1976	1400
1981	1670
1986	2100
1991	2400
1996	2470
2001	3750
2006	4050
2007	4310
2008	4731
2009	5350
2010	5550
2011	5550

Source: Trends in Student Aid, The College Board, 2013.

(a) Find a quadratic regression model for the Pell Grant amounts, and graph it together with a scatter plot of the data.

(b) Find a quartic regression model for the Pell Grant amounts, and graph it together with a scatter plot of the data.

(c) Use each regression equation to predict the amount of a Pell Grant in 2015.

(d) Writing to Learn Determine the end behavior of the two regression models. What does the end behavior say about future Pell Grant amounts?

89. National Institutes of Health Spending Table 2.25 shows the spending at the National Institutes of Health (NIH) for several years. Let $x = 0$ represent 2000, $x = 1$ represent 2001, and so forth.

Table 2.25 Spending at the National Institutes of Health

Year	Amount (billions of dollars)
2003	26.7
2004	28.1
2005	28.6
2006	28.5
2007	29.0
2008	29.3
2009	30.2
2010	31.0
2011	30.6
2012	30.8

Source: National Institutes of Health, 2012.

(a) Find a linear regression model, and graph it together with a scatter plot of the data.

(b) Find a quadratic regression model, and graph it together with a scatter plot of the data.

(c) What do the two models suggest about future NIH spending?

90. Breaking Even Midtown Sporting Goods has determined that it needs to sell its soccer shin guards for \$5.25 a pair in order to be competitive. It costs \$4.32 to produce each pair of shinguards, and the weekly overhead cost is \$4000.

(a) Express the average cost that includes the overhead of producing one shinguard as a function of the number x of shinguards produced each week.

(b) Solve algebraically to find the number of shinguards that must be sold each week to make \$8000 in profit. Support your work graphically.

91. Deer Population The number of deer P at any time t (in years) in a federal game reserve is given by

$$P(t) = \frac{800 + 640t}{20 + 0.8t}.$$

(a) Find the number of deer when t is 15, 70, and 100.

(b) Find the horizontal asymptote of the graph of $y = P(t)$.

(c) According to the model, what is the largest possible deer population?

92. Resistors The total electrical resistance R of two resistors connected in parallel with resistances R_1 and R_2 is given by

$$\frac{1}{R} = \frac{1}{R_1} + \frac{1}{R_2}.$$

The total resistance is 1.2 ohms. Let $x = R_1$.

(a) Express the second resistance R_2 as a function of x.

(b) Find R_2 if x_1 is 3 ohms.

93. Acid Mixture Suppose that x ounces of distilled water are added to 50 oz. of pure acid.

(a) Express the concentration $C(x)$ of the new mixture as a function of x.

(b) Use a graph to determine how much distilled water should be added to the pure acid to produce a new solution that is less than 60% acid.

(c) Solve (b) algebraically.

94. Industrial Design Johnson Cannery will pack peaches in 1-L cylindrical cans. Let x be the radius of the base of the can in centimeters.

(a) Express the surface area S of the can as a function of x.

(b) Find the radius and height of the can if the surface area is 900 cm^2.

(c) What dimensions are possible for the can if the surface area is to be less than 900 cm^2?

95. Industrial Design Gilman Construction is hired to build a rectangular tank with a square base and no top. The tank is to hold 1000 ft^3 of water. Let x be a length of the base.

(a) Express the outside surface area S of the tank as a function of x.

(b) Find the length, width, and height of the tank if the outside surface area is 600 ft^2.

(c) What dimensions are possible for the tank if the outside surface area is to be less than 600 ft^2?

CHAPTER 2 Project

Modeling the Height of a Bouncing Ball

When a ball is bouncing up and down on a flat surface, its height with respect to time can be modeled using a quadratic function. One form of a quadratic function is the vertex form:

$$y = a(x - h)^2 + k$$

In this equation, y represents the height of the ball and x represents the elapsed time. For this project, you will use a motion detection device to collect distance and time data for a bouncing ball, then find a mathematical model that describes the position of the ball with respect to time.

The table shows sample data collected using a Calculator-Based Ranger (CBR™).

Total Elapsed Time (sec)	Height of the Ball (m)	Total Elapsed Time (sec)	Height of the Ball (m)
0.688	0	1.118	0.828
0.731	0.155	1.161	0.811
0.774	0.309	1.204	0.776
0.817	0.441	1.247	0.721
0.860	0.553	1.290	0.650
0.903	0.643	1.333	0.563
0.946	0.716	1.376	0.452
0.989	0.773	1.419	0.322
1.032	0.809	1.462	0.169
1.075	0.828		

Explorations

1. If you collected motion data using a CBL™ or CBR™, a plot of height versus time or distance versus time should be shown on your grapher or computer screen. Either plot will work for this project. If you do not have access to a CBL or CBR, enter the data from the table provided into your grapher or computer. Create a scatter plot for the data.
2. Find values for a, h, and k so that the equation $y = a(x - h)^2 + k$ fits one of the bounces contained in the data plot. Approximate the vertex (h, k) from your data plot and solve for the value of a algebraically.
3. Change the values of a, h, and k in the model found above and observe how the graph of the function is affected on your grapher or computer. Generalize how each of these changes affects the graph.
4. Expand the equation you found in #2 above so that it is in the standard quadratic form: $y = ax^2 + bx + c$.
5. Use your grapher or computer to select the data from the bounce you modeled above and then use quadratic regression to find a model for this data set. (See your grapher's guidebook for instructions on how to do this.) How does this model compare with the standard quadratic form found in #4?
6. Complete the square to transform the regression model to the vertex form of a quadratic, and compare it to the original vertex model found in #2. (Round the values of a, b, and c to the nearest 0.001 before completing the square if desired.)

CHAPTER **3**

3.1 Exponential and Logistic Functions

3.2 Exponential and Logistic Modeling

3.3 Logarithmic Functions and Their Graphs

3.4 Properties of Logarithmic Functions

3.5 Equation Solving and Modeling

3.6 Mathematics of Finance

Exponential, Logistic, and Logarithmic Functions

The loudness of a sound we hear is based on the intensity of the associated sound wave. This sound intensity is the energy per unit time of the wave over a given area, measured in watts per square meter (W/m^2). The intensity is greatest near the source and decreases as you move away, whether the sound is rustling leaves or rock music. Because of the wide range of audible sound intensities, they are generally converted into *decibels*, which are based on logarithms. See page 280.

Chapter 3 Overview

In this chapter, we study three interrelated families of functions: exponential, logistic, and logarithmic functions. Polynomial functions, rational functions, and power functions with rational exponents are **algebraic functions**—functions obtained by adding, subtracting, multiplying, and dividing constants and an independent variable, and raising expressions to integer powers and extracting roots. In this chapter and the next one, we explore **transcendental functions**, which go beyond, or transcend, these algebraic operations.

Just like their algebraic cousins, exponential, logistic, and logarithmic functions have wide application. Exponential functions model growth and decay over time, such as *unrestricted* population growth and the decay of radioactive substances. Logistic functions model *restricted* population growth, certain chemical reactions, and the spread of rumors and diseases. Logarithmic functions are the basis of the Richter scale of earthquake intensity, the pH acidity scale, and the decibel measurement of sound.

The chapter closes with a study of the mathematics of finance, an application of exponential and logarithmic functions often used when making investments.

3.1 Exponential and Logistic Functions

What you'll learn about

- Exponential Functions and Their Graphs
- The Natural Base *e*
- Logistic Functions and Their Graphs
- Population Models

... and why

Exponential and logistic functions model many growth patterns, including the growth of human and animal populations.

Exponential Functions and Their Graphs

The functions $f(x) = x^2$ and $g(x) = 2^x$ each involve a base raised to a power, but the roles are reversed:

- For $f(x) = x^2$, the base is the variable x, and the exponent is the constant 2; f is a familiar monomial and power function.
- For $g(x) = 2^x$, the base is the constant 2, and the exponent is the variable x; g is an *exponential function*. See Figure 3.1.

DEFINITION Exponential Functions

Let a and b be real number constants. An **exponential function** in x is a function that can be written in the form

$$f(x) = a \cdot b^x,$$

where a is nonzero, b is positive, and $b \neq 1$. The constant a is the *initial value* of f (the value at $x = 0$), and b is the **base**.

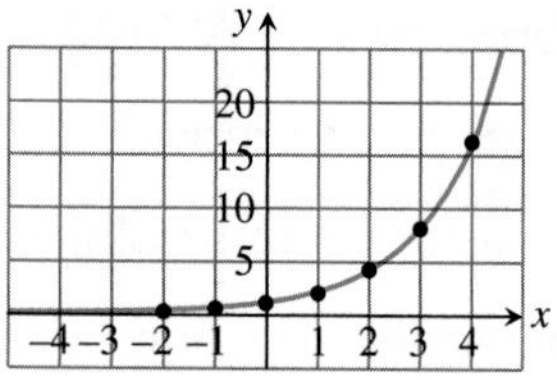

FIGURE 3.1 Sketch of $g(x) = 2^x$.

Exponential functions are defined and continuous for all real numbers. It is important to recognize whether a function is an exponential function.

EXAMPLE 1 Identifying Exponential Functions

(a) $f(x) = 3^x$ is an exponential function, with an initial value of 1 and base of 3.

(b) $g(x) = 6x^{-4}$ is *not* an exponential function because the base x is a variable and the exponent is a constant; g is a power function.

(c) $h(x) = -2 \cdot 1.5^x$ is an exponential function, with an initial value of -2 and base of 1.5.

(d) $k(x) = 7 \cdot 2^{-x}$ is an exponential function, with an initial value of 7 and base of 1/2 because $2^{-x} = (2^{-1})^x = (1/2)^x$.

(e) $q(x) = 5 \cdot 6^{\pi}$ is *not* an exponential function because the exponent π is a constant; q is a constant function. **Now try Exercise 1.**

One way to evaluate an exponential function, when the inputs are rational numbers, is to use the properties of exponents.

EXAMPLE 2 Computing Exponential Function Values for Rational Number Inputs

For $f(x) = 2^x$,

(a) $f(4) = 2^4 = 2 \cdot 2 \cdot 2 \cdot 2 = 16$

(b) $f(0) = 2^0 = 1$

(c) $f(-3) = 2^{-3} = \dfrac{1}{2^3} = \dfrac{1}{8} = 0.125$

(d) $f\left(\dfrac{1}{2}\right) = 2^{1/2} = \sqrt{2} = 1.4142\ldots$

(e) $f\left(-\dfrac{3}{2}\right) = 2^{-3/2} = \dfrac{1}{2^{3/2}} = \dfrac{1}{\sqrt{2^3}} = \dfrac{1}{\sqrt{8}} = 0.35355\ldots$

Now try Exercise 7.

There is no way to use properties of exponents to express an exponential function's value for *irrational* inputs. For example, if $f(x) = 2^x$, $f(\pi) = 2^{\pi}$, but what does 2^{π} mean? Using properties of exponents, $2^3 = 2 \cdot 2 \cdot 2$, $2^{3.1} = 2^{31/10} = \sqrt[10]{2^{31}}$. So we can find meaning for 2^{π} by using successively closer *rational* approximations to π as shown in Table 3.1.

Table 3.1 Values of $f(x) = 2^x$ for Rational Numbers x Approaching $\pi = 3.14159265\ldots$

x	3	3.1	3.14	3.141	3.1415	3.14159
2^x	8	8.5...	8.81...	8.821...	8.8244...	8.82496...

We can conclude that $f(\pi) = 2^{\pi} \approx 8.82$, which could be found directly using a grapher. The methods of calculus permit a more rigorous definition of exponential functions than we give here, a definition that allows for both rational and irrational inputs.

The way exponential functions change makes them useful in applications. This pattern of change can best be observed in tabular form.

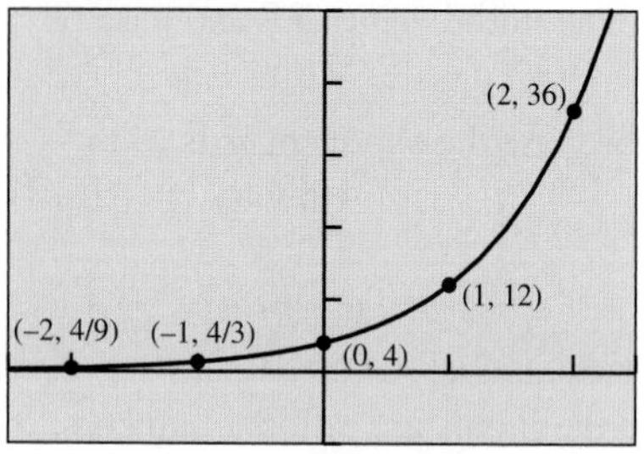

[−2.5, 2.5] by [−10, 50]
(a)

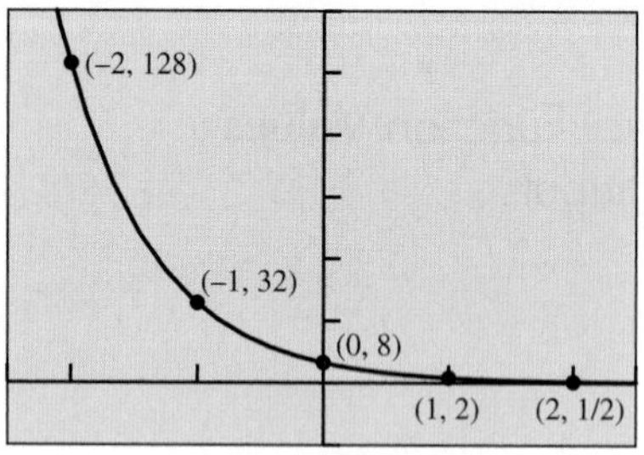

[−2.5, 2.5] by [−25, 150]
(b)

FIGURE 3.2 Graphs of (a) $g(x) = 4 \cdot 3^x$ and (b) $h(x) = 8 \cdot (1/4)^x$. (Example 3)

EXAMPLE 3 Finding an Exponential Function from Its Table of Values

Determine formulas for the exponential functions g and h whose values are given in Table 3.2.

Table 3.2 Values for Two Exponential Functions

x	$g(x)$		$h(x)$	
−2	4/9		128	
		$\times 3$		$\times \frac{1}{4}$
−1	4/3		32	
		$\times 3$		$\times \frac{1}{4}$
0	4		8	
		$\times 3$		$\times \frac{1}{4}$
1	12		2	
		$\times 3$		$\times \frac{1}{4}$
2	36		1/2	

SOLUTION Because g is exponential, $g(x) = a \cdot b^x$. Because $g(0) = 4$, the initial value a is 4. Because $g(1) = 4 \cdot b^1 = 12$, the base b is 3. So,

$$g(x) = 4 \cdot 3^x.$$

Because h is exponential, $h(x) = a \cdot b^x$. Because $h(0) = 8$, the initial value a is 8. Because $h(1) = 8 \cdot b^1 = 2$, the base b is 1/4. So,

$$h(x) = 8 \cdot \left(\frac{1}{4}\right)^x.$$

Figure 3.2 shows the graphs of these functions pass through the points whose coordinates are given in Table 3.2. **Now try Exercise 11.**

Table 3.3 Values for a General Exponential Function $f(x) = a \cdot b^x$

x	$a \cdot b^x$	
−2	ab^{-2}	
		$\times b$
−1	ab^{-1}	
		$\times b$
0	a	
		$\times b$
1	ab	
		$\times b$
2	ab^2	

Observe the patterns in the $g(x)$ and $h(x)$ columns of Table 3.2. The $g(x)$ values increase by a factor of 3 and the $h(x)$ values decrease by a factor of 1/4, as we add 1 to x moving from one row of the table to the next. In each case, the change factor is the base of the exponential function. This pattern generalizes to all exponential functions as illustrated in Table 3.3.

In Table 3.3, as x increases by 1, the function value is multiplied by the base b. This relationship leads to the following *recursive formula.*

Exponential Growth and Decay

For any exponential function $f(x) = a \cdot b^x$ and any real number x,

$$f(x + 1) = b \cdot f(x).$$

If $a > 0$ and $b > 1$, the function f is increasing and is an **exponential growth function**. The base b is its **growth factor**.

If $a > 0$ and $0 < b < 1$, f is decreasing and is an **exponential decay function**. The base b is its **decay factor**.

In Example 3, g is an exponential growth function, and h is an exponential decay function. As x increases by 1, $g(x) = 4 \cdot 3^x$ grows by a factor of 3, and $h(x) = 8 \cdot (1/4)^x$ decays by a factor of 1/4. Whenever the initial value is positive, the base of an exponential function, like the slope of a linear function, tells us whether the function is increasing or decreasing and by how much.

So far, we have focused most of our attention on the algebraic and numerical aspects of exponential functions. We now turn our attention to the graphs of these functions.

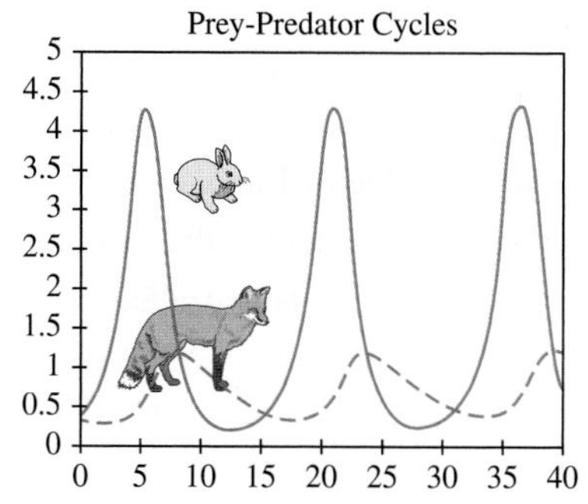

EXPLORATION 1 Graphs of Exponential Functions

1. Graph each function in the viewing window $[-2, 2]$ by $[-1, 6]$.

(a) $y_1 = 2^x$ **(b)** $y_2 = 3^x$ **(c)** $y_3 = 4^x$ **(d)** $y_4 = 5^x$

- Which point is common to all four graphs?
- Analyze the functions for domain, range, continuity, increasing or decreasing behavior, symmetry, boundedness, extrema, asymptotes, and end behavior.

2. Graph each function in the viewing window $[-2, 2]$ by $[-1, 6]$.

(a) $y_1 = \left(\frac{1}{2}\right)^x$ **(b)** $y_2 = \left(\frac{1}{3}\right)^x$

(c) $y_3 = \left(\frac{1}{4}\right)^x$ **(d)** $y_4 = \left(\frac{1}{5}\right)^x$

- Which point is common to all four graphs?
- Analyze the functions for domain, range, continuity, increasing or decreasing behavior, symmetry, boundedness, extrema, asymptotes, and end behavior.

We summarize what we have learned about exponential functions with an initial value of 1.

Exponential Functions $f(x) = b^x$

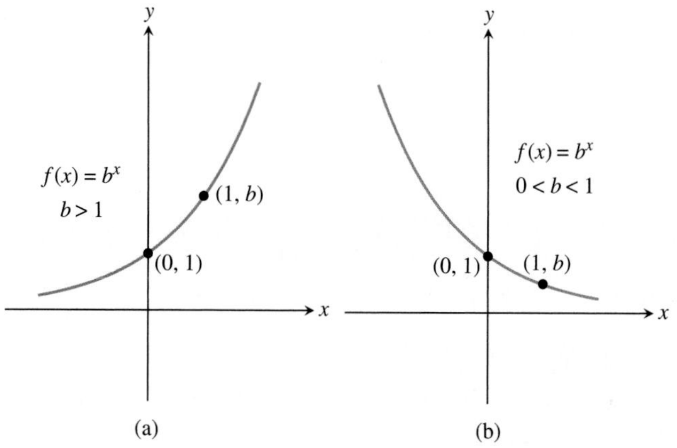

FIGURE 3.3 Graphs of $f(x) = b^x$ for (a) $b > 1$ and (b) $0 < b < 1$.

Domain: $(-\infty, \infty)$
Range: $(0, \infty)$
Continuous
No symmetry: neither even nor odd
Bounded below, but not above
No local extrema
Horizontal asymptote: $y = 0$
No vertical asymptotes
If $b > 1$ (see Figure 3.3a), then

- f is an increasing function,
- $\lim_{x\to-\infty} f(x) = 0$ and $\lim_{x\to\infty} f(x) = \infty$.

If $0 < b < 1$ (see Figure 3.3b), then

- f is a decreasing function,
- $\lim_{x\to-\infty} f(x) = \infty$ and $\lim_{x\to\infty} f(x) = 0$.

The translations, reflections, stretches, and shrinks studied in Section 1.5, together with our knowledge of the graphs of basic exponential functions, allow us to predict the graphs of the functions in Example 4.

EXAMPLE 4 Transforming Exponential Functions

Describe how to transform the graph of $f(x) = 2^x$ into the graph of the given function. Sketch the graphs by hand and support your answer with a grapher.

(a) $g(x) = 2^{x-1}$ **(b)** $h(x) = 2^{-x}$ **(c)** $k(x) = 3 \cdot 2^x$

SOLUTION

(a) The graph of $g(x) = 2^{x-1}$ is obtained by translating the graph of $f(x) = 2^x$ by 1 unit to the right (Figure 3.4a).

(b) We can obtain the graph of $h(x) = 2^{-x}$ by reflecting the graph of $f(x) = 2^x$ across the y-axis (Figure 3.4b). Because $2^{-x} = (2^{-1})^x = (1/2)^x$, we can also think of h as an exponential function with an initial value of 1 and a base of 1/2.

(c) We can obtain the graph of $k(x) = 3 \cdot 2^x$ by vertically stretching the graph of $f(x) = 2^x$ by a factor of 3 (Figure 3.4c). **Now try Exercise 15.**

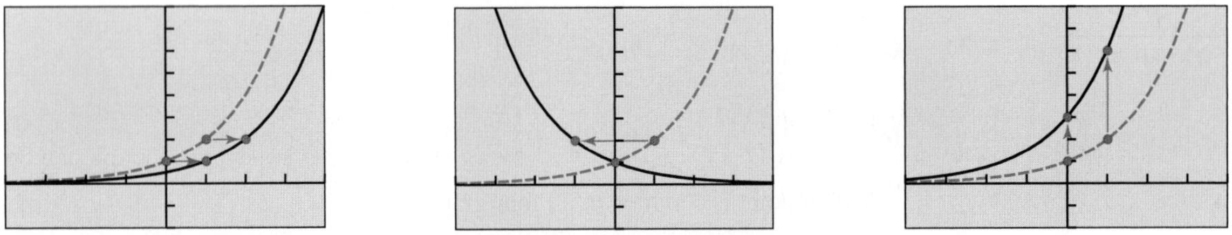

[–4, 4] by [–2, 8]
(a)

[–4, 4] by [–2, 8]
(b)

[–4, 4] by [–2, 8]
(c)

FIGURE 3.4 The graph of $f(x) = 2^x$ shown with (a) $g(x) = 2^{x-1}$, (b) $h(x) = 2^{-x}$, and (c) $k(x) = 3 \cdot 2^x$. (Example 4)

The Natural Base *e*

The function $f(x) = e^x$ is one of the basic functions introduced in Section 1.3, and is an exponential growth function.

BASIC FUNCTION The Exponential Function

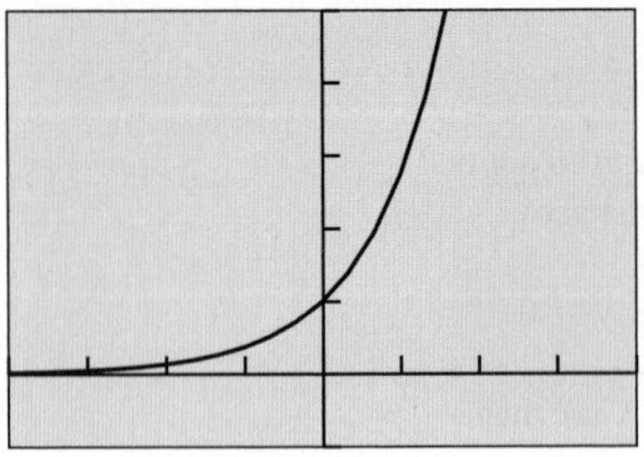

[–4, 4] by [–1, 5]

FIGURE 3.5 The graph of $f(x) = e^x$.

$f(x) = e^x$
Domain: $(-\infty, \infty)$
Range: $(0, \infty)$
Continuous
Increasing for all x
No symmetry
Bounded below, but not above
No local extrema
Horizontal asymptote: $y = 0$
No vertical asymptotes
End behavior: $\lim_{x \to -\infty} e^x = 0$ and $\lim_{x \to \infty} e^x = \infty$

Because $f(x) = e^x$ is increasing, it is an exponential growth function, so $e > 1$. But what is e, and what makes this exponential function *the* exponential function?

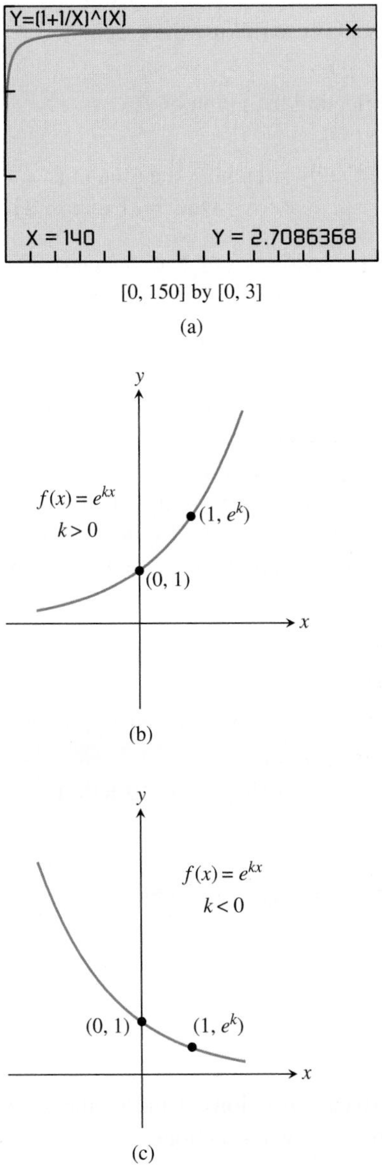

FIGURE 3.6 (a) Graph of $f(x) = (1 + 1/x)^x$ with asymptote in red approximating the number e. Graphs of $f(x) = e^{kx}$ for (b) $k > 0$ and (c) $k < 0$.

The letter e is the initial of the last name of Leonhard Euler (1707–1783), who introduced the notation. Because $f(x) = e^x$ has special calculus properties that simplify many calculations, e is the *natural base* of exponential functions for calculus purposes, and $f(x) = e^x$ is considered the *natural exponential function.*

DEFINITION The Natural Base e

$$e = \lim_{x \to \infty} \left(1 + \frac{1}{x}\right)^x$$

We cannot compute the irrational number e directly, but using this definition we can obtain successively closer approximations to e, as shown in Table 3.4. Continuing the process in Table 3.4 with a sufficiently accurate computer can show that $e \approx 2.718281828459$.

Table 3.4 Approximations Approaching the Natural Base e

x	1	10	100	1000	10,000	100,000
$(1 + 1/x)^x$	2	2.5 . . .	2.70 . . .	2.716 . . .	2.7181 . . .	2.71826 . . .

We are usually more interested in the exponential function $f(x) = e^x$ and variations of this function than in the irrational number e. In fact, *any* exponential function can be expressed in terms of the natural base e.

THEOREM Exponential Functions and the Base e

Any exponential function $f(x) = a \cdot b^x$ can be rewritten as

$$f(x) = a \cdot e^{kx},$$

for an appropriately chosen real number constant k.

If $a > 0$ and $k > 0$, $f(x) = a \cdot e^{kx}$ is an exponential growth function. (See Figure 3.6b.)

If $a > 0$ and $k < 0$, $f(x) = a \cdot e^{kx}$ is an exponential decay function. (See Figure 3.6c.)

In Section 3.3 we will develop some mathematics so that, for any positive number $b \neq 1$, we can easily find the value of k such that $e^{kx} = b^x$. In the meantime, we can use graphical and numerical methods to approximate k, as you will discover in Exploration 2.

EXPLORATION 2 Choosing k so that $e^{kx} = 2^x$

1. Graph $f(x) = 2^x$ in the viewing window $[-4, 4]$ by $[-2, 8]$.
2. One at a time, overlay the graphs of $g(x) = e^{kx}$ for $k = 0.4, 0.5, 0.6, 0.7$, and 0.8. For which of these values of k does the graph of g most closely match the graph of f?
3. Using tables, find the 3-decimal-place value of k for which the values of g most closely approximate the values of f.

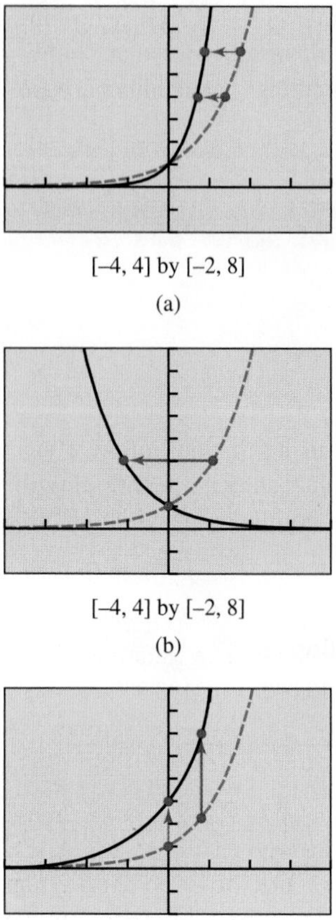

[–4, 4] by [–2, 8]
(a)

[–4, 4] by [–2, 8]
(b)

[–4, 4] by [–2, 8]
(c)

FIGURE 3.7 The graph of $f(x) = e^x$ shown with (a) $g(x) = e^{2x}$, (b) $h(x) = e^{-x}$, and (c) $k(x) = 3e^x$. (Example 5)

EXAMPLE 5 Transforming Exponential Functions

Describe how to transform the graph of $f(x) = e^x$ into the graph of the given function. Sketch the graphs by hand and support your answer with a grapher.

(a) $g(x) = e^{2x}$ **(b)** $h(x) = e^{-x}$ **(c)** $k(x) = 3e^x$

SOLUTION

(a) The graph of $g(x) = e^{2x}$ is obtained by horizontally shrinking the graph of $f(x) = e^x$ by a factor of 2 (Figure 3.7a).

(b) We can obtain the graph of $h(x) = e^{-x}$ by reflecting the graph of $f(x) = e^x$ across the y-axis (Figure 3.7b).

(c) We can obtain the graph of $k(x) = 3e^x$ by vertically stretching the graph of $f(x) = e^x$ by a factor of 3 (Figure 3.7c). Now try Exercise 21.

Logistic Functions and Their Graphs

Exponential growth is *unrestricted.* An exponential growth function increases at an ever-increasing rate and is not bounded above. In many growth situations, however, there is a limit to the possible growth. A plant can only grow so tall. The number of goldfish in an aquarium is limited by the size of the aquarium. In such situations the growth often begins in an exponential manner, but the growth eventually slows and the graph levels out. The associated growth function is bounded both below and above by horizontal asymptotes.

DEFINITION Logistic Growth Functions

Let a, b, c, and k be positive constants, with $b < 1$. A **logistic growth function** in x is a function that can be written in the form

$$f(x) = \frac{c}{1 + a \cdot b^x} \text{ or } f(x) = \frac{c}{1 + a \cdot e^{-kx}}$$

where the constant c is the **limit to growth**.

If $b > 1$ or $k < 0$, these formulas yield **logistic decay functions**. Unless otherwise stated, all *logistic functions* in this book will be logistic growth functions.

By setting $a = c = k = 1$, we obtain the **logistic function**

$$f(x) = \frac{1}{1 + e^{-x}}.$$

This function, though related to the exponential function e^x, *cannot* be obtained from e^x by translations, reflections, and horizontal and vertical stretches and shrinks. So we give the logistic function a formal introduction:

Aliases for Logistic Growth

Logistic growth is also known as *restricted, inhibited,* or *constrained exponential growth.*

BASIC FUNCTION The Logistic Function

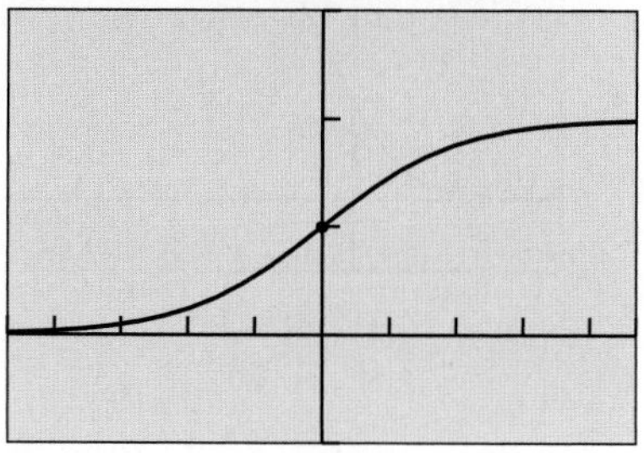

[–4.7, 4.7] by [–0.5, 1.5]

FIGURE 3.8 The graph of $f(x) = 1/(1 + e^{-x})$.

$$f(x) = \frac{1}{1 + e^{-x}}$$

Domain: $(-\infty, \infty)$
Range: $(0, 1)$
Continuous
Increasing for all x
Symmetric about $(0, 1/2)$, but neither even nor odd
Bounded below and above
No local extrema
Horizontal asymptotes: $y = 0$ and $y = 1$
No vertical asymptotes
End behavior: $\lim_{x \to -\infty} f(x) = 0$ and $\lim_{x \to \infty} f(x) = 1$

All logistic growth functions have graphs much like the basic logistic function. Their end behavior is always described by the equations

$$\lim_{x \to -\infty} f(x) = 0 \text{ and } \lim_{x \to \infty} f(x) = c,$$

where c is the limit to growth (see Exercise 74). All logistic functions are bounded by their horizontal asymptotes, $y = 0$ and $y = c$, and have a range of $(0, c)$. Although every logistic function is symmetric about the point of its graph with y-coordinate $c/2$, this point of symmetry is usually not the y-intercept, as we can see in Example 6.

EXAMPLE 6 Graphing Logistic Growth Functions

Graph the function. Find the y-intercept and the horizontal asymptotes.

(a) $f(x) = \dfrac{8}{1 + 3 \cdot 0.7^x}$ **(b)** $g(x) = \dfrac{20}{1 + 2e^{-3x}}$

SOLUTION

(a) The graph of $f(x) = 8/(1 + 3 \cdot 0.7^x)$ is shown in Figure 3.9a on the next page. The y-intercept is

$$f(0) = \frac{8}{1 + 3 \cdot 0.7^0} = \frac{8}{1 + 3} = 2.$$

Because the limit to growth is 8, the horizontal asymptotes are $y = 0$ and $y = 8$.

(b) The graph of $g(x) = 20/(1 + 2e^{-3x})$ is shown in Figure 3.9b. The y-intercept is

$$g(0) = \frac{20}{1 + 2e^{-3 \cdot 0}} = \frac{20}{1 + 2} = 20/3 \approx 6.67.$$

Because the limit to growth is 20, the horizontal asymptotes are $y = 0$ and $y = 20$.

Now try Exercise 41.

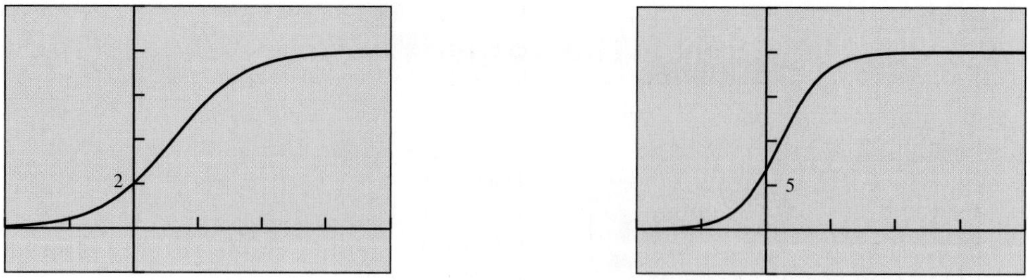

[–10, 20] by [–2, 10]
(a)

[–2, 4] by [–5, 25]
(b)

FIGURE 3.9 The graphs of (a) $f(x) = 8/(1 + 3 \cdot 0.7^x)$ and (b) $g(x) = 20/(1 + 2e^{-3x})$. (Example 6)

Population Models

Exponential and logistic functions have many applications. One area where both types of functions are used is in modeling population. Between 1990 and 2000, both Phoenix and San Antonio passed the 1 million mark. With its Silicon Valley industries, San Jose, California, appears to be the next U.S. city destined to surpass 1 million residents. When a city's population is growing rapidly, as in the case of San Jose, exponential growth is a reasonable model.

EXAMPLE 7 Modeling San Jose's Population

Using the data in Table 3.5 and assuming the growth is exponential, when will the population of San Jose, California, surpass 1 million persons?

SOLUTION

Model Let $P(t)$ be the population of San Jose t years after July 1, 2000. (See margin note.) Because P is exponential, $P(t) = P_0 \cdot b^t$, where P_0 is the initial (2000) population of 898,759. From Table 3.5 we see that $P(7) = 898{,}759b^7 = 939{,}899$. So,

$$b = \sqrt[7]{\frac{939{,}899}{898{,}759}} \approx 1.0064$$

and $P(t) = 898{,}759 \cdot 1.0064^t$.

Solve Graphically Figure 3.10 shows that this population model intersects $y = 1{,}000{,}000$ when the independent variable is about 16.73.

Interpret Because 16.73 years after mid-2000 is in the first half of 2017, according to this model the population of San Jose will surpass the 1 million mark in early 2017.

Now try Exercise 51.

A Note on Population Data

When the U.S. Census Bureau reports a population estimate for a given year, it generally represents the population at the middle of the year, or July 1. We will assume this to be the case when interpreting our results to population problems unless otherwise noted.

Table 3.5 The Population of San Jose, California

Year	Population
2000	898,759
2007	939,899

Source: U.S. Census Bureau.

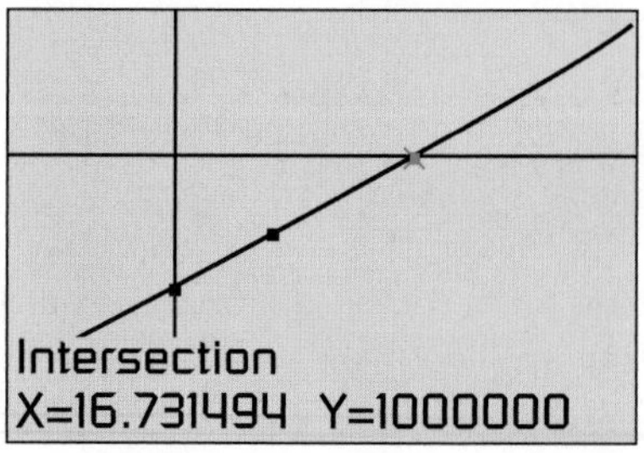

[–10, 30] by [800 000, 1 100 000]

FIGURE 3.10 A population model for San Jose, California. (Example 7)

While San Jose's population is soaring, in other major cities, such as Dallas, the population growth is slowing. The once sprawling Dallas is now *constrained* by its neighboring cities. *A logistic function is often an appropriate model for restricted growth,* such as the growth that Dallas is experiencing.

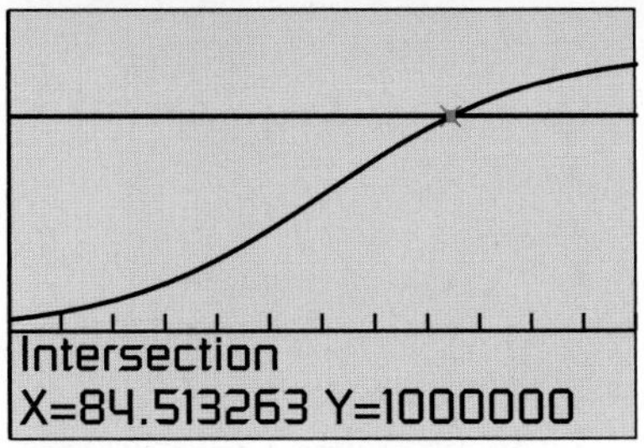

[0, 120] by [−500 000, 1 500 000]

FIGURE 3.11 A population model for Dallas, Texas. (Example 8)

EXAMPLE 8 Modeling Dallas's Population

Based on recent census data, a logistic model for the population of Dallas, t years after 1900, is as follows:

$$P(t) = \frac{1{,}301{,}642}{1 + 21.602e^{-0.05054t}}$$

According to this model, when was the population 1 million?

SOLUTION Figure 3.11 shows that the population model intersects $y = 1{,}000{,}000$ when the independent variable is about 84.51. Because 84.51 years after mid-1900 is at the beginning of 1985, if Dallas's population has followed this logistic model, its population was 1 million then. Now try Exercise 55.

QUICK REVIEW 3.1 *(For help, go to Sections A.1 and P.1.)*

Exercise numbers with a gray background indicate problems that the authors have designed to be solved *without a calculator*.

In Exercises 1–4, evaluate the expression.

1. $\sqrt[3]{-216}$ **2.** $\sqrt[3]{\frac{125}{8}}$

3. $27^{2/3}$ **4.** $4^{5/2}$

In Exercises 5–8, rewrite the expression using a single positive exponent.

5. $(2^{-3})^4$ **6.** $(3^4)^{-2}$

7. $(a^{-2})^3$ **8.** $(b^{-3})^{-5}$

In Exercises 9–10, use a calculator to evaluate the expression.

9. $\sqrt[5]{-5.37824}$ **10.** $\sqrt[4]{92.3521}$

SECTION 3.1 Exercises

In Exercises 1–6, which of the following are exponential functions? For those that are exponential functions, state the initial value and the base. For those that are not, explain why not.

1. $y = x^8$

2. $y = 3^x$

3. $y = 5^x$

4. $y = 4^2$

5. $y = x^{\sqrt{x}}$

6. $y = x^{1.3}$

In Exercises 7–10, compute the exact value of the function for the given x-value without using a calculator.

7. $f(x) = 3 \cdot 5^x$ for $x = 0$

8. $f(x) = 6 \cdot 3^x$ for $x = -2$

9. $f(x) = -2 \cdot 3^x$ for $x = 1/3$

10. $f(x) = 8 \cdot 4^x$ for $x = -3/2$

In Exercises 11 and 12, determine a formula for the exponential function whose values are given in Table 3.6.

11. $f(x)$

12. $g(x)$

Table 3.6 Values for Two Exponential Functions

x	$f(x)$	$g(x)$
−2	6	108
−1	3	36
0	3/2	12
1	3/4	4
2	3/8	4/3

In Exercises 13 and 14, determine a formula for the exponential function whose graph is shown in the figure.

13. $f(x)$

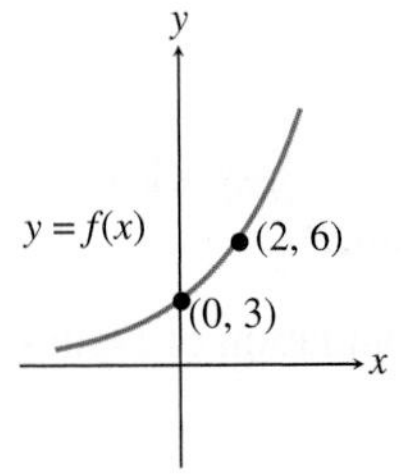

14. $g(x)$

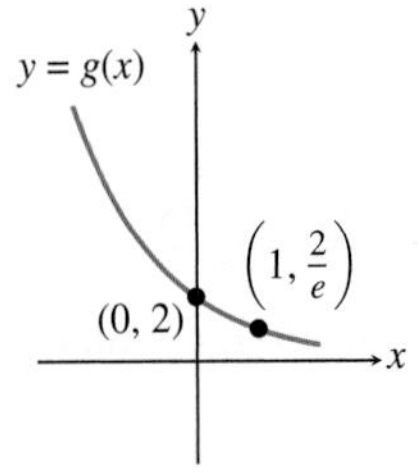

In Exercises 15–24, describe how to transform the graph of f into the graph of g. Sketch the graphs by hand and support your answer with a grapher.

15. $f(x) = 2^x,\ g(x) = 2^{x-3}$

16. $f(x) = 3^x,\ g(x) = 3^{x+4}$

17. $f(x) = 4^x,\ g(x) = 4^{-x}$

18. $f(x) = 2^x,\ g(x) = 2^{5-x}$

19. $f(x) = 0.5^x,\ g(x) = 3 \cdot 0.5^x + 4$

20. $f(x) = 0.6^x,\ g(x) = 2 \cdot 0.6^{3x}$

21. $f(x) = e^x,\ g(x) = e^{-2x}$

22. $f(x) = e^x,\ g(x) = -e^{-3x}$

23. $f(x) = e^x,\ g(x) = 2e^{3-3x}$

24. $f(x) = e^x,\ g(x) = 3e^{2x} - 1$

Writing to Learn In Exercises 25–30, **(a)** match the given function with its graph. **(b)** Explain how to make the choice without using a grapher.

25. $y = 3^x$

26. $y = 2^{-x}$

27. $y = -2^x$

28. $y = -0.5^x$

29. $y = 3^{-x} - 2$

30. $y = 1.5^x - 2$

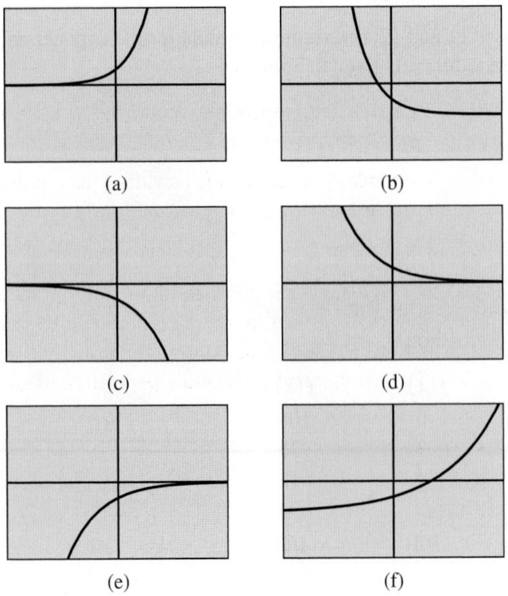

In Exercises 31–34, state whether the function is an exponential growth function or exponential decay function, and describe its end behavior using limits.

31. $f(x) = 3^{-2x}$

32. $f(x) = \left(\frac{1}{e}\right)^x$

33. $f(x) = 0.5^x$

34. $f(x) = 0.75^{-x}$

In Exercises 35–38, solve the inequality graphically.

35. $9^x < 4^x$

36. $6^{-x} > 8^{-x}$

37. $\left(\frac{1}{4}\right)^x > \left(\frac{1}{3}\right)^x$

38. $\left(\frac{1}{3}\right)^x < \left(\frac{1}{2}\right)^x$

Group Activity In Exercises 39 and 40, use the properties of exponents to prove that two of the given three exponential functions are identical. Support graphically.

39. **(a)** $y_1 = 3^{2x+4}$

(b) $y_2 = 3^{2x} + 4$

(c) $y_3 = 9^{x+2}$

40. **(a)** $y_1 = 4^{3x-2}$

(b) $y_2 = 2(2^{3x-2})$

(c) $y_3 = 2^{3x-1}$

In Exercises 41–44, use a grapher to graph the function. Find the y-intercept and the horizontal asymptotes.

41. $f(x) = \dfrac{12}{1 + 2 \cdot 0.8^x}$

42. $f(x) = \dfrac{18}{1 + 5 \cdot 0.2^x}$

43. $f(x) = \dfrac{16}{1 + 3e^{-2x}}$

44. $g(x) = \dfrac{9}{1 + 2e^{-x}}$

In Exercises 45–50, graph the function and analyze it for domain, range, continuity, increasing or decreasing behavior, symmetry, boundedness, extrema, asymptotes, and end behavior.

45. $f(x) = 3 \cdot 2^x$

46. $f(x) = 4 \cdot 0.5^x$

47. $f(x) = 4 \cdot e^{3x}$

48. $f(x) = 5 \cdot e^{-x}$

49. $f(x) = \dfrac{5}{1 + 4 \cdot e^{-2x}}$

50. $f(x) = \dfrac{6}{1 + 2 \cdot e^{-x}}$

51. **Population Growth** Using the midyear data in Table 3.7 and assuming the growth is exponential, when did the population of Austin surpass 800,000 persons?

Table 3.7 Populations of Two Major U.S. Cities

City	1990 Population	2011 Population
Austin, Texas	465,622	820,611
Columbus, Ohio	632,910	797,434

Source: World Almanac and Book of Facts 2013.

52. **Population Growth** Using the data in Table 3.7 and assuming the growth is exponential, when did the population of Columbus surpass 800,000 persons?

53. **Population Growth** Using the data in Table 3.7 and assuming the growth is exponential, when were the populations of Austin and Columbus equal?

54. **Population Growth** Using the data in Table 3.7 and assuming the growth is exponential, which city—Austin or Columbus—would reach a population of 1 million first, and in what year?

55. **Population Growth** Using 20th-century U.S. census data, the population of Ohio can be modeled by

$$P(t) = \frac{12.79}{1 + 2.402e^{-0.0309x}},$$

where P is the population in millions and t is the number of years since April 1, 1900. Based on this model, when was the population of Ohio 10 million?

56. **Population Growth** Using 20th-century U.S. census data, the population of New York state can be modeled by

$$P(t) = \frac{19.875}{1 + 57.993e^{-0.035005t}},$$

where P is the population in millions and t is the number of years since 1800. Based on this model,

(a) What was the population of New York in 1850?

(b) What will New York state's population be in 2015?

(c) What is New York's *maximum sustainable population* (limit to growth)?

57. **Bacterial Growth** The number B of bacteria in a petri dish culture after t hours is given by

$$B = 100e^{0.693t}.$$

(a) What was the initial number of bacteria present?

(b) How many bacteria are present after 6 hr?

58. **Carbon Dating** The amount C in grams of carbon-14 present in a certain substance after t years is given by

$$C = 20e^{-0.0001216t}.$$

(a) What was the initial amount of carbon-14 present?

(b) How much is left after 10,400 years? When will the amount left be 10 g?

Standardized Test Questions

59. **True or False** Every exponential function is strictly increasing. Justify your answer.

60. **True or False** Every logistic growth function has two horizontal asymptotes. Justify your answer.

In Exercises 61–64, solve the problem without using a calculator.

61. **Multiple Choice** Which of the following functions is exponential?

(A) $f(x) = a^2$ **(B)** $f(x) = x^3$

(C) $f(x) = x^{2/3}$ **(D)** $f(x) = \sqrt[3]{x}$

(E) $f(x) = 8^x$

62. **Multiple Choice** What point do all functions of the form $f(x) = b^x (b > 0)$ have in common?

(A) $(1, 1)$ **(B)** $(1, 0)$

(C) $(0, 1)$ **(D)** $(0, 0)$

(E) $(-1, -1)$

63. **Multiple Choice** The growth factor for $f(x) = 4 \cdot 3^x$ is

(A) 3. **(B)** 4.

(C) 12. **(D)** 64.

(E) 81.

64. **Multiple Choice** For $x > 0$, which of the following is true?

(A) $3^x > 4^x$ **(B)** $7^x > 5^x$

(C) $(1/6)^x > (1/2)^x$ **(D)** $9^{-x} > 8^{-x}$

(E) $0.17^x > 0.32^x$

Explorations

65. Graph each function and analyze it for domain, range, increasing or decreasing behavior, boundedness, extrema, asymptotes, and end behavior.

(a) $f(x) = x \cdot e^x$ **(b)** $g(x) = \dfrac{e^{-x}}{x}$

66. Use the properties of exponents to solve each equation. Support graphically.

(a) $2^x = 4^2$ **(b)** $3^x = 27$

(c) $8^{x/2} = 4^{x+1}$

(d) $9^x = 3^{x+1}$

Extending the Ideas

67. **Exponential Growth in Genealogy** In genealogy you are considered generation 0. Your two parents are generation 1. Your four grandparents are generation 2. Your great grandparents are generation 3 and your 2nd great grandparents are generation 4.

(a) How many 4th great grandparents do you have? What generation are they?

(b) Use an exponential function to model the number of nth great grandparents you have.

(c) Use the model you found in (b) and find the number of 6th great grandparents you have.

(d) How many 25th great grandparents do you have?

(e) Discuss how many years it takes to span 25 generations at an average of 30 years between generations. The world's population in 1250 is thought to have been about 400 million. To how much of the world's population might you be related in 1250?

68. Writing to Learn Table 3.8 gives function values for $y = f(x)$ and $y = g(x)$. Also, three different graphs are shown.

Table 3.8 Data for Two Functions

x	$f(x)$	$g(x)$
1.0	5.50	7.40
1.5	5.35	6.97
2.0	5.25	6.44
2.5	5.17	5.76
3.0	5.13	4.90
3.5	5.09	3.82
4.0	5.06	2.44
4.5	5.05	0.71

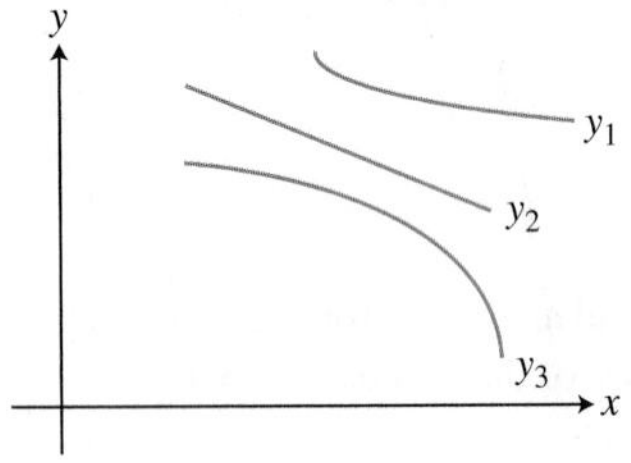

(a) Which curve of those shown in the graph most closely resembles the graph of $y = f(x)$? Explain your choice.

(b) Which curve most closely resembles the graph of $y = g(x)$? Explain your choice.

69. Writing to Learn Let $f(x) = 2^x$. Explain why the graph of $f(ax + b)$ can be obtained by applying one transformation to the graph of $y = c^x$ for an appropriate value of c. What is c?

Exercises 70–73 refer to the expression $f(a, b, c) = a \cdot b^c$. For example, if $a = 2$, $b = 3$, and $c = x$, the expression is $f(2, 3, x) = 2 \cdot 3^x$, an exponential function.

70. If $b = x$, state conditions on a and c under which the expression $f(a, b, c)$ is a quadratic power function.

71. If $b = x$, state conditions on a and c under which the expression $f(a, b, c)$ is a decreasing linear function.

72. If $c = x$, state conditions on a and b under which the expression $f(a, b, c)$ is an increasing exponential function.

73. If $c = x$, state conditions on a and b under which the expression $f(a, b, c)$ is a decreasing exponential function.

74. Prove that $\lim_{x \to -\infty} \frac{c}{1 + a \cdot b^x} = 0$ and $\lim_{x \to \infty} \frac{c}{1 + a \cdot b^x} = c$, for constants a, b, and c, with $a > 0$, $0 < b < 1$, and $c > 0$.

3.2 Exponential and Logistic Modeling

What you'll learn about

- Constant Percentage Rate and Exponential Functions
- Exponential Growth and Decay Models
- Using Regression to Model Population
- Other Logistic Models

... and why

Exponential functions model many types of unrestricted growth; logistic functions model restricted growth, including the spread of disease and the spread of rumors.

Constant Percentage Rate and Exponential Functions

Suppose that a population is changing at a **constant percentage rate *r***, where r is the percent rate of change expressed in decimal form. Then the population follows the pattern shown.

Time in Years	Population
0	$P(0) = P_0 =$ initial population
1	$P(1) = P_0 + P_0 r = P_0(1 + r)$
2	$P(2) = P(1) \cdot (1 + r) = P_0(1 + r)^2$
3	$P(3) = P(2) \cdot (1 + r) = P_0(1 + r)^3$
⋮	⋮
t	$P(t) = P_0(1 + r)^t$

So, in this case, the population is an exponential function of time.

Exponential Population Model

If a population P is changing at a constant percentage rate r each year, then

$$P(t) = P_0(1 + r)^t,$$

where P_0 is the initial population, r is expressed as a decimal, and t is time in years.

If $r > 0$, then $P(t)$ is an exponential growth function, and its *growth factor* is the base of the exponential function, $1 + r$.

On the other hand, if $r < 0$, the base $1 + r < 1$, $P(t)$ is an exponential decay function, and $1 + r$ is the *decay factor* for the population.

EXAMPLE 1 Finding Growth and Decay Rates

Tell whether the population model is an exponential growth function or exponential decay function, and find the constant percentage rate of growth or decay.

(a) San Jose: $P(t) = 898{,}759 \cdot 1.0064^t$

(b) Detroit: $P(t) = 1{,}203{,}368 \cdot 0.9858^t$

SOLUTION

(a) Because $1 + r = 1.0064$, $r = 0.0064 > 0$. So, P is an exponential growth function with a growth rate of 0.64%.

(b) Because $1 + r = 0.9858$, $r = -0.0142 < 0$. So, P is an exponential decay function with a decay rate of 1.42%.

Now try Exercise 1.

EXAMPLE 2 Finding an Exponential Function

Determine the exponential function with initial value $= 12$, increasing at a rate of 8% per year.

SOLUTION Because $P_0 = 12$ and $r = 8\% = 0.08$, the function is $P(t) = 12(1 + 0.08)^t$ or $P(t) = 12 \cdot 1.08^t$. We could write this as $f(x) = 12 \cdot 1.08^x$, where x represents time.

Now try Exercise 7.

Exponential Growth and Decay Models

Exponential growth and decay models are used for populations of animals, bacteria, and even radioactive atoms. Exponential growth and decay apply to any situation where the growth is proportional to the current size of the quantity of interest. Such situations are frequently encountered in biology, chemistry, business, and the social sciences.

Exponential growth models can be developed in terms of the time it takes a quantity to double. On the flip side, exponential decay models can be developed in terms of the time it takes for a quantity to be halved. Examples 3 through 5 use these strategies.

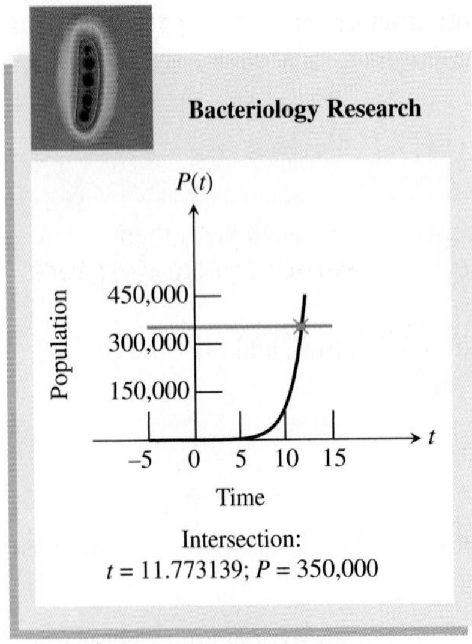

FIGURE 3.12 Rapid growth of a bacterial population. (Example 3)

EXAMPLE 3 Modeling Bacterial Growth

Suppose a culture of 100 bacteria is put into a petri dish and the culture doubles every hour. Predict when the number of bacteria will be 350,000.

SOLUTION

Model

$$
\begin{aligned}
200 &= 100 \cdot 2 && \text{Total bacteria after 1 hr} \\
400 &= 100 \cdot 2^2 && \text{Total bacteria after 2 hr} \\
800 &= 100 \cdot 2^3 && \text{Total bacteria after 3 hr} \\
&\vdots \\
P(t) &= 100 \cdot 2^t && \text{Total bacteria after } t \text{ hr}
\end{aligned}
$$

So the function $P(t) = 100 \cdot 2^t$ represents the bacterial population t hours after it is placed in the petri dish.

Solve Graphically Figure 3.12 shows that the population function intersects $y = 350{,}000$ when $t \approx 11.77$.

Interpret The population of the bacteria in the petri dish will be 350,000 in about 11 hr 46 min.

Now try Exercise 15.

Exponential decay functions model the amount of a radioactive substance present in a sample. The number of atoms of a specific element that change from a radioactive state to a nonradioactive state is a fixed fraction per unit time. The process is called **radioactive decay**, and the time it takes for half of a sample to change its state is the **half-life** of the radioactive substance.

EXAMPLE 4 Modeling Radioactive Decay

Suppose the half-life of a certain radioactive substance is 20 days and there are 5 g (grams) present initially. Find the time when there will be 1 g of the substance remaining.

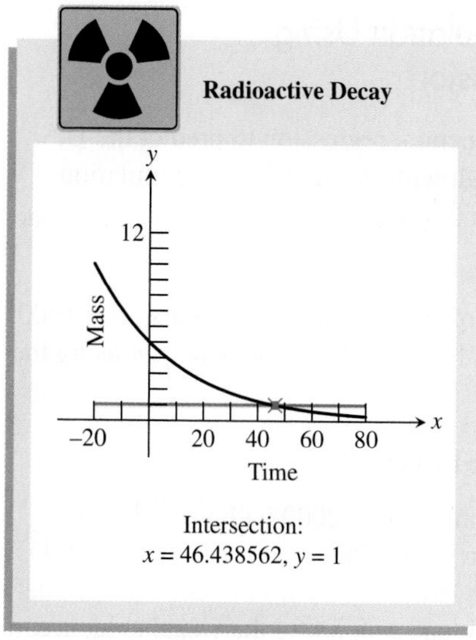

FIGURE 3.13 Radioactive decay. (Example 4)

SOLUTION

Model If t is the time in days, the number of half-lives will be $t/20$.

$$\frac{5}{2} = 5\left(\frac{1}{2}\right)^{20/20} \quad \text{Grams after 20 days}$$

$$\frac{5}{4} = 5\left(\frac{1}{2}\right)^{40/20} \quad \text{Grams after } 2(20) = 40 \text{ days}$$

$$\vdots$$

$$f(t) = 5\left(\frac{1}{2}\right)^{t/20} \quad \text{Grams after } t \text{ days}$$

Thus the function $f(t) = 5 \cdot 0.5^{t/20}$ models the mass in grams of the radioactive substance at time t.

Solve Graphically Figure 3.13 shows that the graph of $f(t) = 5 \cdot 0.5^{t/20}$ intersects $y = 1$ when $t \approx 46.44$.

Interpret There will be 1 g of the radioactive substance left after approximately 46.44 days, or about 46 days, 11 hr. **Now try Exercise 33.**

Scientists have established that atmospheric pressure at sea level is 14.7 lb/in.^{2}, and the pressure is reduced by half for each 3.6 mi above sea level. For example, the pressure 3.6 mi above sea level is $(1/2)(14.7) = 7.35$ lb/in.^{2}. This rule for atmospheric pressure holds for altitudes up to 50 mi above sea level. Though the context is different, the mathematics of atmospheric pressure closely resembles the mathematics of radioactive decay.

EXAMPLE 5 Determining Altitude from Atmospheric Pressure

Find the altitude above sea level at which the atmospheric pressure is 4 lb/in.^{2}.

SOLUTION

Model

$$7.35 = 14.7 \cdot 0.5^{3.6/3.6} \quad \text{Pressure at 3.6 mi}$$

$$3.675 = 14.7 \cdot 0.5^{7.2/3.6} \quad \text{Pressure at } 2(3.6) = 7.2 \text{ mi}$$

$$\vdots$$

$$P(h) = 14.7 \cdot 0.5^{h/3.6} \quad \text{Pressure at } h \text{ mi}$$

So $P(h) = 14.7 \cdot 0.5^{h/3.6}$ models the atmospheric pressure P (in pounds per square inch) as a function of the height h (in miles above sea level). We must find the value of h that satisfies the equation

$$14.7 \cdot 0.5^{h/3.6} = 4.$$

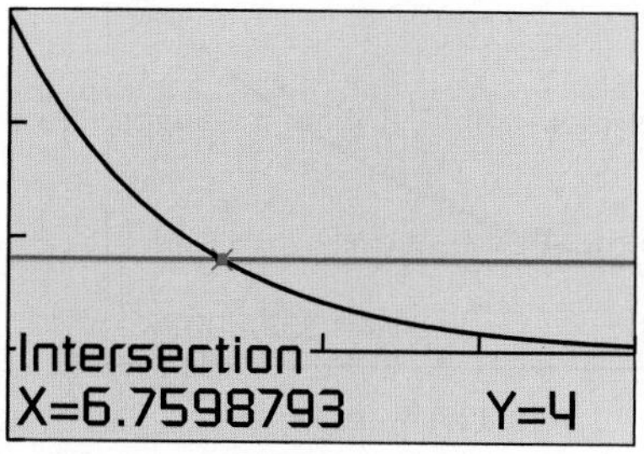

[0, 20] by [–4, 15]

FIGURE 3.14 A model for atmospheric pressure. (Example 5)

Solve Graphically Figure 3.14 shows that the graph of $P(h) = 14.7 \cdot 0.5^{h/3.6}$ intersects $y = 4$ when $h \approx 6.76$.

Interpret The atmospheric pressure is 4 lb/in.^{2} at an altitude of approximately 6.76 mi above sea level. **Now try Exercise 41.**

Using Regression to Model Population

So far, our models have been given to us or developed algebraically. We now use exponential and logistic regression to build models from population data.

Due to the post–World War II baby boom and other factors, exponential growth is not a perfect model for the U.S. population. It does, however, provide a means to make approximate predictions, as illustrated in Example 6.

Table 3.9 U.S. Population (in millions)

Year	Population
1900	76.2
1910	92.2
1920	106.0
1930	123.2
1940	132.2
1950	151.3
1960	179.3
1970	203.3
1980	226.5
1990	248.7
2000	281.4
2011	311.6

Source: World Almanac and Book of Facts 2013.

EXAMPLE 6 Modeling U.S. Population Using Exponential Regression

Use the 1900–2000 data in Table 3.9 and exponential regression to predict the U.S. population for the year 2011. Compare the result with the actual 2011 population shown in the table.

SOLUTION

Model Let $P(t)$ be the population in millions of the United States t years after 1900. Using exponential regression, we find a model for the 1900–2000 data (*not* using the 2011 data):

$$P(t) = 80.5514 \cdot 1.01289^t$$

Figure 3.15a shows a scatter plot of the data from 1900 to 2000 with a graph of the regression equation found. The black "x" on the graph is the data point $(100, 281.4)$ for the year 2000. You can see the curve fits the data very well. The coefficient of determination is $r^2 \approx 0.995$, indicating a close fit and supporting the visual evidence.

Figure 3.15b shows the scatter plot of the data that includes the year 2011 together with a graph of the regression equation based on the data from 1900 to 2000. The black "x" shows the location of the data pair $(111, 311.6)$ for the year 2011.

Solve Numerically and Support Graphically To predict the 2011 U.S. population, we compute $P(111) = 80.5514 \cdot 1.01289^{111} = 333.69027$. In Figure 3.15c the black "x" denotes the *predicted* population for the year 2011 using the regression equation. The open red box below the "x" on the graph is based on the actual U.S. population value for 2011, which is shown in Figure 3.15b. So, using the regression model of the data from year 1900 to 2000, we have *overestimated* the population by about 22 million persons. The visual and numerical evidence in Figure 3.15c supports that the predicted value is an overestimate of the actual population for 2011.

Now try Exercise 43.

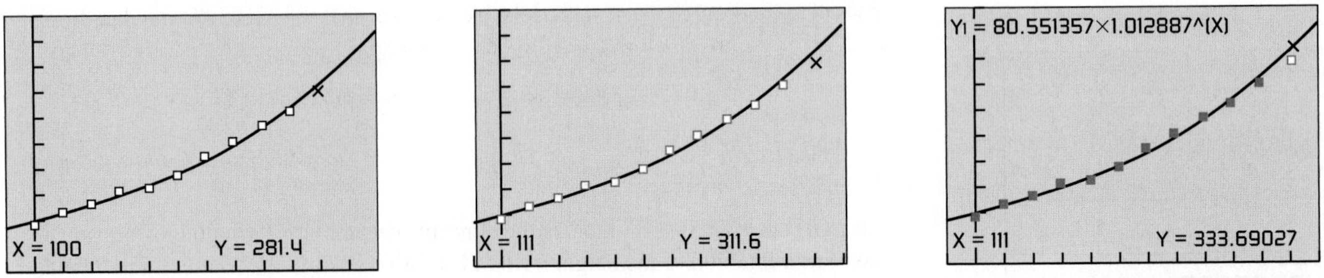

[–10, 120] by [0, 400]
(a)

[–10, 120] by [0, 400]
(b)

[–10, 120] by [0, 400]
(c)

FIGURE 3.15 Graphs of the regression model in Example 6, with scatter plots that (a) exclude the 2011 data point, (b) include the 2011 data point, (c) include the 2011 data point and the predicted value for 2011. In each case, the coordinates shown are those of the black "x".

Exponential growth is unrestricted, but population growth often is not. For many populations, the growth begins exponentially, but eventually slows and approaches a limit to growth called the **maximum sustainable population**.

In Section 3.1 we modeled Dallas's population with a logistic function. We now use logistic regression to do the same for the populations of Florida and Pennsylvania. As the data in Table 3.10 suggest, Florida had rapid growth in the second half of the 20th century, whereas Pennsylvania appears to be approaching its maximum sustainable population.

Table 3.10 Populations of Two U.S. States (in millions)

Year	Florida	Pennsylvania
1900	0.5	6.3
1910	0.8	7.7
1920	1.0	8.7
1930	1.5	9.6
1940	1.9	9.9
1950	2.8	10.5
1960	5.0	11.3
1970	6.8	11.8
1980	9.7	11.9
1990	12.9	11.9
2000	16.0	12.3
2010	18.8	12.7

Source: U.S. Census Bureau.

EXAMPLE 7 Modeling Two States' Populations Using Logistic Regression

Use the data in Table 3.10 and logistic regression to predict the maximum sustainable populations for Florida and Pennsylvania. Graph the logistic models and interpret their significance.

SOLUTION Let $F(t)$ and $P(t)$ be the populations (in millions) of Florida and Pennsylvania, respectively, t years *after 1800*. Figure 3.16a shows a scatter plot of the data for both states; the data for Florida are shown in black, and for Pennsylvania, in red. Using logistic regression, we obtain the models for the two states:

$$F(t) = \frac{26.697023}{1 + 81.903863e^{-0.048030t}} \quad \text{and} \quad P(t) = \frac{12.802428}{1 + 0.957912e^{-0.032184t}}$$

Figure 3.16b shows the scatter plots of the data together with graphs of the two logistic regression population models. You can see that the curves fit the data fairly well. From the numerators of the models we see that

$$\lim_{t \to \infty} F(t) = 26.697 \quad \text{and} \quad \lim_{t \to \infty} P(t) = 12.802.$$

So the maximum sustainable population for Florida is about 26.7 million, and for Pennsylvania is about 12.8 million.

Figure 3.16b also shows a two-century span for the two states. Pennsylvania had rapid growth in the 19th century and first half of the 20th century, and is now approaching its limit to growth. Florida, on the other hand, is currently experiencing extremely rapid growth but should be approaching its maximum sustainable population by the end of the 21st century. **Now try Exercise 50.**

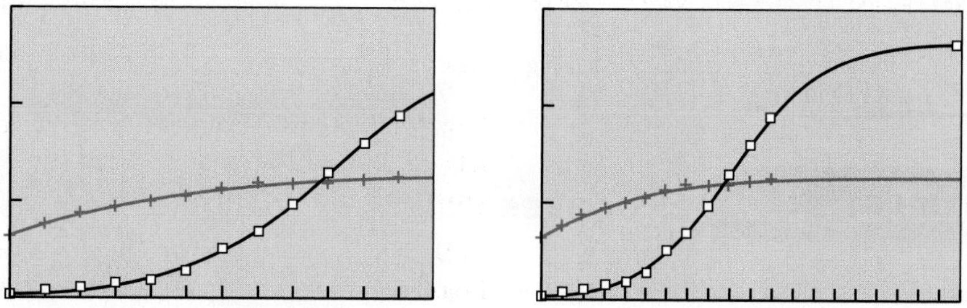

[0, 120] by [0, 30]
(a)

[0, 200] by [0, 30]
(b)

FIGURE 3.16 Scatter plots and graphs for Example 7.

Other Logistic Models

In Example 3, the bacteria cannot continue to grow exponentially forever because they cannot grow beyond the confines of the petri dish. In Example 7, though Florida's population is booming now, it will eventually level off, just as Pennsylvania's has done. Sunflowers and many other plants grow to a natural height following a logistic pattern. Chemical acid-base titration curves are logistic. Yeast cultures grow logistically. Contagious diseases and even rumors spread according to logistic models.

EXAMPLE 8 Modeling a Rumor

Watauga High School has 1200 students. Bob, Carol, Ted, and Alice start a rumor, which spreads logistically so that $S(t) = 1200/(1 + 39e^{-0.9t})$ models the number of students who have heard the rumor by the end of Day t.

(a) How many students have heard the rumor by the end of Day 0?

(b) How long does it take for 1000 students to hear the rumor?

SOLUTION

(a) $S(0) = \dfrac{1200}{1 + 39e^{-0.9 \cdot 0}} = \dfrac{1200}{1 + 39} = 30.$ So, 30 students have heard the rumor by the end of Day 0.

(b) We need to solve $\dfrac{1200}{1 + 39e^{-0.9t}} = 1000.$

Figure 3.17 shows that the graph of $S(t) = 1200/(1 + 39e^{-0.9t})$ intersects $y = 1000$ when $t \approx 5.86$. So toward the end of Day 6 the rumor has reached the ears of 1000 students.

Now try Exercise 45.

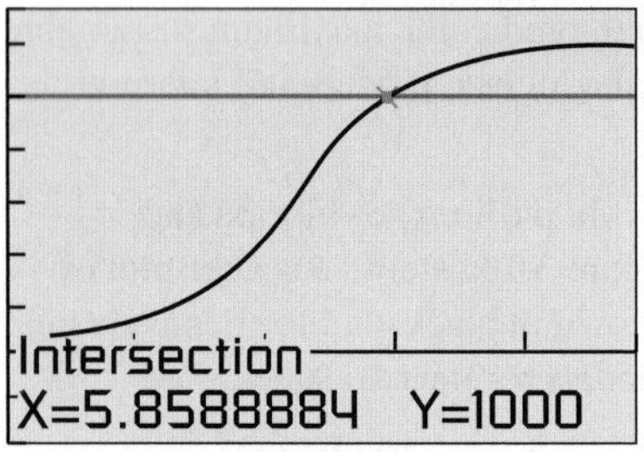

[0, 10] by [−400, 1400]

FIGURE 3.17 The spread of a rumor. (Example 8)

QUICK REVIEW 3.2 *(For help, go to Section P.5.)*

Exercise numbers with a gray background indicate problems that the authors have designed to be solved *without a calculator.*

In Exercises 1 and 2, convert the percent to decimal form or the decimal into a percent.

1. 15%

2. 0.04

3. Show how to increase 23 by 7% using a single multiplication.

4. Show how to decrease 52 by 4% using a single multiplication.

In Exercises 5 and 6, solve the equation algebraically.

5. $40 \cdot b^2 = 160$

6. $243 \cdot b^3 = 9$

In Exercises 7–10, solve the equation numerically.

7. $782b^6 = 838$

8. $93b^5 = 521$

9. $672b^4 = 91$

10. $127b^7 = 56$

SECTION 3.2 Exercises

In Exercises 1–6, tell whether the function is an exponential growth function or exponential decay function, and find the constant percentage rate of growth or decay.

1. $P(t) = 3.5 \cdot 1.09^t$

2. $P(t) = 4.3 \cdot 1.018^t$

3. $f(x) = 78{,}963 \cdot 0.968^x$

4. $f(x) = 5607 \cdot 0.9968^x$

5. $g(t) = 247 \cdot 2^t$

6. $g(t) = 43 \cdot 0.05^t$

In Exercises 7–18, determine the exponential function that satisfies the given conditions.

7. Initial value = 5, increasing at a rate of 17% per year

8. Initial value = 52, increasing at a rate of 2.3% per day

9. Initial value = 16, decreasing at a rate of 50% per month

10. Initial value = 5, decreasing at a rate of 0.59% per week

11. Initial population = 28,900, decreasing at a rate of 2.6% per year

12. Initial population = 502,000, increasing at a rate of 1.7% per year

13. Initial height = 18 cm, growing at a rate of 5.2% per week

14. Initial mass = 15 g, decreasing at a rate of 4.6% per day

15. Initial mass = 0.6 g, doubling every 3 days

16. Initial population = 250, doubling every 7.5 hr

17. Initial mass = 592 g, halving once every 6 years

18. Initial mass = 17 g, halving once every 32 hr

In Exercises 19 and 20, determine a formula for the exponential function whose values are given in Table 3.11.

19. $f(x)$ **20.** $g(x)$

Table 3.11 Values for Two Exponential Functions

x	$f(x)$	$g(x)$
−2	1.472	−9.0625
−1	1.84	−7.25
0	2.3	−5.8
1	2.875	−4.64
2	3.59375	−3.7123

In Exercises 21 and 22, determine a formula for the exponential function whose graph is shown in the figure.

21.

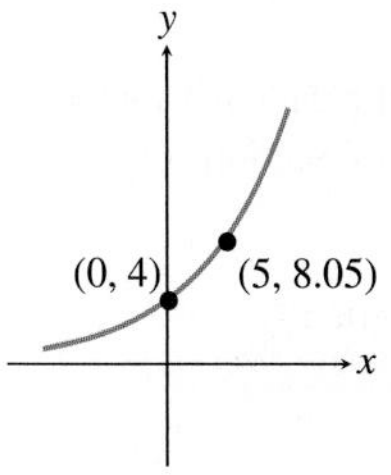

22.

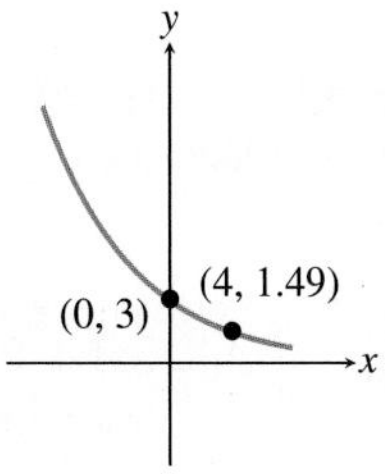

In Exercises 23–26, find the logistic function that satisfies the given conditions.

23. Initial value = 10, limit to growth = 40, passing through $(1, 20)$.

24. Initial value = 12, limit to growth = 60, passing through $(1, 24)$.

25. Initial population = 16, maximum sustainable population = 128, passing through $(5, 32)$.

26. Initial height = 5, limit to growth = 30, passing through $(3, 15)$.

In Exercises 27 and 28, determine a formula for the logistic function whose graph is shown in the figure.

27.

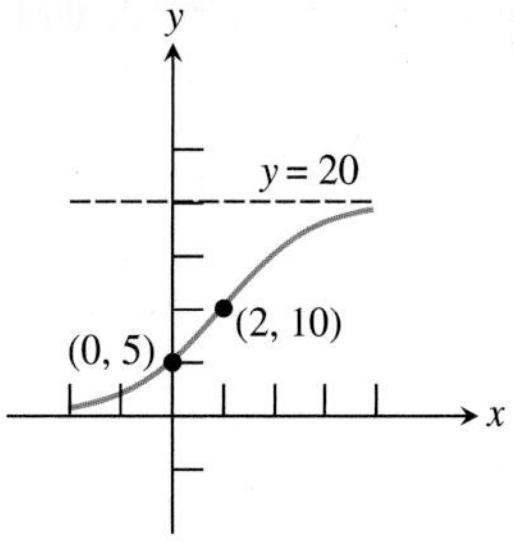

28.

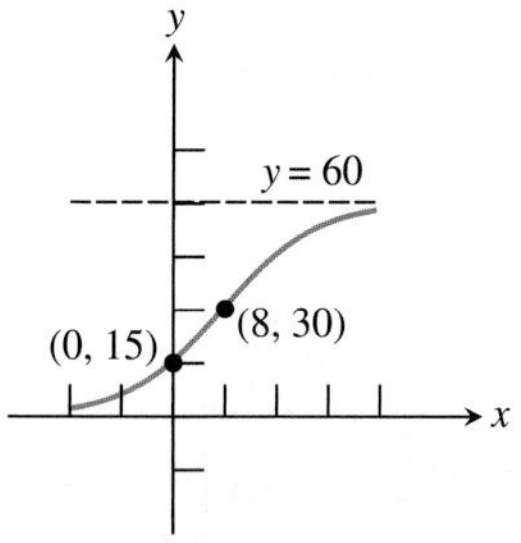

29. Exponential Growth The 2000 population of Jacksonville, Florida, was 736,000 and was increasing at the rate of 1.49% each year. At that rate, when will the population be 1 million?

30. Exponential Growth The 2000 population of Las Vegas, Nevada, was 478,000 and is increasing at the rate of 6.28% each year. At that rate, when will the population be 1 million?

31. Exponential Growth The population of Smallville in the year 1890 was 6250. Assume the population increased at a rate of 2.75% per year.

(a) Estimate the population in 1915 and 1940.

(b) Predict when the population reached 50,000.

32. Exponential Growth The population of River City in the year 1910 was 4200. Assume the population increased at a rate of 2.25% per year.

(a) Estimate the population in 1930 and 1945.

(b) Predict when the population reached 20,000.

33. Radioactive Decay The half-life of a certain radioactive substance is 14 days. There are 6.6 g present initially.

(a) Express the amount of substance remaining as a function of time t.

(b) When will there be less than 1 g remaining?

34. Radioactive Decay The half-life of a certain radioactive substance is 65 days. There are 3.5 g present initially.

(a) Express the amount of substance remaining as a function of time t.

(b) When will there be less than 1 g remaining?

35. Writing to Learn Without using formulas or graphs, compare and contrast exponential functions and linear functions.

36. Writing to Learn Without using formulas or graphs, compare and contrast exponential functions and logistic functions.

37. Writing to Learn Using the population model that is graphed in the figure, explain why the time it takes the population to double (doubling time) is independent of the population size.

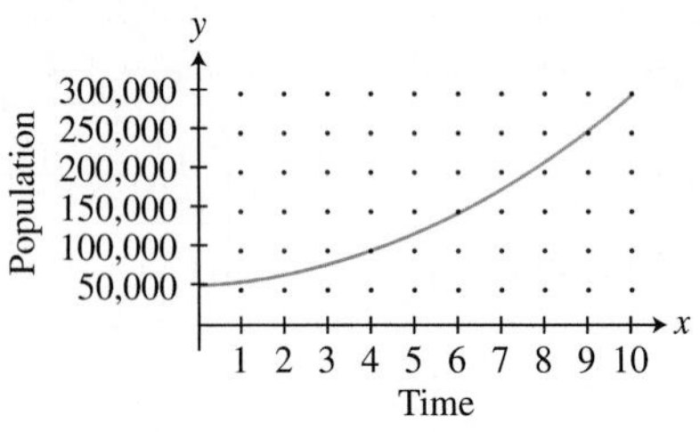

38. Writing to Learn Explain why the half-life of a radioactive substance is independent of the initial amount of the substance that is present.

39. Bacterial Growth The number B of bacteria in a petri dish culture after t hours is given by

$$B = 100e^{0.693t}.$$

When will the number of bacteria be 200? Estimate the doubling time of the bacteria.

40. Radiocarbon Dating The amount C in grams of carbon-14 present in a certain substance after t years is given by

$$C = 20e^{-0.0001216t}.$$

Estimate the half-life of carbon-14.

41. Atmospheric Pressure Determine the atmospheric pressure outside an aircraft flying at 52,800 ft (10 mi above sea level).

42. Atmospheric Pressure Find the altitude above sea level at which the atmospheric pressure is 2.5 lb/in.^{2}.

43. Population Modeling Use the 1950–2000 data in Table 3.12 and exponential regression to predict Los Angeles's population for 2011. Compare the result with the listed value for 2011. [*Hint:* Let 1900 be $t = 0$.]

44. Population Modeling Use the 1950–2000 data in Table 3.12 and exponential regression to predict Phoenix's population for 2011. Compare the result with the listed value for 2011. (*Hint:* Let 1900 be $t = 0$.)

Table 3.12 Populations of Two U.S. Cities (in thousands)

Year	Los Angeles	Phoenix
1950	1970	107
1960	2479	439
1970	2812	584
1980	2969	790
1990	3485	983
2000	3695	1321
2011	3820	1469

Source: World Almanac and Book of Facts 2013.

45. Spread of Flu The number of students infected with flu at Springfield High School after t days is modeled by the function

$$P(t) = \frac{800}{1 + 49e^{-0.2t}}.$$

(a) What was the initial number of infected students?

(b) When will the number of infected students be 200?

(c) The school will close when 300 of the 800-student body are infected. When will the school close?

46. Population of Deer The population of deer after t years in Cedar State Park is modeled by the function

$$P(t) = \frac{1001}{1 + 90e^{-0.2t}}.$$

(a) What was the initial population of deer?

(b) When will the number of deer be 600?

(c) What is the maximum number of deer possible in the park?

47. Population Growth Using all of the data in Table 3.9, compute a logistic regression model, and use it to predict the U.S. population in 2020.

48. Population Growth Using the data in Table 3.13, confirm the model used in Example 8 of Section 3.1.

Table 3.13 Population of Dallas, Texas

Year	Population
1950	434,462
1960	679,684
1970	844,401
1980	904,599
1990	1,006,877
2000	1,188,589

Source: World Almanac and Book of Facts 2013.

49. Population Growth Using the data in Table 3.14, confirm the model used in Exercise 56 of Section 3.1.

Table 3.14 Populations of Two U.S. States (in millions)

Year	Arizona	New York
1900	0.1	7.3
1910	0.2	9.1
1920	0.3	10.3
1930	0.4	12.6
1940	0.5	13.5
1950	0.7	14.8
1960	1.3	16.8
1970	1.8	18.2
1980	2.7	17.6
1990	3.7	18.0
2000	5.1	19.0

Source: World Almanac and Book of Facts 2013.

50 **Population Growth** Using the data in Table 3.14, compute a logistic regression model for Arizona's population for t years since 1900. Based on your model and the New York population model from Exercise 56 of Section 3.1, will the population of Arizona ever surpass that of New York? If so, when?

Standardized Test Questions

51. **True or False** Exponential population growth is constrained with a maximum sustainable population. Justify your answer.

52. **True or False** If the constant percentage rate of an exponential function is negative, then the base of the function is negative. Justify your answer.

In Exercises 53–56, you may use a graphing calculator to solve the problem.

53. **Multiple Choice** What is the constant percentage growth rate of $P(t) = 1.23 \cdot 1.049^t$?

(A) 49% **(B)** 23% **(C)** 4.9% **(D)** 2.3% **(E)** 1.23%

54. **Multiple Choice** What is the constant percentage decay rate of $P(t) = 22.7 \cdot 0.834^t$?

(A) 22.7% **(B)** 16.6% **(C)** 8.34%

(D) 2.27% **(E)** 0.834%

55. **Multiple Choice** A single-cell amoeba doubles every 4 days. About how long will it take one amoeba to produce a population of 1000?

(A) 10 days **(B)** 20 days **(C)** 30 days

(D) 40 days **(E)** 50 days

56. **Multiple Choice** A rumor spreads logistically so that $S(t) = 789/(1 + 16 \cdot e^{-0.8t})$ models the number of persons who have heard the rumor by the end of t days. Based on this model, which of the following is **true**?

(A) After 0 days, 16 persons have heard the rumor.

(B) After 2 days, 439 persons have heard the rumor.

(C) After 4 days, 590 persons have heard the rumor.

(D) After 6 days, 612 persons have heard the rumor.

(E) After 8 days, 769 persons have heard the rumor.

Explorations

57. **Population Growth** **(a)** Use the 1900–1990 data in Table 3.9 and *logistic* regression to predict the U.S. population for 2011.

(b) **Writing to Learn** Compare the prediction with the value listed in the table for 2011.

(c) Noting the results of Example 6, which model—exponential or logistic—makes the better prediction in this case?

58. **Population Growth** Use all of the data in Tables 3.9 and 3.15.

(a) Based on exponential growth models, will Mexico's population surpass that of the United States, and if so, when?

(b) Based on logistic growth models, will Mexico's population surpass that of the United States, and if so, when?

(c) What are the maximum sustainable populations for the two countries?

(d) **Writing to Learn** Which model—exponential or logistic—is more valid in this case? Justify your choice.

Table 3.15 Population of Mexico (in millions)

Year	Population
1900	13.6
1950	25.8
1960	34.9
1970	48.2
1980	66.8
1990	88.1
2001	101.9
2011	115.0

Sources: 1992 Statesman's Yearbook and World Almanac and Book of Facts 2013.

Extending the Ideas

59. The **hyperbolic sine function** is defined by $\sinh(x) = (e^x - e^{-x})/2$. Prove that sinh is an odd function.

60. The **hyperbolic cosine function** is defined by $\cosh(x) = (e^x + e^{-x})/2$. Prove that cosh is an even function.

61. The **hyperbolic tangent function** is defined by $\tanh(x) = (e^x - e^{-x})/(e^x + e^{-x})$.

(a) Prove that $\tanh(x) = \sinh(x)/\cosh(x)$.

(b) Prove that tanh is an odd function.

(c) Prove that $f(x) = 1 + \tanh(x)$ is a logistic function.

3.3 Logarithmic Functions and Their Graphs

What you'll learn about

- Inverses of Exponential Functions
- Common Logarithms—Base 10
- Natural Logarithms—Base *e*
- Graphs of Logarithmic Functions
- Measuring Sound Using Decibels

... and why

Logarithmic functions are used in many applications, including the measurement of the relative intensity of sounds.

Inverses of Exponential Functions

In Section 1.4 we learned that, if a function passes the *horizontal line test*, then the inverse of the function is also a function. Figure 3.18 shows that an exponential function $f(x) = b^x$ would pass the horizontal line test. So it has an inverse that is a function. This inverse is the **logarithmic function with base *b***, denoted $\log_b(x)$, or more simply as $\log_b x$. That is, if $f(x) = b^x$ with $b > 0$ and $b \neq 1$, then $f^{-1}(x) = \log_b x$. See Figure 3.19.

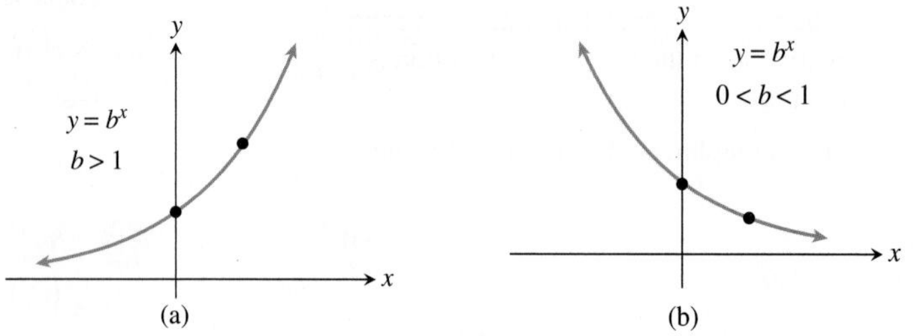

FIGURE 3.18 Exponential functions are either (a) increasing or (b) decreasing.

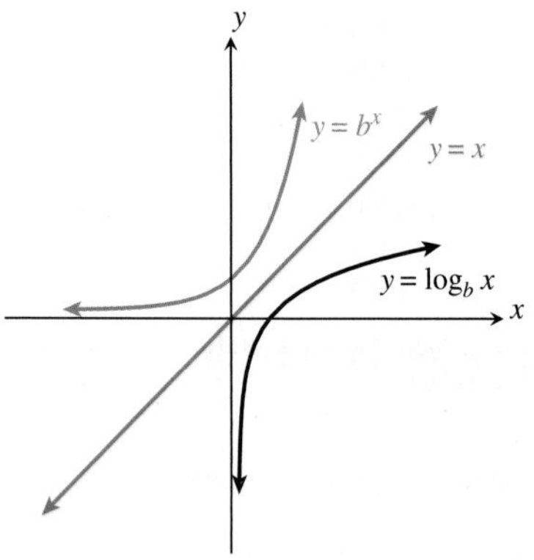

FIGURE 3.19 Because logarithmic functions are inverses of exponential functions, we can obtain the graph of a logarithmic function by the mirror or rotational methods discussed in Section 1.4.

An immediate and useful consequence of this definition is the link between an exponential equation and its logarithmic counterpart.

Changing Between Logarithmic and Exponential Form

If $x > 0$ and $0 < b \neq 1$, then

$$y = \log_b(x) \quad \text{if and only if} \quad b^y = x.$$

This linking statement says that *a logarithm is an exponent*. Because logarithms are exponents, we can evaluate simple logarithmic expressions using our understanding of exponents.

EXAMPLE 1 Evaluating Logarithms

(a) $\log_2 8 = 3$ because $2^3 = 8$.

(b) $\log_3 \sqrt{3} = 1/2$ because $3^{1/2} = \sqrt{3}$.

(c) $\log_5 \dfrac{1}{25} = -2$ because $5^{-2} = \dfrac{1}{5^2} = \dfrac{1}{25}$.

(d) $\log_4 1 = 0$ because $4^0 = 1$.

(e) $\log_7 7 = 1$ because $7^1 = 7$.

Now try Exercise 1.

We can generalize the relationships observed in Example 1.

Basic Properties of Logarithms

For $0 < b \neq 1$, $x > 0$, and any real number y,

- $\log_b 1 = 0$ because $b^0 = 1$.
- $\log_b b = 1$ because $b^1 = b$.
- $\log_b b^y = y$ because $b^y = b^y$.
- $b^{\log_b x} = x$ because $\log_b x = \log_b x$.

A Bit of History

Logarithmic functions were developed around 1614 as computational tools by Scottish mathematician John Napier (1550–1617). He originally called them "artificial numbers," but changed the name to logarithms, which means "reckoning numbers."

Generally $b > 1$

In practice, logarithmic bases are almost always greater than 1.

These properties give us efficient ways to evaluate simple logarithms and some exponential expressions. The first two parts of Example 2 are the same as the first two parts of Example 1.

EXAMPLE 2 Evaluating Logarithmic and Exponential Expressions

(a) $\log_2 8 = \log_2 2^3 = 3.$

(b) $\log_3 \sqrt{3} = \log_3 3^{1/2} = 1/2.$

(c) $6^{\log_6 11} = 11.$

Now try Exercise 5.

Logarithmic functions are inverses of exponential functions. So the inputs and outputs are switched. Table 3.16 illustrates this relationship for $f(x) = 2^x$ and $f^{-1}(x) = \log_2 x$.

Table 3.16 An Exponential Function and Its Inverse

x	$f(x) = 2^x$	x	$f^{-1}(x) = \log_2 x$
-3	1/8	1/8	-3
-2	1/4	1/4	-2
-1	1/2	1/2	-1
0	1	1	0
1	2	2	1
2	4	4	2
3	8	8	3

This relationship can be used to produce both tables and graphs for logarithmic functions, as you will discover in Exploration 1.

EXPLORATION 1 **Comparing Exponential and Logarithmic Functions**

1. Set your grapher to Parametric mode and Simultaneous graphing mode.

Set X1T = T and Y1T = 2^T.

Set X2T = 2^T and Y2T = T.

Creating Tables. Set TblStart = -3 and ΔTbl = 1. Use the Table feature of your grapher to obtain the decimal form of both parts of Table 3.16. Be sure to scroll to the right to see X2T and Y2T.

Drawing Graphs. Set Tmin = -6, Tmax = 6, and Tstep = 0.5. Set the (x, y) window to $[-6, 6]$ by $[-4, 4]$. Use the Graph feature to obtain the simultaneous graphs of $f(x) = 2^x$ and $f^{-1}(x) = \log_2 x$. Use the Trace feature to explore the numerical relationships within the graphs.

2. *Graphing in Function mode.* Graph $y = 2^x$ in the same window. Then use the "draw inverse" command to draw the graph of $y = \log_2 x$.

Common Logarithms—Base 10

Logarithms with base 10 are called **common logarithms**. Because of their connection to our base-ten number system, the metric system, and scientific notation, common logarithms are especially useful. We often drop the subscript of 10 for the base when using common logarithms. The common logarithmic function $\log_{10} x = \log x$ is the inverse of the exponential function $f(x) = 10^x$. So

$$y = \log x \quad \text{if and only if} \quad 10^y = x.$$

Applying this relationship, we can obtain other relationships for logarithms with base 10.

Basic Properties of Common Logarithms

Let x and y be real numbers with $x > 0$.

- $\log 1 = 0$ because $10^0 = 1$.
- $\log 10 = 1$ because $10^1 = 10$.
- $\log 10^y = y$ because $10^y = 10^y$.
- $10^{\log x} = x$ because $\log x = \log x$.

Some Words of Warning

In Figure 3.20, notice we used "10^Ans" instead of "10^1.537819095" to check log (34.5). This is because graphers generally store more digits than they display and so we can obtain a more accurate check. Even so, because log (34.5) is an irrational number, a grapher cannot produce its exact value, thus checks like those shown in Figure 3.20 may not always work out so perfectly.

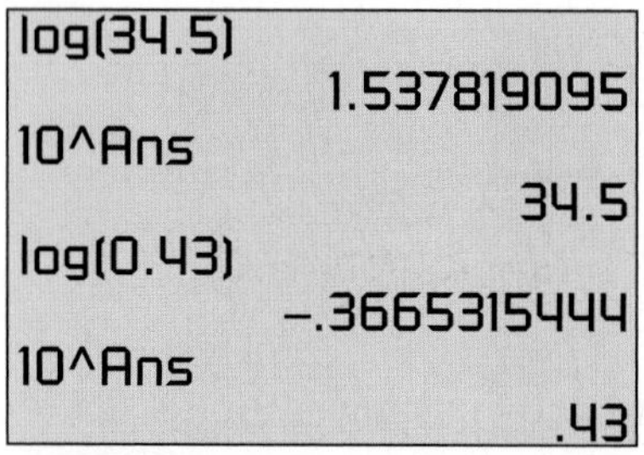

FIGURE 3.20 Doing and checking common logarithmic computations. (Example 4)

Using the definition of common logarithm or these basic properties, we can evaluate expressions involving a base of 10.

EXAMPLE 3 Evaluating Logarithmic and Exponential Expressions—Base 10

(a) $\log 100 = \log_{10} 100 = 2$ because $10^2 = 100$.

(b) $\log \sqrt[5]{10} = \log 10^{1/5} = \dfrac{1}{5}$.

(c) $\log \dfrac{1}{1000} = \log \dfrac{1}{10^3} = \log 10^{-3} = -3$.

(d) $10^{\log 6} = 6$.

Now try Exercise 7.

Common logarithms can be evaluated by using the LOG key on a calculator, as illustrated in Example 4.

EXAMPLE 4 Evaluating Common Logarithms with a Calculator

Use a calculator to evaluate the logarithmic expression if it is defined, and check your result by evaluating the corresponding exponential expression.

(a) $\log 34.5 = 1.537\ldots$ because $10^{1.537\ldots} = 34.5$.

(b) $\log 0.43 = -0.366\ldots$ because $10^{-0.366\ldots} = 0.43$.

See Figure 3.20.

(c) $\log(-3)$ is undefined because there is no real number y such that $10^y = -3$. A grapher will yield either an error message or a complex-number answer for entries such as $\log(-3)$. We shall restrict the domain of logarithmic functions to the set of positive real numbers and ignore such complex-number answers.

Now try Exercise 25.

Changing from logarithmic form to exponential form sometimes is enough to solve an equation involving logarithmic functions.

EXAMPLE 5 Solving Simple Logarithmic Equations

Solve each equation by changing it to exponential form.

(a) $\log x = 3$ **(b)** $\log_2 x = 5$

SOLUTION

(a) Changing to exponential form, $x = 10^3 = 1000$.

(b) Changing to exponential form, $x = 2^5 = 32$.

Now try Exercise 33.

Natural Logarithms—Base e

Because of their special calculus properties, logarithms with the natural base e are used in many situations. Logarithms with base e are **natural logarithms**. We often use the special abbreviation "ln" (without a subscript) to denote a natural logarithm. Thus, the natural logarithmic function $\log_e x = \ln x$. It is the inverse of the exponential function $f(x) = e^x$. So

$$y = \ln x \quad \text{if and only if} \quad e^y = x.$$

Applying this relationship, we can obtain other fundamental relationships for logarithms with the natural base e.

Reading a Natural Log

The expression $\ln x$ is pronounced "el en of ex." The "l" is for logarithm, and the "n" is for natural.

Basic Properties of Natural Logarithms

Let x and y be real numbers with $x > 0$.

- $\ln 1 = 0$ because $e^0 = 1$.
- $\ln e = 1$ because $e^1 = e$.
- $\ln e^y = y$ because $e^y = e^y$.
- $e^{\ln x} = x$ because $\ln x = \ln x$.

Using the definition of natural logarithm or these basic properties, we can evaluate expressions involving the natural base e.

EXAMPLE 6 Evaluating Logarithmic and Exponential Expressions—Base e

(a) $\ln \sqrt{e} = \log_e \sqrt{e} = 1/2$ because $e^{1/2} = \sqrt{e}$.

(b) $\ln e^5 = \log_e e^5 = 5$.

(c) $e^{\ln 4} = 4$.

Now try Exercise 13.

Natural logarithms can be evaluated by using the LN key on a calculator, as illustrated in Example 7.

EXAMPLE 7 Evaluating Natural Logarithms with a Calculator

Use a calculator to evaluate the logarithmic expression, if it is defined, and check your result by evaluating the corresponding exponential expression.

(a) $\ln 23.5 = 3.157\ldots$ because $e^{3.157\ldots} = 23.5$.

(b) $\ln 0.48 = -0.733\ldots$ because $e^{-0.733\ldots} = 0.48$.

See Figure 3.21.

(c) $\ln(-5)$ is undefined because there is no real number y such that $e^y = -5$. A grapher will yield either an error message or a complex-number answer for entries such as $\ln(-5)$. We will continue to restrict the domain of logarithmic functions to the set of positive real numbers and ignore such complex-number answers.

Now try Exercise 29.

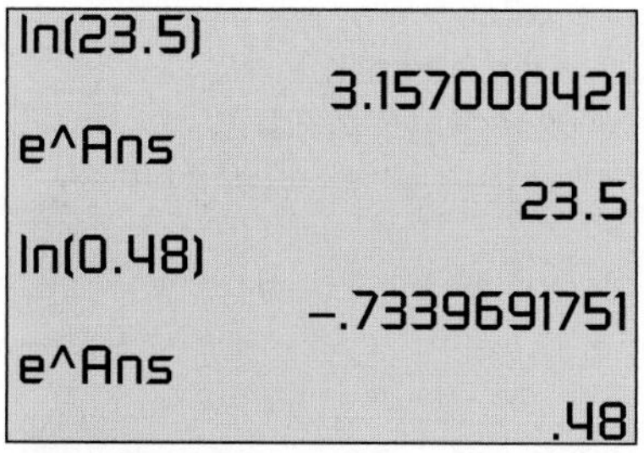

FIGURE 3.21 Doing and checking natural logarithmic computations. (Example 7)

Graphs of Logarithmic Functions

The natural logarithmic function $f(x) = \ln x$ is one of the basic functions introduced in Section 1.3. We now list its properties.

BASIC FUNCTION The Natural Logarithmic Function

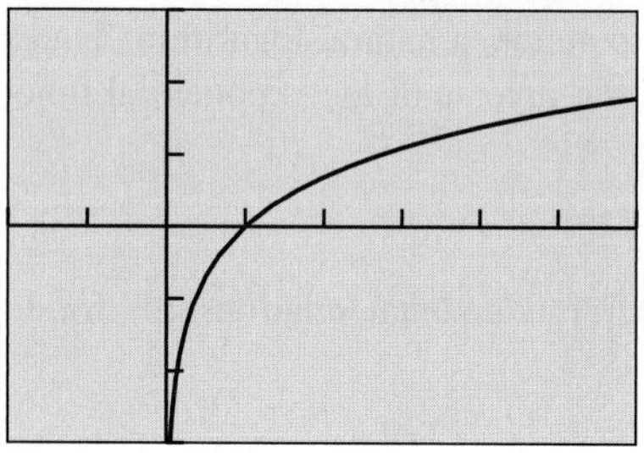

[–2, 6] by [–3, 3]

FIGURE 3.22

$f(x) = \ln x$
Domain: $(0, \infty)$
Range: $(-\infty, \infty)$
Continuous on $(0, \infty)$
Increasing on $(0, \infty)$
No symmetry
Not bounded above or below
No local extrema
No horizontal asymptotes
Vertical asymptote: $x = 0$
End behavior: $\lim_{x \to \infty} \ln x = \infty$

Any logarithmic function $g(x) = \log_b x$ with $b > 1$ has the same domain, range, continuity, increasing behavior, lack of symmetry, and other general behavior as $f(x) = \ln x$. It is rare that we are interested in logarithmic functions $g(x) = \log_b x$ with $0 < b < 1$. So, the graph and behavior of $f(x) = \ln x$ are typical of logarithmic functions.

We now consider the graphs of the common and natural logarithmic functions and their geometric transformations. To understand the graphs of $y = \log x$ and $y = \ln x$, we can compare each to the graph of its inverse, $y = 10^x$ and $y = e^x$, respectively. Figure 3.23a shows that the graphs of $y = \ln x$ and $y = e^x$ are reflections of each other across the line $y = x$. Similarly, Figure 3.23b shows that the graphs of $y = \log x$ and $y = 10^x$ are reflections of each other across this same line.

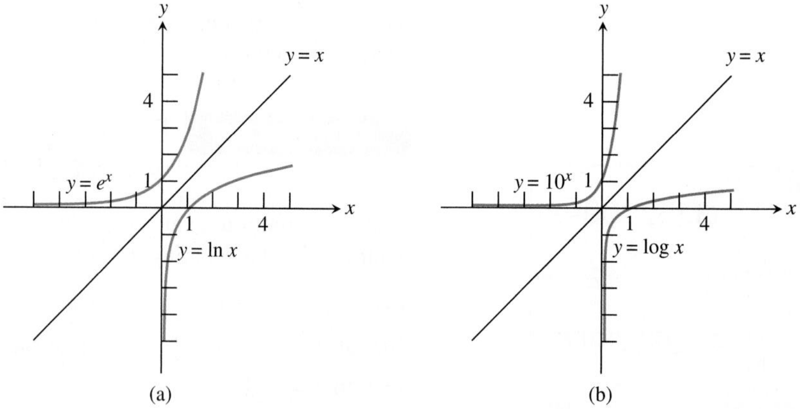

FIGURE 3.23 Two pairs of inverse functions.

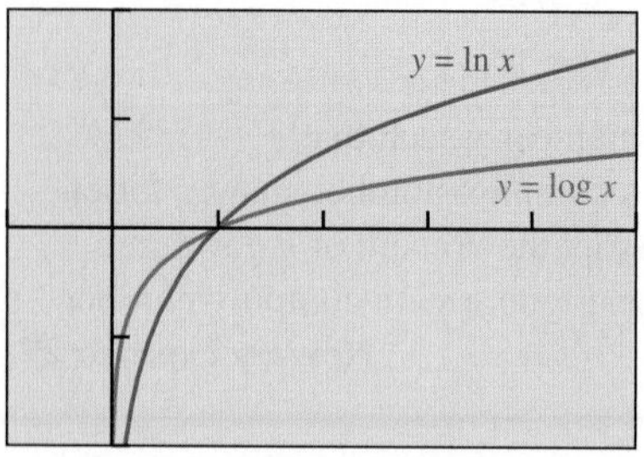

[–1, 5] by [–2, 2]

FIGURE 3.24 The graphs of the common and natural logarithmic functions.

From Figure 3.24 we can see that the graphs of $y = \log x$ and $y = \ln x$ have much in common. Figure 3.24 also shows how they differ.

The geometric transformations studied in Section 1.5, together with our knowledge of the graphs of $y = \ln x$ and $y = \log x$, allow us to predict the graphs of the functions in Example 8.

EXAMPLE 8 Transforming Logarithmic Graphs

Describe how to transform the graph of $y = \ln x$ or $y = \log x$ into the graph of the given function.

(a) $g(x) = \ln(x + 2)$ **(b)** $h(x) = \ln(3 - x)$

(c) $g(x) = 3 \log x$ **(d)** $h(x) = 1 + \log x$

SOLUTION

(a) The graph of $g(x) = \ln(x + 2)$ is obtained by translating the graph of $y = \ln(x)$ 2 units to the left. See Figure 3.25a.

(b) $h(x) = \ln(3 - x) = \ln[-(x - 3)]$. So we obtain the graph of $h(x) = \ln(3 - x)$ from the graph of $y = \ln x$ by applying, in order, a reflection across the y-axis followed by a translation 3 units to the right. See Figure 3.25b.

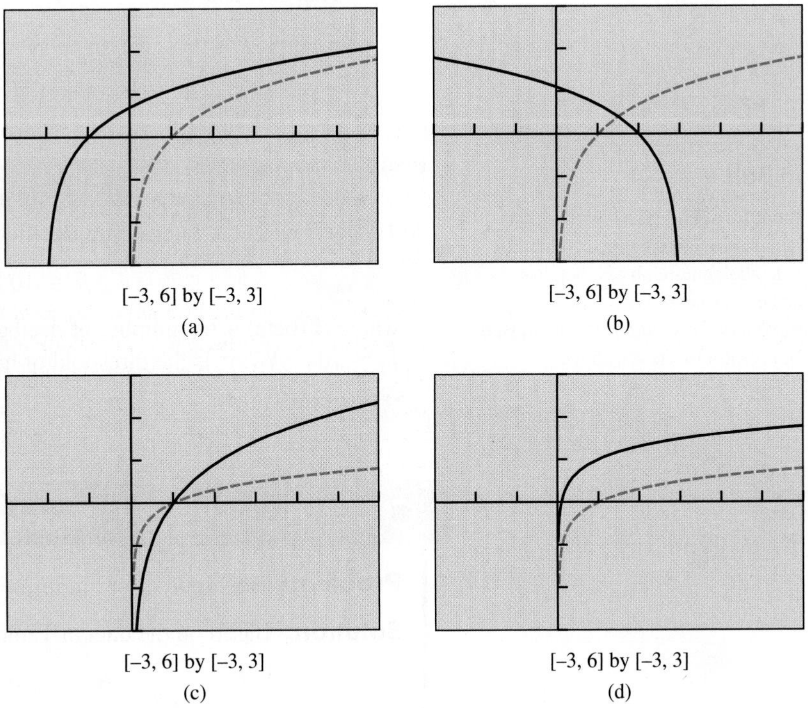

FIGURE 3.25 Transforming $y = \ln x$ to obtain (a) $g(x) = \ln(x + 2)$ and (b) $h(x) = \ln(3 - x)$; and $y = \log x$ to obtain (c) $g(x) = 3 \log x$ and (d) $h(x) = 1 + \log x$. (Example 8)

(c) The graph of $g(x) = 3 \log x$ is obtained by vertically stretching the graph of $f(x) = \log x$ by a factor of 3. See Figure 3.25c.

(d) We can obtain the graph of $h(x) = 1 + \log x$ from the graph of $f(x) = \log x$ by a translation 1 unit up. See Figure 3.25d. **Now try Exercise 41.**

Measuring Sound Using Decibels

Table 3.17 lists assorted sounds. Notice that a jet at takeoff is 100 trillion times as loud as a soft whisper. Because the range of audible sound intensities is so great, common logarithms (powers of 10) are used to compare how loud sounds are.

Sound Intensity

Sound intensity is the energy per unit time of a sound wave over a given area, and is measured in watts per square meter (W/m^2).

Table 3.17 Approximate Intensities of Selected Sounds

Sound	Intensity (W/m^2)
Hearing threshold	10^{-12}
Soft whisper at 5 m	10^{-11}
City traffic	10^{-5}
Subway train	10^{-2}
Pain threshold	10^{0}
Jet at takeoff	10^{3}

Source: Adapted from R. W. Reading, Exploring Physics: Concepts and Applications. Belmont, CA: Wadsworth, 1984.

Bel Is for Bell

The original unit for sound intensity level was the *bel* (B), which proved to be inconveniently large. So the decibel, one-tenth of a bel, has replaced it. The bel was named in honor of Scottish-born American Alexander Graham Bell (1847–1922), inventor of the telephone.

DEFINITION Decibels

The level of sound intensity in **decibels** (dB) is

$$\beta = 10 \log(I/I_0),$$

where β (beta) is the number of decibels, I is the sound intensity in W/m^2, and $I_0 = 10^{-12}$ W/m^2 is the threshold of human hearing (the quietest audible sound intensity).

CHAPTER OPENER Problem (from page 251)

Problem: How loud is a train inside a subway tunnel?

Solution: Based on the data in Table 3.17,

$$\begin{aligned}\beta &= 10 \log(I/I_0)\\ &= 10 \log(10^{-2}/10^{-12})\\ &= 10 \log(10^{10})\\ &= 10 \cdot 10 = 100\end{aligned}$$

So the sound intensity level inside the subway tunnel is 100 dB.

QUICK REVIEW 3.3 *(For help, go to Sections P.1 and A.1.)*

Exercise numbers with a gray background indicate problems that the authors have designed to be solved *without a calculator*.

In Exercises 1–6, evaluate the expression.

1. 5^{-2}
2. 10^{-3}
3. $\frac{4^0}{5}$
4. $\frac{1^0}{2}$
5. $\frac{8^{11}}{2^{28}}$
6. $\frac{9^{13}}{27^8}$

In Exercises 7–10, rewrite as a base raised to a rational number exponent.

7. $\sqrt{5}$
8. $\sqrt[3]{10}$
9. $\frac{1}{\sqrt{e}}$
10. $\frac{1}{\sqrt[3]{e^2}}$

SECTION 3.3 Exercises

In Exercises 1–18, evaluate the logarithmic expression without using a calculator.

1. $\log_4 4$ **2.** $\log_6 1$

3. $\log_2 32$ **4.** $\log_3 81$

5. $\log_5 \sqrt[3]{25}$ **6.** $\log_6 \frac{1}{\sqrt[5]{36}}$

7. $\log 10^3$ **8.** $\log 10{,}000$

9. $\log 100{,}000$ **10.** $\log 10^{-4}$

11. $\log \sqrt[3]{10}$ **12.** $\log \frac{1}{\sqrt{1000}}$

13. $\ln e^3$ **14.** $\ln e^{-4}$

15. $\ln \frac{1}{e}$ **16.** $\ln 1$

17. $\ln \sqrt[4]{e}$ **18.** $\ln \frac{1}{\sqrt{e^7}}$

In Exercises 19–24, evaluate the expression without using a calculator.

19. $7^{\log_7 3}$ **20.** $5^{\log_5 8}$

21. $10^{\log (0.5)}$ **22.** $10^{\log 14}$

23. $e^{\ln 6}$ **24.** $e^{\ln(1/5)}$

In Exercises 25–32, use a calculator to evaluate the logarithmic expression if it is defined, and check your result by evaluating the corresponding exponential expression.

25. $\log 9.43$ **26.** $\log 0.908$

27. $\log(-14)$ **28.** $\log(-5.14)$

29. $\ln 4.05$ **30.** $\ln 0.733$

31. $\ln(-0.49)$ **32.** $\ln(-3.3)$

In Exercises 33–36, solve the equation by changing it to exponential form.

33. $\log x = 2$ **34.** $\log x = 4$

35. $\log x = -1$ **36.** $\log x = -3$

In Exercises 37–40, match the function with its graph.

37. $f(x) = \log(1 - x)$ **38.** $f(x) = \log(x + 1)$

39. $f(x) = -\ln(x - 3)$ **40.** $f(x) = -\ln(4 - x)$

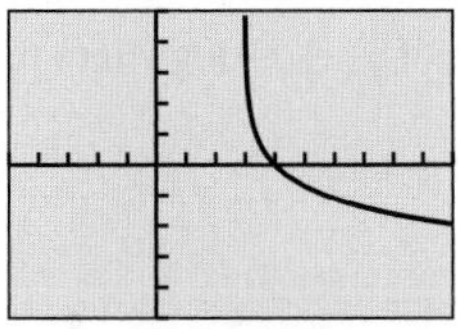
(a)

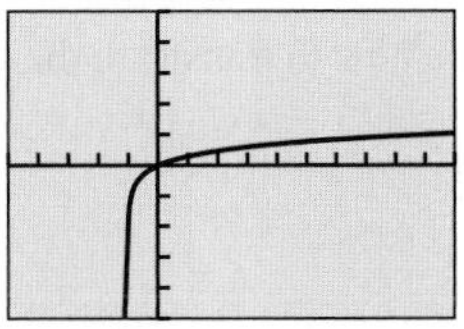
(b)

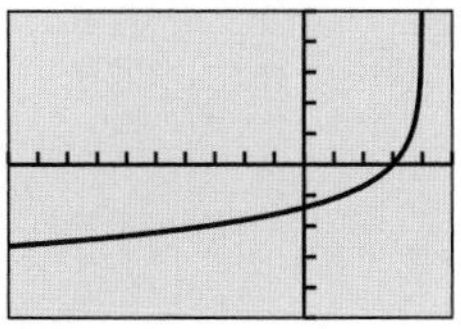
(c)

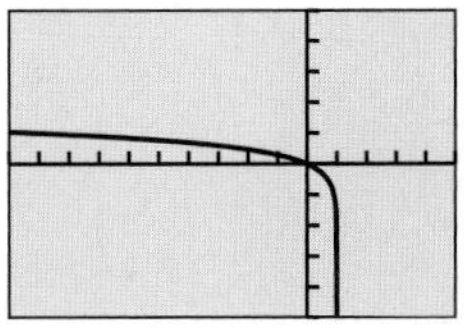
(d)

In Exercises 41–46, describe how to transform the graph of $y = \ln x$ into the graph of the given function. Sketch the graph by hand and support your sketch with a grapher.

41. $f(x) = \ln(x + 3)$ **42.** $f(x) = \ln(x) + 2$

43. $f(x) = \ln(-x) + 3$ **44.** $f(x) = \ln(-x) - 2$

45. $f(x) = \ln(2 - x)$ **46.** $f(x) = \ln(5 - x)$

In Exercises 47–52, describe how to transform the graph of $y = \log x$ into the graph of the given function. Sketch the graph by hand and support with a grapher.

47. $f(x) = -1 + \log(x)$ **48.** $f(x) = \log(x - 3)$

49. $f(x) = -2\log(-x)$ **50.** $f(x) = -3\log(-x)$

51. $f(x) = 2\log(3 - x) - 1$

52. $f(x) = -3\log(1 - x) + 1$

In Exercises 53–58, graph the function, and analyze it for domain, range, continuity, increasing or decreasing behavior, boundedness, extrema, symmetry, asymptotes, and end behavior.

53. $f(x) = \log(x - 2)$ **54.** $f(x) = \ln(x + 1)$

55. $f(x) = -\ln(x - 1)$ **56.** $f(x) = -\log(x + 2)$

57. $f(x) = 3\log(x) - 1$ **58.** $f(x) = 5\ln(2 - x) - 3$

59. Sound Intensity Use the data in Table 3.17 to compute the sound intensity in decibels for **(a)** a soft whisper, **(b)** city traffic, and **(c)** a jet at takeoff.

60. Light Absorption The Beer-Lambert Law of Absorption applied to Lake Erie states that the light intensity I (in lumens), at a depth of x feet, satisfies the equation

$$\log \frac{I}{12} = -0.00235x.$$

Find the intensity of the light at a depth of 30 ft.

61. Earthquake Intensity The Richter magnitude scale was developed in 1935 by Charles F. Richter of the California Institute of Technology as a mathematical device to compare the intensity of earthquakes. Table 3.18 shows the relationship between earthquake magnitude and intensity of an earthquake as measured by ground motion.

Table 3.18 Richter Scale

Magnitude (x)	Ground Motion (y)
1	10
2	100
3	1000
4	10,000
5	100,000
6	1,000,000

(a) What is the difference between an earthquake of magnitude 2 and 3? And of 2 and 5?

(b) Draw the scatter plot of the data in Table 3.18.

(c) Model the data with an exponential function.

(d) Model the data with a logarithmic function

(e) Discuss the advantages of using a logarithmic scale like the Richter Scale.

(f) Are the functions in (c) and (d) inverse functions?

62. Population Decay

(a) Using the data in Table 3.19, find an exponential regression model of the Milwaukee population data.

(b) Graph a scatter plot of the data in Table 3.19 together with its exponential regression model found in (a).

(c) Use the population regression you found in (a) to predict when Milwaukee's population will be 500,000.

(d) Is the model you found realistic? (*Hint:* Graph in the window $[1900, 5000]$ by $[-100{,}000, 1{,}000{,}000]$.) Explain why or why not.

Table 3.19 Population of Milwaukee

Year	Population
1970	717,372
1980	636,297
1990	628,088
2000	596,974
2011	597,867

Source: World Almanac and Book of Facts 2013.

Standardized Test Questions

63. True or False A logarithmic function is the inverse of an exponential function. Justify your answer.

64. True or False Common logarithms are logarithms with base 10. Justify your answer.

In Exercises 65–68, you may use a graphing calculator to solve the problem.

65. Multiple Choice What is the approximate value of the common log of 2?

(A) 0.10523 **(B)** 0.20000

(C) 0.30103 **(D)** 0.69315

(E) 3.32193

66. Multiple Choice Which statement is **false**?

(A) $\log 5 = 2.5 \log 2$ **(B)** $\log 5 = 1 - \log 2$

(C) $\log 5 > \log 2$ **(D)** $\log 5 < \log 10$

(E) $\log 5 = \log 10 - \log 2$

67. Multiple Choice Which statement is **false** about $f(x) = \ln x$?

(A) It is increasing on its domain.

(B) It is symmetric about the origin.

(C) It is continuous on its domain.

(D) It is unbounded.

(E) It has a vertical asymptote.

68. Multiple Choice Which of the following is the inverse of $f(x) = 2 \cdot 3^x$?

(A) $f^{-1}(x) = \log_3 (x/2)$ **(B)** $f^{-1}(x) = \log_2 (x/3)$

(C) $f^{-1}(x) = 2 \log_3 (x)$ **(D)** $f^{-1}(x) = 3 \log_2 (x)$

(E) $f^{-1}(x) = 0.5 \log_3 (x)$

Explorations

69. Writing to Learn Parametric Graphing In the manner of Exploration 1, make tables and graphs for $f(x) = 3^x$ and its inverse $f^{-1}(x) = \log_3 x$. Write a comparative analysis of the two functions regarding domain, range, intercepts, and asymptotes.

70. Writing to Learn Parametric Graphing In the manner of Exploration 1, make tables and graphs for $f(x) = 5^x$ and its inverse $f^{-1}(x) = \log_5 x$. Write a comparative analysis of the two functions regarding domain, range, intercepts, and asymptotes.

71. Group Activity Parametric Graphing In the manner of Exploration 1, find the number $b > 1$ such that the graphs of $f(x) = b^x$ and its inverse $f^{-1}(x) = \log_b x$ have exactly one point of intersection. What is the one point that is in common to the two graphs?

72. Writing to Learn Explain why zero is not in the domain of the logarithmic functions $f(x) = \log_3 x$ and $g(x) = \log_5 x$.

Extending the Ideas

73. Describe how to transform the graph of $f(x) = \ln x$ into the graph of $g(x) = \log_{1/e} x$.

74. Describe how to transform the graph of $f(x) = \log x$ into the graph of $g(x) = \log_{0.1} x$.

3.4 Properties of Logarithmic Functions

What you'll learn about

- Properties of Logarithms
- Change of Base
- Graphs of Logarithmic Functions with Base *b*
- Re-expressing Data

... and why

The applications of logarithms are based on their many special properties, so learn them well.

Properties of Logarithms

Logarithms have special algebraic traits that historically made them indispensable for calculations and that still make them valuable in many areas of application and modeling. In Section 3.3 we learned about the inverse relationship between exponents and logarithms and how to apply some basic properties of logarithms. We now delve deeper into the nature of logarithms to prepare for equation solving and modeling.

Properties of Logarithms

Let b, R, and S be positive real numbers with $b \neq 1$, and c any real number.

- **Product rule:** $\log_b (RS) = \log_b R + \log_b S$
- **Quotient rule:** $\log_b \dfrac{R}{S} = \log_b R - \log_b S$
- **Power rule:** $\log_b R^c = c \log_b R$

Properties of Exponents

Let b, x, and y be real numbers with $b > 0$.

1. $b^x \cdot b^y = b^{x+y}$
2. $\dfrac{b^x}{b^y} = b^{x-y}$
3. $(b^x)^y = b^{xy}$

The properties of exponents in the margin are the basis for these three properties of logarithms. For instance, the first exponent property listed in the margin is used to verify the product rule.

EXAMPLE 1 Proving the Product Rule for Logarithms

Prove $\log_b (RS) = \log_b R + \log_b S$.

SOLUTION Let $x = \log_b R$ and $y = \log_b S$. The corresponding exponential statements are $b^x = R$ and $b^y = S$. Therefore,

$$
\begin{aligned}
RS &= b^x \cdot b^y \\
&= b^{x+y} && \text{First property of exponents} \\
\log_b (RS) &= x + y && \text{Change to logarithmic form.} \\
&= \log_b R + \log_b S && \text{Use the definitions of } x \text{ and } y.
\end{aligned}
$$

Now try Exercise 37.

EXPLORATION 1 Exploring the Arithmetic of Logarithms

Use the 5-decimal-place approximations shown in Figure 3.26 to support the properties of logarithms numerically.

1. Product $\log (2 \cdot 4) = \log 2 + \log 4$
2. Quotient $\log \left(\dfrac{8}{2}\right) = \log 8 - \log 2$
3. Power $\log 2^3 = 3 \log 2$

(continued)

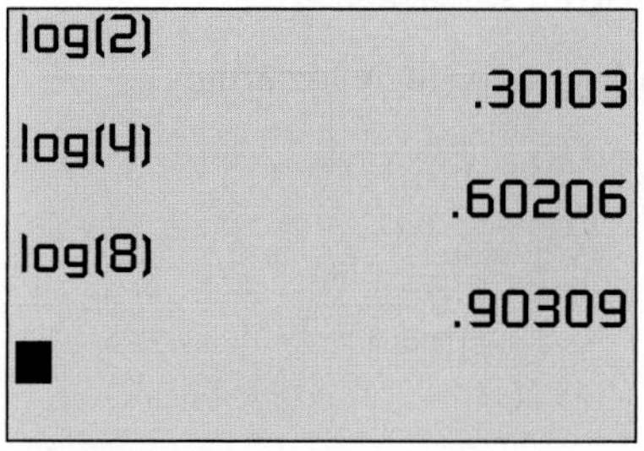

FIGURE 3.26 An arithmetic pattern of logarithms. (Exploration 1)

Now evaluate the common logs of other positive integers using the information given in Figure 3.26 and without using your calculator.

4. Use the fact that $5 = 10/2$ to evaluate log 5.
5. Use the fact that 16, 32, and 64 are powers of 2 to evaluate log 16, log 32, and log 64.
6. Evaluate log 25, log 40, and log 50.

List all of the positive integers less than 100 whose common logs can be evaluated knowing only log 2 and the properties of logarithms and without using a calculator.

A Bit of History: Logarithms and Computation

Logarithms were introduced by John Napier in the early 1600s as a means to simplify tedious calculations. Before logarithms, multidigit multiplication steps were *very* time consuming. A table of values of logarithms was first created by John Briggs in 1624. Using log tables transformed tedious multiplication problems into simple, quick addition problems. The key to using logs for multiplication is to understand that the log of a product is the *sum* of the logs of the factors. That is,

$$\log_b(xy) = \log_b(x) + \log_b(y).$$

(See the Product Rule on page 283). Most precalculus textbooks published before 1900 contained many pages devoted to log tables. With the invention of the electronic calculator log tables became obsolete.

When we solve equations algebraically that involve logarithms, we often have to rewrite expressions using properties of logarithms. Sometimes we need to expand as far as possible, and other times we condense as much as possible. The next three examples illustrate how properties of logarithms can be used to change the form of expressions involving logarithms.

EXAMPLE 2 Expanding the Logarithm of a Product

Assuming x and y are positive, use properties of logarithms to write $\log(8xy^4)$ as a sum of logarithms or multiples of logarithms.

SOLUTION

$\log(8xy^4) = \log 8 + \log x + \log y^4$	Product rule
$= \log 2^3 + \log x + \log y^4$	$8 = 2^3$
$= 3\log 2 + \log x + 4\log y$	Power rule

Now try Exercise 1.

EXAMPLE 3 Expanding the Logarithm of a Quotient

Assuming x is positive, use properties of logarithms to write $\ln(\sqrt{x^2+5}/x)$ as a sum or difference of logarithms or multiples of logarithms.

SOLUTION

$\ln\dfrac{\sqrt{x^2+5}}{x} = \ln\dfrac{(x^2+5)^{1/2}}{x}$	
$= \ln(x^2+5)^{1/2} - \ln x$	Quotient rule
$= \dfrac{1}{2}\ln(x^2+5) - \ln x$	Power rule

Now try Exercise 9.

EXAMPLE 4 Condensing a Logarithmic Expression

Assuming x and y are positive, write $\ln x^5 - 2\ln(xy)$ as a single logarithm.

SOLUTION

$\ln x^5 - 2\ln(xy) = \ln x^5 - \ln(xy)^2$	Power rule
$= \ln x^5 - \ln(x^2y^2)$	
$= \ln\dfrac{x^5}{x^2y^2}$	Quotient rule
$= \ln\dfrac{x^3}{y^2}$	

Now try Exercise 13.

As we have seen, logarithms have some surprising properties. It is easy to overgeneralize and fall into misconceptions about logarithms. Exploration 2 should help you discern what is true and false about logarithmic relationships.

EXPLORATION 2 Discovering Relationships and Nonrelationships

Of the eight relationships suggested here, four are *true* and four are *false* (using values of x within the domains of both sides of the equations). Thinking about the properties of logarithms, make a prediction about the truth of each statement. Then test each with some specific numerical values for x. Finally, compare the graphs of the two sides of the equation.

1. $\ln(x+2) = \ln x + \ln 2$
2. $\log_3(7x) = 7\log_3 x$
3. $\log_2(5x) = \log_2 5 + \log_2 x$
4. $\ln\frac{x}{5} = \ln x - \ln 5$
5. $\log\frac{x}{4} = \frac{\log x}{\log 4}$
6. $\log_4 x^3 = 3\log_4 x$
7. $\log_5 x^2 = (\log_5 x)(\log_5 x)$
8. $\log|4x| = \log 4 + \log|x|$

Which four are true, and which four are false?

Student Investigation

Search the Internet for the term "slide rule."

- What were slide rules?
- When were they invented?
- What were they used for?
- Are they obsolete today?
- Explain why or why not.

Change of Base

When working with a logarithmic expression with an undesirable base, it is possible to change the expression into a quotient of logarithms with a different base. For example, it is hard to evaluate $\log_4 7$ because 7 is not a simple power of 4 and there is no $\boxed{\log_4}$ key on a calculator or grapher.

We can work around this problem with some algebraic trickery. First let $y = \log_4 7$. Then

$$\begin{aligned} 4^y &= 7 && \text{Switch to exponential form.} \\ \ln 4^y &= \ln 7 && \text{Apply ln.} \\ y\ln 4 &= \ln 7 && \text{Power rule} \\ y &= \frac{\ln 7}{\ln 4} && \text{Divide by ln 4.} \end{aligned}$$

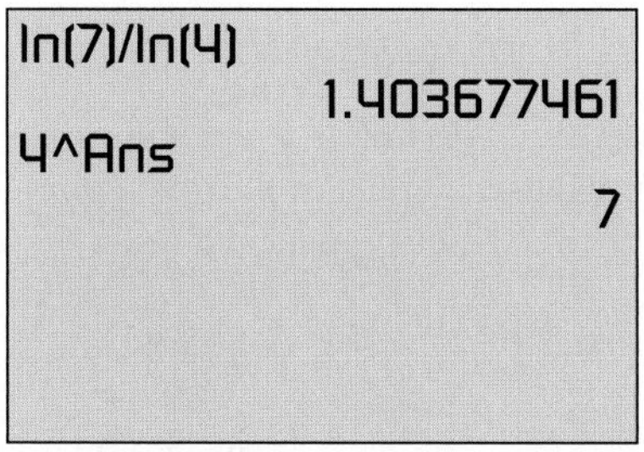

FIGURE 3.27 Evaluating and checking $\log_4 7$.

Using a grapher (Figure 3.27), we see that

$$\log_4 7 = \frac{\ln 7}{\ln 4} = 1.4036\ldots$$

We generalize this useful trickery as the change-of-base formula:

Change-of-Base Formula for Logarithms

For positive real numbers a, b, and x with $a \neq 1$ and $b \neq 1$,

$$\log_b x = \frac{\log_a x}{\log_a b}.$$

Calculators and graphers generally have two logarithm keys—LOG and LN—which correspond to the bases 10 and e, respectively. So we often use the change-of-base formula in one of the following two forms:

$$\log_b x = \frac{\log x}{\log b} \quad \text{or} \quad \log_b x = \frac{\ln x}{\ln b}$$

These two forms are useful in evaluating logarithms and graphing logarithmic functions.

EXAMPLE 5 Evaluating Logarithms by Changing the Base

(a) $\log_3 16 = \dfrac{\ln 16}{\ln 3} = 2.523\ldots \approx 2.52$

(b) $\log_6 10 = \dfrac{\log 10}{\log 6} = \dfrac{1}{\log 6} = 1.285\ldots \approx 1.29$

(c) $\log_{1/2} 2 = \dfrac{\ln 2}{\ln (1/2)} = \dfrac{\ln 2}{\ln 1 - \ln 2} = \dfrac{\ln 2}{-\ln 2} = -1$

Now try Exercise 23.

Graphs of Logarithmic Functions with Base *b*

Using the change-of-base formula we can rewrite any logarithmic function $g(x) = \log_b x$ as

$$g(x) = \frac{\ln x}{\ln b} = \frac{1}{\ln b} \ln x.$$

Therefore, every logarithmic function is a constant multiple of the natural logarithmic function $f(x) = \ln x$. If the base is $b > 1$, the graph of $g(x) = \log_b x$ is a vertical stretch or shrink of the graph of $f(x) = \ln x$ by the factor $1/\ln b$. If $0 < b < 1$, a reflection across the x-axis is required as well.

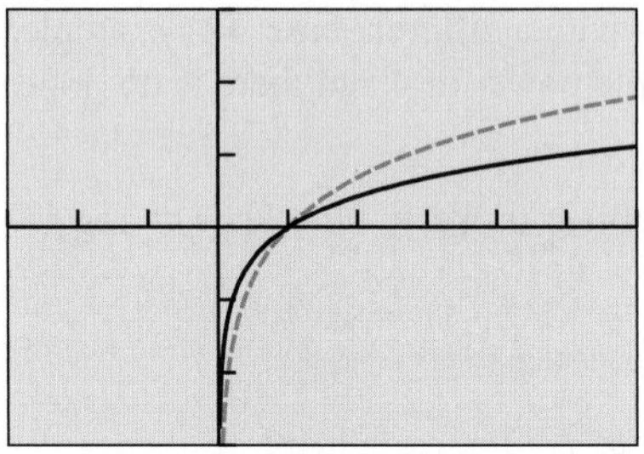

[–3, 6] by [–3, 3]

(a)

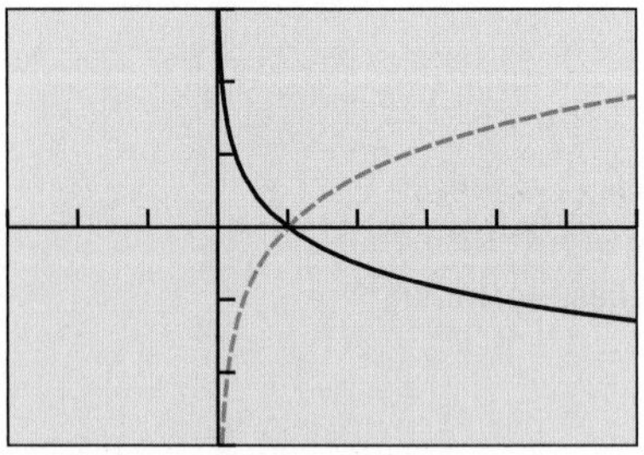

[–3, 6] by [–3, 3]

(b)

FIGURE 3.28 Transforming $f(x) = \ln x$ to obtain (a) $g(x) = \log_5 x$ and (b) $h(x) = \log_{1/4} x$. (Example 6)

EXAMPLE 6 Graphing Logarithmic Functions

Describe how to transform the graph of $f(x) = \ln x$ into the graph of the given function. Sketch the graph by hand and support your answer with a grapher.

(a) $g(x) = \log_5 x$ **(b)** $h(x) = \log_{1/4} x$

SOLUTION

(a) Because $g(x) = \log_5 x = \ln x/\ln 5$, its graph is obtained by vertically shrinking the graph of $f(x) = \ln x$ by a factor of $1/\ln 5 \approx 0.62$. See Figure 3.28a.

(b) $h(x) = \log_{1/4} x = \dfrac{\ln x}{\ln 1/4} = \dfrac{\ln x}{\ln 1 - \ln 4} = \dfrac{\ln x}{-\ln 4} = -\dfrac{1}{\ln 4} \ln x$. We can obtain the graph of h from the graph of $f(x) = \ln x$ by applying, in either order, a reflection across the x-axis and a vertical shrink by a factor of $1/\ln 4 \approx 0.72$. See Figure 3.28b.

Now try Exercise 39.

We can generalize Example 6b in the following way: If $b > 1$, then $0 < 1/b < 1$ and

$$\log_{1/b} x = -\log_b x.$$

So when given a function like $h(x) = \log_{1/4} x$, with a base between 0 and 1, we can immediately rewrite it as $h(x) = -\log_4 x$. Because we can so readily change the base of logarithms with bases between 0 and 1, such logarithms are rarely encountered or used. Instead, we work with logarithms that have bases $b > 1$, which behave much like natural and common logarithms, as we now summarize.

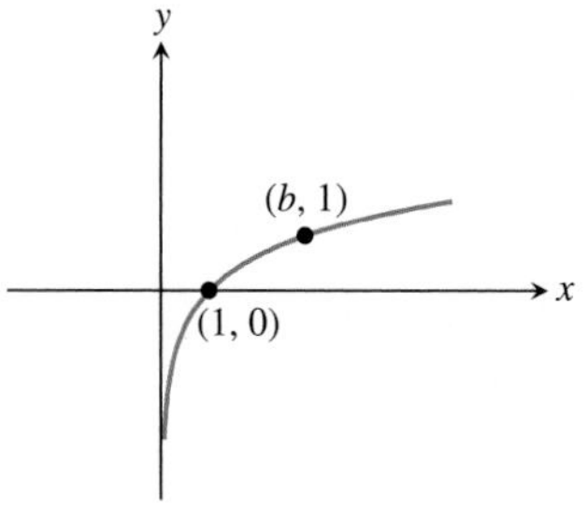

FIGURE 3.29 $f(x) = \log_b x, b > 1$.

Logarithmic Functions $f(x) = \log_b x$, with $b > 1$

Domain: $(0, \infty)$
Range: $(-\infty, \infty)$
Continuous
Increasing on its domain
No symmetry: neither even nor odd
Not bounded above or below
No local extrema
No horizontal asymptotes
Vertical asymptote: $x = 0$
End behavior: $\lim_{x\to\infty} \log_b x = \infty$

Astronomically Speaking

An astronomical unit (AU) is the average distance between the Earth and the Sun, about 149.6 million kilometers (149.6 Gm).

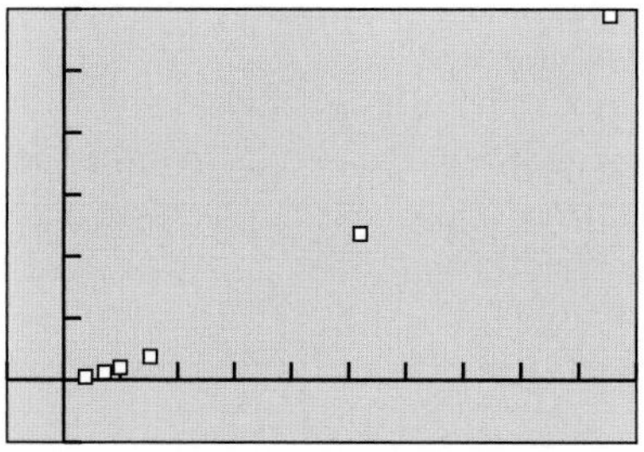

[–1, 10] by [–5, 30]
(a)

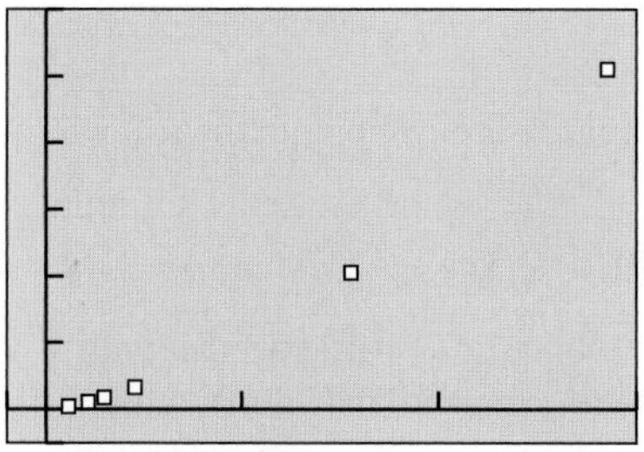

[–100, 1500] by [–1000, 12 000]
(b)

FIGURE 3.30 Scatter plots of the planetary data from (a) Table 3.20 and (b) Table 2.10.

Re-expressing Data

When seeking a model for a set of data, it is often helpful to transform the data by applying a function to one or both of the variables in the data set. We did this already when we treated the years 1900–2000 as 0–100. Such a transformation of a data set is a **re-expression** of the data.

Recall from Section 2.2 that Kepler's Third Law states that the square of the orbit period T for each planet is proportional to the cube of its average distance a from the Sun. If we re-express the Kepler planetary data in Table 2.10 using Earth-based units, the constant of proportion becomes 1 and the "is proportional to" in Kepler's Third Law becomes "equals." We can do this by dividing the "average distance" column by 149.6 Gm/AU and the "period of orbit" column by 365.2 days/yr. The re-expressed data are shown in Table 3.20.

Table 3.20 Average Distances and Orbit Periods for the Six Innermost Planets

Planet	Average Distance from Sun (AU)	Period of Orbit (yr)
Mercury	0.3870	0.2410
Venus	0.7233	0.6161
Earth	1.000	1.000
Mars	1.523	1.881
Jupiter	5.203	11.86
Saturn	9.539	29.46

Source: Re-expression of data from: Shupe, et al., National Geographic Atlas of the World (rev. 6th ed.). Washington, DC: National Geographic Society, 1992, plate 116.

Notice that the pattern in the scatter plot of these re-expressed data, shown in Figure 3.30a, is essentially the same as the pattern in the plot of the original data, shown in Figure 3.30b. What we have done is to make the numerical values of the data more convenient and to guarantee that our plot contains the ordered pair (1, 1) for Earth, which could potentially simplify our model. What we have *not* done and still wish to do is to clarify the relationship between the variables a (distance from the Sun) and T (orbit period).

Logarithms can be used to re-express data and help us clarify relationships and uncover hidden patterns. For the planetary data, if we plot $(\ln a, \ln T)$ pairs instead of (a, T) pairs, the pattern is much clearer. In Example 7, we carry out this re-expression of the data and then use an algebraic *tour de force* to obtain Kepler's Third Law.

EXAMPLE 7 Establishing Kepler's Third Law Using Logarithmic Re-expression

Re-express the (a, T) data pairs in Table 3.20 as $(\ln a, \ln T)$ pairs. Find a linear regression model for the $(\ln a, \ln T)$ pairs. Rewrite the linear regression in terms of a and T, and rewrite the equation in a form with no logs or fractional exponents.

SOLUTION

Model We use grapher list operations to obtain the $(\ln a, \ln T)$ pairs (see Figure 3.31a). We make a scatter plot of the re-expressed data in Figure 3.31b. The $(\ln a, \ln T)$ pairs appear to lie along a straight line.

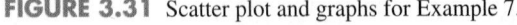

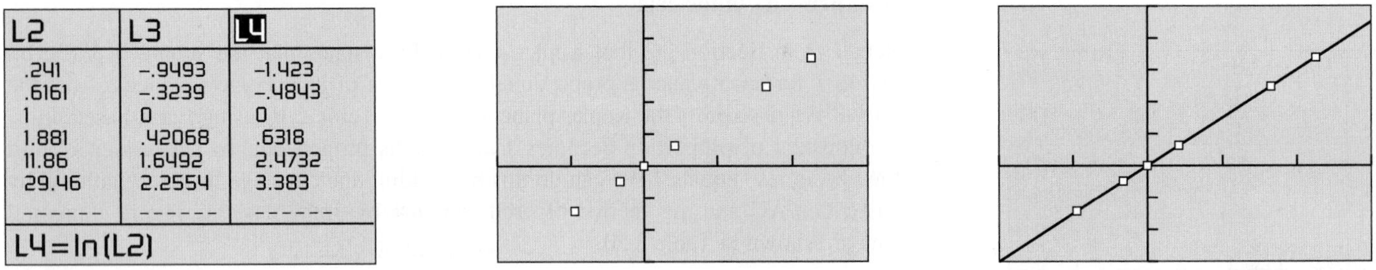

[–2, 3] by [–3, 5]

[–2, 3] by [–3, 5]

(a) (b) (c)

FIGURE 3.31 Scatter plot and graphs for Example 7.

We let $y = \ln T$ and $x = \ln a$. Then using linear regression, we obtain the following model:

$$y = 1.49950x + 0.00070 \approx 1.5x$$

Figure 3.31c shows the scatter plot for the $(x, y) = (\ln a, \ln T)$ pairs together with a graph of $y = 1.5x$. You can see that the line fits the re-expressed data remarkably well.

Remodel Returning to the original variables a and T, we obtain:

$$\ln T = 1.5 \cdot \ln a \qquad y = 1.5x$$

$$\frac{\ln T}{\ln a} = 1.5 \qquad \text{Divide by } \ln a.$$

$$\log_a T = \frac{3}{2} \qquad \text{Change of base}$$

$$T = a^{3/2} \qquad \text{Switch to exponential form.}$$

$$T^2 = a^3 \qquad \text{Square both sides.}$$

Interpret This is Kepler's Third Law!

Now try Exercise 65.

QUICK REVIEW 3.4 *(For help, go to Sections A.1 and 3.3.)*

Exercise numbers with a gray background indicate problems that the authors have designed to be solved *without a calculator*.

In Exercises 1–4, evaluate the expression.

1. $\log 10^2$
2. $\ln e^3$
3. $\ln e^{-2}$
4. $\log 10^{-3}$

In Exercises 5–10, simplify the expression.

5. $\dfrac{x^5 y^{-2}}{x^2 y^{-4}}$
6. $\dfrac{u^{-3}v^7}{u^{-2}v^2}$
7. $(x^6y^{-2})^{1/2}$
8. $(x^{-8}y^{12})^{3/4}$
9. $\dfrac{(u^2v^{-4})^{1/2}}{(27u^6v^{-6})^{1/3}}$
10. $\dfrac{(x^{-2}y^3)^{-2}}{(x^3y^{-2})^{-3}}$

SECTION 3.4 Exercises

In Exercises 1–12, assuming x and y are positive, use properties of logarithms to write the expression as a sum or difference of logarithms or multiples of logarithms.

1. $\ln 8x$
2. $\ln 9y$
3. $\log \dfrac{3}{x}$
4. $\log \dfrac{2}{y}$
5. $\log_2 y^5$
6. $\log_2 x^{-2}$
7. $\log x^3y^2$
8. $\log xy^3$
9. $\ln \dfrac{x^2}{y^3}$
10. $\log 1000x^4$
11. $\log \sqrt[4]{\dfrac{x}{y}}$
12. $\ln \dfrac{\sqrt[3]{x}}{\sqrt[3]{y}}$

In Exercises 13–22, assuming x, y, and z are positive, use properties of logarithms to write the expression as a single logarithm.

13. $\log x + \log y$
14. $\log x + \log 5$
15. $\ln y - \ln 3$
16. $\ln x - \ln y$
17. $\dfrac{1}{3}\log x$
18. $\dfrac{1}{5}\log z$
19. $2\ln x + 3\ln y$
20. $4\log y - \log z$
21. $4\log(xy) - 3\log(yz)$
22. $3\ln(x^3y) + 2\ln(yz^2)$

In Exercises 23–28, use the change-of-base formula and your calculator to evaluate the logarithm.

23. $\log_2 7$
24. $\log_5 19$
25. $\log_8 175$
26. $\log_{12} 259$
27. $\log_{0.5} 12$
28. $\log_{0.2} 29$

In Exercises 29–32, write the expression using only natural logarithms.

29. $\log_3 x$
30. $\log_7 x$
31. $\log_2(a + b)$
32. $\log_5(c - d)$

In Exercises 33–36, write the expression using only common logarithms.

33. $\log_2 x$
34. $\log_4 x$
35. $\log_{1/2}(x + y)$
36. $\log_{1/3}(x - y)$
37. Prove the quotient rule of logarithms.
38. Prove the power rule of logarithms.

In Exercises 39–42, describe how to transform the graph of $g(x) = \ln x$ into the graph of the given function. Sketch the graph by hand and support with a grapher.

39. $f(x) = \log_4 x$
40. $f(x) = \log_7 x$
41. $f(x) = \log_{1/3} x$
42. $f(x) = \log_{1/5} x$

In Exercises 43–46, match the function with its graph. Identify the window dimensions, Xscl, and Yscl of the graph.

43. $f(x) = \log_4(2 - x)$
44. $f(x) = \log_6(x - 3)$
45. $f(x) = \log_{0.5}(x - 2)$
46. $f(x) = \log_{0.7}(3 - x)$

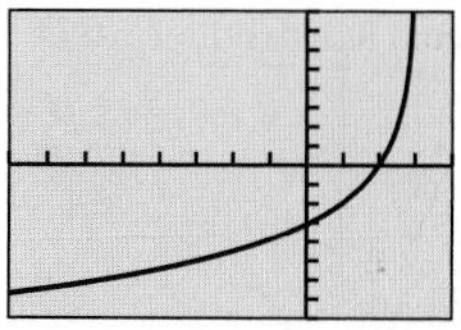
(a)

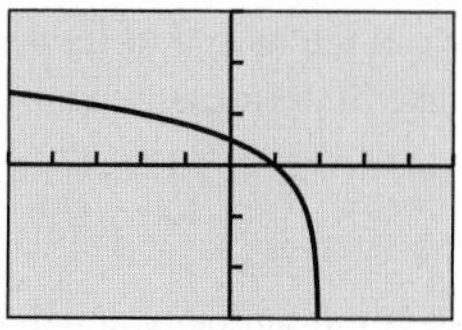
(b)

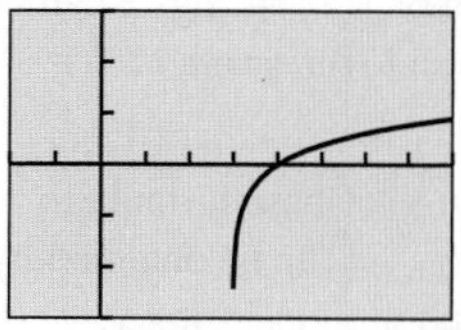
(c)

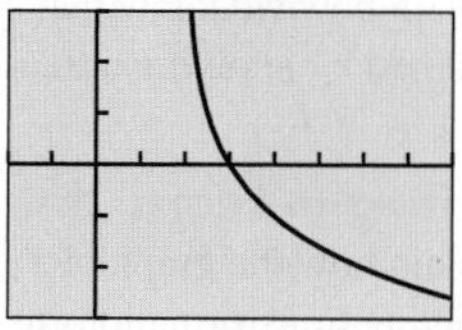
(d)

In Exercises 47–50, graph the function, and analyze it for domain, range, continuity, increasing or decreasing behavior, asymptotes, and end behavior.

47. $f(x) = \log_2(8x)$ **48.** $f(x) = \log_{1/3}(9x)$

49. $f(x) = \log(x^2)$ **50.** $f(x) = \ln(x^3)$

51. Sound Intensity Compute the sound intensity level in decibels for each sound listed in Table 3.21.

Table 3.21 Approximate Intensities for Selected Sounds

Sound	Intensity (Watts/m²)
(a) Hearing threshold	10^{-12}
(b) Rustling leaves	10^{-11}
(c) Conversation	10^{-6}
(d) School cafeteria	10^{-4}
(e) Jack hammer	10^{-2}
(f) Pain threshold	1

Sources: J. J. Dwyer, College Physics. Belmont, CA: Wadsworth, 1984; and E. Connally et al., Functions Modeling Change. New York: Wiley, 2000.

52. Earthquake Intensity The **Richter scale** magnitude R of an earthquake is based on the features of the associated seismic wave and is measured by

$$R = \log(a/T) + B,$$

where a is the amplitude in μm (micrometers), T is the period in seconds, and B accounts for the weakening of the seismic wave due to the distance from the epicenter. Compute the earthquake magnitude R for each set of values.

(a) $a = 250$, $T = 2$, and $B = 4.25$

(b) $a = 300$, $T = 4$, and $B = 3.5$

53. Light Intensity in Lake Erie The relationship between intensity I of light (in lumens) at a depth of x feet in Lake Erie is given by

$$\log \frac{I}{12} = -0.00235x.$$

What is the intensity at a depth of 40 ft?

54. Light Intensity in Lake Superior The relationship between intensity I of light (in lumens) at a depth of x feet in Lake Superior is given by

$$\log \frac{I}{12} = -0.0125x.$$

What is the intensity at a depth of 10 ft?

55. Writing to Learn Use the change-of-base formula to explain how we know that the graph of $f(x) = \log_3 x$ can be obtained by applying a transformation to the graph of $g(x) = \ln x$.

56. Writing to Learn Use the change-of-base formula to explain how the graph of $f(x) = \log_{0.8} x$ can be obtained by applying transformations to the graph of $g(x) = \log x$.

Standardized Test Questions

57. True or False The logarithm of the product of two positive numbers is the sum of the logarithms of the numbers. Justify your answer.

58. True or False The logarithm of a positive number is positive. Justify your answer.

In Exercises 59–62, solve the problem without using a calculator.

59. Multiple Choice $\log 12 =$

(A) $3 \log 4$ **(B)** $\log 3 + \log 4$

(C) $4 \log 3$ **(D)** $\log 3 \cdot \log 4$

(E) $2 \log 6$

60. Multiple Choice $\log_9 64 =$

(A) $5 \log_3 2$ **(B)** $(\log_3 8)^2$

(C) $(\ln 64)/(\ln 9)$ **(D)** $2 \log_9 32$

(E) $(\log 64)/9$

61. Multiple Choice $\ln x^5 =$

(A) $5 \ln x$ **(B)** $2 \ln x^3$

(C) $x \ln 5$ **(D)** $3 \ln x^2$

(E) $\ln x^2 \cdot \ln x^3$

62. Multiple Choice $\log_{1/2} x^2 =$

(A) $-2 \log_2 x$ **(B)** $2 \log_2 x$

(C) $-0.5 \log_2 x$ **(D)** $0.5 \log_2 x$

(E) $-2 \log_2 |x|$

Explorations

63. (a) Compute the power regression model for the following data.

x	4	6.5	8.5	10
y	2816	31,908	122,019	275,000

(b) Predict the y value associated with $x = 7.1$ using the power regression model.

(c) Re-express the data in terms of their natural logarithms and make a scatter plot of $(\ln x, \ln y)$.

(d) Compute the linear regression model $(\ln y) = a(\ln x) + b$ for $(\ln x, \ln y)$.

(e) Confirm that $y = e^b \cdot x^a$ is the power regression model found in (a).

64. (a) Compute the power regression model for the following data.

x	2	3	4.8	7.7
y	7.48	7.14	6.81	6.41

(b) Predict the y value associated with $x = 9.2$ using the power regression model.

(c) Re-express the data in terms of their natural logarithms and make a scatter plot of $(\ln x, \ln y)$.

(d) Compute the linear regression model $(\ln y) = a(\ln x) + b$ for $(\ln x, \ln y)$.

(e) Confirm that $y = e^b \cdot x^a$ is the power regression model found in (a).

65. **Keeping Warm—Revisited** Recall from Exercise 55 of Section 2.2 that scientists have found the pulse rate r of mammals to be a power function of their body weight w.

(a) Re-express the data in Table 3.22 in terms of their *common* logarithms and make a scatter plot of $(\log w, \log r)$.

(b) Compute the linear regression model $(\log r) = a(\log w) + b$ for $(\log w, \log r)$.

(c) Superimpose the regression curve on the scatter plot.

(d) Use the regression equation to predict the pulse rate for a 450-kg horse. Is the result close to the 38 beats/min reported by A. J. Clark in 1927?

(e) **Writing to Learn** Why can we use either common or natural logarithms to re-express data that fit a power regression model?

Table 3.22 Weight and Pulse Rate of Selected Mammals

Mammal	Body Weight (kg)	Pulse Rate (beats/min)
Rat	0.2	420
Guinea pig	0.3	300
Rabbit	2	205
Small dog	5	120
Large dog	30	85
Sheep	50	70
Human	70	72

Source: A. J. Clark, Comparative Physiology of the Heart. New York: Macmillan, 1927.

66. Let $a = \log 2$ and $b = \log 3$. Then, for example, $\log 6 = a + b$ and $\log 15 = 1 - a + b$. List all of the positive integers less than 100 whose common logs can be written as expressions involving a or b or both. (*Hint:* See Exploration 1 on page 283.)

67. **A Bit of History** Explain in the time before electronic calculators and computers how logarithms turned complicated computations into easy arithmetic problems. Use the example $\sqrt[5]{4581}$.

Extending the Ideas

68. Solve $\ln x > \sqrt[3]{x}$.

69. Solve $1.2^x \le \log_{1.2} x$.

70. **Group Activity** Work in groups of three. Have each group member graph and compare the domains for one pair of functions.

(a) $f(x) = 2\ln x + \ln(x - 3)$ and $g(x) = \ln x^2(x - 3)$

(b) $f(x) = \ln(x + 5) - \ln(x - 5)$ and $g(x) = \ln \dfrac{x + 5}{x - 5}$

(c) $f(x) = \log(x + 3)^2$ and $g(x) = 2\log(x + 3)$

Writing to Learn After discussing your findings, write a brief group report that includes your overall conclusions and insights.

71. Prove the change-of-base formula for logarithms.

72. Prove that $f(x) = \log x/\ln x$ is a constant function with restricted domain by finding the exact value of the constant $\log x/\ln x$ expressed as a common logarithm.

73. Graph $f(x) = \ln(\ln(x))$, and analyze it for domain, range, continuity, increasing or decreasing behavior, symmetry, asymptotes, end behavior, and invertibility.

3.5 Equation Solving and Modeling

What you'll learn about

- Solving Exponential Equations
- Solving Logarithmic Equations
- Orders of Magnitude and Logarithmic Models
- Newton's Law of Cooling
- Logarithmic Re-expression

... and why

The Richter scale, pH, and Newton's Law of Cooling are among the most important uses of logarithmic and exponential functions.

Solving Exponential Equations

Some logarithmic equations can be solved by changing to exponential form, as we saw in Example 5 of Section 3.3. For other equations, the properties of exponents or the properties of logarithms are used. A property of both exponential and logarithmic functions that is often helpful for solving equations is that they are one-to-one functions.

One-to-One Properties

For any exponential function $f(x) = b^x$,

- If $b^u = b^v$, then $u = v$.

For any logarithmic function $f(x) = \log_b x$,

- If $\log_b u = \log_b v$, then $u = v$.

Example 1 shows how the one-to-oneness of exponential functions can be used. Examples 3 and 4 use the one-to-one property of logarithms.

EXAMPLE 1 Solving an Exponential Equation Algebraically

Solve $20(1/2)^{x/3} = 5$.

SOLUTION

$$20\left(\frac{1}{2}\right)^{x/3} = 5$$

$$\left(\frac{1}{2}\right)^{x/3} = \frac{1}{4} \quad \text{Divide by 20.}$$

$$\left(\frac{1}{2}\right)^{x/3} = \left(\frac{1}{2}\right)^2 \quad \frac{1}{4} = \left(\frac{1}{2}\right)^2$$

$$\frac{x}{3} = 2 \quad \text{One-to-one property}$$

$$x = 6 \quad \text{Multiply by 3.}$$

Now try Exercise 1.

The equation in Example 2 involves a difference of two exponential functions, which makes it difficult to solve algebraically. So we start with a graphical approach.

EXAMPLE 2 Solving an Exponential Equation

Solve $(e^x - e^{-x})/2 = 5$.

SOLUTION

Solve Graphically Figure 3.32 shows that the graphs of $y = (e^x - e^{-x})/2$ and $y = 5$ intersect when $x \approx 2.31$.

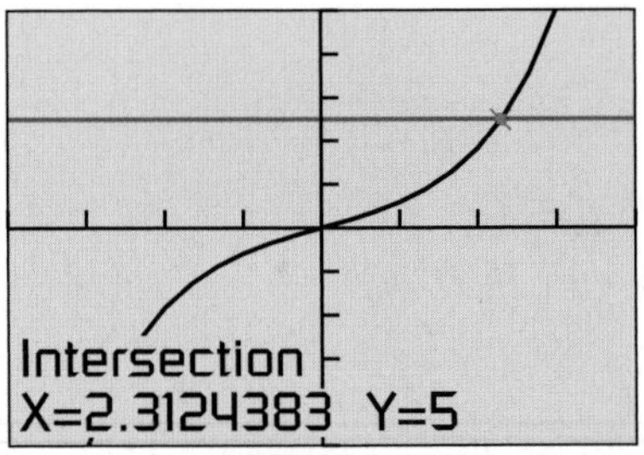

[–4, 4] by [–10, 10]

FIGURE 3.32 $y = (e^x - e^{-x})/2$ and $y = 5$. (Example 2)

Confirm Algebraically The algebraic approach involves some ingenuity. If we multiply each side of the original equation by $2e^x$ and rearrange the terms, we can obtain a quadratic equation in e^x:

$$\frac{e^x - e^{-x}}{2} = 5$$

$$e^{2x} - e^0 = 10e^x \quad \text{Multiply by } 2e^x.$$

$$(e^x)^2 - 10(e^x) - 1 = 0 \quad \text{Subtract } 10e^x.$$

A Cinch?

You may recognize the left-hand side of the equation in Example 2 as the *hyperbolic sine function* that was introduced in Exercise 59 of Section 3.2. This function is often used in calculus. We write $\sinh(x) = (e^x - e^{-x})/2$. "Sinh" is pronounced as if spelled "cinch."

If we let $w = e^x$, this equation becomes $w^2 - 10w - 1 = 0$, and the quadratic formula gives

$$w = e^x = \frac{10 \pm \sqrt{104}}{2} = 5 \pm \sqrt{26}.$$

Because e^x is always positive, we reject the possibility that e^x has the negative value $5 - \sqrt{26}$. Therefore,

$$e^x = 5 + \sqrt{26}$$

$$x = \ln\left(5 + \sqrt{26}\right) \qquad \text{Convert to logarithmic form.}$$

$$x = 2.312\ldots \approx 2.31 \qquad \text{Approximate with a grapher.}$$

Now try Exercise 31.

Solving Logarithmic Equations

When logarithmic equations are solved algebraically, it is important to keep track of the domain of each expression in the equation as it is being solved. A particular algebraic method may introduce extraneous solutions, or worse yet, *lose* some valid solutions, as illustrated in Example 3.

EXAMPLE 3 Solving a Logarithmic Equation

Solve $\log x^2 = 2$.

SOLUTION

Method 1 Use the one-to-one property of logarithms.

$$\log x^2 = 2$$

$$\log x^2 = \log 10^2 \qquad y = \log 10^y$$

$$x^2 = 10^2 \qquad \text{One-to-one property}$$

$$x^2 = 100 \qquad 10^2 = 100$$

$$x = 10 \quad \text{or} \quad x = -10$$

Method 2 Change the equation from logarithmic to exponential form.

$$\log x^2 = 2$$

$$x^2 = 10^2 \qquad \text{Change to exponential form.}$$

$$x^2 = 100 \qquad 10^2 = 100$$

$$x = 10 \quad \text{or} \quad x = -10$$

Method 3 (Incorrect) Use the power rule of logarithms.

$$\log x^2 = 2$$

$$2 \log x = 2 \qquad \text{Power rule, incorrectly applied}$$

$$\log x = 1 \qquad \text{Divide by 2.}$$

$$x = 10 \qquad \text{Change to exponential form.}$$

Support Graphically Figure 3.33 shows that the graphs of $f(x) = \log x^2$ and $y = 2$ intersect when $x = -10$. From the symmetry of the graphs due to f being an even function, we can see that $x = 10$ is also a solution.

Interpret Methods 1 and 2 are correct. Method 3 fails because the domain of $\log x^2$ is all nonzero real numbers, but the domain of $\log x$ is only the positive real numbers. The correct solution includes both 10 and -10 because both of these x values make the original equation true.

Now try Exercise 25.

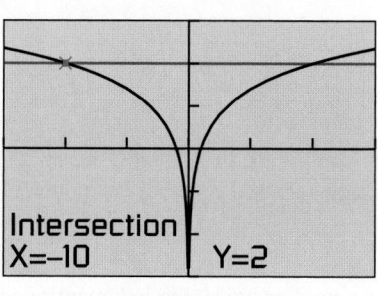

[–15, 15] by [–3, 3]

FIGURE 3.33 Graphs of $f(x) = \log x^2$ and $y = 2$. (Example 3)

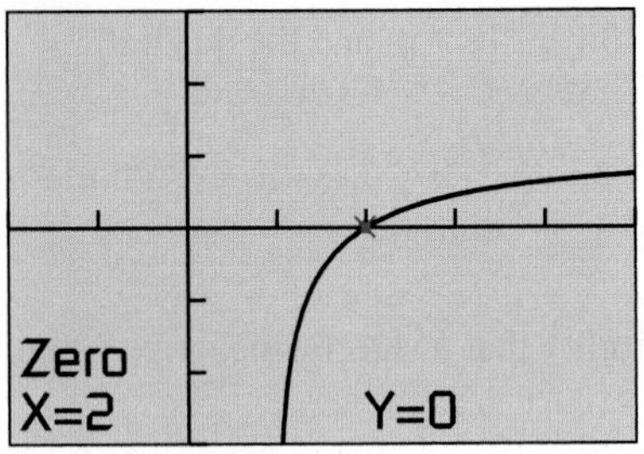

[–2, 5] by [–3, 3]

FIGURE 3.34 The zero of $f(x) = \ln(3x - 2) + \ln(x - 1) - 2\ln x$ is $x = 2$. (Example 4)

Method 3 above violates an easily overlooked condition of the power rule $\log_b R^c = c \log_a R$, namely, that the rule holds *if R is positive*. In the expression $\log x^2$, x plays the role of R, and x can be -10, which is *not* positive. Because algebraic manipulation of a logarithmic equation can produce expressions with different domains, a graphical solution is often less prone to error.

EXAMPLE 4 Solving a Logarithmic Equation

Solve $\ln(3x - 2) + \ln(x - 1) = 2\ln x$.

SOLUTION To use the x-intercept method, we rewrite the equation as

$$\ln(3x - 2) + \ln(x - 1) - 2\ln x = 0,$$

and graph

$$f(x) = \ln(3x - 2) + \ln(x - 1) - 2\ln x,$$

as shown in Figure 3.34. The x-intercept is $x = 2$, which is the solution to the equation. **Now try Exercise 35.**

Orders of Magnitude and Logarithmic Models

When comparing quantities, their sizes sometimes span a wide range. This is why scientific notation was developed.

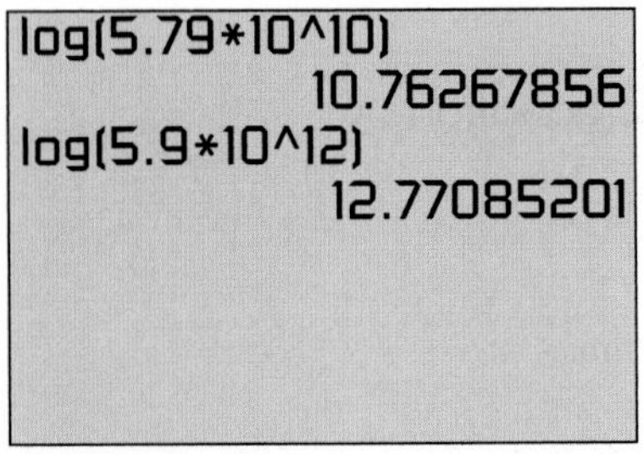

FIGURE 3.35 Pluto is two orders of magnitude farther from the Sun than Mercury.

For instance, the planet Mercury is 57.9 billion meters from the Sun, whereas Pluto is 5900 billion meters from the Sun, roughly 100 times farther! In scientific notation, Mercury is 5.79×10^{10} m from the Sun, and Pluto is 5.9×10^{12} m from the Sun. Pluto's distance is 2 powers of ten greater than Mercury's distance. From Figure 3.35, we see that the difference in the common logs of these two distances is about 2. The common logarithm of a positive quantity is its **order of magnitude**. So we say, Pluto's distance from the Sun is 2 orders of magnitude greater than Mercury's.

Orders of magnitude can be used to compare any like quantities:

- A kilometer is 3 orders of magnitude longer than a meter.
- A dollar is 2 orders of magnitude greater than a penny.
- A horse weighing 400 kg is 4 orders of magnitude heavier than a mouse weighing 40 g.
- New York City with 8 million persons is 6 orders of magnitude bigger than Earmuff Junction with a population of 8.

EXPLORATION 1 **Comparing Scientific Notation and Common Logarithms**

1. Using a calculator compute $\log(4 \cdot 10)$, $\log(4 \cdot 10^2)$, $\log(4 \cdot 10^3), \ldots,$ $\log(4 \cdot 10^{10})$.
2. What is the pattern in the integer parts of these numbers?
3. What is the pattern of their decimal parts?
4. How many orders of magnitude greater is $4 \cdot 10^{10}$ than $4 \cdot 10$?

Orders of magnitude have many applications. For a sound or noise, the *bel*, mentioned in Section 3.3, measures the order of magnitude of its intensity compared to the threshold of hearing. For instance, a sound of 3 bels or 30 dB (decibels) has a sound intensity 3 orders of magnitude above the threshold of hearing.

Orders of magnitude are also used to compare the severity of earthquakes and the acidity of chemical solutions. We now turn our attention to these two applications.

As mentioned in Exercise 52 of Section 3.4, the *Richter scale* magnitude R of an earthquake is

$$R = \log \frac{a}{T} + B,$$

where a is the amplitude in micrometers (μm) of the vertical ground motion at the receiving station, T is the period of the associated seismic wave in seconds, and B accounts for the weakening of the seismic wave with increasing distance from the epicenter of the earthquake.

EXAMPLE 5 Comparing Earthquake Intensities

How many times more severe was the 2001 earthquake in Gujarat, India ($R_1 = 7.9$), than the 1999 earthquake in Athens, Greece ($R_2 = 5.9$)?

SOLUTION

Model The severity of an earthquake is measured by the associated amplitude. Let a_1 be the amplitude for the Gujarat earthquake and a_2 be the amplitude for the Athens earthquake. Then

$$R_1 = \log \frac{a_1}{T} + \mathrm{B} = 7.9$$

$$R_2 = \log \frac{a_2}{T} + B = 5.9$$

Solve Algebraically We seek the ratio of severities a_1/a_2:

$$\left(\log \frac{a_1}{T} + B\right) - \left(\log \frac{a_2}{T} + B\right) = R_1 - R_2$$

$$\log \frac{a_1}{T} - \log \frac{a_2}{T} = 7.9 - 5.9 \qquad B - B = 0$$

$$\log \frac{a_1}{a_2} = 2 \qquad \text{Quotient rule}$$

$$\frac{a_1}{a_2} = 10^2 = 100$$

Interpret A Richter scale difference of 2 corresponds to an amplitude ratio of 2 powers of 10, or $10^2 = 100$. So the Gujarat quake was 100 times as severe as the Athens quake.

Now try Exercise 45.

In chemistry, the acidity of a water-based solution is measured by the concentration of hydrogen ions in the solution (in moles per liter). The hydrogen-ion concentration is written $[H^+]$. Because such concentrations usually involve *negative* powers of ten, *negative* orders of magnitude are used to compare acidity levels. The measure of acidity used is **pH**, the opposite of the common log of the hydrogen-ion concentration:

$$\text{pH} = -\log\,[H^+]$$

More acidic solutions have higher hydrogen-ion concentrations and lower pH values.

EXAMPLE 6 Comparing Chemical Acidity

Some especially sour vinegar has a pH of 2.4, and a box of Leg and Sickle baking soda has a pH of 8.4.

(a) What are their hydrogen-ion concentrations?

(b) How many times greater is the hydrogen-ion concentration of the vinegar than that of the baking soda?

(c) By how many orders of magnitude do the concentrations differ?

SOLUTION

(a)

$$\begin{aligned} \text{Vinegar: } -\log[H^+] &= 2.4 \\ \log[H^+] &= -2.4 \\ [H^+] &= 10^{-2.4} \approx 3.98 \times 10^{-3} \text{ moles per liter} \end{aligned}$$

$$\begin{aligned} \text{Baking soda: } -\log[H^+] &= 8.4 \\ \log[H^+] &= -8.4 \\ [H^+] &= 10^{-8.4} \approx 3.98 \times 10^{-9} \text{ moles per liter} \end{aligned}$$

(b)

$$\frac{[H^+] \text{ of vinegar}}{[H^+] \text{ of baking soda}} = \frac{10^{-2.4}}{10^{-8.4}} = 10^{(-2.4)-(-8.4)} = 10^6$$

(c) The hydrogen-ion concentration of the vinegar is 6 orders of magnitude greater than that of the Leg and Sickle baking soda, exactly the difference in their pH values.

Now try Exercise 47.

Newton's Law of Cooling

An object that has been heated will cool to the temperature of the medium in which it is placed, such as the surrounding air or water. The temperature T of the object at time t can be modeled by

$$T(t) = T_m + (T_0 - T_m)e^{-kt}$$

for an appropriate value of k, where

$$\begin{aligned} T_m &= \text{the temperature of the surrounding medium,} \\ T_0 &= \text{initial temperature of the object.} \end{aligned}$$

This model assumes that the surrounding medium, although taking heat from the object, essentially maintains a constant temperature. In honor of English mathematician and physicist Isaac Newton (1643–1727), this model is called **Newton's Law of Cooling**.

EXAMPLE 7 Applying Newton's Law of Cooling

A hard-boiled egg at temperature 96°C is placed in 16°C water to cool. Four minutes later the temperature of the egg is 45°C. Use Newton's Law of Cooling to determine when the egg will be 20°C.

SOLUTION

Model Because $T_0 = 96$ and $T_m = 16$, $T_0 - T_m = 80$ and

$$T(t) = T_m + (T_0 - T_m)e^{-kt} = 16 + 80e^{-kt}.$$

To find the value of k we use the fact that $T = 45$ when $t = 4$.

$$45 = 16 + 80e^{-4k}$$

$$\frac{29}{80} = e^{-4k} \qquad \text{Subtract 16, then divide by 80.}$$

$$\ln\frac{29}{80} = -4k \qquad \text{Change to logarithmic form.}$$

$$k = -\frac{\ln(29/80)}{4} \qquad \text{Divide by } -4.$$

$$k = 0.253\ldots$$

We save this k value because it is part of our model. (See Figure 3.36.)

Solve Algebraically To find t when $T = 20°C$, we solve the equation:

$$20 = 16 + 80e^{-kt}$$

$$\frac{4}{80} = e^{-kt} \qquad \text{Subtract 16, then divide by 80.}$$

$$\ln\frac{4}{80} = -kt \qquad \text{Change to logarithmic form.}$$

$$t = -\frac{\ln(4/80)}{k} \approx 11.81 \qquad \text{See Figure 3.36.}$$

Interpret The temperature of the egg will be 20°C after about 11.81 min (11 min 49 sec).

Now try Exercise 49.

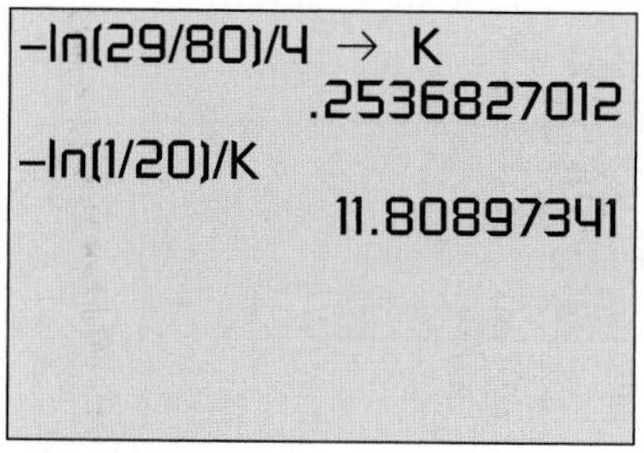

FIGURE 3.36 Storing and using the constant k.

We can rewrite Newton's Law of Cooling in the following form:

$$T(t) - T_m = (T_0 - T_m)e^{-kt}$$

We use this form of Newton's Law of Cooling when modeling temperature using data gathered from an actual experiment. Because the difference $T - T_m$ is an exponential function of time t, we can use exponential regression on $T - T_m$ versus t to obtain a model, as illustrated in Example 8.

EXAMPLE 8 Modeling with Newton's Law of Cooling

In an experiment, a temperature probe connected to a Calculator-Based Laboratory™ (CBL™) device was removed from a cup of hot coffee and placed in a glass of cold water. The first two columns of Table 3.23 show the resulting data for time t (in seconds since the probe was placed in the water) and temperature T (in °C). In the third column, the temperature data have been *re-expressed* by subtracting the temperature of the water, which was 4.5°C.

(a) Estimate the temperature of the coffee.

(b) Estimate the time when the temperature probe reading was 40°C.

Table 3.23 Temperature Data from a CBL™ Experiment

Time t	Temp T	$T - T_m$
2	64.8	60.3
5	49.0	44.5
10	31.4	26.9
15	22.0	17.5
20	16.5	12.0
25	14.2	9.7
30	12.0	7.5

SOLUTION

Model Figure 3.37a shows a scatter plot of the re-expressed temperature data. Using exponential regression, we obtain the following model:

$$T(t) - 4.5 = 61.656 \times 0.92770^t$$

Figure 3.37b shows the graph of this model with the scatter plot of the data. You can see that the curve fits the data fairly well.

(continued)

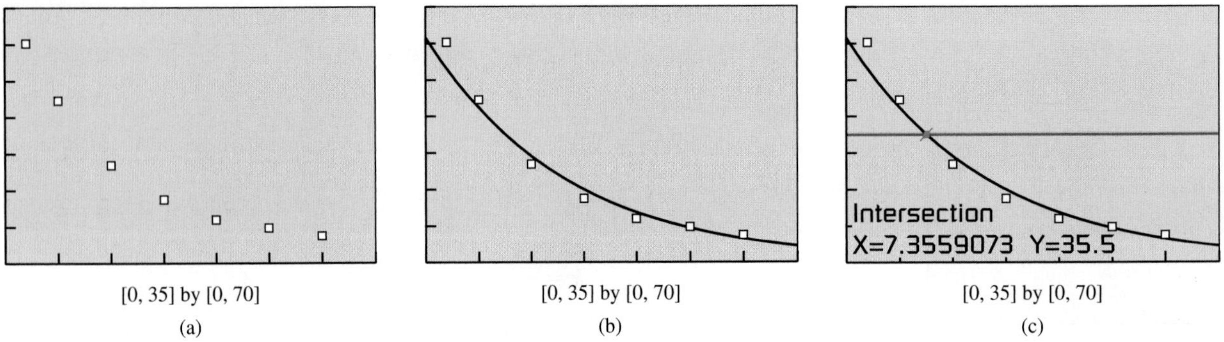

FIGURE 3.37 Scatter plot and graphs for Example 8.

(a) Solve Algebraically From the model we see that $T_0 - T_m \approx 61.656$. So

$$T_0 \approx 61.656 + T_m = 61.656 + 4.5 \approx 66.16.$$

(b) Solve Graphically Figure 3.37c shows that the graph of $T(t) - 4.5 = 61.656 \times 0.92770^t$ intersects $y = 40 - 4.5 = 35.5$ when $t \approx 7.36$.

Interpret The temperature of the coffee was roughly 66.2°C, and the probe reading was 40°C about 7.4 sec after it was placed in the water. **Now try Exercise 51.**

Logarithmic Re-expression

In Example 7 of Section 3.4 we learned that data pairs (x, y) that fit a power model have a linear relationship when re-expressed as $(\ln x, \ln y)$ pairs. We now illustrate that data pairs (x, y) that fit a logarithmic or exponential regression model can also be *linearized* through *logarithmic re-expression.*

Regression Models Related by Logarithmic Re-expression

- **Linear regression:** $y = ax + b$
- **Natural logarithmic regression:** $y = a + b \ln x$
- **Exponential Regression:** $y = a \cdot b^x$
- **Power regression:** $y = a \cdot x^b$

When we examine a scatter plot of data pairs (x, y), we can ask whether one of these four regression models could be the best choice. If the data plot appears to be linear, a linear regression may be the best choice. But when it is visually evident that the data plot is not linear, the best choice may be a natural logarithmic, exponential, or power regression.

Knowing the shapes of logarithmic, exponential, and power function graphs helps us choose an appropriate model. In addition, it is often helpful to re-express the (x, y) data pairs as $(\ln x, y)$, $(x, \ln y)$, or $(\ln x, \ln y)$ and create scatter plots of the re-expressed data. If any of the scatter plots appear to be linear, then we have a likely choice for an appropriate model. See page 299.

The three regression models can be justified algebraically. We give the justification for exponential regression, and leave the other two justifications as exercises.

$$v = ax + b$$

$$\ln y = ax + b \qquad v = \ln y$$

$$y = e^{ax+b} \qquad \text{Change to exponential form.}$$

$$y = e^{ax} \cdot e^b \qquad \text{Use the laws of exponents.}$$

$$y = e^b \cdot (e^a)^x$$

$$y = c \cdot d^x \qquad \text{Let } c = e^b \text{ and } d = e^a.$$

Example 9 illustrates how knowledge about the shapes of logarithmic, exponential, and power function graphs is used in combination with logarithmic re-expression to choose a curve of best fit.

Three Types of Logarithmic Re-expression

1. Natural Logarithmic Regression Re-expressed: $(x, y) \rightarrow (\ln x, y)$

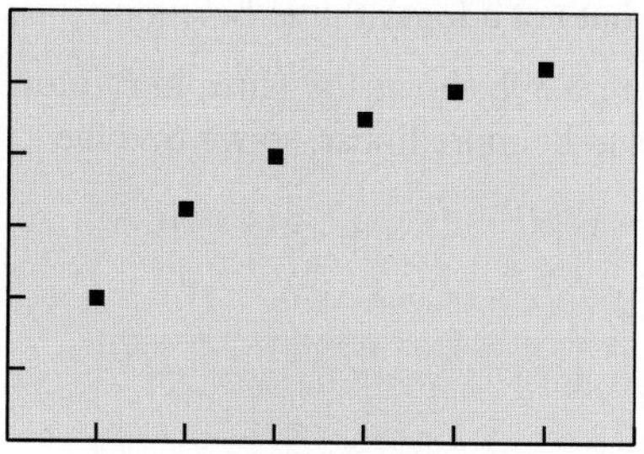

[0, 7] by [0, 30]

(x, y) data

(a)

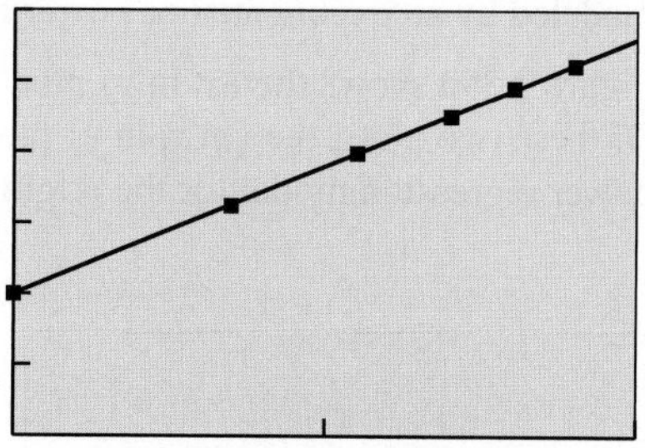

[0, 2] by [0, 30]

$(\ln x, y) = (u, y)$ data with linear regression model $y = au + b$

(b)

Conclusion:

$y = a \ln x + b$ is the logarithmic regression model for the (x, y) data.

2. Exponential Regression Re-expressed: $(x, y) \rightarrow (x, \ln y)$

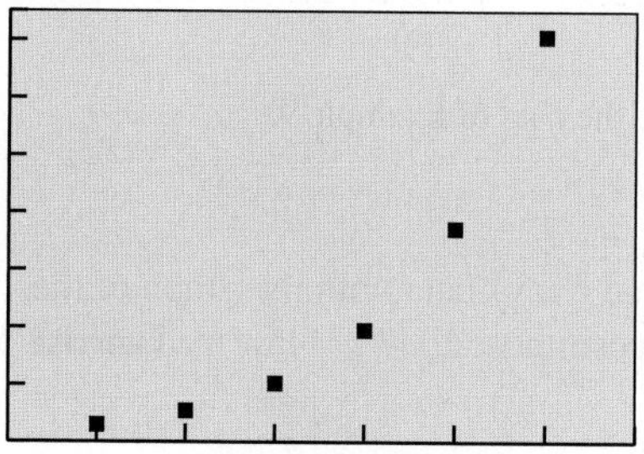

[0, 7] by [0, 75]

(x, y) data

(a)

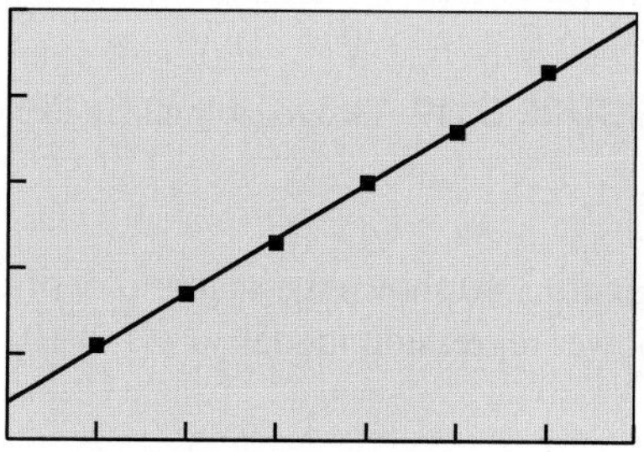

[0, 7] by [0, 5]

$(x, \ln y) = (x, v)$ data with linear regression model $v = ax + b$

(b)

Conclusion:

$y = c(d^x)$, where $c = e^b$ and $d = e^a$, is the exponential regression model for the (x, y) data.

3. Power Regression Re-expressed: $(x, y) \rightarrow (\ln x, \ln y)$

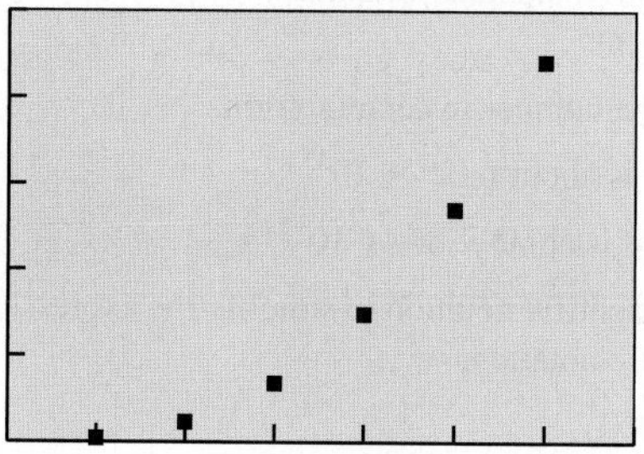

[0, 7] by [0, 50]

(x, y) data

(a)

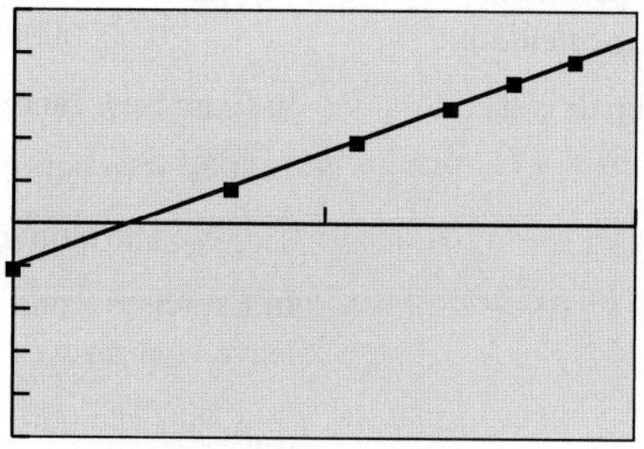

[0, 2] by [–5, 5]

$(\ln x, \ln y) = (u, v)$ data with linear regression model $v = au + b$

(b)

Conclusion:

$y = c(x^a)$, where $c = e^b$, is the power regression model for the (x, y) data.

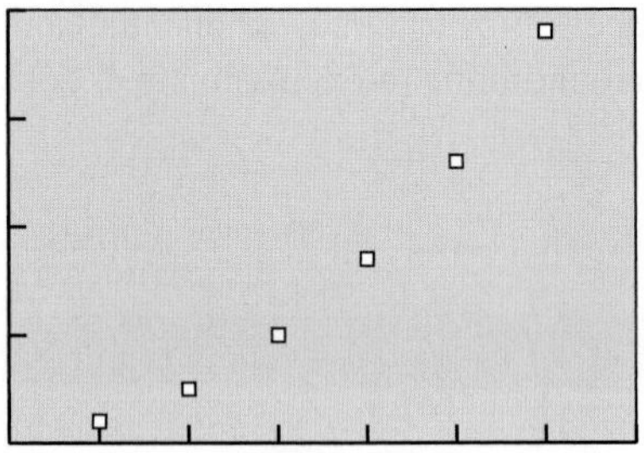

[0, 7] by [0, 40]

FIGURE 3.38 A scatter plot of the original data of Example 9.

EXAMPLE 9 Selecting a Regression Model

Decide whether these data can be best modeled by logarithmic, exponential, or power regression. Find the appropriate regression model.

x	1	2	3	4	5	6
y	2	5	10	17	26	38

SOLUTION The shape of the data plot in Figure 3.38 suggests that the data could be modeled by an exponential or power function, but not a logarithmic function.

Figure 3.39a shows the $(x, \ln y)$ plot, and Figure 3.39b shows the $(\ln x, \ln y)$ plot. Of these two plots, the $(\ln x, \ln y)$ plot appears to be more linear, so we find the power regression model for the original data.

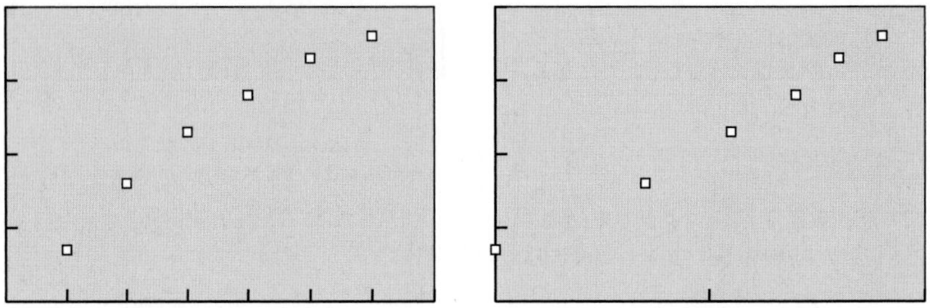

[0, 7] by [0, 4]
$(x, \ln y)$
(a)

[0, 2] by [0, 4]
$(\ln x, \ln y)$
(b)

FIGURE 3.39 Two logarithmic re-expressions of the data of Example 9.

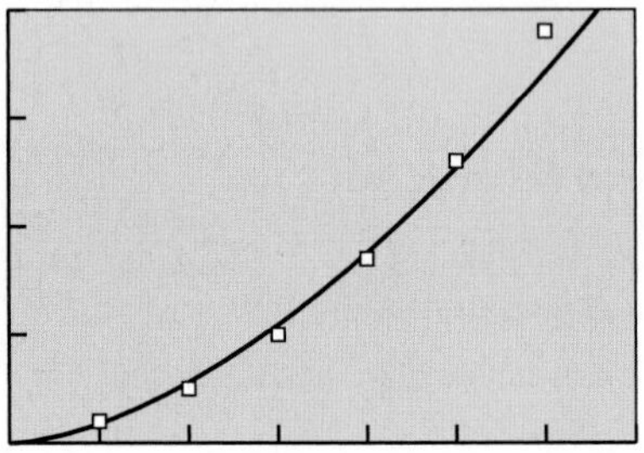

[0, 7] by [0, 40]

FIGURE 3.40 A power regression model fits the data of Example 9.

Figure 3.40 shows the scatter plot of the original (x, y) data with the graph of the power regression model $y = 1.7910x^{1.6472}$ superimposed. **Now try Exercise 55.**

QUICK REVIEW 3.5 *(For help, go to Sections P.1 and 1.4.)*

Exercise numbers with a gray background indicate problems that the authors have designed to be solved *without a calculator*.

In Exercises 1–4, prove that each function in the given pair is the inverse of the other.

1. $f(x) = e^{2x}$ and $g(x) = \ln (x^{1/2})$

2. $f(x) = 10^{x/2}$ and $g(x) = \log x^2, x > 0$

3. $f(x) = (1/3) \ln x$ and $g(x) = e^{3x}$

4. $f(x) = 3 \log x^2, x > 0$ and $g(x) = 10^{x/6}$

In Exercises 5 and 6, write the number in scientific notation.

5. The mean distance from Jupiter to the Sun is about 778,300,000 km.

6. An atomic nucleus has a diameter of about 0.000000000000001 m.

In Exercises 7 and 8, write the number in decimal form.

7. Avogadro's number is about 6.02×10^{23}.

8. The atomic mass unit is about 1.66×10^{-27} kg.

In Exercises 9 and 10, use scientific notation to simplify the expression; leave your answer in scientific notation.

9. $(186{,}000)(31{,}000{,}000)$

10. $\dfrac{0.0000008}{0.000005}$

SECTION 3.5 Exercises

In Exercises 1–10, find the exact solution algebraically, and check it by substituting into the original equation.

1. $36\left(\frac{1}{3}\right)^{x/5} = 4$ **2.** $32\left(\frac{1}{4}\right)^{x/3} = 2$

3. $2 \cdot 5^{x/4} = 250$ **4.** $3 \cdot 4^{x/2} = 96$

5. $2(10^{-x/3}) = 20$ **6.** $3(5^{-x/4}) = 15$

7. $\log x = 4$ **8.** $\log_2 x = 5$

9. $\log_4 (x - 5) = -1$ **10.** $\log_4 (1 - x) = 1$

In Exercises 11–18, solve each equation algebraically. Obtain a numerical approximation for your solution and check it by substituting into the original equation.

11. $1.06^x = 4.1$ **12.** $0.98^x = 1.6$

13. $50e^{0.035x} = 200$ **14.** $80e^{0.045x} = 240$

15. $3 + 2e^{-x} = 6$ **16.** $7 - 3e^{-x} = 2$

17. $3 \ln (x - 3) + 4 = 5$ **18.** $3 - \log (x + 2) = 5$

In Exercises 19–24, state the domain of each function. Then match the function with its graph. (Each graph shown has a window of $[-4.7, 4.7]$ by $[-3.1, 3.1]$).

19. $f(x) = \log [x(x + 1)]$ **20.** $g(x) = \log x + \log (x + 1)$

21. $f(x) = \ln \frac{x}{x + 1}$ **22.** $g(x) = \ln x - \ln (x + 1)$

23. $f(x) = 2 \ln x$ **24.** $g(x) = \ln x^2$

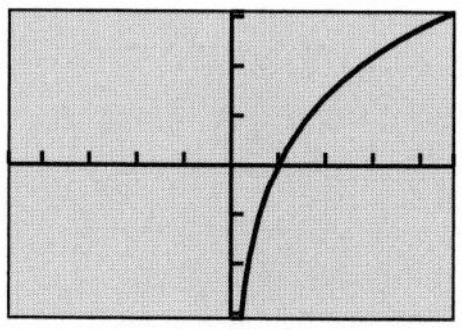
(a)

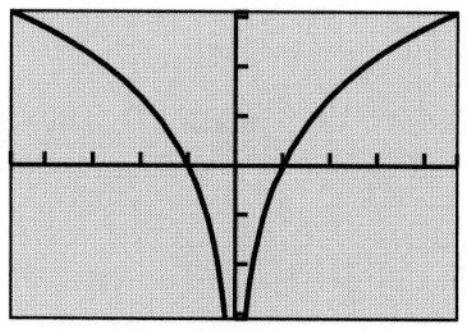
(b)

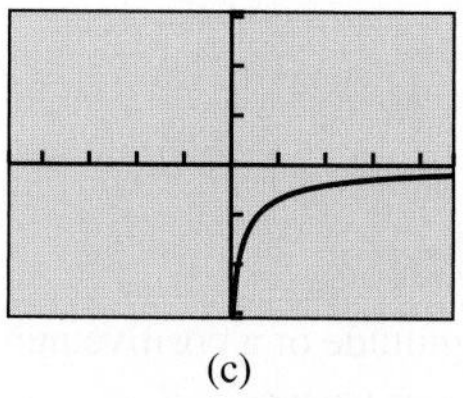
(c)

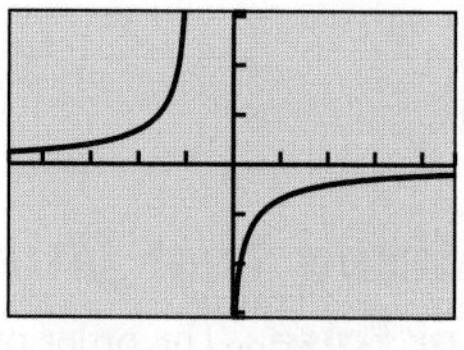
(d)

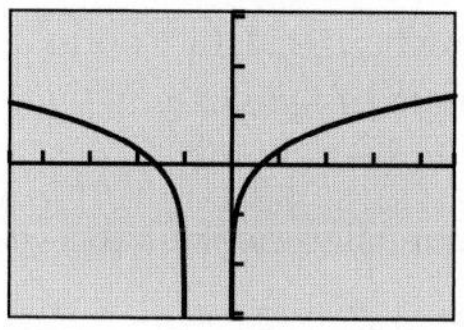
(e)

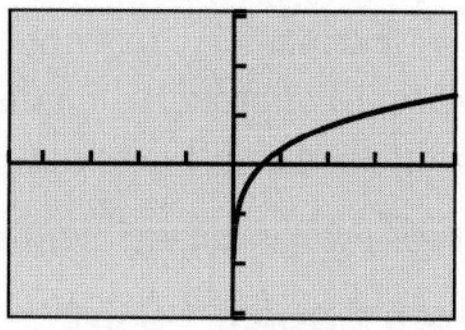
(f)

In Exercises 25–38, solve each equation by the method of your choice. Support your solution by a second method.

25. $\log x^2 = 6$

26. $\ln x^2 = 4$

27. $\log x^4 = 2$ **28.** $\ln x^6 = 12$

29. $\frac{2^x - 2^{-x}}{3} = 4$ **30.** $\frac{2^x + 2^{-x}}{2} = 3$

31. $\frac{e^x + e^{-x}}{2} = 4$ **32.** $2e^{2x} + 5e^x - 3 = 0$

33. $\frac{500}{1 + 25e^{0.3x}} = 200$ **34.** $\frac{400}{1 + 95e^{-0.6x}} = 150$

35. $\frac{1}{2} \ln (x + 3) - \ln x = 0$

36. $\log x - \frac{1}{2} \log (x + 4) = 1$

37. $\ln (x - 3) + \ln (x + 4) = 3 \ln 2$

38. $\log (x - 2) + \log (x + 5) = 2 \log 3$

In Exercises 39–44, determine by how many orders of magnitude the quantities differ.

39. A \$100 bill and a dime

40. A canary weighing 20 g and a hen weighing 2 kg

41. An earthquake rated 7 on the Richter scale and one rated 5.5

42. Lemon juice with pH $= 2.3$ and beer with pH $= 4.1$

43. The sound intensities of a riveter at 95 dB and ordinary conversation at 65 dB

44. The sound intensities of city traffic at 70 dB and rustling leaves at 10 dB

45. Comparing Earthquakes How many times more severe was the 1978 Mexico City earthquake $(R = 7.9)$ than the 1994 Los Angeles earthquake $(R = 6.6)$?

46. Comparing Earthquakes How many times more severe was the 1995 Kobe, Japan, earthquake $(R = 7.2)$ than the 1994 Los Angeles earthquake $(R = 6.6)$?

47. Chemical Acidity The pH of carbonated water is 3.9 and the pH of household ammonia is 11.9.

(a) What are their hydrogen-ion concentrations?

(b) How many times greater is the hydrogen-ion concentration of carbonated water than that of ammonia?

(c) By how many orders of magnitude do the concentrations differ?

48. Chemical Acidity Stomach acid has a pH of about 2.0, and blood has a pH of 7.4.

(a) What are their hydrogen-ion concentrations?

(b) How many times greater is the hydrogen-ion concentration of stomach acid than that of blood?

(c) By how many orders of magnitude do the concentrations differ?

49. Newton's Law of Cooling A cup of coffee has cooled from 92°C to 50°C after 12 min in a room at 22°C. How long will the cup take to cool to 30°C?

50. Newton's Law of Cooling A cake is removed from an oven at 350°F and cools to 120°F after 20 min in a room at 65°F. How long will the cake take to cool to 90°F?

51. Newton's Law of Cooling Experiment A thermometer is removed from a cup of coffee and placed in water with a temperature (T_m) of 10°C. The data in Table 3.24 were collected over the next 30 sec.

Table 3.24 Experimental Data

Time t	Temp T	$T - T_m$
2	80.47	70.47
5	69.39	59.39
10	49.66	39.66
15	35.26	25.26
20	28.15	18.15
25	23.56	13.56
30	20.62	10.62

(a) Draw a scatter plot of the data $T - T_m$.

(b) Find an exponential regression equation for the $T - T_m$ data. Superimpose its graph on the scatter plot.

(c) Estimate the thermometer reading when it was removed from the coffee.

52. Newton's Law of Cooling Experiment A thermometer was removed from a cup of hot chocolate and placed in a saline solution with temperature $T_m = 0$°C. The data in Table 3.25 were collected over the next 30 sec.

(a) Draw a scatter plot of the data $T - T_m$.

(b) Find an exponential regression equation for the $T - T_m$ data. Superimpose its graph on the scatter plot.

(c) Estimate the thermometer reading when it was removed from the hot chocolate.

Table 3.25 Experimental Data

Time t	Temp T	$T - T_m$
2	74.68	74.68
5	61.99	61.99
10	34.89	34.89
15	21.95	21.95
20	15.36	15.36
25	11.89	11.89
30	10.02	10.02

53. Penicillin Use The use of penicillin became so widespread in the 1980s in Hungary that it became practically useless against common sinus and ear infections. Now the use of more effective antibiotics has caused a decline in penicillin resistance. The bar graph shows the use of penicillin in Hungary for selected years.

(a) From the bar graph we read the data pairs to be approximately (1, 11), (8, 6), (15, 4.8), (16, 4), and (17, 2.5), using $t = 1$ for 1976, $t = 8$ for 1983, and so on. Complete a scatter plot for these data.

(b) Writing to Learn Discuss whether the bar graph shown or the scatter plot that you completed best represents the data and why.

Nationwide Consumption of Penicillin

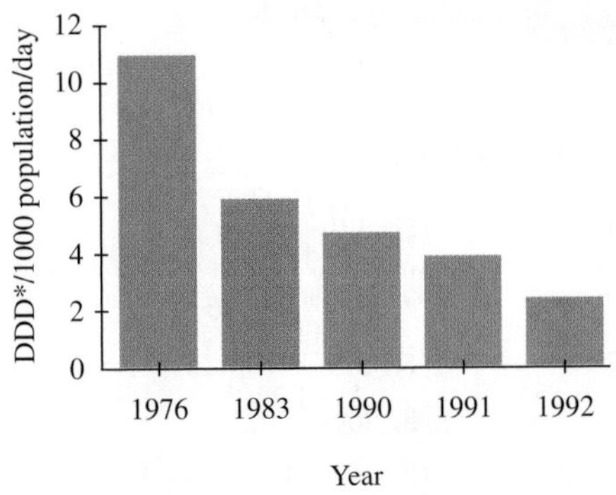

*Defined Daily Dose
Source: Science, vol. 264, April 15, 1994, American Association for the Advancement of Science.

54. Writing to Learn Which regression model would you use for the data in Exercise 53? Discuss various options, and explain why you chose the model you did. Support your writing with tables and graphs as needed.

Writing to Learn In Exercises 55–58, tables of (x, y) data pairs are given. Determine whether a linear, logarithmic, exponential, or power regression equation is the best model for the data. Explain your choice. Support your writing with tables and graphs as needed.

55.

x	1	2	3	4
y	3	4.4	5.2	5.8

56.

x	1	2	3	4
y	6	18	54	162

57.

x	1	2	3	4
y	3	6	12	24

58.

x	1	2	3	4
y	5	7	9	11

Standardized Test Questions

59. True or False The order of magnitude of a positive number is its natural logarithm. Justify your answer.

60. True or False According to Newton's Law of Cooling, an object will approach the temperature of the medium that surrounds it. Justify your answer.

In Exercises 61–64, solve the problem without using a calculator.

61. Multiple Choice Solve $2^{3x-1} = 32$.

(A) $x = 1$ **(B)** $x = 2$ **(C)** $x = 4$

(D) $x = 11$ **(E)** $x = 13$

62. Multiple Choice Solve $\ln x = -1$.

(A) $x = -1$ **(B)** $x = 1/e$ **(C)** $x = 1$

(D) $x = e$ **(E)** No solution is possible.

63. Multiple Choice How many times more severe was the 2001 earthquake in Arequipa, Peru ($R_1 = 8.1$), than the 1998 double earthquake in Takhar province, Afghanistan ($R_2 = 6.1$)?

(A) 2 **(B)** 6.1 **(C)** 8.1

(D) 14.2 **(E)** 100

64. Multiple Choice Newton's Law of Cooling is

(A) an exponential model. **(B)** a linear model.

(C) a logistic model. **(D)** a power model.

Explorations

In Exercises 65 and 66, use the data in Table 3.26. Determine whether a linear, exponential, power, or logistic regression equation is the best model for the data. Explain your choice. Support your writing with tables and graphs as needed.

Table 3.26 Populations of Two U.S. States (in thousands)

Year	Alaska	Hawaii
1900	63.6	154
1910	64.4	192
1920	55.0	256
1930	59.2	368
1940	72.5	423
1950	128.6	500
1960	226.2	633
1970	302.6	770
1980	401.9	965
1990	550.0	1108
2011	722.7	1374

Source: U.S. Census Bureau.

65. Writing to Learn Modeling Population Which regression equation is the best model for Alaska's population?

66. Writing to Learn Modeling Population Which regression equation is the best model for Hawaii's population?

67. Group Activity Normal Distribution The function

$$f(x) = k \cdot e^{-cx^2},$$

where c and k are positive constants, is a bell-shaped curve that is useful in probability and statistics.

(a) Graph f for $c = 1$ and $k = 0.1, 0.5, 1, 2, 10$. Explain the effect of changing k.

(b) Graph f for $k = 1$ and $c = 0.1, 0.5, 1, 2, 10$. Explain the effect of changing c.

68. Group Activity Use the U.S. population data in Table 3.9 on page 268.

(a) Let P be the U.S. population (in millions), and let t be the number of years after 1900. Create a scatter plot of the (t, P) pairs.

(b) Add a plot of the (P, t) pairs to the scatter plot created in part (a).

(c) Be sure to adjust the window to see all the data in both plots and graph in a square window. Add the line $y = x$. Use TRACE to show that the two plots are reflections of each other about the line $y = x$.

(d) Use your grapher to find an exponential model (y_1) for the (t, P) pairs and a logarithmic model (y_2) for the (P, t) pairs.

(e) Graph the exponential regression model as $y_1 = f(x)$ and the logarithmic regression model as $y_2 = g(x)$.

(f) Analytically show how to compute the logarithmic regression value $g(311.6)$ using only the exponential regression equation.

Extending the Ideas

69. Writing to Learn Prove if $u/v = 10^n$ for $u > 0$ and $v > 0$, then $\log u - \log v = n$. Explain how this result relates to powers of ten and orders of magnitude.

70. Potential Energy The potential energy E (the energy stored for use at a later time) between two ions in a certain molecular structure is modeled by the function

$$E = -\frac{5.6}{r} + 10e^{-r/3}$$

where r is the distance separating the nuclei.

(a) Writing to Learn Graph this function in the window $[-10, 10]$ by $[-10, 30]$, and explain which portion of the graph does not represent this potential energy situation.

(b) Identify a viewing window that shows that portion of the graph (with $r \le 10$) which represents this situation, and find the maximum value for E.

71. In Example 8, the Newton's Law of Cooling model was

$$T(t) - T_m = (T_0 - T_m)e^{-kt} = 61.656 \times 0.92770^t.$$

Determine the value of k.

72. Justify the conclusion made about natural logarithmic regression on page 299.

73. Justify the conclusion made about power regression on page 299.

In Exercises 74–79, solve the equation or inequality.

74. $e^x + x = 5$

75. $e^{2x} - 8x + 1 = 0$

76. $e^x < 5 + \ln x$

77. $\ln |x| - e^{2x} \ge 3$

78. $2 \log x - 4 \log 3 > 0$

79. $2 \log (x + 1) - 2 \log 6 < 0$

3.6 Mathematics of Finance

What you'll learn about

- Simple and Compound Interest
- Interest Compounded k Times per Year
- Interest Compounded Continuously
- Annual Percentage Yield
- Annuities—Future Value
- Loans and Mortgages—Present Value

... and why

The mathematics of finance is the science of letting your money work for you—valuable information indeed!

Simple and Compound Interest

In business, as the saying goes, "time is money." We must pay interest for the use of property or money over time. When we borrow money, we pay interest, and when we loan money, we receive interest. When we invest in a savings account, we are actually lending money to the bank.

Simple interest is called "simple" because it ignores the effects of compounding. The interest charge is based solely on the original principal, so interest on interest paid is *not* included. In contrast, for **compound interest** the interest charge is based on the original principal plus interest on accrued interest. Compound interest includes interest earned on the interest itself. Suppose a principal of P dollars is invested in an account bearing interest rate r, expressed in decimals, for n years. The value over time of the investment for both simple interest and interest compounded annually follows the growth pattern in Table 3.27. The nth row of the table gives a formula for simple annual interest and interest compounded annually.

Table 3.27 Interest Computed Using Simple Interest and Compound Interest

	Amount in the Account	
Time (years)	Simple Interest	Compound Interest
0	$A_0 = P =$ principal	$A_0 = P =$ principal
1	$A_1 = P + rP = P(1 + r)$	$A_1 = P + rP = P(1 + r)$
2	$A_2 = P + 2rP = P(1 + 2r)$	$A_2 = A_1(1 + r) = P(1 + r)^2$
3	$A_3 = P + 3rP = P(1 + 3r)$	$A_3 = A_2(1 + r) = P(1 + r)^3$
⋮	⋮	⋮
n	$A = A_n = P(1 + nr)$	$A = A_n = P(1 + r)^n$

Notice in the case of simple interest the growth model is a linear function of time, and in the case of compound interest, the growth model is an exponential function of time. Also notice for the first period simple interest and compound interest are the same. In the exercises, you will show that the graph of the simple interest function is a line and the graph of a compound interest function is an exponential curve.

Interest Compounded Annually

If a principal P is invested at a fixed annual interest rate r, calculated at the end of each year, then the value of the investment after n years is

$$A = P(1 + r)^n,$$

where r is expressed as a decimal.

EXAMPLE 1 Interest Compounded Annually Versus Simple Annual Interest

Suppose George Milligan invests \$5000 at 5% annual simple interest and Quan Li also invests \$5000 at 5% annual compound interest. Compare the values of both investments after 10 years.

SOLUTION Let $P = 5000$, $r = 0.05$, and $n = 10$. Then, using simple interest, $A = 5000(1 + 10 \cdot 0.05) = \$7500$. Using compound interest, $A = 5000(1 + 0.05)^{10} = \$8144.47$. You earn \$644.47 more using compound interest. **Now try Exercise 1.**

Interest Compounded k Times per Year

Suppose a principal P is invested at an annual interest rate r compounded k times a year for t years. Then $\frac{r}{k}$ is the interest rate per compounding period, and kt is the number of compounding periods. The amount A in the account after t years is

$$A = P\left(1 + \frac{r}{k}\right)^{kt}.$$

EXAMPLE 2 Compounding Monthly

Suppose Roberto invests \$500 at 9% annual interest *compounded monthly*, that is, compounded 12 times a year. Find the value of his investment 5 years later.

SOLUTION Letting $P = 500$, $r = 0.09$, $k = 12$, and $t = 5$,

$$A = 500\left(1 + \frac{0.09}{12}\right)^{12(5)} = 782.840\ldots$$

So the value of Roberto's investment after 5 years is \$782.84. **Now try Exercise 5.**

The problems in Examples 1 and 2 required that we calculate A. Examples 3 and 4 illustrate situations that require us to determine the values of other variables in the compound interest formula.

EXAMPLE 3 Finding the Time Period of an Investment

Judy has \$500 to invest at 9% annual interest compounded monthly. How long will it take for her investment to grow to \$3000?

SOLUTION

Model Let $P = 500$, $r = 0.09$, $k = 12$, and $A = 3000$ in the equation

$$A = P\left(1 + \frac{r}{k}\right)^{kt},$$

and solve for t.

Solve Graphically For

$$3000 = 500\left(1 + \frac{0.09}{12}\right)^{12t},$$

we let

$$f(t) = 500\left(1 + \frac{0.09}{12}\right)^{12t} \quad \text{and} \quad y = 3000,$$

and then find the point of intersection of the graphs. Figure 3.41 shows that this occurs at $t \approx 19.98$.

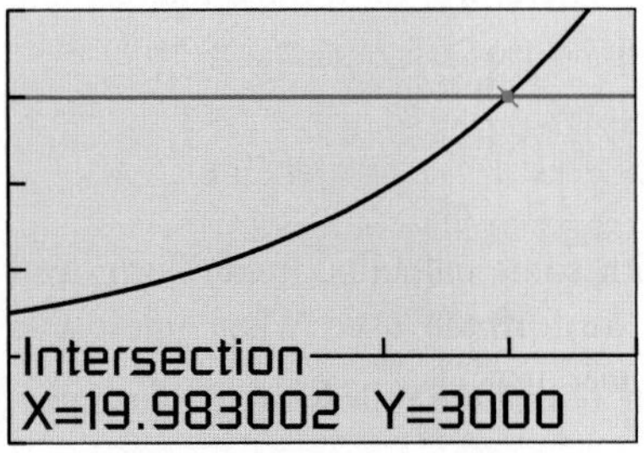

[0, 25] by [−1000, 4000]

FIGURE 3.41 Graph for Example 3.

Confirm Algebraically

$$\begin{aligned}
3000 &= 500(1 + 0.09/12)^{12t} \\
6 &= 1.0075^{12t} && \text{Divide by 500.} \\
\ln 6 &= \ln(1.0075^{12t}) && \text{Apply ln to each side.} \\
\ln 6 &= 12t \ln(1.0075) && \text{Power rule} \\
t &= \frac{\ln 6}{12 \ln 1.0075} && \text{Divide by 12 ln 1.0075.} \\
&= 19.983\ldots && \text{Calculate.}
\end{aligned}$$

Interpret So it will take Judy 20 years for the value of the investment to reach (and slightly exceed) \$3000. **Now try Exercise 21.**

EXAMPLE 4 Finding an Interest Rate

Stephen has \$500 to invest. What annual interest rate *compounded quarterly* (four times per year) is required to double his money in 10 years?

SOLUTION

Model Letting $P = 500$, $k = 4$, $t = 10$, and $A = 1000$ yields the equation

$$1000 = 500\left(1 + \frac{r}{4}\right)^{4(10)}$$

that we solve for r.

Solve Graphically Figure 3.42 shows that $f(r) = 500(1 + r/4)^{40}$ and $y = 1000$ intersect at $r \approx 0.0699$, or $r = 6.99\%$.

Interpret Stephen's investment of \$500 will double in 10 years at an annual interest rate of 6.99% compounded quarterly. Now try Exercise 25.

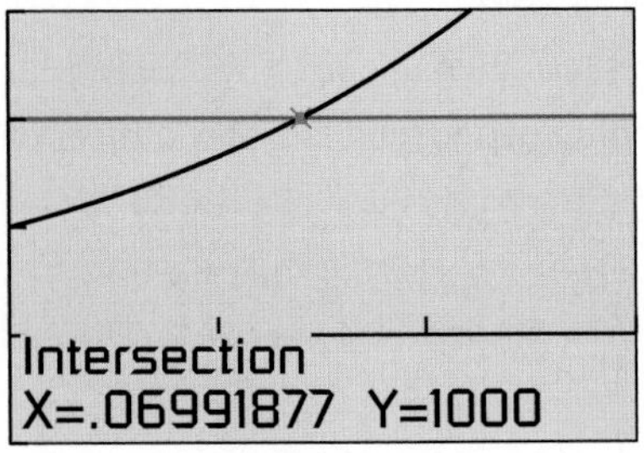

[0, 0.15] by [–500, 1500]

FIGURE 3.42 Graph for Example 4.

Interest Compounded Continuously

In Exploration 1, \$1000 is invested for 1 year at a 10% interest rate. We investigate the value of the investment at the end of 1 year as the number of compounding periods k increases. In other words, we determine the "limiting" value of the expression $1000(1 + 0.1/k)^k$ as k assumes larger and larger integer values.

EXPLORATION 1 Increasing the Number of Compounding Periods Boundlessly

Let $A = 1000\left(1 + \frac{0.1}{k}\right)^k$.

1. Complete a table of values of A for $k = 10, 20, \ldots, 100$. What pattern do you observe?
2. Figure 3.43 shows the graphs of the function $A(k) = 1000(1 + 0.1/k)^k$ and the horizontal line $y = 1000e^{0.1}$. Interpret the meanings of these graphs.

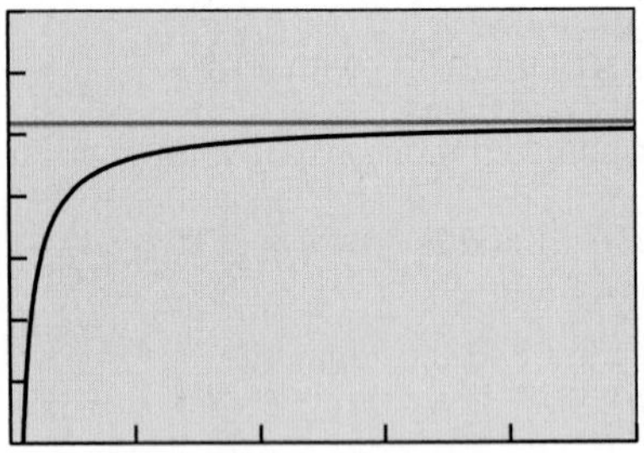

[0, 50] by [1100, 1107]

FIGURE 3.43 Graph for Exploration 1.

Recall from Section 3.1 that $e = \lim_{x\to\infty} (1 + 1/x)^x$. Therefore, for a fixed interest rate r, if we let $x = k/r$,

$$\lim_{k\to\infty}\left(1 + \frac{r}{k}\right)^{k/r} = e.$$

We do not know enough about limits yet, but with some calculus, it can be proved that $\lim_{k\to\infty} P(1 + r/k)^{kt} = Pe^{rt}$. So $A = Pe^{rt}$ is the formula used when interest is **compounded continuously**. In nearly any situation, one of the following two formulas can be used to compute compound interest:

Compound Interest—Value of an Investment

Suppose a principal P is invested at a fixed annual interest rate r. The value of the investment after t years is

- $A = P\left(1 + \frac{r}{k}\right)^{kt}$ when interest compounds k times per year,
- $A = Pe^{rt}$ when interest compounds continuously.

X	Y_1	
1	108.33	
2	117.35	
3	127.12	
4	137.71	
5	149.18	
6	161.61	
7	175.07	

Y_1 = 100e^(0.08X)

FIGURE 3.44 Table of values for Example 5.

EXAMPLE 5 Compounding Continuously

Suppose LaTasha invests \$100 at 8% annual interest compounded continuously. Find the value of her investment at the end of each of the years 1, 2, . . . , 7.

SOLUTION Substituting into the formula for continuous compounding, we obtain $A(t) = 100e^{0.08t}$. Figure 3.44 shows the values of $y_1 = A(x) = 100e^{0.08x}$ for $x = 1, 2, \ldots, 7$. For example, the value of her investment is \$149.18 at the end of 5 years and \$175.07 at the end of 7 years. **Now try Exercise 9.**

Annual Percentage Yield

With so many different interest rates and methods of compounding it is sometimes difficult for a consumer to compare two different options. For example, would you prefer an investment earning 8.75% annual interest compounded quarterly or one earning 8.7% compounded monthly?

A common basis for comparing investments is the **annual percentage yield (APY)**—the percentage rate that, compounded annually, would yield the same return as the given interest rate with the given compounding period.

EXAMPLE 6 Computing Annual Percentage Yield (APY)

Ursula invests \$2000 with Crab Key Bank at 5.15% annual interest compounded quarterly. What is the equivalent APY?

SOLUTION Let $x =$ the equivalent APY. The value of the investment at the end of 1 year using this rate is $A = 2000(1 + x)$. Thus, we have

$$2000(1 + x) = 2000\left(1 + \frac{0.0515}{4}\right)^4$$

$$(1 + x) = \left(1 + \frac{0.0515}{4}\right)^4 \quad \text{Divide by 2000.}$$

$$x = \left(1 + \frac{0.0515}{4}\right)^4 - 1 \quad \text{Subtract 1.}$$

$$\approx 0.0525 \quad \text{Calculate.}$$

The annual percentage yield is 5.25%. In other words, Ursula's \$2000 invested at 5.15% compounded quarterly for 1 year earns the same interest and yields the same value as \$2000 invested elsewhere paying 5.25% interest once at the end of the year. **Now try Exercise 42.**

Example 6 shows that the APY does not depend on the principal P because both sides of the equation were divided by $P = 2000$. So we can assume that $P = 1$ when comparing investments.

EXAMPLE 7 Comparing Annual Percentage Yields (APYs)

Which investment is more attractive, one that pays 8.75% compounded quarterly or another that pays 8.7% compounded monthly?

(continued)

SOLUTION

Let

$$r_1 = \text{ the APY for the 8.75\% rate}$$
$$r_2 = \text{ the APY for the 8.7\% rate}$$

$$1 + r_1 = \left(1 + \frac{0.0875}{4}\right)^4 \qquad 1 + r_2 = \left(1 + \frac{0.087}{12}\right)^{12}$$

$$r_1 = \left(1 + \frac{0.0875}{4}\right)^4 - 1 \qquad r_2 = \left(1 + \frac{0.087}{12}\right)^{12} - 1$$

$$\approx 0.09041 \qquad \approx 0.09055$$

The 8.7% rate compounded monthly is more attractive because its APY is 9.055% compared with 9.041% for the 8.75% rate compounded quarterly.

Now try Exercise 46.

Annuities—Future Value

So far, in all of the investment situations we have considered, the investor has made a single *lump-sum* deposit. But suppose an investor makes regular deposits monthly, quarterly, or yearly—the same amount each time. This is an *annuity* situation.

An **annuity** is a sequence of equal periodic payments. The annuity is **ordinary** if deposits are made at the end of each period at the same time the interest is posted in the account. Figure 3.45 represents this situation graphically. We will consider only ordinary annuities in this textbook.

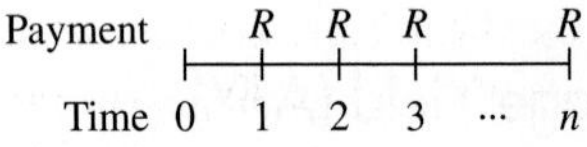

FIGURE 3.45 Payments into an ordinary annuity.

Let's consider an example. Suppose Sarah makes \$500 payments at the end of each quarter into a retirement account that pays 8% interest compounded quarterly. How much will be in Sarah's account at the end of the first year? Notice the pattern.

End of Quarter 1:

$$\$500 = \$500$$

End of Quarter 2:

$$\$500 + \$500(1.02) = \$1010$$

End of Quarter 3:

$$\$500 + \$500(1.02) + \$500(1.02)^2 = \$1530.20$$

End of the year:

$$\$500 + \$500(1.02) + \$500(1.02)^2 + \$500(1.02)^3 \approx \$2060.80$$

Thus the total value of the investment returned from an annuity consists of all the periodic payments together with all the interest. This value is called the **future value** of the annuity because it is typically calculated when projecting into the future.

Future Value of an Annuity

The future value FV of an annuity consisting of n equal periodic payments of R dollars at an interest rate i per compounding period (payment interval) is

$$FV = R\frac{(1 + i)^n - 1}{i}.$$

EXAMPLE 8 Calculating the Value of an Annuity

At the end of each quarter year, Emily makes a \$500 payment into the Lanaghan Mutual Fund. If her investments earn 7.88% annual interest compounded quarterly, what will be the value of Emily's annuity in 20 years?

SOLUTION Let $R = 500$, $i = 0.0788/4$, $n = 20(4) = 80$. Then,

$$FV = R\frac{(1 + i)^n - 1}{i}$$

$$FV = 500 \cdot \frac{(1 + 0.0788/4)^{80} - 1}{0.0788/4}$$

$$FV = 95{,}483.389\ldots$$

So the value of Emily's annuity in 20 years will be \$95,483.39.

Now try Exercise 13.

Loans and Mortgages—Present Value

An annuity is a sequence of equal period payments. The net amount of money put into an annuity is its **present value**. The net amount returned from the annuity is its future value. The periodic and equal payments on a loan or mortgage actually constitute an annuity.

How does the bank determine what the periodic payments should be? It considers what would happen to the present value of an investment with interest compounding over the term of the loan and compares the result to the future value of the loan repayment annuity.

We illustrate this reasoning by assuming that a bank lends you a present value $PV = \$50{,}000$ at 6% to purchase a house with the expectation that you will make a mortgage payment each month (at the monthly interest rate of $0.06/12 = 0.005$).

- The future value of an investment at 6% compounded monthly for n months is

$$PV(1 + i)^n = 50{,}000(1 + 0.005)^n.$$

- The future value of an annuity of R dollars (the loan payments) is

$$R\frac{(1 + i)^n - 1}{i} = R\frac{(1 + 0.005)^n - 1}{0.005}.$$

To find R, we would solve the equation

$$50{,}000(1 + 0.005)^n = R\frac{(1 + 0.005)^n - 1}{0.005}.$$

In general, the monthly payments of R dollars for a loan of PV dollars must satisfy the equation

$$PV(1 + i)^n = R\frac{(1 + i)^n - 1}{i}.$$

Dividing both sides by $(1 + i)^n$ leads to the following formula for the present value of an annuity.

Present Value of an Annuity

The present value PV of an annuity consisting of n equal payments of R dollars earning an interest rate i per period (payment interval) is

$$PV = R\frac{1 - (1 + i)^{-n}}{i}.$$

Student Note

Understand Present Value The concept of "present value" is important in many ways. In many life situations you can have the choice of a lump sum payment "today" or lesser payments over time. Thus it is important to be able to determine the value of future expected payments. The PV formula will tell you the value "today" of a series of equal future payments. However, in choosing an alternative (lump sum or a series of payments over time), other factors may need to be considered, such as risk (how do you know the entity paying you will be solvent in 5 years?) or the need for a large sum of money "today."

The annual interest rate charged on consumer loans is the **annual percentage rate (APR)**. The APY for the lender is higher than the APR. See Exercise 59.

EXAMPLE 9 Calculating Loan Payments

Carlos purchases a pickup truck for \$18,500. What are the monthly payments for a 4-year loan with a \$2000 down payment if the annual interest rate (APR) is 2.9%?

SOLUTION

Model The down payment is \$2000, so the amount borrowed is \$16,500. Since APR = 2.9%, $i = 0.029/12$ and the monthly payment is the solution to

$$16{,}500 = R\frac{1 - (1 + 0.029/12)^{-4(12)}}{0.029/12}.$$

Solve Algebraically

$$R\left[1 - \left(1 + \frac{0.029}{12}\right)^{-4(12)}\right] = 16{,}500\left(\frac{0.029}{12}\right)$$

$$R = \frac{16{,}500(0.029/12)}{1 - (1 + 0.029/12)^{-48}}$$

$$= 364.487\ldots$$

Interpret Carlos will have to pay \$364.49 per month for 47 months, and slightly less the last month. Now try Exercise 19.

QUICK REVIEW 3.6

Exercise numbers with a gray background indicate problems that the authors have designed to be solved *without a calculator*.

1. Find 3.5% of 200.
2. Find 2.5% of 150.
3. What is one-fourth of 7.25%?
4. What is one-twelfth of 6.5%?
5. 78 is what percent of 120?
6. 28 is what percent of 80?
7. 48 is 32% of what number?
8. 176.4 is 84% of what number?
9. How much does Jane have at the end of 1 year if she invests \$300 at 5% simple interest?
10. How much does Reggie have at the end of 1 year if he invests \$500 at 4.5% simple interest?

SECTION 3.6 Exercises

In Exercises 1–4, find the amount A accumulated after investing a principal P for t years at an interest rate r compounded annually, and for simple interest at the same rate r.

1. $P = \$1500, r = 7\%, t = 6$
2. $P = \$3200, r = 8\%, t = 4$
3. $P = \$12{,}000, r = 7.5\%, t = 7$
4. $P = \$15{,}500, r = 9.5\%, t = 12$

In Exercises 5–8, find the amount A accumulated after investing a principal P for t years at an interest rate r compounded k times per year.

5. $P = \$1500, r = 7\%, t = 5, k = 4$
6. $P = \$3500, r = 5\%, t = 10, k = 4$
7. $P = \$40{,}500, r = 3.8\%, t = 20, k = 12$
8. $P = \$25{,}300, r = 4.5\%, t = 25, k = 12$

In Exercises 9–12, find the amount A accumulated after investing a principal P for t years at interest rate r compounded continuously.

9. $P = \$1250, r = 5.4\%, t = 6$
10. $P = \$3350, r = 6.2\%, t = 8$
11. $P = \$21{,}000, r = 3.7\%, t = 10$
12. $P = \$8{,}875, r = 4.4\%, t = 25$

In Exercises 13–15, find the future value FV accumulated in an annuity after investing periodic payments R for t years at an annual interest rate r, with payments made and interest credited k times per year.

13. $R = \$500, r = 7\%, t = 6, k = 4$
14. $R = \$300, r = 6\%, t = 12, k = 4$
15. $R = \$450, r = 5.25\%, t = 10, k = 12$
16. $R = \$610, r = 6.5\%, t = 25, k = 12$

In Exercises 17 and 18, find the present value PV of a loan with an annual interest rate r and periodic payments R for a term of t years, with payments made and interest charged 12 times per year.

17. $r = 4.7\%$, $R = \$815.37$, $t = 5$

18. $r = 6.5\%$, $R = \$1856.82$, $t = 30$

In Exercises 19 and 20, find the periodic payment R of a loan with present value PV and an annual interest rate r for a term of t years, with payments made and interest charged 12 times per year.

19. $PV = \$18{,}000$, $r = 5.4\%$, $t = 6$

20. $PV = \$154{,}000$, $r = 7.2\%$, $t = 15$

21. Finding Time If John invests $2300 in a savings account with a 9% interest rate compounded quarterly, how long will it take until John's account has a balance of $4150?

22. Finding Time If Joelle invests $8000 into a retirement account with a 9% interest rate compounded monthly, how long will it take until this single payment has grown in her account to $16,000?

23. Trust Officer Megan is the trust officer for an estate. If she invests $15,000 into an account that carries an interest rate of 8% compounded monthly, how long will it be until the account has a value of $45,000 for Megan's client?

24. Chief Financial Officer Willis is the financial officer of a private university with the responsibility for managing an endowment. If he invests $1.5 million at an interest rate of 8% compounded quarterly, how long will it be until the account exceeds $3.75 million?

25. Finding the Interest Rate What interest rate compounded daily (365 days/year) is required for a $22,000 investment to grow to $36,500 in 5 years?

26. Finding the Interest Rate What interest rate compounded monthly is required for an $8500 investment to triple in 5 years?

27. Pension Officer Jack is an actuary working for a corporate pension fund. He needs to have $14.6 million grow to $22 million in 6 years. What interest rate compounded annually does he need for this investment?

28. Bank President The president of a bank has $18 million in his bank's investment portfolio that he wants to grow to $25 million in 8 years. What interest rate compounded annually does he need for this investment?

29. Doubling Your Money Determine how much time is required for an investment to double in value if interest is earned at the rate of 5.75% compounded quarterly.

30. Tripling Your Money Determine how much time is required for an investment to triple in value if interest is earned at the rate of 6.25% compounded monthly.

In Exercises 31–34, complete the table about continuous compounding.

	Initial Investment	APR	Time to Double	Amount in 15 Years
31.	$12,500	9%	?	?
32.	$32,500	8%	?	?
33.	$ 9,500	?	4 years	?
34.	$16,800	?	6 years	?

In Exercises 35–41, complete the table about doubling time of an investment.

	APR	Compounding Periods	Time to Double
35.	4%	Quarterly	?
36.	8%	Quarterly	?
37.	7%	Simple annual interest	?
38.	7%	Annually	?
39.	7%	Quarterly	?
40.	7%	Monthly	?
41.	7%	Continuously	?

In Exercises 42–45, find the annual percentage yield (APY) for the investment.

42. $3000 at 6% compounded quarterly

43. $8000 at 5.75% compounded daily

44. P dollars at 6.3% compounded continuously

45. P dollars at 4.7% compounded monthly

46. Comparing Investments Which investment is more attractive, 5% compounded monthly or 5.1% compounded quarterly?

47. Comparing Investments Which investment is more attractive, $5\frac{1}{8}\%$ compounded annually or 5% compounded continuously?

In Exercises 48–51, payments are made and interest is credited at the end of each month.

48. An IRA Account Amy contributes $50 per month into the Lincoln National Bond Fund that earns 7.26% annual interest. What is the value of Amy's investment after 25 years?

49. An IRA Account Andrew contributes $50 per month into the Hoffbrau Fund that earns 15.5% annual interest. What is the value of his investment after 20 years?

50. An Investment Annuity Jolinda contributes to the Celebrity Retirement Fund that earns 12.4% annual interest. What should her monthly payments be if she wants to accumulate $250,000 in 20 years?

51. An Investment Annuity Diego contributes to a Commercial National money market account that earns 4.5% annual interest. What should his monthly payments be if he wants to accumulate $120,000 in 30 years?

52. Car Loan Payment What is Kim's monthly payment for a 4-year $9000 car loan with an APR of 7.95% from Century Bank?

53. Car Loan Payment What is Ericka's monthly payment for a 3-year $4500 car loan with an APR of 10.25% from County Savings Bank?

54. House Mortgage Payment Gendo obtains a 30-year $86,000 house loan with an APR of 8.75% from National City Bank. What is her monthly payment?

55. House Mortgage Payment Roberta obtains a 25-year $100,000 house loan with an APR of 9.25% from NBD Bank. What is her monthly payment?

56. Mortgage Payment Planning An \$86,000 mortgage for 30 years at 12% APR requires monthly payments of \$884.61. Suppose you decided to make monthly payments of \$1050.00.

(a) When would the mortgage be completely paid?

(b) How much do you save with the greater payments compared with the original plan?

57. Mortgage Payment Planning Suppose you make payments of \$884.61 for the \$86,000 mortgage in Exercise 54 for 10 years and then make payments of \$1050 until the loan is paid.

(a) When will the mortgage be completely paid under these circumstances?

(b) How much do you save with the greater payments compared with the original plan?

58. Writing to Learn Explain why computing the APY for an investment does not depend on the actual amount being invested. Give a formula for the APY on a \$1 investment at annual rate r compounded k times a year. How do you extend the result to a \$1000 investment?

59. Writing to Learn Give reasons why banks might not announce their APY on a loan they would make to you at a given APR. What is the bank's APY on a loan that they make at 4.5% APR?

60. Group Activity Work in groups of three or four. Consider population growth of humans or other animals, bacterial growth, radioactive decay, and compounded interest. Explain how these problem situations are similar and how they are different. Give examples to support your point of view.

61. Simple Interest Versus Compounding Annually Steve purchases a \$1000 certificate of deposit and will earn 6% each year. The interest will be mailed to him, so he will not earn interest on his interest.

(a) Writing to Learn Explain why after t years, the total amount of interest he receives from his investment plus the original \$1000 is given by

$$f(t) = 1000(1 + 0.06t).$$

(b) Steve invests another \$1000 at 6% compounded annually. Make a table that compares the value of the two investments for $t = 1, 2, \ldots, 10$ years.

62. Simple Interest Function Let $P(t) = P + P \cdot r \cdot t = P(1 + rt)$, where P is the original amount invested, r is the simple annual interest rate, and t is time in years.

(a) Graph $y = P(t)$ for $P = \$300$ and $r = 3\%$ simple annual interest. What type of function is graphed?

(b) What is the slope of the line graphed?

(c) Describe the effect on the graph of (a) by increasing or decreasing the interest rates.

(d) Evaluate $P(0.5)$ for simple interest and compound interest. Here $t = 0.5$ means one-half year. Use the data in (a) and compare $P_{\text{Compound}}(0.5)$ and $P_{\text{Simple}}(0.5)$. Which one is greater? Explain why and discuss.

Standardized Test Questions

63. True or False If \$100 is invested at 5% annual interest for 1 year, there is no limit to the final value of the investment if it is compounded sufficiently often. Justify your answer.

64. True or False The total interest paid on a 15-year mortgage is less than half of the total interest paid on a 30-year mortgage with the same loan amount and APR. Justify your answer.

In Exercises 65–68, you may use a graphing calculator to solve the problem.

65. Multiple Choice What is the total value after 6 years of an initial investment of \$2250 that earns 7% interest compounded quarterly?

(A) \$3376.64 **(B)** \$3412.00 **(C)** \$3424.41
(D) \$3472.16 **(E)** \$3472.27

66. Multiple Choice The annual percentage yield of an account paying 6% compounded monthly is

(A) 6.03%. **(B)** 6.12%. **(C)** 6.17%.
(D) 6.20%. **(E)** 6.24%.

67. Multiple Choice Mary Jo deposits \$300 each month into her retirement account that pays 4.5% APR (0.375% per month). Use the formula $FV = R((1 + i)^n - 1)/i$ to find the value of her annuity after 20 years.

(A) \$71,625.00 **(B)** \$72,000.00 **(C)** \$72,375.20
(D) \$73,453.62 **(E)** \$116,437.31

68. Multiple Choice To finance their home, Mr. and Mrs. Dass have agreed to a \$120,000 mortgage loan at 7.25% APR. Use the formula $PV = R(1 - (1 + i)^{-n})/i$ to determine their monthly payments if the loan has a term of 15 years.

(A) \$1095.44 **(B)** \$1145.44 **(C)** \$1195.44
(D) \$1245.44 **(E)** \$1295.44

Explorations

69. Loan Payoff Use the information about Carlos's truck loan in Example 9 to make a spreadsheet of the payment schedule. The first few lines of the spreadsheet should look like the following table:

Month No.	Payment	Interest	Principal	Balance
0				\$16,500.00
1	\$364.49	\$39.88	\$324.61	\$16,175.39
2	\$364.49	\$39.09	\$325.40	\$15,849.99

To create the spreadsheet successfully, however, you need to use formulas for many of the cells, as shown in boldface type in the following sample:

Month No.	Payment	Interest	Principal	Balance
0				\$16,500.00
= A2 + 1	\$364.49	**= round(E2*2.9%/12,2)**	**= B3 − C3**	**= E2 − D3**
= A3 + 1	\$364.49	**= round(E3*2.9%/12,2)**	**= B4 − C4**	**= E3 − D4**

Continue the spreadsheet using copy-and-paste techniques, and determine the amount of the 48th and final payment so that the final balance is \$0.00.

70. Writing to Learn Loan Payoff Which of the following graphs is an accurate graph of the loan balance as a function of time, based on Carlos's truck loan in Example 9 and Exercise 69? Explain your choice based on increasing or decreasing behavior and other analytical characteristics. Would you expect the graph of loan balance versus time for a 30-year mortgage loan at twice the interest rate to have the same shape or a different shape as the one for the truck loan? Explain.

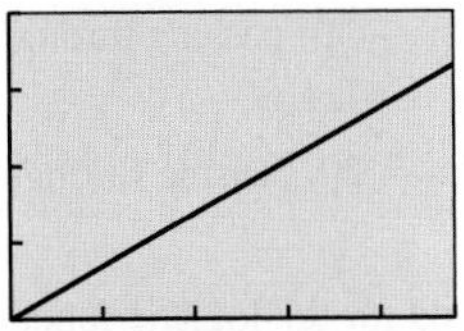

[0, 48] by [0, 20 000]

(a)

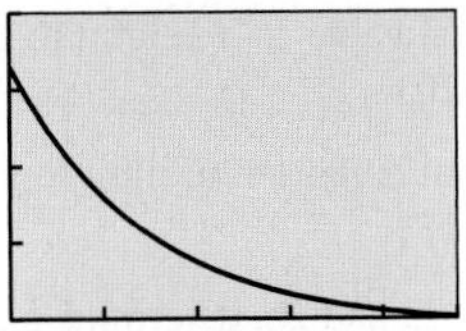

[0, 48] by [0, 20 000]

(b)

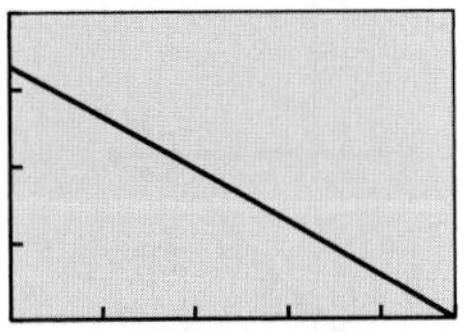

[0, 48] by [0, 20 000]

(c)

71. Graph the present value (PV) function of \$10,000 paid annually for 500 years at 4% annual interest. Compute the PV at 10 years, 50 years, 100 years, 150 years, 200 years, and 300 years by using the TRACE function on a grapher. Discuss why the function seems to have a horizontal asymptote and seems to have a limit of \$250,000.00 to the nearest penny.

Extending the Ideas

72. The function

$$F(x) = 100\frac{\left(1 + \frac{0.08}{12}\right)^x - 1}{\frac{0.08}{12}}, \text{ where } x \text{ is time in months,}$$

describes the future value of a certain annuity.

(a) What is the annual interest rate?

(b) How many payments per year are there?

(c) What is the amount of each payment?

(d) Compute the future value of the annuity paying \$100 monthly for 20 years.

(e) Graph the future value function for x in the interval 0 to 20 years.

73. The function

$$P(x) = 200\frac{1 - \left(1 + \frac{0.08}{12}\right)^{-x}}{\frac{0.08}{12}}, \text{ where } x \text{ is time in months,}$$

describes the present value of a certain annuity.

(a) What is the annual interest rate?

(b) How many payments per year are there?

(c) What is the amount of each payment?

(d) Compute the present value of the annuity paying \$200 monthly after 20 years.

(e) Graph the present value function for x in the interval 0 to 20 years.

(f) Explain why the answer to (d) is much less then the total of all the 240 monthly payments of \$200 each, which is \$48,000. Which would you rather have: \$48,000 by receiving 200 payments each month or \$23,910.86 now? Why?

CHAPTER 3 Key Ideas

Properties, Theorems, and Formulas

Exponential Growth and Decay 254
Exponential Functions $f(x) = b^x$ 255
Exponential Functions and the Base e 257
Exponential Population Model 265
Changing Between Logarithmic and Exponential Form 274
Basic Properties of Logarithms 274
Basic Properties of Common Logarithms 276
Basic Properties of Natural Logarithms 277
Properties of Logarithms 283
Change-of-Base Formula for Logarithms 285
Logarithmic Functions $f(x) = \log_b x$, with $b > 1$ 287
One-to-One Properties 292
Newton's Law of Cooling 296
Interest Compounded Annually 304
Interest Compounded k Times per Year 305–306
Interest Compounded Continuously 306
Future Value of an Annuity 308
Present Value of an Annuity 309

Procedures

Re-expression of Data 287–288
Logarithmic Re-expression of Data 298–299

Gallery of Functions

Exponential

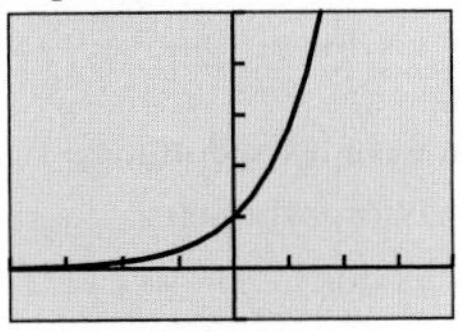

[–4, 4] by [–1, 5]

$f(x) = e^x$

Basic Logistic

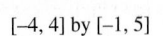

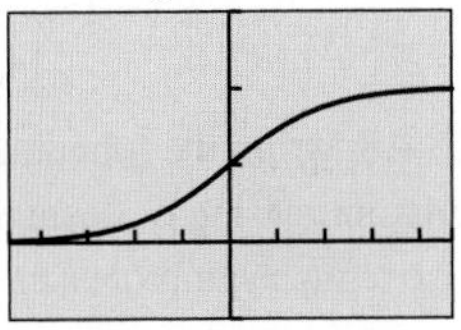

[–4.7, 4.7] by [–0.5, 1.5]

$$f(x) = \frac{1}{1 + e^{-x}}$$

Natural Logarithmic

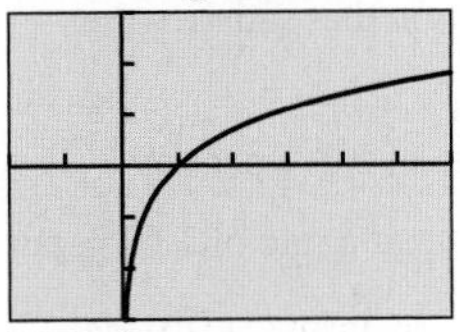

[–2, 6] by [–3, 3]

$f(x) = \ln x$

CHAPTER 3 Review Exercises

Exercise numbers with a gray background indicate problems that the authors have designed to be solved *without a calculator*.

The collection of exercises marked in red could be used as a chapter test.

In Exercises 1 and 2, compute the exact value of the function for the given x value without using a calculator.

1. $f(x) = -3 \cdot 4^x$ for $x = \frac{1}{3}$

2. $f(x) = 6 \cdot 3^x$ for $x = -\frac{3}{2}$

In Exercises 3 and 4, determine a formula for the exponential function whose graph is shown in the figure.

3.

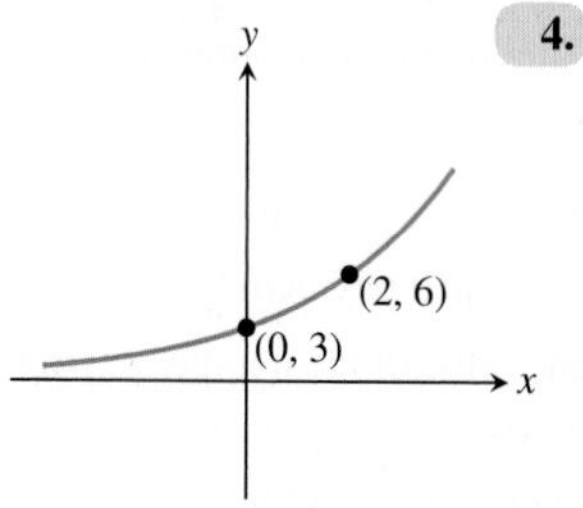

4.

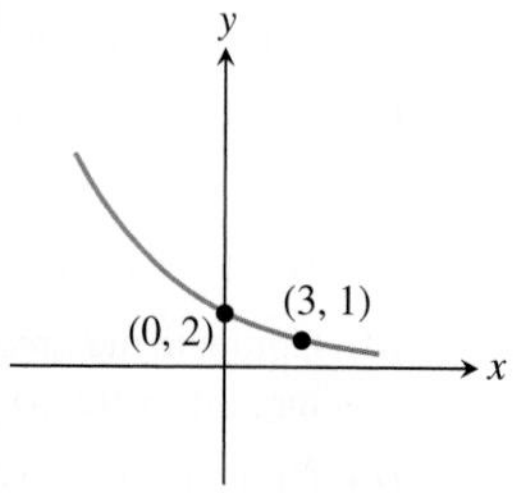

In Exercises 5–10, describe how to transform the graph of f into the graph of $g(x) = 2^x$ or $h(x) = e^x$. Sketch the graph by hand and support your answer with a grapher.

5. $f(x) = 4^{-x} + 3$

6. $f(x) = -4^{-x}$

7. $f(x) = -8^{-x} - 3$

8. $f(x) = 8^{-x} + 3$

9. $f(x) = e^{2x-3}$

10. $f(x) = e^{3x-4}$

In Exercises 11 and 12, find the y-intercept and the horizontal asymptotes.

11. $f(x) = \dfrac{100}{5 + 3e^{-0.05x}}$

12. $f(x) = \dfrac{50}{5 + 2e^{-0.04x}}$

In Exercises 13 and 14, state whether the function is an exponential growth function or an exponential decay function, and describe its end behavior using limits.

13. $f(x) = e^{4-x} + 2$

14. $f(x) = 2(5^{x-3}) + 1$

In Exercises 15–18, graph the function, and analyze it for domain, range, continuity, increasing or decreasing behavior, symmetry, boundedness, extrema, asymptotes, and end behavior.

15. $f(x) = e^{3-x} + 1$

16. $g(x) = 3(4^{x+1}) - 2$

17. $f(x) = \dfrac{6}{1 + 3 \cdot 0.4^x}$

18. $g(x) = \dfrac{100}{4 + 2e^{-0.01x}}$

In Exercises 19–22, find the exponential function that satisfies the given conditions.

19. Initial value = 24, increasing at a rate of 5.3% per day

20. Initial population = 67,000, increasing at a rate of 1.67% per year

21. Initial height = 18 cm, doubling every 3 weeks

22. Initial mass = 117 g, halving once every 262 hr

In Exercises 23 and 24, find the logistic function that satisfies the given conditions.

23. Initial value = 12, limit to growth = 30, passing through (2, 20).

24. Initial height = 6, limit to growth = 20, passing through (3, 15).

In Exercises 25 and 26, determine a formula for the logistic function whose graph is shown in the figure.

25.

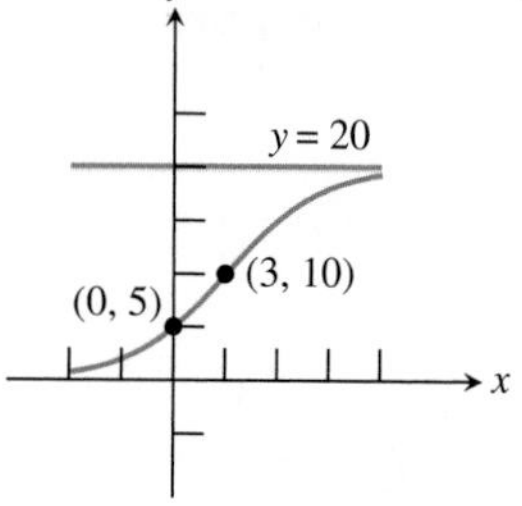

26.

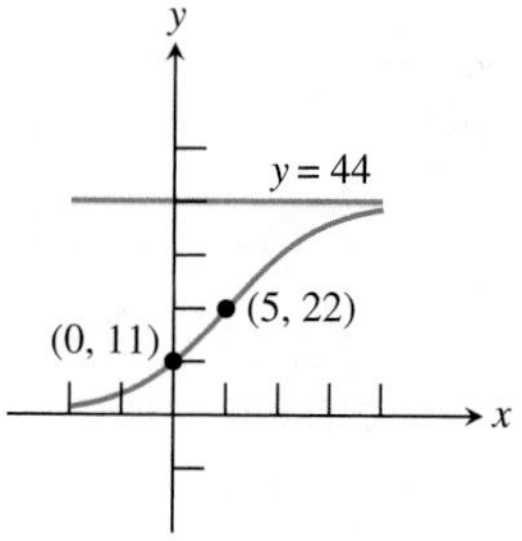

In Exercises 27–30, evaluate the logarithmic expression.

27. $\log_2 32$

28. $\log_3 81$

29. $\log \sqrt[3]{10}$

30. $\ln \dfrac{1}{\sqrt{e^7}}$

In Exercises 31–34, rewrite the equation in exponential form.

31. $\log_3 x = 5$

32. $\log_2 x = y$

33. $\ln \dfrac{x}{y} = -2$

34. $\log \dfrac{a}{b} = -3$

In Exercises 35–38, describe how to transform the graph of $y = \log_2 x$ into the graph of the given function. Sketch the graph by hand and support with a grapher.

35. $f(x) = \log_2 (x + 4)$

36. $g(x) = \log_2 (4 - x)$

37. $h(x) = -\log_2 (x - 1) + 2$

38. $h(x) = -\log_2 (x + 1) + 4$

In Exercises 39–42, graph the function, and analyze it for domain, range, continuity, increasing or decreasing behavior, symmetry, boundedness, extrema, asymptotes, and end behavior.

39. $f(x) = x \ln x$

40. $f(x) = x^2 \ln x$

41. $f(x) = x^2 \ln |x|$

42. $f(x) = \dfrac{\ln x}{x}$

In Exercises 43–54, solve the equation.

43. $10^x = 4$

44. $e^x = 0.25$

45. $1.05^x = 3$

46. $\ln x = 5.4$

47. $\log x = -7$

48. $3^{x-3} = 5$

49. $3 \log_2 x + 1 = 7$

50. $2 \log_3 x - 3 = 4$

51. $\dfrac{3^x - 3^{-x}}{2} = 5$

52. $\dfrac{50}{4 + e^{2x}} = 11$

53. $\log(x + 2) + \log(x - 1) = 4$

54. $\ln(3x + 4) - \ln(2x + 1) = 5$

In Exercises 55 and 56, write the expression using only natural logarithms.

55. $\log_2 x$

56. $\log_{1/6}(6x^2)$

In Exercises 57 and 58, write the expression using only common logarithms.

57. $\log_5 x$

58. $\log_{1/2}(4x^3)$

In Exercises 59–62, match the function with its graph. All graphs are drawn in the window $[-4.7, 4.7]$ by $[-3.1, 3.1]$.

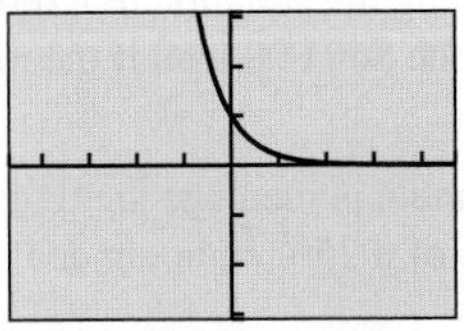

[–4.7, 4.7] by [–3.1, 3.1]

(a)

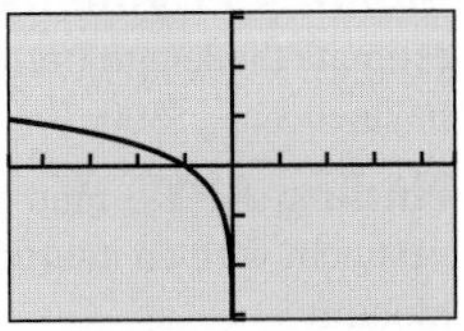

[–4.7, 4.7] by [–3.1, 3.1]

(b)

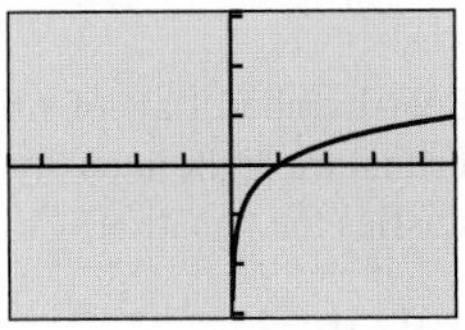

[–4.7, 4.7] by [–3.1, 3.1]

(c)

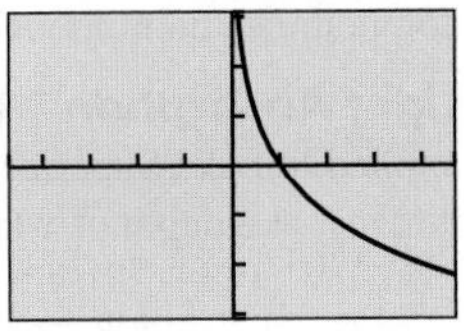

[–4.7, 4.7] by [–3.1, 3.1]

(d)

59. $f(x) = \log_5 x$

60. $f(x) = \log_{0.5} x$

61. $f(x) = \log_5(-x)$

62. $f(x) = 5^{-x}$

63. Compound Interest Find the amount A accumulated after investing a principal $P = \$450$ for 3 years at an interest rate of 4.6% compounded annually.

64. Compound Interest Find the amount A accumulated after investing a principal $P = \$4800$ for 17 years at an interest rate 6.2% compounded quarterly.

65. Compound Interest Find the amount A accumulated after investing a principal P for t years at interest rate r compounded continuously.

66. Future Value Find the future value FV accumulated in an annuity after investing periodic payments R for t years at an annual interest rate r, with payments made and interest credited k times per year.

67. Present Value Find the present value PV of a loan with an annual interest rate $r = 5.5\%$ and periodic payments $R = \$550$ for a term of $t = 5$ years, with payments made and interest charged 12 times per year.

68. Present Value Find the present value PV of a loan with an annual interest rate $r = 7.25\%$ and periodic payments $R = \$953$ for a term of $t = 15$ years, with payments made and interest charged 26 times per year.

In Exercises 69 and 70, determine the value of k so that the graph of f passes through the given point.

69. $f(x) = 20e^{-kx}, (3, 50)$

70. $f(x) = 20e^{-kx}, (1, 30)$

In Exercises 71 and 72, use the data in Table 3.28.

Table 3.28 Populations of Two U.S. States (in millions)

Year	Georgia	Illinois
1900	2.2	4.8
1910	2.6	5.6
1920	2.9	6.5
1930	2.9	7.6
1940	3.1	7.9
1950	3.4	8.7
1960	3.9	10.1
1970	4.6	11.1
1980	5.5	11.4
1990	6.5	11.4
2000	8.2	12.4
2011	9.8	12.9

Source: U.S. Census Bureau as reported in the World Almanac and Book of Facts 2013.

71. Modeling Population Using Table 3.28 find an exponential regression model for Georgia's population, and use it to predict the population in 2020.

72. Modeling Population Using Table 3.28 find a logistic regression model for Illinois's population, and use it to predict the population in 2020.

73. Drug Absorption A drug is administered intravenously for pain. The function $f(t) = 90 - 52 \ln(1 + t)$, where $0 \le t \le 4$ gives the amount of the drug in the body after t hours.

(a) What was the initial $(t = 0)$ number of units of drug administered?

(b) How much is present after 2 hr?

(c) Draw the graph of f.

74. Population Decrease The population of Metroville is 123,000 and is decreasing by 2.4% each year.

(a) Write a function that models the population as a function of time t.

(b) Predict when the population will be 90,000.

75. Population Decrease The population of Preston is 89,000 and is decreasing by 1.8% each year.

(a) Write a function that models the population as a function of time t.

(b) Predict when the population will be 50,000.

76. **Spread of Flu** The number P of students infected with flu at Northridge High School t days after exposure is modeled by

$$P(t) = \frac{300}{1 + e^{4-t}}.$$

(a) What was the initial $(t = 0)$ number of students infected with the flu?

(b) How many students were infected after 3 days?

(c) When will 100 students be infected?

(d) What would be the maximum number of students infected?

77. **Rabbit Population** The number of rabbits in Elkgrove doubles every month. There are 20 rabbits present initially.

(a) Express the number of rabbits as a function of the time t.

(b) How many rabbits were present after 1 year? after 5 years?

(c) When will there be 10,000 rabbits?

78. **Guppy Population** The number of guppies in Susan's aquarium doubles every day. There are four guppies initially.

(a) Express the number of guppies as a function of time t.

(b) How many guppies were present after 4 days? after 1 week?

(c) When will there be 2000 guppies?

79. **Radioactive Decay** The half-life of a certain radioactive substance is 1.5 sec. The initial amount of substance is S_0 grams.

(a) Express the amount of substance S remaining as a function of time t.

(b) How much of the substance is left after 1.5 sec? after 3 sec?

(c) Determine S_0 if there was 1 g left after 1 min.

80. **Radioactive Decay** The half-life of a certain radioactive substance is 2.5 sec. The initial amount of substance is S_0 grams.

(a) Express the amount of substance S remaining as a function of time t.

(b) How much of the substance is left after 2.5 sec? after 7.5 sec?

(c) Determine S_0 if there was 1 g left after 1 min.

81. **Richter Scale** Afghanistan suffered two major earthquakes in 1998. The one on February 4 had a Richter magnitude of 6.1, causing about 2300 deaths, and the one on May 30 measured 6.9 on the Richter scale, killing about 4700 persons. How many times more powerful was the deadlier quake?

82. **Chemical Acidity** The pH of seawater is 7.6, and the pH of milk of magnesia is 10.5.

(a) What are their hydrogen-ion concentrations?

(b) How many times greater is the hydrogen-ion concentration of the seawater than that of milk of magnesia?

(c) By how many orders of magnitude do the concentrations differ?

83. **Annuity Finding Time** If Joenita invests \$1500 into a retirement account with an 8% interest rate compounded quarterly, how long will it take this single payment to grow to \$3750?

84. **Annuity Finding Time** If Juan invests \$12,500 into a retirement account with a 9% interest rate compounded continuously, how long will it take this single payment to triple in value?

85. **Monthly Payments** The time t in months that it takes to pay off a \$60,000 loan at 9% annual interest with monthly payments of x dollars is given by

$$t \approx 133.83 \ln\left(\frac{x}{x - 450}\right).$$

Estimate the length (term) of the \$60,000 loan if the monthly payments are \$700.

86. **Monthly Payments** Using the equation in Exercise 85, estimate the length (term) of the \$60,000 loan if the monthly payments are \$500.

87. **Finding APY** Find the annual percentage yield for an investment with an interest rate of 8.25% compounded monthly.

88. **Finding APY** Find the annual percentage yield that can be used to advertise an account that pays interest at 7.20% compounded continuously.

89. **Light Absorption** The Beer-Lambert Law of Absorption applied to Lake Superior states that the light intensity I (in lumens) at a depth of x feet satisfies the equation

$$\log \frac{I}{12} = -0.0125x.$$

Find the light intensity at a depth of 25 ft.

90. For what values of b is $\log_b x$ a vertical stretch of $y = \ln x$? A vertical shrink of $y = \ln x$?

91. For what values of b is $\log_b x$ a vertical stretch of $y = \log x$? A vertical shrink of $y = \log x$?

92. If $f(x) = ab^x$, $a > 0$, $b > 0$, prove that $g(x) = \ln f(x)$ is a linear function. Find its slope and y-intercept.

93. **Spread of Flu** The number of students infected with flu after t days at Springfield High School is modeled by the function

$$P(t) = \frac{1600}{1 + 99e^{-0.4t}}.$$

(a) What was the initial number of infected students?

(b) When will 800 students be infected?

(c) The school will close when 400 of the 1600 student body are infected. When would the school close?

94. **Population of Deer** The population P of deer after t years in Briggs State Park is modeled by the function

$$P(t) = \frac{1200}{1 + 99e^{-0.4t}}.$$

(a) What was the initial population of deer?

(b) When will there be 1000 deer?

(c) What is the maximum number of deer planned for the park?

95. Newton's Law of Cooling A cup of coffee cooled from 96°C to 65°C after 8 min in a room at 20°C. When will it cool to 25°C?

96. Newton's Law of Cooling A cake is removed from an oven at 220°F and cools to 150°F after 35 min in a room at 75°F. When will it cool to 95°F?

97. The function

$$f(x) = 100\frac{(1 + 0.09/4)^x - 1}{0.09/4}$$

describes the future value of a certain annuity.

(a) What is the annual interest rate?

(b) How many payments per year are there?

(c) What is the amount of each payment?

98. The function

$$g(x) = 200\frac{1 - (1 + 0.11/4)^{-x}}{0.11/4}$$

describes the present value of a certain annuity.

(a) What is the annual interest rate?

(b) How many payments per year are there?

(c) What is the amount of each payment?

99. Simple Interest Versus Compounding Continuously Grace purchases a $1000 certificate of deposit that will earn 5% each year. The interest will be mailed to her, so she will not earn interest on her interest.

(a) Show that after t years, the total amount of interest she receives from her investment plus the original $1000 is given by

$$f(t) = 1000(1 + 0.05t).$$

(b) Grace invests another $1000 at 5% compounded continuously. Make a table that compares the values of the two investments for $t = 1, 2, \ldots, 10$ years.

CHAPTER 3 Project

Analyzing a Bouncing Ball

When a ball bounces up and down on a flat surface, the maximum height of the ball decreases with each bounce. Each rebound is a percentage of the previous height. For most balls, the percentage is a constant. In this project, you will use a motion detection device to collect height data for a ball bouncing underneath the detector, then find a mathematical model that describes the maximum bounce height as a function of bounce number.

Collecting the Data

Set up the Calculator-data Based Laboratory (CBL™) system with a motion detector or a Calculator-Based Ranger (CBR™) system to collect ball bounce data using a ball bounce program for the CBL or the Ball Bounce Application for the CBR. See the CBL or CBR guidebook for specific setup instruction.

Hold the ball at least 2 ft below the detector and release it so that it bounces straight up and down beneath the detector. These programs convert distance versus time data to height from the ground versus time. The graph shows a plot of sample data collected with a racquetball and CBR. The data table below shows each maximum height collected.

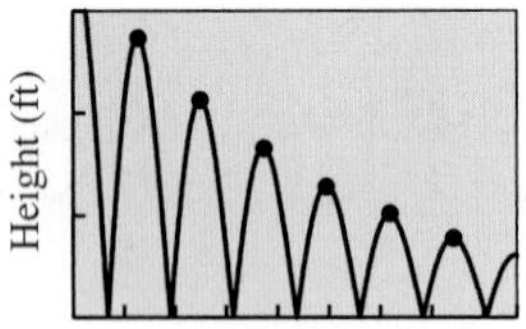

Time (sec)
[0, 4.25] by [0, 3]

Bounce Number	Maximum Height (ft)
0	2.7188
1	2.1426
2	1.6565
3	1.2640
4	0.98309
5	0.77783

Explorations

1. If you collected motion data using a CBL or CBR, a plot of height versus time should be shown on your graphing calculator or computer screen. Trace to the maximum height for each bounce, and record your data in a table. Then use other lists in your grapher to enter these data. If you do not have access to a CBL or CBR, enter the data given in the table into your grapher.
2. Bounce height 1 is what percentage of bounce height 0? Calculate the percentage return for each bounce. The numbers should be fairly constant.
3. Create a scatter plot for maximum height versus bounce number.
4. For bounce 1, the height is predicted by multiplying bounce height 0, or H, by the percentage P. The second height is predicted by multiplying this height HP by P, which gives the HP^2. Explain why $y = HP^x$ is the appropriate model for these data, where x is the bounce number.
5. Enter this equation into your grapher, using your values for H and P. How does the model fit your data?
6. Use the statistical features of the grapher to find the exponential regression for these data. Compare it to the equation that you used as a model.
7. How would your data and equation change if you used a different type of ball?
8. What factors would change the H-value, and what factors affect the P-value?
9. Rewrite your equation using base e instead of using P as the base for the exponential equation.
10. What do you predict the graph of ln (bounce height) versus bounce number to look like?
11. Plot ln (bounce height) versus bounce number. Calculate the linear regression and use the concept of logarithmic re-expression to explain how the slope and y-intercept are related to P and H.

CHAPTER 4

4.1 Angles and Their Measures

4.2 Trigonometric Functions of Acute Angles

4.3 Trigonometry Extended: The Circular Functions

4.4 Graphs of Sine and Cosine: Sinusoids

4.5 Graphs of Tangent, Cotangent, Secant, and Cosecant

4.6 Graphs of Composite Trigonometric Functions

4.7 Inverse Trigonometric Functions

4.8 Solving Problems with Trigonometry

Trigonometric Functions

When the motion of an object causes air molecules to vibrate, we hear a sound. We measure sound according to its pitch and loudness, which are attributes associated with the frequency and amplitude of sound waves. As we shall see, it is the branch of mathematics called trigonometry that enables us to analyze waves of all kinds; indeed, that is only one application of this powerful analytical tool. See page 393 for an application of trigonometry to sound waves.

Chapter 4 Overview

The trigonometric functions arose from the consideration of ratios within right triangles, the ultimate computational tool for engineers in the ancient world. As the great mysteries of civilization progressed from a flat Earth to a world of circles and spheres, trigonometry was soon seen to be the secret to understanding circular phenomena as well. Then circular motion led to harmonic motion and waves, and suddenly trigonometry was the indispensable tool for understanding everything from electrical current to modern telecommunications.

The advent of calculus made the trigonometric functions more important than ever. It turns out that every kind of periodic (recurring) behavior can be modeled to any degree of accuracy by simply combining sine functions. The modeling aspect of trigonometric functions is another focus of our study.

Hipparchus of Nicaea (190–120 BCE)

Hipparchus of Nicaea, the "father of trigonometry," compiled the first trigonometric tables to simplify the study of astronomy more than 2000 years ago. Today, that same mathematics enables us to store sound waves digitally on a compact disc. Hipparchus wrote during the second century BCE, but he was not the first mathematician to "do" trigonometry. Greek mathematicians like Hippocrates of Chois (470–410 BCE) and Eratosthenes of Cyrene (276–194 BCE) had paved the way for using triangle ratios in astronomy, and those same triangle ratios had been used by Egyptian and Babylonian engineers at least 4000 years earlier. The term "trigonometry" itself emerged in the 16th century, although it derives from ancient Greek roots: "tri" (three), "gonos" (side), and "metros" (measure).

4.1 Angles and Their Measures

What you'll learn about

- The Problem of Angular Measure
- Degrees and Radians
- Circular Arc Length
- Angular and Linear Motion

... and why

Angles are the domain elements of the trigonometric functions.

The Problem of Angular Measure

The input variable in a trigonometric function is an angle measure; the output is a real number. Believe it or not, this poses an immediate problem for us if we choose to measure our angles in degrees (as most of us did in our geometry courses).

The problem is that degree units have no mathematical relationship whatsoever to linear units. There are 360 degrees in a circle of radius 1. What relationship does the 360 have to the 1? In what sense is it 360 times as big? Answering these questions isn't possible, because a "degree" is another unit altogether.

Consider the diagrams in Figure 4.1. The ratio of s to h in the right triangle in Figure 4.1a is independent of the size of the triangle. (You may recall this fact about similar triangles from geometry.) This valuable insight enabled early engineers to compute triangle ratios on a small scale before applying them to much larger projects. That was (and still is) trigonometry in its most basic form. For astronomers tracking celestial motion, however, the extended diagram in Figure 4.1b was more interesting. In this picture, s is half a chord in a circle of radius h, and θ is a **central angle** of the circle intercepting a circular arc of length a. If θ were 40°, we might call a a "40-degree arc" because of its direct association with the central angle θ, but notice that a also has a *length* that can be measured in the *same units* as the other lengths in the picture. Over time it became natural to think of the angle being determined by the arc rather than the arc being determined by the angle, and that led to radian measure.

Why 360°?

The idea of dividing a circle into 360 equal pieces dates back to the sexagesimal (60-based) counting system of the ancient Sumerians. The appeal of 60 was that it was evenly divisible by so many numbers (2, 3, 4, 5, 6, 10, 12, 15, 20, and 30). Early astronomical calculations wedded the sexagesimal system to circles, and the rest is history.

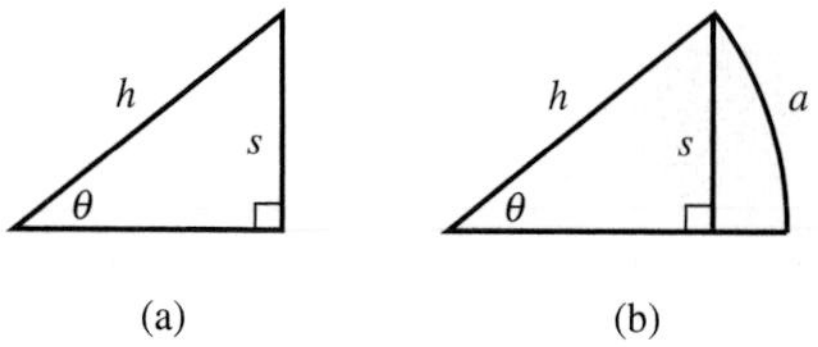

FIGURE 4.1 The pictures that motivated trigonometry.

Degrees and Radians

A **degree**, represented by the symbol °, is a unit of angular measure equal to 1/180th of a straight angle. In the DMS (degree-minute-second) system of angular measure, each degree is subdivided into 60 **minutes** (denoted by ′) and each minute is subdivided into 60 **seconds** (denoted by ″). (Notice that Sumerian influence again.)

Example 1 illustrates how to convert from degrees in decimal form to DMS and vice versa.

EXAMPLE 1 Working with DMS Measure

(a) Convert 37.425° to DMS.

(b) Convert 42°24′36″ to degrees.

SOLUTION

(a) We need to convert the fractional part to minutes and seconds. First we convert 0.425° to minutes:

$$0.425°\left(\frac{60'}{1°}\right) = 25.5'$$

Then we convert 0.5 minute to seconds:

$$0.5'\left(\frac{60''}{1'}\right) = 30''$$

Putting it all together, we find that 37.425° = 37°25′30″.

(b) Each minute is 1/60th of a degree, and each second is 1/3600th of a degree. Therefore,

$$42°24'36'' = 42° + \left(\frac{24}{60}\right)^{\circ} + \left(\frac{36}{3600}\right)^{\circ} = 42.41°.$$

Now try Exercises 3 and 5.

Calculator Conversions

Your calculator probably has built-in functionality to convert degrees to DMS. Consult your owner's manual. Meanwhile, you should try some conversions the "long way" to get a better feel for how DMS works.

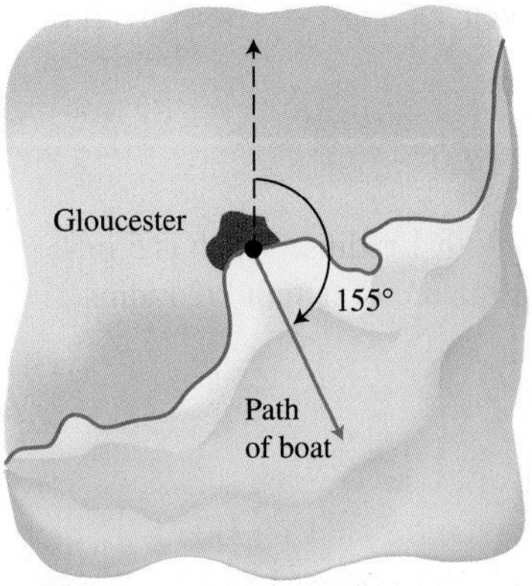

FIGURE 4.2 The course of a fishing boat bearing 155° out of Gloucester.

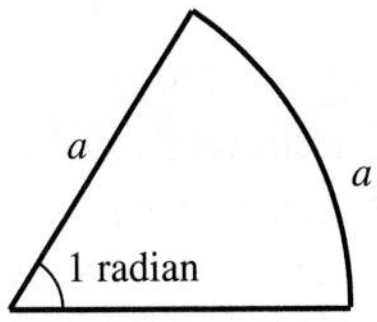

FIGURE 4.3 In a circle, a central angle of 1 radian intercepts an arc of length one radius.

In navigation, the **course** or **bearing** of an object is sometimes given as the angle of the **line of travel** measured clockwise from due north. For example, the line of travel in Figure 4.2 has the bearing of 155°.

In this book we use degrees to measure angles in their familiar geometric contexts, especially when applying trigonometry to real-world problems in surveying, construction, and navigation, where degrees are still the accepted units of measure. When we shift our attention to the trigonometric *functions*, however, we will measure angles in *radians* so that domain and range values can be measured on comparable scales.

DEFINITION Radian

A central angle of a circle has measure 1 **radian** if it intercepts an arc with the same length as the radius. (See Figure 4.3.)

EXPLORATION 1 Constructing a 1-Radian Angle

Carefully draw a large circle on a piece of paper, either by tracing around a circular object or by using a compass. Identify the center of the circle (O) and draw a radius horizontally from O toward the right, intersecting the circle at point A. Then cut a piece of thread or string the same size as the radius. Place one end of the string at A and bend it around the circle counterclockwise, marking the point B on the circle where the other end of the string ends up. Draw the radius from O to B.

The measure of angle AOB is 1 rad.

1. What is the circumference of the circle, in terms of its radius r
2. How many radians must there be in a complete circle?
3. If we cut a piece of thread 3 times as big as the radius, would it extend halfway around the circle? Why or why not?
4. How many radians are in a straight angle?

EXAMPLE 2 Working with Radian Measure

(a) How many radians are in 90 degrees?

(b) How many degrees are in $\pi/3$ radians?

(c) Find the length of an arc intercepted by a central angle of 1/2 radian in a circle of radius 5 in.

(d) Find the radian measure of a central angle that intercepts an arc of length s in a circle of radius r.

SOLUTION

(a) Since π radians and 180° both measure a straight angle, we can use the conversion factor (π radians)/(180°) = 1 to convert degrees to radians:

$$90°\left(\frac{\pi \text{ radians}}{180°}\right) = \frac{90\pi}{180}\text{ rad} = \frac{\pi}{2}\text{ rad}$$

(b) In this case, we use the conversion factor (180°)/(π radians) = 1 to convert radians to degrees:

$$\left(\frac{\pi}{3}\text{ rad}\right)\left(\frac{180°}{\pi \text{ rad}}\right) = \frac{180°}{3} = 60°$$

(c) A central angle of 1 radian intercepts an arc of length 1 radius, which is 5 in. Therefore, a central angle of 1/2 radian intercepts an arc of length 1/2 radius, which is 2.5 in.

(d) We can solve this problem with ratios:

$$\frac{x \text{ radians}}{s \text{ units}} = \frac{1 \text{ radian}}{r \text{ units}}$$
$$xr = s$$
$$x = \frac{s}{r}$$

Now try Exercises 11 and 19.

Degree-Radian Conversion

To convert radians to degrees, multiply by $\dfrac{180°}{\pi \text{ rad}}$.

To convert degrees to radians, multiply by $\dfrac{\pi \text{ rad}}{180°}$.

With practice, you can perform these conversions in your head. The key is to think of a straight angle equaling π rad as readily as you think of it equaling 180°.

Circular Arc Length

Since a central angle of 1 radian always intercepts an arc of one radius in length, it follows that a central angle of θ radians in a circle of radius r intercepts an arc of length θr. This gives us a convenient formula for measuring arc length.

Arc Length Formula (Radian Measure)

If θ is a central angle in a circle of radius r, and if θ is measured in radians, then the length s of the intercepted arc is given by

$$s = r\theta.$$

Does a Radian Have Units?

The formula $s = r\theta$ implies an interesting fact about radians: As long as s and r are measured in the same units, the radian is unit-neutral. For example, if $r = 5$ in. and $\theta = 2$ rad, then $s = 10$ in. (not 10 "in.-rad"). This unusual situation arises from the fact that the definition of the radian is tied to the length of the radius, units and all.

A somewhat less simple formula (which incorporates the degree-radian conversion formula) applies when θ is measured in degrees.

Arc Length Formula (Degree Measure)

If θ is a central angle in a circle of radius r, and if θ is measured in degrees, then the length s of the intercepted arc is given by

$$s = \frac{\pi r \theta}{180}.$$

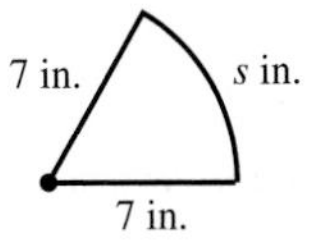

FIGURE 4.4 A 60° slice of a large pizza. (Example 3)

EXAMPLE 3 Perimeter of a Pizza Slice

Find the perimeter of a 60° slice of a large (7-in. radius) pizza.

SOLUTION The perimeter (Figure 4.4) is 7 in. + 7 in. + s in., where s is the arc length of the pizza's curved edge. By the arc length formula,

$$s = \frac{\pi(7)(60)}{180} = \frac{7\pi}{3} \approx 7.3.$$

The perimeter is approximately 21.3 in. **Now try Exercise 35.**

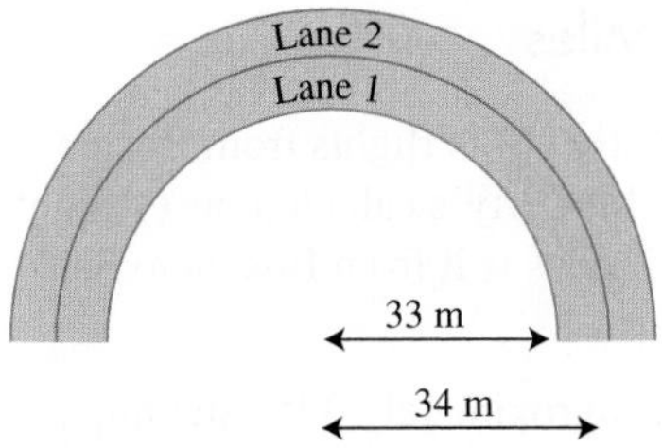

FIGURE 4.5 Two lanes of the track described in Example 4.

EXAMPLE 4 Designing a Running Track

The running lanes at the Emery Sears track at Bluffton College are 1 m wide. The inside radius of lane 1 is 33 m, and the inside radius of lane 2 is 34 m. How much longer is lane 2 than lane 1 around one turn? (See Figure 4.5.)

SOLUTION We think this solution through in radians. Each lane is a semicircle with central angle $\theta = \pi$ and length $s = r\theta = r\pi$. The difference in their lengths, therefore, is $34\pi - 33\pi = \pi$. Lane 2 is about 3.14 m longer than lane 1.

Now try Exercise 37.

Angular and Linear Motion

In applications it is sometimes necessary to connect *angular speed* (measured in units like revolutions per minute) to *linear speed* (measured in units like miles per hour). The connection is usually provided by one of the arc length formulas or by a conversion factor that equates "1 radian" of angular measure to "1 radius" of arc length.

EXAMPLE 5 Using Angular Speed

Albert Juarez's truck has wheels 36 in. in diameter. If the wheels are rotating at 630 rpm (revolutions per minute), find the truck's speed in miles per hour.

SOLUTION We convert revolutions per minute to miles per hour by a series of unit conversion factors. Note that the conversion factor $\frac{18 \text{ in.}}{1 \text{ rad}}$ works for this example because the radius is 18 in.

$$\frac{630 \text{ rev}}{1 \text{ min}} \times \frac{60 \text{ min}}{1 \text{ hr}} \times \frac{2\pi \text{ rad}}{1 \text{ rev}} \times \frac{18 \text{ in.}}{1 \text{ rad}} \times \frac{1 \text{ ft}}{12 \text{ in.}} \times \frac{1 \text{ mi}}{5280 \text{ ft}}$$

$$\approx 67.47 \frac{\text{mi}}{\text{hr}}$$

Now try Exercise 45.

A **nautical mile** (naut mi) is the length of 1 minute of arc along Earth's equator. Figure 4.6 shows, though not to scale, a central angle AOB of Earth that measures 1/60 of a degree. It intercepts an arc 1 naut mi long.

The arc length formula allows us to convert between nautical miles and **statute miles** (stat mi), the familiar "land mile" of 5280 ft.

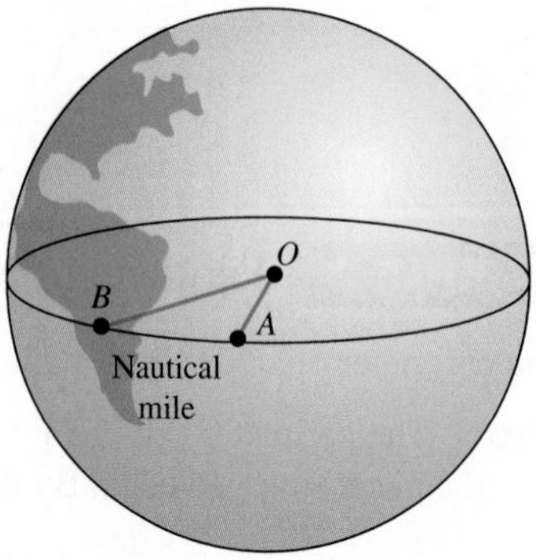

FIGURE 4.6 Although Earth is not a perfect sphere, its diameter is, on average, 7912.18 stat mi. A nautical mile is 1′ of Earth's circumference at the equator.

EXAMPLE 6 Converting to Nautical Miles

Megan McCarty, a pilot for Western Airlines, frequently pilots flights from Boston to San Francisco, a distance of 2698 stat mi. Captain McCarty's calculations of flight time are based on nautical miles. How many nautical miles is it from Boston to San Francisco?

SOLUTION The radius of the Earth at the equator is approximately 3956 stat mi. Convert 1 minute to radians:

$$1' = \left(\frac{1}{60}\right)^{\circ} \times \frac{\pi \text{ rad}}{180^{\circ}} = \frac{\pi}{10{,}800} \text{ rad.}$$

Now we can apply the formula $s = r\theta$:

$$\begin{aligned} 1 \text{ naut mi} &= (3956)\left(\frac{\pi}{10{,}800}\right) \text{ stat mi} \\ &\approx 1.15 \text{ stat mi} \\ 1 \text{ stat mi} &= \left(\frac{10{,}800}{3956\pi}\right) \text{ naut mi} \\ &\approx 0.87 \text{ naut mi} \end{aligned}$$

The distance from Boston to San Francisco is

$$2698 \text{ stat mi} = \frac{2698 \cdot 10{,}800}{3956\pi} \approx 2345 \text{ naut mi.}$$

Now try Exercise 51.

Distance Conversions

$$1 \text{ stat mi} \approx 0.87 \text{ naut mi}$$
$$1 \text{ naut mi} \approx 1.15 \text{ stat mi}$$

QUICK REVIEW 4.1 *(For help, go to Section 1.7.)*

Exercise numbers with a gray background indicate problems that the authors have designed to be solved *without a calculator*.

In Exercises 1 and 2, find the circumference of the circle with the given radius r. State the correct unit.

1. $r = 2.5$ in. **2.** $r = 4.6$ m

In Exercises 3 and 4, find the radius of the circle with the given circumference C.

3. $C = 12$ m **4.** $C = 8$ ft

In Exercises 5 and 6, evaluate the expression for the given values of the variables. State the correct unit.

5. $s = r\theta$

(a) $r = 9.9$ ft $\theta = 4.8$ rad

(b) $r = 4.1$ km $\theta = 9.7$ rad

6. $v = r\omega$

(a) $r = 8.7$ m $\omega = 3.0$ rad/sec

(b) $r = 6.2$ ft $\omega = 1.3$ rad/sec

In Exercises 7–10, convert from miles per hour to feet per second or from feet per second to miles per hour.

7. 60 mph **8.** 45 mph

9. 8.8 ft/sec **10.** 132 ft/sec

SECTION 4.1 Exercises

In Exercises 1–4, convert from DMS to decimal form.

1. 23°12′ **2.** 35°24′

3. 118°44′15″ **4.** 48°30′36″

In Exercises 5–8, convert from decimal form to degrees, minutes, seconds (DMS).

5. 21.2° **6.** 49.7°

7. 118.32° **8.** 99.37°

In Exercises 9–16, convert from DMS to radians.

9. 60° **10.** 90°

11. 120° **12.** 150°

13. 71.72° **14.** 11.83°

15. 61°24′ **16.** 75°30′

In Exercises 17–24, convert from radians to degrees.

17. $\pi/6$ **18.** $\pi/4$

19. $\pi/10$ **20.** $3\pi/5$

21. $7\pi/9$ **22.** $13\pi/20$

23. 2 **24.** 1.3

In Exercises 25–32, use the appropriate arc length formula to find the missing information.

	s	r	θ
25.	?	2 in.	25 rad
26.	?	1 cm	70 rad
27.	1.5 ft	?	$\pi/4$ rad
28.	2.5 cm	?	$\pi/3$ rad
29.	3 m	1 m	?
30.	4 in.	7 in.	?
31.	40 cm	?	20°
32.	?	5 ft	18°

In Exercises 33 and 34, a central angle θ intercepts arcs s_1 and s_2 on two concentric circles with radii r_1 and r_2, respectively. Find the missing information.

	θ	r_1	s_1	r_2	s_2
33.	?	11 cm	9 cm	44 cm	?
34.	?	8 km	36 km	?	72 km

35. To the nearest inch, find the perimeter of a 10° sector cut from a circular disc of radius 11 in.

36. A 100° arc of a circle has a length of 7 cm. To the nearest centimeter, what is the radius of the circle?

37. It takes ten identical pieces to form a circular track for a pair of toy racing cars. If the inside arc of each piece is 3.4 in. shorter than the outside arc, what is the width of the track?

38. The concentric circles on an archery target are 6 in. apart. The inner circle (red) has a perimeter of 37.7 in. What is the perimeter of the next-largest (yellow) circle?

Exercises 39–42 refer to the 16 compass bearings shown. North corresponds to an angle of 0°, and other angles are measured clockwise from north.

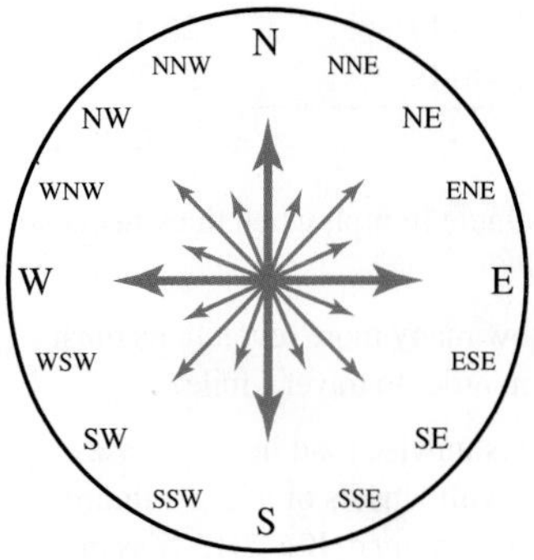

39. **Compass Reading** Find the angle in degrees that describes the compass bearing.

(a) NE (northeast)

(b) NNE (north-northeast)

(c) WSW (west-southwest)

40. **Compass Reading** Find the angle in degrees that describes the compass bearing.

(a) SSW (south-southwest)

(b) WNW (west-northwest)

(c) NNW (north-northwest)

41. **Compass Reading** Which compass direction is closest to a bearing of 121°?

42. **Compass Reading** Which compass direction is closest to a bearing of 219°?

43. **Navigation** Two Coast Guard patrol boats leave Cape May at the same time. One travels with a bearing of 42°30′ and the other with a bearing of 52°12′. If they travel at the same speed, approximately how far apart will they be when they are 25 stat mi from Cape May?

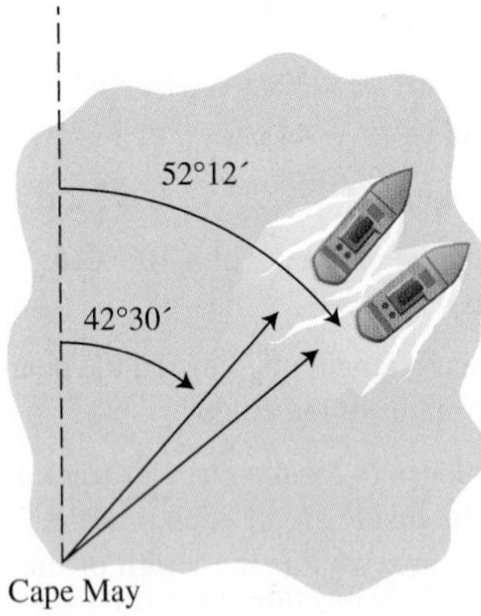

44. **Automobile Design** Table 4.1 shows the size specifications for the tires that come as standard equipment on three 2013 all-electric vehicles.

Table 4.1 Tire Sizes for Electric Vehicles

Vehicle	Tire Type	Tire Diameter (in.)
Nissan Leaf SL	215/50–17	25.5
Chevy Volt	215/55–17	26.3
Tesla S	245/45–19	27.7

Source: Tirerack.com

(a) Find the speed of each vehicle in mph when the wheels are turning at 800 rpm.

(b) Compared to the Tesla, how many more revolutions must the tire of the Leaf make in order to travel a mile?

(c) **Writing to Learn** It is unwise (and in some cases illegal) to equip a vehicle with wheels of a larger diameter than those for which it was designed. If a 2013 Nissan Leaf were equipped with the Tesla's 27.7-in. tires, how would it affect the odometer (which measures mileage) and speedometer readings?

45. **Bicycle Racing** 2012 Olympics gold medalist Mariana Pajón races on a BMX bike with 10-in.-radius wheels. When she is traveling at a speed of 24 ft/sec, how many revolutions per minute are her wheels making?

46. **Tire Sizing** The numbers in the "tire type" column in Exercise 44 give the size of the tire in the P-metric system. Each number is of the form $W/R–D$, where W is the width of the tire in millimeters, $R/100$ is the ratio of the sidewall (S) of the tire to its width W, and D is the diameter (in inches) of the wheel without the tire.

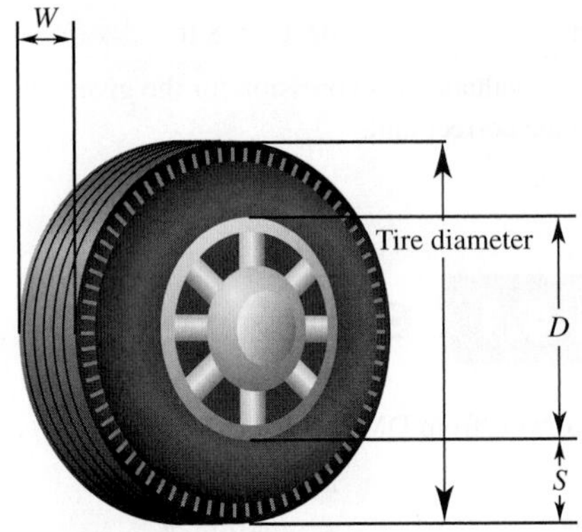

(a) Show that $S = WR/100$ millimeters $= WR/2540$ inches.

(b) The tire diameter is $D + 2S$. Derive a formula for the tire diameter that involves only the variables D, W, and R.

(c) Use the formula in part (b) to verify the tire diameters in Exercise 44. Then find the tire diameter for the 2013 Cadillac Escalade, which comes with 265/65–18 tires.

47. **Tool Design** A radial arm saw has a circular cutting blade with a diameter of 10 in. It spins at 2000 rpm. If there are 12 cutting teeth per inch on the cutting blade, how many teeth cross the cutting surface each second?

48. **Navigation** Sketch a diagram of a ship on the given course.

(a) 35° (b) 128° (c) 310°

49. **Navigation** The captain of the tourist boat *Julia* out of Oak Harbor follows a 38° course for 2 mi and then changes to a 47° course for the next 4 mi. Draw a sketch of this trip.

50. **Navigation** Points A and B are 257 naut mi apart. How far apart are A and B in statute miles?

51. **Navigation** Points C and D are 895 stat mi apart. How far apart are C and D in nautical miles?

52. **Designing a Sports Complex** Example 4 describes how lanes 1 and 2 compare in length around one turn of a track. Find the differences in the lengths of these lanes around one turn of the same track.

(a) Lanes 5 and 6 (b) Lanes 1 and 6

53. **Mechanical Engineering** A simple pulley with the given radius r used to lift heavy objects is positioned 10 ft above ground level. Given that the pulley rotates $\theta°$, determine the height to which the object is lifted.

(a) $r = 4$ in., $\theta = 720°$ (b) $r = 2$ ft, $\theta = 180°$

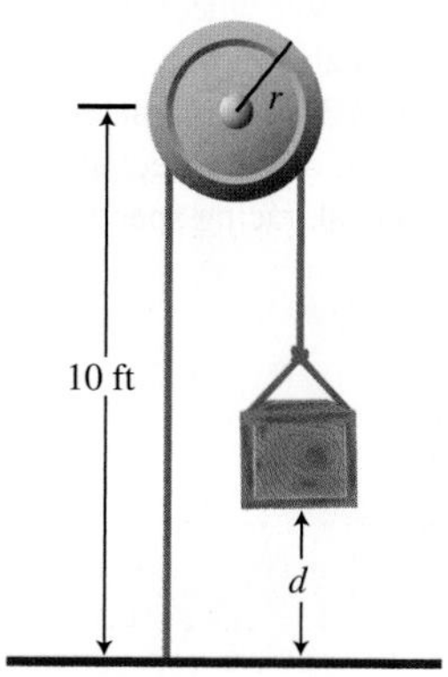

54. **Foucault Pendulum** In 1851 the French physicist Jean Foucault used a pendulum to demonstrate the Earth's rotation. There are now over 30 Foucault pendulum displays in the United States. The Foucault pendulum at the Smithsonian Institution in Washington, DC, consists of a large brass ball suspended by a thin 52-ft cable. If the ball swings through an angle of 1°, how far does it travel?

55. **Group Activity Air Conditioning Belt** The belt on an automobile air conditioner connects metal wheels with radii $r = 4$ cm and $R = 7$ cm. The angular speed of the larger wheel is 120 rpm.

(a) What is the angular speed of the larger wheel in radians per second?

(b) What is the linear speed of the belt in centimeters per second?

(c) What is the angular speed of the smaller wheel in radians per second?

56. **Group Activity Ship's Propeller** The propellers of the *Amazon Paradise* have a radius of 1.2 m. At full throttle the propellers turn at 135 rpm.

(a) What is the angular speed of a propeller blade in radians per second?

(b) What is the linear speed of the tip of the propeller blade in meters per second?

(c) What is the linear speed (in meters per second) of a point on a blade halfway between the center of the propeller and the tip of the blade?

Standardized Test Questions

57. **True or False** If horse A is twice as far as horse B from the center of a merry-go-round, then horse A travels twice as fast as horse B. Justify your answer.

58. **True or False** The radian measure of all three angles in a triangle can be integers. Justify your answer.

You may use a graphing calculator when answering these questions.

59. **Multiple Choice** What is the radian measure of an angle of x degrees?

(A) πx (B) $x/180$

(C) $\pi x/180$ (D) $180x/\pi$

(E) $180/x\pi$

60. **Multiple Choice** If the perimeter of a sector is 4 times its radius, then the radian measure of the central angle of the sector is

(A) 2. (B) 4.

(C) $2/\pi$. (D) $4/\pi$.

(E) impossible to determine without knowing the radius.

61. **Multiple Choice** A bicycle with 26-in.-diameter wheels is traveling at 10 mph. To the nearest whole number, how many revolutions does each wheel make per minute?

(A) 54 (B) 129

(C) 259 (D) 406

(E) 646

62. **Multiple Choice** A central angle in a circle of radius r has a measure of θ radians. If the same central angle were drawn in a circle of radius $2r$, its radian measure would be

(A) $\frac{\theta}{2}$. (B) $\frac{\theta}{2r}$.

(C) θ. (D) 2θ.

(E) $2r\theta$.

Explorations

Table 4.2 shows the latitude-longitude locations of several U.S. cities. Latitude is measured from the equator. Longitude is measured from the Greenwich meridian that passes north-south through London.

Table 4.2 Latitude and Longitude Locations of U.S. Cities

City	Latitude	Longitude
Atlanta	33°45′	84°23′
Chicago	41°51′	87°39′
Detroit	42°20′	83°03′
Los Angeles	34°03′	118°15′
Miami	25°46′	80°12′
Minneapolis	44°59′	93°16′
New Orleans	29°57′	90°05′
New York	40°43′	74°0′
San Diego	32°43′	117°09′
San Francisco	37°47′	122°25′
Seattle	47°36′	122°20′

Source: U.S. Department of the Interior, as reported in The World Almanac and Book of Facts 2009.

In Exercises 63–66, find the difference in longitude between the given cities.

63. Atlanta and San Francisco

64. New York and San Diego

65. Minneapolis and Chicago

66. Miami and Seattle

In Exercises 67–70, assume that the two cities have the same longitude (that is, assume that one is directly north of the other), and find the distance between them in nautical miles.

67. San Diego and Los Angeles

68. Seattle and San Francisco

69. New Orleans and Minneapolis

70. Detroit and Atlanta

71. Group Activity Area of a Sector A *sector of a circle* (shaded in the figure) is a region bounded by a central angle of a circle and its intercepted arc. Use the fact that the areas of sectors are proportional to their central angles to prove that

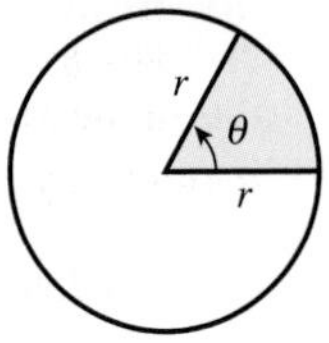

$$A = \frac{1}{2}r^2\theta$$

where r is the radius and θ is in radians.

Extending the Ideas

72. Area of a Sector Use the formula $A = (1/2)r^2\theta$ to determine the area of the sector with given radius r and central angle θ.

(a) $r = 5.9$ ft, $\theta = \pi/5$

(b) $r = 1.6$ km, $\theta = 3.7$

73. Navigation Control tower A is 60 mi east of control tower B. At a certain time an airplane is on bearings of 340° from tower A and 37° from tower B. Use a drawing to model the exact location of the airplane.

74. Bicycle Racing Ben Scheltz's bike wheels are 28 in. in diameter, and for high gear the pedal sprocket is 9 in. in diameter and the wheel sprocket is 3 in. in diameter. Find the angular speed in radians per second of the wheel and of both sprockets when Ben reaches his peak racing speed of 66 ft/sec in high gear.

4.2 Trigonometric Functions of Acute Angles

What you'll learn about

- Right Triangle Trigonometry
- Two Famous Triangles
- Evaluating Trigonometric Functions with a Calculator
- Common Calculator Errors when Evaluating Trig Functions
- Applications of Right Triangle Trigonometry

... and why

The many applications of right triangle trigonometry gave the subject its name.

Right Triangle Trigonometry

Recall that geometric figures are **similar** if they have the same shape even though they may have different sizes. Having the same shape means that the angles of one are congruent to the angles of the other and their corresponding sides are proportional. Similarity is the basis for many applications, including scale drawings, maps, and **right triangle trigonometry**, which is the topic of this section.

Two triangles are similar if the angles of one are congruent to the angles of the other. For two *right* triangles we need only know that an acute angle of one is equal to an acute angle of the other for the triangles to be similar. Thus a single acute angle θ of a right triangle determines six distinct ratios of side lengths. Each ratio can be considered a function of θ as θ takes on values from 0° to 90° or from 0 rad to $\pi/2$ rad. We wish to study these functions of acute angles more closely.

To bring the power of coordinate geometry into the picture, we will often put our acute angles in **standard position** in the xy-plane, with the vertex at the origin, one ray along the positive x-axis, and the other ray extending into the first quadrant. (See Figure 4.7.)

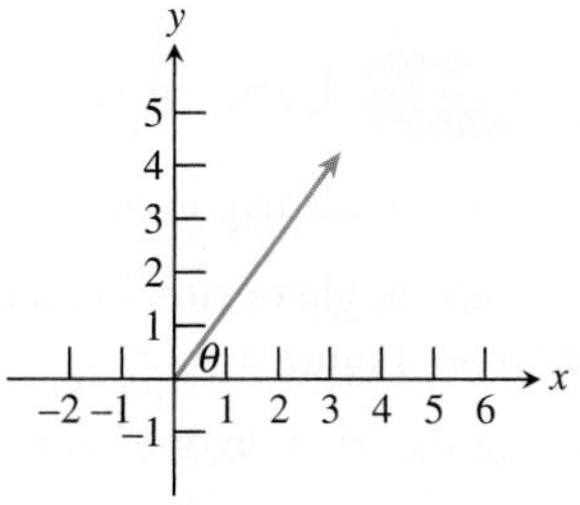

FIGURE 4.7 An acute angle θ in *standard position*, with one ray along the positive x-axis and the other extending into the first quadrant.

The six ratios of side lengths in a right triangle are the six *trigonometric functions* (often abbreviated as *trig functions*) of the acute angle θ. We will define them here with reference to the right ΔABC as labeled in Figure 4.8. The abbreviations *opp*, *adj*, and *hyp* refer to the lengths of the side opposite θ, the side adjacent to θ, and the hypotenuse, respectively.

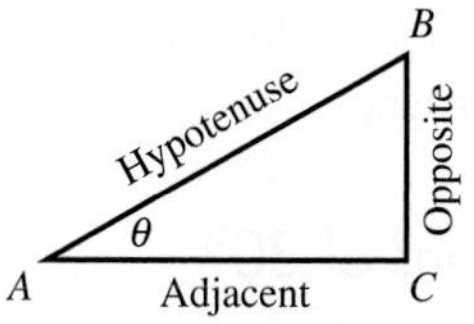

FIGURE 4.8 The triangle referenced in our definition of the trigonometric functions.

DEFINITION Trigonometric Functions

Let θ be an acute angle in the right ΔABC (Figure 4.8). Then

$$\textbf{sine}\ (\theta) = \sin\theta = \frac{opp}{hyp} \qquad \textbf{cosecant}\ (\theta) = \csc\theta = \frac{hyp}{opp}$$

$$\textbf{cosine}\ (\theta) = \cos\theta = \frac{adj}{hyp} \qquad \textbf{secant}\ (\theta) = \sec\theta = \frac{hyp}{adj}$$

$$\textbf{tangent}\ (\theta) = \tan\theta = \frac{opp}{adj} \qquad \textbf{cotangent}\ (\theta) = \cot\theta = \frac{adj}{opp}$$

Function Reminder

Both sin θ and $\sin(\theta)$ represent a function of the variable θ. Neither notation implies multiplication by θ. The notation $\sin(\theta)$ is just like the notation $f(x)$, and the notation sin θ is a widely accepted shorthand. The same note applies to all six trigonometric functions.

EXPLORATION 1 Exploring Trigonometric Functions

There are twice as many trigonometric functions as there are triangle sides that define them, so we can already explore some ways in which the trigonometric functions relate to each other. Doing this Exploration will help you learn which ratios are which.

1. Each of the six trig functions can be paired to another that is its reciprocal. Find the three pairs of reciprocals.
2. Which trig function can be written as the quotient $\sin\theta/\cos\theta$
3. Which trig function can be written as the quotient $\csc\theta/\cot\theta$
4. What is the (simplified) product of all six trig functions multiplied together?
5. Which two trig functions must be less than 1 for any acute angle θ? (*Hint:* What is always the longest side of a right triangle?)

Two Famous Triangles

Evaluating trigonometric functions of particular angles used to require trig tables or slide rules; now it requires only a calculator. However, the side ratios for some angles that appear in right triangles can be found *geometrically*. Every student of trigonometry should be able to find these special ratios without a calculator.

EXAMPLE 1 Evaluating Trigonometric Functions of 45°

Find the values of all six trigonometric functions for an angle of 45°.

SOLUTION A 45° angle occurs in an *isosceles right triangle*, with angles 45°–45°–90° (see Figure 4.9).

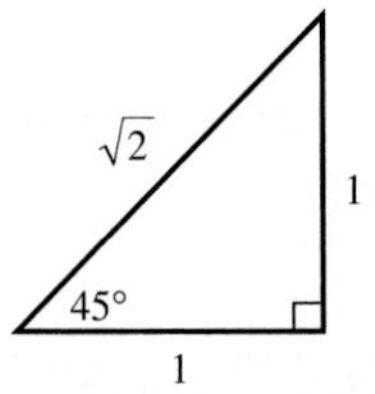

FIGURE 4.9 An isosceles right triangle. (Example 1)

Since the size of the triangle does not matter, we set the length of the two equal legs to 1. The hypotenuse, by the Pythagorean Theorem, is $\sqrt{1+1}=\sqrt{2}$. Applying the definitions, we have

$$\sin 45° = \frac{opp}{hyp} = \frac{1}{\sqrt{2}} = \frac{\sqrt{2}}{2} \qquad \csc 45° = \frac{hyp}{opp} = \frac{\sqrt{2}}{1}$$

$$\cos 45° = \frac{adj}{hyp} = \frac{1}{\sqrt{2}} = \frac{\sqrt{2}}{2} \qquad \sec 45° = \frac{hyp}{adj} = \frac{\sqrt{2}}{1}$$

$$\tan 45° = \frac{opp}{adj} = \frac{1}{1} = 1 \qquad \cot 45° = \frac{adj}{opp} = \frac{1}{1} = 1$$

Now try Exercise 1.

Whenever two sides of a right triangle are known, the third side can be found using the Pythagorean Theorem. All six trigonometric functions of either acute angle can then be found. We illustrate this in Example 2 with another well-known triangle.

EXAMPLE 2 Evaluating Trigonometric Functions of 30°

Find the values of all six trigonometric functions for an angle of 30°.

SOLUTION A 30° angle occurs in a 30°–60°–90° triangle, which can be constructed from an equilateral (60°–60°–60°) triangle by constructing an altitude to any side. Since size does not matter, start with an equilateral triangle with sides 2 units long. The altitude splits it into two congruent 30°–60°–90° triangles, each with hypotenuse 2 and smaller leg 1. By the Pythagorean Theorem, the longer leg has length $\sqrt{2^2-1^2}=\sqrt{3}$. (See Figure 4.10.)

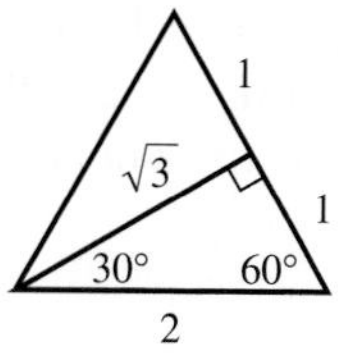

FIGURE 4.10 An altitude to any side of an equilateral triangle creates two congruent 30°–60°–90° triangles. If each side of the equilateral triangle has length 2, then the two 30°–60°–90° triangles have sides of length 2, 1, and $\sqrt{3}$. (Example 2)

We apply the definitions of the trigonometric functions to get:

$$\sin 30^\circ = \frac{opp}{hyp} = \frac{1}{2} \qquad \csc 30^\circ = \frac{hyp}{opp} = \frac{2}{1} = 2$$

$$\cos 30^\circ = \frac{adj}{hyp} = \frac{\sqrt{3}}{2} \qquad \sec 30^\circ = \frac{hyp}{adj} = \frac{2}{\sqrt{3}} = \frac{2\sqrt{3}}{3}$$

$$\tan 30^\circ = \frac{opp}{adj} = \frac{1}{\sqrt{3}} = \frac{\sqrt{3}}{3} \qquad \cot 30^\circ = \frac{adj}{opp} = \frac{\sqrt{3}}{1}$$

Now try Exercise 3.

EXPLORATION 2 Evaluating Trigonometric Functions of 60°

1. Find the values of all six trigonometric functions for an angle of 60°. Note that most of the preliminary work has been done in Example 2.
2. Compare the six function values for 60° with the six function values for 30°. What do you notice?
3. We will eventually learn a rule that relates trigonometric functions of any angle with trigonometric functions of the *complementary* angle. (Recall from geometry that 30° and 60° are complementary because they add up to 90°.) Based on this exploration, can you predict what that rule will be? (*Hint:* The "co" in cosine, cotangent, and cosecant actually comes from "complement.")

Example 3 illustrates that knowing one trigonometric ratio in a right triangle is sufficient for finding all the others.

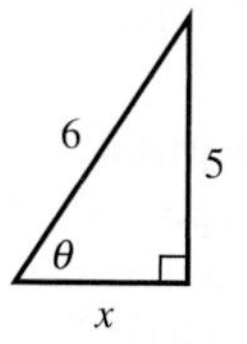

FIGURE 4.11 How to create an acute angle θ such that $\sin \theta = 5/6$. (Example 3)

EXAMPLE 3 Using One Trigonometric Ratio to Find Them All

Let θ be an acute angle such that $\sin \theta = 5/6$. Evaluate the other five trigonometric functions of θ.

SOLUTION Sketch a triangle showing an acute angle θ. Label the opposite side 5 and the hypotenuse 6. (See Figure 4.11.) Since $\sin \theta = 5/6$, this must be our angle! Now we need the other side of the triangle (labeled x in the figure).

From the Pythagorean Theorem it follows that $x^2 + 5^2 = 6^2$, so $x = \sqrt{36 - 25} = \sqrt{11}$. Applying the definitions,

$$\sin \theta = \frac{opp}{hyp} = \frac{5}{6} \qquad \csc \theta = \frac{hyp}{opp} = \frac{6}{5} = 1.2$$

$$\cos \theta = \frac{adj}{hyp} = \frac{\sqrt{11}}{6} \qquad \sec \theta = \frac{hyp}{adj} = \frac{6}{\sqrt{11}} = \frac{6\sqrt{11}}{11}$$

$$\tan \theta = \frac{opp}{adj} = \frac{5}{\sqrt{11}} = \frac{5\sqrt{11}}{11} \qquad \cot \theta = \frac{adj}{opp} = \frac{\sqrt{11}}{5}$$

Now try Exercise 9.

A Word About Radical Fractions

There was a time when $\frac{5\sqrt{11}}{11}$ was considered "simpler" than $\frac{5}{\sqrt{11}}$ because it was easier to approximate, but today they are just equivalent expressions for the same irrational number. With technology, either form leads easily to an approximation of 1.508. We leave the answers in exact form here because we want you to practice problems of this type without a calculator.

Evaluating Trigonometric Functions with a Calculator

Using a calculator for the evaluation step enables you to concentrate all your problem-solving skills on the modeling step, which is where the real trigonometry occurs. The danger is that your calculator will try to evaluate what you ask it to evaluate, even if you ask it to evaluate the wrong thing. If you make a mistake, you might be lucky and see an error message. In most cases you will unfortunately see an answer that you will assume is correct but is actually wrong. We list the most common calculator errors associated with evaluating trigonometric functions.

Common Calculator Errors when Evaluating Trig Functions

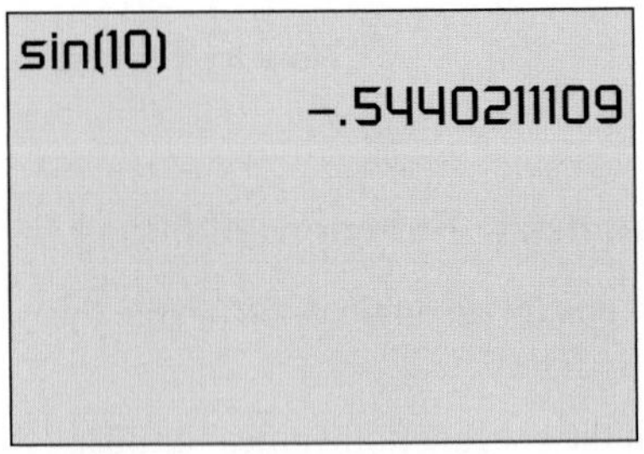

FIGURE 4.12 Wrong mode for finding sin (10°).

1. **Using the Calculator in the Wrong Angle Mode (Degrees/Radians)** This error is so common that everyone encounters it once in a while. You just hope to recognize it when it occurs. For example, suppose we are doing a problem in which we need to evaluate the sine of 10°. Our calculator shows us the screen in Figure 4.12.

   Why is the answer negative? Our first instinct should be to check the mode. Sure enough, it is in Radian mode. Changing to Degree mode, we get $\sin(10) = 0.1736481777$, which is a reasonable answer. (That still leaves open the question of why the sine of 10 rad is negative, but that is a topic for the next section.) We will revisit the mode problem later when we look at trigonometric graphs.

FIGURE 4.13 Finding cot (30°).

2. **Using the Inverse Trig Keys to Evaluate cot, sec, and csc** There are no buttons on most calculators for cotangent, secant, and cosecant. The reason is because they can be easily evaluated by finding reciprocals of tangent, cosine, and sine, respectively. For example, Figure 4.13 shows the correct way to evaluate the cotangent of 30°.

   There is also a key on the calculator for "TAN^{-1}"—but this is *not* the cotangent function! Remember that an exponent of -1 on a *function* is *never* used to denote a reciprocal; it is always used to denote the *inverse function*. We will study the inverse trigonometric functions in a later section, but meanwhile you can see that it is a bad way to evaluate cot (30) (Figure 4.14).

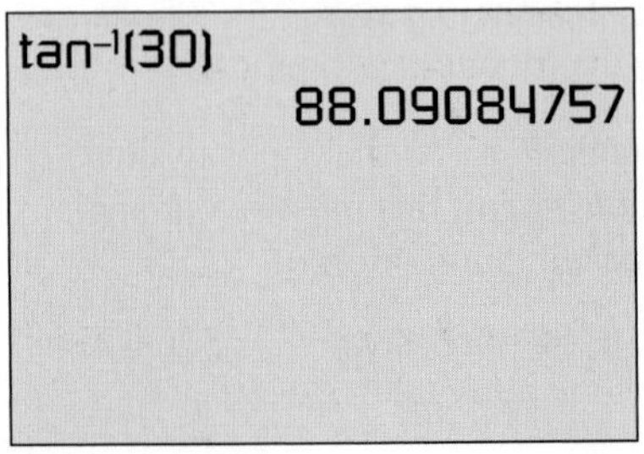

FIGURE 4.14 This is not cot (30°).

3. **Using Function Shorthand That the Calculator Does Not Recognize** This error is less dangerous because it usually results in an error message. We will often abbreviate powers of trig functions, writing (for example) "$\sin^3\theta - \cos^3\theta$" instead of the more cumbersome "$(\sin(\theta))^3 - (\cos(\theta))^3$." The calculator does not recognize the shorthand notation and interprets it as a syntax error.

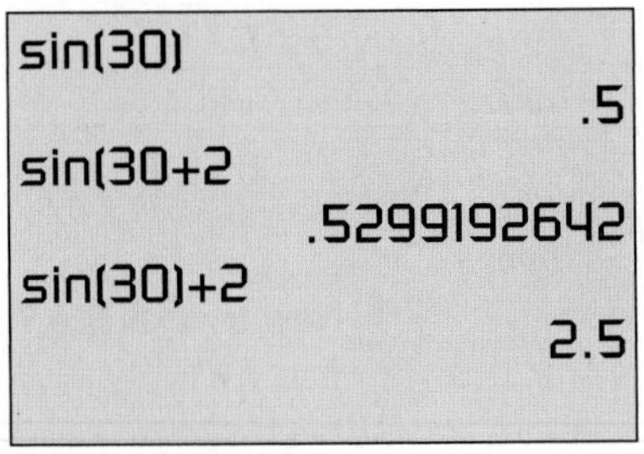

FIGURE 4.15 A correct and incorrect way to find sin (30°) + 2.

4. **Not Closing Parentheses** This general algebraic error is easy to make on calculators that automatically open a parenthesis pair whenever you type a function key. Check your calculator by pressing the SIN key. If the screen displays "sin (" instead of just "sin" then you have such a calculator. The danger arises because the calculator will automatically *close* the parenthesis pair at the end of a command if you have forgotten to do so. That is fine if you *want* the parenthesis at the end of the command, but it is bad if you want it somewhere else. For example, if you want "sin (30)" and you type "sin (30", you will get away with it. But if you want "sin (30) + 2" and you type "sin (30 + 2", you will not (Figure 4.15).

It is usually impossible to find an "exact" answer on a calculator, especially when evaluating trigonometric functions. The actual values are usually irrational numbers with nonterminating, nonrepeating decimal expansions. However, you can find some exact answers if you know what you are looking for, as in Example 4.

EXAMPLE 4 Getting an "Exact Answer" on a Calculator

Find the exact value of cos 30° on a calculator.

SOLUTION As you see in Figure 4.16, the calculator gives the answer 0.8660254038. However, if we recognize 30° as one of our special angles (see Example 2 in this section), we might recall that the exact answer can be written in terms of a square root. We square our answer and get 0.75, which suggests that the exact value of cos 30° is $\sqrt{3/4} = \sqrt{3}/2$.

Now try Exercise 25.

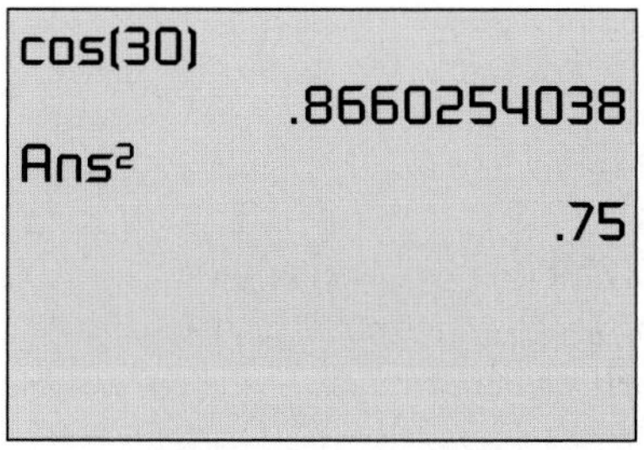

FIGURE 4.16 Calculations for Example 4.

Applications of Right Triangle Trigonometry

A triangle has six "parts," three angles and three sides, but you do not need to know all six parts to determine a triangle up to congruence. In fact, three parts are usually sufficient. The trigonometric functions take this observation a step further by giving us the means for actually *finding* the rest of the parts once we have enough parts to establish congruence. Using some of the parts of a triangle to solve for all the others is **solving a triangle**.

We will learn about solving general triangles in Sections 5.5 and 5.6, but we can already do some right triangle solving just by using the trigonometric ratios.

EXAMPLE 5 Solving a Right Triangle

A right triangle with a hypotenuse of 8 includes a 37° angle (Figure 4.17). Find the measures of the other two angles and the lengths of the other two sides.

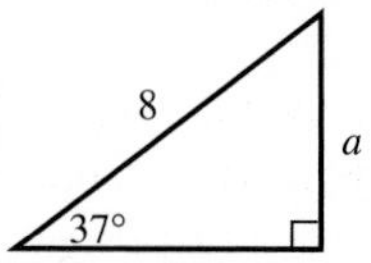

FIGURE 4.17 Diagram for Example 5.

SOLUTION Since it is a right triangle, one of the other angles is 90°. That leaves 180° − 90° − 37° = 53° for the third angle.

Referring to the labels in Figure 4.17, we have

$$\sin 37° = \frac{a}{8} \qquad \cos 37° = \frac{b}{8}$$

$$a = 8 \sin 37° \qquad b = 8 \cos 37°$$

$$a \approx 4.81 \qquad b \approx 6.39$$

Now try Exercise 55.

The real-world applications of triangle solving are many, reflecting the frequency with which one encounters triangular shapes in everyday life.

A Word About Rounding Answers

Notice in Example 6 that we rounded the answer to the nearest integer. In applied problems it is illogical to give answers with more decimal places of accuracy than can be guaranteed for the input values. An answer of 729.132 ft implies razor-sharp accuracy, whereas the reported distance from the building (340 ft) implies a much less precise measurement. (So does the angle of 65°.) Indeed, an engineer following specific rounding criteria based on "significant digits" would probably report the answer to Example 6 as 730 ft. We will not get too picky about rounding, but we will try to be sensible.

EXAMPLE 6 Finding the Height of a Building

From a point 340 ft away from the base of the Peachtree Center Plaza in Atlanta, Georgia, the angle of elevation to the top of the building is 65°. (See Figure 4.18.) Find the height h of the building.

SOLUTION We need a ratio that will relate an angle to its opposite and adjacent sides. The tangent function is the appropriate choice.

$$\tan 65° = \frac{h}{340}$$

$$h = 340 \tan 65°$$

$$h \approx 729 \text{ ft}$$

Now try Exercise 61.

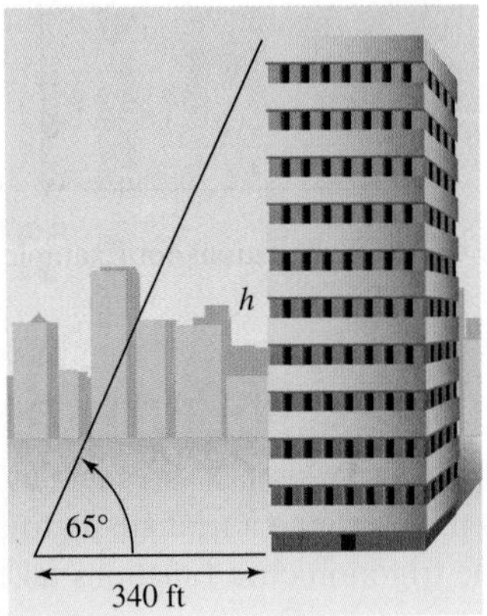

FIGURE 4.18 Labeled diagram for Example 6.

QUICK REVIEW 4.2 *(For help, go to Sections P.2 and 1.7.)*

Exercise numbers with a gray background indicate problems that the authors have designed to be solved *without a calculator*.

In Exercises 1–4, use the Pythagorean Theorem to solve for x.

1.

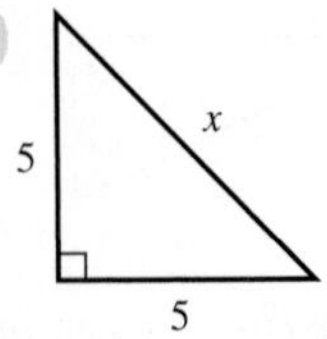

2.

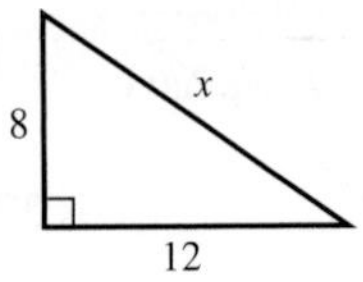

3.

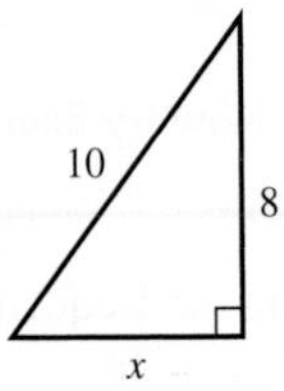

4.

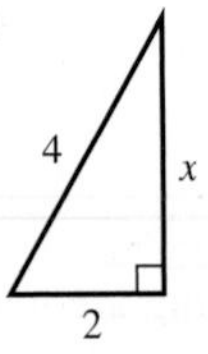

In Exercises 5 and 6, convert units.

5. 8.4 ft to inches

6. 940 ft to miles

In Exercises 7–10, solve the equation. State the correct unit.

7. $0.388 = \dfrac{a}{20.4 \text{ km}}$

8. $1.72 = \dfrac{23.9 \text{ ft}}{b}$

9. $\dfrac{2.4 \text{ in.}}{31.6 \text{ in.}} = \dfrac{a}{13.3}$

10. $\dfrac{5.9}{\beta} = \dfrac{8.66 \text{ cm}}{6.15 \text{ cm}}$

SECTION 4.2 Exercises

In Exercises 1–8, find the values of all six trigonometric functions of the angle θ.

1.

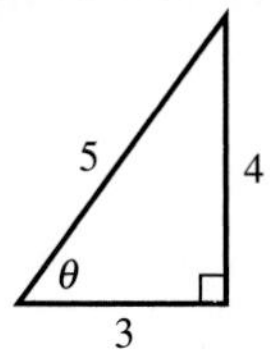

2.

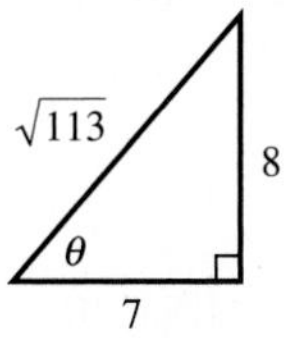

3.

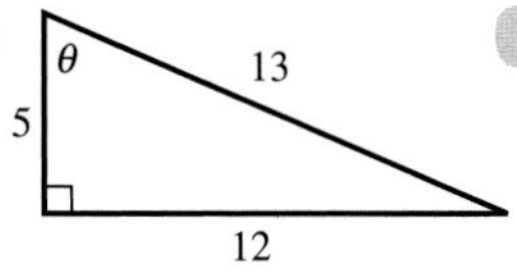

4.

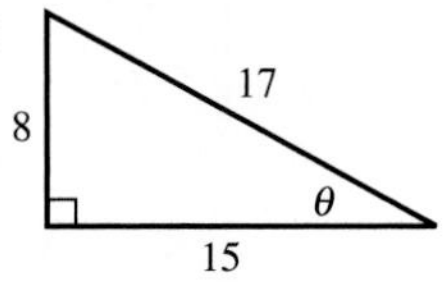

5.

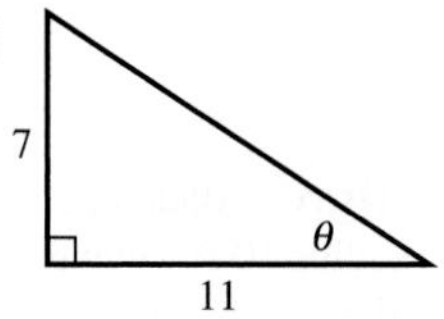

6.

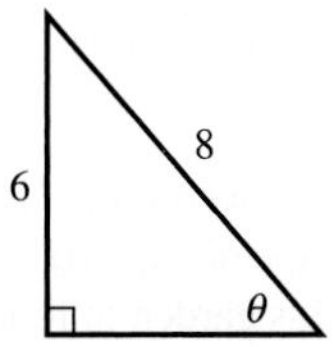

7.

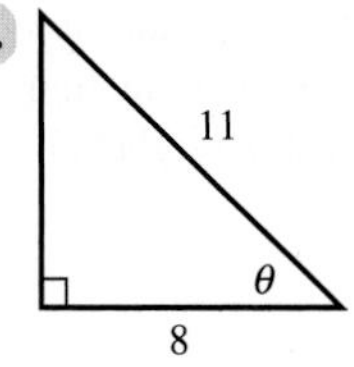

8.

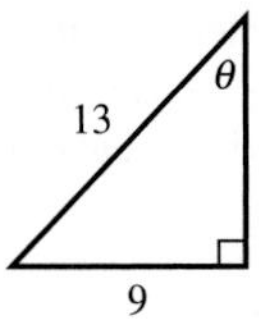

In Exercises 9–18, assume that θ is an acute angle in a right triangle satisfying the given conditions. Evaluate the remaining trigonometric functions.

9. $\sin\theta = \frac{3}{7}$ **10.** $\sin\theta = \frac{2}{3}$

11. $\cos\theta = \frac{5}{11}$ **12.** $\cos\theta = \frac{5}{8}$

13. $\tan\theta = \frac{5}{9}$ **14.** $\tan\theta = \frac{12}{13}$

15. $\cot\theta = \frac{11}{3}$ **16.** $\csc\theta = \frac{12}{5}$

17. $\csc\theta = \frac{23}{9}$ **18.** $\sec\theta = \frac{17}{5}$

In Exercises 19–24, evaluate *without* using a calculator.

19. $\sin\left(\frac{\pi}{3}\right)$ **20.** $\tan\left(\frac{\pi}{4}\right)$

21. $\cot\left(\frac{\pi}{6}\right)$ **22.** $\sec\left(\frac{\pi}{3}\right)$

23. $\cos\left(\frac{\pi}{4}\right)$ **24.** $\csc\left(\frac{\pi}{3}\right)$

In Exercises 25–28, evaluate using a calculator. Give an exact value, not an approximate answer. (See Example 4.)

25. $\sec 45°$ **26.** $\sin 60°$

27. $\csc\left(\frac{\pi}{3}\right)$

28. $\tan\left(\frac{\pi}{3}\right)$

In Exercises 29–40, evaluate using a calculator. Be sure the calculator is in the correct mode. Give answers correct to three decimal places.

29. $\sin 74°$ **30.** $\tan 8°$

31. $\cos 19°23'$ **32.** $\tan 23°42'$

33. $\tan\left(\frac{\pi}{12}\right)$ **34.** $\sin\left(\frac{\pi}{15}\right)$

35. $\sec 49°$ **36.** $\csc 19°$

37. $\cot 0.89$ **38.** $\sec 1.24$

39. $\cot\left(\frac{\pi}{8}\right)$ **40.** $\csc\left(\frac{\pi}{10}\right)$

In Exercises 41–48, find the acute angle θ that satisfies the given equation. Give θ in both degrees and radians. You should do these problems without a calculator.

41. $\sin\theta = \frac{1}{2}$ **42.** $\sin\theta = \frac{\sqrt{3}}{2}$

43. $\cot\theta = \frac{1}{\sqrt{3}}$ **44.** $\cos\theta = \frac{\sqrt{2}}{2}$

45. $\sec\theta = 2$ **46.** $\cot\theta = 1$

47. $\tan\theta = \frac{\sqrt{3}}{3}$ **48.** $\cos\theta = \frac{\sqrt{3}}{2}$

In Exercises 49–54, solve for the variable shown.

49.

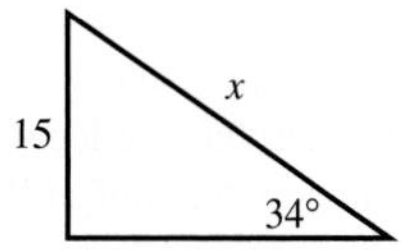

50.

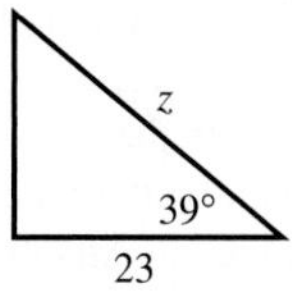

51.

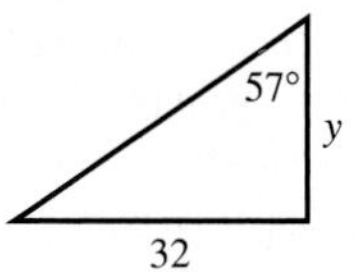

52.

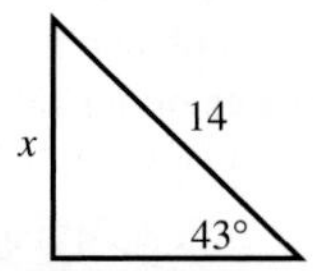

53.

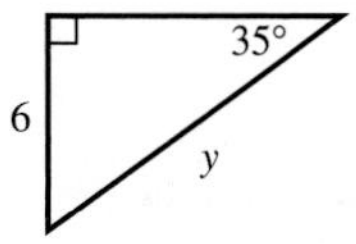

54.

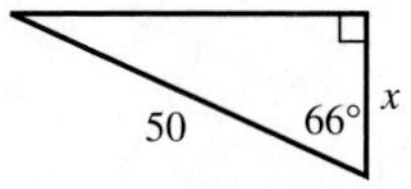

In Exercises 55–58, solve the right ΔABC for all of its unknown parts.

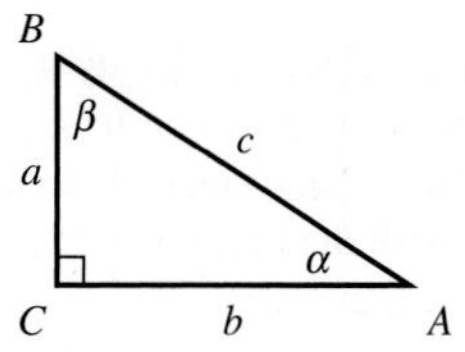

55. $\alpha = 20°; a = 12.3$ **56.** $\alpha = 41°; c = 10$

57. $\beta = 55°; a = 15.58$ **58.** $a = 5; \beta = 59°$

59. Writing to Learn What is $\lim_{\theta \to 0} \sin \theta$? Explain your answer in terms of right triangles in which θ gets smaller and smaller.

60. Writing to Learn What is $\lim_{\theta \to 0} \cos \theta$? Explain your answer in terms of right triangles in which θ gets smaller and smaller.

61. Height A guy wire from the top of the transmission tower at WJBC forms a 75° angle with the ground at a 55-ft distance from the base of the tower. How tall is the tower?

62. Height Kirsten places her surveyor's telescope on the top of a tripod 5 ft above the ground. She measures an 8° elevation above the horizontal to the top of a tree that is 120 ft away. How tall is the tree?

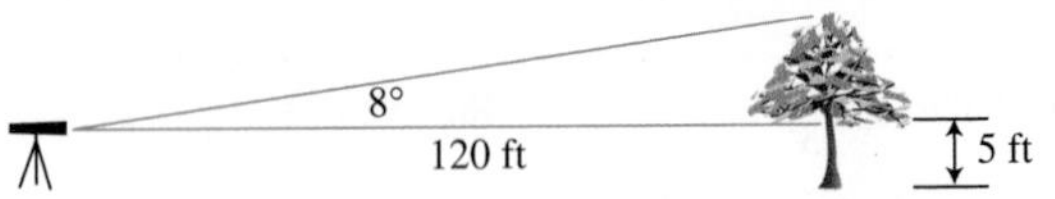

63. Group Activity Area For locations between 20° and 60° north latitude a solar collector panel should be mounted so that its angle with the horizontal is 20 greater than the local latitude. Consequently, the solar panel mounted on the roof of Solar Energy, Inc., in Atlanta (latitude 34°) forms a 54° angle with the horizontal. The bottom edge of the 12-ft-long panel is resting on the roof, and the high edge is 5 ft above the roof. What is the total area of this rectangular collector panel?

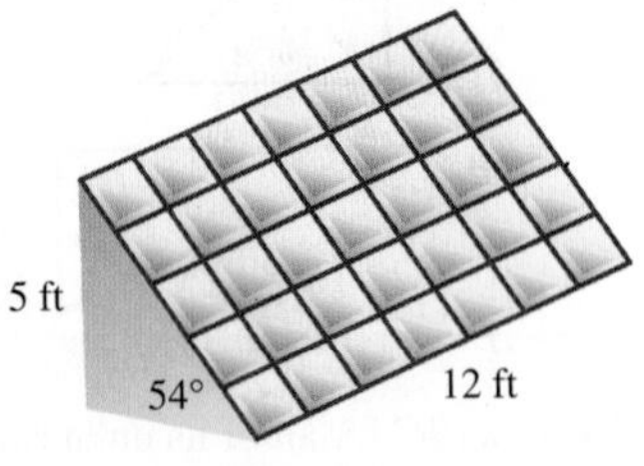

64. Height The Chrysler Building in New York City was the tallest building in the world at the time it was built. It casts a shadow approximately 130 ft long on the street when the Sun's rays form an 82.9° angle with the Earth. How tall is the building?

65. Distance DaShanda's team of surveyors had to find the distance AC across the lake at Montgomery County Park. Field assistants positioned themselves at points A and C while DaShanda set up an angle-measuring instrument at point B, 100 ft from C in a perpendicular direction. DaShanda measured $\angle ABC$ as $75°12'42''$. What is the distance AC?

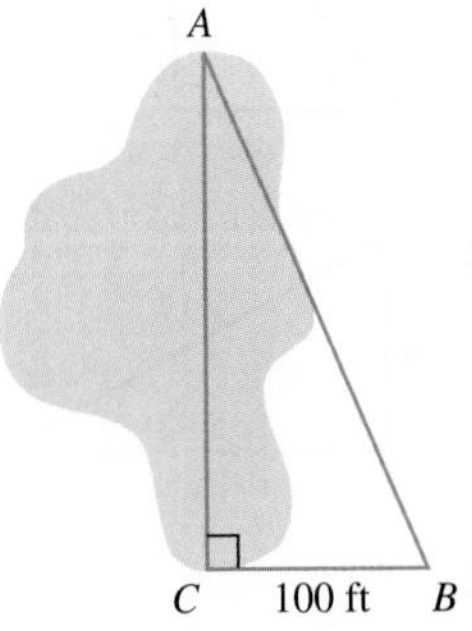

66. Group Activity Garden Design Allen's garden is in the shape of a quarter-circle with radius 10 ft. He wishes to plant his garden in four parallel strips, as shown in the diagram on the left below, so that the four arcs along the circular edge of the garden are all of equal length. After measuring four equal arcs, he carefully measures the widths of the four strips and records his data in the table shown at the right below.

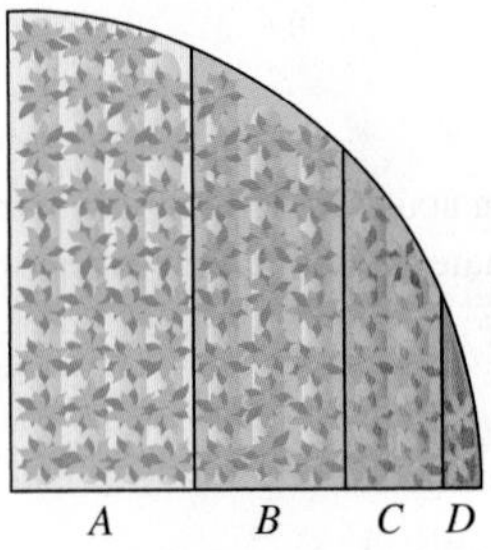

Strip	Width
A	3.827 ft
B	3.344 ft
C	2.068 ft
D	0.761 ft

Alicia sees Allen's data and realizes that he could have saved himself some work by figuring out the strip widths by trigonometry. By checking his data with a calculator she is able to correct two measurement errors he has made. Find Allen's two errors and correct them.

Standardized Test Questions

67. True or False If θ is an angle in any triangle, then $\tan \theta$ is the length of the side opposite θ divided by the length of the side adjacent to θ. Justify your answer.

68. True or False If A and B are angles of a triangle such that $A > B$, then $\cos A > \cos B$. Justify your answer.

You should answer these questions without using a calculator.

69. Multiple Choice Which of the following expressions does not represent a real number?

(A) sin 30° **(B)** tan 45° **(C)** cos 90°

(D) csc 90° **(E)** sec 90°

70. Multiple Choice If θ is the smallest angle in a 3–4–5 right triangle, then $\sin\theta =$

(A) $\frac{3}{5}$. **(B)** $\frac{3}{4}$. **(C)** $\frac{4}{5}$.

(D) $\frac{5}{4}$. **(E)** $\frac{5}{3}$.

71. Multiple Choice If a nonhorizontal line has slope $\sin\theta$, it will be perpendicular to a line with slope

(A) $\cos\theta$. **(B)** $-\cos\theta$. **(C)** $\csc\theta$.

(D) $-\csc\theta$. **(E)** $-\sin\theta$.

72. Multiple Choice Which of the following trigonometric ratios could *not* be π?

(A) $\tan\theta$ **(B)** $\cos\theta$ **(C)** $\cot\theta$

(D) $\sec\theta$ **(E)** $\csc\theta$

73. Trig Tables Before calculators became common classroom tools, students used trig tables to find trigonometric ratios. Below is a simplified trig table for angles between 40° and 50°. *Without using a calculator*, can you determine which column gives sine values, which gives cosine values, and which gives tangent values?

Trig Tables for Sine, Cosine, and Tangent

Angle	?	?	?
40°	0.8391	0.6428	0.7660
42°	0.9004	0.6691	0.7431
44°	0.9657	0.6947	0.7193
46°	1.0355	0.7193	0.6947
48°	1.1106	0.7431	0.6691
50°	1.1917	0.7660	0.6428

74. Trig Tables Below is a simplified trig table for angles between 30° and 40°. *Without using a calculator*, can you determine which column gives cotangent values, which gives secant values, and which gives cosecant values?

Trig Tables for Cotangent, Secant, and Cosecant

Angle	?	?	?
30°	1.1547	1.7321	2.0000
32°	1.1792	1.6003	1.8871
34°	1.2062	1.4826	1.7883
36°	1.2361	1.3764	1.7013
38°	1.2690	1.2799	1.6243
40°	1.3054	1.1918	1.5557

Explorations

75. Mirrors In the figure, a light ray shining from point A to point P on the mirror will bounce to point B in such a way that the *angle of incidence* α will equal the *angle of reflection* β. This is the *law of reflection* derived from physical experiments. Both angles are measured from the *normal line*, which is perpendicular to the mirror at the point of reflection P. If A is 2 m farther from the mirror than is B, and if $\alpha = 30°$ and $AP = 5$ m, what is the length PB?

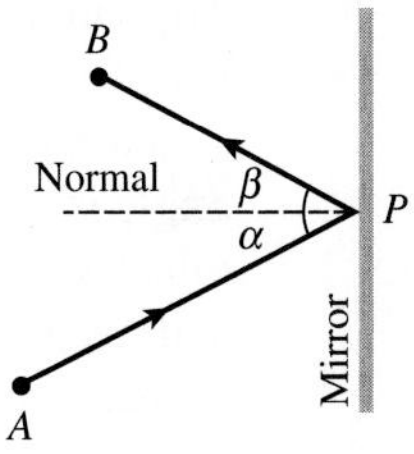

76. Pool On the pool table shown in the figure, where along the portion CD of the railing should you direct ball A so that it will bounce off CD and strike ball B? Assume that A obeys the law of reflection relative to rail CD.

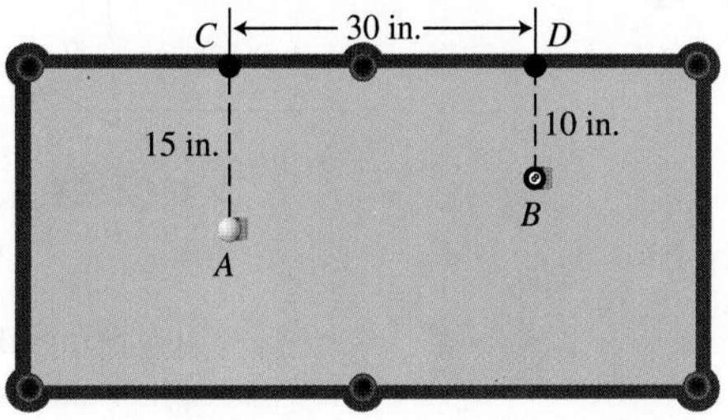

Extending the Ideas

77. Using the labeling of the triangle below, prove that if θ is an acute angle in any right triangle, $(\sin\theta)^2 + (\cos\theta)^2 = 1$.

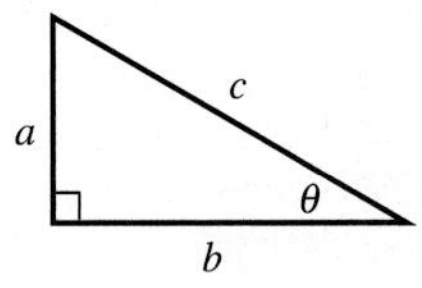

78. Using the labeling of the triangle below, prove that the area of the triangle is equal to $(1/2)\,ab\sin\theta$. (*Hint:* Start by drawing the altitude to side b and finding its length.)

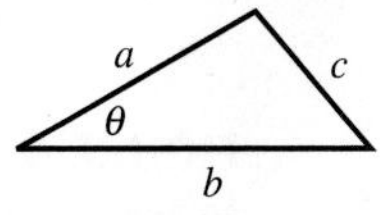

4.3 Trigonometry Extended: The Circular Functions

What you'll learn about

- Trigonometric Functions of Any Angle
- Trigonometric Functions of Real Numbers
- Periodic Functions
- The 16-Point Unit Circle

... and why

Extending trigonometric functions beyond triangle ratios opens up a new world of applications.

Trigonometric Functions of Any Angle

We now extend the definitions of the six basic trigonometric functions beyond triangles so that we do not have to restrict our attention to acute angles, or even to positive angles.

In geometry we think of an angle as a union of two rays with a common vertex. Trigonometry takes a more dynamic view by thinking of an angle in terms of a rotating ray. The beginning position of the ray, the **initial side**, is rotated about its endpoint, called the **vertex**. The final position is called the **terminal side**. The **measure of an angle** is a number that describes the amount of rotation from the initial side to the terminal side of the angle. **Positive angles** are generated by counterclockwise rotations and **negative angles** are generated by clockwise rotations. Figure 4.19 shows an angle of measure α, where α is a positive number.

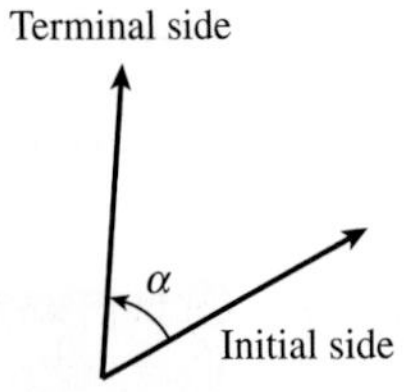

FIGURE 4.19 An angle with positive measure α.

To bring the power of coordinate geometry into the picture (literally), we usually place an angle in **standard position** in the Cartesian plane, with the vertex of the angle at the origin and its initial side lying along the positive x-axis. Figure 4.20 shows two angles in standard position, one with positive measure α and the other with negative measure β.

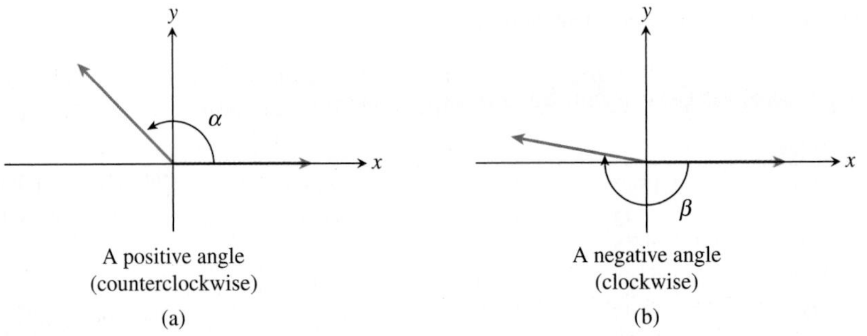

FIGURE 4.20 Two angles in standard position. In (a) the counterclockwise rotation generates an angle with positive measure α. In (b) the clockwise rotation generates an angle with negative measure β.

Two angles in this expanded angle-measurement system can have the same initial side and the same terminal side, yet have different measures. We call such angles **coterminal angles**. (See Figure 4.21 on the next page.) For example, angles of 90°, 450°, and −270° are all coterminal, as are angles of π, 3π, and -99π radians. In fact, angles are coterminal whenever they differ by an integer multiple of 360° or by an integer multiple of 2π rad.

EXAMPLE 1 Finding Coterminal Angles

Find and draw a positive angle and a negative angle that are coterminal with the given angle.

(a) $30°$ **(b)** $-150°$ **(c)** $\frac{2\pi}{3}$ radians

SOLUTION There are infinitely many possible solutions; we will show two for each angle.

(a) Add $360°$: $30° + 360° = 390°$

Subtract $360°$: $30° - 360° = -330°$

Figure 4.22 shows these two angles, which are coterminal with the $30°$ angle.

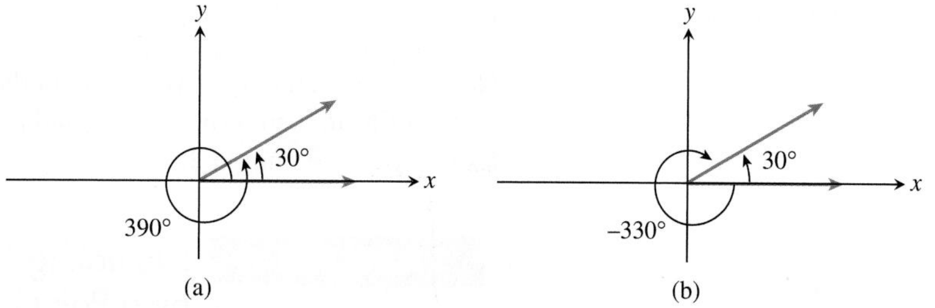

FIGURE 4.22 Two angles coterminal with 30°. (Example 1a)

(b) Add $360°$: $-150° + 360° = 210°$

Subtract $2(360°)$: $-150° - 720° = -870°$

We leave it to you to draw the coterminal angles.

(c) Add 2π: $\frac{2\pi}{3} + 2\pi = \frac{2\pi}{3} + \frac{6\pi}{3} = \frac{8\pi}{3}$

Subtract 2π: $\frac{2\pi}{3} - 2\pi = \frac{2\pi}{3} - \frac{6\pi}{3} = -\frac{4\pi}{3}$

Again, we leave it to you to draw the coterminal angles.

Now try Exercise 1.

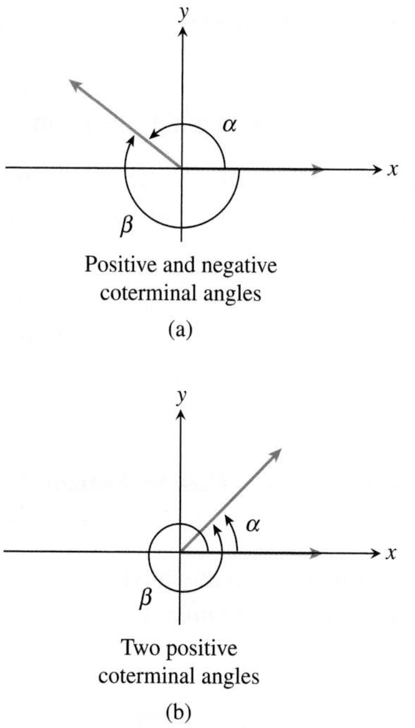

FIGURE 4.21 Coterminal angles. In (a) a positive angle and a negative angle are coterminal, and in (b) both coterminal angles are positive.

Extending the definitions of the six basic trigonometric functions so that they can apply to any angle is surprisingly easy, but first you need to see how our current definitions relate to the (x, y) coordinates in the Cartesian plane. We start in the first quadrant (see Figure 4.23), where the angles are all acute. Work through Exploration 1 before moving on.

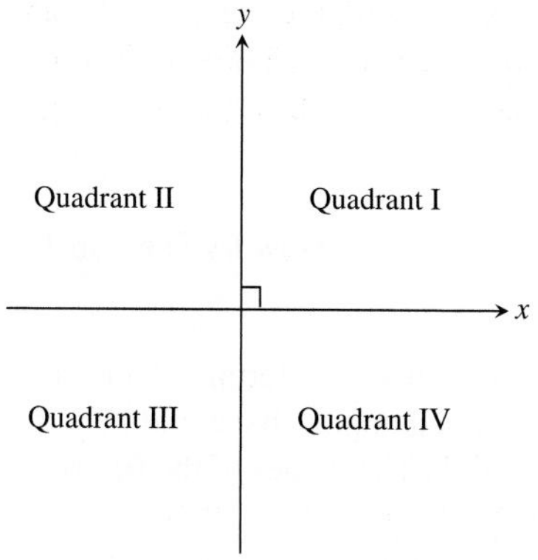

FIGURE 4.23 The four quadrants of the Cartesian plane. Both x and y are positive in QI (Quadrant I). Quadrants, like Super Bowls, are invariably designated by Roman numerals.

EXPLORATION 1 Investigating First Quadrant Trigonometry

Let $P(x, y)$ be any point in the first quadrant (QI), and let r be the distance from P to the origin. (See Figure 4.24.)

1. Use the acute angle definition of the sine function (Section 4.2) to prove that $\sin\theta = y/r$.
2. Express $\cos\theta$ in terms of x and r.
3. Express $\tan\theta$ in terms of x and y.
4. Express the remaining three basic trigonometric functions in terms of x, y, and r.

If you have successfully completed Exploration 1, you should have no trouble verifying the solution to Example 2, which we show without the details.

EXAMPLE 2 Evaluating Trig Functions Determined by a Point in QI

Let θ be the acute angle in standard position whose terminal side contains the point $(5, 3)$. Find the six trigonometric functions of θ.

SOLUTION The distance from $(5, 3)$ to the origin is $\sqrt{34}$.

So
$$\sin\theta = \frac{3}{\sqrt{34}} = \frac{3\sqrt{34}}{34} \qquad \csc\theta = \frac{\sqrt{34}}{3}$$
$$\cos\theta = \frac{5}{\sqrt{34}} = \frac{5\sqrt{34}}{34} \qquad \sec\theta = \frac{\sqrt{34}}{5}$$
$$\tan\theta = \frac{3}{5} \qquad \cot\theta = \frac{5}{3}$$

Now try Exercise 5.

Now we have an easy way to extend the trigonometric functions to any angle: Use the same definitions in terms of x, y, and r—*whether or not x and y are positive*. Compare Example 3 to Example 2.

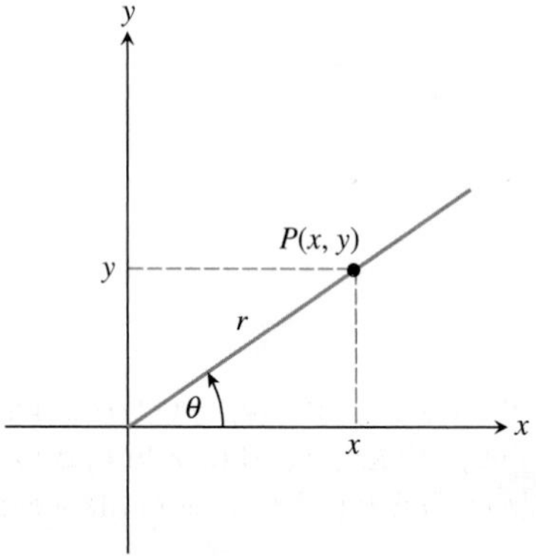

FIGURE 4.24 A point $P(x, y)$ in Quadrant I determines an acute angle θ. The number r denotes the distance from P to the origin. (Exploration 1)

EXAMPLE 3 Evaluating Trig Functions Determined by a Point Not in QI

Let θ be any angle in standard position whose terminal side contains the point $(-5, 3)$. Find the six trigonometric functions of θ.

SOLUTION The distance from $(-5, 3)$ to the origin is $\sqrt{34}$.

So
$$\sin\theta = \frac{3}{\sqrt{34}} = \frac{3\sqrt{34}}{34} \qquad \csc\theta = \frac{\sqrt{34}}{3}$$
$$\cos\theta = \frac{-5}{\sqrt{34}} = \frac{-5\sqrt{34}}{34} \qquad \sec\theta = \frac{\sqrt{34}}{-5}$$
$$\tan\theta = \frac{3}{-5} \qquad \cot\theta = \frac{-5}{3}$$

Now try Exercise 11.

Notice in Example 3 that θ is *any* angle in standard position whose terminal side contains the point $(-5, 3)$. There are infinitely many coterminal angles that could play the role of θ, some of them positive and some of them negative. The values of the six trigonometric functions would be the same for all of them.

We are now ready to state the formal definition.

DEFINITION Trigonometric Functions of Any Angle

Let θ be any angle in standard position and let $P(x, y)$ be any point on the terminal side of the angle (except the origin). Let r denote the distance from $P(x, y)$ to the origin, i.e., let $r = \sqrt{x^2 + y^2}$. (See Figure 4.25.) Then

$$\sin\theta = \frac{y}{r} \qquad \csc\theta = \frac{r}{y}\ (y \neq 0)$$
$$\cos\theta = \frac{x}{r} \qquad \sec\theta = \frac{r}{x}\ (x \neq 0)$$
$$\tan\theta = \frac{y}{x}\ (x \neq 0) \qquad \cot\theta = \frac{x}{y}\ (y \neq 0)$$

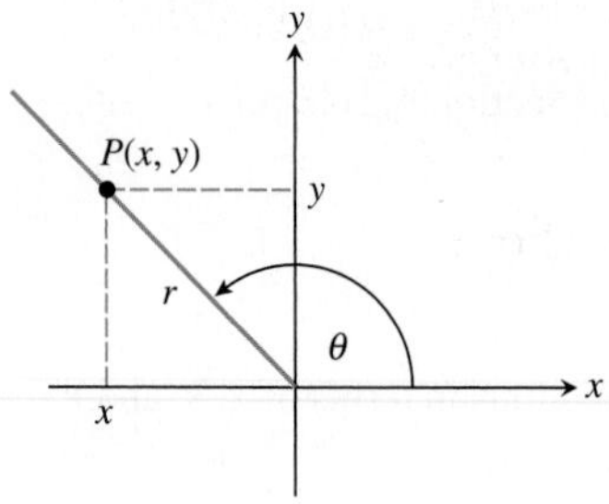

FIGURE 4.25 Defining the six trig functions of θ.

Examples 2 and 3 both began with a point $P(x, y)$ rather than an angle θ. Indeed, the point gave us so much information about the trigonometric ratios that we were able to compute them all without ever finding θ. So what do we do if we start with an angle θ in standard position and we want to evaluate the trigonometric functions? We try to find a point (x, y) on its terminal side. We illustrate this process with Example 4.

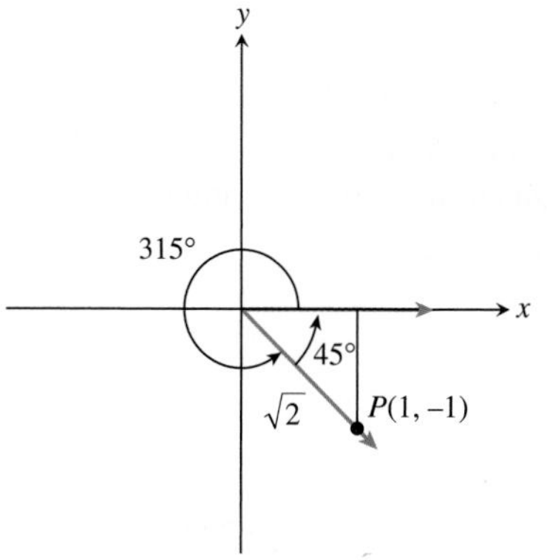

FIGURE 4.26 An angle of 315° in standard position determines a 45°–45°–90° *reference triangle.* (Example 4)

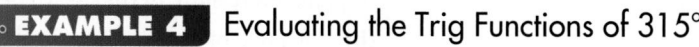

EXAMPLE 4 Evaluating the Trig Functions of 315°

Find the six trigonometric functions of 315°.

SOLUTION First we draw an angle of 315° in standard position. Without declaring a scale, pick a point P on the terminal side and connect it to the x-axis with a perpendicular segment. Notice that the triangle formed (called a **reference triangle**) is a 45°—45°—90° triangle. If we arbitrarily choose the horizontal and vertical sides of the reference triangle to be of length 1, then P has coordinates $(1, -1)$. (See Figure 4.26.)

We can now use the definitions with $x = 1$, $y = -1$, and $r = \sqrt{2}$.

$$\sin 315° = \frac{-1}{\sqrt{2}} = -\frac{\sqrt{2}}{2} \qquad \csc 315° = \frac{\sqrt{2}}{-1} = -\sqrt{2}$$

$$\cos 315° = \frac{1}{\sqrt{2}} = \frac{\sqrt{2}}{2} \qquad \sec 315° = \frac{\sqrt{2}}{1} = \sqrt{2}$$

$$\tan 315° = \frac{-1}{1} = -1 \qquad \cot 315° = \frac{1}{-1} = -1$$

Now try Exercise 25.

The happy fact that the reference triangle in Example 4 was a 45°—45°—90° triangle enabled us to label a point P on the terminal side of the 315° angle and then to find the trigonometric function values. We would also be able to find P if the given angle were to produce a 30°—60°—90° reference triangle.

Evaluating Trig Functions of a Nonquadrantal Angle θ

1. Draw the angle θ in standard position, being careful to place the terminal side in the correct quadrant.
2. Without declaring a scale on either axis, label a point P (other than the origin) on the terminal side of θ.
3. Draw a perpendicular segment from P to the x-axis, determining the *reference triangle*. If this triangle is one of the triangles whose ratios you know, label the sides accordingly. If it is not, then you will need to use your calculator.
4. Use the sides of the triangle to determine the coordinates of point P, making them positive or negative according to the signs of x and y in that particular quadrant.
5. Use the coordinates of point P and the definitions to determine the six trig functions.

EXAMPLE 5 Evaluating More Trig Functions

Find the following without a calculator:

(a) $\sin(-210°)$

(b) $\tan(5\pi/3)$

(c) $\sec(-3\pi/4)$

(continued)

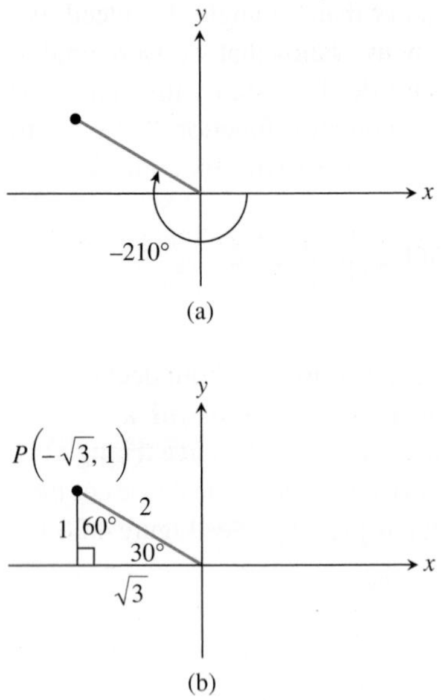

FIGURE 4.27 (Example 5a)

SOLUTION

(a) An angle of $-210°$ in standard position determines a $30°-60°-90°$ reference triangle in the second quadrant (Figure 4.27). We label the sides accordingly, then use the lengths of the sides to determine the point $P(-\sqrt{3}, 1)$. (Note that the x-coordinate is negative in the second quadrant.) The hypotenuse is $r = 2$. Therefore $\sin(-210°) = y/r = 1/2$.

(b) An angle of $5\pi/3$ radians in standard position determines a $30°-60°-90°$ reference triangle in the fourth quadrant (Figure 4.28). We label the sides accordingly, then use the lengths of the sides to determine the point $P(1, -\sqrt{3})$. (Note that the y-coordinate is negative in the fourth quadrant.) The hypotenuse is $r = 2$. Therefore $\tan(5\pi/3) = y/x = -\sqrt{3}/1 = -\sqrt{3}$.

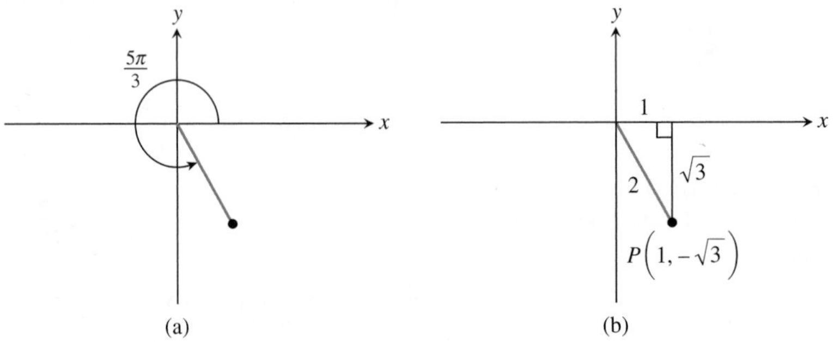

FIGURE 4.28 (Example 5b)

(c) An angle of $-3\pi/4$ radians in standard position determines a $45°-45°-90°$ reference triangle in the third quadrant. (See Figure 4.29.) We label the sides accordingly, then use the lengths of the sides to determine the point $P(-1, -1)$. (Note that both coordinates are negative in the third quadrant.) The hypotenuse is $r = \sqrt{2}$. Therefore $\sec(-3\pi/4) = r/x = \sqrt{2}/(-1) = -\sqrt{2}$.

Now try Exercise 29.

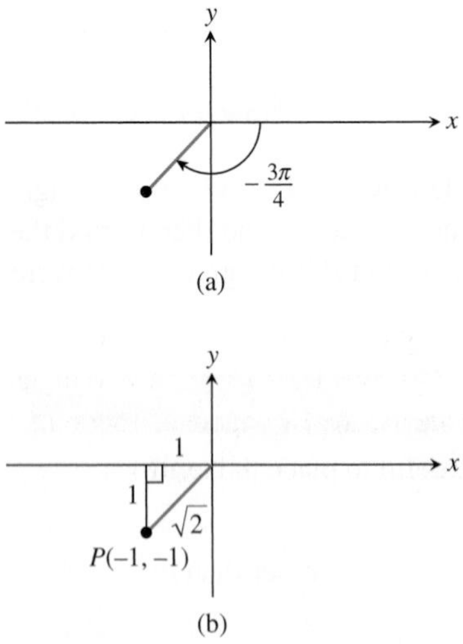

FIGURE 4.29 (Example 5c)

Angles whose terminal sides lie along one of the coordinate axes are called **quadrantal angles**, and although they do not produce reference triangles at all, it is easy to pick a point P along one of the axes.

EXAMPLE 6 Evaluating Trig Functions of Quadrantal Angles

Find each of the following, if it exists. If the value is undefined, write "undefined."

(a) $\sin(-270°)$

(b) $\tan 3\pi$

(c) $\sec \dfrac{11\pi}{2}$

SOLUTION

(a) In standard position, the terminal side of an angle of $-270°$ lies along the positive y-axis (Figure 4.30). A convenient point P along the positive y-axis is the point for which $r = 1$, namely $(0, 1)$. Therefore

$$\sin(-270°) = \frac{y}{r} = \frac{1}{1} = 1.$$

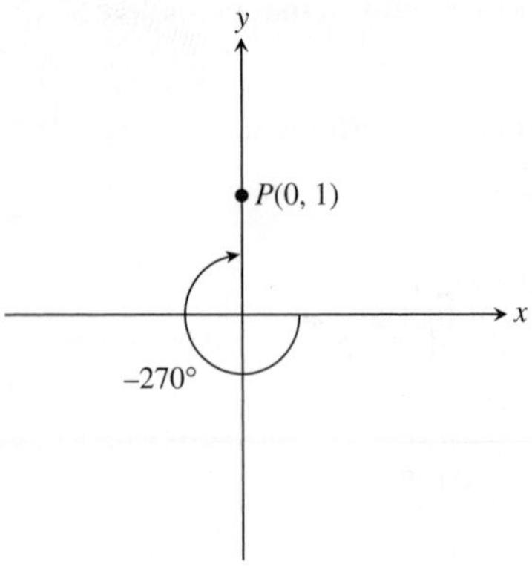

FIGURE 4.30 (Example 6a)

Why Not Use a Calculator?

You might wonder why we would go through this procedure to produce values that could be found so easily with a calculator. The answer is to understand how trigonometry *works* in the coordinate plane. Ironically, technology has made these computational exercises more important than ever, since calculators have eliminated the need for the repetitive evaluations that once gave students their initial insights into the basic trig functions.

(b) In standard position, the terminal side of an angle of 3π lies along the negative x-axis. (See Figure 4.31.) A convenient point P along the negative x-axis is the point for which $r = 1$, namely $(-1, 0)$. Therefore

$$\tan 3\pi = \frac{y}{x} = \frac{0}{-1} = 0.$$

(c) In standard position, the terminal side of an angle of $11\pi/2$ lies along the negative y-axis. (See Figure 4.32.) A convenient point P along the negative y-axis is the point for which $r = 1$, namely $(0, -1)$. Therefore

$$\sec \frac{11\pi}{2} = \frac{r}{x} = \frac{1}{0}. \quad \text{Undefined}$$

Now try Exercise 41.

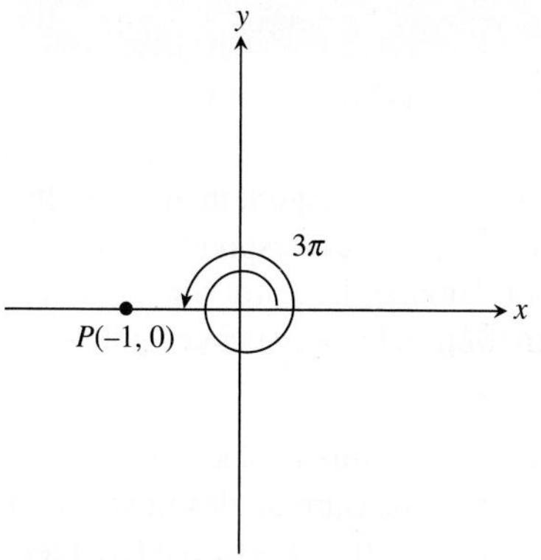

FIGURE 4.31 (Example 6b)

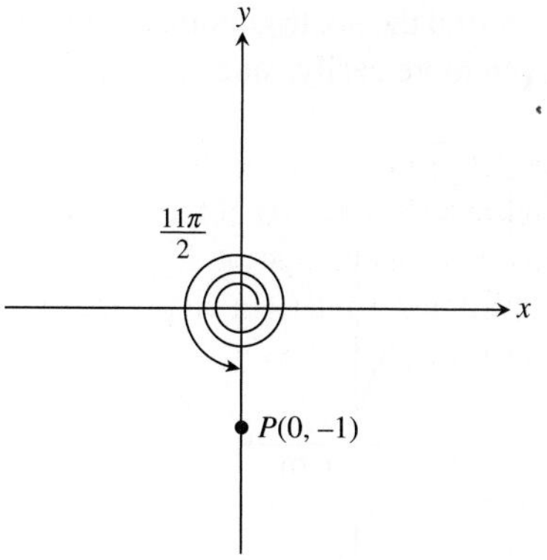

FIGURE 4.32 (Example 6c)

Another good exercise is to use information from one trigonometric ratio to produce the other five. We do not need to know the angle θ, although we do need a hint as to the location of its terminal side so that we can sketch a reference triangle in the correct quadrant (or place a quadrantal angle on the correct side of the origin). Example 7 illustrates how this is done.

EXAMPLE 7 Using One Trig Ratio to Find the Others

Find $\cos\theta$ and $\tan\theta$ by using the given information to construct a reference triangle.

(a) $\sin\theta = \dfrac{3}{7}$ and $\tan\theta < 0$

(b) $\sec\theta = 3$ and $\sin\theta > 0$

(c) $\cot\theta$ is undefined and $\sec\theta$ is negative

SOLUTION

(a) Since $\sin\theta$ is positive, the terminal side is either in QI or in QII. The added fact that $\tan\theta$ is negative means that the terminal side is in QII. We draw a reference triangle in QII with $r = 7$ and $y = 3$ (Figure 4.33); then we use the Pythagorean Theorem to find that $x = -1\sqrt{7^2 - 3^2} = -\sqrt{40}$. (Note that x is negative in QII.)

We then use the definitions to get

$$\cos\theta = \frac{-\sqrt{40}}{7} \text{ and } \tan\theta = \frac{3}{-\sqrt{40}} = \frac{-3\sqrt{10}}{20}.$$

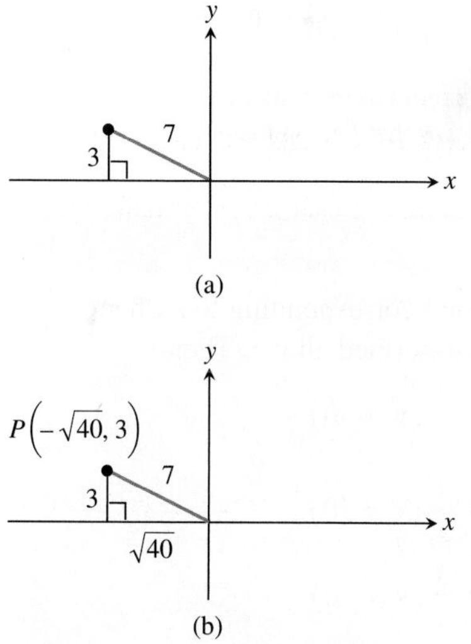

FIGURE 4.33 (Example 7a)

(b) Since $\sec\theta$ is positive, the terminal side is either in QI or in QIV. The added fact that $\sin\theta$ is positive means that the terminal side is in QI. We draw a reference triangle in QI with $r = 3$ and $x = 1$ (Figure 4.34 on the next page); then we use the Pythagorean Theorem to find that $y = \sqrt{3^2 - 1^2} = \sqrt{8}$. (Note that y is positive in QI.)

We then use the definitions to get

$$\cos\theta = \frac{1}{3} \text{ and } \tan\theta = \frac{\sqrt{8}}{1} = \sqrt{8}.$$

(We could also have found $\cos\theta$ directly as the reciprocal of $\sec\theta$.)

(c) Since $\cot\theta$ is undefined, we conclude that $y = 0$ and that θ is a quadrantal angle on the x-axis. The added fact that $\sec\theta$ is negative means that the terminal side is along the negative x-axis. We choose the point $(-1, 0)$ on the terminal side and use the definitions to get

$$\cos\theta = -1 \text{ and } \tan\theta = \frac{0}{-1} = 0.$$

Now try Exercise 43.

Why Not Degrees?

One could actually develop a consistent theory of trigonometric functions based on a rescaled x-axis with "degrees." For example, your graphing calculator will probably produce reasonable-looking graphs in Degree mode. Calculus, however, uses rules that *depend* on radian measure for all trigonometric functions, so it is prudent for precalculus students to become accustomed to that now.

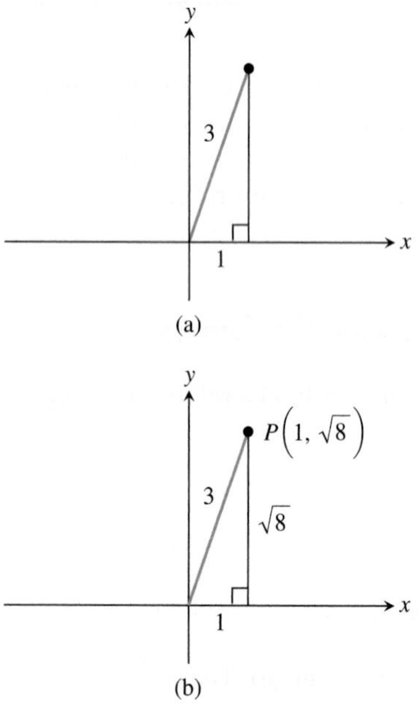

FIGURE 4.34 (Example 7b)

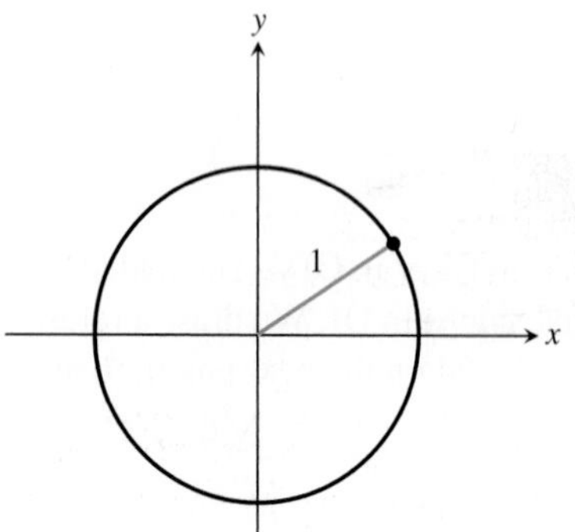

FIGURE 4.35 The unit circle: $x^2 + y^2 = 1$.

Trigonometric Functions of Real Numbers

Now that we have extended the six basic trigonometric functions to apply to any angle, we are ready to appreciate them as functions of real numbers and to study their behavior. First, for reasons discussed in the first section of this chapter, we must agree to measure θ in radian mode so that the real number units of the input will match the real number units of the output.

When considering the trigonometric functions as functions of real numbers, the angles will be measured in radians.

DEFINITION Unit Circle

The **unit circle** is a circle of radius 1 centered at the origin (Figure 4.35).

The unit circle provides an ideal connection between triangle trigonometry and the trigonometric functions. Because arc length along the unit circle corresponds exactly to radian measure, we can use the circle itself as a sort of "number line" for the input values of our functions. This involves the **wrapping function**, which associates points on the number line with points on the circle.

Figure 4.36 shows how the wrapping function works. The real line is placed tangent to the unit circle at the point $(1, 0)$, the point from which we measure angles in standard position. When the line is wrapped around the unit circle in both the positive (counterclockwise) and negative (clockwise) directions, each point t on the real line will fall on a point of the circle that lies on the terminal side of an angle of t radians in standard position. Using the coordinates (x, y) of this point, we can find the six trigonometric ratios for the angle t just as we did in Example 7—except even more easily, since $r = 1$.

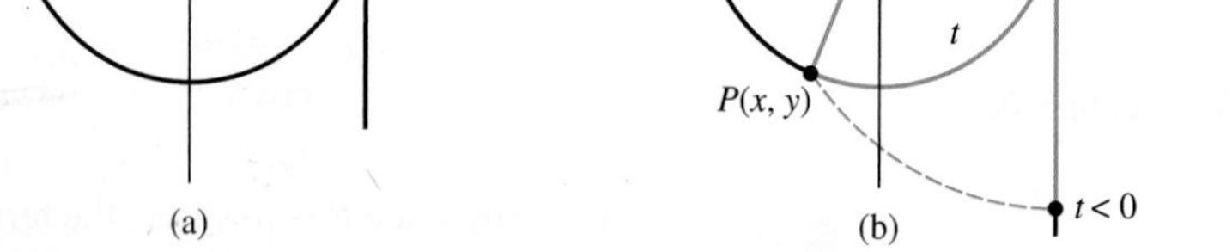

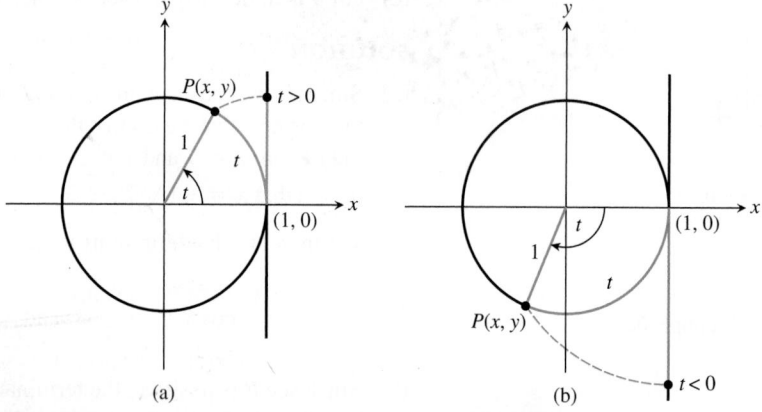

FIGURE 4.36 How the number line is wrapped onto the unit circle. Note that each number t (positive or negative) is "wrapped" to a point P that lies on the terminal side of an angle of t radians in standard position.

DEFINITION Trigonometric Functions of Real Numbers

Let t be any real number, and let $P(x, y)$ be the point corresponding to t when the number line is wrapped onto the unit circle as described above. Then

$$\sin t = y \qquad \csc t = \frac{1}{y}\ (y \neq 0)$$

$$\cos t = x \qquad \sec t = \frac{1}{x}\ (x \neq 0)$$

$$\tan t = \frac{y}{x}\ (x \neq 0) \qquad \cot t = \frac{x}{y}\ (y \neq 0)$$

Therefore, the number t on the number line always wraps onto the point $(\cos t, \sin t)$ on the unit circle (Figure 4.37).

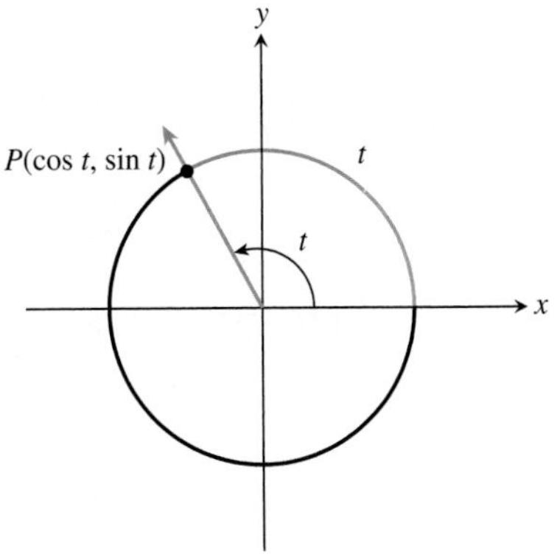

FIGURE 4.37 The real number t always wraps onto the point $(\cos t, \sin t)$ on the unit circle.

Although it is still helpful to draw reference triangles inside the unit circle to see the ratios geometrically, this latest round of definitions does not invoke triangles at all. The real number t determines a point on the unit circle, and the (x, y) coordinates of the point determine the six trigonometric ratios. For this reason, the trigonometric functions when applied to real numbers are usually called the **circular functions**.

EXPLORATION 2 Exploring the Unit Circle

This works well as a group exploration. Get together in groups of two or three and explain to each other why these statements are true. Base your explanations on the unit circle (Figure 4.37). Remember that $-t$ wraps the same distance as t, but in the opposite direction.

1. For any t, the value of $\cos t$ lies between -1 and 1 inclusive.
2. For any t, the value of $\sin t$ lies between -1 and 1 inclusive.
3. The values of $\cos t$ and $\cos(-t)$ are always equal to each other. (Recall that this is the check for an *even* function.)
4. The values of $\sin t$ and $\sin(-t)$ are always opposites of each other. (Recall that this is the check for an *odd* function.)
5. The values of $\sin t$ and $\sin(t + 2\pi)$ are always equal to each other. In fact, that is true of all six trig functions on their domains, and for the same reason.
6. The values of $\sin t$ and $\sin(t + \pi)$ are always opposites of each other. The same is true of $\cos t$ and $\cos(t + \pi)$.
7. The values of $\tan t$ and $\tan(t + \pi)$ are always equal to each other (unless they are both undefined).
8. The sum $(\cos t)^2 + (\sin t)^2$ always equals 1.
9. (Challenge) Can you discover a similar relationship that is not mentioned in our list of eight? There are some to be found.

Periodic Functions

Statements 5 and 7 in Exploration 2 reveal an important property of the circular functions that we need to define for future reference.

DEFINITION Periodic Function

A function $y = f(t)$ is **periodic** if there is a positive number c such that $f(t + c) = f(t)$ for all values of t in the domain of f. The smallest such number c is called the **period** of the function.

Exploration 2 suggests that the sine and cosine functions have period 2π and that the tangent function has period π. We use this periodicity later to model predictably repetitive behavior in the real world, but meanwhile we can also use it to solve little noncalculator training problems such as those found in some of the previous examples in this section.

EXAMPLE 8 Using Periodicity

Find each of the following numbers without a calculator.

(a) $\sin\left(\dfrac{57{,}801\pi}{2}\right)$ **(b)** $\cos(288.45\pi) - \cos(280.45\pi)$

(c) $\tan\left(\dfrac{\pi}{4} - 99{,}999\pi\right)$

(continued)

SOLUTION

(a) $\sin\left(\frac{57{,}801\pi}{2}\right) = \sin\left(\frac{\pi}{2} + \frac{57{,}800\pi}{2}\right) = \sin\left(\frac{\pi}{2} + 28{,}900\pi\right)$

$= \sin\left(\frac{\pi}{2}\right) = 1$

Notice that $28{,}900\pi$ is just a large multiple of 2π, so $\pi/2$ and $((\pi/2) + 28{,}900\pi)$ wrap to the same point on the unit circle, namely $(0, 1)$.

(b) $\cos(288.45\pi) - \cos(280.45\pi) =$

$\cos(280.45\pi + 8\pi) - \cos(280.45\pi) = 0$

Notice that 280.45π and $(280.45\pi + 8\pi)$ wrap to the same point on the unit circle, so the cosine of one is the same as the cosine of the other.

(c) Since the period of the tangent function is π rather than 2π, $99{,}999\pi$ is a large multiple of the period of the tangent function. Therefore,

$$\tan\left(\frac{\pi}{4} - 99{,}999\pi\right) = \tan\left(\frac{\pi}{4}\right) = 1.$$

Now try Exercise 49.

We take a closer look at the properties of the six circular functions in the next two sections.

The 16-Point Unit Circle

At this point you should be able to use reference triangles and quadrantal angles to evaluate trigonometric functions for all integer multiples of 30° or 45° (equivalently, $\pi/6$ radians or $\pi/4$ radians). All of these special values wrap to the 16 special points shown on the unit circle below. Study this diagram until you are confident that you can find the coordinates of these points easily, but avoid using it as a reference when doing problems.

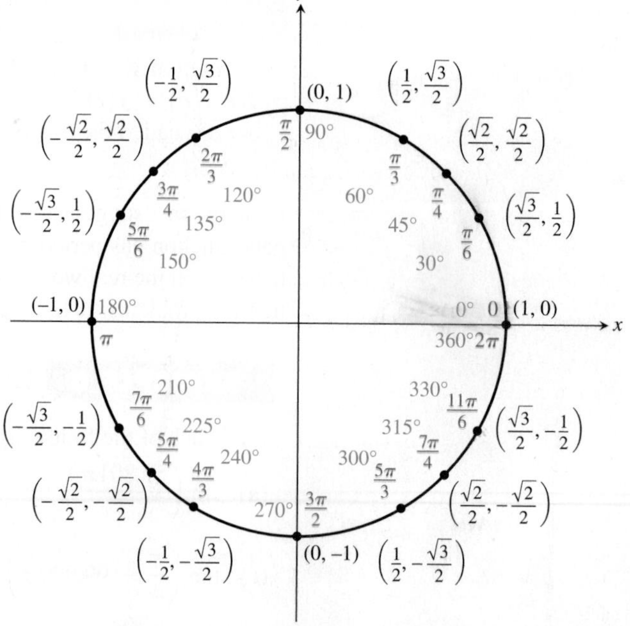

QUICK REVIEW 4.3 *(For help, go to Section 4.1.)*

Exercise numbers with a gray background indicate problems that the authors have designed to be solved *without a calculator*.

In Exercises 1–4, give the value of the angle θ in degrees.

1. $\theta = -\dfrac{\pi}{6}$ **2.** $\theta = -\dfrac{5\pi}{6}$

3. $\theta = \dfrac{25\pi}{4}$ **4.** $\theta = \dfrac{16\pi}{3}$

In Exercises 5–8, use special triangles to evaluate:

5. $\tan \dfrac{\pi}{6}$ **6.** $\cot \dfrac{\pi}{4}$

7. $\csc \dfrac{\pi}{4}$ **8.** $\sec \dfrac{\pi}{3}$

In Exercises 9 and 10, use a right triangle to find the other five trigonometric functions of the *acute* angle θ.

9. $\sin \theta = \dfrac{5}{13}$ **10.** $\cos \theta = \dfrac{15}{17}$

SECTION 4.3 Exercises

In Exercises 1 and 2, identify the one angle that is not coterminal with all the others.

1. 150°, 510°, −210°, 450°, 870°

2. $\dfrac{5\pi}{3}, -\dfrac{5\pi}{3}, \dfrac{11\pi}{3}, -\dfrac{7\pi}{3}, \dfrac{365\pi}{3}$

In Exercises 3–6, evaluate the six trigonometric functions of the angle θ.

3.

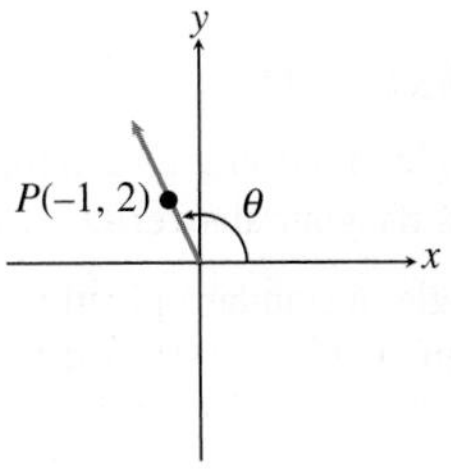

4.

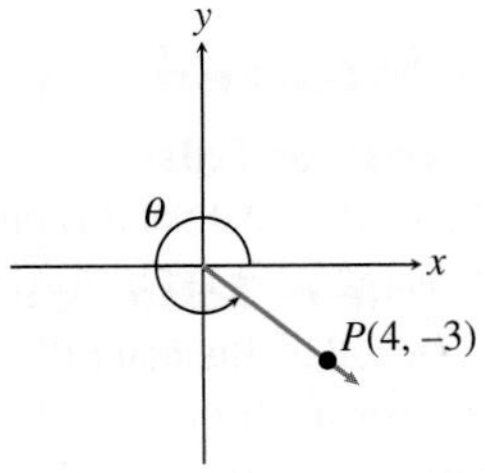

5.

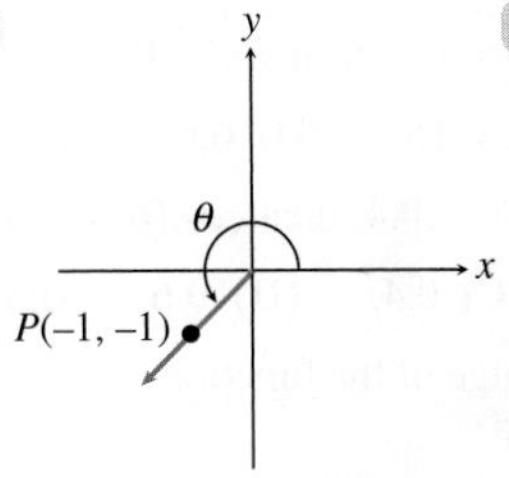

6.

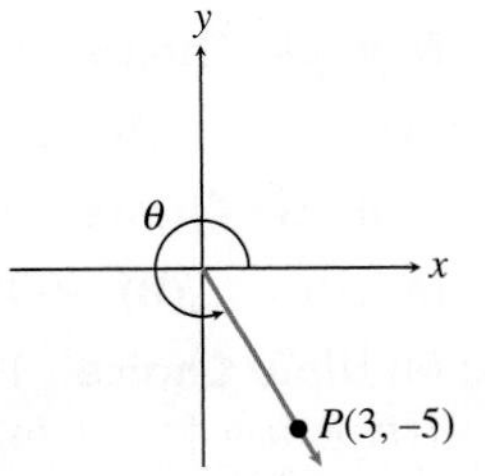

In Exercises 7–12, point P is on the terminal side of angle θ. Evaluate the six trigonometric functions for θ. If the function is undefined, write "undefined."

7. $P(3, 4)$ **8.** $P(-4, -6)$

9. $P(0, 5)$ **10.** $P(-3, 0)$

11. $P(5, -2)$ **12.** $P(22, -22)$

In Exercises 13–16, state the sign (+ or −) of **(a)** $\sin t$, **(b)** $\cos t$, and **(c)** $\tan t$ for values of t in the interval given.

13. $\left(0, \dfrac{\pi}{2}\right)$ **14.** $\left(\dfrac{\pi}{2}, \pi\right)$

15. $\left(\pi, \dfrac{3\pi}{2}\right)$ **16.** $\left(\dfrac{3\pi}{2}, 2\pi\right)$

In Exercises 17–20, determine the sign (+ or −) of the given value without the use of a calculator.

17. $\cos 143°$ **18.** $\tan 192°$

19. $\cos \dfrac{7\pi}{8}$ **20.** $\tan \dfrac{4\pi}{5}$

In Exercises 21–24, choose the point on the terminal side of θ.

21. $\theta = 45°$

(a) $(2, 2)$ **(b)** $(1, \sqrt{3})$ **(c)** $(\sqrt{3}, 1)$

22. $\theta = \dfrac{2\pi}{3}$

(a) $(-1, 1)$ **(b)** $(-1, \sqrt{3})$ **(c)** $(-\sqrt{3}, 1)$

23. $\theta = \dfrac{7\pi}{6}$

(a) $(-\sqrt{3}, -1)$ **(b)** $(-1, \sqrt{3})$ **(c)** $(-\sqrt{3}, 1)$

24. $\theta = -60°$

(a) $(-1, -1)$ **(b)** $(1, -\sqrt{3})$ **(c)** $(-\sqrt{3}, 1)$

In Exercises 25–36, evaluate without using a calculator by using ratios in a reference triangle.

25. $\cos 120°$ **26.** $\tan 300°$

27. $\sec \dfrac{\pi}{3}$ **28.** $\csc \dfrac{3\pi}{4}$

29. $\sin \dfrac{13\pi}{6}$ **30.** $\cos \dfrac{7\pi}{3}$

31. $\tan -\dfrac{15\pi}{4}$ **32.** $\cot \dfrac{13\pi}{4}$

33. $\cos \dfrac{23\pi}{6}$ **34.** $\cos \dfrac{17\pi}{4}$

35. $\sin \dfrac{11\pi}{3}$ **36.** $\cot \dfrac{19\pi}{6}$

In Exercises 37–42, find **(a)** $\sin \theta$, **(b)** $\cos \theta$, and **(c)** $\tan \theta$ for the given quadrantal angle. If the value is undefined, write "undefined."

37. −450° **38.** −270°

39. 7π **40.** $\dfrac{11\pi}{2}$

41. $-\dfrac{7\pi}{2}$ **42.** -4π

In Exercises 43–48, evaluate without using a calculator.

43. Find $\sin\theta$ and $\tan\theta$ if $\cos\theta = \frac{2}{3}$ and $\cot\theta > 0$.

44. Find $\cos\theta$ and $\cot\theta$ if $\sin\theta = \frac{1}{4}$ and $\tan\theta < 0$.

45. Find $\tan\theta$ and $\sec\theta$ if $\sin\theta = -\frac{2}{5}$ and $\cos\theta > 0$.

46. Find $\sin\theta$ and $\cos\theta$ if $\cot\theta = \frac{3}{7}$ and $\sec\theta < 0$.

47. Find $\sec\theta$ and $\csc\theta$ if $\cot\theta = -\frac{4}{3}$ and $\cos\theta < 0$.

48. Find $\csc\theta$ and $\cot\theta$ if $\tan\theta = -\frac{4}{3}$ and $\sin\theta > 0$.

In Exercises 49–52, evaluate by using the period of the function.

49. $\sin\left(\frac{\pi}{6} + 49{,}000\pi\right)$

50. $\tan(1{,}234{,}567\pi) - \tan(7{,}654{,}321\pi)$

51. $\cos\left(\frac{5{,}555{,}555\pi}{2}\right)$

52. $\tan\left(\frac{3\pi - 70{,}000\pi}{2}\right)$

53. Group Activity Use a calculator to evaluate the expressions in Exercises 49–52. Does your calculator give the correct answers? Many calculators miss all four. Give a brief explanation of what probably goes wrong.

54. Writing to Learn Give a convincing argument that the period of $\sin t$ is 2π. That is, show that there is no smaller positive real number p such that $\sin(t + p) = \sin t$ for all real numbers t.

55. Refracted Light Light is *refracted* (bent) as it passes through glass. In the figure, θ_1 is the angle of incidence and θ_2 is the *angle of refraction*. The *index of refraction* is a constant μ that satisfies the equation

$$\sin\theta_1 = \mu\sin\theta_2.$$

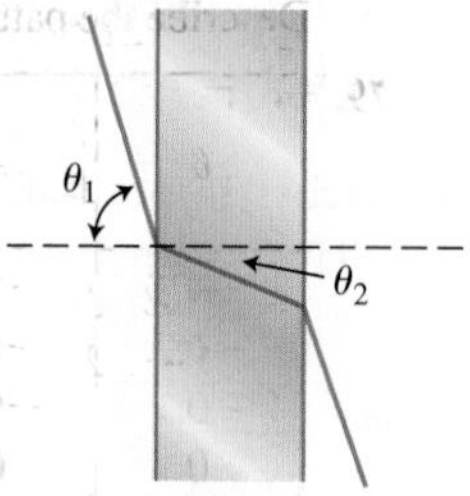

If $\theta_1 = 83°$ and $\theta_2 = 36°$ for a certain piece of flint glass, find the index of refraction.

56. Refracted Light A certain piece of crown glass has an index of refraction of 1.52. If a light ray enters the glass at an angle $\theta_1 = 42°$, what is $\sin\theta_2$?

57. Damped Harmonic Motion A weight suspended from a spring is set into motion. Its displacement d from equilibrium is modeled by the equation

$$d = 0.4e^{-0.2t}\cos 4t,$$

where d is the displacement in inches and t is the time in seconds. Find the displacement at the given time. Use radian mode.

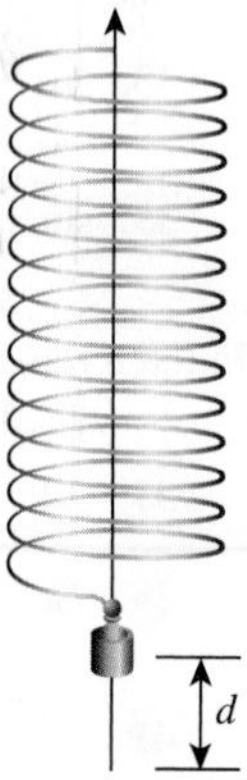

(a) $t = 0$

(b) $t = 3$

58. Swinging Pendulum The Columbus Museum of Science and Industry exhibits a Foucault pendulum 32 ft long that swings back and forth on a cable once in approximately 6 sec. The angle θ (in radians) between the cable and an imaginary vertical line is modeled by the equation

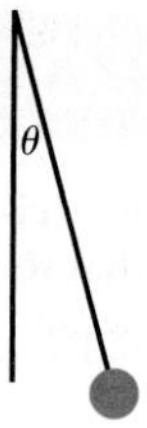

$$\theta = 0.25\cos t.$$

Find the measure of angle θ when $t = 0$ and $t = 2.5$.

59. Too Close for Comfort An F-15 aircraft flying at an altitude of 8000 ft passes directly over a group of vacationers hiking at 7400 ft. If θ is the angle of elevation from the hikers to the F-15, find the distance d from the group to the jet for the given angle.

(a) $\theta = 45°$ **(b)** $\theta = 90°$ **(c)** $\theta = 140°$

60. Manufacturing Swimwear Get Wet, Inc., manufactures swimwear, a seasonal product. The monthly sales x (in thousands) for Get Wet swimsuits are modeled by the equation

$$x = 72.4 + 61.7\sin\frac{\pi t}{6},$$

where $t = 1$ represents January, $t = 2$ February, and so on. Estimate the number of Get Wet swimsuits sold in January, April, June, October, and December. For which two of these months are projected sales the same? Explain why this might be so.

Standardized Test Questions

61. True or False If θ is an angle of a triangle such that $\cos\theta < 0$, then θ is obtuse. Justify your answer.

62. True or False If θ is an angle in standard position determined by the point $(8, -6)$, then $\sin\theta = -0.6$. Justify your answer.

You should answer these questions without using a calculator.

63. Multiple Choice If $\sin\theta = 0.4$, then $\sin(-\theta) + \csc\theta =$

(A) -0.15. **(B)** 0. **(C)** 0.15. **(D)** 0.65. **(E)** 2.1.

64. Multiple Choice If $\cos\theta = 0.4$, then $\cos(\theta + \pi) =$

(A) -0.6. **(B)** -0.4. **(C)** 0.4. **(D)** 0.6. **(E)** 3.54.

65. Multiple Choice The range of the function $f(t) = (\sin t)^2 + (\cos t)^2$ is

(A) $\{1\}$. **(B)** $[-1, 1]$. **(C)** $[0, 1]$.

(D) $[0, 2]$. **(E)** $[0, \infty)$.

66. Multiple Choice If $\cos\theta = -\frac{5}{13}$ and $\tan\theta > 0$, then $\sin\theta =$

(A) $-\frac{12}{13}$. **(B)** $-\frac{5}{12}$. **(C)** $\frac{5}{13}$. **(D)** $\frac{5}{12}$. **(E)** $\frac{12}{13}$.

Explorations

In Exercises 67–70, find the value of the unique real number θ between 0 and 2π that satisfies the two given conditions.

67. $\sin\theta = \frac{1}{2}$ and $\tan\theta < 0$.

68. $\cos\theta = \frac{\sqrt{3}}{2}$ and $\sin\theta < 0$.

69. $\tan\theta = -1$ and $\sin\theta < 0$.

70. $\sin\theta = -\dfrac{\sqrt{2}}{2}$ and $\tan\theta > 0$.

Exercises 71–74 refer to the unit circle in this figure. Point P is on the terminal side of an angle t and point Q is on the terminal side of an angle $t + \pi/2$.

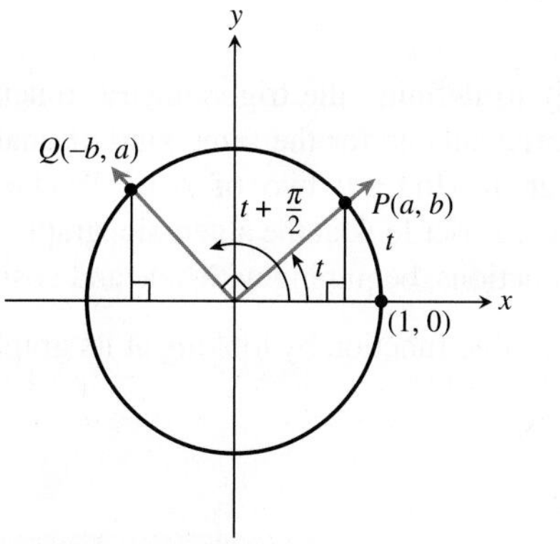

71. Using Geometry in Trigonometry Drop perpendiculars from points P and Q to the x-axis to form two right triangles. Explain how the right triangles are related.

72. Using Geometry in Trigonometry If the coordinates of point P are (a, b), explain why the coordinates of point Q are $(-b, a)$.

73. Explain why $\sin\left(t + \dfrac{\pi}{2}\right) = \cos t$.

74. Explain why $\cos\left(t + \dfrac{\pi}{2}\right) = -\sin t$.

75. Writing to Learn In the figure for Exercises 71–74, t is an angle with radian measure $0 < t < \pi/2$. Draw a similar figure for an angle with radian measure $\pi/2 < t < \pi$ and use it to explain why $\sin(t + \pi/2) = \cos t$.

76. Writing to Learn Use the accompanying figure to explain each of the following.

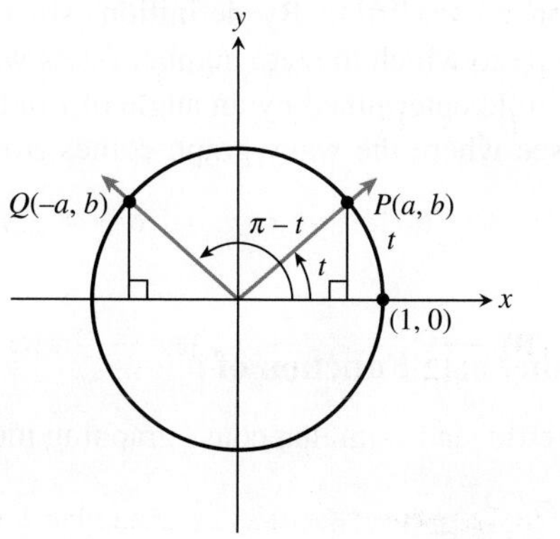

(a) $\sin(\pi - t) = \sin t$ **(b)** $\cos(\pi - t) = -\cos t$

Extending the Ideas

77. Approximation and Error Analysis Use your grapher to complete the table to show that $\sin\theta \approx \theta$ (in radians) when $|\theta|$ is small. Physicists often use the approximation $\sin\theta \approx \theta$ for small values of θ. For what values of θ is the *magnitude of the error* in approximating $\sin\theta$ by θ less than 1% of $\sin\theta$? That is, solve the relation

$$|\sin\theta - \theta| < 0.01|\sin\theta|.$$

(*Hint:* Extend the table to include a column for values of

$$\frac{|\sin\theta - \theta|}{|\sin\theta|}.)$$

θ	$\sin\theta$	$\sin\theta - \theta$
−0.03		
−0.02		
−0.01		
0		
0.01		
0.02		
0.03		

78. Proving a Theorem If t is any real number, prove that $1 + (\tan t)^2 = (\sec t)^2$.

Taylor Polynomials Radian measure allows the trigonometric functions to be approximated by simple polynomial functions. For example, in Exercises 79 and 80, sine and cosine are approximated by Taylor polynomials, named after the English mathematician Brook Taylor (1685–1731). Complete each table showing a Taylor polynomial in the third column. Describe the patterns in the table.

79.

θ	$\sin\theta$	$\theta - \dfrac{\theta^3}{6}$	$\sin\theta - \left(\theta - \dfrac{\theta^3}{6}\right)$
−0.3	−0.295...		
−0.2	−0.198...		
−0.1	−0.099...		
0	0		
0.1	0.099...		
0.2	0.198...		
0.3	0.295...		

80.

θ	$\cos\theta$	$1 - \dfrac{\theta^2}{2} + \dfrac{\theta^4}{24}$	$\cos\theta - \left(1 - \dfrac{\theta^2}{2} + \dfrac{\theta^4}{24}\right)$
−0.3	0.955...		
−0.2	0.980...		
−0.1	0.995...		
0	1		
0.1	0.995...		
0.2	0.980...		
0.3	0.955...		

4.4 Graphs of Sine and Cosine: Sinusoids

What you'll learn about

- The Basic Waves Revisited
- Sinusoids and Transformations
- Modeling Periodic Behavior with Sinusoids

... and why

Sine and cosine gain added significance when used to model waves and periodic behavior.

The Basic Waves Revisited

In the first three sections of this chapter you saw how the trigonometric functions are rooted in the geometry of triangles and circles. It is these connections with geometry that give trigonometric functions their mathematical power and make them widely applicable in many fields.

The unit circle in Section 4.3 was the key to defining the trigonometric functions as functions of real numbers. This makes them available for the same kind of analysis as the other functions introduced in Chapter 1. (Indeed, two of our "Twelve Basic Functions" are trigonometric.) We now take a closer look at the algebraic, graphical, and numerical properties of the trigonometric functions, beginning with sine and cosine.

Recall that we can learn quite a bit about the sine function by looking at its graph. Here is a summary of sine facts:

BASIC FUNCTION The Sine Function

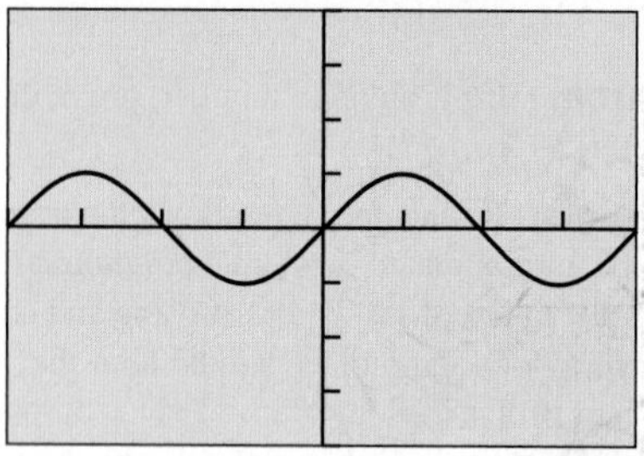

$[-2\pi, 2\pi]$ by $[-4, 4]$

FIGURE 4.38

$f(x) = \sin x$
Domain: $(-\infty, \infty)$
Range: $[-1, 1]$
Continuous
Alternately increasing and decreasing in periodic waves
Symmetric with respect to the origin (odd)
Bounded
Absolute maximum of 1
Absolute minimum of -1
No horizontal asymptotes
No vertical asymptotes
End behavior: $\lim_{x\to-\infty} \sin x$ and $\lim_{x\to\infty} \sin x$ do not exist. (The function values continually oscillate between -1 and 1 and approach no limit.)

We can add to this list that $y = \sin x$ is *periodic*, with period 2π. We can also add understanding of where the sine function comes from: By definition, $\sin t$ is the y-coordinate of the point P on the unit circle to which the real number t gets wrapped (or, equivalently, the point P on the unit circle determined by an angle of t radians in standard position). In fact, now we can see where the wavy graph comes from. Try Exploration 1.

EXPLORATION 1 Graphing sin *t* as a Function of *t*

Set your grapher to radian mode, parametric, and "simultaneous" graphing modes.

Set Tmin $= 0$, Tmax $= 6.3$, Tstep $= \pi/24$.

Set the (x, y) window to $[-1.2, 6.3]$ by $[-2.5, 2.5]$.

Set $X_{1T} = \cos(T)$ and $Y_{1T} = \sin(T)$. This will graph the unit circle. Set $X_{2T} = T$ and $Y_{2T} = \sin(T)$. This will graph $\sin(T)$ as a function of T.

Now start the graph and watch the point go counterclockwise around the unit circle as t goes from 0 to 2π in the positive direction. You will simultaneously see the y-coordinate of the point being graphed as a function of t along the horizontal t-axis. You can clear the drawing and watch the graph as many times as you need to in order to answer the following questions.

1. Where is the point on the unit circle when the wave is at its highest?
2. Where is the point on the unit circle when the wave is at its lowest?
3. Why do both graphs cross the x-axis at the same time?
4. Double the value of Tmax and change the window to $[-2.4, 12.6]$ by $[-5, 5]$. If your grapher can change "style" to show a moving point, choose that style for the unit circle graph. Run the graph and watch how the sine curve tracks the y-coordinate of the point as it moves around the unit circle.
5. Explain from what you have seen why the period of the sine function is 2π.
6. Challenge: Can you modify the grapher settings to show dynamically how the cosine function tracks the x-coordinate as the point moves around the unit circle?

Although a static picture does not do the dynamic simulation justice, Figure 4.39 shows the final screens for the two graphs in Exploration 1.

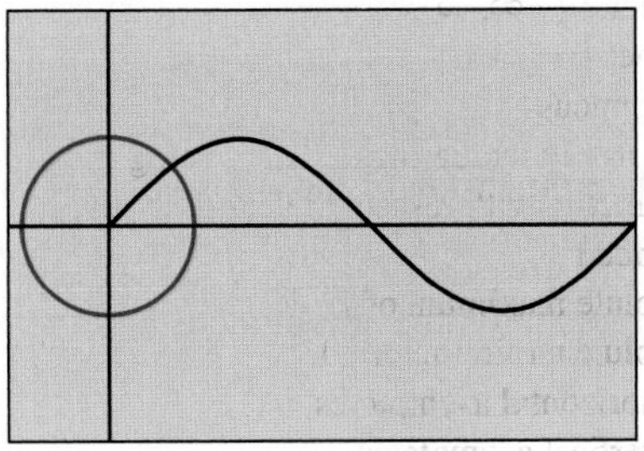

[−1.2, 6.3] by [−2.5, 2.5]
(a)

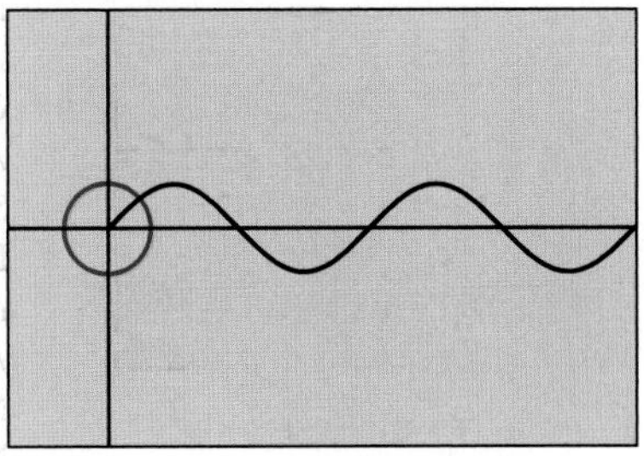

[−2.4, 12.6] by [−5, 5]
(b)

FIGURE 4.39 The graph of $y = \sin t$ tracks the y-coordinate of the point determined by t as it moves around the unit circle.

BASIC FUNCTION The Cosine Function

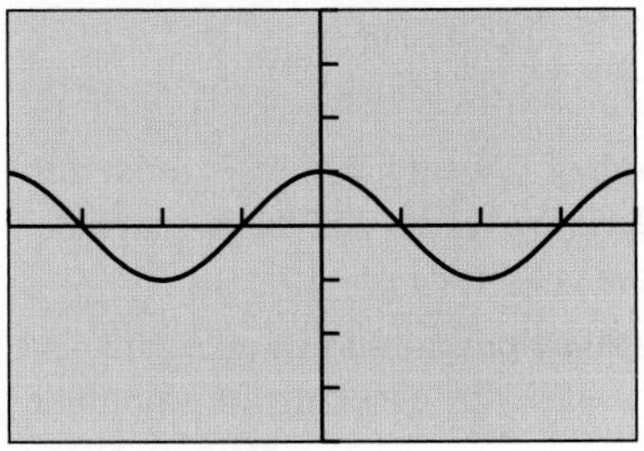

[-2π, 2π] by [−4, 4]

FIGURE 4.40

$f(x) = \cos x$
Domain: $(-\infty, \infty)$
Range: $[-1, 1]$
Continuous
Alternately increasing and decreasing in periodic waves
Symmetric with respect to the y-axis (even)
Bounded
Absolute maximum of 1
Absolute minimum of -1
No horizontal asymptotes
No vertical asymptotes
End behavior: $\lim_{x\to-\infty} \cos x$ and $\lim_{x\to\infty} \cos x$ do not exist. (The function values continually oscillate between -1 and 1 and approach no limit.)

As with the sine function, we can add the observation that it is periodic, with period 2π.

Sinusoids and Transformations

A comparison of the graphs of $y = \sin x$ and $y = \cos x$ suggests that either one can be obtained from the other by a horizontal translation (Section 1.5). In fact, we will prove later in this section that $\cos x = \sin(x + \pi/2)$. Each graph is an example of a *sinusoid*. In general, any transformation of a sine function (or the graph of such a function) is a sinusoid.

DEFINITION Sinusoid

A function is a **sinusoid** if it can be written in the form

$$f(x) = a \sin(bx + c) + d$$

where a, b, c, and d are constants and neither a nor b is 0.

Since cosine functions are themselves translations of sine functions, any transformation of a cosine function is also a sinusoid by the above definition.

There is a special vocabulary used to describe some of our usual graphical transformations when we apply them to sinusoids. Horizontal stretches and shrinks affect the *period* and the *frequency*, vertical stretches and shrinks affect the *amplitude*, and horizontal translations bring about *phase shifts*. All of these terms are associated with waves, and waves are quite naturally associated with sinusoids.

DEFINITION Amplitude of a Sinusoid

The **amplitude** of the sinusoid $f(x) = a \sin(bx + c) + d$ is $|a|$. Similarly, the amplitude of $f(x) = a \cos(bx + c) + d$ is $|a|$.

Graphically, the amplitude is half the height of the wave.

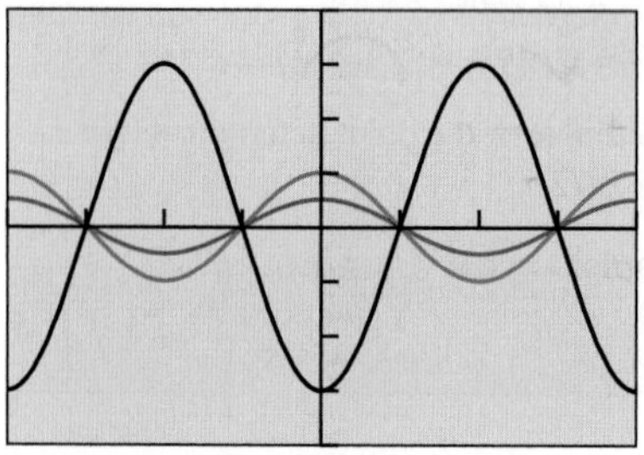

$[-2\pi, 2\pi]$ by $[-4, 4]$

FIGURE 4.41 Sinusoids (in this case, cosine curves) of different amplitudes. (Example 1)

EXAMPLE 1 Vertical Stretch or Shrink and Amplitude

Find the amplitude of each function and use the language of transformations to describe how the graphs are related.

(a) $y_1 = \cos x$ **(b)** $y_2 = \dfrac{1}{2}\cos x$ **(c)** $y_3 = -3\cos x$

SOLUTION

Solve Algebraically The amplitudes are (a) 1, (b) 1/2, and (c) $|-3| = 3$.

The graph of y_2 is a vertical shrink of the graph of y_1 by a factor of 1/2.

The graph of y_3 is a vertical stretch of the graph of y_1 by a factor of 3, and a reflection across the x-axis, performed in either order. (We do not call this a vertical stretch by a factor of -3, nor do we say that the amplitude is -3.)

Support Graphically The graphs of the three functions are shown in Figure 4.41. You should be able to tell which is which quite easily by checking the amplitudes.

Now try Exercise 1.

You learned in Section 1.5 that the graph of $y = f(bx)$ when $|b| > 1$ is a horizontal shrink of the graph of $y = f(x)$ by a factor of $1/|b|$. That is exactly what happens with sinusoids, but we can add the observation that the period shrinks by the same factor. When $|b| < 1$, the effect on both the graph and the period is a horizontal stretch by a factor of $1/|b|$, plus a reflection across the y-axis if $b < 0$.

Period of a Sinusoid

The period of the sinusoid $f(x) = a \sin(bx + c) + d$ is $2\pi/|b|$. Similarly, the period of $f(x) = a \cos(bx + c) + d$ is $2\pi/|b|$.

Graphically, the period is the length of one full cycle of the wave.

EXAMPLE 2 Horizontal Stretch or Shrink and Period

Find the period of each function and use the language of transformations to describe how the graphs are related.

(a) $y_1 = \sin x$ **(b)** $y_2 = -2 \sin\left(\frac{x}{3}\right)$ **(c)** $y_3 = 3 \sin(-2x)$

SOLUTION

Solve Algebraically The periods are (a) 2π, (b) $2\pi/(1/3) = 6\pi$, and (c) $2\pi/|-2| = \pi$.

The graph of y_2 is a horizontal stretch of the graph of y_1 by a factor of 3, a vertical stretch by a factor of 2, and a reflection across the x-axis, performed in any order.

The graph of y_3 is a horizontal shrink of the graph of y_1 by a factor of 1/2, a vertical stretch by a factor of 3, and a reflection across the y-axis, performed in any order. (Note that we do not call this a horizontal shrink by a factor of $-1/2$, nor do we say that the period is $-\pi$.)

Support Graphically The graphs of the three functions are shown in Figure 4.42. You should be able to tell which is which quite easily by checking the periods or the amplitudes. **Now try Exercise 9.**

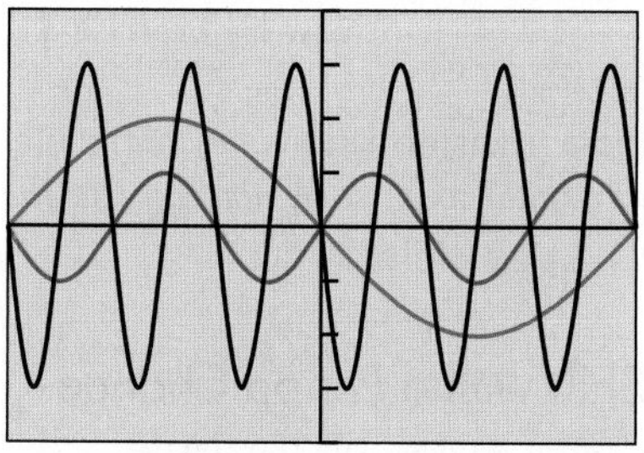

$[-3\pi, 3\pi]$ by $[-4, 4]$

FIGURE 4.42 Sinusoids (in this case, sine curves) of different amplitudes and periods. (Example 2)

In some applications, the *frequency* of a sinusoid is an important consideration. The frequency is simply the reciprocal of the period.

Frequency of a Sinusoid

The **frequency** of the sinusoid $f(x) = a \sin(bx + c) + d$ is $|b|/2\pi$. Similarly, the frequency of $f(x) = a \cos(bx + c) + d$ is $|b|/2\pi$.

Graphically, the frequency is the number of complete cycles the wave completes in a unit interval.

EXAMPLE 3 Finding the Frequency of a Sinusoid

Find the frequency of the function $f(x) = 4 \sin(2x/3)$ and interpret its meaning graphically.

Sketch the graph in the window $[-3\pi, 3\pi]$ by $[-4, 4]$.

SOLUTION The frequency is $(2/3) \div 2\pi = 1/(3\pi)$. This is the reciprocal of the period, which is 3π. The graphical interpretation is that the graph completes 1 full cycle per interval of length 3π. (That, of course, is what having a period of 3π is all about.) The graph is shown in Figure 4.43. **Now try Exercise 17.**

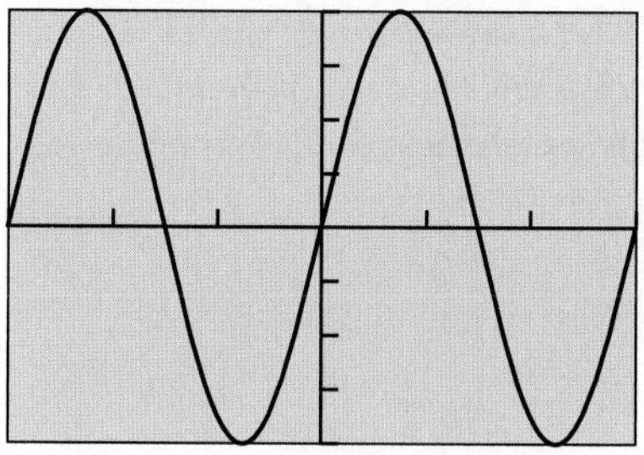

$[-3\pi, 3\pi]$ by $[-4, 4]$

FIGURE 4.43 The graph of the function $f(x) = 4 \sin(2x/3)$. It has frequency $1/(3\pi)$, so it completes 1 full cycle per interval of length 3π. (Example 3)

Recall from Section 1.5 that the graph of $y = f(x + c)$ is a translation of the graph of $y = f(x)$ by c units to the left when $c > 0$. That is exactly what happens with sinusoids, but using terminology with its roots in electrical engineering, we say that the wave undergoes a **phase shift** of $-c$.

EXAMPLE 4 Getting One Sinusoid from Another by a Phase Shift

(a) Write the cosine function as a phase shift of the sine function.

(b) Write the sine function as a phase shift of the cosine function.

SOLUTION

(a) The function $y = \sin x$ has a maximum at $x = \pi/2$, and the function $y = \cos x$ has a maximum at $x = 0$. Therefore, we need to shift the sine curve $\pi/2$ units to the *left* to get the cosine curve:

$$\cos x = \sin(x + \pi/2)$$

(b) It follows from the work in (a) that we need to shift the cosine curve $\pi/2$ units to the right to get the sine curve:

$$\sin x = \cos(x - \pi/2)$$

You can support with your grapher that these statements are true. Incidentally, there are many other translations that would have worked just as well. Adding any integral multiple of 2π to the phase shift would result in the same graph.

Now try Exercise 41.

One note of caution applies when combining these transformations. A horizontal stretch or shrink affects the variable along the horizontal axis, so it *also affects the phase shift*. Consider the transformation in Example 5.

EXAMPLE 5 Combining a Phase Shift with a Period Change

Construct a sinusoid with period $\pi/5$ and amplitude 6 that goes through $(2, 0)$.

SOLUTION To find the coefficient of x, we set $2\pi/|b| = \pi/5$ and solve to find that $b = \pm 10$. We arbitrarily choose $b = 10$. (Either will satisfy the specified conditions.)

For amplitude 6, we have $|a| = 6$. Again, we arbitrarily choose the positive value. The graph of $y = 6 \sin(10x)$ has the required amplitude and period, but it does not go through the point $(2, 0)$. It does, however, go through the point $(0, 0)$, so all that is needed is a phase shift of $+2$ to finish our function. Replacing x by $x - 2$, we get

$$y = 6 \sin(10(x - 2)) = 6 \sin(10x - 20).$$

Notice that we did *not* get the function $y = 6 \sin(10x - 2)$. That function would represent a phase shift of $y = \sin(10x)$, but only by 2/10, not 2. Parentheses are important when combining phase shifts with horizontal stretches and shrinks.

Now try Exercise 59.

Graphs of Sinusoids

The graphs of $y = a \sin(b(x - h)) + k$ and $y = a \cos(b(x - h)) + k$ (where $a \neq 0$ and $b \neq 0$) have the following characteristics:

$$\text{amplitude} = |a|$$

$$\text{period} = \frac{2\pi}{|b|}$$

$$\text{frequency} = \frac{|b|}{2\pi}$$

When compared to the graphs of $y = a \sin bx$ and $y = a \cos bx$, respectively, they also have the following characteristics:

a phase shift of h; a vertical translation of k.

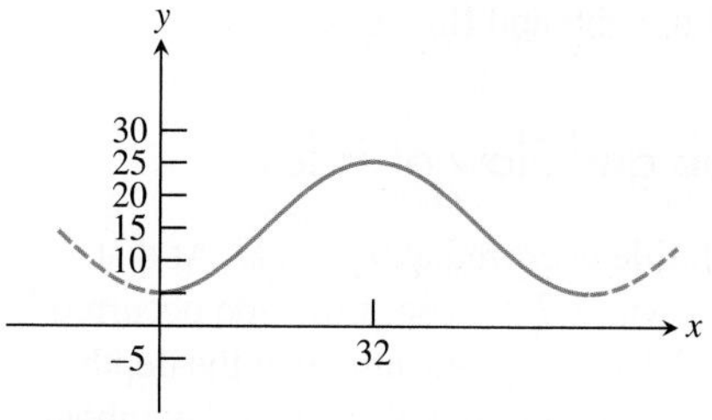

FIGURE 4.44 A sinusoid with specifications. (Example 6)

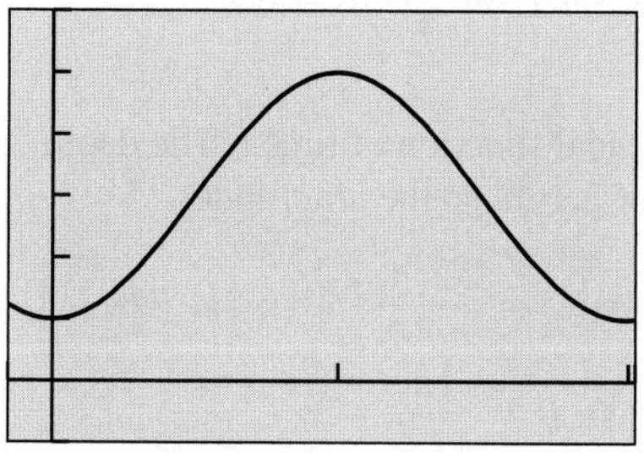

[–5, 65] by [–5, 30]

FIGURE 4.45 The graph of the function $y = -10 \cos((\pi/32)x) + 15$. (Example 8)

EXAMPLE 6 Constructing a Sinusoid by Transformations

Construct a sinusoid $y = f(x)$ that rises from a minimum value of $y = 5$ at $x = 0$ to a maximum value of $y = 25$ at $x = 32$. (See Figure 4.44.)

SOLUTION

Solve Algebraically The amplitude of this sinusoid is half the height of the graph: $(25 - 5)/2 = 10$. So $|a| = 10$. The period is 64 (since a full period goes from minimum to maximum and back down to the minimum). So set $2\pi/|b| = 64$. Solving, we get $|b| = \pi/32$.

We need a sinusoid that takes on its minimum value at $x = 0$. We could shift the graph of sine or cosine horizontally, but it is easier to take the cosine curve (which assumes its *maximum* value at $x = 0$) and turn it upside down. This reflection can be obtained by letting $a = -10$ rather than 10.

So far we have:

$$y = -10 \cos\left(\pm\frac{\pi}{32}x\right)$$

$$= -10 \cos\left(\frac{\pi}{32}x\right) \quad \text{(Since cosine is an even function)}$$

Finally, we note that this function ranges from a minimum of -10 to a maximum of 10. We shift the graph vertically by 15 to obtain a function that ranges from a minimum of 5 to a maximum of 25, as required. Thus

$$y = -10 \cos\left(\frac{\pi}{32}x\right) + 15.$$

Support Graphically We support our answer graphically by graphing the function (Figure 4.45).

Now try Exercise 69.

Modeling Periodic Behavior with Sinusoids

Example 6 was intended as more than just a review of the graphical transformations. Constructing a sinusoid with specific properties is often the key step in modeling physical situations that exhibit periodic behavior over time. The procedure we followed in Example 6 can be summarized as follows:

Constructing a Sinusoidal Model Using Time

1. Determine the maximum value M and minimum value m. The amplitude A of the sinusoid will be $A = \frac{M - m}{2}$, and the vertical shift will be $C = \frac{M + m}{2}$.
2. Determine the period p, the time interval of a single cycle of the periodic function. The horizontal shrink (or stretch) will be $B = \frac{2\pi}{p}$.
3. Choose an appropriate sinusoid based on behavior at some given time T. For example, at time T:

   $f(t) = A \cos(B(t - T)) + C$ attains a maximum value;

   $f(t) = -A \cos(B(t - T)) + C$ attains a minimum value;

   $f(t) = A \sin(B(t - T)) + C$ is halfway between a minimum and a maximum value;

   $f(t) = -A \sin(B(t - T)) + C$ is halfway between a maximum and a minimum value.

We apply the procedure in Example 7 to model the ebb and flow of a tide.

EXAMPLE 7 Calculating the Ebb and Flow of Tides

One particular July 4th in Galveston, TX, high tide occurred at 9:36 A.M. At that time the water at the end of the 61st Street Pier was 2.7 m deep. Low tide occurred at 3:48 P.M., at which time the water was only 2.1 m deep. Assume that the depth of the water is a sinusoidal function of time with a period of half a lunar day (about 12 hr 24 min).

(a) At what time on the 4th of July did the first low tide occur?

(b) What was the approximate depth of the water at 6:00 A.M. and at 3:00 P.M. that day?

(c) What was the first time on July 4th when the water was 2.4 m deep?

SOLUTION

Model We want to model the depth D as a sinusoidal function of time t. The depth varies from a maximum of 2.7 m to a minimum of 2.1 m, so the amplitude $A = \frac{2.7 - 2.1}{2} = 0.3$, and the vertical shift will be $C = \frac{2.7 + 2.1}{2} = 2.4$. The period is 12 hr 24 min, which converts to 12.4 hr, so $B = \frac{2\pi}{12.4} = \frac{\pi}{6.2}$.

We need a sinusoid that assumes its maximum value at 9:36 A.M. (which converts to 9.6 hr after midnight, a convenient time 0). We choose the cosine model. Thus,

$$D(t) = 0.3 \cos\left(\frac{\pi}{6.2}(t - 9.6)\right) + 2.4.$$

Solve Graphically The graph over the 24-hr period of July 4th is shown in Figure 4.46.

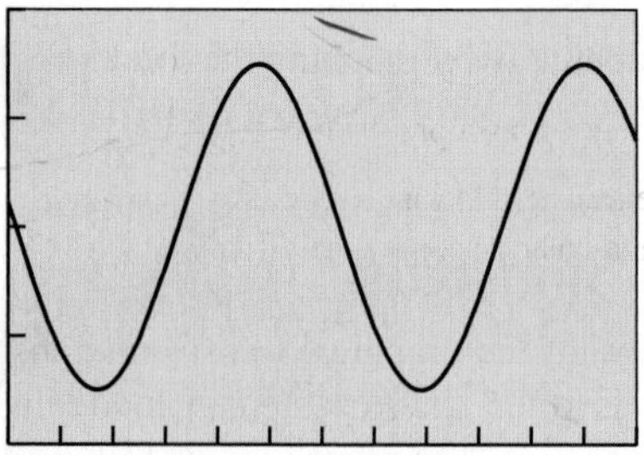

[0, 24] by [2, 2.8]

FIGURE 4.46 The Galveston tide graph. (Example 7)

We now use the graph to answer the questions posed.

(a) The first low tide corresponds to the first local minimum on the graph. We find graphically that this occurs at $t = 3.4$. This translates to $3 + (0.4)(60) = 3{:}24$ A.M.

(b) The depth at 6:00 A.M. is $D(6) \approx 2.32$ m. The depth at 3:00 P.M. is $D(12 + 3) = D(15) \approx 2.12$ m.

(c) The first time the water is 2.4 m deep corresponds to the leftmost intersection of the sinusoid with the line $y = 2.4$. We use the grapher to find that $t = 0.3$. This translates to $0 + (0.3)(60) = 00{:}18$ A.M., which we write as 12:18 A.M.

Now try Exercise 75.

We will see more applications of this kind when we look at *simple harmonic motion* in Section 4.8.

QUICK REVIEW 4.4 *(For help, go to Sections 1.6, 4.1, and 4.2.)*

Exercise numbers with a gray background indicate problems that the authors have designed to be solved *without a calculator*.

In Exercises 1–3, state the sign (positive or negative) of the function in each quadrant.

1. $\sin x$ **2.** $\cos x$

3. $\tan x$

In Exercises 4–6, give the radian measure of the angle.

4. 135° **5.** −150° **6.** 450°

In Exercises 7–10, find a transformation that will transform the graph of y_1 to the graph of y_2.

7. $y_1 = \sqrt{x}$ and $y_2 = 3\sqrt{x}$

8. $y_1 = e^x$ and $y_2 = e^{-x}$

9. $y_1 = \ln x$ and $y_2 = 0.5 \ln x$

10. $y_1 = x^3$ and $y_2 = x^3 - 2$

SECTION 4.4 Exercises

In Exercises 1–6, find the amplitude of the function and use the language of transformations to describe how the graph of the function is related to the graph of $y = \sin x$.

1. $y = 2 \sin x$ **2.** $y = \frac{2}{3} \sin x$

3. $y = -4 \sin x$ **4.** $y = -\frac{7}{4} \sin x$

5. $y = 0.73 \sin x$ **6.** $y = -2.34 \sin x$

In Exercises 7–12, find the period of the function and use the language of transformations to describe how the graph of the function is related to the graph of $y = \cos x$.

7. $y = \cos 3x$ **8.** $y = \cos x/5$

9. $y = \cos(-7x)$ **10.** $y = \cos(-0.4x)$

11. $y = 3 \cos 2x$ **12.** $y = \frac{1}{4} \cos \frac{2x}{3}$

In Exercises 13–16, find the amplitude, period, and frequency of the function and use this information (not your calculator) to sketch a graph of the function in the window $[-3\pi, 3\pi]$ by $[-4, 4]$.

13. $y = 3 \sin \frac{x}{2}$ **14.** $y = 2 \cos \frac{x}{3}$

15. $y = -\frac{3}{2} \sin 2x$ **16.** $y = -4 \sin \frac{2x}{3}$

In Exercises 17–22, graph one period of the function. Use your understanding of transformations, not your graphing calculators. Be sure to show the scale on both axes.

17. $y = 2 \sin x$ **18.** $y = 2.5 \sin x$

19. $y = 3 \cos x$ **20.** $y = -2 \cos x$

21. $y = -0.5 \sin x$ **22.** $y = 4 \cos x$

In Exercises 23–28, graph three periods of the function. Use your understanding of transformations, not your graphing calculators. Be sure to show the scale on both axes.

23. $y = 5 \sin 2x$ **24.** $y = 3 \cos \frac{x}{2}$

25. $y = 0.5 \cos 3x$ **26.** $y = 20 \sin 4x$

27. $y = 4 \sin \frac{x}{4}$ **28.** $y = 8 \cos 5x$

In Exercises 29–34, specify the period and amplitude of each function. Then give the viewing window in which the graph is shown. Use your understanding of transformations, not your graphing calculators.

29. $y = 1.5 \sin 2x$

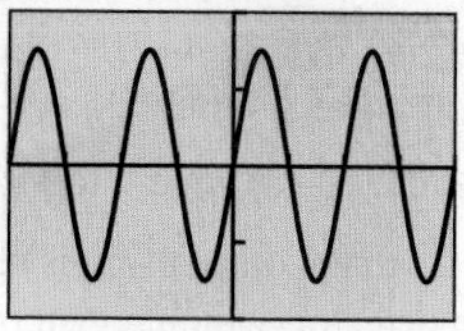

30. $y = 2 \cos 3x$

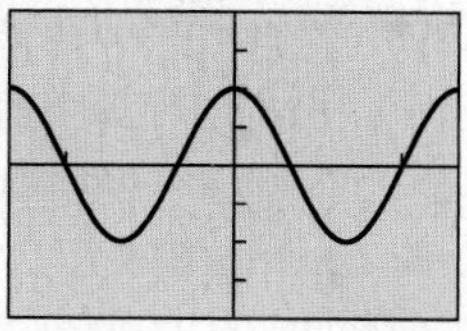

31. $y = -3 \cos 2x$

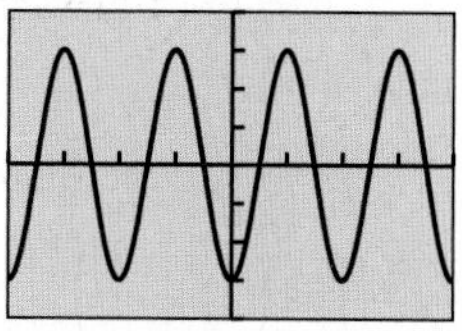

32. $y = 5 \sin \frac{x}{2}$

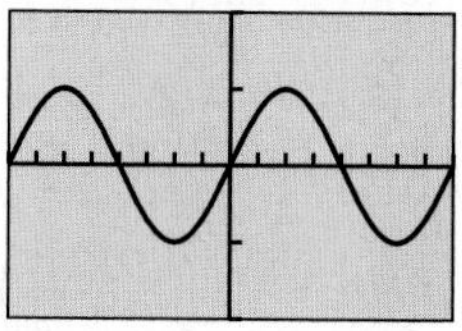

33. $y = -4 \sin \frac{\pi}{3} x$

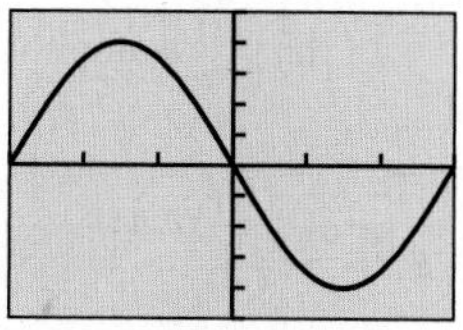

34. $y = 3 \cos \pi x$

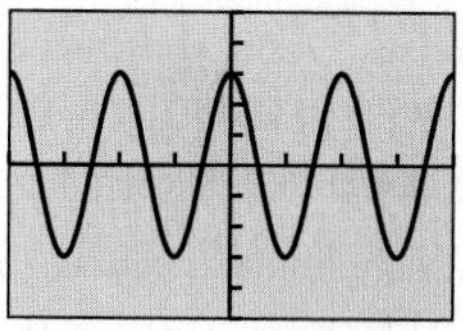

In Exercises 35–40, identify the maximum and minimum values and the zeros of the function in the interval $[-2\pi, 2\pi]$. Use your understanding of transformations, not your graphing calculators.

35. $y = 2 \sin x$ **36.** $y = 3 \cos \frac{x}{2}$

37. $y = \cos 2x$ **38.** $y = \frac{1}{2} \sin x$

39. $y = -\cos 2x$ **40.** $y = -2 \sin x$

41. Write the function $y = -\sin x$ as a phase shift of $y = \sin x$.

42. Write the function $y = -\cos x$ as a phase shift of $y = \sin x$.

In Exercises 43–48, describe the transformations required to obtain the graph of the given function from a basic trigonometric graph.

43. $y = 0.5 \sin 3x$ **44.** $y = 1.5 \cos 4x$

45. $y = -\frac{2}{3} \cos \frac{x}{3}$ **46.** $y = \frac{3}{4} \sin \frac{x}{5}$

47. $y = 3 \cos \frac{2\pi x}{3}$ **48.** $y = -2 \sin \frac{\pi x}{4}$

In Exercises 49–52, describe the transformations required to obtain the graph of y_2 from the graph of y_1.

49. $y_1 = \cos 2x$ and $y_2 = \frac{5}{3} \cos 2x$

50. $y_1 = 2 \cos\left(x + \frac{\pi}{3}\right)$ and $y_2 = \cos\left(x + \frac{\pi}{4}\right)$

51. $y_1 = 2 \cos \pi x$ and $y_2 = 2 \cos 2\pi x$

52. $y_1 = 3 \sin \frac{2\pi x}{3}$ and $y_2 = 2 \sin \frac{\pi x}{3}$

In Exercises 53–56, select the pair of functions that have identical graphs.

53. **(a)** $y = \cos x$ **(b)** $y = \sin\left(x + \frac{\pi}{2}\right)$

(c) $y = \cos\left(x + \frac{\pi}{2}\right)$

54. (a) $y = \sin x$ (b) $y = \cos\left(x - \frac{\pi}{2}\right)$

(c) $y = \cos x$

55. (a) $y = \sin\left(x + \frac{\pi}{2}\right)$ (b) $y = -\cos(x - \pi)$

(c) $y = \cos\left(x - \frac{\pi}{2}\right)$

56. (a) $y = \sin\left(2x + \frac{\pi}{4}\right)$ (b) $y = \cos\left(2x - \frac{\pi}{2}\right)$

(c) $y = \cos\left(2x - \frac{\pi}{4}\right)$

In Exercises 57–60, construct a sinusoid with the given amplitude and period that goes through the given point.

57. Amplitude 3, period π, point $(0, 0)$

58. Amplitude 2, period 3π, point $(0, 0)$

59. Amplitude 1.5, period $\pi/6$, point $(1, 0)$

60. Amplitude 3.2, period $\pi/7$, point $(5, 0)$

In Exercises 61–68, state the amplitude and period of the sinusoid, and (relative to the basic function) the phase shift and vertical translation.

61. $y = -2\sin\left(x - \frac{\pi}{4}\right) + 1$

62. $y = -3.5\sin\left(2x - \frac{\pi}{2}\right) - 1$

63. $y = 5\cos\left(3x - \frac{\pi}{6}\right) + 0.5$

64. $y = 3\cos(x + 3) - 2$

65. $y = 2\cos 2\pi x + 1$

66. $y = 4\cos 3\pi x - 2$

67. $y = \frac{7}{3}\sin\left(x + \frac{5}{2}\right) - 1$

68. $y = \frac{2}{3}\cos\left(\frac{x - 3}{4}\right) + 1$

In Exercises 69 and 70, find values a, b, h, and k so that the graph of the function $y = a\sin(b(x + h)) + k$ is the curve shown.

69.

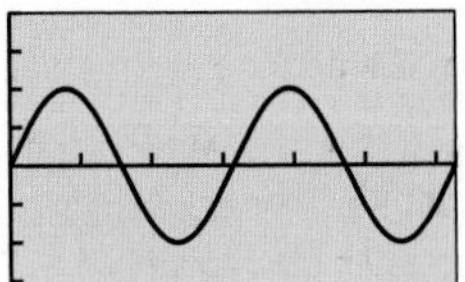

[0, 6.28] by [–4, 4]

70.

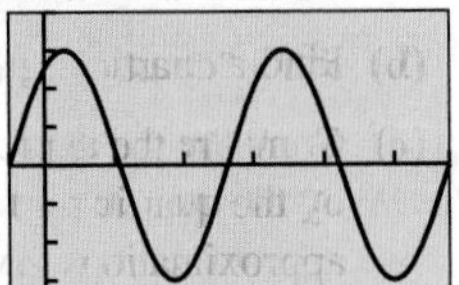

[–0.5, 5.78] by [–4, 4]

71. Points of Intersection Graph $y = 1.3^{-x}$ and $y = 1.3^{-x}\cos x$ for x in the interval $[-1, 8]$.

(a) How many points of intersection do there appear to be?

(b) Find the coordinates of each point of intersection.

72. Motion of a Buoy A signal buoy in the Chesapeake Bay bobs up and down with the height h of its transmitter (in feet) above sea level modeled by $h = a\sin bt + 5$. During a small squall its height varies from 1 ft to 9 ft and there are 3.5 sec from one 9-ft height to the next. What are the values of the constants a and b?

73. Ferris Wheel A Ferris wheel 50 ft in diameter makes one revolution every 40 sec. If the center of the wheel is 30 ft above the ground, how long after reaching the low point is a rider 50 ft above the ground?

74. Tsunami Wave An earthquake occurred at 9:40 A.M. on Nov. 1, 1755, at Lisbon, Portugal, and started a *tsunami* (often called a tidal wave) in the ocean. It produced waves that traveled more than 540 ft/sec (370 mph) and reached a height of 60 ft. If the period of the waves was 30 min or 1800 sec, estimate the length L between the crests.

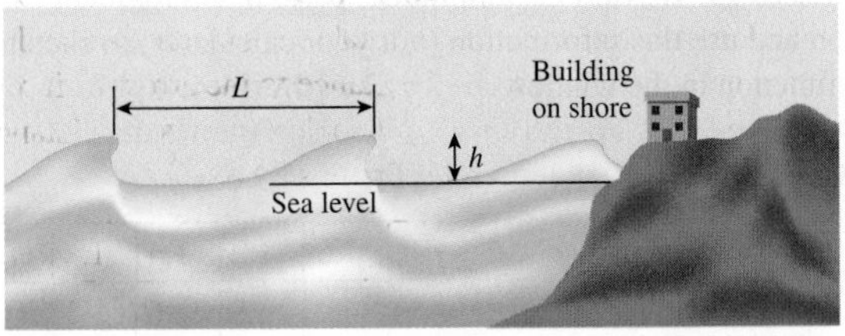

75. Ebb and Flow On a particular Labor Day, the high tide in Southern California occurs at 7:12 A.M. At that time you measure the water at the end of the Santa Monica Pier to be 11 ft deep. At 1:24 P.M. it is low tide, and you measure the water to be only 7 ft deep. Assume the depth of the water is a sinusoidal function of time with a period of 1/2 a lunar day, which is about 12 hr 24 min.

(a) At what time on that Labor Day does the first low tide occur?

(b) What was the approximate depth of the water at 4:00 A.M. and at 9:00 P.M.?

(c) What is the first time on that Labor Day that the water is 9 ft deep?

76. Blood Pressure The function

$$P = 120 + 30\sin 2\pi t$$

models the blood pressure (in millimeters of mercury) for a person who has a (high) blood pressure of 150/90; t represents time in seconds.

(a) What is the period of this function?

(b) How many heartbeats are there each minute?

(c) Graph this function to model a 10-sec time interval.

77. Bouncing Block A block mounted on a spring is set into motion directly above a motion detector, which registers the distance to the block at intervals of 0.1 sec. When the block is released, it is 7.2 cm above the motion detector. The table below shows the data collected by the motion detector during the first 2 sec, with distance d measured in centimeters:

(a) Make a scatter plot of d as a function of t and estimate the maximum d visually. Use this number and the given minimum (7.2) to compute the amplitude of the block's motion.

(b) Estimate the period of the block's motion visually from the scatter plot.

(c) Model the motion of the block as a sinusoidal function $d(t)$.

(d) Graph your function with the scatter plot to support your model graphically.

t	0.1	0.2	0.3	0.4	0.5	0.6	0.7	0.8	0.9	1.0
d	9.2	13.9	18.8	21.4	20.0	15.6	10.5	7.4	8.1	12.1

t	1.1	1.2	1.3	1.4	1.5	1.6	1.7	1.8	1.9	2.0
d	17.3	20.8	20.8	17.2	12.0	8.1	7.5	10.5	15.6	19.9

78. LP Turntable A suction-cup–tipped arrow is secured vertically to the outer edge of a turntable designed for playing LP phonograph records (ask your parents). A motion detector is situated 60 cm away. The turntable is switched on and a motion detector measures the distance to the arrow as it revolves around the turntable. The table below shows the distance d as a function of time during the first 4 sec.

t	0.2	0.4	0.6	0.8	1.0	1.2	1.4	1.6	1.8	2.0
d	63.5	71.6	79.8	84.7	84.7	79.8	71.6	63.5	60.0	63.5

t	2.2	2.4	2.6	2.8	3.0	3.2	3.4	3.6	3.8	4.0
d	71.6	79.8	84.7	84.7	79.8	71.6	63.5	60.0	63.5	71.6

(a) If the turntable is 25.4 cm in diameter, find the amplitude of the arrow's motion.

(b) Find the period of the arrow's motion by analyzing the data.

(c) Model the motion of the arrow as a sinusoidal function $d(t)$.

(d) Graph your function with a scatter plot to support your model graphically.

79. Temperature Data The average monthly Fahrenheit temperatures in Albuquerque, NM, for 2012 are shown in the table below (month 1 = Jan, month 2 = Feb, etc.):

Month	1	2	3	4	5	6	7	8	9	10	11	12
Temp	34	40	47	55	64	74	79	76	69	57	44	35

Source: Climate-zone.com, 2013.

Model the temperature T as a sinusoidal function of time, using 34 as the minimum value and 79 as the maximum value. Support your answer graphically by graphing your function with a scatter plot.

80 Temperature Data The average monthly Fahrenheit temperatures in Helena, MT, for 2012 are shown in the table below (month 1 = Jan, month 2 = Feb, etc.):

Month	1	2	3	4	5	6	7	8	9	10	11	12
Temp	20	26	34	43	53	62	69	67	55	45	32	21

Source: Climate-zone.com, 2013.

Model the temperature T as a sinusoidal function of time, using 20 as the minimum value and 69 as the maximum value. Support your answer graphically by graphing your function with a scatter plot.

Standardized Test Questions

81. True or False The graph of $y = \sin 2x$ has half the period of the graph of $y = \sin 4x$. Justify your answer.

82. True or False Every sinusoid can be written as $y = A \cos (Bx + C)$ for some real numbers A, B, and C. Justify your answer.

You may use a graphing calculator when answering these questions.

83. Multiple Choice A sinusoid with amplitude 4 has a minimum value of 5. Its maximum value is

(A) 7. (B) 9. (C) 11.
(D) 13. (E) 15.

84. Multiple Choice The graph of $y = f(x)$ is a sinusoid with period 45 passing through the point (6, 0). Which of the following can be determined from the given information?

I. $f(0)$ **II.** $f(6)$ **III.** $f(96)$

(A) I only (B) II only
(C) I and III only (D) II and III only
(E) I, II, and III only

85. Multiple Choice The period of the function $f(x) = 210 \sin (420x + 840)$ is

(A) $\pi/840$. (B) $\pi/420$. (C) $\pi/210$.
(D) $210/\pi$. (E) $420/\pi$.

86. Multiple Choice The number of solutions to the equation $\sin (2000x) = 3/7$ in the interval $[0, 2\pi]$ is

(A) 1000. (B) 2000. (C) 4000.
(D) 6000. (E) 8000.

Explorations

87. Approximating Cosine

(a) Draw a scatter plot $(x, \cos x)$ for the 17 special angles x, where $-\pi \le x \le \pi$.

(b) Find a quartic regression for the data.

(c) Compare the approximation to the cosine function given by the quartic regression with the Taylor polynomial approximations given in Exercise 80 of Section 4.3.

88. Approximating Sine

(a) Draw a scatter plot $(x, \sin x)$ for the 17 special angles x, where $-\pi \le x \le \pi$.

(b) Find a cubic regression for the data.

(c) Compare the approximation to the sine function given by the cubic regression with the Taylor polynomial approximations given in Exercise 79 of Section 4.3.

89. Visualizing a Musical Note A piano tuner strikes a tuning fork for the note middle C and creates a sound wave that can be modeled by

$$y = 1.5 \sin 524\pi t,$$

where t is the time in seconds.

(a) What is the period p of this function?

(b) What is the frequency $f = 1/p$ of this note?

(c) Graph the function.

90. Writing to Learn In a certain video game a cursor bounces back and forth horizontally across the screen at a constant rate. Its distance d from the center of the screen varies with time t and hence can be described as a function of t. Explain why this horizontal distance d from the center of the screen *does not vary* according to an equation $d = a \sin bt$, where t represents time in seconds. You may find it helpful to include a graph in your explanation.

91. Group Activity Using only integer values of a and b between 1 and 9 inclusive, look at graphs of functions of the form

$$y = \sin(ax)\cos(bx) - \cos(ax)\sin(bx)$$

for various values of a and b. (A group can look at more graphs at a time than one person can.)

(a) Some values of a and b result in the graph of $y = \sin x$. Find a general rule for such values of a and b.

(b) Some values of a and b result in the graph of $y = \sin 2x$. Find a general rule for such values of a and b.

(c) Can you guess which values of a and b will result in the graph of $y = \sin kx$ for an arbitrary integer k?

92. Group Activity Using only integer values of a and b between 1 and 9 inclusive, look at graphs of functions of the form

$$y = \cos(ax)\cos(bx) + \sin(ax)\sin(bx)$$

for various values of a and b. (A group can look at more graphs at a time than one person can.)

(a) Some values of a and b result in the graph of $y = \cos x$. Find a general rule for such values of a and b.

(b) Some values of a and b result in the graph of $y = \cos 2x$. Find a general rule for such values of a and b.

(c) Can you guess which values of a and b will result in the graph of $y = \cos kx$ for an arbitrary integer k?

Extending the Ideas

In Exercises 93–96, the graphs of the sine and cosine functions are waveforms like the figure below. By correctly labeling the coordinates of points A, B, and C, you will get the graph of the function given.

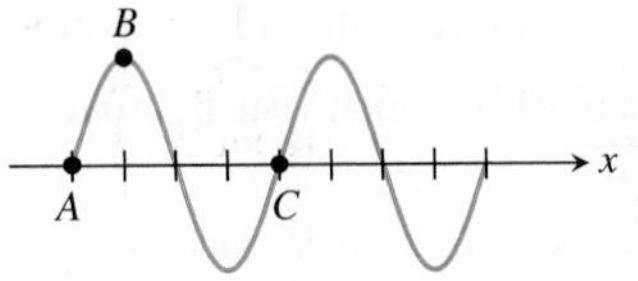

93. $y = 3 \cos 2x$ and $A = \left(-\frac{\pi}{4}, 0\right)$. Find B and C.

94. $y = 4.5 \sin\left(x - \frac{\pi}{4}\right)$ and $A = \left(\frac{\pi}{4}, 0\right)$. Find B and C.

95. $y = 2 \sin\left(3x - \frac{\pi}{4}\right)$ and $A = \left(\frac{\pi}{12}, 0\right)$. Find B and C.

96. $y = 3 \sin(2x - \pi)$, and A is the first x-intercept on the right of the y-axis. Find A, B, and C.

97. The Ultimate Sinusoidal Equation It is an interesting fact that any sinusoid can be written in the form

$$y = a \sin[b(x - H)] + k,$$

where both a and b are positive numbers.

(a) Explain why you can assume b is positive. (*Hint:* Replace b by $-B$ and simplify.)

(b) Use one of the horizontal translation identities to prove that the equation

$$y = a \cos[b(x - h)] + k$$

has the same graph as

$$y = a \sin[b(x - H)] + k$$

for a correctly chosen value of H. Explain how to choose H.

(c) Give a unit circle argument for the identity $\sin(\theta + \pi) = -\sin\theta$. Support your unit circle argument graphically.

(d) Use the identity from (c) to prove that

$$y = -a \sin[b(x - h)] + k, a > 0$$

has the same graph as

$$y = a \sin[b(x - H)] + k, a > 0$$

for a correctly chosen value of H. Explain how to choose H.

(e) Combine your results from (a)–(d) to prove that any sinusoid can be represented by the equation

$$y = a \sin[b(x - H)] + k$$

where a and b are both positive.

4.5 Graphs of Tangent, Cotangent, Secant, and Cosecant

What you'll learn about

- The Tangent Function
- The Cotangent Function
- The Secant Function
- The Cosecant Function

... and why

This will give us functions for the remaining trigonometric ratios.

The Tangent Function

The graph of the tangent function is shown below. As with the sine and cosine graphs, this graph tells us quite a bit about the function's properties. Here is a summary of tangent facts:

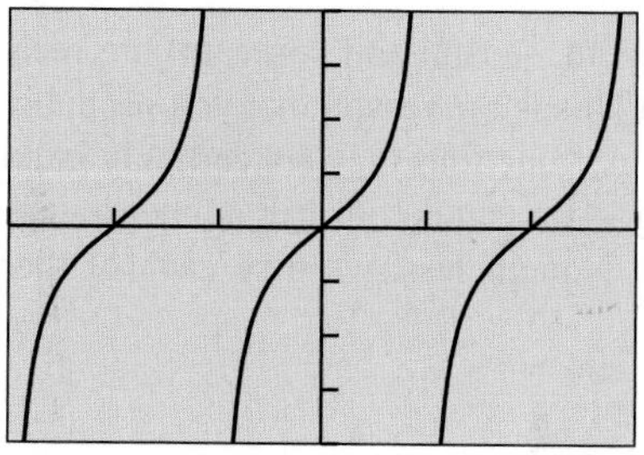

$[-3\pi/2, 3\pi/2]$ by $[-4, 4]$

FIGURE 4.47

The Tangent Function

$f(x) = \tan x$
Domain: All real numbers except odd multiples of $\pi/2$
Range: $(-\infty, \infty)$
Continuous (i.e., continuous on its domain)
Increasing on each interval in its domain
Symmetric with respect to the origin (odd)
Not bounded above or below
No local extrema
No horizontal asymptotes
Vertical asymptotes: $x = k \cdot (\pi/2)$ for all odd integers k
End behavior: $\lim_{x \to -\infty} \tan x$ and $\lim_{x \to \infty} \tan x$ do not exist. (The function values continually oscillate between $-\infty$ and ∞ and approach no limit.)

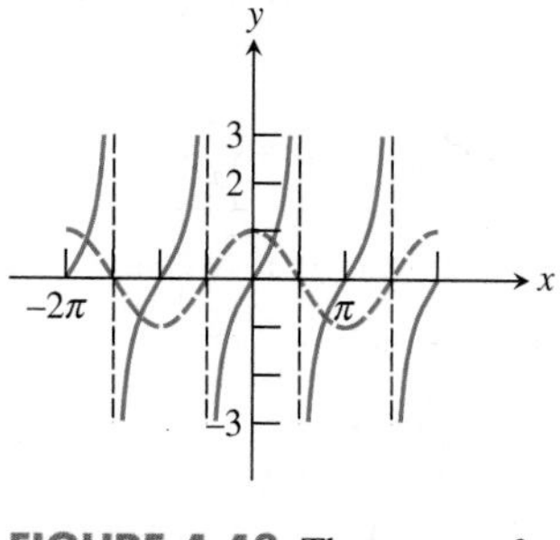

FIGURE 4.48 The tangent function has asymptotes at the zeros of cosine.

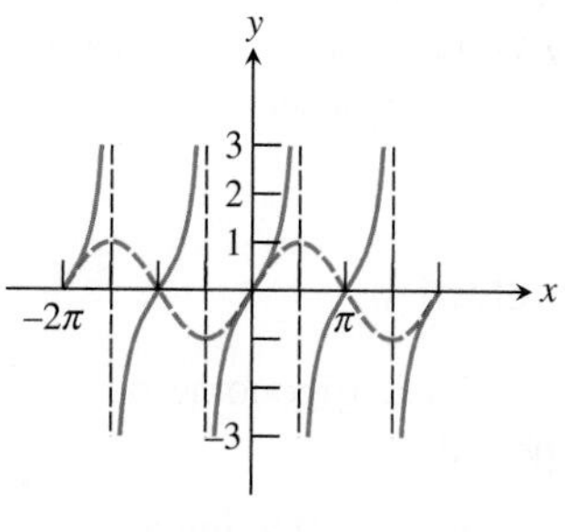

FIGURE 4.49 The tangent function has zeros at the zeros of sine.

We now analyze why the graph of $f(x) = \tan x$ behaves the way it does. It follows from the definitions of the trigonometric functions (Section 4.2) that

$$\tan x = \frac{\sin x}{\cos x}.$$

Unlike the sinusoids, the tangent function has a denominator that might be zero for some values of x, which makes the function undefined at those values. Not only does this actually happen, but it happens an infinite number of times: at all the values of x for which $\cos x = 0$. That is why the tangent function has vertical asymptotes at those values (Figure 4.48). The zeros of the tangent function are the same as the zeros of the sine function: all the integer multiples of π (Figure 4.49).

Because $\sin x$ and $\cos x$ are both periodic with period 2π, you might expect the period of the tangent function to be 2π also. The graph shows, however, that it is π.

The constants a, b, h, and k influence the behavior of $y = a \tan (b(x - h)) + k$ in much the same way that they do for the graph of $y = a \sin (b(x - h)) + k$. The constant a yields a vertical stretch or shrink, b affects the period, h causes a horizontal translation, and k causes a vertical translation. The terms *amplitude* and *phase shift*, however, are not used, as they apply only to sinusoids.

EXAMPLE 1 Graphing a Tangent Function

Describe the graph of the function $y = -\tan 2x$ in terms of a basic trigonometric function. Locate the vertical asymptotes and graph four periods of the function.

SOLUTION The effect of the 2 is a horizontal shrink of the graph of $y = \tan x$ by a factor of 1/2, and the effect of the -1 is a reflection across the x-axis. Since the vertical asymptotes of $y = \tan x$ are all odd multiples of $\pi/2$, the shrink factor causes the vertical asymptotes of $y = \tan 2x$ to be all odd multiples of $\pi/4$ (Figure 4.50a). The reflection across the x-axis (Figure 4.50b) does not change the asymptotes.

Since the period of the function $y = \tan x$ is π, the period of the function $y = -\tan 2x$ is (thanks again to the shrink factor) $\pi/2$. Thus, any window of horizontal length 2π will show four periods of the graph. Figure 4.50b uses the window $[-\pi, \pi]$ by $[-4, 4]$.

Now try Exercise 5.

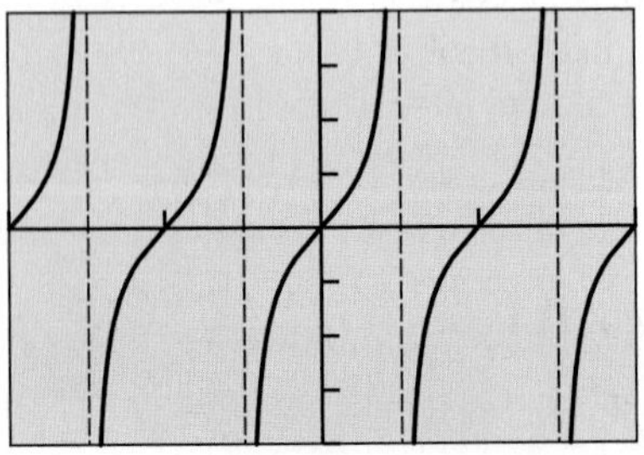

$[-\pi, \pi]$ by $[-4, 4]$

(a)

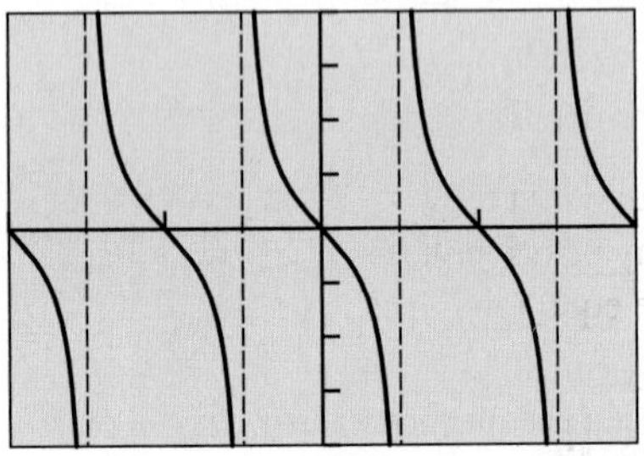

$[-\pi, \pi]$ by $[-4, 4]$

(b)

FIGURE 4.50 The graph of (a) $y = \tan 2x$ is reflected across the x-axis to produce the graph of (b) $y = -\tan 2x$. (Example 1)

The other three trigonometric functions (cotangent, secant, and cosecant) are reciprocals of tangent, cosine, and sine, respectively. (This is the reason that you probably do not have buttons for them on your calculators.) As functions they are certainly interesting, but as basic functions they are unnecessary—we can do our trigonometric modeling and equation solving with the other three. Nonetheless, we give each of them a brief section of its own in this book.

The Cotangent Function

The cotangent function is the reciprocal of the tangent function. Thus,

$$\cot x = \frac{\cos x}{\sin x}.$$

The graph of $y = \cot x$ will have asymptotes at the zeros of the sine function (Figure 4.51) and zeros at the zeros of the cosine function (Figure 4.52).

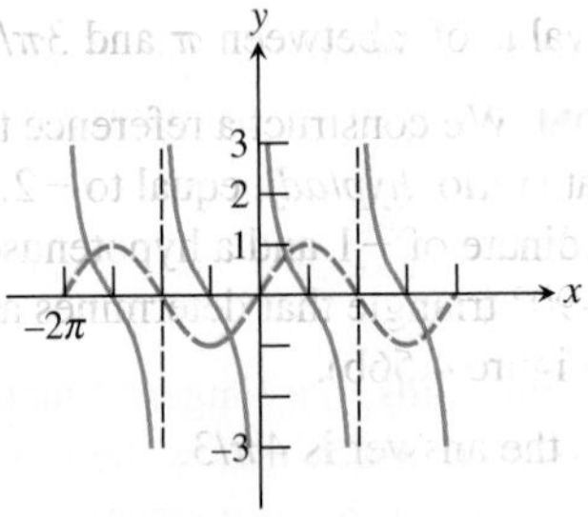

FIGURE 4.51 The cotangent has asymptotes at the zeros of the sine function.

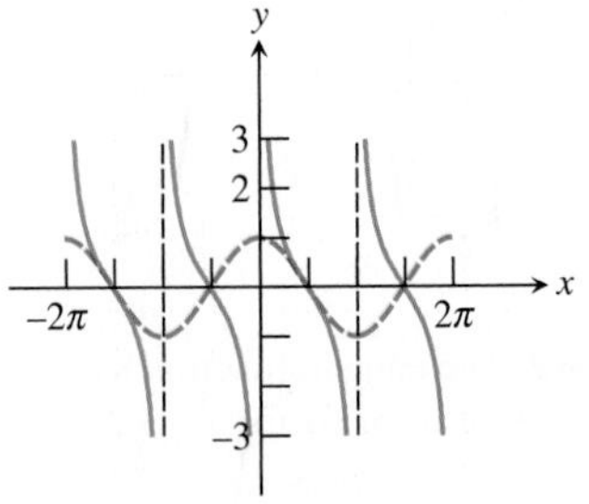

FIGURE 4.52 The cotangent has zeros at the zeros of the cosine function.

Cotangent on the Calculator

If your calculator does not have a "cotan" button, you can use the fact that cotangent and tangent are reciprocals. For example, the function in Example 2 can be entered in a calculator as $y = 3/\tan(x/2) + 1$ or as $y = 3(\tan(x/2))^{-1} + 1$. Remember that it *cannot* be entered as $y = 3\tan^{-1}(x/2) + 1$. (The -1 exponent in that position represents a function *inverse*, not a *reciprocal*.)

EXAMPLE 2 Graphing a Cotangent Function

Describe the graph of $f(x) = 3\cot(x/2) + 1$ in terms of a basic trigonometric function. Locate the vertical asymptotes and graph two periods.

SOLUTION The graph is obtained from the graph of $y = \cot x$ by effecting a horizontal stretch by a factor of 2, a vertical stretch by a factor of 3, and a vertical translation up 1 unit. The horizontal stretch makes the period of the function 2π (twice the period of $y = \cot x$), and the asymptotes are at the even multiples of π. Figure 4.53 shows two periods of the graph of f.

Now try Exercise 9.

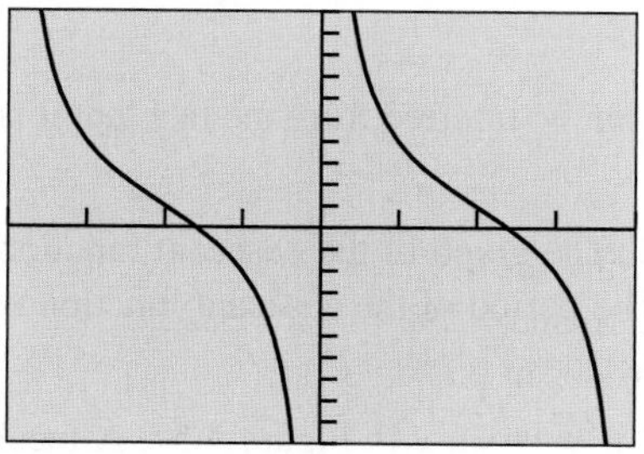

[-2π, 2π] by [-10, 10]

FIGURE 4.53 Two periods of $f(x) = 3\cot(x/2) + 1$. (Example 2)

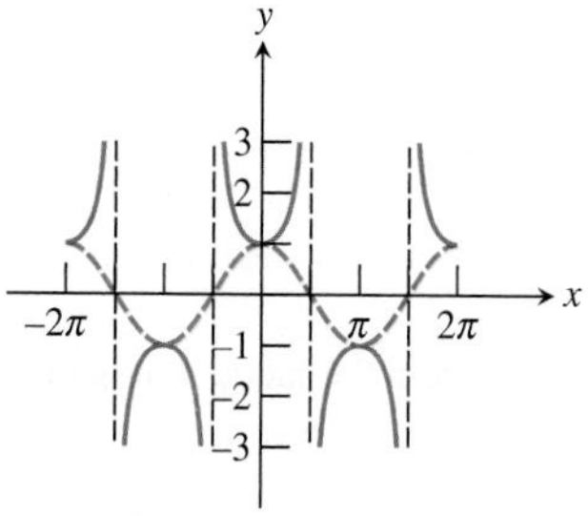

FIGURE 4.54 Characteristics of the secant function are inferred from the fact that it is the reciprocal of the cosine function.

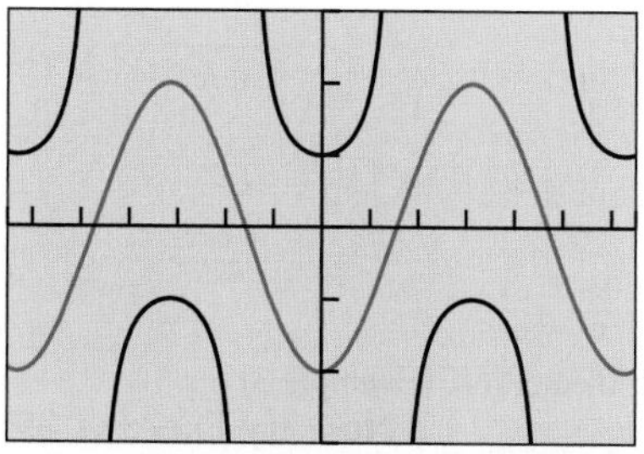

[-6.5, 6.5] by [-3, 3]

FIGURE 4.55 The graphs of $y = \sec x$ and $y = -2\cos x$. (Exploration 1)

The Secant Function

Important characteristics of the secant function can be inferred from the fact that it is the reciprocal of the cosine function.

Whenever $\cos x = 1$, its reciprocal $\sec x$ is also 1. The graph of the secant function has asymptotes at the zeros of the cosine function. The period of the secant function is 2π, the same as its reciprocal, the cosine function.

The graph of $y = \sec x$ is shown with the graph of $y = \cos x$ in Figure 4.54. A local maximum of $y = \cos x$ corresponds to a local minimum of $y = \sec x$, and a local minimum of $y = \cos x$ corresponds to a local maximum of $y = \sec x$.

EXPLORATION 1 Proving a Graphical Hunch

Figure 4.55 shows that the graphs of $y = \sec x$ and $y = -2\cos x$ never seem to intersect.

If we stretch the reflected cosine graph vertically by a large enough number, will it continue to miss the secant graph? Or is there a large enough (positive) value of k so that the graph of $y = \sec x$ *does* intersect the graph of $y = -k\cos x$?

1. Try a few other values of k in your calculator. Do the graphs intersect?
2. Your exploration should lead you to conjecture that the graphs of $y = \sec x$ and $y = -k\cos x$ will *never* intersect for any positive value of k. Verify this conjecture by proving **algebraically** that the equation

$$-k\cos x = \sec x$$

has no real solutions when k is a positive number.

EXAMPLE 3 Solving a Trigonometric Equation Algebraically

Find the value of x between π and $3\pi/2$ that solves $\sec x = -2$.

SOLUTION We construct a reference triangle in the third quadrant that has the appropriate ratio, *hyp/adj*, equal to -2. This is most easily accomplished by choosing an x-coordinate of -1 and a hypotenuse of 2 (Figure 4.56a). We recognize this as a 30°–60°–90° triangle that determines an angle of 240°, which converts to $4\pi/3$ radians (Figure 4.56b).

Therefore the answer is $4\pi/3$.

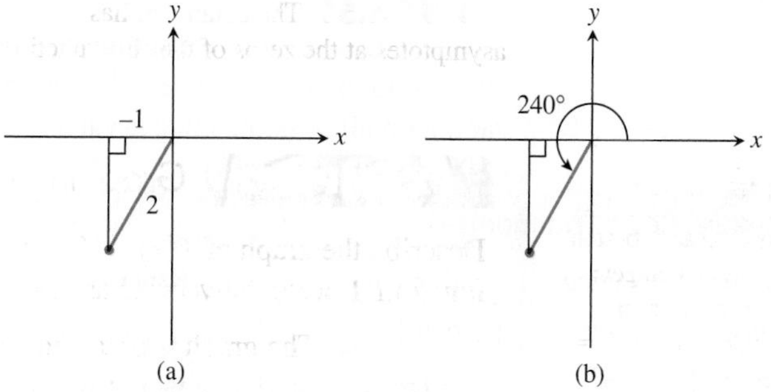

FIGURE 4.56 A reference triangle in the third quadrant (a) with *hyp/adj* $= -2$ determines an angle (b) of 240°, which converts to $4\pi/3$ rad. (Example 3)

Now try Exercise 29.

The Cosecant Function

Important characteristics of the cosecant function can be inferred from the fact that it is the reciprocal of the sine function.

Whenever $\sin x = 1$, its reciprocal $\csc x$ is also 1. The graph of the cosecant function has asymptotes at the zeros of the sine function. The period of the cosecant function is 2π, the same as its reciprocal, the sine function.

The graph of $y = \csc x$ is shown with the graph of $y = \sin x$ in Figure 4.57. A local maximum of $y = \sin x$ corresponds to a local minimum of $y = \csc x$, and a local minimum of $y = \sin x$ corresponds to a local maximum of $y = \csc x$.

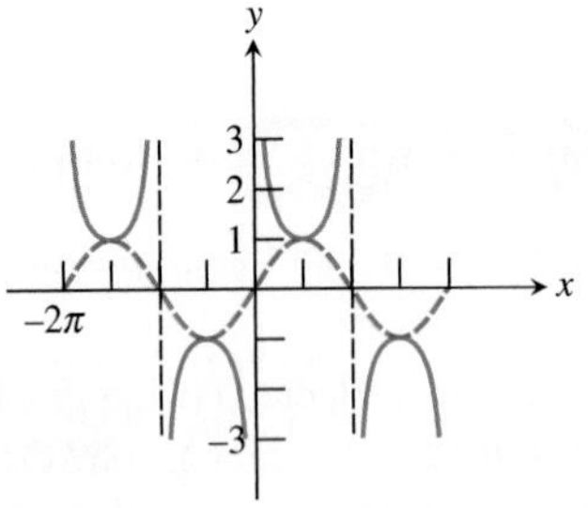

FIGURE 4.57 Characteristics of the cosecant function are inferred from the fact that it is the reciprocal of the sine function.

EXAMPLE 4 Solving a Trigonometric Equation Graphically

Find the smallest positive number x such that $x^2 = \csc x$.

SOLUTION There is no algebraic attack that looks hopeful, so we solve this equation graphically. The intersection point of the graphs of $y = x^2$ and $y = \csc x$ that has the smallest positive x-coordinate is shown in Figure 4.58. We use the grapher to determine that $x \approx 1.068$.

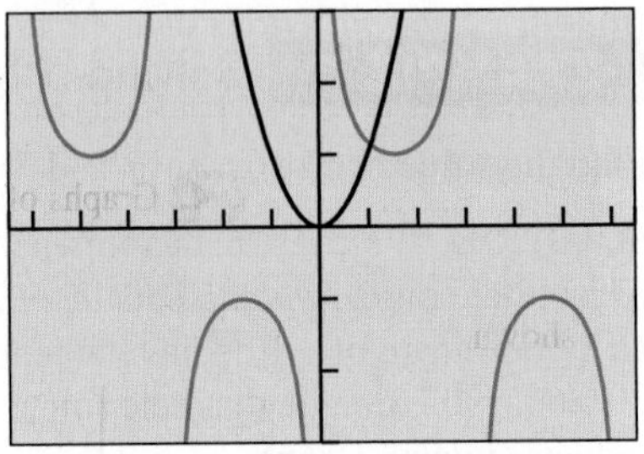

[−6.5, 6.5] by [−3, 3]

FIGURE 4.58 A graphical solution of a trigonometric equation. (Example 3)

Now try Exercise 39.

Are Cosecant Curves Parabolas?

Figure 4.58 shows a parabola intersecting one of the infinite number of U-shaped curves that make up the graph of the cosecant function. In fact, the parabola intersects *all* of those curves that lie above the x-axis, since the parabola must spread out to cover the entire domain of $y = x^2$, which is all real numbers! The cosecant curves do not keep spreading out, as they are hemmed in by asymptotes. That means that the U-shaped curves in the cosecant function are not parabolas.

To close this section, we summarize the properties of the six basic trigonometric functions in tabular form. The "n" that appears in several places should be understood as taking on all possible integer values: $0, \pm 1, \pm 2, \pm 3, \ldots$.

Summary: Basic Trigonometric Functions

Function	Period	Domain	Range	Asymptotes	Zeros	Even/Odd
$\sin x$	2π	$(-\infty, \infty)$	$[-1, 1]$	None	$n\pi$	Odd
$\cos x$	2π	$(-\infty, \infty)$	$[-1, 1]$	None	$\pi/2 + n\pi$	Even
$\tan x$	π	$x \neq \pi/2 + n\pi$	$(-\infty, \infty)$	$x = \pi/2 + n\pi$	$n\pi$	Odd
$\cot x$	π	$x \neq n\pi$	$(-\infty, \infty)$	$x = n\pi$	$\pi/2 + n\pi$	Odd
$\sec x$	2π	$x \neq \pi/2 + n\pi$	$(-\infty, -1] \cup [1, \infty)$	$x = \pi/2 + n\pi$	None	Even
$\csc x$	2π	$x \neq n\pi$	$(-\infty, -1] \cup [1, \infty)$	$x = n\pi$	None	Odd

QUICK REVIEW 4.5 *(For help, go to Sections 1.2, 2.6, and 4.3.)*

Exercise numbers with a gray background indicate problems that the authors have designed to be solved *without a calculator*.

In Exercises 1–4, state the period of the function.

1. $y = \cos 2x$ **2.** $y = \sin 3x$

3. $y = \sin \frac{1}{3}x$ **4.** $y = \cos \frac{1}{2}x$

In Exercises 5–8, find the zeros and vertical asymptotes of the function.

5. $y = \dfrac{x - 3}{x + 4}$ **6.** $y = \dfrac{x + 5}{x - 1}$

7. $y = \dfrac{x + 1}{(x - 2)(x + 2)}$ **8.** $y = \dfrac{x + 2}{x(x - 3)}$

In Exercises 9 and 10, tell whether the function is odd, even, or neither.

9. $y = x^2 + 4$ **10.** $y = \dfrac{1}{x}$

SECTION 4.5 Exercises

In Exercises 1–4, identify the graph of each function. Use your understanding of transformations, not your graphing calculator.

1. Graphs of one period of csc x and 2 csc x are shown.

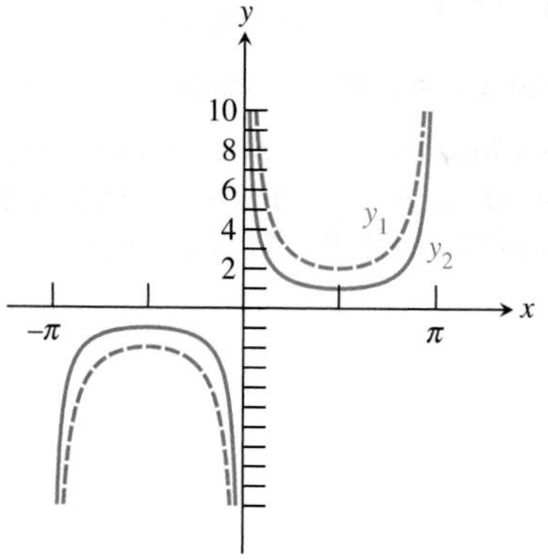

2. Graphs of two periods of 0.5 tan x and 5 tan x are shown.

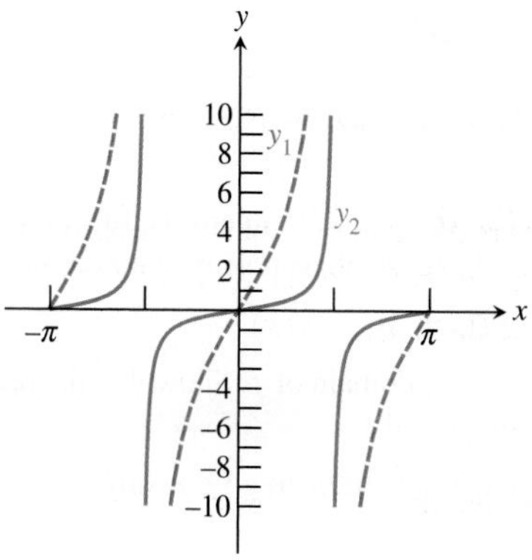

3. Graphs of csc x and 3 csc $2x$ are shown.

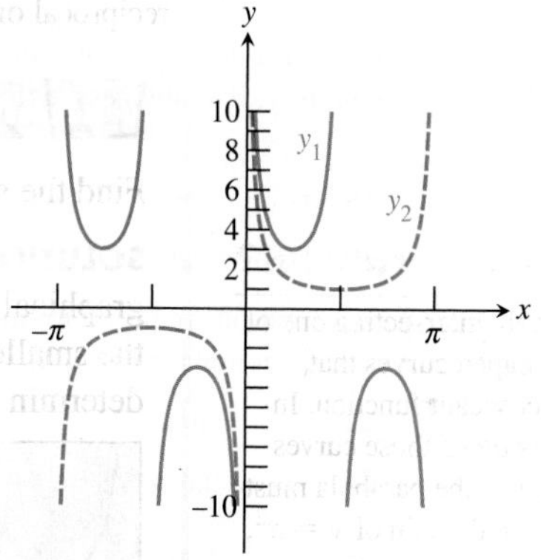

4. Graphs of cot x and cot $(x - 0.5) + 3$ are shown.

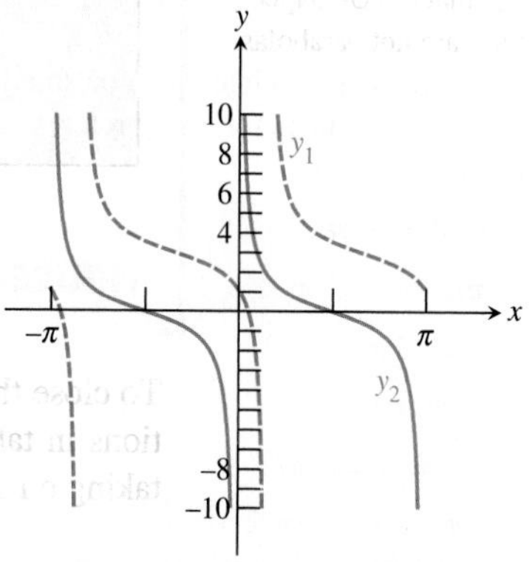

In Exercises 5–12, describe the graph of the function in terms of a basic trigonometric function. Locate the vertical asymptotes and graph two periods of the function.

5. $y = \tan 2x$ **6.** $y = -\cot 3x$

7. $y = \sec 3x$ **8.** $y = \csc 2x$

9. $y = 2 \cot 2x$ **10.** $y = 3 \tan (x/2)$

11. $y = \csc (x/2)$ **12.** $y = 3 \sec 4x$

In Exercises 13–16, match the trigonometric function with its graph. Then give the Xmin and Xmax values for the viewing window in which the graph is shown. Use your understanding of transformations, not your graphing calculator.

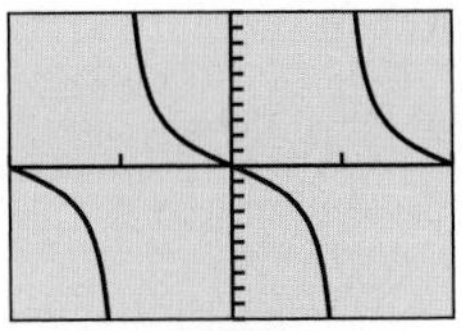

[?, ?] by [–10, 10]
(a)

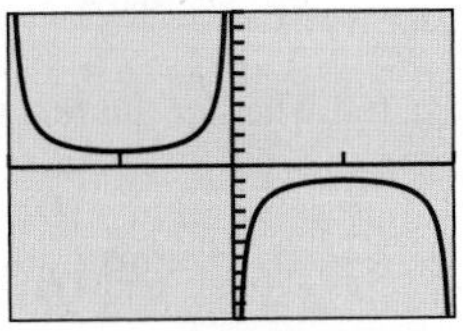

[?, ?] by [–10, 10]
(b)

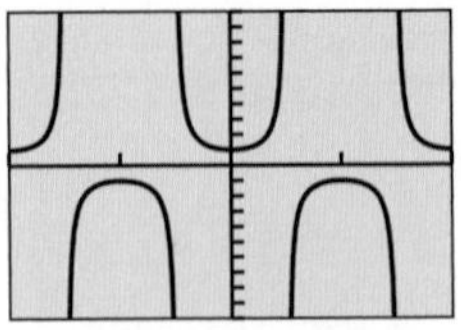

[?, ?] by [–10, 10]
(c)

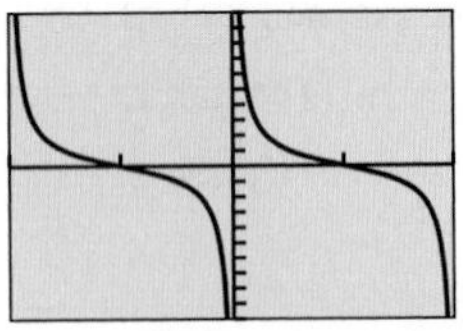

[?, ?] by [–10, 10]
(d)

13. $y = -2 \tan x$

14. $y = \cot x$

15. $y = \sec 2x$

16. $y = -\csc x$

In Exercises 17–20, analyze each function for domain, range, continuity, increasing or decreasing behavior, symmetry, boundedness, extrema, asymptotes, and end behavior.

17. $f(x) = \cot x$

18. $f(x) = \sec x$

19. $f(x) = \csc x$

20. $f(x) = \tan (x/2)$

In Exercises 21–28, describe the transformations required to obtain the graph of the given function from a basic trigonometric graph.

21. $y = 3 \tan x$

22. $y = -\tan x$

23. $y = 3 \csc x$

24. $y = 2 \tan x$

25. $y = -3 \cot \frac{1}{2} x$

26. $y = -2 \sec \frac{1}{2} x$

27. $y = -\tan \frac{\pi}{2} x + 2$

28. $y = 2 \tan \pi x - 2$

In Exercises 29–34, solve for x in the given interval. You should be able to find these numbers without a calculator, using reference triangles in the proper quadrants.

29. $\sec x = 2, \quad 0 \le x \le \pi/2$

30. $\csc x = 2, \quad \pi/2 \le x \le \pi$

31. $\cot x = -\sqrt{3}, \quad \pi/2 \le x \le \pi$

32. $\sec x = -\sqrt{2}, \quad \pi \le x \le 3\pi/2$

33. $\csc x = 1, \quad 2\pi \le x \le 5\pi/2$

34. $\cot x = 1, \quad -\pi \le x \le -\pi/2$

In Exercises 35–40, use a calculator to solve for x in the given interval.

35. $\tan x = 1.3, \quad 0 \le x \le \frac{\pi}{2}$

36. $\sec x = 2.4, \quad 0 \le x \le \frac{\pi}{2}$

37. $\cot x = -0.6, \quad \frac{3\pi}{2} \le x \le 2\pi$

38. $\csc x = -1.5, \quad \pi \le x \le \frac{3\pi}{2}$

39. $\csc x = 2, \quad 0 \le x \le 2\pi$

40. $\tan x = 0.3, \quad 0 \le x \le 2\pi$

41. Writing to Learn The figure shows a unit circle and an angle t whose terminal side is in Quadrant III.

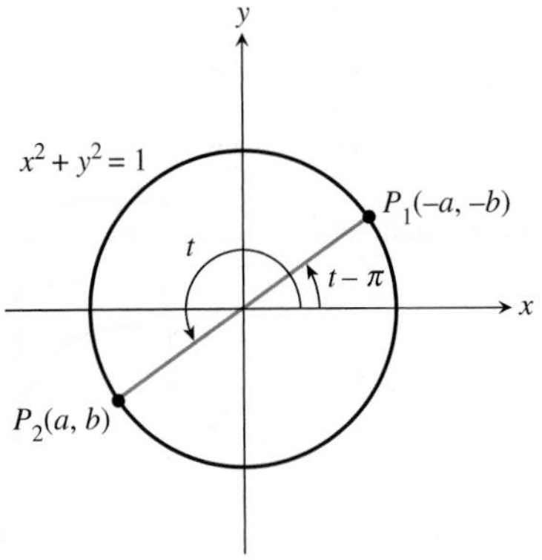

(a) If the coordinates of point P_2 are (a, b), explain why the coordinates of point P_1 on the circle and the terminal side of angle $t - \pi$ are $(-a, -b)$.

(b) Explain why $\tan t = \frac{b}{a}$.

(c) Find $\tan (t - \pi)$, and show that $\tan t = \tan (t - \pi)$.

(d) Explain why the period of the tangent function is π.

(e) Explain why the period of the cotangent function is π.

42. Writing to Learn Explain why it is correct to say $y = \tan x$ is the slope of the terminal side of angle x in standard position. P is on the unit circle.

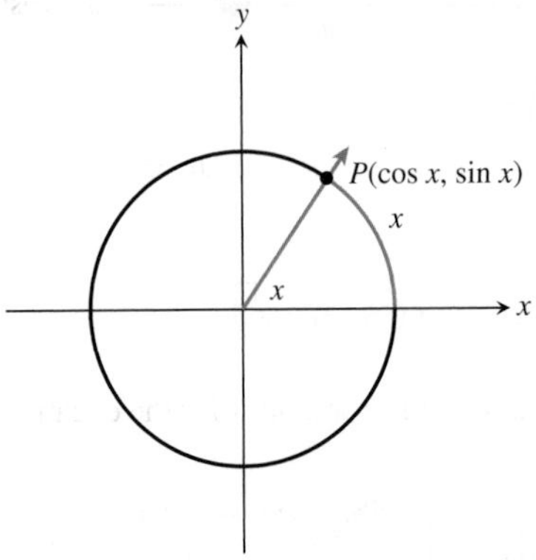

43. Periodic Functions Let f be a periodic function with period p. That is, p is the smallest positive number such that

$$f(x + p) = f(x)$$

for any value of x in the domain of f. Show that the reciprocal $1/f$ is periodic with period p.

44. Identities Use the unit circle to give a convincing argument for the identity.

(a) $\sin (t + \pi) = -\sin t$

(b) $\cos (t + \pi) = -\cos t$

(c) Use (a) and (b) to show that $\tan (t + \pi) = \tan t$. Explain why this is *not* enough to conclude that the period of tangent is π.

45. Lighthouse Coverage The Bolivar Lighthouse is located on a small island 350 ft from the shore of the mainland as shown in the figure.

(a) Express the distance d as a function of the angle x.

(b) If x is 1.55 rad, what is d?

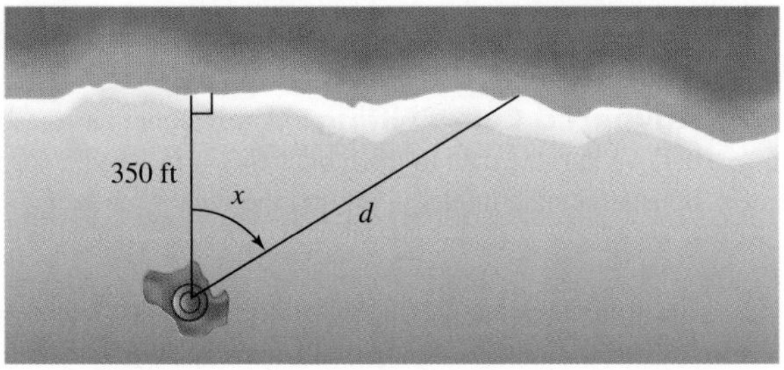

46. Hot-Air Balloon A hot-air balloon over Albuquerque, New Mexico, is being blown due east from point P and traveling at a constant height of 800 ft. The angle y is formed by the ground and the line of vision from P to the balloon. This angle changes as the balloon travels.

(a) Express the horizontal distance x as a function of the angle y.

(b) When the angle is $\pi/20$ rad, what is its horizontal distance from P?

(c) An angle of $\pi/20$ rad is equivalent to how many degrees?

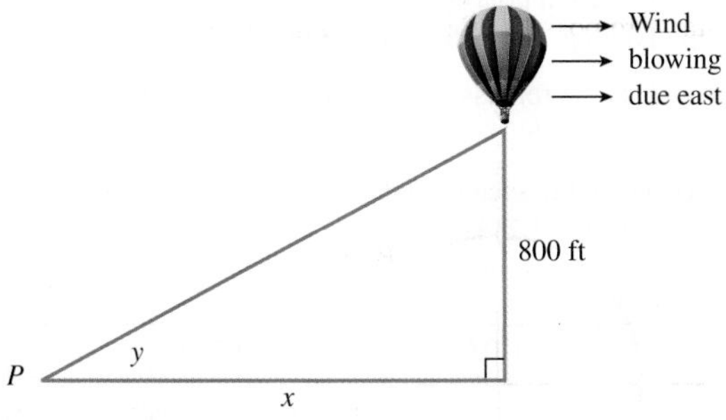

In Exercises 47–50, find approximate solutions for the equation in the interval $-\pi < x < \pi$.

47. $\tan x = \csc x$ **48.** $\sec x = \cot x$

49. $\sec x = 5 \cos x$ **50.** $4 \cos x = \tan x$

Standardized Test Questions

51. True or False The function $f(x) = \tan x$ is increasing on the interval $(-\infty, \infty)$. Justify your answer.

52. True or False If $x = a$ is an asymptote of the secant function, then $\cot a = 0$. Justify your answer.

You should answer these questions without using a calculator.

53. Multiple Choice The graph of $y = \cot x$ can be obtained by a horizontal shift of the graph of $y =$

(A) $-\tan x$. **(B)** $-\cot x$. **(C)** $\sec x$.

(D) $\tan x$. **(E)** $\csc x$.

54. Multiple Choice The graph of $y = \sec x$ never intersects the graph of $y =$

(A) x. **(B)** x^2. **(C)** $\csc x$.

(D) $\cos x$. **(E)** $\sin x$.

55. Multiple Choice If $k \neq 0$, what is the range of the function $y = k \csc x$?

(A) $[-k, k]$ **(B)** $(-k, k)$

(C) $(-\infty, -k) \cup (k, \infty)$ **(D)** $(-\infty, -k] \cup [k, \infty)$

(E) $(-\infty, -1/k] \cup [1/k, \infty)$

56. Multiple Choice The graph of $y = \csc x$ has the same set of asymptotes as the graph of $y =$

(A) $\sin x$. **(B)** $\tan x$. **(C)** $\cot x$.

(D) $\sec x$. **(E)** $\csc 2x$.

Explorations

In Exercises 57 and 58, graph both f and g in the $[-\pi, \pi]$ by $[-10, 10]$ viewing window. Estimate values in the interval $[-\pi, \pi]$ for which $f > g$.

57. $f(x) = 5 \sin x$ and $g(x) = \cot x$

58. $f(x) = -\tan x$ and $g(x) = \csc x$

59. Writing to Learn Graph the function $f(x) = -\cot x$ on the interval $(-\pi, \pi)$. Explain why it is correct to say that f is increasing on the interval $(0, \pi)$, but it is not correct to say that f is increasing on the interval $(-\pi, \pi)$.

60. Writing to Learn Graph functions $f(x) = -\sec x$ and

$$g(x) = \frac{1}{x - (\pi/2)}$$

simultaneously in the viewing window $[0, \pi]$ by $[-10, 10]$. Discuss whether you think functions f and g are equivalent.

61. Write $\csc x$ as a horizontal translation of $\sec x$.

62. Write $\cot x$ as the reflection about the x-axis of a horizontal translation of $\tan x$.

Extending the Ideas

63. Group Activity Television Coverage A television camera is on a platform 30 m from the point on High Street where the Worthington Memorial Day Parade will pass. Express the distance d from the camera to a particular parade float as a function of the angle x, and graph the function over the interval $-\pi/2 < x < \pi/2$.

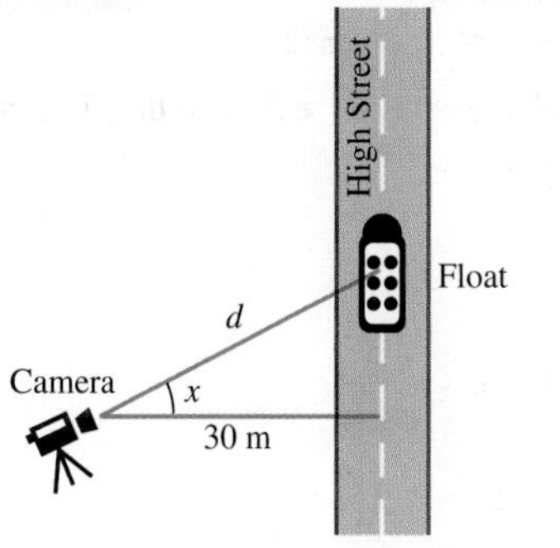

64. What's in a Name? The word *sine* comes from the Latin word *sinus*, which means "bay" or "cove." It entered the language through a mistake (variously attributed to Gerardo of Cremona or Robert of Chester) in translating the Arabic word "jiba" (chord) as if it were "jaib" (bay). This was due to the fact that the Arabs abbreviated their technical terms, much as we do today. Imagine someone unfamiliar with the technical term "cosecant" trying to reconstruct the English word that is abbreviated by "csc." It might well enter their language as their word for "cascade."

The names for the other trigonometric functions can all be explained.

(a) *Cosine* means "*sine* of the *co*mplement." Explain why this is a logical name for cosine.

(b) In the figure below, BC is perpendicular to OC, which is a radius of the unit circle. By a familiar geometry theorem, BC is tangent to the circle. OB is part of a secant that intersects the unit circle at A. It lies along the terminal side of an angle of t radians in standard position. Write the coordinates of A as functions of t.

(c) Use similar triangles to find length BC as a trig function of t.

(d) Use similar triangles to find length OB as a trig function of t.

(e) Use the results from parts (a), (c), and (d) to explain where the names "tangent, cotangent, secant," and "cosecant" came from.

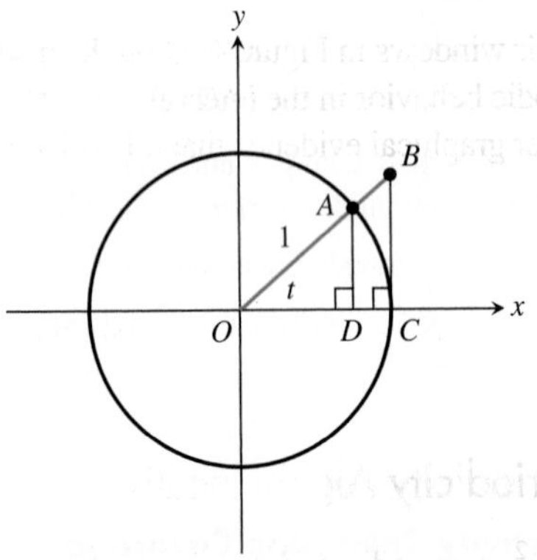

65. Capillary Action A film of liquid in a thin (capillary) tube has surface tension γ (gamma) given by

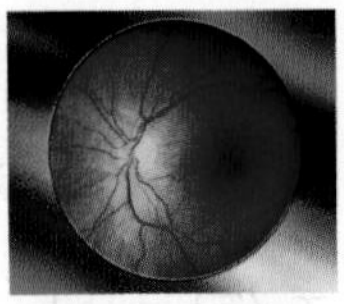

$$\gamma = \frac{1}{2}h\rho gr \sec \phi,$$

where h is the height of the liquid in the tube, ρ (rho) is the density of the liquid, $g = 9.8\ \text{m/sec}^2$ is the acceleration due to gravity, r is the radius of the tube, and ϕ (phi) is the angle of contact between the tube and the liquid's surface. Whole blood has a surface tension of 0.058 N/m (newton per meter) and a density of 1050 kg/m^3. Suppose that blood rises to a height of 1.5 m in a capillary blood vessel of radius 4.7×10^{-6} m. What is the contact angle between the capillary vessel and the blood surface? $[1N = 1(\text{kg} \cdot \text{m})/\text{sec}^2]$

66. Advanced Curve Fitting A researcher has reason to believe that the data in the table below can best be described by an algebraic model involving the secant function:

$$y = a \sec(bx)$$

Unfortunately, her calculator will do only sine regression. She realizes that the following two facts will help her:

$$\frac{1}{y} = \frac{1}{a \sec(bx)} = \frac{1}{a}\cos(bx)$$

and

$$\cos(bx) = \sin\left(bx + \frac{\pi}{2}\right).$$

(a) Use these two facts to show that

$$\frac{1}{y} = \frac{1}{a}\sin\left(bx + \frac{\pi}{2}\right).$$

(b) Store the x-values in the table in L1 in your calculator and the y-values in L2. Store the *reciprocals* of the y-values in L3. Then do a sine regression for L3 $(1/y)$ as a function of L1 (x). Write the regression equation.

(c) Use the regression equation in (b) to determine the values of a and b.

(d) Write the secant model: $y = a \sec(bx)$ Does the curve fit the (L1, L2) scatter plot?

x	1	2	3	4
y	5.0703	5.2912	5.6975	6.3622

x	5	6	7	8
y	7.4359	9.2541	12.716	21.255

4.6 Graphs of Composite Trigonometric Functions

What you'll learn about

- Combining Trigonometric and Algebraic Functions
- Sums and Differences of Sinusoids
- Damped Oscillation

... and why

Function composition extends our ability to model periodic phenomena like heartbeats and sound waves.

Combining Trigonometric and Algebraic Functions

A theme of this text has been "families of functions." We have studied polynomial functions, exponential functions, logarithmic functions, and rational functions (to name a few), and in this chapter we have studied trigonometric functions. Now we consider adding, multiplying, or composing trigonometric functions with functions from these other families.

The notable property that distinguishes the trigonometric function from others we have studied is periodicity. Example 1 shows that when a trigonometric function is combined with a polynomial, the resulting function may or may not be periodic.

EXAMPLE 1 Combining the Sine Function with x^2

Graph each of the following functions for $-2\pi \le x \le 2\pi$, adjusting the vertical window as needed. Which of the functions appear to be periodic?

(a) $y = \sin x + x^2$

(b) $y = x^2 \sin x$

(c) $y = (\sin x)^2$

(d) $y = \sin (x^2)$

SOLUTION We show the graphs and their windows in Figure 4.59 on the next page. Only the graph of $y = (\sin x)^2$ exhibits periodic behavior in the interval $-2\pi \le x \le 2\pi$. (You can widen the window to see further graphical evidence that this is indeed the only periodic function among the four.) **Now try Exercise 5.**

EXAMPLE 2 Verifying Periodicity Algebraically

Verify algebraically that $f(x) = (\sin x)^2$ is periodic and determine its period graphically.

SOLUTION We use the fact that the period of the basic sine function is 2π; that is, $\sin (x + 2\pi) = \sin (x)$ for all x. It follows that

$$\begin{aligned} f(x + 2\pi) &= (\sin (x + 2\pi))^2 \\ &= (\sin (x))^2 \qquad \text{By periodicity of sine} \\ &= f(x) \end{aligned}$$

So $f(x)$ is also periodic, with some period that divides 2π. The graph in Figure 4.59c on the next page shows that the period is actually π. **Now try Exercise 9.**

Exponent Notation

Example 3 introduces a shorthand notation for powers of trigonometric functions: $(\sin \theta)^n$ can be written as $\sin^n \theta$. (*Caution:* This shorthand notation will probably not be recognized by your calculator.)

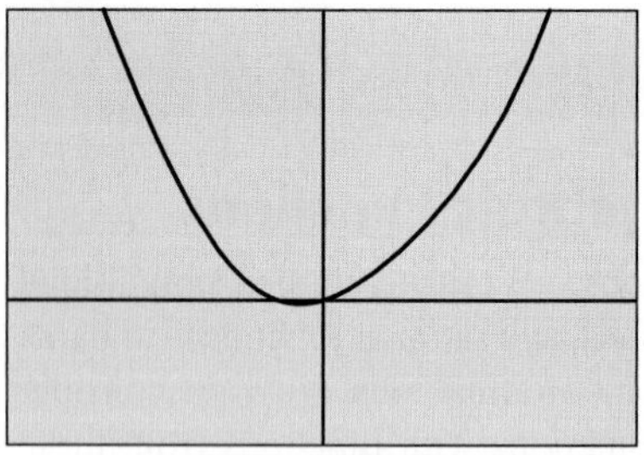

$[-2\pi, 2\pi]$ by $[-10, 20]$
(a)

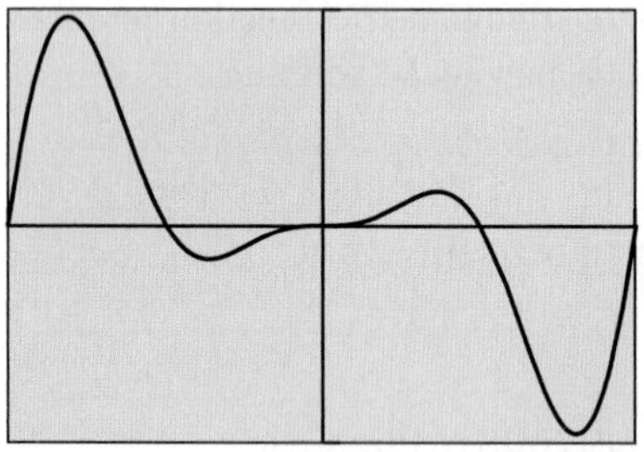

$[-2\pi, 2\pi]$ by $[-25, 25]$
(b)

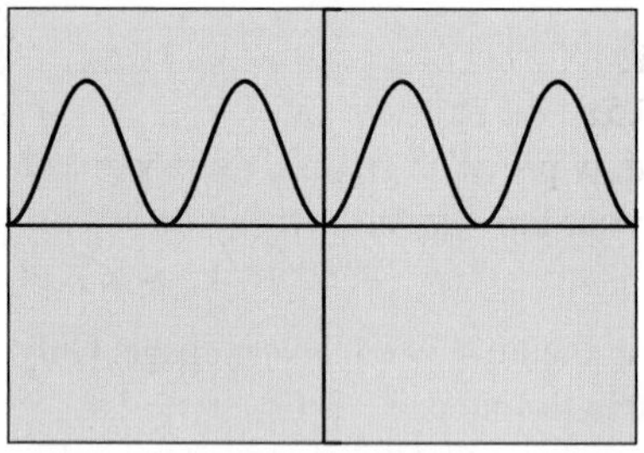

$[-2\pi, 2\pi]$ by $[-1.5, 1.5]$
(c)

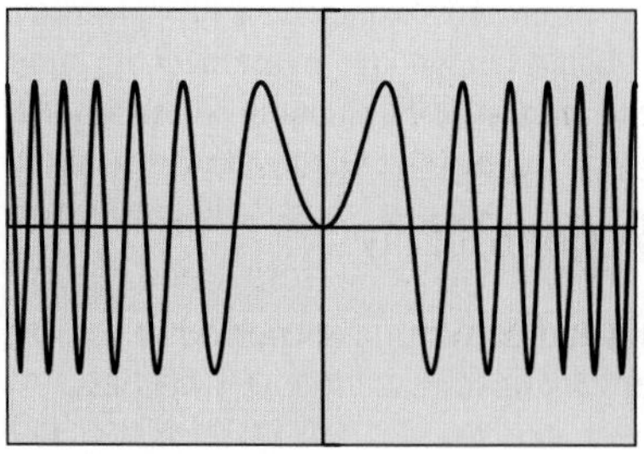

$[-2\pi, 2\pi]$ by $[-1.5, 1.5]$
(d)

FIGURE 4.59 The graphs of the four functions in Example 1. Only graph (c) exhibits periodic behavior over the interval $-2\pi \leq x \leq 2\pi$.

EXAMPLE 3 Composing $y = \sin x$ and $y = x^3$

Prove algebraically that $f(x) = \sin^3 x$ is periodic and find the period graphically. State the domain and range and sketch a graph showing two periods.

SOLUTION To prove that $f(x) = \sin^3 x$ is periodic, we show that $f(x + 2\pi) = f(x)$ for all x.

$$\begin{aligned} f(x + 2\pi) &= \sin^3 (x + 2\pi) \\ &= (\sin (x + 2\pi))^3 && \text{Changing notation} \\ &= (\sin (x))^3 && \text{By periodicity of sine} \\ &= \sin^3 (x) && \text{Changing notation} \\ &= f(x) \end{aligned}$$

Thus $f(x)$ is periodic with a period that divides 2π. Graphing the function over the interval $-2\pi \leq x \leq 2\pi$ (Figure 4.60), we see that the period must be 2π.

Since both functions being composed have domain $(-\infty, \infty)$, the domain of f is also $(-\infty, \infty)$. Since cubing all numbers in the interval $[-1, 1]$ gives all numbers in the interval $[-1, 1]$, the range is $[-1, 1]$ (as supported by the graph).

Now try Exercise 13.

Comparing the graphs of $y = \sin^3 x$ and $y = \sin x$ over a single period (Figure 4.61), we see that the two functions have the same zeros and extreme points, but otherwise the graph of $y = \sin^3 x$ is closer to the x-axis than the graph of $y = \sin x$. This is because $|y^3| < |y|$ whenever y is between -1 and 1. In fact, higher odd powers of $\sin x$ yield graphs that are "sucked in" more and more, but always with the same zeros and extreme points.

The absolute value of a periodic function is also a periodic function. We consider two such functions in Example 4.

EXAMPLE 4 Analyzing Nonnegative Periodic Functions

Find the domain, range, and period of each of the following functions. Sketch a graph showing four periods.

(a) $f(x) = |\tan x|$
(b) $g(x) = |\sin x|$

SOLUTION

(a) Whenever $\tan x$ is defined, so is $|\tan x|$. Therefore, the domain of f is the same as the domain of the tangent function, that is, all real numbers except $\pi/2 + n\pi$, $n = 0, \pm 1, \ldots$. Because $f(x) = |\tan x| \geq 0$ and the range of $\tan x$ is $(-\infty, \infty)$, the range of f is $[0, \infty)$. The period of f, like that of $y = \tan x$, is π. The graph of $y = f(x)$ is shown in Figure 4.62.

(b) Whenever $\sin x$ is defined, so is $|\sin x|$. Therefore, the domain of g is the same as the domain of the sine function, that is, all real numbers. Because $g(x) = |\sin x| \geq 0$ and the range of $\sin x$ is $[-1, 1]$, the range of g is $[0, 1]$.

The period of g is only half the period of $y = \sin x$, for reasons that are apparent from viewing the graph. The negative sections of the sine curve below the x-axis are reflected above the x-axis, where they become repetitions of the positive sections. The graph of $y = g(x)$ is shown in Figure 4.63.

Now try Exercise 15.

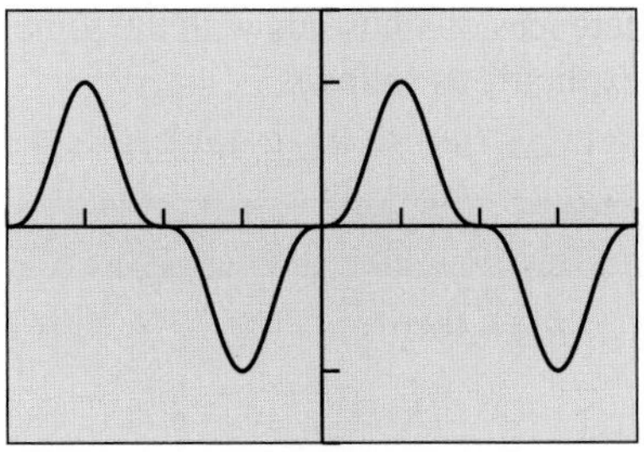

$[-2\pi, 2\pi]$ by $[-1.5, 1.5]$

FIGURE 4.60 The graph of $f(x) = \sin^3 x$. (Example 3)

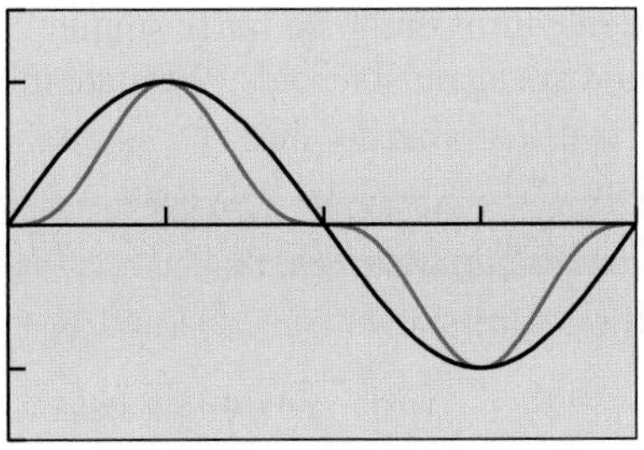

$[0, 2\pi]$ by $[-1.5, 1.5]$

FIGURE 4.61 The graph suggests that $|\sin^3 x| \leq |\sin x|$.

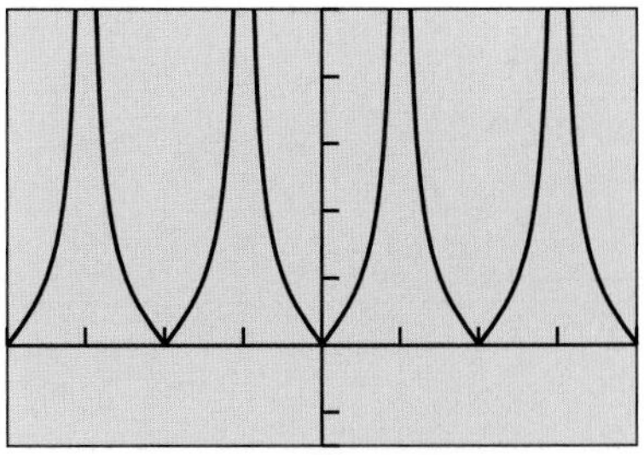

$[-2\pi, 2\pi]$ by $[-1.5, 5]$

FIGURE 4.62 $f(x) = |\tan x|$ has the same period as $y = \tan x$. (Example 4a)

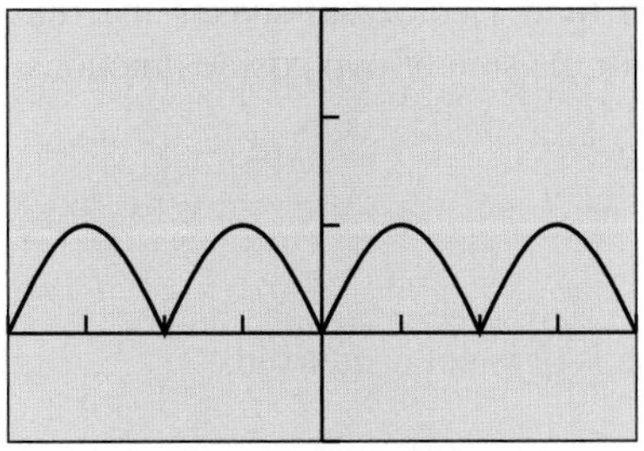

$[-2\pi, 2\pi]$ by $[-1, 3]$

FIGURE 4.63 $g(x) = |\sin x|$ has half the period of $y = \sin x$. (Example 4b)

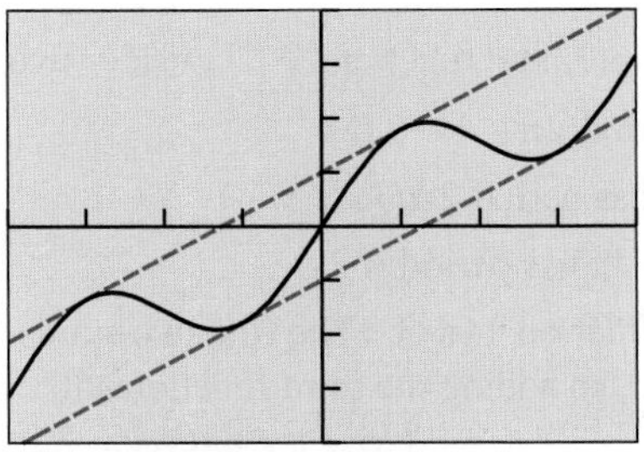

$[-2\pi, 2\pi]$ by $[-4, 4]$

FIGURE 4.64 The graph of $f(x) = 0.5x + \sin x$ oscillates between the lines $y = 0.5x + 1$ and $y = 0.5x - 1$. Although the wave repeats its shape, it is not periodic. (Example 5)

When a sinusoid is added to a (nonconstant) linear function, the result is *not* periodic. The graph repeats its *shape* at regular intervals, but the function takes on different values over those intervals. The graph will show a curve oscillating between two parallel lines, as in Example 5.

EXAMPLE 5 Adding a Sinusoid to a Linear Function

The graph of $f(x) = 0.5x + \sin x$ oscillates between two parallel lines (Figure 4.64). What are the equations of the two lines?

SOLUTION As $\sin x$ oscillates between -1 and 1, $f(x)$ oscillates between $0.5x - 1$ and $0.5x + 1$. Therefore, the two lines are $y = 0.5x - 1$ and $y = 0.5x + 1$. Graphing the two lines and $f(x)$ in the same window provides graphical support. Of course, the graph should resemble Figure 4.61 if your lines are correct.

Now try Exercise 19.

Sums and Differences of Sinusoids

Section 4.4 introduced you to sinusoids, functions that can be written in the form

$$y = a \sin(b(x - h)) + k$$

and therefore have the shape of a sine curve.

Sinusoids model a variety of physical and social phenomena—such as sound waves, voltage in alternating electrical current, the velocity of air flow during the human respiratory cycle, and many others. Sometimes these phenomena interact in an additive fashion. For example, if y_1 models the sound of one tuning fork and y_2 models the sound of a second tuning fork, then $y_1 + y_2$ models the sound when they are both struck simultaneously. So we are interested in whether the sums and differences of sinusoids are again sinusoids.

EXPLORATION 1 Investigating Sinusoids

Graph these functions, one at a time, in the viewing window $[-2\pi, 2\pi]$ by $[-6, 6]$. Which ones appear to be sinusoids?

$y = 3 \sin x + 2 \cos x$ $\qquad$ $y = 2 \sin x - 3 \cos x$

$y = 2 \sin 3x - 4 \cos 2x$ $\qquad$ $y = 2 \sin(5x + 1) - 5 \cos 5x$

$y = \cos\left(\dfrac{7x - 2}{5}\right) + \sin\left(\dfrac{7x}{5}\right)$ $\qquad$ $y = 3 \cos 2x + 2 \sin 7x$

What relationship between the sine and cosine functions ensures that their sum or difference will again be a sinusoid? Check your guess on a graphing calculator by constructing your own examples.

The rule turns out to be fairly simple: Sums and differences of sinusoids with the same period are again sinusoids. We state this rule more explicitly as follows.

Sums That Are Sinusoidal Functions

If $y_1 = a_1 \sin(b(x - h_1))$ and $y_2 = a_2 \cos(b(x - h_2))$, then

$$y_1 + y_2 = a_1 \sin(b(x - h_1)) + a_2 \cos(b(x - h_2))$$

is a sinusoid with period $2\pi/|b|$.

For the sum to be a sinusoid, the two sinusoids being added together must have the same period, and the sum has the same period as both of them. Also, although the rule is stated in terms of a sine function being added to a cosine function, the fact that every cosine function is a translation of a sine function (and vice versa) makes the rule equally applicable to the sum of two sine functions or the sum of two cosine functions. If they have the same period, their sum is a sinusoid.

EXAMPLE 6 Identifying a Sinusoid

Determine whether each of the following functions is or is not a sinusoid.

(a) $f(x) = 5 \cos x + 3 \sin x$

(b) $f(x) = \cos 5x + \sin 3x$

(c) $f(x) = 2 \cos 3x - 3 \cos 2x$

(d) $f(x) = a \cos\left(\frac{3x}{7}\right) - b \cos\left(\frac{3x}{7}\right) + c \sin\left(\frac{3x}{7}\right)$

SOLUTION

(a) Yes, since both functions in the sum have period 2π.

(b) No, since $\cos 5x$ has period $2\pi/5$ and $\sin 3x$ has period $2\pi/3$.

(c) No, since $2 \cos 3x$ has period $2\pi/3$ and $3 \cos 2x$ has period π.

(d) Yes, since all three functions in the sum have period $14\pi/3$. (The first two sum to a sinusoid with the same period as the third, so adding the third function still yields a sinusoid.)

Now try Exercise 25.

EXAMPLE 7 Expressing the Sum of Sinusoids as a Sinusoid

Let $f(x) = 2 \sin x + 5 \cos x$. From the discussion above, you should conclude that $f(x)$ is a sinusoid.

(a) Find the period of f.

(b) Estimate the amplitude and phase shift graphically (to the nearest hundredth).

(c) Give a sinusoid $a \sin(b(x - h))$ that approximates $f(x)$.

SOLUTION

(a) The period of f is the same as the period of $\sin x$ and $\cos x$, namely 2π.

Solve Graphically

(b) We will learn an algebraic way to find the amplitude and phase shift in the next chapter, but for now we will find this information graphically. Figure 4.65 suggests that f is indeed a sinusoid. That is, for some a and b,

$$2 \sin x + 5 \cos x = a \sin(x - h).$$

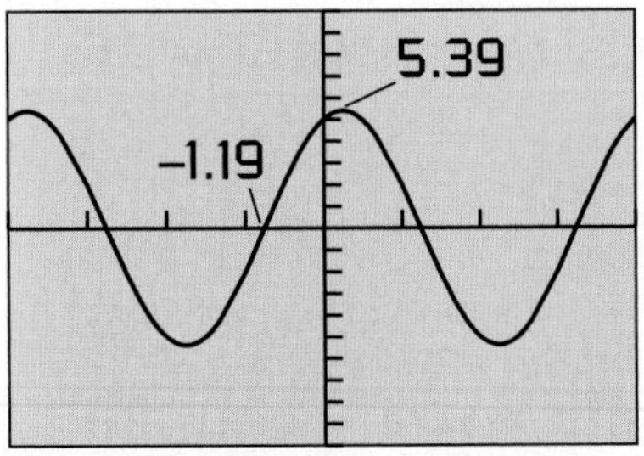

$[-2\pi, 2\pi]$ by $[-10, 10]$

FIGURE 4.65 The sum of two sinusoids: $f(x) = 2 \sin x + 5 \cos x$. (Example 7)

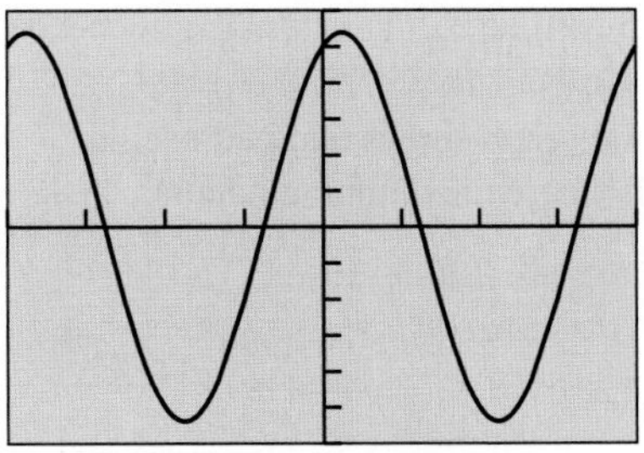

$[-2\pi, 2\pi]$ by $[-6, 6]$

FIGURE 4.66 The graphs of $y = 2 \sin x + 5 \cos x$ and $y = 5.39 \sin (x + 1.19)$ appear to be identical. (Example 7)

The *maximum value*, rounded to the nearest hundredth, is 5.39, so the amplitude of f is about 5.39. The *x-intercept* closest to $x = 0$, rounded to the nearest hundredth, is -1.19, so the phase shift of the sine function is about -1.19. We conclude that

$$f(x) = a \sin (x + h) \approx 5.39 \sin (x + 1.19).$$

(c) We support our answer graphically by showing that the graphs of

$$y = 2 \sin x + 5 \cos x \quad \text{and} \quad y = 5.39 \sin (x + 1.19)$$

are virtually identical (Figure 4.66). **Now try Exercise 29.**

The sum of two sinusoids with different periods, while not a sinusoid, will often be a periodic function. Finding the period of a sum of periodic functions can be tricky. Here is a useful fact to keep in mind. If f is periodic, and if $f(x + s) = f(x)$ for all x in the domain of f, then the period of f divides s exactly. In other words, s is either the period or a multiple of the period.

EXAMPLE 8 Showing a Function Is Periodic but Not a Sinusoid

Show that $f(x) = \sin 2x + \cos 3x$ is periodic but not a sinusoid. Graph one period.

SOLUTION Since $\sin 2x$ and $\cos 3x$ have different periods, the sum is not a sinusoid. Next we show that 2π is a candidate for the period of f, that is, $f(x + 2\pi) = f(x)$ for all x.

$$\begin{aligned} f(x + 2\pi) &= \sin (2(x + 2\pi)) + \cos (3(x + 2\pi)) \\ &= \sin (2x + 4\pi) + \cos (3x + 6\pi) \\ &= \sin 2x + \cos 3x \\ &= f(x) \end{aligned}$$

This means either that 2π is the period of f or that the period is an exact divisor of 2π. Figure 4.67 suggests that the period is not smaller than 2π, so it must be 2π. The graph shows that indeed f is not a sinusoid. **Now try Exercise 35.**

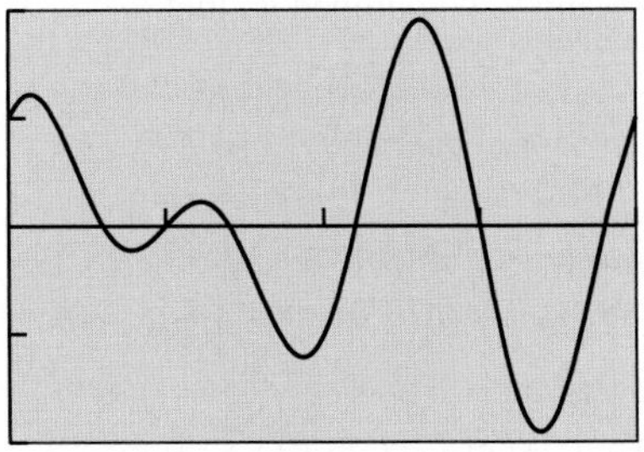

$[0, 2\pi]$ by $[-2, 2]$

FIGURE 4.67 One period of $f(x) = \sin 2x + \cos 3x$. (Example 8)

Damped Oscillation

Because the values of $\sin bt$ and $\cos bt$ oscillate between -1 and 1, something interesting happens when either of these functions is multiplied by another function. For example, consider the function $y = (x^2 + 5) \cos 6x$, graphed in Figure 4.68. The (blue) graph of the function oscillates between the (red) graphs of $y = x^2 + 5$ and $y = -(x^2 + 5)$. The "squeezing" effect that can be seen near the origin is called **damping**.

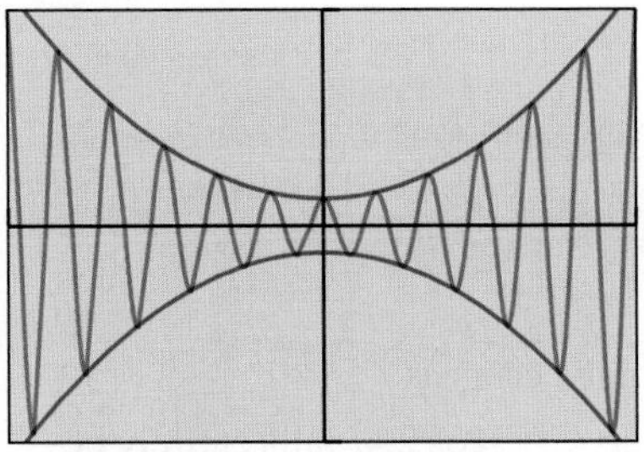

$[-2\pi, 2\pi]$ by $[-40, 40]$

FIGURE 4.68 The graph of $y = (x^2 + 5) \cos 6x$ shows damped oscillation.

DEFINITION Damped Oscillation

The graph of $y = f(x) \cos bx$ (or $y = f(x) \sin bx$) oscillates between the graphs of $y = f(x)$ and $y = -f(x)$. When this reduces the amplitude of the wave, it is called **damped oscillation**. The factor $f(x)$ is called the **damping factor**.

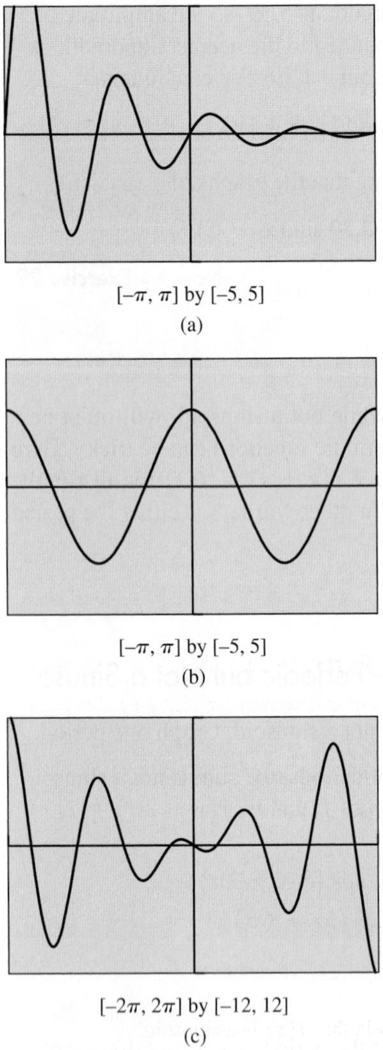

FIGURE 4.69 The graphs of functions (a), (b), and (c) in Example 9. The wave in graph (b) does not exhibit damped oscillation.

EXAMPLE 9 Identifying Damped Oscillation

For each of the following functions, determine if the graph shows damped oscillation. If it does, identify the damping factor and tell where the damping occurs.

(a) $f(x) = 2^{-x} \sin 4x$

(b) $f(x) = 3 \cos 2x$

(c) $f(x) = -2x \cos 2x$

SOLUTION The graphs are shown in Figure 4.69.

(a) This is damped oscillation. The damping factor is 2^{-x} and the damping occurs as $x \to \infty$.

(b) This wave has a constant amplitude of 3. No damping occurs.

(c) This is damped oscillation. The damping factor is $-2x$. The damping occurs as $x \to 0$.

Now try Exercise 43.

EXAMPLE 10 A Damped Oscillating Spring

Dr. Sanchez's physics class collected data for an air table glider that oscillates between two springs. The class determined from the data that the equation

$$y = 0.22e^{-0.065t} \cos 2.4t$$

modeled the displacement y of the spring from its original position as a function of time t.

(a) Identify the damping factor and tell where the damping occurs.

(b) Approximately how long does it take for the spring to be damped so that $-0.1 \le y \le 0.1$?

SOLUTION The graph is shown in Figure 4.70.

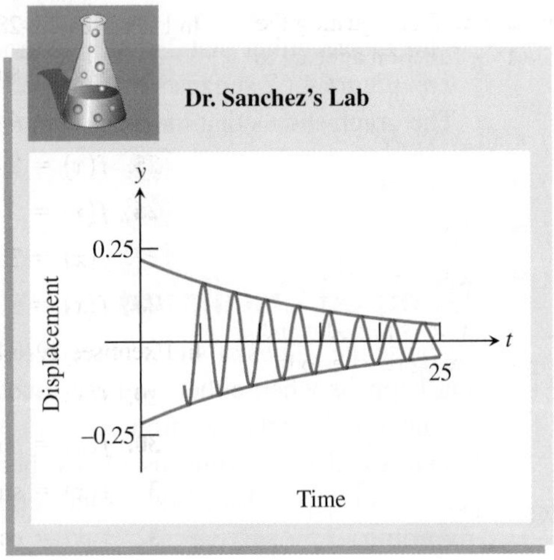

FIGURE 4.70 Damped oscillation in the physics lab. (Example 10)

(a) The damping factor is $0.22e^{-0.065t}$. The damping occurs as $t \to \infty$.

(b) We want to find how soon the curve $y = 0.22e^{-0.065t} \cos 2.4t$ falls entirely between the lines $y = -0.1$ and $y = 0.1$. By zooming in on the region indicated in Figure 4.71a and using grapher methods, we find that it takes approximately 11.86 sec until the graph of $y = 0.22e^{-0.065t} \cos 2.4t$ lies entirely between $y = -0.1$ and $y = 0.1$ (Figure 4.71b).

Now try Exercise 71.

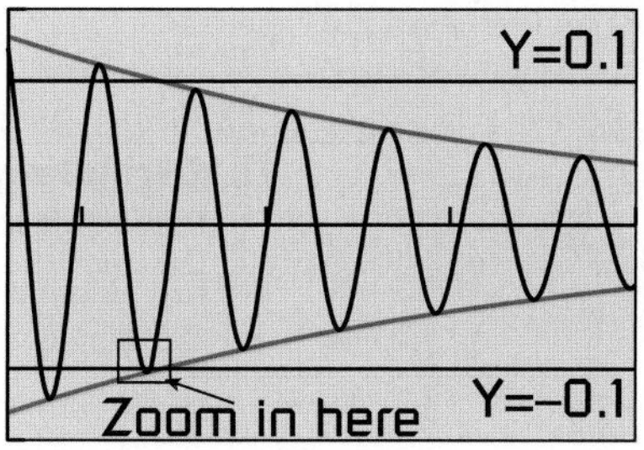

[8, 25] by [–0.15, 0.15]

(a)

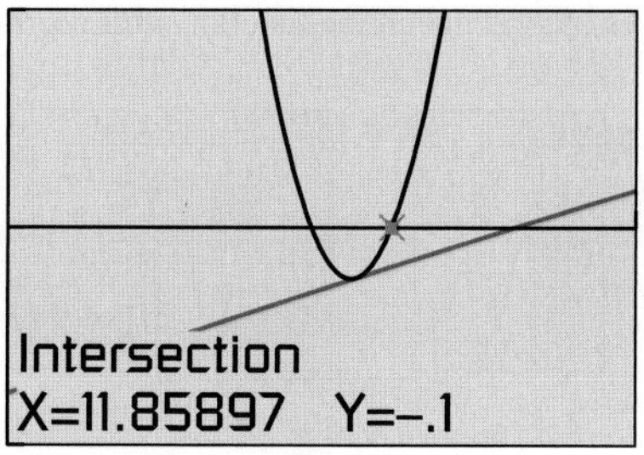

[11, 12.4] by [–0.11, –0.09]

(b)

FIGURE 4.71 The damped oscillation in Example 10 eventually gets to be less than 0.1 in either direction.

QUICK REVIEW 4.6 *(For help, go to Sections 1.2 and 1.4.)*

Exercise numbers with a gray background indicate problems that the authors have designed to be solved *without a calculator*.

In Exercises 1–6, state the domain and range of the function.

1. $f(x) = 3 \sin 2x$
2. $f(x) = -2 \cos 3x$
3. $f(x) = \sqrt{x-1}$
4. $f(x) = \sqrt{x}$
5. $f(x) = |x| - 2$
6. $f(x) = |x+2| + 1$

In Exercises 7 and 8, describe the end behavior of the function, that is, the behavior as $|x| \to \infty$.

7. $f(x) = 5e^{-2x}$
8. $f(x) = -0.2(5^{-0.1x})$

In Exercises 9 and 10, form the compositions $f \circ g$ and $g \circ f$. State the domain of each function.

9. $f(x) = x^2 - 4$ and $g(x) = \sqrt{x}$
10. $f(x) = x^2$ and $g(x) = \cos x$

SECTION 4.6 Exercises

In Exercises 1–8, graph the function for $-2\pi \le x \le 2\pi$, adjusting the vertical window as needed. State whether or not the function appears to be periodic.

1. $f(x) = (\sin x)^2$
2. $f(x) = (1.5 \cos x)^2$
3. $f(x) = x^2 + 2 \sin x$
4. $f(x) = x^2 - 2 \cos x$
5. $f(x) = x \cos x$
6. $f(x) = x^2 \cos x$
7. $f(x) = (\sin x + 1)^3$
8. $f(x) = (2 \cos x - 4)^2$

In Exercises 9–12, verify algebraically that the function is periodic and determine its period graphically. Sketch a graph showing two periods.

9. $f(x) = \cos^2 x$
10. $f(x) = \cos^3 x$
11. $f(x) = \sqrt{\cos^2 x}$
12. $f(x) = |\cos^3 x|$

In Exercises 13–18, state the domain and range of the function and sketch a graph showing four periods.

13. $y = \cos^2 x$
14. $y = |\cos x|$
15. $y = |\cot x|$
16. $y = \cos |x|$
17. $y = -\tan^2 x$
18. $y = -\sin^2 x$

The graph of each function in Exercises 19–22 oscillates between two parallel lines, as in Example 5. Find the equations of the two lines and graph the lines and the function in the same viewing window.

19. $y = 2x + \cos x$
20. $y = 1 - 0.5x + \cos 2x$
21. $y = 2 - 0.3x + \cos x$
22. $y = 1 + x + \cos 3x$

In Exercises 23–28, determine whether $f(x)$ is a sinusoid.

23. $f(x) = \sin x - 3 \cos x$
24. $f(x) = 4 \cos x + 2 \sin x$
25. $f(x) = 2 \cos \pi x + \sin \pi x$
26. $f(x) = 2 \sin x - \tan x$
27. $f(x) = 3 \sin 2x - 5 \cos x$
28. $f(x) = \pi \sin 3x - 4\pi \sin 2x$

In Exercises 29–34, find a, b, and h so that $f(x) \approx a \sin(b(x-h))$.

29. $f(x) = 2 \sin 2x - 3 \cos 2x$
30. $f(x) = \cos 3x + 2 \sin 3x$
31. $f(x) = \sin \pi x - 2 \cos \pi x$
32. $f(x) = \cos 2\pi x + 3 \sin 2\pi x$
33. $f(x) = 2 \cos x + \sin x$
34. $f(x) = 3 \sin 2x - \cos 2x$

In Exercises 35–38, the function is periodic but not a sinusoid. Find the period graphically and sketch a graph showing one period.

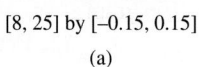

35. $y = 2 \cos x + \cos 3x$
36. $y = 2 \sin 2x + \cos 3x$
37. $y = \cos 3x - 4 \sin 2x$
38. $y = \sin 2x + \sin 5x$

In Exercises 39–42, match the function with its graph.

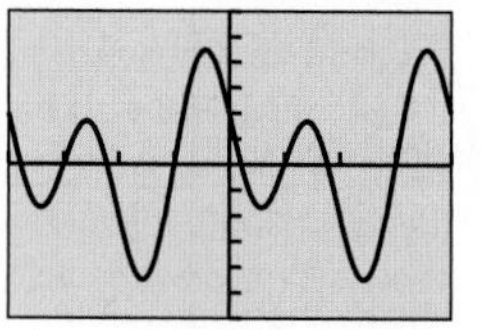

$[-2\pi, 2\pi]$ by $[-6, 6]$

(a)

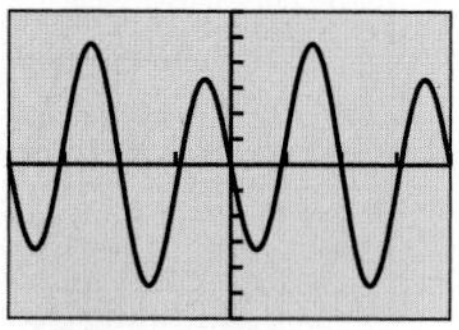

$[-2\pi, 2\pi]$ by $[-6, 6]$

(b)

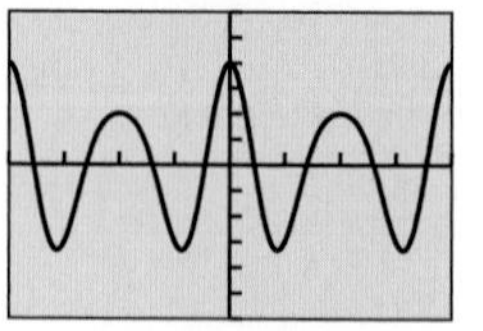

$[-2\pi, 2\pi]$ by $[-6, 6]$

(c)

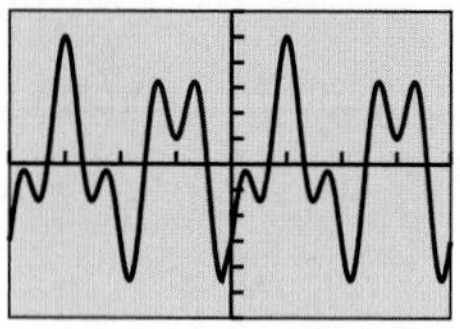

$[-2\pi, 2\pi]$ by $[-6, 6]$

(d)

39. $y = 2\cos x - 3\sin 2x$

40. $y = 2\sin 5x - 3\cos 2x$

41. $y = 3\cos 2x + \cos 3x$

42. $y = \sin x - 4\sin 2x$

In Exercises 43–48, tell whether the function exhibits damped oscillation. If so, identify the damping factor and tell whether the damping occurs as $x \to 0$ or as $x \to \infty$.

43. $f(x) = e^{-x}\sin 3x$ **44.** $f(x) = x\sin 4x$

45. $f(x) = \sqrt{5}\cos 1.2x$ **46.** $f(x) = \pi^2\cos \pi x$

47. $f(x) = x^3\sin 5x$ **48.** $f(x) = \left(\frac{2}{3}\right)^x \sin\left(\frac{2x}{3}\right)$

In Exercises 49–52, graph both f and plus or minus its damping factor in the same viewing window. Describe the behavior of the function f for $x > 0$. What is the end behavior of f?

49. $f(x) = 1.2^{-x}\cos 2x$ **50.** $f(x) = 2^{-x}\sin 4x$

51. $f(x) = x^{-1}\sin 3x$ **52.** $f(x) = e^{-x}\cos 3x$

In Exercises 53–56, find the period and graph the function over two periods.

53. $y = \sin 3x + 2\cos 2x$

54. $y = 4\cos 2x - 2\cos(3x - 1)$

55. $y = 2\sin(3x + 1) - \cos(5x - 1)$

56. $y = 3\cos(2x - 1) - 4\sin(3x - 2)$

In Exercises 57–62, graph f over the interval $[-4\pi, 4\pi]$. Determine whether the function is periodic and, if it is, state the period.

57. $f(x) = \left|\sin \frac{1}{2}x\right| + 2$ **58.** $f(x) = 3x + 4\sin 2x$

59. $f(x) = x - \cos x$ **60.** $f(x) = x + \sin 2x$

61. $f(x) = \frac{1}{2}x + \cos 2x$ **62.** $f(x) = 3 - x + \sin 3x$

In Exercises 63–70, find the domain and range of the function.

63. $f(x) = 2x + \cos x$ **64.** $f(x) = 2 - x + \sin x$

65. $f(x) = |x| + \cos x$ **66.** $f(x) = -2x + |3\sin x|$

67. $f(x) = \sqrt{\sin x}$ **68.** $f(x) = \sin |x|$

69. $f(x) = \sqrt{|\sin x|}$ **70.** $f(x) = \sqrt{\cos x}$

71. Oscillating Spring The oscillations of a spring subject to friction are modeled by the equation $y = 0.43e^{-0.55t}\cos 1.8t$.

(a) Graph y and its two damping curves in the same viewing window for $0 \le t \le 12$.

(b) Approximately how long does it take for the spring to be damped so that $-0.2 \le y \le 0.2$?

72. Predicting Economic Growth The business manager of a small manufacturing company finds that she can model the company's annual growth as roughly exponential, but with cyclical fluctuations. She uses the function $S(t) = 105(1.04)^t + 4\sin(\pi t/3)$ to estimate sales (in millions of dollars), t years after 2012.

(a) What are the company's sales in 2012?

(b) What is the approximate annual growth rate?

(c) What does the model predict for sales in 2020?

(d) How many years are in each economic cycle for this company?

73. Writing to Learn Example 3 shows that the function $y = \sin^3 x$ is periodic. Explain whether you think that $y = \sin x^3$ is periodic and why.

74. Writing to Learn Example 4 shows that $y = |\tan x|$ is periodic. Write a convincing argument that $y = \tan |x|$ is not a periodic function.

In Exercises 75 and 76, select the one correct inequality, **(a)** or **(b).** Give a convincing argument.

75. (a) $x - 1 \le x + \sin x \le x + 1$ for all x.

(b) $x - \sin x \le x + \sin x$ for all x.

76. (a) $-x \le x\sin x \le x$ for all x.

(b) $-|x| \le x\sin x \le |x|$ for all x.

In Exercises 77–80, match the function with its graph. In each case state the viewing window.

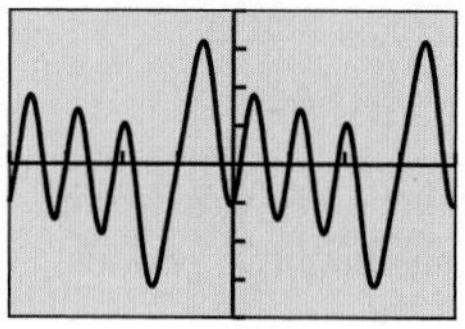

(a)

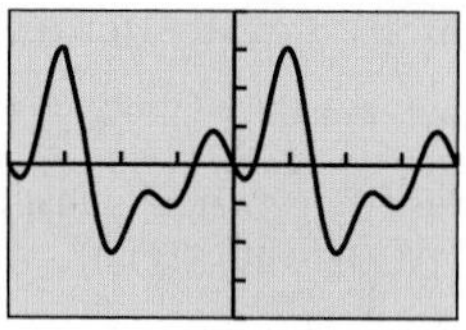

(b)

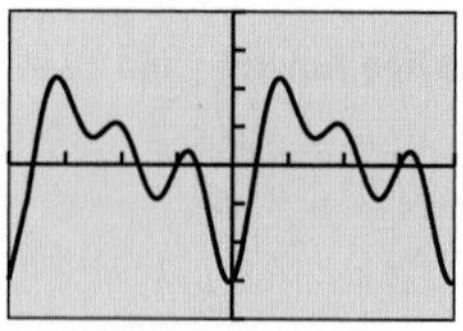

(c)

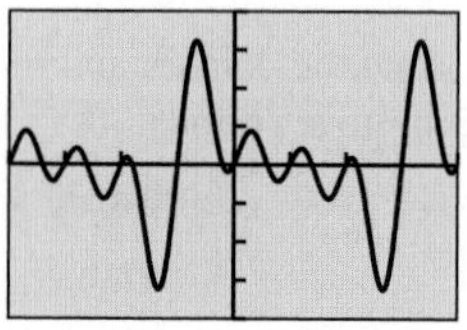

(d)

77. $y = \cos x - \sin 2x - \cos 3x + \sin 4x$

78. $y = \cos x - \sin 2x - \cos 3x + \sin 4x - \cos 5x$

79. $y = \sin x + \cos x - \cos 2x - \sin 3x$

80. $y = \sin x - \cos x - \cos 2x - \cos 3x$

Standardized Test Questions

81. **True or False** The function $f(x) = \sin|x|$ is periodic. Justify your answer.

82. **True or False** The sum of two sinusoids is a sinusoid. Justify your answer.

You may use a graphing calculator when answering these questions.

83. **Multiple Choice** What is the period of the function $f(x) = |\sin x|$?

(A) $\pi/2$ (B) π (C) 2π

(D) 3π (E) None; the function is not periodic.

84. **Multiple Choice** The function $f(x) = x \sin x$ is

(A) discontinuous. (B) bounded. (C) even.

(D) one-to-one. (E) periodic.

85. **Multiple Choice** The function $f(x) = x + \sin x$ is

(A) discontinuous. (B) bounded. (C) even.

(D) odd. (E) periodic.

86. **Multiple Choice** Which of the following functions is *not* a sinusoid?

(A) $2\cos(2x)$ (B) $3\sin(2x)$

(C) $3\sin(2x) + 2\cos(2x)$ (D) $3\sin(3x) + 2\cos(2x)$

(E) $\sin(3x + 3) + \cos(3x + 2)$

Explorations

87. **Group Activity Inaccurate or Misleading Graphs**

(a) Set Xmin = 0 and Xmax = 2π. Move the cursor along the x-axis. What is the distance between one pixel and the next (to the nearest hundredth)?

(b) What is the period of $f(x) = \sin 250x$? Consider that the period is the length of one full cycle of the graph. Approximately how many cycles should there be between two adjacent pixels? Can your grapher produce an accurate graph of this function between 0 and 2π?

88. **Group Activity Length of Days** The graph shows the number of hours of daylight in Boston as a function of the day of the year, from September 22, 2013, to December 15, 2014. Key points are labeled and other critical information is provided. Write a formula for the sinusoidal function and check it by graphing.

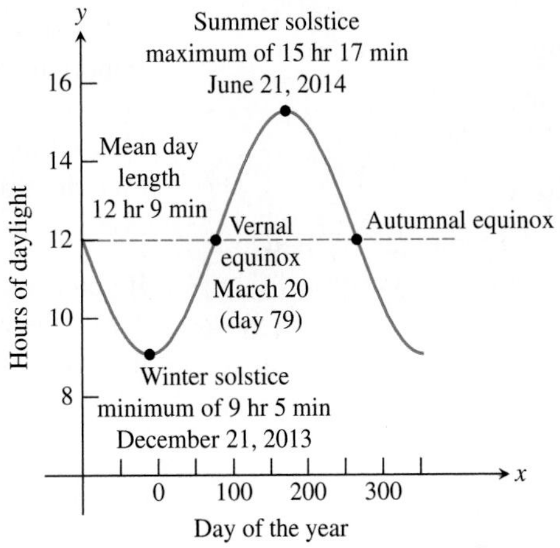

Extending the Ideas

In Exercises 89–96, first try to predict what the graph will look like (without too much effort, that is, just for fun). Then graph the function in one or more viewing windows to determine the main features of the graph, and draw a summary sketch. Where applicable, name the period, amplitude, domain, range, asymptotes, and zeros.

89. $f(x) = \cos e^x$

90. $g(x) = e^{\tan x}$

91. $f(x) = \sqrt{x} \sin x$

92. $g(x) = \sin \pi x + \sqrt{4 - x^2}$

93. $f(x) = \dfrac{\sin x}{x}$

94. $g(x) = \dfrac{\sin x}{x^2}$

95. $f(x) = x \sin \dfrac{1}{x}$

96. $g(x) = x^2 \sin \dfrac{1}{x}$

4.7 Inverse Trigonometric Functions

What you'll learn about

- Inverse Sine Function
- Inverse Cosine and Tangent Functions
- Composing Trigonometric and Inverse Trigonometric Functions
- Applications of Inverse Trigonometric Functions

... and why

Inverse trig functions can be used to solve trigonometric equations.

Inverse Sine Function

You learned in Section 1.4 that each function has an inverse relation, and that this inverse relation is a function only if the original function is one-to-one. The six basic trigonometric functions, being periodic, fail the horizontal line test for one-to-oneness rather spectacularly. However, you also learned in Section 1.4 that some functions are important enough that we want to study their inverse behavior despite the fact that they are not one-to-one. We do this by restricting the domain of the original function to an interval on which it *is* one-to-one, then finding the inverse of the restricted function. (We did this when defining the square root function, which is the inverse of the function $y = x^2$ restricted to a nonnegative domain.)

If you restrict the domain of $y = \sin x$ to the interval $[-\pi/2, \pi/2]$, as shown in Figure 4.72a, the restricted function is one-to-one. The **inverse sine function** $y = \sin^{-1} x$ is the inverse of this restricted portion of the sine function (Figure 4.72b).

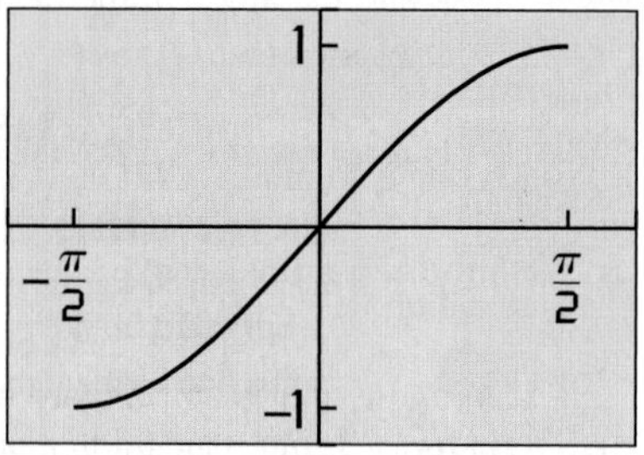

[–2, 2] by [–1.2, 1.2]

(a)

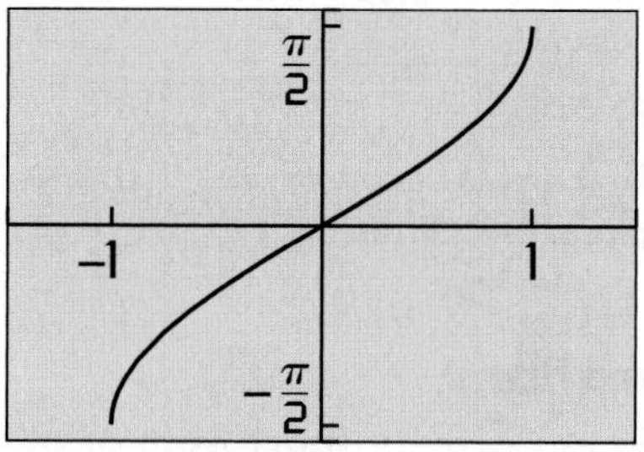

[–1.5, 1.5] by [–1.7, 1.7]

(b)

FIGURE 4.72 The (a) restriction of $y = \sin x$ is one-to-one and (b) has an inverse, $y = \sin^{-1} x$.

By the usual inverse relationship, the statements

$$y = \sin^{-1} x \quad \text{and} \quad x = \sin y$$

are equivalent for y-values in the restricted domain $[-\pi/2, \pi/2]$ and x-values in $[-1, 1]$. This means that $\sin^{-1} x$ can be thought of as *the angle between $-\pi/2$ and $\pi/2$ whose sine is x.* Since angles and directed arcs on the unit circle have the same measure, the angle $\sin^{-1} x$ is also called the **arcsine of x**.

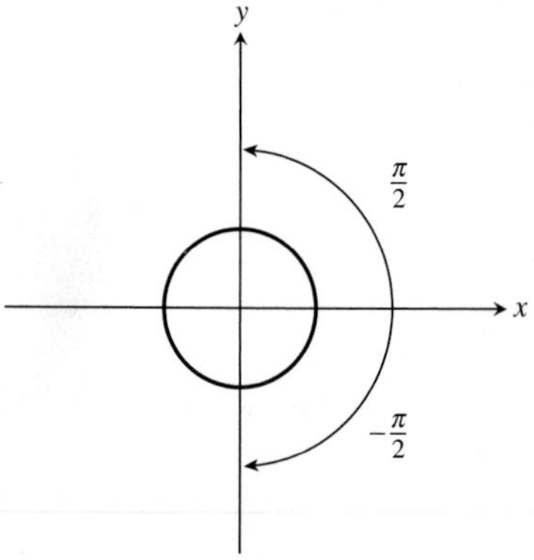

FIGURE 4.73 The values of $y = \sin^{-1} x$ will always be found on the right-hand side of the unit circle, between $-\pi/2$ and $\pi/2$.

DEFINITION Inverse Sine Function (Arcsine Function)

The unique angle y in the interval $[-\pi/2, \pi/2]$ such that $\sin y = x$ is the **inverse sine** (or **arcsine**) of x, denoted $\sin^{-1} x$ or $\arcsin x$.

The domain of $y = \sin^{-1} x$ is $[-1, 1]$ and the range is $[-\pi/2, \pi/2]$.

It helps to think of the range of $y = \sin^{-1} x$ as being along the right-hand side of the unit circle, which is traced out as angles range from $-\pi/2$ to $\pi/2$ (Figure 4.73).

EXAMPLE 1 Evaluating $\sin^{-1} x$ without a Calculator

Find the exact value of each expression without a calculator.

(a) $\sin^{-1}\left(\frac{1}{2}\right)$ **(b)** $\sin^{-1}\left(-\frac{\sqrt{3}}{2}\right)$ **(c)** $\sin^{-1}\left(\frac{\pi}{2}\right)$

(d) $\sin^{-1}\left(\sin\left(\frac{\pi}{9}\right)\right)$ **(e)** $\sin^{-1}\left(\sin\left(\frac{5\pi}{6}\right)\right)$

SOLUTION

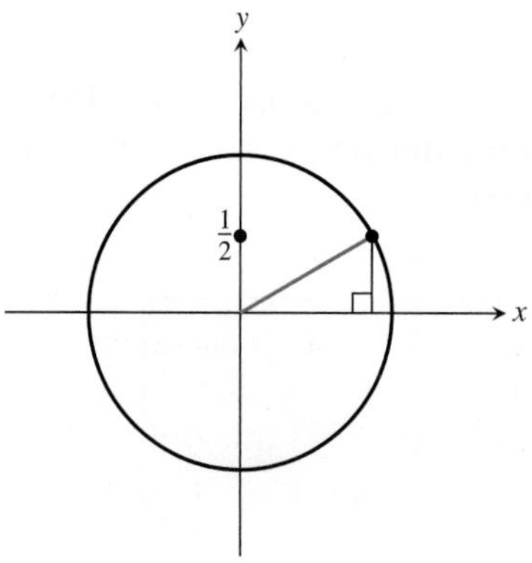

FIGURE 4.74 $\sin^{-1}(1/2) = \pi/6$. (Example 1a)

(a) Find the point on the right half of the unit circle whose y-coordinate is 1/2 and draw a reference triangle (Figure 4.74). We recognize this as one of our special ratios, and the angle in the interval $[-\pi/2, \pi/2]$ whose sine is 1/2 is $\pi/6$. Therefore

$$\sin^{-1}\left(\frac{1}{2}\right) = \frac{\pi}{6}.$$

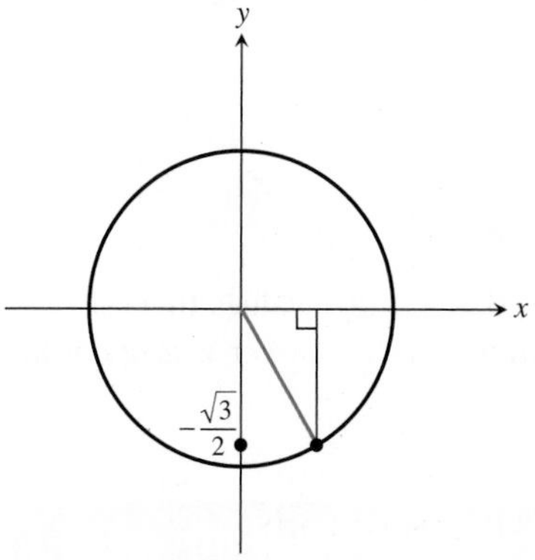

FIGURE 4.75 $\sin^{-1}(-\sqrt{3}/2) = -\pi/3$. (Example 1b)

(b) Find the point on the right half of the unit circle whose y-coordinate is $-\sqrt{3}/2$ and draw a reference triangle (Figure 4.75). We recognize this as one of our special ratios, and the angle in the interval $[-\pi/2, \pi/2]$ whose sine is $-\sqrt{3}/2$ is $-\pi/3$. Therefore

$$\sin^{-1}\left(-\frac{\sqrt{3}}{2}\right) = -\frac{\pi}{3}.$$

(c) $\sin^{-1}(\pi/2)$ does not exist, because the domain of $\sin^{-1}$ is $[-1, 1]$ and $\pi/2 > 1$.

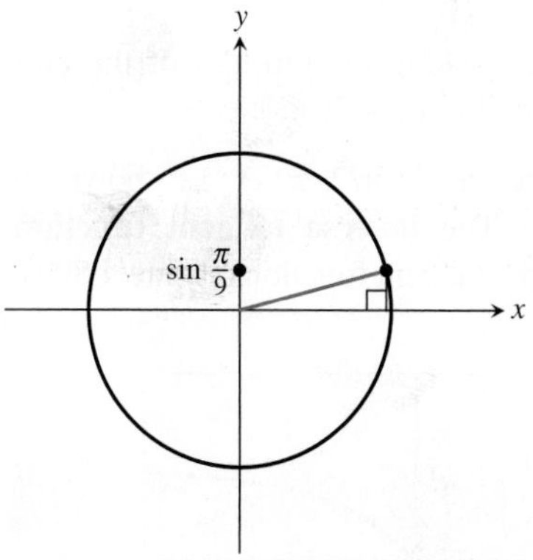

FIGURE 4.76 $\sin^{-1}(\sin(\pi/9)) = \pi/9$. (Example 1d)

(d) Draw an angle of $\pi/9$ in standard position and mark its y-coordinate on the y-axis (Figure 4.76). The angle in the interval $[-\pi/2, \pi/2]$ whose sine is this number is $\pi/9$. Therefore

$$\sin^{-1}\left(\sin\left(\frac{\pi}{9}\right)\right) = \frac{\pi}{9}.$$

(e) Draw an angle of $5\pi/6$ in standard position (notice that this angle is *not* in the interval $[-\pi/2, \pi/2]$) and mark its y-coordinate on the y-axis. (See Figure 4.77 on the next page.) The angle in the interval $[-\pi/2, \pi/2]$ whose sine is this number is $\pi - 5\pi/6 = \pi/6$. Therefore

$$\sin^{-1}\left(\sin\left(\frac{5\pi}{6}\right)\right) = \frac{\pi}{6}.$$

Now try Exercise 1.

EXAMPLE 2 Evaluating $\sin^{-1} x$ with a Calculator

Use a calculator in Radian mode to evaluate these inverse sine values

(a) $\sin^{-1}(-0.81)$ **(b)** $\sin^{-1}(\sin(3.49\pi))$

SOLUTION

(a) $\sin^{-1}(-0.81) = -0.9441521\ldots \approx -0.944$

(b) $\sin^{-1}(\sin(3.49\pi)) = -1.5393804\ldots \approx -1.539$

Although this is a calculator answer, we can use it to get an exact answer if we are alert enough to expect a multiple of π. Divide the answer by π:

$$\text{Ans}/\pi = -0.49$$

Therefore, we conclude that $\sin^{-1}(\sin(3.49\pi)) = -0.49\pi$.

You should also try to work Example 2b without a calculator. It is possible!

Now try Exercise 19.

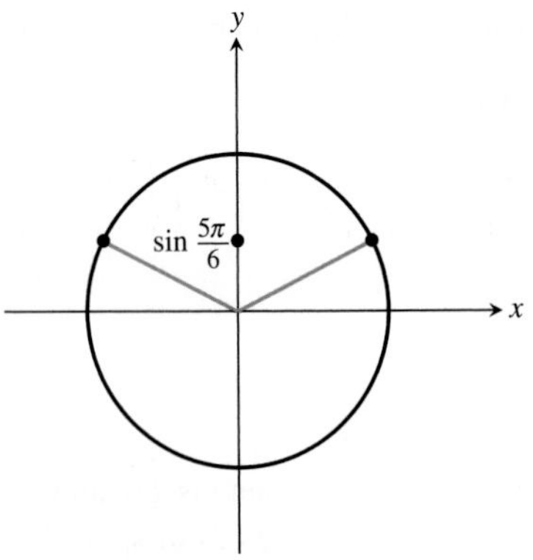

FIGURE 4.77 $\sin^{-1}(\sin(5\pi/6)) = \pi/6$. (Example 1e)

What About the Inverse Composition Rule?

Does Example 1e violate the Inverse Composition Rule of Section 1.4? That rule guarantees that $f^{-1}(f(x)) = x$ for every x in the domain of f. Keep in mind, however, that the domain of f might need to be restricted in order for f^{-1} to exist. That is certainly the case with the sine function. So Example 1e does not violate the Inverse Composition Rule, because that rule *does not apply* at $x = 5\pi/6$. It lies outside the (restricted) domain of sine.

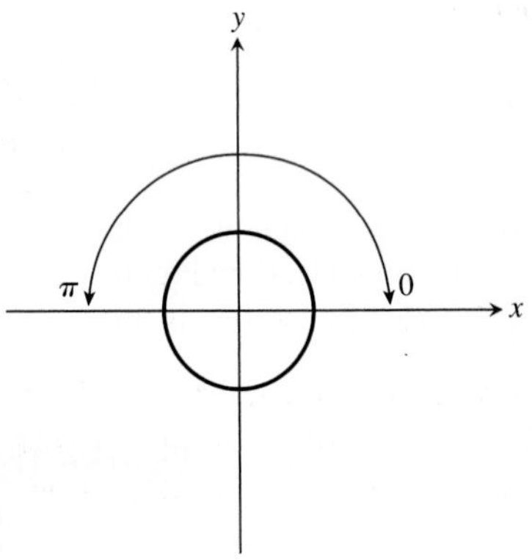

FIGURE 4.79 The values of $y = \cos^{-1} x$ will always be found on the top half of the unit circle, between 0 and π.

Inverse Cosine and Tangent Functions

If you restrict the domain of $y = \cos x$ to the interval $[0, \pi]$, as shown in Figure 4.78a, the restricted function is one-to-one. The **inverse cosine function** $y = \cos^{-1} x$ is the inverse of this restricted portion of the cosine function (Figure 4.78b).

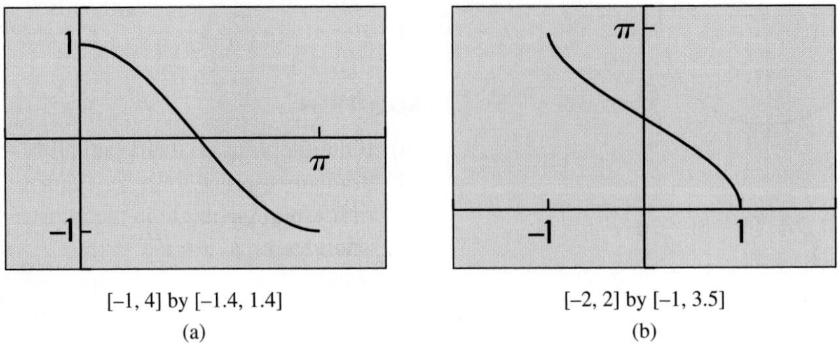

FIGURE 4.78 The (a) restriction of $y = \cos x$ is one-to-one and (b) has an inverse, $y = \cos^{-1} x$.

By the usual inverse relationship, the statements

$$y = \cos^{-1} x \quad \text{and} \quad x = \cos y$$

are equivalent for y-values in the restricted domain $[0, \pi]$ and x-values in $[-1, 1]$. This means that $\cos^{-1} x$ can be thought of as *the angle between* 0 *and* π *whose cosine is* x. The angle $\cos^{-1} x$ is also the **arccosine of *x***.

DEFINITION Inverse Cosine Function (Arccosine Function)

The unique angle y in the interval $[0, \pi]$ such that $\cos y = x$ is the **inverse cosine** (or **arccosine**) of x, denoted $\mathbf{cos^{-1}\, x}$ or **arccos *x***.

The domain of $y = \cos^{-1} x$ is $[-1, 1]$ and the range is $[0, \pi]$.

It helps to think of the range of $y = \cos^{-1} x$ as being along the top half of the unit circle, which is traced out as angles range from 0 to π (Figure 4.79).

If you restrict the domain of $y = \tan x$ to the interval $(-\pi/2, \pi/2)$, as shown in Figure 4.80a, the restricted function is one-to-one. The **inverse tangent function** $y = \tan^{-1} x$ is the inverse of this restricted portion of the tangent function (Figure 4.80b).

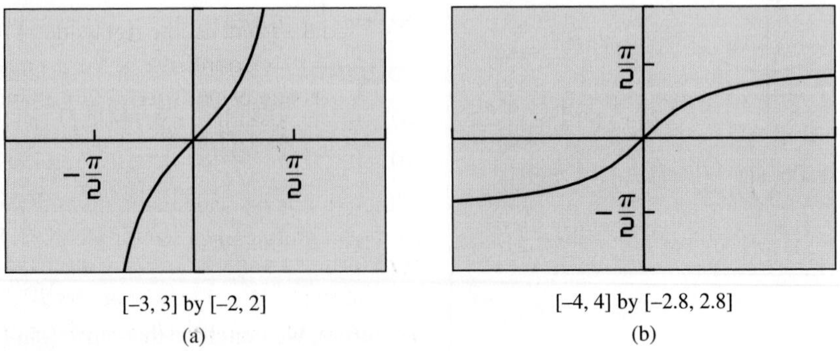

FIGURE 4.80 The (a) restriction of $y = \tan x$ is one-to-one and (b) has an inverse, $y = \tan^{-1} x$.

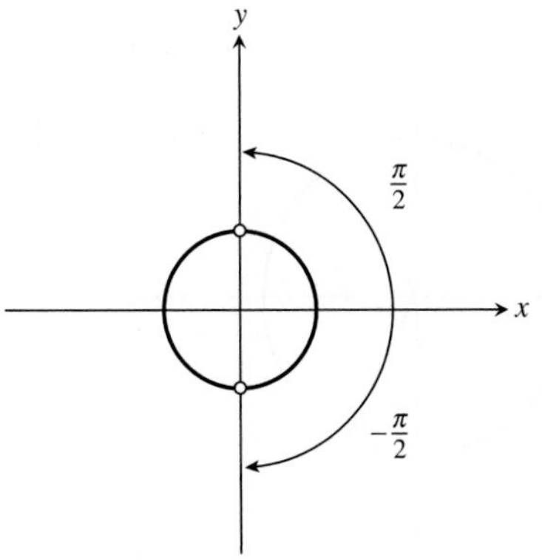

FIGURE 4.81 The values of $y = \tan^{-1} x$ will always be found on the right-hand side of the unit circle, between (but not including) $-\pi/2$ and $\pi/2$.

By the usual inverse relationship, the statements

$$y = \tan^{-1} x \quad \text{and} \quad x = \tan y$$

are equivalent for y-values in the restricted domain $(-\pi/2, \pi/2)$ and x-values in $(-\infty, \infty)$. This means that $\tan^{-1} x$ can be thought of as *the angle between* $-\pi/2$ *and* $\pi/2$ *whose tangent is* x. The angle $\tan^{-1} x$ is also the **arctangent of** $\boldsymbol{x}$.

DEFINITION Inverse Tangent Function (Arctangent Function)

The unique angle y in the interval $(-\pi/2, \pi/2)$ such that $\tan y = x$ is the **inverse tangent** (or **arctangent**) of x, denoted $\mathbf{tan^{-1} x}$ or **arctan** $\boldsymbol{x}$. The domain of $y = \tan^{-1} x$ is $(-\infty, \infty)$ and the range is $(-\pi/2, \pi/2)$.

It helps to think of the range of $y = \tan^{-1} x$ as being along the right-hand side of the unit circle (minus the top and bottom points), which is traced out as angles range from $-\pi/2$ to $\pi/2$ (noninclusive) (Figure 4.81).

EXAMPLE 3 Evaluating Inverse Trig Functions without a Calculator

Find the exact value of the expression without a calculator.

(a) $\cos^{-1}\left(-\frac{\sqrt{2}}{2}\right)$

(b) $\tan^{-1} \sqrt{3}$

(c) $\cos^{-1}(\cos(-1.1))$

SOLUTION

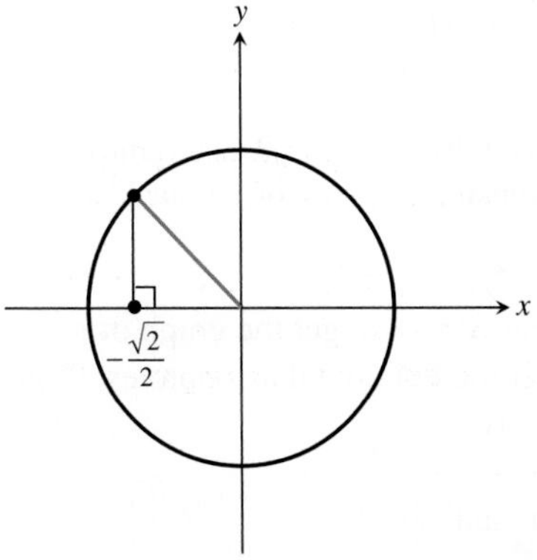

FIGURE 4.82 $\cos^{-1}(-\sqrt{2}/2) = 3\pi/4$. (Example 3a)

(a) Find the point on the top half of the unit circle whose x-coordinate is $-\sqrt{2}/2$ and draw a reference triangle (Figure 4.82). We recognize this as one of our special ratios, and the angle in the interval $[0, \pi]$ whose cosine is $-\sqrt{2}/2$ is $3\pi/4$. Therefore

$$\cos^{-1}\left(-\frac{\sqrt{2}}{2}\right) = \frac{3\pi}{4}.$$

(b) Find the point on the right side of the unit circle whose y-coordinate is $\sqrt{3}$ times its x-coordinate and draw a reference triangle (Figure 4.83). We recognize this as one of our special ratios, and the angle in the interval $(-\pi/2, \pi/2)$ whose tangent is $\sqrt{3}$ is $\pi/3$. Therefore

$$\tan^{-1} \sqrt{3} = \frac{\pi}{3}.$$

(c) Draw an angle of -1.1 in standard position (notice that this angle is *not* in the interval $[0, \pi]$) and mark its x-coordinate on the x-axis (Figure 4.84). The angle in the interval $[0, \pi]$ whose cosine is this number is 1.1. Therefore

$$\cos^{-1}(\cos(-1.1)) = 1.1.$$

Now try Exercises 5 and 7.

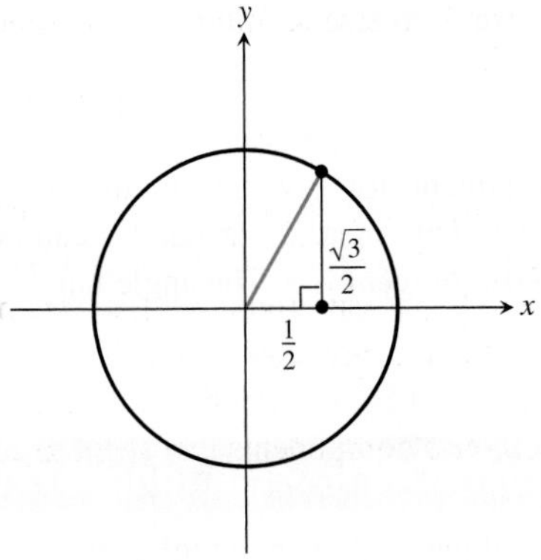

FIGURE 4.83 $\tan^{-1}\sqrt{3} = \pi/3$. (Example 3b)

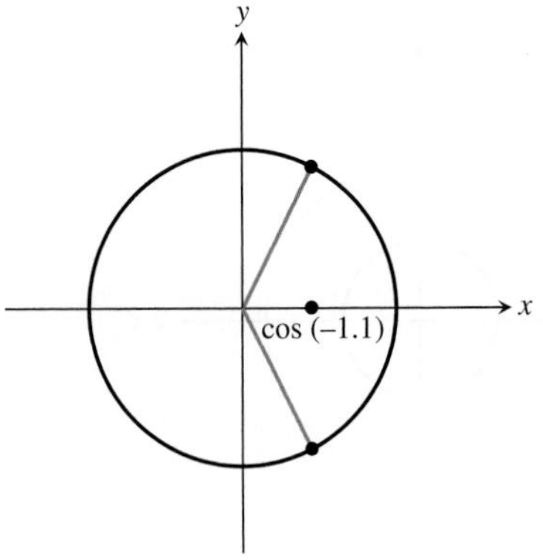

FIGURE 4.84 $\cos^{-1}(\cos(-1.1)) = 1.1$. (Example 3c)

EXAMPLE 4 Describing End Behavior

Describe the end behavior of the function $y = \tan^{-1} x$.

SOLUTION We can get this information most easily by considering the graph of $y = \tan^{-1} x$, remembering how it relates to the restricted graph of $y = \tan x$. (See Figure 4.85.)

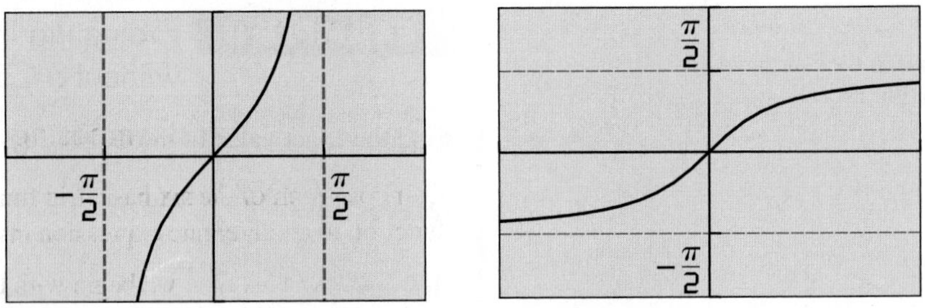

[–3, 3] by [–2, 2]
(a)

[–4, 4] by [–2.8, 2.8]
(b)

FIGURE 4.85 The graphs of (a) $y = \tan x$ (restricted) and (b) $y = \tan^{-1} x$. The vertical asymptotes of $y = \tan x$ are reflected to become the horizontal asymptotes of $y = \tan^{-1} x$. (Example 4)

When we reflect the graph of $y = \tan x$ about the line $y = x$ to get the graph of $y = \tan^{-1} x$, the vertical asymptotes $x = \pm\pi/2$ become horizontal asymptotes $y = \pm\pi/2$. We can state the end behavior accordingly:

$$\lim_{x \to -\infty} \tan^{-1} x = -\frac{\pi}{2} \quad \text{and} \quad \lim_{x \to +\infty} \tan^{-1} x = \frac{\pi}{2}$$

Now try Exercise 21.

Composing Trigonometric and Inverse Trigonometric Functions

We have already seen the need for caution when applying the Inverse Composition Rule to the trigonometric functions and their inverses (Examples 1e and 3c). The following equations are *always* true whenever they are defined:

$$\sin(\sin^{-1}(x)) = x \qquad \cos(\cos^{-1}(x)) = x \qquad \tan(\tan^{-1}(x)) = x$$

What About Arccot, Arcsec, and Arccsc?

Because we already have inverse functions for their reciprocals, we do not really need inverse functions for cot, sec, and csc for computational purposes. Moreover, the decision of how to choose the range of arcsec and arccsc is not as straightforward as with the other functions. See Exercises 63, 71, and 72.

On the other hand, the following equations are true only for x-values in the "restricted" domains of sin, cos, and tan:

$$\sin^{-1}(\sin(x)) = x \quad \cos^{-1}(\cos(x)) = x \quad \tan^{-1}(\tan(x)) = x$$

An even more interesting phenomenon occurs when we compose inverse trigonometric functions of one kind with trigonometric functions of another kind, as in $\sin(\tan^{-1}x)$. Surprisingly, these trigonometric compositions reduce to algebraic functions that involve no trigonometry at all! This curious situation has profound implications in calculus, where it is sometimes useful to decompose nontrigonometric functions into trigonometric components that seem to come out of nowhere. Try Exploration 1.

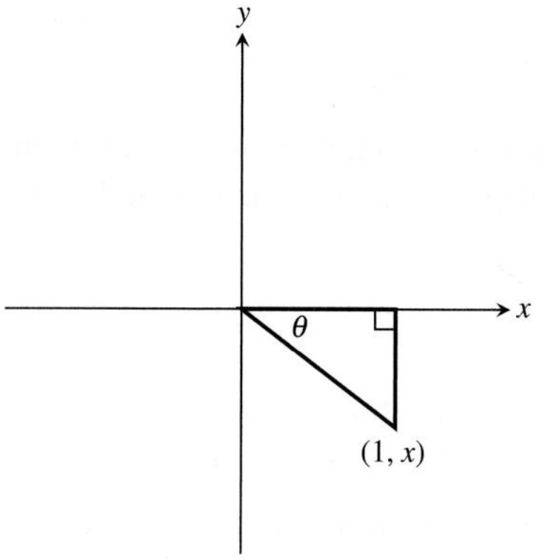

FIGURE 4.86 If $x < 0$, then $\theta = \tan^{-1}x$ is an angle in the fourth quadrant. (Exploration 1)

EXPLORATION 1 Finding Inverse Trig Functions of Trig Functions

In the right triangle shown to the right, the angle θ is measured in radians.

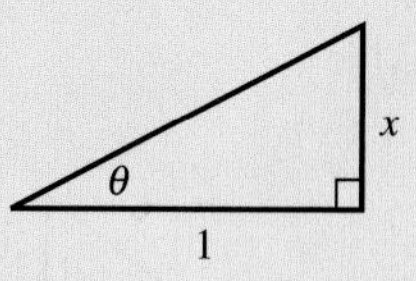

1. Find $\tan\theta$.
2. Find $\tan^{-1}x$.
3. Find the hypotenuse of the triangle as a function of x.
4. Find $\sin(\tan^{-1}(x))$ as a ratio involving no trig functions.
5. Find $\sec(\tan^{-1}(x))$ as a ratio involving no trig functions.
6. If $x < 0$, then $\tan^{-1}x$ is a negative angle in the fourth quadrant (Figure 4.86). Verify that your answers to parts (4) and (5) are still valid in this case.

EXAMPLE 5 Composing Trig Functions with Arcsine

Compose each of the six basic trig functions with $\sin^{-1}x$ and reduce the composite function to an algebraic expression involving no trig functions.

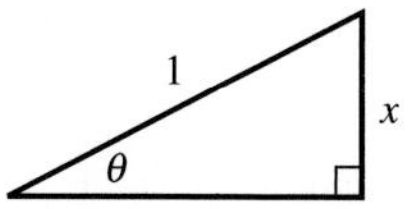

FIGURE 4.87 A triangle in which $\theta = \sin^{-1}x$. (Example 5)

SOLUTION This time we begin with the triangle shown in Figure 4.87, in which $\theta = \sin^{-1}x$. (This triangle could appear in the fourth quadrant if x were negative, but the trig ratios would be the same.)

The remaining side of the triangle (which is $\cos\theta$) can be found by the Pythagorean Theorem. If we denote the unknown side by s, we have

$$s^2 + x^2 = 1$$
$$s^2 = 1 - x^2$$
$$s = \pm\sqrt{1 - x^2}$$

Note the ambiguous sign, which requires a further look. Since $\sin^{-1}x$ is always in Quadrant I or IV, the horizontal side of the triangle can only be positive.

Therefore, we can actually write s unambiguously as $\sqrt{1 - x^2}$, giving us the triangle in Figure 4.88.

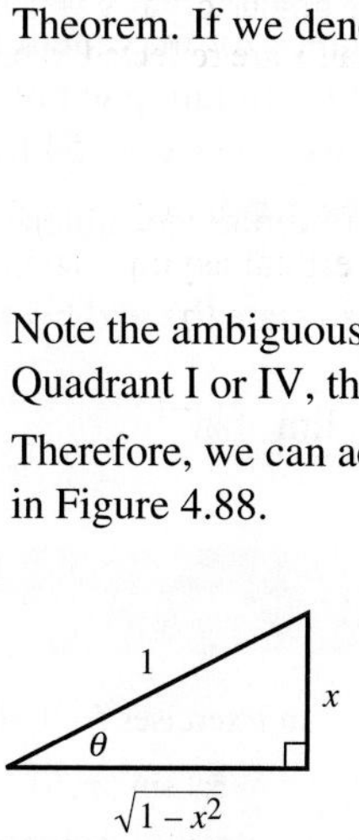

FIGURE 4.88 If $\theta = \sin^{-1}x$, then $\cos\theta = \sqrt{1 - x^2}$. Note that $\cos\theta$ will be positive because $\sin^{-1}x$ can only be in Quadrant I or IV. (Example 5)

(continued)

We can now read all the required ratios straight from the triangle:

$$\sin(\sin^{-1}(x)) = x \qquad \csc(\sin^{-1}(x)) = \frac{1}{x}$$

$$\cos(\sin^{-1}(x)) = \sqrt{1 - x^2} \qquad \sec(\sin^{-1}(x)) = \frac{1}{\sqrt{1 - x^2}}$$

$$\tan(\sin^{-1}(x)) = \frac{x}{\sqrt{1 - x^2}} \qquad \cot(\sin^{-1}(x)) = \frac{\sqrt{1 - x^2}}{x}$$

Now try Exercise 47.

Applications of Inverse Trigonometric Functions

When an application involves an angle as a dependent variable, as in $\theta = f(x)$, then to solve for x, it is natural to use an inverse trigonometric function and find $x = f^{-1}(\theta)$.

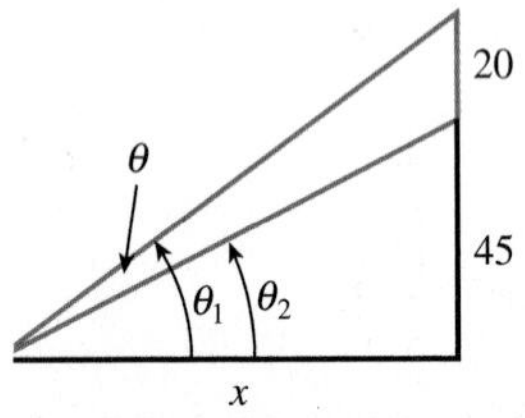

FIGURE 4.89 The diagram for the stadium screen. (Example 6)

EXAMPLE 6 Calculating a Viewing Angle

The bottom of a 20-ft replay screen at Dodger Stadium is 45 ft above the playing field. As you move away from the wall, the angle formed by the screen at your eye changes. There is a distance from the wall at which the angle is the greatest. What is that distance?

SOLUTION

Model The angle subtended by the screen is represented in Figure 4.89 by θ, and $\theta = \theta_1 - \theta_2$. Since $\tan\theta_1 = 65/x$, it follows that $\theta_1 = \tan^{-1}(65/x)$. Similarly, $\theta_2 = \tan^{-1}(45/x)$. Thus,

$$\theta = \tan^{-1}\frac{65}{x} - \tan^{-1}\frac{45}{x}.$$

Solve Graphically Figure 4.90 shows a graph of θ that reflects Degree mode. The question about distance for maximum viewing angle can be answered by finding the x-coordinate of the maximum point of this graph. Using grapher methods we see that this maximum occurs when $x \approx 54$ ft.

Therefore the maximum angle subtended by the replay screen occurs about 54 ft from the wall.

Now try Exercise 55.

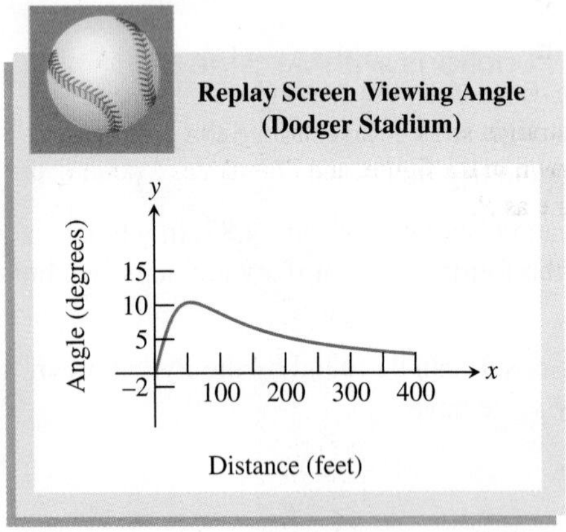

FIGURE 4.90 Viewing angle θ as a function of distance x from the wall. (Example 6)

QUICK REVIEW 4.7 *(For help, go to Section 4.3.)*

Exercise numbers with a gray background indicate problems that the authors have designed to be solved *without a calculator*.

In Exercises 1–4, state the sign (positive or negative) of the sine, cosine, and tangent in the quadrant.

1. Quadrant I
2. Quadrant II
3. Quadrant III
4. Quadrant IV

In Exercises 5–10, find the exact value.

5. $\sin(\pi/6)$
6. $\tan(\pi/4)$
7. $\cos(2\pi/3)$
8. $\sin(2\pi/3)$
9. $\sin(-\pi/6)$
10. $\cos(-\pi/3)$

SECTION 4.7 Exercises

In Exercises 1–12, find the exact value.

1. $\sin^{-1}\left(\frac{\sqrt{3}}{2}\right)$ **2.** $\sin^{-1}\left(-\frac{1}{2}\right)$

3. $\tan^{-1} 0$ **4.** $\cos^{-1} 1$

5. $\cos^{-1}\left(\frac{1}{2}\right)$ **6.** $\tan^{-1} 1$

7. $\tan^{-1}(-1)$ **8.** $\cos^{-1}\left(-\frac{\sqrt{3}}{2}\right)$

9. $\sin^{-1}\left(-\frac{1}{\sqrt{2}}\right)$ **10.** $\tan^{-1}(-\sqrt{3})$

11. $\cos^{-1} 0$ **12.** $\sin^{-1} 1$

In Exercises 13–16, use a calculator to find the approximate value. Express your answer in degrees.

13. $\sin^{-1}(0.362)$ **14.** $\arcsin 0.67$

15. $\tan^{-1}(-12.5)$ **16.** $\cos^{-1}(-0.23)$

In Exercises 17–20, use a calculator to find the approximate value. Express your result in radians.

17. $\tan^{-1}(2.37)$ **18.** $\tan^{-1}(22.8)$

19. $\sin^{-1}(-0.46)$ **20.** $\cos^{-1}(-0.853)$

In Exercises 21 and 22, describe the end behavior of the function.

21. $y = \tan^{-1}(x^2)$ **22.** $y = (\tan^{-1} x)^2$

In Exercises 23–32, find the exact value without a calculator.

23. $\cos(\sin^{-1}(1/2))$ **24.** $\sin(\tan^{-1} 1)$

25. $\sin^{-1}(\cos(\pi/4))$ **26.** $\cos^{-1}(\cos(7\pi/4))$

27. $\cos(2\sin^{-1}(1/2))$ **28.** $\sin(\tan^{-1}(-1))$

29. $\arcsin(\cos(\pi/3))$ **30.** $\arccos(\tan(\pi/4))$

31. $\cos\left(\tan^{-1}\sqrt{3}\right)$ **32.** $\tan^{-1}(\cos\pi)$

In Exercises 33–36, analyze each function for domain, range, continuity, increasing or decreasing behavior, symmetry, boundedness, extrema, asymptotes, and end behavior.

33. $f(x) = \sin^{-1} x$

34. $f(x) = \cos^{-1} x$

35. $f(x) = \tan^{-1} x$

36. $f(x) = \cot^{-1} x$ (See graph in Exercise 67.)

In Exercises 37–40, use transformations to describe how the graph of the function is related to a basic inverse trigonometric graph. State the domain and range.

37. $f(x) = \sin^{-1}(2x)$ **38.** $g(x) = 3\cos^{-1}(2x)$

39. $h(x) = 5\tan^{-1}(x/2)$ **40.** $g(x) = 3\arccos(x/2)$

In Exercises 41–46, find the solution to the equation without a calculator.

41. $\sin(\sin^{-1} x) = 1$ **42.** $\cos^{-1}(\cos x) = 1$

43. $2\sin x = 1$ **44.** $\tan x = -1$

45. $\cos(\cos^{-1} x) = 1/3$ **46.** $\sin^{-1}(\sin x) = \pi/10$

In Exercises 47–52, find an algebraic expression equivalent to the given expression. (*Hint:* Form a right triangle as done in Example 5.)

47. $\sin(\tan^{-1} x)$ **48.** $\cos(\tan^{-1} x)$

49. $\tan(\arcsin x)$ **50.** $\cot(\arccos x)$

51. $\cos(\arctan 2x)$ **52.** $\sin(\arccos 3x)$

53. Group Activity Viewing Angle You are standing in an art museum viewing a picture. The bottom of the picture is 2 ft above your eye level, and the picture is 12 ft tall. Angle θ is formed by the lines of vision to the bottom and to the top of the picture.

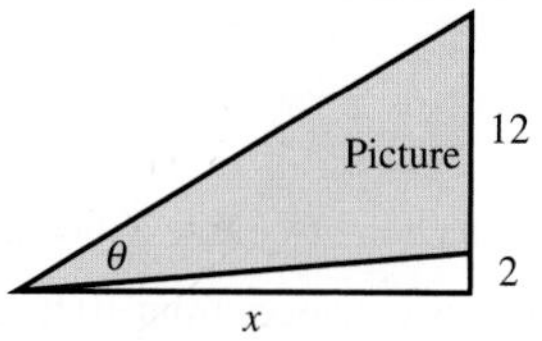

(a) Show that $\theta = \tan^{-1}\left(\frac{14}{x}\right) - \tan^{-1}\left(\frac{2}{x}\right)$.

(b) Graph θ in the $[0, 25]$ by $[0, 55]$ viewing window using Degree mode. Use your grapher to show that the maximum value of θ occurs approximately 5.3 ft from the picture.

(c) How far (to the nearest foot) are you standing from the wall if $\theta = 35°$?

54. Group Activity Analysis of a Lighthouse A rotating beacon L stands 3 m across the harbor from the nearest point P along a straight shoreline. As the light rotates, it forms an angle θ as shown in the figure, and illuminates a point Q on the same shoreline as P.

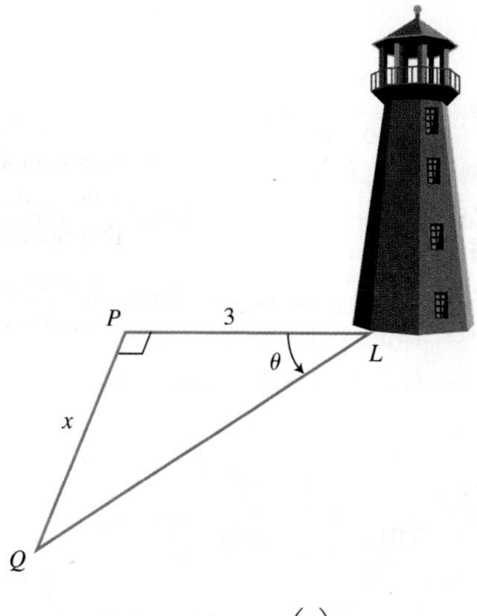

(a) Show that $\theta = \tan^{-1}\left(\frac{x}{3}\right)$.

(b) Graph θ in the viewing window $[-20, 20]$ by $[-90, 90]$ using Degree mode. What do negative values of x represent in the problem? What does a positive angle represent? A negative angle?

(c) Find θ when $x = 15$.

55. Rising Hot-Air Balloon The hot-air balloon festival held each year in Phoenix, Arizona, is a popular event for photographers. Jo Silver, an award-winning photographer at the event, watches a balloon rising from ground level from a point 500 ft away on level ground.

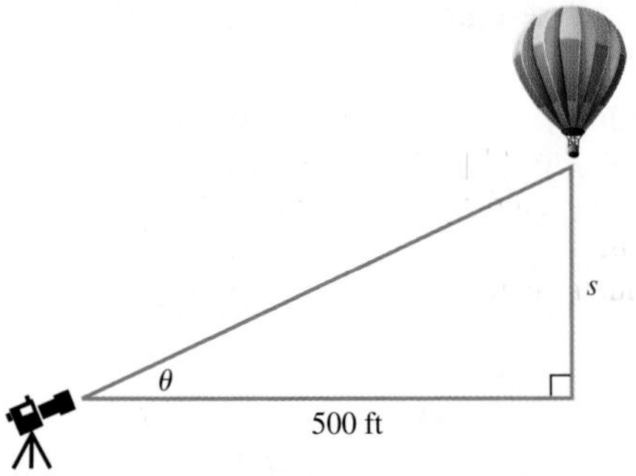

(a) Write θ as a function of the height s of the balloon.

(b) Is the change in θ greater as s changes from 10 ft to 20 ft, or as s changes from 200 ft to 210 ft? Explain.

(c) **Writing to Learn** In the graph of this relationship shown here, do you think that the x-axis represents the height s and the y-axis angle θ, or does the x-axis represent angle θ and the y-axis height s? Explain.

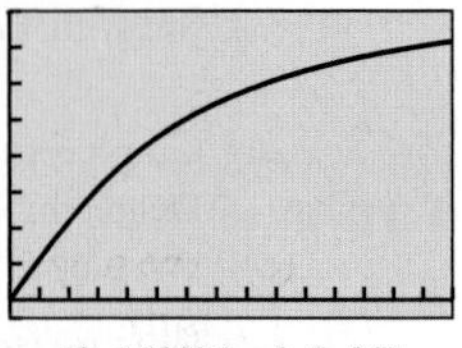

[0, 1500] by [–5, 80]

56. Find the domain and range of each of the following functions.

(a) $f(x) = \sin(\sin^{-1} x)$

(b) $g(x) = \sin^{-1}(x) + \cos^{-1}(x)$

(c) $h(x) = \sin^{-1}(\sin x)$

(d) $k(x) = \sin(\cos^{-1} x)$

(e) $q(x) = \cos^{-1}(\sin x)$

Standardized Test Questions

57. True or False $\sin(\sin^{-1} x) = x$ for all real numbers x. Justify your answer.

58. True or False The graph of $y = \arctan x$ has two horizontal asymptotes. Justify your answer.

You should answer these questions without using a calculator.

59. Multiple Choice $\cos^{-1}\left(-\frac{\sqrt{3}}{2}\right) =$

(A) $-\frac{7\pi}{6}$ (B) $-\frac{\pi}{3}$ (C) $-\frac{\pi}{6}$

(D) $\frac{2\pi}{3}$ (E) $\frac{5\pi}{6}$

60. Multiple Choice $\sin^{-1}(\sin \pi) =$

(A) -2π (B) $-\pi$ (C) 0

(D) π (E) 2π

61. Multiple Choice $\sec(\tan^{-1} x) =$

(A) x (B) $\csc x$ (C) $\sqrt{1 + x^2}$

(D) $\sqrt{1 - x^2}$ (E) $\frac{\sin x}{(\cos x)^2}$

62. Multiple Choice The range of the function $f(x) = \arcsin x$ is

(A) $(-\infty, \infty)$. (B) $(-1, 1)$. (C) $[-1, 1]$.

(D) $[0, \pi]$. (E) $[-\pi/2, \pi/2]$.

Explorations

63. Writing to Learn Using the format demonstrated in this section for the inverse sine, cosine, and tangent functions, give a careful definition of the inverse cotangent function. [*Hint:* The range of $y = \cot^{-1} x$ is $(0, \pi)$.]

64. Writing to Learn Use an appropriately labeled triangle to explain why $\sin^{-1} x + \cos^{-1} x = \pi/2$. For what values of x is the left-hand side of this equation defined?

65. Graph each of the following functions and interpret the graph to find the domain, range, and period of each function. Which of the three functions has points of discontinuity? Are the discontinuities removable or nonremovable?

(a) $y = \sin^{-1}(\sin x)$

(b) $y = \cos^{-1}(\cos x)$

(c) $y = \tan^{-1}(\tan x)$

Extending the Ideas

66. Practicing for Calculus Express each of the following functions as an algebraic expression involving no trig functions.

(a) $\cos(\sin^{-1} 2x)$ (b) $\sec^2(\tan^{-1} x)$

(c) $\sin\left(\cos^{-1} \sqrt{x}\right)$ (d) $-\csc^2(\cot^{-1} x)$

(e) $\tan(\sec^{-1} x^2)$

67. Arccotangent on the Calculator Most graphing calculators do not have a button for the inverse cotangent. The graph is shown below. Find an expression that you can put into your calculator to produce a graph of $y = \cot^{-1} x$.

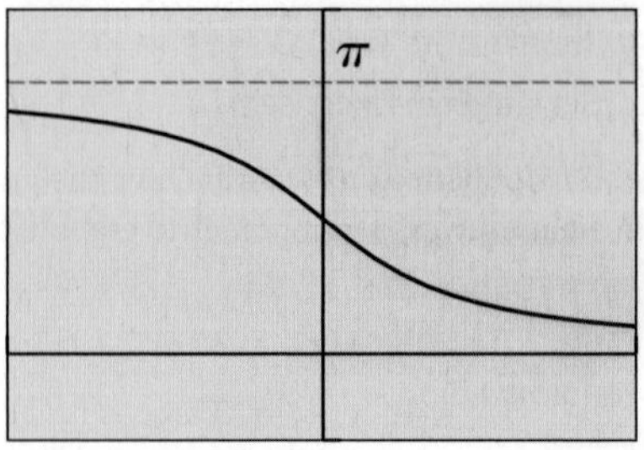

[–3, 3] by [–1, 4]

68. Advanced Decomposition Decompose each of the following algebraic functions by writing it as a trig function of an arctrig function.

(a) $\sqrt{1 - x^2}$ (b) $\dfrac{x}{\sqrt{1 + x^2}}$ (c) $\dfrac{x}{\sqrt{1 - x^2}}$

69. Use elementary transformations and the arctangent function to construct a function with domain all real numbers that has horizontal asymptotes at $y = 24$ and $y = 42$.

70. Avoiding Ambiguities When choosing the right triangle in Example 5, we used a hypotenuse of 1. It is sometimes necessary to use a variable quantity for the hypotenuse, in which case it is a good idea to use x^2 rather than x, just in case x is negative. (All of our definitions of the trig functions have involved triangles in which the hypotenuse is assumed to be positive.)

(a) If we use the triangle below to represent $\theta = \sin^{-1}(1/x)$, explain why side s must be positive regardless of the sign of x.

(b) Use the triangle in part (a) to find $\tan(\sin^{-1}(1/x))$.

(c) Using an appropriate triangle, find $\sin(\cos^{-1}(1/x))$.

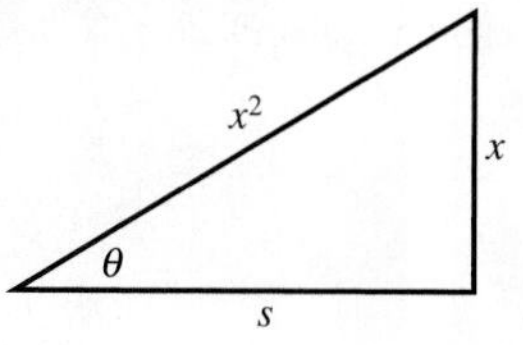

71. Defining Arcsecant The range of the secant function is $(-\infty, -1] \cup [1, \infty)$, which must become the domain of the arcsecant function. The graph of $y = \text{arcsec}\, x$ must therefore be the union of two unbroken curves. Two possible graphs with the correct domain are shown below.

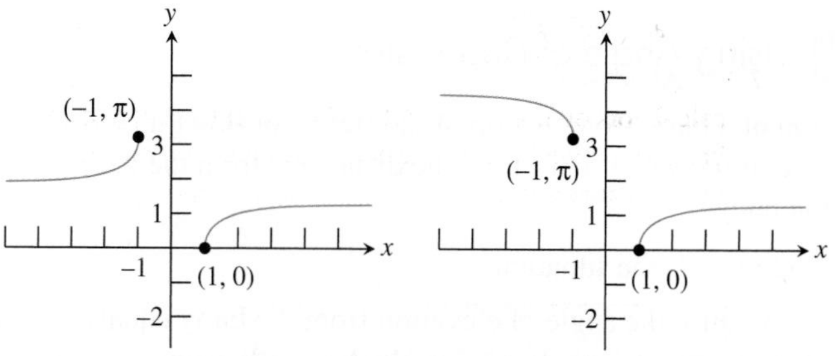

(a) The graph on the left has one horizontal asymptote. What is it?

(b) The graph on the right has two horizontal asymptotes. What are they?

(c) Which of these graphs is also the graph of $y = \cos^{-1}(1/x)$?

(d) Which of these graphs is increasing on both connected intervals?

72. Defining Arccosecant The range of the cosecant function is $(-\infty, -1] \cup [1, \infty)$, which must become the domain of the arccosecant function. The graph of $y = \text{arccsc}\, x$ must therefore be the union of two unbroken curves. Two possible graphs with the correct domain are shown below.

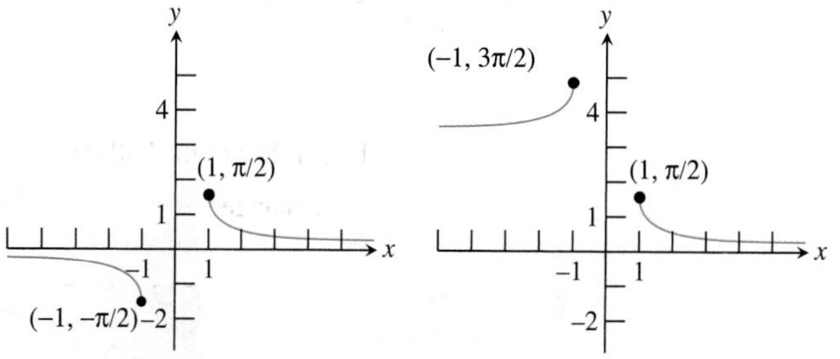

(a) The graph on the left has one horizontal asymptote. What is it?

(b) The graph on the right has two horizontal asymptotes. What are they?

(c) Which of these graphs is also the graph of $y = \sin^{-1}(1/x)$?

(d) Which of these graphs is decreasing on both connected intervals?

4.8 Solving Problems with Trigonometry

What you'll learn about

- More Right Triangle Problems
- Simple Harmonic Motion

... and why

These problems illustrate some of the better-known applications of trigonometry.

More Right Triangle Problems

We close this first of two trigonometry chapters by revisiting some of the applications of Section 4.2 (right triangle trigonometry) and Section 4.4 (sinusoids).

An **angle of elevation** is the angle through which the eye moves up from horizontal to look at something above, and an **angle of depression** is the angle through which the eye moves down from horizontal to look at something below. For two observers at different elevations looking at each other, the angle of elevation for one equals the angle of depression for the other. The concepts are illustrated in Figure 4.91 as they might apply to observers at Mount Rushmore or the Grand Canyon.

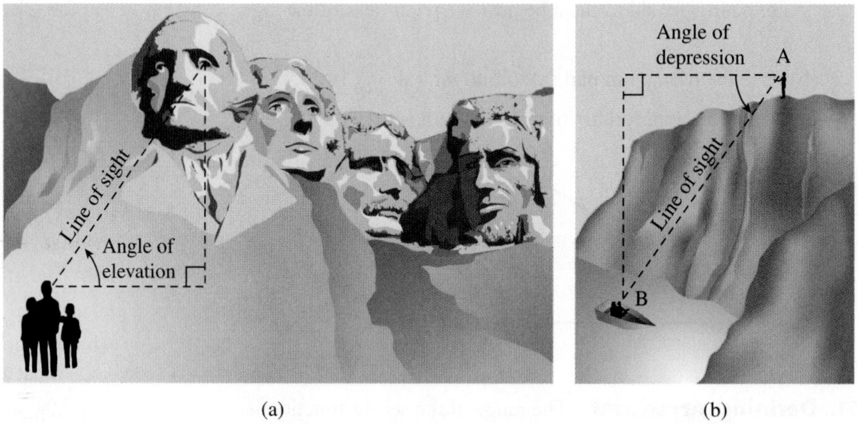

FIGURE 4.91 (a) Angle of elevation at Mount Rushmore. (b) Angle of depression at the Grand Canyon.

EXAMPLE 1 Using Angle of Depression

The angle of depression of a buoy from the top of the Barnegat Bay lighthouse 130 ft above the surface of the water is 6°. Find the distance x from the base of the lighthouse to the buoy.

SOLUTION Figure 4.92 models the situation.

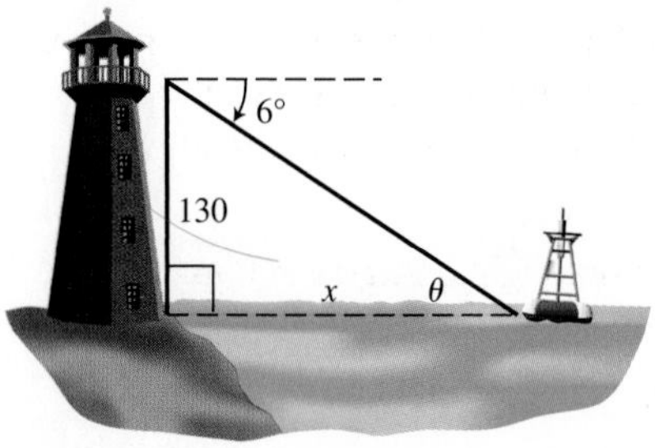

FIGURE 4.92 A big lighthouse and a little buoy. (Example 1)

In the diagram, $\theta = 6°$ because the angle of elevation from the buoy equals the angle of depression from the lighthouse. We **solve algebraically** using the tangent function:

$$\tan \theta = \tan 6° = \frac{130}{x}$$

$$x = \frac{130}{\tan 6°} \approx 1236.9$$

Interpreting We find that the buoy is about 1237 ft from the base of the lighthouse.

Now try Exercise 3.

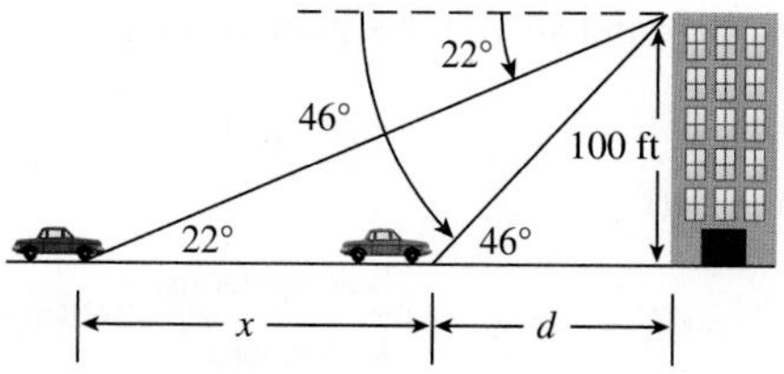

FIGURE 4.93 A car approaches Altgelt Hall. (Example 2)

EXAMPLE 2 Making Indirect Measurements

From the top of the 100-ft-tall Altgelt Hall a man observes a car moving toward the building. If the angle of depression of the car changes from 22° to 46° during the period of observation, how far does the car travel?

SOLUTION

Solve Algebraically Figure 4.93 models the situation. Notice that we have labeled the acute angles at the car's two positions as 22° and 46° (because the angle of elevation from the car equals the angle of depression from the building). Denote the distance the car moves as x. Denote its distance from the building at the second observation as d.

From the smaller right triangle we conclude:

$$\tan 46° = \frac{100}{d}$$

$$d = \frac{100}{\tan 46°}$$

From the larger right triangle we conclude:

$$\tan 22° = \frac{100}{x + d}$$

$$x + d = \frac{100}{\tan 22°}$$

$$x = \frac{100}{\tan 22°} - d$$

$$x = \frac{100}{\tan 22°} - \frac{100}{\tan 46°}$$

$$x \approx 150.9$$

Interpreting We find that the car travels about 151 ft.

Now try Exercise 7.

FIGURE 4.94 A large, helium-filled penguin. (Example 3)

EXAMPLE 3 Finding Height Above Ground

A large, helium-filled penguin is moored at the beginning of a parade route awaiting the start of the parade. Two cables attached to the underside of the penguin make angles of 48° and 40° with the ground and are in the same plane as a perpendicular line from the penguin to the ground. (See Figure 4.94.) If the cables are attached to the ground 10 ft from each other, how high above the ground is the penguin?

SOLUTION We can simplify the drawing to the two right triangles in Figure 4.95 that share the common side h.

Model By the definition of the tangent function,

$$\frac{h}{x} = \tan 48° \quad \text{and} \quad \frac{h}{x + 10} = \tan 40°.$$

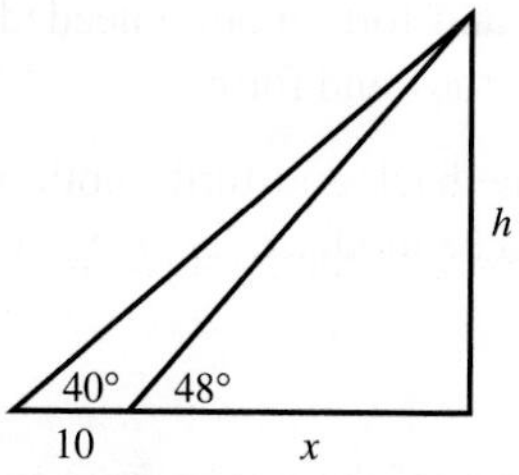

FIGURE 4.95 (Example 3)

Solve Algebraically Solving for h,

$$h = x \tan 48° \quad \text{and} \quad h = (x + 10) \tan 40°.$$

(continued)

Set these two expressions for h equal to each other and solve the equation for x:

$$x \tan 48° = (x + 10) \tan 40° \quad \text{Both equal } h.$$
$$x \tan 48° = x \tan 40° + 10 \tan 40°$$
$$x \tan 48° - x \tan 40° = 10 \tan 40° \quad \text{Isolate } x \text{ terms.}$$
$$x(\tan 48° - \tan 40°) = 10 \tan 40° \quad \text{Factor out } x.$$
$$x = \frac{10 \tan 40°}{\tan 48° - \tan 40°} \approx 30.90459723$$

We retain the full display for x because we are not finished yet; we need to solve for h:

$$h = x \tan 48° = (30.90459723) \tan 48° \approx 34.32$$

The penguin is approximately 34 ft above ground level.

Now try Exercise 15.

EXAMPLE 4 Using Trigonometry in Navigation

A U.S. Coast Guard patrol boat leaves Port Cleveland and averages 35 **knots** (naut mi per hr) traveling for 2 hr on a course of 53° and then 3 hr on a course of 143°. What is the boat's bearing and distance from Port Cleveland?

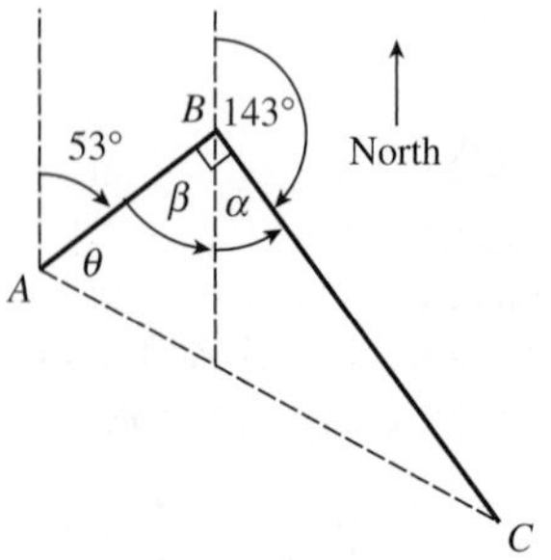

FIGURE 4.96 Path of travel for a Coast Guard boat that corners well at 35 knots. (Example 4)

SOLUTION Figure 4.96 models the situation.

Solve Algebraically In the diagram, line AB is a transversal that cuts a pair of parallel lines. Thus, $\beta = 53°$ because they are alternate interior angles. Angle α, as the supplement of a 143° angle, is 37°. Consequently, $\angle ABC = 90°$ and AC is the hypotenuse of right $\triangle ABC$.

Use distance = rate × time to determine distances AB and BC.

$$AB = (35 \text{ knots})(2 \text{ hr}) = 70 \text{ naut mi}$$
$$BC = (35 \text{ knots})(3 \text{ hr}) = 105 \text{ naut mi}$$

Solve the right triangle for AC and θ.

$$AC = \sqrt{70^2 + 105^2} \quad \text{Pythagorean Theorem}$$
$$AC \approx 126.2$$
$$\theta = \tan^{-1}\left(\frac{105}{70}\right)$$
$$\theta \approx 56.3°$$

Interpreting We find that the boat's bearing from Port Cleveland is $53° + \theta$, or approximately 109.3°. They are about 126 naut mi out.

Now try Exercise 17.

Simple Harmonic Motion

Because of their periodic nature, the sine and cosine functions are helpful in describing the motion of objects that oscillate, vibrate, or rotate. For example, the linkage in Figure 4.97 converts the rotary motion of a motor to the back-and-forth motion needed for some machines. When the wheel rotates, the piston moves back and forth.

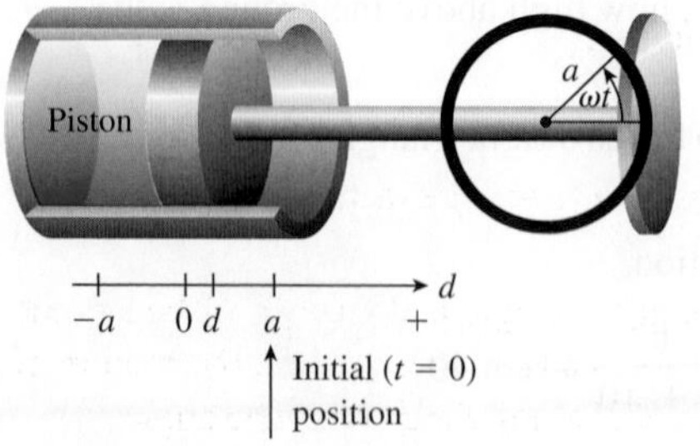

FIGURE 4.97 A piston operated by a wheel rotating at a constant rate demonstrates simple harmonic motion.

If the wheel rotates at a constant rate ω radians per second, the back-and-forth motion of the piston is an example of *simple harmonic motion* and can be modeled by an equation of the form

$$d = a \cos \omega t, \quad \omega > 0,$$

where a is the radius of the wheel and d is the directed distance of the piston from its center of oscillation.

Frequency and Period

Notice that harmonic motion is sinusoidal, with amplitude $|a|$ and period $2\pi/\omega$. The frequency is the reciprocal of the period.

For the sake of simplicity, we will define simple harmonic motion in terms of a point moving along a number line.

DEFINITION Simple Harmonic Motion

A point moving on a number line is in **simple harmonic motion** if its directed distance d from the origin is given by either

$$d = a \sin \omega t \quad \text{or} \quad d = a \cos \omega t,$$

where a and ω are real numbers and $\omega > 0$. The motion has **frequency** $\omega/2\pi$, which is the number of oscillations per unit of time.

EXPLORATION 1 Watching Harmonic Motion

You can watch harmonic motion on your graphing calculator. Set your grapher to parametric mode and set $X_{1T} = \cos(T)$ and $Y_{1T} = \sin(T)$. Set Tmin $= 0$, Tmax $= 25$, Tstep $= 0.2$, Xmin $= -1.5$, Xmax $= 1.5$, Xscl $= 1$, Ymin $= -100$, Ymax $= 100$, Yscl $= 0$.

If your calculator allows you to change style to graph a moving ball, choose that style. When you graph the function, you will see the ball moving along the x-axis between -1 and 1 in simple harmonic motion. If your grapher does not have the moving ball option, wait for the grapher to finish graphing, then press TRACE and keep your finger pressed on the right arrow key to see the tracer move in simple harmonic motion.

1. For each value of T, the parametrization gives the point $(\cos(T), \sin(T))$. What well-known curve should this parametrization produce?
2. Why does the point seem to go back and forth on the x-axis when it should be following the curve identified in part 1? (*Hint:* Check that viewing window again!)
3. Why does the point slow down at the extremes and speed up in the middle? (*Hint:* Remember that the grapher is really following the curve identified in part 1.)
4. How can you tell that this point moves in simple harmonic motion?

EXAMPLE 5 Calculating Harmonic Motion

In a mechanical linkage like the one shown in Figure 4.97, a wheel with an 8-cm radius turns with an angular velocity of 8π rad/sec.

(a) What is the frequency of the piston in cycles per second?

(b) What is the distance from the starting position ($t = 0$) exactly 3.45 sec after starting?

SOLUTION Imagine the wheel to be centered at the origin and let $P(x, y)$ be a point on its perimeter (Figure 4.98). As the wheel rotates and P goes around, the motion of the piston follows the path of the x-coordinate of P along the x-axis. The angle determined by P at any time t is $8\pi t$, so its x-coordinate is $8 \cos 8\pi t$. Therefore, the sinusoid $d = 8 \cos 8\pi t$ models the motion of the piston.

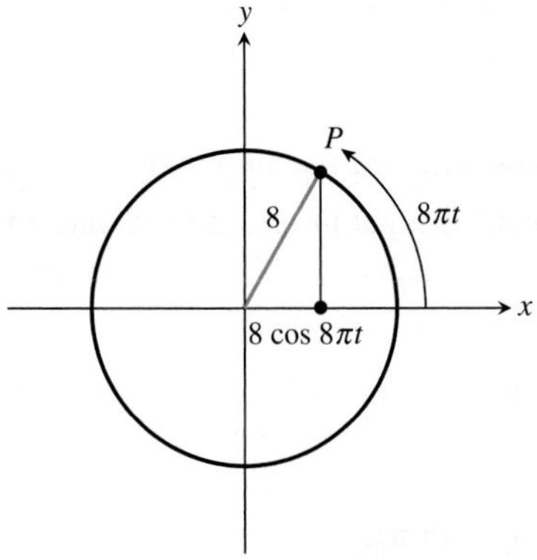

FIGURE 4.98 Modeling the path of a piston by a sinusoid. (Example 5)

(continued)

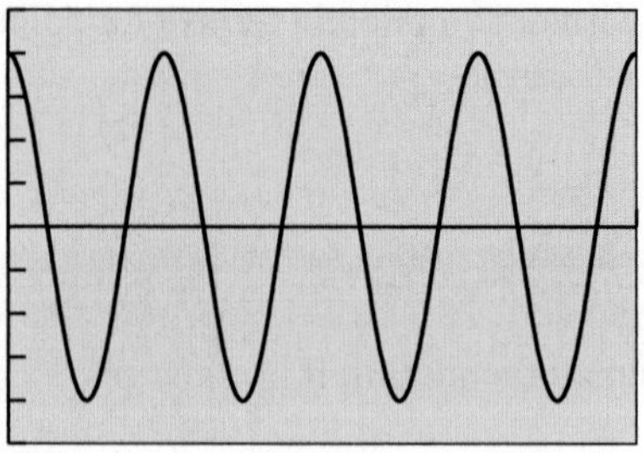

[0, 1] by [–10, 10]

FIGURE 4.99 A sinusoid with frequency 4 models the motion of the piston in Example 5.

(a) The frequency of $d = 8 \cos 8\pi t$ is $8\pi/2\pi$, or 4 cycles/sec. The piston makes four complete back-and-forth strokes per second. The graph of d as a function of t is shown in Figure 4.99. The four cycles of the sinusoidal graph in the interval $[0, 1]$ model the four cycles of the motor or the four strokes of the piston. Note that the sinusoid has a period of 1/4, the reciprocal of the frequency.

(b) We must find the distance between the positions at $t = 0$ and $t = 3.45$. The initial position at $t = 0$ is

$$d(0) = 8.$$

The position at $t = 3.45$ is

$$d(3.45) = 8 \cos (8\pi \cdot 3.45) \approx 2.47.$$

The distance between the two positions is approximately $8 - 2.47 = 5.53$.

Interpreting We conclude that the piston is approximately 5.53 cm from its starting position after 3.45 sec.

Now try Exercise 27.

EXAMPLE 6 Calculating Harmonic Motion

A mass oscillating up and down on the bottom of a spring (assuming perfect elasticity and no friction or air resistance) can be modeled as harmonic motion. If the weight is displaced a maximum of 5 cm, find the modeling equation if it takes 2 sec to complete one cycle. (See Figure 4.100.)

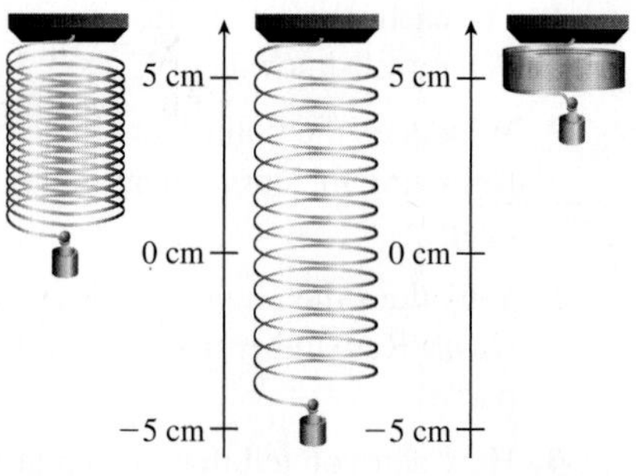

FIGURE 4.100 The mass and spring in Example 6.

SOLUTION We have our choice between the two equations $d = a \sin \omega t$ or $d = a \cos \omega t$. Assuming that the spring is at the origin of the coordinate system when $t = 0$, we choose the equation $d = a \sin \omega t$.

Because the maximum displacement is 5 cm, we conclude that the amplitude $a = 5$.

Because it takes 2 sec to complete one cycle, we conclude that the period is 2 and the frequency is 1/2. Therefore,

$$\frac{\omega}{2\pi} = \frac{1}{2}$$

$$\omega = \pi$$

Putting it all together, our modeling equation is $d = 5 \sin \pi t$.

Now try Exercise 29.

CHAPTER OPENER Problem (from page 319)

Problem: If we know that the musical note A above middle C has a pitch of 440 Hz, how can we model the sound produced by it at 80 dB?

Solution: Sound is modeled by simple harmonic motion, with frequency perceived as pitch and measured in hertz (Hz), and amplitude perceived as loudness and measured in decibels (dB). So for the musical note A with a pitch of 440 Hz, we have frequency $= \omega/2\pi = 440$ and therefore $\omega = 2\pi 440 = 880\pi$.

If this note is played at a loudness of 80 dB, we have $|a| = 80$. Using the simple harmonic motion model $d = a \sin \omega t$, we have

$$d = 80 \sin 880\pi t.$$

QUICK REVIEW 4.8 *(For help, go to Sections 4.1, 4.2, and 4.3.)*

Exercise numbers with a gray background indicate problems that the authors have designed to be solved *without a calculator*.

In Exercises 1–4, find the lengths a, b, and c.

1.

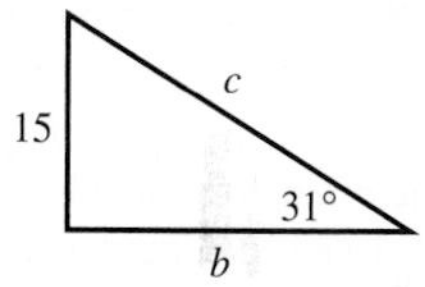

2.

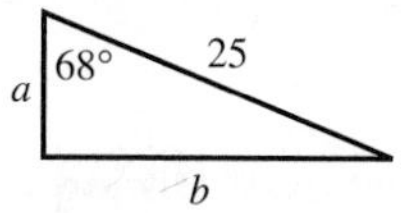

3.

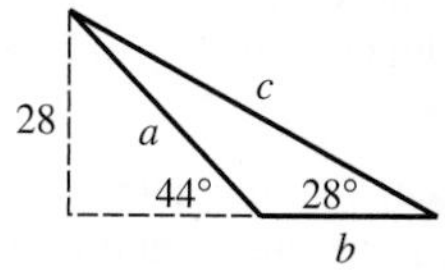

4.

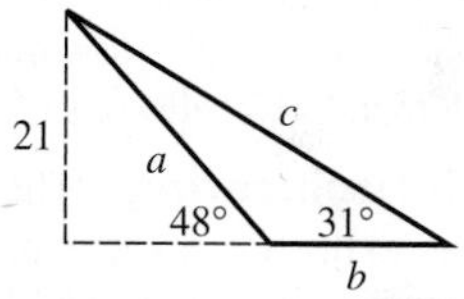

In Exercises 5 and 6, find the complement and supplement of the angle.

5. 32°

6. 73°

In Exercises 7 and 8, state the bearing that describes the direction.

7. NE (northeast)

8. SSW (south-southwest)

In Exercises 9 and 10, state the amplitude and period of the sinusoid.

9. $-3 \sin 2(x - 1)$

10. $4 \cos 4(x + 2)$

SECTION 4.8 Exercises

In Exercises 1–43, solve the problem using your knowledge of geometry and the techniques of this section. Sketch a figure if one is not provided.

1. Finding a Cathedral Height The angle of elevation of the top of the Ulm Cathedral from a point 300 ft away from the base of its steeple on level ground is 60°. Find the height of the cathedral.

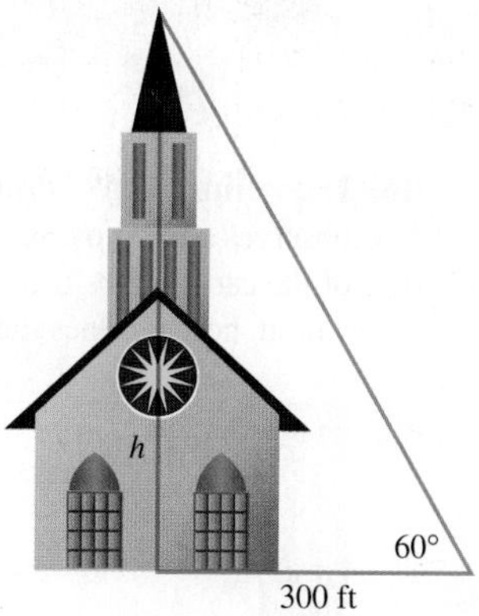

2. Finding a Monument Height From a point 100 ft from its base, the angle of elevation of the top of the Arch of Septimus Severus, in Rome, Italy, is 34°13′12″. How tall is this monument?

3. Finding a Distance The angle of depression from the top of the Smoketown Lighthouse 120 ft above the surface of the water to a buoy is 10°. How far is the buoy from the lighthouse?

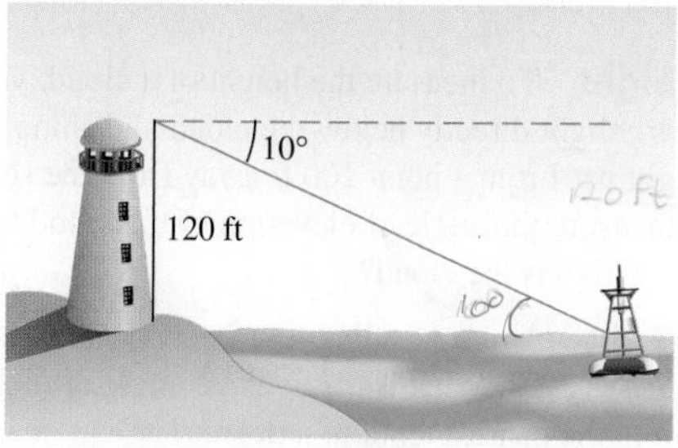

4. **Finding a Baseball Stadium Dimension** The top row of seats behind home plate at Cincinnati's Great American Ball Park is 90 ft above the level of the playing field. The angle of depression to the base of the left field wall is 14°. How far is the base of the left field wall from a point on level ground directly below the top row?

5. **Finding a Guy-Wire Length** A guy wire connects the top of an antenna to a point on level ground 50 ft from the base of the antenna. The angle of elevation formed by this wire is 80°. What are the length of the wire and the height of the antenna?

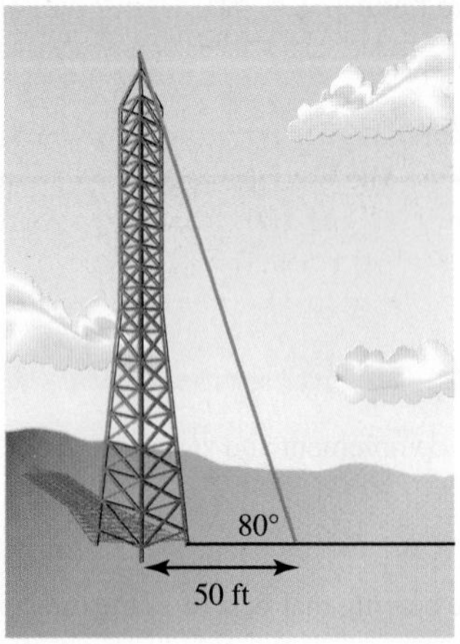

6. **Finding a Length** A wire stretches from the top of a vertical pole to a point on level ground 16 ft from the base of the pole. If the wire makes an angle of 62° with the ground, find the height of the pole and the length of the wire.

7. **Height of Eiffel Tower** The angle of elevation of the top of the TV antenna mounted on top of the Eiffel Tower in Paris is measured to be 80°1′12″ at a point 185 ft from the base of the tower. How tall is the tower plus TV antenna?

8. **Finding the Height of Tallest Chimney** The world's tallest smokestack at the International Nickel Co., Sudbury, Ontario, casts a shadow that is approximately 1580 ft long when the Sun's angle of elevation (measured from the horizon) is 38°. How tall is the smokestack?

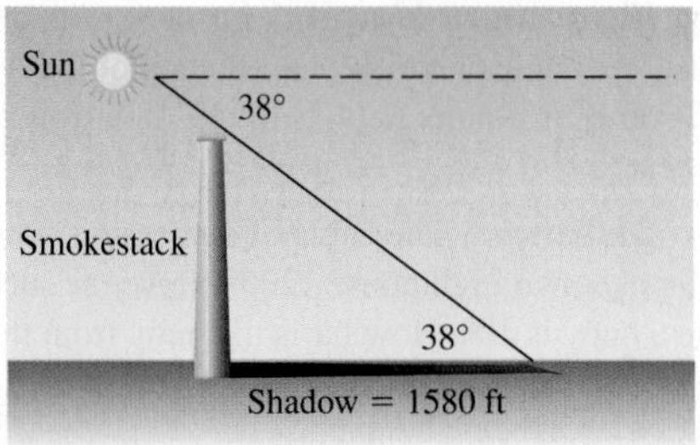

9. **Cloud Height** To measure the height of a cloud, you place a bright searchlight directly below the cloud and shine the beam straight up. From a point 100 ft away from the searchlight, you measure the angle of elevation of the cloud to be 83°12′. How high is the cloud?

10. **Ramping Up** A ramp leading to a freeway overpass is 470 ft long and rises 32 ft. What is the average angle of inclination of the ramp to the nearest tenth of a degree?

11. **Antenna Height** A guy wire attached to the top of the KSAM radio antenna is anchored at a point on the ground 10 m from the antenna's base. If the wire makes an angle of 55° with level ground, how high is the KSAM antenna?

12. **Building Height** To determine the height of the Louisiana-Pacific (LP) Tower, the tallest building in Conroe, Texas, a surveyor stands at a point on the ground, level with the base of the LP building. He measures the point to be 125 ft from the building's base and the angle of elevation to the top of the building to be 29°48′. Find the height of the building.

13. **Navigation** The *Paz Verde*, a whalewatch boat, is located at point P, and L is the nearest point on the Baja California shore. Point Q is located 4.25 mi down the shoreline from L and $\overline{PL} \perp \overline{LQ}$. Determine the distance that the *Paz Verde* is from the shore if $\angle PQL = 35°$.

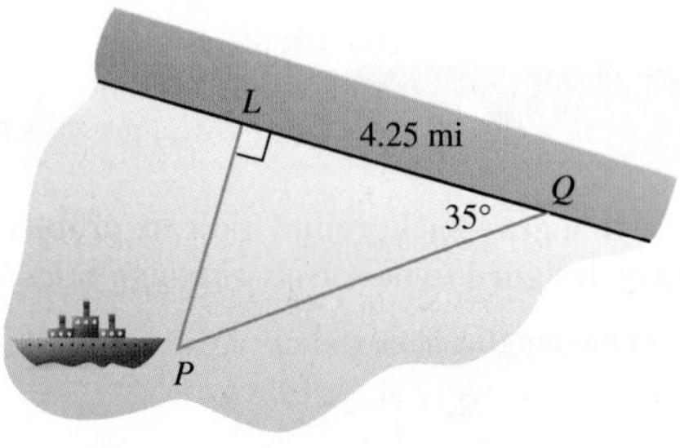

14. **Recreational Hiking** While hiking on a level path toward Colorado's front range, Otis Evans determines that the angle of elevation to the top of Long's Peak is 30°. Moving 1000 ft closer to the mountain, Otis determines the angle of elevation to be 35°. How much higher is the top of Long's Peak than Otis's elevation?

15. **Civil Engineering** The angle of elevation from an observer to the bottom edge of the Delaware River drawbridge observation deck located 200 ft from the observer is 30°. The angle of elevation from the observer to the top of the observation deck is 40°. What is the height of the observation deck?

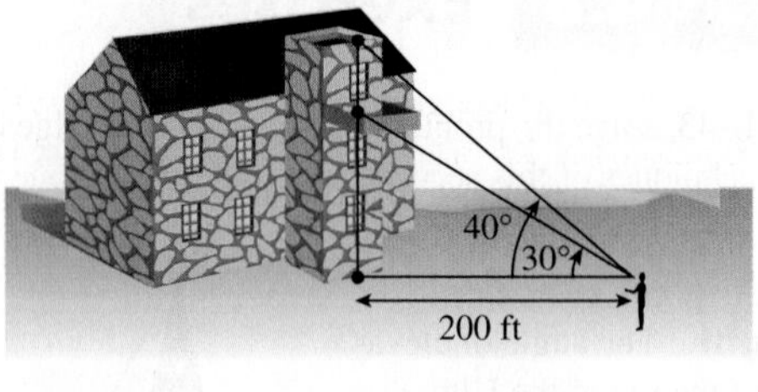

16. **Traveling Car** From the top of a 100-ft building a man observes a car moving toward him. If the angle of depression of the car changes from 15° to 33° during the period of observation, how far does the car travel?

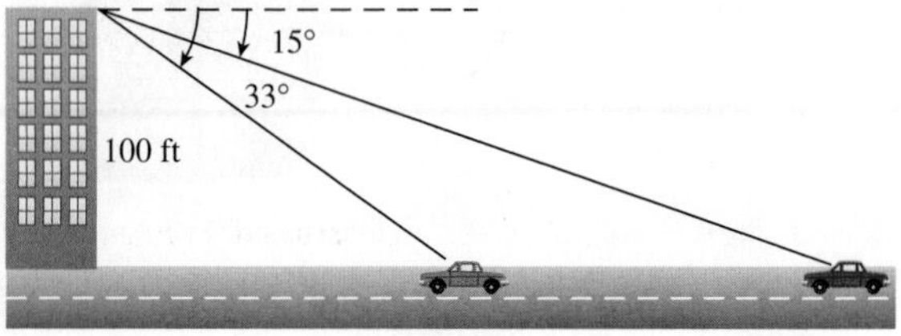

17. Navigation The Coast Guard cutter *Angelica* travels at 30 knots from its home port of Corpus Christi on a course of 95° for 2 hr and then changes to a course of 185° for 2 hr. Find the distance and the bearing from the Corpus Christi port to the boat.

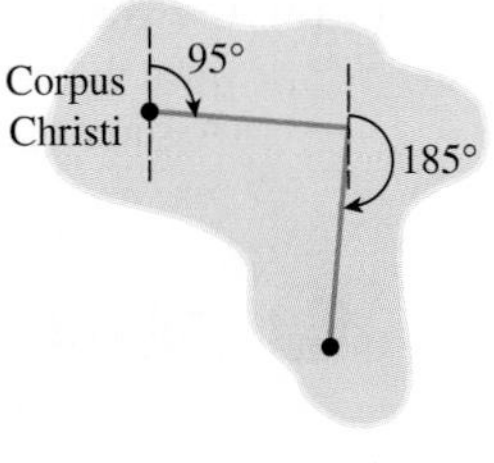

18. Navigation The *Cerrito Lindo* travels at a speed of 40 knots from Fort Lauderdale on a course of 65° for 2 hr and then changes to a course of 155° for 4 hr. Determine the distance and the bearing from Fort Lauderdale to the boat.

19. Land Measure The angle of depression is 19° from a point 7256 ft above sea level on the north rim of the Grand Canyon level to a point 6159 ft above sea level on the south rim. How wide is the canyon at that point?

20. Ranger Fire Watch A ranger spots a fire from a 73-ft tower in Yellowstone National Park. She measures the angle of depression to be 1°20′. How far is the fire from the tower?

21. Civil Engineering The bearing of the line of sight to the east end of the Royal Gorge footbridge from a point 375 ft due north of the west end of the footbridge across the Royal Gorge is 113°. What is the length l of the bridge?

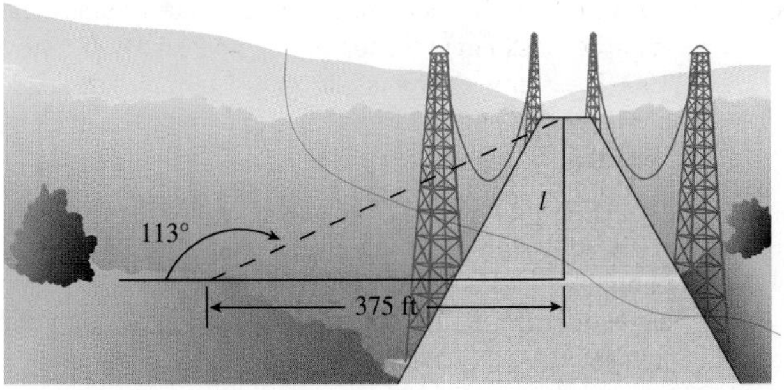

22. Space Flight The angle of elevation of a space shuttle from Cape Canaveral is 17° when the shuttle is directly over a ship 12 mi downrange. What is the altitude of the shuttle when it is directly over the ship?

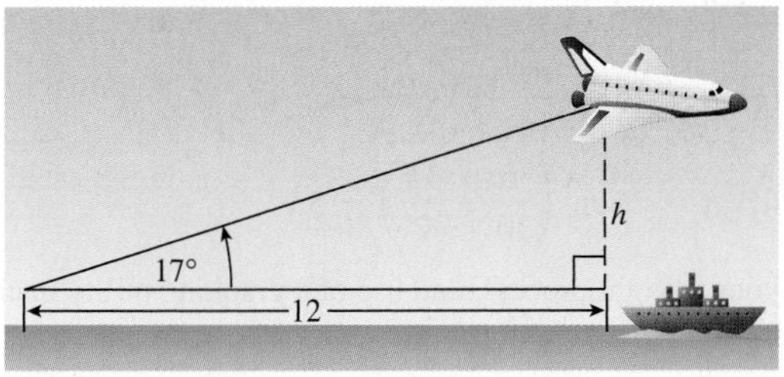

23. Architectural Design A barn roof is constructed as shown in the figure at the top of the next column. What is the height of the vertical center span?

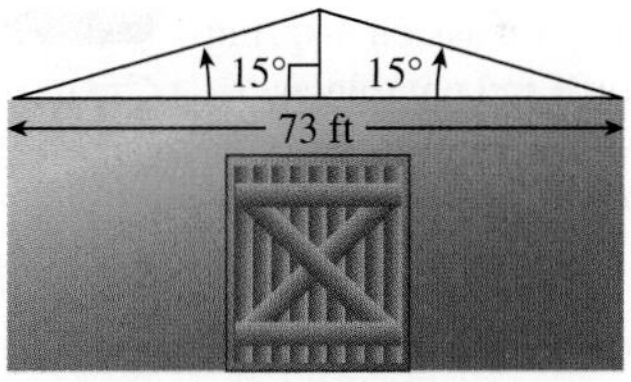

24. Recreational Flying A hot-air balloon over Park City, Utah, is 760 ft above the ground. The angle of depression from the balloon to an observer is 5.25°. Assuming the ground is relatively flat, how far is the observer from a point on the ground directly under the balloon?

25. Navigation A shoreline runs north-south, and a boat is due east of the shoreline. The bearings of the boat from two points on the shore are 110° and 100°. Assume the two points are 550 ft apart. How far is the boat from the shore?

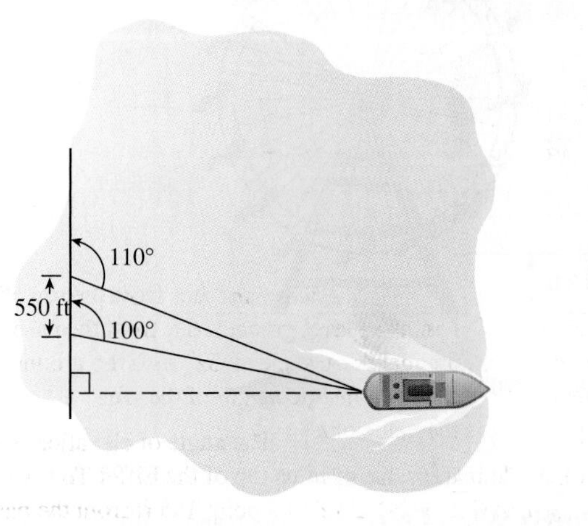

26. Navigation Milwaukee, Wisconsin, is directly west of Grand Haven, Michigan, on opposite sides of Lake Michigan. On a foggy night, a law enforcement boat leaves from Milwaukee on a course of 105° at the same time that a small smuggling craft steers a course of 195° from Grand Haven. The law enforcement boat averages 23 knots and collides with the smuggling craft. What was the smuggling boat's average speed?

27. Mechanical Design *Refer to Figure 4.97.* The wheel in a piston linkage like the one shown in the figure has a radius of 6 in. It turns with an angular velocity of 16π rad/sec. The initial position is the same as that shown in Figure 4.97.

(a) What is the frequency of the piston?

(b) What equation models the motion of the piston?

(c) What is the distance from the initial position 2.85 sec after starting?

28. Mechanical Design Suppose the wheel in a piston linkage like the one shown in Figure 4.97 has a radius of 18 cm and turns with an angular velocity of π rad/sec.

(a) What is the frequency of the piston?

(b) What equation models the motion of the piston?

(c) How many cycles does the piston make in 1 min?

29. Vibrating Spring A mass on a spring oscillates back and forth and completes one cycle in 0.5 sec. Its maximum displacement is 3 cm. Write an equation that models this motion.

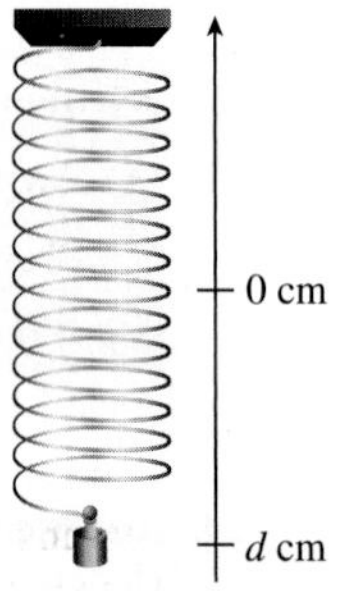

30. Tuning Fork A point on the tip of a tuning fork vibrates in harmonic motion described by the equation $d = 14 \sin \omega t$. Find ω for a tuning fork that has a frequency of 528 vibrations per second.

31. Ferris Wheel Motion The Ferris wheel shown in this figure makes one complete turn every 20 sec. A rider's height, h, above the ground can be modeled by the equation $h = a \sin \omega t + k$, where h and k are given in feet and t is given in seconds.

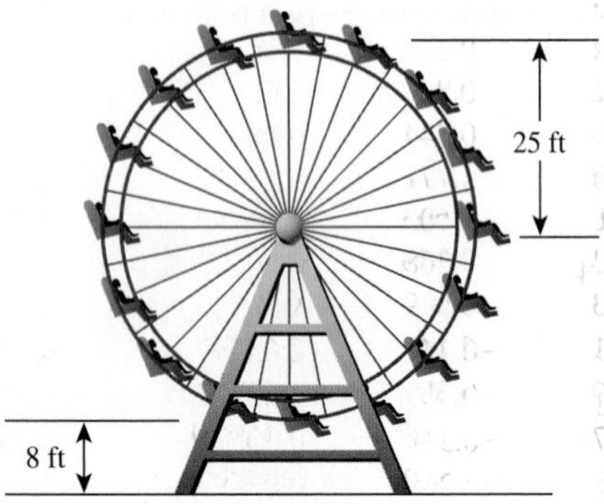

(a) What is the value of a?

(b) What is the value of k?

(c) What is the value of ω?

32. Ferris Wheel Motion Jacob and Emily ride a Ferris wheel at a carnival in Billings, Montana. The wheel has a 16-m diameter and turns at 3 rpm with its lowest point 1 m above the ground. Assume that Jacob and Emily's height h above the ground is a sinusoidal function of time t (in seconds), where $t = 0$ represents the lowest point of the wheel.

(a) Write an equation for h.

(b) Draw a graph of h for $0 \le t \le 30$.

(c) Use h to estimate Jacob and Emily's height above the ground at $t = 4$ and $t = 10$.

33. Monthly Temperatures in Charleston The monthly normal mean temperatures in Charleston, SC, are shown in Table 4.3. A scatter plot suggests that the mean monthly temperatures follow a sinusoidal curve over time. Assume that the sinusoid has equation $y = a \sin (b(t - h)) + k$.

(a) Given that the period is 12 months, find b.

(b) Assuming that the high and low temperatures in the table determine the range of the sinusoid, find a and k.

(c) Find a value of h that will put the minimum at $t = 1$ and the maximum at $t = 7$.

(d) Superimpose a graph of your sinusoid on a scatter plot of the data. How good is the fit?

(e) Use your sinusoidal model to predict dates in the year when the mean temperature in Charleston will be 70°. (Assume that $t = 0$ represents January 1.)

Table 4.3 Average Daily Temperature for Charleston, SC

Month	Temperature (°F)
1	49.2
2	51.3
3	57.7
4	64.9
5	72.7
6	78.8
7	81.5
8	80.7
9	76.3
10	66.8
11	57.9
12	50.9

Source: Southeast Regional Climate Center, 2013.

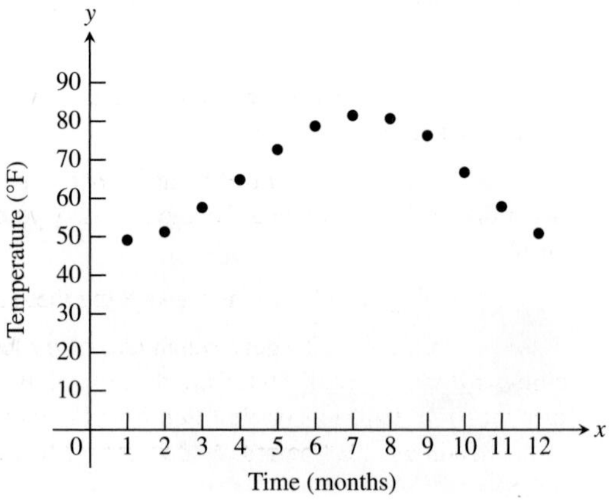

34. Writing to Learn For the Ferris wheel in Exercise 31, which equation correctly models the height of a rider who begins the ride at the bottom of the wheel when $t = 0$?

(a) $h = 25 \sin \dfrac{\pi t}{10}$

(b) $h = 25 \sin \dfrac{\pi t}{10} + 8$

(c) $h = 25 \sin \dfrac{\pi t}{10} + 33$

(d) $h = 25 \sin \left(\dfrac{\pi t}{10} + \dfrac{3\pi}{2}\right) + 33$

Explain your thought process, and use of a graphing utility in choosing the correct modeling equation.

35. **Monthly Sales** Owing to startup costs and seasonal variations, Gina found that the monthly profit in her bagel shop during the first year followed an up-and-down pattern that could be modeled by $P = 2t - 7\sin(\pi t/3)$, where P was measured in hundreds of dollars and t was measured in months after January 1.

(a) In what month did the shop first begin to make money?

(b) In what month did the shop enjoy its greatest profit in that first year?

36. **Weight Loss** Courtney tried several different diets over a two-year period in an attempt to lose weight. She found that her weight W followed a fluctuating curve that could be modeled by the function $W = 220 - 1.5t + 9.81\sin(\pi t/4)$, where t was measured in months after January 1 of the first year and W was measured in pounds.

(a) What was Courtney's weight at the start and at the end of two years?

(b) What was her maximum weight during the two-year period?

(c) What was her minimum weight during the two-year period?

Standardized Test Questions

37. **True or False** Higher frequency sound waves have shorter periods. Justify your answer.

38. **True or False** A car traveling at 30 mph is traveling faster than a ship traveling at 30 knots. Justify your answer.

You may use a graphing calculator when answering these questions.

39. **Multiple Choice** To get a rough idea of the height of a building, John paces off 50 ft from the base of the building, then measures the angle of elevation from the ground to the top of the building at that point to be 58°. About how tall is the building?

(A) 31 ft (B) 42 ft (C) 59 ft

(D) 80 ft (E) 417 ft

40. **Multiple Choice** A boat leaves harbor and travels at 20 knots on a bearing of 90°. After 2 hr, it changes course to a bearing of 150° and continues at the same speed for another hour.

After the entire 3-hr trip, how far is it from the harbor?

(A) 50 naut mi (B) 53 naut mi

(C) 57 naut mi (D) 60 naut mi

(E) 67 naut mi

41. **Multiple Choice** At high tide at 8:15 P.M., the water level on the side of a pier is 9 ft from the top. At low tide 6 hr 12 min later, the water level is 13 ft from the top. At which of the following times in that interval is the water level 10 ft from the top of the pier?

(A) 9:15 P.M. (B) 9:48 P.M. (C) 9:52 P.M.

(d) 10:19 P.M. (E) 11:21 P.M.

42. **Multiple Choice** The loudness of a musical tone is determined by which characteristic of its sound wave?

(A) Amplitude (B) Frequency (C) Period

(D) Phase shift (E) Pitch

Explorations

43. **Group Activity** The data for displacement versus time on a tuning fork, shown in Table 4.4, were collected using a CBL and a microphone.

Table 4.4 Tuning Fork Data

Time	Displacement	Time	Displacement
0.00091	−0.080	0.00362	0.217
0.00108	0.200	0.00379	0.480
0.00125	0.480	0.00398	0.681
0.00144	0.693	0.00416	0.810
0.00162	0.816	0.00435	0.827
0.00180	0.844	0.00453	0.749
0.00198	0.771	0.00471	0.581
0.00216	0.603	0.00489	0.346
0.00234	0.368	0.00507	0.077
0.00253	0.099	0.00525	−0.164
0.00271	−0.141	0.00543	−0.320
0.00289	−0.309	0.00562	−0.354
0.00307	−0.348	0.00579	−0.248
0.00325	−0.248	0.00598	−0.035
0.00344	−0.041		

(a) Graph a scatter plot of the data in the $[0, 0.0062]$ by $[-0.5, 1]$ viewing window.

(b) Select the equation that appears to be the best fit of these data.

i. $y = 0.6\sin(2464x - 2.84) + 0.25$

ii. $y = 0.6\sin(1210x - 2) + 0.25$

iii. $y = 0.6\sin(2440x - 2.1) + 0.15$

(c) What is the approximate frequency of the tuning fork?

44. **Writing to Learn** Human sleep-awake cycles at three different ages are described by the accompanying graphs. The portions of the graphs above the horizontal lines represent times awake, and the portions below represent times asleep.

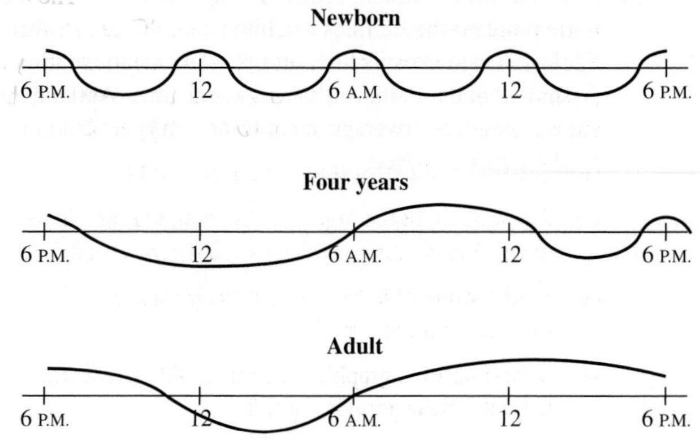

(a) What is the period of the sleep-awake cycle of a newborn? of a four-year-old? of an adult?

(b) Which of these three sleep-awake cycles is the closest to being modeled by a function $y = a \sin bx$?

Using Trigonometry in Geometry In a *regular polygon* all sides have equal length and all angles have equal measure. In Exercises 45 and 46, consider the regular seven-sided polygon whose sides are 5 cm.

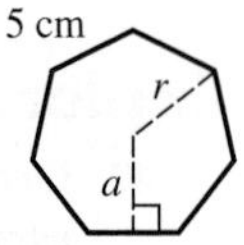

45. Find the length of the *apothem*, the segment from the center of the seven-sided polygon to the midpoint of a side.

46. Find the radius of the circumscribed circle of the regular seven-sided polygon.

47. A *rhombus* is a quadrilateral with all sides equal in length. Recall that a rhombus is also a parallelogram. Find length AC and length BD in the rhombus shown here.

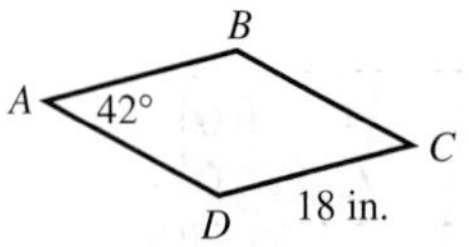

Extending the Ideas

48. A roof has two sections, one with a 50° elevation and the other with a 20° elevation, as shown in the figure.

(a) Find the height BE.

(b) Find the height CD.

(c) Find the length $AE + ED$, and double it to find the length of the roofline.

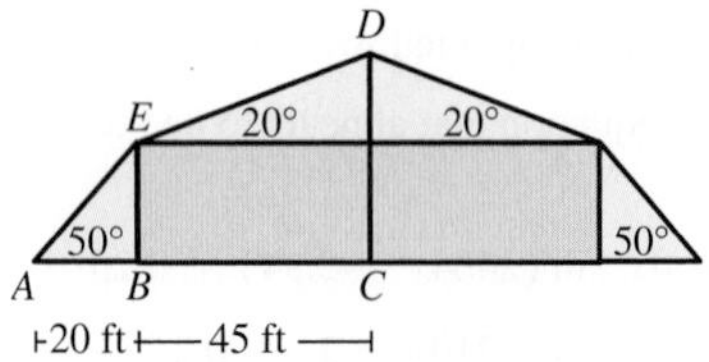

49. Steep Trucking The *percentage grade* of a road is its slope expressed as a percentage. A tractor-trailer rig passes a sign that reads, "6% grade next 7 miles." What is the average angle of inclination of the road?

50. Television Coverage Many satellites travel in *geosynchronous orbits*, which means that the satellite stays over the same point on the Earth. A satellite that broadcasts cable television is in geosynchronous orbit 100 mi above the Earth. Assume that the Earth is a sphere with radius 4000 mi, and find the arc length of coverage area for the cable television satellite on the Earth's surface.

51. Group Activity A musical note like that produced with a tuning fork or pitch meter is a pressure wave. Typically, frequency is measured in hertz (1 Hz = 1 cycle per second). Table 4.5 gives frequency (in Hz) of several musical notes. The time-vs.-pressure tuning fork data in Table 4.6 was collected using a CBL and a microphone.

Table 4.5 Tuning Fork Data

Note	Frequency (Hz)
C	262
C♯ or D♭	277
D	294
D♯ or E♭	311
E	330
F	349
F♯ or G♭	370
G	392
G♯ or A♭	415
A	440
A♯ or B♭	466
B	494
C (next octave)	524

Table 4.6 Tuning Fork Data

Time (sec)	Pressure	Time (sec)	Pressure
0.0002368	1.29021	0.0049024	−1.06632
0.0005664	1.50851	0.0051520	0.09235
0.0008256	1.51971	0.0054112	1.44694
0.0010752	1.51411	0.0056608	1.51411
0.0013344	1.47493	0.0059200	1.51971
0.0015840	0.45619	0.0061696	1.51411
0.0018432	−0.89280	0.0064288	1.43015
0.0020928	−1.51412	0.0066784	0.19871
0.0023520	−1.15588	0.0069408	−1.06072
0.0026016	−0.04758	0.0071904	−1.51412
0.0028640	1.36858	0.0074496	−0.97116
0.0031136	1.50851	0.0076992	0.23229
0.0033728	1.51971	0.0079584	1.46933
0.0036224	1.51411	0.0082080	1.51411
0.0038816	1.45813	0.0084672	1.51971
0.0041312	0.32185	0.0087168	1.50851
0.0043904	−0.97676	0.0089792	1.36298
0.0046400	−1.51971		

(a) Graph a scatter plot of the data.

(b) Determine a, b, and h so that the equation $y = a \sin (b(t - h))$ is a model for the data.

(c) Determine the frequency of the sinusoid in part (b), and use Table 4.5 to identify the musical note produced by the tuning fork.

(d) Identify the musical note produced by the tuning fork used in Exercise 43.

CHAPTER 4 Key Ideas

Properties, Theorems, and Formulas

Arc Length Formulas 322, 323
Distance Conversions 324
Special Angles 330, 331, 346
Period of a Sinusoid 353
Frequency of a Sinusoid 353
Graphs of Sinusoids 354
Sums That Are Sinusoidal Functions 371

Procedures

Degree-Radian Conversion 322
Evaluating Trig Functions of a Nonquadrantal Angle 341
Constructing a Sinusoidal Model Using Time 355

Gallery of Functions

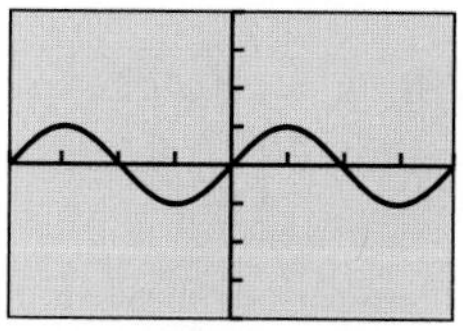

$[-2\pi, 2\pi]$ by $[-4, 4]$
$f(x) = \sin x$

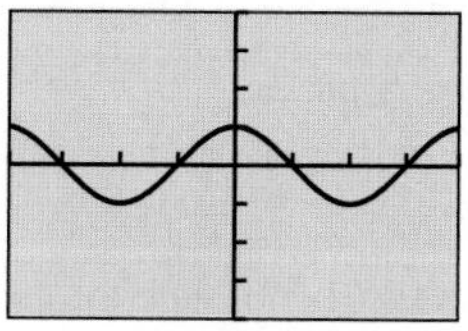

$[-2\pi, 2\pi]$ by $[-4, 4]$
$f(x) = \cos x$

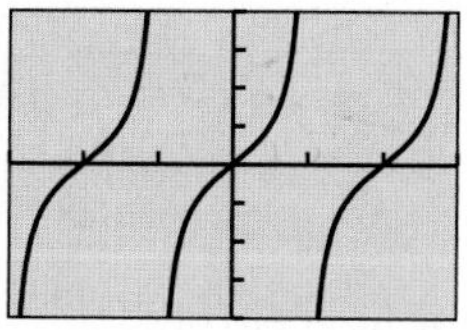

$[-3\pi/2, 3\pi/2]$ by $[-4, 4]$
$f(x) = \tan x$

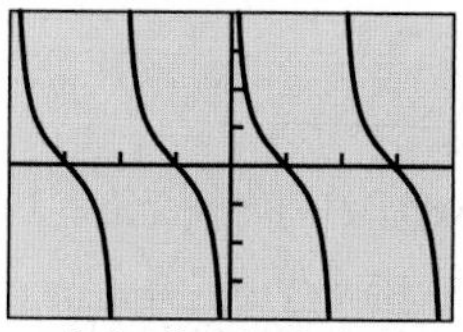

$[-2\pi, 2\pi]$ by $[-4, 4]$
$f(x) = \cot x$

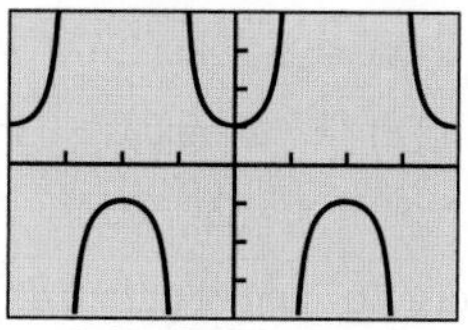

$[-2\pi, 2\pi]$ by $[-4, 4]$
$f(x) = \sec x$

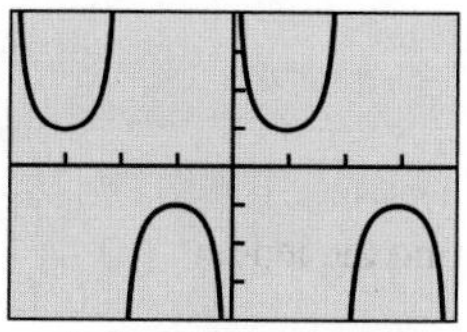

$[-2\pi, 2\pi]$ by $[-4, 4]$
$f(x) = \csc x$

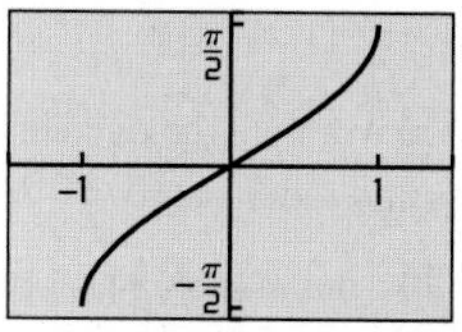

$[-1.5, 1.5]$ by $[-1.7, 1.7]$
$f(x) = \sin^{-1} x$

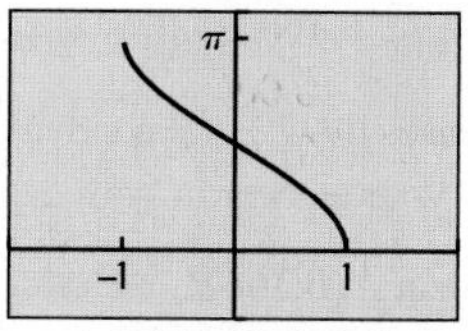

$[-2, 2]$ by $[-1, 3.5]$
$f(x) = \cos^{-1} x$

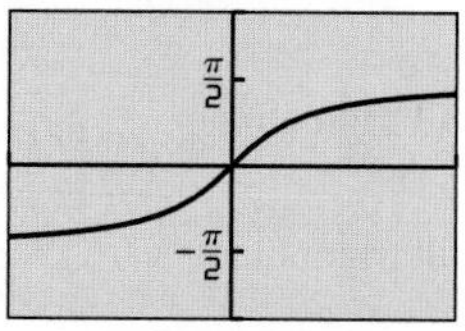

$[-4, 4]$ by $[-2.8, 2.8]$
$f(x) = \tan^{-1} x$

CHAPTER 4 Review Exercises

Exercise numbers with a gray background indicate problems that the authors have designed to be solved *without a calculator*.

The collection of exercises marked in red could be used as a chapter test.

In Exercises 1–8, determine the quadrant of the terminal side of the angle in standard position. Convert degree measures to radians and radian measures to degrees.

1. $\frac{5\pi}{2}$

2. $\frac{3\pi}{4}$

3. $-135°$

4. $-45°$

5. $78°$

6. $112°$

7. $\frac{\pi}{12}$

8. $\frac{7\pi}{10}$

In Exercises 9 and 10, determine the angle measure in both degrees and radians. Draw the angle in standard position if its terminal side is obtained as described.

9. A three-quarters counterclockwise rotation

10. Two and one-half counterclockwise rotations

In Exercises 11–16, the point is on the terminal side of an angle in standard position. Give the smallest positive angle measure in both degrees and radians.

11. $(\sqrt{3}, 1)$ **12.** $(-1, 1)$

13. $(-1, \sqrt{3})$ **14.** $(-3, -3)$

15. $(6, -12)$ **16.** $(2, 4)$

In Exercises 17–28, evaluate the expression exactly without a calculator.

17. $\sin 30°$ **18.** $\cos 330°$

19. $\tan(-135°)$ **20.** $\sec(-135°)$

21. $\sin \dfrac{5\pi}{6}$ **22.** $\csc \dfrac{2\pi}{3}$

23. $\sec\left(-\dfrac{\pi}{3}\right)$ **24.** $\tan\left(-\dfrac{2\pi}{3}\right)$

25. $\csc 270°$ **26.** $\sec 180°$

27. $\cot(-90°)$ **28.** $\tan 360°$

In Exercises 29–32, evaluate exactly all six trigonometric functions of the angle. Use reference triangles and not your calculator.

29. $-\dfrac{\pi}{6}$ **30.** $\dfrac{19\pi}{4}$

31. $-135°$ **32.** $420°$

33. Find all six trigonometric functions of α in $\triangle ABC$.

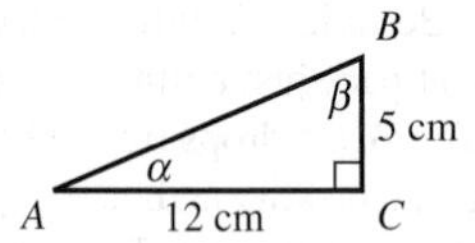

34. Use a right triangle to determine the values of all trigonometric functions of θ, where $\cos\theta = 5/7$.

35. Use a right triangle to determine the values of all trigonometric functions of θ, where $\tan\theta = 15/8$.

36. Use a calculator in Degree mode to solve $\cos\theta = 3/7$ if $0° \le \theta \le 90°$.

37. Use a calculator in Radian mode to solve $\tan x = 1.35$ if $\pi \le x \le 3\pi/2$.

38. Use a calculator in Radian mode to solve $\sin x = 0.218$ if $0 \le x \le 2\pi$.

In Exercises 39–44, solve the right $\triangle ABC$.

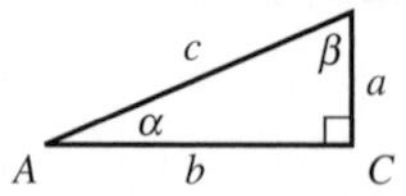

39. $\alpha = 35°, c = 15$ **40.** $b = 8, c = 10$

41. $\beta = 48°, a = 7$ **42.** $\alpha = 28°, c = 8$

43. $b = 5, c = 7$ **44.** $a = 2.5, b = 7.3$

In Exercises 45–48, x is an angle in standard position with $0 \le x \le 2\pi$. Determine the quadrant of x.

45. $\sin x < 0$ and $\tan x > 0$

46. $\cos x < 0$ and $\csc x > 0$

47. $\tan x < 0$ and $\sin x > 0$

48. $\sec x < 0$ and $\csc x > 0$

In Exercises 49–52, point P is on the terminal side of angle θ. Evaluate the six trigonometric functions for θ.

49. $(-3, 6)$ **50.** $(12, 7)$

51. $(-5, -3)$ **52.** $(4, 9)$

In Exercises 53–60, use transformations to describe how the graph of the function is related to a basic trigonometric graph. Graph two periods.

53. $y = \sin(x + \pi)$ **54.** $y = 3 + 2\cos x$

55. $y = -\cos(x + \pi/2) + 4$

56. $y = -2 - 3\sin(x - \pi)$

57. $y = \tan 2x$ **58.** $y = -2\cot 3x$

59. $y = -2\sec \dfrac{x}{2}$ **60.** $y = \csc \pi x$

In Exercises 61–66, state the amplitude, period, phase shift, domain, and range for the sinusoid.

61. $f(x) = 2\sin 3x$ **62.** $g(x) = 3\cos 4x$

63. $f(x) = 1.5\sin(2x - \pi/4)$

64. $g(x) = -2\sin(3x - \pi/3)$

65. $y = 4\cos(2x - 1)$ **66.** $g(x) = -2\cos(3x + 1)$

In Exercises 67 and 68, graph the function. Then estimate the values of a, b, and h so that $f(x) \approx a\sin(b(x - h))$.

67. $f(x) = 2\sin x - 4\cos x$

68. $f(x) = 3\cos 2x - 2\sin 2x$

In Exercises 69–72, use a calculator to evaluate the expression. Express your answer in both degrees and radians.

69. $\sin^{-1}(0.766)$ **70.** $\cos^{-1}(0.479)$

71. $\tan^{-1} 1$ **72.** $\sin^{-1}\left(\dfrac{\sqrt{3}}{2}\right)$

In Exercises 73–76, use transformations to describe how the graph of the function is related to a basic inverse trigonometric graph. State the domain and range.

73. $y = \sin^{-1} 3x$ **74.** $y = \tan^{-1} 2x$

75. $y = \sin^{-1}(3x - 1) + 2$ **76.** $y = \cos^{-1}(2x + 1) - 3$

In Exercises 77–82, find the exact value of x without using a calculator.

77. $\sin x = 0.5, \quad \pi/2 \le x \le \pi$

78. $\cos x = \sqrt{3}/2, \quad 0 \le x \le \pi$

79. $\tan x = -1, \quad 0 \le x \le \pi$

80. $\sec x = 2, \quad \pi \le x \le 2\pi$

81. $\csc x = -1, \quad 0 \le x \le 2\pi$

82. $\cot x = -\sqrt{3}, \quad 0 \le x \le \pi$

In Exercises 83 and 84, describe the end behavior of the function.

83. $\dfrac{\sin x}{x^2}$ **84.** $\dfrac{3}{5}e^{-x/12}\sin(2x - 3)$

In Exercises 85–88, evaluate the expression without a calculator.

85. $\tan(\tan^{-1} 1)$

86. $\cos^{-1}(\cos \pi/3)$

87. $\tan(\sin^{-1} 3/5)$

88. $\cos^{-1}(\cos(-\pi/7))$

In Exercises 89–92, determine whether the function is periodic. State the period (if applicable), the domain, and the range.

89. $f(x) = |\sec x|$

90. $g(x) = \sin|x|$

91. $f(x) = 2x + \tan x$

92. $g(x) = 2\cos 2x + 3\sin 5x$

93. Arc Length Find the length of the arc intercepted by a central angle of $2\pi/3$ rad in a circle with radius 2.

94. Algebraic Expression Find an algebraic expression equivalent to $\tan(\cos^{-1} x)$.

95. Height of Building The angle of elevation of the top of a building from a point 100 m away from the building on level ground is 78°. Find the height of the building.

96. Height of Tree A tree casts a shadow 51 ft long when the angle of elevation of the Sun (measured with the horizon) is 25°. How tall is the tree?

97. Traveling Car From the top of a 150-ft building Flora observes a car moving toward her. If the angle of depression of the car changes from 18° to 42° during the observation, how far does the car travel?

98. Finding Distance A lighthouse L stands 4 mi from the closest point P along a straight shore (see figure). Find the distance from P to a point Q along the shore if $\angle PLQ = 22°$.

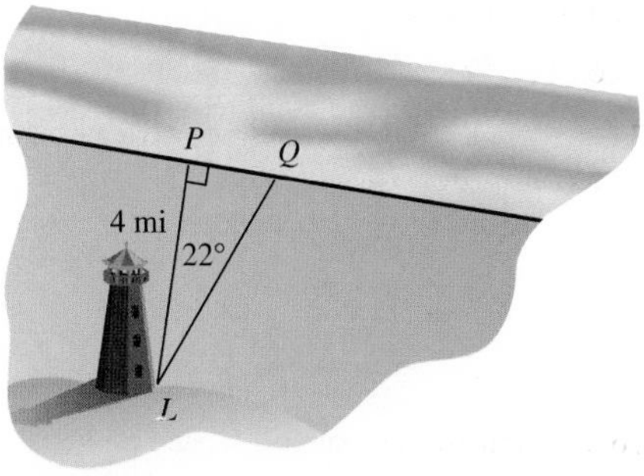

99. Navigation An airplane is flying due east between two signal towers. One tower is due north of the other. The bearing from the plane to the north tower is 23°, and to the south tower is 128°. Use a drawing to show the exact location of the plane.

100. Finding Distance The bearings of two points on the shore from a boat are 115° and 123°. Assume the two points are 855 ft apart. How far is the boat from the nearest point on shore if the shore is straight and runs north-south?

101. Height of Tree Dr. Thom Lawson standing on flat ground 62 ft from the base of a Douglas fir measures the angle of elevation to the top of the tree as 72°24′. What is the height of the tree?

102. Storing Hay A 75-ft-long conveyor is used at the Lovelady Farm to put hay bales up for winter storage. The conveyor is tilted to an angle of elevation of 22°.

(a) To what height can the hay be moved?

(b) If the conveyor is repositioned to an angle of 27°, to what height can the hay be moved?

103. Swinging Pendulum In the Hardy Boys Adventure *While the Clock Ticked*, the pendulum of the grandfather clock at the Purdy place is 44 in. long and swings through an arc of 6°. Find the length of the arc that the pendulum traces.

104. Finding Area A windshield wiper arm on a vintage 1994 Plymouth Acclaim is 20 in. long and has a blade 16 in. long. If the wiper sweeps through an angle of 110°, how large an area does the wiper blade clean? (See Exercise 71 in Section 4.1.)

105. Modeling Low Temperatures The average daily low temperature (°F) in Fairbanks, Alaska, can be modeled by the equation

$$T(x) = 31.2 \sin\left[\frac{2\pi}{365}(x - 106)\right] + 20.4,$$

where x is time in days, with $x = 1$ representing January 1. How many days a year would you expect the low temperature to be above freezing (32°F)?

Source: Weather.com, 2012.

106. Taming The Beast The Beast is a featured roller coaster at the King Island's amusement park just north of Cincinnati. On its first and biggest hill, The Beast drops from a height of 52 ft above the ground along a sinusoidal path to a depth 18 ft underground as it enters a frightening tunnel. The mathematical model for this part of track is

$$h(x) = 35\cos\left(\frac{x}{35}\right) + 17, 0 \le x \le 110,$$

where x is the horizontal distance from the top of the hill and $h(x)$ is the vertical position relative to ground level (both in feet). What is the horizontal distance from the top of the hill to the point where the track reaches ground level?

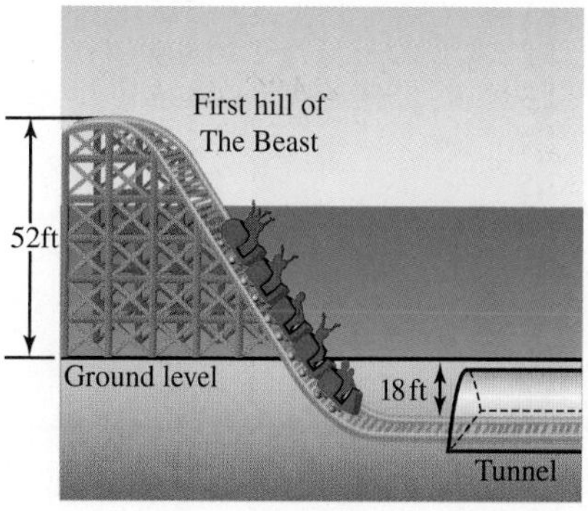

CHAPTER 4 Project

Modeling the Motion of a Pendulum

As a simple pendulum swings back and forth, its displacement can be modeled using a standard sinusoidal equation of the form

$$y = a\cos(b(x - h)) + k$$

where y represents the pendulum's distance from a fixed point and x represents total elapsed time. In this project, you will use a motion detection device to collect distance and time data for a swinging pendulum, then find a mathematical model that describes the pendulum's motion.

Collecting the Data

To start, construct a simple pendulum by fastening about 1 m of string to the end of a ball. Set up the Calculator-Based Laboratory (CBL) system with a motion detector or a Calculator-Based Ranger (CBR) system to collect time and distance readings for between 2 sec and 4 sec (enough time to capture at least one complete swing of the pendulum). See the CBL/CBR guidebook for specific setup instructions. Start the pendulum swinging in front of the detector, then activate the system. The data table below shows a sample set of data collected as a pendulum swung back and forth in front of a CBR.

Total Elapsed Time (sec)	Distance from the CBR (m)
0	0.665
0.1	0.756
0.2	0.855
0.3	0.903
0.4	0.927
0.5	0.931
0.6	0.897
0.7	0.837
0.8	0.753
0.9	0.663
1.0	0.582
1.1	0.525
1.2	0.509
1.3	0.495
1.4	0.521
1.5	0.575
1.6	0.653
1.7	0.741
1.8	0.825
1.9	0.888
2.0	0.921

Explorations

1. If you collected motion data using a CBL or CBR, a plot of distance versus time should be shown on your graphing calculator or computer screen. If you don't have access to a CBL/CBR, enter the data in the sample table into your graphing calculator/computer. Create a scatter plot for the data.
2. Find values for a, b, h, and k so that the equation $y = a\cos(b(x - h)) + k$ fits the distance versus time data plot. Refer to the information box on page 355 in this chapter to review sinusoidal graph characteristics.
3. What are the physical meanings of the constants a and k in the modeling equation $y = a\cos(b(x - h)) + k$? (*Hint:* What distances do a and k measure?)
4. Which, if any, of the values of a, b, h, and/or k would change if you used the equation $y = a\sin(b(x - h)) + k$ to model the data set?
5. Use your calculator or computer to find a sinusoidal regression equation to model this data set (see your grapher's guidebook for instructions on how to do this). If your calculator/computer uses a different sinusoidal form, compare it to the modeling equation you found earlier, $y = a\cos(b(x - h)) + k$.

CHAPTER 5

5.1 Fundamental Identities

5.2 Proving Trigonometric Identities

5.3 Sum and Difference Identities

5.4 Multiple-Angle Identities

5.5 The Law of Sines

5.6 The Law of Cosines

Analytic Trigonometry

It is no surprise that naturalists seeking to estimate wildlife populations must have an understanding of geometry (a word that literally means "earth measurement"). You will learn in this chapter that trigonometry, with its many connections to triangles and circles, enables us to extend the problem-solving tools of geometry significantly. On page 447 we will apply a result called Heron's Formula (which we prove trigonometrically) to estimate the local density of a deer population.

Chapter 5 Overview

Although the title of this chapter suggests that we are now moving into the analytic phase of our study of trigonometric functions, the truth is that we have been in that phase for several sections already. Once the transition is made from triangle ratios to functions and their graphs, one is on analytic soil. But our primary applications of trigonometry so far have been computational; we have not made full use of the properties of the functions to study the connections among the trigonometric functions themselves. In this chapter we will shift our emphasis more toward theory and proof, exploring where the properties of these special functions lead us, often with no immediate concern for real-world relevance at all. We hope in the process to give you an appreciation for the rich and intricate tapestry of interlocking patterns that can be woven from the six basic trigonometric functions—patterns that will take on even greater beauty later on when you can view them through the lens of calculus.

5.1 Fundamental Identities

What you'll learn about

- Identities
- Basic Trigonometric Identities
- Pythagorean Identities
- Cofunction Identities
- Odd-Even Identities
- Simplifying Trigonometric Expressions
- Solving Trigonometric Equations

... and why

Identities are important when working with trigonometric functions in calculus.

Identities

As you probably realize by now, the symbol "=" means several different things in mathematics.

1. $1 + 1 = 2$ means *equality of real numbers.* It is a true sentence.
2. $2(x - 3) = 2x - 6$ signifies *equivalent expressions.* It is a true sentence.
3. $x^2 + 3 = 7$ is an *open sentence,* because it can be true or false, depending on whether x is a solution to the equation.
4. $(x^2 - 1)/(x + 1) = x - 1$ is an *identity.* It is a true sentence (very much like (2) above), but with the important qualification that *x must be in the domain of both expressions.* If either side of the equality is undefined, the sentence is meaningless. Substituting -1 into both sides of the equation in (3) gives a sentence that is mathematically false (i.e., $4 = 7$), whereas substituting -1 into both sides of the identity in (4) gives a sentence that is meaningless.

Statements like "$\tan\theta = \sin\theta/\cos\theta$" and "$\csc\theta = 1/\sin\theta$" are trigonometric **identities** because they are true for all values of the variable for which both sides of the equation are defined. The set of all such values is called the **domain of validity** of the identity. We will spend much of this chapter exploring trigonometric identities, their proofs, their implications, and their applications.

Basic Trigonometric Identities

Some trigonometric identities follow directly from the definitions of the six basic trigonometric functions. These *basic identities* consist of the *reciprocal identities* and the *quotient identities*.

Basic Trigonometric Identities

Reciprocal Identities

$$\csc\theta = \frac{1}{\sin\theta} \qquad \sec\theta = \frac{1}{\cos\theta} \qquad \cot\theta = \frac{1}{\tan\theta}$$

$$\sin\theta = \frac{1}{\csc\theta} \qquad \cos\theta = \frac{1}{\sec\theta} \qquad \tan\theta = \frac{1}{\cot\theta}$$

Quotient Identities

$$\tan\theta = \frac{\sin\theta}{\cos\theta} \qquad \cot\theta = \frac{\cos\theta}{\sin\theta}$$

EXPLORATION 1 Making a Point About Domain of Validity

1. $\theta = 0$ is in the domain of validity of exactly three of the basic identities. Which three?
2. For exactly two of the basic identities, one side of the equation is defined at $\theta = 0$ and the other side is not. Which two?
3. For exactly three of the basic identities, both sides of the equation are undefined at $\theta = 0$. Which three?

Pythagorean Identities

Exploration 2 in Section 4.3 introduced you to the fact that, for any real number t, the numbers $(\cos t)^2$ and $(\sin t)^2$ always sum to 1. This is clearly true for the quadrantal angles that wrap to the points $(\pm 1, 0)$ and $(0, \pm 1)$, and it is true for any other t because $\cos t$ and $\sin t$ are the (signed) lengths of legs of a reference triangle with hypotenuse 1 (Figure 5.1). No matter what quadrant the triangle lies in, the Pythagorean Theorem guarantees the following identity: $(\cos t)^2 + (\sin t)^2 = 1$.

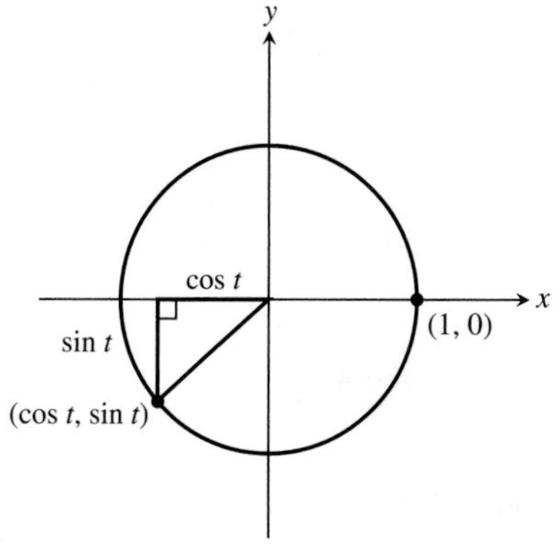

FIGURE 5.1 By the Pythagorean Theorem, $(\cos t)^2 + (\sin t)^2 = 1$.

If we divide each term of the identity by $(\cos t)^2$, we get an identity that involves tangent and secant:

$$\frac{(\cos t)^2}{(\cos t)^2} + \frac{(\sin t)^2}{(\cos t)^2} = \frac{1}{(\cos t)^2}$$

$$1 + (\tan t)^2 = (\sec t)^2$$

If we divide each term of the identity by $(\sin t)^2$, we get an identity that involves cotangent and cosecant:

$$\frac{(\cos t)^2}{(\sin t)^2} + \frac{(\sin t)^2}{(\sin t)^2} = \frac{1}{(\sin t)^2}$$

$$(\cot t)^2 + 1 = (\csc t)^2$$

These three identities are called the *Pythagorean identities*, which we restate using the shorthand notation for powers of trigonometric functions.

Pythagorean Identities

$$\cos^2\theta + \sin^2\theta = 1$$

$$1 + \tan^2\theta = \sec^2\theta$$

$$\cot^2\theta + 1 = \csc^2\theta$$

EXAMPLE 1 Using Identities

Find $\sin\theta$ and $\cos\theta$ if $\tan\theta = 5$ and $\cos\theta > 0$.

SOLUTION We could solve this problem by the reference triangle techniques of Section 4.3 (see Example 7 in that section), but we will show an alternate solution here using only identities.

First, we note that $\sec^2\theta = 1 + \tan^2\theta = 1 + 5^2 = 26$, so $\sec\theta = \pm\sqrt{26}$.

Since $\sec\theta = \pm\sqrt{26}$, we have $\cos\theta = 1/\sec\theta = 1/\pm\sqrt{26}$.

But $\cos\theta > 0$, so $\cos\theta = 1/\sqrt{26}$.

(continued)

Student Note

It would be very inappropriate to use a calculator to approximate the exact numbers in the solution to Example 1.

Finally,

$$\tan\theta = 5$$

$$\frac{\sin\theta}{\cos\theta} = 5$$

$$\sin\theta = 5\cos\theta = 5\left(\frac{1}{\sqrt{26}}\right) = \frac{5}{\sqrt{26}}$$

Therefore, $\sin\theta = \dfrac{5}{\sqrt{26}}$ and $\cos\theta = \dfrac{1}{\sqrt{26}}$. **Now try Exercise 1.**

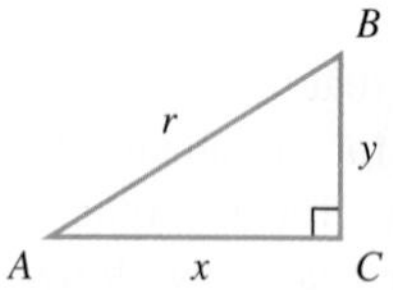

FIGURE 5.2 Angles A and B are complements in right ΔABC.

If you find yourself preferring the reference triangle method, that's fine. Remember that combining the power of geometry and algebra to solve problems is one of the themes of this book, and the instinct to do so will serve you well in calculus.

Cofunction Identities

If C is the right angle in right ΔABC, then angles A and B are complements. Notice what happens if we use the usual triangle ratios to define the six trigonometric functions of angles A and B (Figure 5.2).

Angle A:	$\sin A = \dfrac{y}{r}$	$\tan A = \dfrac{y}{x}$	$\sec A = \dfrac{r}{x}$
	$\cos A = \dfrac{x}{r}$	$\cot A = \dfrac{x}{y}$	$\csc A = \dfrac{r}{y}$
Angle B:	$\sin B = \dfrac{x}{r}$	$\tan B = \dfrac{x}{y}$	$\sec B = \dfrac{r}{y}$
	$\cos B = \dfrac{y}{r}$	$\cot B = \dfrac{y}{x}$	$\csc B = \dfrac{r}{x}$

Do you see what happens? In every case, the value of a function at A is the same as the value of its cofunction at B. This always happens with complementary angles; in fact, it is this phenomenon that gives a "co" function its name. The "co" stands for "complement."

Cofunction Identities

$$\sin\left(\frac{\pi}{2} - \theta\right) = \cos\theta \qquad \cos\left(\frac{\pi}{2} - \theta\right) = \sin\theta$$

$$\tan\left(\frac{\pi}{2} - \theta\right) = \cot\theta \qquad \cot\left(\frac{\pi}{2} - \theta\right) = \tan\theta$$

$$\sec\left(\frac{\pi}{2} - \theta\right) = \csc\theta \qquad \csc\left(\frac{\pi}{2} - \theta\right) = \sec\theta$$

Although our argument on behalf of these equations was based on acute angles in a triangle, these equations are genuine identities, valid for all real numbers for which both sides of the equation are defined. We could extend our acute-angle argument to produce a general proof, but it will be easier to wait and use the identities of Section 5.3. We will revisit this particular set of fundamental identities in that section.

Odd-Even Identities

We have seen that every basic trigonometric function is either odd or even. Either way, the usual function relationship leads to another fundamental identity.

Odd-Even Identities

$\sin(-x) = -\sin x$	$\cos(-x) = \cos x$	$\tan(-x) = -\tan x$
$\csc(-x) = -\csc x$	$\sec(-x) = \sec x$	$\cot(-x) = -\cot x$

EXAMPLE 2 Using More Identities

If $\cos\theta = 0.34$, find $\sin(\theta - \pi/2)$.

SOLUTION This problem can best be solved using identities.

$$\sin\left(\theta - \frac{\pi}{2}\right) = -\sin\left(\frac{\pi}{2} - \theta\right) \quad \text{Sine is odd.}$$
$$= -\cos\theta \quad \text{Cofunction identity}$$
$$= -0.34$$

Now try Exercise 7.

Simplifying Trigonometric Expressions

In calculus it is often necessary to deal with expressions that involve trigonometric functions. Some of those expressions start out looking fairly complicated, but it is often possible to use identities along with algebraic techniques (e.g., factoring or combining fractions over a common denominator) to *simplify* the expressions before dealing with them. In some cases the simplifications can be dramatic.

EXAMPLE 3 Simplifying by Factoring and Using Identities

Simplify the expression $\sin^3 x + \sin x \cos^2 x$.

SOLUTION

Solve Algebraically

$$\sin^3 x + \sin x \cos^2 x = \sin x\,(\sin^2 x + \cos^2 x)$$
$$= \sin x\,(1) \quad \text{Pythagorean identity}$$
$$= \sin x$$

Support Graphically We recognize the graph of $y = \sin^3 x + \sin x \cos^2 x$ (Figure 5.3a) as the same as the graph of $y = \sin x$ (Figure 5.3b). **Now try Exercise 13.**

Student Note

Remember that graphs cannot be used to *prove* identities. However they can provide support of the identity.

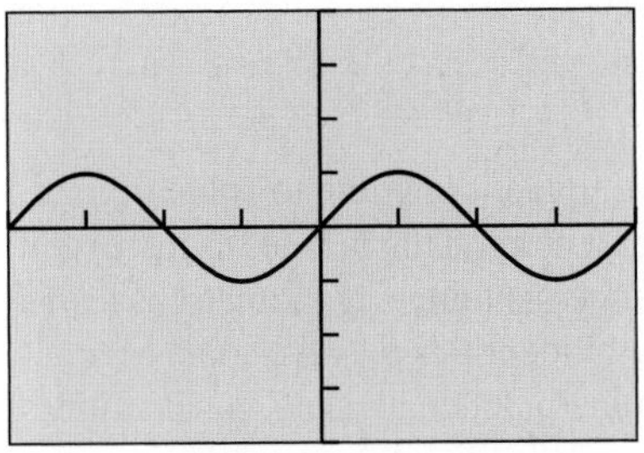

$[-2\pi, 2\pi]$ by $[-4, 4]$

(a)

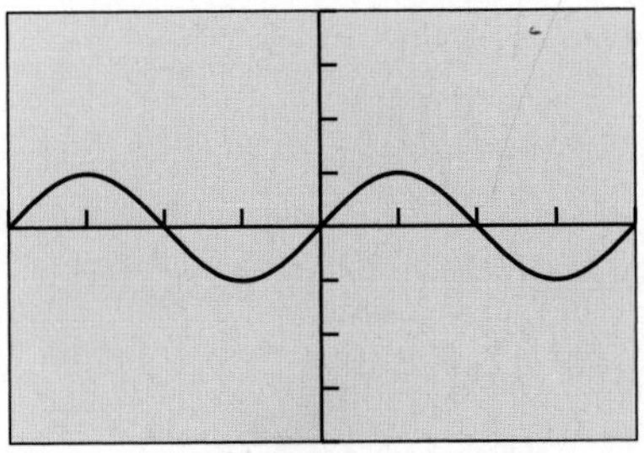

$[-2\pi, 2\pi]$ by $[-4, 4]$

(b)

FIGURE 5.3 Graphical support of the identity $\sin^3 x + \sin x \cos^2 x = \sin x$. (Example 3)

EXAMPLE 4 Simplifying by Expanding and Using Identities

Simplify the expression $[(\sec x + 1)(\sec x - 1)]/\sin^2 x$.

SOLUTION

Solve Algebraically

$$\frac{(\sec x + 1)(\sec x - 1)}{\sin^2 x} = \frac{\sec^2 x - 1}{\sin^2 x} \quad (a + b)(a - b) = a^2 - b^2$$
$$= \frac{\tan^2 x}{\sin^2 x} \quad \text{Pythagorean identity}$$
$$= \frac{\sin^2 x}{\cos^2 x} \cdot \frac{1}{\sin^2 x} \quad \tan x = \frac{\sin x}{\cos x}$$
$$= \frac{1}{\cos^2 x}$$
$$= \sec^2 x$$

(continued)

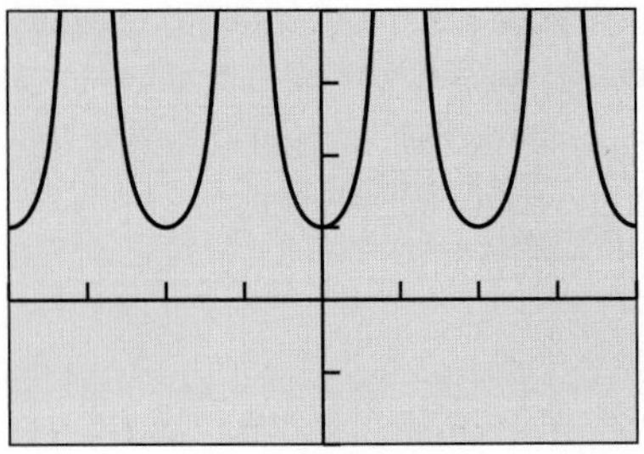

$[-2\pi, 2\pi]$ by $[-2, 4]$

(a)

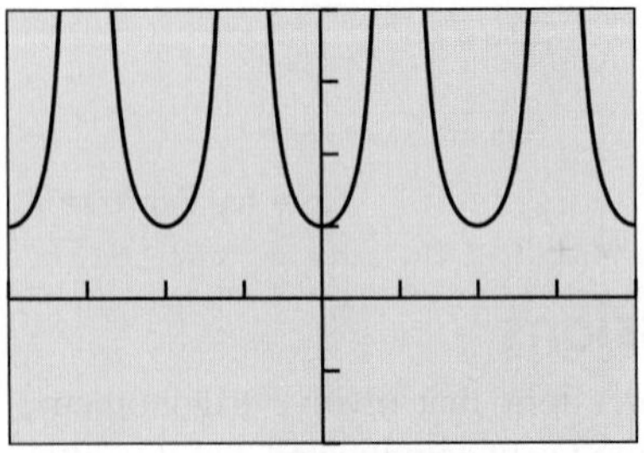

$[-2\pi, 2\pi]$ by $[-2, 4]$

(b)

FIGURE 5.4 Graphical support of the identity $(\sec x + 1)(\sec x - 1)/\sin^2 x = \sec^2 x$. (Example 4)

Support Graphically The graphs of $y = \dfrac{(\sec x + 1)(\sec x - 1)}{\sin^2 x}$ and $y = \sec^2 x$ appear to be identical, as expected (Figure 5.4). **Now try Exercise 25.**

EXAMPLE 5 Simplifying by Combining Fractions and Using Identities

Simplify the expression $\dfrac{\cos x}{1 - \sin x} - \dfrac{\sin x}{\cos x}$.

SOLUTION

$$\frac{\cos x}{1 - \sin x} - \frac{\sin x}{\cos x}$$

$$= \frac{\cos x}{1 - \sin x} \cdot \frac{\cos x}{\cos x} - \frac{\sin x}{\cos x} \cdot \frac{1 - \sin x}{1 - \sin x} \quad \text{Rewrite using common denominator.}$$

$$= \frac{(\cos x)(\cos x) - (\sin x)(1 - \sin x)}{(1 - \sin x)(\cos x)}$$

$$= \frac{\cos^2 x - \sin x + \sin^2 x}{(1 - \sin x)(\cos x)}$$

$$= \frac{1 - \sin x}{(1 - \sin x)(\cos x)} \quad \text{Pythagorean identity}$$

$$= \frac{1}{\cos x}$$

$$= \sec x$$

(We leave it to you to provide the graphical support.) **Now try Exercise 37.**

We will use these same simplifying techniques to prove trigonometric identities in Section 5.2.

Solving Trigonometric Equations

The equation-solving capabilities of calculators have made it possible to solve trigonometric equations without understanding much trigonometry. This is fine, to the extent that solving equations is our goal. However, since understanding trigonometry is also a goal, we will occasionally pause in our development of identities to solve some trigonometric equations with paper and pencil, just to get some practice in using the identities.

EXAMPLE 6 Solving a Trigonometric Equation

Find all values of x in the interval $[0, 2\pi)$ that solve $\cos^3 x/\sin x = \cot x$.

SOLUTION

$$\frac{\cos^3 x}{\sin x} = \cot x$$

$$\frac{\cos^3 x}{\sin x} = \frac{\cos x}{\sin x}$$

$$\cos^3 x = \cos x \quad \text{Multiply both sides by } \sin x.$$

$$\cos^3 x - \cos x = 0$$

$$(\cos x)(\cos^2 x - 1) = 0$$

$$(\cos x)(-\sin^2 x) = 0 \quad \text{Pythagorean identity}$$

$$\cos x = 0 \quad \text{or} \quad \sin x = 0$$

We reject the possibility that $\sin x = 0$ because it would make both sides of the original equation undefined.

The values in the interval $[0, 2\pi)$ that solve $\cos x = 0$ (and therefore $\cos^3 x/\sin x = \cot x$) are $\pi/2$ and $3\pi/2$.

Now try Exercise 51.

EXAMPLE 7 Solving a Trigonometric Equation by Factoring

Find all solutions to the trigonometric equation $2 \sin^2 x + \sin x = 1$.

SOLUTION Let $y = \sin x$. The equation $2y^2 + y = 1$ can be solved by factoring:

$$2y^2 + y = 1$$
$$2y^2 + y - 1 = 0$$
$$(2y - 1)(y + 1) = 0$$
$$2y - 1 = 0 \quad \text{or} \quad y + 1 = 0$$
$$y = \frac{1}{2} \quad \text{or} \quad y = -1$$

So, in the original equation, $\sin x = 1/2$ or $\sin x = -1$. Figure 5.5 shows that the solutions in the interval $[0, 2\pi)$ are $\pi/6$, $5\pi/6$, and $3\pi/2$.

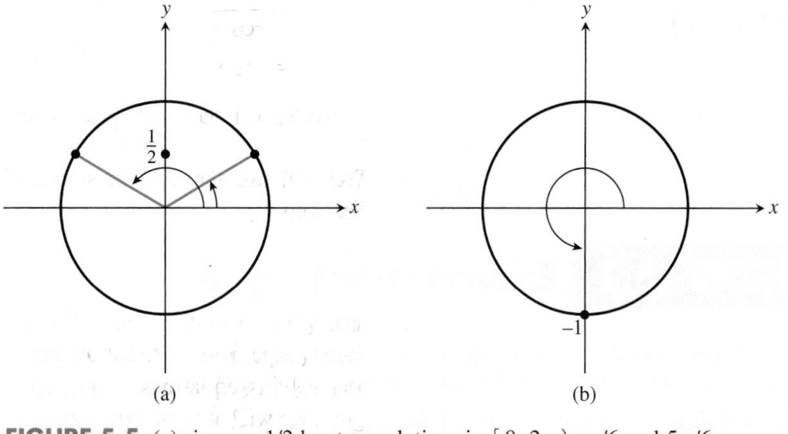

FIGURE 5.5 (a) $\sin x = 1/2$ has two solutions in $[0, 2\pi)$: $\pi/6$ and $5\pi/6$. (b) $\sin x = -1$ has one solution in $[0, 2\pi)$: $3\pi/2$. (Example 7)

To get *all* real solutions, we simply add integer multiples of the period, 2π, of the periodic function $\sin x$:

$$x = \frac{\pi}{6} + 2n\pi \text{ or } x = \frac{5\pi}{6} + 2n\pi \text{ or } x = \frac{3\pi}{2} + 2n\pi$$
$$(n = 0, \pm 1, \pm 2, \ldots)$$

Now try Exercise 57.

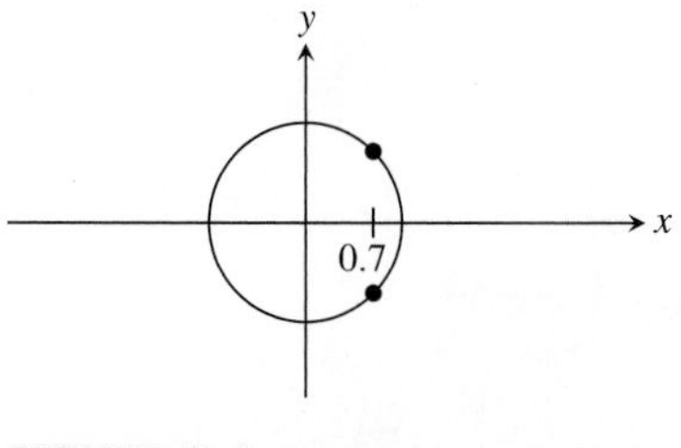

FIGURE 5.6 There are two points on the unit circle with x-coordinate 0.7. (Example 8 on the next page)

You might try solving the equation in Example 7 on your grapher for the sake of comparison. Finding *all* real solutions still requires an understanding of periodicity, and finding *exact* solutions requires the savvy to divide your calculator answers by π. It is likely that anyone who knows that much trigonometry will actually find the algebraic solution to be easier!

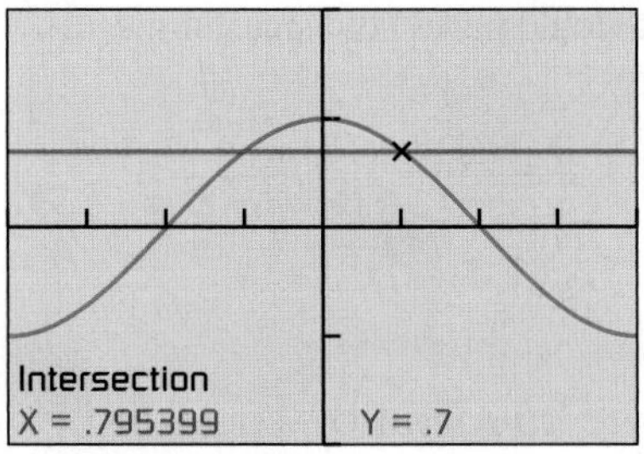

$[-\pi, \pi]$ by $[-2, 2]$

FIGURE 5.7 The two intersections of the graphs of $y = \cos x$ (blue) and $y = 0.7$ (red) give two solutions to the equation $\cos t = 0.7$ in the interval $[-\pi, \pi]$, supporting Example 8.

EXAMPLE 8 Solving a Trig Equation with a Grapher

Find all solutions to the equation $\cos t = 0.7$, using a calculator where needed.

SOLUTION Figure 5.6 on the previous page shows that there are two points on the unit circle with an x-coordinate of 0.7. We do not recognize this value as one of our special triangle ratios, but we can use a graphing calculator to find the smallest positive and negative values for which $\cos x = 0.7$ by intersecting the graphs of $y = \cos x$ and $y = 0.7$ (Figure 5.7).

The two values are predictably opposites of each other: $t \approx \pm 0.80$. Using the period of cosine (which is 2π), we get the complete solution set: $\{\pm 0.80 + 2n\pi \mid n = 0, \pm 1, \pm 2, \pm 3, \ldots\}$.

Now try Exercise 63.

QUICK REVIEW 5.1 *(For help, go to Sections A.2, A.3, and 4.7.)*

Exercise numbers with a gray background indicate problems that the authors have designed to be solved *without a calculator.*

In Exercises 1–4, evaluate the expression.

1. $\sin^{-1}\left(\frac{12}{13}\right)$
2. $\cos^{-1}\left(\frac{3}{5}\right)$
3. $\cos^{-1}\left(-\frac{4}{5}\right)$
4. $\sin^{-1}\left(-\frac{5}{13}\right)$

In Exercises 5–8, factor the expression into a product of linear factors.

5. $a^2 - 2ab + b^2$
6. $4u^2 + 4u + 1$
7. $2x^2 - 3xy - 2y^2$
8. $2v^2 - 5v - 3$

In Exercises 9–12, simplify the expression.

9. $\frac{1}{x} - \frac{2}{y}$
10. $\frac{a}{x} + \frac{b}{y}$
11. $\frac{x + y}{(1/x) + (1/y)}$
12. $\frac{x}{x - y} - \frac{y}{x + y}$

SECTION 5.1 Exercises

In Exercises 1–4, evaluate without using a calculator. Use the Pythagorean identities rather than reference triangles. (See Example 1.)

1. Find $\sin \theta$ and $\cos \theta$ if $\tan \theta = 3/4$ and $\sin \theta > 0$.
2. Find $\sec \theta$ and $\csc \theta$ if $\tan \theta = 3$ and $\cos \theta > 0$.
3. Find $\tan \theta$ and $\cot \theta$ if $\sec \theta = 4$ and $\sin \theta < 0$.
4. Find $\sin \theta$ and $\tan \theta$ if $\cos \theta = 0.8$ and $\tan \theta < 0$.

In Exercises 5–8, use identities to find the value of the expression.

5. If $\sin \theta = 0.45$, find $\cos(\pi/2 - \theta)$.
6. If $\tan(\pi/2 - \theta) = -5.32$, find $\cot \theta$.
7. If $\sin(\theta - \pi/2) = 0.73$, find $\cos(-\theta)$.
8. If $\cot(-\theta) = 7.89$, find $\tan(\theta - \pi/2)$.

In Exercises 9–16, use basic identities to simplify the expression.

9. $\tan x \cos x$
10. $\cot x \tan x$
11. $\sec y \sin(\pi/2 - y)$
12. $\cot u \sin u$
13. $\frac{1 + \tan^2 x}{\csc^2 x}$
14. $\frac{1 - \cos^2 \theta}{\sin \theta}$
15. $\cos x - \cos^3 x$
16. $\frac{\sin^2 u + \tan^2 u + \cos^2 u}{\sec u}$

In Exercises 17–22, simplify the expression to either 1 or -1.

17. $\sin x \csc(-x)$
18. $\sec(-x) \cos(-x)$
19. $\cot(-x) \cot(\pi/2 - x)$
20. $\cot(-x) \tan(-x)$
21. $\sin^2(-x) + \cos^2(-x)$
22. $\sec^2(-x) - \tan^2 x$

In Exercises 23–26, simplify the expression to either a constant or a basic trigonometric function. Support your result graphically.

23. $\frac{\tan(\pi/2 - x) \csc x}{\csc^2 x}$
24. $\frac{1 + \tan x}{1 + \cot x}$
25. $(\sec^2 x + \csc^2 x) - (\tan^2 x + \cot^2 x)$
26. $\frac{\sec^2 u - \tan^2 u}{\cos^2 v + \sin^2 v}$

In Exercises 27–32, use the basic identities to change the expression to one involving only sines and cosines. Then simplify to a basic trigonometric function.

27. $(\sin x)(\tan x + \cot x)$

28. $\sin\theta - \tan\theta\cos\theta + \cos(\pi/2 - \theta)$

29. $\sin x \cos x \tan x \sec x \csc x$

30. $\dfrac{(\sec y - \tan y)(\sec y + \tan y)}{\sec y}$

31. $\dfrac{\tan x}{\csc^2 x} + \dfrac{\tan x}{\sec^2 x}$

32. $\dfrac{\sec^2 x \csc x}{\sec^2 x + \csc^2 x}$

In Exercises 33–38, combine the fractions and simplify to a multiple of a power of a basic trigonometric function (e.g., $3\tan^2 x$).

33. $\dfrac{1}{\sin^2 x} + \dfrac{\sec^2 x}{\tan^2 x}$

34. $\dfrac{1}{1 - \sin x} + \dfrac{1}{1 + \sin x}$

35. $\dfrac{\sin x}{\cot^2 x} - \dfrac{\sin x}{\cos^2 x}$

36. $\dfrac{1}{\sec x - 1} - \dfrac{1}{\sec x + 1}$

37. $\dfrac{\sec x}{\sin x} - \dfrac{\sin x}{\cos x}$

38. $\dfrac{\sin x}{1 - \cos x} + \dfrac{1 - \cos x}{\sin x}$

In Exercises 39–46, write each expression in factored form as an algebraic expression of a single trigonometric function (e.g., $(2\sin x + 3)(\sin x - 1)$).

39. $\cos^2 x + 2\cos x + 1$

40. $1 - 2\sin x + \sin^2 x$

41. $1 - 2\sin x + (1 - \cos^2 x)$

42. $\sin x - \cos^2 x - 1$

43. $\cos x - 2\sin^2 x + 1$

44. $\sin^2 x + \dfrac{2}{\csc x} + 1$

45. $4\tan^2 x - \dfrac{4}{\cot x} + \sin x \csc x$

46. $\sec^2 x - \sec x + \tan^2 x$

In Exercises 47–50, write each expression as an algebraic expression of a single trigonometric function (e.g., $2\sin x + 3$).

47. $\dfrac{1 - \sin^2 x}{1 + \sin x}$

48. $\dfrac{\tan^2\alpha - 1}{1 + \tan\alpha}$

49. $\dfrac{\sin^2 x}{1 + \cos x}$

50. $\dfrac{\tan^2 x}{\sec x + 1}$

In Exercises 51–56, find all solutions to the equation in the interval $[0, 2\pi)$. You do not need a calculator.

51. $2\cos x \sin x - \cos x = 0$

52. $\sqrt{2}\tan x \cos x - \tan x = 0$

53. $\tan x \sin^2 x = \tan x$

54. $\sin x \tan^2 x = \sin x$

55. $\tan^2 x = 3$

56. $2\sin^2 x = 1$

In Exercises 57–62, find all solutions to the equation. You do not need a calculator.

57. $4\cos^2 x - 4\cos x + 1 = 0$

58. $2\sin^2 x + 3\sin x + 1 = 0$

59. $\sin^2\theta - 2\sin\theta = 0$

60. $3\sin t = 2\cos^2 t$

61. $\cos(\sin x) = 1$

62. $2\sin^2 x + 3\sin x = 2$

In Exercises 63–68, find all solutions to the trigonometric equation, using a calculator where needed.

63. $\cos x = 0.37$

64. $\cos x = 0.75$

65. $\sin x = 0.30$

66. $\tan x = 5$

67. $\cos^2 x = 0.4$

68. $\sin^2 x = 0.4$

In Exercises 69–74, make the suggested trigonometric substitution, and then use Pythagorean identities to write the resulting function as a multiple of a basic trigonometric function.

69. $\sqrt{1 - x^2}$, $x = \cos\theta$

70. $\sqrt{x^2 + 1}$, $x = \tan\theta$

71. $\sqrt{x^2 - 9}$, $x = 3\sec\theta$

72. $\sqrt{36 - x^2}$, $x = 6\sin\theta$

73. $\sqrt{x^2 + 81}$, $x = 9\tan\theta$

74. $\sqrt{x^2 - 100}$, $x = 10\sec\theta$

Standardized Test Questions

75. True or False If $\sec(x - \pi/2) = 34$, then $\csc x = 34$. Justify your answer.

76. True or False The domain of validity for the identity $\sin\theta = \tan\theta\cos\theta$ is the set of all real numbers. Justify your answer.

You should answer these questions without using a calculator.

77. Multiple Choice Which of the following could *not* be set equal to $\sin x$ as an identity?

(A) $\cos\left(\dfrac{\pi}{2} - x\right)$ **(B)** $\cos\left(x - \dfrac{\pi}{2}\right)$

(C) $\sqrt{1 - \cos^2 x}$ **(D)** $\tan x \sec x$

(E) $-\sin(-x)$

78. Multiple Choice Exactly four of the six basic trigonometric functions are

(A) odd. **(B)** even.

(C) periodic. **(D)** continuous.

(E) bounded.

79. Multiple Choice A simpler expression for $(\sec\theta + 1)(\sec\theta - 1)$ is

(A) $\sin^2\theta$. **(B)** $\cos^2\theta$.

(C) $\tan^2\theta$. **(D)** $\cot^2\theta$.

(E) $\sec^2\theta$.

80. Multiple Choice How many numbers between 0 and 2π solve the equation $3\cos^2 x + \cos x = 2$?

(A) None **(B)** One

(C) Two **(D)** Three

(E) Four

Explorations

81. Write all six basic trigonometric functions entirely in terms of $\sin x$.

82. Write all six basic trigonometric functions entirely in terms of $\cos x$.

83. Writing to Learn Graph the functions $y = \sin^2 x$ and $y = -\cos^2 x$ in the standard trigonometric viewing window. Describe the apparent relationship between these two graphs and verify it with a trigonometric identity.

84. Writing to Learn Graph the functions $y = \sec^2 x$ and $y = \tan^2 x$ in the standard trigonometric viewing window. Describe the apparent relationship between these two graphs and verify it with a trigonometric identity.

85. Orbit of the Moon Because its orbit is elliptical, the distance from the Moon to the Earth in miles (measured from the center of the Moon to the center of the Earth) varies periodically. On Friday, January 23, 2009, the Moon was at its apogee (farthest from the Earth). The distance of the Moon from the Earth each Friday from January 23 to March 27 is recorded in Table 5.1.

Table 5.1 Distance from Earth to Moon

Date	Day	Distance (mi)
Jan 23	0	251,966
Jan 30	7	238,344
Feb 6	14	225,784
Feb 13	21	240,385
Feb 20	28	251,807
Feb 27	35	236,315
Mar 6	42	226,101
Mar 13	49	242,390
Mar 20	56	251,333
Mar 27	63	234,347

Source: The World Almanac and Book of Facts 2009.

(a) Draw a scatter plot of the data, using "day" as x and "distance" as y.

(b) Use your calculator to do a sine regression of y on x. Find the equation of the best-fit sine curve and superimpose its graph on the scatter plot.

(c) What is the approximate number of days from apogee to apogee? Interpret this number in terms of the orbit of the moon.

(d) Approximately how far is the Moon from the Earth at perigee (closest distance)?

(e) Since the data begin at apogee, perhaps a cosine curve would be a more appropriate model. Use the sine curve in part (b) and a cofunction identity to write a cosine curve that fits the data.

86. Group Activity Divide your class into six groups, each assigned to one of the basic trigonometric functions. With your group, construct a list of five different expressions that can be simplified to your assigned function. When you have finished, exchange lists with your "cofunction" group to check one another for accuracy.

Extending the Ideas

87. Prove that $\sin^4 \theta - \cos^4 \theta = \sin^2 \theta - \cos^2 \theta$.

88. Find all values of k that result in $\sin^2 x + 1 = k \sin x$ having an infinite solution set.

89. Use the cofunction identities and odd-even identities to prove that $\sin(\pi - x) = \sin x$.
[*Hint*: $\sin(\pi - x) = \sin(\pi/2 - (x - \pi/2))$.]

90. Use the cofunction identities and odd-even identities to prove that $\cos(\pi - x) = -\cos x$.
[*Hint*: $\cos(\pi - x) = \cos(\pi/2 - (x - \pi/2))$.]

91. Use the identity in Exercise 89 to prove that in any ΔABC, $\sin(A + B) = \sin C$.

92. Use the identities in Exercises 89 and 90 to find an identity for simplifying $\tan(\pi - x)$.

5.2 Proving Trigonometric Identities

What you'll learn about

- A Proof Strategy
- Proving Identities
- Disproving Non-Identities
- Identities in Calculus

... and why

Proving identities gives you excellent insights into the way mathematical proofs are constructed.

A Proof Strategy

We now arrive at the best opportunity in the precalculus curriculum for you to try your hand at constructing analytic proofs: trigonometric identities. Some are easy and some can be quite challenging, but in every case the *identity itself* frames your work with a beginning and an ending. The proof consists of filling in what lies between.

The strategy for proving an identity is very different from the strategy for solving an equation, most notably in the very first step. Usually the first step in solving an equation is to write down the equation. If you do this with an identity, however, you will have a beginning and an ending—with no proof in between! With an identity, you begin by writing down *one function* and end by writing down *the other*. Example 1 will illustrate what we mean.

EXAMPLE 1 Proving an Algebraic Identity

Prove the algebraic identity $\dfrac{x^2-1}{x-1} - \dfrac{x^2-1}{x+1} = 2$.

SOLUTION We prove this identity by showing a sequence of expressions, each one easily seen to be equivalent to its preceding expression:

$$\begin{aligned}
\frac{x^2-1}{x-1} - \frac{x^2-1}{x+1} &= \frac{(x+1)(x-1)}{x-1} - \frac{(x+1)(x-1)}{x+1} && \text{Factoring difference of squares}\\
&= (x+1)\left(\frac{x-1}{x-1}\right) - (x-1)\left(\frac{x+1}{x+1}\right) && \text{Algebraic manipulation}\\
&= (x+1)(1) - (x-1)(1) && \text{Reducing fractions}\\
&= x+1-x+1 && \text{Algebraic manipulation}\\
&= 2
\end{aligned}$$

Notice that the first thing we wrote down was the expression on the left-hand side (LHS), and the last thing we wrote down was the expression on the right-hand side (RHS). The proof would have been just as legitimate going from RHS to LHS, but it is more natural to move from the more complicated side to the less complicated side. Incidentally, the margin notes on the right, called "floaters," are included here for instructional purposes and are not really part of the proof. A good proof should consist of steps for which a knowledgeable reader could readily supply the floaters.

Now try Exercise 1.

These, then, are our first general strategies for proving an identity:

General Strategies I

1. The proof begins with the expression on one side of the identity.
2. The proof ends with the expression on the other side.
3. The proof in between consists of showing a sequence of expressions, each one easily seen to be equivalent to its preceding expression.

Proving Identities

Trigonometric identity proofs follow General Strategies I. We are told that two expressions are equal, and the object is to prove that they are equal. We do this by changing one expression into the other by a series of intermediate steps that follow the important rule that *every intermediate step yields an expression that is equivalent to the first.*

The changes at every step are accomplished by algebraic manipulations or identities, but the manipulations or identities should be sufficiently obvious as to require no additional justification. Since "obvious" is often in the eye of the beholder, it is usually safer to err on the side of including too many steps than too few.

By working through several examples, we try to give you a sense for what is appropriate as we illustrate some of the algebraic tools that you have at your disposal.

EXAMPLE 2 Proving an Identity

Prove the identity: $\tan x + \cot x = \sec x \csc x$.

SOLUTION We begin by deciding whether to start with the expression on the right or the left. It is usually best to start with the more complicated expression, as it is easier to proceed from the complex toward the simple than to go in the other direction. The expression on the left is slightly more complicated because it involves two terms.

$$\tan x + \cot x = \frac{\sin x}{\cos x} + \frac{\cos x}{\sin x} \qquad \text{Basic identities}$$

$$= \frac{\sin x}{\cos x} \cdot \frac{\sin x}{\sin x} + \frac{\cos x}{\sin x} \cdot \frac{\cos x}{\cos x} \qquad \text{Setting up common denominator}$$

$$= \frac{\sin^2 x + \cos^2 x}{\cos x \cdot \sin x}$$

$$= \frac{1}{\cos x \cdot \sin x} \qquad \text{Pythagorean identity}$$

$$= \frac{1}{\cos x} \cdot \frac{1}{\sin x} \qquad \text{(A step you could choose to omit)}$$

$$= \sec x \csc x \qquad \text{Basic identities}$$

(Remember that the "floaters" are not really part of the proof.)

Now try Exercise 13.

The preceding example illustrates three general strategies that are often useful in proving trigonometric identities.

General Strategies II

1. Begin with the more complicated expression and work toward the less complicated expression.
2. If no other move suggests itself, convert the entire expression to one involving sines and cosines.
3. Combine fractions by combining them over a common denominator.

EXAMPLE 3 Identifying and Proving an Identity

Match the function

$$f(x) = \frac{1}{\sec x - 1} + \frac{1}{\sec x + 1}$$

with one of the following. Then confirm the match with a proof.

(i) $2 \cot x \csc x$ **(ii)** $\dfrac{1}{\sec x}$

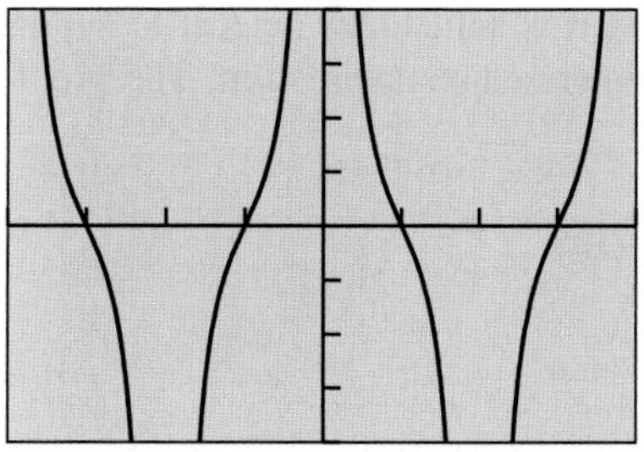

$[-2\pi, 2\pi]$ by $[-4, 4]$

$f(x) = \dfrac{1}{\sec x - 1} + \dfrac{1}{\sec x + 1}$

(a)

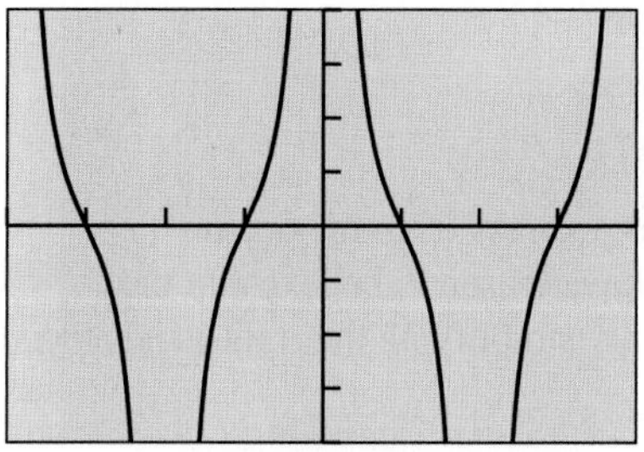

$[-2\pi, 2\pi]$ by $[-4, 4]$

$y = 2 \cot x \csc x$

(b)

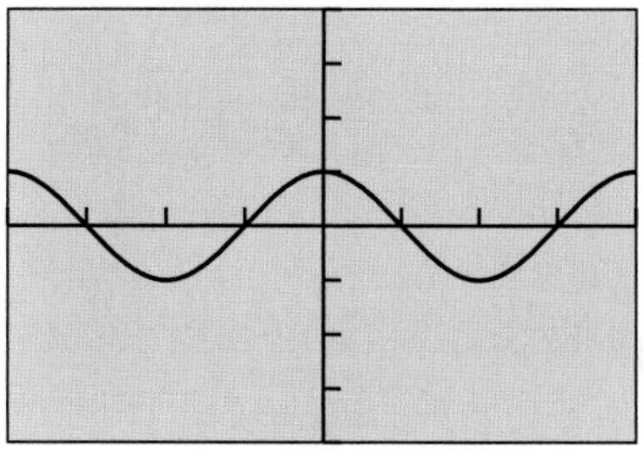

$[-2\pi, 2\pi]$ by $[-4, 4]$

$y = \dfrac{1}{\sec x}$

(c)

FIGURE 5.8 A grapher can be useful for identifying possible identities. (Example 3)

SOLUTION Figures 5.8a, b, and c show the graphs of the functions $y = f(x)$, $y = 2 \cot x \csc x$, and $y = 1/\sec x$, respectively. The graphs in (a) and (c) show that $f(x)$ is not equal to the expression in (ii). From the graphs in (a) and (b), it appears that $f(x)$ is equal to the expression in (i). To confirm algebraically, we begin with the expression for $f(x)$.

$$\frac{1}{\sec x - 1} + \frac{1}{\sec x + 1}$$

$$= \frac{\sec x + 1}{(\sec x - 1)(\sec x + 1)} + \frac{\sec x - 1}{(\sec x - 1)(\sec x + 1)} \quad \text{Common denominator}$$

$$= \frac{\sec x + 1 + \sec x - 1}{\sec^2 x - 1}$$

$$= \frac{2 \sec x}{\tan^2 x} \quad \text{Pythagorean identity}$$

$$= \frac{2}{\cos x} \cdot \frac{\cos^2 x}{\sin^2 x} \quad \text{Basic identities}$$

$$= \frac{2 \cos x}{\sin x} \cdot \frac{1}{\sin x}$$

$$= 2 \cot x \csc x$$

Now try Exercise 55.

The next example illustrates how the algebraic identity $(a + b)(a - b) = a^2 - b^2$ can be used to set up a Pythagorean substitution.

EXAMPLE 4 Setting up a Difference of Squares

Prove the identity: $\dfrac{\cos t}{1 - \sin t} = \dfrac{1 + \sin t}{\cos t}$.

SOLUTION The left-hand expression is slightly more complicated, as we can handle extra terms in a numerator more easily than in a denominator. So we begin with the left.

$$\frac{\cos t}{1 - \sin t} = \frac{\cos t}{1 - \sin t} \cdot \frac{1 + \sin t}{1 + \sin t} \quad \text{Setting up a difference of squares}$$

$$= \frac{(\cos t)(1 + \sin t)}{1 - \sin^2 t}$$

$$= \frac{(\cos t)(1 + \sin t)}{\cos^2 t} \quad \text{Pythagorean identity}$$

$$= \frac{1 + \sin t}{\cos t}$$

Now try Exercise 39.

Notice that we kept $(\cos t)(1 + \sin t)$ in factored form in the hope that we could eventually eliminate the factor $\cos t$ and be left with the numerator we need. It is always a good idea to keep an eye on the "target" expression toward which your proof is aimed.

General Strategies III

1. Use the algebraic identity $(a + b)(a - b) = a^2 - b^2$ to set up applications of the Pythagorean identities.
2. Always be mindful of the "target" expression, and favor manipulations that bring you closer to your goal.

In more complicated identities (as in word ladders) it is sometimes helpful to see if both sides can be manipulated toward a common intermediate expression. The proof can then be reconstructed in a single path.

EXAMPLE 5 Working from Both Sides

Prove the identity: $\dfrac{\cot^2 u}{1 + \csc u} = (\cot u)(\sec u - \tan u)$

SOLUTION Both sides are fairly complicated, but the left-hand side looks as though it needs more work. We start on the left.

$$\begin{aligned} \frac{\cot^2 u}{1 + \csc u} &= \frac{\csc^2 u - 1}{1 + \csc u} && \text{Pythagorean identity} \\ &= \frac{(\csc u + 1)(\csc u - 1)}{\csc u + 1} && \text{Factor.} \\ &= \csc u - 1 \end{aligned}$$

At this point it is not clear how we can get from this expression to the one on the right-hand side of our identity. However, we now have reason to believe that the right-hand side must simplify to $\csc u - 1$, so we try simplifying the right-hand side.

$$\begin{aligned} (\cot u)(\sec u - \tan u) &= \left(\frac{\cos u}{\sin u}\right)\left(\frac{1}{\cos u} - \frac{\sin u}{\cos u}\right) && \text{Basic identities} \\ &= \frac{1}{\sin u} - 1 && \text{Distribute the product.} \\ &= \csc u - 1 \end{aligned}$$

Now we can reconstruct the proof by going through $\csc u - 1$ as an intermediate step.

$$\begin{aligned} \frac{\cot^2 u}{1 + \csc u} &= \frac{\csc^2 u - 1}{1 + \csc u} \\ &= \frac{(\csc u + 1)(\csc u - 1)}{\csc u + 1} \\ &= \csc u - 1 && \text{Intermediate step} \\ &= \frac{1}{\sin u} - 1 \\ &= \left(\frac{\cos u}{\sin u}\right)\left(\frac{1}{\cos u} - \frac{\sin u}{\cos u}\right) \\ &= (\cot u)(\sec u - \tan u) \end{aligned}$$

Now try Exercise 41.

Disproving Non-Identities

Obviously, not every equation involving trigonometric expressions is an identity. How can we spot a non-identity before embarking on a futile attempt at a proof? Try the following exploration.

EXPLORATION 1 Confirming a Non-Identity

Prove or disprove that this is an identity: $\cos 2x = 2 \cos x$.

1. Graph $y = \cos 2x$ and $y = 2 \cos x$ in the same window. Interpret the graphs to make a conclusion about whether or not the equation is an identity.
2. With the help of the graph, find a value of x for which $\cos 2x \neq 2 \cos x$.
3. Does the existence of the x value in part 2 *prove* that the equation is *not* an identity?

4. Graph $y = \cos 2x$ and $y = \cos^2 x - \sin^2 x$ in the same window. Interpret the graphs to make a conclusion about whether or not $\cos 2x = \cos^2 x - \sin^2 x$ is an identity.

5. Do the graphs in part 4 *prove* that $\cos 2x = \cos^2 x - \sin^2 x$ is an identity? Explain your answer.

Exploration 1 suggests that we can use graphers to help confirm a *non-identity*, since we only have to produce a single value of x for which the two compared expressions are defined but unequal. On the other hand, we cannot use graphers to prove that an equation *is* an identity, since, for example, the graphers can never prove that two irrational numbers are equal. Also, graphers cannot show behavior over infinite domains.

Identities in Calculus

In most calculus problems where identities play a role, the object is to make a complicated expression simpler for the sake of computational ease. Occasionally it is actually necessary to make a *simple* expression *more complicated* for the sake of computational ease. Each of the following identities (just a sampling of many) represents a useful substitution in calculus wherein the expression on the right is simpler to deal with (even though it does not look that way). We prove one of these identities in Example 6 and leave the rest for the exercises or for future sections.

1. $\cos^3 x = (1 - \sin^2 x)(\cos x)$

2. $\sec^4 x = (1 + \tan^2 x)(\sec^2 x)$

3. $\sin^2 x = \frac{1}{2} - \frac{1}{2}\cos 2x$

4. $\cos^2 x = \frac{1}{2} + \frac{1}{2}\cos 2x$

5. $\sin^5 x = (1 - 2\cos^2 x + \cos^4 x)(\sin x)$

6. $\sin^2 x \cos^5 x = (\sin^2 x - 2\sin^4 x + \sin^6 x)(\cos x)$

EXAMPLE 6 Proving an Identity Useful in Calculus

Prove the following identity:

$$\sin^2 x \cos^5 x = (\sin^2 x - 2\sin^4 x + \sin^6 x)(\cos x)$$

SOLUTION We begin with the expression on the left.

$$\begin{aligned}
\sin^2 x \cos^5 x &= \sin^2 x \cos^4 x \cos x \\
&= (\sin^2 x)(\cos^2 x)^2 (\cos x) \\
&= (\sin^2 x)(1 - \sin^2 x)^2 (\cos x) \\
&= (\sin^2 x)(1 - 2\sin^2 x + \sin^4 x)(\cos x) \\
&= (\sin^2 x - 2\sin^4 x + \sin^6 x)(\cos x)
\end{aligned}$$

Now try Exercise 51.

QUICK REVIEW 5.2 *(For help, go to Section 5.1.)*

Exercise numbers with a gray background indicate problems that the authors have designed to be solved *without a calculator.*

In Exercises 1–6, write the expression in terms of sines and cosines only. Express your answer as a single fraction.

1. $\csc x + \sec x$
2. $\tan x + \cot x$
3. $\cos x \csc x + \sin x \sec x$
4. $\sin \theta \cot \theta - \cos \theta \tan \theta$
5. $\dfrac{\sin x}{\csc x} + \dfrac{\cos x}{\sec x}$
6. $\dfrac{\sec \alpha}{\cos \alpha} - \dfrac{\sin \alpha}{\csc \alpha \cos^2 \alpha}$

In Exercises 7–12, determine whether or not the equation is an identity. If not, find a single value of x for which the two expressions are not equal.

7. $\sqrt{x^2} = x$
8. $\sqrt[3]{x^3} = x$
9. $\sqrt{1 - \cos^2 x} = \sin x$
10. $\sqrt{\sec^2 x - 1} = \tan x$
11. $\ln \dfrac{1}{x} = -\ln x$
12. $\ln x^2 = 2 \ln x$

SECTION 5.2 Exercises

In Exercises 1–4, prove the algebraic identity by starting with the LHS expression and supplying a sequence of equivalent expressions that ends with the RHS expression.

1. $\dfrac{x^3 - x^2}{x} - (x - 1)(x + 1) = 1 - x$
2. $\dfrac{1}{x} - \dfrac{1}{2} = \dfrac{2 - x}{2x}$
3. $\dfrac{x^2 - 4}{x - 2} - \dfrac{x^2 - 9}{x + 3} = 5$
4. $(x - 1)(x + 2) - (x + 1)(x - 2) = 2x$

In Exercises 5–10, tell whether or not $f(x) = \sin x$ is an identity.

5. $f(x) = \dfrac{\sin^2 x + \cos^2 x}{\csc x}$
6. $f(x) = \dfrac{\tan x}{\sec x}$
7. $f(x) = \cos x \cdot \cot x$
8. $f(x) = \cos (x - \pi/2)$
9. $f(x) = (\sin^3 x)(1 + \cot^2 x)$
10. $f(x) = \dfrac{\sin 2x}{2}$

In Exercises 11–51, prove the identity.

11. $(\cos x)(\tan x + \sin x \cot x) = \sin x + \cos^2 x$
12. $(\sin x)(\cot x + \cos x \tan x) = \cos x + \sin^2 x$
13. $(1 - \tan x)^2 = \sec^2 x - 2 \tan x$
14. $(\cos x - \sin x)^2 = 1 - 2 \sin x \cos x$
15. $\dfrac{(1 - \cos u)(1 + \cos u)}{\cos^2 u} = \tan^2 u$
16. $\tan x + \sec x = \dfrac{\cos x}{1 - \sin x}$
17. $\dfrac{\cos^2 x - 1}{\cos x} = -\tan x \sin x$
18. $\dfrac{\sec^2 \theta - 1}{\sin \theta} = \dfrac{\sin \theta}{1 - \sin^2 \theta}$
19. $(1 - \sin \beta)(1 + \csc \beta) = \csc \beta - \sin \beta$
20. $\dfrac{1}{1 - \cos x} + \dfrac{1}{1 + \cos x} = 2 \csc^2 x$
21. $(\cos t - \sin t)^2 + (\cos t + \sin t)^2 = 2$
22. $\sin^2 \alpha - \cos^2 \alpha = 1 - 2 \cos^2 \alpha$
23. $\dfrac{1 + \tan^2 x}{\sin^2 x + \cos^2 x} = \sec^2 x$
24. $\dfrac{1}{\tan \beta} + \tan \beta = \sec \beta \csc \beta$
25. $\dfrac{\cos \beta}{1 + \sin \beta} = \dfrac{1 - \sin \beta}{\cos \beta}$
26. $\dfrac{\sec x + 1}{\tan x} = \dfrac{\sin x}{1 - \cos x}$
27. $\dfrac{\tan^2 x}{\sec x + 1} = \dfrac{1 - \cos x}{\cos x}$
28. $\dfrac{\cot v - 1}{\cot v + 1} = \dfrac{1 - \tan v}{1 + \tan v}$
29. $\cot^2 x - \cos^2 x = \cos^2 x \cot^2 x$
30. $\tan^2 \theta - \sin^2 \theta = \tan^2 \theta \sin^2 \theta$
31. $\cos^4 x - \sin^4 x = \cos^2 x - \sin^2 x$
32. $\tan^4 t + \tan^2 t = \sec^4 t - \sec^2 t$
33. $(x \sin \alpha + y \cos \alpha)^2 + (x \cos \alpha - y \sin \alpha)^2 = x^2 + y^2$
34. $\dfrac{1 - \cos \theta}{\sin \theta} = \dfrac{\sin \theta}{1 + \cos \theta}$
35. $\dfrac{\tan x}{\sec x - 1} = \dfrac{\sec x + 1}{\tan x}$
36. $\dfrac{\sin t}{1 + \cos t} + \dfrac{1 + \cos t}{\sin t} = 2 \csc t$
37. $\dfrac{\sin x - \cos x}{\sin x + \cos x} = \dfrac{2 \sin^2 x - 1}{1 + 2 \sin x \cos x}$
38. $\dfrac{1 + \cos x}{1 - \cos x} = \dfrac{\sec x + 1}{\sec x - 1}$
39. $\dfrac{\sin t}{1 - \cos t} + \dfrac{1 + \cos t}{\sin t} = \dfrac{2(1 + \cos t)}{\sin t}$
40. $\dfrac{\sin A \cos B + \cos A \sin B}{\cos A \cos B - \sin A \sin B} = \dfrac{\tan A + \tan B}{1 - \tan A \tan B}$

41. $\sin^2 x \cos^3 x = (\sin^2 x - \sin^4 x)(\cos x)$

42. $\sin^5 x \cos^2 x = (\cos^2 x - 2\cos^4 x + \cos^6 x)(\sin x)$

43. $\cos^5 x = (1 - 2\sin^2 x + \sin^4 x)(\cos x)$

44. $\sin^3 x \cos^3 x = (\sin^3 x - \sin^5 x)(\cos x)$

45. $\dfrac{\tan x}{1 - \cot x} + \dfrac{\cot x}{1 - \tan x} = 1 + \sec x \csc x$

46. $\dfrac{\cos x}{1 + \sin x} + \dfrac{\cos x}{1 - \sin x} = 2 \sec x$

47. $\dfrac{2\tan x}{1 - \tan^2 x} + \dfrac{1}{2\cos^2 x - 1} = \dfrac{\cos x + \sin x}{\cos x - \sin x}$

48. $\dfrac{1 - 3\cos x - 4\cos^2 x}{\sin^2 x} = \dfrac{1 - 4\cos x}{1 - \cos x}$

49. $\cos^3 x = (1 - \sin^2 x)(\cos x)$

50. $\sec^4 x = (1 + \tan^2 x)(\sec^2 x)$

51. $\sin^5 x = (1 - 2\cos^2 x + \cos^4 x)(\sin x)$

In Exercises 52–57, match the function with an equivalent expression from the following list. Then confirm the match with a proof. (The matching is not one-to-one.)

(a) $\sec^2 x \csc^2 x$ **(b)** $\sec x + \tan x$ **(c)** $2\sec^2 x$

(d) $\tan x \sin x$ **(e)** $\sin x \cos x$

52. $\dfrac{1 + \sin x}{\cos x}$

53. $(1 + \sec x)(1 - \cos x)$

54. $\sec^2 x + \csc^2 x$

55. $\dfrac{1}{1 + \sin x} + \dfrac{1}{1 - \sin x}$

56. $\dfrac{1}{\tan x + \cot x}$

57. $\dfrac{1}{\sec x - \tan x}$

Standardized Test Questions

58. True or False The equation $\sqrt{x^2} = x$ is an identity. Justify your answer.

59. True or False The equation $(\sqrt{x})^2 = x$ is an identity. Justify your answer.

You should answer these questions without using a calculator.

60. Multiple Choice If $f(x) = g(x)$ is an identity with domain of validity D, which of the following must be true?

I. For any x in D, $f(x)$ is defined.

II. For any x in D, $g(x)$ is defined.

III. For any x in D, $f(x) = g(x)$.

(A) None

(B) I and II only

(C) I and III only

(D) III only

(E) I, II, and III

61. Multiple Choice Which of these is an efficient first step in proving the identity $\dfrac{\sin x}{1 - \cos x} = \dfrac{1 + \cos x}{\sin x}$?

(A) $\dfrac{\sin x}{1 - \cos x} = \dfrac{\cos\left(\frac{\pi}{2} - x\right)}{1 - \cos x}$

(B) $\dfrac{\sin x}{1 - \cos x} = \dfrac{\sin x}{\sin^2 x + \cos^2 x - \cos x}$

(C) $\dfrac{\sin x}{1 - \cos x} = \dfrac{\sin x}{1 - \cos x} \cdot \dfrac{\csc x}{\csc x}$

(D) $\dfrac{\sin x}{1 - \cos x} = \dfrac{\sin x}{1 - \cos x} \cdot \dfrac{1 - \cos x}{1 - \cos x}$

(E) $\dfrac{\sin x}{1 - \cos x} = \dfrac{\sin x}{1 - \cos x} \cdot \dfrac{1 + \cos x}{1 + \cos x}$

62. Multiple Choice Which of the following could be an intermediate expression in a proof of the identity

$\tan\theta + \sec\theta = \dfrac{\cos\theta}{1 - \sin\theta}$?

(A) $\sin\theta + \cos\theta$

(B) $\tan\theta + \csc\theta$

(C) $\dfrac{\sin\theta + 1}{\cos\theta}$

(D) $\dfrac{\cos\theta}{1 + \sin\theta}$

(E) $\cos\theta - \cot\theta$

63. Multiple Choice If $f(x) = g(x)$ is an identity and $\dfrac{f(x)}{g(x)} = k$, which of the following must be *false*?

(A) $g(x) \neq 0$

(B) $f(x) = 0$

(C) $k = 1$

(D) $f(x) - g(x) = 0$

(E) $f(x)g(x) > 0$

Explorations

In Exercises 64–69, identify a simple function that has the same graph. Then confirm your choice with a proof.

64. $\sin x \cot x$

65. $\cos x \tan x$

66. $\dfrac{\sin x}{\csc x} + \dfrac{\cos x}{\sec x}$

67. $\dfrac{\csc x}{\sin x} - \dfrac{\cot x \csc x}{\sec x}$

68. $\dfrac{\sin x}{\tan x}$

69. $(\sec^2 x)(1 - \sin^2 x)$

70. Writing to Learn Let θ be any number that is in the domain of all six trig functions. Explain why the natural logarithms of all six basic trig functions of θ sum to 0.

71. If A and B are complementary angles, prove that $\sin^2 A + \sin^2 B = 1$.

72. Group Activity If your class contains $2n$ students, write the two expressions from n different identities on separate pieces of paper. (If your class contains an odd number of students, invite your teacher to join you for this activity.) You can use the identities from Exercises 11–51 in this section or from other textbooks, but be sure to write them all in the variable x. Mix up the slips of paper and give one to each student in your class. Then see how long it takes you as a class, without looking at the book, to pair yourselves off as identities. (This activity takes on an added degree of difficulty if you try it without calculators.)

Extending the Ideas

In Exercises 73–78, prove the identity.

73. $\sqrt{\dfrac{1 - \sin t}{1 + \sin t}} = \dfrac{1 - \sin t}{|\cos t|}$

74. $\sqrt{\dfrac{1 + \cos t}{1 - \cos t}} = \dfrac{1 + \cos t}{|\sin t|}$

75. $\sin^6 x + \cos^6 x = 1 - 3\sin^2 x \cos^2 x$

76. $\cos^6 x - \sin^6 x = (\cos^2 x - \sin^2 x)(1 - \cos^2 x \sin^2 x)$

77. $\ln|\tan x| = \ln|\sin x| - \ln|\cos x|$

78. $\ln|\sec\theta + \tan\theta| + \ln|\sec\theta - \tan\theta| = 0$

79. Writing to Learn Let $y_1 = [\sin(x + 0.001) - \sin x]/0.001$ and $y_2 = \cos x$.

(a) Use graphs and tables to decide whether $y_1 = y_2$.

(b) Find a value for h so that the graph of $y_3 = y_1 - y_2$ in $[-2\pi, 2\pi]$ by $[-h, h]$ appears to be a sinusoid. Give a convincing argument that y_3 is a sinusoid.

80. Hyperbolic Functions The *hyperbolic trigonometric functions* are defined as follows:

$$\sinh x = \frac{e^x - e^{-x}}{2} \qquad \cosh x = \frac{e^x + e^{-x}}{2} \qquad \tanh x = \frac{\sinh x}{\cosh x}$$

$$\operatorname{csch} x = \frac{1}{\sinh x} \qquad \operatorname{sech} x = \frac{1}{\cosh x} \qquad \coth x = \frac{1}{\tanh x}$$

Confirm the identity.

(a) $\cosh^2 x - \sinh^2 x = 1$

(b) $1 - \tanh^2 x = \operatorname{sech}^2 x$

(c) $\coth^2 x - 1 = \operatorname{csch}^2 x$

81. Writing to Learn Write a paragraph to explain why

$$\cos x = \cos x + \sin(10\pi x)$$

appears to be an identity when the two sides are graphed in a decimal window. Give a convincing argument that it is not an identity.

5.3 Sum and Difference Identities

What you'll learn about

- Cosine of a Difference
- Cosine of a Sum
- Sine of a Difference or Sum
- Tangent of a Difference or Sum
- Verifying a Sinusoid Algebraically

... and why

These identities provide clear examples of how different the algebra of functions can be from the algebra of real numbers.

Cosine of a Difference

There is a powerful instinct in all of us to believe that all functions obey the following law of additivity:

$$f(u + v) = f(u) + f(v)$$

In fact, very few do. If there were a hall of fame for algebraic blunders, the following would probably be the first two inductees:

$$(u + v)^2 = u^2 + v^2$$

$$\sqrt{u + v} = \sqrt{u} + \sqrt{v}$$

So, before we derive the true sum formulas for sine and cosine, let us clear the air with the following exploration.

EXPLORATION 1 Getting Past the Obvious but Incorrect Formulas

1. Let $u = \pi$ and $v = \pi/2$.
   Find $\sin(u + v)$. Find $\sin(u) + \sin(v)$.
   Does $\sin(u + v) = \sin(u) + \sin(v)$?
2. Let $u = 0$ and $v = 2\pi$.
   Find $\cos(u + v)$. Find $\cos(u) + \cos(v)$.
   Does $\cos(u + v) = \cos(u) + \cos(v)$?
3. Find your own values of u and v that will confirm that $\tan(u + v) \neq \tan(u) + \tan(v)$.

We could also show easily that

$$\cos(u - v) \neq \cos(u) - \cos(v) \quad \text{and} \quad \sin(u - v) \neq \sin(u) - \sin(v).$$

As you might expect, there *are* formulas for $\sin(u \pm v)$, $\cos(u \pm v)$, and $\tan(u \pm v)$, but Exploration 1 shows that they are not the ones our instincts would suggest. In a sense, that makes them all the more interesting. We will derive them all, beginning with the formula for $\cos(u - v)$.

Figure 5.9a on the next page shows angles u and v in standard position on the unit circle, determining points A and B with coordinates $(\cos u, \sin u)$ and $(\cos v, \sin v)$, respectively. Figure 5.9b shows the triangle ABO rotated so that the angle $\theta = u - v$ is in standard position. The angle θ determines point C with coordinates $(\cos\theta, \sin\theta)$.

The chord opposite angle θ has the same length in both circles, even though the coordinatization of the endpoints is different. Use the distance formula to find the length in each case, and set the formulas equal to each other:

$$AB = CD$$

$$\sqrt{(\cos v - \cos u)^2 + (\sin v - \sin u)^2} = \sqrt{(\cos\theta - 1)^2 + (\sin\theta - 0)^2}$$

Square both sides to eliminate the radical and expand the binomials to get

$$\cos^2 u - 2\cos u\cos v + \cos^2 v + \sin^2 u - 2\sin u\sin v + \sin^2 v$$
$$= \cos^2\theta - 2\cos\theta + 1 + \sin^2\theta$$

$$(\cos^2 u + \sin^2 u) + (\cos^2 v + \sin^2 v) - 2\cos u\cos v - 2\sin u\sin v$$
$$= (\cos^2\theta + \sin^2\theta) + 1 - 2\cos\theta$$

$$2 - 2\cos u\cos v - 2\sin u\sin v = 2 - 2\cos\theta$$

$$\cos u\cos v + \sin u\sin v = \cos\theta$$

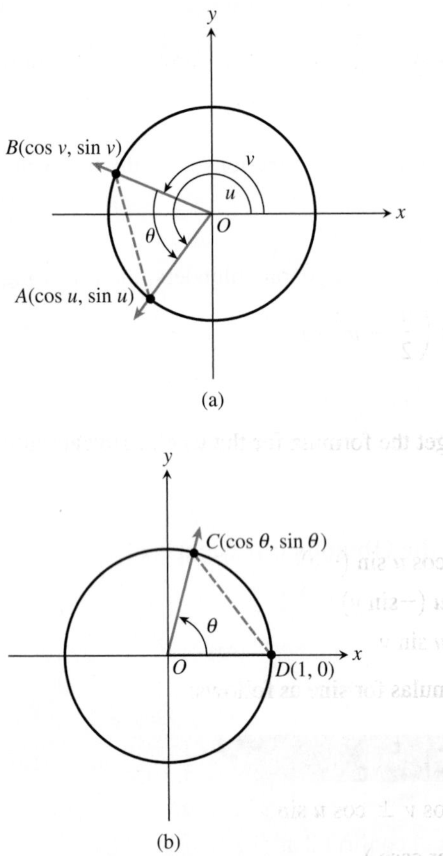

FIGURE 5.9 Angles u and v are in standard position in (a), and angle $\theta = u - v$ is in standard position in (b). The chords shown in the two circles are equal in length.

Finally, since $\theta = u - v$, we can write

$$\cos(u - v) = \cos u \cos v + \sin u \sin v.$$

EXAMPLE 1 Using the Cosine-of-a-Difference Identity

Find the exact value of cos 15° without using a calculator.

SOLUTION The trick is to write cos 15° as cos (45° − 30°); then we can use our knowledge of the special angles.

$$\begin{aligned} \cos 15^\circ &= \cos(45^\circ - 30^\circ) \\ &= \cos 45^\circ \cos 30^\circ + \sin 45^\circ \sin 30^\circ && \text{Cosine difference identity} \\ &= \left(\frac{\sqrt{2}}{2}\right)\left(\frac{\sqrt{3}}{2}\right) + \left(\frac{\sqrt{2}}{2}\right)\left(\frac{1}{2}\right) \\ &= \frac{\sqrt{6} + \sqrt{2}}{4} \end{aligned}$$

Now try Exercise 5.

Cosine of a Sum

Now that we have the formula for the cosine of a difference, we can get the formula for the cosine of a sum almost for free by using the odd-even identities.

$$\begin{aligned} \cos(u + v) &= \cos(u - (-v)) \\ &= \cos u \cos(-v) + \sin u \sin(-v) && \text{Cosine difference identity} \\ &= \cos u \cos v + \sin u(-\sin v) && \text{Odd-even identities} \\ &= \cos u \cos v - \sin u \sin v \end{aligned}$$

We can combine the sum and difference formulas for cosine as follows:

Cosine of a Sum or Difference

$$\cos(u \pm v) = \cos u \cos v \mp \sin u \sin v$$

(Note the sign switch in either case.)

We pointed out in Section 5.1 that the cofunction identities would be easier to prove with the results of Section 5.3. Here is what we mean.

EXAMPLE 2 Confirming Cofunction Identities

Prove the identities **(a)** $\cos((\pi/2) - x) = \sin x$ and **(b)** $\sin((\pi/2) - x) = \cos x$.

SOLUTION

(a)
$$\begin{aligned} \cos\left(\frac{\pi}{2} - x\right) &= \cos\left(\frac{\pi}{2}\right)\cos x + \sin\left(\frac{\pi}{2}\right)\sin x && \text{Cosine sum identity} \\ &= 0 \cdot \cos x + 1 \cdot \sin x \\ &= \sin x \end{aligned}$$

(b)
$$\begin{aligned} \sin\left(\frac{\pi}{2} - x\right) &= \cos\left(\frac{\pi}{2} - \left(\frac{\pi}{2} - x\right)\right) && \sin\theta = \cos((\pi/2) - \theta) \text{ by previous proof} \\ &= \cos(0 + x) \\ &= \cos x \end{aligned}$$

Now try Exercise 41.

Sine of a Difference or Sum

We can use the cofunction identities in Example 2 to get the formula for the sine of a sum from the formula for the cosine of a difference.

$$\begin{aligned}
\sin(u+v) &= \cos\left(\frac{\pi}{2}-(u+v)\right) && \text{Cofunction identity}\\
&= \cos\left(\left(\frac{\pi}{2}-u\right)-v\right) && \text{A little algebra}\\
&= \cos\left(\frac{\pi}{2}-u\right)\cos v + \sin\left(\frac{\pi}{2}-u\right)\sin v && \text{Cosine difference identity}\\
&= \sin u \cos v + \cos u \sin v && \text{Cofunction identities}
\end{aligned}$$

Then we can use the odd-even identities to get the formula for the sine of a difference from the formula for the sine of a sum.

$$\begin{aligned}
\sin(u-v) &= \sin(u+(-v)) && \text{A little algebra}\\
&= \sin u \cos(-v) + \cos u \sin(-v) && \text{Sine sum identity}\\
&= \sin u \cos v + \cos u(-\sin v) && \text{Odd-even identities}\\
&= \sin u \cos v - \cos u \sin v
\end{aligned}$$

We can combine the sum and difference formulas for sine as follows:

Sine of a Sum or Difference

$$\sin(u \pm v) = \sin u \cos v \pm \cos u \sin v$$

(Note that the sign does *not* switch in either case.)

EXAMPLE 3 Using the Sum and Difference Formulas

Write each of the following expressions as the sine or cosine of an angle.

(a) $\sin 22° \cos 13° + \cos 22° \sin 13°$

(b) $\cos\frac{\pi}{3}\cos\frac{\pi}{4} + \sin\frac{\pi}{3}\sin\frac{\pi}{4}$

(c) $\sin x \sin 2x - \cos x \cos 2x$

SOLUTION The key in each case is recognizing which formula applies. (Indeed, the real purpose of such exercises is to help you remember the formulas.)

(a) $\sin 22° \cos 13° + \cos 22° \sin 13°$ — Recognizing sine of sum formula

$= \sin(22° + 13°)$

$= \sin 35°$

(b) $\cos\frac{\pi}{3}\cos\frac{\pi}{4} + \sin\frac{\pi}{3}\sin\frac{\pi}{4}$ — Recognizing cosine of difference formula

$= \cos\left(\frac{\pi}{3}-\frac{\pi}{4}\right)$

$= \cos\frac{\pi}{12}$

(c) $\sin x \sin 2x - \cos x \cos 2x$ — Recognizing *opposite* of cos sum formula

$= -(\cos x \cos 2x - \sin x \sin 2x)$

$= -\cos(x + 2x)$ — Applying formula

$= -\cos 3x$

Now try Exercise 19.

If one of the angles in a sum or difference is a quadrantal angle (that is, a multiple of 90° or of $\pi/2$ rad), then the sum-difference identities yield single-termed expressions. Since the effect is to reduce the complexity, the resulting identity is called a **reduction formula**.

EXAMPLE 4 Proving Reduction Formulas

Prove the reduction formulas:

(a) $\sin(x + \pi) = -\sin x$

(b) $\cos\left(x + \dfrac{3\pi}{2}\right) = \sin x$

SOLUTION

(a)
$$\begin{aligned}\sin(x + \pi) &= \sin x \cos \pi + \cos x \sin \pi \\ &= \sin x \cdot (-1) + \cos x \cdot 0 \\ &= -\sin x\end{aligned}$$

(b)
$$\begin{aligned}\cos\left(x + \frac{3\pi}{2}\right) &= \cos x \cos \frac{3\pi}{2} - \sin x \sin \frac{3\pi}{2} \\ &= \cos x \cdot 0 - \sin x \cdot (-1) \\ &= \sin x\end{aligned}$$

Now try Exercise 23.

Tangent of a Difference or Sum

We can derive a formula for $\tan(u \pm v)$ directly from the corresponding formulas for sine and cosine, as follows:

$$\tan(u \pm v) = \frac{\sin(u \pm v)}{\cos(u \pm v)} = \frac{\sin u \cos v \pm \cos u \sin v}{\cos u \cos v \mp \sin u \sin v}$$

There is also a formula for $\tan(u \pm v)$ that is written entirely in terms of tangent functions:

$$\tan(u \pm v) = \frac{\tan u \pm \tan v}{1 \mp \tan u \tan v}$$

We will leave the proof of the all-tangent formula to the exercises.

EXAMPLE 5 Proving a Tangent Reduction Formula

Prove the reduction formula: $\tan(\theta - (3\pi/2)) = -\cot \theta$.

SOLUTION We can't use the all-tangent formula (Do you see why?), so we convert to sines and cosines.

$$\begin{aligned}\tan\left(\theta - \frac{3\pi}{2}\right) &= \frac{\sin(\theta - (3\pi/2))}{\cos(\theta - (3\pi/2))} \\ &= \frac{\sin\theta \cos(3\pi/2) - \cos\theta \sin(3\pi/2)}{\cos\theta \cos(3\pi/2) + \sin\theta \sin(3\pi/2)} \\ &= \frac{\sin\theta \cdot 0 - \cos\theta \cdot (-1)}{\cos\theta \cdot 0 + \sin\theta \cdot (-1)} \\ &= -\cot\theta\end{aligned}$$

Now try Exercise 39.

Verifying a Sinusoid Algebraically

Example 7 of Section 4.6 asked us to verify that the function $f(x) = 2 \sin x + 5 \cos x$ is a sinusoid. We solved graphically, concluding that $f(x) \approx 5.39 \sin(x + 1.19)$.

We now have a way of solving this kind of problem algebraically, with exact values for the amplitude and phase shift. Example 6 illustrates the technique.

EXAMPLE 6 Expressing a Sum of Sinusoids as a Sinusoid

Express $f(x) = 2\sin x + 5\cos x$ as a sinusoid in the form $f(x) = a\sin(bx + c)$.

SOLUTION Since $a\sin(bx + c) = a(\sin bx\cos c + \cos bx\sin c)$, we have

$$\begin{aligned} 2\sin x + 5\cos x &= a(\sin bx\cos c + \cos bx\sin c) \\ &= (a\cos c)\sin bx + (a\sin c)\cos bx. \end{aligned}$$

Comparing coefficients, we see that $b = 1$ and that $a\cos c = 2$ and $a\sin c = 5$. We can solve for a as follows:

$$\begin{aligned} (a\cos c)^2 + (a\sin c)^2 &= 2^2 + 5^2 \\ a^2\cos^2 c + a^2\sin^2 c &= 29 \\ a^2(\cos^2 c + \sin^2 c) &= 29 \\ a^2 &= 29 \quad \text{Pythagorean identity} \\ a &= \pm\sqrt{29} \end{aligned}$$

If we choose a to be positive, then $\cos c = 2/\sqrt{29}$ and $\sin c = 5/\sqrt{29}$. We can identify an acute angle c with those specifications as either $\cos^{-1}(2/\sqrt{29})$ or $\sin^{-1}(5/\sqrt{29})$, which are equal. So, an exact sinusoid for f is

$$\begin{aligned} f(x) &= 2\sin x + 5\cos x \\ &= a\sin(bx + c) \\ &= \sqrt{29}\sin\left(x + \cos^{-1}(2/\sqrt{29})\right) \text{ or } \sqrt{29}\sin\left(x + \sin^{-1}(5/\sqrt{29})\right) \end{aligned}$$

Now try Exercise 43.

QUICK REVIEW 5.3 *(For help, go to Sections 4.2 and 5.1.)*

Exercise numbers with a gray background indicate problems that the authors have designed to be solved *without a calculator.*

In Exercises 1–6, express the angle as a sum or difference of special angles (multiples of 30°, 45°, $\pi/6$, or $\pi/4$). Answers are not unique.

1. 15°
2. 75°
3. 165°
4. $\pi/12$
5. $5\pi/12$
6. $7\pi/12$

In Exercises 7–10, tell whether or not the identity $f(x + y) = f(x) + f(y)$ holds for the function f.

7. $f(x) = \ln x$
8. $f(x) = e^x$
9. $f(x) = 32x$
10. $f(x) = x + 10$

SECTION 5.3 Exercises

In Exercises 1–10, use a sum or difference identity to find an exact value.

1. $\sin 15°$
2. $\tan 15°$
3. $\sin 75°$
4. $\cos 75°$
5. $\cos\frac{\pi}{12}$
6. $\sin\frac{7\pi}{12}$
7. $\tan\frac{5\pi}{12}$
8. $\tan\frac{11\pi}{12}$
9. $\cos\frac{7\pi}{12}$
10. $\sin\frac{-\pi}{12}$

In Exercises 11–22, write the expression as the sine, cosine, or tangent of an angle.

11. $\sin 42°\cos 17° - \cos 42°\sin 17°$
12. $\cos 94°\cos 18° + \sin 94°\sin 18°$
13. $\sin\frac{\pi}{5}\cos\frac{\pi}{2} + \sin\frac{\pi}{2}\cos\frac{\pi}{5}$
14. $\sin\frac{\pi}{3}\cos\frac{\pi}{7} - \sin\frac{\pi}{7}\cos\frac{\pi}{3}$
15. $\dfrac{\tan 19° + \tan 47°}{1 - \tan 19°\tan 47°}$

16. $\dfrac{\tan(\pi/5) - \tan(\pi/3)}{1 + \tan(\pi/5)\tan(\pi/3)}$

17. $\cos\frac{\pi}{7}\cos x + \sin\frac{\pi}{7}\sin x$

18. $\cos x\cos\frac{\pi}{7} - \sin x\sin\frac{\pi}{7}$

19. $\sin 3x\cos x - \cos 3x\sin x$

20. $\cos 7y\cos 3y - \sin 7y\sin 3y$

21. $\dfrac{\tan 2y + \tan 3x}{1 - \tan 2y\tan 3x}$ **22.** $\dfrac{\tan 3\alpha - \tan 2\beta}{1 + \tan 3\alpha\tan 2\beta}$

In Exercises 23–30, prove the identity.

23. $\sin\left(x - \frac{\pi}{2}\right) = -\cos x$ **24.** $\tan\left(x - \frac{\pi}{2}\right) = -\cot x$

25. $\cos\left(x - \frac{\pi}{2}\right) = \sin x$

26. $\cos\left[\left(\frac{\pi}{2} - x\right) - y\right] = \sin(x + y)$

27. $\sin\left(x + \frac{\pi}{6}\right) = \frac{\sqrt{3}}{2}\sin x + \frac{1}{2}\cos x$

28. $\cos\left(x - \frac{\pi}{4}\right) = \frac{\sqrt{2}}{2}(\cos x + \sin x)$

29. $\tan\left(\theta + \frac{\pi}{4}\right) = \dfrac{1 + \tan\theta}{1 - \tan\theta}$

30. $\cos\left(\theta + \frac{\pi}{2}\right) = -\sin\theta$

In Exercises 31–34, match each graph with a pair of the following equations. Use your knowledge of identities and transformations, not your grapher.

(a) $y = \cos(3 - 2x)$

(b) $y = \sin x\cos 1 + \cos x\sin 1$

(c) $y = \cos(x - 3)$

(d) $y = \sin(2x - 5)$

(e) $y = \cos x\cos 3 + \sin x\sin 3$

(f) $y = \sin(x + 1)$

(g) $y = \cos 3\cos 2x + \sin 3\sin 2x$

(h) $y = \sin 2x\cos 5 - \cos 2x\sin 5$

31.

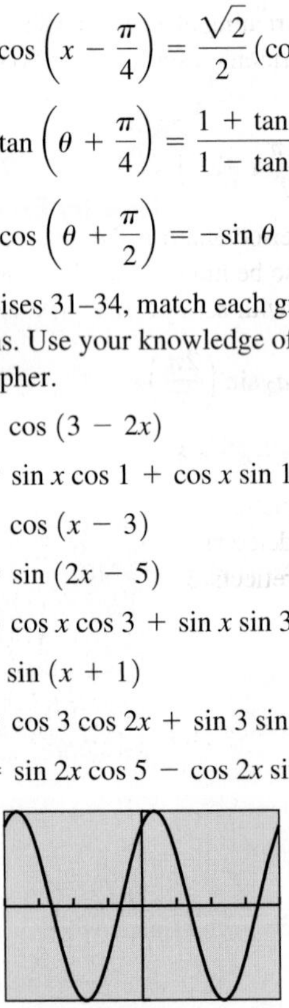

$[-2\pi, 2\pi]$ by $[-1, 1]$

32.

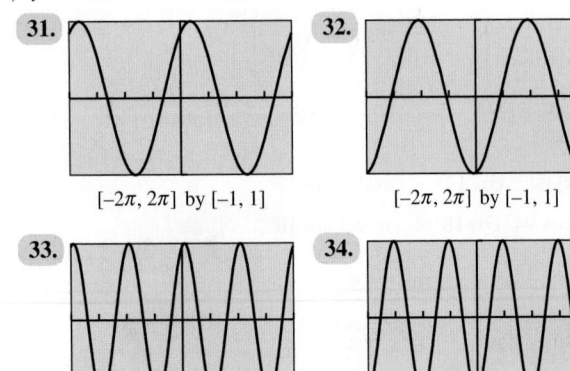

$[-2\pi, 2\pi]$ by $[-1, 1]$

33.

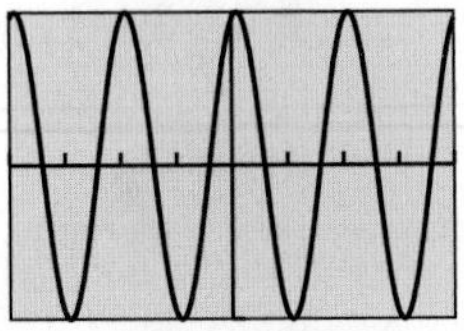

$[-2\pi, 2\pi]$ by $[-1, 1]$

34.

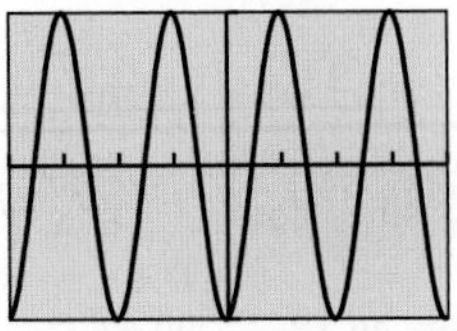

$[-2\pi, 2\pi]$ by $[-1, 1]$

In Exercises 35 and 36, use sum or difference identities (and not your grapher) to solve the equation exactly.

35. $\sin 2x\cos x = \cos 2x\sin x$

36. $\cos 3x\cos x = \sin 3x\sin x$

In Exercises 37–42, prove the reduction formula.

37. $\sin\left(\frac{\pi}{2} - u\right) = \cos u$ **38.** $\tan\left(\frac{\pi}{2} - u\right) = \cot u$

39. $\cot\left(\frac{\pi}{2} - u\right) = \tan u$ **40.** $\sec\left(\frac{\pi}{2} - u\right) = \csc u$

41. $\csc\left(\frac{\pi}{2} - u\right) = \sec u$ **42.** $\cos\left(x + \frac{\pi}{2}\right) = -\sin x$

In Exercises 43–46, express the function as a sinusoid in the form $y = a\sin(bx + c)$.

43. $y = 3\sin x + 4\cos x$

44. $y = 5\sin x - 12\cos x$

45. $y = \cos 3x + 2\sin 3x$

46. $y = 3\cos 2x - 2\sin 2x$

In Exercises 47–55, prove the identity.

47. $\sin(x - y) + \sin(x + y) = 2\sin x\cos y$

48. $\cos(x - y) + \cos(x + y) = 2\cos x\cos y$

49. $\cos 3x = \cos^3 x - 3\sin^2 x\cos x$

50. $\sin 3u = 3\cos^2 u\sin u - \sin^3 u$

51. $\cos 3x + \cos x = 2\cos 2x\cos x$

52. $\sin 4x + \sin 2x = 2\sin 3x\cos x$

53. $\tan(x + y)\tan(x - y) = \dfrac{\tan^2 x - \tan^2 y}{1 - \tan^2 x\tan^2 y}$

54. $\tan 5u\tan 3u = \dfrac{\tan^2 4u - \tan^2 u}{1 - \tan^2 4u\tan^2 u}$

55. $\dfrac{\sin(x + y)}{\sin(x - y)} = \dfrac{(\tan x + \tan y)}{(\tan x - \tan y)}$

Standardized Test Questions

56. True or False If A and B are supplementary angles, then $\cos A + \cos B = 0$. Justify your answer.

57. True or False If $\cos A + \cos B = 0$, then A and B are supplementary angles. Justify your answer.

You should answer these questions without using a calculator.

58. Multiple Choice If $\cos A\cos B = \sin A\sin B$, then $\cos(A + B) =$

(A) 0. **(B)** 1.

(C) $\cos A + \cos B$. **(D)** $\cos B + \cos A$.

(E) $\cos A\cos B + \sin A\sin B$.

59. Multiple Choice The function $y = \sin x\cos 2x + \cos x\sin 2x$ has amplitude

(A) 1. **(B)** 1.5.

(C) 2. **(D)** 3.

(E) 6.

60. Multiple Choice $\sin 15° =$

(A) $\frac{1}{4}$. **(B)** $\frac{\sqrt{3}}{4}$.

(C) $\frac{\sqrt{3}+\sqrt{2}}{4}$. **(D)** $\frac{\sqrt{6}-\sqrt{2}}{4}$.

(E) $\frac{\sqrt{6}+\sqrt{2}}{4}$.

61. Multiple Choice A function with the property

$f(1+2) = \frac{f(1)+f(2)}{1-f(1)f(2)}$ is

(A) $f(x) = \sin x$. **(B)** $f(x) = \tan x$.

(C) $f(x) = \sec x$. **(D)** $f(x) = e^x$.

(E) $f(x) = -1$.

Explorations

62. Prove the identity $\tan(u+v) = \frac{\tan u + \tan v}{1 - \tan u \tan v}$.

63. Prove the identity $\tan(u-v) = \frac{\tan u - \tan v}{1 + \tan u \tan v}$.

64. Writing to Learn Explain why the identity in Exercise 62 cannot be used to prove the reduction formula $\tan(x + \pi/2) = -\cot x$. Then prove the reduction formula.

65. Writing to Learn Explain why the identity in Exercise 63 cannot be used to prove the reduction formula $\tan(x - 3\pi/2) = -\cot x$. Then prove the reduction formula.

66. An Identity for Calculus Prove the following identity, which is used in calculus to prove an important differentiation formula.

$$\frac{\sin(x+h) - \sin x}{h} = \sin x\left(\frac{\cos h - 1}{h}\right) + \cos x\frac{\sin h}{h}$$

67. An Identity for Calculus Prove the following identity, which is used in calculus to prove another important differentiation formula.

$$\frac{\cos(x+h) - \cos x}{h} = \cos x\left(\frac{\cos h - 1}{h}\right) - \sin x\frac{\sin h}{h}$$

68. Group Activity Place 24 points evenly spaced around the unit circle, starting with the point $(1, 0)$. Using only your knowledge of the special angles and the sum and difference identities, work with your group to find the exact coordinates of all 24 points.

Extending the Ideas

In Exercises 69–72, assume that A, B, and C are the three angles of some ΔABC. (Note, then, that $A + B + C = \pi$.) Prove the following identities.

69. $\sin(A+B) = \sin C$

70. $\cos C = \sin A \sin B - \cos A \cos B$

71. $\tan A + \tan B + \tan C = \tan A \tan B \tan C$

72. $\cos A \cos B \cos C - \sin A \sin B \cos C - \sin A \cos B \sin C - \cos A \sin B \sin C = -1$

73. Writing to Learn The figure shows graphs of $y_1 = \cos 5x \cos 4x$ and $y_2 = -\sin 5x \sin 4x$ in one viewing window. Discuss the question, "How many solutions are there to the equation $\cos 5x \cos 4x = -\sin 5x \sin 4x$ in the interval $[-2\pi, 2\pi]$?" Give an algebraic argument that answers the question more convincingly than the graph does. Then support your argument with an *appropriate* graph.

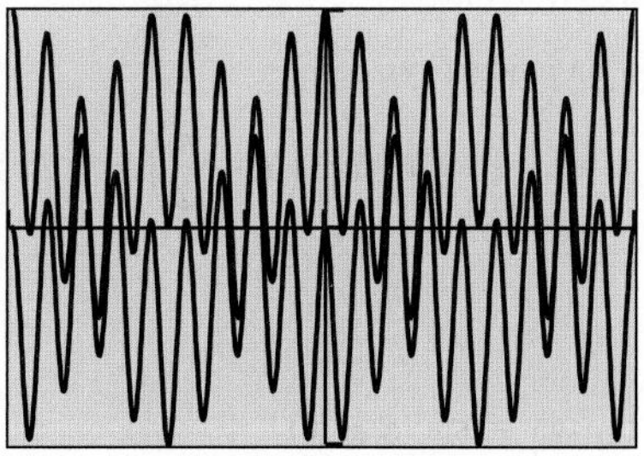

$[-2\pi, 2\pi]$ by $[-1, 1]$

74. Harmonic Motion Alternating electric current, an oscillating spring, or any other harmonic oscillator can be modeled by the equation

$$x = a\cos\left(\frac{2\pi}{T}t + \delta\right),$$

where T is the time for one period and δ is the phase constant. Show that this motion can also be modeled by the following sum of cosine and sine, each with zero phase constant:

$$a_1 \cos\left(\frac{2\pi}{T}\right)t + a_2 \sin\left(\frac{2\pi}{T}\right)t,$$

where $a_1 = a\cos\delta$ and $a_2 = -a\sin\delta$.

75. Magnetic Fields A magnetic field B can sometimes be modeled as the sum of an incident and a reflective field as

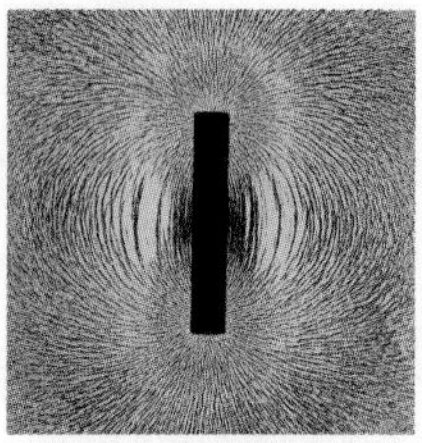

$$B = B_{\text{in}} + B_{\text{ref}},$$

where $B_{\text{in}} = \frac{E_0}{c}\cos\left(\omega t - \frac{\omega x}{c}\right)$, and

$$B_{\text{ref}} = \frac{E_0}{c}\cos\left(\omega t + \frac{\omega x}{c}\right).$$

Show that $B = 2\frac{E_0}{c}\cos \omega t \cos\frac{\omega x}{c}$.

5.4 Multiple-Angle Identities

What you'll learn about

- Double-Angle Identities
- Power-Reducing Identities
- Half-Angle Identities
- Solving Trigonometric Equations

... and why

These identities are useful in calculus courses.

Double-Angle Identities

The formulas that result from letting $u = v$ in the angle sum identities are called the *double-angle identities.* We will state them all and prove one, leaving the rest of the proofs as exercises. (See Exercises 1–4.)

Double-Angle Identities

$$\sin 2u = 2 \sin u \cos u$$

$$\cos 2u = \begin{cases} \cos^2 u - \sin^2 u \\ 2\cos^2 u - 1 \\ 1 - 2\sin^2 u \end{cases}$$

$$\tan 2u = \frac{2 \tan u}{1 - \tan^2 u}$$

There are three identities for $\cos 2u$. This is not unusual; indeed, there are plenty of other identities one could supply for $\sin 2u$ as well, such as $2 \sin u \sin (\pi/2 - u)$. We list the three identities for $\cos 2u$ because they are all *useful* in various contexts and therefore worth memorizing.

EXAMPLE 1 Proving a Double-Angle Identity

Prove the identity: $\sin 2u = 2 \sin u \cos u$.

SOLUTION

$$\begin{aligned} \sin 2u &= \sin (u + u) \\ &= \sin u \cos u + \cos u \sin u \qquad \text{Sine of a sum } (v = u) \\ &= 2 \sin u \cos u \end{aligned}$$

Now try Exercise 1.

Power-Reducing Identities

One immediate use for two of the three formulas for $\cos 2u$ is to derive the *power-reducing identities.* Some simple-looking functions like $y = \sin^2 u$ would be quite difficult to handle in certain calculus contexts were it not for the existence of these identities.

Power-Reducing Identities

$$\sin^2 u = \frac{1 - \cos 2u}{2}$$

$$\cos^2 u = \frac{1 + \cos 2u}{2}$$

$$\tan^2 u = \frac{1 - \cos 2u}{1 + \cos 2u}$$

We will also leave the proofs of these identities as exercises. (See Exercises 37 and 38.)

EXAMPLE 2 Proving an Identity

Prove the identity: $\cos^4 \theta - \sin^4 \theta = \cos 2\theta$.

SOLUTION

$$\begin{aligned}\cos^4 \theta - \sin^4 \theta &= (\cos^2 \theta + \sin^2 \theta)(\cos^2 \theta - \sin^2 \theta) \\ &= 1 \cdot (\cos^2 \theta - \sin^2 \theta) && \text{Pythagorean identity} \\ &= \cos 2\theta && \text{Double-angle identity}\end{aligned}$$

Now try Exercise 15.

EXAMPLE 3 Reducing a Power of 4

Rewrite $\cos^4 x$ in terms of trigonometric functions with no power greater than 1.

SOLUTION

$$\begin{aligned}\cos^4 x &= (\cos^2 x)^2 \\ &= \left(\frac{1 + \cos 2x}{2}\right)^2 && \text{Power-reducing identity} \\ &= \left(\frac{1 + 2\cos 2x + \cos^2 2x}{4}\right) \\ &= \frac{1}{4} + \frac{1}{2}\cos 2x + \frac{1}{4}\left(\frac{1 + \cos 4x}{2}\right) && \text{Power-reducing identity} \\ &= \frac{1}{4} + \frac{1}{2}\cos 2x + \frac{1}{8} + \frac{1}{8}\cos 4x \\ &= \frac{1}{8}(3 + 4\cos 2x + \cos 4x)\end{aligned}$$

Now try Exercise 39.

Half-Angle Identities

The power-reducing identities can be used to extend our stock of "special" angles whose trigonometric ratios can be found without a calculator. As usual, we are not suggesting that this algebraic procedure is any more practical than using a calculator, but we are suggesting that this sort of exercise helps you to understand how the functions behave. In Exploration 1, for example, we use a power-reducing formula to find the exact values of $\sin(\pi/8)$ and $\sin(9\pi/8)$ without a calculator.

EXPLORATION 1 **Finding the Sine of Half an Angle**

Recall the power-reducing formula $\sin^2 u = (1 - \cos 2u)/2$.

1. Use the power-reducing formula to show that $\sin^2(\pi/8) = (2 - \sqrt{2})/4$.
2. Solve for $\sin(\pi/8)$. Do you take the positive or negative square root? Why?
3. Use the power-reducing formula to show that $\sin^2(9\pi/8) = (2 - \sqrt{2})/4$.
4. Solve for $\sin(9\pi/8)$. Do you take the positive or negative square root? Why?

A little alteration of the power-reducing identities results in the *half-angle identities*, which can be used directly to find trigonometric functions of $u/2$ in terms of trigonometric functions of u. As Exploration 1 suggests, there is an unavoidable ambiguity of sign involved with the square root that must be resolved in particular cases by checking the quadrant in which $u/2$ lies.

Did We Miss Two ± Signs?

You might have noticed that all of the half-angle identities have unresolved ± signs except for the last two. The fact that we can omit them on the last two identities for $\tan u/2$ is a fortunate consequence of two facts: (1) $\sin u$ and $\tan(u/2)$ always have the same sign (easily observed from the graphs of the two functions in Figure 5.10), and (2) $1 \pm \cos u$ is never negative.

Half-Angle Identities

$$\sin\frac{u}{2} = \pm\sqrt{\frac{1-\cos u}{2}}$$

$$\cos\frac{u}{2} = \pm\sqrt{\frac{1+\cos u}{2}}$$

$$\tan\frac{u}{2} = \begin{cases} \pm\sqrt{\dfrac{1-\cos u}{1+\cos u}} \\ \dfrac{1-\cos u}{\sin u} \\ \dfrac{\sin u}{1+\cos u} \end{cases}$$

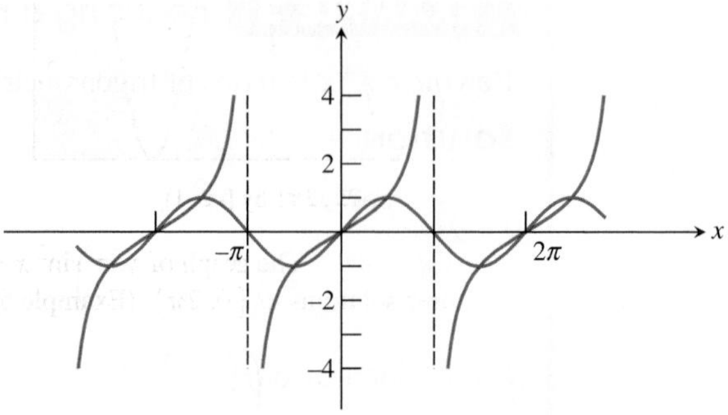

FIGURE 5.10 The functions $\sin u$ and $\tan(u/2)$ always have the same sign.

Solving Trigonometric Equations

New identities always provide new tools for solving trigonometric equations algebraically. Under the right conditions, they even lead to exact solutions. We assert again that we are not presenting these algebraic solutions for their practical value (as the calculator solutions are certainly sufficient for most applications and unquestionably much quicker to obtain), but rather as ways to observe the behavior of the trigonometric functions and their interwoven tapestry of identities.

EXAMPLE 4 Using a Double-Angle Identity

Solve algebraically in the interval $[0, 2\pi)$: $\sin 2x = \cos x$.

SOLUTION

$$\begin{aligned} \sin 2x &= \cos x \\ 2\sin x\cos x &= \cos x \\ 2\sin x\cos x - \cos x &= 0 \\ \cos x\,(2\sin x - 1) &= 0 \\ \cos x = 0 \quad &\text{or} \quad 2\sin x - 1 = 0 \\ \cos x = 0 \quad &\text{or} \quad \sin x = \frac{1}{2} \end{aligned}$$

The two solutions of $\cos x = 0$ are $x = \pi/2$ and $x = 3\pi/2$. The two solutions of $\sin x = 1/2$ are $x = \pi/6$ and $x = 5\pi/6$. Therefore, the solutions of $\sin 2x = \cos x$ are

$$\frac{\pi}{6}, \frac{\pi}{2}, \frac{5\pi}{6}, \frac{3\pi}{2}.$$

We can **support** this result **graphically** by verifying the four x-intercepts of the function $y = \sin 2x - \cos x$ in the interval $[0, 2\pi)$ (Figure 5.11).

Now try Exercise 23.

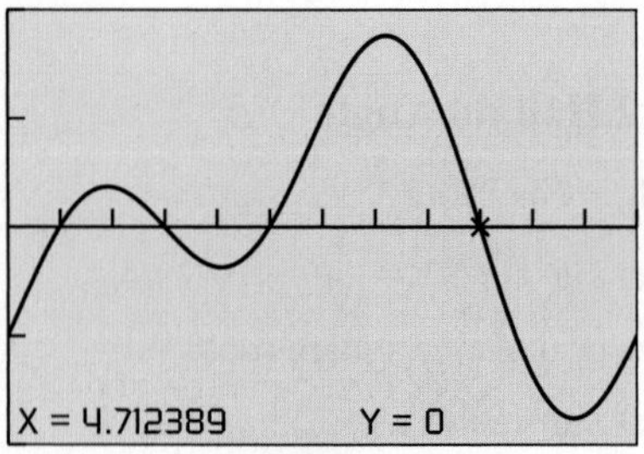

$[0, 2\pi]$ by $[-2, 2]$

FIGURE 5.11 The function $y = \sin 2x - \cos x$ for $0 \le x \le 2\pi$. The scale on the x-axis shows intervals of length $\pi/6$. This graph supports the solution found algebraically in Example 4. For example, using TRACE on a grapher with $x = 3\pi/2$ supports the 4th solution, $3\pi/2$.

EXAMPLE 5 Using Half-Angle Identities

Solve $\sin^2 x = 2 \sin^2 (x/2)$.

SOLUTION The graph of $y = \sin^2 x - 2 \sin^2 (x/2)$ in Figure 5.12 suggests that this function is periodic with period 2π and that the equation $\sin^2 x = 2 \sin^2 (x/2)$ has three solutions in $[0, 2\pi)$.

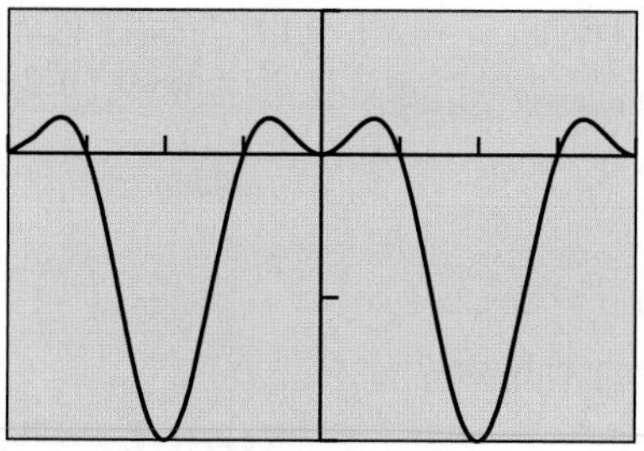

$[-2\pi, 2\pi]$ by $[-2, 1]$

FIGURE 5.12 The graph of $y = \sin^2 x - 2 \sin^2 (x/2)$ suggests that $\sin^2 x = 2 \sin^2 (x/2)$ has three solutions in $[0, 2\pi)$. (Example 5)

Solve Algebraically

$$\sin^2 x = 2 \sin^2 \frac{x}{2}$$

$$\sin^2 x = 2\left(\frac{1 - \cos x}{2}\right) \quad \text{Half-angle identity}$$

$$1 - \cos^2 x = 1 - \cos x \quad \text{Convert to all cosines.}$$

$$\cos x - \cos^2 x = 0$$

$$\cos x (1 - \cos x) = 0$$

$$\cos x = 0 \quad \text{or} \quad \cos x = 1$$

$$x = \frac{\pi}{2} \quad \text{or} \quad \frac{3\pi}{2} \quad \text{or} \quad 0$$

The rest of the solutions are obtained by periodicity:

$$x = 2n\pi, \quad x = \frac{\pi}{2} + 2n\pi, \quad x = \frac{3\pi}{2} + 2n\pi, \quad n = 0, \pm 1, \pm 2, \ldots$$

Now try Exercise 43.

QUICK REVIEW 5.4 *(For help, go to Section 5.1.)*

Exercise numbers with a gray background indicate problems that the authors have designed to be solved *without a calculator*.

In Exercises 1–8, find the general solution of the equation.

1. $\tan x - 1 = 0$

2. $\tan x + 1 = 0$

3. $(\cos x)(1 - \sin x) = 0$

4. $(\sin x)(1 + \cos x) = 0$

5. $\sin x + \cos x = 0$

6. $\sin x - \cos x = 0$

7. $(2 \sin x - 1)(2 \cos x + 1) = 0$

8. $(\sin x + 1)(2 \cos x - \sqrt{2}) = 0$

9. Find the area of the trapezoid.

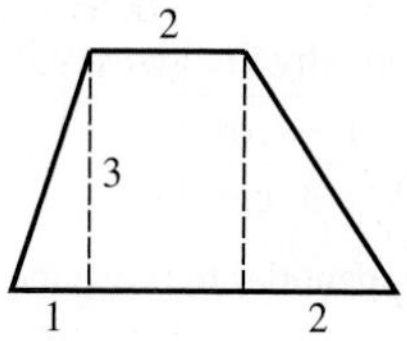

10. Find the height of the isosceles triangle.

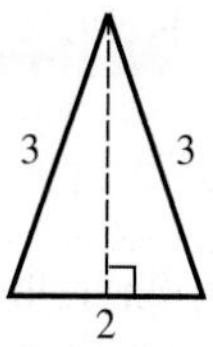

SECTION 5.4 Exercises

In Exercises 1–4, use the appropriate sum or difference identity to prove the double-angle identity.

1. $\cos 2u = \cos^2 u - \sin^2 u$ **2.** $\cos 2u = 2\cos^2 u - 1$

3. $\cos 2u = 1 - 2\sin^2 u$ **4.** $\tan 2u = \dfrac{2\tan u}{1 - \tan^2 u}$

In Exercises 5–10, find all solutions to the equation in the interval $[0, 2\pi)$.

5. $\sin 2x = 2\sin x$ **6.** $\sin 2x = \sin x$

7. $\cos 2x = \sin x$ **8.** $\cos 2x = \cos x$

9. $\sin 2x - \tan x = 0$ **10.** $2\cos^2 x + \cos x = \cos 2x$

In Exercises 11–14, write the expression as one involving only $\sin\theta$ and $\cos\theta$.

11. $\sin 2\theta + \cos\theta$ **12.** $\sin 2\theta + \cos 2\theta$

13. $\sin 2\theta + \cos 3\theta$ **14.** $\sin 3\theta + \cos 2\theta$

In Exercises 15–22, prove the identity.

15. $\sin 4x = 2\sin 2x\cos 2x$ **16.** $\cos 6x = 2\cos^2 3x - 1$

17. $2\csc 2x = \csc^2 x\tan x$ **18.** $2\cot 2x = \cot x - \tan x$

19. $\sin 3x = (\sin x)(4\cos^2 x - 1)$

20. $\sin 3x = (\sin x)(3 - 4\sin^2 x)$

21. $\cos 4x = 1 - 8\sin^2 x\cos^2 x$

22. $\sin 4x = (4\sin x\cos x)(2\cos^2 x - 1)$

In Exercises 23–30, solve algebraically for exact solutions in the interval $[0, 2\pi)$. Use your grapher only to support your algebraic work.

23. $\cos 2x + \cos x = 0$ **24.** $\cos 2x + \sin x = 0$

25. $\cos x + \cos 3x = 0$ **26.** $\sin x + \sin 3x = 0$

27. $\sin 2x + \sin 4x = 0$ **28.** $\cos 2x + \cos 4x = 0$

In Exercises 29 and 30, use a graphing calculator to find all exact solutions in the interval $[0, 2\pi)$. (*Hint*: All solutions are rational multiples of π.)

29. $\sin 2x - \cos 3x = 0$ **30.** $\sin 3x + \cos 2x = 0$

In Exercises 31–36, use half-angle identities to find an exact value without a calculator.

31. $\sin 15°$ **32.** $\tan 195°$

33. $\cos 75°$ **34.** $\sin(5\pi/12)$

35. $\tan(7\pi/12)$ **36.** $\cos(\pi/8)$

37. Prove the power-reducing identities:

(a) $\sin^2 u = \dfrac{1 - \cos 2u}{2}$ (b) $\cos^2 u = \dfrac{1 + \cos 2u}{2}$

38. (a) Use the identities in Exercise 37 to prove the power-reducing identity $\tan^2 u = \dfrac{1 - \cos 2u}{1 + \cos 2u}$.

(b) **Writing to Learn** Explain why the identity in part (a) does not imply that $\tan u = \sqrt{\dfrac{1 - \cos 2u}{1 + \cos 2u}}$.

In Exercises 39–42, use the power-reducing identities to prove the identity.

39. $\sin^4 x = \dfrac{1}{8}(3 - 4\cos 2x + \cos 4x)$

40. $\cos^3 x = \left(\dfrac{1}{2}\cos x\right)(1 + \cos 2x)$

41. $\sin^3 2x = \left(\dfrac{1}{2}\sin 2x\right)(1 - \cos 4x)$

42. $\sin^5 x = \left(\dfrac{1}{8}\sin x\right)(3 - 4\cos 2x + \cos 4x)$

In Exercises 43–46, use the half-angle identities to find all solutions in the interval $[0, 2\pi)$. Then find the general solution.

43. $\cos^2 x = \sin^2\left(\dfrac{x}{2}\right)$ **44.** $\sin^2 x = \cos^2\left(\dfrac{x}{2}\right)$

45. $\tan\left(\dfrac{x}{2}\right) = \dfrac{1 - \cos x}{1 + \cos x}$ **46.** $\sin^2\left(\dfrac{x}{2}\right) = \cos x - 1$

Standardized Test Questions

47. True or False The product of two functions with period 2π has period 2π. Justify your answer.

48. True or False The function $f(x) = \cos^2 x$ is a sinusoid. Justify your answer.

You should answer these questions without using a calculator.

49. Multiple Choice If $f(x) = \sin x$ and $g(x) = \cos x$, then $f(2x) =$

(A) $2\,f(x)$. (B) $f(2)\,f(x)$. (C) $f(x)\,g(x)$.

(D) $2\,f(x)\,g(x)$. (E) $f(2)\,g(x) + g(2)\,f(x)$.

50. Multiple Choice $\sin 22.5° =$

(A) $\dfrac{\sqrt{2}}{4}$. (B) $\dfrac{\sqrt{3}}{4}$. (C) $\dfrac{\sqrt{6} - \sqrt{2}}{4}$.

(D) $\sqrt{\dfrac{2 - \sqrt{2}}{2}}$. (E) $\sqrt{\dfrac{2 - \sqrt{2}}{2}}$.

51. Multiple Choice How many numbers between 0 and 2π satisfy the equation $\sin 2x = \cos x$?

(A) None (B) One (C) Two (D) Three (E) Four

52. Multiple Choice The period of the function $\sin^2 x - \cos^2 x$ is

(A) $\dfrac{\pi}{4}$. (B) $\dfrac{\pi}{2}$. (C) π. (D) 2π. (E) 4π.

Explorations

53. Connecting Trigonometry and Geometry In a regular polygon all sides are the same length and all angles are equal in measure.

(a) If the perpendicular distance from the center of the polygon with n sides to the midpoint of a side is R, and if the length of the side of the polygon is x, show that

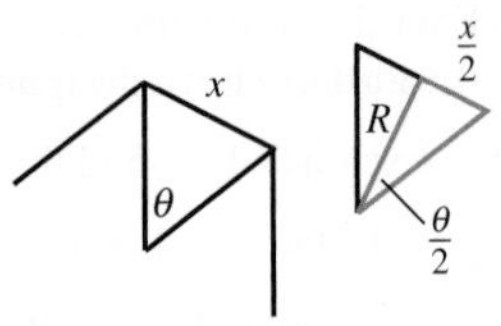

Regular polygon with n sides

$$x = 2R\tan\frac{\theta}{2}$$

where $\theta = 2\pi/n$ is the central angle subtended by one side.

(b) If the length of one side of a regular 11-sided polygon is approximately 5.87 and R is a whole number, what is the value of R?

54. Connecting Trigonometry and Geometry A rhombus is a quadrilateral with equal sides. The diagonals of a rhombus bisect the angles of the rhombus and are perpendicular bisectors of each other. Let $\angle ABC = \theta$, d_1 = length of AC, and d_2 = length of BD.

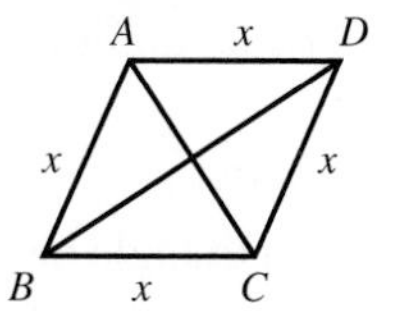

(a) Show that $\cos \frac{\theta}{2} = \frac{d_2}{2x}$ and $\sin \frac{\theta}{2} = \frac{d_1}{2x}$.

(b) Show that $\sin \theta = \frac{d_1 d_2}{2x^2}$.

55. Group Activity Maximizing Volume The ends of a 10-ft-long water trough are isosceles trapezoids as shown in the figure. Find the value of θ that maximizes the volume of the trough and the maximum volume.

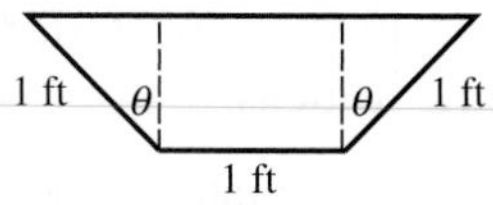

56. Group Activity Tunnel Problem A rectangular tunnel is cut through a mountain to make a road. The upper vertices of the rectangle are on the circle $x^2 + y^2 = 400$, as illustrated in the figure.

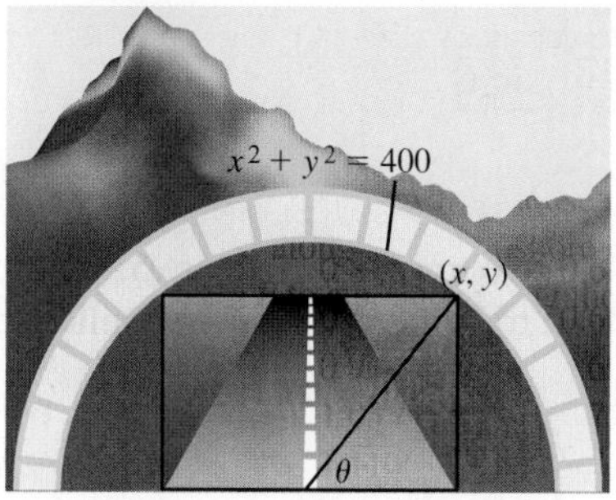

(a) Show that the cross-sectional area of the end of the tunnel is $400 \sin 2\theta$.

(b) Find the dimensions of the rectangular end of the tunnel that maximizes its cross-sectional area.

Extending the Ideas

In Exercises 57–61, prove the double-angle formulas.

57. $\csc 2u = \frac{1}{2} \csc u \sec u$

58. $\cot 2u = \frac{\cot^2 u - 1}{2 \cot u}$

59. $\sec 2u = \frac{\csc^2 u}{\csc^2 u - 2}$

60. $\sec 2u = \frac{\sec^2 u}{2 - \sec^2 u}$

61. $\sec 2u = \frac{\sec^2 u \csc^2 u}{\csc^2 u - \sec^2 u}$

62. Writing to Learn Explain why

$$\sqrt{\frac{1 - \cos 2x}{2}} = |\sin x|$$

is an identity but

$$\sqrt{\frac{1 - \cos 2x}{2}} = \sin x$$

is not an identity.

63. Hawaiian Sunset Table 5.2 gives the time of day for sunset in Honolulu, HI, on the first day of each month of 2009.

Table 5.2 Sunset in Honolulu, 2009

Date	Day	Time	18:30 +
Jan 1	1	18:01	−29
Feb 1	32	18:22	−8
Mar 1	60	18:36	6
Apr 1	91	18:46	16
May 1	121	18:57	27
Jun 1	152	19:10	40
Jul 1	182	19:17	47
Aug 1	213	19:10	40
Sep 1	244	18:47	17
Oct 1	274	18:19	−11
Nov 1	305	17:55	−35
Dec 1	335	17:48	−42

Source: www.timeanddate.com

The second column gives the date as the day of the year, and the fourth column gives the time as the number of minutes past 18:30.

(a) Enter the numbers in column 2 (day) into list L1 and the numbers in column 4 (minutes past 18:30) into list L2. Make a scatter plot with x-coordinates from L1 and y-coordinates from L2.

(b) Using sine regression, find the regression curve through the points and store its equation in Y1. Superimpose the graph of the curve on the scatter plot. Is it a good fit?

(c) Make a new column showing the *residuals* (the difference between the actual y value at each point and the y value predicted by the regression curve) and store them in list L3. Your calculator might have a list called RESID among the NAMES in the LIST menu, in which case the command RESID → L3 will perform this operation. You could also enter L2 − Y1(L1) → L3.

(d) Make a scatter plot with x-coordinates from L1 and y-coordinates from L3. Find the sine regression curve through *these* points and superimpose it on the scatter plot.

(e) **Writing to Learn** Interpret what the two regressions seem to indicate about the periodic behavior of sunset as a function of time. This is not an unusual phenomenon in astronomical data, and it kept astronomers baffled for centuries.

5.5 The Law of Sines

What you'll learn about

- Deriving the Law of Sines
- Solving Triangles (AAS, ASA)
- The Ambiguous Case (SSA)
- Applications

... and why

The Law of Sines is a powerful extension of the triangle congruence theorems of Euclidean geometry.

Deriving the Law of Sines

Recall from geometry that a triangle has six parts (three sides (S), three angles (A)), but that its size and shape can be completely determined by fixing only three of those parts, provided they are the right three. These threesomes that determine *triangle congruence* are known by their acronyms: AAS, ASA, SAS, and SSS. The other two acronyms represent matchups that don't quite work: AAA determines similarity only, but SSA does not even determine similarity.

With trigonometry we can find the other parts of the triangle once congruence is established. The tools we need are the Law of Sines and the Law of Cosines, the subjects of our last two trigonometric sections.

The **Law of Sines** states that the ratio of the sine of an angle to the length of its opposite side is the same for all three angles of any triangle.

Law of Sines

In any ΔABC with angles A, B, and C opposite sides a, b, and c, respectively, the following equation is true:

$$\frac{\sin A}{a} = \frac{\sin B}{b} = \frac{\sin C}{c}.$$

The derivation of the Law of Sines refers to the two triangles in Figure 5.13, in each of which we have drawn an altitude to side c. Right triangle trigonometry applied to either of the triangles in Figure 5.13 tells us that

$$\sin A = \frac{h}{b}.$$

In the acute triangle on the top,

$$\sin B = \frac{h}{a},$$

and in the obtuse triangle on the bottom,

$$\sin(\pi - B) = \frac{h}{a}.$$

But $\sin(\pi - B) = \sin B$, so in either case

$$\sin B = \frac{h}{a}.$$

Solving for h in both equations yields $h = b \sin A = a \sin B$. The equation $b \sin A = a \sin B$ is equivalent to

$$\frac{\sin A}{a} = \frac{\sin B}{b}.$$

If we were to draw an altitude to side a and repeat the same steps as above, we would reach the conclusion that

$$\frac{\sin B}{b} = \frac{\sin C}{c}.$$

Putting the results together,

$$\frac{\sin A}{a} = \frac{\sin B}{b} = \frac{\sin C}{c}.$$

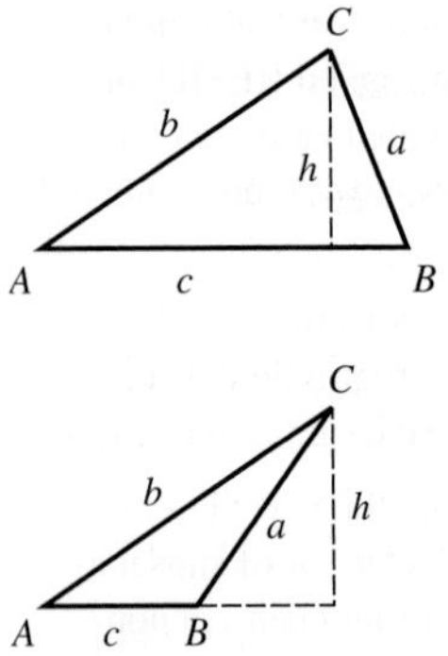

FIGURE 5.13 The Law of Sines.

Solving Triangles (AAS, ASA)

Two angles and a side of a triangle, in any order, determine the size and shape of a triangle completely. Of course, two angles of a triangle determine the third, so we really get one of the missing three parts for free. We solve for the remaining two parts (the unknown sides) with the Law of Sines.

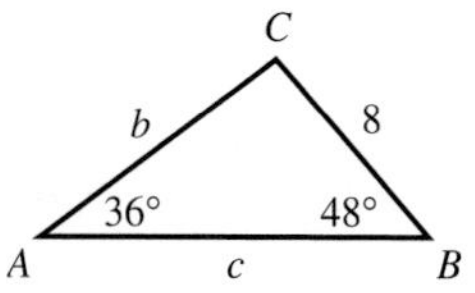

FIGURE 5.14 A triangle determined by AAS. (Example 1)

EXAMPLE 1 Solving a Triangle Given Two Angles and a Side

Solve ΔABC given that $\angle A = 36°, \angle B = 48°$, and $a = 8$. (See Figure 5.14.)

SOLUTION First, we note that $\angle C = 180° - 36° - 48° = 96°$.

We then apply the Law of Sines:

$$\frac{\sin A}{a} = \frac{\sin B}{b} \quad \text{and} \quad \frac{\sin A}{a} = \frac{\sin C}{c}$$

$$\frac{\sin 36°}{8} = \frac{\sin 48°}{b} \qquad \frac{\sin 36°}{8} = \frac{\sin 96°}{c}$$

$$b = \frac{8 \sin 48°}{\sin 36°} \qquad c = \frac{8 \sin 96°}{\sin 36°}$$

$$b \approx 10.11 \qquad c \approx 13.54$$

The six parts of the triangle are:

$$\angle A = 36° \qquad a = 8$$
$$\angle B = 48° \qquad b \approx 10.11$$
$$\angle C = 96° \qquad c \approx 13.54$$

Now try Exercise 1.

The Ambiguous Case (SSA)

Although two angles and a side of a triangle are always sufficient to determine its size and shape, the same cannot be said for two sides and an angle. Perhaps unexpectedly, it depends on where that angle is. If the angle is included between the two sides (the SAS case), then the triangle is uniquely determined up to congruence. If the angle is opposite one of the sides (the SSA case), then there might be one, two, or zero triangles determined.

Solving a triangle in the SAS case involves the Law of Cosines and will be handled in the next section. Solving a triangle in the SSA case is done with the Law of Sines, but with an eye toward the possibilities, as seen in the following Exploration.

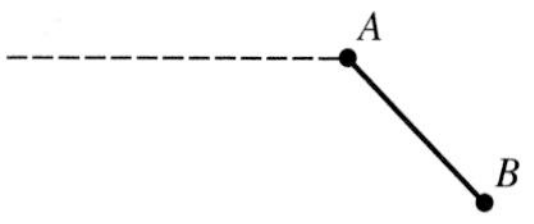

FIGURE 5.15 The diagram for part 1. (Exploration 1)

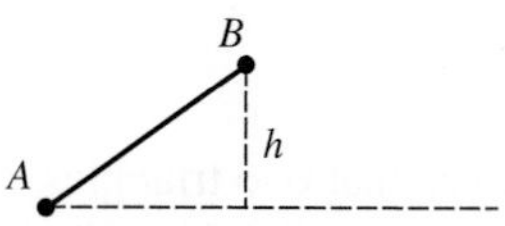

FIGURE 5.16 The diagram for part 2. (Exploration 1)

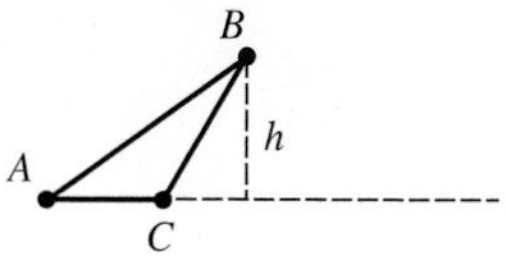

FIGURE 5.17 The diagram for parts 3–5. (Exploration 1)

EXPLORATION 1 Determining the Number of Triangles

We wish to construct ΔABC given angle A, side AB, and side BC.

1. Suppose $\angle A$ is obtuse and that side AB is as shown in Figure 5.15. To complete the triangle, side BC must determine a point on the dotted horizontal line (which extends infinitely to the left). Explain from the picture why a *unique* triangle ΔABC is determined if $BC > AB$, but *no* triangle is determined if $BC \leq AB$.
2. Suppose $\angle A$ is acute and that side AB is as shown in Figure 5.16. To complete the triangle, side BC must determine a point on the dotted horizontal line (which extends infinitely to the right). Explain from the picture why a *unique* triangle ΔABC is determined if $BC = h$, but *no* triangle is determined if $BC < h$.
3. Suppose $\angle A$ is acute and that side AB is as shown in Figure 5.17. If $AB > BC > h$, then we can form a triangle as shown. Find a *second* point C on the dotted horizontal line that gives a side BC of the same length, but determines a different triangle. (This is the "ambiguous case.")
4. Explain why $\sin C$ is the same in both triangles in the ambiguous case. (This is why the Law of Sines is also ambiguous in this case.)
5. Explain from Figure 5.17 why a *unique* triangle is determined if $BC \geq AB$.

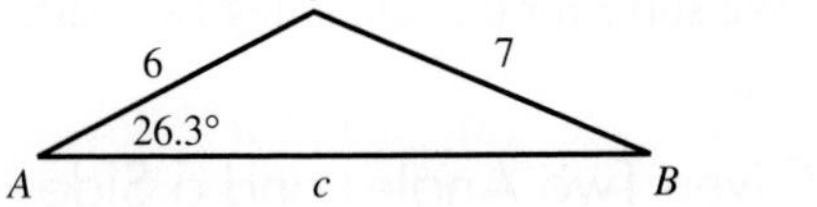

FIGURE 5.18 A triangle determined by SSA. (Example 2)

Now that we know what can happen, let us try the algebra.

EXAMPLE 2 Solving a Triangle Given Two Sides and an Angle

Solve ΔABC given that $a = 7$, $b = 6$, and $\angle A = 26.3°$. (See Figure 5.18.)

SOLUTION By drawing a reasonable sketch (Figure 5.18), we can assure ourselves that this is not the ambiguous case. (In fact, this is the case described in step 5 of Exploration 1.)

Begin by solving for the *acute* angle B, using the Law of Sines:

$$\frac{\sin A}{a} = \frac{\sin B}{b}. \quad \text{Law of Sines}$$

$$\frac{\sin 26.3°}{7} = \frac{\sin B}{6}$$

$$\sin B = \frac{6 \sin 26.3°}{7}$$

$$B = \sin^{-1}\left(\frac{6 \sin 26.3°}{7}\right)$$

$$B = 22.3° \quad \text{Round to match accuracy of given angle.}$$

Then, find the obtuse angle C by subtraction:

$$C = 180° - 26.3° - 22.3°$$
$$= 131.4°$$

Finally, find side c:

$$\frac{\sin A}{a} = \frac{\sin C}{c}$$

$$\frac{\sin 26.3°}{7} = \frac{\sin 131.4°}{c}$$

$$c = \frac{7 \sin 131.4°}{\sin 26.3°}$$

$$c \approx 11.85$$

The six parts of the triangle are:

$\angle A = 26.3°$ $\quad a = 7$

$\angle B = 22.3°$ $\quad b = 6$

$\angle C = 131.4°$ $\quad c \approx 11.85$

Now try Exercise 9.

EXAMPLE 3 Handling the Ambiguous Case

Solve ΔABC given that $a = 6$, $b = 7$, and $\angle A = 30°$.

SOLUTION By drawing a reasonable sketch (Figure 5.19), we see that two triangles are possible with the given information. We keep this in mind as we proceed.

We begin by using the Law of Sines to find angle B.

$$\frac{\sin A}{a} = \frac{\sin B}{b} \quad \text{Law of Sines}$$

$$\frac{\sin 30°}{6} = \frac{\sin B}{7}$$

$$\sin B = \frac{7 \sin 30°}{6}$$

$$B = \sin^{-1}\left(\frac{7 \sin 30°}{6}\right)$$

$$B = 35.7° \quad \text{Round to match accuracy of given angle.}$$

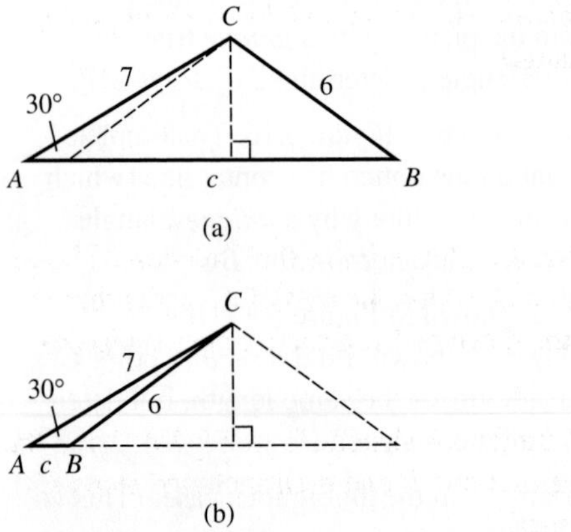

FIGURE 5.19 Two triangles determined by the same SSA values. (Example 3)

Notice that the calculator gave us one value for B, not two. That is because we used the *function* $\sin^{-1}$, which cannot give two output values for the same input value. Indeed, the function $\sin^{-1}$ will *never give an obtuse angle*, which is why we chose to start with the acute angle in Example 2. In this case, the calculator has found the angle B shown in Figure 5.19a.

Find the obtuse angle C by subtraction:

$$C = 180° - 30.0° - 35.7° = 114.3°$$

Finally, find side c:

$$\frac{\sin A}{a} = \frac{\sin C}{c}$$

$$\frac{\sin 30.0°}{6} = \frac{\sin 114.3°}{c}$$

$$c = \frac{6 \sin 114.3°}{\sin 30°}$$

$$c \approx 10.94$$

So, under the assumption that angle B is *acute* (see Figure 5.19a), the six parts of the triangle are:

$$\angle A = 30.0° \qquad a = 6$$
$$\angle B = 35.7° \qquad b = 7$$
$$\angle C = 114.3° \qquad c \approx 10.94$$

If angle B is *obtuse*, then we can see from Figure 5.19b that it has measure $180° - 35.7° = 144.3°$.

By subtraction, the acute angle $C = 180° - 30.0° - 144.3° = 5.7°$. We then recompute c:

$$c = \frac{6 \sin 5.7°}{\sin 30°} \approx 1.19$$ Substitute 5.7° for 114.3° in earlier computation.

So, under the assumption that angle B is *obtuse* (see Figure 5.19b), the six parts of the triangle are:

$$\angle A = 30.0° \qquad a = 6$$
$$\angle B = 144.3° \qquad b = 7$$
$$\angle C = 5.7° \qquad c \approx 1.19$$

Now try Exercise 19.

Applications

Many problems involving angles and distances can be solved by superimposing a triangle onto the situation and solving the triangle.

EXAMPLE 4 Locating a Fire

Forest Ranger Chris Johnson at ranger station A sights a fire in the direction 32° east of north. Ranger Rick Thorpe at ranger station B, 10 mi due east of A, sights the same fire on a line 48° west of north. Find the distance from each ranger station to the fire.

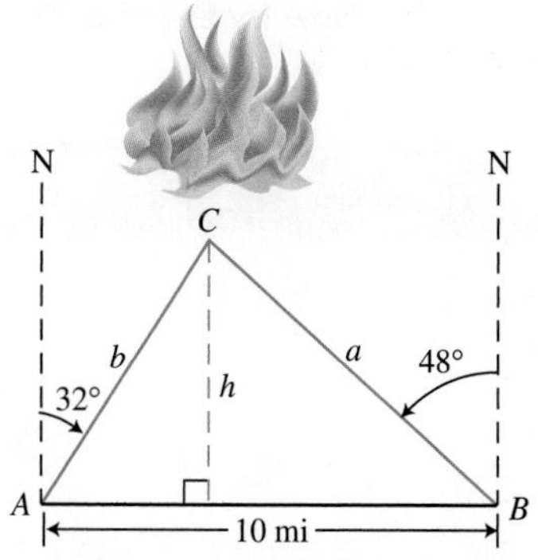

FIGURE 5.20 Determining the location of a fire. (Example 4)

SOLUTION Let C represent the location of the fire. A sketch (Figure 5.20) shows the superimposed triangle, ΔABC, in which angles A and B and their included side (AB) are known. This is a setup for the Law of Sines.

(continued)

Note that $\angle A = 90° - 32° = 58°$ and $\angle B = 90° - 48° = 42°$. By subtraction, we find that $\angle C = 180° - 58° - 42° = 80°$.

$$\frac{\sin A}{a} = \frac{\sin C}{c} \quad \text{and} \quad \frac{\sin B}{b} = \frac{\sin C}{c} \qquad \text{Law of Sines}$$

$$\frac{\sin 58°}{a} = \frac{\sin 80°}{10} \qquad \frac{\sin 42°}{b} = \frac{\sin 80°}{10}$$

$$a = \frac{10 \sin 58°}{\sin 80°} \qquad b = \frac{10 \sin 42°}{\sin 80°}$$

$$a \approx 8.6 \qquad b \approx 6.8 \qquad \text{Round to match accuracy of input.}$$

The fire is about 6.8 mi from ranger station A and about 8.6 mi from ranger station B.

Now try Exercise 45.

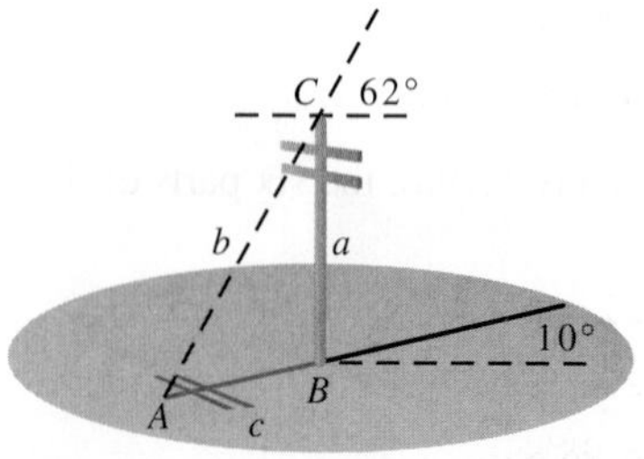

FIGURE 5.21 A telephone pole on a slope. (Example 5)

EXAMPLE 5 Finding the Height of a Pole

A road slopes 10° above the horizontal, and a vertical telephone pole stands beside the road. The angle of elevation of the Sun is 62°, and the pole casts a 14.5-ft shadow downhill along the road. Find the height of the telephone pole.

SOLUTION This is an interesting variation on a typical application of right triangle trigonometry. The slope of the road eliminates the convenient right angle, but we can still solve the problem by solving a triangle.

Figure 5.21 shows the superimposed triangle, ΔABC. A little preliminary geometry is required to find the measure of angles A and C. Due to the slope of the road, angle A is 10° less than the angle of elevation of the Sun and angle B is 10° more than a right angle. That is,

$$\angle A = 62° - 10° = 52°$$
$$\angle B = 90° + 10° = 100°$$
$$\angle C = 180° - 52° - 100° = 28°$$

Therefore,

$$\frac{\sin A}{a} = \frac{\sin C}{c} \qquad \text{Law of Sines}$$

$$\frac{\sin 52°}{a} = \frac{\sin 28°}{14.5}$$

$$a = \frac{14.5 \sin 52°}{\sin 28°}$$

$$a \approx 24.3 \qquad \text{Round to match accuracy of input.}$$

The pole is approximately 24.3 ft high.

Now try Exercise 39.

QUICK REVIEW 5.5 *(For help, go to Sections 4.2 and 4.7.)*

Exercise numbers with a gray background indicate problems that the authors have designed to be solved *without a calculator*.

In Exercises 1–4, solve the equation $a/b = c/d$ for the given variable.

1. a **2.** b **3.** c **4.** d

In Exercises 5 and 6, evaluate the expression.

5. $\dfrac{7 \sin 48°}{\sin 23°}$ **6.** $\dfrac{9 \sin 121°}{\sin 14°}$

In Exercises 7–10, solve for the angle x.

7. $\sin x = 0.3, \quad 0° < x < 90°$

8. $\sin x = 0.3, \quad 90° < x < 180°$

9. $\sin x = -0.7, \quad 180° < x < 270°$

10. $\sin x = -0.7, \quad 270° < x < 360°$

SECTION 5.5 Exercises

In Exercises 1–4, solve the triangle.

1.

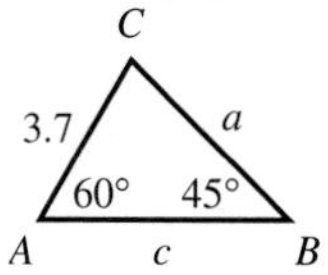

2.

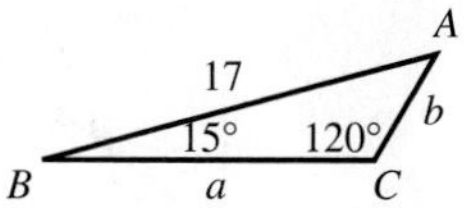

3.

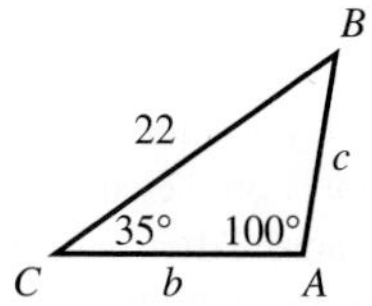

4.

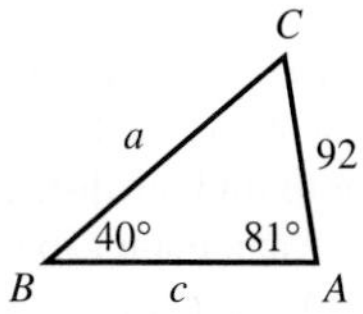

In Exercises 5–8, solve the triangle.

5. $A = 40°$, $B = 30°$, $b = 10$

6. $A = 50°$, $B = 62°$, $a = 4$

7. $A = 33°$, $B = 70°$, $b = 7$

8. $B = 16°$, $C = 103°$, $c = 12$

In Exercises 9–12, solve the triangle.

9. $A = 32°$, $a = 17$, $b = 11$

10. $A = 49°$, $a = 32$, $b = 28$

11. $B = 70°$, $b = 14$, $c = 9$

12. $C = 103°$, $b = 46$, $c = 61$

In Exercises 13–18, state whether the given measurements determine zero, one, or two triangles.

13. $A = 36°$, $a = 2$, $b = 7$

14. $B = 82°$, $b = 17$, $c = 15$

15. $C = 36°$, $a = 17$, $c = 16$

16. $A = 73°$, $a = 24$, $b = 28$

17. $C = 30°$, $a = 18$, $c = 9$

18. $B = 88°$, $b = 14$, $c = 62$

In Exercises 19–22, two triangles can be formed using the given measurements. Solve both triangles.

19. $A = 64°$, $a = 16$, $b = 17$

20. $B = 38°$, $b = 21$, $c = 25$

21. $C = 68°$, $a = 19$, $c = 18$

22. $B = 57°$, $a = 11$, $b = 10$

23. Determine the values of b that will produce the given number of triangles if $a = 10$ and $B = 42°$.

(a) Two triangles **(b)** One triangle **(c)** Zero triangles

24. Determine the values of c that will produce the given number of triangles if $b = 12$ and $C = 53°$.

(a) Two triangles **(b)** One triangle **(c)** Zero triangles

In Exercises 25 and 26, decide whether the triangle can be solved using the Law of Sines. If so, solve it. If not, explain why not.

25.

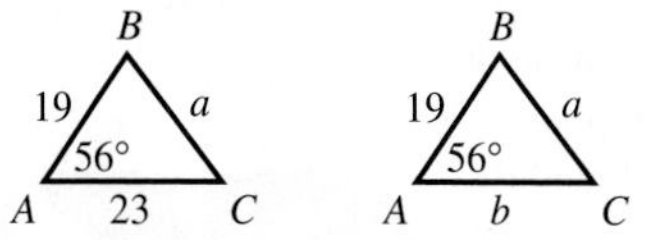

(a) (b)

26.

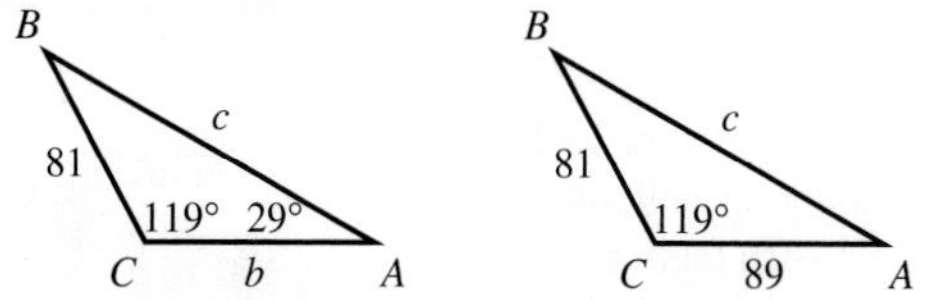

(a) (b)

In Exercises 27–36, respond in one of the following ways:

(a) State, "Cannot be solved with the Law of Sines."

(b) State, "No triangle is formed."

(c) Solve the triangle.

27. $A = 61°$, $a = 8$, $b = 21$

28. $B = 47°$, $a = 8$, $b = 21$

29. $A = 136°$, $a = 15$, $b = 28$

30. $C = 115°$, $b = 12$, $c = 7$

31. $B = 42°$, $c = 18$, $C = 39°$

32. $A = 19°$, $b = 22$, $B = 47°$

33. $C = 75°$, $b = 49$, $c = 48$

34. $A = 54°$, $a = 13$, $b = 15$

35. $B = 31°$, $a = 8$, $c = 11$

36. $C = 65°$, $a = 19$, $b = 22$

37. Surveying a Canyon Two markers A and B on the same side of a canyon rim are 56 ft apart. A third marker C, located across the rim, is positioned so that $\angle BAC = 72°$ and $\angle ABC = 53°$.

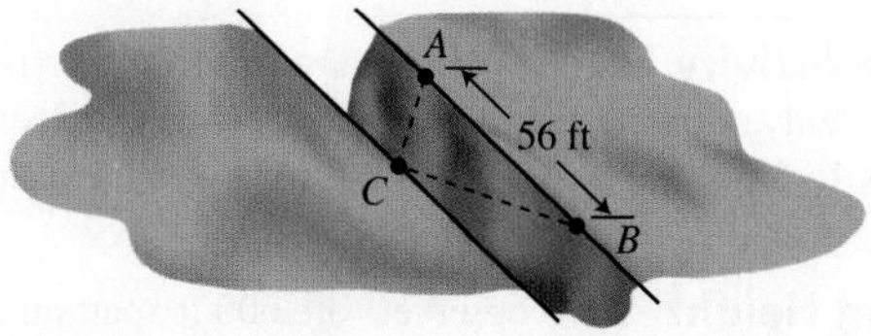

(a) Find the distance between C and A.

(b) Find the distance between the two canyon rims. (Assume they are parallel.)

38. Weather Forecasting Two meteorologists are 25 mi apart located on an east-west road. The meteorologist at point A sights a tornado 38° east of north. The meteorologist at point B sights the same tornado 53° west of north. Find the distance from each meteorologist to the tornado. Also find the distance between the tornado and the road.

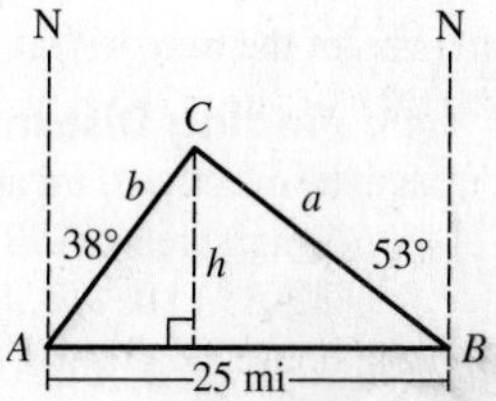

39. **Engineering Design** A vertical flagpole stands beside a road that slopes at an angle of 15° with the horizontal. When the angle of elevation of the Sun is 62°, the flagpole casts a 16-ft shadow downhill along the road. Find the height of the flagpole.

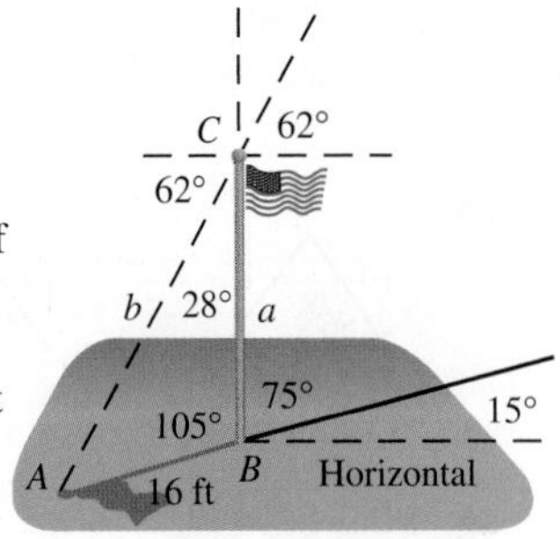

40. **Altitude** Observers 2.32 mi apart see a hot-air balloon directly between them but at the angles of elevation shown in the figure. Find the altitude of the balloon.

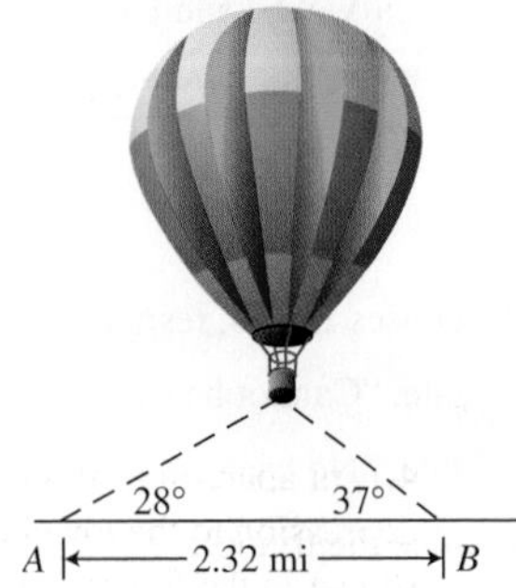

41. **Reducing Air Resistance** A 4-ft airfoil attached to the cab of a truck reduces wind resistance. If the angle between the airfoil and the cab top is 18° and angle B is 10°, find the length of a vertical brace positioned as shown in the figure.

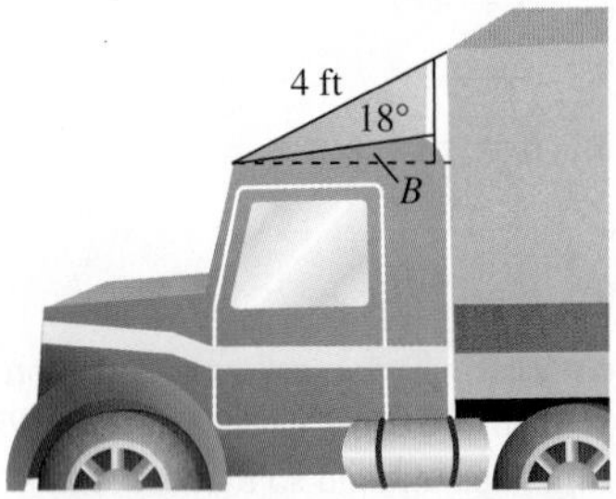

42. **Group Activity Ferris Wheel Design** A Ferris wheel has 16 evenly spaced cars. The distance between adjacent chairs is 15.5 ft. Find the radius of the wheel (to the nearest 0.1 ft).

43. **Finding Height** Two observers are 600 ft apart on opposite sides of a flagpole. The angles of elevation from the observers to the top of the pole are 19° and 21°. Find the height of the flagpole.

44. **Finding Height** Two observers are 400 ft apart on opposite sides of a tree. The angles of elevation from the observers to the top of the tree are 15° and 20°. Find the height of the tree.

45. **Finding Distance** Two lighthouses A and B are known to be exactly 20 mi apart on a north-south line. A ship's captain at S measures $\angle ASB$ to be 33°. A radio operator at B measures $\angle ABS$ to be 52°. Find the distance from the ship to each lighthouse.

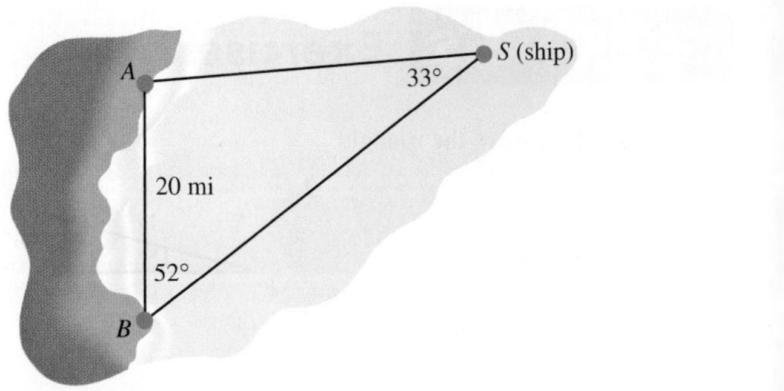

46. **Using Measurement Data** A geometry class is divided into ten teams, each of which is given a yardstick and a protractor to find the distance from a point A on the edge of a pond to a tree at a point C on the opposite shore. After marking points A and B with stakes, each team uses a protractor to measure angles A and B and a yardstick to measure distance AB. Their measurements are given in the table.

A	B	AB
79°	84°	26′ 4″
81°	82°	25′ 5″
79°	83°	26′ 0″
80°	87°	26′ 1″
79°	87°	25′ 11″

A	B	AB
83°	84°	25′ 3″
82°	82°	26′ 5″
78°	85°	25′ 8″
77°	83°	26′ 4″
79°	82°	25′ 7″

Use the data to find the class's best estimate for the distance AC.

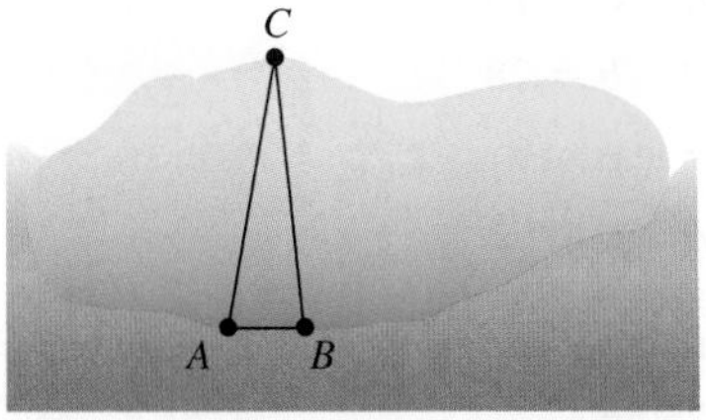

Standardized Test Questions

47. **True or False** The ratio of the sines of any two angles in a triangle equals the ratio of the lengths of their opposite sides. Justify your answer.

48. **True or False** The perimeter of a triangle with two 10-in. sides and two 40° angles is greater than 36 in. Justify your answer.

You may use a graphing calculator when answering these questions.

49. **Multiple Choice** The length x in the triangle shown at the right is

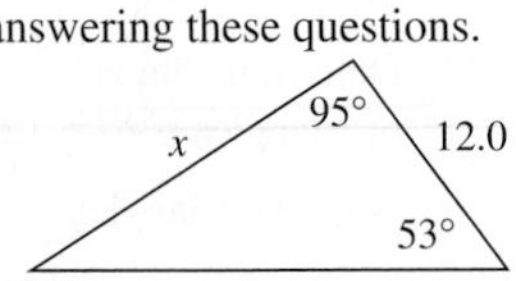

(A) 8.6. **(B)** 15.0. **(C)** 18.1.

(D) 19.2. **(E)** 22.6.

50. Multiple Choice Which of the following three triangle parts do not necessarily determine the other three parts?

(A) AAS (B) ASA (C) SAS

(D) SSA (E) SSS

51. Multiple Choice The shortest side of a triangle with angles 50°, 60°, and 70° has length 9.0. What is the length of the longest side?

(A) 11.0 (B) 11.5 (C) 12.0

(D) 12.5 (E) 13.0

52. Multiple Choice How many noncongruent triangles ABC can be formed if $AB = 5$, $A = 60°$, and $BC = 8$?

(A) None (B) One (C) Two

(D) Three (E) Infinitely many

Explorations

53. Writing to Learn

(a) Show that there are infinitely many triangles with AAA given if the sum of the three positive angles is 180°.

(b) Give three examples of triangles where $A = 30°$, $B = 60°$, and $C = 90°$.

(c) Give three examples where $A = B = C = 60°$.

54. Use the Law of Sines and the cofunction identities to derive the following formulas from right triangle trigonometry:

(a) $\sin A = \dfrac{opp}{hyp}$ (b) $\cos A = \dfrac{adj}{hyp}$ (c) $\tan A = \dfrac{opp}{adj}$

55. Wrapping up Exploration 1 Refer to Figures 5.16 and 5.17 in Exploration 1 of this section.

(a) Express h in terms of angle A and length AB.

(b) In terms of the given angle A and the given length AB, state the conditions on length BC that will result in no triangle being formed.

(c) In terms of the given angle A and the given length AB, state the conditions on length BC that will result in a unique triangle being formed.

(d) In terms of the given angle A and the given length AB, state the conditions on length BC that will result in two possible triangles being formed.

Extending the Ideas

56. Solve this triangle assuming that $\angle B$ is obtuse. (*Hint:* Draw a perpendicular from A to the line through B and C.)

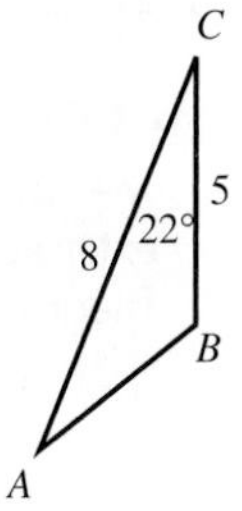

57. Pilot Calculations Towers A and B are known to be 4.1 mi apart on level ground. A pilot measures the angles of depression to the towers to be 36.5° and 25°, respectively, as shown in the figure. Find distances AC and BC and the height of the airplane.

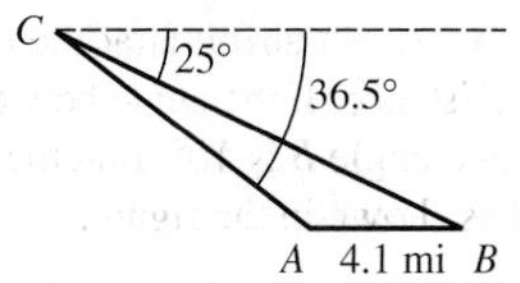

5.6 The Law of Cosines

What you'll learn about

- Deriving the Law of Cosines
- Solving Triangles (SAS, SSS)
- Triangle Area and Heron's Formula
- Applications

... and why

The Law of Cosines is an important extension of the Pythagorean Theorem, with many applications.

Deriving the Law of Cosines

Having seen the Law of Sines, you will probably not be surprised to learn that there is a Law of Cosines. There are many such parallels in mathematics. What you might find surprising is that the Law of Cosines has absolutely no resemblance to the Law of Sines. Instead, it resembles the Pythagorean Theorem. In fact, the Law of Cosines is often called the "generalized Pythagorean Theorem" because it contains that classic theorem as a special case.

Law of Cosines

Let ΔABC be any triangle with sides and angles labeled in the usual way (Figure 5.22).

Then

$$a^2 = b^2 + c^2 - 2bc \cos A$$
$$b^2 = a^2 + c^2 - 2ac \cos B$$
$$c^2 = a^2 + b^2 - 2ab \cos C$$

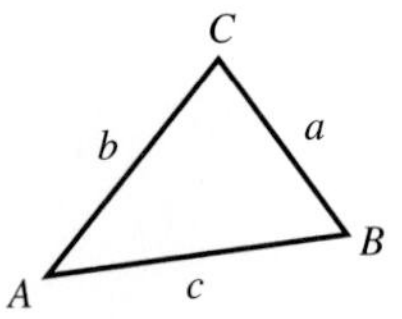

FIGURE 5.22 A triangle with the usual labeling (angles A, B, C; opposite sides a, b, c).

We derive only the first of the three equations, since the other two are derived in exactly the same way. Set the triangle in a coordinate plane so that the angle that appears in the formula (in this case, A) is at the origin in standard position, with side c along the positive x-axis. Depending on whether angle A is right (Figure 5.23a), acute (Figure 5.23b), or obtuse (Figure 5.23c), the point C will be on the y-axis, in QI, or in QII.

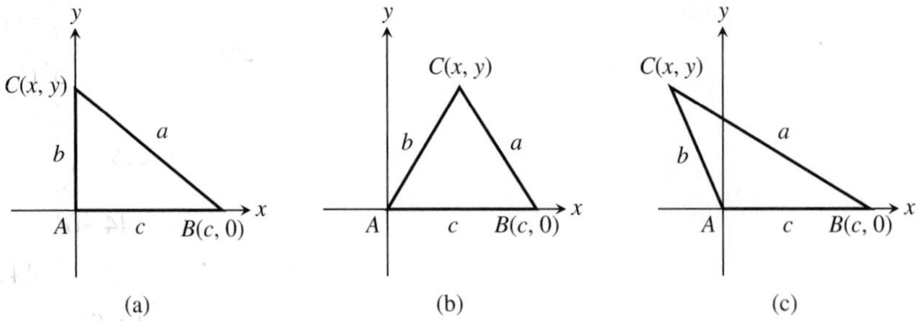

FIGURE 5.23 Three cases for proving the Law of Cosines.

In each of these three cases, C is a point on the terminal side of angle A in standard position, at distance b from the origin. Denote the coordinates of C by (x, y). By our definitions for trigonometric functions of any angle (Section 4.3), we can conclude that

$$\frac{x}{b} = \cos A \quad \text{and} \quad \frac{y}{b} = \sin A,$$

and therefore

$$x = b \cos A \quad \text{and} \quad y = b \sin A.$$

Now set a equal to the distance from C to B using the distance formula:

$$a = \sqrt{(x - c)^2 + (y - 0)^2} \qquad \text{Distance formula}$$
$$a^2 = (x - c)^2 + y^2 \qquad \text{Square both sides.}$$
$$= (b \cos A - c)^2 + (b \sin A)^2 \qquad \text{Substitution}$$
$$= b^2 \cos^2 A - 2bc \cos A + c^2 + b^2 \sin^2 A$$
$$= b^2(\cos^2 A + \sin^2 A) + c^2 - 2bc \cos A$$
$$= b^2 + c^2 - 2bc \cos A \qquad \text{Pythagorean identity}$$

Solving Triangles (SAS, SSS)

Although the Law of Sines is the tool we use to solve triangles in the AAS and ASA cases, the Law of Cosines is the required tool for SAS and SSS. (Both methods can be used in the SSA case, but remember that there might be 0, 1, or 2 triangles.)

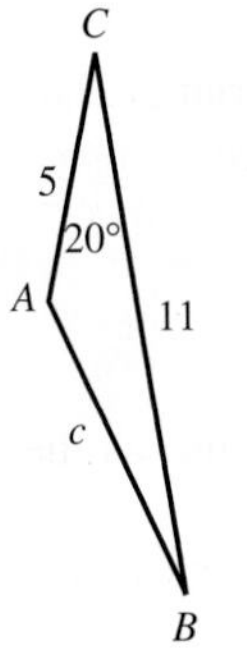

FIGURE 5.24 A triangle with two sides and an included angle known. (Example 1)

EXAMPLE 1 Solving a Triangle (SAS)

Solve ΔABC given that $a = 11$, $b = 5$, and $C = 20°$. (See Figure 5.24.)

SOLUTION

$$c^2 = a^2 + b^2 - 2ab \cos C$$
$$= 11^2 + 5^2 - 2(11)(5) \cos 20°$$
$$\approx 42.63$$
$$c = \sqrt{42.63} \approx 6.53$$

We could now use either the Law of Cosines or the Law of Sines to find one of the two unknown angles. As a general rule, it is better to use the Law of Cosines to find angles, since the arccosine function will distinguish obtuse angles from acute angles.

$$a^2 = b^2 + c^2 - 2bc \cos A$$
$$11^2 = 5^2 + (6.53)^2 - 2(5)(6.53)\cos A$$
$$\cos A = \frac{5^2 + (6.53)^2 - 11^2}{2(5)(6.53)}$$
$$A = \cos^{-1}\left(\frac{5^2 + (6.53)^2 - 11^2}{2(5)(6.53)}\right)$$
$$\approx 144.80°$$
$$B = 180° - 144.80° - 20°$$
$$\approx 15.20°$$

So the six parts of the triangle are:

$$A \approx 144.80° \qquad a = 11$$
$$B \approx 15.20° \qquad b = 5$$
$$C = 20° \qquad c \approx 6.53$$

Now try Exercise 1.

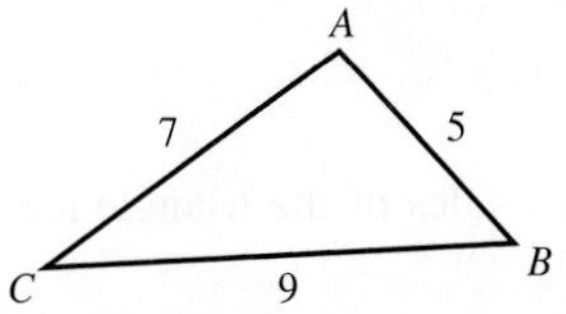

FIGURE 5.25 A triangle with three sides known. (Example 2)

EXAMPLE 2 Solving a Triangle (SSS)

Solve ΔABC if $a = 9$, $b = 7$, and $c = 5$. (See Figure 5.25.)

(continued)

SOLUTION We use the Law of Cosines to find two of the angles. The third angle can be found by subtraction from 180°.

$$\begin{aligned} a^2 &= b^2 + c^2 - 2bc\cos A \\ 9^2 &= 7^2 + 5^2 - 2(7)(5)\cos A \\ 70\cos a &= -7 \\ A &= \cos^{-1}(-0.1) \\ &\approx 95.74° \end{aligned}$$

$$\begin{aligned} b^2 &= a^2 + c^2 - 2ac\cos B \\ 7^2 &= 9^2 + 5^2 - 2(9)(5)\cos B \\ 90\cos b &= 57 \\ B &= \cos^{-1}(57/90) \\ &\approx 50.70° \end{aligned}$$

Then $C = 180° - 95.74° - 50.70° \approx 33.56°$. **Now try Exercise 3.**

Triangle Area and Heron's Formula

The same parts that determine a triangle also determine its area. If the parts happen to be two sides and an included angle (SAS), we get a simple area formula in terms of those three parts that does not require finding an altitude.

Observe in Figure 5.23 (used in explaining the Law of Cosines) that each triangle has base c and altitude $y = b \sin A$. Applying the standard area formula, we have

$$\Delta \text{ Area} = \frac{1}{2}(base)(height) = \frac{1}{2}(c)(b\sin A) = \frac{1}{2}bc\sin A.$$

This is actually three formulas in one, as it does not matter which side we use as the base.

Area of a Triangle

$$\Delta \text{ Area} = \frac{1}{2}bc\sin A = \frac{1}{2}ac\sin B = \frac{1}{2}ab\sin C$$

EXAMPLE 3 Finding the Area of a Regular Polygon

Find the area of a regular octagon (8 equal sides, 8 equal angles) inscribed inside a circle of radius 9 in.

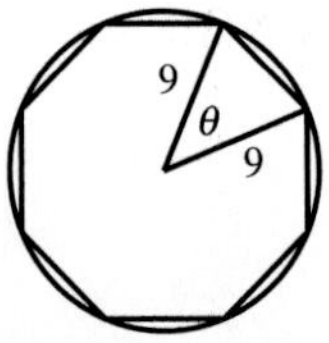

FIGURE 5.26 A regular octagon inscribed inside a circle of radius 9 in. (Example 3)

SOLUTION Figure 5.26 shows that we can split the octagon into 8 congruent triangles. Each triangle has two 9-in. sides with an included angle of $\theta = 360°/8 = 45°$. The area of each triangle is

$$\Delta \text{ Area} = (1/2)(9)(9)\sin 45° = (81/2)\sin 45° = 81\sqrt{2}/4.$$

Therefore, the area of the octagon is

$$\text{Area} = 8\ \Delta \text{ Area} = 162\sqrt{2} \approx 229.10 \text{ in.}^2.$$

Now try Exercise 31.

There is also an area formula that can be used when the three sides of the triangle are known.

Although Heron proved this theorem using only classical geometric methods, we prove it, as most people do today, by using the tools of trigonometry.

THEOREM Heron's Formula

Let a, b, and c be the sides of ΔABC, and let s denote the **semiperimeter**.

$$(a + b + c)/2.$$

Then the area of ΔABC is given by Area $= \sqrt{s(s-a)(s-b)(s-c)}$.

The following proof of Heron's Formula uses both algebra and trigonometry:

$$
\begin{aligned}
\text{Area} &= \frac{1}{2}ab\sin C \\
4(\text{Area}) &= 2ab\sin C \\
16(\text{Area})^2 &= 4a^2b^2\sin^2 C \\
&= 4a^2b^2(1-\cos^2 C) && \text{Pythagorean identity} \\
&= 4a^2b^2 - 4a^2b^2\cos^2 C \\
&= 4a^2b^2 - (2ab\cos C)^2 \\
&= 4a^2b^2 - (a^2+b^2-c^2)^2 && \text{Law of Cosines} \\
&= (2ab-(a^2+b^2-c^2))(2ab+(a^2+b^2-c^2)) && \text{Difference of squares} \\
&= (c^2-(a^2-2ab+b^2))((a^2+2ab+b^2)-c^2) \\
&= (c^2-(a-b)^2)((a+b)^2-c^2) \\
&= (c-(a-b))(c+(a-b))((a+b)-c)((a+b)+c) && \text{Difference of squares} \\
&= (c-a+b)(c+a-b)(a+b-c)(a+b+c) \\
&= (2s-2a)(2s-2b)(2s-2c)(2s) && 2s = a+b+c \\
16(\text{Area})^2 &= 16s(s-a)(s-b)(s-c) \\
(\text{Area})^2 &= s(s-a)(s-b)(s-c) \\
\text{Area} &= \sqrt{s(s-a)(s-b)(s-c)}
\end{aligned}
$$

EXAMPLE 4 Using Heron's Formula

Find the area of a triangle with sides 13, 15, 18.

SOLUTION First we compute the semiperimeter: $s = (13 + 15 + 18)/2 = 23$.

Then we use Heron's Formula

$$
\begin{aligned}
\text{Area} &= \sqrt{23(23-13)(23-15)(23-18)} \\
&= \sqrt{23\cdot 10\cdot 8\cdot 5} = \sqrt{9200} = 20\sqrt{23}.
\end{aligned}
$$

The approximate area is 96 square units.

Now try Exercise 21.

Applications

We end this section with a few applications.

EXAMPLE 5 Measuring a Baseball Diamond

The bases on a baseball diamond are 90 ft apart, and the front edge of the pitcher's rubber is 60.5 ft from the back corner of home plate. Find the distance from the center of the front edge of the pitcher's rubber to the far corner of first base.

(continued)

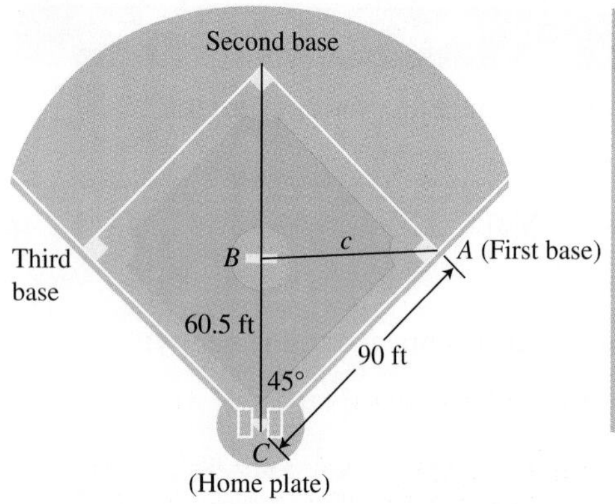

FIGURE 5.27 The diamond-shaped part of a baseball diamond. (Example 5)

SOLUTION Figure 5.27 shows first base as A, the pitcher's rubber as B, and home plate as C. The distance we seek is side c in ΔABC.

By the Law of Cosines,

$$c^2 = 60.5^2 + 90^2 - 2(60.5)(90)\cos 45°$$
$$c = \sqrt{60.5^2 + 90^2 - 2(60.5)(90)\cos 45°}$$
$$\approx 63.72$$

The distance from first base to the pitcher's rubber is about 63.72 ft.

Now try Exercise 37.

Platonic Solids

The regular tetrahedron in Example 6 is one of only 5 *regular* solids (solids with faces that are congruent polygons having equal angles and equal sides). The others are the cube (6 square faces), the octahedron (8 triangular faces), the dodecahedron (12 pentagonal faces), and the icosahedron (20 triangular faces). Referred to as the *Platonic* solids, Plato did not discover them, but they are featured in his cosmology as being the stuff of which everything in the universe is made. The Platonic universe itself is a dodecahedron, a favorite symbol of the Pythagoreans.

EXAMPLE 6 Measuring a Dihedral Angle (Solid Geometry)

A regular tetrahedron is a solid with four faces, each of which is an equilateral triangle. Find the measure of the *dihedral angle* formed along the common edge of two intersecting faces of a regular tetrahedron with edges of length 2.

SOLUTION Figure 5.28 shows the tetrahedron. Point B is the midpoint of edge DE, and A and C are the endpoints of the opposite edge. The measure of $\angle ABC$ is the same as the measure of the dihedral angle formed along edge DE, so we will find the measure of $\angle ABC$.

Because both ΔADB and ΔCDB are $30°-60°-90°$ triangles, AB and BC both have length $\sqrt{3}$. If we apply the Law of Cosines to ΔABC, we obtain

$$2^2 = (\sqrt{3})^2 + (\sqrt{3})^2 - 2\sqrt{3}\sqrt{3}\cos(\angle ABC)$$
$$\cos(\angle ABC) = \frac{1}{3}$$
$$\angle ABC = \cos^{-1}\left(\frac{1}{3}\right) \approx 70.53°$$

The dihedral angle has the same measure as $\angle ABC$, approximately 70.53°. (We chose sides of length 2 for computational convenience, but in fact this is the measure of a dihedral angle in a regular tetrahedron of any size.)

Now try Exercise 43.

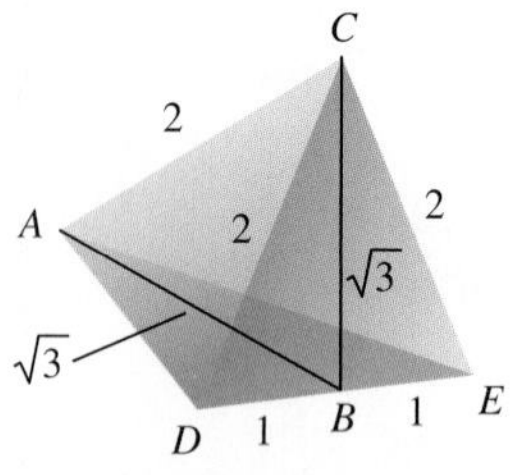

FIGURE 5.28 The measure of $\angle ABC$ is the same as the measure of any dihedral angle formed by two of the tetrahedron's faces. (Example 6)

EXPLORATION 1 Estimating Acreage of a Plot of Land

Jim and Barbara are house hunting and need to estimate the size of an irregular adjacent lot that is described by the owner as "a little more than an acre." With Barbara stationed at a corner of the plot, Jim starts at another corner and walks a straight line toward her, counting his paces. They then shift corners and Jim paces again, until they have recorded the dimensions of the lot (in paces) as in Figure 5.29. They later measure Jim's pace as 2.2 ft. What is the approximate acreage of the lot?

1. Use Heron's Formula to find the area in square paces.
2. Convert the area to square ft, using the measure of Jim's pace.
3. There are 5280 ft in a mile. Convert the area to square miles.
4. There are 640 acres in a square mile. Convert the area to acres.
5. Is there good reason to doubt the owner's estimate of the acreage of the lot?
6. Would Jim and Barbara be able to modify their system to estimate the area of an irregular lot with five straight sides?

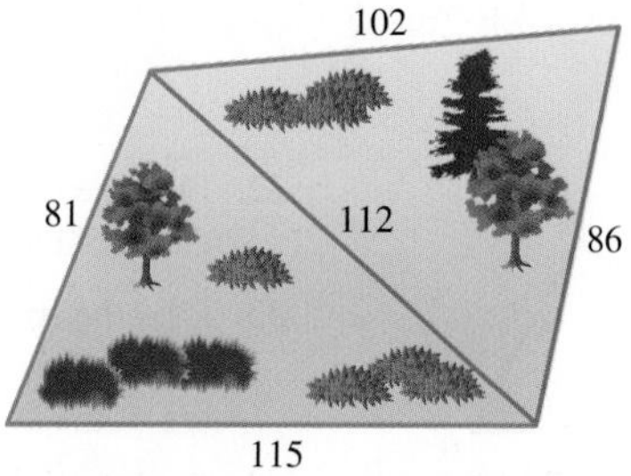

FIGURE 5.29 Dimensions (in paces) of an irregular plot of land. (Exploration 1)

CHAPTER OPENER **Problem** (from page 403)

(from page 403)

Problem: Because deer require food, water, cover for protection from weather and predators, and living space for healthy survival, there are natural limits to the number of deer that a given plot of land can support. Deer populations in national parks average 14 animals per square kilometer. If a triangular region with sides of 3 km, 4 km, and 6 km has a population of 50 deer, how close is the population on this land to the average national park population?

Solution: We can find the area of the land region

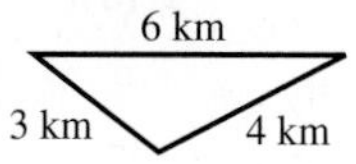

by using Heron's Formula with

$$s = (3 + 4 + 6)/2 = 13/2$$

and

$$\begin{aligned}\text{Area} &= \sqrt{s(s-a)(s-b)(s-c)}\\ &= \sqrt{\frac{13}{2}\left(\frac{13}{2}-3\right)\left(\frac{13}{2}-4\right)\left(\frac{13}{2}-6\right)}\\ &= \sqrt{\frac{13}{2}\left(\frac{7}{2}\right)\left(\frac{5}{2}\right)\left(\frac{1}{2}\right)} \approx 5.33\end{aligned}$$

so the area of the land region is about 5.33 km^2.

If this land were to support 14 deer/km^2, it would have $(5.33\ \text{km}^2)(14\ \text{deer/km}^2) = 74.62$ or about 75 deer. Thus, the land supports about 25 deer less than the average.

QUICK REVIEW 5.6 *(For help, go to Sections 2.4 and 4.7.)*

Exercise numbers with a gray background indicate problems that the authors have designed to be solved *without a calculator*.

In Exercises 1–4, find an angle between 0° and 180° that is a solution to the equation.

1. $\cos A = 3/5$

2. $\cos C = -0.23$

3. $\cos A = -0.68$

4. $3 \cos C = 1.92$

In Exercises 5 and 6, solve the equation (in terms of x and y) for **(a)** $\cos A$ and **(b)** A, $0 \le A \le 180°$.

5. $9^2 = x^2 + y^2 - 2xy \cos A$

6. $y^2 = x^2 + 25 - 10 \cos A$

In Exercises 7–10, find a quadratic polynomial with real coefficients that satisfies the given condition.

7. Has two positive zeros

8. Has one positive and one negative zero

9. Has no real zeros

10. Has exactly one positive zero

SECTION 5.6 Exercises

In Exercises 1–4, solve the triangle.

1.

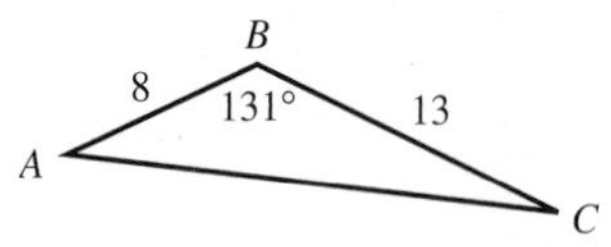

2.

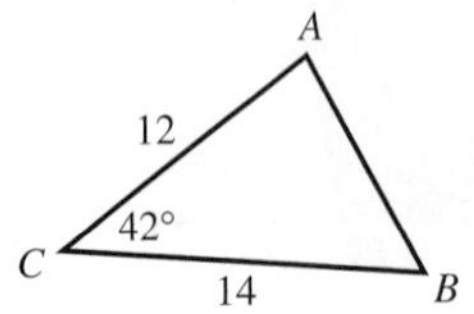

3.

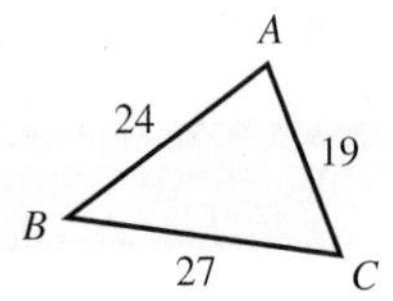

4.

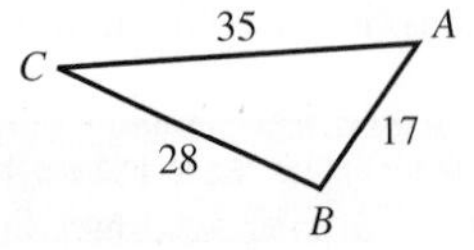

In Exercises 5–16, solve the triangle.

5. $A = 55°$, $b = 12$, $c = 7$

6. $B = 35°$, $a = 43$, $c = 19$

7. $a = 12$, $b = 21$, $C = 95°$

8. $b = 22$, $c = 31$, $A = 82°$

9. $a = 1$, $b = 5$, $c = 4$

10. $a = 1$, $b = 5$, $c = 8$

11. $a = 3.2$, $b = 7.6$, $c = 6.4$

12. $a = 9.8$, $b = 12$, $c = 23$

13. $A = 42°$, $a = 7$, $b = 10$

14. $A = 57°$, $a = 11$, $b = 10$

15. $A = 63°$, $a = 8.6$, $b = 11.1$

16. $A = 71°$, $a = 9.3$, $b = 8.5$

In Exercises 17–20, find the area of the triangle.

17. $A = 47°$, $b = 32$ ft, $c = 19$ ft

18. $A = 52°$, $b = 14$ m, $c = 21$ m

19. $B = 101°$, $a = 10$ cm, $c = 22$ cm

20. $C = 112°$, $a = 1.8$ in., $b = 5.1$ in.

In Exercises 21–28, decide whether a triangle can be formed with the given side lengths. If so, use Heron's Formula to find the area of the triangle.

21. $a = 4$, $b = 5$, $c = 8$

22. $a = 5$, $b = 9$, $c = 7$

23. $a = 3$, $b = 5$, $c = 8$

24. $a = 23$, $b = 19$, $c = 12$

25. $a = 19.3$, $b = 22.5$, $c = 31$

26. $a = 8.2$, $b = 12.5$, $c = 28$

27. $a = 33.4$, $b = 28.5$, $c = 22.3$

28. $a = 18.2$, $b = 17.1$, $c = 12.3$

29. Find the radian measure of the largest angle in the triangle with sides of 4, 5, and 6.

30. A parallelogram has sides of 18 and 26 ft, and an angle of 39°. Find the shorter diagonal.

31. Find the area of a regular hexagon inscribed in a circle of radius 12 in.

32. Find the area of a regular nonagon (9 sides) inscribed in a circle of radius 10 in.

33. Find the area of a regular hexagon circumscribed about a circle of radius 12 in. (*Hint:* Start by finding the distance from a vertex of the hexagon to the center of the circle.)

34. Find the area of a regular nonagon (9 sides) circumscribed about a circle of radius 10 in.

35. Measuring Distance Indirectly Juan wants to find the distance between two points A and B on opposite sides of a building. He locates a point C that is 110 ft from A and 160 ft from B, as illustrated in the figure. If the angle at C is 54°, find distance AB.

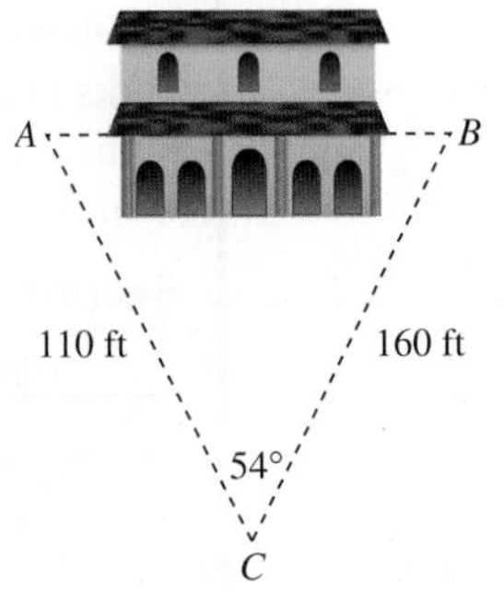

36. Designing a Baseball Field

(a) Find the distance from the center of the front edge of the pitcher's rubber to the far corner of second base. How does this distance compare with the distance from the pitcher's rubber to first base? (See Example 5.)

(b) Find $\angle B$ in ΔABC.

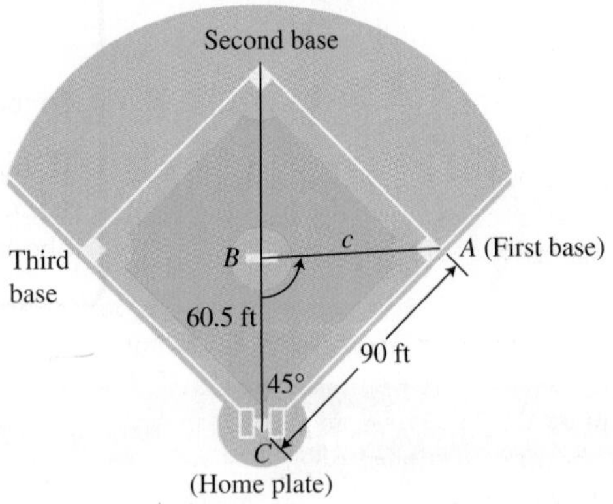

37. Designing a Softball Field In softball, adjacent bases are 60 ft apart. The distance from the center of the front edge of the pitcher's rubber to the far corner of home plate is 40 ft.

(a) Find the distance from the center of the pitcher's rubber to the far corner of first base.

(b) Find the distance from the center of the pitcher's rubber to the far corner of second base.

(c) Find $\angle B$ in ΔABC.

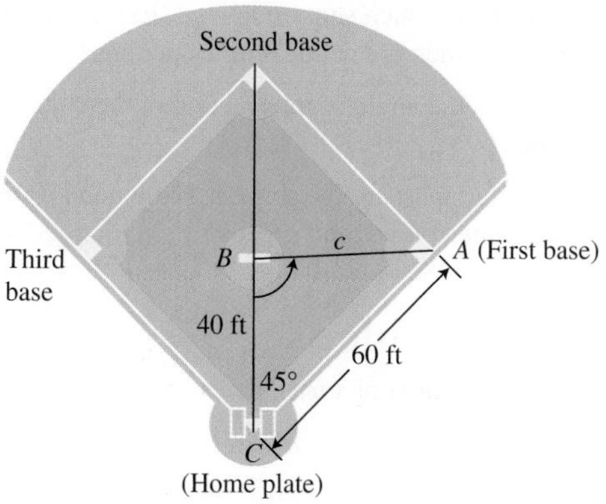

38. Surveyor's Calculations Tony must find the distance from A to B on opposite sides of a lake. He locates a point C that is 860 ft from A and 175 ft from B. He measures the angle at C to be 78°. Find distance AB.

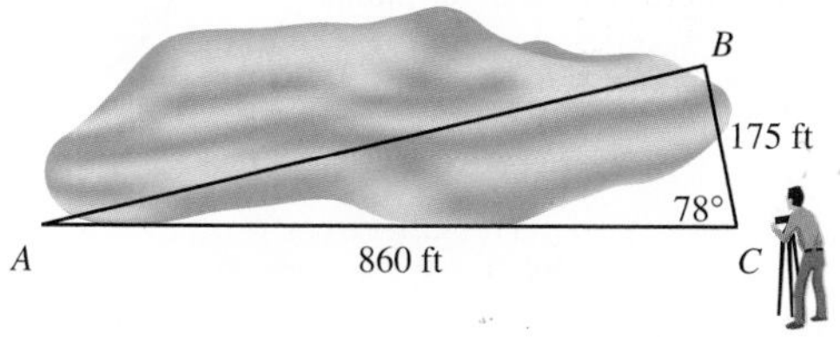

39. Construction Engineering A manufacturer is designing the roof truss that is modeled in the figure shown.

(a) Find the measure of $\angle CAE$.

(b) If $AF = 12$ ft, find the length DF.

(c) Find the length EF.

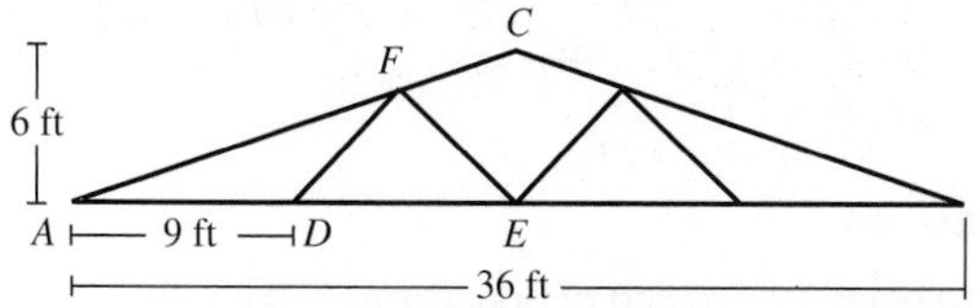

40. Navigation Two airplanes flying together in formation take off in different directions. One flies due east at 350 mph, and the other flies east-northeast at 380 mph. How far apart are the two airplanes 2 hr after they separate, assuming that they fly at the same altitude?

41. Football Kick The player waiting to receive a kickoff stands at the 5-yd line (point A) as the ball is being kicked 65 yd up the field from the opponent's 30-yd line. The kicked ball travels 73 yd at an angle of 8° to the right of the receiver, as shown in the figure (point B). Find the distance the receiver runs to catch the ball.

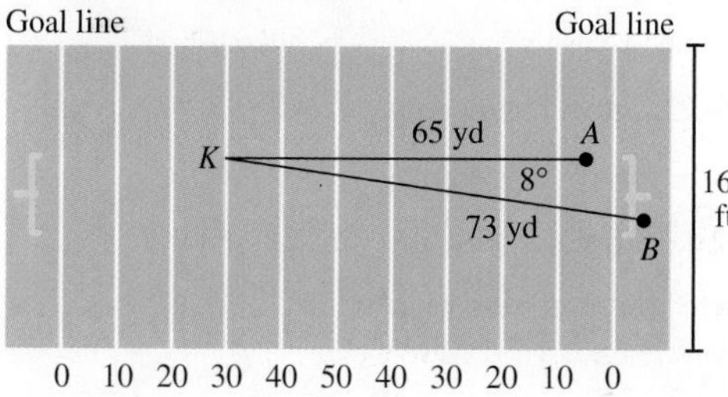

42. Group Activity Architectural Design Building Inspector Julie Wang checks a building in the shape of a regular octagon, each side 20 ft long. She checks that the contractor has located the corners of the foundation correctly by measuring several of the diagonals. Calculate what the lengths of diagonals HB, HC, and HD should be.

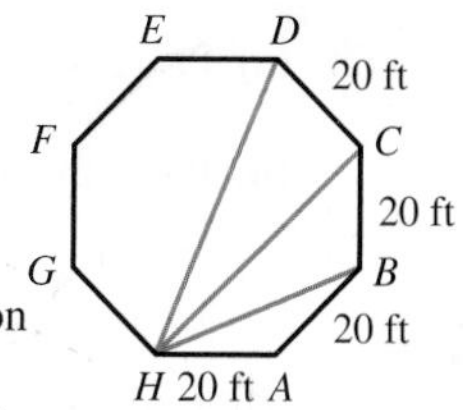

43. Connecting Trigonometry and Geometry $\angle CAB$ is inscribed in a rectangular box whose sides are 1, 2, and 3 ft long as shown. Find the measure of $\angle CAB$.

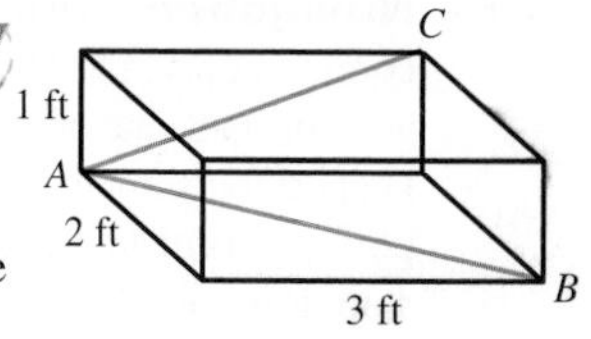

44. Group Activity Connecting Trigonometry and Geometry A cube has edges of length 2 ft. Point A is the midpoint of an edge. Find the measure of $\angle ABC$.

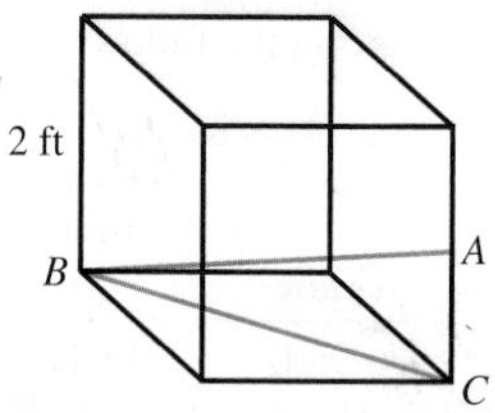

Standardized Test Questions

45. True or False If ΔABC is any triangle with sides and angles labeled in the usual way, then $b^2 + c^2 > 2bc \cos A$. Justify your answer.

46. True or False If a, b, and θ are two sides and an included angle of a parallelogram, the area of the parallelogram is $ab \sin \theta$. Justify your answer.

You may use a graphing calculator when answering these questions.

47. Multiple Choice What is the area of a regular dodecagon (12-sided figure) inscribed in a circle of radius 12?

(A) 427 **(B)** 432 **(C)** 437
(D) 442 **(E)** 447

48. Multiple Choice The area of a triangle with sides 7, 8, and 9 is

(A) $6\sqrt{15}$. **(B)** $12\sqrt{5}$. **(C)** $16\sqrt{3}$.
(D) $17\sqrt{3}$. **(E)** $18\sqrt{3}$.

49. Multiple Choice Two boats start at the same point and speed away along courses that form a 110° angle. If one boat travels at 24 mph and the other boat travels at 32 mph, how far apart are the boats after 30 min?

(A) 21 mi **(B)** 22 mi **(C)** 23 mi
(D) 24 mi **(E)** 25 mi

50. Multiple Choice What is the measure of the smallest angle in a triangle with sides 12, 17, and 25?

(A) 21° **(B)** 22° **(C)** 23°
(D) 24° **(E)** 25°

Explorations

51. Find the area of a regular polygon with n sides inscribed inside a circle of radius r. (Express your answer in terms of n and r.)

52. **(a)** Prove the identity: $\dfrac{\cos A}{a} = \dfrac{b^2 + c^2 - a^2}{2abc}$.

(b) Prove the (tougher) identity:

$$\frac{\cos A}{a} + \frac{\cos B}{b} + \frac{\cos C}{c} = \frac{a^2 + b^2 + c^2}{2abc}.$$

(*Hint:* Use the identity in part (a), along with its other variations.)

53. Navigation Two ships leave a common port at 8:00 A.M. and travel at a constant rate of speed. Each ship keeps a log showing its distance from port and its distance from the other ship. Portions of the logs from later that morning for both ships are shown in the following tables.

Time	Naut mi from Port	Naut mi from Ship B	Time	Naut mi from Port	Naut mi from Ship A
9:00	15.1	8.7	9:00	12.4	8.7
10:00	30.2	17.3	11:00	37.2	26.0

(a) How fast is each ship traveling? (Express your answer in knots, which are nautical miles per hour.)

(b) What is the angle of intersection of the courses of the two ships?

(c) How far apart are the ships at 12:00 noon if they maintain the same courses and speeds?

Extending the Ideas

54. Prove that the area of a triangle can be found with the formula

$$\Delta \text{ Area} = \frac{a^2 \sin B \sin C}{2 \sin A}.$$

55. A **segment** of a circle is the region enclosed between a chord of a circle and the arc intercepted by the chord. Find the area of a segment intercepted by a 7-in. chord in a circle of radius 5 in.

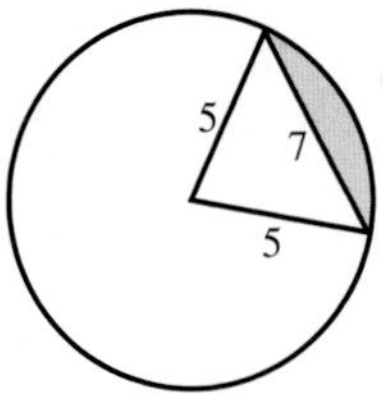

CHAPTER 5 Key Ideas

Properties, Theorems, and Formulas

Reciprocal Identities 404
Quotient Identities 404
Pythagorean Identities 405
Cofunction Identities 406
Odd-Even Identities 407
Sum and Difference Identities 422–424
Double-Angle Identities 428
Power-Reducing Identities 428
Half-Angle Identities 430
Law of Sines 434
Law of Cosines 442
Area of a Triangle 444
Heron's Formula 445

Procedures

Strategies for Proving an Identity 413–415

CHAPTER 5 Review Exercises

Exercise numbers with a gray background indicate problems that the authors have designed to be solved *without a calculator.*

The collection of exercises marked in red could be used as a chapter test.

In Exercises 1 and 2, write the expression as the sine, cosine, or tangent of an angle.

1. $2 \sin 100° \cos 100°$

2. $\dfrac{2 \tan 40°}{1 - \tan^2 40°}$

In Exercises 3 and 4, simplify the expression to a single term. Support your answer graphically.

3. $(1 - 2\sin^2\theta)^2 + 4\sin^2\theta\cos^2\theta$

4. $1 - 4\sin^2 x \cos^2 x$

In Exercises 5–22, prove the identity.

5. $\cos 3x = 4\cos^3 x - 3\cos x$

6. $\cos^2 2x - \cos^2 x = \sin^2 x - \sin^2 2x$

7. $\tan^2 x - \sin^2 x = \sin^2 x \tan^2 x$

8. $2 \sin \theta \cos^3 \theta + 2 \sin^3 \theta \cos \theta = \sin 2\theta$

9. $\csc x - \cos x \cot x = \sin x$

10. $\dfrac{\tan \theta + \sin \theta}{2 \tan \theta} = \cos^2 \left(\dfrac{\theta}{2}\right)$

11. $\dfrac{1 + \tan \theta}{1 - \tan \theta} + \dfrac{1 + \cot \theta}{1 - \cot \theta} = 0$

12. $\sin 3\theta = 3 \cos^2 \theta \sin \theta - \sin^3 \theta$

13. $\cos^2 \left(\dfrac{t}{2}\right) = \dfrac{1 + \sec t}{2 \sec t}$

14. $\dfrac{\tan^3 \gamma - \cot^3 \gamma}{\tan^2 \gamma + \csc^2 \gamma} = \tan \gamma - \cot \gamma$

15. $\dfrac{\cos \phi}{1 - \tan \phi} + \dfrac{\sin \phi}{1 - \cot \phi} = \cos \phi + \sin \phi$

16. $\dfrac{\cos (-z)}{\sec (-z) + \tan (-z)} = 1 + \sin z$

17. $\sqrt{\dfrac{1 - \cos y}{1 + \cos y}} = \dfrac{1 - \cos y}{|\sin y|}$ **18.** $\sqrt{\dfrac{1 - \sin \gamma}{1 + \sin \gamma}} = \dfrac{|\cos \gamma|}{1 + \sin \gamma}$

19. $\tan \left(u + \dfrac{3\pi}{4}\right) = \dfrac{\tan u - 1}{1 + \tan u}$

20. $\dfrac{1}{4} \sin 4\gamma = \sin \gamma \cos^3 \gamma - \cos \gamma \sin^3 \gamma$

21. $\tan \dfrac{1}{2} \beta = \csc \beta - \cot \beta$

22. $\arctan t = \dfrac{1}{2} \arctan \dfrac{2t}{1 - t^2}, -1 < t < 1$

In Exercises 23 and 24, use a grapher to conjecture whether the equation is likely to be an identity. Confirm your conjecture.

23. $\sec x - \sin x \tan x = \cos x$

24. $(\sin^2 \alpha - \cos^2 \alpha)(\tan^2 \alpha + 1) = \tan^2 \alpha - 1$

In Exercises 25–28, write the expression in terms of $\sin x$ and $\cos x$ only.

25. $\sin 3x + \cos 3x$ **26.** $\sin 2x + \cos 3x$

27. $\cos^2 2x - \sin 2x$ **28.** $\sin 3x - 3 \sin 2x$

In Exercises 29–34, find the general solution without using a calculator. Give exact answers.

29. $\sin 2x = 0.5$ **30.** $\cos x = \dfrac{\sqrt{3}}{2}$

31. $\tan x = -1$ **32.** $2 \sin^{-1} x = \sqrt{2}$

33. $\tan^{-1} x = 1$ **34.** $2 \cos 2x = 1$

In Exercises 35–38, solve the equation graphically. Find all solutions in the interval $[0, 2\pi)$.

35. $\sin^2 x - 3 \cos x = -0.5$

36. $\cos^3 x - 2 \sin x - 0.7 = 0$

37. $\sin^4 x + x^2 = 2$

38. $\sin 2x = x^3 - 5x^2 + 5x + 1$

In Exercises 39–44, find all solutions in the interval $[0, 2\pi)$ without using a calculator. Give exact answers.

39. $2 \cos x = 1$ **40.** $\sin 3x = \sin x$

41. $\sin^2 x - 2 \sin x - 3 = 0$ **42.** $\cos 2t = \cos t$

43. $\sin (\cos x) = 1$ **44.** $\cos 2x + 5 \cos x = 2$

In Exercises 45–48, solve the inequality. Use any method, but give exact answers.

45. $2 \cos 2x > 1$ for $0 \leq x < 2\pi$

46. $\sin 2x > 2 \cos x$ for $0 < x \leq 2\pi$

47. $2 \cos x < 1$ for $0 \leq x < 2\pi$

48. $\tan x < \sin x$ for $-\dfrac{\pi}{2} < x < \dfrac{\pi}{2}$

In Exercises 49 and 50, find an equivalent equation of the form $y = a \sin (bx + c)$. Support your work graphically.

49. $y = 3 \sin 3x + 4 \cos 3x$ **50.** $y = 5 \sin 2x - 12 \cos 2x$

In Exercises 51–58, solve ΔABC.

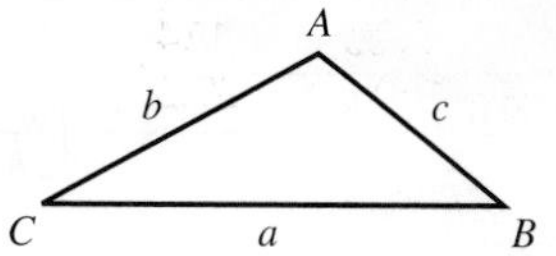

51. $A = 79°, \quad B = 33°, \quad a = 7$

52. $a = 5, \quad b = 8, \quad B = 110°$

53. $a = 8, \quad b = 3, \quad B = 30°$

54. $a = 14.7, \quad A = 29.3°, \quad C = 33°$

55. $A = 34°, \quad B = 74°, \quad c = 5$

56. $c = 41, \quad A = 22.9°, \quad C = 55.1°$

57. $a = 5, \quad b = 7, \quad c = 6$

58. $A = 85°, \quad a = 6, \quad b = 4$

In Exercises 59 and 60, find the area of ΔABC.

59. $a = 3, \quad b = 5, \quad c = 6$

60. $a = 10, \quad b = 6, \quad C = 50°$

61. If $a = 12$ and $B = 28°$, determine the values of b that will produce the indicated number of triangles:

(a) Two **(b)** One **(c)** Zero

62. Surveying a Canyon Two markers A and B on the same side of a canyon rim are 80 ft apart, as shown in the figure. A hiker is located across the rim at point C. A surveyor determines that $\angle BAC = 70°$ and $\angle ABC = 65°$.

(a) What is the distance between the hiker and point A?

(b) What is the distance between the two canyon rims? (Assume they are parallel.)

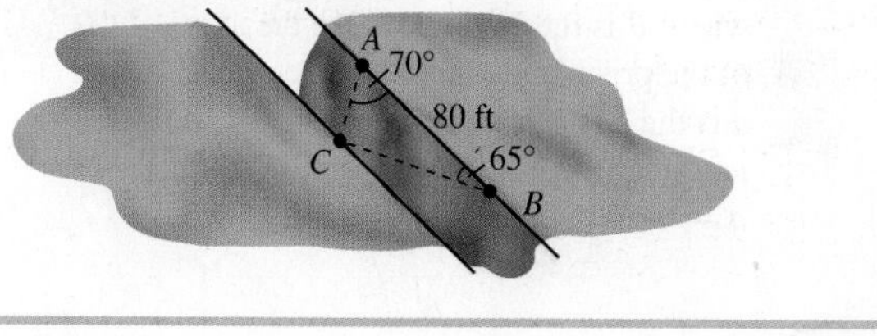

63. **Altitude** A hot-air balloon is seen over Tucson, Arizona, simultaneously by two observers at points A and B that are 1.75 mi apart on level ground and in line with the balloon. The angles of elevation are as shown here. How high above ground is the balloon?

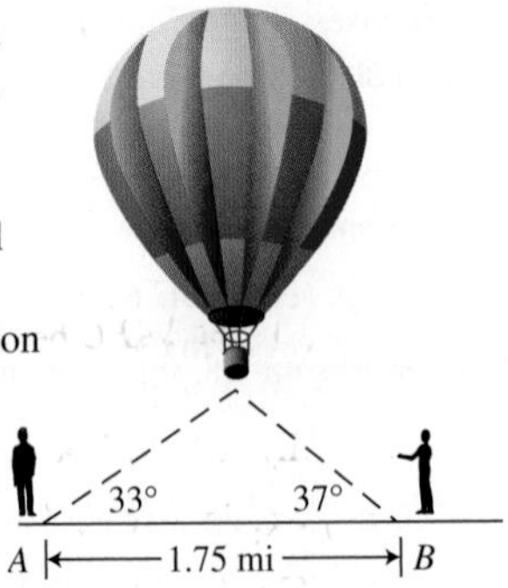

64. **Finding Distance** In order to determine the distance between two points A and B on opposite sides of a lake, a surveyor chooses a point C that is 900 ft from A and 225 ft from B, as shown in the figure. If the measure of the angle at C is 70°, find the distance between A and B.

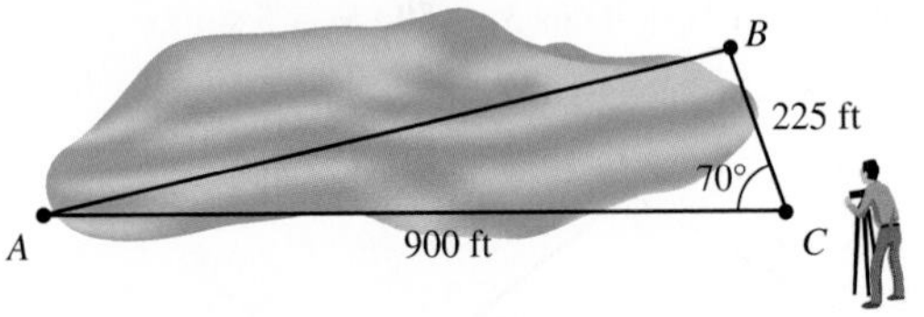

65. **Finding Radian Measure** Find the radian measure of the largest angle of the triangle whose sides have lengths 8, 9, and 10.

66. **Finding a Parallelogram** A parallelogram has sides of 15 and 24 ft, and an angle of 40°. Find the diagonals.

67. **Maximizing Area** A trapezoid is inscribed in the upper half of a unit circle, as shown in the figure.

(a) Write the area of the trapezoid as a function of θ.

(b) Find the value of θ that maximizes the area of the trapezoid and the maximum area.

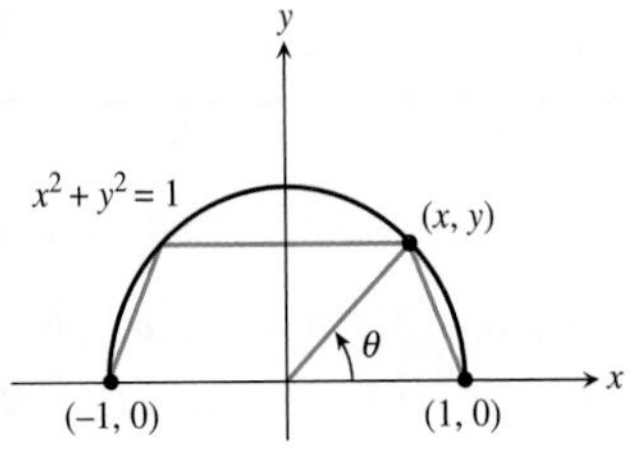

68. **Beehive Cells** A single cell in a beehive is a regular hexagonal prism open at the front with a trihedral cut at the back. Trihedral refers to a vertex formed by three faces of a polyhedron. It can be shown that the surface area of a cell is given by

$$S(\theta) = 6ab + \frac{3}{2}b^2\left(-\cot\theta + \frac{\sqrt{3}}{\sin\theta}\right),$$

where θ is the angle between the axis of the prism and one of the back faces, a is the depth of the prism, and b is the length of the hexagonal front. Assume $a = 1.75$ in. and $b = 0.65$ in.

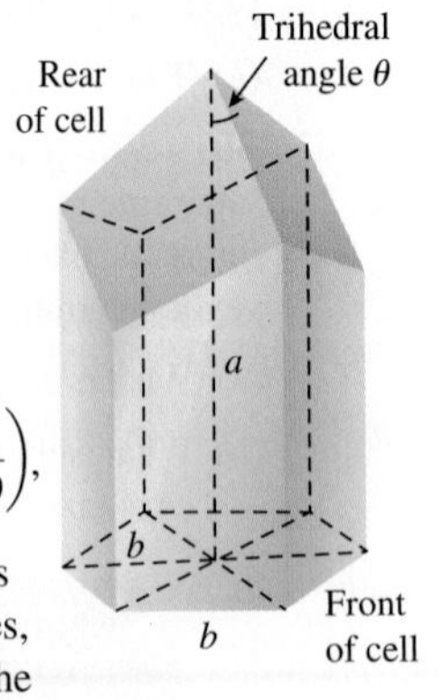

(a) Graph the function $y = S(\theta)$.

(b) What value of θ gives the minimum surface area? (*Note:* This answer is quite close to the observed angle in nature.)

(c) What is the minimum surface area?

69. **Cable Television Coverage** A cable broadcast satellite S orbits the Earth at a height h (in miles) above the Earth's surface, as shown in the figure. The two lines from S are tangent to the Earth's surface. The part of the Earth's surface that is in the broadcast area of the satellite is determined by the central angle θ indicated in the figure.

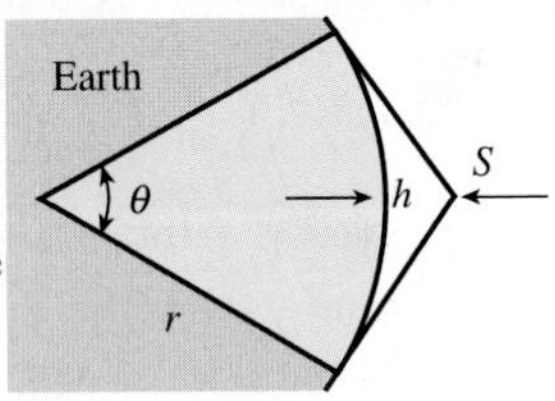

(a) Assuming that the Earth is spherical with a radius of 4000 mi, write h as a function of θ.

(b) Approximate θ for a satellite 200 mi above the surface of the Earth.

70. **Finding Extremum Values** The graph of

$$y = \cos x - \frac{1}{2}\cos 2x + \frac{1}{3}\cos 3x$$

is shown in the figure. The x values that correspond to local maximum and minimum points are solutions of the equation $\sin x - \sin 2x + \sin 3x = 0$. Solve this equation algebraically, and support your solution using the graph of y.

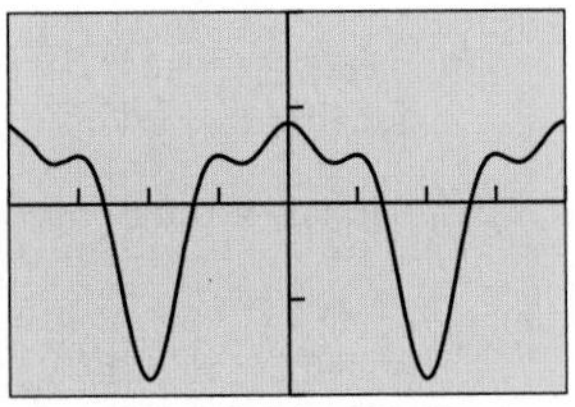

$[-2\pi, 2\pi]$ by $[-2, 2]$

71. **Using Trigonometry in Geometry** A regular hexagon whose sides are 16 cm is inscribed in a circle. Find the area inside the circle and outside the hexagon.

72. **Using Trigonometry in Geometry** A circle is inscribed in a regular pentagon whose sides are 12 cm. Find the area inside the pentagon and outside the circle.

73. **Using Trigonometry in Geometry** A wheel of cheese in the shape of a right circular cylinder is 18 cm in diameter and 5 cm thick. If a wedge of cheese with a central angle of 15° is cut from the wheel, find the volume of the cheese wedge.

74. **Product-to-Sum Formulas** Prove the following identities, which are called the **product-to-sum formulas**.

(a) $\sin u \sin v = \frac{1}{2}(\cos(u - v) - \cos(u + v))$

(b) $\cos u \cos v = \frac{1}{2}(\cos(u - v) + \cos(u + v))$

(c) $\sin u \sin v = \frac{1}{2}(\sin(u + v) + \sin(u - v))$

75. **Sum-to-Product Formulas** Use the product-to-sum formulas in Exercise 74 to prove the following identities, which are called the **sum-to-product formulas**.

(a) $\sin u + \sin v = 2 \sin \dfrac{u+v}{2} \cos \dfrac{u-v}{2}$

(b) $\sin u - \sin v = 2 \sin \dfrac{u-v}{2} \cos \dfrac{u+v}{2}$

(c) $\cos u + \cos v = 2 \cos \dfrac{u+v}{2} \cos \dfrac{u-v}{2}$

(d) $\cos u - \cos v = -2 \sin \dfrac{u+v}{2} \sin \dfrac{u-v}{2}$

76. Catching Students Faking Data Carmen and Pat both need to make up a missed physics lab. They are to measure the total distance $(2x)$ traveled by a beam of light from point A to point B and record it in 20° increments of θ as they adjust the mirror (C) upward vertically. They report the following measurements. However, only one of the students actually did the lab; the other skipped it and faked the data. Who faked the data, and how can you tell?

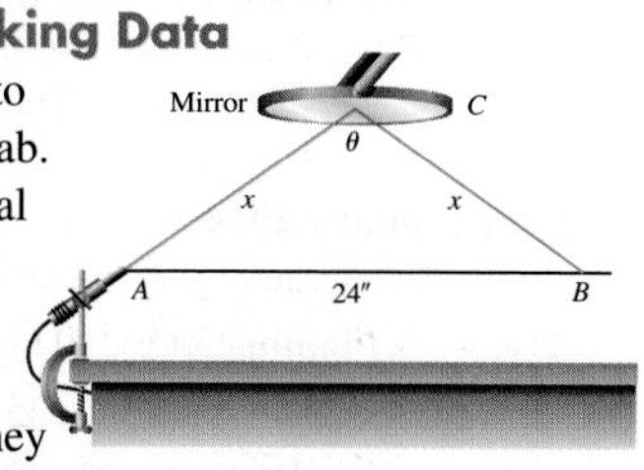

CARMEN		PAT	
θ	$2x$	θ	$2x$
160°	24.4″	160°	24.5″
140°	25.6″	140°	25.2″
120°	28.0″	120°	26.4″
100°	31.2″	100°	30.4″
80°	37.6″	80°	35.2″
60°	48.0″	60°	48.0″
40°	70.4″	40°	84.0″
20°	138.4″	20°	138.4″

77. An Interesting Fact About (sin A)/a The ratio $(\sin A)/a$ that shows up in the Law of Sines shows up another way in the geometry of ΔABC: It is the reciprocal of the radius of the circumscribed circle.

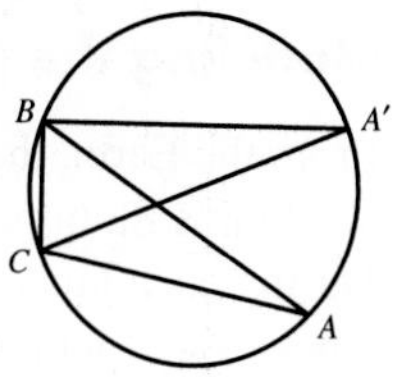

(a) Let ΔABC be circumscribed as shown in the diagram, and construct diameter CA'. Explain why $\angle A'BC$ is a right angle.

(b) Explain why $\angle A'$ and $\angle A$ are congruent.

(c) If a, b, and c are the sides opposite angles A, B, and C as usual, explain why $\sin A' = a/d$, where d is the diameter of the circle.

(d) Finally, explain why $(\sin A)/a = 1/d$.

(e) Do $(\sin B)/b$ and $(\sin C)/c$ also equal $1/d$? Why?

CHAPTER 5 Project

Modeling the Illumination of the Moon

From the Earth, the Moon appears to be a circular disk in the sky that is illuminated to varying degrees by direct sunlight. During each lunar orbit the Moon varies from a status of being a new Moon with no visible illumination to that of a full Moon, which is fully illuminated by direct sunlight. The U.S. Naval Observatory has developed a mathematical model to find the fraction of the Moon's visible disk that is illuminated by the Sun. The data in the table below (obtained from the U.S. Naval Observatory Web site, http://aa.usno.navy.mil/, Astronomical Applications Department) show the fraction of the Moon illuminated at midnight for each day in January 2014.

Fraction of the Moon Illuminated, January 2014

Day	Fraction Illuminated	Day	Fraction Illuminated	Day	Fraction Illuminated	Day	Fraction Illuminated
1	1.00	9	0.32	17	0.03	25	0.68
2	0.97	10	0.23	18	0.07	26	0.78
3	0.92	11	0.15	19	0.13	27	0.87
4	0.84	12	0.09	20	0.20	28	0.94
5	0.74	13	0.04	21	0.28	29	0.98
6	0.64	14	0.01	22	0.38	30	1.00
7	0.53	15	0.00	23	0.48	31	0.99
8	0.42	16	0.01	24	0.58		

Explorations

1. Enter the data in the table above into your grapher or computer. Create a scatter plot of the data.
2. Find values for a, b, h, and k so the equation $y = a \cos(b(x - h)) + k$ models the data in the data plot.
3. Verify graphically the cofunction identity $\sin(\pi/2 - \theta) = \cos\theta$ by substituting $(\pi/2 - \theta)$ for θ in the model above and using sine instead of cosine. (Note $\theta = b(x - h)$.) Observe how well this new model fits the data.
4. Verify graphically the odd-even identity $\cos(\theta) = \cos(-\theta)$ for the model in #2 by substituting $-\theta$ for θ and observing how well the graph fits the data.
5. Find values for a, b, h, and k so the equation $y = a \sin(b(x - h)) + k$ fits the data in the table.
6. Verify graphically the cofunction identity $\cos(\pi/2 - \theta) = \sin\theta$ by substituting $(\pi/2 - \theta)$ for θ in the model above and using cosine rather than sine. (Note $\theta = b(x - h)$.) Observe the fit of this model to the data.
7. Verify graphically the odd-even identity $\sin(-\theta) = -\sin(\theta)$ for the model in #5 by substituting $-\theta$ for θ and graphing $-a \sin(-\theta) + k$. How does this model compare to the original one?

CHAPTER 6

6.1 Vectors in the Plane

6.2 Dot Product of Vectors

6.3 Parametric Equations and Motion

6.4 Polar Coordinates

6.5 Graphs of Polar Equations

6.6 De Moivre's Theorem and *n*th Roots

Applications of Trigonometry

Young salmon migrate from the fresh water they are born in to salt water and live in the ocean for several years. When it's time to spawn, the salmon return from the ocean to the river's mouth, where they follow the organic odors of their homestream to guide them upstream. Researchers believe the fish use currents, salinity, temperature, and the magnetic field of the Earth to guide them. Some fish swim as far as 3500 mi upstream for spawning. See a related problem on page 463.

Jakob Bernoulli (1654–1705)

The first member of the Bernoulli family (driven out of Holland by the Spanish persecutions and settled in Switzerland) to achieve mathematical fame, Jakob defined the numbers now known as Bernoulli numbers. He determined the form (the elastica) taken by an elastic rod acted on at one end by a given force and fixed at the other end.

Chapter 6 Overview

We have used coordinate geometry extensively in previous chapters as a tool for understanding functions (the primary focus of this book), but we have barely begun to make use of the simple fact that we can coordinatize every point on a line with one number, every point in a plane with two numbers, and every point in space with three numbers. We know how to use numerical coordinates to identify position, but with a slight change in interpretation we can also use them to identify *change* in position (i.e., velocity), and thus model problems of motion. This new interpretation involves *vectors*, the subject of most of this chapter.

Modeling coordinates as functions of time leads us in a natural way to revisit parametric equations, which we do in Section 6.3. The last three sections of the chapter introduce *polar* coordinates, with which we can coordinatize the *xy*-plane in a different way. Not only will this lead to some new and interesting graphs, but it will also give us further insight into some graphs we have already studied. The final section uses polar coordinates in the context of complex numbers to explore a theorem about *n*th powers (De Moivre's Theorem) that is truly remarkable—both algebraically and geometrically!

6.1 Vectors in the Plane

What you'll learn about

- Two-Dimensional Vectors
- Vector Operations
- Unit Vectors
- Direction Angles
- Applications of Vectors

... and why

These topics are important in many real-world applications, such as calculating the effect of the wind on an airplane's path.

Two-Dimensional Vectors

Some quantities, like temperature, distance, height, area, and volume, can be represented by a single real number that indicates *magnitude* or *size*. Other quantities, such as force, velocity, and acceleration, have *magnitude* and *direction*. Since the number of possible directions for an object moving in a plane is infinite, you might be surprised to learn that two numbers are all that we need to represent both the magnitude of an object's velocity and its direction of motion. We simply look at ordered pairs of real numbers in a new way. While the pair (a, b) determines a point in the plane, it also determines a **directed line segment** (or **arrow**) with its tail at the origin and its head at (a, b) (Figure 6.1). The length of this arrow represents magnitude, and the direction in which it points represents direction. Because in this context the ordered pair (a, b) represents a mathematical object with both magnitude and direction, we call it the **position vector of (a, b)**, and denote it as $\langle a, b\rangle$ to distinguish it from the point (a, b).

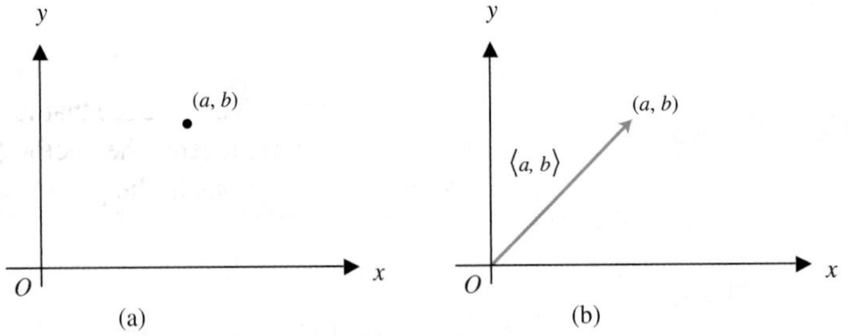

FIGURE 6.1 The point represents the ordered pair (a, b). The arrow (directed line segment) represents the vector $\langle a, b\rangle$.

DEFINITION Two-Dimensional Vector

A **two-dimensional vector v** is an ordered pair of real numbers, denoted in **component form** as $\langle a, b\rangle$. The numbers a and b are the **components** of the vector **v**. The **standard representation** of the vector $\langle a, b\rangle$ is the arrow from the origin to the point (a, b). The **magnitude** of **v** is the length of the arrow, and the **direction** of **v** is the direction in which the arrow is pointing. The vector $\mathbf{0} = \langle 0, 0\rangle$, called the **zero vector**, has zero length and no direction.

Is an Arrow a Vector?

Although an arrow *represents* a vector, it is not a vector itself, since each vector can be represented by an infinite number of equivalent arrows. Still, it is hard to avoid referring to "the vector $\overrightarrow{PQ}$" in practice, and we will often do that ourselves. When we say "the vector $\mathbf{u} = \overrightarrow{PQ}$," we really mean "the vector $\mathbf{u}$ represented by $\overrightarrow{PQ}$."

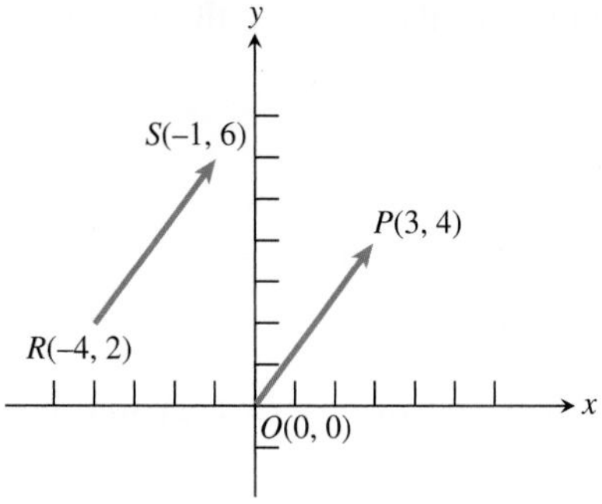

FIGURE 6.2 The arrows $\overrightarrow{RS}$ and $\overrightarrow{OP}$ both represent the vector $\langle 3, 4 \rangle$, as would any arrow with the same length pointing in the same direction. Such arrows are called *equivalent.*

It is often convenient in applications to represent vectors with arrows that begin at points other than the origin. The important thing to remember is that *any two arrows with the same length and pointing in the same direction represent the same vector.* In Figure 6.2, for example, the vector $\langle 3, 4 \rangle$ is shown represented by $\overrightarrow{RS}$, an arrow with **initial point** R and **terminal point** S, as well as by its standard representation $\overrightarrow{OP}$. Two arrows that represent the same vector are called **equivalent**.

The quick way to associate arrows with the vectors they represent is to use the following rule.

Head Minus Tail (HMT) Rule for Vectors

If an arrow has initial point (x_1, y_1) and terminal point (x_2, y_2), it represents the vector $\langle x_2 - x_1, y_2 - y_1 \rangle$.

EXAMPLE 1 Showing Arrows Are Equivalent

Show that the arrow from $R = (-4, 2)$ to $S = (-1, 6)$ is equivalent to the arrow from $P = (2, -1)$ to $Q = (5, 3)$ (Figure 6.3).

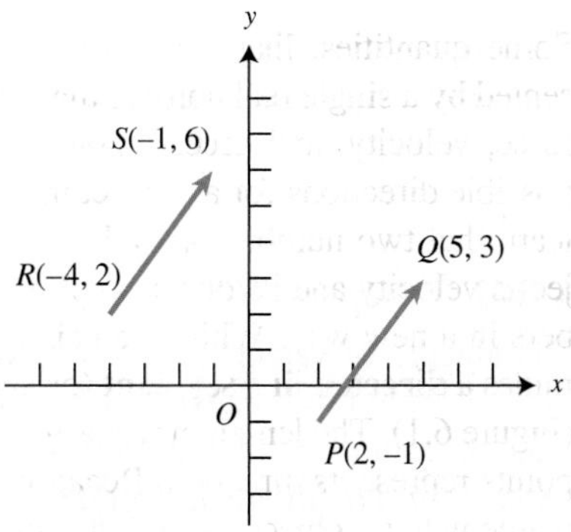

FIGURE 6.3 The arrows $\overrightarrow{RS}$ and $\overrightarrow{PQ}$ appear to have the same magnitude and direction. The Head Minus Tail Rule proves that they represent the same vector. (Example 1)

SOLUTION

Applying the HMT Rule, we see that $\overrightarrow{RS}$ represents the vector $\langle -1 - (-4), 6 - 2 \rangle = \langle 3, 4 \rangle$, and $\overrightarrow{PQ}$ represents the vector $\langle 5 - 2, 3 - (-1) \rangle = \langle 3, 4 \rangle$. Although they have different positions in the plane, these arrows represent the same vector and are therefore equivalent.

Now try Exercise 1.

EXPLORATION 1 Vector Archery

See how well you can direct arrows in the plane using vector information and the HMT Rule.

1. An arrow has initial point $(2, 3)$ and terminal point $(7, 5)$. What vector does it represent?
2. An arrow has initial point $(3, 5)$ and represents the vector $\langle -3, 6 \rangle$. What is the terminal point?
3. If P is the point $(4, -3)$ and $\overrightarrow{PQ}$ represents $\langle 2, -4 \rangle$, find Q.
4. If Q is the point $(4, -3)$ and $\overrightarrow{PQ}$ represents $\langle 2, -4 \rangle$, find P.

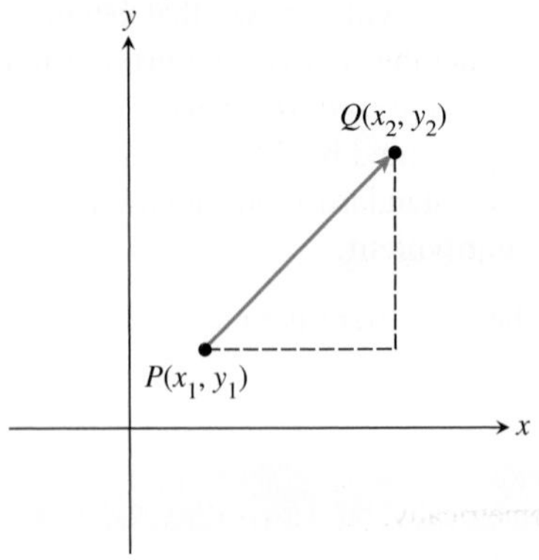

FIGURE 6.4 The magnitude of **v** is the length of the arrow $\overrightarrow{PQ}$, which is found using the distance formula:
$|\mathbf{v}| = \sqrt{(x_2 - x_1)^2 + (y_2 - y_1)^2}$.

What About Direction?

You might expect a quick computational rule for *direction* to accompany the rule for magnitude, but direction is less easily quantified. We will deal with vector direction later in the section.

If you handled Exploration 1 with relative ease, you have a good understanding of how vectors are represented geometrically by arrows. This will help you understand the algebra of vectors, beginning with the concept of magnitude.

The *magnitude* of a vector **v** is also called the *absolute value of* **v**, so it is usually denoted by $|\mathbf{v}|$. (You might see $\|\mathbf{v}\|$ in some textbooks.) Note that it is a nonnegative real number, not a vector. The following computational rule follows directly from the distance formula in the plane (Figure 6.4).

DEFINITION Magnitude (or Length) of a Vector

If **v** is represented by the arrow from (x_1, y_1) to (x_2, y_2), then its **magnitude** is

$|\mathbf{v}| = \sqrt{(x_2 - x_1)^2 + (y_2 - y_1)^2}$.

If $\mathbf{v} = \langle a, b \rangle$, then $|\mathbf{v}| = \sqrt{a^2 + b^2}$.

EXAMPLE 2 Finding Magnitude of a Vector

Find the magnitude of the vector **v** represented by $\overrightarrow{PQ}$, where $P = (-3, 4)$ and $Q = (-5, 2)$.

SOLUTION

Working directly with the arrow, $|\mathbf{v}| = \sqrt{(-5 - (-3))^2 + (2 - 4)^2} = 2\sqrt{2}$. Or, the HMT Rule shows that $\mathbf{v} = \langle -2, -2 \rangle$, so $|\mathbf{v}| = \sqrt{(-2)^2 + (-2)^2} = 2\sqrt{2}$. (See Figure 6.5.)

Now try Exercise 5.

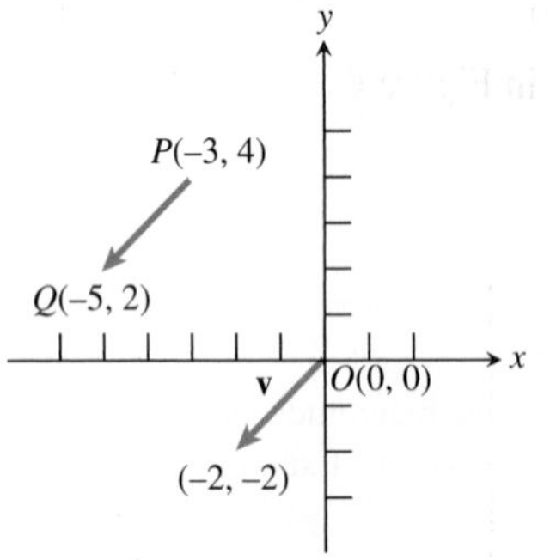

FIGURE 6.5 The vector **v** of Example 2.

Vector Operations

The algebra of vectors sometimes involves working with vectors and numbers at the same time. In this context we refer to the numbers as **scalars**. The two most basic algebraic operations involving vectors are *vector addition* (adding a vector to a vector) and *scalar multiplication* (multiplying a vector by a number). Both operations are easily represented geometrically, and both have immediate applications to many real-world problems.

DEFINITION Vector Addition and Scalar Multiplication

Let $\mathbf{u} = \langle u_1, u_2 \rangle$ and $\mathbf{v} = \langle v_1, v_2 \rangle$ be vectors and let k be a real number (scalar). The **sum** (or **resultant**) **of the vectors u and v** is the vector

$$\mathbf{u} + \mathbf{v} = \langle u_1 + v_1, u_2 + v_2 \rangle.$$

The **product of the scalar k and the vector u** is

$$k\mathbf{u} = k\langle u_1, u_2 \rangle = \langle ku_1, ku_2 \rangle.$$

What About Vector Multiplication?

There is a useful way to define the multiplication of two vectors—in fact, there are two useful ways, but neither one of them is as predictable as vector addition. (You may recall that matrix multiplication was not as predictable as matrix addition either, and for similar reasons.) We will look at the *dot product* in Section 6.2. The *cross product* requires a third dimension, so we will not deal with it in this course.

The sum of the vectors **u** and **v** can be represented geometrically by arrows in two ways.

In the **tail-to-head** representation, the standard representation of **u** points from the origin to (u_1, u_2). The arrow from (u_1, u_2) to $(u_1 + v_1, u_2 + v_2)$ represents **v** (as you can verify by the HMT Rule). The arrow from the origin to $(u_1 + v_1, u_2 + v_2)$ then represents $\mathbf{u} + \mathbf{v}$ (Figure 6.6a).

In the **parallelogram** representation, the standard representations of **u** and **v** determine a parallelogram, the diagonal of which is the standard representation of $\mathbf{u} + \mathbf{v}$ (Figure 6.6b).

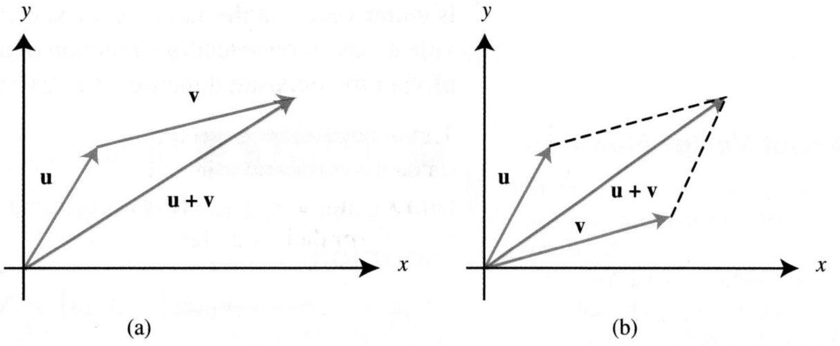

FIGURE 6.6 Two ways to represent vector addition geometrically: (a) tail-to-head, and (b) parallelogram.

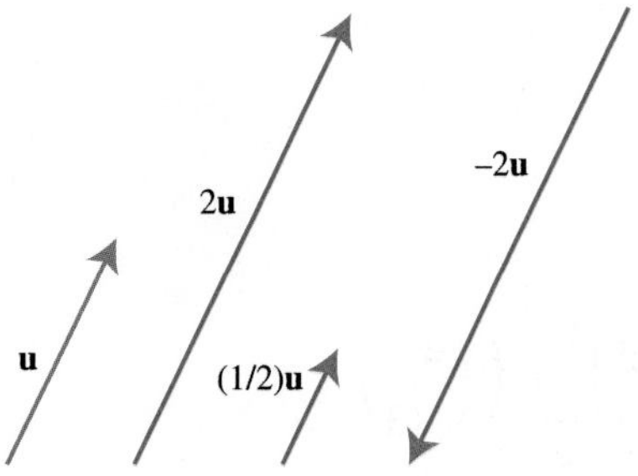

FIGURE 6.7 Representations of **u** and several scalar multiples of **u**.

The product $k\mathbf{u}$ of the scalar k and the vector **u** can be represented by a stretch (or shrink) of **u** by a factor of k. If $k > 0$, then $k\mathbf{u}$ points in the same direction as **u**; if $k < 0$, then $k\mathbf{u}$ points in the opposite direction (Figure 6.7).

EXAMPLE 3 Performing Vector Operations

Let $\mathbf{u} = \langle -1, 3 \rangle$ and $\mathbf{v} = \langle 4, 7 \rangle$. Find the component form of the following vectors:

(a) $\mathbf{u} + \mathbf{v}$ **(b)** $3\mathbf{u}$ **(c)** $2\mathbf{u} + (-1)\mathbf{v}$

SOLUTION Using the vector operations as defined, we have:

(a) $\mathbf{u} + \mathbf{v} = \langle -1, 3 \rangle + \langle 4, 7 \rangle = \langle -1 + 4, 3 + 7 \rangle = \langle 3, 10 \rangle$

(b) $3\mathbf{u} = 3\langle -1, 3 \rangle = \langle -3, 9 \rangle$

(c) $2\mathbf{u} + (-1)\mathbf{v} = 2\langle -1, 3 \rangle + (-1)\langle 4, 7 \rangle = \langle -2, 6 \rangle + \langle -4, -7 \rangle = \langle -6, -1 \rangle$

Geometric representations of $\mathbf{u} + \mathbf{v}$ and $3\mathbf{u}$ are shown in Figure 6.8.

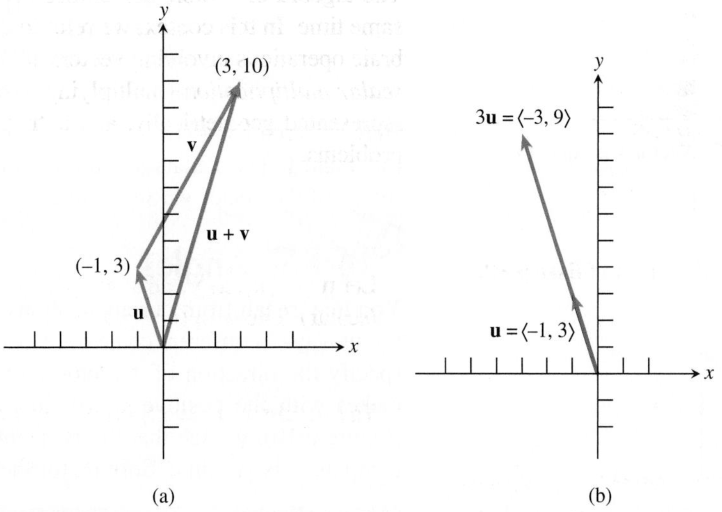

FIGURE 6.8 Given that $\mathbf{u} = \langle -1, 3 \rangle$ and $\mathbf{v} = \langle 4, 7 \rangle$, we can (a) represent $\mathbf{u} + \mathbf{v}$ by the tail-to-head method, and (b) represent $3\mathbf{u}$ as a stretch of **u** by a factor of 3.

Now try Exercise 13.

Unit Vectors

A vector **u** with length $|\mathbf{u}| = 1$ is a **unit vector**. If **v** is not the zero vector $\langle 0, 0 \rangle$, then the vector

$$\mathbf{u} = \frac{\mathbf{v}}{|\mathbf{v}|} = \frac{1}{|\mathbf{v}|}\mathbf{v}$$

is a unit vector in the direction of **v**, often called a **direction vector**. Unit vectors provide a way to represent the direction of any nonzero vector. Any vector in the direction of **v**, or the opposite direction, is a scalar multiple of this unit vector **u**.

A Word About Vector Notation

Both notations, $\langle a, b\rangle$ and $a\mathbf{i} + b\mathbf{j}$, are designed to convey the idea that a single vector **v** has two separate components. This is what makes a two-dimensional vector two-dimensional. You will see both $\langle a, b, c\rangle$ and $a\mathbf{i} + b\mathbf{j} + c\mathbf{k}$ used for three-dimensional vectors, but scientists stick to the $\langle\ \rangle$ notation for dimensions higher than three.

EXAMPLE 4 Finding a Unit Vector

Find a unit vector in the direction of $\mathbf{v} = \langle -3, 2\rangle$, and verify that it has length 1.

SOLUTION

$$|\mathbf{v}| = |\langle -3, 2\rangle| = \sqrt{(-3)^2 + (2)^2} = \sqrt{13}, \text{ so}$$

$$\frac{\mathbf{v}}{|\mathbf{v}|} = \frac{1}{\sqrt{13}}\langle -3, 2\rangle = \left\langle \frac{-3}{\sqrt{13}}, \frac{2}{\sqrt{13}}\right\rangle$$

The magnitude of this vector is

$$\left|\left\langle \frac{-3}{\sqrt{13}}, \frac{2}{\sqrt{13}}\right\rangle\right| = \sqrt{\left(\frac{-3}{\sqrt{13}}\right)^2 + \left(\frac{2}{\sqrt{13}}\right)^2} = \sqrt{\frac{9}{13} + \frac{4}{13}} = \sqrt{\frac{13}{13}} = 1$$

Thus, the magnitude of $\mathbf{v}/|\mathbf{v}|$ is 1. Its direction is the same as **v** because it is a positive scalar multiple of **v**. **Now try Exercise 21.**

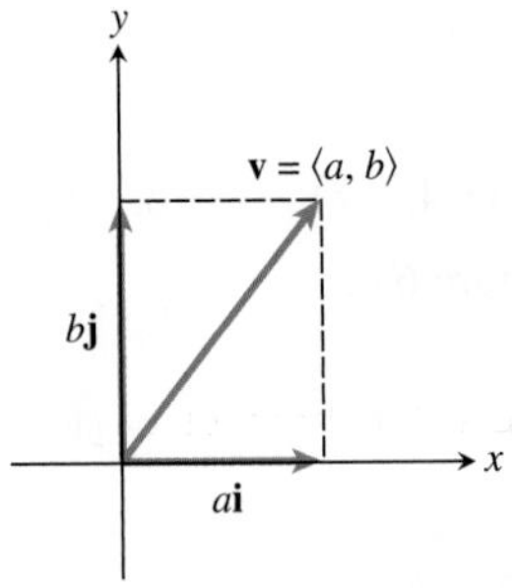

FIGURE 6.9 The vector **v** is equal to $a\mathbf{i} + b\mathbf{j}$.

The two unit vectors $\mathbf{i} = \langle 1, 0\rangle$ and $\mathbf{j} = \langle 0, 1\rangle$ are the **standard unit vectors**. Any vector **v** can be written as an expression in terms of the standard unit vectors:

$$\begin{aligned}\mathbf{v} &= \langle a, b\rangle \\ &= \langle a, 0\rangle + \langle 0, b\rangle \\ &= a\langle 1, 0\rangle + b\langle 0, 1\rangle \\ &= a\mathbf{i} + b\mathbf{j}\end{aligned}$$

Here the vector $\mathbf{v} = \langle a, b\rangle$ is expressed as the **linear combination** $a\mathbf{i} + b\mathbf{j}$ of the vectors **i** and **j**. The scalars a and b are the **horizontal** and **vertical components**, respectively, of the vector **v**. (See Figure 6.9.)

Direction Angles

You may recall from our applications in Section 4.8 that direction is measured in different ways in different contexts, especially in navigation. A simple but precise way to specify the direction of a vector **v** is to state its **direction angle**, the angle θ that **v** makes with the positive x-axis, just as we did in Section 4.3. Using trigonometry (Figure 6.10), we see that the horizontal component of **v** is $|\mathbf{v}| \cos\theta$ and the vertical component is $|\mathbf{v}| \sin\theta$. Solving for these components is called **resolving the vector**.

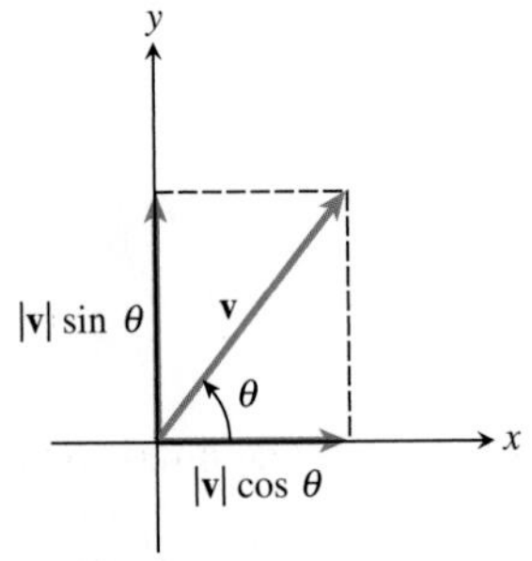

FIGURE 6.10 The horizontal and vertical components of **v**.

Resolving the Vector into Its Components

If **v** has direction angle θ, the components of **v** can be computed using the formula

$$\mathbf{v} = \langle |\mathbf{v}| \cos\theta, |\mathbf{v}| \sin\theta\rangle.$$

From the formula above, it follows that the unit vector in the direction of **v** is

$$\mathbf{u} = \frac{\mathbf{v}}{|\mathbf{v}|} = \langle \cos\theta, \sin\theta\rangle.$$

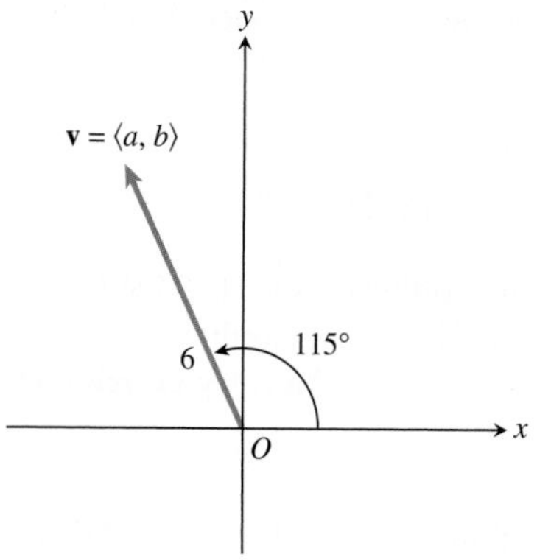

FIGURE 6.11 The direction angle of **v** is 115°. (Example 5)

EXAMPLE 5 Finding the Components of a Vector

Find the components of the vector **v** with direction angle 115° and magnitude 6 (Figure 6.11) using no technology except to approximate the exact trigonometric solution.

SOLUTION If a and b are the horizontal and vertical components, respectively, of **v**, then

$$\mathbf{v} = \langle a, b \rangle = \langle 6 \cos 115°, 6 \sin 115° \rangle.$$

So, $a = 6 \cos 115° \approx -2.54$ and $b = 6 \sin 115° \approx 5.44$.

Now try Exercise 29.

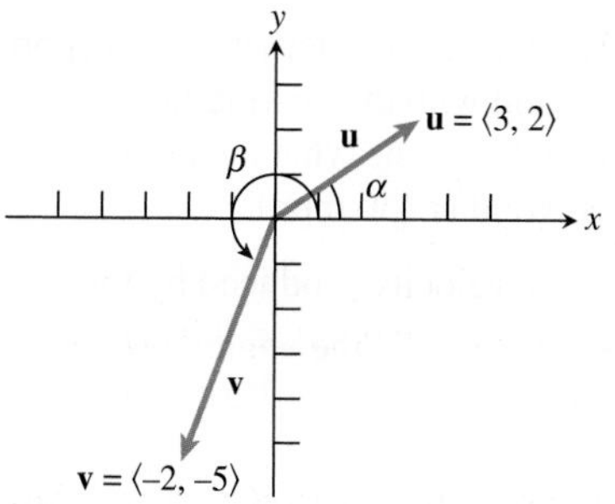

FIGURE 6.12 The two vectors of Example 6.

EXAMPLE 6 Finding the Direction Angle of a Vector

Find the magnitude and direction angle of each vector:

(a) $\mathbf{u} = \langle 3, 2 \rangle$ **(b)** $\mathbf{v} = \langle -2, -5 \rangle$

SOLUTION See Figure 6.12.

(a) $|\mathbf{u}| = \sqrt{3^2 + 2^2} = \sqrt{13}$ is the magnitude. If α is the direction angle of **u**, then $\mathbf{u} = \langle 3, 2 \rangle = \langle |\mathbf{u}| \cos \alpha, |\mathbf{u}| \sin \alpha \rangle$.

$$3 = |\mathbf{u}| \cos \alpha \qquad \text{Horizontal component of } \mathbf{u}$$

$$3 = \sqrt{3^2 + 2^2} \cos \alpha \qquad |\mathbf{u}| = \sqrt{3^2 + 2^2}$$

$$3 = \sqrt{13} \cos \alpha$$

$$\alpha = \cos^{-1}\left(\frac{3}{\sqrt{13}}\right) \approx 33.69° \qquad \alpha \text{ is acute.}$$

So the direction angle is about 33.69°.

(b) $|\mathbf{v}| = \sqrt{(-2)^2 + (-5)^2} = \sqrt{29}$ is the magnitude. If β is the direction angle of **v**, then $\mathbf{v} = \langle -2, -5 \rangle = \langle |\mathbf{v}| \cos \beta, |\mathbf{v}| \sin \beta \rangle$.

$$-2 = |\mathbf{v}| \cos \beta \qquad \text{Horizontal component of } \mathbf{v}$$

$$-2 = \sqrt{(-2)^2 + (-5)^2} \cos \beta \qquad |\mathbf{v}| = \sqrt{(-2)^2 + (-5)^2}$$

$$-2 = \sqrt{29} \cos \beta$$

$$\beta = 360° - \cos^{-1}\left(\frac{-2}{\sqrt{29}}\right) \approx 248.2° \qquad 180° < \beta < 270°$$

So the direction angle is about 248.2°.

Now try Exercise 33.

Applications of Vectors

The **velocity** of a moving object is a vector because velocity has both magnitude and direction. The magnitude of velocity is **speed**.

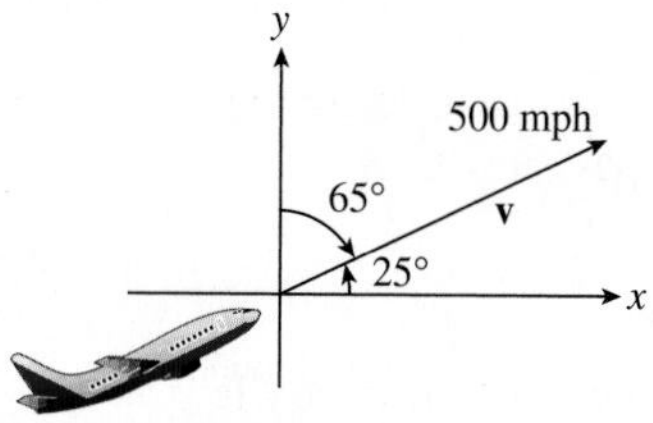

FIGURE 6.13 The airplane's path (bearing) in Example 7.

EXAMPLE 7 Writing Velocity as a Vector

A DC-10 jet aircraft is flying on a bearing of 65° at 500 mph. Find the component form of the velocity of the airplane. Recall that the bearing is the angle that the line of travel makes with due north, measured clockwise (see Section 4.1, Figure 4.2).

SOLUTION Let **v** be the velocity of the airplane. A bearing of 65° is equivalent to a direction angle of 25°. The plane's speed, 500 mph, is the magnitude of vector **v**; that is, $|\mathbf{v}| = 500$. (See Figure 6.13.)

(continued)

The horizontal component of $\mathbf{v}$ is 500 cos 25° and the vertical component is 500 sin 25°, so

$$\mathbf{v} = (500 \cos 25°)\mathbf{i} + (500 \sin 25°)\mathbf{j}$$
$$= \langle 500 \cos 25°, 500 \sin 25° \rangle \approx \langle 453.15, 211.31 \rangle$$

The components of the velocity give the eastward and northward speeds. That is, the airplane travels about 453.15 mph eastward and about 211.31mph northward as it travels at 500 mph on a bearing of 65°. **Now try Exercise 41.**

A typical problem for a navigator involves calculating the effect of wind on the direction and speed of the airplane, as illustrated in Example 8.

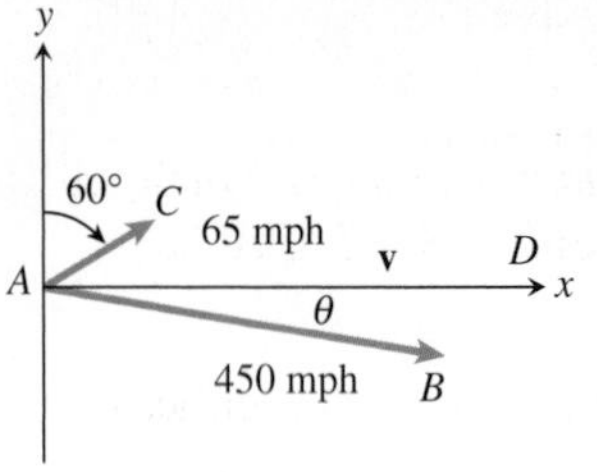

FIGURE 6.14 The x-axis represents the flight path of the plane in Example 8.

EXAMPLE 8 Calculating the Effect of Wind Velocity

Pilot Megan McCarty's flight plan has her leaving San Francisco International Airport and flying a Boeing 727 due east. There is a 65-mph wind with the bearing 60°. Find the compass heading McCarty should follow, and determine what the airplane's ground speed will be (assuming that its speed with no wind is 450 mph).

SOLUTION See Figure 6.14. Vector $\overrightarrow{AB}$ represents the velocity produced by the airplane alone, $\overrightarrow{AC}$ represents the velocity of the wind, and θ is the angle DAB. Vector $\mathbf{v} = \overrightarrow{AD}$ represents the resulting velocity, so

$$\mathbf{v} = \overrightarrow{AD} = \overrightarrow{AC} + \overrightarrow{AB}.$$

We must find the bearing of $\overrightarrow{AB}$ and $|\mathbf{v}|$.

Resolving the vectors, we obtain

$$\overrightarrow{AC} = \langle 65 \cos 30°, 65 \sin 30° \rangle$$
$$\overrightarrow{AB} = \langle 450 \cos \theta, 450 \sin \theta \rangle$$
$$\overrightarrow{AD} = \overrightarrow{AC} + \overrightarrow{AB}$$
$$= \langle 65 \cos 30° + 450 \cos \theta, 65 \sin 30° + 450 \sin \theta \rangle$$

Because the plane is traveling due east, the second component of $\overrightarrow{AD}$ must be zero.

$$65 \sin 30° + 450 \sin \theta = 0$$
$$\theta = \sin^{-1}\left(\frac{-65 \sin 30°}{450}\right)$$
$$\approx -4.14° \qquad \theta < 0$$

Thus, the compass heading McCarty should follow is

$$90° + |\theta| \approx 94.14°. \qquad \text{Bearing} > 90°$$

The ground speed of the airplane is

$$|\mathbf{v}| = |\overrightarrow{AD}| = \sqrt{(65 \cos 30° + 450 \cos \theta)^2 + 0^2}$$
$$= |65 \cos 30° + 450 \cos \theta|$$
$$\approx 505.12$$

Using the unrounded value of θ.

McCarty should use a bearing of approximately 94.14°. The airplane will travel due east at approximately 505.12 mph. **Now try Exercise 43.**

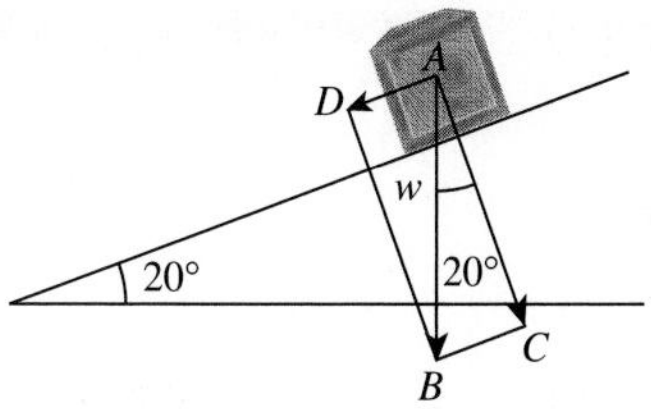

FIGURE 6.15 The force of gravity $\overrightarrow{AB}$ has a component $\overrightarrow{AC}$ that holds the box against the surface of the ramp, and a component $\overrightarrow{AD} = \overrightarrow{CB}$ that tends to push the box down the ramp. (Example 9)

EXAMPLE 9 Finding the Effect of Gravity

A force of 30 lb just keeps the box in Figure 6.15 from sliding down the ramp inclined at 20°. Find the weight of the box.

SOLUTION We are given that $|\overrightarrow{AD}| = 30$. Let $|\overrightarrow{AB}| = w$; then

$$\sin 20° = \frac{|\overrightarrow{CB}|}{w} = \frac{30}{w}.$$

Thus,

$$w = \frac{30}{\sin 20°} \approx 87.71.$$

The weight of the box is about 87.71 lb. **Now try Exercise 47.**

A Reminder About Special Angles

As a reminder, we hope that when confronted with sin 60° you will use your mental skills to write the answer of $\frac{\sqrt{3}}{2}$ *without* the aid *of a calculator*—similarly for all the special angles. However, when a value like sin 20° is needed you will need your calculator to approximate the answer to the agreed-upon accuracy.

CHAPTER OPENER **Problem** (from page 455)

Problem: During one part of its migration, a salmon is swimming at 6 mph, and the current is flowing downstream at 3 mph at an angle of 7°. How fast is the salmon moving upstream?

Solution: Assume the salmon is swimming in a plane parallel to the surface of the water.

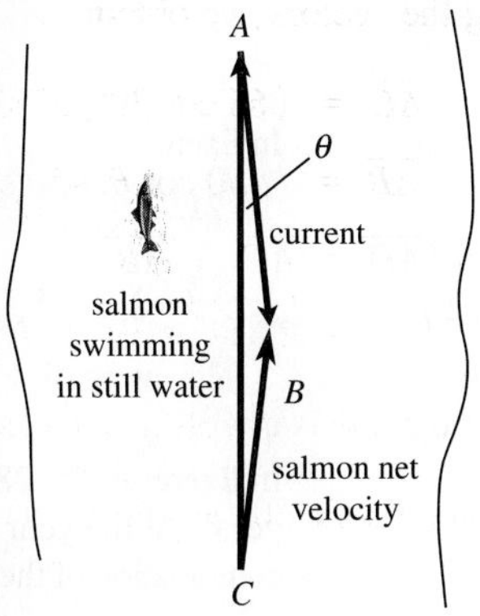

In the figure, vector $\overrightarrow{AB}$ represents the current of 3 mph, θ is the angle CAB, which is 7°, the vector $\overrightarrow{CA}$ represents the velocity of the salmon of 6 mph, and the vector $\overrightarrow{CB}$ is the net velocity at which the fish is moving upstream.

So we have

$$\overrightarrow{AB} = \langle 3\cos(-83°), 3\sin(-83°)\rangle \approx \langle 0.37, -2.98\rangle$$

$$\overrightarrow{CA} = \langle 0, 6\rangle$$

Thus $\overrightarrow{CB} = \overrightarrow{CA} + \overrightarrow{AB} = \langle 3\cos(-83°), 3\sin(-83°) + 6\rangle$

$$\approx \langle 0.37, 3.02\rangle.$$

The speed of the salmon is then $|\overrightarrow{CB}| \approx \sqrt{0.37^2 + 3.02^2} \approx 3.04$ mph upstream.

QUICK REVIEW 6.1 *(For help, go to Sections 4.3 and 4.7.)*

Exercise numbers with a gray background indicate problems that the authors have designed to be solved *without a calculator*.

In Exercises 1–4, find the values of x and y.

1.

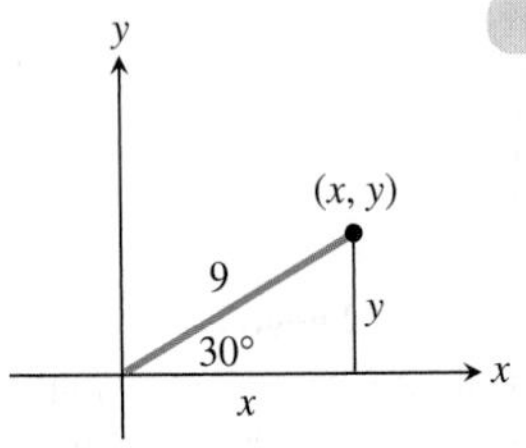

2.

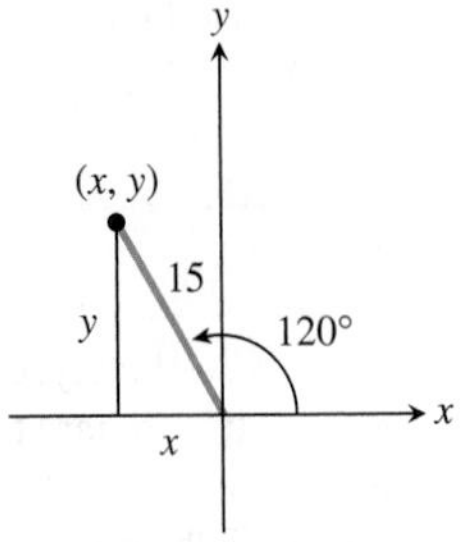

3.

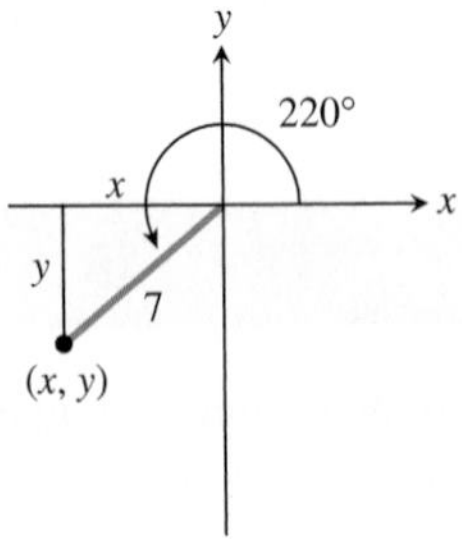

4.

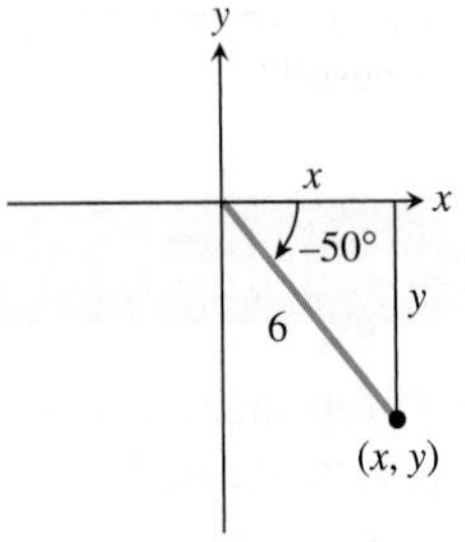

In Exercises 5 and 6, solve for θ in degrees.

5. $\theta = \sin^{-1}\left(\frac{3}{\sqrt{29}}\right)$

6. $\theta = \cos^{-1}\left(\frac{-1}{\sqrt{15}}\right)$

In Exercises 7–9, the point P is on the terminal side of the angle θ. Find the measure of θ if $0° < \theta < 360°$.

7. $P(5, 9)$

8. $P(5, -7)$

9. $P(-2, -5)$

10. A naval ship leaves Port Norfolk and averages 42 knots (naut mi per hr) traveling for 3 hr on a bearing of 40° and then 5 hr on a course of 125°. What is the boat's bearing and distance from Port Norfolk after 8 hr?

SECTION 6.1 Exercises

In Exercises 1–4, prove that $\overrightarrow{RS}$ and $\overrightarrow{PQ}$ are equivalent by showing that they represent the same vector.

1. $R = (-4, 7), S = (-1, 5), P = (0, 0)$, and $Q = (3, -2)$

2. $R = (7, -3), S = (4, -5), P = (0, 0)$, and $Q = (-3, -2)$

3. $R = (2, 1), S = (0, -1), P = (1, 4)$, and $Q = (-1, 2)$

4. $R = (-2, -1), S = (2, 4), P = (-3, -1)$, and $Q = (1, 4)$

In Exercises 5–12, let $P = (-2, 2), Q = (3, 4), R = (-2, 5)$, and $S = (2, -8)$. Find the component form and magnitude of the vector.

5. $\overrightarrow{PQ}$

6. $\overrightarrow{RS}$

7. $\overrightarrow{QR}$

8. $\overrightarrow{PS}$

9. $2\overrightarrow{QS}$

10. $(\sqrt{2})\overrightarrow{PR}$

11. $3\overrightarrow{QR} + \overrightarrow{PS}$

12. $\overrightarrow{PS} - 3\overrightarrow{PQ}$

In Exercises 13–20, let $\mathbf{u} = \langle -1, 3 \rangle, \mathbf{v} = \langle 2, 4 \rangle$, and $\mathbf{w} = \langle 2, -5 \rangle$. Find the component form of the vector.

13. $\mathbf{u} + \mathbf{v}$

14. $\mathbf{u} + (-1)\mathbf{v}$

15. $\mathbf{u} - \mathbf{w}$

16. $3\mathbf{v}$

17. $2\mathbf{u} + 3\mathbf{w}$

18. $2\mathbf{u} - 4\mathbf{v}$

19. $-2\mathbf{u} - 3\mathbf{v}$

20. $-\mathbf{u} - \mathbf{v}$

In Exercises 21–24, find a unit vector in the direction of the given vector.

21. $\mathbf{u} = \langle -2, 4 \rangle$

22. $\mathbf{v} = \langle 1, -1 \rangle$

23. $\mathbf{w} = -\mathbf{i} - 2\mathbf{j}$

24. $\mathbf{w} = 5\mathbf{i} + 5\mathbf{j}$

In Exercises 25–28, find the unit vector in the direction of the given vector. Write your answer in **(a)** component form and **(b)** as a linear combination of the standard unit vectors **i** and **j**.

25. $\mathbf{u} = \langle 2, 1 \rangle$

26. $\mathbf{u} = \langle -3, 2 \rangle$

27. $\mathbf{u} = \langle -4, -5 \rangle$

28. $\mathbf{u} = \langle 3, -4 \rangle$

In Exercises 29–32, find the component form of the vector **v**. Solve algebraically and approximate exact answers with a calculator. Support your solution by estimating the lengths of the components of the vector in each figure and comparing with your answer.

29.

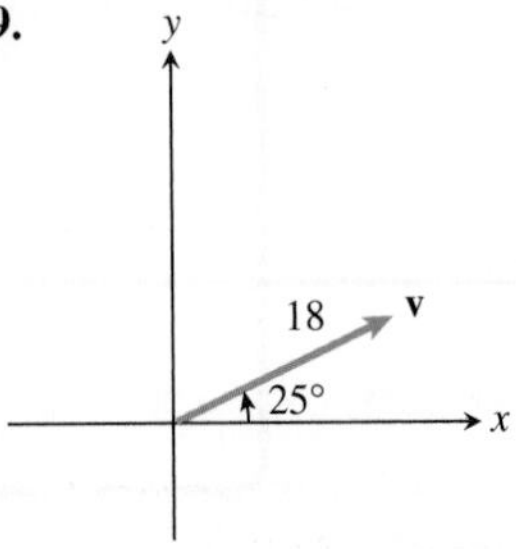

30.

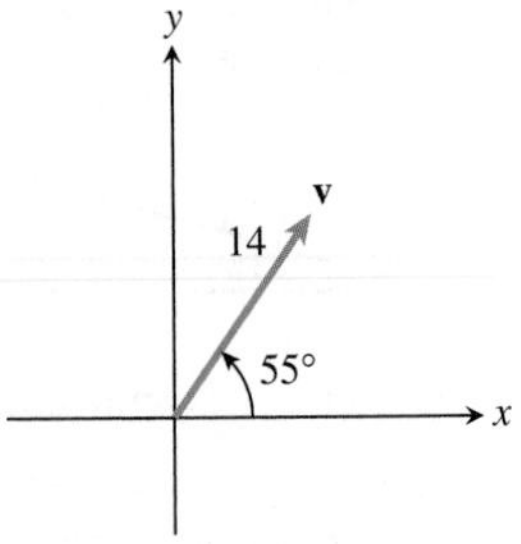

31.

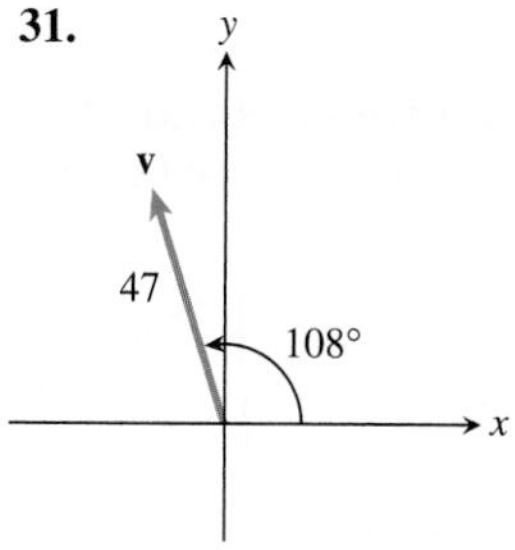

32.

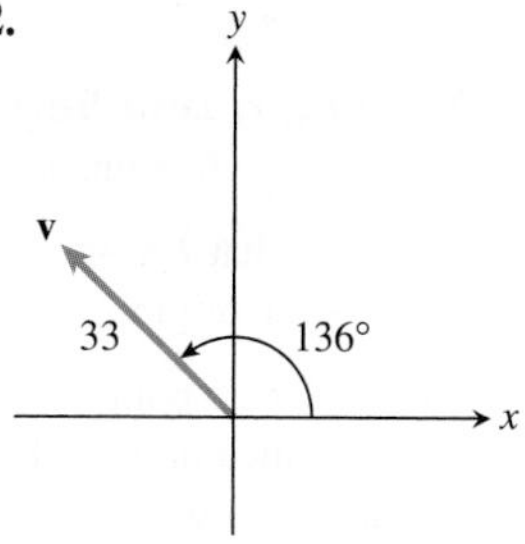

In Exercises 33–38, find the magnitude and direction angle of the vector. Use an algebraic method and approximate exact answers with a calculator when appropriate.

33. $\langle 3, 4\rangle$

34. $\langle -1, 2\rangle$

35. $3\mathbf{i} - 4\mathbf{j}$

36. $-3\mathbf{i} - 5\mathbf{j}$

37. $7(\cos 135°\mathbf{i} + \sin 135°\mathbf{j})$

38. $2(\cos 60°\mathbf{i} + \sin 60°\mathbf{j})$

In Exercises 39 and 40, find the vector **v** with the given magnitude and the same direction as **u**.

39. $|\mathbf{v}| = 2, \mathbf{u} = \langle 3, -3\rangle$

40. $|\mathbf{v}| = 5, \mathbf{u} = \langle -5, 7\rangle$

41. **Navigation** An airplane is flying on a bearing of 335° at 530 mph. Find the component form of the velocity of the airplane.

42. **Navigation** An airplane is flying on a bearing of 170° at 460 mph. Find the component form of the velocity of the airplane.

43. **Flight Engineering** An airplane is flying on a compass heading (bearing) of 340° at 325 mph. A wind is blowing with the bearing 320° at 40 mph.

(a) Find the component form of the velocity of the airplane.

(b) Find the actual ground speed and direction of the plane.

44. **Flight Engineering** An airplane is flying on a compass heading (bearing) of 170° at 460 mph. A wind is blowing with the bearing 200° at 80 mph.

(a) Find the component form of the velocity of the airplane.

(b) Find the actual ground speed and direction of the airplane.

45. **Shooting a Basketball** A basketball is shot at a 70° angle with the horizontal direction with an initial speed of 10 m/sec.

(a) Find the component form of the initial velocity.

(b) **Writing to Learn** Give an interpretation of the horizontal and vertical components of the velocity.

46. **Moving a Heavy Object** In a warehouse a box is being pushed up a 15° inclined plane with a force of 2.5 lb, as shown in the figure.

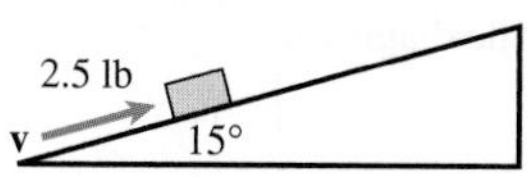

(a) Find the component form of the force.

(b) **Writing to Learn** Give an interpretation of the horizontal and vertical components of the force.

47. **Moving a Heavy Object** Suppose the box described in Exercise 46 is being towed up the inclined plane, as shown in the figure below. Find the force **w** needed in order for the component of the force parallel to the inclined plane to be 2.5 lb. Give the answer in component form.

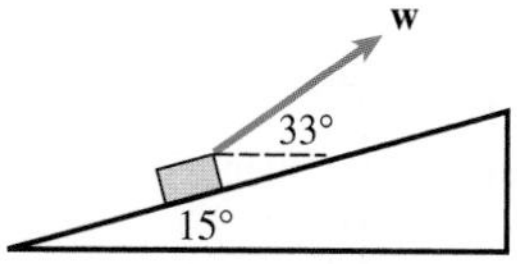

48. **Combining Forces** Juana and Diego Gonzales, ages six and four respectively, own a strong and stubborn puppy named Corporal. It is so hard to take Corporal for a walk that they devise a scheme to use two leashes. If Juana and Diego pull with forces of 23 lb and 27 lb at the angles shown in the figure, how hard is Corporal pulling if the puppy holds the children at a standstill?

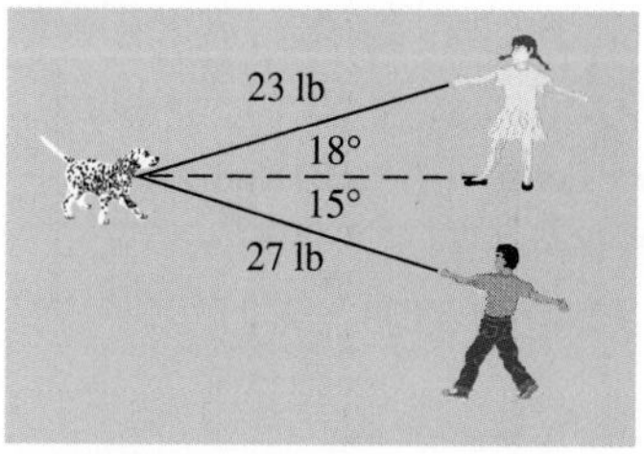

In Exercises 49 and 50, find the direction and magnitude of the resultant force.

49. **Combining Forces** A force of 50 lb acts on an object at an angle of 45°. A second force of 75 lb acts on the object at an angle of −30°.

50. **Combining Forces** Three forces with magnitudes 100, 50, and 80 lb act on an object at angles of 50°, 160°, and −20°, respectively.

51. **Navigation** A ship is heading due north at 12 mph. The current is flowing southwest at 4 mph. Find the actual bearing and speed of the ship.

52. **Navigation** A motor boat capable of 20 mph keeps the bow of the boat pointed straight across a mile-wide river. The current is flowing left to right at 8 mph. Find where the boat meets the opposite shore.

53. **Group Activity** A ship heads due south with the current flowing northwest. Two hours later the ship is 20 mi in the direction 30° west of south from the original starting point. Find the speed with no current of the ship and the rate of the current.

54. **Group Activity** Express each vector in component form to prove the following properties of vectors.

(a) $\mathbf{u} + \mathbf{v} = \mathbf{v} + \mathbf{u}$

(b) $(\mathbf{u} + \mathbf{v}) + \mathbf{w} = \mathbf{u} + (\mathbf{v} + \mathbf{w})$

(c) $\mathbf{u} + \mathbf{0} = \mathbf{u}$, where $\mathbf{0} = \langle 0, 0\rangle$

(d) $\mathbf{u} + (-\mathbf{u}) = \mathbf{0}$, where $-\langle a, b\rangle = \langle -a, -b\rangle$

(e) $a(\mathbf{u} + \mathbf{v}) = a\mathbf{u} + a\mathbf{v}$

(f) $(a + b)\mathbf{u} = a\mathbf{u} + b\mathbf{u}$

(g) $(ab)\mathbf{u} = a(b\mathbf{u})$

(h) $a\mathbf{0} = \mathbf{0}, 0\mathbf{u} = \mathbf{0}$

(i) $(1)\mathbf{u} = \mathbf{u}, (-1)\mathbf{u} = -\mathbf{u}$

(j) $|a\mathbf{u}| = |a|\,|\mathbf{u}|$

Standardized Test Questions

55. True or False If $\mathbf{u}$ is a unit vector, then $-\mathbf{u}$ is also a unit vector. Justify your answer.

56. True or False If $\mathbf{u}$ is a unit vector, then $1/\mathbf{u}$ is also a unit vector. Justify your answer.

In Exercises 57–60, you may use a graphing calculator to solve the problem.

57. Multiple Choice Which of the following is the magnitude of the vector $\langle 2, -1\rangle$?

(A) 1 (B) $\sqrt{3}$ (C) $\frac{\sqrt{5}}{5}$

(D) $\sqrt{5}$ (E) 5

58. Multiple Choice Let $\mathbf{u} = \langle -2, 3\rangle$ and $\mathbf{v} = \langle 4, -1\rangle$. Which of the following is equal to $\mathbf{u} - \mathbf{v}$?

(A) $\langle 6, -4\rangle$ (B) $\langle 2, 2\rangle$ (C) $\langle -2, 2\rangle$

(D) $\langle -6, 2\rangle$ (E) $\langle -6, 4\rangle$

59. Multiple Choice Which of the following represents the vector $\mathbf{v}$ shown in the figure below?

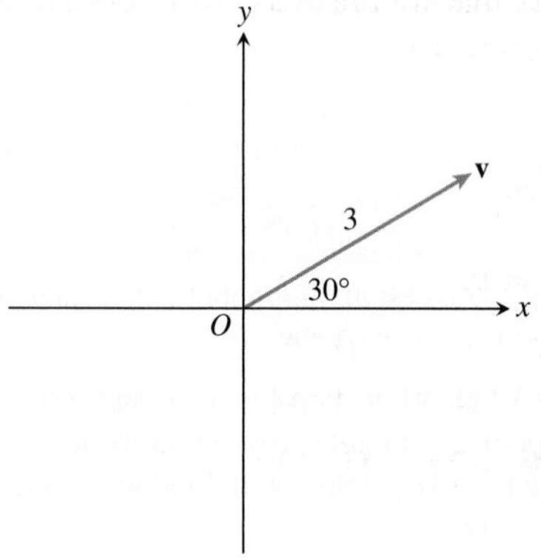

(A) $\langle 3\cos 30°, 3\sin 30°\rangle$ (B) $\langle 3\sin 30°, 3\cos 30°\rangle$

(C) $\langle 3\cos 60°, 3\sin 60°\rangle$ (D) $\langle \sqrt{3}\cos 30°, \sqrt{3}\sin 30°\rangle$

(E) $\langle \sqrt{3}\cos 30°, \sqrt{3}\sin 30°\rangle$

60. Multiple Choice Which of the following is a unit vector in the direction of $\mathbf{v} = -\mathbf{i} + 3\mathbf{j}$?

(A) $-\frac{1}{10}\mathbf{i} + \frac{3}{10}\mathbf{j}$ (B) $\frac{1}{10}\mathbf{i} - \frac{3}{10}\mathbf{j}$ (C) $-\frac{1}{\sqrt{10}}\mathbf{i} + \frac{3}{\sqrt{10}}\mathbf{j}$

(D) $\frac{1}{\sqrt{10}}\mathbf{i} - \frac{3}{\sqrt{10}}\mathbf{j}$ (E) $-\frac{1}{\sqrt{8}}\mathbf{i} + \frac{3}{\sqrt{8}}\mathbf{j}$

Explorations

61. Dividing a Line Segment in a Given Ratio Let A and B be two points in the plane, as shown in the figure.

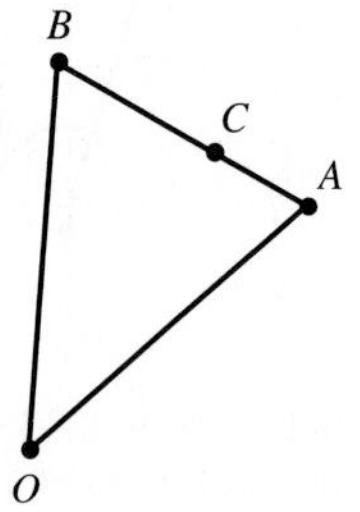

(a) Prove that $\overrightarrow{BA} = \overrightarrow{OA} - \overrightarrow{OB}$, where O is the origin.

(b) Let C be a point on the line segment BA which divides the segment in the ratio $x : y$ where $x + y = 1$. That is,

$$\frac{|\overrightarrow{BC}|}{|\overrightarrow{CA}|} = \frac{x}{y}.$$

Show that $\overrightarrow{OC} = x\overrightarrow{OA} + y\overrightarrow{OB}$.

62. Medians of a Triangle Perform the following steps to use vectors to prove that the medians of a triangle meet at a point O which divides each median in the ratio 1 : 2. M_1, M_2, and M_3 are midpoints of the sides of the triangle shown in the figure.

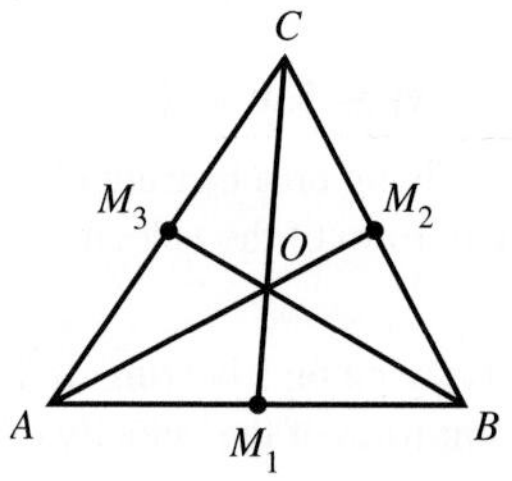

(a) Use Exercise 61 to prove that

$$\overrightarrow{OM_1} = \frac{1}{2}\overrightarrow{OA} + \frac{1}{2}\overrightarrow{OB}$$

$$\overrightarrow{OM_2} = \frac{1}{2}\overrightarrow{OC} + \frac{1}{2}\overrightarrow{OB}$$

$$\overrightarrow{OM_3} = \frac{1}{2}\overrightarrow{OA} + \frac{1}{2}\overrightarrow{OC}$$

(b) Prove that each of $2\overrightarrow{OM_1} + \overrightarrow{OC}$, $2\overrightarrow{OM_2} + \overrightarrow{OA}$, $2\overrightarrow{OM_3} + \overrightarrow{OB}$ is equal to $\overrightarrow{OA} + \overrightarrow{OB} + \overrightarrow{OC}$.

(c) **Writing to Learn** Explain why part (b) establishes the desired result.

Extending the Ideas

63. Vector Equation of a Line Let L be the line through the two points A and B. Prove that $C = (x, y)$ is on the line L if and only if $\overrightarrow{OC} = t\overrightarrow{OA} + (1 - t)\overrightarrow{OB}$, where t is a real number and O is the origin.

64. Connecting Vectors and Geometry Prove that the lines which join one vertex of a parallelogram to the midpoints of the opposite sides trisect the diagonal.

6.2 Dot Product of Vectors

What you'll learn about

- The Dot Product
- Angle Between Vectors
- Projecting One Vector onto Another
- Work

... and why

Vectors are used extensively in mathematics and science applications such as determining the net effect of several forces acting on an object and computing the work done by a force acting on an object.

The Dot Product

Vectors can be multiplied in two different ways, both of which are derived from their usefulness for solving problems in vector applications. The *cross product* (or *vector product* or *outer product*) results in a vector perpendicular to the plane of the two vectors being multiplied, which takes us into a third dimension and outside the scope of this chapter. The *dot product* (or *scalar product* or *inner product*) results in a scalar. In other words, the dot product of two vectors is not a vector but a real number! It is the important information conveyed by that number that makes the dot product so worthwhile, as you will see.

Now that you have some experience with vectors and arrows, we hope we won't confuse you if we occasionally resort to the common convention of using arrows to name the vectors they represent. For example, we might write "$\mathbf{u} = \overrightarrow{PQ}$" as a shorthand for "$\mathbf{u}$ is the vector represented by $\overrightarrow{PQ}$." This greatly simplifies the discussion of concepts like vector projection. Also, we will continue to use both vector notations, $\langle a, b \rangle$ and $a\mathbf{i} + b\mathbf{j}$, so you will get some practice with each.

DEFINITION Dot Product

The **dot product** or **inner product** of $\mathbf{u} = \langle u_1, u_2 \rangle$ and $\mathbf{v} = \langle v_1, v_2 \rangle$ is

$$\mathbf{u} \cdot \mathbf{v} = u_1 v_1 + u_2 v_2.$$

Dot Product and Standard Unit Vectors

$(u_1\mathbf{i} + u_2\mathbf{j}) \cdot (v_1\mathbf{i} + v_2\mathbf{j}) = u_1 v_1 + u_2 v_2$

Dot products have many important properties that we make use of in this section. We prove the first two and leave the rest for the exercises.

Properties of the Dot Product

Let $\mathbf{u}$, $\mathbf{v}$, and $\mathbf{w}$ be vectors and let c be a scalar.

1. $\mathbf{u} \cdot \mathbf{v} = \mathbf{v} \cdot \mathbf{u}$
2. $\mathbf{u} \cdot \mathbf{u} = |\mathbf{u}|^2$
3. $\mathbf{0} \cdot \mathbf{u} = 0$
4. $\mathbf{u} \cdot (\mathbf{v} + \mathbf{w}) = \mathbf{u} \cdot \mathbf{v} + \mathbf{u} \cdot \mathbf{w}$
   $(\mathbf{u} + \mathbf{v}) \cdot \mathbf{w} = \mathbf{u} \cdot \mathbf{w} + \mathbf{v} \cdot \mathbf{w}$
5. $(c\mathbf{u}) \cdot \mathbf{v} = \mathbf{u} \cdot (c\mathbf{v}) = c(\mathbf{u} \cdot \mathbf{v})$

Proof

Let $\mathbf{u} = \langle u_1, u_2 \rangle$ and $\mathbf{v} = \langle v_1, v_2 \rangle$.

Property 1

$\mathbf{u} \cdot \mathbf{v} = u_1 v_1 + u_2 v_2$ — Definition of $\mathbf{u} \cdot \mathbf{v}$

$= v_1 u_1 + v_2 u_2$ — Commutative property of real numbers

$= \mathbf{v} \cdot \mathbf{u}$ — Definition of $\mathbf{u} \cdot \mathbf{v}$

Property 2

$\mathbf{u} \cdot \mathbf{u} = {u_1}^2 + {u_2}^2$ — Definition of $\mathbf{u} \cdot \mathbf{v}$

$= \left(\sqrt{{u_1}^2 + {u_2}^2}\right)^2$

$= |\mathbf{u}|^2$ — Definition of $|\mathbf{u}|$

EXAMPLE 1 Finding Dot Products

Find each dot product.

(a) $\langle 3, 4\rangle \cdot \langle 5, 2\rangle$ **(b)** $\langle 1, -2\rangle \cdot \langle -4, 3\rangle$ **(c)** $(2\mathbf{i} - \mathbf{j}) \cdot (3\mathbf{i} - 5\mathbf{j})$

SOLUTION

(a) $\langle 3, 4\rangle \cdot \langle 5, 2\rangle = (3)(5) + (4)(2) = 23$

(b) $\langle 1, -2\rangle \cdot \langle -4, 3\rangle = (1)(-4) + (-2)(3) = -10$

(c) $(2\mathbf{i} - \mathbf{j}) \cdot (3\mathbf{i} - 5\mathbf{j}) = (2)(3) + (-1)(-5) = 11$

Now try Exercise 3.

Dot Products on Calculators

It is really a waste of time to compute a simple dot product of two-dimensional vectors using a calculator, but it can be done. Some calculators do vector operations outright, and others can do vector operations via matrices. If you have learned about matrix multiplication already, you will know why the matrix product $[u_1, u_2] \cdot \begin{bmatrix} v_1 \\ v_2 \end{bmatrix}$ yields the dot product $\langle u_1, u_2\rangle \cdot \langle v_1, v_2\rangle$ as a 1-by-1 matrix. (The same trick works with vectors of higher dimensions.) This book will cover matrix multiplication in Chapter 7.

Property 2 of the dot product gives us another way to find the length of a vector, as illustrated in Example 2.

EXAMPLE 2 Using Dot Product to Find Length

Use the dot product to find the length of the vector $\mathbf{u} = \langle 4, -3\rangle$.

SOLUTION It follows from property 2 that $|\mathbf{u}| = \sqrt{\mathbf{u} \cdot \mathbf{u}}$. Thus,

$$|\langle 4, -3\rangle| = \sqrt{\langle 4, -3\rangle \cdot \langle 4, -3\rangle} = \sqrt{(4)(4) + (-3)(-3)} = \sqrt{25} = 5.$$

Now try Exercise 9.

Angle Between Vectors

Let $\mathbf{u}$ and $\mathbf{v}$ be two nonzero vectors in standard position as shown in Figure 6.16. The **angle between u and v** is the angle θ, $0 \le \theta \le \pi$ or $0° \le \theta \le 180°$. The angle between any two nonzero vectors is the corresponding angle between their respective standard position representatives.

We can use the dot product to find the angle between nonzero vectors, as we prove in the next theorem.

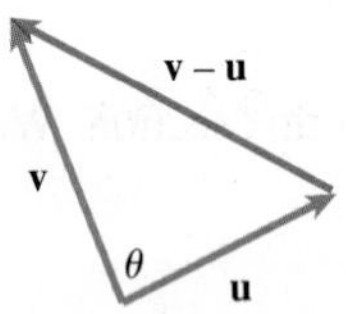

FIGURE 6.16 The angle θ between nonzero vectors **u** and **v**.

THEOREM Angle Between Two Vectors

If θ is the angle between the nonzero vectors **u** and **v**, then

$$\cos\theta = \frac{\mathbf{u} \cdot \mathbf{v}}{|\mathbf{u}|\,|\mathbf{v}|}$$

$$\text{and } \theta = \cos^{-1}\left(\frac{\mathbf{u} \cdot \mathbf{v}}{|\mathbf{u}|\,|\mathbf{v}|}\right)$$

Proof

We apply the Law of Cosines to the triangle determined by **u**, **v**, and **v** − **u** in Figure 6.16, and use the properties of the dot product.

$$\begin{aligned}
|\mathbf{v} - \mathbf{u}|^2 &= |\mathbf{u}|^2 + |\mathbf{v}|^2 - 2|\mathbf{u}|\,|\mathbf{v}|\cos\theta \\
(\mathbf{v} - \mathbf{u}) \cdot (\mathbf{v} - \mathbf{u}) &= |\mathbf{u}|^2 + |\mathbf{v}|^2 - 2|\mathbf{u}|\,|\mathbf{v}|\cos\theta \\
\mathbf{v} \cdot \mathbf{v} - \mathbf{v} \cdot \mathbf{u} - \mathbf{u} \cdot \mathbf{v} + \mathbf{u} \cdot \mathbf{u} &= |\mathbf{u}|^2 + |\mathbf{v}|^2 - 2|\mathbf{u}|\,|\mathbf{v}|\cos\theta \\
|\mathbf{v}|^2 - 2\mathbf{u} \cdot \mathbf{v} + |\mathbf{u}|^2 &= |\mathbf{u}|^2 + |\mathbf{v}|^2 - 2|\mathbf{u}|\,|\mathbf{v}|\cos\theta \\
-2\mathbf{u} \cdot \mathbf{v} &= -2|\mathbf{u}|\,|\mathbf{v}|\cos\theta \\
\cos\theta &= \frac{\mathbf{u} \cdot \mathbf{v}}{|\mathbf{u}|\,|\mathbf{v}|} \\
\theta &= \cos^{-1}\left(\frac{\mathbf{u} \cdot \mathbf{v}}{|\mathbf{u}|\,|\mathbf{v}|}\right)
\end{aligned}$$

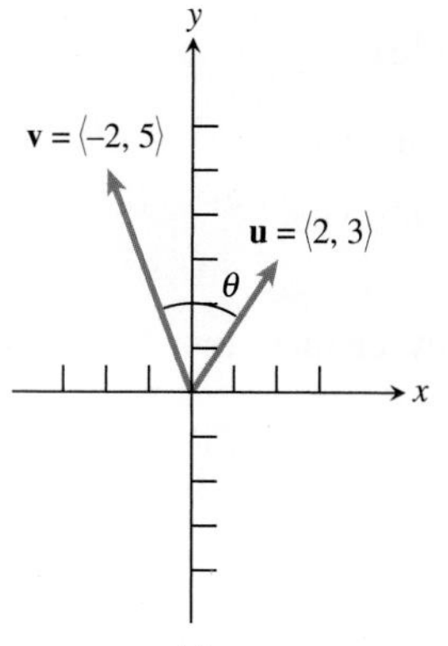

(a)

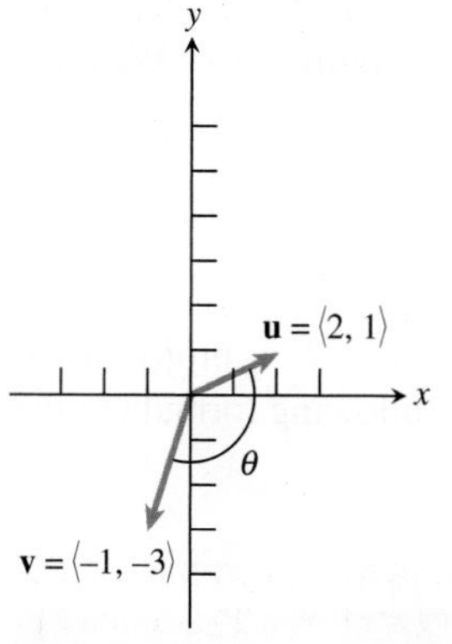

(b)

FIGURE 6.17 The vectors in (a) Example 3a and (b) Example 3b.

EXAMPLE 3 Finding the Angle Between Vectors

Use an algebraic method to find the angle between the vectors **u** and **v**. Use a calculator to approximate the exact answer when appropriate.

(a) $\mathbf{u} = \langle 2, 3 \rangle, \mathbf{v} = \langle -2, 5 \rangle$ **(b)** $\mathbf{u} = \langle 2, 1 \rangle, \mathbf{v} = \langle -1, -3 \rangle$

SOLUTION

(a) See Figure 6.17a. Using the Angle Between Two Vectors Theorem, we have

$$\cos\theta = \frac{\mathbf{u}\cdot\mathbf{v}}{|\mathbf{u}|\,|\mathbf{v}|} = \frac{\langle 2,3\rangle\cdot\langle -2,5\rangle}{|\langle 2,3\rangle|\,|\langle -2,5\rangle|} = \frac{11}{\sqrt{13}\sqrt{29}}.$$

So,

$$\theta = \cos^{-1}\left(\frac{11}{\sqrt{13}\sqrt{29}}\right) \approx 55.5°.$$

(b) See Figure 6.17b. Again using the Angle Between Two Vectors Theorem, we have

$$\cos\theta = \frac{\mathbf{u}\cdot\mathbf{v}}{|\mathbf{u}|\,|\mathbf{v}|} = \frac{\langle 2,1\rangle\cdot\langle -1,-3\rangle}{|\langle 2,1\rangle|\,|\langle -1,-3\rangle|} = \frac{-5}{\sqrt{5}\sqrt{10}} = \frac{-1}{\sqrt{2}}.$$

So,

$$\theta = \cos^{-1}\left(\frac{-1}{\sqrt{2}}\right) = 135°.$$

Now try Exercise 13.

If vectors **u** and **v** are perpendicular, that is, if the angle between them is 90°, then

$$\mathbf{u}\cdot\mathbf{v} = |\mathbf{u}|\,|\mathbf{v}|\cos 90° = 0$$

because $\cos 90° = 0$.

DEFINITION Orthogonal Vectors

The vectors **u** and **v** are **orthogonal** if and only if $\mathbf{u}\cdot\mathbf{v} = 0$.

The terms "perpendicular" and "orthogonal" almost mean the same thing. The zero vector has no direction angle, so technically speaking, the zero vector is not perpendicular to any vector. However, the zero vector is orthogonal to every vector. Except for this special case, orthogonal and perpendicular are the same.

EXAMPLE 4 Proving Vectors Are Orthogonal

Prove that the vectors $\mathbf{u} = \langle 2, 3 \rangle$ and $\mathbf{v} = \langle -6, 4 \rangle$ are orthogonal.

SOLUTION We must prove that their dot product is zero.

$$\mathbf{u}\cdot\mathbf{v} = \langle 2,3\rangle\cdot\langle -6,4\rangle = -12 + 12 = 0$$

The two vectors are orthogonal.

Now try Exercise 23.

EXPLORATION 1 Angles Inscribed in Semicircles

Figure 6.18 shows $\angle ABC$ inscribed in the upper half of the circle $x^2 + y^2 = a^2$.

1. For $a = 2$, find the component form of the vectors $\mathbf{u} = \overrightarrow{BA}$ and $\mathbf{v} = \overrightarrow{BC}$.
2. Find $\mathbf{u} \cdot \mathbf{v}$. What can you conclude about the angle θ between these two vectors?
3. Repeat parts 1 and 2 for arbitrary a.

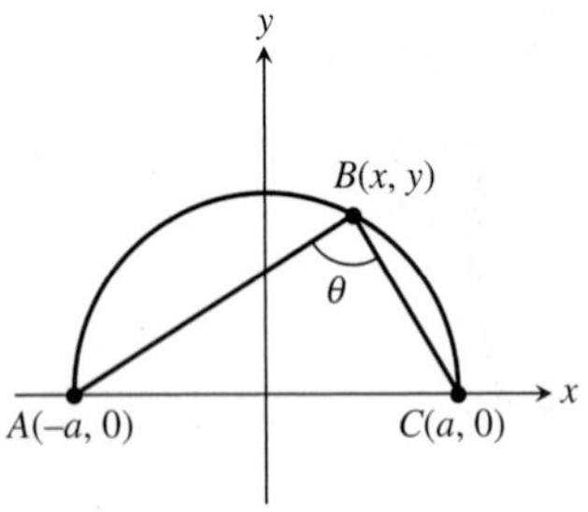

FIGURE 6.18 The angle $\angle ABC$ inscribed in the upper half of the circle $x^2 + y^2 = a^2$. (Exploration 1)

Projecting One Vector onto Another

The **vector projection** of $\mathbf{u} = \overrightarrow{PQ}$ onto a nonzero vector $\mathbf{v} = \overrightarrow{PS}$ is the vector $\overrightarrow{PR}$ determined by dropping a perpendicular from Q to the line PS (Figure 6.19). We have resolved $\mathbf{u}$ into components $\overrightarrow{PR}$ and $\overrightarrow{RQ}$

$$\mathbf{u} = \overrightarrow{PR} + \overrightarrow{RQ}$$

with $\overrightarrow{PR}$ and $\overrightarrow{RQ}$ perpendicular.

The standard notation for $\overrightarrow{PR}$, the vector projection of $\mathbf{u}$ onto $\mathbf{v}$, is $\overrightarrow{PR} = \text{proj}_{\mathbf{v}}\mathbf{u}$. With this notation, $\overrightarrow{RQ} = \mathbf{u} - \text{proj}_{\mathbf{v}}\mathbf{u}$. We ask you to establish the following formula in the exercises (see Exercise 58).

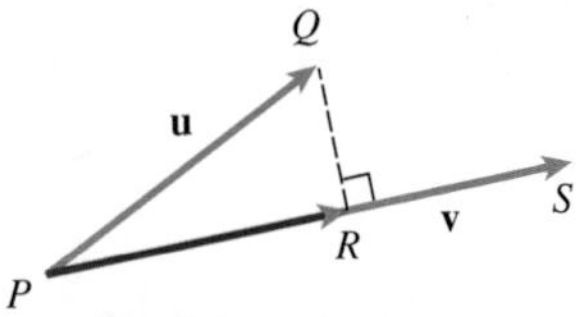

FIGURE 6.19 The vectors $\mathbf{u} = \overrightarrow{PQ}$, $\mathbf{v} = \overrightarrow{PS}$, and the vector projection of $\mathbf{u}$ onto $\mathbf{v}$, $\overrightarrow{PR} = \text{proj}_{\mathbf{v}}\mathbf{u}$.

Projection of the Vector u onto the Vector v

If $\mathbf{u}$ and $\mathbf{v}$ are nonzero vectors, the projection of $\mathbf{u}$ onto $\mathbf{v}$ is

$$\text{proj}_{\mathbf{v}}\mathbf{u} = \left(\frac{\mathbf{u} \cdot \mathbf{v}}{|\mathbf{v}|^2}\right)\mathbf{v}.$$

EXAMPLE 5 Decomposing a Vector into Perpendicular Components

Find the vector projection of $\mathbf{u} = \langle 6, 2 \rangle$ onto $\mathbf{v} = \langle 5, -5 \rangle$. Then write $\mathbf{u}$ as the sum of two orthogonal vectors, one of which is $\text{proj}_{\mathbf{v}}\mathbf{u}$.

SOLUTION We write $\mathbf{u} = \mathbf{u}_1 + \mathbf{u}_2$ where $\mathbf{u}_1 = \text{proj}_{\mathbf{v}}\mathbf{u}$ and $\mathbf{u}_2 = \mathbf{u} - \mathbf{u}_1$ (Figure 6.20).

$$\mathbf{u}_1 = \text{proj}_{\mathbf{v}}\mathbf{u} = \left(\frac{\mathbf{u} \cdot \mathbf{v}}{|\mathbf{v}|^2}\right)\mathbf{v} = \frac{20}{50}\langle 5, -5 \rangle = \langle 2, -2 \rangle$$

$$\mathbf{u}_2 = \mathbf{u} - \mathbf{u}_1 = \langle 6, 2 \rangle - \langle 2, -2 \rangle = \langle 4, 4 \rangle$$

Thus, $\mathbf{u}_1 + \mathbf{u}_2 = \langle 2, -2 \rangle + \langle 4, 4 \rangle = \langle 6, 2 \rangle = \mathbf{u}$.

Now try Exercise 25.

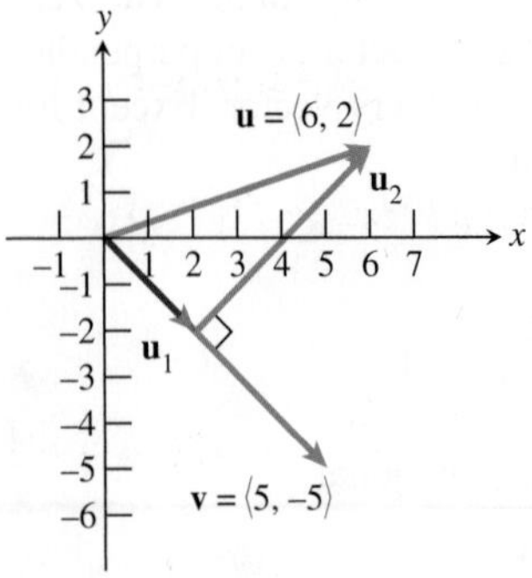

FIGURE 6.20 The vectors $\mathbf{u} = \langle 6, 2 \rangle$, $\mathbf{v} = \langle 5, -5 \rangle$, $\mathbf{u}_1 = \text{proj}_{\mathbf{v}}\mathbf{u}$, and $\mathbf{u}_2 = \mathbf{u} - \mathbf{u}_1$. (Example 5)

If $\mathbf{u}$ is a force, then $\text{proj}_{\mathbf{v}}\mathbf{u}$ represents the effective force in the direction of $\mathbf{v}$ (Figure 6.21).

We can use vector projections to determine the amount of force required in problem situations like Example 6.

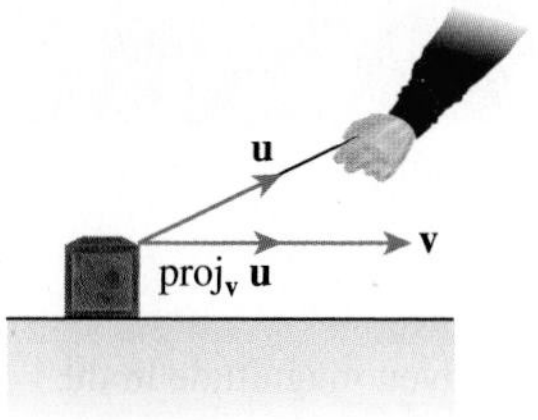

FIGURE 6.21 If we pull on a box with force **u**, the effective force in the direction of **v** is $\text{proj}_{\mathbf{v}}\mathbf{u}$, the vector projection of **u** onto **v**.

EXAMPLE 6 Finding a Force

Juan is sitting on a sled on the side of a hill inclined at 45°. The combined weight of Juan and the sled is 140 lb. What force is required for Rafaela to keep the sled from sliding down the hill? (See Figure 6.22.)

SOLUTION We can represent the force due to gravity as $\mathbf{F} = -140\mathbf{j}$ because gravity acts vertically downward. We can represent the side of the hill with the vector

$$\mathbf{v} = (\cos 45°)\mathbf{i} + (\sin 45°)\mathbf{j} = \frac{\sqrt{2}}{2}\mathbf{i} + \frac{\sqrt{2}}{2}\mathbf{j}.$$

The force required to keep the sled from sliding down the hill is

$$\mathbf{F}_1 = \text{proj}_{\mathbf{v}}\mathbf{F} = \left(\frac{\mathbf{F}\cdot\mathbf{v}}{|\mathbf{v}|^2}\right)\mathbf{v} = (\mathbf{F}\cdot\mathbf{v})\mathbf{v}$$

because $|\mathbf{v}| = 1$. So,

$$\mathbf{F}_1 = (\mathbf{F}\cdot\mathbf{v})\mathbf{v} = (-140)\left(\frac{\sqrt{2}}{2}\right)\mathbf{v} = -70\sqrt{2}(\mathbf{i} + \mathbf{j}).$$

The magnitude of the force that Rafaela must exert to keep the sled from sliding down the hill is $70\sqrt{2} \approx 99$ lb. Now try Exercise 45.

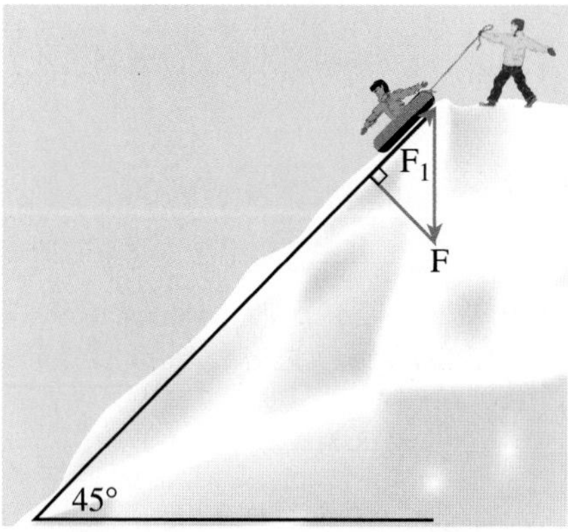

FIGURE 6.22 The sled in Example 6.

Work

If **F** is a constant force whose direction is the same as the direction of $\overrightarrow{AB}$, then the **work** W done by **F** in moving an object from A to B is

$$W = |\mathbf{F}|\,|\overrightarrow{AB}|.$$

If **F** is a constant force in any direction, then the **work** W done by **F** in moving an object from A to B is

$$\begin{aligned} W &= \mathbf{F}\cdot\overrightarrow{AB} \\ &= |\mathbf{F}|\,|\overrightarrow{AB}|\cos\theta \end{aligned}$$

where θ is the angle between **F** and $\overrightarrow{AB}$. Except for the sign, the work is the magnitude of the effective force in the direction of $\overrightarrow{AB}$ times $\overrightarrow{AB}$.

Units for Work

Work is usually measured in foot-pounds or newton-meters. One newton-meter is commonly referred to as one joule (1 J).

EXAMPLE 7 Finding Work

Find the work done by a 10-lb force acting in the direction $\langle 1, 2\rangle$ in moving an object 3 ft from $(0, 0)$ to $(3, 0)$.

SOLUTION The force **F** has magnitude 10 lb and acts in the direction $\langle 1, 2\rangle$, so

$$\mathbf{F} = 10\frac{\langle 1, 2\rangle}{|\langle 1, 2\rangle|} = \frac{10}{\sqrt{5}}\langle 1, 2\rangle.$$

The direction of motion is from $A = (0, 0)$ to $B = (3, 0)$, so $\overrightarrow{AB} = \langle 3, 0\rangle$. Thus, the work done by the force is

$$\mathbf{F}\cdot\overrightarrow{AB} = \frac{10}{\sqrt{5}}\langle 1, 2\rangle\cdot\langle 3, 0\rangle = \frac{30}{\sqrt{5}} \approx 13.42 \text{ ft-lb}.$$

Now try Exercise 53.

QUICK REVIEW 6.2 *(For help, go to Section 6.1.)*

Exercise numbers with a gray background indicate problems that the authors have designed to be solved *without a calculator*.

In Exercises 1–4, find $|\mathbf{u}|$.

1. $\mathbf{u} = \langle 2, -3 \rangle$ **2.** $\mathbf{u} = -3\mathbf{i} - 4\mathbf{j}$

3. $\mathbf{u} = \cos 35°\,\mathbf{i} + \sin 35°\,\mathbf{j}$

4. $\mathbf{u} = 2(\cos 75°\,\mathbf{i} + \sin 75°\,\mathbf{j})$

In Exercises 5–8, the points A and B lie on the circle $x^2 + y^2 = 4$. Find the component form of the vector $\overrightarrow{AB}$.

5. $A = (-2, 0), B = (1, \sqrt{3})$

6. $A = (2, 0), B = (1, \sqrt{3})$

7. $A = (2, 0), B = (1, -\sqrt{3})$

8. $A = (-2, 0), B = (1, -\sqrt{3})$

In Exercises 9 and 10, find a vector **u** with the given magnitude in the direction of **v**.

9. $|\mathbf{u}| = 2, \mathbf{v} = \langle 2, 3 \rangle$

10. $|\mathbf{u}| = 3, \mathbf{v} = -4\mathbf{i} + 3\mathbf{j}$

SECTION 6.2 Exercises

In Exercises 1–8, find the dot product of **u** and **v**.

1. $\mathbf{u} = \langle 5, 3 \rangle, \mathbf{v} = \langle 12, 4 \rangle$

2. $\mathbf{u} = \langle -5, 2 \rangle, \mathbf{v} = \langle 8, 13 \rangle$

3. $\mathbf{u} = \langle 4, 5 \rangle, \mathbf{v} = \langle -3, -7 \rangle$

4. $\mathbf{u} = \langle -2, 7 \rangle, \mathbf{v} = \langle -5, -8 \rangle$

5. $\mathbf{u} = -4\mathbf{i} - 9\mathbf{j}, \mathbf{v} = -3\mathbf{i} - 2\mathbf{j}$

6. $\mathbf{u} = 2\mathbf{i} - 4\mathbf{j}, \mathbf{v} = -8\mathbf{i} + 7\mathbf{j}$

7. $\mathbf{u} = 7\mathbf{i}, \mathbf{v} = -2\mathbf{i} + 5\mathbf{j}$

8. $\mathbf{u} = 4\mathbf{i} - 11\mathbf{j}, \mathbf{v} = -3\mathbf{j}$

In Exercises 9–12, use the dot product to find $|\mathbf{u}|$.

9. $\mathbf{u} = \langle 5, -12 \rangle$ **10.** $\mathbf{u} = \langle -8, 15 \rangle$

11. $\mathbf{u} = -4\mathbf{i}$ **12.** $\mathbf{u} = 3\mathbf{j}$

In Exercises 13–22 use an algebraic method to find the angle between the vectors. Use a calculator to approximate exact answers when appropriate.

13. $\mathbf{u} = \langle -4, -3 \rangle, \mathbf{v} = \langle -1, 5 \rangle$

14. $\mathbf{u} = \langle 2, -2 \rangle, \mathbf{v} = \langle -3, -3 \rangle$

15. $\mathbf{u} = \langle 2, 3 \rangle, \mathbf{v} = \langle -3, 5 \rangle$

16. $\mathbf{u} = \langle 5, 2 \rangle, \mathbf{v} = \langle -6, -1 \rangle$

17. $\mathbf{u} = 3\mathbf{i} - 3\mathbf{j}, \mathbf{v} = -2\mathbf{i} + 2\sqrt{3}\mathbf{j}$

18. $\mathbf{u} = -2\mathbf{i}, \mathbf{v} = 5\mathbf{j}$

19. $\mathbf{u} = \left(2\cos\frac{\pi}{4}\right)\mathbf{i} + \left(2\sin\frac{\pi}{4}\right)\mathbf{j}, \mathbf{v} = \left(\cos\frac{3\pi}{2}\right)\mathbf{i} + \left(\sin\frac{3\pi}{2}\right)\mathbf{j}$

20. $\mathbf{u} = \left(\cos\frac{\pi}{3}\right)\mathbf{i} + \left(\sin\frac{\pi}{3}\right)\mathbf{j}, \mathbf{v} = \left(3\cos\frac{5\pi}{6}\right)\mathbf{i} + \left(3\sin\frac{5\pi}{6}\right)\mathbf{j}$

21.

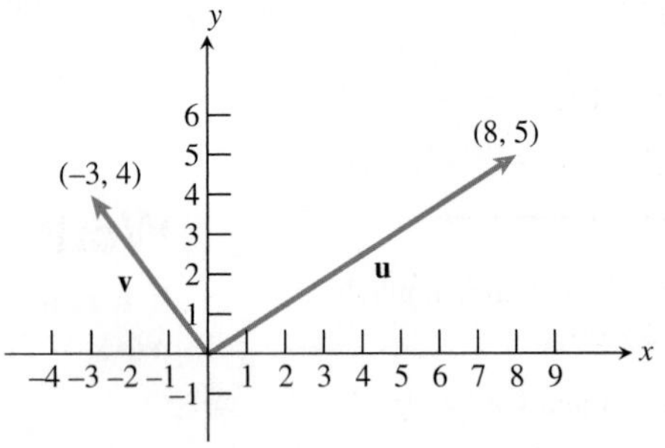

22.

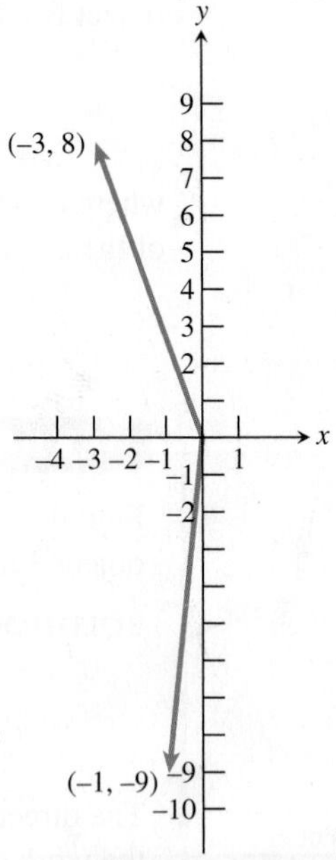

In Exercises 23–24, prove that the vectors **u** and **v** are orthogonal.

23. $\mathbf{u} = \langle 2, 3 \rangle, \mathbf{v} = \langle 3/2, -1 \rangle$

24. $\mathbf{u} = \langle -4, -1 \rangle, \mathbf{v} = \langle 1, -4 \rangle$

In Exercises 25–28, find the vector projection of **u** onto **v**. Then write **u** as a sum of two orthogonal vectors, one of which is $\text{proj}_{\mathbf{v}}\mathbf{u}$.

25. $\mathbf{u} = \langle -8, 3\rangle, \mathbf{v} = \langle -6, -2\rangle$

26. $\mathbf{u} = \langle 3, -7\rangle, \mathbf{v} = \langle -2, -6\rangle$

27. $\mathbf{u} = \langle 8, 5\rangle, \mathbf{v} = \langle -9, -2\rangle$

28. $\mathbf{u} = \langle -2, 8\rangle, \mathbf{v} = \langle 9, -3\rangle$

In Exercises 29 and 30, find the interior angles of the triangle with given vertices.

29. $(-4, 5), (1, 10), (3, 1)$ **30.** $(-4, 1), (1, -6), (5, -1)$

In Exercises 31 and 32, find $\mathbf{u}\cdot\mathbf{v}$ satisfying the given conditions where θ is the angle between **u** and **v**.

31. $\theta = 150°, |\mathbf{u}| = 3, |\mathbf{v}| = 8$

32. $\theta = \frac{\pi}{3}, |\mathbf{u}| = 12, |\mathbf{v}| = 40$

In Exercises 33–38, determine whether the vectors **u** and **v** are parallel, orthogonal, or neither.

33. $\mathbf{u} = \langle 5, 3\rangle, \mathbf{v} = \left\langle -\frac{10}{4}, -\frac{3}{2}\right\rangle$

34. $\mathbf{u} = \langle 2, 5\rangle, \mathbf{v} = \left\langle \frac{10}{3}, \frac{4}{3}\right\rangle$

35. $\mathbf{u} = \langle 15, -12\rangle, \mathbf{v} = \langle -4, 5\rangle$

36. $\mathbf{u} = \langle 5, -6\rangle, \mathbf{v} = \langle -12, -10\rangle$

37. $\mathbf{u} = \langle -3, 4\rangle, \mathbf{v} = \langle 20, 15\rangle$

38. $\mathbf{u} = \langle 2, -7\rangle, \mathbf{v} = \langle -4, 14\rangle$

In Exercises 39–42, find

(a) the x-intercept A and y-intercept B of the line.

(b) the coordinates of the point P so that $\overrightarrow{AP}$ is perpendicular to the line and $|\overrightarrow{AP}| = 1$. (There are two answers.)

39. $3x - 4y = 12$ **40.** $-2x + 5y = 10$

41. $3x - 7y = 21$ **42.** $x + 2y = 6$

In Exercises 43 and 44, find the vector(s) **v** satisfying the given conditions.

43. $\mathbf{u} = \langle 2, 3\rangle, \mathbf{u}\cdot\mathbf{v} = 10, |\mathbf{v}|^2 = 17$

44. $\mathbf{u} = \langle -2, 5\rangle, \mathbf{u}\cdot\mathbf{v} = -11, |\mathbf{v}|^2 = 10$

45. Sliding down a Hill Ojemba is sitting on a sled on the side of a hill inclined at 60°. The combined weight of Ojemba and the sled is 160 lb. What is the magnitude of the force required for Mandisa to keep the sled from sliding down the hill?

46. Revisiting Example 6 Suppose Juan and Rafaela switch positions. The combined weight of Rafaela and the sled is 125 lb. What is the magnitude of the force required for Juan to keep the sled from sliding down the hill?

47. Braking Force A 2000-lb car is parked on a street that makes an angle of 12° with the horizontal (see figure).

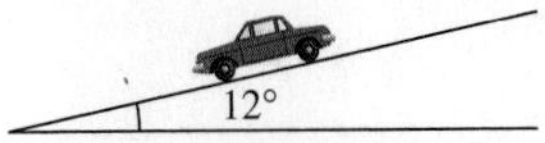

(a) Find the magnitude of the force required to keep the car from rolling down the hill.

(b) Find the force perpendicular to the street.

48. Effective Force A 60-lb force **F** that makes an angle of 25° with an inclined plane is pulling a box up the plane. The inclined plane makes an 18° angle with the horizontal (see figure). What is the magnitude of the effective force pulling the box up the plane?

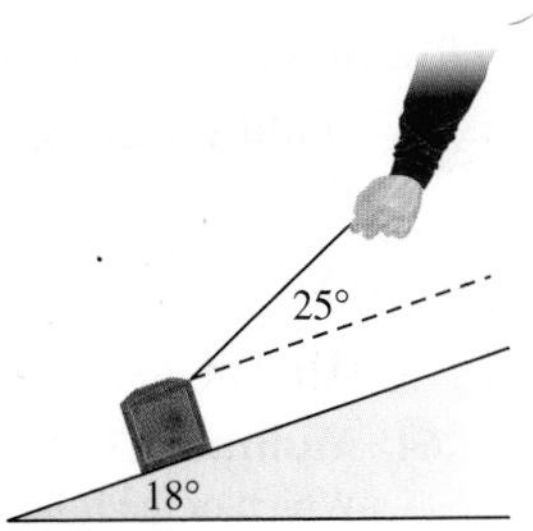

49. Work Find the work done lifting a 2600-lb car 5.5 ft.

50. Work Find the work done lifting a 100-lb bag of potatoes 3 ft.

51. Work Find the work done by a force **F** of 12 N acting in the direction $\langle 1, 2\rangle$ in moving an object 4 m from $(0, 0)$ to $(4, 0)$.

52. Work Find the work done by a force **F** of 24 N acting in the direction $\langle 4, 5\rangle$ in moving an object 5 m from $(0, 0)$ to $(5, 0)$.

53. Work Find the work done by a force **F** of 30 N acting in the direction $\langle 2, 2\rangle$ in moving an object 3 m from $(0, 0)$ to a point in the first quadrant along the line $y = (1/2)x$.

54. Work Find the work done by a force **F** of 50 N acting in the direction $\langle 2, 3\rangle$ in moving an object 5 m from $(0, 0)$ to a point in the first quadrant along the line $y = x$.

55. Work The angle between a 200-lb force **F** and $\overrightarrow{AB} = 2\mathbf{i} + 3\mathbf{j}$ is 30°. Find the work done by **F** in moving an object from A to B.

56. Work The angle between a 75-lb force **F** and $\overrightarrow{AB}$ is 60°, where $A = (-1, 1)$ and $B = (4, 3)$. Find the work done by **F** in moving an object from A to B.

57. Properties of the Dot Product Let **u**, **v**, and **w** be vectors and let c be a scalar. Use the component form of vectors to prove the following properties.

(a) $\mathbf{0}\cdot\mathbf{u} = 0$

(b) $\mathbf{u}\cdot(\mathbf{v} + \mathbf{w}) = \mathbf{u}\cdot\mathbf{v} + \mathbf{u}\cdot\mathbf{w}$

(c) $(\mathbf{u} + \mathbf{v})\cdot\mathbf{w} = \mathbf{u}\cdot\mathbf{w} + \mathbf{v}\cdot\mathbf{w}$

(d) $(c\mathbf{u})\cdot\mathbf{v} = \mathbf{u}\cdot(c\mathbf{v}) = c(\mathbf{u}\cdot\mathbf{v})$

58. Group Activity Projection of a Vector Let **u** and **v** be nonzero vectors. Prove that

(a) $\text{proj}_{\mathbf{v}}\mathbf{u} = \left(\frac{\mathbf{u}\cdot\mathbf{v}}{|\mathbf{v}|^2}\right)\mathbf{v}$ **(b)** $(\mathbf{u} - \text{proj}_{\mathbf{v}}\mathbf{u})\cdot(\text{proj}_{\mathbf{v}}\mathbf{u}) = 0$

59. Group Activity Connecting Geometry and Vectors Prove that the sum of the squares of the diagonals of a parallelogram is equal to the sum of the squares of its sides.

60. If **u** is any vector, prove that we can write **u** as

$$\mathbf{u} = (\mathbf{u}\cdot\mathbf{i})\mathbf{i} + (\mathbf{u}\cdot\mathbf{j})\mathbf{j}.$$

Standardized Test Questions

61. True or False If $\mathbf{u}\cdot\mathbf{v} = 0$, then **u** and **v** are perpendicular. Justify your answer.

62. True or False If **u** is a unit vector, then $\mathbf{u}\cdot\mathbf{u} = 1$. Justify your answer.

In Exercises 63–66, solve the problem without using a calculator.

63. Multiple Choice Let $\mathbf{u} = \langle 1, 1\rangle$ and $\mathbf{v} = \langle -1, 1\rangle$. Which of the following is the angle between **u** and **v**?

(A) 0° **(B)** 45° **(C)** 60°

(D) 90° **(E)** 135°

64. Multiple Choice Let $\mathbf{u} = \langle 4, -5\rangle$ and $\mathbf{v} = \langle -2, -3\rangle$. Which of the following is equal to $\mathbf{u} \cdot \mathbf{v}$?

(A) -23 **(B)** -7 **(C)** 7

(D) 23 **(E)** $\sqrt{7}$

65. Multiple Choice Let $\mathbf{u} = \langle 3/2, -3/2\rangle$ and $\mathbf{v} = \langle 2, 0\rangle$. Which of the following is equal to $\text{proj}_{\mathbf{v}}\mathbf{u}$?

(A) $\langle 3/2, 0\rangle$ **(B)** $\langle 3, 0\rangle$ **(C)** $\langle -3/2, 0\rangle$

(D) $\langle 3/2, 3/2\rangle$ **(E)** $\langle -3/2, -3/2\rangle$

66. Multiple Choice Which of the following vectors describes a 5-lb force acting in the direction of $\mathbf{u} = \langle -1, 1\rangle$?

(A) $5\langle -1, 1\rangle$ **(B)** $\dfrac{5}{\sqrt{2}}\langle -1, 1\rangle$ **(C)** $5\langle 1, -1\rangle$

(D) $\dfrac{5}{\sqrt{2}}\langle 1, -1\rangle$ **(E)** $\dfrac{5}{2}\langle -1, 1\rangle$

Explorations

67. Distance from a Point to a Line Consider the line L with equation $2x + 5y = 10$ and the point $P = (3, 7)$.

(a) Verify that $A = (0, 2)$ and $B = (5, 0)$ are the y- and x-intercepts of L.

(b) Find $\mathbf{w}_1 = \text{proj}_{\overrightarrow{AB}}\overrightarrow{AP}$ and $\mathbf{w}_2 = \overrightarrow{AP} - \text{proj}_{\overrightarrow{AB}}\overrightarrow{AP}$.

(c) Writing to Learn Explain why $|\mathbf{w}_2|$ is the distance from P to L. What is this distance?

(d) Find a formula for the distance of any point $P = (x_0, y_0)$ to L.

(e) Find a formula for the distance of any point $P = (x_0, y_0)$ to the line $ax + by = c$.

Extending the Ideas

68. Writing to Learn Let $\mathbf{w} = (\cos t)\,\mathbf{u} + (\sin t)\,\mathbf{v}$ where **u** and **v** are not parallel.

(a) Can the vector **w** be parallel to the vector **u**? Explain.

(b) Can the vector **w** be parallel to the vector **v**? Explain.

(c) Can the vector **w** be parallel to the vector $\mathbf{u} + \mathbf{v}$? Explain.

69. If the vectors **u** and **v** are not parallel, prove that

$$a\mathbf{u} + b\mathbf{v} = c\mathbf{u} + d\mathbf{v} \Rightarrow a = c,\ b = d.$$

6.3 Parametric Equations and Motion

What you'll learn about

- Parametric Equations
- Parametric Curves
- Eliminating the Parameter
- Lines and Line Segments
- Simulating Motion with a Grapher

... and why

These topics can be used to model the path of an object such as a baseball or a golf ball.

Parametric Equations

When vectors are used to solve problems of motion, the components of the vector vary as functions of time. For example, in the two-dimensional vector $\langle x, y \rangle$ we might have $x = f(t)$ and $y = g(t)$. This is equivalent to using *parametric equations* to write x and y in terms of the *parameter t*, something we have done twice before in this textbook using those very terms (Sections 1.5 and 2.1). You might also recall using the parametric equations $x = \cos(t)$ and $y = \sin(t)$ to define the Unit Circle in Section 4.3. (See Figure 6.23.) Although the study of vector-valued functions is beyond the scope of this course, we can get many of the same insights, and solve many of the same problems, by simply taking a closer look at parametric equations and parametrically defined curves. This section is devoted to taking that closer look.

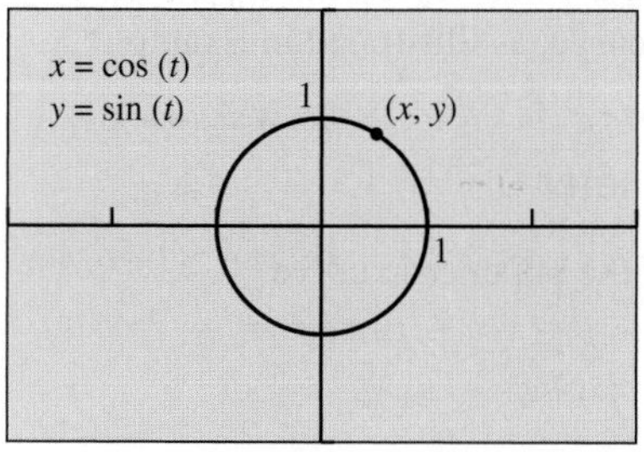

[–3, 3] by [–2, 2]

FIGURE 6.23 In Section 4.3 we defined the Unit Circle parametrically by expressing x and y as trigonometric functions of the parameter t.

Parametric Curves

In this section we study the graphs of *parametric equations* and investigate motion of objects that can be modeled with parametric equations.

DEFINITION Parametric Curve, Parametric Equations

The graph of the ordered pairs (x, y) where

$$x = f(t), \quad y = g(t)$$

are functions defined on an interval I of t values is a **parametric curve**. The equations are **parametric equations** for the curve, the variable t is a **parameter**, and I is the **parameter interval**.

When we give parametric equations and a parameter interval for a curve, we have **parametrized** the curve. A **parametrization** of a curve consists of the parametric equations and the interval of t values. Sometimes parametric equations are used by companies in their design plans. It is then easier for the company to make larger and smaller objects efficiently by just changing the parameter t.

Graphs of parametric equations can be obtained using parametric mode on a grapher.

EXAMPLE 1 Graphing Parametric Equations

For the given parameter interval, graph the parametric equations

$$x = t^2 - 2, \quad y = 3t.$$

(a) $-3 \leq t \leq 1$

(b) $-2 \leq t \leq 3$

(c) $-3 \leq t \leq 3$

SOLUTION In each case, set Tmin equal to the left endpoint of the interval and Tmax equal to the right endpoint of the interval. Figure 6.24 shows a graph of the parametric equations for each parameter interval. The corresponding relations are different because the parameter intervals are different. **Now try Exercise 7.**

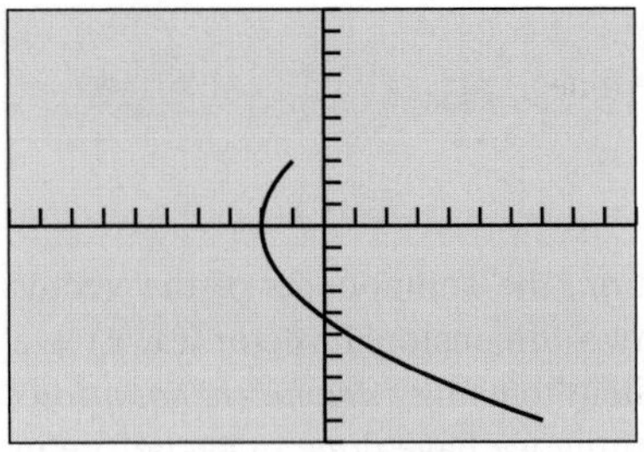

[–10, 10] by [–10, 10]
(a)

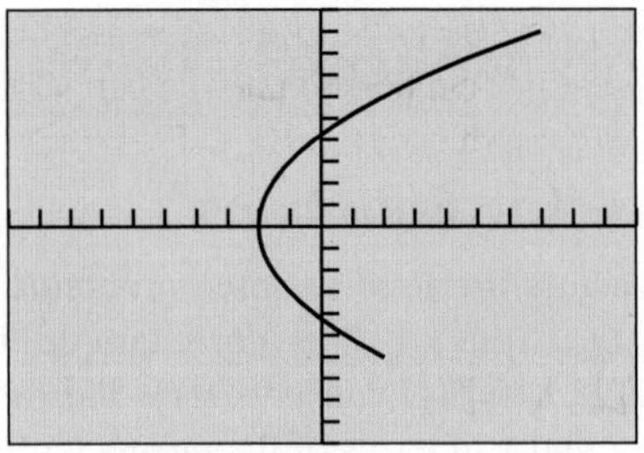

[–10, 10] by [–10, 10]
(b)

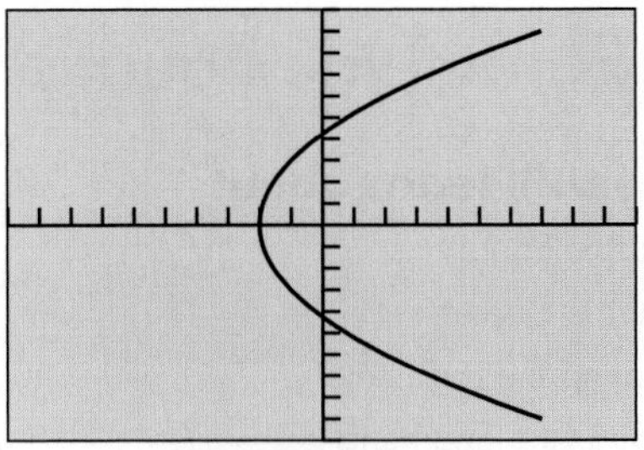

[–10, 10] by [–10, 10]
(c)

FIGURE 6.24 Three different relations defined parametrically. (Example 1)

Eliminating the Parameter

When a curve is defined parametrically it is sometimes possible to *eliminate the parameter* and obtain a rectangular equation in x and y that represents the curve. This often helps us identify the graph of the parametric curve, as illustrated in Example 2.

EXAMPLE 2 Eliminating the Parameter

Eliminate the parameter and identify the graph of the parametric curve

$$x = 1 - 2t, \quad y = 2 - t, \quad -\infty < t < \infty.$$

SOLUTION We solve the first equation for t:

$$x = 1 - 2t$$
$$2t = 1 - x$$
$$t = \frac{1}{2}(1 - x)$$

Then we substitute this expression for t into the second equation:

$$y = 2 - t$$
$$y = 2 - \frac{1}{2}(1 - x)$$
$$y = 0.5x + 1.5$$

The graph of the equation $y = 0.5x + 1.5$ is a line with slope 0.5 and y-intercept 1.5 (Figure 6.25). **Now try Exercise 11.**

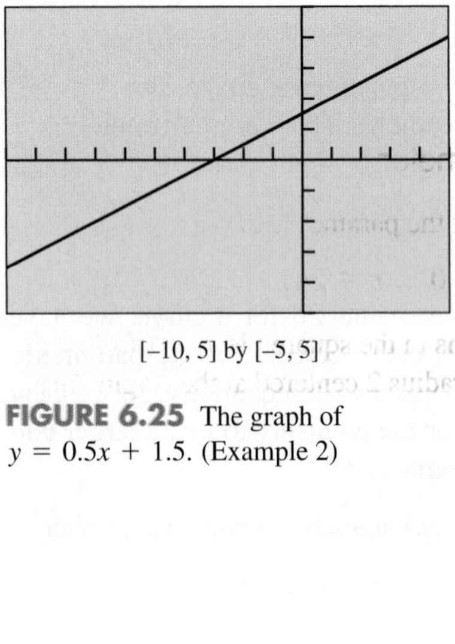

[–10, 5] by [–5, 5]

FIGURE 6.25 The graph of $y = 0.5x + 1.5$. (Example 2)

EXPLORATION 1 Graphing the Curve of Example 2 Parametrically

1. Use the parametric mode of your grapher to reproduce the graph in Figure 6.25. Use -2 for Tmin and 5.5 for Tmax.
2. Prove that the point $(17, 10)$ is on the graph of $y = 0.5x + 1.5$. Find the corresponding value of t that produces this point.
3. Repeat part 2 for the point $(-23, -10)$.
4. Assume that (a, b) is on the graph of $y = 0.5x + 1.5$. Find the corresponding value of t that produces this point.
5. How do you have to choose Tmin and Tmax so that the graph in Figure 6.25 fills the window?

If we do not specify a parameter interval for the parametric equations $x = f(t)$, $y = g(t)$, it is understood that the parameter t can take on all values that produce real numbers for x and y. We use this agreement in Example 3.

EXAMPLE 3 Eliminating the Parameter

Eliminate the parameter and identify the graph of the parametric curve

$$x = t^2 - 2, \quad y = 3t.$$

SOLUTION Here t can be any real number. We solve the second equation for t, obtaining $t = y/3$, and substitute this value for y into the first equation.

$$x = t^2 - 2$$
$$x = \left(\frac{y}{3}\right)^2 - 2$$
$$x = \frac{y^2}{9} - 2$$
$$y^2 = 9(x + 2)$$

Figure 6.24c shows what the graph of these parametric equations looks like. In Chapter 8 we will call this a parabola that opens to the right. Interchanging x and y, we can identify this graph as the inverse of the graph of the parabola $x^2 = 9(y + 2)$.

Now try Exercise 15.

Parabolas

The inverse of a parabola that opens up or down is a parabola that opens left or right. We will investigate these curves in more detail in Chapter 8.

EXAMPLE 4 Eliminating the Parameter

Eliminate the parameter and identify the graph of the parametric curve

$$x = 2\cos t, \quad y = 2\sin t, \quad 0 \le t \le 2\pi.$$

SOLUTION The graph of the parametric equations in the square viewing window of Figure 6.26 suggests that the graph is a circle of radius 2 centered at the origin. We confirm this result algebraically.

$$\begin{aligned} x^2 + y^2 &= 4\cos^2 t + 4\sin^2 t \\ &= 4(\cos^2 t + \sin^2 t) \\ &= 4(1) \qquad \cos^2 t + \sin^2 t = 1 \\ &= 4 \end{aligned}$$

The graph of $x^2 + y^2 = 4$ is a circle of radius 2 centered at the origin. Increasing the length of the interval $0 \le t \le 2\pi$ will cause the grapher to trace all or part of the circle more than once. Decreasing the length of the interval will cause the grapher to only draw a portion of the complete circle. Try it!

Now try Exercise 23.

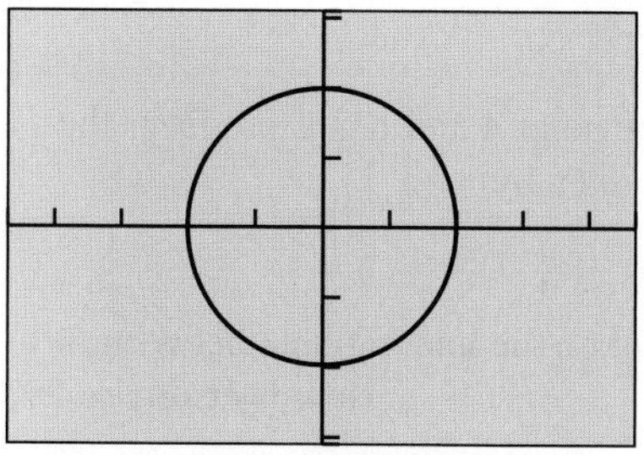

[–4.7, 4.7] by [–3.1, 3.1]

FIGURE 6.26 The graph of the circle of Example 4.

In Exercise 65, you will find parametric equations for any circle in the plane.

Lines and Line Segments

We can use vectors to help us find parametric equations for a line, as illustrated in Example 5.

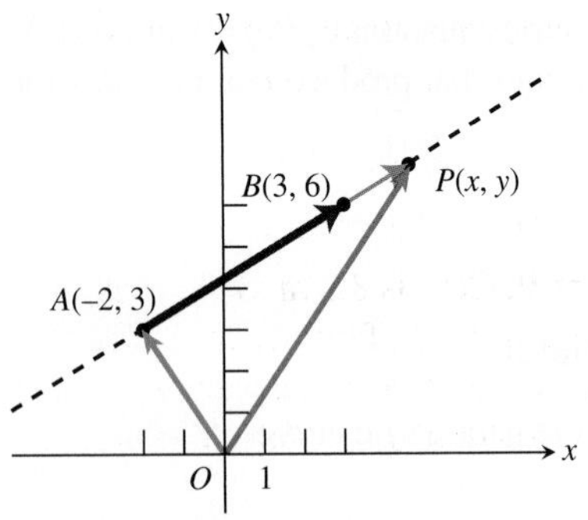

FIGURE 6.27 Example 5 uses vectors to construct a parametrization of the line through A and B.

Lines in 3-Space

You might wonder why anybody would bother to find parametric equations for a line, since a linear equation of the form $ax + by + c = 0$ does the job nicely and is not that hard to find. One reason is that the familiar point-slope method does not work in 3-space, where the graph of the linear equation $ax + by + cz + d = 0$ is a plane, not a line. Happily, the vector/parametric approach of Example 5 works just as well in 3-space as it does in the plane.

EXAMPLE 5 Finding Parametric Equations for a Line

Find a parametrization of the line through the points $A = (-2, 3)$ and $B = (3, 6)$.

SOLUTION Let $P(x, y)$ be an arbitrary point on the line through A and B. As you can see from Figure 6.27, the vector $\overrightarrow{OP}$ is the tail-to-head vector sum of $\overrightarrow{OA}$ and $\overrightarrow{AP}$. You can also see that $\overrightarrow{AP}$ is a scalar multiple of $\overrightarrow{AB}$.

If we let the scalar be t, we have

$$\overrightarrow{OP} = \overrightarrow{OA} + \overrightarrow{AP}$$
$$\overrightarrow{OP} = \overrightarrow{OA} + t \cdot \overrightarrow{AB}$$
$$\langle x, y \rangle = \langle -2, 3 \rangle + t\langle 3 - (-2), 6 - 3 \rangle$$
$$\langle x, y \rangle = \langle -2, 3 \rangle + t\langle 5, 3 \rangle$$
$$\langle x, y \rangle = \langle -2 + 5t, 3 + 3t \rangle$$

This vector equation is equivalent to the parametric equations $x = -2 + 5t$ and $y = 3 + 3t$. Together with the parameter interval $(-\infty, \infty)$, these equations define the line.

We can confirm our work numerically as follows: If $t = 0$, then $x = -2$ and $y = 3$, which gives the point A. Similarly, if $t = 1$, then $x = 3$ and $y = 6$, which gives the point B. **Now try Exercise 27.**

The fact that $t = 0$ yields point A and $t = 1$ yields point B in Example 5 is no accident, as a little reflection on Figure 6.27 and the vector equation $\overrightarrow{OP} = \overrightarrow{OA} + t \cdot \overrightarrow{AB}$ should suggest. We use this fact in Example 6.

EXAMPLE 6 Finding Parametric Equations for a Line Segment

Find a parametrization of the line segment with endpoints $A = (-2, 3)$ and $B = (3, 6)$.

SOLUTION In Example 5 we found parametric equations for the line through A and B:

$$x = -2 + 5t, \quad y = 3 + 3t$$

We also saw in Example 5 that $t = 0$ produces the point A and $t = 1$ produces the point B. A parametrization of the line segment is given by

$$x = -2 + 5t, \quad y = 3 + 3t, \quad 0 \le t \le 1.$$

As t varies between 0 and 1 we pick up every point on the line segment between A and B. **Now try Exercise 29.**

Simulating Motion with a Grapher

Example 7 illustrates several ways to simulate motion along a horizontal line using parametric equations. We use the variable t for the parameter to represent time.

EXAMPLE 7 Simulating Horizontal Motion

Gary walks along a horizontal line (think of it as a number line) with the coordinate of his position (in meters) given by

$$s = -0.1(t^3 - 20t^2 + 110t - 85)$$

where $0 \le t \le 12$. Use parametric equations and a grapher to simulate his motion. Estimate the times when Gary changes direction.

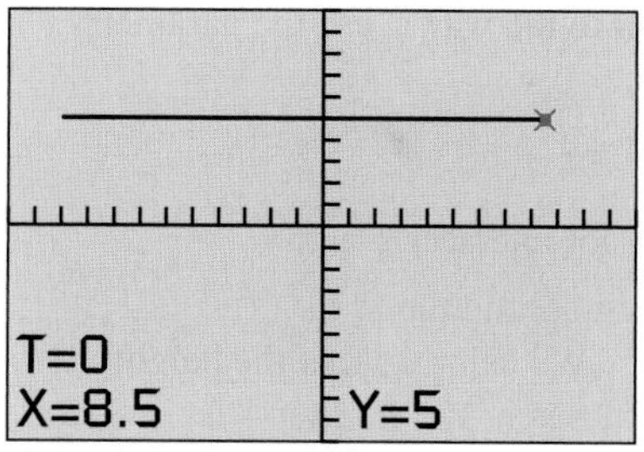

Start, $t = 0$

(a)

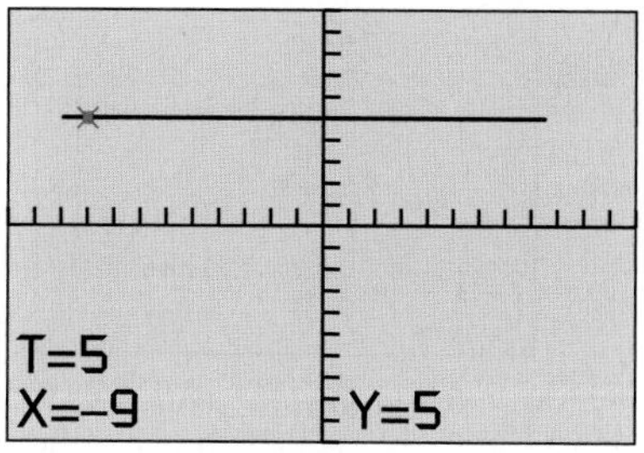

5 sec later, $t = 5$

(b)

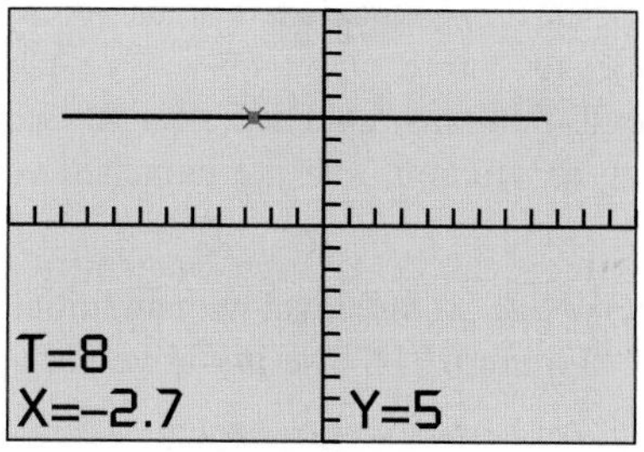

3 sec after that, $t = 8$

(c)

FIGURE 6.28 Three views of the graph C_1: $x_1 = -0.1(t^3 - 20t^2 + 110t - 85)$, $y_1 = 5, 0 \le t \le 12$ in the $[-12, 12]$ by $[-10, 10]$ viewing window. (Example 7)

SOLUTION We arbitrarily choose the horizontal line $y = 5$ to display this motion. The graph C_1 of the parametric equations,

$$C_1: x_1 = -0.1(t^3 - 20t^2 + 110t - 85), y_1 = 5, 0 \le t \le 12,$$

simulates the motion. His position at any time t is given by the point $(x_1(t), 5)$.

Using TRACE in Figure 6.28 we see that when $t = 0$, Gary is 8.5 m to the right of the y-axis at the point $(8.5, 5)$, and that he initially moves left. Five seconds later he is 9 m to the left of the y-axis at the point $(-9, 5)$. And after 8 sec he is only 2.7 m to the left of the y-axis. Gary must have changed direction during the walk. The motion of the trace cursor simulates Gary's motion.

A variation in $y(t)$,

$$C_2: x_2 = -0.1(t^3 - 20t^2 + 110t - 85), y_2 = -t, 0 \le t \le 12,$$

can be used to help visualize where Gary changes direction. The graph C_2 shown in Figure 6.29 suggests that Gary reverses his direction at 3.9 sec and again at 9.5 sec after beginning his walk. **Now try Exercise 37.**

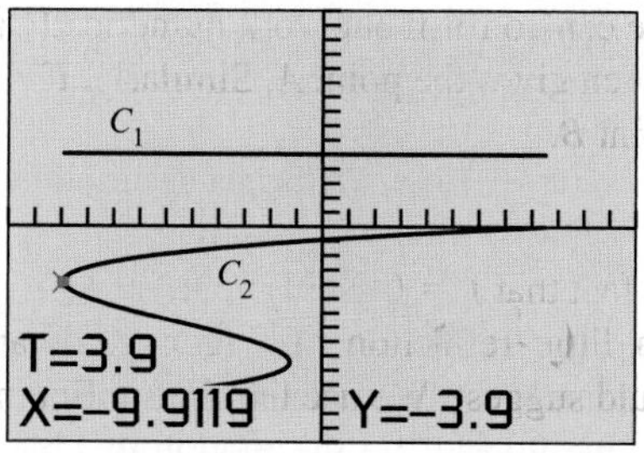

$[-12, 12]$ by $[-15, 15]$

(a)

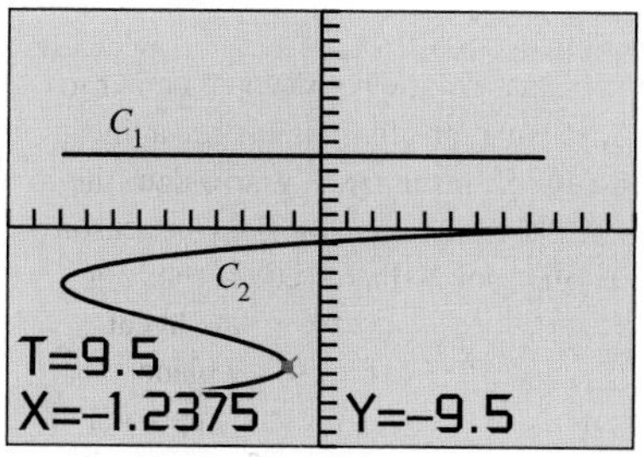

$[-12, 12]$ by $[-15, 15]$

(b)

FIGURE 6.29 Two views of the graph $C_1: x_1 = -0.1(t^3 - 20t^2 + 110t - 85)$, $y_1 = 5, 0 \le t \le 12$ and the graph C_2: $x_2 = -0.1(t^3 - 20t^2 + 110t - 85)$, $y_2 = -t$, $0 \le t \le 12$ in the $[-12, 12]$ by $[-15, 15]$ viewing window. (Example 7)

Example 8 solves a projectile-motion problem. Parametric equations are used in two ways: to find a graph of the modeling equation and to simulate the motion of the projectile.

Grapher Note

The equation $y_2 = t$ is typically used in the parametric equations for the graph C_2 in Figure 6.29. We have chosen $y_2 = -t$ to get two curves in Figure 6.29 that do not overlap. Also notice that the y-coordinates of C_1 are constant $(y_1 = 5)$, and that the y-coordinates of C_2 vary with time t $(y_2 = -t)$.

EXAMPLE 8 Simulating Projectile Motion

A distress flare is launched from a ship's deck 75 ft above the water with an initial vertical velocity of 76 ft/sec. Graph the flare's height against time, give the height of the flare above water at each time, and simulate the flare's motion for each length of time.

(a) 1 sec **(b)** 2 sec **(c)** 4 sec **(d)** 5 sec

SOLUTION An equation that models the flare's height above the water t seconds after launch is

$$y = -16t^2 + 76t + 75.$$

A graph of the flare's height against time can be found using the parametric equations

$$x_1 = t, \quad y_1 = -16t^2 + 76t + 75.$$

(continued)

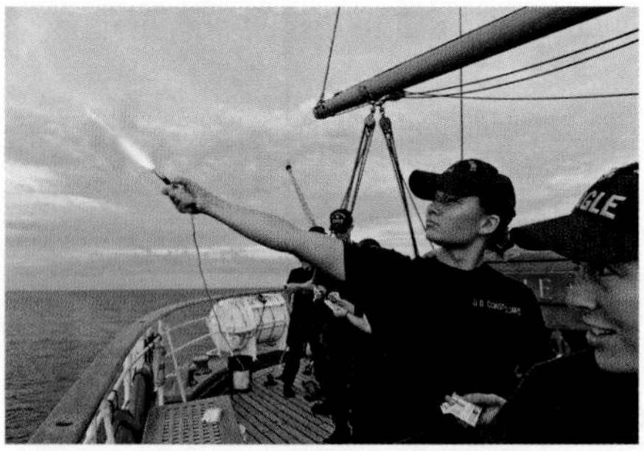

To simulate the flare's flight straight up and its fall to the water, use the parametric equations

$$x_2 = 5.5, \quad y_2 = -16t^2 + 76t + 75.$$

(We chose $x_2 = 5.5$ so that the two graphs would not intersect.)

Figure 6.30 shows the two graphs in simultaneous graphing mode for (a) $0 \le t \le 1$, (b) $0 \le t \le 2$, (c) $0 \le t \le 4$, and (d) $0 \le t \le 5$. We can read that the height of the flare above the water after 1 sec is 135 ft, after 2 sec is 163 ft, after 4 sec is 123 ft, and after 5 sec is 55 ft.

Now try Exercise 39.

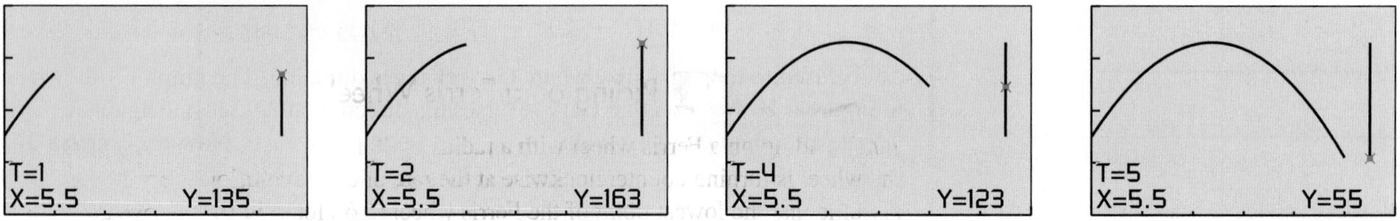

[0, 6] by [0, 200] (a)   [0, 6] by [0, 200] (b)   [0, 6] by [0, 200] (c)   [0, 6] by [0, 200] (d)

FIGURE 6.30 Simultaneous graphing of $x_1 = t$, $y_1 = -16t^2 + 76t + 75$ (height against time) and $x_2 = 5.5$, $y_2 = -16t^2 + 76t + 75$ (the actual path of the flare). (Example 8)

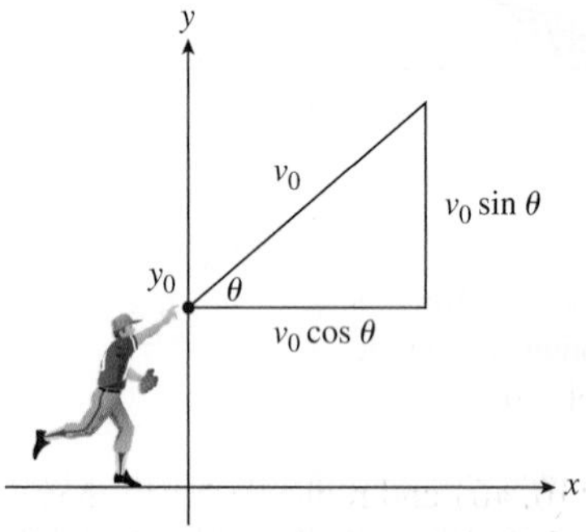

FIGURE 6.31 Throwing a baseball.

In Example 8 we modeled the motion of a projectile that was launched straight up. Now we investigate the motion of objects, ignoring air friction, that are launched at angles other than 90° with the horizontal.

Suppose that a baseball is thrown from a point y_0 feet above ground level with an initial speed of v_0 ft/sec at an angle θ with the horizontal (Figure 6.31). The initial velocity can be represented by the vector

$$\mathbf{v} = \langle v_0 \cos \theta, v_0 \sin \theta \rangle.$$

The path of the object is modeled by the parametric equations

$$x = (v_0 \cos \theta)t, \quad y = -16t^2 + (v_0 \sin \theta)t + y_0.$$

The x-component is simply

$$\text{distance} = (x\text{-component of initial velocity}) \times \text{time}.$$

The y-component is the familiar vertical projectile-motion equation using the y-component of initial velocity.

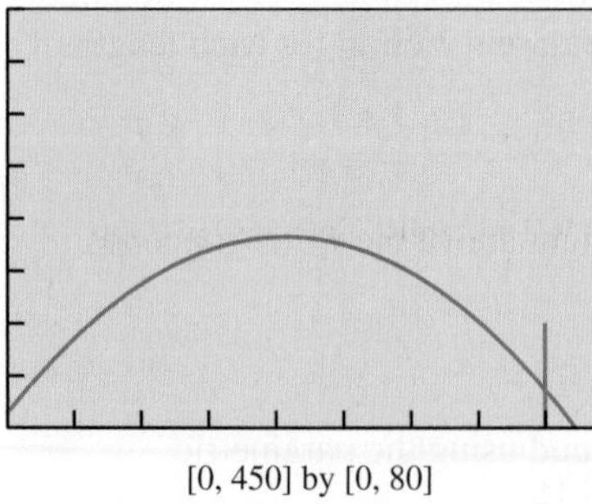

[0, 450] by [0, 80]

FIGURE 6.32 The fence and path of the baseball in Example 9. See Exploration 2 for ways to draw the wall.

EXAMPLE 9 Hitting a Baseball

Kevin hits a baseball at 3 ft above the ground with an initial speed of 150 ft/sec at an angle of 18° with the horizontal. Will the ball clear a 20-ft wall that is 400 ft away?

SOLUTION The path of the ball is modeled by the parametric equations

$$x = (150 \cos 18°)t, \quad y = -16t^2 + (150 \sin 18°)t + 3.$$

A little experimentation will show that the ball will reach the fence in less than 3 sec. Figure 6.32 shows a graph of the path of the ball using the parameter interval $0 \le t \le 3$ and the 20-ft wall. The ball does not clear the wall.

Now try Exercise 43.

EXPLORATION 2 Extending Example 9

1. If your grapher has a line segment feature, draw the fence in Example 9.
2. Describe the graph of the parametric equations

$$x = 400, \quad y = 20(t/3), \quad 0 \le t \le 3.$$

3. Repeat Example 9 for the angles 19°, 20°, 21°, and 22°.

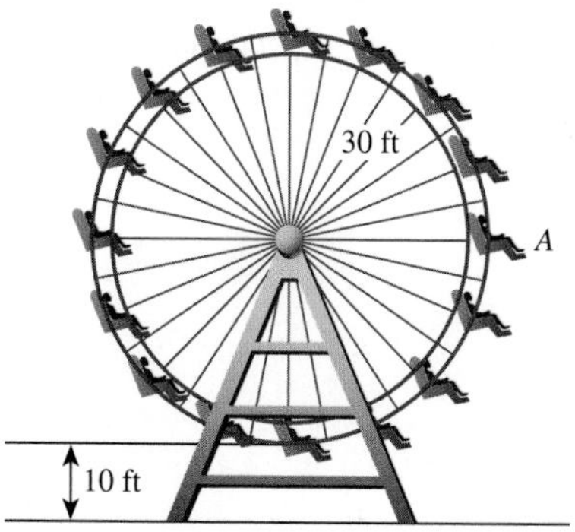

FIGURE 6.33 The Ferris wheel of Example 10.

In Example 10 we see how to write parametric equations for position on a moving Ferris wheel, using time t as the parameter.

EXAMPLE 10 Riding on a Ferris Wheel

Jane is riding on a Ferris wheel with a radius of 30 ft. As we view it in Figure 6.33, the wheel is turning counterclockwise at the rate of one revolution every 30 sec. Assume that the lowest point of the Ferris wheel (6 o'clock) is 10 ft above the ground and that Jane is at the point marked A (3 o'clock) at time $t = 0$. Find parametric equations to model Jane's path and use them to find Jane's position 22 sec into the ride.

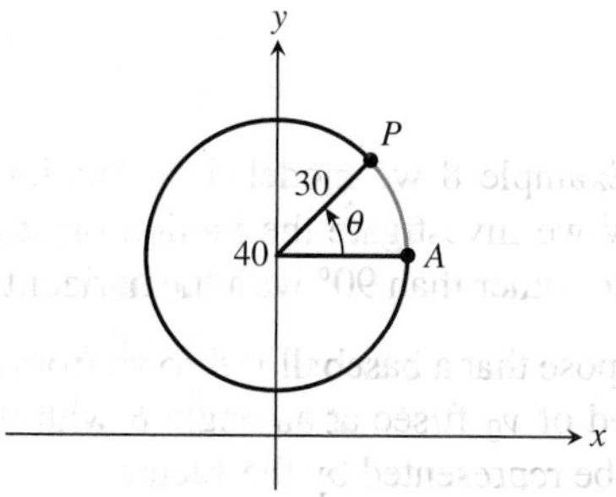

FIGURE 6.34 A model for the Ferris wheel of Example 10.

SOLUTION Figure 6.34 shows a circle with center $(0, 40)$ and radius 30 that models the Ferris wheel. The parametric equations for this circle in terms of the parameter θ, the central angle of the circle determined by the arc AP, are

$$x = 30 \cos \theta, \quad y = 40 + 30 \sin \theta, \quad 0 \le \theta \le 2\pi.$$

To take into account the rate at which the wheel is turning we must describe θ as a function of time t in seconds. The wheel is turning at the rate of 2π rad every 30 sec, or $2\pi/30 = \pi/15$ rad/sec. So, $\theta = (\pi/15)t$. Thus, parametric equations that model Jane's path are given by

$$x = 30 \cos\left(\frac{\pi}{15}t\right), \quad y = 40 + 30 \sin\left(\frac{\pi}{15}t\right), \quad t \ge 0.$$

We substitute $t = 22$ into the parametric equations to find Jane's position at that time:

$$x = 30 \cos\left(\frac{\pi}{15} \cdot 22\right) \qquad y = 40 + 30 \sin\left(\frac{\pi}{15} \cdot 22\right)$$

$$x \approx -3.14 \qquad y \approx 10.16$$

After riding for 22 sec, Jane is approximately 10.16 ft above the ground and approximately 3.14 ft to the left of the y-axis, using the coordinate system of Figure 6.34.

Now try Exercise 51.

QUICK REVIEW 6.3 *(For help, go to Sections P.2, P.4, 1.3, 4.1, and 6.1.)*

Exercise numbers with a gray background indicate problems that the authors have designed to be solved *without a calculator*.

In Exercises 1 and 2, find the component form of the vectors **(a)** $\overrightarrow{OA}$, **(b)** $\overrightarrow{OB}$, and **(c)** $\overrightarrow{AB}$ where O is the origin.

1. $A = (-3, -2), B = (4, 6)$

2. $A = (-1, 3), B = (4, -3)$

In Exercises 3 and 4, write an equation in point-slope form for the line through the two points.

3. $(-3, -2), (4, 6)$ **4.** $(-1, 3), (4, -3)$

In Exercises 5 and 6, find and graph the two functions defined implicitly by each given relation.

5. $y^2 = 8x$ **6.** $y^2 = -5x$

In Exercises 7 and 8, write an equation for the circle with given center and radius.

7. $(0, 0), 2$ **8.** $(-2, 5), 3$

In Exercises 9 and 10, a wheel with radius r spins at the given rate. Find the angular velocity in radians per second.

9. $r = 13$ in., 600 rpm **10.** $r = 12$ in., 700 rpm

SECTION 6.3 Exercises

In Exercises 1–4, match the parametric equation with its graph. Identify the viewing window that seems to have been used.

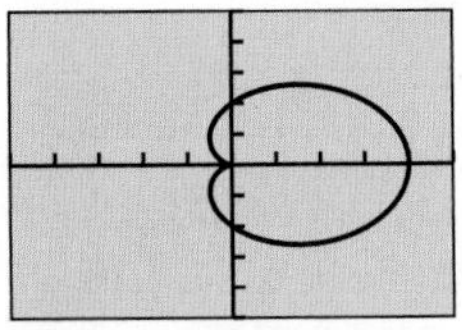
(a)

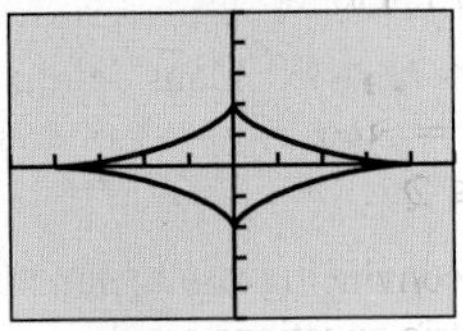
(b)

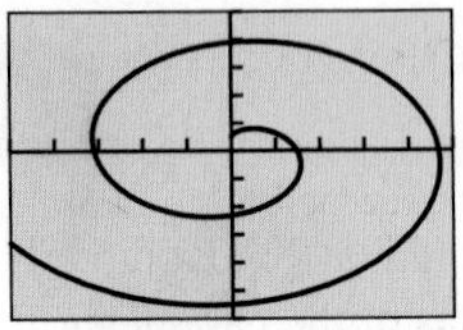
(c)

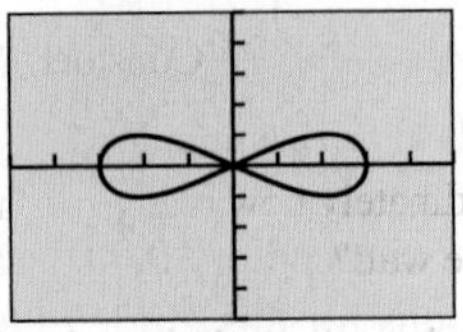
(d)

1. $x = 4\cos^3 t, y = 2\sin^3 t$

2. $x = 3\cos t, y = \sin 2t$

3. $x = 2\cos t + 2\cos^2 t, y = 2\sin t + \sin 2t$

4. $x = \sin t - t\cos t, y = \cos t + t\sin t$

In Exercises 5 and 6, **(a)** complete the table for the parametric equations and **(b)** plot the corresponding points.

5. $x = t + 2, y = 1 + 3/t$

t	-2	-1	0	1	2
x					
y					

6. $x = \cos t, y = \sin t$

t	0	$\pi/2$	π	$3\pi/2$	2π
x					
y					

In Exercises 7–10, graph the parametric equations $x = 3 - t^2, y = 2t$, in the specified parameter interval. Use the standard viewing window.

7. $0 \le t \le 10$ **8.** $-10 \le t \le 0$

9. $-3 \le t \le 3$ **10.** $-2 \le t \le 4$

In Exercises 11–26, use an algebraic method to eliminate the parameter and identify the graph of the parametric curve. Use a grapher to support your answer.

11. $x = 1 + t, y = t$ **12.** $x = 2 - 3t, y = 5 + t$

13. $x = 2t - 3, y = 9 - 4t, \quad 3 \le t \le 5$

14. $x = 5 - 3t, y = 2 + t, \quad -1 \le t \le 3$

15. $x = t^2, y = t + 1$ (*Hint:* Eliminate t and solve for x in terms of y.)

16. $x = t, y = t^2 - 3$

17. $x = t, y = t^3 - 2t + 3$

18. $x = 2t^2 - 1, y = t$ (*Hint:* Eliminate t and solve for x in terms of y.)

19. $x = 4 - t^2, y = t$ (*Hint:* Eliminate t and solve for x in terms of y.)

20. $x = 0.5t, y = 2t^3 - 3, -2 \le t \le 2$

21. $x = t - 3, y = 2/t, -5 \le t \le 5$

22. $x = t + 2, y = 4/t, t \ge 2$

23. $x = 5\cos t, y = 5\sin t$

24. $x = 4\cos t, y = 4\sin t$

25. $x = 2\sin t, y = 2\cos t, 0 \le t \le 3\pi/2$

26. $x = 3\cos t, y = 3\sin t, 0 \le t \le \pi$

In Exercises 27–32 find a parametrization for the curve.

27. The line through the points $(-2, 5)$ and $(4, 2)$

28. The line through the points $(-3, -3)$ and $(5, 1)$

29. The line segment with endpoints $(3, 4)$ and $(6, -3)$

30. The line segment with endpoints $(5, 2)$ and $(-2, -4)$

31. The circle with center $(5, 2)$ and radius 3

32. The circle with center $(-2, -4)$ and radius 2

Exercises 33–36 refer to the graph of the parametric equations

$$x = 2 - |t|, \quad y = t - 0.5, \quad -3 \le t \le 3$$

given below. Find the values of the parameter t that produces the graph in the indicated quadrant.

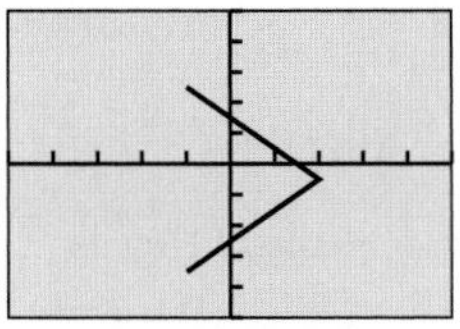

[–5, 5] by [–5, 5]

33. Quadrant I

34. Quadrant II

35. Quadrant III

36. Quadrant IV

37. Simulating a Foot Race Ben can sprint at the rate of 24 ft/sec. Jerry sprints at 20 ft/sec. Ben gives Jerry a 10-ft head start. The parametric equations can be used to model a race.

$$x_1 = 20t, \qquad y_1 = 3$$
$$x_2 = 24t - 10, \quad y_2 = 5$$

(a) Find a viewing window to simulate a 100-yd dash. Graph simultaneously with t starting at $t = 0$ and Tstep $= 0.05$.

(b) Who is ahead after 3 sec and by how much?

38. Capture the Flag Two opposing players in "Capture the Flag" are 100 ft apart. On a signal, they run to capture a flag that is on the ground midway between them. The faster runner, however, hesitates for 0.1 sec. The following parametric equations model the race to the flag:

$$x_1 = 10(t - 0.1), \quad y_1 = 3$$
$$x_2 = 100 - 9t, \qquad y_2 = 3$$

(a) Simulate the game in a $[0, 100]$ by $[-1, 10]$ viewing window with t starting at 0. Graph simultaneously.

(b) Who captures the flag and by how many feet?

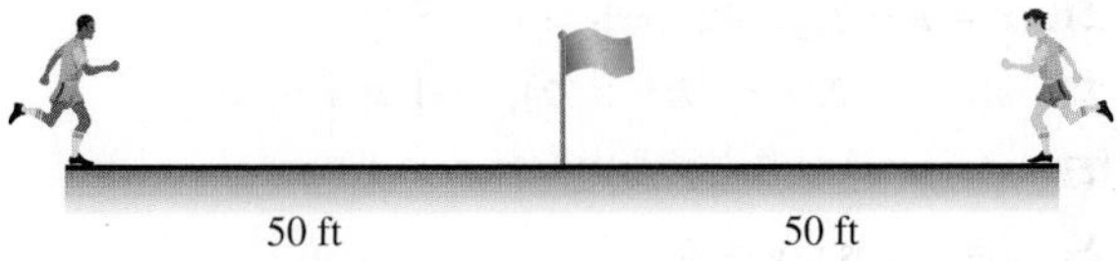

39. Famine Relief Air Drop A relief agency drops food containers from an airplane on a war-torn famine area. The drop was made from an altitude of 1000 ft above ground level.

(a) Use an equation to model the height of the containers (during free fall) as a function of time t.

(b) Use parametric mode to simulate the drop during the first 6 sec.

(c) After 4 sec of free fall, parachutes open. How many feet above the ground are the food containers when the parachutes open?

40. Height of a Pop-up A baseball is hit straight up from a height of 5 ft with an initial velocity of 80 ft/sec.

(a) Write an equation that models the height of the ball as a function of time t.

(b) Use parametric mode to simulate the pop-up.

(c) Use parametric mode to graph height against time. [*Hint:* Let $x(t) = t$.]

(d) How high is the ball after 4 sec?

(e) What is the maximum height of the ball? How many seconds does it take to reach its maximum height?

41. The complete graph of the parametric equations $x = 2 \cos t$, $y = 2 \sin t$ is the circle of radius 2 centered at the origin. Find an interval of values for t so that the graph is the given portion of the circle.

(a) The portion in the first quadrant

(b) The portion above the x-axis

(c) The portion to the left of the y-axis

42. Writing to Learn Consider the two pairs of parametric equations $x = 3 \cos t$, $y = 3 \sin t$ and $x = 3 \sin t$, $y = 3 \cos t$ for $0 \le t \le 2\pi$.

(a) Give a convincing argument that the graphs of the pairs of parametric equations are the same.

(b) Explain how the parametrizations are different.

43. Hitting a Baseball Consider Kevin's hit discussed in Example 9.

(a) Approximately how many seconds after the ball is hit does it hit the wall?

(b) How high up the wall does the ball hit?

(c) Writing to Learn Explain why Kevin's hit might be caught by an outfielder. Then explain why his hit would likely not be caught by an outfielder if the ball had been hit at a 20° angle with the horizontal.

44. Hitting a Baseball Kirby hits a ball when it is 4 ft above the ground with an initial velocity of 120 ft/sec. The ball leaves the bat at a 30° angle with the horizontal and heads toward a 30-ft fence 350 ft from home plate.

(a) Does the ball clear the fence?

(b) If so, by how much does it clear the fence? If not, could the ball be caught?

45. Hitting a Baseball Suppose that the moment Kirby hits the ball in Exercise 44 there is a 5-ft/sec split-second wind gust. Assume the wind acts in the horizontal direction out with the ball.

(a) Does the ball clear the fence?

(b) If so, by how much does it clear the fence? If not, could the ball be caught?

46. **Two-Softball Toss** Chris and Linda warm up in the outfield by tossing softballs to each other. Suppose both tossed a ball at the same time from the same height, as illustrated in the figure. Find the minimum distance between the two balls and when this minimum distance occurs.

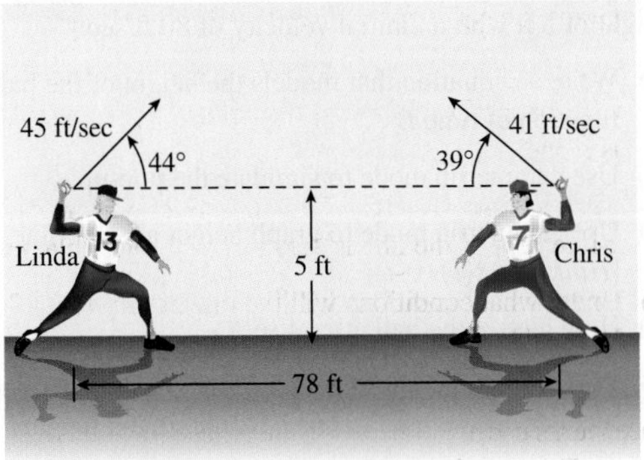

47. **Yard Darts** Tony and Sue are launching yard darts 20 ft from the front edge of a circular target of radius 18 in. on the ground. If Tony throws the dart directly at the target, and releases it 3 ft above the ground with an initial velocity of 30 ft/sec at a 70° angle, will the dart hit the target?

48. **Yard Darts** In the game of darts described in Exercise 47, Sue releases the dart 4 ft above the ground with an initial velocity of 25 ft/sec at a 55° angle. Will the dart hit the target?

49. **Hitting a Baseball** Orlando hits a ball when it is 4 ft above ground level with an initial velocity of 160 ft/sec. The ball leaves the bat at a 20° angle with the horizontal and heads toward a 30-ft fence 400 ft from home plate. How strong must a split-second wind gust be (in feet per second) that acts directly with or against the ball in order for the ball to hit within a few inches of the top of the wall? Estimate the answer graphically and solve algebraically.

50. **Hitting Golf Balls** Nancy hits golf balls off the practice tee with an initial velocity of 180 ft/sec with four different clubs. How far down the fairway does the ball hit the ground if it comes off the club making the specified angle with the horizontal?

(a) 15° (b) 20° (c) 25° (d) 30°

51. **Analysis of a Ferris Wheel** Ron is on a Ferris wheel of radius 35 ft that turns counterclockwise at the rate of one revolution every 12 sec. The lowest point of the Ferris wheel (6 o'clock) is 15 ft above ground level at the point $(0, 15)$ on a rectangular coordinate system. Find parametric equations for the position of Ron as a function of time t (in seconds) if the Ferris wheel starts $(t = 0)$ with Ron at the point $(35, 50)$.

52. **Revisiting Example 5** Eliminate the parameter t from the parametric equations of Example 5 to find an equation in x and y for the line. Verify that the line passes through the points A and B of the example.

53. **Cycloid** The graph of the parametric equations $x = t - \sin t,\ y = 1 - \cos t$ is a *cycloid.*

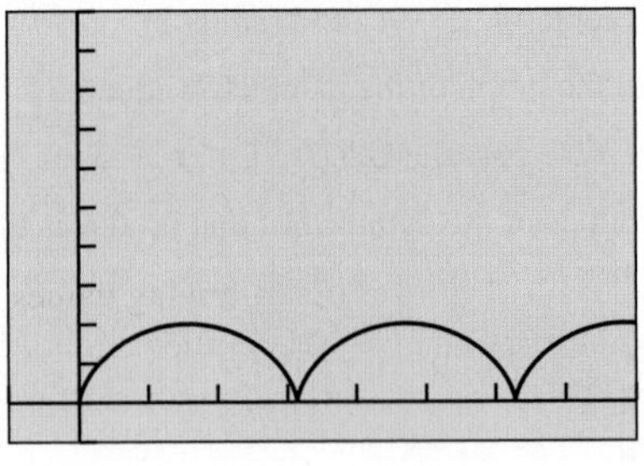

[–2, 16] by [–1, 10]

(a) What is the maximum value of $y = 1 - \cos t$? How is that value related to the graph?

(b) What is the distance between neighboring x-intercepts?

54. **Hypocycloid** The graph of the parametric equations $x = 2\cos t + \cos 2t,\ y = 2\sin t - \sin 2t$ is a *hypocycloid.* The graph is the path of a point P on a circle of radius 1 rolling along the inside of a circle of radius 3, as illustrated in the figure.

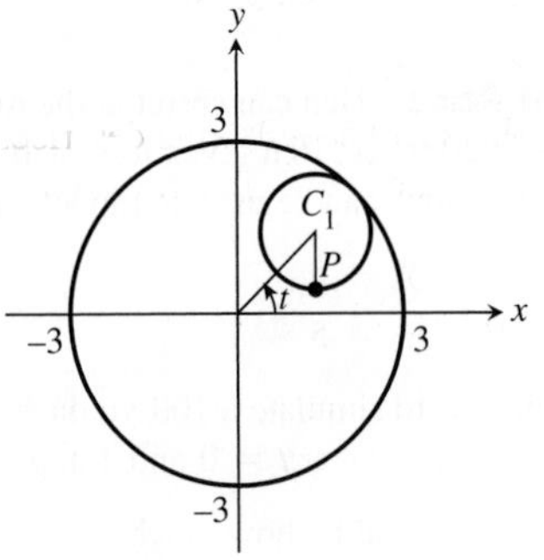

(a) Graph simultaneously this hypocycloid and the circle of radius 3.

(b) Suppose the large circle had a radius of 4. Experiment! How do you think the equations in part (a) should be changed to obtain defining equations? What do you think the hypocycloid would look like in this case? Check your guesses.

Group Activity In Exercises 55–58, a particle moves along a horizontal line so that its position at any time t is given by $s(t)$. Write a description of the motion. (*Hint:* See Example 7.)

55. $s(t) = -t^2 + 3t, \quad -2 \le t \le 4$

56. $s(t) = -t^2 + 4t, \quad -1 \le t \le 5$

57. $s(t) = 0.5(t^3 - 7t^2 + 2t), \quad -1 \le t \le 7$

58. $s(t) = t^3 - 5t^2 + 4t, \quad -1 \le t \le 5$

Standardized Test Questions

59. **True or False** The two sets of parametric equations $x_1 = t - 1,\ y_1 = 3t + 1$ and $x_2 = (2/3)t - 4/3,\ y_2 = 2t$ correspond to the same rectangular equation. Justify your answer.

60. **True or False** The graph of the parametric equations $x = t - 1,\ y = 2t - 1,\ 1 \le t \le 3$ is a line segment with endpoints $(0, 1)$ and $(2, 5)$. Justify your answer.

In Exercises 61–64, solve the problem without using a calculator.

61. Multiple Choice Which of the following points corresponds to $t = -1$ in the parametrization $x = t^2 - 4$, $y = t + \frac{1}{t}$?

(A) $(-3, -2)$ (B) $(-3, 0)$ (C) $(-5, -2)$
(D) $(-5, 0)$ (E) $(3, 2)$

62. Multiple Choice Which of the following values of t produces the same point as $t = 2\pi/3$ in the parametrization $x = 2\cos t$, $y = 2\sin t$?

(A) $t = -\frac{4\pi}{3}$ (B) $t = -\frac{2\pi}{3}$ (C) $t = \frac{-\pi}{3}$
(D) $t = \frac{4\pi}{3}$ (E) $t = \frac{7\pi}{3}$

63. Multiple Choice A rock is thrown straight up from level ground with its position above ground at any time $t \geq 0$ given by $x = 5$, $y = -16t^2 + 80t + 7$. At what time will the rock be 91 ft above ground?

(A) 1.5 sec (B) 2.5 sec
(C) 3.5 sec (D) 1.5 sec and 3.5 sec
(E) The rock never goes that high.

64. Multiple Choice Which of the following describes the graph of the parametric equations $x = 1 - t$, $y = 3t + 2$, $t \geq 0$?

(A) A straight line
(B) A line segment
(C) A ray
(D) A parabola
(E) A circle

Explorations

65. Parametrizing Circles Consider the parametric equations

$$x = a\cos t, \quad y = a\sin t, \quad 0 \leq t \leq 2\pi.$$

(a) Graph the parametric equations for $a = 1, 2, 3, 4$ in the same square viewing window.

(b) Eliminate the parameter t in the parametric equations to verify that they are all circles. What is the radius?

Now consider the parametric equations

$$x = h + a\cos t, \quad y = k + a\sin t, \quad 0 \leq t \leq 2\pi.$$

(c) Graph the equations for $a = 1$ using the following pairs of values for h and k:

h	2	−2	−4	3
k	3	3	−2	−3

(d) Eliminate the parameter t in the parametric equations and identify the graph.

(e) Write a parametrization for the circle with center $(-1, 4)$ and radius 3.

66. Group Activity Parametrization of Lines Consider the parametrization

$$x = at + b, \quad y = ct + d,$$

where a and c are not both zero.

(a) Graph the curve for $a = 2$, $b = 3$, $c = -1$, and $d = 2$.

(b) Graph the curve for $a = 3$, $b = 4$, $c = 1$, and $d = 3$.

(c) Writing to Learn Eliminate the parameter t and write an equation in x and y for the curve. Explain why its graph is a line.

(d) Writing to Learn Find the slope, y-intercept, and x-intercept of the line if they exist. If not, explain why not.

(e) Under what conditions will the line be horizontal? Vertical?

67. Throwing a Ball at a Ferris Wheel A 20-ft Ferris wheel turns counterclockwise one revolution every 12 sec (see figure). Eric stands at point D, 75 ft from the base of the wheel. At the instant Jane is at point A, Eric throws a ball at the Ferris wheel, releasing it from the same height as the bottom of the wheel. If the ball's initial speed is 60 ft/sec and it is released at an angle of 120° with the horizontal, does Jane have a chance to catch the ball? Follow the steps below to obtain the answer.

(a) Assign a coordinate system so that the bottom car of the Ferris wheel is at $(0, 0)$ and the center of the wheel is at $(0, 20)$. Then Eric releases the ball at the point $(75, 0)$. Explain why parametric equations for Jane's path are

$$x_1 = 20\cos\left(\frac{\pi}{6}t\right), \; y_1 = 20 + 20\sin\left(\frac{\pi}{6}t\right), \; t \geq 0.$$

(b) Explain why parametric equations for the path of the ball are

$$x_2 = -30t + 75, \; y_2 = -16t^2 + \left(30\sqrt{3}\right)t, \; t \geq 0.$$

(c) Graph the two paths simultaneously and determine if Jane and the ball arrive at the point of intersection of the two paths at the same time.

(d) Find a formula for the distance $d(t)$ between Jane and the ball at any time t.

(e) Writing to Learn Use the graph of the parametric equations $x_3 = t$, $y_3 = d(t)$, to estimate the minimum distance between Jane and the ball and when it occurs. Do you think Jane has a chance to catch the ball?

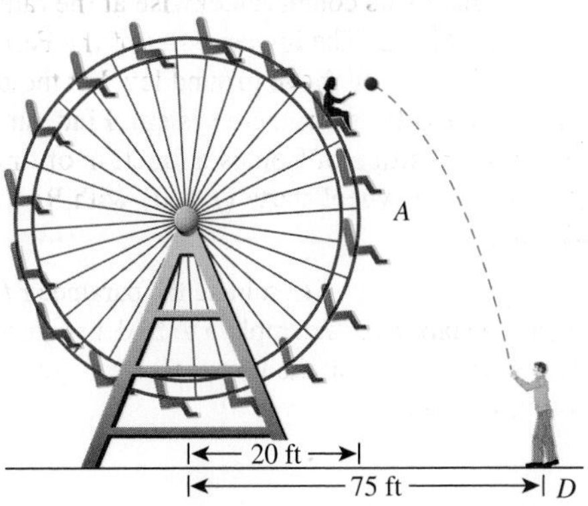

68. Throwing a Ball at a Ferris Wheel A 71-ft-radius Ferris wheel turns counterclockwise one revolution every 20 sec. Tony stands at a point 90 ft to the right of the base of the wheel. At the instant Matthew is at point A (3 o'clock), Tony throws a ball toward the Ferris wheel with an initial velocity of 88 ft/sec at an angle with the horizontal of 100°. Find the minimum distance between the ball and Matthew.

Extending the Ideas

69. Two Ferris Wheels Problem Chang is on a Ferris wheel of center $(0, 20)$ and radius 20 ft turning counterclockwise at the rate of one revolution every 12 sec. Kuan is on a Ferris wheel of center $(15, 15)$ and radius 15 turning counterclockwise at the rate of one revolution every 8 sec. Find the minimum distance between Chang and Kuan if both start out $(t = 0)$ at 3 o'clock.

70. Two Ferris Wheels Problem Chang and Kuan are riding the Ferris wheels described in Exercise 69. Find the minimum distance between Chang and Kuan if Chang starts out $(t = 0)$ at 3 o'clock and Kuan at 6 o'clock.

Exercises 71–73 refer to the graph C of the parametric equations

$$x = tc + (1 - t)a, \quad y = td + (1 - t)b$$

where $P_1(a, b)$ and $P_2(c, d)$ are two fixed points.

71. Using Parametric Equations in Geometry Show that the point $P(x, y)$ on C is equal to

(a) $P_1(a, b)$ when $t = 0$.

(b) $P_2(c, d)$ when $t = 1$.

72. Using Parametric Equations in Geometry Show that if $t = 0.5$, the corresponding point (x, y) on C is the midpoint of the line segment with endpoints (a, b) and (c, d).

73. What values of t will find two points that divide the line segment P_1P_2 into three equal pieces? Four equal pieces?

6.4 Polar Coordinates

What you'll learn about

- Polar Coordinate System
- Coordinate Conversion
- Equation Conversion
- Finding Distance Using Polar Coordinates

... and why

Use of polar coordinates sometimes simplifies complicated rectangular equations and they are useful in calculus.

Polar Coordinate System

A **polar coordinate system** is a plane with a point O, the **pole**, and a ray from O, the **polar axis**, as shown in Figure 6.35. Each point P in the plane is assigned **polar coordinates (r, θ)** as follows: r is the **directed distance** from O to P, and θ is the **directed angle** whose initial side is the polar axis and whose terminal side is the ray $\overrightarrow{OP}$.

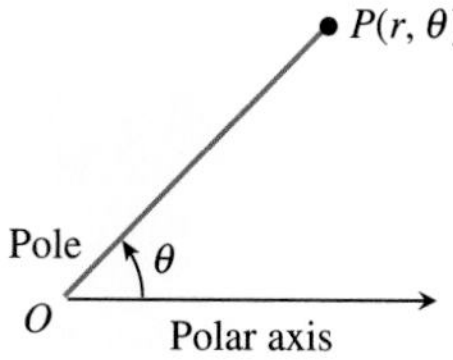

FIGURE 6.35 The polar coordinate system.

As in trigonometry, we measure θ as positive when moving counterclockwise and negative when moving clockwise. If $r > 0$, then P is on the terminal side of θ. If $r < 0$, then P is on the terminal side of $\theta + \pi$. We can use radian or degree measure for the angle θ as illustrated in Example 1.

EXAMPLE 1 Plotting Points in the Polar Coordinate System

Plot the points with the given polar coordinates.

(a) $P(2, \pi/3)$ **(b)** $Q(-1, 3\pi/4)$ **(c)** $R(3, -45°)$

SOLUTION Figure 6.36 shows the three points. Now try Exercise 7.

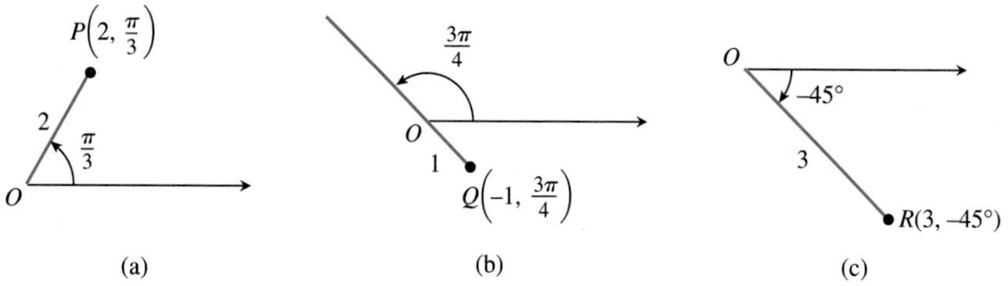

FIGURE 6.36 The three points in Example 1.

Each polar coordinate pair determines a unique point. However, the polar coordinates of a point P in the plane are not unique.

EXAMPLE 2 Finding All Polar Coordinates for a Point

If the point P has polar coordinates $(3, \pi/3)$, find all polar coordinates for P.

SOLUTION Point P is shown in Figure 6.37. Two additional pairs of polar coordinates for P are

$$\left(3, \frac{\pi}{3} + 2\pi\right) = \left(3, \frac{7\pi}{3}\right) \quad \text{and} \quad \left(-3, \frac{\pi}{3} + \pi\right) = \left(-3, \frac{4\pi}{3}\right).$$

We can use these two pairs of polar coordinates for P to write the rest of the possibilities:

$$\left(3, \frac{\pi}{3} + 2n\pi\right) = \left(3, \frac{(6n+1)\pi}{3}\right) \text{ or}$$

$$\left(-3, \frac{\pi}{3} + (2n+1)\pi\right) = \left(-3, \frac{(6n+4)\pi}{3}\right)$$

where n is any integer. Now try Exercise 23.

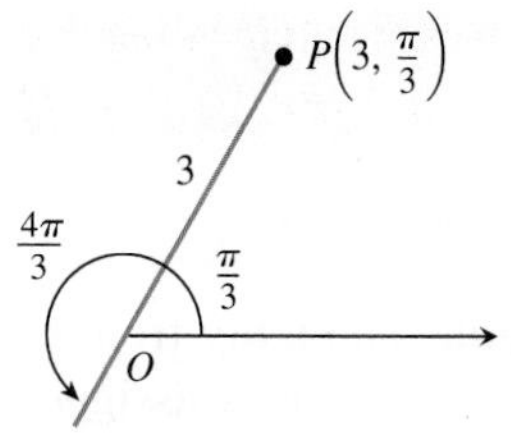

FIGURE 6.37 The point P in Example 2.

The coordinates (r, θ), $(r, \theta + 2\pi)$, and $(-r, \theta + \pi)$ all name the same point. In general, the point with polar coordinates (r, θ) also has the following polar coordinates:

Finding All Polar Coordinates of a Point

Let P have polar coordinates (r, θ). Any other polar coordinate of P must be of the form

$$(r, \theta + 2n\pi) \text{ or } (-r, \theta + (2n + 1)\pi)$$

where n is any integer. In particular, the pole has polar coordinates $(0, \theta)$, where θ is any angle.

Coordinate Conversion

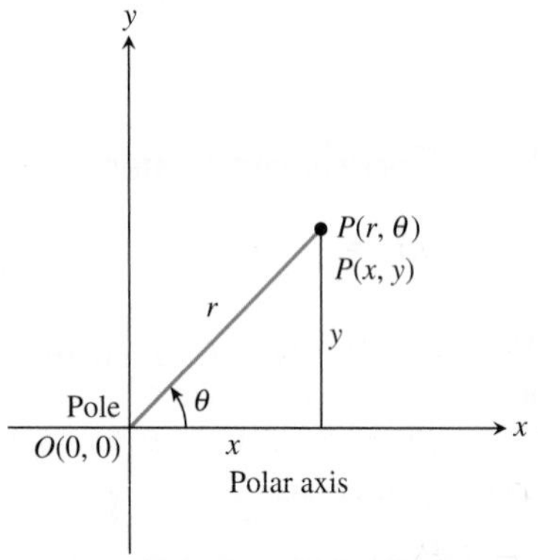

FIGURE 6.38 Polar and rectangular coordinates for P.

When we use both polar coordinates and Cartesian coordinates, the pole is the origin and the polar axis is the positive x-axis as shown in Figure 6.38. By applying trigonometry we can find equations that relate the polar coordinates (r, θ) and the rectangular coordinates (x, y) of a point P.

Coordinate Conversion Equations

Let the point P have polar coordinates (r, θ) and rectangular coordinates (x, y). Then

$$x = r\cos\theta, \qquad r^2 = x^2 + y^2,$$

$$y = r\sin\theta, \quad \tan\theta = \frac{y}{x}.$$

These relationships allow us to convert from one coordinate system to the other.

EXAMPLE 3 Converting from Polar to Rectangular Coordinates

Use an algebraic method to find the rectangular coordinates of the points with given polar coordinates. Approximate exact solution values with a calculator when appropriate.

(a) $P(3, 5\pi/6)$ **(b)** $Q(2, -200°)$

SOLUTION

(a) For $P(3, 5\pi/6)$, $r = 3$ and $\theta = 5\pi/6$:

$$x = r\cos\theta \qquad\qquad y = r\sin\theta$$

$$x = 3\cos\frac{5\pi}{6} \quad \text{and} \quad y = 3\sin\frac{5\pi}{6}$$

$$x = 3\left(-\frac{\sqrt{3}}{2}\right) \approx -2.60 \qquad y = 3\left(\frac{1}{2}\right) = 1.5$$

The rectangular coordinates for P are $\left(-3\sqrt{3}/2, 1.5\right) \approx (-2.60, 1.5)$ (Figure 6.39a).

(b) For $Q(2, -200°)$, $r = 2$ and $\theta = -200°$:

$$x = r\cos\theta \qquad\qquad y = r\sin\theta$$

$$x = 2\cos(-200°) \approx -1.88 \quad \text{and} \quad y = 2\sin(-200°) \approx 0.68$$

The rectangular coordinates for Q are approximately $(-1.88, 0.68)$ (Figure 6.39b).

Now try Exercise 15.

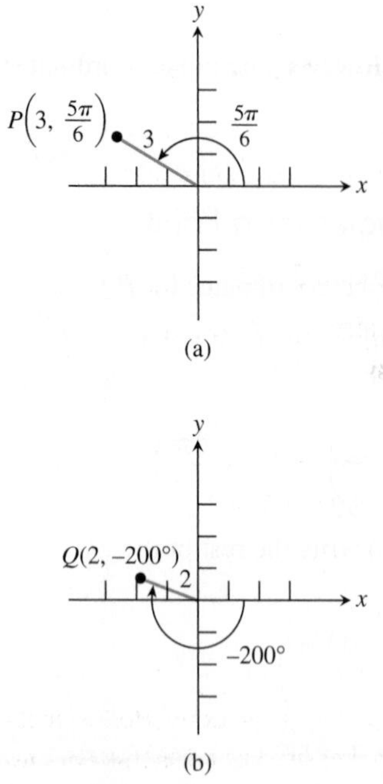

FIGURE 6.39 The points P and Q in Example 3.

When converting rectangular coordinates to polar coordinates, we must remember that there are infinitely many possible polar coordinate pairs. In Example 4 we report two of the possibilities.

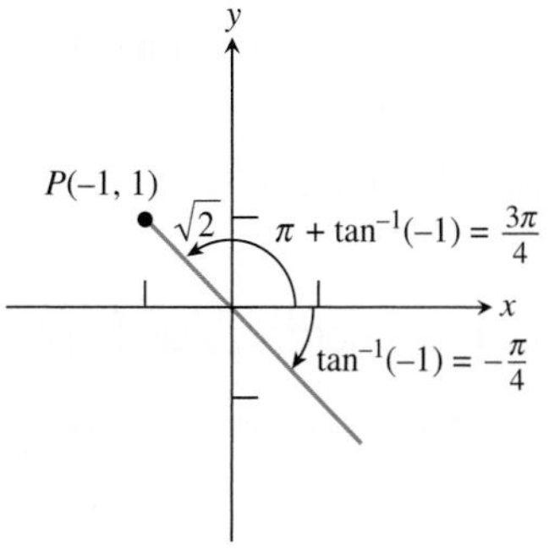

FIGURE 6.40 The point P in Example 4a.

EXAMPLE 4 Converting from Rectangular to Polar Coordinates

Find two polar coordinate pairs for the points with given rectangular coordinates.

(a) $P(-1, 1)$ **(b)** $Q(-3, 0)$

SOLUTION

(a) For $P(-1, 1)$, $x = -1$ and $y = 1$:

$$r^2 = x^2 + y^2 \qquad \tan\theta = \frac{y}{x}$$

$$r^2 = (-1)^2 + (1)^2 \qquad \tan\theta = \frac{1}{-1} = -1$$

$$r = \pm\sqrt{2} \qquad \theta = \tan^{-1}(-1) + n\pi = -\frac{\pi}{4} + n\pi$$

We use the angles $-\pi/4$ and $-\pi/4 + \pi = 3\pi/4$. Because P is on the ray opposite the terminal side of $-\pi/4$, the value of r corresponding to this angle is negative (Figure 6.40). Because P is on the terminal side of $3\pi/4$, the value of r corresponding to this angle is positive. So two polar coordinate pairs of point P are

$$\left(-\sqrt{2}, -\frac{\pi}{4}\right) \text{ and } \left(\sqrt{2}, \frac{3\pi}{4}\right).$$

(b) For $Q(-3, 0)$, $x = -3$ and $y = 0$. Thus, $r = \pm 3$ and $\theta = n\pi$. We use the angles 0 and π. So two polar coordinate pairs for point Q are

$$(-3, 0) \text{ and } (3, \pi).$$

Now try Exercise 27.

EXPLORATION 1 Using a Grapher to Convert Coordinates

Most graphers have the capability to convert polar coordinates to rectangular coordinates and vice versa. Usually they give just one possible polar coordinate pair for a given rectangular coordinate pair.

1. Use your grapher to check the conversions in Examples 3 and 4.
2. Use your grapher to convert the polar coordinate pairs $(2, \pi/3)$, $(-1, \pi/2)$, $(2, \pi)$, $(-5, 3\pi/2)$, $(3, 2\pi)$ to rectangular coordinate pairs.
3. Use your grapher to convert the rectangular coordinate pairs $(-1, -\sqrt{3})$, $(0, 2)$, $(3, 0)$, $(-1, 0)$, $(0, -4)$ to polar coordinate pairs.

Equation Conversion

Simple-looking polar equations can become quite complicated when converted to rectangular form, and vice versa. Certainly, a curve that is expressible as a function in one coordinate system is not always expressible as a function in the other. Nonetheless, all we need to know to attempt such a conversion are the coordinate conversion equations. The trick is to set up the substitutions with a little algebra.

For example, the polar equation $r = 4\cos\theta$ can be converted to rectangular form as follows:

$$
\begin{aligned}
r &= 4\cos\theta \\
r^2 &= 4r\cos\theta && \text{Multiply both sides by } r. \\
x^2 + y^2 &= 4x && r^2 = x^2 + y^2,\ r\cos\theta = x \\
x^2 - 4x + 4 + y^2 &= 4 && \text{Subtract } 4x \text{ and add } 4. \\
(x-2)^2 + y^2 &= 4 && \text{Factor.}
\end{aligned}
$$

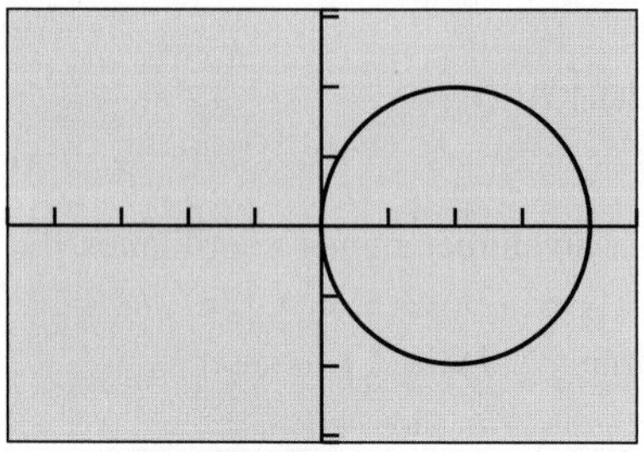

[−4.7, 4.7] by [−3.1, 3.1]

FIGURE 6.41 The graph of the polar equation $r = 4\cos\theta$ in $0 \le \theta \le 2\pi$.

Thus the graph of $r = 4\cos\theta$ is all or part of the circle with center $(2, 0)$ and radius 2.

Figure 6.41 shows the graph of $r = 4\cos\theta$ for $0 \le \theta \le 2\pi$ obtained using the polar graphing mode of our grapher. So, the graph of $r = 4\cos\theta$ is the entire circle.

Just as with parametric equations, the domain of a polar equation in r and θ is understood to be all values of θ for which the corresponding values of r are real numbers. You must also select a value for θ min and θ max to graph in polar mode.

You may be surprised by the polar form for a vertical line in Example 5.

EXAMPLE 5 Converting from Polar Form to Rectangular Form

Convert $r = 4\sec\theta$ to rectangular form and identify the graph. Support your answer with a polar graphing utility.

SOLUTION

$$
\begin{aligned}
r &= 4\sec\theta \\
\frac{r}{\sec\theta} &= 4 && \text{Divide by } \sec\theta. \\
r\cos\theta &= 4 && \cos\theta = \frac{1}{\sec\theta} \\
x &= 4 && r\cos\theta = x
\end{aligned}
$$

The graph is the vertical line $x = 4$ (Figure 6.42). **Now try Exercise 35.**

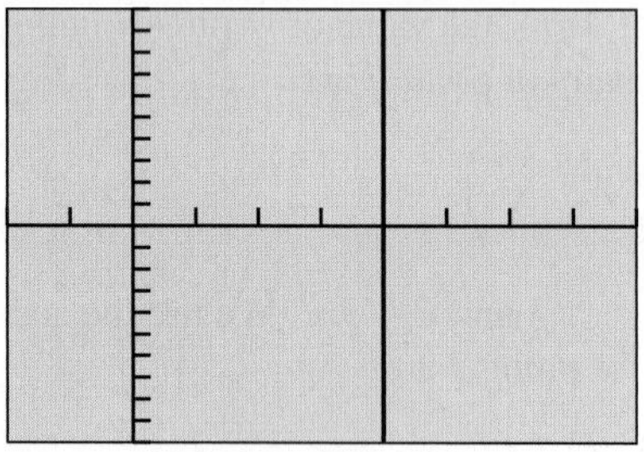

[−2, 8] by [−10, 10]

FIGURE 6.42 The graph of the vertical line $r = 4\sec\theta$ $(x = 4)$. (Example 5)

EXAMPLE 6 Converting from Rectangular Form to Polar Form

Convert $(x-3)^2 + (y-2)^2 = 13$ to polar form.

SOLUTION

$$
\begin{aligned}
(x-3)^2 + (y-2)^2 &= 13 \\
x^2 - 6x + 9 + y^2 - 4y + 4 &= 13 \\
x^2 + y^2 - 6x - 4y &= 0
\end{aligned}
$$

Substituting r^2 for $x^2 + y^2$, $r\cos\theta$ for x, and $r\sin\theta$ for y gives the following:

$$
\begin{aligned}
r^2 - 6r\cos\theta - 4r\sin\theta &= 0 \\
r(r - 6\cos\theta - 4\sin\theta) &= 0 \\
r = 0 \quad \text{or} \quad r - 6\cos\theta - 4\sin\theta &= 0
\end{aligned}
$$

The graph of $r = 0$ consists of a single point, the origin, which is also on the graph of $r - 6\cos\theta - 4\sin\theta = 0$. Thus, the polar form is

$$r = 6\cos\theta + 4\sin\theta.$$

The graph of $r = 6\cos\theta + 4\sin\theta$ for $0 \le \theta \le 2\pi$ is shown in Figure 6.43 and appears to be a circle with center $(3, 2)$ and radius $\sqrt{13}$, as expected.

Now try Exercise 43.

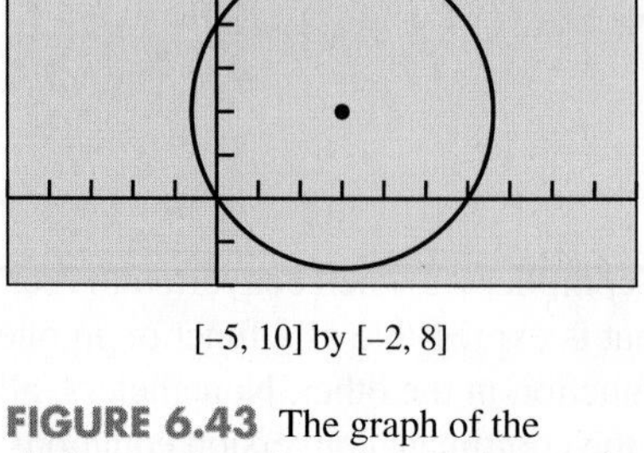

[−5, 10] by [−2, 8]

FIGURE 6.43 The graph of the circle $r = 6\cos\theta + 4\sin\theta$. (Example 6)

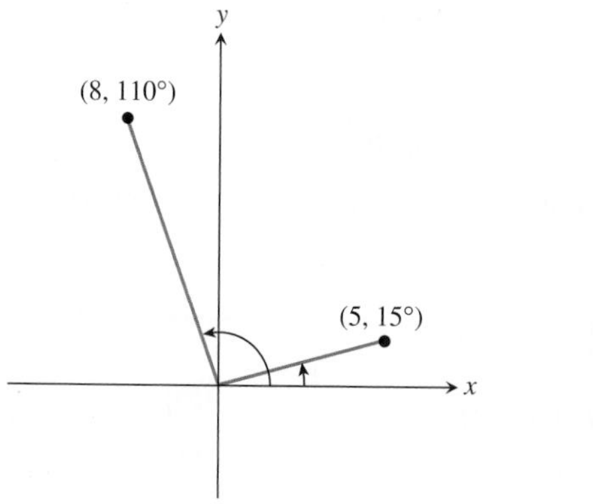

FIGURE 6.44 The distance and direction of two airplanes from a radar source. (Example 7)

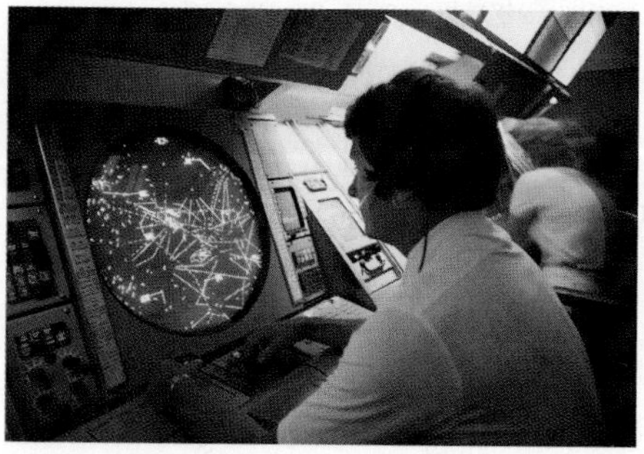

Finding Distance Using Polar Coordinates

A radar tracking system sends out high-frequency radio waves and receives their reflection from an object. The distance and direction of the object from the radar is often given in polar coordinates.

EXAMPLE 7 Using a Radar Tracking System

Radar detects two airplanes at the same altitude. Their polar coordinates are (8 mi, 110°) and (5 mi, 15°). (See Figure 6.44.) How far apart are the airplanes?

SOLUTION By the Law of Cosines (Section 5.6),

$$d^2 = 8^2 + 5^2 - 2 \cdot 8 \cdot 5 \cos (110° - 15°)$$
$$d = \sqrt{8^2 + 5^2 - 2 \cdot 8 \cdot 5 \cos 95°}$$
$$d \approx 9.80$$

The airplanes are about 9.80 mi apart. **Now try Exercise 51.**

We can also use the Law of Cosines to derive a formula for the distance between points in the polar coordinate system. See Exercise 61.

QUICK REVIEW 6.4 *(For help, go to Sections P.2, 4.3, and 5.6.)*

Exercise numbers with a gray background indicate problems that the authors have designed to be solved *without a calculator*.

In Exercises 1 and 2, determine the quadrants containing the terminal side of the angles.

1. **(a)** $5\pi/6$ **(b)** $-3\pi/4$

2. **(a)** $-300°$ **(b)** $210°$

In Exercises 3–6, find a positive and a negative angle coterminal with the given angle.

3. $-\pi/4$ **4.** $\pi/3$

5. $160°$ **6.** $-120°$

In Exercises 7 and 8, write a standard form equation for the circle.

7. Center $(3, 0)$ and radius 2

8. Center $(0, -4)$ and radius 3

In Exercises 9 and 10, use the Law of Cosines to find the measure of the third side of the given triangle.

9.

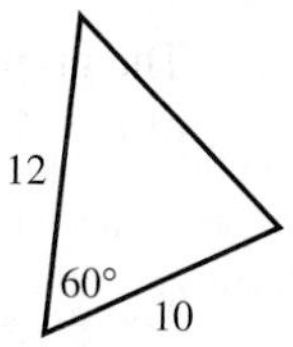

10.

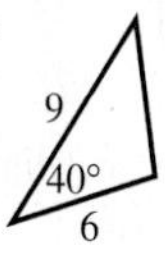

SECTION 6.4 Exercises

In Exercises 1–4, the polar coordinates of a point are given. Find its rectangular coordinates.

1.

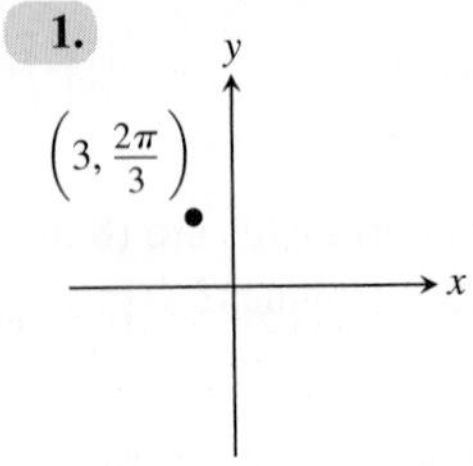

2.

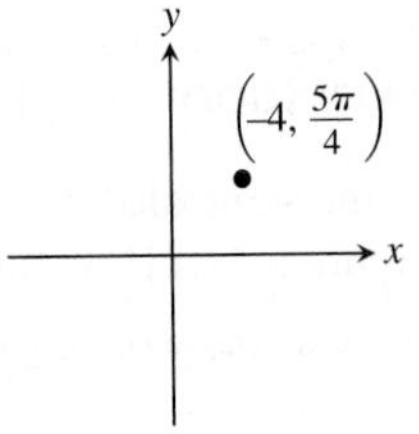

3.

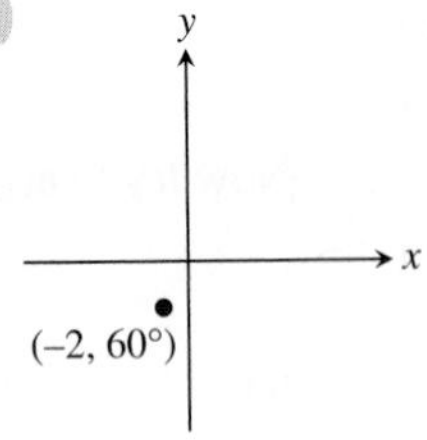

4.

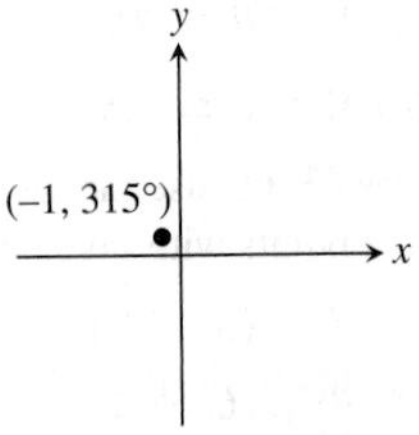

In Exercises 5 and 6, **(a)** complete the table for the polar equation and **(b)** plot the corresponding points.

5. $r = 3 \sin \theta$

θ	$\pi/4$	$\pi/2$	$5\pi/6$	π	$4\pi/3$	2π
r						

6. $r = 2 \csc \theta$

θ	$\pi/4$	$\pi/2$	$5\pi/6$	π	$4\pi/3$	2π
r						

In Exercises 7–14, plot the point with the given polar coordinates.

7. $(3, 4\pi/3)$ **8.** $(2, 5\pi/6)$ **9.** $(-1, 2\pi/5)$

10. $(-3, 17\pi/10)$ **11.** $(2, 30°)$ **12.** $(3, 210°)$

13. $(-2, 120°)$ **14.** $(-3, 135°)$

In Exercises 15–22, use an algebraic method to find the rectangular coordinates of the point with given polar coordinates. Approximate the exact solution values with a calculator when appropriate.

15. $(1.5, 7\pi/3)$ **16.** $(2.5, 17\pi/4)$

17. $(-3, -29\pi/7)$ **18.** $(-2, -14\pi/5)$

19. $(-2, \pi)$ **20.** $(1, \pi/2)$

21. $(2, 270°)$ **22.** $(-3, 360°)$

In Exercises 23–26, polar coordinates of point P are given. Find all of its polar coordinates.

23. $P = (2, \pi/6)$ **24.** $P = (1, -\pi/4)$

25. $P = (1.5, -20°)$ **26.** $P = (-2.5, 50°)$

In Exercises 27–30, rectangular coordinates of a point P are given. Use an algebraic method, and approximate exact solution values with a calculator when appropriate, to find all polar coordinates of P that satisfy

(a) $0 \le \theta \le 2\pi$ **(b)** $-\pi \le \theta \le \pi$ **(c)** $0 \le \theta \le 4\pi$.

27. $P = (1, 1)$ **28.** $P = (1, 3)$

29. $P = (-2, 5)$ **30.** $P = (-1, -2)$

In Exercises 31–34, use your grapher to match the polar equation with its graph.

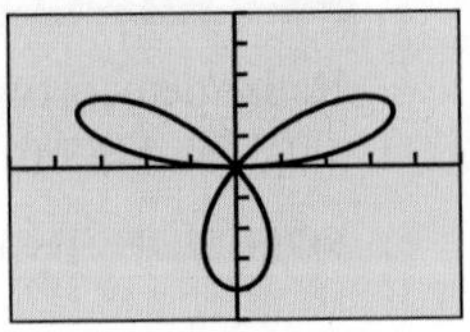

(a)

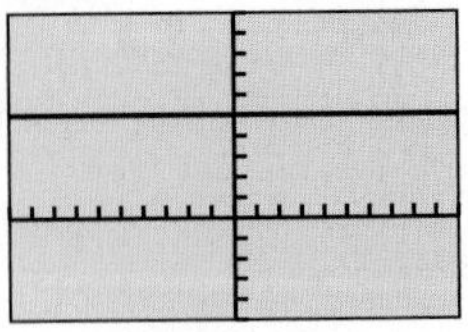

(b)

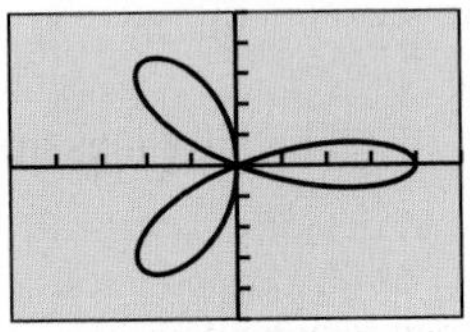

(c)

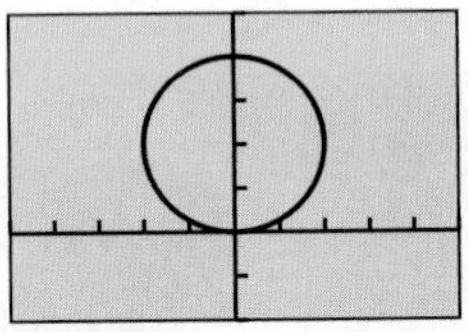

(d)

31. $r = 5 \csc \theta$ **32.** $r = 4 \sin \theta$

33. $r = 4 \cos 3\theta$ **34.** $r = 4 \sin 3\theta$

In Exercises 35–42, convert the polar equation to rectangular form and identify the graph. Support your answer by graphing the polar equation.

35. $r = 3 \sec \theta$ **36.** $r = -2 \csc \theta$

37. $r = -3 \sin \theta$ **38.** $r = -4 \cos \theta$

39. $r \csc \theta = 1$ **40.** $r \sec \theta = 3$

41. $r = 2 \sin \theta - 4 \cos \theta$ **42.** $r = 4 \cos \theta - 4 \sin \theta$

In Exercises 43–50, convert the rectangular equation to polar form. Sketch the graph of the rectangular equation (do *not* use a grapher), then support your hand sketch by graphing the polar equation with your grapher.

43. $x = 2$ **44.** $x = 5$

45. $2x - 3y = 5$ **46.** $3x + 4y = 2$

47. $(x - 3)^2 + y^2 = 9$ **48.** $x^2 + (y - 1)^2 = 1$

49. $(x + 3)^2 + (y + 3)^2 = 18$

50. $(x - 1)^2 + (y + 4)^2 = 17$

51. Tracking Airplanes The location, given in polar coordinates, of two planes approaching the Vicksburg airport are (4 mi, 12°) and (2 mi, 72°). Find the distance between the airplanes.

52. Tracking Ships The locations of two ships from Mays Landing Lighthouse, given in polar coordinates, are (3 mi, 170°) and (5 mi, 150°). Find the distance between the ships.

53. Using Polar Coordinates in Geometry A square with sides of length a and center at the origin has two sides parallel to the x-axis. Find polar coordinates of the vertices.

54. Using Polar Coordinates in Geometry A regular pentagon whose center is at the origin has one vertex on the positive x-axis at a distance a from the center. Find polar coordinates of the vertices.

Standardized Test Questions

55. True or False Every point in the plane has exactly two polar coordinates. Justify your answer.

56. True or False If r_1 and r_2 are not 0, and if (r_1, θ) and $(r_2, \theta + \pi)$ represent the same point in the plane, then $r_1 = -r_2$. Justify your answer.

In Exercises 57–60, solve the problem without using a calculator.

57. Multiple Choice If $r \neq 0$, which of the following polar coordinate pairs represents the same point as the point with polar coordinates (r, θ)?

(A) $(-r, \theta)$ **(B)** $(-r, \theta + 2\pi)$ **(C)** $(-r, \theta + 3\pi)$

(D) $(r, \theta + \pi)$ **(E)** $(r, \theta + 3\pi)$

58. Multiple Choice Which of the following are the rectangular coordinates of the point with polar coordinate $(-2, -\pi/3)$?

(A) $(-\sqrt{3}, 1)$ **(B)** $(-1, -\sqrt{3})$ **(C)** $(-1, \sqrt{3})$

(D) $(1, -\sqrt{3})$ **(E)** $(1, \sqrt{3})$

59. Multiple Choice Which of the following polar coordinate pairs represent the same point as the point with polar coordinates $(2, 110°)$?

(A) $(-2, -70°)$ **(B)** $(-2, 110°)$ **(C)** $(-2, -250°)$

(D) $(2, -70°)$ **(E)** $(2, 290°)$

60. Multiple Choice Which of the following polar coordinate pairs does *not* represent the point with rectangular coordinates $(-2, -2)$?

(A) $(2\sqrt{2}, -135°)$ **(B)** $(2\sqrt{2}, 225°)$

(C) $(-2\sqrt{2}, -315°)$ **(D)** $(-2\sqrt{2}, 45°)$

(E) $(-2\sqrt{2}, 135°)$

Explorations

61. Polar Distance Formula Let P_1 and P_2 have polar coordinates (r_1, θ_1) and (r_2, θ_2), respectively.

(a) If $\theta_1 - \theta_2$ is a multiple of π, write a formula for the distance between P_1 and P_2.

(b) Use the Law of Cosines to prove that the distance between P_1 and P_2 is given by

$$d = \sqrt{r_1^2 + r_2^2 - 2r_1r_2\cos(\theta_1 - \theta_2)}.$$

(c) Writing to Learn Does the formula in part (b) agree with the formula(s) you found in part (a)? Explain.

62. Watching Your θ-Step Consider the polar curve $r = 4\sin\theta$. Describe the graph for each of the following.

(a) $0 \le \theta \le \pi/2$ **(b)** $0 \le \theta \le 3\pi/4$

(c) $0 \le \theta \le 3\pi/2$ **(d)** $0 \le \theta \le 4\pi$

In Exercises 63–66, use the results of Exercise 61 to find the distance between the points with given polar coordinates.

63. $(2, 10°), (5, 130°)$

64. $(4, 20°), (6, 65°)$

65. $(-3, 25°), (-5, 160°)$

66. $(6, -35°), (8, -65°)$

Extending the Ideas

67. Graphing Polar Equations Parametrically Find parametric equations for the polar curve $r = f(\theta)$.

Group Activity In Exercises 68–71, use what you learned in Exercise 67 to write parametric equations for the given polar equation. Support your answers graphically.

68. $r = 2\cos\theta$ **69.** $r = 5\sin\theta$

70. $r = 2\sec\theta$ **71.** $r = 4\csc\theta$

6.5 Graphs of Polar Equations

What you'll learn about

- Polar Curves and Parametric Curves
- Symmetry
- Analyzing Polar Graphs
- Rose Curves
- Limaçon Curves
- Other Polar Curves

... and why

Graphs that have circular or cylindrical symmetry often have simple polar equations, which is very useful in calculus.

Polar Curves and Parametric Curves

Polar curves are actually just special cases of parametric curves. Keep in mind that polar curves are graphed in the xy-plane, despite the fact that they are given in terms of r and θ. That is why the polar graph of $r = 4\cos\theta$ is a circle (see Figure 6.41 in Section 6.4) rather than a cosine curve.

In function mode, points are determined by a vertical coordinate that changes as the horizontal coordinate moves left to right. In polar mode, points are determined by a directed distance from the pole that changes as the angle sweeps around the pole. The connection is provided by the coordinate conversion equations from Section 6.4, which show that the graph of $r = f(\theta)$ is really just the graph of the parametric equations

$$x = f(\theta)\cos\theta$$
$$y = f(\theta)\sin\theta$$

for all values of θ in some parameter interval that suffices to produce a complete graph. (In many of our examples, $0 \le \theta < 2\pi$ will do.)

Since modern graphing calculators produce these graphs so easily in polar mode, we are frankly going to assume that you do not have to sketch them by hand. Instead we will concentrate on analyzing the properties of the curves. In later courses you can discover further properties of the curves using the tools of calculus.

Symmetry

You learned algebraic tests for symmetry for equations in rectangular form in Section 1.2. Algebraic tests also exist for equations in polar form.

Figure 6.45 on the next page shows a rectangular coordinate system superimposed on a polar coordinate system, with the origin and the pole coinciding and the positive x-axis and the polar axis coinciding.

The three types of symmetry figures to be considered will have:

1. The x-axis (polar axis) as a line of symmetry (Figure 6.45a).
2. The y-axis (the line $\theta = \pi/2$) as a line of symmetry (Figure 6.45b).
3. The origin (the pole) as a point of symmetry (Figure 6.45c).

All three algebraic tests for symmetry in polar forms require replacing the pair (r, θ), which satisfies the polar equation, with another coordinate pair and determining whether it also satisfies the polar equation.

Symmetry Tests for Polar Graphs

The graph of a polar equation has the indicated symmetry if either replacement produces an equivalent polar equation.

To Test for Symmetry	Replace	By
1. about the x-axis,	(r, θ)	$(r, -\theta)$ or $(-r, \pi - \theta)$.
2. about the y-axis,	(r, θ)	$(-r, -\theta)$ or $(r, \pi - \theta)$.
3. about the origin,	(r, θ)	$(-r, \theta)$ or $(r, \theta + \pi)$.

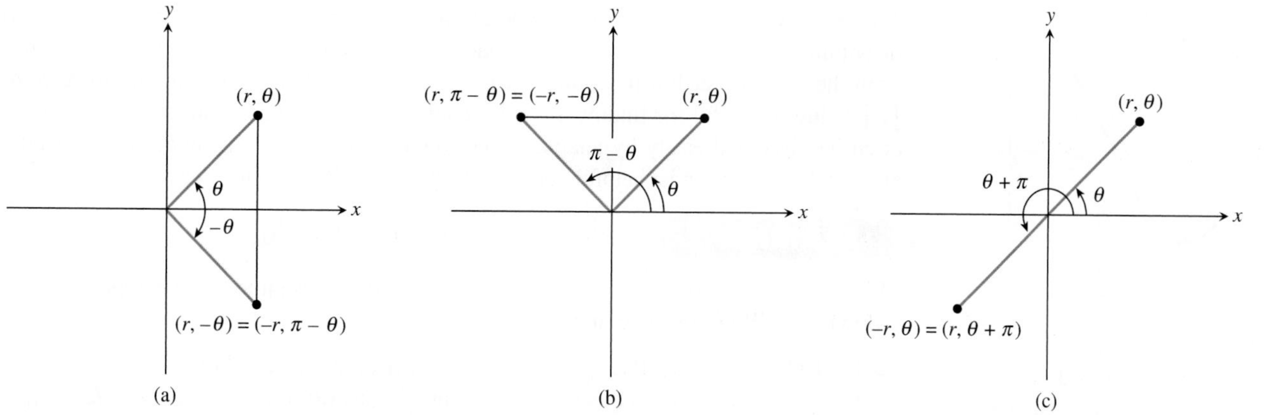

FIGURE 6.45 Symmetry with respect to (a) the x-axis (polar axis), (b) the y-axis (the line $\theta = \pi/2$), and (c) the origin (the pole).

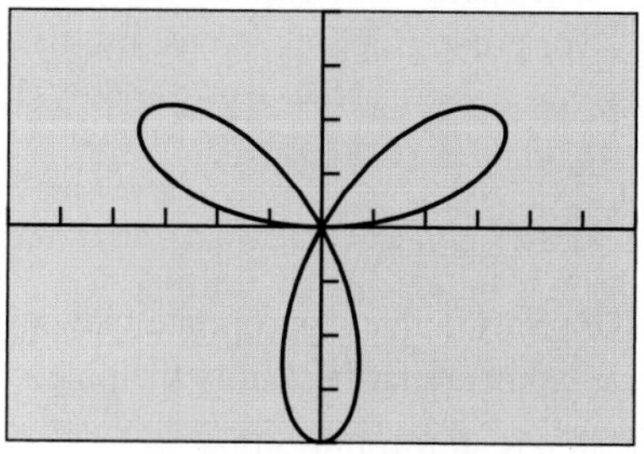

[–6, 6] by [–4, 4]

FIGURE 6.46 The graph of $r = 4 \sin 3\theta$ is symmetric about the y-axis. (Example 1)

EXAMPLE 1 Testing for Symmetry

Use the appropriate symmetry test to prove that the graph of $r = 4 \sin 3\theta$ is symmetric about the y-axis.

SOLUTION Figure 6.46 suggests that the graph of $r = 4 \sin 3\theta$ is symmetric about the y-axis and not symmetric about the x-axis or origin.

$$r = 4 \sin 3\theta$$
$$-r = 4 \sin 3(-\theta) \quad \text{Replace } (r, \theta) \text{ by } (-r, -\theta).$$
$$-r = 4 \sin (-3\theta)$$
$$-r = -4 \sin 3\theta \quad \sin \theta \text{ is an odd function of } \theta.$$
$$r = 4 \sin 3\theta \quad \text{(Same as original)}$$

Because the equations $-r = 4 \sin 3(-\theta)$ and $r = 4 \sin 3\theta$ are equivalent, there is symmetry about the y-axis.

Now try Exercise 13.

Analyzing Polar Graphs

We analyze graphs of polar equations in much the same way that we analyze graphs of rectangular equations. For example, the function r of Example 1 is a continuous function of θ. Also $r = 0$ when $\theta = 0$ and when θ is any integer multiple of $\pi/3$. The domain of this function is the set of all real numbers.

TRACE can be used to help determine the range of this polar function (Figure 6.47), but you should realize that it is the interval $[-4, 4]$.

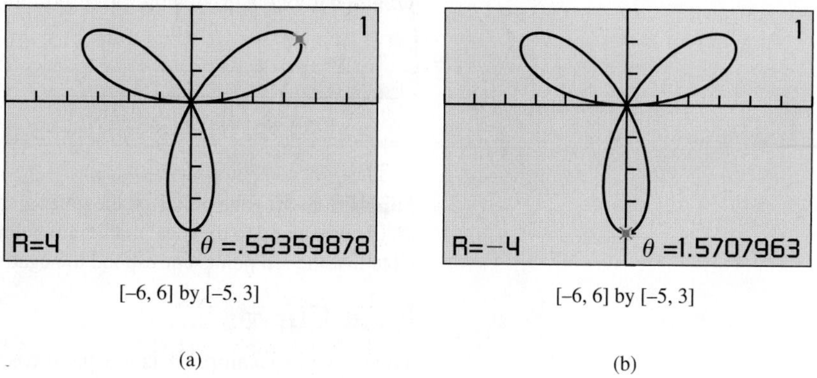

FIGURE 6.47 The values of r in $r = 4 \sin 3\theta$ vary from (a) 4 to (b) −4.

The graphs of most of the interesting polar functions (as you will see later in this section) fit within a bounded (x, y) viewing window that contains the pole $(0, 0)$. The distance from the pole for such a function (that is, $|r|$) is bounded, and knowing the **maximum $|r|$ value** is helpful for finding a good graphing window. The maximum $|r|$ value can often be discerned easily by analyzing the trigonometric functions involved, but a little graphical analysis can be helpful, too. We illustrate this in the following two examples.

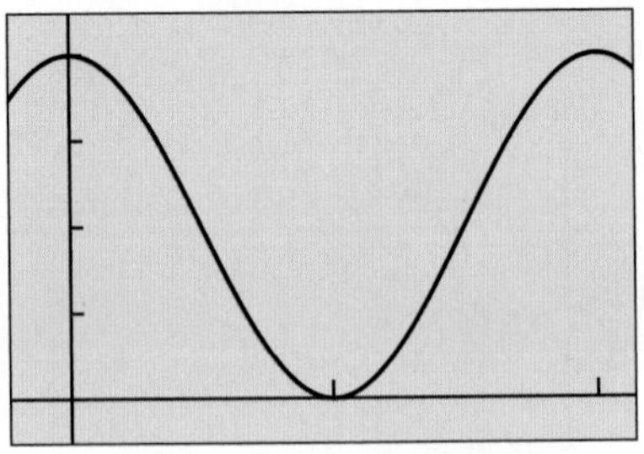

[–0.5, 6.5] by [–0.5, 4.5]

FIGURE 6.48 The graph of $y = |2 + 2\cos x|$ on the interval $[0, 2\pi]$ supports the conclusion that the maximum $|r|$ value for the polar function $r = 2 + 2\cos\theta$ is 4. (Example 2)

EXAMPLE 2 Finding a Maximum $|r|$ Value

Find the maximum $|r|$ value of $r = 2 + 2\cos\theta$ and use it to find an appropriate viewing window for the graph.

SOLUTION We know that $-1 \le \cos\theta \le 1$, from which it follows that $0 \le 2 + 2\cos\theta \le 4$. Thus $0 \le r \le 4$ and the maximum $|r|$ value is 4. A standard rectangular graph of $y = |2 + 2\cos x|$ on the interval $[0, 2\pi]$ shows a maximum value of 4, supporting our algebraic analysis (Figure 6.48).

A "safe" viewing window of $[-4, 4]$ by $[-4, 4]$ will enclose the entire curve, but it is better to keep the $x{:}y$ aspect ratio close to $3{:}2$ for polar graphing, so we use the larger window $[-6, 6]$ by $[-4, 4]$ (Figure 6.49). **Now try Exercise 21.**

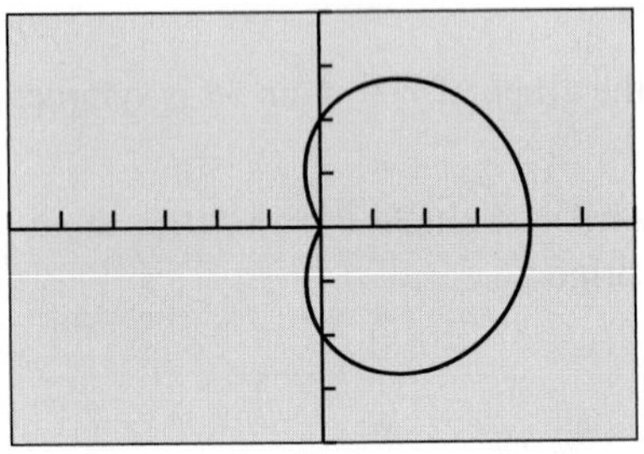

[–6, 6] by [–4, 4]

FIGURE 6.49 The graph of $r = 2 + 2\cos\theta$ on the interval $[-6, 6]$ by $[-4, 4]$. The "safe" window of $[-4, 4]$ by $[-4, 4]$ is stretched to a "square" window with an $x{:}y$ ratio of $3{:}2$. (Example 2)

EXAMPLE 3 Finding Another Maximum $|r|$ Value

Find the maximum $|r|$ value of $r = 3\cos 2\theta$ and use it to find an appropriate viewing window for the graph. Identify the points on the graph that are farthest from the pole.

SOLUTION We know that $-1 \le \cos 2\theta \le 1$, from which it follows that $-3 \le 3\cos 2\theta \le 3$. Thus $-3 \le r \le 3$ and the maximum $|r|$ value is 3. A rectangular graph of $y = |3\cos 2x|$ shows a maximum value of 3, supporting our algebraic analysis (Figure 6.50a).

A "safe" viewing window of $[-3, 3]$ by $[-3, 3]$ will enclose the entire curve, but we use the larger "square" window of $[-4.5, 4.5]$ by $[-3, 3]$ (Figure 6.50b). The four points that are farthest from the pole on the graph in Figure 6.49b correspond to the local maximum points on the graph in Figure 6.50a. Note that the points $(0, 3)$ and $(\pi, 3)$ on the rectangular graph correspond to $(3, 0)$ and $(3, \pi)$ on the polar curve because r is positive at those points, but the points $(\pi/2, 3)$ and $(3\pi/2, 3)$ correspond to points $(-3, \pi/2)$ and $(-3, 3\pi/2)$ because r is negative at those points. Points $(3, 0)$ and $(3, 2\pi)$ are the same in polar coordinates. **Now try Exercise 23.**

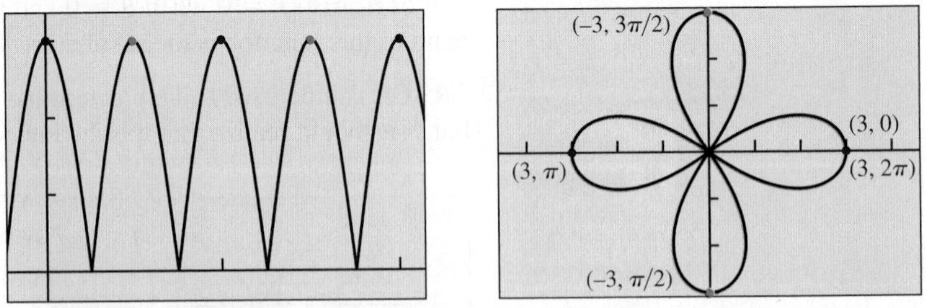

[–0.5, 6.5] by [–0.1, 3.1]
(a)

[–4.5, 4.5] by [–3, 3]
(b)

FIGURE 6.50 The points on the polar graph (b) of $r = 3\cos 2\theta$ that are farthest from the pole (0, 0) correspond to the local maximum points on the rectangular graph (a) of $y = |3\cos 2x|$. Note that the red points correspond to negative values of r on the polar graph. (Example 3)

Rose Curves

The curve in Example 1 is a 3-petal rose curve, and the curve in Example 3 is a 4-petal rose curve. The graphs of the polar equations $r = a\cos n\theta$ and $r = a\sin n\theta$, where n is an integer greater than 1, are **rose curves**. If n is odd there are n petals, and if n is even there are $2n$ petals. (In fact, the graph always produces n petals as θ goes from

0 to π, then n more petals as θ goes from π to 2π. When n is odd, however, the second group of n petals retraces the first set exactly.)

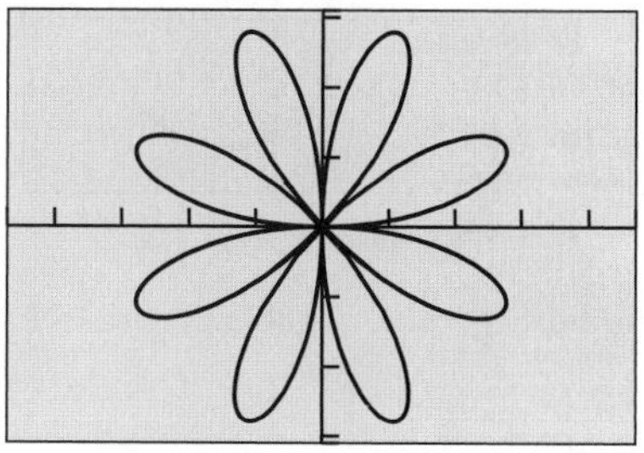

[–4.7, 4.7] by [–3.1, 3.1]

FIGURE 6.51 The graph of 8-petal rose curve $r = 3 \sin 4\theta$. (Example 4)

EXAMPLE 4 Analyzing a Rose Curve

Analyze the graph of the rose curve $r = 3 \sin 4\theta$.

SOLUTION Figure 6.51 shows the graph of the 8-petal rose curve $r = 3 \sin 4\theta$. The maximum $|r|$ value is 3. The graph appears to be symmetric about the x-axis, the y-axis, and the origin. For example, to prove that the graph is symmetric about the x-axis we replace (r, θ) by $(-r, \pi - \theta)$:

$$\begin{aligned} r &= 3 \sin 4\theta \\ -r &= 3 \sin 4(\pi - \theta) \\ -r &= 3 \sin (4\pi - 4\theta) \\ -r &= 3[\sin 4\pi \cos 4\theta - \cos 4\pi \sin 4\theta] && \text{Sine difference identity} \\ -r &= 3[(0) \cos 4\theta - (1) \sin 4\theta] && \sin 4\pi = 0, \cos 4\pi = 1 \\ -r &= -3 \sin 4\theta \\ r &= 3 \sin 4\theta \end{aligned}$$

Because the new polar equation is the same as the original equation, the graph is symmetric about the x-axis. In a similar way, you can prove that the graph is symmetric about the y-axis and the origin. (See Exercise 58.)

Domain: $(-\infty, \infty)$
Range: $[-3, 3]$
Continuous
Symmetric about the x-axis, the y-axis, and the origin
Bounded
Maximum $|r|$ value: 3
No asymptotes

Now try Exercise 29.

Here are the general characteristics of rose curves. You will investigate these curves in more detail in Exercises 67 and 68.

Graphs of Rose Curves

The graphs of $r = a \cos n\theta$ and $r = a \sin n\theta$, where $n > 1$ is an integer, have the following characteristics:

Domain: $(-\infty, \infty)$
Range: $[-|a|, |a|]$
Continuous
Symmetry: n even, symmetric about x-axis, y-axis, and the origin
n odd, $r = a \cos n\theta$ symmetric about x-axis
n odd, $r = a \sin n\theta$ symmetric about y-axis
Bounded
Maximum $|r|$ value: $|a|$
No asymptotes
Number of petals: n, if n is odd
$2n$, if n is even

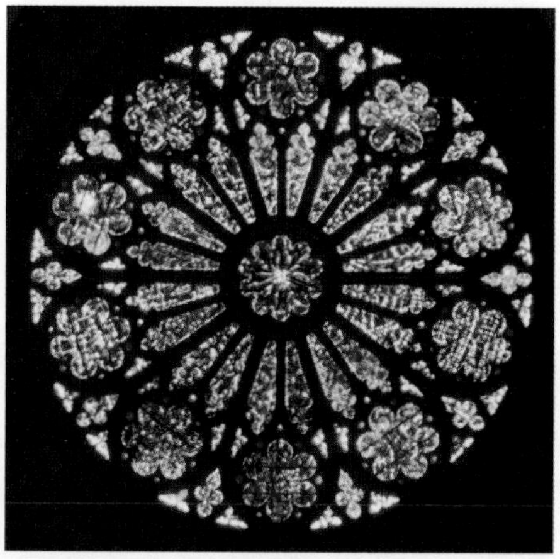

A ROSE IS A ROSE...

Budding botanists like to point out that the rose curve doesn't look much like a rose. However, consider the beautiful stained-glass window shown here, which is a feature of many great cathedrals and is called a "rose window."

Limaçon Curves

The **limaçon curves** are graphs of polar equations of the form

$$r = a \pm b \sin \theta \quad \text{and} \quad r = a \pm b \cos \theta,$$

where $a > 0$ and $b > 0$. *Limaçon*, pronounced "LEE-ma-sohn," is Old French for "snail." There are four different shapes of limaçons, as illustrated in Figure 6.52.

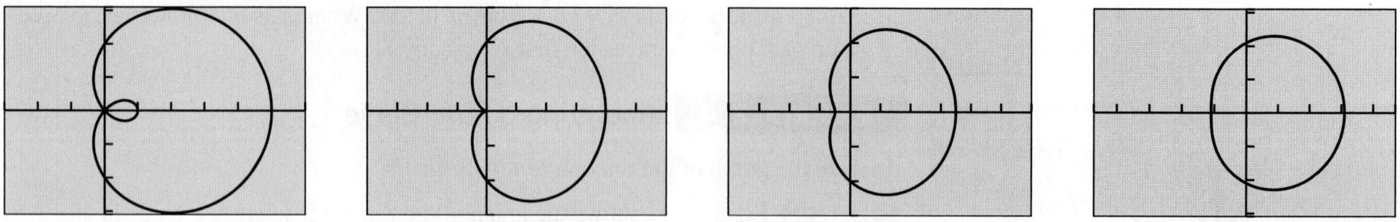

Limaçon with an inner loop: $\frac{a}{b} < 1$ (a)

Cardioid: $\frac{a}{b} = 1$ (b)

Dimpled limaçon: $1 < \frac{a}{b} < 2$ (c)

Convex limaçon: $\frac{a}{b} \geq 2$ (d)

FIGURE 6.52 The four types of limaçons.

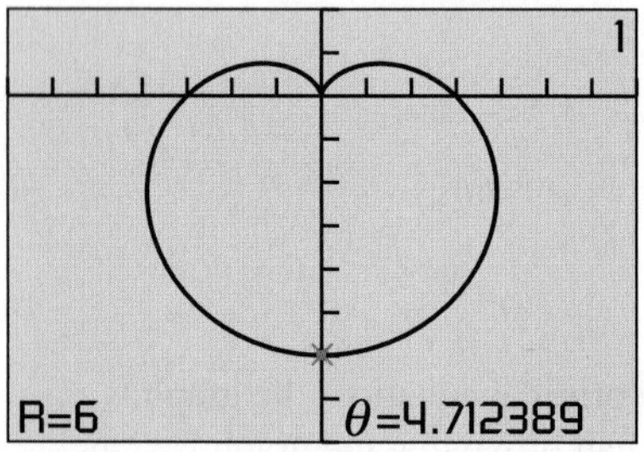

[–7, 7] by [–8, 2]

FIGURE 6.53 The graph of the cardioid of Example 5.

EXAMPLE 5 Analyzing a Limaçon Curve

Analyze the graph of $r = 3 - 3 \sin \theta$.

SOLUTION We can see from Figure 6.53 that the curve is a cardioid with maximum $|r|$ value 6. The graph is symmetric only about the y-axis.

Domain: $(-\infty, \infty)$
Range: $[0, 6]$
Continuous
Symmetric about the y-axis
Bounded
Maximum $|r|$ value: 6
No asymptotes

Now try Exercise 33.

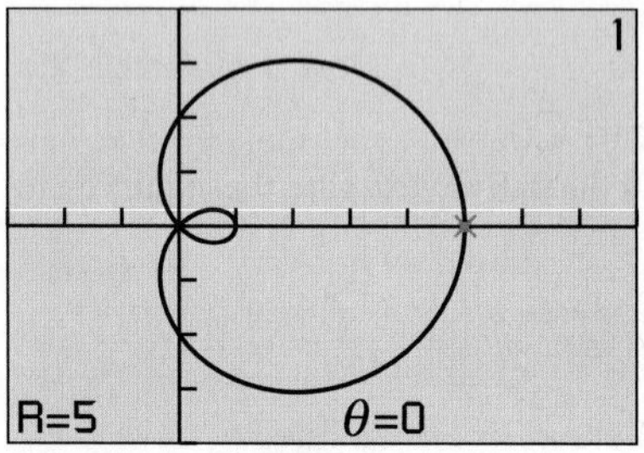

[–3, 8] by [–4, 4]

FIGURE 6.54 The graph of a limaçon with an inner loop. (Example 6)

EXAMPLE 6 Analyzing a Limaçon Curve

Analyze the graph of $r = 2 + 3 \cos \theta$.

SOLUTION We can see from Figure 6.54 that the curve is a limaçon with an inner loop and maximum $|r|$ value 5. The graph is symmetric only about the x-axis.

Domain: $(-\infty, \infty)$
Range: $[-1, 5]$
Continuous
Symmetric about the x-axis
Bounded
Maximum $|r|$ value: 5
No asymptotes

Now try Exercise 39.

Graphs of Limaçon Curves

The graphs of $r = a \pm b \sin \theta$ and $r = a \pm b \cos \theta$, where $a > 0$ and $b > 0$, have the following characteristics:

Domain: $(-\infty, \infty)$
Range: $[a - b, a + b]$
Continuous
Symmetry: $r = a \pm b \sin \theta$, symmetric about y-axis
$r = a \pm b \cos \theta$, symmetric about x-axis
Bounded
Maximum $|r|$ value: $a + b$
No asymptotes

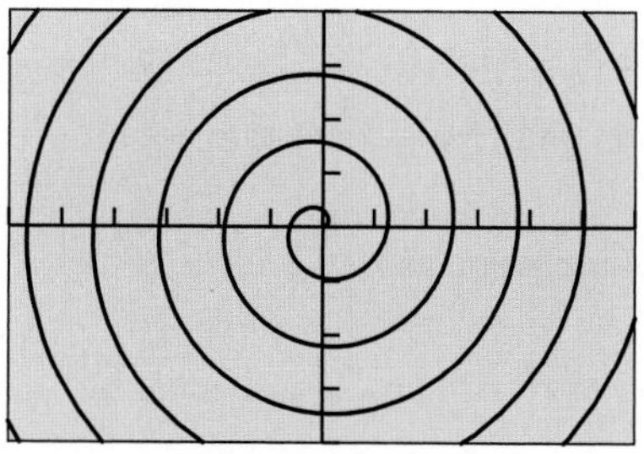

[–30, 30] by [–20, 20]
(a)

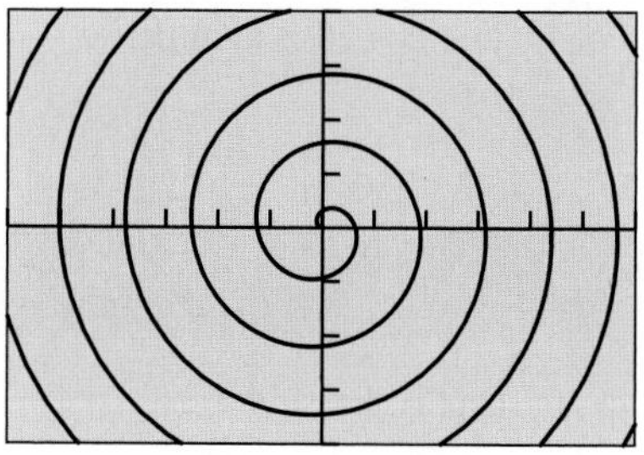

[–30, 30] by [–20, 20]
(b)

FIGURE 6.55 The graph of $r = \theta$ for (a) $\theta \geq 0$ (set θmin $= 0$, θmax $= 45$, θstep $= 0.1$) and (b) $\theta \leq 0$ (set θmin $= -45$, θmax $= 0$, θstep $= 0.1$). (Example 7)

EXPLORATION 1 Limaçon Curves

Try several values for a and b to convince yourself of the characteristics of limaçon curves listed above.

Other Polar Curves

All the polar curves we have graphed so far have been bounded. The spiral in Example 7 is unbounded.

EXAMPLE 7 Analyzing the Spiral of Archimedes

Analyze the graph of $r = \theta$.

SOLUTION

We can see from Figure 6.55 that the curve has no maximum $|r|$ value.

Domain: $(-\infty, \infty)$
Range: $(-\infty, \infty)$
Continuous
No symmetry
Unbounded
No maximum $|r|$ value
No asymptotes

Now try Exercise 41.

The **lemniscate curves** are graphs of polar equations of the form

$$r^2 = a^2 \sin 2\theta \quad \text{and} \quad r^2 = a^2 \cos 2\theta.$$

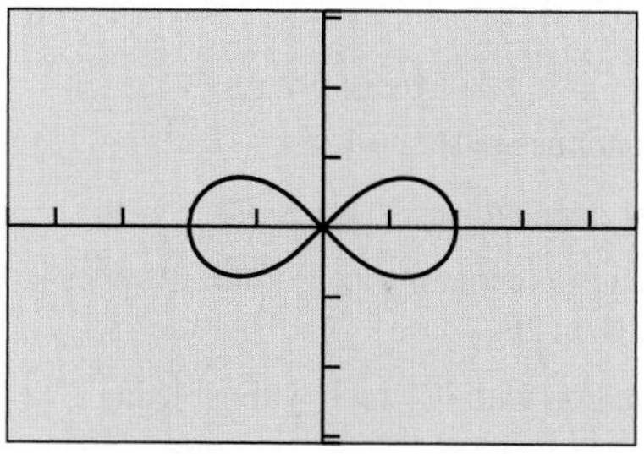

[–4.7, 4.7] by [–3.1, 3.1]

FIGURE 6.56 The graph of the lemniscate $r^2 = 4 \cos 2\theta$. (Example 8)

EXAMPLE 8 Analyzing a Lemniscate Curve

Analyze the graph of $r^2 = 4 \cos 2\theta$ for $[0, 2\pi]$.

SOLUTION It turns out that you can get the complete graph using $r = 2\sqrt{\cos 2\theta}$. You also need to choose a very small θ step to produce the graph in Figure 6.56.

Domain: $[0, \pi/4] \cup [3\pi/4, 5\pi/4] \cup [7\pi/4, 2\pi]$
Range: $[-2, 2]$
Symmetric about the x-axis, the y-axis, and the origin
Continuous (on its domain)
Bounded
Maximum $|r|$ value: 2
No asymptotes

Now try Exercise 43.

EXPLORATION 2 Revisiting Example 8

1. Prove that θ values in the intervals $(\pi/4, 3\pi/4)$ and $(5\pi/4, 7\pi/4)$ are not in the domain of the polar equation $r^2 = 4 \cos 2\theta$.
2. Explain why $r = -2\sqrt{\cos 2\theta}$ produces the same graph as $r = 2\sqrt{\cos 2\theta}$ in the interval $[0, 2\pi]$.
3. Use the symmetry tests to show that the graph of $r^2 = 4 \cos 2\theta$ is symmetric about the x-axis.
4. Use the symmetry tests to show that the graph of $r^2 = 4 \cos 2\theta$ is symmetric about the y-axis.
5. Use the symmetry tests to show that the graph of $r^2 = 4 \cos 2\theta$ is symmetric about the origin.

QUICK REVIEW 6.5 *(For help, go to Sections 1.2 and 5.3.)*

Exercise numbers with a gray background indicate problems that the authors have designed to be solved *without a calculator*.

In Exercises 1–4, find the absolute maximum value and absolute minimum value in $[0, 2\pi]$ and where they occur.

1. $y = 3 \cos 2x$
2. $y = 2 + 3 \cos x$
3. $y = 2\sqrt{\cos 2x}$
4. $y = 3 - 3 \sin x$

In Exercises 5 and 6, determine if the graph of the function is symmetric about the **(a)** x-axis, **(b)** y-axis, and **(c)** origin.

5. $y = \sin 2x$
6. $y = \cos 4x$

In Exercises 7–10, use trig identities to simplify the expression.

7. $\sin(\pi - \theta)$
8. $\cos(\pi - \theta)$
9. $\cos 2(\pi + \theta)$
10. $\sin 2(\pi + \theta)$

SECTION 6.5 Exercises

In Exercises 1 and 2, **(a)** complete the table for the polar equation, and **(b)** plot the corresponding points.

1. $r = 3 \cos 2\theta$

θ	0	$\pi/4$	$\pi/2$	$3\pi/4$	π	$5\pi/4$	$3\pi/2$	$7\pi/4$
r								

2. $r = 2 \sin 3\theta$

θ	0	$\pi/6$	$\pi/3$	$\pi/2$	$2\pi/3$	$5\pi/6$	π
r							

In Exercises 3–6, draw a graph of the rose curve. State the smallest θ-interval $(0 \le \theta \le k)$ that will produce a complete graph.

3. $r = 3 \sin 3\theta$
4. $r = -3 \cos 2\theta$
5. $r = 3 \cos 2\theta$
6. $r = 3 \sin 5\theta$

Exercises 7 and 8 refer to the curves in the given figure.

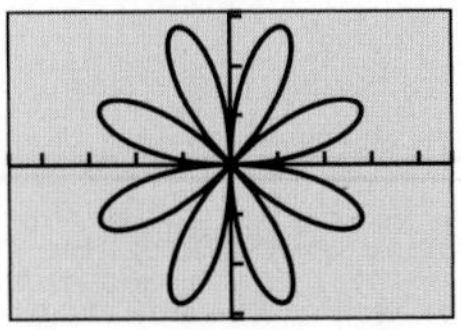

[–4.7, 4.7] by [–3.1, 3.1]

(a)

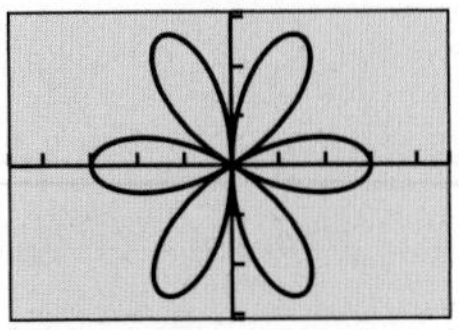

[–4.7, 4.7] by [–3.1, 3.1]

(b)

7. The graphs of which equations are shown?

$r_1 = 3 \cos 6\theta \quad r_2 = 3 \sin 8\theta \quad r_3 = 3|\cos 3\theta|$

8. Use trigonometric identities to explain which of these curves is the graph of $r = 6 \cos 2\theta \sin 2\theta$.

In Exercises 9–12, match the equation with its graph without using your graphing calculator.

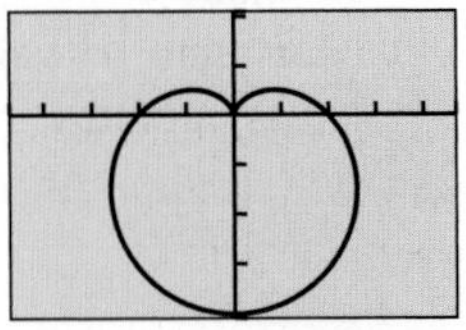

[–4.7, 4.7] by [–4.1, 2.1]

(a)

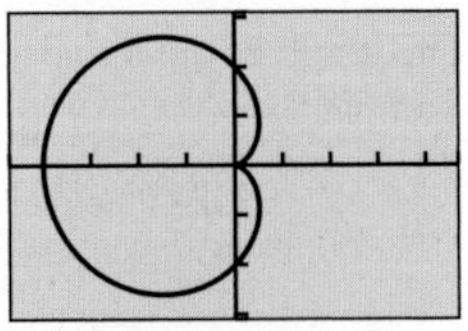

[–4.7, 4.7] by [–3.1, 3.1]

(b)

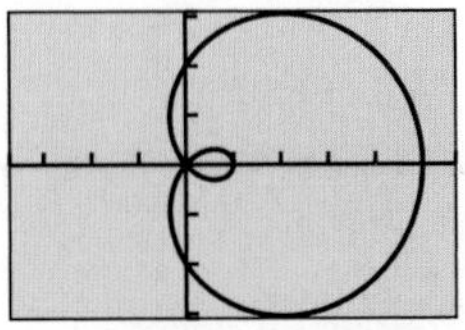

[–3.7, 5.7] by [–3.1, 3.1]

(c)

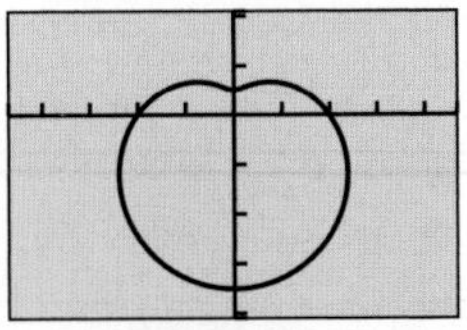

[–4.7, 4.7] by [–4.1, 2.1]

(d)

9. Does the graph of $r = 2 + 2 \sin \theta$ or $r = 2 - 2 \cos \theta$ appear in the figure? Explain.

10. Does the graph of $r = 2 + 3 \cos \theta$ or $r = 2 - 3 \cos \theta$ appear in the figure? Explain.

11. Is the graph in (a) the graph of $r = 2 - 2 \sin \theta$ or $r = 2 + 2 \cos \theta$? Explain.

12. Is the graph in (d) the graph of $r = 2 + 1.5 \cos \theta$ or $r = 2 - 1.5 \sin \theta$? Explain.

In Exercises 13–20, use the polar symmetry tests to determine if the graph is symmetric about the *x*-axis, the *y*-axis, or the origin. Support your algebraic solution with a grapher.

13. $r = 3 + 3 \sin \theta$ **14.** $r = 1 + 2 \cos \theta$

15. $r = 4 - 3 \cos \theta$ **16.** $r = 1 - 3 \sin \theta$

17. $r = 5 \cos 2\theta$ **18.** $r = 7 \sin 3\theta$

19. $r = \dfrac{3}{1 + \sin \theta}$ **20.** $r = \dfrac{2}{1 - \cos \theta}$

In Exercises 21–24, identify the points for $0 \le \theta \le 2\pi$ where maximum $|r|$ values occur on the graph of the polar equation.

21. $r = 2 + 3 \cos \theta$ **22.** $r = -3 + 2 \sin \theta$

23. $r = 3 \cos 3\theta$ **24.** $r = 4 \sin 2\theta$

In Exercises 25–44, analyze the graph of the polar curve.

25. $r = 3$ **26.** $r = -2$

27. $\theta = \pi/3$ **28.** $\theta = -\pi/4$

29. $r = 2 \sin 3\theta$ **30.** $r = -3 \cos 4\theta$

31. $r = 5 + 4 \sin \theta$ **32.** $r = 6 - 5 \cos \theta$

33. $r = 4 + 4 \cos \theta$ **34.** $r = 5 - 5 \sin \theta$

35. $r = 5 + 2 \cos \theta$ **36.** $r = 3 - \sin \theta$

37. $r = 2 + 5 \cos \theta$ **38.** $r = 3 - 4 \sin \theta$

39. $r = 1 - \cos \theta$ **40.** $r = 2 + \sin \theta$

41. $r = 2\theta$ **42.** $r = \theta/4$

43. $r^2 = \sin 2\theta, 0 \le \theta \le 2\pi$

44. $r^2 = 9 \cos 2\theta, 0 \le \theta \le 2\pi$

In Exercises 45–48, find the length of each petal of the polar curve.

45. $r = 2 + 4 \sin 2\theta$ **46.** $r = 3 - 5 \cos 2\theta$

47. $r = 1 - 4 \cos 5\theta$ **48.** $r = 3 + 4 \sin 5\theta$

In Exercises 49–52, select the two equations whose graphs are the same curve. Then, even though the graphs of the equations are identical, describe how the two paths are different as θ increases from 0 to 2π.

49. $r_1 = 1 + 3 \sin \theta,\quad r_2 = -1 + 3 \sin \theta,\quad r_3 = 1 - 3 \sin \theta$

50. $r_1 = 1 + 2 \cos \theta,\quad r_2 = -1 - 2 \cos \theta,\quad r_3 = -1 + 2 \cos \theta$

51. $r_1 = 1 + 2 \cos \theta,\quad r_2 = 1 - 2 \cos \theta,\quad r_3 = -1 - 2 \cos \theta$

52. $r_1 = 2 + 2 \sin \theta,\quad r_2 = -2 + 2 \sin \theta,\quad r_3 = 2 - 2 \sin \theta$

In Exercises 53–56, **(a)** describe the graph of the polar equation, **(b)** state any symmetry that the graph possesses, and **(c)** state its maximum $|r|$ value if it exists.

53. $r = 2 \sin^2 2\theta + \sin 2\theta$ **54.** $r = 3 \cos 2\theta - \sin 3\theta$

55. $r = 1 - 3 \cos 3\theta$ **56.** $r = 1 + 3 \sin 3\theta$

57. Group Activity Analyze the graphs of the polar equations $r = a \cos n\theta$ and $r = a \sin n\theta$ when n is an even integer.

58. Revisiting Example 4 Use the polar symmetry tests to prove that the graph of the curve $r = 3 \sin 4\theta$ is symmetric about the *y*-axis and the origin.

59. Writing to Learn Revisiting Example 5 Confirm the range stated for the polar function $r = 3 - 3 \sin \theta$ of Example 5 by graphing $y = 3 - 3 \sin x$ for $0 \le x \le 2\pi$. Explain why this works.

60. Writing to Learn Revisiting Example 6 Confirm the range stated for the polar function $r = 2 + 3 \cos \theta$ of Example 6 by graphing $y = 2 + 3 \cos x$ for $0 \le x \le 2\pi$. Explain why this works.

Standardized Test Questions

61. True or False A polar curve is always bounded. Justify your answer.

62. True or False The graph of $r = 2 + \cos \theta$ is symmetric about the *x*-axis. Justify your answer.

In Exercises 63–66, solve the problem without using a calculator.

63. Multiple Choice Which of the following gives the number of petals of the rose curve $r = 3 \cos 2\theta$?

(A) 1 **(B)** 2 **(C)** 3 **(D)** 4 **(E)** 6

64. Multiple Choice Which of the following describes the symmetry of the rose graph of $r = 3 \cos 2\theta$?

(A) Only the *x*-axis

(B) Only the *y*-axis

(C) Only the origin

(D) The *x*-axis, the *y*-axis, the origin

(E) Not symmetric about the *x*-axis, the *y*-axis, or the origin

65. Multiple Choice Which of the following is a maximum $|r|$ value for $r = 2 - 3 \cos \theta$?

(A) 6 **(B)** 5 **(C)** 3

(D) 2 **(E)** 1

66. Multiple Choice Which of the following is the number of petals of the rose curve $r = 5 \sin 3\theta$?

(A) 1 **(B)** 3 **(C)** 6

(D) 10 **(E)** 15

Explorations

67. Analyzing Rose Curves Consider the polar equation $r = a \cos n\theta$ for n, an odd integer.

(a) Prove that the graph is symmetric about the *x*-axis.

(b) Prove that the graph is not symmetric about the *y*-axis.

(c) Prove that the graph is not symmetric about the origin.

(d) Prove that the maximum $|r|$ value is $|a|$.

(e) Analyze the graph of this curve.

68. Analyzing Rose Curves Consider the polar equation $r = a \sin n\theta$ for n, an odd integer.

(a) Prove that the graph is symmetric about the y-axis.

(b) Prove that the graph is not symmetric about the x-axis.

(c) Prove that the graph is not symmetric about the origin.

(d) Prove that the maximum $|r|$ value is $|a|$.

(e) Analyze the graph of this curve.

69. Extended Rose Curves The graphs of $r_1 = 3 \sin((7/2)\theta)$ and $r_2 = 3 \sin((7/2)\theta)$ may be called rose curves.

(a) Determine the smallest θ-interval that will produce a complete graph of r_1; of r_2.

(b) How many petals does each graph have?

Extending the Ideas

In Exercises 70–72, graph each polar equation. Describe how they are related to each other.

70. **(a)** $r_1 = 3 \sin 3\theta$ **(b)** $r_2 = 3 \sin 3\left(\theta + \frac{\pi}{12}\right)$

(c) $r_3 = 3 \sin 3\left(\theta + \frac{\pi}{4}\right)$

71. **(a)** $r_1 = 2 \sec \theta$ **(b)** $r_2 = 2 \sec\left(\theta - \frac{\pi}{4}\right)$

(c) $r_3 = 2 \sec\left(\theta - \frac{\pi}{3}\right)$

72. **(a)** $r_1 = 2 - 2 \cos \theta$ **(b)** $r_2 = r_1\left(\theta + \frac{\pi}{4}\right)$

(c) $r_3 = r_1\left(\theta + \frac{\pi}{3}\right)$

73. Writing to Learn Describe how the graphs of $r = f(\theta)$, $r = f(\theta + \alpha)$, and $r = f(\theta - \alpha)$ are related. Explain why you think this generalization is true.

6.6 De Moivre's Theorem and *n*th Roots

What you'll learn about
- The Complex Plane
- Midpoint and Distance Formulas in the Complex Plane
- Polar Form of Complex Numbers
- Multiplication and Division of Complex Numbers
- Powers of Complex Numbers
- Roots of Complex Numbers

... and why
This material extends your equation-solving technique to include equations of the form $z^n = c$, n an integer and c a complex number.

The Complex Plane

You might be curious as to why we addressed complex numbers in Sections P.6 and 2.5, but generally have ignored them elsewhere. (Indeed, after this section we will pretty much ignore them again.) The reason is simply because the key to understanding calculus is the graphing of functions in the Cartesian plane, which consists of two perpendicular real (not complex) lines.

We are not saying that complex numbers are impossible to graph. Just as every real number is associated with a point of the real number line, every complex number can be associated with a point of the **complex plane**. This idea evolved through the work of Caspar Wessel (1745–1818), Jean-Robert Argand (1768–1822), and Carl Friedrich Gauss (1777–1855). Real numbers are placed along the horizontal axis (the **real axis**) and imaginary numbers along the vertical axis (the **imaginary axis**), thus associating the complex number $a + bi$ with the point (a, b). In Figure 6.57 we show the graph of $2 + 3i$ as an example.

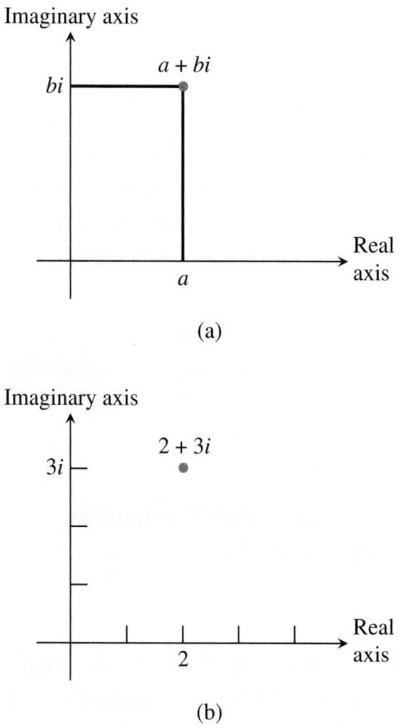

FIGURE 6.57 Plotting points in the complex plane.

EXAMPLE 1 Plotting Complex Numbers

Plot $u = 1 + 3i$, $v = 2 - i$, and $u + v$ in the complex plane. These three points and the origin determine a quadrilateral. Is it a parallelogram?

SOLUTION First notice that $u + v = (1 + 3i) + (2 - i) = 3 + 2i$. The numbers u, v, and $u + v$ are plotted in Figure 6.58a. The quadrilateral is a parallelogram because the arithmetic is exactly the same as in vector addition (Figure 6.58b).

Now try Exercise 1.

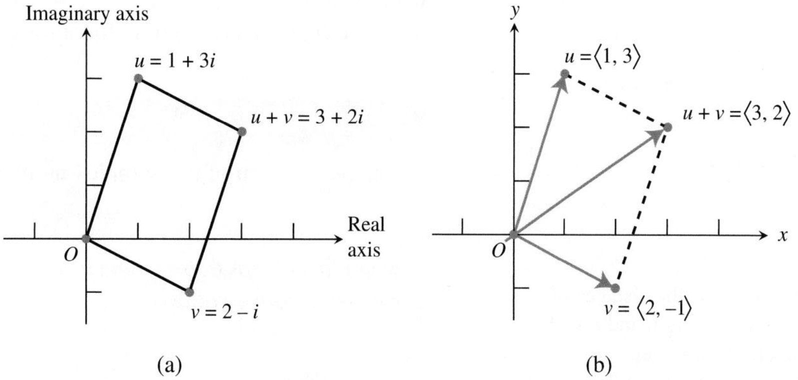

FIGURE 6.58 (a) Two numbers and their sum are plotted in the complex plane. (b) The arithmetic is the same as in vector addition. (Example 1)

Example 1 shows how the complex plane representation of complex number addition is virtually the same as the Cartesian plane representation of vector addition. Another similarity between complex numbers and two-dimensional vectors is the definition of absolute value.

Is There a Calculus of Complex Functions?

There is a calculus of functions of complex variables. If you study it someday, it should only be after acquiring a pretty firm algebraic and geometric understanding of the calculus of functions of real variables.

DEFINITION Absolute Value (Modulus) of a Complex Number

The **absolute value** or **modulus** of a complex number $z = a + bi$ is

$$|z| = |a + bi| = \sqrt{a^2 + b^2}.$$

In the complex plane, $|a + bi|$ is the distance of $a + bi$ from the origin.

Midpoint and Distance Formulas in the Complex Plane

There are some nice parallels between the geometric interpretation of number operations on the real line and the geometric interpretation of number operations in the complex plane. Here are two of them:

Number Operation	Algebraic Interpretation	Geometric Interpretation
Real numbers: $\dfrac{a+b}{2}$	Average (mean) of a and b	Number midway between a and b on the real line
Complex numbers: $\dfrac{z_1+z_2}{2}$	Average (mean) of z_1 and z_2	Number midway between z_1 and z_2 in the complex plane
Real numbers: $\lvert a-b\rvert$	Closeness of a to b	Distance between a and b on the real line
Complex numbers: $\lvert z_1-z_2\rvert$	Closeness of z_1 to z_2	Distance between z_1 and z_2 in the complex plane

These insights are important in future courses for studying functions of complex variables. For example, the idea of closeness is critical for understanding what continuity means!

Polar Form of Complex Numbers

Figure 6.59 shows the graph of $z = a + bi$ in the complex plane. The distance r from the origin is the modulus of z. If we define a direction angle θ for z just as we did with vectors, we see that $a = r\cos\theta$ and $b = r\sin\theta$. Substituting these expressions for a and b gives us the **polar form** (or **trigonometric form**) of the complex number z.

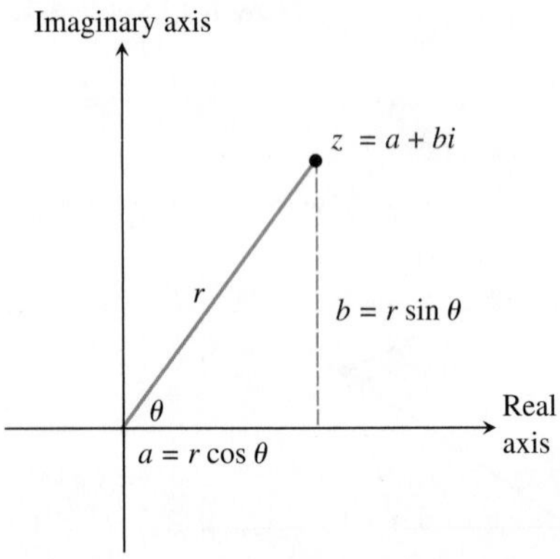

FIGURE 6.59 If r is the distance of $z = a + bi$ from the origin and θ is the directional angle shown, then $z = r(\cos\theta + i\sin\theta)$, which is the polar form of z.

DEFINITION Polar Form of a Complex Number

The **polar form** of the complex number $z = a + bi$ is

$$z = r(\cos\theta + i\sin\theta)$$

where $a = r\cos\theta$, $b = r\sin\theta$, $r = \sqrt{a^2 + b^2}$, and $\tan\theta = b/a$. The number r is the *absolute value* or *modulus* of z, and θ is an **argument** of z.

An angle θ for the polar form of z can always be chosen so that $0 \le \theta \le 2\pi$, although any angle coterminal with θ could be used. Consequently, the *angle* θ and *argument* of a complex number z are not unique. It follows that the polar form of a complex number z is not unique.

Polar Form

What's in a cis?

Polar (or trigonometric) form appears frequently enough in scientific texts to have an abbreviated form. The expression "$\cos\theta + i\sin\theta$" is often shortened to "cis θ" (pronounced "kiss θ"). Thus $z = r\,\text{cis}\,\theta$. Leonhard Euler showed that $re^{i\theta} = r\,cis\,\theta$, giving an exponential version of polar form. (See Exercise 73.)

EXAMPLE 2 Finding Polar Forms

Use an algebraic method to find the polar form with $0 \le \theta < 2\pi$ for the complex number. Approximate exact values with a calculator when appropriate.

(a) $1 - \sqrt{3}i$ **(b)** $-3 - 4i$

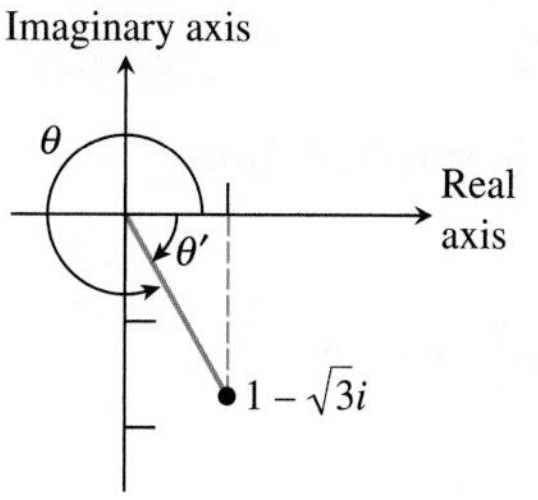

FIGURE 6.60 The complex number for Example 2a.

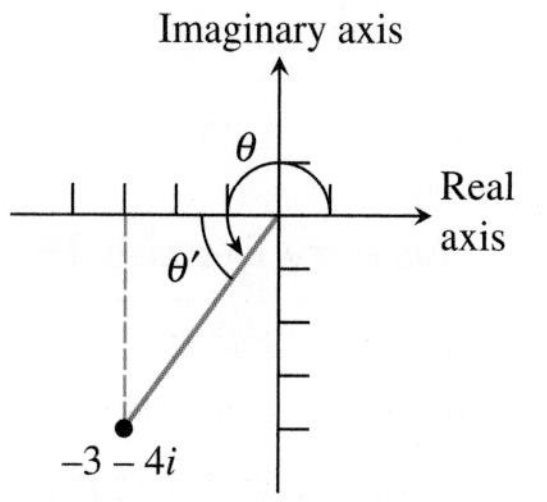

FIGURE 6.61 The complex number for Example 2b.

SOLUTION

(a) For $1 - \sqrt{3}i$,

$$r = |1 - \sqrt{3}i| = \sqrt{(1)^2 + (\sqrt{3})^2} = 2.$$

Because the reference angle θ' for θ is $-\pi/3$ (Figure 6.60),

$$\theta = 2\pi + \left(-\frac{\pi}{3}\right) = \frac{5\pi}{3}.$$

Thus,

$$1 - \sqrt{3}i = 2\cos\frac{5\pi}{3} + 2i\sin\frac{5\pi}{3}.$$

(b) For $-3 - 4i$,

$$|-3 - 4i| = \sqrt{(-3)^2 + (-4)^2} = 5.$$

The reference angle θ' for θ (Figure 6.61) satisfies the equation

$$\tan\theta' = \frac{4}{3}, \quad \text{so}$$

$$\theta' = \tan^{-1}\frac{4}{3} = 0.927\ldots.$$

Because the terminal side of θ is in the third quadrant, we conclude that

$$\theta = \pi + \theta' \approx 4.07.$$

Therefore,

$$-3 - 4i \approx 5(\cos 4.07 + i\sin 4.07).$$

Now try Exercise 5.

Multiplication and Division of Complex Numbers

The polar form for complex numbers is particularly convenient for multiplying and dividing complex numbers. The product involves the product of the moduli and the sum of the arguments. (*Moduli* is the plural of *modulus*.) The quotient involves the quotient of the moduli and the difference of the arguments.

Product and Quotient of Complex Numbers

Let $z_1 = r_1(\cos\theta_1 + i\sin\theta_1)$ and $z_2 = r_2(\cos\theta_2 + i\sin\theta_2)$. Then

1. $z_1 \cdot z_2 = r_1 r_2[\cos(\theta_1 + \theta_2) + i\sin(\theta_1 + \theta_2)]$.

2. $\dfrac{z_1}{z_2} = \dfrac{r_1}{r_2}[\cos(\theta_1 - \theta_2) + i\sin(\theta_1 - \theta_2)], \quad r_2 \neq 0.$

The proof of the product formula relies on two sum identities:

$$\begin{aligned} z_1 \cdot z_2 &= r_1(\cos\theta_1 + i\sin\theta_1) \cdot r_2(\cos\theta_2 + i\sin\theta_2) \\ &= r_1 r_2[(\cos\theta_1\cos\theta_2 - \sin\theta_1\sin\theta_2) + i(\sin\theta_1\cos\theta_2 + \cos\theta_1\sin\theta_2)] \\ &= r_1 r_2[\cos(\theta_1 + \theta_2) + i\sin(\theta_1 + \theta_2)] \end{aligned}$$

You will be asked to prove the quotient formula in Exercise 63.

EXAMPLE 3 Multiplying Complex Numbers

Use an algebraic method to express the product of z_1 and z_2 in standard form. Approximate exact values with a calculator when appropriate.

$$z_1 = 25\sqrt{2}\left(\cos\frac{-\pi}{4} + i\sin\frac{-\pi}{4}\right),\quad z_2 = 14\left(\cos\frac{\pi}{3} + i\sin\frac{\pi}{3}\right)$$

SOLUTION

$$\begin{aligned} z_1 \cdot z_2 &= 25\sqrt{2}\left(\cos\frac{-\pi}{4} + i\sin\frac{-\pi}{4}\right)\cdot 14\left(\cos\frac{\pi}{3} + i\sin\frac{\pi}{3}\right) \\ &= 25\cdot 14\sqrt{2}\left[\cos\left(\frac{-\pi}{4} + \frac{\pi}{3}\right) + i\sin\left(\frac{-\pi}{4} + \frac{\pi}{3}\right)\right] \\ &= 350\sqrt{2}\left(\cos\frac{\pi}{12} + i\sin\frac{\pi}{12}\right) \\ &\approx 478.11 + 128.11i \end{aligned}$$

Now try Exercise 19.

EXAMPLE 4 Dividing Complex Numbers

Use an algebraic method to express the product z_1/z_2 in standard form. Approximate exact values with a calculator when appropriate.

$$z_1 = 2\sqrt{2}(\cos 135° + i\sin 135°),\quad z_2 = 6(\cos 300° + i\sin 300°)$$

SOLUTION

$$\begin{aligned} \frac{z_1}{z_2} &= \frac{2\sqrt{2}\,(\cos 135° + i\sin 135°)}{6(\cos 300° + i\sin 300°)} \\ &= \frac{\sqrt{2}}{3}\left[\cos(135° - 300°) + i\sin(135° - 300°)\right] \\ &= \frac{\sqrt{2}}{3}\left[\cos(-165°) + i\sin(-165°)\right] \\ &\approx -0.46 - 0.12i \end{aligned}$$

Now try Exercise 23.

Powers of Complex Numbers

We can use the product formula to raise a complex number to a power. For example, let $z = r(\cos\theta + i\sin\theta)$. Then

$$\begin{aligned} z^2 &= z\cdot z \\ &= r(\cos\theta + i\sin\theta)\cdot r(\cos\theta + i\sin\theta) \\ &= r^2[\cos(\theta + \theta) + i\sin(\theta + \theta)] \\ &= r^2(\cos 2\theta + i\sin 2\theta) \end{aligned}$$

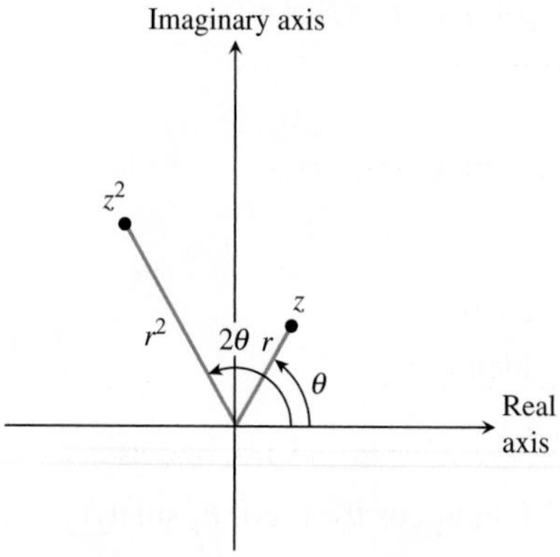

FIGURE 6.62 A geometric interpretation of z^2.

Figure 6.62 gives a geometric interpretation of squaring a complex number: Its argument is doubled and its distance from the origin is multiplied by a factor of r, increased if $r > 1$ or decreased if $r < 1$.

We can find z^3 by multiplying z by z^2:

$$\begin{aligned} z^3 &= z\cdot z^2 \\ &= r(\cos\theta + i\sin\theta)\cdot r^2(\cos 2\theta + i\sin 2\theta) \\ &= r^3[\cos(\theta + 2\theta) + i\sin(\theta + 2\theta)] \\ &= r^3(\cos 3\theta + i\sin 3\theta) \end{aligned}$$

Similarly,

$$z^4 = r^4(\cos 4\theta + i \sin 4\theta)$$
$$z^5 = r^5(\cos 5\theta + i \sin 5\theta)$$
$$\vdots$$

This pattern can be generalized to the following theorem, named after the mathematician Abraham De Moivre (1667–1754), who also made major contributions to the field of probability.

De Moivre's Theorem

Let $z = r(\cos\theta + i\sin\theta)$ and let n be a positive integer. Then

$$z^n = [r(\cos\theta + i\sin\theta)]^n = r^n(\cos n\theta + i \sin n\theta).$$

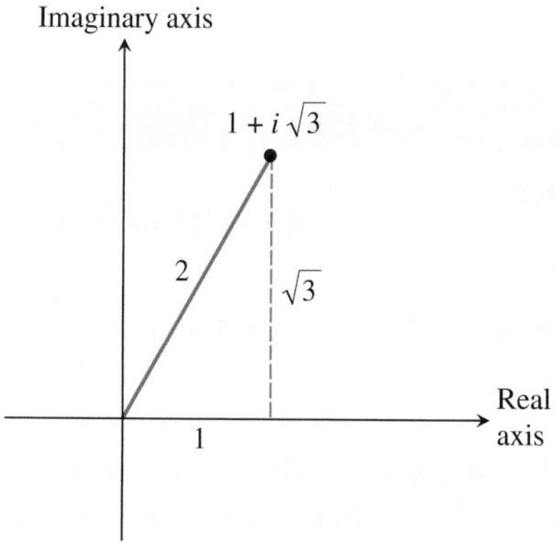

FIGURE 6.63 The complex number in Example 5.

EXAMPLE 5 Using De Moivre's Theorem

Find $(1 + i\sqrt{3})^3$ using De Moivre's Theorem.

SOLUTION

Solve Algebraically See Figure 6.63. The argument of $z = 1 + i\sqrt{3}$ is $\theta = \pi/3$, and its modulus is $|1 + i\sqrt{3}| = \sqrt{1+3} = 2$. Therefore,

$$z = 2\left(\cos\frac{\pi}{3} + i\sin\frac{\pi}{3}\right)$$
$$z^3 = 2^3\left[\cos\left(3\cdot\frac{\pi}{3}\right) + i\sin\left(3\cdot\frac{\pi}{3}\right)\right]$$
$$= 8(\cos\pi + i\sin\pi)$$
$$= 8(-1 + 0i) = -8$$

Support Numerically Figure 6.64a sets the graphing calculator we use in complex number mode. Figure 6.64b supports the result obtained algebraically.

Now try Exercise 31.

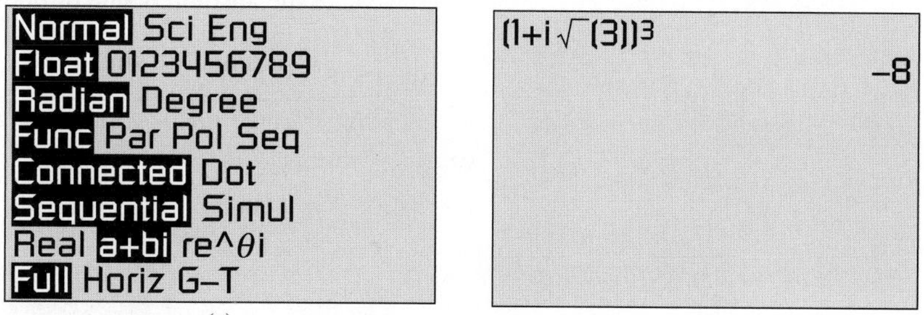

(a) (b)

FIGURE 6.64 (a) Setting a graphing calculator in complex number mode. (b) Computing $(1 + i\sqrt{3})^3$ with a graphing calculator.

EXAMPLE 6 Using De Moivre's Theorem

Find $[(-\sqrt{2}/2) + i(\sqrt{2}/2)]^8$ using De Moivre's Theorem.

SOLUTION The argument of $z = (-\sqrt{2}/2) + i(\sqrt{2}/2)$ is $\theta = 3\pi/4$, and its modulus is

$$\left|\frac{-\sqrt{2}}{2} + i\frac{\sqrt{2}}{2}\right| = \sqrt{\frac{1}{2} + \frac{1}{2}} = 1.$$

(continued)

Therefore,

$$z = \cos\frac{3\pi}{4} + i\sin\frac{3\pi}{4}$$

$$z^8 = \cos\left(8\cdot\frac{3\pi}{4}\right) + i\sin\left(8\cdot\frac{3\pi}{4}\right)$$

$$= \cos 6\pi + i\sin 6\pi$$

$$= 1 + i\cdot 0 = 1$$

Now try Exercise 35.

Roots of Complex Numbers

The complex number $1 + i\sqrt{3}$ in Example 5 is a solution of $z^3 = -8$, and the complex number $\left(-\sqrt{2}/2\right) + i\left(\sqrt{2}/2\right)$ in Example 6 is a solution of $z^8 = 1$. The complex number $1 + i\sqrt{3}$ is a third root of -8, and $\left(-\sqrt{2}/2\right) + i\left(\sqrt{2}/2\right)$ is an eighth root of 1.

DEFINITION *n*th Root of a Complex Number

A complex number $v = a + bi$ is an ***n*th root of *z*** if

$$v^n = z.$$

If $z = 1$, then v is an ***n*th root of unity**.

We use De Moivre's Theorem to develop a general formula for finding the nth roots of a nonzero complex number. Suppose that $v = s(\cos\alpha + i\sin\alpha)$ is an nth root of $z = r(\cos\theta + i\sin\theta)$. Then

$$v^n = z$$

$$\left[s(\cos\alpha + i\sin\alpha)\right]^n = r(\cos\theta + i\sin\theta)$$

$$s^n(\cos n\alpha + i\sin n\alpha) = r(\cos\theta + i\sin\theta) \qquad (1)$$

Next, we take the absolute value of both sides:

$$\left|s^n(\cos n\alpha + i\sin n\alpha)\right| = \left|r(\cos\theta + i\sin\theta)\right|$$

$$\sqrt{s^{2n}(\cos^2 n\alpha + \sin^2 n\alpha)} = \sqrt{r^2(\cos^2\theta + \sin^2\theta)}$$

$$\sqrt{s^{2n}} = \sqrt{r^2}$$

$$s^n = r \qquad s > 0, r > 0$$

$$s = \sqrt[n]{r}$$

Substituting $s^n = r$ into Equation (1), we obtain

$$\cos n\alpha + i\sin n\alpha = \cos\theta + i\sin\theta.$$

Therefore, $n\alpha$ can be any angle coterminal with θ. Consequently, for any integer k, v is an nth root of z if $s = \sqrt[n]{r}$ and

$$n\alpha = \theta + 2\pi k$$

$$\alpha = \frac{\theta + 2\pi k}{n}.$$

The expression for v takes on n different values for $k = 0, 1, \ldots, n - 1$, and the values start to repeat for $k = n, n + 1, \ldots$.

We summarize this result.

Finding *n*th Roots of a Complex Number

If $z = r(\cos\theta + i\sin\theta)$, then the n distinct complex numbers

$$\sqrt[n]{r}\left(\cos\frac{\theta + 2\pi k}{n} + i\sin\frac{\theta + 2\pi k}{n}\right),$$

where $k = 0, 1, 2, \ldots, n - 1$, are the nth roots of the complex number z.

EXAMPLE 7 Finding Fourth Roots

Find the fourth roots of $z = 5(\cos(\pi/3) + i\sin(\pi/3))$.

SOLUTION The fourth roots of z are the complex numbers

$$\sqrt[4]{5}\left(\cos\frac{\pi/3 + 2\pi k}{4} + i\sin\frac{\pi/3 + 2\pi k}{4}\right)$$

for $k = 0, 1, 2, 3$.

Taking into account that $(\pi/3 + 2\pi k)/4 = \pi/12 + \pi k/2$, the list becomes

$$z_1 = \sqrt[4]{5}\left[\cos\left(\frac{\pi}{12} + \frac{0}{2}\right) + i\sin\left(\frac{\pi}{12} + \frac{0}{2}\right)\right]$$

$$= \sqrt[4]{5}\left[\cos\frac{\pi}{12} + i\sin\frac{\pi}{12}\right]$$

$$z_2 = \sqrt[4]{5}\left[\cos\left(\frac{\pi}{12} + \frac{\pi}{2}\right) + i\sin\left(\frac{\pi}{12} + \frac{\pi}{2}\right)\right]$$

$$= \sqrt[4]{5}\left[\cos\frac{7\pi}{12} + i\sin\frac{7\pi}{12}\right]$$

$$z_3 = \sqrt[4]{5}\left[\cos\left(\frac{\pi}{12} + \frac{2\pi}{2}\right) + i\sin\left(\frac{\pi}{12} + \frac{2\pi}{2}\right)\right]$$

$$= \sqrt[4]{5}\left[\cos\frac{13\pi}{12} + i\sin\frac{13\pi}{12}\right]$$

$$z_4 = \sqrt[4]{5}\left[\cos\left(\frac{\pi}{12} + \frac{3\pi}{2}\right) + i\sin\left(\frac{\pi}{12} + \frac{3\pi}{2}\right)\right]$$

$$= \sqrt[4]{5}\left[\cos\frac{19\pi}{12} + i\sin\frac{19\pi}{12}\right]$$

Now try Exercise 45.

EXAMPLE 8 Finding Cube Roots

Find the cube roots of -1 and plot them.

SOLUTION First we write the complex number $z = -1$ in polar form

$$z = -1 + 0i = \cos\pi + i\sin\pi.$$

The third roots of $z = -1 = \cos\pi + i\sin\pi$ are the complex numbers

$$\cos\frac{\pi + 2\pi k}{3} + i\sin\frac{\pi + 2\pi k}{3},$$

(continued)

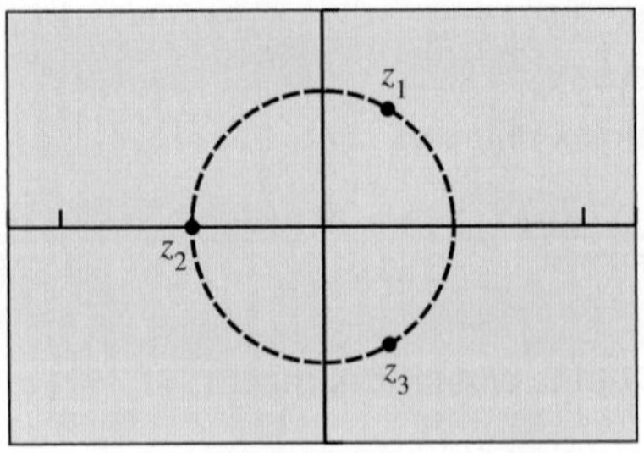

[–2.4, 2.4] by [–1.6, 1.6]

FIGURE 6.65 The three cube roots z_1, z_2, and z_3 of -1 displayed on the unit circle (dashed). (Example 8)

for $k = 0, 1, 2$. The three complex numbers are

$$z_1 = \cos\frac{\pi}{3} + i\sin\frac{\pi}{3} = \frac{1}{2} + \frac{\sqrt{3}}{2}i$$

$$z_2 = \cos\frac{\pi + 2\pi}{3} + i\sin\frac{\pi + 2\pi}{3} = -1 + 0i$$

$$z_3 = \cos\frac{\pi + 4\pi}{3} + i\sin\frac{\pi + 4\pi}{3} = \frac{1}{2} - \frac{\sqrt{3}}{2}i$$

Figure 6.65 shows the graph of the three cube roots z_1, z_2, and z_3. They are evenly spaced ($2\pi/3$ rad apart) around the unit circle.

Now try Exercise 57.

EXAMPLE 9 Finding Roots of Unity

Find the eight eighth roots of unity.

SOLUTION First we write the complex number $z = 1$ in polar form

$$z = 1 + 0i = \cos 0 + i\sin 0.$$

The eighth roots of $z = 1 + 0i = \cos 0 + i\sin 0$ are the complex numbers

$$\cos\frac{0 + 2\pi k}{8} + i\sin\frac{0 + 2\pi k}{8},$$

for $k = 0, 1, 2, \ldots, 7$.

$$z_1 = \cos 0 + i\sin 0 = 1 + 0i$$

$$z_2 = \cos\frac{\pi}{4} + i\sin\frac{\pi}{4} = \frac{\sqrt{2}}{2} + \frac{\sqrt{2}}{2}i$$

$$z_3 = \cos\frac{\pi}{2} + i\sin\frac{\pi}{2} = 0 + i$$

$$z_4 = \cos\frac{3\pi}{4} + i\sin\frac{3\pi}{4} = -\frac{\sqrt{2}}{2} + \frac{\sqrt{2}}{2}i$$

$$z_5 = \cos\pi + i\sin\pi = -1 + 0i$$

$$z_6 = \cos\frac{5\pi}{4} + i\sin\frac{5\pi}{4} = -\frac{\sqrt{2}}{2} - \frac{\sqrt{2}}{2}i$$

$$z_7 = \cos\frac{3\pi}{2} + i\sin\frac{3\pi}{2} = 0 - i$$

$$z_8 = \cos\frac{7\pi}{4} + i\sin\frac{7\pi}{4} = \frac{\sqrt{2}}{2} - \frac{\sqrt{2}}{2}i$$

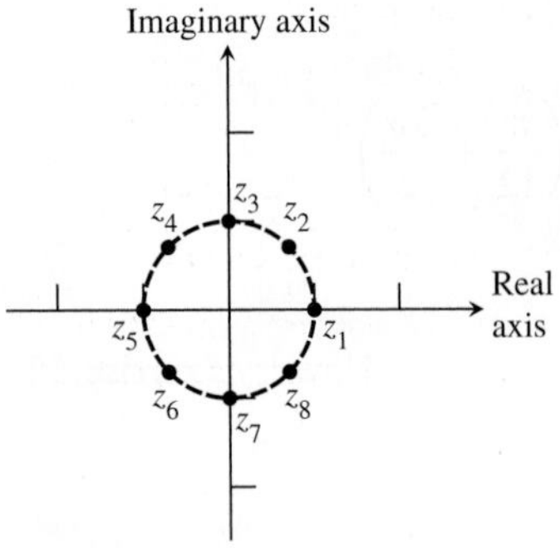

FIGURE 6.66 The eight eighth roots of unity are evenly spaced on a unit circle. (Example 9)

Figure 6.66 shows the eight points. They are spaced $2\pi/8 = \pi/4$ rad apart.

Now try Exercise 59.

QUICK REVIEW 6.6 *(For help, go to Sections P.5, P.6, and 4.3.)*

Exercise numbers with a gray background indicate problems that the authors have designed to be solved *without a calculator*.

In Exercises 1 and 2, write the roots of the equation in $a + bi$ form.

1. $x^2 + 13 = 4x$

2. $5(x^2 + 1) = 6x$

In Exercises 3 and 4, write the complex number in standard form $a + bi$.

3. $(1 + i)^5$

4. $(1 - i)^4$

In Exercises 5–8, find an angle θ in $0 \le \theta < 2\pi$ that satisfies both equations.

5. $\sin \theta = \frac{1}{2}$ and $\cos \theta = -\frac{\sqrt{3}}{2}$

6. $\sin \theta = -\frac{\sqrt{2}}{2}$ and $\cos \theta = \frac{\sqrt{2}}{2}$

7. $\sin \theta = -\frac{\sqrt{3}}{2}$ and $\cos \theta = -\frac{1}{2}$

8. $\sin \theta = -\frac{\sqrt{2}}{2}$ and $\cos \theta = -\frac{\sqrt{2}}{2}$

In Exercises 9 and 10, find all *real* solutions.

9. $x^3 - 1 = 0$

10. $x^4 - 1 = 0$

SECTION 6.6 Exercises

In Exercises 1 and 2, plot all four points in the same complex plan.

1. $1 + 2i, 3 - i, -2 + 2i, i$

2. $2 - 3i, 1 + i, 3, -2 - i$

In Exercises 3–12, find the polar form of the complex number where the argument satisfies $0 \le \theta < 2\pi$.

3. $3i$

4. $-2i$

5. $2 + 2i$

6. $\sqrt{3} + i$

7. $-2 + 2i\sqrt{3}$

8. $3 - 3i$

9. $3 + 2i$

10. $4 - 7i$

11.

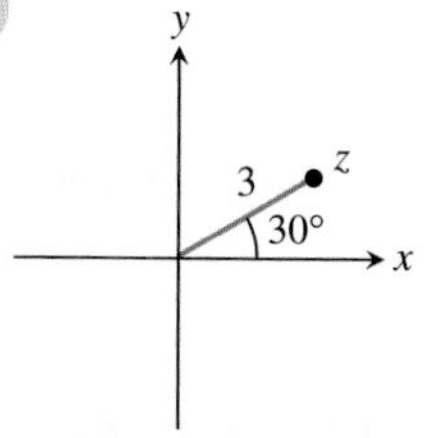

12.

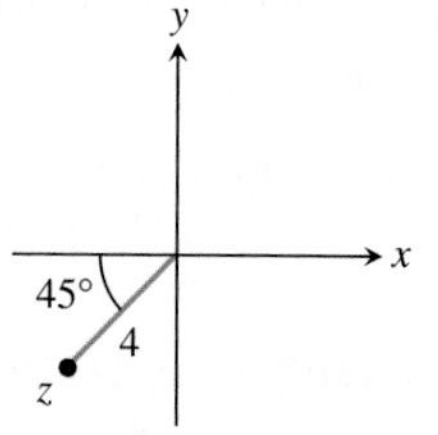

In Exercises 13–18, write the complex number in standard form $a + bi$.

13. $3(\cos 30° - i \sin 30°)$

14. $8(\cos 210° + i \sin 210°)$

15. $5[\cos(-60°) + i \sin(-60°)]$

16. $5\left(\cos \frac{\pi}{4} + i \sin \frac{\pi}{4}\right)$

17. $\sqrt{2}\left(\cos \frac{7\pi}{6} + i \sin \frac{7\pi}{6}\right)$

18. $\sqrt{7}\left(\cos \frac{\pi}{12} + i \sin \frac{\pi}{12}\right)$

In Exercises 19–22, find the product of z_1 and z_2. Leave the answer in polar form.

19. $z_1 = 7(\cos 25° + i \sin 25°)$

$z_2 = 2(\cos 130° + i \sin 130°)$

20. $z_1 = \sqrt{2}(\cos 118° + i \sin 118°)$

$z_2 = 0.5[\cos(-19°) + i \sin(-19°)]$

21. $z_1 = 5\left(\cos \frac{\pi}{4} + i \sin \frac{\pi}{4}\right)$ $z_2 = 3\left(\cos \frac{5\pi}{3} + i \sin \frac{5\pi}{3}\right)$

22. $z_1 = \sqrt{3}\left(\cos \frac{3\pi}{4} + i \sin \frac{3\pi}{4}\right)$ $z_2 = \frac{1}{3}\left(\cos \frac{\pi}{6} + i \sin \frac{\pi}{6}\right)$

In Exercises 23–26, find the polar form of the quotient.

23. $\frac{2(\cos 30° + i \sin 30°)}{3(\cos 60° + i \sin 60°)}$

24. $\frac{5(\cos 220° + i \sin 220°)}{2(\cos 115° + i \sin 115°)}$

25. $\frac{6(\cos 5\pi + i \sin 5\pi)}{3(\cos 2\pi + i \sin 2\pi)}$

26. $\frac{\cos(\pi/2) + i \sin(\pi/2)}{\cos(\pi/4) + i \sin(\pi/4)}$

In Exercises 27–30, find the product $z_1 \cdot z_2$ and quotient z_1/z_2 in two ways, **(a)** using the polar form for z_1 and z_2 and **(b)** using the standard form for z_1 and z_2.

27. $z_1 = 3 - 2i$ and $z_2 = 1 + i$

28. $z_1 = 1 - i$ and $z_2 = \sqrt{3} + i$

29. $z_1 = 3 + i$ and $z_2 = 5 - 3i$

30. $z_1 = 2 - 3i$ and $z_2 = 1 - \sqrt{3}i$

In Exercises 31–38, use De Moivre's Theorem to find the indicated power of the complex number. Write your answer in standard form $a + bi$.

31. $\left(\cos \frac{\pi}{4} + i \sin \frac{\pi}{4}\right)^3$

32. $\left[3\left(\cos \frac{3\pi}{2} + i \sin \frac{3\pi}{2}\right)\right]^5$

33. $\left[2\left(\cos\frac{3\pi}{4} + i\sin\frac{3\pi}{4}\right)\right]^3$

34. $\left[6\left(\cos\frac{5\pi}{6} + i\sin\frac{5\pi}{6}\right)\right]^4$

35. $(1 + i)^5$

36. $(3 + 4i)^{20}$

37. $(1 - \sqrt{3}i)^3$

38. $\left(\frac{1}{2} + i\frac{\sqrt{3}}{2}\right)^3$

Use an algebraic method in Exercises 39–44 to find the cube roots of the complex number. Approximate exact solution values when appropriate.

39. $2(\cos 2\pi + i\sin 2\pi)$

40. $2\left(\cos\frac{\pi}{4} + i\sin\frac{\pi}{4}\right)$

41. $3\left(\cos\frac{4\pi}{3} + i\sin\frac{4\pi}{3}\right)$

42. $27\left(\cos\frac{11\pi}{6} + i\sin\frac{11\pi}{6}\right)$

43. $3 - 4i$

44. $-2 + 2i$

In Exercises 45–50, find the fifth roots of the complex number.

45. $\cos\pi + i\sin\pi$

46. $32\left(\cos\frac{\pi}{2} + i\sin\frac{\pi}{2}\right)$

47. $2\left(\cos\frac{\pi}{6} + i\sin\frac{\pi}{6}\right)$

48. $2\left(\cos\frac{\pi}{4} + i\sin\frac{\pi}{4}\right)$

49. $2i$

50. $1 + \sqrt{3}\,i$

In Exercises 51–56, find the nth roots of the complex number for the specified value of n.

51. $1 + i,\quad n = 4$

52. $1 - i,\quad n = 6$

53. $2 + 2i,\quad n = 3$

54. $-2 + 2i,\quad n = 4$

55. $-2i,\quad n = 6$

56. $32,\quad n = 5$

In Exercises 57–60, express the roots of unity in standard form $a + bi$. Graph each root in the complex plane.

57. Cube roots of unity

58. Fourth roots of unity

59. Sixth roots of unity

60. Square roots of unity

61. Determine z and the three cube roots of z if one cube root of z is $1 + \sqrt{3}i$.

62. Determine z and the four fourth roots of z if one fourth root of z is $-2 - 2i$.

63. Quotient Formula Let $z_1 = r_1(\cos\theta_1 + i\sin\theta_1)$ and $z_2 = r_2(\cos\theta_2 + i\sin\theta_2)$, $r_2 \neq 0$. Verify that $z_1/z_2 = r_1/r_2\left[\cos(\theta_1 - \theta_2) + i\sin(\theta_1 - \theta_2)\right]$.

64. Group Activity *n*th Roots Show that the nth roots of the complex number $r(\cos\theta + i\sin\theta)$ are spaced $2\pi/n$ radians apart on a circle with radius $\sqrt[n]{r}$.

Standardized Test Questions

65. True or False The polar form of a complex number is unique. Justify your answer.

66. True or False The complex number i is a cube root of $-i$. Justify your answer.

In Exercises 67–70, do not use technology to solve the problem.

67. Multiple Choice Which of the following is a polar form of the complex number $-1 + \sqrt{3}i$?

(A) $2\left(\cos\frac{\pi}{3} + i\sin\frac{\pi}{3}\right)$ **(B)** $2\left(\cos\frac{2\pi}{3} + i\sin\frac{2\pi}{3}\right)$

(C) $2\left(\cos\frac{4\pi}{3} + i\sin\frac{4\pi}{3}\right)$ **(D)** $2\left(\cos\frac{5\pi}{3} + i\sin\frac{5\pi}{3}\right)$

(E) $2\left(\cos\frac{7\pi}{3} + i\sin\frac{7\pi}{3}\right)$

68. Multiple Choice Which of the following is the number of distinct complex number solutions of $z^5 = 1 + i$?

(A) 0 **(B)** 1 **(C)** 3 **(D)** 4 **(E)** 5

69. Multiple Choice Which of the following is the standard form for the product of $\sqrt{2}\left(\cos\frac{\pi}{4} + i\sin\frac{\pi}{4}\right)$ and $\sqrt{2}\left(\cos\frac{7\pi}{4} + i\sin\frac{7\pi}{4}\right)$?

(A) 2 **(B)** -2 **(C)** $-2i$ **(D)** $-1 + i$ **(E)** $1 - i$

70. Multiple Choice Which of the following is not a fourth root of 1?

(A) i^2 **(B)** $-i^2$ **(C)** $\sqrt{-1}$ **(D)** $-\sqrt{-1}$ **(E)** $\sqrt{i}$

Explorations

71. Complex Conjugates The complex conjugate of $z = a + bi$ is $\bar{z} = a - bi$. Let $z = r(\cos\theta + i\sin\theta)$.

(a) Prove that $\bar{z} = r\left[\cos(-\theta) + i\sin(-\theta)\right]$.

(b) Use the polar form to find $z \cdot \bar{z}$.

(c) Use the polar form to find $z/\bar{z}$, if $\bar{z} \neq 0$.

(d) Prove that $-z = r\left[\cos(\theta + \pi) + i\sin(\theta + \pi)\right]$.

72. Modulus of Complex Numbers Let $z = r(\cos\theta + i\sin\theta)$.

(a) Prove that $|z| = |r|$.

(b) Use the polar form for the complex numbers z_1 and z_2 to prove that $|z_1 \cdot z_2| = |z_1| \cdot |z_2|$.

Extending the Ideas

73. Using Polar Form on a Graphing Calculator The complex number $r(\cos\theta + i\sin\theta)$ can be entered in polar form on some graphing calculators as $re^{i\theta}$.

(a) Support the result of Example 3 by entering the complex numbers z_1 and z_2 in polar form on your graphing calculator and computing the product with your graphing calculator.

(b) Support the result of Example 4 by entering the complex numbers z_1 and z_2 in polar form on your graphing calculator and computing the quotient with your graphing calculator.

(c) Support the result of Example 5 by entering the complex number in polar form on your graphing calculator and computing the power with your graphing calculator.

74. Visualizing Roots of Unity Set your graphing calculator in parametric mode with $0 \le T \le 8$, Tstep $= 1$, Xmin $= -2.4$, Xmax $= 2.4$, Ymin $= -1.6$, and Ymax $= 1.6$.

(a) Let $x = \cos((2\pi/8)t)$ and $y = \sin((2\pi/8)t)$. Use TRACE to visualize the eight eighth roots of unity. We say that $2\pi/8$ *generates* the eighth roots of unity. (Try both dot mode and connected mode.)

(b) Replace $2\pi/8$ in part (a) by the arguments of other eighth roots of unity. Do any others *generate* the eighth roots of unity?

(c) Repeat parts (a) and (b) for the fifth, sixth, and seventh roots of unity, using appropriate functions for x and y.

(d) What would you conjecture about an nth root of unity that generates all the nth roots of unity in the sense of part (a)?

75. Parametric Graphing Write parametric equations that represent $(\sqrt{2} + i)^n$ for $n = t$. Draw and label an *accurate* spiral representing $(\sqrt{2} + i)^n$ for $n = 0, 1, 2, 3, 4$.

76. Parametric Graphing Write parametric equations that represent $(-1 + i)^n$ for $n = t$. Draw and label an *accurate* spiral representing $(-1 + i)^n$ for $n = 0, 1, 2, 3, 4$.

77. Explain why the triangles formed by 0, 1, and z_1, and by 0, z_2, and $z_1 z_2$ shown in the figure are similar triangles.

78. Compass and Straightedge Construction Using only a compass and straightedge, construct the location of $z_1 z_2$ given the location of 0, 1, z_1, and z_2.

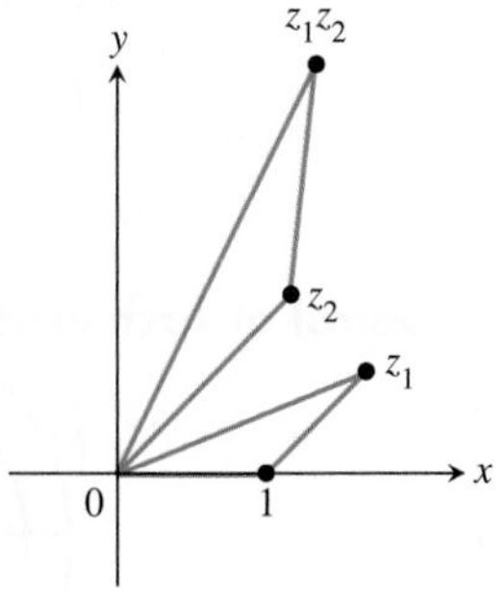

In Exercises 79–84, find all solutions of the equation (real and complex).

79. $x^3 - 1 = 0$ **80.** $x^4 - 1 = 0$

81. $x^3 + 1 = 0$ **82.** $x^4 + 1 = 0$

83. $x^5 + 1 = 0$ **84.** $x^5 - 1 = 0$

CHAPTER 6 KEY IDEAS

Properties, Theorems, and Formulas

Head Minus Tail (HMT) Rule for Vectors 457
Properties of the Dot Product 467
Angle Between Two Vectors 468
Projection of the Vector **u** onto the Vector **v** 470
Coordinate Conversion Equations 488
Product and Quotient of Complex Numbers 505
De Moivre's Theorem 507

Procedures

Resolving a Vector into Its Components 460
Finding All Polar Coordinates of a Point 488
Symmetry Tests for Polar Graphs 494
Finding nth Roots of a Complex Number 509

Gallery of Functions

Rose Curves: $r = a \cos n\theta$ and $r = a \sin n\theta$

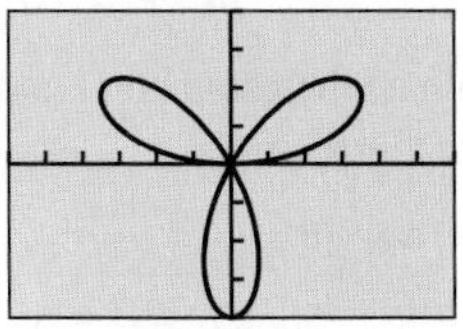

[–6,6] by [–4, 4]
$r = 4 \sin 3\theta$

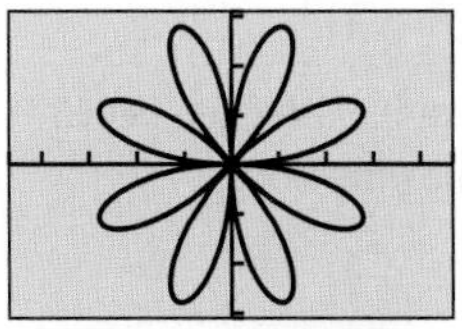

[–4.7, 4.7] by [–3.1, 3.1]
$r = 3 \sin 4\theta$

Limaçon Curves: $r = a \pm b \sin \theta$ and $r = a \pm b \cos \theta$ with $a > 0$ and $b > 0$

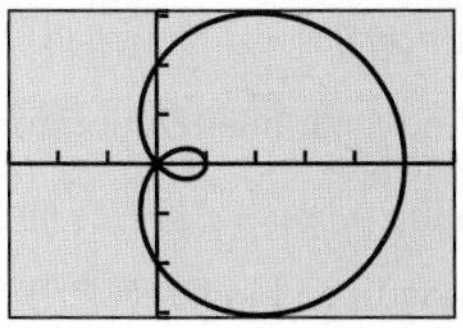

Limaçon with an inner loop: $\frac{a}{b} < 1$

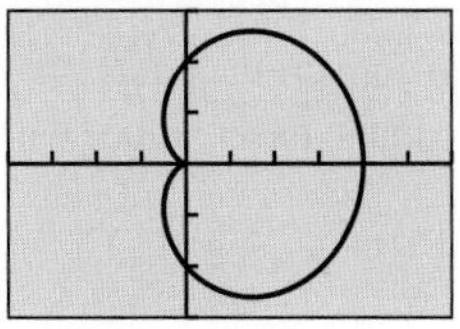

Cardioid: $\frac{a}{b} = 1$

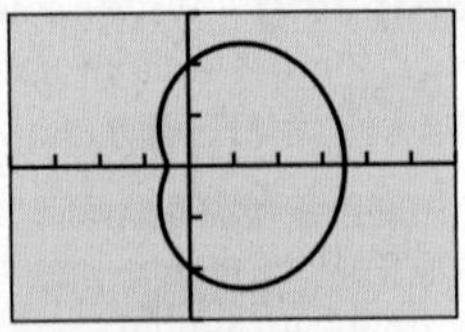

Dimpled limaçon: $1 < \frac{a}{b} < 2$

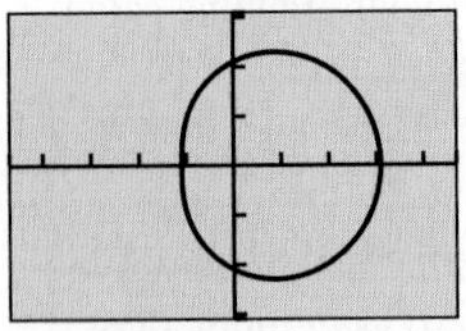

Convex limaçon: $\frac{a}{b} \geq 2$

Spiral of Archimedes:

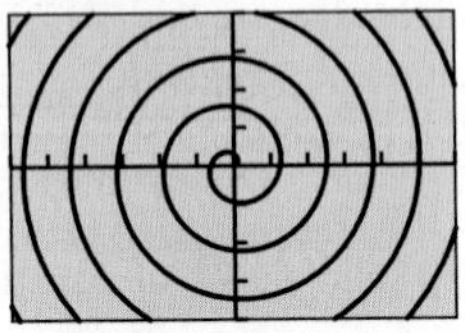

[–30, 30] by [–20, 20]

$r = \theta, 0 \leq \theta \leq 45$

Lemniscate Curves: $r^2 = a^2 \sin 2\theta$ and $r^2 = a^2 \cos 2\theta$

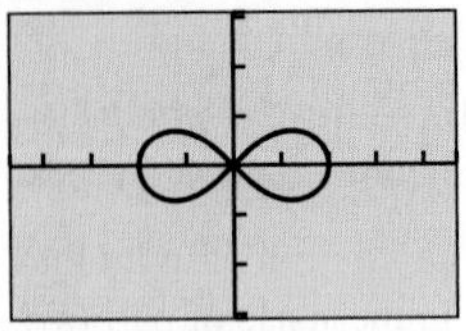

[–4.7, 4.7] by [–3.1, 3.1]

$r^2 = 4 \cos 2\theta$

CHAPTER 6 Review Exercises

Exercise numbers with a gray background indicate problems that the authors have designed to be solved *without a calculator.*

The collection of exercises marked in red could be used as a chapter test.

In Exercises 1–6, let $\mathbf{u} = \langle 2, -1 \rangle$, $\mathbf{v} = \langle 4, 2 \rangle$, and $\mathbf{w} = \langle 1, -3 \rangle$ be vectors. Find the indicated expression.

1. $\mathbf{u} - \mathbf{v}$ **2.** $2\mathbf{u} - 3\mathbf{w}$

3. $|\mathbf{u} + \mathbf{v}|$ **4.** $|\mathbf{w} - 2\mathbf{u}|$

5. $\mathbf{u} \cdot \mathbf{v}$ **6.** $\mathbf{u} \cdot \mathbf{w}$

In Exercises 7–10, let $A = (2, -1)$, $B = (3, 1)$, $C = (-4, 2)$, and $D = (1, -5)$. Find the component form and magnitude of the vector.

7. $3\overrightarrow{AB}$ **8.** $\overrightarrow{AB} + \overrightarrow{CD}$

9. $\overrightarrow{AC} + \overrightarrow{BD}$ **10.** $\overrightarrow{CD} + \overrightarrow{AB}$

In Exercises 11 and 12, find **(a)** a unit vector in the direction of $\overrightarrow{AB}$ and **(b)** a vector of magnitude 3 in the opposite direction.

11. $A = (4, 0), B = (2, 1)$

12. $A = (3, 1), B = (5, 1)$

In Exercises 13 and 14, find **(a)** the direction angles of **u** and **v** and **(b)** the angle between **u** and **v**.

13. $\mathbf{u} = \langle 4, 3 \rangle, \mathbf{v} = \langle 2, 5 \rangle$ **14.** $\mathbf{u} = \langle -2, 4 \rangle, \mathbf{v} = \langle 6, 4 \rangle$

Use an algebraic method in Exercises 15–18 to convert the polar coordinates to rectangular coordinates. Approximate exact values with a calculator when appropriate.

15. $(-2.5, 25°)$

16. $(-3.1, 135°)$

17. $(2, -\pi/4)$

18. $(3.6, 3\pi/4)$

In Exercises 19 and 20, polar coordinates of point P are given. Find all of its polar coordinates.

19. $P = (-1, -2\pi/3)$ **20.** $P = (-2, 5\pi/6)$

Use an algebraic method in Exercises 21–24 to find the polar coordinates of the given rectangular coordinates of point P that satisfy the stated conditions. Approximate exact values with a calculator when appropriate.

(a) $0 \leq \theta \leq 2\pi$ **(b)** $-\pi \leq \theta \leq \pi$ **(c)** $0 \leq \theta \leq 4\pi$

21. $P = (2, -3)$ **22.** $P = (-10, 0)$

23. $P = (5, 0)$ **24.** $P = (0, -2)$

In Exercises 25–30, eliminate the parameter t and identify the graph. Support your answer with a grapher.

25. $x = 3 - 5t, y = 4 + 3t$

26. $x = 4 + t, y = -8 - 5t, -3 \leq t \leq 5$

27. $x = 2t^2 + 3, y = t - 1$

28. $x = 3 \cos t, y = 3 \sin t$

29. $x = e^{2t} - 1, y = e^t$

30. $x = t^3, y = \ln t, t > 0$

In Exercises 31 and 32, find a parametrization for the curve.

31. The line through the points $(-1, -2)$ and $(3, 4)$

32. The line segment with endpoints $(-2, 3)$ and $(5, 1)$

Exercises 33 and 34 refer to the complex number z_1 shown in the figure.

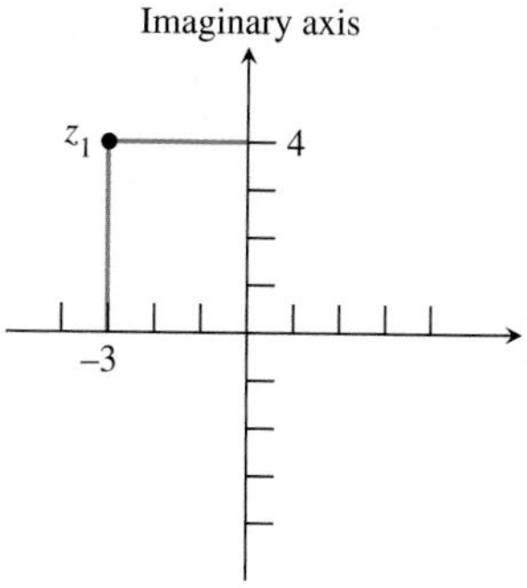

33. If $z_1 = a + bi$, find a, b, and $|z_1|$.

34. Find the polar form of z_1.

In Exercises 35–38, write the complex number in standard form.

35. $6(\cos 30° + i \sin 30°)$ **36.** $3(\cos 150° + i \sin 150°)$

37. $2.5\left(\cos \dfrac{4\pi}{3} + i \sin \dfrac{4\pi}{3}\right)$ **38.** $4(\cos 2.5 + i \sin 2.5)$

Use an algebraic method in Exercises 39–42 to find the polar form with $0 \le \theta < 2\pi$ for the given complex number. Then write three other possible polar forms for the number. Approximate exact values with a calculator when appropriate.

39. $3 - 3i$ **40.** $-1 + i\sqrt{2}$

41. $3 - 5i$ **42.** $-2 - 2i$

In Exercises 43 and 44, write the complex numbers $z_1 \cdot z_2$ and z_1/z_2 in polar form.

43. $z_1 = 3(\cos 30° + i \sin 30°)$ and $z_2 = 4(\cos 60° + i \sin 60°)$

44. $z_1 = 5(\cos 20° + i \sin 20°)$ and $z_2 = -2(\cos 45° + i \sin 45°)$

In Exercises 45–48, use De Moivre's Theorem to find the indicated power of the complex number. Write your answer in **(a)** polar form and **(b)** standard form.

45. $\left[3\left(\cos \dfrac{\pi}{4} + i \sin \dfrac{\pi}{4}\right)\right]^5$

46. $\left[2\left(\cos \dfrac{\pi}{12} + i \sin \dfrac{\pi}{12}\right)\right]^8$

47. $\left[5\left(\cos \dfrac{5\pi}{3} + i \sin \dfrac{5\pi}{3}\right)\right]^3$

48. $\left[7\left(\cos \dfrac{\pi}{24} + i \sin \dfrac{\pi}{24}\right)\right]^6$

In Exercises 49–52, find and graph the nth roots of the complex number for the specified value of n.

49. $3 + 3i, \quad n = 4$ **50.** $8, \quad n = 3$

51. $1, \quad n = 5$ **52.** $-1, \quad n = 6$

In Exercises 53–60, decide whether the graph of the given polar equation appears among the four graphs shown.

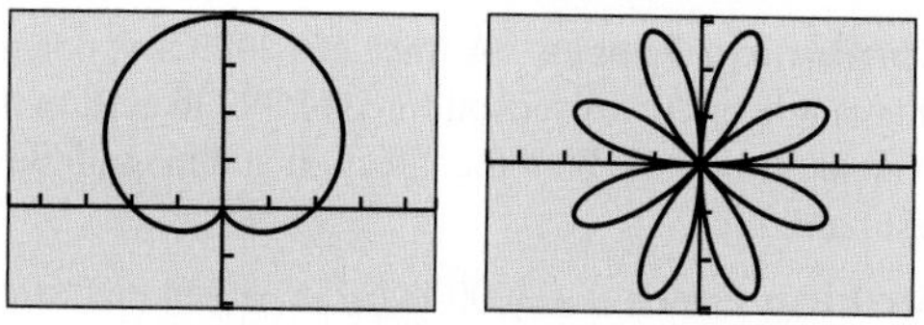

(a) (b)

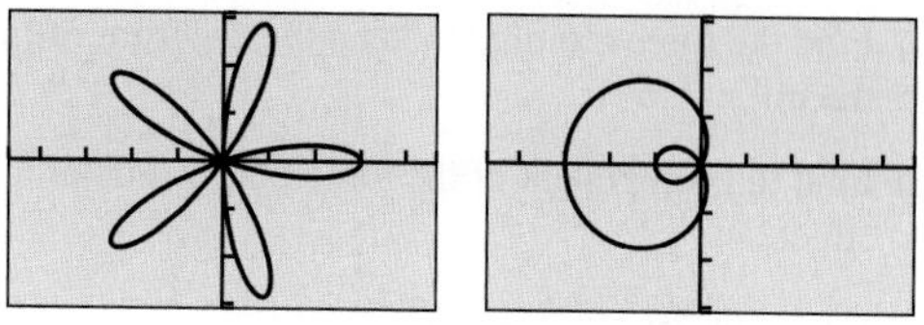

(c) (d)

53. $r = 3 \sin 4\theta$ **54.** $r = 2 + \sin \theta$

55. $r = 2 + 2 \sin \theta$ **56.** $r = 3|\sin 3\theta|$

57. $r = 2 - 2 \sin \theta$ **58.** $r = 1 - 2 \cos \theta$

59. $r = 3 \cos 5\theta$ **60.** $r = 3 - 2 \tan \theta$

In Exercises 61–64, convert the polar equation to rectangular form and identify the graph.

61. $r = -2$ **62.** $r = -2 \sin \theta$

63. $r = -3 \cos \theta - 2 \sin \theta$ **64.** $r = 3 \sec \theta$

In Exercises 65–68, convert the rectangular equation to polar form. Graph the polar equation.

65. $y = -4$ **66.** $x = 5$

67. $(x - 3)^2 + (y + 1)^2 = 10$

68. $2x - 3y = 4$

In Exercises 69–72, analyze the graph of the polar curve.

69. $r = 2 - 5 \sin \theta$ **70.** $r = 4 - 4 \cos \theta$

71. $r = 2 \sin 3\theta$ **72.** $r^2 = 2 \sin 2\theta, 0 \le \theta \le 2\pi$

73. Graphing Lines Using Polar Equations

(a) Explain why $r = a \sec \theta$ is a polar form for the line $x = a$.

(b) Explain why $r = b \csc \theta$ is a polar form for the line $y = b$.

(c) Let $y = mx + b$. Prove that

$$r = \frac{b}{\sin \theta - m \cos \theta}$$

is a polar form for the line. What is the domain of r?

(d) Illustrate the result in part (c) by graphing the line $y = 2x + 3$ using the polar form from part (c).

74. Flight Engineering An airplane is flying on a bearing of 80° at 540 mph. A wind is blowing with the bearing 100° at 55 mph.

(a) Find the component form of the velocity of the airplane.

(b) Find the actual speed and direction of the airplane.

75. Flight Engineering An airplane is flying on a bearing of 285° at 480 mph. A wind is blowing with the bearing 265° at 30 mph.

(a) Find the component form of the velocity of the airplane.

(b) Find the actual speed and direction of the airplane.

76. Combining Forces A force of 120 lb acts on an object at an angle of 20°. A second force of 300 lb acts on the object at an angle of $-5°$. Find the direction and magnitude of the resultant force.

77. Braking Force A 3000-lb car is parked on a street that makes an angle of 16° with the horizontal (see figure).

(a) Find the force required to keep the car from rolling down the hill.

(b) Find the component of the force perpendicular to the street.

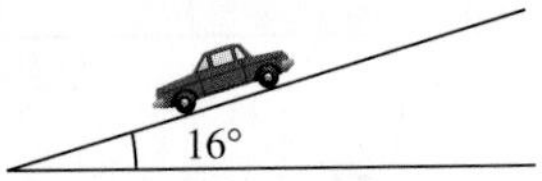

78. Work Find the work done by a force **F** of 36 lb acting in the direction given by the vector $\langle 3, 5 \rangle$ in moving an object 10 ft from $(0, 0)$ to $(10, 0)$.

79. Height of an Arrow Stewart shoots an arrow straight up from the top of a building with initial velocity of 245 ft/sec. The arrow leaves from a point 200 ft above level ground.

(a) Write an equation that models the height of the arrow as a function of time t.

(b) Use parametric equations to simulate the height of the arrow.

(c) Use parametric equations to graph height against time.

(d) How high is the arrow after 4 sec?

(e) What is the maximum height of the arrow? When does it reach its maximum height?

(f) How long will it be before the arrow hits the ground?

80. Ferris Wheel Problem Lucinda is on a Ferris wheel of radius 35 ft that turns at the rate of one revolution every 20 sec. The lowest point of the Ferris wheel (6 o'clock) is 15 ft above ground level at the point $(0, 15)$ of a rectangular coordinate system. Find parametric equations for the position of Lucinda as a function of time t in seconds if Lucinda starts $(t = 0)$ at the point $(35, 50)$.

81. Ferris Wheel Problem The lowest point of a Ferris wheel (6 o'clock) of radius 40 ft is 10 ft above the ground, and the center is on the y-axis. Find parametric equations for Henry's position as a function of time t in seconds if his starting position $(t = 0)$ is the point $(0, 10)$ and the wheel turns at the rate of one revolution every 15 sec.

82. Ferris Wheel Problem Sarah rides the Ferris wheel described in Exercise 81. Find parametric equations for Sarah's position as a function of time t in seconds if her starting position $(t = 0)$ is the point $(0, 90)$ and the wheel turns at the rate of one revolution every 18 sec.

83. Epicycloid The graph of the parametric equations

$$x = 4\cos t - \cos 4t, \quad y = 4\sin t - \sin 4t$$

is an *epicycloid*. The graph is the path of a point P on a circle of radius 1 rolling along the outside of a circle of radius 3, as suggested in the figure.

(a) Graph simultaneously this epicycloid and the circle of radius 3.

(b) Suppose the large circle has a radius of 4. Experiment! How do you think the equations in part (a) should be changed to obtain defining equations? What do you think the epicycloid would look like in this case? Check your guesses.

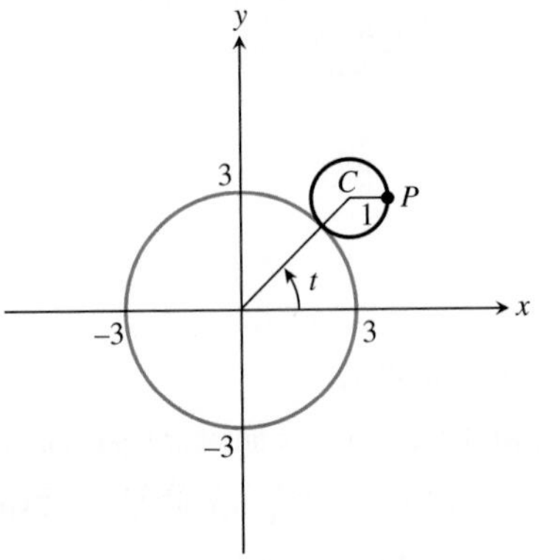

84. Throwing a Baseball Sharon releases a baseball 4 ft above the ground with an initial velocity of 66 ft/sec at an angle of 5° with the horizontal. How many seconds after the ball is thrown will it hit the ground? How far from Sharon will the ball be when it hits the ground?

85. Throwing a Baseball Diego releases a baseball 3.5 ft above the ground with an initial velocity of 66 ft/sec at an angle of 12° with the horizontal. How many seconds after the ball is thrown will it hit the ground? How far from Diego will the ball be when it hits the ground?

86. Field Goal Kicking Spencer practices kicking field goals 40 yd from a goal post with a crossbar 10 ft high. If he kicks the ball with an initial velocity of 70 ft/sec at a 45° angle with the horizontal (see figure), will Spencer make the field goal if the kick sails "true"?

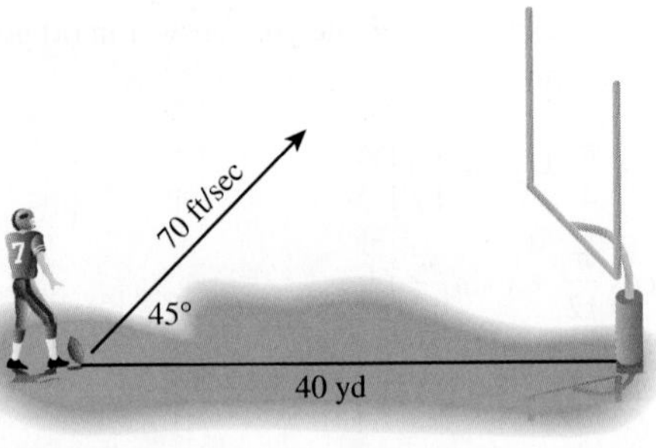

87. Hang Time An NFL place-kicker kicks a football downfield with an initial velocity of 85 ft/sec. The ball leaves his foot at the 15-yd line at an angle of 56° with the horizontal. Determine the following:

(a) The ball's maximum height above the field.

(b) The "hang time" (the total time the football is in the air).

88. Baseball Hitting Brian hits a baseball straight toward a 15-ft-high fence that is 400 ft from home plate. The ball is hit when it is 2.5 ft above the ground and leaves the bat at an angle of 30° with the horizontal. Find the initial velocity needed for the ball to clear the fence.

89. Throwing a Ball at a Ferris Wheel A 60-ft-radius Ferris wheel turns counterclockwise one revolution every 12 sec. Sam stands at a point 80 ft to the left of the bottom (6 o'clock) of the wheel. At the instant Kathy is at 3 o'clock, Sam throws a ball with an initial velocity of 100 ft/sec and an angle with the horizontal of 70°. He releases the ball from the same height as the bottom of the Ferris wheel. Find the minimum distance between the ball and Kathy.

90. Yard Darts Gretta and Lois are launching yard darts 20 ft from the front edge of a circular target of radius 18 in. If Gretta releases the dart 5 ft above the ground with an initial velocity of 20 ft/sec and at a 50° angle with the horizontal, will the dart hit the target?

CHAPTER 6 Project

Parametrizing Ellipses

As you discovered in the Chapter 4 Data Project, it is possible to model the displacement of a swinging pendulum using a sinusoidal equation of the form

$$x = a \sin(b(t - c)) + d$$

where x represents the pendulum's distance from a fixed point and t represents total elapsed time. In fact, a pendulum's velocity behaves sinusoidally as well: $y = ab \cos(b(t - c))$, where y represents the pendulum's velocity and a, b, and c are constants common to both the displacement and velocity equations.

Use a motion detection device to collect distance, velocity, and time data for a pendulum, then determine how a resulting plot of velocity versus displacement (called a phase-space plot) can be modeled using parametric equations.

Collecting the Data

Construct a simple pendulum by fastening about 1 m of string to the end of a ball. Collect time, distance, and velocity readings for between 2 sec and 4 sec (enough time to capture at least one complete swing of the pendulum). Start the pendulum swinging in front of the detector, then activate the system. The data table below shows a sample set of data collected as a pendulum swung back and forth in front of a CBR where t is total elapsed time in seconds, d = distance from the CBR in meters, v = velocity in meters/second.

t	d	v	t	d	v	t	d	v
0	1.021	0.325	0.7	0.621	−0.869	1.4	0.687	0.966
0.1	1.038	0.013	0.8	0.544	−0.654	1.5	0.785	1.013
0.2	1.023	−0.309	0.9	0.493	−0.359	1.6	0.880	0.826
0.3	0.977	−0.598	1.0	0.473	−0.044	1.7	0.954	0.678
0.4	0.903	−0.819	1.1	0.484	0.263	1.8	1.008	0.378
0.5	0.815	−0.996	1.2	0.526	0.573	1.9	1.030	0.049
0.6	0.715	−0.979	1.3	0.596	0.822	2.0	1.020	−0.260

Explorations

1. Create a scatter plot for the data you collected or the data above.
2. With your calculator/computer in function mode, find values for a, b, c, and d so that the equation $y = a \sin(b(x - c)) + d$ (where y is distance and x is time) fits the distance versus time data plot.
3. Make a scatter plot of velocity versus time. Using the same a, b, and c values you found in (2), verify that the equation $y = ab \cos(b(x - c))$ (where y is velocity and x is time) fits the velocity versus time data plot.
4. What do you think a plot of velocity versus distance (with velocity on the vertical axis and distance on the horizontal axis) would look like? Make a rough sketch of your prediction, then create a scatter plot of velocity versus distance. How well did your predicted graph match the actual data plot?
5. With your calculator/computer in parametric mode, graph the parametric curve $x = a \sin(b(t - c)) + d$, $y = ab \cos(b(t - c))$, $0 \le t \le 2$, where x represents distance, y represents velocity, and t is the time parameter. How well does this curve match the scatter plot of velocity versus time?

CHAPTER 7

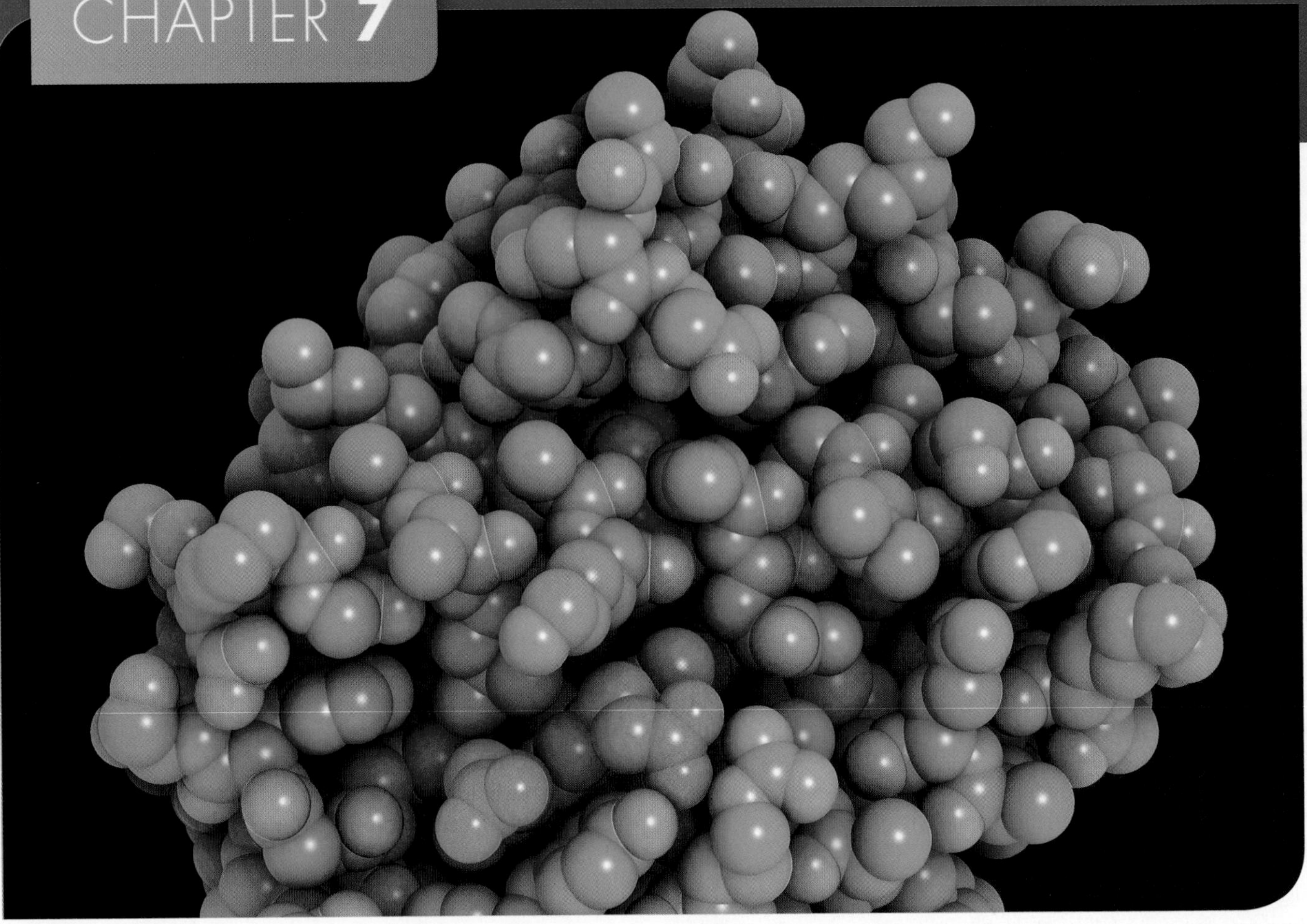

7.1 Solving Systems of Two Equations

7.2 Matrix Algebra

7.3 Multivariate Linear Systems and Row Operations

7.4 Systems of Inequalities in Two Variables

Systems and Matrices

Scientists studying hemoglobin molecules, as represented in the photo, can make new discoveries by viewing the image on a computer. To see all the details, they may need to move the image in some direction (translation), turn it about some line (rotation), or change its size (dilation). In computer graphics these operations are performed using matrix operations. See a related problem involving scaling a triangle on page 538.

Chapter 7 Overview

Many applications in science, engineering, and business use systems of equations or inequalities in two or more variables as models. We investigate several techniques commonly used to solve such systems, and we investigate matrices, which play a central role in several of these techniques. Matrices allow us to handle vast amounts of data. We explore numerous applications, including linear programming, a method used to solve problems in management science.

7.1 Solving Systems of Two Equations

What you'll learn about

- The Method of Substitution
- Solving Systems Graphically
- The Method of Elimination
- Applications

... and why

Many applications in business and science can be modeled using systems of equations.

The Method of Substitution

Here is an example of a system of two linear equations in the two variables x and y:

$$2x - y = 10$$
$$3x + 2y = 1$$

A **solution of a system** of two equations in two variables is an ordered pair of real numbers that is a solution of each equation. For example, the ordered pair $(3, -4)$ is a solution to the above system. We can verify this by showing that $(3, -4)$ is a solution of each equation. Substituting $x = 3$ and $y = -4$ into each equation, we obtain

$$2x - y = 2(3) - (-4) = 6 + 4 = 10$$
$$3x + 2y = 3(3) + 2(-4) = 9 - 8 = 1$$

So, both equations are satisfied.

We **solve a system of equations** when we find all of its solutions.

EXAMPLE 1 Using the Substitution Method

Solve the system

$$2x - y = 10$$
$$3x + 2y = 1$$

SOLUTION

Solve Algebraically Solving the first equation for y yields $y = 2x - 10$. Then substitute the expression for y into the second equation.

$3x + 2y = 1$	Second equation
$3x + 2(2x - 10) = 1$	Replace y by $2x - 10$.
$3x + 4x - 20 = 1$	Distributive property
$7x = 21$	Collect like terms.
$x = 3$	Divide by 7.
$y = -4$	Use $y = 2x - 10$.

(continued)

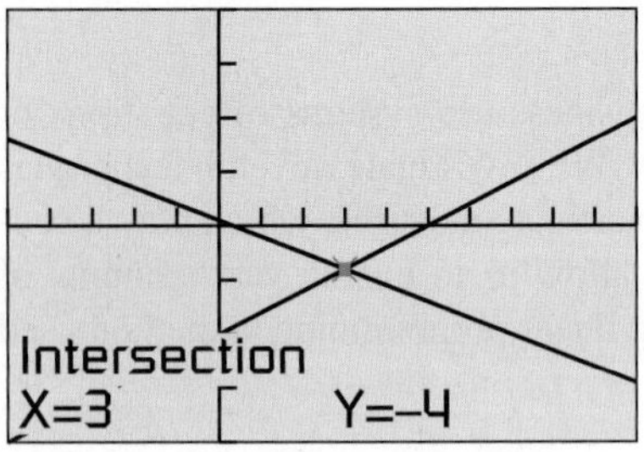

[–5, 10] by [–20, 20]

FIGURE 7.1 The two lines $y = 2x - 10$ and $y = -1.5x + 0.5$ intersect in the point $(3, -4)$. (Example 1)

Support Graphically The graph of each equation is a line. Figure 7.1 shows that the two lines intersect in the single point $(3, -4)$.

Interpret The solution of the system is $x = 3$, $y = -4$, or the ordered pair $(3, -4)$.

Now try Exercise 5.

The method of substitution sometimes can be applied even when the equations in the system are not linear, as illustrated in Examples 2 and 3.

EXAMPLE 2 Solving a Nonlinear System by Substitution

Find the dimensions of a rectangular garden that has a perimeter of 100 ft and an area of 300 ft^2.

SOLUTION

Model Let x and y be the lengths of adjacent sides of the garden (Figure 7.2). Then

$$2x + 2y = 100 \quad \text{Perimeter is 100.}$$
$$xy = 300 \quad \text{Area is 300.}$$

Solve Algebraically Solving the first equation for y yields $y = 50 - x$. Then substitute the expression for y into the second equation.

$$\begin{aligned} xy &= 300 && \text{Second equation} \\ x(50 - x) &= 300 && \text{Replace } y \text{ by } 50 - x. \\ 50x - x^2 &= 300 && \text{Distributive property} \\ x^2 - 50x + 300 &= 0 \\ x &= \frac{50 \pm \sqrt{(-50)^2 - 4(300)}}{2} && \text{Quadratic formula} \\ x \approx 6.97 \quad &\text{or} \quad x \approx 43.03 && \text{Evaluate.} \\ y \approx 43.03 \quad &\text{or} \quad y \approx 6.97 && \text{Use } y = 50 - x. \end{aligned}$$

Support Graphically Figure 7.3 shows that the graphs of $y = 50 - x$ and $y = 300/x$ have two points of intersections.

Interpret The two ordered pairs produce the same rectangle, the dimensions of which are approximately 7 ft by 43 ft.

Now try Exercise 11.

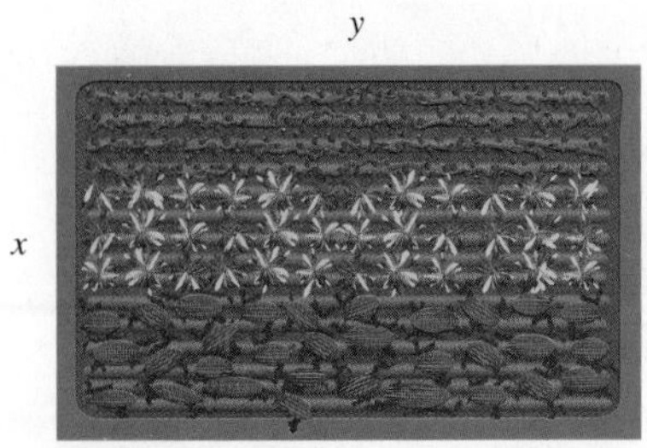

FIGURE 7.2 The rectangular garden in Example 2.

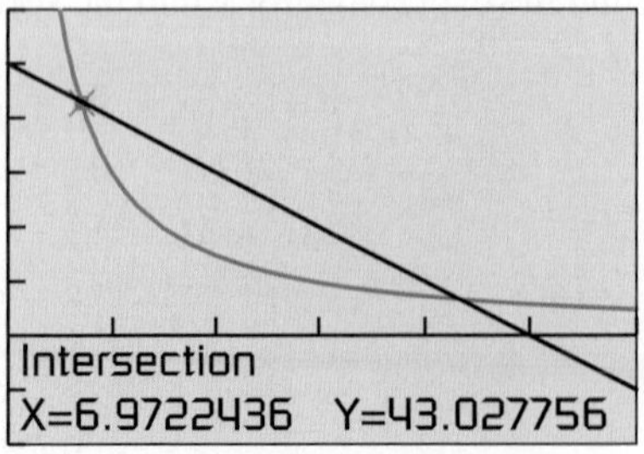

[0, 60] by [–20, 60]

FIGURE 7.3 We can assume $x \geq 0$ and $y \geq 0$ because x and y are lengths. (Example 2)

Rounding at the End

In Example 2, we did *not* round the values found for x until we computed the values for y. For the sake of accuracy, do not round *intermediate results*. Carry all decimals on your calculator computations and then round the final answer(s).

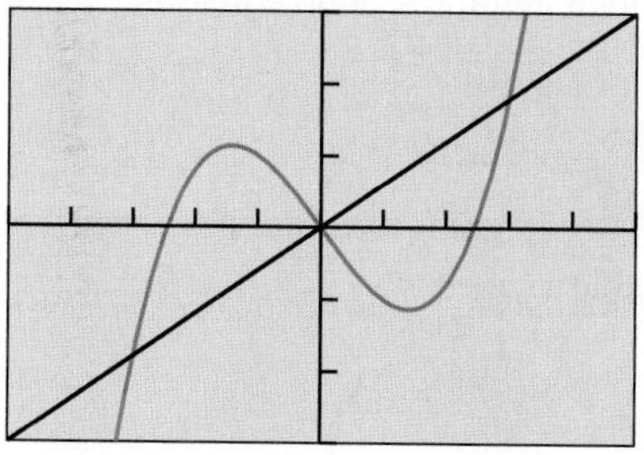

[–5, 5] by [–15, 15]

FIGURE 7.4 The graphs of $y = x^3 - 6x$ and $y = 3x$ have three points of intersection. (Example 3)

EXAMPLE 3 Solving a Nonlinear System Algebraically

Solve the system

$$y = x^3 - 6x$$
$$y = 3x$$

Support your solution graphically.

SOLUTION Substituting the value of y from the first equation into the second equation yields

$$x^3 - 6x = 3x$$
$$x^3 - 9x = 0$$
$$x(x - 3)(x + 3) = 0$$
$$x = 0, x = 3, x = -3 \quad \text{Zero factor property}$$
$$y = 0, y = 9, y = -9 \quad \text{Use } y = 3x.$$

The system of equations has three solutions: $(-3, -9)$, $(0, 0)$, and $(3, 9)$.

Support Graphically The graphs in Figure 7.4 of the two equations suggest that the three solutions found algebraically are correct. **Now try Exercise 13.**

Solving Systems Graphically

Sometimes the method of substitution leads to an equation in one variable that we are not able to solve using the standard algebraic techniques we have studied in this text. In these cases we can solve the system graphically by finding intersections as illustrated in Exploration 1.

EXPLORATION 1 **Solving a System Graphically**

Consider the system:

$$y = \ln x$$
$$y = x^2 - 4x + 2$$

1. Draw the graphs of the two equations in the $[0, 10]$ by $[-5, 5]$ viewing window.
2. Use the graph in part 1 to find the coordinates of the points of intersection shown in the viewing window.
3. Use your knowledge about the graphs of logarithmic and quadratic functions to explain why this system has exactly two solutions.

Substituting the expression for y of the first equation of Exploration 1 into the second equation yields

$$\ln x = x^2 - 4x + 2.$$

We have no standard algebraic technique to solve this equation.

The Method of Elimination

Consider a system of two linear equations in x and y. To **solve by elimination**, we rewrite the two equations as two equivalent equations so that one of the variables has opposite coefficients. Then we add the two equations to eliminate that variable.

EXAMPLE 4 Using the Elimination Method

Solve the system

$$2x + 3y = 5$$
$$-3x + 5y = 21$$

SOLUTION

Solve Algebraically Multiply the first equation by 3 and the second equation by 2 to obtain

$$6x + 9y = 15$$
$$-6x + 10y = 42$$

Then add the two equations to eliminate the variable x.

$$19y = 57$$

Next divide by 19 to solve for y.

$$y = 3$$

Finally, substitute $y = 3$ into either of the two original equations to determine that $x = -2$.

The solution of the original system is $(-2, 3)$. Now try Exercise 19.

EXAMPLE 5 Finding No Solution

Solve the system

$$x - 3y = -2$$
$$2x - 6y = 4$$

SOLUTION We use the elimination method.

Solve Algebraically

$$-2x + 6y = 4 \quad \text{Multiply first equation by } -2.$$
$$2x - 6y = 4 \quad \text{Second equation}$$
$$0 = 8 \quad \text{Add.}$$

The last equation is true for *no* values of x and y. So, the system has no solution.

Support Graphically Figure 7.5 suggests that the two lines that are the graphs of the two equations in the system are parallel. Solving for y in each equation yields

$$y = \frac{1}{3}x + \frac{2}{3}$$

$$y = \frac{1}{3}x - \frac{2}{3}$$

The two lines have the same slope and different y-intercepts, and are therefore parallel. Hence, they do not intersect, which confirms the algebraic findings.

Now try Exercise 23.

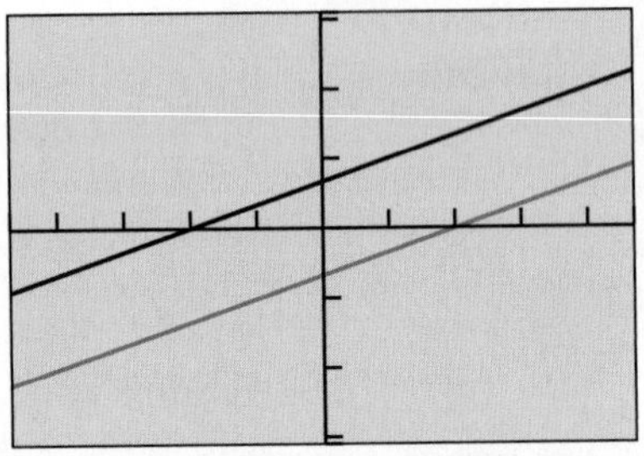

[−4.7, 4.7] by [−3.1, 3.1]

FIGURE 7.5 The graph of the two lines in Example 5 in this square viewing window appear to be parallel.

An easy way to determine the *number of solutions* of a system of two linear equations in two variables is to look at the graphs of the two lines. There are three possibilities. The two lines can intersect in a single point, producing exactly *one* solution as in Examples 1 and 4. The two lines can be parallel, producing *no* solution as in Example 5. The two lines can coincide, producing infinitely many solutions as illustrated in Example 6.

EXAMPLE 6 Finding Infinitely Many Solutions

Solve the system

$$4x - 5y = 2$$
$$-12x + 15y = -6$$

SOLUTION

$$\begin{aligned} 12x - 15y &= 6 && \text{Multiply first equation by 3.} \\ -12x + 15y &= -6 && \text{Second equation} \\ 0 &= 0 && \text{Add.} \end{aligned}$$

The last equation is true for all values of x and y. Thus, every ordered pair that satisfies one equation satisfies the other equation. The system has infinitely many solutions.

Another way to see that there are infinitely many solutions is to solve each equation for y. Both equations yield

$$y = \frac{4}{5}x - \frac{2}{5}.$$

The two equations yield the same line. **Now try Exercise 25.**

Applications

Table 7.1 shows the personal expenditures (in billions of dollars) for dentists and health insurance in the United States for several years.

Table 7.1 Selected U.S. Personal Expenditures

Year	Dentists (billions)	Health Insurance (billions)
2001	66.8	89.4
2002	72.2	96.6
2003	74.6	112.8
2004	80.2	129.5
2005	85.0	141.3
2006	91.1	146.7
2007	95.8	153.2

Source: U.S. Department of Commerce Bureau of Economic Analysis, Table 2.5.5. Personal Consumption Expenditures by Type of Expenditure.

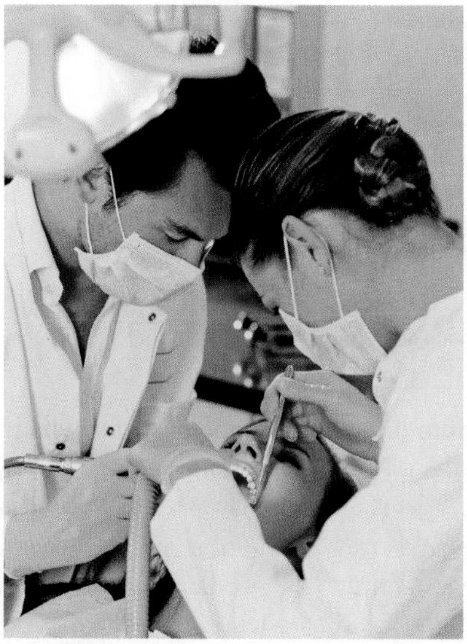

EXAMPLE 7 Estimating Personal Expenditures with Linear Models

(a) Find linear regression equations of the expenditures for dentists and health insurance in Table 7.1. Superimpose their graphs on a scatter plot of the data.

(b) Use the models in part (a) to estimate when the U.S. personal consumption expenditures for dentists was the same as that for health insurance, and find the corresponding amount.

SOLUTION

(a) Let $x = 0$ stand for 2000, $x = 1$ for 2001, and so forth. We use a grapher to find linear regression equations for the amount of expenditures for dentists, y_D, and the amount of expenditures for health insurance, y_{HI}:

$$y_D \approx 4.8286x + 61.5$$
$$y_{HI} \approx 11.4321x + 78.4857$$

(continued)

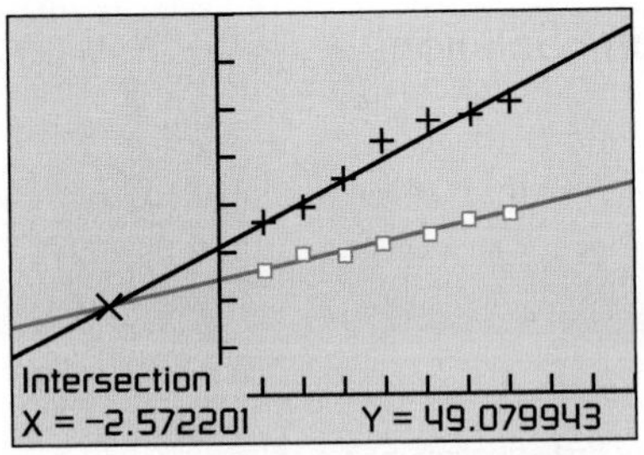

[−5, 10] by [−25, 200]

FIGURE 7.6 The scatter plot and regression equations for the data in Table 7.1. Dentists (□), health insurance (+). (Example 7)

Figure 7.6 shows the two regression equations together with a scatter plot of the two sets of data.

(b) Figure 7.6 shows that the graphs of y_D and y_{HI} intersect at approximately $(-2.57, 49.08)$. $x = -3$ stands for 1997, so Figure 7.6 suggests that the personal expenditures for dentists and for health insurance were both about \$49.1 billion sometime during 1997. **Now try Exercise 45.**

Suppliers will usually increase production, x, if they can get higher prices, p, for their products. So, as one variable increases, the other also increases. Normal mathematical practice would be to use p as the independent variable and x as the dependent variable. However, most economists put x on the horizontal axis and p on the vertical axis. In keeping with this practice, we write $p = f(x)$ for a **supply curve**. On one hand, as the price increases (vertical axis) so does the willingness for suppliers to increase production x (horizontal axis).

On the other hand, the demand, x, for a product by consumers will decrease as the price, p, goes up. So, as one variable increases, the other decreases. Again economists put x (demand) on the horizontal axis and p (price) on the vertical axis. In keeping with this practice, we write $p = g(x)$ for a **demand curve**.

Finally, a point where the supply curve and demand curve intersect is an **equilibrium point**. The corresponding price is the **equilibrium price**.

EXAMPLE 8 Determining the Equilibrium Price

Nibok Manufacturing has determined that production and price of a new tennis shoe should be geared to the equilibrium point for this system of equations.

$$p = 160 - 5x \quad \text{Demand curve}$$
$$p = 35 + 20x \quad \text{Supply curve}$$

The price, p, is in dollars and the number of shoes, x, is in millions of pairs. Find the equilibrium point.

SOLUTION We use substitution to solve the system.

$$160 - 5x = 35 + 20x$$
$$25x = 125$$
$$x = 5$$

Substitute this value of x into the demand curve and solve for p.

$$p = 160 - 5x$$
$$p = 160 - 5(5) = 135$$

The equilibrium point is $(5, 135)$. The equilibrium price is \$135, the price for which supply and demand will be equal at 5 million pairs of tennis shoes.

Now try Exercise 43.

QUICK REVIEW 7.1 *(For help, go to Sections P.4 and P.5.)*

In Exercises 1 and 2, solve for y in terms of x.

1. $2x + 3y = 5$ **2.** $xy + x = 4$

In Exercises 3–6, solve the equation algebraically.

3. $3x^2 - x - 2 = 0$ **4.** $2x^2 + 5x - 10 = 0$

5. $x^3 = 4x$ **6.** $x^3 + x^2 = 6x$

7. Write an equation for the line through the point $(-1, 2)$ and parallel to the line $4x + 5y = 2$.

8. Write an equation for the line through the point $(-1, 2)$ and perpendicular to the line $4x + 5y = 2$.

9. Write an equation equivalent to $2x + 3y = 5$ with the coefficient of x equal to -4.

10. Find the points of intersection of the graphs of $y = 2x$ and $y = x^3 - 2x$ graphically.

SECTION 7.1 Exercises

Exercise numbers with a gray background indicate problems that the authors have designed to be solved *without a calculator.*

In Exercises 1 and 2, determine whether the ordered pair is a solution of the system.

1. $5x - 2y = 8$
$2x - 3y = 1$

(a) $(0, 4)$ **(b)** $(2, 1)$

(c) $(-2, -9)$

2. $y = x^2 - 6x + 5$
$y = 2x - 7$

(a) $(2, -3)$ **(b)** $(1, -5)$

(c) $(6, 5)$

In Exercises 3–12, solve the system by substitution.

3. $x + 2y = 5$
$y = -2$

4. $x = 3$
$x - y = 20$

5. $3x + y = 20$
$x - 2y = 10$

6. $2x - 3y = -23$
$x + y = 0$

7. $2x - 3y = -7$
$4x + 5y = 8$

8. $3x + 2y = -5$
$2x - 5y = -16$

9. $x - 3y = 6$
$-2x + 6y = 4$

10. $3x - y = -2$
$-9x + 3y = 6$

11. $y = x^2$
$y - 9 = 0$

12. $x = y + 3$
$x - y^2 = 3y$

In Exercises 13–18, solve the system algebraically. Support your answer graphically.

13. $y = 6x^2$
$7x + y = 3$

14. $y = 2x^2 + x$
$2x + y = 20$

15. $y = x^3 - x^2$
$y = 2x^2$

16. $y = x^3 + x^2$
$y = -x^2$

17. $x^2 + y^2 = 9$
$x - 3y = -1$

18. $x^2 + y^2 = 16$
$4x + 7y = 13$

In Exercises 19–26, solve the system by elimination.

19. $x - y = 10$
$x + y = 6$

20. $2x + y = 10$
$x - 2y = -5$

21. $3x - 2y = 8$
$5x + 4y = 28$

22. $4x - 5y = -23$
$3x + 4y = 6$

23. $2x - 4y = -10$
$-3x + 6y = -21$

24. $2x - 4y = 8$
$-x + 2y = -4$

25. $2x - 3y = 5$
$-6x + 9y = -15$

26. $2x - y = 3$
$-4x + 2y = 5$

In Exercises 27–30, use the graph to estimate any solutions of the system. Confirm by substitution.

27. $y = 1 + 2x - x^2$
$y = 1 - x$

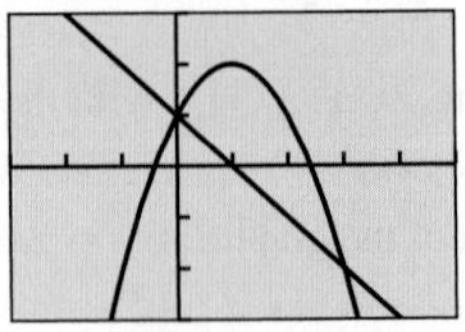

[–3, 5] by [–3, 3]

28. $6x - 2y = 7$
$2x + y = 4$

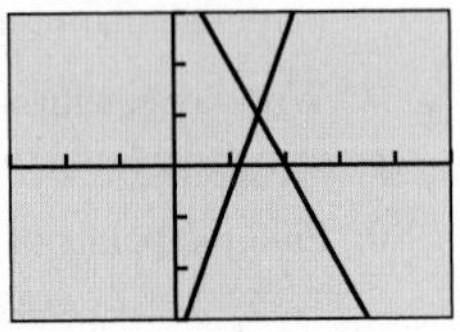

[–3, 5] by [–3, 3]

29. $x + 2y = 0$
$0.5x + y = 2$

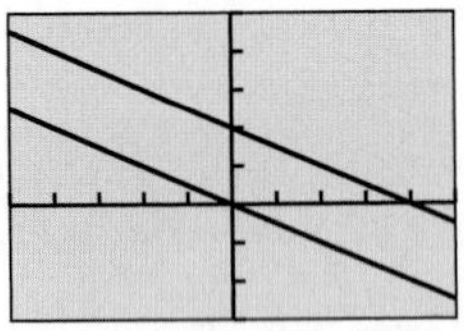

[–5, 5] by [–3, 5]

30. $x^2 + y^2 = 16$
$y + 4 = x^2$

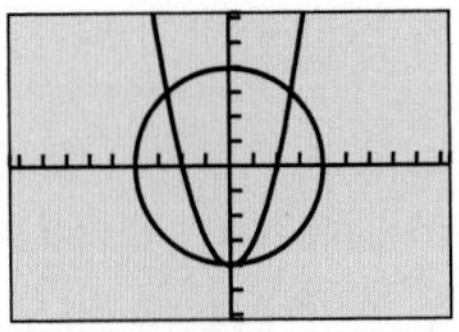

[–9.4, 9.4] by [–6.2, 6.2]

In Exercises 31–34, use graphs of the equations to determine the number of solutions for the system.

31. $3x + 5y = 7$
$4x - 2y = -3$

32. $3x - 9y = 6$
$2x - 6y = 1$

33. $2x - 4y = 6$
$3x - 6y = 9$

34. $x - 7y = 9$
$3x + 4y = 1$

In Exercises 35–42, solve the system graphically. Support your answer numerically using substitution and a calculator.

35. $y = \ln x$
$1 = 2x + y$

36. $y = 3 \cos x$
$1 = 2x - y$

37. $y = x^3 - 4x$
$4 = x - 2y$

38. $y = x^2 - 3x - 5$
$1 = 2x - y$

39. $x^2 + y^2 = 4$
$x + 2y = 2$

40. $x^2 + y^2 = 4$
$x - 2y = 2$

41. $x^2 + y^2 = 9$
$y = x^2 - 2$

42. $x^2 + y^2 = 9$
$y = 2 - x^2$

In Exercises 43 and 44, find the equilibrium point for the given demand and supply curves.

43. $p = 200 - 15x$
$p = 50 + 25x$

44. $p = 15 - \frac{7}{100}x$
$p = 2 + \frac{3}{100}x$

45. Medical Research Expenditures Table 7.2 shows expenditures (in billions of dollars) for public medical research costs for several years. Let x be the number of years since 2000.

(a) Find the quadratic regression equation and superimpose its graph on a scatter plot of the data.

(b) Find the logistic regression equation and superimpose its graph on a scatter plot of the data.

(c) When will the two models predict expenditures of \$42 billion?

(d) Writing to Learn Explain the long-range implications of using the quadratic regression equation to predict future expenditures.

(e) Writing to Learn Explain the long-range implications of using the logistic regression equation to predict future expenditures.

Table 7.2 National Public Expenditures on Medical Research

Year	Expenditures (billions of \$)
2000	23.1
2001	26.0
2002	29.5
2003	32.2
2004	35.4
2005	36.9
2006	37.8

Source: U.S. Government Census Bureau, Table 125. National Health Expenditures by Type: 1990 to 2006.

46. Personal Income Table 7.3 gives the total personal income (in billions of dollars) for residents of the states of Iowa and Nevada for several years. Let x be the number of years since 2000.

(a) Find the linear regression equation for the Iowa data and superimpose its graph on a scatter plot of the Iowa data.

(b) Find the linear regression equation for the Nevada data and superimpose its graph on a scatter plot of the Nevada data.

(c) When was Nevada's personal income about \$14 billion greater than Iowa's?

Table 7.3 Total Personal Income

Year	Iowa (billions of \$)	Nevada (billions of \$)
2002	82.4	66.6
2006	97.2	96.5
2007	104.2	101.8
2008	110.1	104.9

Source: U.S. Total Personal Income by State, U.S. Department of Commerce Bureau of Economic Analysis.

47. Population Table 7.4 gives the populations (in thousands) of Florida and Indiana for selected years. Let x be the number of years since 1980.

(a) Find a linear regression equation for the population of Florida.

(b) Find a linear regression equation for the population of Indiana.

(c) Graph the models from parts (a) and (b), and use these models to determine when the populations of the two states were about the same.

Table 7.4 Populations of Two U.S. States

Year	Florida (thousands)	Indiana (thousands)
1980	9,746	5,490
1990	12,938	5,544
2000	15,982	6,080
2010	18,801	6,483

Source: U.S. Census Bureau, The World Almanac and Book of Facts 2013.

48. Group Activity Describe all possibilities for the number of solutions to a system of two equations in two variables if the graphs of the two equations are **(a)** a line and a circle, and **(b)** a circle and a parabola.

49. Garden Problem Find the dimensions of a rectangle with a perimeter of 200 m and an area of 500 m^2.

50. Cornfield Dimensions Find the dimensions of a rectangular cornfield with a perimeter of 220 yd and an area of 3000 yd^2.

51. Rowing Speed Hank can row a boat 1 mi upstream (against the current) in 24 min. He can row the same distance downstream in 13 min. If both the rowing speed and current speed are constant, find Hank's rowing speed and the speed of the current.

52. Airplane Speed An airplane flying with the wind from Los Angeles to New York City takes 3.75 hr. Flying against the wind, the airplane takes 4.4 hr for the return trip. If the air distance between Los Angeles and New York is 2500 mi and the airplane speed and wind speed are constant, find the airplane speed and the wind speed.

53. Food Prices At Philip's convenience store the total cost of one medium and one large soda is \$1.74. The large soda costs \$0.16 more than the medium soda. Find the cost of each soda.

54. Nut Mixture A 5-lb nut mixture is worth \$3.92 per pound. The mixture contains peanuts worth \$2.38 per pound and cashews worth \$6.37 per pound. How many pounds of each type of nut are in the mixture?

55. Connecting Algebra and Functions Determine a and b so that the graph of $y = ax + b$ contains the two points $(-1, 4)$ and $(2, 6)$.

56. Connecting Algebra and Functions Determine a and b so that the graph of $ax + by = 8$ contains the two points $(2, -1)$ and $(-4, -6)$.

57. Rental Van Pedro has two plans to choose from to rent a van.

Company A: a flat fee of \$40 plus 10 cents a mile.

Company B: a flat fee of \$25 plus 15 cents a mile.

(a) How many miles can Pedro drive in order to be charged the same amount by the two companies?

(b) Writing to Learn Give reasons why Pedro might choose one plan over the other. Explain.

58. Salary Package Stephanie is offered two different salary options to sell major household appliances.

Plan A: a \$300 weekly salary plus 5% of her sales.

Plan B: a \$600 weekly salary plus 1% of her sales.

(a) What must Stephanie's weekly sales be to earn the same amount on the two plans?

(b) Writing to Learn Give reasons why Stephanie might choose one plan over the other. Explain.

Standardized Test Questions

59. True or False Let a and b be real numbers. The following system of equations can have exactly two solutions:

$$2x + 5y = a$$
$$3x - 4y = b$$

Justify your answer.

60. True or False If the resulting equation after using elimination correctly on a system of two linear equations in two variables is $7 = 0$, then the system has infinitely many solutions. Justify your answer.

In Exercises 61–64, solve the problem without using a calculator.

61. Multiple Choice Which of the following is a solution of the system $2x - 3y = 12$
$x + 2y = -1$?

(A) $(-3, 1)$ **(B)** $(-1, 0)$ **(C)** $(3, -2)$
(D) $(3, 2)$ **(E)** $(6, 0)$

62. Multiple Choice Which of the following cannot be the number of solutions of a system of two equations in two variables whose graphs are a circle and a parabola?

(A) 0 **(B)** 1 **(C)** 2
(D) 3 **(E)** 5

63. Multiple Choice Which of the following cannot be the number of solutions of a system of two equations in two variables whose graphs are parabolas?

(A) 1 (B) 2 (C) 4
(D) 5 (E) Infinitely many

64. Multiple Choice Which of the following is the number of solutions of a system of two linear equations in two variables if the resulting equation after using elimination correctly is $4 = 4$?

(A) 0 (B) 1 (C) 2
(D) 3 (E) Infinitely many

Explorations

65. An Ellipse and a Line Consider the system of equations

$$\frac{x^2}{4} + \frac{y^2}{9} = 1$$
$$x + y = 1$$

(a) Solve the equation $x^2/4 + y^2/9 = 1$ for y in terms of x to determine the two functions determined by the equation.

(b) Solve the system of equations graphically.

(c) Use substitution to confirm the solutions found in part (b).

66. A Hyperbola and a Line Consider the system of equations

$$\frac{x^2}{4} - \frac{y^2}{9} = 1$$
$$x - y = 0.$$

(a) Solve the equation $x^2/4 - y^2/9 = 1$ for y in terms of x to determine the two functions determined by the equation.

(b) Solve the system of equations graphically.

(c) Use substitution to confirm the solutions found in part (b).

Extending the Ideas

In Exercises 67 and 68, use the elimination method to solve the system of equations.

67. $x^2 - 2y = -6$
$x^2 + y = 4$

68. $x^2 + y^2 = 1$
$x^2 - y^2 = 1$

In Exercises 69 and 70, $p(x)$ is the demand curve. The total revenue if x units are sold is $R = px$. Find the number of units sold that gives the maximum revenue.

69. $p = 100 - 4x$

70. $p = 80 - x^2$

7.2 Matrix Algebra

What you'll learn about

- Matrices
- Matrix Addition and Subtraction
- Matrix Multiplication
- Identity and Inverse Matrices
- Determinant of a Square Matrix
- Applications

... and why

Matrix algebra provides a powerful technique to manipulate large data sets and solve the related problems that are modeled by the matrices.

Matrices

A *matrix* is a rectangular array of numbers. Matrices provide an efficient way to solve systems of linear equations and to record data. The tables of data presented in this textbook are examples of matrices.

DEFINITION Matrix

Let m and n be positive integers. An $\boldsymbol{m \times n}$ **matrix** (read "m by n matrix") is a rectangular array of m horizontal rows and n vertical columns of real numbers.

$$\begin{bmatrix} a_{11} & a_{12} & \cdots & a_{1n} \\ a_{21} & a_{22} & \cdots & a_{2n} \\ \vdots & \vdots & & \vdots \\ a_{m1} & a_{m2} & \cdots & a_{mn} \end{bmatrix}$$

We also use the shorthand notation $[a_{ij}]$ for this matrix.

Each **element**, or **entry**, a_{ij}, of the matrix uses *double subscript* notation. The **row subscript** is the first subscript i, and the **column subscript** is j. The element a_{ij} is in the ith row and jth column. In general, the **order of an** $\boldsymbol{m \times n}$ **matrix** is $m \times n$. If $m = n$, the matrix is a **square matrix**. Two matrices are **equal matrices** if they have the same order and their corresponding elements are equal.

EXAMPLE 1 Determining the Order of a Matrix

(a) The matrix $\begin{bmatrix} 1 & -2 & 3 \\ 2 & 0 & 4 \end{bmatrix}$ has order 2×3.

(b) The matrix $\begin{bmatrix} 1 & -1 \\ 0 & 14 \\ 2 & -1 \\ 3 & 62 \end{bmatrix}$ has order 4×2.

(c) The matrix $\begin{bmatrix} 1 & 2 & 3 \\ 4 & 5 & 6 \\ 7 & 8 & 9 \end{bmatrix}$ has order 3×3 and is a square matrix.

Now try Exercise 1.

Matrix Addition and Subtraction

We add or subtract two matrices of the same order by adding or subtracting their corresponding entries. Matrices of different orders *cannot* be added or subtracted.

DEFINITION Matrix Addition and Matrix Subtraction

Let $A = [a_{ij}]$ and $B = [b_{ij}]$ be matrices of order $m \times n$.

1. The **sum** $\boldsymbol{A + B}$ is the $m \times n$ matrix
$$A + B = [a_{ij} + b_{ij}].$$
2. The **difference** $\boldsymbol{A - B}$ is the $m \times n$ matrix
$$A - B = [a_{ij} - b_{ij}].$$

Historical Note

Methods used by the Chinese between 200 BCE and 100 BCE to solve problems involving several unknowns were similar to modern methods that use matrices. Matrices were formally developed in the 18th century by several mathematicians, including Leibniz, Cauchy, and Gauss.

EXAMPLE 2 Using Matrix Addition

In 2005 the verbal section of the SAT became critical reading, and a writing section was added. (*Source: The College Board, World Almanac and Book of Facts 2013.*) Matrix V gives the mean SAT verbal (or critical reading) scores for selected years for the six New England states. Matrix M gives the mean SAT mathematics scores for the same years and states. Express the mean combined scores for these years and states.

$$V = \begin{array}{c} \\ \text{CT} \\ \text{ME} \\ \text{MA} \\ \text{NH} \\ \text{RI} \\ \text{VT} \end{array} \begin{array}{c} \begin{array}{cccc} 2000 & 2005 & 2010 & 2012 \end{array} \\ \begin{bmatrix} 508 & 517 & 509 & 506 \\ 504 & 509 & 468 & 470 \\ 511 & 520 & 512 & 513 \\ 520 & 525 & 520 & 521 \\ 505 & 503 & 494 & 490 \\ 513 & 521 & 519 & 519 \end{bmatrix} \end{array} \qquad M = \begin{array}{c} \\ \text{CT} \\ \text{ME} \\ \text{MA} \\ \text{NH} \\ \text{RI} \\ \text{VT} \end{array} \begin{array}{c} \begin{array}{cccc} 2000 & 2005 & 2010 & 2012 \end{array} \\ \begin{bmatrix} 509 & 517 & 514 & 512 \\ 500 & 505 & 467 & 472 \\ 513 & 527 & 526 & 530 \\ 519 & 525 & 524 & 525 \\ 500 & 505 & 495 & 491 \\ 508 & 517 & 521 & 523 \end{bmatrix} \end{array}$$

SOLUTION We obtain the combined scores by adding the two matrices:

$$V + M = \begin{array}{c} \\ \text{CT} \\ \text{ME} \\ \text{MA} \\ \text{NH} \\ \text{RI} \\ \text{VT} \end{array} \begin{array}{c} \begin{array}{cccc} 2000 & 2005 & 2010 & 2012 \end{array} \\ \begin{bmatrix} 1017 & 1034 & 1023 & 1018 \\ 1004 & 1014 & 935 & 942 \\ 1024 & 1047 & 1038 & 1043 \\ 1039 & 1050 & 1044 & 1046 \\ 1005 & 1008 & 989 & 981 \\ 1021 & 1038 & 1040 & 1042 \end{bmatrix} \end{array}$$

Now try Exercise 11.

Power of Matrix Algebra

The result in Example 2 is fairly simple, but it is significant that we found (essentially) 24 pieces of information with a single mathematical operation. That is the power of matrix algebra.

When we work with matrices, real numbers are **scalars**. The product of the real number k and the $m \times n$ matrix $A = [a_{ij}]$ is the $m \times n$ matrix

$$kA = [ka_{ij}].$$

The matrix $kA = [ka_{ij}]$ is a **scalar multiple of A**.

EXAMPLE 3 Using Scalar Multiplication

A consumer advocacy group has computed the mean retail prices for brand name products and generic products at three different stores in a major city. The prices in dollars are shown in the 3×2 matrix.

$$\begin{array}{c} \\ \text{Store A} \\ \text{Store B} \\ \text{Store C} \end{array} \begin{array}{c} \begin{array}{cc} \text{Brand} & \text{Generic} \end{array} \\ \begin{bmatrix} 3.97 & 3.64 \\ 3.78 & 3.69 \\ 3.75 & 3.67 \end{bmatrix} \end{array}$$

The city has a combined sales tax of 7.25%. Construct a matrix showing the comparative prices with sales tax included.

SOLUTION Multiply the original matrix by the scalar 1.0725 to add the sales tax to every price.

$$1.0725 \times \begin{bmatrix} 3.97 & 3.64 \\ 3.78 & 3.69 \\ 3.75 & 3.67 \end{bmatrix} \approx \begin{array}{c} \\ \text{Store A} \\ \text{Store B} \\ \text{Store C} \end{array} \begin{array}{c} \begin{array}{cc} \text{Brand} & \text{Generic} \end{array} \\ \begin{bmatrix} 4.26 & 3.90 \\ 4.05 & 3.96 \\ 4.02 & 3.94 \end{bmatrix} \end{array}$$

Now try Exercise 13.

Matrices inherit many properties possessed by the real numbers. Let $A = [a_{ij}]$ be any $m \times n$ matrix. The $m \times n$ matrix $O = [0]$ consisting entirely of zeros is the **zero matrix** because $A + O = A$. In other words, O is the **additive identity** for the set of all $m \times n$ matrices. The $m \times n$ matrix $B = [-a_{ij}]$ consisting of the *additive inverses* of the entries of A is the **additive inverse of A** because $A + B = O$. We also write $B = -A$.

Subtraction can be defined using such an additive inverse:

$$A - C = [a_{ij} - c_{ij}] = [a_{ij} + (-c_{ij})] = [a_{ij}] + [-c_{ij}] = A + (-C)$$

Thus, subtracting C from A is the same as adding A to the additive inverse of C.

EXPLORATION 1 Computing with Matrices

Let $A = [a_{ij}]$ and $B = [b_{ij}]$ be 2×2 matrices with $a_{ij} = 3i - j$ and $b_{ij} = i^2 + j^2 - 3$ for $i = 1, 2$ and $j = 1, 2$.

1. Determine A and B.
2. Determine the additive inverse $-A$ of A and verify that $A + (-A) = [0]$. What is the order of $[0]$?
3. Determine $3A - 2B$.

Matrix Multiplication

To form the *product AB* of two matrices, the number of columns of the matrix A on the left must be equal to the number of rows of the matrix B on the right. In this case, any row of A has the same number of entries as any column of B. Each entry of the product is obtained by summing the products of the entries of a row of A by the corresponding entries of a column of B.

DEFINITION Matrix Multiplication

Let $A = [a_{ij}]$ be an $m \times r$ matrix and $B = [b_{ij}]$ an $r \times n$ matrix. The **product AB** $= [c_{ij}]$ is the $m \times n$ matrix where $c_{ij} = a_{i1}b_{1j} + a_{i2}b_{2j} + \cdots + a_{ir}b_{rj}$.

The key to understanding how to form the product of any two matrices is first to consider the product of a $1 \times r$ matrix $A = [a_{1j}]$ with an $r \times 1$ matrix $B = [b_{j1}]$. According to the definition, $AB = [c_{11}]$ is the 1×1 matrix where $c_{11} = a_{11}b_{11} + a_{12}b_{21} + \cdots + a_{1r}b_{r1}$. For example, the product AB of the 1×3 matrix A and the 3×1 matrix B, where

$$A = \begin{bmatrix} 1 & 2 & 3 \end{bmatrix} \quad \text{and} \quad B = \begin{bmatrix} 4 \\ 5 \\ 6 \end{bmatrix},$$

is

$$A \cdot B = \begin{bmatrix} 1 & 2 & 3 \end{bmatrix} \cdot \begin{bmatrix} 4 \\ 5 \\ 6 \end{bmatrix} = [1 \cdot 4 + 2 \cdot 5 + 3 \cdot 6] = [32].$$

Then, the ij-entry of the product AB of an $m \times r$ matrix with an $r \times n$ matrix is the product of the ith row of A, considered as a $1 \times r$ matrix, with the jth column of B, considered as an $r \times 1$ matrix, as illustrated in Example 4.

EXAMPLE 4 Finding the Product of Two Matrices

Find the product AB if possible, where

(a) $A = \begin{bmatrix} 2 & 1 & -3 \\ 0 & 1 & 2 \end{bmatrix}$ and $B = \begin{bmatrix} 1 & -4 \\ 0 & 2 \\ 1 & 0 \end{bmatrix}$.

(b) $A = \begin{bmatrix} 2 & 1 & -3 \\ 0 & 1 & 2 \end{bmatrix}$ and $B = \begin{bmatrix} 3 & -4 \\ 2 & 1 \end{bmatrix}$.

SOLUTION

(a) The number of columns of A is 3 and the number of rows of B is 3, so the product AB is defined. The product $AB = [c_{ij}]$ is a 2×2 matrix where

$$c_{11} = [2 \quad 1 \quad -3]\begin{bmatrix} 1 \\ 0 \\ 1 \end{bmatrix} = 2 \cdot 1 + 1 \cdot 0 + (-3) \cdot 1 = -1$$

$$c_{12} = [2 \quad 1 \quad -3]\begin{bmatrix} -4 \\ 2 \\ 0 \end{bmatrix} = 2 \cdot (-4) + 1 \cdot 2 + (-3) \cdot 0 = -6$$

$$c_{21} = [0 \quad 1 \quad 2]\begin{bmatrix} 1 \\ 0 \\ 1 \end{bmatrix} = 0 \cdot 1 + 1 \cdot 0 + 2 \cdot 1 = 2$$

$$c_{22} = [0 \quad 1 \quad 2]\begin{bmatrix} -4 \\ 2 \\ 0 \end{bmatrix} = 0 \cdot (-4) + 1 \cdot 2 + 2 \cdot 0 = 2$$

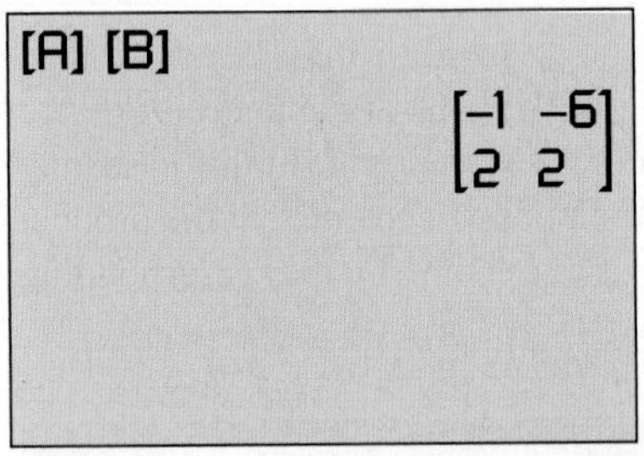

FIGURE 7.7 The matrix product AB of Example 4.

Thus, $AB = \begin{bmatrix} -1 & -6 \\ 2 & 2 \end{bmatrix}$. Figure 7.7 supports this computation.

(b) The number of columns of A is 3 and the number of rows of B is 2, so the product AB is *not* defined.

Now try Exercise 19.

EXAMPLE 5 Using Matrix Multiplication

A florist makes three flower arrangements for Mother's Day (I, II, and III), each including roses, carnations, and lilies. Matrix A shows the number of each type of flower used in each arrangement.

$$A = \begin{array}{r} \\ \text{Roses} \\ \text{Carnations} \\ \text{Lilies} \end{array} \begin{array}{c} \begin{array}{ccc} \text{I} & \text{II} & \text{III} \end{array} \\ \begin{bmatrix} 5 & 8 & 7 \\ 6 & 6 & 7 \\ 4 & 3 & 3 \end{bmatrix} \end{array}$$

The florist can buy flowers from two wholesalers (W1 and W2), but wants to give all of his business to one or the other. The cost of the three flower types from the two wholesalers is shown in matrix B.

$$B = \begin{array}{r} \\ \text{Roses} \\ \text{Carnations} \\ \text{Lilies} \end{array} \begin{array}{c} \begin{array}{cc} \text{W1} & \text{W2} \end{array} \\ \begin{bmatrix} 1.50 & 1.35 \\ 0.95 & 1.00 \\ 1.30 & 1.35 \end{bmatrix} \end{array}$$

Construct a matrix showing the cost of making each of the three flower arrangements from flowers supplied by the two wholesalers.

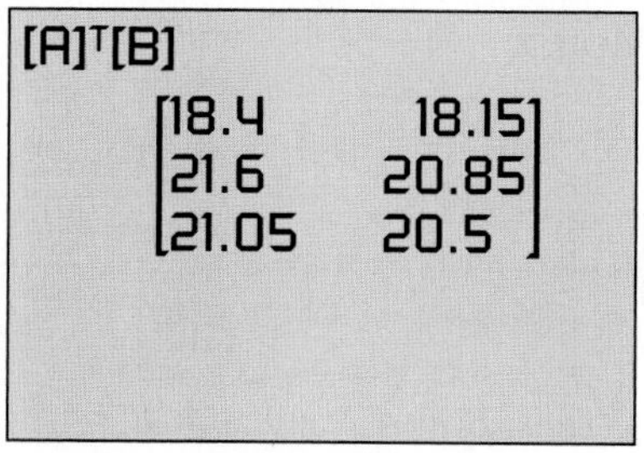

FIGURE 7.8 The product A^TB for the matrices A and B of Example 5.

SOLUTION We can use the labeling of the matrices to help us. We want the columns of A to match up with the rows of B (because that's how the matrix multiplication works). We therefore switch the rows and columns of A to get the flowers along the columns. (The new matrix is called the **transpose** of A, denoted by A^T.) We then find the product A^TB:

$$\begin{array}{c} \\ \text{I} \\ \text{II} \\ \text{III} \end{array} \begin{array}{c} \text{Rose Carn Lily} \\ \begin{bmatrix} 5 & 6 & 4 \\ 8 & 6 & 3 \\ 7 & 7 & 3 \end{bmatrix} \end{array} \times \begin{array}{c} \\ \text{Rose} \\ \text{Carn} \\ \text{Lilly} \end{array} \begin{array}{c} \text{W1} \quad \text{W2} \\ \begin{bmatrix} 1.50 & 1.35 \\ 0.95 & 1.00 \\ 1.30 & 1.35 \end{bmatrix} \end{array} = \begin{array}{c} \\ \text{I} \\ \text{II} \\ \text{III} \end{array} \begin{array}{c} \text{W1} \quad \text{W2} \\ \begin{bmatrix} 18.40 & 18.15 \\ 21.60 & 20.85 \\ 21.05 & 20.50 \end{bmatrix} \end{array}$$

Figure 7.8 shows the product A^TB and supports our computation. The florist should do business with wholesaler W2. **Now try Exercise 47.**

Identity and Inverse Matrices

The $n \times n$ matrix I_n with 1's on the main diagonal (upper left to lower right) and 0's elsewhere is the **identity matrix** of order $n \times n$:

$$I_n = \begin{bmatrix} 1 & 0 & 0 & \cdots & 0 \\ 0 & 1 & 0 & \cdots & 0 \\ 0 & 0 & 1 & \cdots & 0 \\ \vdots & \vdots & \vdots & & \vdots \\ 0 & 0 & 0 & \cdots & 1 \end{bmatrix}$$

For example,

$$I_2 = \begin{bmatrix} 1 & 0 \\ 0 & 1 \end{bmatrix}, \quad I_3 = \begin{bmatrix} 1 & 0 & 0 \\ 0 & 1 & 0 \\ 0 & 0 & 1 \end{bmatrix}, \quad \text{and} \quad I_4 = \begin{bmatrix} 1 & 0 & 0 & 0 \\ 0 & 1 & 0 & 0 \\ 0 & 0 & 1 & 0 \\ 0 & 0 & 0 & 1 \end{bmatrix}.$$

If $A = [a_{ij}]$ is any $n \times n$ matrix, we can prove (see Exercise 56) that

$$AI_n = I_nA = A,$$

that is, I_n is the **multiplicative identity** for the set of $n \times n$ matrices.

If a is a nonzero real number, then $a^{-1} = 1/a$ is the multiplicative inverse of a, that is, $aa^{-1} = a(1/a) = 1$. The definition of the *multiplicative inverse* of a square matrix is similar.

DEFINITION Inverse of a Square Matrix

Let $A = [a_{ij}]$ be an $n \times n$ matrix. If there is a matrix B such that

$$AB = BA = I_n,$$

then B is the **inverse** of A. We write $B = A^{-1}$ (read "A inverse").

We will see in Example 7 that not every square matrix has an inverse. If a square matrix A has no inverse, then A is **singular**. If a square matrix A has an inverse, then A is **nonsingular**, as in Example 6.

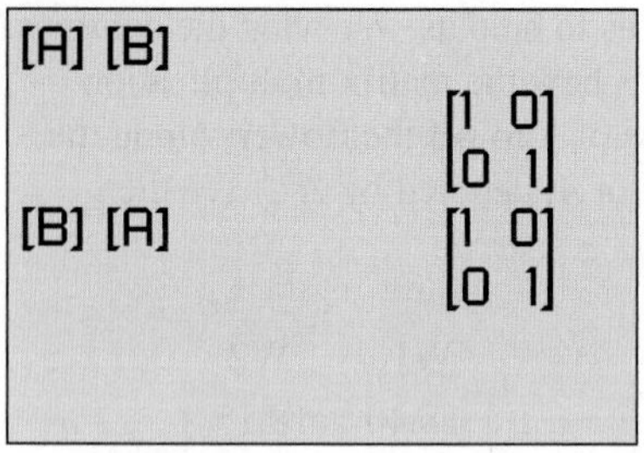

FIGURE 7.9 Showing A and B are inverse matrices. (Example 6)

EXAMPLE 6 Verifying an Inverse Matrix

Prove that

$$A = \begin{bmatrix} 3 & -2 \\ -1 & 1 \end{bmatrix} \quad \text{and} \quad B = \begin{bmatrix} 1 & 2 \\ 1 & 3 \end{bmatrix}$$

are inverse matrices.

SOLUTION Figure 7.9 shows that $AB = BA = I_2$. Thus, $B = A^{-1}$ and $A = B^{-1}$.

Now try Exercise 33.

EXAMPLE 7 Showing a Matrix Has No Inverse

Prove that the matrix $A = \begin{bmatrix} 6 & 3 \\ 2 & 1 \end{bmatrix}$ is singular, that is, A has no inverse.

SOLUTION Suppose A has an inverse $B = \begin{bmatrix} x & y \\ z & w \end{bmatrix}$. Then, $AB = I_2$.

$$AB = \begin{bmatrix} 6 & 3 \\ 2 & 1 \end{bmatrix} \begin{bmatrix} x & y \\ z & w \end{bmatrix} = \begin{bmatrix} 1 & 0 \\ 0 & 1 \end{bmatrix}$$

$$= \begin{bmatrix} 6x + 3z & 6y + 3w \\ 2x + z & 2y + w \end{bmatrix} = \begin{bmatrix} 1 & 0 \\ 0 & 1 \end{bmatrix}$$

Using equality of matrices we obtain:

$$6x + 3z = 1 \quad 6y + 3w = 0$$
$$2x + z = 0 \quad 2y + w = 1$$

Multiplying both sides of the equation $2x + z = 0$ by 3 yields $6x + 3z = 0$. There are no values for x and z for which the value of $6x + 3z$ is both 0 and 1. Thus, A does not have an inverse.

Now try Exercise 37.

Determinant of a Square Matrix

There is a simple test that determines whether a 2×2 matrix has an inverse.

Inverse of a 2×2 Matrix

If $ad - bc \neq 0$, then

$$\begin{bmatrix} a & b \\ c & d \end{bmatrix}^{-1} = \frac{1}{ad - bc} \begin{bmatrix} d & -b \\ -c & a \end{bmatrix}.$$

In Exercise 55 we ask you to prove the theorem above. The number $ad - bc$ is the **determinant** of the 2×2 matrix $A = \begin{bmatrix} a & b \\ c & d \end{bmatrix}$ and is denoted

$$\det A = \begin{vmatrix} a & b \\ c & d \end{vmatrix} = ad - bc.$$

To define the determinant of a higher-order square matrix, we need to introduce the *minors* and *cofactors* associated with the entries of a square matrix. Let $A = [a_{ij}]$ be an $n \times n$ matrix. The **minor** (short for "minor determinant") M_{ij} corresponding to the element a_{ij} is the determinant of the $(n - 1) \times (n - 1)$ matrix obtained by deleting the row and column containing a_{ij}. The **cofactor** corresponding to a_{ij} is $A_{ij} = (-1)^{i+j}M_{ij}$.

Computing Determinants

We expect you to compute the determinant of a 2 × 2 matrix mentally or using paper and pencil. For instance, in Example 8a, $\det A = 3 \cdot 2 - 1 \cdot 4 = 6 - 4 = 2$.

Using a grapher for higher-dimension matrices is appropriate, for instance, in Example 8b:

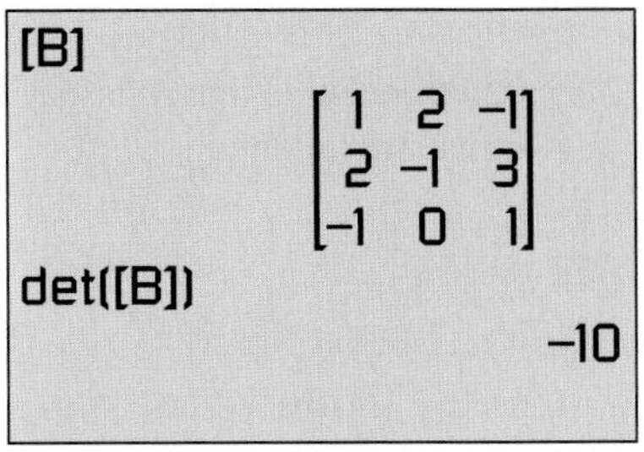

DEFINITION Determinant of a Square Matrix

Let $A = [a_{ij}]$ be a matrix of order $n \times n$ $(n > 2)$. The determinant of A, denoted by $\det A$ or $|A|$, is the sum of the entries in any row or any column multiplied by their respective cofactors. For example, expanding by the ith row gives

$$\det A = |A| = a_{i1}A_{i1} + a_{i2}A_{i2} + \cdots + a_{in}A_{in}.$$

If $A = [a_{ij}]$ is a 3 × 3 matrix, then, using the definition of determinant applied to the second row, we obtain

$$\begin{vmatrix} a_{11} & a_{12} & a_{13} \\ a_{21} & a_{22} & a_{23} \\ a_{31} & a_{32} & a_{33} \end{vmatrix} = a_{21}A_{21} + a_{22}A_{22} + a_{23}A_{23}$$

$$= a_{21}(-1)^3 \begin{vmatrix} a_{12} & a_{13} \\ a_{32} & a_{33} \end{vmatrix} + a_{22}(-1)^4 \begin{vmatrix} a_{11} & a_{13} \\ a_{31} & a_{33} \end{vmatrix}$$

$$+ a_{23}(-1)^5 \begin{vmatrix} a_{11} & a_{12} \\ a_{31} & a_{32} \end{vmatrix}$$

$$= -a_{21}(a_{12}a_{33} - a_{13}a_{32}) + a_{22}(a_{11}a_{33} - a_{13}a_{31})$$

$$- a_{23}(a_{11}a_{32} - a_{12}a_{31})$$

The determinant of a 3 × 3 matrix involves three determinants of 2 × 2 matrices, the determinant of a 4 × 4 matrix involves four determinants of 3 × 3 matrices, and so forth. This is a tedious definition to apply. Most of the time we use a grapher to evaluate determinants in this textbook, as shown in the margin note.

EXPLORATION 2 Investigating the Definition of Determinant

1. Complete the expansion of the determinant of the 3 × 3 matrix $A = [a_{ij}]$ started above. Explain why each term in the expansion contains an element from each row and each column.
2. Use the first row of the 3 × 3 matrix to expand the determinant and compare to the expression in 1.
3. Prove that the determinant of a square matrix with a zero row or a zero column is zero.

We can now state the condition under which square matrices have inverses.

THEOREM Inverses of $n \times n$ Matrices

An $n \times n$ matrix A has an inverse if and only if $\det A \neq 0$.

There are complicated formulas for finding the inverses of nonsingular matrices of order 3 × 3 or higher. We will use a grapher instead of these formulas to find inverses of square matrices.

EXAMPLE 8 Finding Inverse Matrices

Determine whether the matrix has an inverse. If so, find its inverse matrix.

(a) $A = \begin{bmatrix} 3 & 1 \\ 4 & 2 \end{bmatrix}$ **(b)** $B = \begin{bmatrix} 1 & 2 & -1 \\ 2 & -1 & 3 \\ -1 & 0 & 1 \end{bmatrix}$

SOLUTION

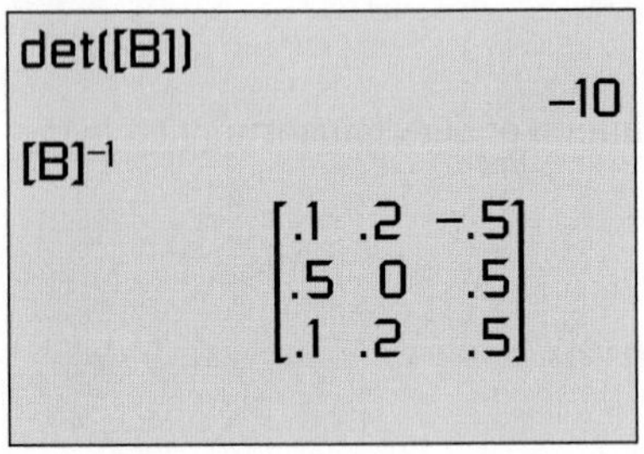

FIGURE 7.10 The matrix B is nonsingular and so has an inverse. (Example 8b)

(a) Because $\det A = ad - bc = 3 \cdot 2 - 1 \cdot 4 = 2 \neq 0$, we conclude that A has an inverse. Using the formula for the inverse of a 2×2 matrix, we obtain

$$A^{-1} = \frac{1}{ad - bc}\begin{bmatrix} d & -b \\ -c & a \end{bmatrix} = \frac{1}{2}\begin{bmatrix} 2 & -1 \\ -4 & 3 \end{bmatrix}$$

$$= \begin{bmatrix} 1 & -0.5 \\ -2 & 1.5 \end{bmatrix}$$

You can check that $A^{-1}A = AA^{-1} = I_2$.

(b) Figure 7.10 shows that $\det B = -10 \neq 0$ and

$$B^{-1} = \begin{bmatrix} 0.1 & 0.2 & -0.5 \\ 0.5 & 0 & 0.5 \\ 0.1 & 0.2 & 0.5 \end{bmatrix}.$$

You can use your grapher to check that $B^{-1}B = BB^{-1} = I_3$.

Now try Exercise 41.

We list five of the important properties of matrices, some of which you will be asked to prove in the exercises.

Properties of Matrices

Let A, B, and C be matrices whose orders are such that the following sums, differences, and products are defined.

1. Commutative property
Addition:
$A + B = B + A$
Multiplication:
(Does not hold in general)

2. Associative property
Addition:
$(A + B) + C = A + (B + C)$
Multiplication:
$(AB)C = A(BC)$

3. Identity property
Addition: $A + O = A$
Multiplication: order of $A = n \times n$
$A \cdot I_n = I_n \cdot A = A$

4. Inverse property
Addition: $A + (-A) = O$
Multiplication: order of $A = n \times n$
$AA^{-1} = A^{-1}A = I_n, \quad |A| \neq 0$

5. Distributive property

Multiplication over addition
$A(B + C) = AB + AC$
$(A + B)C = AC + BC$

Multiplication over subtraction
$A(B - C) = AB - AC$
$(A - B)C = AC - BC$

Applications

Points in the Cartesian coordinate plane can be represented by 1×2 matrices. For example, the point $(2, -3)$ can be represented by the 1×2 matrix $[2 \ \ -3]$. We can calculate the images of points acted upon by some of the transformations studied in Section 1.5 using matrix multiplication as illustrated in Example 9.

EXAMPLE 9 Reflecting with Respect to the x-Axis as Matrix Multiplication

Prove that the image of a point under a reflection across the x-axis can be obtained by multiplying by $\begin{bmatrix} 1 & 0 \\ 0 & -1 \end{bmatrix}$.

SOLUTION The image of the point (x, y) under a reflection across the x-axis is $(x, -y)$. The product

$$\begin{bmatrix} x & y \end{bmatrix}\begin{bmatrix} 1 & 0 \\ 0 & -1 \end{bmatrix} = \begin{bmatrix} x & -y \end{bmatrix}$$

shows that the point (x, y) (in matrix form $\begin{bmatrix} x & y \end{bmatrix}$) is moved to the point $(x, -y)$ (in matrix form $\begin{bmatrix} x & -y \end{bmatrix}$).

Now try Exercise 57.

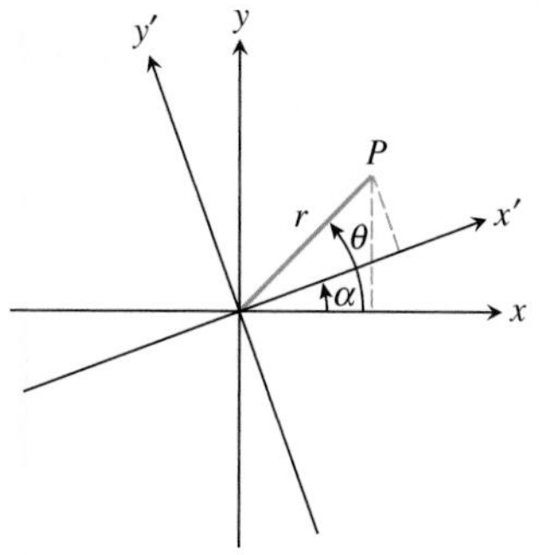

FIGURE 7.11 Rotating the xy-coordinate system through the angle α to obtain the $x'y'$-coordinate system. (Example 10)

Figure 7.11 shows the xy-coordinate system rotated through the angle α to obtain the $x'y'$-coordinate system. In Example 10, we see that the coordinates of a point in the $x'y'$-coordinate system can be obtained by multiplying the coordinates of the point in the xy-coordinate system by an appropriate 2×2 matrix. In Exercise 71, you will see that the reverse is also true.

EXAMPLE 10 Rotating a Coordinate System

Prove that the (x', y') coordinates of P in Figure 7.11 are related to the (x, y) coordinates of P by the equations

$$\begin{aligned} x' &= x \cos \alpha + y \sin \alpha \\ y' &= -x \sin \alpha + y \cos \alpha \end{aligned}$$

Then, prove that the coordinates (x', y') can be obtained from the (x, y) coordinates by matrix multiplication. We use this result in Section 8.4 when we study conic sections.

SOLUTION Using the right triangle formed by P and the $x'y'$-coordinate system, we obtain

$$x' = r \cos (\theta - \alpha) \quad \text{and} \quad y' = r \sin (\theta - \alpha).$$

Expanding the above expressions for x' and y', using trigonometric identities for $\cos (\theta - \alpha)$ and $\sin (\theta - \alpha)$, yields

$$\begin{aligned} x' &= r \cos \theta \cos \alpha + r \sin \theta \sin \alpha \\ y' &= r \sin \theta \cos \alpha - r \cos \theta \sin \alpha \end{aligned}$$

It follows from the right triangle formed by P and the xy-coordinate system that $x = r \cos \theta$ and $y = r \sin \theta$. Substituting these values for x and y into the above pair of equations yields

$$x' = x \cos \alpha + y \sin \alpha \quad \text{and} \quad \begin{aligned} y' &= y \cos \alpha - x \sin \alpha \\ &= -x \sin \alpha + y \cos \alpha, \end{aligned}$$

which is what we were asked to prove. Finally, matrix multiplication shows that

$$\begin{bmatrix} x' & y' \end{bmatrix} = \begin{bmatrix} x & y \end{bmatrix}\begin{bmatrix} \cos \alpha & -\sin \alpha \\ \sin \alpha & \cos \alpha \end{bmatrix}.$$

Now try Exercise 71.

CHAPTER OPENER Problem (from page 518)

Problem: If we have a triangle with vertices at $(0, 0)$, $(1, 1)$, and $(2, 0)$, and we want to double the lengths of the sides of the triangle, where would the vertices of the enlarged triangle be?

Solution: Given a triangle with vertices at $(0, 0)$, $(1, 1)$, and $(2, 0)$, as in Figure 7.12, we can find the vertices of a new triangle whose sides are twice as long by multiplying by the scale matrix.

$$\begin{bmatrix} 2 & 0 \\ 0 & 2 \end{bmatrix}$$

For the point $(0, 0)$, we have

$$[x' \quad y'] = [0 \quad 0]\begin{bmatrix} 2 & 0 \\ 0 & 2 \end{bmatrix} = [0 \quad 0].$$

For the point $(1, 1)$, we have

$$[x' \quad y'] = [1 \quad 1]\begin{bmatrix} 2 & 0 \\ 0 & 2 \end{bmatrix} = [2 \quad 2].$$

And for the point $(2, 0)$, we have

$$[x' \quad y'] = [2 \quad 0]\begin{bmatrix} 2 & 0 \\ 0 & 2 \end{bmatrix} = [4 \quad 0].$$

So the new triangle has vertices $(0, 0)$, $(2, 2)$, and $(4, 0)$, as Figure 7.13 shows.

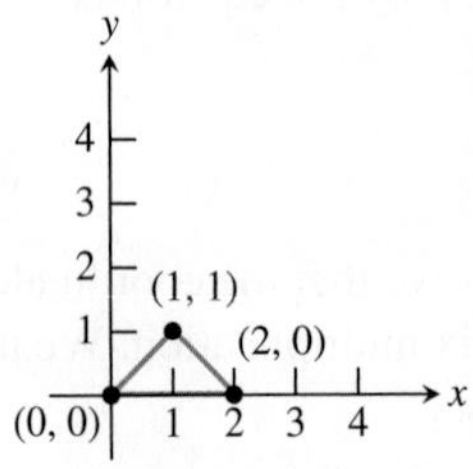

FIGURE 7.12 Original triangle.

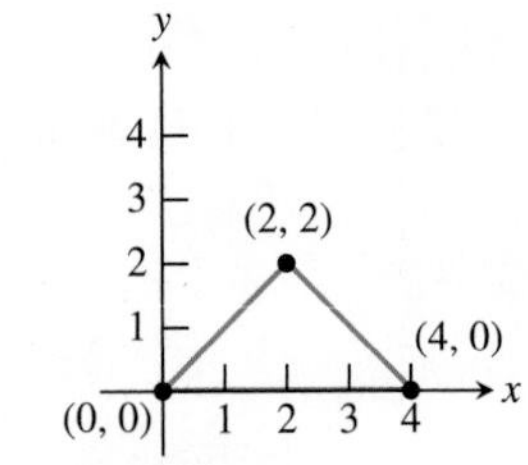

FIGURE 7.13 Transformed triangle.

QUICK REVIEW 7.2 *(For help, go to Sections 1.5, 5.3, and 6.4.)*

In Exercises 1–4, the points **(a)** $(3, -2)$ and **(b)** (x, y) are reflected across the given line. Find the coordinates of the reflected points.

1. The x-axis **2.** The y-axis

3. The line $y = x$ **4.** The line $y = -x$

In Exercises 5 and 6, express the coordinates of P in terms of θ.

5.

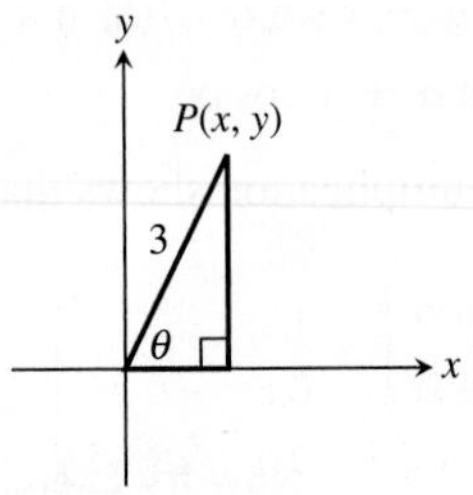

6.

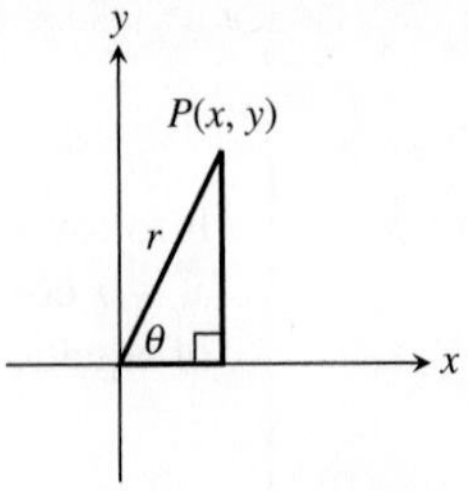

In Exercises 7–10, expand the expression.

7. $\sin(\alpha + \beta)$ **8.** $\sin(\alpha - \beta)$

9. $\cos(\alpha + \beta)$ **10.** $\cos(\alpha - \beta)$

SECTION 7.2 Exercises

Exercise numbers with a gray background indicate problems that the authors have designed to be solved *without a calculator*.

In Exercises 1–6, determine the order of the matrix. Indicate whether the matrix is square.

1. $\begin{bmatrix} 2 & 3 & -1 \\ 1 & 0 & 5 \end{bmatrix}$ **2.** $\begin{bmatrix} 1 & 3 \\ -1 & 2 \end{bmatrix}$ **3.** $\begin{bmatrix} 5 & 6 \\ -1 & 2 \\ 0 & 0 \end{bmatrix}$

4. $[-1 \quad 0 \quad 6]$ **5.** $\begin{bmatrix} 2 \\ -1 \\ 0 \end{bmatrix}$ **6.** $[0]$

In Exercises 7–10, identify the element specified for the following matrix.

$$\begin{bmatrix} -2 & 0 & 3 & 4 \\ 3 & 1 & 5 & -1 \\ 1 & 4 & -1 & 3 \end{bmatrix}$$

7. a_{13} **8.** a_{24} **9.** a_{32} **10.** a_{33}

In Exercises 11–16, find **(a)** $A + B$, **(b)** $A - B$, **(c)** $3A$, and **(d)** $2A - 3B$.

11. $A = \begin{bmatrix} 2 & 3 \\ -1 & 5 \end{bmatrix}, \quad B = \begin{bmatrix} 1 & -3 \\ -2 & -4 \end{bmatrix}$

12. $A = \begin{bmatrix} -1 & 0 & 2 \\ 4 & 1 & -1 \\ 2 & 0 & 1 \end{bmatrix}, \quad B = \begin{bmatrix} 2 & 1 & 0 \\ -1 & 0 & 2 \\ 4 & -3 & -1 \end{bmatrix}$

13. $A = \begin{bmatrix} -3 & 1 \\ 0 & -1 \\ 2 & 1 \end{bmatrix}, \quad B = \begin{bmatrix} 4 & 0 \\ -2 & 1 \\ -3 & -1 \end{bmatrix}$

14. $A = \begin{bmatrix} 5 & -2 & 3 & 1 \\ -1 & 0 & 2 & 2 \end{bmatrix}, \quad B = \begin{bmatrix} -2 & 3 & 1 & 0 \\ 4 & 0 & -1 & -2 \end{bmatrix}$

15. $A = \begin{bmatrix} -2 \\ 1 \\ 0 \end{bmatrix}, \quad B = \begin{bmatrix} -1 \\ 0 \\ 4 \end{bmatrix}$

16. $A = [-1 \quad -2 \quad 0 \quad 3]$, and $B = [1 \quad 2 \quad -2 \quad 0]$

In Exercises 17–22, use the definition of matrix multiplication to find **(a)** AB and **(b)** BA. Support your answer with the matrix feature of your grapher.

17. $A = \begin{bmatrix} 2 & 3 \\ -1 & 5 \end{bmatrix}, \quad B = \begin{bmatrix} 1 & -3 \\ -2 & -4 \end{bmatrix}$

18. $A = \begin{bmatrix} 1 & -4 \\ 2 & 6 \end{bmatrix}, \quad B = \begin{bmatrix} 5 & 1 \\ -2 & -3 \end{bmatrix}$

19. $A = \begin{bmatrix} 2 & 0 & 1 \\ 1 & 4 & -3 \end{bmatrix}, \quad B = \begin{bmatrix} 1 & 2 \\ -3 & 1 \\ 0 & -2 \end{bmatrix}$

20. $A = \begin{bmatrix} 1 & 0 & -2 & 3 \\ 2 & 1 & 4 & -1 \end{bmatrix}, \quad B = \begin{bmatrix} 5 & -1 \\ 0 & 2 \\ -1 & 3 \\ 4 & 2 \end{bmatrix}$

21. $A = \begin{bmatrix} -1 & 0 & 2 \\ 4 & 1 & -1 \\ 2 & 0 & 1 \end{bmatrix}, \quad B = \begin{bmatrix} 2 & 1 & 0 \\ -1 & 0 & 2 \\ 4 & -3 & -1 \end{bmatrix}$

22. $A = \begin{bmatrix} -2 & 3 & 0 \\ 1 & -2 & 4 \\ 3 & 2 & 1 \end{bmatrix}, \quad B = \begin{bmatrix} 4 & -1 & 2 \\ 0 & 2 & 3 \\ -1 & 3 & -1 \end{bmatrix}$

In Exercises 23–28, find **(a)** AB and **(b)** BA, or state that the product is not defined. Support your answer using the matrix feature of your grapher.

23. $A = [2 \quad -1 \quad 3], \quad B = \begin{bmatrix} -5 \\ 4 \\ 2 \end{bmatrix}$

24. $A = \begin{bmatrix} -2 \\ 3 \\ -4 \end{bmatrix}, \quad B = [-1 \quad 2 \quad 4]$

25. $A = \begin{bmatrix} -1 & 2 \\ 3 & 4 \end{bmatrix}, \quad B = [-3 \quad 5]$

26. $A = \begin{bmatrix} -1 & 3 \\ 0 & 1 \\ 1 & 0 \\ -3 & -1 \end{bmatrix}, \quad B = \begin{bmatrix} 5 & -6 \\ 2 & 3 \end{bmatrix}$

27. $A = \begin{bmatrix} 0 & 0 & 1 \\ 0 & 1 & 0 \\ 1 & 0 & 0 \end{bmatrix}, \quad B = \begin{bmatrix} 1 & 2 & 1 \\ 2 & 0 & 1 \\ -1 & 3 & 4 \end{bmatrix}$

28. $A = \begin{bmatrix} 0 & 0 & 1 & 0 \\ 0 & 1 & 0 & 0 \\ 1 & 0 & 0 & 0 \\ 0 & 0 & 0 & 1 \end{bmatrix}, \quad B = \begin{bmatrix} -1 & 2 & 3 & -4 \\ 2 & 1 & 0 & -1 \\ -3 & 2 & 1 & 3 \\ 4 & 0 & 2 & -1 \end{bmatrix}$

In Exercises 29–32, solve for a and b.

29. $\begin{bmatrix} a & -3 \\ 4 & 2 \end{bmatrix} = \begin{bmatrix} 5 & -3 \\ 4 & b \end{bmatrix}$

30. $\begin{bmatrix} 1 & -1 & 0 \\ a & -2 & 1 \end{bmatrix} = \begin{bmatrix} 1 & b & 0 \\ 3 & -2 & 1 \end{bmatrix}$

31. $\begin{bmatrix} 2 & a-1 \\ 2 & 3 \\ -1 & 2 \end{bmatrix} = \begin{bmatrix} 2 & -3 \\ b+2 & 3 \\ -1 & 2 \end{bmatrix}$

32. $\begin{bmatrix} a+3 & 2 \\ 0 & 5 \end{bmatrix} = \begin{bmatrix} 4 & 2 \\ 0 & b-1 \end{bmatrix}$

In Exercises 33 and 34, verify that the matrices are inverses of each other.

33. $A = \begin{bmatrix} 2 & 1 \\ 3 & 4 \end{bmatrix}, \quad B = \begin{bmatrix} 0.8 & -0.2 \\ -0.6 & 0.4 \end{bmatrix}$

34. $A = \begin{bmatrix} -2 & 1 & 3 \\ 1 & 2 & -2 \\ 0 & 1 & -1 \end{bmatrix}, \quad B = \begin{bmatrix} 0 & 1 & -2 \\ 0.25 & 0.5 & -0.25 \\ 0.25 & 0.5 & -1.25 \end{bmatrix}$

In Exercises 35–40, find the inverse of the matrix if it has one, or state that the inverse does not exist.

35. $\begin{bmatrix} 2 & 3 \\ 2 & 2 \end{bmatrix}$ **36.** $\begin{bmatrix} 6 & 3 \\ 10 & 5 \end{bmatrix}$

37. $\begin{bmatrix} 1 & 2 & -1 \\ 2 & -1 & 3 \\ 3 & 1 & 2 \end{bmatrix}$ **38.** $\begin{bmatrix} 2 & 3 & -1 \\ -1 & 0 & 4 \\ 0 & 1 & 1 \end{bmatrix}$

39. $A = [a_{ij}],\ a_{ij} = (-1)^{i+j},\ 1 \le i \le 4,\ 1 \le j \le 4$

40. $B = [b_{ij}],\ b_{ij} = |i - j|,\ 1 \le i \le 3,\ 1 \le j \le 3$

In Exercises 41 and 42, use the definition to evaluate the determinant of the matrix.

41. $\begin{bmatrix} 2 & 1 & 1 \\ -1 & 0 & 2 \\ 1 & 3 & -1 \end{bmatrix}$ **42.** $\begin{bmatrix} 1 & 0 & 2 & 0 \\ 0 & 1 & 2 & 3 \\ 1 & -1 & 0 & 2 \\ 1 & 0 & 0 & 3 \end{bmatrix}$

In Exercises 43 and 44, solve for X.

43. $3X + A = B$, where $A = \begin{bmatrix} 1 \\ 3 \end{bmatrix}$ and $B = \begin{bmatrix} 4 \\ 2 \end{bmatrix}$.

44. $2X + A = B$, where $A = \begin{bmatrix} -1 & 2 \\ 0 & 3 \end{bmatrix}$ and $B = \begin{bmatrix} 1 & 4 \\ 1 & -1 \end{bmatrix}$.

45. Symmetric Matrix The matrix below gives the road mileage between Atlanta (A), Baltimore (B), Cleveland (C), and Denver (D). (*Source: AAA Road Atlas*)

(a) Writing to Learn Explain why the entry in the ith row and jth column is the same as the entry in the jth row and ith column. A matrix with this property is **symmetric**.

(b) Writing to Learn Why are the entries along the diagonal all 0's?

	A	B	C	D
A	0	689	774	1406
B	689	0	371	1685
C	774	371	0	1340
D	1406	1685	1340	0

46. Production Jordan Manufacturing has two factories, each of which manufactures three products. The number of units of product i produced at factory j in one week is represented by a_{ij} in the matrix

$$A = \begin{bmatrix} 120 & 70 \\ 150 & 110 \\ 80 & 160 \end{bmatrix}.$$

If production levels are increased by 10%, write the new production levels as a matrix B. How is B related to A?

47. Egg Production Happy Valley Farms produces three types of eggs: 1 (large), 2 (X-large), 3 (jumbo). The number of dozens of type i eggs sold to grocery store j is represented by a_{ij} in the matrix

$$A = \begin{bmatrix} 100 & 60 \\ 120 & 70 \\ 200 & 120 \end{bmatrix}.$$

The price per dozen that Happy Valley Farms charges for egg type i is represented by b_{i1} in the matrix

$$B = \begin{bmatrix} \$0.80 \\ \$0.85 \\ \$1.00 \end{bmatrix}.$$

(a) Find the product B^TA.

(b) Writing to Learn What does the matrix B^TA represent?

48. Inventory Jekell-Heid, Inc. sells four models of "all-in-one fax, printer, copier, and scanner machine" at three retail stores. The inventory at store i of model j is represented by s_{ij} in the matrix

$$S = \begin{bmatrix} 16 & 10 & 8 & 12 \\ 12 & 0 & 10 & 4 \\ 4 & 12 & 0 & 8 \end{bmatrix}.$$

The wholesale and retail prices of model i are represented by p_{i1} and p_{i2}, respectively, in the matrix

$$P = \begin{bmatrix} \$180 & \$269.99 \\ \$275 & \$399.99 \\ \$355 & \$499.99 \\ \$590 & \$799.99 \end{bmatrix}.$$

(a) Determine the product SP.

(b) Writing to Learn What does the matrix SP represent?

49. Profit Fast Buck Furniture sells four types of 5-piece bedroom sets. The price charged for a bedroom set of type j is represented by a_{1j} in the matrix

$$A = [\$398 \quad \$598 \quad \$798 \quad \$998].$$

The number of sets of type j sold in one period is represented by b_{1j} in the matrix

$$B = [35 \quad 25 \quad 20 \quad 10].$$

The cost to the furniture store for a bedroom set of type j is given by c_{1j} in the matrix

$$C = [\$199 \quad \$268 \quad \$500 \quad \$670].$$

(a) Write a matrix product that gives the total revenue made from the sale of the bedroom sets in the one period.

(b) Write an expression using matrices that gives the profit produced by the sale of the bedroom sets in the one period.

50. Construction A building contractor has agreed to build six ranch-style houses, seven Cape Cod–style houses, and 14 colonial-style houses. The number of units of raw materials that go into each type of house are shown in matrix R:

$R =$

	Steel	Wood	Glass	Paint	Labor
Ranch	5	22	14	7	17
Cape Cod	7	20	10	9	21
Colonial	6	27	8	5	13

Assume that steel costs \$1600 a unit, wood \$900 a unit, glass \$500 a unit, paint \$100 a unit, and labor \$1000 a unit.

(a) Write a 1×3 matrix B that represents the number of each type of house to be built.

(b) Write a matrix product that gives the number of units of each raw material needed to build the houses.

(c) Write a 5×1 matrix C that represents the per unit cost of each type of raw material.

(d) Write a matrix product that gives the cost of each house.

(e) **Writing to Learn** Compute the product BRC. What does this matrix represent?

51. Rotating Coordinate Systems The xy-coordinate system is rotated through the angle 30° to obtain the $x'y'$-coordinate system.

(a) If the coordinates of a point in the xy-coordinate system are $(1, 1)$, what are the coordinates of the rotated point in the $x'y'$-coordinate system?

(b) If the coordinates of a point in the $x'y'$-coordinate system are $(1, 1)$, what are the coordinates of the point in the xy-coordinate system that was rotated to it?

52. Group Activity Let A, B, and C be matrices whose orders are such that the following expressions are defined. Prove that the following properties are true.

(a) $A + B = B + A$

(b) $(A + B) + C = A + (B + C)$

(c) $A(B + C) = AB + AC$

(d) $(A - B)C = AC - BC$

53. Group Activity Let A and B be $m \times n$ matrices and c and d scalars. Prove that the following properties are true.

(a) $c(A + B) = cA + cB$ **(b)** $(c + d)A = cA + dA$

(c) $c(dA) = (cd)A$ **(d)** $1 \cdot A = A$

54. Writing to Learn Explain why the definition given for the determinant of a square matrix agrees with the definition given for the determinant of a 2×2 matrix. (Assume that the determinant of a 1×1 matrix is the entry.)

55. Inverse of a 2 × 2 Matrix Prove that the inverse of the matrix

$$A = \begin{bmatrix} a & b \\ c & d \end{bmatrix} \text{ is } A^{-1} = \frac{1}{ad - bc}\begin{bmatrix} d & -b \\ -c & a \end{bmatrix}$$

provided $ad - bc \neq 0$.

56. Identity Matrix Let $A = [a_{ij}]$ be an $n \times n$ matrix. Prove that $AI_n = I_nA = A$.

In Exercises 57–61, prove that the image of a point under the given transformation of the plane can be obtained by matrix multiplication.

57. A reflection across the y-axis

58. A reflection across the line $y = x$

59. A reflection across the line $y = -x$

60. A vertical stretch or shrink by a factor of a

61. A horizontal stretch or shrink by a factor of c

Standardized Test Questions

62. True or False Every square matrix has an inverse. Justify your answer.

63. True or False The determinant $|A|$ of the square matrix A is greater than or equal to 0. Justify your answer.

In Exercises 64–67, solve the problem without using a calculator.

64. Multiple Choice Which of the following is equal to the determinant of $A = \begin{bmatrix} 2 & 4 \\ -3 & -1 \end{bmatrix}$?

(A) 4 **(B)** −4 **(C)** 10 **(D)** −10 **(E)** −14

65. Multiple Choice Let A be a matrix of order 3×2 and B a matrix of order 2×4. Which of the following gives the order of the product AB?

(A) 2×2 **(B)** 3×4 **(C)** 4×3 **(D)** 6×8

(E) The product is not defined.

66. Multiple Choice Which of the following is the inverse of the matrix $\begin{bmatrix} 2 & 7 \\ 1 & 4 \end{bmatrix}$?

(A) $\begin{bmatrix} -4 & 7 \\ 1 & -2 \end{bmatrix}$ **(B)** $\begin{bmatrix} 2 & -7 \\ -1 & 4 \end{bmatrix}$ **(C)** $\begin{bmatrix} 2 & -1 \\ -7 & 4 \end{bmatrix}$

(D) $\begin{bmatrix} 4 & -1 \\ -7 & 2 \end{bmatrix}$ **(E)** $\begin{bmatrix} 4 & -7 \\ -1 & 2 \end{bmatrix}$

67. Multiple Choice Which of the following is the value of a_{13} in the matrix $[a_{ij}] = \begin{bmatrix} 1 & 2 & 3 \\ 4 & 5 & 6 \\ 7 & 8 & 9 \end{bmatrix}$?

(A) −7 **(B)** 7 **(C)** −3 **(D)** 3 **(E)** 10

Explorations

68. Continuation of Exploration 2 Let $A = [a_{ij}]$ be an $n \times n$ matrix.

(a) Prove that the determinant of A changes sign if two rows or two columns are interchanged. Start with a 3×3 matrix and compare the expansion by expanding by the same row (or column) before and after the interchange. (*Hint:* Compare without expanding the minors.) How can you generalize from the 3×3 case?

(b) Prove that the determinant of a square matrix with two identical rows or two identical columns is zero.

(c) Prove that if a scalar multiple of a row (or column) is added to another row (or column) the value of the determinant of a square matrix is unchanged. (*Hint:* Expand by the row (or column) being added to.)

69. Continuation of Exercise 68 Let $A = [a_{ij}]$ be an $n \times n$ matrix.

(a) Prove that, if every element of a row or column of a matrix is multiplied by the real number c, then the determinant of the matrix is multiplied by c.

(b) Prove that, if all the entries above the main diagonal (or all below it) of a matrix are zero, the determinant is the product of the elements on the main diagonal.

70. Writing Equations for Lines Using Determinants Consider the equation

$$\begin{vmatrix} 1 & x & y \\ 1 & x_1 & y_1 \\ 1 & x_2 & y_2 \end{vmatrix} = 0.$$

(a) Verify that the equation is linear in x and y.

(b) Verify that the two points (x_1, y_1) and (x_2, y_2) lie on the line in part (a).

(c) Use a determinant to state that the point (x_3, y_3) lies on the line in part (a).

(d) Use a determinant to state that the point (x_3, y_3) does not lie on the line in part (a).

71. Continuation of Example 10 The xy-coordinate system is rotated through the angle α to obtain the $x'y'$-coordinate system (see Figure 7.11).

(a) Show that the inverse of the matrix

$$A = \begin{bmatrix} \cos\alpha & -\sin\alpha \\ \sin\alpha & \cos\alpha \end{bmatrix}$$

of Example 10 is

$$A^{-1} = \begin{bmatrix} \cos\alpha & \sin\alpha \\ -\sin\alpha & \cos\alpha \end{bmatrix}.$$

(b) Prove that the (x, y) coordinates of P in Figure 7.11 are related to the (x', y') coordinates of P by the equations

$$x = x'\cos\alpha - y'\sin\alpha$$

$$y = x'\sin\alpha + y'\cos\alpha$$

(c) Prove that the coordinates (x, y) can be obtained from the (x', y') coordinates by matrix multiplication. How is this matrix related to A?

Extending the Ideas

72. Characteristic Polynomial Let $A = [a_{ij}]$ be a 2×2 matrix and define $f(x) = \det(xI_2 - A)$.

(a) Expand the determinant to show that $f(x)$ is a polynomial of degree 2. (The **characteristic polynomial** of A)

(b) How is the constant term of $f(x)$ related to $\det A$?

(c) How is the coefficient of x related to A?

(d) Prove that $f(A) = 0$.

73. Characteristic Polynomial Let $A = [a_{ij}]$ be a 3×3 matrix and define $f(x) = \det(xI_3 - A)$.

(a) Expand the determinant to show that $f(x)$ is a polynomial of degree 3. (The characteristic polynomial of A)

(b) How is the constant term of $f(x)$ related to $\det A$?

(c) How is the coefficient of x^2 related to A?

(d) Prove that $f(A) = 0$.

7.3 Multivariate Linear Systems and Row Operations

What you'll learn about

- Triangular Form for Linear Systems
- Gaussian Elimination
- Elementary Row Operations and Row Echelon Form
- Reduced Row Echelon Form
- Solving Systems Using Inverse Matrices
- Partial Fraction Decomposition
- Other Applications

... and why

Many applications in business and science are modeled by systems of linear equations in three or more variables.

Triangular Form for Linear Systems

The method of elimination used in Section 7.1 can be extended to systems of linear (first-degree) equations in more than two variables. The goal of the elimination method is to rewrite the system as an *equivalent system* of equations whose solution is obvious. Two systems of equations are **equivalent** if they have the same solution.

A *triangular form* of a system is an equivalent form from which the solution is easy to read, as illustrated in Example 1.

EXAMPLE 1 Solving by Substitution

Solve the system

$$\begin{aligned} x - 2y + z &= 7 \\ y - 2z &= -7 \\ z &= 3 \end{aligned}$$

SOLUTION The third equation determines z, namely $z = 3$. Substitute the value of z into the second equation to determine y.

$$\begin{aligned} y - 2z &= -7 && \text{Second equation} \\ y - 2(3) &= -7 && \text{Substitute } z = 3. \\ y &= -1 \end{aligned}$$

Finally, substitute the values for y and z into the first equation to determine x.

$$\begin{aligned} x - 2y + z &= 7 && \text{First equation} \\ x - 2(-1) + 3 &= 7 && \text{Substitute } y = -1, z = 3. \\ x &= 2 \end{aligned}$$

The solution of the system is $x = 2$, $y = -1$, $z = 3$, or the ordered triple $(2, -1, 3)$.

Now try Exercise 1.

Gaussian Elimination

Transforming a system to triangular form is **Gaussian elimination**, named after the famous German mathematician Carl Friedrich Gauss (1777–1855). Here are the operations needed to transform a system of linear equations into triangular form.

Back Substitution

The method of solution used in Example 1 is sometimes referred to as *back substitution*.

Equivalent Systems of Linear Equations

The following operations produce an equivalent system of linear equations:

1. Interchange any two equations of the system.
2. Multiply (or divide) one of the equations by any nonzero real number.
3. Add a multiple of one equation to any other equation in the system.

Observe how we use property 3 to bring the system in Example 2 to triangular form.

EXAMPLE 2 Using Gaussian Elimination

Solve the system

$$\begin{aligned} x - 2y + z &= 7 \\ 3x - 5y + z &= 14 \\ 2x - 2y - z &= 3 \end{aligned}$$

SOLUTION Each step in the following process leads to a system of equations equivalent to the original system.

Multiply the first equation by -3 and add the result to the second equation, replacing the second equation. (Leave the first and third equations unchanged.)

$$\begin{aligned} x - 2y + z &= 7 \\ y - 2z &= -7 \\ 2x - 2y - z &= 3 \end{aligned} \qquad \begin{aligned} -3x + 6y - 3z &= -21 \\ 3x - 5y + z &= 14 \end{aligned}$$

Multiply the first equation by -2 and add the result to the third equation, replacing the third equation.

$$\begin{aligned} x - 2y + z &= 7 \\ y - 2z &= -7 \\ 2y - 3z &= -11 \end{aligned} \qquad \begin{aligned} -2x + 4y - 2z &= -14 \\ 2x - 2y - z &= 3 \end{aligned}$$

Multiply the second equation by -2 and add the result to the third equation, replacing the third equation.

$$\begin{aligned} x - 2y + z &= 7 \\ y - 2z &= -7 \\ z &= 3 \end{aligned} \qquad \begin{aligned} -2y + 4z &= 14 \\ 2y - 3z &= -11 \end{aligned}$$

This is the system that was solved in Example 1. It is a triangular form of the original system. We know from Example 1 that the solution is $(2, -1, 3)$.

Now try Exercise 3.

It Pays to Check

It is easy to make mistakes when performing Gaussian elimination, so check answers by substituting them back into the original equations.

For a system of equations that has exactly one solution, the final system in Example 2 is in **triangular form**. In this case, the leading term of each equation has coefficient 1, the third equation has one variable (z), the second equation has at most two variables including one not in the third equation (y), and the first one has the remaining variable, x in this case.

EXAMPLE 3 Finding No Solution

Solve the system

$$\begin{aligned} x - 3y + z &= 4 \\ -x + 2y - 5z &= 3 \\ 5x - 13y + 13z &= 8 \end{aligned}$$

SOLUTION Use Gaussian elimination.

$$\begin{aligned} x - 3y + z &= 4 \\ -y - 4z &= 7 \\ 5x - 13y + 13z &= 8 \end{aligned}$$ Add first equation to second equation.

$$\begin{aligned} x - 3y + z &= 4 \\ -y - 4z &= 7 \\ 2y + 8z &= -12 \end{aligned}$$ Multiply first equation by -5 and add to third equation.

$$\begin{aligned} x - 3y + z &= 4 \\ -y - 4z &= 7 \\ 0 &= 2 \end{aligned}$$ Multiply second equation by 2 and add to third equation.

Because $0 = 2$ is never true, we conclude that this system has no solution.

Now try Exercise 5.

Elementary Row Operations and Row Echelon Form

When we solve a system of linear equations using Gaussian elimination, all the action is really on the coefficients of the variables. Matrices can be used to record the coefficients as we go through the steps of the Gaussian elimination process. We illustrate with the system of Example 2.

$$\begin{aligned} x - 2y + z &= 7 \\ 3x - 5y + z &= 14 \\ 2x - 2y - z &= 3 \end{aligned}$$

The **augmented matrix** of this system of equations is

$$\begin{bmatrix} 1 & -2 & 1 & 7 \\ 3 & -5 & 1 & 14 \\ 2 & -2 & -1 & 3 \end{bmatrix}.$$

The entries in the last column are the numbers on the right-hand sides of the equations. The **coefficient matrix** of this system is

$$\begin{bmatrix} 1 & -2 & 1 \\ 3 & -5 & 1 \\ 2 & -2 & -1 \end{bmatrix}.$$

Here the entries are the coefficients of the variables. We use coefficient matrices to solve linear systems later in this section.

We repeat the Gaussian elimination process used in Example 2 and record the corresponding action on the augmented matrix.

System of Equations	Augmented Matrix	
$\begin{aligned} x - 2y + z &= 7 \\ y - 2z &= -7 \\ 2x - 2y - z &= 3 \end{aligned}$	$\begin{bmatrix} 1 & -2 & 1 & 7 \\ 0 & 1 & -2 & -7 \\ 2 & -2 & -1 & 3 \end{bmatrix}$	Multiply eq. 1 (row 1) by -3; add result to eq. 2 (row 2), replacing eq. 2 (row 2).
$\begin{aligned} x - 2y + z &= 7 \\ y - 2z &= -7 \\ 2y - 3z &= -11 \end{aligned}$	$\begin{bmatrix} 1 & -2 & 1 & 7 \\ 0 & 1 & -2 & -7 \\ 0 & 2 & -3 & -11 \end{bmatrix}$	Multiply eq. 1 (row 1) by -2; add result to eq. 3 (row 3), replacing eq. 3 (row 3).
$\begin{aligned} x - 2y + z &= 7 \\ y - 2z &= -7 \\ z &= 3 \end{aligned}$	$\begin{bmatrix} 1 & -2 & 1 & 7 \\ 0 & 1 & -2 & -7 \\ 0 & 0 & 1 & 3 \end{bmatrix}$	Multiply eq. 2 (row 2) by -2; add result to eq. 3 (row 3), replacing eq. 3 (row 3).

The augmented matrix above, corresponding to the triangular form of the original system of equations, is a *row echelon form* of the augmented matrix of the original system of equations. In general, the last few rows of a row echelon form of a matrix *can* consist of all 0's.

DEFINITION Row Echelon Form of a Matrix

A matrix is in **row echelon form** if the following conditions are satisfied.

1. Rows consisting entirely of 0's (if there are any) occur at the bottom of the matrix.
2. The first entry in any row with nonzero entries is 1.
3. The column subscript of the leading 1 entries increases as the row subscript increases.

Another way to phrase parts 2 and 3 of the above definition is to say that the leading 1's move to the right as we move down the rows.

Our goal is to take a system of equations, write the corresponding augmented matrix, and transform it to row echelon form without carrying along the equations. From there we can read off the solutions to the system fairly easily.

The operations that we use to transform a linear system to equivalent triangular form correspond to elementary row operations of the corresponding augmented matrix of the linear system.

Elementary Row Operations on a Matrix

A combination of the following operations will transform a matrix to row echelon form:

1. Interchange any two rows.
2. Multiply all elements of a row by a nonzero real number.
3. Add a multiple of one row to any other row.

Example 4 illustrates how we can transform the augmented matrix to row echelon form to solve a system of linear equations.

Notation

1. R_{ij} indicates interchanging the ith and jth row of a matrix.
2. kR_i indicates multiplying the ith row by the nonzero real number k.
3. $kR_i + R_j$ indicates adding k times the ith row to the jth row.

Row Echelon Form

A word of caution! You can use your grapher to find a row echelon form of a matrix. However, row echelon form is *not* unique. Your grapher may produce a row echelon form different from the one you obtained by paper and pencil. Fortunately, all row echelon forms produce the same solution to the system of equations. (Correspondingly, a triangular form of a linear system is also not unique.)

EXAMPLE 4 Finding a Row Echelon Form

Solve the system

$$\begin{aligned} x - y + 2z &= -3 \\ 2x + y - z &= 0 \\ -x + 2y - 3z &= 7 \end{aligned}$$

SOLUTION We apply elementary row operations to find a row echelon form of the augmented matrix. The elementary row operations used are recorded above the arrows, using the notation in the margin.

$$\begin{bmatrix} 1 & -1 & 2 & -3 \\ 2 & 1 & -1 & 0 \\ -1 & 2 & -3 & 7 \end{bmatrix} \xrightarrow{(-2)R_1+R_2} \begin{bmatrix} 1 & -1 & 2 & -3 \\ 0 & 3 & -5 & 6 \\ -1 & 2 & -3 & 7 \end{bmatrix} \xrightarrow{(1)R_1+R_3}$$

$$\begin{bmatrix} 1 & -1 & 2 & -3 \\ 0 & 3 & -5 & 6 \\ 0 & 1 & -1 & 4 \end{bmatrix} \xrightarrow{R_{23}} \begin{bmatrix} 1 & -1 & 2 & -3 \\ 0 & 1 & -1 & 4 \\ 0 & 3 & -5 & 6 \end{bmatrix} \xrightarrow{(-3)R_2+R_3}$$

$$\begin{bmatrix} 1 & -1 & 2 & -3 \\ 0 & 1 & -1 & 4 \\ 0 & 0 & -2 & -6 \end{bmatrix} \xrightarrow{(-1/2)R_3} \begin{bmatrix} 1 & -1 & 2 & -3 \\ 0 & 1 & -1 & 4 \\ 0 & 0 & 1 & 3 \end{bmatrix}$$

The last matrix is in row echelon form. Then we convert each row, beginning with the last, into equation form and complete the solution by substitution.

$$z = 3 \qquad \begin{aligned} y - z &= 4 \\ y - 3 &= 4 \\ y &= 7 \end{aligned} \qquad \begin{aligned} x - y + 2z &= -3 \\ x - 7 + 2(3) &= -3 \\ x &= -2 \end{aligned}$$

The solution of the original system of equations is $(-2, 7, 3)$.

Now try Exercise 33.

Reduced Row Echelon Form

If we continue to apply elementary row operations to a row echelon form of a matrix, we can obtain a matrix in which every column that has a leading 1 has 0's elsewhere. This is the **reduced row echelon form** of the matrix. It is usually easier to read the solution from the reduced row echelon form.

We apply elementary row operations to the row echelon form found in Example 4 until we find the reduced row echelon form.

$$\begin{bmatrix} 1 & -1 & 2 & -3 \\ 0 & 1 & -1 & 4 \\ 0 & 0 & 1 & 3 \end{bmatrix} \xrightarrow{(1)R_2+R_1} \begin{bmatrix} 1 & 0 & 1 & 1 \\ 0 & 1 & -1 & 4 \\ 0 & 0 & 1 & 3 \end{bmatrix} \xrightarrow{(-1)R_3+R_1}$$

$$\begin{bmatrix} 1 & 0 & 0 & -2 \\ 0 & 1 & -1 & 4 \\ 0 & 0 & 1 & 3 \end{bmatrix} \xrightarrow{(1)R_3+R_2} \begin{bmatrix} 1 & 0 & 0 & -2 \\ 0 & 1 & 0 & 7 \\ 0 & 0 & 1 & 3 \end{bmatrix}$$

From this reduced row echelon form, we can immediately read the solution to the system of Example 4: $x = -2$, $y = 7$, $z = 3$. Figure 7.14 shows that the above final matrix is the reduced row echelon form of the augmented matrix of Example 4.

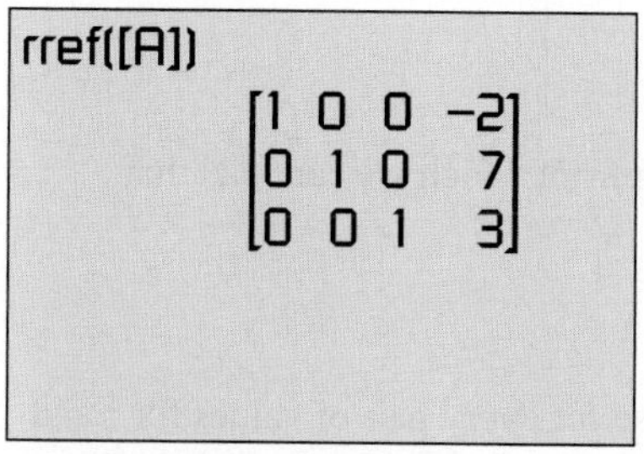

FIGURE 7.14 *A* is the augmented matrix of the system of linear equations in Example 4. "rref" stands for the grapher-produced reduced row echelon form of *A*.

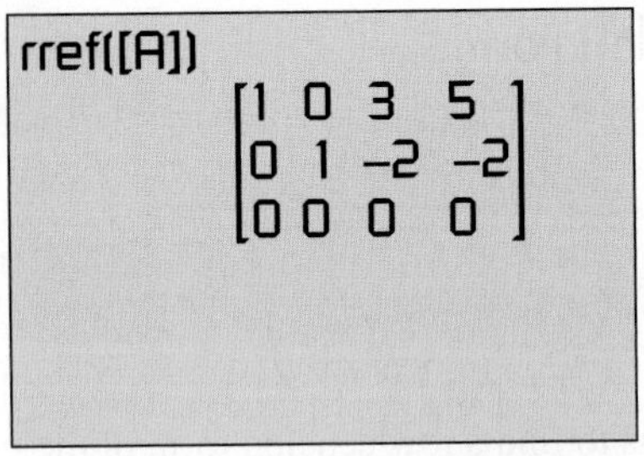

FIGURE 7.15 The reduced row echelon form for the augmented matrix of Example 5.

EXAMPLE 5 Finding Infinitely Many Solutions

Solve the system

$$\begin{aligned} x + y + z &= 3 \\ 2x + y + 4z &= 8 \\ x + 2y - z &= 1 \end{aligned}$$

SOLUTION Figure 7.15 shows the reduced row echelon form for the augmented matrix of the system. So, the following system of equations is equivalent to the original system.

$$\begin{aligned} x + 3z &= 5 \\ y - 2z &= -2 \\ 0 &= 0 \end{aligned}$$

Solving the first two equations for x and y in terms of z yields:

$$\begin{aligned} x &= -3z + 5 \\ y &= 2z - 2 \end{aligned}$$

This system has infinitely many solutions because for every value of z we can use these two equations to find corresponding values for x and y.

Interpret The solution is the set of all ordered triples of the form $(-3z + 5, 2z - 2, z)$ where z is any real number.

Now try Exercise 39.

We can also solve linear systems with more than three variables, or more than three equations, or both, by finding a row (or reduced row) echelon form. The solution set may become more complicated, as illustrated in Example 6.

EXAMPLE 6 Finding Infinitely Many Solutions

Solve the system

$$\begin{aligned} x + 2y - 3z &= -1 \\ 2x + 3y - 4z + w &= -1 \\ 3x + 5y - 7z + w &= -2 \end{aligned}$$

SOLUTION The 3×5 augmented matrix is

$$\begin{bmatrix} 1 & 2 & -3 & 0 & -1 \\ 2 & 3 & -4 & 1 & -1 \\ 3 & 5 & -7 & 1 & -2 \end{bmatrix}.$$

Figure 7.16 shows the reduced row echelon form from which we can read that:

$$\begin{aligned} x &= -z - 2w + 1 \\ y &= 2z + w - 1 \end{aligned}$$

This system has infinitely many solutions because for every pair of values for z and w we can use these two equations to find corresponding values for x and y.

Interpret The solution is the set of all ordered 4-tuples of the form $(-z - 2w + 1, 2z + w - 1, z, w)$ where z and w are any real numbers.

Now try Exercise 43.

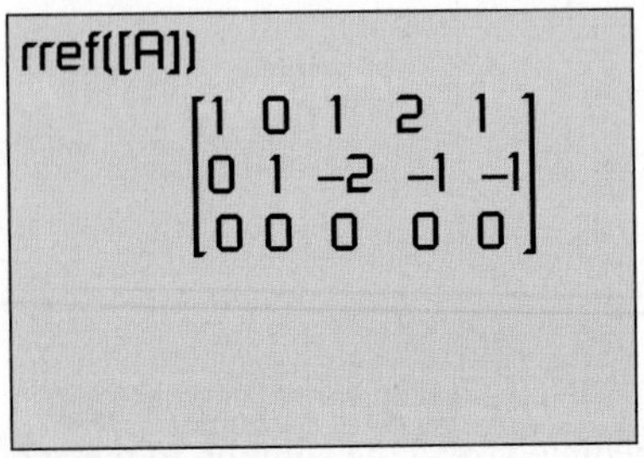

FIGURE 7.16 The reduced row echelon form for the augmented matrix of Example 6.

Solving Systems Using Inverse Matrices

If a linear system consists of the same number of equations as variables, then the coefficient matrix is square. If this matrix is also nonsingular, then we can solve the system using the technique illustrated in Example 7.

EXAMPLE 7 Solving a System Using Inverse Matrices

Solve the system

$$\begin{aligned} 3x - 2y &= 0 \\ -x + y &= 5 \end{aligned}$$

SOLUTION First we write the system as a matrix equation. Let

$$A = \begin{bmatrix} 3 & -2 \\ -1 & 1 \end{bmatrix}, \quad X = \begin{bmatrix} x \\ y \end{bmatrix}, \quad \text{and} \quad B = \begin{bmatrix} 0 \\ 5 \end{bmatrix}.$$

Then

$$A \cdot X = \begin{bmatrix} 3 & -2 \\ -1 & 1 \end{bmatrix} \cdot \begin{bmatrix} x \\ y \end{bmatrix} = \begin{bmatrix} 3x - 2y \\ -x + y \end{bmatrix}$$

so that

$$AX = B,$$

where A is the coefficient matrix of the system. You can easily check that $\det A = 1$, so A^{-1} exists. From Figure 7.17, we obtain

$$X = A^{-1}B = \begin{bmatrix} 10 \\ 15 \end{bmatrix}.$$

The solution of the system is $x = 10$, $y = 15$, or $(10, 15)$. **Now try Exercise 49.**

Linear Equations

If a and b are real numbers with $a \neq 0$, the linear equation $ax = b$ has a unique solution $x = a^{-1}b$. A similar statement holds for the linear matrix equation $AX = B$ when A is a nonsingular square matrix. (See the Invertible Square Linear Systems Theorem.)

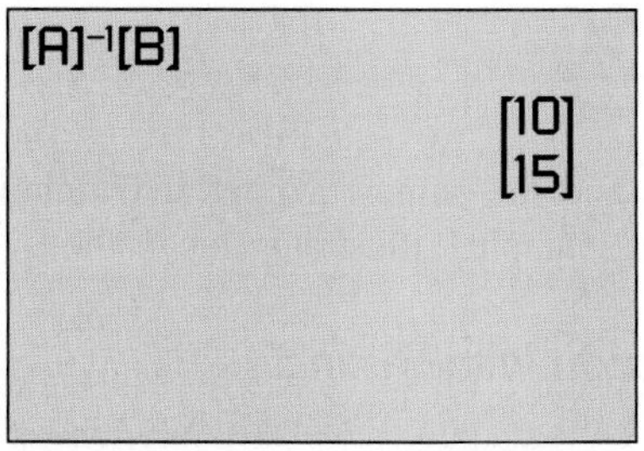

FIGURE 7.17 The solution of the matrix equation of Example 7.

Examples 7 and 8 are two instances of the following theorem.

THEOREM Invertible Square Linear Systems

Let A be the coefficient matrix of a system of n linear equations in n variables given by $AX = B$, where X is the $n \times 1$ matrix of variables and B is the $n \times 1$ matrix of numbers on the right-hand side of the equations. If A^{-1} exists, then the system of equations has the unique solution

$$X = A^{-1}B.$$

EXAMPLE 8 Solving a System Using Inverse Matrices

Solve the system

$$\begin{aligned} 3x - 3y + 6z &= 20 \\ x - 3y + 10z &= 40 \\ -x + 3y - 5z &= 30 \end{aligned}$$

SOLUTION Let

$$A = \begin{bmatrix} 3 & -3 & 6 \\ 1 & -3 & 10 \\ -1 & 3 & -5 \end{bmatrix}, \quad X = \begin{bmatrix} x \\ y \\ z \end{bmatrix}, \quad \text{and} \quad B = \begin{bmatrix} 20 \\ 40 \\ 30 \end{bmatrix}.$$

(continued)

The Order Is Important

Remember that A^{-1} is multiplied on the ***left*** of each expression; matrix multiplication is ***not*** commutative.

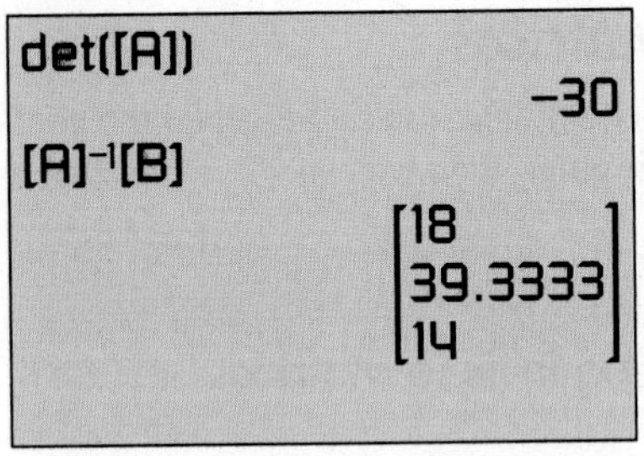

FIGURE 7.18 The solution of the system in Example 8.

The system of equations can be written as

$$A \cdot X = B.$$

Figure 7.18 shows that $\det A = -30 \neq 0$. Thus A^{-1} exists and, as suggested by Figure 7.18,

$$X = A^{-1}B = \begin{bmatrix} 18 \\ 39.\overline{3} \\ 14 \end{bmatrix}.$$

Interpret The solution of the system of equations is $x = 18$, $y = 39\frac{1}{3}$, and $z = 14$, or $(18, 39\frac{1}{3}, 14)$.

Now try Exercise 51.

Partial Fraction Decomposition

In Section 2.6 we saw that a polynomial with real coefficients could be factored into a product of factors with real coefficients, each of which was either a linear factor or an irreducible quadratic factor. Similarly, a rational expression can be written as a sum of rational expression in which each denominator is a power of a linear factor or a power of an irreducible quadratic factor.

For example,

$$\frac{3x - 4}{x^2 - 2x} = \frac{2}{x} + \frac{1}{x - 2}.$$

Each fraction in the sum is a **partial fraction**, and the sum is a **partial fraction decomposition** of the original rational expression. Example 9 illustrates this method.

EXAMPLE 9 Decomposing a Fraction with Distinct Linear Factors

Find the partial fraction decomposition of

$$\frac{5x - 1}{x^2 - 2x - 15}.$$

SOLUTION The denominator factors into $(x + 3)(x - 5)$. We write

$$\frac{5x - 1}{x^2 - 2x - 15} = \frac{A}{x + 3} + \frac{B}{x - 5}$$

and then "clear the fractions" by multiplying both sides of the above equation by $x^2 - 2x - 15$ to obtain

$$5x - 1 = A(x - 5) + B(x + 3)$$
$$5x - 1 = (A + B)x + (-5A + 3B).$$

Comparing coefficients on the left and right side of the above equation, we obtain the following system of two equations in the two variables A and B:

$$A + B = 5$$
$$-5A + 3B = -1$$

We can write this system in matrix form as $CX = D$ where

$$C = \begin{bmatrix} 1 & 1 \\ -5 & 3 \end{bmatrix}, \quad X = \begin{bmatrix} A \\ B \end{bmatrix}, \quad \text{and} \quad D = \begin{bmatrix} 5 \\ -1 \end{bmatrix},$$

and read from Figure 7.19 that

$$X = \begin{bmatrix} 2 \\ 3 \end{bmatrix}.$$

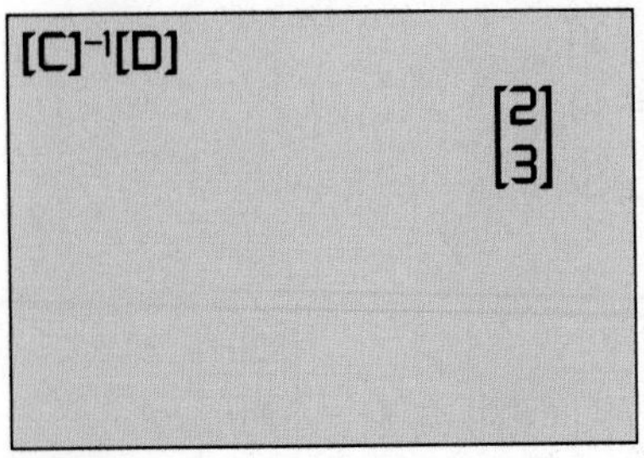

FIGURE 7.19 The solution of the system of equations in Example 9.

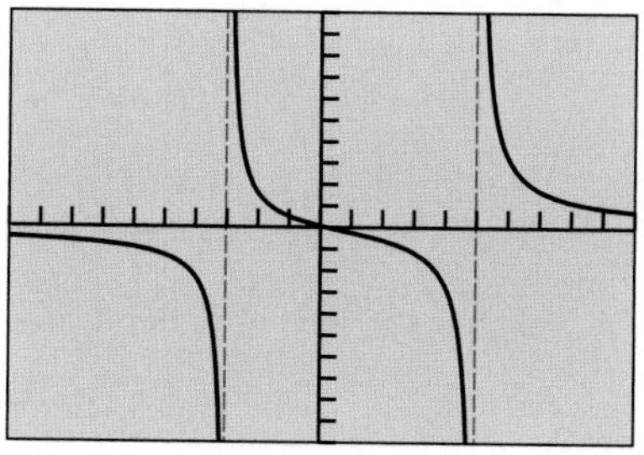

[−10, 10] by [−10, 10]

FIGURE 7.20 The graphs of $y = (5x - 1)/(x^2 - 2x - 15)$ and $y = 2/(x + 3) + 3/(x - 5)$ appear to be the same. (Example 9)

Thus, $A = 2$, $B = 3$, and

$$\frac{5x - 1}{x^2 - 2x - 15} = \frac{2}{x + 3} + \frac{3}{x - 5}.$$

Support Graphically Figure 7.20 suggests that the following two functions are the same:

$$y = \frac{5x - 1}{x^2 - 2x - 15} \quad \text{and} \quad y = \frac{2}{x + 3} + \frac{3}{x - 5}$$

Now try Exercise 67.

Sometimes we can solve for the variables introduced in a partial fraction decomposition by substituting strategic values for x, as illustrated in Exploration 1.

EXPLORATION 1 Revisiting Example 9

When we cleared fractions in Example 9 we obtained the equation $5x - 1 = A(x - 5) + B(x + 3)$.

1. Substitute $x = 5$ into this equation and solve for B.
2. Substitute $x = -3$ into this equation and solve for A.

Other Applications

Any three noncollinear points with distinct x-coordinates determine exactly one second-degree polynomial, as illustrated in Example 10. The graph of a second-degree polynomial is a parabola.

EXAMPLE 10 Fitting a Parabola to Three Points

Determine a, b, and c so that the points $(-1, 5)$, $(2, -1)$, and $(3, 13)$ are on the graph of $f(x) = ax^2 + bx + c$.

SOLUTION

Model We must have $f(-1) = 5$, $f(2) = -1$, and $f(3) = 13$:

$$\begin{aligned} f(-1) &= a - b + c = 5 \\ f(2) &= 4a + 2b + c = -1 \\ f(3) &= 9a + 3b + c = 13 \end{aligned}$$

The above system of three linear equations in the three variables a, b, and c can be written in matrix form $AX = B$, where

$$A = \begin{bmatrix} 1 & -1 & 1 \\ 4 & 2 & 1 \\ 9 & 3 & 1 \end{bmatrix}, \quad X = \begin{bmatrix} a \\ b \\ c \end{bmatrix}, \quad \text{and} \quad B = \begin{bmatrix} 5 \\ -1 \\ 13 \end{bmatrix}.$$

Solve Numerically Figure 7.21a on the next page shows that

$$X = A^{-1}B = \begin{bmatrix} 4 \\ -6 \\ -5 \end{bmatrix}.$$

Thus, $a = 4$, $b = -6$, and $c = -5$. The second-degree polynomial $f(x) = 4x^2 - 6x - 5$ contains the points $(-1, 5)$, $(2, -1)$, and $(3, 13)$ (Figure 7.21b on the next page).

Now try Exercise 83.

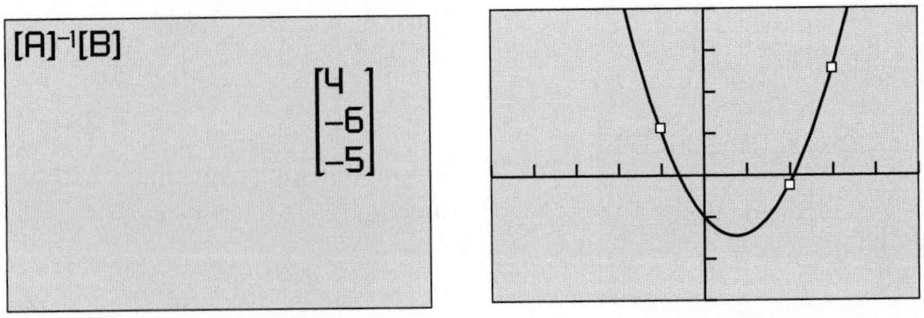

(a)

[–5, 5] by [–15, 20]

(b)

FIGURE 7.21 (a) The solution of the matrix equation in Example 10. (b) A graph of $f(x) = 4x^2 - 6x - 5$ superimposed on a scatter plot of the three points $(-1, 5)$, $(2, -1)$, and $(3, 13)$.

EXPLORATION 2 Mixing Solutions

Aileen's Drugstore needs to prepare a 60-L mixture that is 40% acid, using three concentrations of acid. The first concentration is 15% acid, the second is 35% acid, and the third is 55% acid. Because of the amounts of acid solution on hand, they need to use twice as much of the 35% solution as the 55% solution. How much of each solution should they use?

Let $x =$ the number of liters of 15% solution used, $y =$ the number of liters of 35% solution used, and $z =$ the number of liters of 55% solution used.

1. Explain how the equation $x + y + z = 60$ is related to the problem.
2. Explain how the equation $0.15x + 0.35y + 0.55z = 24$ is related to the problem.
3. Explain how the equation $y = 2z$ is related to the problem.
4. Write the system of three equations obtained from parts 1–3 in the form $AX = B$, where A is the coefficient matrix of the system. What are A, B, and X?
5. Solve the matrix equation in part 4.
6. Interpret the solution in part 5 in terms of the problem situation.

QUICK REVIEW 7.3 *(For help, go to Sections 1.2 and 7.2.)*

In Exercises 1 and 2, find the amount of pure acid in the solution.

1. 40 L of a 32% acid solution

2. 60 mL of a 14% acid solution

In Exercises 3 and 4, find the amount of water in the solution.

3. 50 L of a 24% acid solution

4. 80 mL of a 70% acid solution

In Exercises 5 and 6, determine which points are on the graph of the function.

5. $f(x) = 2x^2 - 3x + 1$

(a) $(-1, 6)$ **(b)** $(2, 1)$

6. $f(x) = x^3 - 4x - 1$

(a) $(0, -1)$ **(b)** $(-2, -17)$

In Exercises 7 and 8, solve for x or y in terms of the other variables.

7. $y + z - w = 1$

8. $x - 2z + w = 3$

In Exercises 9 and 10, find the inverse of the matrix.

9. $\begin{bmatrix} 1 & 3 \\ -2 & -2 \end{bmatrix}$

10. $\begin{bmatrix} 0 & 0 & 2 \\ -2 & 1 & 3 \\ 0 & 2 & -2 \end{bmatrix}$

SECTION 7.3 Exercises

Exercise numbers with a gray background indicate problems that the authors have designed to be solved *without a calculator*.

In Exercises 1 and 2, use substitution to solve the system of equations.

1. $\begin{aligned} x - 3y + z &= 0 \\ 2y + 3z &= 1 \\ z &= -2 \end{aligned}$

2. $\begin{aligned} 3x - y + 2z &= -2 \\ y + 3z &= 3 \\ 2z &= 4 \end{aligned}$

In Exercises 3–8, use Gaussian elimination to solve the system of equations.

3. $\begin{aligned} x - y + z &= 0 \\ 2x - 3z &= -1 \\ -x - y + 2z &= -1 \end{aligned}$

4. $\begin{aligned} 2x - y &= 0 \\ x + 3y - z &= -3 \\ 3y + z &= 8 \end{aligned}$

5. $\begin{aligned} x + y + z &= -3 \\ 4x - y &= -5 \\ -3x + 2y + z &= 4 \end{aligned}$

6. $\begin{aligned} x + y - 3z &= -1 \\ 2x - 3y + z &= 4 \\ 3x - 7y + 5z &= 4 \end{aligned}$

7. $\begin{aligned} x + y - z &= 4 \\ y + w &= -4 \\ x - y &= 1 \\ x + z + w &= 1 \end{aligned}$

8. $\begin{aligned} \frac{1}{2}x - y + z - w &= 1 \\ -x + y + z + 2w &= -3 \\ x - z &= 2 \\ y + w &= 0 \end{aligned}$

In Exercises 9–12, perform the indicated elementary row operation on the matrix

$$\begin{bmatrix} 2 & -6 & 4 \\ 1 & 2 & -3 \\ -3 & 1 & -2 \end{bmatrix}.$$

9. $(3/2)R_1 + R_3$

10. $(1/2)R_1$

11. $(-2)R_2 + R_1$

12. $(1)R_1 + R_2$

In Exercises 13–16, what elementary row operations applied to

$$\begin{bmatrix} -2 & 1 & -1 & 2 \\ 1 & -2 & 3 & 0 \\ 3 & 1 & -1 & 2 \end{bmatrix}$$

will yield the given matrix?

13. $\begin{bmatrix} 1 & -2 & 3 & 0 \\ -2 & 1 & -1 & 2 \\ 3 & 1 & -1 & 2 \end{bmatrix}$

14. $\begin{bmatrix} 0 & -3 & 5 & 2 \\ 1 & -2 & 3 & 0 \\ 3 & 1 & -1 & 2 \end{bmatrix}$

15. $\begin{bmatrix} -2 & 1 & -1 & 2 \\ 1 & -2 & 3 & 0 \\ 0 & 7 & -10 & 2 \end{bmatrix}$

16. $\begin{bmatrix} -2 & 1 & -1 & 2 \\ 1 & -2 & 3 & 0 \\ 0.75 & 0.25 & -0.25 & 0.5 \end{bmatrix}$

In Exercises 17–20, find a row echelon form for the matrix.

17. $\begin{bmatrix} 1 & 2 & -1 \\ 2 & 1 & 4 \\ -3 & 0 & 1 \end{bmatrix}$

18. $\begin{bmatrix} 1 & 2 & -3 \\ -3 & -6 & 10 \\ -2 & -4 & 7 \end{bmatrix}$

19. $\begin{bmatrix} 1 & 2 & 3 & -4 \\ -2 & 6 & -6 & 2 \\ 3 & 12 & 6 & 12 \end{bmatrix}$

20. $\begin{bmatrix} 3 & 6 & 9 & -6 \\ 2 & 5 & 5 & -3 \end{bmatrix}$

In Exercises 21–24, find the reduced row echelon form for the matrix.

21. $\begin{bmatrix} 1 & 0 & 2 & 1 \\ 3 & 2 & 4 & 7 \\ 2 & 1 & 3 & 4 \end{bmatrix}$

22. $\begin{bmatrix} 1 & -2 & 2 & 1 & 1 \\ 3 & -5 & 6 & 3 & -1 \\ -2 & 4 & -3 & -2 & 5 \\ 3 & -5 & 6 & 4 & -3 \end{bmatrix}$

23. $\begin{bmatrix} 1 & 2 & 3 & 1 \\ -3 & -5 & -7 & -4 \end{bmatrix}$

24. $\begin{bmatrix} 3 & -6 & 3 & -3 \\ 2 & -4 & 2 & -2 \\ -3 & 6 & -3 & 3 \end{bmatrix}$

In Exercises 25–28, write the augmented matrix corresponding to the system of equations.

25. $\begin{aligned} 2x - 3y + z &= 1 \\ -x + y - 4z &= -3 \\ 3x - z &= 2 \end{aligned}$

26. $\begin{aligned} 3x - 4y + z - w &= 1 \\ x + z - 2w &= 4 \end{aligned}$

27. $\begin{aligned} 2x - 5y + z - w &= -3 \\ x - 2z + w &= 4 \\ 2y - 3z - w &= 5 \end{aligned}$

28. $\begin{aligned} 3x - 2y &= 5 \\ -x + 5y &= 7 \end{aligned}$

In Exercises 29–32, write the system of equations corresponding to the augmented matrix.

29. $\begin{bmatrix} 3 & 2 & -1 \\ -4 & 5 & 2 \end{bmatrix}$

30. $\begin{bmatrix} 1 & 0 & -1 & 2 & -3 \\ 2 & 1 & 0 & -1 & 4 \\ -1 & 1 & 2 & 0 & 0 \end{bmatrix}$

31. $\begin{bmatrix} 2 & 0 & 1 & 3 \\ -1 & 1 & 0 & 2 \\ 0 & 2 & -3 & -1 \end{bmatrix}$

32. $\begin{bmatrix} 2 & 1 & -2 & 4 \\ -3 & 0 & 2 & -1 \end{bmatrix}$

In Exercises 33–34, solve the system of equations by finding a row echelon form for the augmented matrix.

33. $\begin{aligned} x - 2y + z &= 8 \\ 2x + y - 3z &= -9 \\ -3x + y + 3z &= 5 \end{aligned}$

34. $\begin{aligned} 3x + 7y - 11z &= 44 \\ x + 2y - 3z &= 12 \\ 4x + 9y - 13z &= 53 \end{aligned}$

In Exercises 35–44, solve the system of equations by finding the reduced row echelon form for the augmented matrix.

35. $\begin{aligned} x + 2y - z &= 3 \\ 3x + 7y - 3z &= 12 \\ -2x - 4y + 3z &= -5 \end{aligned}$

36. $\begin{aligned} x - 2y + z &= -2 \\ 2x - 3y + 2z &= 2 \\ 4x - 8y + 5z &= -5 \end{aligned}$

37. $x + y + 3z = 2$
$3x + 4y + 10z = 5$
$x + 2y + 4z = 3$

38. $x - z = 2$
$-2x + y + 3z = -5$
$2x + y - z = 3$

39. $x + z = 2$
$2x + y + z = 5$

40. $x + 2y - 3z = 1$
$-3x - 5y + 8z = -29$

41. $x + 2y = 4$
$3x + 4y = 5$
$2x + 3y = 4$

42. $x + y = 3$
$2x + 3y = 8$
$2x + 2y = 6$

43. $x + y - 3z = 1$
$x - z - w = 2$
$2x + y - 4z - w = 3$

44. $x - y - z + 2w = -3$
$2x - y - 2z + 3w = -3$
$x - 2y - z + 3w = -6$

In Exercises 45 and 46, write the system of equations as a matrix equation $AX = B$, with A as the coefficient matrix of the system.

45. $2x + 5y = -3$
$x - 2y = 1$

46. $5x - 7y + z = 2$
$2x - 3y - z = 3$
$x + y + z = -3$

In Exercises 47 and 48, write the matrix equation as a system of equations.

47. $\begin{bmatrix} 3 & -1 \\ 2 & 4 \end{bmatrix} \begin{bmatrix} x \\ y \end{bmatrix} = \begin{bmatrix} -1 \\ 3 \end{bmatrix}$

48. $\begin{bmatrix} 1 & 0 & -3 \\ 2 & -1 & 3 \\ -2 & 3 & -4 \end{bmatrix} \begin{bmatrix} x \\ y \\ z \end{bmatrix} = \begin{bmatrix} 3 \\ -1 \\ 2 \end{bmatrix}$

In Exercises 49–54, solve the system of equations by using an inverse matrix.

49. $2x - 3y = -13$
$4x + y = -5$

50. $x + 2y = -2$
$3x - 4y = 9$

51. $2x - y + z = -6$
$x + 2y - 3z = 9$
$3x - 2y + z = -3$

52. $x + 4y - 2z = 0$
$2x + y + z = 6$
$-3x + 3y - 5z = -13$

53. $2x - y + z + w = -3$
$x + 2y - 3z + w = 12$
$3x - y - z + 2w = 3$
$-2x + 3y + z - 3w = -3$

54. $2x + y + 2z = 8$
$3x + 2y - z - w = 10$
$-2x + y - 3w = -1$
$4x - 3y + 2z - 5w = 39$

In Exercises 55–66, use a method of your choice to solve the system of equations.

55. $2x - y = 10$
$x - z = -1$
$y + z = -9$

56. $1.25x + z = -2$
$y - 5.5z = -2.75$
$3x - 1.5y = -6$

57. $x + 2y + 2z + w = 5$
$2x + y + 2z = 5$
$3x + 3y + 3z + 2w = 12$
$x + z + w = 1$

58. $x - y + w = -4$
$-2x + y + z = 8$
$2x - 2y - z = -10$
$-2x + z + w = 5$

59. $x - y + z = 6$
$x + y + 2z = -2$

60. $x - 2y + z = 3$
$2x + y - z = -4$

61. $2x + y + z + 4w = -1$
$x + 2y + z + w = 1$
$x + y + z + 2w = 0$

62. $2x + 3y + 3z + 7w = 0$
$x + 2y + 2z + 5w = 0$
$x + y + 2z + 3w = -1$

63. $2x + y + z + 2w = -3.5$
$x + y + z + w = -1.5$

64. $2x + y + 4w = 6$
$x + y + z + w = 5$

65. $x + y - z + 2w = 0$
$y - z + 2w = -1$
$x + y + 3w = 3$
$2x + 2y - z + 5w = 4$

66. $x + y + w = 2$
$x + 4y + z - 2w = 3$
$x + 3y + z - 3w = 2$
$x + y + w = 2$

In Exercises 67 and 68, use inverse matrices to find the partial fraction decomposition.

67. $\dfrac{x + 22}{(x + 4)(x - 2)} = \dfrac{A}{x + 4} + \dfrac{B}{x - 2}$

68. $\dfrac{x - 3}{x(x + 3)} = \dfrac{A}{x + 3} + \dfrac{B}{x}$

In Exercises 69–72, find the partial fraction decomposition. Confirm your answer algebraically by combining the partial fractions.

69. $\dfrac{2}{(x - 5)(x - 3)}$

70. $\dfrac{4}{(x + 3)(x + 7)}$

71. $\dfrac{4}{x^2 - 1}$

72. $\dfrac{6}{x^2 - 9}$

In Exercises 73–76, find the partial fraction decomposition. Support your answer graphically.

73. $\dfrac{2}{x^2 + 2x}$

74. $\dfrac{-6}{x^2 - 3x}$

75. $\dfrac{-x + 10}{x^2 + x - 12}$

76. $\dfrac{7x - 7}{x^2 - 3x - 10}$

Use an algebraic method in Exercises 77 and 78 to find a partial fraction decomposition.

77. $\dfrac{x + 17}{2x^2 + 5x - 3}$

78. $\dfrac{4x - 11}{2x^2 - x - 3}$

In Exercises 79–82, use division to write the rational function in the form $q(x) + r(x)/d(x)$, where the degree of $r(x)$ is less than the degree of $d(x)$. Then find the partial fraction decomposition of $r(x)/d(x)$. Compare the graphs of the rational function with the graphs of its terms in the partial fraction decomposition.

79. $\dfrac{2x^2 + x + 3}{x^2 - 1}$

80. $\dfrac{3x^2 + 2x}{x^2 - 4}$

81. $\dfrac{x^3 - 2}{x^2 + x}$

82. $\dfrac{x^3 + 2}{x^2 - x}$

In Exercises 83–86, determine f so that its graph contains the given points.

83. Curve Fitting $f(x) = ax^2 + bx + c$
$(-1, 3), (1, -3), (2, 0)$

84. Curve Fitting $f(x) = ax^3 + bx^2 + cx + d.$
$(-2, -37), (-1, -11), (0, -5), (2, 19)$

85. Family of Curves $f(x) = ax^2 + bx + c$

$(-1, -4), (1, -2)$

86. Family of Curves $f(x) = ax^3 + bx^2 + cx + d$

$(-1, -6), (0, -1), (1, 2)$

87. Population Table 7.5 gives the populations (in thousands) of two fast-growing suburbs in the Dallas–Forth Worth Metroplex: Garland and Irving. Let x be the number of years since 1980.

(a) Find a linear regression equation for the population of Garland.

(b) Find a linear regression equation for the population of Irving.

(c) Graph the models from parts (a) and (b), and use these models to determine when the populations of the two cities will be about the same.

Table 7.5 Populations of Two Cities

Year	Garland (thousands)	Irving (thousands)
1980	139	110
1990	181	155
2000	216	192
2010	227	216

Source: U.S. Census Bureau, The World Almanac and Book of Facts 2012.

88. Population Table 7.6 gives the populations (in thousands) of Anaheim, CA, and Anchorage, AK, for selected years. Let x be the number of years since 1970.

(a) Find a linear regression equation for the population of Anaheim.

(b) Find a linear regression equation for the population of Anchorage.

(c) Graph the models from parts (a) and (b), and use these models to determine when the populations of the two cities will be about the same.

Table 7.6 Populations of Two Cities

Year	Anaheim (thousands)	Anchorage (thousands)
1970	166	48
1980	219	174
1990	266	226
2000	328	260
2010	336	292

Source: U.S. Census Bureau, The World Almanac and Book of Facts 2012.

89. Train Tickets At the Pittsburgh zoo, children ride a train for 25 cents, adults pay $1.00, and senior citizens 75 cents. On a given day, 1400 passengers paid a total of $740 for the rides. There were 250 more children riders than all other riders. Find the number of children, adult, and senior riders.

90. Manufacturing Stewart's Metals has three silver alloys on hand. One is 22% silver, another is 30% silver, and the third is 42% silver. How many grams of each alloy is required to produce 80 g of a new alloy that is 34% silver if the amount of 30% alloy used is twice the amount of 22% alloy used?

91. Investment Monica receives an $80,000 inheritance. She invests part of it in CDs (certificates of deposit) earning 6.7% APY (annual percentage yield), part in bonds earning 9.3% APY, and the remainder in a growth fund earning 15.6% APY. She invests three times as much in the growth fund as in the other two combined. How much does she have in each investment if she receives $10,843 interest the first year?

92. Investments Oscar invests $20,000 in three investments earning 6% APY, 8% APY, and 10% APY. He invests $9000 more in the 10% investment than in the 6% investment. How much does he have invested at each rate if he receives $1780 interest the first year?

93. Investments Morgan has $50,000 to invest and wants to receive $5000 interest the first year. He puts part in CDs earning 5.75% APY, part in bonds earning 8.7% APY, and the rest in a growth fund earning 14.6% APY. How much should he invest at each rate if he puts the least amount possible in the growth fund?

94. Mixing Acid Solutions Simpson's Drugstore needs to prepare a 40-L mixture that is 32% acid from three solutions: a 10% acid solution, a 25% acid solution, and a 50% acid solution. How much of each solution should be used if Simpson's wants to use as little of the 50% solution as possible?

95. Loose Change Matthew has 74 coins consisting of nickels, dimes, and quarters in his coin box. The total value of the coins is $8.85. If the number of nickels and quarters is four more than the number of dimes, find how many of each coin Matthew has in his coin box.

96. Vacation Money Heather has saved $177 to take with her on the family vacation. She has 51 bills consisting of $1, $5, and $10 bills. If the number of $5 bills is three times the number of $10 bills, find how many of each bill she has.

In Exercises 97 and 98, use inverse matrices to find the equilibrium point for the demand and supply curves.

97. $p = 100 - 5x$ Demand curve

$p = 20 + 10x$ Supply curve

98. $p = 150 - 12x$ Demand curve

$p = 30 + 24x$ Supply curve

99. Writing to Learn Explain why adding one row to another row in a matrix is an elementary row operation.

100. Writing to Learn Explain why subtracting one row from another row in a matrix is an elementary row operation.

Standardized Test Questions

101. True or False Every nonzero square matrix has an inverse. Justify your answer.

102. True or False The reduced row echelon form of the augmented matrix of a system of three linear equations in three variables must be of the form

$$\begin{bmatrix} 1 & 0 & 0 & a \\ 0 & 1 & 0 & b \\ 0 & 0 & 1 & c \end{bmatrix},$$

where a, b, c, are real numbers. Justify your answer.

In Exercises 103–106, you may use a graphing calculator to solve the problem.

103. Multiple Choice Which of the following is the determinant of the matrix $\begin{bmatrix} 2 & 2 \\ -1 & 3 \end{bmatrix}$?

(A) 0 **(B)** 4 **(C)** −4 **(D)** 8 **(E)** −8

104. Multiple Choice Which of the following is the augmented matrix of the system of equations

$$\begin{aligned} x + 2y + z &= -1 \\ 2x - y + 3z &= -4? \\ 3x + y - z &= -2 \end{aligned}$$

(A) $\begin{bmatrix} 1 & 2 & 1 & -1 \\ 2 & -1 & 3 & -4 \\ 3 & 1 & -1 & -2 \end{bmatrix}$ **(B)** $\begin{bmatrix} 1 & 2 & 1 & 1 \\ 2 & -1 & 3 & 4 \\ 3 & 1 & -1 & 2 \end{bmatrix}$

(C) $\begin{bmatrix} 1 & 2 & 1 & 0 \\ 2 & -1 & 3 & 0 \\ 3 & 1 & -1 & 0 \end{bmatrix}$ **(D)** $\begin{bmatrix} 1 & 2 & 1 \\ 2 & -1 & 3 \\ 3 & 1 & -1 \end{bmatrix}$

(E) $\begin{bmatrix} 1 & 2 & -1 \\ 2 & -1 & -3 \\ 3 & 1 & 1 \end{bmatrix}$

105. Multiple Choice The matrix $\begin{bmatrix} 1 & 2 & 3 \\ 2 & 1 & 0 \\ 7 & 8 & 9 \end{bmatrix}$ was obtained from $\begin{bmatrix} 1 & 2 & 3 \\ 4 & 5 & 6 \\ 7 & 8 & 9 \end{bmatrix}$ by an elementary row operation. Which of the following describes the elementary row operation?

(A) $(-2)R_1$ **(B)** $(-2)R_1 + R_2$ **(C)** $(-2)R_2 + R_1$

(D) $(2)R_1 + R_2$ **(E)** $(2)R_2 + R_1$

106. Multiple Choice Which of the following is the reduced row echelon form for the augmented matrix of

$$\begin{aligned} x + 2y - z &= 8 \\ -x + 3y + 2z &= 3 \quad ? \\ 2x - y + 3z &= -19 \end{aligned}$$

(A) $\begin{bmatrix} 1 & 2 & 0 & 4 \\ 0 & 1 & 0 & 3 \\ 0 & 0 & 1 & -4 \end{bmatrix}$ **(B)** $\begin{bmatrix} 1 & 0 & 0 & 2 \\ 0 & 1 & 0 & -3 \\ 0 & 0 & 0 & 4 \end{bmatrix}$

(C) $\begin{bmatrix} 1 & 0 & 0 & -2 \\ 0 & 1 & 0 & 3 \\ 0 & 0 & 0 & -4 \end{bmatrix}$ **(D)** $\begin{bmatrix} 1 & 0 & 0 & 2 \\ 0 & 1 & 0 & -3 \\ 0 & 0 & 1 & 4 \end{bmatrix}$

(E) $\begin{bmatrix} 1 & 0 & 0 & -2 \\ 0 & 1 & 0 & 3 \\ 0 & 0 & 1 & -4 \end{bmatrix}$

Explorations

107. Group Activity Investigating the Solution of a System of 3 Linear Equations in 3 Variables Assume that the graph of a linear equation in three variables is a plane in 3-dimensional space. (You will study these in Chapter 8.)

(a) Explain geometrically how such a system can have a unique solution.

(b) Explain geometrically how such a system can have no solution. Describe several possibilities.

(c) Explain geometrically how such a system can have infinitely many solutions. Describe several possibilities. Construct physical models if you find that helpful.

Extending the Ideas

108. Writing to Learn Explain why a row echelon form of a matrix is not unique. That is, show that a matrix can have two unequal row echelon forms. Give an example.

The roots of the characteristic polynomial $C(x) = \det(xI_n - A)$ of the $n \times n$ matrix A are the **eigenvalues** of A (see Section 7.2, Exercises 72 and 73). Use this information in Exercises 109 and 110.

109. Let $A = \begin{bmatrix} 3 & 2 \\ 1 & 5 \end{bmatrix}$.

(a) Find the characteristic polynomial $C(x)$ of A.

(b) Find the graph of $y = C(x)$.

(c) Find the eigenvalues of A.

(d) Compare $\det A$ with the y-intercept of the graph of $y = C(x)$.

(e) Compare the sum of the main diagonal elements of A $(a_{11} + a_{22})$ with the sum of the eigenvalues.

110. Let $A = \begin{bmatrix} 2 & -1 \\ -5 & 2 \end{bmatrix}$.

(a) Find the characteristic polynomial $C(x)$ of A.

(b) Find the graph of $y = C(x)$.

(c) Find the eigenvalues of A.

(d) Compare $\det A$ with the y-intercept of the graph of $y = C(x)$.

(e) Compare the sum of the main diagonal elements of A $(a_{11} + a_{22})$ with the sum of the eigenvalues.

7.4 Systems of Inequalities in Two Variables

What you'll learn about

- Graph of an Inequality
- Systems of Inequalities
- Linear Programming

... and why

Linear programming is used in business and industry to maximize profits, minimize costs, and help management make decisions.

Graph of an Inequality

An ordered pair (a, b) of real numbers is a **solution of an inequality** in x and y if the substitution $x = a$ and $y = b$ satisfies the inequality. For example, the ordered pair $(2, 5)$ is a solution of $y < 2x + 3$ because

$$5 < 2(2) + 3 = 7.$$

However, the ordered pair $(2, 8)$ is *not* a solution because

$$8 \not< 2(2) + 3 = 7.$$

When we have found all the solutions we have **solved the inequality**.

The **graph of an inequality** in x and y consists of all pairs (x, y) that are solutions of the inequality. The graph of an inequality involving two variables typically is a region of the coordinate plane.

The point $(2, 7)$ is on the graph of the line $y = 2x + 3$ but is not a solution of $y < 2x + 3$. A point $(2, y)$ below the line $y = 2x + 3$ is on the graph of $y < 2x + 3$, and those above it are not. The graph of $y < 2x + 3$ is the set of all points below the line $y = 2x + 3$. The graph of the line $y = 2x + 3$ is the *boundary* of the region (Figure 7.22).

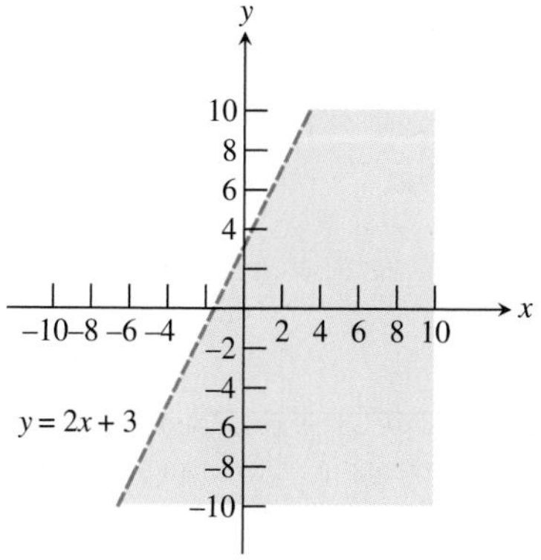

FIGURE 7.22 A graph of $y = 2x + 3$ (dashed line) and $y < 2x + 3$ (shaded area). The line is dashed to indicate it is *not* part of the solution of $y < 2x + 3$.

We can summarize our observations about the graph of an inequality in two variables with the following procedure.

Steps for Drawing the Graph of an Inequality in Two Variables

1. Draw the graph of the equation obtained by replacing the inequality sign by an equals sign. Use a dashed line if the inequality is $<$ or $>$. Use a solid line if the inequality is $\leq$ or $\geq$.
2. Check a point in each of the two regions of the plane determined by the graph of the equation. If the point satisfies the inequality, then shade the region containing the point.

EXAMPLE 1 Graphing a Linear Inequality

Draw the graph of $y \geq 2x + 3$. State the boundary of the region.

SOLUTION

Step 1. Because of "$\geq$," the graph of the line $y = 2x + 3$ is part of the graph of the inequality and should be drawn as a solid line.

Step 2. The point $(0, 4)$ is above the line and satisfies the inequality because

$$4 \geq 2(0) + 3 = 3.$$

Thus, the graph of $y \geq 2x + 3$ consists of all points on or above the line $y = 2x + 3$. The boundary is the graph of $y = 2x + 3$ (Figure 7.23).

Now try Exercise 9.

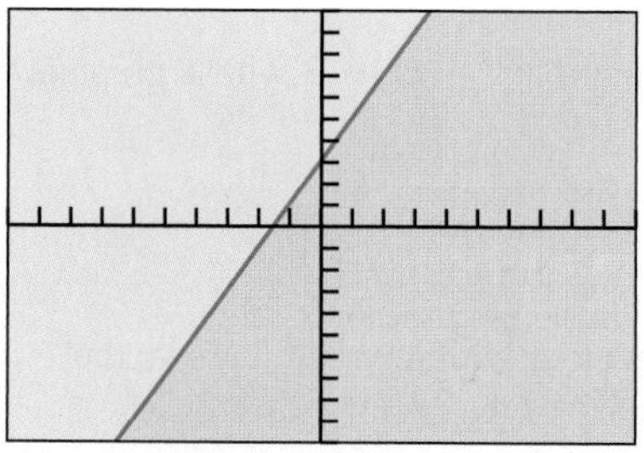

[−10, 10] by [−10, 10]

FIGURE 7.23 The graph of $y \geq 2x + 3$. shown in blue. (Example 1)

The graph of the linear inequality $y \geq ax + b, y > ax + b, y \leq ax + b$, or $y < ax + b$ is a **half-plane**. The graph of the line $y = ax + b$ is the **boundary** of the region.

EXAMPLE 2 Graphing Linear Inequalities

Draw the graph of the inequality. State the boundary of the region.

(a) $x \geq 2$ **(b)** $y < -3$

SOLUTION

(a) Step 1. Replacing "$\geq$" by "$=$" we obtain the equation $x = 2$ whose graph is a vertical line.

Step 2. The graph of $x \geq 2$ is the set of all points on and to the right of the vertical line $x = 2$ (Figure 7.24a). The line $x = 2$ is the boundary of the region.

(b) Step 1. Replacing "$<$" by "$=$" we obtain the equation $y = -3$ whose graph is a horizontal line.

Step 2. The graph of $y < -3$ is the set of all points below the horizontal line $y = -3$ (Figure 7.24b). The line $y = -3$ is the boundary of the region.

Now try Exercise 7.

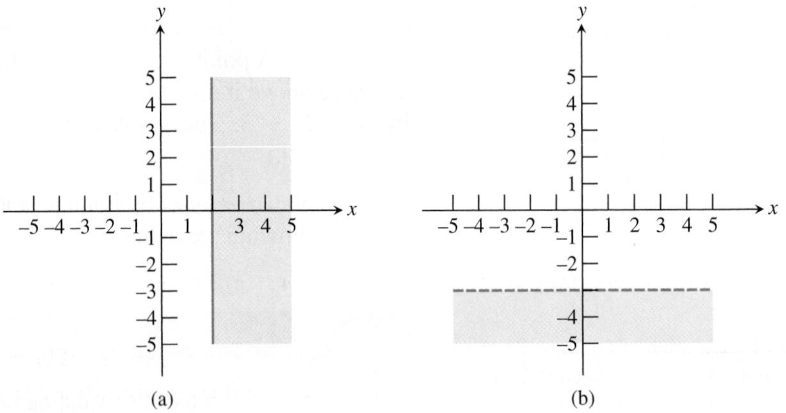

FIGURE 7.24 The graphs of (a) $x \geq 2$ and (b) $y < -3$. (Example 2)

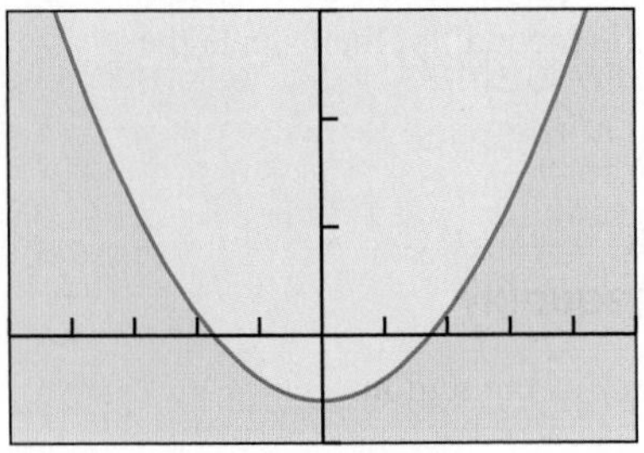

[–5, 5] by [–5, 15]

FIGURE 7.25 The graph of $y \geq x^2 - 3$ shown in blue. (Example 3)

EXAMPLE 3 Graphing a Quadratic Inequality

Draw the graph of $y \geq x^2 - 3$. State the boundary of the region.

SOLUTION

Step 1. Replacing "$\geq$" by "$=$" we obtain the equation $y = x^2 - 3$ whose graph is a parabola.

Step 2. The pair $(0, 1)$ is a solution of the inequality because

$$1 \geq (0)^2 - 3 = -3.$$

Thus, the graph of $y \geq x^2 - 3$ is the parabola together with the region above the parabola (Figure 7.25). The parabola is the boundary of the region.

Now try Exercise 11.

Systems of Inequalities

A **solution** of a system of inequalities in x and y is an ordered pair (x, y) that satisfies each inequality in the system. When we have found all the common solutions we have **solved the system of inequalities**.

The technique for solving a system of inequalities graphically is similar to that for solving a system of equations graphically. We graph each inequality and determine the points common to the individual graphs.

Shading Graphs

Most graphing utilities are capable of shading solutions to inequalities. Check the owner's manual for your grapher.

EXAMPLE 4 Solving a System of Inequalities Graphically

Solve the system:

$$y > x^2$$
$$2x + 3y < 4$$

SOLUTION The graph of $y > x^2$ is shaded in Figure 7.26a. It does not include its boundary $y = x^2$. The graph of $2x + 3y < 4$ is shaded in Figure 7.26b. It does not include its boundary $2x + 3y = 4$. The solution to the system is the intersection of these two graphs, as shaded in Figure 7.26c.

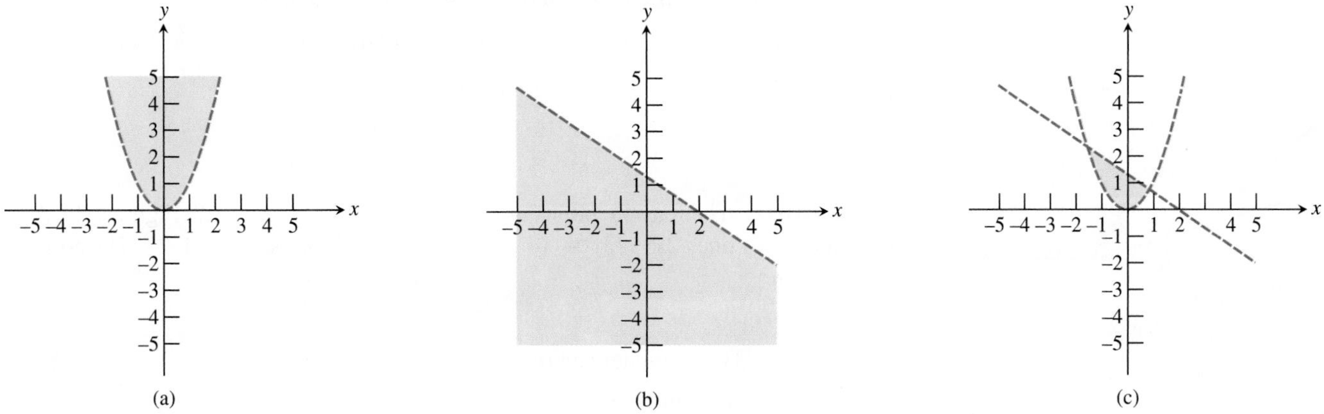

FIGURE 7.26 The graphs of (a) $y > x^2$, (b) $2x + 3y < 4$, and (c) the system which they form (Example 4).

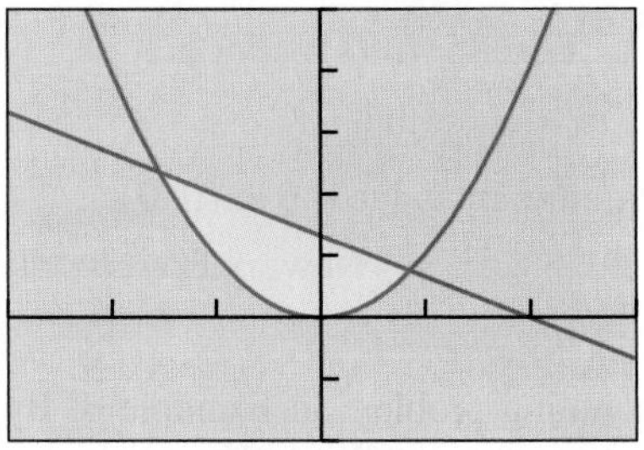

[–3, 3] by [–2, 5]

FIGURE 7.27 The solution (shaded in blue) of the system in Example 4. Most graphers cannot distinguish between dashed and solid boundaries.

Support with a Grapher Figure 7.27 shows what some graphers produce when we shade above the curve $y = x^2$ and below the curve $2x + 3y = 4$. The shaded portion appears to be identical to the shaded portion in Figure 7.26c.

Now try Exercise 19.

EXAMPLE 5 Solving a System of Inequalities

Solve the system:

$$2x + y \le 10$$
$$2x + 3y \le 14$$
$$x \ge 0$$
$$y \ge 0$$

SOLUTION The solution is in the first quadrant because $x \ge 0$ and $y \ge 0$, it lies below each of the two lines $2x + y = 10$ and $2x + 3y = 14$, and includes all of its boundary points (Figure 7.28).

Now try Exercise 23.

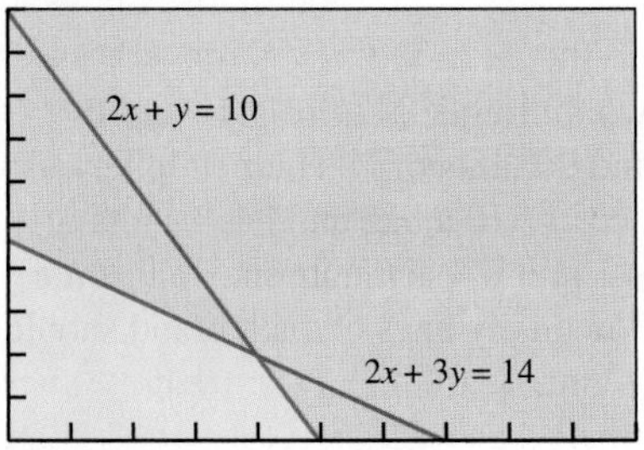

[0, 10] by [0, 10]

FIGURE 7.28 The solution (shaded in blue) of the system in Example 5. The boundary points are included.

Linear Programming

Sometimes decision making in management science requires that we find a minimum or a maximum of a linear function

$$f = a_1x_1 + a_2x_2 + \cdots + a_nx_n,$$

called an **objective function**, over a set of points. Such a problem is a **linear programming problem**. In two dimensions, the function f takes the form $f = ax + by$ and the

Making a Sketch by Hand

It is usually easier to draw a hand sketch of the lines in the examples and exercises in this chapter (by using the x- and y-intercept method) than to use a grapher. You can support your sketches using a grapher.

set of points is the solution of a system of inequalities, called **constraints**, such as the one in Figure 7.30. The solution of the system of inequalities is the set of **feasible points** (x, y) for the optimization problem.

It can be shown that, if a linear programming problem has a solution, it occurs at one of the **vertex points**, or **corner points**, along the boundary of the region of feasible points, or **feasible region**. We use this information in Examples 6 and 7.

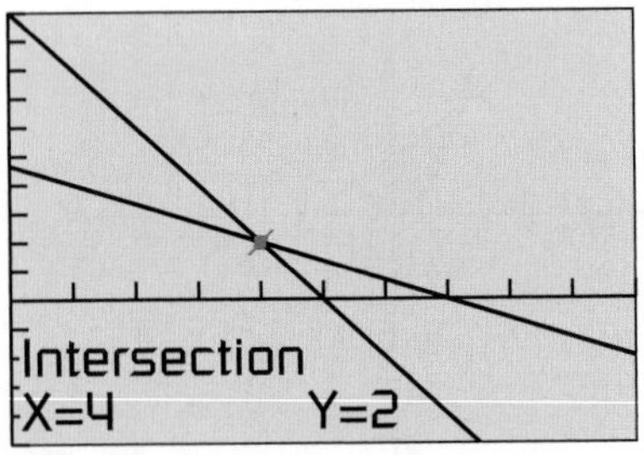

[0, 10] by [–5, 10]

FIGURE 7.29 The lines $2x + 3y = 14$ and $2x + y = 10$ intersect at $(4, 2)$. (Example 6)

EXAMPLE 6 Solving a Linear Programming Problem

Find the maximum and minimum values of the objective function $f = 5x + 8y$, subject to the constraints given by the system of inequalities.

$$\begin{aligned} 2x + y &\le 10 \\ 2x + 3y &\le 14 \\ x &\ge 0 \\ y &\ge 0 \end{aligned}$$

SOLUTION The feasible points are those graphed in Figure 7.28. Figure 7.29 shows that the two lines $2x + 3y = 14$ and $2x + y = 10$ intersect at $(4, 2)$. The corner points are

$(0, 0)$,
$(0, 14/3)$, the y-intercept of $2x + 3y = 14$,
$(5, 0)$, the x-intercept of $2x + y = 10$, and
$(4, 2)$, the point of intersection of $2x + 3y = 14$ and $2x + y = 10$.

The following table evaluates f at the corner points of the region in Figure 7.29.

(x, y)	$(0, 0)$	$(0, 4.\bar{6})$	$(4, 2)$	$(5, 0)$
f	0	$37.\bar{3}$	36	25

The maximum value of f is $37.\bar{3}$ at $(0, 4.\bar{6})$. The minimum value is 0 at $(0, 0)$.

Now try Exercise 31.

There is another way to analyze the linear programming problem in Example 6: By assigning positive values to f in $f = 5x + 8y$, we obtain parallel lines whose distances from the origin increase as f increases. (See Exercise 47.) This family of lines sweeps across the feasible region. Geometrically, we can see that there is a minimum and maximum value for f if the line $f = 5x + 8y$ is to intersect the region of feasible solutions.

EXAMPLE 7 Purchasing Fertilizer

Johnson's Produce is purchasing fertilizer with two nutrients: N (nitrogen) and P (phosphorus). They need at least 180 units of N and 90 units of P. Their supplier has two brands of fertilizer for them to buy. Brand A costs \$10 a bag and has 4 units of N and 1 unit of P. Brand B costs \$5 a bag and has 1 unit of each nutrient. Johnson's Produce can pay at most \$800 for the fertilizer. How many bags of each brand should be purchased to minimize cost?

SOLUTION

Model

Let x = number of bags of Brand A

Let y = number of bags of Brand B

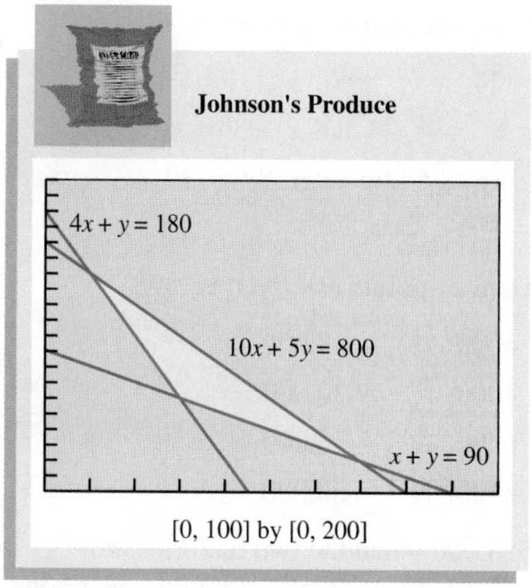

FIGURE 7.30 The feasible region, shaded in blue, for Example 7.

Then $C =$ the total cost $= 10x + 5y$ is the objective function to be minimized. The constraints are:

$$4x + y \geq 180 \quad \text{Amount of N is at least 180.}$$
$$x + y \geq 90 \quad \text{Amount of P is at least 90.}$$
$$10x + 5y \leq 800 \quad \text{Total cost to be at most \$800.}$$
$$x \geq 0, y \geq 0$$

Solve Graphically The feasible region is the intersection of the graphs of $4x + y \geq 180$, $x + y \geq 90$, and $10x + 5y \leq 800$ in the first quadrant (Figure 7.30).

The region has three corner points, which are at the points of intersections of the lines $4x + y = 180$, $x + y = 90$, and $10x + 5y = 800$: $(10, 140)$, $(70, 20)$, and $(30, 60)$. The values of the objective function C at the corner points are as follows:

$$C(10, 140) = 10(10) + 5(140) = 800$$
$$C(70, 20) = 10(70) + 5(20) = 800$$
$$C(30, 60) = 10(30) + 5(60) = 600$$

Interpret The minimum cost for the fertilizer is \$600 when 30 bags of Brand A and 60 bags of Brand B are purchased. For this purchase, Johnson's Produce gets exactly 180 units of nutrient N and 90 units of nutrient P. **Now try Exercise 37.**

The region in Example 8 is unbounded. Using the discussion following Example 6, we can see geometrically that the linear programming problem in Example 8 does not have a maximum value but, fortunately, does have a minimum value.

EXAMPLE 8 Minimizing Operating Cost

Gonza Manufacturing has two factories that produce three grades of paper: low-grade, medium-grade, and high-grade. Gonza needs to supply 24 tons of low-grade, 6 tons of medium-grade, and 30 tons of high-grade paper. Factory A produces 8 tons of low-grade, 1 ton of medium-grade, 2 tons of high-grade paper daily, and costs \$2000 per day to operate. Factory B produces 2 tons of low-grade, 1 ton of medium-grade, 8 tons of high-grade paper daily, and costs \$4000 per day to operate. How many days should each factory operate to fill the orders at minimum cost?

SOLUTION

Model

Let $x =$ the number of days Factory A operates.

Let $y =$ the number of days Factory B operates.

Then $C =$ total operating cost $= 2000x + 4000y$ is the objective function to be minimized. The constraints are:

$$8x + 2y \geq 24 \quad \text{Amount of low-grade is at least 24.}$$
$$x + y \geq 6 \quad \text{Amount of medium-grade is at least 6.}$$
$$2x + 8y \geq 30 \quad \text{Amount of high-grade is at least 30.}$$
$$x \geq 0, y \geq 0$$

Solve Graphically The region of feasible points is the intersection of the graphs of $8x + 2y \geq 24$, $x + y \geq 6$, and $2x + 8y \geq 30$ in the first quadrant.

(continued)

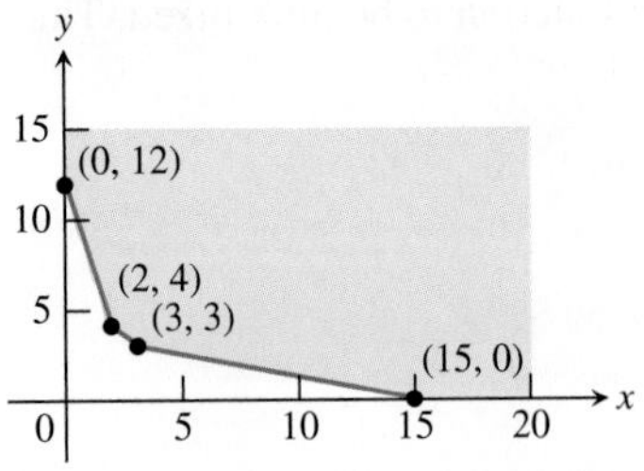

FIGURE 7.31 The graph of the feasible region, showing the coordinates of the four vertex points. (Example 8)

As shown in Figure 7.31, this feasible region has four corner points:

$(0, 12)$, the y-intercept of $8x + 2y = 24$
$(2, 4)$, the point of intersection of $8x + 2y = 24$ and $x + y = 6$
$(3, 3)$, the point of intersection of $x + y = 6$ and $2x + 8y = 30$
$(15, 0)$, the x-intercept of $2x + 8y = 30$

The values of the objective function C at the corner points are given below:

$$\begin{aligned} C(0, 12) &= 2000(0) + 4000(12) = 48{,}000 \\ C(2, 4) &= 2000(2) + 4000(4) = 20{,}000 \\ C(3, 3) &= 2000(3) + 4000(3) = 18{,}000 \\ C(15, 0) &= 2000(15) + 4000(0) = 30{,}000 \end{aligned}$$

Interpret The minimum operational cost is \$18,000 when the two factories are operated for 3 days each. The two factories will produce 30 tons of low-grade, 6 tons of medium-grade, and 30 tons of high-grade. They will have a surplus of 6 tons of low-grade paper.

Now try Exercise 39.

QUICK REVIEW 7.4 *(For help, go to Sections P.4 and 7.1.)*

In Exercises 1–4, find the x- and y-intercepts of the line and draw its graph.

1. $2x - 3y = 6$

2. $5x + 10y = 30$

3. $\dfrac{x}{20} + \dfrac{y}{50} = 1$

4. $\dfrac{x}{30} - \dfrac{y}{20} = 1$

In Exercises 5–9, find the point of intersection of the two lines. (We will use these values in Examples 7 and 8.)

5. $4x + y = 180$ and $x + y = 90$

6. $x + y = 90$ and $10x + 5y = 800$

7. $4x + y = 180$ and $10x + 5y = 800$

8. $8x + 2y = 24$ and $x + y = 6$

9. $x + y = 6$ and $2x + 8y = 30$

10. Solve the system of equations:

$$y = x^2$$
$$2x + 3y = 4$$

SECTION 7.4 Exercises

Exercise numbers with a gray background indicate problems that the authors have designed to be solved *without a calculator*.

In Exercises 1–6, match the inequality with its graph, shaded in blue. Indicate whether the boundary is included in or excluded from the graph. All graphs are drawn in $[-4.7, 4.7]$ by $[-3.1, 3.1]$.

1. $x \le 3$

2. $y > 2$

3. $2x - 5y \ge 2$

4. $y > (1/2)x^2 - 1$

5. $y \ge 2 - x^2$

6. $x^2 + y^2 < 4$

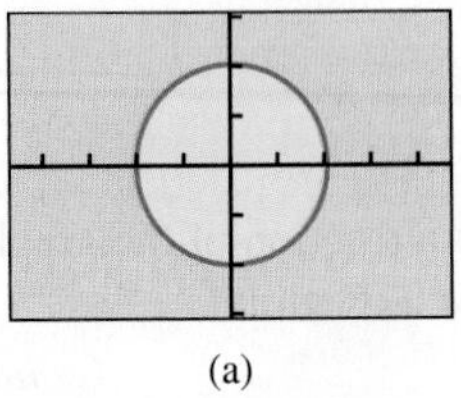

(a)

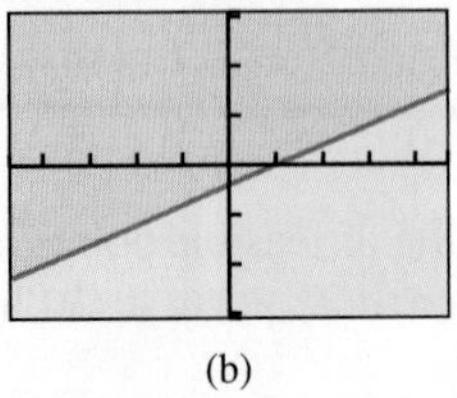

(b)

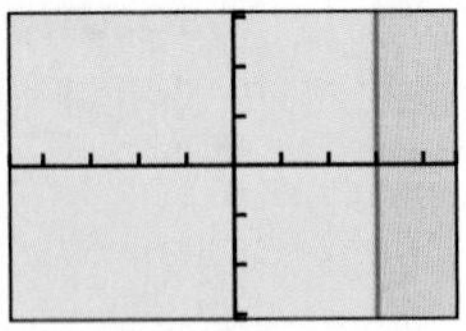

(c)

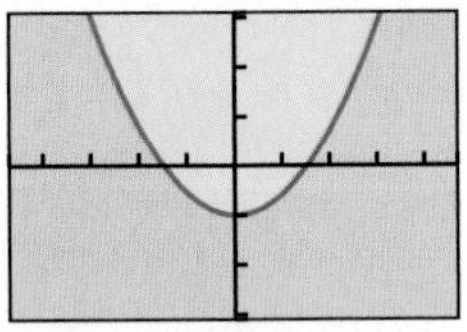

(d)

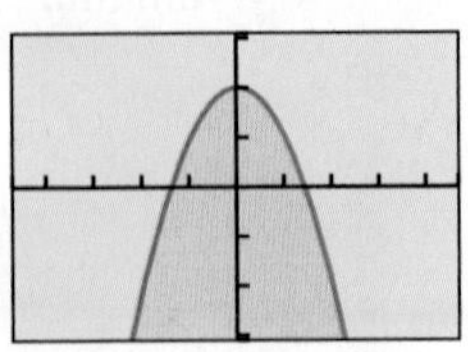

(e)

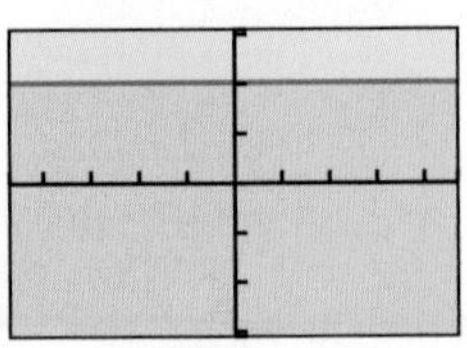

(f)

In Exercises 7–16, draw a hand sketch of the inequality. State the boundary of the region.

7. $x \le 4$

8. $y \ge -3$

9. $2x + 5y \le 7$

10. $3x - y > 4$

11. $y < x^2 + 1$

12. $y \ge x^2 - 3$

13. $x^2 + y^2 < 9$

14. $x^2 + y^2 \ge 4$

15. $y > 2^x$

16. $y < \sin x$

Use an algebraic method in Exercises 17–22 to solve the system of inequalities. Support with a grapher.

17. $5x - 3y > 1$
$3x + 4y \le 18$

18. $4x + 3y \le -6$
$2x - y \le -8$

19. $y \le 2x + 3$
$y \ge x^2 - 2$

20. $x - 3y - 6 < 0$
$y > -x^2 - 2x + 2$

21. $y \ge x^2$
$x^2 + y^2 \le 4$

22. $x^2 + y^2 \le 9$
$y \ge |x|$

Use an algebraic method in Exercises 23–26 to solve the system of inequalities. Support with a grapher.

In Exercises 23–26, solve the system of inequalities.

23. $2x + y \le 80$
$x + 2y \le 80$
$x \ge 0$
$y \ge 0$

24. $3x + 8y \ge 240$
$9x + 4y \ge 360$
$x \ge 6$
$y \ge 0$

25. $5x + 2y \le 20$
$2x + 3y \le 18$
$x + y \ge 2$
$x \ge 0$
$y \ge 0$

26. $7x + 3y \le 210$
$3x + 7y \le 210$
$x + y \ge 30$

In Exercises 27–30, write a system of inequalities whose solution is the region shaded in blue in the given figure. All boundaries are to be included.

27. Group Activity

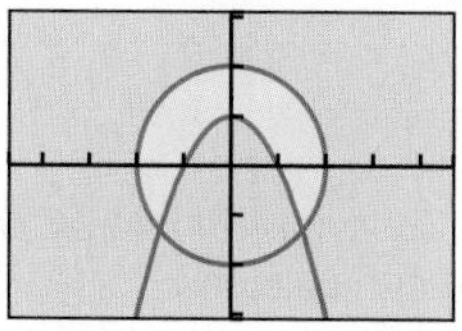

[–4.7, 4.7] by [–3.1, 3.1]

28. Group Activity

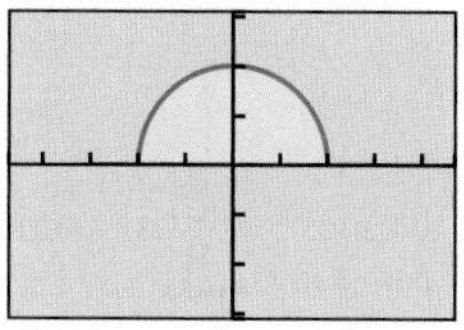

[–4.7, 4.7] by [–3.1, 3.1]

29. Group Activity

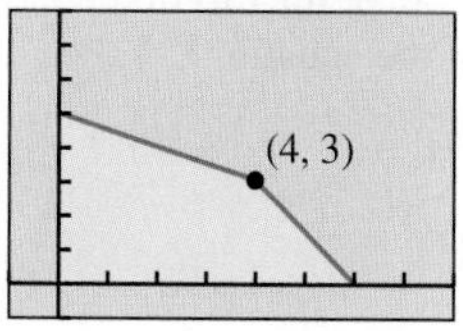

[–1, 8] by [–1, 8]

30. Group Activity

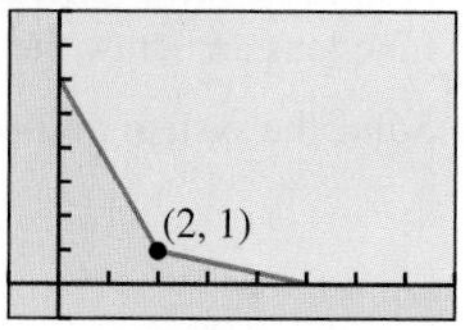

[–1, 8] by [–1, 8]

In Exercises 31–36, find the minimum and maximum, if they exist, of the objective function f, subject to the constraints.

31. Objective function: $f = 4x + 3y$
Constraints:

$$x + y \le 80$$
$$x - 2y \le 0$$
$$x \ge 0,\ y \ge 0$$

32. Objective function: $f = 10x + 11y$
Constraints:

$$x + y \le 90$$
$$3x - y \ge 0$$
$$x \ge 0,\ y \ge 0$$

33. Objective function: $f = 7x + 4y$
Constraints:

$$5x + y \ge 60$$
$$x + 6y \ge 60$$
$$4x + 6y \ge 204$$
$$x \ge 0,\ y \ge 0$$

34. Objective function: $f = 15x + 25y$
Constraints:

$$3x + 4y \ge 60$$
$$x + 8y \ge 40$$
$$11x + 28y \le 380$$
$$x \ge 0,\ y \ge 0$$

35. Objective function: $f = 5x + 2y$
Constraints:

$$2x + y \ge 12$$
$$4x + 3y \ge 30$$
$$x + 2y \ge 10$$
$$x \ge 0,\ y \ge 0$$

36. Objective function: $f = 3x + 5y$
Constraints:

$$3x + 2y \ge 20$$
$$5x + 6y \ge 52$$
$$2x + 7y \ge 30$$
$$x \ge 0,\ y \ge 0$$

37. Mining Ore Pearson's Metals mines two ores: R and S. The company extracts minerals A and B from each type of ore. It costs \$50 per ton to extract 80 lb of A and 160 lb of B from ore R. It costs \$60 per ton to extract 140 lb of A and 50 lb of B from ore S. Pearson's must produce at least 4000 lb of A and 3200 lb of B. How much of each ore should be processed to minimize cost? What is the minimum cost?

38. Planning a Diet Paul's diet is to contain at least 24 units of carbohydrates and 16 units of protein. Food substance A costs \$1.40 per unit and each unit contains 3 units of carbohydrates and 4 units of protein. Food substance B costs \$0.90 per unit and each unit contains 2 units of carbohydrates and 1 unit of protein. How many units of each food substance should be purchased in order to minimize cost? What is the minimum cost?

39. Producing Gasoline Two oil refineries produce three grades of gasoline: A, B, and C. At each refinery, the three grades of gasoline are produced in a single operation in the following proportions: Refinery 1 produces 1 unit of A, 2 units of B, and 1 unit of C; Refinery 2 produces 1 unit of A, 4 units of B, and 4 units of C. For the production of one operation, Refinery 1 charges \$300 and Refinery 2 charges \$600. A customer needs 100 units of A, 320 units of B, and 200 units of C. How should the orders be placed if the customer is to minimize his cost?

40. Maximizing Profit A manufacturer wants to maximize the profit for two products. Product A yields a profit of \$2.25 per unit, and product B yields a profit of \$2.00 per unit. Demand information requires that the total number of units produced be no more than 3000 units, and that the number of units of product B produced be greater than or equal to half the number of units of product A produced. How many of each unit should be produced to maximize profit?

Standardized Test Questions

41. True or False The graph of a linear inequality in x and y is a half-line. Justify your answer.

42. True or False The boundary of the solution of $2x - 3y < 5$ is the graph of $3y = 2x - 5$. Justify your answer.

In Exercises 43–46, you may use a graphing calculator to solve the problem.

For Exercises 43–44, use the figure below, which shows the graphs of the two lines $3x + 4y = 5$ and $2x - 3y = 4$.

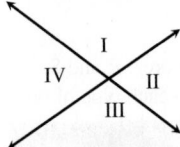

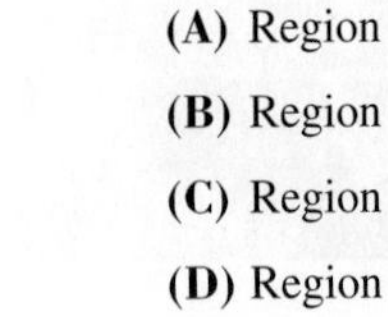

43. Multiple Choice Which of the following represents the solution of the system $3x + 4y \geq 5$
$2x - 3y \leq 4$?

(A) Region I plus its boundary

(B) Region I without its boundary

(C) Region II plus its boundary

(D) Region II without its boundary

(E) Region IV plus its boundary

44. Multiple Choice Which of the following represents the solution of the system $3x + 4y < 5$
$2x - 3y > 4$?

(A) Region II plus its boundary

(B) Region III plus its boundary

(C) Region III without its boundary

(D) Region IV plus its boundary

(E) Region IV without its boundary

Exercises 45–46 refer to the following linear programming problem:

Objective function: $f = 5x + 10y$

Constraints:

$$2x + y \leq 10$$
$$x + 3y \leq 12$$
$$x \geq 0, y \geq 0$$

45. Multiple Choice Which of the following is not a corner point?

(A) $(0, 0)$ **(B)** $(5, 0)$

(C) $(0, 4)$ **(D)** $(3, 4)$

(E) $(3.6, 2.8)$

46. Multiple Choice What is the maximum value of f in the feasible region of the problem?

(A) 0 **(B)** 25 **(C)** 40 **(D)** 46 **(E)** 55

Explorations

47. Revisiting Example 6 Consider the objective function $f = 5x + 8y$ of Example 6.

(a) Prove that for any two real number values for f, the two lines are parallel.

(b) Writing to Learn For $f > 0$, give reasons why the line moves farther away from the origin as the value of f increases.

(c) Writing to Learn Give a geometric explanation of why the region of Example 6 must contain a minimum and a maximum value for f.

48. Writing to Learn Describe all the possible ways that two distinct parabolas of the form $y = f(x)$ can intersect. Give examples.

Extending the Ideas

49. Implicit Functions The equation

$$\frac{x^2}{9} + \frac{y^2}{4} = 1$$

defines y as two functions of x. Solve for y to find the two functions and draw the graph of the equation.

50. Implicit Functions The equation

$$x^2 - y^2 = 4$$

defines y as two functions of x. Solve for y to find the two functions and draw the graph of the equation.

51. Solve the system of inequalities:

$$\frac{x^2}{9} + \frac{y^2}{4} \leq 1$$
$$y \geq x^2 - 1$$

(*Hint*: See Exercise 49.)

52. Graph the inequality $x^2 - y^2 \leq 4$. (*Hint*: See Exercise 50.)

CHAPTER 7 Key Ideas

Properties, Theorems, and Formulas

Inverse of a 2×2 Matrix 534
Inverses of $n \times n$ Matrices 535
Properties of Matrices 536
Invertible Square Linear Systems 549

Procedures

Solving Systems of Equations 519–523, 543–550
Partial Fraction Decomposition 550–551
Graphing Systems of Inequalities 557–559
Linear Programming 559–562

CHAPTER 7 Review Exercises

Exercise numbers with a gray background indicate problems that the authors have designed to be solved *without a calculator.*

The collection of exercises marked in red could be used as a chapter test.

In Exercises 1 and 2, find **(a)** $A + B$, **(b)** $A - B$, **(c)** $-2A$, and **(d)** $3A - 2B$.

1. $A = \begin{bmatrix} -1 & 3 \\ 4 & 0 \end{bmatrix}$, $B = \begin{bmatrix} 2 & -1 \\ 4 & 3 \end{bmatrix}$

2. $A = \begin{bmatrix} 2 & 3 & -1 & 2 \\ 1 & 4 & -2 & -3 \\ 0 & -3 & 2 & 1 \end{bmatrix}$, $B = \begin{bmatrix} -1 & 2 & 0 & 4 \\ 2 & -1 & 3 & 3 \\ -2 & 4 & 1 & 3 \end{bmatrix}$

In Exercises 3–8, find the products AB and BA, or state that a given product is not possible.

3. $A = \begin{bmatrix} -1 & 4 \\ 0 & 6 \end{bmatrix}$, $B = \begin{bmatrix} 3 & -1 & 5 \\ 0 & -2 & 4 \end{bmatrix}$

4. $A = \begin{bmatrix} -1 & 2 \\ 3 & -1 \\ 4 & 3 \end{bmatrix}$, $B = \begin{bmatrix} -2 & 3 & 1 \\ 2 & 1 & 0 \\ -1 & 2 & -3 \end{bmatrix}$

5. $A = [-1 \quad 4]$, $B = \begin{bmatrix} 5 & -3 \\ 2 & 1 \end{bmatrix}$

6. $A = \begin{bmatrix} -1 & 1 \\ 0 & 1 \end{bmatrix}$, $B = \begin{bmatrix} 3 & -4 \\ 1 & 2 \\ 3 & 1 \\ 1 & 1 \end{bmatrix}$

7. $A = \begin{bmatrix} 0 & 1 & 0 \\ 1 & 0 & 0 \\ 0 & 0 & 1 \end{bmatrix}$, $B = \begin{bmatrix} 2 & -3 & 4 \\ 1 & 2 & -3 \\ -2 & 1 & -1 \end{bmatrix}$

8. $A = \begin{bmatrix} 0 & 1 & 0 & 0 \\ 1 & 0 & 0 & 0 \\ 0 & 0 & 0 & 1 \\ 0 & 0 & 1 & 0 \end{bmatrix}$, $B = \begin{bmatrix} -2 & 1 & 0 & 1 \\ 3 & 0 & 2 & 1 \\ -1 & 1 & 2 & -1 \\ 3 & -2 & 1 & 0 \end{bmatrix}$

In Exercises 9 and 10, use multiplication to verify that the matrices are inverses.

9. $A = \begin{bmatrix} 1 & -2 & 1 & 1 \\ 1 & -1 & 0 & 3 \\ 1 & -1 & 2 & 2 \\ 2 & -4 & 2 & 3 \end{bmatrix}$, $B = \begin{bmatrix} 8 & 1.5 & 0.5 & -4.5 \\ 2 & 0.5 & 0.5 & -1.5 \\ -1 & -0.5 & 0.5 & 0.5 \\ -2 & 0 & 0 & 1 \end{bmatrix}$

10. $A = \begin{bmatrix} -1 & 1 & 1 \\ 2 & 1 & 0 \\ -1 & 0 & 2 \end{bmatrix}$, $B = \begin{bmatrix} -0.4 & 0.4 & 0.2 \\ 0.8 & 0.2 & -0.4 \\ -0.2 & 0.2 & 0.6 \end{bmatrix}$

In Exercises 11 and 12, find the inverse of the matrix if it has one. If it does, use multiplication to support your result.

11. $\begin{bmatrix} 1 & 2 & 0 & -1 \\ 2 & -1 & 1 & 2 \\ 2 & 0 & 1 & 2 \\ -1 & 1 & 1 & 4 \end{bmatrix}$

12. $\begin{bmatrix} -1 & 0 & 1 \\ 2 & -1 & 1 \\ 1 & 1 & 1 \end{bmatrix}$

In Exercises 13 and 14, evaluate the determinant of the matrix.

13. $\begin{bmatrix} 1 & -3 & 2 \\ 2 & 4 & -1 \\ -2 & 0 & 1 \end{bmatrix}$

14. $\begin{bmatrix} -2 & 3 & 0 & 1 \\ 3 & 0 & 2 & 0 \\ 5 & 2 & -3 & 4 \\ 1 & -1 & 2 & 3 \end{bmatrix}$

In Exercises 15–18, find the reduced row echelon form of the matrix.

15. $\begin{bmatrix} 1 & 0 & 2 \\ 3 & 1 & 5 \\ 1 & -1 & 3 \end{bmatrix}$

16. $\begin{bmatrix} 2 & 1 & 1 & 1 \\ -3 & -1 & -2 & 1 \\ 5 & 2 & 2 & 3 \end{bmatrix}$

17. $\begin{bmatrix} 1 & 2 & 3 & 1 \\ 2 & 3 & 3 & -2 \\ 1 & 2 & 4 & 6 \end{bmatrix}$

18. $\begin{bmatrix} 1 & -2 & 0 & 4 \\ -2 & 5 & 3 & -6 \\ 2 & -4 & 1 & 9 \end{bmatrix}$

In Exercises 19–22, state whether the system of equations has a solution. If it does, solve the system.

19. $3x - y = 1$
$x + 2y = 5$

20. $x - 2y = -1$
$-2x + y = 5$

21. $x + 2y = 1$
$4y - 4 = -2x$

22. $x - 2y = 9$
$3y - \frac{3}{2}x = -9$

In Exercises 23–28, use Gaussian elimination to solve the system of equations.

23. $\begin{aligned} x + z + w &= 2 \\ x + y + z &= 3 \\ 3x + 2y + 3z + w &= 8 \end{aligned}$

24. $\begin{aligned} x + w &= -2 \\ x + y + z + 2w &= -2 \\ -x - 2y - 2z - 3w &= 2 \end{aligned}$

25. $\begin{aligned} x + y - 2z &= 2 \\ 3x - y + z &= 4 \\ -2x - 2y + 4z &= 6 \end{aligned}$

26. $\begin{aligned} x + y - 2z &= 2 \\ 3x - y + z &= 1 \\ -2x - 2y + 4z &= -4 \end{aligned}$

27. $\begin{aligned} -x - 6y + 4z - 5w &= -13 \\ 2x + y + 3z - w &= 4 \\ 2x + 2y + 2z &= 6 \\ -x - 3y + z - 2w &= -7 \end{aligned}$

28. $\begin{aligned} -x + 2y + 2z - w &= -4 \\ y + z &= -1 \\ -x + 2y + 2z - 2w &= -6 \\ -x + 3y + 3z - w &= -5 \end{aligned}$

In Exercises 29–32, solve the system of equations by using inverse matrices.

29. $\begin{aligned} x + 2y + z &= -1 \\ x - 3y + 2z &= 1 \\ 2x - 3y + z &= 5 \end{aligned}$

30. $\begin{aligned} x + 2y - z &= -2 \\ 2x - y + z &= 1 \\ x + y - 2z &= 3 \end{aligned}$

31. $\begin{aligned} 2x + y + z - w &= 1 \\ 2x - y + z - w &= -2 \\ -x + y - z + w &= -3 \\ x - 2y + z - w &= 1 \end{aligned}$

32. $\begin{aligned} x - 2y + z - w &= 2 \\ 2x + y - z - w &= -1 \\ x - y + 2z - w &= -1 \\ x + 3y - z + w &= 4 \end{aligned}$

In Exercises 33–36, solve the system of equations by finding the reduced row echelon form of the augmented matrix.

33. $\begin{aligned} x + 2y - 2z + w &= 8 \\ 2x + 3y - 3z + 2w &= 13 \end{aligned}$

34. $\begin{aligned} x + 2y - 2z + w &= 8 \\ 2x + 7y - 7z + 2w &= 25 \\ x + 3y - 3z + w &= 11 \end{aligned}$

35. $\begin{aligned} x + 2y + 4z + 6w &= 6 \\ 3x + 4y + 8z + 11w &= 11 \\ 2x + 4y + 7z + 11w &= 10 \\ 3x + 5y + 10z + 14w &= 15 \end{aligned}$

36. $\begin{aligned} x + 2z - 2w &= 5 \\ 2x + y + 4z - 3w &= 7 \\ 4x + y + 7z - 6w &= 15 \\ 2x + y + 5z - 4w &= 9 \end{aligned}$

In Exercises 37 and 38, find the equilibrium point for the demand and supply curves.

37. $\begin{aligned} p &= 100 - x^2 \\ p &= 20 + 3x \end{aligned}$

38. $\begin{aligned} p &= 80 - \frac{1}{10}x^2 \\ p &= 5 + 4x \end{aligned}$

In Exercises 39–44, solve the system of equations graphically.

39. $\begin{aligned} 3x - 2y &= 5 \\ 2x + y &= -2 \end{aligned}$

40. $\begin{aligned} y &= x - 1.5 \\ y &= 0.5x^2 - 3 \end{aligned}$

41. $\begin{aligned} y &= -0.5x^2 + 3 \\ y &= 0.5x^2 - 1 \end{aligned}$

42. $\begin{aligned} x^2 + y^2 &= 4 \\ y &= 2x^2 - 3 \end{aligned}$

43. $\begin{aligned} y &= 2 \sin x \\ y &= 2x - 3 \end{aligned}$

44. $\begin{aligned} y &= \ln 2x \\ y &= 2x^2 - 12x + 15 \end{aligned}$

In Exercises 45 and 46, find the coefficients of the function so that its graph goes through the given points.

45. Curve Fitting $f(x) = ax^3 + bx^2 + cx + d$
$(2, 8),\ (4, 5),\ (6, 3),\ (9, 4)$

46. Curve Fitting $f(x) = ax^4 + bx^3 + cx^2 + dx + e$
$(-2, -4),\ (1, 2),\ (3, 6),\ (4, -2),\ (7, 8)$

In Exercises 47–52, find the partial fraction decomposition of the rational function.

47. $\dfrac{3x - 2}{x^2 - 3x - 4}$

48. $\dfrac{x - 16}{x^2 + x - 2}$

49. $\dfrac{x + 14}{x^2 + 7x + 10}$

50. $\dfrac{-2x - 14}{x^2 + 6x + 8}$

51. $\dfrac{2x^2 - 12x + 12}{x^3 - 6x^2 + 11x - 6}$

52. $\dfrac{4x^2 - 3x - 19}{x^3 - 4x^2 + x + 6}$

In Exercises 53–56 match the inequality with its graph, shaded in blue. Indicate whether the boundary is included in or excluded from the graph. All graphs are drawn in the window $[-4.7, 4.7]$ *by* $[-3.1, 3.1]$.

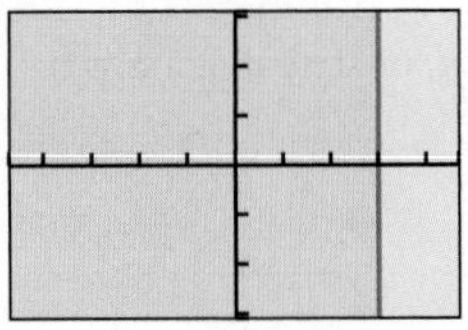

(a)

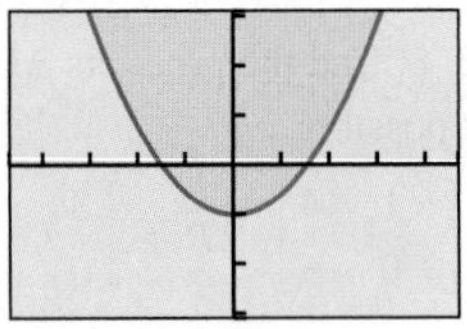

(b)

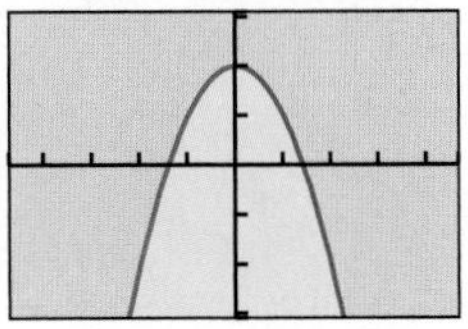

(c)

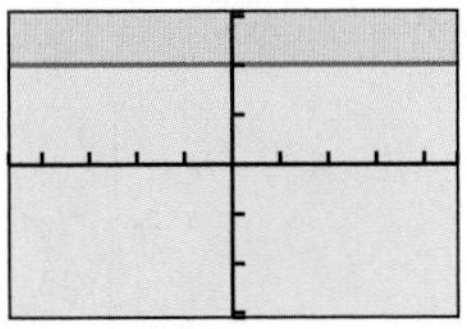

(d)

53. $x > 3$

54. $y \leq 2$

55. $y \leq 2 - x^2$

56. $y < x^2/2 - 1$

In Exercises 57 and 58, graph the inequality.

57. $2x - y \leq 1$

58. $x + 3y < 2$

In Exercises 59–64, solve the system of inequalities. Give the coordinates of any corner points.

59. $4x + 9y \geq 360$
$9x + 4y \geq 360$
$x + y \leq 90$

60. $7x + 10y \leq 70$
$2x + y \leq 10$
$x + y \geq 3$
$x \geq 0$
$y \geq 0$

61. $x - 3y + 6 < 0$
$y > x^2 - 6x + 7$

62. $x + 2y \geq 4$
$y \leq 9 - x^2$

63. $x^2 + y^2 \leq 4$
$y \geq x^2$

64. $y \leq x^2 + 4$
$x^2 + y^2 \geq 4$

In Exercises 65–68, find the minimum and maximum, if they exist, of the objective function f, subject to the constraints.

65. Objective function: $f = 7x + 6y$
Constraints:
$7x + 5y \geq 100$
$2x + 5y \geq 50$
$x \geq 0, y \geq 0$

66. Objective function: $f = 11x + 5y$
Constraints:
$5x + 2y \geq 60$
$5x + 8y \geq 120$
$x \geq 0, y \geq 0$

67. Objective function: $f = 3x + 7y$
Constraints:
$5x + 2y \geq 100$
$x + 4y \geq 110$
$5x + 11y \leq 460$
$x \geq 0, y \geq 0$

68. Objective function: $f = 9x + 14y$
Constraints:
$x + y \leq 120$
$9x + 2y \geq 240$
$3x + 10y \geq 360$

69. Rotating Coordinate Systems The xy-coordinate system is rotated through the angle 45° to obtain the $x'y'$-coordinate system.

(a) If the coordinates of a point in the xy-coordinate system are $(1, 2)$, what are the coordinates of the rotated point in the $x'y'$-coordinate system?

(b) If the coordinates of a point in the $x'y'$-coordinate system are $(1, 2)$, what are the coordinates of the point in the xy-coordinate system that was rotated to it?

70. Medicare Disbursements Table 7.7 shows the total Medicare Disbursements in billions of dollars for several years. Let x be the numbers of years since 1990.

Table 7.7 Total Medicare Disbursements

Year	Disbursements (billions of $)
2000	224.3
2001	247.7
2002	265.1
2003	281.5
2004	309.3
2005	338.0
2006	401.3

Source: U.S. Census Bureau.

(a) Find a linear regression model and superimpose its graph on a scatter plot of the data.

(b) Find a power regression model and superimpose its graph on a scatter plot of the data.

(c) Find when the models in parts (a) and (b) predict the same disbursement amounts.

(d) Writing to Learn Which model appears to be a better fit for the data? Explain.

Which model would you choose to make predictions beyond 2010?

71. Population Table 7.8 gives the populations (in thousands) of Hawaii and Idaho for selected years. Let x be the number of years since 1980.

Table 7.8 Populations of Two U.S. States

Year	Hawaii (thousands)	Idaho (thousands)
1980	965	944
1990	1108	1007
2000	1212	1294
2010	1360	1568

Source: U.S. Census Bureau, The World Almanac and Book of Facts 2012.

(a) Find a linear regression equation for the population of Hawaii.

(b) Find a linear regression equation for the population of Idaho.

(c) Graph the models from parts (a) and (b), and use these models to determine when the populations of the two states were about the same.

(d) Writing to Learn Do these models seem appropriate for these data? Might a nonlinear model be more appropriate in either case? If so, which nonlinear model would you select? Explain.

72. **(a)** The 2010 population data for three states are listed below. Use the 2010 data in Table 7.9 on the next page to create a 3×2 matrix that estimates the number of males and females in these three states.

State	Population (millions)
California	37.3
Florida	18.8
Rhode Island	1.1

(b) Write the data from the 2010 Census table below in the form of a 3×2 matrix.

State	% Pop. Under 18 Years	% Pop. 65 Years or Older
California	25.0	11.4
Florida	21.3	17.3
Rhode Island	22.3	14.3

(c) Multiply your 3×2 matrix in part (b) by the scalar 0.01 to change the values from percentages to decimals.

(d) Use matrix multiplication to multiply the transpose of the matrix from part (c) by the matrix from part (a). What information does the resulting matrix provide?

(e) How many males under age 18 lived in these three states in 2010? How many females age 65 or older lived in these three states?

73. Using Matrices A stockbroker sold a customer 200 shares of stock A, 400 shares of stock B, 600 shares of stock C, and 250 shares of stock D. The price per share of A, B, C, and D is \$80, \$120, \$200, and \$300, respectively.

(a) Write a 1×4 matrix N representing the number of shares of each stock the customer bought.

(b) Write a 1×4 matrix P representing the price per share of each stock.

(c) Write a matrix product that gives the total cost of the stocks that the customer bought.

74. Basketball Attendance At Whetstone High School 452 tickets were sold for the first basketball game. There were two ticket prices: \$0.75 for students and \$2.00 for nonstudents. How many tickets of each type were sold if the total revenue from the sale of tickets was \$429?

75. HDTV Deliveries Brock's Discount Electronics has three sizes of high-definition televisions (HDTVs) on sale: small (19-in.), medium (32-in.), and large (47-in.). Brock's has three types of vehicles that they use for deliveries: minivans, vans, and trucks. The minivans can carry 8 small, 3 medium, and 2 large HDTVs; the vans, 15 small, 10 medium, and 6 large; the trucks, 22 small, 20 medium, and 5 large. On the last day of the sale, they have 115 small, 85 medium, and 35 large HDTVs to deliver. How many vehicles of each type are needed to deliver all of these?

76. Investments Jessica invests \$38,000; part at 7.5% simple interest and the remainder at 6% simple interest. If her annual interest income is \$2600, how much does she have invested at each rate?

77. Business Loans Thompson's Furniture Store borrowed \$650,000 to expand its facilities and extend its product line. Some of the money was borrowed at 4%, some at 6.5%, and the rest at 9%. How much was borrowed at each rate if the annual interest was \$46,250 and the amount borrowed at 9% was twice the amount borrowed at 4%?

78. Home Remodeling Sanchez Remodeling has three painters: Sue, Esther, and Murphy. Working together they can paint a large room in 4 hr. Sue and Murphy can paint the same size room in 6 hr. Esther and Murphy can paint the same size room in 7 hr. How long would it take each of them to paint the room alone?

79. Swimming Pool Three pipes, A, B, and C, are connected to a swimming pool. When all three pipes are running, the pool can be filled in 3 hr. When only A and B are running, the pool can be filled in 4 hr. When only B and C are running, the pool can be filled in 3.75 hr. How long would it take each pipe running alone to fill the pool?

80. Writing to Learn If the products AB and BA are defined for the $n \times n$ matrix A, what can you conclude about the order of matrix B? Explain.

81. Writing to Learn If A is an $m \times n$ matrix and B is a $p \times q$ matrix, and if AB is defined, what can you conclude about their orders? Explain.

CHAPTER 7 Project

Examine the male and female population data from 1990 through 2010 given in Table 7.9.

Table 7.9 U.S. Male and Female Population Data, 1990–2010

Year	Male (millions)	Female (millions)
1990	121.3	127.5
1997	130.8	137.0
1998	132.0	138.3
1999	133.3	139.4
2000	138.1	143.4
2001	140.1	145.1
2002	141.5	146.4
2003	143.0	147.8
2004	144.5	149.2
2005	146.0	150.5
2010	151.8	157.0

Source: U.S. Census Bureau.

1. Plot the data using 1990 as Year 0. Find a linear regression model for each.
2. What do the slope and y-intercept mean in each equation?
3. What conclusions can you draw? According to these models, will the male population ever become greater than the female population? Was the male population ever greater than the female population? Are there enough data to create a model for a hundred years or more? Explain your answers.
4. Notice that the data in Table 7.9 give information for a span of only 20 years. This is not enough information to answer accurately the questions in part 3. Data over a small period often appear to be linear and can be modeled with a linear equation that works well over that limited domain. Table 7.10 gives more data. Now use these data to plot the number of males versus time and females versus time, using 1900 as Year 0.
5. Notice that the data in Table 7.10 do not seem to be linear. Often, you may remember from Chapter 3, a logistic model is used to model population growth. Find the logistic regression model for each data plot. Do these models seem reasonable in the long run? Explain.
6. Go to the U.S. Census Bureau Web site (www.census.gov). How does your model predict the populations for the current year?
7. Use the census data for 2010. What percentage of the population is male and what percentage is female?
8. Go to the U.S. Census Bureau Web site (www.census.gov) and use the most recent data along with the concepts from this chapter to collect and analyze other data.

Table 7.10 U.S. Male and Female Population Data, 1990–2010

Year	Male (millions)	Female (millions)	Year	Male (millions)	Female (millions)
1900	38.8	37.2	1960	88.3	91.0
1910	47.3	44.6	1970	98.9	104.3
1920	53.9	51.8	1980	110.1	116.5
1930	62.1	60.6	1990	121.3	127.5
1940	66.0	65.6	2000	138.1	143.4
1950	75.2	76.1	2010	151.8	157.0

Source: U.S. Census Bureau.

8.1 Conic Sections and a New Look at Parabolas

8.2 Circles and Ellipses

8.3 Hyperbolas

8.4 Quadratic Equations with *xy* Terms

8.5 Polar Equations of Conics

8.6 Three-Dimensional Cartesian Coordinate System

Analytic Geometry in Two and Three Dimensions

The oval lawn behind the White House in Washington, D.C., is called *the Ellipse*. It has views of the Washington Monument, the Jefferson Memorial, the Department of Commerce, and the Old Post Office Building. The Ellipse is 616 ft long and 528 ft wide, and is in the shape of a conic section. Its shape can be modeled using the methods of this chapter. See page 590.

History of Conic Sections

Parabolas, ellipses, and hyperbolas had been studied for many years when Apollonius (c. 250–175 BCE) wrote his eight-volume *Conic Sections*. Apollonius, born in Perga, Asia Minor, was the first to unify these three curves as cross sections of a cone and to view the hyperbola as having two branches. Interest in conic sections was renewed in the 17th century when Galileo proved that projectiles follow parabolic paths and Johannes Kepler (1571–1630) discovered that planets travel in elliptical orbits.

Chapter 8 Overview

By now you should be fairly comfortable thinking about functions graphically and algebraically, since connecting those two representations has been a consistent goal of this textbook. We now want to devote one chapter to a brief look at two interesting extensions of this algebra-geometry connection. The first is an extension from linear equations in x and y (the graphs of which are always lines) to quadratic equations in x and y (the graphs of which are always curves called *conic sections*). The second is an extension from coordinates in 2-space (planar geometry) to coordinates in 3-space (solid geometry). In both cases we will be temporarily abandoning the interpretations that depend on y being a function of x, although you may revisit some other connections to functions in future math courses.

The marriage of algebra and geometry, known as *analytic geometry*, grew out of the work of Frenchmen René Descartes (1596–1650) and Pierre de Fermat (1601–1665). They developed a systematic approach that could be used to solve algebra problems geometrically and geometry problems algebraically—two major themes of this course. It was analytic geometry that opened the door for Newton and Leibniz to develop the fundamental techniques of calculus less than half a century later.

8.1 Conic Sections and a New Look at Parabolas

What you'll learn about

- Conic Sections
- Geometry of a Parabola
- Translations of Parabolas
- Reflective Property of a Parabola

... and why

Conic sections are the paths of nature: Any free-moving object in a gravitational field follows the path of a conic section.

Conic Sections

The connection between algebra and geometry is impressively simple in the case of lines in the xy-plane: Every first-degree (linear) equation $Ax + By + C = 0$ (assuming A and B are not both 0) has a graph that is a line in the plane, and every line in the plane is the graph of some linear equation.

A **second-degree (quadratic) equation in two variables** is an equation of the form $Ax^2 + Bxy + Cy^2 + Dx + Ey + F = 0$ (assuming A, B, and C are not all 0). The geometric connection between these quadratic equations and their graphs is not as simple as it is for linear equations, but it is no less impressive. In fact, every such equation has a graph that is a *conic section*, linking quadratic equations to a class of geometric curves studied extensively by the mathematician Apollonius (c. 250–175 BCE).

Imagine two nonperpendicular lines intersecting at a point V. If we fix one of the lines as an *axis* and rotate the other line (the *generator*) around the axis, then the generator sweeps out a **right circular cone** with **vertex** V, as illustrated in Figure 8.1. Notice that the cone consists of two parts, called **nappes**, joined at the vertex. Each nappe resembles a pointed ice-cream cone, except that these ice-cream cones extend infinitely as you move away from the vertex.

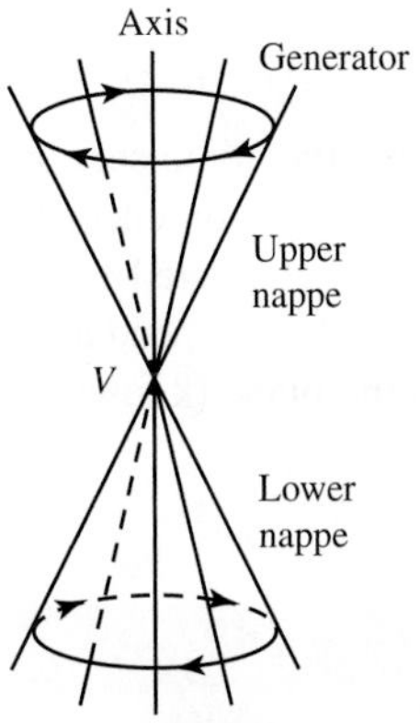

FIGURE 8.1 A right circular cone (of two nappes).

A **conic section** (or **conic** for short) is the intersection of a plane with one of these right circular cones. Figure 8.2 shows some of the possible curves that can result. The three most obvious conic sections (Figure 8.2a) are the *parabola*, the *ellipse* (of which we can consider the *circle* to be a special case), and the *hyperbola*. Less obvious are the so-called **degenerate conic sections** (Figure 8.2b), including a *line*, a *point*, and *two intersecting lines*. (Two other degenerate conics, *two parallel lines* and the *empty set*, can result from intersecting a plane with a cylinder. A cylinder can be viewed as a **degenerate cone** with its vertex at infinity. See Exercise 73.) These eight possibilities appear to be so different from one another that you might have easily concluded that they have nothing in common geometrically, had you not seen that each can be realized as the intersection of a plane with a (possibly degenerate) right circular cone.

We now want to take a deeper look at the geometric properties of the three basic conic sections (parabolas, ellipses, and hyperbolas), using the tools of analytic geometry to

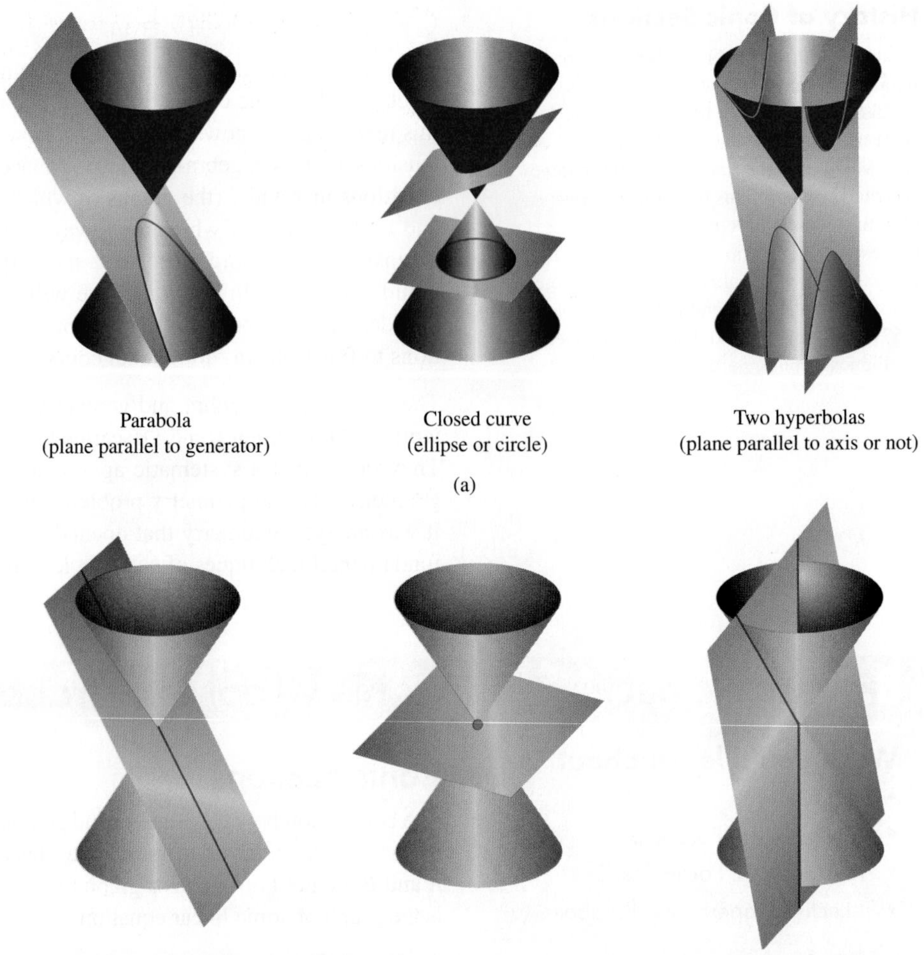

FIGURE 8.2 Conic sections: (a) standard and (b) degenerate. Two additional degenerate cases (two parallel lines and the empty set) are discussed in Exercise 73.

connect them to quadratic equations in x and y. Some of the properties of these curves have real-world applications that have been exploited by scientists and engineers since the days of Apollonius.

Geometry of a Parabola

When the intersecting plane is parallel to the generator of the cone, the conic section formed is a parabola (the leftmost picture in Figure 8.2a). In Section 2.1 you encountered parabolas as the graphs of quadratic functions, a connection that proved to be useful in solving problems involving free fall and projectile motion. What we need now is a purely geometric definition of a parabola as a set of points in the plane (known in geometry courses as a *locus* definition).

A Degenerate Parabola

If the focus F lies on the directrix l, the parabola "degenerates" to the line through F perpendicular to l. Henceforth, we will assume F does not lie on l.

DEFINITION Parabola

A **parabola** is the set of all points in a plane equidistant from a particular line (the **directrix**) and a particular point (the **focus**) in the plane (Figure 8.3).

Locus of a Point

Before the word *set* was used in mathematics, the Latin word *locus*, meaning "place," was often used in geometric definitions. The locus of a point was the set of possible places a point could be and still fit the conditions of the definition. Sometimes, conics are still defined in terms of loci.

As you can see in Figure 8.3, the line passing through the focus and perpendicular to the directrix is the **(focal) axis** of the parabola. The axis is the line of symmetry for the parabola. The point where the parabola intersects its axis is the **vertex** of the parabola. The vertex is located midway between the focus and the directrix and is the point of the parabola that is closest to both the focus and the directrix.

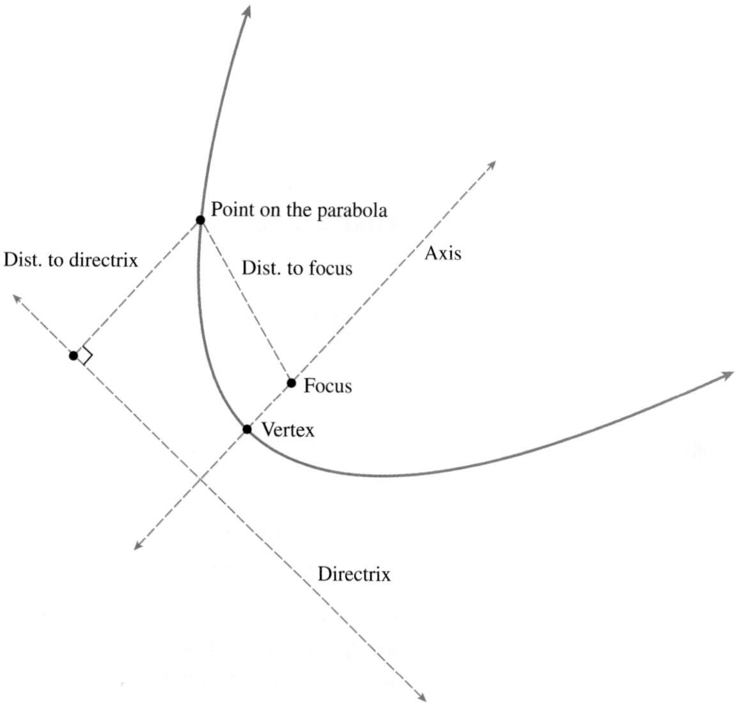

FIGURE 8.3 Structure of a Parabola. The distance from each point on the parabola to both the focus and the directrix is the same.

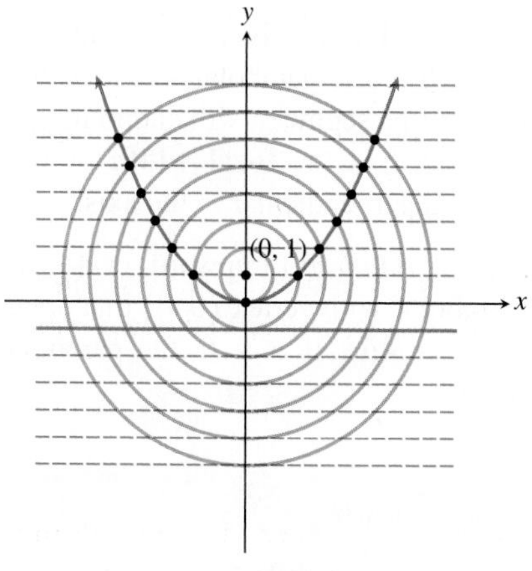

FIGURE 8.4 The geometry of a parabola. Note that each point lies on a circle centered at the focus. The radius of the circle is the distance between the point and the directrix.

EXPLORATION 1 Understanding the Definition of Parabola

1. Prove that the vertex of the parabola with focus $(0, 1)$ and directrix $y = -1$ is $(0, 0)$ (Figure 8.4).
2. Find an equation for the parabola shown in Figure 8.4.
3. Find the coordinates of the points of the parabola that are highlighted in Figure 8.4.

We can generalize the situation in Exploration 1 to show that an equation for the parabola with focus $(0, p)$ and directrix $y = -p$ is $x^2 = 4py$ (Figure 8.5).

We must show first that a point $P(x, y)$ that is equidistant from $F(0, p)$ and the line $y = -p$ satisfies the equation $x^2 = 4py$, and then that a point satisfying the equation $x^2 = 4py$ is equidistant from $F(0, p)$ and the line $y = -p$:

Let $P(x, y)$ be equidistant from $F(0, p)$ and the line $y = -p$. Notice that

$$\sqrt{(x - 0)^2 + (y - p)^2} = \text{distance from } P(x, y) \text{ to } F(0, p), \text{ and}$$

$$\sqrt{(x - x)^2 + (y - (-p))^2} = \text{distance from } P(x, y) \text{ to } y = -p.$$

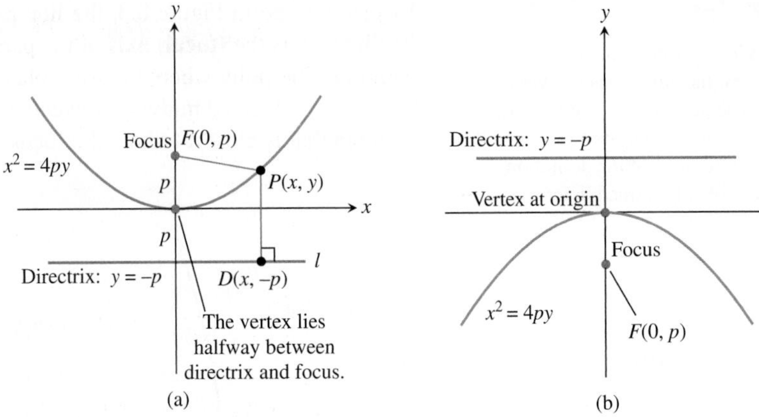

FIGURE 8.5 Graphs of $x^2 = 4py$ with (a) $p > 0$ and (b) $p < 0$.

Name Game

The chord of the parabola that defines the focal width is called the *latus rectum*. Latin students should note that *latus* (side) is the noun, and *rectum* (right) is the adjective. Everyone else can stop snickering. Ellipses and hyperbolas also have latera recta, which are always chords through the foci perpendicular to the axes that pass through the vertices.

Equating these distances and squaring yields:

$$\begin{aligned}(x-0)^2 + (y-p)^2 &= (x-x)^2 + (y-(-p))^2 \\ x^2 + (y-p)^2 &= 0 + (y+p)^2 && \text{Simplify.} \\ x^2 + y^2 - 2py + p^2 &= y^2 + 2py + p^2 && \text{Expand.} \\ x^2 &= 4py && \text{Combine like terms.}\end{aligned}$$

By reversing the above steps, we see that a solution (x, y) of $x^2 = 4py$ is equidistant from $F(0, p)$ and the line $y = -p$.

The equation $x^2 = 4py$ is the **standard form** of the equation of an upward- or downward opening parabola with vertex at the origin. If $p > 0$, the parabola opens upward; if $p < 0$, it opens downward. An alternative algebraic form for such a parabola is $y = ax^2$, where $a = 1/(4p)$. So the graph of $x^2 = 4py$ is also the graph of the quadratic function $f(x) = ax^2$.

When the equation of an upward- or downward-opening parabola is written as $x^2 = 4py$, the value p is interpreted as the **focal length** of the parabola—the *directed* distance from the vertex to the focus of the parabola. A line segment with endpoints on a parabola is a **chord** of the parabola. The value $|4p|$ is the **focal width** of the parabola—the length of the chord through the focus and perpendicular to the axis.

Parabolas that open to the right or to the left are *inverse relations* of upward- or downward opening parabolas. Therefore, equations of parabolas with vertex $(0, 0)$ that open to the right or to the left have the standard form $y^2 = 4px$. If $p > 0$, the parabola opens to the right, and if $p < 0$, to the left (Figure 8.6).

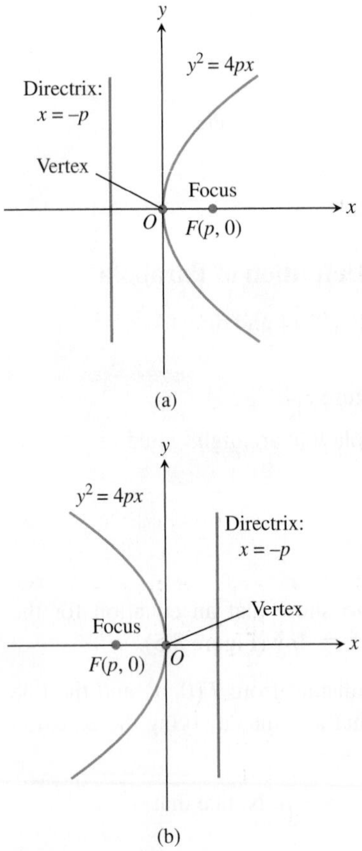

FIGURE 8.6 Graph of $y^2 = 4px$ with (a) $p > 0$ and (b) $p < 0$.

Parabolas with Vertex (0, 0)

• **Standard equation**	$x^2 = 4py$	$y^2 = 4px$
• **Opens**	Upward or downward	To the right or to the left
• **Focus**	$(0, p)$	$(p, 0)$
• **Directrix**	$y = -p$	$x = -p$
• **Axis**	y-axis	x-axis
• **Focal length**	p	p
• **Focal width**	$\lvert 4p \rvert$	$\lvert 4p \rvert$

See Figures 8.5 and 8.6.

EXAMPLE 1 Finding the Focus, Directrix, and Focal Width

Find the focus, the directrix, and the focal width of the parabola $y = -(1/3)x^2$.

SOLUTION Multiplying both sides of the equation by -3 yields the standard form $x^2 = -3y$. The coefficient of y is $4p = -3$; thus $p = -3/4$. So the focus is $(0, p) = (0, -3/4)$. Because $-p = -(-3/4) = 3/4$, the directrix is the line $y = 3/4$. The focal width is $|4p| = |-3| = 3$. **Now try Exercise 1.**

EXAMPLE 2 Finding an Equation of a Parabola

Find an equation in standard form for the parabola whose directrix is the line $x = 2$ and whose focus is the point $(-2, 0)$.

SOLUTION Because the directrix is $x = 2$ and the focus is $(-2, 0)$, the focal length is $p = -2$ and the parabola opens to the left. The equation of the parabola in standard form is $y^2 = 4px$ or, more specifically, $y^2 = -8x$. **Now try Exercise 15.**

Translations of Parabolas

When a parabola with the equation $x^2 = 4py$ or $y^2 = 4px$ is translated horizontally by h units and vertically by k units, the vertex of the parabola moves from $(0, 0)$ to (h, k) (Figure 8.7). Such a translation does not change the focal length, the focal width, or the direction the parabola opens.

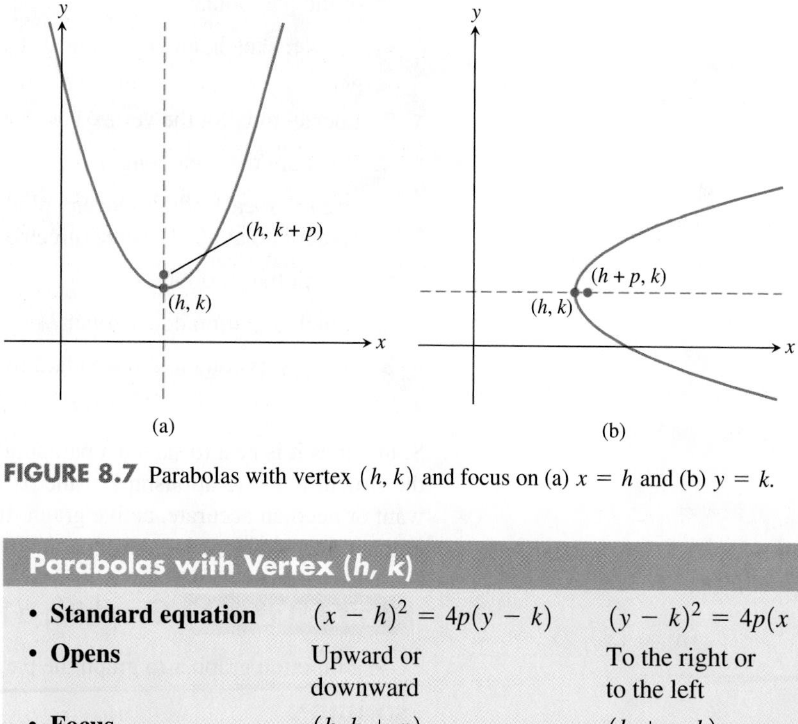

FIGURE 8.7 Parabolas with vertex (h, k) and focus on (a) $x = h$ and (b) $y = k$.

Parabolas with Vertex (h, k)

• **Standard equation**	$(x - h)^2 = 4p(y - k)$	$(y - k)^2 = 4p(x - h)$
• **Opens**	Upward or downward	To the right or to the left
• **Focus**	$(h, k + p)$	$(h + p, k)$
• **Directrix**	$y = k - p$	$x = h - p$
• **Axis**	$x = h$	$y = k$
• **Focal length**	p	p
• **Focal width**	$\|4p\|$	$\|4p\|$

See Figure 8.7.

EXAMPLE 3 Finding an Equation of a Parabola

Find the standard form of the equation for the parabola with vertex $(3, 4)$ and focus $(5, 4)$.

SOLUTION The axis of the parabola is the line passing through the vertex $(3, 4)$ and the focus $(5, 4)$. This is the line $y = 4$. So the equation has the form

$$(y - k)^2 = 4p(x - h).$$

Because the vertex $(h, k) = (3, 4)$, $h = 3$ and $k = 4$. The directed distance from the vertex $(3, 4)$ to the focus $(5, 4)$ is $p = 5 - 3 = 2$, so $4p = 8$. Thus the equation we seek is

$$(y - 4)^2 = 8(x - 3).$$

Now try Exercise 21.

When solving a problem like Example 3, it is a good idea to sketch the vertex, the focus, and other features of the parabola as we solve the problem. This makes it easy to see whether the axis of the parabola is horizontal or vertical and the relative positions of its features. Exploration 2 "walks us through" this process.

EXPLORATION 2 Building a Parabola

Carry out the following steps using a sheet of rectangular graph paper.

1. Let the focus F of a parabola be $(2, -2)$ and its directrix be $y = 4$. Draw the x- and y-axes on the graph paper. Then sketch and label the focus and directrix of the parabola.
2. Locate, sketch, and label the axis of the parabola. What is the equation of the axis?
3. Locate and plot the vertex V of the parabola. Label it by name and coordinates.
4. What are the focal length and focal width of the parabola?
5. Use the focal width to locate, plot, and label the endpoints of a chord of the parabola that parallels the directrix.
6. Sketch the parabola.
7. Which direction does it open?
8. What is its equation in standard form?

Sometimes it is best to sketch a parabola by hand, as in Exploration 2; this helps us see the structure and relationships of the parabola and its features. At other times, we may want or need an accurate, active graph. If we wish to graph a parabola using a function grapher, we need to solve the equation of the parabola for y, as illustrated in Example 4.

EXAMPLE 4 Graphing a Parabola with a Horizontal Axis

Use a function grapher to graph the parabola $(y - 4)^2 = 8(x - 3)$ of Example 3.

SOLUTION

$$(y - 4)^2 = 8(x - 3)$$

$$y - 4 = \pm\sqrt{8(x - 3)} \quad \text{Extract square roots.}$$

$$y = 4 \pm \sqrt{8(x - 3)} \quad \text{Add 4.}$$

Let $Y_1 = 4 + \sqrt{8(x - 3)}$ and $Y_2 = 4 - \sqrt{8(x - 3)}$, and graph the two equations in a window centered at the vertex, as shown in Figure 8.8. **Now try Exercise 45.**

Closing the Gap

In Figure 8.8, we centered the graphing window at the vertex $(3, 4)$ of the parabola to ensure that this point would be plotted. This avoids the common grapher error of a gap between the two upper and lower parts of the conic section being plotted.

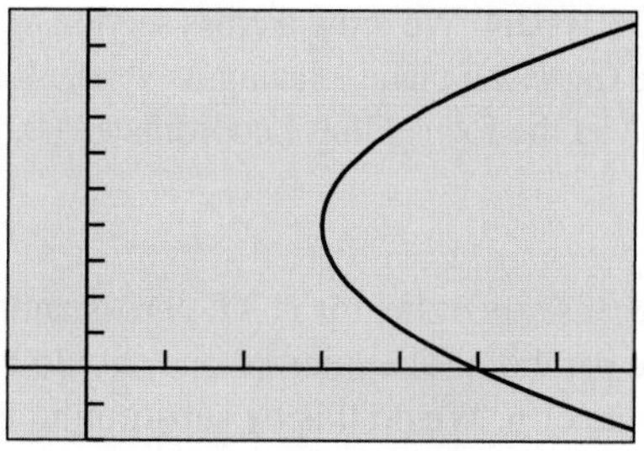

[–1, 7] by [–2, 10]

FIGURE 8.8 The graphs of $Y1 = 4 + \sqrt{8(x - 3)}$ and $Y2 = 4 - \sqrt{8(x - 3)}$ together form the graph of $(y - 4)^2 = 8(x - 3)$. (Example 4)

EXAMPLE 5 Using Standard Forms with a Parabola

Prove that the graph of $y^2 - 6x + 2y + 13 = 0$ is a parabola, and find its vertex, focus, and directrix.

SOLUTION Because this equation is quadratic in the variable y, we complete the square with respect to y to obtain a standard form.

$$y^2 - 6x + 2y + 13 = 0$$
$$y^2 + 2y = 6x - 13 \quad \text{Isolate the } y \text{ terms.}$$
$$y^2 + 2y + 1 = 6x - 13 + 1 \quad \text{Complete the square.}$$
$$(y + 1)^2 = 6x - 12$$
$$(y + 1)^2 = 6(x - 2)$$

This equation is in the standard form $(y - k)^2 = 4p(x - h)$, where $h = 2$, $k = -1$, and $p = 6/4 = 3/2 = 1.5$. It follows that

- the vertex (h, k) is $(2, -1)$;
- the focus $(h + p, k)$ is $(3.5, -1)$, or $(7/2, -1)$;
- the directrix $x = h - p$ is $x = 0.5$, or $x = 1/2$.

Now try Exercise 49.

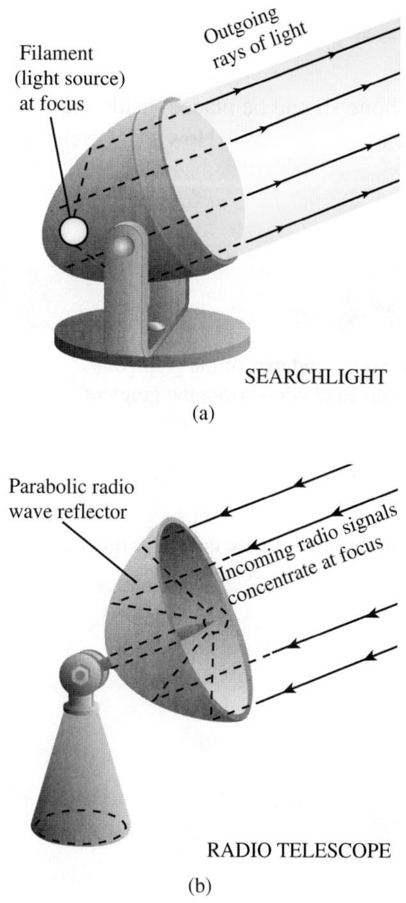

FIGURE 8.9 Examples of parabolic reflectors.

Reflective Property of a Parabola

The main applications of parabolas involve their use as reflectors of sound, light, radio waves, and other electromagnetic waves. If we rotate a parabola in three-dimensional space about its axis, the parabola sweeps out a **paraboloid of revolution**. If we place a signal source at the focus of a reflective paraboloid, the signal reflects off the surface in lines parallel to the axis of symmetry, as illustrated in Figure 8.9a. This property is used by flashlights, headlights, searchlights, microwave relays, and satellite up-links.

The principle works for signals traveling in the reverse direction as well (Figure 8.9b); signals arriving parallel to a parabolic reflector's axis are directed toward the reflector's focus. This property is used to intensify signals picked up by radio telescopes and television satellite dishes, to focus arriving light in reflecting telescopes, to concentrate heat in solar ovens, and to magnify sound for sideline microphones at football games.

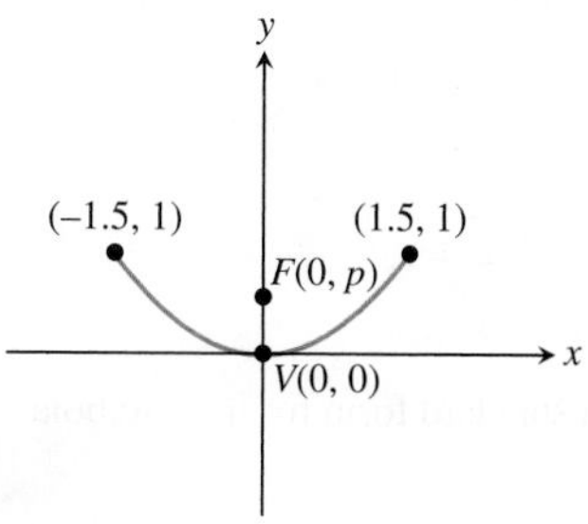

FIGURE 8.10 Cross section of parabolic reflector in Example 6.

EXAMPLE 6 Studying a Parabolic Microphone

On the sidelines of each of its televised football games, the FBTV network uses a parabolic reflector with a microphone at the reflector's focus to capture the conversations among players on the field. If the parabolic reflector is 3 ft across and 1 ft deep, where should the microphone be placed?

(continued)

SOLUTION We draw a cross section of the reflector as an upward-opening parabola in the Cartesian plane, placing its vertex V at the origin (Figure 8.10 on the previous page). We let the focus F have coordinates $(0, p)$ to yield the equation

$$x^2 = 4py.$$

Because the reflector is 3 ft across and 1 ft deep, the points $(\pm 1.5, 1)$ must lie on the parabola. The microphone should be placed at the focus, so we need to find the value of p. We do this by substituting the values we found into the equation:

$$\begin{aligned} x^2 &= 4py \\ (\pm 1.5)^2 &= 4p(1) \\ 2.25 &= 4p \\ p &= \frac{2.25}{4} = 0.5625 \end{aligned}$$

Because $p = 0.5625$ ft, or 6.75 in., the microphone should be placed inside the reflector along its axis and $6\frac{3}{4}$ in. from its vertex. Now try Exercise 59.

QUICK REVIEW 8.1 *(For help, go to Sections P.2, P.5, and 2.1.)*

Exercise numbers with a gray background indicate problems that the authors have designed to be solved *without a calculator*.

In Exercises 1 and 2, find the distance between the given points.

1. $(-1, 3)$ and $(2, 5)$ **2.** $(2, -3)$ and (a, b)

In Exercises 3 and 4, solve for y in terms of x.

3. $2y^2 = 8x$ **4.** $3y^2 = 15x$

In Exercises 5 and 6, complete the square to rewrite the equation in vertex form.

5. $y = -x^2 + 2x - 7$ **6.** $y = 2x^2 + 6x - 5$

In Exercises 7 and 8, find the vertex and axis of the graph of f. Describe how the graph of f can be obtained from the graph of $g(x) = x^2$, and graph f.

7. $f(x) = 3(x - 1)^2 + 5$

8. $f(x) = -2x^2 + 12x + 1$

In Exercises 9 and 10, write an equation for the quadratic function whose graph contains the given vertex and point.

9. Vertex $(-1, 3)$, point $(0, 1)$

10. Vertex $(2, -5)$, point $(5, 13)$

SECTION 8.1 Exercises

In Exercises 1–6, find the vertex, focus, directrix, and focal width of the parabola.

1. $x^2 = 6y$ **2.** $y^2 = -8x$

3. $(y - 2)^2 = 4(x + 3)$ **4.** $(x + 4)^2 = -6(y + 1)$

5. $3x^2 = -4y$ **6.** $5y^2 = 16x$

In Exercises 7–10, match the graph with its equation.

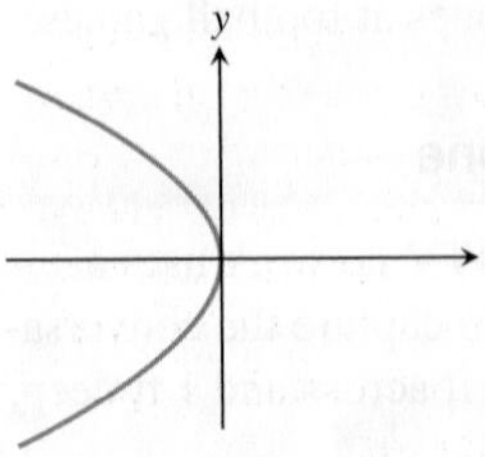

(a)

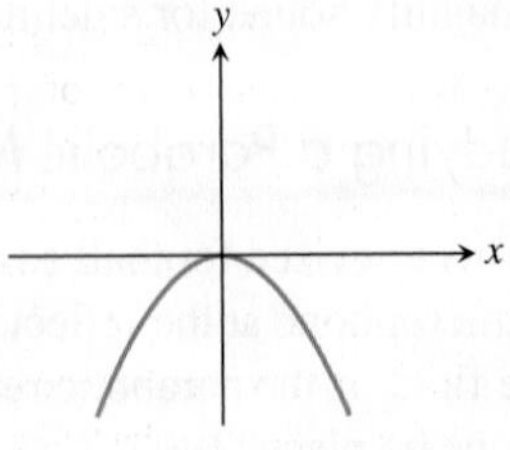

(b)

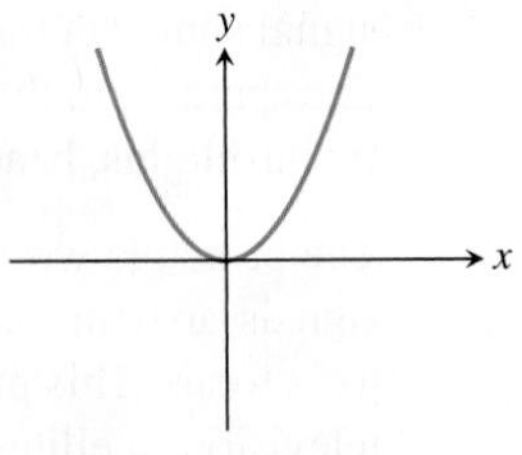

(c)

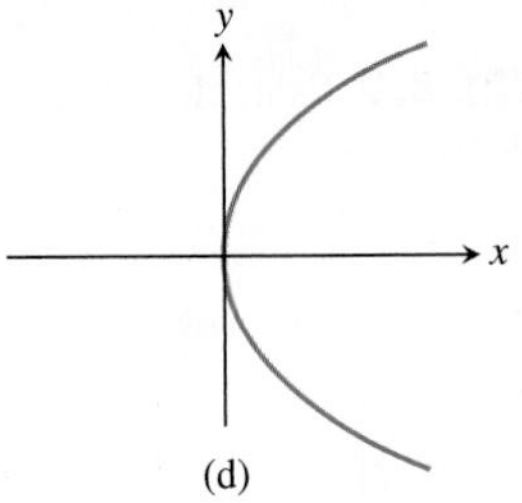

(d)

7. $x^2 = 3y$ **8.** $x^2 = -4y$

9. $y^2 = -5x$ **10.** $y^2 = 10x$

In Exercises 11–30, find an equation in standard form for the parabola that satisfies the given conditions.

11. Vertex $(0, 0)$, focus $(-3, 0)$

12. Vertex $(0, 0)$, focus $(0, 2)$

13. Vertex $(0, 0)$, directrix $y = 4$

14. Vertex $(0, 0)$, directrix $x = -2$

15. Focus $(0, 5)$, directrix $y = -5$

16. Focus $(-4, 0)$, directrix $x = 4$

17. Vertex $(0, 0)$, opens to the right, focal width $= 8$

18. Vertex $(0, 0)$, opens to the left, focal width $= 12$

19. Vertex $(0, 0)$, opens downward, focal width $= 6$

20. Vertex $(0, 0)$, opens upward, focal width $= 3$

21. Focus $(-2, -4)$, vertex $(-4, -4)$

22. Focus $(-5, 3)$, vertex $(-5, 6)$

23. Focus $(3, 4)$, directrix $y = 1$

24. Focus $(2, -3)$, directrix $x = 5$

25. Vertex $(4, 3)$, directrix $x = 6$

26. Vertex $(3, 5)$, directrix $y = 7$

27. Vertex $(2, -1)$, opens upward, focal width $= 16$

28. Vertex $(-3, 3)$, opens downward, focal width $= 20$

29. Vertex $(-1, -4)$, opens to the left, focal width $= 10$

30. Vertex $(2, 3)$, opens to the right, focal width $= 5$

In Exercises 31–36, sketch the graph of the parabola by hand.

31. $y^2 = -4x$ **32.** $x^2 = 8y$

33. $(x + 4)^2 = -12(y + 1)$

34. $(y + 2)^2 = -16(x + 3)$

35. $(y - 1)^2 = 8(x + 3)$

36. $(x - 5)^2 = 20(y + 2)$

In Exercises 37–48, graph the parabola using a function grapher.

37. $y = 4x^2$ **38.** $y = -\frac{1}{6}x^2$

39. $x = -8y^2$ **40.** $x = 2y^2$

41. $12(y + 1) = (x - 3)^2$ **42.** $6(y - 3) = (x + 1)^2$

43. $2 - y = 16(x - 3)^2$ **44.** $(x + 4)^2 = -6(y - 1)$

45. $(y + 3)^2 = 12(x - 2)$ **46.** $(y - 1)^2 = -4(x + 5)$

47. $(y + 2)^2 = -8(x + 1)$ **48.** $(y - 6)^2 = 16(x - 4)$

In Exercises 49–52, prove that the graph of the equation is a parabola, and find its vertex, focus, and directrix.

49. $x^2 + 2x - y + 3 = 0$

50. $3x^2 - 6x - 6y + 10 = 0$

51. $y^2 - 4y - 8x + 20 = 0$

52. $y^2 - 2y + 4x - 12 = 0$

In Exercises 53–56, write an equation for the parabola.

53.

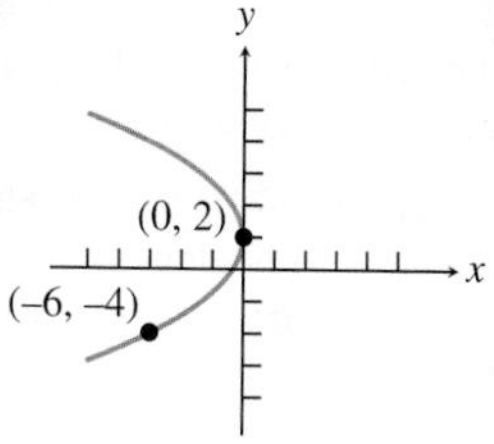

54.

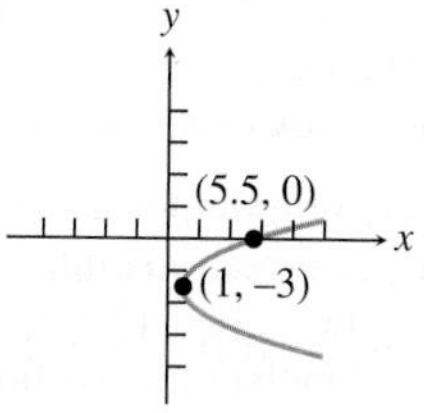

55.

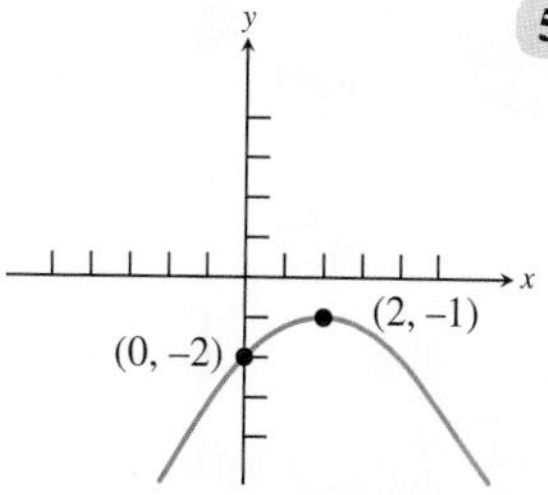

56.

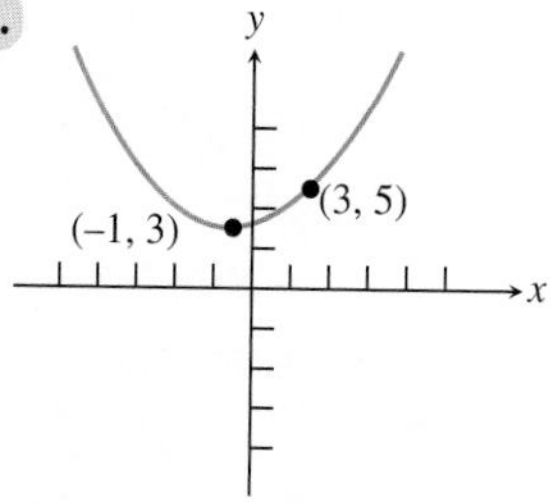

57. Writing to Learn Explain why the derivation of $x^2 = 4py$ is valid regardless of whether $p > 0$ or $p < 0$.

58. Writing to Learn Prove that an equation for the parabola with focus $(p, 0)$ and directrix $x = -p$ is $y^2 = 4px$.

59. Designing a Flashlight Mirror The mirror of a flashlight is a paraboloid of revolution. Its diameter is 6 cm and its depth is 2 cm. How far from the vertex should the filament of the lightbulb be placed for the flashlight to have its beam run parallel to the axis of its mirror?

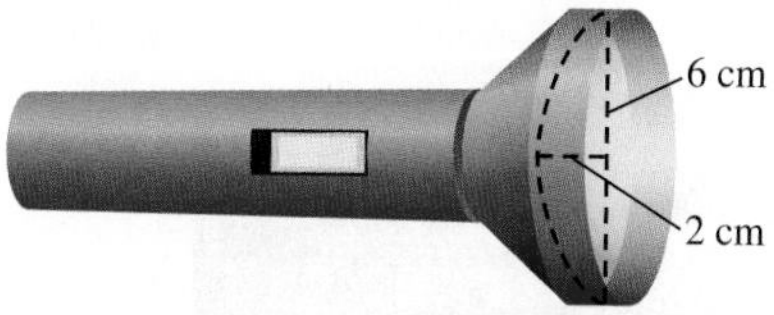

60. Designing a Satellite Dish The reflector of a television satellite dish is a paraboloid of revolution with diameter 5 ft and a depth of 2 ft. How far from the vertex should the receiving antenna be placed?

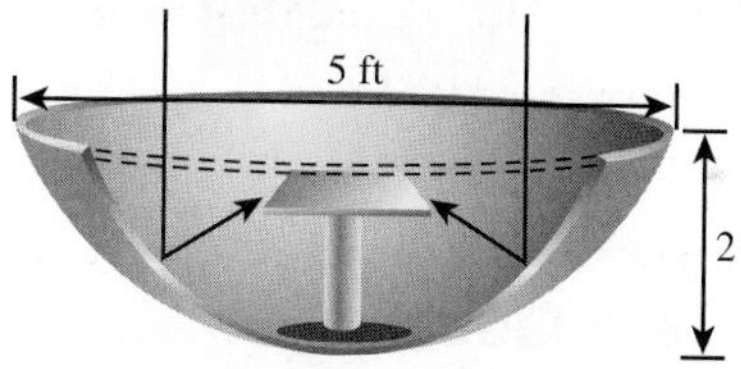

61. Parabolic Microphones The Sports Channel uses a parabolic microphone to capture all the sounds from golf tournaments throughout a season. If one of its microphones has a parabolic surface generated by the parabola $x^2 = 10y$, locate the focus (the electronic receiver) of the parabola.

62. Parabolic Headlights Stein Glass, Inc., makes parabolic headlights for a variety of automobiles. If one of its headlights has a parabolic surface generated by the parabola $x^2 = 12y$, where should its lightbulb be placed?

63. Group Activity Designing a Suspension Bridge The main cables of a suspension bridge uniformly distribute the weight of the bridge when in the form of a parabola. The main cables of a particular bridge are attached to towers that are 600 ft apart. The cables are attached to the towers at a height of 110 ft above the roadway and are 10 ft above the roadway at their lowest points. If vertical support cables are at 50-ft intervals along the level roadway, what are the lengths of these vertical cables? See art on the next page.

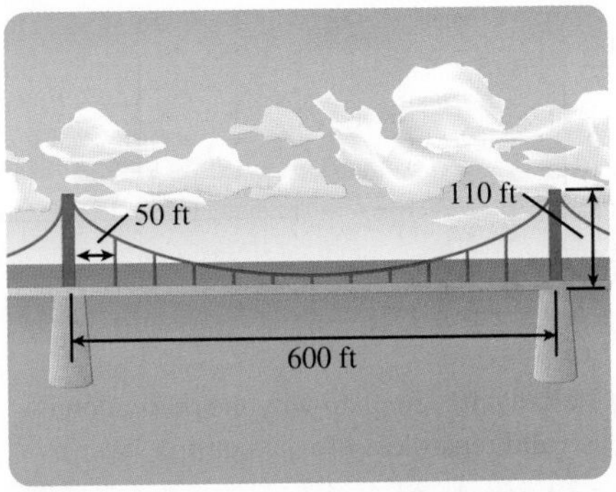

64. Group Activity Designing a Bridge Arch Parabolic arches are known to have greater strength than other arches. A bridge with a supporting parabolic arch spans 60 ft with a 30-ft-wide road passing underneath the bridge. In order to have a minimum clearance of 16 ft, what is the maximum clearance?

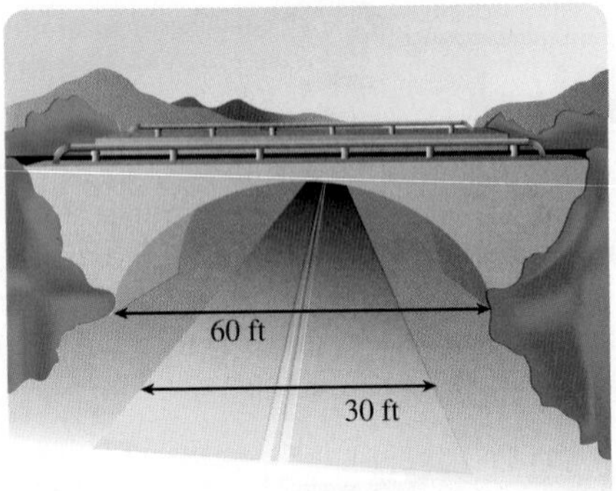

Standardized Test Questions

65. True or False Every point on a parabola is the same distance from its focus and its axis. Justify your answer.

66. True or False The directrix of a parabola is parallel to the parabola's axis. Justify your answer.

In Exercises 67–70, solve the problem without using a calculator.

67. Multiple Choice Which of the following curves is **not** a conic section?

(A) Circle

(B) Ellipse

(C) Hyperbola

(D) Oval

(E) Parabola

68. Multiple Choice Which point do all conics of the form $x^2 = 4py$ have in common?

(A) $(1, 1)$

(B) $(1, 0)$

(C) $(0, 1)$

(D) $(0, 0)$

(E) $(-1, -1)$

69. Multiple Choice The focus of $y^2 = 12x$ is

(A) $(3, 3)$.

(B) $(3, 0)$.

(C) $(0, 3)$.

(D) $(0, 0)$.

(E) $(-3, -3)$.

70. Multiple Choice The vertex of $(y - 3)^2 = -8(x + 2)$ is

(A) $(3, -2)$.

(B) $(-3, -2)$.

(C) $(-3, 2)$.

(D) $(-2, 3)$.

(E) $(-2, -3)$.

Explorations

71. Dynamically Constructing a Parabola Use a geometry software package, such as *Cabri Geometry II*™, *The Geometer's Sketchpad*®, or similar application on a handheld device, to construct a parabola geometrically from its definition. (See Figure 8.3.)

(a) Start by placing a line l (directrix) and a point F (focus) not on the line in the construction window.

(b) Construct a point A on the directrix, and then the segment AF.

(c) Construct a point P where the perpendicular bisector of AF meets the line perpendicular to l through A.

(d) What curve does P trace out as A moves?

(e) Prove that your answer to part (d) is correct.

72. Constructing Points of a Parabola Use a geometry software package, such as *Cabri Geometry II*™, *The Geometer's Sketchpad*®, or similar application on a handheld device, to construct Figure 8.4, associated with Exploration 1.

(a) Start by placing the coordinate axes in the construction window.

(b) Construct the line $y = -1$ as the directrix and the point $(0, 1)$ as the focus.

(c) Construct the horizontal lines and concentric circles shown in Figure 8.4.

(d) Construct the points where these horizontal lines and concentric circles meet.

(e) Prove these points lie on the parabola with directrix $y = -1$ and focus $(0, 1)$.

73. Degenerate Conics from a Degenerate Cone Figure 8.2 shows all the possible graph types that can result from intersecting a plane with a double-napped right circular cone. Still, there are two *more* possibilities for graphs of quadratic equations in two variables that must be considered.

(a) Consider the equation $x^2 - 2xy + y^2 - 1 = 0$. Show algebraically that this equation is equivalent to $(x - y + 1)(x - y - 1) = 0$. What two lines form the graph of the solution set?

(b) Why is the graph in (a) not covered by the cases in Figure 8.2?

(c) Consider the equation $x^2 + y^2 + 1 = 0$. What points (x, y) solve the equation? (Remember that x and y are both real numbers.)

(d) Why is the graph of the solution in (c) not covered by the cases in Figure 8.2?

(e) We may imagine an infinitely extended *cylinder* to be a degenerate "cone" with a vertex at infinity. (This kind of re-imagining is useful in non-Euclidean geometry.) Draw a cylinder and show how each of the graphs in parts (a) and (c) can be realized as the intersection of the cylinder with an appropriate plane.

Extending the Ideas

74. Tangent Lines A **tangent line** of a parabola is a line that intersects but does not cross the parabola. Prove that a line tangent to the parabola $x^2 = 4py$ at the point (a, b) crosses the y-axis at $(0, -b)$.

75. Focal Chords A **focal chord** of a parabola is a chord of the parabola that passes through the focus.

(a) Prove that the x-coordinates of the endpoints of a focal chord of $x^2 = 4py$ are $x = 2p\left(m \pm \sqrt{m^2 + 1}\right)$, where m is the slope of the focal chord.

(b) Using part (a), prove that the minimum length of a focal chord is the focal width $|4p|$.

76. Latus Rectum The focal chord of a parabola perpendicular to the axis of the parabola is the **latus rectum**, which is Latin for "right chord." Using the results from Exercises 74 and 75, prove:

(a) For a parabola the two endpoints of the latus rectum and the point of intersection of the axis and directrix are the vertices of an isosceles right triangle.

(b) The legs of this isosceles right triangle are tangent to the parabola.

8.2 Circles and Ellipses

What you'll learn about
- Transforming the Unit Circle
- Geometry of an Ellipse
- Translations of Ellipses
- Orbits and Eccentricity
- Reflective Property of an Ellipse

... and why

Ellipses are the paths of planets and comets around the Sun, or of moons around planets.

Transforming the Unit Circle

As was noted repeatedly in Chapters 4 and 5, the circle with radius 1 centered at (0, 0) has equation $x^2 + y^2 = 1$. If you stretch this circle horizontally and vertically by the same factor r and shift it h units horizontally and k units vertically, you get a circle of radius r centered at (h, k). Recall from Section 1.6 that we know exactly how these transformations will affect the algebraic equation: They will transform it to $[(x - h)/r]^2 + [(y - k)/r]^2 = 1$. This is equivalent to the "standard form" equation $(x - h)^2 + (y - k)^2 = r^2$ that we reviewed in Section P.2, where it was derived using the distance formula.

Now suppose that you stretch the unit circle horizontally by the factor a and vertically by the factor b before shifting the center to (h, k). If a and b are different numbers, the curve will be elliptical, stretched more in one direction than in the other. The equation, almost the same as for the circle, is $[(x - h)/a]^2 + [(y - k)/b]^2 = 1$, or, equivalently, $(x - h)^2/a^2 + (y - k)^2/b^2 = 1$. That is the equation of an ellipse in standard form.

This quick and simple derivation of the general formula for an ellipse neatly explains why a circle can be thought of as a special case of an ellipse. However, since our goal is to understand the *geometric* properties of ellipses, we need to start with a geometric definition, just as we did with the parabola in Section 8.1. Eventually you will arrive at the same algebraic formula, but with additional geometric understanding of a and b (not to mention a new variable, c).

Geometry of an Ellipse

When a plane intersects one nappe of a right circular cone in a simple closed curve, the curve is an ellipse. (It could also be a circle, which we will consider to be a special case of an ellipse).

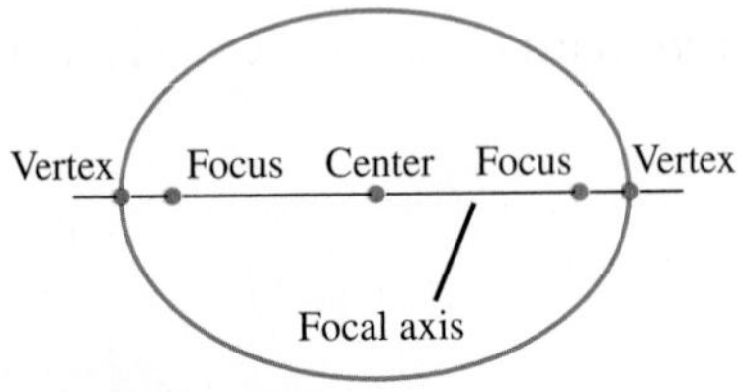

FIGURE 8.11 Key points on the focal axis of an ellipse.

DEFINITION Ellipse

An **ellipse** is the set of all points in a plane whose distances from two fixed points in the plane have a constant sum. The fixed points are the **foci** (plural of focus) of the ellipse. The line through the foci is the **focal axis**. The point on the focal axis midway between the foci is the **center**. The points where the ellipse intersects its axis are the **vertices** of the ellipse (Figure 8.11).

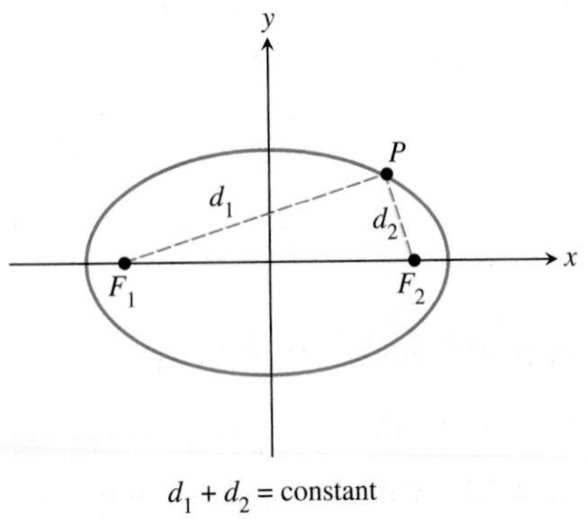

FIGURE 8.12 *Structure of an Ellipse.* The sum of the distances from the foci to each point on the ellipse is a constant.

Figure 8.12 shows a point $P(x, y)$ of an ellipse. The fixed points F_1 and F_2 are the foci of the ellipse, and the distances whose sum is constant are d_1 and d_2. We can construct an ellipse using a pencil, a loop of string, and two pushpins. Put the loop around the two pins placed at F_1 and F_2, pull the string taut with a pencil point P, and move the pencil around to trace out the ellipse (Figure 8.13).

We now use the definition to derive an equation for an ellipse. For some constants a and c with $a > c \geq 0$, let $F_1(-c, 0)$ and $F_2(c, 0)$ be the foci (Figure 8.14). Then an ellipse is defined by the set of points $P(x, y)$ such that

$$PF_1 + PF_2 = 2a.$$

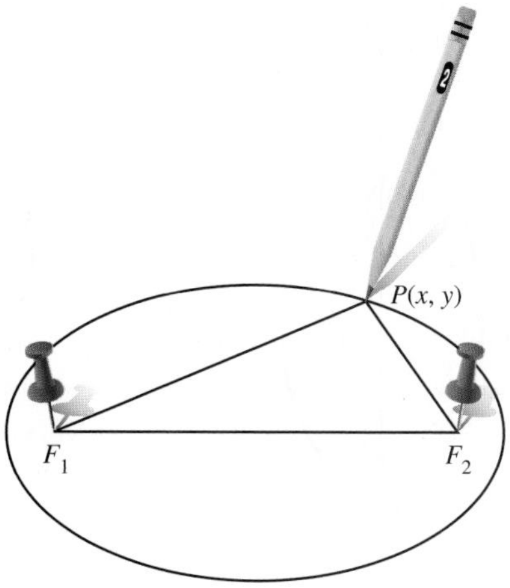

FIGURE 8.13 How to draw an ellipse.

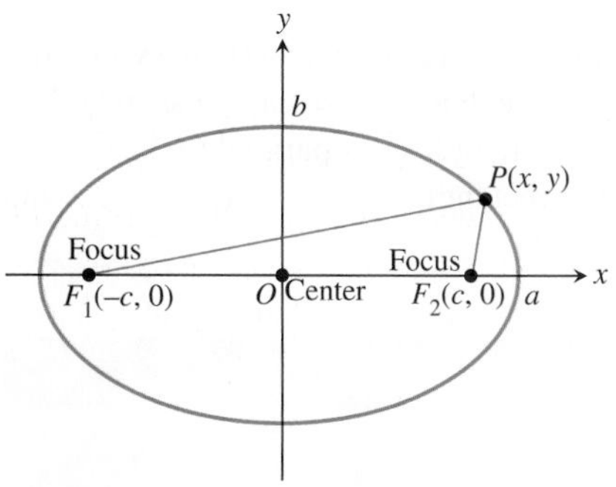

FIGURE 8.14 The ellipse defined by $PF_1 + PF_2 = 2a$ is the graph of the equation $x^2/a^2 + y^2/b^2 = 1$, where $b^2 = a^2 - c^2$.

Axis Alert

For an ellipse, the word *axis* is used in several ways. The focal axis is a *line*. The major and minor axes are *line segments*. The semimajor and semiminor axes are *numbers*.

Using the distance formula, the equation becomes

$$\sqrt{(x+c)^2+(y-0)^2}+\sqrt{(x-c)^2+(y-0)^2}=2a.$$

$$\sqrt{(x-c)^2+y^2}=2a-\sqrt{(x+c)^2+y^2} \quad \text{Rearrange terms.}$$

$$x^2-2cx+c^2+y^2=4a^2-4a\sqrt{(x+c)^2+y^2}+x^2+2cx+c^2+y^2 \quad \text{Square.}$$

$$a\sqrt{(x+c)^2+y^2}=a^2+cx \quad \text{Simplify.}$$

$$a^2(x^2+2cx+c^2+y^2)=a^4+2a^2cx+c^2x^2 \quad \text{Square.}$$

$$(a^2-c^2)x^2+a^2y^2=a^2(a^2-c^2) \quad \text{Simplify.}$$

Letting $b^2 = a^2 - c^2$, we have

$$b^2x^2+a^2y^2=a^2b^2,$$

which is usually written as

$$\frac{x^2}{a^2}+\frac{y^2}{b^2}=1.$$

Because these steps can be reversed, a point $P(x, y)$ satisfies this last equation if and only if the point lies on the ellipse defined by $PF_1 + PF_2 = 2a$, provided that $a > c \geq 0$ and $b^2 = a^2 - c^2$. The *Pythagorean relation* $b^2 = a^2 - c^2$ can be written many ways, including $c^2 = a^2 - b^2$ and $a^2 = b^2 + c^2$.

The equation $x^2/a^2 + y^2/b^2 = 1$ is the **standard form** of the equation of an ellipse centered at the origin with the x-axis as its focal axis. An ellipse centered at the origin with the y-axis as its focal axis is the *inverse* of $x^2/a^2 + y^2/b^2 = 1$, and thus has an equation of the form

$$\frac{y^2}{a^2}+\frac{x^2}{b^2}=1.$$

As with circles and parabolas, a line segment with endpoints on an ellipse is a **chord** of the ellipse. The chord lying on the focal axis is the **major axis** of the ellipse. The chord through the center perpendicular to the focal axis is the **minor axis** of the ellipse. The length of the major axis is $2a$ and of the minor axis is $2b$. The number a is the **semimajor axis**, and b is the **semiminor axis**.

Ellipses with Center (0, 0)

• **Standard equation**	$\frac{x^2}{a^2}+\frac{y^2}{b^2}=1$	$\frac{y^2}{a^2}+\frac{x^2}{b^2}=1$
• **Focal axis**	x-axis	y-axis
• **Foci**	$(\pm c, 0)$	$(0, \pm c)$
• **Vertices**	$(\pm a, 0)$	$(0, \pm a)$
• **Semimajor axis**	a	a
• **Semiminor axis**	b	b
• **Pythagorean relation**	$a^2 = b^2 + c^2$	$a^2 = b^2 + c^2$

See Figure 8.15 on the next page.

EXAMPLE 1 Finding the Vertices and Foci of an Ellipse

Find the vertices and the foci of the ellipse $4x^2 + 9y^2 = 36$.

SOLUTION Dividing both sides of the equation by 36 yields the standard form $x^2/9 + y^2/4 = 1$. Because the larger number is the denominator of x^2, the focal axis is the x-axis. So $a^2 = 9$, $b^2 = 4$, and $c^2 = a^2 - b^2 = 9 - 4 = 5$. Thus the vertices are $(\pm 3, 0)$, and the foci are $(\pm\sqrt{5}, 0)$. **Now try Exercise 1.**

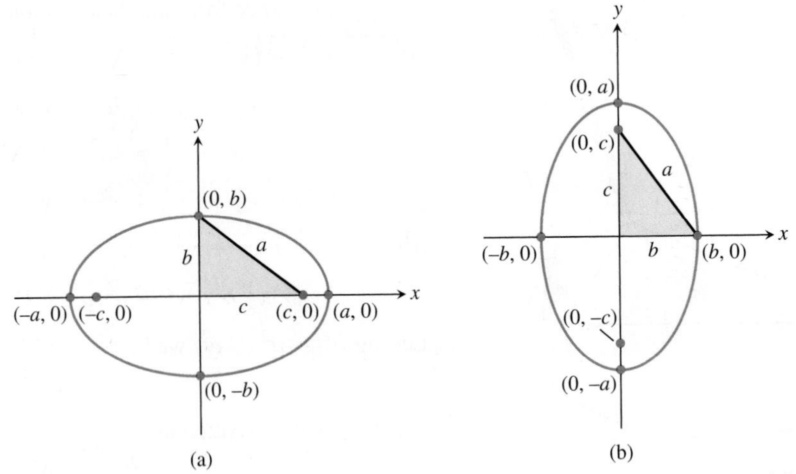

FIGURE 8.15 Ellipses centered at the origin with foci on (a) the x-axis and (b) the y-axis. In each case, a right triangle illustrating the Pythagorean relation is shown.

An ellipse centered at the origin with its focal axis on a coordinate axis is symmetric with respect to the origin and both coordinate axes. Such an ellipse can be sketched by first drawing a *guiding rectangle* centered at the origin with sides parallel to the coordinate axes and then sketching the ellipse inside the rectangle, as shown in the Drawing Lesson.

Drawing Lesson

How to Sketch the Ellipse $x^2/a^2 + y^2/b^2 = 1$

1. Sketch line segments at $x = \pm a$ and $y = \pm b$ and complete the rectangle they determine.
2. Inscribe an ellipse that is tangent to the rectangle at $(\pm a, 0)$ and $(0, \pm b)$.

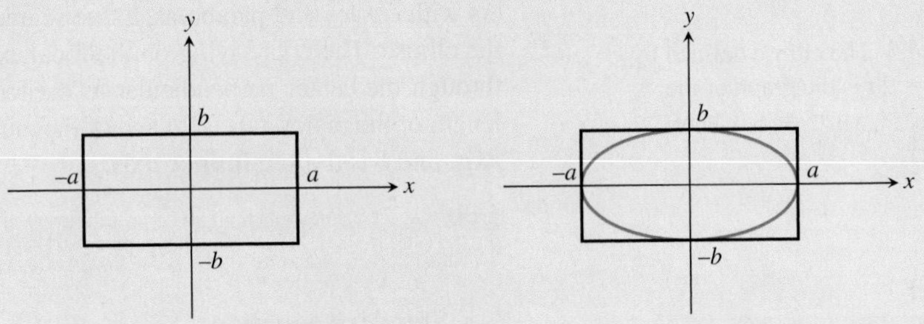

If we wish to graph an ellipse using a function grapher, we need to solve the equation of the ellipse for y, as illustrated in Example 2.

EXAMPLE 2 Finding an Equation and Graphing an Ellipse

Find an equation of the ellipse with foci $(0, -3)$ and $(0, 3)$ whose minor axis has length 4. Sketch the ellipse and support your sketch with a grapher.

SOLUTION The center is $(0, 0)$. The foci are on the y-axis with $c = 3$. The semiminor axis is $b = 4/2 = 2$. Using $a^2 = b^2 + c^2$, we have $a^2 = 2^2 + 3^2 = 13$. So the standard form of the equation for the ellipse is

$$\frac{y^2}{13} + \frac{x^2}{4} = 1.$$

Using $a = \sqrt{13} \approx 3.61$ and $b = 2$, we can sketch a guiding rectangle and then the ellipse itself, as explained in the Drawing Lesson. (Try doing this.) To graph the ellipse using a function grapher, we solve for y in terms of x.

That 3:2 Ratio Again

Notice that we chose square viewing windows in Figure 8.16. A nonsquare window would give a distorted view of an ellipse.

$$\frac{y^2}{13} = 1 - \frac{x^2}{4}$$

$$y^2 = 13(1 - x^2/4)$$

$$y = \pm\sqrt{13(1 - x^2/4)}$$

Figure 8.16 shows three views of the graphs of

$$\text{Y1} = \sqrt{13(1 - x^2/4)} \quad \text{and} \quad \text{Y2} = -\sqrt{13(1 - x^2/4)}.$$

We must select the viewing window carefully to avoid grapher failure.

Now try Exercise 17.

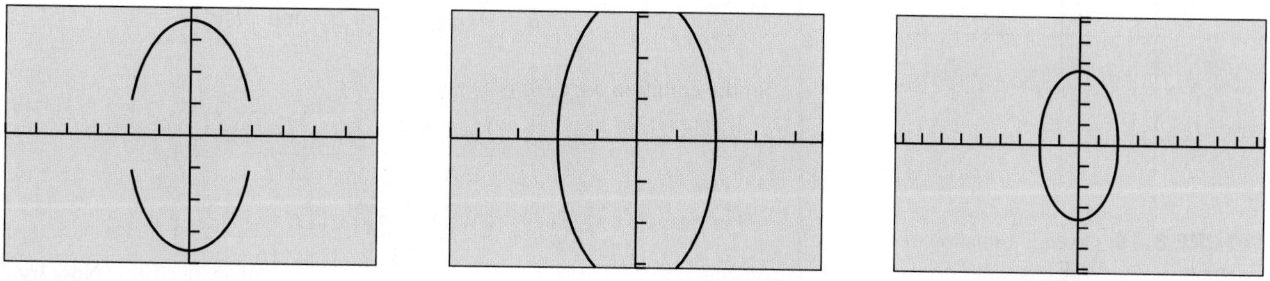

[–6, 6] by [–4, 4]
(a)

[–4.7, 4.7] by [–3.1, 3.1]
(b)

[–9.4, 9.4] by [–6.2, 6.2]
(c)

FIGURE 8.16 Three views of the ellipse $y^2/13 + x^2/4 = 1$. All of the windows are square or approximately square viewing windows so we can see the true shape. Notice that the gaps between the upper and lower function branches do not occur when the grapher window includes columns of pixels whose x-coordinates are ± 2 as in (b) and (c). (Example 2)

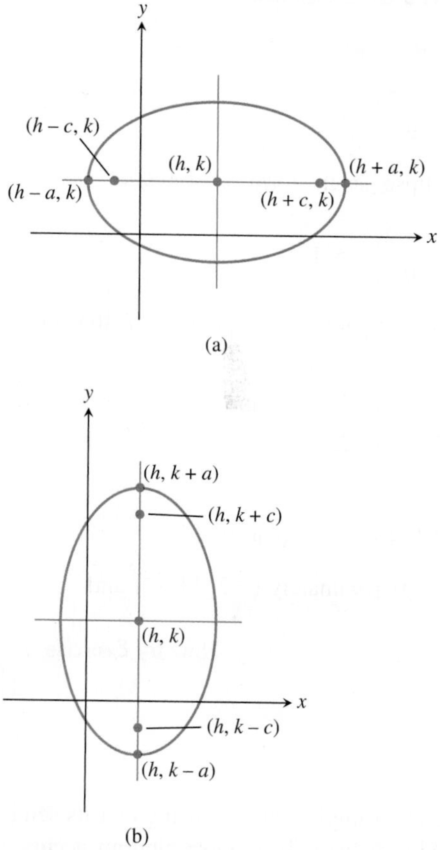

FIGURE 8.17 Ellipses with center (h, k) and foci on (a) $y = k$ and (b) $x = h$.

Translations of Ellipses

When an ellipse with center $(0, 0)$ is translated horizontally by h units and vertically by k units, the center of the ellipse moves from $(0, 0)$ to (h, k), as shown in Figure 8.17. Such a translation does not change the length of the major or minor axis or the Pythagorean relation.

Ellipses with Center (h, k)

• **Standard equation**	$\frac{(x-h)^2}{a^2} + \frac{(y-k)^2}{b^2} = 1$	$\frac{(y-k)^2}{a^2} + \frac{(x-h)^2}{b^2} = 1$
• **Focal axis**	$y = k$	$x = h$
• **Foci**	$(h \pm c, k)$	$(h, k \pm c)$
• **Vertices**	$(h \pm a, k)$	$(h, k \pm a)$
• **Semimajor axis**	a	a
• **Semiminor axis**	b	b
• **Pythagorean relation**	$a^2 = b^2 + c^2$	$a^2 = b^2 + c^2$

See Figure 8.17.

EXAMPLE 3 Finding an Equation of an Ellipse

Find the standard form of the equation for the ellipse whose major axis has endpoints $(-2, -1)$ and $(8, -1)$, and whose minor axis has length 8.

SOLUTION Figure 8.18 shows the major axis endpoints, the minor axis, and the center of the ellipse. The standard equation of this ellipse has the form

$$\frac{(x-h)^2}{a^2} + \frac{(y-k)^2}{b^2} = 1,$$

where the center (h, k) is the midpoint $(3, -1)$ of the major axis. The semimajor axis and semiminor axis are

$$a = \frac{8-(-2)}{2} = 5 \quad \text{and} \quad b = \frac{8}{2} = 4.$$

So the equation we seek is

$$\frac{(x-3)^2}{5^2} + \frac{(y-(-1))^2}{4^2} = 1,$$

$$\frac{(x-3)^2}{25} + \frac{(y+1)^2}{16} = 1.$$

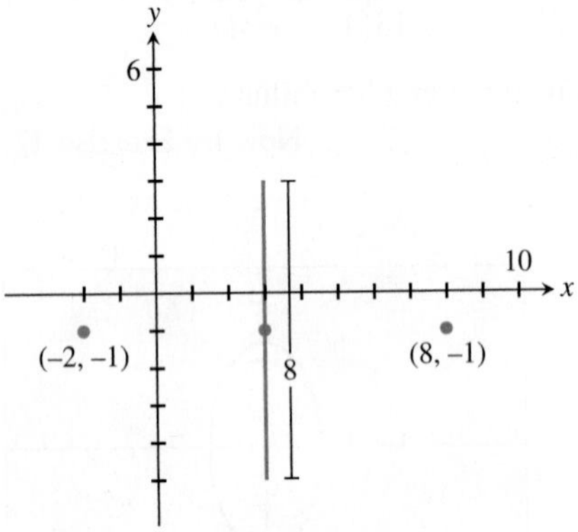

FIGURE 8.18 Given information for Example 3.

Now try Exercise 31.

EXAMPLE 4 Locating Key Points of an Ellipse

Find the center, vertices, and foci of the ellipse

$$\frac{(x+2)^2}{9} + \frac{(y-5)^2}{49} = 1.$$

SOLUTION The standard equation of this ellipse has the form

$$\frac{(y-5)^2}{49} + \frac{(x+2)^2}{9} = 1.$$

The center (h, k) is $(-2, 5)$. Because the semimajor axis $a = \sqrt{49} = 7$, the vertices $(h, k \pm a)$ are

$$(h, k+a) = (-2, 5+7) = (-2, 12) \quad \text{and}$$
$$(h, k-a) = (-2, 5-7) = (-2, -2).$$

Because

$$c = \sqrt{a^2 - b^2} = \sqrt{49-9} = \sqrt{40}$$

the foci $(h, k \pm c)$ are $\left(-2, 5 \pm \sqrt{40}\right)$, or approximately $(-2, 11.32)$ and $(-2, -1.32)$.

Now try Exercise 37.

With the information found about the ellipse in Example 4 and knowing that its semiminor axis $b = \sqrt{9} = 3$, we could easily sketch the ellipse. Obtaining an accurate graph of the ellipse using a function grapher is another matter. Generally, the best way to graph an ellipse using a graphing utility is to use parametric equations.

EXPLORATION 1 Graphing an Ellipse Using Its Parametric Equations

1. Use the Pythagorean trigonometry identity $\cos^2 t + \sin^2 t = 1$ to prove that the *parametrization* $x = -2 + 3 \cos t$, $y = 5 + 7 \sin t$, $0 \le t \le 2\pi$ will produce a graph of the ellipse $(x + 2)^2/9 + (y - 5)^2/49 = 1$.
2. Graph $x = -2 + 3 \cos t$, $y = 5 + 7 \sin t$, $0 \le t \le 2\pi$ in a square viewing window to support part 1 graphically.
3. Create parametrizations for the ellipses in Examples 1, 2, and 3.
4. Graph each of your parametrizations in part 3 and check the features of the obtained graph to see whether they match the expected geometric features of the ellipse. Revise your parametrization and regraph until all features match.
5. Prove that each of your parametrizations is valid.

Orbits and Eccentricity

Kepler's First Law of Planetary Motion, published in 1609, states that the path of a planet's orbit is an ellipse with the Sun at one of the foci. Asteroids, comets, and other bodies that orbit the Sun follow elliptical paths. The closest point to the Sun in such an orbit is the *perihelion*, and the farthest point is the *aphelion* (Figure 8.19). The shape of an ellipse is related to its *eccentricity*.

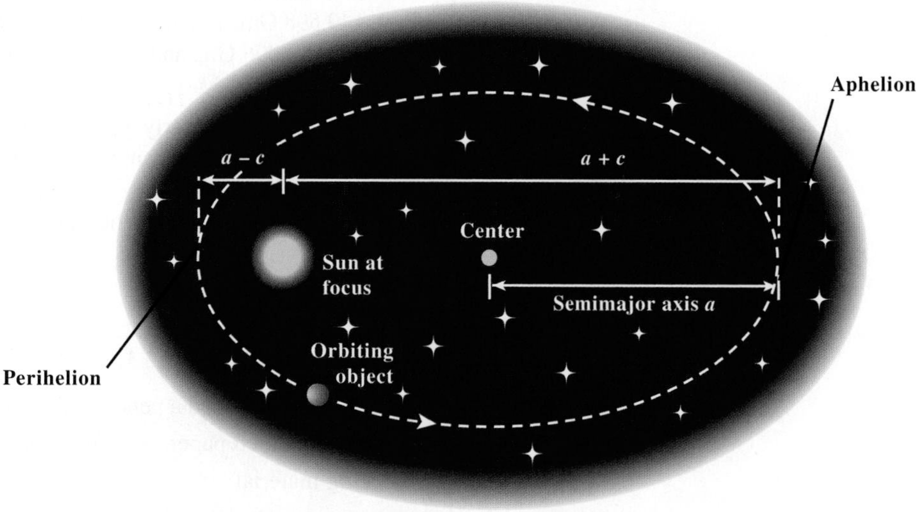

FIGURE 8.19 Many celestial objects have elliptical orbits around the Sun.

A New e

Try not to confuse the eccentricity e with the natural base e used in exponential and logarithmic functions. The context should clarify which meaning is intended.

DEFINITION Eccentricity of an Ellipse

The **eccentricity** of an ellipse is

$$e = \frac{c}{a} = \frac{\sqrt{a^2 - b^2}}{a},$$

where a is the semimajor axis, b is the semiminor axis, and c is the distance from the center of the ellipse to either focus.

The noun *eccentricity* comes from the adjective *eccentric*, which means off-center. Mathematically, the eccentricity is the ratio of c to a. The larger c is, compared to a, the more off-center the foci are.

In any ellipse, $a > c \geq 0$. Dividing this inequality by a shows that $0 \leq e < 1$. So the eccentricity of an ellipse is between 0 and 1.

Ellipses with highly off-center foci are elongated and have eccentricities close to 1; for example, the orbit of Halley's comet has eccentricity $e \approx 0.97$. Ellipses with foci near the center are almost circular and have eccentricities close to 0; for instance, Venus's orbit has an eccentricity of 0.0068.

What happens when the eccentricity $e = 0$? In an ellipse, because a is positive, $e = c/a = 0$ implies that $c = 0$ and thus $a = b$. In this case, the ellipse *degenerates* into a circle. Because the ellipse is a circle when $a = b$, it is customary to denote this common value as r and call it the radius of the circle.

Surprising things happen when an ellipse is nearly but not quite a circle, as in the orbit of our planet, Earth.

EXAMPLE 5 Analyzing Earth's Orbit

Earth's orbit has a semimajor axis $a \approx 149.598$ Gm (gigameters) and an eccentricity of $e \approx 0.0167$. Calculate and interpret b and c.

SOLUTION Because $e = c/a$, $c = ea \approx 0.0167 \times 149.598 = 2.4982866$ and

$$b = \sqrt{a^2 - c^2} \approx \sqrt{149.598^2 - 2.4982866^2} \approx 149.577.$$

The semiminor axis $b \approx 149.577$ Gm is only 0.014% shorter than the semimajor axis $a \approx 149.598$ Gm. The aphelion distance of Earth from the Sun is $a + c \approx 149.598 + 2.498 = 152.096$ Gm, and the perihelion distance is $a - c \approx 149.598 - 2.498 = 147.100$ Gm.

Thus Earth's orbit is nearly a perfect circle, but the distance between the center of the Sun at one focus and the center of Earth's orbit is $c \approx 2.498$ Gm, more than 2 orders of magnitude greater than $a - b$. The eccentricity as a percentage is 1.67%; this measures how far *off-center* the Sun is. **Now try Exercise 53.**

EXPLORATION 2 Constructing Ellipses to Understand Eccentricity

Each group will need a pencil, a centimeter ruler, scissors, some string, several sheets of unlined paper, two pushpins, and a foam board or other appropriate backing material.

1. Make a closed *loop* of string that is 20 cm in circumference.
2. Place a sheet of unlined paper on the backing material, and carefully place the two pushpins 2 cm apart near the center of the paper. Construct an ellipse using the loop of string and a pencil as shown in Figure 8.13. Measure and record the resulting values of a, b, and c for the ellipse, and compute the ratios $e = c/a$ and b/a for the ellipse.
3. On separate sheets of paper repeat step 2 three more times, placing the pushpins 4, 6, and 8 cm apart. Record the values of a, b, c and the ratios e and b/a for each ellipse.
4. Write your observations about the ratio b/a as the eccentricity ratio e increases. Which of these two ratios measures the shape of the ellipse? Which measures how off-center the foci are?
5. Plot the ordered pairs $(e, b/a)$, determine a formula for the ratio b/a as a function of the eccentricity e, and overlay this function's graph on the scatter plot.

Whispering Galleries

In architecture, ceilings in the shape of an ellipsoid are used to create *whispering galleries*. A person whispering at one focus can be heard across the room by a person at the other focus. An ellipsoid is part of the design of the Texas state capitol; a hand clap made in the center of the main vestibule (at one focus of the ellipsoid) bounces off the inner elliptical dome, passes through the other focus, bounces off the dome a second time, and returns to the person as a distinct echo.

Reflective Property of an Ellipse

Because of their shape, ellipses are used to make reflectors of sound, light, and other waves. If we rotate an ellipse in three-dimensional space about its focal axis, the ellipse sweeps out an **ellipsoid of revolution**. If we place a signal source at one focus of a reflective ellipsoid, the signal reflects off the elliptical surface to the other focus, as illustrated in Figure 8.20. This property is used to make mirrors for optical equipment and to study aircraft noise in wind tunnels.

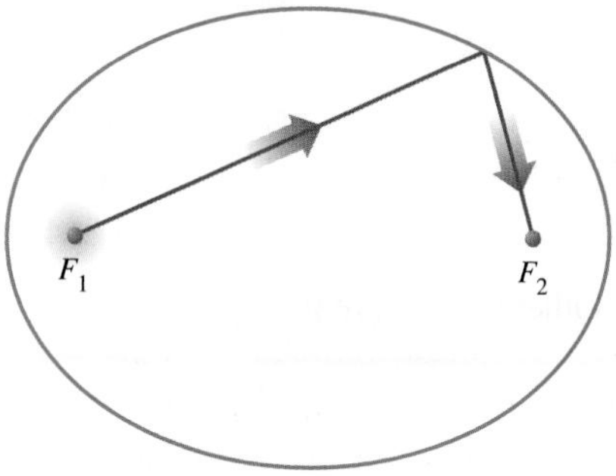

FIGURE 8.20 The reflective property of an ellipse.

Ellipsoids are used in health care to avoid surgery in the treatment of kidney stones. An elliptical *lithotripter* emits underwater ultrahigh-frequency (UHF) shock waves from one focus, with the patient's kidney carefully positioned at the other focus (Figure 8.21).

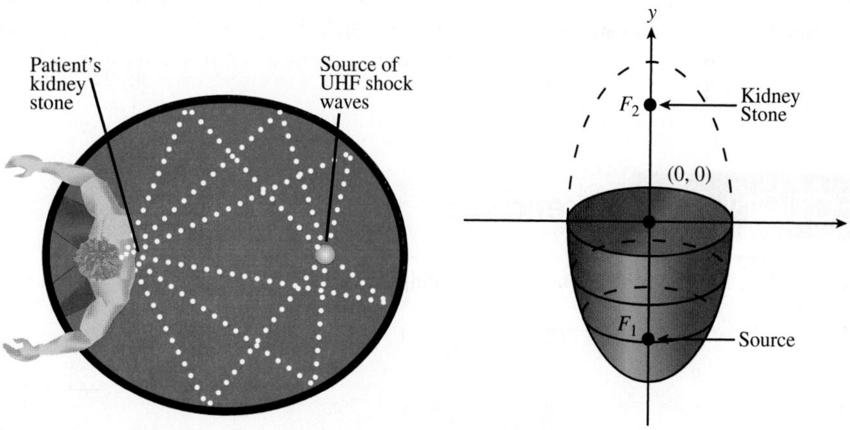

FIGURE 8.21 How a lithotripter breaks up kidney stones.

EXAMPLE 6 Focusing a Lithotripter

The ellipse used to generate the ellipsoid of a lithotripter has a major axis of 12 ft and a minor axis of 5 ft. How far from the center are the foci?

SOLUTION From the given information, we know $a = 12/2 = 6$ and $b = 5/2 = 2.5$. So

$$c = \sqrt{a^2 - b^2} \approx \sqrt{6^2 - 2.5^2} \approx 5.4544.$$

The foci are about 5 ft 5.5 in. from the center of the lithotripter.

Now try Exercise 59.

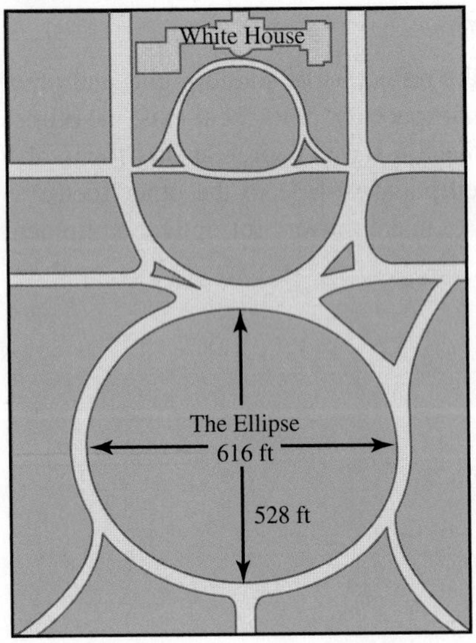

CHAPTER OPENER Problem (from page 570)

Problem: If the Ellipse at the White House is 616 ft long and 528 ft wide, what is its equation?

Solution: For simplicity's sake, we model the Ellipse as centered at (0, 0) with the x-axis as its focal axis. Because the Ellipse is 616 ft long, $a = 616/2 = 308$, and because the Ellipse is 528 ft wide, $b = 528/2 = 264$. Using $x^2/a^2 + y^2/b^2 = 1$, we obtain

$$\frac{x^2}{308^2} + \frac{y^2}{264^2} = 1,$$

$$\frac{x^2}{94{,}864} + \frac{y^2}{69{,}696} = 1.$$

Other models are possible.

QUICK REVIEW 8.2 *(For help, go to Sections P.2 and P.5.)*

Exercise numbers with a gray background indicate problems that the authors have designed to be solved *without a calculator.*

In Exercises 1 and 2, find the distance between the given points.

1. $(-3, -2)$ and $(2, 4)$
2. $(-3, -4)$ and (a, b)

In Exercises 3 and 4, solve for y in terms of x.

3. $\dfrac{y^2}{9} + \dfrac{x^2}{4} = 1$
4. $\dfrac{x^2}{36} + \dfrac{y^2}{25} = 1$

In Exercises 5–8, solve for x algebraically.

5. $\sqrt{3x + 12} + \sqrt{3x - 8} = 10$
6. $\sqrt{6x + 12} - \sqrt{4x + 9} = 1$
7. $\sqrt{6x^2 + 12} + \sqrt{6x^2 + 1} = 11$
8. $\sqrt{2x^2 + 8} + \sqrt{3x^2 + 4} = 8$

In Exercises 9 and 10, find exact solutions by completing the square.

9. $2x^2 - 6x - 3 = 0$
10. $2x^2 + 4x - 5 = 0$

SECTION 8.2 Exercises

In Exercises 1–6, find the vertices and foci of the ellipse.

1. $\dfrac{x^2}{16} + \dfrac{y^2}{7} = 1$
2. $\dfrac{y^2}{25} + \dfrac{x^2}{21} = 1$
3. $\dfrac{y^2}{36} + \dfrac{x^2}{27} = 1$
4. $\dfrac{x^2}{11} + \dfrac{y^2}{7} = 1$
5. $3x^2 + 4y^2 = 12$
6. $9x^2 + 4y^2 = 36$

In Exercises 7–10, match the graph with its equation, given that the tick marks on all axes are 1 unit apart.

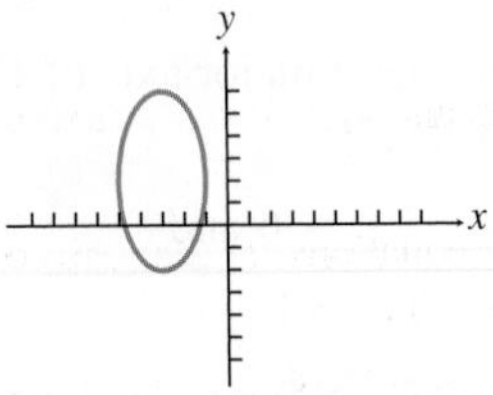

(a)

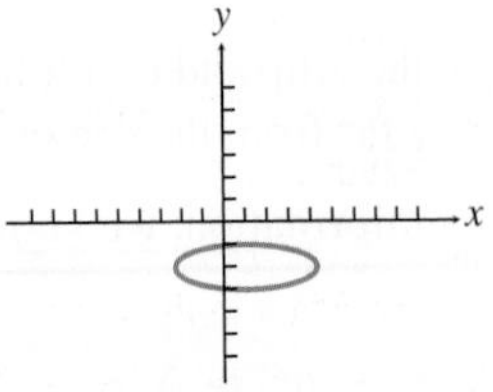

(b)

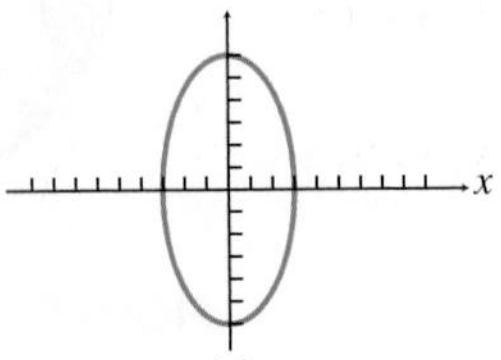

(c)

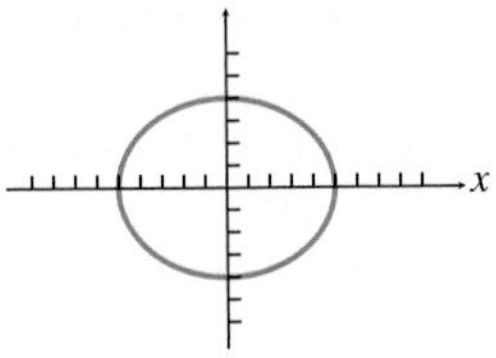

(d)

7. $\dfrac{x^2}{25} + \dfrac{y^2}{16} = 1$
8. $\dfrac{y^2}{36} + \dfrac{x^2}{9} = 1$
9. $\dfrac{(y-2)^2}{16} + \dfrac{(x+3)^2}{4} = 1$
10. $\dfrac{(x-1)^2}{11} + (y+2)^2 = 1$

In Exercises 11–16, sketch the graph of the ellipse by hand.

11. $\dfrac{x^2}{64} + \dfrac{y^2}{36} = 1$
12. $\dfrac{x^2}{81} + \dfrac{y^2}{25} = 1$

13. $\dfrac{y^2}{9} + \dfrac{x^2}{4} = 1$ **14.** $\dfrac{y^2}{49} + \dfrac{x^2}{25} = 1$

15. $\dfrac{(x+3)^2}{16} + \dfrac{(y-1)^2}{4} = 1$ **16.** $\dfrac{(x-1)^2}{2} + \dfrac{(y+3)^2}{4} = 1$

In Exercises 17–20, graph the ellipse using a function grapher.

17. $\dfrac{x^2}{36} + \dfrac{y^2}{16} = 1$ **18.** $\dfrac{y^2}{64} + \dfrac{x^2}{16} = 1$

19. $\dfrac{(x+2)^2}{5} + 2(y-1)^2 = 1$

20. $\dfrac{(x-4)^2}{16} + 16(y+4)^2 = 8$

In Exercises 21–36, find an equation in standard form for the ellipse that satisfies the given conditions.

21. Major axis length 6 on y-axis, minor axis length 4

22. Major axis length 14 on x-axis, minor axis length 10

23. Foci $(\pm 2, 0)$, major axis length 10

24. Foci $(0, \pm 3)$, major axis length 10

25. Endpoints of axes are $(\pm 4, 0)$ and $(0, \pm 5)$

26. Endpoints of axes are $(\pm 7, 0)$ and $(0, \pm 4)$

27. Major axis endpoints $(0, \pm 6)$, minor axis length 8

28. Major axis endpoints $(\pm 5, 0)$, minor axis length 4

29. Minor axis endpoints $(0, \pm 4)$, major axis length 10

30. Minor axis endpoints $(\pm 12, 0)$, major axis length 26

31. Major axis endpoints $(1, -4)$ and $(1, 8)$, minor axis length 8

32. Major axis endpoints $(-2, -3)$ and $(-2, 7)$, minor axis length 4

33. Foci $(1, -4)$ and $(5, -4)$, major axis endpoints $(0, -4)$ and $(6, -4)$

34. Foci $(-2, 1)$ and $(-2, 5)$, major axis endpoints $(-2, -1)$ and $(-2, 7)$

35. Major axis endpoints $(3, -7)$ and $(3, 3)$, minor axis length 6

36. Major axis endpoints $(-5, 2)$ and $(3, 2)$, minor axis length 6

In Exercises 37–40, find the center, vertices, and foci of the ellipse.

37. $\dfrac{(x+1)^2}{25} + \dfrac{(y-2)^2}{16} = 1$

38. $\dfrac{(x-3)^2}{11} + \dfrac{(y-5)^2}{7} = 1$

39. $\dfrac{(y+3)^2}{81} + \dfrac{(x-7)^2}{64} = 1$

40. $\dfrac{(y-1)^2}{25} + \dfrac{(x+2)^2}{16} = 1$

In Exercises 41–44, graph the ellipse using a parametric grapher.

41. $\dfrac{y^2}{25} + \dfrac{x^2}{4} = 1$ **42.** $\dfrac{x^2}{30} + \dfrac{y^2}{20} = 1$

43. $\dfrac{(x+3)^2}{12} + \dfrac{(y-6)^2}{5} = 1$ **44.** $\dfrac{(y+1)^2}{15} + \dfrac{(x-2)^2}{6} = 1$

In Exercises 45–48, prove that the graph of the equation is an ellipse, and find its vertices, foci, and eccentricity.

45. $9x^2 + 4y^2 - 18x + 8y - 23 = 0$

46. $3x^2 + 5y^2 - 12x + 30y + 42 = 0$

47. $9x^2 + 16y^2 + 54x - 32y - 47 = 0$

48. $4x^2 + y^2 - 32x + 16y + 124 = 0$

In Exercises 49 and 50, write an equation for the ellipse.

49.

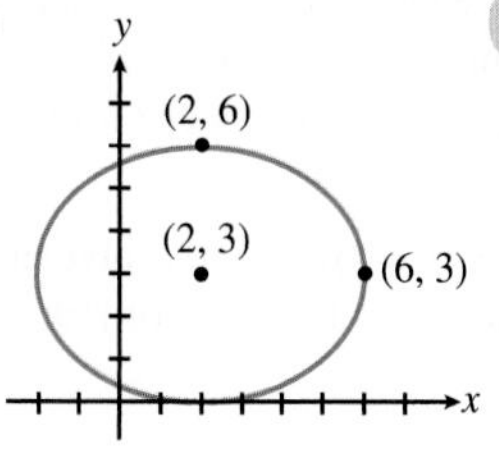

50.

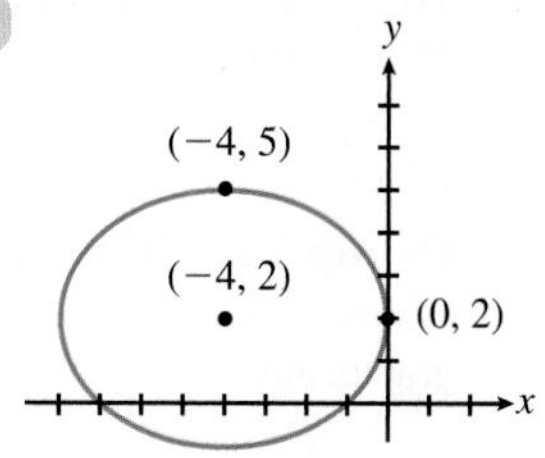

51. Writing to Learn Prove that an equation for the ellipse with center $(0, 0)$, foci $(0, \pm c)$, and semimajor axis $a > c \geq 0$ is $y^2/a^2 + x^2/b^2 = 1$, where $b^2 = a^2 - c^2$. (*Hint*: Refer to derivation at the beginning of the section.)

52. Writing to Learn Dancing Among the Stars Using the data in Table 8.1, prove that the object with the most eccentric orbit sometimes is closer to the Sun than the planet with the least eccentric orbit.

Table 8.1 Semimajor Axes and Eccentricities of the Planets and Pluto

Object	Semimajor Axis (Gm)	Eccentricity
Mercury	57.9	0.2056
Venus	108.2	0.0068
Earth	149.6	0.0167
Mars	227.9	0.0934
Jupiter	778.3	0.0485
Saturn	1427	0.0560
Uranus	2869	0.0461
Neptune	4497	0.0050
Pluto	5900	0.2484

Source: Shupe et al., National Geographic Atlas of the World (rev.6th ed.). Washington, DC: National Geographic Society, 1992, plate 116, and other sources.

53. The Moon's Orbit The Moon's apogee (farthest distance from Earth) is 252,710 mi, and perigee (closest distance to Earth) is 221,463 mi. Assuming the Moon's orbit of Earth is elliptical with Earth at one focus, calculate and interpret a, b, c, and e.

54. Hot Mercury Given that the diameter of the Sun is about 1.392 Gm, how close does Mercury get to the Sun's surface?

55. Saturn Find the perihelion and aphelion distances of Saturn.

56. Venus and Mars Write equations for the orbits of Venus and Mars in the form $x^2/a^2 + y^2/b^2 = 1$.

57. Sungrazers One comet group, known as the sungrazers, passes within a Sun's diameter (1.392 Gm) of the solar surface. What can you conclude about $a - c$ for orbits of the sungrazers?

58. Halley's Comet The orbit of Halley's comet is 36.18 AU long and 9.12 AU wide. What is its eccentricity?

59. Lithotripter For an ellipse that generates the ellipsoid of a lithotripter, the major axis has endpoints $(-8, 0)$ and $(8, 0)$. One endpoint of the minor axis is $(0, 3.5)$. Find the coordinates of the foci.

60. Lithotripter (Refer to Figure 8.21.) A lithotripter's shape is formed by rotating the portion of an ellipse below its minor axis about its major axis. If the length of the major axis is 26 in. and the length of the minor axis is 10 in., where should the shock-wave source and the patient be placed for maximum effect?

Group Activities In Exercises 61 and 62, solve the system of equations algebraically and support your answer graphically.

61. $\dfrac{x^2}{4} + \dfrac{y^2}{9} = 1$
$x^2 + y^2 = 4$

62. $\dfrac{x^2}{9} + y^2 = 1$
$x - 3y = -3$

63. Group Activity Consider the system of equations

$x^2 + 4y^2 = 4$
$y = 2x^2 - 3$

(a) Solve the system graphically.

(b) If you have access to a grapher that also does symbolic algebra, use it to find the exact solutions to the system.

64. Writing to Learn Look up the adjective *eccentric* in a dictionary and read its various definitions. Notice that the word is derived from *ex-centric*, meaning "out-of-center" or "off-center." Explain how this is related to the word's everyday meanings as well as its mathematical meaning for ellipses.

Standardized Test Questions

65. True or False The distance from a focus of an ellipse to the closer vertex is $a(1 + e)$, where a is the semimajor axis and e is the eccentricity. Justify your answer.

66. True or False The distance from a focus of an ellipse to either endpoint of the minor axis is half the length of the major axis. Justify your answer.

In Exercises 67–70, you may use a graphing calculator to solve the problem.

67. Multiple Choice One focus of $x^2 + 4y^2 = 4$ is

(A) $(4, 0)$. **(B)** $(2, 0)$. **(C)** $(\sqrt{3}, 0)$.
(D) $(\sqrt{2}, 0)$. **(E)** $(1, 0)$.

68. Multiple Choice The focal axis of

$$\frac{(x-2)^2}{25} + \frac{(y-3)^2}{16} = 1 \text{ is}$$

(A) $y = 1$. **(B)** $y = 2$. **(C)** $y = 3$.
(D) $y = 4$. **(E)** $y = 5$.

69. Multiple Choice The center of $9x^2 + 4y^2 - 72x - 24y + 144 = 0$ is

(A) $(4, 2)$. **(B)** $(4, 3)$. **(C)** $(4, 4)$.
(D) $(4, 5)$. **(E)** $(4, 6)$.

70. Multiple Choice The perimeter of a triangle with one vertex on the ellipse $x^2/a^2 + y^2/b^2 = 1$ and the other two vertices at the foci of the ellipse would be

(A) $a + b$. **(B)** $2a + 2b$. **(C)** $2a + 2c$.
(D) $2b + 2c$. **(E)** $a + b + c$.

Explorations

71. Area and Perimeter The area of an ellipse is $A = \pi ab$, but the perimeter cannot be expressed so simply:

$$P \approx \pi(a+b)\left(3 - \frac{\sqrt{(3a+b)(a+3b)}}{a+b}\right)$$

(a) Prove that, when $a = b = r$, these become the familiar formulas for the area and perimeter (circumference) of a circle.

(b) Find a pair of ellipses such that the one with greater area has smaller perimeter.

72. Writing to Learn Kepler's Laws We have encountered Kepler's first and third laws (pages 179, 587). Using a library or the Internet,

(a) Read about Kepler's life, and write in your own words how he came to discover his three laws of planetary motion.

(b) What is Kepler's Second Law? Explain it with both pictures and words.

73. Pendulum Velocity vs. Position As a pendulum swings toward and away from a motion detector, its distance (in meters) from the detector is given by the position function $x(t) = 3 + \cos(2t - 5)$, where t represents time (in seconds). The velocity (in m/sec) of the pendulum is given by $y(t) = -2\sin(2t - 5)$.

(a) Using parametric mode on your grapher, plot the (x, y) relation for velocity versus position for $0 \le t \le 2\pi$.

(b) Write the equation of the resulting conic in standard form, in terms of x and y, and eliminating the parameter t.

74. Pendulum Velocity vs. Position A pendulum that swings toward and away from a motion detector has a distance (in feet) from the detector of $x(t) = 5 + 3\sin(\pi t + \pi/2)$ and a velocity (in ft/sec) of $y(t) = 3\pi\cos(\pi t + \pi/2)$, where t represents time (in seconds).

(a) Prove that the plot of velocity versus position (distance) is an ellipse.

(b) Writing to Learn Describe the motion of the pendulum.

Extending the Ideas

75. Prove that a nondegenerate graph of the equation

$$Ax^2 + Cy^2 + Dx + Ey + F = 0$$

is an ellipse if $AC > 0$.

76. Writing to Learn The graph of the equation

$$\frac{(x-h)^2}{a^2} + \frac{(y-k)^2}{b^2} = 0$$

is considered to be a degenerate ellipse. Describe the graph. How is it like a full-fledged ellipse, and how is it different?

8.3 Hyperbolas

What you'll learn about

- Geometry of a Hyperbola
- Translations of Hyperbolas
- Eccentricity and Orbits
- Reflective Property of a Hyperbola
- Long-Range Navigation

... and why

The hyperbola is the least-known conic section, yet it is used in astronomy, optics, and navigation.

Geometry of a Hyperbola

When a plane intersects both nappes of a right circular cone, the intersection is a hyperbola. (Figure 8.2a shows that the intersecting plane does not need to be parallel to the axis of the cone. If the plane intersects both nappes and does not contain the vertex, the intersection is a hyperbola.) The definition, features, and formula derivation for a hyperbola closely resemble those for an ellipse; indeed, it is helpful to make ellipse comparisons as you go through this section in order to notice the differences. As before, we begin with a geometric definition of the curve and use it to derive an algebraic formula.

DEFINITION Hyperbola

A **hyperbola** is the set of all points in a plane whose distances from two fixed points in the plane have a constant *difference*. The fixed points are the **foci** of the hyperbola. The line through the foci is the **focal axis**. The point on the focal axis midway between the foci is the **center**. The points where the hyperbola intersects its focal axis are the **vertices** of the hyperbola (Figure 8.22).

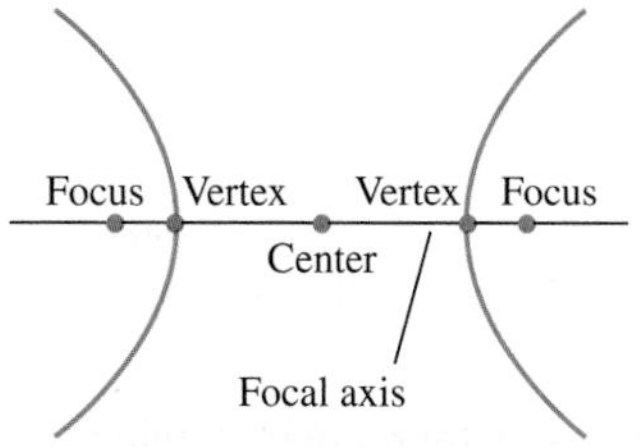

FIGURE 8.22 Key points on the focal axis of a hyperbola.

Figure 8.23 shows a hyperbola centered at the origin with its focal axis on the x-axis. The vertices are at $(-a, 0)$ and $(a, 0)$, where a is some positive constant. The fixed points $F_1(-c, 0)$ and $F_2(c, 0)$ are the foci of the hyperbola, with $c > a$.

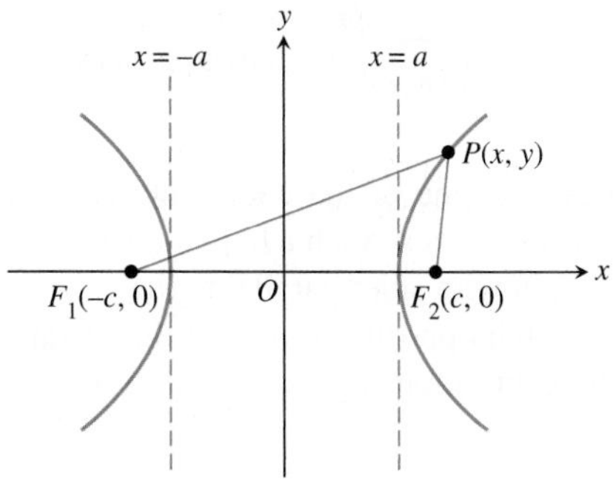

FIGURE 8.23 *Structure of a Hyperbola.* The difference of the distances from the foci to each point on the hyperbola is a constant.

Notice that the hyperbola has two *branches*. For a point $P(x, y)$ on the right-hand branch, $PF_1 - PF_2 = 2a$. On the left-hand branch, $PF_2 - PF_1 = 2a$. Combining these two equations gives us

$$PF_1 - PF_2 = \pm 2a.$$

Using the distance formula, the equation becomes

$$\sqrt{(x+c)^2 + (y-0)^2} - \sqrt{(x-c)^2 + (y-0)^2} = \pm 2a.$$

$$\sqrt{(x-c)^2 + y^2} = \pm 2a + \sqrt{(x+c)^2 + y^2} \quad \text{Rearrange terms.}$$

$$x^2 - 2cx + c^2 + y^2 = 4a^2 \pm 4a\sqrt{(x+c)^2 + y^2} + x^2 + 2cx + c^2 + y^2 \quad \text{Square.}$$

$$\mp a\sqrt{(x+c)^2 + y^2} = a^2 + cx \quad \text{Simplify.}$$

$$a^2(x^2 + 2cx + c^2 + y^2) = a^4 + 2a^2cx + c^2x^2 \quad \text{Square.}$$

$$(c^2 - a^2)x^2 - a^2y^2 = a^2(c^2 - a^2) \quad \text{Multiply by } -1 \text{ and simplify.}$$

Letting $b^2 = c^2 - a^2$, we have

$$b^2x^2 - a^2y^2 = a^2b^2,$$

which is usually written as

$$\frac{x^2}{a^2} - \frac{y^2}{b^2} = 1.$$

Because these steps can be reversed, a point $P(x, y)$ satisfies this last equation if and only if the point lies on the hyperbola defined by $PF_1 - PF_2 = \pm 2a$, provided that

$c > a > 0$ and $b^2 = c^2 - a^2$. The *Pythagorean relation* $b^2 = c^2 - a^2$ can be written many ways, including $a^2 = c^2 - b^2$ and $c^2 = a^2 + b^2$.

The equation $x^2/a^2 - y^2/b^2 = 1$ is the **standard form** of the equation of a hyperbola centered at the origin with the x-axis as its focal axis. A hyperbola centered at the origin with the y-axis as its focal axis is the *inverse relation* of $x^2/a^2 - y^2/b^2 = 1$, and thus has an equation of the form

$$\frac{y^2}{a^2} - \frac{x^2}{b^2} = 1.$$

As with other conics, a line segment with endpoints on a hyperbola is a **chord** of the hyperbola. The chord lying on the focal axis connecting the vertices is the **transverse axis** of the hyperbola. The length of the transverse axis is $2a$. The line segment of length $2b$ that is perpendicular to the focal axis and that has the center of the hyperbola as its midpoint is the **conjugate axis** of the hyperbola. The number a is the **semitransverse axis**, and b is the **semiconjugate axis**.

The hyperbola

$$\frac{x^2}{a^2} - \frac{y^2}{b^2} = 1$$

has two *asymptotes*. These asymptotes are slant lines that can be found by replacing the 1 on the right-hand side of the hyperbola's equation by a 0:

$$\underbrace{\frac{x^2}{a^2} - \frac{y^2}{b^2} = 1}_{\text{hyperbola}} \rightarrow \underbrace{\frac{x^2}{a^2} - \frac{y^2}{b^2} = 0}_{\text{Replace 1 by 0.}} \rightarrow \underbrace{y = \pm\frac{b}{a}x}_{\text{asymptotes}}$$

Naming Axes

The word "transverse" comes from the Latin *trans vertere*: to go across. The transverse axis "goes across" from one vertex to the other. The conjugate axis is the transverse axis for the *conjugate hyperbola*, defined in Exercise 73.

A hyperbola centered at the origin with its focal axis one of the coordinate axes is symmetric with respect to the origin and both coordinate axes. Such a hyperbola can be sketched by drawing a rectangle centered at the origin with sides parallel to the coordinate axes, followed by drawing the asymptotes through opposite corners of the rectangle, and finally sketching the hyperbola using the central rectangle and asymptotes as guides, as shown in the Drawing Lesson.

Drawing Lesson

How to Sketch the Hyperbola $x^2/a^2 - y^2/b^2 = 1$

1. Sketch line segments at $x = \pm a$ and $y = \pm b$, and complete the rectangle they determine.
2. Sketch the asymptotes by extending the rectangle's diagonals.
3. Use the rectangle and asymptotes to guide your drawing.

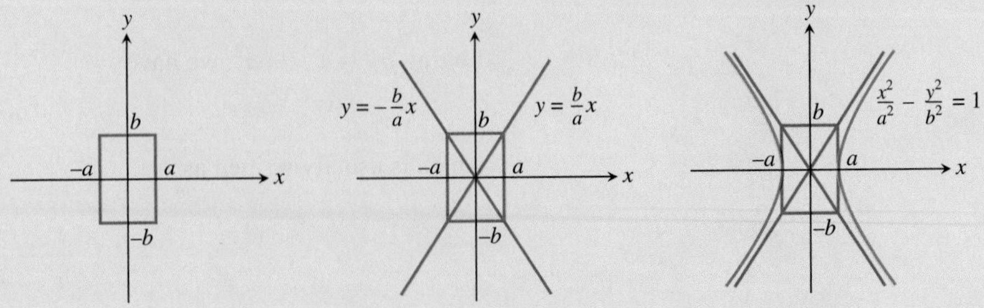

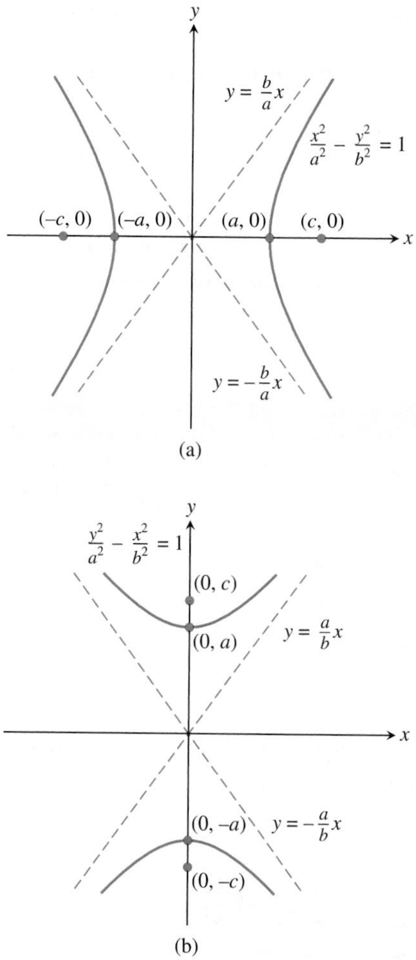

FIGURE 8.24 Hyperbolas centered at the origin with foci on (a) the x-axis and (b) the y-axis.

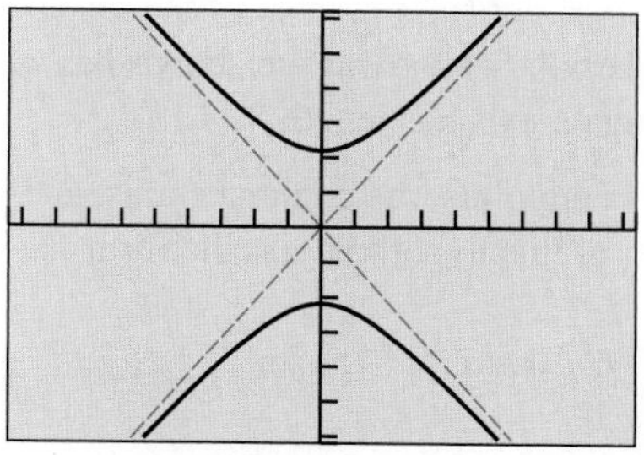

[–9.4, 9.4] by [–6.2, 6.2]

FIGURE 8.25 The hyperbola $y^2/5 - x^2/4 = 1$, shown with its asymptotes. (Example 2)

Hyperbolas with Center (0, 0)

• **Standard equation**	$\dfrac{x^2}{a^2} - \dfrac{y^2}{b^2} = 1$	$\dfrac{y^2}{a^2} - \dfrac{x^2}{b^2} = 1$
• **Focal axis**	x-axis	y-axis
• **Foci**	$(\pm c, 0)$	$(0, \pm c)$
• **Vertices**	$(\pm a, 0)$	$(0, \pm a)$
• **Semitransverse axis**	a	a
• **Semiconjugate axis**	b	b
• **Pythagorean relation**	$c^2 = a^2 + b^2$	$c^2 = a^2 + b^2$
• **Asymptotes**	$y = \pm \dfrac{b}{a} x$	$y = \pm \dfrac{a}{b} x$

See Figure 8.24.

EXAMPLE 1 Finding the Vertices and Foci of a Hyperbola

Find the vertices and the foci of the hyperbola $4x^2 - 9y^2 = 36$.

SOLUTION Dividing both sides of the equation by 36 yields the standard form $x^2/9 - y^2/4 = 1$. So $a^2 = 9$, $b^2 = 4$, and $c^2 = a^2 + b^2 = 9 + 4 = 13$. Thus the vertices are $(\pm 3, 0)$, and the foci are $(\pm\sqrt{13}, 0)$. **Now try Exercise 1.**

If we wish to graph a hyperbola using a function grapher, we need to solve the equation of the hyperbola for y, as illustrated in Example 2.

EXAMPLE 2 Finding an Equation and Graphing a Hyperbola

Find an equation of the hyperbola with foci $(0, -3)$ and $(0, 3)$ whose conjugate axis has length 4. Sketch the hyperbola and its asymptotes, and support your sketch with a grapher.

SOLUTION The center is $(0, 0)$. The foci are on the y-axis with $c = 3$. The semiconjugate axis is $b = 4/2 = 2$. Thus $a^2 = c^2 - b^2 = 3^2 - 2^2 = 5$. The standard form of the equation for the hyperbola is

$$\frac{y^2}{5} - \frac{x^2}{4} = 1.$$

Using $a = \sqrt{5} \approx 2.24$ and $b = 2$, we can sketch the central rectangle, the asymptotes, and the hyperbola itself. Try doing this. To graph the hyperbola using a function grapher, we solve for y in terms of x.

$$\frac{y^2}{5} = 1 + \frac{x^2}{4} \qquad \text{Add } \frac{x^2}{4}.$$

$$y^2 = 5(1 + x^2/4) \qquad \text{Multiply by 5.}$$

$$y = \pm\sqrt{5(1 + x^2/4)} \qquad \text{Extract square roots.}$$

Figure 8.25 shows the graphs of

$$y_1 = \sqrt{5(1 + x^2/4)} \quad \text{and} \quad y_2 = -\sqrt{5(1 + x^2/4)},$$

together with the asymptotes of the hyperbola

$$y_3 = \frac{\sqrt{5}}{2}x \quad \text{and} \quad y_4 = -\frac{\sqrt{5}}{2}x.$$

Now try Exercise 17.

In Example 2, because the hyperbola had a vertical focal axis, selecting a viewing rectangle was easy. When a hyperbola has a horizontal focal axis, we try to select a viewing window to include the two vertices in the plot and thus avoid gaps in the graph of the hyperbola.

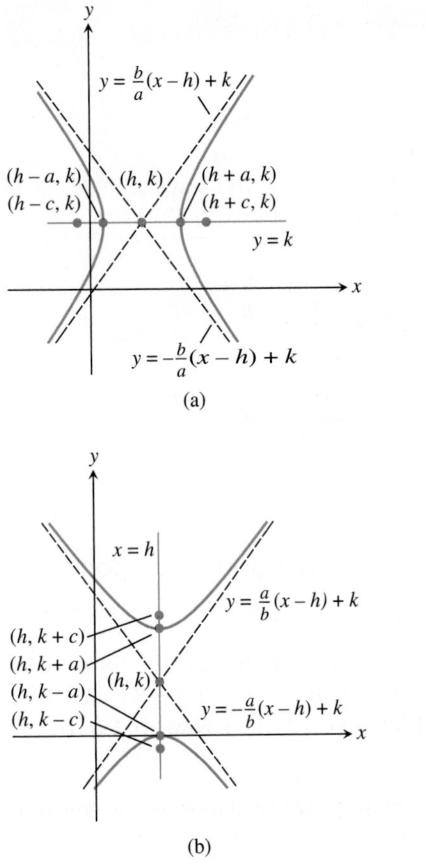

FIGURE 8.26 Hyperbolas with center (h, k) and foci on (a) $y = k$ and (b) $x = h$.

Translations of Hyperbolas

When a hyperbola with center $(0, 0)$ is translated horizontally by h units and vertically by k units, the center of the hyperbola moves from $(0, 0)$ to (h, k), as shown in Figure 8.26. Such a translation does not change the length of the transverse or conjugate axis or the Pythagorean relation.

Hyperbolas with Center (h, k)

• **Standard equation**	$\frac{(x-h)^2}{a^2} - \frac{(y-k)^2}{b^2} = 1$	$\frac{(y-k)^2}{a^2} - \frac{(x-h)^2}{b^2} = 1$
• **Focal axis**	$y = k$	$x = h$
• **Foci**	$(h \pm c, k)$	$(h, k \pm c)$
• **Vertices**	$(h \pm a, k)$	$(h, k \pm a)$
• **Semitransverse axis**	a	a
• **Semiconjugate axis**	b	b
• **Pythagorean relation**	$c^2 = a^2 + b^2$	$c^2 = a^2 + b^2$
• **Asymptotes**	$y = \pm\frac{b}{a}(x - h) + k$	$y = \pm\frac{a}{b}(x - h) + k$

See Figure 8.26.

EXAMPLE 3 Finding an Equation of a Hyperbola

Find the standard form of the equation for the hyperbola whose transverse axis has endpoints $(-2, -1)$ and $(8, -1)$ and whose conjugate axis has length 8.

SOLUTION Figure 8.27 shows the transverse axis endpoints, the conjugate axis, and the center of the hyperbola. The standard equation of this hyperbola has the form

$$\frac{(x-h)^2}{a^2} - \frac{(y-k)^2}{b^2} = 1,$$

where the center (h, k) is the midpoint $(3, -1)$ of the transverse axis. The semitransverse axis and semiconjugate axis are

$$a = \frac{8 - (-2)}{2} = 5 \quad \text{and} \quad b = \frac{8}{2} = 4.$$

So the equation we seek is

$$\frac{(x-3)^2}{5^2} - \frac{(y-(-1))^2}{4^2} = 1,$$

$$\frac{(x-3)^2}{25} - \frac{(y+1)^2}{16} = 1.$$

Now try Exercise 31.

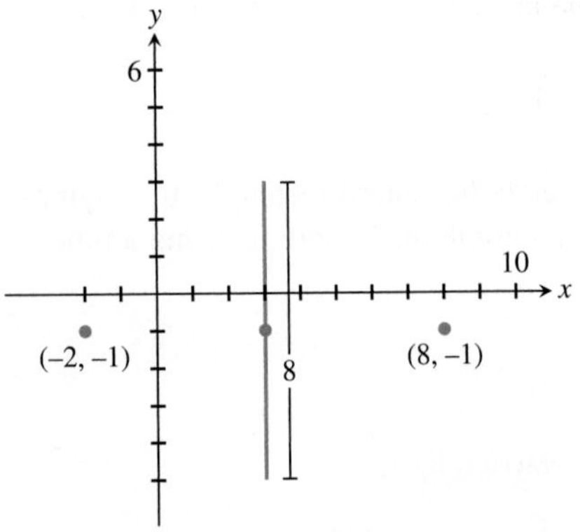

FIGURE 8.27 Given information for Example 3.

EXAMPLE 4 Locating Key Points of a Hyperbola

Find the center, vertices, and foci of the hyperbola

$$\frac{(x+2)^2}{9} - \frac{(y-5)^2}{49} = 1.$$

SOLUTION The center (h, k) is $(-2, 5)$. Because the semitransverse axis $a = \sqrt{9} = 3$, the vertices are

$$(h + a, k) = (-2 + 3, 5) = (1, 5) \quad \text{and}$$
$$(h - a, k) = (-2 - 3, 5) = (-5, 5).$$

Because $c = \sqrt{a^2 + b^2} = \sqrt{9 + 49} = \sqrt{58}$, the foci $(h \pm c, k)$ are $(-2 \pm \sqrt{58}, 5)$, or approximately $(5.62, 5)$ and $(-9.62, 5)$. Now try Exercise 39.

With the information found about the hyperbola in Example 4 and knowing that its semiconjugate axis $b = \sqrt{49} = 7$, we could easily sketch the hyperbola. Obtaining an accurate graph of the hyperbola using a function grapher is another matter. Often, the best way to graph a hyperbola using a graphing utility is to use parametric equations.

EXPLORATION 1 Graphing a Hyperbola Using Its Parametric Equations

1. Use the Pythagorean trigonometry identity $\sec^2 t - \tan^2 t = 1$ to prove that the parametrization $x = -1 + 3/\cos t$, $y = 1 + 2\tan t$ $(0 \le t \le 2\pi)$ will produce a graph of the hyperbola $(x + 1)^2/9 - (y - 1)^2/4 = 1$.
2. Using Dot graphing mode, graph $x = -1 + 3/\cos t$, $y = 1 + 2\tan t$ $(0 \le t \le 2\pi)$ in a square viewing window to support part 1 graphically. Switch to Connected graphing mode, and regraph the equation. What do you observe? Explain.
3. Create parametrizations for the hyperbolas in Examples 1, 2, 3, and 4.
4. Graph each of your parametrizations in part 3 and check the features of the obtained graph to see whether they match the expected geometric features of the hyperbola. If necessary, revise your parametrization and regraph until all features match.
5. Prove that each of your parametrizations is valid.

Eccentricity and Orbits

DEFINITION Eccentricity of a Hyperbola

The **eccentricity** of a hyperbola is

$$e = \frac{c}{a} = \frac{\sqrt{a^2 + b^2}}{a},$$

where a is the semitransverse axis, b is the semiconjugate axis, and c is the distance from the center to either focus.

For a hyperbola, because $c > a$, the eccentricity $e > 1$. In Section 8.2 we learned that the eccentricity of an ellipse satisfies the inequality $0 \le e < 1$ and that, for $e = 0$, the ellipse is a circle. In Section 8.5 we will generalize the concept of eccentricity to all types of conics and learn that the eccentricity of a parabola is $e = 1$.

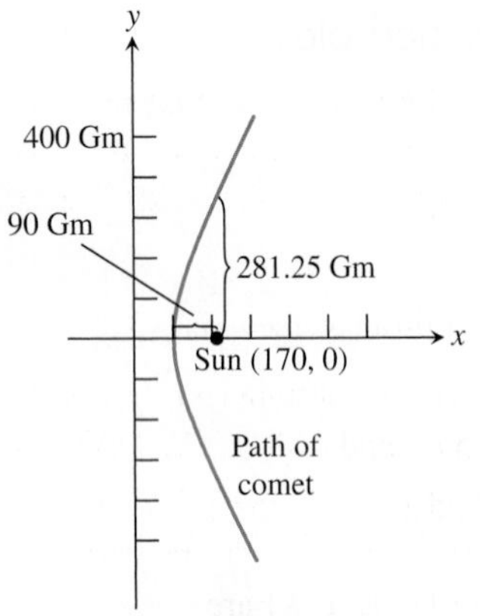

FIGURE 8.28 The graph of one branch of $x^2/6400 - y^2/22{,}500 = 1$. (Example 5)

Kepler's First Law of Planetary Motion says that a planet's orbit is elliptical with the Sun at one focus. Since 1609, astronomers have generalized Kepler's Law; the current theory states: A celestial body that travels within the gravitational field of a much more massive body follows a path that closely approximates a conic section that has the more massive body as a focus. Two bodies that do not differ greatly in mass (such as Earth and the Moon, or Pluto and its moon Charon) actually revolve around their balance point, or *barycenter*. In theory, a comet can approach the Sun from interstellar space, make a partial loop about the Sun, and then leave the solar system, returning to deep space; such a comet follows a path that is one branch of a hyperbola.

EXAMPLE 5 Analyzing a Comet's Orbit

A comet following a hyperbolic path about the Sun has a perihelion distance of 90 Gm. When the line from the comet to the Sun is perpendicular to the focal axis of the orbit, the comet is 281.25 Gm from the Sun. Calculate a, b, c, and e. What are the coordinates of the center of the Sun if we coordinatize space so that the hyperbola is given by

$$\frac{x^2}{a^2} - \frac{y^2}{b^2} = 1?$$

SOLUTION The perihelion distance is $c - a = 90$. When $x = c$, $y = \pm b^2/a$ (see Exercise 74). So $b^2/a = 281.25$, or $b^2 = 281.25a$. Because $b^2 = c^2 - a^2$, we have the system

$$c - a = 90 \quad \text{and} \quad c^2 - a^2 = 281.25a,$$

which yields the equation:

$$\begin{aligned} (a + 90)^2 - a^2 &= 281.25a \\ a^2 + 180a + 8100 - a^2 &= 281.25a \\ 8100 &= 101.25a \\ a &= 80 \end{aligned}$$

Therefore, $a = 80$ Gm, $b = 150$ Gm, $c = 170$ Gm, and $e = 17/8 = 2.125$ (Figure 8.28). If the comet's path is the branch of the hyperbola with positive x-coordinates, then the Sun is at the focus $(c, 0) = (170, 0)$. **Now try Exercise 55.**

Reflective Property of a Hyperbola

Like other conics, a hyperbola can be used to make a reflector of sound, light, and other waves. If we rotate a hyperbola in three-dimensional space about its focal axis, the hyperbola sweeps out a **hyperboloid of revolution**. If a signal is directed toward a focus of a reflective hyperboloid, the signal reflects off the hyperbolic surface to the other focus. In Figure 8.29 light reflects off a primary parabolic mirror toward the mirror's focus $F_P = F_H$, which is also the focus of a small hyperbolic mirror. The light is then reflected off the hyperbolic mirror, toward the hyperboloid's other focus $F_H = F_E$, which is also the focus of an elliptical mirror. Finally the light is reflected into the observer's eye, which is at the second focus of the ellipsoid F_E.

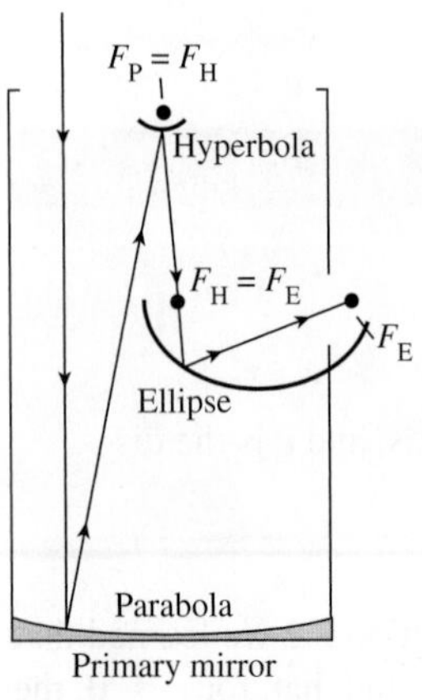

FIGURE 8.29 Cross section of a reflecting telescope.

Reflecting telescopes date back to the 1600s when Isaac Newton used a primary parabolic mirror in combination with a flat secondary mirror, slanted to reflect the light out the side to the eyepiece. French optician G. Cassegrain was the first to use a hyperbolic secondary mirror, which directed the light through a hole at the vertex of the primary mirror (see Exercise 70). Today, reflecting telescopes such as the Hubble Space Telescope have become quite sophisticated and must have nearly perfect mirrors to focus properly.

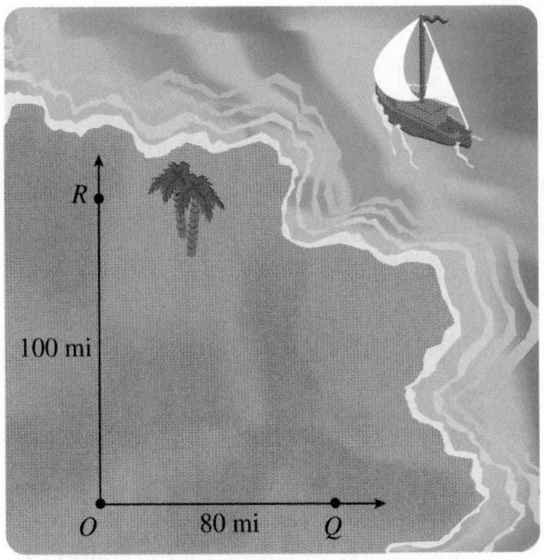

FIGURE 8.30 Strategically located LORAN transmitters O, Q, and R. (Example 6)

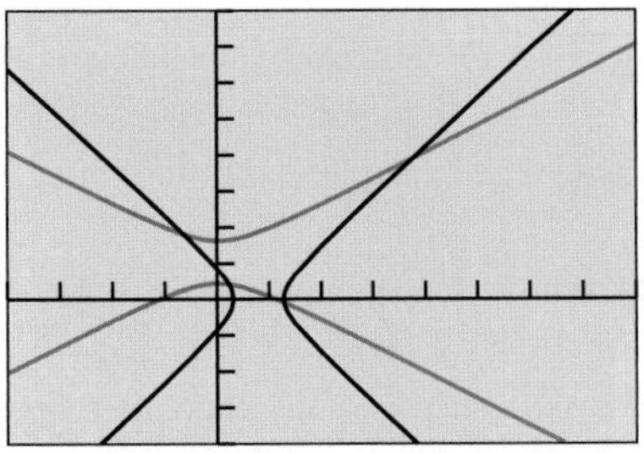

[–200, 400] by [–200, 400]

FIGURE 8.31 Graphs for Example 6.

Long-Range Navigation

Hyperbolas and radio signals are the basis of the LORAN (long-range navigation) system. Example 6 illustrates this system using the definition of hyperbola and the fact that radio signals travel 980 ft per microsecond (1 microsecond $= 1\ \mu\text{sec} = 10^{-6}$ sec).

EXAMPLE 6 Using the LORAN System

Radio signals are sent simultaneously from transmitters located at points O, Q, and R (Figure 8.30). R is 100 mi due north of O, and Q is 80 mi due east of O. The LORAN receiver on sloop *Gloria* receives the signal from O 323.27 μsec after the signal from R, and 258.61 μsec after the signal from Q. What is the sloop's bearing and distance from O?

SOLUTION Denote the *Gloria*'s position as point G. The time difference between the reception of the signals from O and R is 323.27 μsec, so

$$OG - RG = 323.27\ \mu\text{sec} \times \frac{980\text{ ft}}{1\ \mu\text{sec}} \times \frac{1\text{ mi}}{5280\text{ ft}} \approx 60\text{ mi.}$$

The set of possible points G such that $OG - RG = 60$ is a branch of a hyperbola with foci O and R and $2a = 60$. Since $OR = 2c = 100$, we conclude that $a = 30$, $c = 50$, and $b = \sqrt{c^2 - a^2} = 40$. The center of the hyperbola is (0, 50), the midpoint of $\overline{OR}$. Using this information, we know that G lies on a branch of the hyperbola with equation

$$\frac{(y - 50)^2}{30^2} - \frac{x^2}{40^2} = 1.$$

Similarly, G lies on a branch of a hyperbola with foci O and Q. In this hyperbola, the center is (40, 0), $c = 40$, and $2a = 258.61\ \mu\text{sec} \times \frac{980\text{ ft}}{1\ \mu\text{sec}} \times \frac{1\text{ mi}}{5280\text{ ft}} \approx 48$ mi.

Thus $a = 24$ and $b = \sqrt{c^2 - a^2} = 32$. Using this information, we know that G lies on a branch of a hyperbola with equation

$$\frac{(x - 40)^2}{24^2} - \frac{y^2}{32^2} = 1.$$

We use a grapher (Figure 8.31) to find the first-quadrant intersection point of the two hyperbolas: $G \approx (187.09, 193.49)$. The bearing of G from point O is

$$\theta \approx 90° - \tan^{-1}\left(\frac{193.49}{187.09}\right) \approx 44.04°,$$

and the distance of G from point O is $OG \approx \sqrt{187.09^2 + 193.49^2} \approx 269.15$.

So the *Gloria* is about 187.1 mi east and 193.5 mi north of point O on a bearing of roughly 44°, and the sloop is about 269 mi from point O. **Now try Exercise 57.**

QUICK REVIEW 8.3 *(For help, go to Sections P.2, P.5, and 7.1.)*

Exercise numbers with a gray background indicate problems that the authors have designed to be solved *without a calculator*.

In Exercises 1 and 2, find the distance between the given points.

1. $(4, -3)$ and $(-7, -8)$

2. $(a, -3)$ and (b, c)

In Exercises 3 and 4, solve for y in terms of x.

3. $\frac{y^2}{16} - \frac{x^2}{9} = 1$

4. $\frac{x^2}{36} - \frac{y^2}{4} = 1$

In Exercises 5–8, solve for x.

5. $\sqrt{3x + 12} - \sqrt{3x - 8} = 10$

6. $\sqrt{4x + 12} - \sqrt{x + 8} = 1$

7. $\sqrt{6x^2 + 12} - \sqrt{6x^2 + 1} = 1$

8. $\sqrt{2x^2 + 12} - \sqrt{3x^2 + 4} = -8$

In Exercises 9 and 10, solve the system of equations.

9. $c - a = 2$ and $c^2 - a^2 = 16a/3$

10. $c - a = 1$ and $c^2 - a^2 = 25a/12$

SECTION 8.3 Exercises

In Exercises 1–6, find the vertices and foci of the hyperbola.

1. $\dfrac{x^2}{16} - \dfrac{y^2}{7} = 1$

2. $\dfrac{y^2}{25} - \dfrac{x^2}{21} = 1$

3. $\dfrac{y^2}{36} - \dfrac{x^2}{13} = 1$

4. $\dfrac{x^2}{9} - \dfrac{y^2}{16} = 1$

5. $3x^2 - 4y^2 = 12$

6. $9x^2 - 4y^2 = 36$

In Exercises 7–10, match the graph with its equation.

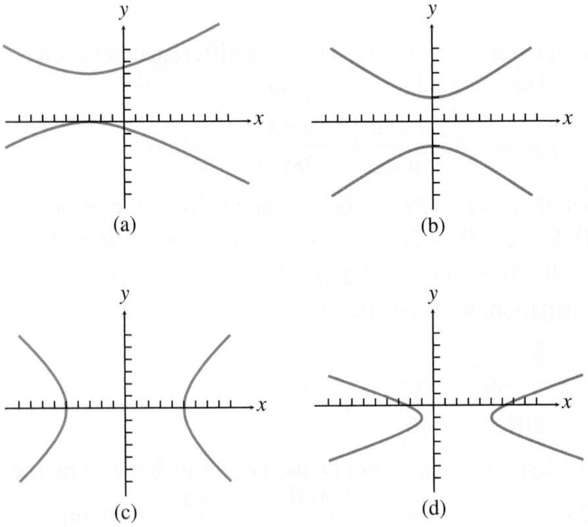

7. $\dfrac{x^2}{25} - \dfrac{y^2}{16} = 1$

8. $\dfrac{y^2}{4} - \dfrac{x^2}{9} = 1$

9. $\dfrac{(y-2)^2}{4} - \dfrac{(x+3)^2}{16} = 1$

10. $\dfrac{(x-2)^2}{9} - (y+1)^2 = 1$

In Exercises 11–16, sketch the graph of the hyperbola by hand.

11. $\dfrac{x^2}{49} - \dfrac{y^2}{25} = 1$

12. $\dfrac{y^2}{64} - \dfrac{x^2}{25} = 1$

13. $\dfrac{y^2}{25} - \dfrac{x^2}{16} = 1$

14. $\dfrac{x^2}{169} - \dfrac{y^2}{144} = 1$

15. $\dfrac{(x+3)^2}{16} - \dfrac{(y-1)^2}{4} = 1$

16. $\dfrac{(x-1)^2}{2} - \dfrac{(y+3)^2}{4} = 1$

In Exercises 17–22, graph the hyperbola using a function grapher.

17. $\dfrac{x^2}{36} - \dfrac{y^2}{16} = 1$

18. $\dfrac{y^2}{64} - \dfrac{x^2}{16} = 1$

19. $\dfrac{x^2}{4} - \dfrac{y^2}{9} = 1$

20. $\dfrac{y^2}{16} - \dfrac{x^2}{9} = 1$

21. $\dfrac{x^2}{4} - \dfrac{(y-3)^2}{5} = 1$

22. $\dfrac{(y-3)^2}{9} - \dfrac{(x+2)^2}{4} = 1$

In Exercises 23–38, find an equation in standard form for the hyperbola that satisfies the given conditions.

23. Foci $(\pm 3, 0)$, transverse axis length 4

24. Foci $(0, \pm 3)$, transverse axis length 4

25. Foci $(0, \pm 15)$, transverse axis length 8

26. Foci $(\pm 5, 0)$, transverse axis length 3

27. Center at $(0, 0)$, $a = 5$, $e = 2$, horizontal focal axis

28. Center at $(0, 0)$, $a = 4$, $e = 3/2$, vertical focal axis

29. Center at $(0, 0)$, $b = 5$, $e = 13/12$, vertical focal axis

30. Center at $(0, 0)$, $c = 6$, $e = 2$, horizontal focal axis

31. Transverse axis endpoints $(2, 3)$ and $(2, -1)$, conjugate axis length 6

32. Transverse axis endpoints $(5, 3)$ and $(-7, 3)$, conjugate axis length 10

33. Transverse axis endpoints $(-1, 3)$ and $(5, 3)$, slope of one asymptote 4/3

34. Transverse axis endpoints $(-2, -2)$ and $(-2, 7)$, slope of one asymptote 4/3

35. Foci $(-4, 2)$ and $(2, 2)$, transverse axis endpoints $(-3, 2)$ and $(1, 2)$

36. Foci $(-3, -11)$ and $(-3, 0)$, transverse axis endpoints $(-3, -9)$ and $(-3, -2)$

37. Center at $(-3, 6)$, $a = 5$, $e = 2$, vertical focal axis

38. Center at $(1, -4)$, $c = 6$, $e = 2$, horizontal focal axis

In Exercises 39–42, find the center, vertices, and the foci of the hyperbola.

39. $\dfrac{(x+1)^2}{144} - \dfrac{(y-2)^2}{25} = 1$

40. $\dfrac{(x+4)^2}{12} - \dfrac{(y+6)^2}{13} = 1$

41. $\dfrac{(y+3)^2}{64} - \dfrac{(x-2)^2}{81} = 1$

42. $\dfrac{(y-1)^2}{25} - \dfrac{(x+5)^2}{11} = 1$

In Exercises 43–46, graph the hyperbola using a parametric grapher in Dot graphing mode.

43. $\dfrac{y^2}{25} - \dfrac{x^2}{4} = 1$

44. $\dfrac{x^2}{30} - \dfrac{y^2}{20} = 1$

45. $\dfrac{(x+3)^2}{12} - \dfrac{(y-6)^2}{5} = 1$

46. $\dfrac{(y+1)^2}{15} - \dfrac{(x-2)^2}{6} = 1$

In Exercises 47–50, graph the hyperbola, and find its vertices, foci, and eccentricity.

47. $4(y-1)^2 - 9(x-3)^2 = 36$

48. $4(x-2)^2 - 9(y+4)^2 = 1$

49. $9x^2 - 4y^2 - 36x + 8y - 4 = 0$

50. $25y^2 - 9x^2 - 50y - 54x - 281 = 0$

In Exercises 51 and 52, write an equation for the hyperbola.

51.

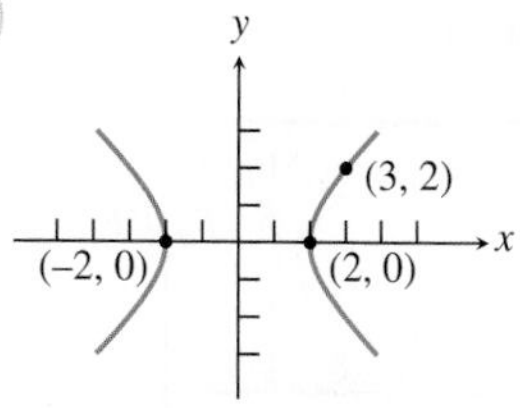

52.

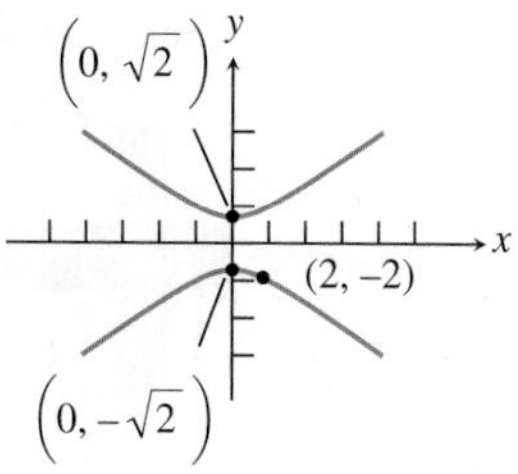

53. Writing to Learn Prove that an equation for the hyperbola with center $(0, 0)$, foci $(0, \pm c)$, and semitransverse axis a is $y^2/a^2 - x^2/b^2 = 1$, where $c > a > 0$ and $b^2 = c^2 - a^2$. (*Hint*: Refer to derivation at the beginning of the section.)

54. Degenerate Hyperbolas Graph the degenerate hyperbola.

(a) $\dfrac{x^2}{4} - y^2 = 0$ **(b)** $\dfrac{y^2}{9} - \dfrac{x^2}{16} = 0$

55. Rogue Comet A comet following a hyperbolic path about the Sun has a perihelion of 120 Gm. When the line from the comet to the Sun is perpendicular to the focal axis of the orbit, the comet is 250 Gm from the Sun. Calculate a, b, c, and e. What are the coordinates of the center of the Sun if the center of the hyperbolic orbit is $(0, 0)$ and the Sun lies on the positive x-axis?

56. Rogue Comet A comet following a hyperbolic path about the Sun has a perihelion of 140 Gm. When the line from the comet to the Sun is perpendicular to the focal axis of the orbit, the comet is 405 Gm from the Sun. Calculate a, b, c, and e. What are the coordinates of the center of the Sun if the center of the hyperbolic orbit is (0, 0) and the Sun lies on the positive x-axis?

57. Long-Range Navigation Three LORAN radio transmitters are positioned as shown in the figure, with R due north of O and Q due east of O. The cruise ship *Princess Ann* receives simultaneous signals from the three transmitters. The signal from O arrives 323.27 μsec after the signal from R, and 646.53 μsec after the signal from Q. Determine the ship's bearing and distance from point O.

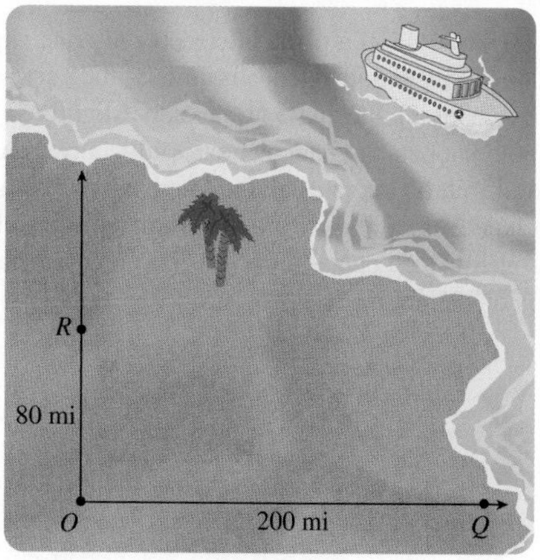

58. Gun Location Observers are located at positions A, B, and C with A due north of B. A cannon is located somewhere in the first quadrant as illustrated in the figure. A hears the sound of the cannon 2 sec before B, and C hears the sound 4 sec before B. Determine the bearing and distance of the cannon from point B. (Assume that sound travels at 1100 ft/sec.)

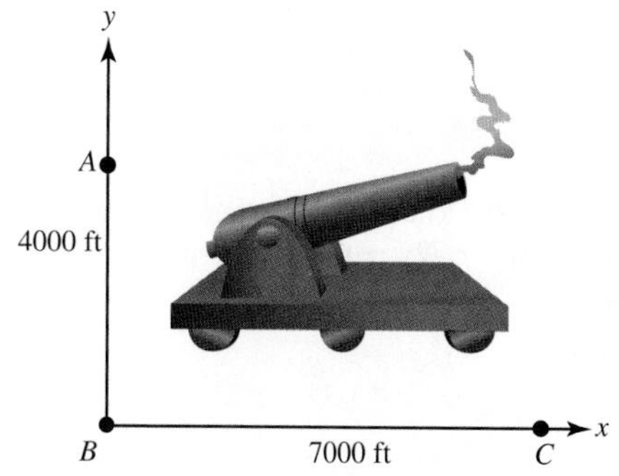

Group Activities In Exercises 59 and 60, solve the system of equations algebraically and support your answer graphically.

59. $\dfrac{x^2}{4} - \dfrac{y^2}{9} = 1$

$x - \dfrac{2\sqrt{3}}{3}y = -2$

60. $\dfrac{x^2}{4} - y^2 = 1$

$x^2 + y^2 = 9$

61. Group Activity Consider the system of equations

$$\frac{x^2}{4} - \frac{y^2}{25} = 1$$

$$\frac{x^2}{25} + \frac{y^2}{4} = 1$$

(a) Solve the system graphically.

(b) If you have access to a grapher that also does symbolic algebra, use it to find the the exact solutions to the system.

62. Writing to Learn Escape of the Unbound When NASA launches a space probe, the probe reaches a speed sufficient for it to become unbound from Earth and escape along a hyperbolic trajectory. Look up *escape speed* in an astronomy textbook or on the Internet, and write a paragraph in your own words about what you find.

Standardized Test Questions

63. True or False The distance from a focus of a hyperbola to the closer vertex is $a(e - 1)$, where a is the semitransverse axis and e is the eccentricity. Justify your answer.

64. True or False Unlike that for an ellipse, the Pythagorean relation for a hyperbola is the usual $a^2 + b^2 = c^2$. Justify your answer.

In Exercises 65–68, you may use a graphing calculator to solve the problem.

65. Multiple Choice One focus of $x^2 - 4y^2 = 4$ is

(A) $(4, 0)$.

(B) $(\sqrt{5}, 0)$.

(C) $(2, 0)$.

(D) $(\sqrt{3}, 0)$.

(E) $(1, 0)$.

66. Multiple Choice The focal axis of $\frac{(x+5)^2}{9} - \frac{(y-6)^2}{16} = 1$ is

(A) $y = 2$.

(B) $y = 3$.

(C) $y = 4$.

(D) $y = 5$.

(E) $y = 6$.

67. Multiple Choice The center of $4x^2 - 12y^2 - 16x - 72y - 44 = 0$ is

(A) $(2, -2)$.

(B) $(2, -3)$.

(C) $(2, -4)$.

(D) $(2, -5)$.

(E) $(2, -6)$.

68. Multiple Choice The slopes of the asymptotes of the hyperbola $\frac{x^2}{4} - \frac{y^2}{3} = 1$ are

(A) ± 1.

(B) $\pm 3/2$.

(C) $\pm \sqrt{3}/2$.

(D) $\pm 2/3$.

(E) $\pm 4/3$.

Explorations

69. Constructing Points of a Hyperbola Use a geometry software package, such as *Cabri Geometry II*™, *The Geometer's Sketchpad*®, or a similar application on a handheld device, to carry out the following construction.

(a) Start by placing the coordinate axes in the construction window.

(b) Construct two points on the x-axis at $(\pm 5, 0)$ as the foci.

(c) Construct concentric circles of radii $r = 1, 2, 3, \ldots, 12$ centered at these two foci.

(d) Construct the points where these concentric circles meet and have a difference of radii of $2a = 6$, and overlay the conic that passes through these points if the software has a conic tool.

(e) Find the equation whose graph includes all of these points.

70. Cassegrain Telescope A Cassegrain telescope as described in the section has the dimensions shown in the figure. Find the standard form for the equation of the hyperbola centered at the origin with the x-axis as the focal axis.

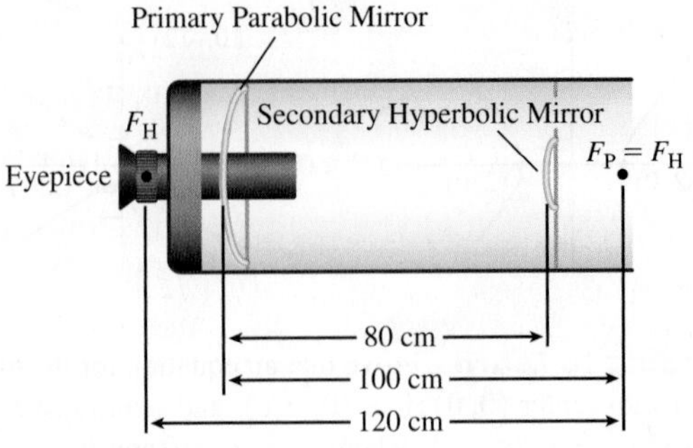

Extending the Ideas

71. Prove that a nondegenerate graph of the equation

$$Ax^2 + Cy^2 + Dx + Ey + F = 0$$

is a hyperbola if $AC < 0$.

72. Writing to Learn The graph of the equation

$$\frac{(x-h)^2}{a^2} - \frac{(y-k)^2}{b^2} = 0$$

is considered to be a degenerate hyperbola. Describe the graph. How is it like a full-fledged hyperbola, and how is it different?

73. Conjugate Hyperbolas The hyperbolas

$$\frac{(x-h)^2}{a^2} - \frac{(y-k)^2}{b^2} = 1 \text{ and } \frac{(y-k)^2}{b^2} - \frac{(x-h)^2}{a^2} = 1$$

obtained by switching the order of subtraction in their standard equations are **conjugate hyperbolas**. Prove that these hyperbolas have the same asymptotes and that the conjugate axis of each of these hyperbolas is the transverse axis of the other hyperbola.

74. Focal Width of a Hyperbola Prove that for the hyperbola

$$\frac{x^2}{a^2} - \frac{y^2}{b^2} = 1,$$

if $x = c$, then $y = \pm b^2/a$. Why is it reasonable to define the **focal width** of such hyperbolas to be $2b^2/a$?

75. Writing to Learn If a plane intersects only one nappe of a cone and the intersection extends infinitely, the curve is a parabola. By changing the angle of the plane slightly so that it intersects both nappes, the intersection results in two curves that extend infinitely, and we call the graph a hyperbola. You might conclude from all this that a hyperbola consists of two parabolas, one on each nappe. Give several good reasons why this conclusion cannot be true.

8.4 Quadratic Equations with *xy* Terms

What you'll learn about

- Quadratic Equations Revisited
- Axis Rotation Formulas
- Discriminant Test

... and why

You will see ellipses, hyperbolas, and parabolas as members of the family of conic sections rather than as separate types of curves.

Quadratic Equations Revisited

Let us now revisit an important fact that we used at the beginning of this chapter to motivate our study of conic sections: Every quadratic equation in two variables, that is, every equation of the form $Ax^2 + Bxy + Cy^2 + Dx + Ey + F = 0$ (assuming A, B, and C are not all 0), has a conic section as its graph, and every conic section in the xy-plane is the graph of a quadratic equation in two variables. If you compare this general quadratic equation with the conic equations we have developed so far, you will probably note that we have not yet encountered an equation with an xy term. There is a reason for that, but before we reveal what it is, see if you can analyze the graph of one such equation. It just happens to be one of our Twelve Basic Functions from Section 1.3.

EXPLORATION 1 A Familiar Conic with an *xy* Term

The equation of the reciprocal function $y = \dfrac{1}{x}$ can be written in the form $xy - 1 = 0$, which is a quadratic equation in x and y. Therefore, its graph is a conic section.

1. Judging from the graph (Figure 8.32), what type of conic section is it?
2. What points are the vertices?
3. What point is the center?
4. What line is the focal axis?
5. What is the distance between the vertices?
6. What is a?
7. What roles do the coordinate axes play?
8. Judging from the graph, what is b?
9. Use a and b to find c.
10. Use c to find the foci.

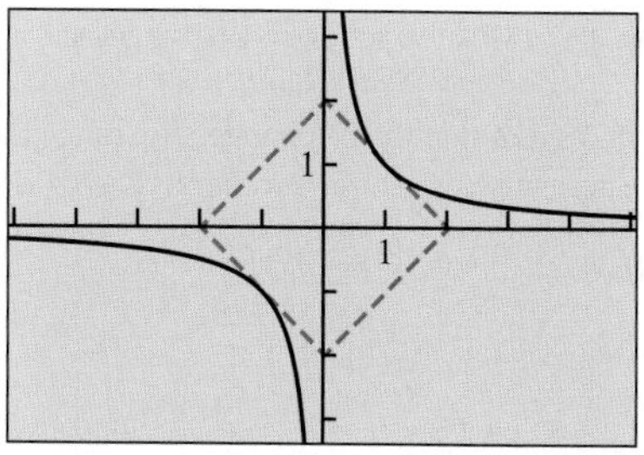

[–5.1, 5.1] by [–3.4, 3.4]

FIGURE 8.32 The graph of the reciprocal function $y = 1/x$, one of our Twelve Basic Functions, is a conic section. (Exploration 1)

What distinguishes the graph of this hyperbola from the ones we encountered in Section 8.3 is its orientation in the plane: It is neither horizontal nor vertical. Still, we hope you were able to find the vertices, asymptotes, and foci of this hyperbola in Exploration 1 by adapting what you know about horizontal hyperbolas to a hyperbola with $y = x$ as its focal axis. In fact, if we were to declare the line $y = x$ to be the "u-axis" and the line $y = -x$ to be the "v-axis," then (using the a and b values from Exploration 1) the equation of the hyperbola would be $u^2/2 - v^2/2 = 1$. (See Figure 8.33 on the next page.)

Notice that the equation in the (u, v) coordinate system has no uv term (known as the **cross-product term**), and so it can be analyzed using techniques from earlier in this chapter. This suggests a general (although possibly tedious) approach to handling quadratic equations with xy terms:

1. Find a rotated (u, v) coordinate system in which the graph is horizontal or vertical.
2. Express the equation in the new (u, v) system by applying rotation formulas.
3. Analyze the graph of the (u, v) quadratic equation as usual.
4. Express the (u, v) points in terms of x and y by applying the same rotation formulas.

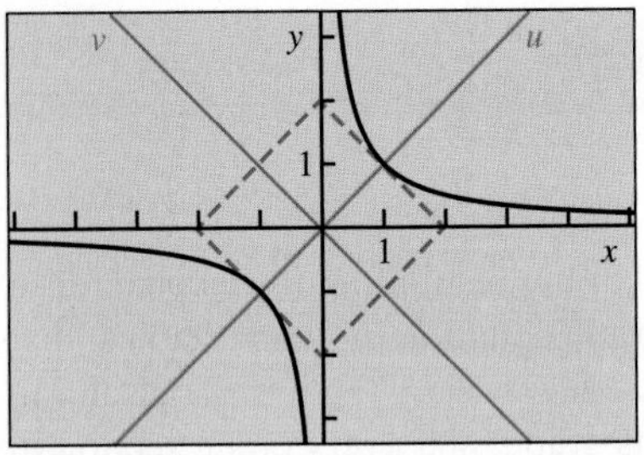

[–5.1, 5.1] by [–3.4, 3.4]

FIGURE 8.33 In the (u, v) coordinate system (the blue coordinate axes), the hyperbola $xy - 1 = 0$ has equation $u^2/2 - v^2/2 = 1$. Example 1 shows how to make this transformation algebraically with a pair of rotation equations.

Before we show you how to determine the rotation formulas, let us see how this strategy plays out with the hyperbola in Exploration 1. In this example we will give you the rotation formulas you will need.

EXAMPLE 1 Revisiting Exploration 1

Use rotation formulas to express the equation $xy - 1 = 0$ in a (u, v) coordinate system in which the u-axis is the focal axis. Find the foci in the (u, v) system and use the same rotation formulas to convert the foci to (x, y) coordinates.

SOLUTION The required rotation formulas are

$$x = \frac{u}{\sqrt{2}} - \frac{v}{\sqrt{2}}$$

$$y = \frac{u}{\sqrt{2}} + \frac{v}{\sqrt{2}}$$

Using these substitutions, $xy - 1 = 0$ becomes

$$\left(u/\sqrt{2} - v/\sqrt{2}\right)\left(u/\sqrt{2} + v/\sqrt{2}\right) - 1 = 0$$

which simplifies to $u^2/2 - v^2/2 = 1$. Analyzing this "horizontal" hyperbola in the usual way, we conclude that $a = \sqrt{2}, b = \sqrt{2}$, and $c = 2$. Therefore, the foci are $(2, 0)$ and $(-2, 0)$. To convert the foci back to (x, y) coordinates, we use the same rotation formulas.

When $(u, v) = (2, 0)$,

$$x = \frac{u}{\sqrt{2}} - \frac{v}{\sqrt{2}} = \frac{2}{\sqrt{2}} - \frac{0}{\sqrt{2}} = \sqrt{2}$$

$$y = \frac{u}{\sqrt{2}} + \frac{v}{\sqrt{2}} = \frac{2}{\sqrt{2}} + \frac{0}{\sqrt{2}} = \sqrt{2}$$

So $(x, y) = \left(\sqrt{2}, \sqrt{2}\right)$.

When $(u, v) = (-2, 0)$,

$$x = \frac{u}{\sqrt{2}} - \frac{v}{\sqrt{2}} = \frac{-2}{\sqrt{2}} - \frac{0}{\sqrt{2}} = -\sqrt{2}$$

$$y = \frac{u}{\sqrt{2}} + \frac{v}{\sqrt{2}} = \frac{-2}{\sqrt{2}} + \frac{0}{\sqrt{2}} = -\sqrt{2}$$

So $(x, y) = \left(-\sqrt{2}, -\sqrt{2}\right)$.

Now try Exercise 1.

Axis Rotation Formulas

The equations we used in Example 1 to write x and y in terms of the "new" coordinates u and v were linear equations with carefully chosen coefficients. If we denote the linear equations by $x = a \cdot u + b \cdot v$ and $y = c \cdot u + d \cdot v$, the coefficients a, b, c, and d are numbers that depend on the first-quadrant angle θ between the x-axis and the u-axis, which we will call the **angle of rotation**.

Look carefully at Figure 8.34. It shows two points on the Unit Circle with their (x, y) coordinates in black and their (u, v) coordinates in blue. Can you see how the (x, y) coordinates were determined? Now, suppose the linear transformations converting (u, v) to (x, y) are brought about by the linear equations $x = a \cdot u + b \cdot v$ and $y = c \cdot u + d \cdot v$.

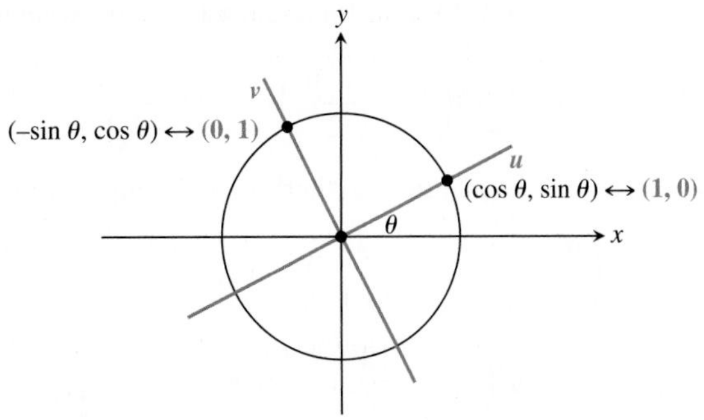

FIGURE 8.34 With a rotation angle of θ, the (u, v) points $(1, 0)$ and $(0, 1)$ correspond to the (x, y) points $(\cos\theta, \sin\theta)$ and $(-\sin\theta, \cos\theta)$, respectively.

Considering the x-coordinates at the two points, we have:

$$\left.\begin{array}{l} \cos\theta = a(1) + b(0) \quad \Rightarrow \quad a = \cos\theta \\ -\sin\theta = a(0) + b(1) \quad \Rightarrow \quad b = -\sin\theta \end{array}\right\} \Rightarrow \quad x = u\cos\theta - v\sin\theta$$

Considering the y-coordinates at the two points, we have:

$$\left.\begin{array}{l} \sin\theta = c(1) + d(0) \quad \Rightarrow \quad c = \sin\theta \\ \cos\theta = c(0) + d(1) \quad \Rightarrow \quad d = \cos\theta \end{array}\right\} \Rightarrow \quad y = u\cos\theta + v\sin\theta$$

We have established the axis rotation formulas.

Axis Rotation Formulas

If the u-axis makes a first-quadrant angle of θ with the x-axis, then the **axis rotation formulas** are

$$x = u\cos\theta - v\sin\theta$$
$$y = u\sin\theta + v\cos\theta$$

EXAMPLE 2 Applying the Rotation Formulas

Consider the conic section with equation $13x^2 - 6\sqrt{3}xy + 7y^2 = 16$.

(a) Use $\dfrac{\pi}{3}$ as the angle of rotation to write the equation in a rotated (u, v) coordinate system with no uv term.

(b) Identify the type of conic and find its vertices, foci, and eccentricity in the (u, v) coordinate system.

(c) Write the information in part (b) in the original coordinate system.

SOLUTION

(a) Since $\cos\dfrac{\pi}{3} = \dfrac{1}{2}$ and $\sin\dfrac{\pi}{3} = \dfrac{\sqrt{3}}{2}$, the rotation formulas are

$$x = \frac{u}{2} - \frac{\sqrt{3}v}{2}$$

$$y = \frac{\sqrt{3}u}{2} + \frac{v}{2}$$

Substituting for x and y in the given equation, we have

$$13\left(\frac{u}{2}-\frac{\sqrt{3}v}{2}\right)^2-6\sqrt{3}\left(\frac{u}{2}-\frac{\sqrt{3}v}{2}\right)\left(\frac{\sqrt{3}u}{2}+\frac{v}{2}\right)+7\left(\frac{\sqrt{3}u}{2}+\frac{v}{2}\right)^2=16.$$

Expanding, gathering, and simplifying, we get:

$$13\left(\frac{u^2}{4}-\frac{\sqrt{3}uv}{2}+\frac{3v^2}{4}\right)-6\sqrt{3}\left(\frac{\sqrt{3}u^2}{4}-\frac{uv}{2}-\frac{\sqrt{3}v^2}{4}\right)+7\left(\frac{3u^2}{4}+\frac{\sqrt{3}uv}{2}+\frac{v^2}{4}\right)=16$$

$$u^2\left(\frac{13}{4}-\frac{18}{4}+\frac{21}{4}\right)+uv\left(-\frac{13\sqrt{3}}{2}+3\sqrt{3}+\frac{7\sqrt{3}}{2}\right)+v^2\left(\frac{39}{4}+\frac{18}{4}+\frac{7}{4}\right)=16$$

$$4u^2+0\cdot uv+16v^2=16$$

$$\frac{u^2}{4}+\frac{v^2}{1}=1$$

(b) In the (u, v) system, the curve is a horizontal ellipse centered at $(0, 0)$.
Since $a = 2$, the vertices are $(\pm 2, 0)$.
Since $c = \sqrt{4-1} = \sqrt{3}$, the foci are $(\pm\sqrt{3}, 0)$.
The eccentricity is $c/a = \sqrt{3}/2$.

(c) We use the same rotation equations to convert the vertices and foci back to (x, y) coordinates.

$$x=\frac{\pm 2}{2}-\frac{\sqrt{3}\cdot 0}{2}=\pm 1$$

$$y=\frac{\sqrt{3}(\pm 2)}{2}+\frac{0}{2}=\pm\sqrt{3}$$

$\Rightarrow$ vertices are $(1, \sqrt{3})$ and $(-1, -\sqrt{3})$

$$x=\frac{\pm\sqrt{3}}{2}-\frac{\sqrt{3}\cdot 0}{2}=\pm\frac{\sqrt{3}}{2}$$

$$y=\frac{\sqrt{3}(\pm\sqrt{3})}{2}+\frac{0}{2}=\pm\frac{3}{2}$$

$\Rightarrow$ foci are $\left(\frac{\sqrt{3}}{2}, \frac{3}{2}\right)$ and $\left(-\frac{\sqrt{3}}{2}, -\frac{3}{2}\right)$

The eccentricity is still $c/a = \sqrt{3}/2$. Since it is a geometric invariant of the ellipse, it is not dependent on which way the ellipse is oriented.

The graph of the ellipse is shown in Figure 8.35. **Now try Exercise 5.**

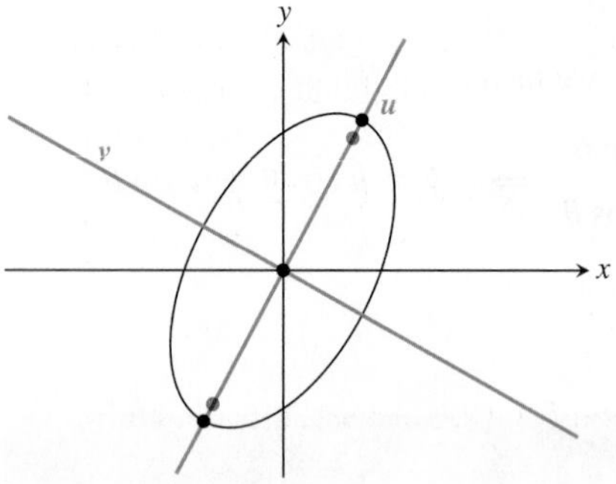

FIGURE 8.35 The graph of the ellipse with equation $13x^2 - 6\sqrt{3}xy + 7y^2 = 16$. The vertices (black dots) are $(1, \sqrt{3})$ and $(-1, -\sqrt{3})$, and the foci (red dots) are $(\sqrt{3}/2, 3/2)$ and $(-\sqrt{3}/2, -3/2)$. The vertices and foci are found by analyzing the curve in a rotated coordinate system (blue axes). (Example 2)

Now that you know how to rotate the axes through an angle of rotation θ, all you need is a way to find the value of θ that will make the graph of a given quadratic equation in two variables either horizontal or vertical. (In Example 2, we told you up front that the angle was $\pi/3$.)

Given the equation $Ax^2 + Bxy + Cy^2 + Dx + Ey + F = 0$ and the substitutions $x = u\cos\theta - v\sin\theta$ and $y = u\sin\theta + v\cos\theta$, what value of θ will result in a (u, v) equation with no uv term? The straightforward (but admittedly tedious) way to find out is to make the substitutions, expand everything, gather and simplify, and see what turns out to be the coefficient of uv. We will leave the details to you as an exercise if you are interested, but suffice it to say that the coefficient of uv turns out to be $B\cos 2\theta + (C - A)\sin 2\theta$. Now all we need to do is find the (acute) angle θ that makes that coefficient zero:

$$B\cos 2\theta + (C-A)\sin 2\theta = 0$$

$$(A-C)\sin 2\theta = B\cos 2\theta$$

$$\cot 2\theta = \frac{A-C}{B}$$

Happily, it is not actually necessary to find θ; what we need are $\sin\theta$ and $\cos\theta$, which can be found directly from $\cot 2\theta$ with a little trigonometry.

Angle of Rotation to Eliminate the Cross-Product Term

Given the equation $Ax^2 + Bxy + Cy^2 + Dx + Ey + F = 0$, a rotation angle θ that will result in a transformed equation with no cross-product term is the angle θ such that

$$\cot 2\theta = \frac{A - C}{B} \text{ and } 0 < \theta < \frac{\pi}{2}.$$

(Actually, the restriction $0 < \theta < \frac{\pi}{2}$ is not necessary, but it is always *possible* to choose an acute angle θ, so we might as well keep it simple.)

EXAMPLE 3 Putting It All Together

Use rotation formulas to eliminate the cross-product term from the quadratic equation $x^2 + 4xy + 4y^2 - 30x - 90y + 450 = 0$. What kind of conic section is its graph?

SOLUTION The angle of rotation must satisfy $\cot 2\theta = \frac{A - C}{B} = \frac{1 - 4}{4} = -\frac{3}{4}$.

Since $0 < 2\theta < \pi$, we know that $\cos 2\theta = -\frac{3}{5}$.

Now apply the half-angle identities from Section 5.4. (We can assume that both $\cos\theta$ and $\sin\theta$ are positive.)

$$\cos\theta = \sqrt{\frac{1 + \cos 2\theta}{2}} = \sqrt{\frac{1 + (-3/5)}{2}} = \frac{1}{\sqrt{5}}$$

$$\sin\theta = \sqrt{\frac{1 - \cos 2\theta}{2}} = \sqrt{\frac{1 - (-3/5)}{2}} = \frac{2}{\sqrt{5}}$$

Thus

$$x = \frac{u}{\sqrt{5}} - \frac{2v}{\sqrt{5}}$$

$$y = \frac{2u}{\sqrt{5}} + \frac{v}{\sqrt{5}}$$

Make the substitutions:

$$\left(\frac{u}{\sqrt{5}} - \frac{2v}{\sqrt{5}}\right)^2 + 4\left(\frac{u}{\sqrt{5}} - \frac{2v}{\sqrt{5}}\right)\left(\frac{2u}{\sqrt{5}} + \frac{v}{\sqrt{5}}\right) + 4\left(\frac{2u}{\sqrt{5}} + \frac{v}{\sqrt{5}}\right)^2$$
$$- 30\left(\frac{u}{\sqrt{5}} - \frac{2v}{\sqrt{5}}\right) - 90\left(\frac{2u}{\sqrt{5}} + \frac{v}{\sqrt{5}}\right) + 450 = 0$$

Expand and gather:

$$u^2\left(\frac{1}{5} + \frac{8}{5} + \frac{16}{5}\right) + uv\left(-\frac{4}{5} - \frac{12}{5} + \frac{16}{5}\right) + v^2\left(\frac{4}{5} - \frac{8}{5} + \frac{4}{5}\right)$$
$$+ u(-6\sqrt{5} - 36\sqrt{5}) + v(12\sqrt{5} - 18\sqrt{5}) + 450 = 0$$

Simplify: $5u^2 - 42\sqrt{5}u - 6\sqrt{5}v + 450 = 0$.

(continued)

Complete the square and write in standard form: $\left(u - 21/\sqrt{5}\right)^2 = 6/\sqrt{5}\left(v - 3\sqrt{5}/10\right)$. This is a horizontal parabola with (u, v) vertex $\left(21/\sqrt{5}, 3\sqrt{5}/10\right)$. Using the same rotation equations, this corresponds to an (x, y) vertex of approximately (3.6, 8.7). Figure 8.36 shows the graph of the parabola with both pairs of coordinate axes.

Now try Exercise 37.

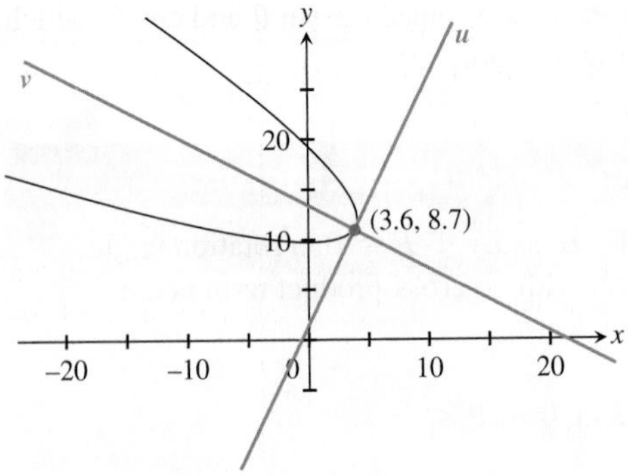

FIGURE 8.36 The graph of the conic with equation $x^2 + 4xy + 4y^2 - 30x - 90y + 450 = 0$. It is a parabola with vertex approximately $(3.6, 8.7)$. The vertex was found by analyzing the curve in a rotated coordinate system (blue axes). (Example 3)

As you can see, it can be quite an algebraic struggle to analyze a quadratic equation with a cross-product term. It can also be very difficult to produce a graph on a graphing calculator, although it is possible with some clever algebra. For example, the equation of the parabola in Example 3 can be written in the form

$$4y^2 + (4x - 90)y + (x^2 - 30x + 450) = 0,$$

which is a quadratic equation in the variable y. The quadratic formula can then be used to show that (in simplified form)

$$y = \frac{45 - 2x \pm \sqrt{225 - 60x}}{4},$$

giving us two equations of the form $y = f(x)$ that can be graphed in the same viewing window to produce a reasonable graph of the tilted parabola (Figure 8.37).

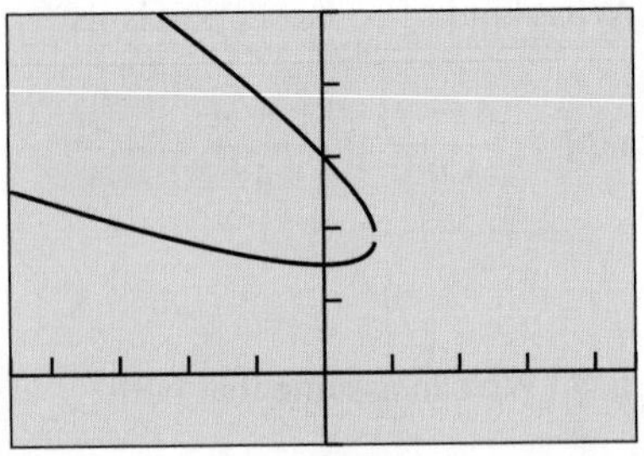

[–23, 23] by [–5, 25]
(a)

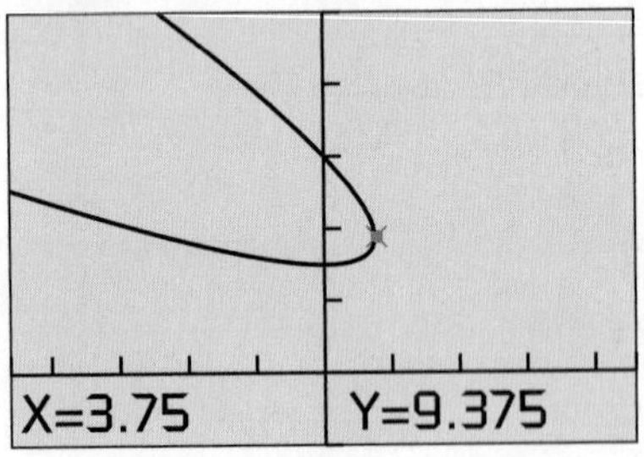

[–23, 23] by [–5, 25]
(b)

FIGURE 8.37 The graph of $x^2 + 4xy + 4y^2 - 30x - 90y + 450 = 0$ (a) with a gap and (b) with the TRACE feature activated at the connecting point. (Example 4)

Discriminant Test

Example 3 demonstrates that the algebra of rotation can get ugly. Fortunately, we can determine which type of conic a second-degree equation represents by looking at the sign of the **discriminant** $B^2 - 4AC$.

Discriminant Test

The second-degree equation $Ax^2 + Bxy + Cy^2 + Dx + Ey + F = 0$ graphs as

- a hyperbola if $B^2 - 4AC > 0$,
- a parabola if $B^2 - 4AC = 0$,
- an ellipse if $B^2 - 4AC < 0$,

except for degenerate cases.

This test hinges on the fact that the discriminant $B^2 - 4AC$ is **invariant under rotation**; in other words, even though A, B, and C do change when we rotate the coordinate axes, the combination $B^2 - 4AC$ maintains its value.

EXAMPLE 4 Revisiting Examples 1, 2, and 3

Use the Discriminant Test to identify the conic sections in Examples 1, 2, and 3.

SOLUTION In Example 1, $B^2 - 4AC = 1^2 - 4(0)(0) = 1$ with the (x, y) coefficients, and (immediately following the rotation) $B^2 - 4AC = 0^2 - 4(1/2)(-1/2) = 1$ with the (u, v) coefficients. The positive discriminant identifies the graph as a hyperbola.

In Example 2, $B^2 - 4AC = \left(-6\sqrt{3}\right)^2 - 4(13)(7) = -256$ with the (x, y) coefficients, and (immediately following the rotation) $B^2 - 4AC = 0^2 - 4(4)(16) = -256$ with the (u, v) coefficients. The negative discriminant identifies the graph as an ellipse.

In Example 3, $B^2 - 4AC = 4^2 - 4(1)(4) = 0$ with the (x, y) coefficients, and (immediately following the rotation) $B^2 - 4AC = 0^2 - 4(5)0 = 0$ with the (u, v) coefficients. The zero discriminant identifies the graph as a parabola.

Now try Exercise 45.

Not only is the discriminant $B^2 - 4AC$ invariant under rotation, but also its *sign* is invariant under translation and under algebraic manipulations that preserve the equivalence of the equation, such as multiplying both sides of the equation by a nonzero constant.

The discriminant test can be applied to degenerate conics. Table 8.2 displays the three basic types of conic sections grouped with their associated degenerate conics. Each conic or degenerate conic is shown with a sample equation and the sign of its discriminant.

Table 8.2 Conics and the Equation $Ax^2 + Bxy + Cy^2 + Dx + Ey + F = 0$

Conic	Sample Equation	A	B	C	D	E	F	Sign of Discriminant
Hyperbola	$x^2 - 2y^2 = 1$	1		−2			−1	Positive
Intersecting lines	$x^2 + xy = 0$	1	1					Positive
Parabola	$x^2 = 2y$	1				−2		Zero
Parallel lines	$x^2 = 4$	1					−4	Zero
One line	$y^2 = 0$			1				Zero
No graph	$x^2 = -1$	1					1	Zero
Ellipse	$x^2 + 2y^2 = 1$	1		2			−1	Negative
Circle	$x^2 + y^2 = 9$	1		1			−9	Negative
Point	$x^2 + y^2 = 0$	1		1				Negative
No graph	$x^2 + y^2 = -1$	1		1			1	Negative

EXAMPLE 5 Identifying Degenerate Conic Sections

Match the quadratic equations 1–5 with the appropriate graph descriptions A–E.

1. $\dfrac{(x-1)^2}{9} + \dfrac{(y-2)^2}{7} = 0$

2. $y^2 - x^2 = 0$

3. $y^2 + 2xy + x^2 = 0$

4. $(y - x)^2 = 1$

5. $(x-1)^2 + (y-3)^2 = -4$

A. The empty set

B. A single line

C. A single point

D. Two parallel lines

E. Two intersecting lines

SOLUTION Equation 1 looks like the standard equation of an ellipse, but notice the 0 on the right-hand side where the 1 should be. The only point that can satisfy this equation is $(1, 2)$, as any other x or y value will make the left-hand side positive. The match (a degenerate ellipse) is C.

Equation 2 would be a hyperbola by the Discriminant Test, but, again, notice the 0 where the 1 should be. The equation factors as $(y - x)(y + x) = 0$, so $y = x$ or $y = -x$. The match (a degenerate hyperbola) is E.

Equation 3 would be a parabola by the Discriminant Test, but it factors as $(y + x)^2 = 0$. The only solution is the line $y = -x$. The match (a degenerate parabola) is B.

Equation 4 would also be a parabola by the Discriminant Test, but it is equivalent to $y - x = \pm 1$. The two possibilities give the lines $y = x + 1$ and $y = x - 1$. The match (a degenerate parabola) is D.

Equation 5 looks almost like a circle, but notice the -4 where the square of the radius should be. In fact, the sum of two squares cannot be negative at all. The match (a degenerate circle, hence a degenerate ellipse) is A. **Now try Exercises 53–56.**

QUICK REVIEW 8.4 *(For help, go to Sections 4.7 and 5.4.)*

Exercise numbers with a gray background indicate problems that the authors have designed to be solved *without a calculator*.

In Exercises 1–10, use trigonometric identities and assume $0 < \theta < \frac{\pi}{2}$.

1. Given that $\cot 2\theta = 5/12$, find $\cos 2\theta$.
2. Given that $\cot 2\theta = 8/15$, find $\cos 2\theta$.
3. Given that $\cot 2\theta = 1/\sqrt{3}$, find $\cos 2\theta$.
4. Given that $\cot 2\theta = 2/\sqrt{5}$, find $\cos 2\theta$.
5. Given that $\cot 2\theta = 0$, find θ.
6. Given that $\cot 2\theta = \sqrt{3}$, find θ.
7. Given that $\cot 2\theta = 3/4$, find $\cos \theta$.
8. Given that $\cot 2\theta = 3/\sqrt{7}$, find $\cos \theta$.
9. Given that $\cot 2\theta = 5/\sqrt{11}$, find $\sin \theta$.
10. Given that $\cot 2\theta = 45/28$, find $\sin \theta$.

SECTION 8.4 Exercises

In Exercises 1–4, find the vertices and foci of the conic section *without* axis rotation by analyzing the graph geometrically in the xy-plane. (Note that all four graphs are symmetric about the lines $y = x$ and $y = -x$.)

1. The hyperbola $xy = 4$
2. The hyperbola $xy = -4$
3. The ellipse $5x^2 - 6xy + 5y^2 = 16$
4. The ellipse $5x^2 + 6xy + 5y^2 = 16$

In Exercises 5–8, use the substitutions $x = u/\sqrt{2} - v/\sqrt{2}$ and $y = u/\sqrt{2} + v/\sqrt{2}$ to convert the equation to a (u, v) equation with no cross-product term.

5. $xy = 4$
6. $xy = -4$
7. $5x^2 - 6xy + 5y^2 = 16$
8. $5x^2 + 6xy + 5y^2 = 16$

In Exercises 9–16, find the linear equations that can be used to convert an (x, y) equation to a (u, v) equation using the given angle of rotation θ.

9. $\theta = \pi/6$
10. $\theta = \pi/3$
11. $\theta = \cos^{-1}(3/5)$
12. $\theta = \tan^{-1}(5/12)$
13. θ such that $\cot 2\theta = 4/3$
14. θ such that $\cot 2\theta = -7/24$
15. $\theta = 40°$
16. $\theta = 70°$

In Exercises 17–24, quadratic equations have been transformed using the rotation equations $x = 0.8u - 0.6v$ and $y = 0.6u + 0.8v$. Convert the given points from (u, v) coordinates back to (x, y) coordinates.

17. parabola vertex $(5, 0)$
18. parabola focus $(4, 2)$
19. ellipse vertices $(0, \pm 4)$
20. hyperbola foci $(\pm 5, 0)$
21. parabola vertex $(3, 7)$
22. parabola focus $(-4, 6)$
23. ellipse center $(12, 5)$
24. hyperbola center $(8, -2)$

In Exercises 25–34, determine axis rotation formulas for x and y that will transform the quadratic equation to an equation in (u, v) coordinates with no cross-product term. (You do not need to find the transformed equation.)

25. $3xy + 17 = 0$
26. $-5xy + 22 = 0$
27. $x^2 + xy + y^2 + 1 = 0$
28. $x^2 - 3xy + y^2 + y = 0$
29. $x^2 - 6xy + y^2 + x + y = 0$
30. $4x^2 + 3xy + 4y^2 - 5x + 30 = 0$
31. $x^2 + 2\sqrt{3}xy + 3y^2 - x + 3 = 0$
32. $2x^2 + \sqrt{3}xy + 3y^2 - 2y + 5 = 0$
33. $24xy - 7y^2 - 4x + 5 = 0$
34. $4x^2 + 24xy - 3y^2 + 17 = 0$

In Exercises 35–40, the graph of each equation is a conic section.

(a) Identify the type of conic section.

(b) Use axis rotation formulas for x and y to transform the quadratic equation to an equation in (u, v) coordinates with no cross-product term.

(c) Find the vertex or vertices in the (u, v) system

(d) Write the vertex or vertices in the (x, y) system.

35. $xy = 8$
36. $3xy + 15 = 0$
37. $2x^2 + \sqrt{3}xy + y^2 - 10 = 0$
38. $x^2 + xy + y^2 = 3$
39. $9x^2 + 24xy + 16y^2 + 20x - 15y + 75 = 0$
40. $9x^2 - 24xy + 16y^2 - 20x - 15y = 50$
41. **Backward Rotation** Find and simplify the axis rotation formulas for a rotation angle of $-\theta$.
42. **Converting Back** If the usual axis rotation formulas are used to convert an (x, y) equation into a (u, v) equation, find two linear equations that will convert the (u, v) equation back into the (x, y) equation. (*Hint:* Look at Exercise 41.)

In Exercises 43–52, use the discriminant $B^2 - 4AC$ to decide whether the equation represents a parabola, an ellipse, or a hyperbola.

43. $x^2 - 4xy + 10y^2 + 2y - 5 = 0$
44. $x^2 - 4xy + 3x + 25y - 6 = 0$
45. $9x^2 - 6xy + y^2 - 7x + 5y = 0$
46. $-xy + 3y^2 - 4x + 2y + 8 = 0$
47. $8x^2 - 4xy + 2y^2 + 6 = 0$
48. $3x^2 - 12xy + 4y^2 + x - 5y - 4 = 0$
49. $x^2 - 3y^2 - y - 22 = 0$
50. $5x^2 + 4xy + 3y^2 + 2x + y = 0$
51. $4x^2 - 2xy + y^2 - 5x + 18 = 0$
52. $6x^2 - 4xy + 9y^2 - 40x + 20y - 56 = 0$

In Exercises 53–56, match the equation on the left with the degenerate conic section on the right that would describe its graph. The matching is one-to-one.

53. $y^2 + 4x^2 = -1$	A. A single point
54. $y^2 - 4x^2 = 0$	B. Two intersecting lines
55. $y^2 + 4x^2 = 0$	C. Two parallel lines
56. $y^2 - 4 = 0$	D. The empty set

57. True or False The graph of the equation $Ax^2 + Cy^2 + Dx + Ey + F = 0$ (A and C not both zero) has a focal axis aligned with the coordinate axes. Justify your answer.

58. True or False The graph of the equation $x^2 + y^2 + Dx + Ey + F = 0$ is a circle or a degenerate circle. Justify your answer.

59. Multiple Choice Which of the following is likely to change when the equation of a hyperbola is transformed by rotation to eliminate the cross-product term?

(A) The length of the transverse axis

(B) The eccentricity

(C) The distance between the foci

(D) The slope of the focal axis

(E) The angle at which the asymptotes intersect

60. Multiple Choice Which of the following is **not** a reason to rotate the axes of a conic?

(A) To simplify its equation

(B) To eliminate the cross-product term

(C) To place its center or vertex at the origin

(D) To make it easier to identify its type

(E) To make it easier to sketch by hand

61. Multiple Choice The graph of the quadratic equation $(y - 2x + 4)^2 = 0$ is degenerate, consisting of

(A) A single point

(B) Two intersecting lines

(C) Two parallel lines

(D) A single line

(E) The empty set

62. Multiple Choice The asymptotes of the hyperbola $xy = 4$ are

(A) $y = x, y = -x.$

(B) $y = 2x, y = -\dfrac{x}{2}.$

(C) $y = -2x, y = \dfrac{x}{2}.$

(D) $y = 4x, y = -\dfrac{x}{4}.$

(E) the coordinate axes.

Explorations

63. The Effect of Rotation on Asymptotes The equation of a hyperbola has been transformed from (x, y) coordinates to (u, v) coordinates using the axis rotation formulas for $\theta = \pi/6$. If the asymptotes in the (u, v) system have slopes ± 1, find the slopes of the asymptotes in the (x, y) system.

64. The Discriminant Determine what happens to the sign of $B^2 - 4AC$ within the equation

$$Ax^2 + Bxy + Cy^2 + Dx + Ey + F = 0$$

when

(a) the axes are translated h units horizontally and k units vertically;

(b) both sides of the equation are multiplied by the same nonzero constant k.

Extending the Ideas

65. Group Activity Use the axis rotation formulas to transform the general quadratic equation $Ax^2 + Bxy + Cy^2 + Dx + Ey + F = 0$ into an equation in (u, v). Working together, show that the new equation can be simplified to form a new quadratic equation $A'u^2 + B'uv + C'v^2 + D'u + E'v + F' = 0$ in which

$$A' = A\cos^2\theta + B\cos\theta\sin\theta + C\sin^2\theta$$

$$B' = B\cos 2\theta + (C - A)\sin 2\theta$$

$$C' = C\cos^2\theta - B\cos\theta\sin\theta + A\sin^2\theta$$

$$D' = D\cos\theta + E\sin\theta$$

$$E' = E\cos\theta - D\sin\theta$$

$$F' = F$$

66. Identifying a Conic Develop a way to decide whether $Ax^2 + Cy^2 + Dx + Ey + F = 0$, with A and C not both 0, represents a parabola, an ellipse, or a hyperbola. Write an example to illustrate each of the three cases.

67. Rotational Invariant Using the formulas in Exercise 65, prove that

$$B'^2 - 4A'C' = B^2 - 4AC,$$

that is, that the discriminant is invariant under axis rotation.

68. Other Rotational Invariants Prove that each of the following are invariants under rotation:

(a) F **(b)** $A + C$ **(c)** $D^2 + E^2$

69. Degenerate Conics Graph all of the degenerate conics listed in Table 8.2. Recall that *degenerate cones* occur when the generator and axis of the cone are parallel or perpendicular. (See Figure 8.2.) Explain the occurrence of all of the degenerate conics listed on the basis of cross sections of typical or degenerate right circular cones.

8.5 Polar Equations of Conics

What you'll learn about

- Eccentricity Revisited
- Writing Polar Equations for Conics
- Analyzing Polar Equations of Conics
- Orbits Revisited

... and why

You will learn the approach to conics used by astronomers.

Eccentricity Revisited

Eccentricity and polar coordinates provide ways to see once again that parabolas, ellipses, and hyperbolas are a unified family of interrelated curves. We can define these three curves simultaneously by generalizing the focus-directrix definition of a parabola given in Section 8.1.

Focus-Directrix Definition of a Conic Section

A **conic section** is the set of all points in a plane whose distances from a particular point (the **focus**) and a particular line (the **directrix**) in the plane have a constant ratio (Figure 8.38). (We assume that the focus does not lie on the directrix.)

The line passing through the focus and perpendicular to the directrix is the **(focal) axis** of the conic section. The axis is a line of symmetry for the conic. Each point where the conic intersects its axis is a **vertex** of the conic. If P is a point of the conic, F is the focus, and D is the point of the directrix closest to P, then the constant ratio PF/PD is the **eccentricity** e of the conic (see Figure 8.38). A parabola has one focus and one directrix. Ellipses and hyperbolas have two focus-directrix pairs, and either focus-directrix pair can be used with the eccentricity to generate the entire conic section.

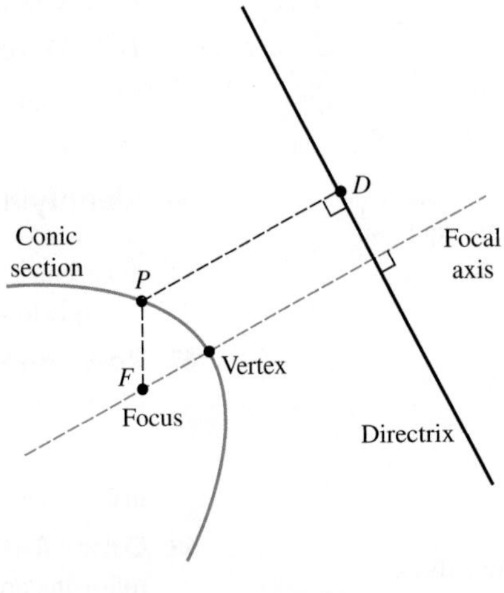

FIGURE 8.38 The geometric structure of a conic section.

Focus-Directrix-Eccentricity Relationship

If P is a point of a conic section, F is the conic's focus, and D is the point of the directrix closest to P, then

$$e = \frac{PF}{PD} \quad \text{and} \quad PF = e \cdot PD,$$

where e is a constant and the eccentricity of the conic. Moreover, the conic is

- a hyperbola if $e > 1$,
- a parabola if $e = 1$,
- an ellipse if $e < 1$.

Remarks

- To be consistent with our work on parabolas, we could use $2p$ for the distance from the focus to the directrix, but following George B. Thomas, Jr., we use k for this distance. This simplifies our polar equations of conics.
- Rather than religiously using polar coordinates and equations, we use a mixture of the polar and Cartesian systems. So, for example, we use $x = k$ for the directrix rather than $r \cos \theta = k$ or $r = k \sec \theta$.

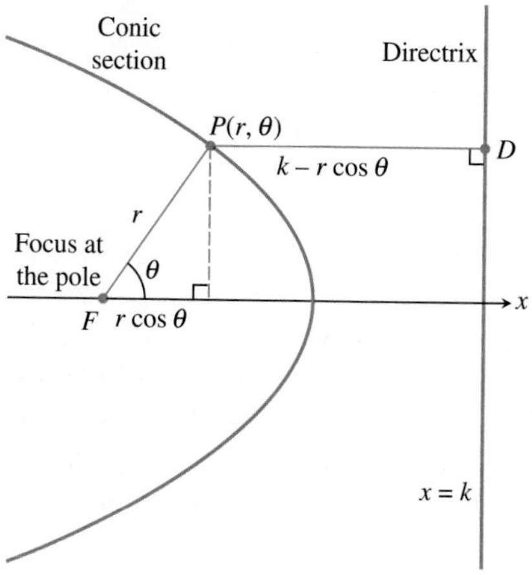

FIGURE 8.39 A conic section in the polar plane.

In this approach to conic sections, the eccentricity e is a strictly positive constant, and there are no circles or other degenerate conics.

Writing Polar Equations for Conics

Our focus-directrix definition of conics works best in combination with polar coordinates. Recall that in polar coordinates the origin is the *pole* and the x-axis is the *polar axis*. To obtain a polar equation for a conic section, we position the pole at the conic's focus and the polar axis along the focal axis with the directrix to the right of the pole (Figure 8.39). If the distance from the focus to the directrix is k, the Cartesian equation of the directrix is $x = k$. From Figure 8.39, we see that

$$PF = r \quad \text{and} \quad PD = k - r\cos\theta.$$

So the equation $PF = e \cdot PD$ becomes

$$r = e(k - r\cos\theta),$$

which when solved for r is

$$r = \frac{ke}{1 + e\cos\theta}.$$

In Exercise 53, you are asked to show that this equation is still valid if $r < 0$ or $r\cos\theta > k$. This one equation can produce all sizes and shapes of nondegenerate conic sections. Figure 8.40 shows three typical graphs for this equation. In Exploration 1, you will investigate how changing the value of e affects the graph of $r = ke/(1 + e\cos\theta)$.

EXPLORATION 1 Graphing Polar Equations of Conics

Set your grapher to Polar and Dot graphing modes, and to Radian mode. Using $k = 3$, an xy window of $[-12, 24]$ by $[-12, 12]$, $\theta\text{min} = 0$, $\theta\text{max} = 2\pi$, and $\theta\text{step} = \pi/48$, graph

$$r = \frac{ke}{1 + e\cos\theta}$$

for $e = 0.7, 0.8, 1, 1.5, 3$. Identify the type of conic section obtained for each e value. Overlay the five graphs, and explain how changing the value of e affects the graph of $r = ke/(1 + e\cos\theta)$. Explain how the five graphs are similar and how they are different.

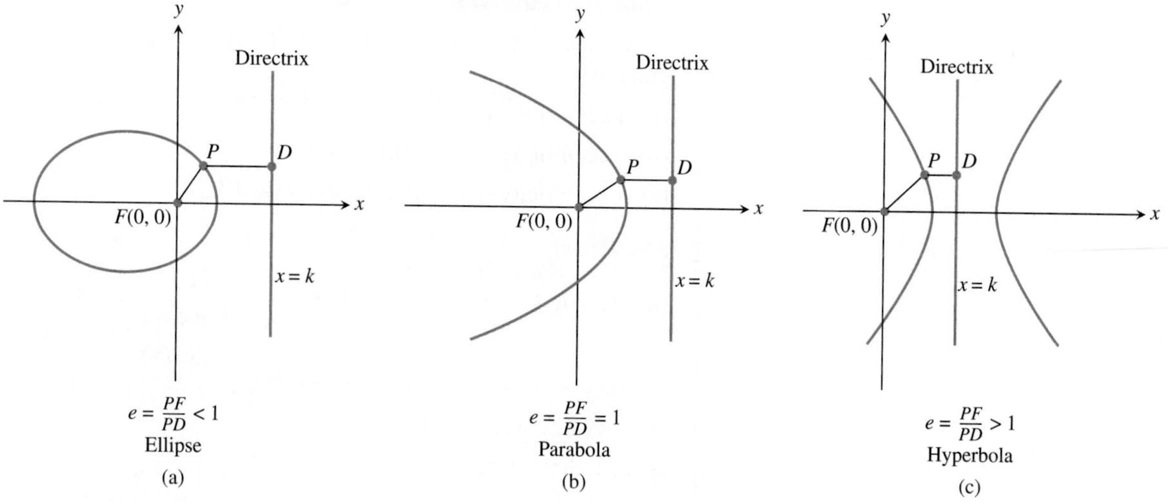

FIGURE 8.40 The three types of conics possible for $r = ke/(1 + e\cos\theta)$.

Polar Equations for Conics

The four standard orientations of a conic in the polar plane are as follows.

(a) $r = \dfrac{ke}{1 + e\cos\theta}$

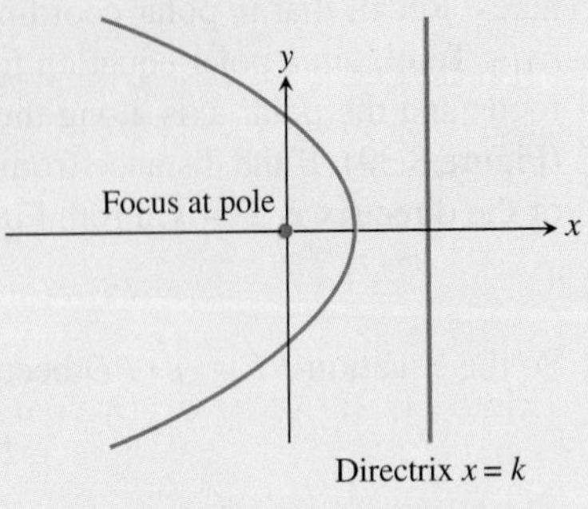

(a)

(b) $r = \dfrac{ke}{1 - e\cos\theta}$

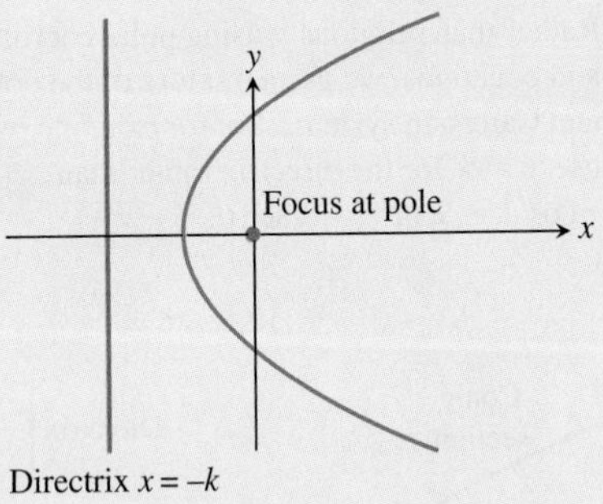

(b)

(c) $r = \dfrac{ke}{1 + e\sin\theta}$

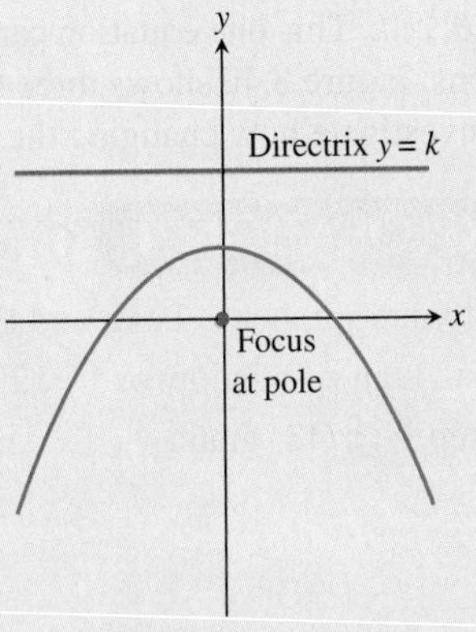

(c)

(d) $r = \dfrac{ke}{1 - e\sin\theta}$

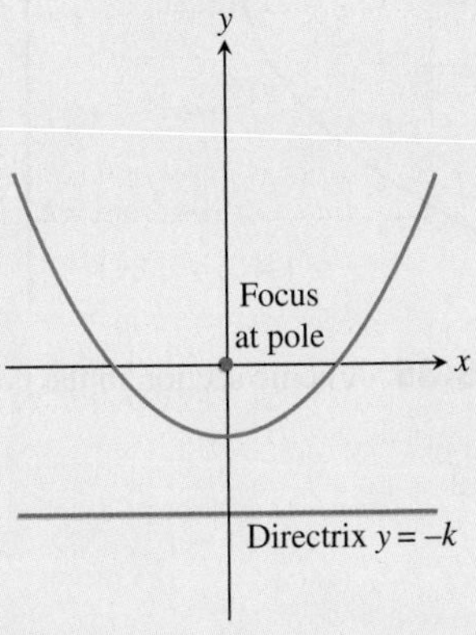

(d)

EXAMPLE 1 Writing and Graphing Polar Equations of Conics

Given that the focus is at the pole, write a polar equation for the specified conic and graph it.

(a) Eccentricity $e = 3/5$, directrix $x = 2$.

(b) Eccentricity $e = 1$, directrix $x = -2$.

(c) Eccentricity $e = 3/2$, directrix $y = 4$.

SOLUTION

(a) Setting $e = 3/5$ and $k = 2$ in $r = \dfrac{ke}{1 + e\cos\theta}$ yields

$$r = \frac{2(3/5)}{1 + (3/5)\cos\theta} = \frac{6}{5 + 3\cos\theta}$$

Figure 8.41a shows this ellipse and the given directrix.

(b) Setting $e = 1$ and $k = 2$ in $r = \dfrac{ke}{1 - e\cos\theta}$ yields

$$r = \frac{2}{1 - \cos\theta}.$$

Figure 8.41b shows this parabola and its directrix.

(c) Setting $e = 3/2$ and $k = 4$ in $r = \dfrac{ke}{1 + e\sin\theta}$ yields

$$r = \frac{4(3/2)}{1 + (3/2)\sin\theta}$$

$$= \frac{12}{2 + 3\sin\theta}$$

Figure 8.41c shows this hyperbola and the given directrix. **Now try Exercise 1.**

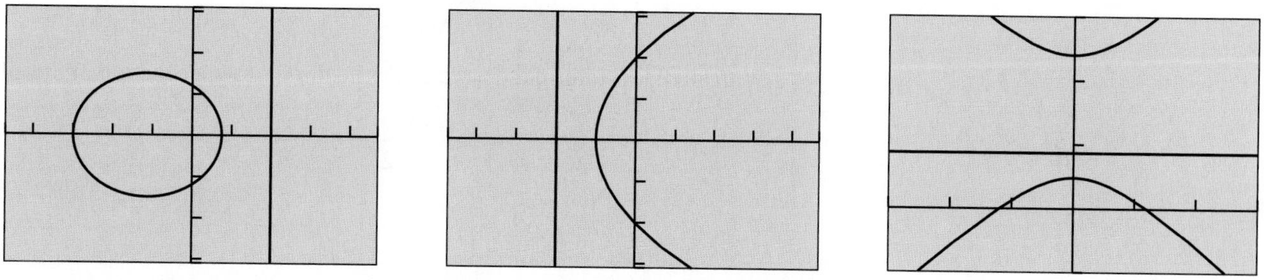

[–4.7, 4.7] by [–3.1, 3.1]
(a)

[–4.7, 4.7] by [–3.1, 3.1]
(b)

[–15,15] by [–5, 15]
(c)

FIGURE 8.41 Graphs for Example 1.

Analyzing Polar Equations of Conics

The first step in analyzing the polar equations of a conic section is to use the eccentricity to identify which type of conic the equation represents. Then we determine the equation of the directrix.

EXAMPLE 2 Identifying Conics from Their Polar Equations

Determine the eccentricity, the type of conic, and the directrix.

(a) $r = \dfrac{6}{2 + 3\cos\theta}$ **(b)** $r = \dfrac{6}{4 - 3\sin\theta}$

SOLUTION

(a) Dividing numerator and denominator by 2 yields $r = 3/(1 + 1.5\cos\theta)$. So the eccentricity $e = 1.5$, and thus the conic is a hyperbola. The numerator $ke = 3$, so $k = 2$, and thus the equation of the directrix is $x = 2$.

(b) Dividing numerator and denominator by 4 yields $r = 1.5/(1 - 0.75\sin\theta)$. So the eccentricity $e = 0.75$, and thus the conic is an ellipse. The numerator $ke = 1.5$, so $k = 2$, and thus the equation of the directrix is $y = -2$. **Now try Exercise 7.**

All of the geometric properties and features of parabolas, ellipses, and hyperbolas developed in Sections 8.1–8.3 still apply in the polar coordinate setting. In Example 3 we use this prior knowledge.

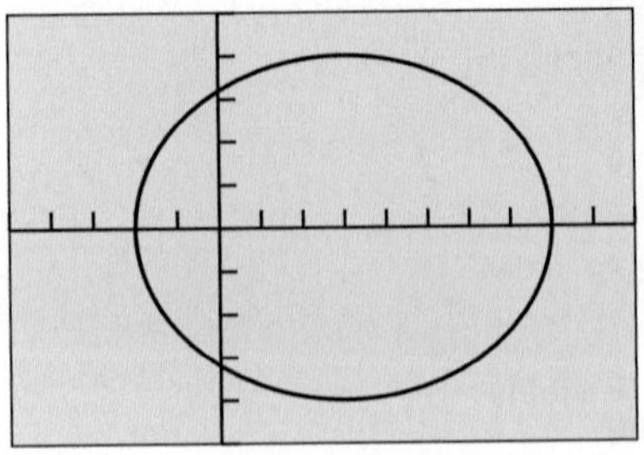

[–5, 10] by [–5, 5]

FIGURE 8.42 Graph of the ellipse $r = 16/(5 - 3\cos\theta)$. (Example 3)

EXAMPLE 3 Analyzing a Conic

Analyze the conic section given by the equation $r = 16/(5 - 3\cos\theta)$. Include in the analysis the values of e, a, b, and c.

SOLUTION Dividing numerator and denominator by 5 yields

$$r = \frac{3.2}{1 - 0.6\cos\theta}.$$

So the eccentricity $e = 0.6$, and thus the conic is an ellipse. Figure 8.42 shows this ellipse. The vertices (endpoints of the major axis) have polar coordinates $(8, 0)$ and $(2, \pi)$. So $2a = 8 + 2 = 10$, and thus $a = 5$.

The vertex $(2, \pi)$ is 2 units to the left of the pole, and the pole is a focus of the ellipse. So $a - c = 2$, and thus $c = 3$. An alternative way to find c is to use the fact that the eccentricity of an ellipse is $e = c/a$, and thus $c = ae = 5 \cdot 0.6 = 3$.

To find b we use the Pythagorean relation of an ellipse:

$$b = \sqrt{a^2 - c^2} = \sqrt{25 - 9} = 4$$

With all of this information, we can write the Cartesian equation of the ellipse:

$$\frac{(x - 3)^2}{25} + \frac{y^2}{16} = 1$$

Now try Exercise 31.

Orbits Revisited

Polar equations for conics are used extensively in celestial mechanics, the branch of astronomy based on the work of Kepler and others who have studied the motion of celestial bodies. The polar equations of conic sections are well suited to the *two-body problem* of celestial mechanics for several reasons. First, the same equations are used for ellipses, parabolas, and hyperbolas—the paths of one body traveling about another. Second, a focus of the conic is always at the pole. This arrangement has two immediate advantages:

- The pole can be thought of as the center of the larger body, such as the Sun, with the smaller body, such as Earth, following a conic path about the larger body.
- The coordinates given by a polar equation are the distance r between the two bodies and the direction θ from the larger body to the smaller body relative to the axis of the conic path of motion.

For these reasons, polar coordinates are preferred over Cartesian coordinates for studying orbital motion.

To use the data in Table 8.3 to create polar equations for the elliptical orbits of the planets, we need to express the equation $r = ke/(1 + e\cos\theta)$ in terms of a and e. We apply the formula $PF = e \cdot PD$ to the ellipse shown in Figure 8.43:

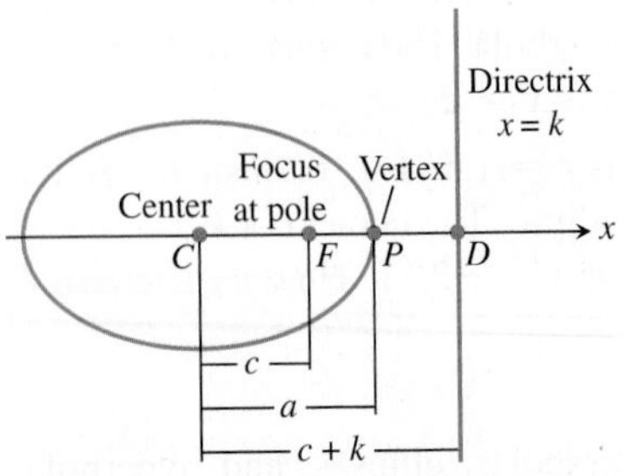

FIGURE 8.43 Geometric relationships within an ellipse.

$e \cdot PD = PF$		
$e(c + k - a) = a - c$	From Figure 8.45
$e(ae + k - a) = a - ae$	Use $e = c/a$.
$ae^2 + ke - ae = a - ae$	Distribute the e.
$ae^2 + ke = a$	Add ae.
$ke = a - ae^2$	Subtract ae^2.
$ke = a(1 - e^2)$	Factor.

Table 8.3 Semimajor Axes and Eccentricities of the Planets

Planet	Semimajor Axis (Gm)	Eccentricity
Mercury	57.9	0.2056
Venus	108.2	0.0068
Earth	149.6	0.0167
Mars	227.9	0.0934
Jupiter	778.3	0.0485
Saturn	1427	0.0560
Uranus	2869	0.0461
Neptune	4497	0.0050

Source: Shupe, et al., National Geographic Atlas of the World (rev. 6th ed.). Washington, DC: National Geographic Society, 1992, plate 116, and other sources.

So the equation $r = ke/(1 + e\cos\theta)$ can be rewritten as follows:

Ellipse with Eccentricity *e* and Semimajor Axis *a*

$$r = \frac{a(1 - e^2)}{1 + e\cos\theta}$$

In this form of the equation, when $e = 0$, the equation reduces to $r = a$, the equation of a circle with radius a.

EXAMPLE 4 Analyzing a Planetary Orbit

Using data from Table 8.3, find a polar equation for the orbit of Mercury, and use it to approximate its aphelion (farthest distance from the Sun) and perihelion (closest distance to the Sun).

SOLUTION Setting $e = 0.2056$ and $a = 57.9$ in

$$r = \frac{a(1 - e^2)}{1 + e\cos\theta} \quad \text{yields} \quad r = \frac{57.9(1 - 0.2056^2)}{1 + 0.2056\cos\theta}.$$

Mercury's aphelion is

$$r = \frac{57.9(1 - 0.2056^2)}{1 - 0.2056} \approx 69.8 \text{ Gm}.$$

Mercury's perihelion is

$$r = \frac{57.9(1 - 0.2056^2)}{1 + 0.2056} \approx 46.0 \text{ Gm}.$$

Now try Exercise 41.

QUICK REVIEW 8.5 *(For help, go to Section 6.4.)*

Exercise numbers with a gray background indicate problems that the authors have designed to be solved *without a calculator*.

In Exercises 1 and 2, solve for r.

1. $(3, \theta) = (r, \theta + \pi)$
2. $(-2, \theta) = (r, \theta + \pi)$

In Exercises 3 and 4, solve for θ.

3. $(1.5, \pi/6) = (-1.5, \theta), -2\pi \le \theta \le 2\pi$
4. $(-3, 4\pi/3) = (3, \theta), -2\pi \le \theta \le 2\pi$

In Exercises 5 and 6, find the focus and directrix of the parabola.

5. $x^2 = 16y$
6. $y^2 = -12x$

In Exercises 7–10, find the foci and vertices of the conic.

7. $\frac{x^2}{9} + \frac{y^2}{4} = 1$
8. $\frac{y^2}{25} + \frac{x^2}{9} = 1$
9. $\frac{x^2}{16} - \frac{y^2}{9} = 1$
10. $\frac{y^2}{36} - \frac{x^2}{4} = 1$

SECTION 8.5 Exercises

In Exercises 1–6, find a polar equation for the conic with a focus at the pole and the given eccentricity and directrix. Identify the conic, and graph it.

1. $e = 1, x = -2$
2. $e = 5/4, x = 4$
3. $e = 3/5, y = 4$
4. $e = 1, y = 2$
5. $e = 7/3, y = -1$
6. $e = 2/3, x = -5$

In Exercises 7–14, determine the eccentricity, type of conic, and directrix.

7. $r = \dfrac{2}{1 + \cos\theta}$
8. $r = \dfrac{6}{1 + 2\cos\theta}$
9. $r = \dfrac{5}{2 - 2\sin\theta}$
10. $r = \dfrac{2}{4 - \cos\theta}$
11. $r = \dfrac{20}{6 + 5\sin\theta}$
12. $r = \dfrac{42}{2 - 7\sin\theta}$
13. $r = \dfrac{6}{5 + 2\cos\theta}$
14. $r = \dfrac{20}{2 + 5\sin\theta}$

In Exercises 15–20, match the polar equation with its graph, and identify the viewing window.

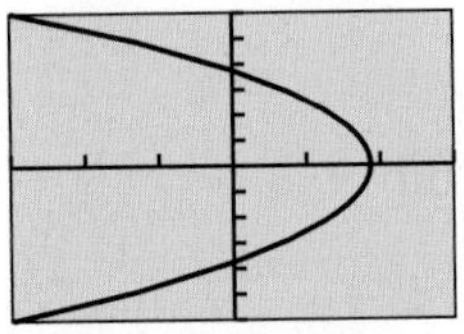
(a)

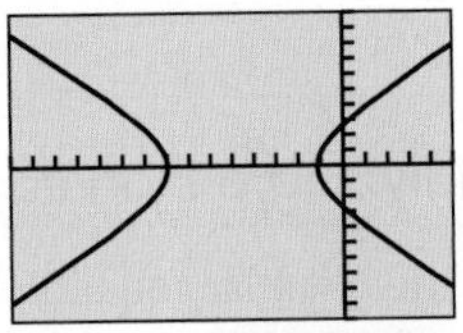
(b)

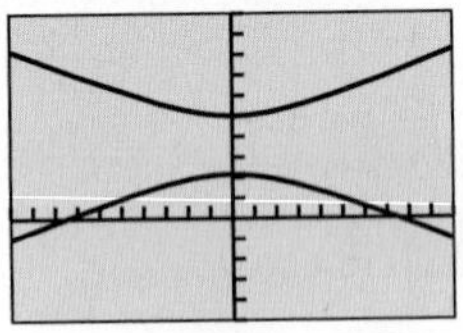
(c)

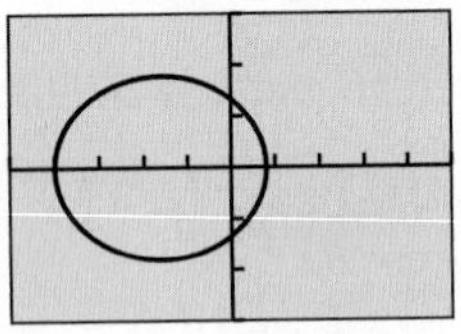
(d)

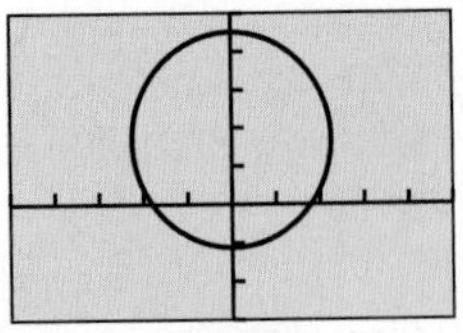
(e)

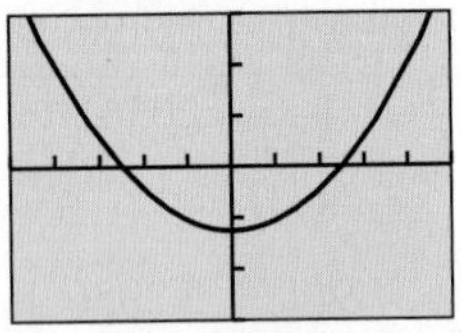
(f)

15. $r = \dfrac{8}{3 - 4\cos\theta}$
16. $r = \dfrac{4}{3 + 2\cos\theta}$
17. $r = \dfrac{5}{2 - 2\sin\theta}$
18. $r = \dfrac{9}{5 - 3\sin\theta}$
19. $r = \dfrac{15}{2 + 5\sin\theta}$
20. $r = \dfrac{15}{4 + 4\cos\theta}$

In Exercises 21–24, find a polar equation for the ellipse with a focus at the pole and the given polar coordinates as the endpoints of its major axis.

21. $(1.5, 0)$ and $(6, \pi)$
22. $(1.5, 0)$ and $(1, \pi)$
23. $(1, \pi/2)$ and $(3, 3\pi/2)$
24. $(3, \pi/2)$ and $(0.75, -\pi/2)$

In Exercises 25–28, find a polar equation for the hyperbola with a focus at the pole and the given polar coordinates as the endpoints of its transverse axis.

25. $(3, 0)$ and $(-15, \pi)$
26. $(-3, 0)$ and $(1.5, \pi)$
27. $\left(2.4, \dfrac{\pi}{2}\right)$ and $\left(-12, \dfrac{3\pi}{2}\right)$
28. $\left(-6, \dfrac{\pi}{2}\right)$ and $\left(2, \dfrac{3\pi}{2}\right)$

In Exercises 29 and 30, find a polar equation for the conic with a focus at the pole.

29.

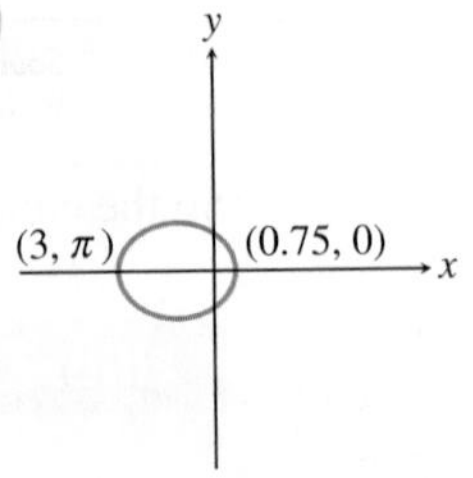

30.

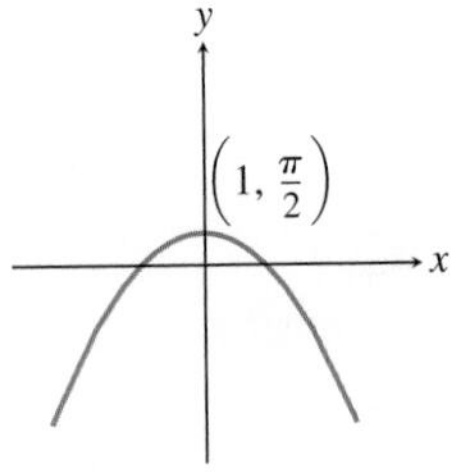

In Exercises 31–36, graph the conic, and find the values of e, a, b, and c.

31. $r = \dfrac{21}{5 - 2\cos\theta}$
32. $r = \dfrac{11}{6 - 5\sin\theta}$
33. $r = \dfrac{24}{4 + 2\sin\theta}$
34. $r = \dfrac{16}{5 + 3\cos\theta}$
35. $r = \dfrac{16}{3 + 5\cos\theta}$
36. $r = \dfrac{12}{1 - 5\sin\theta}$

In Exercises 37 and 38, determine a Cartesian equation for the given polar equation.

37. $r = \dfrac{4}{2 - \sin\theta}$
38. $r = \dfrac{6}{1 + 2\cos\theta}$

In Exercises 39 and 40, use the fact that $k = 2p$ is twice the focal length and half the focal width, to determine a Cartesian equation of the parabola whose polar equation is given.

39. $r = \dfrac{4}{2 - 2\cos\theta}$
40. $r = \dfrac{12}{3 + 3\cos\theta}$

41. Halley's Comet The orbit of Halley's comet has a semimajor axis of 18.09 AU and an orbital eccentricity of 0.97. Compute its perihelion and aphelion distances.

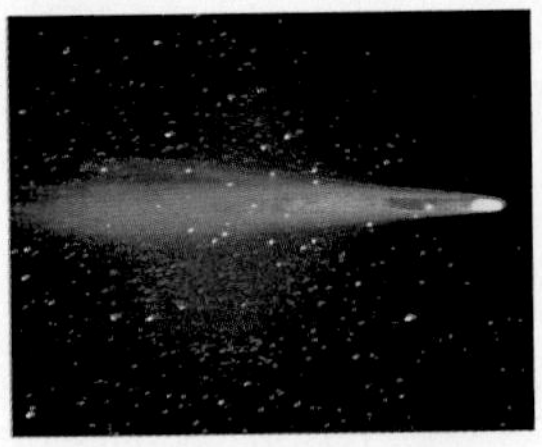

42. Uranus The orbit of the planet Uranus has a semimajor axis of 19.18 AU and an orbital eccentricity of 0.0461. Compute its perihelion and aphelion distances.

In Exercises 43 and 44, the velocity of an object traveling in a circular orbit of radius r (distance from center of planet in meters) around a planet is given by

$$v = \sqrt{\frac{3.99 \times 10^{14}\, k}{r}} \text{ m/sec,}$$

where k is a constant related to the mass of the planet and the orbiting object.

43. **Group Activity Lunar Module** A lunar excursion module is in a circular orbit 250 km above the surface of the Moon. Assume that the Moon's radius is 1740 km and that $k = 0.012$. Find the following.

(a) the velocity of the lunar module

(b) the length of time required for the lunar module to circle the Moon once

44. **Group Activity Mars Satellite** A satellite is in a circular orbit 1000 mi above Mars. Assume that the radius of Mars is 2100 mi and that $k = 0.11$. Find the velocity of the satellite.

Standardized Test Questions

45. **True or False** The equation $r = ke/(1 + e\cos\theta)$ yields no true circles. Justify your answer.

46. **True or False** The equation $r = a(1 - e^2)/(1 + e\cos\theta)$ yields no true parabolas. Justify your answer.

In Exercises 47–50, solve the problem without using a calculator.

47. **Multiple Choice** Which ratio of distances is constant for a point on a nondegenerate conic?

(A) Distance to center : distance to directrix

(B) Distance to focus : distance to vertex

(C) Distance to vertex : distance to directrix

(D) Distance to focus : distance to directrix

(E) Distance to center : distance to vertex

48. **Multiple Choice** Which type of conic section has an eccentricity greater than one?

(A) An ellipse

(B) A parabola

(C) A hyperbola

(D) Two parallel lines

(E) A circle

49. **Multiple Choice** For a conic expressed by $r = ke/(1 + e\sin\theta)$, which point is located at the pole?

(A) The center

(B) A focus

(C) A vertex

(D) An endpoint of the minor axis

(E) An endpoint of the conjugate axis

50. **Multiple Choice** Which of the following is **not** a polar equation of a conic?

(A) $r = 1 + 2\cos\theta$

(B) $r = 1/(1 + \sin\theta)$

(C) $r = 3$

(D) $r = 1/(2 - \cos\theta)$

(E) $r = 1/(1 + 2\cos\theta)$

Explorations

51. **Planetary Orbits** Use the polar equation $r = a(1 - e^2)/(1 + e\cos\theta)$ in completing the following activities.

(a) Use the fact that $-1 \le \cos\theta \le 1$ to prove that the perihelion distance of any planet is $a(1 - e)$ and the aphelion distance is $a(1 + e)$.

(b) Use $e = c/a$ to confirm that $a(1 - e) = a - c$ and $a(1 + e) = a + c$.

(c) Use the formulas $a(1 - e)$ and $a(1 + e)$ to compute the perihelion and aphelion distances of each planet listed in Table 8.4.

(d) For which of these planets is the difference between the perihelion and aphelion distance the greatest?

Table 8.4 Semimajor Axes and Eccentricities of the Six Innermost Planets

Planet	Semimajor Axis (AU)	Eccentricity
Mercury	0.3871	0.206
Venus	0.7233	0.007
Earth	1.0000	0.017
Mars	1.5237	0.093
Jupiter	5.2026	0.048
Saturn	9.5547	0.056

Source: Encrenaz & Bibring. The Solar System (2nd ed.). New York: Springer, p. 5.

52. **Using the Astronomer's Equation for Conics** Using Dot mode, $a = 2$, an xy window of $[-13, 5]$ by $[-6, 6]$, $\theta\min = 0$, $\theta\max = 2\pi$, and $\theta\text{step} = \pi/48$, graph $r = a(1 - e^2)/(1 + e\cos\theta)$ for $e = 0, 0.3, 0.7, 1.5, 3$. Identify the type of conic section obtained for each e value. What happens when $e = 1$?

Extending the Ideas

53. **Revisiting Figure 8.39** In Figure 8.39, if $r < 0$ or $r\cos\theta > k$, then we must use $PD = |k - r\cos\theta|$ and $PF = |r|$. Prove that, even in these cases, the resulting equation is still $r = ke/(1 + e\cos\theta)$.

54. **Deriving Other Polar Forms for Conics** Using Figure 8.39 as a guide, draw an appropriate diagram for and derive the equation.

(a) $r = \dfrac{ke}{1 - e\cos\theta}$

(b) $r = \dfrac{ke}{1 + e\sin\theta}$

(c) $r = \dfrac{ke}{1 - e\sin\theta}$

55. **Revisiting Example 3** Use the formulas $x = r\cos\theta$ and $x^2 + y^2 = r^2$ to transform the polar equation $r = \dfrac{16}{5 - 3\cos\theta}$ into the Cartesian equation $\dfrac{(x-3)^2}{25} + \dfrac{y^2}{16} = 1$.

56. Focal Widths Using polar equations, derive formulas for the focal width of an ellipse and the focal width of a hyperbola. Begin by defining focal width for these conics in a manner analogous to the definition of the focal width of a parabola given in Section 8.1.

57. Prove that for a hyperbola the formula $r = ke/(1 - e\cos\theta)$ is equivalent to $r = a(e^2 - 1)/(1 - e\cos\theta)$, where a is the semitransverse axis of the hyperbola.

58. Connecting Polar to Rectangular Consider the ellipse

$$\frac{x^2}{a^2} + \frac{y^2}{b^2} = 1,$$

where half the length of the major axis is a, and the foci are $(\pm c, 0)$ such that $c^2 = a^2 - b^2$. Let L be the vertical line $x = a^2/c$.

(a) Prove that L is a directrix for the ellipse. (*Hint*: Prove that PF/PD is the constant c/a, where P is a point on the ellipse, and D is the point on L such that PD is perpendicular to L.)

(b) Prove that the eccentricity is $e = c/a$.

(c) Prove that the distance from F to L is $a/e - ea$.

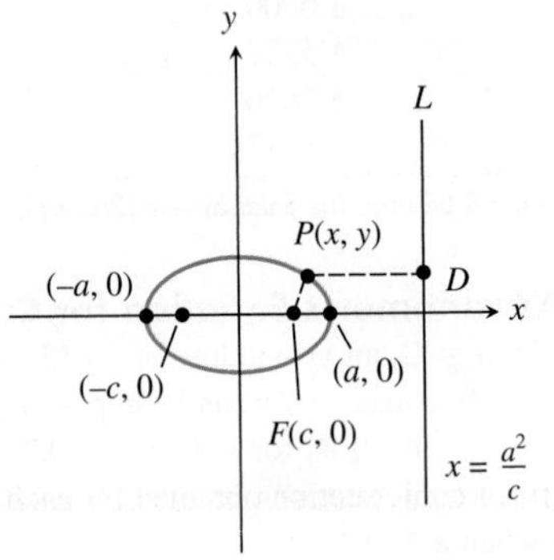

59. Connecting Polar to Rectangular Consider the hyperbola

$$\frac{x^2}{a^2} - \frac{y^2}{b^2} = 1,$$

where half the length of the transverse axis is a, and the foci are $(\pm c, 0)$ such that $c^2 = a^2 + b^2$. Let L be the vertical line $x = a^2/c$.

(a) Prove that L is a directrix for the hyperbola. (*Hint*: Prove that PF/PD is the constant c/a, where P is a point on the hyperbola, and D is the point on L such that PD is perpendicular to L.)

(b) Prove that the eccentricity is $e = c/a$.

(c) Prove that the distance from F to L is $ea - a/e$.

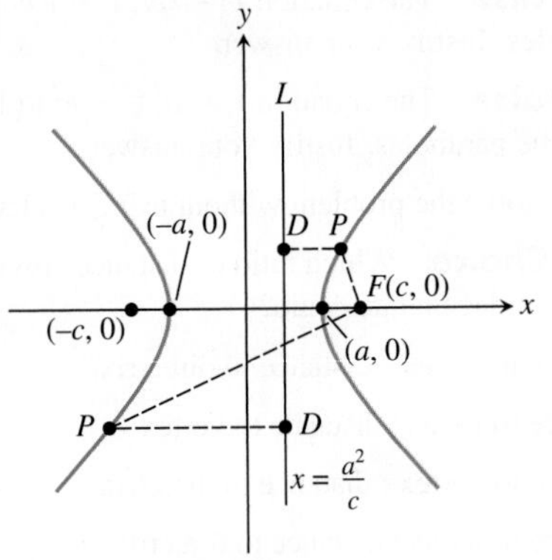

8.6 Three-Dimensional Cartesian Coordinate System

What you'll learn about

- Three-Dimensional Cartesian Coordinates
- Distance and Midpoint Formulas
- Equation of a Sphere
- Planes and Other Surfaces
- Vectors in Space
- Lines in Space

... and why

This is the analytic geometry of our physical world.

Three-Dimensional Cartesian Coordinates

In Sections P.2 and P.4, we studied Cartesian coordinates and the associated basic formulas and equations for the two-dimensional plane; we now extend these ideas to *three-dimensional space*. In the plane, we used two axes and ordered pairs to name points; in space, we use three mutually perpendicular axes and ordered triples of real numbers to name points (Figure 8.44).

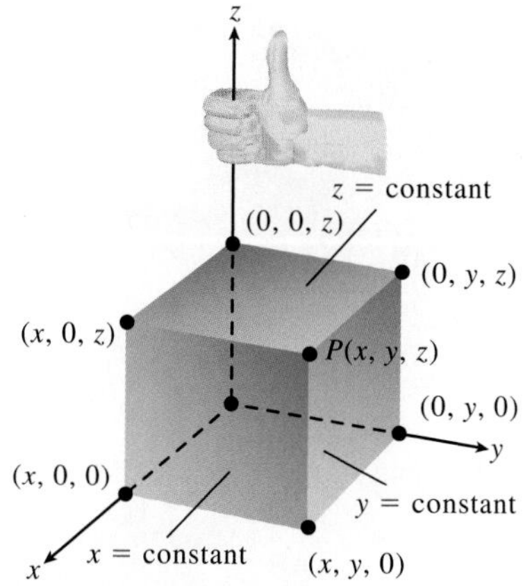

FIGURE 8.44 The point $P(x, y, z)$ in Cartesian space.

Notice that Figure 8.44 exhibits several important features of the *three-dimensional Cartesian coordinate system*:

- The axes are labeled x, y, and z, and these three **coordinate axes** form a **right-handed coordinate frame**: When you hold your right hand with fingers curving from the positive x-axis toward the positive y-axis, your thumb points in the direction of the positive z-axis.
- A point P in space uniquely corresponds to an ordered triple (x, y, z) of real numbers. The numbers x, y, and z are the **Cartesian coordinates of *P***.
- Points on the axes have the form $(x, 0, 0)$, $(0, y, 0)$, or $(0, 0, z)$, with $(x, 0, 0)$ on the x-axis, $(0, y, 0)$ on the y-axis, and $(0, 0, z)$ on the z-axis.

In Figure 8.45, the axes are paired to determine the **coordinate planes**:

- The coordinate planes are the ***xy*-plane**, the ***xz*-plane**, and the ***yz*-plane**, and have equations $z = 0$, $y = 0$, and $x = 0$, respectively.
- Points on the coordinate planes have the form $(x, y, 0)$, $(x, 0, z)$, or $(0, y, z)$, with $(x, y, 0)$ on the xy-plane, $(x, 0, z)$ on the xz-plane, and $(0, y, z)$ on the yz-plane.
- The coordinate planes meet at the **origin**, (0, 0, 0).
- The coordinate planes divide space into eight regions called **octants**, with the **first octant** containing all points in space with three positive coordinates.

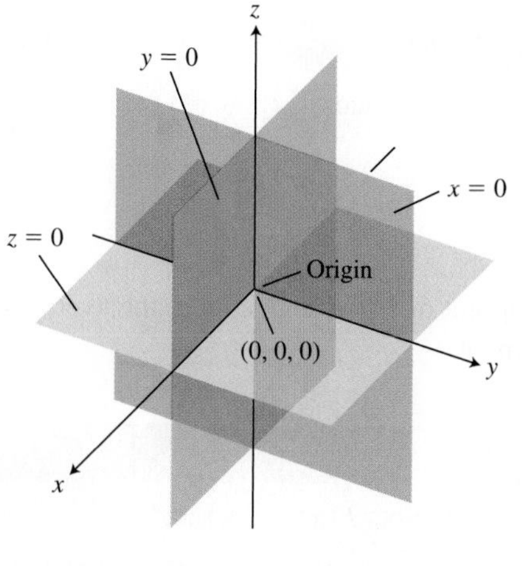

FIGURE 8.45 The coordinate planes divide space into eight octants.

EXAMPLE 1 Locating a Point in Cartesian Space

Draw a sketch that shows the point (2, 3, 5).

SOLUTION To locate the point $(2, 3, 5)$, we first sketch a right-handed three-dimensional coordinate frame. We then draw the planes $x = 2$, $y = 3$, and $z = 5$, which parallel the coordinate planes $x = 0$, $y = 0$, and $z = 0$, respectively. The point $(2, 3, 5)$ lies at the intersection of the planes $x = 2$, $y = 3$, and $z = 5$, as shown in Figure 8.46. **Now try Exercise 1.**

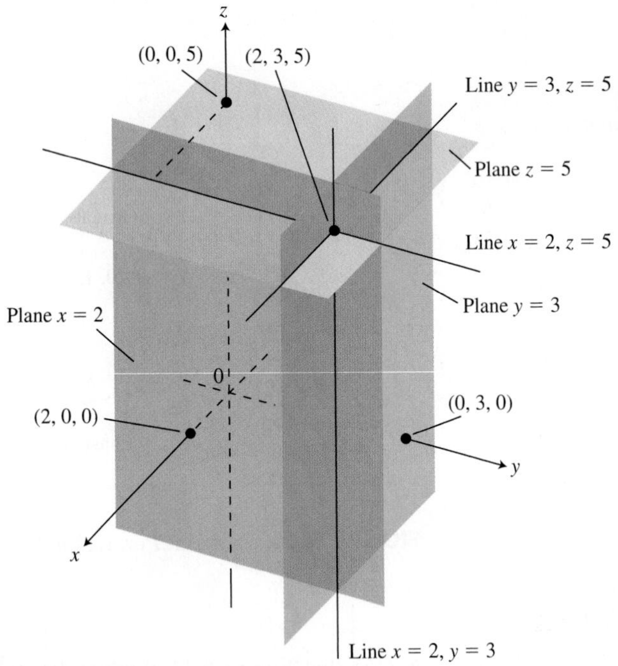

FIGURE 8.46 The planes $x = 2$, $y = 3$, and $z = 5$ determine the point $(2, 3, 5)$. (Example 1)

Distance and Midpoint Formulas

The distance and midpoint formulas for space are natural extensions of the corresponding formulas for the plane.

Distance Formula (Cartesian Space)

The distance $d(P, Q)$ between the points $P(x_1, y_1, z_1)$ and $Q(x_2, y_2, z_2)$ in space is

$$d(P, Q) = \sqrt{(x_1 - x_2)^2 + (y_1 - y_2)^2 + (z_1 - z_2)^2}.$$

Just as in the plane, the coordinates of the midpoint of a line segment are the averages for the coordinates of the endpoints of the segment.

Midpoint Formula (Cartesian Space)

The midpoint M of the line segment PQ with endpoints $P(x_1, y_1, z_1)$ and $Q(x_2, y_2, z_2)$ is

$$M = \left(\frac{x_1 + x_2}{2}, \frac{y_1 + y_2}{2}, \frac{z_1 + z_2}{2}\right).$$

EXAMPLE 2 Calculating a Distance and Finding a Midpoint

Find the distance between the points $P(-2, 3, 1)$ and $Q(4, -1, 5)$, and find the midpoint of line segment PQ.

SOLUTION The distance is given by

$$\begin{aligned} d(P, Q) &= \sqrt{(-2-4)^2 + (3+1)^2 + (1-5)^2} \\ &= \sqrt{36 + 16 + 16} \\ &= \sqrt{68} \approx 8.25 \end{aligned}$$

The midpoint is

$$M = \left(\frac{-2+4}{2}, \frac{3-1}{2}, \frac{1+5}{2}\right) = (1, 1, 3).$$

Now try Exercises 5 and 9.

Drawing Lesson

How to Draw Three-Dimensional Objects to Look Three-Dimensional

1. Make the angle between the positive x-axis and the positive y-axis large enough.

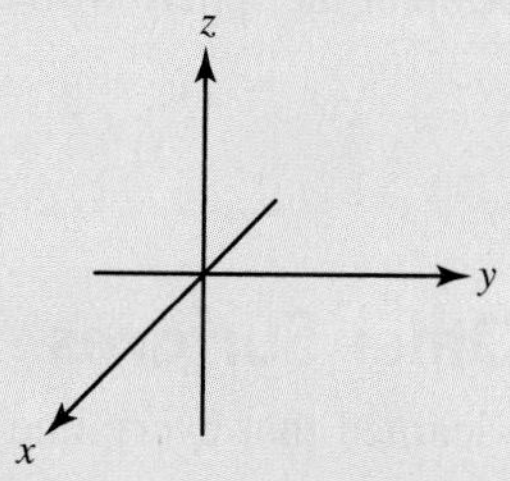

This

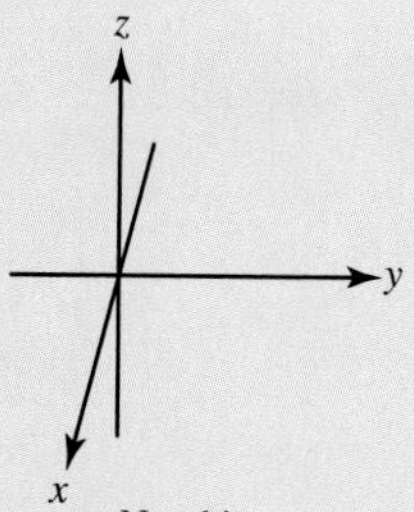

Not this

2. Break lines. When one line passes behind another, break it to show that it doesn't touch and that part of it is hidden.

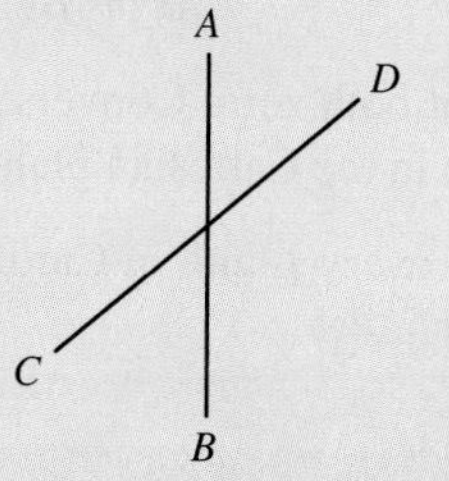

Intersecting

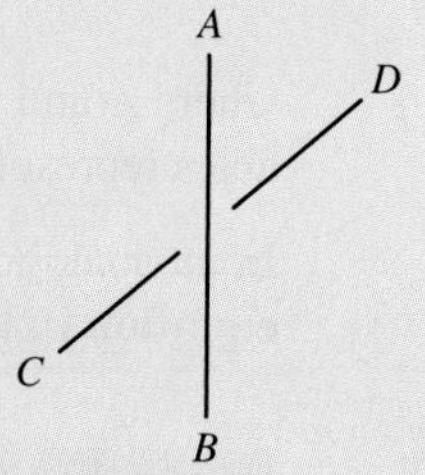

CD behind *AB*

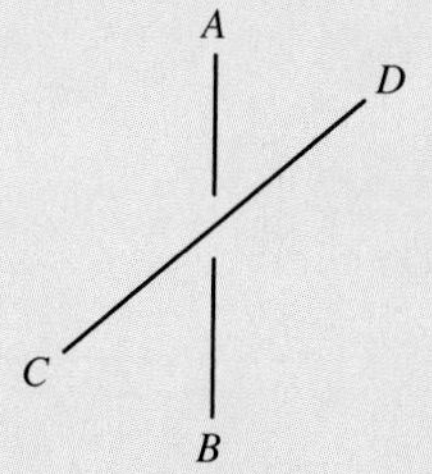

AB behind *CD*

3. Dash or omit hidden portions of lines. Don't let the line touch the boundary of the parallelogram that represents the plane, unless the line lies in the plane.

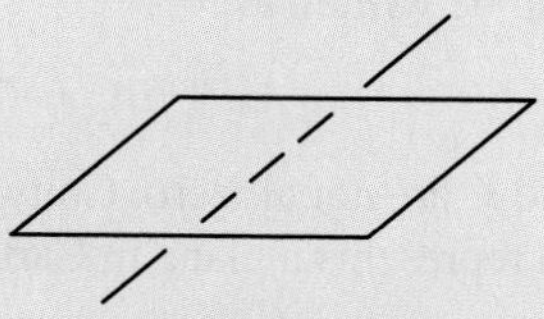
Line below plane

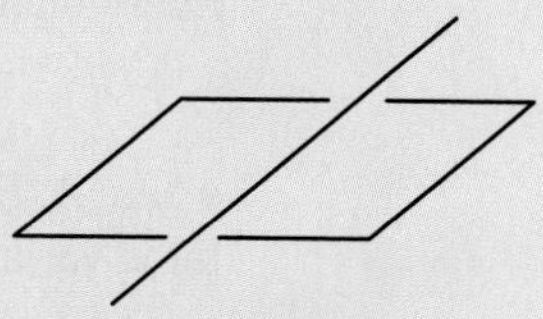
Line above plane

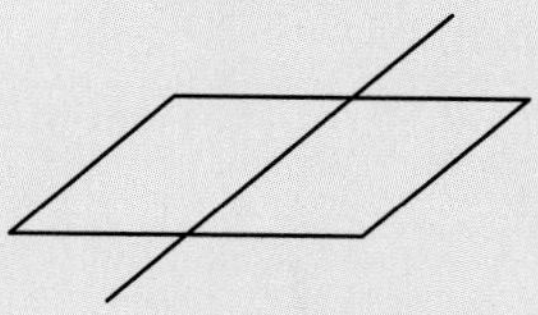
Line *in* plane

4. Spheres: Draw the sphere first (outline and equator); draw axes, if any, later. Use line breaks and dashed lines.

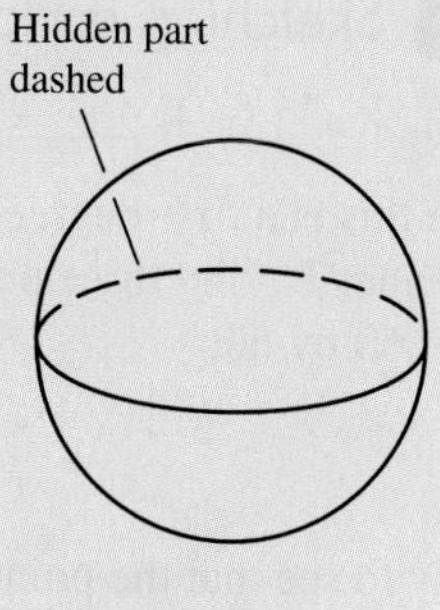

Sphere first

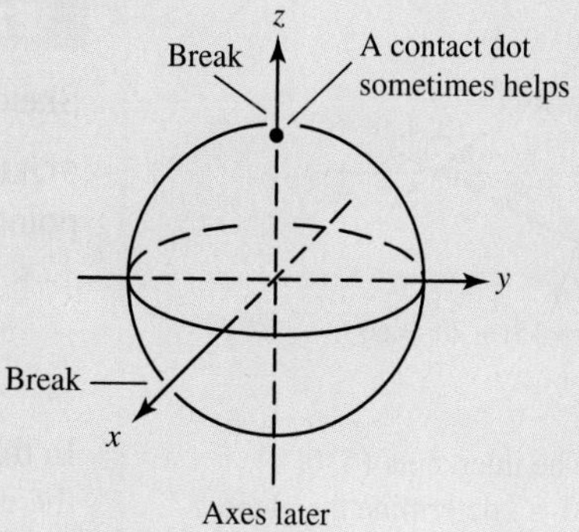

Axes later

Equation of a Sphere

A *sphere* is the three-dimensional analogue of a circle: In space, the set of points that lie a fixed distance from a fixed point is a **sphere**. The fixed distance is the **radius**, and the fixed point is the **center** of the sphere. The point $P(x, y, z)$ is a point of the sphere with center (h, k, l) and radius r if and only if

$$\sqrt{(x-h)^2+(y-k)^2+(z-l)^2}=r.$$

Squaring both sides gives the standard equation shown below.

Standard Equation of a Sphere

A point $P(x, y, z)$ is on the sphere with center (h, k, l) and radius r if and only if

$$(x-h)^2+(y-k)^2+(z-l)^2=r^2.$$

EXAMPLE 3 Finding the Standard Equation of a Sphere

The standard equation of the sphere with center $(2, 0, -3)$ and radius 7 is

$$(x-2)^2+y^2+(z+3)^2=49.$$

Now try Exercise 13.

Planes and Other Surfaces

In Section P.4, we learned that every line in the Cartesian plane can be written as a first-degree (linear) equation in two variables; that is, every line can be written as

$$Ax+By+C=0,$$

where A and B are not both zero. Conversely, every first-degree equation in two variables represents a line in the Cartesian plane.

In an analogous way, every **plane** in Cartesian space can be written as a **first-degree equation in three variables**:

Equation for a Plane in Cartesian Space

Every plane can be written as

$$Ax+By+Cz+D=0,$$

where A, B, and C are not all zero. Conversely, every first-degree equation in three variables represents a plane in Cartesian space.

EXAMPLE 4 Sketching a Plane in Space

Sketch the graph of $12x+15y+20z=60$.

SOLUTION Because this is a first-degree equation, its graph is a plane. Three points determine a plane. To find three points, we first divide both sides of $12x+15y+20z=60$ by 60:

$$\frac{x}{5}+\frac{y}{4}+\frac{z}{3}=1$$

In this form, it is easy to see that the points $(5, 0, 0)$, $(0, 4, 0)$, and $(0, 0, 3)$ satisfy the equation. These are the points where the graph crosses the coordinate axes. Figure 8.47 shows the completed sketch.

Now try Exercise 17.

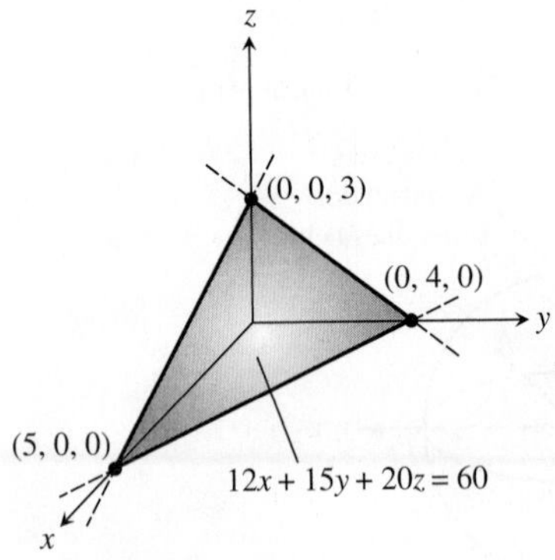

FIGURE 8.47 The intercepts $(5, 0, 0)$, $(0, 4, 0)$, and $(0, 0, 3)$ determine the plane $12x+15y+20z=60$. (Example 4)

Equations in the three variables x, y, and z generally graph as surfaces in three-dimensional space. Just as in the plane, second-degree equations are of particular interest. Recall that second-degree equations in two variables yield conic sections in the Cartesian plane. In space, second-degree equations in *three* variables yield **quadric surfaces**: The paraboloids, ellipsoids, and hyperboloids of revolution that have special reflective properties are all quadric surfaces, as are such exotic-sounding surfaces as hyperbolic paraboloids and elliptic hyperboloids.

Other surfaces of interest include graphs of **functions of two variables**, whose equations have the form $z = f(x, y)$. Some examples are $z = x \ln y$, $z = \sin(xy)$, and $z = \sqrt{1 - x^2 - y^2}$. The last equation graphs as a hemisphere (see Exercise 63). Equations of the form $z = f(x, y)$ can be graphed using some graphing calculators and most computer algebra software. Quadric surfaces and functions of two variables are studied in most university-level calculus course sequences.

Vectors in Space

In space, just as in the plane, the sets of equivalent directed line segments (or arrows) are *vectors*. They are used to represent forces, displacements, and velocities in three dimensions. In space, we use ordered triples to denote vectors:

$$\mathbf{v} = \langle v_1, v_2, v_3 \rangle$$

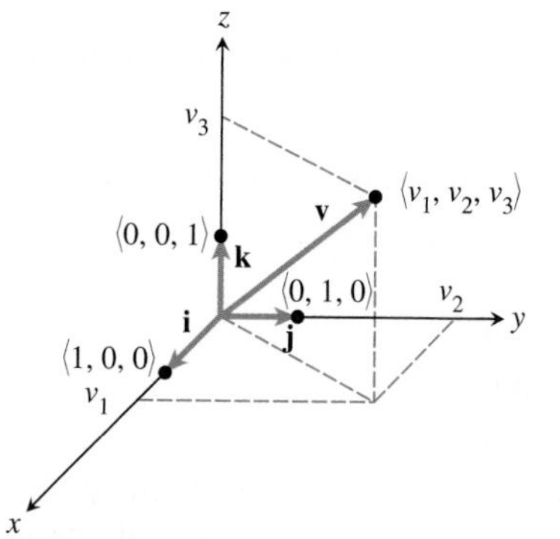

FIGURE 8.48 The vector $\mathbf{v} = \langle v_1, v_2, v_3 \rangle$.

The **zero vector** is $\mathbf{0} = \langle 0, 0, 0 \rangle$, and the **standard unit vectors** are $\mathbf{i} = \langle 1, 0, 0 \rangle$, $\mathbf{j} = \langle 0, 1, 0 \rangle$, and $\mathbf{k} = \langle 0, 0, 1 \rangle$. As shown in Figure 8.48, the vector $\mathbf{v}$ can be expressed in terms of these standard unit vectors:

$$\mathbf{v} = \langle v_1, v_2, v_3 \rangle = v_1\mathbf{i} + v_2\mathbf{j} + v_3\mathbf{k}$$

The vector $\mathbf{v}$ that is represented by the arrow from $P(a, b, c)$ to $Q(x, y, z)$ is

$$\mathbf{v} = \overrightarrow{PQ} = \langle x - a, y - b, z - c \rangle = (x - a)\mathbf{i} + (y - b)\mathbf{j} + (z - c)\mathbf{k}.$$

A vector $\mathbf{v} = \langle v_1, v_2, v_3 \rangle$ can be multiplied by a scalar (real number) c as follows:

$$c\mathbf{v} = c\langle v_1, v_2, v_3 \rangle = \langle cv_1, cv_2, cv_3 \rangle$$

Many other properties of vectors extend in a natural way when we move from two to three dimensions:

Vector Relationships in Space

For vectors $\mathbf{v} = \langle v_1, v_2, v_3 \rangle$ and $\mathbf{w} = \langle w_1, w_2, w_3 \rangle$,

- **Equality:** $\mathbf{v} = \mathbf{w}$ if and only if $v_1 = w_1$, $v_2 = w_2$, and $v_3 = w_3$
- **Addition:** $\mathbf{v} + \mathbf{w} = \langle v_1 + w_1, v_2 + w_2, v_3 + w_3 \rangle$
- **Subtraction:** $\mathbf{v} - \mathbf{w} = \langle v_1 - w_1, v_2 - w_2, v_3 - w_3 \rangle$
- **Magnitude:** $|\mathbf{v}| = \sqrt{v_1^2 + v_2^2 + v_3^2}$
- **Dot product:** $\mathbf{v} \cdot \mathbf{w} = v_1w_1 + v_2w_2 + v_3w_3$
- **Unit vector:** $\mathbf{u} = \mathbf{v}/|\mathbf{v}|$, $\mathbf{v} \neq \mathbf{0}$, is the unit vector in the direction of $\mathbf{v}$.

EXAMPLE 5 Computing with Vectors

(a) $3\langle -2, 1, 4 \rangle = \langle 3 \cdot -2, 3 \cdot 1, 3 \cdot 4 \rangle = \langle -6, 3, 12 \rangle$

(b) $\langle 0, 6, -7 \rangle + \langle -5, 5, 8 \rangle = \langle 0 - 5, 6 + 5, -7 + 8 \rangle = \langle -5, 11, 1 \rangle$

(c) $\langle 1, -3, 4 \rangle - \langle -2, -4, 5 \rangle = \langle 1 + 2, -3 + 4, 4 - 5 \rangle = \langle 3, 1, -1 \rangle$

(d) $|\langle 2, 0, -6 \rangle| = \sqrt{2^2 + 0^2 + 6^2} = \sqrt{40} \approx 6.32$

(e) $\langle 5, 3, -1 \rangle \cdot \langle -6, 2, 3 \rangle = 5 \cdot (-6) + 3 \cdot 2 + (-1) \cdot 3 = -30 + 6 - 3 = -27$

Now try Exercises 23–26.

EXAMPLE 6 Using Vectors in Space

A jet airplane just after takeoff is pointed due east. Its air velocity vector makes an angle of 30° with flat ground with an airspeed of 250 mph. If the wind is out of the southeast at 32 mph, calculate a vector that represents the plane's velocity relative to the point of takeoff.

SOLUTION Let **i** point east, **j** point north, and **k** point up. The plane's air velocity is

$$\mathbf{a} = \langle 250 \cos 30°, 0, 250 \sin 30° \rangle \approx \langle 216.506, 0, 125 \rangle,$$

and the wind velocity, which is pointing northwest, is

$$\mathbf{w} = \langle 32 \cos 135°, 32 \sin 135°, 0 \rangle \approx \langle -22.627, 22.627, 0 \rangle.$$

The velocity relative to the ground is $\mathbf{v} = \mathbf{a} + \mathbf{w}$, so

$$\begin{aligned} \mathbf{v} &\approx \langle 216.506, 0, 125 \rangle + \langle -22.627, 22.627, 0 \rangle \\ &\approx \langle 193.88, 22.63, 125 \rangle \\ &= 193.88\mathbf{i} + 22.63\mathbf{j} + 125\mathbf{k} \end{aligned}$$

Now try Exercise 33.

In Exercise 64, you will be asked to interpret the meaning of the velocity vector obtained in Example 6.

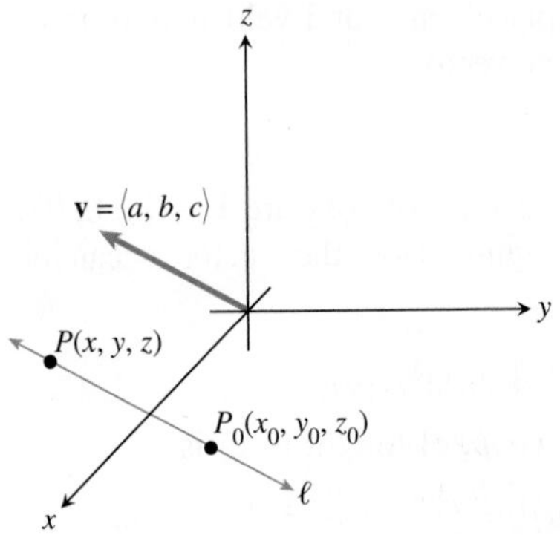

FIGURE 8.49 The line ℓ is parallel to the direction vector $\mathbf{v} = \langle a, b, c \rangle$.

Lines in Space

We have seen that first-degree equations in three variables graph as planes in space. So how do we get lines? There are several ways. First notice that to specify the x-axis, which is a line, we could use the pair of first-degree equations $y = 0$ and $z = 0$. As alternatives to using a pair of Cartesian equations, we can specify any line in space using

- one vector equation, or
- a set of three parametric equations.

Suppose ℓ is a line through the point $P_0(x_0, y_0, z_0)$ and in the direction of a nonzero vector $\mathbf{v} = \langle a, b, c \rangle$ (Figure 8.49). Then for any point $P(x, y, z)$ on ℓ,

$$\overrightarrow{P_0P} = t\mathbf{v}$$

for some real number t. The vector $\mathbf{v}$ is a **direction vector** for line ℓ. If $\mathbf{r} = \overrightarrow{OP} = \langle x, y, z \rangle$ and $\mathbf{r}_0 = \overrightarrow{OP_0} = \langle x_0, y_0, z_0 \rangle$, then $\mathbf{r} - \mathbf{r}_0 = t\mathbf{v}$. So an equation of the line ℓ is $\mathbf{r} = \mathbf{r}_0 + t\mathbf{v}$.

Equations for a Line in Space

If ℓ is a line through the point $P_0(x_0, y_0, z_0)$ in the direction of a nonzero vector $\mathbf{v} = \langle a, b, c \rangle$, then a point $P(x, y, z)$ is on ℓ if and only if

- **Vector form:** $\mathbf{r} = \mathbf{r}_0 + t\mathbf{v}$, where $\mathbf{r} = \langle x, y, z \rangle$ and $\mathbf{r}_0 = \langle x_0, y_0, z_0 \rangle$; or
- **Parametric form:** $x = x_0 + at$, $y = y_0 + bt$, and $z = z_0 + ct$,

where t is a real number.

EXAMPLE 7 Finding Equations for a Line

The line through $P_0(4, 3, -1)$ with direction vector $\mathbf{v} = \langle -2, 2, 7 \rangle$ can be written

- in vector form as $\mathbf{r} = \langle 4, 3, -1 \rangle + t\langle -2, 2, 7 \rangle$; or
- in parametric form as $x = 4 - 2t$, $y = 3 + 2t$, and $z = -1 + 7t$.

Now try Exercise 35.

EXAMPLE 8 Finding Equations for a Line

Using the standard unit vectors **i**, **j**, and **k**, write a vector equation for the line containing the points $A(3, 0, -2)$ and $B(-1, 2, -5)$, and compare it to the parametric equations for the line.

SOLUTION The line is in the direction of

$$\mathbf{v} = \overrightarrow{AB} = \langle -1 - 3, 2 - 0, -5 + 2 \rangle = \langle -4, 2, -3 \rangle.$$

Using $\mathbf{r}_0 = \overrightarrow{OA}$, the vector equation of the line becomes:

$$\begin{aligned} \mathbf{r} &= \mathbf{r}_0 + t\mathbf{v} \\ \langle x, y, z \rangle &= \langle 3, 0, -2 \rangle + t\langle -4, 2, -3 \rangle \\ \langle x, y, z \rangle &= \langle 3 - 4t, 2t, -2 - 3t \rangle \\ x\mathbf{i} + y\mathbf{j} + z\mathbf{k} &= (3 - 4t)\mathbf{i} + 2t\mathbf{j} + (-2 - 3t)\mathbf{k} \end{aligned}$$

The parametric equations are the three component equations

$$x = 3 - 4t,\ y = 2t, \text{ and } z = -2 - 3t.$$

Now try Exercise 41.

QUICK REVIEW 8.6 *(For help, go to Sections 6.1 and 6.3.)*

Exercise numbers with a gray background indicate problems that the authors have designed to be solved *without a calculator*.

In Exercises 1–3, let $P(x, y)$ and $Q(2, -3)$ be points in the xy-plane.

1. Compute the distance between P and Q.

2. Find the midpoint of the line segment PQ.

3. If P is 5 units from Q, describe the position of P.

In Exercises 4–6, let $\mathbf{v} = \langle -4, 5 \rangle = -4\mathbf{i} + 5\mathbf{j}$ be a vector in the xy-plane.

4. Find the magnitude of **v**.

5. Find a unit vector in the direction of **v**.

6. Find a vector 7 units long in the direction of $-\mathbf{v}$.

7. Give a geometric description of the graph of $(x + 1)^2 + (y - 5)^2 = 25$ in the xy-plane.

8. Give a geometric description of the graph of $x = 2 - t$, $y = -4 + 2t$ in the xy-plane.

9. Find the center and radius of the circle $x^2 + y^2 + 2x - 6y + 6 = 0$ in the xy-plane.

10. Find a vector from $P(2, 5)$ to $Q(-1, -4)$ in the xy-plane.

SECTION 8.6 Exercises

In Exercises 1–4, draw a sketch that shows the point.

1. $(3, 4, 2)$ **2.** $(2, -3, 6)$

3. $(1, -2, -4)$ **4.** $(-2, 3, -5)$

In Exercises 5–8, compute the distance between the points.

5. $(-1, 2, 5)$, $(3, -4, 6)$

6. $(2, -1, -8)$, $(6, -3, 4)$

7. (a, b, c), $(1, -3, 2)$

8. (x, y, z), (p, q, r)

In Exercises 9–12, find the midpoint of the segment PQ.

9. $P(-1, 2, 5)$, $Q(3, -4, 6)$

10. $P(2, -1, -8)$, $Q(6, -3, 4)$

11. $P(2x, 2y, 2z)$, $Q(-2, 8, 6)$

12. $P(-a, -b, -c)$, $Q(3a, 3b, 3c)$

In Exercises 13–16, write an equation for the sphere with the given point as its center and the given number as its radius.

13. $(5, -1, -2)$, 8 **14.** $(-1, 5, 8)$, $\sqrt{5}$

15. $(1, -3, 2)$, $\sqrt{a}$, $a > 0$ **16.** (p, q, r), 6

In Exercises 17–22, sketch a graph of the equation. Label all intercepts.

17. $x + y + 3z = 9$ **18.** $x + y - 2z = 8$

19. $x + z = 3$ **20.** $2y + z = 6$

21. $x - 3y = 6$ **22.** $x = 3$

In Exercises 23–32, evaluate the expression using $\mathbf{r} = \langle 1, 0, -3 \rangle$, $\mathbf{v} = \langle -3, 4, -5 \rangle$, and $\mathbf{w} = \langle 4, -3, 12 \rangle$.

23. $\mathbf{r} + \mathbf{v}$ **24.** $\mathbf{r} - \mathbf{w}$

25. $\mathbf{v} \cdot \mathbf{w}$ **26.** $|\mathbf{w}|$

27. $\mathbf{r} \cdot (\mathbf{v} + \mathbf{w})$ **28.** $(\mathbf{r} \cdot \mathbf{v}) + (\mathbf{r} \cdot \mathbf{w})$

29. $\mathbf{w}/|\mathbf{w}|$ **30.** $\mathbf{i} \cdot \mathbf{r}$

31. $\langle \mathbf{i} \cdot \mathbf{v}, \mathbf{j} \cdot \mathbf{v}, \mathbf{k} \cdot \mathbf{v} \rangle$ **32.** $(\mathbf{r} \cdot \mathbf{v})\mathbf{w}$

In Exercises 33 and 34, let **i** point east, **j** point north, and **k** point up.

33. Three-Dimensional Velocity An airplane just after takeoff is headed west and is climbing at a 20° angle relative to flat ground with an airspeed of 200 mph. If the wind is out of the northeast at 10 mph, calculate a vector **v** that represents the plane's velocity relative to the point of takeoff.

34. Three-Dimensional Velocity A rocket soon after takeoff is headed east and is climbing at an 80° angle relative to flat ground with an airspeed of 12,000 mph. If the wind is out of the southwest at 8 mph, calculate a vector **v** that represents the rocket's velocity relative to the point of takeoff.

In Exercises 35–38, write the vector and parametric forms of the line through the point P_0 in the direction of **v**.

35. $P_0(2, -1, 5)$, $\mathbf{v} = \langle 3, 2, -7 \rangle$

36. $P_0(-3, 8, -1)$, $\mathbf{v} = \langle -3, 5, 2 \rangle$

37. $P_0(6, -9, 0)$, $\mathbf{v} = \langle 1, 0, -4 \rangle$

38. $P_0(0, -1, 4)$, $\mathbf{v} = \langle 0, 0, 1 \rangle$

In Exercises 39–48, use the points $A(-1, 2, 4)$, $B(0, 6, -3)$, and $C(2, -4, 1)$.

39. Find the distance from A to the midpoint of BC.

40. Find the vector from A to the midpoint of BC.

41. Write a vector equation of the line through A and B.

42. Write a vector equation of the line through A and the midpoint of BC.

43. Write parametric equations for the line through A and C.

44. Write parametric equations for the line through B and C.

45. Write parametric equations for the line through B and the midpoint of AC.

46. Write parametric equations for the line through C and the midpoint of AB.

47. Is ΔABC equilateral, isosceles, or scalene?

48. If M is the midpoint of BC, what is the midpoint of AM?

Writing to Learn In Exercises 49–52, **(a)** sketch the line defined by the pair of equations, and **(b)** give a geometric description of the line, including its direction and its position relative to the coordinate frame.

49. $x = 0, y = 0$

50. $x = 0, z = 2$

51. $x = -3, y = 0$

52. $y = 1, z = 3$

53. Write a vector equation for the line through the distinct points $P(x_1, y_1, z_1)$ and $Q(x_2, y_2, z_2)$.

54. Write parametric equations for the line through the distinct points $P(x_1, y_1, z_1)$ and $Q(x_2, y_2, z_2)$.

55. Generalizing the Distance Formula Prove that the distance $d(P, Q)$ between the points $P(x_1, y_1, z_1)$ and $Q(x_2, y_2, z_2)$ in space is $\sqrt{(x_1 - x_2)^2 + (y_1 - y_2)^2 + (z_1 - z_2)^2}$ by using the point $R(x_2, y_2, z_1)$, the two-dimensional distance formula within the plane $z = z_1$, the one-dimensional distance formula within the line $\mathbf{r} = \langle x_2, y_2, t \rangle$, and the Pythagorean Theorem. (*Hint:* A sketch may help you visualize the situation.)

56. Generalizing a Property of the Dot Product Prove $\mathbf{u} \cdot \mathbf{u} = |\mathbf{u}|^2$ where **u** is a vector in three-dimensional space.

Standardized Test Questions

57. True or False $x^2 + 4y^2 = 1$ represents a surface in space. Justify your answer.

58. True or False The parametric equations $x = 1 + 0t$, $y = 2 - 0t$, $z = -5 + 0t$ represent a line in space. Justify your answer.

In Exercises 59–62, you may use a graphing calculator to solve the problem.

59. Multiple Choice A first-degree equation in three variables graphs as

(A) a line.

(B) a plane.

(C) a sphere.

(D) a paraboloid.

(E) an ellipsoid.

60. Multiple Choice Which of the following is **not** a quadric surface?

(A) A plane

(B) A sphere

(C) An ellipsoid

(D) An elliptic paraboloid

(E) A hyperbolic paraboloid

61. Multiple Choice If **v** and **w** are vectors and c is a scalar, which of these is a **scalar**?

(A) $\mathbf{v} + \mathbf{w}$

(B) $\mathbf{v} - \mathbf{w}$

(C) $\mathbf{v} \cdot \mathbf{w}$

(D) $c\mathbf{v}$

(E) $|\mathbf{v}|\mathbf{w}$

62. Multiple Choice The parametric form of the line $\mathbf{r} = \langle 2, -3, 0 \rangle + t\langle 1, 0, -1 \rangle$ is

(A) $x = 2 - 3t, y = 0 + 1t, z = 0 - 1t.$

(B) $x = 2t, y = -3 + 0t, z = 0 - 1t.$

(C) $x = 1 + 2t, y = 0 - 3t, z = -1 + 0t.$

(D) $x = 1 + 2t, y = -3, z = -1t.$

(E) $x = 2 + t, y = -3, z = -t.$

Explorations

63. Group Activity Writing to Learn The figure shows a graph of the ellipsoid $x^2/9 + y^2/4 + z^2/16 = 1$ drawn in a *box* using *Mathematica* computer software.

(a) Describe its cross sections in each of the three coordinate planes, that is, for $z = 0$, $y = 0$, and $x = 0$. In your description, include the name of each cross section and its position relative to the coordinate frame.

(b) Explain algebraically why the graph of $z = \sqrt{1 - x^2 - y^2}$ is half of a sphere. What is the equation of the related whole sphere?

(c) By hand, sketch the graph of the hemisphere $z = \sqrt{1 - x^2 - y^2}$. Check your sketch using a 3D grapher if you have access to one.

(d) Explain how the graph of an ellipsoid is related to the graph of a sphere and why a sphere is a *degenerate* ellipsoid.

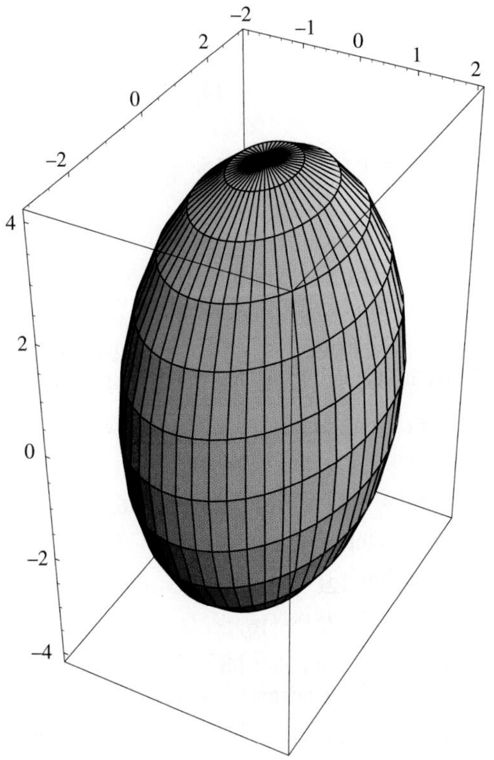

64. Revisiting Example 6 Read Example 6. Then using $\mathbf{v} = 193.88\mathbf{i} + 22.63\mathbf{j} + 125\mathbf{k}$, establish the following:

(a) The plane's compass bearing is 83.34°.

(b) Its speed downrange (that is, ignoring the vertical component) is 195.2 mph.

(c) The plane is climbing at an angle of 32.63°.

(d) The plane's overall speed is 231.8 mph.

Extending the Ideas

The **cross product u × v** of the vectors $\mathbf{u} = u_1\mathbf{i} + u_2\mathbf{j} + u_3\mathbf{k}$ and $\mathbf{v} = v_1\mathbf{i} + v_2\mathbf{j} + v_3\mathbf{k}$ is

$$\mathbf{u} \times \mathbf{v} = \begin{vmatrix} \mathbf{i} & \mathbf{j} & \mathbf{k} \\ u_1 & u_2 & u_3 \\ v_1 & v_2 & v_3 \end{vmatrix}$$

$$= (u_2v_3 - u_3v_2)\mathbf{i} + (u_3v_1 - u_1v_3)\mathbf{j} + (u_1v_2 - u_2v_1)\mathbf{k}.$$

Use this definition in Exercises 65–68.

65. $\langle 1, -2, 3 \rangle \times \langle -2, 1, -1 \rangle$

66. $\langle 4, -1, 2 \rangle \times \langle 1, -3, 2 \rangle$

67. Prove that $\mathbf{i} \times \mathbf{j} = \mathbf{k}$.

68. Assuming the theorem about angles between vectors (Section 6.2) holds for three-dimensional vectors, prove that $\mathbf{u} \times \mathbf{v}$ is perpendicular to both **u** and **v** if they are nonzero.

CHAPTER 8 Key Ideas

Properties, Theorems, and Formulas

Parabolas with Vertex (0, 0) 574
Parabolas with Vertex (h, k) 575
Ellipses with Center (0, 0) 583
Ellipses with Center (h, k) 585
Hyperbolas with Center (0, 0) 595
Hyperbolas with Center (h, k) 596
Axis Rotation Formulas 605
Angle of Rotation to Eliminate the Cross-Product Term 607
Discriminant Test 608
Focus–Directrix–Eccentricity Relationship 612
Polar Equations for Conics 614
Ellipse with Eccentricity e and Semimajor Axis a 617
Distance Formula (Cartesian Space) 622
Midpoint Formula (Cartesian Space) 622
Standard Equation of a Sphere 624
Equation for a Plane in Cartesian Space 624
Vector Relationships in Space 625
Equations for a Line in Space 626

Procedures

How to Sketch the Ellipse $\frac{x^2}{a^2} + \frac{y^2}{b^2} = 1$ 584
How to Sketch the Hyperbola $\frac{x^2}{a^2} - \frac{y^2}{b^2} = 1$ 594
Rotation of Axes 603–608
How to Draw Three-Dimensional Objects to Look Three-Dimensional 623

CHAPTER 8 Review Exercises

Exercise numbers with a gray background indicate problems that the authors have designed to be solved *without a calculator*.

The collection of exercises marked in red could be used as a chapter test.

In Exercises 1–4, find the vertex, focus, directrix, and focal width of the parabola, and sketch the graph.

1. $y^2 = 12x$ **2.** $x^2 = -8y$

3. $(x + 2)^2 = -4(y - 1)$ **4.** $(y + 2)^2 = 16x$

In Exercises 5–12, identify the type of conic. Find the center, vertices, and foci of the conic, and sketch its graph.

5. $\frac{y^2}{8} + \frac{x^2}{5} = 1$ **6.** $\frac{y^2}{16} - \frac{x^2}{49} = 1$

7. $\frac{x^2}{25} - \frac{y^2}{36} = 1$ **8.** $\frac{x^2}{49} - \frac{y^2}{9} = 1$

9. $\frac{(x + 3)^2}{18} - \frac{(y - 5)^2}{28} = 1$

10. $\frac{(y - 3)^2}{9} - \frac{(x - 7)^2}{12} = 1$

11. $\frac{(x - 2)^2}{16} + \frac{(y + 1)^2}{7} = 1$

12. $\frac{y^2}{36} + \frac{(x + 6)^2}{20} = 1$

In Exercises 13–20, match the equation with its graph.

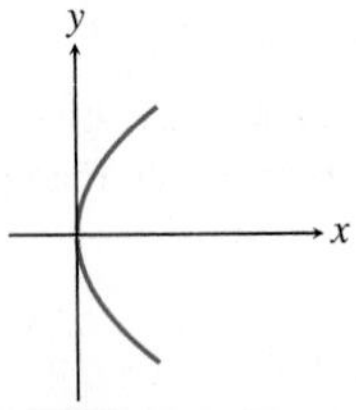

(a)

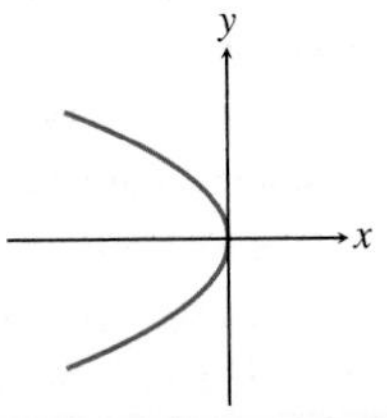

(b)

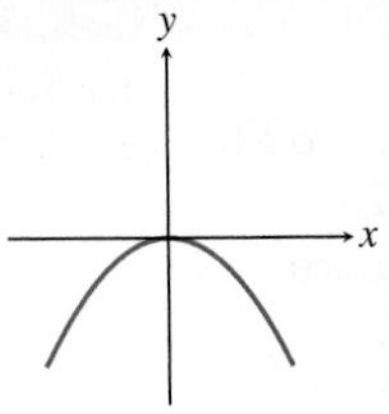

(c)

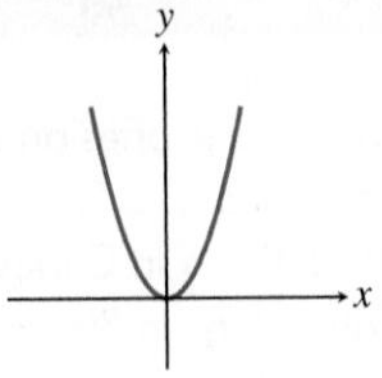

(d)

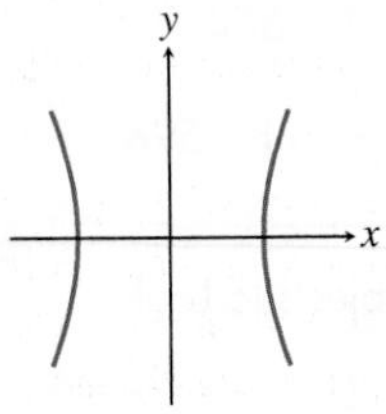

(e)

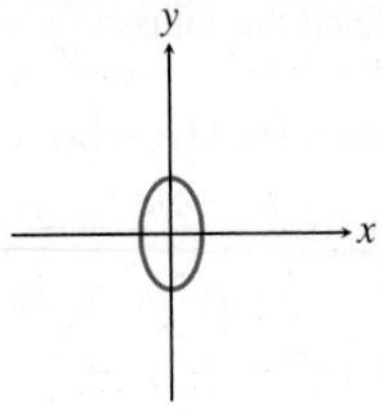

(f)

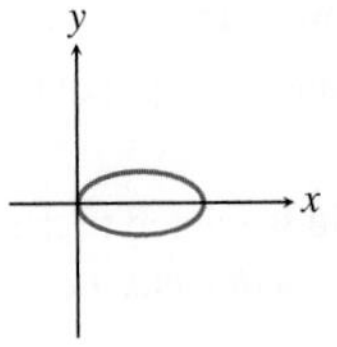

(g)

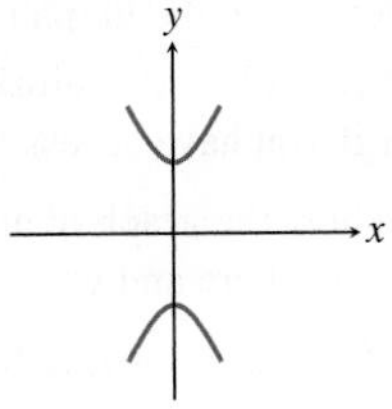

(h)

13. $y^2 = -3x$ **14.** $\frac{(x - 2)^2}{4} + y^2 = 1$

15. $\frac{y^2}{5} - x^2 = 1$ **16.** $\frac{x^2}{9} - \frac{y^2}{25} = 1$

17. $\frac{y^2}{3} + x^2 = 1$ **18.** $x^2 = y$

19. $x^2 = -4y$ **20.** $y^2 = 6x$

In Exercises 21–28, identify the conic. Then complete the square to write the conic in standard form, and sketch the graph.

21. $x^2 - 6x - y - 3 = 0$ **22.** $x^2 + 4x + 3y^2 - 5 = 0$

23. $x^2 - y^2 - 2x + 4y - 6 = 0$

24. $x^2 + 2x + 4y - 7 = 0$ **25.** $y^2 - 6x - 4y - 13 = 0$

26. $3x^2 - 6x - 4y - 9 = 0$

27. $2x^2 - 3y^2 - 12x - 24y + 60 = 0$

28. $12x^2 - 4y^2 - 72x - 16y + 44 = 0$

29. Prove that the parabola with focus $(0, p)$ and directrix $y = -p$ has the equation $x^2 = 4py$.

30. Prove that the equation $y^2 = 4px$ represents a parabola with focus $(p, 0)$ and directrix $x = -p$.

In Exercises 31–34, identify the conic. Solve the equation for y and graph it.

31. $3x^2 - 8xy + 6y^2 - 5x - 5y + 20 = 0$

32. $10x^2 - 8xy + 6y^2 + 8x - 5y - 30 = 0$

33. $3x^2 - 2xy - 5x + 6y - 10 = 0$

34. $5xy - 6y^2 + 10x - 17y + 20 = 0$

In Exercises 35 and 36, (a) identify the type of conic; (b) find rotation formulas that will convert the (x, y) equation to a (u, v) equation with no cross-product term; (c) write the (u, v) equation in standard form; (d) find the vertices and foci in the (u, v) system; and (e) write the vertices and foci in (x, y) coordinates.

35. $x^2 + 2xy + y^2 + 4\sqrt{2}\,x - 4\sqrt{2}\,y + 8 = 0$

36. $91x^2 - 24xy + 84y^2 = 300$

In Exercises 37–48, find the equation for the conic in standard form.

37. Parabola: vertex $(0, 0)$, focus $(2, 0)$

38. Parabola: vertex $(0, 0)$, opens downward, focal width $= 12$

39. Parabola: vertex $(-3, 3)$, directrix $y = 0$

40. Parabola: vertex $(1, -2)$, opens to the left, focal length $= 2$

41. Ellipse: center $(0, 0)$, foci $(\pm 12, 0)$, vertices $(\pm 13, 0)$

42. Ellipse: center $(0, 0)$, foci $(0, \pm 2)$, vertices $(0, \pm 6)$

43. Ellipse: center $(0, 2)$, semimajor axis $= 3$, one focus is $(2, 2)$.

44. Ellipse: center $(-3, -4)$, semimajor axis $= 4$, one focus is $(0, -4)$.

45. Hyperbola: center $(0, 0)$, foci $(0, \pm 6)$, vertices $(0, \pm 5)$

46. Hyperbola: center $(0, 0)$, vertices $(\pm 2, 0)$, asymptotes $y = \pm 2x$

47. Hyperbola: center $(2, 1)$, vertices $(2 \pm 3, 1)$, one asymptote is $y = (4/3)(x - 2) + 1$

48. Hyperbola: center $(-5, 0)$, one focus is $(-5, 3)$, one vertex is $(-5, 2)$.

In Exercises 49–54, find the equation for the conic in standard form.

49. $x = 5 \cos t,\ y = 2 \sin t,\ 0 \le t \le 2\pi$

50. $x = 4 \sin t,\ y = 6 \cos t,\ 0 \le t \le 4\pi$

51. $x = -2 + \cos t,\ y = 4 + \sin t,\ 2\pi \le t \le 4\pi$

52. $x = 5 + 3 \cos t,\ y = -3 + 3 \sin t,\ -2\pi \le t \le 0$

53. $x = 3 \sec t,\ y = 5 \tan t,\ 0 \le t \le 2\pi$

54. $x = 4 \sec t,\ y = 3 \tan t,\ 0 \le t \le 2\pi$

In Exercises 55–62, identify and graph the conic, and rewrite the equation in Cartesian coordinates.

55. $r = \dfrac{4}{1 + \cos\theta}$

56. $r = \dfrac{5}{1 - \sin\theta}$

57. $r = \dfrac{4}{3 - \cos\theta}$

58. $r = \dfrac{3}{4 + \sin\theta}$

59. $r = \dfrac{35}{2 - 7\sin\theta}$

60. $r = \dfrac{15}{2 + 5\cos\theta}$

61. $r = \dfrac{2}{1 + \cos\theta}$

62. $r = \dfrac{4}{4 - 4\cos\theta}$

In Exercises 63–74, use the points $P(-1, 0, 3)$ and $Q(3, -2, -4)$ and the vectors $\mathbf{v} = \langle -3, 1, -2 \rangle$ and $\mathbf{w} = \langle 3, -4, 0 \rangle$.

63. Compute the distance from P to Q.

64. Find the midpoint of segment PQ.

65. Compute $\mathbf{v} + \mathbf{w}$.

66. Compute $\mathbf{v} - \mathbf{w}$.

67. Compute $\mathbf{v} \cdot \mathbf{w}$.

68. Compute the magnitude of $\mathbf{v}$.

69. Write the unit vector in the direction of $\mathbf{w}$.

70. Compute $(\mathbf{v} \cdot \mathbf{w})(\mathbf{v} + \mathbf{w})$.

71. Write an equation for the sphere centered at P with radius 4.

72. Write parametric equations for the line through P and Q.

73. Write a vector equation for the line through P in the direction of $\mathbf{v}$.

74. Write parametric equations for the line in the direction of $\mathbf{w}$ through the midpoint of PQ.

75. Parabolic Microphones B-Ball Network uses a parabolic microphone to capture all the sounds from the basketball players and coaches during each regular season game. If one of its microphones has a parabolic surface generated by the parabola $18y = x^2$, locate the focus (the electronic receiver) of the parabola.

76. Parabolic Headlights Specific Electric makes parabolic headlights for a variety of automobiles. If one of its headlights has a parabolic surface generated by the parabola $y^2 = 15x$ (see figure), where should its lightbulb be placed?

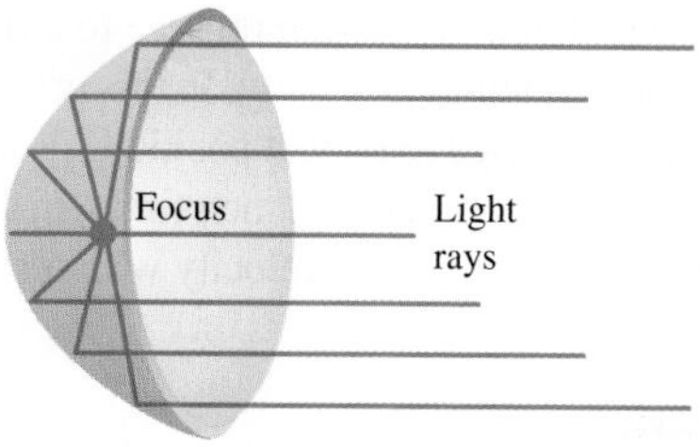

77. Writing to Learn Elliptical Billiard Table Elliptical billiard tables have been constructed with spots marking the foci. Suppose such a table has a major axis of 6 ft and minor axis of 4 ft.

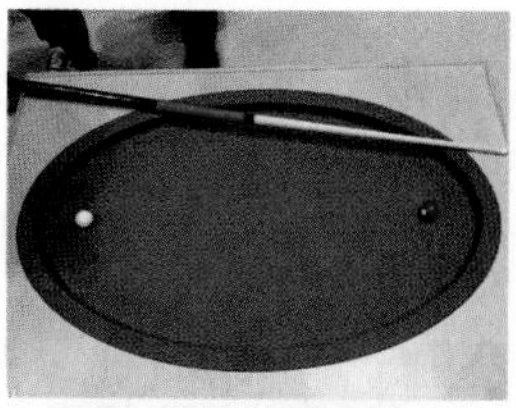

(a) Explain the strategy that a "pool shark" who knows conic geometry would use to hit a blocked spot on this table.

(b) If the table surface is coordinatized so that $(0, 0)$ represents the center of the table and the x-axis is along the focal axis of the ellipse, at which point(s) should the ball be aimed?

78. Weather Satellite The Nimbus weather satellite travels in a north-south circular orbit 500 m above Earth. Find the following. (Assume Earth's radius is 6380 km.)

(a) The velocity of the satellite using the formula for velocity v given for Exercises 43 and 44 in Section 8.5 with $k = 1$

(b) The time required for Nimbus to circle Earth once

79. Elliptical Orbits The velocity of a body in an elliptical Earth orbit at a distance r (in meters) from the focus (the center of Earth) is

$$v = \sqrt{3.99 \times 10^{14}\left(\frac{2}{r} - \frac{1}{a}\right)} \text{ m/sec},$$

where a is the semimajor axis of the ellipse. An Earth satellite has a maximum altitude (at *apogee*) of 18,000 km and has a minimum altitude (at *perigee*) of 170 km. Assuming Earth's radius is 6380 km, find the velocity of the satellite at its apogee and perigee.

80. Icarus The asteroid Icarus is about 1 mi wide. It revolves around the Sun once every 409 Earth days and has an orbital eccentricity of 0.83. Use Kepler's first and third laws to determine Icarus's semimajor axis, perihelion distance, and aphelion distance.

CHAPTER 8 Project

Ellipses as Models of Pendulum Motion

As a simple pendulum swings back and forth, a plot of its velocity versus its position is elliptical in nature and can be modeled using a standard form of the equation of an ellipse,

$$\frac{(x-h)^2}{a^2}+\frac{(y-k)^2}{b^2}=1 \text{ or } \frac{(y-k)^2}{a^2}+\frac{(x-h)^2}{b^2}=1,$$

where x represents the pendulum's position relative to a fixed point and y represents the pendulum's velocity. In this project, you will use a motion detection device to collect position (distance) and velocity data for a swinging pendulum, then find a mathematical model that describes the pendulum's velocity with respect to position.

Collecting the Data

Construct a simple pendulum by fastening about 0.5 m of string to a ball. Set up a Calculator-Based Ranger (CBR) system to collect distance and velocity readings for 4 sec (enough time to capture at least one complete swing of the pendulum). See the printed or online CBR guidebook for specific setup instructions. Start the pendulum swinging toward and away from the detector, then activate the CBR system. The table below is a sample set of data collected in the manner just described.

Explorations

1. If you collected data using a CBR, a plot of distance versus time may be shown on your grapher screen. Go to the plot setup screen and create a scatter plot of velocity versus distance. If you do not have access to a CBR, use the distance and velocity data from the table below to create a scatter plot.
2. Find values for a, b, h, and k so that the equation
$$\frac{(x-h)^2}{a^2}+\frac{(y-k)^2}{b^2}=1 \text{ or } \frac{(y-k)^2}{a^2}+\frac{(x-h)^2}{b^2}=1$$
fits the velocity versus position data plot. To graph this model, you will have to solve the appropriate equation for y and enter it into the calculator in Y1 and Y2.
3. With respect to the ellipse modeled above, what do the variables a, b, h, and k represent?
4. What are the physical meanings of a, b, h, and k with respect to the motion of the pendulum?
5. Set up plots of distance versus time and velocity versus time. Find models for both of these plots and use them to graph the plot of the ellipse using parametric equations.

Time (sec)	Distance from the CBR (m)	Velocity (m/sec)	Time (sec)	Distance from the CBR (m)	Velocity (m/sec)
0	0.682	−0.3	0.647	0.454	0.279
0.059	0.659	−0.445	0.706	0.476	0.429
0.118	0.629	−0.555	0.765	0.505	0.544
0.176	0.594	−0.621	0.824	0.54	0.616
0.235	0.557	−0.638	0.882	0.576	0.639
0.294	0.521	−0.605	0.941	0.612	0.612
0.353	0.489	−0.523	1	0.645	0.536
0.412	0.463	−0.4	1.059	0.672	0.418
0.471	0.446	−0.246	1.118	0.69	0.266
0.529	0.438	−0.071	1.176	0.699	0.094
0.588	0.442	0.106	1.235	0.698	−0.086

CHAPTER 9

9.1 Basic Combinatorics

9.2 Binomial Theorem

9.3 Sequences

9.4 Series

9.5 Mathematical Induction

Discrete Mathematics

As the use of cellular telephones, modems, pagers, and fax machines has grown in recent years, the increasing demand for unique telephone numbers has necessitated the creation of new area codes in many parts of the United States. Counting the number of possible telephone numbers in a given area code is a *combinatorial* problem, and such problems are solved using the techniques of *discrete mathematics*. See page 640 for more on the subject of telephone area codes.

Chapter 9 Overview

The branches of mathematics known broadly as algebra, analysis, and geometry come together so beautifully in calculus that it has been difficult over the years to squeeze other mathematics into the curriculum. Consequently, for many students, worthwhile topics such as probability and statistics, combinatorics, graph theory, and numerical analysis that could easily be introduced in high school are either first seen in college electives or never seen at all. This situation is changing as the applications of noncalculus mathematics become increasingly more important in our modern, computerized, data-driven society. Therefore, besides introducing important topics like sequences and series and the Binomial Theorem, this chapter will touch on some other discrete topics that might prove useful to you in the near future.

9.1 Basic Combinatorics

What you'll learn about

- Discrete Versus Continuous
- The Importance of Counting
- The Multiplication Principle of Counting
- Permutations
- Combinations
- Subsets of an *n*-Set

... and why

Counting large sets is easy if you know the correct formula.

Discrete Versus Continuous

A point has no length and no width, and yet intervals on the real line—which are made up of these dimensionless points—have length! This little mystery illustrates the distinction between *continuous* and *discrete* mathematics. Any interval (a, b) contains a **continuum** of real numbers, which is why you can zoom in on an interval repeatedly and there will still be an interval there. Calculus concepts like limits and continuity depend on the mathematics of the continuum. In *discrete* mathematics, we are concerned with properties of numbers and algebraic systems that do not depend on that kind of analysis. Many of them are related to the first kind of mathematics that most of us ever did, namely counting. Counting is what we will do for the rest of this section.

The Importance of Counting

We begin with a relatively simple counting problem.

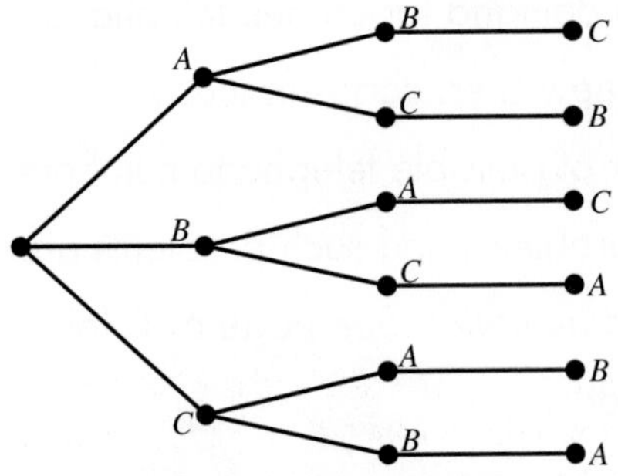

FIGURE 9.1 A tree diagram for ordering the letters *ABC*. (Example 1)

EXAMPLE 1 Arranging Three Objects in Order

In how many different ways can three distinguishable objects be arranged in order?

SOLUTION It is not difficult to list all the possibilities. If we call the objects *A, B,* and *C*, the orderings are *ABC, ACB, BAC, BCA, CAB,* and *CBA*. A good way to visualize the six choices is with a *tree diagram*, as in Figure 9.1. Notice that we have three choices for the first letter. Then, branching off each of those three choices are two choices for the second letter. Finally, branching off each of the $3 \times 2 = 6$ branches formed so far is one choice for the third letter. By beginning at the "root" of the tree, we can proceed to the right along any of the $3 \times 2 \times 1 = 6$ branches and get a different ordering each time. We conclude that there are six ways to arrange three distinguishable objects in order.

Now try Exercise 3.

Scientific studies will usually manipulate one or more **explanatory** variables and observe the effect of that manipulation on one or more **response** variables. The key to understanding the significance of the effect is to know what is likely to occur *by chance alone*, and that often depends on counting. For example, Exploration 1 shows a real-world application of Example 1.

EXPLORATION 1 Questionable Product Claims

A salesman for a copying machine company is trying to convince a client to buy his \$2000 machine instead of his competitor's \$5000 machine. To make his point, he lines up an original document, a copy made by his machine, and a copy made by the more expensive machine on a table and asks 60 office workers to identify which is which. To everyone's surprise, not a single worker identifies all three correctly. The salesman states triumphantly that this proves that all three documents look the same to the naked eye and that therefore the client should buy his company's less expensive machine.

What do you think?

1. Each worker is essentially being asked to put the three documents in the correct order. How many ways can the three documents be ordered?
2. Suppose all three documents really *do* look alike. What fraction of the workers would you expect to put them into the correct order by chance alone?
3. If 0 persons out of 60 put the documents in the correct order, should we conclude that "all three documents look the same to the naked eye"?
4. Can you suggest a more likely conclusion that we might draw from the results of the salesman's experiment?

The Multiplication Principle of Counting

You would not want to draw the tree diagram for ordering five objects (*ABCDE*), but you should be able to see in your mind that it would have $5 \times 4 \times 3 \times 2 \times 1 = 120$ branches. A tree diagram is a geometric visualization of a fundamental counting principle known as the *Multiplication Principle.*

Multiplication Principle of Counting

If a procedure P has a sequence of stages $S_1, S_2, \ldots, S_n$ and if

$$\begin{aligned} &S_1 \text{ can occur in } r_1 \text{ ways,} \\ &S_2 \text{ can occur in } r_2 \text{ ways,} \\ &\vdots \\ &S_n \text{ can occur in } r_n \text{ ways,} \end{aligned}$$

then the number of ways that the procedure P can occur is the product

$$r_1 r_2 \cdots r_n.$$

It is important to be mindful of how the choices at each stage are affected by the choices at preceding stages. For example, when choosing an order for the letters *ABC* we have 3 choices for the first letter, but only 2 choices for the second and 1 for the third.

EXAMPLE 2 Using the Multiplication Principle

A Tennessee license plate consists of three letters of the alphabet followed by three numerical digits (0 through 9). Find the number of different license plates that could be formed

(a) if there is no restriction on the letters or digits that can be used;

(b) if no letter or digit can be repeated.

(continued)

SOLUTION Consider each license plate as having six blanks to be filled in: three letters followed by three numerical digits.

(a) If there are no restrictions on letters or digits, then we can fill in the first blank 26 ways, the second blank 26 ways, the third blank 26 ways, the fourth blank 10 ways, the fifth blank 10 ways, and the sixth blank 10 ways. By the Multiplication Principle, we can fill in all six blanks in $26 \times 26 \times 26 \times 10 \times 10 \times 10 =$ 17,576,000 ways. There are 17,576,000 possible license plates with no restrictions on letters or digits.

(b) If no letter or digit can be repeated, then we can fill in the first blank 26 ways, the second blank 25 ways, the third blank 24 ways, the fourth blank 10 ways, the fifth blank 9 ways, and the sixth blank 8 ways. By the Multiplication Principle, we can fill in all six blanks in $26 \times 25 \times 24 \times 10 \times 9 \times 8 = 11{,}232{,}000$ ways. There are 11,232,000 possible license plates with no letters or digits repeated.

Now try Exercise 5.

License Plate Restrictions

Although prohibiting repeated letters and digits as in Example 2 would make no practical sense (why rule out more than 6 million possible plates for no good reason?), states do impose some restrictions on license plates. They rule out certain letter progressions that could be considered obscene or offensive.

Permutations

One important application of the Multiplication Principle of Counting is to count the number of ways that a set of n objects (called an ***n*-set**) can be arranged in order. Each such ordering is called a **permutation** of the set. Example 1 showed that there are $3! = 6$ permutations of a 3-set. In fact, if you understood the tree diagram in Figure 9.1, you can probably guess how many permutations there are of an n-set.

Permutations of an *n*-set

There are $n!$ permutations of an n-set.

Usually the elements of a set are distinguishable from one another, but we can adjust our counting when they are not, as we see in Example 3.

Factorials

If n is a positive integer, the symbol $n!$ (read "n factorial") represents the product $n(n-1)(n-2)(n-3)\cdots 2\cdot 1$. We also define $0! = 1$.

EXAMPLE 3 Distinguishable Permutations

Count the number of different 9-letter "words" (don't worry about whether they're in the dictionary) that can be formed using the letters in each word.

(a) DRAGONFLY **(b)** BUTTERFLY **(c)** BUMBLEBEE

SOLUTION

(a) Each permutation of the 9 letters forms a different word. There are $9! = 362{,}880$ such permutations.

(b) There are also 9! permutations of these letters, but a simple permutation of the two T's does not result in a new word. We correct for the overcount by dividing by 2!. There are $\frac{9!}{2!} = 181{,}440$ *distinguishable* permutations of the letters in BUTTERFLY.

(c) Again there are 9! permutations, but the three B's are indistinguishable, as are the three E's, so we divide by 3! twice to correct for the overcount. There are $\frac{9!}{3!3!} = 10{,}080$ distinguishable permutations of the letters in BUMBLEBEE.

Now try Exercise 9.

Distinguishable Permutations

There are $n!$ distinguishable permutations of an n-set containing n distinguishable objects.

If an n-set contains n_1 objects of a first kind, n_2 objects of a second kind, and so on, with $n_1 + n_2 + \cdots + n_k = n$, then the number of distinguishable permutations of the n-set is

$$\frac{n!}{n_1!n_2!n_3!\cdots n_k!}.$$

In many counting problems, we are interested in using n objects to fill r blanks in order, where $r < n$. These are called **permutations of *n* objects taken *r* at a time.** The procedure for counting them is the same; only this time we run out of blanks before we run out of objects.

The first blank can be filled in n ways, the second in $n - 1$ ways, and so on until we come to the rth blank, which can be filled in $n - (r - 1)$ ways. By the Multiplication Principle, we can fill in all r blanks in $n(n - 1)(n - 2)\cdots(n - r + 1)$ ways. This expression can be written in a more compact (but less easily computed) way as $n!/(n - r)!$.

Permutations on a Calculator

Most modern calculators have an ${}_nP_r$ selection built in. They also compute factorials, but remember that factorials get very large. If you want to count the number of permutations of 90 objects taken 5 at a time, be sure to use the ${}_nP_r$ function. The expression $90!/85!$ is likely to lead to an overflow error.

Permutation Counting Formula

The number of permutations of n objects taken r at a time is denoted ${}_nP_r$ and is given by

$${}_nP_r = \frac{n!}{(n - r)!} \quad \text{for} \quad 0 \le r \le n.$$

If $r > n$, then ${}_nP_r = 0$.

Notice that ${}_nP_n = n!/(n - n)! = n!/0! = n!/1 = n!$, which we have already seen is the number of permutations of a complete set of n objects. This is why we define $0! = 1$.

EXAMPLE 4 Counting Permutations

Evaluate each expression without a calculator.

(a) ${}_6P_4$ **(b)** ${}_{11}P_3$ **(c)** ${}_nP_3$

SOLUTION

(a) By the formula, ${}_6P_4 = 6!/(6 - 4)! = 6!/2! = (6 \cdot 5 \cdot 4 \cdot 3 \cdot 2!)/2! = 6 \cdot 5 \cdot 4 \cdot 3 = 360$.

(b) Although you could use the formula again, you might prefer to apply the Multiplication Principle directly. We have 11 objects and 3 blanks to fill:

$${}_{11}P_3 = 11 \cdot 10 \cdot 9 = 990$$

(c) This time it is definitely easier to use the Multiplication Principle. We have n objects and 3 blanks to fill; so assuming $n \ge 3$,

$${}_nP_3 = n(n - 1)(n - 2).$$

Now try Exercise 15.

EXAMPLE 5 Applying Permutations

Sixteen actors answer a casting call to try out for roles as dwarfs in a production of *Snow White and the Seven Dwarfs*. In how many different ways can the director cast the seven roles?

SOLUTION The 7 different roles can be thought of as 7 blanks to be filled, and we have 16 actors with which to fill them. The director can cast the roles in ${}_{16}P_7 = 57{,}657{,}600$ ways. **Now try Exercise 12.**

Combinations

When we count permutations of n objects taken r at a time, we consider different orderings of the same r selected objects as being different permutations. In many applications we are interested only in the ways to *select* the r objects, regardless of the order in which we arrange them. These unordered selections are called **combinations of n objects taken r at a time**.

Combination Counting Formula

The number of combinations of n objects taken r at a time is denoted ${}_nC_r$ and is given by

$$ {}_nC_r = \frac{n!}{r!(n-r)!} \quad \text{for} \quad 0 \le r \le n. $$

If $r > n$, then ${}_nC_r = 0$.

We can verify the ${}_nC_r$ formula with the Multiplication Principle. Because every permutation can be thought of as an *unordered* selection of r objects *followed* by a particular *ordering* of the objects selected, the Multiplication Principle gives ${}_nP_r = {}_nC_r \cdot r!$.

Therefore,

$$ {}_nC_r = \frac{{}_nP_r}{r!} = \frac{1}{r!} \cdot \frac{n!}{(n-r)!} = \frac{n!}{r!(n-r)!}. $$

A Word on Notation

Some textbooks use $P(n, r)$ instead of ${}_nP_r$ and $C(n, r)$ instead of ${}_nC_r$. Much more common is the notation $\binom{n}{r}$ for ${}_nC_r$. Both $\binom{n}{r}$ and ${}_nC_r$ are often read "n choose r."

Combinations on a Calculator

Most modern calculators have an ${}_nC_r$ selection built in. As with permutations, it is better to use the ${}_nC_r$ function than to use the formula $\frac{n!}{r!(n-r)!}$, as the individual factorials can get too large for the calculator.

EXAMPLE 6 Distinguishing Combinations from Permutations

In each of the following scenarios, tell whether permutations (ordered) or combinations (unordered) are being described.

(a) A president, vice president, and secretary are chosen from a 25-member garden club.

(b) A cook chooses 5 potatoes from a bag of 12 potatoes to make a potato salad.

(c) A teacher makes a seating chart for 22 students in a classroom with 30 desks.

SOLUTION

(a) Permutations. Order matters because it matters who gets which office.

(b) Combinations. The salad is the same no matter what order the potatoes are chosen.

(c) Permutations. A different ordering of students in the same seats results in a different seating chart.

Notice that once you know what is being counted, getting the correct numerical answer is easy with a calculator. The number of possible choices in the scenarios above are (a) ${}_{25}P_3 = 13{,}800$, (b) ${}_{12}C_5 = 792$, and (c) ${}_{30}P_{22} \approx 6.5787 \times 10^{27}$.

Now try Exercise 19.

EXAMPLE 7 Counting Combinations

In the Miss America pageant, 51 contestants must be narrowed down to 10 finalists who will compete on national television. In how many possible ways can the 10 finalists be selected?

SOLUTION Notice that the *order* of the finalists does not matter at this phase; all that matters is which women are selected. So we count combinations rather than permutations.

$$_{51}C_{10} = \frac{51!}{10!41!} = 12{,}777{,}711{,}870$$

The 10 finalists can be chosen in 12,777,711,870 ways. **Now try Exercise 27.**

EXAMPLE 8 Picking Lottery Numbers

The Georgia Lotto requires winners to pick 6 integers between 1 and 46. The order in which you select them does not matter; indeed, the lottery tickets are always printed with the numbers in ascending order. How many different lottery tickets are possible?

SOLUTION There are $_{46}C_6 = 9{,}366{,}819$ possible lottery tickets of this type. (That's more than enough different tickets for every person in the state of Georgia!) **Now try Exercise 29.**

Subsets of an *n*-Set

As a final application of the counting principle, consider the pizza topping problem.

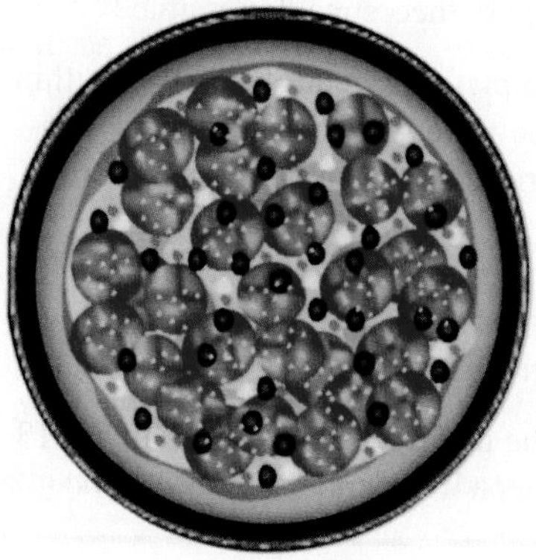

EXAMPLE 9 Selecting Pizza Toppings

Armando's Pizzeria offers patrons any combination of up to 10 different toppings: pepperoni, mushroom, sausage, onion, green pepper, bacon, prosciutto, black olive, green olive, and anchovies. How many different pizzas can be ordered

(a) if we can choose any three toppings?

(b) if we can choose any number of toppings (0 through 10)?

SOLUTION

(a) Order does not matter (for example, the sausage-pepperoni-mushroom pizza is the same as the pepperoni-mushroom-sausage pizza), so the number of possible pizzas is $_{10}C_3 = 120$.

(b) We could add up all the numbers of the form $_{10}C_r$ for $r = 0, 1, \ldots, 10$, but there is an easier way to count the possibilities. Consider the ten options to be lined up as in the statement of the problem. In considering each option, we have two choices: yes or no. (For example, the pepperoni-mushroom-sausage pizza would correspond to the sequence YYYNNNNNNN.) By the Multiplication Principle, the number of such sequences is $2 \cdot 2 \cdot 2 \cdot 2 \cdot 2 \cdot 2 \cdot 2 \cdot 2 \cdot 2 \cdot 2 = 1024$, which is the number of possible pizzas. **Now try Exercise 37.**

The solution to Example 9b suggests a general rule that will be our last counting formula of the section.

Formula for Counting Subsets of an *n*-Set

There are 2^n subsets of a set with n objects (including the empty set and the entire set).

EXAMPLE 10 Analyzing an Advertised Claim

A national hamburger chain used to advertise that it fixed its hamburgers "256 ways," since patrons could order whatever toppings they wanted. How many toppings must have been available?

SOLUTION We need to solve the equation $2^n = 256$ for n. We could solve this easily enough by trial and error, but we will solve it with logarithms just to keep the method fresh in our minds.

$$\begin{aligned} 2^n &= 256 \\ \log 2^n &= \log 256 \\ n \log 2 &= \log 256 \\ n &= \frac{\log 256}{\log 2} \\ n &= 8 \end{aligned}$$

There must have been 8 toppings from which to choose.

Now try Exercise 39.

Why Are There Not 1000 Possible Area Codes?

Although there are 1000 three-digit numbers between 000 and 999, not all of them are available for use as area codes. For example, area codes cannot begin with 0 or 1, and numbers of the form *abb* have been reserved for other purposes.

CHAPTER OPENER Problem (from page 633)

Problem: There are 680 three-digit numbers that are available for use as area codes in North America. As of July 2013, 390 of them were actually being used. How many additional three-digit area codes are available for use? Within a given area code, how many unique telephone numbers are theoretically possible?

Solution: There are $680 - 390 = 290$ additional area codes available. Within a given area code, each telephone number has seven digits chosen from the ten digits 0 through 9. Because each digit can theoretically be any of 10 numbers, there are

$$10 \cdot 10 \cdot 10 \cdot 10 \cdot 10 \cdot 10 \cdot 10 = 10^7 = 10{,}000{,}000$$

different telephone numbers possible within a given area code.

Putting these two results together, we see that the unused area codes in July 2013 represented an additional 2.9 billion possible telephone numbers!

QUICK REVIEW 9.1

In Exercises 1–10, give the number of objects described. In some cases you might have to do a little research or ask a friend.

1. The number of cards in a standard deck
2. The number of cards of each suit in a standard deck
3. The number of faces on a cubical die
4. The number of possible totals when two dice are rolled
5. The number of vertices of a decagon
6. The number of musicians in a string quartet
7. The number of players on a soccer team
8. The number of prime numbers between 1 and 10, inclusive
9. The number of squares on a chessboard
10. The number of cards in a contract bridge hand

SECTION 9.1 Exercises

In Exercises 1–4, count the number of ways that each procedure can be done.

1. Line up three people for a photograph.
2. Prioritize four pending jobs from most to least important.
3. Arrange five books from left to right on a bookshelf.
4. Award ribbons for 1st place to 5th place to the top five dogs in a dog show.
5. **Homecoming King and Queen** There are four candidates for homecoming queen and three candidates for king. How many king-queen pairs are possible?
6. **Possible Routes** There are three roads from town A to town B and four roads from town B to town C. How many different routes are there from A to C by way of B?
7. **Permuting Letters** How many 9-letter "words" (not necessarily in any dictionary) can be formed from the letters of the word LOGARITHM? (Curiously, one such arrangement spells another word related to mathematics. Can you name it?)
8. **Three-Letter Crossword Entries** Excluding J, Q, X, and Z, how many 3-letter crossword puzzle entries can be formed that contain no repeated letters? (It has been conjectured that all of them have appeared in puzzles over the years, sometimes with painfully contrived definitions.)
9. **Permuting Letters** How many distinguishable 11-letter "words" can be formed using the letters in MISSISSIPPI?
10. **Permuting Letters** How many distinguishable 11-letter "words" can be formed using the letters in CHATTANOOGA?
11. **Electing Officers** The 13 members of the East Brainerd Garden Club are electing a President, Vice President, and Secretary from among their members. How many different ways can this be done?
12. **City Government** From among 12 projects under consideration, the mayor must put together a prioritized (that is, ordered) list of 6 projects to submit to the city council for funding. How many such lists can be formed?

In Exercises 13–18, evaluate each expression without a calculator. Then check with your calculator to see if your answer is correct.

13. $4!$
14. $(3!)(0!)$
15. ${}_6P_2$
16. ${}_9P_2$
17. ${}_{10}C_7$
18. ${}_{10}C_3$

In Exercises 19–22, tell whether permutations (ordered) or combinations (unordered) are being described.

19. 13 cards are selected from a deck of 52 to form a bridge hand.
20. 7 digits are selected (without repetition) to form a telephone number.
21. 4 students are selected from the senior class to form a committee to advise the cafeteria director about food.
22. 4 actors are chosen to play the Beatles in a film biography.
23. **License Plates** How many different license plates begin with two digits, followed by two letters and then three digits if no letters or digits are repeated?
24. **License Plates** How many different license plates consist of five symbols, either digits or letters?
25. **Tumbling Dice** Suppose that two dice, one red and one green, are rolled. How many different outcomes are possible for the pair of dice?
26. **Coin Toss** How many different sequences of heads and tails are there if a coin is tossed 10 times?
27. **Forming Committees** A 3-woman committee is to be elected from a 25-member sorority. How many different committees can be elected?
28. **Straight Poker** In the original version of poker known as "straight" poker, a five-card hand is dealt from a standard deck of 52. How many different straight poker hands are possible?
29. **Buying Discs** Juan has money to buy only three of the 48 compact discs available. How many different sets of discs can he purchase?
30. **Coin Toss** A coin is tossed 20 times and the heads and tails sequence is recorded. From among all the possible sequences of heads and tails, how many have exactly seven heads?
31. **Drawing Cards** How many different 13-card hands include the ace and king of spades?
32. **Job Interviews** The head of the personnel department interviews eight people for three identical openings. How many different groups of three can be employed?
33. **Scholarship Nominations** Six seniors at Rydell High School meet the qualifications for a competitive honor scholarship at a major university. The university allows the school to nominate up to three candidates, and the school always nominates at least one. How many different choices could the nominating committee make?
34. **Pu-pu Platters** A Chinese restaurant will make a Pu-pu platter "to order" containing any one, two, or three selections from its appetizer menu. If the menu offers five different appetizers, how many different platters could be made?

35. **Yahtzee** In the game of Yahtzee, five dice are tossed simultaneously. How many outcomes can be distinguished if all the dice are different colors?
36. **Indiana Jones and the Final Exam** Professor Indiana Jones gives his class 20 study questions, from which he will select 8 to be answered on the final exam. How many ways can he select the questions?

37. **Salad Bar** Mary's lunch always consists of a full plate of salad from Ernestine's salad bar. She always takes equal amounts of each salad she chooses, but she likes to vary her selections. If she can choose from among 9 different salads, how many essentially different lunches can she create?

38. **Buying a New Car** A new car customer has to choose from among 3 models, each of which comes in 4 exterior colors, 3 interior colors, and with any combination of up to 6 optional accessories. How many essentially different ways can the customer order the car?

39. **Pizza Possibilities** Luigi sells one size of pizza, but he claims that his selection of toppings allows for "more than 4000 different choices." What is the smallest number of toppings Luigi could offer?

40. **Proper Subsets** A subset of set A is called *proper* if it is neither the empty set nor the entire set A. How many proper subsets does an n-set have?

41. **True-False Tests** How many different answer keys are possible for a 10-question true-false test?

42. **Multiple-Choice Tests** How many different answer keys are possible for a 10-question multiple-choice test in which each question leads to choice a, b, c, d, or e?

Standardized Test Questions

43. **True or False** If a and b are positive integers such that $a + b = n$, then $\binom{n}{a} = \binom{n}{b}$. Justify your answer.

44. **True or False** If a, b, and n are integers such that $a < b < n$, then $\binom{n}{a} < \binom{n}{b}$. Justify your answer.

You may use a graphing calculator when evaluating Exercises 45–48.

45. **Multiple Choice** Lunch at the Gritsy Palace consists of an entrée, two vegetables, and a dessert. If there are four entrées, six vegetables, and six desserts from which to choose, how many essentially different lunches are possible?

(A) 16

(B) 25

(C) 144

(D) 360

(E) 720

46. **Multiple Choice** How many different ways can the judges choose 5th to 1st places from ten Miss America finalists?

(A) 50

(B) 120

(C) 252

(D) 30,240

(E) 3,628,800

47. **Multiple Choice** Assuming r and n are positive integers with $r < n$, which of the following numbers does *not* equal 1?

(A) $(n - n)!$

(B) ${}_nP_n$

(C) ${}_nC_n$

(D) $\binom{n}{n}$

(E) $\binom{n}{r} \div \binom{n}{n-r}$

48. **Multiple Choice** An organization is electing 3 new board members by approval voting. Members are given ballots with the names of 5 candidates and are allowed to check off the names of all candidates whom they would approve (which could be none, or even all five). The three candidates with the most checks overall are elected. In how many different ways can a member fill out the ballot?

(A) 10

(B) 20

(C) 32

(D) 125

(E) 243

Explorations

49. **Group Activity** For each of the following numbers, make up a counting problem that has the number as its answer.

(A) ${}_{52}C_3$

(B) ${}_{12}C_3$

(C) ${}_{25}P_{11}$

(D) 2^5

(E) $3 \cdot 2^{10}$

50. **Writing to Learn** You have a fresh carton containing one dozen eggs and you need to choose two for breakfast. Give a counting argument based on this scenario to explain why ${}_{12}C_2 = {}_{12}C_{10}$.

51. **Factorial Riddle** The number 50! ends in a string of consecutive 0's.

(a) How many 0's are in the string?

(b) How do you know?

52. **Group Activity Diagonals of a Regular Polygon** In Exploration 1 of Section 1.7, you reasoned from data points and quadratic regression that the number of diagonals of a regular polygon with n vertices was $(n^2 - 3n)/2$.

(a) Explain why the number of segments connecting all pairs of vertices is ${}_nC_2$.

(b) Use the result from part (a) to prove that the number of diagonals is $(n^2 - 3n)/2$.

Extending the Ideas

53. **Writing to Learn** Suppose that a chain letter (illegal if money is involved) is sent to five people the first week of the year. Each of these five people sends a copy of the letter to five more people during the second week of the year. Assume that everyone who receives a letter participates. Explain how you know with certainty that someone will receive a second copy of this letter later in the year.

54. **A Round Table** How many different seating arrangements are possible for 4 persons sitting around a round table?

55. **Colored Beads** Four beads—red, blue, yellow, and green—are arranged on a string to make a simple necklace as shown in the figure. How many arrangements are possible?

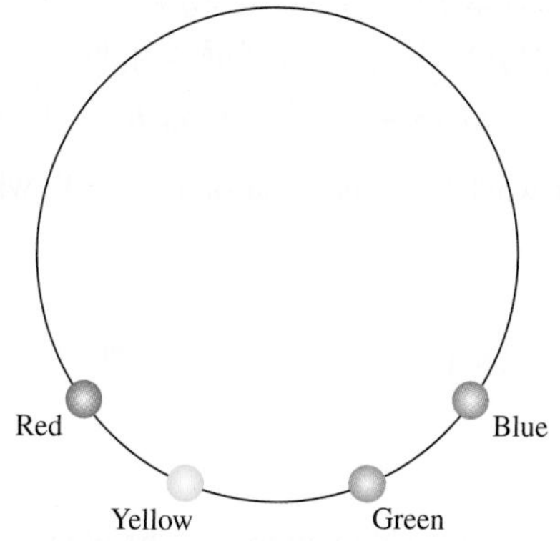

56. **Casting a Play** A director is casting a play with two female leads and wants to have a chance to audition the actresses two at a time to get a feeling for how well they would work together. His casting director and his administrative assistant both prepare charts to show the amount of time that would be required, depending on the number of actresses who come to the audition. Which time chart is more reasonable, and why?

Number Who Audition	Time Required (min)
3	10
6	45
9	110
12	200
15	320

Number Who Audition	Time Required (min)
3	10
6	30
9	60
12	100
15	150

57. **Bridge Around the World** Suppose that a contract bridge hand is dealt somewhere in the world every second. What is the fewest number of years required for every possible bridge hand to be dealt? (See Quick Review Exercise 10.)

58. **Basketball Lineups** Each NBA basketball team has 12 players on its roster. If each coach chooses 5 starters without regard to position, how many different sets of 10 players can start when two given teams play a game?

9.2 Binomial Theorem

What you'll learn about

- Powers of Binomials
- Pascal's Triangle
- Binomial Theorem
- Factorial Identities

... and why

The Binomial Theorem is a marvelous study in combinatorial patterns.

Powers of Binomials

Many important mathematical discoveries have begun with the study of patterns. In this chapter, we want to introduce an important polynomial theorem called the Binomial Theorem, for which we will set the stage by observing some patterns.

If you expand $(a + b)^n$ for $n = 0, 1, 2, 3, 4,$ and 5, here is what you get:

$$
\begin{aligned}
(a + b)^0 &= 1\\
(a + b)^1 &= 1a^1b^0 + 1a^0b^1\\
(a + b)^2 &= 1a^2b^0 + 2a^1b^1 + 1a^0b^2\\
(a + b)^3 &= 1a^3b^0 + 3a^2b^1 + 3a^1b^2 + 1a^0b^3\\
(a + b)^4 &= 1a^4b^0 + 4a^3b^1 + 6a^2b^2 + 4a^1b^3 + 1a^0b^4\\
(a + b)^5 &= 1a^5b^0 + 5a^4b^1 + 10a^3b^2 + 10a^2b^3 + 5a^1b^4 + 1a^0b^5
\end{aligned}
$$

Can you observe the patterns and predict what the expansion of $(a + b)^6$ will look like? You can probably predict the following:

1. The powers of a will decrease from 6 to 0 by 1's.
2. The powers of b will increase from 0 to 6 by 1's.
3. The first two coefficients will be 1 and 6.
4. The last two coefficients will be 6 and 1.

At first you might not see the pattern that would enable you to find the other so-called *binomial coefficients*, but you should see it after the following Exploration.

EXPLORATION 1 Exploring the Binomial Coefficients

1. Compute ${}_3C_0$, ${}_3C_1$, ${}_3C_2$, and ${}_3C_3$. Where can you find these numbers in the binomial expansions above?
2. The expression $4\ {}_nC_r\{0, 1, 2, 3, 4\}$ tells the calculator to compute ${}_4C_r$ for each of the numbers $r = 0, 1, 2, 3, 4$ and display them as a list. Where can you find these numbers in the binomial expansions above?
3. Compute $5\ {}_nC_r\ \{0, 1, 2, 3, 4, 5\}$. Where can you find these numbers in the binomial expansions above?

By now you are probably ready to conclude that the binomial coefficients in the expansion of $(a + b)^n$ are just the values of ${}_nC_r$ for $r = 0, 1, 2, 3, 4, \ldots, n$. We also hope you are wondering *why* this is true.

The expansion of

$$(a + b)^n = \underbrace{(a + b)(a + b)(a + b)\cdots(a + b)}_{n \text{ factors}}$$

consists of all possible products that can be formed by taking one letter (either a or b) from each factor $(a + b)$. The number of ways to form the product a^rb^{n-r} is exactly the same as the number of ways to choose r factors to contribute an a, since the rest of the factors will obviously contribute a b. The number of ways to choose r factors from n factors is ${}_nC_r$.

DEFINITION Binomial Coefficient

The binomial coefficients that appear in the expansion of $(a + b)^n$ are the values of ${}_nC_r$ for $r = 0, 1, 2, 3, \ldots, n$.
A classical notation for ${}_nC_r$, especially in the context of binomial coefficients, is $\binom{n}{r}$. Both notations are read "n choose r."

Table Trick

You can also use the table display to show binomial coefficients. For example, let $Y1 = 5\ {}_nC_r\ X$, and set TblStart $= 0$ and $\Delta\ Tbl = 1$ to display the binomial coefficients for $(a + b)^5$.

EXAMPLE 1 Using ${}_nC_r$ to Expand a Binomial

Expand $(a + b)^5$, using a calculator to compute the binomial coefficients.

SOLUTION Enter $5\ {}_nC_r\ \{0, 1, 2, 3, 4, 5\}$ into the calculator to find the binomial coefficients for $n = 5$. The calculator returns the list $\{1, 5, 10, 10, 5, 1\}$. Using these coefficients, we construct the expansion:

$$(a + b)^5 = 1a^5 + 5a^4b + 10a^3b^2 + 10a^2b^3 + 5ab^4 + 1b^5$$

Now try Exercise 3.

Pascal's Triangle

If we eliminate the plus signs and the powers of the variables a and b in the "triangular" array of binomial coefficients with which we began this section, we get:

1
1 1
1 2 1
1 3 3 1
1 4 6 4 1
1 5 10 10 5 1
⋰ ⋮ ⋱

The Name Game

The fact that Pascal's triangle was not discovered by Pascal is ironic, but hardly unusual in the annals of mathematics. We mentioned in Chapter 5 that Heron did not discover Heron's Formula, and Pythagoras did not even discover the Pythagorean Theorem. The history of calculus is filled with similar injustices.

This is called **Pascal's triangle** in honor of Blaise Pascal (1623–1662), who used it in his work but certainly did not discover it. It appeared in 1303 in a Chinese text, the *Precious Mirror*, by Chu Shih-chieh, who referred to it even then as a "diagram of the old method for finding eighth and lower powers."

For convenience, we refer to the top "1" in Pascal's triangle as row 0. That allows us to associate the numbers along row n with the expansion of $(a + b)^n$.

Pascal's triangle is so rich in patterns that people still write about them today. One of the simplest patterns is the one that we use for getting from one row to the next, as in the following example.

EXAMPLE 2 Extending Pascal's Triangle

Show how row 5 of Pascal's triangle can be used to obtain row 6, and use the information to write the expansion of $(x + y)^6$.

SOLUTION The two outer numbers of every row are 1's. Each number between them is the sum of the two numbers immediately above it. So row 6 can be found from row 5 as follows:

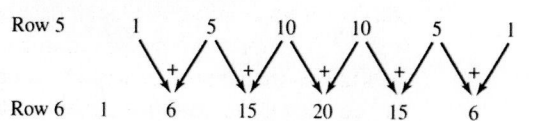

Row 5   1   5   10   10   5   1
(each adjacent pair added: + + + + +)
Row 6 1   6   15   20   15   6   1

(continued)

These are the binomial coefficients for $(x + y)^6$, so

$$(x + y)^6 = x^6 + 6x^5y + 15x^4y^2 + 20x^3y^3 + 15x^2y^4 + 6xy^5 + y^6.$$

Now try Exercise 7.

The technique used in Example 2 to extend Pascal's triangle depends on the following recursive formula.

Recursive Formula for Pascal's Triangle

$$\binom{n}{r} = \binom{n-1}{r-1} + \binom{n-1}{r} \quad \text{or, equivalently, } {}_nC_r = {}_{n-1}C_{r-1} + {}_{n-1}C_r$$

Here's a counting argument to explain why it works. Suppose we are choosing r objects from n objects. As we have seen, this can be done in ${}_nC_r$ ways. Now identify one of the n objects with a special tag. How many ways can we choose r objects if the tagged object is among them? Well, we have $r - 1$ objects yet to be chosen from among the $n - 1$ that are untagged, so ${}_{n-1}C_{r-1}$. How many ways can we choose r objects if the tagged object is *not* among them? This time we must choose all r objects from among the $n - 1$ without tags, so ${}_{n-1}C_r$. Since our selection of r objects must either contain the tagged object or not contain it, ${}_{n-1}C_{r-1} + {}_{n-1}C_r$ counts all the possibilities with no overlap. Therefore, ${}_nC_r = {}_{n-1}C_{r-1} + {}_{n-1}C_r$.

It is not necessary to construct Pascal's triangle to find specific binomial coefficients because we already have a formula for computing them: ${}_nC_r = \binom{n}{r} = \frac{n!}{r!(n-r)!}$. This formula can be used to give an algebraic formula for the recursive formula above, but we will leave that as an exercise for the end of the section.

EXAMPLE 3 Computing Binomial Coefficients

Find the coefficient of x^{10} in the expansion of $(x + 2)^{15}$.

SOLUTION The only term in the expansion that we need to deal with is ${}_{15}C_{10}x^{10}2^5$. This is

$$\frac{15!}{10!5!} \cdot 2^5 \cdot x^{10} = 3003 \cdot 32 \cdot x^{10} = 96{,}096\,x^{10}.$$

The coefficient of x^{10} is 96,096.

Now try Exercise 15.

Binomial Theorem in Σ Notation

For those who are already familiar with summation notation, here is how the Binomial Theorem looks:

$$(a + b)^n = \sum_{r=0}^{n} \binom{n}{r} a^{n-r}b^r$$

Those who are not familiar with this notation will learn about it in Section 9.4.

Binomial Theorem

We now state formally the theorem about expanding powers of binomials, known as the Binomial Theorem. For tradition's sake, we will use the symbol $\binom{n}{r}$ instead of ${}_nC_r$.

Binomial Theorem

For any positive integer n,

$$(a + b)^n = \binom{n}{0}a^n + \binom{n}{1}a^{n-1}b + \cdots + \binom{n}{r}a^{n-r}b^r + \cdots + \binom{n}{n}b^n,$$

where

$$\binom{n}{r} = {}_nC_r = \frac{n!}{r!(n-r)!}.$$

EXAMPLE 4 Expanding a Binomial

Expand $(2x - y^2)^4$.

SOLUTION We use the Binomial Theorem to expand $(a + b)^4$ where $a = 2x$ and $b = -y^2$.

$$\begin{aligned}(a + b)^4 &= a^4 + 4a^3b + 6a^2b^2 + 4ab^3 + b^4\\ (2x - y^2)^4 &= (2x)^4 + 4(2x)^3(-y^2) + 6(2x)^2(-y^2)^2\\ &\quad + 4(2x)(-y^2)^3 + (-y^2)^4\\ &= 16x^4 - 32x^3y^2 + 24x^2y^4 - 8xy^6 + y^8\end{aligned}$$

Now try Exercise 17.

Factorial Identities

Expressions involving factorials combine to give some interesting identities, most of them relying on the basic identities shown below (actually two versions of the same identity).

Basic Factorial Identities

For any integer $n \geq 1$, $n! = n(n - 1)!$

For any integer $n \geq 0$, $(n + 1)! = (n + 1)n!$

EXAMPLE 5 Proving an Identity with Factorials

Prove that $\binom{n+1}{2} - \binom{n}{2} = n$ for all integers $n \geq 2$.

SOLUTION

$$\begin{aligned}\binom{n+1}{2} - \binom{n}{2} &= \frac{(n+1)!}{2!(n+1-2)!} - \frac{n!}{2!(n-2)!} && \text{Combination counting formula}\\ &= \frac{(n+1)(n)(n-1)!}{2(n-1)!} - \frac{n(n-1)(n-2)!}{2(n-2)!} && \text{Basic factorial identities}\\ &= \frac{n^2+n}{2} - \frac{n^2-n}{2}\\ &= \frac{2n}{2}\\ &= n\end{aligned}$$

Now try Exercise 33.

QUICK REVIEW 9.2 *(Prerequisite skill Section A.2)*

In Exercises 1–10, use the distributive property to expand the binomial.

1. $(x + y)^2$
2. $(a + b)^2$
3. $(5x - y)^2$
4. $(a - 3b)^2$
5. $(3s + 2t)^2$
6. $(3p - 4q)^2$
7. $(u + v)^3$
8. $(b - c)^3$
9. $(2x - 3y)^3$
10. $(4m + 3n)^3$

SECTION 9.2 Exercises

In Exercises 1–4, expand the binomial using a calculator to find the binomial coefficients.

1. $(a + b)^4$
2. $(a + b)^6$
3. $(x + y)^7$
4. $(x + y)^{10}$

In Exercises 5–8, expand the binomial using Pascal's triangle to find the coefficients.

5. $(x + y)^3$
6. $(x + y)^5$
7. $(p + q)^8$
8. $(p + q)^9$

In Exercises 9–12, evaluate the expression by hand (using the formula) before checking your answer on a grapher.

9. $\binom{9}{2}$
10. $\binom{15}{11}$
11. $\binom{166}{166}$
12. $\binom{166}{0}$

In Exercises 13–16, find the coefficient of the given term in the binomial expansion.

13. $x^{11}y^3$ term, $(x + y)^{14}$
14. x^5y^8 term, $(x + y)^{13}$
15. x^4 term, $(x - 2)^{12}$
16. x^7 term, $(x - 3)^{11}$

In Exercises 17–20, use the Binomial Theorem to find a polynomial expansion for the function.

17. $f(x) = (x - 2)^5$
18. $g(x) = (x + 3)^6$
19. $h(x) = (2x - 1)^7$
20. $f(x) = (3x + 4)^5$

In Exercises 21–26, use the Binomial Theorem to expand each expression.

21. $(2x + y)^4$
22. $(2y - 3x)^5$
23. $(\sqrt{x} - \sqrt{y})^6$
24. $(\sqrt{x} + \sqrt{3})^4$
25. $(x^{-2} + 3)^5$
26. $(a - b^{-3})^7$

27. Determine the largest integer n for which your calculator will compute $n!$.

28. Determine the largest integer n for which your calculator will compute $\binom{n}{100}$.

29. Prove that $\binom{n}{1} = \binom{n}{n-1} = n$ for all integers $n \geq 1$.

30. Prove that $\binom{n}{r} = \binom{n}{n-r}$ for all integers $n \geq r \geq 0$.

31. Use the formula $\binom{n}{r} = \dfrac{n!}{r!(n-r)!}$ to prove that $\binom{n}{r} = \binom{n-1}{r-1} + \binom{n-1}{r}$. (This is the pattern in Pascal's triangle that appears in Example 2.)

32. Find a counterexample to show that each statement is *false*.

   (a) $(n + m)! = n! + m!$

   (b) $(nm)! = n!m!$

33. Prove that $\binom{n}{2} + \binom{n+1}{2} = n^2$ for all integers $n \geq 2$.

34. Prove that $\binom{n}{n-2} + \binom{n+1}{n-1} = n^2$ for all integers $n \geq 2$.

Standardized Test Questions

35. **True or False** The coefficients in the polynomial expansion of $(x - y)^{50}$ alternate in sign. Justify your answer.

36. **True or False** The sum of any row of Pascal's triangle is an even integer. Justify your answer.

You may use a graphing calculator when evaluating Exercises 37–40.

37. **Multiple Choice** What is the coefficient of x^4 in the expansion of $(2x + 1)^8$?

   (A) 16

   (B) 256

   (C) 1120

   (D) 1680

   (E) 26,680

38. **Multiple Choice** Which of the following numbers does *not* appear on row 10 of Pascal's triangle?

   (A) 1

   (B) 5

   (C) 10

   (D) 120

   (E) 252

39. **Multiple Choice** The *sum* of the coefficients of $(3x - 2y)^{10}$ is

   (A) 1. **(B)** 1024.

   (C) 58,025. **(D)** 59,049.

   (E) 9,765,625.

40. **Multiple Choice** $(x + y)^3 + (x - y)^3 =$

   (A) 0. **(B)** $2x^3$.

   (C) $2x^3 - 2y^3$. **(D)** $2x^3 + 6xy^2$.

   (E) $6x^2y + 2y^3$.

Explorations

41. **Triangular Numbers** Numbers of the form $1 + 2 + \cdots + n$ are called **triangular numbers** because they count numbers in triangular arrays, as shown below:

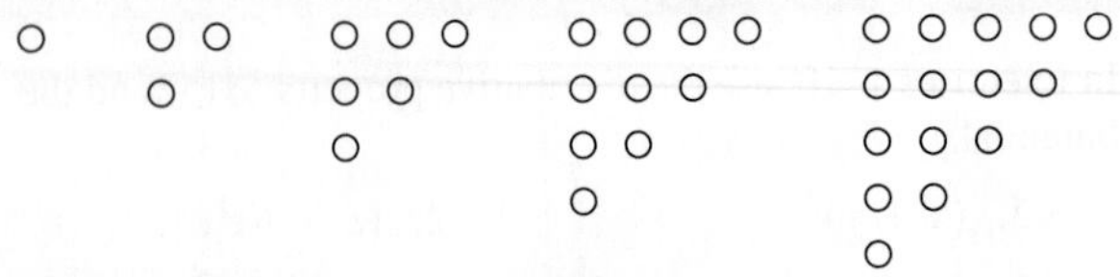

(a) Compute the first 10 triangular numbers.

(b) Where do the triangular numbers appear in Pascal's triangle?

(c) **Writing to Learn** Explain why the diagram below shows that the nth triangular number can be written as $n(n + 1)/2$.

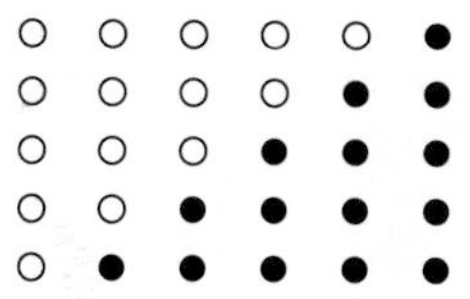

(d) Write the formula in part (c) as a binomial coefficient. (This is why the triangular numbers appear as they do in Pascal's triangle.)

42. Group Activity Exploring Pascal's Triangle Break into groups of two or three. Just by looking at patterns in Pascal's triangle, guess the answers to the following questions. (It is easier to make a conjecture from a pattern than it is to construct a proof!)

(a) What positive integer appears the least number of times?

(b) What number appears the greatest number of times?

(c) Is there any positive integer that does *not* appear in Pascal's triangle?

(d) If you go along any row alternately adding and subtracting the numbers, what is the result?

(e) If p is a prime number, what do all the interior numbers along the pth row have in common?

(f) Which rows have all even interior numbers?

(g) Which rows have all odd numbers?

(h) What other patterns can you find? Share your discoveries with the other groups.

Extending the Ideas

43. Use the Binomial Theorem to prove that the sum of the entries along the nth row of Pascal's triangle is 2^n. That is,

$$\binom{n}{0} + \binom{n}{1} + \binom{n}{2} + \cdots + \binom{n}{n} = 2^n.$$

(*Hint*: Use the Binomial Theorem to expand $(1 + 1)^n$.)

44. Use the Binomial Theorem to prove that the alternating sum along any row of Pascal's triangle is zero. That is,

$$\binom{n}{0} - \binom{n}{1} + \binom{n}{2} - \cdots + (-1)^n\binom{n}{n} = 0.$$

45. Use the Binomial Theorem to prove that

$$\binom{n}{0} + 2\binom{n}{1} + 4\binom{n}{2} + \cdots + 2^n\binom{n}{n} = 3^n.$$

9.3 Sequences

What you'll learn about

- Infinite Sequences
- Limits of Infinite Sequences
- Arithmetic and Geometric Sequences
- Sequences and Technology

... and why

Infinite sequences, especially those with finite limits, are involved in some key concepts of calculus.

Infinite Sequences

One of the most natural ways to study patterns in mathematics is to look at an ordered progression of numbers, called a **sequence**. Here are some examples of sequences:

1. 5, 10, 15, 20, 25
2. $2, 4, 8, 16, 32, \ldots, 2^k, \ldots$
3. $\left\{\frac{1}{k}; k = 1, 2, 3, \ldots\right\}$
4. $\{a_1, a_2, a_3, \ldots, a_k, \ldots\}$, which is sometimes abbreviated $\{a_k\}$

The first of these is a **finite sequence**, and the other three are **infinite sequences**. Notice that in (2) and (3) we were able to give a rule that defines the kth number in the sequence (called the ***k*th term**) as a function of k. In (4) we do not have a rule, but notice how we can use subscript notation (a_k) to identify the kth term of a "general" infinite sequence. In this sense, an infinite sequence can be thought of as a *function* that assigns a unique number (a_k) to each natural number k.

EXAMPLE 1 Defining a Sequence Explicitly

Find the first 6 terms and the 100th term of the sequence $\{a_k\}$ in which $a_k = k^2 - 1$.

SOLUTION Because we know the kth term *explicitly* as a function of k, we need only to evaluate the function to find the required terms:

$$a_1 = 1^2 - 1 = 0,\ a_2 = 3,\ a_3 = 8,\ a_4 = 15,\ a_5 = 24,\ a_6 = 35, \quad \text{and}$$
$$a_{100} = 100^2 - 1 = 9999$$

Now try Exercise 1.

Explicit formulas are the easiest to work with, but there are other ways to define sequences. For example, we can specify values for the first term (or terms) of a sequence, then define each of the following terms **recursively** by a formula relating it to previous terms. Example 2 shows how this is done.

Agreement on Sequences

Because we will be dealing primarily with infinite sequences in this book, the word "sequence" will mean an infinite sequence unless otherwise specified.

EXAMPLE 2 Defining a Sequence Recursively

Find the first 6 terms and the 100th term for the sequence defined recursively by the conditions:

$$b_1 = 3$$
$$b_n = b_{n-1} + 2 \text{ for all } n > 1$$

SOLUTION We proceed one term at a time, starting with $b_1 = 3$ and obtaining each succeeding term by adding 2 to the term just before it:

$$b_1 = 3$$
$$b_2 = b_1 + 2 = 5$$
$$b_3 = b_2 + 2 = 7$$
etc.

Eventually it becomes apparent that we are building the sequence of odd natural numbers beginning with 3:

$$\{3, 5, 7, 9, \ldots\}$$

The 100th term is 99 terms beyond the first, which means that we can get there quickly by adding 99 2's to the number 3:

$$b_{100} = 3 + 99 \times 2 = 201$$

Now try Exercise 5.

Limits of Infinite Sequences

Just as we were interested in the end behavior of functions, we also are interested in the end behavior of sequences.

DEFINITION Limit of a Sequence

Let $\{a_n\}$ be a sequence of real numbers, and consider $\lim_{n\to\infty} a_n$. If the limit is a finite number L, the sequence **converges** and L is the **limit of the sequence**. If the limit is infinite or nonexistent, the sequence **diverges**.

EXAMPLE 3 Finding Limits of Sequences

Determine whether the sequence converges or diverges. If it converges, give the limit.

(a) $\frac{1}{1}, \frac{1}{2}, \frac{1}{3}, \frac{1}{4}, \ldots, \frac{1}{n}, \ldots$

(b) $\frac{2}{1}, \frac{3}{2}, \frac{4}{3}, \frac{5}{4}, \ldots$

(c) $2, 4, 6, 8, 10, \ldots$

(d) $-1, 1, -1, 1, \ldots, (-1)^n, \ldots$

SOLUTION

(a) $\lim_{x\to\infty} \frac{1}{n} = 0$, so the sequence converges to a limit of 0.

(b) Although the nth term is not explicitly given, we can see that $a_n = \frac{n+1}{n}$.

$\lim_{x\to\infty} \frac{n+1}{n} = \lim_{n\to\infty}\left(1 + \frac{1}{n}\right) = 1 + 0 = 1$. The sequence converges to a limit of 1.

(c) This time we see that $a_n = 2n$. Because $\lim_{n\to\infty} 2n = \infty$, the sequence diverges.

(d) This sequence oscillates forever between two values and hence has no limit. The sequence diverges.

Now try Exercise 13.

It might help to review the rules for finding the *end behavior asymptotes* of rational functions (page 221 in Section 2.6) because those same rules apply to sequences that are rational functions of n, as in Example 4.

EXAMPLE 4 Finding Limits of Sequences

Determine whether the sequence converges or diverges. If it converges, give the limit.

(a) $\left\{\frac{3n}{n+1}\right\}$

(b) $\left\{\frac{5n^2}{n^3+1}\right\}$

(c) $\left\{\frac{n^3+2}{n^2+n}\right\}$

(continued)

SOLUTION

(a) Since the degree of the numerator is the same as the degree of the denominator, the limit is the ratio of the leading coefficients.

Thus $\lim_{n \to \infty} \frac{3n}{n+1} = \frac{3}{1} = 3$. The sequence converges to a limit of 3.

(b) Since the degree of the numerator is less than the degree of the denominator, the limit is zero. Thus $\lim_{n \to \infty} \frac{5n^2}{n^3+1} = 0$. The sequence converges to 0.

(c) Since the degree of the numerator is greater than the degree of the denominator, the limit is infinite. Thus $\lim_{n \to \infty} \frac{n^3+2}{n^2+n}$ is infinite. The sequence diverges.

Now try Exercise 15.

Arithmetic and Geometric Sequences

There are all kinds of rules by which we can construct sequences, but two particular types of sequences dominate in mathematical applications: those in which pairs of successive terms all have a common *difference* (*arithmetic sequences*), and those in which pairs of successive terms all have a common *ratio* (*geometric sequences*). We will study these in this section.

Pronunciation Tip

The word "arithmetic" is probably more familiar to you as a noun, referring to the mathematics you studied in elementary school. In this word, the second syllable ("rith") is accented. When used as an adjective, the third syllable ("met") gets the accent. (For the sake of comparison, a similar shift of accent occurs when going from the noun "analysis" to the adjective "analytic.")

DEFINITION Arithmetic Sequence

A sequence $\{a_n\}$ is an **arithmetic sequence** if it can be written in the form

$$\{a, a+d, a+2d, \ldots, a+(n-1)d, \ldots\} \text{ for some constant } d.$$

The number a is the first term, and the number d is the **common difference**. Each term in an arithmetic sequence can be obtained recursively from its preceding term by adding d:

$$a_n = a_{n-1} + d \text{ (for all } n \geq 2)$$

EXAMPLE 5 Defining Arithmetic Sequences

For each of the following arithmetic sequences, find **(a)** the common difference, **(b)** the tenth term, **(c)** a recursive rule for the nth term, and **(d)** an explicit rule for the nth term.

(1) $-6, -2, 2, 6, 10, \ldots$

(2) $\ln 3, \ln 6, \ln 12, \ln 24, \ldots$

SOLUTION

(1) **(a)** The difference between successive terms is 4.

(b) $a_{10} = -6 + (10-1)(4) = 30$

(c) The sequence is defined recursively by $a_1 = -6$ and $a_n = a_{n-1} + 4$ for all $n \geq 2$.

(d) The sequence is defined explicitly by $a_n = -6 + (n-1)(4) = 4n - 10$.

(2) **(a)** This sequence might not look arithmetic at first, but $\ln 6 - \ln 3 = \ln \frac{6}{3} = \ln 2$ (by a law of logarithms) and the difference between successive terms continues to be $\ln 2$.

(b) $a_{10} = \ln 3 + (10-1)\ln 2 = \ln 3 + 9 \ln 2 = \ln (3 \cdot 2^9) = \ln 1536$

(c) The sequence is defined recursively by $a_1 = \ln 3$ and $a_n = a_{n-1} + \ln 2$ for all $n \geq 2$.

(d) The sequence is defined explicitly by $a_n = \ln 3 + (n-1)\ln 2 = \ln (3 \cdot 2^{n-1})$.

Now try Exercise 21.

DEFINITION Geometric Sequence

A sequence $\{a_n\}$ is a **geometric sequence** if it can be written in the form

$$\{a, a \cdot r, a \cdot r^2, \ldots, a \cdot r^{n-1}, \ldots\} \text{ for some nonzero constant } r.$$

The number a is the first term, and the number r is the **common ratio**.
Each term in a geometric sequence can be obtained recursively from its preceding term by multiplying by r:

$$a_n = a_{n-1} \cdot r \text{ (for all } n \geq 2)$$

EXAMPLE 6 Defining Geometric Sequences

For each of the following geometric sequences, find **(a)** the common ratio, **(b)** the tenth term, **(c)** a recursive rule for the nth term, and **(d)** an explicit rule for the nth term.

(1) 3, 6, 12, 24, 48, . . . **(2)** $10^{-3}, 10^{-1}, 10^{1}, 10^{3}, 10^{5}, \ldots$

SOLUTION

(1) **(a)** The ratio between successive terms is 2.

(b) $a_{10} = 3 \cdot 2^{10-1} = 3 \cdot 2^9 = 1536$

(c) The sequence is defined recursively by $a_1 = 3$ and $a_n = 2a_{n-1}$ for $n \geq 2$.

(d) The sequence is defined explicitly by $a_n = 3 \cdot 2^{n-1}$.

(2) **(a)** Applying a law of exponents, $\dfrac{10^{-1}}{10^{-3}} = 10^{-1-(-3)} = 10^2$, and the ratio between successive terms continues to be 10^2.

(b) $a_{10} = 10^{-3} \cdot (10^2)^{10-1} = 10^{-3+18} = 10^{15}$

(c) The sequence is defined recursively by $a_1 = 10^{-3}$ and $a_n = 10^2 a_{n-1}$ for $n \geq 2$.

(d) The sequence is defined explicitly by $a_n = 10^{-3}(10^2)^{n-1} = 10^{-3+2n-2} = 10^{2n-5}$.

Now try Exercise 25.

EXAMPLE 7 Constructing Sequences

The second and fifth terms of a sequence are 3 and 24, respectively. Find explicit and recursive formulas for the sequence if it is **(a)** arithmetic and **(b)** geometric.

SOLUTION

(a) If the sequence is arithmetic, then $a_2 = a_1 + d = 3$ and $a_5 = a_1 + 4d = 24$. Subtracting, we have

$$(a_1 + 4d) - (a_1 + d) = 24 - 3$$
$$3d = 21$$
$$d = 7$$

Then $a_1 + d = 3$ implies $a_1 = -4$.

The sequence is defined explicitly by $a_n = -4 + (n - 1) \cdot 7$, or $a_n = 7n - 11$.

The sequence is defined recursively by $a_1 = -4$ and $a_n = a_{n-1} + 7$ for $n \geq 2$.

(b) If the sequence is geometric, then $a_2 = a \cdot r^1 = 3$ and $a_5 = a \cdot r^4 = 24$. Dividing, we have

$$\frac{a_1 \cdot r^4}{a_1 \cdot r^1} = \frac{24}{3}$$
$$r^3 = 8$$
$$r = 2$$

(continued)

Then $a_1 \cdot r^1 = 3$ implies $a_1 = 1.5$.

The sequence is defined explicitly by $a_n = 1.5(2)^{n-1}$, or $a_n = 3(2)^{n-2}$.

The sequence is defined recursively by $a_1 = 1.5$ and $a_n = 2 \cdot a_{n-1}$.

Now try Exercise 29.

Sequences and Technology

As with other kinds of functions, it helps to be able to represent a sequence graphically. There are at least two ways to obtain a sequence graph using a grapher. One way to graph explicitly defined sequences is as scatter plots of points of the form (k, a_k). A second way is to use the sequence graphing mode.

Sequence Graphing

Most graphers enable you to graph in "sequence mode." Check your owner's manual to see how to use this mode. Your grapher may work differently from the descriptions on this page and the next page.

EXAMPLE 8 Graphing a Sequence Defined Explicitly

Use a grapher to produce a graph of the sequence $\{a_k\}$ in which $a_k = k^2 - 1$.

Method 1 (Scatter Plot) The command seq(K, K, 1, 10) $\rightarrow L_1$ puts the first 10 natural numbers in list L_1. (You could change the 10 if you wanted to graph more or fewer points.)

The command $L_1{}^2 - 1 \rightarrow L_2$ puts the corresponding terms of the sequence in list L_2. A scatter plot of L_1, L_2 produces the graph in Figure 9.2a.

Method 2 (Sequence Mode) With your grapher in Sequence mode, enter the sequence $a_k = k^2 - 1$ in the Y= list as $u(n) = n^2 - 1$ with nMin $= 1$, nMax $= 10$, and $u(n\text{Min}) = 0$. (You could change the 10 if you wanted to graph more or fewer points.) Figure 9.2b shows the graph in the same window as Figure 9.2a.

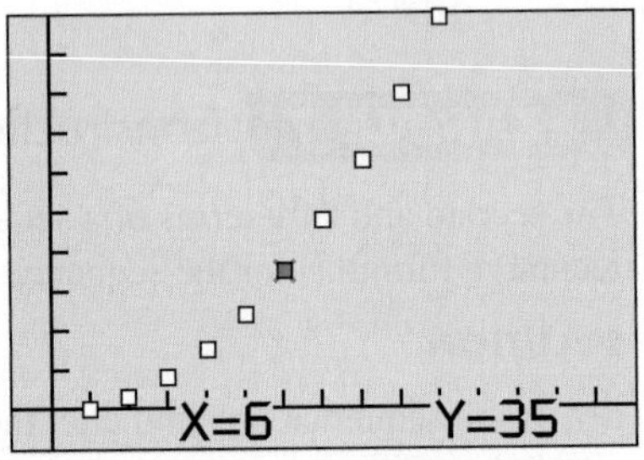

[–1, 15] by [–10, 100]

(a)

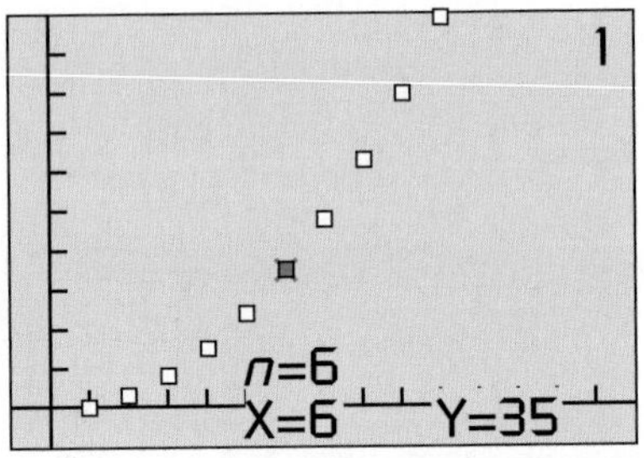

[–1, 15] by [–10, 100]

(b)

FIGURE 9.2 The sequence $a_k = k^2 - 1$ graphed (a) as a scatter plot and (b) using the sequence graphing mode. Tracing along the points gives values of a_k for $k = 1, 2, 3, \ldots$. (Example 8)

Now try Exercise 33.

EXAMPLE 9 Generating Sequences Using a Grapher

Use a grapher to generate the specific terms of the following sequences:

(a) (Explicit) $a_k = 3k - 5$ for $k = 1, 2, 3, \ldots$

(b) (Recursive) $a_1 = -2$ and $a_n = a_{n-1} + 3$ for $n = 2, 3, 4, \ldots$

SOLUTION

(a) On the calculating screen, type the two commands shown in Figure 9.3. The grapher will then generate the terms of the sequence as you press the ENTER key repeatedly.

(b) On the calculating screen, type the two commands shown in Figure 9.4. The first command gives the value of a_1. The grapher will generate the remaining terms of the sequence as you press the ENTER key repeatedly.

Notice that these two definitions generate the very same sequence!

Now try Exercises 1 and 5 on your grapher.

```
0→K
                    0
K+1→K:3K-5
                   -2
                    1
                    4
                    7
```

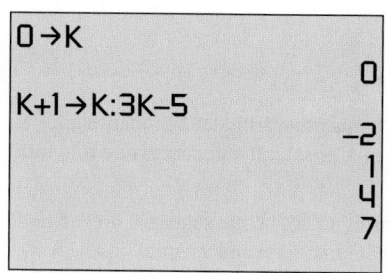

FIGURE 9.3 Typing these two commands (on the left of the viewing screen) will generate the terms of the explicitly defined sequence $a_k = 3k - 5$. (Example 9a)

```
-2
                   -2
Ans+3
                    1
                    4
                    7
                   10
```

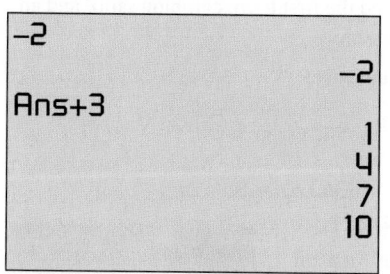

FIGURE 9.4 Typing these two commands (on the left of the viewing screen) will generate the terms of the recursively defined sequence with $a_1 = -2$ and $a_n = a_{n-1} + 3$. (Example 9b)

A recursive definition of a_n can be made in terms of any combination of preceding terms, provided the preceding terms have already been determined. A famous example is the *Fibonacci sequence*, named for Leonardo of Pisa (ca. 1170–1250), who wrote under the name Fibonacci. You can generate it with the two commands shown in Figure 9.5.

```
0→A:1→B
                    1
A+B→C:A→B:C→A
                    1
                    1
                    2
                    3
```

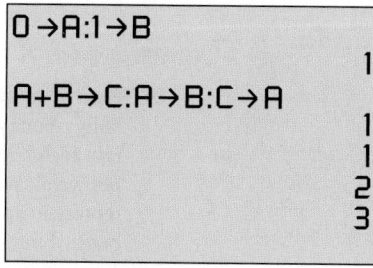

FIGURE 9.5 The two commands on the left will generate the Fibonacci sequence as the ENTER key is pressed repeatedly.

The Fibonacci sequence can be defined recursively using three statements.

DEFINITION The Fibonacci Sequence

The **Fibonacci sequence** can be defined recursively by

$$a_1 = 1$$
$$a_2 = 1$$
$$a_n = a_{n-2} + a_{n-1}$$

for all positive integers $n \geq 3$.

Fibonacci Numbers

The numbers in the Fibonacci sequence have fascinated professional and amateur mathematicians since the 13th century. Not only is the sequence, like Pascal's triangle, a rich source of curious internal patterns, but the Fibonacci numbers seem to appear everywhere in nature. If you count the leaflets on a leaf, the leaves on a stem, the whorls on a pine cone, the rows on an ear of corn, the spirals in a sunflower, or the branches from a trunk of a tree, they tend to be Fibonacci numbers. (Check **phyllotaxy** in a biology book.)

QUICK REVIEW 9.3 *(For help, see Section P.1.)*

Exercise numbers with a gray background indicate problems that the authors have designed to be solved *without a calculator*.

In Exercises 1 and 2, evaluate each expression when $a = 3$, $d = 4$, and $n = 5$.

1. $a + (n - 1)d$

2. $\frac{n}{2}(2a + (n - 1)d)$

In Exercises 3 and 4, evaluate each expression when $a = 5$, $r = 4$, and $n = 3$.

3. $a \cdot r^{n-1}$

4. $\frac{a(1 - r^n)}{1 - r}$

In Exercises 5–10, find a_{10}.

5. $a_k = \frac{k}{k + 1}$

6. $a_k = 5 + (k - 1)3$

7. $a_k = 5 \cdot 2^{k-1}$

8. $a_k = \frac{4}{3}\left(\frac{1}{2}\right)^{k-1}$

9. $a_k = 32 - a_{k-1}$ and $a_9 = 17$

10. $a_k = \frac{k^2}{2^k}$

SECTION 9.3 Exercises

In Exercises 1–4, find the first 6 terms and the 100th term of the explicitly defined sequence.

1. $u_n = \dfrac{n+1}{n}$ **2.** $v_n = \dfrac{4}{n+2}$

3. $c_n = n^3 - n$ **4.** $d_n = n^2 - 5n$

In Exercises 5–10, find the first 4 terms and the eighth term of the recursively defined sequence.

5. $a_1 = 8$ and $a_n = a_{n-1}, - 4$, for $n \ge 2$

6. $u_1 = -3$ and $u_{k+1} = u_k + 10$, for $k \ge 1$,

7. $b_1 = 2$ and $b_{k+1} = 3b_k$, for $k \ge 1$

8. $v_1 = 0.75$ and $v_n = (-2)v_{n-1}$, for $n \ge 2$

9. $c_1 = 2$, $c_2 = -1$, and $c_{k+2} = c_k + c_{k+1}$ 1, for $k \ge 1$

10. $c_1 = -2$, $c_2 = 3$, and $c_k = c_{k-2} + c_{k-1}$, for $k \ge 3$

In Exercises 11–20, determine whether the sequence converges or diverges. If it converges, give the limit.

11. $1, 4, 9, 16, \ldots, n^2, \ldots$

12. $\dfrac{1}{2}, \dfrac{1}{4}, \dfrac{1}{8}, \dfrac{1}{16}, \ldots, \dfrac{1}{2^n}, \ldots$

13. $\dfrac{1}{1}, \dfrac{1}{4}, \dfrac{1}{9}, \dfrac{1}{16}, \ldots$

14. $\{3n - 1\}$

15. $\left\{\dfrac{3n-1}{2-3n}\right\}$

16. $\left\{\dfrac{2n-1}{n+1}\right\}$

17. $\{(0.5)^n\}$

18. $\{(1.5)^n\}$

19. $a_1 = 1$ and $a_{n+1} = a_n + 3$ for $n \ge 1$

20. $u_1 = 1$ and $u_{n+1} = \dfrac{u_n}{3}$ for $n \ge 1$

In Exercises 21–24, the sequences are arithmetic. Find

(a) the common difference,

(b) the tenth term,

(c) a recursive rule for the nth term, and

(d) an explicit rule for the nth term.

21. $6, 10, 14, 18, \ldots$ **22.** $-4, 1, 6, 11, \ldots$

23. $-5, -2, 1, 4, \ldots$ **24.** $-7, 4, 15, 26, \ldots$

In Exercises 25–28, the sequences are geometric. Find

(a) the common ratio,

(b) the eighth term,

(c) a recursive rule for the nth term, and

(d) an explicit rule for the nth term.

25. $2, 6, 18, 54, \ldots$ **26.** $3, 6, 12, 24, \ldots$

27. $1, -2, 4, -8, 16, \ldots$ **28.** $-2, 2, -2, 2, \ldots$

29. The fourth and seventh terms of an arithmetic sequence are -8 and 4, respectively. Find the first term and a recursive rule for the nth term.

30. The fifth and ninth terms of an arithmetic sequence are -5 and -17, respectively. Find the first term and a recursive rule for the nth term.

31. The second and eighth terms of a geometric sequence are 3 and 192, respectively. Find the first term, common ratio, and an explicit rule for the nth term.

32. The third and sixth terms of a geometric sequence are -75 and -9375, respectively. Find the first term, common ratio, and an explicit rule for the nth term.

In Exercises 33–36, graph the sequence.

33. $a_n = 2 - \dfrac{1}{n}$ **34.** $b_n = \sqrt{n} - 3$

35. $c_n = n^2 - 5$ **36.** $d_n = 3 + 2n$

37. Rain Forest Growth The bungy-bungy tree in the Amazon rain forest grows an average 2.3 cm per week. Write a sequence that represents the weekly height of a bungy-bungy over the course of 1 year if it is 7 m tall today. Display the first four terms and the last two terms.

38. Half-Life (See Section 3.2) Thorium-232 has a half-life of 14 billion years. Make a table showing the half-life decay of a sample of thorium-232 from 16 g to 1 g; list the time (in years, starting with $t = 0$) in the first column and the mass (in grams) in the second column. Which type of sequence is each column of the table?

39. Arena Seating The first row of seating in section J of the Athena Arena has 7 seats. In all, there are 25 rows of seats in section J, each row containing two more seats than the row preceding it. How many seats are in section J?

40. Patio Construction Pat designs a patio with a trapezoid-shaped deck consisting of 16 rows of congruent slate tiles. The numbers of tiles in the rows form an arithmetic sequence. The first row contains 15 tiles and the last row contains 30 tiles. How many tiles are used in the deck?

41. Group Activity Pair up with a partner to create a sequence recursively together. Each of you picks five random digits from 1 to 9 (with repetitions, if you wish). Merge your digits to make a list of ten. Now each of you constructs a ten-digit number using exactly the numbers in your list.

Let a_1 = the (positive) difference between your two numbers.

Let a_{n+1} = the sum of the digits of a_n for $n \ge 1$.

This sequence converges because it is eventually constant. What is the limit? (Remember, you can check your answer in the back of the book.)

42. **Group Activity** Here is an interesting recursively defined word sequence. Join up with three or four classmates and, without telling it to the others, pick a word from this sentence. Then, with care, count the letters in your word. Move *ahead* that many words in the text to come to a new word. Count the letters in the new word. Move ahead again, and so on. When you come to a point when your next move would take you out of this problem, stop. Share your last word with your friends. Are they all the same?

Standardized Test Questions

43. **True or False** If the first two terms of a geometric sequence are negative, then so is the third. Justify your answer.

44. **True or False** If the first two terms of an arithmetic sequence are positive, then so is the third. Justify your answer.

You may use a graphing calculator when solving Exercises 45–48.

45. **Multiple Choice** The first two terms of an arithmetic sequence are 2 and 8. The fourth term is

(A) 20. **(B)** 26. **(C)** 64. **(D)** 128. **(E)** 256.

46. **Multiple Choice** Which of the following sequences is divergent?

(A) $\left\{\frac{n+100}{n}\right\}$ **(B)** $\{\sqrt{n}\}$ **(C)** $\{\pi^{-n}\}$ **(D)** $\left\{\frac{2n+2}{n+1}\right\}$ **(E)** $\{n^{-2}\}$

47. **Multiple Choice** A geometric sequence $\{a_n\}$ begins 2, 6, ... What is $\frac{a_6}{a_2}$?

(A) 3 **(B)** 4 **(C)** 9 **(D)** 12 **(E)** 81

48. **Multiple Choice** Which of the following rules for $n \geq 1$ will define a geometric sequence if $a_1 \neq 0$?

(A) $a_{n+1} = a_n + 3$ **(B)** $a_{n+1} = a_n - 3$

(C) $a_{n+1} = a_n \div 3$ **(D)** $a_{n+1} = a_n^3$ **(E)** $a_{n+1} = a_n \cdot 3^{n-1}$

Explorations

49. **Rabbit Populations** Assume that 2 months after birth, each male-female pair of rabbits begins producing one new male-female pair of rabbits each month. Further assume that the rabbit colony begins with one newborn male-female pair of rabbits and no rabbits die for 12 months. Let a_n represent the number of *pairs* of rabbits in the colony after $n - 1$ months.

(a) Writing to Learn Explain why $a_1 = 1$, $a_2 = 1$, and $a_3 = 2$.

(b) Find $a_4, a_5, a_6, \ldots, a_{13}$.

(c) Writing to Learn Explain why the sequence $\{a_n\}$, $1 \leq n \leq 13$, is a model for the size of the rabbit colony for a 1-year period.

50. **Fibonacci Sequence** Compute the first seven terms of the sequence whose nth term is

$$a_n = \frac{1}{\sqrt{5}}\left(\frac{1+\sqrt{5}}{2}\right)^n - \frac{1}{\sqrt{5}}\left(\frac{1-\sqrt{5}}{2}\right)^n.$$

How do these seven terms compare with the first seven terms of the Fibonacci sequence?

51. **Connecting Geometry and Sequences** In the following sequence of diagrams, regular polygons are inscribed in unit circles with at least one side of each polygon perpendicular to the positive x-axis.

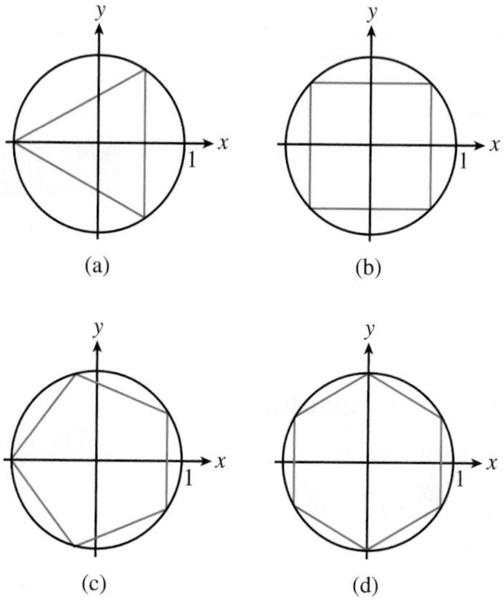

(a) Prove that the perimeter of each polygon in the sequence is given by $a_n = 2n \sin(\pi/n)$, where n is the number of sides in the polygon.

(b) Investigate the value of a_n for $n = 10, 100, 1000$, and 10,000. What conclusion can you draw?

52. **Recursive Sequence** The population of Centerville was 525,000 in 1992 and is growing annually at the rate of 1.75%. Write a recursive sequence $\{P_n\}$ for the population. State the first term P_1 for your sequence.

53. **Writing to Learn** If $\{a_n\}$ is a geometric sequence with all positive terms, explain why $\{\log a_n\}$ must be arithmetic.

54. **Writing to Learn** If $\{b_n\}$ is an arithmetic sequence, explain why $\{10^{b_n}\}$ must be geometric.

Extending the Ideas

55. **A Sequence of Matrices** Write out the first seven terms of the "geometric sequence" with the first term the matrix $[1 \quad 1]$ and the common ratio the matrix $\begin{bmatrix} 0 & 1 \\ 1 & 1 \end{bmatrix}$. How is this sequence of matrices related to the Fibonacci sequence?

56. **Another Sequence of Matrices** Write out the first seven terms of the "geometric sequence" which has for its first term the matrix $[1 \quad a]$ and for its common ratio the matrix $\begin{bmatrix} 1 & d \\ 0 & 1 \end{bmatrix}$. How is this sequence of matrices related to the arithmetic sequence?

9.4 Series

What you'll learn about

- Summation Notation
- Sums of Arithmetic and Geometric Sequences
- Infinite Series
- Convergence of Geometric Series

... and why

Infinite series are at the heart of integral calculus.

Summation Notation

We want to look at the formulas for summing the terms of arithmetic and geometric sequences, but first we need a notation for writing the sum of an indefinite number of terms. The capital Greek letter sigma (Σ) provides our shorthand notation for a "summation."

DEFINITION Summation Notation

In **summation notation**, the sum of the terms of the sequence $\{a_1, a_2, \ldots, a_n\}$ is denoted

$$\sum_{k=1}^{n} a_k$$

which is read "the sum of a_k from $k = 1$ to n."

The variable k is called the **index of summation**.

Summations on a Calculator

If you think of summations as summing sequence values, it is not hard to translate sigma notation into calculator syntax. Here, in calculator syntax, are the first three summations in Exploration 1. (Don't try these on your calculator until you have first figured out the answers with pencil and paper.)

1. sum(seq(3K, K, 1, 5))
2. sum(seq(K^2, K, 5, 8))
3. sum(seq(cos(Nπ), N, 0, 12))

EXPLORATION 1 Summing with Sigma

Sigma notation is actually even more versatile than the definition above suggests. See if you can determine the number represented by each of the following expressions.

1. $\sum_{k=1}^{5} 3k$ **2.** $\sum_{k=5}^{8} k^2$ **3.** $\sum_{n=0}^{12} \cos(n\pi)$ **4.** $\sum_{n=1}^{\infty} \sin(n\pi)$ **5.** $\sum_{k=1}^{\infty} \frac{3}{10^k}$

(If you're having trouble with number 5, here's a hint: Write the sum as a decimal!)

Although you probably computed them correctly, there is more going on in number 4 and number 5 in the above exploration than first meets the eye. We will have more to say about these "infinite" summations toward the end of this section.

Sums of Arithmetic and Geometric Sequences

One of the most famous legends in the lore of mathematics concerns the German mathematician Karl Friedrich Gauss (1777–1855), whose mathematical talent was apparent at a very early age. One version of the story has Gauss, at age ten, being in a class that was challenged by the teacher to add up all the numbers from 1 to 100. While his classmates were still writing down the problem, Gauss walked to the front of the room to present his slate to the teacher. The teacher, certain that Gauss could only be guessing, refused to look at his answer. Gauss simply placed it face down on the teacher's desk, declared "There it is," and returned to his seat. Later, after all the slates had been collected, the teacher looked at Gauss's work, which consisted of a single number: the correct answer. No other student (the legend goes) got it right.

The important feature of this legend for mathematicians is *how* the young Gauss got the answer so quickly. We'll let you reproduce his technique in Exploration 2.

EXPLORATION 2 Gauss's Insight

Your challenge is to find the sum of the natural numbers from 1 to 100 without a calculator.

1. On a wide piece of paper, write the sum

"1 + 2 + 3 + · · · + 98 + 99 + 100."

2. Underneath this sum, write the sum

"100 + 99 + 98 + · · · + 3 + 2 + 1."

3. Add the numbers two-by-two in *vertical* columns and notice that you get the same identical sum 100 times. What is it?
4. What is the sum of the 100 identical numbers referred to in part 3?
5. Explain why half the answer in part 4 is the answer to the challenge. Can you find it without a calculator?

If this story is true, then the youthful Gauss had discovered a fact that his elders knew about arithmetic sequences. If you write a finite arithmetic sequence forward on one line and backward on the line below it, then all the pairs stacked vertically sum to the same number. Multiplying this number by the number of terms n and dividing by 2 gives us a shortcut to the sum of the n terms. We state this result as a theorem.

THEOREM Sum of a Finite Arithmetic Sequence

Let $\{a_1, a_2, \dots, a_n\}$ be a finite arithmetic sequence with common difference d. Then the sum of the terms of the sequence is

$$\begin{aligned}\sum_{k=1}^{n} a_k &= a_1 + a_2 + \cdots + a_n \\ &= n\left(\frac{a_1 + a_n}{2}\right) \\ &= \frac{n}{2}(2a_1 + (n-1)d)\end{aligned}$$

Proof

We can construct the sequence forward by starting with a_1 and *adding d* each time, or we can construct the sequence backward by starting at a_n and *subtracting d* each time. We thus get two expressions for the sum we are looking for:

$$\sum_{k=1}^{n} a_k = a_1 + (a_1 + d) + (a_1 + 2d) + \cdots + (a_1 + (n-1)d)$$

$$\sum_{k=1}^{n} a_k = a_n + (a_n - d) + (a_n - 2d) + \cdots + (a_n - (n-1)d)$$

Summing vertically, we get

$$2\sum_{k=1}^{n} a_k = (a_1 + a_n) + (a_1 + a_n) + \cdots + (a_1 + a_n)$$

$$2\sum_{k=1}^{n} a_k = n(a_1 + a_n)$$

$$\sum_{k=1}^{n} a_k = n\left(\frac{a_1 + a_n}{2}\right)$$

If we substitute $a_1 + (n - 1)d$ for a_n, we get the alternate formula

$$\sum_{k=1}^{n} a_k = \frac{n}{2}(2a_1 + (n - 1)d).$$

EXAMPLE 1 Summing the Terms of an Arithmetic Sequence

A corner section of a stadium has 8 seats along the front row. Each successive row has two more seats than the row preceding it. If the top row has 24 seats, how many seats are in the entire section?

SOLUTION The numbers of seats in the rows form an arithmetic sequence with

$$a_1 = 8, \quad a_n = 24, \quad \text{and} \quad d = 2.$$

Solving $a_n = a_1 + (n - 1)d$, we find that

$$\begin{aligned} 24 &= 8 + (n - 1)(2) \\ 16 &= (n - 1)(2) \\ 8 &= n - 1 \\ n &= 9 \end{aligned}$$

Applying the Sum of a Finite Arithmetic Sequence Theorem, the total number of seats in the section is $9(8 + 24)/2 = 144$.

We can support this answer numerically using technology:

$$\text{sum}(\text{seq}(8+(N-1)2, N, 1, 9) = 144$$

Now try Exercise 7.

Word of Warning

Your grapher may work differently from the descriptions given in Examples 1 and 2.

As you might expect, there is also a convenient formula for summing the terms of a finite geometric sequence.

THEOREM Sum of a Finite Geometric Sequence

Let $\{a_1, a_2, a_3, \ldots, a_n\}$ be a finite geometric sequence with common ratio $r \neq 1$.

Then the sum of the terms of the sequence is

$$\begin{aligned} \sum_{k=1}^{n} a_k &= a_1 + a_2 + \cdots + a_n \\ &= \frac{a_1(1 - r^n)}{1 - r}. \end{aligned}$$

Proof

Because the sequence is geometric, we have

$$\sum_{k=1}^{n} a_k = a_1 + a_1 \cdot r + a_1 \cdot r^2 + \cdots + a_1 \cdot r^{n-1}.$$

Therefore,

$$r \cdot \sum_{k=1}^{n} a_k = a_1 \cdot r + a_1 \cdot r^2 + \cdots + a_1 \cdot r^{n-1} + a_1 \cdot r^n.$$

If we now *subtract* the lower summation from the one above it, we have (after eliminating a lot of zeros):

$$\left(\sum_{k=1}^{n} a_k\right) - r \cdot \left(\sum_{k=1}^{n} a_k\right) = a_1 - a_1 \cdot r^n$$

$$\left(\sum_{k=1}^{n} a_k\right)(1 - r) = a_1(1 - r^n)$$

$$\sum_{k=1}^{n} a_k = \frac{a_1(1 - r^n)}{1 - r}$$

EXAMPLE 2 Summing the Terms of a Geometric Sequence

Find the sum of the geometric sequence $4, -4/3, 4/9, -4/27, \ldots, 4(-1/3)^{10}$.

SOLUTION We can see that $a_1 = 4$ and $r = -1/3$. The nth term is $4(-1/3)^{10}$, which means that $n = 11$. (Remember that the exponent on the nth term is $n - 1$, not n.) Applying the Sum of a Finite Geometric Sequence Theorem, we find that

$$\sum_{n=1}^{11} 4\left(-\frac{1}{3}\right)^{n-1} = \frac{4(1 - (-1/3)^{11})}{1 - (-1/3)} \approx 3.000016935.$$

We can support this answer with technology:
sum(seq(4(−1/3)^(N−1), N, 1, 11) = 3.000016935. **Now try Exercise 13.**

As one practical application of the Sum of a Finite Geometric Sequence Theorem, we will tie up a loose end from Section 3.6, wherein you learned that the future value FV of an ordinary annuity consisting of n equal periodic payments of R dollars at an interest rate i per compounding period (payment interval) is

$$FV = R\frac{(1 + i)^n - 1}{i}.$$

We can now consider the mathematics behind this formula. The n payments remain in the account for different lengths of time and so earn different amounts of interest. The total value of the annuity after n payment periods (see Example 8 in Section 3.6) is

$$FV = R + R(1 + i) + R(1 + i)^2 + \cdots + R(1 + i)^{n-1}.$$

The terms of this sum form a geometric sequence with first term R and common ratio $(1 + i)$. Applying the Sum of a Finite Geometric Sequence Theorem, we find the sum of the n terms to be

$$\begin{aligned} FV &= \frac{R(1 - (1 + i)^n)}{1 - (1 + i)} \\ &= R\frac{1 - (1 + i)^n}{-i} \\ &= R\frac{(1 + i)^n - 1}{i} \end{aligned}$$

Infinite Series

If you change the "11" in the calculator sum in Example 2 to higher and higher numbers, you will find that the sum approaches a value of 3. This is no coincidence. In the language of limits,

$$\begin{aligned} \lim_{x\to\infty} \sum_{k=1}^{n} 4\left(-\frac{1}{3}\right)^{k-1} &= \lim_{x\to\infty} \frac{4(1 - (-1/3)^n)}{1 - (-1/3)} \\ &= \frac{4(1 - 0)}{4/3} \qquad \text{Since } \lim_{n\to\infty}(-1/3)^n = 0 \\ &= 3 \end{aligned}$$

This gives us the opportunity to extend the usual meaning of the word "sum," which always applies to a *finite* number of terms being added together. By using limits, we can make sense of expressions in which an *infinite* number of terms are added together. Such expressions are called *infinite series.*

DEFINITION Infinite Series

An **infinite series** is an expression of the form

$$\sum_{n=1}^{\infty} a_n = a_1 + a_2 + \cdots + a_n + \cdots.$$

The first thing to understand about an infinite series is that it is not a true sum. There are properties of real number addition that allow us to extend the definition of $a + b$ to sums like $a + b + c + d + e + f$, but not to "infinite sums." For example, we can add any finite number of 2's together and get a real number, but if we add an *infinite* number of 2's together we do not get a real number at all. Sums do not behave that way.

What makes series so interesting is that sometimes (as in Example 2) the sequence of **partial sums**, all of which are true sums, approaches a finite limit S:

$$\lim_{n\to\infty} \sum_{k=1}^{n} a_k = \lim_{n\to\infty} (a_1 + a_2 + \cdots + a_n) = S$$

In this case we say that the series **converges** to S, and it makes sense to define S as the **sum of the infinite series**. In sigma notation,

$$\sum_{k=1}^{\infty} a_k = \lim_{n\to\infty} \sum_{k=1}^{n} a_k = S.$$

If the limit of partial sums does not exist, then the series **diverges** and has no sum.

EXAMPLE 3 Looking at Limits of Partial Sums

For each of the following series, find the first five terms in the sequence of partial sums. Which of the series appear to converge?

(a) $0.1 + 0.01 + 0.001 + 0.0001 + \cdots$

(b) $10 + 20 + 30 + 40 + \cdots$

(c) $1 - 1 + 1 - 1 + \cdots$

SOLUTION

(a) The first five partial sums are $\{0.1, 0.11, 0.111, 0.1111, 0.11111\}$. These appear to be approaching a limit of $0.\overline{1} = 1/9$, which would suggest that the series converges to a sum of 1/9.

(b) The first five partial sums are $\{10, 30, 60, 100, 150\}$. These numbers increase without bound and do not approach a limit. The series diverges and has no sum.

(c) The first five partial sums are $\{1, 0, 1, 0, 1\}$. These numbers oscillate and do not approach a limit. The series diverges and has no sum.

Now try Exercise 23.

You might have been tempted to "pair off" the terms in Example 3c to get an infinite summation of 0's (and hence a sum of 0), but you would be applying a rule (namely the *associative property of addition*) that works on *finite* sums but not, in general, on infinite series. The sequence of partial sums does not have a limit, so any manipulation of the series in Example 3c that appears to result in a sum is actually meaningless.

Truncation Error

Since the sum of a convergent series is the limit of the partial sums, the positive *difference between* the partial sums and the infinite sum (called the *truncation error*) will go to zero if and only if the sequence converges. Geometric series are easy to deal with because we have a theorem that tells exactly which of them converge, and, if they converge, exactly what their sums are. There are also many important *non-geometric* series that converge, but finding the actual sums can be quite an adventure. You may encounter some of these in a calculus course.

Representing functions by infinite series was an important step in the history of science. When you enter sin(2) into your calculator, it uses a partial sum approximation of an infinite series to display the value. The finite sum is not the exact value of sin(2), but the truncation error is small enough to make all the displayed digits correct. If you want to explore truncation error a little further, see the last few exercises at the end of this section.

Convergence of Geometric Series

Determining the convergence or divergence of infinite series is an important part of a calculus course, in which series are used to represent functions. Most of the convergence tests are well beyond the scope of this course, but we are in a position to settle the issue completely for geometric series.

THEOREM Sum of an Infinite Geometric Series

The geometric series $\sum_{k=1}^{\infty} a \cdot r^{k-1}$, $a \neq 0$, converges if and only if $|r| < 1$. If it does converge, the sum is $a/(1 - r)$.

Proof

Let $a \neq 0$. If $r = 1$, the series is $a + a + a + \cdots$, which is unbounded and hence diverges. If $r = -1$, the series is $a - a + a - a + \cdots$, which diverges. (See Example 3c.) If $r \neq 1$, then by the Sum of a Finite Geometric Sequence Theorem, the nth partial sum of the series is $\sum_{k=1}^{n} a \cdot r^{k-1} = a(1 - r^n)/(1 - r)$. The limit of the partial sums is $\lim_{n\to\infty}[a(1 - r^n)/(1 - r)]$, which converges if and only if $\lim_{n\to\infty} r^n$ is a finite number. But $\lim_{n\to\infty} r^n$ is 0 when $|r| < 1$ and unbounded when $|r| > 1$. Therefore, the sequence of partial sums converges if and only if $|r| < 1$, in which case the sum of the series is

$$\lim_{n\to\infty}[a(1 - r^n)/(1 - r)] = a(1 - 0)/(1 - r) = a/(1 - r).$$

EXAMPLE 4 Summing Infinite Geometric Series

Determine whether the series converges. If it converges, give the sum.

(a) $\displaystyle\sum_{k=1}^{\infty} 3(0.75)^{k-1}$ **(b)** $\displaystyle\sum_{n=0}^{\infty}\left(-\frac{4}{5}\right)^n$

(c) $\displaystyle\sum_{n=1}^{\infty}\left(\frac{\pi}{2}\right)^n$ **(d)** $1 + \dfrac{1}{2} + \dfrac{1}{4} + \dfrac{1}{8} + \cdots$

SOLUTION

(a) Since $|r| = |0.75| < 1$, the series converges. The first term is $3(0.75)^0 = 3$, so the sum is $a/(1 - r) = 3/(1 - 0.75) = 12$.

(b) Since $|r| = |-4/5| < 1$, the series converges. The first term is $(-4/5)^0 = 1$, so the sum is $a/(1 - r) = 1/(1 - (-4/5)) = 5/9$.

(c) Since $|r| = |\pi/2| > 1$, the series diverges.

(d) Since $|r| = |1/2| < 1$, the series converges. The first term is 1, and so the sum is $a/(1 - r) = 1/(1 - 1/2) = 2$.

Now try Exercise 25.

EXAMPLE 5 Converting a Repeating Decimal to Fraction Form

Express $0.\overline{234} = 0.234234234\ldots$ in fraction form.

SOLUTION We can write this number as a sum: $0.234 + 0.000234 + 0.000000234 + \cdots$.

This is a convergent infinite geometric series in which $a = 0.234$ and $r = 0.001$. The sum is

$$\frac{a}{1 - r} = \frac{0.234}{1 - 0.001} = \frac{0.234}{0.999} = \frac{234}{999} = \frac{26}{111}.$$

Now try Exercise 31.

QUICK REVIEW 9.4 *(For help, see Section 9.4.)*

Exercise numbers with a gray background indicate problems that the authors have designed to be solved *without a calculator.*

In Exercises 1–4, $\{a_n\}$ is arithmetic. Use the given information to find a_{10}.

1. $a_1 = 4; d = 2$

2. $a_1 = 3; a_2 = 1$

3. $a_3 = 6; a_8 = 21$

4. $a_5 = 3; a_{n+1} = a_n + 5$ for $n \geq 1$

In Exercises 5–8, $\{a_n\}$ is geometric. Use the given information to find a_{10}.

5. $a_1 = 1; a_2 = 2$

6. $a_4 = 1; a_6 = 2$

7. $a_7 = 5; r = -2$

8. $a_8 = 10; a_{12} = 40$

9. Find the sum of the first 5 terms of the sequence $\{n^2\}$.

10. Find the sum of the first 5 terms of the sequence $\{2n - 1\}$.

SECTION 9.4 Exercises

In Exercises 1–6, write each sum using summation notation, assuming the suggested pattern continues.

1. $-7 - 1 + 5 + 11 + \cdots + 53$

2. $2 + 5 + 8 + 11 + \cdots + 29$

3. $1 + 4 + 9 + \cdots + (n + 1)^2$

4. $1 + 8 + 27 + \cdots + (n + 1)^3$

5. $6 - 12 + 24 - 48 + \cdots$

6. $5 - 15 + 45 - 135 + \cdots$

In Exercises 7–12, find the sum of the arithmetic sequence.

7. $-7, -3, 1, 5, 9, 13$

8. $-8, -1, 6, 13, 20, 27$

9. $1, 2, 3, 4, \ldots, 80$

10. $2, 4, 6, 8, \ldots, 70$

11. $117, 110, 103, \ldots, 33$

12. $111, 108, 105, \ldots, 27$

In Exercises 13–16, find the sum of the geometric sequence.

13. $3, 6, 12, \ldots, 12{,}288$

14. $5, 15, 45, \ldots, 98{,}415$

15. $42, 7, \frac{7}{6}, \ldots, 42\left(\frac{1}{6}\right)^8$

16. $42, -7, \frac{7}{6}, \ldots, 42\left(-\frac{1}{6}\right)^9$

In Exercises 17–22, find the sum of the first n terms of the sequence. The sequence is either arithmetic or geometric.

17. $2, 5, 8, \ldots; n = 10$

18. $14, 8, 2, \ldots; n = 9$

19. $4, -2, 1, -\frac{1}{2}, \ldots; n = 12$

20. $6, -3, \frac{3}{2}, -\frac{3}{4}, \ldots; n = 11$

21. $-1, 11, -121, \ldots; n = 9$

22. $-2, 24, -288, \ldots; n = 8$

23. Find the first six partial sums of the following infinite series. If the sums have a finite limit, write "convergent." If not, write "divergent."

(a) $0.3 + 0.03 + 0.003 + 0.0003 + \cdots$

(b) $1 - 2 + 3 - 4 + 5 - 6 + \cdots$

24. Find the first six partial sums of the following infinite series. If the sums have a finite limit, write "convergent." If not, write "divergent."

(a) $-2 + 2 - 2 + 2 - 2 + \cdots$

(b) $1 - 0.7 - 0.07 - 0.007 - 0.0007 - \cdots$

In Exercises 25–30, determine whether the infinite geometric series converges. If it does, find its sum.

25. $6 + 3 + \frac{3}{2} + \frac{3}{4} + \cdots$

26. $4 + \frac{4}{3} + \frac{4}{9} + \frac{4}{27} + \cdots$

27. $\frac{1}{64} + \frac{1}{32} + \frac{1}{16} + \frac{1}{8} + \cdots$

28. $\frac{1}{48} + \frac{1}{16} + \frac{3}{16} + \frac{9}{16} + \cdots$

29. $\sum_{j=1}^{\infty} 3\left(\frac{1}{4}\right)^j$

30. $\sum_{n=1}^{\infty} 5\left(\frac{2}{3}\right)^n$

In Exercises 31–34, express the rational number as a fraction of integers.

31. $7.14141414\ldots$

32. $5.93939393\ldots$

33. $-17.268268268\ldots$

34. $-12.876876876\ldots$

35. Savings Account The table below shows the December balance in a fixed-rate compound savings account each year from 1996 to 2000.

Year	1996	1997	1998	1999	2000
Balance	\$20,000	\$22,000	\$24,200	\$26,620	\$29,282

(a) The balances form a geometric sequence. What is r?

(b) Write a formula for the balance in the account n years after December 1996.

(c) Find the sum of the December balances from 1996 to 2006, inclusive.

36. **Savings Account** The table below shows the December balance in a simple interest savings account each year from 1996 to 2000.

Year	1996	1997	1998	1999	2000
Balance	\$18,000	\$20,016	\$22,032	\$24,048	\$26,064

(a) The balances form an arithmetic sequence. What is d?

(b) Write a formula for the balance in the account n years after December 1996.

(c) Find the sum of the December balances from 1996 to 2006, inclusive.

37. **Annuity** Mr. O'Hara deposits \$120 at the end of each month into an account that pays 7% interest compounded monthly. After 10 years, the balance in the account, in dollars, is

$$120\left(1+\frac{0.07}{12}\right)^0 + 120\left(1+\frac{0.07}{12}\right)^1 + \cdots + 120\left(1+\frac{0.07}{12}\right)^{119}.$$

(a) This is a geometric series. What is the first term? What is the common ratio r?

(b) Use the formula for the sum of a finite geometric sequence to find the balance.

38. **Annuity** Ms. Argentieri deposits \$100 at the end of each month into an account that pays 8% interest compounded monthly. After 10 years, the balance in the account, in dollars, is

$$100\left(1+\frac{0.08}{12}\right)^0 + 100\left(1+\frac{0.08}{12}\right)^1 + \cdots + 100\left(1+\frac{0.08}{12}\right)^{119}.$$

(a) This is a geometric series. What is the first term? What is the common ratio r?

(b) Use the formula for the sum of a finite geometric sequence to find the balance.

39. **Group Activity Follow the Bouncing Ball** When "superballs" sprang upon the scene in the 1960s, kids across the United States were amazed that these hard rubber balls could bounce to 90% of the height from which they were dropped. If a superball is dropped from a height of 2 m, how far does it travel until it stops bouncing? (*Hint:* The ball goes down to the first bounce, then up *and* down thereafter.)

40. **Writing to Learn The Trouble with Flubber** In the 1961 movie classic *The Absent Minded Professor*, Prof. Ned Brainard discovers flubber (flying rubber). If a "super duper ball" made of flubber is dropped, it rebounds to an ever greater height with each bounce. How far does it travel if allowed to keep bouncing?

Standardized Test Questions

41. **True or False** If all terms of a series are positive, the series sums to a positive number. Justify your answer.

42. **True or False** If $\sum_{n=1}^{\infty} a_n$ and $\sum_{n=1}^{\infty} b_n$ both diverge, then $\sum_{n=1}^{\infty}(a_n + b_n)$ diverges. Justify your answer.

You should solve Exercises 43–46 without the use of a calculator.

43. **Multiple Choice** The series $3^{-1} + 3^{-2} + 3^{-3} + \cdots + 3^{-n} + \cdots$

(A) converges to 1/2. (B) converges to 1/3.

(C) converges to 2/3. (D) converges to 3/2. (E) diverges.

44. **Multiple Choice** If $\sum_{n=1}^{\infty} x^n = 4$, then $x =$

(A) 0.2. (B) 0.25. (C) 0.4. (D) 0.8. (E) 4.0.

45. **Multiple Choice** The sum of an infinite geometric series with first term 3 and second term 0.75 is

(A) 3.75. (B) 2.4. (C) 4. (D) 5. (E) 12.

46. **Multiple Choice** $\sum_{n=0}^{\infty} 4\left(-\frac{5}{3}\right)^n =$

(A) -6 (B) $-\frac{5}{2}$ (C) $\frac{3}{2}$ (D) 10 (E) Divergent

Explorations

47. **Population Density** The *National Geographic Picture Atlas of Our Fifty States* groups the states into 10 regions. The two largest groupings are the Heartland (Table 9.1) and the Southeast (Table 9.2). Population and area data for the two regions are given in the tables. The populations are official 2010 U.S. Census figures.

(a) What is the total population of each region?

(b) What is the total area of each region?

(c) What is the population density (in persons per square mile) of each region?

(d) **Writing to Learn** For the two regions, compute the population density of each state. What is the average of the seven state population densities for each region? Explain why these answers differ from those found in part (c).

Table 9.1 The Heartland

State	Population	Area (mi^2)
Iowa	3,046,355	56,275
Kansas	2,853,118	82,277
Minnesota	5,303,925	84,402
Missouri	5,988,927	69,697
Nebraska	1,826,341	77,355
North Dakota	672,591	70,703
South Dakota	814,180	77,116

Table 9.2 The Southeast

State	Population	Area (mi^2)
Alabama	4,779,736	51,705
Arkansas	2,915,918	53,187
Florida	18,801,310	58,644
Georgia	9,687,653	58,910
Louisiana	4,533,372	47,751
Mississippi	2,967,297	47,689
South Carolina	4,625,364	31,113

48. Finding a Pattern Write the finite series $-1 + 2 + 7 + 14 + 23 + \cdots + 62$ in summation notation.

Extending the Ideas

49. Fibonacci Sequence and Series Complete the following table, where F_n is the nth term of the Fibonacci sequence and S_n is the nth partial sum of the Fibonacci series. Make a conjecture based on the numerical evidence in the table.

$$S_n = \sum_{k=1}^{n} F_k$$

n	F_n	S_n	$F_{n+2} - 1$
1	1		
2	1		
3	2		
4			
5			
6			
7			
8			
9			

50. Triangular Numbers Revisited Exercise 41 in Section 9.2 introduced **triangular numbers** as numbers that count objects arranged in triangular arrays:

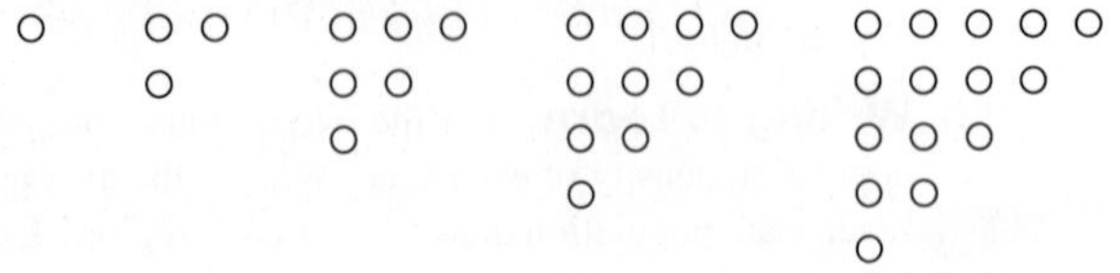

1 3 6 10 15

In that exercise, you gave a geometric argument that the nth triangular number was $n(n + 1)/2$. Prove that formula algebraically using the Sum of a Finite Arithmetic Sequence Theorem.

51. Square Numbers and Triangular Numbers Prove that the sum of two consecutive triangular numbers is a square number; that is, prove

$$T_{n-1} + T_n = n^2$$

for all positive integers $n \geq 2$. Use both a geometric and an algebraic approach.

52. Harmonic Series Graph the sequence of partial sums of the *harmonic series:*

$$1 + \frac{1}{2} + \frac{1}{3} + \frac{1}{4} + \cdots + \frac{1}{n} + \cdots$$

Overlay on it the graph of $f(x) = \ln x$. The resulting picture should support the claim that

$$1 + \frac{1}{2} + \frac{1}{3} + \frac{1}{4} + \cdots + \frac{1}{n} \geq \ln n,$$

for all positive integers n. Make a table of values to further support this claim. Explain why the claim implies that the harmonic series must diverge.

53. Alternating Harmonic Series While the harmonic series (Exercise 52) diverges, the *alternating harmonic series:*

$$1 - \frac{1}{2} + \frac{1}{3} - \frac{1}{4} + \cdots + (-1)^{n+1}\frac{1}{n} + \cdots$$

converges. On a standard number line graph of the interval $[0, 1]$, mark the first several partial sums of the series until you can see what is happening. Using the visual evidence from your graph, explain why the truncation error goes to 0 (thus showing that the series converges).

54. Geometric Series Truncation Error Suppose you estimate the sum of a convergent geometric series by adding up the first n terms. Show that the truncation error is

$$\left|\frac{a_{n+1}}{1 - r}\right|,$$

where a_{n+1} is the $(n + 1)$st term of the series and r is the constant ratio between the terms.

9.5 Mathematical Induction

What you'll learn about

- Tower of Hanoi Problem
- Principle of Mathematical Induction
- Induction and Deduction

... and why

The principle of mathematical induction is a valuable technique for proving combinatorial formulas.

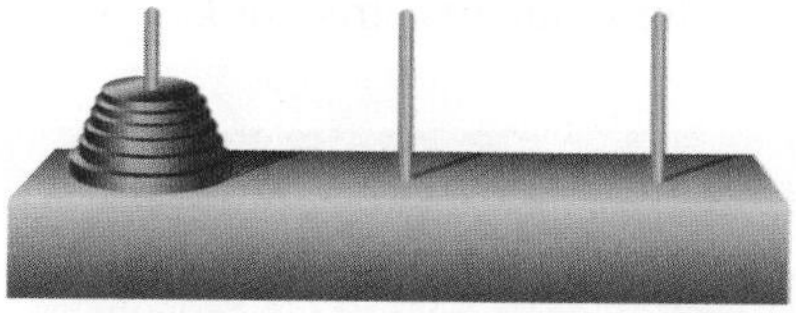

FIGURE 9.6 The Tower of Hanoi Game. The object is to move the entire stack of washers to the rightmost peg, one washer at a time, never placing a larger washer on top of a smaller washer.

Tower of Hanoi History

The legend notwithstanding, the Tower of Hanoi dates back to 1883, when Édouard Lucas marketed the game as "La Tour de Hanoï," brought back from Asia by "Professor N. Claus de Siam"—an anagram of "Professor Lucas d'Amiens." The legend appeared shortly thereafter. The game has been a favorite among computer programmers, so a Web search on "Tower of Hanoi" will bring up multiple sites that allow you to play it on your home computer.

Tower of Hanoi Problem

You might be familiar with a game that is played with a stack of round washers of different diameters and a stand with three vertical pegs (Figure 9.6). The game is not difficult to win once you get the hang of it, but it takes a while to move all the washers even when you know what you are doing. A mathematician, presented with this game, wants to figure out the *minimum* number of moves required to win the game—not because of impatience, but because it is an interesting mathematical problem.

In case mathematics is not sufficient motivation to look into the problem, there is a legend attached to the game that provides a sense of urgency. The legend has it that a game of this sort with 64 golden washers was created at the beginning of time. A special order of Far Eastern monks has been moving the washers at one move per second ever since, always using the minimum number of moves required to win the game. When the final washer is moved, that will be the end of time. The Tower of Hanoi Problem is simply to figure out how much time we have left.

We will solve the problem by proving a general theorem that gives the minimum number of moves for any number of washers. The technique of proof we use is called the principle of mathematical induction, the topic of this section.

THEOREM Tower of Hanoi Solution

The minimum number of moves required to move a stack of n washers in a Tower of Hanoi game is $2^n - 1$.

Proof

(The anchor) First, we note that the assertion is true when $n = 1$. We can certainly move the one washer to the right peg in (minimally) one move, and $2^1 - 1 = 1$.

(The inductive hypothesis) Now let us assume that the assertion holds for $n = k$; that is, the minimum number of moves required to move k washers is $2^k - 1$. (So far the only k we are sure of is 1, but keep reading.)

(The inductive step) We next consider the case when $n = k + 1$ washers. To get at the bottom washer, we must first move the entire stack of k washers sitting on top of it. *By the assumption we just made*, this will take a minimum of $2^k - 1$ moves. We can then move the bottom washer to the free peg (1 move). Finally, we must move the stack of k washers back onto the bottom washer—again, *by our assumption*, a minimum of $2^k - 1$ moves. Altogether, moving $k + 1$ washers requires

$$(2^k - 1) + 1 + (2^k - 1) = 2 \cdot 2^k - 1 = 2^{k+1} - 1$$

moves. Since that agrees with the formula in the statement of the proof, we have shown the assertion to be true for $n = k + 1$ washers—under the assumption that it is true for $n = k$.

Remarkably, we are finished. Recall that we *did* prove the theorem to be true for $n = 1$. Therefore, by the inductive step, it must also be true for $n = 2$. By the inductive step again, it must be true for $n = 3$. And so on, for all positive integers n.

If we apply the Tower of Hanoi Solution to the legendary Tower of Hanoi Problem, the monks will need $2^{64} - 1$ seconds to move the 64 golden washers. The largest current conjecture for the age of the universe is something on the order of 20 billion years. If you convert $2^{64} - 1$ seconds to years, you will find that the end of time (at least according to this particular legend) is not exactly imminent. In fact, you might be surprised at how much time is left!

EXPLORATION 1 Winning the Game

One thing that the Tower of Hanoi Solution does not settle is how to get the stack to finish on the rightmost peg rather than the middle peg. Predictably, it depends on where you move the first washer, but it also depends on the height of the stack. Using a Web site game, or coins of different sizes, or even the real game if you have one, play the game with 1 washer, then 2, then 3, then 4, and so on, keeping track of what your first move must be in order to have the stack wind up on the rightmost peg in $2^n - 1$ moves. What is the general rule for a stack of n washers?

Principle of Mathematical Induction

The proof of the Tower of Hanoi Solution used a general technique known as the Principle of Mathematical Induction. It is a powerful tool for proving all kinds of theorems about positive integers. We *anchor* the proof by establishing the truth of the theorem for 1, then we show the *inductive hypothesis* that "true for k" implies "true for $k + 1$."

FIGURE 9.7 The Principle of Mathematical Induction visualized by dominoes. The toppling of domino #1 guarantees the toppling of domino n for all positive integers n.

Principle of Mathematical Induction

Let P_n be a statement about the integer n. Then P_n is true for all positive integers n provided the following conditions are satisfied:

1. (the anchor) P_1 is true;
2. (the inductive step) if P_k is true, then P_{k+1} is true.

A good way to visualize how the principle works is to imagine an infinite sequence of dominoes stacked upright, each one close enough to its neighbor so that any kth domino, if it falls, will knock over the $(k + 1)$st domino (Figure 9.7). Given that fact (the inductive step), the toppling of domino 1 guarantees the toppling of the entire infinite sequence of dominoes.

Let us use the principle to prove a fact that we already know.

EXAMPLE 1 Using Mathematical Induction

Prove that $1 + 3 + 5 + \cdots + (2n - 1) = n^2$ is true for all positive integers n.

SOLUTION Call the statement P_n. We could verify P_n by using the formula for the sum of an arithmetic sequence, but here is how we prove it by mathematical induction.

(The anchor) For $n = 1$, the equation reduces to P_1: $1 = 1^2$, which is true.

(The inductive hypothesis) Assume that the equation is true for $n = k$. That is, assume

$$P_k: 1 + 3 + \cdots + (2k - 1) = k^2 \text{ is true.}$$

(The inductive step) The next term on the left-hand side would be $2(k + 1) - 1$. We add this to both sides of P_k and get

$$\begin{aligned} 1 + 3 + \cdots + (2k - 1) + (2(k + 1) - 1) &= k^2 + (2(k + 1) - 1) \\ &= k^2 + 2k + 1 \\ &= (k + 1)^2 \end{aligned}$$

This is exactly the statement P_{k+1}, so the equation is true for $n = k + 1$. Therefore, P_n is true for all positive integers, by mathematical induction.

Now try Exercise 1.

Notice that we did *not* plug in $k + 1$ on both sides of the equation P_n in order to verify the inductive step; if we had done that, there would have been nothing to verify. If you find yourself verifying the inductive step without using the inductive hypothesis, you can assume that you have gone astray.

EXAMPLE 2 Using Mathematical Induction

Prove that $1^2 + 2^2 + 3^2 + \cdots + n^2 = [n(n + 1)(2n + 1)]/6$ is true for all positive integers n.

SOLUTION Let P_n be the statement $1^2 + 2^2 + 3^2 + \cdots + n^2 = [n(n + 1)(2n + 1)]/6$.

(The anchor) P_1 is true because $1^2 = [1(2)(3)]/6$.

(The inductive hypothesis) Assume that P_k is true, so that

$$1^2 + 2^2 + \cdots + k^2 = \frac{k(k + 1)(2k + 1)}{6}.$$

(The inductive step) The next term on the left-hand side would be $(k + 1)^2$. We add this to both sides of P_k and get

$$\begin{aligned}1^2 + 2^2 + \cdots + k^2 + (k + 1)^2 &= \frac{k(k + 1)(2k + 1)}{6} + (k + 1)^2 \\ &= \frac{k(k + 1)(2k + 1) + 6(k + 1)^2}{6} \\ &= \frac{(k + 1)(2k^2 + k + 6k + 6)}{6} \\ &= \frac{(k + 1)(k + 2)(2k + 3)}{6} \\ &= \frac{(k + 1)((k + 1) + 1)(2(k + 1) + 1)}{6}\end{aligned}$$

This is exactly the statement P_{k+1}, so the equation is true for $n = k + 1$. Therefore, P_n is true for all positive integers, by mathematical induction.

Now try Exercise 13.

Applications of mathematical induction can be quite different from the first two examples. Here is one involving divisibility.

EXAMPLE 3 Proving Divisibility

Prove that $4^n - 1$ is evenly divisible by 3 for all positive integers n.

SOLUTION Let P_n be the statement that $4^n - 1$ is evenly divisible by 3 for all positive integers n.

(The anchor) P_1 is true because $4^1 - 1 = 3$ is divisible by 3.

(The inductive hypothesis) Assume that P_k is true, so that $4^k - 1$ is divisible by 3.

(The inductive step) We need to prove that $4^{k+1} - 1$ is divisible by 3.

Using a little algebra, we see that $4^{k+1} - 1 = 4 \cdot 4^k - 1 = 4(4^k - 1) + 3$.

By the inductive hypothesis, $4^k - 1$ is divisible by 3. Of course, so is 3. Therefore, $4(4^k - 1) + 3$ is a sum of multiples of 3, and hence divisible by 3. This is exactly the statement P_{k+1}, so P_{k+1} is true. Therefore, P_n is true for all positive integers, by mathematical induction.

Now try Exercise 19.

Induction and Deduction

The words *induction* and *deduction* are usually used to contrast two patterns of logical thought. We reason by **induction** when we use evidence derived from particular examples to draw conclusions about general principles. We reason by **deduction** when we reason from general principles to draw conclusions about specific cases.

When mathematicians prove theorems, they use deduction. In fact, even a "proof by mathematical induction" is a deductive proof, since it consists of applying the general principle to a particular formula. We have been careful to use the term **mathematical induction** in this section to distinguish it from inductive reasoning, which is often good for inspiring conjectures—but not for proving general principles.

Exploration 2 illustrates why mathematicians do not rely on inductive reasoning.

EXPLORATION 2 Is $n^2 + n + 41$ Prime for All n?

1. Plug in the numbers from 1 to 10. Are the results all prime?
2. Repeat for the numbers from 11 to 20.
3. Repeat for the numbers from 21 to 30. (Ready to make your conjecture?)
4. What is the smallest value of n for which $n^2 + n + 41$ is not prime?

The Four-Color Map Theorem

In 1852, Francis Guthrie conjectured that any map on a flat surface could be colored in at most four colors so that no two bordering regions would share the same color. Mathematicians tried unsuccessfully for almost 150 years to prove (or disprove) the conjecture, until Kenneth Appel and Wolfgang Haken finally proved it in 1976.

There is one situation in which (nonmathematical) induction can constitute a proof. In **enumerative induction**, one reasons from specific cases to the general principle by considering *all possible cases*. This is simple enough when proving a theorem like "All one-digit prime numbers are factors of 210," but it can involve some very elegant mathematics when the number of cases is seemingly infinite. Such was the case in the proof of the Four-Color Map Theorem, in which all possible cases were settled with the help of a clever computer program.

QUICK REVIEW 9.5 *(Prerequisite skill Sections A.2 and 1.2)*

Exercise numbers with a gray background indicate problems that the authors have designed to be solved *without a calculator*.

In Exercises 1–3, expand the product.

1. $n(n + 5)$
2. $(n + 2)(n - 3)$
3. $k(k + 1)(k + 2)$

In Exercises 4–6, factor the polynomial.

4. $n^2 + 2n - 3$
5. $k^3 + 3k^2 + 3k + 1$
6. $n^3 - 3n^2 + 3n - 1$

In Exercises 7–10, evaluate the function at the given domain values or variable expressions.

7. $f(x) = x + 4$; $f(1), f(t), f(t + 1)$
8. $f(n) = \dfrac{n}{n + 1}$; $f(1), f(k), f(k + 1)$
9. $P(n) = \dfrac{2n}{3n + 1}$; $P(1), P(k), P(k + 1)$
10. $P(n) = 2n^2 - n - 3$; $P(1), P(k), P(k + 1)$

SECTION 9.5 Exercises

In Exercises 1–4, use mathematical induction to prove that the statement holds for all positive integers.

1. $2 + 4 + 6 + \cdots + 2n = n^2 + n$
2. $8 + 10 + 12 + \cdots + (2n + 6) = n^2 + 7n$
3. $6 + 10 + 14 + \cdots + (4n + 2) = n(2n + 4)$
4. $14 + 18 + 22 + \cdots + (4n + 10) = 2n(n + 6)$

In Exercises 5–8, state an explicit rule for the nth term of the recursively defined sequence. Then use mathematical induction to prove the rule.

5. $a_n = a_{n-1} + 5, a_1 = 3$ **6.** $a_n = a_{n-1} + 2, a_1 = 7$

7. $a_n = 3a_{n-1}, a_1 = 2$ **8.** $a_n = 5a_{n-1}, a_1 = 3$

In Exercises 9–12, write the statements P_1, P_k, and P_{k+1}. (Do not write a proof.)

9. P_n: $1 + 2 + \cdots + n = \dfrac{n(n+1)}{2}$

10. P_n: $1^2 + 3^2 + 5^2 + \cdots + (2n-1)^2 = \dfrac{n(2n-1)(2n+1)}{3}$

11. P_n: $\dfrac{1}{1\cdot 2} + \dfrac{1}{2\cdot 3} + \cdots + \dfrac{1}{n\cdot(n+1)} = \dfrac{n}{n+1}$

12. P_n: $\displaystyle\sum_{k=1}^{n} k^4 = \dfrac{n(n+1)(2n+1)(3n^2+3n-1)}{30}$

In Exercises 13–20, use mathematical induction to prove that the statement holds for all positive integers.

13. $1 + 5 + 9 + \cdots + (4n-3) = n(2n-1)$

14. $1 + 2 + 2^2 + \cdots + 2^{n-1} = 2^n - 1$

15. $\dfrac{1}{1\cdot 2} + \dfrac{1}{2\cdot 3} + \dfrac{1}{3\cdot 4} + \cdots + \dfrac{1}{n(n+1)} = \dfrac{n}{n+1}$

16. $\dfrac{1}{1\cdot 3} + \dfrac{1}{3\cdot 5} + \cdots + \dfrac{1}{(2n-1)(2n+1)} = \dfrac{n}{2n+1}$

17. $2^n \ge 2n$ **18.** $3^n \ge 3n$

19. 3 is a factor of $n^3 + 2n$. **20.** 6 is a factor of $7^n - 1$.

In Exercises 21 and 22, use *mathematical induction* to prove that the statement holds for all positive integers. (We have already seen each proved in another way.)

21. The sum of the first n terms of a geometric sequence with first term a_1 and common ratio $r \ne 1$ is $a_1(1 - r^n)/(1 - r)$.

22. The sum of the first n terms of an arithmetic sequence with first term a_1 and common difference d is

$$S_n = \frac{n}{2}[2a_1 + (n-1)d].$$

In Exercises 23 and 24, use mathematical induction to prove that the formula holds for all positive integers.

23. Triangular Numbers $\displaystyle\sum_{k=1}^{n} k = \dfrac{n(n+1)}{2}$

24. Sum of the First *n* Cubes $\displaystyle\sum_{k=1}^{n} k^3 = \dfrac{n^2(n+1)^2}{4}$

[Note that if you put the results from Exercises 23 and 24 together, you obtain the pleasantly surprising equation

$$1^3 + 2^3 + 3^3 + \cdots + n^3 = (1 + 2 + 3 + \cdots + n)^2.]$$

In Exercises 25–30, use the results of Exercises 21–24 and Example 2 to find the sums.

25. $1 + 2 + 3 + \cdots + 500$ **26.** $1^2 + 2^2 + \cdots + 250^2$

27. $4 + 5 + 6 + \cdots + n$ **28.** $1^3 + 2^3 + 3^3 + \cdots + 75^3$

29. $1 + 2 + 4 + 8 + \cdots + 2^{34}$

30. $1 + 8 + 27 + \cdots + 3375$

In Exercises 31–34, use the results of Exercises 21–24 and Example 2 to find the sum in terms of n.

31. $\displaystyle\sum_{k=1}^{n} (k^2 - 3k + 4)$ **32.** $\displaystyle\sum_{k=1}^{n} (2k^2 + 5k - 2)$

33. $\displaystyle\sum_{k=1}^{n} (k^3 - 1)$ **34.** $\displaystyle\sum_{k=1}^{n} (k^3 + 4k - 5)$

35. Group Activity Here is a proof by mathematical induction that any gathering of n people must all have the same blood type.

(Anchor) If there is 1 person in the gathering, everyone in the gathering obviously has the same blood type.

(Inductive hypothesis) Assume that any gathering of k people must all have the same blood type.

(Inductive step) Suppose $k + 1$ people are gathered. Send one of them out of the room. The remaining k people must all have the same blood type (by the inductive hypothesis). Now bring the first person back and send someone else out of the room. You get another gathering of k people, all of whom must have the same blood type. Therefore all $k + 1$ people must have the same blood type, and we are done by mathematical induction.

This result is obviously false, so there must be something wrong with the proof. Explain where the proof goes wrong.

36. Writing to Learn Kitty is having trouble understanding mathematical induction proofs because she does not understand the inductive hypothesis. If we can assume it is true for k, she asks, why can't we assume it is true for n and be done with it? After all, a variable is a variable! Write a response to Kitty to clear up her confusion.

Standardized Test Questions

37. True or False The goal of mathematical induction is to prove that a statement P_n is true for all real numbers n. Justify your answer.

38. True or False If P_n is the statement "$(n+1)^2 = 4n$," then P_1 is true. Justify your answer.

You may use a graphing calculator when solving Exercises 39–42.

39. Multiple Choice In a proof by mathematical induction that $1 + 2 + 3 + \cdots + n = \dfrac{n(n+1)}{2}$ for all positive integers n, the inductive hypothesis would be to assume that

(A) $n = 1$.

(B) $k = 1$.

(C) $1 = \dfrac{1(1+1)}{2}$.

(D) $1 + 2 + 3 + \cdots + n = \dfrac{n(n+1)}{2}$ for all positive integers n.

(E) $1 + 2 + 3 + \cdots + k = \dfrac{k(k+1)}{2}$ for some positive integer k.

40. Multiple Choice The first step in a proof by mathematical induction is to prove

(A) the anchor.

(B) the inductive hypothesis.

(C) the inductive step.

(D) the inductive principle.

(E) the inductive foundation.

41. Multiple Choice Which of the following could be used to prove that $1 + 3 + 5 + \cdots + (2n - 1) = n^2$ for all positive integers n?

I. Mathematical induction

II. The formula for the sum of a finite arithmetic sequence

III. The formula for the sum of a finite geometric sequence

(A) I only

(B) I and II only

(C) I and III only

(D) II and III only

(E) I, II, and III

42. Multiple Choice Mathematical induction can be used to prove that, for any positive integer n, $\sum_{k=1}^{n} k^3 =$

(A) $\frac{n(n+1)}{2}$. (B) $\frac{n^2(n+1)^2}{2}$.

(C) $\frac{n^2(n+1)^2}{4}$. (D) $\frac{n^3(n+1)^3}{2}$.

(E) $\frac{n^3(n+1)^3}{8}$.

Explorations

43. Use mathematical induction to prove that 2 is a factor of $(n + 1)(n + 2)$ for all positive integers n.

44. Use mathematical induction to prove that 6 is a factor of $n(n + 1)(n + 2)$ for all positive integers n. (You may assume the assertion in Exercise 43 to be true.)

45. Give an alternate proof of the assertion in Exercise 43 based on the fact that $(n + 1)(n + 2)$ is a product of two consecutive integers.

46. Give an alternate proof of the assertion in Exercise 44 based on the fact that $n(n + 1)(n + 2)$ is a product of three consecutive integers.

Extending the Ideas

In Exercises 47 and 48, use mathematical induction to prove that the statement holds for all positive integers.

47. Fibonacci Sequence and Series

$F_{n+2} - 1 = \sum_{k=1}^{n} F_k$, where $\{F_n\}$ is the Fibonacci sequence.

48. If $\{a_n\}$ is the sequence $\sqrt{2}, \sqrt{2 + \sqrt{2}}, \sqrt{2 + \sqrt{2 + \sqrt{2}}}, \ldots$, then $a_n < 2$.

49. Let a be any integer greater than 1. Use mathematical induction to prove that $a - 1$ divides $a^n - 1$ evenly for all positive integers n.

50. Give an alternate proof of the assertion in Exercise 49 based on the Factor Theorem of Section 2.4.

It is not necessary to anchor a mathematical induction proof at $n = 1$; we might be interested only in the integers greater than or equal to some integer c. In this case, we simply modify the anchor and inductive step as follows:

Anchor: P_c is true.

Inductive Step: If P_k is true for some $k \geq c$, then P_{k+1} is true.

(This is sometimes called the **Extended Principle of Mathematical Induction**.) Use this principle to prove the statements in Exercises 51 and 52.

51. $3n - 4 \geq n$, for all $n \geq 2$ **52.** $2^n \geq n^2$, for all $n \geq 4$

53. Proving the Interior Angle Formula Use extended mathematical induction to prove P_n for $n \geq 3$.
P_n: The sum of the interior angles of an n-sided polygon is $180°(n - 2)$.

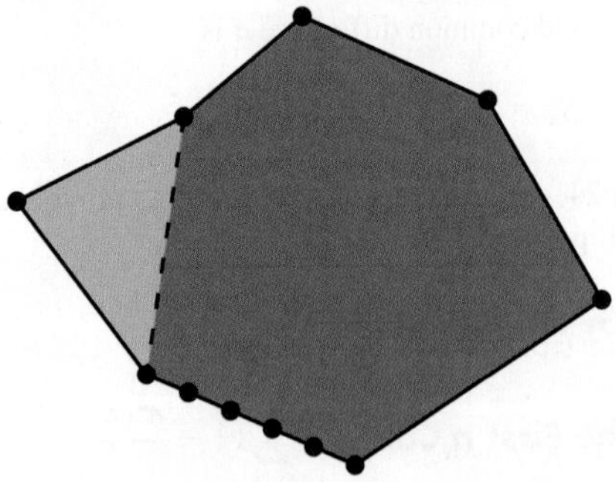

CHAPTER 9 Key Ideas

Properties, Theorems, and Formulas

Multiplication Principle of Counting 635
Permutations of an n-set 636
Distinguishable Permutations 637
Permutation Counting Formula 637
Combination Counting Formula 638
Formula for Counting Subsets of an n-Set 639
Recursive Formula for Pascal's Triangle 646
Binomial Theorem 646
Basic Factorial Identities 647
Sum of a Finite Arithmetic Sequence 659
Sum of a Finite Geometric Sequence 660
Sum of an Infinite Geometric Series 663
Tower of Hanoi Solution 667
Principle of Mathematical Induction 668

CHAPTER 9 Review Exercises

Exercise numbers with a gray background indicate problems that the authors have designed to be solved *without a calculator*.

The collection of exercises marked in red could be used as a chapter test.

In Exercises 1–6, evaluate the expression by hand, then check your result with a calculator.

1. $\binom{12}{5}$
2. $\binom{789}{787}$
3. ${}_{18}C_{12}$
4. ${}_{35}C_{28}$
5. ${}_{12}P_7$
6. ${}_{15}P_8$
7. **Code Words** How many five-character code words are there if the first character is always a letter and the other characters are letters and/or digits?
8. **Scheduling Trips** A travel agent is trying to schedule a client's trip from city A to city B. There are three direct flights, three flights from A to a connecting city C, and four flights from this connecting city C to city B. How many trips are possible?
9. **License Plates** How many license plates begin with two letters followed by four digits or begin with three digits followed by three letters? Assume that no letters or digits are repeated.
10. **Forming Committees** A club has 45 members, and its membership committee has three members. How many different membership committees are possible?
11. **Bridge Hands** How many 13-card bridge hands include the ace, king, and queen of spades?
12. **Bridge Hands** How many 13-card bridge hands include all four aces and exactly one king?
13. **Coin Toss** Suppose that a coin is tossed five times. How many different outcomes include at least two heads?
14. **Forming Committees** A certain small business has 35 employees, 21 women and 14 men. How many different employee representative committees are there if the committee must consist of two women and two men?
15. **Code Words** How many code words of any length can be spelled out using game tiles of five different letters (including single-letter code words)?
16. **A Pocket of Coins** Sean tells Moira that he has less than 50 cents in American coins in his pocket and no two coins of the same denomination. How many possible total amounts could be in Sean's pocket?
17. **Permutations** Find the number of distinguishable permutations that can be made from the letters in

    (a) GERMANY

    (b) PRESBYTERIANS

    In each case, can you find a permutation that spells the first and last name of a female entertainer?
18. **Permutations** Find the number of distinguishable permutations that can be made from the letters in

    (a) FLORIDA **(b)** TALLAHASSEE

In Exercises 19–24, expand each expression.

19. $(2x + y)^5$
20. $(4a - 3b)^7$
21. $(3x^2 + y^3)^5$
22. $\left(1 + \frac{1}{x}\right)^6$
23. $(2a^3 - b^2)^9$
24. $(x^{-2} + y^{-1})^4$
25. Find the coefficient of x^8 in the expansion of $(x - 2)^{11}$.
26. Find the coefficient of x^2y^6 in the expansion of $(2x + y)^8$.

In Exercises 27–30, list the elements of the sample space.

27. **Spinners** A game spinner on a circular region divided into 6 equal sectors numbered 1–6 is spun.
28. **Rolling Dice** A red die and a green die are rolled.
29. **Code Words** A two-digit code is selected from the digits $\{1, 3, 6\}$, where no digits are to be repeated.
30. **Production Line** A product is inspected as it comes off the production line and is classified as either defective or nondefective.

In Exercises 31–34, a penny, a nickel, and a dime are tossed.

31. List all possible outcomes.
32. List all outcomes in the event "two heads or two tails."

33. List all outcomes in the complement of the event in Exercise 32 (i.e., the event "neither two heads nor two tails.")

34. List all outcomes in the event "at least two heads."

In Exercises 35 and 36, find the first 6 terms and the 40th term of the sequence.

35. $a_n = \dfrac{n^2 - 1}{n + 1}$

36. $b_k = \dfrac{(-2)^k}{k + 1}$

In Exercises 37–42, find the first 6 terms and the 12th term of the sequence.

37. $a_1 = -1$ and $a_n = a_{n-1} + 3$, for $n \ge 2$

38. $b_1 = 5$ and $b_k = 2b_{k-1}$, for $k \ge 2$

39. Arithmetic sequence, with $a_1 = -5$ and $d = 1.5$

40. Geometric sequence, with $a_1 = 3$ and $r = 1/3$

41. $v_1 = -3$, $v_2 = 1$, and $v_k = v_{k-2} + v_{k-1}$, for $k \ge 3$

42. $w_1 = -3$, $w_2 = 2$, and $w_k = w_{k-2} + w_{k-1}$, for $k \ge 3$

In Exercises 43–50, the sequences are arithmetic or geometric. Find an explicit formula for the nth term. State the common difference or ratio.

43. $12, 9.5, 7, 4.5, \ldots$

44. $-5, -1, 3, 7, \ldots$

45. $10, 12, 14.4, 17.28, \ldots$

46. $\dfrac{1}{8}, -\dfrac{1}{4}, \dfrac{1}{2}, -1, \ldots$

47. $a_1 = -11$ and $a_n = a_{n-1} + 4.5$ for $n \ge 2$

48. $b_1 = 7$ and $b_n = (1/4)\, b_{n-1}$ for $n \ge 2$

49. The fourth and ninth terms of a geometric sequence are -192 and 196,608, respectively.

50. The third and eighth terms of an arithmetic sequence are 14 and -3.5, respectively.

In Exercises 51–54, find the sum of the terms of the arithmetic sequence.

51. $-11, -8, -5, -2, 1, 4, 7, 10$

52. $13, 9, 5, 1, -3, -7, -11$

53. $2.5, -0.5, -3.5, \ldots, -75.5$

54. $-5, -3, -1, 1, \ldots, 55$

In Exercises 55–58, find the sum of the terms of the geometric sequence.

55. $4, -2, 1, -\dfrac{1}{2}, \dfrac{1}{4}, -\dfrac{1}{8}$

56. $-3, -1, -\dfrac{1}{3}, -\dfrac{1}{9}, -\dfrac{1}{27}$

57. $2, 6, 18, \ldots, 39{,}366$

58. $1, -2, 4, -8, \ldots, -8192$

In Exercises 59 and 60, find the sum of the first 10 terms of the arithmetic or geometric sequence.

59. $2187, 729, 243, \ldots$

60. $94, 91, 88, \ldots$

In Exercises 61 and 62, graph the sequence.

61. $a_n = 1 + \dfrac{(-1)^n}{n}$

62. $a_n = 2n^2 - 1$

63. Annuity Mr. Andalib pays \$150 at the end of each month into an account that pays 8% interest compounded monthly. At the end of 10 years, the balance in the account, in dollars, is

$$150\left(1 + \frac{0.08}{12}\right)^0 + 150\left(1 + \frac{0.08}{12}\right)^1 + \cdots + 150\left(1 + \frac{0.08}{12}\right)^{119}$$

Use the formula for the sum of a finite geometric series to find the balance.

64. Annuity What is the minimum monthly payment at month's end that must be made in an account that pays 8% interest compounded monthly if the balance at the end of 10 years is to be at least \$30,000?

In Exercises 65–70, determine whether the geometric series converges. If it does, find its sum.

65. $\displaystyle\sum_{j=1}^{\infty} 2\left(\frac{3}{4}\right)^j$

66. $\displaystyle\sum_{k=1}^{\infty} 2\left(-\frac{1}{3}\right)^k$

67. $\displaystyle\sum_{j=1}^{\infty} 4\left(-\frac{4}{3}\right)^j$

68. $\displaystyle\sum_{k=1}^{\infty} 5\left(\frac{6}{5}\right)^k$

69. $\displaystyle\sum_{k=1}^{\infty} 3(0.5)^k$

70. $\displaystyle\sum_{k=1}^{\infty} (1.2)^k$

In Exercises 71–74, write the sum in sigma notation.

71. $-8 - 3 + 2 + \cdots + 92$

72. $4 - 8 + 16 - 32 + \cdots - 2048$

73. $1^2 + 3^2 + 5^2 + \cdots$

74. $1 + \dfrac{1}{2} + \dfrac{1}{2^2} + \dfrac{1}{2^3} + \cdots$

In Exercises 75–78, use summation formulas to evaluate the expression.

75. $\displaystyle\sum_{k=1}^{n} (3k + 1)$

76. $\displaystyle\sum_{k=1}^{n} 3k^2$

77. $\displaystyle\sum_{k=1}^{25} (k^2 - 3k + 4)$

78. $\displaystyle\sum_{k=1}^{175} (3k^2 - 5k + 1)$

In Exercises 79–82, use mathematical induction to prove that the statement is true for all positive integers n.

79. $1 + 3 + 6 + \cdots + \dfrac{n(n+1)}{2} = \dfrac{n(n+1)(n+2)}{6}$

80. $1 \cdot 2 + 2 \cdot 3 + 3 \cdot 4 + \cdots + n(n+1) = \dfrac{n(n+1)(n+2)}{3}$

81. $2^{n-1} \le n!$

82. $n^3 + 2n$ is divisible by 3.

83. Find row 9 of Pascal's triangle.

84. Show algebraically that

$${}_nP_k \times {}_{n-k}P_j = {}_nP_{k+j}.$$

CHAPTER 9 Project

Modeling Population Growth with Sequences

In this project, you will use recursive formulas and their associated sequences to explore trends in human and animal populations. In fact, you will use some of the modeling techniques that demographers and wildlife biologists use.

Explorations

1. **U.S. Population.** An arithmetic sequence is the discrete analog to a linear function. Both an arithmetic sequence and a linear function assume a constant *additive* rate of change.
   - **(a)** According to the U.S. Census Bureau, the official U.S. population was 248.7 million in 1990 and 308.7 million in 2010. What was the average change in population per year from 1990 through 2010?
   - **(b)** The official U.S. population was 281.4 million in 2000. Let p_n be an arithmetic sequence that represents the U.S. population (in millions of persons) n years after 2000. Verify that $p_0 = 281.4$. This is the **initial condition** for the sequence.
   - **(c)** Use your answer from part (a) to write a recursive formula for the sequence p_n.
   - **(d)** Write an explicit formula for the arithmetic sequence p_n, assuming that $p_0 = 281.4$.
   - **(e)** Use the information from parts (b), (c), and (d) to estimate the U.S. population in 2010, 2015, and 2020.
2. **World Population.** The following table shows U.S. Census Bureau estimates for key world population benchmarks. A geometric sequence is the discrete analog to an exponent function. Both a geometric sequence and an exponent function assume a constant *multiplicative* rate of change. Let w_n be a geometric sequence that represents the world population (in billions of persons) n years after 1959. Use $w_0 = 3.00$ as the initial condition for the sequence.

Year	Population (in billions)
1959	3
1974	4
1987	5
1999	6
2012	7

   - **(a)** Assuming an annual percentage increase of 1.62% in world population, write a recursive formula for the sequence w_n.
   - **(b)** Write an explicit formula for the geometric sequence w_n in part (a). Use exponential regression to confirm the accuracy of this model.
   - **(c)** Use the formulas from parts (a) and (b) to estimate the world population in 2010, 2015, and 2020.
   - **(d)** The actual annual growth rate in world population has slowed from 2.22% in 1963 to 1.09% in 2013, so a logistic growth model may be more appropriate than an exponential model. Use the data in the table and exponential regression to estimate the world population in 2010, 2015, and 2020.
3. **Bluegill Population.** There are 1100 bluegill in Lake Tawaga at present (Year 0). Each year the population decreases by 25% from the combined effects of deaths and harvesting, and at the end of each year 800 bluegill are added to the lake. Let b_n be a sequence that represents the bluegill population n years into the future.
   - **(a)** Write a recursive formula for the sequence b_n.
   - **(b)** Graph the sequence b_n as a time plot (that is, as a function of time n).
   - **(c)** Use mathematical induction to prove that $b_n = 3200 - 2100(0.75)^n$ is an explicit formula for the sequence.
   - **(d)** What is the population of the bluegill in the long run?
4. **Predator-Prey Population Model.** In the wild, the population of a predator species, such as foxes, depends on prey species, such as rabbits, so the populations are interrelated. Let

$$u_n = \text{rabbit population at time } n = u_{n-1} \cdot (1 + a - b \cdot v_{n-1})$$
$$v_n = \text{fox population at time } n = v_{n-1} \cdot (1 + c \cdot u_{n-1} - d)$$

where

a = growth rate of rabbit population when there are no foxes = 0.047
b = rate at which foxes kill rabbits = 0.0011
c = growth rate of fox population when there are rabbits = 0.00019
d = death rate of foxes when there are no rabbits = 0.032

Let the initial conditions be $u_0 = 288$ rabbits and $w_0 = 63$ foxes.

   - **(a)** Graph the sequences u_n and v_n as functions of time n on the same set of axes.
   - **(b)** Graph the relation v_n versus u_n of the same set of axes. (Either use the uv format in Sequence graphing mode, or use Parametric graphing mode.)
   - **(c)** Write a few sentences to describe your findings. Do you think either population will die out? Why, or why not? If the foxes were to die out, what would happen to the rabbit population? Explain this based on the model.

CHAPTER 10

10.1 Probability

10.2 Statistics (Graphical)

10.3 Statistics (Numerical)

10.4 Random Variables and Probability Models

10.5 Statistical Literacy

Statistics and Probability

"All you need is a dollar and a dream," it's said, and big lottery payouts make big news. Bigger than the payouts, though, are the profits for the states that run these lotteries and apply the millions of dollars spent on losing tickets to fund education and other worthy endeavors. Those profits arise because most people who play a lottery lose.

The games typically require that you pick several numbers (often 6) from a large sequence (sometimes as many as 56). If your picks correctly match a winning set chosen at random, you're a winner. We all know that's unlikely, but just what *are* your chances? See that calculation on page 681.

Many people think they can improve their chances by betting "lucky" numbers. Or numbers that seem to come up more often than others. Or numbers that have not come up recently. See page 727 for an analysis of historical data suggesting that some numbers may be less likely to make you a winner.

Chapter 10 Overview

We live in a world awash in data. Businesses collect information on consumer preferences and spending habits to create and market products. Governments base public policy decisions on census and economic data. Medical researchers run experiments and collect information on patients to determine what treatments may be safe and effective. Weather forecasters tell us the chances for rain on our picnic, and pollsters tell us who'll win the upcoming election. In every walk of life, informed decisions are based on statistical analyses of data coupled with a clear understanding of probability. This chapter will present concepts and skills that provide a useful introduction to these increasingly important ways of understanding the world.

10.1 Probability

What you'll learn about

- Sample Spaces and Probability Functions
- Determining Probabilities
- Venn Diagrams
- Tree Diagrams
- Conditional Probability

... and why

Everyone should know how mathematical the "laws of chance" really are.

Sample Spaces and Probability Functions

Most people have an intuitive sense of probability. Unfortunately, this sense is not often based on a foundation of mathematical principles, so people become victims of scams, misleading statistics, and false advertising. In this section, we want to build on your intuitive sense of probability and give it a mathematical foundation.

Perhaps the simplest example of probability involves tossing a coin. What do we mean when we say there's a 50-50 chance of the coin landing heads (or, the probability of heads is 1/2)? Certainly we are not claiming that a coin tossed twice *must* land heads once and tails once. And we're not saying (even though many people mistakenly believe this) that after 10 tosses that land 8 heads and 2 tails the coin is somehow more likely to land tails until things even out a bit. To understand this better, imagine tossing a coin hundreds of times and keeping a running record of the percent that were heads. A computer simulation of 200 tosses produced the table and graph below.

Toss No.	Outcome	No. of Heads	% So Far
1	H	1	100
2	H	2	100
3	T	2	66.7
4	H	3	75
5	H	4	80
6	H	5	83
7	T	5	71.4
8	H	6	75
9	H	7	77.8
10	H	8	80
⋮			
100	T	54	54
⋮			
200	H	104	52

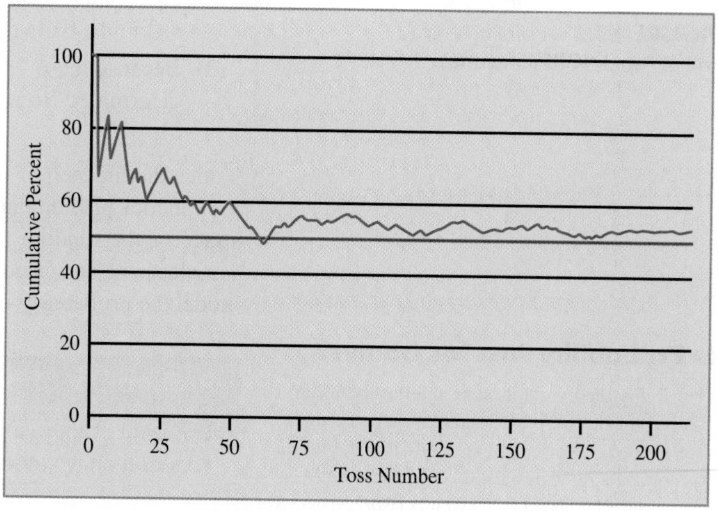

Even though the coin started out with a lot of heads, as the number of tosses increases we see a trend toward 50% heads. That occurs not because the coin starts favoring tails; in fact, tosses 11–100 were $46/90 \approx 51\%$ heads, and tosses 101–200 were 50% heads. Rather, seemingly strange short-term results are simply overwhelmed by the long-term behavior of the coin. It's that long-term behavior that we refer to as the *probability*.

DEFINITION Probability of an Event

The **probability** of an event E is the value that its relative frequency of occurrence approaches in the long run.

Because we can't toss a coin an infinite number of times to see what the probability of heads turns out to be, we use some intuition to say that because there are two possible outcomes $\{H, T\}$ and because it seems they should be equally likely, in the long run heads should occur with a probability of 1/2.

Be careful with this line of reasoning, though. The probability your house will be hit by a meteor before your bedtime is not 1/2, even though there are also two possible outcomes, $\{\text{hit, not hit}\}$. This intuition-based approach works only if the outcomes are *equally likely* to occur.

EXAMPLE 1 Testing Your Intuition About Probability

Find the probability of each of the following events.

(a) Tossing two heads in a row on two tosses of a fair coin

(b) Drawing a queen from a standard deck of 52 cards

(c) Rolling a sum of 4 on a single roll of two fair dice

(d) Having tendonitis, if your doctor tells you that further tests are needed to determine whether your sore knee is caused by tendonitis, bursitis, arthritis, or a torn meniscus.

SOLUTION

(a) There are four equally likely outcomes: $\{TT, TH, HT, HH\}$. The probability is 1/4.

(b) There are 52 equally likely outcomes, 4 of which are queens. The probability is 4/52, or 1/13.

(c) By the Multiplication Principle of Counting (Section 9.1), there are $6 \times 6 = 36$ equally likely outcomes. Of these, three $\{(1, 3), (3, 1), (2, 2)\}$ yield a sum of 4 (Figure 10.1). The probability is 3/36, or 1/12.

(d) Because these 4 outcomes may not be equally likely, the probability cannot be determined from this information.

Now try Exercise 5.

FIGURE 10.1 A sum of 4 on a roll of two dice. (Example 1c)

Notice that in each of these cases we first counted the number of possible outcomes of the random phenomenon in question. The set of all possible outcomes is the **sample space** of the random phenomenon. An **event** is a subset of the sample space. If the sample space consisted of a finite number of **equally likely outcomes**, we were able to model the probability of an event by counting.

Probability of an Event (Equally Likely Outcomes)

If E is an event in a finite, nonempty sample space S of equally likely outcomes, then we model the probability of the event E as

$$P(E) = \frac{\text{the number of outcomes in } E}{\text{the number of outcomes in } S}.$$

The hypothesis of equally likely outcomes is critical here. Many people guess wrongly on the probability in Example 1c because they figure that there are 11 possible outcomes for the sum on two fair dice: $\{2, 3, 4, 5, 6, 7, 8, 9, 10, 11, 12\}$ and that 4 is one of them. (That reasoning is correct so far.) The reason that 1/11 is not the probability of rolling a sum of 4 is that all those sums are *not equally likely*.

Is Probability Just for Games?

Probability theory got its start in letters between Blaise Pascal (1623–1662) and Pierre de Fermat (1601–1665) concerning games of chance, but it has come a long way since then. Modern mathematicians like David Blackwell "(1919–2010)", the first African-American to receive a fellowship to the Institute for Advanced Study at Princeton, have greatly extended both the theory and the applications of probability, especially in the areas of statistics, quantum physics, and information theory. Moreover, the work of John Von Neumann (1903–1957) has led to a separate branch of modern discrete mathematics that really is about games, called game theory.

On the other hand, we can *assign* probabilities to the 11 outcomes in this smaller sample space in a way that is consistent with the number of ways each total can occur. The table below shows a **probability distribution**, in which each sum is assigned a unique probability based on the sample space of 36 equally likely outcomes.

All Outcomes		Second Die 1	2	3	4	5	6
First Die	1	(1,1)	(1,2)	(1,3)	(1,4)	(1,5)	(1,6)
	2	(2,1)	(2,2)	(2,3)	(2,4)	(2,5)	(2,6)
	3	(3,1)	(3,2)	(3,3)	(3,4)	(3,5)	(3,6)
	4	(4,1)	(4,2)	(4,3)	(4,4)	(4,5)	(4,6)
	5	(5,1)	(5,2)	(5,3)	(5,4)	(5,5)	(5,6)
	6	(6,1)	(6,2)	(6,3)	(6,4)	(6,5)	(6,6)

SUM		Second Die 1	2	3	4	5	6
First Die	1	2	3	4	5	6	7
	2	3	4	5	6	7	8
	3	4	5	6	7	8	9
	4	5	6	7	8	9	10
	5	6	7	8	9	10	11
	6	7	8	9	10	11	12

Outcome	Probability
2	1/36
3	2/36
4	3/36
5	4/36
6	5/36
7	6/36
8	5/36
9	4/36
10	3/36
11	2/36
12	1/36

We see that the sums are not equally likely, but we can find the probabilities of events by adding up the probabilities of the outcomes in the event, as in the following example.

EXAMPLE 2 Rolling the Dice

Find the probability of rolling a sum divisible by 3 on a single roll of two fair dice.

SOLUTION The event E consists of the outcomes $\{3, 6, 9, 12\}$. To get the probability of E we add up the probabilities of the outcomes in E (see the table of the probability distribution):

$$P(E) = \frac{2}{36} + \frac{5}{36} + \frac{4}{36} + \frac{1}{36} = \frac{12}{36} = \frac{1}{3}.$$

Now try Exercise 7.

Notice that this method would also have worked just fine with our 36-outcome sample space, in which every outcome has probability 1/36. In general, it is easier to work with sample spaces of equally likely events because it is not necessary to write out the probability distribution. When outcomes do have unequal probabilities, we need to know what probabilities to assign to the outcomes.

Not every function that assigns numbers to outcomes qualifies as a probability function.

DEFINITION Probability Function

A **probability function** is a function P that assigns a real number to each outcome in a sample space S subject to the following conditions:

1. $0 \le P(O) \le 1$ for every outcome O;

2. the sum of the probabilities of all outcomes in S is 1;

3. $P(\varnothing) = 0$.

The probability of any event can then be modeled in terms of the probability function.

Probability of an Event (Generalized)

Let S be a finite, nonempty sample space in which every outcome has a probability assigned to it by a probability function P. If E is any event in S, the probability of the event E is the sum of the probabilities of all the outcomes contained in E.

Empty Set

A set with no elements is the *empty set*, denoted by $\varnothing$.

Random Variables

A more formal treatment of probability would distinguish between an *outcome* in a sample space and a *number* that is associated with that outcome. For example, an outcome in Example 2 is really something like 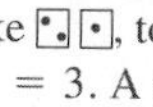, to which we associate the number $2 + 1 = 3$. A function that assigns a number to an outcome is called a **random variable**. A different random variable might assign the number $2 \cdot 1 = 2$ to this outcome of the dice. Another random variable might assign the number 21.

EXAMPLE 3 Testing a Probability Function

Prior to the Final Four basketball tournament, a sports analyst considers the strengths and weaknesses of the 4 remaining teams and suggests the probability distribution shown for their chances of winning the championship.

Team	Probability
1	2/5
2	1/3
3	1/6
4	1/8

SOLUTION

This is not a valid probability function because $2/5 + 1/3 + 1/6 + 1/8 \neq 1$.

Now try Exercise 9a.

Determining Probabilities

It is not always easy to determine probabilities, but the arithmetic involved is fairly simple. It usually comes down to multiplication, addition, and (most importantly) counting. Here is the strategy we will follow:

Strategy for Determining Probabilities

1. Determine the sample space of all possible outcomes. When possible, choose outcomes that are equally likely.
2. If the sample space has equally likely outcomes, the probability of an event E is determined by counting:

$$P(E) = \frac{\text{the number of outcomes in } E}{\text{the number of outcomes in } S}$$

3. If the sample space does not have equally likely outcomes, determine the probability function. (This is not always easy to do.) Check to be sure that the conditions of a probability function are satisfied. Then the probability of an event E is determined by adding up the probabilities of all the outcomes contained in E.

Ordered or Unordered?

Notice that in Example 4 we have a sample space in which order is disregarded, whereas in Example 2 we had a sample space in which order matters. (For example, $3 + 1$ and $1 + 3$ are distinct outcomes.) The order matters in Example 2 because we are considering the probabilities of two events (first die, second die), with distinguishable outcomes. In Example 4, we are simply counting unordered combinations.

EXAMPLE 4 Choosing Chocolates

Sal opens a box of a dozen chocolate-covered cremes and generously offers two of them to Val. Val likes vanilla cremes the best, but all the chocolates look alike on the outside. If four of the twelve cremes are vanilla, what is the probability that both of Val's picks turn out to be vanilla?

SOLUTION The experiment in question is the selection of two chocolates, without regard to order, from a box of 12. There are ${}_{12}C_2 = 66$ outcomes of this experiment, and all of them are equally likely. We can therefore determine the probability by counting.

The event E consists of all possible pairs of 2 vanilla cremes that can be chosen, without regard to order, from 4 vanilla cremes available. There are ${}_4C_2 = 6$ ways to form such pairs.

Therefore, $P(E) = 6/66 = 1/11$.

Now try Exercise 25.

CHAPTER OPENER Problem, part 1 (from page 676)

Problem: Typical of many lotteries, Wisconsin's Megabucks game requires players to correctly choose 6 numbers from 1 to 49. What's the probability of winning the big prize?

Solution: There are ${}_{49}C_6 = 13{,}983{,}816$ equally likely ways that 6 numbers can be chosen from 49 numbers without regard to order. Only one of these wins the grand prize, so the probability of winning is $1/13{,}983{,}816 \approx 0.0000000715$.

Numbers like this are more easily understood by recasting them in more familiar terms. If you stayed at the sales terminal 24 hours a day and purchased one ticket per minute, you could expect to win this Megabucks game about once every 26 1/2 years.

Many probability problems require that we think of events happening in succession, often with the occurrence of one event affecting the probability of the occurrence of another event. In these cases, we use a law of probability called the Multiplication Principle of Probability.

Multiplication Principle of Probability

Suppose an event A has probability p_1 and an event B has probability p_2 under the assumption that A occurs. Then the probability that both A and B occur is p_1p_2.

Independent Events

Two events are **independent** if the occurrence (or non-occurrence) of one does not affect the probability of the other.

If the events A and B are independent, we can omit the phrase "under the assumption that A occurs" because that assumption would not matter.

EXAMPLE 5 Calculating Probabilities

According to the American Red Cross, the distribution of blood types in the U.S. population is 45% Type O, 40% Type A, 11% Type B, and 4% Type AB.

If 2 people are chosen at random, find the probability that

(a) both have Type O;

(b) neither has Type A;

(c) both have the same blood type.

SOLUTION Note that the four blood types are not equally likely, so we need to consider the distribution of probabilities. Also note that choosing people at random makes their blood types independent. Without random selection this would not be true, as family members may share blood types and frequencies vary across ethnic groups.

(a) Applying the Multiplication Principle, $P(OO) = (0.45)(0.45) = 0.2025$.

(b) Let event N = "not A." Since $P(A) = 0.4$, $P(N) = 1 - 0.4 = 0.6$, and thus $P(NN) = (0.6)(0.6) = 0.36$.

(c) The event S = "same blood type" consists of the outcomes $\{OO, AA, BB, AB\_AB\}$, so we find the probability of S by adding up four probabilities.

$$\begin{aligned} P(S) &= P(OO) + P(AA) + P(BB) + P(AB\_AB) \\ &= (0.45)(0.45) + (0.4)(0.4) + (0.11)(0.11) + (0.04)(0.04) \\ &= 0.2025 + 0.16 + 0.0121 + 0.0016 \\ &= 0.3762 \end{aligned}$$

Now try Exercise 45.

John Venn

John Venn (1834–1923) was an English logician and clergyman, just like his contemporary, Charles L. Dodgson. Although both men used overlapping circles to illustrate their logical syllogisms, it is Venn whose name lives on in connection with these diagrams. Dodgson's name barely lives on at all, and yet he is far the more famous of the two: Under the pen name Lewis Carroll, he wrote *Alice's Adventures in Wonderland* and *Through the Looking Glass*.

Venn Diagrams

We have seen many instances in which geometric models help us to understand algebraic models more easily, and probability theory is yet another setting in which this is true. **Venn diagrams**, associated mainly with the world of set theory, are good for visualizing relationships among events within sample spaces.

EXAMPLE 6 Using a Venn Diagram, Part 1

At Pascal High School, 54% of the students are girls and 62% of the students play sports. Half of the girls at the school play sports.

(a) What percentage of the students who play sports are boys?

(b) If a student is chosen at random, what is the probability that it is a boy who does not play sports?

SOLUTION To organize the categories, we draw a large rectangle to represent the sample space (all students at the school) and two overlapping regions to represent "girls" and "sports" (Figure 10.2). We fill in the percentages (Figure 10.3) using the following logic:

- The overlapping (green) region contains half the girls, or $(0.5)(54\%) = 27\%$ of the students.
- The yellow region (the rest of the girls) then contains $(54 - 27)\% = 27\%$ of the students.
- The blue region (the rest of the sports players) then contains $(62 - 27)\% = 35\%$ of the students.
- The white region (the rest of the students) then contains $(100 - 89)\% = 11\%$ of the students. These are boys who do not play sports.

We can now answer the two questions by looking at the Venn diagram.

(a) We see from the diagram that the ratio of *boys* who play sports to *all students* who play sports is $\dfrac{0.35}{0.62}$, which is about 56.45%.

(b) We see that 11% of the students are boys who do not play sports, so 0.11 is the probability.

Now try Exercises 27a–c.

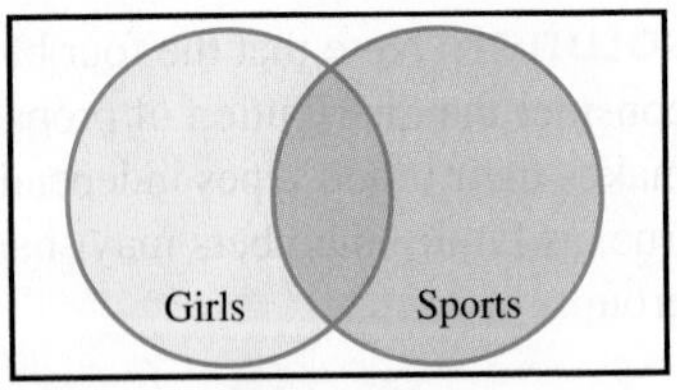

FIGURE 10.2 A Venn diagram for Example 6. The overlapping region common to both circles represents "girls who play sports." The region outside both circles (but inside the rectangle) represents "boys who do not play sports."

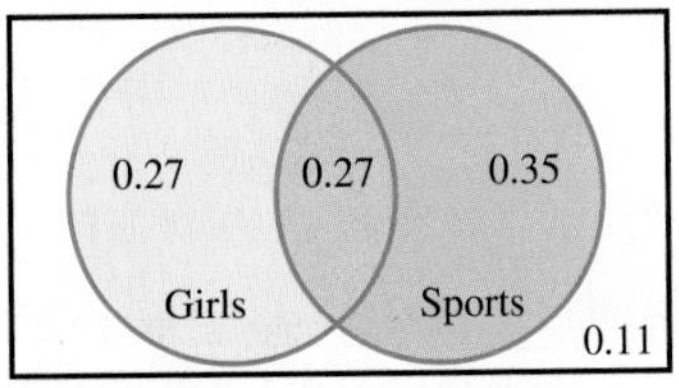

FIGURE 10.3 A Venn diagram for Example 6 with the probabilities filled in.

The Inclusive "Or"

Note that in mathematics we always use the word "or" in the inclusive sense, to mean "one or the other or both." That's often what people mean in everyday speech, too. If you can win a card game by drawing an ace or a heart, getting the ace of hearts counts. But sometimes people use "or" in an exclusive sense, as in "Tonight let's go to a movie or the game." In mathematics we'd need to add "but not both" if that's what we really meant.

What's the probability that a student at Pascal High School (Example 6) is a girl *or* plays sports? To find the probability of rolling a sum of 10 or 11 on two dice, we simply add: $P(11) + P(12) = 2/36 + 1/36 = 3/36$. But adding $P(Girl) + P(Sports) = 0.54 + 0.62 = 1.16$, and that's not a legitimate probability. A careful look at the Venn diagram in

Figure 10.3 reveals the problem: Adding the probabilities represented by the two circles counts the overlapping region twice. To find the correct probability, we need to subtract the amount of overlap from the sum: $P(Girl \text{ or } Sports) = 0.54 + 0.62 - 0.27 = 0.89$.

Addition Principle of Probability

For events A and B in a sample space, $P(A \text{ or } B) = P(A) + P(B) - P(A \text{ and } B)$.

When A and B don't intersect, this reduces to the familiar $P(A \text{ or } B) = P(A) + P(B)$. We call such non-intersecting events **mutually exclusive** (or *disjoint*).

Note also that the Venn diagram in Figure 10.3 suggests the alternative approach of adding the probabilities for the yellow, green, and blue regions:

$$P(Girl \text{ or } Sports) = 0.27 + 0.27 + 0.35 = 0.89.$$

EXAMPLE 7 Using a Venn Diagram, Part 2

At Pascal High School (Example 6), 68% of the students have cell phones, 77% have iPods, and 62% have both a cell phone and an iPod. What's the probability a student has

(a) a cell phone or an iPod?

(b) a cell phone or an iPod, but not both?

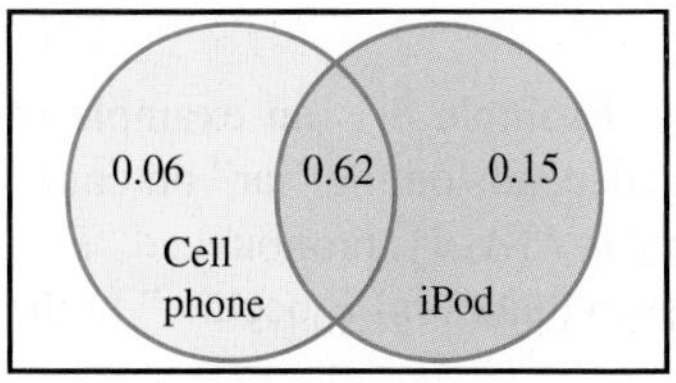

FIGURE 10.4

SOLUTION First we construct a Venn diagram (Figure 10.4). It's usually easiest to start with the overlapping region (62% have both devices) and then determine the percentage that provides the correct total in each circle $(68\% - 62\% = 6\%$ and $77\% - 62\% = 15\%)$.

(a) We let events C = cell phone and I = iPod; then,

$$P(C \text{ or } I) = P(C) + P(I) - P(C \text{ and } I) = 0.68 + 0.77 - 0.62 = 0.83$$

Alternatively, $P(C \text{ or } I) = 0.06 + 0.62 + 0.15 = 0.83$

(b) $P(C \text{ or } I, \text{ but not both}) = 0.06 + 0.15 = 0.21$

Now try Exercise 27d.

Tree Diagrams

Tree diagrams, which we first met in Section 9.1 as a way to visualize the Multiplication Principle of Counting, are good for visualizing the Multiplication Principle of Probability as well.

EXAMPLE 8 Using a Tree Diagram

Two identical cookie jars are on a counter. Jar A contains 2 chocolate chip and 2 peanut butter cookies, and jar B contains 1 chocolate chip cookie. We select a cookie at random. What is the probability that it is a chocolate chip cookie?

SOLUTION It is tempting to say 3/5, since there are 5 cookies in all, 3 of which are chocolate chip. Indeed, this would be the answer if all the cookies were in the same jar. However, the fact that they are in different jars means that the 5 cookies are *not equally likely outcomes*. That lone chocolate chip cookie in jar B has a much better chance of being chosen than any of the cookies in jar A. We need to think of this as a two-step experiment: First pick a jar, then pick a cookie from that jar.

(continued)

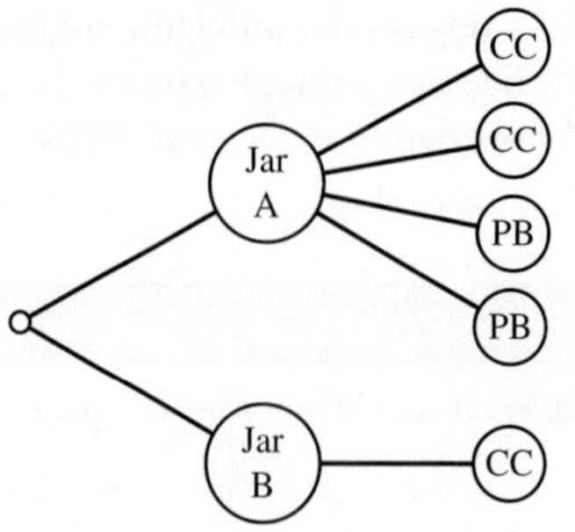

FIGURE 10.5 A tree diagram.

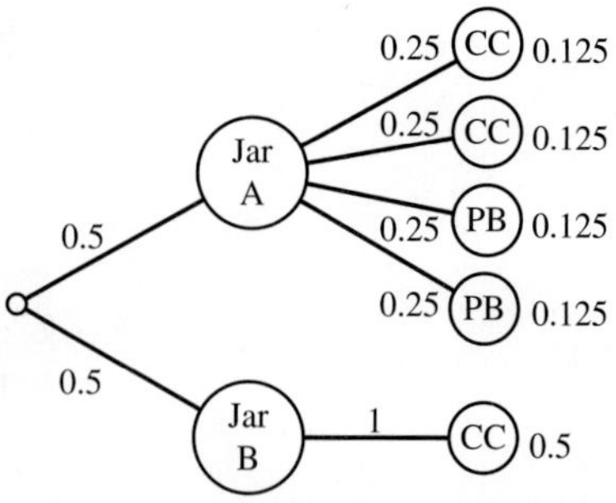

FIGURE 10.6 The tree diagram with the probabilities filled in.

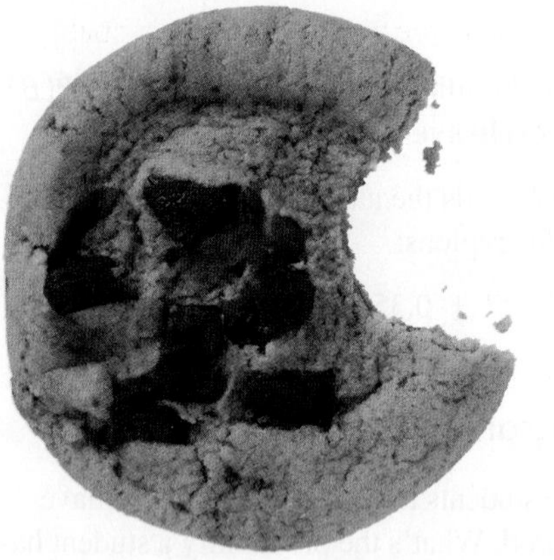

Figure 10.5 gives a visualization of the two-step process. In Figure 10.6, we have filled in the probabilities along each branch, first of picking the jar, then of picking the cookie. The probability at the *end* of each branch is obtained by multiplying the probabilities from the root to the branch. (This is the Multiplication Principle.) Notice that the probabilities of the 5 cookies (as predicted) are not all equal.

The event "chocolate chip" is a set containing three outcomes. We add their probabilities together to get the correct probability:

$P(\text{chocolate chip}) = 0.125 + 0.125 + 0.5 = 0.75$ **Now try Exercise 29.**

Conditional Probability

The probability of drawing a chocolate chip cookie in Example 8 is an example of **conditional probability**, since the "cookie" probability depends on the "jar" outcome. A convenient symbol to use with conditional probability is $P(A|B)$, pronounced "*P* of *A* given *B*," meaning "the probability of the event *A*, given that event *B* occurs." In the cookie jars of Example 8,

$$P(\text{chocolate chip}|\text{jar } A) = \frac{2}{4} \quad \text{and} \quad P(\text{chocolate chip}|\text{jar } B) = 1.$$

(In the tree diagram, these are the probabilities *along* the branches that come out of the two jars, not the probabilities at the *ends* of the branches.)

The Multiplication Principle of Probability can be stated succinctly with this notation as follows:

$$P(A \text{ and } B) = P(A) \cdot P(B|A)$$

This is how we found the numbers at the ends of the branches in Figure 10.6.

In Example 4 we solved a probability problem by counting the number of equally likely events in a sample space. We will now solve the same problem using a tree diagram and conditional probability.

EXAMPLE 9 Choosing Chocolates, Revisited

Sal opens a box of a dozen chocolate-covered cremes and generously offers two of them to Val. Val likes vanilla cremes the best, but all the chocolates look alike on the outside. If four of the twelve cremes are vanilla, what is the probability that both of Val's picks turn out to be vanilla?

SOLUTION As far as Val is concerned, there are two kinds of chocolate cremes: vanilla (V) and nonvanilla (N). When choosing two chocolates, there are four possible outcomes: *VV*, *VN*, *NV*, and *NN*, as shown in the tree diagram, Figure 10.7.

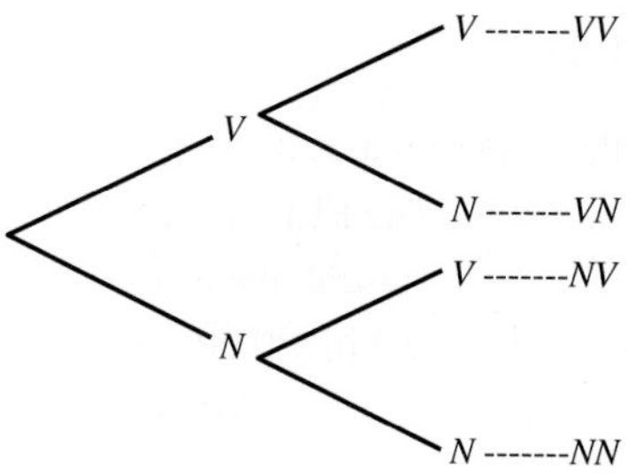

FIGURE 10.7 The tree diagram for Example 9.

We need to determine the probability of the outcome VV. The probability of picking a vanilla creme on the first draw is 4/12. The conditional probability of picking a vanilla creme on the second draw, *given that a vanilla creme was drawn on the first*, is 3/11. By the Multiplication Principle, the probability of drawing a vanilla creme on both draws is

$$\frac{4}{12}\cdot\frac{3}{11}=\frac{1}{11}.$$

Since this is the probability we are looking for, we do not need to compute the probabilities of the other outcomes. However, you should verify that the other probabilities would be:

$$P(VN)=\frac{4}{12}\cdot\frac{8}{11}=\frac{8}{33}$$
$$P(NV)=\frac{8}{12}\cdot\frac{4}{11}=\frac{8}{33}$$
$$P(NN)=\frac{8}{12}\cdot\frac{7}{11}=\frac{14}{33}$$

Notice that $P(VV)+P(VN)+P(NV)+P(NN)=(1/11)+(8/33)+(8/33)+(14/33)=1$, so the probability function checks out. **Now try Exercise 33.**

In Example 9, the probability of getting a second vanilla creme changed, because that outcome depended on what the first candy was. When rolling a die, however, the outcome on the second roll does not depend on the first roll, so the probability does not change.

Independent Events

Events A and B are independent if and only if $P(B|A)=P(B)$.
Therefore, for independent events $P(A \text{ and } B)=P(A)\cdot P(B)$.

As our final example of a probability problem, we will show how to use the Multiplication Principle formula in a different but equivalent form, sometimes called the **conditional probability formula**:

Conditional Probability Formula

If the event B depends on the event A, then $P(B|A)=\dfrac{P(A \text{ and } B)}{P(A)}$.

EXAMPLE 10 Using the Conditional Probability Formula

Suppose we have drawn a cookie at random from one of the jars described in Example 8. Given that it is chocolate chip, what is the probability that it came from jar A?

SOLUTION By the formula,

$$P(\text{jar } A|\text{chocolate chip})=\frac{P(\text{jar } A \text{ and chocolate chip})}{P(\text{chocolate chip})}$$
$$=\frac{(1/2)(2/4)}{0.75}=\frac{0.25}{0.75}=\frac{1}{3}$$

Now try Exercise 31.

EXPLORATION 1 Testing Positive for HIV

As of the year 2009, the probability of an adult in the United States having HIV/AIDS was 0.006 (*Source: 2010 CIA World Factbook*). The ELISA test is used to detect the virus antibody in blood. If the antibody is present, the test reports positive with probability 0.997 and negative with probability 0.003. If the antibody is not present, the test reports positive with probability 0.015 and negative with probability 0.985.

1. Draw a tree diagram with branches to nodes "antibody present" and "antibody absent" branching from the root. Fill in the probabilities for North American adults (age 15–49) along the branches. (Note that these two probabilities must add up to 1.)
2. From the node at the end of each of the two branches, draw branches to "positive" and "negative." Fill in the probabilities along the branches.
3. Use the Multiplication Principle to fill in the probabilities at the ends of the four branches. Check to see that they add up to 1.
4. Find the probability of a positive test result. (Note that this event consists of two outcomes.)
5. Use the conditional probability formula to find the probability that a person with a positive test result actually *has* the antibody; that is, determine $P(\text{antibody present} \mid \text{positive})$.

You might be surprised that the answer to part 5 is so low, but it should be compared with the probability of the antibody being present *before* seeing the positive test result, which was 0.006. Nonetheless, that is why a positive ELISA test is followed by further testing before a diagnosis of HIV/AIDS is made. This is the case with many diagnostic tests.

QUICK REVIEW 10.1 *(Prerequisite skill Section 9.1)*

In Exercises 1–8, tell how many outcomes are possible for the experiment.

1. A single coin is tossed.
2. A single 6-sided die is rolled.
3. Three different coins are tossed.
4. Three different 6-sided dice are rolled.
5. Five different cards are drawn from a standard deck of 52.
6. Two chips are drawn simultaneously from a jar containing 10 chips.
7. Five people are lined up for a photograph.
8. Three-digit numbers are formed from the numbers $\{1, 2, 3, 4, 5\}$ without repetition.

In Exercises 9 and 10, evaluate the expression by pencil and paper. Verify your answer with a calculator.

9. $\dfrac{{}_5C_3}{{}_{10}C_3}$

10. $\dfrac{{}_5C_2}{{}_{10}C_2}$

SECTION 10.1 Exercises

In Exercises 1–8, a red die and a green die have been rolled. What is the probability of the event?

1. The sum is 9.
2. The sum is even.
3. The number on the red die is greater than the number on the green die.
4. The sum is less than 10.
5. Both numbers are odd.
6. Both numbers are even.
7. The sum is prime.
8. The sum is 7 or 11.

9. **Writing to Learn** Alrik's gerbil cage has four compartments, A, B, C, and D. After careful observation, he estimates the proportion of time the gerbil spends in each compartment and constructs the table below.

Compartment	A	B	C	D
Proportion	0.25	0.20	0.35	0.30

(a) Is this a valid probability function? Explain.

(b) Is there a problem with Alrik's reasoning? Explain.

10. (Continuation of Exercise 9) Suppose Alrik determines that his gerbil spends time in the four compartments A, B, C, and D in the ratio 4:3:2:1. What proportions should he fill in the table above? Is this a valid probability function?

The maker of a popular chocolate candy that is covered in a thin colored shell has released information about the overall color proportions in its production of the candy, which is summarized in the following table.

Color	Brown	Red	Yellow	Green	Orange	Tan
Proportion	0.3	0.2	0.2	0.1	0.1	0.1

In Exercises 11–16, a single candy of this type is selected at random from a newly opened bag. What is the probability that the candy has the given color(s)?

11. Brown or tan

12. Red, green, or orange

13. Red

14. Not red

15. Neither orange nor yellow

16. Neither brown nor tan

A peanut version of the same candy has all the same colors except tan. The proportions of the peanut version are given in the following table.

Color	Brown	Red	Yellow	Green	Orange
Proportion	0.3	0.2	0.2	0.2	0.1

In Exercises 17–22, a candy of this type is selected at random from each of two newly opened bags. What is the probability that the two candies have the given color(s)?

17. Both are brown.

18. Both are orange.

19. One is red, and the other is green.

20. The first is brown, and the second is yellow.

21. Neither is yellow.

22. The first is not red, and the second is not orange.

Exercises 23–32 concern a version of the card game "bid Euchre" that uses a pack of 24 cards, consisting of ace, king, queen, jack, 10, and 9 in each of the four suits (spades, hearts, diamonds, and clubs). In bid Euchre, a hand consists of 6 cards. Find the probability of each event.

23. **Euchre** A hand is all spades.

24. **Euchre** All six cards are from the same suit.

25. **Euchre** A hand includes all four aces.

26. **Euchre** A hand includes two jacks of the same color (called the right and left bower).

27. **Using Venn Diagrams** A and B are events in a sample space S such that $P(A) = 0.6$, $P(B) = 0.5$, and $P(A \text{ and } B) = 0.3$.

(a) Draw a Venn diagram showing the overlapping sets A and B and fill in the probabilities of the four regions formed.

(b) Find the probability that A occurs but B does not.

(c) Find the probability that neither A nor B occurs.

(d) Find the probability that A or B occurs.

(e) Are events A and B independent? (That is, does $P(A|B) = P(A)$?)

28. **Using Venn Diagrams** A and B are events in a sample space S such that $P(A) = 0.7$, $P(B) = 0.4$, and $P(A \text{ and } B) = 0.2$.

(a) Draw a Venn diagram showing the overlapping sets A and B and fill in the probabilities of the four regions formed.

(b) Find the probability that A occurs but B does not.

(c) Find the probability that neither A nor B occurs.

(d) Find the probability that A or B occurs.

(e) Are events A and B independent? (That is, does $P(A|B) = P(A)$?)

In Exercises 29 and 30, it will help to draw a tree diagram.

29. **Piano Lessons** If it rains tomorrow, the probability is 0.8 that John will practice his piano lesson. If it does not rain tomorrow, there is only a 0.4 chance that John will practice. Suppose that the chance of rain tomorrow is 60%. What is the probability that John will practice his piano lesson?

30. **Predicting Cafeteria Food** If the school cafeteria serves meat loaf, there is a 70% chance that they will serve peas. If they do not serve meat loaf, there is a 30% chance that they will serve peas anyway. The students know that meat loaf will be served exactly once during the 5-day week, but they do not know which day. If tomorrow is Monday, what is the probability that

(a) The cafeteria serves meat loaf?

(b) The cafeteria serves meat loaf and peas?

(c) The cafeteria serves peas?

31. **Conditional Probability** There are two precalculus sections at West High School. Mr. Abel's class has 12 girls and 8 boys, while Mr. Bonitz's class has 10 girls and 15 boys. If a West High precalculus student chosen at random happens to be a girl, what is the probability she is from Mr. Abel's class? (*Hint*: The answer is not 12/22.)

32. **Group Activity Conditional Probability** Two boxes are on the table. One box contains a normal coin and a two-headed coin; the other box contains three normal coins. A friend reaches into a box, removes a coin, and shows you one side: a head. What is the probability that it came from the box with the two-headed coin?

33. **Renting Cars** Floppy Jalopy Rent-a-Car has 25 cars available for rental—20 big bombs and 5 midsize cars. If two cars are selected at random, what is the probability that both are big bombs?

34. **Defective Calculators** Dull Calculators, Inc., knows that a unit coming off an assembly line has a probability of 0.037 of being defective. If four units are selected at random during the course of a workday, what is the probability that none of the units are defective?

35. **Causes of Death** The government designates a single cause for each death in the United States. The resulting data indicate that 45% of deaths are due to heart and other cardiovascular disease and 22% are due to cancer.

(a) What is the probability that the death of a randomly selected person will be due to cardiovascular disease or cancer?

(b) What is the probability that the death will be due to some other cause?

36. **Yahtzee** In the game of *Yahtzee*, on the first roll five dice are tossed simultaneously. What is the probability of rolling five of a kind (which is Yahtzee!) on the first roll?

37. **Writing to Learn** Explain why the following statement cannot be true. The probabilities that a computer salesperson will sell zero, one, two, or three computers in any one day are 0.12, 0.45, 0.38, and 0.15, respectively.

38. **HIV Testing** A particular test for HIV, the virus that causes AIDS, is 0.7% likely to produce a false positive result—a result indicating that the human subject has HIV when in fact the person is not carrying the virus. If 60 individuals who are HIV-negative are tested, what is the probability of obtaining at least one false result?

39. **Graduate School Survey** The Earmuff Junction College Alumni Office surveys selected members of the class of 2000. Of the 254 who graduated that year, 172 were women, 124 of whom went on to graduate school. Of the male graduates, 58 went on to graduate school. What is the probability of the given event?

(a) The graduate is a woman.

(b) The graduate went on to graduate school.

(c) The graduate was a woman who went on to graduate school.

40. **Indiana Jones and the Final Exam** Professor Indiana Jones gives his class a list of 20 study questions, from which he will select 8 to be answered on the final exam. If a given student knows how to answer 14 of the questions, what is the probability that the student will be able to answer the given number of questions correctly?

(a) All 8 questions

(b) Exactly 5 questions

(c) At least 6 questions

41. **Graduation Requirement** To complete the kinesiology requirement at Palpitation Tech you must pass two classes chosen from aerobics, aquatics, defense arts, gymnastics, racket sports, recreational activities, rhythmic activities, soccer, and volleyball. If you decide to choose your two classes at random by drawing two class names from a box, what is the probability that you will take racket sports and rhythmic activities?

42. **Writing to Learn** During July in Gunnison, Colorado, the probability of at least 1 hr a day of sunshine is 0.78, the probability of at least 30 min of rain is 0.44, and the probability that it will be cloudy all day is 0.22. Write a paragraph explaining whether this statement could be true.

43. **Writing to Learn** After diagnosing an athlete's knee injury, a doctor tells her there is a 90% chance she will make a full recovery after surgery. Where do you think that probability comes from?

44. **Writing to Learn** The TV weather reporter says there is a 30% chance of rain during your picnic tomorrow afternoon. How do you think such predictions are made?

45. **Handedness** By the age of 1 many infants display a tendency to favor one hand over the other. About 60% favor their right hands, about 10% favor their left hands, and the rest show no preference yet. Suppose you observe 3 unrelated year-old infants at a day care center. What's the probability that

(a) none shows any preference?

(b) all show a preference for one hand or the other?

(c) all show a preference for the same hand?

46. **Gumballs** A vending machine contains 12 red gumballs, 5 white ones, and 3 blue ones. You insert 3 coins, getting one gumball after another after another. What's the probability you end up with

(a) no red ones?

(b) 3 of the same color?

(c) a set of red, white, and blue?

47. **Blood** Recall from Example 5 that the distribution of U.S. blood types is 45% Type O, 40% Type A, 11% Type B, and 4% Type AB. At a blood drive, what is the probability that the first Type A donor is the fourth person in line?

48. **Blue Gumball** Suppose your little sister insists on having a blue gumball from the machine described in Exercise 46. You need to buy gumballs until you get a blue one. What's the probability that you have to buy 5 of them?

49. **Homes** A realtor reports that 30% of houses currently listed for sale have garages, 55% have basements, and 25% have both a garage and a basement.

(a) What's the probability that a house for sale has neither a garage nor a basement?

(b) What's the probability that a house for sale has a garage or a basement?

(c) If a house for sale has a basement, what's the probability it has a garage?

(d) What's the probability that a house for sale with a garage has a basement?

(e) Are having a basement and having a garage independent events? Explain.

50. **Courses** At a large high school, 75% of the sophomores take Spanish, 60% take Biology, and 45% take both courses.

(a) What's the probability that a sophomore doesn't take either course?

(b) What's the probability that a sophomore takes Spanish or Biology?

(c) If a sophomore takes Spanish, what's the probability she takes Biology?

(d) What's the probability that a sophomore taking Biology takes Spanish?

(e) Among sophomores, are taking Spanish and taking Biology independent? Explain.

Standardized Test Questions

51. **True or False** A sample space consists of equally likely events. Justify your answer.

52. **True or False** The probability of an event can be greater than 1. Justify your answer.

Evaluate Exercises 53–56 without using a calculator.

53. **Multiple Choice** The probability of rolling a total of 5 on a pair of fair dice is

(a) $\frac{1}{4}$. (b) $\frac{1}{5}$.
(c) $\frac{1}{6}$. (d) $\frac{1}{9}$.
(e) $\frac{1}{11}$.

54. **Multiple Choice** Which of the following numbers could not be the probability of an event?

(a) 0 (b) 0.95
(c) $\frac{\sqrt{3}}{4}$ (d) $\frac{3}{\pi}$
(e) $\frac{\pi}{2}$

55. **Multiple Choice** If A and B are independent events, then $P(A|B) =$

(a) $P(A)$. (b) $P(B)$.
(c) $P(B|A)$. (d) $P(A) \cdot P(B)$.
(e) $P(A) + P(B)$.

56. **Multiple Choice** A fair coin is tossed three times in succession. What is the probability that exactly one of the coins shows heads?

(a) $\frac{1}{8}$ (b) $\frac{1}{3}$
(c) $\frac{3}{8}$ (d) $\frac{1}{2}$
(e) $\frac{2}{3}$

Explorations

57. **Empirical Probability** In real applications, it is often necessary to approximate the probabilities of the various outcomes of an experiment by performing the experiment a large number of times and recording the results. Barney's Bread Basket offers five different kinds of bagels. Barney records the sales of the first 500 bagels in a given week in the table shown below:

Type of Bagel	Number Sold
Plain	185
Onion	60
Rye	55
Cinnamon Raisin	125
Sourdough	75

(a) Use the observed sales number to approximate the probability that a random customer buys a plain bagel. Do the same for each other bagel type and make a table showing the approximate probability distribution.

(b) Assuming independence of the events, find the probability that three customers in a row all order plain bagels.

(c) **Writing to Learn** Do you think it is reasonable to assume that the orders of three consecutive customers actually are independent? Explain.

58. **Straight Poker** In the original version of poker known as "straight" poker, a 5-card hand is dealt from a standard deck of 52 cards. What is the probability of the given event?

(a) A hand will contain at least one king.

(b) A hand will be a "full house" (any three of one kind and a pair of another kind).

59. **Politics** The Republican chairperson of a county legislature had to select 4 of the members to represent the city at a big event. To avoid the appearance of partisanship, she drew names from a hat. When all 4 turned out to be Republicans, the Democrats cried foul. They pointed out that the legislature had 7 Democrats and 10 Republicans, and suggested the choice might have been rigged somehow.

(a) Is it plausible that a fairly chosen group could be all Republicans, or is there a good reason to be suspicious?

(b) What assumption did you base your calculation on? Explain why that assumption may (or may not) be warranted.

60. **Group Activity Investigating Red Lights** The city claims that the new traffic light it just installed at the end of your street is on a cycle that makes it green in your direction for 15 sec/min and yellow for another 5 sec. Nonetheless, this week you hit a red light there all 5 mornings when you left home.

(a) Could this be attributed merely to a run of bad luck, or is there reason to suspect something is wrong with the light? Explain.

(b) What assumption did you base your calculation on? Explain why that assumption is (or is not) warranted here.

Extending the Ideas

61. **Expected Value** If the outcomes of an experiment are given numerical values (such as the total on a roll of two dice, or the payoff on a lottery ticket), we define the **expected value** to be the sum of all the numerical values times their respective probabilities.

For example, suppose we roll a fair die. If we roll a multiple of 3, we win \$3; otherwise we lose \$1. The probabilities of the two possible payoffs are shown in the table below:

Value	Probability
+3	2/6
−1	4/6

The expected value is

$3 \times (2/6) + (-1) \times (4/6) = (6/6) - (4/6) = 1/3.$

We interpret this to mean that we would win an average of 1/3 dollar per game in the long run.

(a) A game is called *fair* if the expected value of the payoff is zero. Assuming that we still win \$3 for a multiple of 3, what should we pay for any other outcome in order to make the game fair?

(b) Suppose we roll *two* fair dice and look at the total under the original rules. That is, we win \$3 for rolling a multiple of 3 and lose \$1 otherwise. What is the expected value of this game?

62. Expected Value (Continuation of Exercise 61) Gladys has a personal rule never to enter the lottery (picking 6 numbers from 1 to 49) until the Megabucks payoff reaches 5 million dollars. When it does reach 5 million, she always buys ten different \$1 tickets.

(a) Assume that the payoff for a winning ticket is 5 million dollars. What is the probability that Gladys holds a winning ticket? (Refer to the Chapter Opener Problem of this section for the probability of any ticket winning.)

(b) Fill in the probability distribution for Gladys's possible payoffs in the table below. (Note that we subtract \$10 from the \$5 million, since Gladys has to pay for her tickets even if she wins.)

Value	Probability
−10	
+4,999,990	

(c) Find the expected value of the game for Gladys.

(d) Writing to Learn In terms of the answer in part (b), explain to Gladys the long-term implications of her strategy.

10.2 Statistics (Graphical)

What you'll learn about

- Statistics
- Categorical Data
- Quantitative Data: Stemplots
- Frequency Tables
- Histograms
- Describing Distributions: Shape
- Time Plots

... and why

Graphical displays of data are increasingly prevalent in professional and popular media. We all need to understand them.

Statistics

The aim of Statistics is to draw meaning from data and communicate it to others. We begin by looking at data.

The objects described by a set of data are **individuals**, which may be people, animals, or things. The characteristic of the individuals being identified or measured is a **variable**. Variables are either *categorical* or *quantitative*. If we use the data to identify each individual as belonging to a distinct class, such as male or female, then the variable is a **categorical variable**. If we use the data as numerical values measuring the characteristic in some kind of units, then the variable is a **quantitative variable**. Examples of quantitative variables are heights of people and weights of lobsters. Variables such as eye color are clearly categorical, but numerical data may also be treated as categorical. Movie reviewers often rate films using a 5-star scale. When we look up the average rating for a film we're thinking of seeing, we're treating ratings as a quantitative variable. When we talk about how many 4-star films we've seen, we're treating the ratings as a categorical variable.

It's hard to understand what the data may tell us about a given situation by just looking at the data themselves. We seek meaning by summarizing the data three ways: graphically, numerically, and verbally. Graphical summaries like pie charts and histograms provide a picture of the data, often revealing important insights or surprising anomalies. Numerical summaries called **statistics** describe properties of the data, such as counts, percentages, or averages. Our verbal summaries communicate to others the meaning we've drawn from the data.

Categorical Data

Government agencies keep detailed records on many aspects of public health, including deaths. A simplified data table might have entries like these:

Name	State	Sex	Age	Cause
Abbott, Aaron	WY	M	63	Cancer
Babbitt, Barbara	ME	F	27	Accident
Cocker, Kirk	AK	M	83	Stroke
⋮				

With over 2,000,000 deaths in the United States annually, it would be nearly impossible to reach any conclusions about the causes of death in America by looking at such a massive table. Instead, each year the Centers for Disease Control and Prevention summarizes the categorical variable cause of death with statistics: the *count* and *proportion* (percentage) of individuals in each category, seen in Table 10.1.

Table 10.1 Leading Causes of Death in the United States in 2010

Cause of death	Number of Deaths	Percentage
Heart disease	595,444	24.1
Cancer	573,855	23.3
Respiratory diseases	137,789	5.6
Stroke	129,180	5.2
Other	1,029,668	41.8

Source: National Vital Statistics Report, Vol. 60, No. 44, Centers for Disease Control and Prevention, January 2012.

The statistics (counts and percentages) communicate information about the categorical variable by telling us the **distribution** of the causes of death in the 2010 population.

We can get that information directly from the numbers, but it is very helpful to see the comparative sizes visually. This is why you will often see categorical data displayed graphically, as a **bar chart** (Figure 10.8a) or a **pie chart** (Figure 10.8b), sometimes called a **circle graph**. For variety, the popular press also makes use of **picture graphs** suited to the categories being displayed. For example, the bars in Figure 10.8a could be made to look like tombstones of different sizes to emphasize that these are causes of death. In each case, the graph provides a visualization of the relative sizes of the categories, with the pie chart providing the added visualization of the categories as parts of a whole population.

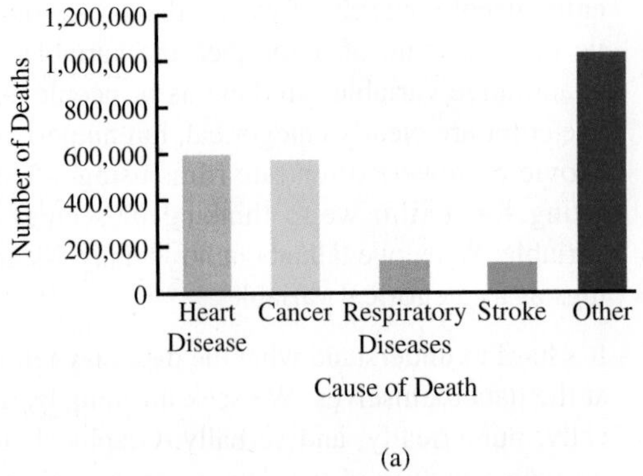

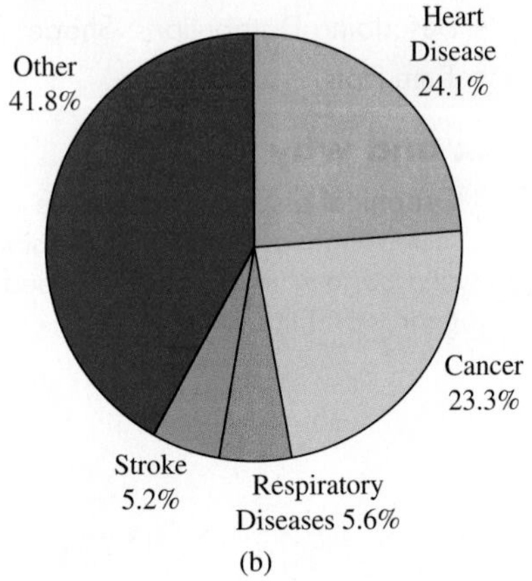

FIGURE 10.8 Causes of death in the United States in 2010 shown in (a) a bar graph and (b) a pie chart (circle graph).

In bar charts of categorical data, the y-axis has a numerical scale, but the x-axis is labeled by category. The rectangular bars are separated by spaces to show that no continuous numerical scale is implied. (In this respect, a bar graph differs from a *histogram*, to be described later in this section.) A circle graph or a pie chart consists of shaded or colored sections of a circle or "pie." The central angles for the sectors are found by multiplying the percentages by 360°. For example, the angle for the sector representing cancer victims in Figure 10.8b is

$$23.3\% \cdot 360^\circ = 83.9^\circ.$$

It used to require time, skill, and mathematical savvy to draw data displays that were both visually appealing and geometrically accurate, but modern spreadsheet programs have made it possible for anyone with a computer to produce high-quality graphs from tabular data with the click of a button.

EXAMPLE 1 Looking for an Association Between Categorical Variables

Table 10.2a summarizes data collected in a survey of 845 college students that asked them which sport was their favorite to attend live.

Table 10.2a Sports Students Prefer to Attend Live

Sport	Number of Males	Number of Females
Baseball	31	21
Basketball	54	96
Football	173	156
Hockey	76	87
Other	64	87
Total	398	447

Source: Sports Culture Among Undergraduates: A Study of Student Athletes and Students at the University of Michigan, MPublishing, University of Michigan, 2007.

Because the students surveyed included so many more females than males, it's hard to tell whether males and females have similar preferences for live sports. This becomes clearer when we express the distribution of preferences for each gender as percentages.

Table 10.2b Sports Students Prefer to Attend Live

Sport	Percent of Males	Percent of Females
Baseball	7.8	4.7
Basketball	13.5	21.5
Football	43.5	34.9
Hockey	19.1	19.5
Other	16.1	19.5
Total	100	100

A higher percentage of male students than female students prefer to attend football or baseball games, and female students were much more likely than males to prefer basketball. Because the distributions of sports preference are different for males and females, we say there is an **association** between sports preference and gender. Two categorical variables are **independent** if the distributions of one variable are the same (or very similar) for all the categories of the other variable.

Now try Exercise 3.

Quantitative Data: Stemplots

A quick way to organize and display a small set of quantitative data is with a **stemplot**, also called a **stem-and-leaf plot**. Each number in the data set is split into a **stem**, consisting of its initial digit or digits, and a **leaf**, which is its final digit.

EXAMPLE 2 Making a Stemplot

Table 10.3 gives the percentage of each state's population that was 65 or older in the 2010 Census. Make a stemplot for the data.

Table 10.3 Percentages of State Residents in 2010 Who Were 65 or Older

AL	13.8	HI	14.3	MA	13.8	NM	13.2	SD	14.3
AK	7.7	ID	12.4	MI	13.8	NY	13.5	TN	13.4
AZ	13.8	IL	12.5	MN	12.9	NC	12.9	TX	10.3
AR	14.4	IN	13.0	MS	12.8	ND	14.5	UT	9.0
CA	11.4	IO	14.9	MO	14.0	OH	14.1	VT	14.6
CO	10.9	KS	13.2	MT	14.8	OK	13.5	VA	12.2
CT	14.2	KY	13.3	NE	13.5	OR	13.9	WA	12.3
DE	14.4	LA	12.3	NV	12.0	PA	15.4	WV	16.0
FL	17.3	ME	15.9	NH	13.5	RI	14.4	WI	13.7
GA	10.7	MD	12.3	NJ	13.5	SC	13.1	WY	12.4

Source: U.S. Census Bureau, 2011.

SOLUTION To form the stem-and-leaf plot, we use the whole number part of each number as the stem and the tenths digit as the leaf. We write the stems in order down

(continued)

the first column and, for each number, write the leaf in the appropriate stem row. It is often helpful to arrange the leaves in each stem row in ascending order. The final plot looks like this:

Stem	Leaf
7	7
8	
9	0
10	3 7 9
11	4
12	0 2 3 3 3 4 4 5 8 9 9
13	0 2 2 3 4 5 5 5 5 5 7 7 8 8 8 8 9
14	0 1 2 3 3 4 4 5 6 8 9
15	4 4 9
16	0
17	3

Notice that we include the "leafless" stem (8) in our plot, as empty gaps are significant features of the distribution. For the same reason, be sure that each "leaf" takes up the same space along the stem. A branch with twice as many leaves should appear to be twice as long.

Now try Exercise 7.

EXPLORATION 1 Using Information from a Stemplot

By looking at both the stemplot and the table, answer the following questions about the distribution of senior citizens among the 50 states.

1. Judging from the stemplot, what was the approximate *average* national percentage of residents who were 65 or older?
2. In how many states were more than 15% of the residents 65 or older?
3. Which states were in the bottom tenth of all states in this statistic?
4. The numbers 7.7 and 17.3 are so far above or below the other numbers in this stemplot that statisticians would call them *outliers*. Quite often there is some special circumstance that sets outliers apart from the other individuals under study and explains the unusual data. What could explain the two outliers in this stemplot?

Sometimes the data are so tightly clustered that a stemplot has too few stems to give a meaningful visualization of the data distribution. In such cases we can spread the data out by splitting the stems, as in Example 3.

EXAMPLE 3 Making a Split-Stem Stemplot

The per capita federal aid to state and local governments for the top 20 states (in this category) in 2007 is shown in Table 10.4. Make a stemplot that provides a good representation of the data distribution. What is the average of the 20 numbers? Why is the stemplot a better summary of the data than the average?

Table 10.4 Federal Aid to State and Local Governments Per Capita (2007) in Dollars

WY	3626	NY	2228	ME	1936	HI	1543
AK	3568	NM	2213	MT	1918	OK	1529
LA	3036	ND	2008	MA	1735	PA	1528
MI	2821	RI	1999	SD	1608	KY	1504
VT	2491	WV	1973	AR	1581	AL	1485

Source: U.S Census Bureau.

SOLUTION

We truncate the data to \$100 units, which does not affect the visualization. Then, to spread out the data a bit, we *split* each stem, putting leaves 0–4 on the lower stem and leaves 5–9 on the upper stem.

Stem	Leaf
1	4
1	5 5 5 5 5 6 7 9 9 9 9
2	0 2 2 4
2	8
3	0
3	5 6

The average of the 20 numbers is \$2117, but this is misleading. The table shows that thirteen of the numbers are lower than that, while only seven are higher. It is better to observe that the distribution is clustered fairly tightly just under \$2000, with the numbers for Wyoming and Alaska being outliers on the high end.

Now try Exercise 11.

Sometimes it is easier to compare two sets of data if we have a visualization that allows us to view both stemplots simultaneously. **Back-to-back stemplots** use the same stems, but leaves from one set of data are added on the left, while leaves from another set are added on the right.

EXAMPLE 4 Making Back-to-Back Stemplots

Babe Ruth and Barry Bonds are two of the great sluggers in the history of baseball. For years "The Babe" held records for hitting 60 home runs in one season and 714 for his legendary career. During the 2001 season Bonds hit 73 home runs, and he retired in 2007 with a career total of 762. Who was the better home run hitter? Table 10.5 shows the number of homers each hit during his productive seasons.

Table 10.5 Season Totals for Home Runs

Babe Ruth								Barry Bonds							
0	1	3	3	11	29	54	59	16	25	24	19	33	25	34	46
35	41	46	25	47	60	54	46	37	33	42	40	37	34	49	73
49	46	41	34	22	6			46	45	45	5	26	28		

Source: www.baseball-reference.com

SOLUTION We form a back-to-back stemplot with Ruth's totals branching off to the left and Bonds's to the right.

Babe Ruth		Barry Bonds
6 3 3 1 0	0	5
1	1	6 9
9 5 2	2	4 5 5 6 8
5 4	3	3 3 4 4 7 7
9 7 6 6 6 1 1	4	0 2 5 5 6 6 9
9 4 4	5	
0	6	
	7	3

(continued)

The low home run years for each player can be explained by fewer times at bat. For the first 4 1/2 years of his career, Babe Ruth was a pitcher and he didn't play much during his final season. Barry Bonds missed most of the 2005 season with an injury. If those years are ignored as anomalies, we see that Bonds's home run totals were typically lower than Ruth's. Bonds often finished a season with fewer than 40 home runs, but Ruth rarely did. Ruth hit over 40 with great consistency, and surpassed 50 home runs four times. Bonds achieved that feat just once. His record-setting 73 in 2001 was (and still is) an outlier of such magnitude that it actually inspired more skepticism than admiration among baseball fans. **Now try Exercise 9.**

Frequency Tables

The visual impact of a stemplot comes from the lengths of the various rows of leaves, which is just a way of seeing *how many* leaves branch off each stem. The number of leaves for a particular stem is the **frequency** of observations within each stem interval. Frequencies are often recorded in a **frequency table**. Table 10.6 shows a frequency table for Babe Ruth's yearly home run totals (see Example 4). The table shows the **frequency distribution**—literally the way that the total frequency of 22 is "distributed" among the various home run intervals. This same information is conveyed visually in a stemplot, but notice that the stemplot has the added numerical advantage of displaying what the numbers in each interval actually are.

Table 10.6 Frequency Table for Babe Ruth's Yearly Home Run Totals

Home Runs	Frequency	Home Runs	Frequency
0–9	5	40–49	7
10–19	1	50–59	3
20–29	3	60–69	1
30–39	2		
		Total	22

Higher frequencies in a table correspond to longer leaf rows in a stemplot. Unlike a stemplot, a frequency table does not display what the numbers in each interval actually are.

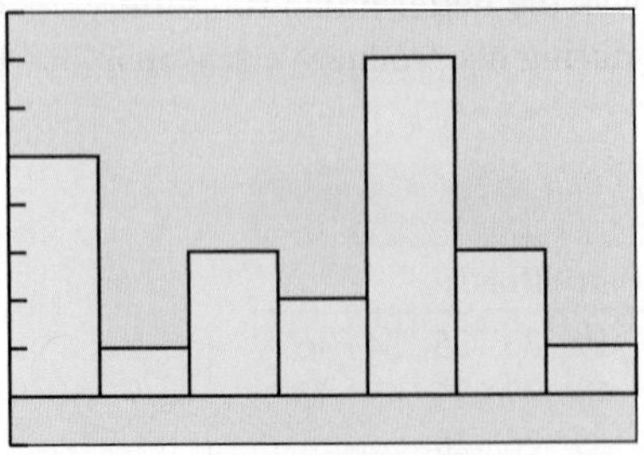

[0, 70] by [–1, 8]
(a)

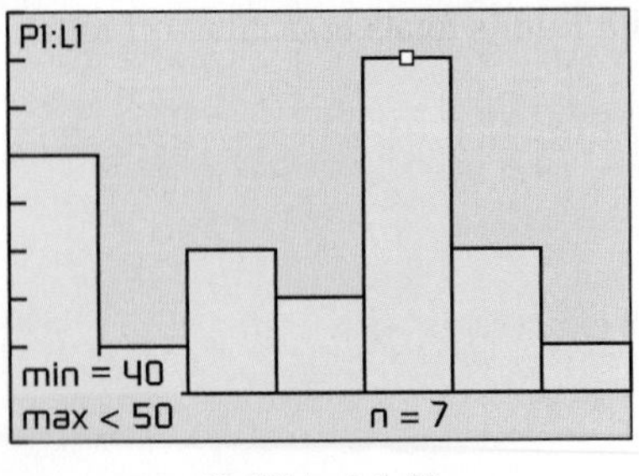

[0, 70] by [–1, 8]
(b)

FIGURE 10.9 Calculator histograms for Babe Ruth's yearly home run totals.

Histograms

A **histogram**, closely related to a stemplot, displays the information of a frequency table. Visually, a histogram is to quantitative data what a bar chart is to categorical data. Unlike a bar chart, however, both axes of a histogram have numerical scales, the rectangular bars cannot be rearranged, and bars on adjacent intervals have no intentional gaps between them.

EXAMPLE 5 Graphing a Histogram on a Calculator

Make a histogram of Babe Ruth's annual home run totals given in Table 10.5, using intervals of width 10.

To scale the x-axis to be consistent with the intervals of the table, let Xmin $= 0$, Xmax $= 70$, and Xscl $= 10$. Notice that the maximum frequency is 7 years (40–49 home runs), so the y-axis ought to extend at least to 8. Enter the data from Table 10.5 into list L1 and plot a histogram in the window $[0, 70]$ by $[-1, 8]$. (See Figure 10.9a.) Tracing along the histogram should reveal the same frequencies as in the frequency table we made. (Compare Figure 10.9b and Table 10.6.) **Now try Exercise 19.**

Describing Distributions: Shape

An important step in drawing meaning from data comes from looking at the shape of the distribution. We describe two aspects of shape: modes and symmetry. A **mode** is an interval where data tend to cluster and to show up as longer leaf rows in a stemplot or higher bars in a histogram. Two examples are shown in Figure 10.10.

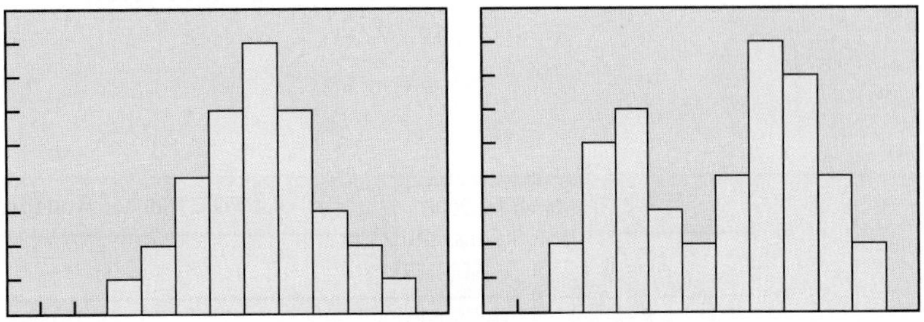

(a) (b)

FIGURE 10.10 Distribution (a) is unimodal and symmetric; distribution (b) is bimodal.

A **unimodal** distribution has one cluster of data. An example is heights of adult males, most within a few inches of the average, with fewer individuals farther from average. A **bimodal** distribution usually indicates individuals from two different groups. An example is the ages of people attending a youth soccer game, where there are players of a certain age with a few younger and older siblings plus many parents there as spectators. For bimodal data like those, subgroups should usually be analyzed separately.

The distribution in Figure 10.10a is **symmetric** because it looks approximately the same when reflected across a vertical line up the middle. The distribution in Figure 10.11a is unimodal and **skewed right** because it has a longer "tail" to the *right*. An example is family incomes, with the great majority of families earning relatively modest amounts and ever smaller numbers of families with increasingly large (very large!) incomes. The distribution in Figure 10.11b is unimodal and **skewed left**, because it has a longer tail extending off to the left. An example could be test scores, where there is an upper limit of 100% and most students do fairly well, but there are also a few low (and even lower) scores.

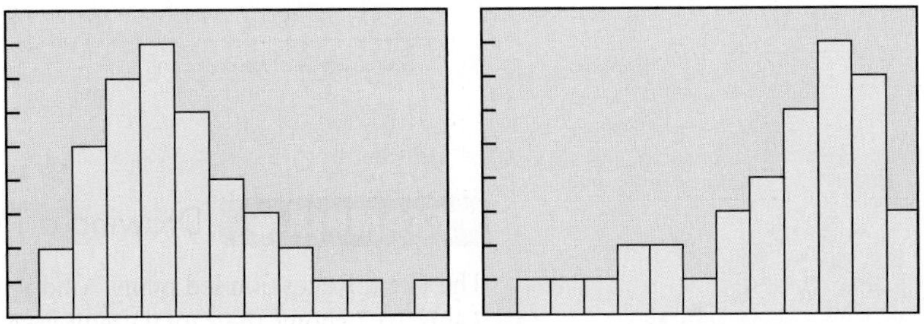

(a) (b)

FIGURE 10.11 Distribution (a) is skewed right; distribution (b) is skewed left.

Time Plots

We have seen in this book many examples of functions in which the input variable is time. It is also quite common to consider quantitative data as a function of time. If we make a scatter plot of the data (y) against the time (x) when they were measured, we can analyze the patterns as the variable changes over time. To help with the visualization, the discrete points from left to right are connected by line segments, just as a grapher would do in connect mode. The resulting **line graph** is called a **time plot**.

Time plots reveal trends in data over time. These plots frequently appear in magazines and newspapers and on the Internet, a typical example being the graph of the Dow Jones Industrial Average (DJIA) during America's Great Recession, in Figure 10.12.

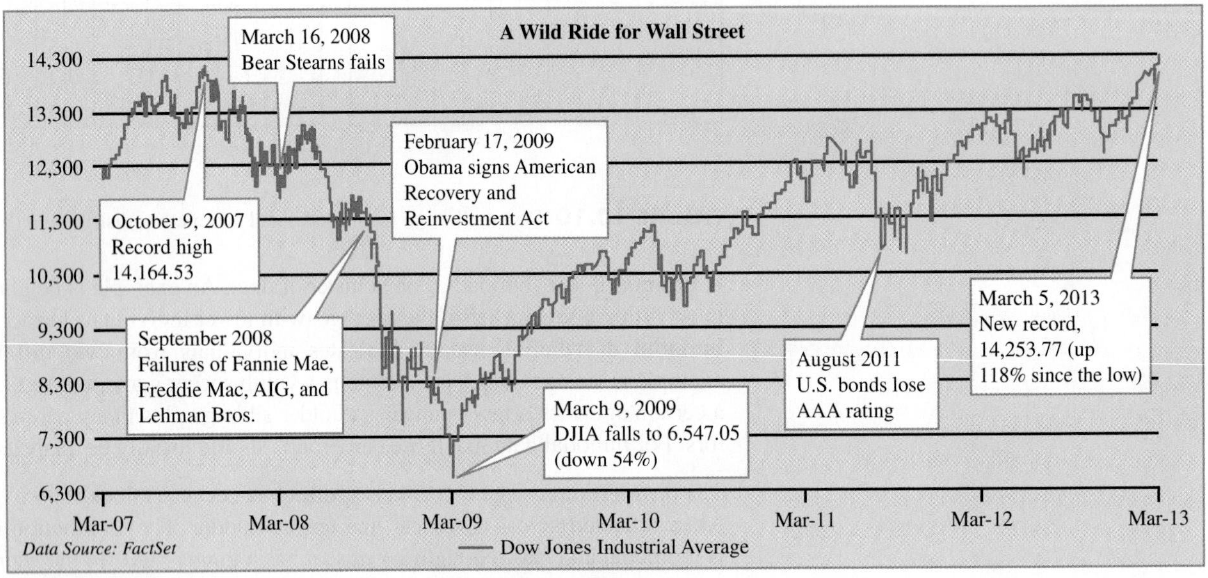

FIGURE 10.12 Time plot of the Dow Jones Industrial Average during the Great Recession. Investors get a good visualization of where the stock market has been, although the trick is to figure out where it is going.

Table 10.7 Total U.S. Credit Card Debt (billions of dollars), 2004–2012

Year	2004	2005	2006	2007	2008	2009	2010	2011	2012
$	707	732	761	839	864	786	725	698	661

Source: Federal Reserve Bank

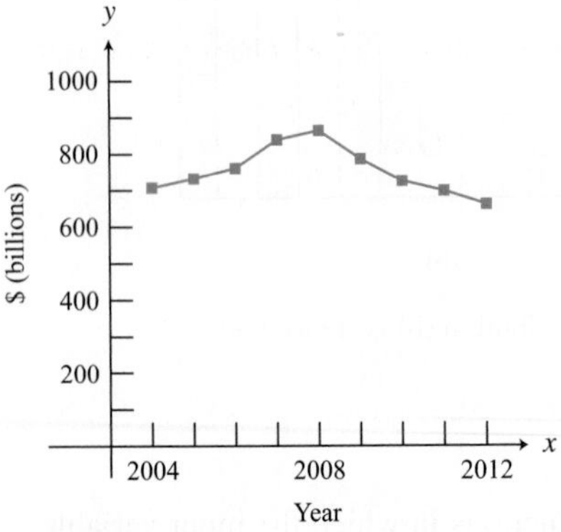

FIGURE 10.13 A time plot of U.S. credit card debt from 2004 through 2012. (Example 6)

EXAMPLE 6 Drawing a Time Plot

The Great Recession led many Americans to make changes in their spending habits. Table 10.7 shows their total credit card debt leading up to and following the recession. Display the data in a time plot and analyze the 9-year trend.

SOLUTION The horizontal axis represents time (in years) from 2004 through 2012. The vertical axis represents total credit card debt in billions of dollars. When working with data like these, it is often best to enter the years as $\{1, 2, 3, \ldots, 10\}$ to make any calculations more manageable. Proper labeling of the x-axis should display the actual years, as shown in Figure 10.13.

The time plot shows that credit card debt rose until the recession began in 2008, and has declined since, falling below its 2004 level by 2012. **Now try Exercise 25.**

Table 10.8 Total Student Loan Debt (billions of dollars), 2004–2012

Year	2004	2005	2006	2007	2008	2009	2010	2011	2012
$	346	392	477	538	631	715	804	865	966

Source: Federal Reserve Bank.

EXAMPLE 7 Comparing Two Time Plots

Table 10.8 shows the total debt for student loans for each year from 2004 to 2012. Compare this debt to the trend in credit card debt by graphing both time plots on the same axes.

SOLUTION Figure 10.14 shows the two time plots. Total student loan debt continues to rise. Much lower than credit card debt in 2003, student loan debt surpassed credit card debt early in 2010, and by the end of 2012 had grown to be nearly 50% larger.

Now try Exercise 27.

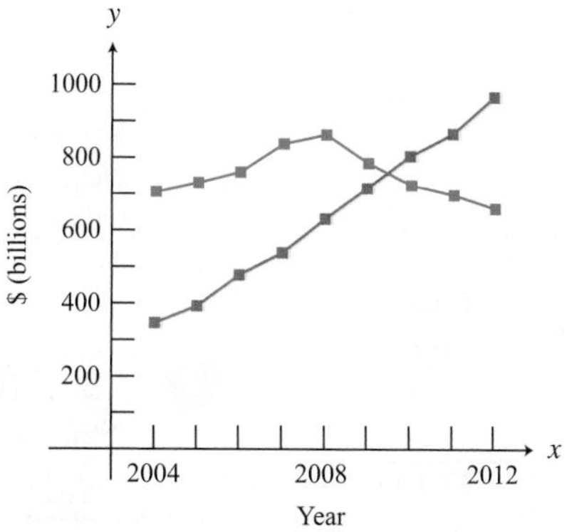

FIGURE 10.14 A time plot comparing student loan debt to credit card debt from 2004 through 2012. (Example 7)

QUICK REVIEW 10.2

In Exercises 1–6, solve for the required value.

1. 457 is what percent of 2953?
2. 827 is what percent of 3950?
3. 52° is what percent of 360°?
4. 98° is what percent of 360°?
5. 734 is 42.6% of what number?
6. 5106 is 55.5% of what number?

In Exercises 7–10, round the given value to the nearest whole number in the specified units.

7. $234,598.43 (thousands of dollars)
8. 237,834,289 (millions)
9. 848.36 thousands (millions)
10. 1432 millions (billions)

SECTION 10.2 Exercises

Table 10.9 Super Bowl Preferences

Primary Interest	Men	Women	Total
The game	279	200	479
The commercials	81	156	237
Won't watch	132	160	292
Total	492	516	1008

Source: Americans and the Super Bowl Phenomenon, Gallup News Service, 2007.

1. A January 2007 Gallup Poll asked people whether they planned to watch the Super Bowl and, if so, whether they were more interested in seeing the game or the commercials. Table 10.9 summarizes the responses by gender.

(a) What percent of the people planned to skip the game?

(b) What percent of the people were men who planned to skip the game?

(c) What percent of the men planned to skip the game?

(d) What percent of the people who planned to skip the game were men?

Table 10.10 Fate of *Titanic* Passengers

Ticket Class	Survived	Died	Total
First class	202	123	325
Second class	118	167	285
Third class	178	528	706
Total	498	818	1316

Source: Titanic Passenger Survival Rates, http://www.dummies.com.

2. On her 1912 maiden voyage across the Atlantic the ocean liner *Titanic* famously struck an iceberg and sank, with great loss of life. Table 10.10 summarizes the fates of the passengers on board by the class of ticket they had purchased.

(a) What percent of the passengers died?

(b) What percent of the passengers were first-class ticket holders who died?

(c) What percent of the passengers who died had first-class tickets?

(d) What percent of the first-class passengers died?

3. Use Table 10.9 to consider whether there is an association between gender and plans for watching the Super Bowl.

(a) Would it be better to compare men's and women's plans with pie charts or a stemplot? Explain.

(b) Are Super Bowl plans and gender independent? Explain.

4. Use Table 10.10 to consider whether there is an association between the type of passenger and their fates on the *Titanic*.

(a) Would it be better to compare fates by passenger class using a bar graph or a histogram? Explain.

(b) Were people's fates and their passenger class independent? Explain.

5. An engineer designing a new electronic device needs to be sure that both left- and right-handed people can operate it successfully. She'll have a test group try it out, and she plans to summarize the results in a table like the one shown here. If ability to operate the device were independent of handedness, what would the counts in the table be?

	Successful	Failed	Total
Lefty	**(a)**	**(b)**	48
Righty	**(c)**	**(d)**	192
Total	210	30	240

6. A pollster plans to survey a group of voters, asking them whether they favor or oppose the death penalty, and will summarize the responses in a table like the one shown here. If opinion turned out to be independent of political party, what would the counts in the table be?

	Favor	Oppose	Total
Republican	**(a)**	**(b)**	220
Democrat	**(c)**	**(d)**	240
Other	**(e)**	**(f)**	140
Total	210	390	600

Table 10.11 Regular Season Home Run Statistics for Roger Maris

Year	Home Runs	Year	Home Runs
1957	14	1963	23
1958	28	1964	26
1959	16	1965	8
1960	39	1966	13
1961	61	1967	9
1962	33	1968	5

Source: The Baseball Encyclopedia, 7th ed., 1988.

7. Make a stemplot of the data in Table 10.11. Are there any outliers?

Table 10.12 Regular Season Home Run Statistics for Mark McGwire

Year	Home Runs	Year	Home Runs
1986	3	1994	9
1987	49	1995	39
1988	32	1996	52
1989	33	1997	58
1990	39	1998	70
1991	22	1999	65
1992	42	2000	32
1993	9	2001	29

Source: www.baseball-reference.com

8. Make a stemplot of the data in Table 10.12. Are there any outliers?

9. Make a back-to-back stemplot comparing the annual home run production of Roger Maris (Table 10.11) with that of Hank Aaron (Table 10.14 in Exercise 20). Write a brief interpretation comparing the distributions.

10. Mark McGwire and Barry Bonds both played during baseball's steroid era. Make a back-to-back stemplot comparing McGwire's annual home run production with that of Bond's (Table 10.5 on page 695). Write a brief interpretation comparing the distributions.

In Exercises 11–14, construct the indicated stemplot from the data in Table 10.13. Then write a brief interpretation of the stemplot.

Table 10.13 Life Expectancy by Gender for 12 Nations of South America

Nation	Male	Female
Argentina	72.5	79.1
Bolivia	65.1	69.2
Brazil	71.0	77.8
Chile	76.2	82.3
Colombia	74.5	80.5
Ecuador	73.2	78.8
Guyana	60.0	67.0
Paraguay	71.7	77.6
Peru	75.1	78.3
Suriname	69.4	75.5
Uruguay	73.0	80.0
Venezuela	72.0	79.3

Source: World Health Organization, 2011.

11. A stemplot showing life expectancies of males in 12 nations of South America (Round to nearest year and use split stems.)

12. A stemplot showing life expectancies of females in the nations of South America (Round to nearest year and use split stems.)

13. A back-to-back stemplot for life expectancies of males and females in the nations of South America (Round to nearest year and use split stems.)

14. A stemplot showing the difference between female and male life expectancies in the nations of South America (Use unrounded data and do not split stems.)

In Exercises 15 and 16, use the data in Table 10.13 to construct the indicated frequency table, using intervals 60.0–64.9, 65.0–69.9, etc.

15. Life expectancies of males in the nations of South America

16. Life expectancies of females in the nations of South America

In Exercises 17–20, draw a histogram for the given table.

17. The frequency table in Exercise 15

18. The frequency table in Exercise 16

19. Table 10.14 of Willie Mays's annual home run totals, using intervals 1–5, 6–10, 11–15, etc.

20. Table 10.14 of Mickey Mantle's annual home run totals, using intervals 0–4, 5–9, 10–14, etc.

Table 10.14 Regular Season Home Run Statistics for Willie Mays, Mickey Mantle, and Hank Aaron

Year	Mays	Mantle	Aaron	Year	Mays	Mantle	Aaron
1951	20	13		1964	52	35	24
1952	4	23		1965	37	19	32
1953	41	21		1966	22	23	44
1954	51	27	13	1967	23	22	39
1955	36	37	27	1968	13	18	29
1956	35	52	26	1969	28		44
1957	29	34	44	1970	18		38
1958	34	42	30	1971	8		47
1959	29	31	39	1972	6		34
1960	40	40	40	1973			40
1961	49	54	34	1974			20
1962	38	30	45	1975			12
1963	47	15	44	1976			10

Source: The Baseball Encyclopedia, 7th ed., 1988.

In Exercises 21–24, make a time plot for the indicated data.

21. Willie Mays's annual home run totals given in Table 10.14

22. Mickey Mantle's annual home run totals given in Table 10.14

23. Mark McGwire's home run totals given in Table 10.12 in Exercise 8.

24. Hank Aaron's home run totals given in Table 10.14.

Table 10.15 shows the first place prize money (in units of $1000) won by the winners of the women's (LPGA) and men's (PGA) professional golf championships for selected years between 1970 and 2012.

Table 10.15 First Place Prize Money (in thousands of dollars) for the PGA and LPGA Championships

Year	Men (PGA)	Women (LPGA)
1970	40	4.5
1975	45	8
1980	60	22.5
1985	125	37.5
1990	225	150
1995	360	180
2000	900	210
2002	990	225
2004	1125	240
2006	1224	270
2008	1350	300
2010	1350	337.5
2012	1445	375

Source: PGA/LPGA Media Guide, 2012.

25. Make a time plot for the men's winnings in Table 10.15. Write a brief interpretation of the time plot.

26. Make a time plot for the women's winnings in Table 10.15. Write a brief interpretation of the time plot.

27. Compare the trends in Table 10.15 by overlaying the time plots. Write a brief interpretation.

28. Writing to Learn (Continuation of Exercise 27) The data in Table 10.15 show that the earnings for the PGA winner rose by a modest 67% in the decade from 1970 to 1980, while the earnings for the top LPGA player rose by a whopping 400%. Although this was, in fact, a strong growth period for women's sports, statisticians would be unlikely to draw any conclusions from a comparison of these two numbers. Use the visualization from the comparative time plot in Exercise 27 to explain why.

In Exercises 29 and 30, compare performances by overlaying time plots.

29. The time plots from Exercises 21 and 22 to compare the performances of Mays and Mantle

30. The time plots from Exercises 23 and 24 to compare the performances of McGwire and Aaron

31. The histogram shows the sugar content (as a percent of weight) of 49 brands of breakfast cereals.

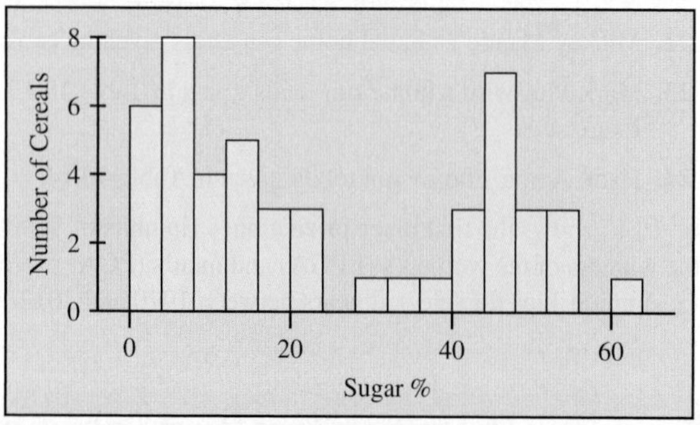

(a) Describe the shape of the distribution.

(b) What do you think might explain why the distribution has this shape?

32. The histogram shows how many points the Super Bowl champions have won the games by, starting with the very first contest in 1967 through Super Bowl XLVII in 2013.

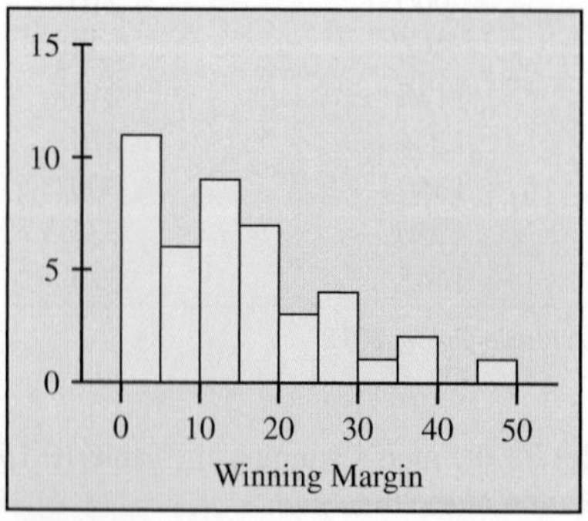

(a) Describe the shape of the distribution.

(b) What does the shape of the distribution say about these football games?

In Exercises 33–36, analyze the data as indicated.

33. The salaries of the workers in one department of the Garcia Brothers Company (given in thousands of dollars) are as follows:

33.5, 35.3, 33.8, 29.3, 36.7, 32.8, 31.7, 36.3, 33.5, 28.2, 34.8, 33.5, 35.3, 29.7, 38.5, 32.7, 34.8, 34.2, 31.6, 35.4

(a) Explain why a bar graph will not work for these data.

(b) Complete a stemplot for this data set.

(c) Describe the shape of the distribution.

34. The average wind speeds for one year at 44 climatic data centers around the United States are as follows:

9.0, 6.9, 9.1, 9.2, 10.2, 12.5, 12.0, 11.2, 12.9, 10.3, 10.6, 10.9, 8.7, 10.3, 11.0, 7.7, 11.4, 7.9, 9.6, 8.0, 10.7, 9.3, 7.9, 6.2, 8.3, 8.9, 9.3, 11.6, 10.6, 9.0, 8.2, 9.4, 10.6, 9.5, 6.3, 9.1, 7.9, 9.7, 8.8, 6.9, 8.7, 9.0, 8.9, 9.3

(a) Explain why a circle graph will not work for these data.

(b) Complete a stemplot for this data set.

(c) Describe the shape of the distribution.

35. Use the Garcia Brothers salary data in Exercise 33.

(a) Create a frequency table for the data.

(b) Draw a histogram for the data.

(c) Describe the shape of the histogram.

36. Use the wind speed data in Exercise 34.

(a) Create a frequency table for the data.

(b) Draw a histogram for the data.

(c) Describe the shape of the histogram.

In Exercises 37 and 38, compare by overlaying time plots for the data in Table 10.16

Table 10.16 Population (in millions) of the Six Most Populous States

Census	CA	FL	IL	NY	PA	TX
1900	1.5	0.5	4.8	7.3	6.3	3.0
1910	2.4	0.8	5.6	9.1	7.7	3.9
1920	3.4	1.0	6.5	10.4	8.7	4.7
1930	5.7	1.5	7.6	12.6	9.6	5.8
1940	6.9	1.9	7.9	13.5	9.9	6.4
1950	10.6	2.8	8.7	14.8	10.5	7.7
1960	15.7	5.0	10.1	16.8	11.3	9.6
1970	20.0	6.8	11.1	18.2	11.8	11.2
1980	23.7	9.7	11.4	17.6	11.9	14.2
1990	29.8	12.9	11.4	18.0	11.9	17.0
2000	33.9	16.0	12.4	19.0	12.3	20.9
2010	37.3	18.9	12.9	19.4	12.7	25.3

Source: U.S. Census Bureau.

37. The populations of California, New York, and Texas from 1990 through 2010.

38. The populations of Florida, Illinois, and Pennsylvania from 1990 through 2010

Standardized Test Questions

39. **True or False** If the percentage of ninth-grade boys who eventually graduate from high school is the same as the percentage of ninth-grade girls who do, then there's an association between graduation and gender. Justify your answer.

40. **True or False** The highest and lowest numbers in a set of data are called outliers. Justify your answer.

Answer Exercises 41–44 without using a calculator.

41. **Multiple Choice** A time plot is an example of a
 (A) Histogram.
 (B) Bar graph.
 (C) Line graph.
 (D) Pie chart.
 (E) Table.

42. **Multiple Choice** A back-to-back stemplot is particularly useful for
 (A) Identifying outliers.
 (B) Comparing two data distributions.
 (C) Merging two sets of data.
 (D) Graphing home runs.
 (E) Distinguishing stems from leaves.

43. **Multiple Choice** The histogram below would most likely result from which set of data?

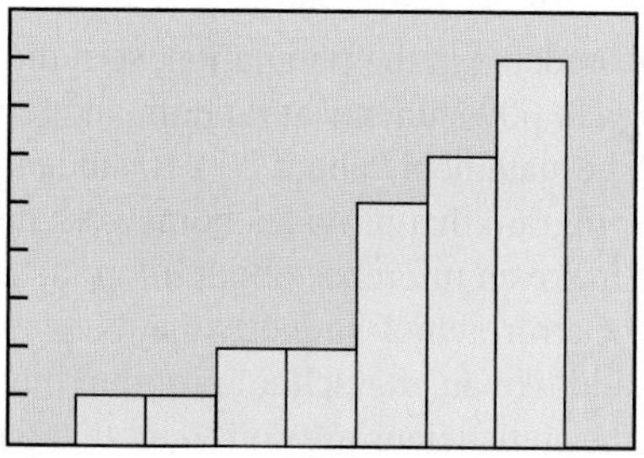

 (A) Scores in an amateur golf tournament
 (B) Weights of children in a third-grade class
 (C) Winning soccer scores for a team over the course of a season
 (D) Ages of all the people visiting the Bronx Zoo at a given point in time
 (E) Prices of all the desserts on the menu at a certain restaurant

44. **Multiple Choice** A sector on a pie chart with a central angle of 45° corresponds to what percentage of the data?
 (A) 8%
 (B) 12.5%
 (C) 15%
 (D) 25%
 (E) 45%

Explorations

45. **Group Activity** Measure the resting pulse rates (beats per min) of the members of your class. Make a stemplot for the data. Are there any outliers? Can they be explained?

46. **Group Activity** Measure the heights (in inches) of the members of your class. Make a back-to-back stemplot to compare the distributions of the male heights and the female heights. Write a brief interpretation of the stemplot.

Extending the Ideas

47. **Time Plot of Periodic Data** Some data are periodic functions of time. If data vary in an annual cycle, the period is 1 year. Use the information in Table 10.17 to overlay the time plots for the average daily high and low temperatures for Beijing, China.

Table 10.17 Average Daily High and Low Temperatures in °C for Beijing, China

Month	High	Low
January	2	−9
February	5	−7
March	12	−1
April	20	7
May	27	13
June	31	18
July	32	22
August	31	21
September	27	14
October	21	7
November	10	−1
December	3	−7

Source: National Geographic Atlas of the World (rev. 6th ed., 1992, Washington, D.C.), plate 132.

48. Find a sinusoidal function that models each time plot in Exercise 47. (See Sections 4.4 and 4.8.)

10.3 Statistics (Numerical)

What you'll learn about

- Parameters and Statistics
- Describing and Comparing Distributions
- Five-Number Summary
- Boxplots
- The Mean (and When to Use It)
- Variance and Standard Deviation
- Normal Distributions

... and why

The language of Statistics is becoming more commonplace in our everyday world.

Statistics and statistics

In this chapter we will capitalize "Statistics" when we refer to the branch of mathematics that collects and analyzes data. When we refer to numerical summaries of those data, we do not capitalize "statistics."

Parameters and Statistics

Numerical summaries of a data set are called **statistics**. They serve to describe the individuals from which the data come, so the gathering and processing of such numerical information is often called **descriptive statistics**. You saw many examples of descriptive statistics in Section 10.2.

The *science* of Statistics comes in when we use descriptive statistics (like the results of a study of 1500 smokers) to make judgments, called *inferences*, about entire *populations* (like all smokers). Statisticians are really interested in numbers called **parameters** that describe entire populations. Since it is usually either impractical or impossible to measure entire populations, statisticians gather statistics from carefully chosen **samples**, then use the science of **inferential statistics** to make inferences about the parameters.

EXAMPLE 1 Distinguishing a Parameter from a Statistic

A 2004 study called *All Work and No Play? Listening to What Kids and Parents Really Want from Out-of-School Time* reported that 32% of middle school and high school students wish there was an afterschool activity that offered homework help. The report was based on a survey of 609 randomly selected students in grades 6–12 and had a margin of error of $\pm 4\%$. (*Source*: *Public Agenda*.) Did the survey measure a parameter or a statistic, and what does that "margin of error" mean?

SOLUTION The survey did not contact all students in the population, so it did not measure a parameter. The researchers *sampled* 609 students at random and reported a statistic based on the data they collected. The statement "about 3 in 10 students say they would very much like an afterschool program that provides homework help (32%)" refers to the sample responses. To make an inference about *all* grade 6–12 students, we must also look at the margin of error, which suggests that between 28% and 36% of all students in grades 6–12 would like an afterschool program that provides homework help. In other words, the statisticians are confident that their sample statistic is within $\pm 4\%$ of the population parameter—even though they sampled only 609 teens, a tiny fraction of the population! That confidence is based on the laws of probability and is scientifically reliable, but we will not go into the mathematics here.

Now try Exercise 1.

Describing and Comparing Distributions

If you wanted to study the effect of chicken feed additives on the thickness of egg shells, you would need to sample many eggs from different hens under various feeding conditions. Suppose you were to gather data from 50 eggs from hens eating feed A and 50 eggs from hens eating feed B. How would you compare the two? To effectively compare sets of data, we must think about 3 aspects of the distributions: shape, center, and spread.

- The **shape** of a distribution may be unimodal or multimodal, symmetric or skewed, as we discussed in Section 10.2.
- The **center** of a distribution indicates what a typical measurement is.
- A **spread** of a distribution tells us whether measurements tend to be close together or vary widely for different individuals.

We can use several statistics to describe center and spread. We start by looking at the *median* and *interquartile range*.

DEFINITION Median

The **median** of a list of n numbers $\{x_1, x_2, \ldots, x_n\}$ arranged in order (either ascending or descending) is

- the middle number if n is odd, and
- the average of the two middle numbers if n is even.

Units = Meaning

Remember that the aim of Statistics is to find meaning in data. Including units in your answers is an important step toward that goal.

EXAMPLE 2 Finding a Median

Find the median of the annual home run totals for Roger Maris's major league career, 1957–1968 (Table 10.11 on page 700).

SOLUTION According to Table 10.11, we are looking for the median of the following list of 12 numbers: $\{14, 28, 16, 39, 61, 33, 23, 26, 8, 13, 9, 5\}$. First, we arrange the list in ascending order: $\{5, 8, 9, 13, 14, 16, 23, 26, 28, 33, 39, 61\}$. Since there are 12 numbers, the median is the average of the 6th and 7th numbers:

$$\frac{16 + 23}{2} = 19.5 \text{ home runs}$$

Now try Exercise 5.

Notice that the median is not affected by the outlier representing Maris's record-breaking season. The median would still have been 19.5 if he had hit only 41 home runs that season, or, indeed, if he had hit 81. We call a statistic **resistant** if it is not strongly affected by outliers. (See Exploration 1, Section 10.2.) While the median is a resistant measure of center, we'll soon see that the mean is not.

Five-Number Summary

Measures of center tell only part of the story of a data set. They do *not* indicate how widely distributed or highly variable the data are. *Measures of spread* do. The simplest and crudest measure of spread is the **range**, which is the difference between the maximum and minimum values in the data set:

$$\text{Range} = \text{maximum} - \text{minimum}$$

For example, the range of numbers in Roger Maris's annual home run production is $61 - 5 = 56$. Like the mean, the range is a statistic that is strongly influenced by outliers, so it can be misleading. A more resistant (and therefore more useful) measure is the *interquartile range*, which is the range of the middle half of the data.

Finding Quartiles

When we are finding the quartiles for a data set with an odd number of values, we do *not* consider the middle value to be included in either the lower or the upper half of the data.

Just as the median separates the data into halves, the **quartiles** separate the data into fourths. The **first quartile** Q_1 is the median of the lower half of the data, the **second quartile** is the median, and the **third quartile** Q_3 is the median of the upper half of the data. The **interquartile range** (IQR) measures the spread between the first and third quartiles, comprising the middle half of the data:

$$IQR = Q_3 - Q_1$$

Taken together, the maximum, the minimum, and the three quartiles give a fairly complete picture of both the center and the spread of a data set.

DEFINITION Five-Number Summary

The **five-number summary** of a data set is the collection

$$\{\text{minimum}, Q_1, \text{median}, Q_3, \text{maximum}\}.$$

Computing Statistics on a Calculator

Modern calculators will usually process lists of data and give statistics like the mean, median, and quartiles with a push of a button. Consult your owner's manual.

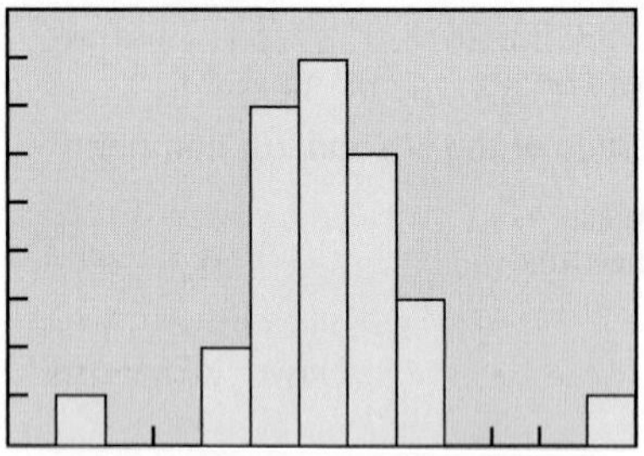

(a)

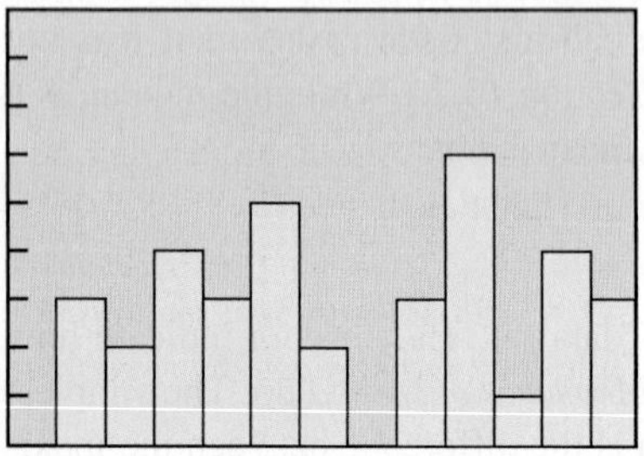

(b)

FIGURE 10.15 Which data set has more variability? (Exploration 1)

EXAMPLE 3 Five-Number Summary and Spread

Find the five-number summaries for the male and female life expectancies in South American nations (Table 10.13 on page 701) and compare the spreads.

SOLUTION Here are the lists in ascending order.

Males:

$\{60.0, 65.1, 69.4, \quad 71.0, 71.7, 72.0, \quad 72.5, 73.0, 73.2, \quad 74.5, 75.1, 76.2\}$

Females:

$\{67.0, 69.2, 75.5, \quad 77.6, 77.8, 78.3, \quad 78.8, 79.1, 79.3, \quad 80.0, 80.5, 82.3\}$

We have spaced the lists to show where the quartiles appear. The median of the 12 values is midway between the 6th and 7th values. The first quartile is the median of the lower 6 values (i.e., midway between the 3rd and 4th), and the third quartile is the median of the upper 6 values (i.e., midway between the 9th and 10th).

The five-number summaries are shown below.

Males: $\{60.0, 70.2, 72.25, 73.85, 76.2\}$

Females: $\{67.0, 76.55, 78.55, 79.65, 82.3\}$

The males have a range of $76.2 - 60.0 = 16.2$ years and an *IQR* of $73.85 - 70.2 = 3.65$ years.

The females have a range of $82.3 - 67.0 = 15.3$ years and an *IQR* of $79.65 - 76.55 = 3.10$ years.

Not only do the women live longer in general, but there is less variability in their life expectancies (as measured by the *IQR*). **Now try Exercise 19.**

EXPLORATION 1 Interpreting Histograms (Part I)

Look at the two histograms shown in Figure 10.15.

1. Which distribution has the greater range?
2. Which distribution has the greater *IQR*?
3. Which histogram displays a data set with more variability?

Boxplots

A **boxplot** (sometimes called a **box-and-whisker plot**) is a graph that depicts the five-number summary of a data set. The plot consists of a central rectangle (box) that extends from the first quartile to the third quartile, with a vertical segment marking the median. Line segments (whiskers) extend at the ends of the box toward the minimum and maximum values, stopping short of any outliers, which are indicated separately.

In fact, a boxplot gives us a convenient way to think of an outlier: a number that makes one of the whiskers noticeably longer than the box. The usual rule of thumb for "noticeably longer" is 1.5 times as long. Since the length of the box is the *IQR*, that leads us to the following numerical check.

Defining Outliers

It must be pointed out that the rule of thumb given here for identifying outliers is not a universal definition. The only sure way to characterize an outlier is that it lies outside the expected range of the data, and that "expected range" can be a judgment call.

Outlier Guideline

A number in a data set may be considered an outlier if it is more than $1.5 \times IQR$ below the first quartile or above the third quartile.

EXAMPLE 4 Identifying an Outlier

When we look at Roger Maris's annual home run totals $\{14, 28, 16, 39, 61, 33, 23, 26, 8, 13, 9, 5\}$, we see that 61 is much larger than any of the others. Is 61 an outlier according to the $1.5 \times IQR$ criterion?

SOLUTION Maris's totals, in order: $\{5, 8, 9, 13, 14, 16, 23, 26, 28, 33, 39, 61\}$

His five-number summary: $\{5, 11, 19.5, 30.5, 61\}$

His IQR: $30.5 - 11 = 19.5$

So,

$$Q_3 + 1.5 \times IQR = 30.5 + 1.5 \times 19.5 = 59.75.$$

Since $61 > 59.75$, the rule of thumb identifies 61 home runs as an outlier.

Now try Exercise 11.

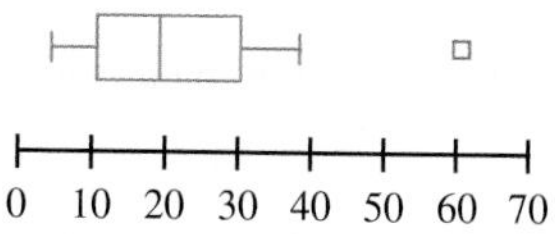

FIGURE 10.16 A boxplot of Roger Maris's annual home run totals.

Figure 10.16 displays a boxplot of Maris's annual home run totals. We see the median indicated by the vertical segment at 19.5. The box representing the middle half of his season totals stands between Q_1 (11) and Q_3 (30.5). The left whisker extends to his minimum, 5 home runs. The right whisker ends at 39 home runs, the highest season total that was not an outlier, and a separate point shows the record-setting 61. Because the whisker and the portion of box to the right of the median are a bit longer than they are to the left, we see that the distribution is slightly skewed to the right.

For a simple plot, then, a boxplot displays a lot of information about a distribution. It offers a hint about the shape, marks the center, shows the spread, and identifies any outliers. It's important to recognize outliers, because they can distort both our overall impression of the data and the values of nonresistant statistics we might calculate. (We'll discuss more about that soon.) For this reason, statisticians need to spot outliers and decide how to handle them. Simply omitting them from your statistical displays and calculations can be risky. While you want to omit a strange laboratory reading that arose through equipment error, you do not want to overlook a discovery that may lead to a Nobel Prize!

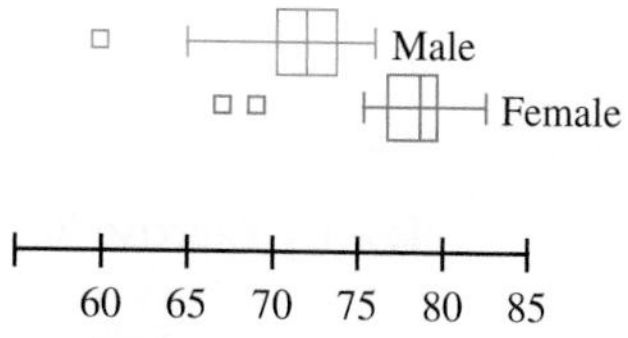

FIGURE 10.17 Parallel boxplots comparing male and female life expectancies in the nations of South America provide a good visualization of the differences in the distributions.

EXAMPLE 5 Comparing Boxplots

Draw the boxplots for the male and female life expectancy data in Example 3 and compare the distributions.

SOLUTION The boxplots are graphed simultaneously in Figure 10.17.

From this graph we see that in general life expectancy for females is much higher than for males. In fact, it exceeds the maximum male life expectancy in at least 75% of the countries. Both distributions are skewed to the left; each has low outliers. Apart from those outliers, life expectancies for females are less variable than for males.

Now try Exercise 17.

EXAMPLE 6 Using a Frequency Table (Part 1)

A teacher gives a 10-point quiz and records the scores in a frequency table (Table 10.18) as shown below. Find the five-number summary and identify any outliers.

Table 10.18 Quiz Scores for Example 6

Score	10	9	8	7	6	5	4	3	2	1	0
Frequency	2	2	3	8	4	3	3	2	1	1	1

(continued)

SOLUTION The total of the frequencies is 30, so there are 30 scores.

The *median* of 30 numbers will be the mean of the 15th and 16th numbers. The table is already arranged in descending order, so we sum the frequencies from left to right until we come to 15. We see that the 15th number is a 7 and the 16th number is a 6. The median, therefore, is 6.5 points.

With 15 values in each half of the data, the *quartiles* will be the eighth numbers in from each end of the distribution. Adding frequencies, we see that the eighth highest score is 7, so $Q_3 = 7$. Similarly, the eighth score from the bottom is $Q_1 = 4$.

The five-number summary of the quiz scores is 0, 4, 6.5, 7, 10.

To check for outliers, we see $1.5(IQR) = 1.5(7 - 4) = 4.5$. Any outliers would have to be scores below $4 - 4.5 = -0.5$ or above $7 + 4.5 = 11.5$. None of the quiz scores is an outlier.

Now try Exercise 13.

The Mean (and When to Use It)

A common way to describe the center of a distribution is the *mean*. This is what most people usually think of as "average."

DEFINITION Mean

The **mean** of a list of n numbers $\{x_1, x_2, \ldots, x_n\}$ is

$$\bar{x} = \frac{x_1 + x_2 + \cdots + x_n}{n} = \frac{1}{n}\sum_{i=1}^{n} x_i.$$

The mean is also called the *arithmetic mean*, *arithmetic average*, or *average value*.

EXAMPLE 7 Computing a Mean

Find the mean annual home run total for Roger Maris's major league career, 1957–1968 (Table 10.11 on page 700).

SOLUTION According to Table 10.11, we are looking for the mean of the following list of 12 numbers: $\{14, 28, 16, 39, 61, 33, 23, 26, 8, 13, 9, 5\}$.

$$\bar{x} = \frac{14 + 28 + 16 + \cdots + 9 + 5}{12} = \frac{275}{12} \approx 22.9 \text{ home runs}$$

Now try Exercise 23.

As common as it is to use the mean as a measure of center, sometimes it can be misleading. For example, if you were to find the mean annual salary of 1984 geography major graduates of the University of North Carolina, it would probably be a number in the millions of dollars. This is because the group being measured, which is not very large, includes an outlier named Michael Jordan.

Unlike the median, the mean is not a resistant measure of center. It can be affected by both outliers and skewness. If you pay attention to news about the economy, you will often encounter reports that refer to "median family income." Why don't they use the mean? The distribution of family incomes is unimodal and strongly skewed to the right. The vast majority of people earn relatively modest incomes, but there is a long upper tail of ever smaller numbers of families who earn ever greater amounts of money. Those high values pull the mean to the right, making it a less meaningful representation of a typical income than the median. Similarly, if a distribution is skewed to the left, the mean will be tugged away from the center toward the very small data values. Before relying on the mean as a description of center, always look at the distribution for signs of skewness or outliers.

EXPLORATION 2 Interpreting Histograms (Part 2)

Of the three histograms in Figure 10.18, which has a median less than its mean? Which has a median greater than its mean? Which has a median approximately equal to its mean?

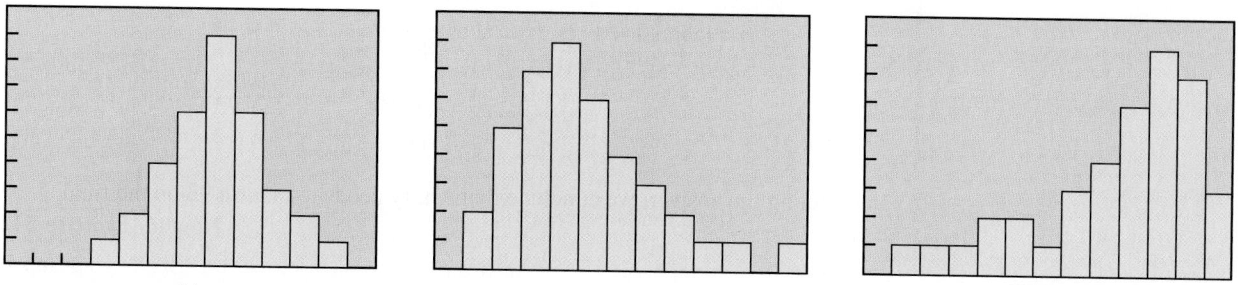

(a) (b) (c)

FIGURE 10.18 Which graph shows a data set in which the mean is less than the median? Greater than the median? Approximately equal to the median? (Exploration 2)

EXAMPLE 8 Using a Frequency Table (Part 2)

Here again is a frequency table for scores on a 10-point quiz. Find the mean score and compare it to the median we found in Example 6 (Table 10.19).

Table 10.19 Quiz Scores for Example 8

Score	10	9	8	7	6	5	4	3	2	1	0
Frequency	2	2	3	8	4	3	3	2	1	1	1

SOLUTION To find the *mean*, we multiply each number by its frequency, add the products, and divide the total by 30, the number of scores:

$$\bar{x} = \frac{\begin{matrix}[10(2) + 9(2) + 8(3) + 7(8) + 6(4) + 5(3) \\ + 4(3) + 3(2) + 2(1) + 1(1) + 0(1)]\end{matrix}}{30}$$

$$= 5.9\overline{3} \text{ points}$$

We found that the median is 6.5. The mean is a bit lower because the tail of low quiz scores extends farther from the center than the upper tail. The distribution is only slightly skewed to the left, so it's okay to use either measure of center.

Now try Exercise 29.

The formula for finding the mean of a list of numbers $\{x_1, x_2, \ldots, x_n\}$ with frequencies $\{f_1, f_2, \ldots, f_n\}$ is

$$\bar{x} = \frac{x_1f_1 + x_2f_2 + \cdots + x_nf_n}{f_1 + f_2 + \cdots + f_n} = \frac{\Sigma x_if_i}{\Sigma f_i}.$$

This same formula can be used to find a **weighted mean**, in which numbers $\{x_1, x_2, \ldots, x_n\}$ are given **weights** before the mean is computed. The weights act the same way as frequencies.

EXAMPLE 9 Working with a Weighted Mean

At Marty's school, it is an administrative policy that the final exam must count 25% of the semester grade. If Marty has an 88.5 average going into the final exam, what is the minimum exam score needed to earn a 90 for the semester?

SOLUTION The preliminary average (88.5) is given a weight of 0.75 and the final exam (x) is given a weight of 0.25. We will assume that a semester average of 89.5 will be rounded to a 90 on the transcript. Therefore,

$$\frac{88.5(0.75) + x(0.25)}{0.75 + 0.25} = 89.5$$

$$0.25x = 89.5(1) - 88.5(0.75)$$

$$x = 92.5$$

Interpreting the answer, we conclude that Marty needs to make a 93 on the final exam.

Now try Exercise 31.

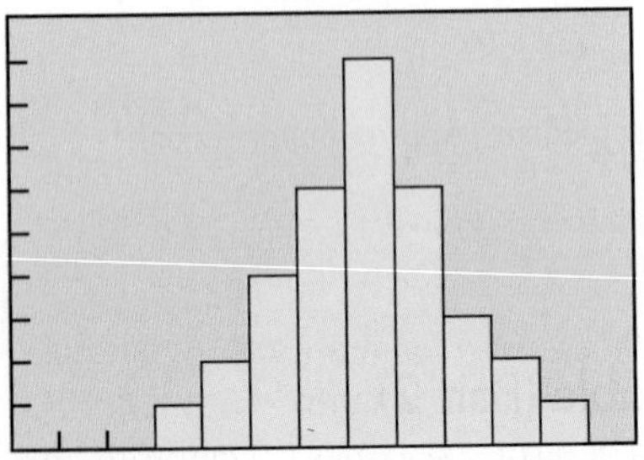

FIGURE 10.19 Histograms based on data gathered from real-world sources are often unimodal and symmetric, without outliers. The mean and standard deviation are most useful for describing the center and spread of distributions like these.

Variance and Standard Deviation

The median and *IQR* are useful statistics for describing the center and spread of any distribution because they are resistant to the presence of skewness and outliers.

On the other hand, the mean is an excellent measure of center when outliers and skewness are not present, which is quite often the case. Indeed, histograms of data from all kinds of real-world sources tend to look something like Figure 10.19, in which frequencies are higher close to the mean and lower as you move away from the mean in either direction.

For roughly symmetric data, the mean is the preferred measure of center. There is also a measure of spread for such data that is better than the *IQR*, called the *standard deviation.* Like the mean, the standard deviation is strongly affected by outliers and can be misleading if outliers are present.

DEFINITION Standard Deviation

The **standard deviation** of a population $\{x_1, x_2, \ldots, x_n\}$ is

$$\sigma = \sqrt{\frac{1}{n}\sum_{i=1}^{n}(x_i - \mu)^2},$$

where μ denotes the population mean. The **variance** is σ^2, the square of the standard deviation.

If we define the "deviation" of a data value to be how much it differs from the mean, then the variance is just the mean of the squared deviations. The standard deviation is the square *root* of the *mean* of the *squared deviations*, which is why it is sometimes called the *root mean square deviation.* The symbol "μ" is the Greek letter mu and "σ" is a lowercase Greek letter sigma.

Calculating a standard deviation by hand can be tedious, but with modern calculators it is usually only necessary to enter the list of data and push a button. In fact, most calculators give you a choice of two standard deviations, one slightly larger than the other. The larger one (usually called s) is based on the formula

$$s = \sqrt{\frac{1}{n-1}\sum_{i=1}^{n}(x_i - \bar{x})^2}.$$

The difference is that the σ formula is for finding the true parameter, which means that it applies only when $\{x_1, x_2, \ldots, x_n\}$ is the whole *population*. If $\{x_1, x_2, \ldots, x_n\}$ is a *sample* from the population, then the s formula actually gives a better estimate of the parameter than the σ formula does. So use the larger standard deviation when your data come from a sample (which is almost always the case).

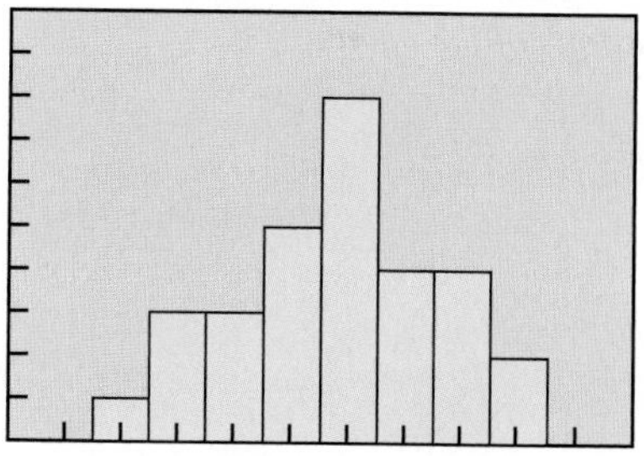

[77, 98] by [0, 10]

FIGURE 10.20 The weights of the loon chicks in Example 10 appear to be normal, with no outliers or strong skewness. We conclude that the mean and standard deviation are appropriate measures of center and variability, respectively.

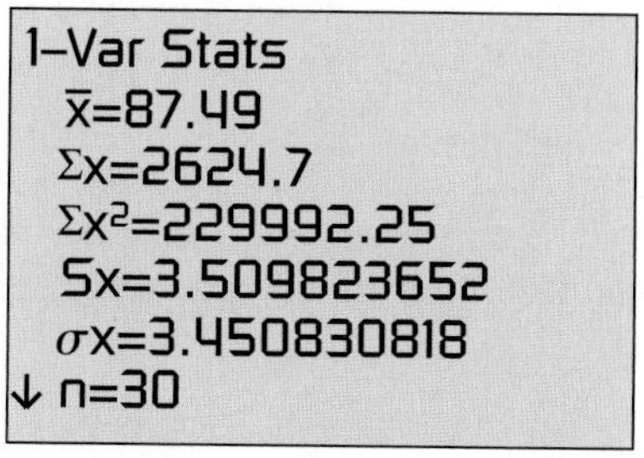

FIGURE 10.21 Single-variable statistics in a typical calculator display. (Example 10)

EXAMPLE 10 Finding Standard Deviation with a Calculator

A researcher measured 30 newly hatched loon chicks and recorded their weights in grams as shown in Table 10.20.

Table 10.20 Weights in Grams of 30 Loon Chicks

79.5	87.5	88.5	89.2	91.6	84.5	82.1	82.3	85.7	89.8
84.0	84.8	88.2	88.2	82.9	89.8	89.2	94.1	88.0	91.1
91.8	87.0	87.7	88.0	85.4	94.4	91.3	86.4	85.7	86.0

Based on the sample, estimate the mean and standard deviation for the weights of newly hatched loon chicks. Are these measures useful in this case, or should we use the five-number summary?

SOLUTION We enter the list of data into a calculator. A histogram (Xscl $= 2$) in the window $[77, 98]$ by $[0, 10]$ shows that the distribution is unimodal (as we would expect from nature), containing no outliers or strong skewness. Therefore, the mean and standard deviation are appropriate measures (Figure 10.20).

We calculate statistics of a single variable. The output from a calculator is shown in Figure 10.21.

The mean is $\bar{x} = 87.49$ grams. For standard deviation, we choose $Sx = 3.51$ grams because the calculations are based on a sample of loon chicks, not the entire population of loon chicks.

Now try Exercise 35.

Normal Distributions

Although we use the word *normal* in many contexts to suggest typical behavior, in the context of statistics and data distributions it is really a technical term. If you graph the function

$$y = e^{-x^2/2}$$

in the window $[-3, 3]$ by $[0, 1]$, you will see what **Normal** means mathematically (Figure 10.22).

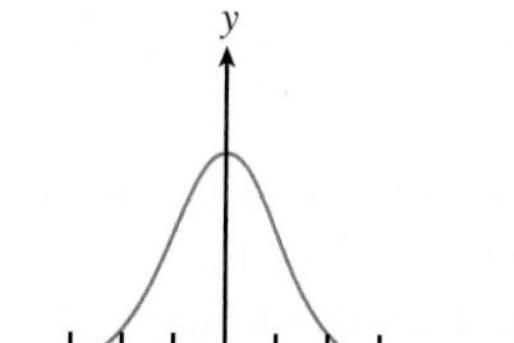

FIGURE 10.22 The graph of $y = e^{-x^2/2}$ This is a Gaussian (or Normal) curve.

The shape models many distributions we commonly encounter. In fact this curve, called a **Gaussian curve** or **Normal curve**, is a *precise mathematical model* for normal behavior. That is where the mean and standard deviation come in.

The standard deviation of the curve in Figure 10.22 is 1. Using calculus, we can find that about 68% of the total area under this curve lies between -1 and 1 (Figure 10.23a). Since any Normal distribution has this shape, *about 68% of the data in any Normal distribution lie within 1 standard deviation of the mean.*

Similarly, we can find that about 95% of the total area under the Gaussian curve lies between -2 and 2 (Figure 10.23b), implying that *about 95% of the data in any Normal distribution lie within 2 standard deviations of the mean.*

Continuous Random Variables

Weights of loon chicks (Example 10) are values of a **continuous random variable**, because, a loon chick can theoretically assume any real-number weight in an interval. Since Normal random variables are continuous, their probability distributions are described using continuous curves.

Similarly, we can find that about 99.7% (nearly all) of the total area under the Gaussian curve lies between −3 and 3 (Figure 10.23c), implying that *about 99.7% of the data in any Normal distribution lie within 3 standard deviations of the mean.*

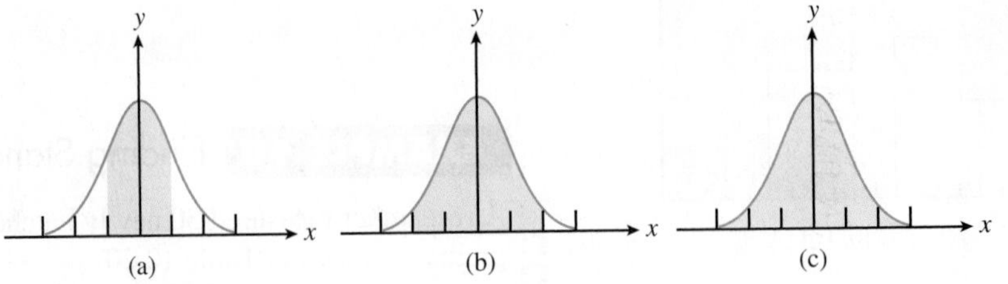

FIGURE 10.23 (a) About 68% of the area under a Gaussian curve lies within 1 unit of the mean. (b) About 95% of the area lies within 2 units of the mean. (c) About 99.7% of the area lies within 3 units of the mean. If we think of the units as standard deviations, this gives us a model for *any* Normal distribution.

It's NOT Normal

No data can actually be normally distributed. Ever. For one thing, the Gaussian curve approaches the x-axis as an asymptote, extending without bound toward $\pm\infty$. Nothing in the real world does that. The Normal model is just that: a model. But it's a spectacularly useful model that describes the behavior of a wide variety of phenomena amazingly well. When data are *approximately* Normal, we can use the 68-95-99.7 Rule to understand *about* what percentage of the values we'd expect to find within ± 1, ± 2, or ± 3 standard deviations of the mean.

The 68–95–99.7 Rule

If the data for a population are approximately normally distributed with mean μ and standard deviation σ, then

- about 68% of the data lie between $\mu - 1\sigma$ and $\mu + 1\sigma$;
- about 95% of the data lie between $\mu - 2\sigma$ and $\mu + 2\sigma$;
- about 99.7% of the data lie between $\mu - 3\sigma$ and $\mu + 3\sigma$.

What makes this rule so useful is that Normal distributions are common in a wide variety of statistical applications. We close the section with a simple application.

EXAMPLE 11 Using the 68–95–99.7 Rule

Based on the research data presented in Example 10, would a loon chick weighing 95 grams be in the top 2.5% of all newly hatched loon chicks?

SOLUTION We assume that the distribution of weights of newly hatched loon chicks in the whole population is approximately Normal. Since we do not know the mean and standard deviation for the whole population (the parameters μ and σ), we use the sample statistics $\bar{x} = 87.49$ and $Sx = 3.51$ as estimates.

Look at Figure 10.23b. The shaded region contains 95% of the area, so the two identical white regions at either end must each contain 2.5% of the area. That is, to be in the top 2.5%, a loon chick will have to weigh at least 2 standard deviations more than the mean:

$$\bar{x} + 2Sx = 87.49 + 2(3.51) = 94.51 \text{ grams}$$

Since $95 > 94.51$, a Normal model suggests that a 95-gram loon chick is probably in the top 2.5%.

Now try Exercise 43.

If you study Statistics more deeply someday, you will learn that more is going on in Example 11 than meets the eye. For starters, we need to know that the chicks are really a random sample of all loon chicks (not, for example, from the same geographical area). Also, we lose some accuracy by using statistics to estimate the true mean and standard deviation. Statisticians have ways of taking that into account.

QUICK REVIEW 10.3 (Prerequisite skill Section 9.4)

In Exercises 1–6, write the sum in expanded form.

1. $\sum_{i=1}^{7} x_i$
2. $\sum_{i=1}^{5} (x_i - \bar{x})$
3. $\frac{1}{7}\sum_{i=1}^{7} x_i$
4. $\frac{1}{5}\sum_{i=1}^{5} (x_i - \bar{x})$
5. $\frac{1}{5}\sum_{i=1}^{5} (x_i - \bar{x})^2$
6. $\sqrt{\frac{1}{5}\sum_{i=1}^{5} (x_i - \bar{x})^2}$

In Exercises 7–10, write the sum in sigma notation.

7. $x_1f_1 + x_2f_2 + x_3f_3 + \cdots + x_8f_8$
8. $(x_1 - \bar{x})^2 + (x_2 - \bar{x})^2 + \cdots + (x_{10} - \bar{x})^2$
9. $\frac{1}{50}\left[(x_1 - \bar{x})^2 + (x_2 - \bar{x})^2 + \cdots + (x_{50} - \bar{x})^2\right]$
10. $\sqrt{\frac{1}{7}\left[(x_1 - \bar{x})^2 + (x_2 - \bar{x})^2 + \cdots + (x_7 - \bar{x})^2\right]}$

SECTION 10.3 Exercises

In Exercises 1 and 2, identify whether the number described is a *parameter* or a *statistic*.

1. **(a)** The average score on last week's quiz was 73.4.
   (b) About 13% of the human population is left-handed.
2. **(a)** The average American is about 23 lb overweight.
   (b) In a study of laboratory rats, 93% became aggressive when deprived of sleep.

In Exercises 3 and 4, identify whether the *average* described is a *mean* or *median*.

3. **(a)** The pitcher's earned run *average* is 2.35.
   (b) The choir lined up with tall people in the back, short people in front, and people of *average* height in the middle.
4. **(a)** As the marathon runners crossed the finish line, the people in the middle of the pack finished in an *average* time of about 3 1/2 hr.
   (b) Bob has a 3.27 grade point *average*.

In Exercises 5 and 6, find the median of the indicated data.

5. The number of satellites (moons), from the data in Table 10.21.

Table 10.21 Planetary Satellites

Planet	Number of Satellites
Mercury	0
Venus	0
Earth	1
Mars	2
Jupiter	67
Saturn	62
Uranus	27
Neptune	14

Source: www.go-astronomy.com, 2013.

6. The area of the continents, from the data in Table 10.22.

Table 10.22 Size of Continent

Continent	Area (km^2)
Africa	30,065,000
Antarctica	13,209,000
Asia	44,579,000
Australia/Oceania	8,112,000
Europe	9,938,000
North America	24,474,000
South America	17,819,000

Source: Worldatlas.com, 2005.

7. Find the median of the following salaries for employees in one department of the Garcia Brothers Company (in thousands of dollars):

   33.5, 35.3, 33.8, 29.3, 36.7, 32.8, 31.7, 37.3, 33.5, 28.2, 34.8, 33.5, 29.7, 38.5, 32.7, 34.8, 34.2, 31.6, 35.4

8. Find the median of the following average annual wind speeds at 44 climatic data centers around the United States:

   9.0, 6.9, 9.1, 9.2, 10.2, 12.5, 12.0, 11.2, 12.9, 10.3, 10.6, 10.9, 8.7, 10.3, 11.0, 7.7, 11.4, 7.9, 9.6, 8.0, 10.7, 9.3, 7.9, 6.2, 8.3, 8.9, 9.3, 11.6, 10.6, 9.0, 8.2, 9.4, 10.6, 9.5, 6.3, 9.1, 7.9, 9.7, 8.8, 6.9, 8.7, 9.0, 8.9, 9.3

9. Find the range and the *IQR* of the salaries in Exercise 7.
10. Find the range and the *IQR* of the wind speeds in Exercise 8.
11. The five-number summary for the weights (in pounds) of the players on a high school soccer team is {113, 143, 152, 165, 210}. Are there any outliers? Explain.
12. The five-number summary for the number of miles (in thousands) on a company's fleet of cars is {7, 49, 66, 72, 98}. Are there any outliers? Explain.
13. The frequency table below summarizes the optical magnification power of lenses on 19 models of cameras classified as "superzoom."

Maximum Zoom	16	18	20	21	22	24	30	40	42
Frequency	4	1	5	2	1	2	2	1	1

Source: Consumer Reports, August 2013.

Find the five-number summary for these data and identify any outliers.

14. National Football League teams play 16 games during the regular season. In 2012–2013 the Broncos and Falcons led the league with 13 wins each. The frequency table below summarizes the number of games won by all 32 teams.

Wins	2	4	5	6	7	8	9	10	11	12	13
Frequency	2	3	2	3	6	2	1	5	4	2	2

Source: www.nfl.com

Find the five-number summary for these data and identify any outliers.

15. Before winning the 2012–2013 NBA championship, the Miami Heat won 62 regular season games, best in the league. The stemplot below displays the number of games won by all 30 NBA teams.

2	0 1 4 5 7 8 9 9
3	1 3 4 4 8
4	1 1 3 4 5 5 5 7 9 9
5	4 6 6 7 8
6	0 6

Source: www.nba.com

Find the five-number summary for these data and identify any outliers.

16. Battery life is an important consideration when purchasing a tablet computer. The stemplot below summarizes test results for 24 highly rated tablets with 9- to 12-in. screens, showing a best-in-class lifespan of 13.4 hr.

8	0 9 9
9	5
10	0 5 6 7 7 9 9
11	0 1 3 4 5 5 6 6 7
12	4 6 9
13	4

Source: Consumer Reports, August 2013

Find the five-number summary for these data and identify any outliers.

17. **Writing to Learn** How important is a home ice advantage in professional hockey? The boxplots below show the number of games the 30 NHL teams won at home and away during the 2011–2012 season. Write a brief comparison of the two distributions.

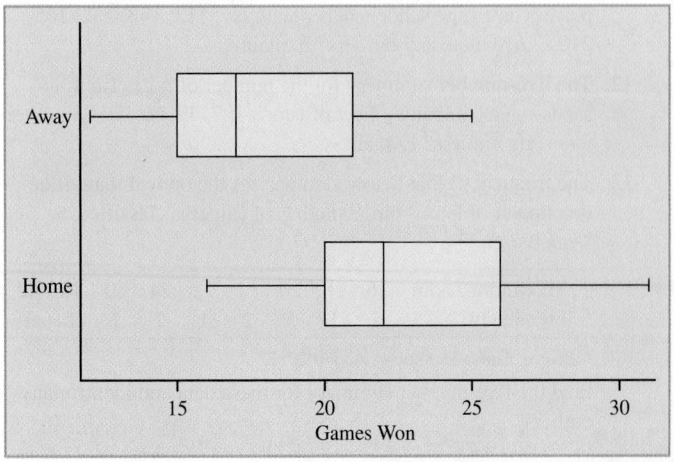

18. **Writing to Learn** How important is a home court advantage in professional basketball? The boxplots below show the number of games the 32 NBA teams won at home and away during the 2012–2013 season. Write a brief comparison of the two distributions.

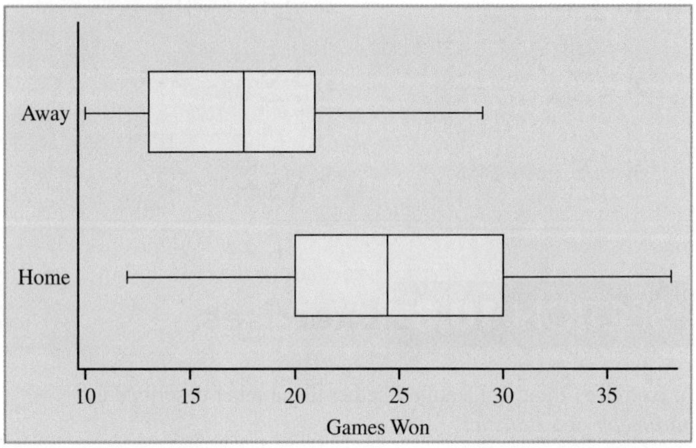

In Exercises 19 and 20, determine the five-number summary, the range, and the interquartile range for the sets of data specified. Identify any outliers.

19. The annual home run production data for Babe Ruth and Barry Bonds in Table 10.5 (page 695)

20. The annual home run production data for Willie Mays and Mickey Mantle in Table 10.14 (page 701)

21. Use your summary statistics from Exercise 19 to create simultaneous boxplots comparing season home run production by Babe Ruth and Barry Bonds.

22. Use your summary statistics from Exercise 20 to create simultaneous boxplots comparing season home run production by Willie Mays and Mickey Mantle.

23. Using the data in Table 10.21, find the mean number of moons for the planets in our solar system. Compare the mean to the median you found in Exercise 5.

24. Using the data in Table 10.22, find the mean area of the continents. Compare the mean to the median you found in Exercise 6.

25. Find the mean salary for the data in Exercise 7. Compare the mean to the median salary you found there.

26. Find the mean wind speed for the data in Exercise 8. Compare the mean to the median weight you found there.

27. In 2012, a total of 266,994 students took the Advanced Placement examination in Calculus AB. From the frequency table below, find the mean score in Calculus AB in 2012. *(Source: The College Board.)*

AP Examination Grade	Number of Students
5	67,394
4	45,523
3	46,526
2	27,216
1	80,335

28. In 2012, a total of 94,403 students took the Advanced Placement examination in Calculus BC. From the frequency table on the next page, find the mean score in Calculus BC in 2012. *(Source: The College Board.)*

AP Examination Grade	Number of Students
5	47,553
4	15,231
3	14,957
2	5,076
1	11,586

29. Find the mean magnification power for the superzoom cameras described in Exercise 13. Which is the better summary of these data, the mean or the median? Explain.

30. Using the frequency table in Exercise 14, find the mean number of games won by NFL teams during the 2012–2013 season. Which is the better summary of these data, the mean or the median? Explain.

In Exercises 31 and 32 use the data in Table 10.17 (page 703).

(a) Find the average (mean) of the indicated temperatures for Beijing.

(b) Find the weighted average using the number of days in the month as the weight. (Assume no leap year.)

(c) Compare your results in parts (a) and (b). Do the weights have an effect on the average? Why or why not? Which average is the better indicator for these temperatures?

31. The monthly high temperatures

32. The monthly low temperatures

In Exercises 33–38, find the standard deviation and variance of the data set. (Since data sets are almost always samples from some larger population, use s in each case, rather than σ.)

33. $\{23, 45, 29, 34, 39, 41, 19, 22\}$

34. $\{28, 84, 67, 71, 92, 37, 45, 32, 74, 96\}$

35. The credit card debt data in Table 10.7 (page 698)

36. The student debt data in Table 10.8 (page 699)

37. The salary data in Exercise 7

38. The wind speeds data in Exercise 8

In Exercises 39 and 40, examine each pair of data sets.

(a) Decide whether you think the standard deviation of the first is smaller, equal to, or larger than the standard deviation of the second. Explain why.

(b) Check your prediction by calculating both standard deviations.

39. $\{4, 6, 8, 8, 8, 10, 12\}$ and $\{4, 5, 7, 8, 9, 11, 12\}$

40. $\{10, 15, 20, 25, 30\}$ and $\{50, 55, 60, 65, 70\}$

41. Writing to Learn Is it possible for the standard deviation of a data set to be zero? Explain your answer.

42. Writing to Learn Is it possible for the standard deviation of a set to be negative? Explain your answer.

43. SAT Scores Scores on the Scholastic Aptitude Tests are scaled to a mean of 500 and a Normal model with a standard deviation of 100.

(a) Approximately what percentage of scores should be between 400 and 600?

(b) Approximately what percentage of scores should be below 300?

(c) In 2012, the national average on the SAT Math section was 514. Is this number a parameter or a statistic?

44. ACT Scores In 2012, the national mean ACT Math score was 21.0, with an approximate standard deviation of 6. ACT scores in the general population approximate a Normal distribution.

(a) Approximately what percentage of the 2012 scores should be higher than 27?

(b) Approximately what ACT Math score would one need to make in 2012 to be ranked among the top 2.5% of all who took the test?

(c) Writing to Learn Mean ACT scores are published state by state. If we add up the 50 state means and divide by 50, will the result be a good estimate for the national mean score? Explain your answer.

45. The weights of eggs laid by mature turkeys are approximately normally distributed with a mean of 90 g and a standard deviation of 5.5 g.

(a) Approximately what percent of all eggs should weigh more than 95.5 g?

(b) Approximately what percent of all eggs should weigh between 79 and 84.5 g?

(c) Approximately what weight would place an egg among the largest 2.5%?

(d) Explain why you'd be very surprised to find a turkey egg that weighed only 72 g.

46. Investors often consider *volatility* before buying stocks. Volatility is essentially the standard deviation of historical price fluctuations, expressed as a percent of the stock's price. Analysts use a Normal model to describe market projections. Suppose a certain stock has an annual volatility of 4% and is projected to increase by 7% this year.

(a) What's the approximate probability the stock's value drops by more than 1%?

(b) What's the approximate probability the stock's value increases by between 3% and 15%?

(c) What change in the stock's price would represent approximately the worst 16% of anticipated outcomes?

(d) Explain why an investor would be foolish to daydream of a 20% increase in this stock's price.

Standardized Test Questions

47. True or False The median is strongly affected by outliers. Justify your answer.

48. True or False The length of the box in a boxplot is the interquartile range. Justify your answer.

You may use a graphing calculator when solving Exercises 49–52.

49. Multiple Choice The plot of a Normal distribution will be

(A) Symmetric.

(B) Skewed left.

(C) Skewed right.

(continued)

(D) Lower in the middle than at the ends.

(E) Of no predictable shape.

50. **Multiple Choice** A frequency table for a set of 25 quiz grades is given below. What is the mean of the data?

Quiz Grade	10	9	8	7	6	5	4	3	2	1
Number of Students	3	3	5	6	4	3	1	0	0	0

(A) 7.00

(B) 7.28

(C) 7.35

(D) 7.60

(E) 7.86

51. **Multiple Choice** Professor Mitchell grades the exams of his 30 students and finds that the scores have a mean of 81.3 and a median of 80.5. He later determines that the top student deserves 9 extra points for a misgraded proof. After the error is corrected, the scores have

(A) A mean of 81.3 and a median of 80.5.

(B) A mean of 81.6 and a median of 80.5.

(C) A mean of 81.6 and a median of 80.8.

(D) A mean of 90.3 and a median of 80.5.

(E) A mean of 90.3 and a median of 89.5.

52. **Multiple Choice** If the calorie contents of robin eggs are normally distributed with a mean of 25 and a standard deviation of 1.2, then approximately 95% of all robin eggs should have a calorie content in the interval

(A) $[21.4, 28.6]$.

(B) $[22, 28]$.

(C) $[22.6, 27.4]$.

(D) $[23, 27]$.

(E) $[23.8, 26.2]$.

Explorations

53. **Group Activity** List a set of data for which the inequality holds.

(a) Median $<$ mean

(b) Mean $<$ median

54. **Group Activity** List a set of data for which the equation or inequality holds.

(a) Standard deviation $<$ interquartile range

(b) Interquartile range $<$ standard deviation

55. Draw a boxplot for which the inequality holds.

(a) Median $<$ mean

(b) $2 \times$ interquartile range $<$ range

(c) Range $< 2 \times$ interquartile range

56. Construct a set of data with median 5 and mean 7.

Extending the Ideas

Weighting Data by Population The average life expectancies for males and females in 12 South American nations were given in Table 10.13 (page 701). To find an overall average life expectancy for males or females in all of these nations, we would need to weight the national data according to the various national populations. Table 10.23 is an extension of Table 10.13, showing the populations (in millions). Assume that males and females appear in roughly equal numbers in each nation.

Table 10.23 Life Expectancy by Gender and Population (in millions) for 12 Nations of South America

Nation	Female	Male	Population
Argentina	79.1	72.5	42.2
Bolivia	69.2	65.1	10.3
Brazil	77.8	71.0	199.3
Chile	82.3	76.2	17.1
Colombia	80.5	74.5	45.2
Ecuador	78.8	73.2	15.2
Guyana	67.0	60.0	0.7
Paraguay	77.6	71.7	6.5
Peru	78.3	75.1	29.5
Suriname	75.5	69.4	0.6
Uruguay	80.0	73.0	3.3
Venezuela	79.3	72.0	28.0

Sources: World Health Organization and www.indexmundi.com, 2012.

In Exercises 57 and 58, use the data in Table 10.23 to find the mean life expectancy for each group.

57. Women living in South American nations

58. Men living in South American nations

59. **Quality Control** A plant manufactures ball bearings to the purchaser's specifications, rejecting any output with a diameter that deviates more than 0.1 mm from the specified value. The variability in such processes can be modeled by a Normal curve. If the ball bearings are produced with the specified mean and a standard deviation of 0.05 mm, what percentage of the output will be rejected?

60. **Quality Control** A machine fills 12-oz. cola cans with a mean of 12.08 oz. of cola and a standard deviation of 0.04 oz. The variability in such processes can be modeled by a Normal curve. Approximately what percentage of the cans will actually contain less than the advertised 12 oz. of cola?

10.4 Random Variables and Probability Models

What you'll learn about

- Probability Models and Expected Values
- Binomial Probability Models
- Normal Model
- Normal Approximation for Binomial Distributions

... and why

These topics allow us to decide on the best course of action when events unfold randomly, and to recognize when an outcome is so unusual that it provides evidence of an important phenomenon.

As we've said before, the aim of Statistics is to find meaning in data. In Sections 10.2 and 10.3 you've taken the first steps: You've learned to describe data. You can now create informative displays, calculate summary statistics, and write appropriate conclusions. One important step remains: comparing what we see in the data to what we expected might happen.

- A medical experiment tests a drug to determine whether it is effective for treating a certain condition. The subjects, 50 people who have the condition, are randomly divided into two groups of 25. One group takes the drug for a month, and the other group takes a placebo (an identical-looking but inert pill). At the end of the month a doctor examines all the patients and finds improved conditions in 16 of the people who took the drug and in 10 of those who were given the placebo. If the drug is not effective, we'd expect the outcomes in the two groups to be the same. Does the somewhat better result among subjects who took the drug provide evidence that the drug actually worked, or could a difference like this just happen by chance? You'll learn how to investigate that question in Section 10.5.
- You watch someone toss a coin 100 times. You expect that fair tossing will yield about 50 heads and 50 tails, but this person tosses 66 heads and only 34 tails. Could the tosser be cheating somehow? If the outcome had been 54 heads and 46 tails, you'd probably think nothing of it. And if it had been 95 heads and 5 tails you'd be pretty sure that the tossing was not being done fairly. But is 66 heads vs. 34 tails evidence of foul play? You'll learn to answer this question in this section.

To think clearly about comparing the results we observe to those we expected, we need to understand probability models.

Probability Models and Expected Values

If you roll a die 5 times and get the results $\{2, 6, 1, 5, 5\}$, the average outcome is $\bar{x} = (2 + 6 + 1 + 5 + 5)/5 = 3.8$. That's the average of the data in this sample, something you already know how to find. Now we ask a different question: For all possible rolls of a die, what do we expect the average result to be in the long run? This question is not about the statistic $\bar{x}$ (the sample mean). It's about the theoretical population mean, the parameter μ. You might find out by simply averaging the faces of the die: $(1 + 2 + 3 + 4 + 5 + 6)/6 = 3.5$. This suggests that in the long run, when rolling a fair die, we expect the result to average 3.5. That's correct, but the calculation works only because the six faces are equally likely to appear. What if the die is "loaded"—weighted so that some faces are more likely to turn up than others? To deal with that, we need a *probability model* for the *random variable* of die outcomes. We start with some definitions.

DEFINITION Random Variable

A **random variable** assumes numerical values as a result of some random event.

We denote random variables by capital letters, such as X, and the values they can assume by subscripted lower case letters, such as $\{x_1, x_2, \ldots, x_n\}$. If the random variable $X =$ the result of rolling a die, then X can have the values $\{1, 2, 3, 4, 5, 6\}$.

DEFINITION Probability Model

A **probability model** is a random variable X together with a function that assigns to each possible outcome a probability $P(x_i)$ such that $\sum_{i=1}^{n} P(x_i) = 1$.

Table 10.24 shows the probability model for a fair die:

Table 10.24 The Probability Model for Rolling a Fair Die

X	1	2	3	4	5	6
$P(X)$	$\frac{1}{6}$	$\frac{1}{6}$	$\frac{1}{6}$	$\frac{1}{6}$	$\frac{1}{6}$	$\frac{1}{6}$

Earlier we calculated the mean by just adding the faces and dividing by 6. Distributing the denominator to each term provides an equivalent way to calculate this long-run average, also known as the **expected value** of the random variable:

$$\frac{1+2+3+4+5+6}{6} = 1\left(\frac{1}{6}\right) + 2\left(\frac{1}{6}\right) + 3\left(\frac{1}{6}\right) + 4\left(\frac{1}{6}\right) + 5\left(\frac{1}{6}\right) + 6\left(\frac{1}{6}\right) = 3.5$$

Expected Value of a Random Variable

If X is a random variable with probability function $P(X)$, the mean of the probability model (μ) is also called the expected value of X, denoted $E(X)$. The expected value is the sum of the values of X times their corresponding probabilities:

$$\mu = E(X) = \sum_{i=1}^{n} x_i \cdot P(x_i)$$

EXAMPLE 1 A Loaded Die

A die is loaded to favor faces 5 and 6, as shown in the probability model below (Table 10.25). Find the expected value for this die and explain what that means.

Table 10.25 A Probability Model for Rolling a Loaded Die

X	1	2	3	4	5	6
$P(X)$	0.1	0.1	0.1	0.1	0.2	0.4

SOLUTION The expected value is

$$E(X) = 1(0.1) + 2(0.1) + 3(0.1) + 4(0.1) + 5(0.2) + 6(0.4)$$
$$= 4.4.$$

If this die were tossed repeatedly, we expect the long-run average of the outcomes to be 4.4.

Now try Exercise 1.

Note that the expected value calculation is simply a weighted average, as seen in Section 10.3.

EXAMPLE 2 A Raffle

As a fundraiser, a student group sells 500 raffle tickets for \$2 each. At the drawing, the top prize will be a gift certificate for \$100. Second prize will be a \$50 gift certificate, and there will be five third prizes, each a \$20 gift certificate.

(a) Create a probability model for the raffle.

(b) Find the expected value of a ticket.

(c) Explain what the expected value means.

SOLUTION

(a) The probability model (Table 10.26) lists all possible outcomes for a ticket-holder (1st, 2nd, 3rd, or no prize), the value of each (taking into account the \$2 paid for the ticket), and the corresponding probabilities.

Check your model

It's easy to overlook the outcome "No prize," but without it the probabilities do not add up to 1. Verifying the sum is an important way to check a probability model.

Table 10.26 The Probability Model for a Raffle

Outcome	1st Prize	2nd Prize	3rd Prize	No prize
X = **value**	98	48	18	−2
$P(X)$	$\frac{1}{500}$	$\frac{1}{500}$	$\frac{5}{500}$	$\frac{493}{500}$

(b) The expected value is

$$E(X) = 98\left(\frac{1}{500}\right) + 48\left(\frac{1}{500}\right) + 18\left(\frac{5}{500}\right) + (-2)\left(\frac{493}{500}\right)$$
$$= -1.5.$$

(c) People who buy these raffle tickets lose an average of \$1.50 each.

Now try Exercise 3.

Sometimes we must do some probability calculations in order to create the model.

EXAMPLE 3 Calculating the Probabilities for the Model

To attract more players on weekdays, a golf course offers players a chance at a Lucky Ace discount. When a player signs in, the manager shuffles the four aces and lays them face down on the counter. The player turns one card over. If it's the ace of diamonds, the player golfs for free, saving the usual \$50 fee. If it's the ace of hearts, the player gets to turn over another card. If this second one is the ace of diamonds, the round of golf is half price. If either the first or second card is one of the black aces, the player gets a \$10 discount. What's the expected cost of a round of weekday golf?

SOLUTION

First we find the probabilities of each outcome, letting D = diamonds, H = hearts, and B = black.

$$P(D) = \frac{1}{4}$$
$$P(H, \text{then } D) = \frac{1}{4}\cdot\frac{1}{3} = \frac{1}{12}$$
$$P(H, \text{then } B) = \frac{1}{4}\cdot\frac{2}{3} = \frac{1}{6}$$
$$P(B) = \frac{2}{4} = \frac{1}{2}$$

We check to be sure the probabilities total 1:

$$\frac{1}{4} + \frac{1}{12} + \frac{1}{6} + \frac{1}{2} = \frac{3}{12} + \frac{1}{12} + \frac{2}{12} + \frac{6}{12} = \frac{12}{12}.$$

(continued)

The probability model is shown in Table 10.27.

Table 10.27 The Probability Model for the Lucky Ace Discount

Outcome	D	H, then D	H, then B	B
X = cost	0	25	40	40
$P(X)$	$\frac{1}{4}$	$\frac{1}{12}$	$\frac{1}{6}$	$\frac{1}{2}$

With the Lucky Ace discount, the expected cost of a round of golf is

$$E(X) = 0\left(\frac{1}{4}\right) + 25\left(\frac{1}{12}\right) + 40\left(\frac{1}{6}\right) + 40\left(\frac{1}{2}\right) = \$28.75.$$

Now try Exercise 13.

EXPLORATION 1 Service Contracts

An office manager must decide whether to purchase a service plan for the copying machine. Past experience suggests this distribution (Table 10.28) for the annual number of service calls:

Table 10.28 The Distribution of Copier Service Calls Annually

No. of calls	0	1	2	3
Probability	0.2	0.4	0.3	0.1

The options are:

- Plan A: Pay a flat fee of \$350 for the year and any service calls are free.
- Plan B: Pay \$250 per service call.
- Plan C: Pay a \$300 fee; then the first service call is free and any others cost \$60 each.

Which of the three plans should the office manager expect would cost the least?

Binomial Probability Models

Some probability models generalize to describe a wide range of random phenomena. In Section 10.3 you saw one example: the Normal curve. We'll explore that model in more detail later in this section. Another common probability model describes random variables that count the number of successes in repeated trials of a simple event.

EXAMPLE 4 Bouncing Dice

We roll a fair die four times. Find the probability that we roll:

(a) All 3's. **(b)** No 3's. **(c)** Exactly two 3's.

SOLUTION

(a) We have a probability 1/6 of rolling a 3 each time. By the Multiplication Principle, the probability of rolling a 3 all four times is $(1/6)^4 \approx 0.00077$.

(b) There is a probability 5/6 of rolling something other than 3 each time. By the Multiplication Principle, the probability of rolling a non-3 all four times is $(5/6)^4 \approx 0.48225$.

Outcomes

33NN
3N3N
3NN3
N33N
N3N3
NN33

(c) The probability of rolling two 3's followed by two non-3's (again by the Multiplication Principle) is $(1/6)^2(5/6)^2 \approx 0.01929$. However, that is not the only outcome we must consider. In fact, the two 3's could occur on any two of the four rolls, in exactly $\binom{4}{2} = 6$ ways (listed at the left). That gives us 6 outcomes, each with probability $(1/6)^2(5/6)^2$. The probability of the event "exactly two 3's" is therefore $\binom{4}{2}(1/6)^2(5/6)^2 \approx 0.11574$. **Now try Exercise 17a.**

The random variable that counts the number of 3's we get when we roll a die repeatedly has a *binomial probability distribution.*

DEFINITION Binomial Probability Distribution

Consider a simple event with these properties:

- Each trial has two possible outcomes, called *success* and *failure.*
- The probability of success on each trial is the same. (We denote the probability of success as p and the probability of failure as q. Note that $q = 1 - p$.)
- The trials are independent.

Let random variable $X =$ the number of successes in n trials. Then the probability model for X is called the **binomial distribution**, and the probability of getting k successes in the n trials is $P(X = k) = \binom{n}{k} p^k q^{n-k}$.

Binomial Probabilities on a Calculator

Your calculator might be programmed to find values for the binomial probability distribution function (binompdf). The solutions to Example 5 in one calculator syntax, for example, could be obtained by:

(a) binompdf(20, .9, 20) (20 repetitions, 0.9 probability, 20 successes)

(b) binompdf(20, .9, 18) (20 repetitions, 0.9 probability, 18 successes)

(c) 1 − binomcdf(20, .9, 17) (1 minus the *cumulative* probability of 17 or fewer successes)

Check your owner's manual for more information.

EXAMPLE 5 Shooting Free Throws

Suppose Michael makes 90% of his free throws. If he shoots 20 free throws, and if his chance of making each one is independent of the other shots (an assumption you might question in a game situation), what is the probability that he makes

(a) All 20? **(b)** Exactly 18? **(c)** At least 18?

SOLUTION We compute three binomial probabilities:

(a) $P(20 \text{ successes}) = (0.9)^{20} \approx 0.12158$

(b) $P(18 \text{ successes}) = \binom{20}{18}(0.9)^{18}(0.1)^2 \approx 0.28518$

(c) $P(\text{at least 18 successes}) = P(18) + P(19) + P(20)$

$$= \binom{20}{18}(0.9)^{18}(0.1)^2 + \binom{20}{19}(0.9)^{19}(0.1) + (0.9)^{20}$$

$$\approx 0.6769$$

Now try Exercise 17b and c.

Recall that when we summarize any distribution, we're always interested in shape, center, and spread. For a binomial random variable, the easiest of these is center. If, overall, Michael makes 90% of his free throws, what's the expected number he'll make when he takes 20 shots? It seems as though 90% of 20 = 18 should be the correct mean, but this simple calculation that multiplies the number of trials by the probability of success is not a formal calculation of the expected value. We investigate this shortcut with an easier example.

EXAMPLE 6 Expected Value and Binomial Distributions

A fair coin is flipped five times. What is the expected number of heads?

SOLUTION

Let random variable X = the number of heads when 5 coins are tossed. Then X takes on the values $\{0, 1, 2, 3, 4, 5\}$, and the probability model is a binomial distribution. For example, $P(X = 2) = \binom{5}{2}(1/2)^2(1/2)^3 = 10/32$. The rest of the model is shown in Table 10.29.

Table 10.29 The Probability Model for the Number of Heads When Tossing 5 Coins

X	0	1	2	3	4	5
$P(X)$	$\frac{1}{32}$	$\frac{5}{32}$	$\frac{10}{32}$	$\frac{10}{32}$	$\frac{5}{32}$	$\frac{1}{32}$

The expected value is

$$E(X) = 0\left(\frac{1}{32}\right) + 1\left(\frac{5}{32}\right) + 2\left(\frac{10}{32}\right) + 3\left(\frac{10}{32}\right) + 4\left(\frac{5}{32}\right) + 5\left(\frac{1}{32}\right) = 2.5.$$

Now try Exercise 54a and b.

Half a Head?

No, we do not "expect" to ever get 2.5 heads when we toss five coins; that's impossible. This is the long-run average. We expect that if we were to toss five coins many, many times, the average number of heads would be about 2.5.

The expected number of heads turns out to be exactly what the shortcut of multiplying the number of trials by the probability of success predicts: $5(0.5) = 2.5$. We state the general case as a theorem.

THEOREM Expected Value for a Binomial Distribution

If the binomial random variable X counts the number of successes in n independent trials with probability of success p, then $\mu = E(X) = np$.

In Example 5, Michael makes 90% of his free throws. If he shoots 20 free throws, we would expect him to make $np = 20(0.90) = 18$ shots.

EXAMPLE 7 A Chimpanzee Takes Organic Chemistry

A chimpanzee is shown 100 questions from a multiple-choice test in organic chemistry. As each question appears on a screen, the chimp randomly presses one of five buttons (A, B, C, D, or E) to indicate its answer. What is the chimp's expected score on the test?

SOLUTION

If we assume the chimpanzee is unfamiliar with the material, does not read well, and is equally likely to press any of the answer buttons, the number of correct answers it will get is a binomial random variable with 100 independent trials and a probability of success $p = 1/5$. The expected number of correct answers is $100(1/5) = 20$. (Of course, this does not mean that the chimp knows 20% of organic chemistry. See the margin note.)

Now try Exercise 25a.

The "Penalty" for Guessing

You may know that some standardized tests are scored with a so-called "guessing penalty formula." For a test with five options, as in Example 7, the formula for R right answers and W wrong answers would be $R - 1/4W$. A chimp with the expected score of 20 would get an adjusted score of $20 - 1/4(80) = 0$. That's a more accurate reflection of what the chimpanzee actually knew about organic chemistry, not really a penalty of any kind.

Note that expected value of 20 correct answers by just guessing is the *mean*, so in a large class of chimpanzees some "students" would get scores higher than 20 (and, of course, some lower). How lucky might some chimp get? To gauge that, we need to know how much variation there is in scores achieved by guessing. We measure spread with standard deviation, and for binomial random variables we can determine that with a shortcut, as we did for the mean. This time, though, it's not so intuitively obvious. We present the theorem here, and will let you investigate further in the exercises.

THEOREM Standard Deviation for a Binomial Distribution

If the binomial random variable X counts the number of successes in n independent trials with probability of success p, then the standard deviation of X is $\sigma = \sqrt{npq}$.

EXAMPLE 8 Variation in Chimpanzee Scores

If many chimpanzees take the 100-question multiple-choice test in organic chemistry, what is the standard deviation of the number of correct answers they would get?

SOLUTION

Again we have the binomial random variable X = the number of correct answers on $n = 100$ questions with probability of success $p = 1/5$. The standard deviation of the number of correct answers is $\sigma = \sqrt{npq} = \sqrt{100(0.2)(0.8)} = 4$.

Now try Exercise 25b.

With a mean of 20 correct answers and a standard deviation of 4, some lucky chimpanzees might be able to get 25 questions (or slightly more) right, but it seems that actually passing the test would be well out of reach.

Normal Model

In Section 10.3, when we first presented the Normal model, you learned the 68-95-99.7 Rule. This handy guideline says that when data appear to be approximately normally distributed, about 68% of observations should fall within ± 1 standard deviation of the mean, 95% within ± 2 standard deviations, and 99.7% within ± 3 standard deviations of the mean. These rules of thumb arise from looking at the area under a Normal curve that describes the probability model for a continuous random variable that is normally distributed. Graphers have a function that calculates such areas more accurately (Figure 10.24).

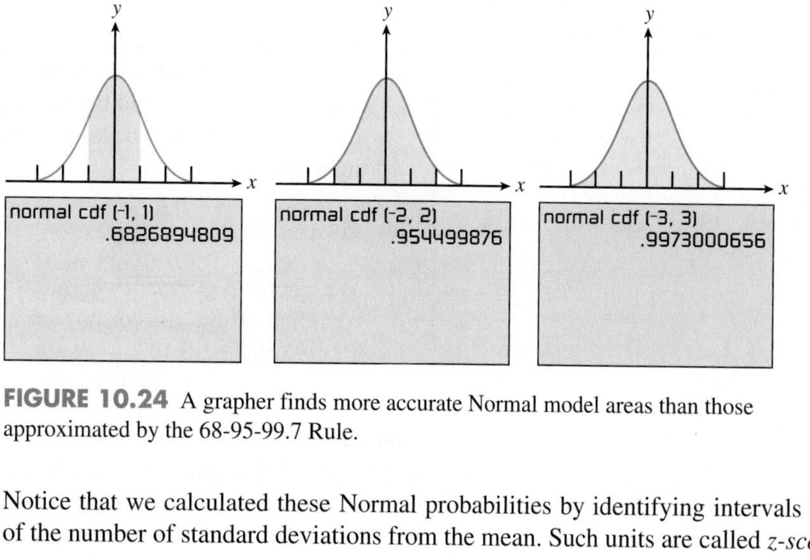

FIGURE 10.24 A grapher finds more accurate Normal model areas than those approximated by the 68-95-99.7 Rule.

Notice that we calculated these Normal probabilities by identifying intervals in terms of the number of standard deviations from the mean. Such units are called *z-scores*.

DEFINITION z-Score

A **z-score** measures distance from the mean in standard deviation units. If a random variable X has mean μ and standard deviation σ, then for a given value of X

$$z = \frac{x - \mu}{\sigma}.$$

If we know that it's appropriate to use a Normal model, we can use z-scores to find probabilities. We identify the interval of interest by specifying the z-score for each endpoint, for example: $1.2 < z < 2.3$. If we're interested in an unbounded interval such as $z > 2$, we simply specify a very extreme value such as 99 for the missing endpoint. With 99.73% of the area in the interval $-3 < z < 3$, there's virtually nothing beyond $z = 99$.

EXAMPLE 9 Finding Normal Probabilities

According to the *National Health and Nutrition Examination Survey* (NHANES), the heights of adult American men can be described by a Normal model with a mean of 69.2 in. and standard deviation of 2.8 in. What's the probability that a randomly selected adult American male is

(a) between 65 and 70 in. tall?

(b) over 6 ft tall?

(c) shorter than 5 ft 6 in. tall?

SOLUTION

(a) First we find the z-scores:

$$z = \frac{65 - 69.2}{2.8} = -1.5$$

$$z = \frac{70 - 69.2}{2.8} \approx 0.29$$

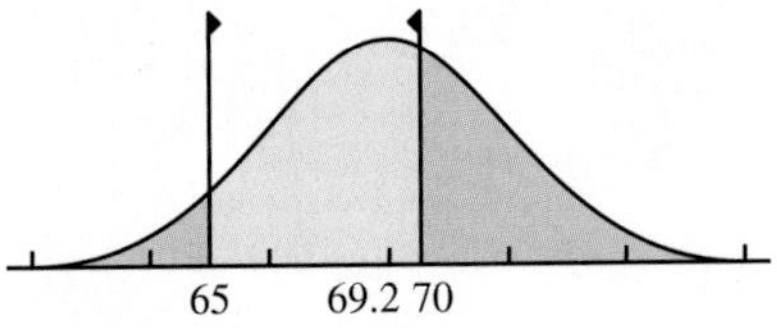

Using the grapher, $P(-1.5 < z < 0.29) = 0.547$.

(b) For $x = 72$, $z = \frac{72 - 69.2}{2.8} = 1.0$

$P(x > 72) = P(z > 1) = 0.159$

(Note that the 68-95-99.7 Rule estimates that about 16% of adult American males should be more than one standard deviation above average in height.)

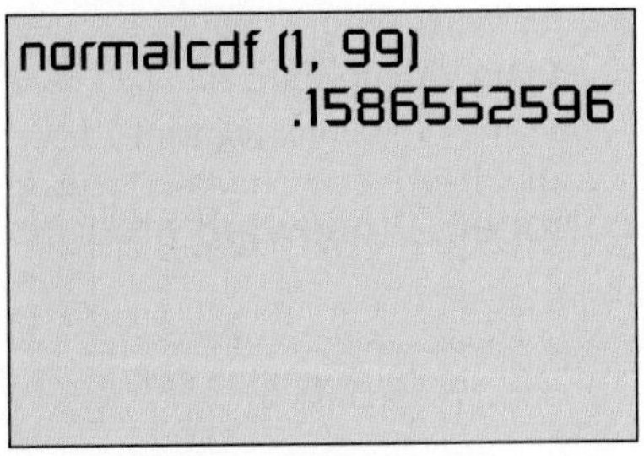

(c) For $x = 66$, $z = \frac{66 - 69.2}{2.8} = -1.14$.

We can use a grapher (without rounding off the z-score) to find $P(x < 66) \approx 0.127$.

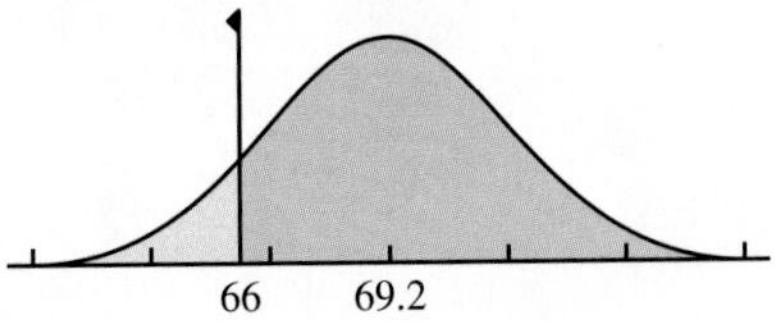

Now try Exercise 33.

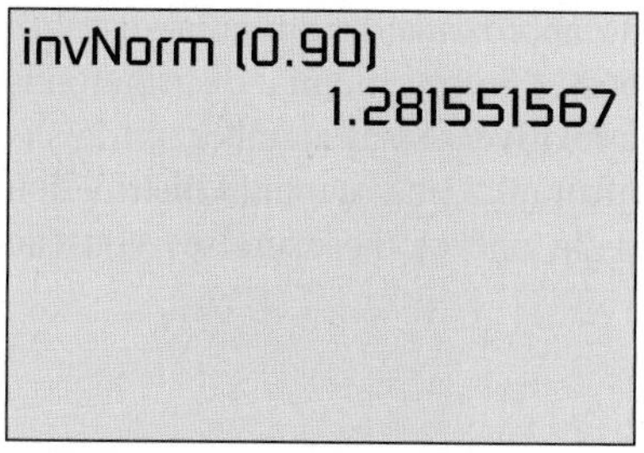

FIGURE 10.25 A grapher finds the z-score for the 90th percentile in a Normal distribution.

In each of the applications above we have used a Normal model to find the probability that certain z-scores occur. It's also possible to reverse the process: We can find the z-score that corresponds to a certain probability using the grapher's inverse Normal function. Figure 10.25 shows that the bottom 90% of a Normal distribution is less than about 1.28 standard deviations above the mean. In other words, $z = 1.28$ represents the 90th *percentile* for a normally distributed random variable. A grapher's inverse Normal function can find the z-score corresponding to any percentile.

EXAMPLE 10 Finding Normal Percentiles

Heights of adult American men can be described by a Normal model with a mean of 69.2 in. and standard deviation of 2.8 in.

(a) The shortest 20% of American men are less than what height?

(b) How big are the tallest 1% of American men?

SOLUTION

(a) Using the grapher's inverse Normal function, we find the 20th percentile at $z = -0.842$. This represents a height that is 0.842 standard deviation below the mean: $69.2 - 0.842(2.8) \approx 66.8$ in. The shortest 20% of adult American men are less than 5 ft 6.8 in tall.

(b) The cutoff height for the tallest 1% of all men is the 99th percentile, and the grapher finds that z-score to be $z = 2.326$. The height that is 2.326 standard deviations above average is $69.2 + 2.326(2.8) \approx 75.7$ in. The tallest 1% of adult American men are over 6 ft 3.7 in. tall. **Now try Exercise 35.**

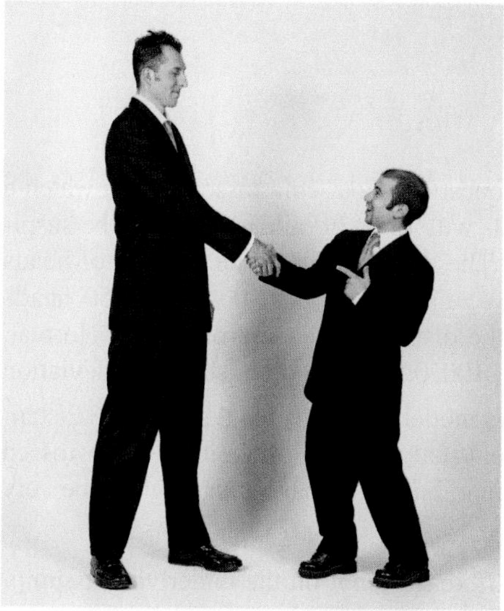

Normal Approximation for Binomial Distributions

A binomial model and a Normal model are quite different. Binomial random variables are discrete: When tossing a coin we can have only a whole number of heads. Normal random variables, on the other hand, are continuous: Heights can (and as you grow, they must) take on any value. Binomial random variables are bounded; in n trials, the number of successes must be between 0 and n. A Normal model extends without bound in either direction. Despite these differences, a binomial probability distribution based on a sufficiently large number of trials is approximately Normal. Figure 10.26a shows the binomial distribution for the number of 3's seen in 10 rolls of a fair die. Because it is skewed to the right, a Normal model is not appropriate. In Figure 10.26b the number of rolls is increased to 100. The superimposed Normal curve is barely discernible, showing how good the approximation is now.

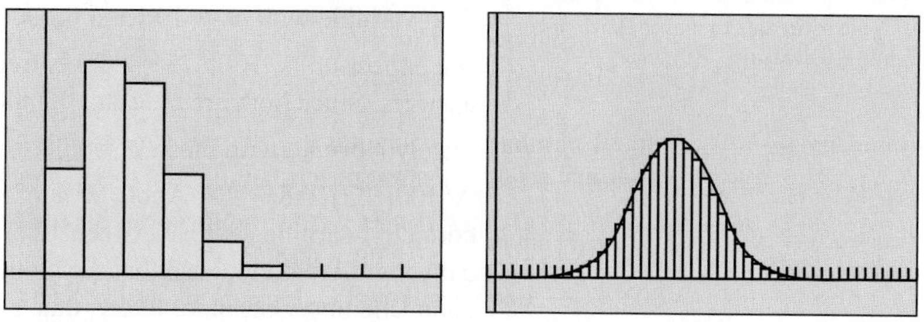

(a) (b)

FIGURE 10.26 The binomial distribution for the number of 3's in 10 rolls of a fair die (a) is skewed to the right, far from Normal. When there are 100 rolls (b), a Normal curve approximates the binomial distribution very well.

The number of trials required for a Normal model to approximate a binomial distribution well depends on the probability of success. If $p = 0.5$, the binomial distribution is symmetric, an important attribute of normality, so a relatively small number of trials is sufficient. But if (as in the case of the die) p is far from 0.5, the binomial distribution will be skewed, and many more trials are needed. In one of the exercises you can explore why the following rule of thumb works.

Success-Failure Rule

If we expect at least 10 successes and 10 failures ($np \geq 10$ and $nq \geq 10$), then the binomial distribution for the number of successes can be approximated by a Normal model with $\mu = np$ and $\sigma = \sqrt{npq}$.

At the beginning of this section we posed a question about coin tosses. If you watch a person toss a coin 100 times and get 66 heads with only 34 tails, should you be suspicious that the tosser is cheating somehow? The distribution of the number of heads is binomial, with $n = 100$ and $p = 0.5$. Because we expect $100(0.5) = 50$ heads and $100(0.5) = 50$ tails, both at least 10, the distribution is approximately Normal. This Normal model has mean $\mu = np = 100(0.5) = 50$ and standard deviation $\sigma = \sqrt{npq} = \sqrt{100(0.5)(0.5)} = 5$. In this model, 66 heads would be $z = 3.2$ standard deviations above the mean, a highly unusual result if the coin is being tossed fairly. This does not *prove* the tosser is cheating, but you'd have every right to be very suspicious.

Outcomes like this, which are so unusual as to cast doubt on the underlying assumptions, are said to be **statistically significant**.

True-False

Suppose that instead of multiple choice, the test given to the chimps consisted of 100 true-false questions. Then the expected number of correct answers would be $np = 100(0.5) = 50$ and the standard deviation would be $\sqrt{npq} = \sqrt{100(0.5)(0.5)} = 5$. A passing score of 60 would be 2 standard deviations above the mean. The 68-95-99.7 Rule estimates that even 2.5% of chimpanzees could pass this true-false test!

EXAMPLE 11 A Lucky Chimpanzee?

In Examples 7 and 8 we imagined chimpanzees choosing 1 of 5 answers at random on a 100-question multiple-choice test in organic chemistry. How high a score might a lucky chimpanzee be likely to get?

SOLUTION

Guessing among five choices makes the number of correct answers a binomial random variable with $n = 100$ and probability of success $p = 1/5$.

We expect $np = 100(1/5) = 20$ correct answers and $nq = 100(0.8) = 80$ wrong answers. Since both are at least 10, this binomial probability distribution is approximately Normal, with mean $\mu = np = 20$ and standard deviation $\sigma = \sqrt{npq} = \sqrt{100(0.2)(0.8)} = 4$. About 95% of all chimps should guess $20 \pm 2(4)$ answers correctly, earning scores between 12 and 28. Any score above 28 could be statistically significant evidence that the chimp actually understood something about organic chemistry or, more likely, that someone was helping with the answers. It would be highly unusual for a chimp to get more than $20 + 3(4) = 32$ answers right, and actually passing is so many standard deviations above average as to be virtually impossible.

Now try Exercise 39.

CHAPTER OPENER Problem, part 2 (from page 676)

Problem: Are some lottery numbers more likely to win than others? That's the impression one might get from historical data for Wisconsin's Megabucks game, in which players choose 6 numbers from 1 to 49. Between the first drawing on June 21, 1992, and the 2205th drawing on July 27, 2013, a total of 13,230 numbers were chosen. The number 9 was a winner more often than any other, 295 times in all. Could this be just random variation, or is there statistically significant evidence that 9 actually *is* a lucky number?

Solution: Let the random variable $X =$ the number of 9's in $n = 13{,}230$ trials. The probability of drawing a 9 is $p = 1/49$, assuming all the lottery numbers are equally likely to appear.

We expect far more than ten of the numbers to be 9's and even more not to be, so we can approximate the distribution of X with a Normal model.

We'd expect a mean of $\mu = np = 13{,}230(1/49) = 270$ nines, with a standard deviation of $\sigma = \sqrt{npq} = \sqrt{13{,}230(1/49)(48/49)} \approx 16.26$.

The z-score for 295 9's is $z = \dfrac{295 - 270}{16.26} \approx 1.54$.

An outcome 1.54 standard deviations above the mean is somewhat higher than expected, but not unusually so. Seeing the number 9 come up 295 times is not statistically significant. This result can be explained simply by random variation, so it is *not* evidence that there are lucky lottery numbers.

QUICK REVIEW 10.4 *(Prerequisite skill Section 9.1)*

In Exercises 1–6, tell how many outcomes are possible for the random event.

1. A single coin is tossed.
2. A coin is tossed 3 times.
3. The number of heads when 3 coins are tossed.
4. A set of 5 winning lottery numbers is chosen from 1–44.
5. Three different cards are drawn from a deck.
6. Five people line up for a photograph.

In Exercises 7–10, evaluate the expression by hand. Verify your answer with a calculator.

7. $\binom{4}{2}$
8. $\binom{10}{0}$
9. $\binom{8}{3}$
10. $\binom{100}{98}$

SECTION 10.4 Exercises

1. Find the expected value of this random variable:

X	10	20	30
$P(X)$	0.5	0.3	0.2

2. Find the expected value of this random variable:

X	2	4	6	8
$P(X)$	0.1	0.2	0.3	0.4

3. A game involves earning points by rolling a special 6-sided die that says 5 on three faces, 10 on two, and 25 on the sixth face. What's the expected number of points a player will earn by rolling this die?

4. A 6-sided die has been weighted so that face 6 will come up 60% of the time. If the remaining faces are all equally likely to appear, what's the expected value of a roll of this die?

5. Many children's games use spinners. Suppose a spinner has 5 regions numbered 1–5. Regions 1, 2, and 3 are each a quarter-circle, and regions 4 and 5 divide the remaining area equally. What's the expected value of a spin?

6. Some fantasy games use a 12-sided die. Suppose one such die has a 1 on seven of the faces, a 5 on three faces, and 10's on the remaining two faces. What's the expected value of a roll of this die?

7. At a carnival, people pay The Great Flubini \$2 each to try to guess their weights within five lb. If he guesses successfully, no prize is given. If he is off from five to ten lb, participants get a prize worth \$5. If he is off by more than ten lb, participants get a \$10 prize. The Great Flubini is correct 72% of the time, off from five to ten lb 21% of the time, and off by more than ten lb 7% of the time. If The Great Flubini guesses the weights of 300 people per day, on average, is the carnival making money or losing money? How much per day?

8. For \$5, a carnival huckster will let you draw a card at random from a fair deck of 52. He will pay you a dollar times the value of your card $(\text{ace} = 1)$, but zero for a face card. What is the expected value of the payoff? Would you pay \$5 to play this game?

9. A bag contains twenty numbered balls. Five of the balls have the number 1 on them. Three balls have the number 2 on them. Seven balls have the number 3 on them. Four balls are marked with a 4. One ball has a 5 on it. Let Y be the random variable that gives the number on a ball chosen at random from the bag.

(a) Show the probability distribution for Y in a table.

(b) Calculate the expected value for Y.

10. A circular spinner consists of eight identical sectors. The spinner is equally likely to land in any of the eight sectors. Four of the eight yield no payout at all if the spinner lands there. Three of the eight pay \$6 if the spinner lands there. One sector pays \$10. It costs \$5 to spin the spinner once. Use expected value to determine if it is a wise idea to play this game.

11. A three-year extended warranty on a laptop computer is offered for \$79. Consumer reports show that typically, 4% of owners incur a \$200 repair in that time and 1% of owners incur a \$300 repair in that time. Does the extended warranty have a positive expected value for the consumer who purchases it?

12. For an extra \$49 you can buy extended warranty protection on your new HDTV for three years. If the set requires maintenance during that time, it will be free. You estimate that there is a 5% chance that you might need a \$300 repair within three years and a 1% chance that you will need a \$500 repair, but the chances are 94% that you will need no repair at all. Based on expected values, should you purchase the extended coverage?

13. A married couple decides to have children until they get a girl, but agree that they will not have more than 3 children even if all are boys.

(a) Assuming that boy and girl children are equally likely, create a probability model for the number of children such couples may have.

(b) Find the expected number of children for these couples.

14. In basketball, a player shooting "one-and-one" takes one foul shot, and if he makes the first one, he gets to shoot a second shot. Suppose that, overall, a certain player makes 80% of his foul shots.

(a) Create a probability model for the number of points the player may score on a one-and-one opportunity.

(b) Find the expected number of points.

15. Games are considered "fair" if the expected value is 0. Suppose you pay \$5 to draw a card from a deck. If it's black, you lose. If it's a heart, you win your \$5 back. If it's any diamond except the ace, you win \$10. What should be the prize for drawing the ace of diamonds in order to make the game fair?

16. A state's Daily Number lottery game offers a \$250 payout on a 50-cent ticket. To play, a person selects a 3-digit number. If the same number is drawn randomly by the state lottery that day, the person wins. On a typical day the state sells a million of these tickets. Find the state's expected daily profit.

17. You roll 4 dice. Find the probability of each of these outcomes.

(a) Exactly one 6

(b) Exactly two 6's

(c) At least two 6's

18. You toss 10 coins. Find the probability of each of these outcomes.

(a) Exactly 5 heads

(b) Exactly 8 heads

(c) At least 8 heads

19. Suppose a basketball player can make 80% of her foul shots. If she gets 6 foul shots during a game, find the probability she makes

(a) exactly 5 of them.

(b) no more than 2 of them.

20. About 12% of people are left-handed. Find the probability of each of these outcomes among a class of 15 students.

(a) Exactly 3 are left-handed.

(b) No more than 2 are left-handed.

21. Suppose that the traffic light at the driveway leading into your school is red for 40% of its green-yellow-red cycle. During a 10-day period, what is the probability your school bus has to stop there

(a) on exactly 5 days?

(b) on at least 8 days?

22 The American Red Cross reports that 11% of people have Type B blood. Find the probability that among 12 donors

(a) exactly one is Type B.

(b) at least two are Type B.

23. Writing to Learn We will roll a single 6-sided die.

(a) Let random variable X = the number that comes up. Explain why the probability model for X is *not* binomial.

(b) In the context of rolling the die, define a random variable that would have a binomial probability model.

24. Writing to Learn We will draw one card from a deck.

(a) We assign 1 point to the aces, 10 points to any face card, and points equal to the number on the card to the rest. Let random variable X = the number of points on the card we draw. Explain why the probability model for X is *not* binomial.

(b) In the context of drawing a card, define a random variable that would have a binomial probability model.

25. A baseball player with a career batting average of .300 comes up to bat 75 times in a month.

(a) What is the mean number of hits such players should get?

(b) What is the standard deviation?

26. The probability a certain archer can hit the bull's-eye is 0.85. If she shoots 50 arrows, what are the mean and standard deviation of number of bull's-eyes she might make?

27. Writing to Learn On your math test students averaged 82, with a standard deviation of 7 points. The English test scores averaged 86, with a standard deviation of only 4 points. You earned a 98 on each test. (Congratulations!) Explain which performance was actually better.

28. Writing to Learn One of the athletes at a track meet ran the 100-m dash in 10.2 sec and had a long jump of 22 ft. Historical records indicate that competitors this age (1) average 10.4 sec for the dash, with a standard deviation of 0.3 sec, and (2) average 20 ft in the long jump, with a standard deviation of 2.5 ft. Which performance by this athlete was better?

29. Find each Normal probability using your grapher.

(a) $P(z > 1.5)$

(b) $P(z < 0.75)$

(c) $P(-1.8 < z < -0.5)$

30. Find each Normal probability using your grapher.

(a) $P(z > -1.2)$

(b) $P(z < -0.6)$

(c) $P(-2 < z < 1.25)$

31. Use your grapher to find the z-score(s) bounding each region in a Normal distribution.

(a) The top 3%

(b) The 60th percentile

(c) The middle 80%

32. Use your grapher to find the z-score(s) bounding each region in a Normal distribution.

(a) The 15th percentile

(b) The top 8%

(c) The middle 90%

33. The distribution of heights of American women is approximately Normal, with a mean of 63.8 in and a standard deviation of 2.8 in. Find the probability of each.

(a) A randomly selected woman is taller than 5 ft 10 in.

(b) A randomly selected woman is shorter than 5 ft 6 in.

34. SAT scores are modeled by a Normal curve with a mean of 500 and standard deviation of 100. Find the probability that a randomly selected student attained these scores.

(a) At least 650

(b) No more than 420

35. Given the distribution of women's heights described in Exercise 33, how tall are the shortest 5% of all adult American women?

36. Given the distribution of SAT scores described in Exercise 34, above what value are the highest 2% of all SAT scores?

37. A certain brand of AAA batteries lasts an average of 13 hr, with a standard deviation of 1.2 hr. The lifespans can be described by a Normal model.

(a) What fraction of these batteries should last at least 10 hr?

(b) What's the probability a battery will last between 14 and 15 hr?

(c) The company wants to guarantee that these batteries will last a certain length of time. What guaranteed lifespan could be achieved by at least 96% of them?

38. A certain type of tire will last for an average of 37,200 mi, with a standard deviation of 2650 mi. Suppose that the tread life can be described by a Normal model.

(a) What percent of these tires wear out before they reach 40,000 mi?

(b) What's the probability a tire will wear out between 32,000 and 35,000 mi?

(c) For how many miles can the company guarantee these tires, if the company wants at least 90% of them to last that long?

39. Many countries of Europe have adopted a common currency. When the "euro" coin was first introduced, several newspapers published an article claiming that it was biased. The story was based on reports that when someone spun a euro it came up heads 140 times in 250 spins. Is this evidence of bias?

(a) Is it okay to use a Normal model to describe the possible number of heads in 250 spins?

(b) Find the mean and standard deviation of the number of heads in 250 spins of a fair coin.

(c) Do you think 140 heads is statistically significant? Explain.

40. Before a blood drive, there was a public call for Type AB donors. Although only 4% of people have Type AB blood, the drive netted 26 Type AB donations among the 320 pints of blood that were collected that day. Is this evidence that the public plea worked?

(a) Is it okay to use a Normal model to describe the possible number of Type AB donors among 320 people?

(b) Find the mean and standard deviation of the number of Type AB donors.

(c) Do you think 26 Type AB donors is statistically significant? Explain.

41. Writing to Learn To test a person who claims to "have ESP," researchers use 4 cards with different symbols (a circle, a square, a triangle, and an X). Someone will pick a card at random and concentrate on it. The "mind reader" must then indicate what picture is on the card. The test will consist of 100 such trials. How many would the person have to get right to convince you that ESP actually exists? Explain.

42. Writing to Learn Biologists are concerned that environmental changes are affecting a certain species of frog. In the past, about 1 frog in 12 exhibited a skin discoloration defect, but the scientists think this may be more prevalent now. To find out, they plan to catch and examine 150 of the frogs. How many frogs in this sample must exhibit the skin discoloration in order to provide statistically significant evidence that this defect is becoming more common? Explain.

Standardized Test Questions

43. True or False The expected value of a random variable is a statistic calculated from sample data. Justify your answer.

44. True or False In any data set, about 16% of the values will be more than one standard deviation larger than the mean. Justify your answer.

45. Multiple Choice Given the probability model below, what is the expected value of the random variable?

X	50	20	5
$P(X)$	0.1	0.3	0.6

(A) 4.67 (B) 5

(C) 7.5 (D) 14

(E) 25

46. Multiple Choice Which of these has a binomial probability model?

(A) The number of aces in a 5-card hand

(B) The number of people with blue eyes in a random sample of 20 people

(C) The total number of spots when two dice are rolled

(D) The number of times you roll a die in order to get a 6

(E) The length of the longest run of heads in 100 tosses of a fair coin

47. Multiple Choice A fair coin is tossed three times in succession. What is the probability that exactly one of the coins shows heads?

(A) $\frac{1}{8}$ (B) $\frac{1}{3}$

(C) $\frac{3}{8}$ (D) $\frac{1}{2}$

(E) $\frac{2}{3}$

48. Multiple Choice In a Normal model, approximately what percent of the observations lie within one half of a standard deviation from the mean?

(A) 31% (B) 34% (C) 38%

(D) 62% (E) 69%

Explorations

49. Married Students Suppose that 23% of all college students are married. Answer the following questions for a random sample of eight college students.

(a) How many would you expect to be married?

(b) Would you regard it as unusual if the sample contained five married students?

(c) What is the probability that five or more of the eight students are married?

50. Investigating an Athletic Program A university widely known for its track and field program claims that 75% of its track athletes get degrees. A journalist investigates what happened to the 32 athletes who began the program over a 6-year period that ended 7 years ago. Of these athletes, 17 have graduated and the remaining 15 are no longer attending any college. If the university's claim is true, the number of athletes who graduate among the 32 examined should have been governed by binomial probability with $p = 0.75$.

(a) What is the probability that exactly 17 athletes should have graduated?

(b) What is the probability that 17 or fewer athletes should have graduated?

(c) If you were the journalist, what would you say in your story on the investigation?

51. Group Activity The Game Show Audience Suppose a TV game show host offers everyone in the audience the choice of three deals:

Deal A: Pick an envelope from five in his hand. The envelopes contain a ten dollar bill, a twenty dollar bill, a fifty dollar bill, a hundred dollar bill, and a check for five thousand dollars.

Deal B: Choose one of three suitcases. Two are empty and the third contains 240 twenty-dollar bills.

Deal C: Take $1000 with no strings attached.

Which deal would you take? Which deal do the show sponsors hope you will take?

52. Group Activity Strange Dice A game is to be played with two 6-sided dice, but not the ordinary kind. The faces of the red die show five 2's and one 6. The faces of the green die show two 1's and four 3's. You will roll one of the dice and your opponent will roll the other.

(a) If the winner will be the person who rolls the higher number, which die do you want? Explain why.

(b) If the winner will be the person who has the highest total after 10 rolls, which die do you want? Explain why.

Extending the Ideas

53. Standard Deviation The mean (μ) of a random variable is the expected value. We can use the probability model to find the standard deviation (σ) in much the same way we found the standard deviation for sample data. We calculate the square root of the expected value of squared deviations from the mean: $\sigma = \sqrt{\sum_{i=1}^{n} (x_i - \mu)^2 \times P(x_i)}$.

(a) Suppose a carnival game offers prizes described by the probability model below. Find the expected value of the game.

X = prize	$0	$10	$50	$100
$P(X)$	0.80	0.15	0.04	0.01

(b) Find the standard deviation of the value of the prizes.

(c) Writing to Learn Explain why the standard deviation is so large.

(d) Writing to Learn Would you be willing to pay \$5 to play this game? Explain.

54. Binomial Standard Deviation Let random variable $X =$ the number of heads in 4 tosses of a fair coin.

(a) The possible values of X are 0, 1, 2, 3, and 4. Create a probability model for X.

(b) Use the expected value formula to find $E(X)$.

(c) Use the formula given in Exercise 53 to calculate the standard deviation of the number of heads.

(d) Verify that these are the same values found by the shortcut formulas $\mu = np$ and $\sigma = \sqrt{npq}$.

55. Margin of Error When pollsters try to predict the outcome of an election, they always report a "margin of error" as a percentage. What they don't say is that the margin of error is based on a Normal model, and they are only 95% confident that the poll correctly predicts the outcome of the election.

Suppose a pollster contacts a random sample of 625 registered voters, and 325 of them express a preference for Candidate A.

(a) What percent of the vote does the sample suggest this candidate will get?

(b) Using that percentage as an estimate of the probability that a voter contacted by a pollster supports Candidate A, what is the standard deviation of the number of supporters pollsters might find in samples this size?

(c) Verify that a Normal model is appropriate here.

(d) What margin of error, expressed as a percent, would provide the pollsters with 95% confidence in their prediction?

(e) Writing to Learn Explain why the pollsters might say this election is "too close to call."

56. The Success-Failure Rule In any binomial probability distribution the number of successes must be at least 0 and less than n, the number of trials. Therefore, when we want to use a Normal model as an approximation, we must cut off the curve's unbounded tails. That will not cause much inaccuracy if what we cut off lies more than 3 standard deviations from the mean. Verify that if np and nq are both at least 10, then 0 and n are both more than 3 standard deviations from the mean of a binomial distribution.

10.5 Statistical Literacy

What you'll learn about

- Uses and Misuses of Statistics
- Correlation Revisited
- Importance of Randomness
- Samples, Surveys, and Observational Studies
- Experimental Design
- Using Randomness
- Simulations

... and why

Statistical literacy is important in today's data-driven world.

Uses and Misuses of Statistics

Just as knowing a little bit about edible wild mushrooms can get you into trouble, so can knowing a little about Statistics. Unfortunately, a lack of true understanding does not stop people from misusing statistics every day to draw conclusions, many of them totally unjustified, and then inflicting those conclusions on you. We will therefore end this chapter with a brief "consumer's guide" to the most common uses and misuses of statistics.

EXPLORATION 1 Test Your Statistical Savvy

Each one of the following scenarios contains at least one common misuse of statistics. How many can you catch?

1. A researcher reported finding a high correlation between aggression in children and gender.
2. Based on a survey of shoppers at the city's busiest mall on two consecutive weekday afternoons, the mayor's staff concluded that 68% of the voters would support his re-election.
3. A doctor recommended vanilla chewing gum to headache sufferers, noting that he had tested it himself on 100 of his patients, 87 of whom reported feeling better within two hours.
4. A school system studied absenteeism in its secondary schools and found a negative correlation between student GPA and student absences. They concluded that absences cause a student's grade to go down.
5. A medical experiment that tested a drug for effectiveness found improvement in 64% of the 25 subjects who took the drug, compared to only 40% of the 25 subjects who were given a placebo (an identical-looking but inert pill). A newspaper article proclaimed that the treatment had been proven effective.

Correlation Revisited

We mentioned r, the correlation coefficient, briefly at the end of Section 1.7, along with R^2, the coefficient of determination. We suggested using the closeness of R^2 to 1 as a measure of how well a curve fits the points of a scatter plot. In Section 2.1 we noted that statisticians commonly use r for a similar purpose, but always as a measure of how close the data points are to a *straight line*. We now want to be a little more emphatic about that point. Consider the following graphs:

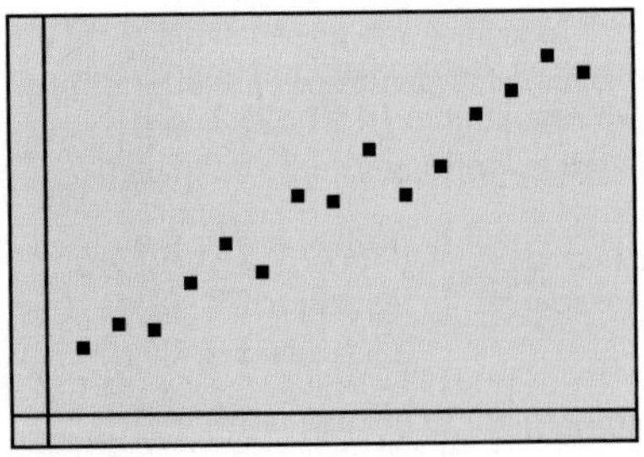

$r = 0.975$

FIGURE 10.27

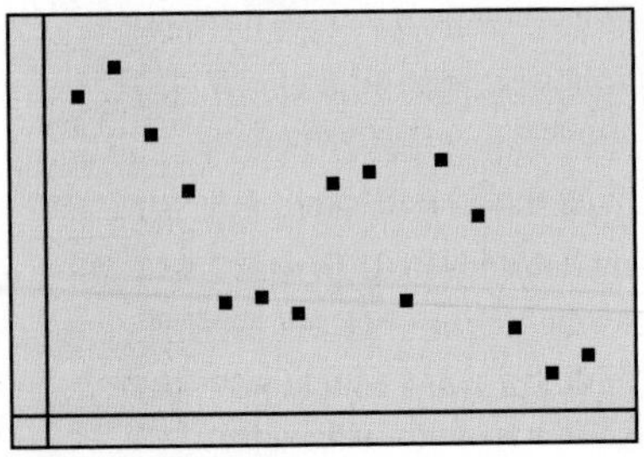

$r = -0.689$

FIGURE 10.28

Figure 10.27 shows a **positive association** between the y-variable and the x-variable, as the higher y values are generally associated with the higher x values. Because the underlying relationship appears to be *linear*, it is appropriate to consider the correlation coefficient, r. It is positive (because the line's slope is positive), and it is close to $+1$ because the points are close to the line. It is appropriate to say that there is a **strong positive correlation**.

Figure 10.28 shows a definite **negative association** between the y-variable and the x-variable, as the higher y values are generally associated with the lower x values. Because the underlying relationship appears to be linear, it is appropriate to consider the correlation coefficient, r. It is negative (because the line's slope is negative), but because the points are quite scattered, it is not very close to -1. It is appropriate to say that there is a **weak negative correlation**.

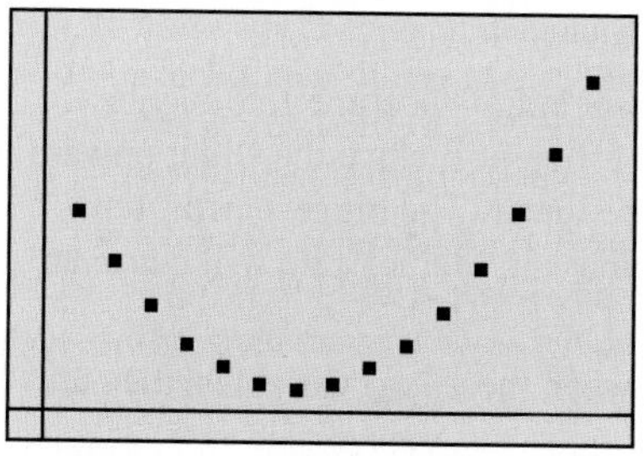

r is not appropriate.

FIGURE 10.29

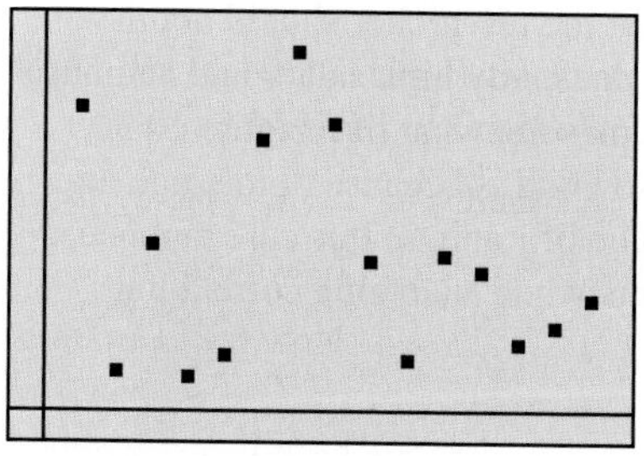

$r = -0.227$

FIGURE 10.30

Transforming Data

Statisticians will often use function techniques to transform data to conform to linear models, precisely so that correlation analysis can be used. (We did this in Example 7 in Section 3.4.) When your calculator reports an r value for a nonlinear regression, it is using the transformed data.

Figure 10.29 shows a **strong association** between the y-variable and the x-variable. The association is negative for the smaller x values and positive for the higher x values. Because the relationship is curved, the correlation coefficient in this case is an *inappropriate measure*. We should not even be talking about correlation here, because that would be treating this relationship as linear!

Figure 10.30 shows a **very weak negative association** between the y-variable and the x-variable. It may not be very useful to model this relationship with a line; indeed, we might wonder whether the down-up-down-up pattern suggests a periodic function. The correlation coefficient happens to be negative and close to 0, but we would be concerned with it only if we believed a *linear model to be appropriate*. If that is our belief, then we could appropriately say that there is a **very weak negative correlation**.

Now that you know the terms, here are the things to watch for.

Correlation Correctness

Be sure both variables are quantitative. If no scatter plot is shown, ask yourself if you could possibly have the data to make one.

Be sure the underlying pattern in the scatter plot is linear. Strength of correlation is not an appropriate measure for data exhibiting nonlinear behavior.

Don't confuse association with correlation. Association is a much broader term (even applicable to categorical data), whereas correlation should be used only for quantitative data to measure the strength of an association that we believe to be linear.

Correlation does not imply causation. If the points line up nicely, it is tempting to conclude that the variable y is reacting to the variable x. It might actually be the other way around, or both variables might be reacting to a third variable not under consideration.

EXAMPLE 1 Being Critical About Correlation

In each case, tell whether correlation is being used correctly. Identify any statistical errors.

(1) A graph of per capita gross domestic product (y) against literacy rate (x) for 17 world countries is shown in Figure 10.31. The strong positive association (with $r = 0.788$) proves that raising a country's literacy rate will improve its per capita gross domestic product.

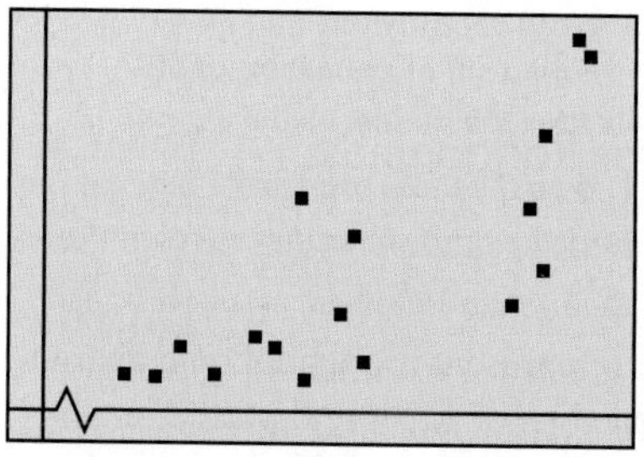

FIGURE 10.31

(2) A research company reported a strong correlation between the religious affiliation of American adults and the types of charitable organizations they support.

(3) A famous study of college freshmen once showed that the high school measure that had the highest positive correlation with freshman GPA in college was "number of mathematics courses taken in high school." It showed that the best preparation for college was to take a lot of math courses.

(4) A researcher in 1990 measured the heights and the annual salaries of all the people named Michael Jordan in Chicago. The correlation was an astounding 0.97, suggesting a strong linear relationship between height and salary among the group.

(continued)

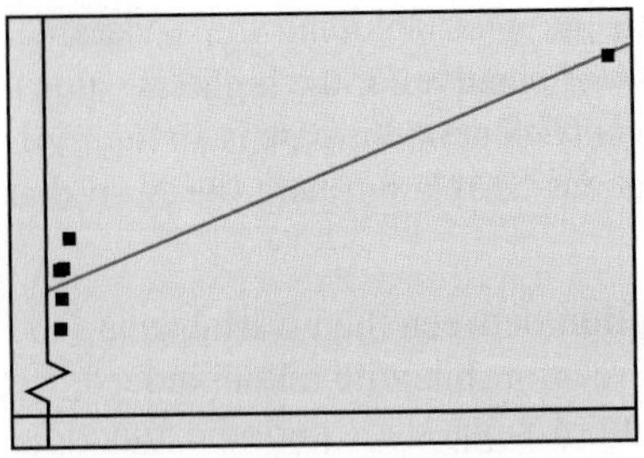

FIGURE 10.32

SOLUTION

(1) While the reference to the "strong positive association" is correct, it may be inappropriate here to report the correlation coefficient r, since the slight upward bend suggests that the best model might not be linear. The biggest error, however, is jumping to the conclusion that an increase in literacy rate will *cause* an increase in per capita GDP.

(2) While there might well be an *association* between these two variables, they are both categorical, so no correlation is possible.

(3) The error is in the conclusion that more math courses will *cause* the GPA to go up. In fact, both variables here are probably responding to a third variable, the academic motivation of the student.

(4) Everything here is actually correct, except that the researcher should not have been astounded. One of the Michaels had an unusually high salary and an unusually high height, causing the scatter plot to look somewhat like Figure 10.32. The *only* reason that the linear model looks so good is because of that one unusual point, which results in a *misleading* value of r and, in this case, a misleading regression equation. Beware of such points when analyzing correlation.

Now try Exercise 5.

Importance of Randomness

As we've said before, the aim of Statistics is to draw meaning from data. The real goal extends beyond the data. **Inferential statistics** uses collected observations to draw conclusions about broader populations and real-world phenomena.

Statistical inference absolutely depends on randomness, ironically because it is so predictable. Take another look at Example 11 in Section 10.3, which used the 68-95-99.7 Rule (a rule based on the laws of probability) to infer a fact about loon chicks in the population. Since the inference was based on a sample of 30 loon chicks, it was vital that we trusted these 30 chicks as being typical of the overall population. Would we have trusted the sample if all 30 chicks had been born in the same wildlife preserve in northern Michigan? Probably not. Suppose a later study were to show that this particular preserve produced unusually large loon chicks. That would render our data useless for making inferences about the *population* of *all* loon chicks.

The *only* reasonable defense the statistician can use against this problem of potentially atypical samples is to ensure that the samples are chosen *randomly* from the population being studied. Then the laws of probability give us a measurable confidence that the atypical cases in the population will, by chance, occur with the same relative frequency in our sample as they do in the population.

EXAMPLE 2 Analyzing Samples for Randomness

Which of the following sampling strategies will result in random samples from the population under consideration?

(1) Your school wants to pick 5 random seniors to represent the school on a district panel. The names of the seniors are written on slips of paper and placed in a bucket. After the class president shakes the bucket, the blindfolded principal draws five names.

(2) A group studying the increasing problem of obesity in America wants to estimate the percentage of American teenagers who order hash browns with their breakfast sandwiches at a particular national fast food chain. They visit each of the chain's restaurants in Idaho and gather data from the first ten teenagers they see ordering breakfast sandwiches in each restaurant.

(3) The Miss America contest wants to choose 8 state representatives at random for a publicity shoot. Each of the 52 women is given a card from a well-shuffled deck of 52 playing cards. The publicity chair then chooses 8 cards at random from another well-shuffled deck, and the women with the matching cards go to the shoot.

SOLUTION

(1) This strategy rarely proves to be random, since the names usually enter the bucket in nonrandom order (for example, alphabetically), and the typical shaking does not usually compensate for that. As a result, the names toward the end of the alphabet are still closer to the top and thus have a better chance of being chosen. (This phenomenon famously affected the 1969 Selective Service draft lottery.)

(2) Not random. Idaho has no guarantee of being typical for the purposes of this study. (For example, people might be more partial to potatoes in Idaho than in, say, Georgia.) Also, sampling the "first ten teenagers" at each site is not random, as friends who arrive together are more apt to be like-minded diners.

(3) This strategy is random. Notice that the method is quite carefully designed, but that is what is usually necessary for attaining randomness. We will return to this point at the end of this section. **Now try Exercise 9.**

Sample Surveys

A well-known example of sampling is a **survey** (such as a political poll), in which data are gathered by questioning a random sample of people.

Samples, Surveys, and Observational Studies

Recall that statisticians use *statistics* (numbers determined from samples) to estimate *parameters* (numbers associated with populations). You might think it would be safer to gather data from the entire population (in which case the study is called a **census**), but that is often impossible, or at least impractical. Instead, statisticians pay attention to their sampling methods so that they can use the laws of probability to make confident inferences about the population parameters. Some sampling designs can be quite complicated, but the most basic requirement is that the sampling be free of **bias**. That is, there should be nothing in the sampling *process* that could *systematically cause* the sample to differ from the population with respect to the statistic being gathered. Sadly, biased sampling is one of the most common flaws in the world of bad statistics.

EXAMPLE 3 Analyzing Samples for Bias

Each of the following studies suffers from a form of sampling bias. Point out the problem, and we will then give it a name for you.

(1) A local talk-show host wanted to assess the opinion of city residents on a proposed tax increase for public education, so he invited listeners to phone in their opinions. According to his poll, 93% opposed the tax increase.

(2) A parent group opposed to the proposed renovation of their high school auditorium passed out a questionnaire at the football games with the question: "Should the school jeopardize vital school programs by diverting funds to improving an auditorium that is needed only a few times a year?" They reported that 89% of the people were opposed to the renovation.

(3) Seeking feedback on their new toothpaste flavor, a company supplied 500 dentists with free samples to give to their clients. With each sample came a stamped postcard with a single box for the user to check: like the new flavor or don't like the new flavor. The company received 897 postcards back, 85% of which reported that the user liked the new flavor.

(4) A biologist wanted to gather evidence that nuclear power plants caused genetic defects in frogs that fed in nearby ponds. He sampled 50 frogs in a pond near his local nuclear power plant and discovered that 8% of them had genetic defects. This was, in fact, considerably higher than the rate in the general frog population, so he considered this to be good evidence supporting his assertion.

(*continued*)

SOLUTION

(1) For starters, the sample was not likely to include anybody not listening to the show. This is called **undercoverage bias**. Also, listeners had to be a little "fired up" about the question to go to the phone, and coming out against a tax increase fires people up. This is called **voluntary response bias**. The 93% number is likely to be much higher than the population parameter.

(2) Sampling at football games (an example of a **convenience sample**) is another source of **undercoverage bias**. (People who are more apt to go to the auditorium might be less apt to go to the stadium.) The question itself also created bias, as the wording was designed to provoke a negative response. Anything (intentional or unintentional) in the study design that influences responses is a form of **response bias**. The 89% is likely to be much higher than the parameter, the true level of support among the whole school community.

(3) Undercoverage was probably not a problem if the 500 dentists were a representative group. (People who did not go to dentists were underrepresented, but they were also less likely to buy toothpaste, so that was probably fine with the toothpaste company.) The real problem here was **response bias**, as the company may have unintentionally elicited a favorable response by giving the people free toothpaste. At the very least, people who slightly disliked it might have been less inclined to send back the card (more **voluntary response** bias). The 85% approval is probably higher than the parameter.

(4) This was another **convenience sample**; the sample of frogs he was able to catch was far from random. These frogs did appear to have a problem, but the fact that they all came from the *same pond* makes **undercoverage bias** a strong possibility. There are many possible explanations other than the nuclear power plant for the high defect rate in the sample.

Now try Exercise 15.

In an **observational study** data are simply observed and recorded. Often this does not involve sampling a population at random. For example, in a wildlife study scientists have access only to the animals they are able to observe or capture. Other researchers may gather information from such available sources as doctors' records, census data, or the Internet. Often observational studies suggest the presence of important associations between variables, but they cannot prove causation, and the lack of random sampling dictates caution when trying to reach conclusions about an entire population.

Observational studies are certainly useful, and it is possible to draw conclusions from the data up to a point. (You can argue about whether baseball standings determine the "best team," but they certainly do reveal which team has won the higher percentage of games.) Observational studies suggested for years that there was a high positive association between smoking rates and lung cancer rates, but did this prove that the smoking variable *caused* an increase in the cancer rate variable? No. Indeed, there was also a high positive association between coffee consumption and lung cancer rates! (Of course, many coffee drinkers used to enjoy a cigarette or two with their coffee.) To actually show *causation*, researchers must control **confounding variables** like coffee drinking. This requires experimentation.

Experimental Design

A well-designed experiment can show there is a cause-and-effect relationship between two variables by accounting for all other possible explanations. In an **experiment**, researchers impose **treatments** on **subjects** who are often volunteers. Each treatment is based on **factors** that might affect the **response variable** under study. By controlling the factors, the researchers hope to identify (with high probability) a causal relationship.

Probability Experiments

The word "experiment" is often used loosely to describe a variety of investigations. The word has quite a different meaning in the context of Statistics (and other sciences), in which it is a procedure specifically designed to test causation.

Four Important Principles of Experimental Design

1. **Control**. Variation between subject groups receiving different treatments should be limited to the factor we are varying, while other conditions are made as similar as possible. For example, if we are testing the effectiveness of a drug, everyone should get a pill, but only one group's pills would contain the drug. (The "fake" pill is called a **placebo**.) The experiment should be **blinded**, meaning that neither group (or anyone in contact with them) should know which treatment they are getting until after the data are collected.
2. **Randomization**. Subjects must be assigned randomly to the different treatments so that variation arising from uncontrolled (often unknown) factors will be randomly spread among the different groups.
3. **Replication**. An effect should be observed in multiple subjects (the more the better) to be statistically convincing, and other researchers should be able to perform the same experiment on subjects from the same population with similar results.
4. **Blocking**. As a form of control, pre-existing differences among the subjects that we think might affect the response variable (like gender or lifestyle) should be deliberately spread evenly (but at random) among the treatment groups. For example, suppose there are 10 women and 40 men among the subjects for a two-treatment experiment, and we believe that men and women might respond differently to the treatments. We could randomly assign 5 women and 20 men to each group. If men and women did react differently to the treatments, we could take that variability into account when assessing the effects of the factor under study.

EXAMPLE 4 Designing an Experiment

A researcher wishes to test the effectiveness of vitamin C supplements for preventing flu. He has a diverse group of 200 men, women, and children who have volunteered for the study. Design an appropriate experiment that takes into account the likelihood that flu will affect different age groups differently.

SOLUTION There is only one factor, requiring two treatment groups (supplements and no supplements). A simple way to control for the age variable is to **match** volunteers of like ages together, then *randomly* assign one from each pair to each of the two treatments. (This spreads the age variability equally among both groups.) One group will take tablets containing vitamin C; the other group will take placebos. The subjects and their caregivers will be blinded as to which tablets they are taking. We monitor each group during flu season and record who gets the flu.

We can show the design of the experiment in a diagram (Figure 10.33):

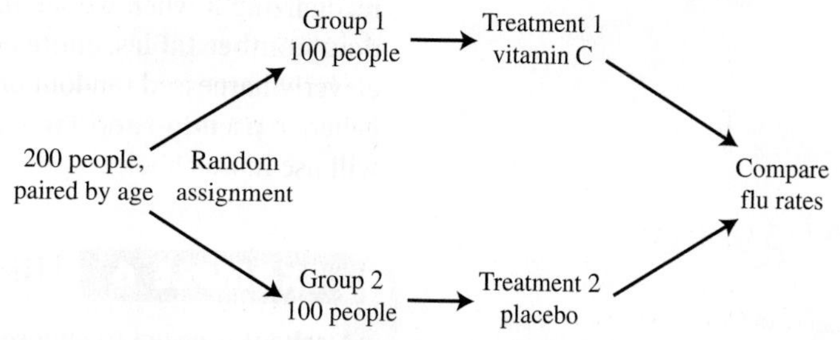

FIGURE 10.33 The flu experiment. (Example 4)

Now try Exercise 25.

EXAMPLE 5 Analyzing Experimental Design

Each of the following experiments suffers from at least one flaw in design. Identify the flaws and tell how each experiment can be improved.

(1) In a study of multitasking, psychologists test the time it takes a group of volunteers to type the same 100-word text message, first listening to loud rock music, then listening to soft classical music, and finally listening to no music at all.

(2) To test the effectiveness of an expensive online homework service, a teacher with two sections of freshman algebra signs up his thirty A period students for the service and teaches his thirty B period students the same lessons, but without the service. At the end of the year he compares the grades of the two sections on the same tests.

(3) A drug company tests a new pain reliever on 100 arthritis patients, which it randomly splits into two groups. Half the patients are given the new pill, while the other half are told to continue their regularly prescribed medications. After 60 days, the two groups are interviewed by representatives of the drug company to report their level of pain relief.

SOLUTION

(1) Since the text message is the same each time, the texters are likely to pick up speed as they gain familiarity with it. The researchers should either work up three different 100-word messages or else randomly assign the music treatments in six different orders (RCN, RNC, CRN, CNR, NCR, NRC) and average the times.

(2) The subjects are not randomly assigned, so any difference in test scores might well be due to other variables related to the two sections. (Is one class in the morning and the other in the afternoon? Were all students taking an advanced science course forced into one of these classes?) The teacher should randomly choose 30 of his 60 freshman algebra students to use the service, without regard to A or B period.

(3) The biggest flaw is that the subjects know which treatment they are getting. Those who get the new pill might feel better simply because they think they ought to feel better. The pills should be made to look the same so that the subjects do not know if they are taking their old medication or the new one. A second flaw is that the drug company representatives might unintentionally treat the subjects differently if they know which ones took the new pill. The person who interviews the subject at the end of the study should not know the subject's treatment. (An experiment in which both the subjects and the data-gatherers are blinded is called a **double-blind** experiment.)

Now try Exercise 29.

Using Randomness

Whether choosing a sample from a population or assigning subjects to treatments for an experiment, we have seen that attention must be paid to randomness. But how do statisticians achieve randomness? It is actually harder than you might think. Not only are people notoriously bad at mimicking true randomness, but we are not even good at recognizing it when we see it. Before the days of computers, researchers relied on **random number tables**, entire books filled with pages of digits (0 through 9) generated by cleverly harnessed random processes in nature. Today we usually rely on computers to generate **pseudo-random** numbers, which work well for most purposes (and which we will use in this book).

EXAMPLE 6 Using Random Numbers in Sampling

A school is asked to choose 10 sixth-graders at random to be part of a multi-year educational study. If there are 135 sixth-graders at the school, how can they be chosen without bias?

SOLUTION

Method 1: Generating Random (actually Pseudo-Random) Numbers Number the students from 1 to 135 (alphabetically, or in any order—this is not the random step). On your calculator, enter the command "randInt(1, 135, 10)." The calculator will return a list of ten random integers between 1 and 135 inclusive (Figure 10.34). If it picks any number more than once, generate more until you have 10 distinct numbers.

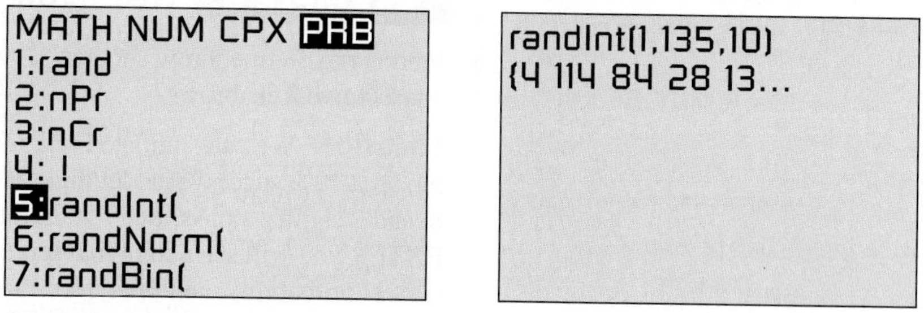

FIGURE 10.34 Use randInt(1, 135, 10) to pick 10 random numbers between 1 and 135 inclusive. (Example 6)

Our list (we had to scroll to the right to see them all) consisted of the students numbered 4, 114, 84, 28, 133, 98, 41, 34, 128, 125. **Now try Exercise 33.**

There's another nice method of randomization that avoids the minor problem of having to skip repeated numbers. We demonstrate that approach in Example 7. Of course, one could use either method for both samples and experiments.

EXAMPLE 7 Using Random Numbers in an Experiment

Fifty people suffering from a certain medical condition have volunteered as subjects in an experiment to test the effectiveness of a new drug. Half will be treated with the drug, and the other half will receive a placebo. Explain how to randomly assign these 50 subjects to the two treatments.

SOLUTION

Method 2: Doing a Random (actually Pseudo-Random) Sort Number the subjects 1 to 50. Enter the numbers 1 to 50 in list L1 using the command "seq(X, X, 1, 50) → L1" and enter 50 random numbers in list L2 using the command "rand(50) → L2." Then sort the random numbers into ascending order, bringing L1 along for the ride, using the command "SortA(L2, L1)." The numbers in list L1 are now in random order. (See Figure 10.35.)

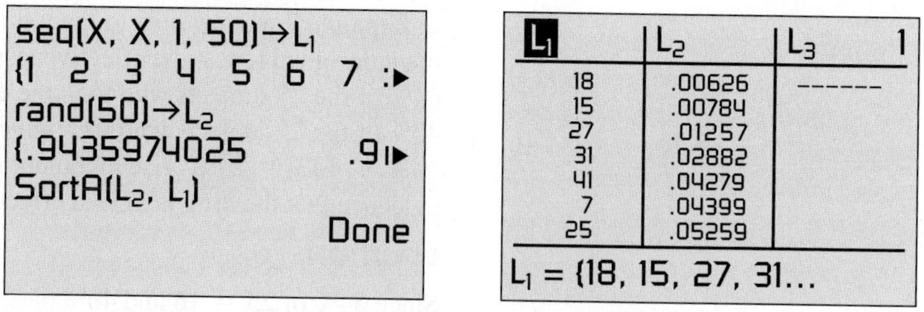

FIGURE 10.35 Randomizing the order of 50 numbers on the calculator. (Example 7)

Assign the people with the first 25 numbers in L1 (here starting with 18, 15, 27, ...) to receive the new drug, and assign the remaining 25 people to get the placebo.

Now try Exercise 37.

Planting the Seed

Your calculator uses a formula to move from one number to the next, so you might be generating the same pseudo-random list as your neighbor. To get a different list, start with a different seed. For example, the command "42 → rand" stores the seed 42. Different seeds will generate different pseudo-random sequences. (You could each enter the last 4 digits of your telephone number.)

Simulations

As long as we have introduced you to random numbers, we will end this section with a quick look at how they can be used to simulate probability and assess statistical significance.

EXAMPLE 8 Free Throw Shooting

Fred is a 70% free throw shooter. How many shots will it typically take him to hit 7 consecutive free throws?

SOLUTION We can simulate Fred's free throw shooting with random digits. To model a 70% success rate, let digits 0 through 6 signify a made free throw, and let 7, 8, and 9 signify a miss. Enter "randInt(0, 9)" in your calculator. Each time you press ENTER you will see a digit that represents the outcome of a free throw with the correct probability, and Fred never has to pick up a ball. The beginning screen is shown below, followed by the first four sequences we (honestly!) generated before getting 7 consecutive "hits."

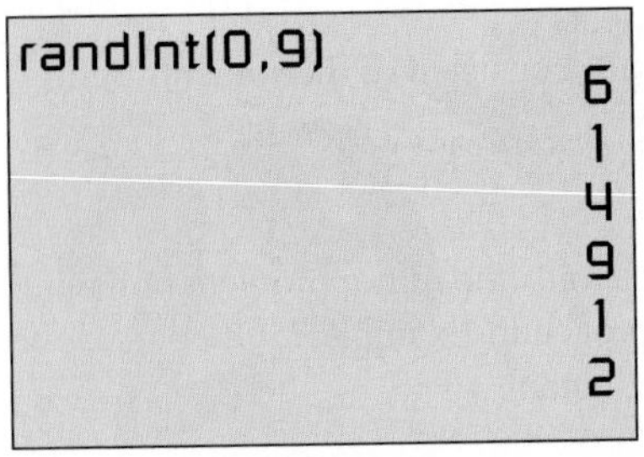

FIGURE 10.36 Keep pressing ENTER to generate random digits.

Sequence #1:
6 1 4 9 1 2 6 0 3 0 8 7 2 4 7 2 0 5 0 3 6 8 6 0 3 4 2 1 8 8 5 5 7 3 6 5 0 9 5 3
8 2 4 9 6 4 1 9 8 5 0 7 7 3 8 3 8 8 6 9 6 1 9 4 1 4 1 4 6 6

Sequence #2:
3 3 3 2 5 5 2

Sequence #3:
6 4 4 5 2 3 4

Sequence #4:
8 8 2 4 6 2 6 6 2

You can verify that poor Fred had to put up 70 shots in the first sequence before achieving the seven consecutive "hits" at the very end: 4141466. In the second and third sequences, though, Fred starts off with seven in a row! By running this simulation many times, you can get an idea of the average number of free throws Fred would need to shoot before he hits seven in a row.

Incidentally, this set of results is pretty good evidence of how difficult it is for us to recognize true randomness. You could ask a thousand people to "make up" free throw sequences for this simulation, and nobody would ever give you a string of 70 followed by two in a row of only 7! **Now try Exercise 41.**

EXAMPLE 9 Assessing Statistical Significance

In Example 7 we saw how to randomly assign 50 subjects to treatments in an experiment designed to test the effectiveness of a new drug. At the end of the experiment 64% of the 25 subjects who took the drug showed improvement, compared to only 40% of the 25 subjects who were given a placebo. Could a difference as large as $64\% - 40\% = 24\%$ arise by random chance, or is that statistically significant evidence that the drug is indeed effective?

SOLUTION

Since 64% of $25 = 16$ and 40% of $25 = 10$, overall $16 + 10 = 26$ subjects showed improvement; the other 24 did not. To simulate this experiment, we create a list of 26 1's and 24 0's to represent the outcomes, and then randomly split them into two groups of 25. We then look at the percentage of 1's in each group to see how big a difference might occur just by chance. The table below shows the results of the first few trials.

Table 10.30 Simulated Differences in the Percentage of Subjects Who Might Show Improvement if the Drug Being Tested Is Not Effective

Trial #	Number (%) of Simulated "Cures": Drug Group	Number (%) of Simulated "Cures": Placebo Group	Difference in Percentages (Drug − Placebo)
1	14 (56%)	12 (48%)	8%
2	11 (44%)	15 (60%)	−16%
3	13 (52%)	13 (52%)	0%
4	18 (72%)	8 (32%)	40%
⋮	⋮	⋮	⋮

In Trial 1, by chance, the difference between the two groups was 8%. That makes the experiment's actual outcome of 24% seem pretty large, perhaps significant. Neither Trial 2 nor Trial 3 produced a very large difference either. But Trial 4 is surprising: Shuffling the outcomes for 50 simulated subjects produced a difference of 40% just at random! Seeing this suggests that maybe the experiment's 24% difference doesn't really signify an effect of the drug. Of course, just as a fair coin *could* show lots of heads now and then by sheer luck, it's possible that Trial 4 is a once-in-a-blue-moon longshot, an outlier we should not be concerned about.

To find out, we ran our simulation for 1000 trials. The histogram below shows the distribution of random differences that arose.

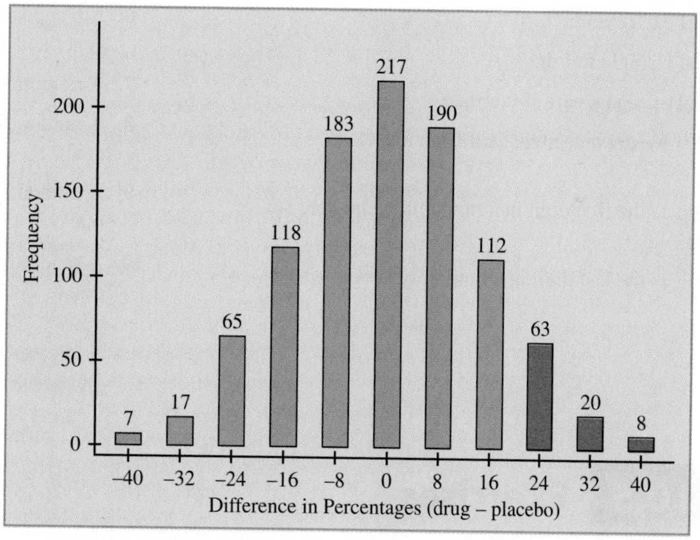

FIGURE 10.37 A histogram of 1000 simulated differences in the percentage of subjects who might show improvement if the drug being tested is not effective.

Notice that differences at least as large as the 24% observed in this experiment are not uncommon. They happened 91 times in the 1000 simulated trials. That suggests there's about a 9% chance that the experiment could have turned out this way even if the drug is worthless. This should make you skeptical that the drug actually works. The simulation shows us that with treatment groups this small a difference of 24% is not large enough to be statistically significant. The experiment does not provide convincing evidence that the drug actually works.

A common rule of thumb is that in order for an observed outcome to be considered statistically significant, there must be less than a 5% chance that it could have occurred randomly. Had the drug group outperformed the placebo group by 32% or more, we'd consider that to be evidence of the drug's effectiveness. In the simulation, differences that large were quite rare; they arose by chance only 2.8% of the time.

(*continued*)

That decision rule can vary with the gravity of the situation. If we're simply curious to see whether the rate of smoking among middle school girls has changed in the past 20 years, we might accept slightly less unusual results as offering some evidence. An outcome with up to a 10% chance of occurring at random might be considered statistically significant. On the other hand, if we're testing the rivets that hold highway bridge girders in place, we'd accept only test results that could occur by chance less than 1 time in 1000 (or 10,000!) to be compelling evidence that they're strong enough.

Now try Exercise 45.

This chapter has offered only a brief glimpse into how statisticians use mathematics. If you are interested in learning more, we urge you to find a good statistics textbook and pursue it!

QUICK REVIEW 10.5 *(Prerequisite skill Sections 10.1 and 10.4)*

Find the probability of each event:

1. Rolling a 6 on a single fair die
2. Rolling a total of 6 on a pair of fair dice
3. Drawing a 6 if we draw a single card from a deck of 52 cards
4. A 6 appearing as the first digit in a randomly generated set of five digits
5. A 6 appearing as the last digit in a randomly generated set of five digits
6. A 6 appearing as the first digit *and* the last digit in a randomly generated set of five digits.
7. A randomly generated set of five digits consisting of all 6's
8. A randomly generated set of five digits containing no 6's at all
9. A randomly generated set of five digits containing at least one 6
10. A randomly generated set of five digits containing at most one 6

SECTION 10.5 Exercises

In Exercises 1–6, tell whether correlation is being used correctly. Identify any statistical errors.

1. A newspaper article reports that a high correlation has been discovered between beauty and intelligence in college women.
2. Because the correlation coefficient for the data in the graph at the right is $r = 0.9$, a student announces that 0.9 is the slope of the regression line that goes through the points.

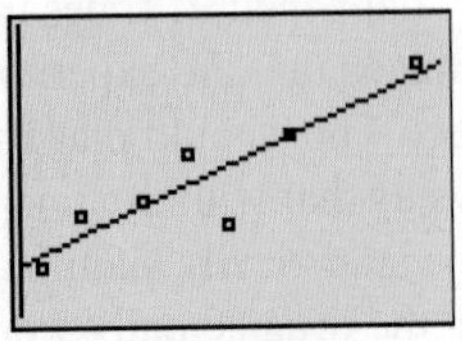

3. Sean has a theory that the average weight of an animal has a high correlation with the number of letters in its name. He checks his theory by gathering data on a rat, a bat, a fox, and a great blue whale. Sure enough, the correlation coefficient is higher than 0.99!
4. Jenna says that the scatter plot at the right shows a negative association between the x-variable and the y-variable. She then adds that the regression line she found does not do a very good job of describing the relationship.

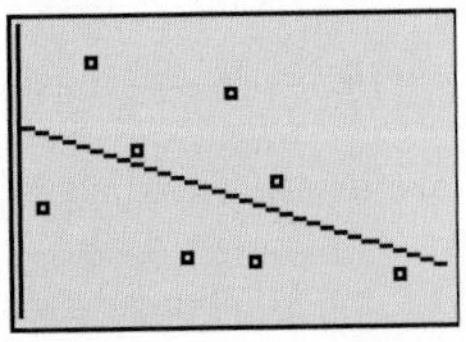

5. Marcus says that the scatter plot at the right shows a weak negative association between the x-variable and the y-variable. He then adds that the low correlation ($r = -0.32$) indicates that there is no significant mathematical relationship.

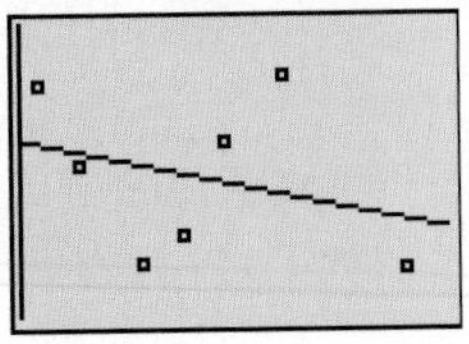

6. A researcher theorizes that sea snakes need to drink rainwater floating on the top of the denser saltwater in order to avoid dehydration. Testing sites around the world, he finds a high correlation between a site's annual rainfall and the size of its sea snake population, thus proving his theory to be true.

In Exercises 7–12, tell whether the sampling strategy will result in a random sample from the population under consideration. If it is not random, explain why it is not.

7. The HR director of a large law firm wants to survey a random sample of the firm's 326 employees. She numbers the employees alphabetically and uses her daughter's calculator to choose 50 distinct numbers with the command "randInt(1, 326, 50)."

8. To get a sample of 50 students for a student services survey, an assistant principal uses the calculator command "randInt(1, 72, 5)" to pick five numbers between 1 and 72. He goes to those five pages in the 72-page student directory and chooses the ten students pictured on each of those five pages.

9. To get a random sample of 50 Reno citizens for a political poll, a worker starts at the beginning of the telephone book. For each name in the book, he flips a fair coin. If it comes up heads, he calls the person; if it comes up tails, he moves on. Once he has called 50 people, he stops.

10. When 50 qualified students audition for the five chorus roles in the school musical, the director decides to choose the five chorus members randomly. He tells them to form five lines of ten people; then he announces that the seventh person in each line will be in the chorus.

11. To choose 9 random campers to represent Camp Pathfinder in a baseball game with Camp Ahmek, the counselors hold tryouts and choose the best 9 players.

12. A talk-show hostess wants to give free automobiles to ten random members of her audience. She instructs the ushers to go into the theater before the show and place special stickers under ten seats of their choice.

Writing to Learn In Exercises 13–18, each sampling method suffers from a form of bias. Identify the bias, tell how the sample statistic might differ significantly from the population parameter, and suggest how the problem might be corrected.

13. To gather feedback on his teaching, Professor Jones invites his students to visit his Web site and respond to a brief questionnaire. He is depressed to discover that 68% of those who respond are dissatisfied with his teaching.

14. To assess consumer response to its new cereal flavor, a cereal company passes out free samples at a local mall and offers shoppers a $5.00 gift certificate if they will fill out a brief form evaluating the new product. They find that 94% of the respondents like the new flavor.

15. The dining committee of the student council surveys students in the lunch hall, asking them, "Do you prefer eating in the lunch hall to eating off campus or bringing your own lunch?" They are mildly surprised to learn that 93% of the students surveyed prefer the lunch hall.

16. A group seeking taxpayer support for a new playground sends pollsters to all the local PTA meetings to ask, "Do you support the use of city taxes to fund a new park that will provide a safe and convenient recreation facility for your children?" A gratifying 88% respond yes.

17. After getting a ticket for running a stop sign, a statistics professor polls a random sample of citizens to ask, "Should the government be allowed to limit the freedom of private citizens to utilize our neighborhood roads as we deem necessary?" He tells the judge at his hearing that 97% of the citizens polled are against stop signs.

18. For twenty years, Mrs. Bohackett has kept a daily log of birds sighted at her backyard feeder. Over the last ten years, the first appearance of a white-throated sparrow in the spring has steadily moved earlier in the year by thirteen days. She attributes the change to global warming.

In Exercises 19–24, identify which are experiments and which are observational studies.

19. A health class analyzes 30 cereal products available at a local food store, recording various nutritional data for a typical 1-cup serving of each.

20. A school counselor looks at ten years of data to see if students in some extracurricular activities make higher grades than students in others.

21. The receptionist at a veterinarian's office keeps a daily log to see if more women than men accompany their pets to the vet.

22. Laboratory rats are fed diets with three different levels of caffeine content, and observers record how much time the rats spend each day running on their treadmills.

23. Members of the school swimming team practice for eight days under simulated racing conditions wearing three different brands of swimsuits to see if any one brand leads to faster times.

24. An English teacher offers extra credit points to students who will keep nightly logs of time spent watching television and time spent reading. She intends to use the data to see if there is an association between either variable and a student's success in her class.

25. A farmer wishes to test the effectiveness of a new kind of fertilizer. He plants his crops on 24 plots of equal size. Design an experiment he can use to test the new fertilizer. (Be sure to explain how randomness will be used.)

26. How would you amend the design in Exercise 25 if the farmer knows that some of his plots have been historically more productive than others?

27. How would you amend the design in Exercise 25 if the farmer can choose between two new fertilizers?

28. How would you amend the design in Exercise 25 if the farmer wants to test the effectiveness of the new fertilizer on each of two different crops?

Writing to Learn In Exercises 29–32, each experiment suffers from at least one flaw in design. Identify each flaw and tell how the experiment can be improved.

29. **Testing Golf Balls** A company tests its new golf ball with 100 experienced golfers. After having each golfer drive 20 of his current favorites to establish his current average distance, the company has him drive 20 of the new balls. The distances are then compared.

30. **Testing Soft Drinks** To test the rumor that her company's bottled soft drinks are "fizzier" than the same drinks in cans, a researcher gives 50 volunteer tasters a can and a bottle of the same soft drink at the same temperature. They then identify which tastes fizzier to them.

31. **Testing Effect of Music** As part of a study to see if music affects babies in the womb, a researcher asks a group of expectant mothers to volunteer to play an hour a day of a particular music type for their babies in their last month of pregnancy. The mothers could choose classical, country, or rock. The researcher plans to gather developmental data on the children at 4-year intervals from birth until adulthood.

32. **Radio Advertising** Researchers wish to test the effects of radio advertising on 100 volunteers between the ages of 18 and 32. They use random numbers to split them into two groups. Both groups hear the same advertisements, but one group hears them delivered with all male voices and the other group hears them with all female voices. Both groups then take the same quiz about the products to see how much information they have retained.

33. **Finding a Random Sample** From 500 people who attend a banquet, the caterer plans to choose 50 at random to fill out a brief questionnaire about the service. The seats at the banquet have already been numbered 1 to 500. How can the caterer use a graphing calculator to select the random sample?

34. **Selecting Customers** Approximately 400 people are in line at a retail store for an advertised chance to buy the season's most anticipated Christmas toy. To avoid unpleasantness, the store has promised that everyone who is in line when the doors open will have an equal chance to buy one of the 100 available toys. How can the store use a graphing calculator to select the lucky 100 customers?

35. **Random Order** A class of 32 students must present their class projects in what the teacher has promised to be "random order." Tell how the teacher can use a graphing calculator to determine the order.

36. **Random Selection** A teacher plans to pair his 28 students randomly to be partners for a collaborative quiz. So they will know it is random, he plans to pair them using a graphing calculator with a student pushing the buttons. Explain how this can be done.

37. **Growing Tomatoes** You plan to use your backyard for an experiment to see if a new plant food offered by your local garden store will help grow tastier tomatoes. You bought 16 tomato plants and will grow half of them with the plant food and half without. Explain how you will assign the plants to the two treatment groups.

38. **Firing Pottery** An artist is trying to decide whether to buy a new kiln for firing her pottery, but wonders whether it will be any better than her current kiln. A friend who already owns one of the newer models offers to let her try it. She creates 10 bowls and plans to fire half of them in her own kiln and the other half in her friend's kiln so she can compare the results. Describe a way to randomly assign the bowls to the two kilns.

39. **Simulating a Spinner** A spinner in a children's game once worked to pick a number between 1 and 8, but the children lost the pointer. Tell how they can use a graphing calculator to simulate the spinner.

40. **Simulate Rolling Two Six-Sided Dice** The Millers want to play Parcheesi, but they have lost the pair of six-sided dice that came with the game. Tell how they can use a graphing calculator to simulate the roll of two dice.

41. **Blood Donors** About 40% of the blood donors at a local facility give O-positive blood. In a typical hour the facility will process twenty donors. Design a simulation to estimate how many hours out of 9 hours of operation the facility will typically see fewer than 4 donors with O-positive blood.

42. **Simulate Drawing Cards** Bapa wants to simulate the drawing of cards for a poker hand using his graphing calculator. Tell how the calculator can simulate the drawing of 5 random cards from a deck of 52.

43. **Rolling Twenty-One** In a game players accumulate points by rolling a die. The first person to reach a total of exactly 21 wins; rolls that would make the total higher than 21 are ignored. Create a simulation to estimate the number of rolls a player might have to make in order to hit 21. Run five trials and report your result.

44. **Spinning Ten** A child's game has a spinner with four equal regions numbered 1, 2, 3, and 4. Players spin until they reach a total of exactly 10. If a spin would push the total above 10, that spin is subtracted from the previous total. For example, if a player with 8 points spins a 3, the player's total is reduced to 5. Create a simulation to estimate the number of spins a player might have to make in order to hit 10. Run five trials and report your result.

45. **Writing to Learn Dice** You just saw someone roll a die 20 times and get eight 6's. That seems like a lot. Is it evidence that something's wrong? Maybe the person has a loaded die or is controlling the roll somehow? To find out, you simulate rolling a fair die 20 times, counting the number of 6's in each trial. Your simulation runs 500 trials and produces the results summarized in the histogram below. Based on these outcomes, would you say there's statistically significant evidence that the die rolling you just witnessed is unfair? Explain.

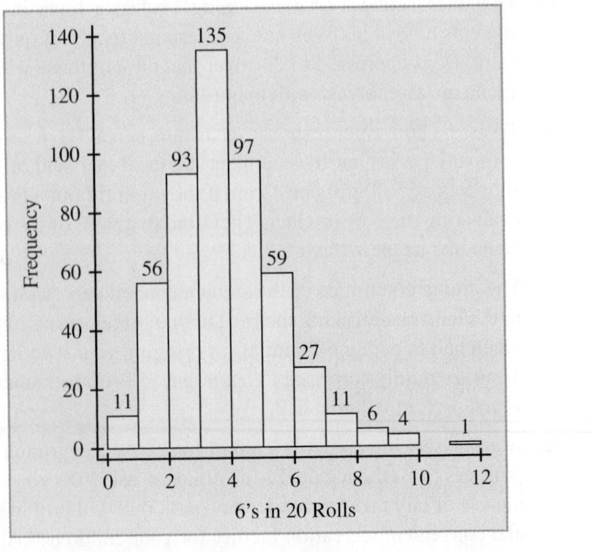

46. Writing to Learn Tasting Tomatoes Exercise 37 describes a backyard experiment to see if a certain plant food can actually produce the tastier tomatoes it promises. After the harvest, a neighbor who did not know about the experiment tasted samples of tomatoes and rated them on a scale from 1 (yuck) to 10 (fantastic). The ratings given tomatoes that grew with the plant food were $\{6, 5, 10, 9, 9, 7, 8, 8\}$, for an average of 7.75. The tomatoes grown without the plant food were rated $\{6, 9, 7, 6, 8, 4, 7, 9\}$, for an average of only 7.0. Do these results provide evidence that the plant food works?

To investigate, a simulation randomly shuffled those taste ratings into two groups to see how large a difference could arise by chance alone. This dotplot displays the random differences in the means seen in 500 trials.

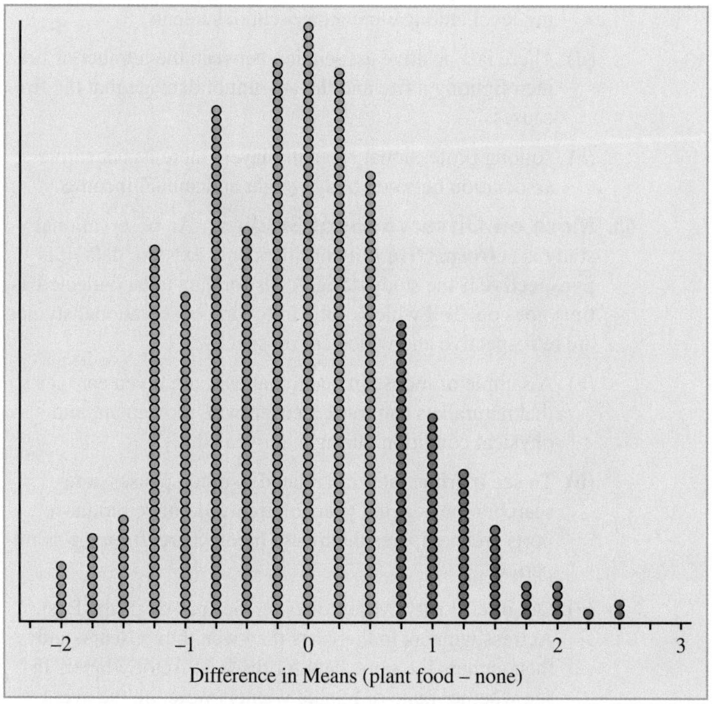

(a) Based on these simulated results, explain why the observed 0.75 difference in mean taste ratings is not statistically significant. Explain.

(b) How large would the difference in means have to be in order for you to believe there to be statistically significant evidence that the plant food caused tomatoes to taste better? Explain.

Standardized Test Questions

47. True or False An observational study, if carefully done, can usually establish causation. Justify your answer.

48. True or False A correlation coefficient of 0.96 would indicate that a linear model for the data is certainly appropriate. Justify your answer.

49. Multiple Choice There could be a positive correlation among high school students between SAT scores and

(A) Colleges of first choice.

(B) Freshman science grades.

(C) Teacher recommendations.

(D) Extracurricular activities.

(E) All of the above.

50. Multiple Choice Randomness is important in experimental design in order to equalize (among the treatment groups)

(A) Unforeseen variation.

(B) Extraneous blocking.

(C) Unnecessary placebos.

(D) Noncausal correlation.

(E) Expected behavior.

51. Multiple Choice A survey question may lead to response bias if it

(A) Causes discomfort among those surveyed.

(B) Is about a particular race or creed.

(C) Has an influence on the response.

(D) Elicits a voluntary response.

(E) Does not fit with the rest of the survey.

52. Multiple Choice Which of the following is *not* a valid way of choosing a random sample of 5 people from 13 people?

(A) Number the people from 1 to 13. Enter "randInt(1, 13)" on your calculator until you have five different numbers.

(B) Number the people from 5 to 17. Enter "randInt(5, 17)" on your calculator until you have five different numbers.

(C) Give each person a playing card from the heart suit (thirteen cards). Shuffle the spade suit thoroughly and draw 5 cards. Pick the people whose hearts match the spades.

(D) Go through the 13 names alphabetically and flip a fair coin for each one. If it is heads, choose the person; otherwise, move on. When you have five people, stop.

(E) Shuffle the thirteen cards from the heart suit thoroughly and have each person draw a card. Line them up in order (ace through king) and pick the middle 5 people in the line.

Explorations

53. Group Activity Attempting Randomness Try this experiment with your classmates. Without a calculator, write down what you think is a sequence of 50 random digits (0 through 9). Count up how many times each digit appears in your sequence, then pool your tallies with those of your classmates. Did any digits appear unusually often?

54. Group Activity Attempting Randomness (Continuation of Exercise 53) Group your random sequence into 25 two-digit numbers. Count how many of them consist of double digits (e.g., 11, 22, 33, etc.). Pool your tally with those of your classmates. By chance, approximately one-tenth of a random collection of two-digit numbers should consist of double digits. Did your class come close?

55. The Effects of Unusual Points Each of the three graphs below has an unusual point. If the unusual point is *removed* and the regression line is recalculated, predict what will happen to the correlation coefficient (increase, decrease, or remain about the same) and the slope of the regression line (increase, decrease, or remain about the same).

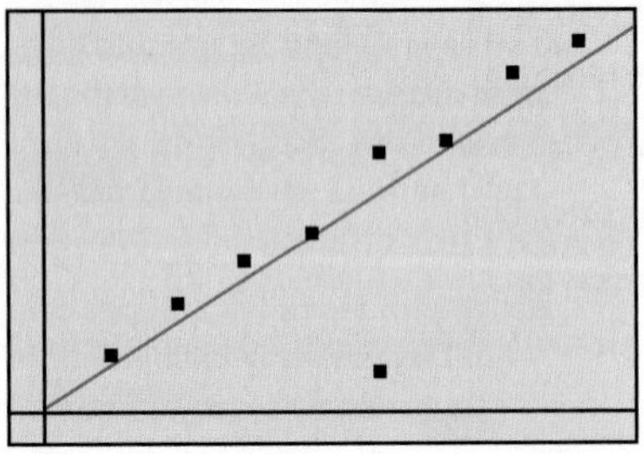

(a)

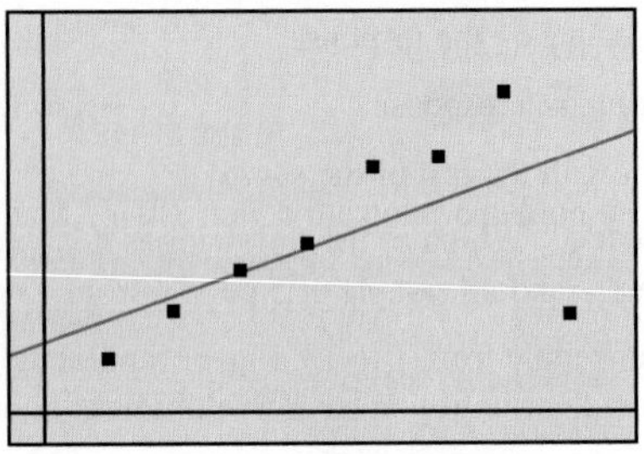

(b)

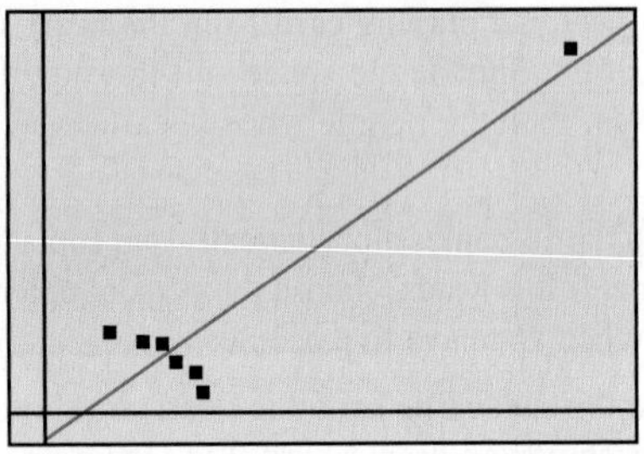

(c)

56. Influential Points An unusual point in a scatter plot is called an **influential point** if the regression model changes significantly when it is removed. Which of the unusual points in Exercise 55 are also influential points?

57. Picture the Possibility Sketch a scatter plot that shows a positive association between x and y for small values of x, a positive association between x and y for large values of x, and a negative association between x and y overall. (*Hint*: Sketch two well-separated clusters of points.)

58. Picture the Possibility Sketch a scatter plot that shows a negative association between x and y for small values of x, a negative association between x and y for large values of x, and a positive association between x and y overall.

Extending the Ideas

59. Lurking Variables One good reason that we should never conclude causation from even the strongest correlation or association is that the association between x and y might be due to a third variable z that influences both x and y. If such a variable is not part of the study model, it is called a **lurking variable**. For each of the following actual associations, identify a possible lurking variable that could be causing it.

(a) There is a positive association between the size of a hospital and the death rate among its patients.

(b) There is a positive association between the number of seats on a commercial jet and the speed at which the aircraft travels.

(c) There is a positive association between shoe size and reading level among elementary school students.

(d) There is a positive association between the number of firemen fighting a fire and the amount of damage that the fire causes.

(e) Among professional football players there is a negative association between body weight and annual income.

60. More on Observational Studies An observational study is **retrospective** if it considers only existing data. It is **prospective** if the study design calls for data to be collected as time goes on. Tell which of the following observational studies are retrospective and which are prospective.

(a) A sample of moose in a national park are given ear tags so that naturalists can track their growth, movement, and physical condition during a ten-year study.

(b) To see if crime rates are related to moon phases, a researcher looks at ten years of archived police blotter reports and compares them with moon charts from the same period.

(c) Out of curiosity, Mimi looks up the ages of all the Best Actress winners in the years they won their Oscars, and then gathers the same data for the Best Actor winners to see whether male or female winners were, on the average, younger.

(d) In a study of post-traumatic stress disorder, soldiers who have been in combat are given biannual physical and psychological tests for five years after they return from active duty.

(e) Paxton devises a complicated formula based on game statistics for rating quarterbacks. He applies it to all NFL quarterbacks who played in the league between 1950 and 2000 and concludes that the quarterbacks from the 1970s were the best overall.

CHAPTER 10 Key Ideas

Properties, Theorems, and Formulas

Probability of an Event (Equally Likely Outcomes) 678
Independent Events 685
Conditional Probability Formula 685
Formula for the Mean 708
Formula for the Standard Deviation 710
The 68-95-99.7 Rule 712
Expected Value of a Random Variable 718
Expected Value for a Binomial Distribution 722
Standard Deviation for a Binomial Distribution 723

Procedures

Probability of an Event (Generalized) 679
Strategy for Determining Probabilities 680
Multiplication Principle of Probability 681
Addition Principle of Probability 683
Outlier Guideline 706
Success Failure Rule 726
Four Important Principles of Experimental Design 737

CHAPTER 10 Review Exercises

The collection of exercises marked in red could be used as a chapter test.

1. The sample space for a random variable X is $\{1, 2, 5, 10\}$. Is $P(1) = 0.45, P(2) = 0.25, P(3) = 0.15$, and $P(10) = 0.05$ a valid probability function? Explain.

2. The sample space for a random variable is $\{A, B, C\}$. Outcome A is only half as likely as B but 3 times as likely as C. Find the probabilities of the three outcomes.

3. Lottery In a state lottery players must pick 5 winning numbers from 1 to 40. What's the probability of winning for a player who buys a dozen different tickets?

4. Poker Hand What's the probability that a 5-card poker hand will contain all hearts?

5. Candy A box of candies contains 6 caramels and 4 buttercreams, all appearing identical. What's the probability that a person who starts eating them one at a time won't get a caramel until the third candy?

6. Celebration A bag contains 10 marbles: 5 red, 3 white, and 2 blue. At a 4th of July carnival there's a nice prize for anyone who can draw out red, white, and blue in that order. What's the probability that a player wins the prize?

7. Seatbelts Highway safety experts estimate that 85% of drivers wear seatbelts. If the police stop 5 cars at a roadblock, what's the probability of each of these outcomes?

(a) All are wearing seatbelts.

(b) The 4th driver is the first one who is not belted in.

(c) At least one driver is not wearing a seatbelt.

8. Bowling A certain bowler makes a strike in 40% of her frames. What's the probability of each of these outcomes?

(a) Her first strike in a game comes in the third frame?

(b) She makes a strike in at least one of the first 5 frames?

(c) She bowls an entire game (10 frames) without a strike?

9. College Amenities On one college campus 66% of dorm rooms have refrigerators, 41% have TVs, and 32% have both.

(a) What's the probability that a randomly selected dorm room has a refrigerator or a TV?

(b) What's the probability that a dorm room with a refrigerator has a TV?

10. At the college described in Exercise 9, are having a refrigerator and having a TV independent? Explain.

11. Chicken In 2007 *Consumer Reports* purchased samples of chicken offered for sale in 23 states and checked them for bacterial contamination. The magazine found that 81% of the packages were contaminated with campylobacter, 15% with salmonella, and 13% with both. What's the probability that a package of chicken was bacteria-free?

12. Given the findings reported in Exercise 11, does it appear that the two kinds of contamination are independent? Explain.

13. Mixed Nuts Two cans of mixed nuts of different brands are open on a table. Brand A consists of 30% cashews, while brand B consists of 40% cashews. A can is chosen at random, and a nut is chosen at random from the can. Find the probability that the nut is

(a) From the brand A can.

(b) A brand A cashew.

(c) A cashew.

(d) From the brand A can, given that it is a cashew.

14. Horse Racing If the track is wet, Mudder Earth has a 70% chance of winning the fifth race at Keeneland. If the track is dry, she only has a 40% chance of winning. Weather forecasts predict an 80% chance that the track will be wet. Find the probability that

(a) The track is wet and Mudder Earth wins.

(b) The track is dry and Mudder Earth wins.

(c) Mudder Earth wins.

(d) (In retrospect) the track was wet, given that Mudder Earth won.

15. Men's Health Workers at a health clinic's walk-in screening event examined 88 men to see if they had high blood pressure and/or high cholesterol. The table summarizes the diagnoses:

		Cholesterol		
		High	OK	Total
Blood Pressure	High	22	12	34
	OK	6	48	54
	Total	28	60	88

(a) What is the probability that a man who was examined had both high blood pressure and high cholesterol?

(b) What is the probability that a man with high blood pressure had high cholesterol?

(c) In this group, were having high blood pressure and having high cholesterol independent? Explain.

16. Reading Pupils at an elementary school took a state proficiency test to see if they were reading at grade level. The results indicated there was no association between reading proficiency and gender. Fill in the missing counts in the table below.

	Boys	Girls	Total
At or above grade level	?	?	150
Below grade level	?	?	30
Total	84	96	180

17. Passing Yardage In 1995, Warren Moon of the Minnesota Vikings became the first pro quarterback to pass for 60,000 total yards. Use intervals of 1000 yd for Moon's regular season passing yards given in Table 10.31.

Table 10.31 Regular Season Passing Yardage Statistics for Warren Moon

Year	Yards	Year	Yards
1978	1112	1987	2806
1979	2382	1988	2327
1980	3127	1989	3631
1981	3959	1990	4689
1982	5000	1991	4690
1983	5648	1992	2521
1984	3338	1993	3485
1985	2709	1994	4264
1986	3489	1995	4228

Source: The Minnesota Vikings, as reported by Julie Stacey in USA Today on September 25, 1995, and www.nfl.com

(a) Create a stemplot of these data.

(b) Describe the shape of the distribution.

18. Based on the data in Exercise 17:

(a) Calculate the median, range, interquartile range, and five-number summary.

(b) Calculate the mean and standard deviation.

(c) Would you summarize these data with the median and *IQR* or the mean and standard deviation? Why?

19. Create a frequency table for the data in Exercise 17.

20. Create a boxplot of the data in Exercise 17 and identify any outliers.

21. Beatles Songs The lengths (in seconds) of 24 randomly selected Beatles songs that appeared on singles are as follows, in order of release date:

143, 120, 120, 139, 124, 144, 131, 132, 148, 163, 140, 177, 136, 124, 179, 131, 180, 137, 156, 202, 191, 197, 230, 190

(Source: Personal collection.)

(a) Create a stemplot of these data.

(b) Describe the shape of the distribution.

22. Based on the data in Exercise 21:

(a) Calculate the median, range, interquartile range, and five-number summary.

(b) Calculate the mean and standard deviation.

(c) Would you summarize these data with the median and *IQR* or the mean and standard deviation? Why?

23. Create a frequency table for the data in Exercise 21.

24. Create a boxplot of the data in Exercise 21 and identify any outliers.

25. Make a back-to-back stemplot of the data in Exercise 21, showing the earlier 12 songs in one plot and the later 12 songs in the other. Write a sentence interpreting the stemplot.

26. Make simultaneous boxplots of the data in Exercise 21, showing the earlier 12 songs in one boxplot and the later 12 songs in the other.

(a) Which set of data has the greater range?

(b) Which set of data has the greater interquartile range?

27. Time Plots Make a time plot for the data in Exercise 21, assuming equal time intervals between songs. Interpret the trend revealed in the time plot.

28. Smoothed Time Plots Statisticians sometimes use a technique called *smoothing* to smooth out random fluctuations in a time plot. Find the mean of the first four numbers in Exercise 21, the mean of the next four numbers, and so on. Then graph the six means as a function of time. Is there a clear trend?

29. Popular Web Sites In 2012 the numbers of unique visitors per month (in millions) to the top 25 Web sites were as follows:

880, 800, 590, 490, 440, 410, 340, 300, 250, 250, 230, 190, 160, 160, 160, 140, 140, 130, 120, 120, 110, 110, 110, 110, 98

(Source: DoubleClick AdPlanner by Google.)

(a) Create a histogram of these data.

(b) Describe the shape of the distribution.

30. Based on the data in Exercise 29:

(a) Calculate the median, range, interquartile range, and five-number summary.

(b) Calculate the mean and standard deviation.

(c) Compare the mean and median and explain why this is true.

(d) Would you summarize these data with the median and *IQR* or the mean and standard deviation? Why?

31. Create a frequency table for the data in Exercise 29.

32. Create a boxplot of the data in Exercise 29 and identify any outliers.

33. Real Estate Prices The median home sale prices (in thousands of $) for 2012 in 30 randomly selected metropolitan areas were as follows:

101.4, 164.3, 156.4, 126.9, 149.1, 175.3, 142.0, 114.6, 252.4, 106.9, 173.9, 126.1, 152.0, 164.8, 119.3, 87.2, 138.2, 203.1, 151.9, 186.9, 145.0, 130.3, 232.9, 189.3, 123.9, 171.9, 149.2, 178.9, 80.4, 144.9

(Source: National Association of Realtors.)

(a) Create a histogram of these data.

(b) Describe the shape of the distribution.

34. Based on the data in Exercise 33:

(a) Calculate the median, range, interquartile range, and five-number summary.

(b) Calculate the mean and standard deviation.

(c) Would you summarize these data with the median and *IQR* or the mean and standard deviation? Why?

35. Create a frequency table for the data in Exercise 33.

36. Create a boxplot of the data in Exercise 33 and identify any outliers.

37. Batters In 2010 the batting averages of the 136 major league baseball players who were eligible for postseason awards (essentially the full season starting players) had a mean of 0.274 and a standard deviation of 0.027. Assuming the distribution of batting averages was approximately Normal, use the 68-95-99.7 Rule to estimate answers for these questions.

(a) About what percent of players hit less than 0.301?

(b) In what interval would the batting averages of approximately 95% of the players be found?

(c) That year Josh Hamilton of the Texas Rangers won the batting title by hitting 0.356. Comment on how unusual that batting average is.

38. Cattle According to the American Angus Association, the distribution of weights of mature Angus steers is approximately Normal with a mean of 1309 lb and a standard deviation of 157 lb. Use the 68-95-99.7 Rule to estimate answers for these questions.

(a) In what interval do the weights of the middle 68% of these cattle fall?

(b) Above what weight are the heaviest 2.5% of these cattle?

(c) In 1988 the Denver champion bull named Dameron Linedrive (seriously) weighed 2527 lb. Why is that so remarkable?

39. Find the expected value of the random variable defined by this probability model:

X	100	150	200	500
$P(X)$	0.4	0.3	0.2	0.1

40. Fair Game The probability model for a game (below) shows some of the prizes a player may win and the losses the player risks. How much must the missing prize be in order to make the game fair?

X	−$10	$1	$5	?
$P(X)$	0.60	0.25	0.10	0.05

41. Car Repair Automobile brakes work when a part called a caliper squeezes brake pads against a spinning disk to slow the rotation of the wheels. The brakes won't work correctly if the caliper is not moving freely. At the brake shop, the simplest fix is to clean and lubricate the caliper. This costs $50 and will work in 35% of cases. In 90% of the cars with a more serious problem, the caliper must be replaced at a cost of $85 for the part and $75 for the labor to install it. The remaining cars need a complete brake overhaul that costs $350. What's the expected cost for car owners who come in for repair of this problem?

42. Contracts A small construction company has bid on two contracts. Preparing the bids cost the company $500. If it gets the smaller job, it expects to earn $8000. The larger job offers $12,000. The manager estimates there's a 60% chance the company will get the smaller contract, a 35% chance it will get the larger one, and a 10% chance it will get both jobs. What's the company's expected profit?

43. Coin Toss A fair coin is tossed five times. Find the probability of obtaining two heads and three tails.

44. Coin Toss A fair coin is tossed four times. Find the probability of obtaining one head and three tails.

45. Baseball Bats Suppose that the probability a bat company produces a defective baseball bat is 0.02. Four bats are selected at random. What is the probability that the lot of four bats contains the following?

(a) No defective bats

(b) One defective bat

46. A professional baseball team purchases 240 bats from the company described in Exercise 45. Find the mean and standard deviation of the number of defective bats the team may get.

47. Would it be appropriate to use a Normal model to describe the distribution of the number of defective bats in batches like the one described in Exercise 46? Explain.

48. Lightbulbs Suppose that the probability a certain manufacturer produces a defective lightbulb is 0.0004. Ten lightbulbs are selected at random. What is the probability that the lot of 10 contains the following?

(a) No defective lightbulbs

(b) Two defective lightbulbs

49. A large hardware supply chain purchases 100,000 lightbulbs from the company described in Exercise 48. Find the mean and standard deviation of the number of defective bulbs in that order.

50. Would it be appropriate to use a Normal model to describe the distribution of the number of defective lightbulbs in orders like the one described in Exercise 49? Explain.

51. Archery At a practice range a bowhunter had been able to hit 65% of his targets. Hoping to improve his accuracy, he bought a new bow. When he tested it, he hit 41 of the first 50 targets.

(a) Find the mean and standard deviation of the number of targets he'd hit if the new bow is no more accurate than the old.

(b) Explain why a Normal model can be used to describe the distribution of targets hit.

(c) Is it plausible that his performance was just a run of good luck, or was this evidence that he is a much better shot with the new bow?

52. Telemarketing A newly hired telemarketer's pay will be based on the number of sales she makes. Her employer predicted she'd successfully sell the product to about 8% of the people she calls. At the end of the first week she had placed 450 phone calls but made only 29 sales. Is this statistically significant evidence that the 8% estimate was misleading?

53. Angus Exercise 38 describes the distribution of weights of mature Angus steers as approximately Normal, with a mean of 1309 lb and a standard deviation of 157 lb.

(a) Approximately what percent of the Angus steers weigh over 1500 lb?

(b) Approximately what percent of the animals weigh only 1100–1250 lb?

(c) What does the model predict the smallest 10% of these animals should weigh?

54. Estimate the interquartile range of the angus steer weights described in Exercise 53.

55. Golf Distances that professional golfers hit the ball off the tee (the "drive") are approximately normally distributed with a mean of 287 yd and a standard deviation of 9 yd.

(a) Approximately what percent of drives carry at least 275 yd?

(b) Approximately what percent of drives travel 290–300 yd?

(c) How far does the model suggest the longest 1% of all drives travel?

56. Estimate the interquartile range of the golf drives described in Exercise 55.

57. Speeding Data collected for a study about traffic tickets revealed that even though white cars account for 25% of the vehicle population, they get only 19% of the speeding tickets. In reporting on these statistics, a newspaper article said there was a high correlation between the color of a car and the risk that the driver gets a speeding ticket. Explain the statistical error in that statement.

58. The newspaper article described in Exercise 57 went on to speculate that the police don't issue as many tickets to white cars because they are less likely to notice the non-flashy color.

(a) Explain the statistical error in that statement.

(b) Propose an alternative explanation for the association between the color and speeding tickets.

59. A student calculates a correlation of $r = -0.96$ between two variables and concludes that the relationship is therefore linear. Explain the statistical error.

60. Another student calculates a correlation of $r = -0.164$ for the data plotted below, and says this indicates there is only a very weak association between the two variables. Explain the statistical error.

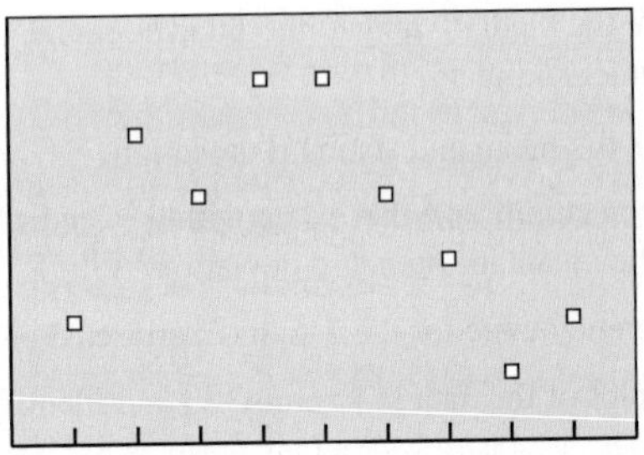

61. The HR director of a large law firm wants to survey a random sample of the firm's 443 employees. She numbers the employees alphabetically and uses her daughter's graphing calculator to choose 60 numbers with the command "randInt(1, 443, 60). Does this result in a random sample?

62. To get a random sample of 50 Atlanta citizens for a political poll, a worker starts at the beginning of the telephone book. For each name in the book, he flips a fair coin. If it comes up heads, he calls the person; if it comes up tails, he moves on. Once he has called 50 people, he stops. Does this result in a random sample?

63. To gather information about the popularity of a radio station, the station staff invites listeners to visit its Web site and respond to a short questionnaire. The station staff was depressed because only 45% of the respondents said they liked the radio station. Identify any bias in this sampling method.

64. A social studies class analyzes 30 canned soups available at a local food store, recording various nutritional data for a typical 1-cup serving. Is this an experiment or an observational study?

65. SAT Prep An organization offers special courses to help students to improve their scores on the SAT exam. It claims that students who participated in this course raised their scores by an average of 50 points when they retook the test. To verify this claim, a researcher has found 40 volunteers who have taken the SAT once and plan to do so again. Explain how you would design an experiment, including specific comments on replication, randomization, and control.

66. Explain how you would randomize the 40 volunteers for the SAT experiment described in Exercise 65.

67. Laundry To test the effectiveness of a new additive for its laundry detergent, a company's researchers have prepared 20 identical swatches of cloth with particularly hard-to-remove dirt and stains. Explain how you would design an experiment to find out whether the additive makes the detergent more effective. Include specific comments on replication, randomization, and control.

68. Explain how you would randomize the experiment described in Exercise 67.

69. Marbles At a 4th of July celebration contestants draw marbles from a bag containing 5 red ones, 3 white ones, and 2 blue ones. Players draw one marble at a time until they have a complete set of red, white, and blue (in any order; for example, one outcome is RBRRBW). The player's prize is $1 for each marble left in the bag. Explain how to use random numbers to create a simulation that would be able to estimate the average amount players might win at this game.

70. Dice In the game Yahtzee a player rolls 5 dice, getting points for various outcomes. One of those outcomes is a "full house": 3 dice showing the same face and the other two showing a different pair—for example, 25525. Explain how to design a simulation to estimate the probability of getting a full house on the first roll.

71. The Hot Hand Many sports fans (and players, too) believe that when a basketball player makes several shots in a row that player is "in the zone," and thus more likely to make the next shot, too. Do long streaks of success provide statistically significant evidence of a "hot hand" phenomenon, or are they just random events? In an NBA season a good player might take 1500 shots and make 55% of them. A simulation modeled a typical season, counting the number of streaks that occurred. The table below shows the distribution of the lengths of the 374 streaks during that simulated season.

Length of Streak	Number of Times
1	173
2	96
3	61
4	23
5	5
6	6
7	4
8	3
9	1
10	1
13	1

(a) Based on these simulated results, do you think that a player who hits 5 shots in a row is "in the zone"? Explain.

(b) How many shots in a row would you consider to be statistically significant? Explain.

72. Pigs A farmer wonders whether a new food supplement does lead to the more rapid weight gain in pigs that it claims. To investigate, he selects 10 of his pigs that are all about the same size and randomly splits them into two groups. One group continues to get the regular feed while the other group gets the feed with the supplement added. After several weeks he weighs the pigs, recording these results:

Regular Feed	194.0	205.5	199.2	172.4	184.0	169.5
Feed w/Supplement	190.7	203.5	203.5	206.5	222.5	209.4

On average, the pigs fed the supplement weighed 18.6 lb more than those fed only the regular food, but is that difference statistically significant? To find out, a simulation divides the 10 weights into two groups at random and calculates the difference in means. Here is a histogram of the random differences for 200 trials.

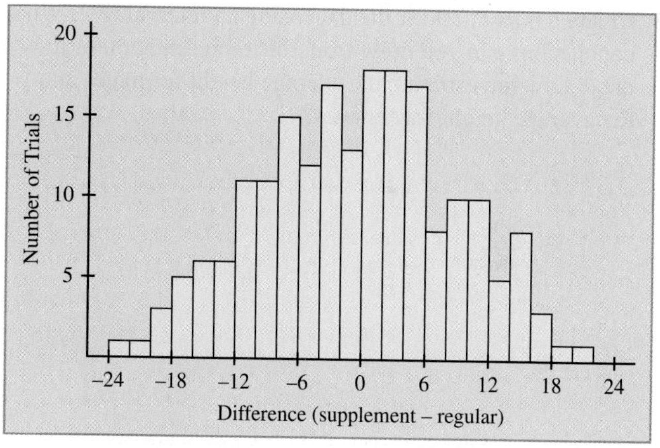

Does this simulation provide evidence that the feed supplement is effective? Explain.

CHAPTER 10 Project

Analyzing Height Data

The set of data below was gathered from a class of 30 precalculus students. Use this set of data or collect the data for your own class and use it for analysis.

Heights of Students (in.)				
66	69	72	64	68
70	71	66	65	63
72	59	64	63	66
68	63	64	71	71
69	62	61	67	69
64	73	75	61	70

Explorations

1. Create a stem-and-leaf plot of the data using split stems. From these data, what is the approximate average height of a student in the class?
2. Create a frequency table for the data using an interval of 2. What information does this give?
3. Create a histogram for the data using an interval of 2. What conclusions can you draw from this representation of the data? Can you estimate the average height for males and the average height for females?
4. Compute the mean and median for the data set. Discuss whether each is a good measure of the average height of a student in the class. Is each a good predictor for average height of students in other precalculus classes?
5. What does it suggest may be true about the distribution of the data if the mean and median values are close?
6. Find the five-number summary for the class heights.
7. Create a boxplot and explain what information it gives about the data set.
8. A new student is now added to the class. He is a 7 ft 2 in. star basketball player. Add his height to the data set. Recalculate the mean, median, and five-number summary. Create a new boxplot and use your calculator to plot it underneath the boxplot for the original class. How does this new student affect the statistics?
9. Explain why this new student would be considered an outlier and the importance of identifying outliers when calculating statistics and making predictions from them.
10. Calculate the standard deviation of the original data. Determine what percent of the data actually lie within ± 1, ± 2, and ± 3 standard deviations of the mean. What does this suggest about modeling these heights with a Normal curve?

11.1 Limits and Motion: The Tangent Problem

11.2 Limits and Motion: The Area Problem

11.3 More on Limits

11.4 Numerical Derivatives and Integrals

An Introduction to Calculus: Limits, Derivatives, and Integrals

Windmills have long been used to pump water from wells, grind grain, and saw wood. More recently they are being used to produce electricity. The propeller radius of these windmills ranges from 1 m to 100 m, and the power output ranges from 100 W to 1000 kW. See page 761 in Section 11.1 for some more information and a question and answer about windmills.

Chapter 11 Overview

By the beginning of the 17th century, algebra and geometry had developed to the point where physical behavior could be modeled both algebraically and graphically, each type of representation providing deeper insights into the other. New discoveries about the solar system had opened up fascinating questions about gravity and its effects on planetary motion, so that finding the mathematical key to studying motion became the scientific quest of the day. The analytic geometry of René Descartes (1596–1650) put the final pieces into place, setting the stage for Isaac Newton (1642–1727) and Gottfried Leibniz (1646–1716) to stand "on the shoulders of giants" and see beyond the algebraic boundaries that had limited their predecessors. With geometry showing them the way, they created the new form of algebra that would come to be known as calculus.

In this chapter we will look at the two central problems of motion much as Newton and Leibniz did, connecting them to geometric problems involving tangent lines and areas. We will see how the obvious geometric solutions to both problems led to algebraic dilemmas, and how the algebraic dilemmas led to the discovery of calculus. The language of limits, which we have used in this book to describe asymptotes, end behavior, and continuity, will serve us well as we make this transition.

11.1 Limits and Motion: The Tangent Problem

What you'll learn about

- Average Velocity
- Instantaneous Velocity
- Limits Revisited
- The Connection to Tangent Lines
- The Derivative

... and why

The derivative allows us to analyze rates of change, which are fundamental to understanding physics, economics, engineering, and even history.

Average Velocity

Average velocity is the change in position divided by the change in time, as in the following familiar-looking example.

EXAMPLE 1 Computing Average Velocity

An automobile travels 120 mi in 2 hr 30 min. What is its average velocity over the entire $2\frac{1}{2}$-hr time interval?

SOLUTION The average velocity is the change in position (120 mi) divided by the change in time (2.5 hr). If we denote position by s and time by t, we have

$$v_{\text{ave}} = \frac{\Delta s}{\Delta t} = \frac{120 \text{ mi}}{2.5 \text{ hr}} = 48 \text{ mph}.$$

Now try Exercise 1.

Notice that the average velocity does not tell us how fast the automobile is traveling at any given moment during the time interval. It could have gone at a constant 48 mph all the way, or it could have sped up, slowed down, or even stopped momentarily several times during the trip. Scientists like Galileo Galilei (1564–1642), who studied motion prior to Newton and Leibniz, were looking for formulas that would give velocity as a *function* of time, that is, formulas that would give *instantaneous* values of v for individual values of t. The step from average to instantaneous sounds simple enough, but there were complications, as we shall see.

Instantaneous Velocity

Galileo experimented with gravity by rolling a ball down an inclined plane and recording its approximate velocity as a function of elapsed time. Here is what he might have asked himself when he began his experiments:

A Velocity Question

A ball rolls a distance of 16 ft in 4 sec. What is the instantaneous velocity of the ball at a moment of time 3 sec after it starts to roll?

You might want to visualize the ball being frozen at that moment, and then try to determine its velocity. Well, then the ball would have zero velocity, because it is frozen! This approach seems foolish, since, of course, the ball is moving.

A Disclaimer

Readers who know a little calculus will recognize that the ball in the Velocity Question really does have a nonzero instantaneous velocity after 3 sec (as we will eventually show). The point of the discussion is to show how difficult it is to demonstrate that fact at a single instant, since both time and the position of the ball appear not to change.

Is this a trick question? On the contrary, it is actually quite profound—it is exactly the question that Galileo (among many others) was trying to answer. Notice how easy it is to find the *average* velocity:

$$v_{\text{ave}} = \frac{\Delta s}{\Delta t} = \frac{16 \text{ ft}}{4 \text{ sec}} = 4 \text{ ft/sec}$$

Now, notice how inadequate our algebra becomes when we try to apply the same formula to *instantaneous* velocity:

$$v_{\text{ave}} = \frac{\Delta s}{\Delta t} = \frac{0 \text{ ft}}{0 \text{ sec}},$$

which involves division by 0—and is therefore undefined!

So Galileo did the best he could by making Δt as small as experimentally possible, measuring the small values of Δs, and then finding the quotients. It only *approximated* the instantaneous velocity, but finding the exact value appeared to be algebraically out of the question, since division by zero was impossible.

Limits Revisited

Newton invented "fluxions" and Leibniz invented "differentials" to explain instantaneous rates of change without resorting to zero denominators. Both involved mysterious quantities that could be infinitesimally small without really being zero. (Their 17th-century colleague Bishop Berkeley called them "ghosts of departed quantities" and dismissed them as nonsense.) Though not well understood by many, the strange quantities modeled the behavior of moving bodies so effectively that most scientists were willing to accept them on faith until a better explanation could be developed. That development, which took about a hundred years, led to our modern understanding of limits.

Since you are already familiar with limit notation, we can show you how this works with a simple example.

EXAMPLE 2 Using Limits to Avoid Zero Division

A ball rolls down a ramp so that its distance s from the top of the ramp after t seconds is exactly t^2 feet. What is its instantaneous velocity after 3 sec?

SOLUTION We might try to answer this question by computing average velocity over smaller and smaller time intervals.

(continued)

On the interval $[3, 3.1]$:

$$\frac{\Delta s}{\Delta t} = \frac{(3.1)^2 - 3^2}{3.1 - 3} = \frac{0.61}{0.1} = 6.1 \text{ ft/sec}$$

On the interval $[3, 3.05]$:

$$\frac{\Delta s}{\Delta t} = \frac{(3.05)^2 - 3^2}{3.05 - 3} = \frac{0.3025}{0.05} = 6.05 \text{ ft/sec}$$

Continuing this process we would eventually conclude that the **instantaneous velocity** must be 6 ft/sec.

However, we can see *directly* what is happening to the quotient by treating it as a *limit* of the average velocity on the interval $[3, t]$ as t approaches 3:

$$\lim_{t\to 3}\frac{\Delta s}{\Delta t} = \lim_{t\to 3}\frac{t^2 - 3^2}{t - 3}$$

$$= \lim_{t\to 3}\frac{(t+3)(t-3)}{t-3} \qquad \text{Factor the numerator.}$$

$$= \lim_{t\to 3}(t+3)\cdot\frac{t-3}{t-3}$$

$$= \lim_{t\to 3}(t+3) \qquad \text{Since } t \neq 3, \frac{t-3}{t-3} = 1.$$

$$= 6$$

Now try Exercise 3.

Notice that t is *not equal to* 3 but is *approaching* 3 as a limit, which allows us to make the crucial cancellation in the second to last line of Example 2. If t were actually equal to 3, the algebra above would lead to the incorrect conclusion that $0/0 = 6$. The difference between equaling 3 and approaching 3 as a limit is a subtle one, but it makes all the difference algebraically.

It is not easy to formulate a rigorous algebraic definition of a limit (which is why Newton and Leibniz never really did). We have used an intuitive approach to limits so far in this book and will continue to do so, deferring the rigorous definitions to your calculus course. For now, we will use the following *informal* definition.

Arbitrarily Close

This definition is useless for mathematical proofs until one defines "arbitrarily close," but if you have a sense of how it applies to the solution in Example 2 above, then you are ready to use limits to study motion problems.

DEFINITION (INFORMAL) Limit at *a*

When we write "$\lim_{x\to a} f(x) = L$," we mean that $f(x)$ gets arbitrarily close to L as x gets arbitrarily close (but not equal) to a.

The Connection to Tangent Lines

What Galileo discovered by rolling balls down ramps was that the distance traveled was proportional to the *square* of the elapsed time. For simplicity, let us suppose that the ramp was tilted just enough so that the relation between s, the distance from the top of the ramp, and t, the elapsed time, was given (as in Example 2) by

$$s = t^2.$$

Graphing s as a function of $t \geq 0$ gives the right half of a parabola (Figure 11.1).

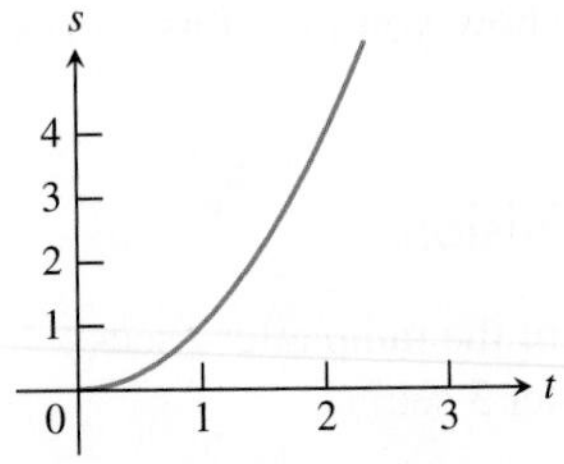

FIGURE 11.1 The graph of $s = t^2$ shows the distance (s) traveled by a ball rolling down a ramp as a function of the elapsed time t.

EXPLORATION 1 Seeing Average Velocity

Copy Figure 11.1 on a piece of paper and connect the points $(1, 1)$ and $(2, 4)$ with a straight line. (This is called a *secant line* because it connects two points on the curve.)

1. Find the slope of the line.
2. Find the average velocity of the ball over the time interval $[1, 2]$.
3. What is the relationship between the numbers that answer questions 1 and 2?
4. In general, how could you represent the average velocity of the ball over the time interval $[a, b]$ geometrically?

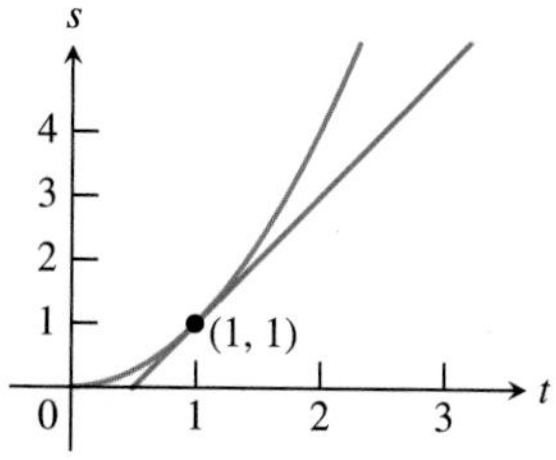

FIGURE 11.2 A line tangent to the graph of $s = t^2$ at the point $(1, 1)$. The slope of this line appears to be the instantaneous velocity at $t = 1$, even though $\Delta s/\Delta t = 0/0$. The geometry succeeds where the algebra fails!

Exploration 1 suggests an important general fact: If $(a, s(a))$ and $(b, s(b))$ are two points on a distance-time graph, then the *average velocity* over the time interval $[a, b]$ can be thought of as the *slope* of the line connecting the two points. In fact, we designate both quantities with the same symbol: $\Delta s/\Delta t$.

Galileo knew this. He also knew that he wanted to find instantaneous velocity by letting the two points become one, resulting in $\Delta s/\Delta t = 0/0$, an algebraic impossibility. The picture, however, told a different story *geometrically*. If, for example, we were to connect pairs of points closer and closer to $(1, 1)$, our secant lines would look more and more like a line that is tangent to the curve at $(1, 1)$ (Figure 11.2).

It seemed obvious to Galileo and the other scientists of his time that the slope of the tangent line was the long-sought-after answer to the quest for instantaneous velocity. They could *see* it, but how could they *compute* it without dividing by zero? That was the "tangent line problem," eventually solved for general functions by Newton and Leibniz in slightly different ways. We will solve it with limits as illustrated in Example 3.

EXAMPLE 3 Finding the Slope of a Tangent Line

Use limits to find the slope of the tangent line to the graph of $s = t^2$ at the point $(1, 1)$ (Figure 11.2).

SOLUTION This will look a lot like the solution to Example 2.

$$\lim_{t\to 1} \frac{\Delta s}{\Delta t} = \lim_{t\to 1} \frac{t^2 - 1^2}{t - 1}$$

$$= \lim_{t\to 1} \frac{(t+1)(t-1)}{t-1} \qquad \text{Factor the numerator.}$$

$$= \lim_{t\to 1} (t+1)\cdot\frac{t-1}{t-1}$$

$$= \lim_{t\to 1} (t+1) \qquad \text{Since } t \neq 1, \frac{t-1}{t-1} = 1.$$

$$= 2$$

Now try Exercise 17(a).

If you compare Example 3 to Example 2 it should be apparent that a method for solving the tangent line problem can be used to solve the instantaneous velocity problem, and vice versa. They are geometric and algebraic versions of the same problem!

The Derivative

Velocity, the rate of change of position with respect to time, is only one application of the general concept of "rate of change." If $y = f(x)$ is *any* function, we can speak of how y changes as x changes.

The Tangent Line Problem

Although we have focused on Galileo's work with motion problems in order to follow a coherent story, it was Pierre de Fermat (1601–1665) who first developed a "method of tangents" for general curves, recognizing its usefulness for finding relative maxima and minima. Fermat is best remembered for his work in number theory, particularly for Fermat's Last Theorem, which states that there are no positive integers x, y, and z that satisfy the equation $x^n + y^n = z^n$ if n is an integer greater than 2. Fermat wrote in the margin of a textbook, "I have a truly marvelous proof that this margin is too narrow to contain," but if he had one, he apparently never wrote it down. Although mathematicians tried for over 330 years to prove (or disprove) Fermat's Last Theorem, nobody succeeded until Andrew Wiles of Princeton University finally proved it in 1994.

DEFINITION Average Rate of Change

If $y = f(x)$, then the **average rate of change** of y with respect to x on the interval $[a, b]$ is

$$\frac{\Delta y}{\Delta x} = \frac{f(b) - f(a)}{b - a}.$$

Geometrically, this is the slope of the **secant line** through $(a, f(a))$ and $(b, f(b))$.

Using limits, we can proceed to a definition of the *instantaneous* rate of change of y with respect to x at the point where $x = a$. This instantaneous rate of change is called the *derivative*.

Differentiability

We say a function is "differentiable" at a if $f'(a)$ exists, because we can find the limit of the "quotient of differences."

DEFINITION Derivative at a Point

The **derivative of the function f at $x = a$**, denoted by $f'(a)$ and read "f prime of a," is

$$f'(a) = \lim_{x \to a} \frac{f(x) - f(a)}{x - a},$$

provided the limit exists.

Geometrically, this is the slope of the **tangent line** to the graph of f at $(a, f(a))$.

A more computationally useful formula for the derivative is obtained by letting $x = a + h$ and looking at the limit as h approaches 0 (equivalent to letting x approach a).

DEFINITION Derivative at a Point (easier for computing)

The **derivative of the function f at $x = a$**, denoted by $f'(a)$ and read "f prime of a," is

$$f'(a) = \lim_{h \to 0} \frac{f(a + h) - f(a)}{h},$$

provided the limit exists.

The fact that the derivative of a function at a point can be viewed geometrically as the slope of the line tangent to the curve $y = f(x)$ at that point provides us with some insight as to how a derivative might fail to exist. Unless a function has a well-defined "slope" when you zoom in on it at a, the derivative at a will not exist. For example, Figure 11.3 shows three cases for which $f(0)$ exists but $f'(0)$ does not.

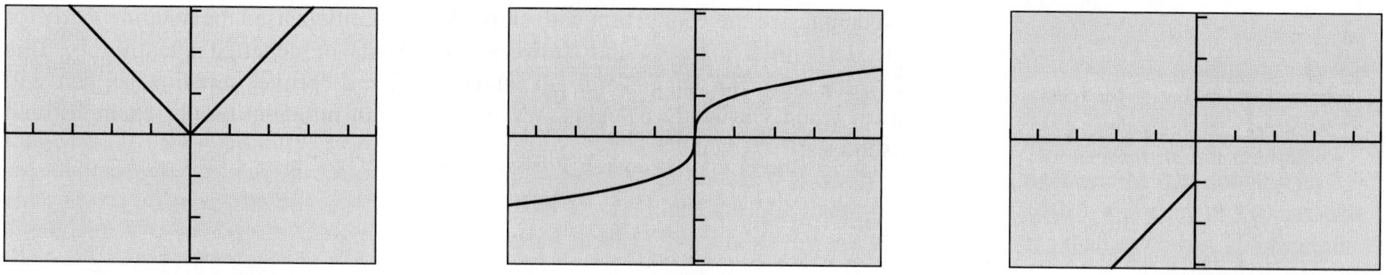

[−4.7, 4.7] by [−3.1, 3.1]
(a)

$f(x) = |x|$ has a graph with no definable slope at $x = 0$.

[−4.7, 4.7] by [−3.1, 3.1]
(b)

$f(x) = \sqrt[3]{x}$ has a graph with a vertical tangent line (no slope) at $x = 0$.

[−4.7, 4.7] by [−3.1, 3.1]
(c)

$f(x) = \begin{cases} x - 1 & \text{for } x < 0 \\ 1 & \text{for } x \geq 0 \end{cases}$
has a graph with no definable slope at $x = 0$.

FIGURE 11.3 Three examples of functions defined at $x = 0$ but not differentiable at $x = 0$.

EXAMPLE 4 Finding a Derivative at a Point

Find $f'(4)$ if $f(x) = 2x^2 - 3$.

SOLUTION

$$\begin{aligned} f'(4) &= \lim_{h \to 0} \frac{f(4 + h) - f(4)}{h} \\ &= \lim_{h \to 0} \frac{[2(4 + h)^2 - 3] - (2 \cdot 4^2 - 3)}{h} \\ &= \lim_{h \to 0} \frac{2(16 + 8h + h^2) - 32}{h} \\ &= \lim_{h \to 0} \frac{16h + 2h^2}{h} \\ &= \lim_{h \to 0} (16 + 2h) \qquad \frac{h}{h} = 1, \text{ since } h \neq 0. \\ &= 16 \end{aligned}$$

Now try Exercise 23.

The derivative can also be thought of as a function of x. Its domain consists of all values in the domain of f for which f is differentiable. The function f' can be defined by adapting the second definition above.

DEFINITION Derivative

If $y = f(x)$, then the **derivative of the function f with respect to x** is the function f' whose value at x is

$$f'(x) = \lim_{h \to 0} \frac{f(x + h) - f(x)}{h},$$

for all values of x where the limit exists.

To emphasize the connection with slope $\Delta y/\Delta x$, Leibniz used the notation dy/dx for the derivative. (The dy and dx were his "ghosts of departed quantities.") This **Leibniz notation** has several advantages over the "prime" notation, as you will learn when you study calculus. We will use both notations in our examples and exercises.

EXAMPLE 5 Finding the Derivative of a Function

(a) Find $f'(x)$ if $f(x) = x^2$.

(b) Find $\dfrac{dy}{dx}$ if $y = \dfrac{1}{x}$.

SOLUTION

(a)

$$\begin{aligned} f'(x) &= \lim_{h \to 0} \frac{f(x+h) - f(x)}{h} \\ &= \lim_{h \to 0} \frac{(x+h)^2 - x^2}{h} \\ &= \lim_{h \to 0} \frac{x^2 + 2xh + h^2 - x^2}{h} \\ &= \lim_{h \to 0} \frac{2xh + h^2}{h} \\ &= \lim_{h \to 0} (2x + h) \\ &= 2x \end{aligned}$$

Since $h \neq 0$, $\dfrac{h}{h} = 1$.

So $f'(x) = 2x$.

(b)

$$\begin{aligned} \frac{dy}{dx} &= \lim_{h \to 0} \frac{f(x+h) - f(x)}{h} \\ &= \lim_{h \to 0} \frac{\dfrac{1}{x+h} - \dfrac{1}{x}}{h} \\ &= \lim_{h \to 0} \frac{\dfrac{x - (x+h)}{x(x+h)}}{h} \\ &= \lim_{h \to 0} \frac{-h}{x(x+h)} \cdot \frac{1}{h} \\ &= \lim_{h \to 0} \frac{-1}{x(x+h)} \\ &= -\frac{1}{x^2} \end{aligned}$$

So $\dfrac{dy}{dx} = -\dfrac{1}{x^2}$.

Now try Exercise 29.

CHAPTER OPENER Problem (from page 753)

Problem: For an efficient windmill, the power generated in watts is given by the equation

$$P = kr^2v^3$$

where r is the radius of the propeller in meters, v is the wind velocity in meters per second, and k is a constant with units of kg/m^3. The exact value of k depends on various characteristics of the windmill.

Suppose a windmill has a propeller with a radius of 5 m and $k = 0.134$ kg/m^3.

(a) Find the function $P(v)$, which gives power as a function of wind velocity.

(b) Find $P'(7)$, the rate of change in power generated with respect to wind velocity, when the wind velocity is 7 m/sec.

Solution:

(a) Since $r = 5$ m and $k = 0.134$ kg/m^3, we have

$$P = kr^2v^3 = (0.134)(5^2)v^3 = 3.35v^3.$$

So, $P(v) = 3.35v^3$, where v is in meters per second and P is in watts.

(b)
$$\begin{aligned}P'(7) &= \lim_{h\to 0} \frac{P(7+h) - P(7)}{h} \\ &= \lim_{h\to 0} \frac{3.35(7+h)^3 - 3.35(7)^3}{h} \\ &= \lim_{h\to 0} \frac{3.35[(7^3 + 147h + 21h^2 + h^3) - 7^3]}{h} \\ &= \lim_{h\to 0} \frac{3.35(147h + 21h^2 + h^3)}{h} \\ &= \lim_{h\to 0} [3.35(147 + 21h + h^2)] \\ &= 3.35(147) \\ &= 492.45\end{aligned}$$

The rate of change in power generated is about 492 watts per m/sec.

QUICK REVIEW 11.1 *(For help, go to Sections P.1 and P.4.)*

Exercise numbers with a gray background indicate problems that the authors have designed to be solved *without a calculator*.

In Exercises 1 and 2, find the slope of the line determined by the points.

1. $(-2, 3), (5, -1)$

2. $(-3, -1), (3, 3)$

In Exercises 3–6, write an equation for the specified line.

3. Through $(-2, 3)$ with slope $= 3/2$

4. Through $(1, 6)$ and $(4, -1)$

5. Through $(1, 4)$ and parallel to $y = (3/4)x + 2$

6. Through $(1, 4)$ and perpendicular to $y = (3/4)x + 2$

In Exercises 7–10, simplify the expression assuming $h \neq 0$.

7. $\dfrac{(2+h)^2 - 4}{h}$

8. $\dfrac{(3+h)^2 + 3 + h - 12}{h}$

9. $\dfrac{1/(2+h) - 1/2}{h}$

10. $\dfrac{1/(x+h) - 1/x}{h}$

SECTION 11.1 Exercises

1. **Average Velocity** A bicyclist travels 21 mi in 1 hr 45 min. What is her average velocity during the entire $1\frac{3}{4}$-hr time interval?

2. **Average Velocity** An automobile travels 540 km in 4 hr 30 min. What is its average velocity over the entire $4\frac{1}{2}$-hr time interval?

In Exercises 3–6, the position of an object at time t is given by $s(t)$. Find the instantaneous velocity at the indicated value of t.

3. $s(t) = 3t - 5$ at $t = 4$

4. $s(t) = \dfrac{2}{t + 1}$ at $t = 2$

5. $s(t) = at^2 + 5$ at $t = 2$

6. $s(t) = \sqrt{t + 1}$ at $t = 1$
(*Hint*: Rationalize the numerator.)

In Exercises 7–10, use the graph to estimate the slope of the tangent line, if it exists, to the graph at the given point.

7. $x = 0$

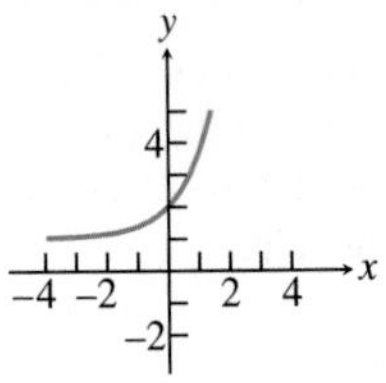

8. $x = 1$

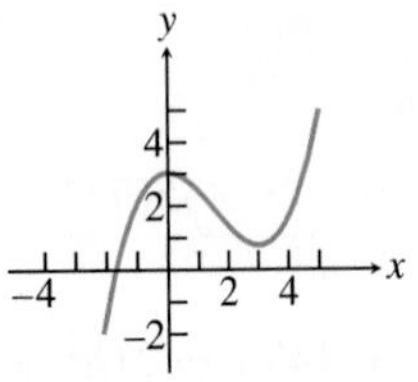

9. $x = 2$

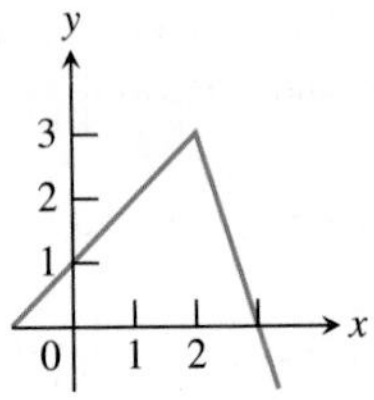

10. $x = 4$

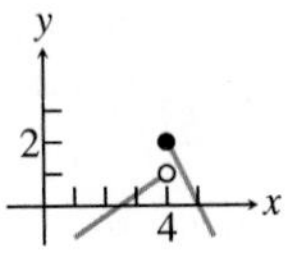

In Exercises 11–14, graph the function in a square viewing window and, without doing any calculations, estimate the derivative of the function at the given point by interpreting it as the tangent line slope, if it exists at the point.

11. $f(x) = x^2 - 2x + 5$ at $x = 3$

12. $f(x) = \frac{1}{2}x^2 + 2x - 5$ at $x = 2$

13. $f(x) = x^3 - 6x^2 + 12x - 9$ at $x = 0$

14. $f(x) = 2 \sin x$ at $x = \pi$

15. **A Rock Toss** A rock is thrown straight up from level ground. The distance (in ft) the ball is above the ground (the position function) is $f(t) = 3 + 48t - 16t^2$ at any time t (in sec). Find

(a) $f'(0)$.

(b) The initial velocity of the rock.

16. **Rocket Launch** A toy rocket is launched straight up in the air from level ground. The distance (in ft) the rocket is above the ground (the position function) is $f(t) = 170t - 16t^2$ at any time t (in sec). Find

(a) $f'(0)$.

(b) The initial velocity of the rocket.

In Exercises 17–20, use the limit definition to find

(a) The slope of the graph of the function at the indicated point.

(b) An equation of the tangent line at the point.

(c) Sketch a graph of the curve near the point without using your graphing calculator.

17. $f(x) = 2x^2$ at $x = -1$

18. $f(x) = 2x - x^2$ at $x = 2$

19. $f(x) = 2x^2 - 7x + 3$ at $x = 2$

20. $f(x) = \dfrac{1}{x + 2}$ at $x = 1$

In Exercises 21 and 22, estimate the slope of the tangent line to the graph of the function, if it exists, at the indicated points.

21. $f(x) = |x|$ at $x = -2, 2,$ and 0.

22. $f(x) = \tan^{-1}(x + 1)$ at $x = -2, 2,$ and 0.

In Exercises 23–28, find the derivative, if it exists, of the function at the specified point.

23. $f(x) = 1 - x^2$ at $x = 2$

24. $f(x) = 2x + \frac{1}{2}x^2$ at $x = 2$

25. $f(x) = 3x^2 + 2$ at $x = -2$

26. $f(x) = x^2 - 3x + 1$ at $x = 1$

27. $f(x) = |x + 2|$ at $x = -2$

28. $f(x) = \dfrac{1}{x + 2}$ at $x = -1$

In Exercises 29–32, find the derivative of f.

29. $f(x) = 2 - 3x$

30. $f(x) = 2 - 3x^2$

31. $f(x) = 3x^2 + 2x - 1$

32. $f(x) = \dfrac{1}{x - 2}$

33. **Average Speed** A lead ball is held at water level and dropped from a boat into a lake. The distance the ball falls at 0.1-sec time intervals is given in Table 11.1.

Table 11.1 Distance Data of the Lead Ball

Time (sec)	Distance (ft)
0	0
0.1	0.1
0.2	0.4
0.3	0.8
0.4	1.5
0.5	2.3
0.6	3.2
0.7	4.4
0.8	5.8
0.9	7.3

(a) Compute the average speed from 0.5 to 0.6 sec and from 0.8 to 0.9 sec.

(b) Find a quadratic regression model for the distance data and overlay its graph on a scatter plot of the data.

(c) Use the model in part (b) to estimate the depth of the lake if the ball hits the bottom after 2 sec.

34. **Finding Derivatives from Data** A ball is dropped from the roof of a two-story building. The distance in feet above ground of the falling ball is given in Table 11.2, where time t is in seconds.

Table 11.2 Distance Data of the Ball

Time (sec)	Distance (ft)
0.2	30.00
0.4	28.36
0.6	25.44
0.8	21.24
1.0	15.76
1.2	9.02
1.4	0.95

(a) Use the data to estimate the average velocity of the ball in the interval $0.8 \le t \le 1$.

(b) Find a quadratic regression model s for the data in Table 11.2 and overlay its graph on a scatter plot of the data.

(c) Find the derivative of the regression equation and use it to estimate the velocity of the ball at time $t = 1$.

In Exercises 35–38, complete the following.

(a) Draw a graph of the function.

(b) Find the derivative of the function at the given point if it exists.

(c) **Writing to Learn** If the derivative does not exist at the point, explain why not.

35. $f(x) = \begin{cases} 4 - x & \text{if } x \le 2 \\ x + 3 & \text{if } x > 2 \end{cases}$ at $x = 2$

36. $f(x) = \begin{cases} 1 + (x - 2)^2 & \text{if } x \le 2 \\ 1 - (x - 2)^2 & \text{if } x > 2 \end{cases}$ at $x = 2$

37. $f(x) = \begin{cases} \dfrac{|x - 2|}{x - 2} & \text{if } x \ne 2 \\ 1 & \text{if } x = 2 \end{cases}$ at $x = 2$

38. $f(x) = \begin{cases} \dfrac{\sin x}{x} & \text{if } x \ne 0 \\ 1 & \text{if } x = 0 \end{cases}$ at $x = 0$

In Exercises 39–42, sketch a possible graph for a function that has the stated properties.

39. The domain of f is $[0, 5]$ and the derivative at $x = 2$ is 3.

40. The domain of f is $[0, 5]$ and the derivative is 0 at both $x = 2$ and $x = 4$.

41. The domain of f is $[0, 5]$ and the derivative at $x = 2$ is undefined.

42. The domain of f is $[0, 5]$, f is nondecreasing on $[0, 5]$, and the derivative at $x = 2$ is 0.

43. **Writing to Learn** Explain why you can find the derivative of $f(x) = ax + b$ without doing any computations. What is $f'(x)$?

44. **Writing to Learn** Use the *first* definition of derivative at a point to express the derivative of $f(x) = |x|$ at $x = 0$ as a limit. Then explain why the limit does not exist. (A graph of the quotient for x values near 0 might help.)

Standardized Test Questions

45. **True or False** When a ball rolls down a ramp, its instantaneous velocity is always zero. Justify your answer.

46. **True or False** If the derivative of the function f exists at $x = a$, then the derivative is equal to the slope of the tangent line at $x = a$. Justify your answer.

In Exercises 47–50, choose the correct answer. You may use a calculator.

47. **Multiple Choice** If $f(x) = x^2 + 3x - 4$, find $f'(x)$.

(A) $x^2 + 3$ (B) $x^2 - 4$ (C) $2x - 1$
(D) $2x + 3$ (E) $2x - 3$

48. **Multiple Choice** If $f(x) = 5x - 3x^2$, find $f'(x)$.

(A) $5 - 6x$ (B) $5 - 3x$ (C) $5x - 6$
(D) $10x - 3$ (E) $5x - 6x^2$

49. **Multiple Choice** If $f(x) = x^3$, find the derivative of f at $x = 2$.

(A) 3 (B) 6 (C) 12
(D) 18 (E) Does not exist

50. **Multiple Choice** If $f(x) = \dfrac{1}{x - 3}$, find the derivative of f at $x = 1$.

(A) $-\dfrac{1}{4}$ (B) $\dfrac{1}{4}$ (C) $-\dfrac{1}{2}$
(D) $\dfrac{1}{2}$ (E) Does not exist

Explorations

Graph each function in Exercises 51–54 and then answer the following questions.

(a) Writing to Learn Does the function have a derivative at $x = 0$? Explain.

(b) Does the function appear to have a tangent line at $x = 0$? If so, what is an equation of the tangent line?

51. $f(x) = |x|$

52. $f(x) = |x^{1/3}|$

53. $f(x) = x^{1/3}$

54. $f(x) = \tan^{-1} x$

55. Free Fall A water balloon dropped from a window will fall a distance of $s = 16t^2$ feet during the first t seconds. Find the balloon's **(a)** average velocity during the first 3 sec of falling and **(b)** instantaneous velocity at $t = 3$.

56. Free Fall on Another Planet It can be established by experimentation that heavy objects dropped from rest free fall near the surface of another planet according to the formula $y = gt^2$, where y is the distance in meters the object falls in t seconds after being dropped. An object falls from the top of a 125-m spaceship that landed on the surface. It hits the surface in 5 sec.

(a) Find the value of g.

(b) Find the average speed for the fall of the object.

(c) With what speed did the object hit the surface?

Extending the Ideas

57. Graphing the Derivative The graph of $f(x) = x^2e^{-x}$ is shown below. Use your knowledge of the geometric interpretation of the derivative to sketch a rough graph of the derivative $y = f'(x)$.

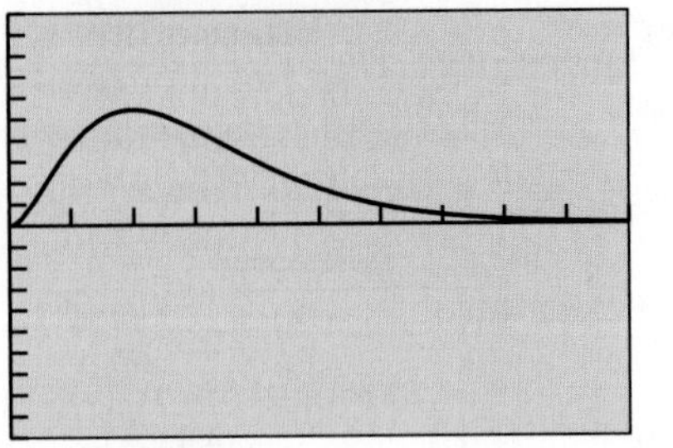

[0, 10] by [–1, 1]

58. Group Activity The graph of $y = f'(x)$ is shown below. Determine a possible graph for the function $y = f(x)$.

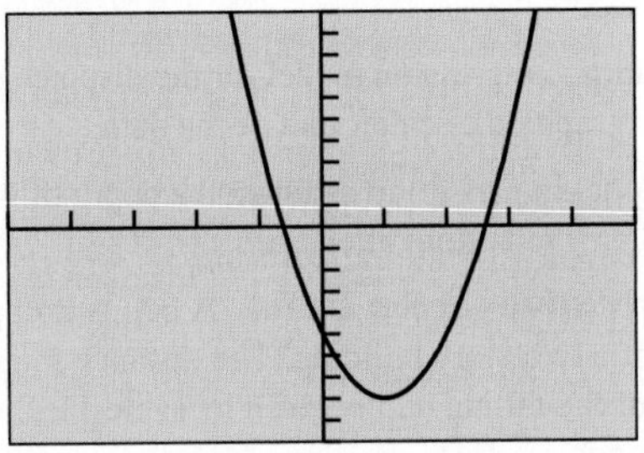

[–5, 5] by [–10, 10]

11.2 Limits and Motion: The Area Problem

What you'll learn about

- Distance from a Constant Velocity
- Distance from a Changing Velocity
- Limits at Infinity
- The Connection to Areas
- The Definite Integral

... and why

Like the tangent line problem, the area problem has many applications in every area of science, as well as historical and economic applications.

Distance from a Constant Velocity

"Distance equals rate times time" is one of the earliest problem-solving formulas that we learn in school mathematics. Given a velocity and a time period, we can use the formula to compute distance traveled—as in the following standard example.

EXAMPLE 1 Computing Distance Traveled

An automobile travels at a constant rate of 48 mph for 2 hr 30 min. How far does the automobile travel?

SOLUTION We apply the formula $d = rt$:

$$d = (48 \text{ mi/hr})(2.5 \text{ hr}) = 120 \text{ mi}$$

Now try Exercise 1.

The similarity to Example 1 in Section 11.1 is intentional. In fact, if we represent distance traveled (i.e., the change in position) by Δs and the time interval by Δt, the formula becomes

$$\Delta s = (48 \text{ mph})\ \Delta t,$$

which is equivalent to

$$\frac{\Delta s}{\Delta t} = 48 \text{ mph}.$$

So the two Example 1's are nearly identical—except that Example 1 of Section 11.1 did not make an assumption about constant velocity. What we computed in that instance was the average velocity over the 2.5-hr interval. This suggests that we could have actually solved the following, slightly different, problem to open this section.

EXAMPLE 2 Computing Distance Traveled

An automobile travels at an *average* rate of 48 mph for 2 hr 30 min. How far does the automobile travel?

SOLUTION The distance traveled is Δs, the time interval has length Δt, and $\Delta s / \Delta t$ is the average velocity.

Therefore,

$$\Delta s = \frac{\Delta s}{\Delta t} \cdot \Delta t = (48 \text{ mph})(2.5 \text{ hr}) = 120 \text{ mi}.$$

Now try Exercise 5.

So, given average velocity over a time interval, we can easily find distance traveled. But suppose we have a velocity function $v(t)$ that gives instantaneous velocity as a changing function of time. How can we use the instantaneous velocity function to find distance traveled over a time interval? This was the other intriguing problem about instantaneous velocity that puzzled the 17th-century scientists—and once again, algebra was inadequate for solving it, as we shall see.

Zeno's Paradoxes

The Greek philosopher Zeno of Elea (490–425 BCE) was noted for presenting paradoxes similar to the Distance Question. One of the most famous concerns the race between Achilles and a slow-but-sure tortoise. Achilles sportingly gives the tortoise a head start, then sets off to catch him. He must first get halfway to the tortoise, but by the time he runs halfway to where the tortoise was when he started, the tortoise has moved ahead. Now Achilles must close half of that distance, but by the time he does, the tortoise has moved ahead again. Continuing this argument forever, we see that Achilles can never even catch the tortoise, let alone pass him, so the tortoise must win the race.

Distance from a Changing Velocity

When Galileo began his experiments, here's what he might have asked himself about using a changing velocity to find distance:

A Distance Question

Suppose a ball rolls down a ramp and its velocity is always $2t$ feet per second, where t is the number of seconds after it started to roll. How far does the ball travel during the first 3 sec?

One might be tempted to offer the following "solution":

Velocity times Δt gives Δs. But instantaneous velocity occurs at an instant of time, so $\Delta t = 0$. That means $\Delta s = 0$. So, at any given instant of time, the ball doesn't move. Since any time interval consists of instants of time, the ball never moves at all! (You might well ask: Is this another trick question?)

As was the case with the Velocity Question in Section 11.1, this foolish-looking example conceals a very subtle algebraic dilemma—and, far from being a trick question, it is exactly the question that needed to be answered in order to compute the distance traveled by an object whose velocity varies as a function of time. The scientists who were working on the tangent line problem realized that the distance-traveled problem must be related to it, but, surprisingly, their geometry led them in another direction. The distance-traveled problem led them not to tangent lines, but to areas.

Limits at Infinity

Before we see the connection to areas, let us revisit another limit concept that will make instantaneous velocity easier to handle, just as in the last section. We will again be content with an informal definition.

DEFINITION (INFORMAL) Limit at Infinity

When we write "$\lim_{x \to \infty} f(x) = L$," we mean that $f(x)$ gets arbitrarily close to L as x gets arbitrarily large.

EXPLORATION 1 An Infinite Limit

A gallon of water is divided equally and poured into teacups. Find the amount in each teacup and the *total amount* in *all* the teacups if there are

1. 10 teacups.
2. 100 teacups.
3. 1 billion teacups.
4. An infinite number of teacups.

The preceding Exploration probably went pretty smoothly until you came to the infinite number of teacups. At that point you were probably pretty comfortable in saying what the *total amount* would be, and probably a little uncomfortable in saying how much would be in each teacup. (Theoretically it would be zero, which is just one reason why the actual experiment cannot be performed.) In the language of limits, the total amount of water in the infinite number of teacups would look like this:

$$\lim_{n\to\infty}\left(n\cdot\frac{1}{n}\right) = \lim_{n\to\infty}\frac{n}{n} = 1 \text{ gal}$$

while the total amount in each teacup would look like this:

$$\lim_{n\to\infty}\frac{1}{n} = 0 \text{ gal}$$

Summing up an infinite number of nothings to get something is mysterious enough when we use limits; *without* limits it seems to be an algebraic impossibility. That is the dilemma that faced the 17th-century scientists who were trying to work with instantaneous velocity. Once again, it was geometry that showed the way when the algebra failed.

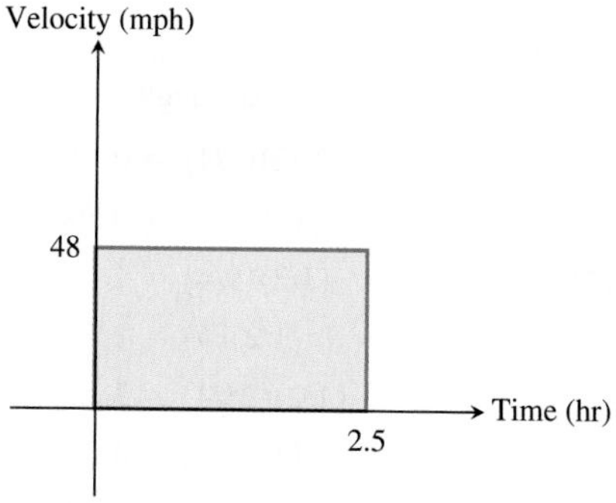

FIGURE 11.4 For constant velocity, the area of the rectangle is the same as the distance traveled, since it represents the product of the same two quantities: $(48 \text{ mph})(2.5 \text{ hr}) = 120$ mi.

The Connection to Areas

If we graph the constant velocity $v = 48$ in Example 1 as a function of time t, we notice that the area of the shaded rectangle is the same as the distance traveled (Figure 11.4). This is no mere coincidence, either, as the area of the rectangle and the distance traveled over the time interval are both computed by multiplying the same two quantities:

$$(48 \text{ mph})(2.5 \text{ hr}) = 120 \text{ mi}$$

Now suppose we graph a velocity function that varies continuously as a function of time (Figure 11.5). Would the area of this irregularly shaped region still give the total distance traveled over the time interval $[a, b]$?

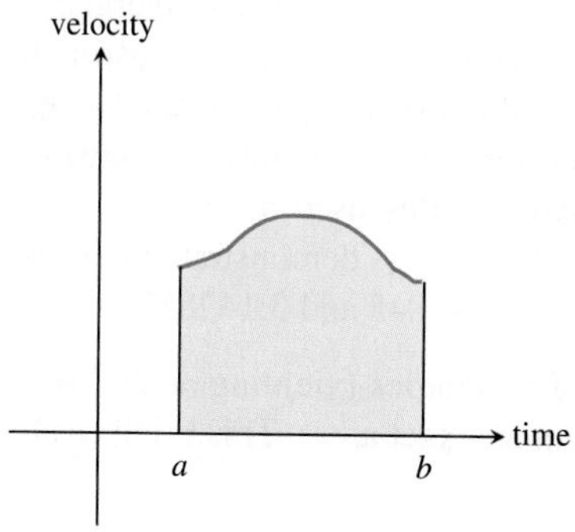

FIGURE 11.5 If the velocity varies over the time interval $[a, b]$, does the shaded region give the distance traveled?

Newton and Leibniz (and, actually, many others who had considered this question) were convinced that it obviously would, and that is why they were interested in a calculus for finding areas under curves. They imagined the time interval being partitioned into many tiny subintervals, each one so small that the velocity over it would essentially be constant. Geometrically, this was equivalent to slicing the area into narrow strips, each one of which would be nearly indistinguishable from a narrow rectangle (Figure 11.6).

The idea of partitioning irregularly shaped areas into approximating rectangles was not new. Indeed, Archimedes had used that very method to approximate the area of a circle with remarkable accuracy. However, it was an exercise in patience and perseverance, as Example 3 will show.

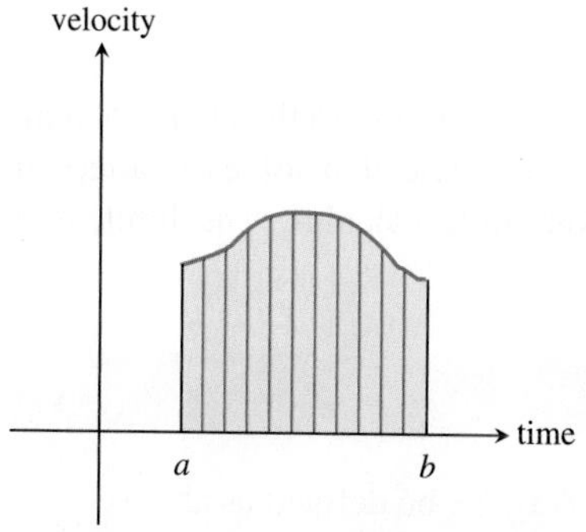

FIGURE 11.6 The region is partitioned into vertical strips. If the strips are narrow enough, they are almost indistinguishable from rectangles. The sum of the areas of these "rectangles" will give the total area and can be interpreted as distance traveled.

EXAMPLE 3 Approximating an Area with Rectangles

Use the six rectangles in Figure 11.7 to approximate the area of the region below the graph of $f(x) = x^2$ over the interval $[0, 3]$.

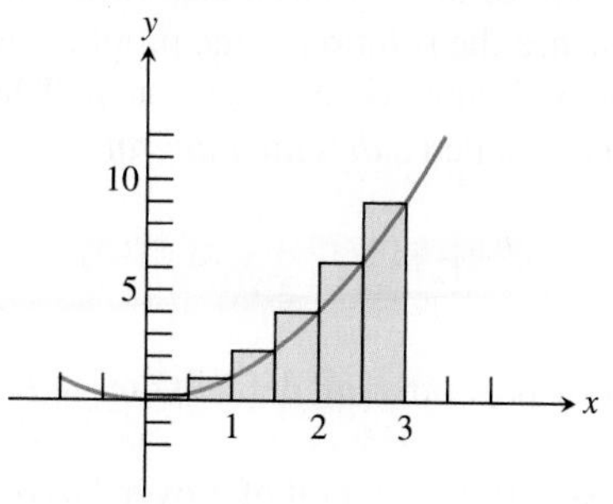

FIGURE 11.7 The area under the graph of $f(x) = x^2$ is approximated by six rectangles, each with base 1/2. The height of each rectangle is the function value at the right-hand endpoint of the subinterval. (Example 3)

(continued)

SOLUTION The base of each approximating rectangle is 1/2. The height is determined by the function value at the right-hand endpoint of each subinterval. The areas of the six rectangles and the total area are computed in the table below:

Subinterval	Base of Rectangle	Height of Rectangle	Area of Rectangle
$[0, 1/2]$	1/2	$f(1/2) = (1/2)^2 = 1/4$	$(1/2)(1/4) = 0.125$
$[1/2, 1]$	1/2	$f(1) = (1)^2 = 1$	$(1/2)(1) = 0.500$
$[1, 3/2]$	1/2	$f(3/2) = (3/2)^2 = 9/4$	$(1/2)(9/4) = 1.125$
$[3/2, 2]$	1/2	$f(2) = (2)^2 = 4$	$(1/2)(4) = 2.000$
$[2, 5/2]$	1/2	$f(5/2) = (5/2)^2 = 25/4$	$(1/2)(25/4) = 3.125$
$[5/2, 3]$	1/2	$f(3) = (3)^2 = 9$	$(1/2)(9) = 4.500$
		Total Area:	11.375

The six rectangles give a (rather crude) approximation of 11.375 square units for the area under the curve from 0 to 3.

Now try Exercise 11.

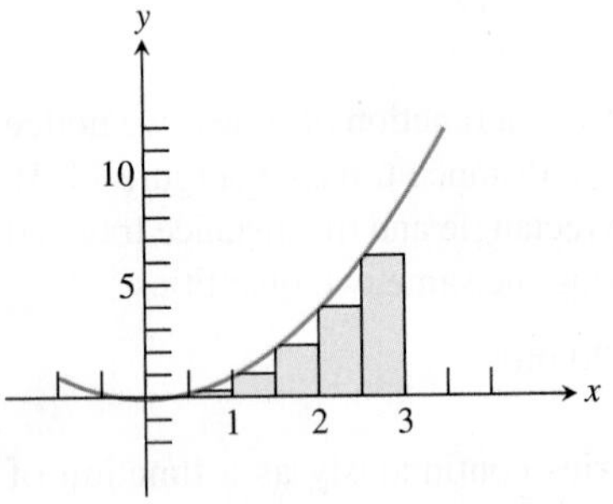

FIGURE 11.8 If we change the rectangles in Figure 11.7 so that their heights are determined by function values at the *left-hand* endpoints, we get an area approximation (6.875 square units) that underestimates the true area.

Figure 11.7 shows that the *right rectangular approximation method* (RRAM) in Example 3 overestimates the true area. If we were to use the function values at the left-hand endpoints of the subintervals (LRAM), we would obtain a rectangular approximation (6.875 square units) that underestimates the true area (Figure 11.8). The average of the two approximations is 9.125 square units, which is actually a pretty good estimate of the true area of 9 square units. If we were to repeat the process with 20 rectangles, the average would be 9.01125. This method of converging toward an unknown area by refining approximations is tedious, but it works—Archimedes used a variation of it 2200 years ago to estimate the area of a circle, and in the process demonstrated that the ratio of the circumference to the diameter was between 3.140845 and 3.142857.

The calculus step is to move from a finite number of rectangles (yielding an approximate area) to an infinite number of rectangles (yielding an exact area). This brings us to the definite integral.

The Definite Integral

In general, begin with a continuous function $y = f(x)$ over an interval $[a, b]$. Divide $[a, b]$ into n subintervals of length $\Delta x = (b - a)/n$. Choose any value x_1 in the first subinterval, x_2 in the second, and so on. Compute $f(x_1), f(x_2), f(x_3), \ldots, f(x_n)$, multiply each value by Δx, and sum up the products. In sigma notation, the sum of the products is

$$\sum_{i=1}^{n} f(x_i)\Delta x.$$

The *limit* of this sum as n approaches infinity is the exact solution to the area problem, and hence the solution to the problem of distance traveled. Indeed, it solves a variety of other problems as well, as you will learn when you study calculus. The limit, if it exists, is called a *definite integral.*

Riemann Sums

A sum of the form $\sum_{i=1}^{n} f(x_i)\Delta x$ in which x_1 is in the first subinterval, x_2 is in the second, and so on, is called a **Riemann sum**, in honor of Georg Riemann (1826–1866), who determined the functions for which such sums had limits as $n \to \infty$.

Definite Integral Notation

Notice that the notation for the definite integral (another legacy of Leibniz) parallels the sigma notation of the sum for which it is a limit. The "Σ" in the limit becomes a stylized "S," for "sum." The "Δx" becomes "dx" (as it did in the derivative), and the "$f(x_i)$" becomes simply "$f(x)$" because we are effectively summing up *all* the $f(x)$-values along the interval (times an arbitrarily small change in x), rendering the subscripts unnecessary.

DEFINITION Definite Integral

Let f be a function defined on $[a, b]$ and let $\sum_{i=1}^{n} f(x_i)\Delta x$ be defined as above.

The definite integral of f over $[a, b]$, denoted $\int_a^b f(x)\,dx$, is given by

$$\int_a^b f(x)\,dx = \lim_{n\to\infty} \sum_{i=1}^{n} f(x_i)\Delta x,$$

provided the limit exists. If the limit exists, we say f is **integrable** on $[a, b]$.

The solution to Example 3 shows that it can be tedious to approximate a definite integral by working out the sum for a large value of n. One of the crowning achievements of calculus was to demonstrate how the exact value of a definite integral could be obtained without summing up any products at all. You will have to wait until calculus to see how that is done; meanwhile, you will learn in Section 11.4 how to use a calculator to take the tedium out of finding definite integrals by summing.

You can also use the area connection to your advantage, as shown in these next two examples.

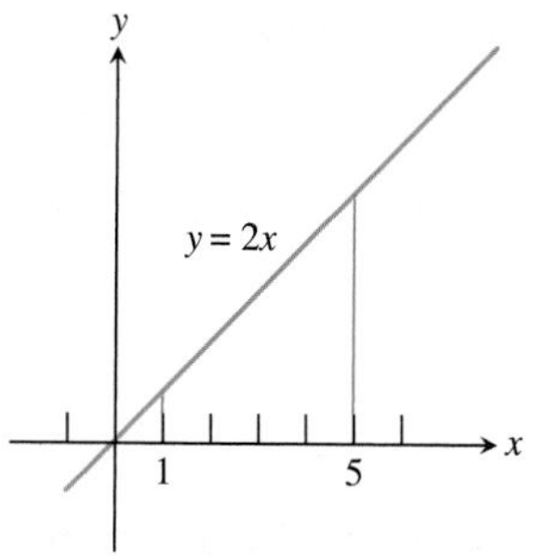

FIGURE 11.9 The area of the trapezoid equals $\int_1^5 2x\,dx$. (Example 4)

EXAMPLE 4 Computing an Integral

Find $\int_1^5 2x\,dx$.

SOLUTION This will be the area under the line $y = 2x$ over the interval $[1, 5]$. The graph in Figure 11.9 shows that this is the area of a trapezoid.

Using the formula $A = h\left(\frac{b_1 + b_2}{2}\right)$, we find that

$$\int_1^5 2x\,dx = 4\left(\frac{2(1) + 2(5)}{2}\right) = 24.$$

Now try Exercise 23.

EXAMPLE 5 Computing an Integral

Suppose a ball rolls down a ramp so that its velocity after t seconds is always $2t$ feet per second. How far does it fall during the first 3 sec?

SOLUTION The distance traveled will be the same as the area under the velocity graph, $v(t) = 2t$, over the interval $[0, 3]$. The graph is shown in Figure 11.10. Since the region is triangular, we can find its area: $A = (1/2)(3)(6) = 9$. The distance traveled in the first 3 sec, therefore, is $\Delta s = (1/2)(3 \text{ sec})(6 \text{ ft/sec}) = 9$ ft.

Now try Exercise 45.

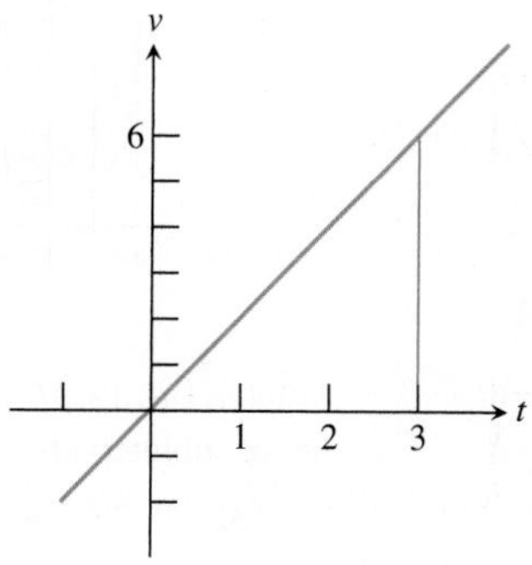

FIGURE 11.10 The area under the velocity graph $v(t) = 2t$, over the interval $[0, 3]$, is the distance traveled by the ball in Example 5 during the first 3 sec.

QUICK REVIEW 11.2 *(For help, go to Sections 1.1 and 9.4.)*

In Exercises 1 and 2, list the elements of the sequence.

1. $a_k = \frac{1}{2}\left(\frac{1}{2}k\right)^2$ for $k = 1, 2, 3, 4, \ldots, 9, 10$

2. $a_k = \frac{1}{4}\left(2 + \frac{1}{4}k\right)^2$ for $k = 1, 2, 3, 4, \ldots, 9, 10$

In Exercises 3–6, find the sum.

3. $\sum_{k=1}^{10} \frac{1}{2}(k + 1)$

4. $\sum_{k=1}^{n} (k + 1)$

5. $\sum_{k=1}^{10} \frac{1}{2}(k + 1)^2$

6. $\sum_{k=1}^{n} \frac{1}{2}k^2$

7. A truck travels at an average speed of 57 mph for 4 hr. How far does it travel?

8. A pump working at 5 gal/min pumps for 2 hr. How many gallons are pumped?

9. Water flows over a spillway at a steady rate of 200 cubic feet per second. How many cubic feet of water pass over the spillway in 6 hr?

10. A county has a population density of 560 persons per square mile in an area of 35,000 mi^2. What is the population of the county?

SECTION 11.2 Exercises

In Exercises 1–6, explain how to represent the problem situation as an area question and then solve the problem.

1. A train travels at 65 mph for 3 hr. How far does it travel?

2. A pump working at 15 gal/min pumps for 0.5 hr. How many gallons are pumped?

3. Water flows over a spillway at a steady rate of 150 cubic feet per second. How many cubic feet of water pass over the spillway in 1 hr?

4. A city has a population density of 650 persons per square mile in an area of 20 mi^2. What is the population of the city?

5. An airplane travels at an average velocity of 640 km/hr for 3 hr 24 min. How far does the airplane travel?

6. A train travels at an average velocity of 24 mph for 4 hr 50 min. How far does the train travel?

In Exercises 7–10, estimate the area of the region above the x-axis and under the graph of the function from $x = 0$ to $x = 5$.

7.

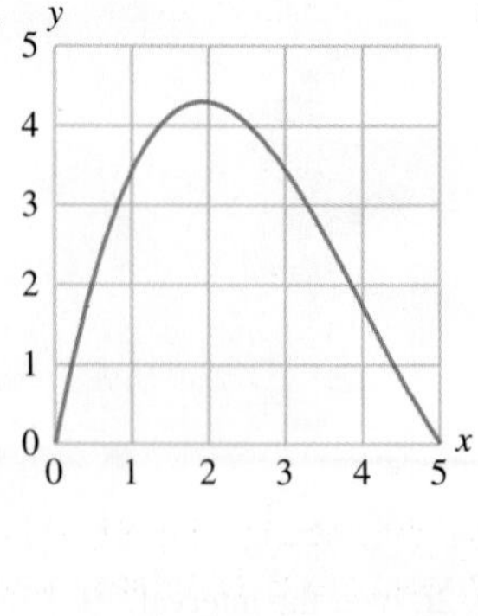

8.

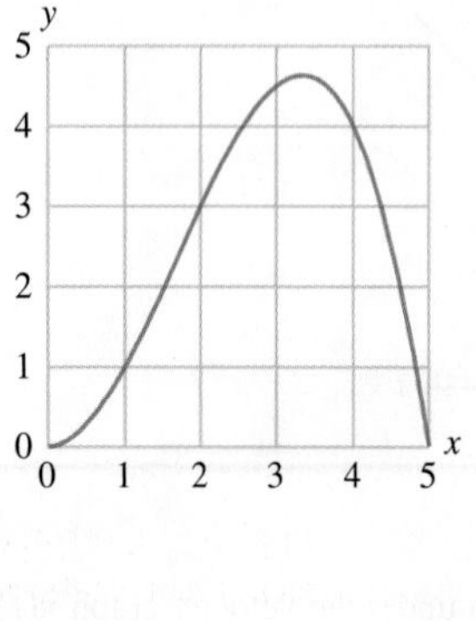

9.

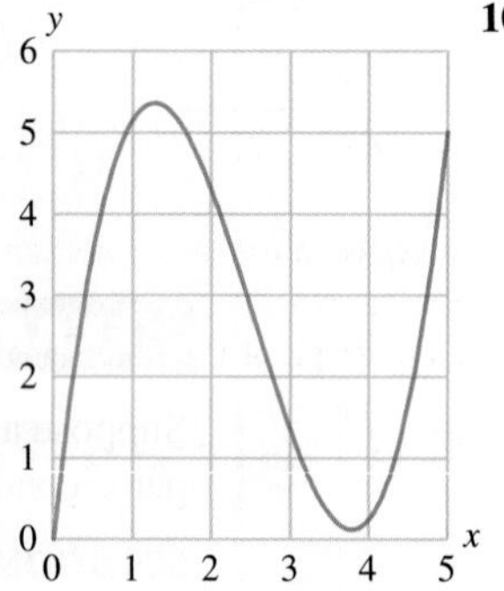

10.

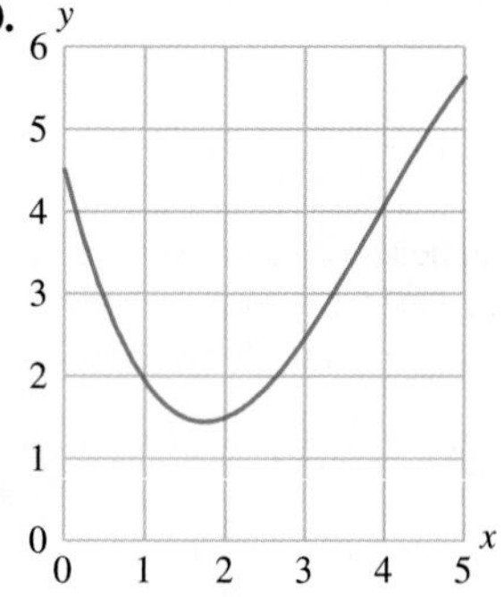

In Exercises 11 and 12, use the 8 rectangles shown to approximate the area of the region below the graph of $f(x) = 10 - x^2$ over the interval $[-1, 3]$.

11.

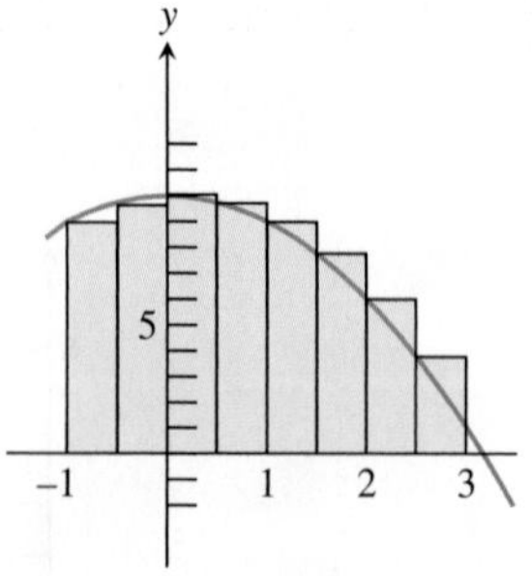

12.

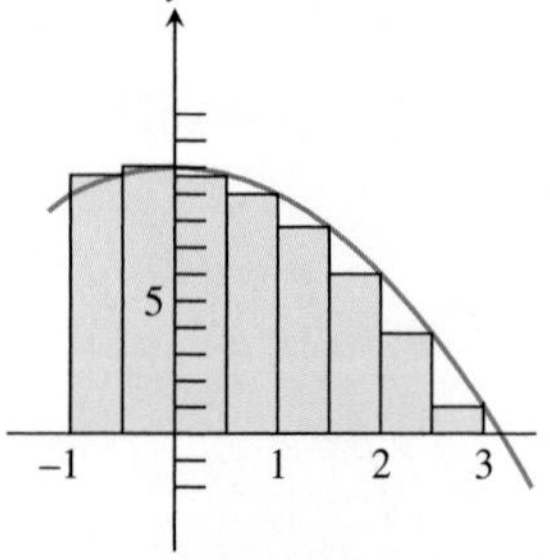

In Exercises 13–16, partition the given interval into the indicated number of subintervals.

13. $[0, 2]$; 4

14. $[0, 2]$; 8

15. $[1, 4]$; 6

16. $[1, 5]$; 8

In Exercises 17–20, complete the following.

(a) Draw the graph of the function for x in the specified interval. Verify that the function is nonnegative in that interval.

(b) On the graph in part (a), draw and shade the approximating rectangles for the RRAM using the specified partition. Compute the RRAM area estimate without using a calculator.

(c) Repeat part (b) using the LRAM.

(d) Average the RRAM and LRAM approximations from parts (b) and (c) to find an average estimate of the area.

17. $f(x) = x^2;\ [0, 4];$ 4 subintervals

18. $f(x) = x^2 + 2;\ [0, 6];$ 6 subintervals

19. $f(x) = 4x - x^2;\ [0, 4];$ 4 subintervals

20. $f(x) = x^3;\ [0, 3];$ 3 subintervals

In Exercises 21–28, find the definite integral by computing an area. (It may help to look at a graph of the function.)

21. $\int_3^7 5\,dx$

22. $\int_{-1}^4 6\,dx$

23. $\int_0^5 3x\,dx$

24. $\int_1^7 0.5x\,dx$

25. $\int_1^4 (x + 3)\,dx$

26. $\int_1^4 (3x - 2)\,dx$

27. $\int_{-2}^2 \sqrt{4 - x^2}\,dx$

28. $\int_0^6 \sqrt{36 - x^2}\,dx$

It can be shown that the area enclosed between the x-axis and one arch of the sine curve is 2. Use this fact in Exercises 29–38 to compute the definite integral. (It may help to look at a graph of the function.)

29. $\int_0^\pi \sin x\,dx$

30. $\int_0^\pi (\sin x + 2)\,dx$

31. $\int_2^{\pi+2} \sin(x - 2)\,dx$

32. $\int_{-\pi/2}^{\pi/2} \cos x\,dx$

33. $\int_0^{\pi/2} \sin x\,dx$

34. $\int_0^{\pi/2} \cos x\,dx$

35. $\int_0^\pi 2\sin x\,dx$ (*Hint:* All the rectangles are twice as tall.)

36. $\int_0^{2\pi} \sin\left(\frac{x}{2}\right) dx$ (*Hint:* All the rectangles are twice as wide.)

37. $\int_0^{2\pi} |\sin x|\,dx$

38. $\int_{-\pi}^{3\pi/2} |\cos x|\,dx$

In Exercises 39–42, find the integral, assuming that k is a number between 0 and 4.

39. $\int_0^4 (kx + 3)\,dx$

40. $\int_0^k (4x + 3)\,dx$

41. $\int_0^4 (3x + k)\,dx$

42. $\int_k^4 (4x + 3)\,dx$

43. Writing to Learn Let $g(x) = -f(x)$ where f has nonnegative function values on an interval $[a, b]$. Explain why the area above the graph of g is the same as the area under the graph of f in the same interval.

44. Writing to Learn Explain how you can find the area under the graph of $f(x) = \sqrt{16 - x^2}$ from $x = 0$ to $x = 4$ by mental computation only.

45. Falling Ball Suppose a ball is dropped from a tower and its velocity after t seconds is always $32t$ feet per second. How far does the ball fall during the first 2 sec?

46. Accelerating Automobile Suppose an automobile accelerates so that its velocity after t seconds is always $6t$ feet per second. How far does the car travel in the first 7 sec?

47. Rock Toss A rock is thrown straight up from level ground. The velocity of the rock at any time t (sec) is $v(t) = 48 - 32t$ ft/sec.

(a) Graph the velocity function.

(b) At what time does the rock reach its maximum height?

(c) Find how far the rock has traveled at its maximum height.

48. Rocket Launch A toy rocket is launched straight up from level ground. Its velocity function is $f(t) = 170 - 32t$ feet per second, where t is the number of seconds after launch.

(a) Graph the velocity function.

(b) At what time does the rocket reach its maximum height?

(c) Find how far the rocket has traveled at its maximum height.

49. Finding Distance Traveled as Area A ball is pushed off the roof of a three-story building. Table 11.3 gives the velocity (in feet per second) of the falling ball at 0.2-sec intervals until it hits the ground 1.4 sec later.

Table 11.3 Velocity Data of the Ball

Time (sec)	Velocity (ft/sec)
0.2	−5.05
0.4	−11.43
0.6	−17.46
0.8	−24.21
1.0	−30.62
1.2	−37.06
1.4	−43.47

(a) Draw a scatter plot of the data.

(b) Find the approximate building height using RRAM areas as in Example 4. Use the fact that if the velocity function is always negative the distance traveled will be the same as if the absolute value of the velocity values were used.

50. Work Work is defined as force times distance. A full water barrel weighing 1250 lb has a significant leak and must be lifted 35 ft. Table 11.4 displays the weight of the barrel measured after each 5 ft of movement. Find the approximate work in foot-pounds done in lifting the barrel 35 ft.

Table 11.4 Weight of a Leaking Water Barrel

Distance (ft)	Weight (lb)
0	1250
5	1150
10	1050
15	950
20	850
25	750
30	650

Standardized Test Questions

51. True or False When estimating the area under a curve using LRAM, the accuracy typically improves as the number n of subintervals is increased.

52. True or False The statement $\lim_{x\to\infty} f(x) = L$ means that $f(x)$ gets arbitrarily large as x gets arbitrarily close to L.

It can be shown that the area of the region enclosed by the curve $y = \sqrt{x}$, the x-axis, and the line $x = 9$ is 18. Use this fact in Exercises 53–56 to choose the correct answer. Do not use a calculator.

53. Multiple Choice $\int_0^9 2\sqrt{x}\,dx$

(A) 36 **(B)** 27 **(C)** 18 **(D)** 9 **(E)** 6

54. Multiple Choice $\int_0^9 \left(\sqrt{x} + 5\right) dx$

(A) 14 **(B)** 23 **(C)** 33 **(D)** 45 **(E)** 63

55. Multiple Choice $\int_5^{14} \left(\sqrt{x - 5}\right) dx$

(A) 9 **(B)** 13 **(C)** 18 **(D)** 23 **(E)** 28

56. Multiple Choice $\int_0^3 \sqrt{3x}\,dx$

(A) 54 **(B)** 18 **(C)** 9 **(D)** 6 **(E)** 3

Explorations

57. Group Activity You may have erroneously assumed that the function f had to be positive in the definition of the definite integral. It is a fact that $\int_0^{2\pi} \sin x\,dx = 0$. Use the definition of the definite integral to explain why this is so. What does this imply about $\int_0^1 (x - 1)\,dx$?

58. Area Under a Discontinuous Function Let

$$f(x) = \begin{cases} 1 & \text{if } x < 2 \\ x & \text{if } x > 2. \end{cases}$$

(a) Draw a graph of f. Determine its domain and range.

(b) Writing to Learn How would you define the area under f from $x = 0$ to $x = 4$? Does it make a difference if the function has no value at $x = 2$?

Extending the Ideas

Group Activity From what you know about definite integrals, decide whether each of the following statements is true or false for integrable functions (in general). Work with your classmates to justify your answers.

59. $\int_a^b f(x)\,dx + \int_a^b g(x)\,dx = \int_a^b (f(x) + g(x))\,dx$

60. $\int_a^b 8 \cdot f(x)\,dx = 8 \cdot \int_a^b f(x)\,dx$

61. $\int_a^b f(x) \cdot g(x)\,dx = \int_a^b f(x)\,dx \cdot \int_a^b g(x)\,dx$

62. $\int_a^c f(x)\,dx + \int_c^b f(x)\,dx = \int_a^b f(x)\,dx$ for $a < c < b$

63. $\int_a^b f(x) = \int_b^a f(x)$

64. $\int_a^a f(x)\,dx = 0$

11.3 More on Limits

What you'll learn about

- A Little History
- Defining a Limit Informally
- Properties of Limits
- Limits of Continuous Functions
- One-Sided and Two-Sided Limits
- Limits Involving Infinity

... and why

Limits are essential concepts in the development of calculus.

A Little History

Progress in mathematics occurs gradually and without much fanfare in the early stages. The fanfare occurs much later, after the discoveries and innovations have been cleaned up and put into perspective. Calculus is certainly a case in point. Most of the ideas in this chapter predate Newton and Leibniz. Others were solving calculus problems as far back as Archimedes of Syracuse (ca. 287–212 BCE), long before calculus was "discovered." What Newton and Leibniz did was to develop the rules of the game so that derivatives and integrals could be computed algebraically. Most importantly, they developed what has come to be called the Fundamental Theorem of Calculus, which explains the connection between the "tangent line problem" and the "area problem."

But the methods of Newton and Leibniz depended on mysterious "infinitesimal" quantities that were small enough to vanish and yet were not zero. Jean Le Rond d'Alembert (1717–1783) was a strong proponent of replacing infinitesimals with limits (the strategy that would eventually work), but these concepts were not well understood until Karl Weierstrass (1815–1897) and his student Eduard Heine (1821–1881) introduced the formal, unassailable definitions that are used in our higher mathematics courses today. By that time, Newton and Leibniz had been dead for over 150 years.

Defining a Limit Informally

There is nothing difficult about the following limit statements:

$$\lim_{x \to 3} (2x - 1) = 5 \qquad \lim_{x \to \infty} (x^2 + 3) = \infty \qquad \lim_{n \to \infty} \frac{1}{n} = 0$$

That is why we have used limit notation throughout this book. Particularly when electronic graphers are available, analyzing the limiting behavior of functions algebraically, numerically, and graphically can tell us much of what we need to know about the functions.

What *is* difficult is to come up with an airtight definition of what a limit really is. If it had been easy, it would not have taken 150 years. The subtleties of the "epsilon-delta" definition of Weierstrass and Heine are as beautiful as they are profound, but they are not the stuff of a precalculus course. Therefore, even as we look more closely at limits and their properties in this section, we will continue to refer to our "informal" definition of limit (essentially that of d'Alembert). We repeat it here for ready reference:

DEFINITION (INFORMAL) Limit at *a*

When we write "$\lim_{x \to a} f(x) = L$," we mean that $f(x)$ gets arbitrarily close to L as x gets arbitrarily close (but not equal) to a.

EXPLORATION 1 What's the Limit?

As a class, discuss the following two limit statements until you really understand why they are true. Look at them every way you can. Use your calculators. Do you see how the definition on the previous page verifies that they are true? In particular, can you defend your position against the challenges that follow the statements? (This exploration is intended to be free-wheeling and philosophical. You can't *prove* these statements without a stronger definition.)

1. $\lim_{x \to 2} 7x \neq 14.000000000000000001$

Challenges:

- Isn't $7x$ getting "arbitrarily close" to that number as x approaches 2?
- How can you tell that 14 is the limit and 14.000000000000000001 is not?

2. $\lim_{x \to 0} \dfrac{x^2 + 2x}{x} = 2$

Challenges:

- How can the limit be 2 when the quotient isn't even defined at 0?
- Won't there be an asymptote at $x = 0$? The denominator equals 0 there.
- How can you *tell* that 2 is the limit and 1.9999999999999999999 is not?

EXAMPLE 1 Finding a Limit

Find $\lim_{x \to 1} \dfrac{x^3 - 1}{x - 1}$.

Solve Graphically

The graph in Figure 11.11a suggests that the limit exists and is about 3.

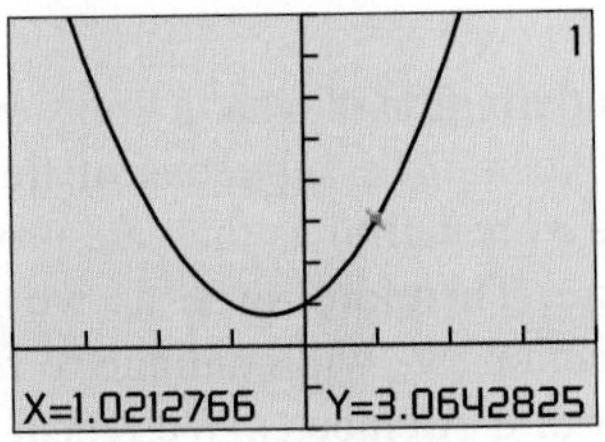

[−4, 4] by [−2, 8]

FIGURE 11.11a A graph of $f(x) = (x^3 - 1)/(x - 1)$.

Solve Numerically

The table also gives compelling evidence that the limit is 3.

X	Y1	
.997	2.991	
.998	2.994	
.999	2.997	
1	ERROR	
1.001	3.003	
1.002	3.006	
1.003	3.009	

Y1 ■ (X³−1)/(X−1)

FIGURE 11.11b A table of values for $f(x) = (x^3 - 1)/(x - 1)$.

Solve Algebraically

$$\lim_{x \to 1} \frac{x^3 - 1}{x - 1}$$

$$= \lim_{x \to 1} \frac{(x - 1)(x^2 + x + 1)}{x - 1}$$

$$= \lim_{x \to 1} (x^2 + x + 1)$$

$$= 1 + 1 + 1$$

$$= 3$$

As convincing as the graphical and numerical evidence is, the best evidence is algebraic. The limit is 3.

Now try Exercise 11.

Properties of Limits

When limits exist, there is nothing unusual about the way they interact algebraically with each other. You could easily predict that the following properties would hold. These are all theorems that one could prove with a rigorous definition of limit, but we must state them without proof here.

Properties of Limits

If $\lim_{x \to c} f(x)$ and $\lim_{x \to c} g(x)$ both exist, then

1. **Sum Rule** $\lim_{x \to c} (f(x) + g(x)) = \lim_{x \to c} f(x) + \lim_{x \to c} g(x)$
2. **Difference Rule** $\lim_{x \to c} (f(x) - g(x)) = \lim_{x \to c} f(x) - \lim_{x \to c} g(x)$
3. **Product Rule** $\lim_{x \to c} (f(x) \cdot g(x)) = \lim_{x \to c} f(x) \cdot \lim_{x \to c} g(x)$
4. **Constant Multiple Rule** $\lim_{x \to c} (k \cdot g(x)) = k \cdot \lim_{x \to c} g(x)$
5. **Quotient Rule** $\lim_{x \to c} \dfrac{f(x)}{g(x)} = \dfrac{\lim_{x \to c} f(x)}{\lim_{x \to c} g(x)}$, provided $\lim_{x \to c} g(x) \neq 0$
6. **Power Rule** $\lim_{x \to c} (f(x))^n = (\lim_{x \to c} f(x))^n$ for n a positive integer
7. **Root Rule** $\lim_{x \to c} \sqrt[n]{f(x)} = \sqrt[n]{\lim_{x \to c} f(x)}$ for $n \geq 2$ a positive integer, provided $\sqrt[n]{\lim_{x \to c} f(x)}$ and $\lim_{x \to c} \sqrt[n]{f(x)}$ are real numbers.

EXAMPLE 2 Using the Limit Properties

You will learn in Example 10 that $\lim_{x \to 0} \dfrac{\sin x}{x} = 1$. Use this fact, along with the limit properties, to find the following limits:

(a) $\lim_{x \to 0} \dfrac{x + \sin x}{x}$ **(b)** $\lim_{x \to 0} \dfrac{1 - \cos^2 x}{x^2}$ **(c)** $\lim_{x \to 0} \dfrac{\sqrt[3]{\sin x}}{\sqrt[3]{x}}$

SOLUTION

(a)
$$\begin{aligned}
\lim_{x \to 0} \frac{x + \sin x}{x} &= \lim_{x \to 0} \left(\frac{x}{x} + \frac{\sin x}{x}\right) \\
&= \lim_{x \to 0} \frac{x}{x} + \lim_{x \to 0} \frac{\sin x}{x} && \text{Sum Rule} \\
&= 1 + 1 \\
&= 2
\end{aligned}$$

(b)
$$\begin{aligned}
\lim_{x \to 0} \frac{1 - \cos^2 x}{x^2} &= \lim_{x \to 0} \frac{\sin^2 x}{x^2} && \text{Pythagorean identity} \\
&= \lim_{x \to 0} \left(\frac{\sin x}{x}\right)\left(\frac{\sin x}{x}\right) \\
&= \lim_{x \to 0} \left(\frac{\sin x}{x}\right) \cdot \lim_{x \to 0} \left(\frac{\sin x}{x}\right) && \text{Product Rule} \\
&= 1 \cdot 1 \\
&= 1
\end{aligned}$$

(c)
$$\begin{aligned}
\lim_{x \to 0} \frac{\sqrt[3]{\sin x}}{\sqrt[3]{x}} &= \lim_{x \to 0} \sqrt[3]{\frac{\sin x}{x}} \\
&= \sqrt[3]{\lim_{x \to 0} \frac{\sin x}{x}} && \text{Root Rule} \\
&= \sqrt[3]{1} \\
&= 1
\end{aligned}$$

Now try Exercise 19.

Limits of Continuous Functions

Recall from Section 1.2 that a function is continuous at a if $\lim_{x \to a} f(x) = f(a)$. This means that the limit (at a) of a function can be found by "plugging in a" provided the function is continuous at a. (The condition of continuity is essential when employing this strategy. For example, plugging in 0 does not work on any of the limits in Example 2.)

EXAMPLE 3 Finding Limits by Substitution

Find the limits.

(a) $\lim_{x \to 0} \dfrac{e^x - \tan x}{\cos^2 x}$ **(b)** $\lim_{n \to 16} \dfrac{\sqrt{n}}{\log_2 n}$

SOLUTION You might not recognize these functions as being continuous, but you can use the limit properties to write the limits in terms of limits of basic functions.

(a)

$$\lim_{x \to 0} \frac{e^x - \tan x}{\cos^2 x} = \frac{\lim_{x \to 0} (e^x - \tan x)}{\lim_{x \to 0} (\cos^2 x)} \quad \text{Quotient Rule}$$

$$= \frac{\lim_{x \to 0} e^x - \lim_{x \to 0} \tan x}{\left(\lim_{x \to 0} \cos x\right)^2} \quad \text{Difference and Power Rules}$$

$$= \frac{e^0 - \tan 0}{(\cos 0)^2} \quad \text{Limits of continuous functions}$$

$$= \frac{1 - 0}{1}$$

$$= 1$$

(b)

$$\lim_{n \to 16} \frac{\sqrt{n}}{\log_2 n} = \frac{\lim_{n \to 16} \sqrt{n}}{\lim_{n \to 16} \log_2 n} \quad \text{Quotient Rule}$$

$$= \frac{\sqrt{16}}{\log_2 16} \quad \text{Limits of continuous functions}$$

$$= \frac{4}{4}$$

$$= 1$$

Now try Exercise 23.

Example 3 hints at some important properties of continuous functions that follow from the properties of limits. If f and g are both continuous at $x = a$, then so are $f + g, f - g, fg$, and f/g (with the assumption that $g(a)$ does not create a zero denominator in the quotient). Also, the nth power and nth root of a function that is continuous at a will also be continuous at a (with the assumption that $\sqrt{f(a)}$ is real).

One-Sided and Two-Sided Limits

We can see that the limit of the function in Figure 11.11 is 3 whether x approaches 1 from the left or right. Sometimes the values of a function f can approach different values as x approaches a number c from opposite sides. When this happens, the limit of f as x approaches c from the left is the **left-hand limit** of f at c and the limit of f as x approaches c from the right is the **right-hand limit** of f at c. Here is the notation we use:

left-hand: $\lim_{x \to c^-} f(x)$ *The limit of f as x approaches c from the left.*

right-hand: $\lim_{x \to c^+} f(x)$ *The limit of f as x approaches c from the right.*

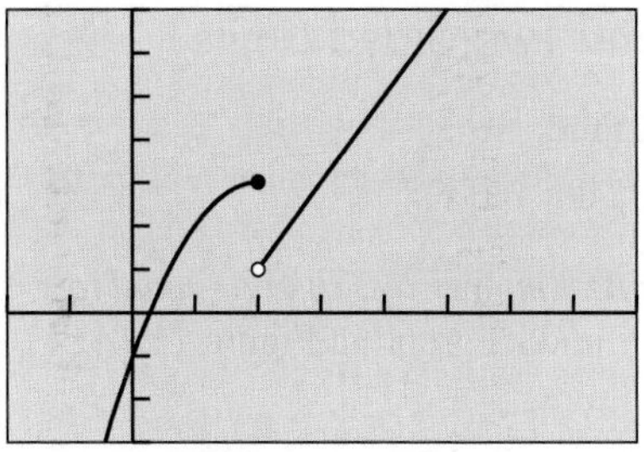

[–2, 8] by [–3, 7]

FIGURE 11.12 A graph of the piecewise-defined function.

$$f(x) = \begin{cases} -x^2 + 4x - 1 & x \le 2 \\ 2x - 3 & x > 2 \end{cases}$$

(Example 4)

EXAMPLE 4 Finding Left- and Right-Hand Limits

Find $\lim_{x \to 2^-} f(x)$ and $\lim_{x \to 2^+} f(x)$ where $f(x) = \begin{cases} -x^2 + 4x - 1 & \text{if } x \le 2 \\ 2x - 3 & \text{if } x > 2. \end{cases}$

SOLUTION Figure 11.12 suggests that the left- and right-hand limits of f exist but are not equal. Using algebra we find:

$$\begin{aligned} \lim_{x \to 2^-} f(x) &= \lim_{x \to 2^-} (-x^2 + 4x - 1) && \text{Definition of } f \\ &= -2^2 + 4 \cdot 2 - 1 \\ &= 3 \end{aligned}$$

$$\begin{aligned} \lim_{x \to 2^+} f(x) &= \lim_{x \to 2^+} (2x - 3) && \text{Definition of } f \\ &= 2 \cdot 2 - 3 \\ &= 1 \end{aligned}$$

You can use TRACE or tables to support the above results.

Now try Exercise 27, parts (a) and (b).

The limit $\lim_{x \to c} f(x)$ is sometimes called the **two-sided limit** of f at c to distinguish it from the *one-sided* left-hand and right-hand limits of f at c. The following theorem indicates how these limits are related.

THEOREM One-Sided and Two-Sided Limits

A function $f(x)$ has a limit as x approaches c if and only if the left-hand and right-hand limits at c exist and are equal. That is,

$$\lim_{x \to c} f(x) = L \Leftrightarrow \lim_{x \to c^-} f(x) = L \text{ and } \lim_{x \to c^+} f(x) = L.$$

The limit of the function f of Example 4 as x approaches 2 does not exist, so f is discontinuous at $x = 2$. However, discontinuous functions can have a limit at a point of discontinuity. The function f of Example 1 is discontinuous at $x = 1$ because $f(1)$ does not exist, but it has the limit 3 as x approaches 1. Example 5 illustrates another way a function can have a limit and still be discontinuous.

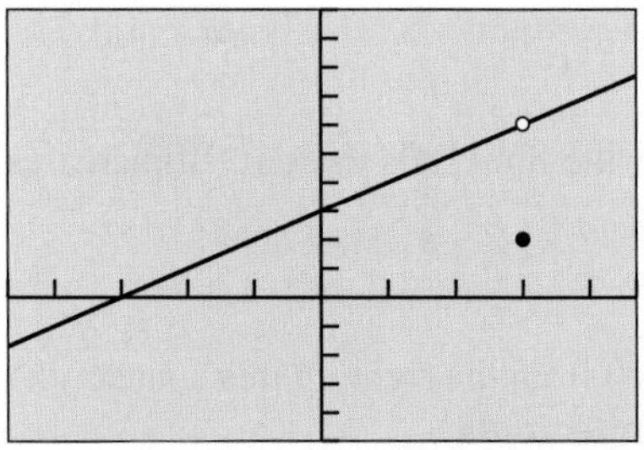

[–4.7, 4.7] by [–5, 10]

FIGURE 11.13 A graph of the function in Example 5.

EXAMPLE 5 Finding a Limit at a Point of Discontinuity

Let

$$f(x) = \begin{cases} \dfrac{x^2 - 9}{x - 3} & \text{if } x \ne 3 \\ 2 & \text{if } x = 3. \end{cases}$$

Find $\lim_{x \to 3} f(x)$ and prove that f is discontinuous at $x = 3$.

SOLUTION Figure 11.13 suggests that the limit of f as x approaches 3 exists. Using algebra we find

$$\begin{aligned} \lim_{x \to 3} \frac{x^2 - 9}{x - 3} &= \lim_{x \to 3} \frac{(x - 3)(x + 3)}{x - 3} \\ &= \lim_{x \to 3} (x + 3) && \text{We can assume } x \ne 3. \\ &= 6 \end{aligned}$$

Because $f(3) = 2 \ne \lim_{x \to 3} f(x)$, f is discontinuous at $x = 3$.

Now try Exercise 37.

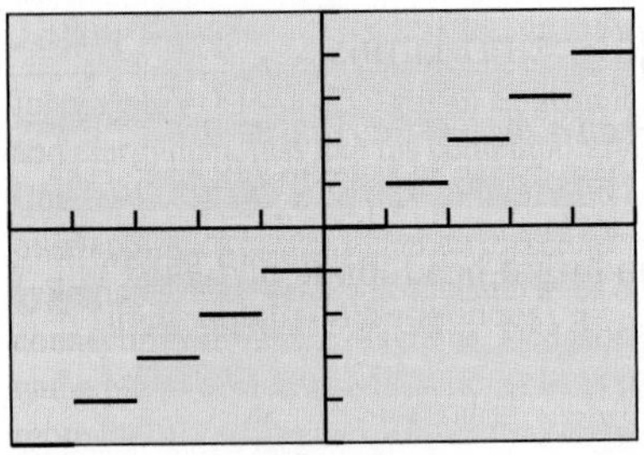

[–5, 5] by [–5, 5]

FIGURE 11.14 The graph of $f(x) = \text{int}(x)$. (Example 6)

EXAMPLE 6 Finding One-Sided and Two-Sided Limits

Let $f(x) = \text{int}(x)$ (the greatest integer function). Find:

(a) $\lim_{x\to 3^-} \text{int}(x)$ **(b)** $\lim_{x\to 3^+} \text{int}(x)$ **(c)** $\lim_{x\to 3} \text{int}(x)$

SOLUTION Recall that $\text{int}(x)$ is equal to *the greatest integer less than or equal to x.* For example, $\text{int}(3) = 3$. From the definition of f and its graph in Figure 11.14 we can see that

(a) $\lim_{x\to 3^-} \text{int}(x) = 2$

(b) $\lim_{x\to 3^+} \text{int}(x) = 3$

(c) $\lim_{x\to 3} \text{int}(x)$ does not exist.

Now try Exercise 41.

Limits Involving Infinity

The informal definition that we have for a limit refers to $\lim_{x\to a} f(x) = L$ where both a and L are real numbers. In Section 11.2 we adapted the definition to apply to limits of the form $\lim_{x\to\infty} f(x) = L$ so that we could use this notation in describing definite integrals. This is one type of "limit at infinity." Notice that the limit itself (L) is a finite real number, assuming the limit exists, but that the values of x are approaching infinity.

Infinite Limits Are Not Limits

It is important to realize that *an infinite limit is not a limit*, despite what the name might imply. It describes a special case of a limit that does not exist. Recall that a sawhorse is not a horse and a badminton bird is not a bird.

DEFINITION (INFORMAL) Limits at Infinity

When we write "$\lim_{x\to\infty} f(x) = L$," we mean that $f(x)$ gets arbitrarily close to L as x gets arbitrarily large. We say that **f has a limit L as x approaches ∞.**

When we write "$\lim_{x\to-\infty} f(x) = L$," we mean that $f(x)$ gets arbitrarily close to L as $-x$ gets arbitrarily large. We say that **f has a limit L as x approaches $-\infty$.**

Archimedes (c. 287–212 BCE)

The Greek mathematician Archimedes found the area of a circle using a method involving infinite limits. See Exercise 89 for a modern version of his method.

Notice that limits, whether at a or at infinity, are always finite real numbers; otherwise, the limits do not exist. For example, it is correct to write

$$\lim_{x\to 0} \frac{1}{x^2} \text{ does not exist,}$$

since it approaches no real number L. In this case, however, it is also convenient to write

$$\lim_{x\to 0} \frac{1}{x^2} = \infty,$$

which gives us a little more information about *why* the limit fails to exist. (It increases without bound.) Similarly, it is convenient to write

$$\lim_{x\to 0^+} \ln x = -\infty,$$

since $\ln x$ decreases without bound as x approaches 0 from the right. In this context, the symbols "∞" and "$-\infty$" are sometimes called **infinite limits**.

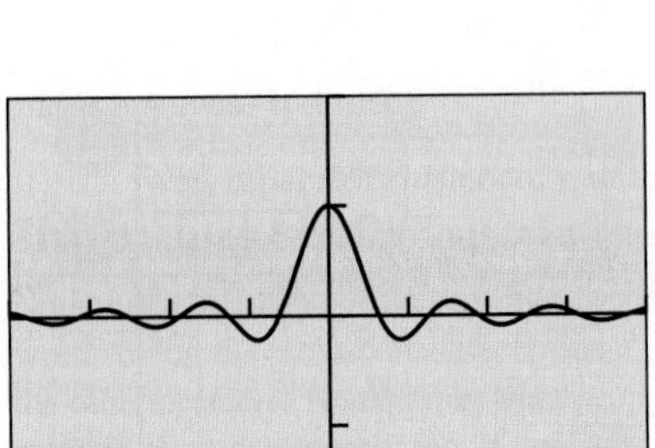

[–20, 20] by [–2, 2]

FIGURE 11.15 The graph of $f(x) = (\sin x)/x$. (Example 7)

EXAMPLE 7 Investigating Limits as $x \to \pm\infty$

Let $f(x) = (\sin x)/x$. Find $\lim_{x\to\infty} f(x)$ and $\lim_{x\to-\infty} f(x)$.

SOLUTION The graph of f in Figure 11.15 suggests that

$$\lim_{x\to\infty} \frac{\sin x}{x} = \lim_{x\to-\infty} \frac{\sin x}{x} = 0.$$

Now try Exercise 47.

In Section 1.3, we used limits to describe the *unbounded behavior* of the function $f(x) = x^3$ as $x \to \pm\infty$:

$$\lim_{x\to\infty} x^3 = \infty \quad \text{and} \quad \lim_{x\to-\infty} x^3 = -\infty$$

The behavior of the function $g(x) = e^x$ as $x \to \pm\infty$ can be described by the following two limits:

$$\lim_{x\to\infty} e^x = \infty \quad \text{and} \quad \lim_{x\to-\infty} e^x = 0$$

The function $g(x) = e^x$ has unbounded behavior as $x \to \infty$ and has a finite limit as $x \to -\infty$.

EXAMPLE 8 Using Tables to Investigate Limits as $x \to \pm\infty$

Let $f(x) = xe^{-x}$. Find $\lim_{x\to\infty} f(x)$ and $\lim_{x\to-\infty} f(x)$.

SOLUTION The tables in Figure 11.16 suggest that

$$\lim_{x\to\infty} xe^{-x} = 0 \quad \text{and} \quad \lim_{x\to-\infty} xe^{-x} = -\infty.$$

X	Y2	
0	0	
10	4.5E–4	
20	4.1E–8	
30	3E–12	
40	2E–16	
50	1E–20	
60	5E–25	

Y2 = Xe^(–X)

(a)

X	Y2	
0	0	
–10	–2.2E5	
–20	–9.7E9	
–30	–3E14	
–40	–9E18	
–50	–3E23	
–60	–7E27	

Y2 = Xe^(–X)

(b)

FIGURE 11.16 The table in (a) suggests that the values of $f(x) = xe^{-x}$ approach 0 as $x \to \infty$ and the table in (b) suggests that the values of $f(x) = xe^{-x}$ approach $-\infty$ as $x \to -\infty$. (Example 8)

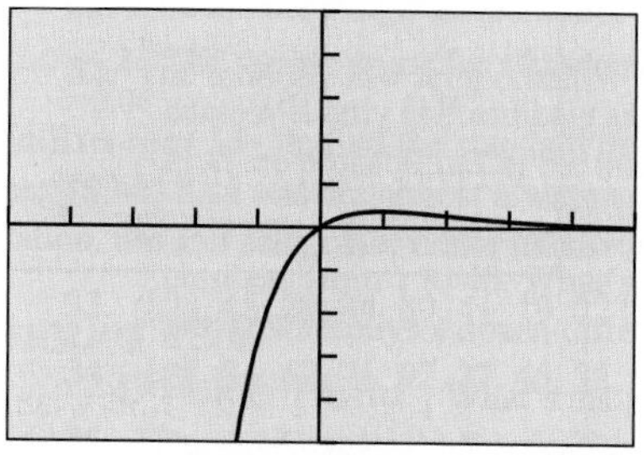

[–5, 5] by [–5, 5]

FIGURE 11.17 The graph of the function $f(x) = xe^{-x}$. (Example 8)

The graph of f in Figure 11.17 supports these results.

Now try Exercise 49.

EXPLORATION 2 Investigating a Logistic Function

Let $f(x) = \dfrac{50}{1 + 2^{3-x}}$.

1. Use tables and graphs to find $\lim_{x\to\infty} f(x)$ and $\lim_{x\to-\infty} f(x)$.
2. Identify any horizontal asymptotes.
3. How is the numerator of the fraction for f related to part 2?

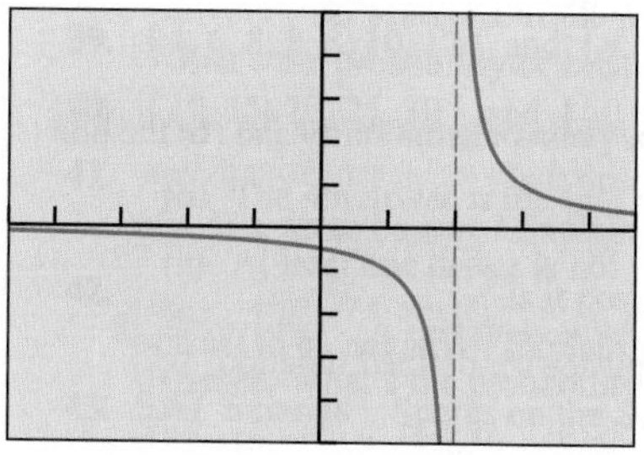

[–4.7, 4.7] by [–5, 5]

FIGURE 11.18 The graph of $f(x) = 1/(x - 2)$ with its vertical asymptote overlaid.

In Section 2.6 we used the graph of $f(x) = 1/(x - 2)$ to state that

$$\lim_{x\to2^-} \frac{1}{x-2} = -\infty \quad \text{and} \quad \lim_{x\to2^+} \frac{1}{x-2} = \infty.$$

Either one of these unbounded limits allows us to conclude that the vertical line $x = 2$ is a vertical asymptote of the graph of f (Figure 11.18).

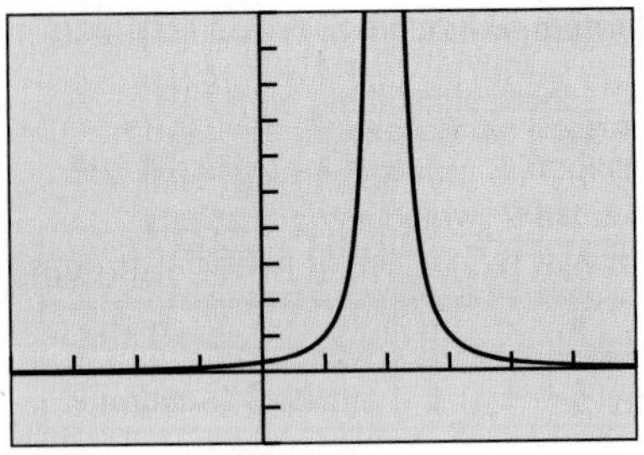

[−4, 6] by [−2, 10]

FIGURE 11.19 The graph of $f(x) = 1/(x - 2)^2$ in Example 9.

EXAMPLE 9 Investigating Unbounded Limits

Find $\lim_{x \to 2} 1/(x - 2)^2$.

SOLUTION The graph of $f(x) = 1/(x - 2)^2$ in Figure 11.19 suggests that

$$\lim_{x \to 2^-} \frac{1}{(x-2)^2} = \infty \quad \text{and} \quad \lim_{x \to 2^+} \frac{1}{(x-2)^2} = \infty.$$

This means that the limit of f as x approaches 2 does not exist. The table of values in Figure 11.20 agrees with this conclusion. The graph of f has a vertical asymptote at $x = 2$.

Now try Exercise 55.

X	Y1	
1.9	100	
1.99	10000	
1.999	1E6	
2	ERROR	
2.001	1E6	
2.01	10000	
2.1	100	

Y1 = 1/(X−2)²

FIGURE 11.20 A table of values for $f(x) = \dfrac{1}{(x-2)^2}$. (Example 9)

Not all zeros of denominators correspond to vertical asymptotes as illustrated in Examples 5 and 7.

EXAMPLE 10 Investigating a Limit at $x = 0$

Find $\lim_{x \to 0} (\sin x)/x$.

SOLUTION The graph of $f(x) = (\sin x)/x$ in Figure 11.15 suggests this limit exists. The table of values in Figure 11.21 suggests that

$$\lim_{x \to 0} \frac{\sin x}{x} = 1.$$

X	Y1	
−.03	.99985	
−.02	.99993	
−.01	.99998	
0	ERROR	
.01	.99998	
.02	.99993	
.03	.99985	

Y1 = sin(X)/X

FIGURE 11.21 A table of values for $f(x) = (\sin x)/x$. (Example 10)

Now try Exercise 63.

QUICK REVIEW 11.3 *(For help, go to Sections 1.2 and 1.3.)*

In Exercises 1 and 2, find **(a)** $f(-2)$, **(b)** $f(0)$, and **(c)** $f(2)$.

1. $f(x) = \dfrac{2x + 1}{(2x - 4)^2}$ **2.** $f(x) = \dfrac{\sin x}{x}$

In Exercises 3 and 4, find the **(a)** vertical asymptotes and **(b)** horizontal asymptotes of the graph of f, if any.

3. $f(x) = \dfrac{2x^2 + 3}{x^2 - 4}$ **4.** $f(x) = \dfrac{x^3 + 1}{2 - x - x^2}$

In Exercises 5 and 6, the end behavior asymptote of the function f is one of the following. Which one is it?

(a) $y = 2x^2$ **(b)** $y = -2x^2$ **(c)** $y = x^3$ **(d)** $y = -x^3$

5. $f(x) = \dfrac{2x^3 - 3x^2 + 1}{3 - x}$

6. $f(x) = \dfrac{x^4 + 2x^2 + x + 1}{x - 3}$

In Exercises 7 and 8, find **(a)** the points of continuity and **(b)** the points of discontinuity of the function.

7. $f(x) = \sqrt{x + 2}$ **8.** $g(x) = \dfrac{2x + 1}{x^2 - 4}$

Exercises 9 and 10 refer to the piecewise-defined function

$$f(x) = \begin{cases} 3x + 1 & \text{if } x \le 1 \\ 4 - x^2 & \text{if } x > 1 \end{cases}.$$

9. Draw the graph of f.

10. Find the points of continuity and the points of discontinuity of f.

SECTION 11.3 Exercises

Exercise numbers with a gray background indicate problems that the authors have designed to be solved *without a calculator*.

In Exercises 1–10, find the limit by direct substitution if it exists.

1. $\lim_{x\to -1} x(x-1)^2$ **2.** $\lim_{x\to 3} (x-1)^{12}$

3. $\lim_{x\to 2} (x^3 - 2x + 3)$ **4.** $\lim_{x\to -2} (x^3 - x + 5)$

5. $\lim_{x\to 2} \sqrt{x+5}$ **6.** $\lim_{x\to -2} (x-4)^{2/3}$

7. $\lim_{x\to 0} (e^x \sin x)$ **8.** $\lim_{x\to \pi} \ln\left(\sin \frac{x}{2}\right)$

9. $\lim_{x\to a} (x^2 - 2)$ **10.** $\lim_{x\to a} \frac{x^2-1}{x^2+1}$

In Exercises 11–18, **(a)** explain why you cannot use substitution to find the limit and **(b)** find the limit algebraically if it exists.

11. $\lim_{x\to -3} \frac{x^2+7x+12}{x^2-9}$ **12.** $\lim_{x\to 3} \frac{x^2-9}{x^2+2x-15}$

13. $\lim_{x\to -1} \frac{x^3+1}{x+1}$ **14.** $\lim_{x\to 2} \frac{x^3-2x^2+x-2}{x-2}$

15. $\lim_{x\to -2} \frac{x^2-4}{x+2}$ **16.** $\lim_{x\to -2} \frac{|x^2-4|}{x+2}$

17. $\lim_{x\to 0} \sqrt{x-3}$ **18.** $\lim_{x\to 0} \frac{x-2}{x^2}$

In Exercises 19–22, use the fact that $\lim_{x\to 0} \frac{\sin x}{x} = 1$, along with the limit properties, to find the following limits.

19. $\lim_{x\to 0} \frac{\sin x}{2x^2 - x}$ **20.** $\lim_{x\to 0} \frac{\sin 3x}{x}$

21. $\lim_{x\to 0} \frac{\sin^2 x}{x}$ **22.** $\lim_{x\to 0} \frac{x+\sin x}{2x}$

In Exercises 23–26, find the limits.

23. $\lim_{x\to 0} \frac{e^x - \sqrt{x}}{\log_4(x+2)}$ **24.** $\lim_{x\to 0} \frac{3\sin x - 4\cos x}{5\sin x + \cos x}$

25. $\lim_{x\to \pi/2} \frac{\ln(2x)}{\sin^2 x}$ **26.** $\lim_{x\to 27} \frac{\sqrt{x+9}}{\log_3 x}$

In Exercises 27–30, use the given graph to find the limits or to explain why the limits do not exist.

27. **(a)** $\lim_{x\to 2^-} f(x)$
(b) $\lim_{x\to 2^+} f(x)$
(c) $\lim_{x\to 2} f(x)$

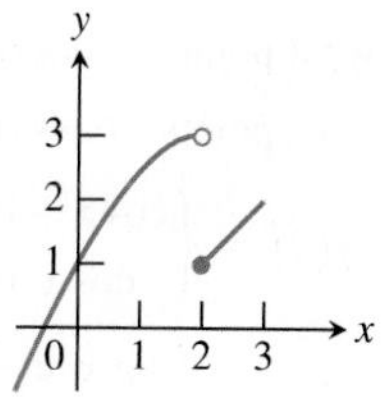

28. **(a)** $\lim_{x\to 3^-} f(x)$
(b) $\lim_{x\to 3^+} f(x)$
(c) $\lim_{x\to 3} f(x)$

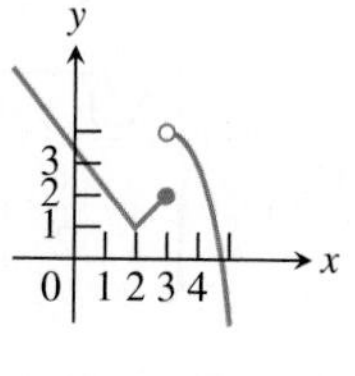

29. **(a)** $\lim_{x\to 3^-} f(x)$
(b) $\lim_{x\to 3^+} f(x)$
(c) $\lim_{x\to 3} f(x)$

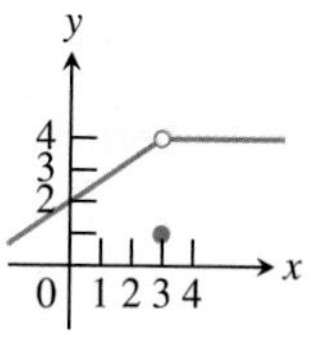

30. **(a)** $\lim_{x\to 1^-} f(x)$
(b) $\lim_{x\to 1^+} f(x)$
(c) $\lim_{x\to 1} f(x)$

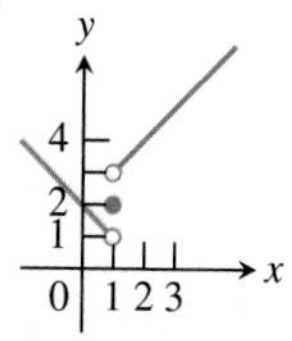

In Exercises 31 and 32, the graph of a function $y = f(x)$ is given. Which of the statements about the function are true and which are false?

31. **(a)** $\lim_{x\to -1^+} f(x) = 1$
(b) $\lim_{x\to 0^-} f(x) = 0$
(c) $\lim_{x\to 0^-} f(x) = 1$
(d) $\lim_{x\to 0^-} f(x) = \lim_{x\to 0^+} f(x)$
(e) $\lim_{x\to 0} f(x)$ exists. **(f)** $\lim_{x\to 0} f(x) = 0$
(g) $\lim_{x\to 0} f(x) = 1$ **(h)** $\lim_{x\to 1} f(x) = 1$
(i) $\lim_{x\to 1} f(x) = 0$ **(j)** $\lim_{x\to 2^-} f(x) = 2$

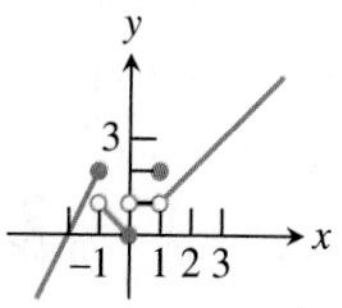

32. **(a)** $\lim_{x\to -1^+} f(x) = 1$
(b) $\lim_{x\to 2} f(x)$ does not exist.
(c) $\lim_{x\to 2} f(x) = 2$
(d) $\lim_{x\to 1^-} f(x) = 2$
(e) $\lim_{x\to 1^+} f(x) = 1$
(f) $\lim_{x\to 1} f(x)$ does not exist.
(g) $\lim_{x\to 0^+} f(x) = \lim_{x\to 0^-} f(x)$
(h) $\lim_{x\to c} f(x)$ exists for every c in $(-1, 1)$.
(i) $\lim_{x\to c} f(x)$ exists for every c in $(1, 3)$.

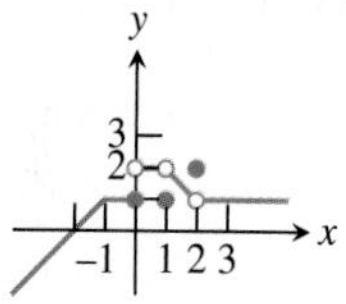

In Exercises 33 and 34, use a graph of f to find **(a)** $\lim_{x\to 0^-} f(x)$, **(b)** $\lim_{x\to 0^+} f(x)$, and **(c)** $\lim_{x\to 0} f(x)$ if they exist.

33. $f(x) = (1+x)^{1/x}$

34. $f(x) = (1+x)^{1/(2x)}$

35. Group Activity Assume that $\lim_{x\to 4} f(x) = -1$ and $\lim_{x\to 4} g(x) = 4$. Find the limit.

(a) $\lim_{x\to 4} (g(x) + 2)$ **(b)** $\lim_{x\to 4} xf(x)$

(c) $\lim_{x\to 4} g^2(x)$ **(d)** $\lim_{x\to 4} \frac{g(x)}{f(x) - 1}$

36. Group Activity Assume that $\lim_{x\to a} f(x) = 2$ and $\lim_{x\to a} g(x) = -3$. Find the limit.

(a) $\lim_{x\to a} (f(x) + g(x))$ **(b)** $\lim_{x\to a} (f(x) \cdot g(x))$

(c) $\lim_{x\to a} (3g(x) + 1)$ **(d)** $\lim_{x\to a} \frac{f(x)}{g(x)}$

In Exercises 37–40, complete the following for the given piecewise-defined function f.

(a) Draw the graph of f.

(b) Determine $\lim_{x\to a^+} f(x)$ and $\lim_{x\to a^-} f(x)$.

(c) Writing to Learn Does $\lim_{x\to a} f(x)$ exist? If it does, give its value. If it does not exist, give an explanation.

37. $a = 2,\ f(x) = \begin{cases} 2 - x & \text{if } x < 2 \\ 1 & \text{if } x = 2 \\ x^2 - 4 & \text{if } x > 2 \end{cases}$

38. $a = 1,\ f(x) = \begin{cases} 2 - x & \text{if } x < 1 \\ x + 1 & \text{if } x \ge 1 \end{cases}$

39. $a = 0,\ f(x) = \begin{cases} |x - 3| & \text{if } x < 0 \\ x^2 - 2x & \text{if } x \ge 0 \end{cases}$

40. $a = -3,\ f(x) = \begin{cases} 1 - x^2 & \text{if } x \ge -3 \\ 8 - x & \text{if } x < -3 \end{cases}$

In Exercises 41–46, find the limit.

41. $\lim_{x\to 2^+} \operatorname{int}(x)$

42. $\lim_{x\to 2^-} \operatorname{int}(x)$

43. $\lim_{x\to 0.0001} \operatorname{int}(x)$

44. $\lim_{x\to 5/2^-} \operatorname{int}(2x)$

45. $\lim_{x\to -3^+} \dfrac{x + 3}{|x + 3|}$

46. $\lim_{x\to 0^-} \dfrac{5x}{|2x|}$

In Exercises 47–54, find **(a)** $\lim_{x\to\infty} y$ and **(b)** $\lim_{x\to-\infty} y$.

47. $y = \dfrac{\cos x}{1 + x}$

48. $y = \dfrac{x + \sin x}{x}$

49. $y = 1 + 2^x$

50. $y = \dfrac{x}{1 + 2x}$

51. $y = x + \sin x$

52. $y = e^{-x} + \sin x$

53. $y = -e^x \sin x$

54. $y = e^{-x} \cos x$

In Exercises 55–60, use graphs and tables to find the limit and identify any vertical asymptotes.

55. $\lim_{x\to 3^-} \dfrac{1}{x - 3}$

56. $\lim_{x\to 3^+} \dfrac{x}{x - 3}$

57. $\lim_{x\to -2^+} \dfrac{1}{x + 2}$

58. $\lim_{x\to -2^-} \dfrac{x}{x + 2}$

59. $\lim_{x\to 5} \dfrac{1}{(x - 5)^2}$

60. $\lim_{x\to 2} \dfrac{1}{x^2 - 4}$

In Exercises 61–64, determine the limit algebraically if possible. Support your answer graphically.

61. $\lim_{x\to 0} \dfrac{(1 + x)^3 - 1}{x}$

62. $\lim_{x\to 0} \dfrac{1/(3 + x) - 1/3}{x}$

63. $\lim_{x\to 0} \dfrac{\tan x}{x}$

64. $\lim_{x\to 2} \dfrac{x - 4}{x^2 - 4}$

In Exercises 65–72, find the limit.

65. $\lim_{x\to 0} \dfrac{|x|}{x^2}$

66. $\lim_{x\to 0} \dfrac{x^2}{|x|}$

67. $\lim_{x\to 0} \left[x \sin\left(\dfrac{1}{x}\right)\right]$

68. $\lim_{x\to 27} \cos\left(\dfrac{1}{x}\right)$

69. $\lim_{x\to 1} \dfrac{x^2 + 1}{x - 1}$

70. $\lim_{x\to\infty} \dfrac{\ln x^2}{\ln x}$

71. $\lim_{x\to\infty} \dfrac{\ln x}{\ln x^2}$

72. $\lim_{x\to\infty} 3^{-x}$

Standardized Test Questions

73. True or False If $f(x) = \begin{cases} x + 2 & \text{if } x \le 3 \\ 8 - x & \text{if } x > 3 \end{cases}$, then $\lim_{x\to 3} f(x)$ is undefined. Justify your answer.

74. True or False If $f(x)$ and $g(x)$ are two functions and $\lim_{x\to 0} f(x)$ does not exist, then $\lim_{x\to 0} [f(x) \cdot g(x)]$ cannot exist. Justify your answer.

Multiple Choice In Exercises 75–78, match the function $y = f(x)$ with the table. Do not use a calculator.

X	Y1
2.7	-52.3
2.8	-82.2
2.9	-172.1
3	ERROR
3.1	188.1
3.2	98.2
3.3	68.3

X=2.7

(A)

X	Y1
2.7	3.7
2.8	3.8
2.9	3.9
3	ERROR
3.1	4.1
3.2	4.2
3.3	4.3

X=2.7

(B)

X	Y1
2.7	23.7
2.8	33.8
2.9	63.9
3	ERROR
3.1	-55.9
3.2	-25.8
3.3	-15.7

X=2.7

(C)

X	Y1
2.7	24.39
2.8	25.24
2.9	26.11
3	ERROR
3.1	27.91
3.2	28.84
3.3	29.79

X=2.7

(D)

X	Y1
2.7	3.7
2.8	3.8
2.9	3.9
3	4
3.1	4.1
3.2	4.2
3.3	4.3

X=2.7

(E)

75. $y = \dfrac{x^2 - 2x - 3}{x - 3}$

76. $y = \dfrac{x^2 + 2x + 3}{x - 3}$

77. $y = \dfrac{x^2 - 2x - 9}{x - 3}$

78. $y = \dfrac{x^3 - 27}{x - 3}$

Explorations

In Exercises 79–82, complete the following for the given piecewise-defined function f.

(a) Draw the graph of f.

(b) At what points c in the domain of f does $\lim_{x\to c} f(x)$ exist?

(c) At what points c does only the left-hand limit exist?

(d) At what points c does only the right-hand limit exist?

79. $f(x) = \begin{cases} \cos x & \text{if } -\pi \le x < 0 \\ -\cos x & \text{if } 0 \le x \le \pi \end{cases}$

80. $f(x) = \begin{cases} \sin x & \text{if } -\pi \le x < 0 \\ \csc x & \text{if } 0 \le x \le \pi \end{cases}$

81. $f(x) = \begin{cases} \sqrt{1 - x^2} & \text{if } -1 \le x < 0 \\ x & \text{if } 0 \le x < 1 \\ 2 & \text{if } x = 1 \end{cases}$

82. $f(x) = \begin{cases} x^2 \text{ if } -2 \le x < 0 \text{ or } 0 < x \le 2 \\ 1 \text{ if } x = 0 \\ 2x \text{ if } x < -2 \text{ or } x > 2 \end{cases}$

83. Rabbit Population The population of rabbits over a 2-year period in a certain county is given in Table 11.5.

Table 11.5 Rabbit Population

Beginning of Month	Number (in thousands)
0	10
2	12
4	14
6	16
8	22
10	30
12	35
14	39
16	44
18	48
20	50
22	51

(a) Draw a scatter plot of the data in Table 11.5.

(b) Find a logistic regression model for the data. Find the limit of that model as time approaches infinity.

(c) What can you conclude about the limit of the rabbit population growth in the county?

(d) Provide a reasonable explanation for the population growth limit.

Group Activity In Exercises 84–87, sketch a graph of a function $y = f(x)$ that satisfies the stated conditions. Include any asymptotes.

84. $\lim_{x \to 0} f(x) = \infty$, $\lim_{x \to \infty} f(x) = \infty$, $\lim_{x \to -\infty} f(x) = 2$

85. $\lim_{x \to 4} f(x) = -\infty$, $\lim_{x \to \infty} f(x) = -\infty$, $\lim_{x \to -\infty} f(x) = 2$

86. $\lim_{x \to \infty} f(x) = 2$, $\lim_{x \to -2^+} f(x) = -\infty$,
$\lim_{x \to -2^-} f(x) = -\infty$, $\lim_{x \to -\infty} f(x) = \infty$

87. $\lim_{x \to 1} f(x) = \infty$, $\lim_{x \to 2^+} f(x) = -\infty$,
$\lim_{x \to 2^-} f(x) = -\infty$, $\lim_{x \to -\infty} f(x) = \infty$

Extending the Ideas

88. Properties of Limits Find the limits of f, g, and fg as x approaches c.

(a) $f(x) = \dfrac{2}{x^2}$, $g(x) = x^2$, $c = 0$

(b) $f(x) = \left|\dfrac{1}{x}\right|$, $g(x) = \sqrt[3]{x}$, $c = 0$

(c) $f(x) = \left|\dfrac{3}{x-1}\right|$, $g(x) = (x-1)^2$, $c = 1$

(d) $f(x) = \dfrac{1}{(x-1)^4}$, $g(x) = (x-1)^2$, $c = 1$

(e) Writing to Learn Suppose that $\lim_{x \to c} f(x) = \infty$ and $\lim_{x \to c} g(x) = 0$. Based on your results in parts (a)–(d), what can you say about $\lim_{x \to a} (f(x) \cdot g(x))$?

89. Limits and the Area of a Circle Consider an n-sided regular polygon made up of n congruent isosceles triangles, each with height h and base b. The figure shows an 8-sided regular polygon.

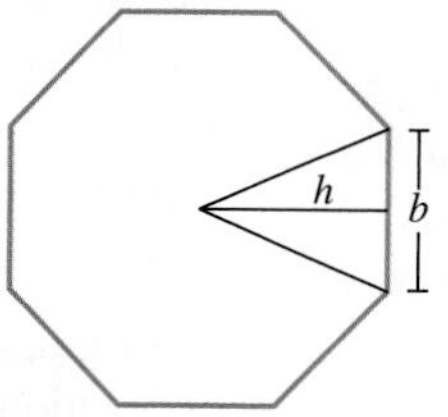

(a) Show that the area of an 8-sided regular polygon is $A = 4hb$ and the area of the n-sided regular polygon is $A = (1/2)nhb$.

(b) Show that the base b of the n-sided regular polygon is $b = 2h \tan (180/n)°$.

(c) Show that the area A of the n-sided regular polygon is $A = nh^2 \tan (180/n)°$.

(d) Let $h = 1$. Construct a table of values for n and A for $n = 4, 8, 16, 100, 500, 1000, 5000, 10{,}000, 100{,}000$. Does A have a limit as $n \to \infty$?

(e) Repeat part (d) with $h = 3$.

(f) Give a convincing argument that $\lim_{n \to \infty} A = \pi h^2$, the area of a circle of radius h.

90. Continuous Extension of a Function Let

$$f(x) = \begin{cases} x^2 - 3x + 3 & \text{if } x \ne 2 \\ a & \text{if } x = 2. \end{cases}$$

(a) Sketch several possible graphs for f.

(b) Find a value for a so that the function is continuous at $x = 2$.

In Exercises 91–93, **(a)** graph the function, **(b)** verify that the function has one removable discontinuity, and **(c)** give a formula for a continuous extension of the function. (*Hint*: See Exercise 90.)

91. $y = \dfrac{2x+4}{x+2}$

92. $y = \dfrac{x-5}{5-x}$

93. $y = \dfrac{x^3-1}{x-1}$

11.4 Numerical Derivatives and Integrals

What you'll learn about

- Derivatives on a Calculator
- Definite Integrals on a Calculator
- Computing a Derivative from Data
- Computing a Definite Integral from Data

... and why

The numerical capabilities of a graphing calculator make it easy to perform many calculations that would have been exceedingly difficult in the past.

Derivatives on a Calculator

As computers and sophisticated calculators have become indispensable tools for modern engineers and mathematicians (and, ultimately, for modern students of mathematics), numerical techniques of differentiation and integration have re-emerged as primary methods of problem solving. This is no small irony, as it was precisely to avoid the tedious computations inherent in such methods that calculus was invented in the first place. Although nothing can diminish the magnitude of calculus as a significant human achievement, and although nobody can get far in mathematics or science without it, the modern fact is that applying the old-fashioned methods of limiting approximations—with the help of a calculator—is often the most efficient way to solve a calculus problem.

Most graphing calculators have built-in algorithms that will approximate derivatives of functions numerically with good accuracy at most points of their domains. We will use the notation NDER $f(a)$ to denote such a calculator derivative approximation to $f'(a)$.

For small values of h, the regular difference quotient

$$\frac{f(a+h)-f(a)}{h}$$

is often a good approximation of $f'(a)$. However, the same value of h will usually produce a better approximation of $f'(a)$ if we use the **symmetric difference quotient**

$$\frac{f(a+h)-f(a-h)}{2h},$$

as illustrated in Figure 11.22.

Many graphing utilities use the symmetric difference quotient with a default value of $h = 0.001$ for computing NDER $f(a)$. When we refer to the numerical derivative in this book, we will assume that it is the symmetric difference quotient with $h = 0.001$.

DEFINITION Numerical Derivative

In this book, we define the **numerical derivative of *f* at *a*** to be

$$\text{NDER } f(a) = \frac{f(a+0.001)-f(a-0.001)}{0.002}.$$

Similarly, we define the **numerical derivative of *f*** to be the function

$$\text{NDER } f(x) = \frac{f(x+0.001)-f(x-0.001)}{0.002}.$$

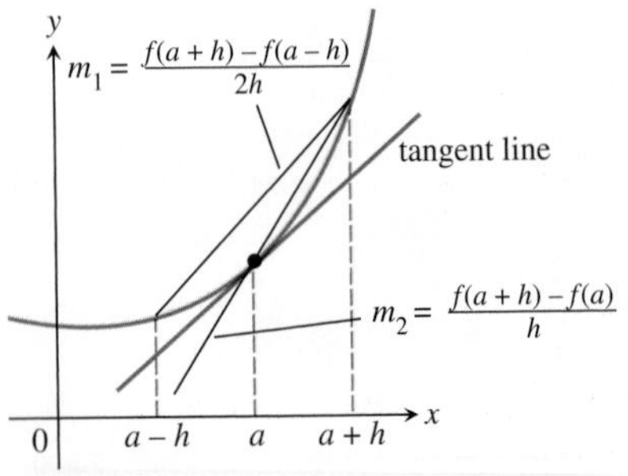

FIGURE 11.22 The symmetric difference quotient (slope m_1) usually gives a better approximation of the derivative for a given value of h than does the regular difference quotient (slope m_2).

EXAMPLE 1 Computing a Numerical Derivative

Let $f(x) = x^3$. Compute NDER $f(2)$ by calculating the symmetric difference quotient with $h = 0.001$. Compare it to the actual value of $f'(x)$.

SOLUTION

$$\text{NDER } f(x) = \frac{f(2 + 0.001) - f(2 - 0.001)}{0.002}$$
$$= \frac{f(2.001) - f(1.999)}{0.002}$$
$$= 12.000001$$

The actual value is

$$f'(2) = \lim_{h \to 0} \frac{(2 + h)^3 - 2^3}{h}$$
$$= \lim_{h \to 0} \frac{8 + 12h + 6h^2 + h^3 - 8}{h}$$
$$= \lim_{h \to 0} (12 + 6h + h^2)$$
$$= 12$$

Now try Exercise 3.

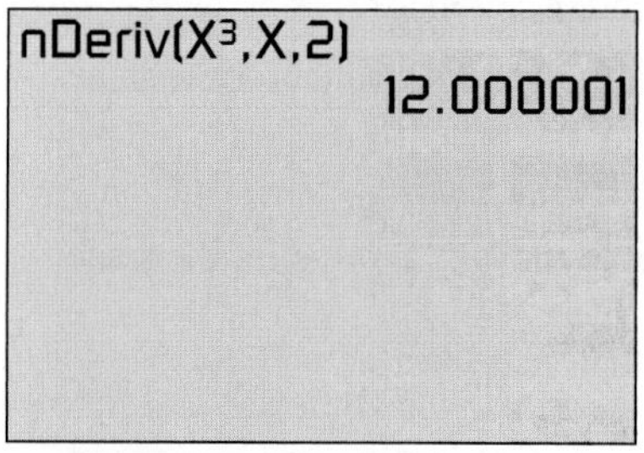

FIGURE 11.23 Applying the numerical derivative command on a graphing calculator. (Example 1)

The numerical derivative in this case is obviously quite accurate. In practice, it is not necessary to key in the symmetric difference quotient, as it is done by the calculator with its built-in algorithm. Figure 11.23 shows the command that would be used on one such calculator to find the numerical derivative in Example 1.

If $f'(a)$ exists, then NDER $f(a)$ usually gives a good approximation to the actual value. On the other hand, the algorithm will sometimes return a value for NDER $f(a)$ when $f'(a)$ does not exist. (See Exercise 51.)

Definite Integrals on a Calculator

Recall from the history of the area problem (Section 11.2) that the strategy of summing up thin rectangles to approximate areas is ancient. The thinner the rectangles, the better the approximation—and, of course, the more tedious the computation. Today, thanks to technology, we can employ the ancient strategy without the tedium.

Many graphing calculators have built-in algorithms to compute definite integrals with great accuracy. We use the notation NINT $(f(x), x, a, b)$ to denote such a calculator approximation to $\int_a^b f(x)\,dx$. Unlike NDER, which uses a fixed value of Δx, NINT will vary the value of Δx until the numerical integral gets close to a limiting value, often resulting in an exact answer (at least to the number of digits in the calculator display). Because the algorithm for NINT finds the definite integral by Riemann sum approximation rather than by calculus, we call it a **numerical integral**.

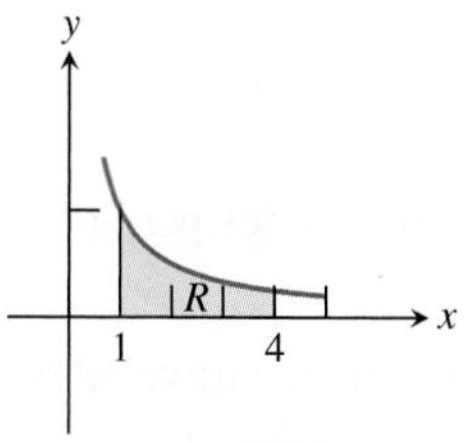

FIGURE 11.24 The graph of $f(x) = 1/x$ with the area under the curve between $x = 1$ and $x = 4$ shaded. (Example 2)

EXAMPLE 2 Finding a Numerical Integral

Use NINT to find the area of the region R enclosed between the x-axis and the graph of $y = 1/x$ from $x = 1$ to $x = 4$.

SOLUTION The region is shown in Figure 11.24.

The area can be written as the definite integral $\int_1^4 \frac{1}{x}\,dx$, which we find on a graphing calculator: NINT $(1/x, x, 1, 4) = 1.386294361$. The exact answer (as you will learn in a calculus course) is ln 4, which agrees in every displayed digit with the NINT value! Figure 11.25 shows the syntax for numerical integration on one type of calculator.

Now try Exercise 13.

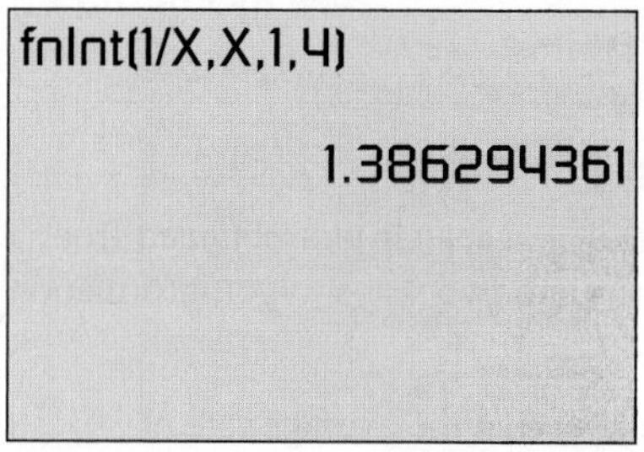

FIGURE 11.25 A numerical integral approximation for $\int_1^4 \frac{1}{x}\,dx$. (Example 2)

EXPLORATION 1 A Do-It-Yourself Numerical Integrator

Recall that a definite integral is the limit at infinity of a Riemann sum—that is, a sum of the form $\sum_{k=1}^{n} f(x_k)\Delta x$. You can use your calculator to evaluate sums of sequences using LIST commands. (It is not as accurate as NINT, and certainly not as easy, but at least you can see the summing that takes place.)

1. The integral in Example 2 can be computed using the command

   $$\text{sum}(\text{seq}(1/(1 + K \cdot 3/50) \cdot 3/50, K, 1, 50)).$$

   This uses 50 RRAM rectangles, each with width $\Delta x = 3/50$. Find the sum on your calculator and compare it to the NINT value.
2. Study the command until you see how it works. Adapt the command to find the RRAM approximation for 100 rectangles and compute it on your calculator. Does the approximation get better?
3. What definite integral is approximated by the command

   $$\text{sum}(\text{seq}(\sin(0 + K \cdot \pi/50) \cdot \pi/50, K, 1, 50))?$$

   Compute it on your calculator and compare it to the NINT value for the same integral.
4. Write a command that uses 50 RRAM rectangles to approximate $\int_4^9 \sqrt{x}\, dx$. Compute it on your calculator and compare it to the NINT value for the same integral.

Remember that we were originally motivated to find areas because of their connection to the problem of distance traveled. To show just one of the many applications of integration, we use the numerical integral to solve a distance problem in Example 3.

EXAMPLE 3 Finding Distance Traveled

An automobile is driven at a variable rate along a test track for 2 hr so that its velocity at any time t, $0 \le t \le 2$, is given by $v(t) = 30 + 10 \sin 6t$ miles per hour. How far does the automobile travel during the 2-hr test?

SOLUTION According to the analysis found in Section 11.2, the distance traveled is given by $\int_0^2 (30 + 10 \sin (6t))\, dt$. We use a calculator to find the numerical integral:

$$\text{NINT}\ (30 + 10 \sin (6t), t, 0, 2) \approx 60.26$$

Interpreting the answer, we conclude that the automobile travels about 60.26 mi.

Now try Exercise 21.

Computing a Derivative from Data

Sometimes all we are given about a problem situation is a scatter plot obtained from a set of data—a numerical model of the problem. There are two ways to get information about the derivative of the model.

1. To approximate the derivative at a point: Remember that the average rate of change over a small interval, $\Delta y/\Delta x$, approximates the derivative at points in that interval. (Generally, the approximation is better near the middle of the interval than it is near the endpoints.) The average rate of change on an interval between two data points can be computed directly from the data.

Who's Driving???

We will candidly admit that the conditions in Example 3 would be virtually impossible to replicate in a real setting, even if one could imagine a reason for doing so. Mathematics textbooks are filled with such unreal real-world problems, but they do serve a purpose when students are being exposed to new material. Real real-world problems are often either too easy to illustrate the concept or too hard for beginners to solve.

Table 11.6 Falling Ball

Time (sec)	Position (ft)
0.04	6.80
0.08	6.40
0.12	5.95
0.16	5.45
0.20	4.90
0.24	4.30
0.28	3.60
0.32	2.90
0.36	2.15
0.40	1.30
0.44	0.40

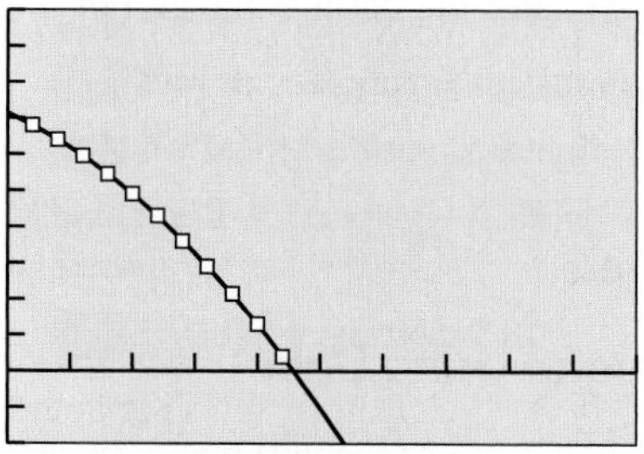

[0, 1] by [–2, 10]

FIGURE 11.26 A scatter plot of the data in Table 11.6 together with its quadratic regression model. (Example 4)

Table 11.7 Change over Intervals from Table 11.6

Midpoint	$\Delta s/\Delta t$
0.06	−10.00
0.10	−11.25
0.14	−12.50
0.18	−13.75
0.22	−15.00
0.26	−17.50
0.30	−17.50
0.34	−18.75
0.38	−21.25
0.42	−22.50

2. To approximate the derivative function: Regression techniques can be used to fit a curve to the data, and then NDER can be applied to the regression model to approximate the derivative. Alternatively, the values of $\Delta y/\Delta x$ can be plotted, and then regression techniques can be used to fit a derivative approximation through those points.

EXAMPLE 4 Finding Derivatives from Data

Table 11.6 shows the height (in feet) of a falling ball above ground level as measured by a motion detector at time intervals of 0.04 sec.

(a) Estimate the instantaneous speed of the ball at $t = 0.2$ sec.

(b) Draw a scatter plot of the data and use quadratic regression to model the height s of the ball above the ground as a function of t.

(c) Use NDER to approximate $s'(0.2)$ and compare it to the value found in (a).

SOLUTION

(a) Since 0.2 is the midpoint of the time interval $[0.16, 0.24]$, the average rate of change $\Delta s/\Delta t$ on the interval $[0.16, 0.24]$ should give a good approximation to $s'(0.2)$.

$$s'(0.2) \approx \frac{\Delta s}{\Delta t} = \frac{4.30 - 5.45}{0.24 - 0.16} = -14.375.$$

The speed is about 14.38 ft/sec at $t = 0.2$.

(b) The scatter plot is shown in Figure 11.26, along with the quadratic regression curve. A graphing calculator gives $s(t) \approx -17.12t^2 - 7.74t + 7.13$ as the equation of the quadratic regression model.

(c) The calculator computes NDER $s(0.2)$ to be about -14.59, which agrees quite well with the approximation in (a). In fact, the difference is only 0.213 ft/sec, less than 1.5% of the speed of the ball. **Now try Exercise 23.**

EXAMPLE 5 Finding Derivatives from Data

This example also uses the falling ball data in Table 11.6.

(a) Compute the average velocity, $\Delta s/\Delta t$, on each subinterval of length 0.04. Make a table showing the midpoints of the subintervals in one column and the values of $\Delta s/\Delta t$ in the second column.

(b) Make a scatter plot showing the numbers in the second column as a function of the numbers in the first column and find a linear regression model to model the data.

(c) Use the linear regression model in (b) to approximate the velocity of the ball at $t = 0.2$ and compare it to the values found in Example 4.

SOLUTION

(a) The first subinterval, which begins at 0.04 and ends at 0.08, has a midpoint of $(0.04 + 0.08)/2 = 0.06$. On that interval, $\Delta s/\Delta t = (6.40 - 6.80)/(0.08 - 0.04) = -10.00$. The rest of the midpoints and values of $\Delta s/\Delta t$ are computed similarly and are shown in Table 11.7.

(b) The scatter plot and the regression line are shown in Figure 11.27. A graphing calculator gives $v(t) = -34.470t - 7.727$ as the regression line.

(continued)

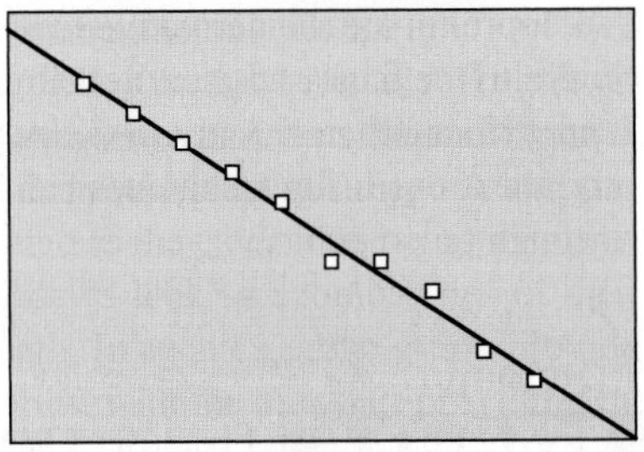

FIGURE 11.27 A scatter plot of the data in Table 11.7 together with its linear regression model. (Example 5)

(c) The linear regression model gives $v(0.2) \approx -14.62$, which is close to the values found in Example 4.

Now try Exercise 25.

Computing a Definite Integral from Data

If we are given a set of data points, the x-coordinates of the points define subintervals between the smallest and largest x-values in the data. We can form a Riemann sum

$$\sum_{k=1}^{n} f(x_k)\,\Delta x$$

using the lengths of the subintervals for Δx and either the left or right endpoints of the intervals as the x_k's. The Riemann sum then approximates the definite integral of the function over the interval.

Example 6 illustrates how this is done.

Table 11.8 Velocity of the Moving Body

Time (sec)	Velocity (m/sec)
0.00	0.00
0.25	0.28
0.50	0.53
0.75	0.73
1.00	0.90
1.25	1.01
1.50	1.11
1.75	1.18
2.00	1.21
2.25	1.17
2.50	1.13
2.75	1.05
3.00	0.91
3.25	0.72
3.50	0.55
3.75	0.26

EXAMPLE 6 Finding a Definite Integral Using Data

Table 11.8 shows the velocity of a moving body (in meters per second) measured at regular 0.25-sec intervals. Estimate the distance traveled by the body from $t = 0$ to $t = 3.75$.

SOLUTION Figure 11.28 gives a scatter plot of the velocity data.

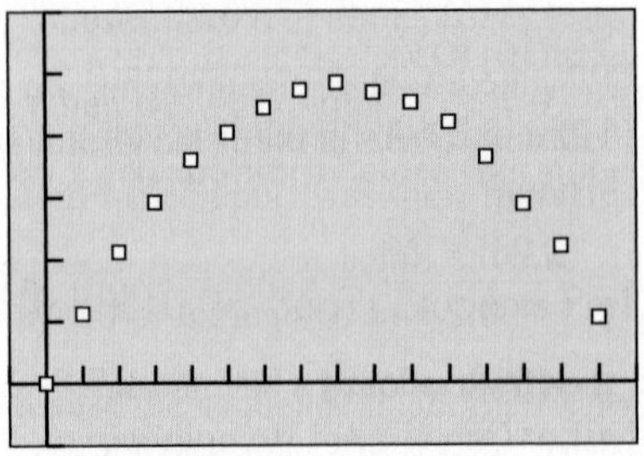

FIGURE 11.28 A scatter plot of the velocity data in Table 11.8. (Example 6)

The distance traveled is $\int_0^{3.75} v(t)\,dx$, which we approximate with a Riemann sum constructed directly from the data. We sum up 15 products of the form $v(t_k)\,\Delta t$, using the right endpoint for t_k each time. (This is the RRAM approximation in the notation of Section 11.2.) Note that $\Delta t = 0.25$ for every subinterval.

$$\begin{aligned}\int_0^{3.75} v(t)\,dx &= \sum_{k=1}^{15} v(t_k)\,\Delta t \\ &= 0.25(0.28 + 0.53 + 0.73 + 0.90 + 1.01 + 1.11 \\ &\quad + 1.18 + 1.21 + 1.17 + 1.13 + 1.05 + 0.91 + 0.72 \\ &\quad + 0.55 + 0.26) \\ &= 3.185\end{aligned}$$

So the distance traveled by the body is about 3.2 m. **Now try Exercise 27.**

QUICK REVIEW 11.4 *(For help, go to Sections P.4 and 1.3.)*

In Exercises 1–6, find $\Delta y/\Delta x$ on the interval $[1, 4]$ under the given conditions.

1. $y = x^2$

2. $y = \sqrt{x}$

3. $y = \log_2 x$

4. $y = 3^x$

5. $y = 2$ when $x = 1$ and $y = 11$ when $x = 4$.

6. The graph of $y = f(x)$ passes through points $(4, 10)$ and $(1, -2)$.

In Exercises 7–10, compute the quotient $(f(1 + h) - f(1 - h))/2h$ with the given f and h.

7. $f(x) = \sin x, h = 0.01$

8. $f(x) = x^4, h = 0.001$

9. $f(x) = \ln x, h = 0.001$

10. $f(x) = e^x, h = 0.0001$

SECTION 11.4 Exercises

In Exercises 1–10, use NDER on a calculator to find the numerical derivative of the function at the specified point.

1. $f(x) = 1 - x^2$ at $x = 2$

2. $f(x) = 2x + \dfrac{1}{2}x^2$ at $x = 2$

3. $f(x) = 3x^2 + 2$ at $x = -2$

4. $f(x) = x^2 - 3x + 1$ at $x = 1$

5. $f(x) = |x + 2|$ at $x = -2$

6. $f(x) = \dfrac{1}{x + 2}$ at $x = -1$

7. $f(x) = \ln 2x$ at $x = 1$

8. $f(x) = 2 \ln x$ at $x = 1$

9. $f(x) = 3 \sin x$ at $x = \pi$

10. $f(x) = \sin 3x$ at $x = \pi$

In Exercises 11–20, use NINT on a calculator to find the numerical integral of the function over the specified interval.

11. $f(x) = x^2, [0, 4]$

12. $f(x) = x^2, [-4, 0]$

13. $f(x) = \sin x, [0, \pi]$

14. $f(x) = \sin x, [\pi, 2\pi]$

15. $f(x) = \cos x, [0, \pi]$

16. $f(x) = |\cos x|, [0, \pi]$

17. $f(x) = 1/x, [1, e]$

18. $f(x) = 1/x, [e, 2e]$

19. $f(x) = \dfrac{2}{1 + x^2}, [0, 10^8]$

20. $f(x) = \sec^2 x - \tan^2 x, [0, 10]$

21. Travel Time A truck is driven at a variable rate for 3 hr so that its velocity at any time t $(0 \le t \le 3)$ is given by $v(t) = 35 - 12 \cos 4t$ mph. How far does the truck travel during the 3 hr? Round your answer to the nearest hundredth.

22. Travel Time A bicyclist rides for 90 min, and her velocity at any time t hours $(0 \le t \le 1.5)$ is given by $v(t) = 12 - 8 \sin 5t$ mph. How far does she travel during the 90 min? Round your answer to the nearest hundredth.

23. Finding Derivatives from Data A ball is dropped from the roof of a 30-story building. The height in feet above the ground of the falling ball is measured at 1/2-sec intervals and recorded in the table on the next page.

Time (sec)	Height (ft)
0	500
0.5	495
1.0	485
1.5	465
2.0	435
2.5	400
3.0	355
3.5	305
4.0	245
4.5	175
5.0	100
5.5	15

(a) Use the average velocity on the interval $[1, 2]$ to estimate the velocity of the ball at $t = 1.5$ sec.

(b) Draw a scatter plot of the data.

(c) Find a quadratic regression model for the data.

(d) Use NDER of the model in part (c) to estimate the velocity of the ball at $t = 1.5$ sec.

(e) Use the model to estimate how fast the ball is going when it hits the ground.

24. Estimating Average Rate of Change from Data Table 11.9 gives the U.S. gross domestic product data in billions of dollars for the years 2004–2012.

Table 11.9 U.S. Gross Domestic Product Data

Year	Amount (billions of dollars)
2004	10,908
2005	11,631
2006	12,364
2007	13,362
2008	13,995
2009	14,297
2010	14,044
2011	14,582
2012	15,094

Source: http://www.tradingeconomics.com/united-states/gdp

(a) Find the average rate of change of the gross domestic product from 2006 to 2007 and then from 2010 to 2011.

(b) Find a quadratic regression model for the data in Table 11.9 and overlay its graph on a scatter plot of the data. Let $x = 0$ stand for 2004, $x = 1$ stand for 2005, and so forth.

(c) Use the model in part (b), and a calculator NDER computation to estimate the rate of change of the gross domestic product in 2006 and in 2010.

(d) Writing to Learn Use the model in part (b) to predict the gross domestic product in 2016. Is this reasonable? Why or why not?

25. Estimating Velocity Refer to the data in Exercise 23.

(a) Compute the average velocity, $\Delta y/\Delta x$, on each subinterval of length 0.5. Make a table showing the midpoints of the subintervals in one column and the average velocities in the second column.

(b) Make a scatter plot showing the numbers in the second column as a function of the numbers in the first column and find a linear regression model to model the data.

(c) Use the linear regression model in part (b) to approximate the velocity of the ball at $t = 1.5$ sec, and compare your result to the value found in Exercise 23(d).

26. Approximating Rate of Change Refer to the data in Exercise 24.

(a) Compute the average rate of change, $\Delta y/\Delta x$, on each subinterval. Make a table showing the midpoints of the subintervals in one column and the average rates of change in the second column.

(b) Make a scatter plot showing the numbers in the second column as a function of the numbers in the first column and find a linear regression model to model the data.

(c) Use the linear regression model in part (b) to approximate the rate of change in 2006 and in 2010, and compare your results to the values found in Exercise 24(c).

27. Estimating Distance A stone is dropped from a cliff and its velocity (in feet per second) at regular 0.5-sec intervals is shown in Table 11.10. Estimate the distance that the stone travels from $t = 0$ to $t = 2.5$.

Table 11.10 Velocity of the Stone

Time (sec)	Velocity (ft/sec)
0	0
0.5	16
1	32
1.5	48
2	64
2.5	80

28. Estimating Distance Table 11.11 shows the velocity of a moving object in meters per second, measured at regular 0.2-sec intervals. Estimate the distance traveled by the body from $t = 0$ to $t = 1.6$.

Table 11.11 Velocity of the Object

Time (sec)	Velocity (m/sec)
0	1.20
0.2	0.98
0.4	0.72
0.6	0.50
0.8	0.34
1.0	0.30
1.2	0.44
1.4	0.79
1.6	1.40

29. **Writing to Learn** Analyze the following program, which produces an LRAM approximation for the function entered in Y1 in the calculator. Then write a short paragraph explaining how it works.

```
PROGRAM:LRAM
:Input "A",A
:Input "B",B
:Input "N",N
:sum(seq(((B–A)/
N)Y1(A+K((B–A)/N
)),K,0,N–1))→C
:Disp "AREA =",C
```

30. **Writing to Learn** Analyze the following program, which produces an RRAM approximation for the function entered in Y1 in the calculator. Then write a short paragraph explaining how it works.

```
PROGRAM:RRAM
:Input "A",A
:Input "B",B
:Input "N",N
:sum(seq(((B–A)/
N)Y1(A+K((B–A)/N
)),K,1,N))→C
:Disp "AREA =",C
```

In Exercises 31–42, complete the following for the indicated interval $[a, b]$.

(a) Verify the given function is nonnegative.

(b) Use a calculator to find the LRAM, RRAM, and average approximations for the area under the graph of the function from $x = a$ to $x = b$ with 10, 20, 50, and 100 approximating rectangles. (You may want to use the programs in Exercises 29 and 30.)

(c) Compare the average area estimate in part (b) using 100 approximating rectangles with the calculator NINT area estimate, if your calculator has this feature.

31. $f(x) = x^2 - x + 1;\ [0, 4]$
32. $f(x) = 2x^2 - 2x + 1;\ [0, 6]$
33. $f(x) = x^2 - 2x + 1;\ [0, 4]$
34. $f(x) = x^2 + x + 5;\ [0, 6]$
35. $f(x) = x^2 + x + 5;\ [2, 6]$
36. $f(x) = x^3 + 1;\ [1, 5]$
37. $f(x) = \sqrt{x + 2};\ [0, 4]$
38. $f(x) = \sqrt{x - 2};\ [3, 6]$
39. $f(x) = \cos x;\ [0, \pi/2]$
40. $f(x) = x \cos x;\ [0, \pi/2]$
41. $f(x) = xe^{-x};\ [0, 2]$
42. $f(x) = \dfrac{1}{x - 2};\ [3, 5]$

Standardized Test Questions

43. **True or False** The numerical derivative algorithm NDER always uses the same value of Δx (or h) to complete its calculations. Justify your answer.

44. **True or False** The numerical integral algorithm always uses the same value of Δx to complete its calculations. Justify your answer.

In Exercises 45–48, choose the correct answer. Do not use a calculator.

45. **Multiple Choice** Which of the following will typically produce the most accurate estimate of an area under a curve?

(A) NDER **(B)** NINT

(C) LRAM, 10 rectangles **(D)** RRAM, 25 rectangles

(e) LRAM, 60 rectangles

46. **Multiple Choice** Given a continuous function f, which of the following expressions will typically produce the most accurate estimate of $f'(a)$?

(A) $\dfrac{f(a + 0.05) - f(a - 0.05)}{0.05}$

(B) $\dfrac{f(a + 0.05) - f(a - 0.05)}{0.1}$

(C) $\dfrac{f(a + 0.01) - f(a)}{0.01}$

(D) $\dfrac{f(a + 0.01) - f(a - 0.01)}{0.01}$

(E) $\dfrac{f(a + 0.01) - f(a - 0.01)}{0.02}$

47. **Multiple Choice** Which of the following *cannot* be estimated using a numerical integral?

(A) The area under a curve that represents some function $f(x)$

(B) The distance traveled, when the velocity function is known

(C) The instantaneous velocity of an object, when the position function is known

(D) The change of a city's population over a 10-year period, when the rate-of-change function is known

(E) The change of a child's height over a 4-year period, when the rate-of-change function is known

48. **Multiple Choice** Which of the following *cannot* be estimated using a numerical derivative?

(A) The instantaneous velocity of an object, when the position function is known

(B) The slope of a curve that represents some function $g(x)$

(C) The growth rate of a city's population, when the population is known as a function of time

(D) The area under a curve that represents some function $f(x)$

(E) The rate of change of an airplane's altitude, when the altitude is known as a function of time

Explorations

49. Let $f(x) = 2x^2 + 3x + 1$ and $g(x) = x^3 + 1$.

(a) Compute the derivative of f.

(b) Compute the derivative of g.

(c) Using $x = 2$ and $h = 0.001$, compute the standard difference quotient

$$\frac{f(x + h) - f(x)}{h}$$

and the symmetric difference quotient

$$\frac{f(x + h) - f(x - h)}{2h}.$$

(d) Using $x = 2$ and $h = 0.001$, compare the approximations to $f'(2)$ in part (c). Which is the better approximation?

(e) Repeat parts (c) and (d) for g.

50. When Are Derivatives and Areas Equal? Let $f(x) = 1 + e^x$.

(a) Draw a graph of f for $0 \leq x \leq 1$.

(b) Use NDER on your calculator to compute the derivative of f at 1.

(c) Use NINT on your calculator to compute the area under f from $x = 0$ to $x = 1$ and compare it to the answer in part (b).

(d) Group Activity What do you think the exact answers to parts (b) and (c) are?

51. Calculator Failure Many calculators report that NDER of $f(x) = |x|$ evaluated at $x = 0$ is equal to 0. Explain why this is incorrect. Explain why this error occurs.

52. Grapher Failure Graph the function $f(x) = |x|/x$ in the window $[-5, 5]$ by $[-3, 3]$ and explain why $f'(0)$ does not exist. Find the value of NDER $f(0)$ on the calculator and explain why it gives an incorrect answer.

Extending the Ideas

53. Group Activity Finding Total Area The **total area** bounded by the graph of the function $y = f(x)$ and the x-axis from $x = a$ to $x = b$ is the area below the graph of $y = |f(x)|$ from $x = a$ to $x = b$.

(a) Find the total area bounded by the graph of $f(x) = \sin x$ and the x-axis from $x = 0$ to $x = 2\pi$.

(b) Find the total area bounded by the graph of $f(x) = x^2 - 2x - 3$ and the x-axis from $x = 0$ to $x = 5$.

54. Writing to Learn If a function is *unbounded* in an interval $[a, b]$ it may have *finite* area. Use your knowledge of limits at infinity to explain why this might be the case.

55. Writing to Learn Let f and g be two continuous functions with $f(x) \geq g(x)$ on an interval $[a, b]$. Devise a limit of sums definition of the area of the region between the two curves. Explain how to compute the area if the area under both curves is already known.

56. Area as a Function Consider the function $f(t) = 2t$.

(a) Use NINT on a calculator to compute $A(x)$ where A is the area under the graph of f from $t = 0$ to $t = x$ for $x = 0.25$, 0.5, 1, 1.5, 2, 2.5, and 3.

(b) Make a table of pairs $(x, A(x))$ for the values of x given in part (a) and plot them using graph paper. Connect the plotted points with a smooth curve.

(c) Use a quadratic regression equation to model the data in part (b) and overlay its graph on a scatter plot of the data.

(d) Make a conjecture about the exact value of $A(x)$ for any x greater than zero.

(e) Find the derivative of the $A(x)$ found in part (d). Record any observations.

57. Area as a Function Consider the function $f(t) = 3t^2$.

(a) Use NINT on a calculator to compute $A(x)$ where A is the area under the graph of f from $t = 0$ to $t = x$ for $x = 0.25$, 0.5, 1, 1.5, 2, 2.5, and 3.

(b) Make a table of pairs $(x, A(x))$ for the values of x given in part (a) and plot them using graph paper. Connect the plotted points with a smooth curve.

(c) Use a cubic regression equation to model the data in part (b) and overlay its graph on a scatter plot of the data.

(d) Make a conjecture about the exact value of $A(x)$ for any x greater than zero.

(e) Find the derivative of the $A(x)$ found in part (d). Record any observations.

58. Group Activity Based on Exercises 56 and 57, discuss how derivatives (slope functions) and integrals (area functions) may be connected.

CHAPTER 11 Key Ideas

Properties, Theorems, and Formulas

Properties of Limits 775
One-Sided and Two-Sided Limits 777

Procedures

Computing a Derivative from Data 786–788
Computing a Definite Integral from Data 788–789

CHAPTER 11 Review Exercises

The collection of exercises marked in red could be used as a chapter test.

In Exercises 1–4, use the graph of the function $y = f(x)$ to find **(a)** $\lim_{x \to 1^-} f(x)$ and **(b)** $\lim_{x \to 1} f(x)$.

1.

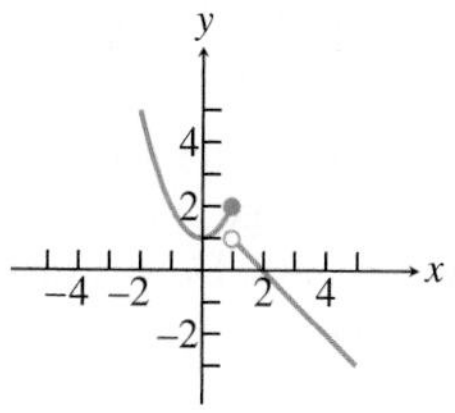

2.

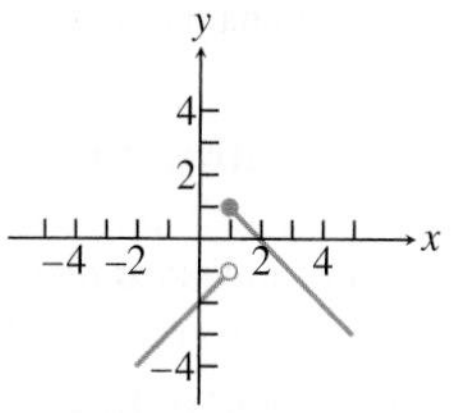

3.

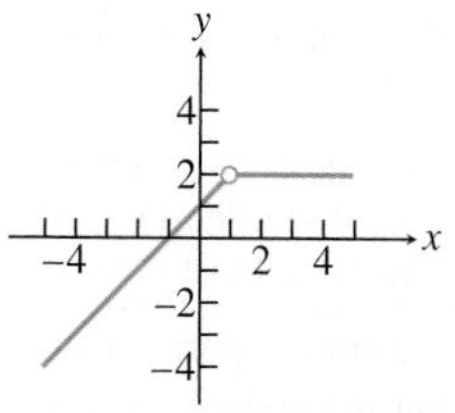

4.

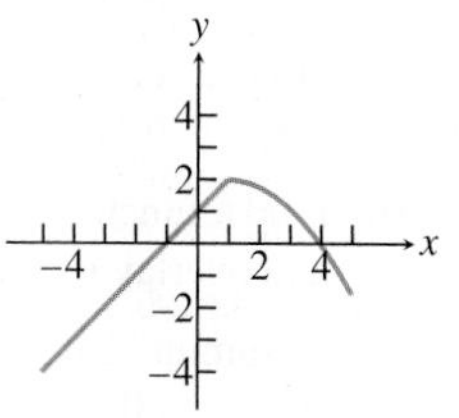

In Exercises 5–10, find the limit at the indicated point, if it exists. Support your answer graphically.

5. $f(x) = \dfrac{x - 1}{x^2 + 1}, x = -1$

6. $f(x) = \dfrac{\sin 5x}{x}, x = 0$

7. $f(x) = \dfrac{x^2 - 3x - 10}{x + 2}, x = -2$

8. $f(x) = |x - 1|, x = 1$

9. $f(x) = 2 \tan^{-1} x, x = 0$

10. $f(x) = \dfrac{2}{1 - 2^x}, x = 0$

In Exercises 11–14, find the limit. Support your answer with an appropriate table.

11. $\lim_{x \to -\infty} \dfrac{-1}{(x + 2)^2}$

12. $\lim_{x \to \infty} \dfrac{x + 5}{x - 3}$

13. $\lim_{x \to \infty} \dfrac{2 - x^2}{x}$

14. $\lim_{x \to -\infty} \dfrac{x^2}{x - 2}$

In Exercises 15–18, find the limit.

15. $\lim_{x \to 2^+} \dfrac{1}{x - 2}$

16. $\lim_{x \to 2^-} \dfrac{1}{x^2 - 4}$

17. $\lim_{x \to 0} \dfrac{1/(2 + x) - 1/2}{x}$

18. $\lim_{x \to 0} \dfrac{(2 + x)^3 - 8}{x}$

In Exercises 19–20, find the vertical and horizontal asymptotes, if any.

19. $f(x) = \dfrac{x - 5}{x^2 + 6x + 5}$

20. $f(x) = \dfrac{x^2 + 1}{2x - 4}$

In Exercises 21–26, find the limit algebraically.

21. $\lim_{x \to 3} \dfrac{x^2 + 2x - 15}{3 - x}$

22. $\lim_{x \to 1} \dfrac{x^2 - 4x + 3}{x - 1}$

23. $\lim_{x \to 0} \dfrac{1/(-3 + x) + 1/3}{x}$

24. $\lim_{x \to 2} \dfrac{\tan(3x - 6)}{x - 2}$

25. $\lim_{x \to 2} \dfrac{x^2 - 5x + 6}{x^2 - 3x + 2}$

26. $\lim_{x \to 3} \dfrac{(x - 3)^2}{x - 3}$

In Exercises 27 and 28, state a formula for the continuous extension of the function. (See Exercise 90, Section 11.3.)

27. $f(x) = \dfrac{x^3 - 1}{x - 1}$

28. $f(x) = \dfrac{x^2 - 6x + 5}{x - 5}$

In Exercises 29 and 30, use the limit definition to find the derivative of the function at the specified point, if it exists. Support your answer numerically with an NDER calculator estimate.

29. $f(x) = 1 - x - 2x^2$ at $x = 2$

30. $f(x) = (x + 3)^2$ at $x = 2$

In Exercises 31 and 32, find **(a)** the average rate of change of the function over the interval $[3, 3.01]$ and **(b)** the instantaneous rate of change at $x = 3$.

31. $f(x) = x^2 + 2x - 3$

32. $f(x) = \dfrac{3}{x + 2}$

In Exercises 33 and 34, find **(a)** the slope and **(b)** an equation of the line tangent to the graph of the function at the indicated point.

33. $f(x) = x^3 - 2x + 1$ at $x = 1$

34. $f(x) = \sqrt{x - 4}$ at $x = 7$

In Exercises 35 and 36, find the derivative of f.

35. $f(x) = 5x^2 + 7x - 1$ **36.** $f(x) = 2 - 8x + 3x^2$

In Exercises 37 and 38, complete the following for the indicated interval $[a, b]$.

(a) Verify the given function is nonnegative.

(b) Use a calculator to find the LRAM, RRAM, and average approximations for the area under the graph of the function from $x = a$ to $x = b$ with 50 approximating rectangles.

37. $f(x) = (x - 5)^2$; $[0, 4]$

38. $f(x) = 2x^2 - 3x + 1$; $[1, 5]$

39. Gasoline Prices The annual average retail price for unleaded regular gasoline in the United States for the years 1995–2012 is given in Table 11.12.

Table 11.12 U.S. Average Unleaded Regular Gasoline Price

Year	Price ($/gal)
1995	1.15
1996	1.23
1997	1.23
1998	1.06
1999	1.17
2000	1.51
2001	1.46
2002	1.36
2003	1.59
2004	1.88
2005	2.30
2006	2.59
2007	2.80
2008	3.27
2009	2.35
2010	2.79
2011	3.53
2012	3.71

Source: World Almanac and Book of Facts 2013

(a) Draw a scatter plot of the data in Table 10.12. Use $x = 0$ for 1995, $x = 1$ for 1996, and so forth.

(b) Find the average rate of change from 2000 to 2001 and from 2010 to 2011.

(c) From what year to the next consecutive year does the average rate exhibit the greatest increase?

(d) From what year to the next consecutive year does the average rate exhibit the greatest decrease?

(e) Find a linear regression model for the data and overlay its graph on a scatter plot of the data.

(f) **Group Activity** Find a cubic regression model for the data and overlay its graph on a scatter plot of the data. Discuss pro and con arguments that this is a good model for the gasoline price data. Compare the cubic model with the linear model. Which one does your group think is best? Why?

(g) **Writing to Learn** Use the cubic regression model found in part (f) and NDER to find the instantaneous rate of change in 2000, 2004, 2007, and 2011. How could the cubic model give some misleading information?

(h) **Writing to Learn** Use the cubic regression model found in part (f) to predict the average price of a gallon of unleaded regular gasoline in 2018. Do you think this is a reasonable estimate? Give reasons.

40. An Interesting Connection

Let $A(x) = \text{NINT}(\cos t, t, 0, x)$

(a) Draw a scatter plot of the pairs $(x, A(x))$ for $x = 0$, 0.4, 0.8, 0.12, . . . , 6.0, 6.4.

(b) Find a function that seems to model the data in part (a) and overlay its graph on a scatter plot of the data.

(c) Assuming that the function found in part (b) agrees with $A(x)$ for all values of x, what is the derivative of $A(x)$?

(d) **Writing to Learn** Describe what seems to be true about the derivative of $\text{NINT}(f(t), t, 0, x)$.

CHAPTER 11 Project

Estimating Population Growth Rates

Austin, Texas, has been experiencing rapid growth in recent years. In the past 20 years the population of Austin has almost doubled. Below are actual population figures for Austin for the years specified.

Year	Actual Population
1950	132,459
1970	253,539
1980	345,890
1990	465,622
2000	656,562
2011	820,611

Source: The World Almanac and Book of Facts, 2013, page 615.

Explorations

1. Create a scatter plot of the above data. Let $t = 1950$ represent 1950, $t = 1970$ represent 1970, and so forth.
2. Find the average population growth rates for Austin from 1950 to 2011, from 1970 to 2011, from 1980 to 2011, and from 2000 to 2011.
3. Find a logistic regression model for the population data of Austin from 1950 to 2011 given in the table and graph it, together with the scatter plot of the data. Explain why the logistics growth model of the population growth of Austin is more realistic than an exponential growth model of the population.
4. Use the logistic regression model you found in question 3 and NDER to estimate the instantaneous population growth rate in 2011.
5. Use the logistic regression model you found in question 3 to predict the population of Austin in the years 2013, 2020, and 2040. Compare your predictions with estimated Austin populations below from a local newspaper. Which predictions seem more reasonable? Explain.

Year	Estimated Population
2013	850,000
2020	950,000
2040	1,250,000

Source: http://impactnews.com/austin-metro/central-austin/city-expects-continued-population-increase/

Appendices Overview

Appendix A extends Chapter P: Prerequisites and supports the methods used in Chapter 2 to solve equations and inequalities. (Please read Section P.1 before reading Section A.1, and review Sections A.2 and A.3 before working through Sections 2.4–2.8.) Appendix B extends the ideas of Chapter 9: Discrete Mathematics. Appendix C provides a summary of key information and useful formulas for the entire book.

A.1 Radicals and Rational Exponents

What you'll learn about

- Radicals
- Simplifying Radical Expressions
- Rationalizing the Denominator
- Rational Exponents

... and why

You need to review these basic algebraic skills if you don't remember them.

Radicals

If $b^2 = a$, then b is a **square root** of a. For example, both 2 and -2 are square roots of 4 because $2^2 = (-2)^2 = 4$. Similarly, b is a **cube root** of a if $b^3 = a$. For example, 2 is a cube root of 8 because $2^3 = 8$.

DEFINITION Real *n*th Root of a Real Number

Let n be an integer greater than 1 and a and b real numbers.

1. If $b^n = a$, then b is an ***n*th root** of a.
2. If a has an nth root, the **principal *n*th root** of a is the nth root having the same sign as a.

The principal nth root of a is denoted by the **radical expression** $\sqrt[n]{a}$. The positive integer n is the **index** of the radical and a is the **radicand**.

Every real number has exactly one real *nth root* whenever n is odd. For instance, 2 is the only real cube root of 8. When n is even, positive real numbers have two real *nth roots* and negative real numbers have no real nth roots. For example, the real fourth roots of 16 are ± 2, and -16 has no real fourth roots. The *principal* fourth root of 16 is 2.

When $n = 2$, special notation is used for roots. We omit the index and write $\sqrt{a}$ instead of $\sqrt[2]{a}$. If a is a positive real number and n a positive even integer, its two nth roots are denoted by $\sqrt[n]{a}$ and $-\sqrt[n]{a}$.

EXAMPLE 1 Finding Principal *n*th Roots

(a) $\sqrt{36} = 6$ because $6^2 = 36$.

(b) $\sqrt[3]{\frac{27}{8}} = \frac{3}{2}$ because $\left(\frac{3}{2}\right)^3 = \frac{27}{8}$.

(c) $\sqrt[3]{-\frac{27}{8}} = -\frac{3}{2}$ because $\left(-\frac{3}{2}\right)^3 = -\frac{27}{8}$.

(d) $\sqrt[4]{-625}$ is not a real number because the index 4 is even and the radicand -625 is negative. (There is *no* real number whose fourth power is negative.)

Now try Exercises 7 and 9.

Principal *n*th Roots and Calculators

Most calculators have a key for the principal nth root. Use this feature of your calculator to check the computations in Example 1.

Here are some properties of radicals together with examples that help illustrate their meaning.

Caution

Without the restriction that preceded the list, property 5 would need special attention. For example,

$$\sqrt{(-3)^2} \neq (\sqrt{-3})^2$$

because $\sqrt{-3}$ on the right is not a real number.

Properties of Radicals

Let u and v be real numbers, variables, or algebraic expressions, and m and n be positive integers greater than 1. We assume that all of the roots are real numbers and all of the denominators are nonzero.

Property	Example
1. $\sqrt[n]{uv} = \sqrt[n]{u} \cdot \sqrt[n]{v}$	$\sqrt{75} = \sqrt{25 \cdot 3} = \sqrt{25} \cdot \sqrt{3} = 5\sqrt{3}$
2. $\sqrt[n]{\dfrac{u}{v}} = \dfrac{\sqrt[n]{u}}{\sqrt[n]{v}}$	$\dfrac{\sqrt[4]{96}}{\sqrt[4]{6}} = \sqrt[4]{\dfrac{96}{6}} = \sqrt[4]{16} = 2$
3. $\sqrt[m]{\sqrt[n]{u}} = \sqrt[m \cdot n]{u}$	$\sqrt{\sqrt[3]{7}} = \sqrt[2 \cdot 3]{7} = \sqrt[6]{7}$
4. $(\sqrt[n]{u})^n = u$	$(\sqrt[4]{5})^4 = 5$
5. $\sqrt[n]{u^m} = (\sqrt[n]{u})^m$	$\sqrt[3]{27^2} = (\sqrt[3]{27})^2 = 3^2 = 9$
6. $\sqrt[n]{u^n} = \begin{cases} \lvert u \rvert & n \text{ even} \\ u & n \text{ odd} \end{cases}$	$\sqrt{(-6)^2} = \lvert -6 \rvert = 6$ $\sqrt[3]{(-6)^3} = -6$

Simplifying Radical Expressions

Many simplifying techniques for roots of real numbers have been rendered obsolete because of calculators. For example, when determining the decimal form of $1/\sqrt{2}$, it was once common first to change the fraction so that the radical was in the numerator:

$$\frac{1}{\sqrt{2}} = \frac{1}{\sqrt{2}} \cdot \frac{\sqrt{2}}{\sqrt{2}} = \frac{\sqrt{2}}{2}$$

Using paper and pencil, it was then easier to divide a decimal approximation for $\sqrt{2}$ by 2 than to divide that decimal into 1. Now either form is quickly computed with a calculator. However, these techniques are still valid for radicals involving algebraic expressions and for numerical computations when you need exact answers. Example 2 illustrates the technique of *removing factors from radicands.*

Properties of Exponents

Check the Properties of Exponents on page 7 of Section P.1 to see why

$$16 = 2^4 \text{ and } 9x^4 = (3x^2)^2.$$

EXAMPLE 2 Removing Factors from Radicands

(a) $\sqrt[4]{80} = \sqrt[4]{16 \cdot 5}$ — Finding greatest fourth-power factor

$= \sqrt[4]{2^4 \cdot 5}$ — $16 = 2^4$

$= \sqrt[4]{2^4} \cdot \sqrt[4]{5}$ — Property 1

$= 2\sqrt[4]{5}$ — Property 6

(b) $\sqrt{18x^5} = \sqrt{9x^4 \cdot 2x}$ — Finding greatest square factor

$= \sqrt{(3x^2)^2 \cdot 2x}$ — $9x^4 = (3x^2)^2$

$= 3x^2\sqrt{2x}$ — Properties 1 and 6

(c) $\sqrt[4]{x^4y^4} = \sqrt[4]{(xy)^4}$ — Finding greatest fourth-power factor

$= \lvert xy \rvert$ — Property 6

(d) $\sqrt[3]{-24y^6} = \sqrt[3]{(-2y^2)^3 \cdot 3}$ — Finding greatest cube factor

$= -2y^2\sqrt[3]{3}$ — Properties 1 and 6

Now try Exercises 29 and 33.

Rationalizing the Denominator

The process of rewriting fractions containing radicals so that the denominator is free of radicals is **rationalizing the denominator**. When the denominator has the form $\sqrt[n]{u^k}$, multiplying numerator and denominator by $\sqrt[n]{u^{n-k}}$ and using property 6 will eliminate the radical from the denominator because

$$\sqrt[n]{u^k} \cdot \sqrt[n]{u^{n-k}} = \sqrt[n]{u^{k+n-k}} = \sqrt[n]{u^n}.$$

Example 3 illustrates the process.

EXAMPLE 3 Rationalizing the Denominator

(a) $\sqrt{\dfrac{2}{3}} = \dfrac{\sqrt{2}}{\sqrt{3}} = \dfrac{\sqrt{2}}{\sqrt{3}} \cdot \dfrac{\sqrt{3}}{\sqrt{3}} = \dfrac{\sqrt{6}}{3}$

(b) $\dfrac{1}{\sqrt[4]{x}} = \dfrac{1}{\sqrt[4]{x}} \cdot \dfrac{\sqrt[4]{x^3}}{\sqrt[4]{x^3}} = \dfrac{\sqrt[4]{x^3}}{\sqrt[4]{x^4}} = \dfrac{\sqrt[4]{x^3}}{|x|}$

(c) $\sqrt[5]{\dfrac{x^2}{y^3}} = \dfrac{\sqrt[5]{x^2}}{\sqrt[5]{y^3}} = \dfrac{\sqrt[5]{x^2}}{\sqrt[5]{y^3}} \cdot \dfrac{\sqrt[5]{y^2}}{\sqrt[5]{y^2}} = \dfrac{\sqrt[5]{x^2y^2}}{\sqrt[5]{y^5}} = \dfrac{\sqrt[5]{x^2y^2}}{y}$

Now try Exercise 37.

Rational Exponents

We know how to handle exponential expressions with integer exponents (see Section P.1). For example, $x^3 \cdot x^4 = x^7$, $(x^3)^2 = x^6$, $x^5/x^2 = x^3$, $x^{-2} = 1/x^2$, and so forth. But exponents can also be rational numbers. How should we define $x^{1/2}$? If we assume that the same rules that apply for integer exponents also apply for rational exponents we get a clue. For example, we want

$$x^{1/2} \cdot x^{1/2} = x^1.$$

This equation suggests that $x^{1/2} = \sqrt{x}$. In general, we have the following definition.

DEFINITION Rational Exponents

Let u be a real number, variable, or algebraic expression, and n an integer greater than 1. Then

$$u^{1/n} = \sqrt[n]{u}.$$

If m is a positive integer, m/n is in reduced form, and all roots are real numbers, then

$$u^{m/n} = (u^{1/n})^m = \left(\sqrt[n]{u}\right)^m \quad \text{and} \quad u^{m/n} = (u^m)^{1/n} = \sqrt[n]{u^m}.$$

The numerator of a rational exponent is the *power* to which the base is raised, and the denominator is the *root* to be taken. The fraction m/n needs to be in reduced form because, for instance,

$$u^{2/3} = \left(\sqrt[3]{u}\right)^2$$

is defined for all real numbers u (every real number has a cube root), but

$$u^{4/6} = \left(\sqrt[6]{u}\right)^4$$

is defined only for $u \geq 0$ (only nonnegative real numbers have sixth roots).

Simplifying Radicals

If you also want the radical form in Example 4d to be simplified, then continue as follows:

$$\frac{1}{\sqrt{z^3}} = \frac{1}{\sqrt{z^3}} \cdot \frac{\sqrt{z}}{\sqrt{z}} = \frac{\sqrt{z}}{z^2}$$

EXAMPLE 4 Converting Radicals to Exponentials and Vice Versa

(a) $\sqrt{(x+y)^3} = (x+y)^{3/2}$

(b) $3x\sqrt[5]{x^2} = 3x \cdot x^{2/5} = 3x^{7/5}$

(c) $x^{2/3}y^{1/3} = (x^2y)^{1/3} = \sqrt[3]{x^2y}$

(d) $z^{-3/2} = \frac{1}{z^{3/2}} = \frac{1}{\sqrt{z^3}}$

Now try Exercises 43 and 47.

An expression involving powers is *simplified* if each factor appears only once and all exponents are positive. Example 5 illustrates.

EXAMPLE 5 Simplifying Exponential Expressions

(a) $(x^2y^9)^{1/3}(xy^2) = (x^{2/3}y^3)(xy^2) = x^{5/3}y^5$

(b) $\left(\frac{3x^{2/3}}{y^{1/2}}\right)\left(\frac{2x^{-1/2}}{y^{2/5}}\right) = \frac{6x^{1/6}}{y^{9/10}}$

Now try Exercise 61.

Example 6 suggests how to simplify a sum or difference of radicals.

EXAMPLE 6 Combining Radicals

(a) $2\sqrt{80} - \sqrt{125} = 2\sqrt{16 \cdot 5} - \sqrt{25 \cdot 5}$ Find greatest square factors.

$= 8\sqrt{5} - 5\sqrt{5}$ Remove factors from radicands.

$= 3\sqrt{5}$ Distributive property

(b) $\sqrt{4x^2y} - \sqrt{y^3} = \sqrt{(2x)^2y} - \sqrt{y^2y}$ Find greatest square factors.

$= 2|x|\sqrt{y} - |y|\sqrt{y}$ Remove factors from radicands.

$= (2|x| - |y|)\sqrt{y}$ Distributive property

Now try Exercise 71.

Here's a summary of the procedures we use to *simplify expressions* involving radicals.

Simplifying Radical Expressions

1. Remove factors from the radicand (see Example 2).
2. Eliminate radicals from denominators and denominators from radicands (see Example 3).
3. Combine sums and differences of radicals, if possible (see Example 6).

APPENDIX A.1 Exercises

In Exercises 1–6, find the indicated real roots.

1. Square roots of 81 **2.** Fourth roots of 81

3. Cube roots of 64 **4.** Fifth roots of 243

5. Square roots of 16/9 **6.** Cube roots of −27/8

In Exercises 7–12, evaluate the expression without using a calculator.

7. $\sqrt{144}$ **8.** $\sqrt{-16}$ **9.** $\sqrt[3]{-216}$

10. $\sqrt[3]{216}$ **11.** $\sqrt[3]{-\frac{64}{27}}$ **12.** $\sqrt{\frac{64}{25}}$

In Exercises 13–22, use a calculator to evaluate the expression.

13. $\sqrt[4]{256}$ **14.** $\sqrt[5]{3125}$

15. $\sqrt[3]{15.625}$ **16.** $\sqrt{12.25}$

17. $81^{3/2}$ **18.** $16^{5/4}$

19. $32^{-2/5}$ **20.** $27^{-4/3}$

21. $\left(-\frac{1}{8}\right)^{-1/3}$ **22.** $\left(-\frac{125}{64}\right)^{-1/3}$

In Exercises 23–26, use the information from the grapher screens below to evaluate the expression.

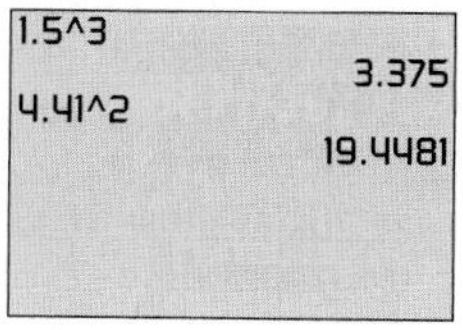

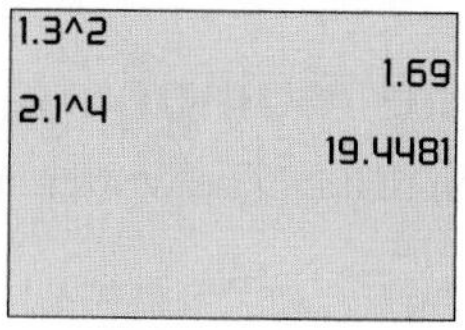

23. $\sqrt{1.69}$

24. $\sqrt{19.4481}$

25. $\sqrt[4]{19.4481}$

26. $\sqrt[3]{3.375}$

In Exercises 27–36, simplify by removing factors from the radicand.

27. $\sqrt{288}$

28. $\sqrt[3]{500}$

29. $\sqrt[3]{-250}$

30. $\sqrt[4]{192}$

31. $\sqrt{2x^3y^4}$

32. $\sqrt[3]{-27x^3y^6}$

33. $\sqrt[4]{3x^8y^6}$

34. $\sqrt[3]{8x^6y^4}$

35. $\sqrt[5]{96x^{10}}$

36. $\sqrt{108x^4y^9}$

In Exercises 37–42, rationalize the denominator.

37. $\frac{4}{\sqrt[3]{2}}$

38. $\frac{1}{\sqrt{5}}$

39. $\frac{1}{\sqrt[5]{x^2}}$

40. $\frac{2}{\sqrt[4]{y}}$

41. $\sqrt[3]{\frac{x^2}{y}}$

42. $\sqrt[5]{\frac{a^3}{b^2}}$

In Exercises 43–46, convert to exponential form.

43. $\sqrt[3]{(a+2b)^2}$

44. $\sqrt[5]{x^2y^3}$

45. $2x\sqrt[3]{x^2y}$

46. $xy\sqrt[4]{xy^3}$

In Exercises 47–50, convert to radical form.

47. $a^{3/4}b^{1/4}$

48. $x^{2/3}y^{1/3}$

49. $x^{-5/3}$

50. $(xy)^{-3/4}$

In Exercises 51–56, write using a single radical.

51. $\sqrt{\sqrt{2x}}$

52. $\sqrt{\sqrt[3]{3x^2}}$

53. $\sqrt[4]{\sqrt{xy}}$

54. $\sqrt[3]{\sqrt{ab}}$

55. $\frac{\sqrt[5]{a^2}}{\sqrt[3]{a}}$

56. $\sqrt{a}\sqrt[3]{a^2}$

In Exercises 57–64, simplify the exponential expression.

57. $\frac{a^{3/5}a^{1/3}}{a^{3/2}}$

58. $(x^2y^4)^{1/2}$

59. $(a^{5/3}b^{3/4})(3a^{1/3}b^{5/4})$

60. $\left(\frac{x^{1/2}}{y^{2/3}}\right)^6$

61. $\left(\frac{-8x^6}{y^{-3}}\right)^{2/3}$

62. $\frac{(p^2q^4)^{1/2}}{(27q^3p^6)^{1/3}}$

63. $\frac{(x^9y^6)^{-1/3}}{(x^6y^2)^{-1/2}}$

64. $\left(\frac{2x^{1/2}}{y^{2/3}}\right)\left(\frac{3x^{-2/3}}{y^{1/2}}\right)$

In Exercises 65–74, simplify the radical expression.

65. $\sqrt{9x^{-6}y^4}$

66. $\sqrt{16y^8z^{-2}}$

67. $\sqrt[4]{\frac{3x^8y^2}{8x^2}}$

68. $\sqrt[5]{\frac{4x^6y}{9x^3}}$

69. $\sqrt[3]{\frac{4x^2}{y^2}} \cdot \sqrt[3]{\frac{2x^2}{y}}$

70. $\sqrt[5]{9ab^6} \cdot \sqrt[5]{27a^2b^{-1}}$

71. $3\sqrt{48} - 2\sqrt{108}$

72. $2\sqrt{175} - 4\sqrt{28}$

73. $\sqrt{x^3} - \sqrt{4xy^2}$

74. $\sqrt{18x^2y} + \sqrt{2y^3}$

In Exercises 75–82, replace ○ with <, =, or > to make a true statement.

75. $\sqrt{2+6}$ ○ $\sqrt{2} + \sqrt{6}$

76. $\sqrt{4} + \sqrt{9}$ ○ $\sqrt{4+9}$

77. $(3^{-2})^{-1/2}$ ○ 3

78. $(2^{-3})^{1/3}$ ○ 2

79. $\sqrt[4]{(-2)^4}$ ○ -2

80. $\sqrt[3]{(-2)^3}$ ○ -2

81. $2^{2/3}$ ○ $3^{3/4}$

82. $4^{-2/3}$ ○ $3^{-3/4}$

83. The time t (in seconds) that it takes for a pendulum to complete one cycle is approximately $t = 1.1\sqrt{L}$, where L is the length (in feet) of the pendulum. How long is the period of a pendulum of length 10 ft?

84. The time t (in seconds) that it takes for a rock to fall a distance d (in meters) is approximately $t = 0.45\sqrt{d}$. How long does it take for the rock to fall a distance of 200 m?

85. Writing to Learn Explain why $\sqrt[n]{a}$ and a real nth root of a need not have the same value.

A.2 Polynomials and Factoring

What you'll learn about

- Adding, Subtracting, and Multiplying Polynomials
- Special Products
- Factoring Polynomials Using Special Products
- Factoring Trinomials
- Factoring by Grouping

... and why

You need to review these basic algebraic skills if you don't remember them.

Adding, Subtracting, and Multiplying Polynomials

A **polynomial in** x is any expression that can be written in the form

$$a_n x^n + a_{n-1}x^{n-1} + \cdots + a_1 x + a_0,$$

where n is a nonnegative integer, and $a_n \neq 0$. The numbers $a_{n-1}, \ldots, a_1, a_0$ are real numbers called **coefficients**. The **degree of the polynomial** is n and the **leading coefficient** is a_n. Polynomials with one, two, or three terms are **monomials**, **binomials**, or **trinomials**, respectively. A polynomial written with powers of x in *descending order* is in **standard form**.

To add or subtract polynomials, we add or subtract *like terms* using the distributive property. Terms of polynomials that have the same variable each raised to the same power are **like terms**.

EXAMPLE 1 Adding and Subtracting Polynomials

(a) $(2x^3 - 3x^2 + 4x - 1) + (x^3 + 2x^2 - 5x + 3)$

(b) $(4x^2 + 3x - 4) - (2x^3 + x^2 - x + 2)$

SOLUTION

(a) We group like terms and then combine them as follows:

$$(2x^3 + x^3) + (-3x^2 + 2x^2) + (4x + (-5x)) + (-1 + 3)$$
$$= 3x^3 - x^2 - x + 2$$

(b) We group like terms and then combine them as follows:

$$(0 - 2x^3) + (4x^2 - x^2) + (3x - (-x)) + (-4 - 2)$$
$$= -2x^3 + 3x^2 + 4x - 6$$

Now try Exercises 9 and 11.

To *expand the product* of two polynomials we use the distributive property. Here is what the procedure looks like when we multiply the binomials $3x + 2$ and $4x - 5$.

$$(3x + 2)(4x - 5)$$
$$= 3x(4x - 5) + 2(4x - 5) \qquad \text{Distributive property}$$
$$= (3x)(4x) - (3x)(5) + (2)(4x) - (2)(5) \qquad \text{Distributive property}$$
$$= \underbrace{12x^2}_{\text{Product of First terms}} - \underbrace{15x}_{\text{Product of Outer terms}} + \underbrace{8x}_{\text{Product of Inner terms}} - \underbrace{10}_{\text{Product of Last terms}}$$

In the above *FOIL method* for products of binomials, the outer (O) and inner (I) terms are like terms and can be added to give

$$(3x + 2)(4x - 5) = 12x^2 - 7x - 10.$$

Multiplying two polynomials requires multiplying each term of one polynomial by every term of the other polynomial. A convenient way to compute a product is to arrange the polynomials in standard form one on top of another so their terms align vertically, as illustrated in Example 2.

EXAMPLE 2 Multiplying Polynomials in Vertical Form

Write $(x^2 - 4x + 3)(x^2 + 4x + 5)$ in standard form.

SOLUTION

$$\begin{array}{lr}
\quad x^2 - 4x + 3 & \\
\quad x^2 + 4x + 5 & \\
\hline
x^4 - 4x^3 + 3x^2 & = x^2(x^2 - 4x + 3) \\
\quad 4x^3 - 16x^2 + 12x & = 4x(x^2 - 4x + 3) \\
\quad\quad 5x^2 - 20x + 15 & = 5(x^2 - 4x + 3) \\
\hline
x^4 + 0x^3 - 8x^2 - 8x + 15 & \text{Add.}
\end{array}$$

Thus,

$$(x^2 - 4x + 3)(x^2 + 4x + 5) = x^4 - 8x^2 - 8x + 15.$$

Now try Exercise 33.

Special Products

Certain products provide patterns that will be useful when we factor polynomials. Here is a list of some special products for binomials.

Special Binomial Products

Let u and v be real numbers, variables, or algebraic expressions.

1. **Product of a sum and a difference:** $(u + v)(u - v) = u^2 - v^2$
2. **Square of a sum:** $(u + v)^2 = u^2 + 2uv + v^2$
3. **Square of a difference:** $(u - v)^2 = u^2 - 2uv + v^2$
4. **Cube of a sum:** $(u + v)^3 = u^3 + 3u^2v + 3uv^2 + v^3$
5. **Cube of a difference:** $(u - v)^3 = u^3 - 3u^2v + 3uv^2 - v^3$

EXAMPLE 3 Using Special Products

Expand the products.

(a) $(3x + 8)(3x - 8) = (3x)^2 - 8^2$
$\quad = 9x^2 - 64$

(b) $(5y - 4)^2 = (5y)^2 - 2(5y)(4) + 4^2$
$\quad = 25y^2 - 40y + 16$

(c) $(2x - 3y)^3 = (2x)^3 - 3(2x)^2(3y)$
$\quad + 3(2x)(3y)^2 - (3y)^3$
$\quad = 8x^3 - 36x^2y + 54xy^2 - 27y^3$

Now try Exercises 23, 25, and 27.

Factoring Polynomials Using Special Products

When we write a polynomial as a product of two or more *polynomial factors* we are **factoring a polynomial**. Unless specified otherwise, we factor polynomials into factors of lesser degree and with integer coefficients in this appendix. A polynomial that cannot be factored using integer coefficients is a **prime polynomial**.

A polynomial is *completely factored* if it is written as a product of its prime factors. For example,

$$2x^2 + 7x - 4 = (2x - 1)(x + 4)$$

and

$$x^3 + x^2 + x + 1 = (x + 1)(x^2 + 1)$$

are completely factored (it can be proved that $x^2 + 1$ is prime). However,

$$x^3 - 9x = x(x^2 - 9)$$

is not completely factored because $(x^2 - 9)$ is *not* prime. In fact, $x^2 - 9 = (x - 3)(x + 3)$ and

$$x^3 - 9x = x(x - 3)(x + 3)$$

is completely factored.

The first step in factoring a polynomial is to remove common factors from its terms using the distributive property as illustrated by Example 4.

EXAMPLE 4 Removing Common Factors

(a) $2x^3 + 2x^2 - 6x = 2x(x^2 + x - 3)$ — 2x is the common factor.

(b) $u^3v + uv^3 = uv(u^2 + v^2)$ — uv is the common factor.

Now try Exercise 43.

Recognizing the expanded form of the five special binomial products will help us factor them. The special form that is easiest to identify is the difference of two squares. The two binomial factors have opposite signs:

Two squares — Square roots

$$u^2 - v^2 = (u + v)(u - v).$$

Difference — Opposite signs

EXAMPLE 5 Factoring the Difference of Two Squares

(a) $25x^2 - 36 = (5x)^2 - 6^2$ — Difference of two squares

$= (5x + 6)(5x - 6)$ — Factors are prime.

(b) $4x^2 - (y + 3)^2 = (2x)^2 - (y + 3)^2$ — Difference of two squares

$= [2x + (y + 3)][2x - (y + 3)]$ — Factors are prime.

$= (2x + y + 3)(2x - y - 3)$ — Simplify.

Now try Exercise 45.

A perfect square trinomial is the square of a binomial and has one of the two forms shown here. The first and last terms are squares of u and v, and the middle term is twice the product of u and v. The operation signs before the middle term and in the binomial factor are the same.

Perfect square (sum)

$$u^2 + 2uv + v^2 = (u + v)^2$$

Same signs

Perfect square (difference)

$$u^2 - 2uv + v^2 = (u - v)^2$$

Same signs

EXAMPLE 6 Factoring Perfect Square Trinomials

(a) $9x^2 + 6x + 1 = (3x)^2 + 2(3x)(1) + 1^2$ $\qquad u = 3x, v = 1$

$= (3x + 1)^2$

(b) $4x^2 - 12xy + 9y^2 = (2x)^2 - 2(2x)(3y) + (3y)^2$ $\qquad u = 2x, v = 3y$

$= (2x - 3y)^2$

Now try Exercise 49.

In the sum and difference of two cubes, notice the pattern of the signs.

Same signs / Opposite signs

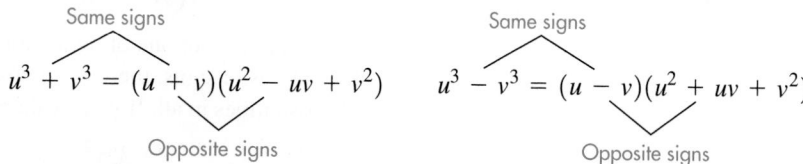

$$u^3 + v^3 = (u + v)(u^2 - uv + v^2) \qquad u^3 - v^3 = (u - v)(u^2 + uv + v^2)$$

Same signs / Opposite signs

EXAMPLE 7 Factoring the Sum and Difference of Two Cubes

(a) $x^3 - 64 = x^3 - 4^3$ $\qquad$ Difference of two cubes

$= (x - 4)(x^2 + 4x + 16)$ $\qquad$ Factors are prime.

(b) $8x^3 + 27 = (2x)^3 + 3^3$ $\qquad$ Sum of two cubes

$= (2x + 3)(4x^2 - 6x + 9)$ $\qquad$ Factors are prime.

Now try Exercise 55.

Factoring Trinomials

Factoring the trinomial $ax^2 + bx + c$ into a product of binomials with integer coefficients requires factoring the integers a and c.

Factors of a

$$ax^2 + bx + c = (\square x + \square)(\square x + \square)$$

Factors of c

Because the number of integer factors of a and c is finite, we can list all possible binomial factors. Then we begin checking each possibility until we find a pair that works. (If no pair works, then the trinomial is prime.) Example 8 illustrates.

EXAMPLE 8 Factoring a Trinomial with Leading Coefficient = 1

Factor $x^2 + 5x - 14$.

SOLUTION The only factor pair of the leading coefficient is 1 and 1. The factor pairs of 14 are 1 and 14, and 2 and 7. Here are the four possible factorizations of the trinomial:

$$(x + 1)(x - 14) \qquad (x - 1)(x + 14)$$
$$(x + 2)(x - 7) \qquad (x - 2)(x + 7)$$

If you check the middle term from each factorization you will find that

$x^2 + 5x - 14 = (x - 2)(x + 7)$.

Now try Exercise 59.

With practice you will find that it usually is not necessary to list all possible binomial factors. Often you can test the possibilities mentally.

EXAMPLE 9 Factoring a Trinomial with Leading Coefficient $\neq 1$

Factor $35x^2 - x - 12$.

SOLUTION The factor pairs of the leading coefficient are 1 and 35, and 5 and 7. The factor pairs of 12 are 1 and 12, 2 and 6, and 3 and 4. The possible factorizations must be of the form

$$(x - *)(35x + ?), \quad (x + *)(35x - ?),$$
$$(5x - *)(7x + ?), \quad (5x + *)(7x - ?),$$

where * and ? are one of the factor pairs of 12. Because the two binomial factors have opposite signs, there are 6 possibilities for each of the four forms—a total of 24 possibilities in all. If you try them, mentally and systematically, you should find that

$$35x^2 - x - 12 = (5x - 3)(7x + 4).$$

Now try Exercise 63.

We can extend the technique of Examples 8 and 9 to trinomials in two variables as illustrated in Example 10.

EXAMPLE 10 Factoring Trinomials in x and y

Factor $3x^2 - 7xy + 2y^2$.

SOLUTION The only way to get $-7xy$ as the middle term is with $3x^2 - 7xy + 2y^2 = (3x - *y)(x - ?y)$.

The signs in the binomials must be negative because the coefficient of y^2 is positive *and* the coefficient of the middle term is negative. Checking the two possibilities, $(3x - y)(x - 2y)$ and $(3x - 2y)(x - y)$, proves that

$$3x^2 - 7xy + 2y^2 = (3x - y)(x - 2y).$$

Now try Exercise 67.

Factoring by Grouping

Notice that $(a + b)(c + d) = ac + ad + bc + bd$. If a polynomial with four terms is the product of two binomials, we can group terms to factor. There are only three ways to group the terms and two of them work. So, if two of the possibilities fail, then it is not factorable.

EXAMPLE 11 Factoring by Grouping

(a) $3x^3 + x^2 - 6x - 2$

$= (3x^3 + x^2) - (6x + 2)$ Group terms.

$= x^2(3x + 1) - 2(3x + 1)$ Factor each group.

$= (3x + 1)(x^2 - 2)$ Distributive property

(b) $2ac - 2ad + bc - bd$

$= (2ac - 2ad) + (bc - bd)$ Group terms.

$= 2a(c - d) + b(c - d)$ Factor each group.

$= (c - d)(2a + b)$ Distributive property

Now try Exercise 69.

Here is a checklist for factoring polynomials.

Factoring Polynomials

1. Look for common factors.
2. Look for special polynomial forms.
3. Use factor pairs.
4. If there are four terms, try grouping.

APPENDIX A.2 Exercises

In Exercises 1–4, write the polynomial in standard form and state its degree.

1. $2x - 1 + 3x^2$ **2.** $x^2 - 2x - 2x^3 + 1$

3. $1 - x^7$ **4.** $x^2 - x^4 + x - 3$

In Exercises 5–8, state whether the expression is a polynomial.

5. $x^3 - 2x^2 + x^{-1}$ **6.** $\dfrac{2x - 4}{x}$

7. $(x^2 + x + 1)^2$ **8.** $1 - 3x + x^4$

In Exercises 9–18, simplify the expression. Write your answer in standard form.

9. $(x^2 - 3x + 7) + (3x^2 + 5x - 3)$

10. $(-3x^2 - 5) - (x^2 + 7x + 12)$

11. $(4x^3 - x^2 + 3x) - (x^3 + 12x - 3)$

12. $-(y^2 + 2y - 3) + (5y^2 + 3y + 4)$

13. $2x(x^2 - x + 3)$ **14.** $y^2(2y^2 + 3y - 4)$

15. $-3u(4u - 1)$ **16.** $-4v(2 - 3v^3)$

17. $(2 - x - 3x^2)(5x)$ **18.** $(1 - x^2 + x^4)(2x)$

In Exercises 19–40, expand the product. Use vertical alignment in Exercises 33 and 34.

19. $(x - 2)(x + 5)$

20. $(2x + 3)(4x + 1)$

21. $(3x - 5)(x + 2)$

22. $(2x - 3)(2x + 3)$

23. $(3x - y)(3x + y)$

24. $(3 - 5x)^2$

25. $(3x + 4y)^2$ **26.** $(x - 1)^3$

27. $(2u - v)^3$ **28.** $(u + 3v)^3$

29. $(2x^3 - 3y)(2x^3 + 3y)$ **30.** $(5x^3 - 1)^2$

31. $(x^2 - 2x + 3)(x + 4)$ **32.** $(x^2 + 3x - 2)(x - 3)$

33. $(x^2 + x - 3)(x^2 + x + 1)$

34. $(2x^2 - 3x + 1)(x^2 - x + 2)$

35. $(x - \sqrt{2})(x + \sqrt{2})$

36. $(x^{1/2} - y^{1/2})(x^{1/2} + y^{1/2})$

37. $(\sqrt{u} + \sqrt{v})(\sqrt{u} - \sqrt{v})$

38. $(x^2 - \sqrt{3})(x^2 + \sqrt{3})$

39. $(x - 2)(x^2 + 2x + 4)$

40. $(x + 1)(x^2 - x + 1)$

In Exercises 41–44, factor out the common factor.

41. $5x - 15$ **42.** $5x^3 - 20x$

43. $yz^3 - 3yz^2 + 2yz$ **44.** $2x(x + 3) - 5(x + 3)$

In Exercises 45–48, factor the difference of two squares.

45. $z^2 - 49$

46. $9y^2 - 16$

47. $64 - 25y^2$

48. $16 - (x + 2)^2$

In Exercises 49–52, factor the perfect square trinomial.

49. $y^2 + 8y + 16$ **50.** $36y^2 + 12y + 1$

51. $4z^2 - 4z + 1$ **52.** $9z^2 - 24z + 16$

In Exercises 53–58, factor the sum or difference of two cubes.

53. $y^3 - 8$

54. $z^3 + 64$

55. $27y^3 - 8$

56. $64z^3 + 27$

57. $1 - x^3$

58. $27 - y^3$

In Exercises 59–68, factor the trinomial.

59. $x^2 + 9x + 14$

60. $y^2 - 11y + 30$

61. $z^2 - 5z - 24$

62. $6t^2 + 5t + 1$

63. $14u^2 - 33u - 5$ **64.** $10v^2 + 23v + 12$

65. $12x^2 + 11x - 15$ **66.** $2x^2 - 3xy + y^2$

67. $6x^2 + 11xy - 10y^2$ **68.** $15x^2 + 29xy - 14y^2$

In Exercises 69–74, factor by grouping.

69. $x^3 - 4x^2 + 5x - 20$ **70.** $2x^3 - 3x^2 + 2x - 3$

71. $x^6 - 3x^4 + x^2 - 3$ **72.** $x^6 + 2x^4 + x^2 + 2$

73. $2ac + 6ad - bc - 3bd$

74. $3uw + 12uz - 2vw - 8vz$

In Exercises 75–90, factor completely.

75. $x^3 + x$

76. $4y^3 - 20y^2 + 25y$

77. $18y^3 + 48y^2 + 32y$

78. $2x^3 - 16x^2 + 14x$

79. $16y - y^3$

80. $3x^4 + 24x$

81. $5y + 3y^2 - 2y^3$

82. $z - 8z^4$

83. $2(5x + 1)^2 - 18$

84. $5(2x - 3)^2 - 20$

85. $12x^2 + 22x - 20$

86. $3x^2 + 13xy - 10y^2$

87. $2ac - 2bd + 4ad - bc$

88. $6ac - 2bd + 4bc - 3ad$

89. $x^3 - 3x^2 - 4x + 12$

90. $x^4 - 4x^3 - x^2 + 4x$

91. Writing to Learn Prove that the grouping

$$(2ac + bc) - (2ad + bd)$$

leads to the same factorization as in Example 11b. Explain why the third possibility,

$$(2ac - bd) + (-2ad + bc)$$

does not lead to a factorization.

A.3 Fractional Expressions

What you'll learn about

- Algebraic Expressions and Their Domains
- Reducing Rational Expressions
- Operations with Rational Expressions
- Compound Rational Expressions

... and why

You need to review these basic algebraic skills if you don't remember them.

Algebraic Expressions and Their Domains

A quotient of two algebraic expressions, besides being another algebraic expression, is a **fractional expression**, or simply a fraction. If the quotient can be written as the ratio of two polynomials, the fractional expression is a **rational expression**. Following are examples of each:

$$\frac{x^2 - 5x + 2}{\sqrt{x^2 + 1}} \qquad \frac{2x^3 - x^2 + 1}{5x^2 - x - 3}$$

The one on the left is a fractional expression but not a rational expression. The other is both a fractional expression and a rational expression.

Unlike polynomials, which are defined for all real numbers, some algebraic expressions are not defined for some real numbers. The set of real numbers for which an expression is defined is the **domain of the expression**.

EXAMPLE 1 Finding Domains of Algebraic Expressions

(a) $3x^2 - x + 5$ **(b)** $\sqrt{x - 1}$ **(c)** $\dfrac{x}{x - 2}$

SOLUTION

(a) The domain of $3x^2 - x + 5$, like that of any polynomial, is the set of all real numbers.

(b) Because only nonnegative numbers have square roots, $x - 1 \geq 0$, or $x \geq 1$. In interval notation, the domain is $[1, \infty)$.

(c) Because division by zero is undefined, $x - 2 \neq 0$, or $x \neq 2$. The domain is the set of all real numbers except 2.

Now try Exercises 11 and 13.

Reducing Rational Expressions

Let u, v, and z be real numbers, variables, or algebraic expressions. We can write rational expressions in simpler form using

$$\frac{uz}{vz} = \frac{u}{v}$$

provided $z \neq 0$. This requires that we first factor the numerator and denominator into prime factors. When all factors common to numerator and denominator have been removed, the rational expression (or rational number) is in **reduced form**.

EXAMPLE 2 Reducing Rational Expressions

Write $(x^2 - 3x)/(x^2 - 9)$ in reduced form.

SOLUTION

$$\frac{x^2 - 3x}{x^2 - 9} = \frac{x(x - 3)}{(x + 3)(x - 3)} \qquad \text{Factor completely.}$$

$$= \frac{x}{x + 3}, \quad x \neq 3 \qquad \text{Remove common factors.}$$

We include $x \neq 3$ as part of the reduced form because 3 is not in the domain of the original rational expression and thus should not be in the domain of the final rational expression.

Now try Exercise 35.

Two rational expressions are **equivalent expressions** if they have the same domain and have the same value for all numbers in the domain. The reduced form of a rational expression must have the same domain as the original rational expression. This is why we attached the restriction $x \neq 3$ to the reduced form in Example 2.

Operations with Rational Expressions

Two fractions are equal, $u/v = z/w$, if and only if $uw = vz$. Here is how we operate with fractions.

Operations with Fractions

Let u, v, w, and z be real numbers, variables, or algebraic expressions. All of the denominators are assumed to be different from zero.

Operation	Example
1. $\frac{u}{v} + \frac{w}{v} = \frac{u + w}{v}$	$\frac{2}{3} + \frac{5}{3} = \frac{2 + 5}{3} = \frac{7}{3}$
2. $\frac{u}{v} + \frac{w}{z} = \frac{uz + vw}{vz}$	$\frac{2}{3} + \frac{4}{5} = \frac{2 \cdot 5 + 3 \cdot 4}{3 \cdot 5} = \frac{22}{15}$
3. $\frac{u}{v} \cdot \frac{w}{z} = \frac{uw}{vz}$	$\frac{2}{3} \cdot \frac{4}{5} = \frac{2 \cdot 4}{3 \cdot 5} = \frac{8}{15}$
4. $\frac{u}{v} \div \frac{w}{z} = \frac{u}{v} \cdot \frac{z}{w} = \frac{uz}{vw}$	$\frac{2}{3} \div \frac{4}{5} = \frac{2}{3} \cdot \frac{5}{4} = \frac{10}{12} = \frac{5}{6}$

5. For subtraction, replace "+" by "−" in 1 and 2.

Invert and Multiply

The division step shown in 4 is often referred to as invert the divisor (the fraction following the division symbol) and multiply the result times the numerator (the first fraction).

EXAMPLE 3 Multiplying and Dividing Rational Expressions

(a) $\frac{2x^2 + 11x - 21}{x^3 + 2x^2 + 4x} \cdot \frac{x^3 - 8}{x^2 + 5x - 14}$

$= \frac{(2x - 3)(x + 7)}{x(x^2 + 2x + 4)} \cdot \frac{(x - 2)(x^2 + 2x + 4)}{(x - 2)(x + 7)}$ Factor completely.

$= \frac{2x - 3}{x}, \quad x \neq 2, \quad x \neq -7$ Remove common factors.

(b) $\frac{x^3 + 1}{x^2 - x - 2} \div \frac{x^2 - x + 1}{x^2 - 4x + 4}$

$= \frac{(x^3 + 1)(x^2 - 4x + 4)}{(x^2 - x - 2)(x^2 - x + 1)}$ Invert and multiply.

$= \frac{(x + 1)(x^2 - x + 1)(x - 2)^{2}}{(x + 1)(x - 2)(x^2 - x + 1)}$ Factor completely.

$= x - 2, \quad x \neq -1, \quad x \neq 2$ Remove common factors.

Now try Exercises 49 and 55.

Note on Example

The numerator, $x^2 + 4x - 6$, of the final expression in Example 4 is a prime polynomial. Thus, there are no common factors.

EXAMPLE 4 Adding Rational Expressions

$\frac{x}{3x - 2} + \frac{3}{x - 5} = \frac{x(x - 5) + 3(3x - 2)}{(3x - 2)(x - 5)}$ Definition of addition

$= \frac{x^2 - 5x + 9x - 6}{(3x - 2)(x - 5)}$ Distributive property

$= \frac{x^2 + 4x - 6}{(3x - 2)(x - 5)}$ Combine like terms.

Now try Exercise 59.

If the denominators of fractions have common factors, then it is often more efficient to find the LCD before adding or subtracting the fractions. The **LCD (least common denominator)** is the product of all the prime factors in the denominators, where each factor is raised to the greatest power found in any one denominator for that factor.

EXAMPLE 5 Using the LCD

Write the following expression as a fraction in reduced form.

$$\frac{2}{x^2 - 2x} + \frac{1}{x} - \frac{3}{x^2 - 4}$$

SOLUTION The factored denominators are $x(x - 2)$, x, and $(x - 2)(x + 2)$, respectively. The LCD is $x(x - 2)(x + 2)$.

$$\frac{2}{x^2 - 2x} + \frac{1}{x} - \frac{3}{x^2 - 4}$$

$$= \frac{2}{x(x - 2)} + \frac{1}{x} - \frac{3}{(x - 2)(x + 2)} \quad \text{Factor.}$$

$$= \frac{2(x + 2)}{x(x - 2)(x + 2)} + \frac{(x - 2)(x + 2)}{x(x - 2)(x + 2)} - \frac{3x}{x(x - 2)(x + 2)} \quad \text{Equivalent fractions}$$

$$= \frac{2(x + 2) + (x - 2)(x + 2) - 3x}{x(x - 2)(x + 2)} \quad \text{Combine numerators.}$$

$$= \frac{2x + 4 + x^2 - 4 - 3x}{x(x - 2)(x + 2)} \quad \text{Expand terms.}$$

$$= \frac{x^2 - x}{x(x - 2)(x + 2)} \quad \text{Simplify.}$$

$$= \frac{x(x - 1)}{x(x - 2)(x + 2)} \quad \text{Factor.}$$

$$= \frac{x - 1}{(x - 2)(x + 2)}, \quad x \neq 0 \quad \text{Reduce.}$$

Now try Exercise 61.

Compound Rational Expressions

Sometimes a complicated algebraic expression needs to be changed to a more familiar form before we can work on it. A **compound fraction** (sometimes called a **complex fraction**), in which the numerators and denominators may themselves contain fractions, is such an example. One way to simplify a compound fraction is to write both the numerator and denominator as single fractions and then invert and multiply. If the fraction then takes the form of a rational expression, we write the expression in reduced or simplest form.

EXAMPLE 6 Simplifying a Compound Fraction

$$\frac{3 - \frac{7}{x + 2}}{1 - \frac{1}{x - 3}} = \frac{\frac{3(x + 2) - 7}{x + 2}}{\frac{(x - 3) - 1}{x - 3}} \quad \text{Combine fractions.}$$

$$= \frac{\frac{3x - 1}{x + 2}}{\frac{x - 4}{x - 3}} \quad \text{Simplify.}$$

$$= \frac{(3x - 1)(x - 3)}{(x + 2)(x - 4)}, \quad x \neq 3 \quad \text{Invert and multiply.}$$

Now try Exercise 63.

A second way to simplify a compound fraction is to multiply the numerator and denominator by the LCD of all fractions in the numerator and denominator as illustrated in Example 7.

EXAMPLE 7 Simplifying Another Compound Fraction

Use the LCD to simplify the compound fraction

$$\frac{\frac{1}{a^2} - \frac{1}{b^2}}{\frac{1}{a} - \frac{1}{b}}.$$

SOLUTION The LCD of the four fractions in the numerator and denominator is a^2b^2.

$$\frac{\frac{1}{a^2} - \frac{1}{b^2}}{\frac{1}{a} - \frac{1}{b}} = \frac{\left(\frac{1}{a^2} - \frac{1}{b^2}\right)a^2b^2}{\left(\frac{1}{a} - \frac{1}{b}\right)a^2b^2} \qquad \text{Multiply numerator and denominator by LCD.}$$

$$= \frac{b^2 - a^2}{ab^2 - a^2b} \qquad \text{Simplify}$$

$$= \frac{(b + a)(b - a)}{ab(b - a)} \qquad \text{Factor.}$$

$$= \frac{b + a}{ab}, \quad a \neq b \qquad \text{Reduce.}$$

Now try Exercise 69.

APPENDIX A.3 Exercises

In Exercises 1–8, rewrite as a single fraction.

1. $\frac{5}{9} + \frac{10}{9}$

2. $\frac{17}{32} - \frac{9}{32}$

3. $\frac{20}{21} \cdot \frac{9}{22}$

4. $\frac{33}{25} \cdot \frac{20}{77}$

5. $\frac{2}{3} \div \frac{4}{5}$

6. $\frac{9}{4} \div \frac{15}{10}$

7. $\frac{1}{14} + \frac{4}{15} - \frac{5}{21}$

8. $\frac{1}{6} + \frac{6}{35} - \frac{4}{15}$

In Exercises 9–18, find the domain of the algebraic expression.

9. $5x^2 - 3x - 7$

10. $2x - 5$

11. $\sqrt{x - 4}$

12. $\frac{2}{\sqrt{x + 3}}$

13. $\frac{2x + 1}{x^2 + 3x}$

14. $\frac{x^2 - 2}{x^2 - 4}$

15. $\frac{x}{x - 1}, \quad x \neq 2$

16. $\frac{3x - 1}{x - 2}, \quad x \neq 0$

17. $x^2 + x^{-1}$

18. $x(x + 1)^{-2}$

In Exercises 19–26, find the missing numerator or denominator so that the two rational expressions are equal.

19. $\frac{2}{3x} = \frac{?}{12x^3}$

20. $\frac{5}{2y} = \frac{15y}{?}$

21. $\frac{x - 4}{x} = \frac{x^2 - 4x}{?}$

22. $\frac{x}{x + 2} = \frac{?}{x^2 - 4}$

23. $\frac{x + 3}{x - 2} = \frac{?}{x^2 + 2x - 8}$

24. $\frac{x - 4}{x + 5} = \frac{x^2 - x - 12}{?}$

25. $\frac{x^2 - 3x}{?} = \frac{x - 3}{x^2 + 2x}$

26. $\frac{?}{x^2 - 9} = \frac{x^2 + x - 6}{x - 3}$

Writing to Learn In Exercises 27–32, consider the original fraction and its reduced form from the specified example. Explain why the given restriction is needed on the reduced form.

27. Example 3a, $x \neq 2, x \neq -7$

28. Example 3b, $x \neq -1, x \neq 2$

29. Example 4, none

30. Example 5, $x \neq 0$

31. Example 6, $x \neq 3$

32. Example 7, $a \neq b$

In Exercises 33–44, write the expression in reduced form.

33. $\dfrac{18x^3}{15x}$ **34.** $\dfrac{75y^2}{9y^4}$

35. $\dfrac{x^3}{x^2 - 2x}$ **36.** $\dfrac{2y^2 + 6y}{4y + 12}$

37. $\dfrac{z^2 - 3z}{9 - z^2}$ **38.** $\dfrac{x^2 + 6x + 9}{x^2 - x - 12}$

39. $\dfrac{y^2 - y - 30}{y^2 - 3y - 18}$

40. $\dfrac{y^3 + 4y^2 - 21y}{y^2 - 49}$

41. $\dfrac{8z^3 - 1}{2z^2 + 5z - 3}$ **42.** $\dfrac{2z^3 + 6z^2 + 18z}{z^3 - 27}$

43. $\dfrac{x^3 + 2x^2 - 3x - 6}{x^3 + 2x^2}$ **44.** $\dfrac{y^2 + 3y}{y^3 + 3y^2 - 5y - 15}$

In Exercises 45–62, simplify.

45. $\dfrac{3}{x - 1} \cdot \dfrac{x^2 - 1}{9}$

46. $\dfrac{x + 3}{7} \cdot \dfrac{14}{2x + 6}$

47. $\dfrac{x + 3}{x - 1} \cdot \dfrac{1 - x}{x^2 - 9}$ **48.** $\dfrac{18x^2 - 3x}{3xy} \cdot \dfrac{12y^2}{6x - 1}$

49. $\dfrac{x^3 - 1}{2x^2} \cdot \dfrac{4x}{x^2 + x + 1}$ **50.** $\dfrac{y^3 + 2y^2 + 4y}{y^3 + 2y^2} \cdot \dfrac{y^2 - 4}{y^3 - 8}$

51. $\dfrac{2y^2 + 9y - 5}{y^2 - 25} \cdot \dfrac{y - 5}{2y^2 - y}$ **52.** $\dfrac{y^2 + 8y + 16}{3y^2 - y - 2} \cdot \dfrac{3y^2 + 2y}{y + 4}$

53. $\dfrac{1}{2x} \div \dfrac{1}{4}$ **54.** $\dfrac{4x}{y} \div \dfrac{8y}{x}$

55. $\dfrac{x^2 - 3x}{14y} \div \dfrac{2xy}{3y^2}$ **56.** $\dfrac{7x - 7y}{4y} \div \dfrac{14x - 14y}{3y}$

57. $\dfrac{\dfrac{2x^2y}{(x - 3)^2}}{\dfrac{8xy}{x - 3}}$ **58.** $\dfrac{\dfrac{x^2 - y^2}{2xy}}{\dfrac{y^2 - x^2}{4x^2y}}$

59. $\dfrac{2x + 1}{x + 5} - \dfrac{3}{x + 5}$

60. $\dfrac{3}{x - 2} + \dfrac{x + 1}{x - 2}$

61. $\dfrac{3}{x^2 + 3x} - \dfrac{1}{x} - \dfrac{6}{x^2 - 9}$

62. $\dfrac{5}{x^2 + x - 6} - \dfrac{2}{x - 2} + \dfrac{4}{x^2 - 4}$

In Exercises 63–70, simplify the compound fraction.

63. $\dfrac{\dfrac{x}{y^2} - \dfrac{y}{x^2}}{\dfrac{1}{y^2} - \dfrac{1}{x^2}}$ **64.** $\dfrac{\dfrac{1}{x} + \dfrac{1}{y}}{\dfrac{1}{x^2} - \dfrac{1}{y^2}}$

65. $\dfrac{2x + \dfrac{13x - 3}{x - 4}}{2x + \dfrac{x + 3}{x - 4}}$ **66.** $\dfrac{2 - \dfrac{13}{x + 5}}{2 + \dfrac{3}{x - 3}}$

67. $\dfrac{\dfrac{1}{(x + h)^2} - \dfrac{1}{x^2}}{h}$ **68.** $\dfrac{\dfrac{x + h}{x + h + 2} - \dfrac{x}{x + 2}}{h}$

69. $\dfrac{\dfrac{b}{a} - \dfrac{a}{b}}{\dfrac{1}{a} - \dfrac{1}{b}}$ **70.** $\dfrac{\dfrac{1}{a} + \dfrac{1}{b}}{\dfrac{b}{a} - \dfrac{a}{b}}$

In Exercises 71–74, write with positive exponents and simplify.

71. $\left(\dfrac{1}{x} + \dfrac{1}{y}\right)(x + y)^{-1}$ **72.** $\dfrac{(x + y)^{-1}}{(x - y)^{-1}}$

73. $x^{-1} + y^{-1}$ **74.** $(x^{-1} + y^{-1})^{-1}$

B.1 Logic: An Introduction

What you'll learn about

- Statements
- Compound Statements

... and why

These topics are important in the study of logic.

Statements

Logic is a tool used in mathematical thinking and problem solving. In logic, a **statement** *is a sentence that is either true or false, but not both.*

The following sentences are not statements because their truth values cannot be determined without more information.

1. She has blue eyes.
2. $x + 7 = 18$
3. $2y + 7 > 1$

The sentences above become statements if, for (1), "she" is identified, and for (2) and (3), values are assigned to x and y, respectively. However, a sentence involving *he* or *she* or x or y may already be a statement. For example, "If he is over 210 cm tall, then he is over 2 m tall," and "$2(x + y) = 2x + 2y$" are both statements because they are true no matter who he is or what the numerical values of x and y are.

From a given statement, it is possible to create a new statement by forming a **negation**. The negation of a statement is a statement with the opposite truth value of the given statement. If a statement is true, its negation is false, and if a statement is false, its negation is true. Consider the statement "It is snowing." The negation of this statement may be stated simply as "It is not snowing."

EXAMPLE 1 Negation of Statements

Negate each of the following statements:

(a) $2 + 3 = 5$

(b) A hexagon has six sides.

(c) Today is not Monday.

SOLUTION

(a) $2 + 3 \neq 5$

(b) A hexagon does not have six sides.

(c) Today is Monday.

Now try Exercise 5, parts (a), (b), and (c).

The statements "The shirt is blue" and "The shirt is green" are not negations of each other. A statement and its negation must have opposite truth values. If the shirt is actually red, then both of the above statements are false and, hence, cannot be negations of each other. However, the statements "The shirt is blue" and "The shirt is not blue" are negations of each other because they have opposite truth values no matter what color the shirt really is.

Some statements involve **quantifiers** and are more complicated to negate. Quantifiers include words such as *all, some, every*, and *there exists.*

The quantifiers *all, every*, and *no* refer to each and every element in a set and are **universal quantifiers**. The quantifiers *some* and *there exists at least one* refer to one or more, or possibly all, of the elements in a set. *Some* and *there exists* are called **existential quantifiers**. Examples with universal and existential quantifiers follow:

1. All roses are red. [universal]
2. Every student is important. [universal]
3. For each counting number x, $x + 0 = x$. [universal]
4. Some roses are red. [existential]
5. There exists at least one even counting number less than 3. [existential]
6. There are women who are taller than 200 cm. [existential]

Venn diagrams can be used to picture statements involving quantifiers. For example, Figures B.1a and B.1b picture statements (1) and (4). The x in Figure B.1b is used to show that there must be at least one element of the set of roses that is red.

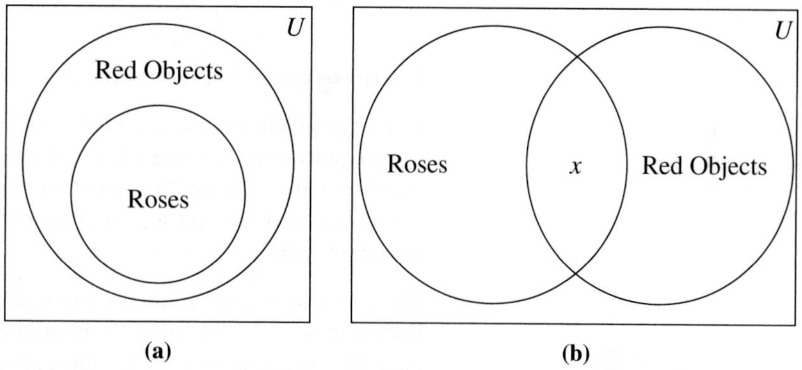

FIGURE B.1 **(a)** All roses are red. **(b)** Some roses are red.

Consider the following statement involving the existential quantifier *some*. "Some professors at Paxson University have blue eyes." This means that at least one professor at Paxson University has blue eyes. It does not rule out the possibilities that all the Paxson professors have blue eyes or that some of the Paxson professors do not have blue eyes. Because the negation of a true statement is false, neither "Some professors at Paxson University do not have blue eyes" nor "All professors at Paxson have blue eyes" are negations of the original statement. One possible negation of the original statement is "No professors at Paxson University have blue eyes."

Statement	**Negation**
Some *a* are *b*.	No *a* is *b*.
Some *a* are not *b*.	All *a* are *b*.
All *a* are *b*.	Some *a* are not *b*.
No *a* is *b*.	Some *a* are *b*.

EXAMPLE 2 Negation with Quantifiers

Negate each of the following statements:

(a) All students like hamburgers.

(b) Some people like mathematics.

(c) There exists a counting number x such that $3x = 6$.

(d) For all counting numbers x, $3x = 3x$.

SOLUTION

(a) Some students do not like hamburgers.

(b) No people like mathematics.

(c) For all counting numbers x, $3x \neq 6$.

(d) There exists a counting number x such that $3x \neq 3x$.

Now try Exercise 5, parts (e) and (f).

There is a symbolic system defined to help in the study of logic. If p represents a statement, the negation of the statement p is denoted by $\sim p$. **Truth tables** are often used to show all possible true-false patterns for statements. Table B.1 summarizes the truth tables for p and $\sim p$.

Table B.1 Negation

p	$\sim p$
T	F
F	T

Observe that p and $\sim p$ are analogous to sets P and $\overline{P}$. If x is an element of P, then x is not an element of $\overline{P}$.

Compound Statements

From two given statements, it is possible to create a new, **compound statement** by using a connective such as *and*. For example, "It is snowing" and "the ski run is open" together with *and* give "It is snowing and the ski run is open." Other compound statements can be obtained by using the connective *or*. For example, "It is snowing or the ski run is open."

The symbols $\wedge$ and $\vee$ are used to represent the connectives *and* and *or*, respectively. For example, if p represents "It is snowing," and if q represents "The ski run is open," then "It is snowing and the ski run is open" is denoted by $p \wedge q$. Similarly, "It is snowing or the ski run is open" is denoted by $p \vee q$.

Table B.2 Conjunction

p	q	$p \wedge q$
T	T	T
T	F	F
F	T	F
F	F	F

The truth value of any compound statement, such as $p \wedge q$, is defined using the truth table of each of the simple statements. Because each of the statements p and q may be either true or false, there are four distinct possibilities for the truth values of p and q, as shown in Table B.2. The compound statement $p \wedge q$ is the **conjunction** of p and q and is defined to be true if, and only if, both p and q are true. Otherwise, it is false.

The compound statement $p \vee q$—that is, p or q—is a **disjunction**. In everyday language, *or* is not always interpreted in the same way. In logic, we use an *inclusive or*. The statement "I will go to a movie or I will read a book" means that I will either go to a movie, or read a book, or do both. Hence, in logic, p or q, symbolized as $p \vee q$, is defined to be false if both p and q are false and true in all other cases. This is summarized in Table B.3.

Table B.3 Disjunction

p	q	$p \vee q$
T	T	T
T	F	T
F	T	T
F	F	F

EXAMPLE 3 Conjunction and Disjunction

Given the following statements, classify each of the conjunctions and disjunctions as true or false:

p: $2 + 3 = 5$ $\quad$ r: $5 + 3 = 9$

q: $2 \cdot 3 = 6$ $\quad$ s: $2 \cdot 4 = 9$

(a) $p \wedge q$ **(b)** $p \wedge r$ **(c)** $s \wedge q$ **(d)** $r \wedge s$

(e) $\sim p \wedge q$ **(f)** $\sim(p \wedge q)$ **(g)** $p \vee q$ **(h)** $p \vee r$

(i) $s \vee q$ **(j)** $r \vee s$ **(k)** $\sim p \vee q$ **(l)** $\sim(p \vee q)$

(continued)

SOLUTION

(a) p is true and q is true, so $p \wedge q$ is true.

(b) p is true and r is false, so $p \wedge r$ is false.

(c) s is false and q is true, so $s \wedge q$ is false.

(d) r is false and s is false, so $r \wedge s$ is false.

(e) $\sim p$ is false and q is true, so $\sim p \wedge q$ is false.

(f) $p \wedge q$ is true [part (a)], so $\sim(p \wedge q)$ is false.

(g) p is true and q is true, so $p \vee q$ is true.

(h) p is true and r is false, so $p \vee r$ is true.

(i) s is false and q is true, so $s \vee q$ is true.

(j) r is false and s is false, so $r \vee s$ is false.

(k) $\sim p$ is false and q is true, so $\sim p \vee q$ is true.

(l) $p \vee q$ is true [part (g)], so $\sim(p \vee q)$ is false.

Now try Exercise 7, parts (a) and (f).

There is an analogy between the connectives $\wedge$ and $\vee$ and the set operations of intersection $(\cap)$ and union $(\cup)$. Just as the statement $p \wedge q$ is true only when p and q are both true, so an element x belongs to the set $P \cap Q$ only when x belongs to both P and Q. Similarly, the statement $p \vee q$ is true when either p or q is true, and an element x belongs to the set $P \cup Q$ when x belongs to either P or Q.

EXAMPLE 4 Statements and Sets

Use set operations to construct a set that corresponds, by analogy, to each of the following statements:

(a) $p \wedge r$ **(b)** $\sim r \vee q$ **(c)** $\sim(p \wedge q)$ **(d)** $\sim(p \vee \sim r)$

SOLUTION

(a) $P \cap R$ **(b)** $\overline{R} \cup Q$ **(c)** $\overline{P \cap Q}$ **(d)** $\overline{P \cup R}$

Now try Exercise 9.

Table B.4

p	q	$p \wedge q$	$q \wedge p$
T	T	T	T
T	F	F	F
F	T	F	F
F	F	F	F

Not only are truth tables used to summarize the truth values of compound statements, they also are used to determine if two statements are logically equivalent. Two statements are **logically equivalent** if, and only if, they have the same truth values. For example, we could prove that $p \wedge q$ is logically equivalent to $q \wedge p$ by using a truth table as in Table B.4.

EXAMPLE 5 Logical Equivalence

Use a truth table to determine if $\sim p \vee \sim q$ and $\sim(p \wedge q)$ are logically equivalent.

SOLUTION Table B.5 shows headings and the four distinct possibilities for p and q. In the column headed $\sim p$, we write the negations of the p column. In the $\sim q$ column, we write the negations of the q column. Next, we use the values in the $\sim p$ and the $\sim q$ columns to construct the $\sim p \vee \sim q$ column. To find the truth values for $\sim(p \wedge q)$, we use the p and q columns to find the truth values for $p \wedge q$ and then negate $p \wedge q$.

Table B.5

p	q	$\sim p$	$\sim q$	$\sim p \vee \sim q$	$p \wedge q$	$\sim(p \wedge q)$
T	T	F	F	F	T	F
T	F	F	T	T	F	T
F	T	T	F	T	F	T
F	F	T	T	T	F	T

Since the values in the columns for $\sim p \vee \sim q$ and $\sim(p \wedge q)$ are identical, the statements are equivalent.

Now try Exercise 4, parts (b) and (d).

APPENDIX B.1 Exercises

1. Determine which of the following are statements, and then classify each statement as true or false:

   (a) $2 + 4 = 8$ **(b)** Shut the window.

   (c) Los Angeles is a state. **(d)** He is in town.

   (e) What time is it? **(f)** $5x = 15$

   (g) $3 \cdot 2 = 6$ **(h)** $2x^2 > x$

   (i) This statement is false. **(j)** Stay put!

2. **Writing to Learn** Use quantifiers to make each of the following true where x is a natural number:

   (a) $x + 8 = 11$ **(b)** $x + 0 = x$

   (c) $x^2 = 4$ **(d)** $x + 1 = x + 2$

3. **Writing to Learn** Use quantifiers to make each equation in Exercise 2 false.

4. Complete each of the following truth tables:

   (a)

p	$\sim p$	$\sim(\sim p)$
T		
F		

   (b)

p	$\sim p$	$p \vee \sim p$	$p \wedge \sim p$
T			
F			

   (c) Based on part (a), is p logically equivalent to $\sim(\sim p)$?

   (d) Based on part (b), is $p \vee \sim p$ logically equivalent to $p \wedge \sim p$?

5. Write the negation for each of the following statements:

   (a) The book has 500 pages.

   (b) Six is less than eight.

   (c) $3 \cdot 5 = 15$

   (d) Some people have blond hair.

   (e) All dogs have four legs.

   (f) Some cats do not have nine lives.

   (g) All squares are rectangles.

   (h) Not all rectangles are squares.

   (i) For all natural numbers x, $x + 3 = 3 + x$.

   (j) There exists a natural number x such that $3 \cdot (x + 2) = 12$.

   (k) Every counting number is divisible by itself and 1.

   (l) Not all natural numbers are divisible by 2.

   (m) For all natural numbers x, $5x + 4x = 9x$.

6. If q stands for "This course is easy" and r stands for "Lazy students do not study," write each of the following in symbolic form:

   (a) This course is easy and lazy students do not study.

   (b) Lazy students do not study or this course is not easy.

   (c) It is false that both this course is easy and lazy students do not study.

   (d) This course is not easy.

7. If p is false and q is true, find the truth values for each of the following:

   (a) $p \wedge q$ **(b)** $p \vee q$

   (c) $\sim p$ **(d)** $\sim q$

   (e) $\sim(\sim p)$ **(f)** $\sim p \vee q$

   (g) $p \wedge \sim q$ **(h)** $\sim(p \vee q)$

   (i) $\sim(\sim p \wedge q)$ **(j)** $\sim q \wedge \sim p$

8. Find the truth value for each statement in Exercise 7 if p is false and q is false.

9. Use set operations to construct a set that corresponds, by analogy, to each of the following statements.

   (a) $r \vee s$ **(b)** $q \wedge \sim q$

   (c) $\sim(r \vee q)$ **(d)** $p \wedge (r \vee s)$

10. For each of the following, is the pair of statements logically equivalent?

(a) $\sim(p \vee q)$ and $\sim p \vee \sim q$

(b) $\sim(p \vee q)$ and $\sim p \wedge \sim q$

(c) $\sim(p \wedge q)$ and $\sim p \wedge \sim q$

(d) $\sim(p \wedge q)$ and $\sim p \vee \sim q$

11. **(a)** Write two logical equivalences discovered in parts 10(a)–(d). These equivalences are called DeMorgan's Laws for "*and*" and "*or*."

(b) Write an explanation of the analogy between DeMorgan's Laws for sets and those found in part (a).

12. Complete the following truth table:

p	q	$\sim p$	$\sim q$	$\sim p \vee q$
T	T			
T	F			
F	T			
F	F			

13. **Writing to Learn** Restate the following in a logically equivalent form:

(a) It is not true that both today is Wednesday and the month is June.

(b) It is not true that yesterday I both ate breakfast and watched television.

(c) It is not raining or it is not July.

B.2 Conditionals and Biconditionals

What you'll learn about

- Forms of Statements
- Valid Reasoning

... and why

These topics are important in the study of logic.

Forms of Statements

Statements expressed in the form "if p, then q" are called **conditionals**, or **implications**, and are denoted by $p \rightarrow q$. Such statements also can be read "p implies q." The "if" part of a conditional is called the **hypotheses** of the implication and the "then" part is called the **conclusion**.

Many types of statements can be put in "if-then" form; an example follows:

Statement: All first graders are 6 years old.

If-then form: If a child is a first grader, then the child is 6 years old.

An implication may also be thought of as a promise. Suppose Betty makes the promise, "If I get a raise, then I will take you to dinner." If Betty keeps her promise, the implication is true; if Betty breaks her promise, the implication is false. Consider the following four possibilities:

	p	q	
(1)	T	T	Betty gets the raise; she takes you to dinner.
(2)	T	F	Betty gets the raise; she does not take you to dinner.
(3)	F	T	Betty does not get the raise; she takes you to dinner.
(4)	F	F	Betty does not get the raise; she does not take you to dinner.

The only case in which Betty breaks her promise is when she gets her raise and fails to take you to dinner, case (2). If she does not get the raise, she can either take you to dinner or not without breaking her promise. The definition of implication is summarized in Table B.6. Observe that the only cause for which the implication is false is when p is true and q is false.

Table B.6 Implication

p	q	$p \rightarrow q$
T	T	T
T	F	F
F	T	T
F	F	T

An implication may be worded in several equivalent ways, as follows:

1. If the sun shines, then the swimming pool is open. (If p, then q.)
2. If the sun shines, the swimming pool is open. (If p, q.)
3. The swimming pool is open if the sun shines. (q if p.)
4. The sun shines implies the swimming pool is open. (p implies q.)
5. The sun is shining only if the pool is open. (p only if q.)
6. The sun's shining is a sufficient condition for the swimming pool to be open. (p is a sufficient condition for q.)
7. The swimming pool's being open is a necessary condition for the sun to be shining. (q is a necessary condition for p.)

Any implication $p \rightarrow q$ has three related implication statements, as follows:

Statement:	If p, then q.	$p \rightarrow q$
Converse:	If q, then p.	$q \rightarrow p$
Inverse:	If not p, then not q.	$\sim p \rightarrow \sim q$
Contrapositive:	If not q, then not p.	$\sim q \rightarrow \sim p$

EXAMPLE 1 Converse, Inverse, Contrapositive

Write the converse, the inverse, and the contrapositive for each of the following statements:

(a) If $2x = 6$, then $x = 3$.

(b) If I am in San Francisco, then I am in California.

SOLUTION

(a) *Converse:* If $x = 3$, then $2x = 6$.
Inverse: If $2x \neq 6$, then $x \neq 3$.
Contrapositive: If $x \neq 3$, then $2x \neq 6$.

(b) *Converse:* If I am in California, then I am in San Francisco.
Inverse: If I am not in San Francisco, then I am not in California.
Contrapositive: If I am not in California, then I am not in San Francisco.

Now try Exercise 3, parts (a) and (b).

Table B.7 shows that an implication and its converse do not always have the same truth value. However, an implication and its contrapositive always have the same truth value. Also, the converse and inverse of a conditional statement are logically equivalent.

Table B.7 Converse, Inverse, Contrapositive

				Implication	Converse	Inverse	Contra-positive
p	q	$\sim p$	$\sim q$	$p \to q$	$q \to p$	$\sim p \to \sim q$	$\sim q \to \sim p$
T	T	F	F	T	T	T	T
T	F	F	T	F	T	T	F
F	T	T	F	T	F	F	T
F	F	T	T	T	T	T	T

Connecting a statement and its converse with the connective *and* gives $(p \to q) \wedge (q \to p)$. This compound statement can be written as $p \leftrightarrow q$ and usually is read "p if and only if q." The statement "p if and only if q" is a **biconditional**. A truth table for $p \leftrightarrow q$ is given in Table B.8. Observe that $p \leftrightarrow q$ is true if and only if both statements are true or both are false.

Table B.8 Biconditional

				Biconditional $(p \to q) \wedge (q \to p)$ or
p	q	$p \to q$	$q \to p$	$p \leftrightarrow q$
T	T	T	T	T
T	F	F	T	F
F	T	T	F	F
F	F	T	T	T

EXAMPLE 2 Biconditionals

Given the following statements, classify each of the biconditionals as true or false:

p: $2 = 2$ r: $2 = 1$

q: $2 \neq 1$ s: $2 + 3 = 1 + 3$

(a) $p \leftrightarrow q$ **(b)** $p \leftrightarrow r$

(c) $s \leftrightarrow q$ **(d)** $r \leftrightarrow s$

SOLUTION

(a) $p \rightarrow q$ is true and $q \rightarrow p$ is true, so $p \leftrightarrow q$ is true.

(b) $p \rightarrow r$ is false and $r \rightarrow p$ is true, so $p \leftrightarrow r$ is false.

(c) $s \rightarrow q$ is true and $q \rightarrow s$ is false, so $s \leftrightarrow q$ is false.

(d) $r \rightarrow s$ is true and $s \rightarrow r$ is true, so $r \leftrightarrow s$ is true.

Now try Exercise 5, parts (a) and (f).

Now consider the following statement:

It is raining or it is not raining.

This statement, which can be modeled as $p \vee (\sim p)$, is always true, as shown in Table B.9. A statement that is always true is called a **tautology**. One way to make a tautology is to take two logically equivalent statements such as $p \rightarrow q$ and $\sim q \rightarrow \sim p$ (from Table B.7) and form them into a biconditional as follows:

$$p \rightarrow q \leftrightarrow (\sim q \rightarrow \sim p)$$

Because $p \rightarrow q$ and $\sim q \rightarrow \sim p$ have the same truth values, $(p \rightarrow q) \leftrightarrow (\sim q \rightarrow \sim p)$ is a tautology.

Table B.9 A Tautology

p	$\sim p$	$p \vee (\sim p)$
T	F	T
F	T	T

Valid Reasoning

In problem solving, the reasoning used is said to be **valid** if the conclusion follows unavoidably from the hypotheses. Consider the following example:

Hypotheses: All roses are red.
This flower is a rose.

Conclusion: Therefore, this flower is red.

The statement "All roses are red" can be written as the implication "If a flower is a rose, then it is red" and pictured with the Venn diagram in Figure B.2a.

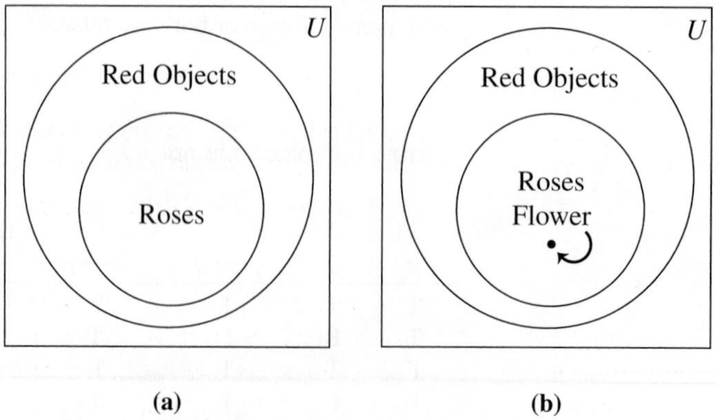

FIGURE B.2 (a) All roses are red. (b) This flower is a rose.

The information "This flower is a rose" implies that this flower must belong to the circle containing roses, as pictured in Figure B.2b. This flower also must belong to the circle containing red objects. Thus the reasoning is valid because it is impossible to draw a picture satisfying the hypotheses and contradicting the conclusion.

Consider the following argument:

Hypotheses:	All elementary school teachers are mathematically literate. Some mathematically literate people are not children.
Conclusion:	Therefore, no elementary school teacher is a child.

Let E be the set of elementary school teachers, M be the set of mathematically literate people, and C be the set of children. Then the statement "All elementary school teachers are mathematically literate" can be pictured as in Figure B.3a. The statement "Some mathematically literate people are not children" can be pictured in several ways. Three of these are illustrated in Figure B.3b–d.

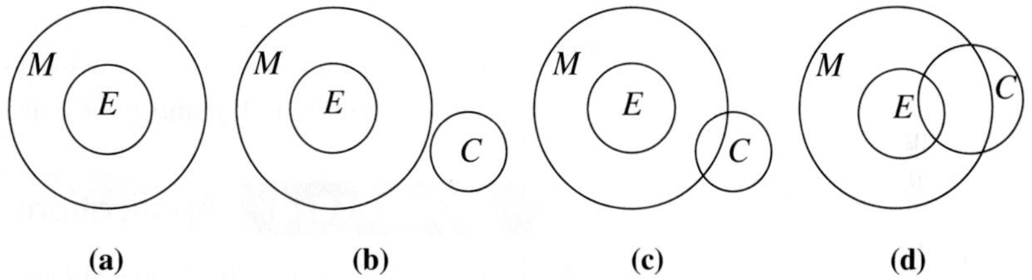

FIGURE B.3 (a) All elementary school teachers are mathematically literate. (b)–(d) Some mathematically literate people are not children.

According to Figure B.3d, it is possible that some elementary school teachers are children, and yet the given statements are satisfied. Therefore, the conclusion that "No elementary school teacher is a child" does not follow from the given hypotheses. Hence, the reasoning is not valid.

If a single picture can be drawn to satisfy the hypotheses of an argument and contradict the conclusion, the argument is not valid. However, to demonstrate that an argument is valid, all possible pictures must be considered to prove that there are no contradictions. There must be no way to satisfy the hypotheses and contradict the conclusion if the argument is valid.

EXAMPLE 3 Argument Validity

Determine if the following argument is valid:

Hypotheses:	In Washington, D.C., all senators wear power ties. No one in Washington, D.C., over 6 ft tall wears a power tie.
Conclusion:	Persons over 6 ft tall are not senators in Washington, D.C.

SOLUTION

If S represents the set of senators and P represents the set of people who wear power ties, the first hypothesis is pictured as shown in Figure B.4a. If T represents the set of people in Washington, D.C., over 6 ft tall, the second hypothesis is pictured in Figure B.4b. Because people over 6 ft tall are outside the circle representing power tie wearers and senators are inside the circle P, the conclusion is valid and no person over 6 ft tall can be a senator in Washington, D.C.

Now try Exercise 14(a).

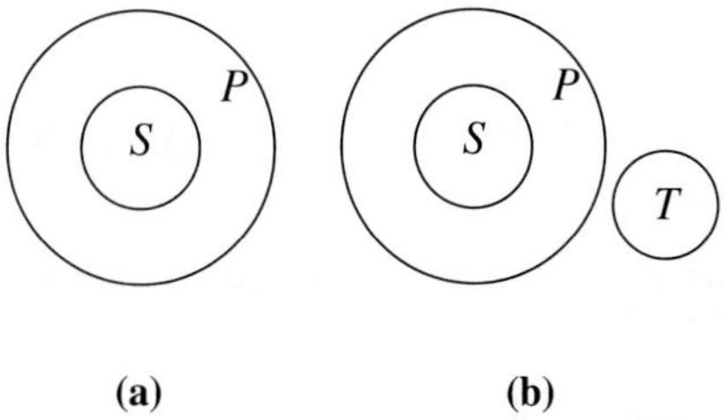

FIGURE B.4 (a) In Washington, D.C., all senators wear power ties. (b) No one in Washington, D.C., over 6 ft tall wears a power tie.

A different method for determining if an argument is valid uses **direct reasoning** and a form of argument called the *Law of Detachment* (or **Modus Ponens**). For example, consider the following true statements:

If the sun is shining, then we shall take a trip.

The sun is shining.

Using these two statements, we can conclude that we shall take a trip. In general, the **Law of Detachment** is stated as follows:

If a statement in the form "if p, then q" is true, and p is true, then q must also be true.

The Law of Detachment is sometimes described schematically as follows, where all statements above the horizontal line are true and the statement below the horizontal line is the conclusion.

$$\begin{array}{l} p \rightarrow q \\ \underline{p \qquad\quad} \\ \qquad\quad q \end{array}$$

The Law of Detachment follows from the truth table for $p \rightarrow q$ given in Table B.6. The only case in which both p and $p \rightarrow q$ are true is when q is true (line 1 in the table).

EXAMPLE 4 Applications of the Law of Detachment

Determine if each of the following arguments is valid:

Hypotheses:	If you eat spinach, then you will be strong. You eat spinach.
Conclusion:	Therefore, you will be strong.
Hypotheses:	If Claude goes skiing, he will break his leg. If Claude breaks his leg, he cannot enter the dance contest. Claude goes skiing.
Conclusion:	Therefore, Claude cannot enter the dance contest.

SOLUTION

(a) Using the Law of Detachment, we see that the conclusion is valid.

(b) By using the Law of Detachment twice, we see that the conclusion is valid.

Now try Exercise 14(d).

A different type of reasoning, **indirect reasoning**, uses a form of argument called **Modus Tollens**. For example, consider the following true statements:

If Chicken Little had been hit by a jumping frog, he would have thought Earth was rising.

Chicken Little did not think Earth was rising.

What is the conclusion? The conclusion is that Chicken Little did not get hit by a jumping frog. This leads us to the general form of Modus Tollens:

If we have a conditional accepted as true, and we know the conclusion is false, then the hypothesis must be false.

Modus Tollens is sometimes schematically described as follows:

$$\begin{array}{l} \quad p \rightarrow q \\ \underline{\sim q \qquad\quad} \\ \qquad\quad \sim p \end{array}$$

The validity of Modus Tollens also follows from the truth table for $p \rightarrow q$ given in Table B.6. The only case in which both $p \rightarrow q$ is true and q is false is when p is false (line 4 in the table). The validity of Modus Tollens also can be established from the fact that an implication and its contrapositive are equivalent.

EXAMPLE 5 Applications of Modus Tollens

Determine conclusions for each of the following sets of true statements:

(a) If an old woman lives in a shoe, then she does not know what to do. Mrs. Pumpkin Eater, an old woman, knows what to do.

(b) If Jack is nimble, he will not get burned. Jack was burned.

SOLUTION

(a) Mrs. Pumpkin Eater does not live in a shoe.

(b) Jack was not nimble.

Now try Exercise 13(a).

People often make invalid conclusions based on advertising or other information. Consider, for example, the statement "Healthy people eat Super-Bran cereal." Are the following conclusions valid?

If a person eats Super-Bran cereal, then the person is healthy.
If a person is not healthy, the person does not eat Super-Bran cereal.

If the original statement is denoted by $p \rightarrow q$, where p is "a person is healthy" and q is "a person eats Super-Bran cereal," then the first conclusion is the converse of $p \rightarrow q$—that is, $q \rightarrow p$—and the second conclusion is the inverse of $p \rightarrow q$—that is, $\sim p \rightarrow \sim q$. Table B.7 points out that neither the converse nor the inverse is logically equivalent to the original statement, and consequently the conclusions are not necessarily true.

The final form of argument to be considered here involves the **Chain Rule**. Consider the following statements:

If my wife works, I will retire early.
If I retire early, I will become lazy.

What is the conclusion? The conclusion is that if my wife works, I will become lazy. In general, the Chain Rule can be stated as follows:

If "if p, then q," and "if q, then r" are true, then "if p, then r" is true.

The Chain Rule is sometimes schematically described as follows:

$$\begin{array}{c} p \rightarrow q \\ \underline{q \rightarrow r} \\ p \rightarrow r \end{array}$$

Notice that the Chain Rule states that implication is a transitive relation.

EXAMPLE 6 Applications of the Chain Rule

Remark

Note that in Example 6, the Chain Rule can be extended to contain several implications.

Determine conclusions for each of the following sets of true statements:

(a) If Alice follows the White Rabbit, she falls into a hole. If she falls into a hole, she goes to a tea party.

(b) If Chicken Little is hit by an acorn, we think the sky is falling. If we think the sky is falling, we will go to a fallout shelter. If we go to a fallout shelter, we will stay there a month.

SOLUTION

(a) If Alice follows the White Rabbit, she goes to a tea party.

(b) If Chicken Little is hit by an acorn, we will stay in a fallout shelter for a month.

Now try Exercise 13(c).

APPENDIX B.2 Exercises

1. Write each of the following in symbolic form if p is the statement "It is raining" and q is the statement "The grass is wet."
   (a) If it is raining, then the grass is wet.
   (b) If it is not raining, then the grass is wet.
   (c) If it is raining, then the grass is not wet.
   (d) The grass is wet if it is raining.
   (e) The grass is not wet implies that it is not raining.
   (f) The grass is wet if, and only if, it is raining.
2. Construct a truth table for each of the following:
   (a) $p \to (p \vee q)$ (b) $(p \wedge q) \to q$
   (c) $p \leftrightarrow \sim(\sim p)$ (d) $\sim(p \to q)$
3. **Writing to Learn** For each of the following implications, state the converse, inverse, and contrapositive.
   (a) If you eat Meaties, then you are good in sports.
   (b) If you do not like this book, then you do not like mathematics.
   (c) If you do not use Ultra Brush toothpaste, then you have cavities.
   (d) If you are good at logic, then your grades are high.
4. Can an implication and its converse both be false? Explain your answer.
5. If p is true and q is false, find the truth values for each of the following:
   (a) $\sim p \to \sim q$ (b) $\sim(p \to q)$
   (c) $(p \vee q) \to (p \wedge q)$ (d) $p \to \sim p$
   (e) $(p \vee \sim p) \to p$ (f) $(p \vee q) \leftrightarrow (p \wedge q)$
6. If p is false and q is false, find the truth values for each of the statements in Exercise 5.
7. Iris makes the true statement, "If it rains, then I am going to the movies." Does it follow logically that if it does not rain, then Iris does not go to the movies?
8. Consider the statement "If every digit of a number is 6, then the number is divisible by 3." Determine whether each of the following is logically equivalent to the statement.
   (a) If every digit of a number is not 6, then the number is not divisible by 3.
   (b) If a number is not divisible by 3, then some digit of the number is not 6.
   (c) If a number is divisible by 3, then every digit of the number is 6.
9. Write a statement logically equivalent to the statement "If a number is a multiple of 8, then it is a multiple of 4."
10. Use truth tables to prove that the following are tautologies:
   (a) $(p \to q) \to [(p \wedge r) \to q]$ Law of Added Hypothesis
   (b) $[(p \to q) \wedge p] \to q$ Law of Detachment
   (c) $[(p \to q) \wedge \sim q] \to \sim p$ Modus Tollens
   (d) $[(p \to q) \wedge (q \to r)] \to (p \to r)$ Chain Rule
11. (a) Suppose that $p \to q$, $q \to r$, and $r \to s$ are all true, but s is false. What can you conclude about the truth value of p?
   (b) Suppose that $(p \wedge q) \to r$ is true, r is false, and q is true. What can you conclude about the truth value of p?
   (c) **Writing to Learn** Suppose that $p \to q$ is true and $q \to p$ is false. Can q be true? Why or why not?
12. **Writing to Learn** Translate the following statements into symbolic form. Give the meanings of the symbols that you use.
   (a) If Mary's little lamb follows her to school, then its appearance there will break the rules and Mary will be sent home.
   (b) If it is not the case that Jack is nimble and quick, then Jack will not make it over the candlestick.
   (c) If the apple had not hit Isaac Newton on the head, then the laws of gravity would not have been discovered.
13. **Writing to Learn** For each of the following, form a conclusion that follows logically from the given statements:
   (a) All college students are poor.
   Helen is a college student.
   (b) Some freshmen like mathematics.
   All people who like mathematics are intelligent.
   (c) If I study for the final, then I will pass the final.
   If I pass the final, then I will pass the course.
   If I pass the course, then I will look for a teaching job.
   (d) Every equilateral triangle is isosceles.
   There exist triangles that are equilateral.
14. Investigate the validity of each of the following arguments:
   (a) All women are mortal.
   Hypatia was a woman.
   Therefore, Hypatia was mortal.
   (b) All squares are quadrilaterals.
   All quadrilaterals are polygons.
   Therefore, all squares are polygons.
   (c) All teachers are intelligent.
   Some teachers are rich.
   Therefore, some intelligent people are rich.
   (d) If a student is a freshman, then she takes mathematics.
   Jane is a sophomore.
   Therefore, Jane does not take mathematics.
15. **Writing to Learn** Write the following in if-then form:
   (a) Every figure that is a square is a rectangle.
   (b) All integers are rational numbers.
   (c) Figures with exactly three sides may be triangles.
   (d) It rains only if it is cloudy.

C Key Formulas

C.1 Formulas from Algebra

Exponents

If all bases are nonzero,

$u^m u^n = u^{m+n}$ $\qquad \dfrac{u^m}{u^n} = u^{m-n}$

$u^0 = 1$ $\qquad u^{-n} = \dfrac{1}{u^n}$

$(uv)^m = u^m v^m$ $\qquad (u^m)^n = u^{mn}$

$\left(\dfrac{u}{v}\right)^m = \dfrac{u^m}{v^m}$

Radicals and Rational Exponents

If all roots are real numbers,

$\sqrt[n]{uv} = \sqrt[n]{u} \cdot \sqrt[n]{v}$ $\qquad \sqrt[n]{\dfrac{u}{v}} = \dfrac{\sqrt[n]{u}}{\sqrt[n]{v}} \; (v \neq 0)$

$\sqrt[m]{\sqrt[n]{u}} = \sqrt[mn]{u}$ $\qquad \left(\sqrt[n]{u}\right)^n = u$

$\sqrt[n]{u^m} = \left(\sqrt[n]{u}\right)^m$ $\qquad \sqrt[n]{u^n} = \begin{cases} |u| & n \text{ even} \\ u & n \text{ odd} \end{cases}$

$u^{1/n} = \sqrt[n]{u}$ $\qquad u^{m/n} = \left(u^{1/n}\right)^m = \left(\sqrt[n]{u}\right)^m$

$u^{m/n} = (u^m)^{1/n} = \sqrt[n]{u^m}$

Special Products

$(u + v)(u - v) = u^2 - v^2$

$(u + v)^2 = u^2 + 2uv + v^2$

$(u - v)^2 = u^2 - 2uv + v^2$

$(u + v)^3 = u^3 + 3u^2v + 3uv^2 + v^3$

$(u - v)^3 = u^3 - 3u^2v + 3uv^2 - v^3$

Factoring Polynomials

$u^2 - v^2 = (u + v)(u - v)$

$u^2 + 2uv + v^2 = (u + v)^2$

$u^2 - 2uv + v^2 = (u - v)^2$

$u^3 + v^3 = (u + v)(u^2 - uv + v^2)$

$u^3 - v^3 = (u - v)(u^2 + uv + v^2)$

Inequalities

If $u < v$ and $v < w$, then $u < w$.

If $u < v$, then $u + w < v + w$.

If $u < v$ and $c > 0$, then $uc < vc$.

If $u < v$ and $c < 0$, then $uc > vc$.

If $c > 0$, $|u| < c$ is equivalent to $-c < u < c$.

If $c > 0$, $|u| > c$ is equivalent to $u < -c$ or $u > c$.

Quadratic Formula

If $a \neq 0$, the solutions of the equation $ax^2 + bx + c = 0$ are given by

$$x = \frac{-b \pm \sqrt{b^2 - 4ac}}{2a}.$$

Logarithms

If $0 < b \neq 1$, $0 < a \neq 1$, and $x, R, S > 0$,

$y = \log_b x$ if and only if $b^y = x$

$\log_b 1 = 0$ $\qquad \log_b b = 1$

$\log_b b^y = y$ $\qquad b^{\log_b x} = x$

$\log_b RS = \log_b R + \log_b S$ $\qquad \log_b \dfrac{R}{S} = \log_b R - \log_b S$

$\log_b R^c = c \log_b R$ $\qquad \log_b x = \dfrac{\log_a x}{\log_a b}$

Determinants

$$\begin{vmatrix} a & b \\ c & d \end{vmatrix} = ad - bc$$

Arithmetic Sequences and Series

$a_n = a_1 + (n - 1)d$

$S_n = n\left(\dfrac{a_1 + a_n}{2}\right)$ or $S_n = \dfrac{n}{2}\left[2a_1 + (n - 1)d\right]$

Geometric Sequences and Series

$a_n = a_1 \cdot r^{n-1}$

$S_n = \dfrac{a_1(1 - r^n)}{1 - r} \; (r \neq 1)$

$S = \dfrac{a_1}{1 - r} \; (|r| < 1)$ infinite geometric series

Factorial

$n! = n \cdot (n - 1) \cdot (n - 2) \cdot \cdots \cdot 3 \cdot 2 \cdot 1$

$n \cdot (n - 1)! = n!$, $0! = 1$

Binomial Coefficient

$$\binom{n}{r} = \frac{n!}{r!(n - r)!} \quad (\text{integers } n \text{ and } r,\ n \geq r \geq 0)$$

Binomial Theorem

If n is a positive integer,

$$(a + b)^n = \binom{n}{0}a^n + \binom{n}{1}a^{n-1}b + \cdots + \binom{n}{r}a^{n-r}b^r + \cdots + \binom{n}{n}b^n.$$

C.2 Formulas from Geometry

Triangle

$h = a \sin \theta$

$\text{Area} = \frac{1}{2}bh$

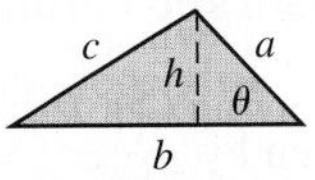

Trapezoid

$\text{Area} = \frac{h}{2}(a + b)$

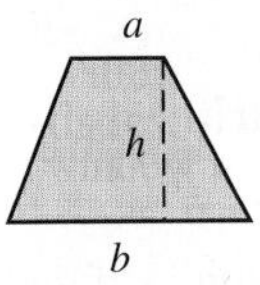

Circle

$\text{Area} = \pi r^2$

$\text{Circumference} = 2\pi r$

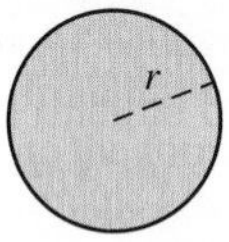

Sector of Circle

$\text{Area} = \frac{\theta r^2}{2}$ (θ in radians)

$s = r\theta$ (θ in radians)

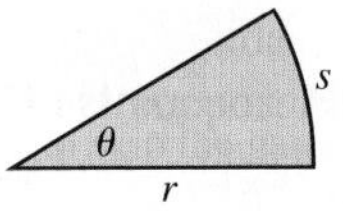

Right Circular Cone

$\text{Volume} = \frac{\pi r^2 h}{3}$

$\text{Lateral surface area} = \pi r\sqrt{r^2 + h^2}$

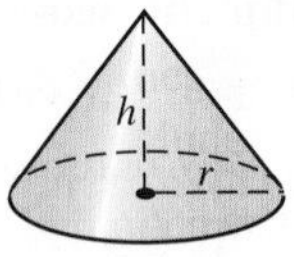

Right Circular Cylinder

$\text{Volume} = \pi r^2 h$

$\text{Lateral surface area} = 2\pi rh$

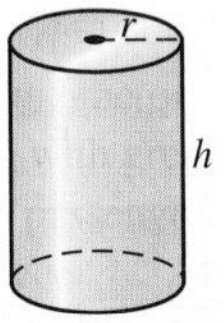

Right Triangle

Pythagorean Theorem:
$c^2 = a^2 + b^2$

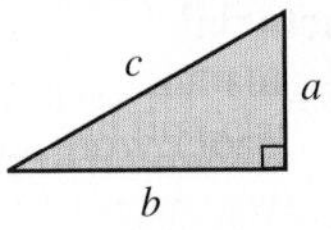

Parallelogram

$\text{Area} = bh$

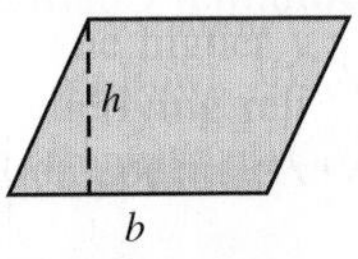

Circular Ring

$\text{Area} = \pi(R^2 - r^2)$

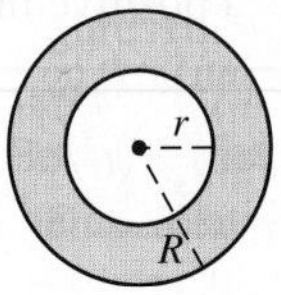

Ellipse

$\text{Area} = \pi ab$

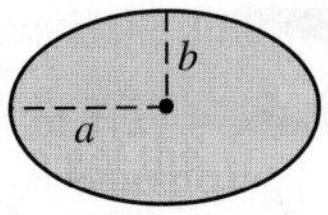

Cone

$\text{Volume} = \frac{Ah}{3}$ (A = Area of base)

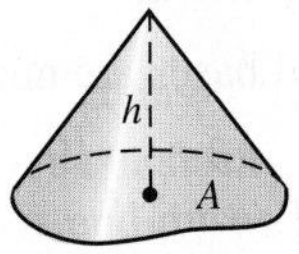

Sphere

$\text{Volume} = \frac{4}{3}\pi r^3$

$\text{Surface area} = 4\pi r^2$

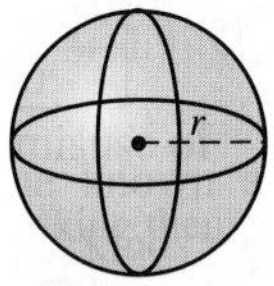

C.3 Formulas from Trigonometry

Angular Measure

π radians $= 180°$

So, 1 radian $= \frac{180}{\pi}$ degrees,

and 1 degree $= \frac{\pi}{180}$ radians.

Reciprocal Identities

$\sin x = \frac{1}{\csc x}$ $\qquad$ $\csc x = \frac{1}{\sin x}$

$\cos x = \frac{1}{\sec x}$ $\qquad$ $\sec x = \frac{1}{\cos x}$

$\tan x = \frac{1}{\cot x}$ $\qquad$ $\cot x = \frac{1}{\tan x}$

Quotient Identities

$\tan x = \frac{\sin x}{\cos x}$ $\qquad$ $\cot x = \frac{\cos x}{\sin x}$

Pythagorean Identities

$\sin^2 x + \cos^2 x = 1$

$1 + \tan^2 x = \sec^2 x$

$1 + \cot^2 x = \csc^2 x$

Odd-Even Identities

$\sin(-x) = -\sin x$ $\qquad$ $\csc(-x) = -\csc x$

$\cos(-x) = \cos x$ $\qquad$ $\sec(-x) = \sec x$

$\tan(-x) = -\tan x$ $\qquad$ $\cot(-x) = -\cot x$

Sum and Difference Identities

$\sin(u+v) = \sin u \cos v + \cos u \sin v$

$\sin(u-v) = \sin u \cos v - \cos u \sin v$

$\cos(u+v) = \cos u \cos v - \sin u \sin v$

$\cos(u-v) = \cos u \cos v + \sin u \sin v$

$\tan(u+v) = \dfrac{\tan u + \tan v}{1 - \tan u \tan v}$

$\tan(u-v) = \dfrac{\tan u - \tan v}{1 + \tan u \tan v}$

Cofunction Identities

$\cos\left(\dfrac{\pi}{2} - u\right) = \sin u$

$\sin\left(\dfrac{\pi}{2} - u\right) = \cos u$

$\tan\left(\dfrac{\pi}{2} - u\right) = \cot u$

$\cot\left(\dfrac{\pi}{2} - u\right) = \tan u$

$\sec\left(\dfrac{\pi}{2} - u\right) = \csc u$

$\csc\left(\dfrac{\pi}{2} - u\right) = \sec u$

Double-Angle Identities

$\sin 2u = 2 \sin u \cos u$

$\cos 2u = \cos^2 u - \sin^2 u$

$= 2\cos^2 u - 1$

$= 1 - 2\sin^2 u$

$\tan 2u = \dfrac{2 \tan u}{1 - \tan^2 u}$

Power-Reducing Identities

$\sin^2 u = \dfrac{1 - \cos 2u}{2}$

$\cos^2 u = \dfrac{1 + \cos 2u}{2}$

$\tan^2 u = \dfrac{1 - \cos 2u}{1 + \cos 2u}$

Half-Angle Identities

$\sin \dfrac{u}{2} = \pm\sqrt{\dfrac{1 - \cos u}{2}}$

$\cos \dfrac{u}{2} = \pm\sqrt{\dfrac{1 + \cos u}{2}}$

$\tan \dfrac{u}{2} = \pm\sqrt{\dfrac{1 - \cos u}{1 + \cos u}}$

$= \dfrac{1 - \cos u}{\sin u} = \dfrac{\sin u}{1 + \cos u}$

Triangles

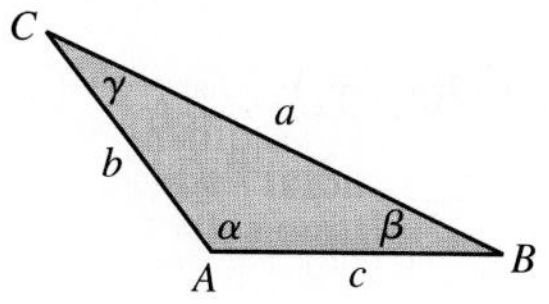

Law of sines:

$\dfrac{\sin A}{a} = \dfrac{\sin B}{b} = \dfrac{\sin C}{c}$

Law of cosines:

$a^2 = b^2 + c^2 - 2bc \cos A$

$b^2 = a^2 + c^2 - 2ac \cos B$

$c^2 = a^2 + b^2 - 2ab \cos C$

Area:

$\text{Area} = \dfrac{1}{2} bc \sin A$

$= \dfrac{1}{2} ac \sin B = \dfrac{1}{2} ab \sin C$

$\text{Area} = \sqrt{s(s-a)(s-b)(s-c)}$

where $s = \dfrac{1}{2}(a + b + c)$

Trigonometric Form of a Complex Number

$z = a + bi = (r\cos\theta) + (r\sin\theta)i$

$= r(\cos\theta + i\sin\theta)$

De Moivre's Theorem

$z^n = [r(\cos\theta + i\sin\theta)]^n$

$= r^n(\cos n\theta + i\sin n\theta)$

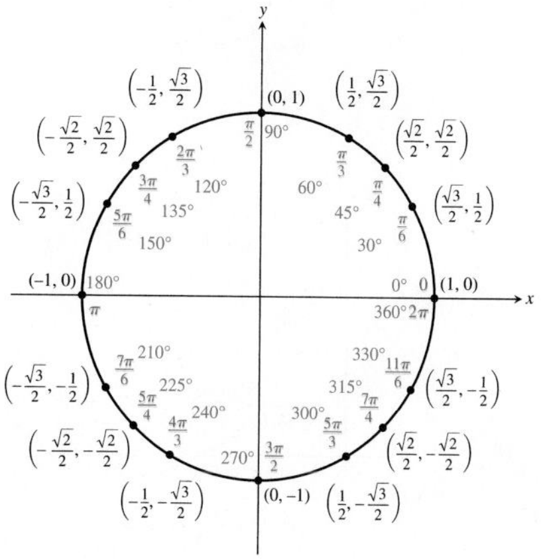

C.4 Formulas from Analytic Geometry

Basic Formulas

Distance d between points $P(x_1, y_1)$ and $Q(x_2, y_2)$:

$d = \sqrt{(x_1 - x_2)^2 + (y_1 - y_2)^2}$

Midpoint:

$\left(\frac{x_1 + x_2}{2}, \frac{y_1 + y_2}{2}\right)$

Slope of a line: $m = \frac{y_2 - y_1}{x_2 - x_1}$

Condition for parallel lines: $m_1 = m_2$

Condition for perpendicular lines: $m_2 = \frac{-1}{m_1}$

Equations of a Line

The point-slope form, slope m and through (x_1, y_1):

$y - y_1 = m(x - x_1)$

The slope-intercept form, slope m and y-intercept b:

$y = mx + b$

Equation of a Circle

The circle with center (h, k) and radius r:

$(x - h)^2 + (y - k)^2 = r^2$

Parabolas with Vertex (h, k)

Standard equation	$(x - h)^2 = 4p(y - k)$	$(y - k)^2 = 4p(x - h)$
Opens	Upward or downward	To the right or to the left
Focus	$(h, k + p)$	$(h + p, k)$
Directrix	$y = k - p$	$x = h - p$
Axis	$x = h$	$y = k$

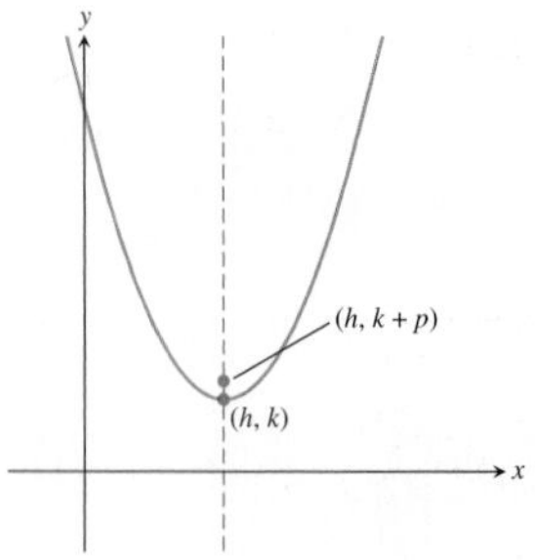

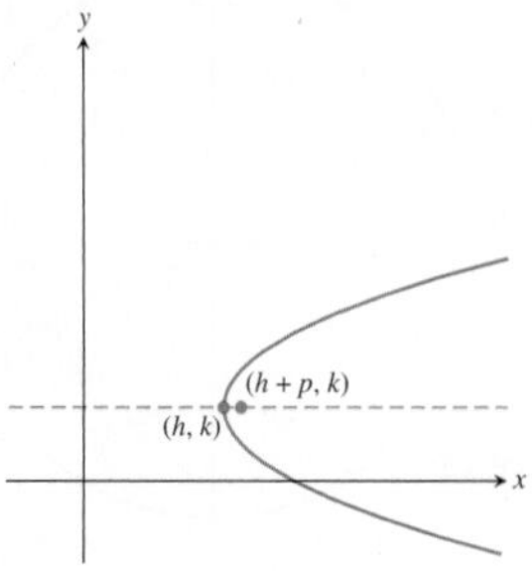

Ellipses with Center (h, k) and $a > b > 0$

Standard equation	$\frac{(x - h)^2}{a^2} + \frac{(y - k)^2}{b^2} = 1$	$\frac{(y - k)^2}{a^2} + \frac{(x - h)^2}{b^2} = 1$
Focal axis	$y = k$	$x = h$
Foci	$(h \pm c, k)$	$(h, k \pm c)$
Vertices	$(h \pm a, k)$	$(h, k \pm a)$
Pythagorean relation	$a^2 = b^2 + c^2$	$a^2 = b^2 + c^2$

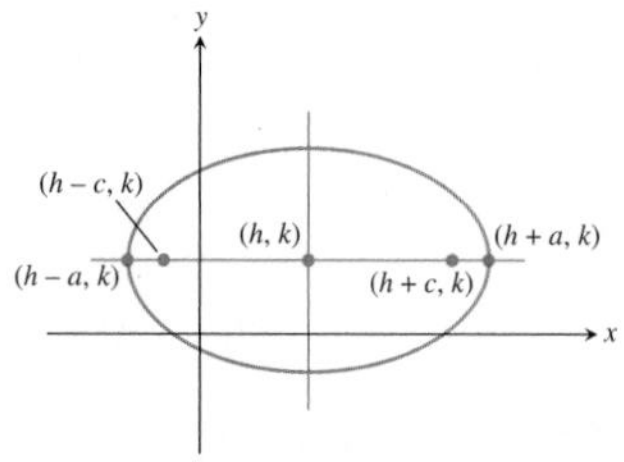

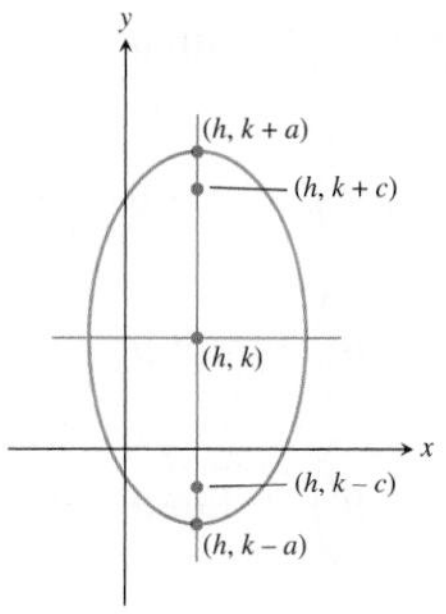

Hyperbolas with Center (h, k)

Standard equation	$\frac{(x - h)^2}{a^2} - \frac{(y - k)^2}{b^2} = 1$	$\frac{(y - k)^2}{a^2} - \frac{(x - h)^2}{b^2} = 1$
Focal axis	$y = k$	$x = h$
Foci	$(h \pm c, k)$	$(h, k \pm c)$
Vertices	$(h \pm a, k)$	$(h, k \pm a)$
Pythagorean relation	$c^2 = a^2 + b^2$	$c^2 = a^2 + b^2$
Asymptotes	$y = \pm\frac{b}{a}(x - h) + k$	$y = \pm\frac{a}{b}(x - h) + k$

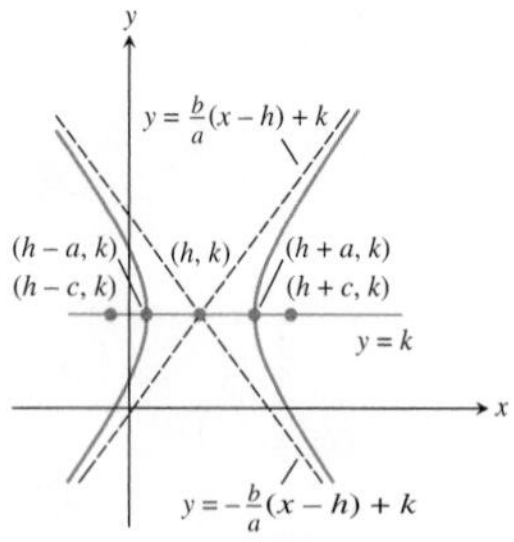

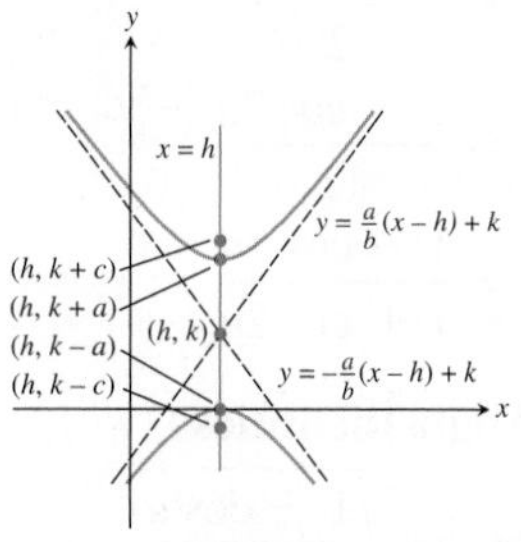

C.5 Gallery of Basic Functions

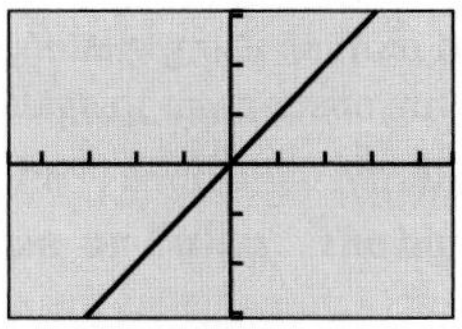

[–4.7, 4.7] by [–3.1, 3.1]

Identity Function

$f(x) = x$

Domain $= (-\infty, \infty)$

Range $= (-\infty, \infty)$

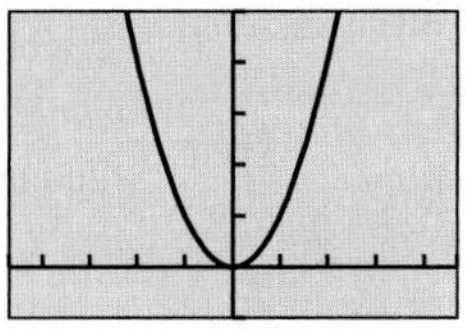

[–4.7, 4.7] by [–1, 5]

Squaring Function

$f(x) = x^2$

Domain $= (-\infty, \infty)$

Range $= [0, \infty)$

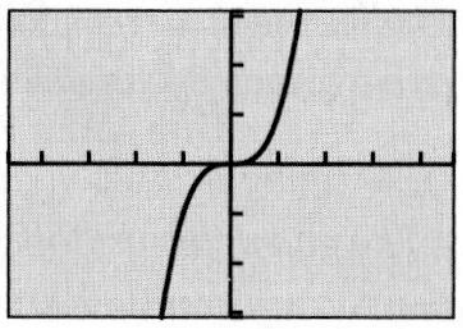

[–4.7, 4.7] by [–3.1, 3.1]

Cubing Function

$f(x) = x^3$

Domain $= (-\infty, \infty)$

Range $= (-\infty, \infty)$

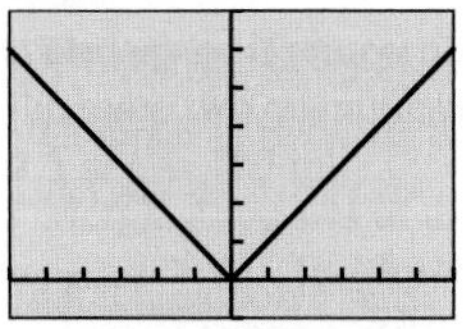

[–6, 6] by [–1, 7]

Absolute Value Function

$f(x) = |x| = \text{abs}\,(x)$

Domain $= (-\infty, \infty)$

Range $= [0, \infty)$

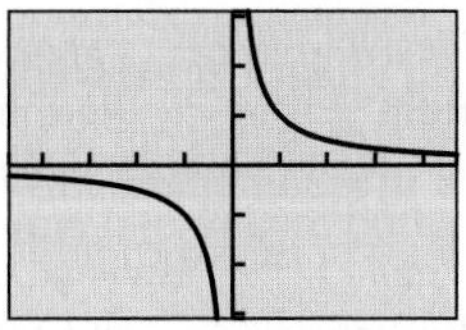

[–4.7, 4.7] by [–3.1, 3.1]

Reciprocal Function

$f(x) = \dfrac{1}{x}$

Domain $= (-\infty, 0) \cup (0, \infty)$

Range $= (-\infty, 0) \cup (0, \infty)$

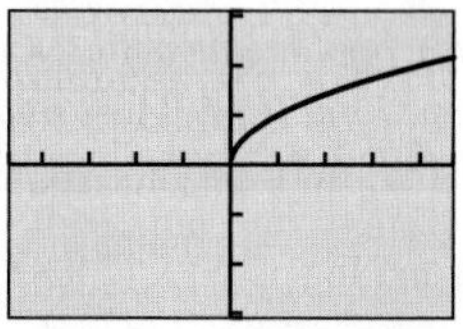

[–4.7, 4.7] by [–3.1, 3.1]

Square Root Function

$f(x) = \sqrt{x}$

Domain $= [0, \infty)$

Range $= [0, \infty)$

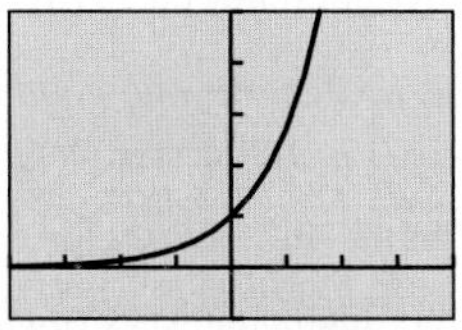

[–4, 4] by [–1, 5]

Exponential Function

$f(x) = e^x$

Domain $= (-\infty, \infty)$

Range $= (0, \infty)$

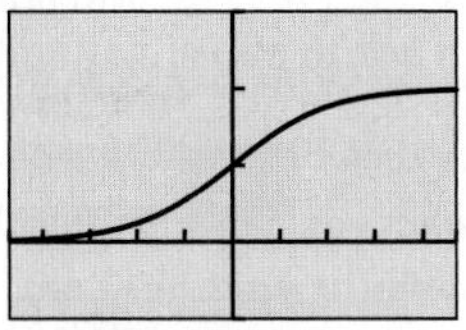

[–4.7, 4.7] by [–0.5, 1.5]

Logistic Function

$f(x) = \dfrac{1}{1 + e^{-x}}$

Domain $= (-\infty, \infty)$

Range $= (0, 1)$

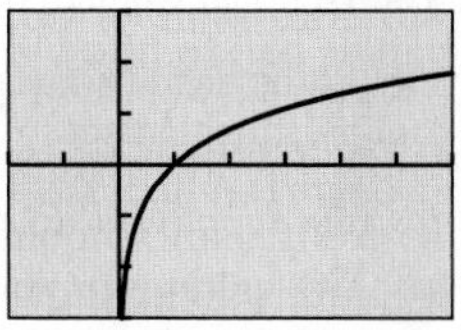

[–2, 6] by [–3, 3]

Natural Logarithmic Function

$f(x) = \ln x$

Domain $= (0, \infty)$

Range $= (-\infty, \infty)$

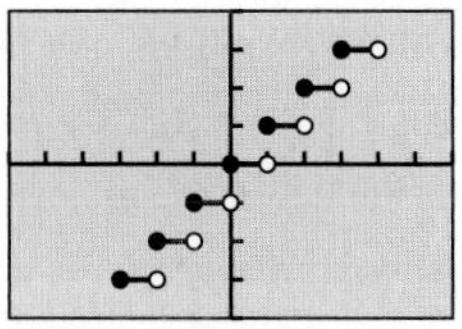

[–6, 6] by [–4, 4]

Greatest Integer Function

$f(x) = \text{int}\,(x)$

Domain $= (-\infty, \infty)$

Range = all integers

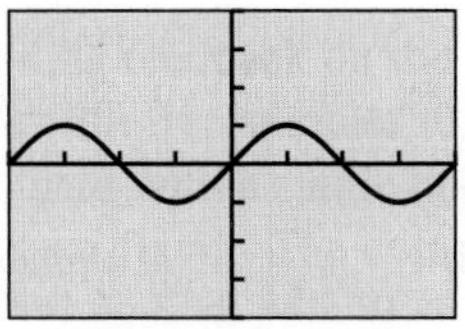

[-2π, 2π] by [–4, 4]

Sine Function

$f(x) = \sin\,(x)$

Domain $= (-\infty, \infty)$

Range $= [-1, 1]$

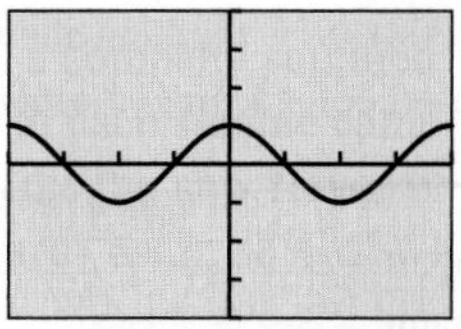

[-2π, 2π] by [–4, 4]

Cosine Function

$f(x) = \cos\,(x)$

Domain $= (-\infty, \infty)$

Range $= [-1, 1]$

Bibliography

Achieve. (2012, September). *Closing the expectations gap 2012: 50-state progress report on the alignment of high school policies with the demands of college and careers.* Washington, DC: Author. Retrieved from http://www.achieve.org/closingtheexpectationsgap2012

ACT. (2007). *Aligning postsecondary expectations and high school practice: The gap defined. Policy implications of the ACT National Curriculum Survey results 2005–2006.* Iowa City, IA: Author.

American Diploma Project. (2004). *Ready or not: Creating a high school diploma that counts.* Washington, DC: Achieve. Retrieved from http://www.achieve.org/files/ReadyorNot.pdf

American Mathematical Association of Two-Year Colleges. (2006). *Beyond crossroads: Implementing mathematics standards in the first two years of college.* Blair, R. (Ed.). Memphis, TN: Author.

Batanero, C., Burrill, G., and Reading, C. (Eds.). (2011). *Teaching statistics in school mathematics: Challenges for teaching and teacher education. Joint ICMI/IASE study: 18th ICMI study.* Dordrecht, Netherlands: Springer. doi:10.1007/978-94-007-1131-0

Bozick, R., and Ingels, S. J. (2008). *Mathematics coursetaking and achievement at the end of high school: Evidence from the Education Longitudinal Study of 2002 (ELS:2002)* (NCES 2008-319). Washington, DC: National Center for Education Statistics, Institute of Education Science, U.S. Department of Education.

College Board. (2006). *College Board standards for college success: Mathematics and statistics.* New York: Author.

Dick, T. P., and Hollebrands, K. F. (Eds.). (2011). *Focus in high school mathematics: Technology to support reasoning and sense making.* Reston, VA: National Council of Teachers of Mathematics.

Franklin, C., Kader, G., Mewborn, D., Moreno, J., Peck, R., Perry, M., and Scheaffer, R. (2007). *Guidelines for assessment and instruction in statistics education (GAISE) report: A pre-K–12 curriculum framework.* Alexandria, VA: American Statistical Association.

Garfield, J. B., and Ben-Zvi, D., with Chance, B. L., Medina, E., Roseth, C., and Zieffer, A. (2008). *Developing students' statistical reasoning: Connecting research and teaching practice.* New York, NY: Springer.

Graham, K., Cuoco, A., and Zimmermann, G. (2010). *Focus in high school mathematics: Reasoning and sense making—Algebra.* Reston, VA: National Council of Teachers of Mathematics.

Marcus, R., Fukawa-Connelly, T., Conklin, M., and Fey, J. T. (2007/2008). New thinking about college mathematics: Implications for high school teaching. *Mathematics Teacher, 101,* 354–358.

Murray, M. (2004). *Teaching mathematics vocabulary in context: Windows, doors, and secret passageways.* Portsmouth, NH: Heinemann.

National Council of Teachers of Mathematics. (2000). *Principles and standards for school mathematics.* Reston, VA: Author.

National Council of Teachers of Mathematics. (2007). *Mathematics teaching today: Improving practice, improving student learning* (2nd ed.). T. S. Martin (Ed.). Reston, VA: Author. (Originally published in 1991 as *Professional standards for teaching mathematics*)

National Council of Teachers of Mathematics. (2009, June). *Guiding principles for mathematics curriculum and assessment.* Reston, VA: Author. Retrieved from http://www.nctm.org/standards/content.aspx?id=23273

National Council of Teachers of Mathematics. (2009). *Focus in high school mathematics: Reasoning and sense making.* Reston, VA: Author.

National Council of Teachers of Mathematics. (2012). *Closing the opportunity gap in mathematics education* [Position statement]. Reston, VA: Author. Retrieved from http://www.nctm.org/about/content.aspx?id=6350

National Governors Association Center for Best Practices and Council of Chief State School Officers. (2010). *Common core state standards for mathematics.* Washington, DC: Author. Retrieved from http://corestandards.org/assets/CCSSI_Math%20Standards.pdf

Peck, R., and Starnes, D., with Kranendonk, H., and Morita, J. (2009). *Making sense of statistical studies.* Alexandria, VA: American Statistical Association.

Pólya, G. (1957). *How to solve it: A new aspect of mathematical method* (2nd ed.). Princeton, NJ: Princeton University Press. (Reprinted in paperback 1971; copyright renewed 1973)

Ramirez, N., and Celedon-Pattichis, S. (2012). *Beyond good teaching: Advancing mathematics education for ELLs.* Reston, VA: National Council of Teachers of Mathematics.

Seeley, C. (2004, October). 21st century mathematics. *Principal Leadership,* 22–26.

Steen, L. A. (Executive Ed.). (2001). *Mathematics and democracy: The case for quantitative literacy.* Princeton, NJ: National Council on Education and the Disciplines, and Woodrow Wilson National Fellowship Foundation.

Steen, L. A. (2006). Twenty questions about precalculus. In N. Baxter Hastings, F. S. Gordon, S. P. Gordon, and J. Narayan (Eds.). *A fresh start for collegiate mathematics: Rethinking the courses below calculus* (pp. 8–12). Washington, DC: Mathematical Association of America.

Waits, B. K., and Demana, F. (1988). Is three years enough? *Mathematics Teacher, 81,* 11–14.

Glossary

Absolute maximum A value $f(c)$ is an absolute maximum value of f if $f(c) \geq f(x)$ for all x in the domain of f, p. 90.

Absolute minimum A value $f(c)$ is an absolute minimum value of f if $f(c) \leq f(x)$ for all x in the domain of f, p. 90.

Absolute value of a complex number The absolute value of the complex number $z = a + bi$ is $\sqrt{a^2 + b^2}$, which is the length of the segment from the origin to z in the complex plane, p. 504.

Absolute value of a real number Denoted by $|a|$, represents the number a if $a \geq 0$ or the positive number $-a$ if $a < 0$, p. 13.

Absolute value of a vector See *Magnitude of a vector.*

Acceleration due to gravity $g \approx 32\ \text{ft/sec}^2 \approx 9.8\ \text{m/sec}^2$, p. 167.

Acute angle An angle whose measure is between 0° and 90°, p. 329.

Acute triangle A triangle in which all angles measure less than 90°, p. 434.

Addition principle of probability $P(A \text{ or } B) = P(A) + P(B) - P(A \text{ and } B)$. If A and B are mutually exclusive events, then $P(A \text{ or } B) = P(A) + P(B)$, p. 683.

Addition property of equality If $u = v$ and $w = z$, then $u + w = v + z$, p. 21.

Addition property of inequality If $u < v$, then $u + w < v + w$, p. 23.

Additive identity for the complex numbers $0 + 0i$ is the complex number zero, p. 50.

Additive inverse of a real number The opposite of b, or $-b$, p. 6.

Additive inverse of a complex number The opposite of $a + bi$, or $-a - bi$, p. 50.

Algebraic expression A combination of variables and constants involving addition, subtraction, multiplication, division, powers, and roots, p. 5.

Algebraic model An equation that relates variable quantities associated with phenomena being studied, p. 65.

Ambiguous case The case in which two sides and a nonincluded angle of a triangle are given, p. 435.

Amplitude See *Sinusoid.*

Anchor See *Mathematical induction.*

Angle Union of two rays with a common endpoint (the vertex). The beginning ray (the initial side) can be rotated about its endpoint to obtain the final position (the terminal side), p. 338.

Angle between vectors The angle formed by two nonzero vectors sharing a common initial point, p. 468.

Angle of depression The acute angle formed by the line of sight (downward) and the horizontal, p. 388.

Angle of elevation The acute angle formed by the line of sight (upward) and the horizontal, p. 388.

Angular speed Speed of rotation, typically measured in radians per unit time or revolutions per unit time, p. 323.

Annual percentage rate (APR) The annual interest rate expressed as a percent, p. 310.

Annual percentage yield (APY) The rate that would give the same return if interest were computed just once a year, p. 307.

Annuity A sequence of equal periodic payments, p. 308.

Aphelion The farthest point from the Sun in an object's orbit, p. 587.

Arc length formula The length of an arc in a circle of radius r intercepted by a central angle of θ radians is $s = r\theta$, p. 322.

Arccosecant function See *Inverse cosecant function.*

Arccosine function See *Inverse cosine function.*

Arccotangent function See *Inverse cotangent function.*

Arcsecant function See *Inverse secant function.*

Arcsine function See *Inverse sine function.*

Arctangent function See *Inverse tangent function.*

Argument of a complex number The argument of $a + bi$ is the direction angle of the vector $\langle a, b \rangle$, p. 504.

Arithmetic sequence A sequence $\{a_n\}$ in which $a_n = a_{n-1} + d$ for every integer $n \geq 2$. The number a_1 is the first term, and the number d is the common difference, p. 652.

Arrow The notation $\overrightarrow{PQ}$ denoting the directed line segment with initial point P and terminal point Q.

Association A relationship between two variables in which the distribution of one varies across values of the other, pp. 161, 692–693.

Associative properties $a + (b + c) = (a + b) + c$, $a(bc) = (ab)c$, p. 6.

Augmented matrix A matrix that represents a system of equations, p. 545.

Average rate of change of *f* over [*a*, *b*] The number $\dfrac{f(b) - f(a)}{b - a}$, provided $a \neq b$, p. 160.

Average velocity The change in position divided by the change in time, p. 754.

Axis of symmetry See *Line of symmetry.*

Back-to-back stemplot A stemplot with leaves on either side used to compare two distributions, p. 695.

Bar chart A rectangular graphical display of categorical data, p. 692.

Base See *Exponential function, Logarithmic function, nth power of a.*

Basic logistic function The function $f(x) = \dfrac{1}{1 + e^{-x}}$, p. 259.

Bearing Measure of the clockwise angle that the line of travel makes with due north, p. 321.

Bias A flaw in the design of a sampling process that systematically causes the sample to differ from the population with respect to the statistic being measured. **Undercoverage bias** results when the sample systematically excludes one or more segments of the population. **Voluntary response bias** results when a sample consists only of those who volunteer their responses. **Response bias** results when the sampling design intentionally or unintentionally influences the responses, pp. 735, 736.

Biconditional (statement) An if-and-only-if statement, p. 821.

Binomial A polynomial with exactly two terms, pp. 644, 802.

Binomial coefficients The numbers in Pascal's triangle:

$${}_nC_r = \binom{n}{r} = \frac{n!}{r!(n-r)!}, \text{ p. 645.}$$

Binomial probability For a random event with two possible outcomes, the probability of one outcome occurring k times in n independent trials is $P(E) = \dfrac{n!}{k!(n-k)!}p^k(1-p)^{n-k}$, where p is the probability of the outcome occurring once, p. 721.

Binomial theorem A theorem that gives an expansion formula for $(a + b)^n$, p. 644.

Blind experiment An experiment in which subjects do not know which treatment they have been given, p. 737.

Blocking A feature of some experimental designs that controls for potential differences between subject groups by applying treatments randomly within homogeneous blocks of subjects, p. 737.

Boundary (of a region) The set of points on the "edge" of a region, p. 557.

Bounded A function f is bounded if there are numbers b and B such that $b \le f(x) \le B$ for all x in the domain of f, p. 89.

Bounded above A function f is bounded above if there is a number B such that $f(x) \le B$ for all x in the domain of f, p. 89.

Bounded below A function f is bounded below if there is a number b such that $b \le f(x)$ for all x in the domain of f, p. 89.

Bounded interval An interval that has finite length (does not extend to ∞ or $-\infty$), p. 4.

Boxplot (or box-and-whisker plot) A graph that displays a five-number summary and identifies outliers, p. 706.

Branches (of a hyperbola) The two separate curves that make up a hyperbola, p. 593.

Cardioid A limaçon whose polar equation is $r = a \pm a \sin\theta$, or $r = a \pm a\cos\theta$, where $a > 0$, p. 498.

Cartesian coordinate system An association between the points in a plane and ordered pairs of real numbers; or an association between the points in three-dimensional space and ordered triples of real numbers, pp. 12, 621.

Categorical variable In statistics, a variable such as gender, hair color, or ZIP code that identifies an individual as having a certain characteristic, p. 691.

Causation A relationship between two variables in which the values of the response variable are directly affected by the values of the explanatory variable, p. 162.

Census A survey or observational study that gathers data from an entire population, p. 735.

Center The central point of a geometric figure, pp. 15, 582, 593, 624.

Central angle An angle whose vertex is the center of a circle, p. 320.

Chain rule A rule of logic that states: An implication is a transitive relation, p. 824.

Characteristic polynomial of a square matrix *A* $\det(xI_n - A)$, where A is an $n \times n$ matrix, p. 556.

Chord of a conic A line segment with endpoints on the conic, pp. 574, 583, 594.

Circle A set of points in a plane equally distant from a fixed point called the center, p. 15.

Circle graph A circular graphical display of categorical data, p. 692.

Circular functions Trigonometric functions when applied to real numbers are circular functions, p. 344.

Closed interval An interval that includes its endpoint(s), pp. 4–5.

Coefficient The real number multiplied by the variable(s) in a polynomial term, pp. 185, 802.

Coefficient of determination The number r^2 or R^2 that measures how well a regression curve fits the data, p. 146.

Coefficient matrix A matrix whose elements are the coefficients in a system of linear equations, p. 545.

Cofunction identity An identity that relates the sine, secant, or tangent to the cosine, cosecant, or cotangent, respectively, p. 406.

Combination A collection of elements of a set, in which order is not important, p. 638.

Combinations of *n* objects taken *r* at a time There are ${}_nC_r = \dfrac{n!}{r!(n-r)!}$ such combinations, p. 638.

Combinatorics A branch of mathematics related to determining the number of elements of a set or the number of ways objects can be arranged or combined, p. 634.

Common difference See *Arithmetic sequence.*

Common logarithm A logarithm with base 10, p. 275.

Common ratio See *Geometric sequence.*

Commutative properties $a + b = b + a$, $ab = ba$, p. 6.

Complements or complementary angles Two angles of positive measure whose sum is 90°, p. 406.

Completing the square A method of adding a constant to an expression in order to form a perfect square, p. 41.

Complex conjugates Complex numbers $a + bi$ and $a - bi$, p. 51.

Complex fraction See *Compound fraction*, p. 810.

Complex number An expression $a + bi$, where a (the real part) and b (the imaginary part) are real numbers, p. 49.

Complex plane A coordinate plane used to represent the set of complex numbers. The x-axis of the complex plane is the real axis, and the y-axis is the imaginary axis, p. 503.

Component form of a vector If a vector's representative in standard position has a terminal point (a, b) (or (a, b, c)), then $\langle a, b\rangle$ (or $\langle a, b, c\rangle$) is the component form of the vector, and a and b are the horizontal and vertical components of the vector (or a, b, and c are the x-, y-, and z-components of the vector), respectively, p. 456.

Components of a vector See *Component form of a vector.*

Composition of functions $(f \circ g)(x) = f(g(x))$, p. 111.

Compound fraction A fractional expression in which the numerator or denominator contains fractions, p. 811.

Compound interest Interest that becomes part of the investment, p. 304.

Compound statement A statement created by combining two or more statements using connectives such as *and* or *or*, p. 816.

Compounded annually See *Compounded k times per year.*

Compounded continuously Interest compounded using the formula $A = Pe^{rt}$, p. 306.

Compounded *k* times per year Interest compounded using the formula $A = P\left(1 + \frac{r}{k}\right)^{kt}$, where $k = 1$ is compounded annually, $k = 4$ is compounded quarterly, $k = 12$ is compounded monthly, etc., pp. 304–306.

Compounded monthly See *Compounded k times per year.*

Compounded quarterly See *Compounded k times per year.*

Conclusion The "then" portion of a conditional statement, p. 820.

Conditional probability The probability of an event B given that an event A has already occurred $(P(B|A))$, p. 684.

Conditional (statement) An if-then statement, p. 820.

Cone See *Right circular cone.*

Confounding variable A third variable that affects either of two variables being studied, making inferences about causation unreliable, p. 736.

Conic section (or conic) A curve obtained by intersecting a double-napped right circular cone with a plane, p. 571.

Conjugate axis of a hyperbola The line segment of length $2b$ that is perpendicular to the focal axis and has the center of the hyperbola as its midpoint, p. 594.

Conjunction A statement created by combining two statements using the connective *and*, p. 816.

Constant A letter or symbol that stands for a specific number, p. 5.

Constant function (on an interval) $f(x_1) = f(x_2)$ for any x_1 and x_2 (in the interval), pp. 87, 93, 159.

Constant term See *Polynomial function*, p. 161.

Constant of variation See *Power function.*

Constraints See *Linear programming problem.*

Continuous function A function that is continuous on its entire domain, p. 102.

Continuous at x = a $\lim_{x\to a} f(x) = f(a)$, p. 85.

Control The principle of experimental design that makes it possible to rule out other factors when making inferences about a particular explanatory variable, p. 737.

Convenience sample A sample that sacrifices randomness for convenience, p. 736.

Convergence of a sequence A sequence $\{a_n\}$ converges to a if $\lim_{n\to\infty} a_n = a$, p. 651.

Convergence of a series A series $\sum_{k=1}^{\infty} a_k$ converges to a sum S if $\lim_{n\to\infty} \sum_{k=1}^{n} a_k = S$, p. 662.

Conversion factor A ratio equal to 1, used for unit conversion, p. 143.

Coordinate(s) of a point The number associated with a point on a number line, the ordered pair associated with a point in the Cartesian coordinate plane, or the ordered triple associated with a point in the Cartesian three-dimensional space, pp. 3, 12, 621.

Coordinate plane See *Cartesian coordinate system.*

Correlation A measure of the strength and direction of a linear association between two quantitative variables, pp. 161, 732–734.

Cosecant The function $y = \csc x$, p. 364.

Cosine The function $y = \cos x$, p. 351.

Cotangent The function $y = \cot x$, p. 362.

Coterminal angles Two angles having the same initial side and the same terminal side, p. 338.

Course See *Bearing.*

Cube root nth root, where $n = 3$ (see *Principal* n*th root*), pp. 509, 797.

Cubic A degree 3 polynomial function, p. 185.

Cycloid The graph of the parametric equations $x = t - \sin t$, $y = 1 - \cos t$, p. 484.

Damping factor The factor Ae^{-at} in an equation such as $y = Ae^{-at} \cos bt$, p. 373.

Data Facts collected for statistical purposes (singular form is *datum*), p. 691.

De Moivre's theorem $(r(\cos \theta + i \sin \theta))^n = r^n(\cos n\theta + i \sin n\theta)$, p. 507.

Decreasing on an interval A function f is decreasing on an interval I if, for any two points in I, a positive change in x results in a negative change in $f(x)$, pp. 87, 93.

Deductive reasoning The process of using general information to prove a specific hypothesis, pp. 73, 80.

Definite integral The definite integral of the function f over $[a, b]$ is $\int_a^b f(x)\,dx = \lim_{n\to\infty} \sum_{i=1}^{n} f(x_i)\,\Delta x$ provided the limit exists, p. 768.

Degree Unit of measurement (represented by the symbol °) for angles or arcs, equal to 1/360 of a complete revolution, p. 320.

Degree of a polynomial (function) The largest exponent on the variable in any of the terms of the polynomial (function), pp. 158, 784.

Demand curve $p = g(x)$, where x represents demand and p represents price, p. 524.

Dependent variable Variable representing the range value of a function (usually y), p. 80.

Derivative of *f* The function f' defined by $f'(x) = \lim_{h\to 0} \dfrac{f(x+h) - f(x)}{h}$ for all x's for which the limit exists, p. 758.

Derivative of *f* at $x = a$ $f'(a) = \lim_{x\to a} \dfrac{f(x) - f(a)}{x - a}$, provided the limit exists, p. 758.

Descriptive statistics The process of gathering and analyzing data, p. 704.

Determinant A number calculated using the minors of a square matrix, pp. 534–535.

Difference identity An identity involving a trigonometric function of $u - v$, pp. 422–424.

Difference of complex numbers $(a + bi) - (c + di) = (a - c) + (b - d)i$, p. 49.

Difference of functions $(f - g)(x) = f(x) - g(x)$, p. 110.

Difference of two vectors $\langle u_1, u_2 \rangle - \langle v_1, v_2 \rangle = \langle u_1 - v_1, u_2 - v_2 \rangle$ or $\langle u_1, u_2, u_3 \rangle - \langle v_1, v_2, v_3 \rangle = \langle u_1 - v_1, u_2 - v_2, u_3 - v_3 \rangle$, p. 625.

Differentiable at $x = a$ $f'(a)$ exists, p. 740.

Dihedral angle An angle formed by two intersecting planes, p. 446.

Direct reasoning Reasoning based on the Law of Detachment (Modus Ponens), p. 824.

Direct variation See *Power function.*

Directed angle See *Polar coordinates.*

Directed distance See *Polar coordinates.*

Directed line segment See *Arrow.*

Direction angle of a vector in the plane The angle that the vector makes with the positive x-axis, p. 460.

Direction vector for a line A vector in the direction of a line, p. 626.

Direction of an arrow in the plane The angle the arrow makes with the positive x-axis, p. 456.

Directrix of a parabola, ellipse, or hyperbola A line used to determine the conic, pp. 572, 612.

Discriminant For the equation $ax^2 + bx + c = 0$, the expression $b^2 - 4ac$; for the equation $Ax^2 + Bxy + Cy^2 + Dx + Ey + F = 0$, the expression $B^2 - 4AC$, pp. 51, 608.

Disjunction A statement created by combining two statements using the connective *or*, p. 816.

Distance (in a coordinate plane) The distance $d(P, Q)$ between $P(x, y)$ and $Q(x, y)$: $d(P, Q) = \sqrt{(x_1 - x_2)^2 + (y_1 - y_2)^2}$, p. 14.

Distance (on a number line) The distance between real numbers a and b: $|a - b|$, p. 13.

Distance (in Cartesian space) The distance $d(P, Q)$ between $P(x, y, z)$ and $Q(x, y, z)$: $d(P, Q)\sqrt{(x_1 - x_2)^2 + (y_1 - y_2)^2 + (z_1 - z_2)^2}$, p. 622.

Distributive property $a(b + c) = ab + ac$ and related properties, p. 6.

Divergence A sequence or series diverges if it does not converge, pp. 651, 662.

Division $\dfrac{a}{b} = a\left(\dfrac{1}{b}\right)$, $b \neq 0$, p. 6.

Division algorithm for polynomials Given $f(x)$ and divisor $d(x) \neq 0$, there are unique polynomials $q(x)$ (quotient) and $r(x)$ (remainder) such that $f(x) = d(x)q(x) + r(x)$ with either $r(x) = 0$ or degree of $r(x) <$ degree of $d(x)$, p. 197.

Divisor of a polynomial *See Division algorithm for polynomials.*

DMS measure The measure of an angle in degrees, minutes, and seconds, p. 320.

Domain of a function The set of all input values for a function, p. 80.

Domain of an expression The set of all real numbers for which an expression is defined, p. 791.

Domain of validity of an identity The set of values of the variable for which both sides of the identity are defined, p. 404.

Dot product The sum found when the corresponding components of two vectors are multiplied and then added, p. 467.

Double-angle identity An identity involving a trigonometric function of $2u$, p. 428.

Double-blind experiment A blind experiment in which the researcher gathering data from the subjects is not told which subjects have received which treatment, p. 738.

Double inequality A statement that describes a bounded interval, such as $3 \le x < 5$, p. 25.

Eccentricity A nonnegative number that specifies how off-center the focus of a conic is, pp. 587, 597, 612.

Elementary row operations The following three row operations: Multiply all elements of a row by a nonzero constant; interchange two rows; and add a multiple of one row to another row, p. 546.

Elements of a matrix See *Matrix element.*

Elimination method A method of solving a system of linear equations, p. 521.

Ellipse The set of all points in the plane such that the sum of the distances from a pair of fixed points (the foci) is a constant, p. 582.

Ellipsoid of revolution A surface generated by rotating an ellipse about one of its axes, p. 589.

Empty set A set with no elements, p. 679.

End behavior The behavior of a graph of a function as $|x| \to \infty$, p. 187.

End behavior asymptote of a rational function A polynomial that the function approaches as $|x| \to \infty$, p. 221.

Endpoint of an interval A real number that is a bounding value of the interval, p. 5.

Equal complex numbers Complex numbers whose real parts are equal and whose imaginary parts are equal, p. 49.

Equal matrices Matrices that have the same order and equal corresponding elements, p. 529.

Equally likely outcomes Outcomes of a random phenomenon that have the same probability of occurring, p. 678.

Equation A statement of equality between two expressions, p. 21.

Equilibrium point A point where the supply curve and demand curve intersect. The corresponding price is the equilibrium price, p. 524.

Equilibrium price See *Equilibrium point.*

Equivalent arrows Arrows that have the same magnitude and direction, p. 457.

Equivalent expressions Expressions with a common value throughout a common domain, p. 810.

Equivalent equations (inequalities) Equations (inequalities) that have the same solutions, pp. 21, 24.

Equivalent systems of equations Systems of equations that have the same solution, p. 543.

Equivalent vectors Vectors with the same magnitude and direction, p. 457.

Even function A function whose graph is symmetric about the y-axis ($f(-x) = f(x)$ for all x in the domain of f), p. 90.

Event A subset of a sample space, p. 678.

Existential quantifier A word or phrase, such as *some* or *there is*, that refers to a subset of a set that is cited in a statement, p. 814.

Expanded form of a series A series written explicitly as a sum of terms (not in summation notation), p. 662.

Expected value The mean, μ, of a random variable X with probability function $P(X)$, denoted $E(X)$, pp. 689–690, 718.

Experiment A controlled study in which one or more treatments are imposed, pp. 678, 736.

Explanatory variable A variable that is associated with a response variable, p. 634.

Explicitly defined sequence A sequence in which the kth term is given as a function of k, p. 650.

Exponent See *nth power of a*, p. 7.

Exponential decay function Decay modeled by $f(x) = a \cdot b^x$, $a > 0$ with $0 < b < 1$, p. 254.

Exponential form An equation written with exponents instead of logarithms, p. 274.

Exponential function A function of the form $f(x) = a \cdot b^x$, where $a \neq 0, b > 0, b \neq 1$, p. 252.

Exponential growth function Growth modeled by $f(x) = a \cdot b^x$, $a > 0, b > 1$, p. 254.

Exponential regression A procedure for fitting an exponential function to a set of data, p. 145.

Extracting square roots A method for solving equations in the form $x^2 = k$, p. 41.

Extraneous solution Any solution of a resulting equation that is not a solution of the original equation, p. 228.

Factor (in algebra) A quantity being multiplied in a product, p. 199.

Factor (in statistics) A variable that is controlled or manipulated in an experiment, p. 736.

Factor theorem $x - c$ is a factor of a polynomial if and only if c is a zero of the polynomial, p. 199.

Factoring (a polynomial) Writing a polynomial as a product of two or more polynomial factors, p. 43.

Feasible points Points that satisfy the constraints in a linear programming problem, p. 560.

Feasible region Region of feasible points, p. 560

Fibonacci numbers The terms of the Fibonacci sequence, p. 655.

Fibonacci sequence The sequence 1, 1, 2, 3, 5, 8, 13, . . . , p. 655.

Finite sequence A function whose domain is the first n positive integers for some fixed integer n, p. 650.

Finite series Sum of a finite number of terms, p. 658.

First-degree equation in *x*, *y*, and *z* An equation that can be written in the form $Ax + By + Cz + D = 0$, p. 624.

First octant The points (x, y, z) in space with $x > 0$, $y > 0$, and $z > 0$, p. 621.

First quartile See *Quartile.*

Fitting a line or curve to data Finding a line or curve that comes close to passing through all the points in a scatter plot, p. 143.

Five-number summary The minimum, first quartile, median, third quartile, and maximum of a data set, p. 705.

Focal axis The line through the focus (foci) and perpendicular to the directrix (directrices) of a conic, pp. 573, 582, 593, 612.

Focal length of a parabola The *directed distance* from the vertex to the focus, p. 574.

Focal width of a parabola The length of the chord through the focus and perpendicular to the axis, p. 574.

Focus, foci See *Ellipse, Hyperbola, Parabola.*

Fractional expression A quotient of two algebraic expressions, p. 809.

Frequency Reciprocal of the period of a sinusoid, p. 353.

Frequency (in statistics) The number of individuals or observations within a category or interval, p. 696.

Frequency distribution See *Frequency table.*

Frequency table (in statistics) A table showing frequencies, p. 696.

Function A relation that associates each value in the domain with exactly one value in the range, pp. 80, 625.

Fundamental Theorem of Algebra A polynomial function of degree $n > 0$ has n complex zeros (counting multiplicity), p. 210.

Future value of an annuity The total value of an annuity at some point in the future, p. 308.

Gaussian curve See *Normal curve.*

Gaussian elimination A method of solving a system of n linear equations in n unknowns, p. 543.

General form (of a line) $Ax + By + C = 0$, where A and B are not both zero, p. 30.

Geometric sequence A sequence $\{a_n\}$ in which $a_n = a_{n-1} \cdot r$ for every positive integer $n \geq 2$. The number a_1 is the first term, and the nonzero number r is the common ratio, p. 653.

Geometric series A series whose terms form a geometric sequence, p. 663.

Graph of a function *f* The set of all points in the coordinate plane corresponding to the pairs $(x, f(x))$ for x in the domain of f, p. 81.

Graph of a polar equation The set of all points in the polar coordinate system corresponding to the ordered pairs (r, θ) that are solutions of the polar equation, p. 494.

Graph of a relation The set of all points in the coordinate plane corresponding to the ordered pairs of the relation, p. 116.

Graph of an equation in x and y The set of all points in the coordinate plane corresponding to the pairs (x, y) that are solutions of the equation, p. 31.

Graph of an inequality in x and y The set of all points in the coordinate plane corresponding to the solutions (x, y) of the inequality, p. 557.

Graph of parametric equations The set of all points in the coordinate plane corresponding to the ordered pairs determined by the parametric equations, p. 475.

Grapher or graphing utility Graphing calculator or a computer with graphing software, p. 31.

Graphical model A visual representation of a numerical or algebraic model, p. 66.

Half-angle identity Identity involving a trigonometric function of $u/2$, p. 430.

Half-life The amount of time required for half of a radioactive substance to decay, p. 266.

Half-plane The graph of the linear inequality $y \geq ax + b$, $y > ax + b$, $y \leq ax + b$, or $y < ax + b$, p. 557.

Head minus tail (HMT) rule An arrow with initial point (x_1, y_1) and terminal point (x_2, y_2) represents the vector $\langle x_2 - x_1, y_2 - y_1 \rangle$, p. 457.

Heron's formula The area of ΔABC with semiperimeter s is given by $\sqrt{s(s-a)(s-b)(s-c)}$, p. 445.

Higher-degree polynomial function A polynomial function whose degree is ≥ 3, p. 185.

Histogram A graph that displays the distribution of a quantitative variable using rectangular areas proportional to the frequencies, p. 696.

Horizontal asymptote The line $y = b$ is a horizontal asymptote of the graph of a function f if $\lim_{x\to-\infty} f(x) = b$ or $\lim_{x\to\infty} f(x) = b$, p. 93.

Horizontal component See *Component form of a vector.*

Horizontal line $y = b$, p. 30.

Horizontal Line Test A test for determining whether the inverse of a relation is a function, p. 122.

Horizontal shrink or stretch See *Shrink, Stretch.*

Horizontal translation A shift of a graph to the left or right, p. 130.

Hyperbola A set of points in a plane, the absolute value of the difference of whose distances from two fixed points (the foci) is a constant, p. 593.

Hyperboloid of revolution A surface generated by rotating a hyperbola about one of its axes, p. 598.

Hypotenuse Side opposite the right angle in a right triangle, p. 329.

Hypothesis The "if" portion of a conditional statement, p. 817.

Identity An equation that is always true throughout its domain, p. 404.

Identity function The function $f(x) = x$, p. 99.

Identity matrix A square matrix with 1's in the main diagonal and 0's elsewhere, p. 534.

Identity properties $a + 0 = a$, $a \cdot 1 = a$, p. 6.

Imaginary axis See *Complex plane.*

Imaginary part of a complex number See *Complex number.*

Imaginary unit The complex number $i = \sqrt{-1}$, p. 49.

Implication An if-then statement, p. 820.

Implicitly defined function A function that is a subset of a relation defined by an equation in x and y, p. 115.

Implied domain The domain of a function's algebraic expression, p. 82.

Increasing on an interval A function f is increasing on an interval I if, for any two points in I, a positive change in x results in a positive change in $f(x)$, p. 87.

Independent events Events A and B such that $P(B|A) = P(B)$, pp. 681, 685.

Independent variable Variable representing the domain value of a function (usually x), p. 80.

Index See *Radical.*

Index of summation See *Summation notation.*

Indirect reasoning Reasoning based on Modus Tollens, p. 824.

Inductive step See *Mathematical induction.*

Inequality A statement that compares two quantities using an inequality symbol, p. 3.

Inequality symbol $<, >, \leq$, or $\geq$, p. 3.

Inferential statistics Using the science of statistics to make inferences about the parameters in a population from information about a sample, p. 704.

Infinite discontinuity at $x = a$ $\lim_{x\to a^+} f(x) = \pm\infty$ or $\lim_{x\to a^-} f(x) = \pm\infty$, p. 85.

Infinite limit A special case of a limit that does not exist, p. 778.

Infinite sequence A function whose domain is the set of all natural numbers, p. 650.

Initial point See *Arrow.*

Initial side of an angle See *Angle.*

Initial value of a function $f(0)$, p. 161.

Instantaneous rate of change See *Derivative f at $x = a$.*

Instantaneous velocity The instantaneous rate of change of a position function with respect to time, p. 755.

Integers The numbers $\ldots, -3, -2, -1, 0, 1, 2, \ldots$, p. 2.

Integrable over [a, b] $\int_a^b f(x)\,dx$ exists, p. 768.

Intercept Point where a curve crosses the x-, y-, or z-axis in a graph, pp. 31, 70, 621.

Intercepted arc Arc of a circle from the initial side through the terminal side of a central angle, p. 320.

Intermediate Value Theorem If f is a polynomial function and $a < b$, then f assumes every value between $f(a)$ and $f(b)$, p. 190.

Interquartile range (*IQR*) A measure of spread, found by subtracting the first quartile from the third quartile, p. 705.

Interval Connected subset of the real number line with at least two points, p. 4.

Inverse composition rule The composition of a one-to-one function with its inverse results in the identity function, p. 124.

Interval notation Notation used to specify intervals, pp. 4, 5.

Inverse cosecant function The function $y = \csc^{-1} x$.

Inverse cosine function The function $y = \cos^{-1} x$, p. 380.

Inverse cotangent function The function $y = \cot^{-1} x$.

Inverse function The inverse relation of a one-to-one function, p. 122.

Inverse of a matrix The inverse of a square matrix A, if it exists, is a matrix B, such that $AB = BA = I$, where I is an identity matrix, p. 533.

Inverse properties $a + (-a) = 0$, $a \cdot \frac{1}{a} = 1 (a \neq 0)$, p. 6.

Inverse reflection principle If the graph of a relation is reflected across the line $y = x$, the graph of the inverse relation results, p. 123.

Inverse relation (of the relation *R*) A relation that consists of all ordered pairs (b, a) for which (a, b) belongs to R, p. 121.

Inverse secant function The function $y = \sec^{-1} x$.

Inverse sine function The function $y = \sin^{-1} x$, p. 378.

Inverse tangent function The function $y = \tan^{-1} x$, p. 381.

Inverse variation See *Power function.*

Invertible linear system A system of n linear equations in n variables whose coefficient matrix has a nonzero determinant, p. 549.

Irrational numbers Real numbers that are not rational, p. 2.

Irrational zeros Zeros of a function that are irrational numbers, p. 201.

Irreducible quadratic over the reals A quadratic polynomial with real coefficients that cannot be factored using real coefficients, p. 214.

Jump discontinuity at $x = a$ $\lim_{x \to a^-} f(x)$ and $\lim_{x \to a^+} f(x)$ exist but are not equal, p. 85.

***k*th term of a sequence** The *k*th expression in the sequence, p. 650.

Law of cosines $a^2 = b^2 + c^2 - 2bc \cos A$, $b^2 = a^2 + c^2 - 2ac \cos B$, $c^2 = a^2 + b^2 - 2ab \cos C$, p. 442.

Law of Detachment A law of logic that states: If an implication and its hypothesis are both true, then the conclusion must be true, p. 824.

Law of sines $\dfrac{\sin A}{a} = \dfrac{\sin B}{b} = \dfrac{\sin C}{c}$, p. 434.

Leading coefficient See *Polynomial function in x.*

Leading term See *Polynomial function in x,* p. 693.

Leaf The last significant digit of a number in a stemplot, p. 694.

Least common denominator (LCD) The product of the prime factors across a set of denominators, for which each factor in the product is raised to the greatest power found in any one denominator for that factor, p. 811.

Least-squares line See *Linear regression line.*

Leibniz notation The notation *dy/dx* for the derivative of *f*, p. 760.

Left-hand limit of *f* at $x = a$ The limit of *f* as *x* approaches *a* from the left, p. 776.

Lemniscate A graph of a polar equation of the form $r^2 = a^2 \sin 2\theta$ or $r^2 = a^2 \cos 2\theta$, p. 499.

Length of an arrow See *Magnitude of an arrow.*

Length of a vector See *Magnitude of a vector.*

Like terms Terms of a polynomial that have the same variables raised to the same powers, p. 802.

Limaçon A graph of a polar equation $r = a \pm b \sin \theta$ or $r = a \pm b \cos \theta$ with $a > 0$, $b > 0$, p. 497.

Limit $\lim_{x \to a} f(x) = L$ means that $f(x)$ gets arbitrarily close to L as x gets arbitrarily close (but not equal) to a, p. 773.

Limit to growth See *Logistic growth function.*

Limit at infinity $\lim_{x \to \infty} f(x) = L$ means that $f(x)$ gets arbitrarily close to L as x gets arbitrarily large; $\lim_{x \to -\infty} f(x) = L$ means that $f(x)$ gets arbitrarily close to L as $-x$ gets arbitrarily large, pp. 766, 778.

Line graph A graph of data in which consecutive data points are connected by line segments, p. 698.

Line of symmetry A line over which a graph is the mirror image of itself, p. 164.

Line of travel The path along which an object travels, p. 321.

Linear combination of vectors u and v An expression $a\mathbf{u} + b\mathbf{v}$, where a and b are real numbers, p. 460.

Linear equation in x An equation that can be written in the form $ax + b = 0$, where a and b are real numbers and $a \neq 0$, p. 21.

Linear factorization theorem A polynomial $f(x)$ of degree $n > 0$ has the factorization $f(x) = a(x - z_1)(x - z_2) \cdots (x - z_n)$ where the z_i are the (complex) zeros of f, p. 210.

Linear function A function that can be written in the form $f(x) = mx + b$, where $m \neq 0$ and b are real numbers, p. 159.

Linear inequality in two variables x and y An inequality that can be written in one of the following forms: $y < mx + b$, $y \leq mx + b$, $y > mx + b$, or $y \geq mx + b$ with $m \neq 0$, p. 557.

Linear inequality in x An inequality that can be written in the form $ax + b < 0$, $ax + b \leq 0$, $ax + b > 0$, or $ax + b \geq 0$, where a and b are real numbers and $a \neq 0$, p. 23.

Linear programming problem A method of solving certain problems involving maximizing or minimizing a function of two variables (called an objective function) subject to restrictions (called constraints), p. 559.

Linear regression A procedure for finding the straight line that is the best fit for the association between two variables, p. 145.

Linear regression equation Equation of a linear regression line, p. 144.

Linear regression line The line for which the sum of the squares of the residuals is the smallest possible, p. 144.

Linear system A system of linear equations, p. 519.

Local extremum A local maximum or a local minimum, p. 90.

Local maximum A value $f(c)$ is a local maximum of f if there is an open interval I containing c such that $f(x) \leq f(c)$ for all values of x in I, p. 90.

Local minimum A value $f(c)$ is a local minimum of f if there is an open interval I containing c such that $f(x) \geq f(c)$ for all values of x in I, p. 90.

Logarithm An expression of the form $\log_b x$ (see *Logarithmic function*), p. 274.

Logarithmic form An equation written with logarithms instead of exponents, p. 274.

Logarithmic function with base *b* The inverse of the exponential function $y = b^x$, denoted by $y = \log_b x$, pp. 274, 287.

Logarithmic re-expression of data Transformation of a data set using logarithms, p. 298.

Logarithmic regression See *Natural logarithmic regression.*

Logically equivalent (statements) Statements that have the same truth value, p. 817.

Logistic growth function A model of population growth: $f(x) = \frac{c}{1 + a \cdot b^x}$ or $f(x) = \frac{c}{1 + ae^{-kx}}$, where a, b, c, and k are positive with $b < 1$. c is the limit to growth, p. 258.

Logistic regression A procedure for fitting a logistic curve to a set of data, p. 145.

Lower bound of *f* Any number b for which $b \le f(x)$ for all x in the domain of f, p. 89.

Lower bound for real zeros A number c is a lower bound for the set of real zeros of f if $f(x) \neq 0$ whenever $x < c$, p. 202.

Lower bound test for real zeros A test for finding a lower bound for the real zeros of a polynomial, p. 202.

LRAM A Riemann sum approximation of the area under a curve $f(x)$ from $x = a$ to $x = b$ using x_i as the left-hand endpoint of each subinterval, p. 768.

Magnitude of an arrow The magnitude of $\overrightarrow{PQ}$ is the distance between P and Q, p. 456.

Magnitude of a real number See *Absolute value of a real number.*

Magnitude of a vector The magnitude of $\langle a, b \rangle$ is $\sqrt{a^2 + b^2}$. The magnitude of $\langle a, b, c \rangle$ is $\sqrt{a^2 + b^2 + c^2}$, pp. 458, 625.

Main diagonal The diagonal from the top left to the bottom right of a square matrix, p. 533.

Major axis The line segment through the foci of an ellipse with endpoints on the ellipse, p. 583.

Mapping A function viewed as a correspondence of the elements of the domain onto the elements of the range, p. 80.

Mathematical model A mathematical structure that approximates a phenomenon for the purpose of studying or predicting its behavior, p. 64.

Mathematical induction A process for proving that a statement is true for all natural numbers n by showing that it is true for $n = 1$ (the anchor) and that, if it is true for $n = k$, then it must be true for $n = k + 1$ (the inductive step), p. 667.

Matrix, $m \times n$ A rectangular array of m rows and n columns of numbers, p. 529.

Matrix element Any of the numbers in a matrix, p. 529.

Maximum *r*-value The value of $|r|$ at the point on the graph of a polar equation that has the maximum distance from the pole, p. 496.

Mean (of a set of data) The sum of all the data divided by the total number of items, p. 708.

Measure of an angle The number of degrees or radians in an angle, p. 320.

Measure of center A measure of the typical, middle, or average value for a data set, p. 704.

Measure of spread A measure of variability in quantitative data, p. 705.

Median (of a data set) The middle number (or the mean of the two middle numbers) if the data are listed in order, p. 705.

Midpoint (in a coordinate plane) For the line segment with endpoints (a, b) and (c, d), $\left(\frac{a + c}{2}, \frac{b + d}{2}\right)$, p. 15.

Midpoint (on a number line) For the line segment with endpoints a and b, $\frac{a + b}{2}$, p. 14.

Midpoint (in Cartesian space) For the line segment with endpoints (x_1, y_1, z_1) and (x_2, y_2, z_2), $\left(\frac{x_1 + x_2}{2}, \frac{y_1 + y_2}{2}, \frac{z_1 + z_2}{2}\right)$, p. 622.

Minor axis The perpendicular bisector of the major axis of an ellipse with endpoints on the ellipse, p. 583.

Minute Angle measure equal to 1/60 of a degree, p. 320.

Modulus See *Absolute value of a complex number.*

Modus Ponens See *Law of Detachment*, p. 824.

Modus Tollens A law of logic that states: If an implication is true and its conclusion is false, then the hypothesis must be false, p. 824.

Monomial (function) A polynomial with exactly one term, pp. 175, 802.

Multiplication principle of counting A principle used to find the number of ways an event can occur, p. 635.

Multiplication principle of probability If A and B are independent events, then $P(A \text{ and } B) = P(A) \cdot P(B)$. If B depends on A, then $P(A \text{ and } B) = P(A) \cdot P(B|A)$, p. 681.

Multiplication property of equality If $u = v$ and $w = z$, then $uw = vz$, p. 21.

Multiplication property of inequality If $u < v$ and $c > 0$, then $uc < vc$. If $u < v$ and $c < 0$, then $uc > vc$, p. 23.

Multiplicative identity for matrices See *Identity matrix.*

Multiplicative inverse of a complex number The reciprocal of $a + bi$, or $\frac{1}{a + bi} = \frac{a}{a^2 + b^2} - \frac{b}{a^2 + b^2}i$, $ab \neq 0$, p. 51.

Multiplicative inverse of a matrix See *Inverse of a matrix.*

Multiplicative inverse of a real number The reciprocal of b, or $1/b$, $b \neq 0$, p. 6.

Multiplicity The multiplicity of a zero c of a polynomial $f(x)$ of degree $n > 0$ is the number of times the factor $(x - c)$ occurs in the linear factorization $f(x) = a(x - z_i)(x - z_2) \cdots (x - z_n)$, p. 189.

Nappe See *Right circular cone.*

Natural exponential function The function $f(x) = e^x$, p. 256.

Natural logarithm A logarithm with base *e*, p. 277.

Natural logarithmic function The inverse of the exponential function $y = e^x$, denoted by $y = \ln x$, p. 278.

Natural logarithmic regression A procedure for fitting a natural logarithmic curve to a set of data, p. 145.

Natural numbers The numbers 1, 2, 3, . . . , p. 2.

Nautical mile Length of 1 minute of arc along the Earth's equator, p. 324.

NDER $f(a)$ See *Numerical derivative of f at $x = a$*, p. 784.

Negation A statement with a truth value opposite that of a given statement, p. 801.

Negative association A relationship between two quantitative variables in which greater values of one variable are generally associated with lesser values of the other variable, p. 732.

Negative angle Angle generated by clockwise rotation, p. 338.

Negative numbers Real numbers less than zero (shown to the left of the origin on a number line), p. 3.

Newton's law of cooling $T(t) = T_m + (T_0 - T_m)e^{-kt}$, p. 296.

***n* factorial** For any positive integer *n*, *n* factorial is $n! = n \cdot (n-1) \cdot (n-2) \cdot \cdots \cdot 3 \cdot 2 \cdot 1$; zero factorial is $0! = 1$, p. 636.

NINT $(f(x), x, a, b)$ A calculator approximation to $\int_a^b f(x)dx$, p. 785.

Nonsingular matrix A square matrix with nonzero determinant, p. 533.

Normal curve The graph of $f(x) = e^{-x^2/2}$, p. 711.

Normal distribution See *Normal model.*

Normal model A theoretical probability distribution shaped like a *Normal curve*, p. 723.

***n*-set** A set of *n* objects, p. 636.

n*th power of *a The number $a^n = a \cdot a \cdot \cdots \cdot a$ (with *n* factors of *a*), where *n* is the exponent and *a* is the base, p. 7.

***n*th root** See *Principal nth root.*

n*th root of a complex number *z A complex number *v* such that $v^n = z$, p. 508.

***n*th root of unity** A complex number *v* such that $v^n = 1$, p. 508.

Number line graph of a linear inequality The graph of the solutions of a linear inequality (in *x*) on a number line, p. 24.

Numerical derivative of *f* at *a* NDER $f(a) = \dfrac{f(a + 0.001) - f(a - 0.001)}{0.002}$, p. 784.

Numerical model A model determined by analyzing numbers or data in order to gain insight into a phenomenon, p. 64.

Objective function See *Linear programming problem.*

Observational study A process for gathering data from a subset of a population through current or past observations. This differs from a sample survey in that the subset may not be randomly chosen and from an experiment in that no treatment is imposed, p. 735.

Obtuse triangle A triangle in which one angle is greater than 90°, p. 434.

Octants The eight regions of space determined by the coordinate planes, p. 621.

Odd-even identity For a basic trigonometric function *f*, an identity relating $f(x)$ to $f(-x)$ p. 407.

Odd function A function whose graph is symmetric about the origin ($f(-x) = -f(x)$ for all *x* in the domain of *f*), p. 91.

One-to-one function A function in which each element of the range corresponds to exactly one element in the domain, p. 122.

One-to-one rule of exponents $x = y$ if and only if $b^x = b^y$, p. 292.

One-to-one rule of logarithms For $x > 0$ and $y > 0$, $x = y$ if and only if $\log_b x = \log_b y$, p. 292.

Open interval An interval that does not include its endpoints, p. 5.

Opens upward or downward A parabola $y = ax^2 + bx + c$ opens upward if $a > 0$ and opens downward if $a < 0$, p. 575.

Opposite See *Additive inverse of a real number and Additive inverse of a complex number.*

Order of magnitude (of *n*) $\log n$, $n > 0$, p. 294.

Order of an $m \times n$ matrix The order of an $m \times n$ matrix is $m \times n$, p. 529.

Ordered pair A pair of real numbers (x, y), p. 12.

Ordinary annuity An annuity in which deposits are made at the same time interest is posted, p. 308.

Origin The number zero on a number line, or the point where the *x*- and *y*-axes cross in the Cartesian plane, or the point where the *x*-, *y*-, and *z*-axes cross in Cartesian three-dimensional space, p. 3.

Orthogonal vectors Two vectors **u** and **v** such that $\mathbf{u} \cdot \mathbf{v} = 0$, p. 469.

Outcomes The various possible results of an experiment, p. 678.

Outliers Data values more than 1.5 times the *IQR* below the first quartile or above the third quartile, p. 706.

Parabola The graph of a quadratic function, or the set of points in a plane that are equidistant from a fixed point (the focus) and a fixed line (the directrix), p. 572.

Paraboloid of revolution A surface generated by rotating a parabola about its line of symmetry, p. 577.

Parallel lines Two lines that are both vertical or that have equal slopes, p. 32.

Parallelogram representation of vector addition Geometric representation of vector addition using the parallelogram determined by the position vectors.

Parameter See *Parametric equations.*

Parameter (in statistics) A numerical description of a population or model, p. 704.

Parameter interval See *Parametric equations.*

Parametric curve The graph of a set of parametric equations, p. 475.

Parametric equations Equations of the form $x = f(t)$ and $y = g(t)$ for all t in an interval I. The variable t is the parameter and I is the parameter interval, pp. 119, 475.

Parametric equations for a line in space The line through $P_0(x_0, y_0, z_0)$ in the direction of the nonzero vector $\mathbf{v} = \langle a, b, c \rangle$ has parametric equations $x = x_0 + at$, $y = y_0 + bt$, $z = z_0 + ct$, p. 626.

Parametrization A set of parametric equations for a curve, p. 475.

Partial fraction decomposition See *Partial fractions.*

Partial fractions The process of expanding a fraction into a sum of fractions. The sum is called the partial fraction decomposition of the original fraction, p. 550.

Partial sums See *Sequence of partial sums.*

Pascal's triangle A number pattern in which row n (beginning with $n = 0$) consists of the coefficients of the expanded form of $(a + b)^n$, p. 645.

Perihelion The closest point to the Sun in an object's orbit, p. 587.

Period See *Periodic function.*

Periodic function A function f for which there is a positive number c such that $f(t + c) = f(t)$ for every value t in the domain of f. The smallest such number c is the period of the function, p. 345.

Permutation An arrangement of elements of a set, in which order is important, p. 636.

Permutations of *n* objects taken *r* at a time There are $_nP_r = \dfrac{n!}{(n - r)!}$ such permutations, p. 637.

Perpendicular lines Two lines that are at right angles to each other, p. 31.

pH A logarithmic measure of acidity, p. 295.

Phase shift See *Sinusoid.*

Piecewise-defined function A function whose domain is divided into several parts with a different function rule applied to each part, p. 104.

Pie chart See *Circle graph.*

Placebo In an experimental study, an inactive treatment that is equivalent to the active treatment in every respect except for the factor about which an inference is to be made. Subjects in a blind experiment do not know whether they have been given the active treatment or the placebo, p. 737.

Plane in Cartesian space The graph of $Ax + By + Cz + D = 0$, where A, B, and C are not all zero, p. 624.

Point-slope form (of a line) $y - y_1 = m(x - x_1)$, p. 29.

Polar axis See *Polar coordinate system.*

Polar coordinate system A coordinate system whose ordered pair is based on the directed distance from a central point (the pole) and the angle measured from a ray from the pole (the polar axis), p. 487.

Polar coordinates The numbers (r, θ) that determine a point's location in a polar coordinate system. The number r is the directed distance and θ is the directed angle, p. 487.

Polar distance formula The distance between the points with polar coordinates (r_1, θ_1) and (r_2, θ_2) $= \sqrt{r_1^2 + r_2^2 - 2r_1r_2 \cos(\theta_1 - \theta_2)}$, p. 493.

Polar equation An equation in r and θ, p. 490.

Polar form of a complex number See *Trigonometric form of a complex number.*

Pole See *Polar coordinate system.*

Polynomial function A function in which $f(x)$ is a polynomial in x, p. 158.

Polynomial in x An expression that can be written in the form $a_nx^n + a_{n-1}x^{n-1} + \cdots + a_1x + a_0$, where n is a nonnegative integer, the coefficients are real numbers, and $a_n \neq 0$. The degree of the polynomial is n, the leading coefficient is a_n, the leading term is a_nx^n, and the constant term is a_0. (The number 0 is the zero polynomial), pp. 158, 802.

Polynomial interpolation The process of fitting a polynomial of degree n to $(n + 1)$ points, p. 192.

Position vector of the point (*a*, *b*) The vector $\langle a, b \rangle$, p. 456.

Positive angle Angle generated by a counterclockwise rotation, p. 338.

Positive association A relationship between two quantitative variables in which greater values of one variable are generally associated with greater values of the other variable, p. 732.

Positive numbers Real numbers greater than zero (shown to the right of the origin on a number line), p. 3.

Power function A function of the form $f(x) = k \cdot x^a$, where k and a are nonzero constants. k is the constant of variation and a is the power, p. 174.

Power-reducing identity A trigonometric identity that reduces the power to which the trigonometric functions are raised, p. 454.

Power regression A procedure for fitting a curve $y = a \cdot x^b$ to a set of data, p. 145.

Power rule of logarithms $\log_b R^c = c \log_b R, R > 0$, p. 283.

Present value of an annuity The current value of an investment put into an annuity, p. 309.

Prime polynomial A polynomial that cannot be factored using integer coefficients, p. 803.

Principal *n*th root If $b^n = a$, then b is an nth root of a. If $b^n = a$ and a and b have the same sign, b is the principal nth root of a (see *Radical*), pp. 508, 797.

Principle of mathematical induction A principle related to mathematical induction, p. 668.

Probability distribution The collection of probabilities of outcomes in a sample space assigned by a probability function, p. 679.

Probability of an event The value that an event's relative frequency of occurrence approaches in the long run, p. 678.

Probability of an event modeled by a finite sample space of equally likely outcomes The number of outcomes in the event divided by the number of outcomes in the sample space, p. 678.

Probability model A random variable X together with a function P that assigns a real number to each outcome x_i satisfying: $0 \le P(x_i) \le 1$, $P(\varnothing) = 0$, and the sum of the probabilities of all outcomes is 1, pp. 717–718.

Product of complex numbers $(a + bi)(c + di) = (ac - bd) + (ad + bc)i$, pp. 50, 505.

Product of a scalar and a vector The product of scalar k and vector $\mathbf{u} = \langle u_1, u_2 \rangle$ (or $\mathbf{u} = \langle u_1, u_2, u_3 \rangle$) is $k \cdot \mathbf{u} = \langle ku_1, ku_2 \rangle$ (or $k \cdot \mathbf{u} = \langle ku_1, ku_2, ku_3 \rangle$), pp. 458, 625.

Product of functions $(fg)(x) = f(x)g(x)$, p. 110.

Product of matrices *A* and *B* The matrix in which each entry is obtained by multiplying the entries of a row of A by the corresponding entries of a column of B and then adding, p. 530.

Product rule of logarithms $\log_b (RS) = \log_b R + \log_b S$, $R > 0, S > 0$, p. 283.

Projection of u onto v The vector $\text{proj}_{\mathbf{v}}\, \mathbf{u} = \left(\dfrac{\mathbf{u} \cdot \mathbf{v}}{|\mathbf{v}|}\right)^2 \mathbf{v}$, p. 470.

Proportional See *Power function.*

Pseudo-random numbers Computer-generated numbers that can be used to approximate true randomness in scientific studies. Because they depend on iterative computer algorithms, they are not truly random, p. 738.

Pythagorean identities $\sin^2 \theta + \cos^2 \theta = 1$, $1 + \tan^2 \theta = \sec^2 \theta$, and $1 + \cot^2 \theta = \csc^2 \theta$, p. 405.

Pythagorean Theorem In a right triangle with legs a and b and hypotenuse c, $c^2 = a^2 + b^2$, p. 14.

Quadrant Any one of the four parts into which a plane is divided by the perpendicular coordinate axes, p. 12.

Quadrantal angle An angle in standard position whose terminal side lies on an axis, p. 342.

Quadratic equation in *x* An equation that can be written in the form $ax^2 + bx + c = 0 (a \ne 0)$, p. 41.

Quadratic formula The formula $x = \dfrac{-b \pm \sqrt{b^2 - 4ac}}{2a}$ used to solve $ax^2 + bx + c = 0$, $a \ne 0$, p. 42.

Quadratic function A function that can be written in the form $f(x) = ax^2 + bx + c$, where a, b, and c are real numbers, and $a \ne 0$, p. 164.

Quadratic regression A procedure for fitting a quadratic function to a set of data, p. 145.

Quadric surface The graph in three dimensions of a second-degree equation in three variables, p. 625.

Quantifier A word or phrase, such as *all*, *some*, or *for any*, that specifies all or part of a set that is referenced in a statement, p. 814.

Quantitative variable A variable (in statistics) that takes on numerical values for an attribute being measured, p. 691.

Quartic function A degree 4 polynomial function, p. 185.

Quartic regression A procedure for fitting a quartic function to a set of data, p. 145.

Quartile The first quartile is the median of the lesser half of a set of quantitative data, the second quartile is the median, and the third quartile is the median of the greater half of the data, p. 705.

Quotient identities $\tan \theta = \dfrac{\sin \theta}{\cos \theta}$ and $\cot \theta = \dfrac{\cos \theta}{\sin \theta}$, p. 404.

Quotient of complex numbers
$\dfrac{a + bi}{c + di} = \dfrac{ac + bd}{c^2 + d^2} + \dfrac{bc - ad}{c^2 + d^2} i$, pp. 51, 505.

Quotient of functions $\left(\dfrac{f}{g}\right)(x) = \dfrac{f(x)}{g(x)}$, $g(x) \ne 0$, p. 110.

Quotient rule of logarithms $\log_b \left(\dfrac{R}{S}\right) = \log_b R - \log_b S$, $R > 0, S > 0$, p. 283.

Quotient polynomial See *Division algorithm for polynomials.*

Radian The measure of a central angle whose intercepted arc has a length equal to the circle's radius, p. 321.

Radian measure The measure of an angle in radians, or, for a central angle, the ratio of the length of the intercepted arc to the radius of the circle, p. 322.

Radical expression An expression of the form $\sqrt[n]{a}$; a is the *radicand*, p. 797.

Radicand See *Radical expression*, p. 797.

Radius The distance from a point on a circle (or a sphere) to the center of the circle (or the sphere), pp. 15, 624.

Random behavior Behavior that is determined only by the laws of probability, p. 734.

Random numbers Numbers that can be used by researchers to simulate randomness in scientific studies (they are usually obtained from lengthy tables of decimal digits that have been generated by verifiably random natural phenomena), p. 738.

Random sample A subset of a population chosen in a way that makes each member of the population equally likely to be included in the sample, pp. 735, 738.

Random variable A variable that assumes numerical values as the result of a random event, pp. 679, 712, 717.

Randomization The principle of experimental design that makes it possible to use the laws of probability when making inferences, p. 737.

Range of a function The set of all output values corresponding to elements in the domain, p. 80.

Range (in statistics) The difference *maximum – minimum* in a data set, p. 705.

Range screen See *Viewing window.*

Rational expression An expression that can be written as a ratio of two polynomials, p. 809.

Rational function Function of the form $\frac{f(x)}{g(x)}$, where $f(x)$ and $g(x)$ are polynomials and $g(x)$ is not the zero polynomial, p. 218.

Rational numbers Numbers that can be written as a/b, where a and b are integers, and $b \neq 0$, p. 2.

Rational zeros Zeros of a function that are rational numbers, p. 201.

Rational zeros theorem A procedure for finding the possible rational zeros of a polynomial, p. 201.

Rationalizing the denominator A method to remove radicals from the denominator of a fractional expression, p. 799.

Real axis See *Complex plane.*

Real number Any number that can be written as a decimal, p. 2.

Real number line A line that represents the set of real numbers, p. 3.

Real part of a complex number See *Complex number.*

Real zeros Zeros of a function that are real numbers, p. 189.

Reciprocal function The function $f(x) = \frac{1}{x}$, p. 100.

Reciprocal identity An identity that equates a trigonometric function with the reciprocal of another trigonometric function, p. 404.

Reciprocal of a real number See *Multiplicative inverse of a real number.*

Rectangular coordinate system See *Cartesian coordinate system.*

Recursively defined sequence A sequence defined by giving the first term (or the first few terms) along with a procedure for finding the subsequent terms, p. 650.

Reduced form A rational number or rational expression with no factors common to numerator and denominator, p. 809.

Reduced row echelon form A matrix in row echelon form with every column that has a leading 1 having 0's in all other positions, p. 547.

Re-expression of data A transformation of a data set, p. 287.

Reference angle See *Reference triangle.*

Reference triangle For an angle θ in standard position, a reference triangle is a triangle formed by the terminal side of angle θ, the x-axis, and a perpendicular dropped from a point on the terminal side to the x-axis. The angle in a reference triangle at the origin is the reference angle, p. 341.

Reflection Two points that are symmetric with respect to a line or a point, p. 131.

Reflection across the *x*-axis (x, y) and $(x, -y)$ are reflections of each other across the x-axis, p. 131.

Reflection across the *y*-axis (x, y) and $(-x, y)$ are reflections of each other across the y-axis, p. 131.

Reflection through the origin (x, y) and $(-x, -y)$ are reflections of each other through the origin, p. 131.

Reflexive property of equality $a = a$, p. 21.

Regression model An equation found by regression that can be used to predict unknown output values, p. 144.

Relation A set of ordered pairs of real numbers, p. 115.

Relevant domain The portion of the domain applicable to the situation being modeled, p. 82.

Remainder polynomial See *Division algorithm for polynomials.*

Remainder theorem If a polynomial $f(x)$ is divided by $x - c$, the remainder is $f(c)$, p. 198.

Removable discontinuity at $x = a$ $\lim_{x \to a^-} f(x) = \lim_{x \to a^+} f(x)$ but either the common limit is not equal to $f(a)$ or $f(a)$ is not defined, p. 84.

Repeated zeros Zeros of multiplicity ≥ 2 (see *Multiplicity*), p. 190.

Replication The principle of experimental design that minimizes the effects of chance variation by repeating the experiment multiple times, p. 737.

Residual The difference $y_1 - (ax_1 + b)$, where (x_1, y_1) is a point in a scatter plot and $y = ax + b$ is a line that fits the set of data, p. 173.

Resistant measure A statistical measure that does not change much in response to outliers, p. 705.

Resolving a vector Finding the horizontal and vertical components of a vector, p. 460.

Response variable A variable that is associated with an explanatory variable, p. 634.

Richter scale A logarithmic scale used to measure the intensity of an earthquake, pp. 290, 295.

Riemann sum A sum $\sum_{i=1}^{n} f(x_i)\Delta x$ where the interval $[a, b]$ is divided into n subintervals of equal length Δx and x_i is in the ith subinterval, p. 768.

Right angle A 90° angle, p. 329.

Right circular cone The surface created when a line is rotated in three-dimensional space about a second line in the same plane, p. 571.

Right-hand limit of *f* at $x = a$ The limit of f as x approaches a from the right, p. 776.

Right triangle A triangle with a 90° angle, p. 14.

Rigid transformation A transformation that leaves the size and shape of a graph unchanged, p. 129.

Root of a number See *Principal nth root.*

Root of an equation A solution, p. 70.

Rose curve A graph of a polar equation $r = a \cos n\theta$ or $r = a \sin n\theta$, p. 496.

Row echelon form A matrix in which rows consisting of all 0's occur only at the bottom of the matrix, the first nonzero entry in any row with nonzero entries is 1, and the leading 1's move to the right as we move down the rows, p. 546.

Row operations See *Elementary row operations.*

RRAM A Riemann sum approximation of the area under a curve $f(x)$ from $x = a$ to $x = b$ using x_i as the right-hand end point of each subinterval, p. 768.

Sample A subset of a population, usually random, from which data are collected, p. 704.

Sample space Set of all possible outcomes of a random phenomenon, p. 678.

Sample standard deviation The *standard deviation* computed using only a sample of the entire population, p. 710.

Scalar A real number, p. 458.

Scatter plot A plot of all the ordered pairs of a two-variable data set on a coordinate plane, p. 12.

Scientific notation A positive number written as $c \times 10^m$, where $1 \le c < 10$ and m is an integer, p. 8.

Secant The function $y = \sec x$, p. 363.

Secant line of *f* A line joining two points of the graph of f, p. 758.

Second Angle measure equal to 1/60 of a minute, p. 320.

Second-degree equation in two variables $Ax^2 + Bxy + Cy^2 + Dx + Ey + F = 0$, where A, B, and C are not all zero, p. 571.

Semimajor axis The distance from the center to a vertex of an ellipse, p. 583.

Semiminor axis The distance from the center of an ellipse to a point on the ellipse along a line perpendicular to the major axis, p. 583.

Semiperimeter of a triangle One-half of the sum of the lengths of the sides of a triangle, p. 445.

Sequence *See Finite sequence, Infinite sequence.*

Sequence of partial sums The sequence $\{S_n\}$, where S_n is the nth partial sum of the series, that is, the sum of the first n terms of the series, p. 662.

Series A finite or infinite sum of terms, p. 658.

Shrink of factor *c* A transformation of a graph obtained by multiplying all the x-coordinates (horizontal shrink) or all of the y-coordinates (vertical shrink) by the constant c, $0 < c < 1$, p. 134.

Simple harmonic motion Motion described by $d = a \sin \omega t$ or $d = a \cos \omega t$, p. 391.

Simulation A process that uses random numbers to model a real-world phenomenon, p. 740.

Sine The function $y = \sin x$, p. 350.

Singular matrix A square matrix for which the determinant is zero, p. 533.

Sinusoid A function that can be written in the form $f(x) = a \sin (b(x - h)) + k$ or $f(x) = a \cos (b(x - h)) + k$. The number $|a|$ is the amplitude, and the number h is the phase shift, p. 352.

Sinusoidal regression A procedure for fitting a curve $y = a \sin (bx + c) + d$ to a set of data, p. 145.

68-95-99.7 Rule An approximation of the percent of data values that lie within ± 1, ± 2, and ± 3 standard deviations of the mean in a Normal distribution, p. 712.

Skewed distribution A distribution that is not symmetric; distributions with a longer tail to the left are *skewed left*, and those with a longer tail on the right are *skewed right*, pp. 697, 704.

Slant asymptote An end behavior asymptote that is a slant line, p. 221.

Slant line A line that is neither horizontal nor vertical, p. 159.

Slope The ratio $\dfrac{\text{change in } y}{\text{change in } x}$, p. 28.

Slope-intercept form (of a line) $y = mx + b$, p. 30.

Solution set of an inequality The set of all solutions of an inequality, p. 23.

Solution of a system in two variables An ordered pair of real numbers that satisfies all of the equations or inequalities in the system, p. 519.

Solution of an equation or inequality A value of the variable (or values of the variables) for which the equation or inequality is true, pp. 21, 23.

Solve algebraically Use an algebraic method, including paper and pencil manipulation and obvious mental work, with no calculator or grapher use. When appropriate, the final exact solution may be approximated by a calculator, p. 69.

Solve graphically Use a graphical method, including use of a hand sketch or use of a grapher. When appropriate, the approximate solution should be confirmed algebraically, p. 69.

Solve an equation or inequality To find all solutions of the equation or inequality, pp. 21, 23.

Solve a triangle To find all unknown sides or angles of a triangle, p. 333.

Solve a system To find all solutions of a system, p. 519.

Solve by elimination A method for solving systems of linear equations, p. 521.

Solve by substitution A method for solving systems of linear equations, p. 519.

Speed The magnitude of a velocity vector, p. 461.

Sphere A set of points in Cartesian space equally distant from a fixed point called the center, p. 624.

Spiral of Archimedes The graph of the polar curve $r = \theta$, p. 499.

Square matrix A matrix for which the number of rows equals the number of columns, p. 529.

Square root nth root, where $n = 2$ (see *Principle* n*th root*), p. 797.

Standard deviation A measure of spread in a quantitative data set or population model, p. 710.

Standard form:

equation of a circle $(x - h)^2 + (y - k)^2 = r^2$, p. 15.

equation of an ellipse $\frac{(x - h)^2}{a^2} + \frac{(y - k)^2}{b^2} = 1$ or $\frac{(y - k)^2}{a^2} + \frac{(x - h)^2}{b^2} = 1$, p. 583.

equation of a hyperbola $\frac{(x - h)^2}{a^2} - \frac{(y - k)^2}{b^2} = 1$ or $\frac{(y - k)^2}{a^2} - \frac{(x - h)^2}{b^2} = 1$, p. 594.

equation of a parabola $(x - h)^2 = 4p(y - k)$ or $(y - k)^2 = 4p(x - h)$, p. 574.

equation of a quadratic function $f(x) = ax^2 + bx + c (a \neq 0)$, p. 164.

Standard form of a complex number $a + bi$, where a and b are real numbers, p. 49.

Standard form of a polar equation of a conic $r = \frac{ke}{1 \pm e \cos \theta}$ or $r = \frac{ke}{1 \pm e \sin \theta}$, pp. 612–613.

Standard form of a polynomial function $f(x) = a_n x^n + a_{n-1}x^{n-1} + \cdots + a_1 x + a_0$, p. 185.

Standard position (angle) An angle positioned on a rectangular coordinate system with its vertex at the origin and its initial side on the positive x-axis, pp. 329, 338.

Standard representation of a vector A representative arrow with its initial point at the origin, p. 456.

Standard unit vectors In the plane $\mathbf{i} = \langle 1, 0 \rangle$ and $\mathbf{j} = \langle 0, 1 \rangle$; in space $\mathbf{i} = \langle 1, 0, 0 \rangle$, $\mathbf{j} = \langle 0, 1, 0 \rangle$, and $\mathbf{k} = \langle 0, 0, 1 \rangle$, pp. 460, 625.

Statement A sentence that is either true or false, but not both, p. 801.

Statistic A numerical description of a variable in a sample from a population, p. 704.

Statistically significant An outcome so unlikely to have arisen by chance alone as to cast doubt on underlying assumptions, p. 726.

Statute mile 5280 ft, p. 324.

Stem The initial digit or digits of a number in a stemplot, p. 693.

Stemplot (or stem-and-leaf plot) An arrangement of a numerical data set into a specific tabular format, p. 693.

Stretch of factor c A transformation of a graph obtained by multiplying all the x-coordinates (horizontal stretch) or all of the y-coordinates (vertical stretch) of the points by a constant c, $c > 1$, p. 134.

Subtraction $a - b = a + (-b)$, p. 6.

Sum identity An identity involving a trigonometric function of $u + v$, p. 421.

Sum of a finite arithmetic series $S_n = n\left(\frac{a_1 + a_n}{2}\right) = \frac{n}{2}[2a_1 + (n - 1)d]$, p. 659.

Sum of a finite geometric series $S_n = \frac{a_1(1 - r^n)}{1 - r}$, p. 660.

Sum of an infinite geometric series $S_n = \frac{a}{1 - r}$, $|r| < 1$, p. 663.

Sum of an infinite series See *Convergence of a series.*

Sum of complex numbers $(a + bi) + (c + di) = (a + c) + (b + d)i$, p. 49.

Sum of functions $(f + g)(x) = f(x) + g(x)$, p. 110.

Sum of two vectors $\langle u_1, u_2 \rangle + \langle v_1, v_2 \rangle = \langle u_1 + v_1, u_2 + v_2 \rangle$ or $\langle u_1, u_2, u_3 \rangle + \langle v_1, v_2, v_3 \rangle = \langle u_1 + v_1, u_2 + v_2, u_3 + v_3 \rangle$, pp. 458, 625.

Summation notation The series $\sum_{k=1}^{n} a_k$, where n is a natural number (or ∞) is in summation notation and is read "the sum of a_k from $k = 1$ to n (or infinity)." k is the index of summation, and a_k is the kth term of the series, p. 658.

Supply curve $p = f(x)$, where x represents production and p represents price, p. 524.

Symmetric difference quotient of *f* at *a*
$\frac{f(a+h) - f(a-h)}{2h}$, p. 784.

Symmetric matrix A matrix $A = [a_{ij}]$ with the property $a_{ij} = a_{ji}$ for all i and j, p. 540.

Symmetric about the origin A graph in which $(-x, -y)$ is on the graph whenever (x, y) is; or a graph in which $(-r, \theta)$ or $(r, \theta + \pi)$ is on the graph whenever (r, θ) is, pp. 91, 494.

Symmetric about the *x*-axis A graph in which $(x, -y)$ is on the graph whenever (x, y) is; or a graph in which $(r, -\theta)$ or $(-r, \pi - \theta)$ is on the graph whenever (r, θ) is, pp. 91, 494.

Symmetric about the *y*-axis A graph in which $(-x, y)$ is on the graph whenever (x, y) is; or a graph in which $(-r, -\theta)$ or $(r, \pi - \theta)$ is on the graph whenever (r, θ) is, pp. 90, 494.

Symmetric property of equality If $a = b$, then $b = a$, p. 21.

Synthetic division A procedure used to divide a polynomial by a linear expression of the form, $x - a$, p. 199.

System A set of equations or inequalities, p. 519.

Tangent The function $y = \tan x$, p. 361.

Tangent line to the graph of *f* at *x* = *a* The line through $(a, f(a))$ with slope $f'(a)$ provided $f'(a)$ exists, p. 758.

Tautology A statement that is always true, p. 822.

Terminal point See *Arrow.*

Terminal side of an angle See *Angle.*

Term of a polynomial (function) An expression of the form $a_n x^n$ in a polynomial (function), p. 185.

Term of a sequence A range element of a sequence, p. 650.

Third quartile See *Quartile.*

Time plot A line graph in which time is measured on the horizontal axis, p. 698.

Transformation A function that maps points to points, p. 129.

Transitive property If $a = b$ and $b = c$, then $a = c$. Similar properties hold for inequalities: $<, \leq, >, \geq$, pp. 21, 23.

Translation See *Horizontal translation, Vertical translation.*

Transpose of a matrix The matrix A^T obtained by interchanging the rows and columns of A, p. 533.

Transverse axis The line segment whose endpoints are the vertices of a hyperbola, p. 594.

Tree diagram A pictorial representation of the *Multiplication Principle of Probability*, p. 683.

Triangular form A special form for a system of linear equations that facilitates finding the solution, p. 543.

Triangular number A number that is a sum of the arithmetic series $1 + 2 + 3 + \cdots + n$ for some natural number n, p. 666.

Trichotomy property For real numbers a and b, exactly one of the following is true: $a < b$, $a = b$, or $a > b$, p. 4.

Trigonometric form of a complex number $r(\cos\theta + i\sin\theta)$, p. 504.

Trinomial A polynomial that has exactly three terms, p. 802.

Truth table A table used to show the true-false patterns of a statement, p. 816.

Unbounded interval An interval that extends to $-\infty$ or ∞ (or both), p. 5.

Union of two sets *A* and *B* The set of all elements that belong to A or B or both, p. 55.

Unit circle A circle with radius 1 centered at the origin, p. 344.

Unit ratio See *Conversion factor.*

Unit vector Vector of length 1, pp. 459, 625.

Unit vector in the direction of a vector A unit vector that has the same direction as the given vector, pp. 460, 625.

Universal quantifier A word or phrase, such as *all*, *none*, or *for any*, that refers to an entire set that is cited in a statement, p. 814.

Upper bound for *f* Any number B for which $f(x) \leq B$ for all x in the domain of f, p. 89.

Upper bound for real zeros A number d is an upper bound for the set of real zeros of f if $f(x) \neq 0$ whenever $x > d$, p. 202.

Upper bound test for real zeros A test for finding an upper bound for the real zeros of a polynomial, p. 202.

Valid A form of argument or reasoning for which the conclusion follows unavoidably from the hypothesis, p. 822.

Variable A letter that represents an unspecified number, p. 5.

Variable (in statistics) A characteristic that is being identified or measured, p. 691.

Variance The square of the standard deviation, p. 710.

Variation See *Power function.*

Vector An ordered pair $\langle a, b\rangle$ of real numbers in the plane, or an ordered triple $\langle a, b, c\rangle$ of real numbers in space. A vector has both magnitude and direction, pp. 456, 625.

Vector equation for a line in space The line through $P_0(x_0, y_0, z_0)$ in the direction of the nonzero vector $\mathbf{v} = \langle a, b, c\rangle$ has vector equation $\mathbf{r} = \mathbf{r}_0 + t\mathbf{v}$, where $\mathbf{r}_0 = \langle x_0, y_0, z_0\rangle$, p. 626.

Velocity A vector that specifies an object's speed and direction, p. 461.

Venn diagram A pictorial representation of the relationships among events within a sample space, p. 682.

Vertex of a cone See *Right circular cone.*

Vertex of a parabola The point of intersection of a parabola and its line of symmetry, pp. 164, 573.

Vertex of an angle See *Angle.*

Vertex form for a quadratic function $f(x) = a(x - h)^2 + k$, p. 165.

Vertical asymptote The line $x = a$ is a vertical asymptote of the graph of the function f if $\lim_{x \to a^+} f(x) = \pm\infty$ or $\lim_{x \to a^-} f(x) = \pm\infty$, pp. 93, 221.

Vertical component See *Component form of a vector.*

Vertical line $x = a$, p. 30.

Vertical line test A test for determining whether a graph is a function, p. 81.

Vertical stretch or shrink See *Stretch, Shrink.*

Vertical translation A shift of a graph up or down, p. 131.

Vertices of an ellipse The points where an ellipse intersects its focal axis, p. 582.

Vertices of a hyperbola The points where a hyperbola intersects its focal axis, p. 593.

Viewing window The rectangular portion of the coordinate plane specified by the dimensions [Xmin, Xmax] by [Ymin, Ymax], p. 31.

Weighted mean A mean calculated in such a way that some elements of the data set have higher weights (that is, are counted more strongly in determining the mean) than others, p. 709.

Whole numbers The numbers 0, 1, 2, 3, . . . , p. 2.

Window dimensions The restrictions on x and y that specify a viewing window. See *Viewing window.*

Work The dot product of a force and direction vector $W = \mathbf{F} \cdot \overrightarrow{AB}$, p. 471.

Wrapping function The function that maps points on a real number line to points on the unit circle, p. 344.

x-axis Usually the horizontal coordinate line in a Cartesian coordinate system with positive direction to the right, pp. 12, 621.

x-coordinate The directed distance from the y-axis (yz-plane) to a point in a plane (space), or the first number in an ordered pair (triple), pp. 12, 621.

x-intercept A point that lies on both the graph and the x-axis, pp. 31, 70, 199, 624.

Xmax The x-value of the right side of the viewing window, p. 31.

Xmin The x-value of the left side of the viewing window, p. 31.

Xscl The scale of the tick marks on the x-axis in a viewing window, p. 31.

xy-plane The points $(x, y, 0)$ in Cartesian space, p. 621.

xz-plane The points $(x, 0, z)$ in Cartesian space, p. 621.

y-axis Usually the vertical coordinate line in a Cartesian coordinate system with positive direction up, pp. 12, 629.

y-coordinate The directed distance from the x-axis (xz-plane) to a point in a plane (space), or the second number in an ordered pair (triple), pp. 12, 621.

y-intercept A point that lies on both the graph and the y-axis, pp. 29, 624.

Ymax The y-value of the top of the viewing window, p. 31.

Ymin The y-value of the bottom of the viewing window, p. 31.

Yscl The scale of the tick marks on the y-axis in a viewing window, p. 31.

yz-plane The points $(0, y, z)$ in Cartesian space, p. 621.

z-axis Usually the third dimension in Cartesian space, p. 621.

z-coordinate The directed distance from the xy-plane to a point in space, or the third number in an ordered triple, p. 621.

z-intercept A point that lies on both the graph and the z-axis, p. 624.

z-score The number of standard deviations a value lies from the mean of its distribution, pp. 723–724.

Zero factor property If $ab = 0$, then $a = 0$ or $b = 0$, pp. 41, 69.

Zero factorial $0! = 1$. See *n factorial.*

Zero of a function A value in the domain of a function that makes the function value zero, pp. 70, 199.

Zero matrix A matrix consisting entirely of zeros, p. 531.

Zero vector The vector $\langle 0, 0 \rangle$ or $\langle 0, 0, 0 \rangle$, pp. 456, 625.

Zoom out A procedure of a graphing utility used to view more of the coordinate plane (used, for example, to find the end behavior of a function), p. 188.

Selected Answers

SECTION P.1

Quick Review P.1

1. $\{1, 2, 3, 4, 5, 6\}$ **3.** $\{-3, -2, -1\}$ **5. (a)** 1187.75 **(b)** -4.72 **7.** -3; 1.375 **9.** 0, 1, 2, 3, 4, 5, 6

Exercises P.1

1. -4.625 (terminates) **3.** $-2.1\overline{6}$ (repeats) **5.** ; All real numbers less than or equal to 2

7. 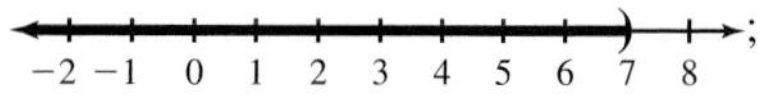; All real numbers less than 7 **9.** ; All real numbers less than 0

11. $-1 \le x < 1$ **13.** $-\infty < x < 5$, or $x < 5$ **15.** $-1 < x < 2$ **17.** $(-3, \infty)$ **19.** $(-2, -1)$ **21.** $(-3, 4]$

23. The real numbers greater than 4 and less than or equal to 9 **25.** The real numbers greater than or equal to -3, or the real numbers which are at least -3 **27.** The real numbers greater than -1 **29.** $-3 < x \le 4$; endpoints -3 and 4; bounded; half-open **31.** $x < 5$; endpoint 5; unbounded; open **33.** $x \ge 29$ or $[29, \infty)$; x = Bill's age in years **35.** $3.099 \le x \le 4.399$ or $[3.099, 4.399]$; x = dollars per gallon of gasoline **37.** $ax^2 + ab$ **39.** $(a + d)x^2$ **41.** $\pi - 6$ **43.** 5 **45. (a)** Associative property of multiplication **(b)** Commutative property of multiplication **(c)** Addition inverse property **(d)** Addition identity property **(e)** Distributive property of multiplication over addition **47.** $\frac{x^2}{y^2}$ **49.** $\frac{16}{x^4}$ **51.** x^4y^4 **53.** $\$5.18997 \times 10^{11}$ **55.** $\$1.6691 \times 10^{10}$

57. 4.839×10^8 mi **59.** 0.000 000 033 3 **61.** 5,870,000,000,000 mi **63.** 2.4×10^{-8}

65. (a) Because $a^m \ne 0$, $a^m a^0 = a^{m+0} = a^m$ implies that $a^0 = 1$. **(b)** Because $a^m \ne 0$, $a^m a^{-m} = a^{m-m} = a^0 = 1$ implies that $a^{-m} = \frac{1}{a^m}$.

67. False. For example, the additive inverse of -5 is 5, which is positive. **69.** E **71.** B **73.** 0, 1, 2, 3, 4, 5, 6
75. $-6, -5, -4, -3, -2, -1, 0, 1, 2, 3, 4, 5, 6$

SECTION P.2

Quick Review P.2

1. 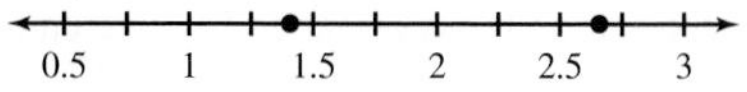 **3.** 

Distance: $\sqrt{7} - \sqrt{2} \approx 1.232$

5. 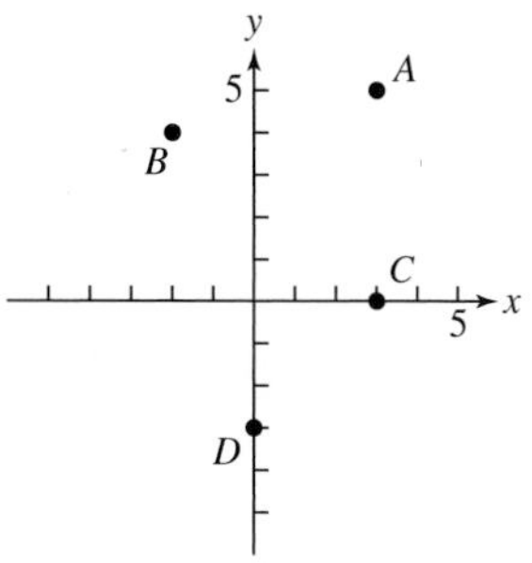

7. 5.5 **9.** 10

Exercises P.2

1. $A(1, 0)$, $B(2, 4)$, $C(-3, -2)$, $D(0, -2)$ **3. (a)** First quadrant **(b)** On the y-axis, between quadrants I and II **(c)** Second quadrant **(d)** Third quadrant **5.** 6 **7.** 6 **9.** $4 - \pi$ **11.** 19.9 **13.** 8 **15.** 5 **17.** 7
19. Perimeter $= 2\sqrt{41} + \sqrt{82} \approx 21.86$; area $= 20.5$ **21.** Perimeter $= 2\sqrt{20} + 16 \approx 24.94$; area $= 32$

23. 0.65 **25.** $(2, 6)$ **27.** $\left(-\frac{1}{3}, -\frac{3}{4}\right)$

29.

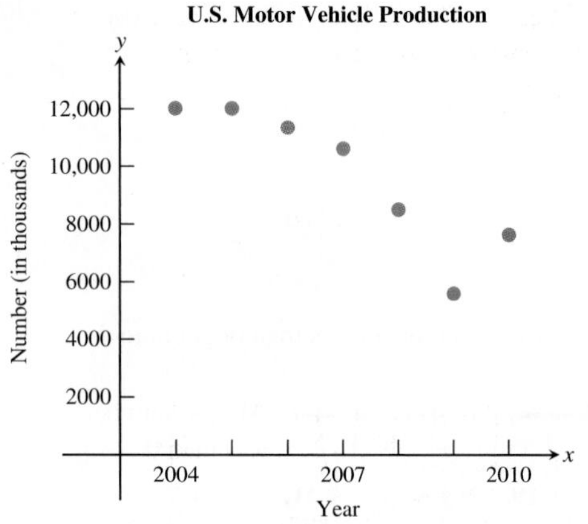

31.

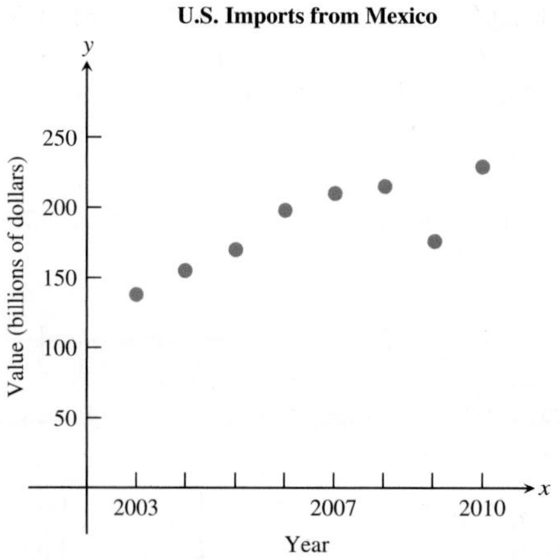

33.

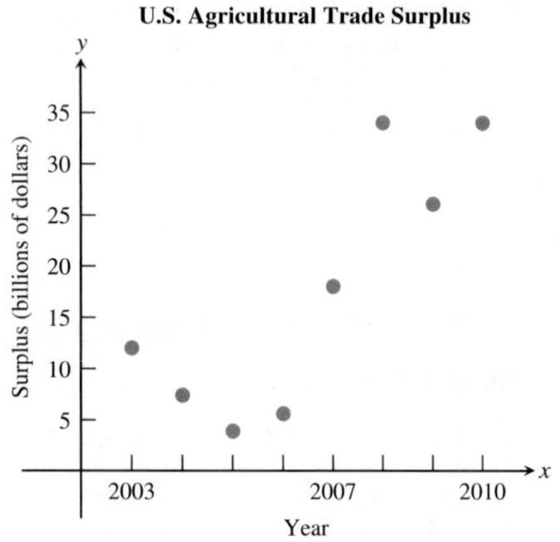

35. **(a)** About \$1.40 **(b)** About \$2.80 **(c)** About \$2.80 **37.** The three sides have lengths 5, 5, and $5\sqrt{2}$. Because two sides have the same length, the triangle is isosceles.
39. **(a)** 8; 5; $\sqrt{89}$ **(b)** $8^2 + 5^2 = 64 + 25 = 89 = (\sqrt{89})^2$
41. $(x - 1)^2 + (y - 2)^2 = 25$
43. $(x + 1)^2 + (y + 4)^2 = 9$ **45.** Center: $(3, 1)$; radius: 6
47. Center: $(0, 0)$; radius: $\sqrt{5}$
49. $|x - 4| = 3$ **51.** $|x - c| < d$ **53.** 7; 6
55. Midpoint is $\left(\frac{5}{2}, \frac{7}{2}\right)$. Distances from this point to vertices are equal to $\sqrt{18.5}$.
57. $x \le -8$ or $x \ge 2$

59. True. $\frac{\text{length of } AM}{\text{length of } AB} = \frac{1}{2}$ because M is the midpoint of AB. By similar triangles, $\frac{\text{length of } AM'}{\text{length of } AC} = \frac{\text{length of } AM}{\text{length of } AB} = \frac{1}{2}$, so M' is the midpoint of AC. **61.** C **63.** E **65.** If the legs have lengths a and b, and the hypotenuse is c units long, then without loss of generality, we can assume the vertices are $(0, 0)$, $(a, 0)$, and $(0, b)$. Then the midpoint of the hypotenuse is $\left(\frac{a+0}{2}, \frac{b+0}{2}\right) = \left(\frac{a}{2}, \frac{b}{2}\right)$. The distance to the other vertices is $\sqrt{\left(\frac{a}{2}\right)^2 + \left(\frac{b}{2}\right)^2} = \sqrt{\frac{a^2}{4} + \frac{b^2}{4}} = \frac{c}{2} = \frac{1}{2}c$. **67.** $Q(a, -b)$ **69.** $Q(-a, -b)$

SECTION P.3

Quick Review P.3

1. $4x + 5y + 9$ **3.** $3x + 2y$ **5.** $\frac{5}{y}$ **7.** $\frac{2x + 1}{x}$ **9.** $\frac{11x + 18}{10}$

Exercises P.3

1. (a) and (c) **3.** (b) **5.** Yes **7.** No **9.** No **11.** $x = 8$ **13.** $t = 4$ **15.** $x = 1$ **17.** $y = -\frac{4}{5} = -0.8$
19. $x = \frac{7}{4} = 1.75$ **21.** $x = \frac{4}{3}$ **23.** $z = \frac{8}{19}$ **25.** $x = \frac{17}{10} = 1.7$ **27.** $t = \frac{31}{9}$ **29.** **(a)** The figure shows that $x = -2$ is a solution of the equation $2x^2 + x - 6 = 0$. **(b)** The figure shows that $x = \frac{3}{2}$ is a solution of the equation $2x^2 + x - 6 = 0$. **31.** (a)
33. (b) and (c) **35.** 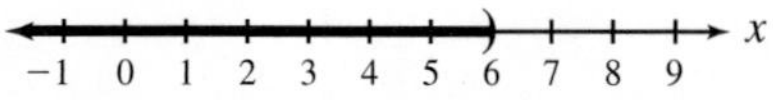

$x < 6$ **37.** 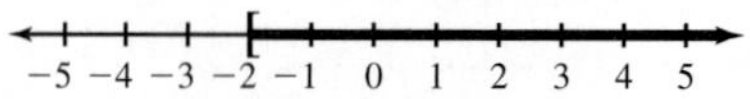

$x \ge -2$
39. 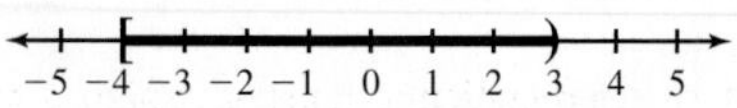

$-4 \le x < 3$ **41.** 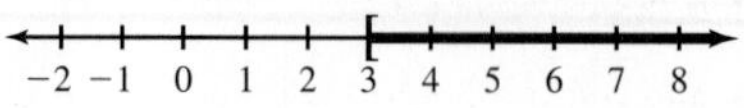

$x \ge 3$
43. $x \le -\frac{19}{5}$ **45.** $-\frac{1}{2} \le y \le \frac{17}{2}$ **47.** $-\frac{5}{2} \le z < \frac{3}{2}$ **49.** $x > \frac{21}{5}$ **51.** $y < \frac{7}{6}$ **53.** $x \le \frac{34}{7}$ **55.** $x = 1$

57. $x = 3, 4, 5, 6$ **59.** Multiply both sides of the first equation by 2. **61.** **(a)** No **(b)** Yes **63.** False. $-6 < -2$ because -6 lies to the left of -2 on the number line. **65.** E **67.** A **69.** **(a)** (no answer) **(b)** (no answer) **(c)** $800/801 > 799/800$ **(d)** $-103/102 > -102/101$ **(e)** The value stored in x is not a solution of the inequality $2x + 1 < 4$.
71. $b_1 = \frac{2A}{h} - b_2$ **73.** $F = \frac{9}{5}C + 32$

SECTION P.4

Exploration 1

1. The graphs of $y = mx + b$ and $y = mx + c$ have the same slope but different y-intercepts.

3.

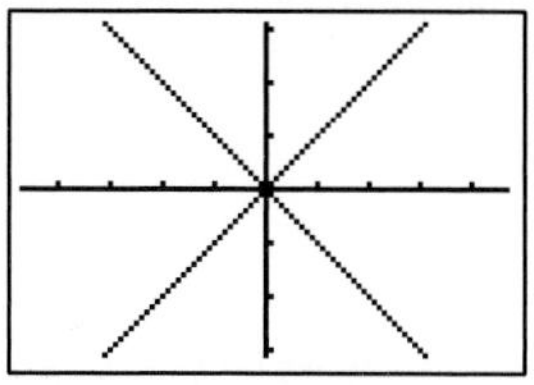

$m = 1$

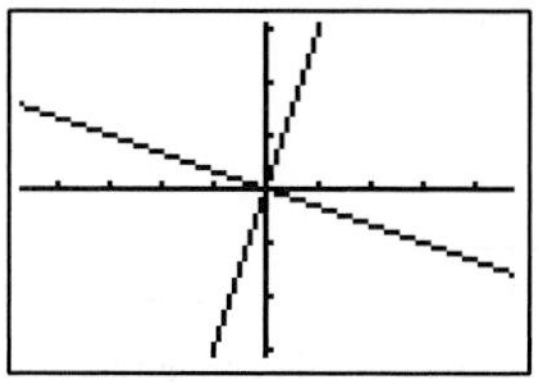

$m = 3$

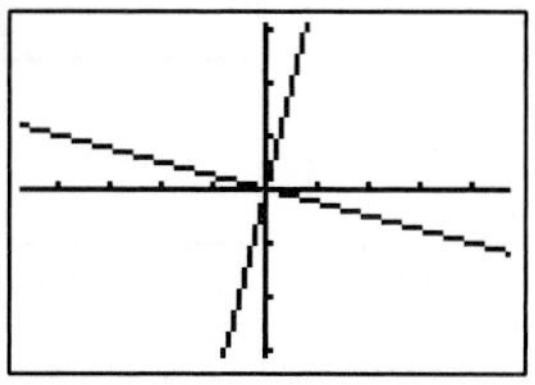

$m = 4$

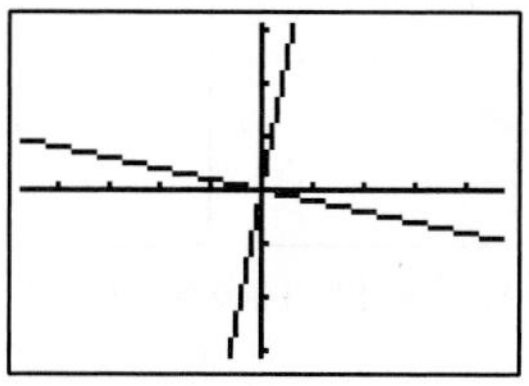

$m = 5$

In each case, the two lines appear to be at right angles (90°) to one another.

Quick Review P.4

1. $x = -\frac{7}{3}$ **3.** $x = 12$ **5.** $y = \frac{2}{5}x - \frac{21}{5}$ **7.** $y = \frac{17}{5}$ **9.** $\frac{2}{3}$

Exercises P.4

1. -2 **3.** $\frac{4}{7}$ **5.** 8 **7.** $x = 2$ **9.** $y = 16$ **11.** $y - 4 = 2(x - 1)$ **13.** $y + 4 = -2(x - 5)$ **15.** $x - y + 5 = 0$

17. $y + 3 = 0$ **19.** $x - y + 3 = 0$ **21.** $y = -3x + 5$ **23.** $y = -\frac{1}{4}x + 4$ **25.** $y = -\frac{2}{5}x + \frac{12}{5}$

27.

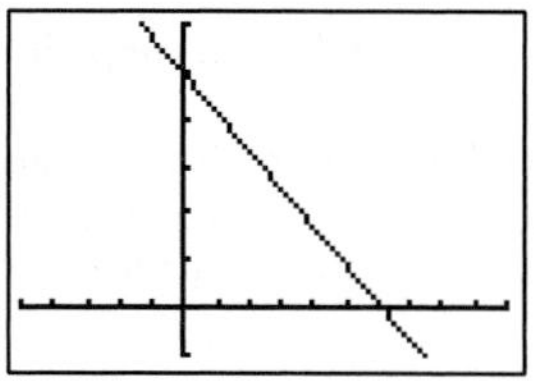

[–5, 10] by [–10, 60]

29.

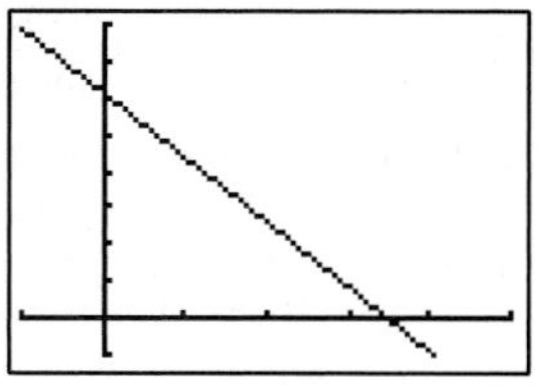

[–1, 5] by [–10, 80]

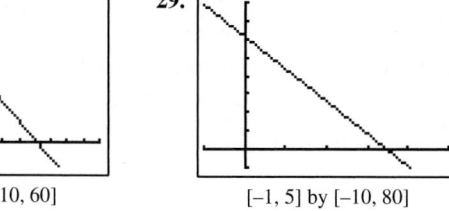

31. (a): the slope is 1.5, compared to 1 in (b). **33.** $x = 4; y = 21$ **35.** $x = -10; y = -7$ **37.** Ymin $= -30$, Ymax $= 30$, Yscl $= 3$ **39.** Ymin $= -20/3$, Ymax $= 20/3$, Yscl $= 2/3$
41. **(a)** $y = 3x - 1$ **(b)** $y = -\frac{1}{3}x + \frac{7}{3}$
43. **(a)** $y = -\frac{2}{3}x + 3$ **(b)** $y = \frac{3}{2}x - \frac{7}{2}$

45. **(a)** 3187.5; 42,000 **(b)** 9.57 years **(c)** $3187.5t + 42{,}000 = 74{,}000$; $t = 10.04$ **(d)** 12 years **47.** 32,000 ft
49. $m = \frac{3}{8} > \frac{4}{12}$, so asphalt shingles are acceptable.
51. **(a)** $y = 0.32(x - 1990) + 3.8$ **(b)** \$8.6 trillion **(c)** \$11.6 trillion **(d)**

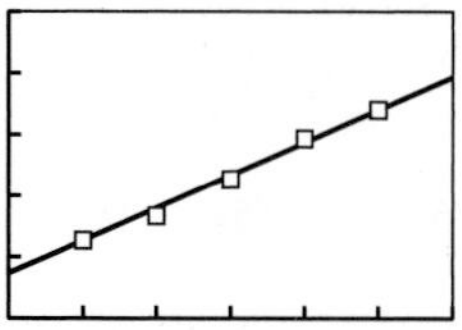

[1985, 2015] by [0, 15]

53. **(a)**

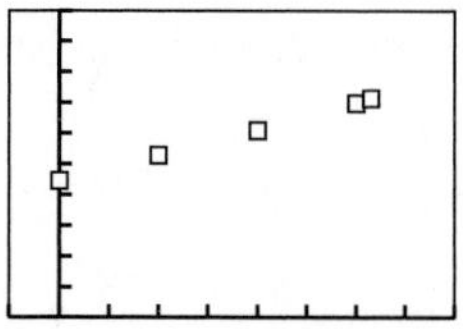

[–5, 40] by [0, 10,000]

(b)

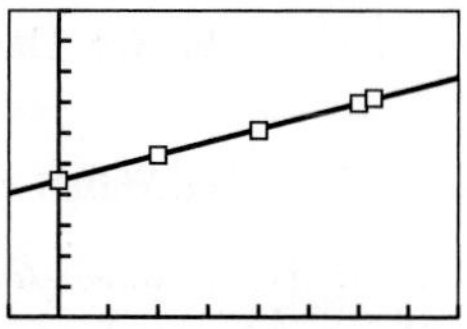

[–5, 40] by [0, 10,000]

$y = 84x + 4453$ **(c)** 7.813 billion

55. 9 **57.** $b = 5; a = 6$ **59.** **(a)** No; perpendicular lines have slopes with opposite signs. **(b)** No; perpendicular lines have slopes with opposite signs. **61.** False. The slope of a vertical line is undefined. For example, the vertical line through (3, 1) and (3, 6) would have slope $(6 - 1)/(3 - 3) = 5/0$, which is undefined. **63.** A **65.** E

67. (a)

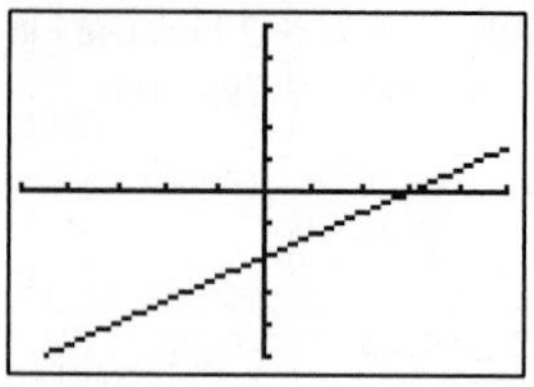

[–5, 5] by [–5, 5]

(b)

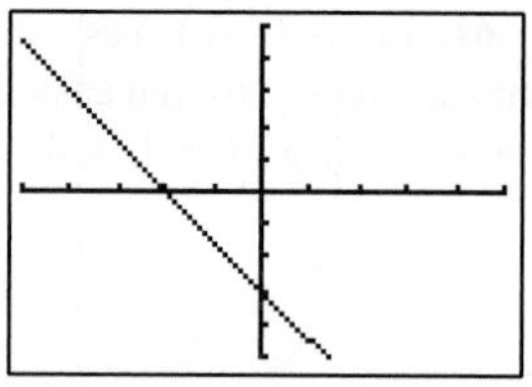

[–5, 5] by [–5, 5]

(c)

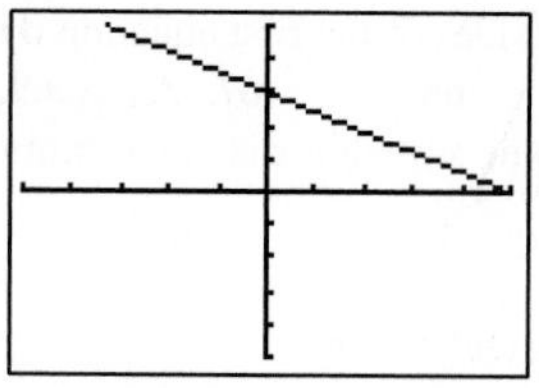

[–5, 5] by [–5, 5]

(d) a is the x-intercept and b is the y-intercept when $c = 1$.

(e)

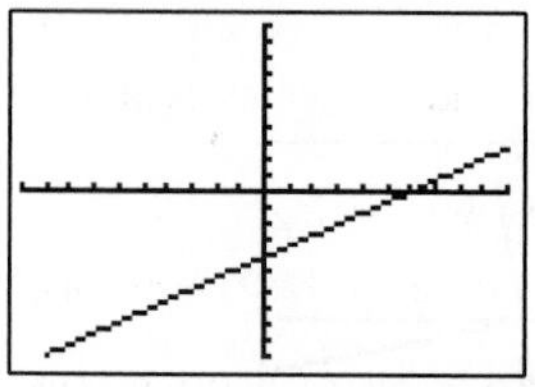

[–10, 10] by [–10, 10]

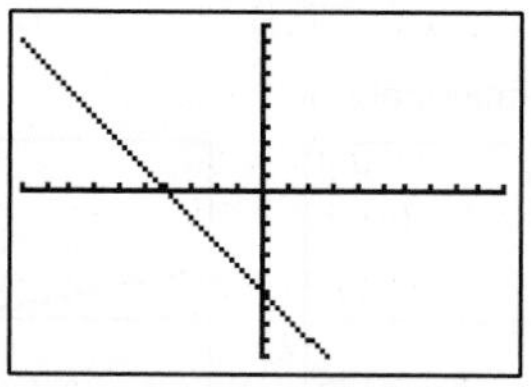

[–10, 10] by [–10, 10]

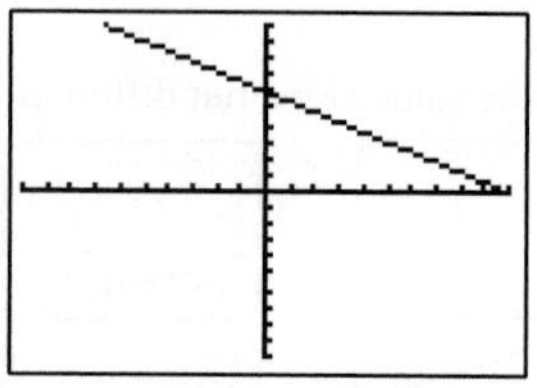

[–10, 10] by [–10, 10]

(f) When $c = -1$, a is the opposite of the x-intercept and b is the opposite of the y-intercept.

a is half the x-intercept and b is half the y-intercept when $c = 2$.

69. As in the diagram, we can choose one point to be the origin, and another to be on the x-axis. The midpoints of the sides, starting from the origin and working around counterclockwise in the diagram, are then $A\left(\frac{a}{2}, 0\right)$, $B\left(\frac{a+b}{2}, \frac{c}{2}\right)$, $C\left(\frac{b+d}{2}, \frac{c+e}{2}\right)$, and $D\left(\frac{d}{2}, \frac{e}{2}\right)$. The opposite sides are therefore parallel, since the slopes of the four lines connecting those points are: $m_{AB} = \frac{c}{b}$; $m_{BC} = \frac{e}{d-a}$; $m_{CD} = \frac{c}{b}$; $m_{DA} = \frac{e}{d-a}$.

71. A has coordinates $\left(\frac{b}{2}, \frac{c}{2}\right)$, while B is $\left(\frac{a+b}{2}, \frac{c}{2}\right)$, so the line containing A and B is the horizontal line $y = \frac{c}{2}$, and the distance from A to B is $\left|\frac{a+b}{2} - \frac{b}{2}\right| = \frac{a}{2}$.

SECTION P.5

Exploration 1

1.

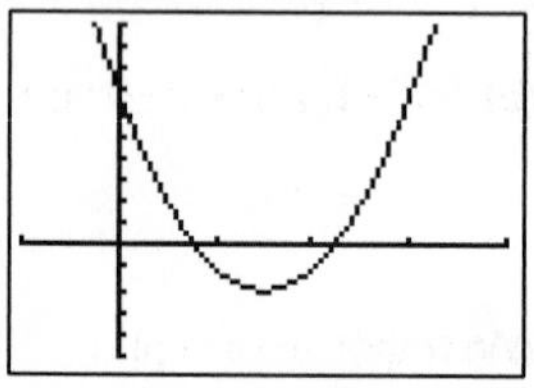

[–1, 4] by [–5, 10]

3.

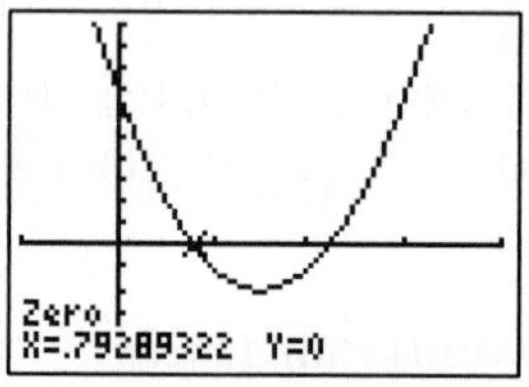

[–1, 4] by [–5, 10]

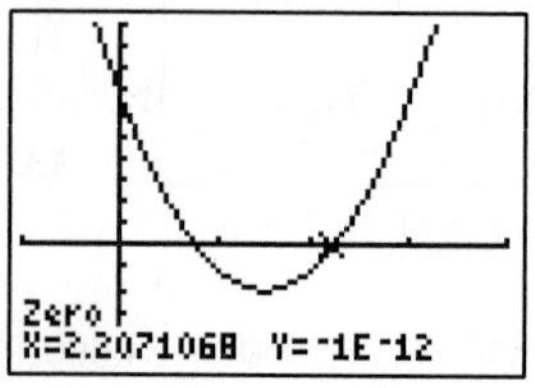

[–1, 4] by [–5, 10]

By this method, we have zeros at 0.79 and 2.21.

5. The answers in parts 2, 3, and 4 are the same. **7.** 0.792893; 2.207107

Quick Review P.5

1. $9x^2 - 24x + 16$ **3.** $6x^2 - 7x - 5$ **5.** $(5x - 2)^2$ **7.** $(3x + 1)(x^2 - 5)$ **9.** $\frac{(x-2)(x+1)}{(2x+1)(x+3)}$

Exercises P.5

1. $x = -4$ or $x = 5$ **3.** $x = 0.5$ or $x = 1.5$ **5.** $x = -\frac{2}{3}$ or $x = 3$ **7.** $x = \pm\frac{5}{2}$ **9.** $x = -4 \pm \sqrt{\frac{8}{3}}$ **11.** $y = \pm\sqrt{\frac{7}{2}}$

13. $x = -7$ or $x = 1$ **15.** $x = \frac{7}{2} - \sqrt{11} \approx 0.18$ or $x = \frac{7}{2} + \sqrt{11} \approx 6.82$ **17.** $x = 2$ or $x = 6$

19. $x = -4 - 3\sqrt{2} \approx -8.24$ or $x = -4 + 3\sqrt{2} \approx 0.24$ **21.** $x = -1$ or $x = 4$ **23.** $-\frac{5}{2} + \frac{\sqrt{73}}{2} \approx 1.77$ or $-\frac{5}{2} - \frac{\sqrt{73}}{2} \approx -6.77$

25. x-intercept: 3; y-intercept: -2 **27.** x-intercepts: $-2, 0, 2$; y-intercept: 0

29.

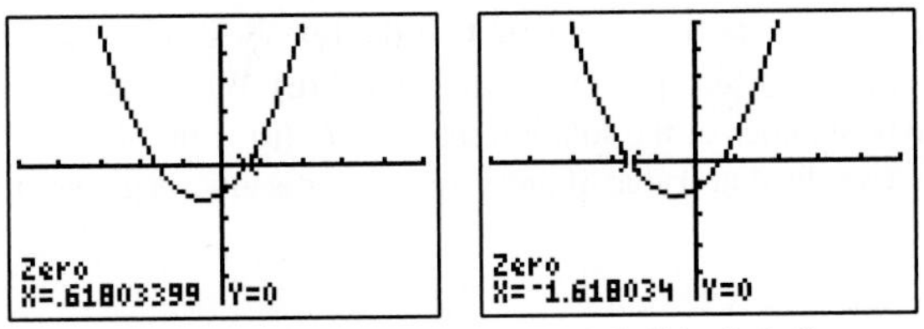

[–5, 5] by [–5, 5] [–5, 5] by [–5, 5]

31.

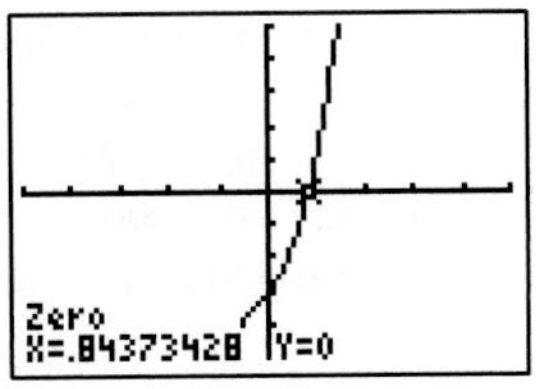

[–5, 5] by [–5, 5]

33.

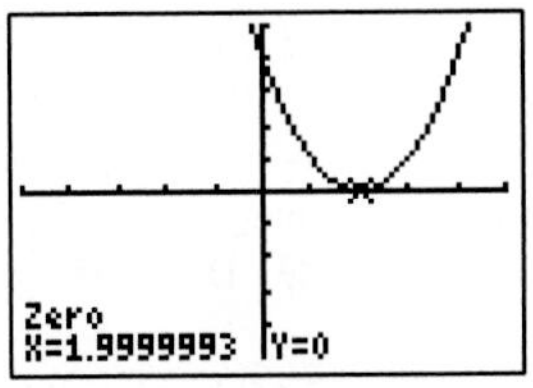

[–5, 5] by [–5, 5]

35. $x^2 + 2x - 1 = 0; x \approx 0.4$ **37.** 1.62; −0.62 **39.** $t = 6$ or $t = 10$ **41.** $x = 1$ or $x = -6$ **43.** $x = -3$ or $x = 1$ **45. (a)** $y_1 = 3\sqrt{x+4}$ (the one that begins on the x-axis) and $y_2 = x^2 - 1$ **(b)** $y = 3\sqrt{x+4} - x^2 + 1$ **(c)** The x-coordinates of the intersections in the first picture are the same as the x-coordinates where the second graph crosses the x-axis. **47.** $x = -2$ or $x = 1$ **49.** $x = 3$ or $x = -2$ **51.** $x \approx -4.56$ or $x \approx -0.44$ or $x = 1$ **53.** $x = -2 \pm 2\sqrt{3}$ **55.** $x \approx -2.41$ or $x \approx 2.91$ **57. (a)** There must be 2 distinct real zeros, because $b^2 - 4ac > 0$ implies that $\pm\sqrt{b^2 - 4ac}$ are 2 distinct real numbers. **(b)** There must be 1 real zero, because $b^2 - 4ac = 0$ implies that $\pm\sqrt{b^2 - 4ac} = 0$, so the root must be $x = -\frac{b}{2a}$. **(c)** There must be no real zeros, because $b^2 - 4ac < 0$ implies that $\pm\sqrt{b^2 - 4ac}$ are not real numbers. **59.** 80 yd wide; 110 yd long **61.** $\approx$ 11.98 ft **63.** False. Notice that $2(-3)^2 = 18$, so x could also be −3. **65.** B **67.** E **69. (a)** $c = 2$ **(b)** $c = 4$ **(c)** $c = 5$ **(d)** $c = -1$ **(e)** There is no other possible number of solutions of this equation. For any c, the solution involves solving two quadratic equations, each of which can have 0, 1, or 2 solutions. **71.** $2.5 + \frac{1}{2}\sqrt{13}$, or approximately 0.697 and 4.303

SECTION P.6

Quick Review P.6

1. $x + 9$ **3.** $a + 2d$ **5.** $x^2 - x - 6$ **7.** $x^2 - 2$ **9.** $x^2 - 2x - 1$

Exercises P.6

1. $8 + 2i$ **3.** $13 - 4i$ **5.** $5 - (1 + \sqrt{3})i$ **7.** $-5 + i$ **9.** $7 + 4i$ **11.** $-5 - 14i$ **13.** $-48 - 4i$ **15.** $5 - 10i$ **17.** $4i$ **19.** $\sqrt{3}i$ **21.** $x = 2, y = 3$ **23.** $x = 1, y = 2$ **25.** $5 + 12i$ **27.** $-1 + 0i$ **29.** 13 **31.** 25 **33.** $2/5 - 1/5i$ **35.** $3/5 + 4/5i$ **37.** $1/2 - 7/2i$ **39.** $7/5 - 1/5i$ **41.** $x = -1 \pm 2i$ **43.** $x = \frac{7}{8} \pm \frac{\sqrt{15}}{8}i$ **45.** False. Any complex number bi has this property. **47.** E **49.** A **51. (a)** $i; -1; -i; 1; i; -1; -i; 1$ **(b)** $-i; -1; i; 1; -i; -1; i; 1$ **(c)** 1 **(d)** (no answer) **53.** $(a + bi) - (a - bi) = 2bi$, real part is zero

55. $\overline{(a + bi)\cdot(c + di)} = \overline{(ac - bd) - (ad + bc)i} = (ac - bd) - (ad + bc)i$ and

$\overline{(a + bi)}\cdot\overline{(c + di)} = (a - bi)\cdot(c - di) = (ac - bd) - (ad + bc)i$ are equal

57. $(-i)^2 - i(-i) + 2 = 0$ but $(i)^2 - i(i) + 2 \neq 0$. Because the coefficient of x in $x^2 - ix + 2 = 0$ is not a real number, the complex conjugate, i, of $-i$, need not be a solution.

SECTION P.7

Quick Review P.7

1. $-2 < x < 5$ **3.** $x = 1$ or $x = -5$ **5.** $x(x - 2)(x + 2)$ **7.** $\frac{z + 5}{z}$ **9.** $\frac{4x^2 - 4x - 1}{(x - 1)(3x - 4)}$

Exercises P.7

1. 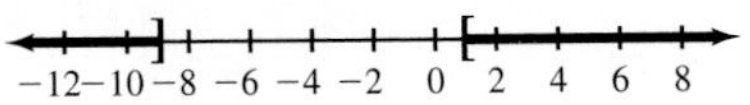

$(-\infty, -9] \cup [1, \infty)$

3. 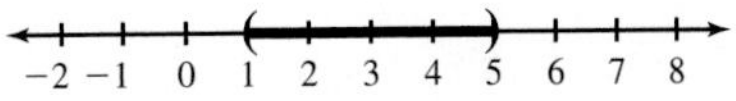

$(1, 5)$

5. 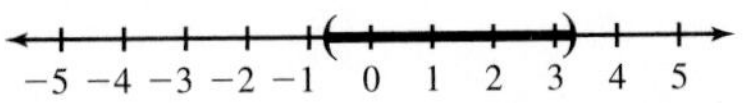

$\left(-\frac{2}{3}, \frac{10}{3}\right)$

7. 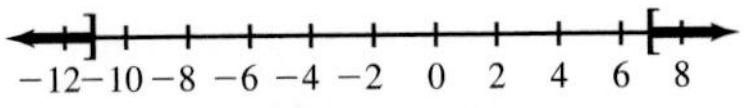

$(-\infty, -11] \cup [7, \infty)$

9. $[-7, -3/2]$ **11.** $(-\infty, -5) \cup \left(\frac{3}{2}, \infty\right)$ **13.** $(-\infty, -2) \cup \left(\frac{1}{3}, \infty\right)$ **15.** $[-1, 0] \cup [1, \infty)$ **17.** $(-0.24, 4.24)$ **19.** $\left(-\infty, -\frac{1}{2}\right) \cup \left(\frac{4}{3}, \infty\right)$ **21.** $(-\infty, -1.41] \cup [0.08, \infty)$ **23.** $\left(-\infty, \frac{1}{2}\right) \cup \left(\frac{1}{2}, \infty\right)$ **25.** No solution

27. $[-2.08, 0.17] \cup [1.19, \infty)$ **29.** $(1.11, \infty)$ **31.** Answers can vary: **(a)** $x^2 + 1 > 0$ **(b)** $x^2 + 1 < 0$ **(c)** $x^2 \leq 0$ **(d)** $(x + 2)(x - 5) \leq 0$ **(e)** $(x + 1)(x - 4) > 0$ **(f)** $x(x - 4) \geq 0$ **33. (a)** $t = 4$ sec up; $t = 12$ sec down **(b)** When t is in the interval $[4, 12]$ **(c)** When t is in the interval $(0, 4]$ or $[12, 16)$ **35.** Reveals the boundaries of the solution set **37. (a)** 1 in. $< x <$ 34 in. **(b)** When x is in the interval $(1, 25]$ **39.** No more than \$100,000 **41.** True. The absolute value of any real number is always greater than or equal to zero. **43.** D **45.** D **47.** $(-5.69, -4.11) \cup (0.61, 2.19)$

CHAPTER P REVIEW EXERCISES

1. Endpoints 0 and 5; bounded **3.** $2x^2 - 2x$ **5.** v^4 **7.** 3.68×10^9 **9.** 5,000,000,000 **11. (a)** 1.45×10^{10} **(b)** 5.456×10^8 **(c)** 1.15×10^{10} **(d)** 9.7×10^7 **(e)** 7.1×10^9 **13.** 19; 4.5 **15.** $5\sqrt{5}; 2\sqrt{5}; \sqrt{145}$ **17.** $(x - 0)^2 + (y - 0)^2 = 2^2$, or $x^2 + y^2 = 4$ **19.** Center: $(-5, -4)$; radius: 3 **21. (a)** $\sqrt{20} \approx 4.47, \sqrt{80} \approx 8.94, 10$ **(b)** $(\sqrt{20})^2 + (\sqrt{80})^2 = 20 + 80 = 100 = 10^2$ **23.** $a = 7; b = 9$

25. $y + 1 = -\frac{2}{3}(x - 2)$ **27.** $y = \frac{4}{5}x - 4.4$

29. $y = 4$ **31.** $y = -\frac{2}{5}x - \frac{11}{5}$

33. (a)

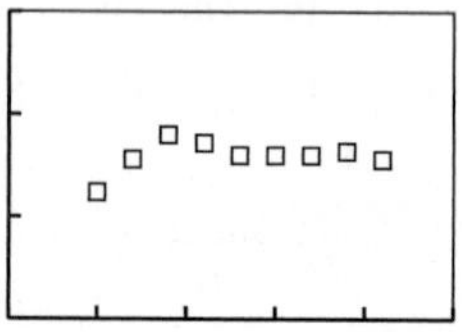

[1990, 2015] by [475, 550]

(b) $y = 1.4(x - 1995) + 506$

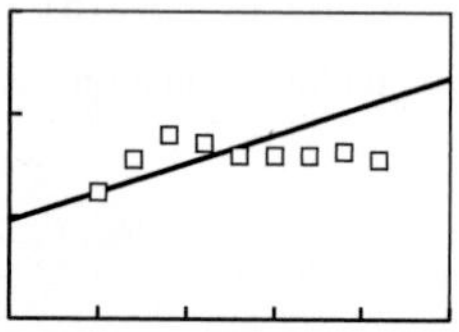

[1990, 2015] by [475, 550]

(c) 534 **(d)** No; the superimposed graphs in (b) show that data for 2005–2011 do not follow the pattern for 1995–2005, so it is unlikely that the average SAT math score in 2015 will be nearly as high as 534.

35. 2.5 **37.** $x = -3$ **39.** $y = 3$ **41.** $x = 2 - \sqrt{7} \approx -0.65; x = 2 + \sqrt{7} \approx 4.65$ **43.** $x = \frac{1}{3}$ or $x = -\frac{3}{2}$

45. $x = \frac{7}{2}$ or $x = -4$ **47.** $x = \frac{5}{2}$ **49.** $x = 0, x = 3$ **51.** $3 \pm 2i$ **53.** $x = \frac{3}{4} - \frac{\sqrt{17}}{4} \approx -0.28$ or $x = \frac{3}{4} + \frac{\sqrt{17}}{4} \approx 1.78$

55. $x = 0$ or $x = -\frac{2}{3}$ or $x = 7$ **57.** $x \approx 2.36$ **59.** $(-6, 3]$

61. $\left(-\infty, \frac{1}{3}\right]$

63. $(-\infty, -2] \cup \left[-\frac{2}{3}, \infty\right)$ **65.** $(-\infty, -0.37) \cup (1.37, \infty)$ **67.** $(-\infty, -2.82] \cup [-0.34, 3.15]$ **69.** $(-\infty, -17) \cup (3, \infty)$

71. $(-\infty, \infty)$ **73.** $1 + 3i$ **75.** $7 + 4i$ **77.** $25 + 0i$ **79.** $0 + 4i$ **81. (a)** $t \approx 8$ sec up; $t \approx 12$ sec down **(b)** When t is in the interval $(0, 8]$ or $[12, 20]$ (approximately) **(c)** When t is in the interval $[8, 12]$ (approximately) **83. (a)** When w is in the interval $(0, 18.5]$ **(b)** When w is in the interval $(22.19, \infty)$ (approximately)

SECTION 1.1

Exploration 1

1. 0.75 **3.** 0.8025 **5.** \$124.61

Exploration 2

1. The linear model increases without bound, whereas there is a finite limit to human life expectancy.
3. More than 58,000 Americans, predominantly males, died prematurely in the Vietnam War, which would have affected the point in 1970 but not the point on either side. The male data in Figure 1.3 suggest that this explains the dip.

Quick Review 1.1

1. $(x + 4)(x - 4)$ **3.** $(9y + 2)(9y - 2)$ **5.** $(4h^2 + 9)(2h + 3)(2h - 3)$ **7.** $(x + 4)(x - 1)$ **9.** $(2x - 1)(x - 5)$

Exercises 1.1

1. (d)(q) **3.** (a)(p) **5.** (e)(l) **7.** (g)(t) **9.** (i)(m) **11. (a)** Increasing steadily until 2000, then leveling off. It has slightly decreased since 2000. **(b)** 1975 to 1980 **13.** Women (+), Men (□)

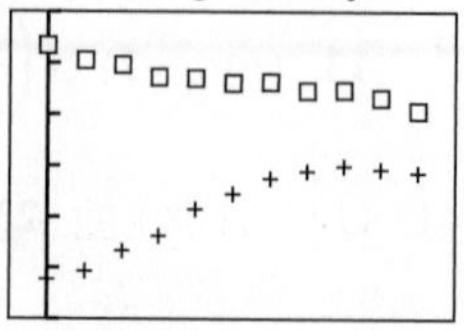

[−5, 55] by [30, 90]

15. Males: $y = -0.254x + 83.4$; females: $y = 0.418x + 37.7$ **17.** The female data follow a linear model fairly well until 1995, but the lack of growth from 1995 to 2010 makes a linear model less likely. Possible explanations for the obvious change in the trend after 1995 will vary.

19. (a) and **(b)**

L3
3.7975
4.375
5.5405
5.8986
6.657

21. Square stones **23.** $y = 1.2t^2$ **25.** The lower line shows the minimum salaries, since they are lower than the average salaries. **27.** Year 15. There is a clear drop in the average salary right after the 1994 strike.

29. $\pm\sqrt{\frac{13}{3}}$ **31.** $-1; 4$ **33.** $-1.5; 4$ **35.** $-\frac{7}{2} \pm \frac{\sqrt{105}}{2}$ **37.** 5

39.
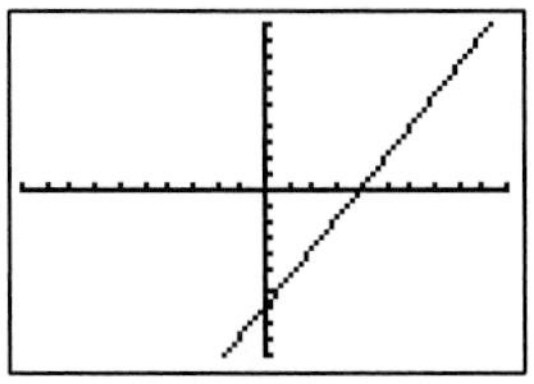
[−10, 10] by [−10, 10]
$x \approx 3.91$

41.
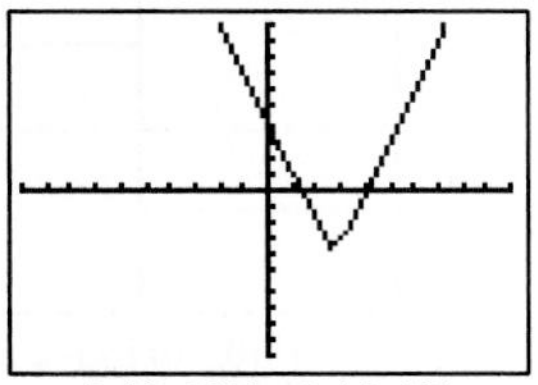
[−10, 10] by [−10, 10]
$x \approx 1.33$ or $x = 4$

43.
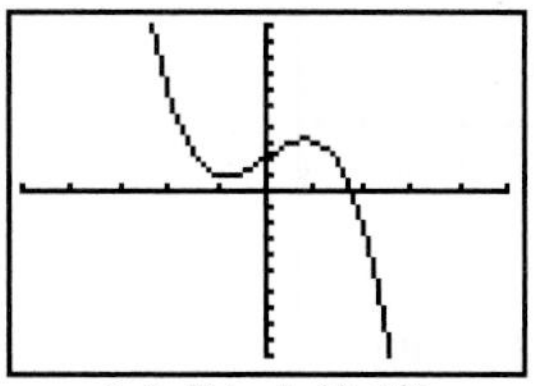
[−5, 5] by [−10, 10]
$x \approx 1.77$

45.
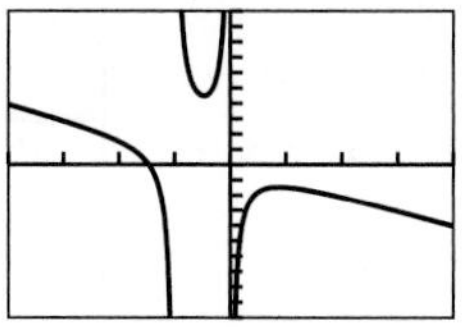
[−4, 4] by [−10, 10]
$x \approx -1.47$

47. (a) \$46.94 **(b)** 210 mi

49. (a) $y = (x^{200})^{1/200} = x^{200/200} = x^1 = x$ for all $x \geq 0$. **(b)**
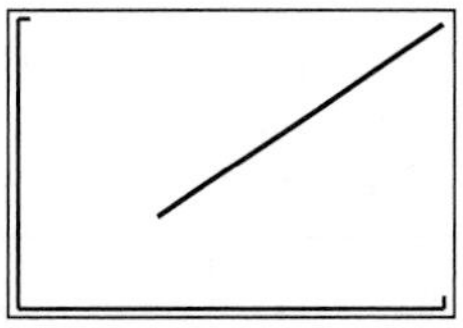
[0, 1] by [0, 1]

(c) Yes **(d)** For values of x close to 0, x^{200} is so small that the calculator is unable to distinguish it from zero. It returns a value of $0^{1/200} = 0$ rather than x.

51. (a) -3 or 1.1 or 1.15 **(b)** -3 **53.** Let n be any integer. $n^2 + 2n = n(n + 2)$, which is either the product of two odd integers or the product of two even integers. The product of two odd integers is odd. The product of two even integers is a multiple of 4, since each even integer in the product contributes a factor of 2 to the product. Therefore, $n^2 + 2n$ is either odd or a multiple of 4. **55.** False; a product is zero if at least one factor is zero. **57.** C **59.** B **61. (a)** March **(b)** \$120 **(c)** June, after three months of poor performance **(d)** About \$2000 **(e)** After reaching a low in June, the stock climbed back to a price near \$140 by December. LaToya's shares had gained \$2000 by that point. **(f)** Any graph that decreases steadily from March to December would favor Ahmad's strategy over LaToya's.

63. (a)

Subscribers
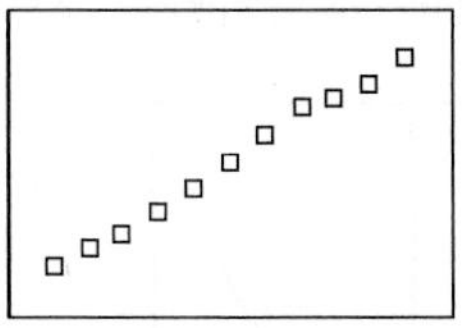
[9, 21] by [77, 335]

Average Monthly Bill
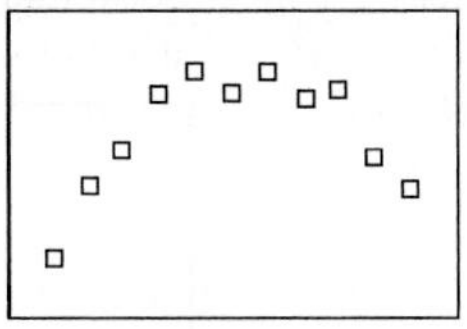
[9, 21] by [44, 52]

(b) $y = 19.1x - 81.5$. **(c)** The linear model fits very well.
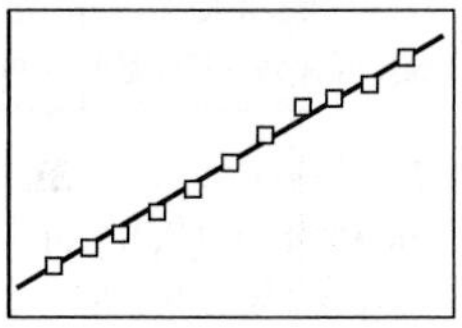
[9, 21] by [77, 335]

(d) The monthly bill scatter plot has an obvious curved shape that could be modeled more effectively over this time period by a parabola. (We will learn about more nonlinear models later in the book.)

(e)

Subscribers
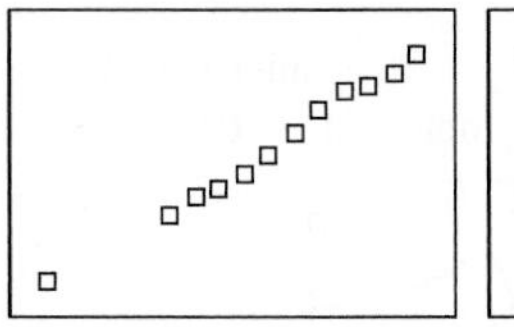
[4, 21] by [5, 345]

Average Monthly Bill
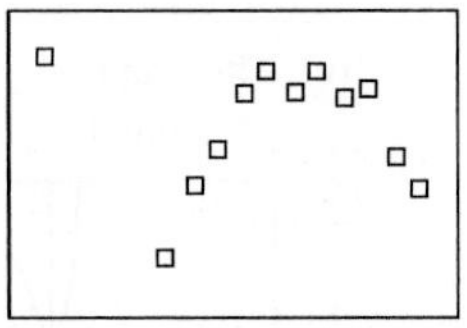
[4, 21] by [44, 52]

The 1995 point fits well with the linear trend in the subscriber data, but the point in the local monthly bill scatter plot does not fit the trend at all.

(f) Cellular phone technology was still emerging in 1995, and early subscribers paid premium prices for the service. Competition brought prices down sharply after more companies entered the market.

SECTION 1.2

Exploration 1

1. From left to right, the tables are (c) constant, (b) decreasing, and (a) increasing. **3.** Positive; negative; 0

Quick Review 1.2

1. $x = \pm 4$ **3.** $x < 10$ **5.** $x = 16$ **7.** $x < 16$ **9.** $x < -2, x \geq 3$

Exercises 1.2

1. Function **3.** Not a function; y has two values for each positive value of x. **5.** Yes **7.** No **9.** $(-\infty, \infty)$
11. $(-\infty, -3) \cup (-3, 1) \cup (1, \infty)$ **13.** $(-\infty, 0) \cup (0, 5) \cup (5, \infty)$ **15.** $(-\infty, -1) \cup (-1, 4]$
17. $(-\infty, 10]$ **19.** $(-\infty, -1) \cup [0, \infty)$
21. Yes, nonremovable

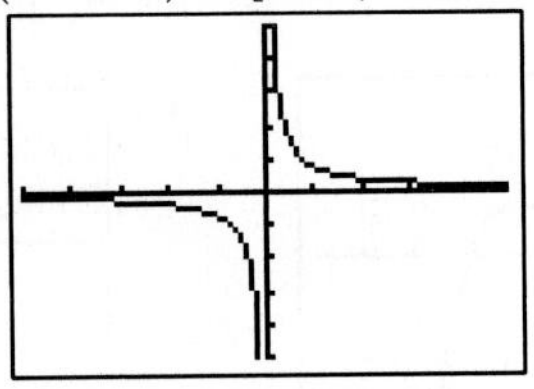

[–10, 10] by [–10, 10]

23. Yes, nonremovable

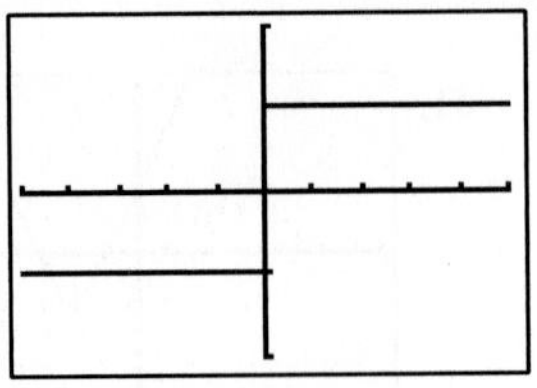

[–10, 10] by [–2, 2]

25. Local maxima at $(-1, 4)$ and $(5, 5)$, local minimum at $(2, 2)$. The function increases on $(-\infty, -1]$, decreases on $[-1, 2]$, increases on $[2, 5]$, and decreases on $[5, \infty)$. **27.** $(-1, 3)$ and $(3, 3)$ are neither, $(1, 5)$ is a local maximum, and $(5, 1)$ is a local minimum. The function increases on $(-\infty, 1]$, decreases on $[1, 5]$, and increases on $[5, \infty)$.

29. Decreasing on $(-\infty, -2]$; increasing on $[-2, \infty)$

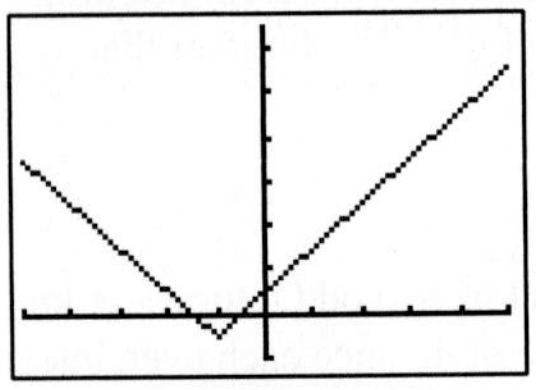

[–10, 10] by [–2, 18]

31. Decreasing on $(-\infty, -2]$; constant on $[-2, 1]$; increasing on $[1, \infty)$

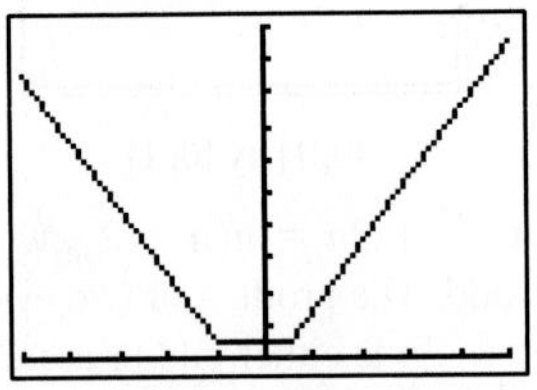

[–10, 10] by [0, 20]

33. Increasing on $(-\infty, 1]$; decreasing on $[1, \infty)$

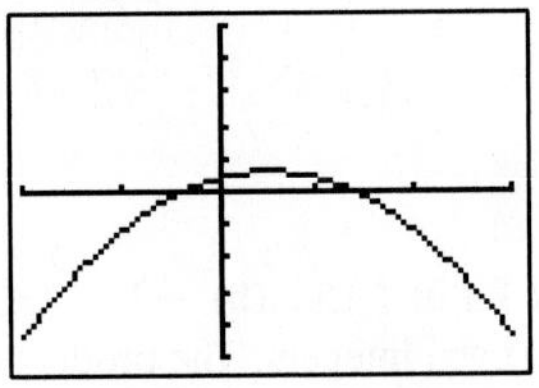

[–4, 6] by [–25, 25]

35. Bounded **37.** Bounded below **39.** Bounded

41. f has a local minimum of $y = 3.75$ at $x = 0.5$. It has no maximum.

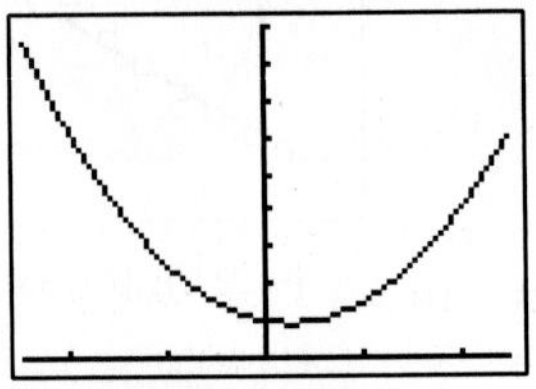

[–5, 5] by [0, 36]

43. Local minimum: $y \approx -4.09$ at $x \approx -0.82$. Local maximum: $y \approx -1.91$ at $x \approx 0.82$.

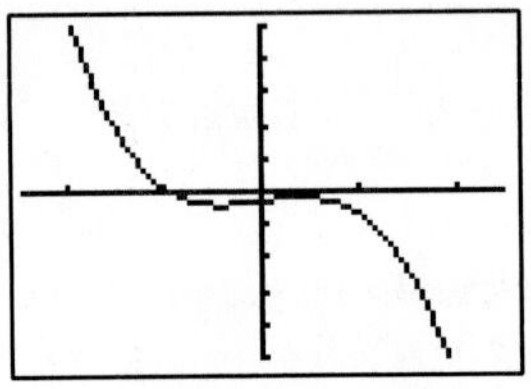

[–5, 5] by [–50, 50]

45. Local maximum: $y \approx 9.16$ at $x \approx -3.20$. Local minimum: $y = 0$ at $x = 0$ and $y = 0$ at $x = -4$.

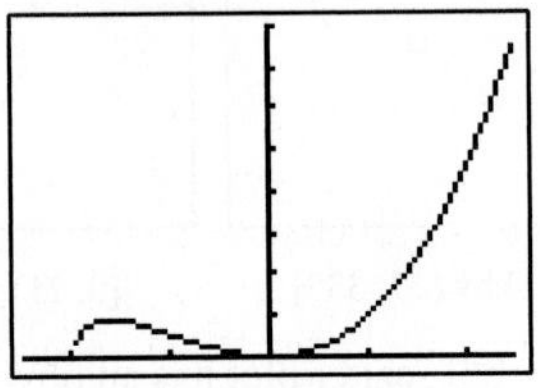

[–5, 5] by [0, 80]

47. Even **49.** Even **51.** Neither **53.** Odd **55.** $y = 1; x = 1$ **57.** $y = -1; x = 3$ **59.** $y = 1; x = 1; x = -1$
61. $y = 0; x = 2$ **63.** (b) **65.** (a)

67. (a) The graph of $y = f(x)$ crosses the horizontal asymptote at $(0, 0)$.

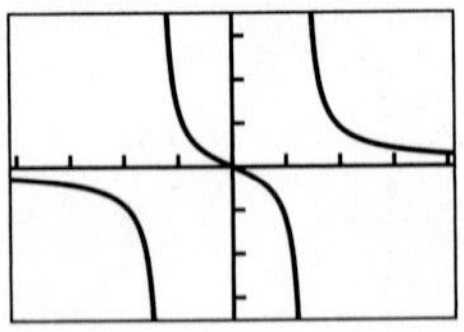

[–4.7, 4.7] by [–3.1, 3.1]

(b) The graph of $y = g(x)$ crosses the horizontal asymptote at $(0, 0)$.

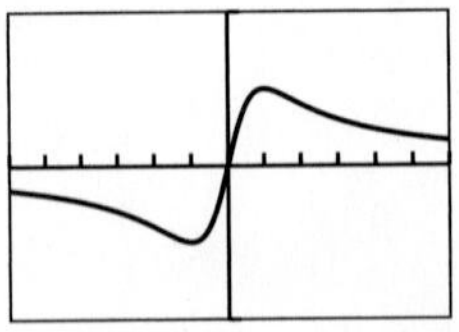

[–6, 6] by [–1, 1]

(c) The graph of $y = h(x)$ intersects the horizontal asymptote at $(0, 0)$.

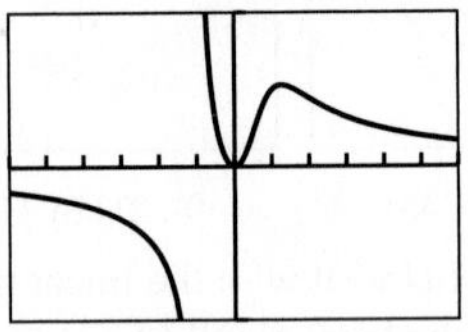

[–6, 6] by [–1, 1]

69. **(a)** The vertical asymptote is $x = 0$, and this function is undefined at $x = 0$ (because a denominator can't be zero).
(b)

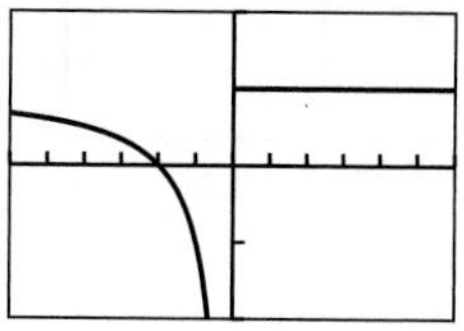

[–6, 6] by [–2, 2]

Add the point $(0, 0)$.
(c) Yes

71. True; this is the definition of the graph of a function.
73. B **75.** C
77. **(a)**

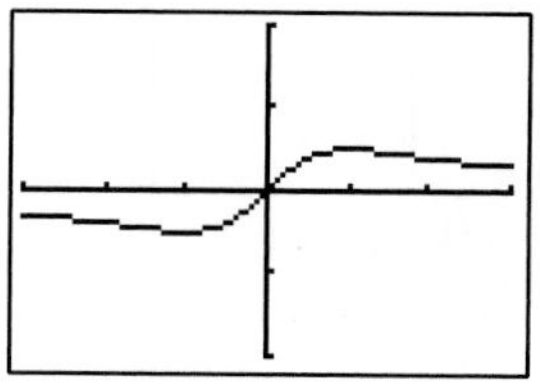

[–3, 3] by [–2, 2]

$k = 1$

(b) $\frac{x}{1 + x^2} < 1 \Leftrightarrow x < 1 + x^2 \Leftrightarrow x^2 - x + 1 > 0$; but the discriminant of $x^2 - x + 1$ is negative (-3), so the graph never crosses the x-axis on the interval $(0, \infty)$.
(c) $k = -1$
(d) $\frac{x}{1 + x^2} > -1 \Leftrightarrow x > -1 - x^2 \Leftrightarrow x^2 + x + 1 > 0$; but the discriminant of $x^2 - x + 1$ is negative (-3), so the graph never crosses the x-axis on the interval $(-\infty, 0)$.

79. **(a)** **(b)** **(c)** **(d)** **(e)** Answers vary. **81.** **(a)** **(b)** **(c)** **(d)** Answers vary. **83.** **(a)** 2. It is in the range. **(b)** 3. It is not in the range. **(c)** $h(x)$ is not bounded above. **(d)** 2. It is in the range. **(e)** 1. It is in the range.
85. Since f is odd, $f(-x) = -f(x)$ for all x. In particular, $f(-0) = -f(0)$. This is equivalent to saying that $f(0) = -f(0)$, and the only number which equals its opposite is 0. Therefore $f(0) = 0$, which means the graph must pass through the origin.
87. **(a)** f is continuous on $[-2, 4]$; the maximum value is 13, which occurs at $x = 4$, and the minimum value is -3, which occurs at $x = 0$.
(b) f is continuous on $[1, 5]$; the maximum value is 1, which occurs at $x = 1$, and the minimum value is 0.2, which occurs at $x = 5$.
(c) f is continuous on $[-4, 1]$; the maximum value is 5, which occurs at $x = -4$, and the minimum value is 2, which occurs at $x = -1$.
(d) f is continuous on $[-4, 4]$; the maximum value is 5, which occurs at both $x = -4$ and $x = 4$, and the minimum value is 3, which occurs at $x = 0$.

SECTION 1.3

Exploration 1

1. $f(x) = 1/x, f(x) = \ln x$ **3.** $f(x) = 1/x, f(x) = e^x, f(x) = 1/(1 + e^{-x})$ **5.** No. There is a removable discontinuity at $x = 0$.

Quick Review 1.3

1. 59.34 **3.** $7 - \pi$ **5.** 0 **7.** 3 **9.** -4

Exercises 1.3

1. (e) **3.** (j) **5.** (i) **7.** (k) **9.** (d) **11.** (l) **13.** Ex. 8 **15.** Ex. 7, 8 **17.** Ex. 2, 4, 6, 10, 11, 12
19. $y = x, y = x^3, y = 1/x, y = \sin x$ **21.** $y = x^2, y = 1/x, y = |x|$ **23.** $y = 1/x, y = e^x, y = 1/(1 + e^{-x})$
25. $y = 1/x, y = \sin x, y = \cos x, y = 1/(1 + e^{-x})$ **27.** $y = x, y = x^3, y = 1/x, y = \sin x$ **29.** Domain: $(-\infty, \infty)$; Range: $[-5, \infty)$
31. Domain: $(-6, \infty)$; Range: $(-\infty, \infty)$ **33.** Domain: $(-\infty, \infty)$; Range: all integers **35.** **(a)** Increasing on $[10, \infty)$ **(b)** Neither **(c)** Minimum value of 0 at $x = 10$ **(d)** Square root function, shifted 10 units right **37.** **(a)** Increasing on $(-\infty, \infty)$ **(b)** Neither **(c)** None **(d)** Logistic function, stretched vertically by a factor of 3 **39.** **(a)** Increasing on $[0, \infty)$; decreasing on $(-\infty, 0]$ **(b)** Even **(c)** Minimum of -10 at $x = 0$ **(d)** Absolute value function, shifted 10 units down **41.** **(a)** Increasing on $[2, \infty)$; decreasing on $(-\infty, 2]$ **(b)** Neither **(c)** Minimum of 0 at $x = 2$ **(d)** Absolute value function, shifted 2 units right **43.** $y = 2, y = -2$

45.

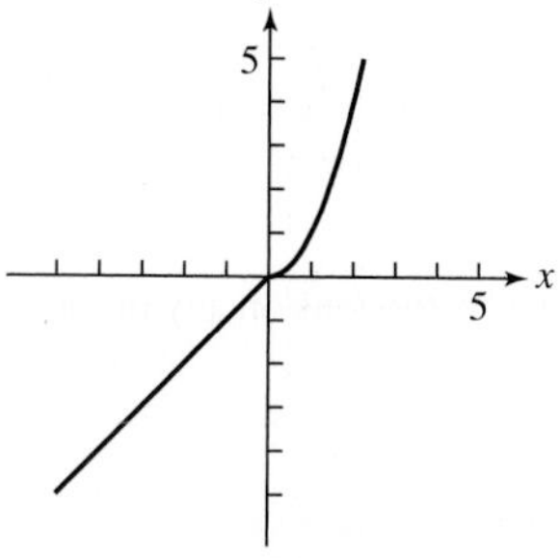

No points of discontinuity

47.

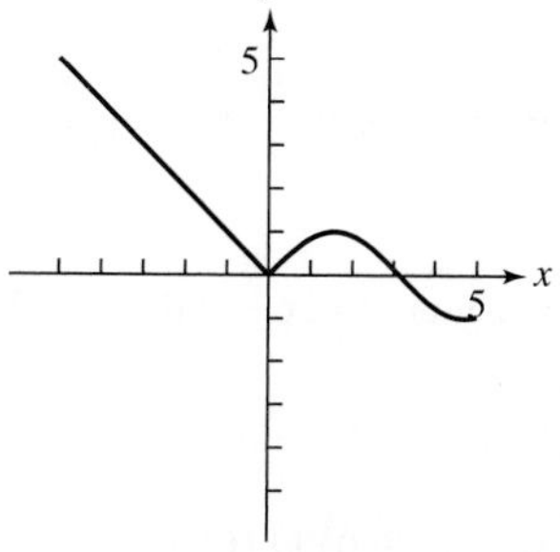

No points of discontinuity

49.

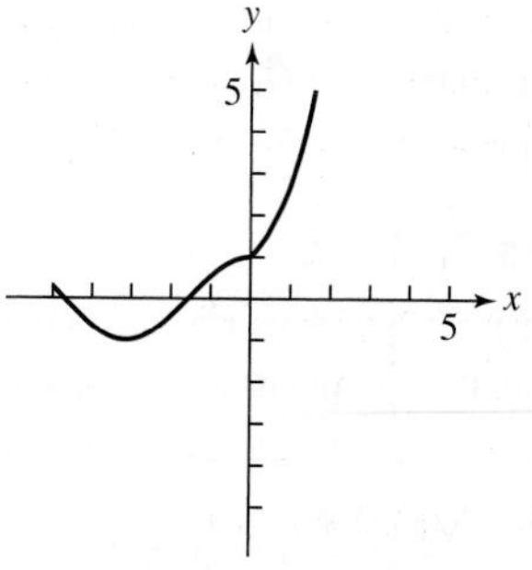

No points of discontinuity

51.

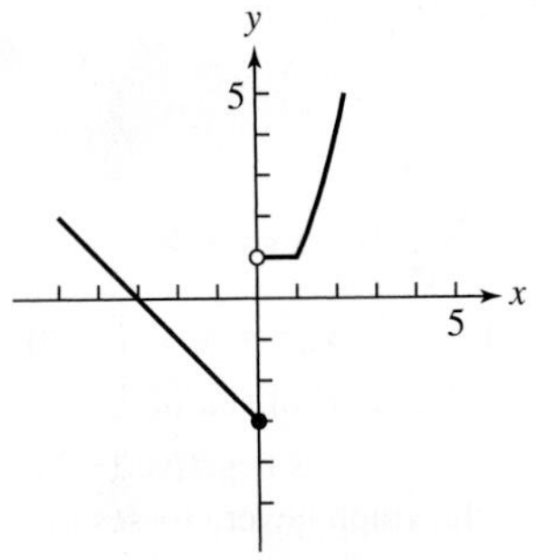

$x = 0$

53. (a)

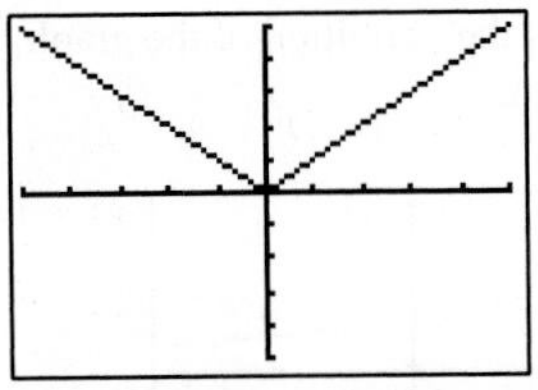

[–5, 5] by [–5, 5]

$g(x) = |x|$

(b) $f(x) = \sqrt{x^2} = \sqrt{|x|^2} = |x| = g(x)$

55. (a)

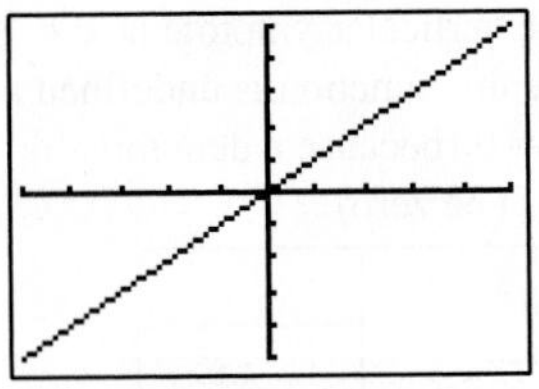

[–5, 5] by [–5, 5]

$f(x) = x$

(b) The fact that $\ln(e^x) = x$ shows that the natural logarithm function takes on arbitrarily large values. In particular, it takes on the value L when $x = e^L$.

57. Domain: $(-\infty, \infty)$; Range: all integers; Continuity: There is a discontinuity at each integer value of x; Increasing/decreasing behavior: constant on intervals of the form $[k, k + 1)$, where k is an integer; Symmetry: none; Boundedness: not bounded; Local extrema: every noninteger is both a local minimum and local maximum; Horizontal asymptotes: none; Vertical asymptotes: none; End behavior: $\text{int}(x) \to -\infty$ as $x \to -\infty$ and $\text{int}(x) \to \infty$ as $x \to \infty$. **59.** True; the asymptotes are $x = 0$ and $x = 1$. **61.** D **63.** E **65. (a)** Even **(b)** Even **(c)** Odd **67. (a)** Pepperoni count ought to be proportional to the area of the pizza, which is proportional to the square of the radius. **(b)** 0.75 **(c)** Yes, very well **(d)** The fact that the pepperoni count fits the expected quadratic model so perfectly suggests that the pizzeria uses such a chart. If repeated observations produced the same results, there would be little doubt. **69. (a)** $f(x) = 1/x$, $f(x) = e^x$, $f(x) = \ln x$, $f(x) = \cos x$, $f(x) = 1/(1 + e^{-x})$ **(b)** $f(x) = x$ **(c)** $f(x) = e^x$ **(d)** $f(x) = \ln x$ **(e)** The odd functions: x, x^3, $1/x$, $\sin x$

SECTION 1.4

Exploration 1

f	g	$f \circ g$
$2x - 3$	$\frac{x+3}{2}$	x
$\lvert 2x + 4 \rvert$	$\frac{(x-2)(x+2)}{2}$	x^2
$\sqrt{x}$	x^2	$\lvert x \rvert$
x^5	$x^{0.6}$	x^3
$x - 3$	$\ln(e^3 x)$	$\ln x$
$2 \sin x \cos x$	$\frac{x}{2}$	$\sin x$
$1 - 2x^2$	$\sin\left(\frac{x}{2}\right)$	$\cos x$

Quick Review 1.4

1. $(-\infty, -3) \cup (-3, \infty)$ **3.** $(-\infty, 5]$ **5.** $[1, \infty)$ **7.** $(-\infty, \infty)$ **9.** $(-1, 1)$

Exercises 1.4

1. $(f + g)(x) = 2x - 1 + x^2$; $(f - g)(x) = 2x - 1 - x^2$; $(fg)(x) = (2x - 1)(x^2) = 2x^3 - x^2$. There are no restrictions on any of the domains, so all three domains are $(-\infty, \infty)$.

3. $(f + g)(x) = \sqrt{x} + \sin x$; $(f - g)(x) = \sqrt{x} - \sin x$; $(fg)(x) = \sqrt{x} \sin x$. Domain in each case is $[0, \infty)$.

5. $(f/g)(x) = \dfrac{\sqrt{x+3}}{x^2}$; $x + 3 \geq 0$ and $x \neq 0$, so the domain is $[-3, 0) \cup (0, \infty)$.

$(g/f)(x) = \dfrac{x^2}{\sqrt{x+3}}$; $x + 3 > 0$, so the domain is $(-3, \infty)$.

7. $(f/g)(x) = x^2/\sqrt{1 - x^2}$; $1 - x^2 > 0$, so $x^2 < 1$; the domain is $(-1, 1)$.

$(g/f)(x) = \sqrt{1 - x^2}/x^2$; $1 - x^2 \geq 0$ and $x \neq 0$; the domain is $[-1, 0) \cup (0, 1]$.

9.

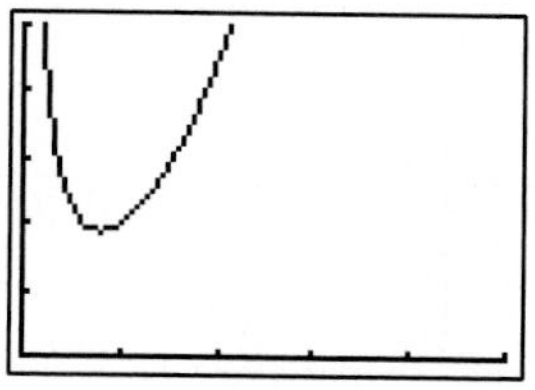

[0, 5] by [0, 5]

11. 5; −6 **13.** 8; 3 **15.** $f(g(x)) = 3x - 1; (-\infty, \infty); g(f(x)) = 3x + 1; (-\infty, \infty)$
17. $f(g(x)) = x - 1; [-1, \infty); g(f(x)) = \sqrt{x^2 - 1}; (-\infty, -1] \cup [1, \infty)$
19. $f(g(x)) = 1 - x^2; [-1, 1]; g(f(x)) = \sqrt{1 - x^4}; [-1, 1]$
21. $f(g(x)) = \frac{3x}{2}; (-\infty, 0) \cup (0, \infty); g(f(x)) = \frac{2x}{3}; (-\infty, 0) \cup (0, \infty)$
23. One possibility: $f(x) = \sqrt{x}$ and $g(x) = x^2 - 5x$
25. One possibility: $f(x) = |x|$ and $g(x) = 3x - 2$ **27.** One possibility: $f(x) = x^5 + 2$ and $g(x) = x - 3$
29. One possibility: $f(x) = \cos x$ and $g(x) = \sqrt{x}$ **31.** $V = \frac{4}{3}\pi r^3 = \frac{4}{3}\pi(48 + 0.03t)^3$; 775,734.6 in.^{3} **33.** $t \approx 3.63$ sec
35. $(3, -1)$ **37.** $y = \sqrt{25 - x^2}$ and $y = -\sqrt{25 - x^2}$ **39.** $y = \sqrt{x^2 - 25}$ and $y = -\sqrt{x^2 - 25}$ **41.** $y = 1 - x$ and $y = x - 1$
43. $y = x$ and $y = -x$ or $y = |x|$ and $y = -|x|$ **45.** False; x is not in the domain of $(f/g)(x)$ if $g(x) = 0$. **47.** C **49.** E

51.

f	g	D
e^x	$2 \ln x$	$(0, \infty)$
$(x^2 + 2)^2$	$\sqrt{x - 2}$	$[2, \infty)$
$(x^2 - 2)^2$	$\sqrt{2 - x}$	$(-\infty, 2]$
$\frac{1}{(x - 1)^2}$	$\frac{x + 1}{x}$	$x \neq 0$
$x^2 - 2x + 1$	$x + 1$	$(-\infty, \infty)$
$\left(\frac{x + 1}{x}\right)^2$	$\frac{1}{x - 1}$	$x \neq 1$

53. **(a)** $g(x) = 0$ **(b)** $g(x) = 1$ **(c)** $g(x) = x$

55.

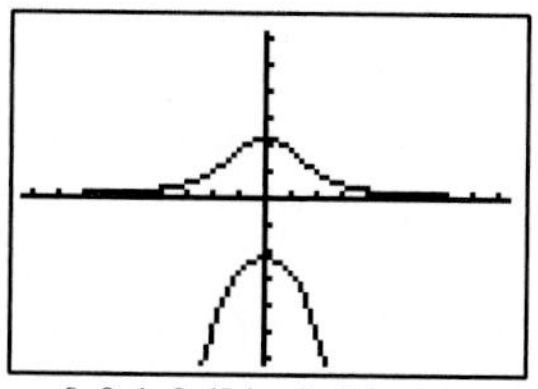

[−9.4, 9.4] by [−6.2, 6.2]

$y = \frac{-x^2 \pm \sqrt{x^4 + 20}}{2}$

SECTION 1.5

Exploration 1

1. T starts at −4, at the point $(-8, -3)$. It stops at T = 2, at the point $(8, 3)$. 61 points are computed.
3. The graph is less smooth because the plotted points are further apart. **5.** The grapher skips directly from the point $(0, -1)$ to the point $(0, 1)$, corresponding to the T-values T = −2 and T = 0. The two points are connected by a straight line, hidden by the Y-axis.
7. Leave everything else the same, but change Tmin back to −4 and Tmax to −1.

Quick Review 1.5

1. $y = \frac{1}{3}x + 2$ **3.** $y = \pm\sqrt{x - 4}$ **5.** $y = \frac{3x + 2}{1 - x}$ **7.** $y = \frac{4x + 1}{x - 2}$ **9.** $y = x^2 - 3, y \geq -3$, and $x \geq 0$

Exercises 1.5

1. $(6, 9)$ **3.** $(15, 2)$

5. **(a)** $(-6, -10), (-4, -7), (-2, -4), (0, -1), (2, 2), (4, 5), (6, 8)$
(b) $1.5x - 1$; It is a function.
(c)

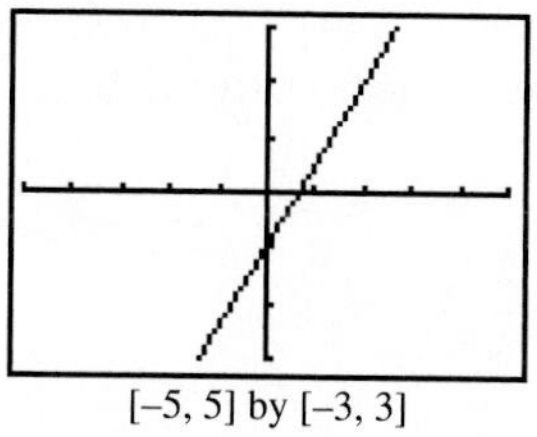

[−5, 5] by [−3, 3]

7. **(a)** $(9, -5), (4, -4), (1, -3), (0, -2), (1, -1), (4, 0), (9, 1)$
(b) $x = (y + 2)^2$; It is not a function.
(c)

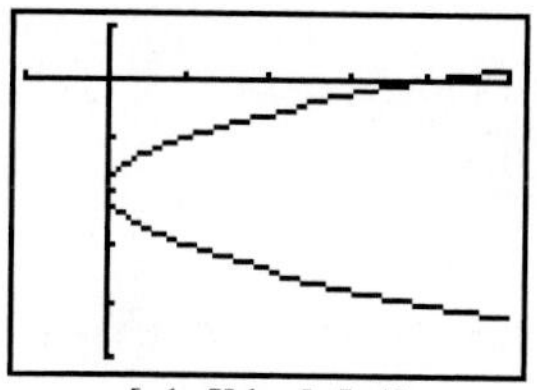

[−1, 5] by [−5, 1]

9. **(a)** No **(b)** Yes **11.** **(a)** Yes **(b)** Yes **13.** $f^{-1}(x) = \frac{1}{3}x + 2, (-\infty, \infty)$ **15.** $f^{-1}(x) = \frac{x + 3}{2 - x}, (-\infty, 2) \cup (2, \infty)$
17. $f^{-1}(x) = x^2 + 3, [0, \infty)$ **19.** $f^{-1}(x) = \sqrt[3]{x}, (-\infty, \infty)$ **21.** $f^{-1}(x) = x^3 - 5, (-\infty, \infty)$

23. One-to-one

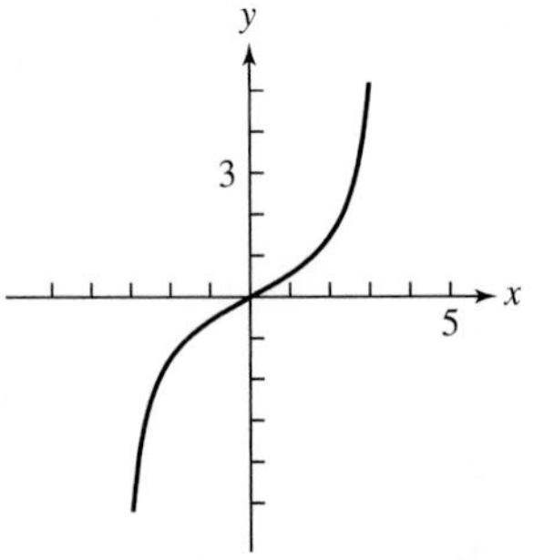

25. One-to-one

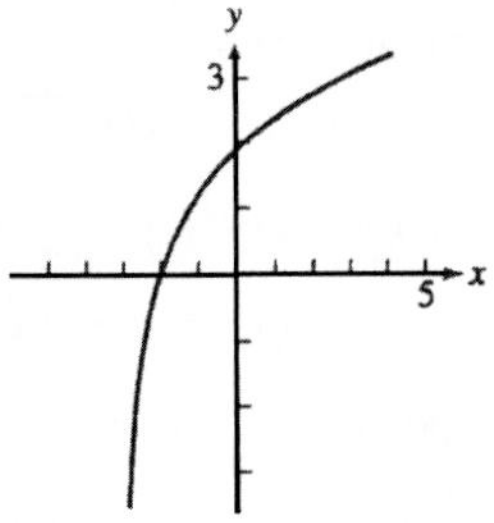

27. $f(g(x)) = 3\left[\frac{1}{3}(x+2)\right] - 2 = x + 2 - 2 = x$; $g(f(x)) = \frac{1}{3}[(3x-2)+2] = \frac{1}{3}(3x) = x$

29. $f(g(x)) = [(x-1)^{1/3}]^3 + 1 = (x-1)^1 + 1 = x - 1 + 1 = x$; $g(f(x)) = [(x^3+1) - 1]^{1/3} = (x^3)^{1/3} = x^1 = x$

31. $f(g(x)) = \dfrac{\frac{1}{x-1}+1}{\frac{1}{x-1}} = (x-1)\left(\frac{1}{x-1}+1\right) = 1 + x - 1 = x$; $g(f(x)) = \dfrac{1}{\frac{x+1}{x}-1} = \left(\dfrac{1}{\frac{x+1}{x}-1}\right)\cdot\dfrac{x}{x} = \dfrac{x}{x+1-x} = \dfrac{x}{1} = x$

33. **(a)** 76 euros **(b)** $y = \frac{25}{19}x$. This converts euros (x) to dollars (y). **(c)** \$63.16 **35.** $y = e^x$ and $y = \ln x$ are inverses. If we restrict the domain of the function $y = x^2$ to the interval $[0, \infty)$, then the restricted function and $y = \sqrt{x}$ are inverses. **37.** $y = |x|$ **39.** True. All the ordered pairs swap domain and range values. **41.** E **43.** C **45.** Answers can vary: **(a)** If the graph of f is unbroken, its reflection in the line $y = x$ will be also. **(b)** Both f and its inverse must be one-to-one in order to be inverse functions. **(c)** Since f is odd, $(-x, -y)$ is on the graph whenever (x, y) is. This implies that $(-y, -x)$ is on the graph of f^{-1} whenever (x, y) is. That implies that f^{-1} is odd. **(d)** Let $y = f(x)$. Since the ratio of Δy to Δx is positive, the ratio of Δx to Δy is positive. Any ratio of Δy to Δx on the graph of f^{-1} is the same as some ratio of Δx to Δy on the graph of f, hence positive. This implies that f^{-1} is increasing.

47. **(a)** $y = 0.75x + 31$ **(b)** $y = \frac{4}{3}(x - 31)$. It converts scaled scores to raw scores. **49.** **(a)** No **(b)** No **(c)** 45°; yes

51. When $k = 1$, the scaling function is linear. Opinions will vary as to which is the best value of k.

SECTION 1.6

Exploration 1

1. They raise or lower the parabola along the y-axis. **3.** Yes

Exploration 2

1. Graph C. Points with positive y-coordinates remain unchanged, while points with negative y-coordinates are reflected across the x-axis.
3. Graph F. The graph will be a reflection across the y-axis of graph C.

Exploration 3

1.

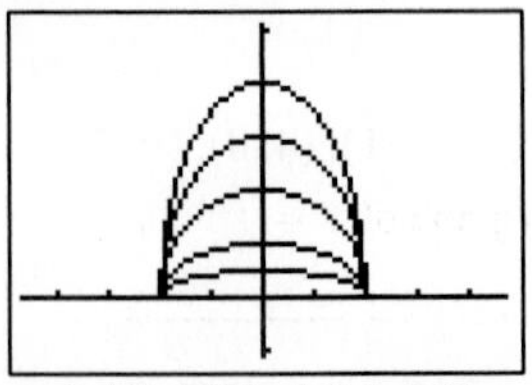

[–4.7, 4.7] by [–1.1, 5.1]

The 1.5 and the 2 stretch the graph vertically; the 0.5 and the 0.25 shrink the graph vertically.

Quick Review 1.6

1. $(x+1)^2$ **3.** $(x+6)^2$ **5.** $(x - 5/2)^2$ **7.** $x^2 - x + 2$ **9.** $x^3 - 6x + 5$

Exercises 1.6

1. Vertical translation down 3 units **3.** Horizontal translation left 4 units **5.** Horizontal translation to the right 100 units **7.** Horizontal translation to the right 1 unit, and Vertical translation up 3 units **9.** Reflection across x-axis **11.** Reflection across y-axis **13.** Vertically stretch by 2 **15.** Horizontally stretch by $\frac{1}{0.2} = 5$, or vertically shrink by $0.2^3 = 0.008$ **17.** Translate right 6 units to get g

19. Translate left 4 units, and reflect across the x-axis to get g

21.

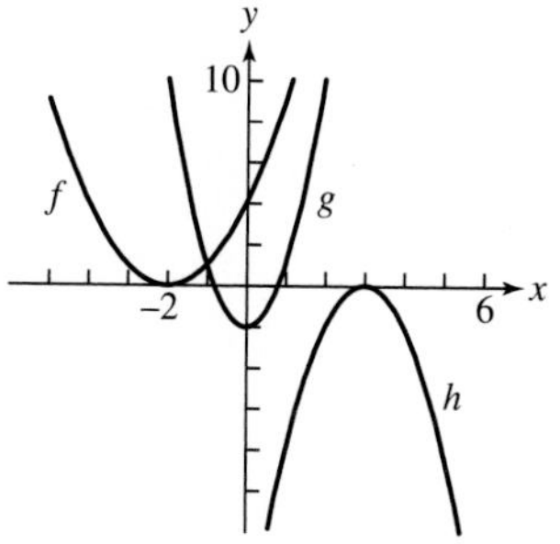

23.

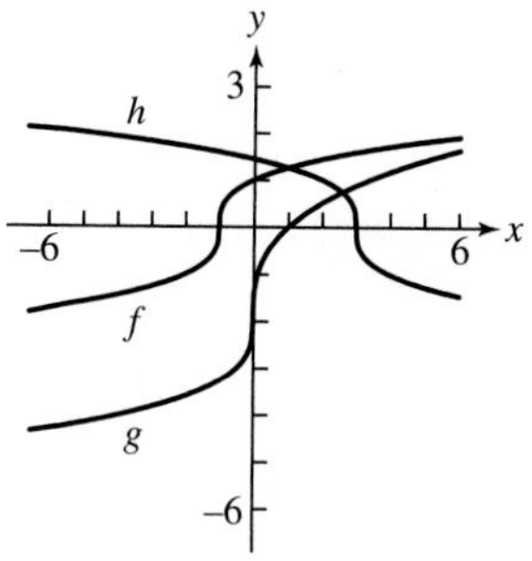

25. $f(x) = \sqrt{x + 5}$ **27.** $f(x) = -\sqrt{x + 2} + 3 = 3 - \sqrt{x + 2}$ **29.** **(a)** $-x^3 + 5x^2 + 3x - 2$ **(b)** $-x^3 - 5x^2 + 3x + 2$

31. **(a)** $y = -f(x) = -(\sqrt[3]{8x}) = -2\sqrt[3]{x}$ **(b)** $y = f(-x) = \sqrt[3]{8(-x)} = -2\sqrt[3]{x}$

33. Let f be an odd function; that is, $f(-x) = -f(x)$ for all $x$in the domain of f. To reflect the graph of $y = f(x)$ across the y-axis, we make the tranformation $y = f(-x)$. But $f(-x) = -f(x)$ for all x in the domain of f, so this transformation results in $y = -f(x)$. That is exactly the translation that reflects the graph of f across the x-axis, so the two reflections yield the same graph.

35.

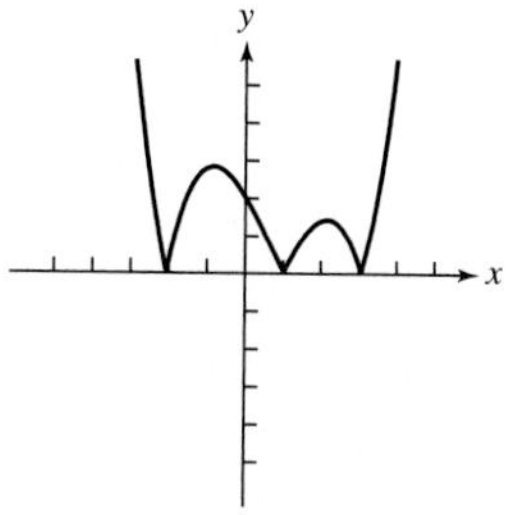

37.

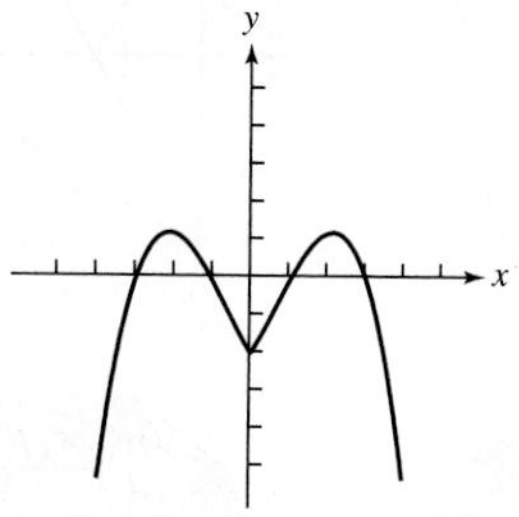

39. **(a)** $2x^3 - 8x$ **(b)** $27x^3 - 12x$

41. **(a)** $2x^2 + 2x - 4$ **(b)** $9x^2 + 3x - 2$

43. Starting with $y = x^2$, translate right 3 units, vertically stretch by 2, and translate down 4 units.

45. Starting with $y = x^2$, horizontally shrink by $\frac{1}{3}$ and translate down 4 units.

47. $y = 3(x - 4)^2$ **49.** $y = 2|x + 2| - 4$

51.

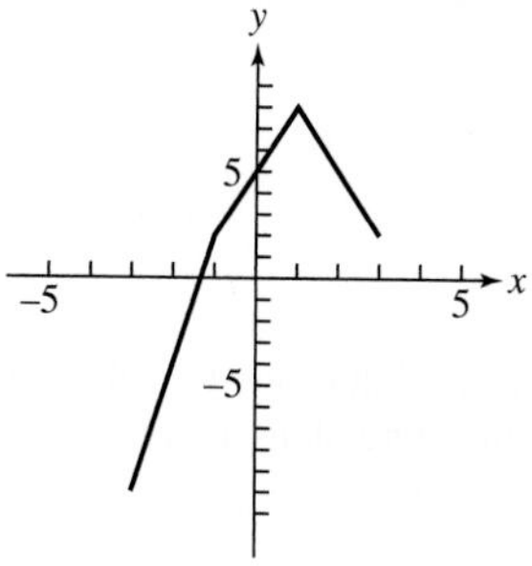

53.

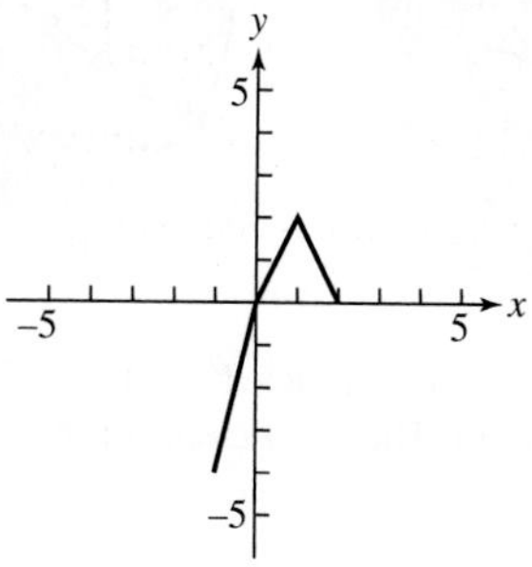

55. Reflections have more effect on points that are farther away from the line of reflection. Translations affect the distance of points from the axes, and hence change the effect of the reflections.

57. First vertically stretch by $\frac{9}{5}$, then translate up 32 units.

59. False; it is translated left. **61.** C **63.** A

65. **(a)**

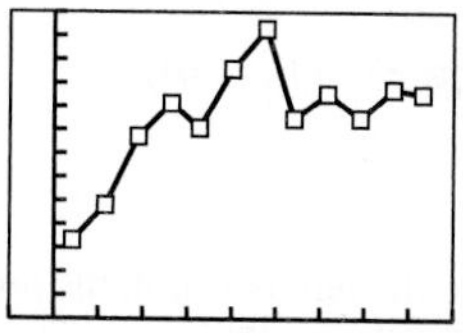

[–1, 13] by [30, 40]

(b) Change the y-value by multiplying by the conversion rate from dollars to yen, a number that changes according to international market conditions. This results in a vertical stretch by the conversion rate.

67. **(a)** The original graph is on the top; the graph of $y = |f(x)|$ is on the bottom.

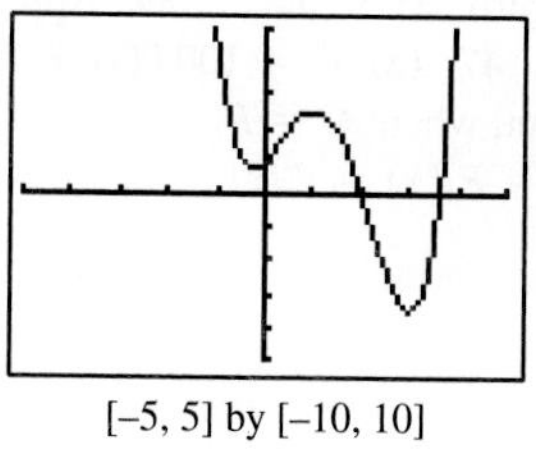

[–5, 5] by [–10, 10]

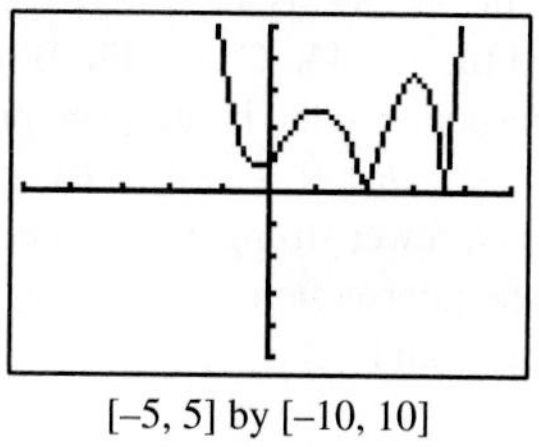

[–5, 5] by [–10, 10]

(b) The original graph is on the top; the graph of $y = f(|x|)$ is on the bottom.

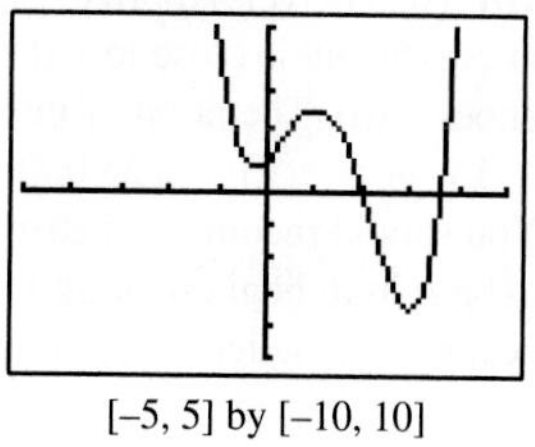

[–5, 5] by [–10, 10]

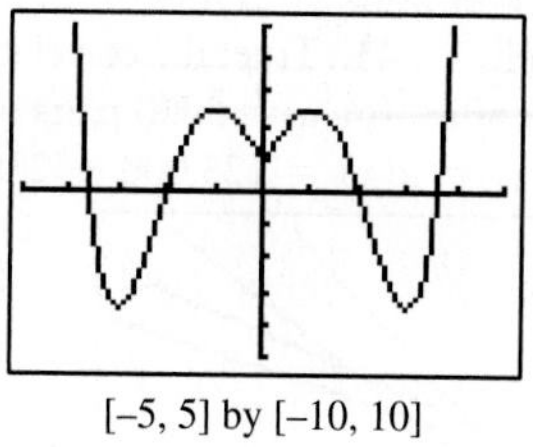

[–5, 5] by [–10, 10]

(c)

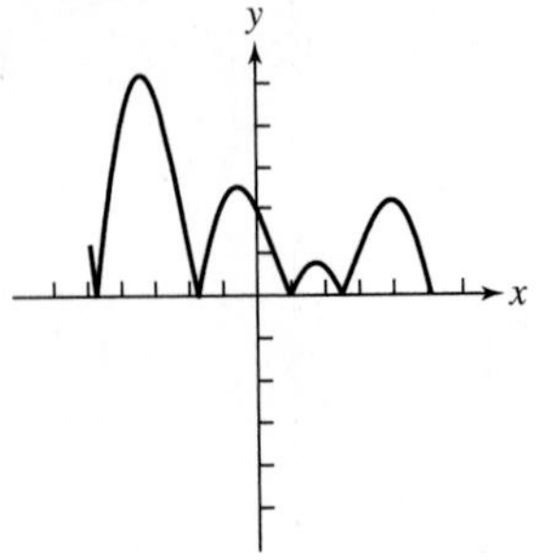

(d)

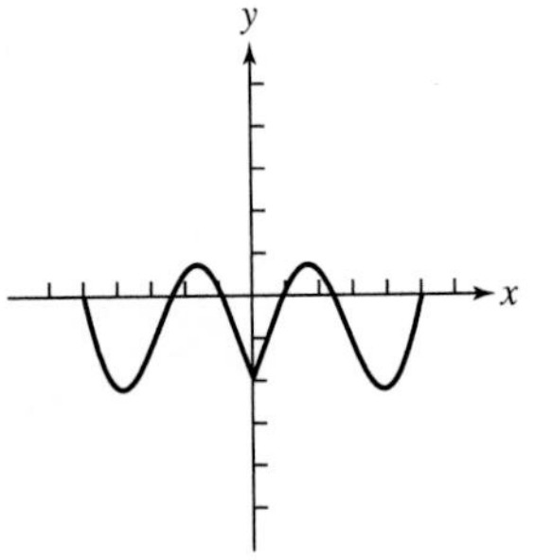

SECTION 1.7

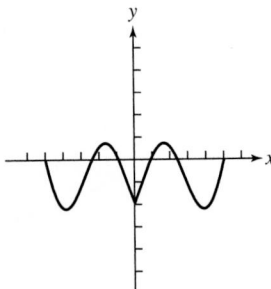

Exploration 1

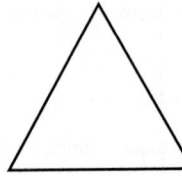

1.

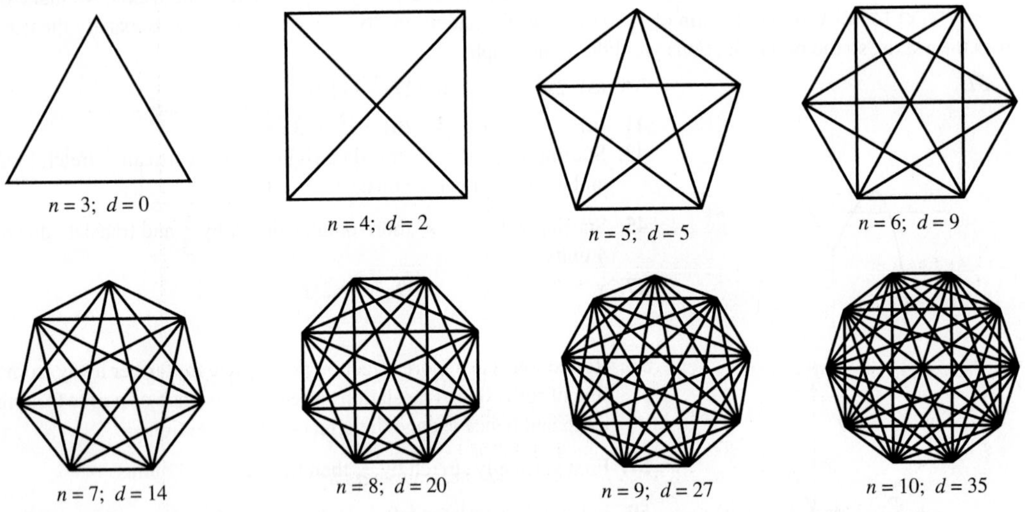

2, 5, 9, 14, 20, 27

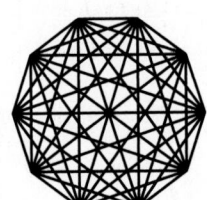

3. Linear: $r^2 = 0.9758$; power: $r^2 = 0.9903$; quadratic: $R^2 = 1$; cubic: $R^2 = 1$; quartic: $R^2 = 1$ **5.** Since the quadratic curve fits the points perfectly, there is nothing to be gained by adding a cubic term or a quartic term. The coefficients of these terms in the regressions are zero.

Quick Review 1.7

1. $h = 2(A/b)$ **3.** $h = V/(\pi r^2)$ **5.** $r = \sqrt[3]{\dfrac{3V}{4\pi}}$ **7.** $h = \dfrac{A - 2\pi r^2}{2\pi r} = \dfrac{A}{2\pi r} - r$ **9.** $P = \dfrac{A}{(1 + r/n)^{nt}} = A(1 + r/n)^{-nt}$

Exercises 1.7

1. $3x + 5$ **3.** $0.17x$ **5.** $(x + 12)(x)$ **7.** $1.045x$ **9.** $0.60\,x$ **11.** Let C be the total cost and n be the number of items produced; $C = 34{,}500 + 5.75n$. **13.** Let R be the revenue and n be the number of items sold: $R = 3.75n$. **15.** $V = 2\pi r^3$ **17.** $A = a^2\sqrt{15}/4$ **19.** $A = 24r^2$ **21.** $x + 4x = 620$; $x = 124$; $4x = 496$ **23.** $1.035x = 36{,}432$; $x = 35{,}200$ **25.** $182 = 52t$, so $t = 3.5$ hr **27.** $0.60(33) = 19.8$, $0.75(27) = 20.25$; The \$33 shirt is a better bargain, because the sale price is cheaper. **29.** 9% **31. (a)** $0.10x + 0.45(100 - x) = 0.25(100)$ **(b)** Use about 57.14 gal of the 10% solution and about 42.86 gal of the 45% solution. **33. (a)** $V = x(10 - 2x)(18 - 2x)$ **(b)** $(0, 5)$ **(c)** Approx. 2.06 in. by 2.06 in. **35.** 6 in. **37.** Approx. 21.36 in. **39.** Approx. 11.42 mph **41.** True; the correlation coefficient is close to 1 if there is a good fit. **43.** C **45.** B **47. (a)** $C = 100{,}000 + 30x$ **(b)** $R = 50x$ **(c)** $x = 5000$ pairs of shoes **(d)** The point of intersection corresponds to the break-even point, where $C = R$. **49. (a)** $y_1 = u(x) = 125{,}000 + 23x$ **(b)** $y_2 = s(x) = 125{,}000 + 31x$ **(c)** $y_3 = R_u(x) = 56x$ **(d)** $y_4 = R_s(x) = 79x$

(e)

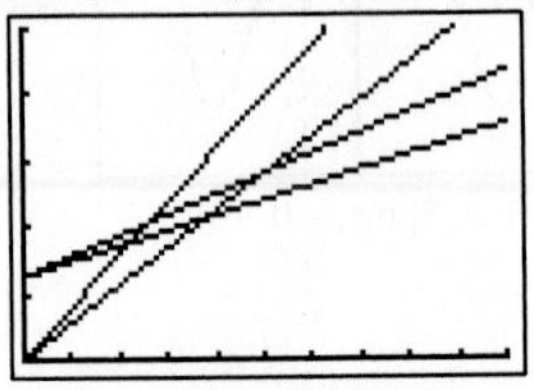

[0, 10,000] by [0, 500,000]

(f) You should recommend stringing the rackets; fewer strung rackets need to be sold to begin making a profit (since the intersection of y_2 and y_4 occurs for smaller x than the intersection of y_1 and y_3).

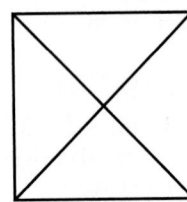

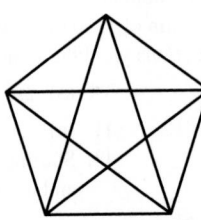

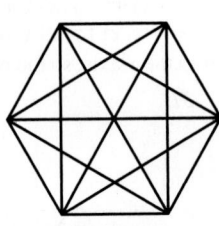

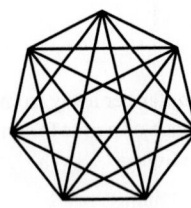

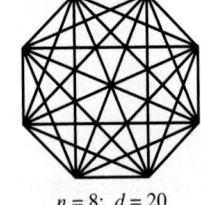

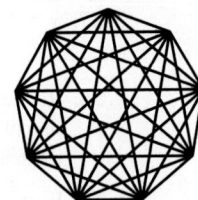

51. (a)

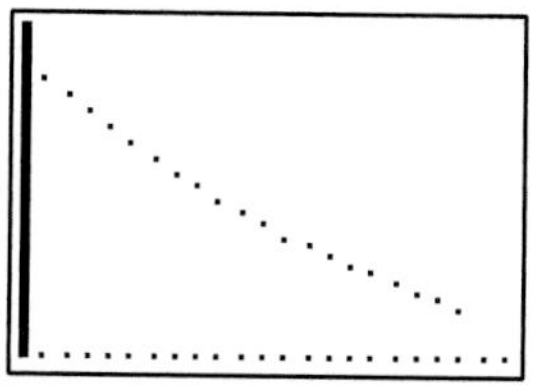

[0, 22] by [100, 200]

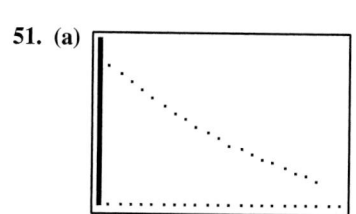

(b) List $L3 = \{112.3, 106.5, 101.5, 96.6, 92.0, 87.2, 83.1, 79.8, 75.0, 71.7, 68, 64.1, 61.5, 58.5, 55.9, 53.0, 50.8, 47.9, 45.2, 43.2\}$
(c) $y = 118.07 \times 0.951^x$. It fits the data extremely well.

CHAPTER 1 REVIEW EXERCISES

1. d **3.** i **5.** b **7.** g **9.** a **11. (a)** All real numbers **(b)** All real numbers **13. (a)** All real numbers **(b)** $[0, \infty)$ **15. (a)** All real numbers **(b)** $[8, \infty)$ **17. (a)** All real numbers except 0 and 2 **(b)** All real numbers except 0 **19.** Continuous **21. (a)** Vertical asymptotes at $x = 0$ and $x = 5$ **(b)** $y = 0$ **23. (a)** none **(b)** $y = 7$ and $y = -7$ **25.** $(-\infty, \infty)$ **27.** $(-\infty, -1), (-1, 1), (1, \infty)$ **29.** Not bounded **31.** Bounded above **33. (a)** None **(b)** -7, at $x = -1$ **35. (a)** -1, at $x = 0$ **(b)** None **37.** Even **39.** Neither **41.** $(x - 3)/2$ **43.** $2/x$

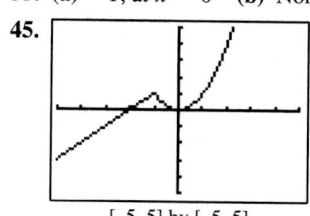

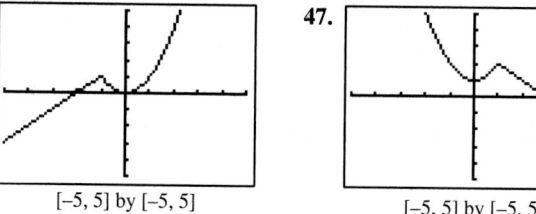

45.

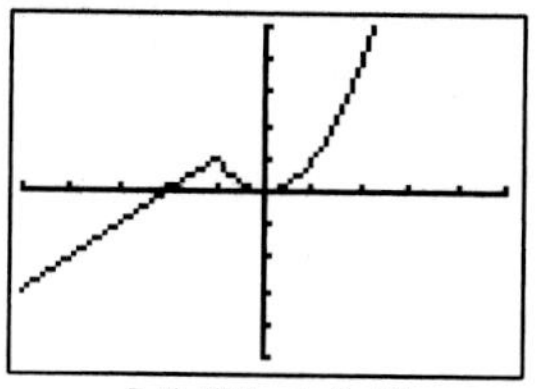

[–5, 5] by [–5, 5]

47.

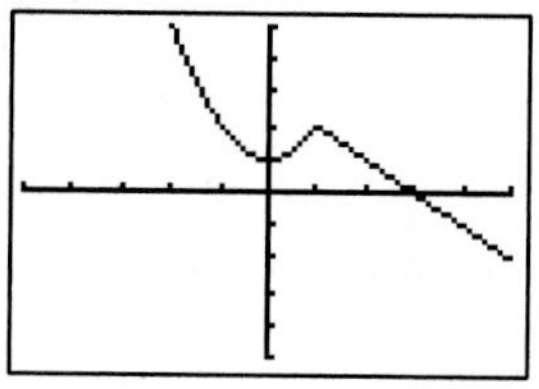

[–5, 5] by [–5, 5]

49.

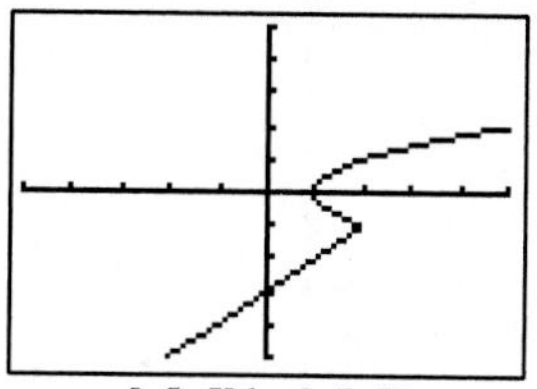

[–5, 5] by [–5, 5]

51.

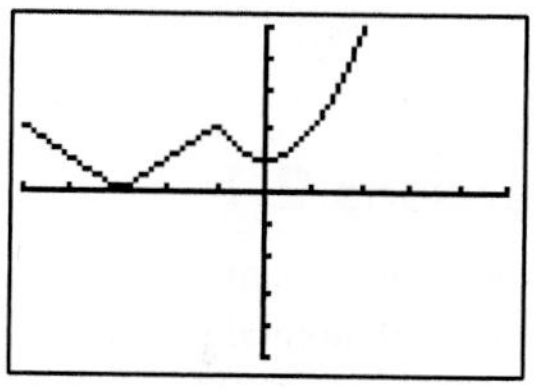

[–5, 5] by [–5, 5]

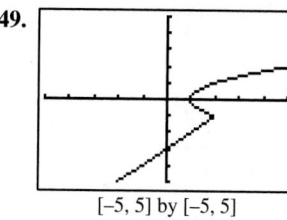

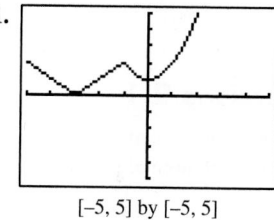

53. $(f \circ g)(x) = \sqrt{x^2 - 4}$; $(-\infty, -2] \cup [2, \infty)$ **55.** $(f \circ g)(x) = \sqrt{x}(x^2 - 4)$; $[0, \infty)$
57. $\lim_{x \to \infty} \sqrt{x} = \infty$ **59.** $\pi s^2/2$ **61.** $100\pi h$ **63.** $40 - t/(50\pi)$

65. (a)

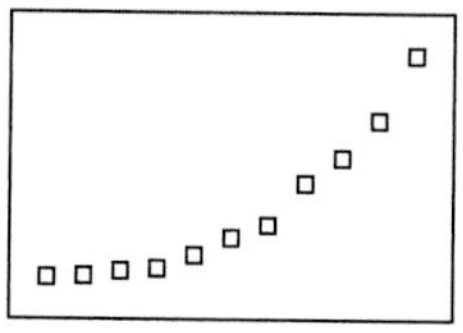

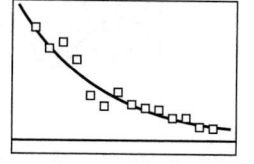

[0, 12] by [640, 3260]

(b) The regression curve is $y = 25.34x^2 - 122.22x + 1117.38$.

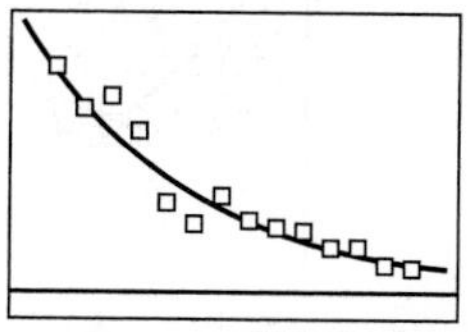

[55, 117] by [–0.5, 11]

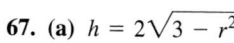

The fit is almost unbelievably good for an economic time graph based on real data. We did not, however, make these numbers up.

(c) 5650

67. (a) $h = 2\sqrt{3 - r^2}$

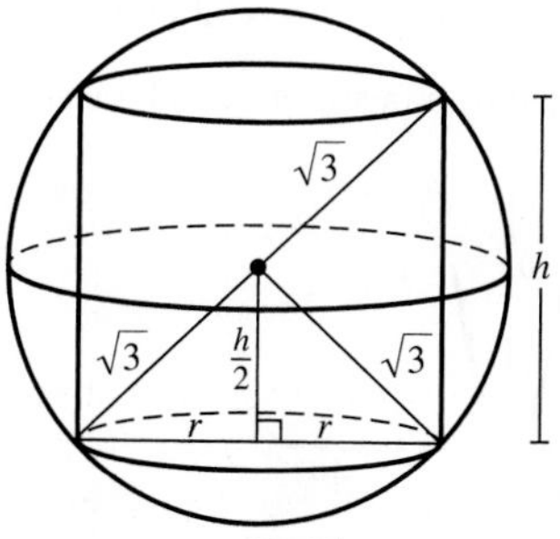

$h = 2\sqrt{3 - r^2}$

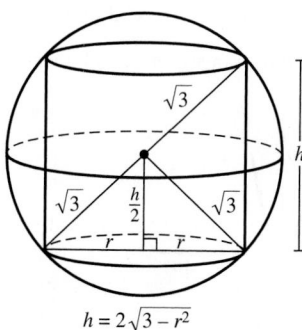

(b) $2\pi r^2\sqrt{3 - r^2}$ **(c)** $[0, \sqrt{3}]$ **(d)**

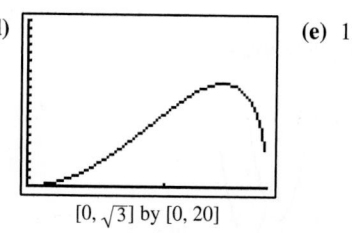

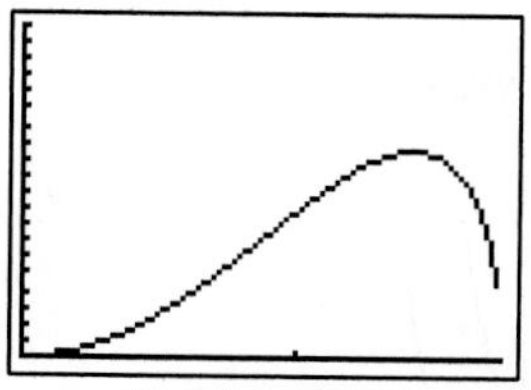

[0, √3] by [0, 20]

(e) 12.57 in.^{3}

CHAPTER 1 PROJECT

1.

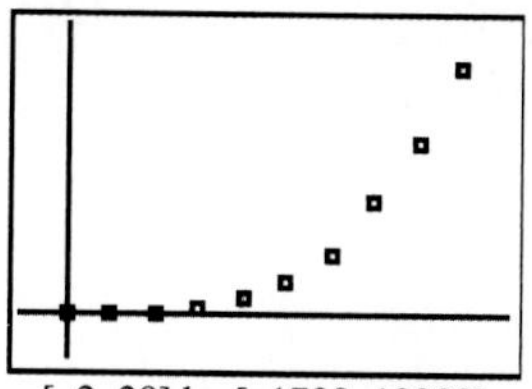

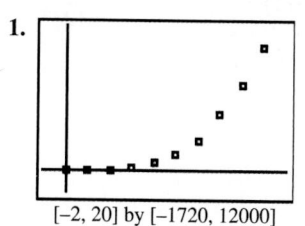

[–2, 20] by [–1720, 12000]

3. 34.854; 49.785 **5.** The logistic model for the data is $y = \dfrac{16{,}098}{1 + 432.23e^{-0.3678x}}$.

SECTION 2.1

Exploration 1

1. −$2000/year **3.** $50,000; $18,000

Quick Review 2.1

1. $y = 8x + 3.6$ **3.** $y = -0.6x + 2.8$ **5.** $x^2 + 6x + 9$ **7.** $3x^2 - 36x + 108$ **9.** $2(x - 1)^2$

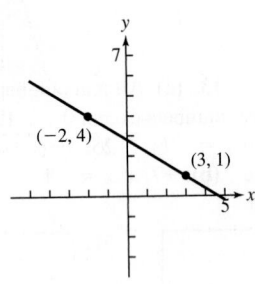

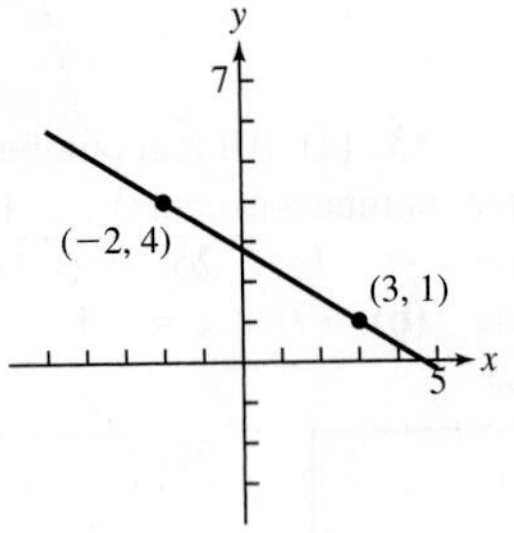

Exercises 2.1

1. Not a polynomial function because of the exponent −5 **3.** Polynomial of degree 5 with leading coefficient 2

5. Not a polynomial function because of the radical

7. $f(x) = \frac{5}{7}x + \frac{18}{7}$

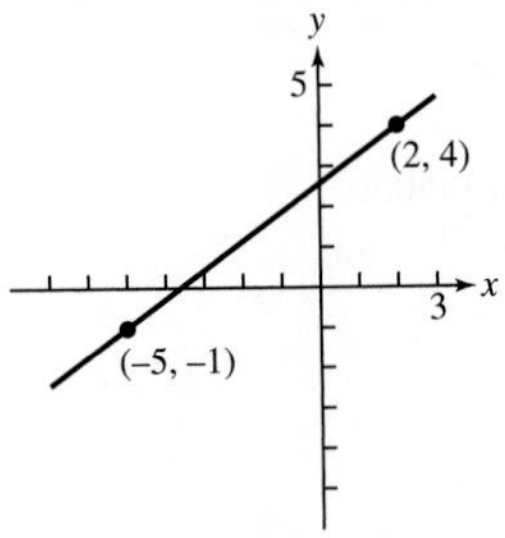

9. $f(x) = -\frac{4}{3}x + \frac{2}{3}$

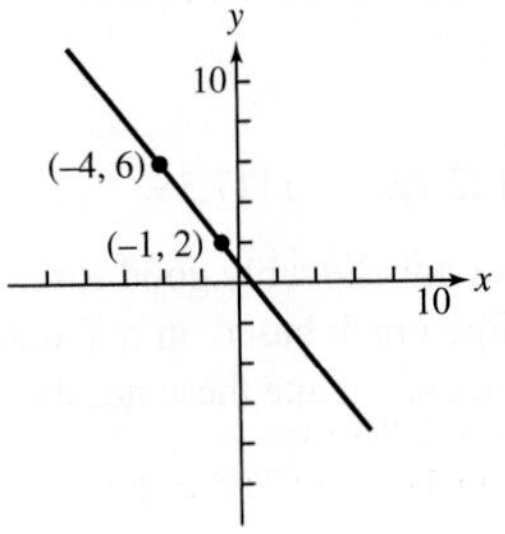

11. $f(x) = -x + 3$

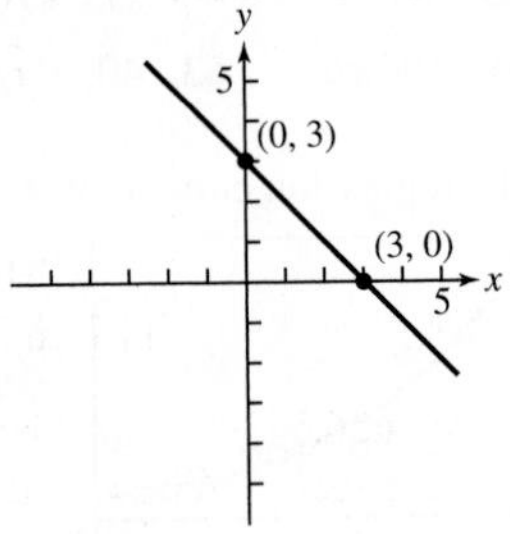

13. (a) **15.** (b) **17.** (e)

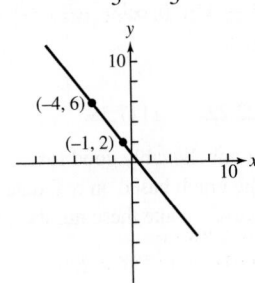

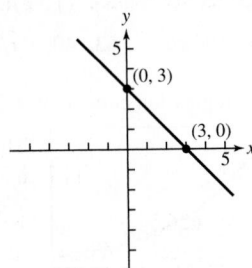

19. Translate the graph of $y = x^2$ 3 units right and the result 2 units down.

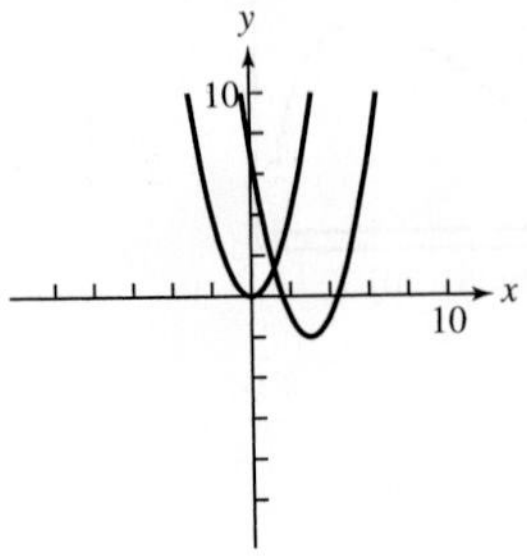

21. Translate the graph of $y = x^2$ 2 units left, vertically shrink the resulting graph by a factor of $\frac{1}{2}$, and translate that graph 3 units down.

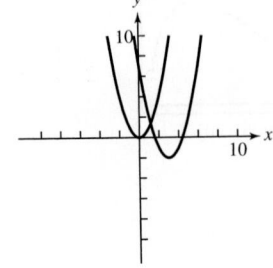

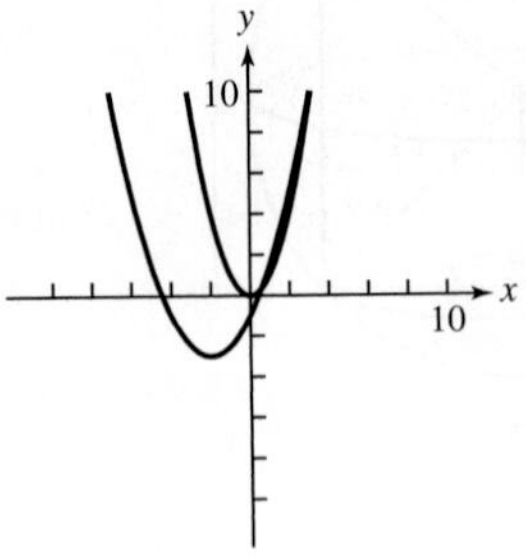

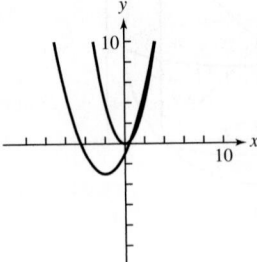

23. Vertex: $(1, 5)$; axis: $x = 1$ **25.** Vertex: $(1, -7)$; axis: $x = 1$ **27.** Vertex: $\left(-\frac{5}{6}, -\frac{73}{12}\right)$; axis: $x = -\frac{5}{6}$; $f(x) = 3\left(x + \frac{5}{6}\right)^2 - \frac{73}{12}$

29. Vertex: $(4, 19)$; axis: $x = 4$; $f(x) = -(x - 4)^2 + 19$ **31.** Vertex: $\left(\frac{3}{5}, \frac{11}{5}\right)$; axis: $x = \frac{3}{5}$; $g(x) = 5\left(x - \frac{3}{5}\right)^2 + \frac{11}{5}$

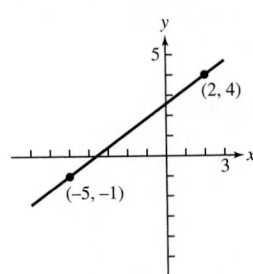

33. $f(x) = (x - 2)^2 + 2$; Vertex: $(2, 2)$; axis: $x = 2$; opens upward; does not intersect x-axis

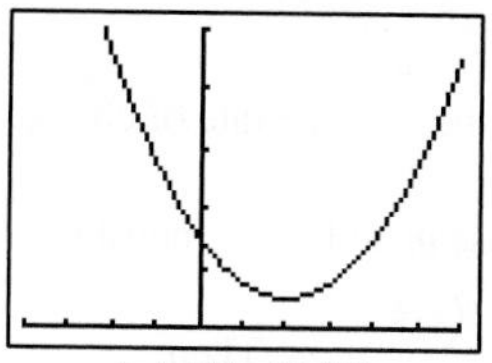

[−4, 6] by [0, 20]

35. $f(x) = -(x + 8)^2 + 74$; Vertex: $(-8, 74)$; axis: $x = -8$; opens downward; intersects x-axis at about -16.602 and 0.602, or $\left(-8 \pm \sqrt{74}\right)$

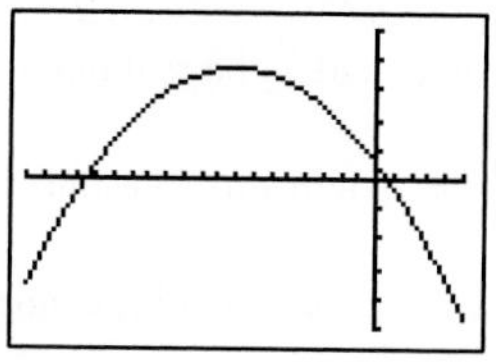

[−20, 5] by [−100, 100]

37. $f(x) = 2\left(x + \frac{3}{2}\right)^2 + \frac{5}{2}$; Vertex: $\left(-\frac{3}{2}, \frac{5}{2}\right)$; axis: $x = -\frac{3}{2}$; opens upward; does not intersect x-axis; vertically stretched by 2

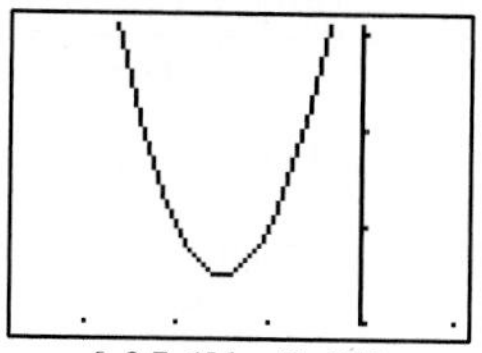

[−3.7, 1] by [2, 5.1]

39. $y = 2(x + 1)^2 - 3$ **41.** $y = -2(x - 1)^2 + 11$ **43.** $y = 2(x - 1)^2 + 3$ **45.** Strong positive **47.** Weak positive
49. (a)

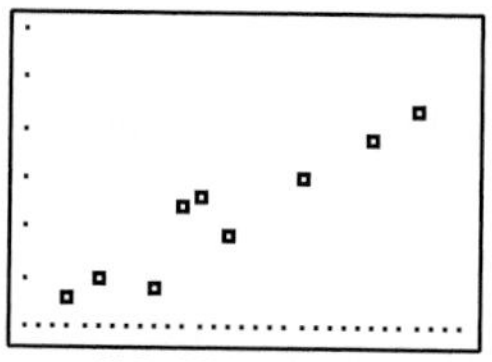

[15, 45] by [20, 50]

(b) Strong positive **51.** \$940 **53. (a)** $y \approx 19.56 + 0.148x$. The slope says that average fuel economy for light duty vehicles increases about 0.15 mpg per year. **(b)** 24 mpg **55. (a)** $[0, 100]$ by $[0, 1000]$ is one possibility. **(b)** Either 107,335 units or 372,665 units **57.** 3.5 ft

59. (a) $R(x) = (26{,}000 - 1000x)(0.50 + 0.05x)$ **(b)**

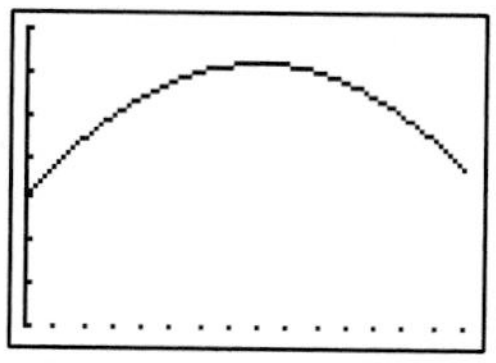

[0, 15] by [10,000, 17,000]

(c) 90 cents per can; \$16,200

61. (a) About 215 ft above the field **(b)** About 6.54 sec **(c)** About 117 ft/sec downward
63. (a) $h = -16t^2 + 80t - 10$ **(b)** 90 ft, 2.5 sec

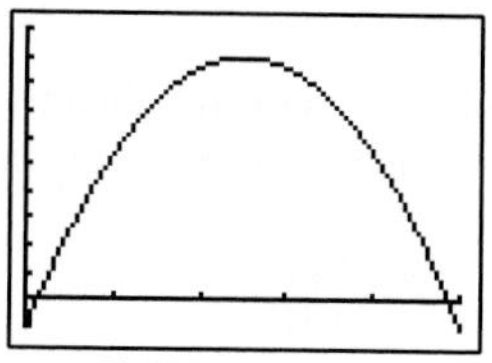

[0, 5] by [−10, 100]

65. $y \approx 0.3721x^2 + 2.697x + 111.264$; 2002

67. (a)

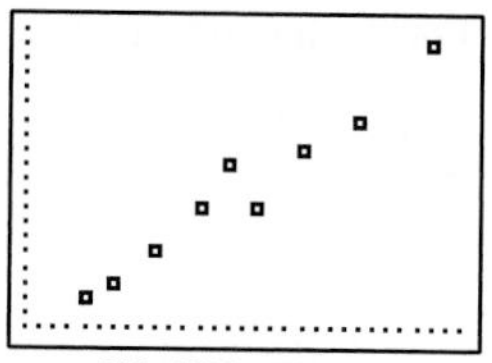

[15, 45] by [20, 40]

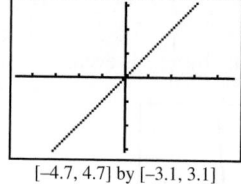

(b) $y \approx 0.68x + 9.01$
(c) On average, the children gained 0.68 pounds per month.

(d)

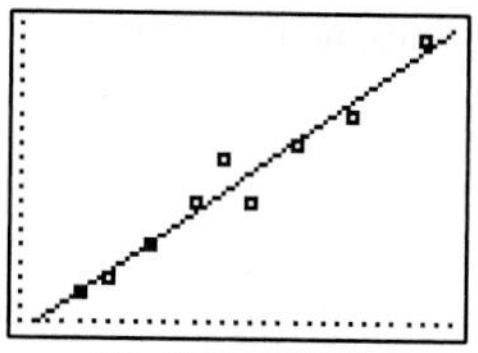

[15, 45] by [20, 40]

(e) ≈ 29.41 lb

69. The Identity Function $f(x) = x$

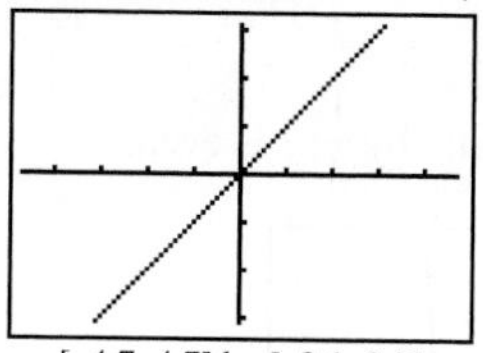

[−4.7, 4.7] by [−3.1, 3.1]

Domain: $(-\infty, \infty)$; Range: $(-\infty, \infty)$; Continuous; Increasing for all x; Symmetric about the origin; Not bounded; No local extrema; No horizontal or vertical asymptotes; End behavior: $\lim_{x \to -\infty} f(x) = -\infty$, $\lim_{x \to \infty} f(x) = \infty$

71. False. The initial value is $f(0) = -3$.
73. E **75.** B **77. (a)** (i), (iii), and (v) because they are slant lines **(b)** (i), (iii), (iv), (v), and (vi) because they are not vertical **(c)** (ii) is not a function because it is vertical.

79. The line that minimizes the sum of the squares of vertical distances is nearly always different from the line that minimizes the sum of the squares of horizontal distances to the points in a scatter plot. For the data in Table 2.2, the regression line obtained from reversing the ordered pairs has a slope of $-1/15{,}974.90$; whereas, the inverse of the function in Example 3 has a slope of $-1/15{,}358.93$—close but not the same slope.

81. **(a)** The two solutions are $\dfrac{-b+\sqrt{b^2-4ac}}{2a}$ and $\dfrac{-b-\sqrt{b^2-4ac}}{2a}$; their sum is $2\left(-\dfrac{b}{2a}\right)=-\dfrac{b}{a}$.
(b) The product of the two solutions given above is $\dfrac{b^2-(b^2-4ac)}{4a^2}=\dfrac{c}{a}$. **83.** $\left(\dfrac{a+b}{2}, -\dfrac{(a-b)^2}{4}\right)$

85. Suppose $f(x)=mx+b$ with m and b constants and $m\neq 0$. Let x_1 and x_2 be real numbers with $x_1\neq x_2$. Then the average rate of change of f is $\dfrac{f(x_2)-f(x_1)}{x_2-x_1}=\dfrac{(mx_2+b)-(mx_1+b)}{x_2-x_1}=\dfrac{m(x_2-x_1)}{x_2-x_1}=m$, a nonzero constant. On the other hand, suppose m and x_1 are constants and $m\neq 0$. Let x be a real number with $x\neq x_1$ and let f be a function defined on all real numbers such that $\dfrac{f(x)-f(x_1)}{x-x_1}=m$. Then $f(x)-f(x_1)=m(x-x_1)$ and $f(x)=mx+(f(x_1)-mx_1)$. Notice that the expression $f(x_1)-mx_1$ is a constant; call it b. Then $f(x_1)-mx_1=b$; so, $f(x_1)=mx_1+b$ and $f(x)=mx+b$ for all $x\neq x_1$. Thus f is a linear function.

SECTION 2.2

Exploration 1

1.

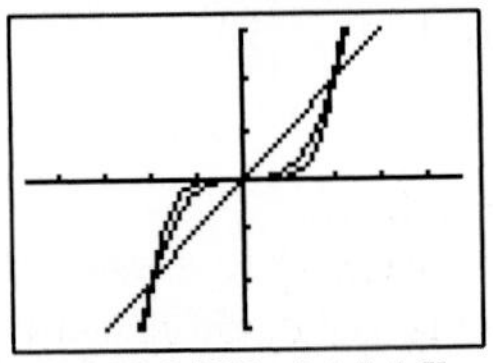

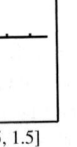

[–2.35, 2.35] by [–1.5, 1.5]

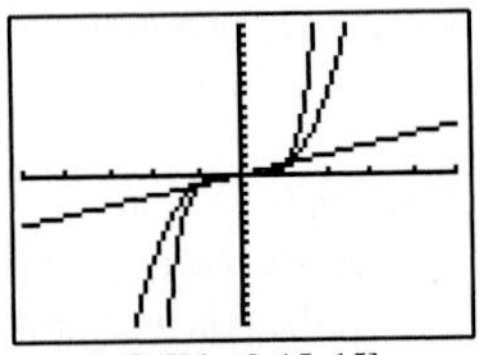

[–5, 5] by [–15, 15]

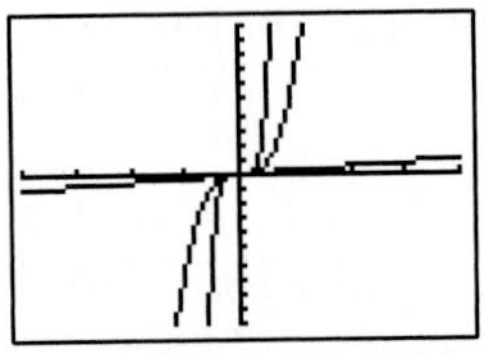

[–20, 20] by [–200, 200]

The pairs $(0, 0)$, $(1, 1)$ and $(-1, -1)$ are common to all three graphs.

Quick Review 2.2

1. $\sqrt[3]{x^2}$ **3.** $1/d^2$ **5.** $1/\sqrt[5]{q^4}$ **7.** $3x^{3/2}$ **9.** $\approx 1.71x^{-4/3}$

Exercises 2.2

1. Power $=5$, constant $=-\dfrac{1}{2}$ **3.** Not a power function **5.** Power $=1$, constant $=c^2$ **7.** Power $=2$, constant $=\dfrac{g}{2}$
9. Power $=-2$, constant $=k$ **11.** Degree $=0$, coefficient $=-4$ **13.** Degree $=7$, coefficient $=-6$
15. Degree $=2$, coefficient $=4\pi$ **17.** $A=ks^2$ **19.** $I=V/R$ **21.** $E=mc^2$ **23.** The weight w of an object varies directly with its mass m, with the constant of variation g. **25.** The refractive index n of a medium is inversely proportional to v, the velocity of light in the medium, with constant of variation c, the constant velocity of light in free space.

27. Power $=4$, constant $=2$; Domain: $(-\infty, \infty)$; Range: $[0, \infty)$; Continuous; Decreasing on $(-\infty, 0)$. Increasing on $(0, \infty)$; Even. Symmetric with respect to y-axis; Bounded below, but not above; Local minimum at $x=0$; Asymptotes: none; End Behavior: $\lim_{x\to-\infty} 2x^4=\infty$, $\lim_{x\to\infty} 2x^4=\infty$

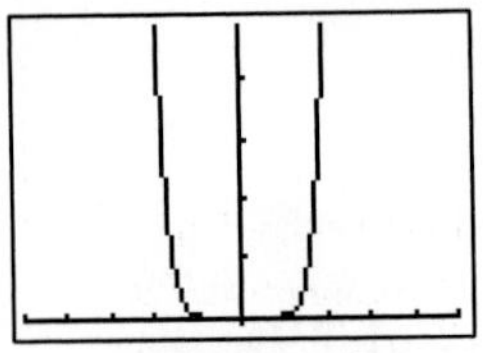

[–5, 5] by [–1, 49]

29. Power $=\dfrac{1}{4}$, constant $=\dfrac{1}{2}$; Domain: $[0, \infty)$; Range: $[0, \infty)$; Continuous; Increasing on $[0, \infty)$; Bounded below; Neither even nor odd; Local minimum at $(0, 0)$; Asymptotes: none; End Behavior: $\lim_{x\to\infty} \dfrac{1}{2}\sqrt[4]{x}=\infty$

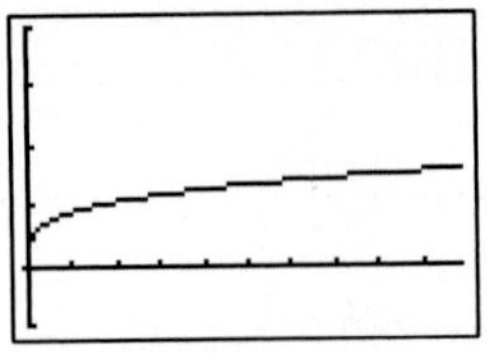

[–1, 99] by [–1, 4]

31. Shrink vertically by $\dfrac{2}{3}$; f is even.

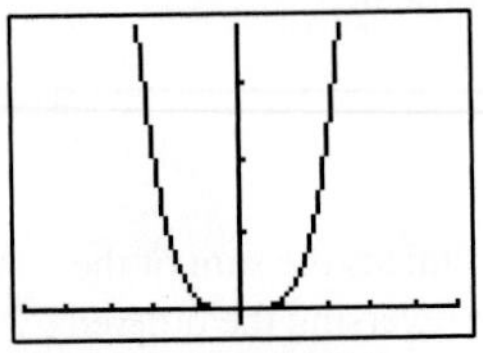

[–5, 5] by [–1, 19]

33. Stretch vertically by 1.5 and reflect over the x-axis; f is odd.

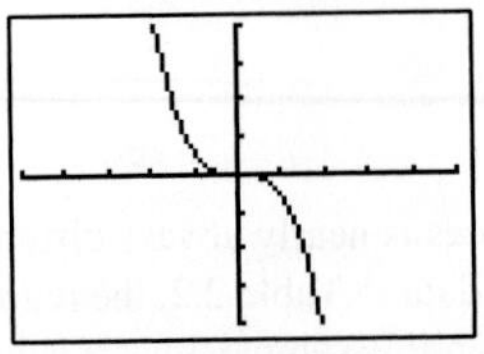

[–5, 5] by [–20, 20]

35. Shrink vertically by $\dfrac{1}{4}$; f is even.

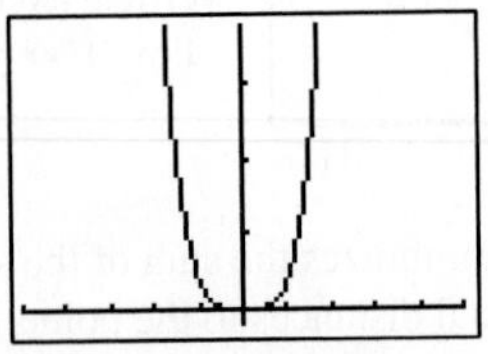

[–5, 5] by [–1, 49]

37. (g) **39.** (d) **41.** (h) **43.** $k = 3, a = \frac{1}{4}$. f is increasing in Quadrant I. f is undefined for $x < 0$.

45. $k = -2, a = \frac{4}{3}$. f is decreasing in Quadrant IV. f is even. **47.** $k = \frac{1}{2}, a = -3$. f is decreasing in Quadrant I. f is odd.

49. $y = \frac{8}{x^2}$, power $= -2$, constant $= 8$ **51.** 2.21 L **53.** 1.24×10^8 m/sec

55. (a)

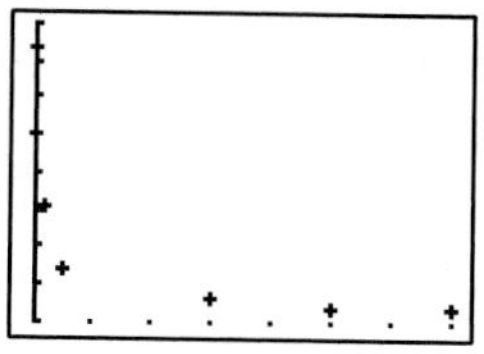

[–2, 71] by [50, 450]

(b) $r \approx 231.204 \cdot w^{-0.297}$ **(c)**

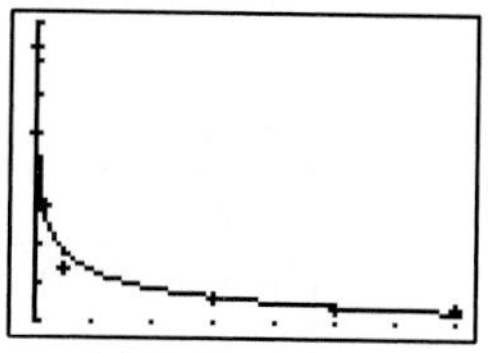

[–2, 71] by [50, 450]

(d) Approximately 37.67 beats/min, which is very close to Clark's observed value

57. (a)

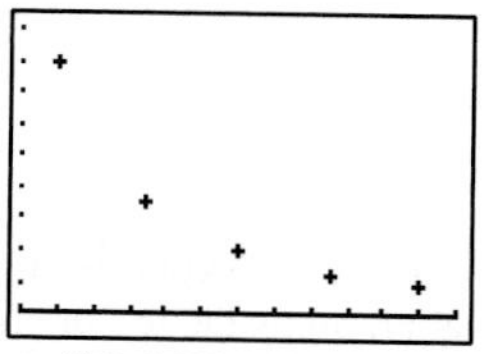

[0.8, 3.2] by [–0.3, 9.2]

(b) $y \approx 7.932 \cdot x^{-1.987}$; yes **(c)**

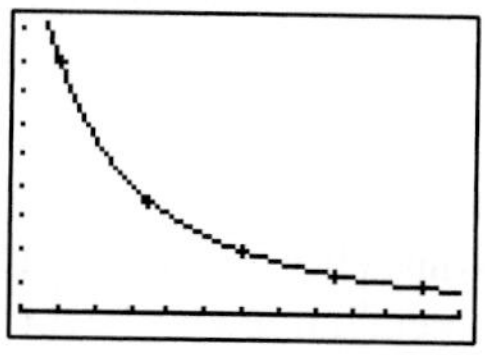

[0.8, 3.2] by [−0.3, 9.2]

(d) Approximately 2.76 W/m^2 and 0.697 W/m^2, respectively

59. False, because $f(-x) = (-x)^{1/3} = -x^{1/3} = -f(x)$. The graph of f is symmetric about the origin. **61.** E **63.** B

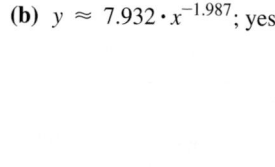

65. (a)

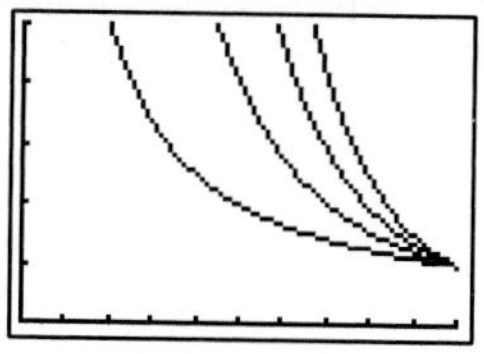

[0, 1] by [0, 5]

[0, 3] by [0, 3]

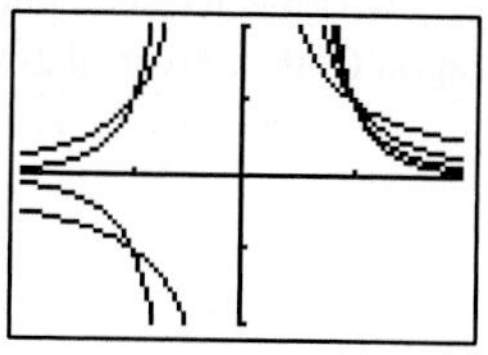

[–2, 2] by [–2, 2]

The graphs of $f(x) = x^{-1}$ and $h(x) = x^{-3}$ are similar and appear in the first and third quadrants only. The graphs of $g(x) = x^{-2}$ and $k(x) = x^{-4}$ are similar and appear in the first and second quadrants only. The pair $(1, 1)$ is common to all four functions.

	f	g	h	k
Domain	$x \neq 0$	$x \neq 0$	$x \neq 0$	$x \neq 0$
Range	$y \neq 0$	$y > 0$	$y \neq 0$	$y > 0$
Continuous	yes	yes	yes	yes
Increasing		$(-\infty, 0)$		$(-\infty, 0)$
Decreasing	$(-\infty, 0), (0, \infty)$	$(0, \infty)$	$(-\infty, 0), (0, \infty)$	$(0, \infty)$
Symmetry	w.r.t. origin	w.r.t. y-axis	w.r.t. origin	w.r.t. y-axis
Bounded	not	below	not	below
Extrema	none	none	none	none
Asymptotes	x-axis, y-axis	x-axis, y-axis	x-axis, y-axis	x-axis, y-axis
End behavior	$\lim_{x\to\pm\infty} f(x) = 0$	$\lim_{x\to\pm\infty} g(x) = 0$	$\lim_{x\to\pm\infty} h(x) = 0$	$\lim_{x\to\pm\infty} k(x) = 0$

(b)

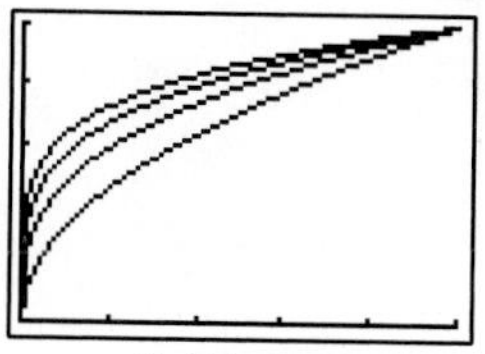

[0, 1] by [0, 1]

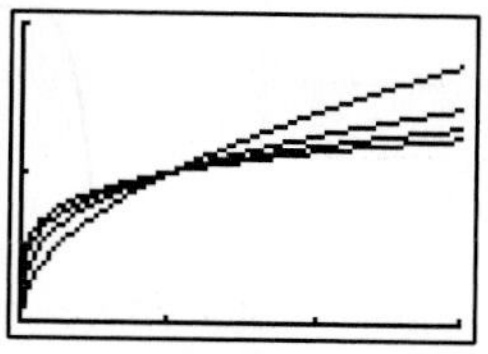

[0, 3] by [0, 2]

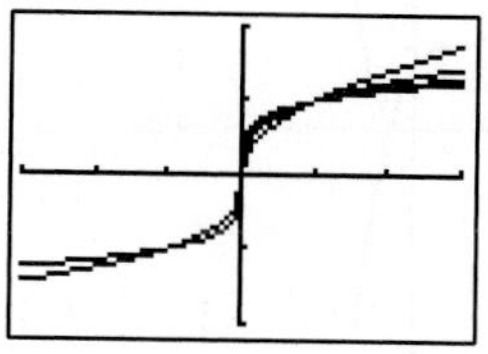

[–3, 3] by [–2, 2]

The graphs of $f(x) = x^{1/2}$ and $h(x) = x^{1/4}$ are similar and appear in the first quadrant only. The graphs of $g(x) = x^{1/3}$ and $k(x) = x^{1/5}$ are similar and appear in the first and third quadrants only. The pairs $(0, 0)$, $(1, 1)$ are common to all four functions.

	f	g	h	k
Domain	$[0, \infty)$	$(-\infty, \infty)$	$[0, \infty)$	$[-\infty, \infty)$
Range	$[0, \infty)$	$(-\infty, \infty)$	$[0, \infty)$	$(-\infty, \infty)$
Continuous	yes	yes	yes	yes
Increasing	$[0, \infty)$	$(-\infty, \infty)$	$[0, \infty)$	$(-\infty, \infty)$
Decreasing				
Symmetry	none	w.r.t. origin	none	w.r.t. origin
Bounded	below	not	below	not
Extrema	min at $(0, 0)$	none	min at $(0, 0)$	none
Asymptotes	none	none	none	none
End behavior	$\lim_{x\to\infty} f(x) = \infty$	$\lim_{x\to\infty} g(x) = \infty$ $\lim_{x\to-\infty} g(x) = -\infty$	$\lim_{x\to\infty} h(x) = \infty$	$\lim_{x\to\infty} k(x) = \infty$ $\lim_{x\to-\infty} k(x) = -\infty$

67. $T \approx a^{1.5}$. Squaring both sides shows that approximately $T^2 = a^3$. **69.** If $f(x)$ is even, $g(-x) = 1/f(-x) = 1/f(x) = g(x)$. If $f(x)$ is odd, $g(-x) = 1/f(-x) = 1/(-f(x)) = -1/f(x) = -g(x)$. If $g(x) = 1/f(x)$, then $f(x) \cdot g(x) = 1$ and $f(x) = 1/g(x)$. So by the reasoning used above, if $g(x)$ is even, so is $f(x)$, and if $g(x)$ is odd, so is $f(x)$. **71. (a)** The force F acting on an object varies jointly as the mass m of the object and the acceleration a of the object. **(b)** The kinetic energy KE of an object varies jointly as the mass m of the object and the square of the velocity v of the object. **(c)** The force of gravity F acting on two objects varies jointly as their masses m_1 and m_2 and inversely as the square of the distance r between their centers, with the constant of variation G, the universal gravitational constant.

SECTION 2.3

Exploration 1

1. (a) $\infty; -\infty$ **(b)** $-\infty; \infty$ **(c)** $\infty; -\infty$ **(d)** $-\infty; \infty$ **3. (a)** $-\infty; \infty$ **(b)** $-\infty; -\infty$ **(c)** $\infty; \infty$ **(d)** $\infty; -\infty$

Exploration 2

1. $y = 0.0061x^3 + 0.0177x^2 - 0.5007x + 0.9769$

Quick Review 2.3

1. $(x - 4)(x + 3)$ **3.** $(3x - 2)(x - 3)$ **5.** $x(3x - 2)(x - 1)$ **7.** $x = 0, x = 1$ **9.** $x = -6, x = -3, x = 1.5$

Exercises 2.3

1. Shift $y = x^3$ to the right by 3 units, stretch vertically by 2. y-intercept: $(0, -54)$

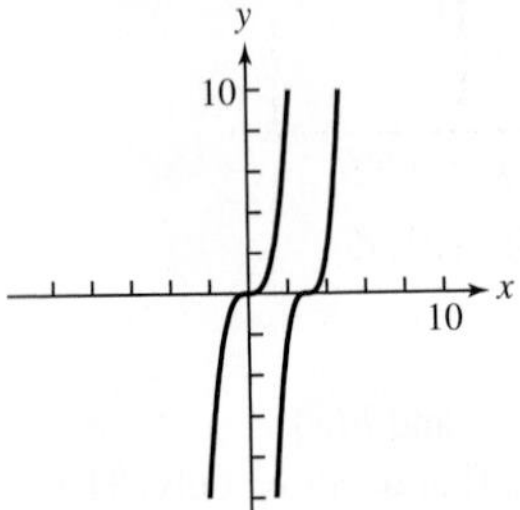

3. Shift $y = x^3$ to the left by 1 unit, vertically shrink by $\frac{1}{2}$, reflect over the x-axis, and then vertically shift up 2 units. y-intercept: $\left(0, \frac{3}{2}\right)$

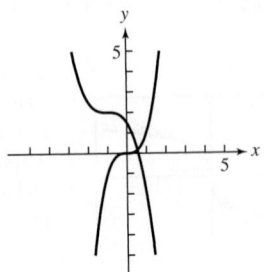

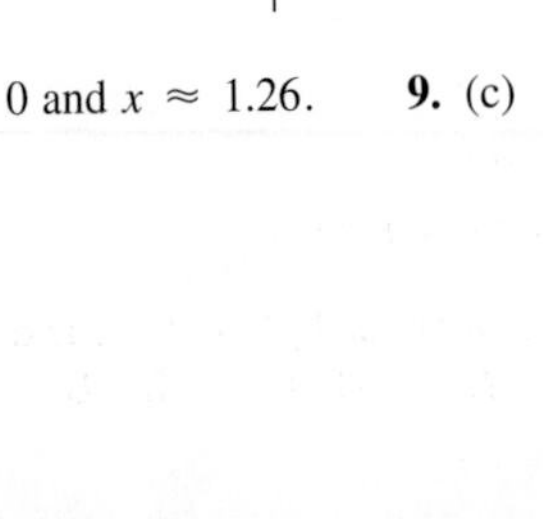

5. Shift $y = x^4$ to the left 2 units, vertically stretch by 2, reflect over the x-axis, and vertically shift down 3 units. y-intercept: $(0, -35)$

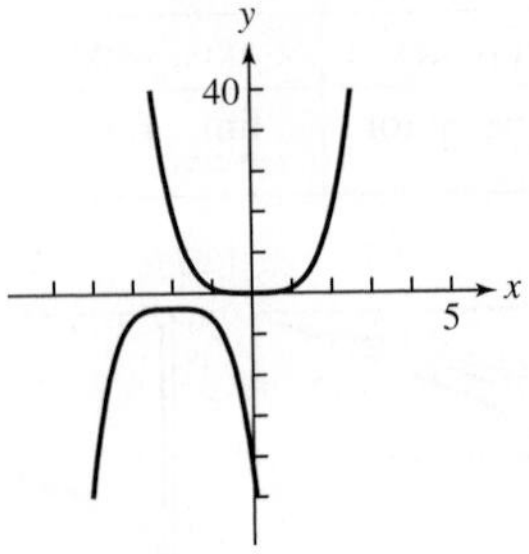

7. Local maximum: $\approx (0.79, 1.19)$; zeros: $x = 0$ and $x \approx 1.26$. **9.** (c) **11.** (a)

13. One possibility:

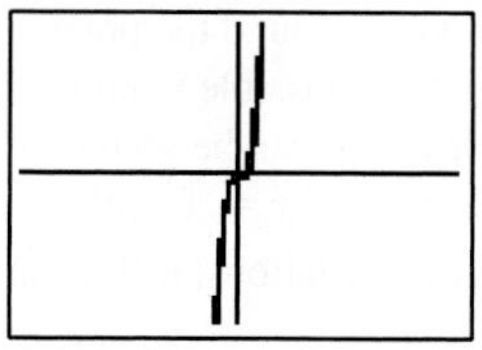

[–100, 100] by [–1000, 1000]

15. One possibility:

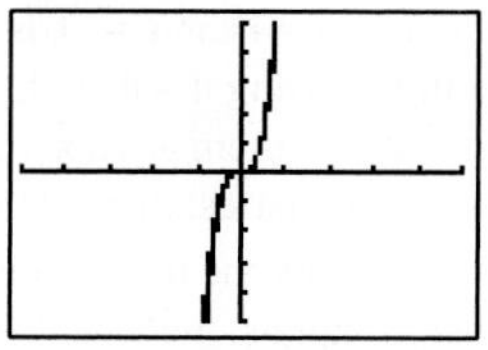

[–50, 50] by [–1000, 1000]

17.

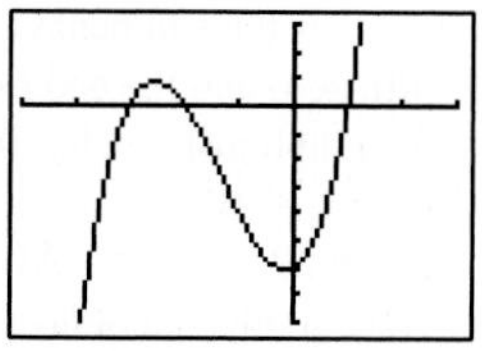

[–5, 3] by [–8, 3]

$\lim_{x\to\infty} f(x) = \infty;\ \lim_{x\to-\infty} f(x) = -\infty$

19.

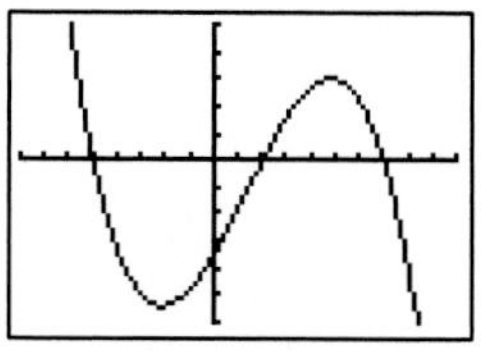

[–8, 10] by [–120, 100]

$\lim_{x\to\infty} f(x) = -\infty;\ \lim_{x\to-\infty} f(x) = \infty$

21.

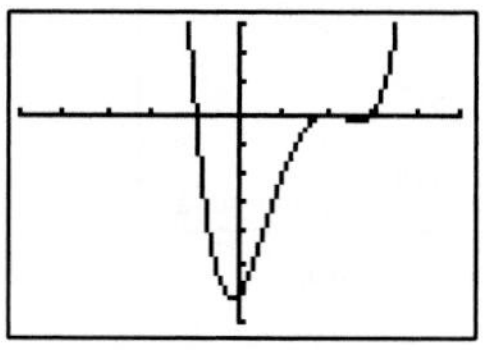

[–5, 5] by [–14, 6]

$\lim_{x\to\infty} f(x) = \infty;\ \lim_{x\to-\infty} f(x) = \infty$

23.

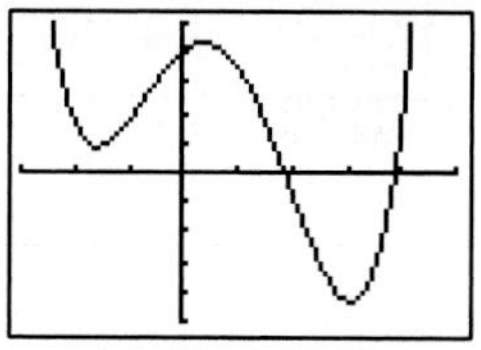

[–3, 5] by [–50, 50]

$\lim_{x\to\infty} f(x) = \infty;\ \lim_{x\to-\infty} f(x) = \infty$

25. ∞, ∞ **27.** $-\infty, \infty$ **29.** **(a)** There are 3 zeros: they are -2.5, 1, and 1.1.

31. **(c)** There are 3 zeros: approximately -0.273 $\left(\text{actually } -\frac{3}{11}\right)$, -0.25, and 1. **33.** -4 and 2 **35.** $\frac{2}{3}$ and $-\frac{1}{3}$ **37.** $0, -\frac{2}{3}$, and 1

39. Degree 3; zeros: $x = 0$ (mult. 1, graph crosses x-axis), $x = 3$ (mult. 2, graph is tangent)

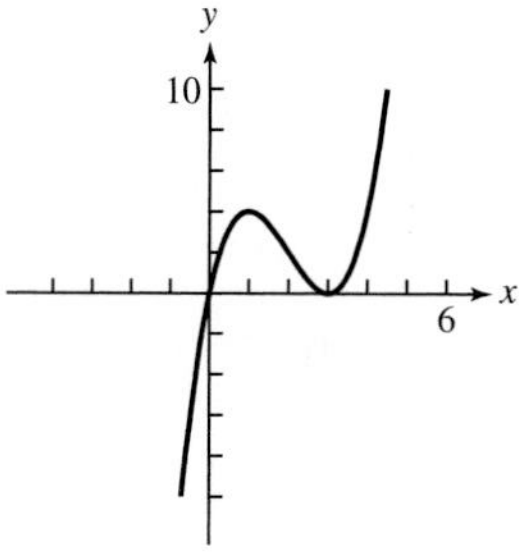

41. Degree 5; zeros: $x = 1$ (mult. 3, graph crosses x-axis), $x = -2$ (mult. 2, graph is tangent)

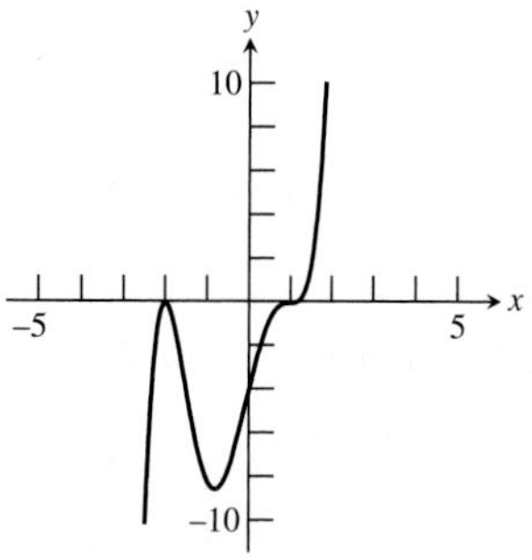

43.

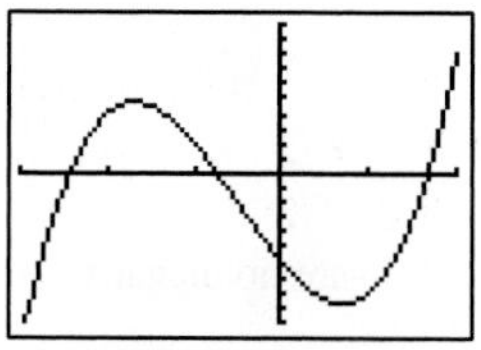

[–3, 2] by [–10, 10]

$-2.43, -0.74, 1.67$

45.

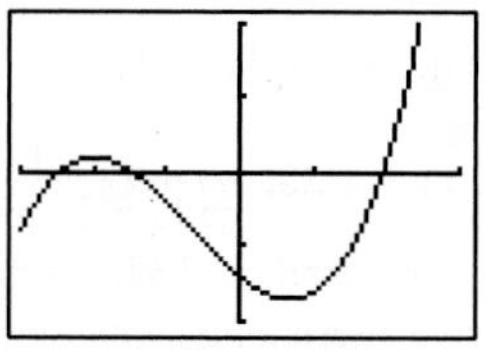

[–3, 3] by [–10, 10]

$-2.47, -1.46, 1.94$

47.

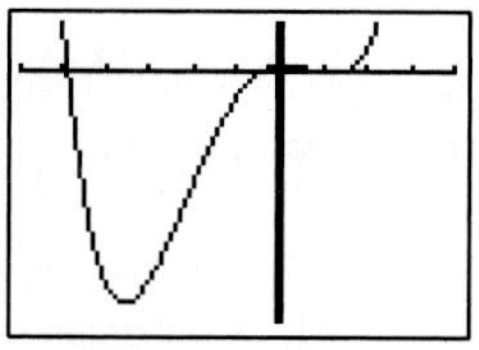

[–6, 4] by [–100, 20]

$-4.90, -0.45, 1, 1.35$

49. $0, -6$, and 6 **51.** $-5, 1, 11$ **53.** $f(x) = x^3 - 5x^2 - 18x + 72$ **55.** $f(x) = x^3 - 4x^2 - 3x + 12$
57. $y = 0.25x^3 - 1.25x^2 - 6.75x + 19.75$ **59.** $y = -2.21x^4 + 45.75x^3 - 339.79x^2 + 1075.25x - 1231$
61. It follows from the Intermediate Value Theorem.

63. **(a)**

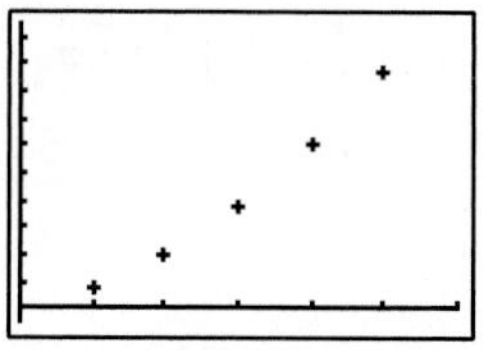

[0, 60] by [–10, 210]

(b) $y = 0.051x^2 + 0.97x + 0.26$

(c)

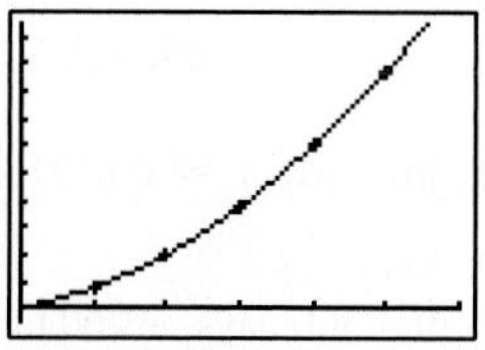

[0, 60] by [–10, 210]

(d) ≈ 56.39 ft **(e)** 67.74 mph

65. **(a)**

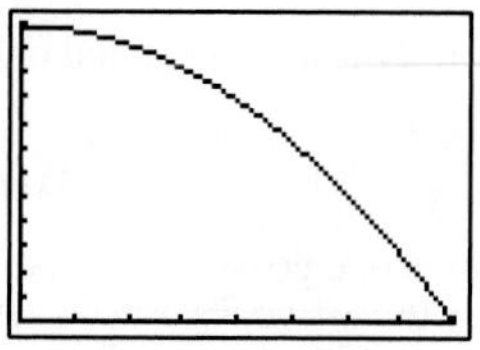

[0, 0.8] by [0, 1.20]

(b) 0.3391 cm

67. $0 < x \le 0.929$ or $3.644 \le x < 5$ **69.** True. Because f is continuous and $f(1) = -2$ and $f(2) = 2$, the Intermediate Value Theorem assures us that the graph of f crosses the x-axis between $x = 1$ and $x = 2$. **71.** C **73.** B

75. The figure at left shows the end behavior and a zero of $x \approx 9$, but hides the other four zeros. The figure at right shows zeros near $-2, -1, 1$, and 3, but hides the fifth zero and the end behavior. **77.** The exact behavior near $x = 1$ is hard to see. A zoomed-in view around the point $(1, 0)$ suggests that the graph just touches the x-axis at 0 without actually crossing it—that is, $(1, 0)$ is alocal maximum. One possible window is $[0.9999, 1.0001]$ by $[-1 \times 10^{-7}, 1 \times 10^{-7}]$. **79.** A maximum and minimum are not visible in the standard window, but can be seen on the window $[0.2, 0.4]$ by $[5.29, 5.3]$. **81.** The graph of $y = 3(x^3 - x)$ increases, then decreases, then increases; the graph of $y = x^3$ only increases. Therefore, this graph cannot be obtained from the graph of $y = x^3$ by the transformations studied in Chapter 1 (translations, reflections, and stretching/shrinking). Since the right side includes only these transformations, there can be no solution.

83. (a) Substituting $x = 2$, $y = 7$, we find that $7 = 5(2 - 2) + 7$, so Q is on line L, and also $f(2) = -8 + 8 + 18 - 11 = 7$, so Q is on the graph of $f(x)$.

(b)

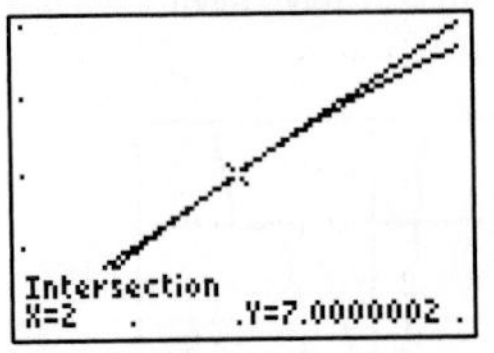

[1.8, 2.2] by [6, 8]

(c) The line L also crosses the graph of $f(x)$ at $(-2, -13)$.

85. (a) $\frac{8}{D - u} = \frac{x}{D}$ and $\frac{8}{u} = \frac{y}{D}$ imply $D - u = \frac{uy}{x} \cdot \frac{8}{u} = \frac{y - 8}{D - u}$ implies $D - u = \frac{u(y - 8)}{8}$. Combining these yields $\frac{uy}{x} = \frac{u(y - 8)}{8}$, which implies $\frac{8}{x} = \frac{y - 8}{y}$. **(b)** Equation (a) says $\frac{8}{x} = 1 - \frac{8}{y}$. So, $\frac{8}{y} = 1 - \frac{8}{x} = \frac{x - 8}{x}$. Thus $y = \frac{8x}{x - 8}$. **(c)** By the Pythagorean Theorem, $y^2 + D^2 = 900$ and $x^2 + D^2 = 400$. Subtracting equal quantities yields $y^2 - x^2 = 500$. So, $500 = \left(\frac{8x}{x - 8}\right)^2 - x^2$. Thus, $500\,(x - 8)^2 = 64x^2 - x^2(x - 8)^2$, or $500x^2 - 8000x + 32000 = 64x^2 - x^4 + 16x^3 - 64x^2$. This is equivalent to $x^4 - 16x^3 + 500x^2 - 8000x + 32{,}000 = 0$. **(d)** Notice that $8 < x < 20$. So, the solution we seek is $x \approx 11.71$, which yields $y \approx 25.24$ and $D \approx 16.21$.

SECTION 2.4

Quick Review 2.4

1. $x^2 - 4x + 7$ **3.** $7x^3 + x^2 - 3$ **5.** $x(x + 2)(x - 2)$ **7.** $4(x + 5)(x - 3)$ **9.** $(x + 2)(x + 1)(x - 1)$

Exercises 2.4

1. $f(x) = (x - 1)^2 + 2; \frac{f(x)}{x - 1} = x - 1 + \frac{2}{x - 1}$ **3.** $f(x) = (x^2 + x + 4)(x + 3) - 21; \frac{f(x)}{x + 3} = x^2 + x + 4 - \frac{21}{x + 3}$

5. $f(x) = (x^2 - 4x + 12)(x^2 + 2x - 1) - 32x + 18; \frac{f(x)}{x^2 + 2x - 1} = x^2 - 4x + 12 + \frac{-32x + 18}{x^2 + 2x - 1}$

7. $x^2 - 6x + 9 + \frac{-11}{x + 1}$ **9.** $9x^2 + 97x + 967 + \frac{9670}{x - 10}$ **11.** $-5x^3 - 20x^2 - 80x - 317 + \frac{-1269}{4 - x}$ **13.** 3 **15.** -43

17. 5 **19.** Yes **21.** No **23.** Yes **25.** $f(x) = (x + 3)(x - 1)(5x - 17)$ **27.** $2x^3 - 6x^2 - 12x + 16$

29. $2x^3 - 8x^2 + \frac{19}{2}x - 3$ **31.** $f(x) = 3(x + 4)(x - 3)(x - 5)$ **33.** $\frac{\pm 1}{\pm 1, \pm 2, \pm 3, \pm 6}; 1$ **35.** $\frac{\pm 1, \pm 3, \pm 9}{\pm 1, \pm 2}; \frac{3}{2}$

37. Numbers in last line—2, 2, 7, 19—are nonnegative, so 3 is an upper bound. **39.** Numbers in last line—1, 1, 3, 7, 2—are nonnegative, so 2 is an upper bound. **41.** Numbers in last line—3, −7, 8, −5—have alternating signs, so −1 is a lower bound. **43.** Numbers in last line—1, −4, 7, −2—are nonnegative, so 0 is a lower bound. **45.** No zeros outside window **47.** There *are* zeros not shown (approx. −11.002 and 12.003). **49.** Rational zero: $\frac{3}{2}$; irrational zeros: $\pm\sqrt{2}$ **51.** Rational: -3; irrational: $1 \pm \sqrt{3}$

53. Rational: -1 and 4; irrational: $\pm\sqrt{2}$ **55.** Rational: $-\frac{1}{2}$ and 4; irrational: none **57.** \$36.27; 54 **59.** -2 **61. (b)** 2 is a zero of $f(x)$ **(c)** $(x - 2)(x^3 + 4x^2 - 3x - 19)$ **(d)** One irrational zero is $x \approx 2.04$. **(e)** $f(x) \approx (x - 2)(x - 2.04)(x^2 + 6.04x + 9.3116)$

63. False. $(x + 2)$ is a factor if and only if $f(-2) = 0$. **65.** A **67.** B **69. (a)** Volume of buoy $= \frac{4}{3}\pi r^3 = \frac{4}{3}\pi \cdot (1)^3 = \frac{4}{3}\pi$

(b) $\frac{4}{3}\pi \cdot \frac{d}{4} = \frac{\pi d}{3}$ **(c)** $V \cdot d = \frac{\pi x}{6} \cdot (3r^2 + x^2) \cdot d = \pi d \cdot x(3r^2 + x^2)/6$ **(d)** $x \approx 0.6527$ m

71. (a) Shown is one possible view, on the window $[0, 600]$ by $[0, 500]$.

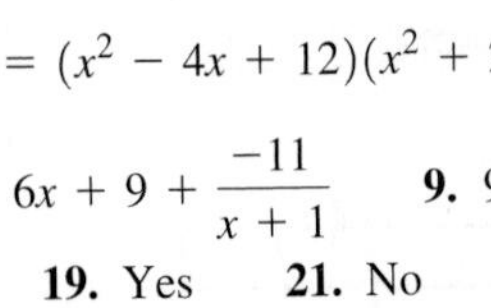

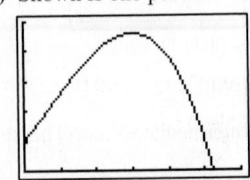

[0, 600] by [0, 500]

(b) The maximum population, after 300 days, is 460 turkeys.
(c) $P = 0$ when $t \approx 523.22$—about 523 days after release.
(d) Answers will vary.

73. **(a)** 0 or 2 positive zeros, 1 negative zero **(b)** No positive zeros, 1 or 3 negative zeros **(c)** 1 positive zero, no negative zeros **(d)** 1 positive zero, 1 negative zero

75. Answers will vary, but should include a diagram of the synthetic division and a summary:

$$4x^3 - 5x^2 + 3x + 1 = \left(x - \frac{1}{2}\right)\left(4x^2 - 3x + \frac{3}{2}\right) + \frac{7}{4}$$
$$= (2x - 1)\left(2x^2 - \frac{3}{2}x + \frac{3}{4}\right) + \frac{7}{4}$$

77. **(a)** **(b)** $-\frac{7}{3}, \frac{1}{2}$, and 3 **(c)** There are no rational zeros. **79.** **(a)** Approximate zeros: $-3.126, -1.075, 0.910, 2.291$. **(b)** $f(x) \approx g(x) = (x + 3.126)(x + 1.075)(x - 0.910)(x - 2.291)$ **(c)** Graphically: graph the original function and the approximate factorization on a variety of windows and observe their similarity. Numerically: Compute $f(c)$ and $g(c)$ for several values of c.

SECTION 2.5

Exploration 1

1. $f(2i) = (2i)^2 - i(2i) + 2 = -4 + 2 + 2 = 0$; $f(-i) = (-i)^2 - i(-i) + 2 = -1 - 1 + 2 = 0$; no.
3. The Complex Conjugate Zeros Theorem does not necessarily hold true for a polynomial function with *complex* coefficients.

Quick Review 2.5

1. $1 + 3i$ **3.** $7 + 4i$ **5.** $(2x - 3)(x + 1)$ **7.** $\frac{5}{2} \pm \frac{\sqrt{19}}{2}i$ **9.** $\pm 1, \pm 2, \pm 1/3, \pm 2/3$

Exercises 2.5

1. $x^2 + 9$; zeros: $\pm 3i$; x-intercepts: none **3.** $x^4 - 2x^3 + 5x^2 - 8x + 4$; zeros: 1 (mult. 2), $\pm 2i$; x-intercept: $x = 1$ **5.** $x^2 + 1$ **7.** $x^3 - x^2 + 9x - 9$ **9.** $x^4 - 5x^3 + 7x^2 - 5x + 6$ **11.** $x^3 - 11x^2 + 43x - 65$ **13.** $x^5 + 4x^4 + x^3 - 10x^2 - 4x + 8$ **15.** $x^4 - 10x^3 + 38x^2 - 64x + 40$ **17.** (b) **19.** (d) **21.** 2 complex zeros; none real **23.** 3 complex zeros; 1 real

25. 4 complex zeros; 2 real **27.** Zeros: $x = 1, x = -\frac{1}{2} \pm \frac{\sqrt{19}}{2}i$; $f(x) = \frac{1}{4}(x - 1)(2x + 1 + \sqrt{19}i)(2x + 1 - \sqrt{19}i)$

29. Zeros: $x = \pm 1, x = -\frac{1}{2} \pm \frac{\sqrt{23}}{2}i$; $f(x) = \frac{1}{4}(x - 1)(x + 1)(2x + 1 + \sqrt{23}i)(2x + 1 - \sqrt{23}i)$

31. Zeros: $x = -\frac{7}{3}, x = \frac{3}{2}, x = 1 \pm 2i$; $f(x) = (3x + 7)(2x - 3)(x - 1 + 2i)(x - 1 - 2i)$

33. Zeros: $x = \pm\sqrt{3}, x = 1 \pm i$; $f(x) = (x - \sqrt{3})(x + \sqrt{3})(x - 1 + i)(x - 1 - i)$

35. Zeros: $x = \pm\sqrt{2}, x = 3 \pm 2i$; $f(x) = (x - \sqrt{2})(x + \sqrt{2})(x - 3 + 2i)(x - 3 - 2i)$

37. $(x - 2)(x^2 + x + 1)$ **39.** $(x - 1)(2x^2 + x + 3)$ **41.** $(x - 1)(x + 4)(x^2 + 1)$ **43.** $h \approx 3.776$ ft

45. Yes, $f(x) = (x + 2)^2 = x^3 + 6x^2 + 12x + 8$. **47.** No, either the degree would be at least 5 or some of the coefficients would be nonreal.

49. $f(x) = -2x^4 + 12x^3 - 20x^2 - 4x + 30$

51. **(a)** $D \approx -0.0820t^3 + 0.9162t^2 - 2.5126t + 3.3779$

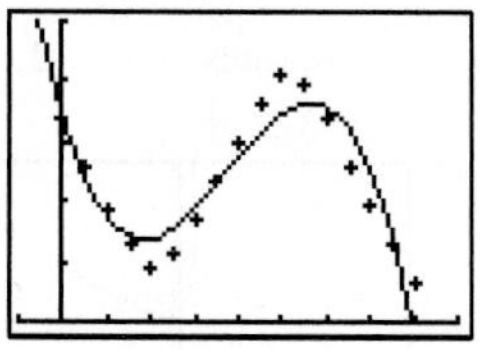

[–1, 9] by [0, 5]

(b) Sally walks toward the detector, turns and walks away (or walks backward), then walks toward the detector again.
(c) $t \approx 1.81$ sec ($D \approx 1.35$ m) and $t \approx 5.64$ sec ($D \approx 3.65$ m).

53. False. If $1 - 2i$ is a zero, then $1 + 2i$ must also be a zero. **55.** E **57.** C

59. **(a)**

Power	Real Part	Imaginary Part
7	8	−8
8	16	0
9	16	16
10	0	32

(b) $(1 + i)^7 = 8 - 8i$
$(1 + i)^8 = 16$
$(1 + i)^9 = 16 + 16i$
$(1 + i)^{10} = 32i$

(c) Reconcile as needed.

61. $f(i) = i^3 - i(i)^2 + 2i(i) + 2 = -i + i - 2 + 2 = 0$ **63.** Synthetic division shows that $f(i) = 0$ (the remainder), and at the same time gives $f(x) \div (x - i) = x^2 + 3x - i = h(x)$, so $f(x) = (x - i)(x^2 + 3x - i)$. **65.** $2, -1 + \sqrt{3}i, -1 - \sqrt{3}i$

SECTION 2.6

Exploration 1

1. $g(x) = \dfrac{1}{x - 2}$

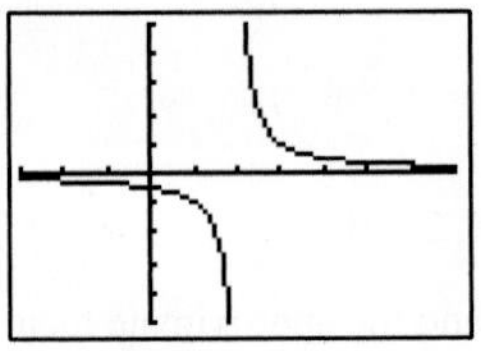

[–3, 7] by [–5, 5]

3. $k(x) = \dfrac{3}{x + 4} - 2$

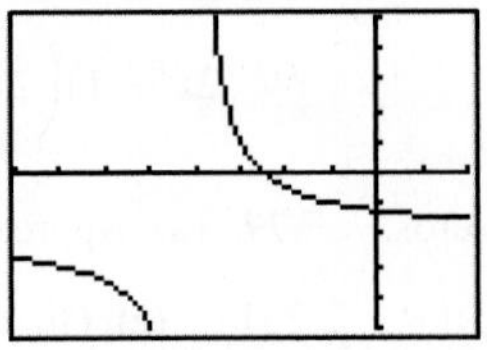

[–8, 2] by [–5, 5]

Quick Review 2.6

1. $x = -3, x = \dfrac{1}{2}$ **3.** $x = \pm 2$ **5.** $x = 1$ **7.** 2; 7 **9.** 3; −5

Exercises 2.6

1. Domain: all $x \neq -3$; $\lim_{x \to -3^-} f(x) = -\infty$, $\lim_{x \to -3^+} f(x) = \infty$

3. Domain: all $x \neq -2, 2$; $\lim_{x \to -2^-} f(x) = -\infty$, $\lim_{x \to -2^+} f(x) = \infty$, $\lim_{x \to 2^-} f(x) = \infty$, $\lim_{x \to 2^+} f(x) = -\infty$

5. Translate right 3 units.
Asymptotes: $x = 3$, $y = 0$

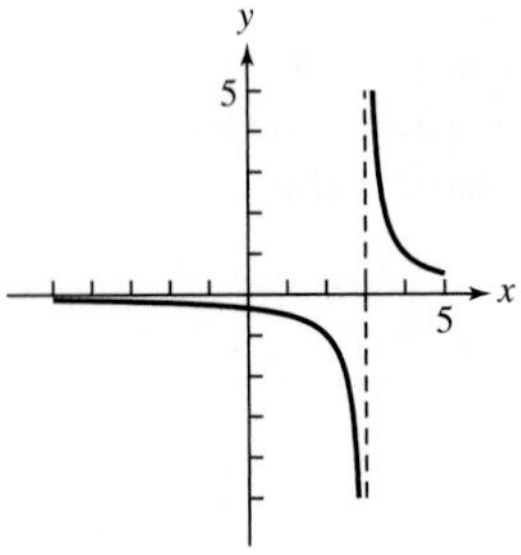

7. Translate left 3 units, reflect across x-axis, vertically stretch by 7, translate up 2 units.
Asymptotes: $x = -3$, $y = 2$

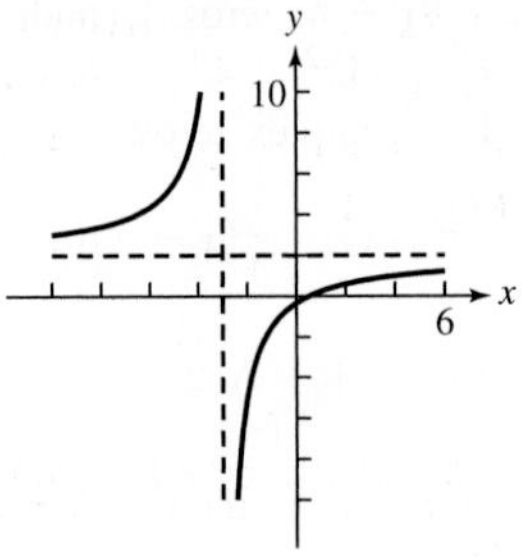

9. Translate left 4 units, vertically stretch by 13, translate down 2 units. Asymptotes: $x = -4$, $y = -2$

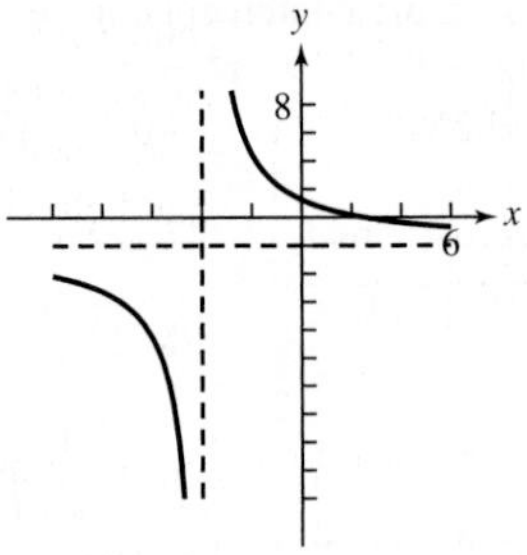

11. ∞ **13.** 0 **15.** ∞ **17.** 5 **19.** Vertical asymptote: none; horizontal asymptote: $y = 2$; $\lim_{x \to -\infty} f(x) = \lim_{x \to \infty} f(x) = 2$

21. Vertical asymptotes: $x = 0$, $x = 1$; horizontal asymptote: $y = 0$; $\lim_{x \to 0^-} f(x) = \infty$, $\lim_{x \to 0^+} f(x) = -\infty$, $\lim_{x \to 1^-} f(x) = -\infty$, $\lim_{x \to 1^+} f(x) = \infty$, $\lim_{x \to -\infty} f(x) = \lim_{x \to \infty} f(x) = 0$

23. Intercepts: $\left(0, \dfrac{2}{3}\right)$ and $(2, 0)$; asymptotes: $x = -1$, $x = 3$, and $y = 0$

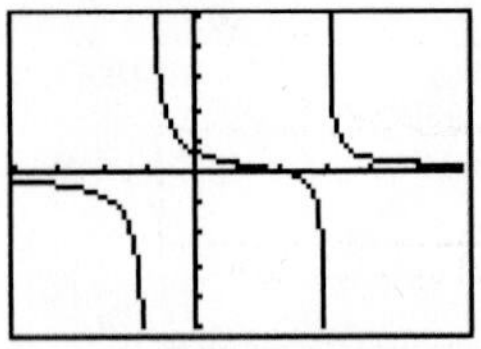

[–4, 6] by [–5, 5]

25. No intercepts; asymptotes: $x = -1$, $x = 0$, $x = 1$, and $y = 0$

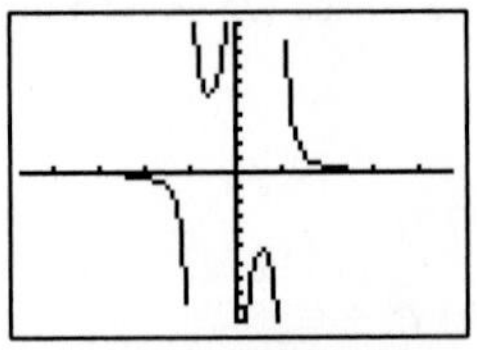

[–4.7, 4.7] by [–10, 10]

27. Intercepts: $(0, 2)$, $(-1.28, 0)$, and $(0.78, 0)$; asymptotes: $x = 1$, $x = -1$, and $y = 2$

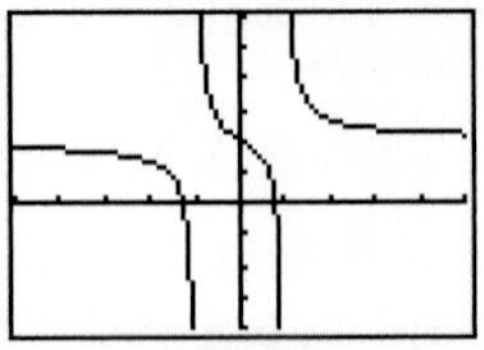

[–5, 5] by [–4, 6]

29. Intercept: $\left(0, \dfrac{3}{2}\right)$; asymptotes: $x = -2$, $y = x - 4$

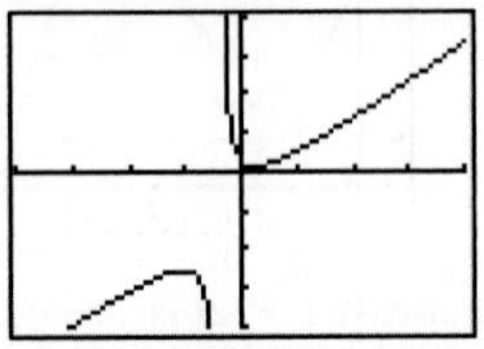

[–20, 20] by [–20, 20]

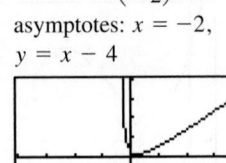

31. (d); Xmin = −2, Xmax = 8, Xscl = 1, and Ymin = −3, Ymax = 3, Yscl = 1
33. (a); Xmin = −3, Xmax = 5, Xscl = 1, and Ymin = −5, Ymax = 10, Yscl = 1
35. (e); Xmin = −2, Xmax = 8, Xscl = 1, and Ymin = −3, Ymax = 3, Yscl = 1

37. Intercept: $\left(0, -\frac{2}{3}\right)$; asymptotes: $x = -1, x = \frac{3}{2}, y = 0$; $\lim_{x\to -1^-} f(x) = \infty$, $\lim_{x\to -1^+} f(x) = -\infty$, $\lim_{x\to (3/2)^-} f(x) = -\infty$, $\lim_{x\to (3/2)^+} f(x) = \infty$;

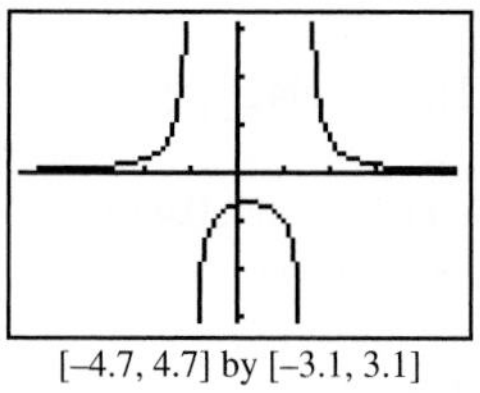

[−4.7, 4.7] by [−3.1, 3.1]

Domain: $x \neq -1, \frac{3}{2}$; Range: $\left(-\infty, -\frac{16}{25}\right] \cup (0, \infty)$; Continuity: all $x \neq -1, \frac{3}{2}$;
Increasing: $(-\infty, -1)$, $\left(-1, \frac{1}{4}\right]$; Decreasing: $\left[\frac{1}{4}, \frac{3}{2}\right)$, $\left(\frac{3}{2}, \infty\right)$; Unbounded;
Local Maximum at $\left(\frac{1}{4}, -\frac{16}{25}\right)$; Horizontal asymptote: $y = 0$;
Vertical asymptotes: $x = -1, x = \frac{3}{2}$; End behavior: $\lim_{x\to -\infty} f(x) = \lim_{x\to\infty} f(x) = 0$

39. Intercepts: $\left(0, \frac{1}{12}\right)$, $(1, 0)$; asymptotes: $x = -3, x = 4, y = 0$; $\lim_{x\to -3^-} h(x) = -\infty$, $\lim_{x\to -3^+} h(x) = \infty$, $\lim_{x\to 4^-} h(x) = -\infty$, $\lim_{x\to 4^+} h(x) = \infty$

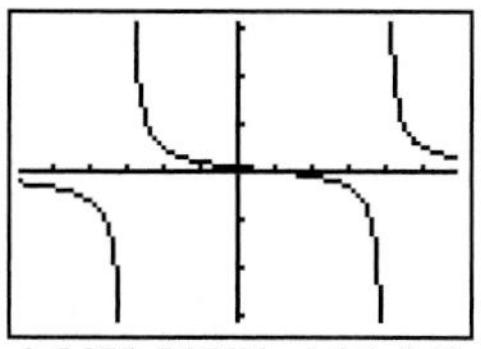

[−5.875, 5.875] by [−3.1, 3.1]

Domain: $x \neq -3, 4$; Range: $(-\infty, \infty)$;
Continuity: all $x \neq -3, 4$;
Decreasing: $(-\infty, -3)$, $(-3, 4)$, $(4, \infty)$;
No symmetry; Unbounded; No extrema;
Horizontal asymptote: $y = 0$; Vertical asymptotes: $x = -3, x = 4$;
End behavior: $\lim_{x\to -\infty} h(x) = \lim_{x\to\infty} h(x) = 0$

41. Intercepts: $(-2, 0)$, $(1, 0)$, $\left(0, \frac{2}{9}\right)$; asymptotes: $x = -3, x = 3, y = 1$; $\lim_{x\to -3^-} f(x) = \infty$, $\lim_{x\to -3^+} f(x) = -\infty$, $\lim_{x\to 3^-} f(x) = -\infty$, $\lim_{x\to 3^+} f(x) = \infty$

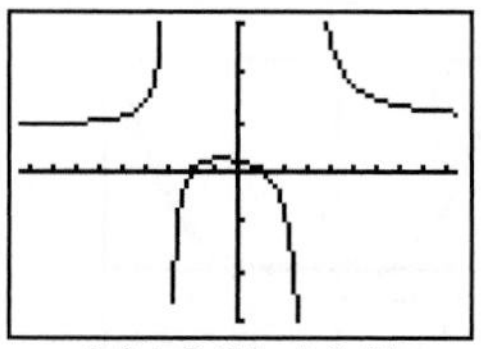

[−9.4, 9.4] by [−3, 3]

Domain: $x \neq -3, 3$; Range: $(-\infty, 0.260] \cup (1, \infty)$; Continuity: all $x \neq -3, 3$;
Increasing: $(-\infty, -3)$, $(-3, -0.675)$; Decreasing: $(-0.675, 3)$, $(3, \infty)$;
No symmetry; Unbounded; Local maximum at $(-0.675, 0.260)$;
Horizontal asymptote: $y = 1$; Vertical asymptotes: $x = -3, x = 3$;
End behavior: $\lim_{x\to -\infty} f(x) = \lim_{x\to\infty} f(x) = 1$

43. Intercepts: $(-3, 0)$, $(1, 0)$, $\left(0, -\frac{3}{2}\right)$; asymptotes: $x = -2, y = x$; $\lim_{x\to -2^-} h(x) = \infty$, $\lim_{x\to -2^+} h(x) = -\infty$

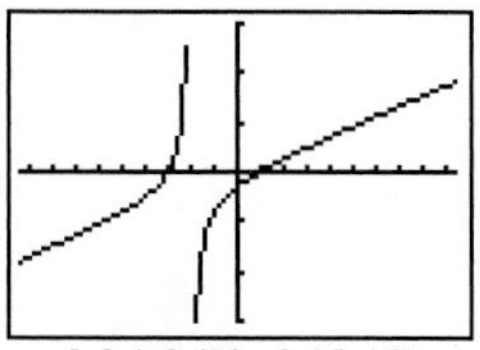

[−9.4, 9.4] by [−15, 15]

Domain: $x \neq -2$, Range: $(-\infty, \infty)$;
Continuity: all $x \neq -2$;
Increasing: $(-\infty, -2)$, $(-2, \infty)$;
No symmetry; Unbounded; No extrema;
Horizontal asymptote: none; Vertical asymptote: $x = -2$;
Slant asymptote: $y = x$; End behavior: $\lim_{x\to -\infty} h(x) = -\infty$, $\lim_{x\to\infty} h(x) = \infty$

45. $y = x + 3$

(a)

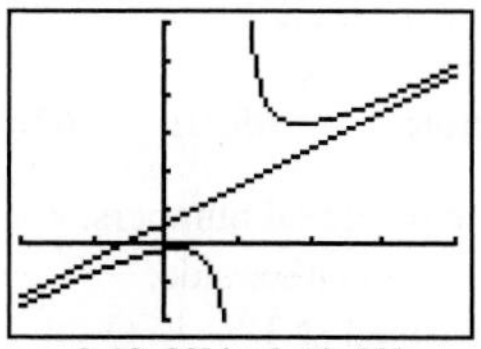

[−10, 20] by [−10, 30]

(b)

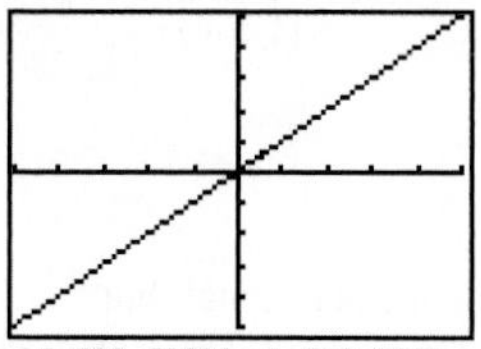

[−500, 500] by [−500, 500]

47. $y = x^2 - 3x + 6$

(a)

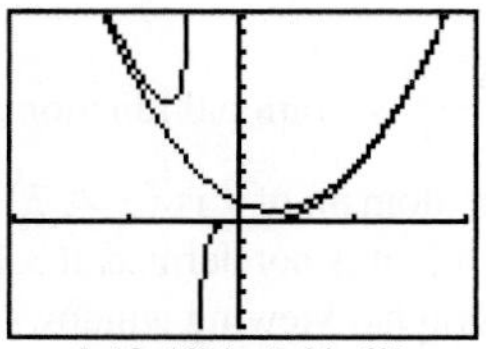

[−10, 10] by [−30, 60]

(b)

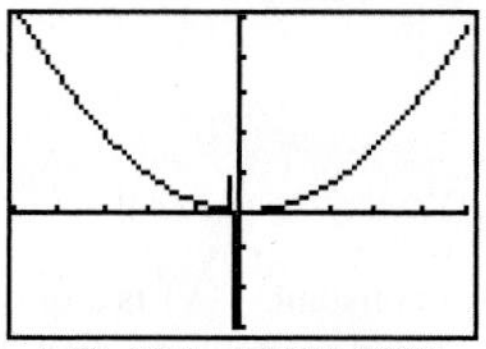

[−50, 50] by [−1500, 2500]

49. $y = x^3 + 2x^2 + 4x + 6$

(a)

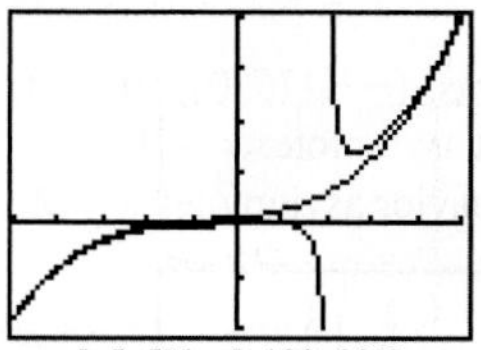

[−5, 5] by [−100, 200]

(b)

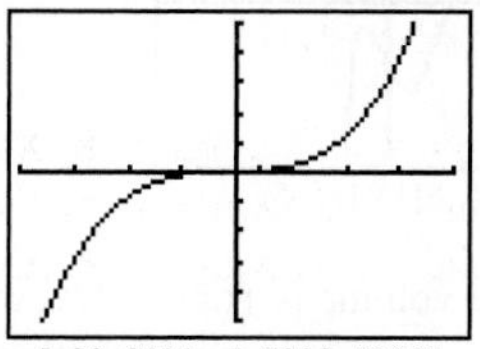

[−20, 20] by [−5000, 5000]

51. Intercept: $\left(0, \frac{4}{5}\right)$;
Domain: $(-\infty, \infty)$; Range: $[0.773, 14.227]$;
Continuity: $(-\infty, \infty)$; Increasing: $[-0.245, 2.445]$;
Decreasing: $(-\infty, -0.245]$, $[2.445, \infty)$;
No symmetry; Bounded;
Local max at $(2.445, 14.227)$, local min at $(-0.245, 0.773)$;
Horizontal asymptote: $y = 3$; Vertical asymptote: none;
End behavior: $\lim_{x \to -\infty} f(x) = \lim_{x \to \infty} f(x) = 3$

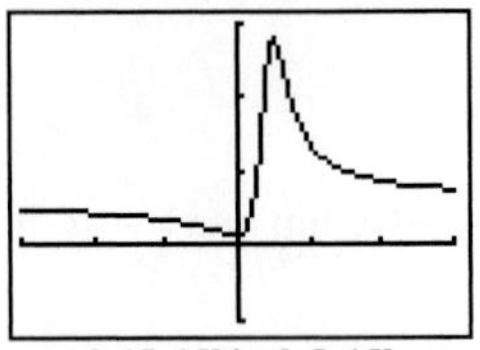

[–15, 15] by [–5, 15]

53. Intercepts: $(1, 0)$, $\left(0, \frac{1}{2}\right)$;
Domain: $x \neq 2$; Range: $(-\infty, \infty)$; Continuity: $x \neq 2$;
Increasing: $[-0.384, 0.442]$, $[2.942, \infty)$;
Decreasing: $(-\infty, -0.384]$, $[0.442, 2)$, $(2, 2.942]$;
No symmetry; Not bounded;
Local max at $(0.442, 0.586)$, local min at $(-0.384, 0.443)$ and $(2.942, 25.970)$; Horizontal asymptote: none;
Vertical asymptote: $x = 2$;
End behavior: $\lim_{x \to -\infty} h(x) = \lim_{x \to \infty} h(x) = \infty$;
End behavior asymptote: $y = x^2 + 2x + 4$

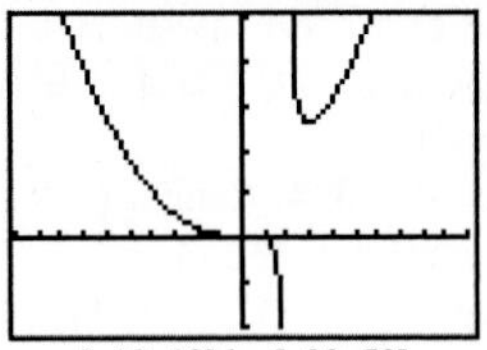

[–10, 10] by [–20, 50]

55. Intercepts: $(1.755, 0)$, $(0, 1)$;
Domain: $x \neq \frac{1}{2}$; Range: $(-\infty, \infty)$; Continuity: $x \neq \frac{1}{2}$;
Increasing: $\left[-0.184, \frac{1}{2}\right)$, $\left(\frac{1}{2}, \infty\right)$;
Decreasing: $(-\infty, -0.184]$;
No symmetry; Not bounded;
Local min at $(-0.184, 0.920)$;
Horizontal asymptote: none;
Vertical asymptote: $x = \frac{1}{2}$;
End behavior: $\lim_{x \to -\infty} f(x) = \lim_{x \to \infty} f(x) = \infty$;
End behavior asymptote: $y = \frac{1}{2}x^2 - \frac{3}{4}x + \frac{1}{8}$

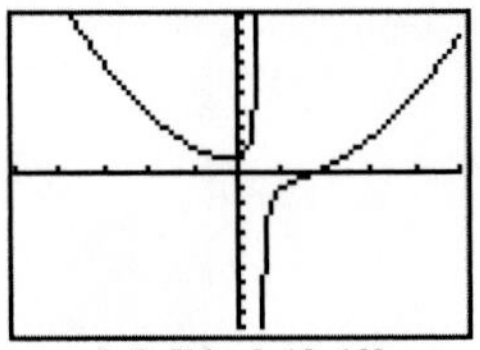

[–5, 5] by [–10, 10]

57. Intercept: $(0, 1)$;
Vertical asymptote: $x = -1$;
End behavior asymptote:
$y = x^3 - x^2 + x - 1$

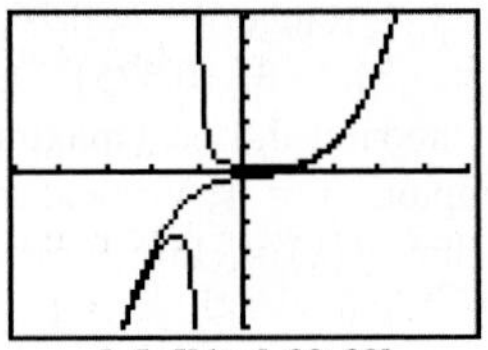

[–5, 5] by [–30, 30]

59. Intercepts: $(1, 0)$, $\left(0, -\frac{1}{2}\right)$;
Vertical asymptote: $x = -2$;
End behavior asymptote:
$y = x^4 - 2x^3 + 4x^2 - 8x + 16$

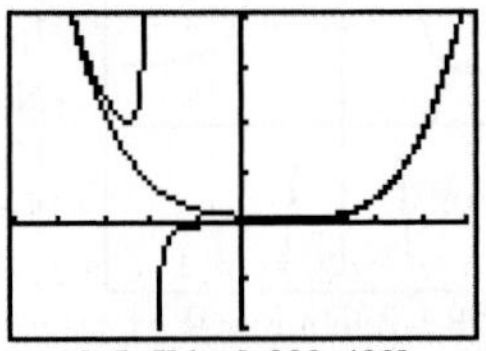

[–5, 5] by [–200, 400]

61. Intercepts: $(-1.476, 0)$, $(0, -2)$;
Vertical asymptote: $x = 1$;
End behavior asymptote: $y = 2$

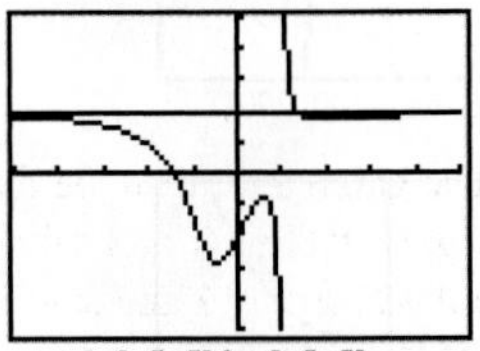

[–5, 5] by [–5, 5]

63. False. $\frac{1}{x^2 + 1}$ is a rational function and has no vertical asymptotes. **65.** E **67.** D

69. **(a)** No: the domain of f is $(-\infty, 3) \cup (3, \infty)$; the domain of g is all real numbers.
(b) No: while it is not defined at 3, it does not tend toward $\pm\infty$ on either side.
(c) Most grapher viewing windows do not reveal that f is undefined at 3.
(d) Almost—but not quite; they are equal for all $x \neq 3$.

71. **(a)** The volume is $f(x) = k/x$, where x is pressure and k is a constant. $f(x)$ is a quotient of polynomials and hence is rational, but $f(x) = k \cdot x^{-1}$, so is a power function with constant of variation k and power -1. **(b)** If $f(x) = kx^a$, where a is a negative integer, then the power function f is also a rational function. **(c)** 4.22 L

73. Horizontal asymptotes: $y = -2$ and $y = 2$;

intercepts: $\left(0, -\frac{3}{2}\right), \left(\frac{3}{2}, 0\right)$;

$$h(x) = \begin{cases} \dfrac{2x - 3}{x + 2} & x \geq 0 \\ \dfrac{2x - 3}{-x + 2} & x < 0 \end{cases}$$

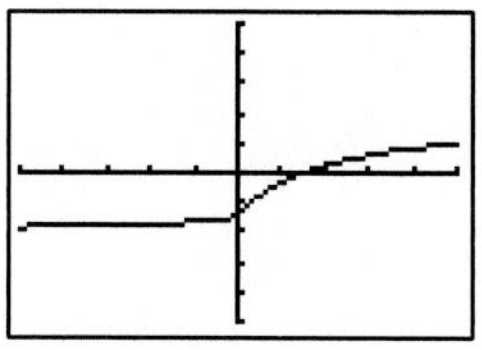

[–5, 5] by [–5, 5]

75. Horizontal asymptotes: $y = \pm 3$;

intercepts: $\left(0, \frac{5}{4}\right), \left(\frac{5}{3}, 0\right)$;

$$f(x) = \begin{cases} \dfrac{5 - 3x}{x + 4} & x \geq 0 \\ \dfrac{5 - 3x}{-x + 4} & x < 0 \end{cases}$$

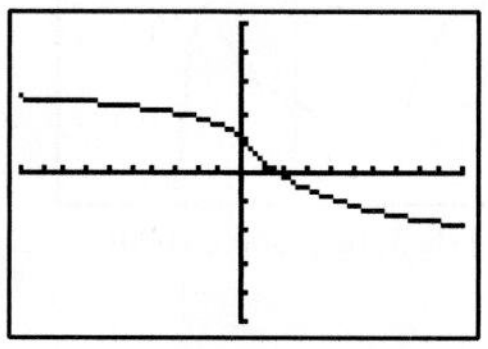

[–10, 10] by [–5, 5]

77. The graph of f is the graph $y = \frac{1}{x}$ shifted horizontally $-d/c$ units, stretched vertically by a factor of $|bc - ad|/c^2$, reflected across x-axis if and only if $bc - ad < 0$, and then shifted vertically by a/c.

SECTION 2.7

Quick Review 2.7

1. $2x^2 + 8x$ **3.** LCD: 36; $-\frac{1}{36}$ **5.** LCD: $(2x + 1)(x - 3)$; $\frac{x^2 - 7x - 2}{(2x + 1)(x - 3)}$ **7.** $\frac{3 \pm \sqrt{17}}{4}$ **9.** $\frac{-1 \pm \sqrt{7}}{3}$

Exercises 2.7

1. $x = -1$ **3.** $x = 2$ or $x = -7$ **5.** $x = -4$ or $x = 3$; the latter is extraneous. **7.** $x = 2$ or $x = 5$ **9.** $x = 3$ or $x = 4$ **11.** $x = \frac{1}{2}$ or $x = -1$; the latter is extraneous. **13.** $x = -\frac{1}{3}$ or $x = 2$; the latter is extraneous. **15.** $x = 5$ or $x = 0$; the latter is extraneous. **17.** $x = -2$ or $x = 0$; both are extraneous. The equation has no solutions. **19.** $x = -2$ **21.** Both **23.** $x = 3 + \sqrt{2} \approx 4.414$ or $x = 3 - \sqrt{2} \approx 1.586$ **25.** $x = 1$ **27.** No real solutions **29.** $x \approx -3.100$ or $x \approx 0.661$ or $x \approx 2.439$

31. (a) The total amount of solution is $(125 + x)$ mL; of this, the amount of acid is x plus 60% of the original amount, or $x + 0.6(125)$. **(b)** $y = 0.83$ **(c)** $C(x) = \frac{x + 75}{x + 125} = 0.83$; $x \approx 169.12$ mL

33. (a) $C(x) = \frac{3000 + 2.12x}{x}$ **(b)** 4762 hats per week **(c)** 6350 hats per week

35. (a) $P(x) = 2x + \frac{364}{x}$ **(b)** $x \approx 13.49$ (a square); $P \approx 53.96$

37. (a) $S = \frac{2\pi x^3 + 1000}{x}$ **(b)** Either $x \approx 1.12$ cm and $h \approx 126.88$ cm or $x \approx 11.37$ cm and $h \approx 1.23$ cm.

39. (a) $R(x) = \frac{2.3x}{x + 2.3}$ **(b)** $x \approx 6.52$ ohms **41. (a)** $D(t) = \frac{4.75 + t}{4.75t}$ **(b)** $t \approx 5.74$ h

43. (a)

[60, 110] by [0, 50]

(b) About 10.6 years

45. False. An extraneous solution is a solution of the equation cleared of fractions that is *not* a solution of the original equation.

47. D **49.** E **51. (a)** $f(x) = \frac{x^2 + 2x}{x^2 + 2x}$ **(b)** $x \neq 0, -2$ **(c)** $f(x) = \begin{cases} 1, x \neq -2, 0 \\ \text{undefined}, x = -2 \text{ or } x = 0 \end{cases}$ **(d)** The graph appears to be the horizontal line $y = 1$ with holes at $x = -2$ and $x = 0$.

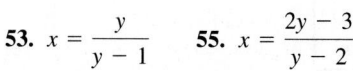

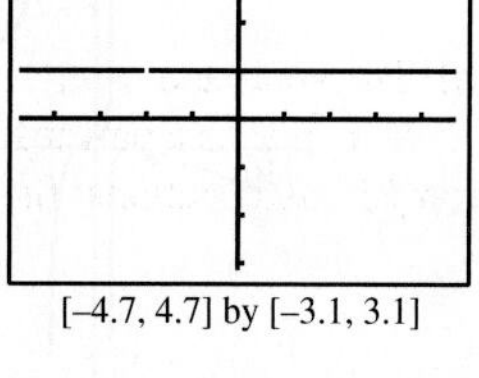

[–4.7, 4.7] by [–3.1, 3.1]

53. $x = \frac{y}{y - 1}$ **55.** $x = \frac{2y - 3}{y - 2}$

SECTION 2.8

Exploration 1

1. (a)

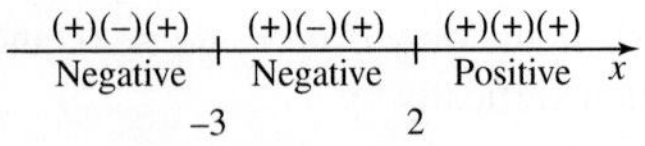

(b)

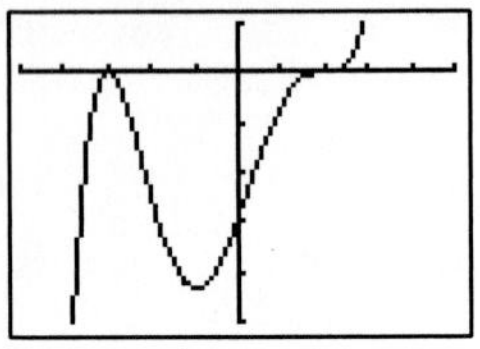

[–5, 5] by [–250, 50]

3. (a)

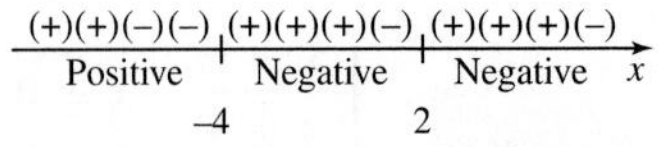

(b)

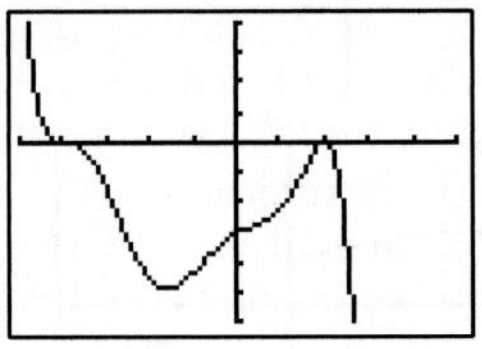

[–5, 5] by [–3000, 2000]

Quick Review 2.8

1. $\lim_{x\to\infty} f(x) = \infty$; $\lim_{x\to-\infty} f(x) = -\infty$ **3.** $\lim_{x\to\infty} g(x) = \infty$, $\lim_{x\to-\infty} g(x) = \infty$ **5.** $(x^3 + 5)/x$ **7.** $\dfrac{x^2 - 7x - 2}{2x^2 - 5x - 3}$

9. (a) $\pm 1, \pm\frac{1}{2}, \pm 3, \pm\frac{3}{2}$ **(b)** $(x + 1)(2x - 3)(x + 1)$

Exercises 2.8

1. (a) $x = -2, -1, 5$ **(b)** $-2 < x < -1$ or $x > 5$ **(c)** $x < -2$ or $-1 < x < 5$
3. (a) $x = -7, -4, 6$ **(b)** $x < -7$ or $-4 < x < 6$ or $x > 6$ **(c)** $-7 < x < -4$
5. (a) $x = 8, -1$ **(b)** $-1 < x < 8$ or $x > 8$ **(c)** $x < -1$ **7.** $(-1, 3) \cup (3, \infty)$ **9.** $(-\infty, -1) \cup (1, 2)$ **11.** $[-2, 1/2] \cup [3, \infty)$
13. $[-1, 0] \cup [2, \infty)$ **15.** $(-1, 3/2) \cup (2, \infty)$ **17.** $[-1.15, \infty)$ **19.** $(3/2, 2)$ **21. (a)** $(-\infty, \infty)$ **(b)** $(-\infty, \infty)$
(c) There are no solutions. **(d)** There are no solutions. **23. (a)** $x \neq \frac{4}{3}$ **(b)** $(-\infty, \infty)$ **(c)** There are no solutions. **(d)** $x = \frac{4}{3}$
25. (a) $x = 1$ **(b)** $x = -\frac{3}{2}, 4$ **(c)** $-\frac{3}{2} < x < 1$ or $x > 4$ **(d)** $x < -\frac{3}{2}$, or $1 < x < 4$
27. (a) $x = 0, -3$ **(b)** $x < -3$ **(c)** $x > 0$ **(d)** $-3 < x < 0$ **29. (a)** $x = -5$ **(b)** $x = -\frac{1}{2}, x = 1, x < -5$
(c) $-5 < x < -\frac{1}{2}$ or $x > 1$ **(d)** $-\frac{1}{2} < x < 1$ **31. (a)** $x = 3$ **(b)** $x = 4, x < 3$ **(c)** $3 < x < 4$ or $x > 4$
(d) $f(x)$ is never negative. **33.** $(-\infty, -2) \cup (1, 2)$ **35.** $[-1, 1]$ **37.** $(-\infty, -4) \cup (3, \infty)$ **39.** $[-1, 0] \cup [1, \infty)$
41. $(0, 2) \cup (2, \infty)$ **43.** $\left(-4, \frac{1}{2}\right)$ **45.** $(0, 2)$ **47.** $(-\infty, 0) \cup (\sqrt[3]{2}, \infty)$ **49.** $(-\infty, -1) \cup [1, 3)$
51. $[-3, \infty)$ **53.** $[5, \infty)$ **55.** Answers will vary; can be solved graphically, numerically, and algebraically: $[-3.5, \infty)$
57. 1 in. $< x <$ 34 in. **59.** 0 in. $\leq x \leq$ 0.69 in. or 4.20 in. $\leq x \leq$ 6 in. **61. (a)** $S = 2\pi x^2 + 1000/x$
(b) 1.12 cm $\leq x \leq$ 11.37 cm, 1.23 cm $\leq h \leq$ 126.88 cm **(c)** About 348.73 cm^2 **63. (a)** $y \approx 2.723x + 282.043$ **(b)** Shortly before July 1, 2014 **65.** False, because the factor x^4 does not change sign at $x = 0$ **67.** C **69.** D

71. Vertical asymptotes: $x = -1$, $x = 3$; x-intercepts: $(-2, 0)$, $(1, 0)$; y-intercept: $\left(0, \frac{4}{3}\right)$

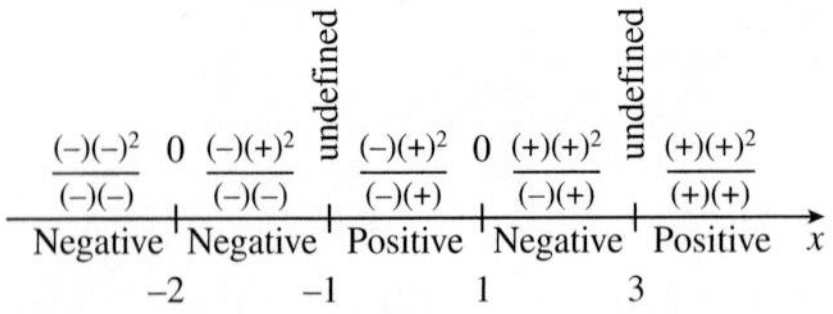

By hand:

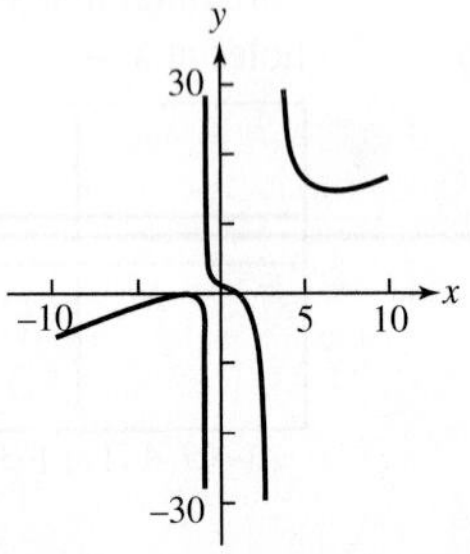

Grapher support:

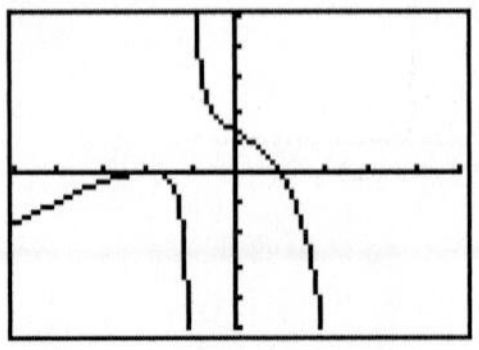

[–5, 5] by [–5, 5]

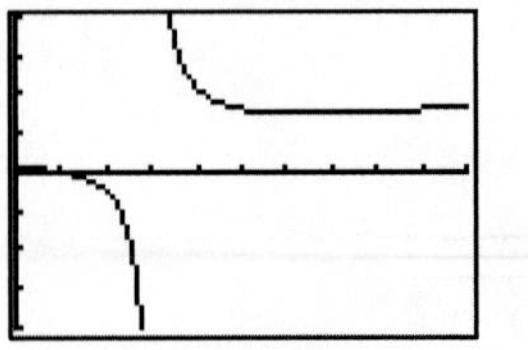

[0, 10] by [–40, 40]

73. **(a)** $|x - 3| < 1/3 \Rightarrow |3x - 9| < 1 \Rightarrow |3x - 5 - 4| < 1 \Rightarrow |f(x) - 4| < 1.$
(b) If x stays within the dashed vertical lines, $f(x)$ will stay within the dashed horizontal lines.
(c) $|x - 3| < 0.01 \Rightarrow |3x - 9| < 0.03 \Rightarrow |3x - 5 - 4| < 0.03 \Rightarrow |f(x) - 4| < 0.03.$
The dashed lines would be closer when $x = 3$ and $y = 4$.
75. $0 < a < b \Rightarrow a^2 < ab$ and $ab < b^2$; so, $a^2 < b^2$.

REVIEW EXERCISES

1. $y = -x - 5$

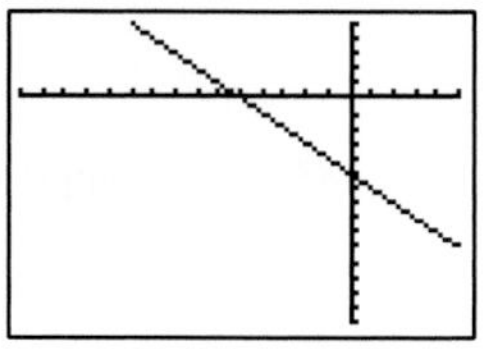

[–15, 5] by [–15, 5]

3. Starting from $y = x^2$, translate right 2 units and vertically stretch by 3 (either order), then translate up 4 units.

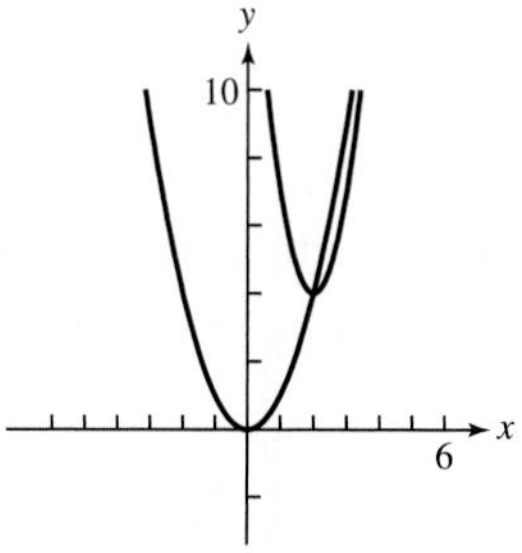

5. Vertex: $(-3, 5)$; axis: $x = -3$ **7.** Vertex: $(-4, 1)$; axis: $x = -4$ **9.** $y = (5/9)(x + 2)^2 - 3$ **11.** $y = \frac{1}{2}(x - 3)^2 - 2$

13.

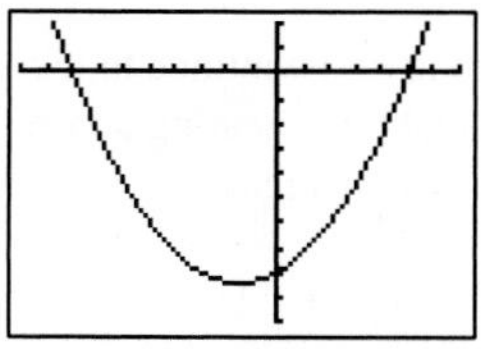

[–10, 7] by [–50, 10]

15.

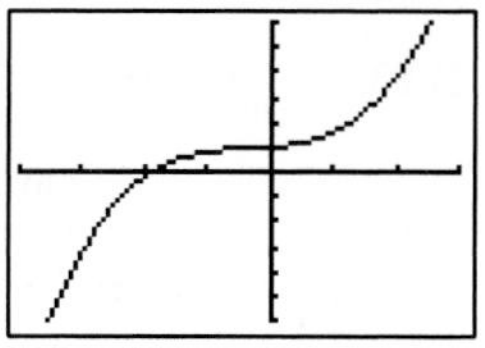

[–4, 3] by [–30, 30]

17. $S = kr^2$

19. The force F needed varies directly with the distance x from its resting position, with constant of variation k.

21. $k = 4$, $a = \frac{1}{3}$, f is increasing in the first quadrant, f is odd. **23.** $k = -2$, $a = -3$, f is increasing in the fourth quadrant, f is odd.

25. $2x^2 - x + 1 - \frac{2}{x - 3}$ **27.** $2x^2 - 3x + 1 + \frac{-2x + 3}{x^2 + 4}$ **29.** -39 **31.** Yes **37.** $\pm 1, \pm 2, \pm 3, \pm 6, \pm\frac{1}{2}, \pm\frac{3}{2}$; $-\frac{3}{2}$ and 2 are zeros. **39.** $-2 + 2i$ **41.** i **43.** $3 \pm 2i$ **45.** (c) **47.** (b) **49.** Rational: 0. Irrational: $5 \pm \sqrt{2}$. No nonreal zeros.

51. Rational: none. Irrational: approximately $-2.34, 0.57, 3.77$. No nonreal zeros. **53.** $-\frac{3}{2}, 3 \pm i$; $f(x) = (2x + 3)(x - 3 + i)(x - 3 - i)$

55. $1, -1, \frac{2}{3}$, and $-\frac{5}{2}$; $f(x) = (3x - 2)(2x + 5)(x - 1)(x + 1)$ **57.** $f(x) = (x - 2)(x^2 + x + 1)$

59. $f(x) = (2x - 3)(x - 1)(x^2 - 2x + 5)$ **61.** $x^3 - 3x^2 - 5x + 15$ **63.** $6x^4 - 5x^3 - 38x^2 - 5x + 6$

65. $x^4 - 4x^3 - 12x^2 + 32x + 64$ **67.** Translate right 5 units and vertically stretch by 2 (either order), then translate down 1 unit. Horizontal asymptote: $y = -1$; vertical asymptote: $x = 5$.

69. Asymptotes: $y = 1$, $x = -1$, and $x = 1$; intercept: $(0, -1)$

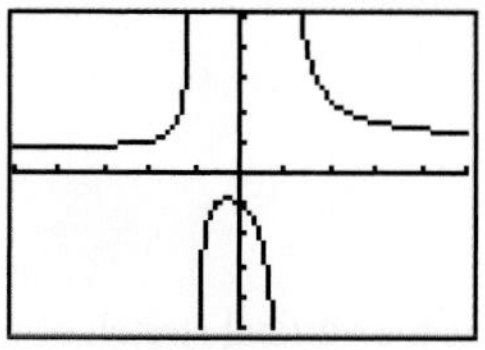

[–5, 5] by [–5, 5]

71. End behavior asymptote: $y = x - 7$; vertical asymptote: $x = -3$; intercept: $\left(0, \frac{5}{3}\right)$

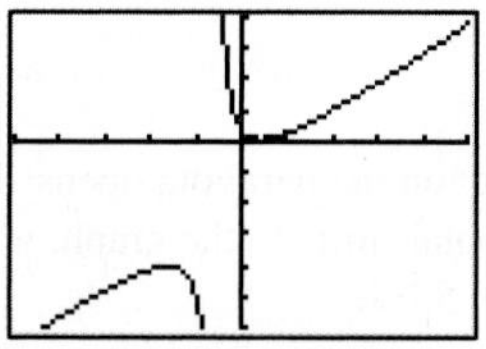

[–25, 25] by [–30, 20]

73. y-intercept: $\left(0, \frac{5}{2}\right)$, x-intercept: $(-2.55, 0)$;
Domain: $x \neq -2$; Range: $(-\infty, \infty)$;
Continuity: all $x \neq -2$;
Decreasing: $(-\infty, -2)$, $(-2, 0.82]$;
Increasing: $[0.82, \infty)$;
Unbounded; Local minimum: $(0.82, 1.63)$;
Vertical asymptote: $x = -2$;
End behavior asymptote: $y = x^2 - x$;
$\lim_{x \to -\infty} f(x) = \lim_{x \to \infty} f(x) = \infty$

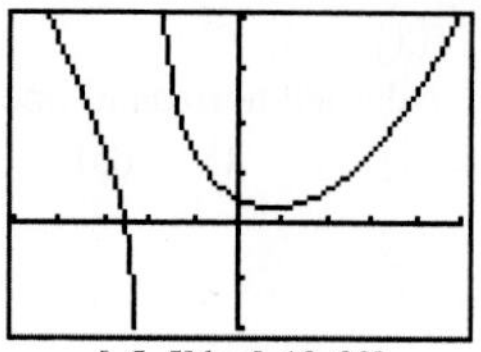

[–5, 5] by [–10, 20]

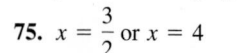

75. $x = \frac{3}{2}$ or $x = 4$ **77.** $(-\infty, -5/2) \cup (-2, 3)$ **79.** $[-3, -2) \cup (2, \infty)$ **81.** $x = -3, x = \frac{1}{2}$ **83.** Yes; at approximately 10.0002

85. **(a)** $V = x(30 - 2x)(70 - 2x)$ in.^{3} **(b)** Either $x \approx 4.57$ or $x \approx 8.63$ in.

87. **(a)** $V = \frac{4}{3}\pi x^3 + \pi x^2(140 - 2x)$ **(b)**

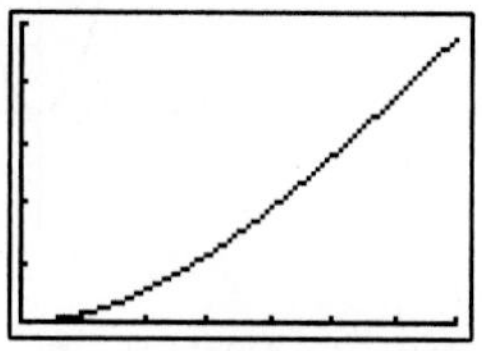

[0, 70] by [0, 1,500,000]

(c) The largest volume occurs when $x = 70$ (so it is actually a sphere). This volume is $\frac{4}{3}\pi(70)^3 \approx 1{,}436{,}755$ ft^3.

89. **(a)** $y \approx 0.435x + 26.016$

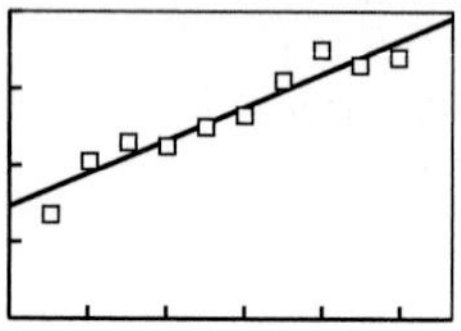

[2, 13] by [24, 32]

(b) $y \approx -0.246x^2 + 0.804x + 24.835$

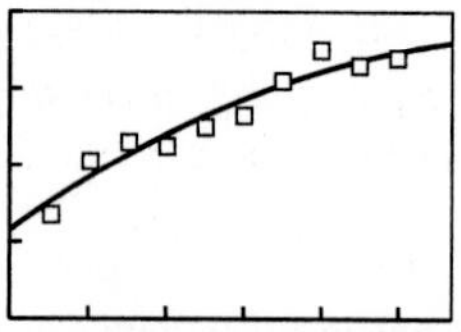

[2, 13] by [24, 32]

(c) Linear: Spending will continue to increase. Quadratic: Spending is nearing a maximum and then will decrease.

91. **(a)** $P(15) = 325$, $P(70) = 600$, $P(100) = 648$ **(b)** $y = \frac{640}{0.8} = 800$ **(c)** The deer population approaches (but never equals) 800.

93. **(a)** $C(x) = \frac{50}{50 + x}$ **(b)** About 33.33 oz. of distilled water **(c)** $x = \frac{100}{3} \approx 33.33$

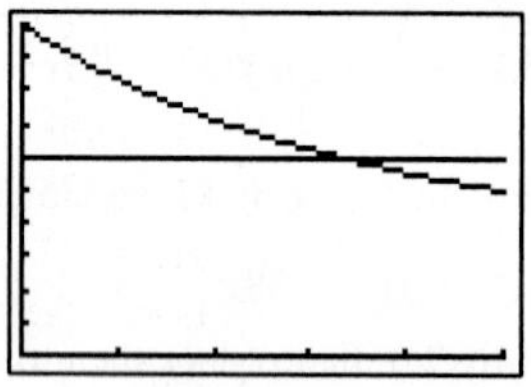

[0, 50] by [0, 1]

95. **(a)** $S = x^2 + \frac{4000}{x}$ **(b)** 20 ft by 20 ft by 2.5 ft or $x \approx 7.32$, giving approximate dimensions 7.32 by 7.32 by 18.66.
(c) $7.32 < x < 20$ (lower bound approximate), so y must be between 2.5 and about 18.66.

CHAPTER 2 PROJECT

Answers are based on the sample date shown in the table.

1.

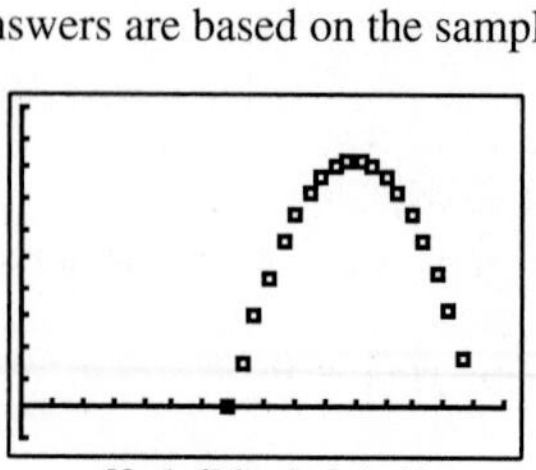

[0, 1.6] by [–0.1, 1]

3. The sign of a affects the direction the parabola opens. The magnitude of a affects the vertical stretch of the graph.Changes to h cause horizontal shifts to the graph, while changes to k cause vertical shifts.
5. $y \approx -4.968x^2 - 10.913x - 5.160$

SECTION 3.1

Exploration 1

1. (0, 1) is in common; Domain: $(-\infty, \infty)$; Range: $(0, \infty)$; Continuous; Always increasing; Not symmetric; No local extrema; Bounded below by $y = 0$, which is also the only asymptote; $\lim_{x\to\infty} f(x) = \infty$. $\lim_{x\to-\infty} f(x) = 0$

Exploration 2

1.

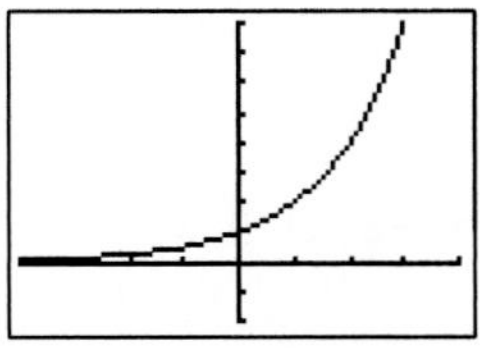

[–4, 4] by [–2, 8]

3. $k \approx 0.693$

Quick Review 3.1

1. -6 **3.** 9 **5.** $1/2^{12}$ **7.** $1/a^6$ **9.** -1.4

Exercises 3.1

1. Not exponential, a monomial function **3.** Exponential function, initial value of 1 and base of 5 **5.** Not exponential, variable base **7.** 3 **9.** $-2\sqrt[3]{3}$ **11.** $3/2 \cdot (1/2)^x$ **13.** $3 \cdot 2^{x/2}$ **15.** Translate $f(x) = 2^x$ by 3 units to the right. **17.** Reflect $f(x) = 4^x$ over the y-axis. **19.** Vertically stretch $f(x) = 0.5^x$ by a factor of 3 and then shift 4 units up. **21.** Reflect $f(x) = e^x$ across the y-axis and horizontally shrink by a factor of 2. **23.** Reflect $f(x) = e^x$ across the y-axis, horizontally shrink by a factor of 3, translate 1 unit to the right, and vertically stretch by a factor of 2. **25.** Graph (a) is the only graph shaped and positioned like the graph of $y = b^x$, $b > 1$. **27.** Graph (c) is the reflection of $y = 2^x$ across the x-axis. **29.** Graph (b) is the graph of $y = 3^{-x}$ translated down 2 units. **31.** Exponential decay; $\lim_{x\to\infty} f(x) = 0$, $\lim_{x\to-\infty} f(x) = \infty$ **33.** Exponential decay; $\lim_{x\to\infty} f(x) = 0$, $\lim_{x\to-\infty} f(x) = \infty$

35. $x < 0$ **37.** $x < 0$ **39.** $y_1 = y_3$ since $3^{2x+4} = 3^{2(x+2)} = (3^2)^{x+2} = 9^{x+2}$

41.

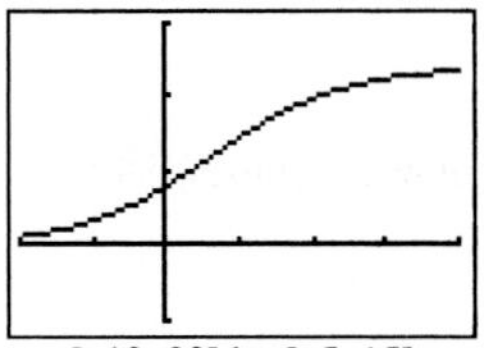

[–10, 20] by [–5, 15]

y-intercept: (0, 4)
Horizontal asymptotes:
$y = 0$, $y = 12$

43.

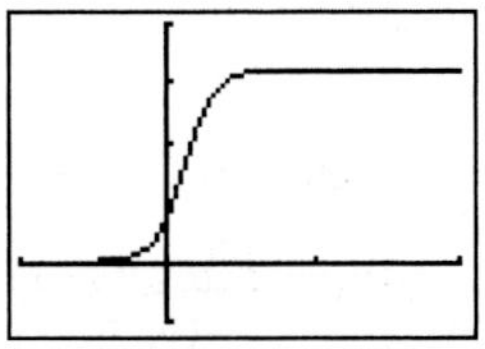

[–5, 10] by [–5, 20]

y-intercept: (0, 4)
Horizontal asymptotes:
$y = 0$, $y = 16$

45.

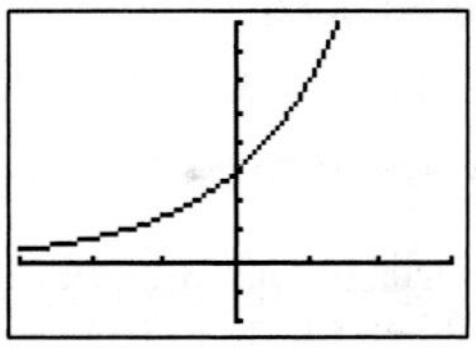

[–3, 3] by [–2, 8]

Domain: $(-\infty, \infty)$; Range: $(0, \infty)$; Continuous; Always increasing; Not symmetric; Bounded below by $y = 0$, which is also the only asymptote; No local extrema; $\lim_{x\to\infty} f(x) = \infty$; $\lim_{x\to-\infty} f(x) = 0$

47.

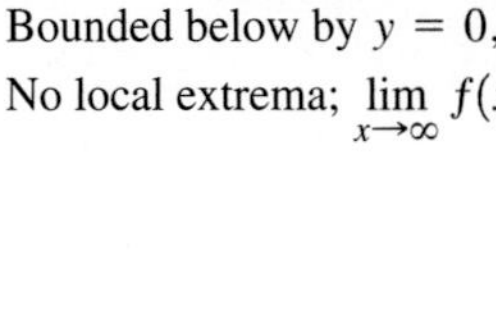

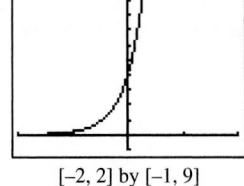

[–2, 2] by [–1, 9]

Domain: $(-\infty, \infty)$; Range: $(0, \infty)$; Continuous;
Always increasing; Not symmetric;
Bounded below by $y = 0$, which is the only asymptote;
No local extrema; $\lim_{x\to\infty} f(x) = \infty$; $\lim_{x\to-\infty} f(x) = 0$

49.

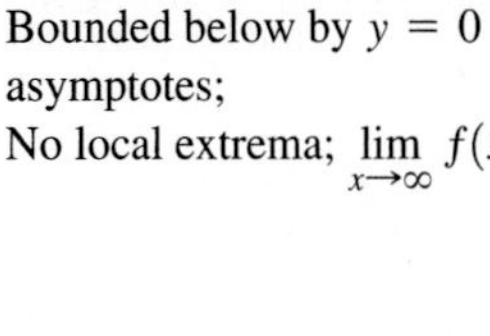

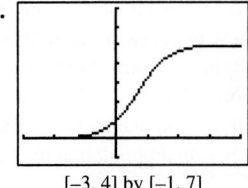

[–3, 4] by [–1, 7]

Domain: $(-\infty, \infty)$; Range: $(0, 5)$; Continuous;
Always increasing; Symmetric about (0.69, 2.5);
Bounded below by $y = 0$ and above by $y = 5$, both of which are asymptotes;
No local extrema; $\lim_{x\to\infty} f(x) = 5$; $\lim_{x\to-\infty} f(x) = 0$

51. In 2010 **53.** In 2009 **55.** Late in 1969 **57. (a)** 100 **(b)** ≈ 6394
59. False. If $a > 0$ and $0 < b < 1$, then $f(x) = a \cdot b^x$ is decreasing. **61.** E **63.** A

65. (a)

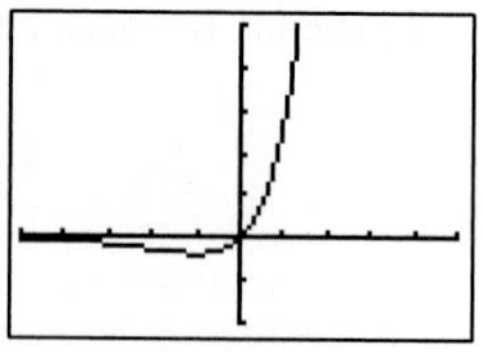

[–5, 5] by [–2, 5]

Domain: $(-\infty, \infty)$; Range: $\left[-\frac{1}{e}, \infty\right)$;

Decreasing on $(-\infty, -1]$; Increasing on $[-1, \infty)$;

Bounded below by $y = -\frac{1}{e}$; Local minimum at $\left(-1, -\frac{1}{e}\right)$;

Asymptote: $y = 0$; $\lim_{x\to\infty} f(x) = \infty$; $\lim_{x\to-\infty} f(x) = 0$

(b)

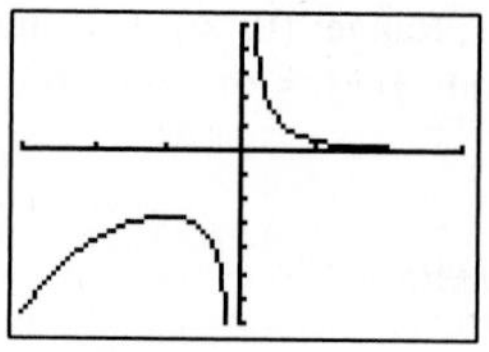

[–3, 3] by [–7, 5]

Domain: $(-\infty, 0) \cup (0, \infty)$; Range: $(-\infty, -e] \cup (0, \infty)$;
Increasing on $(-\infty, -1]$; Decreasing on $[-1, 0) \cup (0, \infty)$;
Not bounded; Local maximum at $(-1, -e)$;
Asymptotes: $x = 0$, $y = 0$; $\lim_{x\to\infty} g(x) = 0$; $\lim_{x\to-\infty} g(x) = -\infty$

67. (a) You have $2^4 = 16$ 4th great grandparents and they are the 4th generation. **(b)** $y = 2^x$ with domain 0, 1, 2, 3, etc. **(c)** $2^6 = 64$ 6th great grandparents. **(d)** $2^{25} = 33{,}554{,}432$ 25th great grandparents. **(e)** About 750 years to span 25 generations. You would be directly related to about 8.4% of the world's population in the year 1250.
69. $c = 2^a$: to the graph of $(2^a)^x$ apply a vertical stretch by 2^b because $f(ax + b) = 2^{ax+b} = 2^{ax}2^b = (2^b)(2^a)^x$.
71. $a < 0$, $c = 1$ **73.** $a > 0$ and $0 < b < 1$, or $a < 0$ and $b > 1$

SECTION 3.2

Quick Review 3.2

1. 0.15 **3.** $23 \cdot 1.07$ **5.** ± 2 **7.** 1.01 **9.** 0.61

Exercises 3.2

1. Exponential growth, 9% **3.** Exponential decay, 3.2% **5.** Exponential growth, 100% **7.** $5 \cdot 1.17^x$ **9.** $16 \cdot 0.5^x$
11. $28{,}900 \cdot 0.974^x$ **13.** $18 \cdot 1.052^x$ **15.** $0.6 \cdot 2^{x/3}$ **17.** $592 \cdot 2^{-x/6}$ **19.** $2.3 \cdot 1.25^x$ **21.** $\approx 4 \cdot 1.15^x$

23. $40/[1 + 3 \cdot (1/3)^x]$ **25.** $\approx 128/(1 + 7 \cdot 0.844^x)$ **27.** $\dfrac{20}{1 + 3 \cdot 0.58^x}$ **29.** In 2021

31. (a) 12,315; 24,265 **(b)** 1966 **33. (a)** $y = 6.6\left(\frac{1}{2}\right)^{t/14}$, where t is time in days **(b)** After 38.11 days

35. One possible answer: Exponential and linear functions are similar in that they are always increasing or always decreasing. However, the two functions vary in how *quickly* they increase or decrease. Although a linear function will increase or decrease at a steady rate over a given interval, the rate at which exponential functions increase or decrease over a given interval will vary.
37. One possible answer: From the graph, we see that doubling time for this model is 4 yr. This is the time required to grow from 50,000 to 100,000, from 100,000 to 200,000, or from any population size to twice that size. Regardless of the population size, it takes 4 yr for it to double.
39. when $t = 1$; every hour **41.** 2.14 lb/in.^{2} **43.** About 4,384,500, overestimated by 565,000, 15% error
45. (a) 16 **(b)** About 14 days **(c)** In about 17 days **47.** About 339.3 million **49.** Model matches.
51. False. This holds true for *logistic* growth, not exponential. **53.** C **55.** D
57. (a) $\approx$308,900,000 **(b)** Underestimates actual population by 2.7 million. **(c)** Logistic model

59. $\sinh(-x) = \dfrac{e^{-x} - e^{-(-x)}}{2} = -\dfrac{e^x - e^{-x}}{2} = -\sinh(x)$

61. (a) $\dfrac{\sinh(x)}{\cosh(x)} = \dfrac{(e^x - e^{-x})/2}{(e^x + e^{-x})/2} = \dfrac{e^x - e^{-x}}{e^x + e^{-x}} = \tanh(x)$ **(b)** $\tanh(-x) = \dfrac{\sinh(-x)}{\cosh(-x)} = \dfrac{-\sinh(x)}{\cosh(x)} = -\tanh(x)$

(c) $f(x) = 1 + \tanh(x) = 1 + \dfrac{e^x - e^{-x}}{e^x + e^{-x}} = \dfrac{e^x + e^{-x}}{e^x + e^{-x}} + \dfrac{e^x - e^{-x}}{e^x + e^{-x}} = \dfrac{2e^x}{e^x + e^{-x}} \cdot \dfrac{e^{-x}}{e^{-x}} = \dfrac{2}{1 + e^{-2x}}$, which is logistic.

SECTION 3.3

Exploration 1

1.

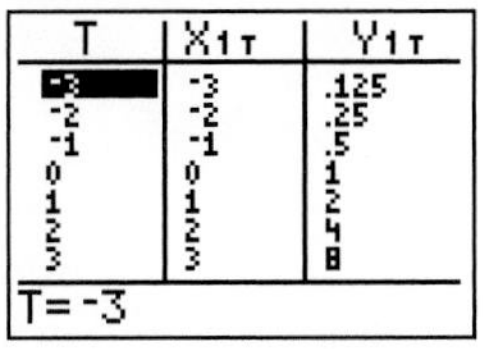

T	X_{1T}	Y_{1T}
-3	-3	.125
-2	-2	.25
-1	-1	.5
0	0	1
1	1	2
2	2	4
3	3	8

T=-3

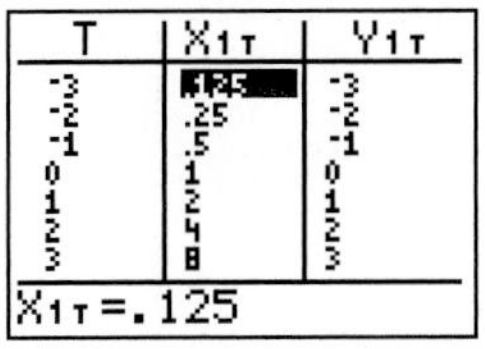

T	X_{1T}	Y_{1T}
-3	.125	-3
-2	.25	-2
-1	.5	-1
0	1	0
1	2	1
2	4	2
3	8	3

X_{1T}=.125

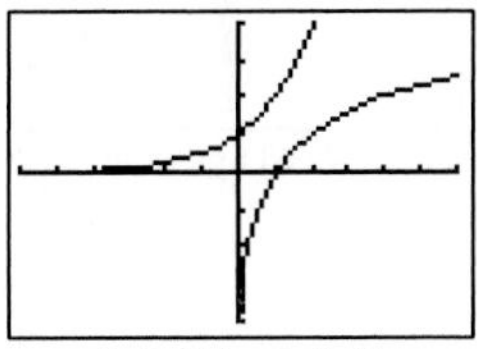

[–6, 6] by [–4, 4]

Quick Review 3.3

1. $1/25 = 0.04$ **3.** $1/5 = 0.2$ **5.** 32 **7.** $5^{1/2}$ **9.** $e^{-1/2}$

Exercises 3.3

1. 1 **3.** 5 **5.** $2/3$ **7.** 3 **9.** 5 **11.** $1/3$ **13.** 3 **15.** -1 **17.** $1/4$ **19.** 3 **21.** 0.5 **23.** 6

25. ≈ 0.975 **27.** Undefined **29.** ≈ 1.399 **31.** Undefined **33.** 100 **35.** 0.1 **37.** (d) **39.** (a)

41.

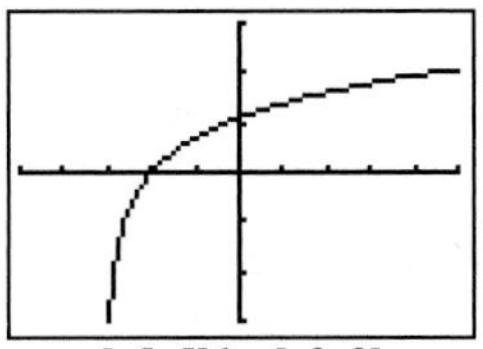

[–5, 5] by [–3, 3]

Starting with $y = \ln x$: shift left 3 units.

43.

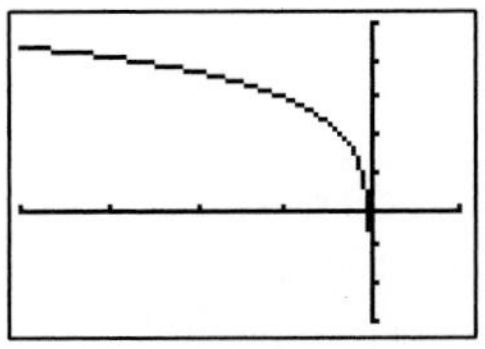

[–4, 1] by [–3, 5]

Starting from $y = \ln x$: reflect across the y-axis and translate up 3 units.

45.

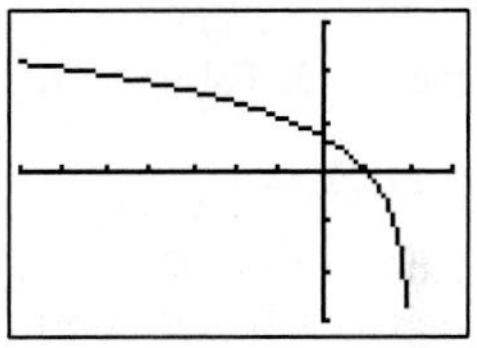

[–7, 3] by [–3, 3]

Starting from $y = \ln x$: reflect across the y-axis and translate right 2 units.

47. Starting with $y = \log x$: shift down 1 unit.

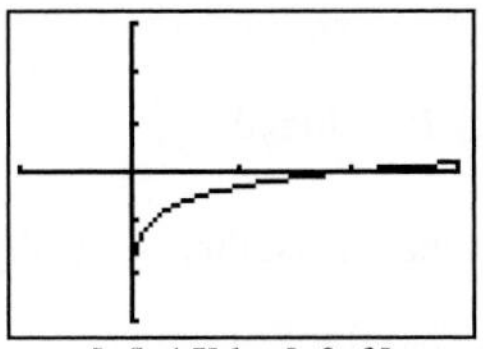

[–5, 15] by [–3, 3]

49. Starting from $y = \log x$: reflect across both axes and vertically stretch by 2.

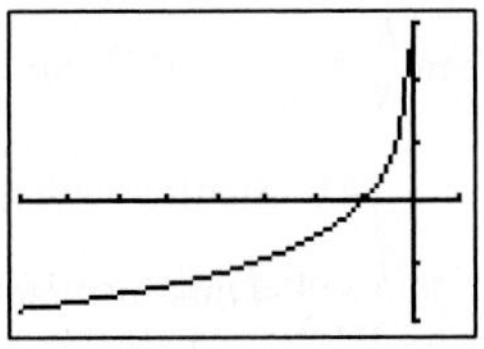

[–8, 1] by [–2, 3]

51. Starting from $y = \log x$: reflect across the y-axis, translate right 3 units, vertically stretch by 2, translate down 1 unit.

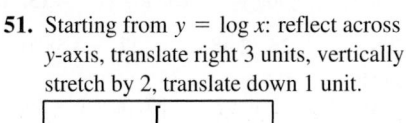

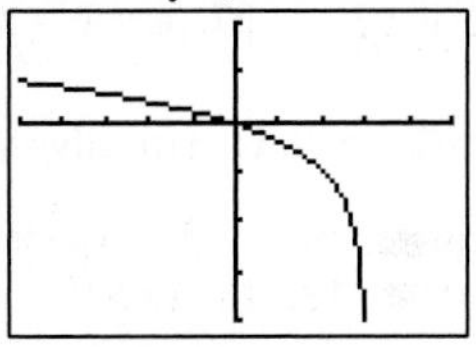

[–5, 5] by [–4, 2]

53.

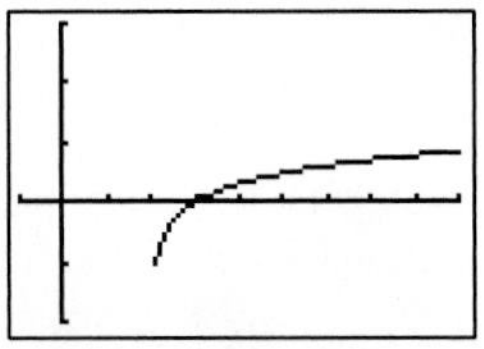

[–1, 9] by [–3, 3]

Domain: $(2, \infty)$; Range: $(-\infty, \infty)$; Continuous; Always increasing; Not symmetric; Not bounded; No local extrema; Asymptote at $x = 2$; $\lim_{x\to\infty} f(x) = \infty$

55.

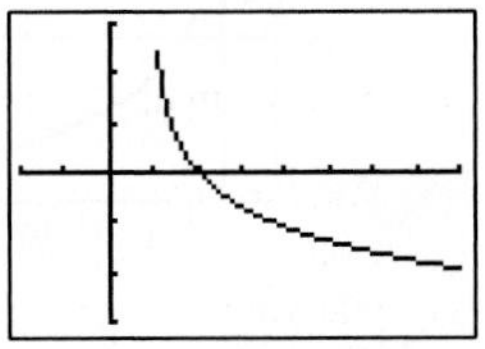

[–2, 8] by [–3, 3]

Domain: $(1, \infty)$; Range: $(-\infty, \infty)$; Continuous; Always decreasing; Not symmetric; Not bounded; No local extrema; Asymptote: $x = 1$; $\lim_{x\to\infty} f(x) = -\infty$

57.

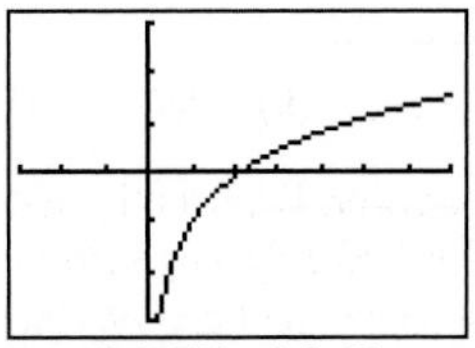

[–3, 7] by [–3, 3]

Domain: $(0, \infty)$; Range: $(-\infty, \infty)$; Continuous; Always increasing on its domain; Not symmetric; Not bounded; No local extrema; Asymptote: $x = 0$; $\lim_{x\to\infty} x = \infty$

59. **(a)** 10 dB **(b)** 70 dB **(c)** 150 dB

61. **(a)** Magnitude 3 is 10 times more powerful than magnitude 2; magnitude 5 is 1000 times more powerful than magnitude 2.

(b)

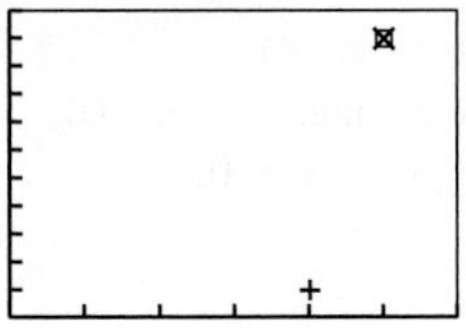

[0, 6] by [0, 110000]

(c) Ground motion $y = 10^x$, where x is the magnitude of the earthquake.
(d) Magnitude $x = \log y$, where y is the ground motion of the earthquake.
(f) Yes

63. True, by definition. **65.** C **67.** B

69.

$f(x)$	3^x	$\log_3 x$
Domain	$(-\infty, \infty)$	$(0, \infty)$
Range	$(0, \infty)$	$(-\infty, \infty)$
Intercepts	$(0, 1)$	$(1, 0)$
Asymptotes	$y = 0$	$x = 0$

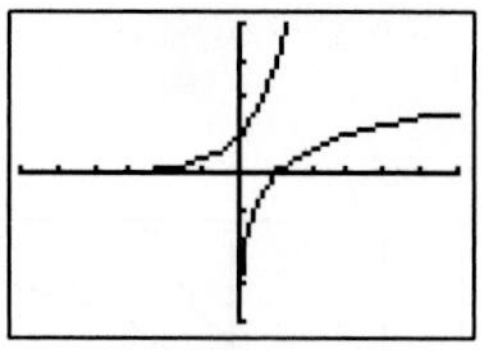

[–6, 6] by [–4, 4]

71. $b = \sqrt[e]{e}$: (e, e) **73.** Reflect across the x-axis

SECTION 3.4

Exploration 1

1. $0.90309 = 0.30103 + 0.60206$ **3.** $0.90309 = 3 \times 0.30103$ **5.** 1.20412; 1.50515; 1.80618

Exploration 2

1. False **3.** True **5.** False **7.** False

Quick Review 3.4

1. 2 **3.** -2 **5.** x^3y^2 **7.** $|x|^3/|y|$ **9.** $1/(3|u|)$

Exercises 3.4

1. $3 \ln 2 + \ln x$ **3.** $\log 3 - \log x$ **5.** $5 \log_2 y$ **7.** $3 \log x + 2 \log y$ **9.** $2 \ln x - 3 \ln y$

11. $\frac{1}{4} \log x - \frac{1}{4} \log y$ **13.** $\log xy$ **15.** $\ln (y/3)$ **17.** $\log \sqrt[3]{x}$ **19.** $\ln (x^2 y^3)$ **21.** $\log (x^4 y/z^3)$

23. 2.8074 **25.** 2.4837 **27.** -3.5850 **29.** $\ln x/\ln 3$ **31.** $\ln (a + b)/\ln 2$ **33.** $\log x/\log 2$

35. $-\log (x + y)/\log 2$ **37.** Let $x = \log_b R$ and $y = \log_b S$. Then $\frac{R}{S} = \frac{b^x}{b^y} = b^{x-y}$. So $\log_b \left(\frac{R}{S}\right) = x - y = \log_b R - \log_b S$.

39. Starting with $g(x) = \ln x$: vertically shrink by a factor of $\frac{1}{\ln 4} \approx 0.72$.

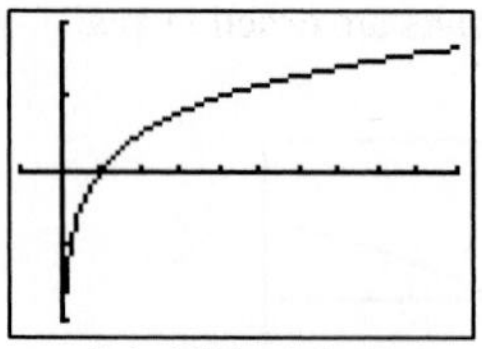

[–1, 10] by [–2, 2]

41. Starting with $g(x) = \ln x$: reflect across the x-axis, then vertically shrink by a factor of $\frac{1}{\ln 3} \approx 0.91$.

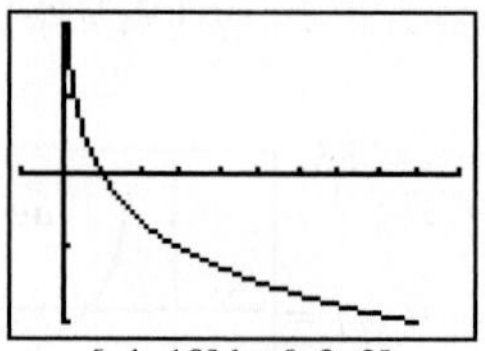

[–1, 10] by [–2, 2]

43. (b): $[-5, 5]$ by $[-3, 3]$, with Xscl = 1 and Yscl = 1
45. (d): $[-2, 8]$ by $[-3, 3]$, with Xscl = 1 and Yscl = 1

47.

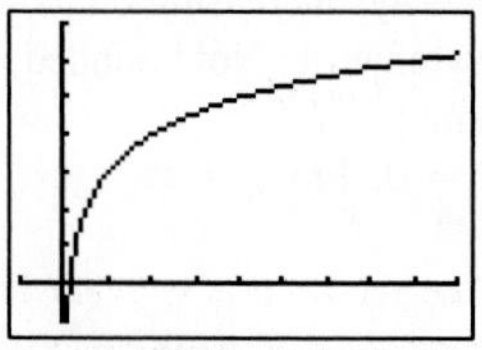

[–1, 9] by [–1, 7]

Domain: $(0, \infty)$; Range: $(-\infty, \infty)$; Continuous; Always increasing; Asymptote: $x = 0$; $\lim_{x\to\infty} f(x) = \infty$

49.

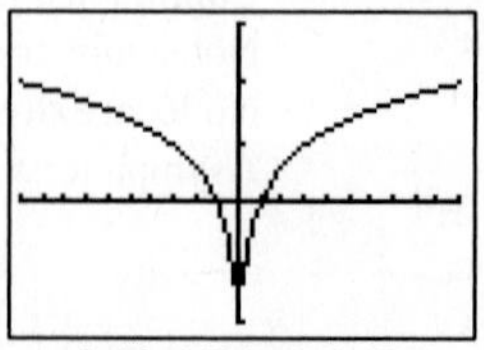

[–10, 10] by [–2, 3]

Domain: $(-\infty, 0) \cup (0, \infty)$; Range: $(-\infty, \infty)$; Discontinuous at $x = 0$; Decreasing on interval $(-\infty, 0)$; Increasing on interval $(0, \infty)$; Asymptote: $x = 0$; $\lim_{x\to\infty} f(x) = \infty$; $\lim_{x\to-\infty} f(x) = \infty$

51. **(a)** 0 **(b)** 10 **(c)** 60 **(d)** 80 **(e)** 100 **(f)** 120 **53.** ≈ 9.6645 lumens **55.** Vertical stretch by a factor of ≈ 0.9102
57. True, by the product rule for logarithms **59.** B **61.** A

63. **(a)** $y = 2.75x^{5.0}$ **(b)** 49,616

(c)

$\ln(x)$	1.39	1.87	2.14	2.30
$\ln(y)$	7.94	10.37	11.71	12.52

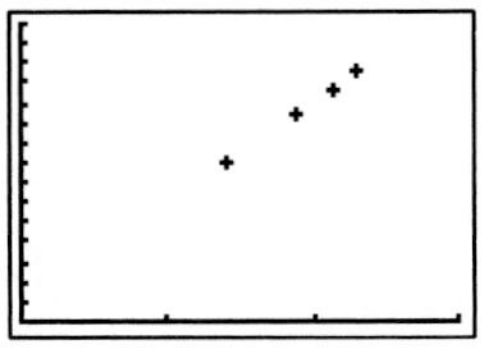

[0, 3] by [0, 15]

(d) $(\ln y) = 5.00\ (\ln x) + 1.01$

(e) $a \approx 5, b \approx 1$ so $f(x) = e^1x^5 = ex^5 \approx 2.72x^5$: The two equations are almost the same.

65. **(a)**

$\log(w)$	−0.70	−0.52	0.30	0.70	1.48	1.70	1.85
$\log(r)$	2.62	2.48	2.31	2.08	1.93	1.85	1.86

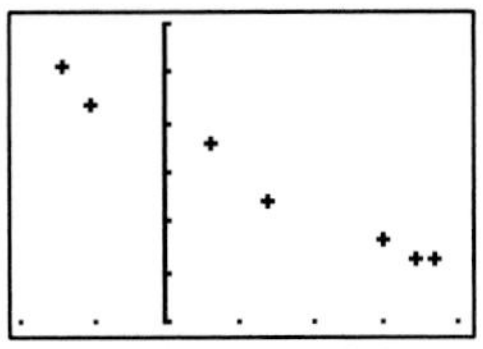

[−1, 2] by [1.6, 2.8]

(b) $\log r = (-0.30)\log w + 2.36$

(c)

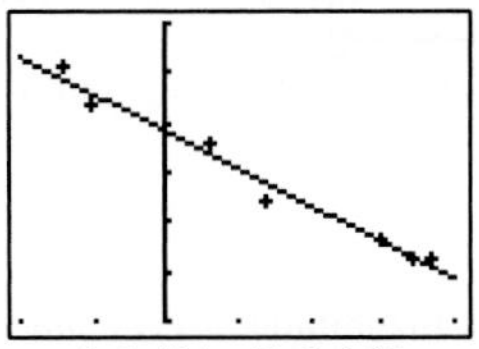

[−1, 2] by [1.6, 2.8]

(d) One possible answer: Consider the power function $y = a \cdot x^b$. Then $\log y = \log (a \cdot x^b) = \log a + \log x^b = \log a + b \log x = b(\log x) + \log a$, which is clearly a linear function of the form $f(t) = mt + c$ where $m = b$, $c = \log a$, $f(t) = \log y$ and $t = \log x$. As a result, there is a linear relationship between log y and log x.

67. Let $x = \sqrt[5]{4581}$. Then $\log x = \frac{1}{5}\log 4581 \approx 3.666 \div 5 = 0.7322$. Thus, $x \approx 5.40$. **69.** $[1.26, 14.77]$

71. Given a, b, $x > 0$, $a \neq 1$, $b \neq 1$, $\log_a x = \log_a b^{\log_b x} = \log_b x \cdot \log_a b$, which yields the desired formula: $\log_b x = \dfrac{\log_a x}{\log_a b}$.

73.

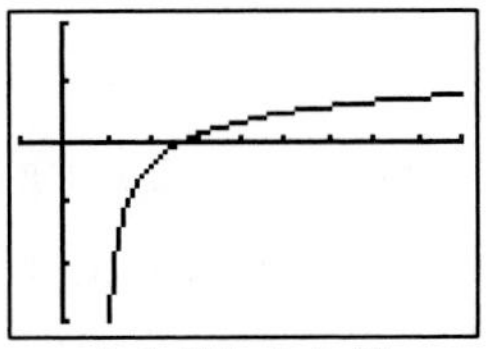

[−1, 9] by [−3, 2]

Domain: $(1, \infty)$; Range: $(-\infty, \infty)$;
Continuous; Increasing; Not symmetric;
Vertical asymptote: $x = 1$; $\lim_{x\to\infty} f(x) = \infty$;
One-to-one, hence invertible: $f^{-1}(x) = e^{e^x}$

SECTION 3.5

Exploration 1

1. 1.60206, 2.60206, 3.60206, 4.60206, 5.60206, 6.60206, 7.60206, 8.60206, 9.60206, 10.60206 **3.** The decimal parts are exactly equal.

Quick Review 3.5

1. $f(g(x)) = e^{2\ln(x^{1/2})} = e^{\ln x} = x$ and $g(f(x)) = \ln(e^{2x})^{1/2} = \ln(e^x) = x$

3. $f(g(x)) = \frac{1}{3}\ln(e^{3x}) = \frac{1}{3}(3x) = x$ and $g(f(x)) = e^{3(1/3 \ln x)} = e^{\ln x} = x$

5. 7.783×10^8 km **7.** 602,000,000,000,000,000,000,000 **9.** 5.766×10^{12}

Exercises 3.5

1. 10 **3.** 12 **5.** −3 **7.** 10,000 **9.** 5.25 **11.** ≈ 24.2151 **13.** ≈ 39.6084 **15.** ≈ -0.4055
17. ≈ 4.3956 **19.** Domain: $(-\infty, -1) \cup (0, \infty)$; graph (e) **21.** Domain: $(-\infty, -1) \cup (0, \infty)$; graph (d)
23. Domain: $(0, \infty)$; graph (a) **25.** $x = 1000$ or $x = -1000$ **27.** $\pm\sqrt{10}$ **29.** $x \approx 3.5949$ **31.** $x \approx \pm 2.0634$
33. $x \approx -9.3780$ **35.** $x \approx 2.3028$ **37.** 4 **39.** 3 **41.** 1.5 **43.** 3 **45.** About 20 times greater
47. **(a)** 1.26×10^{-4}; 1.26×10^{-12} **(b)** 10^8 **(c)** 8 **49.** ≈ 28.41 min

51. **(a)**

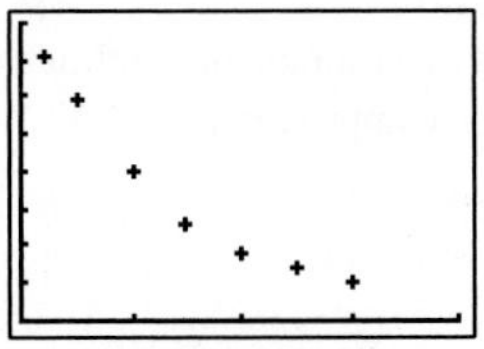

[0, 40] by [0, 80]

(b)

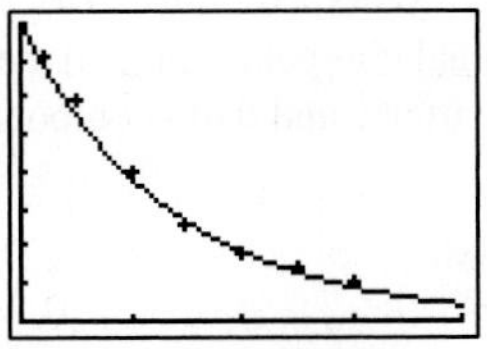

[0, 40] by [0, 80]

$T(x) \approx 79.47 \cdot 0.93^x$

(c) 89.47°C

53. **(a)**

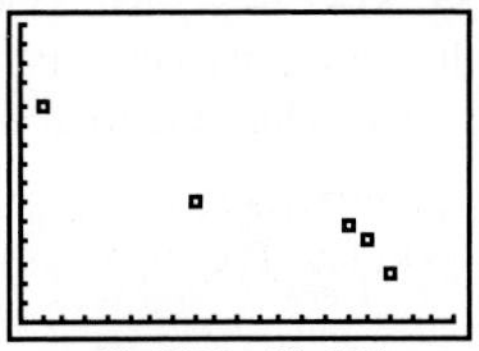

[0, 20] by [0, 15]

(b) The scatter plot is better because it accurately represents the times between the measurements. The equal spacing on the bar graph suggests that the measurements were taken at equally spaced intervals, which distorts our perception of how the consumption has changed over time.

55. Logarithmic seems best—the scatterplot of (x, y) looks most logarithmic. (The data can be modeled by $y = 3 + 2 \ln x$.)

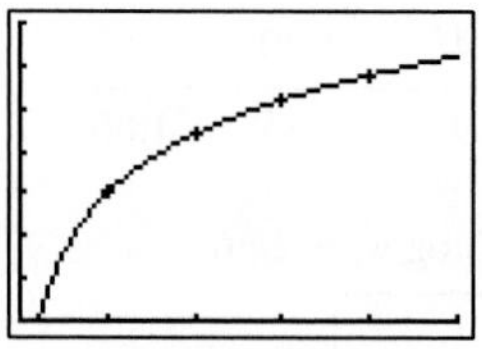

[0, 5] by [0, 7]

57. Exponential—the scatterplot of (x, y) is *exactly* exponential. $\left(\text{The data can be modeled by } y = \frac{3}{2} \cdot 2^x.\right)$

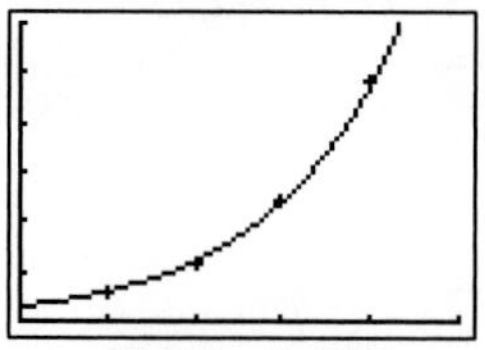

[0, 5] by [0, 30]

59. False. The order of magnitude is its *common* logarithm. **61.** B **63.** E **65.** Logistic regression
67. **(a)** As k increases, the bell curve stretches vertically. **(b)** As c increases, the bell curve compresses horizontally.
69. $n = \log 10^n = \log\left(\frac{u}{v}\right) = \log u - \log v$; u is n orders of magnitude greater than v.
71. $k \approx 0.075$ **73.** $y = cx^a$, where $c = e^b$, exactly the power regression **75.** $x \approx 0.4073$ or $x \approx 0.9333$
77. $x \leq -20.0855$ (approx.) **79.** $-1 < x < 5$

SECTION 3.6

Exploration 1

1.

k	A
10	1104.6
20	1104.9
30	1105
40	1105
50	1105.1
60	1105.1
70	1105.1
80	1105.1
90	1105.1
100	1105.1

A approaches a limit of about 1105.1.

Quick Review 3.6

1. 7 **3.** 1.8125% **5.** 65% **7.** 150 **9.** $315

Exercises 3.6

1. $2251.10, $2130 **3.** $19,908.59, $18,300 **5.** $2122.17 **7.** $86,496.26 **9.** $1728.31
11. $30,402.43 **13.** $14,755.51 **15.** $70,819.63 **17.** $43,523.31 **19.** $293.24
21. 6.63 years — round to 6 years 9 months **23.** 13.78 years — round to 13 years 10 months
25. ≈ 10.13% **27.** 7.07% **29.** 12.14 — round to 12 years 3 months **31.** 7.7016 years; $48,217.82
33. 17.33%; $127,816.26 **35.** 17.42 — round to 17 years 6 months **37.** 15 years
39. 9.99 — round to 10 years **41.** 9.90 years **43.** ≈ 5.92% **45.** ≈ 4.80% **47.** 5% continuously
49. $80,367.73 **51.** $158.03 **53.** $145.74 **55.** $856.39 **57.** **(a)** 22 years 2 months **(b)** $59,006.40
59. One possible answer: The APR will be lower than the APY (except under annual compounding), so the bank's offer looks more attractive when the APR is given. Assuming monthly compounding, the APY is about 4.594%; quarterly and daily compounding give approximately 4.577% and 4.602%, respectively.

61. **(a)** Steve's balance will always remain \$1000, since interest is not added to it. Every year he receives 6% of that \$1000 in interest: 6% in the first year, then another 6% in the second year (for a total of $2 \cdot 6\% = 12\%$), then another 6% (totaling $3 \cdot 6\% = 18\%$), etc. After t years, he has earned $6t\%$ of the \$1000 investment, meaning that altogether he has $1000 + 1000 \cdot 0.06t = 1000(1 + 0.06t)$.

(b)

Years	Not Compounded	Compounded
0	1000.00	1000.00
1	1060.00	1060.00
2	1120.00	1123.60
3	1180.00	1191.02
4	1240.00	1262.48
5	1300.00	1338.23
6	1360.00	1418.52
7	1420.00	1503.63
8	1480.00	1593.85
9	1540.00	1689.48
10	1600.00	1790.85

63. False. The limit is $A = Pe^{rt} = 100e^{0.05} \approx \$105.13$. **65.** B **67.** E **69.** \$364.38

71.

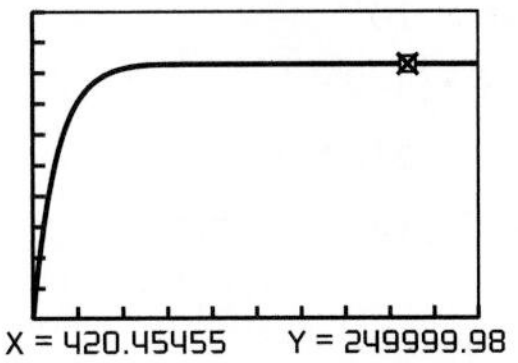

[0, 500] by [0, 300000]

73. **(a)** 8% **(b)** 12 **(c)** \$200 **(d)** \$23,910.86 **(e)**

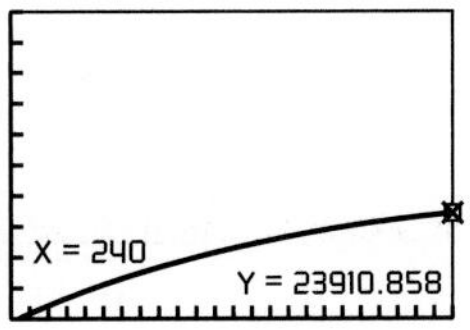

[0, 240] by [0, 70000]

CHAPTR 3 REVIEW EXERCISES

1. $-3\sqrt[3]{4}$ **3.** $y = 3 \cdot 2^{x/2}$ **5.** $f(x) = 2^{-2x} + 3$ — starting from 2^x, horizontally shrink by $\frac{1}{2}$, reflect across y-axis, and translate up 3 units.

7. $f(x) = -2^{-3x} - 3$ — starting from 2^x, horizontally shrink by $\frac{1}{3}$, reflect across y-axis, reflect across x-axis, translate down 3 units.

9. Starting from e^x, horizontally shrink by $\frac{1}{2}$, then translate right $\frac{3}{2}$ units — or translate right 3 units, then horizontally shrink by $\frac{1}{2}$.

11. y-intercept: $(0, 12.5)$; Asymptotes: $y = 0$, $y = 20$ **13.** Exp. decay; $\lim_{x\to\infty} f(x) = 2$, $\lim_{x\to-\infty} f(x) = \infty$

15.

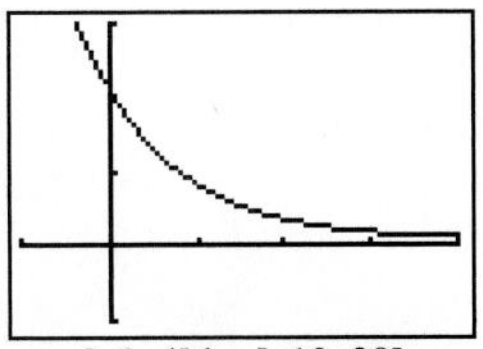

[–1, 4] by [–10, 30]

Domain: $(-\infty, \infty)$; Range: $(1, \infty)$; Continuous;
Always decreasing; Not symmetric;
Bounded below by $y = 1$, which is also the only asymptote;
No local extrema; $\lim_{x\to\infty} f(x) = 1$; $\lim_{x\to-\infty} f(x) = \infty$

17.

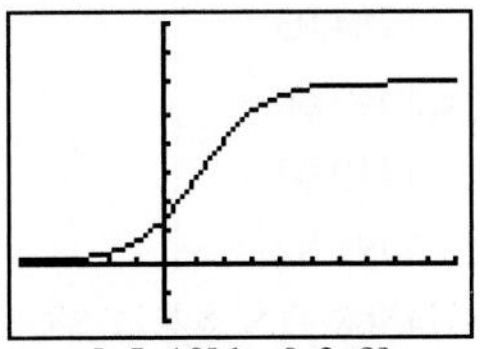

[–5, 10] by [–2, 8]

Domain: $(-\infty, \infty)$; Range: $(0, 6)$; Continuous;
Increasing; Symmetric about $(1.20, 3)$;
Bounded by the asymptotes $y = 0$, $y = 6$;
No extrema; $\lim_{x\to\infty} f(x) = 6$; $\lim_{x\to-\infty} f(x) = 0$

19. $f(x) = 24 \cdot 1.053^x$ **21.** $f(x) = 18 \cdot 2^{x/21}$ **23.** $f(x) \approx 30/(1 + 1.5e^{-0.55x})$ **25.** $f(x) \approx \dfrac{20}{1 + 3e^{-0.37x}}$

27. 5 **29.** 1/3 **31.** $3^5 = x$ **33.** $y = xe^2$ **35.** Translate left 4 units.
37. Translate right 1 unit, reflect across x-axis, and translate up 2 units.

39. 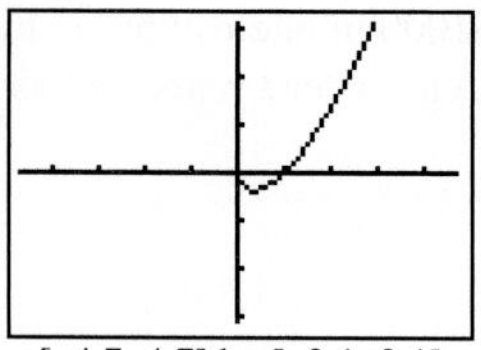

[–4.7, 4.7] by [–3.1, 3.1]

Domain: $(0, \infty)$; Range: $\left[-\frac{1}{e}, \infty\right) \approx [-0.37, \infty)$;
Continuous; Decreasing on $(0, 0.37]$;
Increasing on $[0.37, \infty)$; Not symmetric;
Bounded below;

Local minimum at $\left(\frac{1}{e}, -\frac{1}{e}\right)$; $\lim_{x\to\infty} f(x) = \infty$

41. 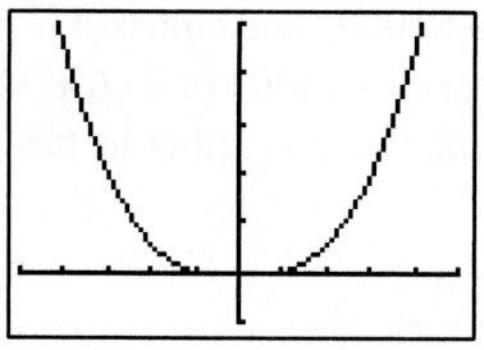

[–5, 5] by [–5, 25]

Domain: $(-\infty, 0) \cup (0, \infty)$; Range: $[-0.18, \infty)$;
Discontinuous at $x = 0$;
Decreasing on $(-\infty, -0.61]$, $(0, 0.61]$;
Increasing on $[-0.61, 0)$, $[0.61, \infty)$;
Symmetric across y-axis; Bounded below;
Local minima at $(-0.61, -0.18)$ and $(0.61, -0.18)$;
No asymptotes; $\lim_{x\to\infty} f(x) = \infty$; $\lim_{x\to-\infty} f(x) = \infty$

43. $\log 4 \approx 0.6021$ **45.** ≈ 22.5171 **47.** 0.0000001 **49.** 4 **51.** ≈ 2.1049 **53.** ≈ 99.5112 **55.** ln *x*/ln 2
57. log *x*/log 5 **59.** (c) **61.** (b) **63.** \$515.00 **65.** Pe^{rt} **67.** \$28,794.06 **69.** -0.3054
71. $P(t) \approx 2.041979 \cdot 1.012960^t$, $P(120) \approx 9.57$ million
73. **(a)** 90 units **(b)** 32.8722 units **(c)** 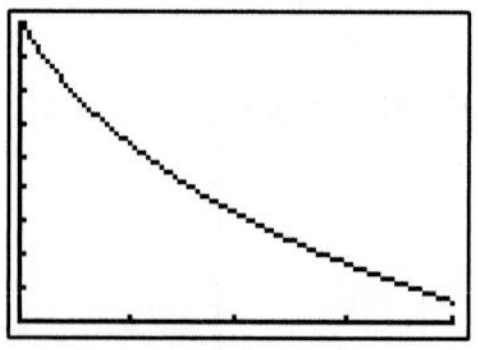

[0, 4] by [0, 90]

75. **(a)** $P(t) = 89{,}000(0.982)^t$ **(b)** 31.74 years
77. **(a)** $P(t) = 20 \cdot 2^t$ **(b)** 81,920; 2.3058×10^{19}
(c) ≈ 8.9658 months **79.** **(a)** $S(t) = S_0 \cdot (1/2)^{t/1.5}$
(b) $S_0/2$; $S_0/4$ **(c)** 1,099,500 metric tons **81.** 6.31
83. 11.75 years **85.** 137.7940 — about 11 years 6 months

87. $\approx 8.57\%$ **89.** ≈ 5.84 lumens **91.** $\frac{1}{10} < b < 10$; $0 < b < \frac{1}{10}$ or $b > 10$ **93.** **(a)** 16**(b)** About $11\frac{1}{2}$ days
(c) 8.7413 — about 8 or 9 days **95.** ≈ 41.54 min **97.** **(a)** 9% **(b)** 4 **(c)** \$100
99. **(a)** Grace's balance will always remain \$1000 since interest is not added to it. Every year she receives 5% of that \$1000, in interest; after t years, she has been paid $5t\%$ of the \$1000 investment, meaning that altogether she has $1000 + 1000 \cdot 0.05t = 1000(1 + 0.05t)$.

(b)

Years	Not Compounded	Compounded
0	1000.00	1000.00
1	1050.00	1051.27
2	1100.00	1105.17
3	1150.00	1161.83
4	1200.00	1221.40
5	1250.00	1284.03
6	1300.00	1349.86
7	1350.00	1419.07
8	1400.00	1491.82
9	1450.00	1568.31
10	1500.00	1648.72

CHAPTER 3 PROJECT

Answers are based on the sample data shown in the table.

3. 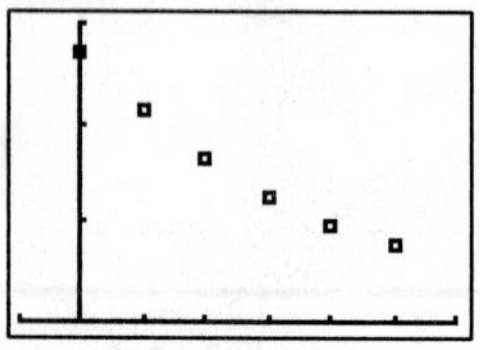

[–1, 6] by [0, 3]

5. $y \approx 2.7188 \cdot 0.788^x$ **7.** A different ball would change the rebound percentage P.
9. $y = He^{\ln(P)x}$ so $y = 2.7188e^{-0.238x}$
11. The linear regression is $y \approx -0.253x + 1.005$. Because $\ln y = (\ln P)x + \ln H$, the slope is $\ln P$ and the y-intercept is $\ln H$.

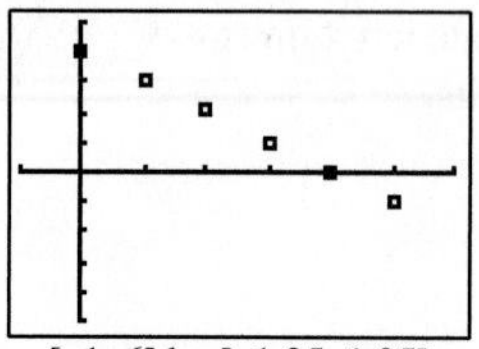

[–1, 6] by [–1.25, 1.25]

SECTION 4.1

Exploration 1

1. $2\pi r$ **3.** No, not quite, since the distance πr would require a piece of thread π times as long, and $\pi > 3$

Quick Review 4.1

1. 5π in. **3.** $\frac{6}{\pi}$ m **5.** **(a)** 47.52 ft **(b)** 39.77 km **7.** 88 ft/sec **9.** 6 mph

Exercises 4.1

1. 23.2° **3.** 118.7375° **5.** 21°12′ **7.** 118°19′ 12″ **9.** $\pi/3$ **11.** $2\pi/3$ **13.** $\approx$ 1.2518 rad **15.** $\approx$1.0716 rad **17.** 30° **19.** 18° **21.** 140° **23.** $\approx$ 114.59° **25.** 50 in. **27.** $6/\pi$ ft **29.** 3 (rad) **31.** $360/\pi$ cm **33.** $\theta = \frac{9}{11}$ rad and $s_2 = 36$ cm **35.** 24 in. **37.** $\approx$ 5.4 in. **39.** **(a)** 45° **(b)** 22.5° **(c)** 247.5° **41.** ESE is closest at 112.5°. **43.** $\approx$ 4.23 stat mi **45.** $\approx$ 275.2 rpm **47.** $\approx$ 12,566 teeth

49.

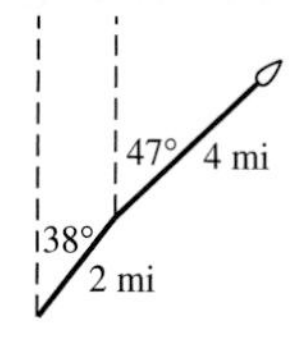

51. $\approx$ 778 naut mi **53.** **(a)** $16\pi \approx 50.265$ in. **(b)** $2\pi \approx 6.283$ ft **55.** **(a)** 4π rad/sec **(b)** 28π cm/sec **(c)** 7π rad/sec **57.** True. Horse *A* travels $2\pi(2r) = 2(2\pi r)$ units of distance in the same amount of time as horse *B* travels $2\pi r$ units of distance, and so is moving twice as fast. **59.** C **61.** B **63.** 38°02′ **65.** 5°37′ **67.** 80 naut mi **69.** 902 naut mi

71. The whole circle's area is πr^2; the sector with central angle θ makes up $\frac{\theta}{2\pi}$ of that area, or $\frac{\theta}{2\pi} \cdot \pi r^2 = \frac{1}{2} r^2 \theta$.

73.

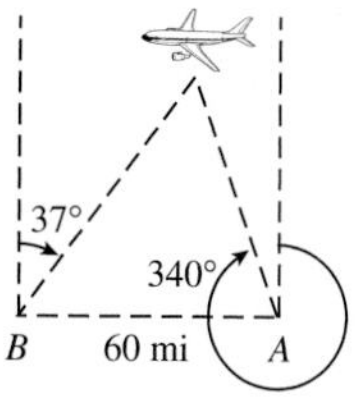

SECTION 4.2

Exploration 1

1. sin and csc, cos and sec, and tan and cot **3.** sec θ **5.** sin θ and cos θ

Exploration 2

1. Let $\theta = 60°$. Then

$$\sin\theta = \frac{\sqrt{3}}{2} \qquad \csc\theta = \frac{2}{\sqrt{3}} = \frac{2\sqrt{3}}{3}$$

$$\cos\theta = \frac{1}{2} \qquad \sec\theta = 2$$

$$\tan\theta = \sqrt{3} \approx 1.732 \qquad \cot\theta = \frac{1}{\sqrt{3}} = \frac{\sqrt{3}}{3}$$

3. The value of a trig function at θ is the same as the value of its cofunction at $90° - \theta$.

Quick Review 4.2

1. $5\sqrt{2}$ **3.** 6 **5.** 100.8 in. **7.** 7.9152 km **9.** $\approx$1.0101 (no units)

Exercises 4.2

1. $\sin\theta = \frac{4}{5}$, $\cos\theta = \frac{3}{5}$, $\tan\theta = \frac{4}{3}$, $\csc\theta = \frac{5}{4}$, $\sec\theta = \frac{5}{3}$, $\cot\theta = \frac{3}{4}$ **3.** $\sin\theta = \frac{12}{13}$, $\cos\theta = \frac{5}{13}$, $\tan\theta = \frac{12}{5}$, $\csc\theta = \frac{13}{12}$, $\sec\theta = \frac{13}{5}$, $\cot\theta = \frac{5}{12}$ **5.** $\sin\theta = \frac{7}{\sqrt{170}}$, $\cos\theta = \frac{11}{\sqrt{170}}$, $\tan\theta = \frac{7}{11}$, $\csc\theta = \frac{\sqrt{170}}{7}$, $\sec\theta = \frac{\sqrt{170}}{11}$, $\cot\theta = \frac{11}{7}$ **7.** $\sin\theta = \frac{\sqrt{57}}{11}$, $\cos\theta = \frac{8}{11}$, $\tan\theta = \frac{\sqrt{57}}{8}$, $\csc\theta = \frac{11}{\sqrt{57}}$, $\sec\theta = \frac{11}{8}$, $\cot\theta = \frac{8}{\sqrt{57}}$ **9.** $\cos\theta = \frac{2\sqrt{10}}{7}$, $\tan\theta = \frac{3}{2\sqrt{10}}$, $\csc\theta = \frac{7}{3}$, $\sec\theta = \frac{7}{2\sqrt{10}}$, $\cot\theta = \frac{2\sqrt{10}}{3}$

11. $\sin\theta = \frac{4\sqrt{6}}{11}$, $\tan\theta = \frac{4\sqrt{6}}{5}$, $\csc\theta = \frac{11}{4\sqrt{6}}$, $\sec\theta = \frac{11}{5}$, $\cot\theta = \frac{5}{4\sqrt{6}}$ **13.** $\sin\theta = \frac{5}{\sqrt{106}}$, $\cos\theta = \frac{9}{\sqrt{106}}$, $\csc\theta = \frac{\sqrt{106}}{5}$, $\sec\theta = \frac{\sqrt{106}}{9}$, $\cot\theta = \frac{9}{5}$ **15.** $\sin\theta = \frac{3}{\sqrt{130}}$, $\cos\theta = \frac{11}{\sqrt{130}}$, $\tan\theta = \frac{3}{11}$, $\csc\theta = \frac{\sqrt{130}}{3}$, $\sec\theta = \frac{\sqrt{130}}{11}$, $\cot\theta = \frac{11}{3}$

17. $\sin\theta = \frac{9}{23}$, $\cos\theta = \frac{8\sqrt{7}}{23}$, $\tan\theta = \frac{9}{8\sqrt{7}}$, $\sec\theta = \frac{23}{8\sqrt{7}}$, $\cot\theta = \frac{8\sqrt{7}}{9}$ **19.** $\frac{\sqrt{3}}{2}$ **21.** $\sqrt{3}$ **23.** $\frac{\sqrt{2}}{2}$ **25.** $\sqrt{2}$

27. $\sqrt{4/3} = 2/\sqrt{3} = 2\sqrt{3}/3$ **29.** 0.961 **31.** 0.943 **33.** 0.268 **35.** 1.524 **37.** 0.810 **39.** 2.414

41. $30° = \frac{\pi}{6}$ **43.** $60° = \frac{\pi}{3}$ **45.** $60° = \frac{\pi}{3}$ **47.** $30° = \frac{\pi}{6}$ **49.** $\frac{15}{\sin 34°} \approx 26.82$ **51.** $\frac{32}{\tan 57°} \approx 20.78$

53. $\frac{6}{\sin 35°} \approx 10.46$ **55.** $b \approx 33.79$, $c \approx 35.96$, $\beta = 70°$ **57.** $b \approx 22.25$, $c \approx 27.16$, $\alpha = 35°$

59. As θ gets smaller and smaller, the side opposite θ gets smaller and smaller, so its ratio to the hypotenuse approaches 0 as a limit.

61. ≈ 205.26 ft **63.** ≈ 74.16 ft^2 **65.** ≈ 378.80 ft **67.** False. This is true only if θ is an acute angle in a right triangle.

69. E **71.** D **73.** Sine values should be increasing, cosine values should be decreasing, and only tangent values can be greater than 1. Therefore, the first column is tangent, the second column is sine, and the third column is cosine.

75. The distance d_A from A to the mirror is 5 cos 30°; the distance from B to the mirror is $d_B = d_A - 2$.

Then $PB = \frac{d_B}{\cos\beta} = \frac{d_A - 2}{\cos 30°} = 5 - \frac{2}{\cos 30°} = 5 - \frac{4}{\sqrt{3}} \approx 2.69$ m

77. One possible proof: $(\sin\theta)^2 + (\cos\theta)^2 = \left(\frac{a}{c}\right)^2 + \left(\frac{b}{c}\right)^2 = \frac{a^2}{c^2} + \frac{b^2}{c^2} = \frac{a^2 + b^2}{c^2} = \frac{c^2}{c^2} = 1$ (Pythagorean Theorem: $a^2 + b^2 = c^2$.)

SECTION 4.3

Exploration 1

1. The side opposite θ in the triangle has length y and the hypotenuse has length r. Therefore $\sin\theta = \frac{opp}{hyp} = \frac{y}{r}$. **3.** $\tan\theta = y/x$

Exploration 2

1. The x-coordinates on the unit circle lie between -1 and 1, and cos t is always an x-coordinate on the unit circle. **3.** The points corresponding to t and $-t$ on the number line are wrapped to points above and below the x-axis with the same x-coordinates. Therefore cos t and cos $(-t)$ are equal. **5.** Since 2π is the distance around the unit circle, both t and $t + 2\pi$ get wrapped to the same point. **7.** By the observation in (6), tan t and tan $(t + \pi)$ are ratios of the form $\frac{y}{x}$ and $\frac{-y}{-x}$, which are either equal to each other or both undefined. **9.** Answers will vary. For example, similar statements can be made about the functions cot, sec, and csc.

Quick Review 4.3

1. $-30°$ **3.** 1125° **5.** $\sqrt{3}/3$ **7.** $\sqrt{2}$ **9.** $\cos\theta = \frac{12}{13}$, $\tan\theta = \frac{5}{12}$, $\csc\theta = \frac{13}{5}$, $\sec\theta = \frac{13}{12}$, $\cot\theta = \frac{12}{5}$

Exercises 4.3

1. 450° **3.** $\sin\theta = \frac{2}{\sqrt{5}}$, $\cos\theta = -\frac{1}{\sqrt{5}}$, $\tan\theta = -2$, $\csc\theta = \frac{\sqrt{5}}{2}$, $\sec\theta = -\sqrt{5}$, $\cot\theta = -\frac{1}{2}$

5. $\sin\theta = -\frac{1}{\sqrt{2}}$, $\cos\theta = -\frac{1}{\sqrt{2}}$, $\tan\theta = 1$, $\csc\theta = -\sqrt{2}$, $\sec\theta = -\sqrt{2}$, $\cot\theta = 1$ **7.** $\sin\theta = \frac{4}{5}$, $\cos\theta = \frac{3}{5}$, $\tan\theta = \frac{4}{3}$, $\csc\theta = \frac{5}{4}$, $\sec\theta = \frac{5}{3}$, $\cot\theta = \frac{3}{4}$ **9.** $\sin\theta = 1$, $\cos\theta = 0$, $\tan\theta$ undefined, $\csc\theta = 1$, $\sec\theta$ undefined, $\cot\theta = 0$

11. $\sin\theta = -\frac{2}{\sqrt{29}}$, $\cos\theta = \frac{5}{\sqrt{29}}$, $\tan\theta = -\frac{2}{5}$, $\csc\theta = -\frac{\sqrt{29}}{2}$, $\sec\theta = \frac{\sqrt{29}}{5}$, $\cot\theta = -\frac{5}{2}$ **13.** +, +, + **15.** −, −, +

17. − **19.** − **21.** (a) **23.** (a) **25.** $-1/2$ **27.** 2 **29.** $\frac{1}{2}$ **31.** 1 **33.** $\frac{\sqrt{3}}{2}$ **35.** $-\frac{\sqrt{3}}{2}$

37. (a) -1 (b) 0 (c) Undefined **39.** (a) 0 (b) -1 (c) 0 **41.** (a) 1 (b) 0 (c) Undefined

43. $\sin\theta = \frac{\sqrt{5}}{3}$; $\tan\theta = \frac{\sqrt{5}}{2}$ **45.** $\tan\theta = -\frac{2}{\sqrt{21}}$; $\sec\theta = \frac{5}{\sqrt{21}}$ **47.** $\sec\theta = -\frac{5}{4}$; $\csc\theta = \frac{5}{3}$ **49.** 1/2 **51.** 0

53. The calculator's value of the irrational number π is necessarily an approximation. When multiplied by a very large number, the slight error of the original approximation is magnified sufficiently to throw the trigonometric functions off.

55. $\mu = \dfrac{\sin 83°}{\sin 36°} \approx 1.69$ **57.** **(a)** 0.4 in. **(b)** ≈ 0.1852 in. **59.** The difference in the elevations is 600 ft, so $d = 600/\sin\theta$. Then: **(a)** ≈ 848.53 ft **(b)** 600 ft **(c)** ≈ 933.43 ft **61.** True. Acute angles determine reference triangles in QI, where cosine is positive, and obtuse angles determine reference triangles in QII, where cosine is negative. **63.** E **65.** A **67.** $5\pi/6$ **69.** $7\pi/4$

71. The two triangles are congruent: Both have hypotenuse 1, and the corresponding angles are congruent—the smaller acute angle has measure t in both triangles, and the two acute angles in a right triangle add up to $\pi/2$. **73.** One possible answer: Starting from the point (a, b) on the unit circle—at an angle of t, so that $\cos t = a$—then measuring a quarter of the way around the circle (which corresponds to adding $\pi/2$ to the angle), we end at $(-b, a)$, so that $\sin(t + \pi/2) = a$. For (a, b) in Quadrant I, this is shown in the figure; similar illustrations can be drawn for the other quadrants.

75. Starting from the point (a, b) on the unit circle—at an angle of t, so that $\cos t = a$—then measuring a quarter of the way around the circle (which corresponds to adding $\pi/2$ to the angle), we end at $(-b, a)$, so that $\sin(t + \pi/2) = a$. This holds true when (a, b) is in Quadrant II, just as it did for Quadrant I.

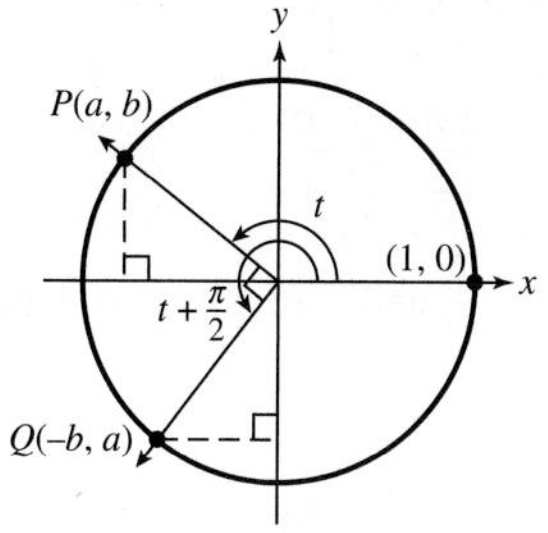

77. $|\theta| < 0.2441$ (approximately) **79.** This Taylor polynomial is generally a very good approximation for $\sin\theta$—in fact, the relative error is less than 1% for $|\theta| < 1$ (approx.). It is better for θ close to 0; it is slightly larger than $\sin\theta$ when $\theta < 0$ and slightly smaller when $\theta > 0$.

SECTION 4.4

Exploration 1

1. $\pi/2$ (at the point $(0, 1)$) **3.** Both graphs cross the x-axis when the y-coordinate on the unit circle is 0. **5.** The sine function tracks the y-coordinate of the point as it moves around the unit circle. After the point has gone completely around the unit circle (a distance of 2π), the same pattern of y-coordinates starts over again.

Quick Review 4.4

1. In order: $+, +, -, -$ **3.** In order: $+, -, +, -$ **5.** $-5\pi/6$ **7.** Vertically stretch by 3. **9.** Vertically shrink by 0.5.

Exercises 4.4

1. Amplitude 2; vertical stretch by a factor of 2 **3.** Amplitude 4; vertical stretch by a factor of 4, reflection across x-axis

5. Amplitude 0.73; vertical shrink by a factor of 0.73 **7.** Period $\dfrac{2\pi}{3}$; horizontal shrink by a factor of $\dfrac{1}{3}$ **9.** Period $\dfrac{2\pi}{7}$; horizontal shrink by a factor of $\dfrac{1}{7}$, reflection across y-axis **11.** Period π; horizontal shrink by a factor of $\dfrac{1}{2}$, vertical stretch by a factor of 3

13. Amplitude 3, period 4π, frequency $\dfrac{1}{4\pi}$

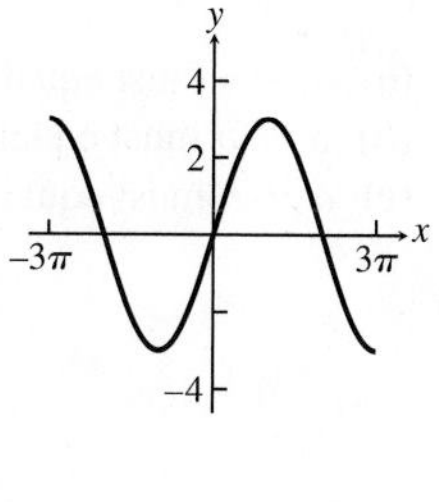

15. Amplitude $\dfrac{3}{2}$, period π, frequency $\dfrac{1}{\pi}$

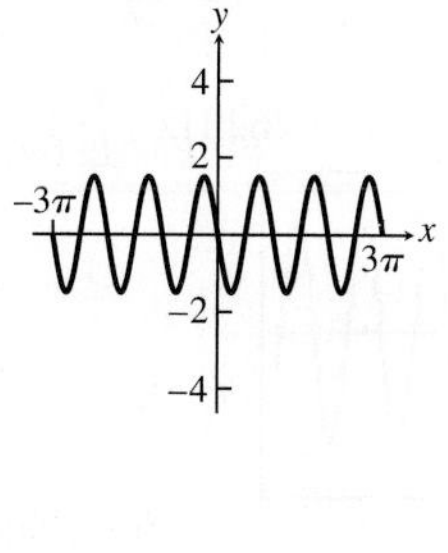

17.

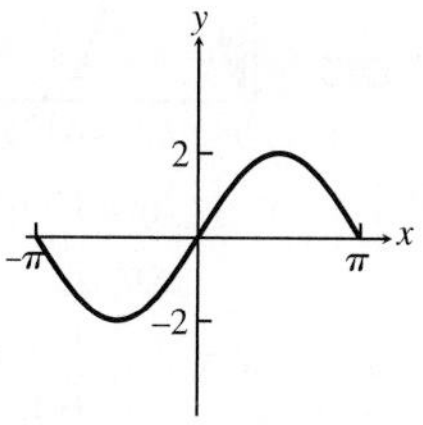

19.

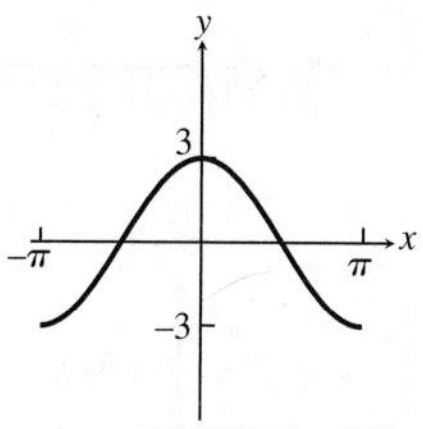

21.

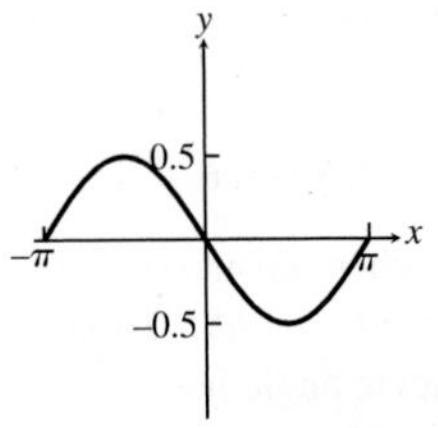

23.

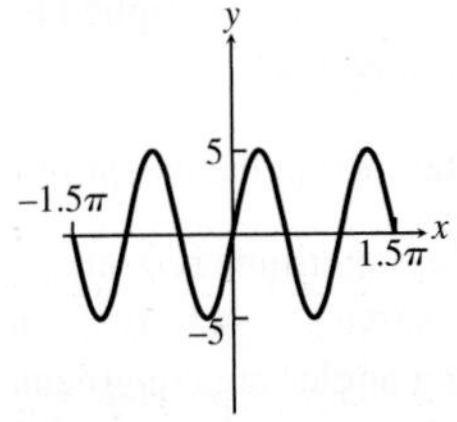

25.

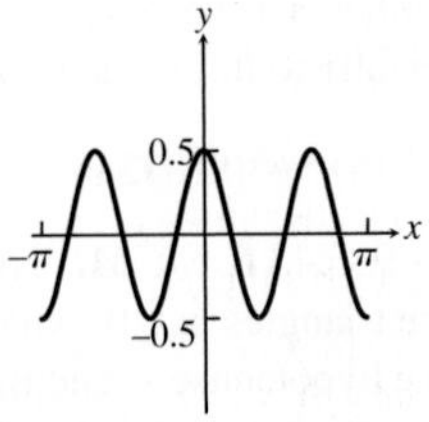

27.

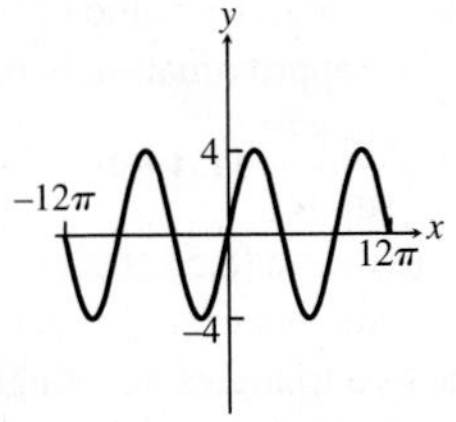

29. Period π; amplitude 1.5; $[-2\pi, 2\pi]$ by $[-2, 2]$ **31.** Period π; amplitude 3; $[-2\pi, 2\pi]$ by $[-4, 4]$

33. Period 6; amplitude 4; $[-3, 3]$ by $[-5, 5]$ **35.** Maximum: 2 $\left(\text{at } -\frac{3\pi}{2} \text{ and } \frac{\pi}{2}\right)$; minimum: -2 $\left(\text{at } -\frac{\pi}{2} \text{ and } \frac{3\pi}{2}\right)$; zeros: 0, $\pm\pi$, $\pm 2\pi$

37. Maximum: 1 (at 0, $\pm\pi$, $\pm 2\pi$); minimum: -1 $\left(\text{at } \pm\frac{\pi}{2} \text{ and } \pm\frac{3\pi}{2}\right)$; zeros: $\pm\frac{\pi}{4}, \pm\frac{3\pi}{4}, \pm\frac{5\pi}{4}, \pm\frac{7\pi}{4}$

39. Maximum: 1 $\left(\text{at } \pm\frac{\pi}{2} \text{ and } \pm\frac{3\pi}{2}\right)$; minimum: -1 (at 0, $\pm\pi$, $\pm 2\pi$); zeros: $\pm\frac{\pi}{4}, \pm\frac{3\pi}{4}, \pm\frac{5\pi}{4}, \pm\frac{7\pi}{4}$

41. One possibility is $y = \sin(x + \pi)$. **43.** Starting from $y = \sin x$, horizontally shrink by $\frac{1}{3}$ and vertically shrink by 0.5.

45. Starting from $y = \cos x$, horizontally stretch by 3, vertically shrink by $\frac{2}{3}$, reflect across x-axis.

47. Starting from $y = \cos x$, horizontally shrink by $\frac{3}{2\pi}$ and vertically stretch by 3. **49.** Starting with y_1, vertically stretch by $\frac{5}{3}$.

51. Starting with y_1, horizontally shrink by $\frac{1}{2}$. **53.** (a) and (b) **55.** (a) and (b) **57.** One possibility is $y = 3 \sin 2x$.

59. One possibility is $y = 1.5 \sin 12(x - 1)$. **61.** Amplitude 2, period 2π, phase shift $\frac{\pi}{4}$, vertical translation 1 unit up

63. Amplitude 5, period $\frac{2\pi}{3}$, phase shift $\frac{\pi}{18}$, vertical translation 0.5 unit up **65.** Amplitude 2, period 1, phase shift 0, vertical translation 1 unit up

67. Amplitude $\frac{7}{3}$, period 2π, phase shift $-\frac{5}{2}$, vertical translation 1 unit down **69.** $y = 2 \sin 2x$ ($a = 2, b = 2, h = 0, k = 0$)

71. **(a)** two **(b)** $(0, 1)$ and $(2\pi, 1.3^{-2\pi}) \approx (6.28, 0.19)$ **73.** ≈ 15.90 sec **75.** **(a)** 1:00 A.M. **(b)** 8.90 ft; 10.52 ft **(c)** 4:06 A.M.

77. **(a)** The maximum d is approximately 21.4 cm. The amplitude is 7.1 cm; scatter plot:

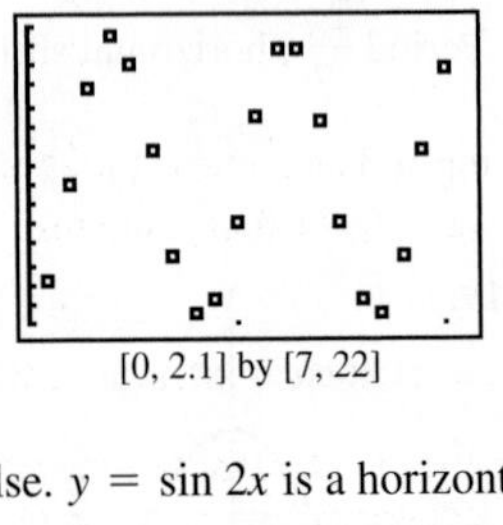

[0, 2.1] by [7, 22]

(b) ≈ 0.83 sec

(c) $d(t) = -7.1 \cos\left(\frac{2\pi x}{0.83}\right) + 14.3$

(d)

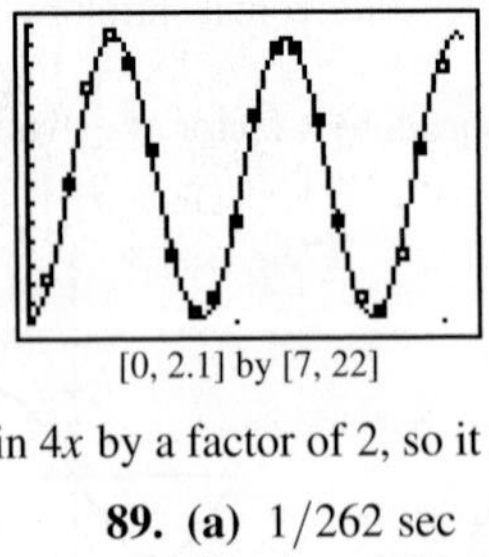

[0, 2.1] by [7, 22]

79. One possible solution is

$$T = 22.5 \cos\left[\frac{\pi}{6}(x - 7)\right] + 56.5$$

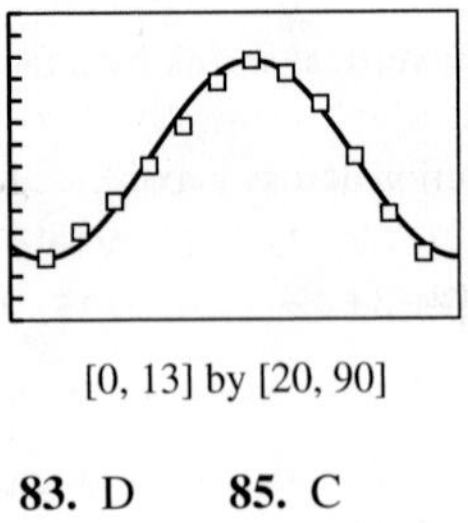

[0, 13] by [20, 90]

81. False. $y = \sin 2x$ is a horizontal *stretch* of $y = \sin 4x$ by a factor of 2, so it has twice the period. **83.** D **85.** C

87. **(a)**

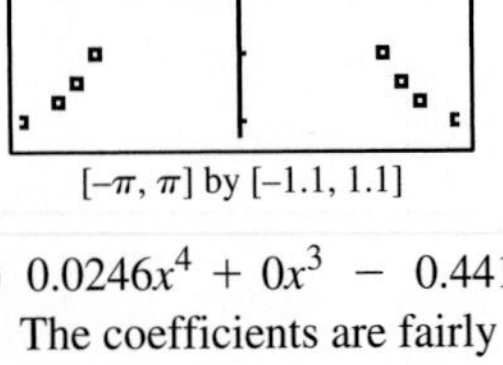

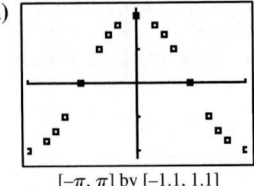

[−π, π] by [−1.1, 1.1]

(b) $0.0246x^4 + 0x^3 - 0.4410x^2 + 0x + 0.9703$

(c) The coefficients are fairly similar.

89. **(a)** 1/262 sec

(b) f = 262 cycle/sec = 262 Hz

(c)

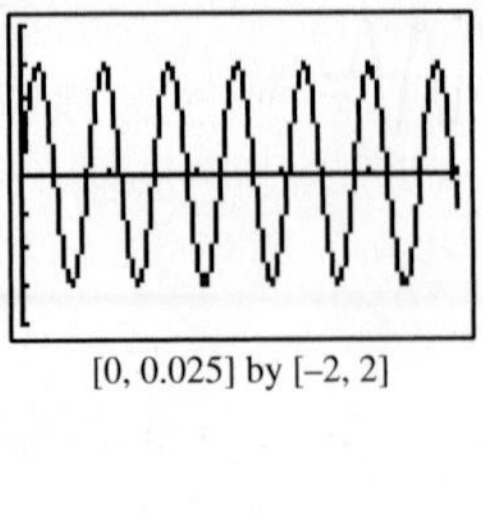

[0, 0.025] by [−2, 2]

91. **(a)** $a - b$ must equal 1.

(b) $a - b$ must equal 2.

(c) $a - b$ must equal k.

93. $B = (0, 3); C = \left(\frac{3\pi}{4}, 0\right)$ **95.** $B = \left(\frac{\pi}{4}, 2\right); C = \left(\frac{3\pi}{4}, 0\right)$

97. (a) If b is negative, then $b = -B$, where B is positive. Then $y = a \sin\left[-B(x - H)\right] + k = -a \sin\left[B(x - H)\right] + k$, since sine is an odd function. We will see in part (d) what to do if the number out front is negative.
(b) A sine graph can be translated a quarter of a period to the left to become a cosine graph of the same sinusoid. Thus

$y = a \sin\left[b\left((x - h) + \frac{1}{4}\cdot\frac{2\pi}{b}\right)\right] + k = a \sin\left[b\left(x - \left(h - \frac{\pi}{2b}\right)\right)\right] + k$ has the same graph as $y = a \cos\left[b(x - h)\right] + k$.

We therefore choose $H = h - \frac{\pi}{2b}$. **(c)** The angles $\theta + \pi$ and θ determine diametrically opposite points on the unit circle, so they have point symmetry with respect to the origin. The y-coordinates are therefore opposites, so $\sin(\theta + \pi) = -\sin\theta$. **(d)** By the identity in (c), $y = a \sin\left[b(x - h) + \pi\right] + k = -a \sin\left[b(x - h)\right] + k$. We therefore choose $H = h - \frac{\pi}{b}$. **(e)** Part (b) shows how to convert $y = a \cos\left[b(x - h)\right] + k$ to $y = a \sin\left[b(x - H)\right] + k$, and parts (a) and (d) show how to ensure that a and b are positive.

SECTION 4.5

Exploration 1

1. The graphs do not seem to intersect.

Quick Review 4.5

1. π **3.** 6π **5.** Zero: 3; asymptote: $x = -4$ **7.** Zero: -1; asymptotes: $x = 2$ and $x = -2$ **9.** Even

Exercises 4.5

1. The graph of $y = 2 \csc x$ must be vertically stretched by 2 compared to $y = \csc x$, so $y_1 = 2 \csc x$ and $y_2 = \csc x$.
3. $y_1 = 3 \csc 2x$, $y_2 = \csc x$

5.

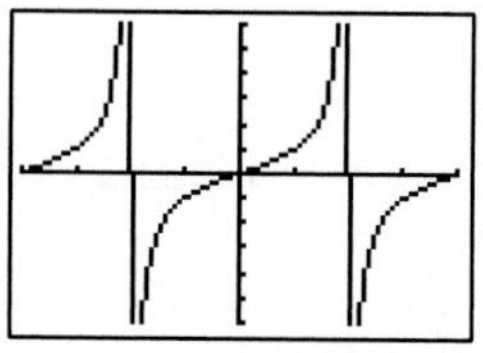

$[-\frac{\pi}{2}, \frac{\pi}{2}]$ by $[-6, 6]$

Horizontal shrink of $y = \tan x$ by factor 1/2; asymptotes at multiples of $\pi/4$

7.

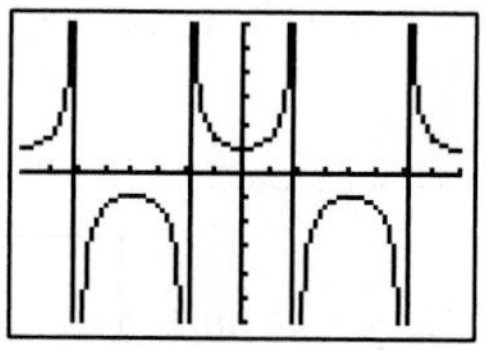

$[-\frac{2\pi}{3}, \frac{2\pi}{3}]$ by $[-6, 6]$

Horizontal shrink of $y = \sec x$ by factor 1/3; asymptotes at odd multiples of $\pi/6$

9.

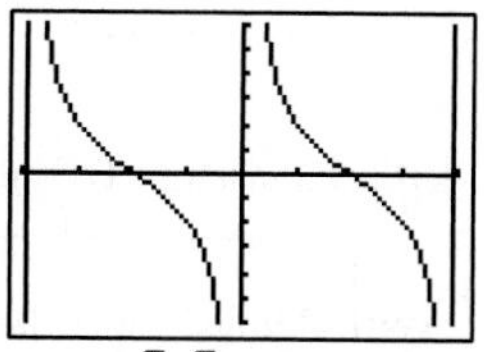

$[-\frac{\pi}{2}, \frac{\pi}{2}]$ by $[-6, 6]$

Horizontal shrink of $y = \cot x$ by factor 1/2; vertical stretch by factor 2; asymptotes at multiples of $\pi/2$

11.

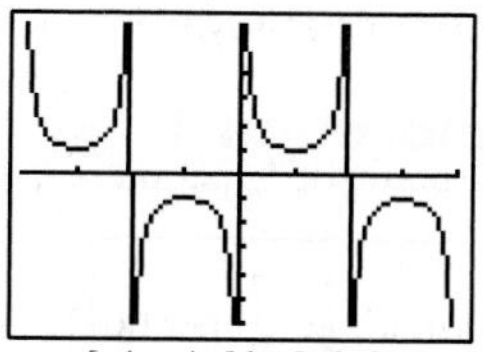

$[-4\pi, 4\pi]$ by $[-6, 6]$

Horizontal stretch of $y = \csc x$ by factor 2; asymptotes at multiples of 2π

13. Graph (a); Xmin $= -\pi$ and Xmax $= \pi$ **15.** Graph (c); Xmin $= -\pi$ and Xmax $= \pi$
17. Domain: All real numbers except integer multiples of π; Range: $(-\infty, \infty)$; Continuous on its domain; Decreasing on each interval in its domain; Symmetry with respect to the origin (odd); Not bounded above or below; No local extrema; No horizontal asymptotes; Vertical asymptotes: $x = k\pi$ for all integers k; End behavior: $\lim_{x\to\infty} \cot x$ and $\lim_{x\to-\infty} \cot x$ do not exist.
19. Domain: All real numbers except integer multiples of π; Range: $(-\infty, -1] \cup [1, \infty)$; Continuous on its domain; On each interval centered at $x = \frac{\pi}{2} + 2k\pi$, where k is an integer, decreasing on the left half of the interval and increasing on the right; for $x = \frac{3\pi}{2} + 2k\pi$, increasing on the first half of the interval and decreasing on the second half; Symmetric with respect to the origin (odd); Not bounded above or below; Local minimum 1 at each $x = \frac{\pi}{2} + 2k\pi$ and local maximum -1 at each $x = \frac{3\pi}{2} + 2k\pi$, where k is an integer; No horizontal asymptotes; Vertical asymptotes: $x = k\pi$ for all integers k; End behavior: $\lim_{x\to\infty} \csc x$ and $\lim_{x\to-\infty} \csc x$ do not exist.
21. Starting with $y = \tan x$, vertically stretch by 3. **23.** Starting with $y = \csc x$, vertically stretch by 3.
25. Starting with $y = \cot x$, horizontally stretch by 2, vertically stretch by 3, and reflect across x-axis.
27. Starting with $y = \tan x$, horizontally shrink by $\frac{2}{\pi}$, reflect across x-axis, and shift up by 2 units.
29. $\pi/3$ **31.** $5\pi/6$ **33.** $5\pi/2$ **35.** $x \approx 0.92$ **37.** $x \approx 5.25$ **39.** $x \approx 0.52$ or $x \approx 2.62$

41. **(a)** The reflection of (a, b) across the origin is $(-a, -b)$. **(b)** Definition of tangent **(c)** $\tan t = \frac{b}{a} = \frac{-b}{-a} = \tan(t - \pi)$

(d) Since points on opposite sides of the unit circle determine the same tangent ratio, $\tan(t \pm \pi) = \tan t$ for all numbers t in the domain. Other points on the unit circle yield triangles with different tangent ratios, so no smaller period is possible. **(e)** The same argument that uses the ratio $\frac{b}{a}$ above can be repeated using the ratio $\frac{a}{b}$, which is the cotangent ratio. **43.** For any x, $\left(\frac{1}{f}\right)(x + p) = \frac{1}{f(x + p)} = \frac{1}{f(x)} = \left(\frac{1}{f}\right)(x)$. This is not true for any smaller value of p, since this is the smallest value that works for f. **45.** **(a)** $d = 350 \sec x$ **(b)** $\approx 16{,}831$ ft **47.** $\approx \pm 0.905$ **49.** $\approx \pm 1.107$ or $\approx \pm 2.034$ **51.** False. It is increasing only over intervals on which it is defined; i.e., intervals bounded by consecutive asymptotes. **53.** A **55.** D

57. About $(-0.44, 0) \cup (0.44, \pi)$

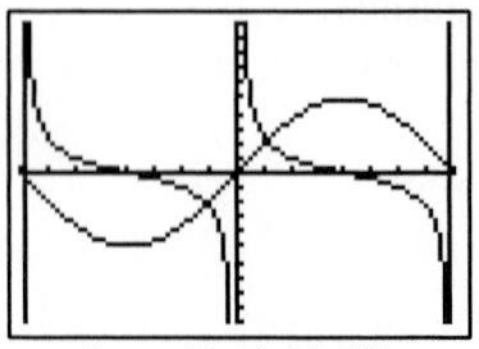

$[-\pi, \pi]$ by $[-10, 10]$

59. $\cot x$ is not defined at 0; the definition of "increasing on (a, b)" requires that the function be defined everywhere in (a, b). Also, choosing $a = -\pi/4$ and $b = \pi/4$, we have $a < b$ but $f(a) = 1 > f(b) = -1$.

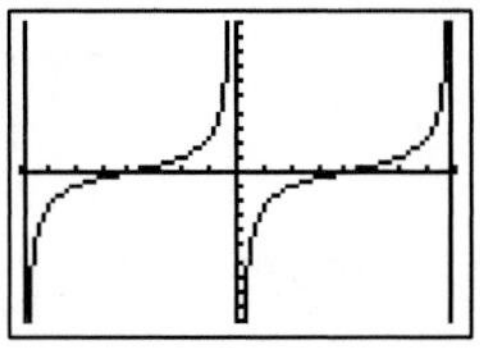

$[-\pi, \pi]$ by $[-10, 10]$

61. $\csc x = \sec\left(x - \frac{\pi}{2}\right)$

63. $d = \frac{30}{\cos x} = 30 \sec x$

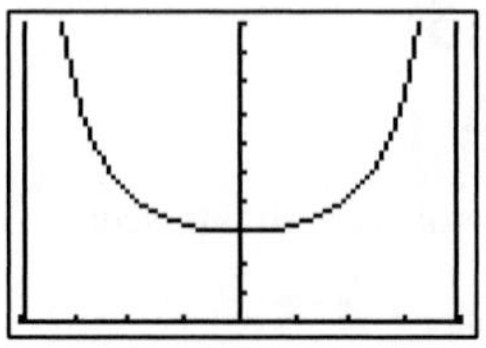

$[-0.5\pi, 0.5\pi]$ by $[0, 100]$

65. ≈ 0.8952 rad $\approx 51.29°$

SECTION 4.6

Exploration 1

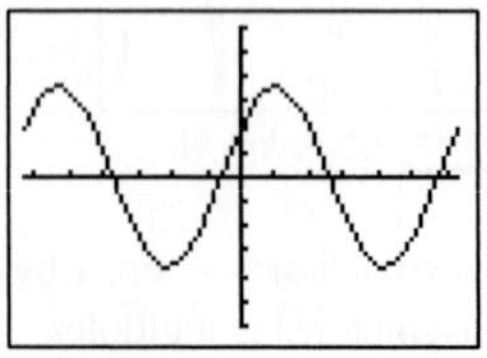

$[-2\pi, 2\pi]$ by $[-6, 6]$

Sinusoid

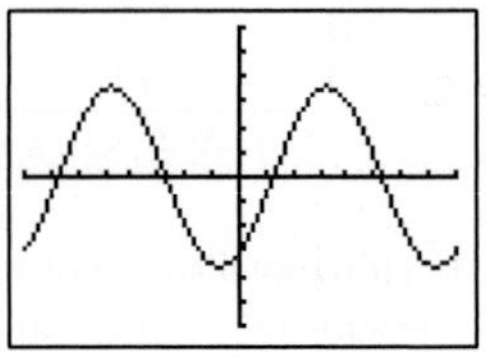

$[-2\pi, 2\pi]$ by $[-6, 6]$

Sinusoid

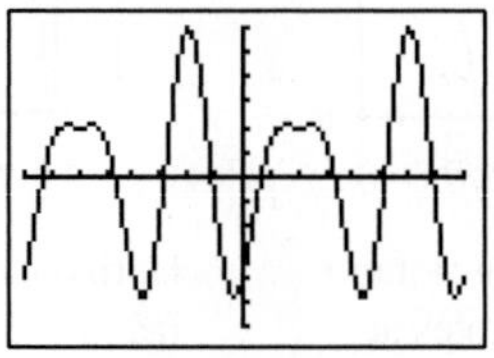

$[-2\pi, 2\pi]$ by $[-6, 6]$

Not a Sinusoid

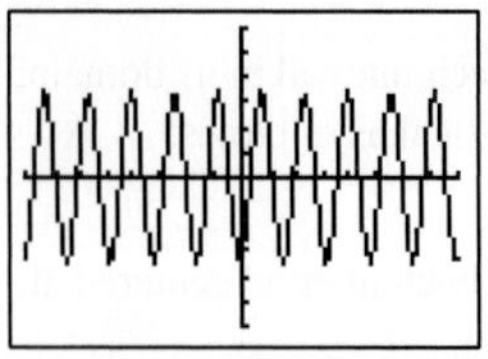

$[-2\pi, 2\pi]$ by $[-6, 6]$

Sinusoid

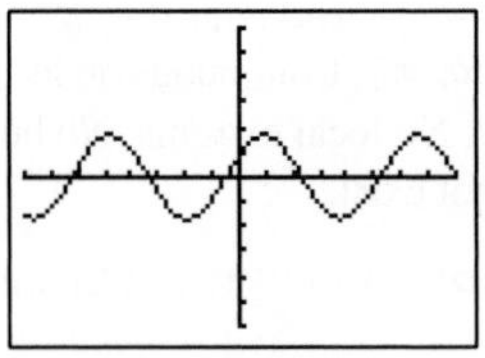

$[-2\pi, 2\pi]$ by $[-6, 6]$

Sinusoid

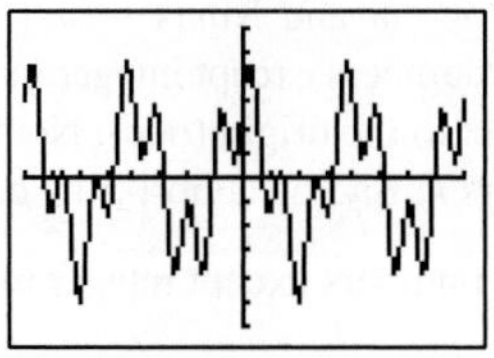

$[-2\pi, 2\pi]$ by $[-6, 6]$

Not a Sinusoid

Quick Review 4.6

1. Domain: $(-\infty, \infty)$; range: $[-3, 3]$ **3.** Domain: $[1, \infty)$; range: $[0, \infty)$
5. Domain: $(-\infty, \infty)$; range: $[-2, \infty)$ **7.** As $x \to -\infty$, $f(x) \to \infty$; as $x \to \infty$, $f(x) \to 0$.
9. $(f \circ g)(x) = x - 4$, domain: $[0, \infty)$; $(g \circ f)(x) = \sqrt{x^2 - 4}$, domain: $(-\infty, -2] \cup [2, \infty)$

Exercises 4.6

1. Periodic

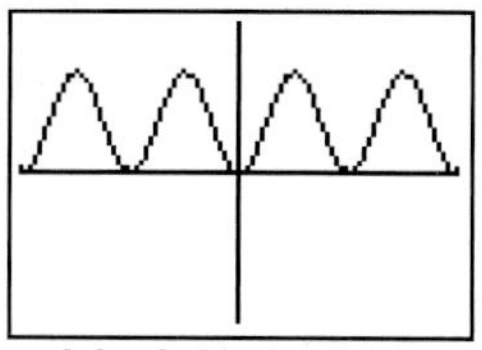

[-2π, 2π] by [-1.5, 1.5]

3. Not periodic

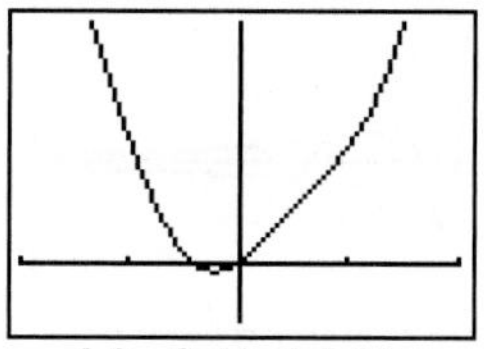

[-2π, 2π] by [-5, 20]

5. Not periodic

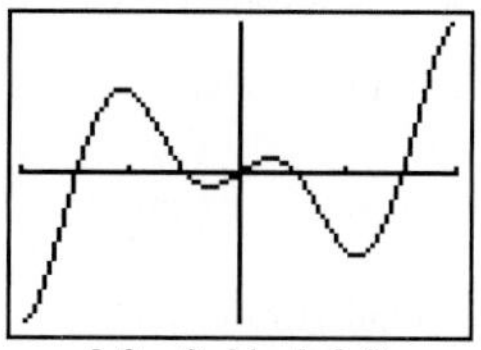

[-2π, 2π] by [-6, 6]

7. Periodic

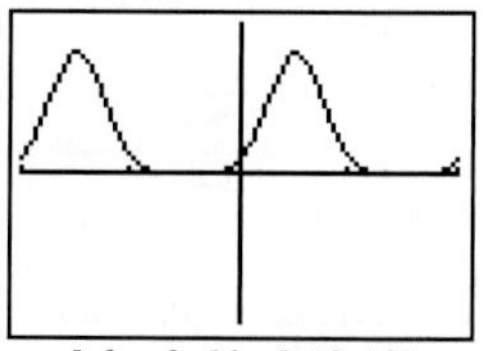

[-2π, 2π] by [-10, 10]

9. Since the period of $\cos x$ is 2π, we have $\cos^2(x + 2\pi) = (\cos(x + 2\pi))^2 = (\cos x)^2 = \cos^2 x$. The period is therefore an exact divisor of 2π, and we see graphically that it is π. A graph for $-\pi \le x \le \pi$ is shown:

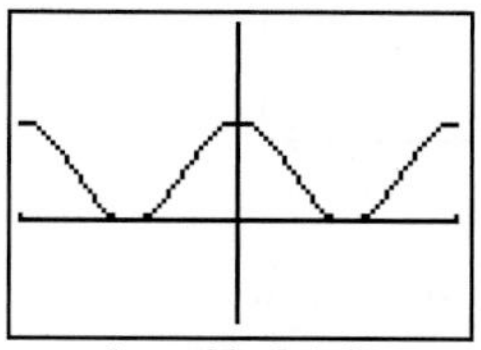

[$-\pi$, π] by [-1, 2]

11. Since the period of $\cos x$ is 2π, we have $\sqrt{\cos^2(x + 2\pi)} = \sqrt{(\cos(x + 2\pi))^2} = \sqrt{(\cos x)^2} = \sqrt{\cos^2 x}$. The period is therefore an exact divisor of 2π, and we see graphically that it is π. A graph for $-\pi \le x \le \pi$ is shown:

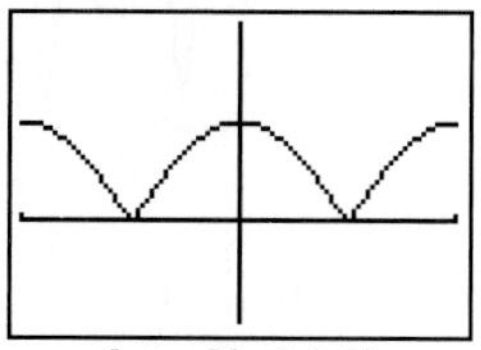

[$-\pi$, π] by [-1, 2]

13. Domain: $(-\infty, \infty)$; range: $[0, 1]$;

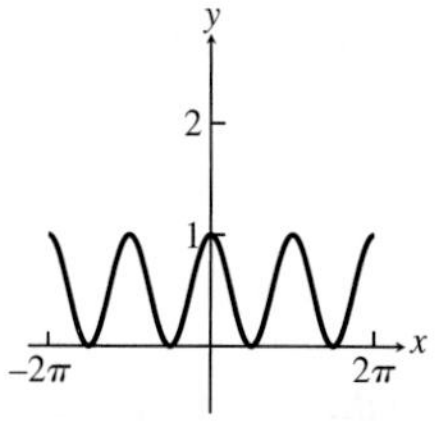

15. Domain: all $x \ne n\pi$, n an integer; range: $[0, \infty)$;

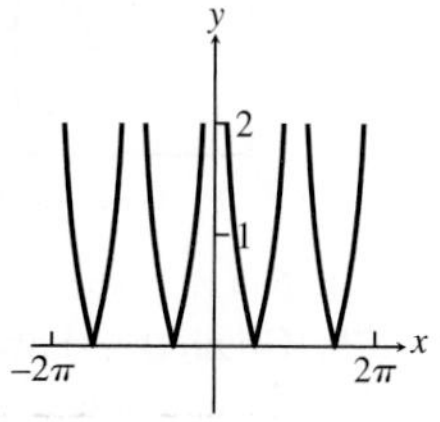

17. Domain: all $x \ne \dfrac{\pi}{2} + n\pi$, n an integer; range: $(-\infty, 0]$;

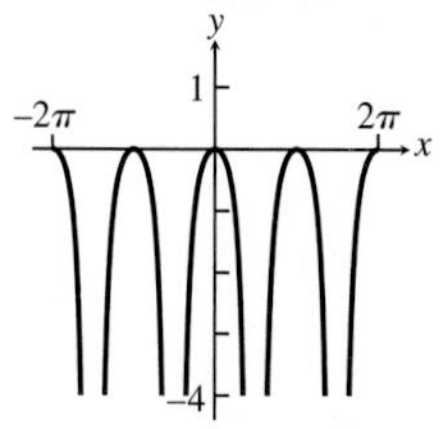

19. $y = 2x - 1$; $y = 2x + 1$

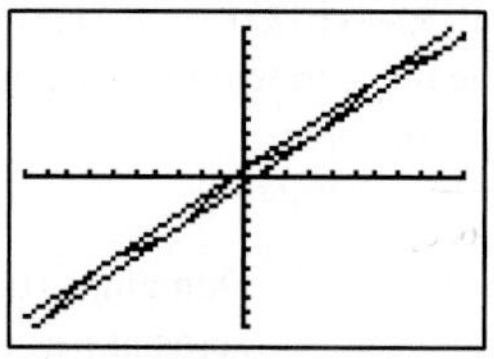

[-10, 10] by [-20, 20]

21. $y = 1 - 0.3x$; $y = 3 - 0.3x$

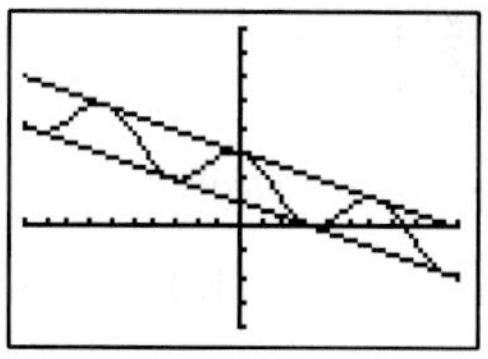

[-10, 10] by [-4, 8]

23. Yes **25.** Yes **27.** No **29.** $a \approx 3.61$, $b = 2$, $h \approx 0.49$
31. $a \approx 2.24$, $b = \pi$, $h \approx 0.35$ **33.** $a \approx 2.24$, $b = 1$, $h \approx -1.11$

35.

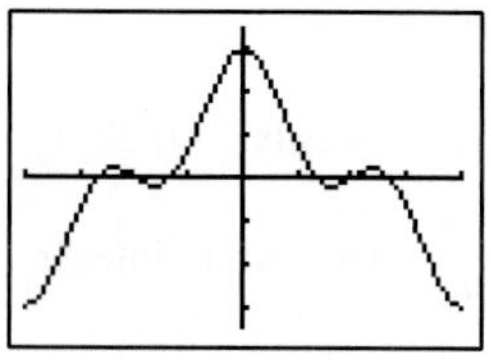

[$-\pi$, π] by [-3.5, 3.5]

37.

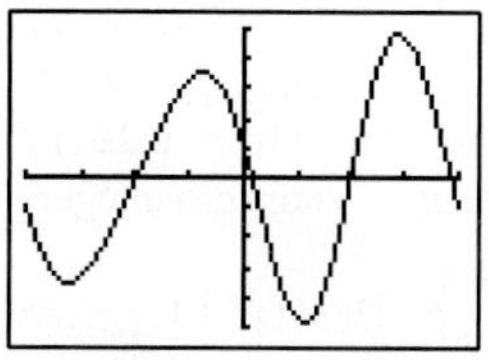

[$-\pi$, π] by [-5, 5]

39. (a) **41.** (c)
43. The damping factor is e^{-x}, and the damping occurs as $x \to \infty$. **45.** No damping
47. The damping factor is x^3, and the damping occurs as $x \to 0$.

49. f oscillates between 1.2^{-x} and -1.2^{-x}.
As $x \to \infty$, $f(x) \to 0$.

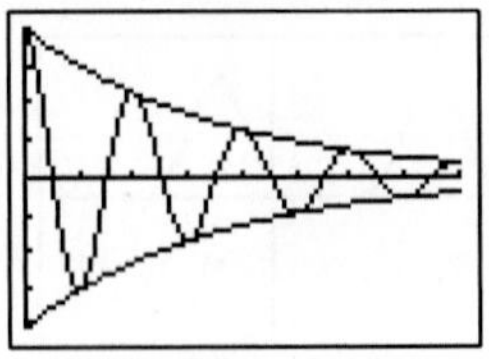

$[0, 4\pi]$ by $[-1, 1]$

51. f oscillates between $\frac{1}{x}$ and $-\frac{1}{x}$.
As $x \to \infty$, $f(x) \to 0$.

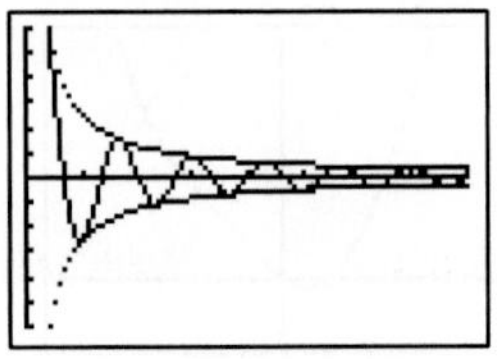

$[0, 4\pi]$ by $[-1.5, 1.5]$

53. 2π

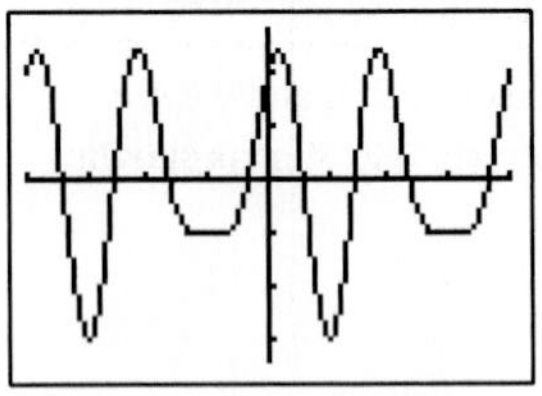

$[-2\pi, 2\pi]$ by $[-3.4, 2.8]$

55. 2π

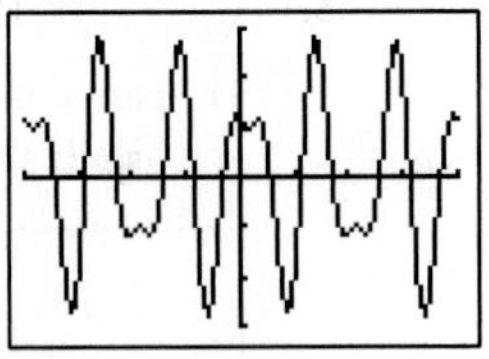

$[-2\pi, 2\pi]$ by $[-3, 3]$

57. Period 2π

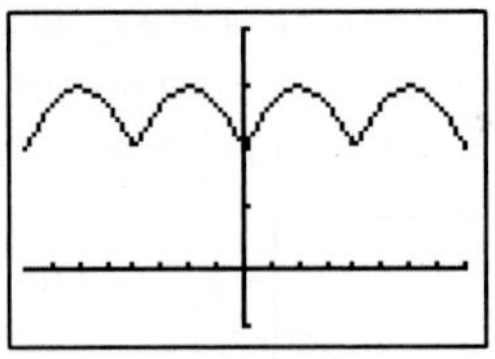

$[-4\pi, 4\pi]$ by $[-1, 4]$

59. Not periodic

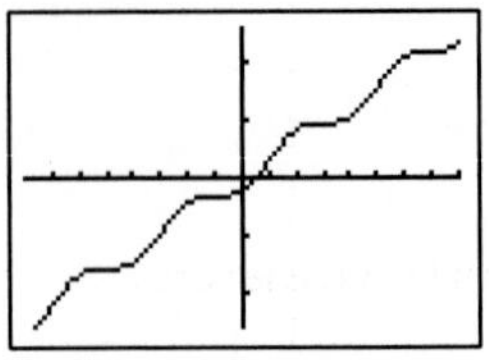

$[-4\pi, 4\pi]$ by $[-13, 13]$

61. Not periodic

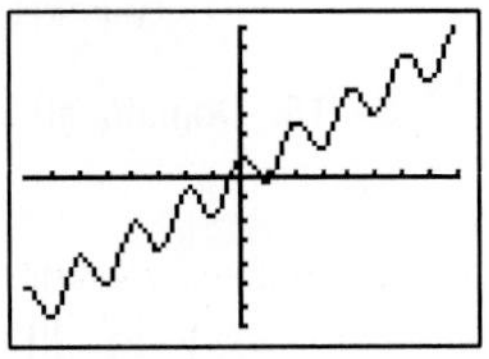

$[-4\pi, 4\pi]$ by $[-7, 7]$

63. Domain: $(-\infty, \infty)$; range: $(-\infty, \infty)$
65. Domain: $(-\infty, \infty)$; range: $[1, \infty)$
67. Domain: all real numbers x with $2n\pi \leq x \leq (2n + 1)\pi$, n an integer; range: $[0, 1]$
69. Domain: $(-\infty, \infty)$; range: $[0, 1]$

71. **(a)**

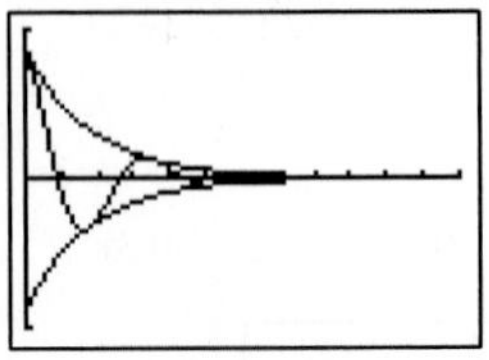

$[0, 12]$ by $[-0.5, 0.5]$

(b) For $t > 0.51$ (approximately).

73. Not periodic **75.** **(a)** **77.** Graph (d), shown on $[-2\pi, 2\pi]$ by $[-4, 4]$
79. Graph (b), shown on $[-2\pi, 2\pi]$ by $[-4, 4]$
81. False. For example, the function has a relative minimum of 0 at $x = 0$ that is not repeated anywhere else. **83.** B **85.** D
87. **(a)** Answers will vary depending on the grapher used.
(b) Period: $\pi/125 = 0.0251\ldots$. Answers will vary for the last two questions. The graph produced will be inaccurate on many graphers because multiple cycles of the true graph occur from one pixel to the next.

89. Domain: $(-\infty, \infty)$; range: $[-1, 1]$; horizontal asymptote: $y = 1$; zeros at $\ln\left(\frac{\pi}{2} + n\pi\right)$, n a nonnegative integer

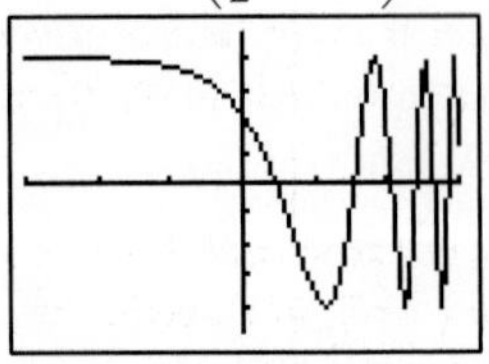

$[-3, 3]$ by $[-1.2, 1.2]$

91. Domain: $[0, \infty)$; range: $(-\infty, \infty)$; zeros at $n\pi$, n a nonnegative integer

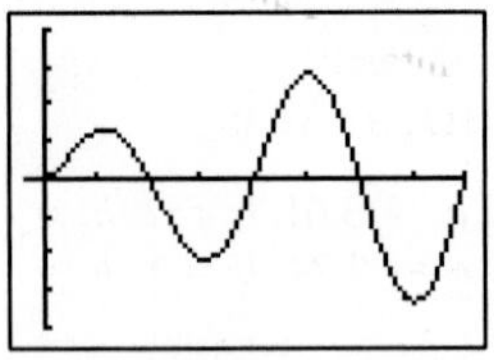

$[-0.5, 4\pi]$ by $[-4, 4]$

93. Domain: $(-\infty, 0) \cup (0, \infty)$; range: approximately $[-0.22, 1)$; horizontal asymptote: $y = 0$; zeros at $n\pi$, n a nonzero integer

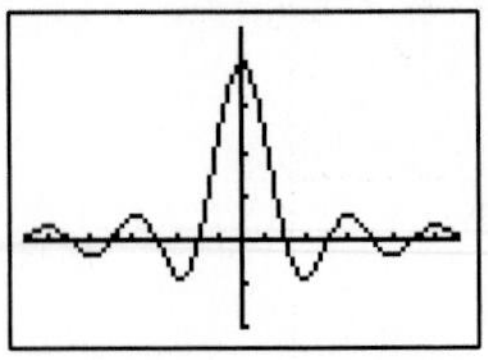

$[-5\pi, 5\pi]$ by $[-0.5, 1.2]$

95. Domain: $(-\infty, 0) \cup (0, \infty)$; range: approximately $[-0.22, 1)$; horizontal asymptote: $y = 1$; zeros at $\frac{1}{n\pi}$, n a nonzero integer

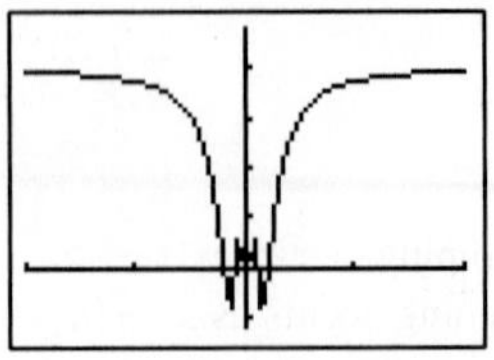

$[-\pi, \pi]$ by $[-0.3, 1.2]$

SECTION 4.7

Exploration 1

1. x **3.** $\sqrt{1 + x^2}$ **5.** $\sqrt{1 + x^2}$

Quick Review 4.7

1. $\sin x$: positive; $\cos x$: positive; $\tan x$: positive **3.** $\sin x$: negative; $\cos x$: negative; $\tan x$: positive **5.** $1/2$ **7.** $-1/2$ **9.** $-1/2$

Exercises 4.7

1. $\pi/3$ **3.** 0 **5.** $\pi/3$ **7.** $-\pi/4$ **9.** $-\pi/4$ **11.** $\pi/2$ **13.** 21.22° **15.** −85.43° **17.** 1.172 **19.** −0.478

21. $\lim_{x\to\infty} \tan^{-1}(x^2) = \frac{\pi}{2}$ and $\lim_{x\to-\infty} \tan^{-1}(x^2) = \frac{\pi}{2}$ **23.** $\sqrt{3}/2$ **25.** $\pi/4$ **27.** $1/2$ **29.** $\pi/6$ **31.** $1/2$

33. Domain: $[-1, 1]$; Range: $\left[-\frac{\pi}{2}, \frac{\pi}{2}\right]$; Continuous; Increasing; Symmetric with respect to the origin (odd); Bounded; Absolute maximum of $\frac{\pi}{2}$, absolute minimum of $-\frac{\pi}{2}$; No asymptotes; No end behavior (bounded domain)

35. Domain: $(-\infty, \infty)$; Range: $\left(-\frac{\pi}{2}, \frac{\pi}{2}\right)$; Continuous; Increasing; Symmetric with respect to the origin (odd); Bounded; No local extrema; Horizontal asymptotes: $y = \frac{\pi}{2}$ and $y = -\frac{\pi}{2}$; End behavior: $\lim_{x\to\infty} \tan^{-1} x = \frac{\pi}{2}$ and $\lim_{x\to-\infty} \tan^{-1} x = -\frac{\pi}{2}$

37. Domain: $\left[-\frac{1}{2}, \frac{1}{2}\right]$; Range: $\left[-\frac{\pi}{2}, \frac{\pi}{2}\right]$; Starting from $y = \sin^{-1} x$, horizontally shrink by $\frac{1}{2}$.

39. Domain: $(-\infty, \infty)$; Range: $\left(-\frac{5\pi}{2}, \frac{5\pi}{2}\right)$; Starting from $y = \tan^{-1} x$, horizontally stretch by 2 and vertically stretch by 5 (either order).

41. 1 **43.** $\frac{\pi}{6} + 2n\pi$ and $\frac{5\pi}{6} + 2n\pi$, for all integers n **45.** $\frac{1}{3}$ **47.** $x/\sqrt{1 + x^2}$ **49.** $x/\sqrt{1 - x^2}$ **51.** $\frac{1}{\sqrt{1 + 4x^2}}$

53. (b)

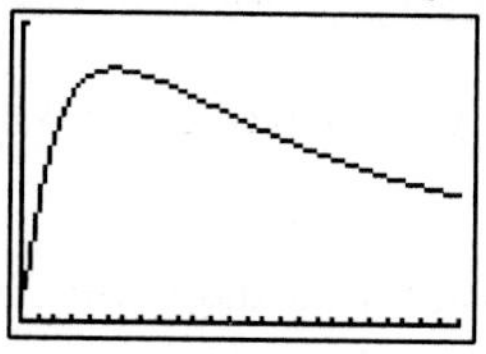

[0, 25] by [0, 55]

(c) 2 or 15 ft.

55. (a) $\theta = \tan^{-1}\frac{s}{500}$ **(b)** As s changes from 10 to 20 ft, θ changes from about 1.1458° to 2.2906°—it almost exactly doubles (a 99.92% increase). As s changes from 200 to 210 ft, θ changes from about 21.80° to 22.78°—an increase of less than 1°, and a very small relative change (only about 4.5%). **(c)** The x-axis represents the height and the y-axis represents the angle: The angle cannot grow past 90° (in fact, it *approaches* but never exactly equals 90°). **57.** False. This is only true for $-1 \le x \le 1$, the domain of $\sin^{-1} x$. **59.** E **61.** C
63. The cotangent function restricted to the interval $(0, \pi)$ is one-to-one and has an inverse. The unique angle y between 0 and π (noninclusive) such that $\cot y = x$ is called the inverse cotangent (or arccotangent) of x, denoted $\cot^{-1} x$ or $\text{arccot}\, x$. The domain of $y = \cot^{-1} x$ is $(-\infty, \infty)$ and the range is $(0, \pi)$.

65. (a) Domain all reals, range $[-\pi/2, \pi/2]$, period 2π

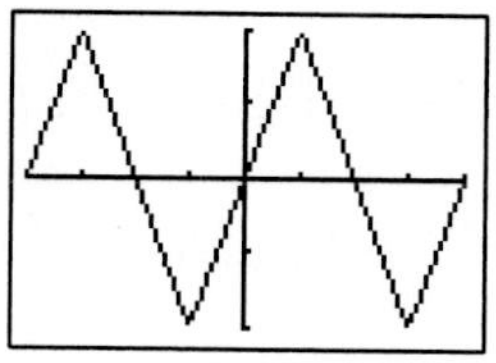

$[-2\pi, 2\pi]$ by $[-0.5\pi, 0.5\pi]$

(b) Domain all reals, range $[0, \pi]$, period 2π

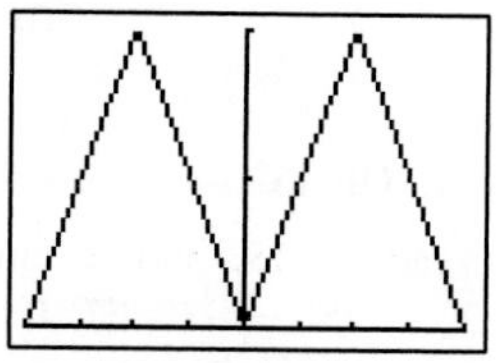

$[-2\pi, 2\pi]$ by $[0, \pi]$

(c) Domain all reals except $\pi/2 + n\pi$, n an integer, range $(-\pi/2, \pi/2)$, period π. Discontinuity is not removable.

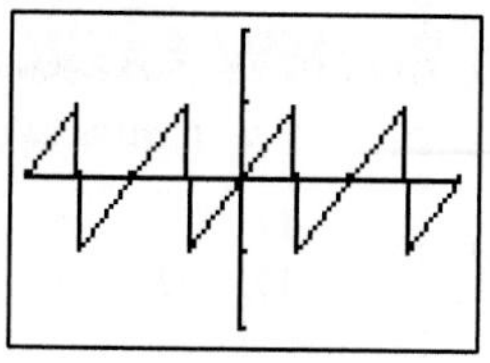

$[-2\pi, 2\pi]$ by $[-\pi, \pi]$

67. $y = \frac{\pi}{2} - \tan^{-1} x$ **69.** $\frac{18}{\pi}\tan^{-1} x + 33$

71. (a) $y = \pi/2$ **(b)** $y = \pi/2, y = 3\pi/2$ **(c)** The graph on the left **(d)** The graph on the left

SECTION 4.8

Exploration 1

1. The unit circle **3.** Since the grapher is plotting points along the unit circle, it covers the circle at a constant speed. Toward the extremes its motion is mostly vertical, so not much horizontal progress (which is all that we see) occurs. Toward the middle, the motion is mostly horizontal, so it moves faster.

Quick Review 4.8

1. $b = 15 \cot 31° \approx 24.964$, $c = 15 \csc 31° \approx 29.124$ **3.** $b = 28 \cot 28° - 28 \cot 44° \approx 23.665$, $c = 28 \csc 28° \approx 59.642$, $a = 28 \csc 44° \approx 40.308$ **5.** Complement: 58°; supplement: 148° **7.** 45° **9.** Amplitude: 3; period: π

Exercises 4.8

1. $300\sqrt{3} \approx 519.62$ ft **3.** $120 \cot 10° \approx 680.55$ ft **5.** wire length $= 50 \sec 80° \approx 288$ ft; tower height $= 50 \tan 80° \approx 284$ ft **7.** $185 \tan 80°1'12'' \approx 1051$ ft **9.** $100 \tan 83°12' \approx 839$ ft **11.** $10 \tan 55° \approx 14.3$ ft **13.** $4.25 \tan 35° \approx 2.98$ mi **15.** $200(\tan 40° - \tan 30°) \approx 52.35$ ft **17.** Distance: $60\sqrt{2} \approx 84.85$ naut mi; bearing is 140°. **19.** $1097 \cot 19° \approx 3186$ ft **21.** $375 \tan 67° \approx 883$ ft **23.** $36.5 \tan 15° \approx 9.8$ ft **25.** $\dfrac{550}{\cot 70° - \cot 80°} \approx 2931$ ft

27. (a) 8 cycles/sec **(b)** $d = 6 \cos 16\pi t$ **(c)** About 4.1 in. left of the starting position **29.** $d = 3 \cos 4\pi t$ cm
31. (a) 25 ft **(b)** 33 ft **(c)** $\pi/10$ rad/sec
33. (a) $\pi/6$ **(b)** $a = (81.5 - 49.2)/2 = 16.15$ and $k = 81.5 - 16.15 = 65.35$ **(c)** $(3 + 1 = 4)$
(d)

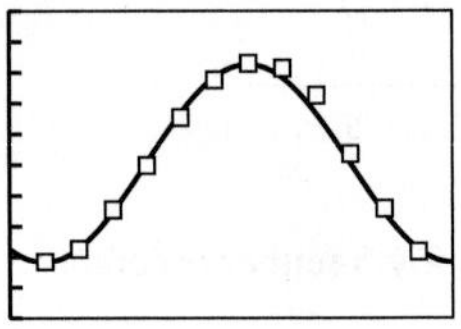

[0, 13] by [40, 90]

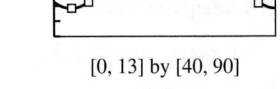

(e) Setting $16.15 \sin(\pi/6(t - 4)) + 65.35 = 70$, we get $t = 4.57$ or $t = 9.44$. These represent (approximately) days #139 and #287 of a 365-day year, namely May 19 and October 14.

35. (a) March **(b)** November **37.** True. Since the frequency and the period are reciprocals, the higher the frequency, the shorter the period.
39. D **41.** D **43. (a)**

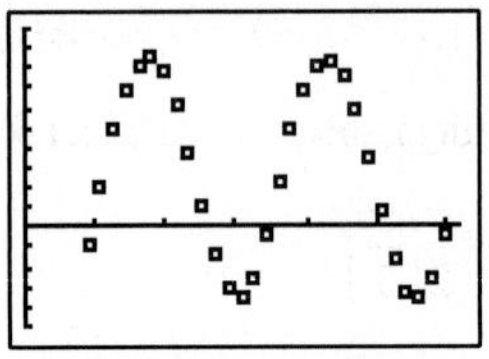

[0, 0.0062] by [–0.5, 1]

(b) The first is the best.
(c) About $\dfrac{2464}{2\pi} = \dfrac{1232}{\pi} \approx 392$ oscillations/sec.

45. $2.5 \cot \dfrac{\pi}{7} \approx 5.2$ cm **47.** $AC \approx 33.6$ in.; $BD \approx 12.9$ in **49.** $\tan^{-1} 0.06 \approx 3.4°$

51. (a)

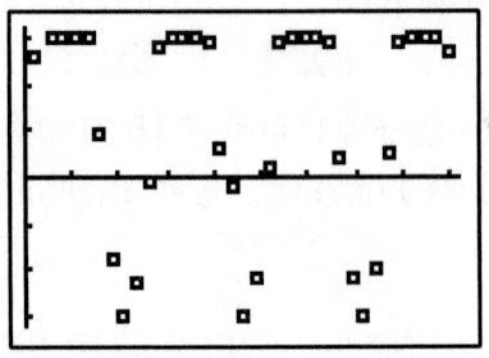

[0, 0.0092] by [–1.6, 1.6]

(b) One pretty good match is $y = 1.51971 \sin[2467(t - 0.0002)]$ (that is, $a = 1.51971$, $b = 2467$, $h = 0.0002$). Answers will vary but should be close to these values.
(c) Frequency: about $\dfrac{2467}{2\pi} \approx 393$ Hz; It appears to be a G.
(d) G

CHAPTER 4 REVIEW EXERCISES

1. positive y-axis; 450° **3.** QIII; $-3\pi/4$ **5.** QI; $13\pi/30$ **7.** QI; 15°
9. 270° or $\dfrac{3\pi}{2}$ radians

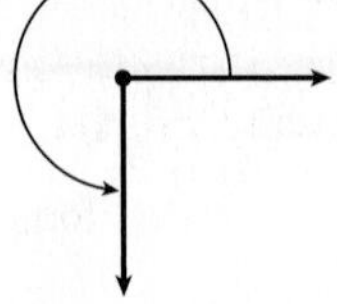

11. $30° = \pi/6$ rad **13.** $120° = 2\pi/3$ rad **15.** $360° + \tan^{-1}(-2) \approx 296.565° \approx 5.176$ radians
17. 1/2 **19.** 1 **21.** 1/2 **23.** 2 **25.** -1 **27.** 0

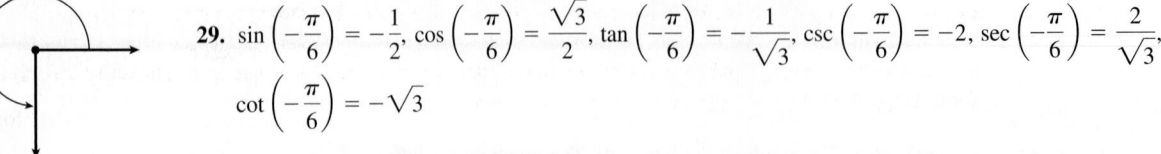

29. $\sin\left(-\dfrac{\pi}{6}\right) = -\dfrac{1}{2}$, $\cos\left(-\dfrac{\pi}{6}\right) = \dfrac{\sqrt{3}}{2}$, $\tan\left(-\dfrac{\pi}{6}\right) = -\dfrac{1}{\sqrt{3}}$, $\csc\left(-\dfrac{\pi}{6}\right) = -2$, $\sec\left(-\dfrac{\pi}{6}\right) = \dfrac{2}{\sqrt{3}}$, $\cot\left(-\dfrac{\pi}{6}\right) = -\sqrt{3}$

31. $\sin(-135°) = -\frac{1}{\sqrt{2}}$, $\cos(-135°) = -\frac{1}{\sqrt{2}}$, $\tan(-135°) = 1$, $\csc(-135°) = -\sqrt{2}$, $\sec(-135°) = -\sqrt{2}$, $\cot(-135°) = 1$

33. $\sin\alpha = \frac{5}{13}$, $\cos\alpha = \frac{12}{13}$, $\tan\alpha = \frac{5}{12}$, $\csc\alpha = \frac{13}{5}$, $\sec\alpha = \frac{13}{12}$, $\cot\alpha = \frac{12}{5}$ **35.** $\sin\theta = \frac{15}{17}$, $\cos\theta = \frac{8}{17}$, $\tan\theta = \frac{15}{8}$, $\csc\theta = \frac{17}{15}$, $\sec\theta = \frac{17}{8}$, $\cot\theta = \frac{8}{15}$ **37.** ≈ 4.075 rad **39.** $a = 15\sin 35° \approx 8.604$, $b = 15\cos 35° \approx 12.287$, $\beta = 55°$

41. $b = 7\tan 48° \approx 7.774$, $c = \frac{7}{\cos 48°} \approx 10.461$, $\alpha = 42°$ **43.** $a = 2\sqrt{6} \approx 4.90$, $\alpha \approx 44.42°$, $\beta \approx 45.58°$

45. QIII **47.** QII **49.** $\sin\theta = \frac{2}{\sqrt{5}}$, $\cos\theta = -\frac{1}{\sqrt{5}}$, $\tan\theta = -2$, $\csc\theta = \frac{\sqrt{5}}{2}$, $\sec\theta = -\sqrt{5}$, $\cot\theta = -\frac{1}{2}$

51. $\sin\theta = -\frac{3}{\sqrt{34}}$, $\cos\theta = -\frac{5}{\sqrt{34}}$, $\tan\theta = \frac{3}{5}$, $\csc\theta = -\frac{\sqrt{34}}{3}$, $\sec\theta = -\frac{\sqrt{34}}{5}$, $\cot\theta = \frac{5}{3}$

53. Starting from $y = \sin x$, translate left π units.

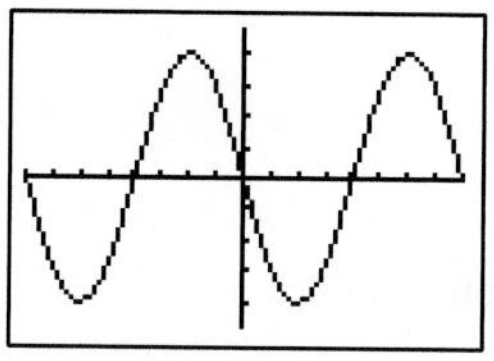

$[-2\pi, 2\pi]$ by $[-1.2, 1.2]$

55. Starting from $y = \cos x$, translate left $\frac{\pi}{2}$ units, reflect across x-axis, and translate up 4 units.

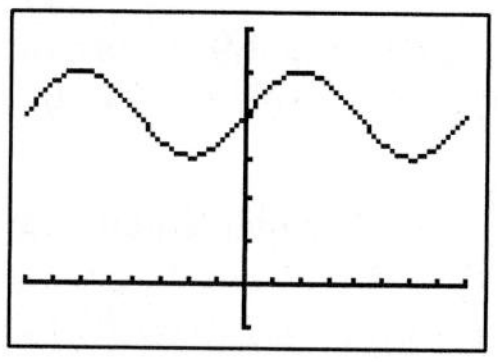

$[-2\pi, 2\pi]$ by $[-1, 6]$

57. Starting from $y = \tan x$, horizontally shrink by $\frac{1}{2}$.

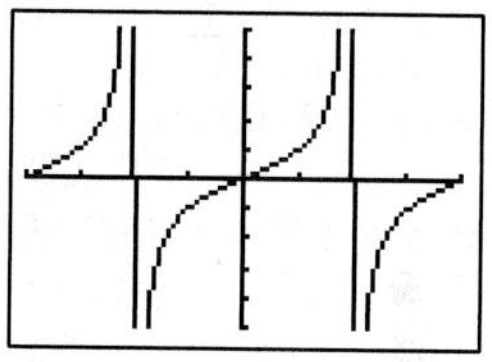

$[-0.5\pi, 0.5\pi]$ by $[-5, 5]$

59. Starting from $y = \sec x$, horizontally stretch by 2, vertically stretch by 2, and reflect across x-axis (in any order).

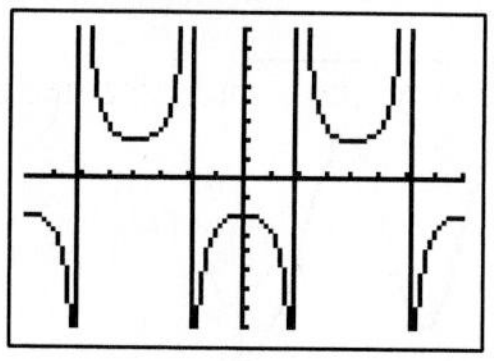

$[-4\pi, 4\pi]$ by $[-8, 8]$

61. Amplitude: 2; period: $\frac{2\pi}{3}$; phase shift: 0; domain: $(-\infty, \infty)$; range: $[-2, 2]$

63. Amplitude: 1.5; period: π; phase shift: $\frac{\pi}{8}$; domain: $(-\infty, \infty)$; range: $[-1.5, 1.5]$

65. Amplitude: 4; period: π; phase shift: $\frac{1}{2}$; domain: $(-\infty, \infty)$; range: $[-4, 4]$

67. $a \approx 4.47$, $b = 1$, and $h \approx 1.11$ **69.** $\approx 49.996° \approx 0.873$ radians

71. $45° = \pi/4$ rad **73.** Starting from $y = \sin^{-1} x$, horizontally shrink by $\frac{1}{3}$. Domain: $\left[-\frac{1}{3}, \frac{1}{3}\right]$; range: $\left[-\frac{\pi}{2}, \frac{\pi}{2}\right]$

75. Starting from $y = \sin^{-1} x$, translate right 1 unit, horizontally shrink by $\frac{1}{3}$, translate up 2 units. Domain: $\left[0, \frac{2}{3}\right]$; range: $\left[2 - \frac{\pi}{2}, 2 + \frac{\pi}{2}\right]$

77. $5\pi/6$ **79.** $3\pi/4$ **81.** $3\pi/2$ **83.** As $|x| \to \infty$, $\frac{\sin x}{x^2} \to 0$. **85.** 1 **87.** 3/4 **89.** Periodic; period: π; domain: $x \neq \frac{\pi}{2} + n\pi$, n an integer; range: $[1, \infty)$ **91.** Not periodic; domain: $x \neq \frac{\pi}{2} + n\pi$, n an integer; range: $(-\infty, \infty)$ **93.** $4\pi/3$ **95.** $100\tan 78° \approx 470$ m

97. $150(\cot 18° - \cot 42°) \approx 295$ ft **99.**

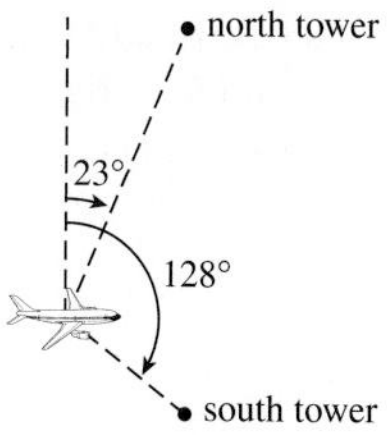

101. $62\tan 72°24' \approx 195.4$ ft **103.** $22\pi/15 \approx 4.6$ in.
105. About 139 days, from day 128 (May 8) through day 266 (Sept. 23).

PROJECT

Answers are based on the sample data shown in the table.

1.

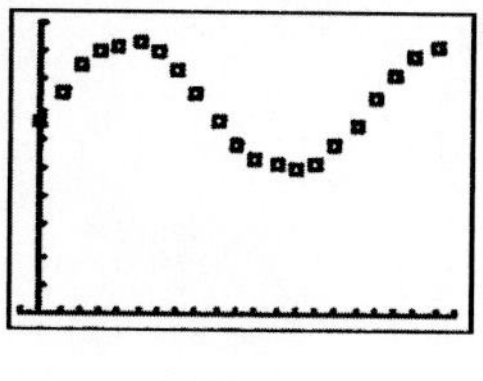

3. The constant a represents half the distance the pendulum bob swings as it moves from its highest point to its lowest point; k represents the distance from the detector to the pendulum bob when it is in midswing.

5. $y \approx 0.22\sin(3.87x - 0.16) + 0.71$; Most calculator/computer regression models are expressed in the form $y = a\sin(bx + p) + k$, where $-p/b = h$ in the equation $y = a\sin(b(x - h)) + k$. The latter equation form differs from $y = a\cos(b(x - h)) + k$ only in h.

SECTION 5.1

Exploration 1

1. $\cos\theta = \dfrac{1}{\sec\theta}$, $\sec\theta = \dfrac{1}{\cos\theta}$, and $\tan\theta = \dfrac{\sin\theta}{\cos\theta}$ **3.** $\csc\theta = \dfrac{1}{\sin\theta}$, $\cot\theta = \dfrac{1}{\tan\theta}$, and $\cot\theta = \dfrac{\cos\theta}{\sin\theta}$

Quick Review 5.1

1. 1.1760 rad = 67.380° **3.** 2.4981 rad = 143.130° **5.** $(a - b)^2$ **7.** $(2x + y)(x - 2y)$ **9.** $\dfrac{y - 2x}{xy}$ **11.** xy

Exercises 5.1

1. $\sin\theta = 3/5$ and $\cos\theta = 4/5$ **3.** $\tan\theta = -\sqrt{15}$ and $\cot\theta = -1/\sqrt{15} = -\sqrt{15}/15$ **5.** 0.45 **7.** −0.73 **9.** $\sin x$
11. 1 **13.** $\tan^2 x$ **15.** $\cos x \sin^2 x$ **17.** −1 **19.** −1 **21.** 1 **23.** $\cos x$ **25.** 2 **27.** $\sec x$ **29.** $\tan x$
31. $\tan x$ **33.** $2\csc^2 x$ **35.** $-\sin x$ **37.** $\cot x$ **39.** $(\cos x + 1)^2$ **41.** $(1 - \sin x)^2$ **43.** $(2\cos x - 1)(\cos x + 1)$
45. $(2\tan x - 1)^2$ **47.** $1 - \sin x$ **49.** $1 - \cos x$ **51.** $\pi/6, \pi/2, 5\pi/6, 3\pi/2$ **53.** $0, \pi$ **55.** $\pi/3, 2\pi/3, 4\pi/3, 5\pi/3$

57. $\pm\dfrac{\pi}{3} + 2n\pi,\ n = 0, \pm 1, \pm 2, \ldots$ **59.** $n\pi,\ n = 0, \pm 1, \pm 2, \ldots$ **61.** $n\pi,\ n = 0, \pm 1, \pm 2, \ldots$

63. $\{\pm 1.1918 + 2n\pi \mid n = 0, \pm 1, \pm 2, \ldots\}$ **65.** $\{0.3047 + 2n\pi$ or $.8369 + 2n\pi \mid n = 0, \pm 1, \pm 2, \ldots\}$
67. $\{\pm 0.8861 + n\pi \mid n = 0, \pm 1, \pm 2, \ldots\}$ **69.** $|\sin\theta|$ **71.** $3\,|\tan\theta|$ **73.** $9\,|\sec\theta|$

75. True. Since secant is an even function, $\sec\left(x - \dfrac{\pi}{2}\right) = \sec\left(\dfrac{\pi}{2} - x\right)$, which equals $\csc x$ by one of the cofunction identities.

77. D **79.** C

81. $\sin x$, $\cos x = \pm\sqrt{1 - \sin^2 x}$, $\tan x = \pm\dfrac{\sin x}{\sqrt{1 - \sin^2 x}}$, $\csc x = \dfrac{1}{\sin x}$, $\sec x = \pm\dfrac{1}{\sqrt{1 - \sin^2 x}}$, $\cot x = \pm\dfrac{\sqrt{1 - \sin^2 x}}{\sin x}$

83. The two functions are parallel to each other, separated by 1 unit for every x. At any x, the distance between the two graphs is $\sin^2 x - (-\cos^2 x) = \sin^2 x + \cos^2 x = 1$.

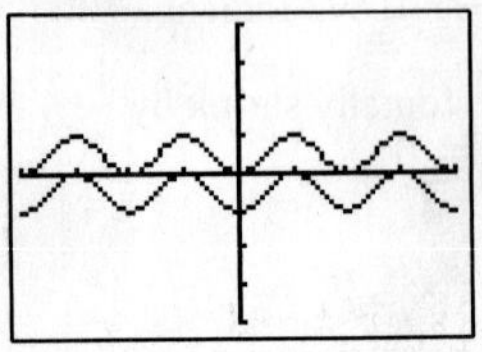

$[-2\pi, 2\pi]$ by $[-4, 4]$

85. (a)

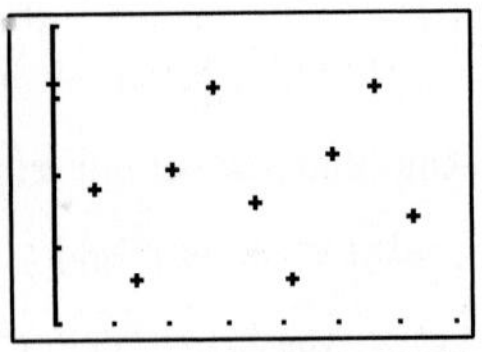

[−6, 70] by [220000, 260000]

(b) The equation is $y = 13{,}111 \sin(0.22997x + 1.571) + 238{,}855$.

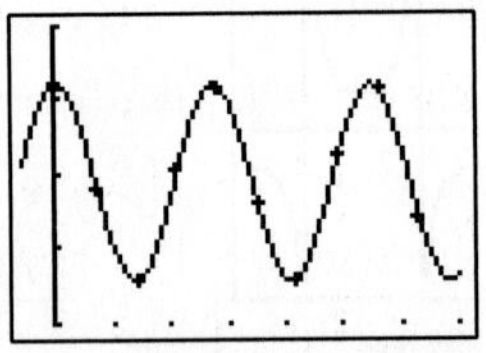

[−6, 70] by [220000, 260000]

(c) $(2\pi)/0.22998 \approx 27.32$ days. This is the number of days that it takes the moon to make one complete orbit of the Earth (known as the moon's sidereal period). **(d)** 225,744 mi **(e)** $y = 13{,}111\cos(-0.22997x) + 238855$, or $y = 13{,}111\cos(0.22997x) + 238855$.

87. Factor the left-hand side: $\sin^4\theta - \cos^4\theta = (\sin^2\theta - \cos^2\theta)(\sin^2\theta + \cos^2\theta) = (\sin^2\theta - \cos^2\theta)\cdot 1 = \sin^2\theta - \cos^2\theta$

89. Use the hint: $\sin(\pi - x) = \sin(\pi/2 - (x - \pi/2))$
$= \cos(x - \pi/2)$ Cofunction identity
$= \cos(\pi/2 - x)$ Since cos is even
$= \sin x$ Cofunction identity

91. Since A, B, and C are angles of a triangle, $A + B = \pi - C$. So: $\sin(A + B) = \sin(\pi - C) = \sin C$.

SECTION 5.2

Exploration 1

1. The graphs lead us to conclude that this is not an identity.

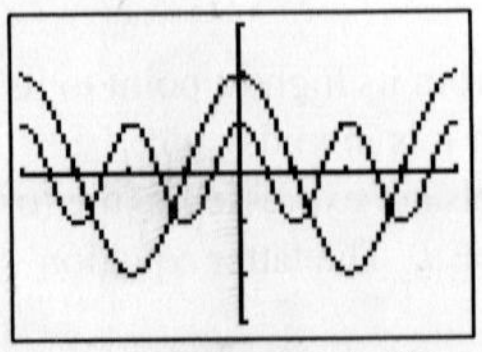

$[-2\pi, 2\pi]$ by $[-4, 4]$

3. Yes **5.** No. The graph window cannot show the full graphs, so they could differ outside the viewing window. Also, the function values could be so close that the graphs *appear* to coincide.

Quick Review 5.2

1. $\dfrac{\sin x + \cos x}{\sin x \cos x}$ **3.** $\dfrac{1}{\sin x \cos x}$ **5.** 1 **7.** No. Any negative x. **9.** No. Any x for which $\sin x < 0$, e.g., $x = -\pi/2$. **11.** Yes

Exercises 5.2

1. One possible proof: $\dfrac{x^3 - x^2}{x} - (x - 1)(x + 1) = \dfrac{x(x^2 - x)}{x} - (x^2 - 1) = x^2 - x - (x^2 - 1) = -x + 1 = 1 - x$

3. One possible proof: $\dfrac{x^2 - 4}{x - 2} - \dfrac{x^2 - 9}{x + 3} = \dfrac{(x + 2)(x - 2)}{x - 2} - \dfrac{(x + 3)(x - 3)}{x + 3} = x + 2 - (x - 3) = 5$ **5.** Yes **7.** No **9.** Yes

11. $(\cos x)(\tan x + \sin x \cot x) = \cos x \cdot \dfrac{\sin x}{\cos x} + \cos x \sin x \cdot \dfrac{\cos x}{\sin x} = \sin x + \cos^2 x$

13. $(1 - \tan x)^2 = 1 - 2\tan x + \tan^2 x = (1 + \tan^2 x) - 2\tan x = \sec^2 x - 2\tan x$

15. $\dfrac{(1 - \cos u)(1 + \cos u)}{\cos^2 u} = \dfrac{1 - \cos^2 u}{\cos^2 u} = \dfrac{\sin^2 u}{\cos^2 u} = \tan^2 u$ **17.** $\dfrac{\cos^2 x - 1}{\cos x} = \dfrac{-\sin^2 x}{\cos x} = -\dfrac{\sin x}{\cos x} \sin x = -\tan x \sin x$

19. Multiply out the expression on the left side.

21. $(\cos t - \sin t)^2 + (\cos t + \sin t)^2 = \cos^2 t - 2\cos t \sin t + \sin^2 t + \cos^2 t + 2\cos t \sin t + \sin^2 t = 2\cos^2 t + 2\sin^2 t = 2$

23. $\dfrac{1 + \tan^2 x}{\sin^2 x + \cos^2 x} = \dfrac{\sec^2 x}{1} = \sec^2 x$ **25.** $\dfrac{\cos \beta}{1 + \sin \beta} = \dfrac{\cos^2 \beta}{\cos \beta(1 + \sin \beta)} = \dfrac{1 - \sin^2 \beta}{\cos \beta(1 + \sin \beta)} = \dfrac{(1 - \sin \beta)(1 + \sin \beta)}{\cos \beta(1 + \sin \beta)} = \dfrac{1 - \sin \beta}{\cos \beta}$

27. $\dfrac{\tan^2 x}{\sec x + 1} = \dfrac{\sec^2 x - 1}{\sec x + 1} = \sec x - 1 = \dfrac{1}{\cos x} - 1 = \dfrac{1 - \cos x}{\cos x}$

29. $\cot^2 x - \cos^2 x = \left(\dfrac{\cos x}{\sin x}\right)^2 - \cos^2 x = \dfrac{\cos^2 x(1 - \sin^2 x)}{\sin^2 x} = \cos^2 x \cdot \dfrac{\cos^2 x}{\sin^2 x} = \cos^2 x \cot^2 x$

31. $\cos^4 x - \sin^4 x = (\cos^2 x + \sin^2 x)(\cos^2 x - \sin^2 x) = 1(\cos^2 x - \sin^2 x) = \cos^2 x - \sin^2 x$

33. $(x \sin \alpha + y \cos \alpha)^2 + (x \cos \alpha - y \sin \alpha)^2 = (x^2 \sin^2 \alpha + 2xy \sin \alpha \cos \alpha + y^2 \cos^2 \alpha) + (x^2 \cos^2 \alpha - 2xy \cos \alpha \sin \alpha$
$+ y^2 \sin^2 \alpha) = x^2 \sin^2 \alpha + y^2 \cos^2 \alpha + x^2 \cos^2 \alpha + y^2 \sin^2 \alpha = (x^2 + y^2)(\sin^2 \alpha + \cos^2 \alpha) = x^2 + y^2$

35. $\dfrac{\tan x}{\sec x - 1} = \dfrac{\tan x(\sec x + 1)}{\sec^2 x + 1} = \dfrac{\tan x(\sec x + 1)}{\tan^2 x} = \dfrac{\sec x + 1}{\tan x}$. See also #26.

37. $\dfrac{\sin x - \cos x}{\sin x + \cos x} = \dfrac{(\sin x - \cos x)(\sin x + \cos x)}{(\sin x + \cos x)^2} = \dfrac{\sin^2 x - \cos^2 x}{\sin^2 x + 2\sin x \cos x + \cos^2 x} = \dfrac{\sin^2 x - (1 - \sin^2 x)}{1 + 2\sin x \cos x} = \dfrac{2\sin^2 x - 1}{1 + 2\sin x \cos x}$

39. $\dfrac{\sin t}{1 - \cos t} + \dfrac{1 + \cos t}{\sin t} = \dfrac{\sin^2 t + (1 + \cos t)(1 - \cos t)}{(\sin t)(1 - \cos t)} = \dfrac{1 - \cos^2 t + 1 - \cos^2 t}{(\sin t)(1 - \cos t)} = \dfrac{2(1 - \cos^2 t)}{(\sin t)(1 - \cos t)} = \dfrac{2(1 + \cos t)}{\sin t}$

41. $\sin^2 x \cos^3 x = \sin^2 x \cos^2 x \cos x = \sin^2 x(1 - \sin^2 x)(\cos x) = (\sin^2 x - \sin^4 x)(\cos x)$

43. $\cos^5 x = \cos^4 x \cos x = (\cos^2 x)^2(\cos x) = (1 - \sin^2 x)^2 (\cos x) = (1 - 2\sin^2 x + \sin^4 x)(\cos x)$

45. $\dfrac{\tan x}{1 - \cot x} + \dfrac{\cot x}{1 - \tan x} = \dfrac{\tan x}{1 - \cot x} \cdot \dfrac{\sin x}{\sin x} + \dfrac{\cot x}{1 - \tan x} \cdot \dfrac{\cos x}{\cos x} = \left(\dfrac{\sin^2 x/\cos x}{\sin x - \cos x} + \dfrac{\cos^2 x/\sin x}{\cos x - \sin x}\right)\dfrac{\sin x \cos x}{\sin x \cos x}$

$= \dfrac{\sin^3 x - \cos^3 x}{\sin x \cos x(\sin x - \cos x)} = \dfrac{\sin^2 x + \sin x \cos x + \cos^2 x}{\sin x \cos x} = \dfrac{1 + \sin x \cos x}{\sin x \cos x} = \dfrac{1}{\sin x \cos x} + 1 = \csc x \sec x + 1$

47. $\dfrac{2\tan x}{1 - \tan^2 x} + \dfrac{1}{2\cos^2 x - 1} = \dfrac{2\tan x}{1 - \tan^2 x} \cdot \dfrac{\cos^2 x}{\cos^2 x} + \dfrac{1}{\cos^2 x - \sin^2 x} = \dfrac{2\sin x \cos x}{\cos^2 x - \sin^2 x} + \dfrac{\cos^2 x + \sin^2 x}{\cos^2 x - \sin^2 x}$

$= \dfrac{2\sin x \cos x + \cos^2 x + \sin^2 x}{(\cos x - \sin x)(\cos x + \sin x)} = \dfrac{(\cos x + \sin x)^2}{(\cos x - \sin x)(\cos x + \sin x)} = \dfrac{\cos x + \sin x}{\cos x - \sin x}$

49. $\cos^3 x = (\cos^2 x)(\cos x) = (1 - \sin^2 x)(\cos x)$

51. $\sin^5 x = (\sin^4 x)(\sin x) = (\sin^2 x)^2(\sin x) = (1 - \cos^2 x)^2(\sin x) = (1 - 2\cos^2 x + \cos^4 x)(\sin x)$ **53.** (d) **55.** (c) **57.** (b)

59. True. If x is in the domain of both sides of the equation, then $x \geq 0$. The equation holds for all $x \geq 0$, so it is an identity.

61. E **63.** B **65.** $\sin x$ **67.** 1 **69.** 1

71. If A and B are complementary angles, then $\sin^2 A + \sin^2 B = \sin^2 A + \sin^2(\pi/2 - A) = \sin^2 A + \cos^2 A = 1$.

73. Multiply and divide by $1 - \sin t$ under the radical: $\sqrt{\dfrac{1-\sin t}{1+\sin t}\cdot\dfrac{1-\sin t}{1-\sin t}} = \sqrt{\dfrac{(1-\sin t)^2}{1-\sin^2 t}} = \sqrt{\dfrac{(1-\sin t)^2}{\cos^2 t}} = \dfrac{|1-\sin t|}{|\cos t|}$ since $\sqrt{a^2} = |a|$. Now, since $1 - \sin t \geq 0$, we can dispense with the absolute value in the numerator, but it must stay in the denominator.

75. $\sin^6 x + \cos^6 x = (\sin^2 x)^3 + \cos^6 x = (1 - \cos^2 x)^3 + \cos^6 x = (1 - 3\cos^2 x + 3\cos^4 x - \cos^6 x) + \cos^6 x$
$= 1 - 3\cos^2 x(1 - \cos^2 x) = 1 - 3\cos^2 x \sin^2 x$

77. $\ln|\tan x| = \ln\dfrac{|\sin x|}{|\cos x|} = \ln|\sin x| - \ln|\cos x|$

79. **(a)** They are not equal. Shown is the window $[-2\pi, 2\pi]$ by $[-2, 2]$; graphing on nearly any viewing window does not show any apparent difference—but using TRACE, one finds that the y-coordinates are not identical. Likewise, a table of values will show slight differences; for example, when $x = 1$, $y_1 = 0.53988$ and $y_2 = 0.54030$.

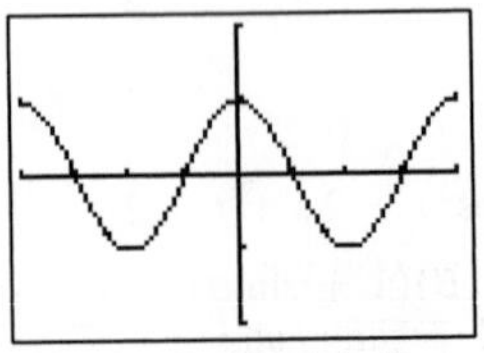

$[-2\pi, 2\pi]$ by $[-2, 2]$

(b) One choice for h is 0.001 (shown). The function y_3 is a combination of three sinusoidal functions ($1000 \sin(x + 0.001)$, $1000 \sin x$, and $\cos x$), all with period 2π.

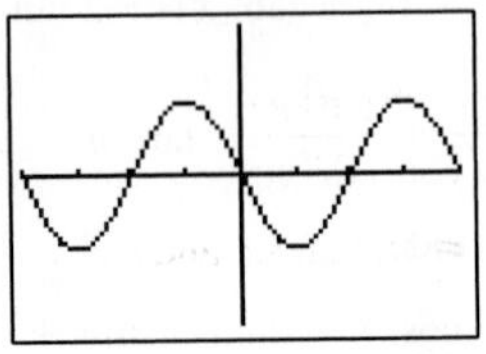

$[-2\pi, 2\pi]$ by $[-0.001, 0.001]$

81. In the decimal window, the x-coordinates used to plot the graph on the calculator are (e.g.) 0, 0.1, 0.2, 0.3, etc.—that is, $x = n/10$, where n is an integer. Then $10\pi x = \pi n$, and the sine of integer multiples of π is 0; therefore, $\cos x + \sin 10\pi x = \cos x + \sin \pi n = \cos x + 0 = \cos x$. However, for other choices of x, such as $x = \dfrac{1}{\pi}$, we have $\cos x + \sin 10\pi x = \cos x + \sin 10 \neq \cos x$.

SECTION 5.3

Exploration 1

1. No **3.** $\tan\left(\dfrac{\pi}{3} + \dfrac{\pi}{3}\right) = -\sqrt{3}$, $\tan\dfrac{\pi}{3} + \tan\dfrac{\pi}{3} = 2\sqrt{3}$. (Many other answers are possible.)

Quick Review 5.3

1. $45° - 30°$ **3.** $210° - 45°$ **5.** $2\pi/3 - \pi/4$ **7.** No **9.** Yes

Exercises 5.3

1. $(\sqrt{6} - \sqrt{2})/4$ **3.** $(\sqrt{6} + \sqrt{2})/4$ **5.** $(\sqrt{2} + \sqrt{6})/4$ **7.** $2 + \sqrt{3}$ **9.** $(\sqrt{2} - \sqrt{6})/4$ **11.** $\sin 25°$ **13.** $\sin 7\pi/10$

15. $\tan 66°$ **17.** $\cos\left(x - \dfrac{\pi}{7}\right)$ **19.** $\sin 2x$ **21.** $\tan(2y + 3x)$

23. $\sin\left(x - \dfrac{\pi}{2}\right) = \sin x \cos\dfrac{\pi}{2} - \cos x \sin\dfrac{\pi}{2} = \sin x \cdot 0 - \cos x \cdot 1 = -\cos x$

25. $\cos\left(x - \dfrac{\pi}{2}\right) = \cos x \cos\dfrac{\pi}{2} + \sin x \sin\dfrac{\pi}{2} = \cos x \cdot 0 + \sin x \cdot 1 = \sin x$

27. $\sin\left(x + \dfrac{\pi}{6}\right) = \sin x \cos\dfrac{\pi}{6} + \cos x \sin\dfrac{\pi}{6} = \sin x \cdot \dfrac{\sqrt{3}}{2} + \cos x \cdot \dfrac{1}{2}$

29. $\tan\left(\theta + \dfrac{\pi}{4}\right) = \dfrac{\tan\theta + \tan(\pi/4)}{1 - \tan\theta\tan(\pi/4)} = \dfrac{\tan\theta + 1}{1 - \tan\theta \cdot 1} = \dfrac{1 + \tan\theta}{1 - \tan\theta}$

31. Equations (b) and (f) **33.** Equations (d) and (h). **35.** $x = n\pi, n = 0, \pm 1, \pm 2, \ldots$

37. $\sin\left(\dfrac{\pi}{2} - u\right) = \sin\dfrac{\pi}{2}\cos u - \cos\dfrac{\pi}{2}\sin u = 1 \cdot \cos u - 0 \cdot \sin u = \cos u$

39. $\cot\left(\dfrac{\pi}{2} - u\right) = \dfrac{\cos(\pi/2 - u)}{\sin(\pi/2 - u)} = \dfrac{\sin u}{\cos u} = \tan u$, using the first two cofunction identities.

41. $\csc\left(\frac{\pi}{2} - u\right) = \frac{1}{\sin(\pi/2 - u)} = \frac{1}{\cos u} = \sec u$, using the second cofunction identity.

43. $y \approx 5\sin(x + 0.9273)$ **45.** $y \approx 2.236\sin(3x + 0.4636)$

47. $\sin(x - y) + \sin(x + y) = (\sin x\cos y - \cos x\sin y) + (\sin x\cos y + \cos x\sin y) = 2\sin x\cos y$

49. $\cos 3x = \cos[(x + x) + x] = \cos(x + x)\cos x - \sin(x + x)\sin x = (\cos x\cos x - \sin x\sin x)\cos x$
$- (\sin x\cos x + \cos x\sin x)\sin x = \cos^3 x - \sin^2 x\cos x - 2\cos x\sin^2 x = \cos^3 x - 3\sin^2 x\cos x$

51. $\cos 3x + \cos x = \cos(2x + x) + \cos(2x - x)$; use Exercise 48 with x replaced with $2x$ and y replaced with x.

53. $\tan(x + y)\tan(x - y) = \left(\frac{\tan x + \tan y}{1 - \tan x\tan y}\right)\cdot\left(\frac{\tan x - \tan y}{1 + \tan x\tan y}\right) = \frac{\tan^2 x - \tan^2 y}{1 - \tan^2 x\tan^2 y}$ since both the numerator and denominator are factored forms for differences of squares.

55. $\frac{\sin(x + y)}{\sin(x - y)} = \frac{\sin x\cos y + \cos x\sin y}{\sin x\cos y - \cos x\sin y} = \frac{\sin x\cos y + \cos x\sin y}{\sin x\cos y - \cos x\sin y}\cdot\frac{1/(\cos x\cos y)}{1/(\cos x\cos y)}$
$= \frac{(\sin x\cos y)/(\cos x\cos y) + (\cos x\sin y)/(\cos x\cos y)}{(\sin x\cos y)/(\cos x\cos y) - (\cos x\sin y)/(\cos x\cos y)} = \frac{(\sin x/\cos x) + (\sin y/\cos y)}{(\sin x/\cos x) - (\sin y/\cos y)} = \frac{\tan x + \tan y}{\tan x - \tan y}$

57. False. For example, $\cos 3\pi + \cos 4\pi = 0$, but 3π and 4π are not supplementary. **59.** A **61.** B

63. $\tan(u - v) = \frac{\sin(u - v)}{\cos(u - v)} = \frac{\sin u\cos v - \cos u\sin v}{\cos u\cos v + \sin u\sin v} = \frac{\frac{\sin u\cos v}{\cos u\cos v} - \frac{\cos u\sin v}{\cos u\cos v}}{\frac{\cos u\cos v}{\cos u\cos v} + \frac{\sin u\sin v}{\cos u\cos v}} = \frac{\frac{\sin u}{\cos u} - \frac{\sin v}{\cos v}}{1 + \frac{\sin u\sin v}{\cos u\cos v}} = \frac{\tan u - \tan v}{1 + \tan u\tan v}$

65. The identity would involve $\tan\left(\frac{3\pi}{2}\right)$, which does not exist. $\tan\left(x - \frac{3\pi}{2}\right) = \frac{\sin\left(x - \frac{3\pi}{2}\right)}{\cos\left(x - \frac{3\pi}{2}\right)} = \frac{\sin x\cos\frac{3\pi}{2} - \cos x\sin\frac{3\pi}{2}}{\cos x\cos\frac{3\pi}{2} + \sin x\sin\frac{3\pi}{2}}$
$= \frac{\sin x\cdot 0 - \cos x\cdot(-1)}{\cos x\cdot 0 + \sin x\cdot(-1)} = -\cot x$

67. $\frac{\cos(x + h) - \cos x}{h} = \frac{\cos x\cos h - \sin x\sin h - \cos x}{h} = \frac{\cos x(\cos h - 1) - \sin x\sin h}{h} = \cos x\left(\frac{\cos h - 1}{h}\right) - \sin x\frac{\sin h}{h}$

69. $\sin(A + B) = \sin(\pi - C) = \sin\pi\cos C - \cos\pi\sin C = 0\cdot\cos C - (-1)\sin C = \sin C$

71. $\tan A + \tan B + \tan C = \frac{\sin A}{\cos A} + \frac{\sin B}{\cos B} + \frac{\sin C}{\cos C} = \frac{\sin A(\cos B\cos C) + \sin B(\cos A\cos C)}{\cos A\cos B\cos C} + \frac{\sin C(\cos A\cos B)}{\cos A\cos B\cos C}$
$= \frac{\cos C(\sin A\cos B + \cos A\sin B) + \sin C(\cos A\cos B)}{\cos A\cos B\cos C} = \frac{\cos C\sin(A + B) + \sin C(\cos(A + B) + \sin A\sin B)}{\cos A\cos B\cos C}$
$= \frac{\cos C\sin(\pi - C) + \sin C(\cos(\pi - C) + \sin A\sin B)}{\cos A\cos B\cos C} = \frac{\cos C\sin C + \sin C(-\cos C) + \sin C\sin A\sin B}{\cos A\cos B\cos C}$
$= \frac{\sin A\sin B\sin C}{\cos A\cos B\cos C} = \tan A\tan B\tan C$

73. This equation is easier to deal with after rewriting it as $\cos 5x\cos 4x + \sin 5x\sin 4x = 0$. The left side of this equation is the expanded form of $\cos(5x - 4x)$, which of course equals $\cos x$; the graph shown is simply $y = \cos x$. The equation $\cos x = 0$ is easily solved on the interval $[-2\pi, 2\pi]$: $x = \pm\frac{\pi}{2}$ or $x = \pm\frac{3\pi}{2}$. The original graph is so crowded that one cannot see where crossings occur. The window shown is $[-2\pi, 2\pi]$ by $[-1.1, 1.1]$.

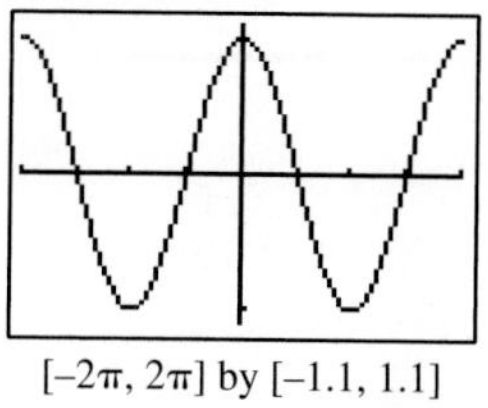

$[-2\pi, 2\pi]$ by $[-1.1, 1.1]$

75. $B = B_{in} + B_{ref} = \frac{E_0}{c}\cos\left(\omega t - \frac{\omega x}{c}\right) + \frac{E_0}{c}\cos\left(\omega t + \frac{\omega x}{c}\right) = \frac{E_0}{c}\left[\cos\left(\omega t - \frac{\omega x}{c}\right) + \cos\left(\omega t + \frac{\omega x}{c}\right)\right]$

$= \frac{E_0}{c}\left(2\cos\omega t\cos\frac{\omega x}{c}\right) = 2\frac{E_0}{c}\cos\omega t\cos\frac{\omega x}{c}$

The next-to-last step follows by the identity in Exercise 48.

SECTION 5.4

Exploration 1

1. $\sin^2\frac{\pi}{8} = \frac{1-\cos(\pi/4)}{2} = \frac{1-(\sqrt{2}/2)}{2}\cdot\frac{2}{2} = \frac{2-\sqrt{2}}{4}$ **3.** $\sin^2\frac{9\pi}{8} = \frac{1-\cos(9\pi/4)}{2} = \frac{1-(\sqrt{2}/2)}{2}\cdot\frac{2}{2} = \frac{2-\sqrt{2}}{4}$

Quick Review 5.4

1. $x = \frac{\pi}{4} + n\pi, n = 0, \pm1, \pm2, \ldots$ **3.** $x = \frac{\pi}{2} + n\pi, n = 0, \pm1, \pm2, \ldots$ **5.** $x = -\frac{\pi}{4} + n\pi, n = 0, \pm1, \pm2, \ldots$

7. $x = \frac{\pi}{6} + 2n\pi$ or $x = \frac{5\pi}{6} + 2n\pi$ or $x = \pm\frac{2\pi}{3} + 2n\pi, n = 0, \pm1, \pm2, \ldots$ **9.** 10 1/2 sq units

Exercises 5.4

1. $\cos 2u = \cos(u + u) = \cos u\cos u - \sin u\sin u = \cos^2 u - \sin^2 u$

3. Starting with the result of Exercise 1: $\cos 2u = \cos^2 u - \sin^2 u = (1 - \sin^2 u) - \sin^2 u = 1 - 2\sin^2 u$ **5.** $0, \pi$

7. $\frac{\pi}{6}, \frac{5\pi}{6}, \frac{3\pi}{2}$ **9.** $0, \frac{\pi}{4}, \frac{3\pi}{4}, \pi, \frac{5\pi}{4}, \frac{7\pi}{4}$ **11.** $2\sin\theta\cos\theta + \cos\theta$ or $(\cos\theta)(2\sin\theta + 1)$

13. $2\sin\theta\cos\theta + 4\cos^3\theta - 3\cos\theta$ or $2\sin\theta\cos\theta + \cos^3\theta - 3\sin^2\theta\cos\theta$ **15.** $\sin 4x = \sin 2(2x) = 2\sin 2x\cos 2x$

17. $2\csc 2x = \frac{2}{\sin 2x} = \frac{2}{2\sin x\cos x} = \frac{1}{\sin^2 x}\cdot\frac{\sin x}{\cos x} = \csc^2 x\tan x$

19. $\sin 3x = \sin 2x\cos x + \cos 2x\sin x = 2\sin x\cos^2 x + (2\cos^2 x - 1)\sin x = (\sin x)(4\cos^2 x - 1)$

21. $\cos 4x = \cos 2(2x) = 1 - 2\sin^2 2x = 1 - 2(2\sin x\cos x)^2 = 1 - 8\sin^2 x\cos^2 x$

23. $\frac{\pi}{3}, \pi, \frac{5\pi}{3}$ **25.** $\frac{\pi}{4}, \frac{\pi}{2}, \frac{3\pi}{4}, \frac{5\pi}{4}, \frac{3\pi}{2}, \frac{7\pi}{4}$ **27.** $0, \frac{\pi}{3}, \frac{\pi}{2}, \frac{2\pi}{3}, \pi, \frac{4\pi}{3}, \frac{3\pi}{2}, \frac{5\pi}{3}$

29.

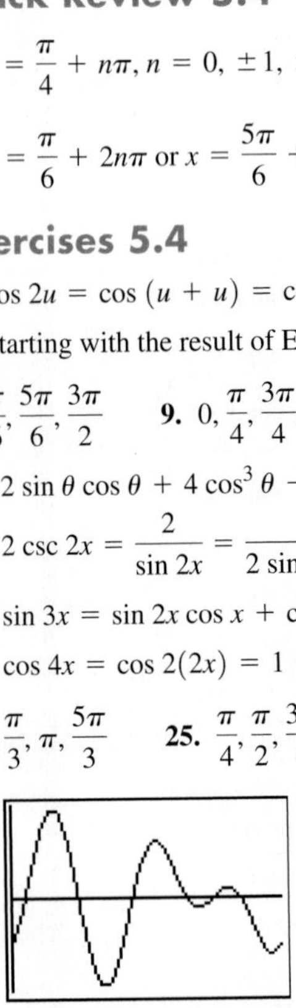

[0, 2π] by [−2, 2]

$\{0.1\pi, 0.5\pi, 0.9\pi, 1.3\pi, 1.5\pi, 1.7\pi\}$

31. $(1/2)\sqrt{2-\sqrt{3}}$ **33.** $(1/2)\sqrt{2-\sqrt{3}}$ **35.** $-2-\sqrt{3}$

37. (a) Starting from the right side:
$\frac{1}{2}(1 - \cos 2u) = \frac{1}{2}[1 - (1 - 2\sin^2 u)] = \frac{1}{2}(2\sin^2 u) = \sin^2 u$

(b) Starting from the right side:
$\frac{1}{2}(1 + \cos 2u) = \frac{1}{2}[1 + (2\cos^2 u - 1)] = \frac{1}{2}(2\cos^2 u) = \cos^2 u$

39. $\sin^4 x = (\sin^2 x)^2 = \left[\frac{1}{2}(1 - \cos 2x)\right]^2 = \frac{1}{4}(1 - 2\cos 2x + \cos^2 2x) = \frac{1}{4}\left[1 - 2\cos 2x + \frac{1}{2}(1 + \cos 4x)\right]$

$= \frac{1}{8}(2 - 4\cos 2x + 1 + \cos 4x) = \frac{1}{8}(3 - 4\cos 2x + \cos 4x)$

41. $\sin^3 2x = \sin 2x\sin^2 2x = \sin 2x\cdot\frac{1}{2}(1 - \cos 4x) = \frac{1}{2}(\sin 2x)(1 - \cos 4x)$

43. $\frac{\pi}{3}, \pi, \frac{5\pi}{3}$; general solution: $\pm\frac{\pi}{3} + 2n\pi$ or $\pi + 2n\pi, n = 0, \pm1, \pm2, \ldots$

45. $0, \frac{\pi}{2}$; general solution: $2n\pi$ or $\frac{\pi}{2} + 2n\pi, n = 0, \pm1, \pm2, \ldots$

47. False. For example, $f(x) = 2\sin x$ has period 2π and $g(x) = \cos x$ has period 2π, but the product $f(x)\,g(x) = 2\sin x\cos x = \sin 2x$ has period π. **49.** D **51.** E

53. **(a)** In the figure, the triangle with side lengths $x/2$ and R is a right triangle, since R is given as the perpendicular distance. Then the tangent of the angle $\theta/2$ is the ratio "opposite over adjacent": $\tan\frac{\theta}{2} = \frac{x/2}{R}$. Solving for x gives the desired equation. The central angle θ is $2\pi/n$ since one full revolution of 2π rad is divided evenly into n sections.

(b) $5.87 \approx 2R\tan\frac{\theta}{2}$, where $\theta = \frac{2\pi}{11}$, so $R \approx 5.87/\left(2\tan\frac{\pi}{11}\right) \approx 9.9957$, $R = 10$.

55. $\theta = \frac{\pi}{6}$; the maximum value is about 12.99 ft^3.

57. $\csc 2u = \frac{1}{\sin 2u} = \frac{1}{2\sin u\cos u} = \frac{1}{2}\cdot\frac{1}{\sin u}\cdot\frac{1}{\cos u} = \frac{1}{2}\csc u\sec u$

59. $\sec 2u = \frac{1}{\cos 2u} = \frac{1}{1 - 2\sin^2 u} = \left(\frac{1}{1 - 2\sin^2 u}\right)\left(\frac{\csc^2 u}{\csc^2 u}\right) = \frac{\csc^2 u}{\csc^2 u - 2}$

61. $\sec 2u = \frac{1}{\cos 2u} = \frac{1}{\cos^2 u - \sin^2 u} = \left(\frac{1}{\cos^2 u - \sin^2 u}\right)\left(\frac{\sec^2 u\csc^2 u}{\sec^2 u\csc^2 u}\right) = \frac{\sec^2 u\csc^2 u}{\csc^2 u - \sec^2 u}$

63. **(a)**

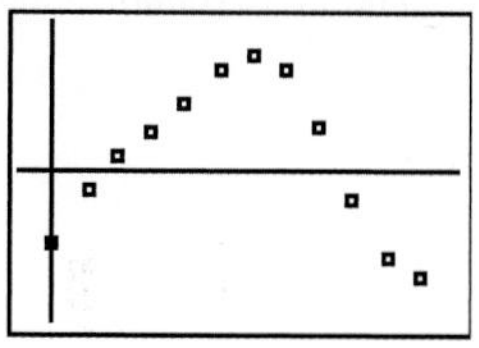

[–30, 370] by [–60, 60]

(b)

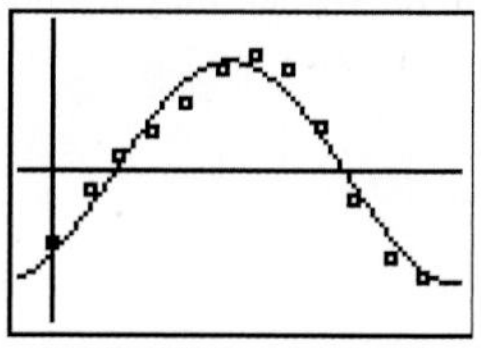

[–30, 370] by [–60, 60]

$Y1 = 44.52\sin(0.015x - 0.82) - 0.59$
This is a fairly good fit, but not really as good as one might expect from data generated by a sinusoidal physical model.

(c) The residual list:
$\{3.73, 7.48, 3.05, -6.50, -9.77, -3.67, 5.63, 9.79, 4.50, -3.66, -8.62, -2.39\}$.

(d)

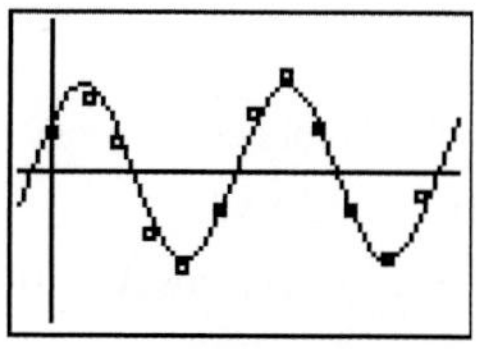

[–30, 370] by [–15, 15]

$Y2 = 8.73\sin(0.034x + 0.62) - 0.05$
This is another fairly good fit, which indicates that the residuals are not random. There is a periodic variation that is most probably due to physical causes.

(e) The first regression indicates that the data are nearly sinusoidal. The second regression indicates that the *variation* of the data around the predicted values is also nearly sinusoidal. Periodic variation around periodic models is a predictable consequence of bodies orbiting bodies, but ancient astronomers had a difficult time reconciling their data with their simpler models of the universe.

SECTION 5.5

Exploration 1

1. If $BC \le AB$, the segment will not reach from point B to the dotted line. On the other hand, if $BC > AB$, then a circle of radius BC will intersect the dotted line in a unique point. (Note that the line only extends to the left of point A.)
3. The second point (C_2) is the reflection of the first point (C_1) on the other side of the altitude.
5. If $BC \ge AB$, then BC can only extend to the right of the altitude, thus determining a unique triangle.

Quick Review 5.5

1. bc/d **3.** ad/b **5.** 13.314 **7.** 17.458° **9.** 224.427°

Exercises 5.5

1. $C = 75°$; $a \approx 4.5$; $c \approx 5.1$ **3.** $B = 45°$; $b \approx 15.8$; $c \approx 12.8$ **5.** $C = 110°$; $a \approx 12.9$; $c \approx 18.8$
7. $C = 77°$; $a \approx 4.1$; $c \approx 7.3$ **9.** $B \approx 20.1°$; $C \approx 127.9°$; $c \approx 25.3$ **11.** $C \approx 37.2°$; $A \approx 72.8°$; $a \approx 14.2$
13. zero **15.** two **17.** one **19.** $B_1 \approx 72.7°$; $C_1 \approx 43.3°$; $c_1 \approx 12.2$; $B_2 \approx 107.3°$; $C_2 \approx 8.7°$; $c_2 \approx 2.7$
21. $A_1 \approx 78.2°$; $B_1 \approx 33.8°$; $b_1 \approx 10.8$; $A_2 \approx 101.8°$; $B_2 \approx 10.2°$; $b_2 \approx 3.4$
23. **(a)** $6.69 < b < 10$ **(b)** $b \approx 6.69$ or $b \ge 10$ **(c)** $b < 6.69$

25. **(a)** No: this is an SAS case. **(b)** No: only two pieces of information given **27.** No triangle is formed **29.** No triangle is formed **31.** $A = 99°$; $a \approx 28.3$; $b \approx 19.1$ **33.** $A_1 \approx 24.6°$; $B_1 \approx 80.4°$; $a_1 \approx 20.7$; $A_2 \approx 5.4°$; $B_2 \approx 99.6°$; $a_2 \approx 4.7$ **35.** Cannot be solved with Law of Sines (an SAS case). **37.** **(a)** 54.6 ft **(b)** 51.9 ft **39.** $\approx$ 24.9 ft **41.** 1.9 ft **43.** $\approx$ 108.9 ft **45.** 36.6 mi to A; 28.9 mi to B **47.** True. By the Law of Sines, $\frac{\sin A}{a} = \frac{\sin B}{b}$, which is equivalent to $\frac{\sin A}{\sin B} = \frac{a}{b}$. **49.** C **51.** A **53.** **(b)** Possible answers: $a = 1, b = \sqrt{3}, c = 2$ (or any set of three numbers proportional to these). **(c)** Any set of three identical numbers. **55.** **(a)** $h = AB \sin A$ **(b)** $BC < AB \sin A$ **(c)** $BC \geq AB$ or $BC = AB \sin A$ **(d)** $AB \sin A < BC < AB$ **57.** $AC \approx 8.7$ mi; $BC \approx 12.2$ mi; $h \approx 5.2$ mi

SECTION 5.6

Exploration 1

1. 8475.742818 paces^2 **3.** 0.0014714831 mi^2
5. The estimate of "a little over an acre" seems questionable, but the roughness of their measurement system does not provide firm evidence that it is incorrect. If Jim and Barbara wish to make an issue of it with the owner, they would be well advised to get some more reliable data.

Quick Review 5.6

1. $A \approx 53.130°$ **3.** $A \approx 132.844°$

5. **(a)** $\cos A = \frac{x^2 + y^2 - 81}{2xy}$ **(b)** $A = \cos^{-1}\left(\frac{x^2 + y^2 - 81}{2xy}\right)$

7. One answer: $(x - 1)(x - 2)$ **9.** One answer $(x - i)(x + i) = x^2 + 1$

Exercises 5.6

1. $A \approx 30.7°$; $C \approx 18.3°$, $b \approx 19.2$ **3.** $A \approx 76.8°$; $B \approx 43.2°$, $C \approx 60°$ **5.** $B \approx 89.3°$; $C \approx 35.7°$, $a \approx 9.8$ **7.** $A \approx 28.5°$; $B \approx 56.5°$, $c \approx 25.1$ **9.** No triangles possible **11.** $A \approx 24.6°$; $B \approx 99.2°$, $C \approx 56.2$ **13.** $B_1 \approx 72.9°$; $C_1 \approx 65.1°$, $c_1 \approx 9.5$; $B_2 \approx 107.1°$; $C_2 \approx 30.9°$, $c_2 \approx 5.4$ **15.** No triangle **17.** ≈ 222.33 ft^2 **19.** ≈ 107.98 cm^2 **21.** ≈ 8.18 **23.** No triangle is formed **25.** ≈ 216.15 **27.** ≈ 314.05 **29.** ≈ 1.445 rad **31.** ≈ 374.1 in.^{2} **33.** ≈ 498.8 in.^{2} **35.** ≈ 130.42 ft **37.** **(a)** ≈ 42.5 ft **(b)** The home-to-second segment is the hypotenuse of a right triangle, so the distance from the pitcher's rubber to second base is $60\sqrt{2} - 40 \approx 44.9$ ft. **(c)** $\approx 93.3°$ **39.** **(a)** $\tan^{-1}(1/3) \approx 18.4°$ **(b)** ≈ 4.5 ft **(c)** ≈ 7.6 ft **41.** ≈ 12.5 yd **43.** $\approx 37.9°$ **45.** True. By the Law of Cosines, $b^2 + c^2 - 2bc \cos A = a^2$, which is a positive number. Since $b^2 + c^2 - 2bc \cos A > 0$, it follows that $b^2 + c^2 > 2bc \cos A$. **47.** B **49.** C **51.** Area $= (nr^2/2) \sin(360°/n)$

53. **(a)** Ship A: $\frac{30.2 - 15.1}{1 \text{ hr}} = 15.1$ knots; Ship B: $\frac{37.2 - 12.4}{2 \text{ hrs}} = 12.4$ knots **(b)** 35.18° **(c)** 34.8 naut mi **55.** 6.9 in.^{2}

CHAPTER 5 REVIEW EXERCISES

1. $\sin 200°$ **3.** 1

5. $\cos 3x = \cos(2x + x) = \cos 2x \cos x - \sin 2x \sin x = (\cos^2 x - \sin^2 x)\cos x - (2 \sin x \cos x)\sin x$
$= \cos^3 x - 3\sin^2 x \cos x = \cos^3 x - 3(1 - \cos^2 x)\cos x = \cos^3 x - 3\cos x + 3\cos^3 x = 4\cos^3 x - 3\cos x$

7. $\tan^2 x - \sin^2 x = \sin^2 x\left(\frac{1 - \cos^2 x}{\cos^2 x}\right) = \sin^2 x \cdot \frac{\sin^2 x}{\cos^2 x} = \sin^2 x \tan^2 x$

9. $\csc x - \cos x \cot x = \frac{1}{\sin x} - \cos x \cdot \frac{\cos x}{\sin x} = \frac{1 - \cos^2 x}{\sin x} = \frac{\sin^2 x}{\sin x} = \sin x$

11. $\frac{1 + \tan\theta}{1 - \tan\theta} + \frac{1 + \cot\theta}{1 - \cot\theta} = \frac{(1 + \tan\theta)(1 - \cot\theta) + (1 + \cot\theta)(1 - \tan\theta)}{(1 - \tan\theta)(1 - \cot\theta)}$
$= \frac{(1 + \tan\theta - \cot\theta - 1) + (1 + \cot\theta - \tan\theta - 1)}{(1 - \tan\theta)(1 - \cot\theta)} = \frac{0}{(1 - \tan\theta)(1 - \cot\theta)} = 0$

13. $\cos^2 \frac{t}{2} = \left[\pm\sqrt{\frac{1}{2}(1 + \cos t)}\right]^2 = \frac{1}{2}(1 + \cos t) = \left(\frac{1 + \cos t}{2}\right)\left(\frac{\sec t}{\sec t}\right) = \frac{1 + \sec t}{2 \sec t}$

15. $\frac{\cos\phi}{1 - \tan\phi} + \frac{\sin\phi}{1 - \cot\phi} = \left(\frac{\cos\phi}{1 - \tan\phi}\right)\left(\frac{\cos\phi}{\cos\phi}\right) + \left(\frac{\sin\phi}{1 - \cot\phi}\right)\left(\frac{\sin\phi}{\sin\phi}\right) = \frac{\cos^2\phi}{\cos\phi - \sin\phi} + \frac{\sin^2\phi}{\sin\phi - \cos\phi}$
$= \frac{\cos^2\phi - \sin^2\phi}{\cos\phi - \sin\phi} = \cos\phi + \sin\phi$

17. $\sqrt{\frac{1-\cos y}{1+\cos y}} = \sqrt{\frac{(1-\cos y)^2}{(1+\cos y)(1-\cos y)}} = \sqrt{\frac{(1-\cos y)^2}{1-\cos^2 y}} = \sqrt{\frac{(1-\cos y)^2}{\sin^2 y}} = \frac{|1-\cos y|}{|\sin y|}$

$= \frac{1-\cos y}{|\sin y|}$; since $1-\cos y \geq 0$, we can drop that absolute value.

19. $\tan\left(u+\frac{3\pi}{4}\right) = \frac{\tan u + \tan(3\pi/4)}{1-\tan u \tan(3\pi/4)} = \frac{\tan u + (-1)}{1-\tan u(-1)} = \frac{\tan u - 1}{1+\tan u}$

21. $\tan\frac{1}{2}\beta = \frac{1-\cos\beta}{\sin\beta} = \frac{1}{\sin\beta} - \frac{\cos\beta}{\sin\beta} = \csc\beta - \cot\beta$

23. Yes: $\sec x - \sin x \tan x = \frac{1}{\cos x} - \frac{\sin^2 x}{\cos x} = \frac{1-\sin^2 x}{\cos x} = \frac{\cos^2 x}{\cos x} = \cos x$

25. Many answers are possible, for example, $(\cos x - \sin x)(1 + 4\sin x \cos x)$.

27. Many answers are possible, for example, $1 - 4\sin^2 x \cos^2 x - 2\sin x \cos x$. **29.** $\frac{\pi}{12} + n\pi$ or $\frac{5\pi}{12} + n\pi, n = 0, \pm 1, \pm 2, \ldots$

31. $-\frac{\pi}{4} + n\pi$ **33.** tan 1 **35.** $x \approx 1.12$ or $x \approx 5.16$ **37.** $x \approx 1.15$ **39.** $\pi/3, 5\pi/3$

41. $\frac{3\pi}{2}$ **43.** No solutions **45.** $\left[0, \frac{\pi}{6}\right) \cup \left(\frac{5\pi}{6}, \frac{7\pi}{6}\right) \cup \left(\frac{11\pi}{6}, 2\pi\right)$ **47.** $(\pi/3, 5\pi/3)$ **49.** $y \approx 5\sin(3x + 0.93)$

51. $C = 68°, b \approx 3.9, c \approx 6.6$ **53.** No triangle is formed. **55.** $C = 72°, a \approx 2.9, b \approx 5.1$ **57.** $A \approx 44.4°, B \approx 78.5°, C \approx 57.1°$

59. ≈ 7.5 **61.** **(a)** $\approx 5.6 < b < 12$ **(b)** $b \approx 5.6$ or $b \geq 12$ **(c)** $b < 5.6$ **63.** ≈ 0.6 mi **65.** 1.25 rad

67. **(a)** $\sin\theta + \frac{1}{2}\sin 2\theta$ **(b)** $\theta = 60°$; ≈ 1.30 square units **69.** **(a)** $h = 4000\sec\frac{\theta}{2} - 4000$ mi **(b)** $\approx 35.51°$

71. Area of circle − area of hexagon $= 256\pi - 6 \cdot 64\sqrt{3} \approx 139.14$ cm^2 **73.** $405\pi/24 \approx 53.01$ cm^3

75. **(a)** By the product-to-sum formula in Exercise 74c, $2\sin\frac{u+v}{2}\cos\frac{u-v}{2}$

$= 2\cdot\frac{1}{2}\left(\sin\frac{u+v+u-v}{2} + \sin\frac{u+v-(u-v)}{2}\right) = \sin u + \sin v.$

(b) By the product-to-sum formula in Exercise 74c, $2\sin\frac{u-v}{2}\cos\frac{u+v}{2} = 2\cdot\frac{1}{2}\left(\sin\frac{u-v+u+v}{2} + \sin\frac{u-v-(u+v)}{2}\right)$

$= \sin u + \sin(-v) = \sin u - \sin v.$ **(c)** By the product-to-sum formula in Exercise 74b, $2\cos\frac{u+v}{2}\cos\frac{u-v}{2}$

$= 2\cdot\frac{1}{2}\left(\cos\frac{u+v-(u-v)}{2} + \cos\frac{u+v+u-v}{2}\right) = \cos v + \cos u = \cos u + \cos v.$ **(d)** By the product-to-sum formula in Exercise 74a, $-2\sin\frac{u+v}{2}\sin\frac{u-v}{2} = -2\cdot\frac{1}{2}\left(\cos\frac{u+v(u-v)}{2} - \cos\frac{u+v+u-v}{2}\right) = -(\cos v - \cos u) = \cos u - \cos v.$

77. **(a)** Any inscribed angle that intercepts an arc of 180° is a right angle. **(b)** Two inscribed angles that intercept the same arc are congruent.

(c) In right $\Delta A'BC$, $\sin A' = \frac{\text{opp}}{\text{hyp}} = \frac{a}{d}$. **(d)** Because $\angle A'$ and $\angle A$ are congruent, $\frac{\sin A}{a} = \frac{\sin A'}{a} = \frac{a/d}{a} = \frac{1}{d}$.

(e) Of course. They both equal $\frac{\sin A}{a}$ by the Law of Sines.

Chapter 5 Project

1.

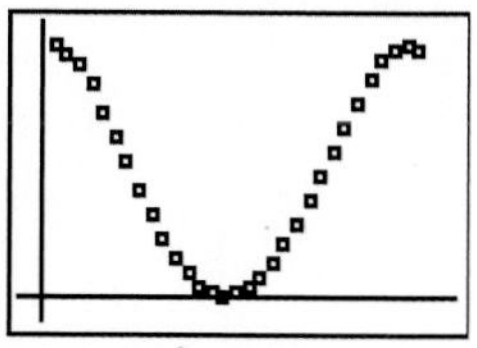

[−2, 34] by [−0.1, 1.1]

5. One possible model is $y = 0.5\sin(0.2(x + 8.4)) + 0.5$.

SECTION 6.1

Exploration 1

1. $\langle 5, 2\rangle$ **3.** $(6, -7)$

Quick Review 6.1

1. $\frac{9\sqrt{3}}{2}$; 4.5 **3.** -5.36; -4.50 **5.** 33.85° **7.** 60.95° **9.** $180° + \tan^{-1}(5/2) \approx 248.20°$

Exercises 6.1

1. Both vectors represent $\langle 3, -2\rangle$ by the HMT Rule. **3.** Both vectors represent $\langle -2, -2\rangle$ by the HMT Rule.
5. $\langle 5, 2\rangle$; $\sqrt{29}$ **7.** $\langle -5, 1\rangle$; $\sqrt{26}$ **9.** $\langle -2, -24\rangle$; $2\sqrt{145}$ **11.** $\langle -11, -7\rangle$; $\sqrt{170}$ **13.** $\langle 1, 7\rangle$
15. $\langle -3, 8\rangle$ **17.** $\langle 4, -9\rangle$ **19.** $\langle -4, -18\rangle$

21. $-\frac{1}{\sqrt{5}}\mathbf{i} + \frac{2}{\sqrt{5}}\mathbf{j}$ **23.** $-\frac{1}{\sqrt{5}}\mathbf{i} - \frac{2}{\sqrt{5}}\mathbf{j}$ **25. (a)** $\left\langle \frac{2}{\sqrt{5}}, \frac{1}{\sqrt{5}}\right\rangle$ **(b)** $\frac{2}{\sqrt{5}}\mathbf{i} + \frac{1}{\sqrt{5}}\mathbf{j}$

27. (a) $\left\langle -\frac{4}{\sqrt{41}}, -\frac{5}{\sqrt{41}}\right\rangle$ **(b)** $-\frac{4}{\sqrt{41}}\mathbf{i} - \frac{5}{\sqrt{41}}\mathbf{j}$ **29.** $\approx \langle 16.31, 7.61\rangle$ **31.** $\approx \langle -14.52, 44.70\rangle$

33. 5; $\approx 53.13°$ **35.** 5; $\approx 306.87°$ **37.** 7; 135° **39.** $\langle \sqrt{2}, -\sqrt{2}\rangle$ **41.** $\approx \langle -223.99, 480.34\rangle$

43. (a) $\approx \langle -111.16, 305.40\rangle$ **(b)** $\approx$ 362.85 mph; bearing $\approx$ 337.84°

45. (a) $\approx \langle 3.42, 9.40\rangle$ **(b)** The horizontal component is the (constant) horizontal speed of the basketball as it travels toward the basket. The vertical component is the vertical velocity of the basketball, affected by both the initial speed and the downward pull of gravity.

47. $\approx \langle 2.20, 1.43\rangle$ **49.** $|\mathbf{F}| \approx 100.33$ lb and $\theta \approx -1.22°$

51. $\approx 342.86°$; ≈ 9.6 mph **53.** ≈ 13.66 mph; ≈ 7.07 mph

55. True. $\mathbf{u}$ and $-\mathbf{u}$ have the same length but opposite directions. Thus, the length of $-\mathbf{u}$ is also 1.

57. D **59.** A

SECTION 6.2

Exploration 1

1. $\langle -2 - x, -y\rangle$, $\langle 2 - x, -y\rangle$ **3.** Answers will vary.

Quick Review 6.2

1. $\sqrt{13}$ **3.** 1 **5.** $\langle 3, \sqrt{3}\rangle$ **7.** $\langle -1, -\sqrt{3}\rangle$ **9.** $\left\langle \frac{4}{\sqrt{13}}, \frac{6}{\sqrt{13}}\right\rangle$

Exercises 6.2

1. 72 **3.** -47 **5.** 30 **7.** -14 **9.** 13 **11.** 4 **13.** $\approx 115.6°$ **15.** $\approx 64.65°$ **17.** 165° **19.** 135°

21. $\approx 94.86°$ **23.** $(2)(3/2) + (3)(-1) = 0$, so $\mathbf{u}$ is orthogonal to $\mathbf{v}$. **25.** $-\frac{21}{10}\langle 3, 1\rangle$; $-\frac{21}{10}\langle 3, 1\rangle + \frac{17}{10}\langle -1, 3\rangle$

27. $\frac{82}{85}\langle 9, 2\rangle$; $\frac{82}{85}\langle 9, 2\rangle + \frac{29}{85}\langle -2, 9\rangle$ **29.** 47.73°, 74.74°, 57.53° **31.** ≈ -20.78 **33.** Parallel **35.** Neither **37.** Orthogonal

39. (a) $(4, 0)$ and $(0, -3)$ **(b)** $(4.6, -0.8)$ or $(3.4, 0.8)$ **41. (a)** $(7, 0)$ and $(0, -3)$ **(b)** $\approx (7.39, -0.92)$ or $(6.61, 0.92)$

43. $\langle -1, 4\rangle$ or $\left\langle \frac{53}{13}, \frac{8}{13}\right\rangle$ **45.** ≈ 138.56 lb **47. (a)** ≈ 415.82 lb **(b)** ≈ 1956.30 lb

49. 14,300 ft-lb **51.** ≈ 21.47 ft-lb **53.** ≈ 85.38 J **55.** $100\sqrt{39} \approx 624.5$ ft-lb

61. False. If one of $\mathbf{u}$ or $\mathbf{v}$ is the zero vector, then $\mathbf{u} \cdot \mathbf{v} = 0$ but $\mathbf{u}$ and $\mathbf{v}$ are not perpendicular.

63. D **65.** A **67. (a)** $2 \cdot 0 + 5 \cdot 2 = 10$ and $2 \cdot 5 + 5 \cdot 0 = 10$

(b) $\frac{5}{29}\langle 5, -2\rangle$; $\frac{1}{29}\langle 62, 155\rangle$ **(c)** $|\mathbf{w}_2| = \frac{31\sqrt{29}}{29}$ **(d)** $d = \frac{|2x_0 + 5y_0 - 10|}{\sqrt{29}}$ **(e)** $d = \frac{|ax_0 + by_0 - c|}{\sqrt{a^2 + b^2}}$

SECTION 6.3

Exploration 1

1.

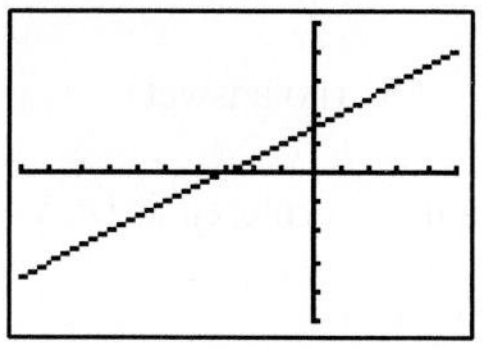

[–10, 5] by [–5, 5]

3. $t = 12$ **5.** Tmin ≤ -2 and Tmax ≥ 5.5

Exploration 2

3.

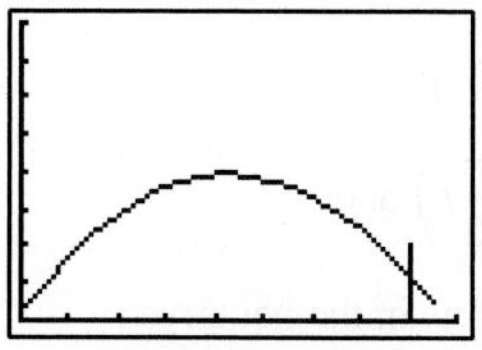

[0, 450] by [0, 80]

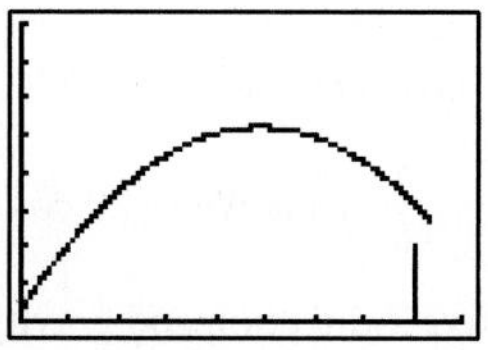

[0, 450] by [0, 80]

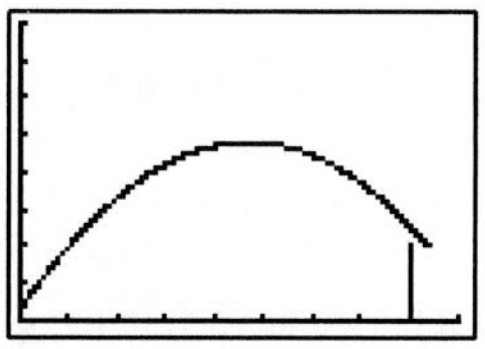

[0, 450] by [0, 80]

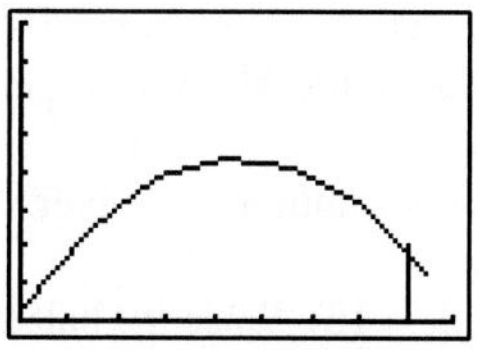

[0, 450] by [0, 80]

Quick Review 6.3

1. **(a)** $\langle -3, -2 \rangle$ **(b)** $\langle 4, 6 \rangle$ **(c)** $\langle 7, 8 \rangle$ **3.** $y + 2 = \frac{8}{7}(x + 3)$ or $y - 6 = \frac{8}{7}(x - 4)$

5.

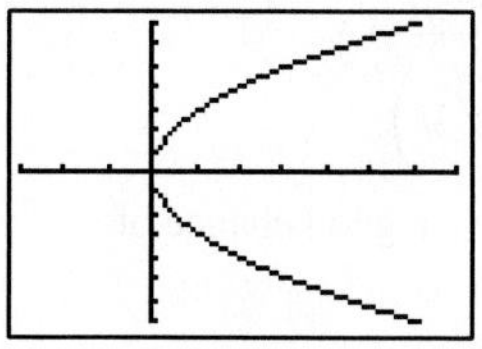

[–3, 7] by [–7, 7]

7. $x^2 + y^2 = 4$ **9.** 20π rad/sec

Exercises 6.3

1. **(b)** $[-5, 5]$ by $[-5, 5]$ **3.** **(a)** $[-5, 5]$ by $[-5, 5]$

5. **(a)**

t	−2	−1	0	1	2
x	0	1	2	3	4
y	1/2	−2	und.	4	5/2

(b)

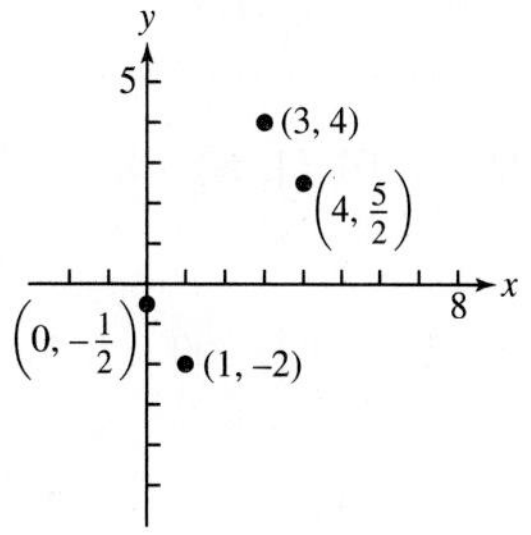

7.

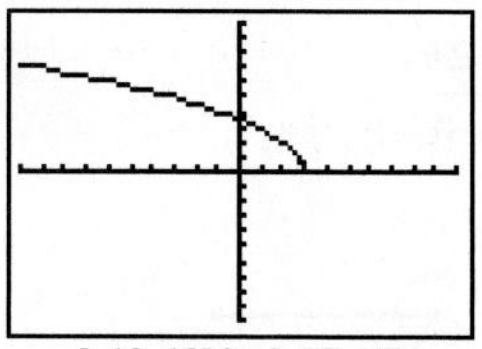

[–10, 10] by [–10, 10]

9.

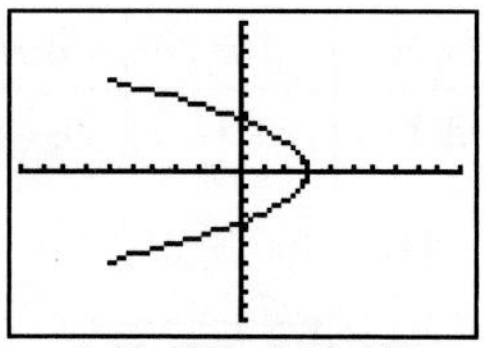

[–10, 10] by [–10, 10]

11. $y = x - 1$: line through $(0, -1)$ and $(1, 0)$ **13.** $y = -2x + 3, 3 \leq x \leq 7$: line segment with endpoints $(3, -3)$ and $(7, -11)$ **15.** $x = (y - 1)^2$: parabola that opens to right with vertex at $(0, 1)$ **17.** $y = x^3 - 2x + 3$: cubic polynomial **19.** $x = 4 - y^2$ parabola that opens to left with vertex at $(4, 0)$ **21.** $t = x + 3$, so $y = \frac{2}{x + 3}, -8 \leq x \leq 2, x \neq -3$ **23.** $x^2 + y^2 = 25$, circle of radius 5 centered at $(0, 0)$ **25.** $x^2 + y^2 = 4$, three-fourths of a circle of radius 2 centered at $(0, 0)$ (not in Quadrant II) **27.** $x = 6t - 2; y = -3t + 5$ **29.** $x = 3t + 3, y = 4 - 7t, 0 \leq t \leq 1$ **31.** $x = 5 + 3\cos t, y = 2 + 3\sin t, 0 \leq t \leq 2\pi$ **33.** $0.5 < t < 2$ **35.** $-3 \leq t < -2$ **37.** **(b)** Ben is ahead by 2 ft.

39. **(a)** $y = -16t^2 + 1000$ **(c)** 744 ft **41.** **(a)** $0 < t < \pi/2$ **(b)** $0 < t < \pi$ **(c)** $\pi/2 < t < 3\pi/2$
43. **(a)** About 2.80 sec **(b)** ≈ 7.18 ft **45.** **(a)** Yes **(b)** 1.59 ft **47.** No
49. $v \approx -10.00$ ft/sec or 551.20 ft/sec **51.** $x = 35\cos\left(\frac{\pi}{6}t\right)$ and $y = 50 + 35\sin\left(\frac{\pi}{6}t\right)$
53. **(a)** When $t = \pi$ (or 3π, or 5π, etc.), $y = 2$. This corresponds to the highest points on the graph. **(b)** 2π units **55.** (no answer)
57. (no answer) **59.** True. Both correspond to the rectangular equation $y = 3x + 4$. **61.** A **63.** D
65. **(a)**

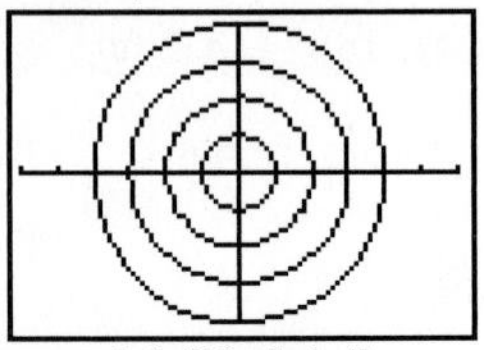

[–6, 6] by [–4, 4]

(b) a **(c)**

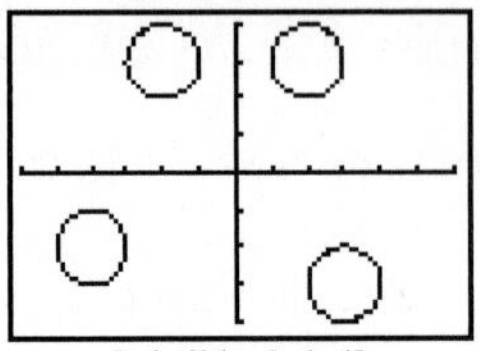

[–6, 6] by [–4, 4]

(d) $(x - h)^2 + (y - k)^2 = a^2$; circle of radius a centered at (h, k)
(e) $x = 3\cos t - 1$; $y = 3\sin t + 4$

67. **(a)** Jane is traveling in a circle of radius 20 ft and origin $(0, 20)$, which yields $x_1 = 20\cos(nt)$ and $y_1 = 20 + 20\sin(nt)$. Since the Ferris wheel is making one revolution (2π) every 12 sec, $2\pi = 12n$, so $n = \frac{2\pi}{12} = \frac{\pi}{6}$. Thus, $x_1 = 20\cos\left(\frac{\pi}{6}t\right)$ and $y_1 = 20 + 20\sin\left(\frac{\pi}{6}t\right)$, in rad mode. **(b)** Since the ball was released at 75 ft in the positive x-direction and gravity acts in the negative y-direction at 16 ft/s^2, we have $x_2 = at + 75$ and $y_2 = -16t^2 + bt$, where a is the initial speed of the ball in the x-direction and b is the initial speed of the ball in the y-direction. The initial velocity vector of the ball is $60\langle\cos 120°, \sin 120°\rangle = \langle -30, 30\sqrt{3}\rangle$, so $a = -30$ and $b = 30\sqrt{3}$. As a result, $x_2 = -30t + 75$ and $y_2 = -16t^3 + (30\sqrt{3})t$ are the parametric equations for the ball.

(c)

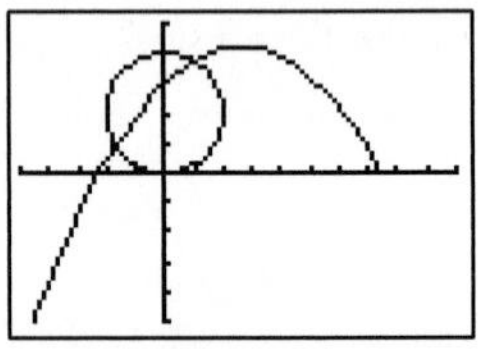

[–50, 100] by [–50, 50]

Jane and the ball will be close to each other but not at the exact same point at $t = 2.2$ sec.

(d) $d(t) = \sqrt{\left(20\cos\left(\frac{\pi}{6}t\right) + 30t - 75\right)^2 + \left(20 + 20\sin\left(\frac{\pi}{6}t\right) + 16t^2 - 30\sqrt{3}t\right)^2}$

(e) The minimum distance occurs at $t \approx 2.2$ sec, when $d(t) \approx 1.64$ ft. Jane will have a good chance of catching the ball.

69. About 4.11 ft **71.** **(a)** (no answer) **(b)** (no answer) **73.** $t = \frac{1}{3}, \frac{2}{3}$; $t = \frac{1}{4}, \frac{1}{2}, \frac{3}{4}$

SECTION 6.4

Exploration 1

1. (no answer) **3.** $(-2, \pi/3), (2, \pi/2), (3, 0), (1, \pi), (4, 3\pi/2)$

Quick Review 6.4

1. **(a)** II **(b)** III **3.** $7\pi/4, -9\pi/4$ **5.** $500°, -200°$ **7.** $(x - 3)^2 + y^2 = 4$ **9.** ≈ 11.14

Exercises 6.4

1. $\left(-\frac{3}{2}, \frac{3\sqrt{3}}{2}\right)$ **3.** $(-1, -\sqrt{3})$ **5.** **(a)**

θ	$\pi/4$	$\pi/2$	$5\pi/6$	π	$4\pi/3$	2π
r	$3\sqrt{2}/2$	3	3/2	0	$-3\sqrt{3}/2$	0

(b)

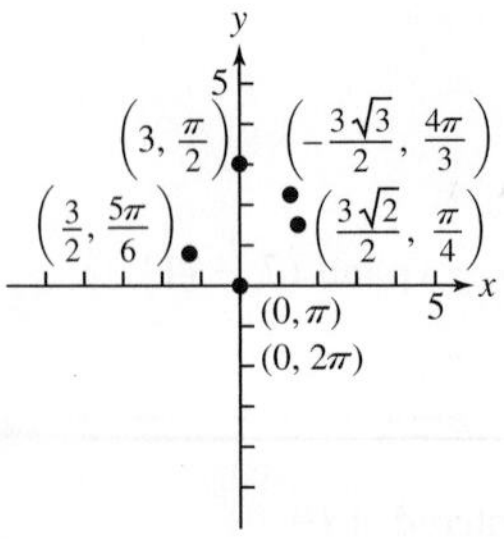

7.

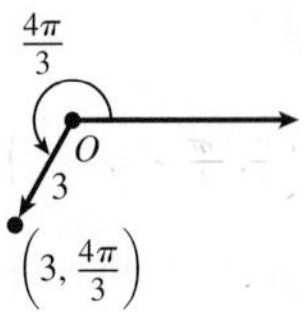

9.

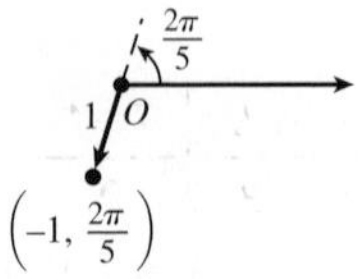

11.

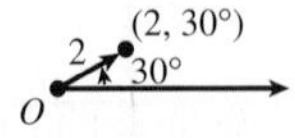

13.

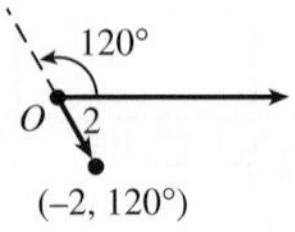

15. $\left(\frac{3}{4}, \frac{3}{4}\sqrt{3}\right)$ **17.** $(-2.70, 1.30)$ **19.** $(2, 0)$ **21.** $(0, -2)$ **23.** $\left(2, \frac{\pi}{6} + 2n\pi\right)$ and $\left(-2, \frac{\pi}{6} + (2n+1)\pi\right)$, n an integer

25. $(1.5, -20° + 360n°)$ and $(-1.5, 160° + 360n°)$, n an integer **27. (a)** $\left(\sqrt{2}, \frac{\pi}{4}\right)$ or $\left(-\sqrt{2}, \frac{5\pi}{4}\right)$ **(b)** $\left(\sqrt{2}, \frac{\pi}{4}\right)$ or $\left(-\sqrt{2}, -\frac{3\pi}{4}\right)$ **(c)** The answers from part (a), and also $\left(\sqrt{2}, \frac{9\pi}{4}\right)$ or $\left(-\sqrt{2}, \frac{13\pi}{4}\right)$. **29. (a)** $(\sqrt{29}, 1.95)$ or $(-\sqrt{29}, 5.09)$ **(b)** $(-\sqrt{29}, -1.19)$ or $(\sqrt{29}, 1.95)$ **(c)** The answers from part (a), plus $(\sqrt{29}, 8.23)$ or $(-\sqrt{29}, 11.38)$ **31.** (b) **33.** (c)

35. $x = 3$ — a vertical line **37.** $x^2 + \left(y + \frac{3}{2}\right)^2 = \frac{9}{4}$ — a circle centered at $\left(0, -\frac{3}{2}\right)$ with radius $\frac{3}{2}$ **39.** $x^2 + \left(y - \frac{1}{2}\right)^2 = \frac{1}{4}$ — a circle centered at $\left(0, \frac{1}{2}\right)$ with radius $\frac{1}{2}$ **41.** $(x+2)^2 + (y-1)^2 = 5$ — a circle centered at $(-2, 1)$ with radius $\sqrt{5}$

43. $r = 2/\cos\theta = 2\sec\theta$

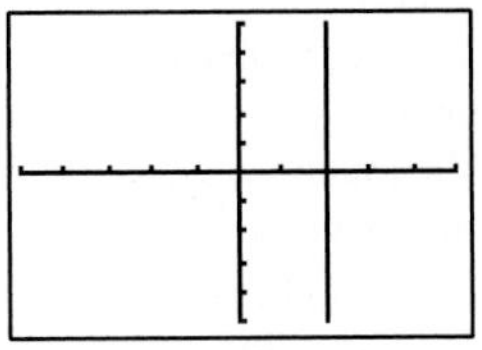

[–5, 5] by [–5, 5]

45. $r = \dfrac{5}{2\cos\theta - 3\sin\theta}$

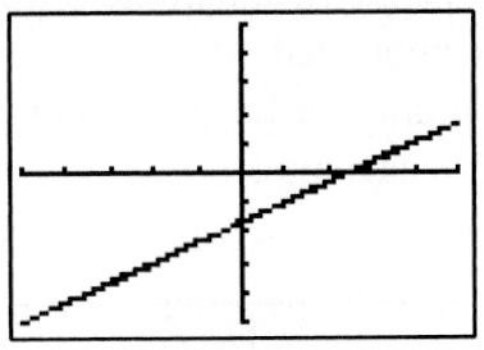

[–5, 5] by [–5, 5]

47. $r^2 - 6r\cos\theta = 0$, so $r = 6\cos\theta$

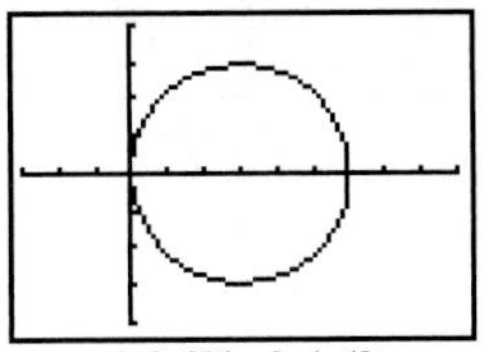

[–3, 9] by [–4, 4]

49. $r^2 + 6r\cos\theta + 6r\sin\theta = 0$, so $r = -6\cos\theta - 6\sin\theta$

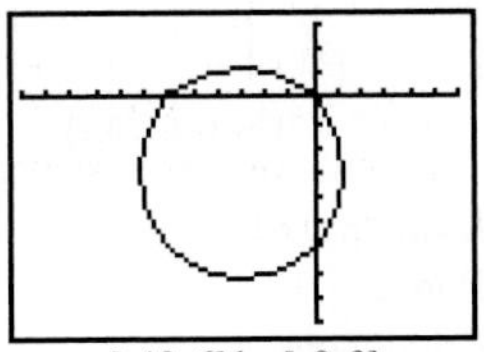

[–12, 6] by [–9, 3]

51. $2\sqrt{3} \approx 3.46$ mi **53.** $\left(\frac{a}{\sqrt{2}}, \frac{\pi}{4}\right), \left(\frac{a}{\sqrt{2}}, \frac{3\pi}{4}\right), \left(\frac{a}{\sqrt{2}}, \frac{5\pi}{4}\right)$, and $\left(\frac{a}{\sqrt{2}}, \frac{7\pi}{4}\right)$

55. False. $(r, \theta) = (r, \theta + 2n\pi)$ for any integer n. These are all distinct polar coordinates. **57.** C **59.** A **61. (a)** If $\theta_1 - \theta_2$ is an odd integer multiple of π, then the distance is $|r_1 + r_2|$. If $\theta_1 - \theta_2$ is an even integer multiple of π, then the distance is $|r_1 - r_2|$. **(b)** **63.** ≈ 6.24 **65.** ≈ 7.43 **67.** $x = f(\theta)\cos(\theta), y = f(\theta)\sin(\theta)$ **69.** $x = 5(\cos\theta)(\sin\theta), y = 5\sin^2\theta$ **71.** $x = 4\cot\theta, y = 4$

SECTION 6.5

Quick Review 6.5

1. Minimum: −3 at $x = \left\{\frac{\pi}{2}, \frac{3\pi}{2}\right\}$; Maximum: 3 at $x = \{0, \pi, 2\pi\}$ **3.** Minimum: 0 at $x = \left\{\frac{\pi}{4}, \frac{3\pi}{4}, \frac{5\pi}{4}, \frac{7\pi}{4}\right\}$; Maximum: 2 at $x = \{0, \pi, 2\pi\}$ **5.** No; no; yes **7.** $\sin\theta$ **9.** $\cos^2\theta - \sin^2\theta$

Exercises 6.5

1. (a)

θ	0	$\pi/4$	$\pi/2$	$3\pi/4$	π	$5\pi/4$	$3\pi/2$	$7\pi/4$
r	3	0	−3	0	3	0	−3	0

(b)

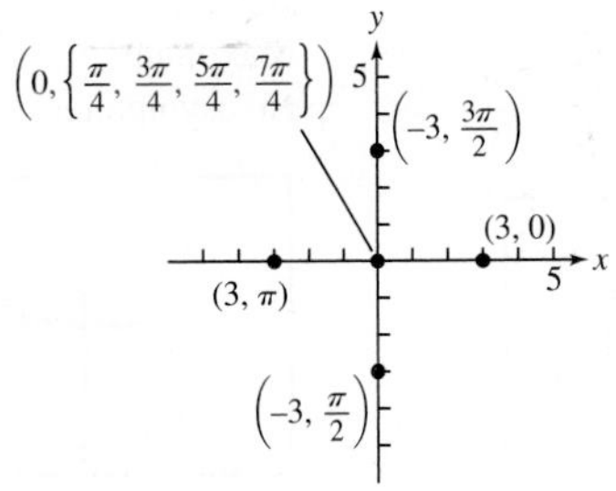

3. $k = \pi$

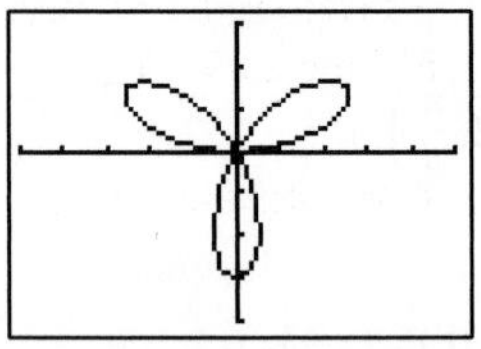

[–5, 5] by [–4, 3]

5. $k = 2\pi$

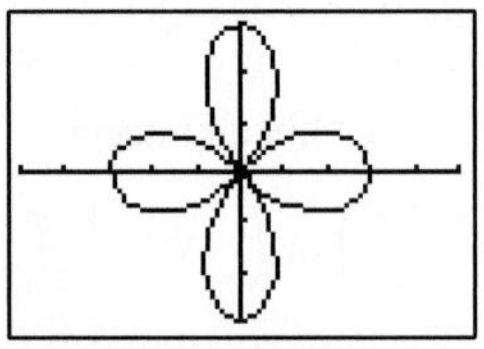

[–5, 5] by [–3, 3]

7. r_3 is graph (b). **9.** Graph (b) is $r = 2 - 2\cos\theta$. **11.** Graph (a) is $r = 2 - 2\sin\theta$.
13. Symmetric about the y-axis **15.** Symmetric about the x-axis **17.** All three symmetries **19.** Symmetric about the y-axis
21. Maximum $|r|$ is 5—when $\theta = 2n\pi$ for any integer n. **23.** Maximum $|r|$ is 3 (when $r = \pm 3$)—when $\theta = 2n\pi/3$ for any integer n.

25. Domain: $(-\infty, \infty)$
Range: $r = 3$
Continuous
Symmetric about the x-axis, y-axis, and origin
Bounded
Maximum $|r|$ value: 3
No asymptotes

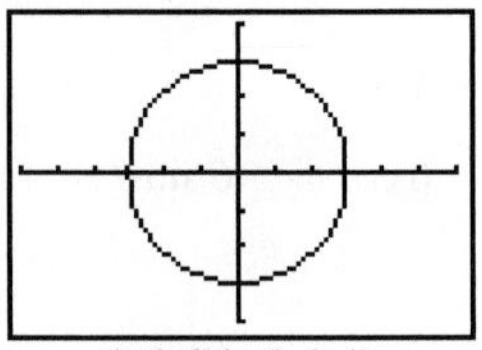

[–6, 6] by [–4, 4]

27. Domain: $\theta = \pi/3$
Range: $(-\infty, \infty)$
Continuous
Symmetric about the origin
Unbounded
Maximum $|r|$ value: none
No asymptotes

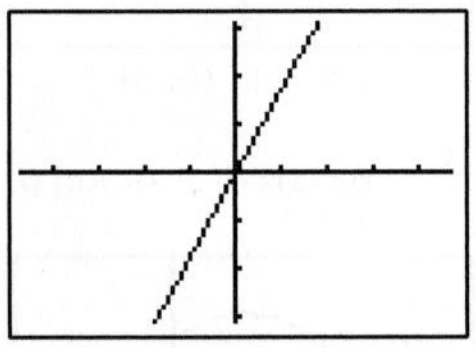

[–4.7, 4.7] by [–3.1, 3.1]

29. Domain: $(-\infty, \infty)$
Range: $[-2, 2]$
Continuous
Symmetric about the y-axis
Bounded
Maximum $|r|$ value: 2
No asymptotes

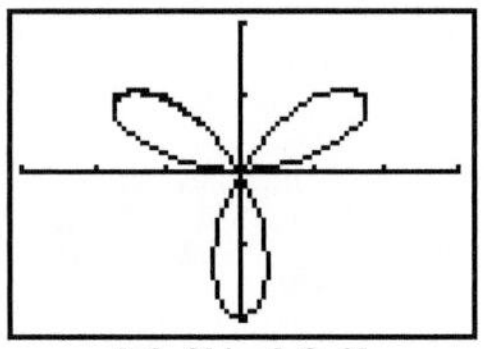

[–3, 3] by [–2, 2]

31. Domain: $(-\infty, \infty)$
Range: $[1, 9]$
Continuous
Symmetric about the y-axis
Bounded
Maximum $|r|$ value: 9
No asymptotes

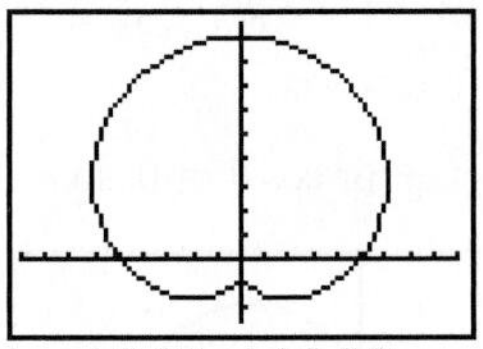

[–9, 9] by [–2.5, 9.5]

33. Domain: $(-\infty, \infty)$
Range: $[0, 8]$
Continuous
Symmetric about the x-axis
Bounded
Maximum $|r|$ value: 8
No asymptotes

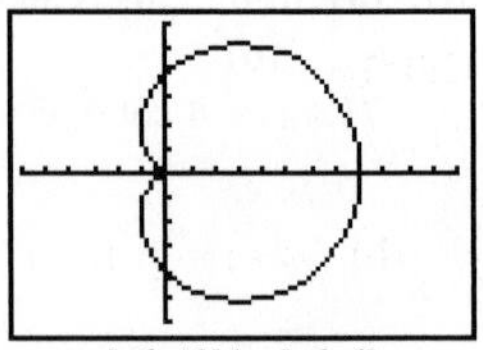

[–6, 12] by [–6, 6]

35. Domain: $(-\infty, \infty)$
Range: $[3, 7]$
Continuous
Symmetric about the x-axis
Bounded
Maximum $|r|$ value: 7
No asymptotes

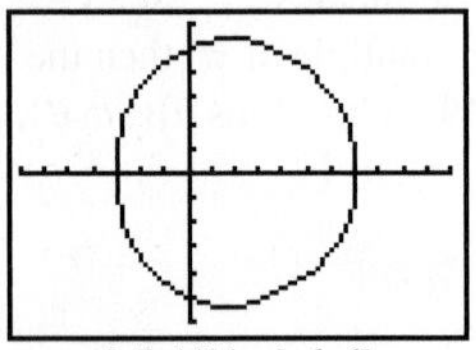

[–7, 11] by [–6, 6]

37. Domain: $(-\infty, \infty)$
Range: $[-3, 7]$
Continuous
Symmetric about the x-axis
Bounded
Maximum $|r|$ value: 7
No asymptotes

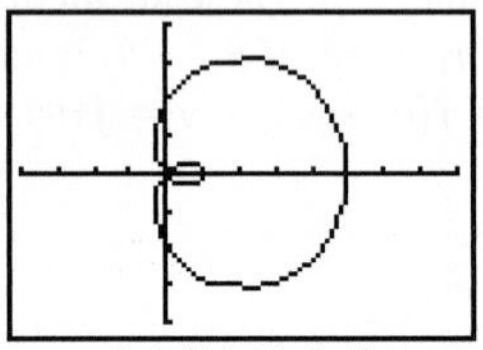

[–4, 8] by [–4, 4]

39. Domain: $(-\infty, \infty)$
Range: $[0, 2]$
Continuous
Symmetric about the x-axis
Bounded
Maximum $|r|$ value: 2
No asymptotes

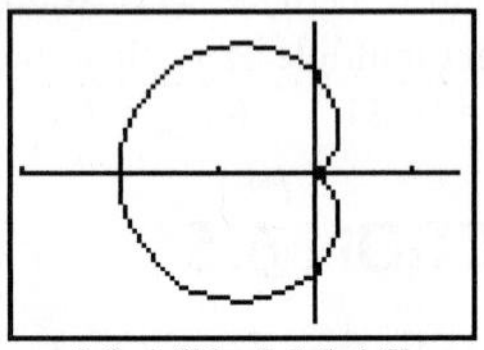

[–3, 1.5] by [–1.5, 1.5]

41. Domain: $(-\infty, \infty)$
Range: $[0, \infty)$
Continuous
No symmetry
Unbounded
Maximum $|r|$ value: none
No asymptotes
Graph for $\theta \geq 0$:

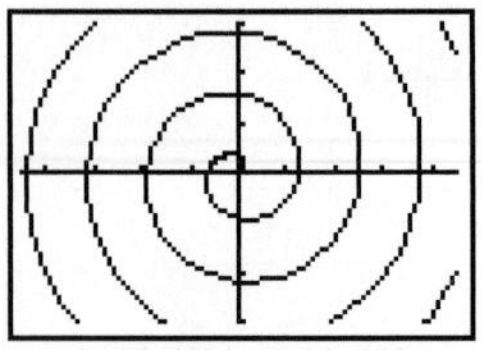

[–45, 45] by [–30, 30]

43. Domain: $\left[0, \frac{\pi}{2}\right] \cup \left[\pi, \frac{3\pi}{2}\right]$
Range: $[0, 1]$
Continuous on domain
Symmetric about the origin
Bounded
Maximum $|r|$ value: 1
No asymptotes

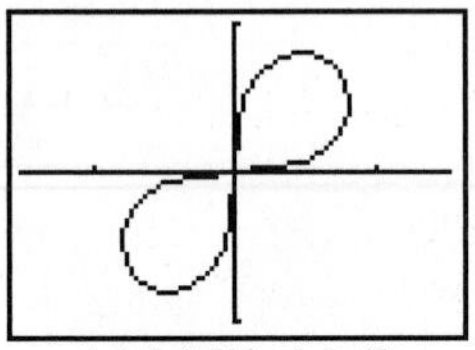

[–1.5, 1.5] by [–1, 1]

45. $\{6, 2, 6, 2\}$ **47.** $\{3, 5, 3, 5, 3, 5, 3, 5, 3, 5\}$ **49.** r_1 and r_2 **51.** r_2 and r_3 **53.** **(a)** A 4-petal rose curve with 2 short petals of length 1 and 2 long petals of length 3 **(b)** Symmetric about the origin **(c)** Maximum $|r|$ value: 3 **55.** **(a)** A 6-petal rose curve with 3 short petals of length 2, and 3 long petals of length 4 **(c)** Symmetric about the x-axis **(d)** Maximum $|r|$-value: 4 **57.** (no answer) **59.** (no answer) **61.** False. The spiral $r = \theta$ is unbounded. **63.** D **65.** B

67. **(e)** Domain: $(-\infty, \infty)$
Range: $[-|a|, |a|]$
Continuous
Symmetric about the x-axis
Bounded
Maximum $|r|$ value: $|a|$
No asymptotes

69. **(a)** For r_1: $0 \le \theta \le 4\pi$ (or any interval that is 4π units long). For r_2: same answer.
(b) r_1: 10 (overlapping) petals; r_2: 14 (overlapping) petals

71. Starting with the graph of r_1, if we rotate counterclockwise (centered at the origin) by $\pi/4$ rad $(45°)$, we get the graph of r_2; rotating r_1 counterclockwise by $\pi/3$ rad $(60°)$ gives the graph of r_3.

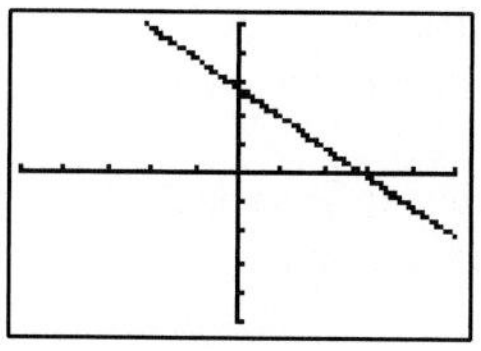
[–5, 5] by [–5, 5]

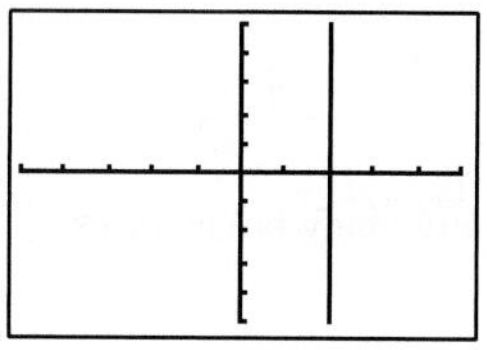
[–5, 5] by [–5, 5]

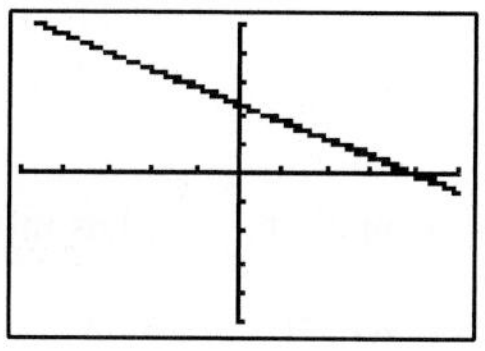
[–5, 5] by [–5, 5]

73. (no answer)

SECTION 6.6

Quick Review 6.6

1. $2 + 3i, 2 - 3i$ **3.** $-4 - 4i$ **5.** $\theta = \frac{5\pi}{6}$ **7.** $\frac{4\pi}{3}$ **9.** 1

Exercises 6.6

1.

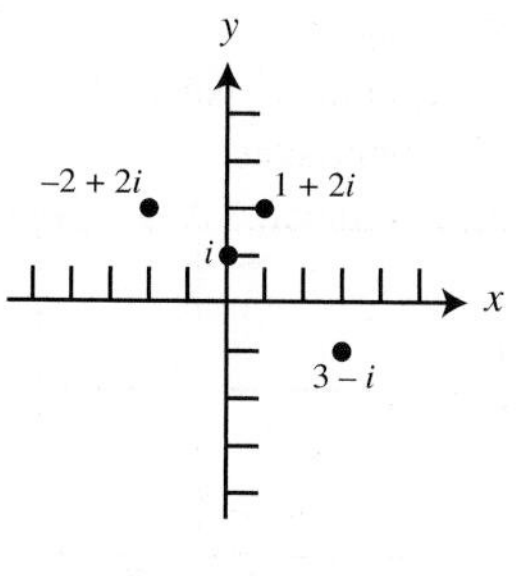

3. $3\left(\cos\frac{\pi}{2} + i\sin\frac{\pi}{2}\right)$ **5.** $2\sqrt{2}\left(\cos\frac{\pi}{4} + i\sin\frac{\pi}{4}\right)$ **7.** $4\left(\cos\frac{2\pi}{3} + i\sin\frac{2\pi}{3}\right)$

9. $\approx \sqrt{13}(\cos 0.59 + i\sin 0.59)$ **11.** $3\left(\cos\frac{\pi}{6} + i\sin\frac{\pi}{6}\right)$ **13.** $\frac{3\sqrt{3}}{2} - \frac{3}{2}i$ **15.** $5/2 - (5/2)\sqrt{3}i$

17. $\frac{\sqrt{6}}{2} - \frac{\sqrt{2}}{2}i$ **19.** $14(\cos 155° + i\sin 155°)$ **21.** $15\left(\cos\frac{23\pi}{12} + i\sin\frac{23\pi}{12}\right)$

23. $\frac{2}{3}(\cos 30° - i\sin 30°)$ **25.** $2(\cos\pi + i\sin\pi)$ **27.** **(a)** $5 + i; \frac{1}{2} - \frac{5}{2}i$ **(b)** Same as part (a)

29. **(a)** $18 - 4i; \approx 0.35 + 0.41i$ **(b)** Same as part (a) **31.** $-\frac{\sqrt{2}}{2} + i\frac{\sqrt{2}}{2}$ **33.** $4\sqrt{2} + 4\sqrt{2}i$ **35.** $-4 - 4i$ **37.** -8

39. $\frac{-1 + \sqrt{3}i}{\sqrt[3]{4}}; \frac{-1 - \sqrt{3}i}{\sqrt[3]{4}}; \sqrt[3]{2}$ **41.** $\sqrt[3]{3}\left(\cos\frac{4\pi}{9} + i\sin\frac{4\pi}{9}\right), \sqrt[3]{3}\left(\cos\frac{10\pi}{9} + i\sin\frac{10\pi}{9}\right), \sqrt[3]{3}\left(\cos\frac{16\pi}{9} + i\sin\frac{16\pi}{9}\right)$

43. $\approx \sqrt[3]{5}(\cos 1.79 + i\sin 1.79), \approx \sqrt[3]{5}(\cos 3.88 + i\sin 3.88), \approx \sqrt[3]{5}(\cos 5.97 + i\sin 5.97)$

45. $\cos\frac{\pi}{5} + i\sin\frac{\pi}{5}, \cos\frac{3\pi}{5} + i\sin\frac{3\pi}{5}, -1, \cos\frac{7\pi}{5} + i\sin\frac{7\pi}{5}, \cos\frac{9\pi}{5} + i\sin\frac{9\pi}{5}$

47. $\sqrt[5]{2}\left(\cos\frac{\pi}{30} + i\sin\frac{\pi}{30}\right), \sqrt[5]{2}\left(\cos\frac{13\pi}{30} + i\sin\frac{13\pi}{30}\right), \sqrt[5]{2}\left(\cos\frac{5\pi}{6} + i\sin\frac{5\pi}{6}\right), \sqrt[5]{2}\left(\cos\frac{37\pi}{30} + i\sin\frac{37\pi}{30}\right), \sqrt[5]{2}\left(\cos\frac{49\pi}{30} + i\sin\frac{49\pi}{30}\right)$

49. $\sqrt[5]{2}\left(\cos\frac{\pi}{10} + i\sin\frac{\pi}{10}\right), \sqrt[5]{2}i, \sqrt[5]{2}\left(\cos\frac{9\pi}{10} + i\sin\frac{9\pi}{10}\right), \sqrt[5]{2}\left(\cos\frac{13\pi}{10} + i\sin\frac{13\pi}{10}\right), \sqrt[5]{2}\left(\cos\frac{17\pi}{10} + i\sin\frac{17\pi}{10}\right)$

51. $\sqrt[8]{2}\left(\cos\frac{\pi}{16} + i\sin\frac{\pi}{16}\right), \sqrt[8]{2}\left(\cos\frac{9\pi}{16} + i\sin\frac{9\pi}{16}\right), \sqrt[8]{2}\left(\cos\frac{17\pi}{16} + i\sin\frac{17\pi}{16}\right), \sqrt[8]{2}\left(\cos\frac{25\pi}{16} + i\sin\frac{25\pi}{16}\right)$

53. $\sqrt{2}\left(\cos\frac{\pi}{12} + i\sin\frac{\pi}{12}\right), -1 + i, \sqrt{2}\left(\cos\frac{17\pi}{12} + i\sin\frac{17\pi}{12}\right)$

55. $\frac{1+i}{\sqrt[6]{4}}$, $\sqrt[6]{2}\left(\cos\frac{7\pi}{12}+i\sin\frac{7\pi}{12}\right)$, $\sqrt[6]{2}\left(\cos\frac{11\pi}{12}+i\sin\frac{11\pi}{12}\right)$, $\sqrt[6]{2}\left(\cos\frac{5\pi}{4}+i\sin\frac{5\pi}{4}\right)$, $\sqrt[6]{2}\left(\cos\frac{19\pi}{12}+i\sin\frac{19\pi}{12}\right)$, $\sqrt[6]{2}\left(\cos\frac{23\pi}{12}+i\sin\frac{23\pi}{12}\right)$

57. $1, -\frac{1}{2} \pm \frac{\sqrt{3}}{2}i$

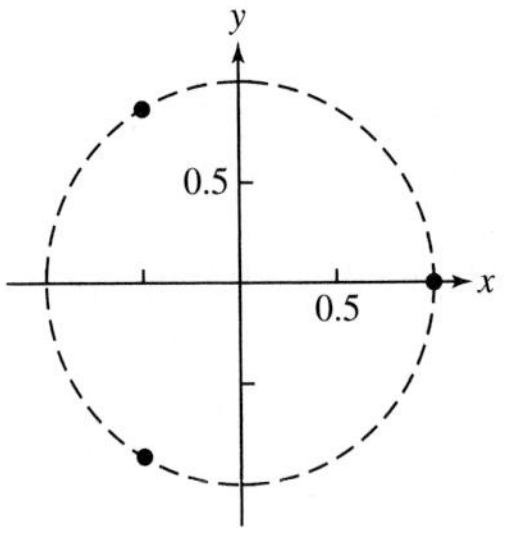

59. $\pm 1, \frac{1}{2} \pm \frac{\sqrt{3}}{2}i, -\frac{1}{2} \pm \frac{\sqrt{3}}{2}i$

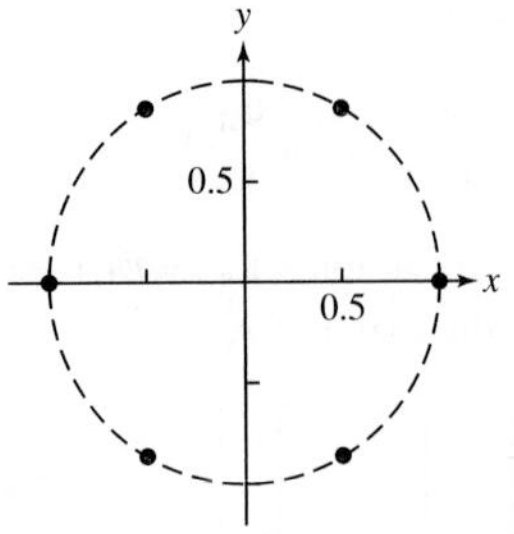

61. -8; -2 and $1 \pm \sqrt{3}i$ **65.** False. For example, the complex number $1 + i$ has infinitely many polar forms. Here are two: $\sqrt{2}\left(\cos\frac{\pi}{4}+i\sin\frac{\pi}{4}\right)$, $\sqrt{2}\left(\cos\frac{9\pi}{4}+i\sin\frac{9\pi}{4}\right)$. **67.** B **69.** A

71. **(a)** (no answer) **(b)** r^2 **(c)** $\cos(2\theta) + i\sin(2\theta)$ **(d)** (no answer)

73. Set the calculator for rounding to 2 decimal places. In part (b), use Degree mode.

(a)
```
25√(2)*(cos(-π/4
)+isin(-π/4))*14
*(cos(π/3)+isin(
π/3))
  478.11+128.11i
```
(b)
```
2√(2)*(cos(135)+
isin(135))/(6*(c
os(300)+isin(300
)))
         -.46-.12i
```
(c)
```
(2(cos(π/3)+isin(
π/3)))^3
                -8
```

75. $x(t) = (\sqrt{3})^t\cos(0.62t)\, y(t) = (\sqrt{3})^t\sin(0.62t)$

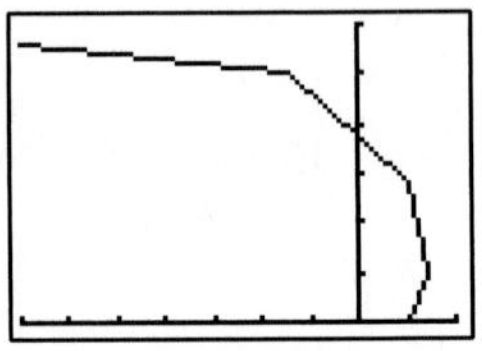

[–7, 2] by [0, 6]

79. $1, -\frac{1}{2} + \frac{\sqrt{3}}{2}i, -\frac{1}{2} - \frac{\sqrt{3}}{2}i$

81. $-1, \frac{1}{2} + \frac{\sqrt{3}}{2}i, \frac{1}{2} - \frac{\sqrt{3}}{2}i$

83. $-1, \approx 0.81 + 0.59i, 0.81 - 0.59i, -0.31 + 0.95i, -0.31 - 0.95i$

CHAPTER 6 REVIEW EXERCISES

1. $\langle -2, -3\rangle$ **3.** $\sqrt{37}$ **5.** 6 **7.** $\langle 3, 6\rangle$; $3\sqrt{5}$ **9.** $\langle -8, -3\rangle$; $\sqrt{73}$ **11.** **(a)** $\left\langle -\frac{2}{\sqrt{5}}, \frac{1}{\sqrt{5}}\right\rangle$ **(b)** $\left\langle \frac{6}{\sqrt{5}}, -\frac{3}{\sqrt{5}}\right\rangle$

13. **(a)** $\tan^{-1}\left(\frac{3}{4}\right) \approx 0.64$, $\tan^{-1}\left(\frac{5}{2}\right) \approx 1.19$ **(b)** ≈ 0.55 **15.** $\approx (-2.27, -1.06)$ **17.** $(\sqrt{2}, -\sqrt{2})$

19. $\left(1, -\frac{2\pi}{3} + (2n+1)\pi\right)$ and $\left(-1, -\frac{2\pi}{3} + 2n\pi\right)$, n an integer

21. **(a)** $\left(-\sqrt{13}, \pi + \tan^{-1}\left(-\frac{3}{2}\right)\right) \approx (-\sqrt{13}, 2.16)$ or $\left(\sqrt{13}, 2\pi + \tan^{-1}\left(-\frac{3}{2}\right)\right) \approx (\sqrt{13}, 5.30)$

(b) $\left(\sqrt{13}, \tan^{-1}\left(-\frac{3}{2}\right)\right) \approx (\sqrt{13}, -0.98)$ or $\left(-\sqrt{13}, \pi + \tan^{-1}\left(-\frac{3}{2}\right)\right) \approx (-\sqrt{13}, 2.16)$

(c) The answers from part (a), and also $\left(-\sqrt{13}, 3\pi + \tan^{-1}\left(-\frac{3}{2}\right)\right) \approx (-\sqrt{13}, 8.44)$ or $\left(\sqrt{13}, 4\pi + \tan^{-1}\left(-\frac{3}{2}\right)\right) \approx (\sqrt{13}, 11.58)$

23. **(a)** $(5, 0)$ or $(-5, \pi)$ or $(5, 2\pi)$ **(b)** $(-5, -\pi)$ or $(5, 0)$ or $(-5, \pi)$ **(c)** The answers from part (a), and also $(-5, 3\pi)$ or $(5, 4\pi)$

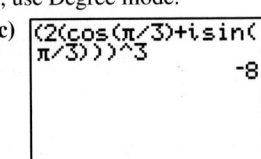

25. $y = -\frac{3}{5}x + \frac{29}{5}$: line through $\left(0, \frac{29}{5}\right)$ with slope $m = -\frac{3}{5}$ **27.** $x = 2(y+1)^2 + 3$: parabola that opens to right with vertex at $(3, -1)$

29. $y = \sqrt{x+1}$: square root function starting at $(-1, 0)$ **31.** $x = 2t + 3, y = 3t + 4$ **33.** $a = -3, b = 4, |z_1| = 5$ **35.** $3\sqrt{3} + 3i$

37. $-1.25 - 1.25\sqrt{3}i$ **39.** $3\sqrt{2}\left(\cos \frac{7\pi}{4} + i \sin \frac{7\pi}{4}\right)$. Other representations would use angles $\frac{7\pi}{4} + 2n\pi$, n an integer.

41. $\approx \sqrt{34}[\cos(5.25) + i \sin(5.25)]$. Other representations would use angles $\approx 5.25 + 2n\pi$, n an integer.

43. $12(\cos 90° + i \sin 90°); \frac{3}{4}(\cos 330° + i \sin 330°)$ **45.** **(a)** $243\left(\cos \frac{5\pi}{4} + i \sin \frac{5\pi}{4}\right)$ **(b)** $-\frac{243\sqrt{2}}{2} - \frac{243\sqrt{2}}{2}i$

47. **(a)** $125(\cos \pi + i \sin \pi)$ **(b)** -125

49. $\sqrt[8]{18}\left(\cos \frac{\pi}{16} + i \sin \frac{\pi}{16}\right), \sqrt[8]{18}\left(\cos \frac{9\pi}{16} + i \sin \frac{9\pi}{16}\right),$
$\sqrt[8]{18}\left(\cos \frac{17\pi}{16} + i \sin \frac{17\pi}{16}\right), \sqrt[8]{18}\left(\cos \frac{25\pi}{16} + i \sin \frac{25\pi}{16}\right)$

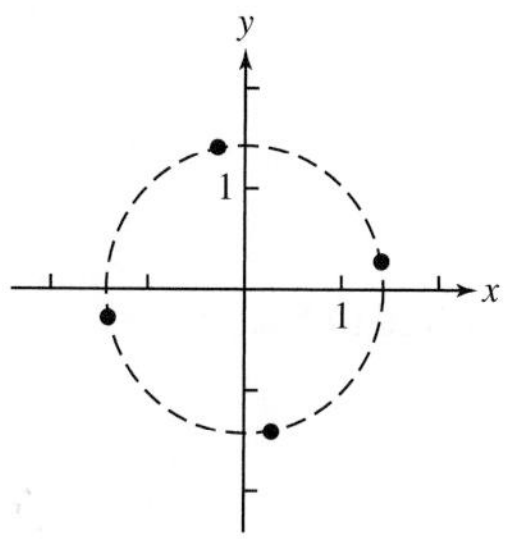

51. $1, \cos \frac{2\pi}{5} + i \sin \frac{2\pi}{5}, \cos \frac{4\pi}{5} + i \sin \frac{4\pi}{5},$
$\cos \frac{6\pi}{5} + i \sin \frac{6\pi}{5}, \cos \frac{8\pi}{5} + i \sin \frac{8\pi}{5}$

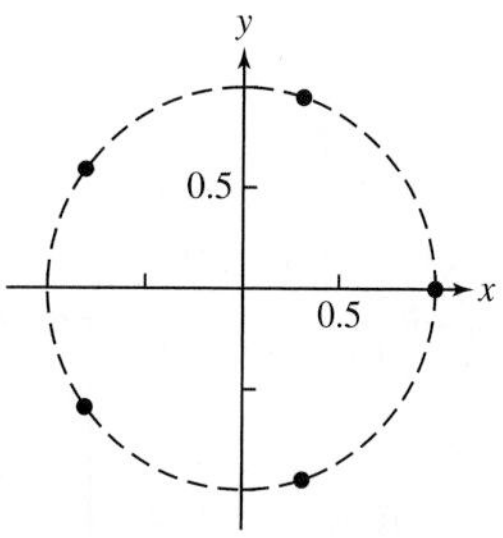

53. (b) **55.** (a) **57.** Not shown **59.** (c) **61.** $x^2 + y^2 = 4$—a circle with center $(0, 0)$ and radius 2

63. $\left(x + \frac{3}{2}\right)^2 + (y + 1)^2 = \frac{13}{4}$—a circle of radius $\frac{\sqrt{13}}{2}$ with center $\left(-\frac{3}{2}, -1\right)$

65. $r = -\frac{4}{\sin \theta} = -4 \csc \theta$

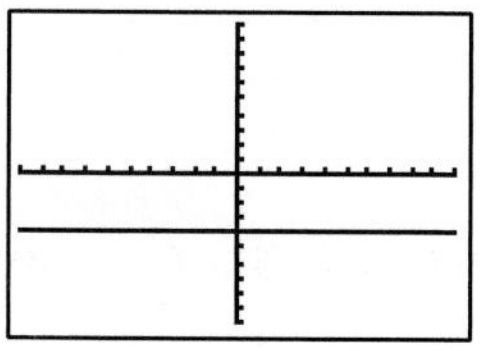

[–10, 10] by [–10, 10]

67. $r = 6 \cos \theta - 2 \sin \theta$

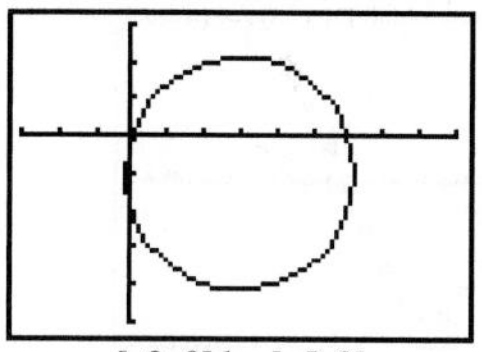

[–3, 9] by [–5, 3]

69.

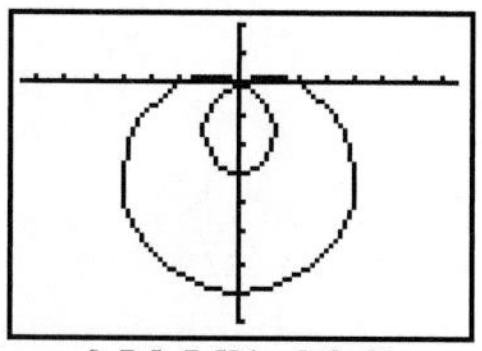

[–7.5, 7.5] by [–8, 2]

Domain: $(-\infty, \infty)$
Range: $[-3, 7]$
Continuous
Symmetric about the y-axis
Bounded
Maximum $|r|$ value: 7
No asymptotes

71.

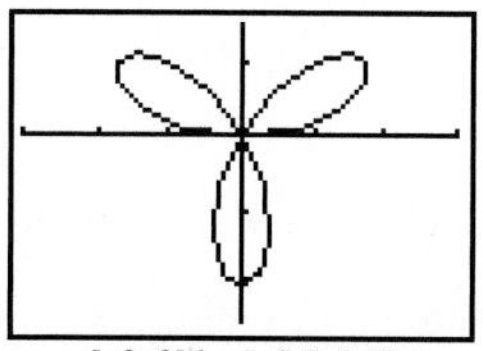

[–3, 3] by [–2.5, 1.5]

Domain: $(-\infty, \infty)$
Range: $[-2, 2]$
Continuous
Symmetric about the y-axis
Bounded
Maximum $|r|$ value: 2
No asymptotes

73. **(a)** (no answer) **(b)** (no answer) **(c)** (no answer)
(d)

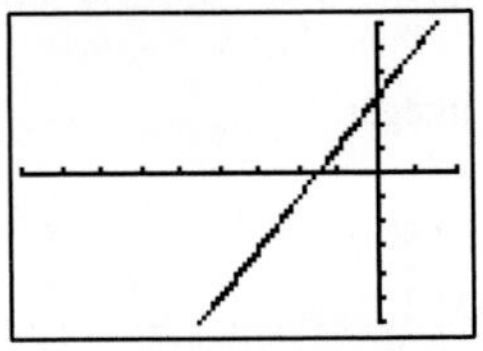

[–9, 2] by [–6, 6]

75. **(a)** $\approx \langle -463.64, 124.23 \rangle$ **(b)** $\approx$ 508.29 mph; bearing $\approx 283.84°$
77. **(a)** $\approx$ 826.91 lb **(b)** 2883.79 lb

79. **(a)** $h = -16t^2 + 245t + 200$ **(b)** Graph and trace: $x = 17$ and $y = -16t^2 + 245t + 200$ with $0 \le t \le 16.1$ (upper limit may vary) on $[0, 18]$ by $[0, 1200]$. This graph will appear as a vertical line from about (17, 0) to about (17, 1138). Tracing shows how the arrow begins at a height of 200 ft, rises to over 1000 ft, then falls back to the ground.
(c)

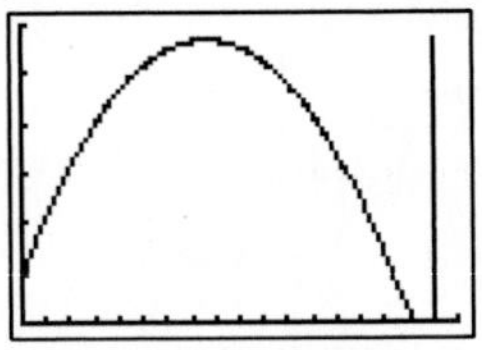

[0, 18] by [0, 1200]

(d) 924 ft **(e)** $\approx$ 1138 ft; $t \approx 7.66$ **(f)** About 16.09 sec

81. $x = 40 \sin\left(\frac{2\pi}{15}t\right)$, $y = 50 - 40 \cos\left(\frac{2\pi}{15}t\right)$ **83.** **(a)**

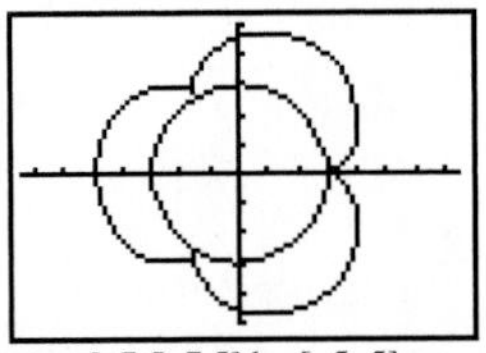

[–7.5, 7.5] by [–5, 5]

(b) All 4's should be changed to 5's.

85. $t \approx 1.06$ sec, $x \approx 68.65$ ft **87.** **(a)** $\approx$ 77.59 ft **(b)** $\approx$ 4.404 sec **89.** $\approx$ 17.65 ft

Chapter 6 Project

Answers are based on the sample data shown in the table.

1.

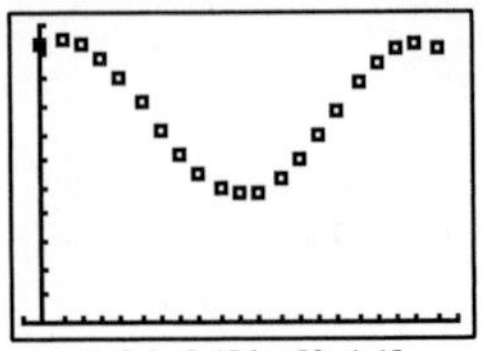

[–0.1, 2.1] by [0, 1.1]

3.

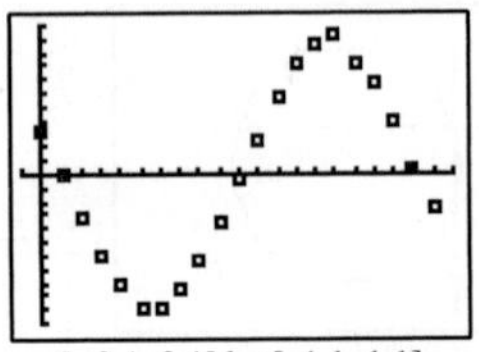

[–0.1, 2.1] by [–1.1, 1.1]

5.

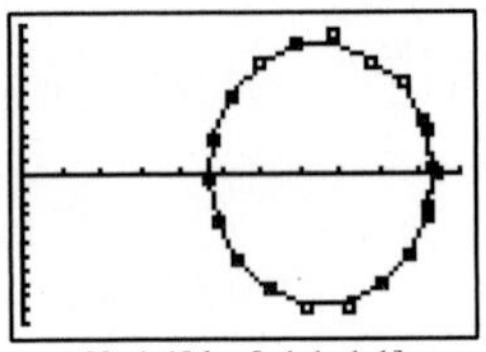

[0, 1.1] by [–1.1, 1.1]

SECTION 7.1

Exploration 1

1.

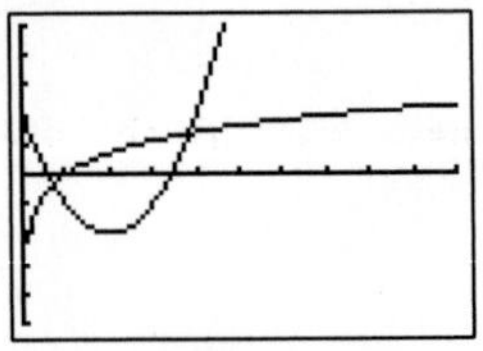

[0, 10] by [–5, 5]

Quick Review 7.1

1. $y = \frac{5}{3} - \frac{2}{3}x$ **3.** $x = -\frac{2}{3}, x = 1$ **5.** 0, 2, –2 **7.** $4x + 5y = 6$ **9.** $-4x - 6y = -10$

Exercises 7.1

1. **(a)** No **(b)** Yes **(c)** No **3.** (9, –2) **5.** (50/7, –10/7) **7.** (–1/2, 2) **9.** No solution **11.** $(\pm 3, 9)$

13. (–3/2, 27/2) and (1/3, 2/3) **15.** (0, 0) and (3, 18) **17.** $\left(\frac{-1 + 3\sqrt{89}}{10}, \frac{3 + \sqrt{89}}{10}\right)$ and $\left(\frac{-1 - 3\sqrt{89}}{10}, \frac{3 - \sqrt{89}}{10}\right)$

19. $(8, -2)$ **21.** $(4, 2)$ **23.** No solution **25.** Infinitely many solutions **27.** $(0, 1)$ and $(3, -2)$ **29.** No solution
31. One solution **33.** Infinitely many solutions **35.** $\approx (0.69, -0.37)$ **37.** $\approx (-2.32, -3.16)$, $(0.47, -1.77)$ and $(1.85, -1.08)$
39. $(-1.2, 1.6)$ and $(2, 0)$ **41.** $\approx (2.05, 2.19)$ and $(-2.05, 2.19)$ **43.** Demand curve Supply curve $(3.75, 143.75)$

45. **(a)** $Y \approx -0.2262x^2 + 3.9214x + 22.7333$

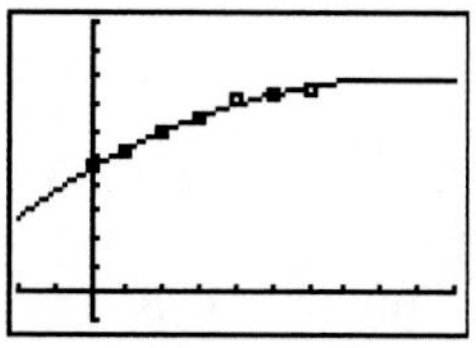

[–2, 10] by [–5, 50]

(b) $y \approx \dfrac{42.2253}{1 + 0.8520 + e^{-0.3467x}}$

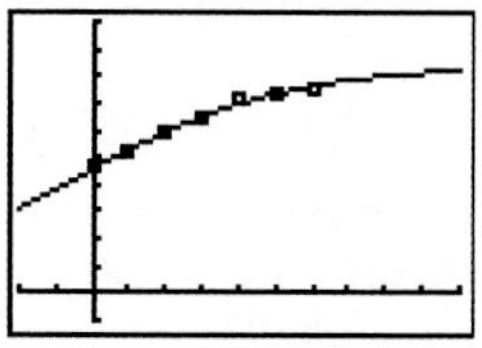

[–2, 10] by [–5, 50]

(c) Quadratic: Never. Logistic: About 2015.
(d) The quadratic regression predicts that the expenditures will eventually be zero.
(e) The logistic regression predicts that the expenditures will eventually level off at about \$42.2 billion.

47. **(a)** $y \approx 302.09x + 9835.4$
(b) $y \approx 35.15x + 5372$
(c) About 1963

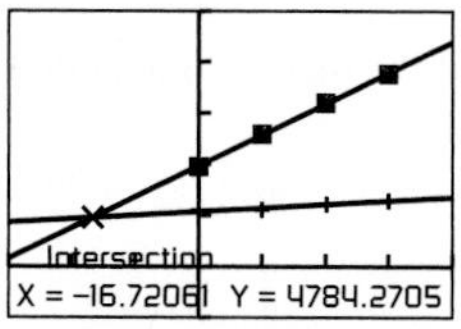

[–30, 40] by [–5000, 25000]

49. ≈ 5.28 m $\times \approx 94.72$ m **51.** Current speed ≈ 1.06 mph; rowing speed ≈ 3.56 mph
53. Medium: \$0.79; large: \$0.95 **55.** $a = 2/3$ and $b = 14/3$
57. **(a)** 300 mi
59. False. A system of two linear equations in two variables has either 0, 1, or infinitely many solutions.
61. C **63.** D

65. **(a)** $y = (3/2)\sqrt{4 - x^2}$, $y = -(3/2)\sqrt{4 - x^2}$
(b)

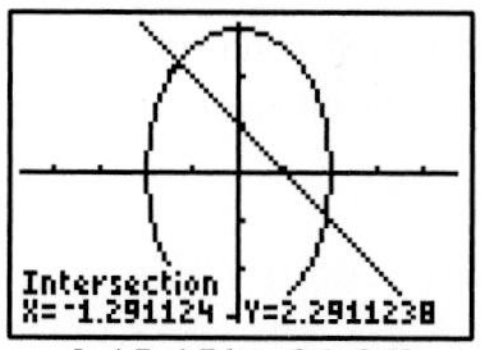

[–4.7, 4.7 by –3.1, 3.1]

$\approx (-1.29, 2.29)$ or $(1.91, -0.91)$

67. $\left(\pm\sqrt{2/3}, 10/3\right)$
69. 12.5 units

SECTION 7.2

Exploration 1

1. $A = \begin{bmatrix} 2 & 1 \\ 5 & 4 \end{bmatrix}$; $B = \begin{bmatrix} -1 & 2 \\ 2 & 5 \end{bmatrix}$ **3.** $\begin{bmatrix} 8 & -1 \\ 11 & 2 \end{bmatrix}$

Exploration 2

1. $\det A = -a_{12}a_{21}a_{33} + a_{13}a_{21}a_{32} + a_{11}a_{22}a_{33} - a_{13}a_{22}a_{31} - a_{11}a_{23}a_{32} + a_{12}a_{23}a_{31}$
3. Since each term in the expansion contains an element from each row and each column, at least one factor in each term is a zero. Therefore, the expansion will be the sum of n zero terms, or zero.

Quick Review 7.2

1. $(3, 2)$; $(x, -y)$ **3.** $(-2, 3)$; (y, x) **5.** $(3 \cos \theta, 3 \sin \theta)$ **7.** $\sin \alpha \cos \beta + \cos \alpha \sin \beta$ **9.** $\cos \alpha \cos \beta - \sin \alpha \sin \beta$

Exercises 7.2

1. 2×3; not square **3.** 3×2; not square **5.** 3×1; not square **7.** 3 **9.** 4

11. **(a)** $\begin{bmatrix} 3 & 0 \\ -3 & 1 \end{bmatrix}$ **(b)** $\begin{bmatrix} 1 & 6 \\ 1 & 9 \end{bmatrix}$ **(c)** $\begin{bmatrix} 6 & 9 \\ -3 & 15 \end{bmatrix}$ **(d)** $\begin{bmatrix} 1 & 15 \\ 4 & 22 \end{bmatrix}$ **13.** **(a)** $\begin{bmatrix} 1 & 1 \\ -2 & 0 \\ -1 & 0 \end{bmatrix}$ **(b)** $\begin{bmatrix} -7 & 1 \\ 2 & -2 \\ 5 & 2 \end{bmatrix}$ **(c)** $\begin{bmatrix} -9 & 3 \\ 0 & -3 \\ 6 & 3 \end{bmatrix}$

(d) $\begin{bmatrix} -18 & 2 \\ 6 & -5 \\ 13 & 5 \end{bmatrix}$ **15.** **(a)** $\begin{bmatrix} -3 \\ 1 \\ 4 \end{bmatrix}$ **(b)** $\begin{bmatrix} -1 \\ 1 \\ -4 \end{bmatrix}$ **(c)** $\begin{bmatrix} -6 \\ 3 \\ 0 \end{bmatrix}$ **(d)** $\begin{bmatrix} -1 \\ 2 \\ -12 \end{bmatrix}$ **17.** **(a)** $\begin{bmatrix} -4 & -18 \\ -11 & -17 \end{bmatrix}$ **(b)** $\begin{bmatrix} 5 & -12 \\ 0 & -26 \end{bmatrix}$

19. **(a)** $\begin{bmatrix} 2 & 2 \\ -11 & 12 \end{bmatrix}$ **(b)** $\begin{bmatrix} 4 & 8 & -5 \\ -5 & 4 & -6 \\ -2 & -8 & 6 \end{bmatrix}$ **21.** **(a)** $\begin{bmatrix} 6 & -7 & -2 \\ 3 & 7 & 3 \\ 8 & -1 & -1 \end{bmatrix}$ **(b)** $\begin{bmatrix} 2 & 1 & 3 \\ 5 & 0 & 0 \\ -18 & -3 & 10 \end{bmatrix}$ **23.** **(a)** $[-8]$

(b) $\begin{bmatrix} -10 & 5 & -15 \\ 8 & -4 & 12 \\ 4 & -2 & 6 \end{bmatrix}$ **25.** Not possible; $[18\ 14]$ **27.** **(a)** $\begin{bmatrix} -1 & 3 & 4 \\ 2 & 0 & 1 \\ 1 & 2 & 1 \end{bmatrix}$ **(b)** $\begin{bmatrix} 1 & 2 & 1 \\ 1 & 0 & 2 \\ 4 & 3 & -1 \end{bmatrix}$ **29.** $a = 5, b = 2$

31. $a = -2, b = 0$ **33.** $AB = BA = I_2$ **35.** $\begin{bmatrix} -1 & 1.5 \\ 1 & -1 \end{bmatrix}$ **37.** No inverse **39.** No inverse **41.** -14 **43.** $\begin{bmatrix} 1 \\ -1/3 \end{bmatrix}$

45. **(a)** The distance from city *X* to city *Y* is the same as the distance from city *Y* to city *X*.
(b) Each entry represents the distance from city *X* to city *X*. **47.** **(a)** $[382\ 227.50]$ **49.** **(a)** AB^T or BA^T **(b)** $(A - C)B^T$

51. **(a)** $\approx (1.37\ 0.37)$ **(b)** $\approx (0.37\ 1.37)$ **55.** $A \cdot A^{-1} = I_2$ **57.** $[x\ y]\begin{bmatrix} -1 & 0 \\ 0 & 1 \end{bmatrix}$ **59.** $[x\ y]\begin{bmatrix} 0 & -1 \\ -1 & 0 \end{bmatrix}$ **61.** $[x\ y]\begin{bmatrix} c & 0 \\ 0 & 1 \end{bmatrix}$

63. False. It can be negative. For example, the determinant of $A = \begin{bmatrix} 1 & 0 \\ 2 & -1 \end{bmatrix}$ is -1. **65.** B **67.** D

71. **(a)** $A \cdot A^{-1} = A^{-1} \cdot A = I_2$ **(c)** It is the inverse of *A*.
73. **(b)** The constant term equals $-\det A$. **(c)** The coefficient of x^2 is the opposite of the sum of the elements of the main diagonal in *A*.

SECTION 7.3

Exploration 1

1. $B = 3$

Exploration 2

1. $x + y + z$ must equal 60 L. **3.** The number of liters of 35% solution must equal twice the number of liters of 55% solution.

5. $\begin{bmatrix} 3.75 \\ 37.5 \\ 18.75 \end{bmatrix}$

Quick Review 7.3

1. 12.8 L **3.** 38 L **5.** $(-1, 6)$ **7.** $y = -z + w + 1$ **9.** $\begin{bmatrix} -0.5 & -0.75 \\ 0.5 & 0.25 \end{bmatrix}$

Exercises 7.3

1. $\left(\frac{25}{2}, \frac{7}{2}, -2\right)$ **3.** $(1, 2, 1)$ **5.** No solution **7.** $\left(\frac{9}{2}, \frac{7}{2}, 4, -\frac{15}{2}\right)$

9. $\begin{bmatrix} 2 & -6 & 4 \\ 1 & 2 & -3 \\ 0 & -8 & 4 \end{bmatrix}$ **11.** $\begin{bmatrix} 0 & -10 & 10 \\ 1 & 2 & -3 \\ -3 & 1 & -2 \end{bmatrix}$ **13.** R_{12} **15.** $(-3)R_2 + R_3$

For Exercises 17–20, possible answers are given.

17. $\begin{bmatrix} 1 & 3 & -1 \\ 0 & 1 & -1.2 \\ 0 & 0 & 1 \end{bmatrix}$ **19.** $\begin{bmatrix} 1 & 2 & 3 & -4 \\ 0 & 1 & 0 & -0.6 \\ 0 & 0 & 1 & -9.2 \end{bmatrix}$ **21.** $\begin{bmatrix} 1 & 0 & 2 & 1 \\ 0 & 1 & -1 & 2 \\ 0 & 0 & 0 & 0 \end{bmatrix}$ **23.** $\begin{bmatrix} 1 & 0 & -1 & 3 \\ 0 & 1 & 2 & -1 \end{bmatrix}$

25. $\begin{bmatrix} 2 & -3 & 1 & 1 \\ -1 & 1 & -4 & -3 \\ 3 & 0 & -1 & 2 \end{bmatrix}$ **27.** $\begin{bmatrix} 2 & -5 & 1 & -1 & -3 \\ 1 & 0 & -2 & 1 & 4 \\ 0 & 2 & -3 & -1 & 5 \end{bmatrix}$

In Exercises 29–32, the variable names are arbitrary.

29. $3x + 2y = -1$
$-4x + 5y = 2$

31. $2x + z = 3$
$-x + y = 2$
$2y - 3z = -1$

33. $(2, -1, 4)$; $\begin{bmatrix} 1 & -2 & 1 & 8 \\ 0 & 1 & -1 & -5 \\ 0 & 0 & 1 & 4 \end{bmatrix}$ **35.** $(-2, 3, 1)$ **37.** No solution **39.** $(2 - z, 1 + z, z)$ **41.** No solution

43. $(z + w + 2, 2z - w - 1, z, w)$ **45.** $\begin{bmatrix} 2 & 5 \\ 1 & -2 \end{bmatrix}\begin{bmatrix} x \\ y \end{bmatrix} = \begin{bmatrix} -3 \\ 1 \end{bmatrix}$ **47.** $\begin{aligned} 3x - y &= -1 \\ 2x + 4y &= 3 \end{aligned}$ **49.** $(-2, 3)$

51. $(-2, -5, -7)$ **53.** $(-1, 2, -2, 3)$ **55.** $(0, -10, 1)$ **57.** $(3, 3, -2, 0)$ **59.** $\left(2 - \frac{3}{2}z, -\frac{1}{2}z - 4, z\right)$

61. $(-2w - 1, w + 1, -w, w)$ **63.** $(-w - 2, -z + 0.5, z, w)$ **65.** No solution **67.** $\frac{-3}{x + 4} + \frac{4}{x - 2}$ **69.** $\frac{1}{x - 5} + \frac{-1}{x - 3}$

71. $\frac{2}{x - 1} + \frac{-2}{x - 1}$ **73.** $\frac{1}{x} - \frac{1}{x + 2}$ **75.** $\frac{1}{x - 3} + \frac{-2}{x + 4}$ **77.** $\frac{-2}{x + 3} + \frac{5}{2x - 1}$

79. $2 + \frac{x + 5}{x^2 - 1}; \frac{3}{x - 1} + \frac{-2}{x + 1}$

Graph of $(2x^2 + x + 3)/(x^2 - 1)$:

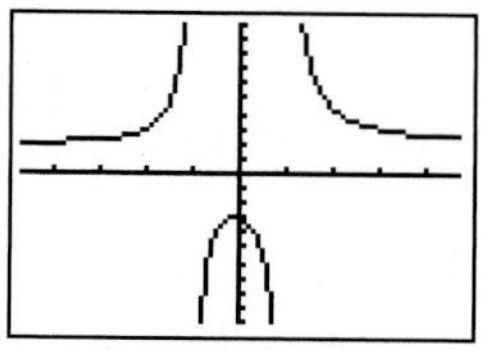

[–4.7, 4.7] by [–10, 10]

Graph of $3/(x - 1)$:

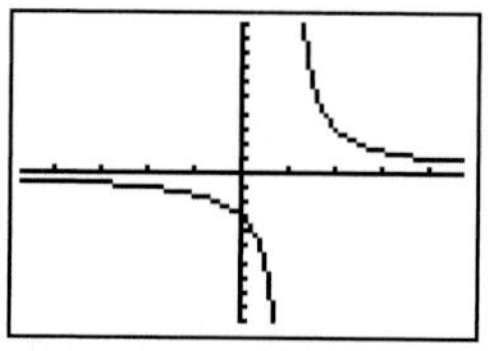

[–4.7, 4.7] by [–10, 10]

Graph of $-2/(x + 1)$:

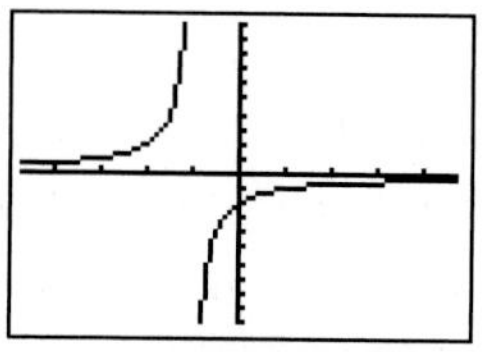

[–4.7, 4.7] by [–10, 10]

81. $x - 1 + \frac{x - 2}{x^2 + x}; \frac{3}{x + 1} + \frac{-2}{x}$

Graph of $y = (x^3 - 2)/(x^2 + x)$:

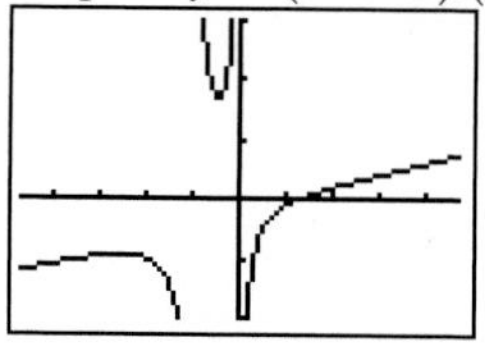

[–4.7, 4.7] by [–10, 15]

Graph of $y = x - 1$:

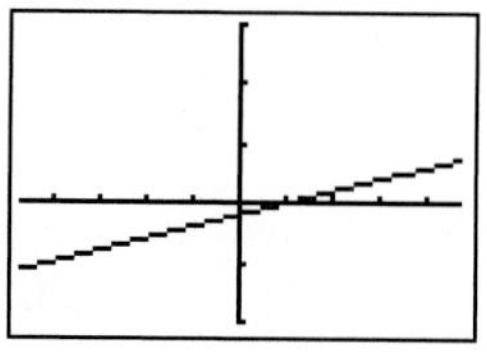

[–4.7, 4.7] by [–10, 15]

Graph of $y = 3/(x + 1)$:

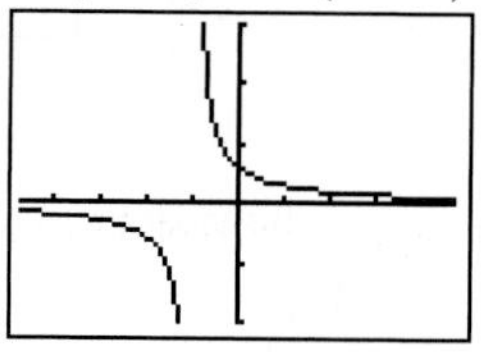

[–4.7, 4.7] by [–10, 15]

Graph of $y = -2/x$:

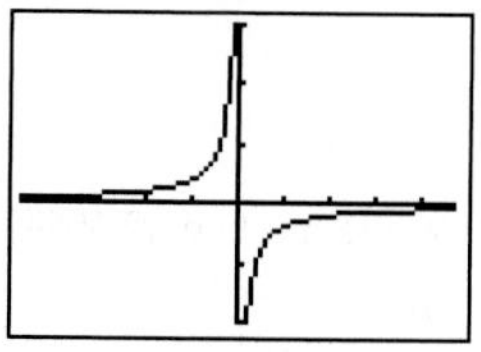

[–4.7, 4.7] by [–10, 15]

83. $f(x) = 2x^2 - 3x - 2$ **85.** $f(x) = (-c - 3)x^2 + x + c$, for any c **87.** **(a)** $y \approx 2.99x + 145.9$ **(b)** $y \approx 3.55x + 115$
(c) About 2035

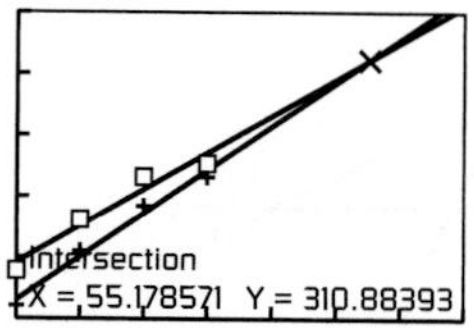

[0, 70] by [100, 350]

89. 825 children, 410 adults, 165 senior citizens **91.** \$14,500 CDs, \$5500 bonds, \$60,000 growth funds **93.** \$0 CDs, \$38,983.05 bonds, \$11,016.95 growth fund **95.** 22 nickels, 35 dimes, and 17 quarters **97.** $(16/3, 220/3)$ **101.** False. The determinant of the matrix must be not equal to zero. **103.** D **105.** D **109.** **(a)** $C(x) = x^2 - 8x + 13$
(b)

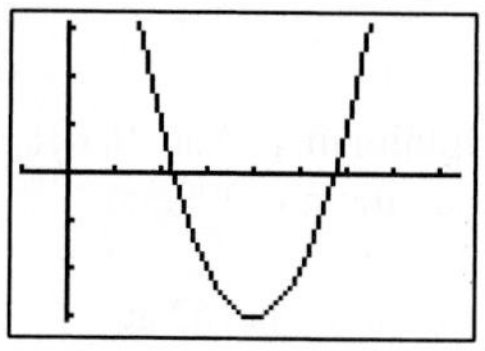

[–1, 8.4] by [–3.1, 3.1]

(c) $4 \pm \sqrt{3}$ **(d)** $\det A = C(0) = 13$
(e) $a_{11} + a_{22} = (4 - \sqrt{3}) + (4 + \sqrt{3}) = 8$

SECTION 7.4

Quick Review 7.4

1. $(3, 0); (0, -2)$

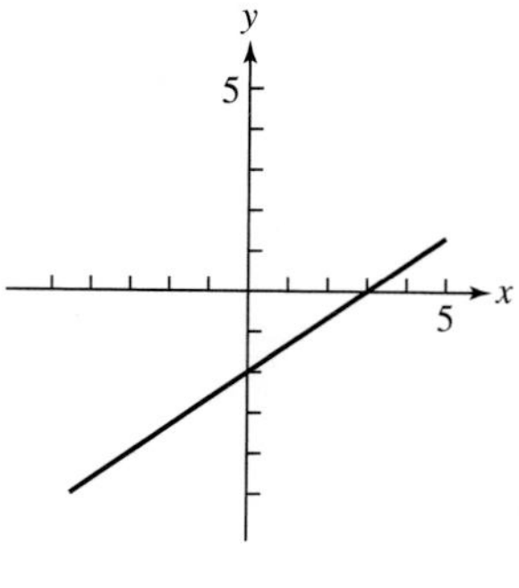

3. $(20, 0); (0, 50)$

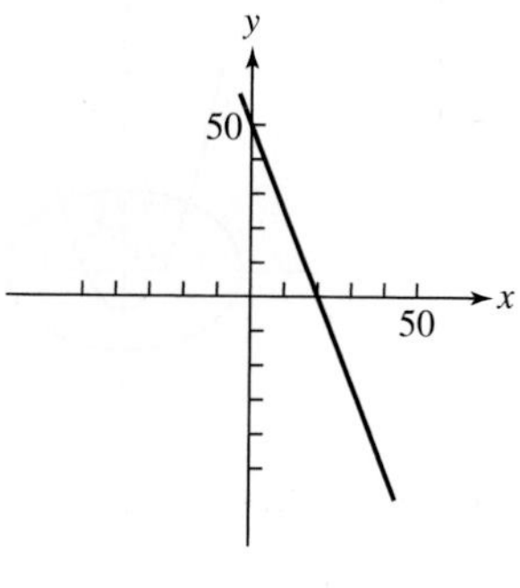

5. $(30, 60)$
7. $(10, 140)$
9. $(3, 3)$

Exercises 7.4

1. Graph (c); boundary included **3.** Graph (b); boundary included **5.** Graph (e); boundary included

7.

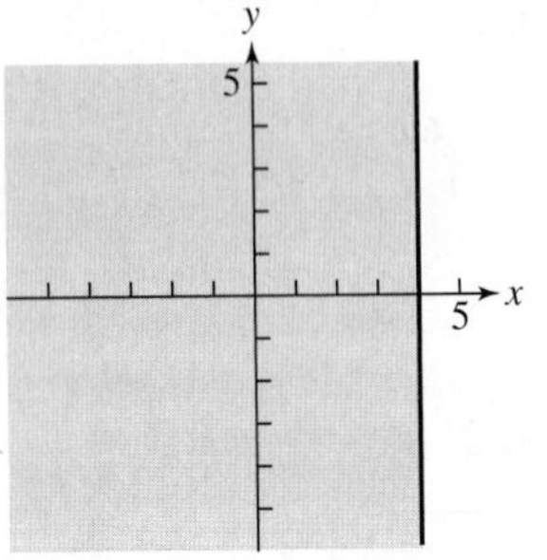

Boundary $x = 4$ included

9.

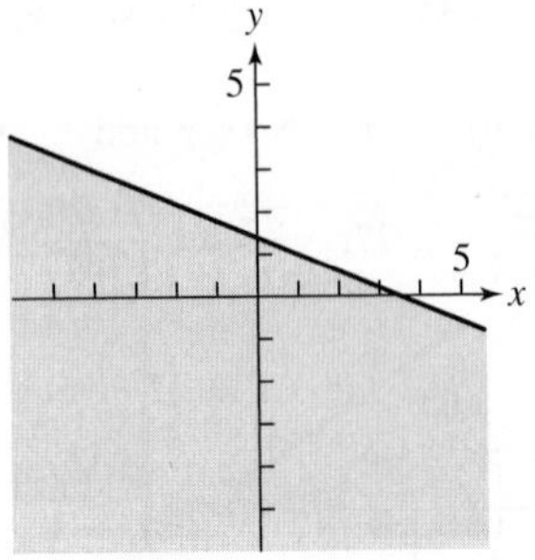

Boundary $2x + 5y = 7$ included

11.

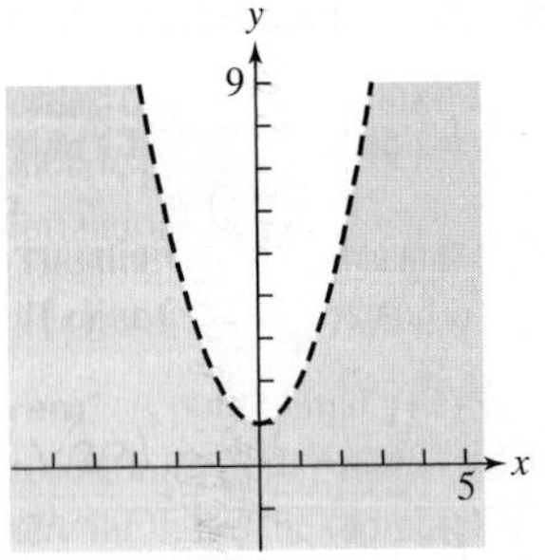

Boundary $y = x^2 + 1$ excluded

13.

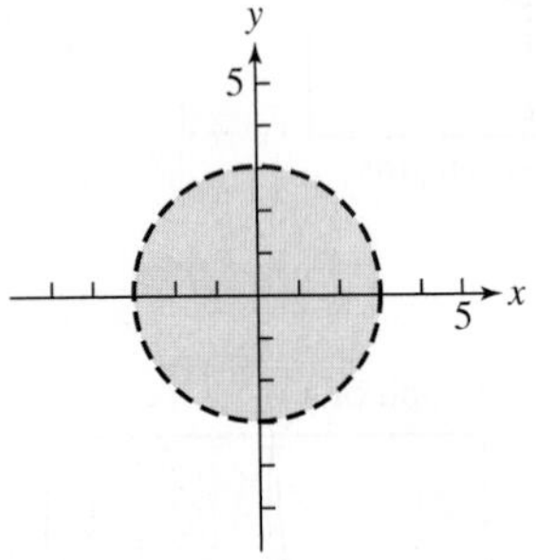

Boundary $x^2 + y^2 = 9$ excluded

15.

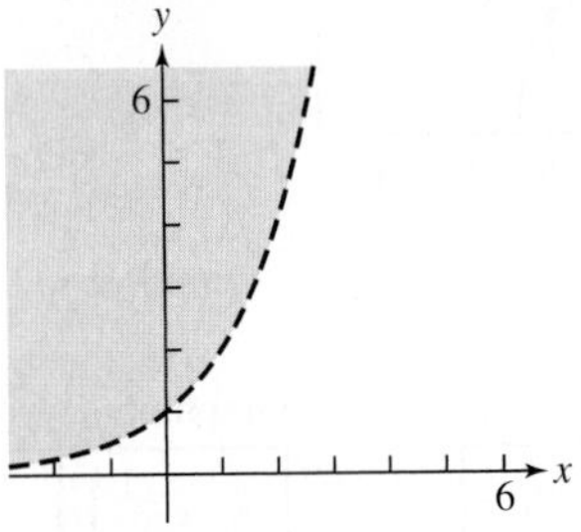

Boundary $y = 2^x$ excluded

17.

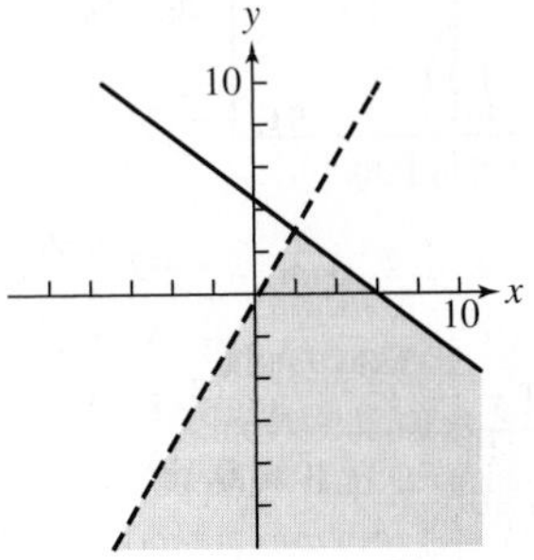

19.

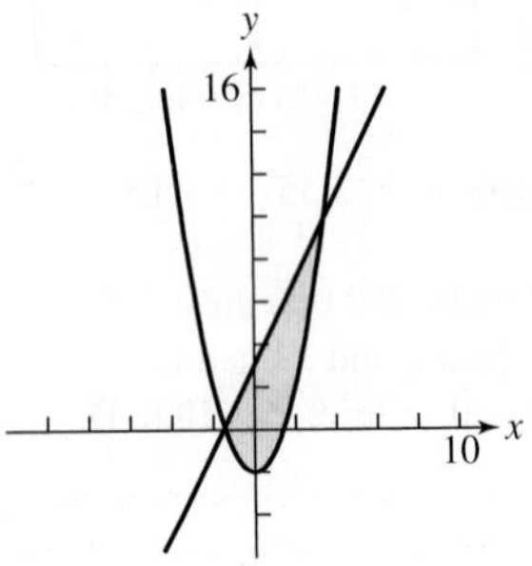

21.

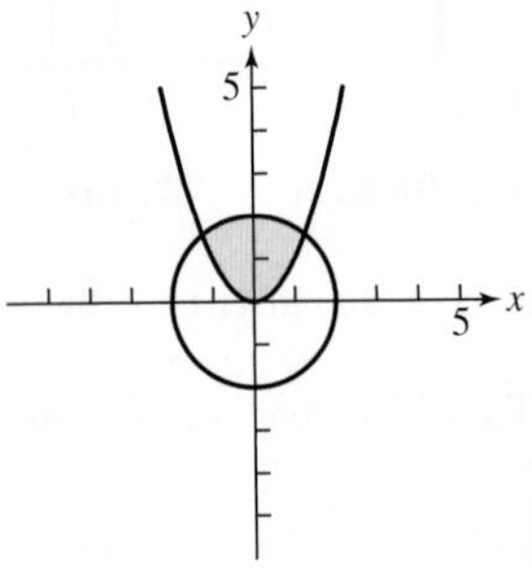

23.

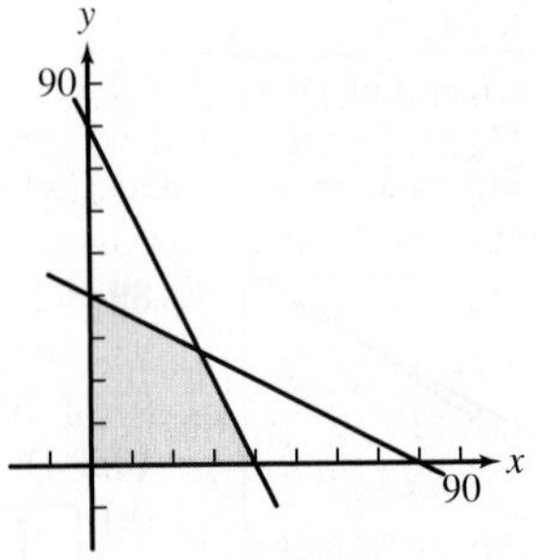

25.

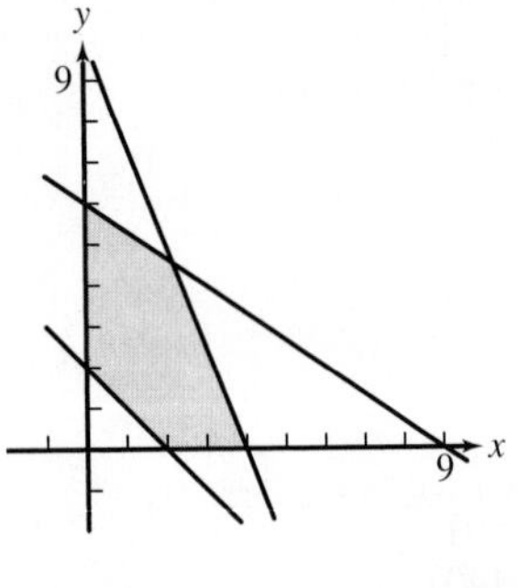

27. $x^2 + y^2 \leq 4$
$y \geq -x^2 + 1$

29. $y \leq -\frac{1}{2}x + 5$
$y \leq -\frac{3}{2}x + 9$
$x \geq 0$
$y \geq 0$

31. The minimum is 0 at $(0, 0)$; the maximum is $293.\overline{3}$ at $(53.\overline{3}, 26.\overline{6})$.

33. The minimum is 162 at $(6, 30)$; there is no maximum.

35. The minimum is 24 at $(0, 12)$; there is no maximum.

37. ≈ 13.48 tons of ore R and ≈ 20.87 tons of ore S; \$1926.20

39. x operations at Refinery 1 and y operations at Refinery 2 such that $2x + 4y = 320$ with $40 \leq x \leq 120$.

41. False. It is a half-plane.

43. A **45.** D **49.** $y_1 = 2\sqrt{1 - \frac{x^2}{9}}$; $y_2 = -2\sqrt{1 - \frac{x^2}{9}}$ **51.**

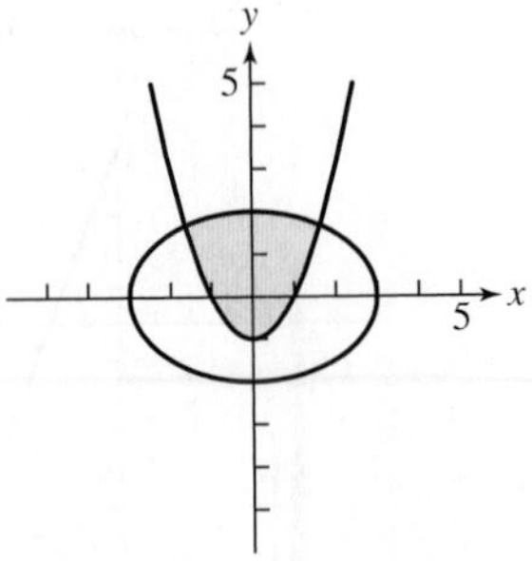

CHAPTER 7 REVIEW EXERCISES

1. **(a)** $\begin{bmatrix} 1 & 2 \\ 8 & 3 \end{bmatrix}$ **(b)** $\begin{bmatrix} -3 & 4 \\ 0 & -3 \end{bmatrix}$ **(c)** $\begin{bmatrix} 2 & -6 \\ -8 & 0 \end{bmatrix}$ **(d)** $\begin{bmatrix} -7 & 11 \\ 4 & -6 \end{bmatrix}$ **3.** $\begin{bmatrix} -3 & -7 & 11 \\ 0 & -12 & 24 \end{bmatrix}$; not possible **5.** $[3\ 7]$; not possible

7. $\begin{bmatrix} 1 & 2 & -3 \\ 2 & -3 & 4 \\ -2 & 1 & -1 \end{bmatrix}$; $\begin{bmatrix} -3 & 2 & 4 \\ 2 & 1 & -3 \\ 1 & -2 & -1 \end{bmatrix}$ **9.** $AB = BA = I_4$ **11.** $\begin{bmatrix} -2 & -5 & 6 & -1 \\ 0 & -1 & 1 & 0 \\ 10 & 24 & -27 & 4 \\ -3 & -7 & 8 & -1 \end{bmatrix}$ **13.** 20 **15.** $\begin{bmatrix} 1 & 0 & 2 \\ 0 & 1 & -1 \\ 0 & 0 & 0 \end{bmatrix}$

17. $\begin{bmatrix} 1 & 0 & 0 & 8 \\ 0 & 1 & 0 & -11 \\ 0 & 0 & 1 & 5 \end{bmatrix}$ **19.** $(1, 2)$ **21.** No solution **23.** $(-z - w + 2, w + 1, z, w)$ **25.** No solution

27. $(-2z + w + 1, z - w + 2, z, w)$ **29.** $(9/4, -3/4, -7/4)$ **31.** No solution **33.** $(-w + 2, z + 3, z, w)$ **35.** $(-2, 1, 3, -1)$
37. Demand curve Supply curve $(7.57, 42.71)$ **39.** $(x, y) \approx (0.14, -2.29)$ **41.** $(x, y) = (-2, 1)$ or $(x, y) = (2, 1)$

43. $(x, y) \approx (2.27, 1.53)$ **45.** $(a, b, c, d) = (17/840, -33/280, -571/420, 386/35)$ **47.** $\dfrac{1}{x+1} + \dfrac{2}{x-4}$

49. $\dfrac{4}{x+2} + \dfrac{3}{x+5}$ **51.** $\dfrac{4}{x-2} + \dfrac{1}{x-1} - \dfrac{3}{x-3}$ **53.** Graph (a); boundary excluded **55.** Graph (c); boundary included

57.

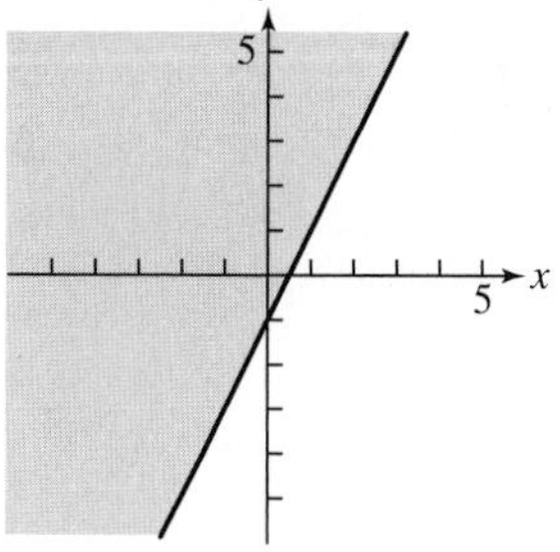

59.

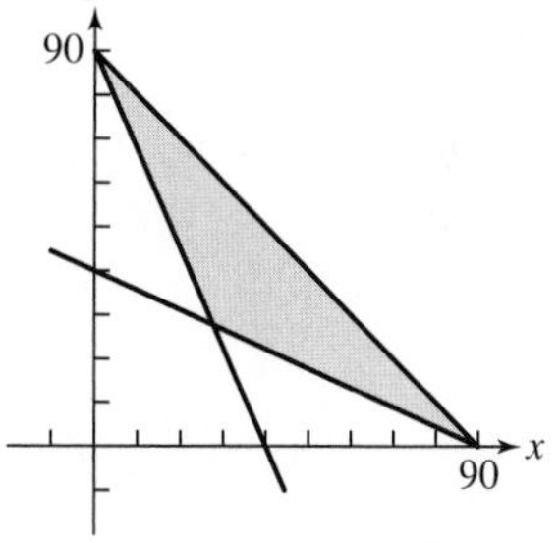

Corners at $(0, 90)$, $(90, 0)$, $(360/13, 360/13)$ Boundaries included

61.

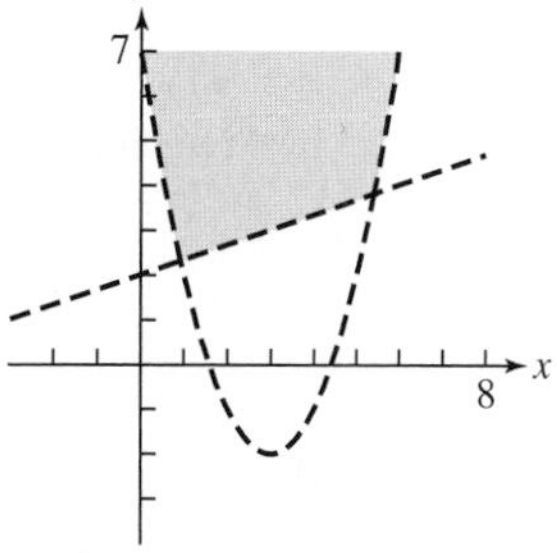

Corners at $\approx (0.92, 2.31)$ and $\approx (5.41, 3.80)$ Boundaries included

63.

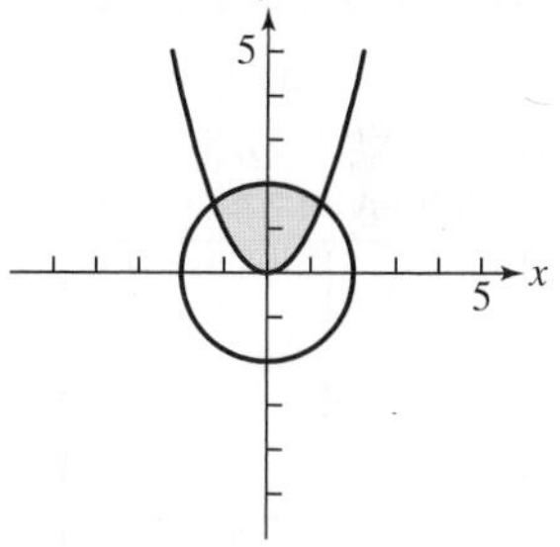

Corners at $\approx (-1.25, 1.56)$ and $\approx (1.25, 1.56)$ Boundaries included

65. The minimum is 106 at $(10, 6)$; there is no maximum. **67.** The minimum is 205 at $(10, 25)$; the maximum is 292 at $(4, 40)$.
69. **(a)** $\approx (2.12, 0.71)$ **(b)** $\approx (-0.71, 2.12)$ **71.** **(a)** $y \approx 12.89x + 967.9$ **(b)** $y \approx 21.59x + 879.4$
(c) About 1989

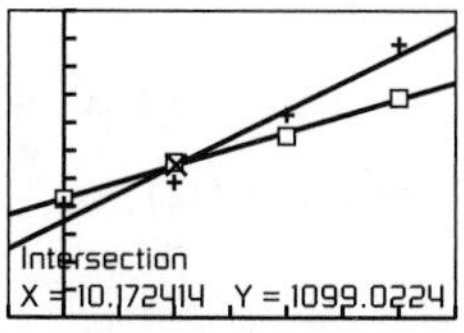

[–5, 35] by [500, 1700]

(d) A linear model seems appropriate for Hawaii's population due to its fairly steady increases over this span of three decades. An exponential or logistic model might be a better fit for Idaho's population, which made big jumps from 1990 to 2000 and from 2000 to 2010 relative to the modest increase of 63,000 persons from 1980 to 1990.

73. **(a)** $N = [200\ 400\ 600\ 250]$ **(b)** $P = [\$80\ \$120\ \$200\ \$300]$ **(c)** $NP^T = \$259{,}000$ **75.** Answers will vary.
77. \$160,000 at 4%, \$170,000 at 6.5%, \$320,000 at 9% **79.** Pipe A: 15 hr; Pipe B: $\approx$ 5.45 hr; Pipe C: 12 hr **81.** n must be equal to p.

Chapter 7 Project

1.

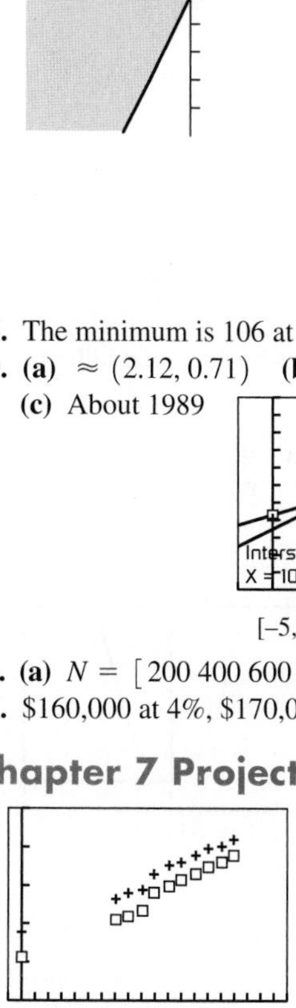

[–2, 22] by [110, 160]

Males population: $y \approx 1.643x + 120.63$
Females population: $y \approx 1.546x + 127.05$

3. The population gap appears to be closing, but the data span only two decades.

5. Male population: $y \approx \dfrac{514.14}{1 + 11.74e^{-0.0144x}}$

Female population: $y \approx \dfrac{315.75}{1 + 7.473e^{-0.0182x}}$

7. Male: 49.16%; female: 50.84%

SECTION 8.1

Exploration 1

1. The axis of the parabola with focus (0, 1) and directrix $y = -1$ is the y-axis because it is perpendicular to $y = -1$ and passes through (0, 1). The vertex lies on this axis midway between the directrix and the focus, so the vertex is the point (0, 0).

3. $\{(-2\sqrt{6}, 6), (-2\sqrt{5}, 5), (-4, 4), (-2\sqrt{3}, 3), (-2\sqrt{2}, 2), (-2, 1), (0, 0), (2, 1), (2\sqrt{2}, 2), (2\sqrt{3}, 3), (4, 4), (2\sqrt{5}, 5), (2\sqrt{6}, 6)\}$

Exploration 2

1.

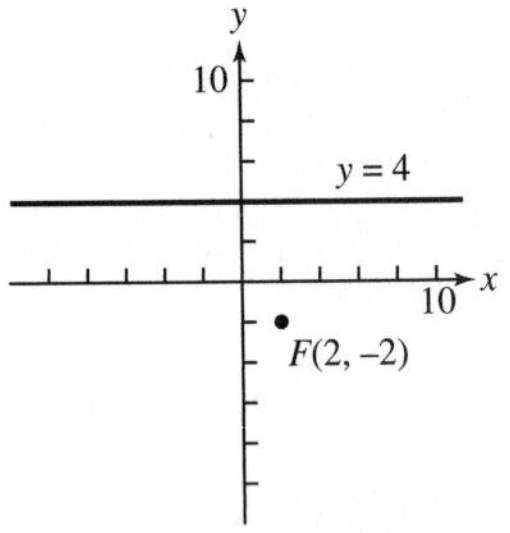

3.

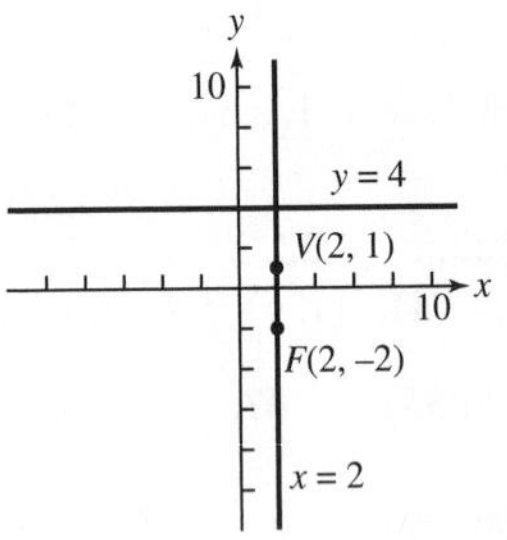

5.

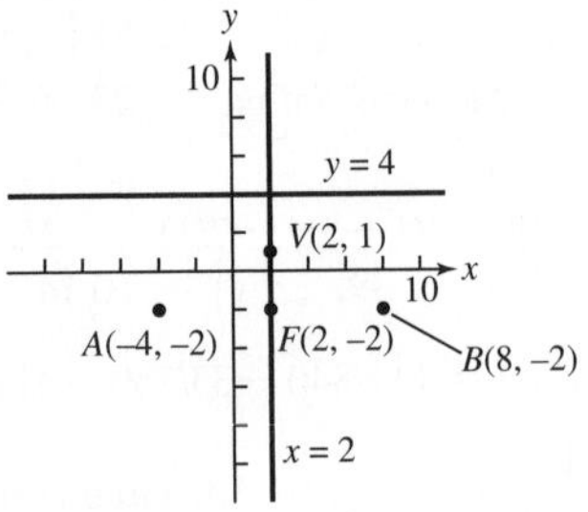

7. Downward

Quick Review 8.1

1. $\sqrt{13}$ **3.** $y = \pm 2\sqrt{x}$ **5.** $y + 6 = -(x - 1)^2$

7. Vertex: (1, 5); $f(x)$ can be obtained from $g(x)$ by stretching x^2 by 3, shifting up 5 units, and shifting right 1 unit.

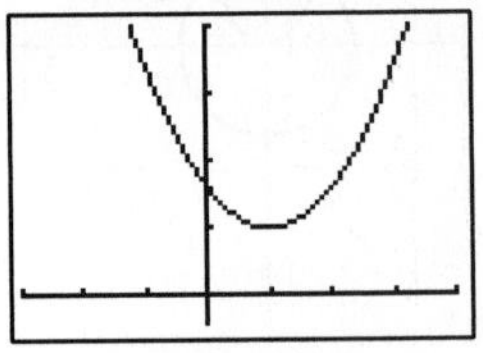

[–3, 4] by [–2, 20]

9. $f(x) = -2(x + 1)^2 + 3$

Exercises 8.1

1. Vertex: (0, 0); focus: $\left(0, \frac{3}{2}\right)$; directrix: $y = -\frac{3}{2}$; focal width: 6 **3.** Vertex: $(-3, 2)$; focus: $(-2, 2)$; directrix: $x = -4$; focal width: 4

5. Vertex: (0, 0); focus: $\left(0, -\frac{1}{3}\right)$; directrix: $y = \frac{1}{3}$; focal width: $\frac{4}{3}$ **7.** (c) **9.** (a) **11.** $y^2 = -12x$ **13.** $x^2 = -16y$

15. $x^2 = 20y$ **17.** $y^2 = 8x$ **19.** $x^2 = -6y$ **21.** $(y + 4)^2 = 8(x + 4)$ **23.** $(x - 3)^2 = 6(y - 5/2)$

25. $(y - 3)^2 = -8(x - 4)$ **27.** $(x - 2)^2 = 16(y + 1)$ **29.** $(y + 4)^2 = -10(x + 1)$

31.

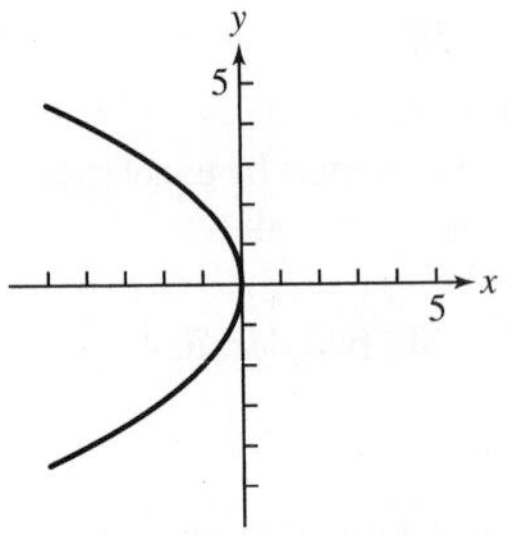

33.

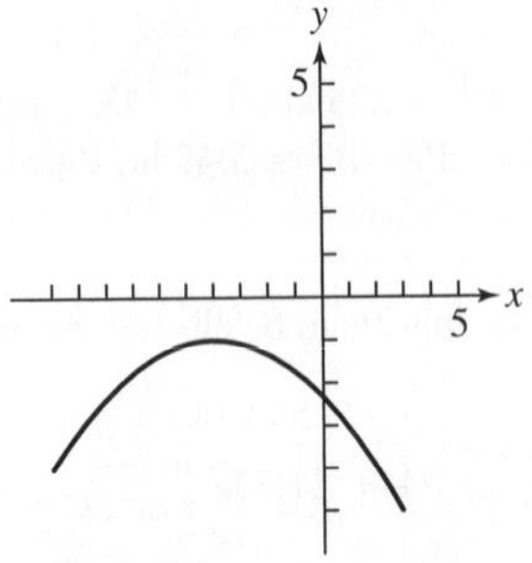

35.

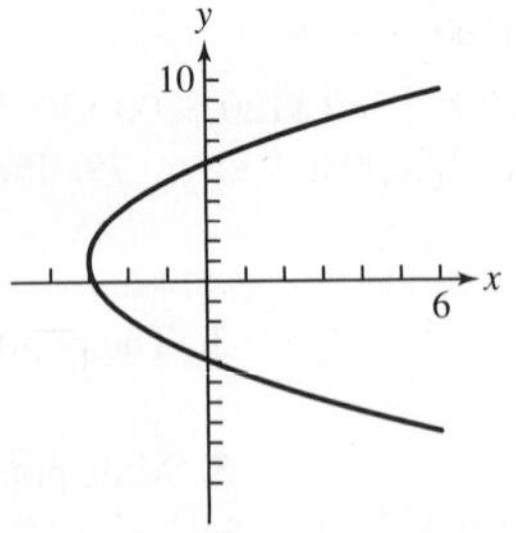

37.

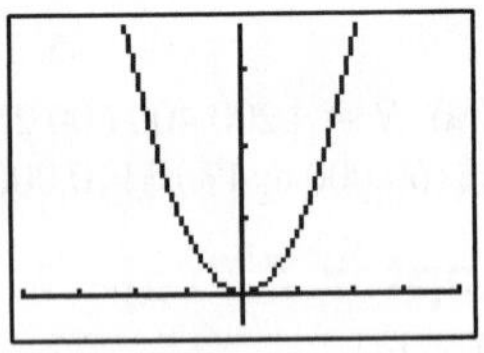

[–4, 4] by [–2, 18]

39.

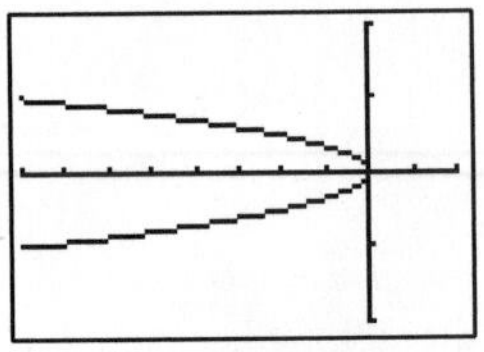

[–8, 2] by [–2, 2]

41.

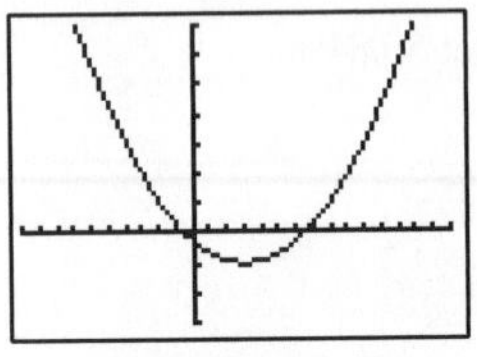

[–10, 15] by [–3, 7]

43.

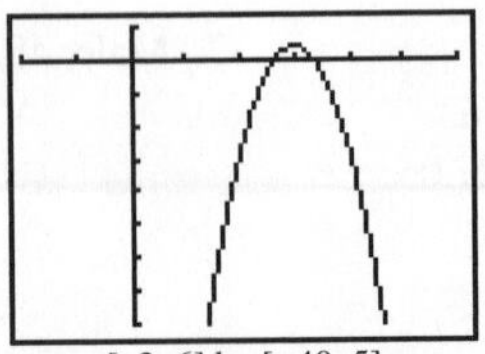

[–2, 6] by [–40, 5]

45.

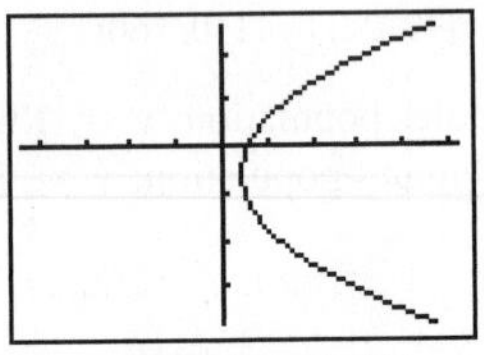

[–22, 26] by [–19, 13]

47.

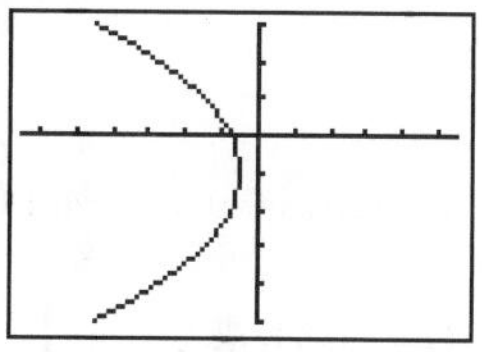

[−13, 11] by [−10, 6]

49. Completing the square, the equation becomes $(x + 1)^2 = y - 2$, a parabola with vertex $(-1, 2)$, focus $(-1, 9/4)$, and directrix $y = 7/4$.

51. Completing the square, the equation becomes $(y - 2)^2 = 8(x - 2)$, a parabola with vertex $(2, 2)$, focus $(4, 2)$, and directrix $x = 0$.

53. $(y - 2)^2 = -6x$ **55.** $(x - 2)^2 = -4(y + 1)$

57. The derivation only requires that p is a fixed real number.

59. The filament should be placed 1.125 cm from the vertex along the axis of the mirror.

61. The electronic receiver is located 2.5 units from the vertex along the axis of the parabolic microphone.

63. Starting at the leftmost tower, the lengths of the cables are roughly 79.44, 54.44, 35, 21.11, 12.78, 10, 12.78, 21.11, 35, 54.44, and 79.44.

65. False. Every point on a parabola is the same distance from its focus and its directrix. **67.** D **69.** B

71. (a)–(c)

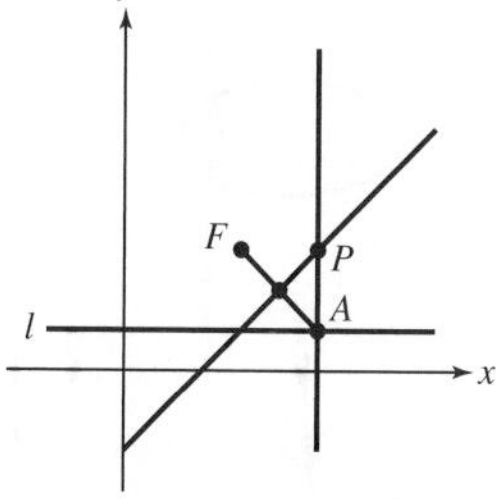

(d) Parabola

73. (a) The lines $y = x + 1$ and $y = x - 1$.

(b) These lines are parallel, not intersecting. There is no way to intersect the cone in Figure 8.2 with a plane and get two parallel lines.

(c) There are no pairs that solve the equation, since $x^2 + y^2$ cannot be negative.

(d) The graph would be the empty set. Because the cone in Figure 8.2 extends infinitely, it has a nonempty intersection with every possible plane.

(e)

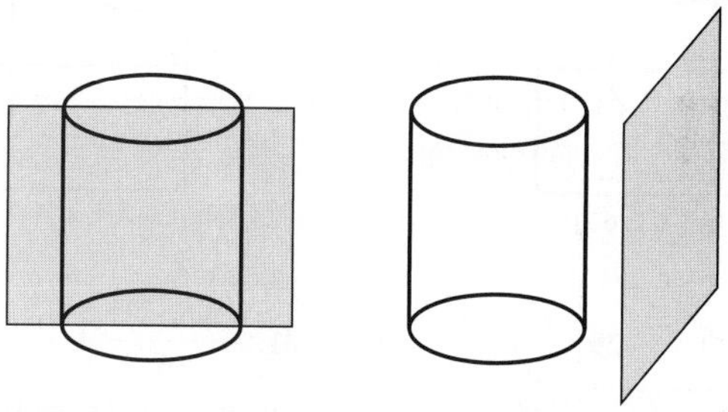

Two parallel lines Empty set

SECTION 8.2

Exploration 1

1. $x = -2 + 3\cos t$ and $y = 5 + 7\sin t$; $\cos t = \dfrac{x+2}{3}$ and $\sin t = \dfrac{y-5}{7}$; $\cos^2 t + \sin^2 t = 1$ yields the equation $\dfrac{(x+2)^2}{9} + \dfrac{(y-5)^2}{49} = 1$.

3. Example 1: $x = 3\cos t$ and $y = 2\sin t$
Example 2: $x = 2\cos t$ and $y = \sqrt{13}\sin t$
Example 3: $x = 3 + 5\cos t$ and $y = -1 + 4\sin t$
(Other parametrizations are possible. For example, sin and cos can be swapped.)

5. Example 1: $x = 3\cos t$, $y = 2\sin t$; $\cos t = \dfrac{x}{3}$, $\sin t = \dfrac{y}{2}$; $\cos^2 t + \sin^2 t = 1$ yields $\dfrac{x^2}{9} + \dfrac{y^2}{4} = 1$, or $4x^2 + 9y^2 = 36$.

Example 2: $x = 2\cos t$, $y = \sqrt{13}\sin t$; $\cos t = \dfrac{x}{2}$, $\sin t = \dfrac{y}{\sqrt{13}}$; $\sin^2 t + \cos^2 t = 1$ yields $\dfrac{y^2}{13} + \dfrac{x^2}{4} = 1$.

Example 3: $x = 3 + 5\cos t$, $y = -1 + 4\sin t$; $\cos t = \dfrac{x-3}{5}$, $\sin t = \dfrac{y+1}{4}$; $\cos^2 t + \sin^2 t = 1$ yields $\dfrac{(x-3)^2}{25} + \dfrac{(y+1)^2}{16} = 1$.

Exploration 2

3. $a = 8$ cm, $b \approx 7.75$ cm, $c = 2$ cm, $e = 0.25$, $b/a \approx 0.97$; $a = 7$ cm, $b \approx 6.32$ cm, $c = 3$ cm, $e \approx 0.43$, $b/a \approx 0.90$; $a = 6$ cm, $b \approx 4.47$ cm, $c = 4$ cm, $e \approx 0.67$, $b/a \approx 0.75$

5.

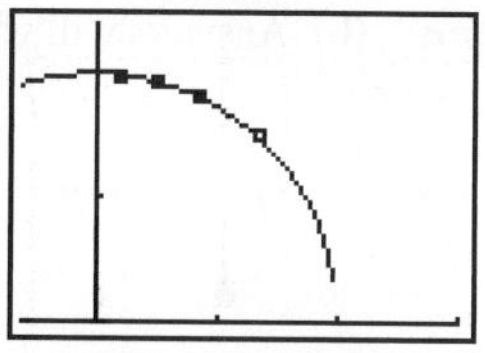

[−0.3, 1.5] by [0, 1.2]

$b/a = \sqrt{1 - e^2}$

Quick Review 8.2

1. $\sqrt{61}$ **3.** $y = \pm\frac{3}{2}\sqrt{4 - x^2}$ **5.** $x = 8$ **7.** $x = 2, x = -2$ **9.** $x = \frac{3 \pm \sqrt{15}}{2}$

Exercises 8.2

1. Vertices: $(4, 0), (-4, 0)$; Foci: $(3, 0), (-3, 0)$ **3.** Vertices: $(0, 6), (0, -6)$; Foci: $(0, 3), (0, -3)$
5. Vertices: $(2, 0), (-2, 0)$; Foci: $(1, 0), (-1, 0)$ **7.** (d) **9.** (a)

11.
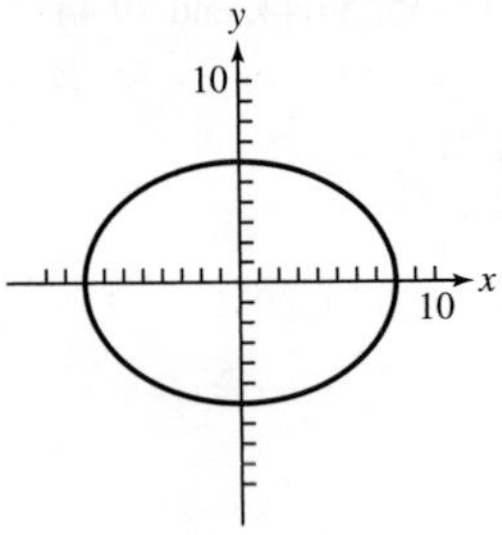

13.
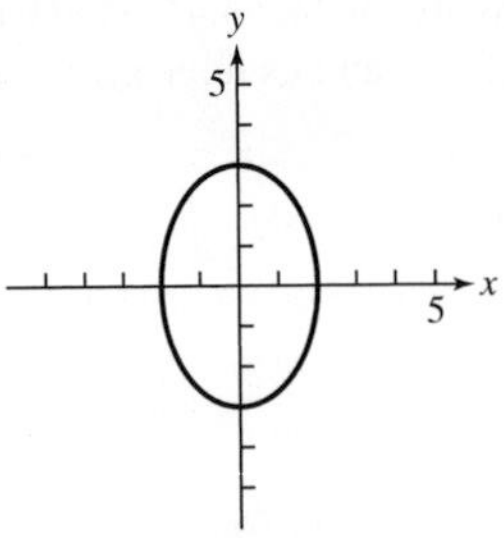

15.
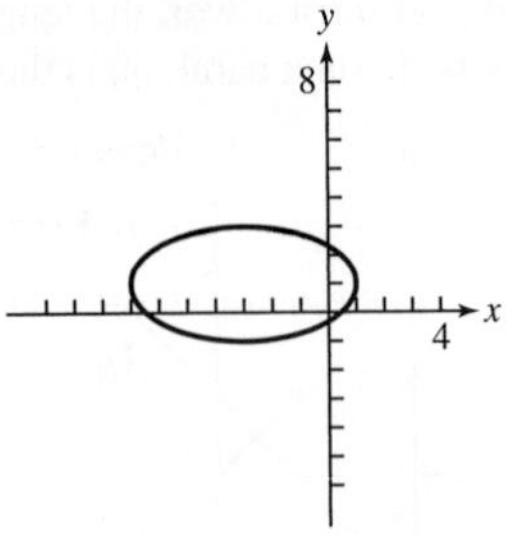

17.
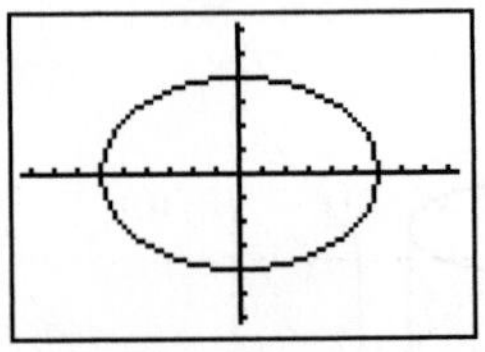
[–9.4, 9.4] by [–6.2, 6.2]

19.
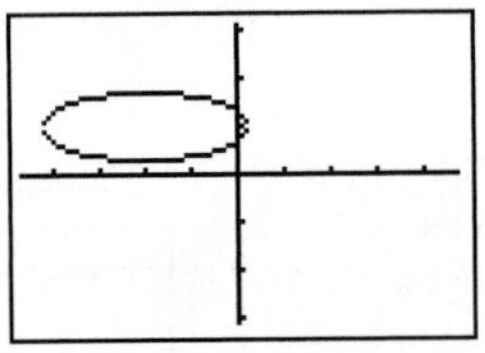
[–17, 4.7] by [–3.1, 3.1]

21. $\frac{y^2}{9} + \frac{x^2}{4} = 1$
23. $x^2/25 + y^2/21 = 1$
25. $\frac{y^2}{25} + \frac{x^2}{16} = 1$

27. $\frac{y^2}{36} + \frac{x^2}{16} = 1$ **29.** $\frac{x^2}{25} + \frac{y^2}{16} = 1$ **31.** $\frac{(y-2)^2}{36} + \frac{(x-1)^2}{16} = 1$ **33.** $(x - 3)^2/9 + (y + 4)^2/5 = 1$
35. $(y + 2)^2/25 + (x - 3)^2/9 = 1$ **37.** Center: $(-1, 2)$; vertices: $(-6, 2), (4, 2)$; foci: $(-4, 2), (2, 2)$
39. Center: $(7, -3)$; vertices: $(7, 6), (7, -12)$; foci: $(7, -3 \pm \sqrt{17})$

41.
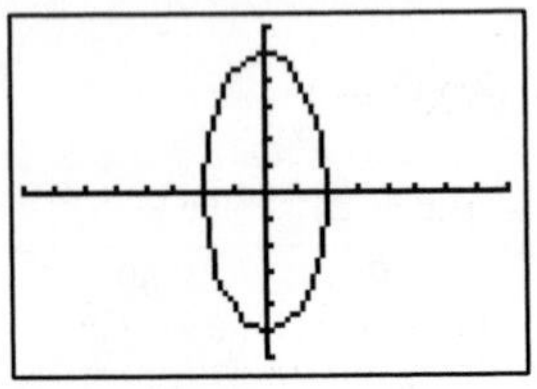
[–8, 8] by [–6, 6]

$x = 2\cos t, y = 5\sin t$

43.
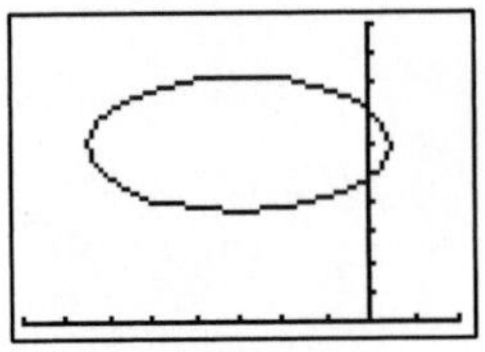
[–8, 2] by [0, 10]

$x = 2\sqrt{3}\cos t - 3,$
$y = \sqrt{5}\sin t + 6$

45. Vertices: $(1, -4), (1, 2)$;
foci: $(1, -1 \pm \sqrt{5})$;
eccentricity: $\frac{\sqrt{5}}{3}$

47. Vertices: $(-7, 1), (1, 1)$; foci: $(-3 \pm \sqrt{7}, 1)$; eccentricity: $\frac{\sqrt{7}}{4}$
49. $\frac{(x-2)^2}{16} + \frac{(y-3)^2}{9} = 1$ **51.** $\frac{y^2}{a^2} + \frac{x^2}{b^2} = 1$ **53.** $a = 237{,}086.5, b \approx 236{,}571, c = 15{,}623.5, e \approx 0.066$
55. ≈ 1347 Gm, ≈ 1507 Gm **57.** $a - c < 1.5(1.392) = 2.088$ Gm **59.** $(\pm\sqrt{51.75}, 0) \approx (\pm 7.19, 0)$ **61.** $(-2, 0), (2, 0)$
63. **(a)** Approximate solutions: $(\pm 1.04, -0.86), (\pm 1.37, 0.73)$
(b) $\left(\pm\frac{\sqrt{94 - 2\sqrt{161}}}{8}, -\frac{1 + \sqrt{161}}{16}\right), \left(\pm\frac{\sqrt{94 + 2\sqrt{161}}}{8}, \frac{-1 + \sqrt{161}}{16}\right)$
65. False. The distance is $a(1 - e)$. **67.** C **69.** B
71. **(a)** When $a = b = r, A = \pi ab = \pi rr = \pi r^2$ and $P \approx \pi(2r)\cdot(3 - \sqrt{(4r)(4r)/(2r)}) = \pi(2r)\cdot(3 - 2) = 2\pi r$. **(b)** Answers will vary.

73. **(a)**
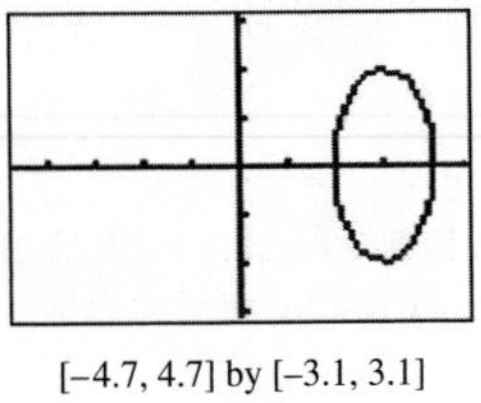
[–4.7, 4.7] by [–3.1, 3.1]

(b) $y^2/4 + (x - 3)^2 = 1$ **75.** Answers vary.

SECTION 8.3

Exploration 1

1. $x = -1 + 3/\cos t = -1 + 3 \sec t$ and $y = 1 + 2 \tan t$; $\sec t = \dfrac{x+1}{3}$ and $\tan t = \dfrac{y-1}{2}$; $\sec^2 t - \tan^2 t = 1$ yields the equation $\dfrac{(x+1)^2}{9} - \dfrac{(y-1)^2}{4} = 1$.

3. Example 1: $x = 3/\cos t$, $y = 2 \tan t$; Example 2: $x = 2 \tan t$, $y = \sqrt{5}/\cos t$; Example 3: $x = 3 + 5/\cos t$, $y = -1 + 4 \tan t$; Example 4: $x = -2 + 3/\cos t$, $y = 5 + 7 \tan t$

5. Example 1: $x = 3/\cos t = 3 \sec t$, $y = 2 \tan t$; $\sec t = x/3$, $\tan t = y/2$; $\sec^2 t - \tan^2 t = 1$ yields $x^2/9 - y^2/4 = 1$, or $4x^2 - 9y^2 = 36$. Example 2: $x = 2 \tan t$, $y = \sqrt{5}/\cos t = \sqrt{5} \sec t$; $\tan t = x/2$, $\sec t = y/\sqrt{5}$; $\sec^2 t - \tan^2 t = 1$ yields $y^2/5 - x^2/4 = 1$. Example 3: $x = 3 + 5/\cos t = 3 + 5 \sec t$, $y = -1 + 4 \tan t$; $\sec t = (x-3)/5$, $\tan t = (y+1)/4$; $\sec^2 t - \tan^2 t = 1$ yields $(x-3)^2/25 - (y+1)^2/16 = 1$. Example 4: $x = -2 + 3/\cos t = -2 + 3 \sec t$, $y = 5 + 7 \tan t$; $\sec t = (x+2)/3$, $\tan t = (y-5)/7$; $\sec^2 t - \tan^2 t = 1$ yields $(x+2)^2/9 - (y-5)^2/49 = 1$.

Quick Review 8.3

1. $\sqrt{146}$ **3.** $y = \pm\frac{4}{3}\sqrt{x^2 + 9}$ **5.** No solution **7.** $x = 2, x = -2$ **9.** $a = 3, c = 5$

Exercises 8.3

1. Vertices: $(\pm 4, 0)$; foci: $(\pm\sqrt{23}, 0)$ **3.** Vertices: $(0, \pm 6)$; foci: $(0, \pm 7)$ **5.** Vertices: $(\pm 2, 0)$; foci: $(\pm\sqrt{7}, 0)$ **7.** (c) **9.** (a)

11.

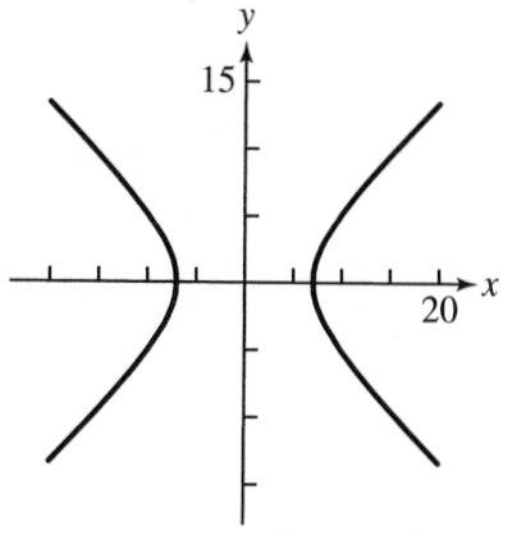

13.

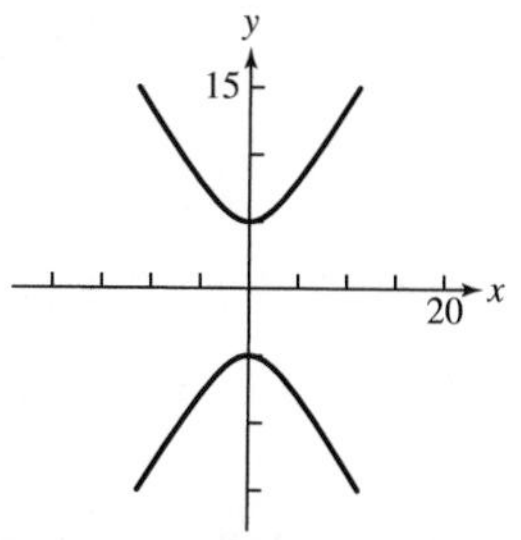

15.

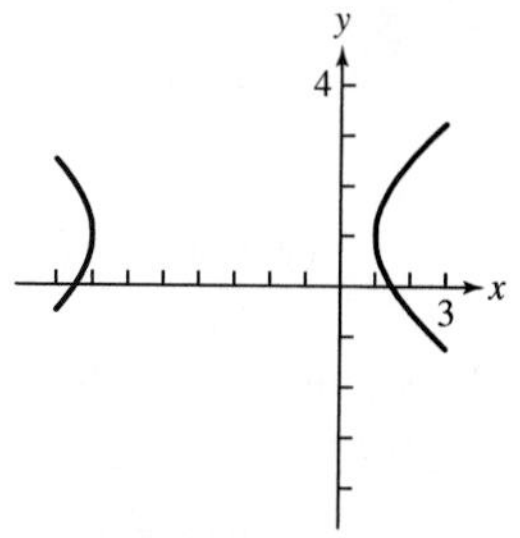

17.

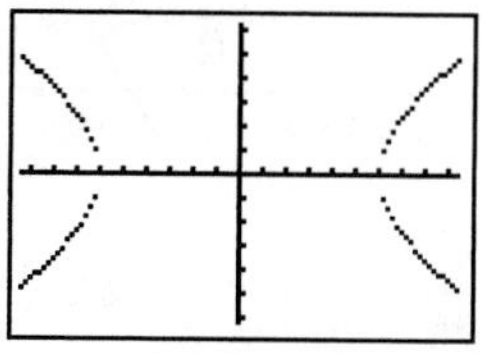

[–9.4, 9.4] by [–6.2, 6.2]

19.

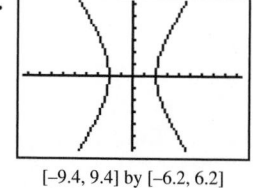

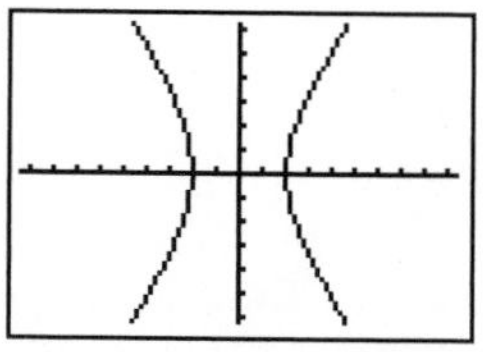

[–9.4, 9.4] by [–6.2, 6.2]

21.

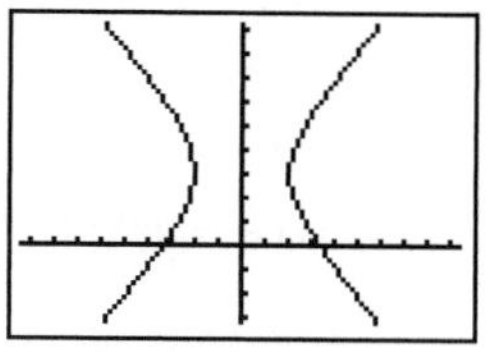

[–9.4, 9.4] by [–3.2, 9.2]

23. $x^2/4 - y^2/5 = 1$ **25.** $y^2/16 - x^2/209 = 1$

27. $\dfrac{x^2}{25} - \dfrac{y^2}{75} = 1$ **29.** $\dfrac{y^2}{144} - \dfrac{x^2}{25} = 1$

31. $\dfrac{(y-1)^2}{4} - \dfrac{(x-2)^2}{9} = 1$ **33.** $\dfrac{(x-2)^2}{9} - \dfrac{(y-3)^2}{16} = 1$

35. $\dfrac{(x+1)^2}{4} - \dfrac{(y-2)^2}{5} = 1$ **37.** $\dfrac{(y-6)^2}{25} - \dfrac{(x+3)^2}{75} = 1$

39. Center: $(-1, 2)$; vertices: $(11, 2)$, $(-13, 2)$; foci: $(12, 2)$, $(-14, 2)$ **41.** Center: $(2, -3)$; vertices: $(2, 5)$, $(2, -11)$; foci: $(2, -3 \pm \sqrt{145})$

43.

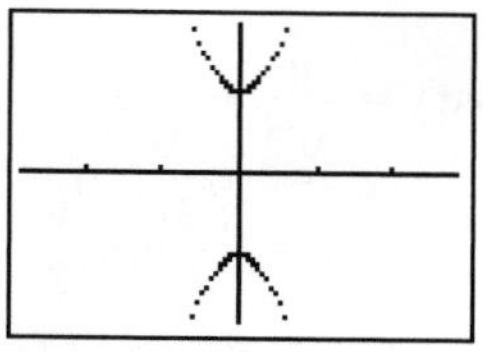

[–14.1, 14.1] by [–9.3, 9.3]

45.

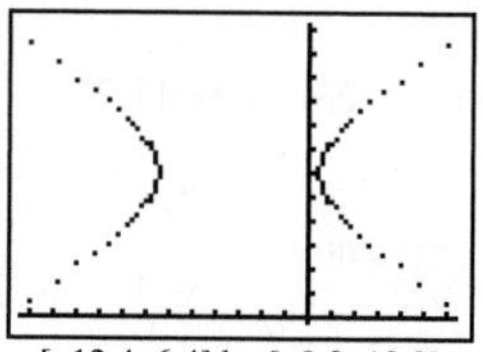

[–12.4, 6.4] by [–0.2, 12.2]

47.

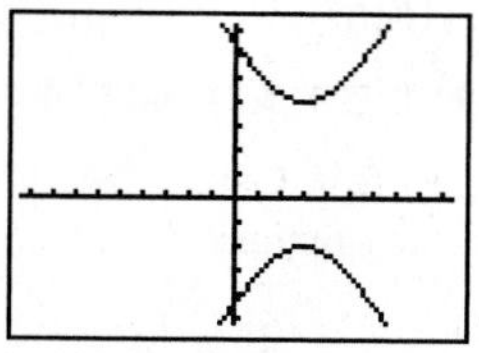

[–9.4, 9.4] by [–5.2, 7.2]

Vertices: $(3, -2)$, $(3, 4)$; foci: $(3, 1 \pm \sqrt{13})$; $e = \dfrac{\sqrt{13}}{3}$

49.

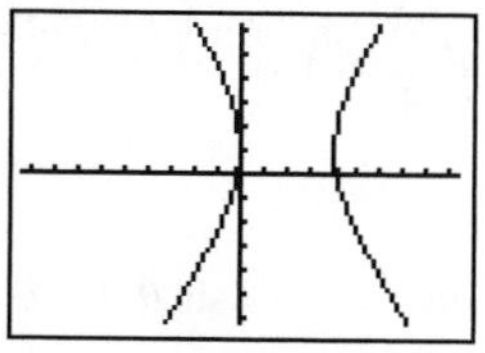

[–9.4, 9.4] by [–6.2, 6.2]

Vertices: $(0, 1)$, $(4, 1)$; foci: $(2 \pm \sqrt{13}, 1)$; $e = \dfrac{\sqrt{13}}{2}$

51. $\dfrac{x^2}{4} - \dfrac{5y^2}{16} = 1$ **53.** $\dfrac{y^2}{a^2} - \dfrac{x^2}{b^2} = 1$

55. $a = 1440$, $b = 600$, $c = 1560$, $e = 13/12$; The Sun is centered at $(1560, 0)$.

57. A bearing and distance of about 40.29° and 1371.11 mi, respectively

59. $(-2, 0), (4, 3\sqrt{3})$

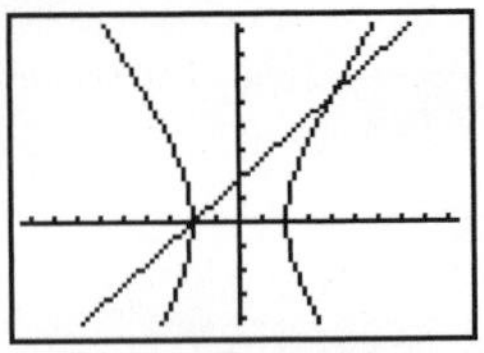

[–9.4, 9.4] by [–4.2, 8.2]

61. (a)

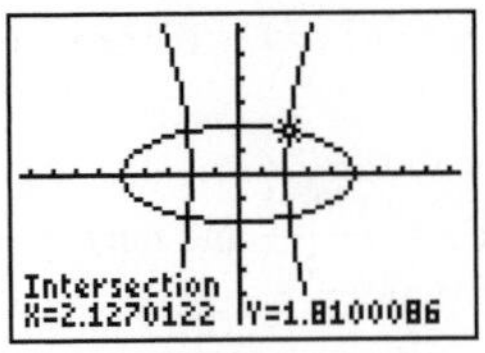

[–9.4, 9.4] by [–6.2, 6.2]

Four solutions: $(\pm 2.13, \pm 1.81)$

(b) $\left(\pm 10\sqrt{\dfrac{29}{641}}, \pm 10\sqrt{\dfrac{21}{641}}\right)$

63. True, because $c - a = ae - a$. **65.** B **67.** B

69. (a–d)

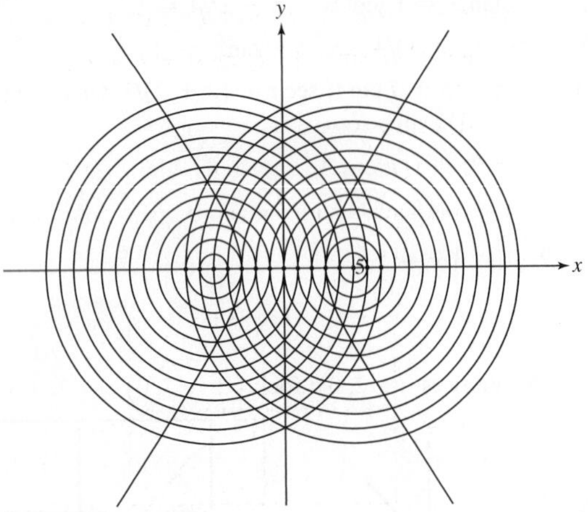

(e) $x^2/9 - y^2/16 = 1$

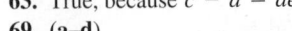

71. Answers will vary.
73. Answers will vary.
75. Answers will vary.

SECTION 8.4

Exploration 1

1. It is a hyperbola. **3.** The origin, (0, 0) **5.** $\sqrt{(1-(-1))^2 + (1-(-1))^2} = \sqrt{8} = 2\sqrt{2}$
7. They are the asymptotes of the hyperbola. **9.** $c = \sqrt{a^2 + b^2} = 2$

Quick Review 8.4

1. $\cos 2\theta = 5/13$ **3.** $\cos 2\theta = 1/2$ **5.** $\theta = \pi/4$ **7.** $\cos\theta = 2/\sqrt{5}$ **9.** $\sin\theta = 1/\sqrt{12}$

Exercises 8.4

1. Vertices: $(\pm 2, \pm 2)$; foci: $(\pm 2\sqrt{2}, \pm 2\sqrt{2})$ **3.** Vertices: $(\pm 2, \pm 2)$; foci: $(\pm\sqrt{3}, \pm\sqrt{3})$

5. $u^2 - v^2 = 8$ **7.** $4u^2 + v^2 = 8$ **9.** $x = \dfrac{\sqrt{3}u}{2} - \dfrac{v}{2}$ and $y = \dfrac{u}{2} + \dfrac{\sqrt{3}v}{2}$ **11.** $x = \dfrac{3u}{5} - \dfrac{4v}{5}$ and $y = \dfrac{4u}{5} + \dfrac{3v}{5}$

13. $x = \dfrac{3u}{\sqrt{10}} - \dfrac{v}{\sqrt{10}}$ and $y = \dfrac{u}{\sqrt{10}} + \dfrac{3v}{\sqrt{10}}$ **15.** $x = 0.766u - 0.643v$ and $y = 0.643u + 0.766v$

17. (4, 3) **19.** (−2.4, 3.2) and (2.4, −3.2) **21.** (−1.8, 7.4) **23.** (6.6, 11.2) **25.** $x = \dfrac{u}{\sqrt{2}} - \dfrac{v}{\sqrt{2}}$ and $y = \dfrac{u}{\sqrt{2}} + \dfrac{v}{\sqrt{2}}$

27. $x = \dfrac{u}{\sqrt{2}} - \dfrac{v}{\sqrt{2}}$ and $y = \dfrac{u}{\sqrt{2}} + \dfrac{v}{\sqrt{2}}$ **29.** $x = \dfrac{u}{\sqrt{2}} - \dfrac{v}{\sqrt{2}}$ and $y = \dfrac{u}{\sqrt{2}} + \dfrac{v}{\sqrt{2}}$ **31.** $x = \dfrac{u}{2} - \dfrac{\sqrt{3}v}{2}$ and $y = \dfrac{\sqrt{3}u}{2} + \dfrac{v}{2}$

33. $x = \dfrac{4u}{5} - \dfrac{3v}{5}$ and $y = \dfrac{3u}{5} + \dfrac{4v}{5}$ **35. (a)** Hyperbola **(b)** $x = \dfrac{u}{\sqrt{2}} - \dfrac{v}{\sqrt{2}}$ and $y = \dfrac{u}{\sqrt{2}} + \dfrac{v}{\sqrt{2}} \Rightarrow \dfrac{u^2}{16} - \dfrac{v^2}{16} = 1$ **(c)** (4, 0) and (−4, 0)

(d) $(2\sqrt{2}, 2\sqrt{2})$ and $(-2\sqrt{2}, -2\sqrt{2})$ **37. (a)** Ellipse **(b)** $x = \dfrac{\sqrt{3}u}{2} - \dfrac{v}{2}$ and $y = \dfrac{u}{2} + \dfrac{\sqrt{3}v}{2} \Rightarrow \dfrac{v^2}{20} + \dfrac{u^2}{4} = 1$

(c) $(0, 2\sqrt{5})$ and $(0, -2\sqrt{5})$ **(d)** $(-\sqrt{5}, \sqrt{15})$ and $(\sqrt{5}, -\sqrt{15})$ **39. (a)** Parabola

(b) $x = \dfrac{3}{5}u - \dfrac{4}{5}v$ and $y = \dfrac{4}{5}u + \dfrac{3}{5}v \Rightarrow v = u^2 + 3$ **(c)** (0, 3) **(d)** (−2.4, 1.8) **41.** $x = u\cos\theta + v\sin\theta$ and $y = -u\sin\theta + v\cos\theta$

43. $-24 < 0$; ellipse **45.** 0; parabola **47.** $-48 < 0$; ellipse **49.** $12 > 0$; hyperbola **51.** $-12 < 0$; ellipse
53. D **55.** A **57.** True, because there is no xy term **59.** D **61.** D

63. The asymptote with slope 1 in the (x, y) system forms an angle of $\pi/4$ with the x-axis. After the axes are rotated through an angle of $\pi/6$, the same asymptote makes an angle of $\pi/4 - \pi/6 = \pi/12$ with the u-axis. The new slope will be $\tan(\pi/12) = \dfrac{\sqrt{6}-\sqrt{2}}{\sqrt{6}+\sqrt{2}} \approx 0.268$. The second asymptote, which is perpendicular to the first, must therefore have slope $-\dfrac{\sqrt{6}+\sqrt{2}}{\sqrt{6}-\sqrt{2}} \approx -3.732$.

65. With the substitutions, $Ax^2 + Bxy + Cy^2 + Dx + Ey + F = 0$ becomes

$$A(u\cos\theta - v\sin\theta)^2 + B(u\cos\theta - v\sin\theta)(u\sin\theta + v\cos\theta) + C(u\sin\theta + v\cos\theta)^2$$
$$+ D(u\cos\theta - v\sin\theta) + E(u\sin\theta + v\cos\theta) + F = 0$$

Expanding, we have

$$u^2 A\cos^2\theta - uv2A\cos\theta\sin\theta + v^2 A\sin^2\theta + u^2 B\cos\theta\sin\theta + uvB\cos^2\theta - uvB\sin^2\theta - v^2 B\cos\theta\sin\theta$$
$$+ u^2C\sin^2\theta + uv2C\sin\theta\cos\theta + v^2C\cos^2\theta + uD\cos\theta - vD\sin\theta + uE\sin\theta + vE\cos\theta + F = 0$$

Combining like terms, we have

$$u^2(A\cos^2\theta + B\cos\theta\sin\theta + C\sin^2\theta) + uv(\cos^2\theta - \sin^2\theta + (C - A)2\sin\theta\cos\theta)$$
$$+ v^2(C\cos^2\theta - B\cos\theta\sin\theta + A\sin^2\theta) + u(D\cos\theta + E\sin\theta) + v(E\cos\theta - D\sin\theta) + F = 0$$

Substitute $\cos 2\theta$ for $\cos^2\theta - \sin^2\theta$ and $\sin 2\theta$ for $2\sin\theta\cos\theta$. Then, comparing these coefficients with the (u, v) coefficients, the formulas follow. **67.** Answers will vary.

69. Intersecting lines:

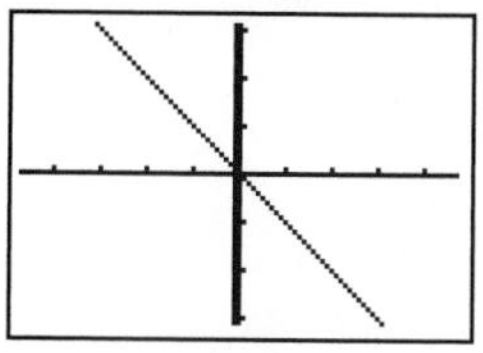

[−4.7, 4.7] by [−3.1, 3.1]

A plane containing the axis of a cone intersects the cone.

Parallel lines:

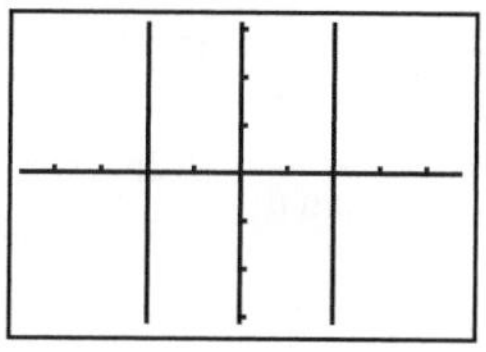

[−4.7, 4.7] by [−3.1, 3.1]

A degenerate cone is created by a generator that is parallel to the axis, producing a cylinder. A plane parallel to a generator of the cylinder intersects the cylinder and its interior.

One line:

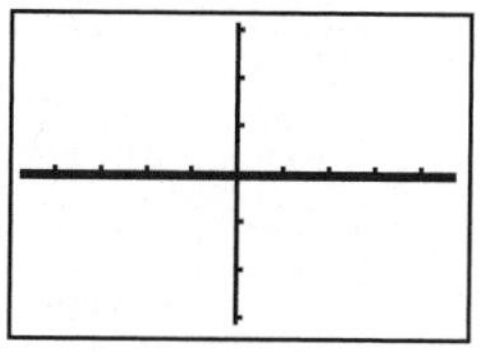

[−4.7, 4.7] by [−3.1, 3.1]

A plane containing a generator of a cone intersects the cone.

No graph:

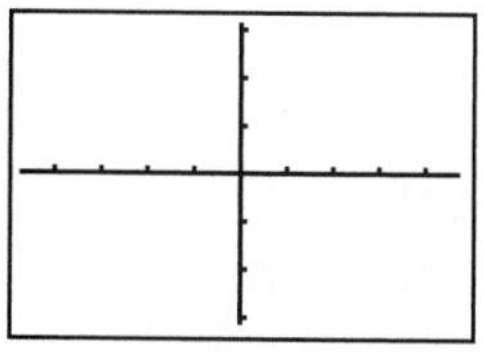

[−4.7, 4.7] by [−3.1, 3.1]

A plane parallel to a generator of a cylinder fails to intersect the cylinder. Also, a degenerate cone is created by a generator that is perpendicular to the axis, producing a plane. A second plane perpendicular to the axis of this degenerate cone fails to intersect it.

Circle:

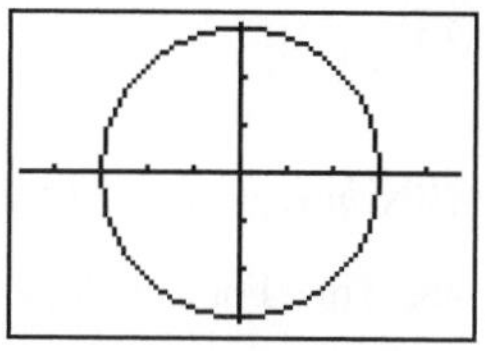

[−4.7, 4.7] by [−3.1, 3.1]

A plane perpendicular to the axis of a cone intersects the cone but not its vertex.

Point:

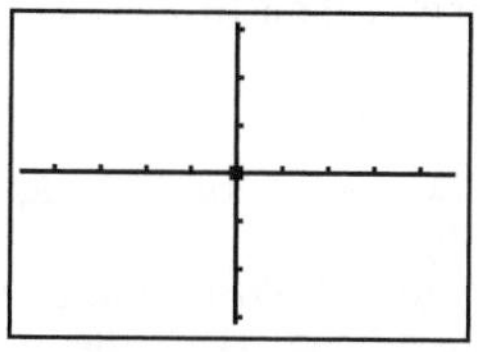

[−4.7, 4.7] by [−3.1, 3.1]

A plane perpendicular to the axis of a cone intersects the vertex of the cone.

SECTION 8.5

Exploration 1

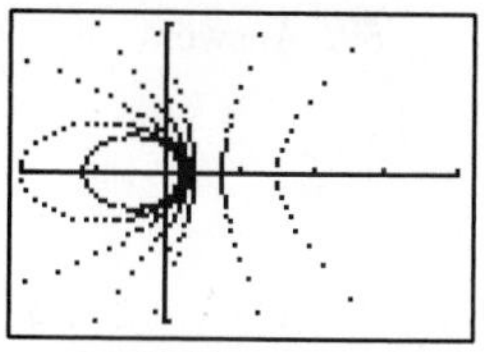

[−12, 24] by [−12, 12]

$e = 0.7$, $e = 0.8$: an ellipse; $e = 1$: a parabola; $e = 1.5$, $e = 3$: a hyperbola
The graphs have a common focus, $(0, 0)$, and a common directrix, the line $x = 3$. As e increases, the graphs move away from the focus and toward the directrix.

Quick Review 8.5

1. $r = -3$ **3.** $\theta = \frac{7\pi}{6}, \theta = -\frac{5\pi}{6}$ **5.** The focus is $(0, 4)$, and the directrix is $y = -4$.
7. Foci: $(\pm\sqrt{5}, 0)$; vertices: $(\pm 3, 0)$ **9.** Foci: $(\pm 5, 0)$; vertices: $(\pm 4, 0)$

Exercises 8.5

1. $r = \frac{2}{1 - \cos\theta}$; parabola

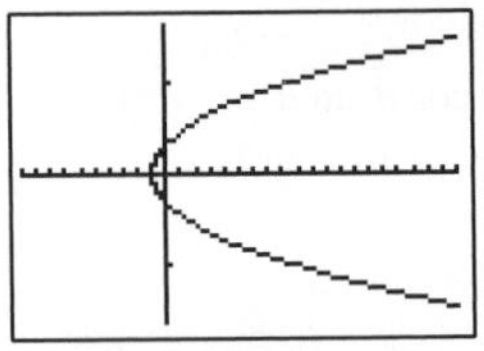

[–10, 20] by [–10, 10]

3. $r = \frac{12}{5 + 3\sin\theta}$; ellipse

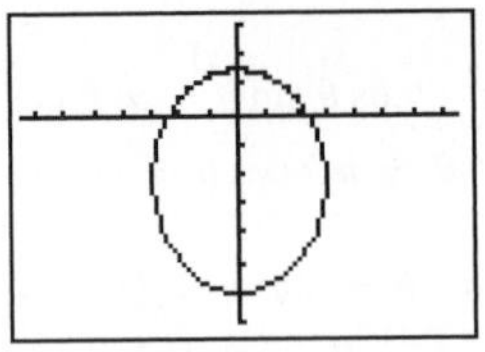

[–7.5, 7.5] by [–7, 3]

5. $r = \frac{7}{3 - 7\sin\theta}$; hyperbola

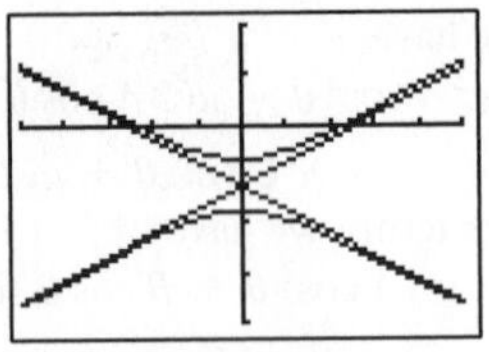

[–5, 5] by [–4, 2]

7. $e = 1$, parabola; directrix: $x = 2$ **9.** $e = 1$, parabola; directrix: $y = -\frac{5}{2} = -2.5$ **11.** $e = \frac{5}{6}$, ellipse; directrix: $y = 4$
13. $e = \frac{2}{5} = 0.4$, ellipse; directrix: $x = 3$ **15.** (b); $[-15, 5]$ by $[-10, 10]$ **17.** (f); $[-5, 5]$ by $[-3, 3]$
19. (c); $[-10, 10]$ by $[-5, 10]$ **21.** $r = \frac{12}{5 + 3\cos\theta}$ **23.** $r = \frac{3}{2 + \sin\theta}$ **25.** $r = \frac{15}{2 + 3\cos\theta}$
27. $r = \frac{12}{2 + 3\sin\theta}$ **29.** $r = \frac{6}{5 + 3\cos\theta}$

31.

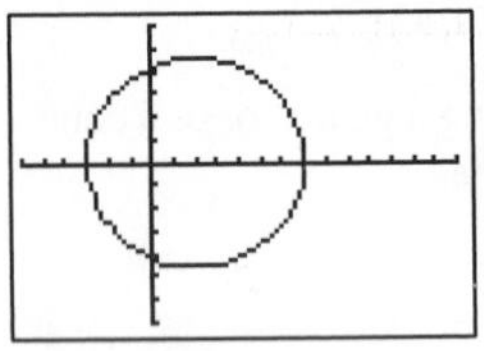

[–6, 14] by [–7, 6]

$e = 0.4, a = 5, b = \sqrt{21}, c = 2$

33.

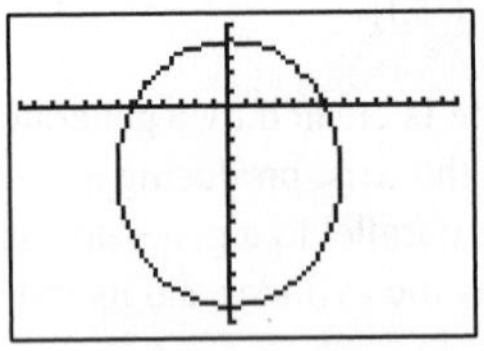

[–13, 14] by [–13, 5]

$e = \frac{1}{2}, a = 8, b = 4\sqrt{3}, c = 4$

35.

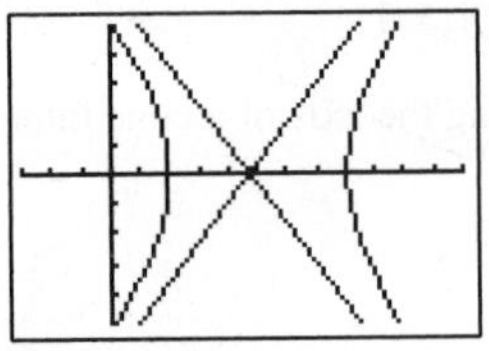

[–3, 12] by [–5, 5]

$e = \frac{5}{3}, a = 3, b = 4, c = 5$

37. $\frac{9(y - 4/3)^2}{64} + \frac{3x^2}{16} = 1$ **39.** $y^2 = 4(x + 1)$ **41.** Perihelion distance $\approx$ 0.54 AU; aphelion distance $\approx$ 35.64 AU
43. **(a)** $v \approx$ 1551 m/sec = 1.551 km/sec **(b)** About 2 hr 14 min **45.** True. For a circle, $e = 0$, so the equation is $r = 0$, which graphs as a point. **47.** D **49.** B
51. **(c)**

Planet	Perihelion Distance (AU)	Aphelion Distance (AU)
Mercury	0.307	0.467
Venus	0.718	0.728
Earth	0.983	1.017
Mars	1.382	1.665
Jupiter	4.953	5.452
Saturn	9.020	10.090

(d) The difference is greatest for Saturn.

53. Answers vary. **55.** $5r - 3r\cos\theta = 16 \Rightarrow 5r = 3x + 16$. So, $25r^2 = 25(x^2 + y^2) = (3x + 16)^2$. $25x^2 + 25y^2 = 9x^2 + 96x + 256 \Rightarrow 16x^2 - 96x + 25y^2 = 256$. Completing the square yields $\frac{(x - 3)^2}{25} + \frac{y^2}{16} = 1$, the desired result. **57.** Answers vary. **59.** Answers vary.

SECTION 8.6

Quick Review 8.6

1. $\sqrt{(x - 2)^2 + (y + 3)^2}$ **3.** P lies on the circle of radius 5 centered at $(2, -3)$. **5.** $\left\langle \frac{-4}{\sqrt{41}}, \frac{5}{\sqrt{41}} \right\rangle$
7. Circle of radius 5 centered at $(-1, 5)$ **9.** Center: $(-1, 3)$; radius: 2

Exercises 8.6

1.

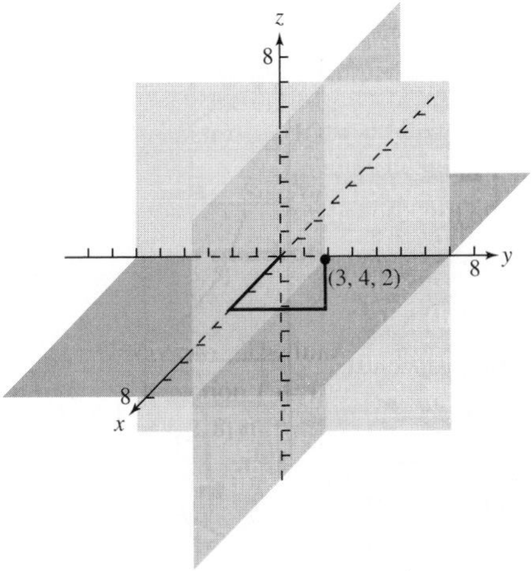

3.

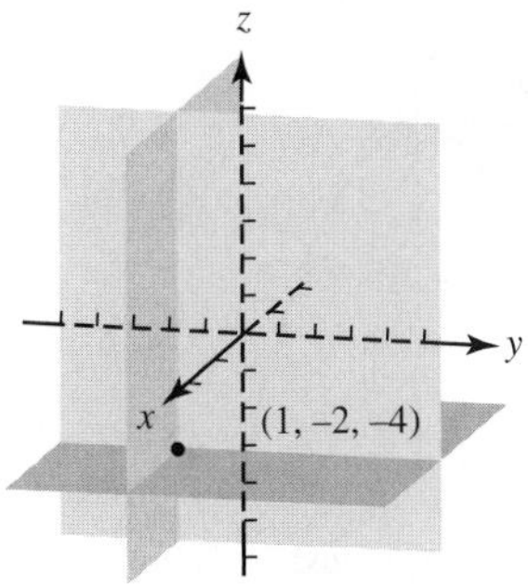

5. $\sqrt{53}$

7. $\sqrt{(a-1)^2+(b+3)^2+(c-2)^2}$

9. $(1, -1, 11/2)$

11. $(x-1, y+4, z+3)$

13. $(x-5)^2+(y+1)^2+(z+2)^2=64$

15. $(x-1)^2+(y+3)^2+(z-2)^2=a$

17.

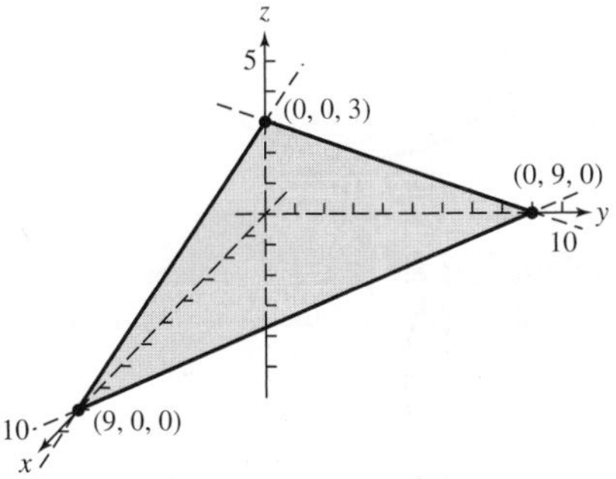

19.

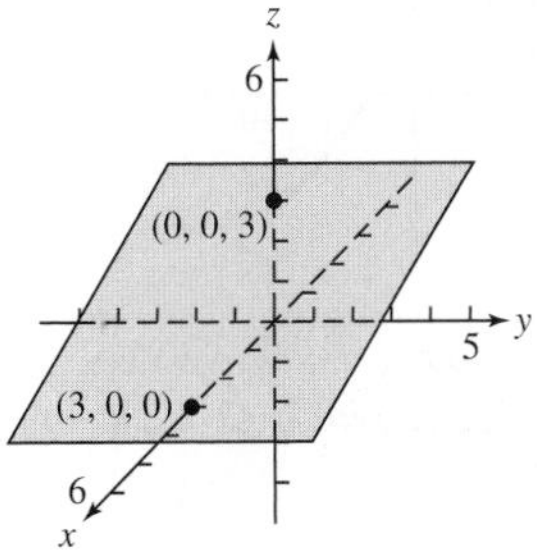

21.

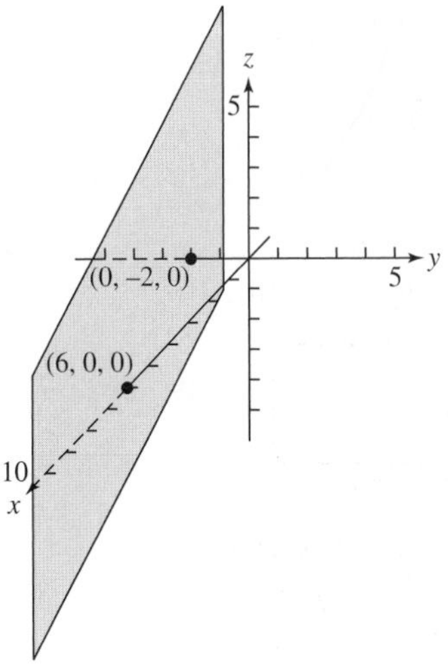

23. $\langle -2, 4, -8\rangle$ **25.** -84 **27.** -20 **29.** $\left\langle \frac{4}{13}, -\frac{3}{13}, \frac{12}{13}\right\rangle$ **31.** $\langle -3, 4, -5\rangle$ **33.** $\mathbf{v} = -195.01\mathbf{i} - 7.07\mathbf{j} + 68.40\mathbf{k}$

35. $\mathbf{r} = \langle 2, -1, 5\rangle + t\langle 3, 2, -7\rangle$; $x = 2 + 3t$, $y = -1 + 2t$, $z = 5 - 7t$ **37.** $\mathbf{r} = \langle 6, -9, 0\rangle + t\langle 1, 0, -4\rangle$; $x = 6 + t$, $y = -9$, $z = -4t$

39. $\sqrt{30}$ **41.** $\mathbf{r} = \langle -1, 2, 4\rangle + t\langle 1, 4, -7\rangle$ **43.** $x = -1 + 3t$, $y = 2 - 6t$, $z = 4 - 3t$ **45.** $x = \frac{1}{2}t$, $y = 6 - 7t$, $z = -3 + \frac{11}{2}t$

47. Scalene

49. (a)

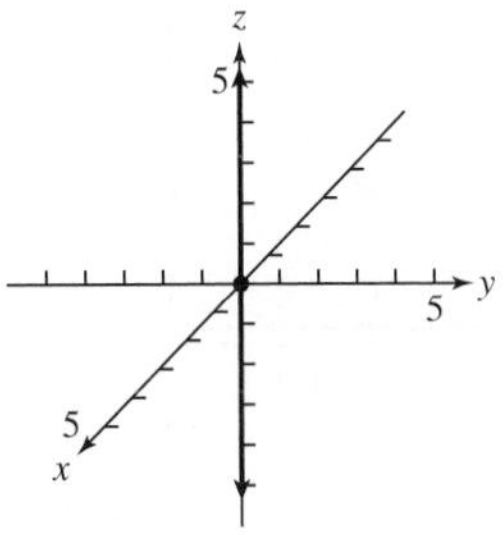

(b) The z-axis; a line through the origin in the direction $\mathbf{k}$

51. (a)

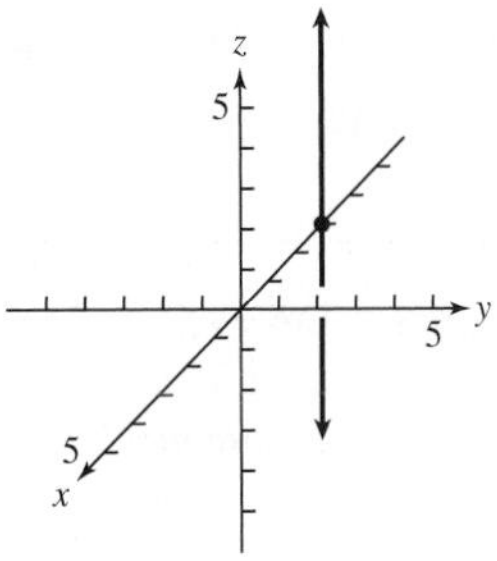

(b) The intersection of the xz-plane $(y = 0)$ and the plane $x = -3$; a line parallel to the z-axis through $(-3, 0, 0)$

53. $\mathbf{r} = \langle x_1 + (x_2 - x_1)t, y_1 + (y_2 - y_1)t, z_1 + (z_2 - z_1)t\rangle$ **55.** Answers vary.

57. True. The equation can be viewed as an equation in three variables, where the coefficient of z is zero. The surface is an elliptical cylinder.

59. B **61.** C **65.** $\langle -1, -5, -3\rangle$ **67.** $\mathbf{i} \times \mathbf{j} = \langle 1, 0, 0\rangle \times \langle 0, 1, 0\rangle = \langle 0 - 0, 0 - 0, 1 - 0\rangle = \langle 0, 0, 1\rangle = \mathbf{k}$

CHAPTER 8 REVIEW EXERCISES

1.

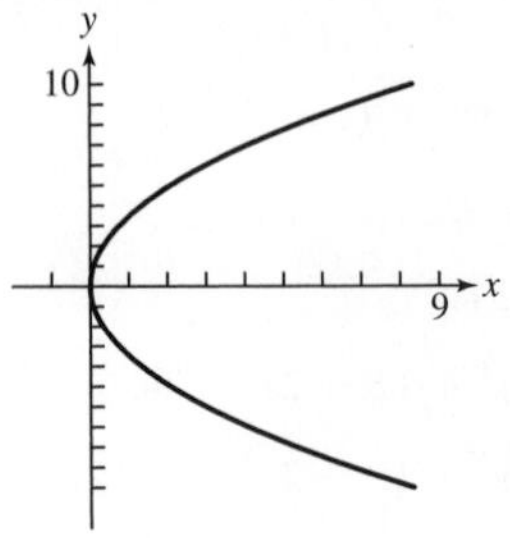

Vertex: $(0, 0)$; focus: $(3, 0)$; directrix: $x = -3$; focal width: 12

3.

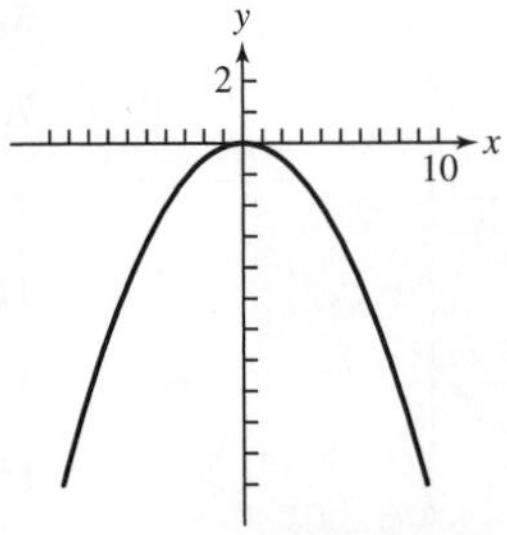

Vertex: $(-2, 1)$; focus: $(-2, 0)$; directrix: $y = 2$; focal width: 4

5.

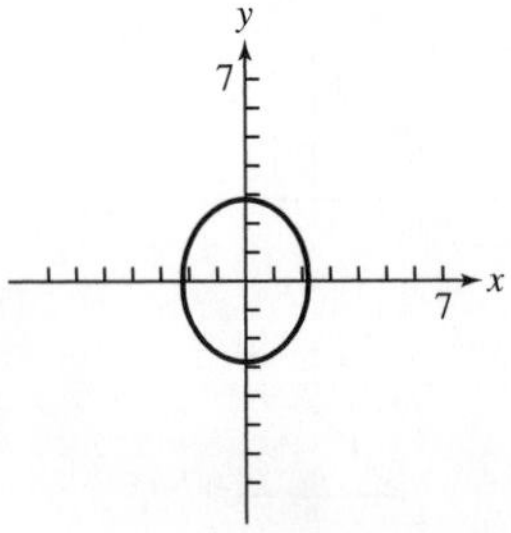

Ellipse; center: $(0, 0)$; vertices: $\left(0, \pm 2\sqrt{2}\right)$; foci: $\left(0, \pm\sqrt{3}\right)$

7.

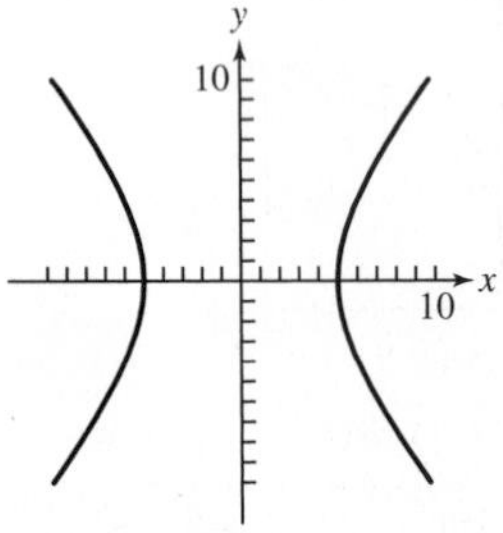

Hyperbola; center: $(0, 0)$; vertices: $(\pm 5, 0)$; foci: $\left(\pm\sqrt{61}, 0\right)$

9.

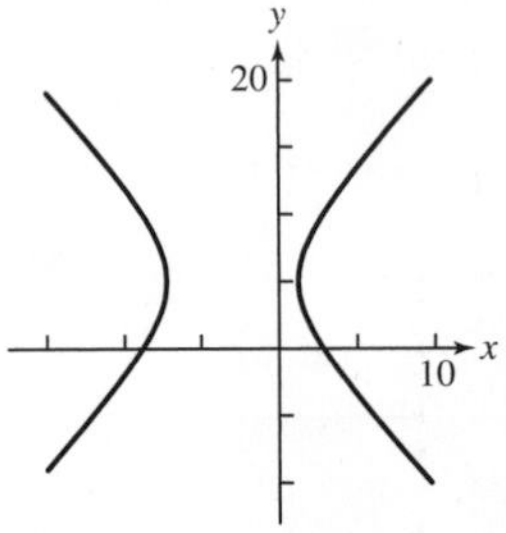

Hyperbola; center: $(-3, 5)$; vertices: $\left(-3 \pm 3\sqrt{2}, 5\right)$; foci: $\left(-3 \pm \sqrt{46}, 5\right)$

11.

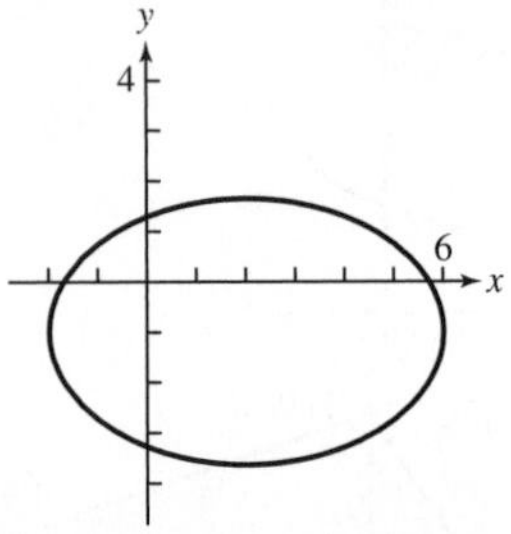

Ellipse; center: $(2, -1)$; vertices: $(6, -1)$, $(-2, -1)$; foci: $(5, -1)$, $(-1, -1)$

13. (b) **15.** (h) **17.** (f) **19.** (c)

21.

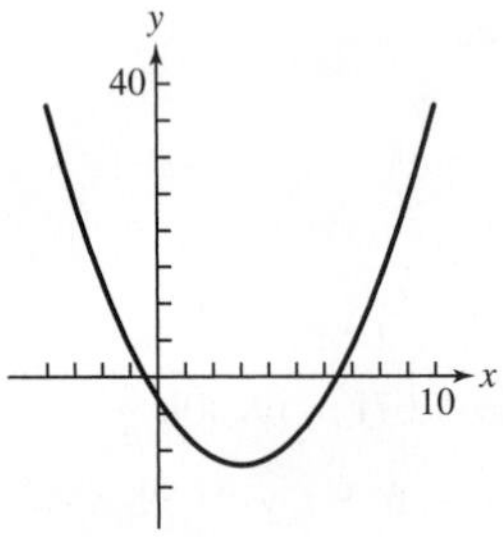

Parabola; $(x - 3)^2 = y + 12$

23.

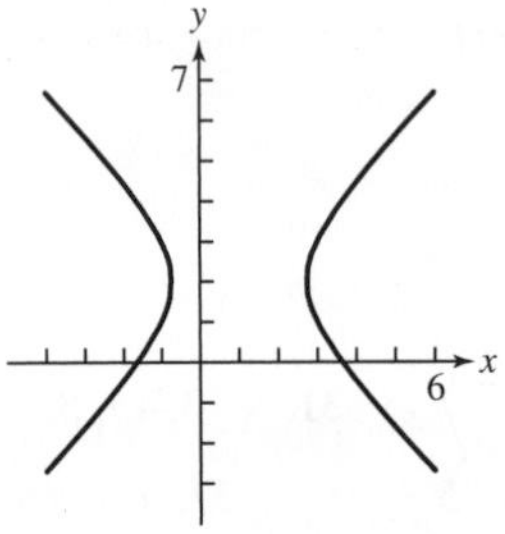

Hyperbola; $\dfrac{(x-1)^2}{3} - \dfrac{(y-2)^2}{3} = 1$

25.

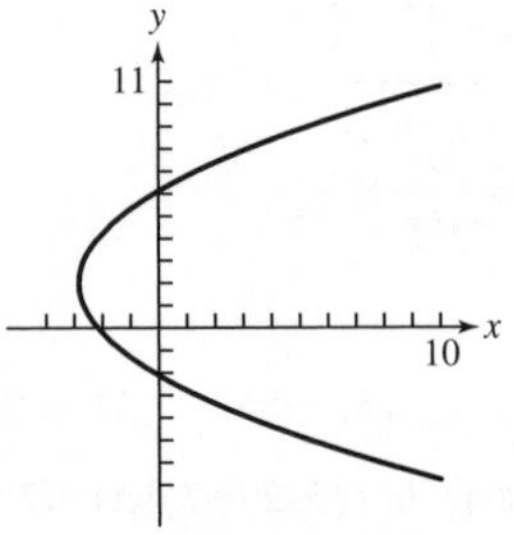

Parabola; $(y - 2)^2 = 6\left(x + \dfrac{17}{6}\right)$

27.

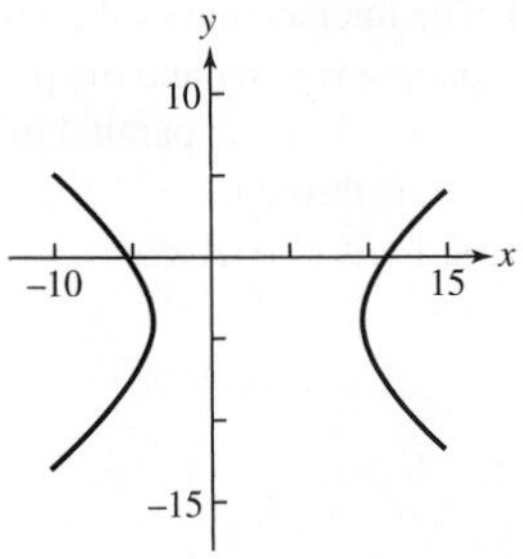

Hyperbola; $\dfrac{(y+4)^2}{30} - \dfrac{(x-3)^2}{45} = 1$

29. See proof on pages 573–574.

31.

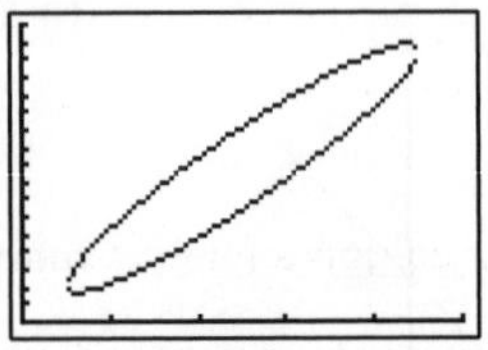

[0, 25] by [0, 17]

Ellipse;

$$y = \frac{1}{12}\left[8x + 5 \pm \sqrt{-8x^2 + 200x - 455}\right]$$

33.

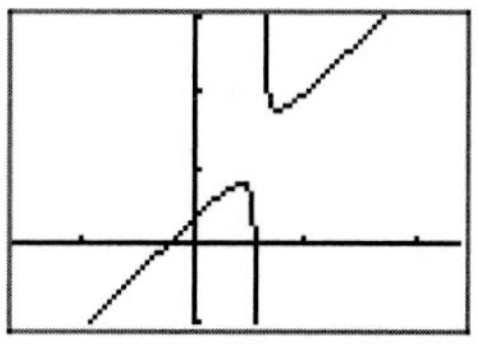

[–8, 12] by [–5, 15]

Hyperbola; $y = \dfrac{3x^2 - 5x - 10}{2x - 6}$

35. **(a)** Parabola

(b) $x = \dfrac{u}{\sqrt{2}} - \dfrac{v}{\sqrt{2}}$ and $y = \dfrac{u}{\sqrt{2}} + \dfrac{v}{\sqrt{2}}$

(c) $u^2 = 4(v - 1)$

(d) Vertex (0, 1); focus (0, 2)

(e) Vertex $\left(-\dfrac{\sqrt{2}}{2}, \dfrac{\sqrt{2}}{2}\right)$; focus $\left(-\sqrt{2}, \sqrt{2}\right)$

37. $y^2 = 8x$

39. $(x + 3)^2 = 12(y - 3)$

41. $\dfrac{x^2}{169} + \dfrac{y^2}{25} = 1$

43. $x^2/9 + (y - 2)^2/5 = 1$

45. $\dfrac{y^2}{25} - \dfrac{x^2}{11} = 1$

47. $\dfrac{(x - 2)^2}{9} - \dfrac{(y - 1)^2}{16} = 1$ **49.** $x^2/25 + y^2/4 = 1$ **51.** $(x + 2)^2 + (y - 4)^2 = 1$ **53.** $x^2/9 - y^2/25 = 1$

55.

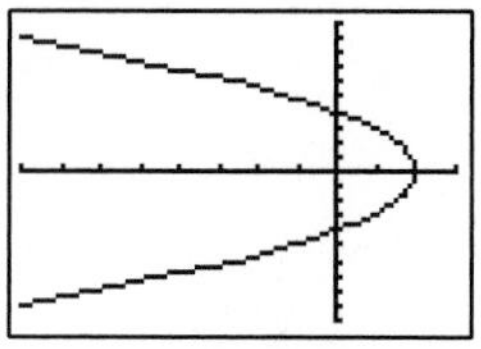

[–8, 3] by [–10, 10]

Parabola; $y^2 = -8(x - 2)$

57.

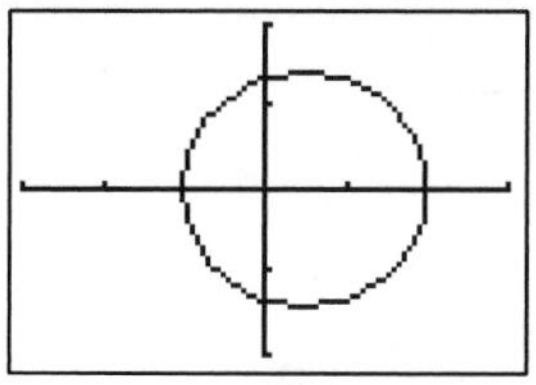

[–3, 3] by [–2, 2]

Ellipse; $\dfrac{4(x - 1/2)}{9} + \dfrac{y^2}{2} = 1$

59.

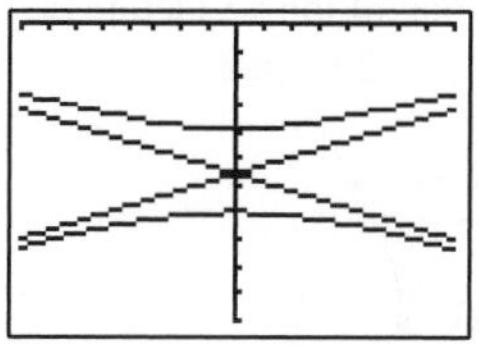

[–8, 8] by [–11, 0]

Hyperbola; $\dfrac{81(y + 49/9)^2}{196} - \dfrac{9x^2}{245} = 1$

61.

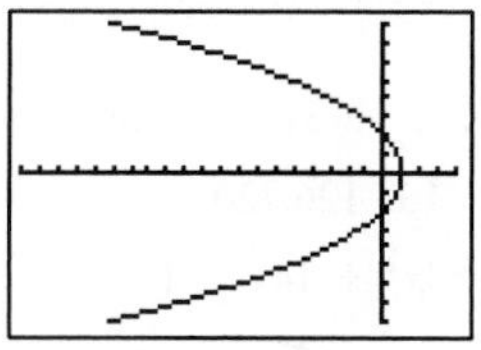

[–20, 4] by [–8, 8]

Parabola; $y^2 = -4(x - 1)$

63. $\sqrt{69}$ **65.** $\langle 0, -3, -2 \rangle$ **67.** -13 **69.** $\langle 3/5, -4/5, 0 \rangle$

71. $(x + 1)^2 + y^2 + (z - 3)^2 = 16$ **73.** $r = \langle -1, 0, 3 \rangle + t\langle -3, 1, -2 \rangle$

75. (0, 4.5) **77.** Answers vary.

79. At apogee, $v \approx 2633$ m/sec; at perigee, $v \approx 9800$ m/sec

Chapter 8 Project

Answers are based on the sample data provided.

1.

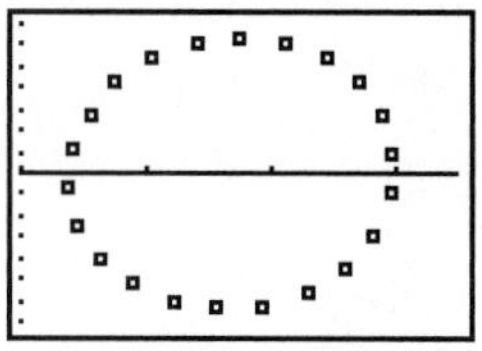

[0.4, 0.75] by [–0.7, 0.7]

3. With respect to the graph of the ellipse, the point (h, k) represents the center of the ellipse. The value a is the semimajor axis, and b is the semiminor axis.

5. The parametric equations for the sample data set are $x_{1T} \approx 0.131 \sin(4.80T + 2.10) + 0.569$ and $y_{1T} \approx 0.639 \sin(4.80T - 2.65)$.

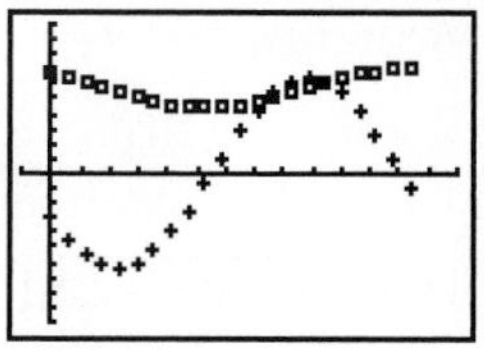

[–0.1, 1.4] by [–1, 1]

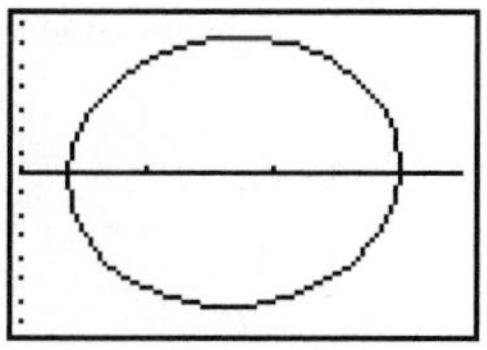

[0.4, 0.75] by [–0.7, 0.7]

SECTION 9.1

Exploration 1

1. 6 **3.** No

Quick Review 9.1

1. 52 **3.** 6 **5.** 10 **7.** 11 **9.** 64 (or 204 if you count squares of all sizes)

Exercises 9.1

1. 6 **3.** 120 **5.** 12 **7.** 362,880 (ALGORITHM) **9.** 34,650 **11.** 1716 **13.** 24 **15.** 30 **17.** 120 **19.** Combinations **21.** Combinations **23.** 19,656,000 **25.** 36 **27.** 2300 **29.** 17,296 **31.** 37,353,738,800 **33.** 41 **35.** 7776 **37.** 511 **39.** 12 **41.** 1024 **43.** True. Both equal $\frac{n!}{a!\,b!}$. **45.** D **47.** B

51. (a) 12 **(b)** There are 12 factors of 5 in 50!, one in each of 5, 10, 15, 20, 30, 35, 40, and 45 and two in each of 25 and 50. Each factor of 5, when paired with one of the 47 factors of 2, yields a factor of 10 and consequently a 0 at the end of 50!

55. 3 **57.** $\approx$ 20,123 years

SECTION 9.2

Exploration 1

1. 1, 3, 3, 1; These are (in order) the coefficients in the expansion of $(a + b)^3$.

3. $\{1\ 5\ 10\ 10\ 5\ 1\}$; These are (in order) the coefficients in the expansion of $(a + b)^5$.

Quick Review 9.2

1. $x^2 + 2xy + y^2$ **3.** $25x^2 - 10xy + y^2$ **5.** $9s^2 + 12st + 4t^2$ **7.** $u^3 + 3u^2v + 3uv^2 + v^3$ **9.** $8x^3 - 36x^2y + 54xy^2 - 27y^3$

Exercises 9.2

1. $a^4 + 4a^3b + 6a^2b^2 + 4ab^3 + b^4$ **3.** $x^7 + 7x^6y + 21x^5y^2 + 35x^4y^3 + 35x^3y^4 + 21x^2y^5 + 7xy^6 + y^7$ **5.** $x^3 + 3x^2y + 3xy^2 + y^3$ **7.** $p^8 + 8p^7q + 28p^6q^2 + 56p^5q^3 + 70p^4q^4 + 56p^3q^5 + 28p^2q^6 + 8pq^7 + q^8$ **9.** 36 **11.** 1 **13.** 364 **15.** 126,720 **17.** $f(x) = x^5 - 10x^4 + 40x^3 - 80x^2 + 80x - 32$ **19.** $h(x) = 128x^7 - 448x^6 + 672x^5 - 560x^4 + 280x^3 - 84x^2 + 14x - 1$ **21.** $16x^4 + 32x^3y + 24x^2y^2 + 8xy^3 + y^4$ **23.** $x^3 - 6x^{5/2}y^{1/2} + 15x^2y - 20x^{3/2}y^{3/2} + 15xy^2 - 6x^{1/2}y^{5/2} + y^3$ **25.** $x^{-10} + 15x^{-8} + 90x^{-6} + 270x^{-4} + 405x^{-2} + 243$ **29.** If $n \geq 1$, $\binom{n}{1} = \frac{n!}{1!(n-1)!} = \frac{n!}{(n-1)!\,1!} = \binom{n}{n-1}$.

31.
$$\begin{aligned}\binom{n-1}{r-1} + \binom{n-1}{r} &= \frac{(n-1)!}{(r-1)!(n-r)!} + \frac{(n-1)!}{r!(n-r-1)!}\\ &= \frac{r(n-1)!}{r(r-1)!(n-r)!} + \frac{(n-r)(n-1)!}{r!(n-r)(n-r-1)!}\\ &= \frac{[r + (n-r)](n-1)!}{r!(n-r)!}\\ &= \frac{n(n-1)!}{r!(n-r)!}\\ &= \frac{n!}{r!(n-r)!}\end{aligned}$$

33. Let $n \geq 2$.
$$\begin{aligned}\binom{n}{2} + \binom{n+1}{2} &= \frac{n!}{2(n-2)!} + \frac{(n+1)!}{2(n-1)!}\\ &= \frac{n(n-1)(n-2)!}{2(n-2)!} + \frac{(n+1)n(n-1)!}{2(n-1)!}\\ &= \frac{n^2 - n}{2} + \frac{n^2 + n}{2}\\ &= \frac{2n^2}{2}\\ &= n^2\end{aligned}$$

35. True. The signs of the coefficients are determined by the powers of the $(-y)$. **37.** C **39.** A
41. **(a)** 1, 3, 6, 10, 15, 21, 28, 36, 45, 55 **(b)** They appear diagonally down the triangle, starting with either of the 1's in row 2.
(c) The number of dots in the array is $n(n + 1)$. Half of the dots are open circles and half are black disks. So, the number of dots in each of these triangular numbers is $n(n + 1)/2$.

(d) $\binom{n+1}{2}$ **43.** $2^n = (1 + 1)^n = \binom{n}{0}1^n1^0 + \binom{n}{1}1^{n-1}1^1 + \binom{n}{2}1^{n-2}1^2 + \cdots + \binom{n}{n}1^01^n = \binom{n}{0} + \binom{n}{1} + \binom{n}{2} + \cdots + \binom{n}{n}$

SECTION 9.3

Quick Review 9.3

1. 19 **3.** 80 **5.** 10/11 **7.** 2560 **9.** 15

Exercises 9.3

1. $2, \frac{3}{2}, \frac{4}{3}, \frac{5}{4}, \frac{6}{5}, \frac{7}{6}; \frac{101}{100}$ **3.** 0, 6, 24, 60, 120, 210; 999,900 **5.** 8, 4, 0, −4; −20 **7.** 2, 6, 18, 54; 4374 **9.** 2, −1, 1, 0; 3
11. Diverges **13.** Converges to 0 **15.** Converges to −1 **17.** Converges to 0 **19.** Diverges
21. **(a)** 4 **(b)** 42 **(c)** $a_1 = 6$ and $a_n = a_{n-1} + 4$ for $n \geq 2$ **(d)** $a_n = 6 + 4(n - 1)$
23. **(a)** 3 **(b)** 22 **(c)** $a_1 = -5$ and $a_n = a_{n-1} + 3$ for $n \geq 2$ **(d)** $a_n = -5 + 3(n - 1)$
25. **(a)** 3 **(b)** 4374 **(c)** $a_1 = 2$ and $a_n = 3a_{n-1}$ for $n \geq 2$ **(d)** $a_n = 2 \cdot 3^{n-1}$
27. **(a)** −2 **(b)** −128 **(c)** $a_1 = 1$ and $a_n = -2a_{n-1}$ for $n \geq 2$ **(d)** $a_n = (-2)^{n-1}$
29. $a_1 = -20$; $a_n = a_{n-1} + 4$ for $n \geq 2$ **31.** $a_1 = \pm\frac{3}{2}$, $r = \pm 2$, and $a_n = 3(\pm 2)^{n-2}$

33.
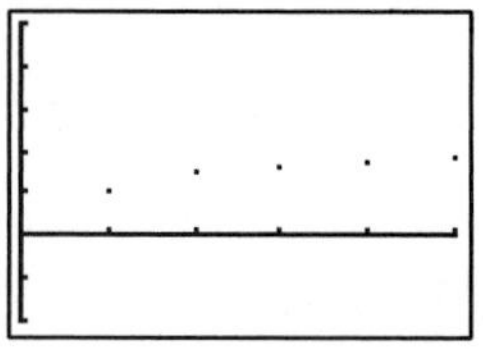
[0, 5] by [–2, 5]

35.
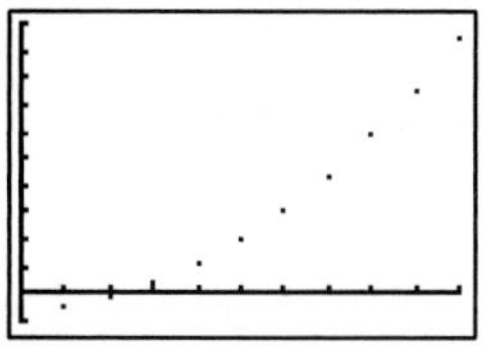
[0, 10] by [–10, 100]

37. 700, 702.3, 704.6, 706.9, . . . , 815, 817.3 **39.** 775 **41.** 9
43. True. The common ratio r must be positive, so the sign of the first term determines the sign of every number in the sequence.
45. A **47.** E **49.** **(b)** 3, 5, 8, 13, 21, 34, 55, 89, 144, 233 **51.** **(b)** $a_n \to 2\pi$ as $n \to \infty$
55. $a_1 = [1 \;\; 1]$, $a_2 = [1 \;\; 2]$, $a_3 = [2 \;\; 3]$, $a_4 = [3 \;\; 5]$, $a_5 = [5 \;\; 8]$, $a_6 = [8 \;\; 13]$, $a_7 = [13 \;\; 21]$. The entries in the terms of this sequence are successive pairs of terms from the Fibonacci sequence.

SECTION 9.4

Exploration 1

1. 45 **3.** 1 **5.** $\frac{1}{3}$

Exploration 2

1. $1 + 2 + 3 + \cdots + 99 + 100$ **3.** 101
5. The sum in 4 involves two copies of the same progression, so it doubles the sum of the progression. The answer is 5050.

Quick Review 9.4

1. 22 **3.** 27 **5.** 512 **7.** −40 **9.** 55

Exercises 9.4

1. $\sum_{k=1}^{11} (6k - 13)$ **3.** $\sum_{k=1}^{n+1} k^2$ **5.** $\sum_{k=0}^{\infty} 6(-2)^k$ **7.** 18 **9.** 3240 **11.** 975 **13.** 24,573 **15.** $50.4(1 - 6^{-9}) \approx 50.4$
17. 155 **19.** $\frac{8}{3}(1 - 2^{-12}) \approx 2.666$ **21.** −196,495,641
23. **(a)** 0.3, 0.33, 0.333, 0.3333, 0.33333, 0.333333; convergent **(b)** 1, −1, 2, −2, 3, −3; divergent **25.** Yes; 12 **27.** No
29. Yes; 1 **31.** 707/99 **33.** $-\frac{17{,}251}{999}$ **35.** **(a)** 1.1 **(b)** $20{,}000(1.1)^n$ **(c)** \$370,623.34

37. **(a)** 120; $1 + 0.07/12$ **(b)** \$20,770.18 **39.** 38 m **41.** **False**. The series might well diverge. **43.** A **45** C
47. **(a)** Heartland: 20,505,437 persons; Southeast: 48,310,650 persons **(b)** Heartland: 517,825 mi^2; Southeast: 348,999 mi^2
(c) Heartland: $\approx$ 39.60 persons/mi^2; Southeast: $\approx$ 138.43 persons/mi^2 **(d)** Heartland: 40.18 persons/mi^2; Southeast: 134.02 persons/mi^2
The answers in part (c) are the proper population densities; they account for the varying areas of the states. In general, rates should be average, as in part (c) and not as in part (d).

49.

n	F_n	S_n	$F_{n+2} - 1$
1	1	1	1
2	1	2	2
3	2	4	4
4	3	7	7
5	5	12	12
6	8	20	20
7	13	33	33
8	21	54	54
9	34	88	88

Conjecture: $S_n = F_{n+2} - 1$

SECTION 9.5

Exploration 1

Start with the rightmost peg if n is odd and the middle peg if n is even.

Exploration 2

1. Yes **3.** Still all prime

Quick Review 9.5

1. $n^2 + 5n$ **3.** $k^3 + 3k^2 + 2k$ **5.** $(k+1)^3$ **7.** $5; t+4; t+5$ **9.** $\frac{1}{2}; \frac{2k}{3k+1}; \frac{2k+2}{3k+4}$

Exercises 9.5

1. P_n: $2 + 4 + 6 + \cdots + 2n = n^2 + n$. P_1 is true: $2(1) = 1^2 + 1$. Now assume P_k is true: $2 + 4 + 6 + \cdots + 2k = k^2 + k$. Add $2(k+1)$ to both sides: $2 + 4 + 6 + \cdots + 2k + 2(k+1) = k^2 + k + 2(k+1) = k^2 + 3k + 2 = k^2 + 2k + 1 + k + 1 = (k+1)^2 + (k+1)$, so P_{k+1} is true. Therefore, P_n is true for all $n \ge 1$.

3. P_n: $6 + 10 + 14 + \cdots + (4n+2) = n(2n+4)$. P_1 is true: $4(1) + 2 = 1(2(1) + 4)$.
Now assume P_k is true: $6 + 10 + 14 + \cdots + (4k+2) = k(2k+4)$. Add $4(k+1) + 2 = 4k + 6$ to both sides:
$6 + 10 + 14 + \cdots + (4k+2) + [4(k+1) + 2] = k(2k+4) + 4k + 6 = 2k^2 + 8k + 6 = (k+1)(2k+6) = (k+1)[2(k+1) + 4]$, so P_{k+1} is true. Therefore, P_n is true for all $n \ge 1$.

5. P_n: $a_n = 5n - 2$. P_1 is true: $a_1 = 5 \cdot 1 - 2 = 3$. Now assume P_k is true: $a_k = 5k - 2$. To get a_{k+1}, add 5 to a_k; that is, $a_{k+1} = (5k - 2) + 5 = 5(k+1) - 2$. This shows that P_{k+1} is true. Therefore, P_n is true for all $n \ge 1$.

7. P_n: $a_n = 2 \cdot 3^{n-1}$. P_1 is true: $a_1 = 2 \cdot 3^{1-1} = 2 \cdot 3^0 = 2$. Now assume P_k is true: $a_k = 2 \cdot 3^{k-1}$. To get a_{k+1}, multiply a_k by 3; that is, $a_{k+1} = 3 \cdot 2 \cdot 3^{k-1} = 2 \cdot 3k = 2 \cdot 3^{(k+1)-1}$. This shows that P_{k+1} is true. Therefore, P_n is true for all $n \ge 1$.

9. P_1: $1 = \frac{1(1+1)}{2}$. P_k: $1 + 2 + \cdots + k = \frac{k(k+1)}{2}$. P_{k+1}: $1 + 2 + \cdots + k + (k+1) = \frac{(k+1)(k+2)}{2}$.

11. P_1: $\frac{1}{1 \cdot 2} = \frac{1}{1+1}$. P_k: $\frac{1}{1 \cdot 2} + \frac{1}{2 \cdot 3} + \cdots + \frac{1}{k(k+1)} = \frac{k}{k+1}$.

P_{k+1}: $\frac{1}{1 \cdot 2} + \frac{1}{2 \cdot 3} + \cdots + \frac{1}{k(k+1)} + \frac{1}{(k+1)(k+2)} = \frac{k+1}{k+2}$.

13. P_n: $1 + 5 + 9 + \cdots + (4n - 3) = n(2n - 1)$. P_1 is true: $4(1) - 3 = 1 \cdot (2 \cdot 1 - 1)$.
Now assume P_k is true: $1 + 5 + 9 + \cdots + (4k - 3) = k(2k - 1)$. Add $4(k+1) - 3 = 4k + 1$ to both sides: $1 + 5 + 9 + \cdots + (4k - 3) + [4(k+1) - 3] = k(2k - 1) + 4k + 1 = 2k^2 + 3k + 1 = (k+1)(2k+1) = (k+1)[2(k+1) - 1]$, so P_{k+1} is true. Therefore, P_n is true for all $n \ge 1$.

15. P_n: $\frac{1}{1\cdot 2}+\frac{1}{2\cdot 3}+\cdots+\frac{1}{n(n+1)}=\frac{n}{n+1}$. P_1 is true: $\frac{1}{1\cdot 2}=\frac{1}{1+1}$.

Now assume P_k is true: $\frac{1}{1\cdot 2}+\frac{1}{2\cdot 3}+\cdots+\frac{1}{k(k+1)}=\frac{k}{k+1}$.

Add $\frac{1}{(k+1)(k+2)}$ to both sides: $\frac{1}{1\cdot 2}+\frac{1}{2\cdot 3}+\cdots+\frac{1}{k(k+1)}+\frac{1}{(k+1)(k+2)}=\frac{k}{k+1}+\frac{1}{(k+1)(k+2)}$

$=\frac{k(k+2)+1}{(k+1)(k+2)}=\frac{(k+1)(k+1)}{(k+1)(k+2)}=\frac{k+1}{k+2}=\frac{k+1}{(k+1)+1}$, so P_{k+1} is true. Therefore, P_n is true for all $n\geq 1$.

17. P_n: $2^n\geq 2n$. P_1 is true: $2^1\geq 2\cdot 1$ (in fact, they are equal). Now assume P_k is true: $2^k\geq 2k$.
Then $2^{k+1}=2\cdot 2^k\geq 2\cdot 2k=2\cdot(k+k)\geq 2(k+1)$, so P_{k+1} is true. Therefore, P_n is true for all $n\geq 1$.

19. P_n: 3 is a factor of n^3+2n. P_1 is true: 3 is a factor of $1^3+2\cdot 1=3$. Now assume P_k is true: 3 is a factor of k^3+2k.
Then $(k+1)^3+2(k+1)=(k^3+3k^2+3k+1)+(2k+2)=(k^3+2k)+3(k^2+k+1)$. Since 3 is a factor of both terms, it is a factor of the sum, so P_{k+1} is true. Therefore, P_n is true for all $n\geq 1$.

21. P_n: The sum of the first n terms of a geometric sequence with first term a_1 and common ratio $r\neq 1$ is $\frac{a_1(1-r^n)}{1-r}$.

P_1 is true: $a_1=\frac{a_1(1-r^1)}{1-r}$. Now assume P_k is true so that $a_1+a_1r+\cdots+a_1r^{k-1}=\frac{a_1(1-r^k)}{(1-r)}$.

Add a_1r^k to both sides: $a_1+a_1r+\cdots+a_1r^{k-1}+a_1r^k=\frac{a_1(1-r^k)}{(1-r)}+a_1r^k=\frac{a_1(1-r^k)+a_1r^k(1-r)}{1-r}$

$=\frac{a_1-a_1r^k+a_1r^k-a_1r^{k+1}}{1-r}=\frac{a_1-a_1r^{k+1}}{1-r}$, so P_{k+1} is true. Therefore, P_n is true for all positive integers n.

23. P_n: $\sum_{k=1}^{n}k=\frac{n(n+1)}{2}$. P_1 is true: $\sum_{k=1}^{1}k=1=\frac{1\cdot 2}{2}$. Now assume P_k is true: $\sum_{i=1}^{k}i=\frac{k(k+1)}{2}$.

Add $(k+1)$ to both sides, and we have $\sum_{i=1}^{k+1}i=\frac{k(k+1)}{2}+(k+1)=\frac{k(k+1)}{2}+\frac{2(k+1)}{2}=\frac{(k+1)(k+2)}{2}$

$=\frac{(k+1)((k+1)+1)}{2}$, so P_{k+1} is true. Therefore, P_n is true for all $n\geq 1$.

25. 125,250 **27.** $\frac{(n-3)(n+4)}{2}$ **29.** $\approx 3.44\times 10^{10}$ **31.** $\frac{n(n^2-3n+8)}{3}$ **33.** $\frac{n(n-1)(n^2+3n+4)}{4}$

35. The inductive step does not work for 2 persons. Sending them alternately out of the room leaves 1 person (and one blood type) each time, but we cannot conclude that their blood types will match *each other.*

37. False. Mathematical induction is used to show that a statement P_n is true for all positive integers. **39.** E **41.** B

43. P_n: 2 is a factor of $(n+1)(n+2)$. P_1 is true because 2 is a factor of $(2)(3)$. Now assume P_k is true so that 2 is a factor of $(k+1)(k+2)$. Then $[(k+1)+1][(k+1)+2]=(k+2)(k+3)=k^2+5k+6=k^2+3k+2+2k+4=(k+1)(k+2)+2(k+2)$. Since 2 is a factor of both terms of this sum, it is a factor of the sum, and so P_{k+1} is true. Therefore, P_n is true for all positive integers n.

45. Given any two consecutive integers, one of them must be even. Therefore, their product is even. Since $n+1$ and $n+2$ are consecutive integers, their product is even. Therefore, 2 is a factor of $(n+1)(n+2)$.

47. P_n: $F_{n+2}-1=\sum_{k=1}^{n}F_k$. P_1 is true since $F_{1+2}-1=F_3-1=2-1=1$, which equals $\sum_{k=1}^{1}F_k=1$.

Now assume that P_k is true: $F_{k+2}-1=\sum_{i=1}^{k}F_i$. Then $F_{(k+1)+2}-1=F_{k+3}-1=F_{k+1}+F_{k+2}-1$

$=(F_{k+2}-1)+F_{k+1}=\left(\sum_{i=1}^{k}F_i\right)+F_{k+1}=\sum_{i=1}^{k+1}F_i$, so P_{k+1} is true. Therefore, P_n is true for all $n\geq 1$.

49. P_n: $a-1$ is a factor of a^n-1. P_1 is true because $a-1$ is a factor of $a-1$. Now assume P_k is true so that $a-1$ is a factor of a^k-1. Then $a^{k+1}-1=a\cdot a^k-1=a(a^k-1)+(a-1)$. Since $a-1$ is a factor of both terms in the sum, it is a factor of the sum, and so P_{k+1} is true. Therefore, P_n is true for all positive integers n.

51. P_n: $3n-4\geq n$ for $n\geq 2$. P_2 is true, since $3\cdot 2-4\geq 2$. Now assume that P_k is true: $3k-4\geq k$.
Then $3(k+1)-4=3k+3-4=(3k-4)+3\geq k+3\geq k+1$, so P_{k+1} is true. Therefore, P_n is true for all $n\geq 2$.

53. Use P_3 as the anchor and obtain the inductive step by representing any n-gon as the union of a triangle and an $(n-1)$-gon.

CHAPTER 9 REVIEW EXERCISES

1. 792 **3.** 18,564 **5.** 3,991,680 **7.** 43,670,016 **9.** 14,508,000 **11.** 8,217,822,536 **13.** 26 **15.** 325
17. (a) 5040; Meg Ryan **(b)** 778,377,600; Britney Spears **19.** $32x^5 + 80x^4y + 80x^3y^2 + 40x^2y^3 + 10xy^4 + y^5$
21. $243x^{10} + 405x^8y^3 + 270x^6y^6 + 90x^4y^9 + 15x^2y^{12} + y^{15}$
23. $512a^{27} - 2304a^{24}b^2 + 4608a^{21}b^4 - 5376a^{18}b^6 + 4032a^{15}b^8 - 2016a^{12}b^{10} + 672a^9b^{12} - 144a^6b^{14} + 18a^3b^{16} - b^{18}$
25. -1320 **27.** {1, 2, 3, 4, 5, 6} **29.** {13, 16, 31, 36, 61, 63} **31.** {HHH, HHT, HTH, HTT, THH, THT, TTH, TTT}
33. {HHH, TTT} **35.** 0, 1, 2, 3, 4, 5; 39 **37.** $-1, 2, 5, 8, 11, 14; 32$ **39.** $-5, -3.5, -2, -0.5, 1, 2.5; 11.5$
41. $-3, 1, -2, -1, -3, -4; -76$ **43.** Arithmetic with $d = -2.5$; $a_n = 14.5 - 2.5n$ **45.** Geometric with $r = 1.2$; $a_n = 10 \cdot (1.2)^{n-1}$
47. Arithmetic with $d = 4.5$; $a_n = 4.5n - 15.5$ **49.** $a_n = 3(-4)^{n-1}$; $r = -4$ **51.** -4
53. -985.5 **55.** 21/8 **57.** 59,048 **59.** $3280.\overline{4}$
61.

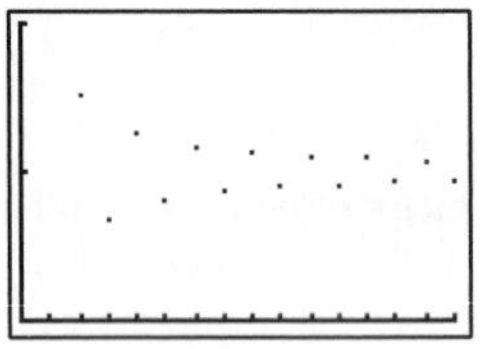

[0, 15] by [0, 2]

63. \$27,441.91 **65.** converges; 6 **67.** diverges **69.** converges; 3
71. $\sum_{k=1}^{2k}(5k - 13)$ **73.** $\sum_{k=0}^{\infty}(2k + 1)^2$ or $\sum_{k=1}^{\infty}(2k - 1)^2$ **75.** $\dfrac{n(3n + 5)}{2}$ **77.** 4650

79. P_n: $1 + 3 + 6 + \cdots + \dfrac{n(n + 1)}{2} = \dfrac{n(n + 1)(n + 2)}{6}$. P_1 is true: $\dfrac{1(1 + 1)}{2} = \dfrac{1(1 + 1)(1 + 2)}{6}$.

Now assume P_k is true: $1 + 3 + 6 + \cdots + \dfrac{k(k + 1)}{2} = \dfrac{k(k + 1)(k + 2)}{6}$.

Add $\dfrac{(k + 1)(k + 2)}{2}$ to both sides: $1 + 3 + 6 + \cdots + \dfrac{k(k + 1)}{2} + \dfrac{(k + 1)(k + 2)}{2} = \dfrac{k(k + 1)(k + 2)}{6} + \dfrac{(k + 1)(k + 2)}{2}$

$= (k + 1)(k + 2)\left(\dfrac{k}{6} + \dfrac{1}{2}\right) = (k + 1)(k + 2)\left(\dfrac{k + 3}{6}\right) = \dfrac{(k + 1)((k + 1) + 1)((k + 1) + 2)}{6}$, so P_{k+1} is true.

Therefore, P_n is true for all $n \geq 1$.

81. P_n: $2^{n-1} \leq n!$. P_1 is true: $2^{1-1} \leq 1!$ (they are equal). Now assume P_k is true: $2^{k-1} \leq k!$.
Then $2^{(k+1)-1} = 2 \cdot 2^{k-1} \leq 2 \cdot k! \leq (k + 1)k! = (k + 1)!$, so P_{k+1} is true. Therefore, P_n is true for all $n \geq 1$.
83. 1 9 36 84 126 126 84 36 9 1

Chapter 9 Project

Answers are based on the sample data shown in the table.

1. (a) 3.0 million persons/year **(b)** The year 2000 was 0 years after 2000. **(c)** $p_n = p_{n-1} + 3$ **(d)** $p_n = 3n + 281.4$
(e) 311.4 million, 326.4 million, 341.4 million
3. (a) $b_n = 0.75 \cdot b_{n-1} + 800$
(b)

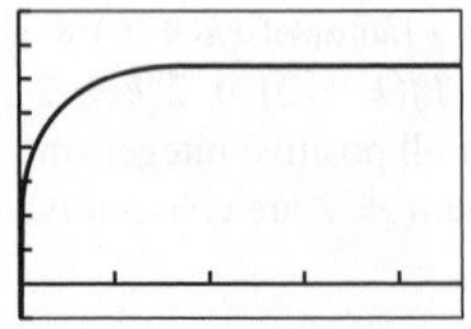

[0, 47] by [–500, 4000]

(c) P_n: $b_n = 3200 - 2100(0.75)^n$. P_0 is true because $b_0 = 3200 - 2100 = 1100$. Now assume P_k is true: $b_k = 3200 - 2100(0.75)^k$. By the recursive formula, $b_{k+1} = 0.75 \cdot (3200 - 2100(0.75)^k) + 800 = 2400 - 2100(0.75)^{k+1} + 800 = 3200 - 2100(0.75)^{k+1}$. So, P_{k+1} is true. Thus, P_n is true for all $n \geq 0$.
(d) 3200

SECTION 10.1

Exploration 1

1.

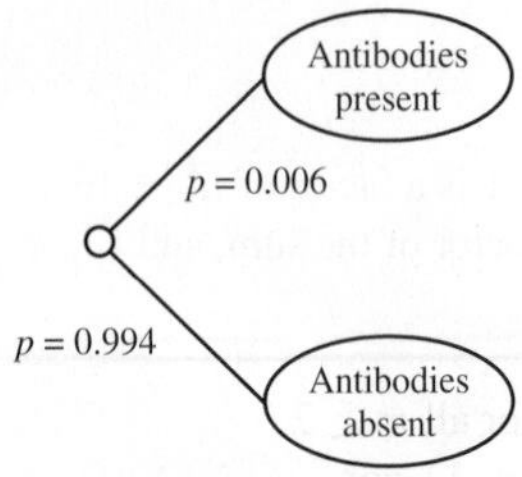

3.

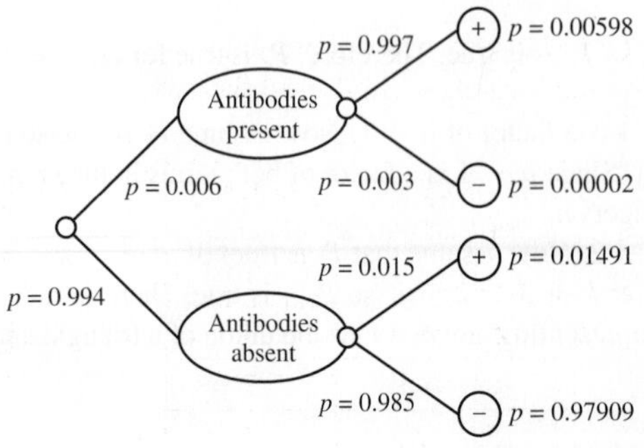

5. ≈ 0.286

Quick Review 10.1

1. 2 **3.** 8 **5.** 2,598,960 **7.** 120 **9.** $\frac{1}{12}$

Exercises 10.1

1. 1/9 **3.** 5/12 **5.** 1/4 **7.** 5/12

9. **(a)** No; the numbers do not add up to 1.
(b) Yes; assuming the gerbil cannot be in more than one compartment at a time, the proportions cannot sum to more than 1.

11. 0.4 **13.** 0.2 **15.** 0.7 **17.** 0.09 **19.** 0.08 **21.** 0.64 **23.** 1/134,596 **25.** 5/3542

27. **(a)**

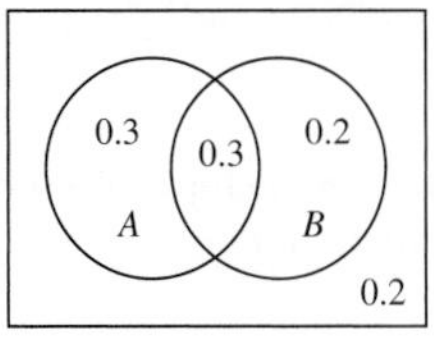

(b) 0.3 **(c)** 0.2 **(d)** 0.8 **(e)** Yes

29. 0.64 **31.** 3/5 **33.** 19/30 **35.** **(a)** 0.67 **(b)** 0.33 **39.** **(a)** 86/127 **(b)** 91/127 **(c)** 62/127
41. 1/36 **45.** **(a)** 0.027 **(b)** 0.343 **(c)** 0.217 **47.** 0.0864 **49.** **(a)** 0.4 **(b)** 0.6 **(c)** 0.455 **(d)** 0.833 **(e)** No
51. False. A sample space consists of outcomes, which are not necessarily equally likely. **53.** D **55.** A

57. **(a)**

Type of Bagel	Probability
Plain	0.37
Onion	0.12
Rye	0.11
Cinnamon Raisin	0.25
Sourdough	0.15

(b) ≈ 0.051

59. **(a)** 8.8% chance; plausible but unlikely **(b)** All names equally likely to be chosen.
61. **(a)** $1.50 **(b)** 1/3

SECTION 10.2

Exploration 1

1. The average is about 12.8. **3.** Alaska, Colorado, Georgia, Texas, and Utah

Quick Review 10.2

1. $\approx$ 15.48% **3.** $\approx$ 14.44% **5.** $\approx$ 1723 **7.** $235 thousand **9.** 1 million

Exercises 10.2

1. **(a)** 29% **(b)** 13.1% **(c)** 26.8% **(d)** 45.2%
3. **(a)** Pie charts; stemplots are not appropriate for categorical variables. **(b)** No. Men are more likely than women to be interested in the game (57% to 39%); women are more likely than men to be interested in the commercials (30% to 16.5%) or to not watch at all (31% to 27%).
5. **(a)** 42 **(b)** 6 **(c)** 168 **(d)** 24

7.

0	5 8 9
1	3 4 6
2	3 6 8
3	3 9
4	
5	
6	1

61 is an outlier.

9.

Maris		Aaron
9 8 5	0	
6 4 3	1	0 2 3
8 6 3	2	0 4 6 7 9
9 3	3	0 2 4 4 8 9 9
	4	0 0 4 4 4 4 5 7
	5	
1	6	

Except for Maris's one record-breaking year, his home run output falls well short of Aaron's.

11. Males

6	0
6	5 9
7	1 2 2 3 3 3
7	5 5 6

13.

Males		Females
3 0	6	
9 5	6	7 9
3 3 3 2 2 1	7	
6 5 5	7	6 8 8 8 9 9 9
	8	0 1 2

15.

Life expectancy (years)	Frequency
60.0–64.9	1
65.0–69.9	2
70.0–74.9	7
75.0–79.9	2

17.

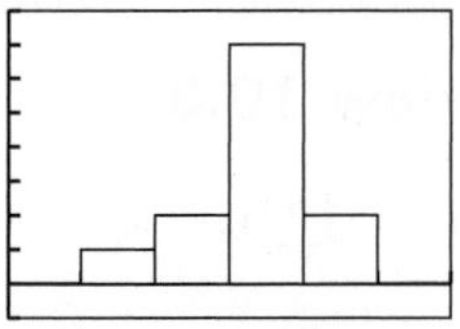

[55, 85] by [–1, 8]

19.

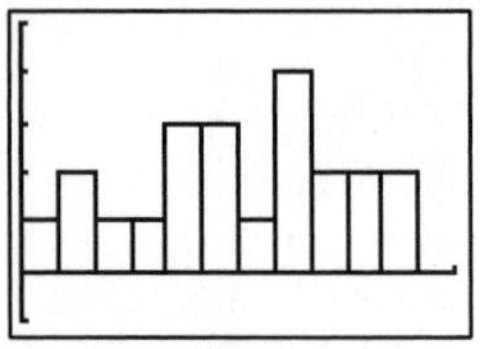

[0, 60] by [–1, 5]

21.

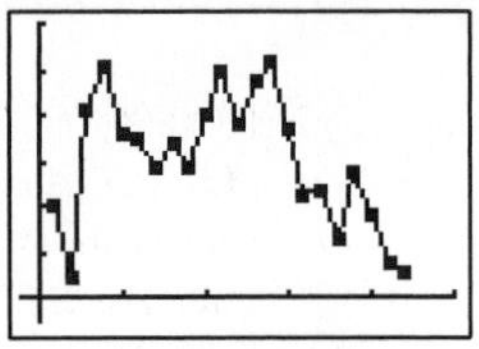

[–1, 25] by [–5, 60]

23.

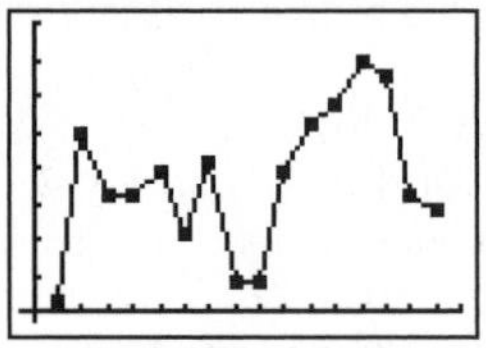

[–1.5, 17] by [–2, 80]

25.

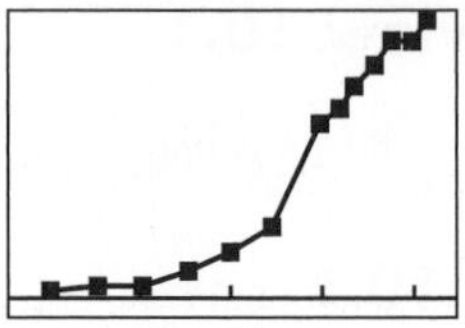

[1965, 2015] by [–100, 1500]

The winner's prize money for the PGA Championship appears to be growing exponentially, though the rate of growth may be slowing down in recent years.

27.

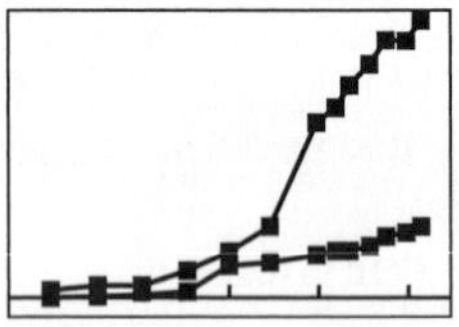

[1965, 2015] by [–100, 1500]

The women's winner's prize money grew at a rate similar to that of the men's until 1995; then the men's PGA purse began increasing rapidly, leaving the LPGA purse far behind.

29.

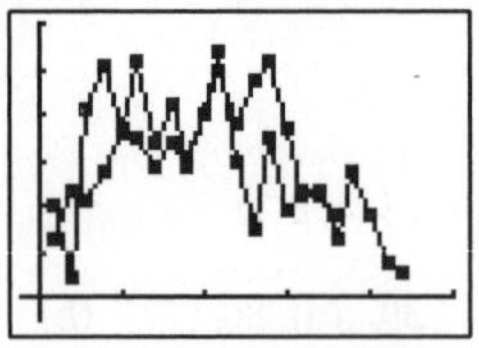

[–1, 25] by [–5, 60]

The two home run hitters enjoyed similar success.

31. **(a)** Bimodal
(b) Healthier cereals for adults and sugary cereals for children

33. **(a)** The data are quantitative.

(b)

28	2
29	3 7
30	
31	6 7
32	7 8
33	5 5 5 8
34	2 8 8
35	3 3 4
36	3 7
37	
38	5

(c) Unimodal and symmetric

35 **(a)**

Interval	Frequency
25.0–29.9	3
30.0–34.9	11
35.0–39.9	6

(b)

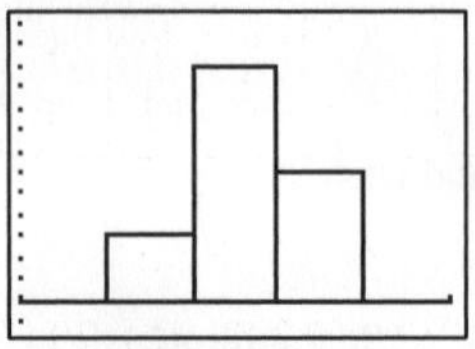

[20, 45] by [–1, 13]

(c) The distribution is unimodal and symmetric, with most salaries in the interval \$30,000–34,999.

37.

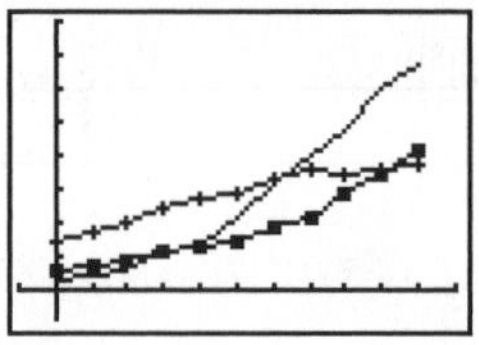

[1890, 2010] by [–4, 40]

_____ = CA; + = NY, ■ = TX

39. False. If the graduation rates are the same, then the likelihood of graduating is independent of a student's gender; there is no association.

41. C **43.** A **47.**

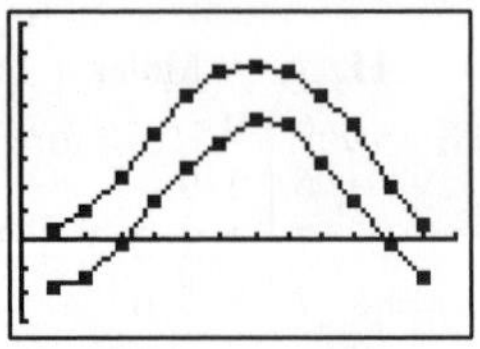

[0, 13] by [–15, 40]

SECTION 10.3

Exploration 1

1. Approximately the same **3.** Figure (b)

Quick Review 10.3

1. $x_1 + x_2 + x_3 + x_4 + x_5 + x_6 + x_7$ **3.** $\frac{1}{7}(x_1 + x_2 + x_3 + x_4 + x_5 + x_6 + x_7)$ **5.** $\frac{1}{5}[(x_1 - \bar{x})^2 + (x_2 - \bar{x})^2 + \cdots + (x_5 - \bar{x})^2]$

7. $\sum_{i=1}^{8} x_i f_i$ **9.** $\frac{1}{50}\sum_{i=1}^{50}(x_i - \bar{x})^2$

Exercises 10.3

1. **(a)** Statistic **(b)** Parameter **3.** **(a)** Mean **(b)** Median **5.** 8 satellites **7.** $33,500 **9.** Range: $10,300; $IQR = \$3,600$ **11.** 210 lb (and any other weights over 198) **13.** $\{16, 18, 20, 24, 42\}$; outliers: 40, 42 **15.** $\{20, 29, 42, 49, 66\}$; no outliers **17.** In general, NHL teams do win more games at home than on the road; the median for home wins is about 5 games higher than the median for away games, and greater than the third quartile for away wins. Although the variability is comparable, over 25% of the teams won more home games than even the best team won on the road. **19.** Babe Ruth: Five-number summary: $\{0, 11, 38, 47, 60\}$; Range: 60; *IQR*: 36; No outliers; Barry Bonds: Five-number summary: $\{5, 25, 34, 45, 73\}$; Range: 68; *IQR*: 20; Outlier: 73 (perhaps)

21.

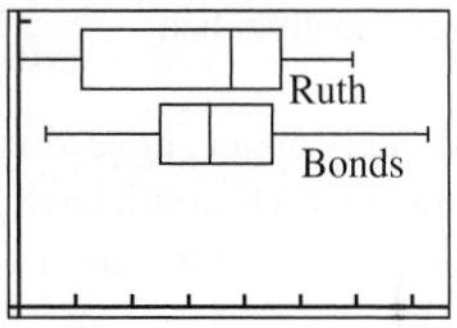

23. 21.625; much larger than the median because 2 planets have over 60 moons **25.** $\approx$ \$33,542; almost exactly the same **27.** ≈ 2.97 **29.** ≈ 22.95; median; mean is pulled high because the distribution is skewed to the right. **31.** **(a)** ≈ 18.42°C **(b)** ≈ 18.49°C **(c)** The weighted average is the better indicator. **33.** $s \approx 9.71$; $s^2 \approx 94.3$ **35.** $s \approx$ \$66.8 billion; $s^2 \approx 4462$

37. $s \approx \$2673$; $s^2 \approx 7{,}150{,}000$ **39.** **(a)** smaller; values closer to the mean **(b)** $2.58 < 2.94$ **43.** **(a)** 68% **(b)** 2.5% **(c)** A statistic **45.** **(a)** 16% **(b)** 13.5% **(c)** Over 101 g **(d)** Individuals more than 3 standard deviations below the mean are very rare. **47.** False. The median is a resistant measure. **49.** A **51.** B **53.** There are many possible answers; examples are given. **(a)** $\{2, 2, 2, 3, 6, 8, 20\}$ **(b)** $\{-20, 1, 1, 1, 2, 3, 4, 5, 6\}$ **57.** 75.9 years **59.** 5%

SECTION 10.4

Exploration 1

Plan B

Quick Review 10.4

1. 2 **3.** 4 **5.** 22,100 **7.** 6 **9.** 56

Exercises 10.4

1. 17 **3.** 10 **5.** 2.625 **7.** Making \$75/day **9.** **(a)**

Y	1	2	3	4	5
$P(Y)$	$\frac{5}{20}$	$\frac{3}{20}$	$\frac{3}{20}$	$\frac{4}{20}$	$\frac{1}{20}$

(b) $53/20 = 2.65$

11. No; $-\$79$, compared to $-\$11$ without the warranty.

13. **(a)**

Family	G	BG	BBG	BBB
X = children	1	2	3	3
$P(X)$	$\frac{1}{2}$	$\frac{1}{4}$	$\frac{1}{8}$	$\frac{1}{8}$

(b) 1.75

15. \$75 **17.** **(a)** 0.386 **(b)** 0.116 **(c)** 0.132 **19.** **(a)** 0.393 **(b)** 0.017 **21.** **(a)** 0.200 **(b)** 0.012 **25.** **(a)** 22.5 **(b)** 3.97 **29.** **(a)** 0.067 **(b)** 0.773 **(c)** 0.083 **31.** **(a)** 1.88 **(b)** 0.25 **(c)** ± 1.28 **33.** **(a)** 0.013 **(b)** 0.784 **35.** Under 59.2 in. **37.** **(a)** 99.4% **(b)** 0.155 **(c)** 10.9 hr **39.** **(a)** Yes; $125 > 10$ **(b)** 125, 7.91 **(c)** No; $z < 2$

43. False; expected value is a parameter calculated from the theoretical probability distribution. **45.** D **47.** C **49.** **(a)** ≈ 2 **(b)** Yes **(c)** $\approx 1.913\%$ **51.** Highest expected winnings = \$1600 for Deal B; lowest = \$1000 for Deal C. **53.** **(a)** \$4.50 **(b)** \$13.96 **55.** **(a)** 52% **(b)** 12.49 **(c)** 325 & 300 are both > 10 **(d)** $\pm 4\%$

SECTION 10.5

Exploration 1

1. Correlation begins with a scatter plot, which requires numerical data from two quantitative variables (like height and weight). "Gender" is categorical.
3. The doctor's "experiment" proves nothing about the effect of vanilla gum on headache pain unless we can compare these subjects with a similar group that does not use vanilla gum. Many headaches are gone in two hours anyway.
5. The percentages make the difference seem large, but it actually amounts to only 6 of the 50 people. The results may not be statistically significant.

Quick Review 10.5

1. $\frac{1}{6}$ **3.** $\frac{4}{52} = \frac{1}{13}$ **5.** $\frac{1}{10}$ **7.** $\left(\frac{1}{10}\right)^5 = 0.00001$ **9.** $1 - \left(\frac{9}{10}\right)^5 = 0.40951$

Exercises 10.5

1. Incorrect. Intelligence might be associated with some quantitative variable, but beauty is categorical.
3. Incorrect. The high correlation coefficient does nothing to support Sean's crazy theory, because the great blue whale (with a long name and a huge weight) is an unusual point that lies far away from the other three.
5. Incorrect. Marcus is OK with his first observation, but not with his second. While his *linear* model is a bad fit, he should not conclude that "there is no significant mathematical relationship." In fact, check out this sinusoidal fit:

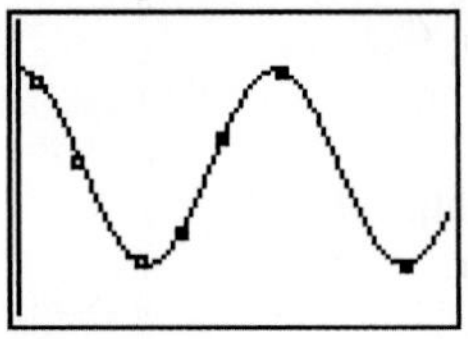

7. This is random (technically pseudo-random, but it should suffice).
9. This is not a random sample of all Reno citizens. All fifty are likely to be from early in the alphabet.
11. This is not random, nor does it apparently try to be.
13. Voluntary response bias. The students most likely to respond were those who felt strongly about suggestions for improvement, so the rate of negative responses was probably higher than the parameter. He could have gotten a less biased response with an in-class census of all his students (ideally in multiple-choice form so that their handwriting would not betray their identities).
15. Undercoverage bias. The survey systematically excluded the students who were not actually eating in the dining hall, so the sample statistic was bound to be higher than the population parameter. A better method would have been to choose a random sample from the student body first, then seek them out for the survey (perhaps in their homerooms).
17. Response bias. The question was designed to elicit a negative response, and it never even mentioned stop signs. The 97% was much higher than it would have been with a simple question like, "Should citizens be allowed to ignore stop signs?"
19. Observational study. No treatment imposed. **21.** Observational study. No treatment imposed. **23.** Experiment.
25. Using random numbers, select 12 of the 24 plots to get the new fertilizer. Use the original fertilizer on the other 12 plots. Compare the yields at harvest time.
27. This requires three treatments. Split the 24 plots randomly into three groups of 8: new fertilizer 1, new fertilizer 2, and original fertilizer.
29. Fatigue may be a factor after they have driven 20 golf balls. They could gather the data on different days, or they could randomly choose half the golfers to drive the new ball first.
31. The music assignment should be randomized, not left to the choice of the mother. Otherwise, the mother's music preference (with possible lifestyle implications) becomes a potentially significant confounding variable.
33. One possible solution: Use the command "randInt(1, 500, 50)" to choose 50 random numbers from 1 to 500. If there are any repeat numbers in the list, use "randInt(1, 500)" to pick additional numbers until you have a sample of 50.
35. One possible solution: Enter the numbers 1 to 32 in list L1 using the command "seq(X, X, 1, 32) → L1" and enter 32 random numbers in list L2 using the command "rand(32) → L2." Then sort the random numbers into ascending order, bringing L1 along for the ride, using the command "SortA(L2, L1)." The numbers in list L1 are now in random order.
37. Number the plants 1–16. Use "randInt(1, 16)" to generate 8 distinct random numbers. Grow those 8 plants with the plant food and the other 8 without it. **39.** One possible solution: Use the command "randInt(1, 8)" to generate random numbers between 1 and 8.
41. One possible solution: Use the command "randInt(1, 5, 20)" to generate 20 random numbers from 1 to 5. Let 1 and 2 designate donors with O-positive blood. Do this nine times and keep track of how many strings have fewer than four numbers that are 1 or 2.
43. Use "randInt(1, 6)" to generate a series of random rolls of the die. Keep a running total, but don't add rolls that would make the sum greater than 21. Stop when the total equals 21. Report the number of rolls.
45. Yes, there is enough evidence to warrant suspicion. In only 10 of the 500 simulated trials did 8 or more 6's show up. There's only a 2% chance that rolling a die fairly would produce a result like this.
47. False. Observational studies can show strong associations, but experiments would be required to establish causation.
49. **B** (Note that this is the only quantitative variable among the choices.) **51.** **C**
53. Answers will vary. Note that you should not expect all the counts to be exactly the same (that would suggest nonrandomness in itself), but "randomness" would predict an approximately equal distribution, especially for a large class.
55. **(a)** Correlation coefficient will increase; slope will remain about the same. **(b)** Correlation coefficient will increase; slope will increase. **(c)** Correlation coefficient will decrease; slope will decrease.
57. One possible picture:

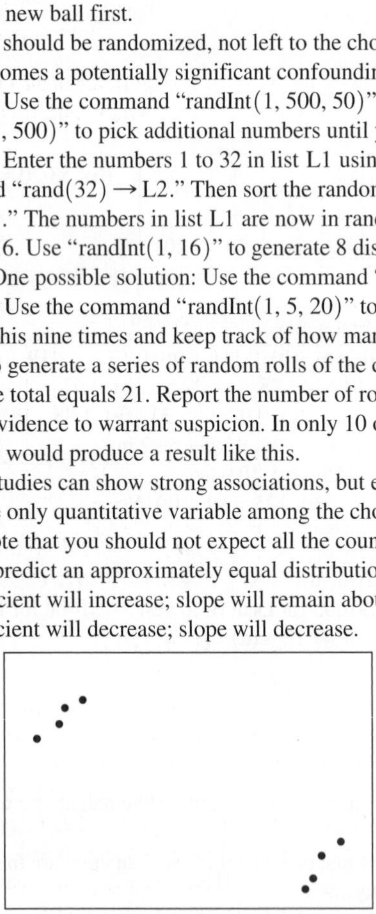

59. **(a)** The size of the hospital is not affecting the death rates of the patients. The lurking variable is the patient's condition. Bigger hospitals tend to get the more critical cases, and critical cases have a higher death rate.
(b) The number of seats is not affecting the speed of the jet. The lurking variable is the size of the aircraft. Larger jets generally have more seats and go faster.
(c) The size of the shoe does not affect reading ability. The lurking variable is the age of the student. In general, older students have larger feet and read at a higher level.
(d) The extra firemen are not causing more damage. The lurking variable is the size of the fire. Larger fires cause more damage and require more firefighters. **(e)** The salary is not generally affected by the player's weight. The lurking variable is the player's position on the team. Linemen weigh more and tend to earn less money than the (usually lighter) players in the so-called "skill" positions (e.g., quarterbacks, running backs, receivers, and defensive backs).

CHAPTER 10 REVIEW EXERCISES

1. No; sum $\neq 1$ **3.** ≈ 0.00001824 **5.** 1/10 **7. (a)** 0.444 **(b)** 0.092 **(c)** 0.556 **9. (a)** 0.75 **(b)** 0.48 **11.** 0.17 **13. (a)** 0.5 **(b)** 0.15 **(c)** 0.35 **(d)** ≈ 0.43 **15. (a)** 0.25 **(b)** 0.647 **(c)** No; the overall rate of high cholesterol was 31.8%, but it was 64.7% among men with high blood pressure.

17. (a)

1	1
2	3 3 5 7 8
3	1 3 4 4 6 9
4	2 2 6 6
5	0 6

(b) Unimodal and symmetric

19.

Yardage	Frequency
1000–1999	1
2000–2999	5
3000–3999	6
4000–4999	4
5000–5999	2

21. (a)

12	0 0 4 4
13	1 1 2 6 7 9
14	0 3 4 8
15	6
16	3
17	7 9
18	0
19	0 1 7
20	2
21	
22	
23	0

(b) Unimodal, skewed to the right

23.

Length (in seconds)	Frequency
120–129	4
130–139	6
140–149	4
150–159	1
160–169	1
170–179	2
180–189	1
190–199	3
200–209	1
210–219	0
220–229	0
230–239	1

25.

Earlier		Later
4 0 0	12	4
9 2 1	13	1 6 7
8 4 3 0	14	
	15	6
3	16	
7	17	9
	18	0
	19	0 1 7
	20	2
	21	
	22	
	23	0

The songs released in the earlier years tended to be shorter.

27.

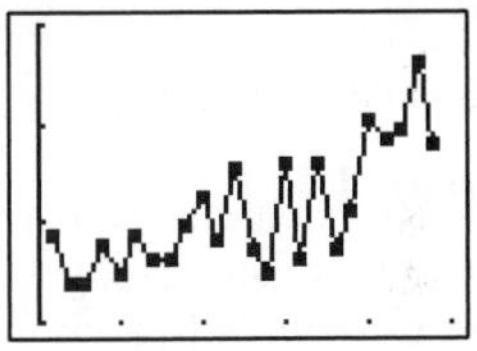

[–1, 25] by [100, 250]

Again, the data demonstrate that songs appearing later tended to be longer.

29. (a)

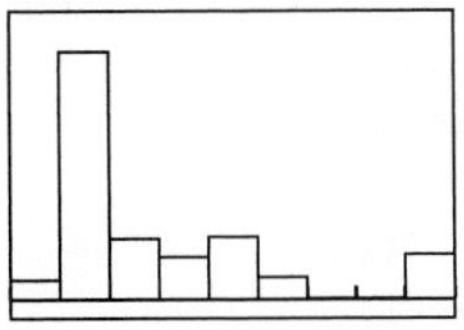

[0, 900] by [–1, 15]

(b) Unimodal and skewed to the right

31.

Visitors (in millions)	Frequency
0–99	1
100–199	13
200–299	3
300–399	2
400–499	3
500–599	1
600–699	0
700–799	0
800–899	1

33. (a)

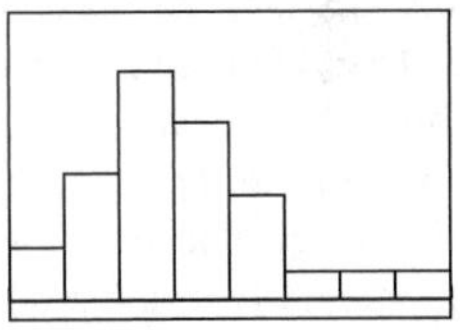

[75, 275] by [–1, 11]

(b) Unimodal and slightly skewed

35.

Median price (in $1000s)	Frequency
75–99	2
100–124	5
125–149	9
150–174	7
175–199	4
200–224	1
225–249	1
250–274	1

37. (a) 84% **(b)** 0.222, 0.328 **39.** 175 **41.** $133.85 **43.** 5/16 **45. (a)** ≈ 0.922 **(b)** ≈ 0.075 **47.** No. The number of expected defective bats is less than 10 (only 4.8). **49.** 40, 6.32 **51. (a)** 32.5, 3.37 **(b)** Yes; the number of expected hits (32.5) and misses (17.5) are both greater than 10. **(c)** The z-score for 41 hits is 2.52, statistically significant evidence of improvement. **53. (a)** 11% **(b)** 26% **(c)** less than 1108 lb **55. (a)** 91% **(b)** 29.5% **(c)** over 308 yd **57.** "Correlation" is incorrect, because color is categorical. **59.** Correlation does not measure straightness. **61.** This will work. **63.** Voluntary response bias. **65.** Randomly divide the students into two groups of 20 (replication). Have one group take the course and the other group study independently (control). Compare improvement in scores.

67. Randomly divide the swatches into two groups of 10 (replication). Wash one group with the old detergent and the other with the new additive. Wash each in the same machine for the same length of time using the same temperature water (control). Compare the cleanliness of the swatches. **69.** Use "randInt(0, 9)" to generate random digits, letting 0–4 = red, 5–7 = white, and 8–9 = blue. Generate a series of digits until one marble of each color is seen. Record the number of marbles not drawn. Repeat many times and find the mean. **71. (a)** No; 21 of the 374 streaks were runs of 5 or more in a row—not unusual. **(b)** (Answers will vary) 8 or more in a row seems unusual, but even very long runs can occur by chance.

CHAPTER 10 PROJECT

Answers are based on the sample data shown in the table.

1.

5	
5	9
6	1 1 2 3 3 3 4 4 4 4
6	5 6 6 6 7 8 8 9 9 9
7	0 0 1 1 1 2 2 3
7	5

66 or 67 inches

3.

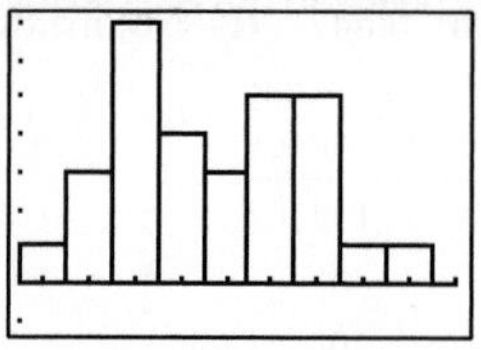

[59, 78] by [–1, 7]

5. The distribution is somewhat symmetric and probably does not have outliers.

7.

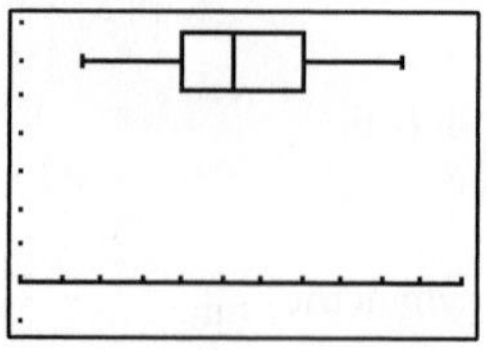

[56, 78] by [–1, 7]

SECTION 11.1

Exploration 1

1. 3 **3.** They are the same.

Quick Review 11.1

1. $-4/7$ **3.** $y = (3/2)x + 6$ **5.** $y - 4 = \frac{3}{4}(x - 1)$ **7.** $h + 4$ **9.** $-\frac{1}{2(h + 2)}$

Exercises 11.1

1. 12 mph **3.** 3 **5.** $4a$ **7.** 1 **9.** No tangent

11. 4

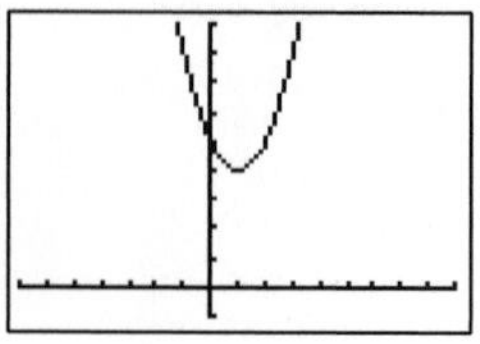

[–7, 9] by [–1, 9]

13. 12

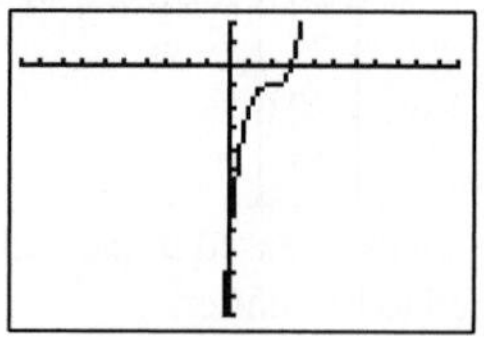

[–10, 11] by [–12, 2]

15. **(a)** 48 **(b)** 48 ft/sec

17. **(a)** -4
(b) $y - 2 = -4(x + 1)$
(c)

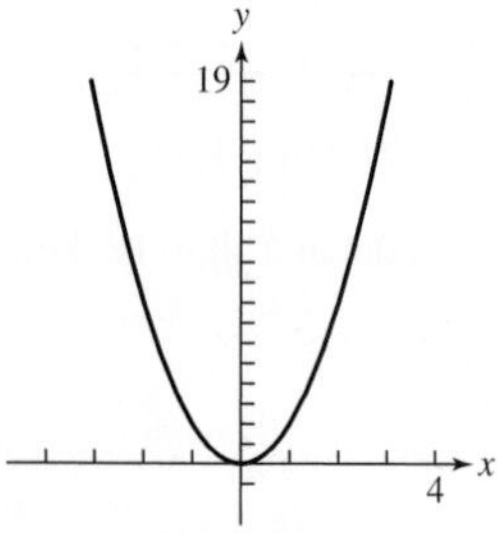

19. **(a)** 1
(b) $y = x - 5$
(c)

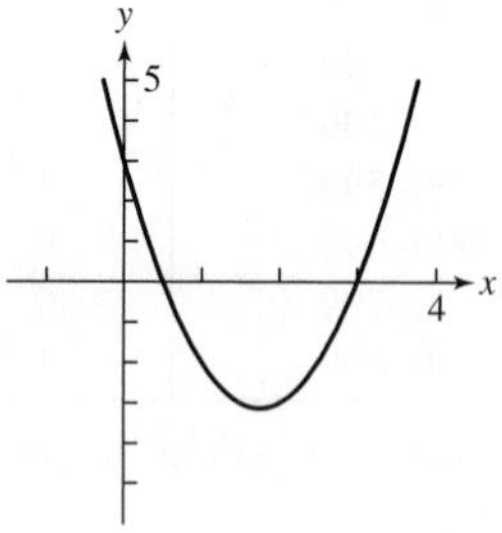

21. -1; 1; none
23. -4
25. -12
27. Does not exist
29. -3
31. $6x + 2$

33. **(a)** 9 ft/sec; 15 ft/sec
(b) $f(x) = 8.94x^2 + 0.05x + 0.01$, x = time in seconds

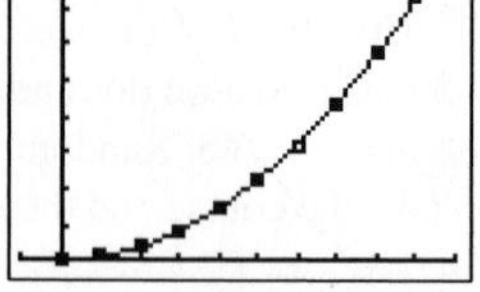

[–0.1, 1] by [–0.1, 8]

(c) ≈ 35.9 ft

35. **(a)**

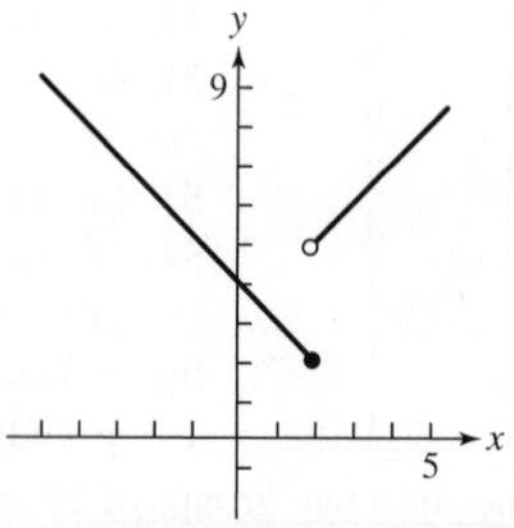

(b) Since the graph of the function does not have a definable slope at $x = 2$, the derivative of f does not exist at $x = 2$.
(c) Derivatives do not exist at points where functions have discontinuities.

37. **(a)**

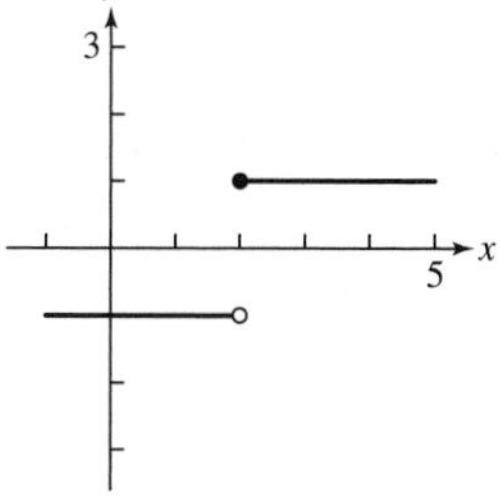

(b) Since the graph of the function does not have a definable slope at $x = 2$, the derivative of f does not exist at $x = 2$.

(c) Derivatives do not exist at points where functions have discontinuities.

39. Possible answer:

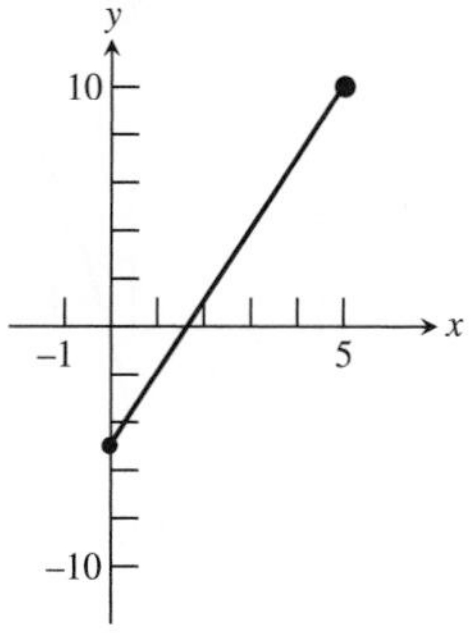

41. Possible answer:

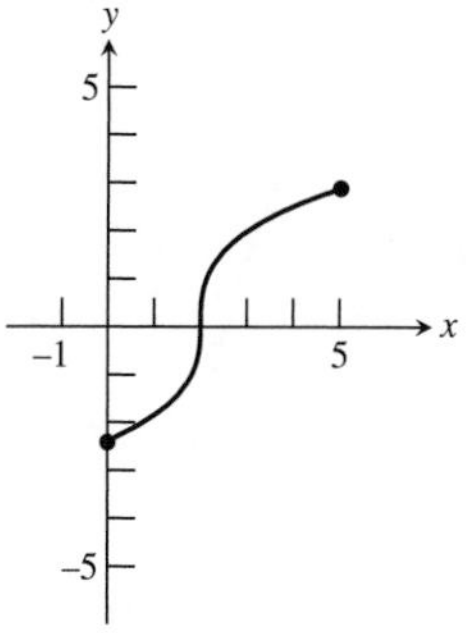

43. The slope of the line is a; $f'(x) = a$.

45. False; the instantaneous velocity is a limit of average velocities. It is nonzero when the ball is moving. **47.** D **49.** C

51.

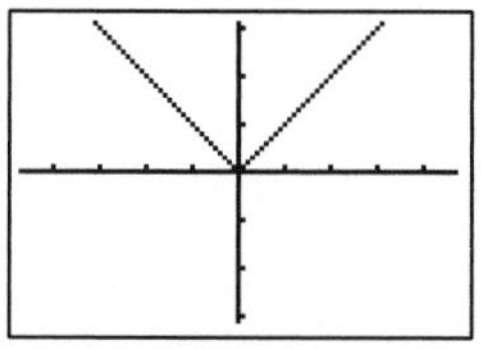

[–4.7, 4.7] by [–3.1, 3.1]

(a) No, there is no derivative because the graph has a corner at $x = 0$.

(b) No

53.

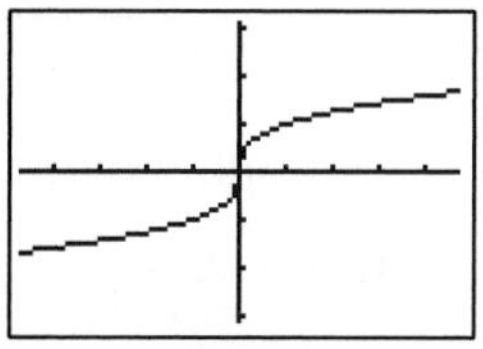

[–4.7, 4.7] by [–3.1, 3.1]

(a) No, there is no derivative because the graph has a vertical tangent (no slope) at $x = 0$.

(b) Yes, $x = 0$.

55. **(a)** 48 ft/sec

(b) 96 ft/sec

57.

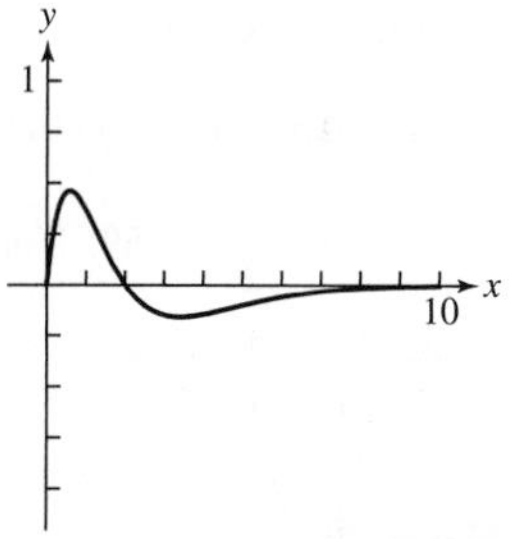

SECTION 11.2

Exploration 1

1. 0.1 gal; 1 gal **3.** 0.000000001 gal; 1 gal

Quick Review 11.2

1. $\frac{1}{8}, \frac{1}{2}, \frac{9}{8}, 2, \frac{25}{8}, \frac{9}{2}, \frac{49}{8}, 8, \frac{81}{8}, \frac{25}{2}$ **3.** $\frac{65}{2}$ **5.** $\frac{505}{2}$ **7.** 228 mi **9.** 4,320,000 ft^3

Exercises 11.2

1. 195 mi **3.** 540,000 ft^3 **5.** 2176 km **7.** 13; Answers will vary. **9.** 13; Answers will vary. **11.** 32.5

13. $\left[0, \frac{1}{2}\right], \left[\frac{1}{2}, 1\right], \left[1, \frac{3}{2}\right], \left[\frac{3}{2}, 2\right]$ **15.** $\left[1, \frac{3}{2}\right], \left[\frac{3}{2}, 2\right], \left[2, \frac{5}{2}\right], \left[\frac{5}{2}, 3\right], \left[3, \frac{7}{2}\right], \left[\frac{7}{2}, 4\right]$

17. **(a)**

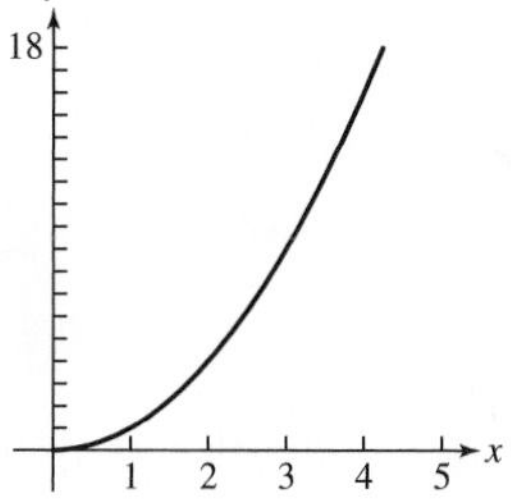

(b)

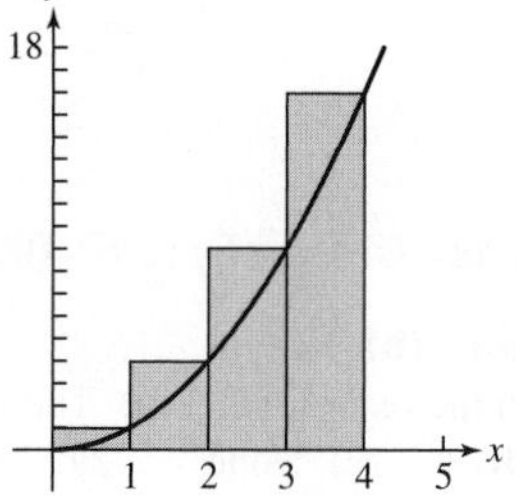

RRAM: 30

(c)

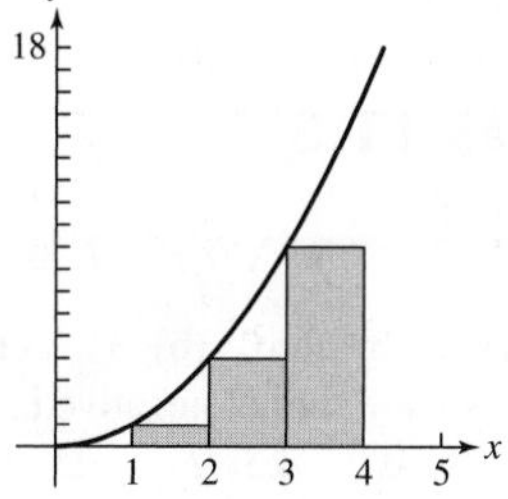

LRAM: 14

(d) Average: 22

19. (a)

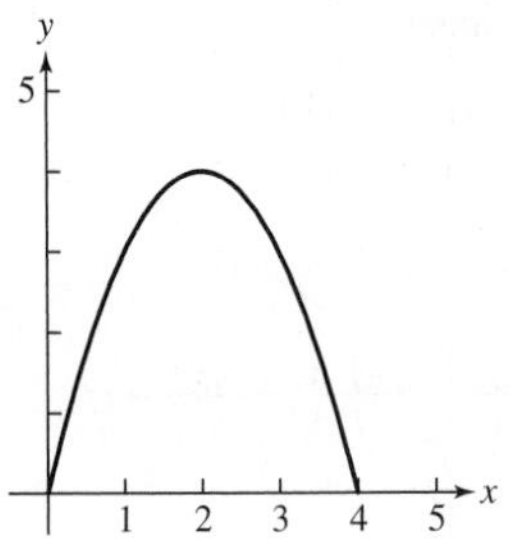

(b)

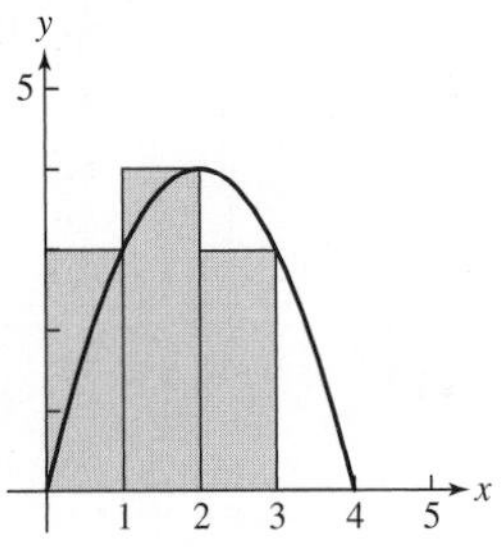

RRAM: 10

(c)

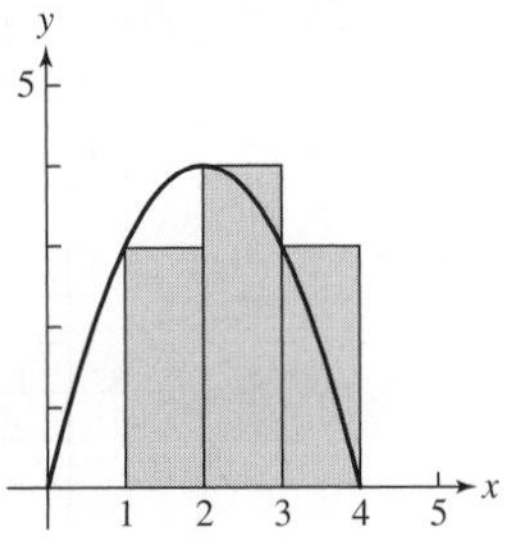

LRAM: 10

(d) Average:10

21. 20 **23.** 37.5 **25.** 16.5 **27.** 2π **29.** 2 **31.** 2 **33.** 1 **35.** 4 **37.** 4
39. $8k + 12$ **41.** $24 + 4k$ **45.** 64 ft

47. (a)

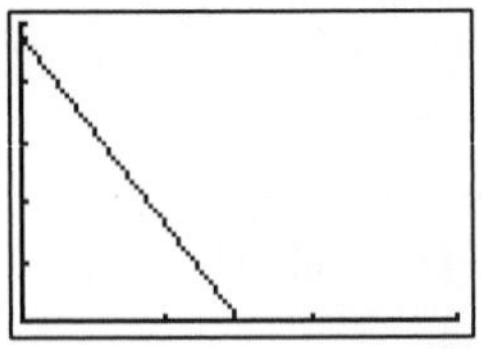

[0, 3] by [0, 50]

(b) $t = 1.5$ sec **(c)** 36 ft

49. (a)

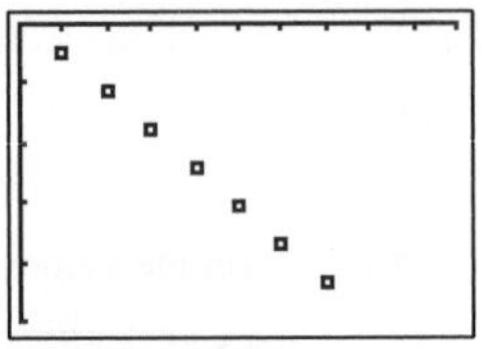

[0, 2] by [–50, 0]

(b) 33.86 ft

51. True; the exact area is given by the limit as $n \to \infty$. **53.** A **55.** C

57. $\int_0^1 (x - 1)\,dx = -\frac{1}{2}$ **59.** True **61.** False **63.** False

SECTION 11.3

Exploration 2

1. 50; 0

Quick Review 11.3

1. (a) $-\frac{3}{64}$ **(b)** $\frac{1}{16}$ **(c)** Undefined **3. (a)** $x = -2$ and $x = 2$ **(b)** $y = 2$
5. (b) **7. (a)** $[-2, \infty)$ **(b)** None

9.

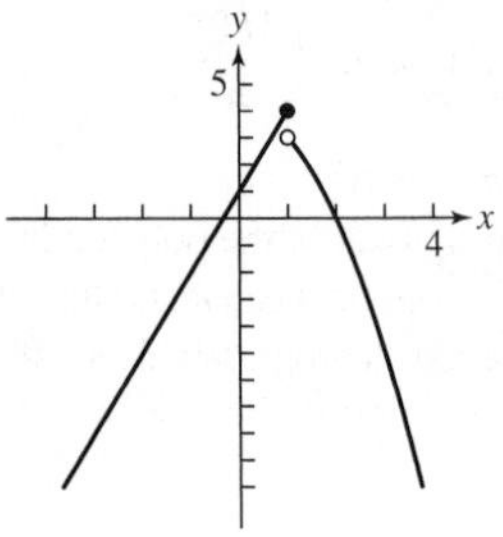

Exercises 11.3

1. -4 **3.** 7 **5.** $\sqrt{7}$ **7.** 0 **9.** $a^2 - 2$ **11. (a)** Division by zero **(b)** $-\frac{1}{6}$
13. (a) Division by zero **(b)** 3 **15. (a)** Division by zero **(b)** -4
17. (a) The square root of negative numbers is not defined in the real plane. **(b)** The limit does not exist.
19. -1 **21.** 0 **23.** 2 **25.** $\ln \pi$ **27. (a)** 3 **(b)** 1 **(c)** None **29. (a)** 4 **(b)** 4 **(c)** 4
31. (a) True **(b)** True **(c)** False **(d)** False **(e)** False **(f)** False **(g)** False **(h)** True **(i)** False **(j)** True
33. (a) ≈ 2.72 **(b)** ≈ 2.72 **(c)** ≈ 2.72 **35. (a)** 6 **(b)** -4 **(c)** 16 **(d)** -2

37. **(a)**

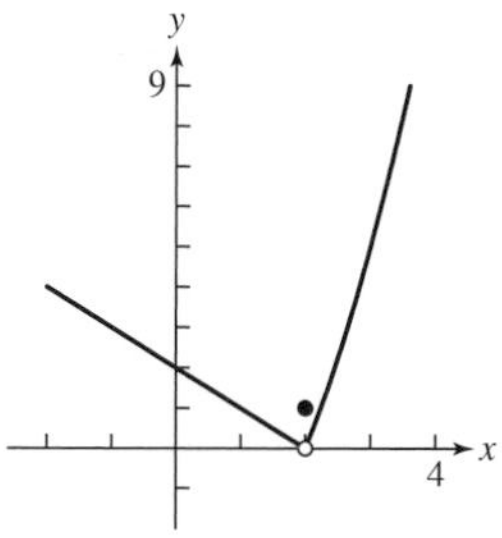

(b) 0; 0
(c) 0

39. **(a)**

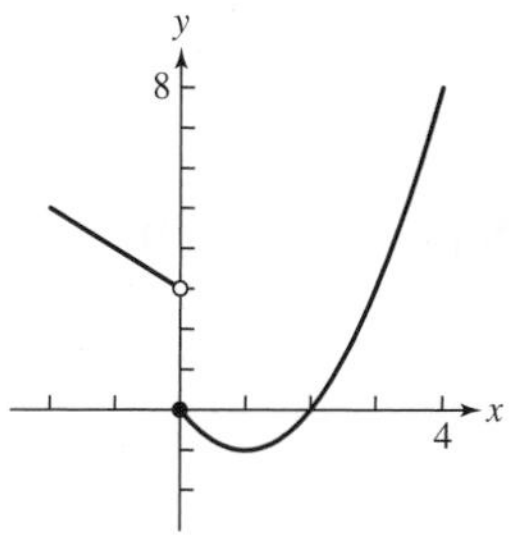

(b) 0; 3
(c) Does not exist; $\lim_{x\to 0^-} f(x) \neq \lim_{x\to 0^+} f(x)$

41. 2 **43.** 0 **45.** 1 **47.** 0; 0
49. ∞; 1 **51.** **(a)** ∞ **(b)** $-\infty$
53. **(a)** Undefined **(b)** 0
55. $-\infty$; $x = 3$ **57.** ∞; $x = -2$
59. ∞; $x = 5$ **61.** 3 **63.** 1
65. ∞ **67.** 0 **69.** Undefined
71. $\dfrac{1}{2}$ **73.** False; $\lim_{x\to 3} f(x) = 5$
75. B **77.** C

79. **(a)**

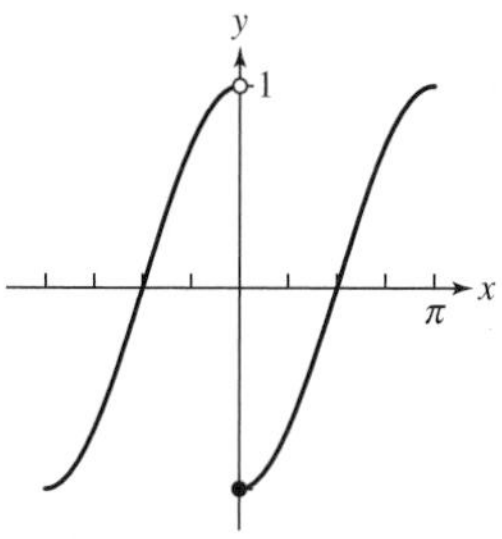

(b) $(-\pi, 0) \cup (0, \pi)$
(c) $x = \pi$
(d) $x = -\pi$

81. **(a)**

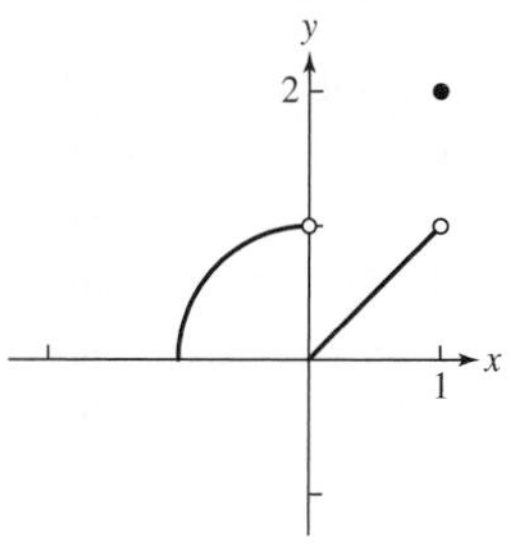

(b) $(-1, 0) \cup (0, 1)$
(c) $x = 1$
(d) $x = -1$

83. **(a)**

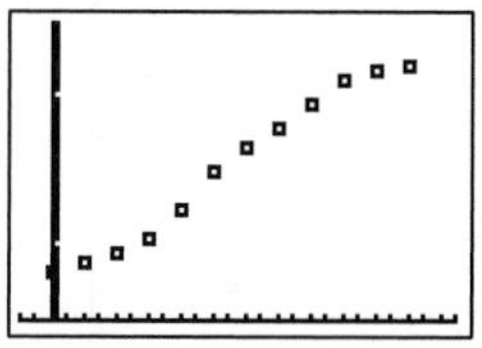

[–2, 25] by [0, 60]

(b) $f(x) \approx \dfrac{57.71}{1 + 6.39e^{-0.19x}}$, where $x =$ the number of months;
$\lim_{x\to\infty} f(x) \approx 57.71$
(c) It's about 58,000.

85.

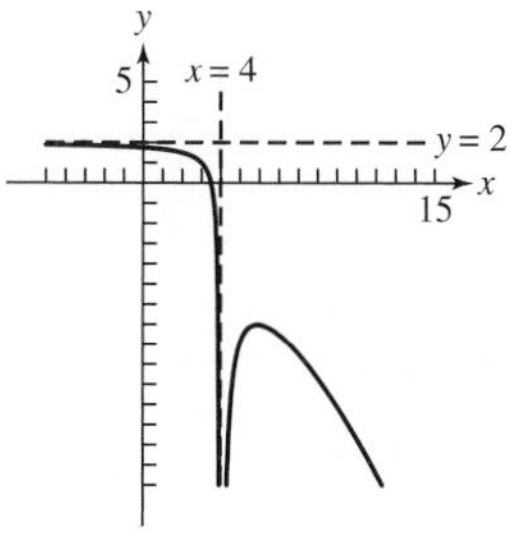

87.

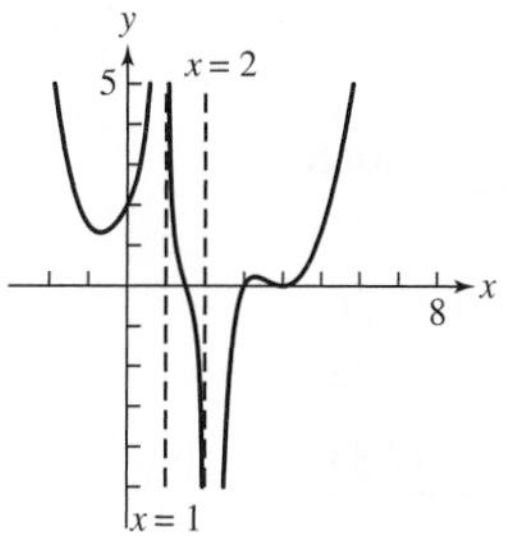

89. **(d)**

n	A
4	4
8	3.3137
16	3.1826
100	3.1426
500	3.1416
1,000	3.1416
5,000	3.1416
10,000	3.1416
100,000	3.1416

Yes, $A \to \pi$ as $n \to \infty$.

(e)

n	A
4	36
8	29.823
16	28.643
100	28.284
500	28.275
1,000	28.274
5,000	28.274
10,000	28.274
100,000	28.274

As $n \to \infty$, $A \to 9\pi$.

(f) One possible answer:

$$\lim_{n\to\infty} A = \lim_{n\to\infty} nh^2 \tan\left(\frac{180°}{n}\right) = h^2 \lim_{n\to\infty} n \tan\left(\frac{180°}{n}\right) = h^2\pi = \pi h^2$$

As the number of sides of the polygon increases, the distance between h and the edge of the circle becomes progressively smaller. As $n \to \infty$, $h \to$ radius of the circle.

91. **(a)**

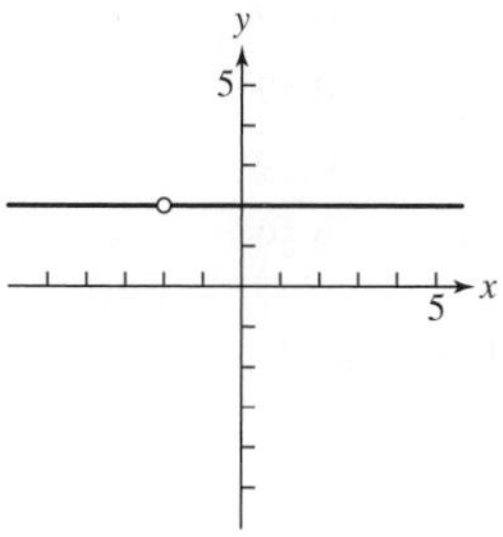

(b) $y = \dfrac{2x + 4}{x + 2} = \dfrac{2(x + 2)}{x + 2} = 2$
(c) $y = 2$

93. **(a)**

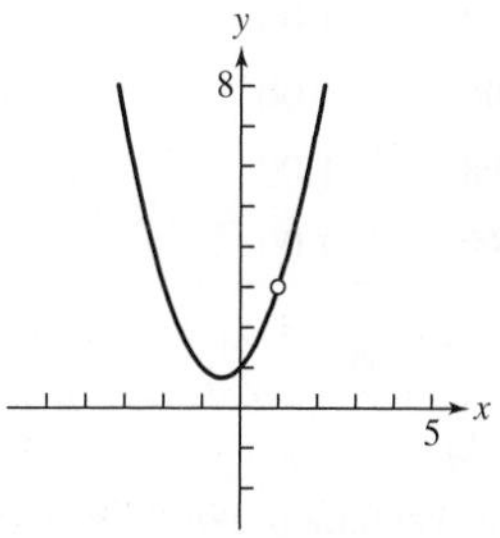

(b) $y = \dfrac{x^3 - 1}{x - 1} = \dfrac{(x - 1)(x^2 + x + 1)}{x - 1} = x^2 + x + 1$
(c) $y = x^2 + x + 1$

SECTION 11.4

Exploration 1

1. 1.364075504 **3.** $\int_0^{\pi} \sin x\, dx$; sum(seq(sin(0 + K*π/50)*π/50, K, 1, 50)) = 1.999341983; fnInt(sin(X), X, 0, π) = 2

Quick Review 11.4

1. 5 **3.** 2/3 **5.** 3 **7.** ≈ 0.5403 **9.** ≈ 1.000

Exercises 11.4

1. −4 **3.** −12 **5.** 0 **7.** ≈ 1.0000 **9.** ≈ −3.0000 **11.** 64/3 **13.** 2 **15.** ≈ 0 **17.** 1 **19.** ≈ 3.1416
21. 106.61 mi

23. **(a)** −50 ft/sec
(b)

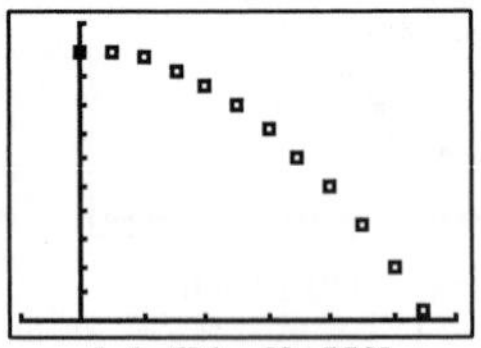

[−1, 6] by [0, 550]

(c) $s(t) = -16.08t^2 + 0.36t + 499.77$
(d) ≈ −47.88 ft/sec
(e) ≈ 179.28 ft/sec

25. **(a)**

Midpoint	$\Delta s/\Delta t$
0.25	−10
0.75	−20
1.25	−40
1.75	−60
2.25	−70
2.75	−90
3.25	−100
3.75	−120
4.25	−140
4.75	−150
5.25	−170

(b)

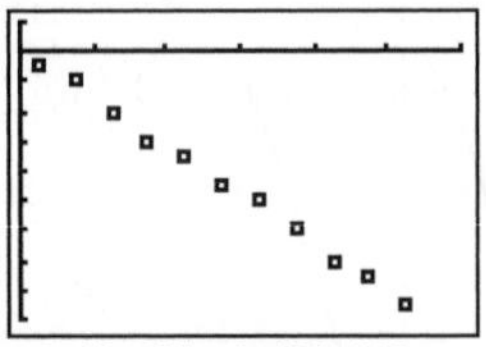

[0, 6] by [−180, 20]

(c) Approximately −47.95 ft/sec; this is close to the results in Exercise 23.

27. 100 ft

31. **(b)**

N	LRAM	RRAM	Average
10	15.04	19.84	17.44
20	16.16	18.56	17.36
50	16.86	17.82	17.34
100	17.09	17.57	17.33

(c) fnInt gives 17.33; at N_{100}, the average is 17.3344.

33. **(b)**

N	LRAM	RRAM	Average
10	7.84	11.04	9.44
20	8.56	10.16	9.36
50	9.02	9.66	9.34
100	9.17	9.49	9.33

(c) fnInt gives 9.33; at N_{100}, the average is 9.3344.

35. **(b)**

N	LRAM	RRAM	Average
10	98.24	112.64	105.44
20	101.76	108.96	105.36
50	103.90	106.78	105.34
100	104.61	106.05	105.33

(c) fnInt gives 105.33; at N_{100}, the average is 105.3344.

37. **(b)**

N	LRAM	RRAM	Average
10	7.70	8.12	7.91
20	7.81	8.02	7.91
50	7.87	7.95	7.91
100	7.89	7.93	7.91

(c) fnInt gives 7.91, the same result as N_{100}.

39. **(b)**

N	LRAM	RRAM	Average
10	1.08	0.92	1.00
20	1.04	0.96	1.00
50	1.02	0.98	1.00
100	1.01	0.99	1.00

(c) fnInt gives 1, the same result as N_{100}.

41. **(b)**

N	LRAM	RRAM	Average
10	0.56	0.62	0.59
20	0.58	0.61	0.59
50	0.59	0.60	0.59
100	0.59	0.60	0.59

(c) fnInt = 0.59, the same result as N_{100}.

43. True; the notation NDER refers to a symmetric difference quotient using $h = 0.001$. **45.** B **47.** C
49. **(a)** $4x + 3$ **(b)** $3x^2$ **(c)** 11.002, 11 **(d)** The symmetric method provides a closer approximation to $f'(2) = 11$.
(e) 12.006001; 12.000001; symmetric **51.** The values of $f(0 + h)$ and $f(0 - h)$ are the same. **53.** **(a)** 4 **(b)** ≈ 19.67

57. (b)

x	$A(x)$
0.25	0.0156
0.5	0.125
1	1
1.5	3.375
2	8
2.5	15.625
3	27

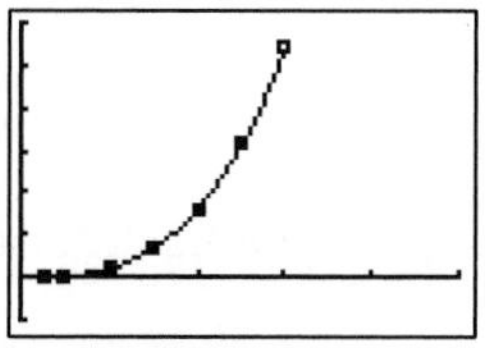

[0, 5] by [–5, 30]

(c) $y \approx x^3$

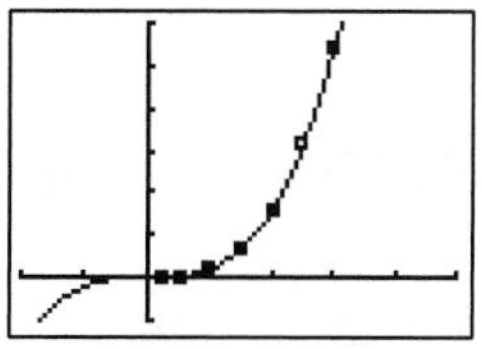

[–2, 5] by [–5, 30]

(d) The exact value of $A(x)$ for any x greater than zero appears to be x^3.

(e) $A'(x) = 3x^2$

CHAPTER 11 REVIEW EXERCISES

1. (a) 2 **(b)** Does not exist **3. (a)** 2 **(b)** 2 **5.** -1 **7.** -7 **9.** 0 **11.** 0 **13.** $-\infty$ **15.** ∞

17. $-\frac{1}{4}$ **19.** f has vertical asymptotes at $x = -1$ and $x = -5$; f has a horizontal asymptote at $y = 0$. **21.** -8

23. $-\frac{1}{9}$ **25.** -1 **27.** $y = \begin{cases} \frac{x^3 - 1}{x - 1} & \text{if } x \neq 1 \\ 3 & \text{if } x = 1 \end{cases}$ **29.** -9 **31. (a)** 8.01 **(b)** 8 **33.** 1; $y = x - 1$ **35.** $10x + 7$

37. LRAM: 42.2976; RRAM: 40.3776; 41.3376

39. (a)

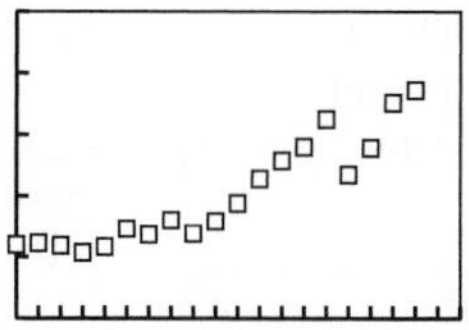

[0, 20] by [0, 5]

(b) 2000 to 2001: 5 cents per year; 2010 to 2011: 74 cents per year
(c) 2010 to 2011 **(d)** 2008 to 2009
(e) $y = 0.143x + 0.738$

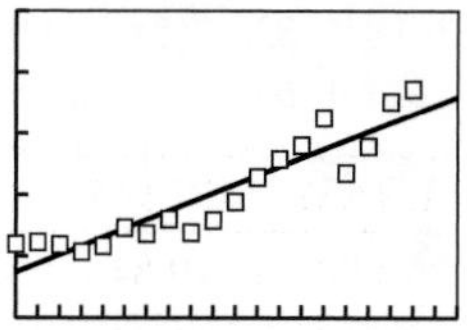

[0, 20] by [0, 5]

(f) $y = -4.467x^3 + 0.019x^2 - 0.073x + 1.220$

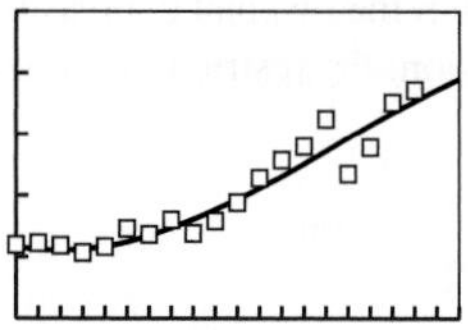

[0, 20] by [0, 5]

(g) 2000: 0.087; 2004: 0.167; 2007: 0.199; 2011: 0.204
(h) \$4.35/gal. Eventually too low.

Chapter 11 Project

1.

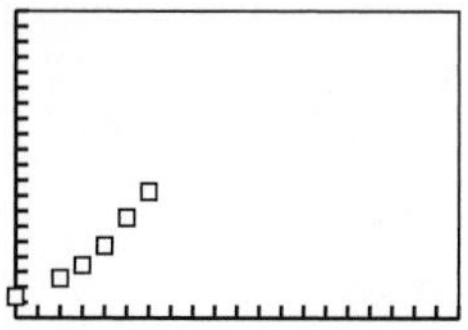

[1950, 2150] by [0, 2000000]

3. $1{,}872{,}804/3.9657959/(1 + 4.9170440783521\text{E}34e^{-0.03960589662546})$

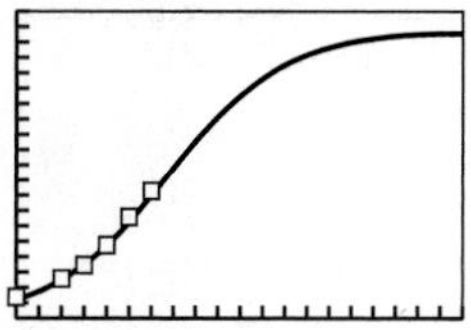

[1950, 2150] by [0, 2000000]

5. 2013: 864,474 persons; 2020: 994,063 persons; 2040: 1,337,387 persons. Both predictions seem reasonable.

APPENDIX A

Appendix A.1

1. 9 or -9 **3.** 4 **5.** $\frac{4}{3}$ or $-\frac{4}{3}$ **7.** 12 **9.** -6 **11.** $-\frac{4}{3}$ **13.** 4 **15.** 2.5 **17.** 729 **19.** 0.25 **21.** -2
23. 1.3 **25.** 2.1 **27.** $12\sqrt{2}$ **29.** $-5\sqrt[3]{2}$ **31.** $|x|y^2\sqrt{2x}$ **33.** $x^2|y|\sqrt[4]{3y^2}$ **35.** $2x^2\sqrt[5]{3}$ **37.** $2\sqrt[3]{4}$ **39.** $\frac{\sqrt[5]{x^3}}{x}$
41. $\frac{\sqrt[3]{x^2y^2}}{y}$ **43.** $(a + 2b)^{2/3}$ **45.** $2x^{5/3}y^{1/3}$ **47.** $\sqrt[4]{a^3b}$ **49.** $\sqrt[3]{x^{-5}} = 1/\sqrt[3]{x^5}$ **51.** $\sqrt[4]{2x}$ **53.** $\sqrt[8]{xy}$
55. $\sqrt[5]{a}$ **57.** $a^{-17/30}$ **59.** $3a^2b^2(b \geq 0)$ **61.** $4x^4y^2$ **63.** $\frac{|x|}{x|y|}$ **65.** $3y^2/|x^3|$ **67.** $|x|\sqrt[4]{6x^2y^2}/2$ **69.** $\frac{2x\sqrt[3]{x}}{y}$
71. 0 **73.** $(x - 2|y|)\sqrt{x}$ **75.** $<$ **77.** $=$ **79.** $>$ **81.** $<$ **83.** ≈ 3.48 sec
85. If n is even, then there are two real nth roots of a (when $a > 0$): $\sqrt[n]{a}$ and $-\sqrt[n]{a}$.

Appendix A.2

1. $3x^2 + 2x - 1$; degree 2 **3.** $-x^7 + 1$; degree 7 **5.** No **7.** Yes **9.** $4x^2 + 2x + 4$ **11.** $3x^3 - x^2 - 9x + 3$
13. $2x^3 - 2x^2 + 6x$ **15.** $-12u^2 + 3u$ **17.** $-15x^3 - 5x^2 + 10x$ **19.** $x^2 + 3x - 10$ **21.** $3x^2 + x - 10$
23. $9x^2 - y^2$ **25.** $9x^2 + 24xy + 16y^2$ **27.** $8u^3 - 12u^2v + 6uv^2 - v^3$ **29.** $4x^6 - 9y^2$ **31.** $x^3 + 2x^2 - 5x + 12$
33. $x^4 + 2x^3 - x^2 - 2x - 3$ **35.** $x^2 - 2$ **37.** $u - v, u \geq 0$ and $v \geq 0$ **39.** $x^3 - 8$ **41.** $5(x - 3)$
43. $yz(z^2 - 3z + 2)$ **45.** $(z + 7)(z - 7)$ **47.** $(8 + 5y)(8 - 5y)$ **49.** $(y + 4)^2$ **51.** $(2z - 1)^2$
53. $(y - 2)(y^2 + 2y + 4)$ **55.** $(3y - 2)(9y^2 + 6y + 4)$ **57.** $(1 - x)(1 + x + x^2)$ **59.** $(x + 2)(x + 7)$
61. $(z - 8)(z + 3)$ **63.** $(2u - 5)(7u + 1)$ **65.** $(3x + 5)(4x - 3)$ **67.** $(2x + 5y)(3x - 2y)$ **69.** $(x - 4)(x^2 + 5)$
71. $(x^2 - 3)(x^4 + 1)$ **73.** $(c + 3d)(2a - b)$ **75.** $x(x^2 + 1)$ **77.** $2y(3y + 4)^2$ **79.** $y(4 + y)(4 - y)$
81. $y(1 + y)(5 - 2y)$ **83.** $2(5x + 4)(5x - 2)$ **85.** $2(2x + 5)(3x - 2)$ **87.** $(2a - b)(c + 2d)$ **89.** $(x - 3)(x + 2)(x - 2)$
91. $(2ac + bc) - (2ad + bd) = c\,(2a + b) - d\,(2a + b) = (2a + b)(c - d)$;
Neither of the groupings $(2ac - bd)$ and $(-2ad + bc)$ have a common factor to remove.

Appendix A.3

1. $\frac{5}{3}$ **3.** $\frac{30}{77}$ **5.** $\frac{5}{6}$ **7.** $\frac{1}{10}$ **9.** All real numbers **11.** $x \geq 4$ or $[4, \infty)$ **13.** $x \neq 0$ and $x \neq -3$
15. $x \neq 2$ and $x \neq 1$ **17.** $x \neq 0$ **19.** $8x^2$ **21.** x^2 **23.** $x^2 + 7x + 12$ **25.** $x^3 + 2x^2$
27. $(x - 2)(x + 7)$ cancels out during simplification; the restriction indicates that the values 2 and -7 were not valid in the original expression.
29. No factors were removed from the expression. **31.** $(x - 3)$ ends up in the numerator of the simplified expression; the restriction reminds us that it began in the denominator so that 3 is not allowed.
33. $\frac{6x^2}{5}, x \neq 0$ **35.** $\frac{x^2}{x - 2}, x \neq 0$ **37.** $-\frac{z}{z + 3}, z \neq 3$ **39.** $\frac{y + 5}{y + 3}, y \neq 6$ **41.** $\frac{4z^2 + 2z + 1}{z + 3}, z \neq \frac{1}{2}$
43. $\frac{x^2 - 3}{x^2}, x \neq -2$ **45.** $\frac{x + 1}{3}, x \neq 1$ **47.** $-\frac{1}{x - 3}, x \neq 1$ and $x \neq -3$ **49.** $\frac{2(x - 1)}{x}$ **51.** $\frac{1}{y}, y \neq 5, y \neq -5$, and $y \neq \frac{1}{2}$
53. $\frac{2}{x}$ **55.** $\frac{3(x - 3)}{28}, x \neq 0$ and $y \neq 0$ **57.** $\frac{x}{4(x - 3)}, x \neq 0$ and $y \neq 0$ **59.** $\frac{2x - 2}{x + 5}$ **61.** $\frac{1}{3 - x}, x \neq 0$ and $x \neq -3$
63. $\frac{x^2 + xy + y^2}{x + y}, x \neq y, x \neq 0$, and $y \neq 0$ **65.** $\frac{x + 3}{x - 3}, x \neq 4$ and $x \neq \frac{1}{2}$ **67.** $-\frac{2x + h}{x^2(x + h)^2}, h \neq 0$
69. $a + b, a \neq 0, b \neq 0$, and $a \neq b$ **71.** $\frac{1}{xy}, x \neq -y$ **73.** $\frac{x + y}{xy}$

APPENDIX B

Appendix B.1

1. **(a)** False statement **(b)** Not a statement **(c)** False statement **(d)** Not a statement **(e)** Not a statement **(f)** Not a statement **(g)** True statement **(h)** Not a statement **(i)** Not a statement **(j)** Not a statement
3. **(a)** There is no natural number x such that $x + 8 = 11$. **(b)** There exists a natural number x such that $x + 0 \neq x$. **(c)** There is no natural x such that $x^2 = 4$. **(d)** There exists a natural number x such that $x + 1 = x + 2$.
5. **(a)** The book does not have 500 pages. **(b)** Six is not less than eight. **(c)** $3 \cdot 5 \neq 15$ **(d)** No people have blond hair. **(e)** Some dogs do not have four legs. **(f)** All cats have nine lives. **(g)** Some squares are not rectangles. **(h)** All rectangles are squares.

(i) There exists a natural number x such that $x + 3 \neq 3 + x$. **(j)** For all natural numbers x, $3 \cdot (x + 2) \neq 12$.
(k) Not every counting number is divisible by itself and 1. **(l)** All natural numbers are divisible by 2.
(m) For some natural number x, $5x + 4x \neq 9x$. **7. (a)** F **(b)** T **(c)** T **(d)** F **(e)** F **(f)** T **(g)** F **(h)** F **(i)** F **(j)** F
9. (a) $R \cup S$ **(b)** $Q \cap \overline{Q}$ **(c)** $\overline{R \cup Q}$ **(d)** $P \cap (R \cup S)$
11. (a) The statements $\sim(p \vee q)$ and $\sim p \wedge \sim q$ are equivalent, and the statements $\sim(p \wedge q)$ and $\sim p \vee \sim q$ are equivalent.
(b) The corresponding DeMorgan's Laws for sets are $\overline{P \cup Q} = \overline{P} \cap \overline{Q}$ and $\overline{P \cap Q} = \overline{P} \cup \overline{Q}$. The analogy comes from letting p mean "x is a member of P" and letting q mean "x is a member of Q." Then, for the first law, $\sim(p \vee q)$ means "x is a member of $\overline{P \cup Q}$," which is equivalent to "x is a member of $\overline{P} \cap \overline{Q}$," which translates into $\sim p \wedge \sim q$.
13. (a) Today is not Wednesday or the month is not June. **(b)** I did not eat breakfast yesterday, or I did not watch television yesterday.
(c) It is not true that both it is raining and it is July.

Appendix B.2

1. (a) $p \rightarrow q$ **(b)** $\sim p \rightarrow q$ **(c)** $p \rightarrow \sim q$ **(d)** $p \rightarrow q$ **(e)** $\sim q \rightarrow \sim p$ **(f)** $q \leftrightarrow p$
3. (a) Converse: If you're good in sports, then you eat Meaties; Inverse: If you don't eat Meaties, then you're not good in sports; Contrapositive: If you're not good in sports, then you don't eat Meaties.
(b) Converse: If you don't like mathematics, then you don't like this book; Inverse: If you like this book, then you like mathematics; Contrapositive: If you like mathematics, then you like this book.
(c) Converse: If you have cavities, then you don't use Ultra Brush toothpaste; Inverse: If you use Ultra Brush toothpaste, then you don't have cavities; Contrapositive: If you don't have cavities, then you use Ultra Brush toothpaste.
(d) Converse: If your grades are high, then you're good at logic; Inverse: If you're not good at logic, then your grades aren't high; Contrapositive: If your grades aren't high, then you're not good at logic.
5. (a) T **(b)** T **(c)** F **(d)** F **(e)** T **(f)** F **7.** No **9.** If a number is not a multiple of 4, then it is not a multiple of 8.
11. (a) p is false. **(b)** p is false. **(c)** q can be true, and in fact q true and p false makes $p \rightarrow q$ true and is the only way for $q \rightarrow p$ to be false.
13. (a) Helen is poor. **(b)** Some freshmen are intelligent. **(c)** If I study for the final, then I will look for a teaching job.
(d) There exist triangles that are isosceles.
15. (a) If a figure is a square, then it is a rectangle. **(b)** If a number is an integer, then it is a rational number.
(c) If a figure has exactly three sides, then it may be a triangle. **(d)** If it rains, then it is cloudy.

Photo Credits

About the Authors **Page xiv,** Franklin D. Demana, courtesy of the author; **Page xiv,** Bert K. Waits, courtesy of the author; **Page xiv,** Gregory D. Foley, courtesy of the author; **Page xv,** Daniel Kennedy, © Lifetouch, Inc. Used with permission; **Page xv,** David E. Bock, courtesy of the author.

Chapter P **Page 1,** Don Figer/Space Telescope Science Institute/NASA; **Page 8,** Photodisc/Getty Images; **Page 27,** Photodisc/Getty Images; Ignatius Wooster/Fotolia; **Page 38,** Martin Barraud/Getty Images.

Chapter 1 **Page 63,** Maxim Loskutnikov/Fotolia; **Page 65,** Scott Houston/Alamy; **Page 108,** Photodisc/Getty Images; **Page 117,** NOAA; **Page 138,** xPACIFICA/Alamy.

Chapter 2 **Page 157,** Stéphane Bidouze/Fotolia; **Page 162,** Photos.com; **Page 179,** iDesign2000/Fotolia; **Page 244,** Getty Images.

Chapter 3 **Page 251,** Andrey Armyagov/Fotolia; **Page 261,** Photodisc/Getty Images; **Page 270,** Photodisc/Getty Images; **Page 281,** Kerenby/Fotolia; **Page 288,** Robert Llewellyn/Corbis; **Page 291,** Byrdyak/Fotolia; **Page 295,** FrankBirds/Fotolia; **Page 312,** Digital Vision/Getty Images.

Chapter 4 **Page 319,** Pavel Losevsky/Fotolia; **Page 356,** Sablin/Fotolia; **Page 368,** Getty Images Inc.

Chapter 5 **Page 403,** Alexandra Giese/Fotolia; **Page 412,** Digital Vision/Getty Images; **Page 427,** NASA; **Page 450,** Digital Vision/Getty Images.

Chapter 6 **Page 455,** Mark Caunt/Shutterstock; **Page 462,** Digital Vision/Getty Images; **Page 480,** Petty Officer 2nd Class Rob Simpson/US Coast Guard; **Page 491,** PhotoDisc/Getty Images; **Page 497,** Hisham Ibrahim/Getty Images.

Chapter 7 **Page 518,** Pawel Szczesny/Fotolia; **Page 520,** John Wang/Getty Images; **Page 523,** Kzenon/Fotolia; **Page 530,** Stockbyte/Getty Images; **Page 555,** Paul Fleet/Fotolia; **Page 563,** Jess Alford/Photodisc/Getty Images; **Page 568,** Yalcinsonat/Fotolia.

Chapter 8 **Page 570,** Meyerbroeker/Alamy; **Page 579,** Tony Roberts/Corbis; **Page 598,** NASA; **Page 618,** National Optical Astronomy Observatories; **Page 631,** Gregory D. Foley.

Chapter 9 **Page 633,** Fotolia; **Page 641,** FoodCollection/SuperStock; **Page 656,** PNC/Photodisc/Getty Images.

Chapter 10 **Page 676,** Jon Boyes/Getty Images; **Page 684,** PhotoDisc/Getty Images; **Page 702,** Photos.com; **Page 716,** Harald Richter/Thinkstock; **Page 723,** Eric Isselée/Fotolia; **Page 725,** Radius Images/Alamy.

Chapter 11 **Page 753,** Michael Rosskothen/Fotolia; **Page 762,** CLEO Photo/Alamy; **Page 783,** Subbotina Anna/Fotolia; **Page 786,** Hans Dieter Seufert/culture-images GmbH/Alamy.

Application Index

Note: Numbers in parentheses refer to exercises on the pages indicated.

Biology, Medicine, and Life Sciences

Analyzing height data, 752
Angioplasty, 114
Average deer population, 403, 447
Bacterial growth, 263(57), 266, 272(39)
Bear population, 234(34)
Biological research, 208(71)
Blood donors, 681, 688(47), 728(22), 729(40), 744(41)
Blood pressure, 358(76)
Blood type, 688(47)
Capillary action, 368(65)
Cattle weights, 749(38), 750(53)
Circulation of blood, 194(65)
Comparing age and weight, 170(49), 172(67)
Deer population, 249(91), 272(46), 316(94)
Drug absorption, 315(73)
Focusing a lithotripter, 589, 592(59, 60)
Galileo's gravity experiment, 66–67
Genealogy, 263(67)
Handedness, 688(45), 728(20)
Heights of American adults, 725, 729(33,35)
HIV testing, 688(38)
Humidity, 157, 224
Life expectancy, 170(50), 234(43), 701(11-14), 716(57, 58)
Medicare expenditures, 526(45), 567(70)
National Institutes of Health spending, 249(89)
Penicillin use, 302(53, 54)
Predatory-prey cycles, 255
Rabbit population, 316(77), 657(49)
Rain forest growth, 656(37)
Salmon migration, 455, 463
Sleep cycles, 397(44)
Spread of flu, 272(45), 316(76, 93)
Testing positive for HIV, 686
Weight and pulse rate of selected mammals, 183(55), 291(65)
Weight loss, 397(36)
Weights of loon chicks, 711, 712, 734

Business, Economics, Industry, and Labor

Agriculture exports, 18(32)
Agriculture surplus, 18(33)
Americans' income, 34, 37(50)
Americans' spending, 37(51)
Analyzing an advertised claim, 640
Analyzing profit, 194(64)
Analyzing the stock market, 78(61)
Annual housing cost, 244(64)
Beverage business, 171(59)
Breaking even, 233(33), 249(90)
Buying a new car, 642(38)
Cash-flow planning, 58(39)
Cell phone antennas, 149(29, 30)
City government, 641(12)
Company wages, 243(56)
Comparing prices, 65, 530
Construction worker's compensation, 170(53)
Costly doll making, 170(52)
Defective baseball bats, 749(45)
Defective calculators, 687(34)
Defective light bulbs, 750(48)
Depreciation of real estate, 33
Electing officers, 641(11)
Equilibrium price, 524
Estimating personal expenditures with linear, 523
Exports to Canada, 18(34), 38(54)
Exports to Mexico, 12
Food prices, 527(53)
Forming committees, 641(27)
Great Recession, 698–699
Gross domestic product data, 790(24)
HDTV deliveries, 568(75)
Imports from Mexico, 18(31), 37(52)
Interior design, 149(36)
Inventory, 540(48)
Job interviews, 641(32)
Job offers, 149(28)
Labor force participation, 75(11–18)
Linear programming problems, 557–561
Management planning, 171(58, 60)
Manufacturing, 150(47, 49), 555(74)
Manufacturing swimwear, 348(60)
Maximizing profit, 564(40)
Maximum revenue, 528(69, 70)
Median women's income, 172(68)
Minimizing operating cost, 561
Minimum hourly wage, 64
Mining ore, 563(37)
Modeling depreciation, 160
Monthly sales, 397(35)
Nut mixture, 527(54)
Percent discounts, 66
Petroleum exports, 154(65)
Pizza toppings, 639, 642(39)
Planning a diet, 564(38)
Planning for profit, 243(58)
Predicting cafeteria food, 687(30)
Predicting demand, 163
Predicting economic growth, 376(72)
Predicting revenue, 166–167
Price of gasoline, 18–19(35–36)
Producing gasoline, 564(39)
Production, 541(46, 47), 673(30)
Profit, 540(49)
Purchasing fertilizer, 560
Quality control, 716(59, 60)
Questionable product claims, 635
Radio advertising, 744(32)
Real estate appreciation, 37(45)
Real estate prices, 749(33–36)
Rental van, 527(57)
Renting cars, 687(33)
Salad bar, 642(37)
Salary package, 527(58)
Sale prices, 149(27)
Selecting customers, 744(34)
Starbucks Coffee, 156
Stocks and matrices, 568(73)
Straight-line depreciation, 170(51)
Supply and demand, 206(57), 207(58)
Total revenue, 170(55), 171(59)
Train tickets, 555(89)
U.S. motor vehicle production, 18(29)
World motor vehicle production, 18(30)
Yearly average gasoline prices, 794(39)

Chemistry and Physical Sciences

Accelerating automobile, 771(46)
Acid mixtures, 230-231, 233(31, 32), 249(93)
Airplane speed, 527(52)
Angular speed, 323
Archimedes' Principle, 207-208(69, 70)
Atmospheric C02, 151(50)
Atmospheric pressure, 267, 272(41, 42)
Boyle's Law, 58(38), 183(51), 227(71)
Carbon dating, 263(58), 272(40)
Charles's Law, 183(52)
Chemical acidity (pH), 296, 301(47, 48), 316(82)
Combining forces, 465(48–50), 516(76)
Diamond refraction, 183(53)
Earthquake intensity, 281(61), 290(52), 295, 301(45, 46)
Ebb and flow of tides, 356, 358(75)
Escape velocity, 601(62)
Finding a force, 471, 473(45-47, 48), 516(77)
Finding the effect of gravity, 463
Flight engineering, 465(43, 44), 515(74, 75)
Galileo's gravity experiment, 66–67
Half-life, 656(38)
Latitude and longitude, 327–328(63–70)
Length of days, 377(88)
Light absorption, 281(60), 316(89)
Light intensity, 227(72), 290(53, 54)
Magnetic fields, 427(75)
Melting snowball, 117(32)
Mixing solutions, 149(31, 32), 552, 555(94)
Modeling a musical note, 393
Modeling illumination of the Moon, 454
Modeling temperature, 396(33), 401(105)
Moving a heavy object, 465(46, 47)
Newton's Law of Cooling, 151(51, 52), 296-298, 301-302(49-52), 317(95, 96)
Oscillating spring, 374, 376(71), 396(29)
Physics experiment, 76(23)
Potential energy, 303(69)
Radioactive decay, 266, 271(33, 34), 316(79, 80)
Reflective property of a hyperbola, 598
Reflective property of an ellipse, 589
Reflective property of a parabola, 577
Refracted light, 348(55, 56)
Resistors, 234(39), 244(62), 249(92)
Rowing speed, 527(51)
Sound intensity, 251, 280, 281(59), 290(51)
Sound waves, 319, 393
Speed of light, 1, 35
Temperature, 359(79, 80)
Temperature conversion, 126(34)
Temperatures of selected cities, 144
Tuning fork, 396(30)
Visualizing a musical note, 360(89)
Weather balloons, 117(31)

Weather forecasting, 439(38)
Windmill power, 183(54)
Wind speeds, 702(36), 713(8,10), 714(26), 715(38)
Work, 471, 473(49–56), 516(78), 772(50)

Computer Science, Geometry, and Other Mathematics

Angle of depression, 388, 395(24)
Angle of elevation, 348(59), 395(22)
Approximation and error analysis, 349(77)
Area of a sector, 328(71,72)
Area of regular polygon, 444
Beehive cells, 452(68)
Calculating a viewing angle, 384, 385(53)
Characteristic polynomial, 542(72, 73)
Computer graphics, 518
Computer imaging, 117(34)
Computing definite integrals from data, 788–789
Cone-shaped pile of grain, 142
Conical tank of water, 149(37)
Connecting algebra and geometry, 39(69–71), 77(50–52), 173(82–84), 243(57)
Connecting geometry and sequences, 657(51)
Constructing a cone, 243(60)
Curve fitting, 554(83–84), 566(45–46)
Cycloid, 484(53)
Designing a juice can, 231, 241, 243(61)
Designing an experiment, 737
Designing a running track, 323, 327(52)
Designing a swimming pool, 234(38)
Designing rectangles, 234(40)
Diagonals of a parallelogram, 16–17
Diagonals of a regular polygon, 146, 642(52)
Distance from a point to a line, 474(67)
Dividing a line segment in a given ratio, 466(61)
Dividing a line segment into thirds, 20(64)
Eigenvalues of a matrix, 556(108–110)
Epicycloid, 516(83)
Expected value, 689–690(61–62)
Finding a maximum area, 170(54)
Finding a minimum perimeter, 231
Finding a random sample, 744(33)
Finding area, 401(104)
Finding derivatives from data, 763(34), 787–788, 789(23)
Finding dimensions of a painting, 170(56)
Finding dimensions of a rectangular cornfield, 527(50)
Finding dimensions of a rectangular garden, 520, 527(49)
Finding distance, 336(65), 393(3), 394(4, 13), 395(19, 20, 24), 401(97, 98, 100), 439(38), 440(41, 45, 46), 452(64)
Finding distance from a velocity, 790(27–28), 791(47)
Finding distance traveled as area, 765, 767, 771(49), 786
Finding faked data, 453(76)
Finding height, 334, 336(61, 62, 64), 389, 393(1, 2), 394(5-9, 11, 12, 14), 395(22), 401(95, 96, 101, 102), 440(40, 44), 452(63)
Finding height of pole, 438, 440(39, 43)
Fitting a parabola to three points, 551
Graphing polar equations parametrically, 493(67–71)
Harmonic series, 666(52)
Hyperbolic functions, 420(80)
Hypocycloid, 484(54)
Indirect measurements, 389, 448(35)
Inscribing a cylinder inside a sphere, 155(67)
Inscribing a rectangle under a parabola, 155(68)
Limits and area of a circle, 783(89)
Locating a fire, 437
Lottery, 639, 680–681, 689(61), 690(62), 727, 728(16)
Maximizing area of an inscribed trapezoid, 452(67)
Maximizing volume, 433(55)
Measuring a dihedral angle, 446
Medians of a triangle, 466(62)
Minimizing perimeter, 234(35)
Mirrors, 337(75)
Modeling a rumor, 270
Normal distribution, 303(67)
Observational studies, 735, 746(60)
Page design, 234(36)
Parametrizing circles, 476, 485(65)
Parametrizing lines, 478, 485(66)
Permuting letters, 641(7, 9, 10)
Polar distance formula, 493(61)
Pool table, 337(76)
Reflecting graphs with matrices, 537
Rotating with matrices, 538, 542(51), 567(69)
Scaling triangles with matrices, 538
Simulate a spinner, 744(39)
Simulate drawing cards, 744(42)
Simulate rolling two six-sided dice, 744(40)
Solving triangles, 435, 443–444, 489
Symmetric matrices, 540(45)
Taylor polynomials, 349(79, 80)
Telephone area codes, 633, 640
Television coverage, 367(63)
Testing effects of music, 744(31)
Testing golf balls, 743(29)
Testing inequalities on a calculator, 27(69)
Testing soft drinks, 744(30)
Time-rate problem, 234(42)
Tire sizing, 326(46)
Tower of Hanoi, 667–668
Transformations and matrices, 541(57–61)
Tunnel problem, 433(56)

Construction and Engineering

Architectural design, 395(23), 449(42)
Architectural engineering, 208(72), 248(86)
Box with maximum volume, 140, 149(33)
Box with no top, 59(46), 194–195(66–68)
Building construction, 540(50)
Building specifications, 37(49)
Cassegrain telescope, 602(70)
Civil engineering, 394(15), 395(21)
Construction engineering, 449(39)
Designing a baseball field, 448(36)
Designing a box, 191, 241, 243(59)
Designing a bridge arch, 580(64)
Designing a flashlight mirror, 579(59)
Designing a satellite dish, 579(60)
Designing a softball field, 448(37)
Designing a suspension bridge, 579(63)
Dimensions of a Norman window, 47(61)
Draining a cylindrical tank, 154(62–64)
Elliptical pool table, 631(77)
Ferris wheel design, 440(42)
Garden design, 336(66)
Grade of a highway, 37(48), 398(49)
Gun location, 601(58)
Height of a ladder, 47(60)
Industrial design, 234(37), 249(94, 95)
Landscape design, 170(57)
Lighthouse coverage, 367(45), 385(54)
Measuring a baseball diamond, 445–446
Packaging a satellite dish, 141, 149(35)
Parabolic headlights, 579(62), 631(76)
Parabolic microphone, 577, 579(61), 631(75)
Patio construction, 656(40)
Radar tracking system, 491
Residential construction, 149(34)
Solar collection panel, 336(63)
Storage container, 248(87)
Surveying a canyon, 439(37), 451(62)
Surveyor's calculations, 449(38)
Swimming pool drainage, 234(41)
Using the LORAN system, 599, 601(57)
Volume of a box, 194 (66–68), 248(85)

Data Collection (CBR™, CBL™)

Analyzing a bouncing ball, 318
Ellipses as models of pendulum motion, 517, 592(73, 74), 632
Free fall with a ball, 167, 180, 764(55, 56), 787
Height of a bouncing ball, 250
Light intensity, 183(57), 227(72)
Motion detector distance, 207(62), 216(51, 52), 402
Tuning fork, 396(30), 397(43), 398(51)

Education

ACT scores, 715(44)
Advanced Placement Calculus exam scores, 714(28)
Baylor School grade point averages, 97(78)
Education budget, 10(53–56), 60(11)
Graduate requirement, 688(41)
Graduate school survey, 688(39)
Indiana Jones and the final exam, 641(36), 688(40)
Pell Grants, 248(88)
SAT Math scores, 61(33)
SAT scores, 530, 715(43)
Scaling grades, 127(47)

Finance and Investments

Annual percentage rate (APR), 310
Annual percentage yield (APY), 307, 312(57), 316(87, 88)

Annuities, 308-309, 311(50, 51), 313(72, 73), 316(83, 84), 317(98, 99), 666(37-38), 674(63-64)
Business loans, 568(77)
Car loan, 310, 312(51, 52)
Cellular phone subscribers, 79(63)
Comparing salaries, 702(33), 713(7)
Comparing simple and compound interest, 312(60), 317(99)
Compound interest, 304, 311(21–30), 315(63–65)
Consumer price index, 63, 146, 149(24)
Credit card debt, 698–699, 715(36)
Currency conversion, 126(33)
Employee benefits, 150(48)
Future value, 315(66), 661
House mortgage, 312(53-56), 316(85, 86)
International finance, 138(65)
Investment planning, 37(46)
Investment returns, 149(38), 150(40)
Investments, 555(91–93), 568(76)
IRA account, 311(47, 48)
Loan payoff (spreadsheet), 312-313(67, 68)
Major League Baseball salaries, 76(25–28)
Present value, 315(67, 68)
Savings account, 664-665(35, 36)
Simple interest, 304, 312(61, 62), 317(99)
Stock market volatility, 715(46)
Student loan debt, 699, 715(36)

Motion

Air conditioning belt, 327(55)
Airplane velocity as a vector, 461–462
Automobile design, 326(44)
Average speed, 763(33)
Baseball throwing machine, 171(62)
Bicycle racing, 326(45), 328(74)
Bouncing ball, 665(39)
Bouncing block, 358(77)
Calculating the effect of wind velocity, 462
Capture the flag, 483(38)
Current affecting ship's path, 465(53)
Damped harmonic motion, 348(57)
Falling rock, 475
Famine relief air drop, 483(39)
Ferris wheel motion, 358(73), 396(31, 32, 34)
Field goal kicking, 516(86)
Fireworks planning, 171(63)
Foucault Pendulum, 327(54)
Free-fall motion, 76(22), 167, 171(61), 184(68), 764(55, 56)
Hang time, 516(87)
Harmonic motion, 390–392, 427(74)
Height of an arrow, 516(79)
Hitting a baseball, 480, 483(40, 43-45), 484(49), 516(88)
Hitting golf balls, 484(50)
Hot-air balloon, 367(46), 386(55)
Landscape engineering, 171(64)
Launching a rock, 248(84)
LP turntable, 359(78)
Mechanical design, 395(27, 28)
Mechanical engineering, 327(53)
Motion of a buoy, 358(72)
Navigation, 37(47), 62(82), 326(43, 48-50), 328(73), 390, 395(17, 18, 25, 26), 449(40), 450(53), 465(41, 42, 51, 52)
Path of a baseball, 127(49)
Pendulum, 348(58), 401(103), 402
Pilot calculations, 441(57)
Projectile motion, 57, 58(33, 34), 62(81)
Riding on a Ferris Wheel, 481, 484(51), 516(80-82)
Rocket launch, 762(16), 771(48)
Rock toss, 485(63), 762(15), 771(47)
Rotating tire, 143, 149(39)
Salmon swimming, 463
Ship's propeller, 327(56)
Shooting a basketball, 465(45)
Simulating a foot race, 483(37)
Simulating horizontal motion, 478
Simulating projectile motion, 479, 483(37, 38)
Stopping distance, 194(63)
Taming The Beast, 401(106)
Throwing a ball at a Ferris Wheel, 485(67), 486(68), 517(89)
Throwing a baseball, 516(84, 85)
Tool design, 326(47)
Travel time, 149(25, 26), 789(21, 22)
Tsunami wave, 358(74)
Two Ferris Wheel problem, 486(69, 70)
Two softball toss, 484(46)
Velocity in 3d-space, 625, 628(33, 34)
Windmill, 753, 761
Yard darts, 484(47, 48), 517(90)

Planets and Satellites

Analyzing a comet's orbit, 587–588
Analyzing a planetary orbit, 617
Analyzing the Earth's orbit, 588
Dancing planets, 591(52)
Elliptical orbits, 589, 631(79)
Halley's Comet, 592(58), 618(41)
Icarus, 631(80)
Kepler's Third Law, 288
Lunar Module, 619(43)
Mars satellite, 619(44)
Mercury, 591(54)
Modeling planetary data, 179, 184(67)
Orbit of the Moon, 412(85)
Orbit periods, 287
Planetary orbits, 619(51)
Planetary satellites, 713(5)
Rogue comet, 601(55, 56)
Saturn, 591(55)
Sungrazers, 591(57)
Television coverage, 398(50), 452(69)
The Moon's orbit, 591(53)
Uranus, 618(42)
Venus and Mars, 591(56)
Weather satellite, 631(78)

Population and Demographics

Alaska, 303(65)
Anaheim, CA, 555(88)
Anchorage, AK, 555(88)
Arizona, 273(50)
Austin, TX, 262–263(51–54)
Causes of death, 688(35), 691–692
Columbus, OH, 262–263(51–54)
Comparing populations, 702(37, 38)
Dallas, TX, 261, 272(48)
Detroit, MI, 265
Employment statistics, 75(11–18)
Estimating population growth rates from data, 795
Florida, 269, 527(47)
Garland, TX, 555(87)
Georgia, 315(71)
Guppy population, 316(78)
Hawaii, 303(66), 567(71)
Idaho, 567(71)
Illinois, 315(72)
Indiana, 527(47)
Los Angeles, CA, 272(43)
Mexico, 273(58)
Milwaukee, WI, 282(62)
New York State, 263(56), 272(49)
Ohio, 263(55)
Pennsylvania, 269
Per capita federal aid, 694
Per capita income, 244(63)
Personal income, 526(46)
Phoenix, AZ, 272(44)
Population and matrices, 568(72)
Population decrease, 315(74, 75)
Population density, 665(47)
Population growth, 271(29–32)
Prison population, 65
Residents 65 or older, 693
San Jose, CA, 260
Size of continents, 713(6)
Temperatures in Beijing, China, 703(47), 715(31, 32)
U.S. population, 244(63), 268, 272(47, 49), 273(57, 58)
World population, 37(53)

Sports

Arena seating, 656(39)
Babe Ruth home runs, 695-696, 714(19, 21)
Barry Bonds home runs, 695-696, 714(19, 21)
Basketball attendance, 568(74)
Basketball lineups, 643(58)
Batting averages, 749(37)
Football kick, 449(41)
Hank Aaron home runs, 701(24), 702(30)
Horse racing, 747(14)
Investigating an athletic program, 730(50)
LPGA golf, 701–702(25–28)
Mark McGwire home runs, 700(7), 701(10, 23), 702(30)
Mickey Mantle home runs, 701(20, 22), 702(29), 714(20, 22)
NBA wins, 714(15, 18)
NFL wins, 714(14), 715(30)
NHL wins, 714(17)
PGA golf, 701–702(25–28)

Roger Maris home runs, 700(7), 701(9), 705, 707–708
Shooting free throws, 721–722, 740
Soccer field dimensions, 47(59)
Super Bowl, 700(1,3), 702(32)
Warren Moon passing yardage, 748(17)
Willie Mays home runs, 701(19, 21), 702(29), 714(20, 22)
Women's 100-meter freestyle, 154(66)

Miscellaneous

Acreage, 446
Air travel, 76(24)
Beatles songs, 748(21–28)
Casting a play, 638, 643(56)
Chain letter, 643(53)
Choosing chocolates, 680–681
Comparing cakes, 76(20)
Computer battery life, 714(16)
Converting to nautical miles, 324
Flower arrangements, 532, 533
Home remodeling, 568(78)
License plate restrictions, 636, 641(23, 24), 673(9)
Loose change, 555(95)
Pageant finalists, 639
Page design, 234(36)
Patent applications, 171(65)
Pepperoni pizza, 108(67)
Piano lessons, 687(29)
Picking lottery numbers, 639
Popular Web sites, 748(29), 750(63)
Radio advertising, 744(27)
Real estate prices, 749(33)
Satellite photography, 117(33)
Stepping stones, 76(21)
Sugar in breakfast cereal, 702(31)
Swimming pool, 568(79)
Swimming pool drainage, 234(41)
Telephone numbers, 641
Titanic survival rates, 700(2, 4)
Tracking airplanes, 492(51)
Tracking ships, 492(52)
Travel planning, 58(36)
Vacation money, 555(96)
Visualizing a musical note, 360(89)
Website visitors, 748–749(29–32)
White House Ellipse, 570, 590
Yahtzee, 641(35), 688(36)
Zoom cameras, 713(13), 715(29)

Index

A

Absolute extrema, 89–90
Absolute maximum, 90
Absolute minimum, 90
Absolute value
  of complex numbers, 504
  compositions, 133
  function, 101, 105
  graphing compositions, 132
  inequality involving, 54–55
  properties of, 13
  of real numbers, 12–13
Acceleration due to gravity, 167
Acute angles, 329–337
Addition
  associative property of, 662
  function, 110
  of inequality, 23
  matrix, 529–531
  of polynomial, 801–802
  principle of probability, 683
  property of equality, 21
  of vector, 458, 625
Additive identity, 50
  of algebraic expression, 6
Additive inverse, 50
  of algebraic expression, 6
  of matrix, 531
Agreement About Approximate Solutions, 43
Algebra
  associative property, 6
  basic properties of, 5–6
  commutative property, 6
  distributive property, 6
  formulas from, 826
  Fundamental Theorem of, 210
  identity property, 6
  inverse property, 6
Algebraically solving equations, 69
Algebraic expression, 5. *See also* Rational expression; Real number
  domain of, 808
  multiplicative inverse of, 6
Algebraic function, 252
  trigonometric function and, 369–371
Algebraic models, 65–66
Algebraic relations in function, 114
Algorithms
  common, 294
  division, 197–198
Amplitude, 352
Analytic geometry, 571, 829–830
Anchor, 667
Angle
  acute, 329–337
  central, 320
  complementary, 331
  coterminal, 338
  degrees of, 320–322
  of depression, 388
  directed, 487
  direction, 460–461
  of elevation, 388
  initial side of, 338
  measures of, 320–328, 338
  negative, 338
  nonquadrantal, 341
  positive, 338
  quadrantal, 342
  radians of, 320–322
  of rotation, 607
  in semicircles, 470
  special, 463
  standard position of, 338
  terminal side of, 338
  trigonometric function of any, 338–343
  between vector, 468–470
  vertex of, 338
Angular measure, 320
Angular motion, 323–324
Annual percentage rate (APR), 309
Annual percentage yield (APY), 307
Annuities
  future value of, 308
  ordinary, 308
  present value of, 309
Aphelion, 587
APOLLONIUS OF PERGA (c. 250–175 BCE), 571
Approximately equal, 3
Approximate solution(s)
  agreement about, 43
  inequalities, 56
  with table feature, 43–44
APR. *See* Annual percentage rate
APY. *See* Annual percentage yield
Arbitrarily close, 756
Arc, circular, 322–323
Arccosine, 380
ARCHIMEDES OF SYRACUSE (287–212 BCE), 767–768, 773, 778
Arc length formula, 322
Arcsine, 378
Arctangent, 381
Area
  limits and, 765–772
  motion and, 765–772
  problem, 765–772
  total, 792
  of triangle, 444–445
ARGAND, JEAN ROBERT (1768–1822), 503
Arguments, 504
Arithmetic sequence, 652–654
  finite, 659
  sum of, 658–661
Association
  between categorical variables, 693
  linear, 161
  negative linear, 161
  positive linear, 161
  between quantities, 161
Associative property, 6
  of addition, 662
  for complex numbers, 50
  of matrix, 536
Asymptote
  end behavior, 221
  in function, 92–93
  horizontal, 92–93, 105
  of hyperbola, 595
  of rational function, 220–222
  slant (oblique), 221
  vertical, 92–93, 221
Augmented matrix, 545
Average rate of change, 160–161
Average velocity, 754, 757
Axis
  of conic section, 612
  conjugate, 594
  coordinate, 621
  of ellipse, 582, 583, 585
  of hyperbola, 593–595
  imaginary, 503
  major, 583
  minor, 583
  of parabola, 573, 575
  polar, 487
  real, 503
  rotation formulas, 604–608
  semimajor, 583, 585, 617
  semiminor, 583, 585
  of symmetry, 164
  transverse, 594
  x-axis, 12, 91, 123, 131
  y-axis, 12, 90, 123, 131

B

Back substitution, 543
Back-to-back stemplot, 695
Bar chart, 692
Base
  change of, 285–286
  of exponential function, 252, 257
  of logarithmic function, 274
  natural, 256–258
Basic combinatorics, 634–643
Basic wave, 350–351
Bearing, 321
BELL, ALEXANDER GRAHAM (1847–1922), 280
BERNOULLI, JAKOB (1654–1705), 456
Bias, 735, 736
Bimodal distribution, 697
Binomial, 801
  coefficients, 644
  normal approximation, 725–727
  powers of, 644–645
  theorem, 644–649
Binomial probability distribution, 721
  expected value for, 722
  standard deviation for, 723
Binomial probability models, 720–723
BLACKWELL, DAVID (1919–2010), 678
Blind experiment, 737
Blocking, 737
Boundary, 557
Bounded function
  above, 88, 89
  below, 88, 89
  on intervals, 89
  not above, 88
  not below, 88

Bounded interval, 4
Boxplot, 706–708
Boyle's Law, 174
Branches of hyperbola, 593
BRIGGS, JOHN, 284

C

Calculator evaluation
  of cotangent function, 362
  of definite integrals, 785–786
  of derivatives, 784–785
  of dot products, 468
  errors in, 332–333
  of permutations, 637
  of trigonometric function, 331–332
Cartesian coordinate, 12, 621
Cartesian plane, 12
CASSEGRAIN, G., 598
Categorical data, 691–693
Categorical variable, 691
Causation, 162
Census, 735
Center, 15
  of ellipse, 583, 585
  of hyperbola, 593
Central angle, 320
Chain Rule, 824
Change-of-base formula, 285
Characteristic polynomial, 542
Chord, 574
  of ellipse, 583
  of hyperbola, 594
CHU SHIH-CHIEH, 645
Circle graph, 692
Circles, 581–592
  equations of, 15–16
  16-point unit, 346
  standard form equation of, 15
  unit, 344, 345
Circular arc length, 322–323
Circular function, 338–349
Closed curve, 572
Closed interval, 4
Coefficient
  binomial, 644
  correlation, 146
  of determination, 146
  leading, 158, 801
  in polynomial function, 185
  real numbers, 214–215
Coefficient matrix, 545
Cofactor, 534
Cofunction identity, 406
Column subscript of matrix, 529
Combinations, 638–639
Combinatorics, basic, 634–643
Common algorithm, 294
Common difference, 652
Common factor, 201
Common logarithms, 275–276
Common ratios, 653
Commutative property, 6
  of complex numbers, 50
  of matrix, 536
Complementary angle, 331
Completing the square, 41
Complex conjugates, 51
  zeros, 211–212
Complex fraction, 810
Complex function, 503
Complex number, 49–53
  absolute value of, 504
  associative property for, 50
  commutative property of, 50
  distributive property for, 50
  division of, 51, 505–506
  equal, 49
  modulus of, 504
  multiplication of, 50, 505–506
  multiplicative identity for, 51
  multiplicative inverse of, 51
  nonreal, 210
  *n*th root of, 508
  operations with, 49–51
  polar form of, 504–505
  power of, 506–508
  root of, 508–510
  solution(s), 210–214
  zero of function, 210–214
Complex plane, 503–504
Complex root, 211
Complex solution, 211
Complex zero, 210–217
Component form of two-dimensional vector, 456
Composite trigonometric function, 369–377
Compound interest, 304
  continuous, 306–307
  per year, 305–306
Compound rational expression, 810–811
Compound statement, 815–817
Conclusion, 819
Conditional probability, 684–686
  formula, 685
Conditionals, 819
Cone
  degenerate, 571
  nappe of, 571
  right circular, 571
Confounding variable, 736
Conic section, 571–572
  analyzing, 615–616
  degenerate, 571
  eccentricity of, 612
  focal axis of, 612
  focus-directrix of, 612
  geometric section of, 612
  orbits, 616–617
  polar equations of, 612–620
  vertex of, 612
Conjugate axis of hyperbola, 594
Conjugate hyperbola, 602
Conjunction, 815
Constant, 5
  of proportion, 174
  of variation, 174
Constant function, 86, 87, 159, 160
Constant of proportion, 174
Constant percentage rate, 265–266
Constant rate of change, 160
Constant term, 161
Constant velocity, 765
Constraint, 560
Continuity of function, 84–86
Continuous, 634
Continuous function, 102
  limits of, 776
Continuously compounded interest, 306–307
Continuum, 634
Control, 737
Convenience sample, 736
Convergence, 651
  of geometric sequences, 663
  of series, 662
Conversion factor, 143
Coordinate axis, 621
Coordinate of point, 3
Coordinate plane
  determining, 621
  midpoint formulas, 15
Corner point, 560
Correlation, 161, 732–734
Correlation coefficient, 146
  causation and, 162
  properties of, 162
Cosecant function, 340
  graphs of, 364
  trigonometric, 329
Cosine function, 101, 340
  of difference, 421–422
  graphs of, 350–360
  inverse, 380–382
  of sums, 422
  trigonometric, 329
Cotangent function, 340
  on calculator, 362
  graphs of, 362
  trigonometric, 329
Coterminal angles, 338
Counting
  combinations, 638
  importance of, 634–635
  multiplication principle of, 635–636
  permutation, 637
  subsets of *n*-sets, 639–640
Course, 321
Cross product, 467, 629
  elimination of, 607
  term, 603, 607
Cube root, 796–797
  function, 175
Cubic regression, 145
Cubing function, 100, 176

D

d'ALEMBERT, JEAN LE ROND (1717–1783), 773
Damped oscillation, 373–375
Damping, 373
Damping factor, 373
Data
  categorical, 691–693
  constant, 86

definite integrals computed with, 788–789
derivatives computed with, 786–788
function from, 143–147
quantitative, 693–696
re-expressing, 287–288
Decay factor, 254
Decibel, 280
Decomposing function, 113
Decreasing function, 86–88
Deduction, 670
Deductive reasoning, 73
Definite integral, 768–769, 785–786
data computation of, 788–789
notation, 768
Degenerate conic sections, 571
Degenerate parabola, 572
Degree-radian conversion, 322
Degrees of angles, 320–322
DE MOIVRE, ABRAHAM (1667–1754), 507
De Moivre's Theorem, 503, 507
Denominator
least common, 810
rationalizing, 798
Dependent variable, 80
Depreciation, linear function modeling, 160
Depression, angle of, 388
Derivative, 758–761
on calculators, 784–785
data computation of, 786–788
of function, 759
numerical, 784–792
at points, 758
DESCARTES, RENÉ (1596–1650), 49, 571, 754
Descriptive statistics, 704
Determinant
minor, 534
of square matrix, 534–536
Determination, coefficient of, 146
Difference
common, 652
cosine function of, 421–422
of function, 110
sine function of, 423–424
of sinusoids, 371–375
tangent function of, 424
Difference identity, 421–428
Difference rule, 775
Differentiability, 758
Directed angle, 487
Directed line segment, 456
Direction, 458
Direction angle, 460–461
Direction vector, 460, 626
Direct reasoning, 823
Directrix, 572
of parabola, 575
Direct variation, 174
Discontinuity
infinite, 85
jump, 85
points of, 84
removable, 84–85
Discrete, 634
Discriminant, 46
of quadratic equation, 52
test, 608–609
Disjunction, 815
Disproving non-identity, 416–417
Distance
from changing velocity, 766
from constant velocity, 765
polar coordinates in finding, 490
Distance conversion, 324
Distance formula, 13–14
three-dimensional Cartesian coordinate system, 622–623
Distinguishable permutations, 637
Distributions
bimodal, 697
binomial probability, 721, 722, 723
center of, 704
comparing, 704–705
describing, 704–705
normal, 711–712
shape of, 697, 704
skewed, 697
spread of, 704
unimodal, 697
Distributive property, 6
for complex numbers, 50
of matrix, 536
Divergence
of infinite sequences, 651
of series, 662
Dividend, 197
Division, 6
algorithm, 197–198
of complex numbers, 51, 505–506
long, 197–198
synthetic, 199–200
Divisor, 197
Domain
of function, 80–84
implied, 82
relevant, 82
of validity of identities, 404
Dot product
on calculators, 468
of vectors, 467–474, 625
Double-angle identity, 428
Double-blind experiment, 738
Double inequality, 25

E

Eccentricity
of conic section, 612
of ellipse, 587, 617
focus-directrix and, 612
of hyperbola, 597–598
Elements, 2
of matrices, 529
Elevation, angle of, 388
Ellipse, 581–592
center of, 583, 585
chord of, 583
eccentricity of, 587, 617
focal axis of, 582, 583, 585
focus of, 582, 585
geometry of, 582–585
major axis of, 583
minor axis of, 583
orbits of, 587–588
Pythagorean relation of, 583, 585
reflective property of, 589–590
semimajor axis of, 583, 585, 617
semiminor axis of, 583, 585
sketching, 584
standard equation of, 585
standard form of, 583
structure of, 582
translations, 585–586
vertex of, 582, 583, 585
Ellipsoid of revolution, 589
End behavior
asymptote, 221
of function, 93–94
leading term test for, 188
of polynomial function, 187–189
zooming on, 94
Endpoint, 4
Entry of matrix, 529
Enumerative induction, 670
Equal
approximately, 3
complex numbers, 49
fractions, 809
matrices, 529
vectors, 625
Equality, properties of, 21
Equally likely outcomes, 678
Equation. *See also* Linear first degree equation; Polar equation; Solving equations; Solving systems of equations
conversion, 489–490
equivalent, 22
exponential, 292–293
logarithmic, 293–294
parametric, 475–486
polynomial, 41
quadratic, 51–52
rational, 228–229
real zeros of, 44
second-degree, in two variables, 571
solving in one variable, 228–235
solving trigonometric, 408–410, 430–431
trigonometric, 408–410, 430–431
Equivalent, 21
Equivalent arrows, 457
Equivalent equation, 22
Equivalent expression, 809
Equivalent inequality, 24
Equivalent systems of linear equations, 543
ERATOSTHENES OF CYRENE (276–194 BCE), 320
EULER, LEONHARD (1707–1783), 80, 100, 257
Even function, 90
Even multiplicity, 190
Events
independent, 681, 685
mutually exclusive, 683
probability of, 678, 679

Existence theorems, 210
Existential quantifiers, 814
Expanded form, 6
Expected value
  for binomial probability distribution, 722
  of random variable, 718
Experimental design, 736–738
Explanatory variable, 634
Exponent
  integer, 7–9
  notation, 369
  properties of, 7
  rational, 798–799
Exponential decay function, 254
Exponential decay models, 266–267
Exponential equation
  one-to-one properties of, 292
  solving, 292–293
Exponential form, 274
Exponential function, 100. *See also* Logarithmic function
  base of, 252, 257
  constant percentage rate and, 265–266
  graphs, 252–256
  inverses of, 274–275
  logarithmic function and, 275
  natural, 257
  natural base in, 256–258
Exponential growth function, 254
Exponential growth model, 266–267
Exponential modeling, 265–273
Exponential notation, 7
Exponential population model, 265
Exponential regression, 145, 298, 299
Extended Principle of Mathematical Induction, 672
Extraneous solutions, 228–230

F

Factor
  common, 201
  in statistics, 736
Factored form, 6
Factorial identities, 647
Factorials, 636
Factoring, 41
  by grouping, 805–806
  polynomials, 802–804
  with real number coefficients, 214–215
  trinomials, 804–805
Factor Theorem
  polynomial function and, 198–199
  proof of, 199
Feasible point, 560
Feasible region, 560
FERMAT, PIERRE DE (1601–1665), 571, 678
Fibonacci. *See* LEONARDO OF PISA
Fibonacci numbers, 655
Fibonacci sequence, 655
Finance, mathematics of, 304–313
Finite sequence, 650, 659–660
First quadrant trigonometry, 339
Five-number summary, 705–706
Focal axis
  of conic section, 612
  of ellipse, 582, 583, 585
  of hyperbola, 593, 595
  of parabola, 573
Focal chord, 581
Focal length, 574, 575
Focal width
  of hyperbola, 602
  of parabola, 574, 575
Focus, 572
  of ellipse, 582, 585
  of hyperbola, 593, 595
  of parabola, 575
Focus-directrix
  of conic section, 612
  eccentricity and, 612
FOUCAULT, JEAN BERNARD LÉON (1819–1868), 35
Four-Color Map Theorem, 670
Fraction
  complex, 810
  equal, 809
  operation with, 808–809
  partial, 550–551
  reduced form of, 808
Fractional expression, 808
Free-fall motion, vertical, 167
Frequency
  distribution, 696
  of simple harmonic motion, 391
  of sinusoids, 353
  tables, 696
Function. *See also* Graph
  absolute extrema of, 89–90
  absolute value, 101, 105
  addition, 110
  agreement, 82
  algebraic, 252, 369–371
  algebraic combination of, 110–111
  algebraic relation in, 114
  asymptote in, 92–93
  basic, 99–109
  bounded above, 89
  bounded below, 89
  boundedness of, 88–89
  building, 110–118
  circular, 338–349
  complex, 503
  composite trigonometric, 369–377
  composition of, 111–114
  constant, 86, 87, 159, 160
  constant data and, 86
  continuity of, 84–86
  continuous, 102
  cosecant, 329, 340, 364
  cosine, 101, 329, 340, 350–360, 380–382, 421–422
  cotangent, 329, 340, 362
  cube root, 175
  cubing, 100, 176
  from data, 143–147
  decomposing, 113
  decreasing, 86–88
  derivative, 759
  difference of, 110
  domains, 80–84
  end behavior, 93–94
  even, 90
  exponential, 100, 252–256, 265–266, 274–275
  exponential decay, 254
  exponential growth, 254
  from formulas, 140–141
  geometric relations in, 114
  graphical analysis of, 103–105
  graph of, 81
  from graphs, 141–142
  greatest integer, 101, 107
  hyperbolic sine, 293
  identity, 99, 172
  implicitly defined, 114–116
  increasing, 86–88
  intervals, 87
  inverse, 121–125
  inverse cosine, 380–382
  inverse sine, 378–379
  inverse tangent, 380–382
  inverse trigonometric, 378–387
  linear, 158–173, 159
  local extrema of, 89–90
  logarithmic, 274–291
  logistic, 101, 258–259, 779
  logistic decay, 258
  logistic growth, 258
  lower bounds of, 89
  mapping, 80
  modeling with, 140–151
  monomial, 175–177
  natural exponential, 257
  natural logarithmic, 100
  notation, 80–81, 122
  odd, 91
  one-to-one, 122
  periodic, 345–346
  piecewise-defined, 104
  polynomial, 158–159, 185–209, 214
  power, 174–184
  product of, 110
  properties of, 80–98
  quadratic, 158–173, 214
  quotient of, 110
  range, 80–84
  rational, 218–227
  reciprocal, 100, 219–220
  relations, 114–116
  secant, 329, 340, 363
  sine, 100, 329, 340, 350–360, 378–379, 423–424
  sinusoid, 372
  square root, 100, 124, 179
  squaring, 99
  step, 101
  sum of, 110
  symmetry of, 90–92
  tangent, 329, 340, 361–362, 424
  transcendental, 252
  trigonometric, 329–346, 364, 369–377
  upper bound of, 89
  of variable, 625
  from verbal descriptions, 142–143

wrapping, 344
zero of, 70, 159, 210–214
Fundamental connection, 70
Fundamental identity0, 404–412
Fundamental polynomial connection, 211
Fundamental Theorem of Algebra, 210
Future value of annuity, 308

G

GALILEI, GALILEO (1564–1642), 35, 167, 754, 756, 757
GAUSS, CARL FRIEDRICH (1777–1855), 49, 503, 657, 659
Gaussian curve, 711
Gaussian elimination, 543–544
General form of equation of line, 30
Geometric formula, 827
Geometric relation in function, 114
Geometric sequence, 652–654
convergence of, 663
finite, 660
sums of, 658–661
Graph, 99. *See also* Function
bounded above, 88
bounded below, 88
circle, 692
of cosecant function, 364
of cosine function, 350–360
of cotangent function, 362
of exponential functions, 252–256
functions from, 141–142
of limaçon curves, 498
line, 698
of linear functions, 159–160
of logarithmic functions, 274–282, 277–279
of monomial functions, 175–177
not bounded above, 88
not bounded below, 88
picture, 692
of polar equations, 494–502
of polynomial functions, 185–187
of power functions, 177–178
quadratic functions, 164–166
of rational functions, 218–227
of rose curves, 497
of secant function, 363
shading, 559
from sign charts, 237
of sine function, 350–360
of sinusoids, 354
for solving systems of equations, 521
of tangent function, 361–362
Grapher, 559
failure, 71–73
hidden behavior, 71–73
for local extrema, 90
motion simulated with, 478–481
polar coordinate conversion with, 489
sequences and, 654–655
Graphically solving equation, 40–41, 43–44, 69
Graphical model, 66–68
Graphical statistics, 691–703
Graphical transformation, 129–139
Graphing utility. *See* Grapher
Gravity, acceleration due to, 167
Greatest integer function, 101, 107
Growth, limits to, 258
Growth factor, 254
Guessing penalty, 722
Guiding rectangle, 584

H

Half-angle identity, 429–430
Half-life, 266
Half-plane, 557
Hand sketching, 560
Head minus tail (HMT) rule, 457
Height, 167
HEINE, HEINRICH EDUARD (1821–1881), 773
Heron's formula, 444–445, 645
Hidden behavior, 71–73
HIPPARCHUS OF NICAEA (190–120 BCE), 320
HIPPOCRATES OF CHOIS (470–410 BCE), 320
Histograms, 696
HMT rule. *See* Head minus tail rule
Horizontal asymptote, 92–93, 105
Horizontal line, 30
Horizontal line test, 121, 122
Horizontal shrink, 133–135
Horizontal stretch, 133–135
Horizontal translation, 129–131
Humidity, relative, 224
Hyperbola, 572
asymptote of, 595
branches of, 593
center of, 593
chord of, 594
conjugate, 602
conjugate axis of, 594
eccentricity of, 597–598
focal axis of, 593, 595
focal width of, 602
focus of, 593, 595
geometry of, 593–596
graphing, 597
in long-range navigation, 599
orbits of, 597–598
Pythagorean relation of, 595
reflective property of, 598
semiconjugate axis of, 594, 595
semitransverse axis of, 594, 595
sketch of, 594
standard equation of, 595
standard form of, 594
translation of, 596–597
transverse axis of, 594
vertex of, 593, 595
Hyperbolic sine function, 293
Hyperboloid of revolution, 598
Hypotheses, 819

I

Identity
in calculus, 417
cofunction, 406
difference, 421–428
domain of validity of, 404
double-angle, 428
function, 99, 172
fundamental, 404–412
half-angle, 429–430
matrix, 533–534
multiple-angle, 428–433
odd-even, 406–407
power-reducing, 428–429
property, 6
property of matrix, 536
Pythagorean, 405–406
quotient, 404
reciprocal, 404
sum, 421–428
trigonometric, 404–405, 413–420
Imaginary axis, 503
Imaginary number, 49
Imaginary part, 49
Imaginary unit, 49
Implications, 819
Implicitly defined function, 114–116
Implied domain, 82
Increasing function, 86–88
Independent events, 681, 685
Independent variable, 80, 693
Index, 796
Index of summation, 658
Induction
deduction and, 670
enumerative, 670
mathematical, 670
Inductive hypothesis, 667
Inductive step, 667
Inequality
absolute value, 54–55
addition of, 23
approximate solutions to, 56
double, 25
equivalent, 24
graphs of, 557–558
linear, 21–27
multiplication of, 23
polynomial, 236–239
properties of, 23
quadratic, 55–56
rational, 239–240
solving, 23, 54–55
solving in one variable, 236–245
symbol, 3
systems of, 558–559
transitive property, 23
Inferential statistics, 704
Infinite discontinuity, 85
Infinitely repeating, 2
Infinite sequence, 650
convergence of, 651
divergence of, 651
limits of, 651–652
Infinite series, 661–662
Infinity limit, 766–767, 778–780
Inflection, point of, 100
Informal limit, 756
Initial conduction, 675

Initial point of vector, 457
Initial side of angle, 338
Inner product. *See* Dot product
Instantaneous velocity, 755
Integer, 2
Integer exponent, 7–9
Interest
  compound, 304
  compounded per year, 305–306
  continuously compounded, 306–307
  simple, 304
Intermediate value theorem, 190–191
Interquartile range, 705
Interval
  bounded, 4
  closed, 4
  endpoints, 4
  function bounded on, 89
  function on, 87
  half-open, 4
  notation, 3–5
  open, 5
  unbounded, 5
Invariant under rotation, 608
Inverse composition rule, 124, 380
Inverse cosine function, 380–382
Inverse exponential function, 274–275
Inverse function, 121–125
Inverse matrix, 533–535
  solving systems with, 549–550
Inverse property, 6
  of matrix, 536
Inverse reflection principle, 123
Inverse relations, 121–125, 574
Inverse sine function, 378–379
Inverse-square law, 175
Inverse tangent function, 380–382
Inverse trigonometric function, 378–387
Inverse variation, 174
Invertible square linear systems, 549
Irrational number, 2
Irrational zero of polynomial, 201
Irreducible quadratic function, 214
Isosceles right triangle, 330

J

Jump discontinuity, 85

K

KEPLER, JOHANNES (1571–1630), 179, 571, 616
Kepler's First Law, 598
Kepler's Third Law, 287
Knot, 390

L

Latus rectum, 574, 581
Law of Cosines, 442–450
  solving triangles with, 443–444
Law of Detachment, 823
Law of Sines, 434–441
  solving triangles with, 435–437
LCD. *See* Least common denominator
Leading coefficient, 158, 801
Leading term
  in polynomial function, 185
  test for end behavior, 188
Leaf, 693
Least common denominator (LCD), 810
Left-hand limit, 776
Left rectangular approximation method (LRAM), 768
LEIBNIZ, GOTTFRIED WILHELM (1646–1716), 80, 754, 766, 773
Leibniz notation, 760
Lemniscate curve, 499
LEONARDO OF PISA (c. 1170–1250), 655
Light, speed of, 35
Like terms, 801
Limaçon curve
  graphs of, 498
  in polar equations, 497–499
Limit. *See also* Asymptote; Continuous function; End behavior
  area and, 765–772
  of continuous function, 776
  defining, 773–774
  Difference Rule for, 775
  to growth, 258
  history of, 773
  infinite, 766–767, 778–780
  informal, 756
  left-hand, 776
  motion and, 756–772
  notation of, 85
  one-sided, 776–778
  Power Rule for, 775
  Product Rule for, 775
  properties of, 774–775
  Quotient Rule for, 775
  rational function, 220–222
  right-hand, 776
  Root Rule for, 775
  Sum Rule for, 775
  two-sided, 776–778
Line
  of best fit, 144–145, 173
  general form of equation of, 30
  graph, 698
  horizontal, 30
  parallel, 31–32
  in parametric equation, 477–478
  perpendicular, 31–32
  point-slope form, 29
  regression, 144
  secant, 758
  segment, 477–478
  single, 572
  slant, 159
  slope-intercept form, 29–30
  slope of, 28–29
  tangent, 581, 756–758
  in three-dimensional Cartesian coordinate system, 626–627
  of travel, 321
  vertical, 30
Linear association, 161
Linear combination, 460
Linear correlation, 161–163
Linear Factorization Theorem, 210
Linear first degree equation, 21–27
  equivalent system of, 543
  graphing in two variables, 30–31
  in one variable, 21
  in two variables, 33–35
Linear function, 158–173
  depreciation modeling with, 160
  graphs of, 159–160
  nature of, 161
Linear inequality, 21–27
  in one variable, 23–25
  solution set of, 23
  solving, 23
  in *x*, 23
Linear modeling, 161–163
Linear motion, 323–324
Linear programming, 559–562
Linear regression, 145, 298
Linear system
  invertible square, 549
  multivariate, 543–556
  triangular form for, 543, 544
Lithotripter, 589
Loans, 309–310
Local extrema, 89–90
  grapher for, 90
  of polynomial function, 187
Local maximum, 90
Logarithm
  basic properties of, 274
  change of base, 285–286
  history of, 284
  natural, 276
  power rule for, 283
  product rule for, 283
  properties of, 283–285
  quotient rule for, 283
Logarithmic equation, 293–294
Logarithmic form, 274
Logarithmic function. *See also* Exponential function
  base of, 274, 286–287
  exponential function and, 275
  graph of, 274–282
  properties of, 283–291
Logarithmic model, 294–296
Logarithmic re-expression, 298–299
Logic, 813–818
Logically equivalent, 816
Logistic decay function, 258
Logistic function, 101, 258–259, 779
Logistic growth function, 258
Logistic modeling, 265–273
Logistic regression, 145
Long division, 197–198
Long-range navigation, 599
Loop, 587
LORAN, 599
Lower bound
  of function, 89
  of real zero, 202–204
  test, 202

LRAM. *See* Left rectangular approximation method
Lurking variable, 746

M

Magnitude, 11, 12–13
order of, 294–296
in two-dimensional vector, 456
of vector, 625
Major axis of ellipse, 583
Mapping, 80
Match, 736
Mathematical induction, 667–672
principle of, 668–669
Mathematical model, 64
Matrix, 529–542. *See also* Solving systems of equations
addition, 529–531
additive identity, 531
additive inverse, 531
associative property of, 536
augmented, 545
coefficient, 545
column subscript of, 529
commutative property of, 536
distributive property of, 536
elements of, 529
entries of, 529
equal, 529
identity, 533–534
identity property of, 536
inverse, 533–535
inverse property of, 536
multiplication, 531–533
multiplicative identity for, 533
nonsingular, 533
order of, 529
properties of, 536
row echelon form of, 546
row subscript of, 529
scalar, 530
singular, 533
square, 529, 534–536
subtraction, 529–531
symmetry in, 540
systems of equations solved with, 549–550
transpose, 533
zero, 531
Maximum $|r|$ value, 496
Maximum sustainable population, 268
Maximum value, 373
Mean, 708–710
weighted, 709
Measure of angle, 320–328
Median, 705
Midpoint formula
coordinate plane, 15
number line, 14
in three-dimensional Cartesian coordinate system, 622–623
Minimum, local, 90
Minor axis of ellipse, 583
Minor determinant, 534
Minutes, 320
Mirror method, 124
Model
algebraic, 65–66
exponential decay, 266–267
exponential growth, 266–267
with function, 140–151
graphical, 66–68
interpreting, 68
logarithmic, 294–296
mathematical, 64
numerical, 64–65
regression, 298
sinusoidal, 355
Modeling, 145
exponential, 265–273
logistic, 265–273
periodic behavior, 355–356
with power function, 179–181
Modulus of complex number, 504
Modus Ponens, 823
Monomial, 801
Monomial function, 175–177
Mortgage, 309–310
Motion, 475–486
angular, 323–324
area and, 765–772
grapher simulation of, 478–481
limits and, 756–772
linear, 323–324
Multiple-angle identity, 428–433
Multiplication
of complex numbers, 50, 505–506
of inequality, 23
matrix, 531–533
of polynomials, 801–802
principle of probability, 681
property of equality, 21
scalar, 458
vector, 458
Multiplication Principle of Counting, 635–636
Multiplicative identity
for complex number, 51
for matrix, 533
Multiplicative inverse
of algebraic expression, 6, 175
of complex numbers, 51
Multiplicity of zero of function, 189
Multivariate linear system, 543–556
Mutually exclusive event, 683

N

NAPIER, JOHN (1550–1617), 284
Nappe of cone, 571
Natural base, 256–258
Natural exponential function, 257
Natural logarithm, 276
Natural logarithmic function, 100, 278
Natural logarithmic regression, 145, 298, 299
Natural number, 2
Nautical mile, 324
Negation, 813
Negative angle, 338
Negative association, 732, 733
Negative linear association, 161
Negative number, 3
NEWTON, ISAAC (1642–1727), 167, 754–757, 767, 773
Newton's Law of Cooling, 296–298
Non-identities, disproving, 416–417
Nonquadrantal angle, 341
Nonreal zero of polynomial, 210
Nonrelationship, 285
Nonrigid transformation, 129
Nonsingular matrix, 533
Normal approximation for binomial, 725–727
Normal curve, 711
Normal distribution, 711–712
Normal model, 723–725
Notation
definite integral, 768
exponent, 369
function, 80–81, 122
Leibniz, 760
of limits, 85
set, 111
set-builder, 2
n-set, 636, 639–640
*n*th power of *a*, 7
*n*th root, 796
of complex numbers, 508
Number line, real, 3, 14
Numerical derivative, 784–792
Numerical integral, 785
Numerically solving equations, 69
Numerical model, 64–65
Numerical statistics, 704–716
Numerical support, 71

O

Objects, 2
Observational studies, 735–736
Octant, 621
Odd degree, polynomial function of, 214
Odd-even identity, 406–407
Odd function, 91
Off-center, 588
One-sided limit, 776–778
One-to-one function, 122
One-to-one property, 292
Open intervals, 5
Open of parabola, 575
Opposite, 6
Orbits
conic section, 616–617
of ellipses, 587–588
of hyperbola, 597–598
Order, 3–5
Ordered pair, 12
Order of magnitude, 294–296
Ordinary annuity, 308
Origin, 3, 621
of Cartesian plane, 12
reflections through, 131
symmetry with respect to, 91
Orthogonal vector, 469
Outcomes, equally likely, 678
Outer product. *See* Cross product
Outlier, 706

P

Parabola, 477
  axis of, 575
  degenerate, 572
  directrix of, 575
  focal axis of, 573
  focal chord of, 581
  focal length of, 574, 575
  focal width of, 574, 575
  focus of, 575
  geometry of, 572–575
  graph of, 121
  open of, 575
  reflective property of, 577–578
  standard equation, 575
  transformations of, 575
  translations of, 575–577
  vertex of, 573, 574
Paraboloid of revolution, 577
Parallel line, 31–32
Parallelogram representation of vector, 458
Parameter
  elimination of, 476–477
  interval, 475
  in statistics, 704
Parametrically defined relation, 119–121
Parametric curves, 475
  in polar equations, 494
Parametric equation, 475–486
  lines in, 477–478
Parametric form, 626
Parametric mode, 120
Parametrization, 475
Partial fraction, 550–551
Partial sums, 662
PASCAL, BLAISE (1623–1662), 645, 678
Pascal's triangle, 645–646
Pendulum motion, 663
Perihelion, 587
Periodic behavior modeling, 355–356
Periodic function, 345–346
Period of sinusoids, 353
Permutation, 636–638
  on calculator, 637
  counting formula, 637
  distinguishable, 637
Perpendicular line, 31–32
pH, 295
Phase shift, 352, 353
Picture graph, 692
Piecewise definition, 104
Pie chart, 692
Placebo, 737
Plane
  complex, 503–504
  equation of, 624
  vector in, 456–466
Pluto, 591
Point
  coordinates of, 3
  corner, 560
  derivatives at, 758
  of discontinuity, 84
  feasible, 560
  of inflection, 100
  of intersection, 45
  single, 572
  vertex, 560
Point-slope form, 29
Polar axis, 487
Polar coordinate, 487–493
  conversion, 488–489
  for distance, 490
  equation conversion, 489–490
  finding all, 488
  grapher conversion of, 489
Polar curve, 494
Polar equation
  analyzing, 495–497, 615–616
  of conic, 612–620
  graphs of, 494–502, 613
  lemniscate curve in, 499
  limaçon curve in, 497–499
  parametric curve in, 494
  polar curve in, 494
  symmetry in, 494–495
  writing, 613–615
Polar form. *See* Trigonometric form
Pole, 487
PÓLYA, GEORGE (1887–1985), 70
Polygon, regular, 146
Polynomial. *See also* Rational function
  addition of, 801–802
  characteristic, 542
  degree of, 801
  equation, 41
  factoring, 802–804
  inequality, 236–239
  irrational zero of, 201
  multiplication of, 801–802
  nonreal zero of, 210
  prime, 802
  rational zero of, 201
  subtraction of, 801–802
  vocabulary of, 185
Polynomial function, 158, 185–196
  coefficient in, 185
  division algorithm and, 197–198
  end behavior of, 187–189
  Factor Theorem, 198–199
  fundamental connection, 199
  graph of, 185–187
  leading term in, 185
  local extrema of, 187
  long division and, 197–198
  of low degree, 159
  multiplicity of zero of, 189
  of no degree, 159
  of odd degree, 214
  Rational Zeros Theorem, 201–202
  real zero of, 197–209
  Remainder Theorem and, 198–199
  standard form of, 185
  synthetic division and, 199–200
  term, 185
  zero of, 187, 189–190
Population model, 260–261, 711
  exponential, 265
  regression in, 267–269
Position vector, 456
Positive angle, 338
Positive association, 732
Positive linear association, 161
Positive number, 3
Positive x-axis, 123
Positive y-axis, 123
Power function, 174–184
  graphs of, 177–178
  modeling with, 179–181
Power-reducing identity, 428–429
Power regression, 145, 298, 299
Power rule
  for limits, 775
  for logarithms, 283
Powers of complex number, 506–508
Precious Mirror (Chu), 645
Present value of annuity, 309–310
Prime polynomial, 802
Principal nth root, 796
Probability, 677–690
  addition principle of, 683
  conditional, 684–686
  determining, 680–681
  distribution, 679
  of events, 678, 679
  function, 677–680
  multiplication principle of, 681
Probability model
  binomial, 720–723
  random variable, 717–731
Problem solving, 70–71
Product rule
  for limits, 775
  for logarithms, 283
Products of function, 110
Projectile motion, 57
Projection of vector, 470–471
Proof, 73–74
  of identities, 413–420
  of product formula, 505
  reflections, 132
  strategies, 413
Prospective, 746
Pseudo-random numbers, 738
Pythagoras, 645
Pythagorean identity, 405–406
Pythagorean relation
  of ellipse, 583, 585
  of hyperbola, 595
Pythagorean Theorem, 14, 330, 645

Q

Quadrant, 12
Quadrantal angle, 342
Quadratic equation
  complex solution of, 51–52
  discriminant of, 52
  solving, 41–43
  with xy terms, 603–611

Quadratic formula, 42
Quadratic function, 158–173
  applications of, 166–168
  characterizing, 166
  graphs, 164–166
  irreducible, 214
  standard form of, 164
  vertex form of, 165
Quadratic inequality, 55–56
Quadratic regression, 145
Quadric surfaces, 625
Quantitative data, 693–696
Quantitative variable, 691
Quartic regression, 145
Quartiles, 705
Quotient, 197
  of function, 110
  identities, 404
Quotient rule
  for limits, 775
  for logarithms, 283

R

Radian, 321
Radians of angles, 320–322
Radical, 796–797
Radical expression, 796
  simplifying, 797–798
Radicand, 796
Radioactive decay, 266
Radius, 15
Randomization, 737
Randomness, 734–735, 738–739
Random number tables, 738
Random sample, 712, 734–735
Random variable, 679
  continuous, 712
  expected value of, 718
  probability model, 717–731
Range, 705
  of function, 80–84
Rate of change, 160–161
Ratio, 160
  common, 653
Rational equation, 228–229
Rational exponent, 798–799
Rational expression, 808–809
  compound, 810–811
  operations with, 809–810
Rational function, 218. *See also* Polynomial
  asymptote of, 220–222
  end behavior asymptote of, 221
  graph of, 218–227
  limit, 220–222
  vertical asymptote, 221
  x-intercept of, 221
  y-intercept of, 221
Rational inequality, 239–240
Rationalizing denominator, 798
Rational number, 2
Rational zero of polynomial, 201
Rational Zeros Theorem, 201–202
Real axis, 503
Real number, 2–11. *See also* Algebraic expression; Rational expression
  absolute value of, 12–13
  bounded interval of, 4
  coefficient, 214–215
  representing, 2–3
  trigonometric function of, 344–346
Real part, 49
Real root, 199
Real solution, 199
Real zero of polynomial, 197–209
  lower bound test, 202
  rational and irrational zeros, 201
  upper bound test, 202
Real zeros of equation, 44
Reciprocal, 6
Reciprocal function, 100
  transformation of, 219–220
Reciprocal identity, 404
Rectangle, guiding, 584
Rectangular approximation method, 768
Rectangular coordinate system, 12
Recursion formula for Pascal's triangle, 646
Recursive, 650
Recursive formula, 254
Reduced form, 808
Reduced row echelon form, 547–548
Reduction formula, 424
Re-expressing data, 287–288
Re-expression, logarithmic, 298–299
Reference triangle, 341
Reflection, 123
  across axes, 131–133
  origin, 131
  proof, 132
  x-axis, 131
  y-axis, 131
Reflective property
  of ellipse, 589–590
  of hyperbola, 598
  of parabola, 577–578
Reflexive property, 21
Regression, 181
  analysis, 163
  cubic, 145
  exponential, 145, 298, 299
  line, 144
  linear, 145, 298
  logistic, 145
  model, 298
  natural logarithmic, 145, 298, 299
  in population model, 267–269
  power, 145, 298, 299
  quadratic, 145
  quartic, 145
  sinusoidal, 145
Regular polygon, 146
Relation
  algebraic, 114
  discovering, 285
  function, 114–116
  geometric, 114
  graphing, 114
  inverse, 121–125
  parametrically defined, 119–121
Relative extrema, 90
Relative humidity, 224
Relevant domain, 82
Remainder, 197
Remainder Theorem, 198–199
Removable discontinuity, 84–85
Repeated zeros, 190
Replacing translation, 130
Replication, 737
Resistant statistics, 705
Response bias, 736
Response variable, 634
Retrospective, 746
RIEMANN, GEORG BERNHARD (1826–1866), 768
Riemann sum, 768
Right circular cone, 571
Right-handed coordinate frame, 621
Right-hand limit, 776
Right rectangular approximation method (RRAM), 768
Right triangle, 329
  isosceles, 330
  problems, 388–390
Right triangle trigonometry, 329–330
  solving, 333
  standard position, 329
Rigid transformation, 129
Root, 70
  of complex number, 508–510
  cube, 175, 796–797
  nth root, 508, 796
  real, 199
  rule, 775
  square, 41, 100, 124, 179, 796
Rose curve
  graphs of, 497
  in polar equation, 496–497
Row echelon form, 545
  of matrix, 546
  reduced, 547–548
Row operation, 543–556
  elementary, 545–547
Row subscript of matrix, 529
RRAM. *See* Right rectangular approximation method

S

Sample, 704, 735–736
  random, 712, 734–735
Sample space, 677–680
Scalar
  matrix, 530
  multiplication, 458
Scientific notation, 8–9
  common algorithms, 294
Secant function, 340
  graph of, 363
  trigonometric, 329
Secant line, 758
Second-degree equation, 571
Seconds, 320

Semicircle, 470
Semiconjugate axis of hyperbola, 594, 595
Semimajor axis of ellipse, 583, 585, 617
Semiminor axis of ellipse, 583, 585
Semiperimeter, 445
Semitransverse axis of hyperbola, 594, 595
Sequence
  arithmetic, 652–654, 658–661
  finite, 650, 659–660
  geometric, 652–654, 658–661, 663
  graphing calculators and, 654–655
  infinite, 650–652
Series, 658–666
  convergence of, 662
  divergence of, 662
  infinite, 661–662
Set-builder notation, 2
Set notation, 111
Shrink, 133–135
Sigma notation, 658
Sign chart, 236, 237
Simple harmonic motion, 390–393
  frequency of, 391
  watching, 391
Simple interest, 304
Simulation, 740–742
Sine function, 100, 340
  of difference, 423–424
  graph of, 350–360
  inverse, 378–379
  of sum, 423–424
  trigonometric, 329
Single line, 572
Single point, 572
Singular matrix, 533
Sinusoidal model, 355
Sinusoidal regression, 145
Sinusoid function, 372
Sinusoids
  amplitude of, 352
  frequency of, 353
  graph of, 354
  investigating, 371
  periodic behavior modeling with, 355–356
  period of, 353
  sum and difference of, 371–375
  transformation and, 352–355
  verifying, 424–425
16-point unit circle, 346
68-95-99.7 rule, 712
Skewed distribution, 697
Slant asymptote, 221
Slant line, 159
Slope-intercept form, 29–30
Slope of line, 28–29
Solution. *See also* Approximate solution(s)
  to complex numbers, 210–214
Solving equations, 21
  algebraically, 69
  approximating, graphically, 43–44
  exponential, 292–293
  extraneous solutions in, 228–230
  graphically, 40–41, 69
  with intersections, 45
  logarithmic, 293–294
  numerically, 69
  in one variable, 228–235
  quadratic, 41–43
  rational, 228–229
  with tables, 43–44
  trigonometric, 408–410, 430–431
Solving inequalities, 557
  absolute value, 54–55
  approximating, 56–57
  linear, 23
  in one variable, 236–245
Solving systems of equations, 519–528. *See also* Matrix
  by elimination, 521–523
  graphically, 521
  with inverse matrices, 549–550
  substitution method of, 519–521
Solving triangles, 333
  with Law of Cosines, 443–444
  with Law of Sines, 435–437
Sound, measuring, 280
Special angle, 463
Special product, 802–804
Speed
  of light, 35
  of vector, 461
Sphere, 624
Spiral of Archimedes, 499
Square matrix, 529
  determinants of, 534–536
  inverse of, 533
Square root, 796
Square root function, 100, 124, 179
Square root principle, 41
Square viewing window, 31
Squaring function, 99
Standard deviation, 710–712
  for binomial probability distribution, 723
Standard equation
  of ellipse, 585
  of hyperbola, 595
  of parabola, 575
Standard form equation, 49, 574, 801
  of circles, 15
  of ellipse, 583
  of hyperbola, 594
  of polynomial function, 185
  quadratic, 164
Standard position
  of angles, 338
  in right triangle trigonometry, 329
Standard representation of vector, 456
Standard unit vector, 460, 467, 625
Standard vector in three-dimensional Cartesian coordinate system, 625
Statements, 813–815
  forms of, 819–825
Statistically significant, 726, 740
Statistics
  descriptive, 704
  factors in, 736
  graphical, 691–703
  inferential, 704
  literacy in, 732–746
  numerical, 704–716
  parameters and, 704
  resistant, 705
  uses and misuses, 732
Statute mile, 324
Stem-and-leaf plot, 693
Stemplot, 693
  back-to-back, 695
Step function, 101
Stretch, 133–135
Strong association, 733
Strong positive correlation, 732
Subject, 736
Substitution method, 519–521
Subtraction, 6
  matrix, 529–531
  of polynomials, 801–802
  of vectors, 625
Success-failure rate, 726
Sum
  of arithmetic sequences, 658–661
  cosine function of, 422
  of function, 110
  of geometric sequences, 658–661
  identity, 421–428
  partial, 662
  rule, 775
  sine function of, 423–424
  of sinusoids, 371–375
  tangent function of, 424
Summation, index of, 658
Summation notation, 658
Supported numerically, 71
Survey, 735–736
Symmetric difference quotient, 784
Symmetric property, 21
Symmetry, 697
  axis, 164
  of function, 90–92
  in matrix, 540
  origin, 91
  in polar equation, 494–495
  x-axis, 91
  y-axis, 90
Synthetic division, 199–200
Systems of equations, solving, 519–528
Systems of inequalities, 558–559

T

Tables, 43–44
Tail-to-head representation of vector, 458
Tangent function, 340
  of difference, 424
  graphs of, 361–362
  inverse, 380–382
  of sum, 424
  trigonometric, 329
Tangent line, 581, 756–757
  problem, 758
Tangent problem, 756–764
Tautology, 821

Term
constant, 161
cross product, 603, 607
leading, 185
in polynomial function, 185
Terminal point of a vector, 457
Terminal side of angle, 338
Terminate, 2
Three-dimensional Cartesian coordinate system, 621–629
distance formula, 622–623
drawing, 623
lines in, 626–627
midpoint formulas, 622–623
standard vector in, 625
vector in, 625–626
zero vector, 625
Time plot, 697–699
Total area, 792
Tower of Hanoi problem, 667, 668
TRACE feature, 40, 102
Transcendental function, 252
Transformation
combining, 135–136
graphical, 129–139
nonrigid, 129
of reciprocal function, 219–220
rigid, 129
sinusoid and, 352–355
of unit circle, 582
Transitive property
of equality, 21
of inequality, 23
Translation
ellipses, 585–586
horizontal, 129–131
introducing, 129
of parabola, 575–577
replacing, 130
vertical, 129–131
Transpose of matrix, 533
Transverse axis of hyperbola, 594
Tree diagram, 683–684
Triangle
area of, 444–445
determining number, 435
famous, 330–331
isosceles right, 330
reference, 341
right, 329–330, 388–390
solving, 435–437, 443–444
Triangular form for linear systems, 543, 544
Triangular number, 666
Trichotomy property, 4
Trigonometric equation, 408–410, 430–431
Trigonometric expression, 407–408
Trigonometric form, 504–505
Trigonometric function, 340
of acute angle, 329–337
algebraic function and, 369–371
of any angle, 338–343
basic, 364
calculation errors, 332–333
calculator evaluation of, 331–332
composing, 382–384
composite, 369–377
damped oscillation, 373–375
exploring, 330
inverse, 378–387
of nonquadrantal angle, 341
of real numbers, 344–346
sine, 329
standard position in, 329
Trigonometric identities, 404–405, 413–420
Trigonometry
first quadrant, 339
formulas from, 827–828
right triangle, 329–330
solving problems with, 388–397
Trinomial, 801
factoring, 804–805
Truth table, 815
Tstep, 121
Two-dimensional vector, 456–458
component form of, 456
initial point, 457
magnitude in, 456
standard representation of, 456
terminal point, 457
Two-sided limit, 776–778

U

Unbounded behavior, 779
Unbounded interval, 5
Undercoverage bias, 736
Unimodal distribution, 697
Union of two sets, 55
Unit circle, 344, 345
16-point, 346
transforming, 582
Unit vector, 459–460, 625
standard, 460, 467
Unordered system, 3
Upper bound
of functions, 89
of real zeros, 202–204
tests, 202

V

Validity, domain of, 404
Variable, 5
categorical and quantitative, 691
confounding, 736
continuous random, 712
dependent and independent, 80
explanatory and response, 634
lurking, 746
random, 679, 717–731
Variance, 710–712
Variation
constant of, 174
direct, 174
inverse, 174
Varies, 174
Vector
addition of, 458, 625
angles between, 468–470
archery, 457
direction, 460, 626
dot product of, 467–474, 625
equality, 625
form, 626
as linear combinations, 460
magnitude of, 625
multiplication, 458
operations, 458–459
orthogonal, 469
parallelogram representation, 458
in plane, 456–466
position, 456
projecting, 470–471
relationships, 625
resolving, 460
speed of, 461
standard unit, 460, 467, 625
subtraction of, 625
tail-to-head representation, 458
in three-dimensional Cartesian coordinate system, 625–626
two-dimensional, 456–458
unit, 459–460, 625
velocity of, 461
zero, 456
Velocity
average, 754, 757
changing, 766
constant, 765
instantaneous, 755
questions of, 755
of vector, 461
vertical, 167
VENN, JOHN (1834–1923), 682
Venn diagrams, 682–684
Verbal descriptions, 142–143
Vertex, 164
of angle, 338
of conic section, 612
of ellipse, 582, 583
form of quadratic function, 165
of hyperbola, 593, 595
of parabola, 573, 574
points, 560
in right circular cone, 571
Vertical asymptote, 92–93
rational function of, 221
Vertical free-fall motion, 167
Vertical line, 30
Vertical line test, 81
Vertical shrink, 133–135
Vertical stretch, 133–135
Vertical translation, 129–131
Vertical velocity, 167
Very weak negative association, 733
Very weak negative correlation, 733
Viewing windows, 31
Voluntary response bias, 736
VON NEUMANN, JOHN (1903–1957), 678

W

Weak negative correlation, 732
WEIERSTRASS, KARL (1815–1897), 773
Weighted mean, 709

WESSEL, CASPAR (1745–1818), 503
Whispering galleries, 589
Whole number, 2
Work, 471
Wrapping function, 344

X

x-axis, 12
  positive, 123
  reflections across, 131
  symmetry with respect to, 91
x-coordinate, 12
x-intercept, 31, 70
  of rational functions, 221
xy-plane, 621
xz-plane, 621

Y

y-axis, 12
  positive, 123
  reflections across, 131
  symmetry with respect to, 90
y-coordinate, 12
y-intercept, 29
  of rational functions, 221
yz-plane, 621

Z

z-axis, 621
ZENO OF ELEA (c. 490–425 BCE), 766
Zeno's paradoxes, 766
Zero
  complex, 210–217
  complex conjugates, 211–212
  of even multiplicity, 190
  factor property, 68–70
  of function, 70, 159, 210–214
  irrational, 201
  matrix, 531
  multiplicity of, 189
  nonreal, 210
  of odd multiplicity, 190
  of polynomial functions, 187, 189–190
  real, 44, 197–209
  repeated, 190
Zero vector, 456
  in three-dimensional space, 625
Zooming on end behavior, 94
z-score, 724